Laxton's
NRM Building Price Book

2017 edition

Edited by V B Johnson LLP

Published by Laxton's Publishing Limited

Laxton's NRM Building Price Book – 2017

Published by
Laxton's Publishing Limited
St John's House
304 – 310 St Albans Road
Watford
Hertfordshire WD24 6PW

First published 2016

ISBN...978-0-9571386-9-8

Laxton's 2017 has up to date prices in an easy to follow and comprehensive format. The main contents page and the index section will enable quick identification of specific items. There is also a contents list to the Alterations, Repairs and Conservation section to assist those pricing alteration works.

Inflation is generally low with some prices of materials unchanged or below last year's prices, other imported materials are subject to fluctuations in exchange rates and care must be taken when pricing these.. Products that are no longer in great demand also are subject to higher prices, for example clay drainage pipes and alternative materials may be considered to offset these increases in costs

Many products come and go each year and Laxton's endeavours to ensure that new products, changes in name and specification are included in the pricing guide, users are welcome to make suggestions for items that they consider should be included.

The Price Book is structured generally in accordance with the RICS New Rules of Measurement (NRM2: Detailed measurement for building works - first edition). The detailed rates can easily be applied to items measured in accordance with the Standard Method of Measurement of Building Works (SMM7).

NRM 2 replaces the *Standard Method of Measurement for Building Works* ('SMM'), The RICS advises that NRM2 became operative from 1st January 2013

The user is urged to read the relevant sections of NRM2 which requires certain measurement rules, definition rules and coverage rules to be observed. Certain work items are deemed to be included in the measured items (and prices) without mention and care should be taken to ascertain precisely the extent of these labours by reference to the Coverage Rules in NRM2

The pricing data herein is prepared from "first principles" with the results detailed indicating labour allowances and the cost of materials required for each item of work.

The information detailed will be of interest and may be used by Quantity Surveyors, Contractors, Sub-contractors, Self Builders, Architects, Developers, Students, Government and Public Authorities and those interested in the cost and price of Building Works.

Users should understand the limitations of any price book and should familiarise themselves with these introductory sections including "How to Use", "Essential Information including Regional Variations" and the "Preliminaries Section".

The "Build-up" of the "All-in" labour rates has been calculated to include the basic Labour Rates and National Insurance contributions and is illustrated in the Preliminaries/General Conditions pages. All-in labour rates are shown at the head of each page in each of the Work sections. The National agreed wage award included in this edition is applicable from 30th June 2016. The plumbing and electrical wage rates included herein are current from January 2016

Material prices used are those current during the third quarter of 2016. The internet has made it possible to easily obtain very competitive prices and check that quotations received are competitive even where large quantities are required so it is still worthwhile searching the internet for pricing information. The prices used have been obtained from a mixture of online prices, quotations from suppliers and suppliers' price lists with an allowance for delivery costs included where applicable.

The analysis of materials as included herein has shown that an item may consist of a number of materials with different waste factors. The waste % indicated within the book is for the predominant material.

The Composite Prices section has the prices linked to those included in the detailed priced sections and is useful for those who wish to produce a quick estimate as a guide perhaps for budget purposes, although care must be taken to include items not described.

The Basic Prices of Materials sections are linked to the built up prices in the delivered to site column of the priced sections.

The Approximate Estimating section has been costed to enable complete buildings to be priced following the format of the Standard Form of Cost Analysis as published by BCIS. The section gives examples of a detailed analysis for a detached house and an office block together with alternative composite prices enabling quick and comprehensive estimates to be produced and amended as detail design continues. Users should note that alternative forms of construction and compliance with the latest Building Regulations and the requirements to achieve reductions in carbon emissions will need to be incorporated in any cost estimate. This section has been tailored to the needs of the Contractor and Quantity Surveyor for approximate estimating data in detail but in a simpler format than can be obtained from other sources. Descriptions in the approximate estimating section have been abbreviated and care must be taken to understand the extent of works included. Allowances should be added for any changes from the norm

Regional Variations in construction costs occur and allowances should be made for the location where the works are to be carried out. Not all materials incur regional variations, such as ironmongery, with some resources being subject to greater regional variation than others although differences in delivery costs will need to be considered. Further information is included in the Essential Information chapter. The prices in this book are based on national average prices of 1.00.

The Landfill Tax (Qualifying Material) Order 1996 and 2011 gives a lower rate for inactive (or inert) waste for materials listed and a standard rate for other waste.
Rates are:-

	Standard Rate	Lower Rate
01.04.08 to 31.03.09 –	£32.00 per tonne	2.50
01.04.09 to 31.03.10 –	£40.00 per tonne	2.50
01.04.10 to 31.03.11 –	£48.00 per tonne	2.50
01.04.11 to 31.03.12 –	£56.00 per tonne	2.50
01.04.12 to 31.03.13 -	£64.00 per tonne	2.50
01.04.13 to 31.03.14 -	£72.00 per tonne	2.50
01.04.14 to 31.03.15 -	£80.00 per tonne	2.50
01.04.15 to 31.03.16	£82.60 per tonne	2.60
From 01.04.16-	£84.40 per tonne	2.65

Users of Laxton's should note that the Landfill Tax has been included within Disposal items with alternatives for Inert, Active and Contaminated materials. In this edition the standard rate of tax included is £84.40 per tonne and lower rate of £2.65.

The Construction (Design and Management) Regulations 2015 must be observed and it is a criminal offence not to observe the regulations, and places specific requirements upon Clients, Principal Designers, Designers, Principal Contractors, Contractors and Workers

The Aggregates Levy is charged at a single flat rate of £2.00p per tonne, less on smaller amounts The levy applies to any sand, gravel or crushed rock commercially exploited in the UK or its territorial waters or imported into the UK. The levy will not apply to coal, clay, slate, metals and metal ores, gemstones or semi precious stones, industrial minerals and recycled aggregates.

The London Congestion Charge and other local congestion charges are not included within the rates. Where applicable due allowances should be made to preliminaries and rates, these costs could be considerable.

The Company Information Section contains details of internet links to many company and useful websites and will enable the reader to find further information on materials, products and services and where to source, search or obtain further details.

INTRODUCTION

Every endeavour has been made to ensure the accuracy of the information printed in the book but the Publishers and Editors cannot accept liability for any loss occasioned by the use of the information given.

Further information and cost information with updated links is added to the website on an on-going basis and users may wish to check latest developments on a regular basis.

The contents list and comprehensive index should be used to assist in locating specific items and subjects. **The section headings** are as follows:

Introduction
How to use
Essential Information
Contents
Index
Regional Variations
Preliminaries/General Conditions

Measured Rates
    Demolitions
    Alterations, Repairs and Conservation
    Excavation and Filling
    Piling
    Underpinning
        Diaphragm Walls and Embedded Retaining Walls
    Crib Walls Gabions and Reinforced Earth
    In-Situ Concrete Works
    Precast/Composite Concrete
    Precast Concrete
    Masonry
    Structural Metalwork
    Carpentry
    Sheet Roof Coverings
    Tile and Slate Roof and Wall Coverings
    Waterproofing
    Proprietary Linings and Partitions
    Cladding and Covering
    General Joinery
    Windows, Screens and Lights
    Doors, Shutters and Hatches

Stairs, Walkways and Balustrades
Metalwork
Glazing
Floor, Wall, Ceiling and Roof Finishings
Decoration
Suspended Ceilings
Insulation, Fire Stopping and Fire Protection
Furniture, Fittings and Equipment
Drainage Above Ground
Drainage Below Ground
Site Works
Fencing
Soft Landscaping
Mechanical Services
Electrical Services
Transportation
Builders Work In Connection with Services

Basic Prices of Materials
Composite Prices for Approximate Estimating -
        Cost Plan for an office development
        Cost Plan for a detached house
        Composite Prices

GENERAL INFORMATION
Standard Rates of Wages
Builders' and Contractors Plant
Guide Prices to Building Types per Square Metre
Capital Allowances
Housing Association
National Working Rule Agreement
Daywork Charges
Fees
Construction (Design and Management) Regulations 2015
Construction Procurement
Sustainable Construction
Tables and Memoranda
Metric System

BRANDS AND TRADE NAMES
COMPANY INFORMATION
PRODUCTS AND SERVICES

The Editor wishes to thank the Professional bodies, Trade Organisations and firms who have kindly given help and provided information for this edition. For addresses, websites and details see company information section.

ACO Technologies plc
Akzo Nobel Decorative Coatings Ltd
Alumasc Interior Building Products
Ancon Building Products
Angle Ring Co. Ltd.
Ariel Plastics
Birtley Building Products Ltd.
Blucher UK Ltd.
Boddingtons Ltd.
BRC Ltd.
British Gypsum Ltd.
British Sisalkraft Ltd.
Building and Engineering Services
  Association (was HVCA)
Burlington Slate
Cannock Gates
Carter Concrete Ltd.
Crendon Timber Engineering
Celuform Ltd.
Cementation SKANSKA Ltd.
Civil Engineering Contractors Association.
Clay Pipe Development Association Ltd.
Crendon Timber Engineering Ltd.
Crittall Steel Windows Ltd.
Crowthorne Fencing
Dow Construction Products
Durable Ltd.
E H Smith (Builders Merchants) Ltd
Electrical Contractors' Association
Fastcall
Fixatrad Ltd.
Forterra
Fosroc Ltd.
Grace Construction Products Ltd.
Grass Concrete Ltd.

Halfen Ltd
Hanson Concrete Products
Heidelberg Cement Group
Hepworth Building Products Ltd.
H.S.S. Hire Shops
Ibstock Brick Ltd
Jackson H.S. and Son (Fencing) Ltd.
Jeld-wen UK Ltd
Joint Industry Board for the Electrical
  Contracting Industry
Keyline Builders Merchants Ltd
Keller Ground Engineering
Kee Klamp Ltd.
Kingston Craftsmen Structural Timber
Kirkstone
Klargester
London Concrete
Marley Eternit
Marshalls plc
Masterbill Micro Systems
Mastic Asphalt Council Ltd.
Milton Pipes Ltd.
National Domelight Company
National Federation of Terrazzo, Marble
and Mosaic Specialists
Painting & Decorating Association
PFC Corofil
Polypipe Terrain Ltd
Promat UK Limited
Ramsay & Sons (Forfar) Ltd.
R.C. Cutting & Co Ltd
Rentokil Initial plc
Rockwool Ltd.
Royal Institute of British Architects
Royal Institution of Chartered Surveyors

Ruberoid Building Products Ltd.
Saint-Gobain PAM UK
Sealmaster
SIG Roofing Supplies
Smyth Composites Limited
Solaglas Ltd.
Stoakes Systems Ltd.
Stowell Concrete Ltd.
Swish Building Products Ltd
Syston Rolling Shutters Ltd.
Tarkett
Tarmac Ltd.
Tata Steel
Thorn Lighting Ltd.
Timloc Building Products
Townscape Products Ltd.
Travis Perkins
Tremco Illbruck Ltd.
Velux Co.Ltd.
Wavin Ltd.
Welco
Zarges (UK) Ltd

V.B.JOHNSON LLP,
Chartered Quantity Surveyors,
St John's House,
304-310 St Albans Road,
Watford   Herts
WD24 6PW
Telephone : 01923 227236
Facsimile : 03332 407363

Web :   www.vbjohnson.com

HOW TO USE AND ESSENTIAL INFORMATION

Laxton's Pricing Book is divided into 4 main sections;
The sections are:

SECTIONS

1   INTRODUCTION AND PRELIMINARIES

Important information to be used in conjunction with the price book, including How to Use, Allowances to be made for Regional Factors, Tender Values and Preliminary items.

2   MEASURED RATES

Measured rates are given for all descriptions of building work generally in accordance with SMM7 Revised 1998. A complete breakdown is shown for all measured items, under 10 headings. This approach provides the reader with the price build-up at a glance enabling adjustments to be made easily. Net and gross rates are given for each item.

Details of Basic Prices of materials, Composite Prices for Approximate Estimating and examples of a cost plan for an Office Development and a Detached House are included.

Prices reflect national average costs and represent target costs that should be achievable on projects in excess of £50,000. Estimators and purchasers should prepare material schedules and obtain quotations as appropriate to check that prices are achieved or improved and adjustments made to the rates.

3   GENERAL INFORMATION

A useful reference section covering: Standard Rates of Wages, Builders' and Contractors' Plant, Guide Prices to Building Types, National Working Rule Agreement, Daywork Charges, Construction (Design and Management) Regulations 2015, Fees, Tables and Memoranda and Metric System.

4   BRANDS AND TRADE NAMES - COMPANY INFORMATION – PRODUCTS AND SERVICES

A unique list of Brands and Trade Names, with names and addresses given in the adjacent Company Information Section. The Product and Services section provides details of manufacturers and suppliers. These lists are useful for locating a particular branded item. Every company listed has been written to and amendments received entered within the database. The majority of the companies now have websites and links to these are on our website at **www.Laxton-s.co.uk**, and a useful route should you wish to make frequent visits to these company websites. Simply add our website to your list of favourites.
A Product and Services Works Sections index is provided following this section at the end of the book

RATES GENERALLY

Rates given are within the context of a complete contract of construction where all trades work is involved. When pricing sub-contract work careful consideration must be given to the special circumstances pertaining to the pricing of such work and the very different labour outputs that can be achieved by specialists.

The comprehensive index will take the reader straight to any particular measured item.

SCHEDULES OF RATES

In the event that Laxton's is to be used as a Schedule of Rates the employer should at tender stage clearly indicate whether rates are to be adjusted for regional factors, tender values and for overheads and profit.

It is suggested that the employer should state that works are to be valued in accordance with Laxton's Price Book 2016. Net Rate column with the addition of a percentage to be stated by the contractor to include for regional factor adjustment, tender value adjustment, overheads, profit and preliminary items not covered elsewhere. A procedure for adjustment of fluctuations should be stated (e.g. by the use of indices) and dates applicable.

Contractors must recognise that the net rates indicated within Laxton's works sections are national average costs. For large or small items of work, works in listed buildings, work of high quality or with difficult access, rates are likely to require adjustment . Minor items of maintenance works may require considerable adjustment.

PRICING AND FORMAT

Materials

The delivered to site price is given together with a percentage allowance for waste on site, where appropriate, the resultant materials cost being shown in the shaded column. The prices used are those included in the Basic Prices of Materials section. Note that where small quantities are required additional delivery charges may be incurred or allowances should be included in preliminaries for collection costs.

Labour

Hours for craft operative and labourer (General Building Operative) are given, these being applied to the labour rates shown in the top left hand shaded panel, the resultant total labour cost being shown in the shaded column.

Sundries

These are incidental costs not included in the main materials or labour cost columns..

Mechanical Plant

Items of Mechanical Plant not included in Preliminaries/General conditions have been allowed for in the measured rates.

Plant and Transport costs are indicated separately in the measured rates for items where these are a major factor and material costs are insignificant as in the case of some groundwork items.

Rates

Net rates shown in the shaded column are exclusive of overheads and profit.

Gross rates are inclusive of 10% Main Contractor's overheads and profit. Readers can, of course, adjust the rates for any other percentages that they may wish to apply.

HOW TO USE AND ESSENTIAL INFORMATION

PERCENTAGE ADJUSTMENTS FOR TENDER VALUES

Care must be taken when applying percentage additions and omissions to Laxton's prices and it should be noted that discounts can not be achieved on some items of work, whilst considerable discounts can be achieved on other items, however as an overall guide to pricing works of lower or larger values and for cost planning or budgetary purposes, the following adjustments may be applied to overall contract values

Contract value

£	15,000 to £	25,000 .... add 10.0%
£	25,000 to £	50,000 .... add 5.0%
£	50,000 to £	200,000 .... rates as shown
£	200,000 to £	500,000 .... deduct 5.0%
£	500,000 to £ 1,000,000 .... deduct 7.5%
£ 1,000,000 to £ 5,000,000 .... deduct 10.0%

MINOR WORKS, RENOVATION WORKS AND WORK IN EXISTING UNOCCUPIED PREMISES

The rates may be used for orders of lower value, renovation works and works in existing unoccupied premises however due allowances must be added for disproportionate waste factors, preparation works, difficulty of access and any double handling and additional preliminary items.

WORKING IN OCCUPIED PREMISES

Nearly all work of this nature is more costly in execution in that it has to be organised to suit the specific working conditions due to occupation and allowances should be made for the following:
    (a) Reduction in output arising there from.
    (b) Moving tradesmen on and off site as the progress of the contract demands.
    (c) Suppression of noise and dust.
As a guide, the extra labour involved in carrying out work in occupied premises could add between 50% and 100% to the labour cost of the same work in unoccupied premises.

REGIONAL VARIATIONS

The cost of a building is affected by its location. Many localised variables combine to produce a unique cost, including market factors such as demand and supply of labour and materials, workload, taxation and grants. The physical characteristics of a particular site, its size, accessibility and topography also contribute. Not even identical buildings built at the same time but in different localities will obtain identical tenders.

While all these factors are particular to a time and place, certain areas of the country tend to have different tender levels than others.

Laxton's Building Price Book is based upon national average prices.

The following Location Factors have been extracted from National Statistics for 'Regional factors for the public sector building (non-housing) tender price index' as at 2nd Quarter 2014.
    Great Britain –

North -	0.94
Wales -	0.98
Midlands -	0.94
East -	1.05
South West -	1.04
South East -	1.07
London –	1.09
Scotland -	0.94

The regions chosen are administrative areas and are not significant cost boundaries as far as the building industry is concerned. It should be stressed that even within counties or large conurbations great variations in tender levels are evident and that in many cases these will outweigh the effect of general regional factors
Further information on the effect of location factors may be derived from the BCIS Tender Price Index.

USE OF FACTORS AND ADJUSTMENTS

Regional factors and tender value adjustments should only be applied to adjust overall pricing or total contract values. They should not be applied to individual rates or trades. Users will need to allow for individual local fluctuations in prices by obtaining quotations from suppliers for specific items.

VALUE ADDED TAX

Prices throughout exclude Value Added Tax.
Note:
    VAT increased from 17.5% to 20% on 4th January 2011

CONTENTS

The greatest care has been taken to ensure accuracy but the publishers can accept no responsibility for errors or omission.

This page left blank intentionally

This page left blank intentionally

GENERALLY

The prices throughout this section offer in a convenient form the means of arriving at approximate rates for the various operations commonly met in connection with normal types of buildings. The basis on which the rates are estimated is given below and should be examined and understood before making any adjustments necessary to adapt the prices to a specific contract or to rates of wages other than those given.

The basic rates of wages used throughout are those which come into force on 25th July 2016.

Note: From 1st January 2017 Industry Holiday entitlement increases to 22 days and from Monday 26th June 2017 a pay increase of 2.75% is agreed.

BASIC RATES OF WAGES – BUILDING WORKERS

ALL-IN LABOUR RATES

Effective Dates of rates used:
Wages..July 25th, 2016
National Insurances ....April 6th, 2016
Holidays with Pay (Template) .... June 30th, 2004
(Note: The National Insurance concession ceased from October 2012)
EasyBuild Stakeholder Pension..................................June 30th, 2004
(Note: Pension arrangements changed from October 2011)
For earlier effective dates of the above see Standard Rates of Wages Section

BUILDING

Calculation of labour rates used throughout the book:-

		Craft Operative £		General Building Operative £
Guaranteed minimum weekly earnings		452.79		340.47
- 39 hours		11.61		8.73
2,023 hours (includes inclement weather allowance)	at £11.61	23,487.03	at £8.73	17,660.79
Productivity Payments		2,348.70		1,766.08
Non-Productive Overtime 131 hours	at £11.61	1,520.91	at £8.73	1,143.63
Sick Pay as WR.20 (three days unpaid, three days paid)		116.32		116.32
Holidays with Pay (Template) - 226 hours	at £11.61	2,623.86	at £8.73	1,972.98
		30,096.82		22,659.80
National Insurance (Earnings threshold 52 weeks at £156.01)....13.8%	of £21,984.30	3,033.83	of £14,547.28	2,007.52
Training 0.50% of Payroll (CITB Levy)	of £30,096.82	150.48	of £24,632.78	123.16
Retirement Benefit - 52 weeks	at £10.90	566.80	at £10.90	566.80
		33,847.94		23,357.28
Severance Pay 1.5% (Including loss of production during notice period absenteeism and turnover of labour)		507.72		380.36
		34,847.94		25,737.64
Employers Liability and Third Party Insurance	2.5%	858.89	2.5%	643.44
Trade Supervision	3%	1,030.67	3%	772.13
Cost per annum		36,245.22		27,153.21
Cost per hour - 1,973 working hours		£ 18.37		£ 13.76

Calculation of other skill rates as above would be as follows:-

Skilled Operative Rate:-	Skill Rate 1.	£17.49
	Skill Rate 2.	£16.83
	Skill Rate 3.	£15.73
	Skill Rate 4.	£14.85

Note: Public Holidays included with Holidays with Pay

Calculation of Hours Worked used above.

Summer: based on average 45 hours per week working

		Hours
	40 weeks	1800
Less Holidays:		
Summer	2 weeks	(90)
Other (Four days plus Easter Monday)		
	1 week	(45)
	37 weeks at 45 hours	1665

Less Bank Holidays (Good Friday, May, Spring and Summer Bank holidays)   4 days at 8 hours.

	(32)
	1633

Winter: based on average 43 hours per week working

		Hours
	12 weeks	516
Less Holidays: Winter (includes Christmas, Boxing and New Years days)	2 weeks	(86)
	10 weeks at 43 hours	430
Less Sick leave	5 days at 8 hours	(40)
		390
Add Summer Hours		1633
Total Hours		2023
(Less Inclement Weather time)		50
Total Actual Hours Worked		1973

PRELIMINARIES/GENERAL CONDITIONS

BASIC RATES OF WAGES (Cont'd)

Calculation of Non-Productive Overtime hours included above

Based on a 45 hour working week in Summer and a 43 hour working week in Winter, the calculation is as follows:-

				Hours
Summer	6 hours overtime per week at time and a half	=	3 hours x 37 weeks	111
Winter	4 hours overtime per week at time and a half	=	2 hours x 10 weeks	20
	Non-Productive hours per annum		....	131

Holidays with Pay as per W.R. 18

Bank Holidays	....	71 hours	(9 days)
Winter, Spring, and Summer Holidays	....	163 hours	(21 days)
		234 hours	(30 days)

Note From January 2017 the rule becomes:-

The holiday year runs from the 1st January for each year with an annual (52 weeks) entitlement of 22 days of Industry plus 8 days of Public/Bank holidays. The above calculations will require adjusting accordingly

PLUMBING

The rates of wages used in the Plumbing and Mechanical Engineering Installations Section are those approved by The Joint Industry Board for Plumbing and Mechanical Engineering Services in England and Wales.

The All-in Labour rate calculations are as follows:

Effective dates of rates used:

Wages..................................	January 4th 2016
National Insurance................................	April 6th 2016
Holiday and Welfare contributions..............	January 4th 2016

		Trained Plumber £		Advanced Plumber £		Technical Plumber £
46 weeks at 37.5 hours = 1725 hours ....at £12.61		21,752.25	at £14.41	24,857.25	at £16.00	27,600.00
Welding Supplement 1725 hours (Gas or Arc)		-	at £0.31	534.75	at £0.31	534.75
Travel time		- .		- .		- ..
		21,752.25		25,392.00		28,134.75
Allowance for Incentive Pay	15%	3,262.84	15%	3,808.80	15%	4,220.21
	(a)	25,015.09		29,200.80		32,354.96
Holiday and Welfare Benefit Credit 60 No...................	at £55.34	3,320.40	at £62.45	3,747.00	at £68.73	4,123.80
		28,335.49		32,947.80		36,478.76
National Insurance 13.8% of............................... 20,222.97		2,790.77	24,835.28	3,427.27	28,366.24	3,914.54
(Earnings Threshold 52 weeks at £156.01 = £8,112.52)						
less rebate on earnings LEL to UEL 3.50% 20,222.97		-707.00	24,835.28	-869.23	28,366.24	-992.62
Pension Contribution ....of (a)... 7.50%		1,876.13	7.50%	2,190.06	7.50%	2,426.62
		32,294.58		37,695.89		41,827.11
Redundancy Payments	1.5%	484.42	1.50%	565.44	1.50%	627.41
Employer's Liability Insurance	2.5%	807.36	2.50%	942.40	2.50%	1,045.68
Fares say 4720 miles @ 43p...		2,029.60		2,029.60		2,029.60
9Cost per annum		35,615.97		41,233.33		45,529.79
Cost per hour – 1725 hours		20.65		23.90		26.39
Inclement Weather time	1%	0.21	1%	0.24	1%	0.26
		£ 20.85		£ 24.14		£ 26.66

The Plumber's all-in wage rate used throughout the book is an average of 1 Trained, 3 Advanced and 1 Technical Plumbers rates giving an average rate of **£ 23.99** per hour.

Note : 1. Total Weekly Credit Value is the sum of Holiday Contribution and Welfare Credit.
2. Travel time – Discontinued for distances up to 20 miles
3. Tool Allowance – Discontinued

ELECTRICAL

The rates of wages used in the Electrical Engineering Installation Section are those approved by the Joint Industry Board for the Electrical Contracting Industry. www.jib.org.uk

The All-in Labour rate calculations using Job reporting operatives with own transport is as follows:

Effective dates of rates used:

| Wages.. | January 4th 2016 |
| National Insurance.................... | April 6th 2016 |

BASIC RATES OF WAGES (Cont'd)
ELECTRICAL (Cont'd)

Job reporting; own transport

	dayssweekshours	APPROVED + 50p		APPROVED		ELECT		TECH.	
		rate		rate		rate		rate	
HOURS WORKED (LESS 1 WK SICK)	45.0045.00	16.11	32,622.75	15.61	31,610.25	14.39	29,139.75	17.57	35,579.25
SICK PAY	2.00	120.83	241.65	117.08	234.15	107.93	215.85	131.78	263.55
NPOT	45.00 3.75	16.11	2,718.56	15.61	2,634.19	14.39	2,428.31	17.57	2,964.94
INCENTIVE PAYMENT 15.00%		15.00%	4,893.41	15.00%	4,741.54	15.00%	4,370.96	15.00%	5,336.89
TRAVEL TIME– up to 15 miles - Nil									
TRAVEL ALLOWANCE. – up to 15 miles - Nil									
STAT HOLS	8.00 7.50	16.11	966.60	15.61	936.60	14.39	863.40	17.57	1,054.20
sub total A			41,442.98		40,156.73		37,018.28		45,198.83
JIB STAMP	52.00	57.65	2,997.80	57.65	2,997.80	53.90	2,802.80	63.52	3,303.04
			44,440.78		43,154.53		39,821.08		48,501.87
NATIONAL INS		13.80%	6,132.83	13.80%	5,955.32	13.80%	5,495.31	13.80%	6,693.26
(Lower earnings allowance)	52	£156.01	-1,119.53	£156.01	-1,119.53	£156.01	-1,119.53	£156.01	-1,119.53
sub total B			49,454.07		47,990.32		44,196.86		54,075.59
TRAINING 2.50%of A		2.50%	1036.07	2.50%	1,003.92	2.50%	925.46	2.50%	1,129.97
TRAVEL ALLOWANCE– up to 15 miles - Nil. (TA TAX & NI EXEMPT ELEMENT)			0.00		0.00		0.00		0.00
sub total			50,490.15		48,994.24		45,122.31		55,205.57
SEVERANCE PAY 2.00%		2.00%	1,009.80	2.00%	979.88	2.00%	902.45	2.00%	1,104.11
sub total			51,499.95		49,974.12		46,024.76		56,309.68
EMP. LIAB & 3rd PARTY		2.50%	1,287.50	2.50%	1,249.35	2.50%	1,150.62	2.50%	1,407.74
COST PER YEAR			52,787.45		51,223.48		47,175.38		57,717.42

GANG RATE						INDIVIDUAL RATES
TECH.	57,717.42 x	1.00	NON PRODUCTIVE		57,717.42	£28.50
APPROVED ELEC. +	52,787.45 x	1.00	100% PRODUCTIVE		52,787.45	£26.07
APPROVED ELEC.	51,223.48 x	2.00	100% PRODUCTIVE		102,446.96	£25.30
ELEC.	47,175.38 x	4.00	100% PRODUCTIVE		188,701.51	£23.30
					390,580.41	
AVERAGE MAN HOURS	45.00	x 45.00	2025	hrs		
AVERAGE COST PER MAN		7.00 Men working		£28.34 Per hour		

Note:

Based on National JIB rates

a Basic week= 37.5hrs
b Hours to be worked before overtime paid= 37.5
c The above JIB hourly rates are from the 6/1/2016.
d Hours remain at 45 as this seems to be the minimum paid in the industry at the moment

The Electricians all in wage rate used throughout the book is an average 1 Approved Electrician +, 2 Approved Electricians and 4 Electricians rates giving an average rate of **£28.34** per hour.

JIB Categories for Operatives Hourly Pay

All-in wage costs

1.   Shop Reporting
     - payable to an operative who is required to start and finish at the shop:
        -   Technician £25.28
        -   Approved Electrician Plus £23.10
        -   Approved Electrician £22.13
        -   Electrician £20.14
     - Average Cost per Man £24.74 (as above with Technician Non Productive)

PRELIMINARIES/GENERAL CONDITIONS

BASIC RATES OF WAGES (Cont'd)
ELECTRICAL (Cont'd)
All-in wage costs (Cont'd)

- Electrician £21.76
- Average Cost per Man £26.61 (Technician Non Productive)

2. Job Reporting (transport provided but not included within these calculations)
    - payable to an operative who is required to start and finish at normal starting and finishing times on jobs, travelling in their own time in transport provided by the employer. The operative shall also be entitled to payment for Travel Time, where eligible as detailed in the appropriate scale:
        - Technician £26.92
        - Approved Electrician Plus £24.95
        - Approved Electrician £23.77

3. Job Reporting - (own transport) as used in the example above
    - payable to an operative who is required to start and finish at normal starting and finishing times on jobs, travelling in their own time by their own means; the operative shall be entitled to payment for travel time and travel allowance where eligible, as detailed in the appropriate scale.

For basic London rates within the M25 see the Wages section at the rear of this book.

The wages and All-in Labour rate effective from January 2nd 2017 will be:-

Job reporting; own transport	days	weeks	hours	APPROVED + 50p rate		APPROVED rate		ELECT rate		TECH. rate	
HOURS WORKED (LESS 1 WK SICK)		45.00	45.00	16.42	33,250.50	15.92	32,238.00	14.68	29,727.00	17.92	36,288.00
SICK PAY	2.00			123.15	246.30	119.40	238.80	110.10	220.20	134.40	268.60
NPOT	45.00	3.75		16.42	2,770.88	15.92	2,686.50	14.68	2,477.25	17.92	3,024.00
INCENTIVE PAYMENT	15.00%			15.00%	4,987.58	15.00%	4,835.70	15.00%	4,459.05	15.00%	5,443.20
TRAVEL TIME– up to 15 miles - Nil											
TRAVEL ALLOWANCE. – up to 15 miles - Nil											
STAT HOLS	8.00		7.50	16.42	985.20	15.92	955.20	14.68	880.80	17.92	1,075.20
sub total A					42,240.45		40,954.20		37,764.30		46,099.20
JIB STAMP	52.00			57.65	2,997.80	57.65	2,997.80	53.90	2,802.80	63.52	3,303.04
					45,238.25		43,952.00		40,567.10		49,402.24
NATIONAL INS				13.80%	6,242.88	13.80%	6,065.38	13.80%	5,598.26	13.80%	6,817.51
(Lower earnings allowance)		52		£156.01	-1,119.53	£156.01	-1,119.53	£156.01	-1,119.53	£156.01	-1,119.53
sub total B					50,361.60		48,897.85		45,045.83		55,100.22
TRAINING	2.50%of A			2.50%	1056.01	2.50%	1,023.86	2.50%	944.11	2.50%	1,152.48
TRAVEL ALLOWANCE– up to 15 miles - Nil. (TA TAX & NI EXEMPT ELEMENT)					0.00		0.00		0.00		0.00
sub total					51,417.61		49,921.70		45,989.94		56,252.70
SEVERANCE PAY	2.00%			2.00%	1,028.35	2.00%	998.43	2.00%	919.80	2.00%	1,125.05
sub total					52,445.96		50,920.14		46,909.74		57,377.76
EMP. LIAB & 3rd PARTY				2.50%	1,311.15	2.50%	1,273.00	2.50%	1,172.74	2.50%	1,434.44
COST PER YEAR					53,757.11		52,193.14		48,082.48		58,812.20

GANG RATE						INDIVIDUAL RATES
TECH.	58,812.20	x	1.00	NON PRODUCTIVE	58,812.20	£29.04
APPROVED ELEC. +	53,757.11	x	1.00	100% PRODUCTIVE	53,757.11	£26.55
APPROVED ELEC.	52,193.14	x	2.00	100% PRODUCTIVE	104,386.28	£25.77
ELEC.	48,082.48	x	4.00	100% PRODUCTIVE	192,329.93	£23.74
					409,285.52	
AVERAGE MAN HOURS	45.00	x	45.00	2025 hrs		
AVERAGE COST PER MAN			7.00 Men working	£28.87 Per hour		

SITE PRELIMINARIES

GENERAL NOTES

The calculation of the costs to be included for Site Preliminaries requires careful consideration of the specific conditions relating to the particular project. No pre-conceived percentage addition can be adopted with any degree of accuracy. The cost of Site Preliminaries is dependent on the type of contract under consideration, its value, its planned programme time, its location and also on the anticipated requirements for the administration and management of the actual site. Alteration works may require consideration of special supervision, whether water and power are available locally, additional allowances for transport and time required to obtain and dispose of small quantities, any requirements relating to working in occupied premises, are the measured rates adequate for the project under consideration.

Two factors which must be established prior to commencement of the detailed calculations are (i) the value of the main contractor's work (from which is derived the value of the contractor's labour force) and (ii) the programme time for completion of the Works. If the latter is not stated in the contract conditions, it must be decided by the contractor. If it is stated in the contract conditions, it must be checked and confirmed by the contractor as being reasonable.

All figures quoted are exclusive of VAT and due allowances should be made and identified on all quotations as appropriate

NOTES ON SITE PRELIMINARIES

Section 1.1 – Employers requirements

Section 1.1.1 - Site accommodation

Site accommodation for the employer and the employer's representatives where separate from main contractor's site accommodation such as may be required for the Architect, Clerk of Works or Resident Engineer.

Attendance on site

Provision of chainmen and any special attendance required by the Clerk of Works or employer's representatives.

Section 1.1.2 - Site records

Operation and maintenance manuals and compilation of health and safety file

Section 1.1.2 - Completion and post-completion requirements

Handover requirements, to include training of building user's staff, provision of spare parts tools and portable indicating instruments, operation and maintenance of building engineering services installations, mechanical plant and equipment and the like and landscape management services

Section 1.2 - Main Contractor's cost items

Section 1.2.1 - Management and Staff

Project-specific management and staff

The numbers, types and grades of Site Staff to be included are matters for judgement by the Contractor.

A selection from the following list will provide the range of staff required to administer a normal building project:

Project manager/director	Supervisor
Construction manager/Site Agent	General Foreman
Engineer	Checker
Storeman	Cashier/Wages Clerk
Materials Purchasing Clerk	Plant Clerk
Quantity Surveyor	Resources Manager
Productivity Controller	Health and Safety Officer
Plant Fitter	Administrative Manager
Planner	Typist/Telephonist

Allowance for Trades Supervision has been made in the Gross Wage build up.

On small contracts, individual staff members may combine the duties of several of the foregoing grades. On larger contracts more than one member may be required for certain duties. For example, two or more Engineers may be required in the initial stages for setting out the Works and for checking the frame construction.

Part 4 of CDM 2015 contains the duties to control specific worksite health and safety risks and is equivalent to the duties contained in the old Construction (Health, Safety and Welfare) Regulations 1996. The regulations apply to the health, safety and welfare of workers on construction sites - the associated costs of which are not specifically identified in this section and acquaintance with these regulations is necessary.

Visiting management and staff

As noted above staff to be included is a matter of judgement.

Extraordinary support costs

Legal advice, recruitment, team building, associated transport and other extraordinary support costs.

Special Conditions of Contract

These can refer to any particular point required by the Employer, and a careful perusal of the Conditions and the Preambles to the Bills of Quantities or Specification is necessary to ensure that no requirement is overlooked. Typical items are: the suppression of noisy mechanical equipment; regulations dealing with the restricted use of tower cranes; or limitations to ground disturbance during piling operations

Travelling Time and Fares

Payments for travelling time and fares can be a major expense and since all contracts, and particularly the JCT Conditions of Contract, place on the Contractor the onus of including in the tender sum the cost of importing labour, investigation should be made into the prevailing conditions regarding the availability of local labour. When all enquiries have been made, the calculations of anticipated expenditure must be based on the Contractor's judgement of the labour position. Rates of payment for travelling time and fares are laid down in the relevant national and local Working Rule Agreements.

Defects

Allowance for the cost of labour and materials in rectifying minor defects prior to handover of the Works and during the Defects Liability Period.

Section 1.2.2 – Site establishment

Site accommodation

The size of the office required to accommodate the contractor's site staff will depend on the number of staff envisaged. An allowance of 10m2 per staff member plus 25% for circulation space and toilets is normal.

The overall cost will vary according to the contract period if the hutting is to be hired. Otherwise a use and waste allowance can be made if the hutting is supplied from the contractors own resources. The cost should include for erection, maintenance, adaptations/alterations during the works and eventual dismantling and, in the case of contracts extending over several years, for labour and materials in repairs and redecoration

In works of renovation and rehabilitation, office accommodation and space for storage of materials is often available in the existing premises. Otherwise a simple lock up hut is all that is normally required for small works contracts.

Accommodation priced in this Section could include offices, stores, welfare facilities, site latrines, canteen, drying rooms and sanitary facilities, first aid cabin, plant fitters sheds, reinforcement bending sheds.

Consideration should also be given to any off-site rented temporary accommodation, maintenance and associated costs.

Storage of Materials

Consideration should be given to the provision of the following items:

Cement storage shed.
Aggregate storage bins.
Use and waste of tarpaulins and plastic sheeting.
Lockups and toolboxes for tradesmens' tools.

Temporary works in connection with site establishment

Allowance should be made for foundations to site accommodation, temporary drainage, temporary services, intruder alarms, heating, lighting, furniture, equipment and any attendance required.

PRELIMINARIES

NOTES ON SITE PRELIMINARIES (Cont'd)
Section 1.2 - Main Contractor's cost items (Cont'd)
Section 1.2.2 – Site establishment (Cont'd)

Temporary Roads

This section includes providing access to the site for heavy excavating equipment and for site roads and hardstandings constructed in ash, hardcore or timber sleepers. Allowance should be made for drainage, maintenance over the contract period and for eventual breaking up and removal. Crossovers, planked footways and any automatic stop/go lights required should be included. If the base courses of permanent roads can be utilised as temporary roads, an allowance should be made for making good the base prior to final surfacing.

Allow also for any cleaning and maintenance which may be required on Public or Private roads, especially if surplus excavated material has to be carted off site.

Furniture and equipment

General office furniture, including canteen equipment, floor coverings, maintenance and removal

IT systems

Computer hardware, software and consumables. Line rental, website and internet services, support and maintenance.

Consumables and services

Office consumables including stationery, postage, tea, coffee, water bottles and the like

Brought-in services

Catering, maintenance, printing, off-site parking etc.

Sundries

Signboards and notices

General Attendance on Nominated Subcontractors

General Attendance on Nominated Subcontractors by the Main Contractor is deemed to include: the use of the Contractor's temporary roads, pavings and paths; standing scaffolding not required to be altered or retained; standing power-operated hoisting plant; the provision of temporary lighting and water supplies; clearing away rubbish; provision of space for the subcontractor's own offices; the storage of his plant and materials; and the use of messrooms, sanitary accommodation and welfare facilities provided by the Contractor for his own use.

The estimated cost of such attendance is generally added by the Contractor in the Bills of Quantities as a percentage of the Prime Cost Sum shown for each Nominated Subcontractor.

Alternatively it can be included as a lump sum according to the Contractor's experience and judgement. The actual percentage can vary from as little as 0.1%, up to 2.5% or more, depending on the anticipated or known requirements of the Nominated Subcontractor

Section 1.2.3 – Temporary services

Water Company charges

Charges made by various Water Companies for the provision of water and drainage differ considerably throughout the country and the actual rate to be paid should be ascertained regionally. Some water companies meter the water required for building purposes. Charges to be considered include:

Water infrastructure charges are £347.00 per connection.

Wastewater infrastructure charges are £347.00 per connection

Water connection charges

Boundary Box	£433.00
Temporary Building Supply	£852.00

Off site connections

Single/double connection charge up to 5 metres	£778.00
Single/double connection charge over 5 metres	£1,057.00
4 port manifold connection	£1,580.00
6 port manifold connection charge	£2,723.00

Additional charges for excavation – per linear metre

Thrust bore	£25.00

Part made roads and footpaths	£100.00
Highway/carriage	£130.00
Verge, unmade and open ground	£60.00
Use duct supplied by developer or lay additional pipes in same trench	£10.00

Additional charges for excavations longer than 2 metres – per linear metre

Part made roads and footpaths	£99.00
Highway/carriage	£128.00
Verge, unmade and open ground	£56.00

Design deposit

1 to 100 properties	£1,447.20

ON COSTS

To these costs must be added the cost of all temporary plumbing required on the site such as trenching and builder's work in connection, distribution piping, standpipes, hoses and tanks, alterations, repairs and removals.

Any reinstatement costs would also need to be added.

Volume Charges per cubic metre

Water Supply	£1.60p
Wastewater Services	£1.70p

Drainage charges

Fees for new sewer connection – vetting and administration:-

Connection type	Price per application
Direct Connection -- Public sewer via lateral drain	£501.60
Indirect connection to public sewer	£213.60
Adoption of a lateral drain	£328.80
Highway drainage connection by highway Authority	£501.80 per street
Building near to public sewer	£475.20 per agreement

Note.
New Build Standards and Mandatory Adoption

Section 42 will introduce mandatory adoption and new build standards for new sewers and lateral drains.

In the near future it will not be possible to apply for a sewer connection unless there is an adoption agreement in place for all of the new sewers and lateral drains on the site

Temporary gas supply

Include for connection, distribution, charges and bottled gas

Temporary electricity supply

Lighting

This item provides for any lighting required inside the buildings during construction or for external floodlighting installed during the winter months to maintain progress. To the cost of wiring, lights and other equipment, should be added the charge for electricity and electrician's time for the installation and removal of the system.

On small works contracts the contractor, by agreement with the client, may be permitted to use existing services. A nominal amount should be included for the use of electricity and any temporary lighting required.

Temporary telecommunication systems

This item provides for line connections and rental, equipment, maintenance and support

length of time it is thought necessary to keep pumps working. A typical build up of cost is as follows:

NOTES ON SITE PRELIMINARIES (Cont'd)
Section 1.2 - Main Contractor's cost items (Cont'd)
Section 1.2.3 – Temporary services (Cont'd)
Temporary drainage

Pumping of ground water - Provision for the pumping of ground water (as opposed to water from normal rainfall) is entirely dependent upon the conditions applying to the particular site in question. Under the JCT Standard Form of Contract this is usually covered by a Provisional or Prime Cost sum, but in contracts where the onus is placed on the contractor it is most important that a visit should be paid during the tendering period to the site of the proposed works, to ascertain as far as possible the nature of the ground and the likelihood of encountering water. When, as a result of such an inspection, it is decided that pumping will be necessary, say during excavation of foundations and until the foundation walls have reached a certain height, the calculations will be based upon the

Assuming one 75mm pump will be kept running continuously for five weeks and for three weeks a second 75mm pump will be required during working hours with a standby pump of 50mm diameter available for emergencies:

(a)	Hire of 75mm pump & hose 5 weeks at £104.00	520.00
	Operator's time attending, fuelling, etc. –	
	28 hours at £14.48 per hour = £405.44 x 5 weeks	2,027.20
	Hire of extra hose 5 weeks at £20.00	100.00
(b)	Hire of 75mm pump & hose 3 weeks at £104.00	312.00
	Operator's time attending, fuelling, etc. –	
	12 hours at £14.48 per hour = £173.76 x 3 weeks	521.28
	Hire of extra hose 3 weeks at £20.00	60.00
(c)	Standby pump 5 weeks at £63.00	315.00
	Total cost of pumping	£3,855.48

Section 1.2.4 - Security

Security of Site - Allow for the provision of a watchman if required, or for periodic visits by a security organisation; electronic surveillance, and protection of scaffolds, security equipment, hoardings, fences and gates. Some allowance may also be required in particular localities for loss of materials due to theft.

Section 1.2.5 – Safety and environmental protection

Safety programme

Personal protective equipment (PPE), including for employer and consultants

Protective helmets, boots, goggles and protective clothing for operatives; wet weather clothing, drying sheds; provision of site latrines and items of a similar nature.

The requirement to provide welfare facilities on construction sites is defined in the Construction (Design and Management) Regulations 2015 and applies to the health, safety and welfare of workers on construction sites - the associated costs are not specifically identified in this section and acquaintance with these regulations is necessary.

Barriers and safety scaffolding

Guard rails, barriers, debris netting etc (see also section 1.2.8 scaffolding)

Environmental protection measures

Control of pollution and noise, management and reporting

Other safety measures

Provision for winter working.

This can range from heating aggregates in the stockpiles to the provision of polythene sheeting to protect the operatives and/or the Works during construction

Section 1.2.6 – Control and protection

Surveys, inspections and monitoring

Setting out

Setting out equipment including the use of theodolites, levels or other items and provision of pegs, paint etc.

Protection of Works

The protection of completed work during the contract period is usually priced in the relevant bill rates. Types of work requiring special protection include expensive stonework or precast concrete features, stair treads, floor finishes, wall linings, high class joinery and glazing to windows.

Glass breakage during unloading or after fixing.

Samples

Sample panels.

Environmental control of building

Dry out building

Allowance should be made in this Section for drying out the building by the use of blow heaters or, if permitted, by the provision of fuel to the heating system as installed.

Section 1.2.7 - Mechanical Plant

This section comprises the cost of all mechanical plant which has not been priced in the Bills of Quantities. It includes site transport, Tower Cranes, Mobile Cranes, Goods and Passenger hoists, pumping plant for dewatering, small plan and tools.

The apportioning of the cost of tower cranes and hoists to separate items of the Bills of Quantities is made difficult and unreliable by the fact that very often they deal with various materials in differing trades. As such, the correct time allocation to the individual trades is virtually impossible. Moreover, although not always working to capacity, they cannot be removed from site entirely and a certain proportion of their time is wasted in standing idle. It is therefore normal practice to view the contract as a whole, decide the overall period for which the particular tower crane or hoist will be required, and include the whole cost in the Preliminaries.

Site Transport

Include for all requisite site transport not priced in the rates, such as site lorries, dumpers, vans, forklift trucks, tractors and trailers. The cost of transporting mechanical plant, scaffolding and temporary accommodation, is also included in this section. The cost of double handling materials during unloading may also be considered for inclusion in this section.

Tower Cranes

There is a large variety of cranes available in the market and the most suitable crane for each particular contract must be selected. The transport of the crane from the plant yard to the site can be a costly item, and the installation of track, the erection, testing and final dismantling of the crane can, on occasion, constitute a greater expenditure than the hire charge for the crane.

Mobile Tower Crane - A typical build up of cost of mobile tower crane is as follows:

	£
Installation	
Prepare ground and lay crane track 5,000	
Supply of track materials...3,000	
Transport erect & dismantle....................................15,000	
	23,000
Running Costs per week	£
Crane hire & driver 2,200	
Running costs, electricity 150	
Banksman... 730	
	3,080
x 36 weeks	110,880
	£133,880

Static Tower Crane - A typical build up of cost of small static tower crane is as follows:

	£
Installation	
Transport, erect & dismantle......................................10,000	
Expendable base and foundation 12,600	
	22,600
Running Costs per week	£
Crane hire & driver... 2200	
Running costs, electricity 70	
Banksman... 730	
	3,000
x 15 weeks	45,000
	£67,600

NOTES ON SITE PRELIMINARIES (Cont'd)
Section 1.2 - Main Contractor's cost items (Cont'd)
Section 1.2.3 – Temporary services (Cont'd)
Section 1.2.7 - Mechanical Plant (Cont'd)

Mobile Cranes

A mobile crane may be required during the frame construction for lifting heavy items such as precast concrete floors or cladding units. The weekly hire rates will vary according to the lifting capacity of the crane required but a typical build up of cost is as follows:

	£
Crane hire & driver	1,800
Banksman	730
	2,530
x 4 weeks	10,120
Transporting to and from site	400
	£13,050

Hoists

Provision of hoist, base, erection, testing, operator, maintenance and removal

Access plant

Fork lifts, scissor lifts, platforms etc

Concrete Plant

Mixers, batching etc

Other plant

Small plant and tools

Small plant and tools comprise ladders, wheelbarrows, trestles, picks, shovels, spades, sledgehammers, cold chisels and similar items. The cost of these minor items is usually allowed for by an addition to the overall cost of labour of between 1% and 2%.

Section 1.2.8 – Temporary works

Access scaffolding

The cost of scaffolding varies not only with the superficial area of the structure but also with the length of time the scaffolding is required to stand and the nature of the building. When calculating the cost, the girth of the building should be multiplied by the height and the area priced at the appropriate rate as shown below, to which should be added the additional charge for standing time exceeding the basic 12 weeks as shown. Provision must also be made for any internal scaffolding required to lift and stair wells and also for fixing suspended ceilings. Debris netting along with boards and ladder access are included with the rates.

Erected

Scaffolding for brickwork		Putlog £	Independent £
Average 6m high	m2	8.20	12.80
Average 12m high	m2	9.40	13.30
Average 18m high	m2	9.60	13.90
Add for each additional week	m2	0.50	0.55

Hire Charges

Mobile towers for one week

Height	2.20m £	4.20m £	6.20m £	8.20m £
	81.20	117.30	153.40	189.50

Support Scaffolding

Hire Charges

		£
Scaffolding for suspended ceilings for one week	m2	7.20
Lift shaft tower to 20m high for one week	item	100.00

Hoardings, fans and fencing

The nature of these items is so dependent upon the site, type of contract, and the frontage of the building area on the public highway that it is difficult to offer any general method of calculating the anticipated cost. Consideration must be given to the means of stabilizing hoardings on sites where basements or other deep excavations come right up to the line of the hoarding, and gates must be sited to give easy access from the road. Gantries may require artificial lighting throughout their length in certain conditions, particularly in narrow streets in city centres. Any police requirements for offloading materials must be ascertained. Allowance must also be made for giving notices and paying any licence fees required by Local Authorities.

Hire Charges

		£
Mesh fence 2m high per week – minimum 13 weeks	m	1.50

Section 1.2.9 – Site records

Photography, works records, manuals, as built drawings, CDM and Health and Safety File

Section 1.2.10 – Completion and post-completion requirements

Testing, commissioning, handover and post completion services

Section 1.2.11 – Cleaning

Site tidy

This Section includes keeping the site tidy throughout the construction period, clearing out after subcontract trades and the final cleaning up on completion of the Works prior to handover.

Rubbish disposal

Waste management including removal of rubbish during construction and clearing out after subcontract trades.

Maintenance of roads, paths and pavings

Maintenance of public, private and temporary roads, paths and pavings

Building clean

Final builder's clean

Section 1.2.12 – Fees and charges

Rates and Taxes on Temporary Buildings. The charges made by the Local Authority for rates on site accommodation should be ascertained locally.

Section 1.2.13 – Site services

Temporary non specific works

Section 1.2.14 – Insurance, bonds, guarantees and warranties

Works Insurance

The Contract Bills of Quantities will normally state whether Fire Insurance is the responsibility of the Contractor or of the Employer. Rates of insurance vary according to the nature of the work but are usually in the range £0.10% to £0.30% of the total value to be insured.

The total insurable value should include for full reinstatement of any damage caused, all cost escalation, any demolition of damaged work required and an appropriate percentage addition for professional fees.

Note: allowance should be made for Terrorism insurance if required.

Bonds

Sureties or bonds. The cost of any surety or performance bond for due completion of the Works required by the Contract Documents. Rates for this vary depending on the financial status of the Contractor.

SPECIMEN SITE PRELIMINARIES

The examples of the build up of Site Preliminaries which follow are based on alternative building contract values of £100,000 and £1,500,000 (excluding Preliminaries) made up as follows:

	£100,000 project Minor Works £	£1,500,000 project Major Works £
Specialist Subcontractors	-	300,000
Named/Specialist Suppliers		100,000
Provisional Sums	20,000	100,000
Main Contractors Labour	25,000	300,000
Main Contractors Materials	25,000	300,000
Subcontractors	30,000	400,000
	£100,000	£1,500,000

The Contract Periods have been programmed as 18 weeks (Minor Works) and 40 weeks (Major Works).

Section 1.1 – Employers requirements

Attendance on site

	Minor Works	Major Works
Chainman attending Engineer	-	400
Carried to Summary		£ 400

Section 1.2 - Main Contractor's cost items

Section 1.2.1 - Management and Staff

	Minor Works			Major Works		
	Per Week £	No. of Weeks	£	Per Week £	No. of Weeks	£
Project Manager/Site Agent	-	-	-	1250	24	30,000
General Foreman	1000 x 40%	18	7,200	1000	20	20,000
Site Engineer	-	-	-	1050	4	4,200
Wages Clerk/Productivity Controller	-	-	-	550	20	11,000
Typist/Telephonist	-	-	-	450	20	9,000
			£			£

Travelling Time and Fares

		Minor Works		Major Works	
Main Contractor's Labour	300,000				
Labour in Preliminaries	40,000				
	£340,000				
- Average £500 per week = 680 man/weeks in contract.					
680 man/weeks x 50% of men local		-		-	
x 20% of men receive £15.00 per week		-		2,040	
x 20% of men receive £20.00 per week		-		2,720	
x 10% of men receive £25.00 per week		-		1,700	
Travelling time and fares:					
Allow 40 man weeks at average £20.00 per week	800	800		-	6,460

Defects

	Minor Works		Major Works	
Defects Liability period:				
Allow 2 tradesmen and 2 labourers for 1 week	-		2,500	
Materials	-		600	
Handover and defects liability:				
Labour	500		-	
Materials	150	650	-	3,100
Carried to Summary		£8,650		£83,760

Section 1.2.2 - Site Accommodation

	Minor Works		Major Works	
	£	£	£	£
Contractors Office: 50m² at £26.00 per m²		-		1,300
Office furniture and equipment		225		425
Clerk of Works Office: 15m² at £26.00 per m²		-		390
Stores: 30m2 at £16.00 per m²		-		480
Canteen/Drying Room: 50m² at £26.00 per m²		-		1,300
Canteen equipment		-		365
Cabins or site huts		-		225
Lockups and toolboxes		-		130
Office: Use of existing accommodation. Allow for making good on completion		150		-
Stores: Use existing. Allow for shelving and lockups for materials storage		225		-
Contractors signboards	-		250	
Labour attendance in offices	-		1,000	

Temporary Roads and Hardstandings

	Minor Works		Major Works	
Temporary Roads and Hardstandings:				
Hardcore road: 100m2 at £12.50 per m2	-		1,250	
Hardstandings: 100m2 at £12.50 per m2	-		1,250	
Temporary Roads - allow for access to site	200	200	-	2,500
Carried to Summary		£ 800		£8,365

SPECIMEN SITE PRELIMINARIES (Cont'd)

Section 1.2.3 – Temporary services

	Minor Works		Major Works	
	£	£	£	£
Water				
Site Water connection	50		850	
Water Company charges	220		2200	
Service piping: 100m at £6.00 per m	-		600	
Stand pipes: 3 no. at £40	-		120	
Hoses tanks and sundry materials	100	370	220	3,990
Power and Lighting				
Connection to mains	-		500	
Distribution unit, cables, floodlights, lamps, plugs and sundry materials	-		1,450	
Electrician in attendance	-		1,530	
Cost of electricity: 40 weeks at £24 per week	-		960	
Heating and lighting offices: 40 weeks at £22.00 per week	-		880	
Power and lighting: 18 weeks at £20.00 per week	360	360	-	5,320
Telephone and Administration				
Installation	-		220	
Rental	-		280	
Calls: 40 weeks at £25 per week	-		1,000	
Telephones: 18 weeks at £20.00 per week	360	360	-	1,500
Carried to Summary		£1,090		£10,810

Section 1.2.4 – Security

	Minor Works		Major Works	
	£	£	£	£
Weekend site supervision: 40 weeks at average £50 per week	-		2,000	
Security: allow for loss of materials	200	200	-	2,000
Carried to Summary		£ 200		£2,000

Section 1.2.5 - Safety and environmental protection

	Minor Works		Major Works	
	£	£	£	£
Site Latrines: 2 no. at £300.00	-		600	
Latrine materials: 40 weeks at £6.00 per week	-		240	
Latrine materials 18 weeks at £2.50 per week	45		-	
Sewer connection	-		300	
Drainage; installation and removal	-		600	
First Aid	35		200	
Protective clothing	120		750	
CDM	200		650	
Winter working	-	400	1,250	4,590
Carried to Summary		£ 400		£4,590

Section 1.2.6 – Control and protection

	Minor Works		Major Works	
	£	£	£	£
Samples				
Sample panels	-		300	
Testing concrete cubes	-		350	
Drying Out				
Heaters and Fuel	100	100	500	1,150
Carried to Summary		£ 100		£1,150

Section 1.2.7 – Mechanical plant -

	Minor Works		Major Works	
	£	£	£	£
Transport				
Site tractor and trailer	-		750	
Site van and driver	-		1,000	
Transport Plant to Site:				
Machine excavator	-		300	
Mechanical equipment including concrete mixers, hoists, and pumps	-		400	
Site offices and storage hutting	-		300	
Scaffolding	-		300	
Sundries	-		300	
Double handling materials on site	250		-	
Transport scaffolding and small plant to site	300	550	-	3,350

SPECIMEN SITE PRELIMINARIES (Cont'd)

Mechanical Plant

Hoists: Assuming a 300kg hoist is required for 12 weeks:

		per week	
Hire of hoist per week	-	200	
Fuel, oil and grease	-	40	
Operator (part-time)	-	160	
	-	400	
x 12 hoist weeks			4,800
Scaffold Hoist: Allow 4 weeks at £100 per week	400		-

Pumping and Dewatering. It is assumed that a Provisional Sum is included in the Bills of Quantities:

A typical build up is shown in the notes	-		-	
Pumps: Allow 50mm pump at £80 per week x 2 weeks	160	560	-	4,800

Small Plant and Tools

Labour value £300,000 x 1.5%	-		4,500	
Labour value £ 25,000 x 1.5%	375	375	-	4,500
Carried to Summary		£1,485		£12,650

Section 1.2.8 – Temporary works

	Minor Works		Major Works	
	£	£	£	£

Scaffolding

Assuming a three storey office building 60m x 10m x 10m high in traditional brick construction:

External putlog scaffolding				
1400m2 at £8.20 per m2	-		11,480	
Mobile tower 8 weeks at £175 each including erection and dismantling	-		1,400	
Internal scaffolding for suspended ceilings				
1800m2 at £7.20 per m2	-		12,960	
Mobile towers				
4 weeks at £180 each including erection and dismantling.	720	720	-	25,840

Hoardings, Fans, Fencing etc

Hoarding 2m high: 100m at £25 per metre	-		2,500	
Gate	-	-	200	2,700
Carried to summary		£720		£28,540

Note;- Where quotations are used for scaffolding, plant and the like allowances must be made for additional hire and changes to anticipated requirements; scaffolding is notorious for costing more than anticipated with actual cost being possibly more than double the original quotation used at tender stage.

Section 1.2.11 – Cleaning

	Minor Works		Major Works	
	£	£	£	£
Rubbish Disposal and Cleaning Site				
Site clearance during construction: 18 weeks at average £50 per week	900		-	
Site clearance during construction: 40 weeks at average £65 per week	-		2,600	
Final cleaning	220	1,120	450	3,050
Carried to summary		£1,120		£3,050

Section 1.2.12 – Fees and charges

	Minor Works	Major Works
	£	£
Rates and Taxes on temporary buildings	250	1,200
Carried to summary	250	1,200

SPECIMEN SITE PRELIMINARIES (Cont'd)

Section 1.2.14 – Insurance, bonds, guarantees and warranties

	Minor Works		Major Works	
	£	£	£	£
Insurance of works				
Contract Sum say	100,000		1,500,000	
Allow for part demolition and clearing up	1,600		25,000	
	101,600		1,525,000	
Increased costs over contract period: say 2%	2,032		-	
Increased costs over contract period: say 4%	-		61,000	
	103,632		1,586,000	
Increased costs over reconstruction period: say 2%	2,073		-	
Increased costs over reconstruction period: say 5%	-		79,300	
	105,705		1,665,300	
Professional fees: 15%	15,856		249,795	
	£121,561		£ 1,915,095	
Cost of Insurance: £0.35%		425	£0.25%	4,788
Carried to summary		£ 425		£4,788

SITE PRELIMINARIES - SUMMARY

NRM Section		Minor Works		Major Works	
		£	£	£	£
1.1.1	Employers requirements	-	-	400	400
1.2.1	Management and staff	8,650	8,650	83,760	83,760
1.2.2	Site accommodation	800	800	8,365	8,365
1.2.3	Temporary services				
	Water	370		3,990	
	Power and lighting	360		5,320	
	Telephone and administration	360	1,090	1,500	10,810
1.2.4	Security	200	200	2,000	2,000
1.2.5	Safety and environmental protection	400	400	4.590	4,590
1.2.6	Control and protection				
	Samples and drying out	100	100	1,150	1,150
1.2.7	Mechanical Plant				
	Transport	550		3,350	
	Mechanical Plant	560		4,800	
	Small Plant and Tools	375	1,485	4,500	12,650
1.2.8	Temporary works				
	Scaffolding	720		25,840	
	Hoardings	-	720	2,700	28,540
1.2.11	Cleaning				
	Rubbish Disposal and Cleaning Site	1,120	1,120	3,050	3,050
1.2.12	Fees and charges				
	Rates and taxes	250	250	1,200	1,200
1.2.14	Insurance, bonds, guarantees and warranties				
	Insurance of Works	425	425	4,788	4,788
			£15,240		£161,303
	Overheads and Profit: 12.5%		£1,905	7.5%	£12,097
	Total of Site Preliminaries		£ 17,145		£ 173,400

SPECIMEN SITE PRELIMINARIES (Cont'd)

FIRM PRICE ADDITION (Major Works)

The amount to be added for a firm price tender is dependent upon the overall contract period of the project, the known or anticipated dates of wage awards and the estimated percentage amount of the award. Separate calculations must be carried out for anticipated labour increased costs, materials increases, plant increases, staff salary awards and the increased costs required by direct subcontractors, if any. The total of these represents the addition to be made for a firm price tender.

A typical calculation is as follows:

		£	£
Labour:			
Assume a 4% wage award half way through the 40 week contract period			
The value of labour affected by the award is calculated as £150,000			
The firm price addition is 4% on £150,000	....		6,000
Materials:			
The value of materials affected by the award is estimated as £150,000			
Firm price addition for materials: allow average 2% on £150,000	....		3,000
Plant: Required mainly at the start of project: allow	....		500
Staff: Allow 4% increase on £25,000	....		1,200
Subcontractors:			
Net value affected by increased costs, say £150,000			
Allow labour 4% on £75,000	....	3,000	
Allow materials 2% on £75,000	....	1,500	4,500
	Net addition for firm price:		£15,200

FIRM PRICE ADDITION (Minor Works)

Firm Price: Most small works contracts are based on firm price tenders. This is due in the main to the relatively short contract periods normally involved. It is therefore the usual practice to make a lump sum allowance in the Preliminaries for any anticipated cost increases in labour and materials. (or maybe ignored altogether should the contractor consider adequate provision is included within the rates)

	£
Allowance for firm price...	200
Net addition for firm price:	£ 200

LANDFILL TAX

The Landfill Tax is included as indicated within the Disposal items of the Groundworks Section in this edition.

The two rates of Tax applicable from 1st April 2016 and as used in this edition are:-
        £2.65 per tonne for inactive waste.
        £84.40 per tonne for all other waste.

Additional allowances should also be added for contaminated waste and specific quotations obtained as required.

The cost of excavation will be affected by the nature of the soil and its subsequent bulking factor in handling and disposal.

A typical conversion rate is 2.10 tonnes per cubic metre. This could result in an addition of £5.56 per cubic metre for inactive waste and £177.24 per cubic metre for all other waste to basic disposal costs.

Allowances for disposal and Landfill tax must be added to all items of demolitions, alterations and rubbish disposal as appropriate.

Recycling of redundant building materials is now essential, not only on environmental grounds but also to achieve savings in costs.

AGGREGATES LEVY (Major and Minor Works)

The Aggregates Levy came into effect on 1st April 2002. It applies to any sand, gravel or crushed rock commercially exploited in the UK or its territorial waters, or imported into the UK.

The levy will not apply to coal, clay and slate, metals and metal ores, gemstones or semi-precious stones, industrial minerals and recycled aggregates.

The levy will be charged at a single flat rate of £2.00 per tonne. (£5m3 approx) from 1 April 2009. The effect of this levy on building materials has generally been included within the rates.

There is one rate per whole tonne. This is apportioned on amounts less than one tonne. So, for example, the levy due on half a tonne from 1 April 2009 is £1.00.

Date	£ per tonne
01.04.02 to 31.03.08 (inclusive)	£1.60
01.04.08 to 31.03.09 (inclusive)	£1.95
From 01.04.09	£2.00

Further details of the Aggregates Levy may be obtained from the www.hmrc.gov.uk website

This page left blank intentionally

Laxton's NRM Building Price Book

2017 edition

WORK SECTIONS

-

MEASURED RATES

The Measured rates in the following guide give
an indication of prices applied to works in the
range of £50,000 to £200,000 based on
National Average Prices

RATES GENERALLY

Rates given are within the context of a complete contract of construction where all trades work is involved.

PERCENTAGE ADJUSTMENTS FOR TENDER VALUES

As a guide to pricing works of lower or larger values and for cost planning or budgetary purposes, the following adjustments may be applied to overall contract values

Contract value

£15,000 to £25,000....................add 10.0%
£25,000 to £50,000.....................add 5.0%
£50,000 to £200,000............rates as shown
£200,000 to £500,000..............deduct 5.0%
£500,000 to £1,000,000...........deduct 7.5%
£1,000,000 to £5,000,000......deduct 10.0%

MINOR WORKS, RENOVATION WORKS AND WORK IN EXISTING UNOCCUPIED PREMISES

The rates may be used for orders of lower value, renovation works and works in existing unoccupied premises however due allowances must be added for disproportionate waste factors, preparation works, difficulty of access and any double handling and additional preliminary items.

REGIONAL VARIATIONS

The cost of a building is affected by its location; Laxton's Building Price Book is based upon national average prices.

The following Location Factors have been extracted from National Statistics for 'Regional factors for the public sector building (non-housing) tender price index' as at 2nd Quarter 2014.

Great Britain –

North -	0.94
Wales -	0.98
Midlands -	0.94
East -	1.05
South West -	1.04
South East -	1.07
London –	1.09
Scotland -	0.94

USE OF FACTORS AND ADJUSTMENTS

Regional factors and tender value adjustments should only be applied to adjust overall pricing or total contract values. They should not be applied to individual rates or trades. Users will need to allow for individual local fluctuations in prices by obtaining quotations from suppliers for specific items.

VALUE ADDED TAX

Prices throughout exclude Value Added Tax.

Labour hourly rates: (except Specialists) Craft Operatives 18.37 Labourer 13.76 Rates are national average prices. Refer to REGIONAL VARIATIONS for indicative levels of overall pricing in regions	MATERIALS			LABOUR				RATES		
	Del to Site	Waste	Material Cost	Craft Optve	Lab	Labour Cost	Sunds	Nett Rate		Gross rate (10%)
	£	%	£	Hrs	Hrs	£	£	£	Unit	£
DEMOLITIONS										
Pulling down outbuildings										
Demolishing individual structures										
timber outbuilding; 2.50 x 2.00 x 3.00m maximum high	-	-	-	-	11.00	151.36	89.25	240.61	nr	264.67
outbuilding with half brick walls, one brick piers and felt covered timber roof; 2.50 x 2.00 x 3.00m maximum high	-	-	-	-	22.00	302.72	178.50	481.22	nr	529.34
greenhouse with half brick dwarf walls and timber framing to upper walls and roof; 3.00 x 2.00 x 3.00m maximum high	-	-	-	-	18.00	247.68	153.00	400.68	nr	440.75
Demolishing individual structures; setting aside materials for re-use										
metal framed greenhouse; 3.00 x 2.00 x 3.00m maximum high	-	-	-	4.00	7.00	169.80	-	169.80	nr	186.78
prefabricated concrete garage; 5.40 x 2.60 x 2.40m maximum high	-	-	-	6.00	11.00	261.58	-	261.58	nr	287.74
Demolishing individual structures; making good structures										
timber framed lean-to outbuilding and remove flashings and make good facing brick wall at ridge and vertical abutments										
3.00 x 1.90 x 2.40m maximum high	-	-	-	1.50	12.00	192.68	94.50	287.17	nr	315.89
Unit rates for pricing the above and similar work										
pull down building; 3.00 x 1.90 x 2.40m maximum high	-	-	-	-	10.00	137.60	89.25	226.85	nr	249.54
remove flashing	-	-	-	-	0.12	1.65	0.51	2.16	m	2.38
make good facing brick wall at ridge	-	-	-	0.20	0.20	6.43	0.47	6.89	m	7.58
make good facing brick wall at vertical abutment	-	-	-	0.17	0.17	5.46	0.47	5.93	m	6.52
make good rendered wall at ridge	-	-	-	0.20	0.20	6.43	1.08	7.51	m	8.26
make good rendered wall at vertical abutment	-	-	-	0.20	0.20	6.43	1.08	7.51	m	8.26
Demolishing individual structures; making good structures										
lean-to outbuilding with half brick walls and slate covered timber roof and remove flashings, hack off plaster to house wall and make good with rendering to match existing										
1.50 x 1.00 x 2.00m maximum high	-	-	-	4.00	12.50	245.48	60.06	305.54	nr	336.09
Unit rates for pricing the above and similar work										
pull down building with half brick walls; 1.50 x 1.00 x 2.00m maximum high	-	-	-	-	5.30	72.93	40.80	113.73	nr	125.10
pull down building with one brick walls; 1.50 x 1.00 x 2.00m maximum high	-	-	-	-	10.00	137.60	76.50	214.10	nr	235.51
hack off plaster or rendering and make good to match existing brick facings	-	-	-	1.00	2.00	45.89	1.53	47.42	m²	52.16
hack off plaster or rendering and make good to match existing rendering	-	-	-	1.30	2.30	55.53	6.16	61.69	m²	67.86
Removal of old work										
Demolishing parts of structures										
in-situ plain concrete bed										
75mm thick	-	-	-	-	0.90	12.38	3.83	16.21	m²	17.83
100mm thick	-	-	-	-	1.20	16.51	5.10	21.61	m²	23.77
150mm thick	-	-	-	-	1.80	24.77	7.65	32.42	m²	35.66
200mm thick	-	-	-	-	2.40	33.02	10.20	43.22	m²	47.55
in-situ reinforced concrete flat roof										
100mm thick	-	-	-	-	1.90	26.14	5.10	31.24	m²	34.37
150mm thick	-	-	-	-	2.85	39.22	7.65	46.87	m²	51.55
200mm thick	-	-	-	-	3.80	52.29	10.20	62.49	m²	68.74
225mm thick	-	-	-	-	4.28	58.89	11.48	70.37	m²	77.40
in-situ reinforced concrete upper floor										
100mm thick	-	-	-	-	1.90	26.14	5.10	31.24	m²	34.37
150mm thick	-	-	-	-	2.85	39.22	7.65	46.87	m²	51.55
200mm thick	-	-	-	-	3.80	52.29	10.20	62.49	m²	68.74
225mm thick	-	-	-	-	4.28	58.89	11.48	70.37	m²	77.40
in-situ reinforced concrete beam	-	-	-	-	19.00	261.44	51.00	312.44	m³	343.68
in-situ reinforced concrete column	-	-	-	-	18.00	247.68	51.00	298.68	m³	328.55
in-situ reinforced concrete wall										
100mm thick	-	-	-	-	1.80	24.77	5.10	29.87	m²	32.85
150mm thick	-	-	-	-	2.70	37.15	7.65	44.80	m²	49.28
200mm thick	-	-	-	-	3.60	49.54	10.20	59.74	m²	65.71
225mm thick	-	-	-	-	4.05	55.73	11.48	67.20	m²	73.92
in-situ reinforced concrete casing to beam	-	-	-	-	16.00	220.16	51.00	271.16	m³	298.28
in-situ reinforced concrete casing to column	-	-	-	-	15.25	209.84	51.00	260.84	m³	286.92
brick internal walls in lime mortar										
102mm thick	-	-	-	-	0.63	8.67	5.20	13.87	m²	15.26
215mm thick	-	-	-	-	1.30	17.89	10.97	28.85	m²	31.74
327mm thick	-	-	-	-	2.00	27.52	16.68	44.20	m²	48.62
brick internal walls in cement mortar										
102mm thick	-	-	-	-	0.95	13.07	5.20	18.28	m²	20.10
215mm thick	-	-	-	-	1.94	26.69	10.97	37.66	m²	41.43
327mm thick	-	-	-	-	2.98	41.00	16.68	57.68	m²	63.45
reinforced brick internal walls in cement lime mortar										
102mm thick	-	-	-	-	0.95	13.07	5.20	18.28	m²	20.10
215mm thick	-	-	-	-	1.94	26.69	10.97	37.66	m²	41.43
hollow clay block internal walls in cement-lime mortar										
50mm thick	-	-	-	-	0.30	4.13	2.55	6.68	m²	7.35
75mm thick	-	-	-	-	0.40	5.50	3.83	9.33	m²	10.26
100mm thick	-	-	-	-	0.52	7.16	5.10	12.26	m²	13.48

DEMOLITIONS

	MATERIALS			LABOUR				RATES		
Labour hourly rates: (except Specialists) Craft Operatives 18.37 Labourer 13.76. Rates are national average prices. Refer to REGIONAL VARIATIONS for indicative levels of overall pricing in regions	Del to Site £	Waste %	Material Cost £	Craft Optve Hrs	Lab Hrs	Labour Cost £	Sunds £	Nett Rate £	Unit	Gross rate (10%) £

DEMOLITIONS (Cont'd)

Removal of old work (Cont'd)

Demolishing parts of structures (Cont'd)

	Del to Site	Waste	Material Cost	Craft Optve	Lab	Labour Cost	Sunds	Nett Rate	Unit	Gross rate
hollow clay block internal walls in cement mortar										
50mm thick	-	-	-	-	0.32	4.40	2.55	6.95	m²	7.65
75mm thick	-	-	-	-	0.43	5.92	3.83	9.74	m²	10.72
100mm thick	-	-	-	-	0.57	7.84	5.10	12.94	m²	14.24
concrete block internal walls in cement lime mortar										
75mm thick	-	-	-	-	0.35	4.82	3.83	8.64	m²	9.51
100mm thick	-	-	-	-	0.53	7.29	5.10	12.39	m²	13.63
190mm thick	-	-	-	-	0.70	9.63	9.69	19.32	m²	21.25
215mm thick	-	-	-	-	1.41	19.40	10.97	30.37	m²	33.40
concrete block internal walls in cement mortar										
75mm thick	-	-	-	-	0.56	7.71	3.83	11.53	m²	12.68
100mm thick	-	-	-	-	0.73	10.04	5.10	15.14	m²	16.66
190mm thick	-	-	-	-	0.91	12.52	9.69	22.21	m²	24.43
215mm thick	-	-	-	-	1.67	22.98	10.97	33.94	m²	37.34
if internal walls plastered, add per side	-	-	-	-	0.12	1.65	1.02	2.67	m²	2.94
if internal walls rendered, add per side	-	-	-	-	0.18	2.48	1.02	3.50	m²	3.85
brick external walls in lime mortar										
102mm thick	-	-	-	-	0.48	6.60	5.20	11.81	m²	12.99
215mm thick	-	-	-	-	0.98	13.48	10.97	24.45	m²	26.89
327mm thick	-	-	-	-	1.51	20.78	16.68	37.46	m²	41.20
brick external walls in cement mortar										
102mm thick	-	-	-	-	0.63	8.67	5.20	13.87	m²	15.26
215mm thick	-	-	-	-	1.30	17.89	10.97	28.85	m²	31.74
327mm thick	-	-	-	-	2.00	27.52	16.68	44.20	m²	48.62
reinforced brick external walls in cement lime mortar										
102mm thick	-	-	-	-	0.63	8.67	5.20	13.87	m²	15.26
215mm thick	-	-	-	-	1.30	17.89	10.97	28.85	m²	31.74
if external walls plastered, add per side	-	-	-	-	0.09	1.24	1.02	2.26	m²	2.48
if external walls rendered or rough cast, add per side	-	-	-	-	0.12	1.65	1.02	2.67	m²	2.94
clean old bricks in lime mortar and stack for re-use, per thousand	-	-	-	-	15.00	206.40	-	206.40	nr	227.04
clean old bricks in cement mortar and stack for re-use, per thousand	-	-	-	-	25.00	344.00	-	344.00	nr	378.40
rough rubble walling, in lime mortar; 600mm thick	-	-	-	-	3.75	51.60	30.60	82.20	m²	90.42
random rubble walling, in lime mortar; 350mm thick	-	-	-	-	2.20	30.27	17.85	48.12	m²	52.93
random rubble walling, in lime mortar; 500mm thick	-	-	-	-	3.10	42.66	25.50	68.16	m²	74.97
dressed stone walling, in gauged mortar; 100mm thick	-	-	-	2.00	2.30	68.39	5.10	73.49	m²	80.84
dressed stone walling, in gauged mortar; 200mm thick	-	-	-	3.90	4.50	133.56	10.20	143.76	m²	158.14
stone copings; 300 x 50mm	-	-	-	0.20	0.26	7.25	0.76	8.02	m	8.82
stone copings; 375 x 100mm	-	-	-	0.40	0.52	14.50	1.92	16.42	m	18.07
stone staircases	-	-	-	-	15.00	206.40	51.00	257.40	m³	283.14
stone steps	-	-	-	-	14.75	202.96	51.00	253.96	m³	279.36
structural timbers										
50 x 75mm	-	-	-	-	0.13	1.79	0.19	1.98	m	2.18
50 x 100mm	-	-	-	-	0.15	2.06	0.25	2.32	m	2.55
50 x 150mm	-	-	-	-	0.18	2.48	0.39	2.87	m	3.15
50 x 225mm	-	-	-	-	0.24	3.30	0.57	3.87	m	4.26
75 x 100mm	-	-	-	-	0.18	2.48	0.39	2.87	m	3.15
75 x 150mm	-	-	-	-	0.20	2.75	0.57	3.32	m	3.65
75 x 225mm	-	-	-	-	0.31	4.27	0.86	5.12	m	5.63
steelwork										
steel beams, joists and lintels; not exceeding 10 kg/m	-	-	-	50.00	50.00	1606.50	51.00	1657.50	t	1823.25
steel beams, joists and lintels 10 - 20 kg/m	-	-	-	46.00	46.00	1477.98	51.00	1528.98	t	1681.88
steel beams, joists and lintels 20 - 50 kg/m	-	-	-	40.00	40.00	1285.20	51.00	1336.20	t	1469.82
steel columns and stanchions; not exceeding 10 kg/m	-	-	-	60.00	60.00	1927.80	51.00	1978.80	t	2176.68
steel columns and stanchions; 10 - 20 kg/m	-	-	-	56.00	56.00	1799.28	51.00	1850.28	t	2035.31
steel columns and stanchions; 20 - 50 kg/m	-	-	-	50.00	50.00	1606.50	51.00	1657.50	t	1823.25
steel purlins and rails; not exceeding 10 kg/m	-	-	-	50.00	50.00	1606.50	51.00	1657.50	t	1823.25
steel purlins and rails; 10 - 20 kg/m	-	-	-	50.00	50.00	1606.50	51.00	1657.50	t	1823.25

SHORING, FAÇADE RETENTION AND TEMPORARY WORKS

Temporary support of structures, roads and the like

	Del to Site	Waste	Material Cost	Craft Optve	Lab	Labour Cost	Sunds	Nett Rate	Unit	Gross rate
Provide and erect timber raking shore complete with sole piece, cleats, needles and 25mm brace boarding										
two 150 x 150mm rakers (total length 8.00m) and 50 x 175mm wall piece	128.87	10.00	141.76	4.00	4.00	128.52	29.28	299.56	nr	329.51
weekly cost of maintaining last	-	-	-	1.00	1.00	32.13	1.47	33.60	nr	36.96
three 225 x 225mm rakers (total length 21.00m) and 50 x 250mm wall piece	492.14	10.00	541.35	23.20	23.20	745.42	127.31	1414.08	nr	1555.48
weekly cost of maintaining last	-	-	-	1.50	1.50	48.19	6.28	54.48	nr	59.93
Provide and erect timber flying shore of 100 x 150mm main member, 50 x 175mm wall pieces and 50 x 100mm straining pieces and 75 x 100mm struts, distance between wall faces										
4.00m	132.12	10.00	145.33	15.00	15.00	481.95	6.28	633.57	nr	696.92
weekly cost of maintaining last	-	-	-	2.50	2.50	80.33	1.47	81.79	nr	89.97
5.00m	154.50	10.00	169.95	17.80	17.80	571.91	7.32	749.18	nr	824.10
weekly cost of maintaining last	-	-	-	3.00	3.00	96.39	1.68	98.07	nr	107.88
6.00m	170.38	10.00	187.42	5.10	20.60	377.14	8.11	572.68	nr	629.94
weekly cost of maintaining last	-	-	-	3.50	3.50	112.46	1.92	114.38	nr	125.81

ALTERATIONS, REPAIRS AND CONSERVATION

Section Contents
The main headings in this section are:-

Labour hourly rates: (except Specialists) Craft Operatives 18.37 Labourer 13.76 Rates are national average prices. Refer to REGIONAL VARIATIONS for indicative levels of overall pricing in regions	MATERIALS			LABOUR				RATES		
	Del to Site £	Waste %	Material Cost £	Craft Optve Hrs	Lab Hrs	Labour Cost £	Sunds £	Nett Rate £	Unit	Gross rate (10%) £
ALTERATION WORK TO EXISTING BUILDINGS										
The following items are usually specified as spot items where all work in all trades is included in one item; to assist in adapting these items for varying circumstances, unit rates have been given for the component operation where possible while these items may not be in accordance with NRM2 either by description or measurement, it is felt that the pricing information given will be of value to the reader										
Works of alteration; strutting for forming openings										
Strutting generally the cost of strutting in connection with forming openings is to be added to the cost of forming openings; the following are examples of costs for strutting for various sizes and locations of openings. Costs are based on three uses of timber										
Strutting for forming small openings in internal load bearing walls										
opening size 900 x 2000mm..............	10.55	5.00	11.07	2.00	0.33	41.28	0.99	53.34	nr	58.68
opening size 2000 x 2300mm..............	18.30	5.00	19.21	5.00	0.83	103.27	2.14	124.63	nr	137.09
Strutting for forming openings in external one brick walls										
opening size 900 x 2000mm..............	10.55	5.00	11.07	3.00	0.50	61.99	1.71	74.77	nr	82.25
opening size 1500 x 1200mm..............	12.48	5.00	13.10	3.00	0.50	61.99	1.71	76.80	nr	84.48
Strutting for forming openings in external 280mm brick cavity walls										
opening size 900 x 2000mm..............	12.45	5.00	13.07	4.00	0.66	82.56	1.92	97.55	nr	107.31
opening size 1500 x 1200mm..............	19.08	5.00	20.03	4.00	0.66	82.56	1.92	104.52	nr	114.97

	MATERIALS			LABOUR				RATES		
Labour hourly rates: (except Specialists) Craft Operatives 18.37 Labourer 13.76 Rates are national average prices. Refer to REGIONAL VARIATIONS for indicative levels of overall pricing in regions	Del to Site £	Waste %	Material Cost £	Craft Optve Hrs	Lab Hrs	Labour Cost £	Sunds £	Nett Rate £	Unit	Gross rate (10%) £
ALTERATION WORK TO EXISTING BUILDINGS (Cont'd)										
Works of alteration; strutting for forming openings (Cont'd)										
Strutting for forming large openings in internal load bearing walls on the ground floor of an average two storey building with timber floors and pitched roof; with load bearing surface 450mm below ground floor; including cutting holes and making good										
opening size 3700 x 2300mm.....................	134.01	5.00	140.71	20.00	20.00	642.60	22.74	806.05	nr	886.65
Unit rates for pricing the above and similar work										
plates, struts, braces or wedges in supports to floor and roof	2.11	5.00	2.21	0.28	0.28	9.00	0.33	11.54	m	12.70
dead shore and needle, sole plates, braces and wedges including cutting holes and making good.....................	67.61	5.00	70.99	10.00	10.00	321.30	11.32	403.62	nr	443.98
Strutting for forming large openings in external one brick walls on the ground floor of an average two storey building as described above										
opening size 6000 x 2500mm.....................	253.06	5.00	265.72	40.00	40.00	1285.20	45.10	1596.02	nr	1755.62
Unit rates for pricing the above and similar work										
strutting to window opening over new opening....................	12.08	5.00	12.68	2.00	2.00	64.26	2.23	79.18	nr	87.10
plates, struts, braces or wedges in supports to floor and roof	1.68	5.00	1.76	0.28	0.28	9.00	0.33	11.09	m	12.20
dead shore and needle, sole plates, braces and wedges including cutting holes and making good....................	92.79	5.00	97.43	15.70	15.70	504.44	17.77	619.64	nr	681.61
set of two raking shores with 50mm wall piece, wedges and dogs.	30.36	5.00	31.88	4.00	4.00	128.52	4.49	164.89	nr	181.38
Strutting for forming large openings in external 280mm brick cavity walls on the ground floor of an average two storey building as above described										
opening size 6000 x 2500mm.....................	259.30	5.00	272.27	43.90	43.90	1410.51	49.59	1732.36	nr	1905.60
Unit rates for pricing the above and similar work										
strutting to window opening over new opening....................	25.16	5.00	26.42	4.00	4.00	128.52	4.49	159.42	nr	175.37
plates, struts, braces and wedges in supports to floor and roof......	1.68	5.00	1.76	0.28	0.28	9.00	0.33	11.09	m	12.20
dead shore and needle, sole plates, braces and wedges including cutting holes and making good....................	100.67	5.00	105.70	16.70	16.70	536.57	18.94	661.22	nr	727.34
set of two raking shores with 50mm wall piece, wedges and dogs.	30.36	5.00	31.88	4.00	4.00	128.52	4.49	164.89	nr	181.38
Works of alteration; work to chimney stacks										
Seal and ventilate top of flue - remove chimney pot and flaunching and provide 50mm precast concrete, BS 8500, designed mix C20, 20mm aggregate, weathered and throated capping and bed in cement mortar (1:3) to seal top of stack; cut opening through brick side of stack and build in 225 x 225mm clay air brick to ventilate flue and make good facings										
stack size 600 x 600mm with one flue.....................	43.47	10.00	46.67	4.90	4.90	157.44	6.48	210.59	nr	231.65
stack size 825 x 600mm with two flues.....................	66.94	10.00	71.34	9.00	9.00	289.17	11.94	372.45	nr	409.70
stack size 1050 x 600mm with three flues.....................	90.41	10.00	96.01	13.10	13.10	420.90	16.36	533.28	nr	586.61
Renew defective chimney pot - remove chimney pot and provide new clay chimney pot set and flaunched in cement mortar (1:3)										
150mm diameter x 600mm high	41.29	-	41.29	2.50	2.50	80.33	5.86	127.48	nr	140.23
Rebuild defective chimney stack - pull down defective stack to below roof level for a height of 2.00m; prepare for raising and rebuild in common brickwork faced with picked stock facings in gauged mortar (1:2:9) pointed to match existing; parge and core flues; provide No. 4 lead flashings and soakers; provide 150mm diameter clay chimney pots 600mm high set and flaunched in cement mortar (1:3); make good roof tiling and all other work disturbed										
600 x 600mm with one flue......................	470.73	10.00	513.67	35.30	37.55	1165.15	88.50	1767.32	nr	1944.05
825 x 600mm with two flues	611.24	10.00	664.11	47.15	47.90	1525.25	123.00	2312.36	nr	2543.59
Remove defective chimney stack - pull down stack to below roof level; remove all flashings and soakers; piece in 50 x 100mm rafters; extend roof tiling with machine made plain tiles to 100mm gauge nailed every fourth course with galvanised nails to and with 25 x 19mm battens; bond new tiling to existing										
850 x 600mm for a height of 2.00m.....................	26.76	10.00	29.44	26.00	22.30	784.47	54.00	867.90	nr	954.69
Works of alteration; work to chimney breasts and fireplaces										
Take out fireplace and fill in opening - remove fire surround, fire back and hearth; hack up screed to hearth and extend 25mm tongued and grooved softwood floor boarding over hearth on and with bearers; fill in opening where fireplace removed with 75mm concrete blocks in gauged mortar (1:2:9) bonded to existing at jambs and wedged and pinned up to soffits; form 225 x 225mm opening and provide and build in 225 x 225mm fibrous plaster louvered air vent; plaster filling with 12mm two coat lightweight plaster on 12mm (average) dubbing plaster; extend 19 x 100mm softwood chamfered skirting over filling and join with existing; make good all new finishings to existing										
fireplace opening 570 x 685mm, hearth 1030 x 405mm..................	45.80	-	45.80	8.00	7.45	249.47	15.45	310.73	nr	341.80
fireplace opening 800 x 760mm, hearth 1260 x 405mm..................	56.00	-	56.00	9.65	8.95	300.42	18.00	374.43	nr	411.87

Labour hourly rates: (except Specialists)
Craft Operatives 18.37   Labourer 13.76
Rates are national average prices.
Refer to REGIONAL VARIATIONS for indicative levels of overall pricing in regions

	MATERIALS			LABOUR				RATES		
	Del to Site	Waste	Material Cost	Craft Optve	Lab	Labour Cost	Sunds	Nett Rate		Gross rate (10%)
	£	%	£	Hrs	Hrs	£	£	£	Unit	£

ALTERATION WORK TO EXISTING BUILDINGS (Cont'd)

Works of alteration; work to chimney breasts and fireplaces (Cont'd)

Remove chimney breast - pull down chimney breast for full height from ground to roof level (two storeys) including removing two fire surrounds and hearths complete; make out brickwork where flues removed and make out brickwork where breasts removed; extend 50 x 175mm floor joists and 25mm tongued and grooved softwood floor boarding (ground and first floors); extend 50 x 100mm ceiling joists (first floor); plaster walls with 12mm two coat lightweight plaster on 12mm (average) plaster dubbing; extend ceiling plaster with expanded metal lathing and 19mm three coat lightweight plaster; run 19 x 100mm softwood chamfered skirting to walls to join with existing; make good all new finishings up to existing chimney breast 2.00m wide and 7.50m high including gathering in roof space..........	188.87	-	188.87	62.00	84.00	2294.78	216.00	2699.65	nr	2969.61

Works of alteration; form dormer in roof

Form opening through existing roof and covering for and construct dormer window; remove covering, cut out rafters, insert 75 x 100mm rafters trimmers; construct dormer cheeks of 50 x 100mm framing covered with 19mm tongue and grooved boarding and three layer fibre based felt roofing; construct flat dormer roof of 50 x 100mm joists covered with 19mm plywood on firrings and three layer fibre based felt roofing dressed under existing rood covering; form No. 4 lead secret gutter at cheeks with tilting fillets and insert No. 4 lead apron at sill dressed over existing roof coverings; make good existing roof covering all round, insulate cheeks and roof with 100mm glass fibre insulation quilt and line internally with 12.5mm plasterboard with filled and scrimmed joints finished with 5mm setting coat of plaster and painted one thinned and two coats of emulsion paint; run 25 x 150mm fascia at edge of roof on three sides and finish edge of roof covering with and seal to anodised aluminium alloy edge trim; knot, prime, stop and paint two undercoats and one gloss finishing coat on fascia complete with pre-finished softwood casement window, double glazed with clear glass

Flat roofed dormer with softwood window in plain tile roof; size 631 x 1050mm; type 110 V..............	500.82	10.00	529.39	47.25	22.00	1170.70	25.95	1726.05	nr	1898.65
Dormer with softwood window in plain tile roof; size 1200 x 1050mm; type 210 CV..............	735.32	10.00	771.98	53.30	23.00	1295.60	28.95	2096.53	nr	2306.19
Dormer with softwood window in slated roof; size 631 x 1050mm; 110 V..............	563.28	10.00	599.83	60.25	23.50	1430.15	28.50	2058.49	nr	2264.33
Dormer with softwood window type in slated roof; size 1200 x 1050mm; type 210 CV..............	821.88	10.00	869.56	66.80	26.00	1584.88	31.20	2485.64	nr	2734.20

Works of alteration; form trap door in ceiling

Form new trap door in ceiling with 100mm deep joists and lath and plaster finish; cut away and trim ceiling joists around opening and insert new trimming and trimmer joists; provide 32x 127mm softwood lining with 15 x 25mm stop and 20 x 70mm architrave; provide and place in position 18mm blockboard trap door, lipped all round; make good existing ceiling plaster around openings

600 x 900mm..............	52.80	-	52.80	21.85	5.60	478.44	13.65	544.89	nr	599.38
Unit rates for pricing the above and similar work										
cut away plastered plasterboard ceiling..............	-	-	-	1.50	0.33	32.10	1.27	33.37	m²	36.71
cut away lath and plaster ceiling..............	-	-	-	2.00	0.41	42.38	1.53	43.91	m²	48.30
cut away and trim existing 100mm deep joists around opening										
664 x 964mm..............	-	-	-	0.95	0.16	19.65	1.07	20.72	nr	22.79
964 x 1444mm..............	-	-	-	1.90	0.33	39.44	2.14	41.59	nr	45.75
cut away and trim existing 150mm deep joists around opening										
664 x 964mm..............	-	-	-	1.43	0.24	29.57	1.63	31.21	nr	34.33
964 x 1444mm..............	-	-	-	2.86	0.48	59.14	3.21	62.35	nr	68.59
softwood trimming or trimmer joists										
75 x 100mm..............	2.22	10.00	2.44	0.95	0.16	19.65	0.54	22.64	m	24.90
75 x 150mm..............	3.13	10.00	3.44	1.00	0.17	20.71	0.58	24.74	m	27.21
softwood lining										
32 x 125mm..............	1.92	15.00	2.21	1.50	0.25	31.00	0.17	33.37	m	36.70
32 x 150mm..............	3.00	15.00	3.45	1.60	0.27	33.11	0.12	36.68	m	40.34
15 x 25mm softwood stop..............	0.36	10.00	0.40	0.33	0.05	6.75	0.08	7.22	m	7.94
19 x 50mm softwood twice rounded architrave to Patt. 18	0.38	10.00	0.42	0.33	0.05	6.75	0.08	7.24	m	7.97
18mm blockboard trap door lipped all round										
600 x 900mm..............	16.00	10.00	17.60	2.00	0.33	41.28	0.25	59.14	nr	65.05
900 x 1280mm..............	35.00	10.00	38.50	3.00	0.50	61.99	0.45	100.94	nr	111.03
75mm steel butt hinge to softwood..............	0.12	2.50	0.12	0.33	0.05	6.82	0.15	7.09	nr	7.80
make good plastered plasterboard ceiling around opening	-	-	-	0.75	0.75	24.10	1.85	25.94	m	28.54
make good lath and plaster ceiling around opening..............	-	-	-	0.83	0.83	26.67	2.27	28.93	m	31.83

Removing; fittings and fixtures

Removing fittings and fixtures										
cupboard with doors, frames, architraves, etc	-	-	-	-	0.59	8.12	1.56	9.68	m²	10.65
worktop with legs and bearers..............	-	-	-	-	0.56	7.71	2.09	9.79	m²	10.77
shelving exceeding 300mm wide with bearers and brackets..........	-	-	-	-	0.45	6.19	1.32	7.51	m²	8.26
shelving not exceeding 300mm wide with bearers and brackets....	-	-	-	-	0.20	2.75	0.42	3.17	m	3.49
draining board and bearers										
500 x 600mm..............	-	-	-	-	0.28	3.85	0.80	4.65	nr	5.11
600 x 900mm..............	-	-	-	-	0.37	5.09	1.41	6.50	nr	7.15
wall cupboard unit										
510 x 305 x 305mm	-	-	-	-	0.47	6.47	1.26	7.73	nr	8.50
510 x 305 x 610mm..............	-	-	-	-	0.65	8.94	2.44	11.39	nr	12.53

Labour hourly rates: (except Specialists) Craft Operatives 18.37 Labourer 13.76 Rates are national average prices. Refer to REGIONAL VARIATIONS for indicative levels of overall pricing in regions	MATERIALS			LABOUR				RATES		
	Del to Site	Waste	Material Cost	Craft Optve	Lab	Labour Cost	Sunds	Nett Rate	Unit	Gross rate (10%)
	£	%	£	Hrs	Hrs	£	£	£		£
ALTERATION WORK TO EXISTING BUILDINGS (Cont'd)										
Removing; fittings and fixtures (Cont'd)										
Removing fittings and fixtures (Cont'd)										
wall cupboard unit (Cont'd)										
1020 x 305 x 610mm	-	-	-	-	0.94	12.93	4.97	17.90	nr	19.69
1220 x 305 x 610mm	-	-	-	-	1.15	15.82	5.91	21.73	nr	23.91
floor cupboard unit										
510 x 510 x 915mm	-	-	-	-	0.92	12.66	6.21	18.87	nr	20.76
610 x 510 x 915mm	-	-	-	-	1.21	16.65	7.41	24.06	nr	26.47
1020 x 510 x 915mm	-	-	-	-	1.65	22.70	12.42	35.12	nr	38.64
1220 x 510 x 915mm	-	-	-	-	1.93	26.56	14.88	41.44	nr	45.58
sink base unit										
585 x 510 x 895mm	-	-	-	-	0.92	12.66	7.01	19.66	nr	21.63
1070 x 510 x 895mm	-	-	-	-	1.65	22.70	12.75	35.45	nr	39.00
tall cupboard unit										
510 x 510 x 1505mm	-	-	-	-	1.56	21.47	10.23	31.70	nr	34.87
610 x 510 x 1505mm	-	-	-	-	1.77	24.36	12.23	36.58	nr	40.24
bath panel and bearers...........................	-	-	-	-	0.47	6.47	1.32	7.79	nr	8.57
steel or concrete clothes line post with concrete base..............	-	-	-	-	0.50	6.88	2.61	9.49	nr	10.44

This section continues
on the next page

ALTERATION WORK TO EXISTING BUILDINGS (Cont'd)

Labour hourly rates: (except Specialists) Craft Operatives 23.99 Labourer 13.76 Rates are national average prices. Refer to REGIONAL VARIATIONS for indicative levels of overall pricing in regions	MATERIALS			LABOUR				RATES		
	Del to Site	Waste	Material Cost	Craft Optve	Lab	Labour Cost	Sunds	Nett Rate		Gross rate (10%)
	£	%	£	Hrs	Hrs	£	£	£	Unit	£
ALTERATION WORK TO EXISTING BUILDINGS (Cont'd)										
Removing; plumbing items or installations										
Removing plumbing items or installations										
cast iron rainwater gutters										
100mm	-	-	-	0.20	-	4.80	0.25	5.05	m	5.56
cast iron rainwater pipes										
75mm	-	-	-	0.25	-	6.00	0.28	6.28	m	6.91
cast iron rainwater pipes										
100mm	-	-	-	0.25	-	6.00	0.48	6.48	m	7.13
cast iron rainwater head	-	-	-	0.30	-	7.20	0.80	7.99	nr	8.79
cast iron soil and vent pipe with caulked joints										
50mm	-	-	-	0.35	-	8.40	0.15	8.55	m	9.40
75mm	-	-	-	0.35	-	8.40	0.28	8.68	m	9.55
100mm	-	-	-	0.35	-	8.40	0.48	8.88	m	9.76
150mm	-	-	-	0.50	-	11.99	1.14	13.14	m	14.45
asbestos cement soil and vent pipe										
100mm	-	-	-	0.25	-	6.00	0.48	6.48	m	7.13
150mm	-	-	-	0.25	-	6.00	1.14	7.14	m	7.85
p.v.c. soil and vent pipe										
100mm	-	-	-	0.25	-	6.00	0.48	6.48	m	7.13
150mm	-	-	-	0.25	-	6.00	1.14	7.14	m	7.85
lead soil and vent pipe										
100mm	-	-	-	0.35	-	8.40	0.48	8.88	m	9.76
150mm	-	-	-	0.35	-	8.40	1.14	9.54	m	10.49
sanitary fittings										
w.c. suites	-	-	-	1.55	-	37.18	4.89	42.07	nr	46.28
lavatory basins	-	-	-	1.35	-	32.39	2.94	35.33	nr	38.86
baths	-	-	-	1.55	-	37.18	9.80	46.98	nr	51.68
glazed ware sinks	-	-	-	1.35	-	32.39	4.89	37.28	nr	41.00
stainless steel sinks	-	-	-	1.35	-	32.39	2.94	35.33	nr	38.86
stainless steel sinks with single drainer	-	-	-	1.35	-	32.39	3.93	36.32	nr	39.95
stainless steel sinks with double drainer	-	-	-	1.35	-	32.39	4.89	37.28	nr	41.00
hot and cold water, waste pipes, etc. 8 - 25mm diameter										
copper	-	-	-	0.15	-	3.60	0.06	3.66	m	4.02
lead	-	-	-	0.15	-	3.60	0.06	3.66	m	4.02
polythene or p.v.c.	-	-	-	0.15	-	3.60	0.06	3.66	m	4.02
stainless steel	-	-	-	0.25	-	6.00	0.06	6.06	m	6.66
steel	-	-	-	0.25	-	6.00	0.06	6.06	m	6.66
hot and cold water, waste pipes, etc. 32 - 50mm diameter										
copper	-	-	-	0.20	-	4.80	0.15	4.95	m	5.44
lead	-	-	-	0.20	-	4.80	0.15	4.95	m	5.44
polythene or p.v.c.	-	-	-	0.20	-	4.80	0.15	4.95	m	5.44
stainless steel	-	-	-	0.30	-	7.20	0.15	7.35	m	8.08
steel	-	-	-	0.30	-	7.20	0.15	7.35	m	8.08
fittings										
cold water cisterns up to 454 litres capacity	-	-	-	2.00	-	47.98	22.23	70.21	nr	77.23
hot water tanks or cylinders up to 227 litre capacity	-	-	-	2.00	-	47.98	11.10	59.08	nr	64.99
solid fuel boilers up to 66000 BTU output	-	-	-	3.75	-	89.96	-	89.96	nr	98.96
gas or oil fired boilers up to 66000 BTU output	-	-	-	3.75	-	89.96	-	89.96	nr	98.96
wall mounted water heaters	-	-	-	2.00	-	47.98	-	47.98	nr	52.78
wall type radiators; up to 914mm long, up to 610mm high	-	-	-	1.00	-	23.99	0.99	24.98	nr	27.48
wall type radiators; up to 914mm long, 610 - 914mm high	-	-	-	1.35	-	32.39	1.97	34.35	nr	37.79
wall type radiators; 914 - 1829mm long, up to 610mm high	-	-	-	1.55	-	37.18	3.00	40.18	nr	44.20
wall type radiators; 914 - 1829mm long, 610 - 914mm high	-	-	-	1.95	-	46.78	3.93	50.71	nr	55.78
column type radiators; up to 914mm long, up to 610mm high	-	-	-	1.00	-	23.99	5.14	29.14	nr	32.05
column type radiators; up to 914mm long, 610 - 914mm high	-	-	-	1.35	-	32.39	10.27	42.66	nr	46.93
column type radiators; 914 - 1829mm long, up to 610mm high	-	-	-	1.55	-	37.18	15.42	52.60	nr	57.86
column type radiators; 914 - 1829mm long, 610 - 914mm high	-	-	-	1.95	-	46.78	20.55	67.33	nr	74.06
insulation from										
pipes; 8 - 25mm diameter	-	-	-	0.10	-	2.40	0.15	2.55	m	2.80
pipes; 32 - 50mm diameter	-	-	-	0.13	-	3.12	0.28	3.40	m	3.74
water tank or calorifier from tanks; up to 227 litres	-	-	-	0.50	-	11.99	11.10	23.10	nr	25.40
water tank or calorifier from tanks; 227 - 454 litres	-	-	-	0.75	-	17.99	22.23	40.22	nr	44.24
Removing plumbing and electrical installations; setting aside for re-use										
cast iron soil and vent pipe with caulked joints										
50mm	-	-	-	0.55	-	13.19	-	13.19	m	14.51
75mm	-	-	-	0.60	-	14.39	-	14.39	m	15.83
100mm	-	-	-	0.80	-	19.19	-	19.19	m	21.11
150mm	-	-	-	1.00	-	23.99	-	23.99	m	26.39

This section continues on the next page

Labour hourly rates: (except Specialists) Craft Operatives 28.34 Labourer 13.76 Rates are national average prices. Refer to REGIONAL VARIATIONS for indicative levels of overall pricing in regions	MATERIALS			LABOUR				RATES		
	Del to Site	Waste	Material Cost	Craft Optve	Lab	Labour Cost	Sunds	Nett Rate	Unit	Gross rate (10%)
	£	%	£	Hrs	Hrs	£	£	£		£
ALTERATION WORK TO EXISTING BUILDINGS (Cont'd)										
Removing; electrical items or installations										
Removing electrical items or installations										
wall mounted electric fire	-	-	-	2.00	-	56.68	2.61	59.29	nr	65.22
night storage heater; 1.5 kW	-	-	-	2.50	-	70.85	7.83	78.68	nr	86.55
night storage heater; 2 KW	-	-	-	2.50	-	70.85	10.44	81.29	nr	89.42
night storage heater; 3 KW	-	-	-	2.50	-	70.85	15.66	86.51	nr	95.16
Removing electrical items or installations; extending and making good finishings										
lighting points	-	-	-	1.25	-	35.42	2.13	37.56	nr	41.31
flush type switches	-	-	-	1.25	-	35.42	2.13	37.56	nr	41.31
surface mounted type switches	-	-	-	1.00	-	28.34	1.71	30.05	nr	33.06
flush type socket outlets	-	-	-	1.25	-	35.42	2.13	37.56	nr	41.31
surface mounted type socket outlets	-	-	-	1.00	-	28.34	1.71	30.05	nr	33.06
flush type fitting points	-	-	-	1.25	-	35.42	2.13	37.56	nr	41.31
surface mounted type fitting points	-	-	-	1.00	-	28.34	1.71	30.05	nr	33.06
surface mounted p.v.c. insulated and sheathed cables	-	-	-	0.04	-	1.13	0.09	1.22	m	1.35
surface mounted mineral insulated copper sheathed cables	-	-	-	0.10	-	2.83	0.18	3.01	m	3.32
conduits up to 25mm diameter with junction boxes	-	-	-	0.15	-	4.25	0.25	4.51	m	4.96
surface mounted cable trunking up to 100 x 100mm	-	-	-	0.55	-	15.59	0.93	16.52	m	18.17

This section continues
on the next page

Labour hourly rates: (except Specialists) Craft Operatives 18.37 Labourer 13.76 Rates are national average prices. Refer to REGIONAL VARIATIONS for indicative levels of overall pricing in regions	MATERIALS			LABOUR				RATES		
	Del to Site £	Waste %	Material Cost £	Craft Optve Hrs	Lab Hrs	Labour Cost £	Sunds £	Nett Rate £	Unit	Gross rate (10%) £
ALTERATION WORK TO EXISTING BUILDINGS (Cont'd)										
Removing; finishes										
Removing finishings										
cement and sand to floors	-	-	-	-	0.86	11.83	2.61	14.44	m²	15.89
granolithic to floors	-	-	-	-	0.92	12.66	2.61	15.27	m²	16.80
granolithic to treads and risers	-	-	-	-	1.43	19.68	2.61	22.29	m²	24.52
plastic or similar tiles to floors	-	-	-	-	0.20	2.75	0.27	3.02	m²	3.32
plastic or similar tiles to floors; cleaning off for new	-	-	-	-	0.47	6.47	0.27	6.74	m²	7.41
plastic or similar tiles to floors and screed under	-	-	-	-	1.05	14.45	2.88	17.33	m²	19.06
linoleum and underlay to floors	-	-	-	-	0.07	0.96	2.61	3.57	m²	3.93
ceramic or quarry tiles to floors	-	-	-	-	1.21	16.65	1.56	18.21	m²	20.03
ceramic or quarry tiles to floors and screed under	-	-	-	-	2.39	32.89	4.17	37.06	m²	40.76
extra; cleaning and keying surface of concrete under	-	-	-	-	0.53	7.29	0.06	7.35	m²	8.09
asphalt to floors on loose underlay	-	-	-	-	0.35	4.82	1.05	5.87	m²	6.45
asphalt to floors keyed to concrete or screed	-	-	-	-	0.76	10.46	1.05	11.51	m²	12.66
granolithic skirtings	-	-	-	-	0.33	4.54	0.27	4.81	m	5.29
plastic or similar skirtings; cleaning off for new	-	-	-	-	0.13	1.79	0.06	1.85	m	2.03
timber skirtings	-	-	-	-	0.12	1.65	0.27	1.92	m	2.11
ceramic or quarry tile skirtings	-	-	-	-	0.43	5.92	0.30	6.22	m	6.84
plaster to walls	-	-	-	-	0.60	8.26	1.05	9.31	m²	10.24
rendering to walls	-	-	-	-	0.87	11.97	1.05	13.02	m²	14.32
rough cast to walls	-	-	-	-	0.90	12.38	1.32	13.70	m²	15.07
match boarding linings to walls with battens	-	-	-	-	0.38	5.23	1.56	6.79	m²	7.47
plywood or similar sheet linings to walls with battens	-	-	-	-	0.31	4.27	1.56	5.83	m²	6.41
insulating board linings to walls with battens	-	-	-	-	0.31	4.27	1.56	5.83	m²	6.41
plasterboard to walls	-	-	-	-	0.31	4.27	0.80	5.06	m²	5.57
plasterboard dry linings to walls	-	-	-	-	0.50	6.88	0.80	7.68	m²	8.44
plasterboard and skim to walls	-	-	-	-	0.60	8.26	1.05	9.31	m²	10.24
lath and plaster to walls	-	-	-	-	0.44	6.05	1.32	7.37	m²	8.11
metal lath and plaster to walls	-	-	-	-	0.33	4.54	1.32	5.86	m²	6.45
ceramic tiles to walls	-	-	-	-	0.75	10.32	0.80	11.11	m²	12.23
ceramic tiles to walls and backing under	-	-	-	-	1.11	15.27	1.32	16.59	m²	18.25
asphalt coverings to walls keyed to concrete or brickwork	-	-	-	-	0.66	9.08	1.05	10.13	m²	11.14
plaster to ceilings	-	-	-	-	0.74	10.18	1.05	11.23	m²	12.36
match boarding linings to ceilings with battens	-	-	-	-	0.54	7.43	1.56	8.99	m²	9.89
plywood or similar sheet linings to ceilings with battens	-	-	-	-	0.38	5.23	1.56	6.79	m²	7.47
insulating board linings to ceilings with battens	-	-	-	-	0.26	3.58	1.56	5.14	m²	5.65
plasterboard to ceilings	-	-	-	-	0.33	4.54	0.80	5.34	m²	5.87
plasterboard and skim to ceilings	-	-	-	-	0.40	5.50	1.05	6.55	m²	7.21
lath and plaster to ceilings	-	-	-	-	0.59	8.12	1.32	9.44	m²	10.38
metal lath and plaster to ceilings	-	-	-	-	0.47	6.47	1.32	7.79	m²	8.57
Extending and making good finishings										
plaster cornices; up to 100mm girth on face	0.63	10.00	0.69	0.78	0.78	25.06	0.10	25.86	m	28.44
plaster cornices; 100 - 200mm girth on face	1.03	10.00	1.13	0.88	0.88	28.27	0.27	29.67	m	32.64
plaster cornices; 200 - 300mm girth on face	1.81	10.00	1.99	0.97	0.97	31.17	0.80	33.95	m	37.35
plaster ceiling roses; up to 300mm diameter	0.31	10.00	0.34	0.41	0.41	13.17	0.21	13.73	nr	15.10
plaster ceiling roses; 300 - 450mm diameter	0.63	10.00	0.69	0.64	0.64	20.56	0.42	21.67	nr	23.84
plaster ceiling roses; 450 - 600mm diameter	1.23	10.00	1.35	0.90	0.90	28.92	0.82	31.09	nr	34.20
Carefully handling and disposing toxic or other special waste by approved method										
asbestos cement sheet linings to walls	-	-	-	2.00	2.00	64.26	4.17	68.43	m²	75.27
asbestos cement sheet linings to ceilings	-	-	-	3.00	3.00	96.39	4.17	100.56	m²	110.62
Removing; roof coverings										
Removing coverings										
felt to roofs	-	-	-	-	0.28	3.85	1.05	4.90	m²	5.39
felt skirtings to roofs	-	-	-	-	0.33	4.54	0.10	4.65	m	5.11
asphalt to roofs	-	-	-	-	0.35	4.82	1.32	6.14	m²	6.75
asphalt skirtings to roofs; on expanded metal reinforcement	-	-	-	-	0.20	2.75	0.21	2.96	m	3.26
asphalt skirtings to roofs; keyed to concrete, brickwork, etc.	-	-	-	-	0.30	4.13	0.21	4.34	m	4.77
asphalt coverings to roofs; on expanded metal reinforcement; per 100mm of width	-	-	-	-	0.13	1.79	0.17	1.95	m	2.15
asphalt coverings to roofs; keyed to concrete, brickwork, etc.; per 100mm of width	-	-	-	-	0.20	2.75	0.17	2.92	m	3.21
slate to roofs	-	-	-	-	0.34	4.68	1.05	5.73	m²	6.30
tiles to roofs	-	-	-	-	0.31	4.27	1.56	5.83	m²	6.41
extra; removing battens	-	-	-	-	0.28	3.85	1.05	4.90	m²	5.39
extra; removing counter battens	-	-	-	-	0.10	1.38	0.21	1.59	m²	1.74
extra; removing underfelt	-	-	-	-	0.20	2.75	0.42	3.17	m²	3.49
lead to roofs	-	-	-	-	0.48	6.60	1.05	7.65	m²	8.42
lead flashings to roofs; per 25mm of girth	-	-	-	-	0.05	0.69	0.06	0.75	m	0.82
zinc to roofs	-	-	-	-	0.34	4.68	1.05	5.73	m²	6.30
zinc flashings to roofs; per 25mm of girth	-	-	-	-	0.03	0.41	0.06	0.47	m	0.52
copper to roofs	-	-	-	-	0.34	4.68	1.05	5.73	m²	6.30
copper flashings to roofs; per 25mm of girth	-	-	-	-	0.03	0.41	0.06	0.47	m	0.52
corrugated metal sheeting to roofs	-	-	-	-	0.31	4.27	1.56	5.83	m²	6.41
corrugated translucent sheeting to roofs	-	-	-	-	0.31	4.27	1.56	5.83	m²	6.41
board roof decking	-	-	-	-	0.31	4.27	1.56	5.83	m²	6.41
woodwool roof decking	-	-	-	-	0.39	5.37	1.56	6.93	m²	7.62
Removing coverings; setting aside for re-use										
slate to roofs	-	-	-	-	0.60	8.26	-	8.26	m²	9.08
tiles to roofs	-	-	-	-	0.60	8.26	-	8.26	m²	9.08
clean and stack 405 x 205mm slates; per 100	-	-	-	-	1.50	20.64	-	20.64	nr	22.70
clean and stack 510 x 255mm slates; per 100	-	-	-	-	1.89	26.01	-	26.01	nr	28.61
clean and stack 610 x 305mm slates; per 100	-	-	-	-	2.24	30.82	-	30.82	nr	33.90
clean and stack concrete tiles; per 100	-	-	-	-	2.24	30.82	-	30.82	nr	33.90
clean and stack plain tiles; per 100	-	-	-	-	1.50	20.64	-	20.64	nr	22.70

ALTERATIONS, REPAIRS AND CONSERVATION

Labour hourly rates: (except Specialists) Craft Operatives 18.37 Labourer 13.76 Rates are national average prices. Refer to REGIONAL VARIATIONS for indicative levels of overall pricing in regions	MATERIALS			LABOUR				RATES		
	Del to Site	Waste	Material Cost	Craft Optve	Lab	Labour Cost	Sunds	Nett Rate	Unit	Gross rate (10%)
	£	%	£	Hrs	Hrs	£	£	£		£
ALTERATION WORK TO EXISTING BUILDINGS (Cont'd)										
Removing; roof coverings (Cont'd)										
Removing coverings; carefully handling and disposing toxic or other special waste by approved method										
asbestos cement sheeting to roofs	-	-	-	2.00	2.00	64.26	4.17	68.43	m²	75.27
asbestos cement roof decking	-	-	-	2.00	2.00	64.26	4.17	68.43	m²	75.27
Removing; wall cladding										
Removing cladding										
corrugated metal sheeting to walls	-	-	-	-	0.25	3.44	1.56	5.00	m²	5.50
corrugated translucent sheeting to walls	-	-	-	-	0.25	3.44	1.56	5.00	m²	5.50
Timber weather boarding	-	-	-	-	0.31	4.27	1.56	5.83	m²	6.41
Removing coverings; carefully handling and disposing toxic or other special waste by approved method										
asbestos cement sheeting to walls	-	-	-	1.75	1.75	56.23	4.17	60.40	m²	66.44
Removing; woodwork										
Removing										
stud partitions plastered both sides	-	-	-	-	0.66	9.08	3.93	13.01	m²	14.31
roof boarding	-	-	-	-	0.19	2.61	1.32	3.93	m²	4.33
roof boarding; prepare joists for new	-	-	-	-	0.45	6.19	1.32	7.51	m²	8.26
gutter boarding and the like	-	-	-	-	0.19	2.61	1.32	3.93	m²	4.33
weather boarding and battens	-	-	-	-	0.24	3.30	1.56	4.86	m²	5.35
tilting fillets, angle fillets and the like	-	-	-	-	0.09	1.24	0.10	1.34	m	1.48
fascia boards 150 mm wide	-	-	-	-	0.10	1.38	0.21	1.59	m	1.74
barge boards 200mm wide	-	-	-	-	0.10	1.38	0.27	1.65	m	1.81
soffit boards 300mm wide	-	-	-	-	0.10	1.38	0.42	1.80	m	1.98
floor boarding	-	-	-	-	0.33	4.54	1.32	5.86	m²	6.45
handrails and brackets	-	-	-	-	0.10	1.38	0.82	2.20	m	2.42
balustrades complete down to and with cappings to aprons or strings	-	-	-	0.58	0.10	12.03	2.09	14.12	m	15.53
newel posts; cut off flush with landing or string	-	-	-	0.48	0.08	9.92	0.52	10.44	nr	11.49
ends of treads projecting beyond face of cut outer string including scotia under; make good treads where balusters removed	-	-	-	0.90	0.15	18.60	1.56	20.16	m	22.17
dado or picture rails with grounds	-	-	-	-	0.10	1.38	0.21	1.59	m	1.74
architrave	-	-	-	-	0.06	0.83	0.21	1.04	m	1.14
window boards and bearers	-	-	-	-	0.12	1.65	0.42	2.07	m	2.28
Removing; setting aside for reuse										
handrails and brackets	-	-	-	0.25	0.04	5.14	-	5.14	m	5.66
Removing; windows and doors										
Removing										
metal windows with internal and external sills; in conjunction with demolition										
997 x 923mm	-	-	-	-	0.40	5.50	2.51	8.01	nr	8.81
1486 x 923mm	-	-	-	-	0.50	6.88	3.60	10.48	nr	11.53
1486 x 1513mm	-	-	-	-	0.60	8.26	5.91	14.17	nr	15.58
1994 x 1513mm	-	-	-	-	0.70	9.63	7.89	17.52	nr	19.27
metal windows with internal and external sills; preparatory to filling openings; cut out lugs										
997 x 923mm	-	-	-	-	1.30	17.89	2.51	20.39	nr	22.43
1486 x 923mm	-	-	-	-	1.60	22.02	3.60	25.62	nr	28.18
1486 x 1513mm	-	-	-	-	2.00	27.52	5.91	33.43	nr	36.77
1994 x 1513mm	-	-	-	-	2.30	31.65	7.89	39.54	nr	43.49
wood single or multi-light casements and frames with internal and external sills; in conjunction with demolition										
440 x 920mm	-	-	-	-	0.28	3.85	1.62	5.47	nr	6.02
1225 x 1070mm	-	-	-	-	0.29	3.99	5.11	9.11	nr	10.02
2395 x 1225mm	-	-	-	-	0.30	4.13	11.49	15.62	nr	17.18
wood single or multi-light casements and frames with internal and external sills; preparatory to filling openings; cut out fixing cramps										
440 x 920mm	-	-	-	-	1.25	17.20	1.62	18.82	nr	20.70
1225 x 1070mm	-	-	-	-	1.70	23.39	5.11	28.51	nr	31.36
2395 x 1225mm	-	-	-	-	2.52	34.68	11.49	46.17	nr	50.78
wood cased frames and sashes with internal and external sills complete with accessories and weights; in conjunction with demolition										
610 x 1225mm	-	-	-	-	0.61	8.39	3.93	12.32	nr	13.56
915 x 1525mm	-	-	-	-	0.66	9.08	7.30	16.39	nr	18.03
1370 x 1525mm	-	-	-	-	0.75	10.32	10.90	21.23	nr	23.35
wood cased frames and sashes with internal and external sills complete with accessories and weights; preparatory to filling openings; cut out fixing cramps										
610 x 1225mm	-	-	-	-	1.79	24.63	3.93	28.56	nr	31.42
915 x 1525mm	-	-	-	-	2.56	35.23	7.30	42.53	nr	46.78
1370 x 1525mm	-	-	-	-	4.11	56.55	10.90	67.46	nr	74.20
wood doors										
single internal doors	-	-	-	-	0.23	3.16	3.13	6.30	nr	6.93
single internal doors and frames; in conjunction with demolition	-	-	-	-	0.18	2.48	7.05	9.53	nr	10.48
single internal doors and frames; preparatory to filling openings; cut out fixing cramps	-	-	-	-	1.00	13.76	7.05	20.81	nr	22.89
double internal doors	-	-	-	-	0.47	6.47	6.27	12.74	nr	14.01
double internal doors and frames; in conjunction with demolition	-	-	-	-	0.19	2.61	10.71	13.32	nr	14.66
double internal doors and frames; preparatory to filling openings; cut out fixing cramps	-	-	-	-	1.20	16.51	10.71	27.22	nr	29.94
single external doors	-	-	-	-	0.25	3.44	5.22	8.66	nr	9.53
single external doors and frames; in conjunction with demolition	-	-	-	-	0.18	2.48	9.39	11.87	nr	13.05
single external doors and frames; preparatory to filling openings; cut out fixing cramps	-	-	-	-	1.00	13.76	9.39	23.15	nr	25.47

Labour hourly rates: (except Specialists) Craft Operatives 18.37 Labourer 13.76 Rates are national average prices. Refer to REGIONAL VARIATIONS for indicative levels of overall pricing in regions	MATERIALS			LABOUR				RATES		
	Del to Site	Waste	Material Cost	Craft Optve	Lab	Labour Cost	Sunds	Nett Rate		Gross rate (10%)
	£	%	£	Hrs	Hrs	£	£	£	Unit	£
ALTERATION WORK TO EXISTING BUILDINGS (Cont'd)										
Removing; windows and doors (Cont'd)										
Removing (Cont'd)										
wood doors (Cont'd)										
double external doors	-	-	-	-	0.47	6.47	6.27	12.74	nr	14.01
double external doors and frames; in conjunction with demolition	-	-	-	-	0.19	2.61	10.71	13.32	nr	14.66
double external doors and frames; preparatory to filling openings; cut out fixing cramps	-	-	-	-	1.20	16.51	10.71	27.22	nr	29.94
single door frames; in conjunction with demolition	-	-	-	-	0.17	2.34	3.93	6.27	nr	6.90
single door frames; preparatory to filling openings; cut out fixing cramps	-	-	-	-	0.77	10.60	3.93	14.53	nr	15.98
double door frames; in conjunction with demolition	-	-	-	-	0.18	2.48	4.44	6.92	nr	7.61
double door frames; preparatory to filling openings; cut out fixing cramps	-	-	-	-	0.82	11.28	4.44	15.72	nr	17.30
Removing; setting aside for reuse										
metal windows with internal and external sills; in conjunction with demolition										
997 x 923mm	-	-	-	0.70	0.50	19.74	-	19.74	nr	21.71
1486 x 923mm	-	-	-	0.85	0.65	24.56	-	24.56	nr	27.01
1486 x 1513mm	-	-	-	0.90	0.75	26.85	-	26.85	nr	29.54
1994 x 1513mm	-	-	-	1.00	0.85	30.07	-	30.07	nr	33.07
metal windows with internal and external sills; preparatory to filling openings; cut out lugs										
997 x 923mm	-	-	-	0.70	1.40	32.12	-	32.12	nr	35.34
1486 x 923mm	-	-	-	0.85	1.75	39.69	-	39.69	nr	43.66
1486 x 1513mm	-	-	-	0.90	2.15	46.12	-	46.12	nr	50.73
1994 x 1513mm	-	-	-	1.00	2.45	52.08	-	52.08	nr	57.29
wood single or multi-light casements and frames with internal and external sills; in conjunction with demolition										
440 x 920mm	-	-	-	0.50	0.36	14.14	-	14.14	nr	15.55
1225 x 1070mm	-	-	-	0.83	0.43	21.16	-	21.16	nr	23.28
2395 x 1225mm	-	-	-	1.00	0.46	24.70	-	24.70	nr	27.17
wood single or multi-light casements and frames with internal and external sills; preparatory to filling openings; cut out fixing cramps										
440 x 920mm	-	-	-	0.50	1.33	27.49	-	27.49	nr	30.23
1225 x 1070mm	-	-	-	0.83	1.84	40.57	-	40.57	nr	44.62
2395 x 1225mm	-	-	-	1.00	2.68	55.25	-	55.25	nr	60.77
wood cased frames and sashes with internal and external sills complete with accessories and weights; in conjunction with demolition										
610 x 1225mm	-	-	-	0.97	0.78	28.55	-	28.55	nr	31.41
915 x 1525mm	-	-	-	1.63	0.93	42.74	-	42.74	nr	47.01
1370 x 1525mm	-	-	-	1.94	1.07	50.36	-	50.36	nr	55.40
wood cased frames and sashes with internal and external sills complete with accessories and weights; preparatory to filling openings; cut out fixing cramps										
610 x 1225mm	-	-	-	0.97	1.96	44.79	-	44.79	nr	49.27
915 x 1525mm	-	-	-	1.63	2.83	68.88	-	68.88	nr	75.77
1370 x 1525mm	-	-	-	1.94	4.43	96.59	-	96.59	nr	106.25
note: the above rates assume that windows and sashes will be re-glazed when re-used										
wood doors										
single internal doors	-	-	-	0.50	0.31	13.45	-	13.45	nr	14.80
single internal doors and frames; in conjunction with demolition .	-	-	-	1.00	0.34	23.05	-	23.05	nr	25.35
single internal doors and frames; preparatory to filling openings; cut out fixing cramps	-	-	-	1.00	1.16	34.33	-	34.33	nr	37.76
double internal doors	-	-	-	1.00	0.63	27.04	-	27.04	nr	29.74
double internal doors and frames; in conjunction with demolition	-	-	-	1.47	0.43	32.92	-	32.92	nr	36.21
double internal doors and frames; preparatory to filling openings; cut out fixing cramps	-	-	-	1.47	1.45	46.96	-	46.96	nr	51.65
single external doors	-	-	-	0.50	0.33	13.73	-	13.73	nr	15.10
single external doors and frames; in conjunction with demolition	-	-	-	1.00	0.34	23.05	-	23.05	nr	25.35
single external doors and frames; preparatory to filling openings; cut out fixing cramps	-	-	-	1.00	1.16	34.33	-	34.33	nr	37.76
double external doors	-	-	-	1.00	0.63	27.04	-	27.04	nr	29.74
double external doors and frames; in conjunction with demolition	-	-	-	1.47	0.44	33.06	-	33.06	nr	36.36
double external doors and frames; preparatory to filling openings; cut out fixing cramps	-	-	-	1.47	1.45	46.96	-	46.96	nr	51.65
single door frames; in conjunction with demolition	-	-	-	0.50	0.25	12.62	-	12.62	nr	13.89
single door frames; preparatory to filling openings; cut out fixing cramps	-	-	-	0.50	0.85	20.88	-	20.88	nr	22.97
double door frames; in conjunction with demolition	-	-	-	0.50	0.26	12.76	-	12.76	nr	14.04
double door frames; preparatory to filling openings; cut out fixing cramps	-	-	-	0.50	0.90	21.57	-	21.57	nr	23.73
Removing; ironmongery										
Removing door ironmongery (piecing in doors and frames included elsewhere)										
butt hinges	-	-	-	-	0.06	0.83	0.01	0.84	nr	0.92
tee hinges up to 300mm	-	-	-	-	0.20	2.75	0.01	2.77	nr	3.04
floor spring hinges and top centres	-	-	-	-	0.92	12.66	0.30	12.96	nr	14.26
barrel bolts up to 200mm	-	-	-	-	0.13	1.79	0.01	1.80	nr	1.98
flush bolts up to 200mm	-	-	-	-	0.17	2.34	0.01	2.35	nr	2.59
indicating bolts	-	-	-	-	0.23	3.16	0.01	3.18	nr	3.50
double panic bolts	-	-	-	-	0.69	9.49	0.17	9.66	nr	10.63
single panic bolts	-	-	-	-	0.58	7.98	0.10	8.09	nr	8.89
Norfolk or Suffolk latches	-	-	-	-	0.23	3.16	0.01	3.18	nr	3.50
cylinder rim night latches	-	-	-	-	0.23	3.16	0.01	3.18	nr	3.50
rim locks or latches and furniture	-	-	-	-	0.23	3.16	0.01	3.18	nr	3.50

Labour hourly rates: (except Specialists) Craft Operatives 18.37 Labourer 13.76 Rates are national average prices. Refer to REGIONAL VARIATIONS for indicative levels of overall pricing in regions	MATERIALS			LABOUR				RATES		
	Del to Site	Waste	Material Cost	Craft Optve	Lab	Labour Cost	Sunds	Nett Rate	Unit	Gross rate (10%)
	£	%	£	Hrs	Hrs	£	£	£		£

ALTERATION WORK TO EXISTING BUILDINGS (Cont'd)

Removing; ironmongery (Cont'd)

Removing door ironmongery (piecing in doors and frames included elsewhere) (Cont'd)

	Del to Site	Waste	Material Cost	Craft Optve	Lab	Labour Cost	Sunds	Nett Rate	Unit	Gross rate
mortice dead locks	-	-	-	-	0.17	2.34	0.01	2.35	nr	2.59
mortice locks or latches and furniture	-	-	-	-	0.29	3.99	0.01	4.01	nr	4.41
Bales catches	-	-	-	-	0.12	1.65	0.01	1.67	nr	1.83
overhead door closers, surface fixed	-	-	-	-	0.29	3.99	0.17	4.16	nr	4.57
pull handles	-	-	-	-	0.12	1.65	0.01	1.67	nr	1.83
push plates	-	-	-	-	0.12	1.65	0.01	1.67	nr	1.83
kicking plates	-	-	-	-	0.17	2.34	0.06	2.40	nr	2.64
letter plates	-	-	-	-	0.12	1.65	0.01	1.67	nr	1.83
Removing window ironmongery										
sash centres	-	-	-	-	0.12	1.65	0.01	1.67	nr	1.83
sash fasteners	-	-	-	-	0.17	2.34	0.01	2.35	nr	2.59
sash lifts	-	-	-	-	0.07	0.96	0.01	0.98	nr	1.08
sash screws	-	-	-	-	0.17	2.34	0.01	2.35	nr	2.59
casement fasteners	-	-	-	-	0.12	1.65	0.01	1.67	nr	1.83
casement stays	-	-	-	-	0.12	1.65	0.01	1.67	nr	1.83
quadrant stays	-	-	-	-	0.12	1.65	0.01	1.67	nr	1.83
fanlight catches	-	-	-	-	0.13	1.79	0.01	1.80	nr	1.98
curtain tracks	-	-	-	-	0.15	2.06	0.10	2.17	m	2.39
Removing sundry ironmongery										
hat and coat hooks	-	-	-	-	0.07	0.96	0.01	0.98	nr	1.08
cabin hooks and eyes	-	-	-	-	0.12	1.65	0.01	1.67	nr	1.83
shelf brackets	-	-	-	-	0.07	0.96	0.06	1.02	nr	1.13
toilet roll holders	-	-	-	-	0.09	1.24	0.06	1.30	nr	1.43
towel rollers	-	-	-	-	0.12	1.65	0.10	1.76	nr	1.93

Removing; metalwork

	Del to Site	Waste	Material Cost	Craft Optve	Lab	Labour Cost	Sunds	Nett Rate	Unit	Gross rate
Removing										
balustrades 1067mm high	-	-	-	-	0.81	11.15	1.35	12.50	m	13.75
wire mesh screens with timber or metal beads screwed on										
generally	-	-	-	-	0.62	8.53	1.32	9.85	m²	10.84
225 x 225mm	-	-	-	-	0.20	2.75	0.10	2.86	nr	3.14
305 x 305mm	-	-	-	-	0.22	3.03	0.17	3.19	nr	3.51
guard bars of vertical bars at 100mm centres welded to horizontal fixing bars at 450mm centres, fixed with bolts	-	-	-	-	0.62	8.53	1.32	9.85	m²	10.84
guard bars of vertical bars at 100mm centres welded to horizontal fixing bars at 450mm centres, fixed with screws	-	-	-	-	0.39	5.37	1.32	6.69	m²	7.36
extra; piecing in softwood after removal of bolts	-	-	-	0.25	0.04	5.14	0.14	5.28	nr	5.81
small bracket	-	-	-	-	0.18	2.48	0.10	2.58	nr	2.84
Removing; making good finishings										
balustrades 1067mm high; making good mortices in treads	-	-	-	0.97	0.97	31.17	1.71	32.88	m	36.16
guard bars of vertical bars at 100mm centres welded to horizontal fixing bars at 450mm centres fixed in mortices; making good concrete or brickwork and plaster	-	-	-	0.60	1.42	30.56	1.78	32.35	m²	35.58
small bracket built or cast in; making good concrete or brickwork and plaster	-	-	-	0.25	0.25	8.03	0.70	8.74	nr	9.61

Removing; manholes

	Del to Site	Waste	Material Cost	Craft Optve	Lab	Labour Cost	Sunds	Nett Rate	Unit	Gross rate
Removing										
remove cover and frame; break up brick sides and concrete bottom; fill in void with hardcore										
overall size 914 x 1067mm and 600mm deep to invert	4.76	30.00	6.19	-	10.00	137.60	5.22	149.01	nr	163.91
overall size 914 x 1219mm and 900mm deep to invert	9.52	30.00	12.38	-	11.50	158.24	5.74	176.37	nr	194.00
manhole covers and frames; clean off brickwork	-	-	-	-	0.50	6.88	0.93	7.81	nr	8.59
fresh air inlets; clean out pipe socket	-	-	-	-	0.50	6.88	0.30	7.18	nr	7.90

Removing; fencing

	Del to Site	Waste	Material Cost	Craft Optve	Lab	Labour Cost	Sunds	Nett Rate	Unit	Gross rate
Removing										
chestnut pale fencing with posts										
610mm high	-	-	-	-	0.30	4.13	0.82	4.95	m	5.45
914mm high	-	-	-	-	0.34	4.68	1.26	5.94	m	6.53
1219mm high	-	-	-	-	0.38	5.23	1.68	6.91	m	7.60
close boarded fencing with posts										
1219mm high	-	-	-	-	0.68	9.36	1.92	11.28	m	12.40
1524mm high	-	-	-	-	0.81	11.15	2.40	13.55	m	14.90
1829mm high	-	-	-	-	0.91	12.52	2.88	15.40	m	16.94
chain link fencing with posts										
914mm high	-	-	-	-	0.29	3.99	0.93	4.92	m	5.41
1219mm high	-	-	-	-	0.34	4.68	3.87	8.55	m	9.40
1524mm high	-	-	-	-	0.38	5.23	1.56	6.79	m	7.47
1829mm high	-	-	-	-	0.42	5.78	1.92	7.70	m	8.47
timber gate and posts	-	-	-	-	0.98	13.48	3.93	17.41	nr	19.16

Cutting or forming openings/recesses/chases etc; openings through concrete

Form opening for door frame through 150mm reinforced concrete wall plastered both sides and with skirting both sides; take off skirtings, cut opening through wall, square up reveals, make good existing plaster up to new frame both sides and form fitted ends on existing skirtings up to new frame; extend cement and sand floor screed and vinyl floor covering through opening and make good up to existing

	Del to Site	Waste	Material Cost	Craft Optve	Lab	Labour Cost	Sunds	Nett Rate	Unit	Gross rate
838 x 2032mm	21.12	-	21.12	22.50	28.00	798.61	45.30	865.03	nr	951.53

Labour hourly rates: (except Specialists) Craft Operatives 18.37 Labourer 13.76 Rates are national average prices. Refer to REGIONAL VARIATIONS for indicative levels of overall pricing in regions	MATERIALS			LABOUR				RATES		
	Del to Site	Waste	Material Cost	Craft Optve	Lab	Labour Cost	Sunds	Nett Rate		Gross rate (10%)
	£	%	£	Hrs	Hrs	£	£	£	Unit	£
ALTERATION WORK TO EXISTING BUILDINGS (Cont'd)										
Cutting or forming openings/recesses/chases etc; openings through concrete (Cont'd)										
Unit rates for pricing the above and similar work										
cut opening through wall										
100mm thick	-	-	-	-	5.30	72.93	10.66	83.59	m²	91.95
150mm thick	-	-	-	-	7.90	108.70	15.94	124.65	m²	137.11
200mm thick	-	-	-	-	10.50	144.48	21.23	165.71	m²	182.28
make good fair face around opening	-	-	-	-	0.80	11.01	0.75	11.76	m	12.93
square up reveals to opening										
100mm wide	-	-	-	0.75	0.75	24.10	1.86	25.96	m	28.55
150mm wide	-	-	-	1.07	1.07	34.38	2.62	37.00	m	40.70
200mm wide	-	-	-	1.38	1.38	44.34	4.17	48.51	m	53.36
make good existing plaster up to new frame	0.58	10.00	0.64	0.55	0.55	17.67	-	18.31	m	20.14
13mm two coat hardwall plaster to reveal not exceeding										
300mm wide	1.16	10.00	1.28	0.64	0.64	20.56	-	21.84	m	24.02
take off old skirting	-	-	-	0.22	-	4.04	0.15	4.19	m	4.61
form fitted end on existing skirting up to new frame	-	-	-	0.38	-	6.98	-	6.98	nr	7.68
short length of old skirting up to 100mm long with mitre with existing one end	-	-	-	0.58	-	10.65	0.06	10.71	nr	11.79
short length of old skirting up to 200mm long with mitres with existing both ends	-	-	-	1.15	-	21.13	0.09	21.22	nr	23.34
38mm cement and sand (1:3) screeded bed in opening	5.18	10.00	5.70	1.54	1.54	49.48	-	55.18	m²	60.70
vinyl or similar floor covering to match existing, fixed with adhesive, not exceeding 300mm wide	7.59	10.00	8.35	0.15	0.15	4.82	0.28	13.45	m	14.80
iroko threshold; twice oiled; plugged and screwed										
15 x 100mm	4.11	10.00	4.52	0.90	-	16.53	0.08	21.13	m	23.25
15 x 150mm	6.17	10.00	6.79	1.20	-	22.04	0.09	28.92	m	31.81
15 x 200mm	8.23	10.00	9.05	1.55	-	28.47	0.14	37.66	m	41.42
Form opening for staircase through 150mm reinforced concrete suspended floor plastered on soffit and with screed and vinyl floor covering on top; take up floor covering and screed, cut opening through floor, square up edges of slab, make good ceiling plaster up to lining and make good screed and floor covering to lining										
900 x 2134mm	83.05	-	83.05	19.00	39.00	885.67	58.50	1027.22	nr	1129.94
Unit rates for pricing the above and similar work										
take up vinyl or similar floor covering	-	-	-	-	0.55	7.57	0.21	7.78	m²	8.56
hack up screed	-	-	-	-	1.50	20.64	1.93	22.58	m²	24.83
cut opening through slab; 150mm thick	-	-	-	-	7.90	108.70	15.94	124.65	m²	137.11
cut opening through slab; 200mm thick	-	-	-	-	10.50	144.48	21.23	165.71	m²	182.28
make good fair face around opening	-	-	-	-	0.80	11.01	0.70	11.71	m	12.88
square up edges of slab; 150mm thick	-	-	-	1.07	1.07	34.38	2.78	37.15	m	40.87
square up edges of slab; 200mm thick	-	-	-	1.38	1.38	44.34	3.70	48.04	m	52.85
make good existing plaster up to new lining	0.58	10.00	0.64	0.55	0.55	17.67	-	18.31	m	20.14
make good existing 38mm floor screed up to new lining	0.82	10.00	0.90	0.45	0.45	14.46	-	15.36	m	16.89
make good existing vinyl or similar floor covering up to new lining	10.12	10.00	11.13	0.67	0.67	21.53	0.45	33.11	m	36.42
Cutting or forming openings/recesses/chases etc; openings through brickwork or blockwork										
Form openings for door frames through 100mm block partition plastered both sides and with skirting both sides; take off skirtings, cut opening through partition, insert 100 x 150mm precast concrete lintel and wedge and pin up over, quoin up jambs, extend plaster to faces of lintel and make good junction with existing plaster; make good existing plaster up to new frame both sides and form fitted ends on existing skirtings up to new frame; extend softwood board flooring through opening on and with bearers and make good up to existing										
838 x 2032mm	25.10	-	25.10	20.00	17.20	604.07	19.05	648.22	nr	713.04
Unit rates for pricing the above and similar work										
cut opening through 75mm block partition plastered both sides	-	-	-	1.10	1.30	38.10	5.86	43.96	m²	48.36
cut opening through 100mm block partition plastered both sides	-	-	-	1.30	1.56	45.35	6.89	52.23	m²	57.45
make good fair face around opening	-	-	-	0.25	0.25	8.03	0.15	8.18	m	9.00
precast concrete, BS 8500, designed mix C20, 20mm aggregate lintel; 75 x 150 x 1200mm, reinforced with 1.01 kg of 12mm mild steel bars	20.25	10.00	22.27	0.80	0.80	25.70	0.38	48.35	nr	53.19
precast concrete, BS 8500, designed mix C20, 20mm aggregate lintel; 100 x 150 x 1200mm, reinforced with 1.01 kg of 12mm mild steel bars	24.63	10.00	27.09	1.00	1.00	32.13	0.50	59.72	nr	65.69
precast concrete, BS 8500 designed mix C20, 20mm aggregate lintel; fair finish all faces; 75 x 150 x 1200mm, reinforced with 1.01 kg of 12mm mild steel bars	20.25	10.00	22.27	0.80	0.80	25.70	0.38	48.35	nr	53.19
precast concrete, BS 8500 designed mix C20, 20mm aggregate lintel; fair finish all faces; 100 x 150 x 1200mm, reinforced with 1.01 kg of 12mm mild steel bars	24.63	10.00	27.09	1.00	1.00	32.13	0.50	59.72	nr	65.69
wedge and pin up over lintel; 75mm wide	-	-	-	0.50	0.50	16.07	0.93	17.00	m	18.69
wedge and pin up over lintel; 100mm wide	-	-	-	0.66	0.66	21.21	1.23	22.44	m	24.68
quoin up 75mm wide jambs	-	-	-	0.52	0.52	16.71	0.93	17.64	m	19.40
quoin up 100mm wide jambs	-	-	-	0.68	0.68	21.85	1.23	23.08	m	25.39
extend 13mm hardwall plaster to face of lintel	3.89	10.00	4.27	2.13	2.13	68.44	0.09	72.80	m²	80.08
make good existing plaster up to new frame	0.58	10.00	0.64	0.55	0.55	17.67	-	18.31	m	20.14
take off old skirting	-	-	-	0.22	-	4.04	0.15	4.19	m	4.61
form fitted end on existing skirting up to new frame	-	-	-	0.38	-	6.98	-	6.98	nr	7.68
38mm cement and sand (1:3) screeded bed in opening	5.18	10.00	5.70	1.54	1.54	49.48	-	55.18	m²	60.70
extend 25mm softwood board flooring through opening on and with bearers	25.36	20.00	28.75	16.20	2.70	334.75	0.81	364.31	m²	400.74
iroko threshold; twice oiled; plugged and screwed										
19 x 75mm	3.91	10.00	4.30	0.80	-	14.70	0.10	19.10	m	21.01
19 x 100mm	5.21	10.00	5.73	0.90	-	16.53	0.12	22.38	m	24.62

ALTERATIONS, REPAIRS AND CONSERVATION

Labour hourly rates: (except Specialists) Craft Operatives 18.37 Labourer 13.76 Rates are national average prices. Refer to REGIONAL VARIATIONS for indicative levels of overall pricing in regions	MATERIALS			LABOUR				RATES		
	Del to Site £	Waste %	Material Cost £	Craft Optve Hrs	Lab Hrs	Labour Cost £	Sunds £	Nett Rate £	Unit	Gross rate (10%) £
ALTERATION WORK TO EXISTING BUILDINGS (Cont'd)										
Cutting or forming openings/recesses/chases etc; openings through brickwork or blockwork (Cont'd)										
Form openings for door frames through one brick wall plastered one side and with external rendering other side; take off skirting, cut opening through wall, insert 215 x 150mm precast concrete lintel and wedge and pin up over; quoin up jambs; extend plaster to face of lintel and make good junction with existing plaster; make good existing plaster up to new frame one side; extend external rendering to face of lintel and to reveals and make good up to existing rendering and new frame the other side; form fitted ends on existing skirting up to new frame										
914 x 2082mm	46.01	-	46.01	28.00	25.40	863.86	42.15	952.02	nr	1047.22
Unit rates for pricing the above and similar work										
cut opening through half brick wall	-	-	-	1.95	2.40	68.85	7.14	75.99	m²	83.58
cut opening through one brick wall	-	-	-	3.80	4.70	134.48	13.52	147.99	m²	162.79
make good fair face around opening	-	-	-	0.50	0.50	16.07	0.93	17.00	m	18.69
make good facings to match existing around opening	2.78	10.00	3.06	0.50	0.50	16.07	0.93	20.06	m	22.06
precast concrete, BS 8500, designed mix C20, 20mm aggregate lintel; 102 x 150 x 1200mm, reinforced with 1.07 kg of 12mm mild steel bars	24.63	10.00	27.09	1.00	1.00	32.13	0.54	59.76	nr	65.74
precast concrete, BS 8500, designed mix C20, 20mm aggregate lintel; 215 x 150 x 1200mm, reinforced with 2.13 kg of 12mm mild steel bars	67.47	10.00	74.22	1.50	1.50	48.19	1.11	123.52	nr	135.87
precast concrete, BS 8500, designed mix C20, 20mm aggregate lintel, fair finish all faces; 102 x 150 x 1200mm, reinforced with 1.07 kg of 12mm mild steel bars	24.63	10.00	27.09	1.00	1.00	32.13	0.54	59.76	nr	65.74
precast concrete, BS 8500, designed mix C20, 20mm aggregate lintel, fair finish all faces; 215 x 150 x 1200mm, reinforced with 2.13 kg of 12mm mild steel bars	67.47	10.00	74.22	1.50	1.50	48.19	1.08	123.49	nr	135.84
wedge and pin up over lintel; 102mm wide	-	-	-	0.66	0.66	21.21	1.23	22.44	m	24.68
wedge and pin up over lintel; 215mm wide	-	-	-	1.00	1.00	32.13	2.62	34.76	m	38.23
quoin up half brick jambs	-	-	-	0.70	0.70	22.49	1.23	23.72	m	26.09
quoin up one brick jambs	-	-	-	1.27	1.27	40.81	2.62	43.43	m	47.77
facings to match existing to margin	2.78	10.00	3.06	0.67	0.67	21.53	1.23	25.82	m	28.40
make good existing plaster up to new frame	0.58	10.00	0.64	0.55	0.55	17.67	-	18.31	m	20.14
13mm two coat hardwall plaster to reveal not exceeding 300mm wide	1.16	10.00	1.28	0.78	0.78	25.06	0.03	26.37	m	29.00
make good existing external rendering up to new frame	0.23	10.00	0.26	0.47	0.47	15.10	-	15.36	m	16.89
13mm two coat cement and sand (1:3) external rendering to reveal not exceeding 300mm wide	0.47	10.00	0.51	0.74	0.74	23.78	0.03	24.32	m	26.75
take off old skirting	-	-	-	0.23	-	4.23	0.14	4.36	m	4.80
short length of old skirting up to 100mm long with mitre with existing one end	-	-	-	0.58	-	10.65	0.06	10.71	nr	11.79
short length of old skirting up to 300mm long with mitres with existing both ends	-	-	-	1.15	-	21.13	0.08	21.20	nr	23.32
38mm cement and sand (1:3) screeded bed in opening	5.18	10.00	5.70	1.54	1.54	49.48	-	55.18	m²	60.70
extend 25mm softwood board flooring through opening on and with bearers	5.28	15.00	5.90	3.65	0.61	75.44	0.21	81.56	nr	89.71
iroko threshold, twice oiled, plugged and screwed										
15 x 113mm	5.89	10.00	6.48	0.90	-	16.53	0.08	23.08	m	25.39
15 x 225mm	11.72	10.00	12.89	1.65	-	30.31	0.14	43.34	m	47.67
Form opening for window through 275mm hollow wall with two half brick skins, plastered one side and faced with picked stock facings the other; cut opening through wall, insert 275 x 225mm precast concrete boot lintel and wedge and pin up over; insert damp proof course and cavity gutter; quoin up jambs, close cavity at jambs with brickwork bonded to existing and with vertical damp proof course and close cavity at sill with one course of slates; face margin externally to match existing and extend plaster to face of lintel and to reveals and make good up to existing plaster and new frame										
900 x 900mm	67.52	-	67.52	15.90	16.80	523.25	21.90	612.67	nr	673.94
Unit rates for pricing the above and similar work										
cut opening through wall with two half brick skins	-	-	-	3.90	4.80	137.69	12.24	149.93	m²	164.92
cut opening through wall with one half brick and one 100mm block skins	-	-	-	3.25	3.96	114.19	12.24	126.43	m²	139.08
precast concrete, BS 8500, designed mix C20, 20mm aggregate boot lintel; 275 x 225 x 1200mm long reinforced with 3.60 kg of 12mm mild steel bars and 0.67 kg of 6mm mild steel links	79.00	10.00	86.90	2.30	2.30	73.90	1.23	162.03	nr	178.23
close cavity at jambs with blockwork bonded to existing and with 112.5mm wide Hyload pitch polymer damp proof course	2.41	10.00	2.65	0.50	0.50	16.07	0.31	19.03	m	20.94
close cavity at jambs with brickwork bonded to existing and with 112.5mm wide Hyload pitch polymer damp proof course	2.76	10.00	3.03	0.70	0.70	22.49	1.23	26.75	m	29.43
close cavity at jambs with slates in cement mortar (1:3) set vertically	3.12	10.00	3.43	0.44	0.44	14.14	0.31	17.88	m	19.67
close cavity at sill with one course of slates in cement mortar (1:3)	3.12	10.00	3.43	0.44	0.44	14.14	0.31	17.88	m	19.67
polythene DPC damp proof course and cavity tray; BS 6515; width exceeding 300mm	0.79	10.00	0.87	0.80	0.80	25.70	0.31	26.89	m²	29.57
precast concrete, BS 8500, designed mix C20, 20mm aggregate lintel; 160 x 225 x 1200mm long reinforced with 2.13 kg of 12mm mild steel bars	52.62	10.00	57.88	1.50	1.50	48.19	0.93	107.01	nr	117.71
mild steel angle arch bar, primed; 76 x 76 x 6mm	22.69	5.00	23.82	0.50	0.50	16.07	0.93	40.82	m	44.90
take out three courses of facing bricks and build brick-on-end flat arch in picked stock facing bricks	10.60	10.00	11.66	1.00	1.00	32.13	2.71	46.50	m	51.15
For unit rates for finishes, quoining up jambs, etc see previous items										

Labour hourly rates: (except Specialists) Craft Operatives 18.37 Labourer 13.76 Rates are national average prices. Refer to REGIONAL VARIATIONS for indicative levels of overall pricing in regions	MATERIALS			LABOUR				RATES		
	Del to Site	Waste	Material Cost	Craft Optve	Lab	Labour Cost	Sunds	Nett Rate	Unit	Gross rate (10%)
	£	%	£	Hrs	Hrs	£	£	£		£

ALTERATION WORK TO EXISTING BUILDINGS (Cont'd)

Cutting or forming openings/recesses/chases etc; openings through brickwork or blockwork (Cont'd)

	Del to Site £	Waste %	Material Cost £	Craft Optve Hrs	Lab Hrs	Labour Cost £	Sunds £	Nett Rate £	Unit	Gross rate (10%) £
Form opening for window where door removed, with window head at same level as old door head, through one brick wall plastered one side and faced the other with picked stock facings; remove old lintel and flat arch, cut away for and insert 102 x 225mm precast concrete lintel and wedge and pin up over; build brick-on-end flat arch in facing bricks on 76 x 76 x 6mm mild steel angle arch bar and point to match existing; cut away jambs of old door opening to extend width; quoin up jambs and face margins externally to match existing; fill in lower part of old opening with brickwork in gauged mortar (1:2:9), bond to existing and face externally and point to match existing 1200 x 900mm; old door opening 914 x 2082mm; extend plaster to face of lintel, reveals and filling and make good up to existing plaster and new frame; extend skirting one side and make good junction with existing...............	146.05	-	146.05	22.00	21.50	699.98	37.50	883.53	nr	971.88
Unit rates for pricing the above and similar work take out old lintel...............	-	-	-	1.00	1.30	36.26	5.10	41.36	m	45.49
cut away one brick wall to increase width of opening...............	-	-	-	3.80	4.70	134.48	12.75	147.23	m²	161.95
cut away 275mm hollow wall of two half brick skins to increase width of opening...............	-	-	-	3.90	4.80	137.69	12.75	150.44	m²	165.49

For other unit rates for pricing the above and similar work see previous items and items in "Filling openings in Brickwork and Blockwork" section

	Del to Site £	Waste %	Material Cost £	Craft Optve Hrs	Lab Hrs	Labour Cost £	Sunds £	Nett Rate £	Unit	Gross rate (10%) £
Form opening for door frame where old window removed, with door head at same level as old window head, through one brick wall plastered one side and faced the other with picked stock facings; remove old lintel and flat arch, cut away for and insert 102 x 225mm precast concrete lintel and wedge and pin up over; build brick-on-end flat arch in facing bricks on 76 x 76 x 6mm mild steel angle arch bar and point, fill old opening at sides of new opening to reduce width with brickwork in gauged mortar (1:2:9), bond to existing, wedge and pin up at soffit, face externally and to margins and point to match existing; take off skirtings, cut away wall below sill of old opening, quoin up jambs and face margins externally 914 x 2082mm, old window opening 1200 x 900mm; extend plaster to face of lintel and filling and make good up to existing plaster and new frame; form fitted ends on existing skirting up to new frame...............	88.32	-	88.32	24.00	23.80	768.37	42.75	899.44	nr	989.39

For unit rates for pricing the above and similar work see previous items and items in "Filling openings in Brickwork and Blockwork" section

Cutting or forming openings/recesses/chases etc; openings through rubble walling

	Del to Site £	Waste %	Material Cost £	Craft Optve Hrs	Lab Hrs	Labour Cost £	Sunds £	Nett Rate £	Unit	Gross rate (10%) £
Form opening for door frame through 600mm rough rubble wall faced both sides; cut opening through wall, insert 600 x 150mm precast concrete lintel finished fair on all exposed faces and wedge and pin up over; quoin and face up jambs and extend softwood board flooring through opening on and with bearers and make good up to existing 838 x 2032mm...............	94.54	-	94.54	41.30	43.70	1359.99	91.65	1546.18	nr	1700.80
Unit rates for pricing the above and similar work cut opening through 350mm random rubble wall...............	-	-	-	6.00	7.00	206.54	17.85	224.39	m²	246.83
cut opening through 600mm rough rubble wall...............	-	-	-	10.00	12.00	348.82	30.60	379.42	m²	417.36
precast concrete, BS 8500, designed mix C20, 20mm aggregate lintel; fair finish all faces; 350 x 150 x 1140mm reinforced with 3.04 kg of 12mm mild steel bars...............	79.00	10.00	86.90	2.25	2.25	72.29	1.08	160.27	nr	176.30
precast concrete, BS 8500, designed mix C20, 20mm aggregate lintel; fair finish all faces; 600 x 150 x 1140mm reinforced with 5.00 kg of 12mm mild steel bars...............	135.00	10.00	148.50	3.65	3.65	117.27	1.70	267.47	nr	294.22
wedge and pin up over lintel; 350mm wide...............	-	-	-	2.00	2.00	64.26	3.09	67.35	m	74.09
wedge and pin up over lintel; 600mm wide...............	-	-	-	3.00	3.00	96.39	5.41	101.81	m	111.99
quoin up; 350mm wide jambs...............	-	-	-	2.00	2.00	64.26	3.09	67.35	m	74.09
quoin up; 600mm wide jambs...............	-	-	-	3.00	3.00	96.39	5.41	101.81	m	111.99
extend 25mm softwood board flooring through opening on and with bearers...............	13.20	15.00	14.76	9.70	1.60	200.21	0.58	215.55	nr	237.11

Cutting or forming openings/recesses/chases etc; openings through stone faced walls

	Del to Site £	Waste %	Material Cost £	Craft Optve Hrs	Lab Hrs	Labour Cost £	Sunds £	Nett Rate £	Unit	Gross rate (10%) £
Form opening for window through wall of half brick backing, plastered, and 100mm stone facing; cut opening through wall, insert 102 x 150mm precast concrete lintel and wedge and pin up over and build in 76 x 76 x 6mm mild steel angle arch bar to support stone facing, clean soffit of stone and point; quoin up jambs and face up and point externally; extend plaster to face of lintel and to reveals and make good up to existing plaster and new frame 900 x 900mm...............	33.28	-	33.28	19.20	19.80	625.15	28.35	686.78	nr	755.46
Unit rates for pricing the above and similar work cut opening through wall...............	-	-	-	5.15	5.85	175.10	12.24	187.34	m²	206.08
precast concrete, BS 8500, designed mix C20, 20mm aggregate lintel; 102 x 150 x 1200mm reinforced with 1.01 kg of 12mm mild steel bars...............	24.63	10.00	27.09	1.05	1.05	33.74	0.47	61.29	nr	67.42
mild steel angle arch bar, primed; 76 x 76 x 6mm...............	22.69	5.00	23.82	0.50	0.50	16.07	0.93	40.82	m	44.90

Labour hourly rates: (except Specialists) Craft Operatives 18.37 Labourer 13.76 Rates are national average prices. Refer to REGIONAL VARIATIONS for indicative levels of overall pricing in regions	MATERIALS			LABOUR				RATES		
	Del to Site	Waste	Material Cost	Craft Optve	Lab	Labour Cost	Sunds	Nett Rate	Unit	Gross rate (10%)
	£	%	£	Hrs	Hrs	£	£	£		£
ALTERATION WORK TO EXISTING BUILDINGS (Cont'd)										
Cutting or forming openings/recesses/chases etc; openings through stone faced walls (Cont'd)										
Unit rates for pricing the above and similar work (Cont'd)										
quoin up jambs	-	-	-	0.70	0.70	22.49	1.39	23.89	m	26.27
face stone jambs and point	-	-	-	2.10	2.10	67.47	3.87	71.34	m	78.48
clean stone head and point	-	-	-	2.10	2.10	67.47	0.47	67.94	m	74.73
For unit rates for finishes, etc. see previous items										
Cutting or forming openings/recesses/chases etc; openings through stud partitions										
Form opening for door frame through lath and plaster finished 50 x 100mm stud partition; take off skirtings, cut away plaster, cut away and trim studding around opening and insert new studding at head and jambs of opening; make good existing lath and plaster up to new frame both sides and form fitted ends on existing skirtings up to new frame; extend softwood board flooring through opening on and with bearers and make good up to existing 838 x 2032mm	18.15	-	18.15	18.10	6.10	416.43	24.15	458.73	nr	504.61
Unit rates for pricing the above and similar work										
cut away plastered plasterboard	-	-	-	1.00	0.25	21.81	1.27	23.09	m²	25.39
cut away lath and plaster	-	-	-	1.10	0.26	23.78	1.53	25.31	m²	27.85
cut away and trim existing 75mm studding around opening										
838 x 2032mm	-	-	-	2.05	0.36	42.61	3.27	45.88	nr	50.47
914 x 2032mm	-	-	-	2.05	0.36	42.61	3.27	45.88	nr	50.47
cut away and trim existing 100mm studding around opening										
838 x 2032mm	-	-	-	2.55	0.45	53.04	4.28	57.31	nr	63.04
914 x 2032mm	-	-	-	2.55	0.45	53.04	4.28	57.31	nr	63.04
softwood studding; 50 x 75mm	1.13	10.00	1.24	0.50	0.08	10.29	0.17	11.69	m	12.86
softwood studding; 50 x 100mm	1.42	10.00	1.56	0.60	0.10	12.40	0.18	14.14	m	15.55
make good plastered plasterboard around opening	-	-	-	0.75	0.75	24.10	1.85	25.94	m	28.54
make good lath and plaster around opening	-	-	-	0.83	0.83	26.67	2.27	28.93	m	31.83
take off skirting	-	-	-	0.21	-	3.86	0.27	4.13	m	4.54
form fitted end on existing skirting	-	-	-	0.57	-	10.47	-	10.47	nr	11.52
extend 25mm softwood tongued and grooved board flooring through opening on and with bearers	2.54	20.00	2.88	1.50	0.25	31.00	0.09	33.96	nr	37.36
Filling in openings/recesses etc; filling holes in concrete										
Fill holes in concrete structure where pipes removed with concrete to BS 8500, designed mix C20, 20mm aggregate; formwork										
wall; thickness 100mm; pipe diameter										
50mm	-	-	-	0.60	0.30	15.15	2.30	17.45	nr	19.19
100mm	-	-	-	0.78	0.39	19.70	3.03	22.73	nr	25.00
150mm	-	-	-	0.98	0.49	24.75	3.64	28.39	nr	31.23
wall; thickness 200mm; pipe diameter										
50mm	-	-	-	0.78	0.39	19.70	3.03	22.73	nr	25.00
100mm	-	-	-	0.98	0.49	24.75	3.64	28.39	nr	31.23
150mm	-	-	-	1.00	0.77	28.97	4.68	33.65	nr	37.01
floor; thickness 200mm; pipe diameter										
50mm	-	-	-	0.74	0.37	18.69	2.74	21.43	nr	23.57
100mm	-	-	-	0.93	0.46	23.41	3.51	26.92	nr	29.62
150mm	-	-	-	0.95	0.48	24.06	4.44	28.50	nr	31.35
Make good fair face to filling and surrounding work										
to one side of wall; pipe diameter										
50mm	-	-	-	0.10	0.10	3.21	0.31	3.53	nr	3.88
100mm	-	-	-	0.20	0.20	6.43	0.48	6.91	nr	7.60
150mm	-	-	-	0.25	0.25	8.03	0.58	8.62	nr	9.48
to soffit of floor; pipe diameter										
50mm	-	-	-	0.10	0.10	3.21	0.31	3.53	nr	3.88
100mm	-	-	-	0.20	0.20	6.43	0.48	6.91	nr	7.60
150mm	-	-	-	0.25	0.25	8.03	0.58	8.62	nr	9.48
Make good plaster to filling and surrounding work										
to one side of wall; pipe diameter										
50mm	-	-	-	0.19	0.19	6.10	0.31	6.42	nr	7.06
100mm	-	-	-	0.36	0.36	11.57	0.48	12.05	nr	13.25
150mm	-	-	-	0.45	0.45	14.46	0.58	15.04	nr	16.55
to soffit of floor; pipe diameter										
50mm	-	-	-	0.20	0.20	6.43	0.31	6.74	nr	7.42
100mm	-	-	-	0.40	0.40	12.85	0.48	13.33	nr	14.67
150mm	-	-	-	0.50	0.50	16.07	0.58	16.65	nr	18.32
Make good floor screed and surrounding work; pipe diameter										
50mm	-	-	-	0.19	0.19	6.10	0.14	6.24	nr	6.86
100mm	-	-	-	0.36	0.36	11.57	0.28	11.85	nr	13.04
150mm	-	-	-	0.45	0.45	14.46	0.39	14.85	nr	16.33
Make good vinyl or similar floor covering and surrounding work; pipe diameter										
50mm	-	-	-	0.50	0.50	16.07	2.34	18.41	nr	20.25
100mm	-	-	-	1.00	1.00	32.13	3.13	35.26	nr	38.79
150mm	-	-	-	1.20	1.20	38.56	4.39	42.95	nr	47.25
Fill holes in concrete structure where metal sections removed with concrete to BS 8500, designed mix C20, 20mm aggregate; formwork										
wall; thickness 100mm; section not exceeding 250mm deep	-	-	-	0.60	0.60	19.28	2.79	22.07	nr	24.27
wall; thickness 200mm; section not exceeding 250mm deep	-	-	-	0.75	0.75	24.10	4.07	28.16	nr	30.98
wall; thickness 100mm; section 250 - 500mm deep	-	-	-	0.75	0.75	24.10	4.07	28.16	nr	30.98
wall; thickness 200mm; section 250 - 500mm deep	-	-	-	0.90	0.90	28.92	4.91	33.82	nr	37.20

Labour hourly rates: (except Specialists) Craft Operatives 18.37 Labourer 13.76. Rates are national average prices. Refer to REGIONAL VARIATIONS for indicative levels of overall pricing in regions	MATERIALS			LABOUR				RATES		
	Del to Site £	Waste %	Material Cost £	Craft Optve Hrs	Lab Hrs	Labour Cost £	Sunds £	Nett Rate £	Unit	Gross rate (10%) £
ALTERATION WORK TO EXISTING BUILDINGS (Cont'd)										
Filling in openings/recesses etc; filling holes in concrete (Cont'd)										
Make good fair face to filling and surrounding work to one side of wall; section not exceeding 250mm deep										
section not exceeding 250mm deep	-	-	-	0.22	0.22	7.07	0.27	7.34	nr	8.07
to one side of wall; section 250 - 500mm deep	-	-	-	0.26	0.26	8.35	0.42	8.77	nr	9.65
Make good plaster to filling and surrounding work to one side of wall; section not exceeding 250mm deep	-	-	-	0.35	0.35	11.25	0.74	11.98	nr	13.18
to one side of wall; section 250 - 500mm deep	-	-	-	0.43	0.43	13.82	1.11	14.93	nr	16.42
Filling in openings/recesses etc; filling openings in concrete										
Fill opening where door removed in 150mm reinforced concrete wall with concrete reinforced with 12mm mild steel bars at 300mm centres both ways tied to existing structure; hack off plaster to reveals, hack up floor covering and screed in opening, prepare edges of opening to form joint with filling, drill edges of opening and tie in new reinforcement, plaster filling both sides and extended skirting both sides; making good junction of new and existing plaster and skirtings and make good floor screed and vinyl covering up to filling both sides										
838 x 2032mm	139.15	-	139.15	22.50	34.50	888.05	14.44	1041.64	nr	1145.80
Unit rates for pricing the above and similar work										
hack off plaster to reveal; not exceeding 100mm wide	-	-	-	-	0.24	3.30	0.10	3.41	m	3.75
hack off plaster to reveal; 100 - 200mm wide	-	-	-	-	0.33	4.54	0.21	4.75	m	5.23
take up vinyl or similar floor covering and screed	-	-	-	-	2.00	27.52	2.14	29.67	m²	32.63
prepare edge of opening wide to form joint with filling; 100mm wide	-	-	-	0.25	0.25	8.03	-	8.03	m	8.84
prepare edge of opening wide to form joint with filling; 200mm wide	-	-	-	0.35	0.35	11.25	-	11.25	m	12.37
concrete to BS 8500, design mix C 16/20 - 20 N/mm²; 20mm aggregate in walls; 150 - 450m thick	96.25	5.00	101.06	-	12.00	165.12	0.62	266.79	m³	293.47
concrete to BS 8500, design mix C 16/20 - 20 N/mm²; 20mm aggregate in walls; not exceeding 150mm thick	96.25	5.00	101.06	-	12.00	165.12	0.62	266.79	m³	293.47
12mm mild steel bar reinforcement; straight	1070.00	10.00	1177.00	85.00	-	1561.45	6.76	2745.22	t	3019.74
drill edge of opening and tie in new reinforcement	-	-	-	-	1.00	13.76	0.62	14.38	m	15.81
formwork and basic finish to wall, per side	8.08	10.00	8.89	2.50	2.50	80.33	1.47	90.69	m²	99.76
formwork and fine formed finish to wall, per side	8.08	10.00	8.89	2.75	2.75	88.36	1.65	98.90	m²	108.79
junction between new and existing fair face	-	-	-	-	0.30	4.13	-	4.13	m	4.54
13mm two coat hardwall plaster over 300mm wide	3.89	10.00	4.27	1.30	1.30	41.77	-	46.04	m²	50.65
25 x 100mm softwood chamfered skirting, primed all round	0.96	10.00	1.06	0.56	0.09	11.53	0.17	12.75	m	14.02
junction with existing	-	-	-	0.50	0.08	10.29	-	10.29	nr	11.31
make good floor screed and vinyl or similar floor covering up to filling not exceeding 300mm wide	9.09	10.00	10.00	1.12	1.12	35.99	0.48	46.47	m	51.11
Fill opening where staircase removed in 150mm reinforced concrete floor with concrete reinforced with 12mm mild steel bars at 225mm centres both ways tied to existing structure; remove timber lining, prepare edges of opening to form joint with filling, drill edges of opening tie in new reinforcement, plaster filling on soffit and extend cement and sand screed and vinyl floor covering to match existing; make good junction of new and existing plaster and vinyl floor covering										
900 x 2134mm	157.50	-	157.50	14.75	22.30	577.81	12.00	747.31	nr	822.04
Unit rates for pricing the above and similar work										
take off timber lining	-	-	-	-	0.24	3.30	0.15	3.45	m	3.80
prepare edge of opening to form joint with filling; 150mm wide	-	-	-	0.30	0.30	9.64	-	9.64	m	10.60
prepare edge of opening to form joint with filling; 200mm wide	-	-	-	0.35	0.35	11.25	-	11.25	m	12.37
concrete to BS 8500, design mix C 16/20 - 20 N/mm²; 20mm aggregate in slabs; 150 - 450mm thick	96.25	5.00	101.06	-	12.00	165.12	0.62	266.79	m³	293.47
concrete to BS 8500, design mix C 16/20 - 20 N/mm²; 20mm aggregate in slabs; not exceeding 150mm thick	96.25	5.00	101.06	-	12.00	165.12	0.62	266.79	m³	293.47
12mm mild steel bar reinforcement; straight	1070.00	10.00	1177.00	85.00	-	1561.45	6.76	2745.22	t	3019.74
drill edge of opening and tie in new reinforcement	-	-	-	-	1.00	13.76	0.62	14.38	m	15.81
formwork and basic finish to soffit; slab thickness not exceeding 200mm; height to soffit 1.50 - 3.00m	11.29	10.00	12.42	2.25	2.25	72.29	1.46	86.17	m²	94.78
formwork and basic finish to soffit; slab thickness 200 - 300mm; height to soffit 1.50 - 3.00m	12.42	10.00	13.66	2.50	2.50	80.33	1.65	95.64	m²	105.20
formwork and fine formed finish to soffit; slab thickness not exceeding 200mm; height to soffit 1.50 - 3.00m	11.29	10.00	12.42	2.50	2.50	80.33	1.68	94.42	m²	103.87
formwork and fine formed finish to soffit; slab thickness 200 - 300mm; height to soffit 1.50 - 3.00m	12.42	10.00	13.66	2.75	2.75	88.36	1.74	103.76	m²	114.14
junction between new and existing fair face	-	-	-	-	0.30	4.13	-	4.13	m	4.54
10mm two coat lightweight plaster over 300mm wide	3.89	10.00	4.27	1.30	1.30	41.77	-	46.04	m²	50.65
38mm cement and sand (1:3) trowelled bed	5.18	10.00	5.70	0.63	0.63	20.24	-	25.94	m²	28.54
vinyl or similar floor covering to match existing fixed with adhesive exceeding 300mm wide	25.30	10.00	27.83	1.25	0.50	29.84	0.84	58.51	m²	64.36
Filling openings/recesses etc; filling openings in brickwork or blockwork										
Fill opening where door removed in 100mm block wall with concrete blocks in gauged mortar (1:2:9); provide 50mm preservative treated softwood sole plate on timber floor in opening, bond to blockwork at jambs and wedge and pin up at head; plaster filling both sides and extend skirting both sides; make good junction of new and existing plaster and skirtings										
838 x 2032mm	65.85	-	65.85	14.10	8.15	371.16	12.72	449.73	nr	494.71
Unit rates for pricing the above and similar work										
blockwork in filling; 75mm	11.20	10.00	12.32	1.00	1.25	35.57	1.70	49.58	m²	54.54

ALTERATIONS, REPAIRS AND CONSERVATION

Labour hourly rates: (except Specialists) Craft Operatives 18.37 Labourer 13.76 Rates are national average prices. Refer to REGIONAL VARIATIONS for indicative levels of overall pricing in regions	MATERIALS			LABOUR				RATES		
	Del to Site	Waste	Material Cost	Craft Optve	Lab	Labour Cost	Sunds	Nett Rate	Unit	Gross rate (10%)
	£	%	£	Hrs	Hrs	£	£	£		£
ALTERATION WORK TO EXISTING BUILDINGS (Cont'd)										
Filling openings/recesses etc; filling openings in brickwork or blockwork (Cont'd)										
Unit rates for pricing the above and similar work (Cont'd) blockwork in filling; 100mm	12.44	10.00	13.68	1.20	1.50	42.68	2.33	58.69	m²	64.56
blockwork in filling and fair face and flush smooth pointing one side; 75mm 75mm	11.20	10.00	12.32	1.25	1.50	43.60	2.01	57.93	m²	63.73
blockwork in filling and fair face and flush smooth pointing one side; 100mm	12.44	10.00	13.68	1.45	1.75	50.72	2.47	66.88	m²	73.56
preservative treated softwood sole plate 50 x 75mm	1.13	10.00	1.24	0.30	0.05	6.20	0.31	7.76	m	8.53
50 x 100mm	1.42	10.00	1.56	0.33	0.06	6.89	0.42	8.87	m	9.76
Hyload pitch polymer damp proof course, width not exceeding 225mm	4.83	10.00	5.31	0.80	0.80	25.70	0.62	31.63	m²	34.79
wedge and pin up at head; 75mm wide	-	-	-	0.10	0.10	3.21	0.62	3.83	m	4.21
wedge and pin up at head; 100mm wide	-	-	-	0.12	0.12	3.86	0.93	4.79	m	5.26
cut pockets and bond to existing blockwork at jambs; 75mm filling to existing blockwork at jambs	-	-	-	0.40	0.40	12.85	1.08	13.93	m	15.33
cut pockets and bond to existing blockwork at jambs; 100mm filling	-	-	-	0.50	0.50	16.07	1.53	17.60	m	19.35
12mm (average) cement and sand dubbing over 300mm wide on filling	1.64	10.00	1.80	0.54	0.54	17.35	-	19.15	m²	21.07
13mm two coat lightweight plaster over 300mm wide on filling	3.89	10.00	4.27	1.27	1.27	40.81	0.03	45.11	m²	49.62
19 x 100mm softwood chamfered skirting, primed all round	0.73	10.00	0.80	0.50	0.08	10.29	0.17	11.25	m	12.38
junction with existing	-	-	-	0.50	0.08	10.29	-	10.29	nr	11.31
make good vinyl or similar floor covering up to filling	8.42	10.00	9.27	0.50	0.50	16.07	0.21	25.54	m	28.10
Fill opening where door removed in half brick wall with brickwork in gauged mortar (1:2:9); provide lead cored damp proof course in opening, lapped with existing, bond to existing brickwork at jambs and wedge and pin up at head; plaster filling both sides and extend skirting both sides; make good junction of new and existing plaster and skirtings 838 x 2032mm	69.84	-	69.84	15.70	12.75	463.85	15.60	549.29	nr	604.22
Unit rates for pricing the above and similar work half brick filling in common bricks	18.96	10.00	20.86	2.00	2.00	64.26	3.87	88.99	m²	97.88
half brick filling in common bricks fair faced and flush pointed one side	18.96	10.00	20.86	2.33	2.33	74.86	4.17	99.89	m²	109.88
50 x 100mm softwood sole plate	1.42	10.00	1.56	0.33	0.06	6.89	0.38	8.82	m	9.71
Hyload pitch polymer damp proof course, width not exceeding 225mm	4.83	10.00	5.31	0.80	0.80	25.70	0.62	31.63	m²	34.79
wedge and pin up at head 102mm wide	-	-	-	0.12	0.12	3.86	0.93	4.79	m	5.26
cut pockets and bond half brick filling to existing brickwork at jambs	-	-	-	0.50	0.50	16.07	1.54	17.61	m	19.37
For unit rates for finishes, etc see previous items										
Fill opening where door removed in one brick external wall with brickwork in gauged mortar (1:2:9) faced externally with picked stock facings pointed to match existing; remove old lintel and arch, provide lead cored damp proof course in opening lapped with existing, bond to existing brickwork at jambs and wedge and pin up at head; plaster filling one side and extend skirting; make good junction of new and existing plaster and skirting 914 x 2082mm	229.17	-	229.17	19.65	18.85	620.35	38.55	888.07	nr	976.87
Unit rates for pricing the above and similar work remove old lintel and arch	-	-	-	1.46	1.64	49.39	3.06	52.45	m	57.69
one brick filling in common bricks	37.92	10.00	41.71	4.00	4.00	128.52	9.27	179.50	m²	197.45
one brick filling in common bricks faced one side with picked stock facings and point to match existing	81.03	10.00	89.13	5.00	5.00	160.65	9.27	259.05	m²	284.96
one brick filling in common bricks fair faced and flush smooth pointing one side	37.92	10.00	41.71	4.33	4.33	139.12	9.27	190.11	m²	209.12
two course slate damp proof course, width not exceeding 225mm	24.96	10.00	27.46	1.02	1.02	32.77	1.08	61.31	m²	67.44
Hyload pitch polymer damp proof course, BS 6398, Class F, width not exceeding 225mm	4.83	10.00	5.31	0.80	0.80	25.70	0.47	31.48	m²	34.63
wedge and pin up at head 215mm wide	-	-	-	0.20	0.20	6.43	2.16	8.59	m	9.44
cut pockets and bond one brick filling to existing brickwork at jambs	-	-	-	1.00	1.00	32.13	3.56	35.69	m	39.25
12mm cement and sand (1:3) two coat external rendering over 300mm wide on filling	1.64	10.00	1.80	1.25	1.25	40.16	0.06	42.02	m²	46.23
For unit rates for internal finishes, etc. see previous items										
Fill opening where door removed in 275mm external hollow wall with inner skin in common bricks and outer skin in picked stock facings pointed to match existing, in gauged mortar (1:2:9); remove old lintel and arch, provide lead cored combined damp proof course and cavity gutter lapped with existing; form cavity with ties; bond to existing brickwork at jambs and wedge and pin up at head; plaster filling one side and extend skirting; make good junction of new and existing plaster and skirting 914 x 2082mm	144.73	-	144.73	19.90	19.10	628.38	39.38	812.49	nr	893.74
Unit rates for pricing the above and similar work half brick filling in common bricks	18.96	10.00	20.86	2.00	2.00	64.26	3.87	88.99	m²	97.88
half brick filling in common bricks and fair face and flush smooth pointing one side	18.96	10.00	20.86	2.33	2.33	74.86	4.17	99.89	m²	109.88
half brick filling in picked stock facings and point to match existing	47.70	10.00	52.47	2.95	2.95	94.78	4.17	151.42	m²	166.57

Labour hourly rates: (except Specialists) Craft Operatives 18.37 Labourer 13.76 Rates are national average prices. Refer to REGIONAL VARIATIONS for indicative levels of overall pricing in regions	MATERIALS			LABOUR				RATES		
	Del to Site	Waste	Material Cost	Craft Optve	Lab	Labour Cost	Sunds	Nett Rate	Unit	Gross rate (10%)
	£	%	£	Hrs	Hrs	£	£	£		£
ALTERATION WORK TO EXISTING BUILDINGS (Cont'd)										
Filling openings/recesses etc; filling openings in brickwork or blockwork (Cont'd)										
Unit rates for pricing the above and similar work (Cont'd)										
form 50mm wide cavity with stainless steel housing type 4 wall ties built in	0.27	10.00	0.30	0.20	0.20	6.43	-	6.72	m²	7.40
Hyload "Original", pitch polymer damp proof course and cavity tray, width exceeding 225mm	4.83	10.00	5.31	0.80	0.80	25.70	0.47	31.48	m²	34.63
wedge and pin up at head 102mm wide	-	-	-	0.12	0.12	3.86	0.93	4.79	m	5.26
cut pockets and bond half brick wall in common bricks to existing brickwork at jambs	-	-	-	0.50	0.50	16.07	1.54	17.61	m	19.37
cut pockets and bond half brick wall in facing bricks to existing brickwork at jambs and make good facings	-	-	-	0.60	0.60	19.28	3.24	22.52	m	24.77
For unit rates for internal finishes, etc. see previous items										
Filling in openings/recesses etc; fillings openings in rubble walling										
Fill opening where window removed in 600mm rough rubble wall with rubble walling in lime mortar (1:3) pointed to match existing; bond to existing walling at jambs and wedge and pin up at head; plaster filling one side and make good junction of new and existing plaster										
900 x 900mm	130.29	-	130.29	12.00	12.00	385.56	23.85	539.70	nr	593.67
Unit rates for pricing the above and similar work										
rough rubble walling in filling										
600mm thick	133.00	10.00	146.30	6.00	6.00	192.78	18.54	357.62	m²	393.38
random rubble walling in filling										
300mm thick	97.00	10.00	106.70	5.80	5.80	186.35	9.27	302.32	m²	332.56
500mm thick	143.00	10.00	157.30	10.90	10.90	350.22	15.45	522.97	m²	575.26
wedge and pin up at head										
300mm wide	-	-	-	0.28	0.28	9.00	3.09	12.09	m	13.30
500mm wide	-	-	-	0.47	0.47	15.10	5.25	20.35	m	22.39
600mm wide	-	-	-	0.56	0.56	17.99	6.18	24.17	m	26.59
cut pockets and bond walling to existing at jambs										
300mm	7.28	10.00	8.00	0.83	0.83	26.67	0.93	35.60	m	39.16
500mm	10.73	10.00	11.80	1.38	1.38	44.34	1.39	57.53	m	63.29
600mm	9.98	10.00	10.97	1.75	1.75	56.23	1.70	68.90	m	75.78
Filling openings/recesses etc; filling openings in stud partitions										
Fill opening where door removed in stud partition with 50 x 100mm studding and sole plate, covered on both sides with 9.5mm plasterboard baseboard with 5mm one coat board finish plaster; extend skirting on both sides; make good junction of new and existing plaster and skirtings										
838 x 2032mm	36.30	-	36.30	16.00	5.36	367.67	4.50	408.47	nr	449.32
Unit rates for pricing the above and similar work										
softwood sole plate; 50 x 75mm	1.13	10.00	1.24	0.50	0.08	10.29	0.17	11.69	m	12.86
softwood sole plate; 50 x 100mm	1.42	10.00	1.56	0.60	0.10	12.40	0.21	14.17	m	15.59
softwood studding; 50 x 75mm	1.13	10.00	1.24	0.50	0.08	10.29	0.17	11.69	m	12.86
softwood studding; 50 x 100mm	1.42	10.00	1.56	0.60	0.10	12.40	0.21	14.17	m	15.59
9.5mm plasterboard baseboard to filling	1.41	10.00	1.55	0.50	0.10	10.56	0.42	12.54	m²	13.79
9.5mm plasterboard baseboard to filling including packing out not exceeding 10mm	1.41	10.00	1.55	1.00	0.17	20.71	0.56	22.82	m²	25.10
5mm one coat board finish plaster on plasterboard to filling	1.18	10.00	1.29	0.91	0.91	29.24	0.05	30.58	m²	33.63
10mm two coat hardwall plaster on plasterboard to filling	1.94	10.00	2.14	1.27	1.27	40.81	0.05	42.99	m²	47.29
softwood chamfered skirting, primed all round; 19 x 100mm	0.73	10.00	0.80	0.54	0.09	11.16	0.27	12.23	m	13.45
junction with existing; 19 x 100mm	-	-	-	0.57	-	10.47	-	10.47	nr	11.52
softwood chamfered skirting, primed all round; 25 x 150mm	2.00	10.00	2.20	0.55	0.09	11.34	0.36	13.90	m	15.29
junction with existing; 25 x 150mm	-	-	-	0.57	-	10.47	-	10.47	nr	11.52
For unit rates for internal finishes, etc. see previous items										
Removing existing and replacing; masonry										
Cut out decayed bricks and replace with new bricks in gauged mortar (1:2:9) and make good surrounding work										
Fair faced common bricks										
singly	0.32	10.00	0.35	0.17	0.17	5.46	0.21	6.02	nr	6.62
small patches	18.96	10.00	20.86	5.80	5.80	186.35	8.97	216.18	m²	237.80
Picked stock facing bricks										
singly	0.80	10.00	0.87	0.17	0.17	5.46	0.21	6.55	nr	7.20
small patches	47.70	10.00	52.47	5.80	5.80	186.35	8.97	247.79	m²	272.57
Cut out defective wall and re-build in gauged mortar (1:2:9) and tooth and bond to surrounding work (25% new bricks allowed)										
Half brick wall; stretcher bond										
common bricks	4.74	10.00	5.21	4.00	4.00	128.52	5.14	138.88	m²	152.77
One brick wall; English bond										
common bricks	9.48	10.00	10.43	7.90	7.90	253.83	11.06	275.31	m²	302.84
common bricks faced one side with picked stock facings	20.26	10.00	22.28	8.30	8.30	266.68	11.06	300.02	m²	330.02
picked stock facing bricks faced both sides	23.85	10.00	26.24	8.80	8.80	282.74	11.06	320.03	m²	352.04
One brick wall; Flemish bond										
common bricks	9.48	10.00	10.43	7.90	7.90	253.83	5.14	269.40	m²	296.34
common bricks faced one side with picked stock facings	20.26	10.00	22.28	8.30	8.30	266.68	11.06	300.02	m²	330.02
picked stock facing bricks faced both sides	23.85	10.00	26.24	8.80	8.80	282.74	11.06	320.03	m²	352.04
275mm hollow wall; skins in stretcher bond										
two half brick skins in common bricks	9.48	10.00	10.43	8.16	8.16	262.18	11.27	283.87	m²	312.26

Labour hourly rates: (except Specialists) Craft Operatives 18.37 Labourer 13.76 Rates are national average prices. Refer to REGIONAL VARIATIONS for indicative levels of overall pricing in regions	MATERIALS			LABOUR				RATES		
	Del to Site	Waste	Material Cost	Craft Optve	Lab	Labour Cost	Sunds	Nett Rate	Unit	Gross rate (10%)
	£	%	£	Hrs	Hrs	£	£	£		£
ALTERATION WORK TO EXISTING BUILDINGS (Cont'd)										
Removing existing and replacing; masonry (Cont'd)										
Cut out defective wall and re-build in gauged mortar (1:2:9) and tooth and bond to surrounding work (25% new bricks allowed) (Cont'd)										
275mm hollow wall; skins in stretcher bond (Cont'd)										
one half brick skin in common bricks and one half brick skin in picked stock facings	16.67	10.00	18.33	8.60	8.60	276.32	11.27	305.91	m²	336.51
one 100mm skin in concrete blocks and one half brick skin in picked stock facings	15.03	10.00	16.54	8.26	8.26	265.39	10.53	292.46	m²	321.71
Cut out crack in brickwork and stitch across with new bricks in gauged mortar(1:2:9); average 450mm wide										
Half brick wall										
common bricks	8.22	10.00	9.04	1.90	1.90	61.05	4.99	75.08	m	82.59
common bricks fair faced one side	8.22	10.00	9.04	2.30	2.30	73.90	4.99	87.93	m	96.72
picked stock facing bricks faced both sides	20.67	10.00	22.74	2.40	2.40	77.11	4.99	104.84	m	115.33
One brick wall										
common bricks	16.43	10.00	18.08	3.90	3.90	125.31	10.76	154.14	m	169.55
common bricks fair faced one side	16.43	10.00	18.08	4.40	4.40	141.37	10.76	170.20	m	187.22
common bricks faced one side with picked stock facings	35.11	10.00	38.62	4.60	4.60	147.80	10.76	197.18	m	216.90
picked stock facing bricks faced both sides	41.34	10.00	45.47	4.70	4.70	151.01	10.76	207.24	m	227.96
Cut out defective arch; re-build in picked stock facing bricks in gauged mortar (1:2:9) including centering										
Brick-on-edge flat arch										
102mm on soffit	5.57	10.00	6.12	1.00	1.00	32.13	1.26	39.51	m	43.46
215mm on soffit	10.60	10.00	11.66	0.95	0.95	30.52	2.20	44.39	m	48.82
Brick-on-end flat arch										
102mm on soffit	10.60	10.00	11.66	0.95	0.95	30.52	2.20	44.39	m	48.82
215mm on soffit	21.19	10.00	23.31	1.90	1.90	61.05	4.41	88.77	m	97.65
Segmental arch in two half brick rings										
102mm on soffit	10.60	10.00	11.66	2.90	2.90	93.18	2.20	107.04	m	117.74
215mm on soffit	21.19	10.00	23.31	2.90	2.90	93.18	4.41	120.90	m	132.99
Semi-circular arch in two half brick rings										
102mm on soffit	10.60	10.00	11.66	2.90	2.90	93.18	2.20	107.04	m	117.74
215mm on soffit	21.19	10.00	23.31	2.90	2.90	93.18	4.41	120.90	m	132.99
Cut out defective arch; replace with precast concrete, BS 8500, designed mix C20, 20mm aggregate lintel; wedge and pin up to brickwork over with slates in cement mortar (1:3) and extend external rendering over										
Remove defective brick-on-end flat arch										
replace with 102 x 215mm precast concrete lintel reinforced with one 12mm mild steel bar	57.72	10.00	63.49	1.90	1.90	61.05	3.46	128.00	m	140.80
Cut out defective external sill and re-build										
Shaped brick-on-edge sill in picked stock facing bricks in gauged mortar (1:2:9)										
225mm wide	10.60	10.00	11.66	1.00	1.00	32.13	3.08	46.86	m	51.55
Roofing tile sill set weathering and projecting in cement mortar (1:3)										
two courses	4.32	10.00	4.75	1.20	1.20	38.56	1.59	44.90	m	49.39
Take down defective brick-on-edge coping and re-build in picked stock facing bricks in gauged mortar (1:2:9)										
Coping										
to one brick wall	10.60	10.00	11.66	0.60	0.60	19.28	3.08	34.01	m	37.41
Coping with cement fillets both sides										
to one brick wall; with oversailing course	15.90	10.00	17.49	0.67	0.67	21.53	4.62	43.64	m	48.00
to one brick wall; with single course tile creasing	12.76	10.00	14.03	1.09	1.09	35.02	3.87	52.92	m	58.22
to one brick wall; with double course tile creasing	14.92	10.00	16.41	1.38	1.38	44.34	5.46	66.21	m	72.83
Cut out defective air brick and replace with new; bed and point in gauged mortar (1:2:9)										
Cast iron, light, square hole										
225 x 75mm	12.49	-	12.49	0.45	0.45	14.46	0.24	27.19	nr	29.91
225 x 150mm	16.66	-	16.66	0.45	0.45	14.46	0.39	31.51	nr	34.66
Terra cotta										
215 x 65mm	3.08	-	3.08	0.45	0.45	14.46	0.24	17.78	nr	19.56
215 x 140mm	4.57	-	4.57	0.45	0.45	14.46	0.39	19.42	nr	21.36
Air bricks in old walls										
Cut opening through old one brick wall, render all round in cement mortar (1:3); build in clay air brick externally and fibrous plaster ventilator internally and make good facings and plaster										
215 x 140mm	7.90	-	7.90	1.75	1.75	56.23	0.82	64.95	nr	71.45
225 x 225mm	15.47	-	15.47	2.33	2.33	74.86	1.03	91.37	nr	100.50
Cut opening through old 275mm hollow brick wall and seal cavity with slates; build in clay air brick externally and fibrous plaster ventilator internally and make good facings and plaster										
215 x 140mm	11.02	10.00	11.33	2.14	2.14	68.76	0.82	80.92	nr	89.01
225 x 225mm	18.59	10.00	18.90	2.81	2.81	90.29	1.03	110.22	nr	121.24
Removing existing and replacing; concrete										
Re-bed loose copings										
Take off loose concrete coping and clean and remove old bedding mortar										
re-bed, joint and point in cement mortar (1:3)	-	-	-	1.00	1.00	32.13	1.50	33.63	m	36.99
Repairing cracks in concrete										
Repair cracks; clean out all dust and debris and fill with mortar mixed with a bonding agent										
up to 5mm wide	-	-	-	0.50	0.50	16.07	0.72	16.78	m	18.46

Labour hourly rates: (except Specialists) Craft Operatives 18.37 Labourer 13.76 Rates are national average prices. Refer to REGIONAL VARIATIONS for indicative levels of overall pricing in regions	MATERIALS			LABOUR				RATES		
	Del to Site	Waste	Material Cost	Craft Optve	Lab	Labour Cost	Sunds	Nett Rate		Gross rate (10%)
	£	%	£	Hrs	Hrs	£	£	£	Unit	£
ALTERATION WORK TO EXISTING BUILDINGS (Cont'd)										
Removing existing and replacing; concrete (Cont'd)										
Repairing cracks in concrete (Cont'd)										
Repair cracks; cutting out to form groove, treat with bonding agent and fill with fine concrete mixed with a bonding agent										
25 x 25mm deep	-	-	-	0.50	0.50	16.07	1.30	17.37	m	19.11
40 x 40mm deep	-	-	-	0.60	0.60	19.28	3.09	22.37	m	24.60
Breaking up and reinstatement of concrete floors for excavations										
Break up concrete bed and 150mm hardcore bed under for a width of 760mm for excavation of trench and reinstate with new hardcore and in-situ concrete, BS 8500, design mix C20P; 20mm aggregate; make good up to new wall and existing concrete bed both sides										
with 100mm plain concrete bed	13.01	7.50	14.74	-	2.70	37.15	9.69	61.59	m	67.75
with 150mm plain concrete bed	17.81	7.50	19.91	-	5.00	68.80	11.73	100.44	m	110.49
with 100mm concrete bed reinforced with steel fabric to BS 4483, Reference A193	15.28	7.50	17.24	-	4.05	55.73	9.69	82.66	m	90.93
with 150mm concrete bed reinforced with steel fabric to BS 4483, Reference A193	20.08	7.50	22.41	-	7.33	100.86	11.73	135.00	m	148.50
Removing existing and replacing; stone										
Note										
restoration and repair of stonework is work for a specialist; prices being obtained for specific projects. Some firms that specialise in this class of work are included within the list at the end of section F-Masonry										
Repairs in Portland stone work and set in gauged mortar (1:2:9) and point to match existing										
Cut out decayed stones in facings of walls, piers or the like, prepare for and set new 75mm thick stone facings										
single stone	400.00	-	400.00	15.00	15.00	481.95	10.12	892.08	m²	981.28
areas of two or more adjacent stones	400.00	-	400.00	12.00	12.00	385.56	10.12	795.68	m²	875.25
Take out decayed stone sill, prepare for and set new sunk weathered and throated sill										
175 x 100mm	210.00	-	210.00	1.80	1.80	57.83	1.37	269.20	m	296.12
250 x 125mm	286.00	-	286.00	2.50	2.50	80.33	2.31	368.63	m	405.50
Take out sections of decayed stone coping, prepare for and set new weathered and twice throated coping										
300 x 75mm	255.00	-	255.00	1.40	1.40	44.98	1.71	301.69	m	331.86
300 x 100mm	290.00	-	290.00	1.70	1.70	54.62	2.27	346.89	m	381.57
Repairs in York stone work and set in gauged mortar (1:2:9) and point to match existing										
Cut out decayed stones in facings of walls, piers or the like, prepare for and set new 75mm thick stone facings										
single stone	302.00	-	302.00	17.00	17.00	546.21	10.12	858.34	m²	944.17
areas of two or more adjacent stones	302.00	-	302.00	13.75	13.75	441.79	10.12	753.91	m²	829.30
Take out decayed stone sill, prepare for and set new sunk weathered and throated sill										
175 x 100mm	131.00	-	131.00	1.80	1.80	57.83	1.71	190.54	m	209.60
250 x 125mm	203.00	-	203.00	2.50	2.50	80.33	2.27	285.59	m	314.15
Take out sections of decayed stone coping, prepare for and set new weathered and twice throated coping										
300 x 75mm	143.00	-	143.00	1.40	1.40	44.98	1.71	189.69	m	208.66
300 x 100mm	170.00	-	170.00	1.70	1.70	54.62	2.27	226.89	m	249.57
Rake out decayed mortar joints and re-point in gauged mortar (1:2:9)										
Re-point rubble walling										
coursed	-	-	-	1.25	1.25	40.16	1.68	41.84	m²	46.03
uncoursed	-	-	-	1.32	1.32	42.41	1.68	44.09	m²	48.50
squared	-	-	-	1.25	1.25	40.16	1.68	41.84	m²	46.03
Re-point stonework										
ashlar	-	-	-	1.20	1.20	38.56	1.68	40.24	m²	44.26
coursed block	-	-	-	1.20	1.20	38.56	1.68	40.24	m²	44.26
Removing existing and replacing; timber										
Cut out rotten or infected structural timber members including cutting away wall, etc. and shoring up adjacent work and replace with new pressure impregnated timber										
Plates and bed in cement mortar (1:3)										
50 x 75mm	1.13	10.00	1.24	0.30	0.30	9.64	0.25	11.14	m	12.25
75 x 100mm	2.22	10.00	2.44	0.50	0.50	16.07	0.50	19.00	m	20.90
Floor or roof joists										
50 x 175mm	2.27	10.00	2.50	0.57	0.57	18.31	0.57	21.38	m	23.52
50 x 225mm	2.96	10.00	3.26	0.67	0.67	21.53	0.74	25.52	m	28.07
75 x 175mm	3.50	10.00	3.85	0.79	0.79	25.38	0.87	30.10	m	33.11
Rafters										
38 x 100mm	1.20	10.00	1.32	0.35	0.35	11.25	0.25	12.82	m	14.10
50 x 100mm	1.42	10.00	1.56	0.38	0.38	12.21	0.33	14.10	m	15.51
Ceiling joists and collars										
50 x 100mm	1.42	10.00	1.56	0.38	0.38	12.21	0.33	14.10	m	15.51
50 x 150mm	2.04	10.00	2.24	0.52	0.52	16.71	0.50	19.45	m	21.39
Purlins, ceiling beams and struts										
75 x 100mm	2.22	10.00	2.44	0.55	0.55	17.67	0.50	20.61	m	22.67
100 x 225mm	7.13	10.00	7.84	1.23	1.23	39.52	1.49	48.85	m	53.73
Roof trusses										
75 x 100mm	2.22	10.00	2.44	0.92	0.92	29.56	0.50	32.50	m	35.75
75 x 150mm	3.13	10.00	3.44	1.37	1.37	44.02	0.74	48.20	m	53.02

ALTERATIONS, REPAIRS AND CONSERVATION

	MATERIALS			LABOUR				RATES		
Labour hourly rates: (except Specialists) Craft Operatives 18.37 Labourer 13.76 Rates are national average prices. Refer to REGIONAL VARIATIONS for indicative levels of overall pricing in regions	Del to Site £	Waste %	Material Cost £	Craft Optve Hrs	Lab Hrs	Labour Cost £	Sunds £	Nett Rate £	Unit	Gross rate (10%) £
ALTERATION WORK TO EXISTING BUILDINGS (Cont'd)										
Removing existing and replacing; timber (Cont'd)										
Cut out rotten or infected structural timber members including cutting away wall, etc. and shoring up adjacent work and replace with new pressure impregnated timber (Cont'd)										
Hangers in roof										
25 x 75mm	0.50	10.00	0.55	0.30	0.30	9.64	0.14	10.32	m	11.36
38 x 100mm	1.20	10.00	1.32	0.45	0.45	14.46	0.25	16.03	m	17.64
Cut out rotten or infected timber and prepare timber under to receive new										
Roof boarding; provide new 25mm pressure impregnated softwood boarding										
flat boarding; areas not exceeding 0.50m²	10.40	10.00	11.44	1.96	1.96	62.97	1.63	76.05	m²	83.65
flat boarding; areas 0.50 - 5.00m²	10.40	10.00	11.44	1.63	1.63	52.37	1.63	65.45	m²	71.99
sloping boarding; areas not exceeding 0.50m²	10.40	10.00	11.44	2.34	2.34	75.18	1.63	88.26	m²	97.09
sloping boarding; areas 0.50 - 5.00m²	10.40	10.00	11.44	1.89	1.89	60.73	1.63	73.80	m²	81.18
Gutter boarding and bearers; provide new 25mm pressure impregnated softwood boarding and 50 x 50mm bearers										
over 300mm wide	13.76	10.00	15.14	2.06	2.06	66.19	1.97	83.29	m²	91.62
200mm	3.76	10.00	4.14	0.55	0.55	17.67	0.66	22.47	m	24.71
225mm	4.02	10.00	4.42	0.62	0.62	19.92	0.72	25.06	m	27.57
Gutter sides; provide new 19mm pressure impregnated softwood boarding										
150mm	1.19	15.00	1.37	0.46	0.46	14.78	0.18	16.33	m	17.96
200mm	1.86	15.00	2.14	0.50	0.50	16.07	0.25	18.46	m	20.30
225mm	2.26	15.00	2.60	0.57	0.57	18.31	0.28	21.20	m	23.32
Eaves or verge soffit and bearers and provide new 25mm pressure impregnated softwood boarding and 25 x 50mm bearers										
over 300mm wide	11.14	15.00	12.77	2.10	2.10	67.47	1.97	82.21	m²	90.43
225mm	3.10	15.00	3.53	0.67	0.67	21.53	0.72	25.78	m	28.35
Fascia; provide new pressure impregnated softwood										
25 x 150mm	1.47	15.00	1.69	0.46	0.46	14.78	0.25	16.73	m	18.40
25 x 200mm	2.08	15.00	2.39	0.51	0.51	16.39	0.33	19.11	m	21.02
Barge board; provide new pressure impregnated softwood										
25 x 225mm	2.36	15.00	2.71	0.55	0.55	17.67	0.38	20.76	m	22.84
25 x 250mm	2.65	15.00	3.05	0.57	0.57	18.31	0.42	21.78	m	23.96
Take up floor boarding, remove all nails and clean joists under, re-lay boarding, clear away and replace damaged boards and sand surface (20% new softwood boards allowed)										
Square edged boarding										
25mm	1.92	15.00	2.21	1.73	1.73	55.58	0.66	58.45	m²	64.30
32mm	2.60	15.00	2.98	1.76	1.76	56.55	0.82	60.36	m²	66.39
Tongued and grooved boarding										
25mm	1.71	20.00	2.05	1.87	1.87	60.08	0.66	62.80	m²	69.08
32mm	2.60	20.00	3.11	1.89	1.89	60.73	0.82	64.67	m²	71.13
Take up floor boarding, remove all nails and clean joists under, re-lay boarding, destroy and replace infected boards with new pressure impregnated boards and sand surface (20% new softwood boards allowed)										
Square edged boarding										
25mm	2.08	15.00	2.39	1.81	1.81	58.16	0.66	61.21	m²	67.33
32mm	2.65	15.00	3.05	1.83	1.83	58.80	0.82	62.68	m²	68.94
Tongued and grooved boarding										
25mm	1.87	20.00	2.25	1.94	1.94	62.33	0.66	65.24	m²	71.76
32mm	2.67	20.00	3.21	1.95	1.95	62.65	0.82	66.68	m²	73.35
Take up worn or damaged floor boarding, remove all nails and clean joists under and replace with new softwood boarding										
Square edged boarding										
25mm in areas not exceeding 0.50m²	9.60	15.00	11.04	1.90	1.90	61.05	1.63	73.72	m²	81.09
32mm in areas not exceeding 0.50m²	12.98	15.00	14.92	1.94	1.94	62.33	2.09	79.34	m²	87.28
25mm in areas 0.50 - 3.00m²	9.60	15.00	11.04	1.57	1.57	50.44	1.63	63.12	m²	69.43
32mm in areas 0.50 - 3.00m²	12.98	15.00	14.92	1.61	1.61	51.73	2.09	68.74	m²	75.61
Tongued and grooved boarding										
25mm in areas not exceeding 0.50m²	8.56	20.00	10.27	2.00	2.00	64.26	1.63	76.17	m²	83.78
32mm in areas not exceeding 0.50m²	12.98	20.00	15.57	2.06	2.06	66.19	2.09	83.85	m²	92.23
25mm in areas 0.50 - 3.00m²	8.56	20.00	10.27	1.82	1.82	58.48	1.63	70.38	m²	77.42
32mm in areas 0.50 - 3.00m²	12.98	20.00	15.57	1.87	1.87	60.08	2.09	77.74	m²	85.52
Punch down all nails, remove all tacks, etc. and fill nail holes										
Surface of existing flooring										
generally	-	-	-	0.50	0.10	10.56	0.05	10.61	m²	11.67
generally and machine sand in addition	-	-	-	1.00	0.30	22.50	1.77	24.27	m²	26.69
Note removal of old floor coverings included elsewhere										
Easing and overhauling doors										
Ease and adjust door and oil ironmongery										
762 x 1981mm	-	-	-	0.33	0.06	6.89	0.03	6.92	nr	7.61
914 x 2134mm	-	-	-	0.39	0.06	7.99	0.03	8.02	nr	8.82
Take off door, shave 12mm off bottom edge and re-hang										
762 x 1981mm	-	-	-	0.66	0.11	13.64	-	13.64	nr	15.00
914 x 2134mm	-	-	-	0.79	0.13	16.30	-	16.30	nr	17.93
Adapting doors										
Take off door and re-hang on opposite hand, piece out as necessary and oil ironmongery										
762 x 1981mm	-	-	-	2.50	0.25	49.37	0.15	49.52	nr	54.47
914 x 2134mm	-	-	-	2.65	0.44	54.73	0.15	54.88	nr	60.37

Labour hourly rates: (except Specialists) Craft Operatives 18.37 Labourer 13.76 Rates are national average prices. Refer to REGIONAL VARIATIONS for indicative levels of overall pricing in regions	MATERIALS			LABOUR				RATES		
	Del to Site	Waste	Material Cost	Craft Optve	Lab	Labour Cost	Sunds	Nett Rate		Gross rate (10%)
	£	%	£	Hrs	Hrs	£	£	£	Unit	£
ALTERATION WORK TO EXISTING BUILDINGS (Cont'd)										
Removing existing and replacing; timber (Cont'd)										
Refixing removed doors and frames										
Doors										
762 x 1981mm	-	-	-	2.00	0.34	41.42	0.21	41.63	nr	45.79
914 x 2134mm	-	-	-	2.25	0.38	46.56	0.21	46.77	nr	51.45
Doors and frames; oil existing ironmongery										
762 x 1981mm	-	-	-	4.00	0.66	82.56	0.52	83.09	nr	91.40
914 x 2134mm	-	-	-	4.50	0.75	92.99	0.52	93.51	nr	102.86
Repairing doors and frames										
Piece in door where ironmongery removed										
rim lock	-	-	-	1.00	0.15	20.43	0.24	20.67	nr	22.74
rim lock or latch and furniture	-	-	-	1.50	0.25	31.00	0.58	31.58	nr	34.74
mortice lock	-	-	-	1.50	0.25	31.00	0.58	31.58	nr	34.74
mortice lock or latch and furniture	-	-	-	2.00	0.32	41.14	0.69	41.83	nr	46.02
butt hinge	-	-	-	0.50	0.08	10.29	0.33	10.62	nr	11.68
Piece in frame where ironmongery removed										
lock or latch keep	-	-	-	0.50	0.08	10.29	0.33	10.62	nr	11.68
butt hinge	-	-	-	0.50	0.08	10.29	0.33	10.62	nr	11.68
Take off door, dismantle as necessary, cramp and re-wedge with new glued wedges, pin and re-hang										
panelled; 762 x 1981mm	-	-	-	5.00	0.84	103.41	0.24	103.65	nr	114.01
panelled; 914 x 2134mm	-	-	-	6.00	1.00	123.98	0.24	124.22	nr	136.64
glazed panelled; 762 x 1981mm	-	-	-	5.50	0.90	113.42	0.24	113.66	nr	125.02
glazed panelled; 914 x 2134mm	-	-	-	6.50	1.08	134.27	0.24	134.51	nr	147.96
Take off softwood panelled door 762 x 1981mm, dismantle as necessary and replace the following damaged members and re-hang										
top rail; 38 x 125mm	1.33	10.00	1.47	2.15	0.36	44.45	0.68	46.59	nr	51.25
hanging stile; 38 x 125mm	3.47	10.00	3.81	2.36	0.39	48.72	1.71	54.24	nr	59.67
locking stile; 38 x 125mm	3.47	10.00	3.81	2.36	0.39	48.72	1.71	54.24	nr	59.67
middle rail; 38 x 200mm	2.29	10.00	2.52	4.23	0.71	87.47	1.11	91.11	nr	100.22
bottom rail; 38 x 200mm	2.29	10.00	2.52	2.20	0.37	45.51	1.11	49.14	nr	54.05
Take off ash panelled door 762 x 1981mm, dismantle as necessary and replace the following damaged members and re-hang										
top rail; 38 x 125mm	13.41	10.00	14.75	5.20	0.87	107.50	0.68	122.92	nr	135.22
top rail; 38 x 150mm	16.09	10.00	17.70	5.20	0.87	107.50	0.68	125.87	nr	138.46
hanging stile; 38 x 125mm	34.87	10.00	38.35	5.36	0.89	110.71	1.71	150.77	nr	165.85
locking stile; 38 x 125mm	34.87	10.00	38.35	5.36	0.89	110.71	1.71	150.77	nr	165.85
middle rail; 38 x 150mm	16.09	10.00	17.70	10.40	1.73	214.85	1.11	233.67	nr	257.03
middle rail; 38 x 200mm	21.46	10.00	23.60	10.40	1.73	214.85	1.11	239.57	nr	263.52
bottom rail; 38 x 175mm	16.09	10.00	17.70	5.30	0.88	109.47	1.11	128.28	nr	141.11
bottom rail; 38 x 200mm	21.46	10.00	23.60	5.30	0.88	109.47	1.11	134.18	nr	147.60
Take off damaged softwood stop and replace with new										
12 x 38mm	0.32	10.00	0.35	0.35	0.06	7.26	0.15	7.76	m	8.53
25 (fin) x 50mm glue and screw on	0.48	10.00	0.53	0.45	0.08	9.37	0.18	10.08	m	11.08
Take off damaged ash stop and replace with new										
12 x 38mm	1.69	10.00	1.86	0.45	0.08	9.37	0.15	11.38	m	12.51
25 (fin) x 50mm glue and screw on	4.63	10.00	5.09	0.69	0.12	14.33	0.18	19.60	m	21.56
Flush facing old doors										
Take off panelled door, remove ironmongery, cover both sides with 3.2mm hardboard; refix ironmongery, adjust stops and re-hang										
762 x 1981mm	11.36	10.00	12.49	3.10	0.50	63.83	0.38	76.69	nr	84.36
838 x 1981mm	11.36	10.00	12.49	3.15	0.51	64.88	0.38	77.75	nr	85.53
Take off panelled door, remove ironmongery, cover one side with 3.2mm hardboard and other side with 6mm fire resistant sheeting screwed on; refix ironmongery, adjust stops and re-hang										
762 x 1981mm	69.25	10.00	76.17	4.00	0.67	82.70	1.23	160.10	nr	176.11
838 x 1981mm	69.25	10.00	76.17	4.05	0.68	83.76	1.23	161.16	nr	177.27
Take off panelled door, remove ironmongery, cover one side with 3.2mm hardboard, fill panels on other side with 6mm fire resistant sheeting and cover with 6mm fire resistant sheeting screwed on; refix ironmongery, adjust stops and re-hang										
762 x 1981mm	101.03	10.00	111.14	5.00	0.83	103.27	1.58	215.98	nr	237.58
838 x 1981mm	101.03	10.00	111.14	5.05	0.84	104.33	1.58	217.04	nr	238.74
Take off flush door, remove ironmongery, cover one side with 6mm fire resistant sheeting screwed on; refix ironmongery and re-hang										
762 x 1981mm	63.57	10.00	69.93	2.50	0.42	51.70	0.80	122.43	nr	134.67
838 x 1981mm	63.57	10.00	69.93	2.50	0.43	51.84	0.80	122.56	nr	134.82
Easing and overhauling windows										
Ease and adjust and oil ironmongery										
casement opening light	-	-	-	0.41	0.07	8.49	0.03	8.52	nr	9.38
double hung sash and renew sash lines with Flax cord	0.98	10.00	1.08	1.64	0.27	33.84	0.08	35.00	nr	38.50
Refixing removed windows										
Windows										
casement opening light and oil ironmongery	-	-	-	1.03	0.17	21.26	0.06	21.32	nr	23.45
double hung sash and provide new Flax cord	0.98	10.00	1.08	1.70	0.28	35.08	0.08	36.24	nr	39.86
Single or multi-light timber casements and frames										
440 x 920mm	-	-	-	4.00	0.66	82.56	0.06	82.62	nr	90.88
1225 x 1070mm	-	-	-	6.00	1.00	123.98	0.08	124.06	nr	136.46
2395 x 1225mm	-	-	-	8.00	1.33	165.26	0.09	165.35	nr	181.89
Repairing windows										
Remove damaged members of cased frame and replace with new softwood										
parting bead	0.36	10.00	0.40	1.16	0.19	23.92	0.08	24.39	m	26.83
pulley stile	2.40	10.00	2.64	1.68	0.28	34.71	0.21	37.56	m	41.32

ALTERATIONS, REPAIRS AND CONSERVATION

	MATERIALS			LABOUR				RATES		
Labour hourly rates: (except Specialists) Craft Operatives 18.37 Labourer 13.76 Rates are national average prices. Refer to REGIONAL VARIATIONS for indicative levels of overall pricing in regions	Del to Site	Waste	Material Cost	Craft Optve	Lab	Labour Cost	Sunds	Nett Rate	Unit	Gross rate (10%)
	£	%	£	Hrs	Hrs	£	£	£		£
ALTERATION WORK TO EXISTING BUILDINGS (Cont'd)										
Removing existing and replacing; timber (Cont'd)										
Repairing windows (Cont'd)										
Remove damaged members of cased frame and replace with new softwood (Cont'd)										
stop bead	0.89	10.00	0.98	0.62	0.10	12.77	0.08	13.82	m	15.20
Take out sash not exceeding 50mm thick, dismantle as necessary, replace the following damaged members in softwood and re-hang with new Flax sash cord										
glazing bar	2.75	10.00	3.03	3.14	0.52	64.84	0.08	67.94	m	74.73
stile	7.69	10.00	8.46	3.03	0.50	62.54	0.14	71.14	m	78.25
rail	5.08	10.00	5.59	2.80	0.45	57.63	0.38	63.59	m	69.95
Take out casement opening light not exceeding 50mm thick, dismantle as necessary, replace the following damaged members in softwood and re-hang										
glazing bar	1.77	10.00	1.95	1.97	0.32	40.59	0.08	42.61	m	46.88
hanging stile	6.71	10.00	7.38	1.85	0.30	38.11	0.15	45.64	m	50.21
shutting stile	6.71	10.00	7.38	1.85	0.30	38.11	0.15	45.64	m	50.21
rail	4.10	10.00	4.51	1.61	0.27	33.29	0.39	38.19	m	42.01
Take out sash, cramp and re-wedge with new glued wedges, pin and re-hang										
sliding sash	-	-	-	5.65	0.95	116.86	1.30	118.17	nr	129.98
opening casement light	-	-	-	5.00	0.85	103.55	1.30	104.85	nr	115.34
Repair corner of sash with 100 x 100mm angle repair plate										
let in flush	0.45	-	0.45	1.00	0.17	20.71	0.25	21.41	nr	23.56
Carefully rake out and dry crack in sill										
and fill with two part wood repair system	-	-	-	0.50	0.08	10.29	2.25	12.54	m	13.79
Repairs to softwood skirtings, picture rails, architraves, etc.										
Take off and refix										
skirting	-	-	-	0.40	0.07	8.31	0.06	8.37	m	9.21
dado or picture rail	-	-	-	0.35	0.06	7.26	0.05	7.30	m	8.03
architrave	-	-	-	0.26	0.05	5.46	0.05	5.51	m	6.06
Cut out damaged section and piece in new										
square skirting; 19 x 100mm	0.73	10.00	0.80	0.69	0.11	14.19	0.30	15.29	m	16.82
square skirting; 25 x 150mm	1.37	10.00	1.51	0.71	0.12	14.69	0.36	16.56	m	18.22
moulded skirting; 19 x 125mm	0.91	10.00	1.00	0.70	0.11	14.37	0.34	15.72	m	17.29
moulded dado or picture rail; 19 x 50mm	0.87	10.00	0.96	0.74	0.12	15.25	0.15	16.35	m	17.99
moulded or splayed architrave; 25 x 50mm	0.87	10.00	0.96	0.51	0.09	10.61	0.18	11.74	m	12.92
moulded or splayed architrave; 38 x 75mm	1.68	10.00	1.85	0.52	0.09	10.79	0.34	12.98	m	14.28
Cut out rotten or infected skirting and grounds and replace with new treated softwood										
square skirting; 19 x 100mm	0.83	10.00	0.91	1.00	0.17	20.71	0.41	22.03	m	24.23
square skirting; 25 x 150mm	1.41	10.00	1.55	1.07	0.18	22.13	0.56	24.24	m	26.66
moulded skirting; 19 x 125mm	1.01	10.00	1.11	1.03	0.17	21.26	0.47	22.84	m	25.12
Repairs to staircase members in softwood										
Cut out damaged members of staircase and piece in new										
tread; 32 x 275mm	4.23	10.00	4.65	4.33	0.72	89.45	0.69	94.79	m	104.27
tread; 38 x 275mm	4.47	10.00	4.92	4.85	0.81	100.24	0.80	105.95	m	116.55
riser; 19 x 200mm	1.70	10.00	1.87	2.25	0.38	46.56	0.27	48.70	m	53.57
riser; 25 x 200mm	1.98	10.00	2.18	3.03	0.51	62.68	0.41	65.26	m	71.79
Cut back old tread for a width of 125mm and provide and fix new nosing with glued and dowelled joints to tread										
with rounded nosing; 32mm	1.35	10.00	1.49	2.16	0.36	44.63	1.14	47.26	m	51.98
with rounded nosing; 38mm	1.40	10.00	1.54	2.55	0.43	52.76	1.27	55.58	m	61.13
with moulded nosing; 32mm	1.35	10.00	1.49	2.54	0.42	52.44	1.14	55.06	m	60.57
with moulded nosing; 38mm	1.40	10.00	1.54	3.06	0.51	63.23	1.35	66.12	m	72.73
Replace missing square bar baluster 914mm high; clean out old mortices										
25 x 25mm	0.36	-	0.36	0.56	0.09	11.53	-	11.89	nr	13.07
38 x 38mm	0.60	-	0.60	0.75	0.13	15.57	-	16.17	nr	17.78
Cut out damaged square bar baluster 914mm high and piece in new										
25 x 25mm	0.36	-	0.36	0.75	0.03	14.19	0.05	14.60	nr	16.05
38 x 38mm	0.60	-	0.60	0.91	0.15	18.78	0.08	19.46	nr	21.40
Cut out damaged handrail and piece in new										
50mm mopstick	1.10	10.00	1.21	1.95	0.33	40.36	4.80	46.37	m	51.01
75 x 100mm moulded	3.60	10.00	3.96	3.97	0.66	82.01	5.20	91.18	m	100.29
Refixing removed handrail										
to new balustrade	-	-	-	0.69	0.12	14.33	-	14.33	m	15.76
Repairs to staircase members in oak										
Cut out damaged members of staircase and piece in new										
tread; 32 x 275mm	45.40	10.00	49.94	6.09	1.02	125.91	0.69	176.54	m	194.19
tread; 38 x 275mm	54.71	10.00	60.18	7.18	1.20	148.41	0.80	209.38	m	230.32
riser; 19 x 200mm	24.45	10.00	26.90	3.34	0.56	69.06	0.27	96.23	m	105.85
riser; 25 x 200mm	30.27	10.00	33.30	4.56	0.76	94.22	0.44	127.96	m	140.75
Cut back old tread for a width of 125mm and provide and fix new nosing with glued and dowelled joints to tread										
with rounded nosing; 32mm	26.76	10.00	29.44	2.23	0.37	46.06	1.14	76.63	m	84.30
with rounded nosing; 38mm	30.27	10.00	33.30	2.92	0.49	60.38	1.35	95.03	m	104.53
with moulded nosing; 32mm	27.94	10.00	30.73	3.20	0.53	66.08	1.14	97.95	m	107.75
with moulded nosing; 38mm	31.43	10.00	34.57	3.95	0.66	81.64	1.35	117.57	m	129.32
Replace missing square bar baluster 914mm high; clean out old mortices										
25 x 25mm	4.50	-	4.50	0.94	0.16	19.47	-	23.97	nr	26.37
38 x 38mm	5.00	-	5.00	0.99	0.17	20.53	-	25.53	nr	28.08
Cut out damaged square bar baluster 914mm high and piece in new										
25 x 25mm	4.50	-	4.50	1.21	0.20	24.98	0.05	29.52	nr	32.48
38 x 38mm	5.00	-	5.00	1.32	0.22	27.28	0.08	32.35	nr	35.59

Labour hourly rates: (except Specialists) Craft Operatives 18.37 Labourer 13.76 Rates are national average prices. Refer to REGIONAL VARIATIONS for indicative levels of overall pricing in regions	MATERIALS			LABOUR				RATES		
	Del to Site	Waste	Material Cost	Craft Optve	Lab	Labour Cost	Sunds	Nett Rate		Gross rate (10%)
	£	%	£	Hrs	Hrs	£	£	£	Unit	£
ALTERATION WORK TO EXISTING BUILDINGS (Cont'd)										
Removing existing and replacing; timber (Cont'd)										
Repairs to staircase members in oak (Cont'd)										
Cut out damaged handrail and piece in new										
32mm mopstick	18.20	10.00	20.02	3.64	0.61	75.26	4.80	100.08	m	110.09
75 x 100mm moulded	92.47	10.00	101.72	5.50	0.92	113.69	5.20	220.62	m	242.68
Refixing removed handrail										
to new balustrade	-	-	-	1.35	0.23	27.96	-	27.96	m	30.76
Remove broken ironmongery and fix only new to softwood										
Door ironmongery										
75mm butt hinges	-	-	-	0.23	0.04	4.78	0.01	4.79	nr	5.27
100mm butt hinges	-	-	-	0.23	0.04	4.78	0.01	4.79	nr	5.27
tee hinges up to 300mm	-	-	-	0.62	0.10	12.77	0.01	12.78	nr	14.06
single action floor spring hinges	-	-	-	3.00	0.50	61.99	1.95	63.94	nr	70.33
barrel bolt up to 300mm	-	-	-	0.41	0.07	8.49	0.01	8.51	nr	9.36
lock or latch furniture	-	-	-	0.44	0.07	9.05	0.01	9.06	nr	9.97
rim lock	-	-	-	0.80	0.13	16.48	0.01	16.50	nr	18.15
rim latch	-	-	-	0.80	0.13	16.48	0.01	16.50	nr	18.15
mortice lock	-	-	-	1.22	0.20	25.16	0.01	25.18	nr	27.70
mortice latch	-	-	-	1.08	0.18	22.32	0.01	22.33	nr	24.56
pull handle	-	-	-	0.40	0.07	8.31	0.05	8.36	nr	9.19
Window ironmongery										
sash fastener	-	-	-	1.09	0.18	22.50	0.01	22.52	nr	24.77
sash lift	-	-	-	0.23	0.04	4.78	0.01	4.79	nr	5.27
casement stay	-	-	-	0.40	0.07	8.31	0.01	8.33	nr	9.16
casement fastener	-	-	-	0.40	0.07	8.31	0.01	8.33	nr	9.16
Sundry ironmongery										
hat and coat hook	-	-	-	0.29	0.05	6.02	0.01	6.03	nr	6.63
toilet roll holder	-	-	-	0.29	0.05	6.02	0.01	6.03	nr	6.63
Remove broken ironmongery and fix only new to hardwood										
Door ironmongery										
75mm butt hinges	-	-	-	0.30	0.05	6.20	0.01	6.21	nr	6.84
100mm butt hinges	-	-	-	0.30	0.05	6.20	0.01	6.21	nr	6.84
tee hinges up to 300mm	-	-	-	0.86	0.14	17.72	0.01	17.74	nr	19.51
single action floor spring hinges	-	-	-	4.20	0.70	86.79	1.95	88.74	nr	97.61
barrel bolt up to 300mm	-	-	-	0.60	0.10	12.40	0.01	12.41	nr	13.65
lock or latch furniture	-	-	-	0.61	0.10	12.58	0.01	12.60	nr	13.86
rim lock	-	-	-	1.15	0.19	23.74	0.01	23.75	nr	26.13
rim latch	-	-	-	1.15	0.19	23.74	0.01	23.75	nr	26.13
mortice lock	-	-	-	1.77	0.30	36.64	0.01	36.66	nr	40.32
mortice latch	-	-	-	1.58	0.26	32.60	0.01	32.62	nr	35.88
pull handle	-	-	-	0.57	0.10	11.85	0.05	11.89	nr	13.08
Window ironmongery										
sash fastener	-	-	-	1.57	0.26	32.42	0.01	32.43	nr	35.68
sash lift	-	-	-	0.31	0.05	6.38	0.01	6.40	nr	7.04
casement stay	-	-	-	0.56	0.09	11.53	0.01	11.54	nr	12.69
casement fastener	-	-	-	0.56	0.09	11.53	0.01	11.54	nr	12.69
Sundry ironmongery										
hat and coat hook	-	-	-	0.42	0.07	8.68	0.01	8.69	nr	9.56
toilet roll holder	-	-	-	0.42	0.07	8.68	0.01	8.69	nr	9.56
Removing existing and replacing; metal										
Replace defective arch bar										
Take out defective arch bar including cutting away brickwork as necessary, replace with new mild steel arch bar primed all round and make good all work disturbed										
flat arch bar; 30 x 6mm	5.70	-	5.70	1.03	1.03	33.09	1.07	39.86	m	43.84
flat arch bar; 50 x 6mm	9.54	-	9.54	1.12	1.12	35.99	1.56	47.09	m	51.79
angle arch bar; 50 x 50 x 6mm	14.78	-	14.78	1.27	1.27	40.81	2.78	58.36	m	64.20
angle arch bar; 75 x 50 x 6mm	18.69	-	18.69	1.44	1.44	46.27	3.49	68.45	m	75.29
Replace defective baluster										
Cut out defective baluster, clean core and bottom rails, provide new mild steel baluster and weld on										
12mm diameter x 686mm long	3.31	-	3.31	0.92	0.92	29.56	2.22	35.09	nr	38.60
12mm diameter x 762mm long	4.41	-	4.41	0.94	0.94	30.20	2.22	36.84	nr	40.52
12 x 23 x 686mm long	3.05	-	3.05	0.94	0.94	30.20	2.62	35.88	nr	39.46
12 x 12 x 762mm long	4.07	-	4.07	0.95	0.95	30.52	2.62	37.22	nr	40.94
Repair damaged weld										
Clean off damaged weld including removing remains of weld and cleaning off surrounding paint back to bright metal and re-weld connection between 10mm rail and										
13mm diameter rod	-	-	-	0.50	0.50	16.07	3.27	19.33	nr	21.27
13 x 13mm bar	-	-	-	0.53	0.53	17.03	3.88	20.91	nr	23.01
25 x 25mm bar	-	-	-	0.60	0.60	19.28	4.88	24.15	nr	26.57
Removing existing and replacing; corrugated fibre cement roofing repairs										
Remove damaged sheets and provide and fix new with screws and washers to timber purlins and ease and adjust edges of adjoining sheets as necessary										
one sheet; 2450mm long Profile 3 natural grey	24.79	10.00	27.27	1.23	1.23	39.52	2.98	69.77	nr	76.75
one sheet; 2450mm long Profile 3 standard colour	26.77	10.00	29.45	1.23	1.23	39.52	2.98	71.95	nr	79.15
one sheet; 1825mm long Profile 6 natural grey	17.16	10.00	18.88	1.78	1.78	57.19	4.41	80.48	nr	88.53
one sheet; 1825mm long Profile 6 standard colour	18.98	10.00	20.88	1.78	1.78	57.19	4.41	82.48	nr	90.73
over one sheet; Profile 3 natural grey	15.59	10.00	17.15	0.89	0.89	28.60	2.82	48.57	m²	53.42
over one sheet; Profile 3 standard colour	10.43	10.00	11.48	0.89	0.89	28.60	2.82	42.89	m²	47.18
over one sheet; Profile 6 natural grey	5.23	10.00	5.76	0.89	0.89	28.60	4.16	38.51	m²	42.36
over one sheet; Profile 6 standard colour	6.03	10.00	6.63	0.89	0.89	28.60	4.16	39.38	m²	43.32

Labour hourly rates: (except Specialists) Craft Operatives 18.37 Labourer 13.76 Rates are national average prices. Refer to REGIONAL VARIATIONS for indicative levels of overall pricing in regions	MATERIALS			LABOUR				RATES		
	Del to Site	Waste	Material Cost	Craft Optve	Lab	Labour Cost	Sunds	Nett Rate	Unit	Gross rate (10%)
	£	%	£	Hrs	Hrs	£	£	£		£
ALTERATION WORK TO EXISTING BUILDINGS (Cont'd)										
Removing existing and replacing; corrugated fibre cement roofing repairs (Cont'd)										
Remove damaged sheets and provide and fix new with hook bolts and washers to steel purlins and ease and adjust edges of adjoining sheets as necessary										
one sheet; 2450mm long Profile 3 natural grey	24.79	10.00	27.27	1.37	1.37	44.02	2.98	74.27	nr	81.70
one sheet; 2450mm long Profile 3 standard colour	26.77	10.00	29.45	1.37	1.37	44.02	2.98	76.45	nr	84.10
one sheet; 1825mm long Profile 6 natural grey	17.16	10.00	18.88	1.98	1.98	63.62	4.41	86.90	nr	95.59
one sheet; 1825mm long Profile 6 standard colour	18.98	10.00	20.88	1.98	1.98	63.62	4.41	88.91	nr	97.80
over one sheet; Profile 3 natural grey	15.59	10.00	17.15	1.03	1.03	33.09	2.82	53.06	m²	58.37
over one sheet; Profile 3 standard colour	10.43	10.00	11.48	1.03	1.03	33.09	2.82	47.39	m²	52.13
over one sheet; Profile 6 natural grey	5.23	10.00	5.76	1.03	1.03	33.09	4.16	43.01	m²	47.31
over one sheet; Profile 6 standard colour	6.03	10.00	6.63	1.03	1.03	33.09	4.16	43.88	m²	48.27
Removing existing and replacing; felt roofing repairs for roofs covered with bituminous felt roofing, BS EN 13707										
Sweep clean, cut out defective layer, re-bond adjacent felt and cover with one layer of felt bonded in hot bitumen; patches not exceeding 1.00m²										
fibre based roofing felt type 1B	1.47	15.00	1.69	0.88	0.88	28.27	3.57	33.54	m²	36.89
for each defective underlayer cut out and replaced, add	1.47	15.00	1.69	0.88	0.88	28.27	3.57	33.54	m²	36.89
fibre based mineral surfaced roofing felt type 1E	2.01	15.00	2.31	1.05	1.05	33.74	3.57	39.62	m²	43.58
high performance polyester based roofing felt type 5B	6.65	15.00	7.65	1.02	1.02	32.77	3.57	43.99	m²	48.39
for each defective underlayer cut out and replaced, add	6.65	15.00	7.65	1.02	1.02	32.77	3.57	43.99	m²	48.39
high performance polyester based mineral surfaced roofing felt type 5E	6.65	15.00	7.65	1.13	1.13	36.31	3.57	47.52	m²	52.28
Sweep clean, cut out defective layer, re-bond adjacent felt and cover with one layer of felt bonded in hot bitumen; patches 1.00 - 3.00m²										
fibre based roofing felt type 1B	1.47	15.00	1.69	0.43	0.43	13.82	3.57	19.08	m²	20.99
for each defective underlayer cut out and replaced, add	1.47	15.00	1.69	0.43	0.43	13.82	3.57	19.08	m²	20.99
fibre based mineral surfaced roofing felt type 1E	2.01	15.00	2.31	0.56	0.56	17.99	3.57	23.88	m²	26.26
high performance polyester based roofing felt type 5B	6.65	15.00	7.65	0.53	0.53	17.03	3.57	28.25	m²	31.07
for each defective underlayer cut out and replaced, add	6.65	15.00	7.65	0.53	0.53	17.03	3.57	28.25	m²	31.07
high performance polyester based mineral surfaced roofing felt type 5E	6.65	15.00	7.65	0.65	0.65	20.88	3.57	32.10	m²	35.31
Sweep clean surface of existing roof, prime with hot bitumen and dress with										
13mm layer of limestone or granite chippings	1.22	10.00	1.34	-	0.23	3.16	2.00	6.50	m²	7.15
13mm layer of pea shingle	0.80	10.00	0.88	-	0.36	4.95	2.00	7.83	m²	8.61
extra; removing existing chippings or shingle	-	-	-	-	0.24	3.30	0.60	3.90	m²	4.29

This section continues
on the next page

Labour hourly rates: (except Specialists) Craft Operatives 23.99 Labourer 13.76 Rates are national average prices. Refer to REGIONAL VARIATIONS for indicative levels of overall pricing in regions	MATERIALS			LABOUR				RATES		
	Del to Site £	Waste %	Material Cost £	Craft Optve Hrs	Lab Hrs	Labour Cost £	Sunds £	Nett Rate £	Unit	Gross rate (10%) £
ALTERATION WORK TO EXISTING BUILDINGS (Cont'd)										
Removing existing and replacing; lead flashings, etc.; replace with new lead										
Remove flashing and provide new code Nr4 lead; wedge into groove										
150mm girth	5.61	5.00	5.89	1.06	-	25.43	0.39	31.71	m	34.88
240mm girth	8.80	5.00	9.24	1.49	-	35.75	0.39	45.37	m	49.91
300mm girth	11.22	5.00	11.78	1.82	-	43.66	0.39	55.84	m	61.42
Remove stepped flashing and provide new code Nr 4 lead; wedge into groove										
180mm girth	6.67	5.00	7.01	1.54	-	36.94	0.39	44.34	m	48.78
240mm girth	8.80	5.00	9.24	1.94	-	46.54	0.39	56.17	m	61.78
300mm girth	11.22	5.00	11.78	2.34	-	56.14	0.39	68.31	m	75.14
Remove apron flashing and provide new code Nr 5 lead; wedge into groove										
150mm girth	6.98	5.00	7.33	1.06	-	25.43	0.39	33.14	m	36.46
300mm girth	13.95	5.00	14.65	1.82	-	43.66	0.39	58.70	m	64.57
450mm girth	20.93	5.00	21.98	2.58	-	61.89	0.60	84.47	m	92.92
Remove lining to valley gutter and provide new code Nr 5 lead										
240mm girth	11.07	5.00	11.63	1.49	-	35.75	-	47.37	m	52.11
300mm girth	13.95	5.00	14.65	1.82	-	43.66	-	58.31	m	64.14
450mm girth	20.93	5.00	21.98	2.58	-	61.89	-	83.87	m	92.26
Remove gutter lining and provide new code Nr 6 lead										
360mm girth	19.72	5.00	20.70	2.12	-	50.86	-	71.56	m	78.72
390mm girth	21.39	5.00	22.45	2.43	-	58.30	-	80.75	m	88.82
450mm girth	24.72	5.00	25.96	2.58	-	61.89	-	87.85	m	96.64
Removing existing and replacing; zinc flashings, etc.; replace with 0.8mm zinc										
Remove flashing and provide new; wedge into groove										
150mm girth	5.68	5.00	5.97	1.12	-	26.87	0.27	33.10	m	36.42
240mm girth	9.09	5.00	9.55	1.60	-	38.38	0.27	48.20	m	53.02
300mm girth	11.36	5.00	11.93	1.94	-	46.54	0.39	58.86	m	64.75
Remove stepped flashing and provide new; wedge into groove										
180mm girth	6.82	5.00	7.16	1.65	-	39.58	0.27	47.01	m	51.71
240mm girth	9.09	5.00	9.55	2.08	-	49.90	0.27	59.71	m	65.69
300mm girth	11.36	5.00	11.93	2.52	-	60.45	0.39	72.78	m	80.05
Remove apron flashing and provide new; wedge into groove										
150mm girth	5.68	5.00	5.97	1.12	-	26.87	0.27	33.10	m	36.42
300mm girth	11.36	5.00	11.93	1.94	-	46.54	0.39	58.86	m	64.75
450mm girth	17.05	5.00	17.90	2.76	-	66.21	0.48	84.59	m	93.05
Remove lining to valley gutter and provide new										
240mm girth	9.09	5.00	9.55	1.60	-	38.38	-	47.93	m	52.72
300mm girth	11.36	5.00	11.93	1.94	-	46.54	-	58.47	m	64.32
450mm girth	17.05	5.00	17.90	2.76	-	66.21	-	84.11	m	92.52
Remove gutter lining and provide new										
360mm girth	13.64	5.00	14.32	2.27	-	54.46	-	68.78	m	75.65
420mm girth	15.91	5.00	16.70	2.61	-	62.61	-	79.32	m	87.25
450mm girth	17.05	5.00	17.90	2.76	-	66.21	-	84.11	m	92.52
Removing existing and replacing; copper flashings, etc.; replace with 0.70mm copper										
Remove flashing and provide new; wedge into groove										
150mm girth	11.25	5.00	11.81	1.06	-	25.43	0.54	37.78	m	41.56
240mm girth	18.00	5.00	18.90	1.49	-	35.75	0.54	55.19	m	60.70
300mm girth	22.50	5.00	23.62	1.82	-	43.66	0.54	67.83	m	74.61
Remove stepped flashing and provide new; wedge into groove										
180mm girth	13.50	5.00	14.17	1.54	-	36.94	0.54	51.66	m	56.83
240mm girth	18.00	5.00	18.90	1.94	-	46.54	0.54	65.98	m	72.58
300mm girth	22.50	5.00	23.62	2.34	-	56.14	0.54	80.30	m	88.33
Remove apron flashing and provide new; wedge into groove										
150mm girth	11.25	5.00	11.81	1.06	-	25.43	0.54	37.78	m	41.56
300mm girth	22.50	5.00	23.62	1.82	-	43.66	0.54	67.83	m	74.61
450mm girth	33.75	5.00	35.44	2.58	-	61.89	0.81	98.14	m	107.96
Remove lining to valley gutter and provide new										
240mm girth	18.00	5.00	18.90	1.49	-	35.75	-	54.65	m	60.11
300mm girth	22.50	5.00	23.62	1.82	-	43.66	-	67.29	m	74.02
450mm girth	33.75	5.00	35.44	2.58	-	61.89	-	97.33	m	107.06
Remove gutter lining and provide new										
360mm girth	27.00	5.00	28.35	2.12	-	50.86	-	79.21	m	87.13
420mm girth	31.50	5.00	33.07	2.43	-	58.30	-	91.37	m	100.51
450mm girth	33.75	5.00	35.44	2.58	-	61.89	-	97.33	m	107.06
Removing existing and replacing; aluminium flashings etc; replace with 0.60mm aluminium										
Remove flashings and provide new; wedge into groove										
150mm girth	2.06	5.00	2.17	1.06	-	25.43	0.24	27.84	m	30.62
240mm girth	3.30	5.00	3.46	1.49	-	35.75	0.24	39.45	m	43.40
300mm girth	4.12	5.00	4.33	1.82	-	43.66	0.24	48.23	m	53.06
Remove stepped flashing and provide new; wedge into groove										
180mm girth	2.47	5.00	2.60	1.54	-	36.94	0.24	39.78	m	43.76
240mm girth	3.30	5.00	3.46	1.94	-	46.54	0.24	50.25	m	55.27

27

Labour hourly rates: (except Specialists) Craft Operatives 23.99 Labourer 13.76 Rates are national average prices. Refer to REGIONAL VARIATIONS for indicative levels of overall pricing in regions	MATERIALS			LABOUR				RATES		
	Del to Site	Waste	Material Cost	Craft Optve	Lab	Labour Cost	Sunds	Nett Rate	Unit	Gross rate (10%)
	£	%	£	Hrs	Hrs	£	£	£		£
ALTERATION WORK TO EXISTING BUILDINGS (Cont'd)										
Removing existing and replacing; aluminium flashings etc; replace with 0.60mm aluminium (Cont'd)										
Remove stepped flashing and provide new; wedge into groove (Cont'd)										
300mm girth	4.12	5.00	4.33	2.34	-	56.14	0.24	60.71	m	66.78
Remove apron flashing and provide new; wedge into groove										
150mm girth	2.06	5.00	2.17	1.06	-	25.43	0.24	27.84	m	30.62
300mm girth	4.12	5.00	4.33	1.82	-	43.66	0.24	48.23	m	53.06
450mm girth	6.19	5.00	6.50	2.56	-	61.41	4.50	72.41	m	79.65
Remove lining to valley gutter and provide new										
240mm girth	3.30	5.00	3.46	1.49	-	35.75	-	39.21	m	43.13
300mm girth	4.12	5.00	4.33	1.82	-	43.66	-	47.99	m	52.79
450mm girth	6.19	5.00	6.50	2.58	-	61.89	-	68.39	m	75.23
Remove gutter lining and provide new										
360mm girth	4.95	5.00	5.20	2.12	-	50.86	-	56.06	m	61.66
420mm girth	5.78	5.00	6.06	2.43	-	58.30	-	64.36	m	70.80
450mm girth	6.19	5.00	6.50	2.58	-	61.89	-	68.39	m	75.23

This section continues
on the next page

ALTERATION WORK TO EXISTING BUILDINGS (Cont'd)

Labour hourly rates: (except Specialists) Craft Operatives 18.37 Labourer 13.76 Rates are national average prices. Refer to REGIONAL VARIATIONS for indicative levels of overall pricing in regions	MATERIALS			LABOUR				RATES		
	Del to Site £	Waste %	Material Cost £	Craft Optve Hrs	Lab Hrs	Labour Cost £	Sunds £	Nett Rate £	Unit	Gross rate (10%) £
ALTERATION WORK TO EXISTING BUILDINGS (Cont'd)										
Removing existing and replacing; wall and ceiling finishings										
Cut out damaged two coat plaster to wall in patches and make out in new plaster to match existing										
under 1.00m²	3.88	10.00	4.27	2.13	2.13	68.44	0.78	73.49	m²	80.83
1.00 - 2.00m²	3.88	10.00	4.27	1.76	1.76	56.55	0.78	61.60	m²	67.76
2.00 - 4.00m²	3.88	10.00	4.27	1.39	1.39	44.66	0.78	49.71	m²	54.68
Cut out damaged three coat plaster to wall in patches and make out in new plaster to match existing										
under 1.00m²	5.06	10.00	5.57	2.50	2.50	80.33	1.20	87.09	m²	95.80
1.00 - 2.00m²	5.06	10.00	5.57	2.06	2.06	66.19	1.20	72.95	m²	80.25
2.00 - 4.00m²	5.06	10.00	5.57	1.62	1.62	52.05	1.20	58.82	m²	64.70
Cut out damaged two coat rendering to wall in patches and make out in new rendering to match existing										
under 1.00m²	1.56	10.00	1.72	1.93	1.93	62.01	0.78	64.51	m²	70.96
1.00 - 2.00m²	1.56	10.00	1.72	1.60	1.60	51.41	0.78	53.90	m²	59.30
2.00 - 4.00m²	1.56	10.00	1.72	1.26	1.26	40.48	0.78	42.98	m²	47.28
Cut out damaged three coat rendering to wall in patches and make out in new rendering to match existing										
under 1.00m²	2.40	10.00	2.64	2.28	2.28	73.26	1.20	77.10	m²	84.81
1.00 - 2.00m²	2.40	10.00	2.64	1.87	1.87	60.08	1.20	63.92	m²	70.32
2.00 - 4.00m²	2.40	10.00	2.64	1.47	1.47	47.23	1.20	51.07	m²	56.18
Cut out damaged two coat plaster to ceiling in patches and make out in new plaster to match existing										
under 1.00m²	3.88	10.00	4.27	2.38	2.38	76.47	0.78	81.52	m²	89.67
1.00 - 2.00m²	3.88	10.00	4.27	1.96	1.96	62.97	0.78	68.02	m²	74.83
2.00 - 4.00m²	3.88	10.00	4.27	1.54	1.54	49.48	0.78	54.53	m²	59.98
Cut out damaged three coat plaster to ceiling in patches and make out in new plaster to match existing										
under 1.00m²	5.06	10.00	5.57	2.79	2.79	89.64	1.20	96.41	m²	106.05
1.00 - 2.00m²	5.06	10.00	5.57	2.30	2.30	73.90	1.20	80.67	m²	88.73
2.00 - 4.00m²	5.06	10.00	5.57	1.81	1.81	58.16	1.20	64.92	m²	71.41
Cut out damaged 9.5mm plasterboard and two coat plaster to wall in patches and make out in new plasterboard and plaster to match existing										
under 1.00m²	7.39	10.00	8.13	2.23	2.23	71.65	1.34	81.11	m²	89.22
1.00 - 2.00m²	7.39	10.00	8.13	1.78	1.78	57.19	1.34	66.65	m²	73.32
2.00 - 4.00m²	7.39	10.00	8.13	1.36	1.36	43.70	1.34	53.16	m²	58.47
Cut out damaged 12.5mm plasterboard and two coat plaster to wall in patches and make out in new plasterboard and plaster to match existing										
under 1.00m²	6.42	10.00	7.07	2.28	2.28	73.26	1.49	81.81	m²	89.99
1.00 - 2.00m²	6.42	10.00	7.07	1.83	1.83	58.80	1.49	67.35	m²	74.08
2.00 - 4.00m²	6.42	10.00	7.07	1.38	1.38	44.34	1.49	52.89	m²	58.18
Cut out damaged 9.5mm plasterboard and two coat plaster to ceiling in patches and make out in new plasterboard and plaster to match existing										
under 1.00m²	7.39	10.00	8.13	2.54	2.54	81.61	1.34	91.07	m²	100.18
1.00 - 2.00m²	7.39	10.00	8.13	2.03	2.03	65.22	1.34	74.68	m²	82.15
2.00 - 4.00m²	7.39	10.00	8.13	1.55	1.55	49.80	1.34	59.26	m²	65.19
Cut out damaged 12.5mm plasterboard and two coat plaster to ceiling in patches and make out in new plasterboard and plaster to match existing										
under 1.00m²	6.42	10.00	7.07	2.56	2.56	82.25	1.49	90.80	m²	99.88
1.00 - 2.00m²	6.42	10.00	7.07	2.06	2.06	66.19	1.49	74.74	m²	82.21
2.00 - 4.00m²	6.42	10.00	7.07	1.57	1.57	50.44	1.49	58.99	m²	64.89
Cut out damaged lath and plaster to wall in patches and make out with new metal lathing and three coat plaster to match existing										
under 1.00m²	9.44	10.00	10.38	2.53	2.53	81.29	1.49	93.16	m²	102.47
1.00 - 2.00m²	9.44	10.00	10.38	2.14	2.14	68.76	1.49	80.63	m²	88.69
2.00 - 4.00m²	9.44	10.00	10.38	1.73	1.73	55.58	1.49	67.45	m²	74.20
Cut out damaged lath and plaster to ceiling in patches and make out with new metal lathing and three coat plaster to match existing										
under 1.00m²	9.44	10.00	10.38	2.87	2.87	92.21	1.49	104.08	m²	114.49
1.00 - 2.00m²	9.44	10.00	10.38	2.43	2.43	78.08	1.49	89.94	m²	98.94
2.00 - 4.00m²	9.44	10.00	10.38	1.95	1.95	62.65	1.49	74.52	m²	81.97
Cut out crack in plaster, form dovetailed key and make good with plaster										
not exceeding 50mm wide	0.23	10.00	0.26	0.20	0.20	6.43	0.05	6.73	m	7.40
Make good plaster where the following removed										
small pipe	-	-	-	0.16	0.16	5.14	0.05	5.19	nr	5.70
small steel section	-	-	-	0.21	0.21	6.75	0.06	6.81	nr	7.49
Cut out damaged plaster moulding or cornice and make out with new to match existing										
up to 100mm girth on face	1.74	10.00	1.91	1.25	0.63	31.63	0.31	33.86	m	37.25
100 - 150mm girth on face	2.90	10.00	3.19	1.86	0.93	46.97	0.62	50.77	m	55.85
150 - 200mm girth on face	5.80	10.00	6.38	2.25	1.13	56.88	1.11	64.37	m	70.81
Remove damaged fibrous plaster louvered ventilator, replace with new and make good										
229 x 75mm	2.73	5.00	2.87	0.50	0.50	16.07	0.15	19.08	nr	20.99
229 x 150mm	3.33	-	3.33	0.60	0.60	19.28	0.25	22.86	nr	25.15

	MATERIALS			LABOUR				RATES		
Labour hourly rates: (except Specialists) Craft Operatives 18.37 Labourer 13.76 Rates are national average prices. Refer to REGIONAL VARIATIONS for indicative levels of overall pricing in regions	Del to Site	Waste	Material Cost	Craft Optve	Lab	Labour Cost	Sunds	Nett Rate	Unit	Gross rate (10%)
	£	%	£	Hrs	Hrs	£	£	£		£
ALTERATION WORK TO EXISTING BUILDINGS (Cont'd)										
Removing existing and replacing; wall and ceiling finishings (Cont'd)										
Take out damaged 152 x 152 x 5.5mm white glazed wall tiles and renew to match existing										
isolated tile	0.45	5.00	0.47	0.32	0.32	10.28	0.09	10.84	nr	11.93
patch of 5 tiles	2.25	5.00	2.36	0.60	0.60	19.28	0.45	22.09	nr	24.30
patch 0.50 - 1.00m²	19.80	5.00	20.79	2.32	2.32	74.54	3.53	98.86	m²	108.74
patch 1.00 - 2.00m²	19.80	5.00	20.79	1.66	1.66	53.34	3.53	77.65	m²	85.42
Removing existing and replacing; floor finishings										
Take out damaged 150 x 150 x 12.5mm clay floor tiles and renew to match existing										
isolated tile	0.50	5.00	0.52	0.32	0.21	8.77	0.09	9.38	nr	10.32
patch of 5 tiles	2.50	5.00	2.62	0.65	0.45	18.13	0.45	21.21	nr	23.33
patch 0.50 - 1.00m²	22.50	5.00	23.62	2.40	1.61	66.24	3.53	93.39	m²	102.73
patch 1.00 - 2.00m²	22.50	5.00	23.62	1.90	1.27	52.38	3.53	79.53	m²	87.48
Cut out damaged 25mm granolithic paving in patches and make out in new granolithic to match existing										
under 1.00m²	4.12	5.00	4.32	1.50	1.50	48.19	1.65	54.17	m²	59.59
1.00 - 2.00m²	4.12	5.00	4.32	1.25	1.25	40.16	1.65	46.14	m²	50.75
2.00 - 4.00m²	4.12	5.00	4.32	1.07	1.07	34.38	1.65	40.35	m²	44.39
Cut out damaged 32mm granolithic paving in patches and make out in new granolithic to match existing										
under 1.00m²	5.27	5.00	5.54	1.68	1.68	53.98	1.93	61.45	m²	67.59
1.00 - 2.00m²	5.27	5.00	5.54	1.38	1.38	44.34	1.93	51.81	m²	56.99
2.00 - 4.00m²	5.27	5.00	5.54	1.15	1.15	36.95	1.93	44.42	m²	48.86
Cut out damaged 38mm granolithic paving in patches and make out in new granolithic to match existing										
under 1.00m²	6.26	5.00	6.57	1.80	1.80	57.83	2.28	66.69	m²	73.36
1.00 - 2.00m²	6.26	5.00	6.57	1.50	1.50	48.19	2.28	57.05	m²	62.75
2.00 - 4.00m²	6.26	5.00	6.57	1.25	1.25	40.16	2.28	49.02	m²	53.92
Cut out damaged 50mm granolithic paving in patches and make out in new granolithic to match existing										
under 1.00m²	8.24	5.00	8.65	2.12	2.12	68.12	2.78	79.54	m²	87.49
1.00 - 2.00m²	8.24	5.00	8.65	1.73	1.73	55.58	2.78	67.01	m²	73.71
2.00 - 4.00m²	8.24	5.00	8.65	1.45	1.45	46.59	2.78	58.01	m²	63.81
Cut out damaged granolithic paving to tread 275mm wide and make out in new granolithic to match existing										
25mm thick	1.15	5.00	1.21	0.44	0.44	14.14	0.47	15.81	m	17.39
32mm thick	1.48	5.00	1.56	0.48	0.48	15.42	0.52	17.50	m	19.25
38mm thick	1.65	5.00	1.73	0.50	0.50	16.07	0.62	18.41	m	20.25
50mm thick	2.31	5.00	2.42	0.59	0.59	18.96	0.76	22.14	m	24.36
Cut out damaged granolithic covering to plain riser 175mm wide and make out in new granolithic to match existing										
13mm thick	0.33	5.00	0.35	0.80	0.80	25.70	4.23	30.28	m	33.31
19mm thick	0.49	5.00	0.52	0.84	0.84	26.99	4.26	31.77	m	34.95
25mm thick	0.66	5.00	0.69	0.88	0.88	28.27	4.30	33.27	m	36.60
32mm thick	0.99	5.00	1.04	0.93	0.93	29.88	4.34	35.25	m	38.78
Cut out damaged granolithic covering to undercut riser 175mm wide and make out in new granolithic to match existing										
13mm thick	0.33	5.00	0.35	0.87	0.87	27.95	4.23	32.53	m	35.78
19mm thick	0.49	5.00	0.52	0.93	0.93	29.88	4.26	34.66	m	38.13
25mm thick	0.66	5.00	0.69	0.98	0.98	31.49	4.30	36.48	m	40.13
32mm thick	0.99	5.00	1.04	1.04	1.04	33.42	4.34	38.79	m	42.67
Cut out damaged granolithic skirting 150mm wide and make out in new granolithic to match existing										
13mm thick	0.33	5.00	0.35	0.72	0.72	23.13	3.62	27.09	m	29.80
19mm thick	0.49	5.00	0.52	0.76	0.76	24.42	3.66	28.60	m	31.46
25mm thick	0.66	5.00	0.69	0.80	0.80	25.70	3.69	30.09	m	33.09
32mm thick	0.82	5.00	0.86	0.82	0.82	26.35	3.73	30.95	m	34.04
Cut out crack in granolithic paving not exceeding 50mm wide, form dovetailed key and make good to match existing										
25mm thick	0.16	5.00	0.17	0.60	0.60	19.28	0.08	19.53	m	21.48
32mm thick	0.33	5.00	0.35	0.66	0.66	21.21	0.09	21.64	m	23.81
38mm thick	0.33	5.00	0.35	0.69	0.69	22.17	0.12	22.64	m	24.90
50mm thick	0.49	5.00	0.52	0.79	0.79	25.38	0.14	26.04	m	28.64

This section continues
on the next page

Labour hourly rates: (except Specialists) Craft Operatives 23.99 Labourer 13.76 Rates are national average prices. Refer to REGIONAL VARIATIONS for indicative levels of overall pricing in regions	MATERIALS			LABOUR				RATES		
	Del to Site	Waste	Material Cost	Craft Optve	Lab	Labour Cost	Sunds	Nett Rate		Gross rate (10%)
	£	%	£	Hrs	Hrs	£	£	£	Unit	£
ALTERATION WORK TO EXISTING BUILDINGS (Cont'd)										
Removing existing and replacing; balloon gratings										
Remove remains of old balloon grating and provide new plastic grating in outlet or end of pipe										
50mm diameter	1.78	2.00	1.82	0.50	-	11.99	-	13.81	nr	15.19
82mm diameter	2.68	2.00	2.73	0.50	-	11.99	-	14.73	nr	16.20
110mm diameter	2.72	2.00	2.77	0.50	-	11.99	-	14.77	nr	16.25
Removing existing and replacing; stopcocks, valves, etc.										
Turn off water supply, drain down as necessary, take out old valves and re-new with joints to copper										
brass stopcock; 13mm	4.39	2.00	4.48	2.00	-	47.98	-	52.46	nr	57.70
chromium plated sink pillar cock; 13mm	14.75	2.00	15.05	2.00	-	47.98	-	63.03	nr	69.33
brass bib cock; 13mm	13.15	2.00	13.41	2.00	-	47.98	-	61.39	nr	67.53
drain tap (BS 2879) type 2; 13mm	4.39	2.00	4.48	2.00	-	47.98	-	52.46	nr	57.70
drain tap (BS 2879) type 2; 19mm	17.22	2.00	17.56	2.00	-	47.98	-	65.54	nr	72.10
high pressure ball valve (BS 1212) piston type with copper ball; 13mm	11.49	2.00	11.72	2.00	-	47.98	-	59.70	nr	65.67
high pressure ball valve (BS 1212) piston type with plastic ball; 13mm	5.73	2.00	5.84	2.00	-	47.98	-	53.82	nr	59.21
low pressure ball valve (BS 1212) piston type with copper ball; 13mm	11.49	2.00	11.72	2.00	-	47.98	-	59.70	nr	65.67
low pressure ball valve (BS 1212) piston type with plastic ball; 13mm	5.90	2.00	6.02	2.00	-	47.98	-	54.00	nr	59.40
high pressure ball valve (BS 1212) piston type with copper ball; 19mm	20.30	2.00	20.71	2.50	-	59.97	-	80.68	nr	88.75
high pressure ball valve (BS 1212) piston type with plastic ball; 19mm	13.95	2.00	14.23	2.50	-	59.97	-	74.20	nr	81.62
Removing existing and replacing; traps										
Take out old plastic trap and replace with new 76mm seal trap (BS EN 274) with "O"" ring joint outlet										
P trap; 36mm diameter	3.77	2.00	3.85	1.25	-	29.99	-	33.83	nr	37.22
P trap; 42mm diameter	4.36	2.00	4.45	1.50	-	35.99	-	40.43	nr	44.48
S trap; 36mm diameter	4.78	2.00	4.88	1.25	-	29.99	-	34.86	nr	38.35
S trap; 42mm diameter	5.59	2.00	5.70	1.50	-	35.99	-	41.69	nr	45.86
bath trap with overflow connection; 42mm diameter	8.07	2.00	8.23	1.75	-	41.98	-	50.21	nr	55.24
Take out old copper trap and replace with new 76mm seal two piece trap with compression joint										
P trap; 35mm diameter	32.96	2.00	33.62	1.25	-	29.99	-	63.61	nr	69.97
P trap; 42mm diameter	38.82	2.00	39.60	1.50	-	35.99	-	75.58	nr	83.14
S trap; 35mm diameter	34.91	2.00	35.61	1.25	-	29.99	-	65.60	nr	72.16
S trap; 42mm diameter	42.00	2.00	42.84	1.50	-	35.99	-	78.82	nr	86.71
bath trap with male iron overflow connection; 42mm diameter	117.68	2.00	120.03	1.75	-	41.98	-	162.02	nr	178.22
Removing existing and replacing; preparation of pipe for insertion of new pipe or fitting										
Cut into cast iron pipe and take out a length up to 1.00m for insertion of new pipe or fitting										
50mm diameter	-	-	-	3.70	-	88.76	-	88.76	nr	97.64
64 and 76mm diameter	-	-	-	4.30	-	103.16	-	103.16	nr	113.47
89 and 102mm diameter	-	-	-	4.90	-	117.55	-	117.55	nr	129.31
Cut into cast iron pipe with caulked joint and take out a length up to 1.00m for insertion of new pipe or fitting										
50mm diameter	-	-	-	3.70	-	88.76	-	88.76	nr	97.64
64 and 76mm diameter	-	-	-	4.30	-	103.16	-	103.16	nr	113.47
89 and 102mm diameter	-	-	-	4.90	-	117.55	-	117.55	nr	129.31
Cut into asbestos cement pipe and take out a length up to 1.00m for insertion of new pipe or fitting										
50mm diameter	-	-	-	3.15	-	75.57	-	75.57	nr	83.13
64 and 76mm diameter	-	-	-	3.75	-	89.96	-	89.96	nr	98.96
89 and 102mm diameter	-	-	-	4.35	-	104.36	-	104.36	nr	114.79
Cut into polythene or p.v.c. pipe and take out a length up to 1.00m for insertion of new pipe or fitting										
up to 25mm diameter	-	-	-	2.50	-	59.97	-	59.97	nr	65.97
32 - 50mm diameter	-	-	-	3.15	-	75.57	-	75.57	nr	83.13
64 and 76mm diameter	-	-	-	3.75	-	89.96	-	89.96	nr	98.96
89 and 102mm diameter	-	-	-	4.40	-	105.56	-	105.56	nr	116.11
Cut into steel pipe and take out a length up to 1.00m for insertion of new pipe or fitting										
up to 25mm diameter	-	-	-	3.05	-	73.17	-	73.17	nr	80.49
32 - 50mm diameter	-	-	-	3.70	-	88.76	-	88.76	nr	97.64
64 and 76mm diameter	-	-	-	4.30	-	103.16	-	103.16	nr	113.47
89 and 102mm diameter	-	-	-	4.90	-	117.55	-	117.55	nr	129.31
Cut into stainless steel pipe and take out a length up to 1.00m for insertion of new pipe or fitting										
up to 25mm diameter	-	-	-	2.90	-	69.57	-	69.57	nr	76.53
32 - 50mm diameter	-	-	-	3.50	-	83.96	-	83.96	nr	92.36
64 and 76mm diameter	-	-	-	4.10	-	98.36	-	98.36	nr	108.19
89 and 102mm diameter	-	-	-	4.75	-	113.95	-	113.95	nr	125.35
Cut into copper pipe and take out a length up to 1.00m for insertion of new pipe or fitting										
up to 25mm diameter	-	-	-	2.90	-	69.57	-	69.57	nr	76.53
32 - 50mm diameter	-	-	-	3.50	-	83.96	-	83.96	nr	92.36
64 and 76mm diameter	-	-	-	4.10	-	98.36	-	98.36	nr	108.19
89 and 102mm diameter	-	-	-	4.75	-	113.95	-	113.95	nr	125.35

ALTERATIONS, REPAIRS AND CONSERVATION

Labour hourly rates: (except Specialists) Craft Operatives 23.99 Labourer 13.76 Rates are national average prices. Refer to REGIONAL VARIATIONS for indicative levels of overall pricing in regions	MATERIALS			LABOUR				RATES		
	Del to Site £	Waste %	Material Cost £	Craft Optve Hrs	Lab Hrs	Labour Cost £	Sunds £	Nett Rate £	Unit	Gross rate (10%) £
ALTERATION WORK TO EXISTING BUILDINGS (Cont'd)										
Removing existing and replacing; preparation of pipe for insertion of new pipe or fitting (Cont'd)										
Cut into lead pipe and take out a length up to 1.00m for insertion of new pipe or fitting										
up to 25mm diameter	-	-	-	3.05	-	73.17	-	73.17	nr	80.49
32 - 50mm diameter	-	-	-	3.70	-	88.76	-	88.76	nr	97.64
64 and 76mm diameter	-	-	-	4.30	-	103.16	-	103.16	nr	113.47
89 and 102mm diameter	-	-	-	4.90	-	117.55	-	117.55	nr	129.31
Removing existing and replacing; sanitary fittings										
Disconnect trap, valves and services pipes, remove old fitting and provide and fix new including re-connecting trap, valves and services pipes										
vitreous china lavatory basin with 32mm chromium plated waste outlet, plug, chain and stay and pair of cast iron brackets; 560 x 405mm	71.00	5.00	74.55	6.00	-	143.94	-	218.49	nr	240.34
stainless steel sink with single drainer with 38mm chromium plated waste outlet, plug, chain and stay; 600 x 1000mm	116.00	5.00	121.80	6.00	-	143.94	-	265.74	nr	292.31
stainless steel sink with double drainer with 38mm chromium plated waste outlet, overflow plug, chain and stay; 600 x 1500mm	140.00	5.00	147.00	6.50	-	155.93	-	302.93	nr	333.23
pressed steel vitreous enamelled rectangular top bath with cradles and with 38mm chromium plated waste outlet, plug, chain and stay and 32mm chromium plated overflow and including removing old cast iron bath; 1500mm	119.00	5.00	124.95	8.00	-	191.92	-	316.87	nr	348.56
pressed steel vitreous enamelled rectangular top bath with cradles and with 38mm chromium plated waste outlet, plug, chain and stay and 32mm chromium plated overflow and including removing old cast iron bath; 1700mm	119.00	5.00	124.95	8.00	-	191.92	-	316.87	nr	348.56
Take off old W.C. seat and provide and fix new ring pattern black plastic seat and cover	14.23	5.00	14.94	1.50	-	35.99	-	50.93	nr	56.02
Disconnect supply pipe, overflow and flush pipe and take out old W.C. cistern and provide and fix new 9 litre black plastic W.C. cistern with cover and connect pipe and ball valve										
low level	63.00	5.00	66.15	5.00	-	119.95	-	186.10	nr	204.71
high level	75.00	5.00	78.75	5.00	-	119.95	-	198.70	nr	218.57
Disconnect flush pipe, take off seat and remove old W.C. pan, provide and fix new white china pan, connect flush pipe, form cement joint with drain and refix old seat										
S or P trap pan	41.00	5.00	43.05	5.00	-	119.95	-	163.00	nr	179.30
Disconnect and remove old high level W.C. suite complete and provide and fix new suite with P or S trap pan, plastic ring seat and cover, plastic flush pipe or bend, 9 litre plastic W.C. cistern with cover, chain and pull or lever handle and 13mm low pressure ball valve; adapt supply pipe and overflow as necessary and connect to new suite and cement joint to drain										
new high level suite	124.00	5.00	130.20	8.50	-	203.91	-	334.11	nr	367.53
new low level suite (vitreous china cistern)	116.00	5.00	121.80	9.00	-	215.91	-	337.71	nr	371.48
Disconnect and remove old low level W.C. suite complete and provide and fix new suite with P or S trap pan, plastic ring seat and cover, plastic flush bend, 9 litre vitreous china W.C. cistern with cover, lever handle and 13mm low pressure ball valve with plastic ball; re-connect supply pipe and overflow and cement joint to drain										
new low level suite	116.00	5.00	121.80	8.50	-	203.91	-	325.71	nr	358.29
Removing existing and replacing; water tanks										
Disconnect all pipework, set aside ball valve and take out cold water storage cistern in roof space; replace with new GRP water storage tank (BS EN 13280); reconnect all pipework and ball valve; remove metal filings and clean inside of tank										
size 915 x 610 x 580mm (227 litre); type SC70	264.00	2.00	269.28	9.00	-	215.91	0.75	485.94	nr	534.53

This section continues
on the next page

Labour hourly rates: (except Specialists) Craft Operatives 28.34 Labourer 13.76 Rates are national average prices. Refer to REGIONAL VARIATIONS for indicative levels of overall pricing in regions	MATERIALS			LABOUR				RATES		
	Del to Site	Waste	Material Cost	Craft Optve	Lab	Labour Cost	Sunds	Nett Rate		Gross rate (10%)
	£	%	£	Hrs	Hrs	£	£	£	Unit	£
ALTERATION WORK TO EXISTING BUILDINGS (Cont'd)										
Removing existing and replacing; electrical installations										
Note the following approximate estimates of costs are dependent on the number and disposition of the points. Lamps and fittings together with cutting and making good are excluded										
Removing existing and replacing; strip out and re-wire complete										
Provide new lamp holders, flush switches, flush socket outlets, fitting outlets and consumer unit with circuit breakers; a three bedroom house with fourteen lighting points and associated switches, nineteen 13 amp socket outlets, four 13 amp fitting outlets, one 30 amp cooker control panel, three earth bonding points and one fifteen way consumer unit										
PVC cables in existing conduit	335.83	2.50	360.36	126.38	-	3581.61	100.00	4041.97	nr	4446.17
PVC cables and conduit	373.54	2.50	400.19	178.42	-	5056.42	100.00	5556.62	nr	6112.28
Removing existing and replacing; strip out and re-wire with PVC insulated and sheathed cable in existing conduit										
Point to junction box, re-using existing lamp holder, socket outlet etc										
5 amp socket outlet	1.30	15.00	1.49	3.93	-	111.38	-	112.87	nr	124.15
13 amp socket or fitting outlet	1.87	15.00	2.15	3.31	-	93.81	-	95.95	nr	105.55
lighting point or switch	0.70	15.00	0.80	2.92	-	82.75	-	83.55	nr	91.91
Removing existing and replacing; strip out and re-wire with PVC insulated and sheathed cable and new conduit										
Point to junction box, re-using existing lamp holder, socket outlet etc										
5 amp socket outlet	2.56	15.00	2.86	5.56	-	157.57	-	160.43	nr	176.47
13 amp socket outlet	3.23	15.00	3.62	4.68	-	132.63	-	136.25	nr	149.87
lighting point or switch	1.92	15.00	2.12	4.13	-	117.04	-	119.17	nr	131.08
Removing existing and replacing; luminaires and accessories										
Take out damaged and renew										
flourescent lamp fitting 1500mm, batten type, single tube	12.50	2.50	12.81	0.56	-	15.87	0.38	29.06	nr	31.96
flourescent lamp fitting 1500mm, batten type, twin tube	23.40	2.50	23.98	0.56	-	15.87	0.38	40.23	nr	44.25
flourescent lamp fitting 1500mm, plastic diffused type, single tube	21.50	2.50	22.04	0.56	-	15.87	0.38	38.28	nr	42.11
flourescent lamp fitting 1500mm, plastic diffused type, twin tube	36.40	2.50	37.31	0.56	-	15.87	0.38	53.56	nr	58.91
lamp holder, batten type	3.34	2.50	3.43	0.56	-	15.87	0.38	19.67	nr	21.64
lamp holder, pendant type with rose	1.45	2.50	1.49	1.06	-	30.04	0.38	31.90	nr	35.09
switch, flush type	0.93	2.50	0.95	0.75	-	21.26	0.68	22.88	nr	25.17
switch, surface mounted type	1.67	2.50	1.71	0.75	-	21.26	0.68	23.64	nr	26.00
13 amp socket outlet, flush type	2.80	2.50	2.87	0.75	-	21.26	0.68	24.80	nr	27.28
13 amp socket outlet, surface mounted type	4.32	2.50	4.43	0.88	-	24.94	0.68	30.05	nr	33.05
5 amp socket outlet, flush type	5.14	2.50	5.27	0.70	-	19.84	0.68	25.78	nr	28.36
5 amp socket outlet, surface mounted type	3.62	2.50	3.71	0.88	-	24.94	0.68	29.32	nr	32.25

This section continues
on the next page

Labour hourly rates: (except Specialists) Craft Operatives 18.37 Labourer 13.76 Rates are national average prices. Refer to REGIONAL VARIATIONS for indicative levels of overall pricing in regions	MATERIALS			LABOUR				RATES		
	Del to Site	Waste	Material Cost	Craft Optve	Lab	Labour Cost	Sunds	Nett Rate	Unit	Gross rate (10%)
	£	%	£	Hrs	Hrs	£	£	£		£
ALTERATION WORK TO EXISTING BUILDINGS (Cont'd)										
Removing existing and replacing; work to existing drains										
Break up concrete paving for excavating drain trench and reinstate - break up 100mm concrete paving and 150mm hardcore bed under for excavation of drain trench and reinstate with new hardcore and concrete, BS 8500, design C20P; 20mm aggregate and make good up to existing paving both sides										
400mm wide	6.85	30.00	7.76	-	3.27	45.00	5.10	57.86	m	63.64
Take up flag paving for excavating drain trench and reinstate - carefully take up 50mm precast concrete flag paving and set aside for re-use and break up 150mm hardcore bed under for excavation of drain trench and reinstate with new hardcore and re-lay salvaged flag paving on and with 25mm sand bed and grout in lime sand (1:3) and make good up to existing paving on both sides										
400mm wide	2.82	30.00	3.64	1.47	1.47	47.23	4.70	55.57	m	61.13
Insert new junction in existing drain - excavate for and trace and expose existing drain, break into glazed vitrified clay drain and insert junction with 100mm branch, short length of new pipe and double collar and joint to existing drain; support earthwork, make good concrete bed and haunching and backfill										
existing 100mm diameter drain, invert depth										
600mm	32.09	5.00	33.69	3.00	12.00	220.23	12.98	266.90	72	293.59
750mm	32.09	5.00	33.69	3.00	14.25	251.19	14.02	298.91	nr	328.80
900mm	32.09	5.00	33.69	3.00	16.50	282.15	15.38	331.22	nr	364.34
existing 150mm diameter drain invert depth										
750mm	66.14	5.00	69.45	4.50	15.75	299.39	16.20	385.04	nr	423.54
900mm	66.14	5.00	69.45	4.50	18.00	330.35	18.38	418.17	nr	459.99
1050mm	66.14	5.00	69.45	4.50	20.25	361.31	20.10	450.86	nr	495.94
1200mm	66.14	5.00	69.45	4.50	22.50	392.27	21.98	483.69	nr	532.06
Repair defective drain - excavate for and trace and expose existing drain, break out fractured glazed vitrified clay drain pipe, replace with new pipe and double collars and joint to existing drain; support earthwork, make good concrete bed and haunching and backfill										
existing 100mm diameter drain; single pipe length; invert depth										
450mm	17.08	5.00	17.93	3.00	11.50	213.35	9.82	241.11	nr	265.22
600mm	17.08	5.00	17.93	3.00	13.50	240.87	11.85	270.65	nr	297.72
750mm	17.08	5.00	17.93	3.00	16.00	275.27	14.48	307.68	nr	338.45
add for each additional pipe length; invert depth										
450mm	11.99	5.00	12.59	2.00	6.75	129.62	9.38	151.58	nr	166.74
600mm	11.99	5.00	12.59	2.00	7.60	141.32	11.32	165.23	nr	181.75
750mm	11.99	5.00	12.59	2.00	8.80	157.83	13.57	183.99	nr	202.39
Removing existing and replacing; work to existing manholes										
Raise top of existing manhole - take off cover and frame and set aside for re-use, prepare level bed on existing one brick sides of manhole internal size 610 x 457mm and raise with Class B engineering brickwork in cement mortar (1:3) finished with a fair face and flush pointed; refix salvaged cover and frame, bed frame in cement mortar and cover in grease and sand										
raising 150mm high	19.01	10.00	20.91	5.30	5.30	170.29	4.76	195.95	nr	215.55
raising 225mm high	28.51	10.00	31.36	6.75	6.75	216.88	7.05	255.29	nr	280.82
raising 300mm high	38.02	10.00	41.82	8.50	8.50	273.11	9.03	323.95	nr	356.35
Insert new branch bend in bottom of existing manhole - break into bottom and one brick side of manhole, insert new glazed vitrified clay three quarter section branch bend to discharge over existing main channel, build in end of new drain and make good benching and side of manhole to match existing										
100mm diameter branch bend	32.67	10.00	35.94	5.75	5.75	184.75	3.40	224.09	nr	246.50
Extend existing manhole - take off cover and frame and set aside; break up one brick end wall of 457mm wide (internal) manhole and excavate for and extend manhole 225mm; support earthwork, level and compact bottom, part backfill and remove surplus spoil from site; extend bottom with 150mm concrete, BS 8500, ordinary prescribed mix C15P, 20mm aggregate bed; extend one brick sides and end with Class B engineering brickwork in cement mortar (1:3) fair faced and flush pointed and bond to existing; extend 100mm main channel and insert 100mm three quarter section branch channel bend; extend benching and build in end of new 100mm drain to one brick side, corbel over end of manhole										
manholes of the following invert depths; refix salvaged cover and frame, bed frame in cement mortar and cover in grease and sand										
450mm	60.00	10.00	66.00	17.60	17.90	569.62	13.14	648.76	nr	713.63
600mm	65.00	10.00	71.50	19.80	20.20	641.68	16.89	730.07	nr	803.07
750mm	72.00	10.00	79.20	22.00	22.50	713.74	20.64	813.58	nr	894.94
900mm	80.00	10.00	88.00	24.20	24.80	785.80	24.39	898.19	nr	988.01
Removing existing and replacing; softwood fencing										
Remove defective timber fence post and replace with new 127 x 102mm post; letting post 450mm into ground; backfilling around in concrete mix C20P; 20mm aggregate; securing ends of existing arris rails and gravel boards to new post										
to suit fencing; 914mm high	14.79	7.50	15.22	2.04	2.04	65.55	0.62	81.38	nr	89.52
to suit fencing; 1219mm high	15.89	7.50	16.32	2.06	2.06	66.19	0.70	83.21	nr	91.53
to suit fencing; 1524mm high	16.94	7.50	17.37	2.08	2.08	66.83	0.82	85.02	nr	93.53
to suit fencing; 1829mm high	18.19	7.50	18.62	2.10	2.10	67.47	0.96	87.05	nr	95.76

Labour hourly rates: (except Specialists) Craft Operatives 18.37 Labourer 13.76 Rates are national average prices. Refer to REGIONAL VARIATIONS for indicative levels of overall pricing in regions	MATERIALS			LABOUR				RATES		
	Del to Site	Waste	Material Cost	Craft Optve	Lab	Labour Cost	Sunds	Nett Rate	Unit	Gross rate (10%)
	£	%	£	Hrs	Hrs	£	£	£		£
ALTERATION WORK TO EXISTING BUILDINGS (Cont'd)										
Removing existing and replacing; softwood fencing (Cont'd)										
Remove defective pale from palisade fence and replace with new 19 x 75mm pointed pale; fixing with galvanised nails										
764mm long	0.95	-	0.95	0.55	0.09	11.34	0.27	12.56	nr	13.82
1069mm long	1.25	-	1.25	0.56	0.09	11.53	0.33	13.11	nr	14.42
1374mm long	1.60	-	1.60	0.58	0.10	12.03	0.38	14.01	nr	15.41
Remove defective pales from close boarded fence and replace with new 100mm wide feather edged pales; fixing with galvanised nails										
singly	1.75	-	1.75	0.25	0.04	5.14	0.31	7.21	m	7.93
areas up to 1.00m²	10.90	20.00	13.08	1.32	0.22	27.28	2.62	42.98	m²	47.28
areas over 1.00m²	10.90	20.00	13.08	0.98	0.16	20.20	2.62	35.91	m²	39.50
Remove defective gravel board and centre stump and replace with new 25 x 150mm gravel board and 50 x 50mm centre stump; fixing with galvanised nails										
securing to fence posts	3.00	5.00	3.15	0.53	0.09	10.97	0.31	14.44	m	15.88
Removing existing and replacing; oak fencing										
Remove defective timber fence post and replace with new 127 x 102mm post; letting post 450mm into ground; backfilling around in concrete mix C20P; 20mm aggregate; securing ends of existing arris rails and gravel boards to new post										
to suit fencing; 914mm high	23.40	7.50	23.83	2.17	2.17	69.72	0.62	94.17	nr	103.58
to suit fencing; 1219mm high	27.21	7.50	27.64	2.21	2.21	71.01	0.70	99.35	nr	109.29
to suit fencing; 1524mm high	31.01	7.50	31.44	2.25	2.25	72.29	0.82	104.56	nr	115.01
to suit fencing; 1829mm high	24.82	7.50	25.25	2.29	2.29	73.58	0.96	99.79	nr	109.77
Remove defective pale from palisade fence and replace with new 19 x 75mm pointed pale; fixing with galvanised nails										
764mm long	2.10	-	2.10	0.70	0.12	14.51	0.27	16.88	nr	18.57
1069mm long	2.88	-	2.88	0.76	0.13	15.75	0.33	18.96	nr	20.86
1374mm long	3.75	-	3.75	0.83	0.14	17.17	0.38	21.30	nr	23.43
Remove defective pales from close boarded fence and replace with new 100mm wide feather edged pales; fixing with galvanised nails										
singly	4.65	-	4.65	0.27	0.05	5.65	0.31	10.61	m	11.67
areas up to 1.00m²	28.20	20.00	33.84	2.56	0.43	52.94	2.62	89.41	m²	98.35
areas over 1.00m²	28.20	20.00	33.84	2.07	0.35	42.84	2.62	79.31	m²	87.24
Remove defective gravel board and centre stump and replace with new 25 x 150mm gravel board and 50 x 50mm centre stump; fixing with galvanised nails										
securing to fence posts	7.20	5.00	7.56	1.10	0.18	22.68	0.31	30.56	m	33.61
Repairs to posts 100 x 100mm precast concrete spur 1219mm long; setting into ground; backfilling around in concrete mix C20P; 20mm aggregate										
bolting to existing timber post	10.54	7.50	10.97	0.76	0.76	24.42	2.49	37.88	nr	41.67
Removing existing and replacing; precast concrete flag paving										
Take up uneven paving and set aside for re-use; level up and consolidate existing hardcore and relay salvaged paving										
bedding on 25mm sand bed; grouting in lime mortar (1:3)	0.78	33.00	1.02	0.67	0.93	25.10	0.18	26.30	m²	28.93
Removing existing and replacing; York stone paving										
Take up old paving; re-square; relay random in random sizes										
bedding, jointing and pointing in cement mortar (1:3)	4.19	10.00	4.61	1.10	1.10	35.34	1.35	41.30	m²	45.43
Removing existing and replacing; granite sett paving										
Take up, clean and stack										
old paving	-	-	-	-	0.90	12.38	-	12.38	m²	13.62
Take up old paving, clean and re-lay										
bedding and grouting in cement mortar (1:3)	-	-	-	1.65	2.55	65.40	1.80	67.20	m²	73.92
Clean out joints of old paving and re-grout										
in cement mortar (1:3)	-	-	-	-	0.55	7.57	0.78	8.35	m²	9.18
Removing existing and replacing; bitumen paving										
Cut out damaged 50mm paving to footpath										
and fill with bitumen macadam under 1.00m²	66.00	10.00	72.60	1.65	2.55	65.40	1.80	139.80	m²	153.78
Cut out damaged 75mm paving to road fill with macadam and seal all around with hot bitumen under 1.00m²	99.00	10.00	108.90	1.65	2.55	65.40	-	174.30	m²	191.73
Preparing existing structures for connection or attachment of new works; Hacking surfaces										
of concrete as key for new finishings										
ceiling	-	-	-	-	0.67	9.22	0.06	9.28	m²	10.21
floor	-	-	-	-	0.53	7.29	0.06	7.35	m²	8.09
wall	-	-	-	-	0.53	7.29	0.06	7.35	m²	8.09
of brick wall and raking out joints as key for new finishings										
wall	-	-	-	-	0.60	8.26	0.06	8.32	m²	9.15

Labour hourly rates: (except Specialists) Craft Operatives 18.37 Labourer 13.76 Rates are national average prices. Refer to REGIONAL VARIATIONS for indicative levels of overall pricing in regions	MATERIALS			LABOUR				RATES		
	Del to Site	Waste	Material Cost	Craft Optve	Lab	Labour Cost	Sunds	Nett Rate	Unit	Gross rate (10%)
	£	%	£	Hrs	Hrs	£	£	£		£
ALTERATION WORK TO EXISTING BUILDINGS (Cont'd)										
Repointing joints										
Rake out joints of old brickwork 20mm deep and re-point in cement mortar (1:3)										
For turned in edge of lead flashing										
flush pointing; horizontal	-	-	-	0.50	0.50	16.07	0.36	16.43	m	18.07
flush pointing; stepped	-	-	-	0.85	0.85	27.31	0.51	27.82	m	30.60
weathered pointing; horizontal	-	-	-	0.55	0.55	17.67	0.36	18.03	m	19.83
weathered pointing; stepped	-	-	-	0.90	0.90	28.92	0.51	29.43	m	32.37
ironed in pointing; horizontal	-	-	-	0.55	0.55	17.67	0.36	18.03	m	19.83
ironed in pointing; stepped	-	-	-	0.90	0.90	28.92	0.51	29.43	m	32.37
Repointing										
Rake out joints of old brickwork 20mm deep and re-point in cement-lime mortar(1:1:6)										
Generally; English bond										
flush pointing	-	-	-	1.00	0.66	27.45	1.03	28.49	m²	31.34
weathered pointing	-	-	-	1.00	0.66	27.45	1.03	28.49	m²	31.34
ironed in pointing	-	-	-	1.20	0.80	33.05	1.03	34.09	m²	37.50
Generally; Flemish bond										
flush pointing	-	-	-	1.00	0.66	27.45	1.03	28.49	m²	31.34
weathered pointing	-	-	-	1.00	0.66	27.45	1.03	28.49	m²	31.34
ironed in pointing	-	-	-	1.20	0.80	33.05	1.03	34.09	m²	37.50
Isolated areas not exceeding 1.00m²; English bond										
flush pointing	-	-	-	1.50	1.00	41.32	1.03	42.35	m²	46.59
weathered pointing	-	-	-	1.50	1.00	41.32	1.03	42.35	m²	46.59
ironed in pointing	-	-	-	1.80	1.20	49.58	1.03	50.61	m²	55.67
Isolated areas not exceeding 1.00m²; Flemish bond										
flush pointing	-	-	-	1.50	1.00	41.32	1.03	42.35	m²	46.59
weathered pointing	-	-	-	1.50	1.00	41.32	1.03	42.35	m²	46.59
ironed in pointing	-	-	-	1.80	1.20	49.58	1.03	50.61	m²	55.67
For using coloured mortar										
add	-	-	-	-	-	-	0.75	0.75	m²	0.83
Rake out joints of old brickwork 20mm deep and re-point in cement mortar (1:3)										
Generally; English bond										
flush pointing	-	-	-	1.20	0.80	33.05	1.03	34.09	m²	37.50
weathered pointing	-	-	-	1.20	0.80	33.05	1.03	34.09	m²	37.50
Generally; Flemish bond										
flush pointing	-	-	-	1.20	0.80	33.05	1.03	34.09	m²	37.50
weathered pointing	-	-	-	1.20	0.80	33.05	1.03	34.09	m²	37.50
Isolated areas not exceeding 1.00m²; English bond										
flush pointing	-	-	-	1.80	1.20	49.58	1.03	50.61	m²	55.67
weathered pointing	-	-	-	1.80	1.20	49.58	1.03	50.61	m²	55.67
Isolated areas not exceeding 1.00m²; Flemish bond										
flush pointing	-	-	-	1.80	1.20	49.58	1.03	50.61	m²	55.67
weathered pointing	-	-	-	1.80	1.20	49.58	1.03	50.61	m²	55.67
REPAIRS/CLEANING/RENOVATING AND CONSERVING										
Repairing; asphalt work										
Cut out crack in old covering and make good with new material to match existing										
Floor or tanking										
in two coats, horizontal; 20mm thick	-	-	-	0.66	0.66	21.21	0.75	21.96	m	24.15
in three coats, horizontal; 30mm thick	-	-	-	0.90	0.90	28.92	1.12	30.04	m	33.05
in three coats, vertical; 30mm thick	-	-	-	1.19	1.19	38.23	1.12	39.36	m	43.30
Cut out detached blister in old covering and make good with new material to match existing										
Floor or tanking										
in two coats, horizontal; 20mm thick	-	-	-	0.30	0.30	9.64	0.38	10.01	nr	11.02
in three coats, horizontal; 30mm thick	-	-	-	0.35	0.35	11.25	0.52	11.77	nr	12.95
in three coats, vertical; 30mm thick	-	-	-	0.48	0.48	15.42	0.52	15.95	nr	17.54
Roof covering										
in two coats; 20mm thick	-	-	-	0.30	0.30	9.64	0.38	10.01	nr	11.02
Cut out crack in old covering and make good with new material to match existing										
Roof covering										
in two coats; 20mm thick	-	-	-	0.66	0.66	21.21	0.75	21.96	m	24.15
Repairing slate roofing; repairs to roofs covered with 610 x 300mm slates										
Remove damaged slates and replace with new; sloping or vertical										
one slate	8.40	5.00	8.82	0.32	0.32	10.28	0.14	19.24	nr	21.16
patch of 10 slates	84.00	5.00	88.20	1.10	1.10	35.34	1.05	124.59	nr	137.05
patch of slates 1.00 - 3.00m²	102.90	5.00	108.04	1.25	1.25	40.16	1.14	149.35	m²	164.28
Examine battens, remove defective and provide 20% new										
19 x 38mm	0.18	10.00	0.20	0.09	0.09	2.89	0.08	3.16	m²	3.48
25 x 50mm	0.27	10.00	0.30	0.09	0.09	2.89	0.10	3.29	m²	3.62
Re-cover roof with slates previously removed and stacked and fix with slate nails										
with 75mm lap	-	-	-	0.77	0.77	24.74	0.14	24.88	m²	27.36
extra for providing 20% new slates	20.58	5.00	21.61	-	-	-	0.27	21.88	m²	24.07
Remove double course at eaves and										
refix	-	-	-	0.24	0.24	7.71	0.47	8.18	m	8.99
extra for providing 20% new slates	7.14	5.00	7.50	-	-	-	0.14	7.63	m	8.40

Labour hourly rates: (except Specialists) Craft Operatives 18.37 Labourer 13.76 Rates are national average prices. Refer to REGIONAL VARIATIONS for indicative levels of overall pricing in regions	MATERIALS			LABOUR				RATES		
	Del to Site	Waste	Material Cost	Craft Optve	Lab	Labour Cost	Sunds	Nett Rate	Unit	Gross rate (10%)
	£	%	£	Hrs	Hrs	£	£	£		£
REPAIRS/CLEANING/RENOVATING AND CONSERVING (Cont'd)										
Repairing slate roofing; repairs to roofs covered with 610 x 300mm slates (Cont'd)										
Remove double course at verge and replace with new slates and bed and point in mortar......	31.42	5.00	32.99	0.29	0.29	9.32	0.87	43.17	m	47.49
Repairing slate roofing; repairs to roofs covered with 510 x 250mm slates										
Remove damaged slates and replace with new; sloping or vertical										
one slate........................	3.83	5.00	4.02	0.32	0.32	10.28	0.14	14.44	nr	15.88
patch of 10 slates........................	38.30	5.00	40.21	1.00	1.00	32.13	0.88	73.23	nr	80.55
patch of slates 1.00 - 3.00m²........................	68.94	5.00	72.39	1.49	1.49	47.87	1.12	121.39	m²	133.52
Examine battens, remove defective and provide 20% new										
19 x 38mm........................	0.22	10.00	0.24	0.11	0.11	3.53	0.10	3.88	m²	4.27
25 x 50mm........................	0.33	10.00	0.36	0.11	0.11	3.53	0.14	4.03	m²	4.44
Re-cover roof with slates previously removed and stacked and fix with slate nails										
with 75mm lap........................	-	-	-	1.00	1.00	32.13	0.19	32.33	m²	35.56
extra for providing 20% new slates........................	13.79	5.00	14.48	-	-	-	0.27	14.75	m²	16.22
Remove double course at eaves and										
refix........................	-	-	-	0.25	0.25	8.03	0.80	8.83	m	9.71
extra for providing 20% new slates........................	4.29	5.00	4.50	-	-	-	0.12	4.62	m	5.09
Remove double course at verge and										
replace with new slates and bed and point in mortar......	17.62	5.00	18.50	0.29	0.29	9.32	0.86	28.67	m	31.54
Repairing slate roofing; repairs to roofs covered with 405 x 200mm slates										
Remove damaged slates and replace with new; sloping or vertical										
one slate........................	1.74	5.00	1.83	0.32	0.32	10.28	0.12	12.23	nr	13.45
patch of 10 slates........................	17.40	5.00	18.27	0.90	0.90	28.92	0.76	47.95	nr	52.75
patch of slates 1.00 - 3.00m²........................	52.20	5.00	54.81	2.07	2.07	66.51	1.25	122.56	m²	134.82
Examine battens, remove defective and provide 20% new										
19 x 38mm........................	0.30	10.00	0.32	0.13	0.13	4.18	0.12	4.62	m²	5.08
25 x 50mm........................	0.44	10.00	0.49	0.13	0.13	4.18	0.14	4.80	m²	5.28
Re-cover roof with slates previously removed and stacked and fix with slate nails										
with 75mm lap........................	-	-	-	1.14	1.14	36.63	0.28	36.91	m²	40.60
extra for providing 20% new slates........................	10.44	5.00	10.96	-	-	-	0.27	11.23	m²	12.36
Remove double course at eaves and										
refix........................	-	-	-	0.26	0.26	8.35	0.78	9.13	m	10.05
extra for providing 20% new slates........................	2.37	5.00	2.48	-	-	-	0.12	2.60	m	2.87
Remove double course at verge and										
replace with new slates and bed and point in mortar......	10.14	5.00	10.65	0.30	0.30	9.64	0.94	21.24	m	23.36
Repairing tile roofing; repairs to roofs covered with machine made clay plain tiles laid to 100mm gauge										
Remove damaged tiles and replace with new, sloping or vertical										
one tile........................	0.35	10.00	0.38	0.32	0.32	10.28	0.05	10.71	nr	11.78
patch of 10 tiles........................	3.50	10.00	3.85	0.90	0.90	28.92	0.38	33.14	nr	36.46
patch of tiles 1.00 - 3.00m²........................	21.35	10.00	23.49	1.72	1.72	55.26	2.40	81.15	m²	89.26
Examine battens, remove defective and provide 20% new										
19 x 25mm........................	0.48	10.00	0.53	0.13	0.13	4.18	0.12	4.82	m²	5.31
19 x 38mm........................	0.48	10.00	0.53	0.14	0.14	4.50	0.14	5.16	m²	5.68
19 x 50mm........................	0.72	10.00	0.79	0.14	0.14	4.50	0.17	5.46	m²	6.00
Remove defective and provide twin ply polypropylene underlay; 150 laps; fixing with galvanized steel clout nails										
horizontal or vertical	0.65	15.00	0.74	0.25	0.25	8.03	0.08	8.85	m2	9.74
Re-cover roof with tiles previously removed and stacked										
nail every fourth course with galvanised nails	-	-	-	1.00	1.00	32.13	0.25	32.39	m²	35.62
extra for nailing every course	-	-	-	0.20	0.20	6.43	0.68	7.10	m²	7.81
extra for providing 20% new tiles........................	4.27	10.00	4.70	-	-	-	0.27	4.97	m²	5.46
Repairing tile roofing; repairs to roofs covered with machine made clay plain tiles laid to 90mm gauge										
Remove damaged tiles and replace with new, sloping or vertical										
one tile........................	0.35	10.00	0.38	0.32	0.32	10.28	0.05	10.71	nr	11.78
patch of 10 tiles	3.50	10.00	3.85	0.90	0.90	28.92	0.38	33.14	nr	36.46
patch of tiles 1.00 - 3.00m²........................	23.80	10.00	26.18	2.05	2.05	65.87	2.67	94.72	m²	104.19
Examine battens, remove defective and provide 20% new										
19 x 25mm........................	0.53	10.00	0.59	0.14	0.14	4.50	0.14	5.22	m²	5.74
19 x 38mm........................	0.53	10.00	0.58	0.15	0.15	4.82	0.17	5.57	m²	6.13
19 x 50mm........................	0.80	10.00	0.88	0.15	0.15	4.82	0.18	5.88	m²	6.47
Re-cover roof with tiles previously removed and stacked										
nail every fourth course with galvanised nails	-	-	-	1.15	1.15	36.95	0.28	37.23	m²	40.96
extra for nailing every course	-	-	-	0.23	0.23	7.39	0.70	8.09	m²	8.90
extra for providing 20% new tiles........................	4.76	10.00	5.24	-	-	-	0.27	5.51	m²	6.06

ALTERATIONS, REPAIRS AND CONSERVATION

	MATERIALS			LABOUR				RATES		
Labour hourly rates: (except Specialists) Craft Operatives 18.37 Labourer 13.76 Rates are national average prices. Refer to REGIONAL VARIATIONS for indicative levels of overall pricing in regions	Del to Site	Waste	Material Cost	Craft Optve	Lab	Labour Cost	Sunds	Nett Rate		Gross rate (10%)
	£	%	£	Hrs	Hrs	£	£	£	Unit	£
REPAIRS/CLEANING/RENOVATING AND CONSERVING (Cont'd)										
Repairing tile roofing; repairs to roofs covered with hand made clay plain tiles laid to 100mm gauge										
Remove damaged tiles and replace with new, sloping or vertical										
one tile..	0.92	10.00	1.01	0.32	0.32	10.28	0.05	11.34	nr	12.47
patch of 10 tiles..	9.20	10.00	10.12	0.90	0.90	28.92	0.38	39.41	nr	43.35
patch of tiles 1.00 - 3.00m²	56.12	10.00	61.73	1.72	1.72	55.26	2.40	119.40	m²	131.34
Examine battens, remove defective and provide 20% new										
19 x 25mm..	0.48	10.00	0.53	0.13	0.13	4.18	0.12	4.82	m²	5.31
19 x 38mm..	0.48	10.00	0.53	0.14	0.14	4.50	0.14	5.16	m²	5.68
19 x 50mm..	0.72	10.00	0.79	0.14	0.14	4.50	0.17	5.46	m²	6.00
Remove defective and provide twin ply polypropylene; 150 laps; fixing with galvanized steel clout nails										
horizontal or vertical	0.65	15.00	0.74	0.25	0.25	8.03	0.08	8.85	m2	9.74
Re-cover roof with tiles previously removed and stacked										
nail every fourth course with galvanised nails	-	-	-	1.00	1.00	32.13	0.25	32.39	m²	35.62
extra for nailing every course	-	-	-	0.20	0.20	6.43	0.68	7.10	m²	7.81
extra for providing 20% new tiles..................................	11.22	10.00	12.35	-	-	-	0.27	12.62	m²	13.88
Repairing tile roofing; repairs to roofs covered with hand made clay plain tiles laid to 90mm gauge										
Remove damaged tiles and replace with new, sloping or vertical										
one tile..	0.92	10.00	1.01	0.32	0.32	10.28	0.05	11.34	nr	12.47
patch of 10 tiles..	9.20	10.00	10.12	0.90	0.90	28.92	0.38	39.41	nr	43.35
patch of tiles 1.00 - 3.00m²...	62.56	10.00	68.82	2.05	2.05	65.87	2.67	137.35	m²	151.09
Examine battens, remove defective and provide 20% new										
19 x 25mm..	0.53	10.00	0.59	0.14	0.14	4.50	0.14	5.22	m²	5.74
19 x 38mm..	0.53	10.00	0.58	0.15	0.15	4.82	0.17	5.57	m²	6.13
19 x 50mm..	0.80	10.00	0.88	0.15	0.15	4.82	0.18	5.88	m²	6.47
Remove defective and provide twin ply polypropylene; 150 laps; fixing with galvanized steel clout nails										
horizontal or vertical	0.65	15.00	0.74	0.25	0.25	8.03	0.08	8.85	m2	9.74
Re-cover roof with tiles previously removed and stacked										
nail every fourth course with galvanised nails	-	-	-	1.15	1.15	36.95	0.28	37.23	m²	40.96
extra for nailing every course	-	-	-	0.23	0.23	7.39	0.70	8.09	m²	8.90
extra for providing 20% new tiles..................................	12.51	10.00	13.76	-	-	-	0.27	14.03	m²	15.44
Repairing tile roofing; repairs to roofs covered with concrete granular faced plain tiles laid to 100mm gauge										
Remove damaged tiles and replace with new, sloping or vertical										
one tile..	0.39	10.00	0.43	0.32	0.32	10.28	0.05	10.76	nr	11.83
patch of 10 tiles..	3.90	10.00	4.29	0.90	0.90	28.92	0.38	33.58	nr	36.94
patch of tiles 1.00 - 3.00m²	23.79	10.00	26.17	1.72	1.72	55.26	2.40	83.83	m²	92.22
Examine battens, remove defective and provide 20% new										
19 x 25mm..	0.48	10.00	0.53	0.13	0.13	4.18	0.12	4.82	m²	5.31
19 x 38mm..	0.48	10.00	0.53	0.14	0.14	4.50	0.14	5.16	m²	5.68
19 x 50mm..	0.72	10.00	0.79	0.14	0.14	4.50	0.17	5.46	m²	6.00
Remove defective and provide twin ply polypropylene; 150 laps; fixing with galvanized steel clout nails										
horizontal or vertical	0.65	15.00	0.74	0.25	0.25	8.03	0.08	8.85	m2	9.74
Re-cover roof with tiles previously removed and stacked										
nail every fourth course with galvanised nails	-	-	-	1.00	1.00	32.13	0.25	32.39	m²	35.62
extra for nailing every course	-	-	-	0.20	0.20	6.43	0.68	7.10	m²	7.81
extra for providing 20% new tiles..................................	4.76	10.00	5.23	-	-	-	0.27	5.50	m²	6.05
Repairing tile roofing; repairs to roofs covered with 381 x 227mm concrete interlocking tiles laid to 75mm laps										
Remove damaged tiles and replace with new, sloping or vertical										
one tile..	1.15	10.00	1.26	0.32	0.32	10.28	0.10	11.65	nr	12.82
patch of 10 tiles..	11.50	10.00	12.65	0.90	0.90	28.92	1.10	42.66	nr	46.93
patch of tiles 1.00 - 3.00m²...	19.55	10.00	21.50	0.85	0.85	27.31	1.74	50.56	m²	55.61
Examine battens, remove defective and provide 20% new										
19 x 25mm..	0.17	10.00	0.19	0.05	0.05	1.61	0.10	1.90	m²	2.09
19 x 38mm..	0.17	10.00	0.19	0.06	0.06	1.93	0.10	2.22	m²	2.45
19 x 50mm..	0.26	10.00	0.29	0.07	0.07	2.25	0.12	2.65	m²	2.92
Re-cover roof with tiles previously removed and stacked										
nail every fourth course with galvanised nails	-	-	-	0.50	0.50	16.07	0.03	16.10	m²	17.70
extra for nailing every course	-	-	-	0.05	0.05	1.61	0.09	1.70	m²	1.87
extra for providing 20% new tiles..................................	4.60	10.00	5.06	-	-	-	0.33	5.39	m²	5.93
Repairing tile roofing; repairs generally										
Remove damaged clay tiles and provide new machine made tiles										
half round ridge or hip and bed in mortar......................	15.94	5.00	16.74	0.60	0.60	19.28	1.63	37.65	m	41.41
bonnet hip...	42.32	5.00	44.44	0.75	0.75	24.10	0.45	68.98	m	75.88
trough valley	48.80	5.00	51.24	0.75	0.75	24.10	0.45	75.79	m	83.37
vertical angle	44.55	5.00	46.78	0.75	0.75	24.10	0.45	71.32	m	78.46
Remove double course and replace with new machine made clay tiles and bed and point in mortar										
at eaves..	4.50	10.00	4.95	0.23	0.23	7.39	0.68	13.01	m	14.32
at verge...	10.75	10.00	11.82	0.38	0.38	12.21	2.09	26.12	m	28.73

Labour hourly rates: (except Specialists) Craft Operatives 18.37 Labourer 13.76 Rates are national average prices. Refer to REGIONAL VARIATIONS for indicative levels of overall pricing in regions	MATERIALS			LABOUR				RATES		
	Del to Site	Waste	Material Cost	Craft Optve	Lab	Labour Cost	Sunds	Nett Rate	Unit	Gross rate (10%)
	£	%	£	Hrs	Hrs	£	£	£		£
REPAIRS/CLEANING/RENOVATING AND CONSERVING (Cont'd)										
Repairing tile roofing; repairs generally (Cont'd)										
Rake out defective pointing and re-point in cement mortar (1:3) ridge or hip tiles	-	-	-	0.34	0.34	10.92	0.25	11.18	m	12.30
Hack off defective cement mortar fillet and renew in cement mortar (1:3)	-	-	-	0.34	0.34	10.92	0.51	11.43	m	12.58
Remove defective hip hook and replace with new	0.80	5.00	0.84	0.34	0.34	10.92	0.10	11.87	nr	13.06
Remove damaged clay tiles and provide new hand made tiles										
half round ridge or hip and bed in mortar	27.42	5.00	28.79	0.60	0.60	19.28	1.63	49.71	m	54.68
bonnet hip	44.80	5.00	47.04	0.75	0.75	24.10	0.45	71.59	m	78.75
trough valley	44.80	5.00	47.04	0.75	0.75	24.10	0.45	71.59	m	78.75
vertical angle	51.30	5.00	53.86	0.75	0.75	24.10	0.45	78.41	m	86.25
Remove double course and replace with hand made new clay tiles and bed and point in mortar										
at eaves	7.92	10.00	8.71	0.23	0.23	7.39	0.68	16.78	m	18.45
at verge	22.15	10.00	24.37	0.38	0.38	12.21	2.09	38.66	m	42.53
Rake out defective pointing and re-point in cement mortar (1:3) ridge or hip tiles	-	-	-	0.34	0.34	10.92	0.25	11.18	m	12.30
Hack off defective cement mortar fillet and renew in cement mortar (1:3)	-	-	-	0.34	0.34	10.92	0.51	11.43	m	12.58
Remove defective hip hook and replace with new	0.80	5.00	0.84	0.34	0.34	10.92	0.10	11.87	nr	13.06
Remove damaged clay tiles and provide new concrete tiles										
half round ridge or hip and bed in mortar	8.98	5.00	9.43	0.60	0.60	19.28	1.63	30.34	m	33.37
bonnet hip	33.68	5.00	35.36	0.75	0.75	24.10	0.45	59.91	m	65.90
trough valley	29.60	5.00	31.08	0.75	0.75	24.10	0.45	55.63	m	61.19
vertical angle	37.35	5.00	39.22	0.75	0.75	24.10	0.45	63.76	m	70.14
Remove double course and replace with new concrete tiles and bed and point in mortar										
at eaves	4.44	10.00	4.88	0.23	0.23	7.39	0.68	12.95	m	14.24
at verge	10.90	10.00	11.99	0.38	0.38	12.21	2.09	26.28	m	28.91
Rake out defective pointing and re-point in cement mortar (1:3) ridge or hip tiles	-	-	-	0.34	0.34	10.92	0.25	11.18	m	12.30
Hack off defective cement mortar fillet and renew in cement mortar (1:3)	-	-	-	0.34	0.34	10.92	0.51	11.43	m	12.58
Remove defective hip hook and replace with new	0.80	5.00	0.84	0.34	0.34	10.92	0.10	11.87	nr	13.06
Repairing liquid bitumen; proofing (black) brush applied on old roof covering including cleaning old covering										
Corrugated asbestos roofing										
one coat	2.85	10.00	3.14	-	0.21	2.89	-	6.03	m²	6.63
two coats	2.85	10.00	3.14	-	0.25	3.44	-	6.58	m²	7.23
Felt roofing										
one coat	2.85	10.00	3.14	-	0.17	2.34	-	5.48	m²	6.02
two coats	2.85	10.00	3.14	-	0.21	2.89	-	6.03	m²	6.63
For top coat in green in lieu black add	1.00	10.00	1.10	-	-	-	-	1.10	m²	1.21

This section continues
on the next page

Labour hourly rates: (except Specialists) Craft Operatives 23.99 Labourer 13.76 Rates are national average prices. Refer to REGIONAL VARIATIONS for indicative levels of overall pricing in regions	MATERIALS			LABOUR				RATES		
	Del to Site	Waste	Material Cost	Craft Optve	Lab	Labour Cost	Sunds	Nett Rate	Unit	Gross rate (10%)
	£	%	£	Hrs	Hrs	£	£	£		£
REPAIRS/CLEANING/RENOVATING AND CONSERVING (Cont'd)										
Repairing; lead roofing										
Repair crack clean out and fill with copper bit solder............................	-	-	-	0.80	-	19.19	0.45	19.64	m	21.61
Turn back flashing and re-dress	-	-	-	0.48	-	11.52	-	11.52	m	12.67

This section continues
on the next page

Labour hourly rates: (except Specialists) Craft Operatives 18.37 Labourer 13.76 Rates are national average prices. Refer to REGIONAL VARIATIONS for indicative levels of overall pricing in regions	MATERIALS			LABOUR				RATES		
	Del to Site £	Waste %	Material Cost £	Craft Optve Hrs	Lab Hrs	Labour Cost £	Sunds £	Nett Rate £	Unit	Gross rate (10%) £
REPAIRS/CLEANING/RENOVATING AND CONSERVING (Cont'd)										
Repairing; resecure glass										
Remove decayed putties, paint one coat on edge of rebate and re-putty										
wood window	0.15	10.00	0.17	0.69	-	12.68	0.03	12.87	m	14.16
metal window	0.15	10.00	0.17	0.76	-	13.96	0.03	14.16	m	15.57
Remove beads, remove decayed bedding materials, paint one coat on edge of rebate, re-bed glass and re-fix beads										
wood window	0.15	10.00	0.17	1.22	-	22.41	0.03	22.61	m	24.87
metal window	0.15	10.00	0.17	1.32	-	24.25	0.03	24.44	m	26.89

This section continues
on the next page

Labour hourly rates: (except Specialists) Craft Operatives 23.99 Labourer 13.76 Rates are national average prices. Refer to REGIONAL VARIATIONS for indicative levels of overall pricing in regions	MATERIALS			LABOUR				RATES		
	Del to Site	Waste	Material Cost	Craft Optve	Lab	Labour Cost	Sunds	Nett Rate		Gross rate (10%)
	£	%	£	Hrs	Hrs	£	£	£	Unit	£
REPAIRS/CLEANING/RENOVATING AND CONSERVING (Cont'd)										
Repairing; rainwater goods										
Clean out and make good loose fixings										
eaves gutter..........	-	-	-	0.40	-	9.60	-	9.60	m	10.56
rainwater pipe..........	-	-	-	0.40	-	9.60	-	9.60	m	10.56
rainwater head..........	-	-	-	0.75	-	17.99	-	17.99	nr	19.79
Clean out defective joint to existing eaves gutter and re-make joint with jointing compound and new bolt..........	0.10	-	0.10	0.75	-	17.99	-	18.09	nr	19.90
Repairing; stopcocks, valves, etc.										
Turn off water supply drain down as necessary and re-new washer to the following up to 19mm										
main stopcock..........	0.25	-	0.25	2.00	-	47.98	-	48.23	nr	53.05
service stopcock..........	0.25	-	0.25	2.00	-	47.98	-	48.23	nr	53.05
bib or pillar cock..........	0.25	-	0.25	2.00	-	47.98	-	48.23	nr	53.05
supatap..........	0.25	-	0.25	2.00	-	47.98	-	48.23	nr	53.05
draining tap..........	0.25	-	0.25	2.00	-	47.98	-	48.23	nr	53.05
Repairing; removal of traps, valves, etc., for re-use										
Take out the following, including unscrewing or uncoupling, for re-use and including draining down as necessary										
trap; up to 25mm diameter..........	-	-	-	0.70	-	16.79	-	16.79	nr	18.47
trap; 32 - 50mm diameter..........	-	-	-	0.85	-	20.39	-	20.39	nr	22.43
stop valve; up to 25mm diameter..........	-	-	-	2.00	-	47.98	-	47.98	nr	52.78
stop valve; 32 - 50mm diameter..........	-	-	-	2.25	-	53.98	-	53.98	nr	59.38
radiator valve; up to 25mm diameter..........	-	-	-	2.00	-	47.98	-	47.98	nr	52.78
radiator valve; 32 - 50mm diameter..........	-	-	-	2.25	-	53.98	-	53.98	nr	59.38
tap; up to 25mm diameter..........	-	-	-	2.00	-	47.98	-	47.98	nr	52.78
tap; 32 - 50mm diameter..........	-	-	-	2.25	-	53.98	-	53.98	nr	59.38
extra; burning out one soldered joint up to 25mm diameter..........	-	-	-	0.50	-	11.99	-	11.99	nr	13.19
extra; burning out one soldered joint 32 - 50mm diameter..........	-	-	-	0.80	-	19.19	-	19.19	nr	21.11

This section continues
on the next page

Labour hourly rates: (except Specialists) Craft Operatives 18.37 Labourer 13.76 Rates are national average prices. Refer to REGIONAL VARIATIONS for indicative levels of overall pricing in regions	MATERIALS			LABOUR				RATES		
	Del to Site £	Waste %	Material Cost £	Craft Optve Hrs	Lab Hrs	Labour Cost £	Sunds £	Nett Rate £	Unit	Gross rate (10%) £
REPAIRS/CLEANING/RENOVATING AND CONSERVING (Cont'd)										
Damp-proof course renewal/insertion										
Insert damp proof course in old wall by cutting out one course of brickwork in alternate lengths not exceeding 1.00m and replace old bricks with 50mm bricks in cement mortar (1:3)										
Hyload, pitch polymer in										
one brick wall	1.09	10.00	1.20	3.00	3.00	96.39	8.30	105.88	m	116.47
one and a half brick wall	1.63	10.00	1.80	4.00	4.00	128.52	12.47	142.78	m	157.06
254mm hollow wall	1.09	10.00	1.20	3.50	3.50	112.46	8.30	121.95	m	134.14
Two courses slates in cement mortar (1:3) in										
one brick wall	6.24	10.00	6.86	3.00	3.00	96.39	8.30	111.55	m	122.70
one and a half brick wall	9.36	10.00	10.30	4.00	4.00	128.52	12.47	151.28	m	166.41
254mm hollow wall	6.24	10.00	6.86	3.50	3.50	112.46	8.30	127.61	m	140.38
0.6mm bitumen coated copper in										
one brick wall	22.48	10.00	24.73	3.00	3.00	96.39	8.30	129.41	m²	142.35
one and a half brick wall	33.77	10.00	37.14	4.00	4.00	128.52	12.47	178.13	m	195.94
254mm hollow wall	22.48	10.00	24.73	3.50	3.50	112.46	8.30	145.48	m	160.02
Insert damp proof course in old wall by hand sawing in 600mm lengths										
Hyload, pitch polymer in										
one brick wall	1.09	10.00	1.20	2.50	2.50	80.33	12.40	93.93	m	103.32
one and a half brick wall	1.63	10.00	1.80	3.50	3.50	112.46	18.61	132.87	m	146.15
254mm hollow wall	1.09	10.00	1.20	3.00	3.00	96.39	12.40	109.99	m	120.99
Two courses slates in cement mortar (1:3) in										
one brick wall	6.24	10.00	6.86	2.50	2.50	80.33	12.40	99.59	m	109.55
one and a half brick wall	9.36	10.00	10.30	3.50	3.50	112.46	18.61	141.37	m	155.50
254mm hollow wall	3.12	10.00	3.43	3.00	3.00	96.39	12.40	112.23	m	123.45
0.6mm bitumen coated copper in										
one brick wall	22.48	10.00	24.73	2.50	2.50	80.33	12.40	117.46	m	129.20
one and a half brick wall	33.77	10.00	37.14	3.50	3.50	112.46	18.61	168.21	m	185.03
254mm hollow wall	22.48	10.00	24.73	3.00	3.00	96.39	12.40	133.52	m	146.87
Insert damp proof course in old wall by machine sawing in 600mm lengths										
Hyload, pitch polymer in										
one brick wall	1.09	10.00	1.20	2.00	2.00	64.26	16.47	81.93	m	90.12
one and a half brick wall	1.63	10.00	1.80	3.00	3.00	96.39	24.69	122.88	m	135.16
254mm hollow wall	1.09	10.00	1.20	2.50	2.50	80.33	16.47	97.99	m	107.79
Two courses slates in cement mortar (1:3) in										
one brick wall	6.24	10.00	6.86	2.00	2.00	64.26	16.47	87.59	m	96.35
one and a half brick wall	9.36	10.00	10.30	3.00	3.00	96.39	24.69	131.38	m	144.51
254mm hollow wall	3.12	10.00	3.43	2.50	2.50	80.33	16.47	100.23	m	110.25
0.6mm bitumen coated copper in										
one brick wall	22.48	10.00	24.73	2.00	2.00	64.26	16.47	105.46	m	116.00
one and a half brick wall	33.77	10.00	37.14	3.00	3.00	96.39	24.69	158.22	m	174.05
254mm hollow wall	22.48	10.00	24.73	2.50	2.50	80.33	16.47	121.52	m	133.67
Insert cavity tray in old wall by cutting out by hand in short lengths; insert individual trays and make good with picked stock facing bricks in gauged mortar (1:2:9) including pinning up with slates										
Type E Cavitray by Cavity Trays Ltd in										
half brick skin of cavity wall	10.50	10.00	11.55	1.75	1.75	56.23	10.53	78.31	m	86.14
external angle	6.87	10.00	7.56	1.00	1.00	32.13	9.68	49.36	nr	54.30
internal angle	6.87	10.00	7.56	1.25	1.25	40.16	9.68	57.39	nr	63.13
Type X Cavity Trays Ltd to suit 40 degree pitched roof complete with attached code 4 lead flashing and dress over tiles in										
half brick skin of cavity wall	64.09	10.00	70.50	4.25	4.25	136.55	21.08	228.13	m	250.94
ridge tray	11.98	10.00	13.18	1.00	1.00	32.13	4.91	50.21	nr	55.23
catchment tray; short	6.57	10.00	7.23	0.50	0.50	16.07	2.58	25.87	nr	28.46
corner catchment tray	15.70	10.00	17.27	1.00	1.00	32.13	4.91	54.31	nr	59.74
Damp proofing old brick wall by silicone injection method (excluding removal and reinstatement of plaster, etc.)										
Damp proofing										
one brick wall	1.15	-	1.15	1.00	1.00	32.13	1.12	34.41	m	37.85
one and a half brick wall	1.73	-	1.73	1.50	1.50	48.19	1.68	51.60	m	56.76
254mm hollow wall	1.15	-	1.15	1.35	1.35	43.38	1.12	45.65	m	50.22
Cleaning surfaces										
Thoroughly clean existing concrete surfaces prior to applying damp proof membrane										
floors	-	-	-	-	0.19	2.61	0.03	2.64	m²	2.91
walls	-	-	-	-	0.21	2.89	0.03	2.92	m²	3.21
Thoroughly clean existing concrete surfaces, fill in nail holes and small surface imperfections and leave smooth										
walls	-	-	-	0.35	0.20	9.18	0.06	9.24	m²	10.17
soffits	-	-	-	0.50	0.29	13.18	0.06	13.24	m²	14.56
Thoroughly clean existing brick or block surfaces prior to applying damp proof membrane										
walls	-	-	-	-	0.25	3.44	-	3.44	m²	3.78
Clean out air brick										
225 x 150mm	-	-	-	-	0.20	2.75	-	2.75	nr	3.03
DECONTAMINATION										
Infestation removal/eradication; protective treatment of existing timbers										
Treat with two coats of spray applied preservative										
boarding	0.59	10.00	0.65	-	0.15	2.06	0.06	2.77	m²	3.05

ALTERATIONS, REPAIRS AND CONSERVATION

Labour hourly rates: (except Specialists) Craft Operatives 18.37 Labourer 13.76 Rates are national average prices. Refer to REGIONAL VARIATIONS for indicative levels of overall pricing in regions	MATERIALS			LABOUR				RATES		
	Del to Site £	Waste %	Material Cost £	Craft Optve Hrs	Lab Hrs	Labour Cost £	Sunds £	Nett Rate £	Unit	Gross rate (10%) £
DECONTAMINATION (Cont'd)										
Infestation removal/eradication; protective treatment of existing timbers (Cont'd)										
Treat with two coats of spray applied preservative (Cont'd) structural timbers	0.88	10.00	0.97	-	0.20	2.75	0.08	3.80	m²	4.18
Treat with two coats of brush applied preservative boarding	0.88	10.00	0.97	-	0.26	3.58	0.05	4.59	m²	5.05
structural timbers	1.35	10.00	1.49	-	0.80	11.01	0.08	12.57	m²	13.82
Infestation removal/eradication; insecticide treatment of existing timbers										
Treat worm infected timbers with spray applied proprietary insecticide										
boarding	0.88	10.00	0.97	-	0.24	3.30	0.08	4.35	m²	4.78
structural timbers	1.17	10.00	1.29	-	0.28	3.85	0.09	5.23	m²	5.76
joinery timbers	1.17	10.00	1.29	-	0.32	4.40	0.09	5.78	m²	6.36
Treat worm infected timbers with brush applied proprietary insecticide										
boarding	1.17	10.00	1.29	-	0.36	4.95	0.05	6.29	m²	6.92
structural timbers	1.41	10.00	1.55	-	0.42	5.78	0.08	7.40	m²	8.14
joinery timbers	1.41	10.00	1.55	-	0.48	6.60	0.08	8.23	m²	9.05
Infestation removal/eradication; treatment of wall surfaces										
Treat surfaces of concrete and brickwork adjoining areas where infected timbers removed with a blow lamp	-	-	-	-	0.25	3.44	0.05	3.49	m²	3.83
TEMPORARY WORKS										
Temporary weatherproof coverings										
Providing and erecting and clearing away on completion; for scaffold tube framework see preliminaries										
corrugated iron sheeting	7.78	10.00	8.56	0.15	0.50	9.64	1.18	19.38	m²	21.31
flexible reinforced plastic sheeting	0.71	10.00	0.78	0.10	0.25	5.28	0.96	7.02	m²	7.72
tarpaulins	0.89	10.00	0.98	0.10	0.25	5.28	0.96	7.22	m²	7.94
Temporary internal screens										
Providing and erecting and clearing away on completion; temporary screen of 50 x 50mm softwood framing and cover with										
reinforced building paper	1.93	10.00	2.13	0.43	0.30	12.03	0.33	14.49	m²	15.93
heavy duty polythene sheeting	1.73	10.00	1.90	0.43	0.30	12.03	0.30	14.23	m²	15.65
3mm hardboard	3.12	10.00	3.43	0.53	0.32	14.14	0.50	18.07	m²	19.88
Providing and erecting and clearing away on completion; temporary dustproof screen of 50 x 75mm softwood framing securely fixed to walls, floor and ceiling, the joints and edges of lining sealed with masking tape and cover with										
3mm hardboard lining one side	3.74	10.00	4.12	1.10	0.18	22.68	0.66	27.46	m²	30.21
3mm hardboard lining both sides	5.30	10.00	5.83	1.30	0.22	26.91	0.82	33.56	m²	36.92
6mm plywood lining one side	5.21	10.00	5.73	1.15	0.19	23.74	0.74	30.20	m²	33.22
6mm plywood lining both sides	8.23	10.00	9.05	1.40	0.23	28.88	1.15	39.09	m²	43.00
providing 35mm hardboard faced flush door size 838 x 1981mm in dustproof screen with 25 x 87mm softwood frame, 18 x 25mm softwood stop, pair of 100mm butts, pull handle and ball catch	114.50	10.00	125.95	1.50	0.25	31.00	4.17	161.12	nr	177.23
Temporary dustproof corridors										
Note for prices of walls and doors, see dustproof screens above										
Providing and erecting and clearing away on completion; ceiling of 50 x 75mm softwood joists at 450mm centres, the joints and edges sealed with masking tape and cover with										
3mm hardboard lining to soffit	4.15	10.00	4.56	1.10	0.18	22.68	0.66	27.91	m²	30.70
6mm plywood lining to soffit	5.61	10.00	6.17	1.15	0.19	23.74	0.82	30.74	m²	33.81
Temporary timber balustrades										
Providing and erecting and clearing away on completion; temporary softwood balustrade consisting of 50 x 75mm plate, 50 x 50mm standards at 900mm centres, four 25 x 150mm intermediate rails and 50 x 50mm handrail										
1150mm high	8.65	10.00	9.51	1.00	0.26	21.95	1.89	33.35	m	36.69
Temporary steel balustrades										
Providing and erecting and clearing away on completion; temporary steel balustrade constructed of 50mm diameter galvanised scaffold tubing and fittings with standards at 900mm centres intermediate rail, handrail, plastic mesh infill and with plates on ends of standards fixed to floor										
1150mm high	25.63	-	25.63	0.18	0.90	15.69	0.45	41.77	m	45.95
weekly cost of hire and maintenance	6.30	-	6.30	0.02	0.10	1.74	0.06	8.10	m	8.91
Temporary fillings to openings in external walls										
Providing and erecting and clearing away on completion; temporary filling to window or door opening of 50 x 100mm softwood framing and cover with										
corrugated iron sheeting	12.88	10.00	14.17	0.45	0.45	14.46	1.97	30.59	m²	33.65
25mm softwood boarding	13.62	10.00	14.98	0.60	0.60	19.28	2.30	36.55	m²	40.21
		10.00	11.43	0.50	0.50	16.07	1.49	28.98	m²	31.88

Labour hourly rates: (except Specialists) Craft Operatives 18.37 Labourer 13.76 Rates are national average prices. Refer to REGIONAL VARIATIONS for indicative levels of overall pricing in regions	PLANT AND TRANSPORT			LABOUR				RATES		
	Plant Cost	Trans Cost	P and T Cost	Craft Optve	Lab	Labour Cost	Sunds	Nett Rate		Gross rate (10%)
	£	£	£	Hrs	Hrs	£	£	£	Unit	£
SITE CLEARANCE/PREPARATION										
Removing trees										
Removing trees; filling voids left by removal of roots with selected material arising from excavation										
girth 0.60 - 1.50m	153.50	-	153.50	-	11.00	151.36	-	304.86	nr	335.35
girth 1.50 - 3.00m	283.38	-	283.38	-	23.00	316.48	-	599.86	nr	659.85
girth exceeding 3.00m	413.27	-	413.27	-	35.00	481.60	-	894.87	nr	984.36
Site clearance										
Clearing site vegetation; filling voids left by removal of roots with selected material arising from excavation										
bushes, scrub, undergrowth, hedges, trees and tree stumps not exceeding 600mm girth.	1.20	-	1.20	-	-	-	-	1.20	m²	1.32
Site preparation										
Lifting turf for preservation										
stacking on site average 100m distant for future use; watering	-	-	-	-	0.60	8.26	-	8.26	m²	9.08
Remove top soil for preservation - by machine; (depositing on site measured separately)										
average 150mm deep	0.66	-	0.66	-	-	-	-	0.66	m²	0.73
Remove top soil for preservation - by hand; (depositing on site measured separately)										
average 150mm deep	-	-	-	-	0.33	4.54	-	4.54	m²	4.99
Breaking out existing hard pavings; extra over any types of excavating irrespective of depth - by machine										
concrete, 150mm thick	2.76	-	2.76	-	-	-	-	2.76	m²	3.04
reinforced concrete, 200mm thick	4.61	-	4.61	-	-	-	-	4.61	m²	5.07
coated macadam or asphalt, 75mm thick	0.55	-	0.55	-	-	-	-	0.55	m²	0.61
Breaking out existing hard pavings; extra over any types of excavating irrespective of depth - by hand										
concrete, 150mm thick	-	-	-	-	1.65	22.70	-	22.70	m²	24.97
reinforced concrete, 200mm thick	-	-	-	-	3.30	45.41	-	45.41	m²	49.95
coated macadam or asphalt, 75mm thick	-	-	-	-	0.30	4.13	-	4.13	m²	4.54
EXCAVATIONS										
Excavating - by machine										
Bulk excavation; to reduce levels										
maximum depth not exceeding										
0.25m	5.12	-	5.12	-	-	-	-	5.12	m³	5.64
1.00m	2.69	-	2.69	-	-	-	-	2.69	m³	2.96
2.00m	3.09	-	3.09	-	-	-	-	3.09	m³	3.40
4.00m	2.86	-	2.86	-	-	-	-	2.86	m³	3.15
6.00m	3.51	-	3.51	-	-	-	-	3.51	m³	3.87
Bulk excavation; basements and the like										
maximum depth not exceeding										
0.25m	5.12	-	5.12	-	0.45	6.19	-	11.32	m³	12.45
1.00m	3.24	-	3.24	-	0.50	6.88	-	10.12	m³	11.13
2.00m	3.78	-	3.78	-	0.55	7.57	-	11.35	m³	12.48
4.00m	4.41	-	4.41	-	0.70	9.63	-	14.04	m³	15.45
6.00m	6.19	-	6.19	-	0.85	11.70	-	17.89	m³	19.68
Foundation excavation; trenches width not exceeding 0.30m										
maximum depth not exceeding										
0.25m	13.22	-	13.22	-	2.00	27.52	-	40.74	m³	44.82
1.00m	13.70	-	13.70	-	2.10	28.90	-	42.59	m³	46.85
Foundation excavation; trenches width exceeding 0.30m										
maximum depth not exceeding										
0.25m	13.20	-	13.20	-	0.65	8.94	-	22.14	m³	24.36
1.00m	11.31	-	11.31	-	0.70	9.63	-	20.94	m³	23.04
2.00m	11.31	-	11.31	-	0.80	11.01	-	22.32	m³	24.55
4.00m	11.03	-	11.03	-	1.00	13.76	-	24.79	m³	27.27
6.00m	11.03	-	11.03	-	1.20	16.51	-	27.54	m³	30.30
commencing 1.50m below existing ground level, maximum depth not exceeding										
0.25m	11.34	-	11.34	-	1.00	13.76	-	25.10	m³	27.60
1.00m	11.34	-	11.34	-	1.05	14.45	-	25.78	m³	28.36
2.00m	9.29	-	9.29	-	1.30	17.89	-	27.18	m³	29.89
commencing 3.00m below existing ground level, maximum depth not exceeding										
0.25m	9.29	-	9.29	-	1.45	19.95	-	29.24	m³	32.16
1.00m	11.03	-	11.03	-	1.55	21.33	-	32.36	m³	35.60
2.00m	11.03	-	11.03	-	1.75	24.08	-	35.11	m³	38.62

Labour hourly rates: (except Specialists) Craft Operatives 18.37 Labourer 13.76 Rates are national average prices. Refer to REGIONAL VARIATIONS for indicative levels of overall pricing in regions	PLANT AND TRANSPORT			LABOUR				RATES		
	Plant Cost	Trans Cost	P and T Cost	Craft Optve	Lab	Labour Cost	Sunds	Nett Rate	Unit	Gross rate (10%)
	£	£	£	Hrs	Hrs	£	£	£		£
EXCAVATIONS (Cont'd)										
Excavating - by machine (Cont'd)										
Foundation excavation; pile caps and ground beams between piles maximum depth not exceeding										
0.25m	13.72	-	13.72	-	0.95	13.07	-	26.79	m³	29.47
1.00m	13.46	-	13.46	-	1.00	13.76	-	27.22	m³	29.94
2.00m	13.46	-	13.46	-	1.15	15.82	-	29.28	m³	32.21
Foundation excavation; pits maximum depth not exceeding										
0.25m	11.57	-	11.57	-	0.95	13.07	-	24.64	m³	27.11
1.00m	14.26	-	14.26	-	1.00	13.76	-	28.02	m³	30.83
2.00m	14.26	-	14.26	-	1.15	15.82	-	30.09	m³	33.10
4.00m	13.24	-	13.24	-	1.45	19.95	-	33.19	m³	36.51
Extra over any types of excavating irrespective of depth excavating below ground water level	11.71	-	11.71	-	-	-	-	11.71	m³	12.88
excavating in unstable ground; running silt, running sand or liquid mud	20.33	-	20.33	-	-	-	-	20.33	m³	22.37
excavating below ground water level in running silt, running sand or liquid mud	27.96	-	27.96	-	-	-	-	27.96	m³	30.76
excavating in heavy soil or clay	2.60	-	2.60	-	-	-	-	2.60	m³	2.86
excavating in gravel	4.06	-	4.06	-	-	-	-	4.06	m³	4.47
excavating in brash, loose rock or chalk	4.86	-	4.86	-	-	-	-	4.86	m³	5.35
Extra over any types of excavating irrespective of depth Breaking out existing materials										
sandstone	15.67	-	15.67	-	-	-	-	15.67	m³	17.23
hard rock	23.59	-	23.59	-	-	-	-	23.59	m³	25.95
concrete	20.27	-	20.27	-	-	-	-	20.27	m³	22.30
reinforced concrete	29.49	-	29.49	-	-	-	-	29.49	m³	32.44
brickwork, blockwork or stonework	11.24	-	11.24	-	-	-	-	11.24	m³	12.37
Breaking out existing materials; drain and concrete bed under										
100mm diameter	1.84	-	1.84	-	-	-	-	1.84	m	2.03
150mm diameter	2.21	-	2.21	-	-	-	-	2.21	m	2.43
225mm diameter	2.58	-	2.58	-	-	-	-	2.58	m	2.84
Excavating - by hand										
Bulk excavation; to reduce levels maximum depth not exceeding										
0.25m	-	-	-	-	2.75	37.84	-	37.84	m³	41.62
1.00m	-	-	-	-	2.90	39.90	-	39.90	m³	43.89
2.00m	-	-	-	-	3.30	45.41	-	45.41	m³	49.95
4.00m	-	-	-	-	4.30	59.17	-	59.17	m³	65.08
6.00m	-	-	-	-	5.35	73.62	-	73.62	m³	80.98
Bulk excavation; basements and the like maximum depth not exceeding										
0.25m	-	-	-	-	2.85	39.22	-	39.22	m³	43.14
1.00m	-	-	-	-	3.05	41.97	-	41.97	m³	46.16
2.00m	-	-	-	-	3.40	46.78	-	46.78	m³	51.46
4.00m	-	-	-	-	4.40	60.54	-	60.54	m³	66.60
6.00m	-	-	-	-	5.35	73.62	-	73.62	m³	80.98
Foundation excavation; trenches width not exceeding 0.30m maximum depth not exceeding										
0.25m	-	-	-	-	4.70	64.67	-	64.67	m³	71.14
1.00m	-	-	-	-	4.95	68.11	-	68.11	m³	74.92
Foundation excavation; trenches width exceeding 0.30m maximum depth not exceeding										
0.25m	-	-	-	-	3.14	43.21	-	43.21	m³	47.53
1.00m	-	-	-	-	3.30	45.41	-	45.41	m³	49.95
2.00m	-	-	-	-	3.85	52.98	-	52.98	m³	58.27
4.00m	-	-	-	-	4.80	66.05	-	66.05	m³	72.65
6.00m	-	-	-	-	6.05	83.25	-	83.25	m³	91.57
commencing 1.50m below existing ground level, maximum depth not exceeding										
0.25m	-	-	-	-	4.70	64.67	-	64.67	m³	71.14
1.00m	-	-	-	-	4.95	68.11	-	68.11	m³	74.92
2.00m	-	-	-	-	6.20	85.31	-	85.31	m³	93.84
commencing 3.00m below existing ground level, maximum depth not exceeding										
0.25m	-	-	-	-	7.00	96.32	-	96.32	m³	105.95
1.00m	-	-	-	-	7.30	100.45	-	100.45	m³	110.49
2.00m	-	-	-	-	8.40	115.58	-	115.58	m³	127.14
Foundation excavation; pile caps and ground beams between piles maximum depth not exceeding										
0.25m	-	-	-	-	3.70	50.91	-	50.91	m³	56.00
1.00m	-	-	-	-	3.85	52.98	-	52.98	m³	58.27
2.00m	-	-	-	-	4.40	60.54	-	60.54	m³	66.60
Foundation excavation; pits maximum depth not exceeding										
0.25m	-	-	-	-	3.70	50.91	-	50.91	m³	56.00
1.00m	-	-	-	-	3.85	52.98	-	52.98	m³	58.27
2.00m	-	-	-	-	4.40	60.54	-	60.54	m³	66.60
4.00m	-	-	-	-	5.60	77.06	-	77.06	m³	84.76
Extra over any types of excavating irrespective of depth excavating below ground water level	-	-	-	-	1.65	22.70	-	22.70	m³	24.97
excavating in running silt, running sand or liquid mud	-	-	-	-	4.95	68.11	-	68.11	m³	74.92
excavating below ground water level in running silt, running sand or liquid mud	-	-	-	-	6.60	90.82	-	90.82	m³	99.90

Labour hourly rates: (except Specialists) Craft Operatives 18.37 Labourer 13.76 Rates are national average prices. Refer to REGIONAL VARIATIONS for indicative levels of overall pricing in regions	PLANT AND TRANSPORT			LABOUR				RATES		
	Plant Cost	Trans Cost	P and T Cost	Craft Optve	Lab	Labour Cost	Sunds	Nett Rate	Unit	Gross rate (10%)
	£	£	£	Hrs	Hrs	£	£	£		£
EXCAVATIONS (Cont'd)										
Excavating - by hand (Cont'd)										
Extra over any types of excavating irrespective of depth (Cont'd)										
excavating in heavy soil or clay	-	-	-	-	0.70	9.63	-	9.63	m³	10.60
excavating in gravel	-	-	-	-	1.80	24.77	-	24.77	m³	27.24
excavating in brash, loose rock or chalk	-	-	-	-	3.70	50.91	-	50.91	m³	56.00
Breaking out existing materials; extra over any types of excavating irrespective of depth										
sandstone	-	-	-	-	7.30	100.45	-	100.45	m³	110.49
hard rock	-	-	-	-	14.70	202.27	-	202.27	m³	222.50
concrete	-	-	-	-	11.00	151.36	-	151.36	m³	166.50
reinforced concrete	-	-	-	-	16.50	227.04	-	227.04	m³	249.74
brickwork, blockwork or stonework	-	-	-	-	7.70	105.95	-	105.95	m³	116.55
drain and concrete bed under										
100mm diameter	-	-	-	-	0.90	12.38	-	12.38	m	13.62
150mm diameter	-	-	-	-	1.10	15.14	-	15.14	m	16.65
225mm diameter	-	-	-	-	1.25	17.20	-	17.20	m	18.92
Excavating inside existing building - by hand										
Bulk excavation; to reduce levels										
maximum depth not exceeding										
0.25m	-	-	-	-	3.85	52.98	-	52.98	m³	58.27
1.00m	-	-	-	-	4.10	56.42	-	56.42	m³	62.06
2.00m	-	-	-	-	4.65	63.98	-	63.98	m³	70.38
4.00m	-	-	-	-	6.00	82.56	-	82.56	m³	90.82
Foundation excavation; trenches width not exceeding 0.30m										
maximum depth not exceeding										
0.25m	-	-	-	-	6.65	91.50	-	91.50	m³	100.65
1.00m	-	-	-	-	7.00	96.32	-	96.32	m³	105.95
Foundation excavation; trenches width exceeding 0.30m										
maximum depth not exceeding										
0.25m	-	-	-	-	4.45	61.23	-	61.23	m³	67.36
1.00m	-	-	-	-	4.70	64.67	-	64.67	m³	71.14
2.00m	-	-	-	-	5.50	75.68	-	75.68	m³	83.25
4.00m	-	-	-	-	6.80	93.57	-	93.57	m³	102.92
Foundation excavation; pits										
maximum depth not exceeding										
0.25m	-	-	-	-	5.20	71.55	-	71.55	m³	78.71
1.00m	-	-	-	-	5.40	74.30	-	74.30	m³	81.73
2.00m	-	-	-	-	6.20	85.31	-	85.31	m³	93.84
4.00m	-	-	-	-	7.85	108.02	-	108.02	m³	118.82
Extra over any types of excavating irrespective of depth										
excavating below ground water level	-	-	-	-	2.35	32.34	-	32.34	m³	35.57
Breaking out existing materials; extra over any types of excavating irrespective of depth										
concrete	-	-	-	-	12.00	165.12	-	165.12	m³	181.63
reinforced concrete	-	-	-	-	17.30	238.05	-	238.05	m³	261.85
brickwork, blockwork or stonework	-	-	-	-	7.80	107.33	-	107.33	m³	118.06
drain and concrete bed under										
100mm diameter	-	-	-	-	1.25	17.20	-	17.20	m	18.92
150mm diameter	-	-	-	-	1.50	20.64	-	20.64	m	22.70
Breaking out existing hard pavings; extra over any types of excavating irrespective of depth										
concrete 150mm thick	-	-	-	-	1.65	22.70	-	22.70	m²	24.97
reinforced concrete 200mm thick	-	-	-	-	3.30	45.41	-	45.41	m²	49.95

This section continues
on the next page

Labour hourly rates: (except Specialists) Craft Operatives 18.37 Labourer 13.76 Rates are national average prices. Refer to REGIONAL VARIATIONS for indicative levels of overall pricing in regions	MATERIALS			LABOUR				RATES		
	Del to Site	Waste	Material Cost	Craft Optve	Lab	Labour Cost	Sunds	Nett Rate	Unit	Gross rate (10%)
	£	%	£	Hrs	Hrs	£	£	£		£

EXCAVATIONS (Cont'd)

Earthwork support

Note

Not withstanding the requirements of NRM rates for earthwork support are given below; appropriate allowances should be added to items of excavtion where earthwork support is deemed to be included

	Del to Site £	Waste %	Material Cost £	Craft Optve Hrs	Lab Hrs	Labour Cost £	Sunds £	Nett Rate £	Unit	Gross rate (10%) £
Earthwork support										
distance between opposing faces not exceeding 2.00m;										
maximum depth not exceeding										
1.00m	1.02	-	1.02	-	0.21	2.89	0.05	3.96	m²	4.35
2.00m	1.28	-	1.28	-	0.26	3.58	0.06	4.92	m²	5.41
4.00m	1.46	-	1.46	-	0.32	4.40	0.08	5.93	m²	6.53
6.00m	1.65	-	1.65	-	0.37	5.09	0.08	6.82	m²	7.50
distance between opposing faces 2.00 - 4.00m; maximum depth not exceeding										
1.00m	1.11	-	1.11	-	0.22	3.03	0.06	4.19	m²	4.61
2.00m	1.36	-	1.36	-	0.27	3.72	0.06	5.13	m²	5.65
4.00m	1.61	-	1.61	-	0.34	4.68	0.08	6.36	m²	7.00
6.00m	1.81	-	1.81	-	0.39	5.37	0.09	7.27	m²	8.00
distance between opposing faces exceeding 4.00m; maximum depth not exceeding										
1.00m	1.26	-	1.26	-	0.23	3.16	0.06	4.48	m²	4.93
2.00m	1.47	-	1.47	-	0.29	3.99	0.08	5.53	m²	6.08
4.00m	1.74	-	1.74	-	0.35	4.82	0.09	6.64	m²	7.31
6.00m	1.95	-	1.95	-	0.41	5.64	0.09	7.69	m²	8.45
Earthwork support; unstable ground										
distance between opposing faces not exceeding 2.00m;										
maximum depth not exceeding										
1.00m	1.74	-	1.74	-	0.35	4.82	0.09	6.64	m²	7.31
2.00m	2.09	-	2.09	-	0.44	6.05	0.10	8.24	m²	9.07
4.00m	2.41	-	2.41	-	0.53	7.29	0.12	9.82	m²	10.81
6.00m	2.75	-	2.75	-	0.62	8.53	0.14	11.41	m²	12.56
distance between opposing faces 2.00 - 4.00m; maximum depth not exceeding										
1.00m	1.91	-	1.91	-	0.37	5.09	0.09	7.09	m²	7.80
2.00m	2.28	-	2.28	-	0.45	6.19	0.10	8.58	m²	9.44
4.00m	2.66	-	2.66	-	0.56	7.71	0.12	10.49	m²	11.53
6.00m	3.05	-	3.05	-	0.65	8.94	0.14	12.13	m²	13.34
distance between opposing faces exceeding 4.00m; maximum depth not exceeding										
1.00m	2.09	-	2.09	-	0.37	5.09	0.10	7.28	m²	8.01
2.00m	2.45	-	2.45	-	0.49	6.74	0.12	9.32	m²	10.25
4.00m	2.92	-	2.92	-	0.58	7.98	0.14	11.04	m²	12.14
6.00m	3.31	-	3.31	-	0.68	9.36	0.15	12.82	m²	14.10
Earthwork support; next to roadways										
distance between opposing faces not exceeding 2.00m;										
maximum depth not exceeding										
1.00m	2.02	-	2.02	-	0.42	5.78	0.09	7.89	m²	8.68
2.00m	2.45	-	2.45	-	0.53	7.29	0.12	9.87	m²	10.85
4.00m	2.92	-	2.92	-	0.63	8.67	0.14	11.73	m²	12.90
6.00m	3.31	-	3.31	-	0.74	10.18	0.15	13.64	m²	15.01
distance between opposing faces 2.00 - 4.00m; maximum depth not exceeding										
1.00m	2.28	-	2.28	-	0.44	6.05	0.10	8.44	m²	9.28
2.00m	2.73	-	2.73	-	0.55	7.57	0.14	10.43	m²	11.47
4.00m	3.21	-	3.21	-	0.67	9.22	0.15	12.58	m²	13.84
6.00m	3.63	-	3.63	-	0.78	10.73	0.18	14.54	m²	15.99
distance between opposing faces exceeding 4.00m; maximum depth not exceeding										
1.00m	2.45	-	2.45	-	0.46	6.33	0.12	8.90	m²	9.79
2.00m	2.98	-	2.98	-	0.59	8.12	0.14	11.23	m²	12.35
4.00m	3.48	-	3.48	-	0.69	9.49	0.17	13.13	m²	14.45
6.00m	3.97	-	3.97	-	0.82	11.28	0.19	15.45	m²	17.00
Earthwork support; left in										
distance between opposing faces not exceeding 2.00m;										
maximum depth not exceeding										
1.00m	3.67	-	3.67	-	0.17	2.34	0.09	6.10	m²	6.71
2.00m	3.67	-	3.67	-	0.21	2.89	0.10	6.67	m²	7.33
4.00m	3.67	-	3.67	-	0.25	3.44	0.12	7.23	m²	7.95
6.00m	3.67	-	3.67	-	0.29	3.99	0.14	7.80	m²	8.58
distance between opposing faces 2.00 - 4.00m; maximum depth not exceeding										
1.00m	3.67	-	3.67	-	0.18	2.48	0.09	6.24	m²	6.86
2.00m	3.67	-	3.67	-	0.22	3.03	0.12	6.82	m²	7.50
4.00m	3.67	-	3.67	-	0.27	3.72	0.14	7.52	m²	8.27
6.00m	3.67	-	3.67	-	0.32	4.40	0.17	8.24	m²	9.06
distance between opposing faces exceeding 4.00m; maximum depth not exceeding										
1.00m	4.89	-	4.89	-	0.19	2.61	0.10	7.61	m²	8.37
2.00m	4.89	-	4.89	-	0.23	3.16	0.12	8.17	m²	8.99
4.00m	4.89	-	4.89	-	0.27	3.72	0.15	8.75	m²	9.63
6.00m	4.89	-	4.89	-	0.33	4.54	0.17	9.59	m²	10.55
Earthwork support; next to roadways; left in										
distance between opposing faces not exceeding 2.00m;										
maximum depth not exceeding										
1.00m	12.22	-	12.22	-	0.34	4.68	0.17	17.06	m²	18.77
2.00m	12.22	-	12.22	-	0.42	5.78	0.21	18.21	m²	20.03

Labour hourly rates: (except Specialists) Craft Operatives 18.37 Labourer 13.76 Rates are national average prices. Refer to REGIONAL VARIATIONS for indicative levels of overall pricing in regions	MATERIALS			LABOUR				RATES		
	Del to Site	Waste	Material Cost	Craft Optve	Lab	Labour Cost	Sunds	Nett Rate		Gross rate (10%)
	£	%	£	Hrs	Hrs	£	£	£	Unit	£
EXCAVATIONS (Cont'd)										
Note (Cont'd)										

Earthwork support; next to roadways; left in (Cont'd)
  distance between opposing faces not exceeding 2.00m;
  maximum depth not exceeding (Cont'd)										

4.00m	12.22	-	12.22	-	0.51	7.02	0.24	19.47	m²	21.42
6.00m	12.22	-	12.22	-	0.59	8.12	0.27	20.61	m²	22.67
distance between opposing faces 2.00 - 4.00m; maximum depth not exceeding										
1.00m	14.66	-	14.66	-	0.36	4.95	0.19	19.81	m²	21.79
2.00m	14.66	-	14.66	-	0.44	6.05	0.24	20.95	m²	23.05
4.00m	14.66	-	14.66	-	0.54	7.43	0.30	22.39	m²	24.63
6.00m	14.66	-	14.66	-	0.62	8.53	0.33	23.52	m²	25.87
distance between opposing faces exceeding 4.00m; maximum depth not exceeding										
1.00m	18.33	-	18.33	-	0.37	5.09	0.22	23.65	m²	26.01
2.00m	18.33	-	18.33	-	0.47	6.47	0.25	25.05	m²	27.56
4.00m	18.33	-	18.33	-	0.56	7.71	0.31	26.35	m²	28.99
6.00m	18.33	-	18.33	-	0.65	8.94	0.38	27.65	m²	30.41

Earthwork support; unstable ground; left in
  distance between opposing faces not exceeding 2.00m;
  maximum depth not exceeding										

1.00m	8.56	-	8.56	-	0.27	3.72	0.15	12.42	m²	13.66
2.00m	8.56	-	8.56	-	0.36	4.95	0.18	13.69	m²	15.06
4.00m	8.56	-	8.56	-	0.42	5.78	0.19	14.53	m²	15.98
6.00m	8.56	-	8.56	-	0.49	6.74	0.22	15.52	m²	17.08
distance between opposing faces 2.00 - 4.00m; maximum depth not exceeding										
1.00m	9.77	-	9.77	-	0.29	3.99	0.17	13.93	m²	15.32
2.00m	9.77	-	9.77	-	0.36	4.95	0.19	14.92	m²	16.41
4.00m	9.77	-	9.77	-	0.44	6.05	0.22	16.05	m²	17.66
6.00m	9.77	-	9.77	-	0.53	7.29	0.25	17.32	m²	19.05
distance between opposing faces exceeding 4.00m; maximum depth not exceeding										
1.00m	12.22	-	12.22	-	0.32	4.40	0.18	16.80	m²	18.48
2.00m	12.22	-	12.22	-	0.40	5.50	0.21	17.93	m²	19.72
4.00m	12.22	-	12.22	-	0.46	6.33	0.24	18.79	m²	20.67
6.00m	12.22	-	12.22	-	0.55	7.57	0.27	20.06	m²	22.06

Earthwork support; inside existing building
  distance between opposing faces not exceeding 2.00m;
  maximum depth not exceeding										

1.00m	1.06	-	1.06	-	0.23	3.16	0.05	4.27	m²	4.70
2.00m	1.32	-	1.32	-	0.29	3.99	0.06	5.38	m²	5.91
4.00m	1.63	-	1.63	-	0.32	4.40	0.08	6.11	m²	6.72
distance between opposing faces 2.00 - 4.00m; maximum depth not exceeding										
1.00m	1.17	-	1.17	-	0.24	3.30	0.06	4.54	m²	4.99
2.00m	1.46	-	1.46	-	0.30	4.13	0.08	5.66	m²	6.22
4.00m	1.65	-	1.65	-	0.37	5.09	0.08	6.82	m²	7.50
distance between opposing faces exceeding 4.00m; maximum depth not exceeding										
1.00m	1.31	-	1.31	-	0.25	3.44	0.06	4.81	m²	5.30
2.00m	1.57	-	1.57	-	0.33	4.54	0.08	6.19	m²	6.81
4.00m	1.81	-	1.81	-	0.38	5.23	0.09	7.13	m²	7.85

Earthwork support; unstable ground; inside existing building
  distance between opposing faces not exceeding 2.00m;
  maximum depth not exceeding										

1.00m	1.81	-	1.81	-	0.38	5.23	0.09	7.13	m²	7.85
2.00m	2.19	-	2.19	-	0.48	6.60	0.10	8.90	m²	9.79
4.00m	2.54	-	2.54	-	0.58	7.98	0.12	10.64	m²	11.71
distance between opposing faces 2.00 - 4.00m; maximum depth not exceeding										
1.00m	2.00	-	2.00	-	0.41	5.64	0.09	7.73	m²	8.50
2.00m	2.39	-	2.39	-	0.49	6.74	0.10	9.24	m²	10.16
4.00m	2.80	-	2.80	-	0.61	8.39	0.14	11.33	m²	12.46
distance between opposing faces exceeding 4.00m; maximum depth not exceeding										
1.00m	2.19	-	2.19	-	0.43	5.92	0.09	8.20	m²	9.02
2.00m	2.58	-	2.58	-	0.55	7.57	0.12	10.27	m²	11.30
4.00m	3.05	-	3.05	-	0.64	8.81	0.14	11.99	m²	13.19

This section continues
on the next page

Labour hourly rates: (except Specialists) Craft Operatives 18.37 Labourer 13.76 Rates are national average prices. Refer to REGIONAL VARIATIONS for indicative levels of overall pricing in regions	PLANT AND TRANSPORT			LABOUR				RATES		
	Plant Cost	Trans Cost	P and T Cost	Craft Optve	Lab	Labour Cost	Sunds	Nett Rate	Unit	Gross rate (10%)
	£	£	£	Hrs	Hrs	£	£	£		£
DISPOSAL										
Disposal - by machine										
Disposal of preserved top soil										
depositing on site in temporary spoil heaps where directed										
25m	-	3.66	3.66	-	0.12	1.65	-	5.31	m³	5.84
50m	-	3.82	3.82	-	0.12	1.65	-	5.47	m³	6.02
100m	-	4.17	4.17	-	0.12	1.65	-	5.82	m³	6.41
Disposal of excavated material										
depositing on site in temporary spoil heaps where directed										
25m	-	3.66	3.66	-	0.12	1.65	-	5.31	m³	5.84
50m	-	3.82	3.82	-	0.12	1.65	-	5.47	m³	6.02
100m	-	4.17	4.17	-	0.12	1.65	-	5.82	m³	6.41
400m	-	4.94	4.94	-	0.12	1.65	-	6.59	m³	7.25
800m	-	6.42	6.42	-	0.12	1.65	-	8.07	m³	8.87
1200m	-	6.85	6.85	-	0.12	1.65	-	8.50	m³	9.35
1600m	-	7.26	7.26	-	0.12	1.65	-	8.91	m³	9.80
extra for each additional 1600m	-	1.13	1.13	-	-	-	-	1.13	m³	1.24
removing from site to tip a) Inert	-	34.58	34.58	-	-	-	-	34.58	m³	38.04
removing from site to tip b) Active	-	71.46	71.46	-	-	-	-	71.46	m³	78.61
removing from site to tip c) Contaminated (Guide price - always seek a quotation for specialist disposal costs.)	-	207.47	207.47	-	-	-	-	207.47	m³	228.22
DISPOSAL										
Disposal - by hand										
Disposal of preserved top soil										
depositing on site in temporary spoil heaps where directed										
25m	-	-	-	-	1.40	19.26	-	19.26	m³	21.19
50m	-	-	-	-	1.65	22.70	-	22.70	m³	24.97
100m	-	-	-	-	2.20	30.27	-	30.27	m³	33.30
Disposal of excavated material										
depositing on site in temporary spoil heaps where directed										
25m	-	-	-	-	1.40	19.26	-	19.26	m³	21.19
50m	-	-	-	-	1.65	22.70	-	22.70	m³	24.97
100m	-	-	-	-	2.20	30.27	-	30.27	m³	33.30
400m	-	4.94	4.94	-	1.72	23.67	-	28.61	m³	31.47
800m	-	6.42	6.42	-	1.72	23.67	-	30.08	m³	33.09
1200m	-	6.85	6.85	-	1.72	23.67	-	30.52	m³	33.57
1600m	-	7.26	7.26	-	1.72	23.67	-	30.93	m³	34.02
extra for each additional 1600m	-	1.12	1.12	-	-	-	-	1.12	m³	1.23
removing from site to tip a) Inert	-	34.58	34.58	-	1.50	20.64	-	55.22	m³	60.74
removing from site to tip b) Active	-	71.46	71.46	-	1.50	20.64	-	92.10	m³	101.31
removing from site to tip c) Contaminated (Guide price - always seek a quotation for specialist disposal costs.)	-	207.47	207.47	-	1.50	20.64	-	228.11	m³	250.92
DISPOSAL										
Disposal inside existing building - by hand										
Disposal of excavated material										
depositing on site in temporary spoil heaps where directed										
25m	-	-	-	-	1.95	26.83	-	26.83	m³	29.52
50m	-	-	-	-	2.15	29.58	-	29.58	m³	32.54
removing from site to tip a) Inert	-	34.58	34.58	-	2.50	34.40	-	68.98	m³	75.88
removing from site to tip b) Active	-	71.46	71.46	-	2.50	34.40	-	105.86	m³	116.45
removing from site to tip c) Contaminated (Guide price - always seek a quotation for specialist disposal costs.)	-	207.47	207.47	-	2.50	34.40	-	241.87	m³	266.06
FILLINGS										
Note										

Not withstanding the requirements of NRM rates for filling for thicknesses not exceeding 500mm thick are given in M2

Not withstanding the requirements of NRM rates for trimming vertical or battered sides are given below; appropriate allowances should be added to items of filling where trimming vertical or battered sides is deemed to be included; compacting layers and surfaces generally are included in the rates

Excavated material arising from excavation - by machine										
Filling to excavation										
average thickness exceeding 500mm	-	-	-	-	-	-	4.11	4.11	m³	4.52
Selected excavated material obtained from on site spoil heaps 25m distant - by machine										
Filling to make up levels; depositing in layers 150mm maximum thickness										
finished thickness 250mm	-	-	-	-	-	-	1.88	1.88	m²	2.06
average thickness exceeding 500mm	-	-	-	-	-	-	5.11	5.11	m³	5.63

Labour hourly rates: (except Specialists) Craft Operatives 18.37 Labourer 13.76 Rates are national average prices. Refer to REGIONAL VARIATIONS for indicative levels of overall pricing in regions	PLANT AND TRANSPORT			LABOUR				RATES		
	Plant Cost	Trans Cost	P and T Cost	Craft Optve	Lab	Labour Cost	Sunds	Nett Rate		Gross rate (10%)
	£	£	£	Hrs	Hrs	£	£	£	Unit	£
FILLINGS (Cont'd)										
Note										
Not withstanding the requirements of NRM rates for trimming vertical or battered sides are given below; appropriate allowances should be added to items of filling where trimming vertical or battered sides is deemed to be included; compacting layers and surfaces generally are included in the rates										
Trimming										
sides of cuttings; vertical or battered	-	-	-	-	-	-	0.86	0.86	m²	0.94
sides of embankments; vertical or battered	-	-	-	-	-	-	0.86	0.86	m²	0.94
Selected excavated material obtained from on site spoil heaps 50m distant - by machine										
Filling to make up levels; depositing in layers 150mm maximum thickness										
finished thickness 250mm	-	-	-	-	-	-	2.10	2.10	m²	2.31
average thickness exceeding 500mm	-	-	-	-	-	-	6.28	6.28	m³	6.91
Note										
Not withstanding the requirements of NRM rates for trimming vertical or battered sides are given below; appropriate allowances should be added to items of filling where trimming vertical or battered sides is deemed to be included; compacting layers and surfaces generally are included in the rates										
Trimming										
sides of cuttings; vertical or battered	-	-	-	-	-	-	0.86	0.86	m²	0.94
sides of embankments; vertical or battered	-	-	-	-	-	-	0.86	0.86	m²	0.94
Preserved topsoil obtained from on site spoil heaps not exceeding 100m distant - by machine										
Filling to make up levels										
finished thickness 250mm	-	-	-	-	-	-	2.46	2.46	m²	2.71

This section continues
on the next page

FILLINGS (Cont'd)										

Labour hourly rates: (except Specialists) Craft Operatives 18.37 Labourer 13.76 Rates are national average prices. Refer to REGIONAL VARIATIONS for indicative levels of overall pricing in regions	MATERIALS			LABOUR				RATES		
	Del to Site	Waste	Material Cost	Craft Optve	Lab	Labour Cost	Sunds	Nett Rate	Unit	Gross rate (10%)
	£	%	£	Hrs	Hrs	£	£	£		£
FILLINGS (Cont'd)										
Imported topsoil; wheeling not exceeding 50m - by machine										
Filling to make up levels finished thickness 250mm	5.00	10.00	5.50	-	-	-	3.00	8.50	m²	9.35
Hard, dry, broken brick or stone to be obtained off site; wheeling not exceeding 25m - by machine										
Filling to make up levels; depositing in layers 150mm maximum thickness										
finished thickness 250mm	6.26	25.00	7.83	-	-	-	1.29	9.12	m²	10.03
average thickness exceeding 500mm	25.06	25.00	31.32	-	-	-	4.24	35.56	m³	39.12
Note										
Not withstanding the requirements of NRM; rates for surface treatments are given below; appropriate allowances should be added to items of filling where surface treatments is deemed to be included										
Surface packing to filling to vertical or battered faces	-	-	-	-	-	-	0.86	0.86	m²	0.94
Compacting with 680 kg vibratory roller filling; blinding with sand, ashes or similar fine material	0.34	33.00	0.45	-	0.05	0.69	0.31	1.46	m²	1.60
Compacting with 6 - 8 tonnes smooth wheeled roller filling; blinding with sand, ashes or similar fine material	0.34	33.00	0.45	-	0.05	0.69	0.62	1.76	m²	1.93
Hard, dry, broken brick or stone to be obtained off site; wheeling not exceeding 50m - by machine										
Filling to make up levels; depositing in layers 150mm maximum thickness										
finished thickness 250mm	6.26	25.00	7.83	-	-	-	1.50	9.33	m²	10.26
average thickness exceeding 500mm	25.06	25.00	31.32	-	-	-	5.11	36.43	m³	40.08
Note										
Not withstanding the requirements of NRM; rates for surface treatments are given below; appropriate allowances should be added to items of filling where surface treatments is deemed to be included										
Surface packing to filling to vertical or battered faces	-	-	-	-	-	-	0.86	0.86	m²	0.94
Compacting with 680 kg vibratory roller filling; blinding with sand, ashes or similar fine material	0.34	33.00	0.45	-	0.05	0.69	0.31	1.46	m²	1.60
Compacting with 6 - 8 tonnes smooth wheeled roller filling; blinding with sand, ashes or similar fine material	0.34	33.00	0.45	-	0.05	0.69	0.62	1.76	m²	1.93
MOT Type 1 to be obtained off site; wheeling not exceeding 25m - by machine										
Filling to make up levels; depositing in layers 150mm maximum thickness										
finished thickness 250mm	8.64	25.00	10.80	-	-	-	1.29	12.09	m²	13.30
average thickness exceeding 500mm	34.56	25.00	43.20	-	-	-	4.24	47.44	m³	52.19
Note										
Not withstanding the requirements of NRM; rates for surface treatments are given below; appropriate allowances should be added to items of filling where surface treatments is deemed to be included										
Surface packing to filling to vertical or battered faces	-	-	-	-	-	-	0.86	0.86	m²	0.94
Compacting with 680 kg vibratory roller filling; blinding with sand, ashes or similar fine material	0.34	33.00	0.45	-	0.05	0.69	0.31	1.46	m²	1.60
Compacting with 6 - 8 tonnes smooth wheeled roller filling; blinding with sand, ashes or similar fine material	0.34	33.00	0.45	-	0.05	0.69	0.62	1.76	m²	1.93
MOT Type 1 to be obtained off site; wheeling not exceeding 50m - by machine										
Filling to make up levels; depositing in layers 150mm maximum thickness										
finished thickness 250mm	8.64	25.00	10.80	-	-	-	1.50	12.30	m²	13.53
average thickness exceeding 500mm	34.56	25.00	43.20	-	-	-	5.11	48.32	m³	53.15
Note										
Not withstanding the requirements of NRM; rates for surface treatments are given below; appropriate allowances should be added to items of filling where surface treatments is deemed to be included										
Surface packing to filling to vertical or battered faces	-	-	-	-	-	-	0.86	0.86	m²	0.94
Compacting with 680 kg vibratory roller filling; blinding with sand, ashes or similar fine material	0.34	33.00	0.45	-	0.05	0.69	0.31	1.46	m²	1.60

Labour hourly rates: (except Specialists) Craft Operatives 18.37 Labourer 13.76 Rates are national average prices. Refer to REGIONAL VARIATIONS for indicative levels of overall pricing in regions	MATERIALS			LABOUR				RATES			
	Del to Site	Waste	Material Cost	Craft Optve	Lab	Labour Cost	Sunds	Nett Rate		Unit	Gross rate (10%)
	£	%	£	Hrs	Hrs	£	£	£			£
FILLINGS (Cont'd)											
Note (Cont'd)											
Compacting with 6 - 8 tonnes smooth wheeled roller											
filling; blinding with sand, ashes or similar fine material	0.34	33.00	0.45	-	0.05	0.69	0.62	1.76		m²	1.93
MOT Type 2 to be obtained off site; wheeling not exceeding 25m - by machine											
Filling to make up levels; depositing in layers 150mm maximum thickness											
finished thickness 250mm	8.64	25.00	10.80	-	-	-	1.29	12.09		m²	13.30
average thickness exceeding 500mm	34.56	25.00	43.20	-	-	-	4.24	47.44		m³	52.19
Note											
Not withstanding the requirements of NRM; rates for surface treatments are given below; appropriate allowances should be added to items of filling where surface treatments is deemed to be included											
Surface packing to filling											
to vertical or battered faces	-	-	-	-	-	-	0.86	0.86		m²	0.94
Compacting with 680 kg vibratory roller											
filling; blinding with sand, ashes or similar fine material	0.34	33.00	0.45	-	0.05	0.69	0.31	1.46		m²	1.60
Compacting with 6 - 8 tonnes smooth wheeled roller											
filling; blinding with sand, ashes or similar fine material	0.34	33.00	0.45	-	0.05	0.69	0.62	1.76		m²	1.93
MOT Type 2 to be obtained off site; wheeling not exceeding 50m - by machine											
Filling to make up levels; depositing in layers 150mm maximum thickness											
finished thickness 250mm	8.64	25.00	10.80	-	-	-	1.50	12.30		m²	13.53
average thickness exceeding 500mm	34.56	25.00	43.20	-	-	-	5.11	48.32		m³	53.15
Note											
Not withstanding the requirements of NRM; rates for surface treatments are given below; appropriate allowances should be added to items of filling where surface treatments is deemed to be included											
Surface packing to filling											
to vertical or battered faces	-	-	-	-	-	-	0.86	0.86		m²	0.94
Compacting with 680 kg vibratory roller											
filling; blinding with sand, ashes or similar fine material	0.34	33.00	0.45	-	0.05	0.69	0.31	1.46		m²	1.60
Compacting with 6 - 8 tonnes smooth wheeled roller											
filling; blinding with sand, ashes or similar fine material	0.34	33.00	0.45	-	0.05	0.69	0.62	1.76		m²	1.93
Sand to be obtained off site; wheeling not exceeding 25m - by machine											
Filling to make up levels; depositing in layers 150mm maximum thickness											
finished thickness 250mm	5.70	33.00	7.58	-	-	-	1.88	9.46		m²	10.40
average thickness exceeding 500mm	22.80	33.00	30.32	-	-	-	5.10	35.42		m³	38.97
Sand to be obtained off site; wheeling not exceeding 50m - by machine											
Filling to make up levels; depositing in layers 150mm maximum thickness											
finished thickness 250mm	5.70	33.00	7.58	-	-	-	2.10	9.68		m²	10.65
average thickness exceeding 500mm	22.80	33.00	30.32	-	-	-	5.94	36.26		m³	39.89
Hoggin to be obtained off site; wheeling not exceeding 25m - by machine											
Filling to make up levels; depositing in layers 150mm maximum thickness											
finished thickness 250mm	7.44	33.00	9.90	-	-	-	1.65	11.55		m²	12.70
average thickness exceeding 500mm	29.76	33.00	39.58	-	-	-	4.24	43.83		m³	48.21
Hoggin to be obtained off site; wheeling not exceeding 50m - by machine											
Filling to make up levels; depositing in layers 150mm maximum thickness											
finished thickness 250mm	7.44	33.00	9.90	-	-	-	1.88	11.77		m²	12.95
average thickness exceeding 500mm	29.76	33.00	39.58	-	-	-	5.10	44.68		m³	49.15
Excavated material arising from excavation - by hand											
Filling to excavation											
average thickness exceeding 500mm	-	-	-	-	1.44	19.81	-	19.81		m³	21.80
Selected excavated material obtained from on site spoil heaps 25m distant - by hand											
Filling to make up levels; depositing in layers 150mm maximum thickness											
finished thickness 250mm	-	-	-	-	0.50	6.88	-	6.88		m²	7.57
average thickness exceeding 500mm	-	-	-	-	1.60	22.02	-	22.02		m³	24.22

EXCAVATION AND FILLING

Labour hourly rates: (except Specialists) Craft Operatives 18.37 Labourer 13.76 Rates are national average prices. Refer to REGIONAL VARIATIONS for indicative levels of overall pricing in regions	MATERIALS			LABOUR				RATES		
	Del to Site	Waste	Material Cost	Craft Optve	Lab	Labour Cost	Sunds	Nett Rate	Unit	Gross rate (10%)
	£	%	£	Hrs	Hrs	£	£	£		£
FILLINGS (Cont'd)										
Note										
Not withstanding the requirements of NRM rates for trimming vertical or battered sides are given below; appropriate allowances should be added to items of filling where trimming vertical or battered sides is deemed to be included										
Trimming										
sides of cuttings; vertical or battered	-	-	-	-	0.22	3.03	-	3.03	m²	3.33
sides of embankments; vertical or battered	-	-	-	-	0.22	3.03	-	3.03	m²	3.33
Selected excavated material obtained from on site spoil heaps 50m distant - by hand										
Filling to make up levels; depositing in layers 150mm maximum thickness										
finished thickness 250mm	-	-	-	-	0.80	11.01	-	11.01	m²	12.11
average thickness exceeding 500mm	-	-	-	-	1.95	26.83	-	26.83	m³	29.52
Note										
Not withstanding the requirements of NRM rates for trimming vertical or battered sides are given below; appropriate allowances should be added to items of filling where trimming vertical or battered sides is deemed to be included										
Trimming										
sides of cuttings; vertical or battered	-	-	-	-	0.22	3.03	-	3.03	m²	3.33
sides of embankments; vertical or battered	-	-	-	-	0.22	3.03	-	3.03	m²	3.33
Preserved topsoil obtained from on site spoil heaps not exceeding 100m distant - by hand										
Filling to make up levels										
finished thickness 250mm	-	%	-	-	1.10	15.14	-	15.14	m²	16.65
Filling to external planters										
finished thickness 250mm	-	-	-	-	1.40	19.26	-	19.26	m²	21.19
Imported topsoil; wheeling not exceeding 50m - by hand										
Filling to make up levels										
finished thickness 250mm	5.00	10.00	5.50	-	1.00	13.76	-	19.26	m²	21.19
Filling to external planters										
finished thickness 250mm	5.00	10.00	5.50	-	1.25	17.20	-	22.70	m²	24.97
Hard, dry, broken brick or stone to be obtained off site; wheeling not exceeding 25m - by hand										
Filling to make up levels; depositing in layers 150mm maximum thickness										
finished thickness 250mm	6.26	25.00	7.83	-	0.65	8.94	-	16.77	m²	18.45
average thickness exceeding 500mm	25.06	25.00	31.32	-	1.31	18.03	-	49.35	m³	54.28
Note										
Not withstanding the requirements of NRM; rates for surface treatments are given below; appropriate allowances should be added to items of filling where surface treatments is deemed to be included										
Surface packing to filling										
to vertical or battered faces	-	-	-	-	0.22	3.03	-	3.03	m²	3.33
Compacting with 680 kg vibratory roller										
filling; blinding with sand, ashes or similar fine material	0.34	33.00	0.45	-	0.05	0.69	0.31	1.46	m²	1.60
Compacting with 6 - 8 tonnes smooth wheeled roller										
filling; blinding with sand, ashes or similar fine material	0.34	33.00	0.45	-	0.05	0.69	0.62	1.76	m²	1.93
Hard, dry, broken brick or stone to be obtained off site; wheeling not exceeding 50m - by hand										
Filling to make up levels; depositing in layers 150mm maximum thickness										
finished thickness 250mm	6.26	25.00	7.83	-	0.75	10.32	-	18.15	m²	19.97
average thickness exceeding 500mm	25.06	25.00	31.32	-	1.58	21.74	-	53.06	m³	58.37
Note										
Not withstanding the requirements of NRM; rates for surface treatments are given below; appropriate allowances should be added to items of filling where surface treatments is deemed to be included										
Surface packing to filling										
to vertical or battered faces	-	-	-	-	0.22	3.03	-	3.03	m²	3.33
Compacting with 680 kg vibratory roller										
filling; blinding with sand, ashes or similar fine material	0.34	33.00	0.45	-	0.05	0.69	0.31	1.46	m²	1.60
Compacting with 6 - 8 tonnes smooth wheeled roller										
filling; blinding with sand, ashes or similar fine material	0.34	33.00	0.45	-	0.05	0.69	0.62	1.76	m²	1.93

EXCAVATION AND FILLING

Labour hourly rates: (except Specialists) Craft Operatives 18.37 Labourer 13.76 Rates are national average prices. Refer to REGIONAL VARIATIONS for indicative levels of overall pricing in regions	MATERIALS			LABOUR				RATES		
	Del to Site	Waste	Material Cost	Craft Optve	Lab	Labour Cost	Sunds	Nett Rate		Gross rate (10%)
	£	%	£	Hrs	Hrs	£	£	£	Unit	£
FILLINGS (Cont'd)										
MOT Type 1 to be obtained off site; wheeling not exceeding 25m - by hand										
Filling to make up levels; depositing in layers 150mm maximum thickness										
finished thickness 250mm	8.64	25.00	10.80	-	0.65	8.94	-	19.74	m²	21.72
average thickness exceeding 500mm	34.56	25.00	43.20	-	1.31	18.03	-	61.23	m³	67.35
Note										
Not withstanding the requirements of NRM; rates for surface treatments are given below; appropriate allowances should be added to items of filling where surface treatments is deemed to be included										
Surface packing to filling										
to vertical or battered faces	-	-	-	-	0.22	3.03	-	3.03	m²	3.33
Compacting with 680 kg vibratory roller										
filling; blinding with sand, ashes or similar fine material	0.34	33.00	0.45	-	0.05	0.69	0.31	1.46	m²	1.60
Compacting with 6 - 8 tonnes smooth wheeled roller										
filling; blinding with sand, ashes or similar fine material	0.34	33.00	0.45	-	0.05	0.69	0.62	1.76	m²	1.93
MOT Type 1 to be obtained off site; wheeling not exceeding 50m - by hand										
Filling to make up levels; depositing in layers 150mm maximum thickness										
finished thickness 250mm	8.64	25.00	10.80	-	0.75	10.32	-	21.12	m²	23.23
average thickness exceeding 500mm	34.56	25.00	43.20	-	1.58	21.74	-	64.94	m³	71.43
Note										
Not withstanding the requirements of NRM; rates for surface treatments are given below; appropriate allowances should be added to items of filling where surface treatments is deemed to be included										
Surface packing to filling										
to vertical or battered faces	-	-	-	-	0.22	3.03	-	3.03	m²	3.33
Compacting with 680 kg vibratory roller										
filling; blinding with sand, ashes or similar fine material	0.34	33.00	0.45	-	0.05	0.69	0.31	1.46	m²	1.60
Compacting with 6 - 8 tonnes smooth wheeled roller										
filling; blinding with sand, ashes or similar fine material	0.34	33.00	0.45	-	0.05	0.69	0.62	1.76	m²	1.93
MOT Type 2 to be obtained off site; wheeling not exceeding 25m - by hand										
Filling to make up levels; depositing in layers 150mm maximum thickness										
finished thickness 250mm	8.64	25.00	10.80	-	0.65	8.94	-	19.74	m²	21.72
average thickness exceeding 500mm	34.56	25.00	43.20	-	1.31	18.03	-	61.23	m³	67.35
Note										
Not withstanding the requirements of NRM; rates for surface treatments are given below; appropriate allowances should be added to items of filling where surface treatments is deemed to be included										
Surface packing to filling										
to vertical or battered faces	-	-	-	-	0.22	3.03	-	3.03	m²	3.33
Compacting with 680 kg vibratory roller										
filling; blinding with sand, ashes or similar fine material	0.34	33.00	0.45	-	0.05	0.69	0.31	1.46	m²	1.60
Compacting with 6 - 8 tonnes smooth wheeled roller										
filling; blinding with sand, ashes or similar fine material	0.34	33.00	0.45	-	0.05	0.69	0.62	1.76	m²	1.93
MOT Type 2 to be obtained off site; wheeling not exceeding 50m - by hand										
Filling to make up levels; depositing in layers 150mm maximum thickness										
finished thickness 250mm	8.64	25.00	10.80	-	0.75	10.32	-	21.12	m²	23.23
average thickness exceeding 500mm	34.56	25.00	43.20	-	1.58	21.74	-	64.94	m³	71.43
Note										
Not withstanding the requirements of NRM; rates for surface treatments are given below; appropriate allowances should be added to items of filling where surface treatments is deemed to be included										
Surface packing to filling										
to vertical or battered faces	-	-	-	-	0.22	3.03	-	3.03	m²	3.33
Compacting with 680 kg vibratory roller										
filling; blinding with sand, ashes or similar fine material	0.34	33.00	0.45	-	0.05	0.69	0.31	1.46	m²	1.60
Compacting with 6 - 8 tonnes smooth wheeled roller										
filling; blinding with sand, ashes or similar fine material	0.34	33.00	0.45	-	0.05	0.69	0.62	1.76	m²	1.93

Labour hourly rates: (except Specialists) Craft Operatives 18.37 Labourer 13.76 Rates are national average prices. Refer to REGIONAL VARIATIONS for indicative levels of overall pricing in regions	MATERIALS			LABOUR				RATES		
	Del to Site	Waste	Material Cost	Craft Optve	Lab	Labour Cost	Sunds	Nett Rate	Unit	Gross rate (10%)
	£	%	£	Hrs	Hrs	£	£	£		£
FILLINGS (Cont'd)										
Sand to be obtained off site; wheeling not exceeding 25m - by hand										
Filling to make up levels; depositing in layers 150mm maximum thickness										
finished thickness 250mm	5.70	33.00	7.58	-	0.75	10.32	-	17.90	m²	19.69
average thickness exceeding 500mm	22.80	33.00	30.32	-	1.58	21.74	-	52.06	m³	57.27
Sand to be obtained off site; wheeling not exceeding 50m - by hand										
Filling to make up levels; depositing in layers 150mm maximum thickness										
finished thickness 250mm	5.70	33.00	7.58	-	0.80	11.01	-	18.59	m²	20.45
average thickness exceeding 500mm	22.80	33.00	30.32	-	1.84	25.32	-	55.64	m³	61.21
Hoggin to be obtained off site; wheeling not exceeding 25m - by hand										
Filling to make up levels; depositing in layers 150mm maximum thickness										
finished thickness 250mm	7.44	33.00	9.90	-	0.65	8.94	-	18.84	m²	20.72
average thickness exceeding 500mm	29.76	33.00	39.58	-	1.30	17.89	-	57.47	m³	63.22
Hoggin to be obtained off site; wheeling not exceeding 50m - by hand										
Filling to make up levels; depositing in layers 150mm maximum thickness										
finished thickness 250mm	7.44	33.00	9.90	-	0.70	9.63	-	19.53	m²	21.48
average thickness exceeding 500mm	29.76	33.00	39.58	-	1.55	21.33	-	60.91	m³	67.00
Excavated material arising from excavation; work inside existing building - by hand										
Filling to excavation										
average thickness exceeding 500mm	-	-	-	-	2.80	38.53	-	38.53	m³	42.38
Selected excavated material obtained from on site spoil heaps 25m distant; work inside existing building - by hand										
Filling to make up levels; depositing in layers 150mm maximum thickness										
finished thickness 250mm	-	-	-	-	0.82	11.28	-	11.28	m²	12.41
average thickness exceeding 500mm	-	-	-	-	3.08	42.38	-	42.38	m³	46.62
Selected excavated material obtained from on site spoil heaps 50m distant; work inside existing building - by hand										
Filling to make up levels; depositing in layers 150mm maximum thickness										
finished thickness 250mm	-	-	-	-	1.15	15.82	-	15.82	m²	17.41
average thickness exceeding 500mm	-	-	-	-	3.36	46.23	-	46.23	m³	50.86
Hard, dry, broken brick or stone to be obtained off site; wheeling not exceeding 25m; work inside existing building - by hand										
Filling to make up levels; depositing in layers 150mm maximum thickness										
finished thickness 250mm	6.26	25.00	7.83	-	1.00	13.76	-	21.59	m²	23.75
average thickness exceeding 500mm	25.06	25.00	31.32	-	2.81	38.67	-	69.99	m³	76.98
Note										
Not withstanding the requirements of NRM; rates for surface treatments are given below; appropriate allowances should be added to items of filling where surface treatments is deemed to be included										
Surface packing to filling										
to vertical or battered faces	-	-	-	-	0.22	3.03	-	3.03	m²	3.33
Compacting										
filling; blinding with sand, ashes or similar fine material	0.34	33.00	0.45	-	0.05	0.69	0.31	1.46	m²	1.60
Hard, dry, broken brick or stone to be obtained off site; wheeling not exceeding 50m; work inside existing building - by hand										
Filling to make up levels; depositing in layers 150mm maximum thickness										
finished thickness 250mm	6.26	25.00	7.83	-	1.07	14.72	-	22.55	m²	24.81
average thickness exceeding 500mm	25.06	25.00	31.32	-	2.99	41.14	-	72.46	m³	79.71
Note										
Not withstanding the requirements of NRM; rates for surface treatments are given below; appropriate allowances should be added to items of filling where surface treatments is deemed to be included										
Surface packing to filling										
to vertical or battered faces	-	%	-	-	0.22	3.03	-	3.03	m²	3.33

Labour hourly rates: (except Specialists) Craft Operatives 18.37 Labourer 13.76 Rates are national average prices. Refer to REGIONAL VARIATIONS for indicative levels of overall pricing in regions	MATERIALS			LABOUR				RATES		
	Del to Site	Waste	Material Cost	Craft Optve	Lab	Labour Cost	Sunds	Nett Rate		Gross rate (10%)
	£	%	£	Hrs	Hrs	£	£	£	Unit	£
FILLINGS (Cont'd)										
Note (Cont'd)										
Compacting										
filling; blinding with sand, ashes or similar fine material	0.34	33.00	0.45	-	0.05	0.69	0.31	1.46	m²	1.60
Sand to be obtained off site; wheeling not exceeding 25m; work inside existing building - by hand										
Filling to make up levels; depositing in layers 150mm maximum thickness										
finished thickness 250mm	5.70	33.00	7.58	-	1.07	14.72	-	22.30	m²	24.53
average thickness exceeding 500mm	22.80	33.00	30.32	-	2.99	41.14	-	71.47	m³	78.61
Sand to be obtained off site; wheeling not exceeding 50m; work inside existing building - by hand										
Filling to make up levels; depositing in layers 150mm maximum thickness										
finished thickness 250mm	5.70	33.00	7.58	-	1.13	15.55	-	23.13	m²	25.44
average thickness exceeding 500mm	22.80	33.00	30.32	-	3.26	44.86	-	75.18	m³	82.70
MEMBRANES										
Waterproof building paper										
laying on hardcore to receive concrete										
Grade B1F	0.46	25.00	0.57	-	0.06	0.83	-	1.40	m²	1.54
Grade B2	1.32	25.00	1.65	-	0.06	0.83	-	2.47	m²	2.72
Polythene sheeting										
laying on hardcore to receive concrete										
300mu	0.34	20.00	0.41	-	0.06	0.83	-	1.23	m²	1.36
Claymaster low density expanded polystyrene permanent formwork; fixing with Clayfix hooks - 3/m²										
To underside of foundations; horizontal; laid on earth or hardcore; 75mm thick										
over 500mm wide	5.19	5.00	5.45	-	0.05	0.69	-	6.14	m²	6.75
To underside of foundations; horizontal; laid on earth or hardcore; 100mm thick										
over 500mm wide	6.93	5.00	7.27	-	0.07	0.96	-	8.24	m²	9.06
To underside of foundations; horizontal; laid on earth or hardcore; 150mm thick										
over 500mm wide	10.42	5.00	10.94	-	0.08	1.10	-	12.04	m²	13.24
Sides of foundations; vertical 75mm thick										
height not exceeding 250mm	1.30	5.00	1.36	-	0.02	0.28	0.31	1.95	m	2.15
height 250-500mm	2.60	5.00	2.73	-	0.08	1.10	0.31	4.14	m	4.56
height exceeding 500mm	5.19	5.00	5.45	-	0.07	0.96	0.94	7.36	m²	8.09
Sides of foundations; vertical 100mm thick										
height not exceeding 250mm	1.73	5.00	1.82	-	0.03	0.41	0.31	2.55	m	2.80
height 250-500mm	3.46	5.00	3.64	-	0.05	0.69	0.31	4.64	m	5.10
height exceeding 500mm	6.93	5.00	7.27	-	0.09	1.24	0.94	9.46	m²	10.40
Sides of foundations; vertical 150mm thick										
height not exceeding 250mm	2.60	5.00	2.73	-	0.03	0.41	0.31	3.46	m	3.81
height 250-500mm	5.21	5.00	5.47	-	0.07	0.96	0.31	6.75	m	7.42
height exceeding 500mm	10.42	5.00	10.94	-	0.11	1.51	0.94	13.40	m²	14.74

This page left blank intentionally

Labour hourly rates: (except Specialists) Craft Operatives 18.37 Labourer 13.76 Rates are national average prices. Refer to REGIONAL VARIATIONS for indicative levels of overall pricing in regions	MATERIALS			LABOUR				RATES		
	Del to Site	Waste	Material Cost	Craft Optve	Lab	Labour Cost	Sunds	Nett Rate		Gross rate (10%)
	£	%	£	Hrs	Hrs	£	£	£	Unit	£

BORED PILING

Prices include for a 21 N/mm² concrete mix, nominal reinforcement and a minimum number of 50 piles on any one contract. The working loads sizes and lengths given below will depend on the nature of the soils in which the piles will be founded as well as structure to be supported (KELLER GROUND ENGINEERING)

Description	Del to Site	Waste	Material Cost	Craft Optve	Lab	Labour Cost	Sunds	Nett Rate	Unit	Gross rate
On/Off site charge in addition to the following prices add approximately	-	-	Specialist	-	-	Specialist	-	8611.20	Sm	9472.32
Short auger piles up to 6m long; 450mm nominal diameter; 8 tonnes normal working load	-	-	Specialist	-	-	Specialist	-	50.43	m	55.47
up to 6m long; 610mm nominal diameter; 30 tonnes normal working load	-	-	Specialist	-	-	Specialist	-	76.28	m	83.91
Bored cast in-situ piles up to 15m long; 450mm nominal diameter; 40 tonnes normal working load	-	-	Specialist	-	-	Specialist	-	46.53	m	51.19
up to 15m long; 610mm nominal diameter; 120 tonnes normal working load	-	-	Specialist	-	-	Specialist	-	72.41	m	79.65
Auger piles up to 20m long; 450mm nominal diameter	-	-	Specialist	-	-	Specialist	-	47.20	m	51.92
up to 30m long; 1200mm nominal diameter	-	-	Specialist	-	-	Specialist	-	234.03	m	257.44
Large diameter auger piles (plain shaft) up to 20m long; 610mm nominal diameter; 150 tonnes normal working load	-	-	Specialist	-	-	Specialist	-	71.81	m	78.99
up to 30m long; 1525mm nominal diameter; 600 tonnes normal working load	-	-	Specialist	-	-	Specialist	-	364.09	m	400.50
Large diameter auger piles (belled base) up to 20m long; 610mm nominal diameter; 150 tonnes normal working load	-	-	Specialist	-	-	Specialist	-	98.28	m	108.11
up to 30m long; 1525mm nominal diameter; 1000 tonnes normal working load	-	-	Specialist	-	-	Specialist	-	468.21	m	515.04
Boring through obstructions, rock like formations, etc; undertaken on a time basis at a rate per piling rig per hour Cutting off tops of piles; including preparation and integration of reinforcement into pile cap or ground beam and disposal 450-610mm nominal diameter	-	-	Specialist	-	-	Specialist	-	17.28	m	19.01
610-1525mm nominal diameter	-	-	Specialist	-	-	Specialist	-	66.53	m	73.18

DRIVEN PILING

Mild steel Universal Bearing Piles "H" section, to BS EN 10025 Grade S275 and high yield steel Grade S275JR in lengths 9 - 15m supplied, handled, pitched and driven vertically with landbased plant (KELLER GROUND ENGINEERING)

Note; the following prices are based on quantities of 25 - 150 tonnes

Description	Del to Site	Waste	Material Cost	Craft Optve	Lab	Labour Cost	Sunds	Nett Rate	Unit	Gross rate
On/Off site charge in addition to the following prices, add approximately	-	-	Specialist	-	-	Specialist	-	7300.60	Sm	8030.66
Mild steel piles 203 x 203 x 45 kg/m, SWL 40 tonnes	-	-	Specialist	-	-	Specialist	-	70.60	m	77.66
203 x 203 x 54 kg/m, SWL 50 tonnes	-	-	Specialist	-	-	Specialist	-	79.63	m	87.60
254 x 254 x 63 kg/m, SWL 60 tonnes	-	-	Specialist	-	-	Specialist	-	90.01	m	99.01
254 x 254 x 71 kg/m, SWL 70 tonnes	-	-	Specialist	-	-	Specialist	-	96.93	m	106.63
305 x 305 x 79 kg/m, SWL 75 tonnes	-	-	Specialist	-	-	Specialist	-	114.15	m	125.57
254 x 254 x 85 kg/m, SWL 80 tonnes	-	-	Specialist	-	-	Specialist	-	117.48	m	129.23
305 x 305 x 88 kg/m, SWL 85 tonnes	-	-	Specialist	-	-	Specialist	-	123.77	m	136.15
305 x 305 x 95 kg/m, SWL 90 tonnes	-	-	Specialist	-	-	Specialist	-	131.60	m	144.76
356 x 368 x 109 kg/m, SWL 105 tonnes	-	-	Specialist	-	-	Specialist	-	157.68	m	173.45
305 x 305 x 110 kg/m, SWL 105 tonnes	-	-	Specialist	-	-	Specialist	-	148.33	m	163.17
305 x 305 x 126 kg/m, SWL 120 tonnes	-	-	Specialist	-	-	Specialist	-	169.22	m	186.14
305 x 305 x 149 kg/m, SWL 140 tonnes	-	-	Specialist	-	-	Specialist	-	186.76	m	205.43
356 x 368 x 133 kg/m, SWL 130 tonnes	-	-	Specialist	-	-	Specialist	-	195.47	m	215.02
356 x 368 x 152 kg/m, SWL 140 tonnes	-	-	Specialist	-	-	Specialist	-	207.46	m	228.20
356 x 368 x 174 kg/m, SWL 165 tonnes	-	-	Specialist	-	-	Specialist	-	238.58	m	262.44
305 x 305 x 186 kg/m, SWL 175 tonnes	-	-	Specialist	-	-	Specialist	-	259.40	m	285.34
305 x 305 x 223 kg/m, SWL 210 tonnes	-	-	Specialist	-	-	Specialist	-	283.90	m	312.29
Extra for high yield steel add	-	-	Specialist	-	-	Specialist	-	77.71	T	85.48
For quiet piling, add to terminal charge, lump sum	-	-	Specialist	-	-	Specialist	-	1550.28	Sm	1705.31
plus for all piles	-	-	Specialist	-	-	Specialist	-	3.76	m	4.13

PILING

Labour hourly rates: (except Specialists) Craft Operatives 18.37 Labourer 13.76 Rates are national average prices. Refer to REGIONAL VARIATIONS for indicative levels of overall pricing in regions	MATERIALS			LABOUR				RATES		
	Del to Site	Waste	Material Cost	Craft Optve	Lab	Labour Cost	Sunds	Nett Rate	Unit	Gross rate (10%)
	£	%	£	Hrs	Hrs	£	£	£		£
INTERLOCKING PILING										
Mild steel sheet piling to BS EN 10025 Grade S275 or high yield steel Grade S275JR in lengths 4.5-15m supplied, handled, pitched and driven by land based plant in one visit (KELLER GROUND ENGINEERING)										
On/Off site charge										
in addition to the following prices, add approximately	-	-	Specialist	-	-	Specialist	-	6963.00	Sm	7659.30
Mild steel sheet piling										
Larssen Section 6W	-	-	Specialist	-	-	Specialist	-	153.04	m²	168.34
extra for high yield steel	-	-	Specialist	-	-	Specialist	-	8.38	m²	9.21
extra for one coat L.B.V.(Lowca Varnish to BS 1070 Type 2)										
before driving	-	-	Specialist	-	-	Specialist	-	6.12	m²	6.73
extra for corners	-	-	Specialist	-	-	Specialist	-	61.17	m	67.28
extra for junctions	-	-	Specialist	-	-	Specialist	-	83.41	m	91.75
Frodingham Section 1N and Larssen Section 9W	-	-	Specialist	-	-	Specialist	-	169.03	m²	185.94
extra for high yield steel	-	-	Specialist	-	-	Specialist	-	6.96	m²	7.66
extra for one coat L.B.V.(Lowca Varnish to BS 1070 Type 2)										
before driving	-	-	Specialist	-	-	Specialist	-	6.12	m²	6.73
extra for corners	-	-	Specialist	-	-	Specialist	-	61.17	m	67.28
extra for junctions	-	-	Specialist	-	-	Specialist	-	83.41	m	91.75
Frodingham Section 2N and Larssen Section 12W	-	-	Specialist	-	-	Specialist	-	188.15	m²	206.96
extra for high yield steel	-	-	Specialist	-	-	Specialist	-	8.90	m²	9.79
extra for one coat L.B.V.(Lowca Varnish to BS 1070 Type 2)										
before driving	-	-	Specialist	-	-	Specialist	-	6.12	m²	6.73
extra for corners	-	-	Specialist	-	-	Specialist	-	61.17	m	67.28
extra for junctions	-	-	Specialist	-	-	Specialist	-	83.41	m	91.75
Frodingham Section 1BXN	-	-	Specialist	-	-	Specialist	-	208.53	m²	229.39
extra for high yield steel	-	-	Specialist	-	-	Specialist	-	13.80	m²	15.18
extra for one coat L.B.V.(Lowca Varnish to BS 1070 Type 2)										
before driving	-	-	Specialist	-	-	Specialist	-	5.57	m²	6.13
extra for corners	-	-	Specialist	-	-	Specialist	-	66.74	m	73.41
extra for junctions	-	-	Specialist	-	-	Specialist	-	83.41	m	91.75
Frodingham Section 3N and Larssen Section 16W	-	-	Specialist	-	-	Specialist	-	210.39	m²	231.43
extra for high yield steel	-	-	Specialist	-	-	Specialist	-	11.42	m²	12.56
extra for one coat L.B.V.(Lowca Varnish to BS 1070 Type 2)										
before driving	-	-	Specialist	-	-	Specialist	-	5.57	m²	6.13
extra for corners	-	-	Specialist	-	-	Specialist	-	66.74	m	73.41
extra for junctions	-	-	Specialist	-	-	Specialist	-	83.41	m	91.75
Larssen Section 20W	-	-	Specialist	-	-	Specialist	-	238.05	m²	261.86
extra for high yield steel	-	-	Specialist	-	-	Specialist	-	12.51	m²	13.76
extra for one coat L.B.V.(Lowca Varnish to BS 1070 Type 2)										
before driving	-	-	Specialist	-	-	Specialist	-	6.27	m²	6.89
extra for corners	-	-	Specialist	-	-	Specialist	-	72.29	m	79.52
extra for junctions	-	-	Specialist	-	-	Specialist	-	97.29	m	107.02
Frodingham Section 4N and Larssen Section 25W	-	-	Specialist	-	-	Specialist	-	263.60	m²	289.96
extra for high yield steel	-	-	Specialist	-	-	Specialist	-	13.63	m²	14.99
extra for one coat L.B.V.(Lowca Varnish to BS 1070 Type 2)										
before driving	-	-	Specialist	-	-	Specialist	-	6.96	m²	7.66
extra for corners	-	-	Specialist	-	-	Specialist	-	77.84	m	85.62
extra for junctions	-	-	Specialist	-	-	Specialist	-	97.29	m	107.02
Larssen Section 32W	-	-	Specialist	-	-	Specialist	-	311.23	m²	342.35
extra for high yield steel	-	-	Specialist	-	-	Specialist	-	15.15	m²	16.67
extra for one coat L.B.V.(Lowca Varnish to BS 1070 Type 2)										
before driving	-	-	Specialist	-	-	Specialist	-	7.51	m²	8.26
extra for corners	-	-	Specialist	-	-	Specialist	-	83.41	m	91.75
extra for junctions	-	-	Specialist	-	-	Specialist	-	102.97	m	113.26
Frodingham Section 5	-	-	Specialist	-	-	Specialist	-	351.21	m²	386.33
extra for high yield steel	-	-	Specialist	-	-	Specialist	-	20.02	m²	22.03
extra for one coat L.B.V.(Lowca Varnish to BS 1070 Type 2)										
before driving	-	-	Specialist	-	-	Specialist	-	13.06	m²	14.37
extra for corners	-	-	Specialist	-	-	Specialist	-	83.41	m	91.75
extra for junctions	-	-	Specialist	-	-	Specialist	-	111.22	m	122.34
For quiet piling, add										
lump sum	-	-	Specialist	-	-	Specialist	-	1413.70	Sm	1555.07
plus for all sections, from	-	-	Specialist	-	-	Specialist	-	6.46	m²	7.10
to	-	-	Specialist	-	-	Specialist	-	10.78	m²	11.86

Labour hourly rates: (except Specialists) Craft Operatives 18.37 Labourer 13.76 Rates are national average prices. Refer to REGIONAL VARIATIONS for indicative levels of overall pricing in regions	MATERIALS			LABOUR				RATES		
	Del to Site	Waste	Material Cost	Craft Optve	Lab	Labour Cost	Sunds	Nett Rate		Gross rate (10%)
	£	%	£	Hrs	Hrs	£	£	£	Unit	£
UNDERPINNING										
Information										
The work of underpinning in this section comprises work to be carried out in short lengths; prices are exclusive of shoring and other temporary supports										
Excavating										
Preliminary trenches maximum depth not exceeding										
1.00m	-	-	-	-	5.00	68.80	-	68.80	m³	75.68
2.00m	-	-	-	-	6.00	82.56	-	82.56	m³	90.82
4.00m	-	-	-	-	7.00	96.32	-	96.32	m³	105.95
Underpinning pits maximum depth not exceeding										
1.00m	-	-	-	-	5.85	80.50	-	80.50	m³	88.55
2.00m	-	-	-	-	7.30	100.45	-	100.45	m³	110.49
4.00m	-	-	-	-	8.80	121.09	-	121.09	m³	133.20
Earthwork support preliminary trenches; distance between opposing faces not exceeding 2.00m; maximum depth not exceeding										
1.00m	1.47	-	1.47	-	0.53	7.29	0.03	8.79	m²	9.67
2.00m	1.85	-	1.85	-	0.67	9.22	0.05	11.11	m²	12.22
4.00m	2.11	-	2.11	-	0.72	9.91	0.05	12.06	m²	13.26
preliminary trenches; distance between opposing faces 2.00 - 4.00m; maximum depth not exceeding										
1.00m	1.60	-	1.60	-	0.55	7.57	0.03	9.19	m²	10.11
2.00m	2.02	-	2.02	-	0.70	9.63	0.05	11.70	m²	12.87
4.00m	2.22	-	2.22	-	0.84	11.56	0.06	13.83	m²	15.22
underpinning pits; distance between opposing faces not exceeding 2.00m; maximum depth not exceeding										
1.00m	1.73	-	1.73	-	0.61	8.39	0.03	10.15	m²	11.17
2.00m	2.14	-	2.14	-	0.77	10.60	0.05	12.78	m²	14.06
4.00m	2.40	-	2.40	-	0.83	11.42	0.06	13.88	m²	15.27
underpinning pits; distance between opposing faces 2.00 - 4.00m; maximum depth not exceeding										
1.00m	1.82	-	1.82	-	0.64	8.81	0.03	10.66	m²	11.73
2.00m	2.31	-	2.31	-	0.80	11.01	0.05	13.37	m²	14.70
4.00m	2.66	-	2.66	-	0.95	13.07	0.06	15.79	m²	17.37
Cutting away existing projecting foundations										
masonry; maximum width 103mm; maximum depth 150mm	-	-	-	-	0.90	12.38	-	12.38	m	13.62
masonry; maximum width 154mm; maximum depth 225mm	-	-	-	-	1.07	14.72	-	14.72	m	16.20
concrete; maximum width 253mm; maximum depth 190mm	-	-	-	-	1.29	17.75	-	17.75	m	19.53
concrete; maximum width 304mm; maximum depth 300mm	-	-	-	-	1.52	20.92	-	20.92	m	23.01
Preparing the underside of the existing work to receive the pinning up of the new work										
350mm wide	-	-	-	-	0.45	6.19	-	6.19	m	6.81
500mm wide	-	-	-	-	0.56	7.71	-	7.71	m	8.48
1000mm wide	-	-	-	-	1.13	15.55	-	15.55	m	17.10
Compacting bottoms of excavations	-	-	-	-	0.45	6.19	-	6.19	m²	6.81
Disposal of excavated material removing from site to tip (including tipping charges but excluding landfill tax).	-	-	-	-	3.38	46.51	17.55	64.06	m³	70.46
Excavated material arising from excavations										
Filling to excavations average thickness exceeding 0.25m	-	-	-	-	2.25	30.96	-	30.96	m³	34.06
Compacting filling	-	-	-	-	0.23	3.16	-	3.16	m²	3.48
Plain in-situ concrete; BS 8500, ordinary prescribed mix ST3, 20mm aggregate										
Foundations; poured on or against earth or unblinded hardcore generally	97.99	7.50	105.34	-	5.34	73.48	-	178.82	m³	196.70
Plain in-situ concrete; BS 8500, ordinary prescribed mix ST4, 20mm aggregate										
Foundations; poured on or against earth or unblinded hardcore generally	99.73	7.50	107.21	-	5.34	73.48	-	180.69	m³	198.76
Formwork and basic finish										
Sides of foundations; plain vertical										
height exceeding 1.00m	9.45	10.00	10.40	3.94	0.79	83.25	2.31	95.95	m²	105.55
height not exceeding 250mm	3.31	10.00	3.64	1.19	0.25	25.30	0.69	29.63	m	32.59

Labour hourly rates: (except Specialists) Craft Operatives 18.37 Labourer 13.76 Rates are national average prices. Refer to REGIONAL VARIATIONS for indicative levels of overall pricing in regions	MATERIALS			LABOUR				RATES		
	Del to Site £	Waste %	Material Cost £	Craft Optve Hrs	Lab Hrs	Labour Cost £	Sunds £	Nett Rate £	Unit	Gross rate (10%) £
UNDERPINNING (Cont'd)										
Formwork and basic finish (Cont'd)										
Sides of foundations; plain vertical (Cont'd)										
height 250 - 500mm	5.20	10.00	5.72	2.16	0.43	45.60	1.27	52.59	m	57.85
height 0.50m - 1.00m	10.40	10.00	11.43	4.03	0.81	85.18	2.34	98.95	m	108.85
Common bricks, BS EN 772, Category M, 215 x 102.5 x 65mm, compressive strength not less than 5.2 N/mm²; in cement mortar (1:3)										
Walls; vertical										
215mm thick; English bond	37.60	5.00	39.48	4.70	5.00	155.14	10.02	204.64	m²	225.11
327mm thick; English bond	56.56	5.00	59.39	5.70	6.15	189.33	15.75	264.48	m²	290.92
440mm thick; English bond	75.21	5.00	78.97	6.30	6.90	210.68	21.47	311.11	m²	342.22
Bonding to existing including extra material										
thickness of new work 215mm	4.11	5.00	4.31	1.32	1.32	42.41	1.08	47.81	m	52.59
thickness of new work 327mm	6.32	5.00	6.64	1.90	1.90	61.05	1.62	69.30	m	76.23
thickness of new work 440mm	8.22	5.00	8.63	2.50	2.50	80.33	2.16	91.11	m	100.22
Second Hard Stock bricks, BS EN 772, Category M, 215 x 102.5 x 65mm, in cement mortar (1:3)										
Walls; vertical										
215mm thick; English bond	94.61	5.00	99.34	4.70	5.00	155.14	10.02	264.49	m²	290.94
327mm thick; English bond	142.31	5.00	149.42	5.70	6.15	189.33	15.75	354.50	m²	389.95
440mm thick; English bond	189.21	5.00	198.67	6.30	6.90	210.68	21.47	430.81	m²	473.89
Bonding to existing including extra material										
thickness of new work 215mm	10.34	5.00	10.85	1.90	1.90	61.05	1.08	72.98	m	80.28
thickness of new work 327mm	15.90	5.00	16.69	1.90	1.90	61.05	1.62	79.36	m	87.30
thickness of new work 440mm	20.67	5.00	21.70	2.50	2.50	80.33	2.16	104.19	m	114.61
Engineering bricks, BS EN 772, Category F, 215 x 102.5 x 65mm, class A; in cement mortar (1:3)										
Walls; vertical										
215mm thick; English bond	89.25	5.00	93.71	5.20	5.50	171.20	10.02	274.94	m²	302.43
327mm thick; English bond	134.25	5.00	140.96	6.30	6.75	208.61	15.75	365.32	m²	401.86
440mm thick; English bond	178.50	5.00	187.42	6.95	7.55	231.56	21.47	440.45	m²	484.49
Bonding to existing including extra material										
thickness of new work 215mm	9.75	5.00	10.24	1.45	1.45	46.59	1.08	57.91	m	63.70
thickness of new work 327mm	15.00	5.00	15.75	2.10	2.10	67.47	1.62	84.84	m	93.33
thickness of new work 440mm	19.50	5.00	20.47	2.75	2.75	88.36	2.16	110.99	m	122.09
Engineering bricks, BS EN 772, Category F, 215 x 102.5 x 65mm, Class B; in cement mortar (1:3)										
Walls; vertical										
215mm thick; English bond	38.44	5.00	40.36	5.20	5.50	171.20	10.02	221.58	m²	243.74
327mm thick; English bond	57.82	5.00	60.71	6.30	6.75	208.61	15.75	285.07	m²	313.58
440mm thick; English bond	76.87	5.00	80.72	6.95	7.55	231.56	21.47	333.74	m²	367.12
Bonding to existing including extra material										
thickness of new work 215mm	4.20	5.00	4.41	1.45	1.45	46.59	1.08	52.08	m	57.29
thickness of new work 327mm	6.46	5.00	6.78	2.10	2.10	67.47	1.62	75.88	m	83.46
thickness of new work 440mm	8.40	5.00	8.82	2.75	2.75	88.36	2.16	99.34	m	109.27
Sundry items										
Wedging and pinning up to underside of existing construction with two courses slates in cement mortar (1:3)										
215mm walls	4.54	15.00	5.22	0.60	0.60	19.28	0.28	24.78	m	27.26
327mm walls	6.81	15.00	7.83	0.87	0.87	27.95	0.34	36.13	m	39.74
440mm walls	9.14	15.00	10.51	1.14	1.14	36.63	0.38	47.52	m	52.27

DIAPHRAGM WALLS AND EMBEDDED RETAINING WALLS

Labour hourly rates: (except Specialists) Craft Operatives 18.37 Labourer 13.76 Rates are national average prices. Refer to REGIONAL VARIATIONS for indicative levels of overall pricing in regions	MATERIALS			LABOUR				RATES		
	Del to Site	Waste	Material Cost	Craft Optve	Lab	Labour Cost	Sunds	Nett Rate		Gross rate (10%)
	£	%	£	Hrs	Hrs	£	£	£	Unit	£
WALLS										
Embedded retaining walls; contiguous panel construction; panel lengths not exceeding 5m. Note:- the following prices are indicative only: firm quotations should always be obtained (CEMENTATION FOUNDATIONS SKANSKA LIMITED)										
On/Off site charge in addition to the following prices, add for bringing plant to site, erecting and dismantling, maintaining and removing from site, approximately.	-	-	Specialist	-	-	Specialist	-	142425.00	Sm	156667.50
Excavation and Bentonite slurry and disposal										
600mm thick wall; maximum depth										
5m	-	-	Specialist	-	-	Specialist	-	428.33	m³	471.16
10m	-	-	Specialist	-	-	Specialist	-	428.33	m³	471.16
15m	-	-	Specialist	-	-	Specialist	-	428.33	m³	471.16
20m	-	-	Specialist	-	-	Specialist	-	428.33	m³	471.16
800mm thick wall; maximum depth										
5m	-	-	Specialist	-	-	Specialist	-	356.59	m³	392.25
10m	-	-	Specialist	-	-	Specialist	-	356.59	m³	392.25
15m	-	-	Specialist	-	-	Specialist	-	356.59	m³	392.25
20m	-	-	Specialist	-	-	Specialist	-	356.59	m³	392.25
1000mm thick wall; maximum depth										
5m	-	-	Specialist	-	-	Specialist	-	284.85	m³	313.34
10m	-	-	Specialist	-	-	Specialist	-	284.85	m³	313.34
15m	-	-	Specialist	-	-	Specialist	-	284.85	m³	313.34
20m	-	-	Specialist	-	-	Specialist	-	284.85	m³	313.34
Excavating through obstructions, rock like formations, etc undertaken on a time basis at a rate per rig per hour	-	-	Specialist	-	-	Specialist	-	831.34	hr	914.47
Reinforced concrete; BS 8500, designed mix C25, 20mm aggregate, minimum cement content 400 kg/m³										
600mm thick wall	-	-	Specialist	-	-	Specialist	-	177.24	m³	194.96
800mm thick wall	-	-	Specialist	-	-	Specialist	-	177.24	m³	194.96
1000mm thick wall	-	-	Specialist	-	-	Specialist	-	177.24	m³	194.96
Reinforced in-situ concrete; sulphate resisting; BS 8500, designed mix C25, 20mm aggregate, minimum cement content 400 kg/m³										
600mm thick wall	-	-	Specialist	-	-	Specialist	-	202.56	m³	222.82
800mm thick wall	-	-	Specialist	-	-	Specialist	-	202.56	m³	222.82
1000mm thick wall	-	-	Specialist	-	-	Specialist	-	202.56	m³	222.82
Reinforcement bars; BS 4449 hot rolled plain round mild steel; including hooks tying wire, and spacers and chairs which are at the discretion of the Contractor										
16mm; straight	-	-	Specialist	-	-	Specialist	-	1270.22	t	1397.24
20mm; straight	-	-	Specialist	-	-	Specialist	-	1270.22	t	1397.24
25mm; straight	-	-	Specialist	-	-	Specialist	-	1270.22	t	1397.24
32mm; straight	-	-	Specialist	-	-	Specialist	-	1270.22	t	1397.24
40mm; straight	-	-	Specialist	-	-	Specialist	-	1270.22	t	1397.24
16mm; bent	-	-	Specialist	-	-	Specialist	-	1270.22	t	1397.24
20mm; bent	-	-	Specialist	-	-	Specialist	-	1270.22	t	1397.24
25mm; bent	-	-	Specialist	-	-	Specialist	-	1270.22	t	1397.24
32mm; bent	-	-	Specialist	-	-	Specialist	-	1268.22	t	1395.04
40mm; bent	-	-	Specialist	-	-	Specialist	-	1268.22	t	1395.04
Reinforcement bars; BS 4449 hot rolled deformed high yield steel; including hooks tying wire, and spacers and chairs which are at the discretion of the Contractor										
16mm; straight	-	-	Specialist	-	-	Specialist	-	1268.22	t	1395.04
20mm; straight	-	-	Specialist	-	-	Specialist	-	1268.22	t	1395.04
25mm; straight	-	-	Specialist	-	-	Specialist	-	1268.22	t	1395.04
32mm; straight	-	-	Specialist	-	-	Specialist	-	1268.22	t	1395.04
40mm; straight	-	-	Specialist	-	-	Specialist	-	1268.22	t	1395.04
16mm; bent	-	-	Specialist	-	-	Specialist	-	1268.22	t	1395.04
20mm; bent	-	-	Specialist	-	-	Specialist	-	1268.22	t	1395.04
25mm; bent	-	-	Specialist	-	-	Specialist	-	1268.22	t	1395.04
32mm; bent	-	-	Specialist	-	-	Specialist	-	1268.22	t	1395.04
40mm; bent	-	-	Specialist	-	-	Specialist	-	1268.22	t	1395.04
Guide walls; excavation, disposal and support; reinforced in-situ concrete; BS 8500, designed mix C25, 20mm aggregate, minimum cement content 290 kg/m³; reinforced with one layer fabric BS 4483 reference A252, 3.95 kg/m² including laps, tying wire, all cutting and bending, and spacers and chairs which are at the discretion of the Contractor; formwork both sides										
600mm apart; propped top and bottom at 2000mm centres; both sides, 1000mm high	-	-	Specialist	-	-	Specialist	-	438.88	m	482.77
800mm apart; propped top and bottom at 2000mm centres; both sides, 1000mm high	-	-	Specialist	-	-	Specialist	-	438.88	m	482.77
1000mm apart; propped top and bottom at 2000mm centres; both sides, 1000mm high	-	-	Specialist	-	-	Specialist	-	438.88	m	482.77

Labour hourly rates: (except Specialists) Craft Operatives 18.37 Labourer 13.76 Rates are national average prices. Refer to REGIONAL VARIATIONS for indicative levels of overall pricing in regions	MATERIALS			LABOUR				RATES		
	Del to Site	Waste	Material Cost	Craft Optve	Lab	Labour Cost	Sunds	Nett Rate	Unit	Gross rate (10%)
	£	%	£	Hrs	Hrs	£	£	£		£
WALLS (Cont'd)										
Embedded retaining walls; contiguous panel construction; panel lengths not exceeding 5m. Note:- the following prices are indicative only: firm quotations should always be obtained (CEMENTATION FOUNDATIONS SKANSKA LIMITED) (Cont'd)										
Guide walls; excavation, disposal and support; reinforced in-situ concrete; BS 8500, designed mix C25, 20mm aggregate, minimum cement content 290 kg/m³; reinforced with one layer fabric BS 4483 reference A252, 3.95 kg/m² including laps, tying wire, all cutting and bending, and spacers and chairs which are at the discretion of the Contractor; formwork both sides (Cont'd)										
600mm apart; propped top and bottom at 2000mm centres; both sides, 1500mm high	-	-	Specialist	-	-	Specialist	-	675.20	m	742.72
800mm apart; propped top and bottom at 2000mm centres; both sides, 1500mm high										
both sides, 1500mm high	-	-	Specialist	-	-	Specialist	-	675.20	m	742.72
1000mm apart; propped top and bottom at 2000mm centres; both sides, 1500mm high	-	-	Specialist	-	-	Specialist	-	675.20	m	742.72

Labour hourly rates: (except Specialists) Craft Operatives 18.37 Labourer 13.76 Rates are national average prices. Refer to REGIONAL VARIATIONS for indicative levels of overall pricing in regions	MATERIALS			LABOUR				RATES		
	Del to Site	Waste	Material Cost	Craft Optve	Lab	Labour Cost	Sunds	Nett Rate		Gross rate (10%)
	£	%	£	Hrs	Hrs	£	£	£	Unit	£
CRIB WALLS										
Retaining Walls										
Betoflor precast concrete landscape retaining walls including soil filling to pockets but excluding excavation, concrete foundations, stone backfill to rear of wall and planting which are all deemed measured separately										
Betoflor interlocking units 500mm long x 250mm wide x 200mm modular deep in wall 250mm wide	63.45	5.00	66.62	-	5.00	68.80	-	135.42	m²	148.96
Extra over for colours	11.29	5.00	11.85	-	0.50	6.88	-	18.73	m²	20.61
Betoatlas interlocking units 250mm long x 500mm wide x 200mm modular deep in wall 500mm wide	109.10	5.00	114.55	-	7.00	96.32	-	210.88	m²	231.96
Extra over for colours	19.41	5.00	20.38	-	0.10	1.38	-	21.76	m²	23.93
Betonap 150/50 woven mesh geotextile as reinforcement to backfill	4.11	5.00	4.32	-	0.30	4.13	-	8.44	m²	9.29
graded stone filling behind Betoflor or Betoatlas.	41.98	5.00	44.08	-	4.00	55.04	-	99.12	m³	109.04
concrete haunching to base of blocks	7.58	5.00	7.96	-	0.45	6.19	-	14.15	m	15.56

This page left blank intentionally

Labour hourly rates: (except Specialists) Craft Operatives 18.37 Labourer 13.76 Rates are national average prices. Refer to REGIONAL VARIATIONS for indicative levels of overall pricing in regions	MATERIALS			LABOUR				RATES		
	Del to Site	Waste	Material Cost	Craft Optve	Lab	Labour Cost	Sunds	Nett Rate		Gross rate (10%)
	£	%	£	Hrs	Hrs	£	£	£	Unit	£
IN-SITU CONCRETE; PLAIN IN-SITU CONCRETE; READY MIXED										
Plain in-situ mass concrete; BS 8500; ordinary prescribed mix ST3, 20mm aggregate										
In filling voids	97.99	7.50	105.34	-	2.30	31.65	-	136.99	m³	150.69
In trench filling	97.99	7.50	105.34	-	2.30	31.65	-	136.99	m³	150.69
Poured on or against earth or unblinded hardcore										
In filling voids	97.99	7.50	105.34	-	2.30	31.65	-	136.99	m³	150.69
In trench filling	97.99	7.50	105.34	-	2.30	31.65	-	136.99	m³	150.69
Plain in-situ concrete in horizontal work; BS 8500; ordinary prescribed mix ST3, 20mm aggregate										
In blinding	97.99	7.50	105.34	-	2.30	31.65	-	136.99	m³	150.69
thickness not exceeding 300mm	97.99	7.50	105.34	-	3.10	42.66	-	148.00	m³	162.79
In structures	97.99	7.50	105.34	-	3.30	45.41	-	150.75	m³	165.82
thickness not exceeding 300mm	97.99	7.50	105.34	-	3.60	49.54	-	154.88	m³	170.36
Poured on or against earth or unblinded hardcore										
In blinding	97.99	7.50	105.34	-	2.30	31.65	-	136.99	m³	150.69
thickness not exceeding 300mm	97.99	7.50	105.34	-	3.10	42.66	-	148.00	m³	162.79
In structures	97.99	7.50	105.34	-	3.30	45.41	-	150.75	m³	165.82
thickness not exceeding 300mm	97.99	7.50	105.34	-	3.60	49.54	-	154.88	m³	170.36
Plain in-situ concrete in sloping work less than 15o; BS 8500; ordinary prescribed mix ST3, 20mm aggregate										
In blinding	97.99	7.50	105.34	-	2.30	31.65	-	136.99	m³	150.69
thickness not exceeding 300mm	97.99	7.50	105.34	-	3.10	42.66	-	148.00	m³	162.79
In structures	97.99	7.50	105.34	-	3.30	45.41	-	150.75	m³	165.82
thickness not exceeding 300mm	97.99	7.50	105.34	-	3.60	49.54	-	154.88	m³	170.36
In staircases	97.99	7.50	105.34	-	3.30	45.41	-	150.75	m³	165.82
thickness not exceeding 300mm	97.99	7.50	105.34	-	5.50	75.68	-	181.02	m³	199.12
Poured on or against earth or unblinded hardcore										
In blinding	97.99	7.50	105.34	-	2.30	31.65	-	136.99	m³	150.69
thickness not exceeding 300mm	97.99	7.50	105.34	-	3.10	42.66	-	148.00	m³	162.79
In structures	97.99	7.50	105.34	-	3.30	45.41	-	150.75	m³	165.82
thickness not exceeding 300mm	97.99	7.50	105.34	-	3.60	49.54	-	154.88	m³	170.36
In staircases	97.99	7.50	105.34	-	3.30	45.41	-	150.75	m³	165.82
thickness not exceeding 300mm	97.99	7.50	105.34	-	5.50	75.68	-	181.02	m³	199.12
Plain in-situ concrete in sloping work greater than 15o; BS 8500; ordinary prescribed mix ST3, 20mm aggregate										
In blinding	97.99	7.50	105.34	-	2.30	31.65	-	136.99	m³	150.69
thickness not exceeding 300mm	97.99	7.50	105.34	-	3.10	42.66	-	148.00	m³	162.79
In structures	97.99	7.50	105.34	-	3.30	45.41	-	150.75	m³	165.82
thickness not exceeding 300mm	97.99	7.50	105.34	-	3.60	49.54	-	154.88	m³	170.36
In staircases	97.99	7.50	105.34	-	3.30	45.41	-	150.75	m³	165.82
thickness not exceeding 300mm	97.99	7.50	105.34	-	5.50	75.68	-	181.02	m³	199.12
Poured on or against earth or unblinded hardcore										
In blinding	97.99	7.50	105.34	-	2.30	31.65	-	136.99	m³	150.69
thickness not exceeding 300mm	97.99	7.50	105.34	-	3.10	42.66	-	148.00	m³	162.79
In structures	97.99	7.50	105.34	-	3.30	45.41	-	150.75	m³	165.82
thickness not exceeding 300mm	97.99	7.50	105.34	-	3.60	49.54	-	154.88	m³	170.36
In staircases	97.99	7.50	105.34	-	3.30	45.41	-	150.75	m³	165.82
thickness not exceeding 300mm	97.99	7.50	105.34	-	5.50	75.68	-	181.02	m³	199.12
Plain in-situ concrete in vertical work; BS 8500; ordinary prescribed mix ST3, 20mm aggregate										
In structures	97.99	7.50	105.34	-	3.30	45.41	-	150.75	m³	165.82
thickness not exceeding 300mm	97.99	5.00	102.89	-	5.50	75.68	-	178.57	m³	196.43
Plain sundry in-situ concrete work; BS 8500; ordinary prescribed mix ST3, 20mm aggregate										
Horizontal	97.99	7.50	105.34	-	2.30	31.65	-	136.99	m³	150.69
width or thickness not exceeding 300mm	97.99	5.00	102.89	-	5.50	75.68	-	178.57	m³	196.43
Sloping	97.99	7.50	105.34	-	2.30	31.65	-	136.99	m³	150.69
width or thickness not exceeding 300mm	97.99	5.00	102.89	-	5.50	75.68	-	178.57	m³	196.43
Vertical	97.99	7.50	105.34	-	3.30	45.41	-	150.75	m³	165.82
width or thickness not exceeding 300mm	97.99	5.00	102.89	-	5.50	75.68	-	178.57	m³	196.43
Plain in-situ mass concrete; BS 8500; ordinary prescribed mix ST4, 20mm aggregate										
In filling voids	99.73	7.50	107.21	-	2.30	31.65	-	138.86	m³	152.75
In trench filling	99.73	7.50	107.21	-	2.30	31.65	-	138.86	m³	152.75
Poured on or against earth or unblinded hardcore										
In filling voids	99.73	7.50	107.21	-	2.30	31.65	-	138.86	m³	152.75
In trench filling	99.73	7.50	107.21	-	2.30	31.65	-	138.86	m³	152.75

Labour hourly rates: (except Specialists) Craft Operatives 18.37 Labourer 13.76 Rates are national average prices. Refer to REGIONAL VARIATIONS for indicative levels of overall pricing in regions	MATERIALS			LABOUR				RATES		
	Del to Site	Waste	Material Cost	Craft Optve	Lab	Labour Cost	Sunds	Nett Rate	Unit	Gross rate (10%)
	£	%	£	Hrs	Hrs	£	£	£		£
IN-SITU CONCRETE; PLAIN IN-SITU CONCRETE; READY MIXED (Cont'd)										
Plain in-situ concrete in horizontal work; BS 8500; ordinary prescribed mix ST4, 20mm aggregate										
In blinding	99.73	7.50	107.21	-	2.30	31.65	-	138.86	m³	152.75
thickness not exceeding 300mm	99.73	7.50	107.21	-	3.10	42.66	-	149.87	m³	164.86
In structures	99.73	7.50	107.21	-	3.30	45.41	-	152.62	m³	167.88
thickness not exceeding 300mm	99.73	7.50	107.21	-	3.60	49.54	-	156.75	m³	172.42
Poured on or against earth or unblinded hardcore										
In blinding	99.73	7.50	107.21	-	2.30	31.65	-	138.86	m³	152.75
thickness not exceeding 300mm	99.73	7.50	107.21	-	3.10	42.66	-	149.87	m³	164.86
In structures	99.73	7.50	107.21	-	3.30	45.41	-	152.62	m³	167.88
thickness not exceeding 300mm	99.73	7.50	107.21	-	3.60	49.54	-	156.75	m³	172.42
Plain in-situ concrete in sloping work less than 15o; BS 8500; ordinary prescribed mix ST4, 20mm aggregate										
In blinding	99.73	7.50	107.21	-	2.30	31.65	-	138.86	m³	152.75
thickness not exceeding 300mm	99.73	7.50	107.21	-	3.10	42.66	-	149.87	m³	164.86
In structures	99.73	7.50	107.21	-	3.30	45.41	-	152.62	m³	167.88
thickness not exceeding 300mm	99.73	7.50	107.21	-	3.60	49.54	-	156.75	m³	172.42
In staircases	99.73	7.50	107.21	-	3.30	45.41	-	152.62	m³	167.88
thickness not exceeding 300mm	99.73	7.50	107.21	-	5.50	75.68	-	182.89	m³	201.18
Poured on or against earth or unblinded hardcore										
In blinding	99.73	7.50	107.21	-	2.30	31.65	-	138.86	m³	152.75
thickness not exceeding 300mm	99.73	7.50	107.21	-	3.10	42.66	-	149.87	m³	164.86
In structures	99.73	7.50	107.21	-	3.30	45.41	-	152.62	m³	167.88
thickness not exceeding 300mm	99.73	7.50	107.21	-	3.60	49.54	-	156.75	m³	172.42
In staircases	99.73	7.50	107.21	-	3.30	45.41	-	152.62	m³	167.88
thickness not exceeding 300mm	99.73	7.50	107.21	-	5.50	75.68	-	182.89	m³	201.18
Plain in-situ concrete in sloping work greater than 15o; BS 8500; ordinary prescribed mix ST4, 20mm aggregate										
In blinding	99.73	7.50	107.21	-	2.30	31.65	-	138.86	m³	152.75
thickness not exceeding 300mm	99.73	7.50	107.21	-	3.10	42.66	-	149.87	m³	164.86
In structures	99.73	7.50	107.21	-	3.30	45.41	-	152.62	m³	167.88
thickness not exceeding 300mm	99.73	7.50	107.21	-	3.60	49.54	-	156.75	m³	172.42
In staircases	99.73	7.50	107.21	-	3.30	45.41	-	152.62	m³	167.88
thickness not exceeding 300mm	99.73	7.50	107.21	-	5.50	75.68	-	182.89	m³	201.18
Poured on or against earth or unblinded hardcore										
In blinding	99.73	7.50	107.21	-	2.30	31.65	-	138.86	m³	152.75
thickness not exceeding 300mm	99.73	7.50	107.21	-	3.10	42.66	-	149.87	m³	164.86
In structures	99.73	7.50	107.21	-	3.30	45.41	-	152.62	m³	167.88
thickness not exceeding 300mm	99.73	7.50	107.21	-	3.60	49.54	-	156.75	m³	172.42
In staircases	99.73	7.50	107.21	-	3.30	45.41	-	152.62	m³	167.88
thickness not exceeding 300mm	99.73	7.50	107.21	-	5.50	75.68	-	182.89	m³	201.18
Plain in-situ concrete in vertical work; BS 8500; ordinary prescribed mix ST4, 20mm aggregate										
In structures	99.73	7.50	107.21	-	3.30	45.41	-	152.62	m³	167.88
thickness not exceeding 300mm	101.68	5.00	106.76	-	5.50	75.68	-	182.44	m³	200.69
Plain sundry in-situ concrete work; BS 8500; ordinary prescribed mix ST4, 20mm aggregate										
Horizontal	99.73	7.50	107.21	-	3.10	42.66	-	149.87	m³	164.86
width or thickness not exceeding 300mm	101.68	5.00	106.76	-	5.50	75.68	-	182.44	m³	200.69
Sloping	99.73	7.50	107.21	-	3.10	42.66	-	149.87	m³	164.86
width or thickness not exceeding 300mm	101.68	5.00	106.76	-	5.50	75.68	-	182.44	m³	200.69
Vertical	99.73	7.50	107.21	-	3.30	45.41	-	152.62	m³	167.88
width or thickness not exceeding 300mm	101.68	5.00	106.76	-	5.50	75.68	-	182.44	m³	200.69
Plain in-situ mass concrete; BS 8500; ordinary prescribed mix ST5, 20mm aggregate										
In filling voids	101.68	7.50	109.31	-	2.30	31.65	-	140.95	m³	155.05
In trench filling	101.68	7.50	109.31	-	2.30	31.65	-	140.95	m³	155.05
Poured on or against earth or unblinded hardcore										
In filling voids	101.68	7.50	109.31	-	2.30	31.65	-	140.95	m³	155.05
In trench filling	101.68	7.50	109.31	-	2.30	31.65	-	140.95	m³	155.05
Plain in-situ concrete in horizontal work; BS 8500; ordinary prescribed mix ST5, 20mm aggregate										
In blinding	101.68	7.50	109.31	-	2.30	31.65	-	140.95	m³	155.05
thickness not exceeding 300mm	101.68	7.50	109.31	-	3.10	42.66	-	151.96	m³	167.16
In structures	101.68	7.50	109.31	-	3.30	45.41	-	154.71	m³	170.19
thickness not exceeding 300mm	101.68	7.50	109.31	-	3.60	49.54	-	158.84	m³	174.73
Poured on or against earth or unblinded hardcore										
In blinding	101.68	7.50	109.31	-	2.30	31.65	-	140.95	m³	155.05
thickness not exceeding 300mm	101.68	7.50	109.31	-	3.10	42.66	-	151.96	m³	167.16
In structures	101.68	7.50	109.31	-	3.30	45.41	-	154.71	m³	170.19
thickness not exceeding 300mm	101.68	7.50	109.31	-	3.60	49.54	-	158.84	m³	174.73
Plain in-situ concrete in sloping work less than 15o; BS 8500; ordinary prescribed mix ST5, 20mm aggregate										
In blinding	101.68	7.50	109.31	-	2.30	31.65	-	140.95	m³	155.05
thickness not exceeding 300mm	101.68	7.50	109.31	-	3.10	42.66	-	151.96	m³	167.16
In structures	101.68	7.50	109.31	-	3.30	45.41	-	154.71	m³	170.19
thickness not exceeding 300mm	101.68	7.50	109.31	-	3.60	49.54	-	158.84	m³	174.73
In staircases	101.68	7.50	109.31	-	3.30	45.41	-	154.71	m³	170.19
thickness not exceeding 300mm	101.68	7.50	109.31	-	5.50	75.68	-	184.99	m³	203.48
Poured on or against earth or unblinded hardcore										
In blinding	101.68	7.50	109.31	-	2.30	31.65	-	140.95	m³	155.05

Labour hourly rates: (except Specialists) Craft Operatives 18.37 Labourer 13.76 Rates are national average prices. Refer to REGIONAL VARIATIONS for indicative levels of overall pricing in regions	MATERIALS			LABOUR				RATES		
	Del to Site	Waste	Material Cost	Craft Optve	Lab	Labour Cost	Sunds	Nett Rate		Gross rate (10%)
	£	%	£	Hrs	Hrs	£	£	£	Unit	£
IN-SITU CONCRETE; PLAIN IN-SITU CONCRETE; READY MIXED (Cont'd)										
Plain in-situ concrete in sloping work less than 15o; BS 8500; ordinary prescribed mix ST5, 20mm aggregate (Cont'd)										
Poured on or against earth or unblinded hardcore (Cont'd)										
thickness not exceeding 300mm	101.68	7.50	109.31	-	3.10	42.66	-	151.96	m³	167.16
In structures	101.68	7.50	109.31	-	3.30	45.41	-	154.71	m³	170.19
thickness not exceeding 300mm	101.68	7.50	109.31	-	3.60	49.54	-	158.84	m³	174.73
In staircases	101.68	7.50	109.31	-	3.30	45.41	-	154.71	m³	170.19
thickness not exceeding 300mm	101.68	7.50	109.31	-	5.50	75.68	-	184.99	m³	203.48
Plain in-situ concrete in sloping work greater than 15o; BS 8500; ordinary prescribed mix ST5, 20mm aggregate										
In blinding	101.68	7.50	109.31	-	2.30	31.65	-	140.95	m³	155.05
thickness not exceeding 300mm	101.68	7.50	109.31	-	3.10	42.66	-	151.96	m³	167.16
In structures	101.68	7.50	109.31	-	3.30	45.41	-	154.71	m³	170.19
thickness not exceeding 300mm	101.68	7.50	109.31	-	3.60	49.54	-	158.84	m³	174.73
In staircases	101.68	7.50	109.31	-	3.30	45.41	-	154.71	m³	170.19
thickness not exceeding 300mm	101.68	7.50	109.31	-	5.50	75.68	-	184.99	m³	203.48
Poured on or against earth or unblinded hardcore										
In blinding	101.68	7.50	109.31	-	2.30	31.65	-	140.95	m³	155.05
thickness not exceeding 300mm	101.68	7.50	109.31	-	3.10	42.66	-	151.96	m³	167.16
In structures	101.68	7.50	109.31	-	3.30	45.41	-	154.71	m³	170.19
thickness not exceeding 300mm	101.68	7.50	109.31	-	3.60	49.54	-	158.84	m³	174.73
In staircases	101.68	7.50	109.31	-	3.30	45.41	-	154.71	m³	170.19
thickness not exceeding 300mm	101.68	7.50	109.31	-	5.50	75.68	-	184.99	m³	203.48
Plain in-situ concrete in vertical work; BS 8500; ordinary prescribed mix ST5, 20mm aggregate										
In structures	101.68	7.50	109.31	-	3.30	45.41	-	154.71	m³	170.19
thickness not exceeding 300mm	101.68	5.00	106.76	-	5.50	75.68	-	182.44	m³	200.69
Plain sundry in-situ concrete work; BS 8500; ordinary prescribed mix ST5, 20mm aggregate										
Horizontal	101.68	7.50	109.31	-	3.10	42.66	-	151.96	m³	167.16
width or thickness not exceeding 300mm	101.68	5.00	106.76	-	5.50	75.68	-	182.44	m³	200.69
Sloping	101.68	7.50	109.31	-	3.10	42.66	-	151.96	m³	167.16
width or thickness not exceeding 300mm	101.68	5.00	106.76	-	5.50	75.68	-	182.44	m³	200.69
Vertical	101.68	7.50	109.31	-	3.30	45.41	-	154.71	m³	170.19
width or thickness not exceeding 300mm	101.68	5.00	106.76	-	5.50	75.68	-	182.44	m³	200.69
IN-SITU CONCRETE; REINFORCED IN-SITU CONCRETE; READY MIXED										
Reinforced in-situ mass concrete; BS 8500; designed mix C12/15, 20mm aggregate, minimum cement content 220 kg/m³; vibrated										
In filling voids	95.12	5.00	99.88	-	2.60	35.78	0.75	136.40	m³	150.04
In trench filling	95.12	5.00	99.88	-	2.60	35.78	0.75	136.40	m³	150.04
Poured on or against earth or unblinded hardcore										
In filling voids	95.12	5.00	99.88	-	2.60	35.78	0.75	136.40	m³	150.04
In trench filling	95.12	5.00	99.88	-	2.60	35.78	0.75	136.40	m³	150.04
Reinforced in-situ concrete in horizontal work; BS 8500; designed mix C12/15, 20mm aggregate, minimum cement content 220 kg/m³; vibrated										
In blinding	95.12	5.00	99.88	-	2.60	35.78	0.75	136.40	m³	150.04
thickness not exceeding 300mm	95.12	5.00	99.88	-	3.50	48.16	0.75	148.79	m³	163.66
In structures	95.12	5.00	99.88	-	3.60	49.54	0.75	150.16	m³	165.18
thickness not exceeding 300mm	95.12	5.00	99.88	-	3.90	53.66	0.75	154.29	m³	169.72
Poured on or against earth or unblinded hardcore										
In blinding	95.12	5.00	99.88	-	2.60	35.78	0.75	136.40	m³	150.04
thickness not exceeding 300mm	95.12	5.00	99.88	-	3.50	48.16	0.75	148.79	m³	163.66
In structures	95.12	5.00	99.88	-	3.60	49.54	0.75	150.16	m³	165.18
thickness not exceeding 300mm	95.12	5.00	99.88	-	3.90	53.66	0.75	154.29	m³	169.72
Reinforced in-situ concrete in sloping work less than 15o; BS 8500; designed mix C12/15, 20mm aggregate, minimum cement content 220 kg/m³; vibrated										
In blinding	95.12	5.00	99.88	-	2.60	35.78	0.75	136.40	m³	150.04
thickness not exceeding 300mm	95.12	5.00	99.88	-	3.50	48.16	0.75	148.79	m³	163.66
In structures	95.12	5.00	99.88	-	3.60	49.54	0.75	150.16	m³	165.18
thickness not exceeding 300mm	95.12	5.00	99.88	-	3.90	53.66	0.75	154.29	m³	169.72
In staircases	95.12	5.00	99.88	-	3.60	49.54	0.75	150.16	m³	165.18
thickness not exceeding 300mm	95.12	2.50	97.50	-	5.50	75.68	0.75	173.93	m³	191.32
Poured on or against earth or unblinded hardcore										
In blinding	95.12	5.00	99.88	-	2.60	35.78	0.75	136.40	m³	150.04
thickness not exceeding 300mm	95.12	5.00	99.88	-	3.50	48.16	0.75	148.79	m³	163.66
In structures	95.12	5.00	99.88	-	3.60	49.54	0.75	150.16	m³	165.18
thickness not exceeding 300mm	95.12	5.00	99.88	-	3.90	53.66	0.75	154.29	m³	169.72
In staircases	95.12	5.00	99.88	-	3.60	49.54	0.75	150.16	m³	165.18
thickness not exceeding 300mm	95.12	2.50	97.50	-	5.50	75.68	0.75	173.93	m³	191.32
Reinforced in-situ concrete in sloping work greater than 15o; BS 8500; designed mix C12/15, 20mm aggregate, minimum cement content 220 kg/m³; vibrated										
In blinding	95.12	5.00	99.88	-	2.60	35.78	0.75	136.40	m³	150.04
thickness not exceeding 300mm	95.12	5.00	99.88	-	3.50	48.16	0.75	148.79	m³	163.66
In structures	95.12	5.00	99.88	-	3.60	49.54	0.75	150.16	m³	165.18
thickness not exceeding 300mm	95.12	5.00	99.88	-	3.90	53.66	0.75	154.29	m³	169.72
In staircases	95.12	5.00	99.88	-	3.60	49.54	0.75	150.16	m³	165.18

69

Labour hourly rates: (except Specialists) Craft Operatives 18.37 Labourer 13.76 Rates are national average prices. Refer to REGIONAL VARIATIONS for indicative levels of overall pricing in regions	MATERIALS			LABOUR				RATES		
	Del to Site	Waste	Material Cost	Craft Optve	Lab	Labour Cost	Sunds	Nett Rate	Unit	Gross rate (10%)
	£	%	£	Hrs	Hrs	£	£	£		£
IN-SITU CONCRETE; REINFORCED IN-SITU CONCRETE; READY MIXED (Cont'd)										
Reinforced in-situ concrete in sloping work greater than 15o; BS 8500; designed mix C12/15, 20mm aggregate, minimum cement content 220 kg/m³; vibrated (Cont'd)										
thickness not exceeding 300mm	95.12	2.50	97.50	-	5.50	75.68	0.75	173.93	m³	191.32
Poured on or against earth or unblinded hardcore										
In blinding	95.12	5.00	99.88	-	2.60	35.78	0.75	136.40	m³	150.04
thickness not exceeding 300mm	95.12	5.00	99.88	-	3.50	48.16	0.75	148.79	m³	163.66
In structures	95.12	5.00	99.88	-	3.60	49.54	0.75	150.16	m³	165.18
thickness not exceeding 300mm	95.12	5.00	99.88	-	3.90	53.66	0.75	154.29	m³	169.72
In staircases	95.12	5.00	99.88	-	3.60	49.54	0.75	150.16	m³	165.18
thickness not exceeding 300mm	95.12	2.50	97.50	-	5.50	75.68	0.75	173.93	m³	191.32
Reinforced in-situ concrete in vertical work; BS 8500; designed mix C12/15, 20mm aggregate, minimum cement content 220 kg/m³; vibrated										
In structures	95.12	5.00	99.88	-	3.60	49.54	0.75	150.16	m³	165.18
thickness not exceeding 300mm	95.12	2.50	97.50	-	5.00	68.80	0.75	167.05	m³	183.75
Reinforced sundry in-situ concrete work; BS 8500; designed mix C12/15, 20mm aggregate, minimum cement content 220 kg/m³; vibrated										
Horizontal	95.12	5.00	99.88	-	2.60	35.78	0.75	136.40	m³	150.04
width or thickness not exceeding 300mm	95.12	5.00	99.88	-	3.50	48.16	0.75	148.79	m³	163.66
Sloping	95.12	5.00	99.88	-	2.60	35.78	0.75	136.40	m³	150.04
width or thickness not exceeding 300mm	95.12	5.00	99.88	-	3.50	48.16	0.75	148.79	m³	163.66
Vertical	95.12	5.00	99.88	-	3.60	49.54	0.75	150.16	m³	165.18
width or thickness not exceeding 300mm	95.12	2.50	97.50	-	5.50	75.68	0.75	173.93	m³	191.32
Reinforced in-situ mass concrete; BS 8500; designed mix C16/20, 20mm aggregate, minimum cement content 240 kg/m³; vibrated										
In filling voids	96.25	5.00	101.06	-	2.60	35.78	0.75	137.59	m³	151.34
In trench filling	96.25	5.00	101.06	-	2.60	35.78	0.75	137.59	m³	151.34
Poured on or against earth or unblinded hardcore										
In filling voids	96.25	5.00	101.06	-	2.60	35.78	0.75	137.59	m³	151.34
In trench filling	96.25	5.00	101.06	-	2.60	35.78	0.75	137.59	m³	151.34
Reinforced in-situ concrete in horizontal work; BS 8500; designed mix C16/20, 20mm aggregate, minimum cement content 240 kg/m³; vibrated										
In blinding	96.25	5.00	101.06	-	2.60	35.78	0.75	137.59	m³	151.34
thickness not exceeding 300mm	96.25	5.00	101.06	-	3.50	48.16	0.75	149.97	m³	164.97
In structures	96.25	5.00	101.06	-	3.60	49.54	0.75	151.35	m³	166.48
thickness not exceeding 300mm	96.25	5.00	101.06	-	3.90	53.66	0.75	155.47	m³	171.02
Poured on or against earth or unblinded hardcore										
In blinding	96.25	5.00	101.06	-	2.60	35.78	0.75	137.59	m³	151.34
thickness not exceeding 300mm	96.25	5.00	101.06	-	3.50	48.16	0.75	149.97	m³	164.97
In structures	96.25	5.00	101.06	-	3.60	49.54	0.75	151.35	m³	166.48
thickness not exceeding 300mm	96.25	5.00	101.06	-	3.90	53.66	0.75	155.47	m³	171.02
Reinforced in-situ concrete in sloping work less than 15o; BS 8500; designed mix C16/20, 20mm aggregate, minimum cement content 240 kg/m³; vibrated										
In blinding	96.25	5.00	101.06	-	2.60	35.78	0.75	137.59	m³	151.34
thickness not exceeding 300mm	96.25	5.00	101.06	-	3.50	48.16	0.75	149.97	m³	164.97
In structures	96.25	5.00	101.06	-	3.60	49.54	0.75	151.35	m³	166.48
thickness not exceeding 300mm	96.25	5.00	101.06	-	3.90	53.66	0.75	155.47	m³	171.02
In staircases	96.25	5.00	101.06	-	3.60	49.54	0.75	151.35	m³	166.48
thickness not exceeding 300mm	96.25	2.50	98.65	-	5.50	75.68	0.75	175.08	m³	192.59
Poured on or against earth or unblinded hardcore										
In blinding	96.25	5.00	101.06	-	2.60	35.78	0.75	137.59	m³	151.34
thickness not exceeding 300mm	96.25	5.00	101.06	-	3.50	48.16	0.75	149.97	m³	164.97
In structures	96.25	5.00	101.06	-	3.60	49.54	0.75	151.35	m³	166.48
thickness not exceeding 300mm	96.25	5.00	101.06	-	3.90	53.66	0.75	155.47	m³	171.02
In staircases	96.25	5.00	101.06	-	3.60	49.54	0.75	151.35	m³	166.48
thickness not exceeding 300mm	96.25	2.50	98.65	-	5.50	75.68	0.75	175.08	m³	192.59
Reinforced in-situ concrete in sloping work greater than 15o; BS 8500; designed mix C16/20, 20mm aggregate, minimum cement content 240 kg/m³; vibrated										
In blinding	96.25	5.00	101.06	-	2.60	35.78	0.75	137.59	m³	151.34
thickness not exceeding 300mm	96.25	5.00	101.06	-	3.50	48.16	0.75	149.97	m³	164.97
In structures	96.25	5.00	101.06	-	3.60	49.54	0.75	151.35	m³	166.48
thickness not exceeding 300mm	96.25	5.00	101.06	-	3.90	53.66	0.75	155.47	m³	171.02
In staircases	96.25	5.00	101.06	-	3.60	49.54	0.75	151.35	m³	166.48
thickness not exceeding 300mm	96.25	2.50	98.65	-	5.50	75.68	0.75	175.08	m³	192.59
Poured on or against earth or unblinded hardcore										
In blinding	96.25	5.00	101.06	-	2.60	35.78	0.75	137.59	m³	151.34
thickness not exceeding 300wmm	96.25	5.00	101.06	-	3.50	48.16	0.75	149.97	m³	164.97
In structures	96.25	5.00	101.06	-	3.60	49.54	0.75	151.35	m³	166.48
thickness not exceeding 300mm	96.25	5.00	101.06	-	3.90	53.66	0.75	155.47	m³	171.02
In staircases	96.25	5.00	101.06	-	3.60	49.54	0.75	151.35	m³	166.48
thickness not exceeding 300mm	96.25	2.50	98.65	-	5.50	75.68	0.75	175.08	m³	192.59
Reinforced in-situ concrete in vertical work; BS 8500; designed mix C16/20, 20mm aggregate, minimum cement content 240 kg/m³; vibrated										
In structures	96.25	5.00	101.06	-	3.60	49.54	0.75	151.35	m³	166.48
thickness not exceeding 300mm	96.25	2.50	98.65	-	5.00	68.80	0.75	168.20	m³	185.02

Labour hourly rates: (except Specialists) Craft Operatives 18.37 Labourer 13.76 Rates are national average prices. Refer to REGIONAL VARIATIONS for indicative levels of overall pricing in regions	MATERIALS			LABOUR				RATES		
	Del to Site	Waste	Material Cost	Craft Optve	Lab	Labour Cost	Sunds	Nett Rate	Unit	Gross rate (10%)
	£	%	£	Hrs	Hrs	£	£	£		£
IN-SITU CONCRETE; REINFORCED IN-SITU CONCRETE; READY MIXED (Cont'd)										
Reinforced sundry in-situ concrete work; BS 8500; designed mix C16/20, 20mm aggregate, minimum cement content 240 kg/m³; vibrated										
Horizontal	96.25	5.00	101.06	-	2.60	35.78	0.75	137.59	m³	151.34
width or thickness not exceeding 300mm	96.25	5.00	101.06	-	3.50	48.16	0.75	149.97	m³	164.97
Sloping	96.25	5.00	101.06	-	2.60	35.78	0.75	137.59	m³	151.34
width or thickness not exceeding 300mm	96.25	5.00	101.06	-	3.50	48.16	0.75	149.97	m³	164.97
Vertical	96.25	5.00	101.06	-	3.60	49.54	0.75	151.35	m³	166.48
width or thickness not exceeding 300mm	96.25	2.50	98.65	-	5.50	75.68	0.75	175.08	m³	192.59
Reinforced in-situ mass concrete; BS 8500; designed mix C20/25, 20mm aggregate, minimum cement content 290 kg/m³; vibrated										
In filling voids	98.61	5.00	103.54	-	2.60	35.78	0.75	140.06	m³	154.07
In trench filling	98.61	5.00	103.54	-	2.60	35.78	0.75	140.06	m³	154.07
Poured on or against earth or unblinded hardcore										
In filling voids	98.61	5.00	103.54	-	2.60	35.78	0.75	140.06	m³	154.07
In trench filling	98.61	5.00	103.54	-	2.60	35.78	0.75	140.06	m³	154.07
Reinforced in-situ concrete in horizontal work; BS 8500; designed mix C20/25, 20mm aggregate, minimum cement content 290 kg/m³; vibrated										
In blinding	98.61	5.00	103.54	-	2.60	35.78	0.75	140.06	m³	154.07
thickness not exceeding 300mm	98.61	5.00	103.54	-	3.50	48.16	0.75	152.45	m³	167.69
In structures	98.61	5.00	103.54	-	3.60	49.54	0.75	153.82	m³	169.20
thickness not exceeding 300mm	98.61	5.00	103.54	-	3.90	53.66	0.75	157.95	m³	173.74
Poured on or against earth or unblinded hardcore										
In blinding	98.61	5.00	103.54	-	2.60	35.78	0.75	140.06	m³	154.07
thickness not exceeding 300mm	98.61	5.00	103.54	-	3.50	48.16	0.75	152.45	m³	167.69
In structures	98.61	5.00	103.54	-	3.60	49.54	0.75	153.82	m³	169.20
thickness not exceeding 300mm	98.61	5.00	103.54	-	3.90	53.66	0.75	157.95	m³	173.74
Reinforced in-situ concrete in sloping work less than 15o; BS 8500; designed mix C20/25, 20mm aggregate, minimum cement content 290 kg/m³; vibrated										
In blinding	98.61	5.00	103.54	-	2.60	35.78	0.75	140.06	m³	154.07
thickness not exceeding 300mm	98.61	5.00	103.54	-	3.50	48.16	0.75	152.45	m³	167.69
In structures	98.61	5.00	103.54	-	3.60	49.54	0.75	153.82	m³	169.20
thickness not exceeding 300mm	98.61	5.00	103.54	-	3.90	53.66	0.75	157.95	m³	173.74
In staircases	98.61	5.00	103.54	-	3.60	49.54	0.75	153.82	m³	169.20
thickness not exceeding 300mm	98.61	2.50	101.07	-	5.50	75.68	0.75	177.50	m³	195.25
Poured on or against earth or unblinded hardcore										
In blinding	98.61	5.00	103.54	-	2.60	35.78	0.75	140.06	m³	154.07
thickness not exceeding 300mm	98.61	5.00	103.54	-	3.50	48.16	0.75	152.45	m³	167.69
In structures	98.61	5.00	103.54	-	3.60	49.54	0.75	153.82	m³	169.20
thickness not exceeding 300mm	98.61	5.00	103.54	-	3.90	53.66	0.75	157.95	m³	173.74
In staircases	98.61	5.00	103.54	-	3.60	49.54	0.75	153.82	m³	169.20
thickness not exceeding 300mm	98.61	2.50	101.07	-	5.50	75.68	0.75	177.50	m³	195.25
Reinforced in-situ concrete in sloping work greater than 15o; BS 8500; designed mix C20/25, 20mm aggregate, minimum cement content 290 kg/m³; vibrated										
In blinding	98.61	5.00	103.54	-	2.60	35.78	0.75	140.06	m³	154.07
thickness not exceeding 300mm	98.61	5.00	103.54	-	3.50	48.16	0.75	152.45	m³	167.69
In structures	98.61	5.00	103.54	-	3.60	49.54	0.75	153.82	m³	169.20
thickness not exceeding 300mm	98.61	5.00	103.54	-	3.90	53.66	0.75	157.95	m³	173.74
In staircases	98.61	5.00	103.54	-	3.60	49.54	0.75	153.82	m³	169.20
thickness not exceeding 300mm	98.61	2.50	101.07	-	5.50	75.68	0.75	177.50	m³	195.25
Poured on or against earth or unblinded hardcore										
In blinding	98.61	5.00	103.54	-	2.60	35.78	0.75	140.06	m³	154.07
thickness not exceeding 300mm	98.61	5.00	103.54	-	3.50	48.16	0.75	152.45	m³	167.69
In structures	98.61	5.00	103.54	-	3.60	49.54	0.75	153.82	m³	169.20
thickness not exceeding 300mm	98.61	5.00	103.54	-	3.90	53.66	0.75	157.95	m³	173.74
In staircases	98.61	5.00	103.54	-	3.60	49.54	0.75	153.82	m³	169.20
thickness not exceeding 300mm	98.61	2.50	101.07	-	5.50	75.68	0.75	177.50	m³	195.25
Reinforced in-situ concrete in vertical work; BS 8500; designed mix C20/25, 20mm aggregate, minimum cement content 290 kg/m³; vibrated										
In structures	98.61	5.00	103.54	-	3.60	49.54	0.75	153.82	m³	169.20
thickness not exceeding 300mm	98.61	2.50	101.07	-	5.00	68.80	0.75	170.62	m³	187.68
Reinforced sundry in-situ concrete work; BS 8500; designed mix C20/25, 20mm aggregate, minimum cement content 290 kg/m³; vibrated										
Horizontal	98.61	5.00	103.54	-	2.60	35.78	0.75	140.06	m³	154.07
width or thickness not exceeding 300mm	98.61	5.00	103.54	-	3.50	48.16	0.75	152.45	m³	167.69
Sloping	98.61	5.00	103.54	-	2.60	35.78	0.75	140.06	m³	154.07
width or thickness not exceeding 300mm	98.61	5.00	103.54	-	3.50	48.16	0.75	152.45	m³	167.69
Vertical	98.61	5.00	103.54	-	3.60	49.54	0.75	153.82	m³	169.20
width or thickness not exceeding 300mm	98.61	2.50	101.07	-	5.50	75.68	0.75	177.50	m³	195.25
Reinforced in-situ mass concrete; BS 8500; designed mix C25/30, 20mm aggregate, minimum cement content 290 kg/m³; vibrated										
In filling voids	101.27	5.00	106.33	-	2.60	35.78	0.75	142.86	m³	157.15
In trench filling	101.27	5.00	106.33	-	2.60	35.78	0.75	142.86	m³	157.15
Poured on or against earth or unblinded hardcore										
In filling voids	101.27	5.00	106.33	-	2.60	35.78	0.75	142.86	m³	157.15

Labour hourly rates: (except Specialists) Craft Operatives 18.37 Labourer 13.76 Rates are national average prices. Refer to REGIONAL VARIATIONS for indicative levels of overall pricing in regions	MATERIALS			LABOUR				RATES		
	Del to Site	Waste	Material Cost	Craft Optve	Lab	Labour Cost	Sunds	Nett Rate		Gross rate (10%)
	£	%	£	Hrs	Hrs	£	£	£	Unit	£

IN-SITU CONCRETE; REINFORCED IN-SITU CONCRETE; READY MIXED (Cont'd)

Reinforced in-situ mass concrete; BS 8500; designed mix C25/30, 20mm aggregate, minimum cement content 290 kg/m³; vibrated (Cont'd)

Poured on or against earth or unblinded hardcore (Cont'd)

In trench filling	101.27	5.00	106.33	-	2.60	35.78	0.75	142.86	m³	157.15

Reinforced in-situ concrete in horizontal work; BS 8500; designed mix C25/30, 20mm aggregate, minimum cement content 290 kg/m³; vibrated

designed mix C25/30, 20mm aggregate, minimum cement content 290 kg/m³; vibrated

In blinding	101.27	5.00	106.33	-	2.60	35.78	0.75	142.86	m³	157.15
thickness not exceeding 300mm	101.27	5.00	106.33	-	3.50	48.16	0.75	155.24	m³	170.77
In structures	101.27	5.00	106.33	-	3.60	49.54	0.75	156.62	m³	172.28
thickness not exceeding 300mm	101.27	5.00	106.33	-	3.90	53.66	0.75	160.75	m³	176.82

Poured on or against earth or unblinded hardcore

In blinding	101.27	5.00	106.33	-	2.60	35.78	0.75	142.86	m³	157.15
thickness not exceeding 300mm	101.27	5.00	106.33	-	3.50	48.16	0.75	155.24	m³	170.77
In structures	101.27	5.00	106.33	-	3.60	49.54	0.75	156.62	m³	172.28
thickness not exceeding 300mm	101.27	5.00	106.33	-	3.90	53.66	0.75	160.75	m³	176.82

Reinforced in-situ concrete in sloping work less than 15o; BS 8500; designed mix C25/30, 20mm aggregate, minimum cement content 290 kg/m³; vibrated

In blinding	101.27	5.00	106.33	-	2.60	35.78	0.75	142.86	m³	157.15
thickness not exceeding 300mm	101.27	5.00	106.33	-	3.50	48.16	0.75	155.24	m³	170.77
In structures	101.27	5.00	106.33	-	3.60	49.54	0.75	156.62	m³	172.28
thickness not exceeding 300mm	101.27	5.00	106.33	-	3.90	53.66	0.75	160.75	m³	176.82
In staircases	101.27	2.50	103.80	-	3.60	49.54	0.75	154.09	m³	169.50
thickness not exceeding 300mm	101.27	2.50	103.80	-	5.50	75.68	0.75	180.23	m³	198.25

Poured on or against earth or unblinded hardcore

In blinding	101.27	5.00	106.33	-	2.60	35.78	0.75	142.86	m³	157.15
thickness not exceeding 300mm	101.27	5.00	106.33	-	3.50	48.16	0.75	155.24	m³	170.77
In structures	101.27	5.00	106.33	-	3.60	49.54	0.75	156.62	m³	172.28
thickness not exceeding 300mm	101.27	5.00	106.33	-	3.90	53.66	0.75	160.75	m³	176.82
In staircases	101.27	5.00	106.33	-	3.60	49.54	0.75	156.62	m³	172.28
thickness not exceeding 300mm	101.27	2.50	103.80	-	5.50	75.68	0.75	180.23	m³	198.25

Reinforced in-situ concrete in sloping work greater than 15o; BS 8500; designed mix C25/30, 20mm aggregate, minimum cement content 290 kg/m³; vibrated

In blinding	101.27	5.00	106.33	-	2.60	35.78	0.75	142.86	m³	157.15
thickness not exceeding 300mm	101.27	5.00	106.33	-	3.50	48.16	0.75	155.24	m³	170.77
In structures	101.27	5.00	106.33	-	3.60	49.54	0.75	156.62	m³	172.28
thickness not exceeding 300mm	101.27	5.00	106.33	-	3.90	53.66	0.75	160.75	m³	176.82
In staircases	101.27	5.00	106.33	-	3.60	49.54	0.75	156.62	m³	172.28
thickness not exceeding 300mm	101.27	2.50	103.80	-	5.50	75.68	0.75	180.23	m³	198.25

Poured on or against earth or unblinded hardcore

In blinding	101.27	5.00	106.33	-	2.60	35.78	0.75	142.86	m³	157.15
thickness not exceeding 300mm	101.27	5.00	106.33	-	3.50	48.16	0.75	155.24	m³	170.77
In structures	101.27	5.00	106.33	-	3.60	49.54	0.75	156.62	m³	172.28
thickness not exceeding 300mm	101.27	5.00	106.33	-	3.90	53.66	0.75	160.75	m³	176.82
In staircases	101.27	5.00	106.33	-	3.60	49.54	0.75	156.62	m³	172.28
thickness not exceeding 300mm	101.27	2.50	103.80	-	5.50	75.68	0.75	180.23	m³	198.25

Reinforced in-situ concrete in vertical work; BS 8500; designed mix C25/30, 20mm aggregate, minimum cement content 290 kg/m³; vibrated

In structures	101.27	5.00	106.33	-	3.60	49.54	0.75	156.62	m³	172.28
thickness not exceeding 300mm	101.27	2.50	103.80	-	5.00	68.80	0.75	173.35	m³	190.69

Reinforced sundry in-situ concrete work; BS 8500; designed mix C25/30, 20mm aggregate, minimum cement content 290 kg/m³; vibrated

Horizontal	101.27	5.00	106.33	-	2.60	35.78	0.75	142.86	m³	157.15
width or thickness not exceeding 300mm	101.27	5.00	106.33	-	3.50	48.16	0.75	155.24	m³	170.77
Sloping	101.27	5.00	106.33	-	2.60	35.78	0.75	142.86	m³	157.15
width or thickness not exceeding 300mm	101.27	5.00	106.33	-	3.50	48.16	0.75	155.24	m³	170.77
Vertical	101.27	5.00	106.33	-	3.60	49.54	0.75	156.62	m³	172.28
width or thickness not exceeding 300mm	101.27	5.00	106.33	-	5.50	75.68	0.75	182.76	m³	201.04

IN-SITU CONCRETE; PLAIN IN-SITU CONCRETE; SITE MIXED

Plain in-situ mass concrete; mix 1:6, all in aggregate

In filling voids	137.06	7.50	147.34	-	3.40	46.78	2.70	196.82	m³	216.50
In trench filling	137.06	7.50	147.34	-	3.40	46.78	2.70	196.82	m³	216.50

Poured on or against earth or unblinded hardcore

In filling voids	137.06	7.50	147.34	-	3.40	46.78	2.70	196.82	m³	216.50
In trench filling	137.06	7.50	147.34	-	3.40	46.78	2.70	196.82	m³	216.50

Plain in-situ concrete in horizontal work; mix 1:6, all in aggregate

In blinding	137.06	7.50	147.34	-	3.40	46.78	2.70	196.82	m³	216.50
thickness not exceeding 300mm	137.06	7.50	147.34	-	4.45	61.23	2.70	211.27	m³	232.40
In structures	137.06	7.50	147.34	-	4.40	60.54	2.70	210.58	m³	231.64
thickness not exceeding 300mm	137.06	7.50	147.34	-	4.70	64.67	2.70	214.71	m³	236.18

Labour hourly rates: (except Specialists) Craft Operatives 18.37 Labourer 13.76 Rates are national average prices. Refer to REGIONAL VARIATIONS for indicative levels of overall pricing in regions	MATERIALS			LABOUR				RATES		
	Del to Site	Waste	Material Cost	Craft Optve	Lab	Labour Cost	Sunds	Nett Rate		Gross rate (10%)
	£	%	£	Hrs	Hrs	£	£	£	Unit	£
IN-SITU CONCRETE; PLAIN IN-SITU CONCRETE; SITE MIXED (Cont'd)										
Plain in-situ concrete in horizontal work; mix 1:6, all in aggregate (Cont'd)										
Poured on or against earth or unblinded hardcore										
In blinding	137.06	7.50	147.34	-	3.40	46.78	2.70	196.82	m³	216.50
thickness not exceeding 300mm	137.06	7.50	147.34	-	4.45	61.23	2.70	211.27	m³	232.40
In structures	137.06	7.50	147.34	-	4.40	60.54	2.70	210.58	m³	231.64
thickness not exceeding 300mm	137.06	7.50	147.34	-	4.70	64.67	2.70	214.71	m³	236.18
Plain in-situ concrete in sloping work less than 15o; mix 1:6, all in aggregate										
In blinding	137.06	7.50	147.34	-	3.40	46.78	2.70	196.82	m³	216.50
thickness not exceeding 300mm	137.06	7.50	147.34	-	4.45	61.23	2.70	211.27	m³	232.40
In structures	137.06	7.50	147.34	-	4.40	60.54	2.70	210.58	m³	231.64
thickness not exceeding 300mm	137.06	7.50	147.34	-	4.70	64.67	2.70	214.71	m³	236.18
Poured on or against earth or unblinded hardcore										
In blinding	137.06	7.50	147.34	-	3.40	46.78	2.70	196.82	m³	216.50
thickness not exceeding 300mm	137.06	7.50	147.34	-	4.45	61.23	2.70	211.27	m³	232.40
In structures	137.06	7.50	147.34	-	4.40	60.54	2.70	210.58	m³	231.64
thickness not exceeding 300mm	137.06	7.50	147.34	-	4.70	64.67	2.70	214.71	m³	236.18
Plain in-situ concrete in sloping work greater than 15o; mix 1:6, all in aggregate										
In blinding	137.06	7.50	147.34	-	3.40	46.78	2.70	196.82	m³	216.50
thickness not exceeding 300mm	137.06	7.50	147.34	-	4.45	61.23	2.70	211.27	m³	232.40
In structures	137.06	7.50	147.34	-	4.40	60.54	2.70	210.58	m³	231.64
thickness not exceeding 300mm	137.06	7.50	147.34	-	4.70	64.67	2.70	214.71	m³	236.18
Poured on or against earth or unblinded hardcore										
In blinding	137.06	7.50	147.34	-	3.40	46.78	2.70	196.82	m³	216.50
thickness not exceeding 300mm	137.06	7.50	147.34	-	4.45	61.23	2.70	211.27	m³	232.40
In structures	603.05	7.50	648.28	-	4.40	60.54	2.70	711.53	m³	782.68
thickness not exceeding 300mm	137.06	7.50	147.34	-	4.70	64.67	2.70	214.71	m³	236.18
Plain in-situ concrete in vertical work; mix 1:6, all in aggregate										
In structures	137.06	7.50	147.34	-	4.40	60.54	2.70	210.58	m³	231.64
thickness not exceeding 300mm	137.06	7.50	147.34	-	6.60	90.82	2.70	240.85	m³	264.94
Plain sundry in-situ concrete work; mix 1:6, all in aggregate										
Horizontal	137.06	7.50	147.34	-	3.40	46.78	2.70	196.82	m³	216.50
width or thickness not exceeding 300mm	137.06	7.50	147.34	-	4.45	61.23	2.70	211.27	m³	232.40
Sloping	137.06	7.50	147.34	-	3.40	46.78	2.70	196.82	m³	216.50
width or thickness not exceeding 300mm	137.06	7.50	147.34	-	4.45	61.23	2.70	211.27	m³	232.40
Plain in-situ mass concrete; mix 1:8, all in aggregate										
In filling voids	130.10	7.50	139.86	-	3.40	46.78	2.70	189.34	m³	208.28
In trench filling	130.10	7.50	139.86	-	3.40	46.78	2.70	189.34	m³	208.28
Poured on or against earth or unblinded hardcore										
In filling voids	130.10	7.50	139.86	-	3.40	46.78	2.70	189.34	m³	208.28
In trench filling	130.10	7.50	139.86	-	3.40	46.78	2.70	189.34	m³	208.28
Plain in-situ concrete in horizontal work; mix 1:8, all in aggregate										
In blinding	130.10	7.50	139.86	-	3.40	46.78	2.70	189.34	m³	208.28
thickness not exceeding 300mm	130.10	7.50	139.86	-	3.40	46.78	2.70	189.34	m³	208.28
In structures	130.10	7.50	139.86	-	4.40	60.54	2.70	203.10	m³	223.41
thickness not exceeding 300mm	130.10	7.50	139.86	-	4.70	64.67	2.70	207.23	m³	227.96
Poured on or against earth or unblinded hardcore										
In blinding	130.10	7.50	139.86	-	3.40	46.78	2.70	189.34	m³	208.28
thickness not exceeding 300mm	130.10	7.50	139.86	-	4.25	58.48	2.70	201.04	m³	221.14
In structures	130.10	7.50	139.86	-	4.40	60.54	2.70	203.10	m³	223.41
thickness not exceeding 300mm	130.10	7.50	139.86	-	4.70	64.67	2.70	207.23	m³	227.96
Plain in-situ concrete in sloping work less than 15o; mix 1:8, all in aggregate										
In blinding	130.10	7.50	139.86	-	3.40	46.78	2.70	189.34	m³	208.28
thickness not exceeding 300mm	130.10	7.50	139.86	-	4.25	58.48	2.70	201.04	m³	221.14
In structures	130.10	7.50	139.86	-	4.40	60.54	2.70	203.10	m³	223.41
thickness not exceeding 300mm	130.10	7.50	139.86	-	4.70	64.67	2.70	207.23	m³	227.96
Poured on or against earth or unblinded hardcore										
In blinding	130.10	7.50	139.86	-	3.40	46.78	2.70	189.34	m³	208.28
thickness not exceeding 300mm	130.10	7.50	139.86	-	4.25	58.48	2.70	201.04	m³	221.14
In structures	130.10	7.50	139.86	-	4.40	60.54	2.70	203.10	m³	223.41
thickness not exceeding 300mm	130.10	7.50	139.86	-	4.70	64.67	2.70	207.23	m³	227.96
Plain in-situ concrete in sloping work greater than 15o; mix 1:8, all in aggregate										
In blinding	130.10	7.50	139.86	-	3.40	46.78	2.70	189.34	m³	208.28
thickness not exceeding 300mm	130.10	7.50	139.86	-	4.25	58.48	2.70	201.04	m³	221.14
In structures	130.10	7.50	139.86	-	4.40	60.54	2.70	203.10	m³	223.41
thickness not exceeding 300mm	130.10	7.50	139.86	-	4.70	64.67	2.70	207.23	m³	227.96
Poured on or against earth or unblinded hardcore										
In blinding	130.10	7.50	139.86	-	3.40	46.78	2.70	189.34	m³	208.28
thickness not exceeding 300mm	130.10	7.50	139.86	-	4.25	58.48	2.70	201.04	m³	221.14
In structures	130.10	7.50	139.86	-	4.40	60.54	2.70	203.10	m³	223.41
thickness not exceeding 300mm	130.10	7.50	139.86	-	4.70	64.67	2.70	207.23	m³	227.96
Plain in-situ concrete in vertical work; mix 1:8, all in aggregate										
In structures	130.10	7.50	139.86	-	4.40	60.54	2.70	203.10	m³	223.41
thickness not exceeding 300mm	130.10	7.50	139.86	-	6.60	90.82	2.70	233.38	m³	256.71

IN-SITU CONCRETE WORKS

Labour hourly rates: (except Specialists) Craft Operatives 18.37 Labourer 13.76 Rates are national average prices. Refer to REGIONAL VARIATIONS for indicative levels of overall pricing in regions	MATERIALS			LABOUR				RATES		
	Del to Site	Waste	Material Cost	Craft Optve	Lab	Labour Cost	Sunds	Nett Rate	Unit	Gross rate (10%)
	£	%	£	Hrs	Hrs	£	£	£		£
IN-SITU CONCRETE; PLAIN IN-SITU CONCRETE; SITE MIXED (Cont'd)										
Plain sundry in-situ concrete work; mix 1:8, all in aggregate										
Horizontal	130.10	7.50	139.86	-	3.40	46.78	2.70	189.34	m³	208.28
width or thickness not exceeding 300mm	130.10	7.50	139.86	-	4.25	58.48	2.70	201.04	m³	221.14
Sloping	130.10	7.50	139.86	-	3.40	46.78	2.70	189.34	m³	208.28
width or thickness not exceeding 300mm	130.10	7.50	139.86	-	4.25	58.48	2.70	201.04	m³	221.14
Plain in-situ mass concrete; mix 1:12, all in aggregate										
In filling voids	126.27	7.50	135.74	-	3.40	46.78	2.70	185.22	m³	203.74
In trench filling	126.27	7.50	135.74	-	3.40	46.78	2.70	185.22	m³	203.74
Poured on or against earth or unblinded hardcore										
In filling voids	126.27	7.50	135.74	-	3.40	46.78	2.70	185.22	m³	203.74
In trench filling	126.27	7.50	135.74	-	3.40	46.78	2.70	185.22	m³	203.74
Plain in-situ concrete in horizontal work; mix 1:12, all in aggregate										
In blinding	126.27	7.50	135.74	-	3.40	46.78	2.70	185.22	m³	203.74
thickness not exceeding 300mm	126.27	7.50	135.74	-	4.25	58.48	2.70	196.92	m³	216.61
In structures	126.27	7.50	135.74	-	4.40	60.54	2.70	198.98	m³	218.88
thickness not exceeding 300mm	126.27	7.50	135.74	-	4.70	64.67	2.70	203.11	m³	223.42
Poured on or against earth or unblinded hardcore										
In blinding	126.27	7.50	135.74	-	3.40	46.78	2.70	185.22	m³	203.74
thickness not exceeding 300mm	126.27	7.50	135.74	-	4.25	58.48	2.70	196.92	m³	216.61
In structures	126.27	7.50	135.74	-	4.40	60.54	2.70	198.98	m³	218.88
thickness not exceeding 300mm	126.27	7.50	135.74	-	4.70	64.67	2.70	203.11	m³	223.42
Plain in-situ concrete in sloping work less than 15o; mix 1:12, all in aggregate										
In blinding	126.27	7.50	135.74	-	3.40	46.78	2.70	185.22	m³	203.74
thickness not exceeding 300mm	126.27	7.50	135.74	-	4.25	58.48	2.70	196.92	m³	216.61
In structures	126.27	7.50	135.74	-	4.40	60.54	2.70	198.98	m³	218.88
thickness not exceeding 300mm	126.27	7.50	135.74	-	4.70	64.67	2.70	203.11	m³	223.42
Poured on or against earth or unblinded hardcore										
In blinding	126.27	7.50	135.74	-	3.40	46.78	2.70	185.22	m³	203.74
thickness not exceeding 300mm	126.27	7.50	135.74	-	4.25	58.48	2.70	196.92	m³	216.61
In structures	126.27	7.50	135.74	-	4.40	60.54	2.70	198.98	m³	218.88
thickness not exceeding 300mm	126.27	7.50	135.74	-	4.70	64.67	2.70	203.11	m³	223.42
Plain in-situ concrete in sloping work greater than 15o; mix 1:12, all in aggregate										
In blinding	126.27	7.50	135.74	-	3.40	46.78	2.70	185.22	m³	203.74
thickness not exceeding 300mm	126.27	7.50	135.74	-	4.25	58.48	2.70	196.92	m³	216.61
In structures	126.27	7.50	135.74	-	4.40	60.54	2.70	198.98	m³	218.88
thickness not exceeding 300mm	126.27	7.50	135.74	-	4.70	64.67	2.70	203.11	m³	223.42
Poured on or against earth or unblinded hardcore										
In blinding	126.27	7.50	135.74	-	3.40	46.78	2.70	185.22	m³	203.74
thickness not exceeding 300mm	126.27	7.50	135.74	-	4.25	58.48	2.70	196.92	m³	216.61
In structures	126.27	7.50	135.74	-	4.40	60.54	2.70	198.98	m³	218.88
thickness not exceeding 300mm	126.27	7.50	135.74	-	4.70	64.67	2.70	203.11	m³	223.42
Plain in-situ concrete in vertical work; mix 1:12, all in aggregate										
In structures	126.27	7.50	135.74	-	4.40	60.54	2.70	198.98	m³	218.88
thickness not exceeding 300mm	126.27	7.50	135.74	-	6.60	90.82	2.70	229.25	m³	252.18
Plain sundry in-situ concrete work; mix 1:12, all in aggregate										
Horizontal	126.27	7.50	135.74	-	3.40	46.78	2.70	185.22	m³	203.74
width or thickness not exceeding 300mm	126.27	7.50	135.74	-	4.25	58.48	2.70	196.92	m³	216.61
Sloping	126.27	7.50	135.74	-	3.40	46.78	2.70	185.22	m³	203.74
width or thickness not exceeding 300mm	126.27	7.50	135.74	-	4.25	58.48	2.70	196.92	m³	216.61
Plain in-situ mass concrete; BS 8500; ordinary prescribed mix C15P, 20mm aggregate										
In filling voids	124.45	7.50	133.79	-	3.40	46.78	2.70	183.27	m³	201.60
In trench filling	124.45	7.50	133.79	-	3.40	46.78	2.70	183.27	m³	201.60
Poured on or against earth or unblinded hardcore										
In filling voids	124.45	7.50	133.79	-	3.40	46.78	2.70	183.27	m³	201.60
In trench filling	124.45	7.50	133.79	-	3.40	46.78	2.70	183.27	m³	201.60
Plain in-situ concrete in horizontal work; BS 8500; ordinary prescribed mix C15P, 20mm aggregate										
In blinding	124.45	7.50	133.79	-	3.40	46.78	2.70	183.27	m³	201.60
thickness not exceeding 300mm	124.45	7.50	133.79	-	3.40	46.78	2.70	183.27	m³	201.60
In structures	124.45	7.50	133.79	-	4.40	60.54	2.70	197.03	m³	216.73
thickness not exceeding 300mm	124.45	7.50	133.79	-	4.70	64.67	2.70	201.16	m³	221.27
Poured on or against earth or unblinded hardcore										
In blinding	124.45	7.50	133.79	-	3.40	46.78	2.70	183.27	m³	201.60
thickness not exceeding 300mm	124.45	7.50	133.79	-	4.25	58.48	2.70	194.97	m³	214.46
In structures	124.45	7.50	133.79	-	4.40	60.54	2.70	197.03	m³	216.73
thickness not exceeding 300mm	124.45	7.50	133.79	-	4.70	64.67	2.70	201.16	m³	221.27
Plain in-situ concrete in sloping work less than 15o; BS 8500; ordinary prescribed mix C15P, 20mm aggregate										
In blinding	124.45	7.50	133.79	-	3.40	46.78	2.70	183.27	m³	201.60
thickness not exceeding 300mm	124.45	7.50	133.79	-	4.25	58.48	2.70	194.97	m³	214.46
In structures	124.45	7.50	133.79	-	4.40	60.54	2.70	197.03	m³	216.73
thickness not exceeding 300mm	124.45	7.50	133.79	-	4.70	64.67	2.70	201.16	m³	221.27
Poured on or against earth or unblinded hardcore										
In blinding	124.45	7.50	133.79	-	3.40	46.78	2.70	183.27	m³	201.60

Labour hourly rates: (except Specialists) Craft Operatives 18.37 Labourer 13.76 Rates are national average prices. Refer to REGIONAL VARIATIONS for indicative levels of overall pricing in regions	MATERIALS			LABOUR				RATES		
	Del to Site	Waste	Material Cost	Craft Optve	Lab	Labour Cost	Sunds	Nett Rate	Unit	Gross rate (10%)
	£	%	£	Hrs	Hrs	£	£	£		£
IN-SITU CONCRETE; PLAIN IN-SITU CONCRETE; SITE MIXED (Cont'd)										
Plain in-situ concrete in sloping work less than 15o; BS 8500; ordinary prescribed mix C15P, 20mm aggregate (Cont'd)										
Poured on or against earth or unblinded hardcore (Cont'd)										
thickness not exceeding 300mm	124.45	7.50	133.79	-	4.25	58.48	2.70	194.97	m³	214.46
In structures	124.45	7.50	133.79	-	4.40	60.54	2.70	197.03	m³	216.73
thickness not exceeding 300mm	124.45	7.50	133.79	-	4.70	64.67	2.70	201.16	m³	221.27
Plain in-situ concrete in sloping work greater than 15o; BS 8500; ordinary prescribed mix C15P, 20mm aggregate										
In blinding	124.45	7.50	133.79	-	3.40	46.78	2.70	183.27	m³	201.60
thickness not exceeding 300mm	124.45	7.50	133.79	-	4.25	58.48	2.70	194.97	m³	214.46
In structures	124.45	7.50	133.79	-	4.40	60.54	2.70	197.03	m³	216.73
thickness not exceeding 300mm	124.45	7.50	133.79	-	4.70	64.67	2.70	201.16	m³	221.27
Poured on or against earth or unblinded hardcore										
In blinding	124.45	7.50	133.79	-	3.40	46.78	2.70	183.27	m³	201.60
thickness not exceeding 300mm	124.45	7.50	133.79	-	4.25	58.48	2.70	194.97	m³	214.46
In structures	124.45	7.50	133.79	-	4.40	60.54	2.70	197.03	m³	216.73
thickness not exceeding 300mm	124.45	7.50	133.79	-	4.70	64.67	2.70	201.16	m³	221.27
Plain in-situ concrete in vertical work; BS 8500; ordinary prescribed mix C15P, 20mm aggregate										
In structures	124.45	7.50	133.79	-	4.40	60.54	2.70	197.03	m³	216.73
thickness not exceeding 300mm	124.45	7.50	133.79	-	6.60	90.82	2.70	227.30	m³	250.03
Plain sundry in-situ concrete work; BS 8500; ordinary prescribed mix C15P, 20mm aggregate										
Horizontal	124.45	7.50	133.79	-	3.40	46.78	2.70	183.27	m³	201.60
width or thickness not exceeding 300mm	124.45	7.50	133.79	-	4.25	58.48	2.70	194.97	m³	214.46
Sloping	124.45	7.50	133.79	-	3.40	46.78	2.70	183.27	m³	201.60
width or thickness not exceeding 300mm	124.45	7.50	133.79	-	4.25	58.48	2.70	194.97	m³	214.46
Plain in-situ mass concrete; BS 8500; ordinary prescribed mix C20P, 20mm aggregate										
In filling voids	126.49	7.50	135.98	-	3.40	46.78	2.70	185.46	m³	204.01
In trench filling	126.49	7.50	135.98	-	3.40	46.78	2.70	185.46	m³	204.01
Poured on or against earth or unblinded hardcore										
In filling voids	126.49	7.50	135.98	-	3.40	46.78	2.70	185.46	m³	204.01
In trench filling	126.49	7.50	135.98	-	3.40	46.78	2.70	185.46	m³	204.01
Plain in-situ concrete in horizontal work; BS 8500; ordinary prescribed mix C20P, 20mm aggregate										
In blinding	126.49	7.50	135.98	-	3.40	46.78	2.70	185.46	m³	204.01
thickness not exceeding 300mm	126.49	7.50	135.98	-	4.25	58.48	2.70	197.16	m³	216.88
In structures	126.49	7.50	135.98	-	4.40	60.54	2.70	199.22	m³	219.15
thickness not exceeding 300mm	126.49	7.50	135.98	-	4.70	64.67	2.70	203.35	m³	223.69
Poured on or against earth or unblinded hardcore										
In blinding	126.49	7.50	135.98	-	3.40	46.78	2.70	185.46	m³	204.01
thickness not exceeding 300mm	126.49	7.50	135.98	-	4.25	58.48	2.70	197.16	m³	216.88
In structures	126.49	7.50	135.98	-	4.40	60.54	2.70	199.22	m³	219.15
thickness not exceeding 300mm	126.49	7.50	135.98	-	4.70	64.67	2.70	203.35	m³	223.69
Plain in-situ concrete in sloping work less than 15o; BS 8500; ordinary prescribed mix C20P, 20mm aggregate										
In blinding	126.49	7.50	135.98	-	3.40	46.78	2.70	185.46	m³	204.01
thickness not exceeding 300mm	126.49	7.50	135.98	-	4.25	58.48	2.70	197.16	m³	216.88
In structures	126.49	7.50	135.98	-	4.40	60.54	2.70	199.22	m³	219.15
thickness not exceeding 300mm	126.49	7.50	135.98	-	4.70	64.67	2.70	203.35	m³	223.69
Poured on or against earth or unblinded hardcore										
In blinding	126.49	7.50	135.98	-	3.40	46.78	2.70	185.46	m³	204.01
thickness not exceeding 300mm	126.49	7.50	135.98	-	4.25	58.48	2.70	197.16	m³	216.88
In structures	126.49	7.50	135.98	-	4.40	60.54	2.70	199.22	m³	219.15
thickness not exceeding 300mm	126.49	7.50	135.98	-	4.70	64.67	2.70	203.35	m³	223.69
Plain in-situ concrete in sloping work greater than 15o; BS 8500; ordinary prescribed mix C20P, 20mm aggregate										
In blinding	126.49	7.50	135.98	-	3.40	46.78	2.70	185.46	m³	204.01
thickness not exceeding 300mm	126.49	7.50	135.98	-	4.25	58.48	2.70	197.16	m³	216.88
In structures	126.49	7.50	135.98	-	4.40	60.54	2.70	199.22	m³	219.15
thickness not exceeding 300mm	126.49	7.50	135.98	-	4.70	64.67	2.70	203.35	m³	223.69
Poured on or against earth or unblinded hardcore										
In blinding	126.49	7.50	135.98	-	3.40	46.78	2.70	185.46	m³	204.01
thickness not exceeding 300mm	126.49	7.50	135.98	-	4.25	58.48	2.70	197.16	m³	216.88
In structures	126.49	7.50	135.98	-	4.40	60.54	2.70	199.22	m³	219.15
thickness not exceeding 300mm	126.49	7.50	135.98	-	4.70	64.67	2.70	203.35	m³	223.69
Plain in-situ concrete in vertical work; BS 8500; ordinary prescribed mix C20P, 20mm aggregate										
In structures	126.49	7.50	135.98	-	4.40	60.54	2.70	199.22	m³	219.15
thickness not exceeding 300mm	126.49	7.50	135.98	-	6.60	90.82	2.70	229.50	m³	252.45
Plain sundry in-situ concrete work; BS 8500; ordinary prescribed mix C20P, 20mm aggregate										
Horizontal	126.49	7.50	135.98	-	3.40	46.78	2.70	185.46	m³	204.01
width or thickness not exceeding 300mm	126.49	7.50	135.98	-	4.25	58.48	2.70	197.16	m³	216.88
Sloping	126.49	7.50	135.98	-	3.40	46.78	2.70	185.46	m³	204.01
width or thickness not exceeding 300mm	126.49	7.50	135.98	-	4.25	58.48	2.70	197.16	m³	216.88

IN-SITU CONCRETE WORKS

Labour hourly rates: (except Specialists) Craft Operatives 18.37 Labourer 13.76 Rates are national average prices. Refer to REGIONAL VARIATIONS for indicative levels of overall pricing in regions	MATERIALS			LABOUR				RATES		
	Del to Site £	Waste %	Material Cost £	Craft Optve Hrs	Lab Hrs	Labour Cost £	Sunds £	Nett Rate £	Unit	Gross rate (10%) £
IN-SITU CONCRETE; PLAIN IN-SITU CONCRETE; SITE MIXED (Cont'd)										
Plain in-situ mass concrete; BS 8500; ordinary prescribed mix C25P, 20mm aggregate										
In filling voids	134.44	7.50	144.52	-	3.40	46.78	2.70	194.01	m³	213.41
In trench filling	134.44	7.50	144.52	-	3.40	46.78	2.70	194.01	m³	213.41
Poured on or against earth or unblinded hardcore										
In filling voids	134.44	7.50	144.52	-	3.40	46.78	2.70	194.01	m³	213.41
In trench filling	134.44	7.50	144.52	-	3.40	46.78	2.70	194.01	m³	213.41
Plain in-situ concrete in horizontal work; BS 8500; ordinary prescribed mix C25P, 20mm aggregate										
In blinding	134.44	7.50	144.52	-	3.40	46.78	2.70	194.01	m³	213.41
thickness not exceeding 300mm	134.44	7.50	144.52	-	4.25	58.48	2.70	205.70	m³	226.28
In structures	134.44	7.50	144.52	-	4.40	60.54	2.70	207.77	m³	228.55
thickness not exceeding 300mm	134.44	7.50	144.52	-	4.70	64.67	2.70	211.90	m³	233.09
Poured on or against earth or unblinded hardcore										
In blinding	134.44	7.50	144.52	-	3.40	46.78	2.70	194.01	m³	213.41
thickness not exceeding 300mm	134.44	7.50	144.52	-	4.25	58.48	2.70	205.70	m³	226.28
In structures	134.44	7.50	144.52	-	4.40	60.54	2.70	207.77	m³	228.55
thickness not exceeding 300mm	134.44	7.50	144.52	-	4.70	64.67	2.70	211.90	m³	233.09
Plain in-situ concrete in sloping work less than 15o; BS 8500; ordinary prescribed mix C25P, 20mm aggregate										
In blinding	134.44	7.50	144.52	-	3.40	46.78	2.70	194.01	m³	213.41
thickness not exceeding 300mm	134.44	7.50	144.52	-	4.25	58.48	2.70	205.70	m³	226.28
In structures	134.44	7.50	144.52	-	4.40	60.54	2.70	207.77	m³	228.55
thickness not exceeding 300mm	134.44	7.50	144.52	-	4.70	64.67	2.70	211.90	m³	233.09
Poured on or against earth or unblinded hardcore										
In blinding	134.44	7.50	144.52	-	3.40	46.78	2.70	194.01	m³	213.41
thickness not exceeding 300mm	134.44	7.50	144.52	-	4.25	58.48	2.70	205.70	m³	226.28
In structures	134.44	7.50	144.52	-	4.40	60.54	2.70	207.77	m³	228.55
thickness not exceeding 300mm	134.44	7.50	144.52	-	4.70	64.67	2.70	211.90	m³	233.09
Plain in-situ concrete in sloping work greater than 15o; BS 8500; ordinary prescribed mix C25P, 20mm aggregate										
In blinding	134.44	7.50	144.52	-	3.40	46.78	2.70	194.01	m³	213.41
thickness not exceeding 300mm	134.44	7.50	144.52	-	4.25	58.48	2.70	205.70	m³	226.28
In structures	134.44	7.50	144.52	-	4.40	60.54	2.70	207.77	m³	228.55
thickness not exceeding 300mm	134.44	7.50	144.52	-	4.70	64.67	2.70	211.90	m³	233.09
Poured on or against earth or unblinded hardcore										
In blinding	134.44	7.50	144.52	-	3.40	46.78	2.70	194.01	m³	213.41
thickness not exceeding 300mm	134.44	7.50	144.52	-	4.25	58.48	2.70	205.70	m³	226.28
In structures	134.44	7.50	144.52	-	4.40	60.54	2.70	207.77	m³	228.55
thickness not exceeding 300mm	134.44	7.50	144.52	-	4.70	64.67	2.70	211.90	m³	233.09
Plain in-situ concrete in vertical work; BS 8500; ordinary prescribed mix C25P, 20mm aggregate										
In structures	134.44	7.50	144.52	-	4.40	60.54	2.70	207.77	m³	228.55
thickness not exceeding 300mm	134.44	7.50	144.52	-	6.60	90.82	2.70	238.04	m³	261.84
Plain sundry in-situ concrete work; BS 8500; ordinary prescribed mix C25P, 20mm aggregate										
Horizontal	134.44	7.50	144.52	-	3.40	46.78	2.70	194.01	m³	213.41
width or thickness not exceeding 300mm	134.44	7.50	144.52	-	4.25	58.48	2.70	205.70	m³	226.28
Sloping	134.44	7.50	144.52	-	3.40	46.78	2.70	194.01	m³	213.41
width or thickness not exceeding 300mm	134.44	7.50	144.52	-	4.25	58.48	2.70	205.70	m³	226.28
IN-SITU CONCRETE; REINFORCED IN-SITU CONCRETE; SITE MIXED										
The foregoing concrete is based on the use of Portland cement. For other cements and waterproofers ADD as follows										
Concrete 1:6										
rapid hardening cement	90.85	-	90.85	-	-	-	-	90.85	m³	99.94
sulphate resisting cement	20.59	-	20.59	-	-	-	-	20.59	m³	22.65
waterproofing powder	23.36	-	23.36	-	-	-	-	23.36	m³	25.70
waterproofing liquid	3.55	-	3.55	-	-	-	-	3.55	m³	3.90
Concrete 1:8										
rapid hardening cement	72.22	-	72.22	-	-	-	-	72.22	m³	79.44
sulphate resisting cement	16.37	-	16.37	-	-	-	-	16.37	m³	18.00
waterproofing powder	18.57	-	18.57	-	-	-	-	18.57	m³	20.43
waterproofing liquid	2.82	-	2.82	-	-	-	-	2.82	m³	3.10
Concrete 1:12										
rapid hardening cement	50.09	-	50.09	-	-	-	-	50.09	m³	55.10
sulphate resisting cement	11.35	-	11.35	-	-	-	-	11.35	m³	12.49
waterproofing powder	12.88	-	12.88	-	-	-	-	12.88	m³	14.17
waterproofing liquid	1.96	-	1.96	-	-	-	-	1.96	m³	2.15
Concrete ST3 and C15										
rapid hardening cement	81.54	-	81.54	-	-	-	-	81.54	m³	89.69
sulphate resisting cement	18.48	-	18.48	-	-	-	-	18.48	m³	20.33
waterproofing powder	20.97	-	20.97	-	-	-	-	20.97	m³	23.06
waterproofing liquid	3.18	-	3.18	-	-	-	-	3.18	m³	3.50
Concrete ST4 and C20										
rapid hardening cement	93.18	-	93.18	-	-	-	-	93.18	m³	102.50
sulphate resisting cement	21.12	-	21.12	-	-	-	-	21.12	m³	23.23
waterproofing powder	23.96	-	23.96	-	-	-	-	23.96	m³	26.36
waterproofing liquid	3.64	-	3.64	-	-	-	-	3.64	m³	4.00

Labour hourly rates: (except Specialists) Craft Operatives 18.37 Labourer 13.76 Rates are national average prices. Refer to REGIONAL VARIATIONS for indicative levels of overall pricing in regions	MATERIALS			LABOUR				RATES		
	Del to Site	Waste	Material Cost	Craft Optve	Lab	Labour Cost	Sunds	Nett Rate		Gross rate (10%)
	£	%	£	Hrs	Hrs	£	£	£	Unit	£
IN-SITU CONCRETE; REINFORCED IN-SITU CONCRETE; SITE MIXED (Cont'd)										
The foregoing concrete is based on the use of Portland cement. For other cements and waterproofers ADD as follows (Cont'd)										
Concrete ST5 and C25										
rapid hardening cement	104.83	-	104.83	-	-	-	-	104.83	m³	115.32
sulphate resisting cement	23.76	-	23.76	-	-	-	-	23.76	m³	26.14
waterproofing powder	26.96	-	26.96	-	-	-	-	26.96	m³	29.65
waterproofing liquid	4.09	-	4.09	-	-	-	-	4.09	m³	4.50
Reinforced in-situ lightweight concrete in horizontal work; 20.5 N/mm²; vibrated										
In structures	131.25	2.50	134.53	-	4.40	60.54	3.45	198.53	m³	218.38
thickness not exceeding 300mm	131.25	2.50	134.53	-	5.60	77.06	3.45	215.04	m³	236.54
Reinforced in-situ lightweight concrete in sloping work less than 15o; 20.5 N/mm²; vibrated										
In structures	131.25	2.50	134.53	-	4.40	60.54	3.45	198.53	m³	218.38
thickness not exceeding 300mm	131.25	2.50	134.53	-	5.60	77.06	3.45	215.04	m³	236.54
Reinforced in-situ lightweight concrete in sloping work greater than 15o; 20.5 N/mm²; vibrated										
In structures	131.25	2.50	134.53	-	4.40	60.54	3.45	198.53	m³	218.38
thickness not exceeding 300mm	131.25	2.50	134.53	-	5.60	77.06	3.45	215.04	m³	236.54
Reinforced in-situ lightweight concrete in vertical work; 20.5 N/mm²; vibrated										
In structures	131.25	2.50	134.53	-	4.85	66.74	3.45	204.72	m³	225.19
thickness not exceeding 300mm	131.25	2.50	134.53	-	5.60	77.06	3.45	215.04	m³	236.54
Reinforced sundry in-situ lightweight concrete work; 20.5 N/mm²; vibrated										
Horizontal	131.25	2.50	134.53	-	4.40	60.54	3.45	198.53	m³	218.38
width or thickness not exceeding 300mm	131.25	2.50	134.53	-	5.60	77.06	3.45	215.04	m³	236.54
Sloping	131.25	2.50	134.53	-	4.40	60.54	3.45	198.53	m³	218.38
width or thickness not exceeding 300mm	131.25	2.50	134.53	-	5.60	77.06	3.45	215.04	m³	236.54
Reinforced in-situ lightweight concrete; 26.0 N/mm²; vibrated										
In structures	135.45	2.50	138.84	-	4.40	60.54	3.45	202.83	m³	223.11
thickness not exceeding 300mm	135.45	2.50	138.84	-	5.60	77.06	3.45	219.34	m³	241.28
Reinforced in-situ lightweight concrete in sloping work less than 15o; 26.0 N/mm²; vibrated										
In structures	135.45	2.50	138.84	-	4.40	60.54	3.45	202.83	m³	223.11
thickness not exceeding 300mm	135.45	2.50	138.84	-	5.60	77.06	3.45	219.34	m³	241.28
Reinforced in-situ lightweight concrete in sloping work greater than 15o; 26.0 N/mm²; vibrated										
In structures	135.45	2.50	138.84	-	4.40	60.54	3.45	202.83	m³	223.11
thickness not exceeding 300mm	135.45	2.50	138.84	-	5.60	77.06	3.45	219.34	m³	241.28
Reinforced in-situ lightweight concrete in vertical work; 26.0 N/mm²; vibrated										
In structures	135.45	2.50	138.84	-	4.85	66.74	3.45	209.02	m³	229.92
thickness not exceeding 300mm	135.45	2.50	138.84	-	5.60	77.06	3.45	219.34	m³	241.28
Reinforced sundry in-situ lightweight concrete work; 26.0 N/mm²; vibrated										
Horizontal	135.45	2.50	138.84	-	4.40	60.54	3.45	202.83	m³	223.11
width or thickness not exceeding 300mm	135.45	2.50	138.84	-	5.60	77.06	3.45	219.34	m³	241.28
Sloping	135.45	2.50	138.84	-	4.40	60.54	3.45	202.83	m³	223.11
width or thickness not exceeding 300mm	135.45	2.50	138.84	-	5.60	77.06	3.45	219.34	m³	241.28
Reinforced in-situ lightweight concrete; 41.5 N/mm²; vibrated										
In structures	135.45	2.50	138.84	-	4.40	60.54	3.45	202.83	m³	223.11
thickness not exceeding 300mm	135.45	2.50	138.84	-	5.60	77.06	3.45	219.34	m³	241.28
Reinforced in-situ lightweight concrete in sloping work less than 15o; 41.5 N/mm²; vibrated										
In structures	143.85	2.50	147.45	-	4.40	60.54	3.45	211.44	m³	232.58
thickness not exceeding 300mm	143.85	2.50	147.45	-	5.60	77.06	3.45	227.95	m³	250.75
Reinforced in-situ lightweight concrete in sloping work greater than 15o; 41.5 N/mm²; vibrated										
In structures	143.85	2.50	147.45	-	4.40	60.54	3.45	211.44	m³	232.58
thickness not exceeding 300mm	143.85	2.50	147.45	-	5.60	77.06	3.45	227.95	m³	250.75
Reinforced in-situ lightweight concrete in vertical work; 41.5 N/mm²; vibrated										
In structures	143.85	2.50	147.45	-	4.85	66.74	3.45	217.63	m³	239.40
thickness not exceeding 300mm	143.85	2.50	147.45	-	5.60	77.06	3.45	227.95	m³	250.75
Reinforced sundry in-situ lightweight concrete work; 41.5 N/mm²; vibrated										
Horizontal	143.85	2.50	147.45	-	4.40	60.54	3.45	211.44	m³	232.58
width or thickness not exceeding 300mm	143.85	2.50	147.45	-	5.60	77.06	3.45	227.95	m³	250.75
Sloping	143.85	2.50	147.45	-	4.40	60.54	3.45	211.44	m³	232.58
width or thickness not exceeding 300mm	143.85	2.50	147.45	-	5.60	77.06	3.45	227.95	m³	250.75
SURFACE FINISHES TO INSITU CONCRETE										
Trowelling										
Trowelling to top surfaces	-	-	-	-	0.36	4.95	-	4.95	m²	5.45

Labour hourly rates: (except Specialists) Craft Operatives 18.37 Labourer 13.76 Rates are national average prices. Refer to REGIONAL VARIATIONS for indicative levels of overall pricing in regions	MATERIALS			LABOUR				RATES		
	Del to Site	Waste	Material Cost	Craft Optve	Lab	Labour Cost	Sunds	Nett Rate	Unit	Gross rate (10%)
	£	%	£	Hrs	Hrs	£	£	£		£
SURFACE FINISHES TO INSITU CONCRETE (Cont'd)										
Trowelling (Cont'd)										
Trowelling (Cont'd) to top surfaces to falls.............................	-	-	-	-	0.39	5.37	-	5.37	m²	5.90
Power floating										
Vacuum dewatering; power floating and power trowelling to top surfaces	-	-	-	-	0.22	3.03	0.33	3.36	m²	3.69
Hacking										
Hacking; by hand to top surfaces	-	-	-	-	0.55	7.57	-	7.57	m²	8.32
to faces..	-	-	-	-	0.60	8.26	-	8.26	m²	9.08
to soffits	-	-	-	-	0.72	9.91	-	9.91	m²	10.90
Hacking; by machine to top surfaces	-	-	-	-	0.22	3.03	0.34	3.37	m²	3.71
to faces..	-	-	-	-	0.24	3.30	0.34	3.65	m²	4.01
to soffits	-	-	-	-	0.28	3.85	0.34	4.20	m²	4.62
Other surface treatments										
Tamping unset concrete to top surfaces	-	-	-	-	0.22	3.03	0.44	3.46	m²	3.81
to top surfaces to falls............................	-	-	-	-	0.45	6.19	0.57	6.76	m²	7.44
Bush hammering to top surfaces	-	-	-	1.43	-	26.27	0.98	27.24	m²	29.97
to faces..	-	-	-	1.60	-	29.39	1.12	30.52	m²	33.57
to soffits	-	-	-	2.20	-	40.41	1.29	41.70	m²	45.87
Polythene sheeting laying as temporary protection to surface of concrete (use and waste) 125mu	0.28	10.00	0.30	-	0.03	0.41	-	0.72	m²	0.79
FORMWORK										
Plain formwork; Sides of foundations and bases										
plain vertical height not exceeding 250mm	1.16	12.50	1.30	0.53	0.11	11.25	0.33	12.88	m	14.17
height 250 - 500mm.................................	2.04	12.50	2.30	0.96	0.19	20.25	0.62	23.16	m	25.48
height exceeding 500mm	3.73	12.50	4.20	1.75	0.35	36.96	1.07	42.22	m²	46.45
plain vertical; left in height not exceeding 250mm	5.16	-	5.16	0.30	0.06	6.34	0.22	11.72	m	12.89
height 250 - 500mm.................................	9.53	-	9.53	0.55	0.11	11.62	0.36	21.51	m	23.66
height exceeding 500mm	17.31	-	17.31	1.00	0.20	21.12	0.64	39.08	m²	42.98
Plain formwork; Edges of horizontal work										
plain vertical height not exceeding 250mm	1.16	12.50	1.30	0.53	0.11	11.25	0.33	12.88	m	14.17
height 250 - 500mm.................................	2.04	12.50	2.29	0.96	0.19	20.25	0.62	23.16	m	25.47
height exceeding 500mm	3.73	12.50	4.20	1.75	0.35	36.96	1.07	42.22	m²	46.45
plain vertical; curved 10m radius height not exceeding 250mm	2.10	12.50	2.36	1.00	0.20	21.12	0.64	24.13	m	26.54
plain vertical; curved 1m radius height not exceeding 250mm	3.55	12.50	3.99	1.65	0.33	34.85	1.02	39.87	m	43.85
plain vertical; curved 20m radius height not exceeding 250mm	1.58	12.50	1.78	0.74	0.15	15.66	0.42	17.86	m	19.64
plain vertical; left in height not exceeding 250mm	5.16	-	5.16	0.30	0.06	6.34	0.22	11.72	m	12.89
height 250 - 500mm.................................	9.53	-	9.53	0.55	0.11	11.62	0.36	21.51	m	23.66
height exceeding 500mm	17.31	-	17.31	1.00	0.20	21.12	0.62	39.05	m²	42.95
Plain formwork; Soffits of horizontal work										
horizontal concrete thickness not exceeding 300mm; height to soffit not exceeding 3.00m.................................	11.29	10.00	12.42	1.65	0.33	34.85	1.68	48.95	m²	53.85
3.00 - 4.50m..	11.29	10.00	12.42	1.87	0.38	39.58	2.20	54.20	m²	59.63
4.50 - 6.00m..	11.29	10.00	12.42	2.09	0.43	44.31	2.70	59.43	m²	65.37
concrete thickness 300 - 450mm; height to soffit not exceeding 3.00m.................................	13.91	10.00	15.30	1.82	0.38	38.66	1.85	55.81	m²	61.39
3.00 - 4.50m..	13.91	10.00	15.30	2.06	0.42	43.62	2.43	61.35	m²	67.49
4.50 - 6.00m..	13.91	10.00	15.30	2.30	0.47	48.72	3.00	67.02	m²	73.72
Soffits of landings; horizontal concrete thickness not exceeding 300mm; height to soffit not exceeding 1.50m.................................	12.42	10.00	13.66	1.98	0.40	41.88	1.65	57.19	m²	62.90
1.50 - 3.00m..	12.42	10.00	13.66	2.22	0.45	46.97	2.27	62.90	m²	69.19
3.00 - 4.50m..	12.42	10.00	13.66	2.46	0.51	52.21	2.82	68.69	m²	75.55
Soffits of landings; horizontal; left in concrete thickness not exceeding 300mm; height to soffit not exceeding 3.00m.................................	34.23	-	34.23	1.10	0.22	23.23	1.20	58.67	m²	64.53
3.00 - 4.50m..	37.66	-	37.66	1.21	0.24	25.53	1.35	64.54	m²	70.99
Soffits of landings; horizontal; with frequent uses of prefabricated panels concrete thickness not exceeding 300mm; height to soffit not exceeding 3.00m.................................	3.35	15.00	3.85	1.65	0.33	34.85	1.63	40.34	m²	44.37

Labour hourly rates: (except Specialists) Craft Operatives 18.37 Labourer 13.76 Rates are national average prices. Refer to REGIONAL VARIATIONS for indicative levels of overall pricing in regions	MATERIALS			LABOUR				RATES		
	Del to Site	Waste	Material Cost	Craft Optve	Lab	Labour Cost	Sunds	Nett Rate		Gross rate (10%)
	£	%	£	Hrs	Hrs	£	£	£	Unit	£
FORMWORK (Cont'd)										
Plain formwork; Soffits of horizontal work (Cont'd)										
Soffits of landings; horizontal; with frequent uses of prefabricated panels (Cont'd)										
concrete thickness not exceeding 300mm; height to soffit (Cont'd)										
3.00 - 4.50m	3.35	15.00	3.85	1.87	0.38	39.58	2.20	45.64	m²	50.20
4.50 - 6.00m	3.35	15.00	3.85	2.09	0.43	44.31	2.70	50.86	m²	55.95
concrete thickness 300 - 450mm; height to soffit										
not exceeding 3.00m	4.04	15.00	4.64	1.82	0.36	38.39	1.80	44.83	m²	49.31
3.00 - 4.50m	4.04	15.00	4.64	2.06	0.42	43.62	2.43	50.69	m²	55.76
4.50 - 6.00m	4.04	15.00	4.64	3.00	0.47	61.58	3.00	69.22	m²	76.14
Plain formwork; Sides and soffits of isolated beams										
rectangular; height to soffit										
1.50 - 3.00m	8.27	10.00	9.10	2.35	0.47	49.64	1.32	60.06	m²	66.06
3.00 - 4.50m	8.27	10.00	9.10	2.35	0.47	49.64	1.61	60.34	m²	66.37
Plain formwork; Sides and soffits of attached beams										
rectangular; height to soffit										
not exceeding 3.00m	7.89	10.00	8.68	2.25	0.45	47.52	1.32	57.52	m²	63.27
3.00 - 4.50m	7.89	10.00	8.68	2.25	0.45	47.52	1.61	57.81	m²	63.59
Note										
not withstanding the requirements of NRM works to 500mm high are shown in lineal metres.										
Plain formwork; Sides of upstand beams										
rectangular; height to soffit										
not exceeding 3.00m										
height not exceeding 250mm	1.66	12.50	1.87	0.75	0.15	15.84	0.42	18.13	m	19.94
height 250 - 500mm	1.94	12.50	2.18	0.88	0.18	18.64	0.56	21.38	m	23.52
height exceeding 500mm	3.73	12.50	4.20	1.75	0.35	36.96	1.07	42.22	m²	46.45
3.00 - 4.50m										
height not exceeding 250mm	1.66	12.50	1.87	0.83	0.17	17.59	0.50	19.95	m	21.94
height 250 - 500mm	1.94	12.50	2.18	0.97	0.20	20.57	0.68	23.43	m	25.77
height exceeding 500mm	3.73	12.50	4.20	1.93	0.39	40.82	1.27	46.29	m²	50.92
Plain formwork; Sides of isolated columns										
Columns										
rectangular, height not exceeding 3.00m above floor level	5.45	10.00	5.99	2.25	0.45	47.52	1.25	54.76	m²	60.24
rectangular, height exceeding 3.00m above floor level	5.45	10.00	5.99	2.30	0.45	48.44	1.50	55.93	m²	61.53
Column casings										
rectangular, height not exceeding 3.00m above floor level	5.86	10.00	6.45	2.25	0.45	47.52	1.25	55.22	m²	60.74
rectangular, height exceeding 3.00m above floor level	5.86	10.00	6.45	2.30	0.45	48.44	1.50	56.39	m²	62.03
Plain formwork; Sides of attached columns										
Columns										
rectangular, height not exceeding 3.00m above floor level	5.48	10.00	6.03	2.30	0.46	48.58	1.25	55.85	m²	61.44
rectangular, height exceeding 3.00m above floor level	5.09	10.00	5.60	2.40	0.46	50.42	1.50	57.52	m²	63.27
Column casings										
rectangular, height not exceeding 3.00m above floor level	5.45	10.00	5.99	2.30	0.46	48.58	1.25	55.82	m²	61.40
rectangular, height exceeding 3.00m above floor level	5.45	10.00	5.99	2.40	0.46	50.42	1.50	57.91	m²	63.70
Plain formwork; Faces of walls and other vertical work										
plain	9.14	15.00	10.51	1.60	0.32	33.80	1.38	45.69	m²	50.26
plain; height exceeding 5.00m above floor level	8.43	15.00	9.70	1.60	0.32	33.80	1.54	45.04	m²	49.54
interrupted	8.43	15.00	9.70	1.75	0.35	36.96	1.51	48.17	m²	52.99
interrupted; height exceeding 5.00m above floor level	9.00	15.00	10.35	1.75	0.35	36.96	1.68	48.99	m²	53.89
Walls; vertical; curved 1m radius										
plain	18.70	15.00	21.51	4.00	0.80	84.49	3.43	109.43	m²	120.37
Walls; vertical; curved 10m radius										
plain	11.25	15.00	12.94	2.40	0.48	50.69	2.10	65.73	m²	72.30
Walls; vertical; curved 20m radius										
plain	8.42	15.00	9.68	1.80	0.36	38.02	1.54	49.24	m²	54.17
Walls; battered one face										
plain	8.96	15.00	10.30	1.92	0.38	40.50	1.68	52.48	m²	57.73
Plain formwork; Extra over										
Openings in walls for doors or the like; wall thickness not exceeding 250mm										
not exceeding 5m²	19.90	15.00	22.89	4.77	0.99	101.25	3.78	127.91	nr	140.71
5m² to 10m²	28.75	15.00	33.06	6.89	1.43	146.25	5.46	184.77	nr	203.24
Plain formwork; Wall ends, soffits and steps in walls										
Wall ends, soffits and steps in walls; plain										
width not exceeding 250mm	2.21	15.00	2.54	0.53	0.11	11.25	0.42	14.21	m	15.63
Plain formwork; Soffits of sloping work										
sloping one way not exceeding 15 degrees										
concrete thickness not exceeding 300mm; height to soffit										
not exceeding 3.00m	13.93	10.00	15.33	1.82	0.38	38.66	1.85	55.83	m²	61.42
3.00 - 4.50m	13.93	10.00	15.33	2.06	0.42	43.62	2.43	61.38	m²	67.52
4.50 - 6.00m	13.93	10.00	15.33	2.30	0.47	48.72	3.00	67.04	m²	73.75

Labour hourly rates: (except Specialists) Craft Operatives 18.37 Labourer 13.76 Rates are national average prices. Refer to REGIONAL VARIATIONS for indicative levels of overall pricing in regions	MATERIALS			LABOUR				RATES		
	Del to Site	Waste	Material Cost	Craft Optve	Lab	Labour Cost	Sunds	Nett Rate	Unit	Gross rate (10%)
	£	%	£	Hrs	Hrs	£	£	£		£
FORMWORK (Cont'd)										
Plain formwork; Soffits of sloping work (Cont'd)										
sloping exceeding 15 degrees concrete thickness not exceeding 300mm; height to soffit										
not exceeding 3.00m	16.00	10.00	17.60	2.00	0.42	42.52	2.03	62.15	m²	68.36
3.00 - 4.50m	16.00	10.00	17.60	2.27	0.47	48.17	2.67	68.44	m²	75.28
4.50 - 6.00m	16.00	10.00	17.60	2.53	0.52	53.63	3.30	74.53	m²	81.99
Plain formwork; Staircase strings, risers and the like										
Stairflights; 1000mm wide; 155mm thick waist; 178mm risers; includes formwork to soffits, risers and strings										
strings 300mm wide	16.85	10.00	18.53	6.05	1.21	127.79	3.12	149.44	m	164.39
string 300mm wide; junction with wall	16.85	10.00	18.53	6.05	1.21	127.79	3.12	149.44	m	164.39
Stairflights; 1500mm wide; 180mm thick waist; 178mm risers; includes formwork to soffits, risers and strings										
strings 325mm wide	21.09	10.00	23.20	7.56	1.51	159.65	3.90	186.75	m	205.43
string 325mm wide; junction with wall	21.09	10.00	23.20	7.56	1.51	159.65	3.90	186.75	m	205.43
Stairflights; 1000mm wide; 155mm thick waist; 178mm undercut risers; includes formwork to soffits, risers and strings										
strings 300mm wide	16.85	10.00	18.53	6.05	1.21	127.79	3.12	149.44	m	164.39
string 300mm wide; junction with wall	16.85	10.00	18.53	6.05	1.21	127.79	3.12	149.44	m	164.39
Stairflights; 1500mm wide; 180mm thick waist; 178mm undercut risers; includes formwork to soffits, risers and strings										
strings 325mm wide	21.08	10.00	23.19	7.56	1.51	159.65	3.90	186.74	m	205.42
string 325mm wide; junction with wall	21.08	10.00	23.19	7.56	1.51	159.65	3.90	186.74	m	205.42
Plain formwork; Sloping top surfaces										
Top formwork sloping exceeding 15 degrees	4.09	10.00	4.50	1.50	0.30	31.68	0.86	37.04	m²	40.74
Plain formwork; Steps in top surfaces										
plain vertical										
height not exceeding 250mm	1.66	12.50	1.87	0.75	0.15	15.84	0.42	18.13	m	19.94
height 250 - 500mm	1.94	12.50	2.18	0.88	0.18	18.64	0.56	21.38	m	23.52
Wall kickers										
Plain										
straight	1.09	15.00	1.25	0.25	0.05	5.28	0.27	6.80	m	7.48
curved 2m radius	2.67	15.00	3.07	0.66	0.15	14.19	0.54	17.80	m	19.58
curved 10m radius	1.62	15.00	1.86	0.40	0.08	8.45	0.34	10.66	m	11.72
curved 20m radius	1.19	15.00	1.36	0.30	0.06	6.34	0.30	8.00	m	8.80
Suspended										
straight	1.17	15.00	1.35	0.28	0.06	5.97	0.28	7.61	m	8.37
curved 2m radius	2.96	15.00	3.40	0.73	0.17	15.75	0.58	19.73	m	21.71
curved 10m radius	1.74	15.00	2.00	0.44	0.09	9.32	0.38	11.70	m	12.87
curved 20m radius	1.35	15.00	1.55	0.33	0.07	7.03	0.33	8.91	m	9.80
Special finish formwork										
Extra over a basic finish for a fine formed finish										
slabs	-	-	-	-	0.35	4.82	-	4.82	m²	5.30
walls	-	-	-	-	0.35	4.82	-	4.82	m²	5.30
beams	-	-	-	-	0.35	4.82	-	4.82	m²	5.30
columns	-	-	-	-	0.35	4.82	-	4.82	m²	5.30
Formwork and basic finish; coating with retarding agent; Sides of foundations and bases										
plain vertical										
height not exceeding 250mm	1.24	12.50	1.39	0.53	0.14	11.66	0.33	13.38	m	14.72
height 250 - 500mm	2.34	12.50	2.63	0.96	0.28	21.49	0.62	24.73	m	27.21
height exceeding 500mm	4.51	12.50	5.08	1.75	0.59	40.27	1.07	46.41	m²	51.05
Formwork and basic finish; coating with retarding agent; Edges of horizontal work										
plain vertical										
height not exceeding 250mm	1.24	12.50	1.39	0.53	0.14	11.66	0.33	13.38	m	14.72
height 250 - 500mm	2.34	12.50	2.63	0.96	0.28	21.49	0.62	24.73	m	27.21
height exceeding 500mm	4.54	12.50	5.10	1.75	0.59	40.27	1.07	46.43	m²	51.08
plain vertical; curved 10m radius										
height not exceeding 250mm	3.63	12.50	4.09	1.65	0.36	35.26	1.02	40.37	m	44.41
plain vertical; curved 1m radius										
height not exceeding 250mm	2.20	12.50	2.48	1.00	0.23	21.53	0.64	24.66	m	27.12
plain vertical; curved 20m radius										
height not exceeding 250mm	1.67	12.50	1.88	0.74	0.18	16.07	0.41	18.35	m	20.19
Formwork and basic finish; coating with retarding agent; Soffits of horizontal work										
horizontal slab thickness not exceeding 300mm; height to soffit										
not exceeding 3.00m	12.08	10.00	13.28	1.65	0.57	38.15	1.63	53.07	m²	58.38
3.00 - 4.50m	12.08	10.00	13.28	1.87	0.62	42.88	2.20	58.37	m²	64.21
4.50 - 6.00m	12.08	10.00	13.28	2.09	0.67	47.61	2.70	63.60	m²	69.96
slab thickness 300 - 450mm; height to soffit										
not exceeding 3.00m	14.54	10.00	15.99	1.82	0.63	42.10	1.80	59.89	m²	65.88
3.00 - 4.50m	14.54	10.00	15.99	2.06	0.68	47.20	2.43	65.62	m²	72.18

Labour hourly rates: (except Specialists) Craft Operatives 18.37 Labourer 13.76 Rates are national average prices. Refer to REGIONAL VARIATIONS for indicative levels of overall pricing in regions	MATERIALS			LABOUR				RATES		
	Del to Site £	Waste %	Material Cost £	Craft Optve Hrs	Lab Hrs	Labour Cost £	Sunds £	Nett Rate £	Unit	Gross rate (10%) £
FORMWORK (Cont'd)										
Formwork and basic finish; coating with retarding agent; Soffits of horizontal work (Cont'd)										
horizontal (Cont'd) slab thickness 300 - 450mm; height to soffit (Cont'd)										
4.50 - 6.00m..........	14.54	10.00	15.99	2.30	0.74	52.43	3.00	71.42	m²	78.57
horizontal; with frequent uses of prefabricated panels slab thickness not exceeding 300mm; height to soffit										
not exceeding 3.00m	4.14	15.00	4.76	1.65	0.78	41.04	1.63	47.44	m²	52.19
3.00 - 4.50m..........	4.14	15.00	4.76	1.87	0.83	45.77	2.20	52.74	m²	58.02
4.50 - 6.00m..........	4.14	15.00	4.76	2.09	0.88	50.50	2.70	57.97	m²	63.76
slab thickness 300 - 450mm; height to soffit										
not exceeding 3.00m	5.43	15.00	6.24	1.82	0.86	45.27	1.80	53.31	m²	58.64
3.00 - 4.50m..........	5.43	15.00	6.24	2.06	0.91	50.36	2.43	59.04	m²	64.94
4.50 - 6.00m..........	5.43	15.00	6.24	2.30	0.97	55.60	2.97	64.81	m²	71.29
Soffits of landings; horizontal slab thickness not exceeding 300mm; height to soffit										
not exceeding 1.50m	13.28	15.00	15.28	2.00	0.66	45.82	1.65	62.75	m²	69.02
1.50 - 3.00m..........	13.28	15.00	15.28	2.22	0.72	50.69	2.27	68.23	m²	75.05
3.00 - 4.50m..........	13.28	15.00	15.28	2.46	0.77	55.79	2.57	73.63	m²	80.99
Formwork and basic finish; coating with retarding agent; Sides and soffits of isolated beams										
regular shaped; rectangular; height to soffit										
not exceeding - 3.00m	9.11	10.00	10.03	2.35	0.71	52.94	1.32	64.29	m²	70.71
3.00 - 4.50m..........	9.11	10.00	10.03	2.35	0.71	52.94	1.61	64.57	m²	71.03
Formwork and basic finish; coating with retarding agent; Sides and soffits of attached beams										
regular shaped; rectangular; height to soffit										
not exceeding - 3.00m	8.77	10.00	9.65	2.25	0.69	50.83	1.32	61.80	m²	67.98
3.00 - 4.50m..........	8.77	10.00	9.65	2.25	0.69	50.83	1.61	62.08	m²	68.29
Formwork and basic finish; coating with retarding agent; Sides of isolated columns										
Columns										
rectangular, height not exceeding 3.00m above floor level.............	6.41	10.00	7.05	2.25	0.70	50.96	1.25	59.26	m²	65.18
rectangular, height exceeding 3.00m above floor level..................	6.41	10.00	7.05	2.48	0.70	55.19	1.50	63.74	m²	70.11
Column casings										
regular shaped; rectangular..................	6.41	10.00	7.05	2.38	0.69	53.22	1.25	61.51	m²	67.66
regular shaped; rectangular, height exceeding 3.00m above floor level..................	6.41	10.00	7.05	2.48	0.69	55.05	1.50	63.60	m²	69.96
Formwork and basic finish; coating with retarding agent; Sides of attached columns										
Columns										
rectangular, height not exceeding 3.00m above floor level.............	6.41	10.00	7.05	2.30	0.70	51.88	1.25	60.17	m²	66.19
rectangular, height exceeding 3.00m above floor level..................	6.41	10.00	7.05	2.43	0.70	54.27	1.50	62.82	m²	69.10
Column casings										
regular shaped; rectangular..................	6.41	10.00	7.05	2.30	0.70	51.88	1.50	60.43	m²	66.47
regular shaped; rectangular, height exceeding 3.00m above floor level..................	6.41	10.00	7.05	2.40	0.70	53.72	1.25	62.01	m²	68.21
Formwork and basic finish; coating with retarding agent; Faces of walls and other vertical work										
plain..................	8.47	15.00	9.74	1.60	0.56	37.10	1.38	48.22	m²	53.04
plain; height exceeding 3.00m above floor level..........................	9.34	15.00	10.74	1.60	0.56	37.10	1.54	49.39	m²	54.33
interrupted	9.14	15.00	10.51	1.75	0.59	40.27	1.51	52.30	m²	57.52
interrupted; height exceeding 3.00m above floor level	9.79	15.00	11.25	1.75	0.59	40.27	1.68	53.20	m²	58.52
Walls; vertical; curved 1m radius										
plain..................	19.57	15.00	22.51	4.00	1.04	87.79	3.43	113.73	m²	125.10
Walls; vertical; curved 10m radius										
plain..................	12.18	15.00	14.01	2.40	0.72	54.00	2.10	70.11	m²	77.12
Walls; vertical; curved 20m radius										
plain..................	9.33	15.00	10.73	1.80	0.60	41.32	1.54	53.59	m²	58.95
Walls; battered										
plain..................	9.87	15.00	11.35	1.92	0.62	43.80	1.68	56.83	m²	62.52
Formwork and basic finish; coating with retarding agent; Extra over										
Openings in walls; wall thickness not exceeding 250mm										
not exceeding 5m2	22.71	15.00	26.11	4.77	12.24	256.05	3.78	285.94	m	314.54
5m2 to 10m2	32.80	15.00	37.72	6.89	2.21	156.98	5.46	200.16	m	220.18
Formwork and basic finish; coating with retarding agent; Wall ends, soffits and steps in walls										
Plain										
width not exceeding 250mm..................	2.52	15.00	2.90	0.53	0.17	12.08	0.42	15.40	m	16.94
Formwork and basic finish; coating with retarding agent; Soffits of sloping work										
Sloping one way not exceeding 15 degrees slab thickness not exceeding 300mm; height to soffit										
not exceeding 3.00m	14.58	15.00	16.77	1.82	0.63	42.10	1.80	60.67	m²	66.73
3.00 - 4.50m..................	14.58	15.00	16.77	2.06	0.68	47.20	2.43	66.39	m²	73.03

Labour hourly rates: (except Specialists) Craft Operatives 18.37 Labourer 13.76 Rates are national average prices. Refer to REGIONAL VARIATIONS for indicative levels of overall pricing in regions	MATERIALS			LABOUR				RATES		
	Del to Site	Waste	Material Cost	Craft Optve	Lab	Labour Cost	Sunds	Nett Rate	Unit	Gross rate (10%)
	£	%	£	Hrs	Hrs	£	£	£		£
FORMWORK (Cont'd)										
Formwork and basic finish; coating with retarding agent; Soffits of sloping work (Cont'd)										
Sloping one way not exceeding 15 degrees (Cont'd) slab thickness not exceeding 300mm; height to soffit (Cont'd)										
4.50 - 6.00m.............	14.58	15.00	16.77	2.30	0.74	52.43	3.00	72.20	m²	79.42
Soffits of slabs; sloping exceeding 15 degrees										
sloping exceeding 15 degrees slab thickness not exceeding 300mm; height to soffit										
not exceeding 3.00m..........	15.19	15.00	17.47	1.90	0.65	43.85	2.10	63.42	m²	69.76
3.00 - 4.50m..........	15.19	15.00	17.47	2.15	0.71	49.27	2.67	69.40	m²	76.34
4.50 - 6.00m..........	15.19	15.00	17.47	2.40	0.76	54.55	3.27	75.28	m²	82.81
Formwork and basic finish; coating with retarding agent; Staircase strings, risers and the like										
Stairflights; 1000mm wide; 155mm thick waist; 178mm risers; includes formwork to soffits, risers and strings										
strings 300mm wide..........	18.59	10.00	20.45	6.05	1.74	135.08	3.12	158.65	m	174.52
string 300mm wide; junction with wall..........	18.59	10.00	20.45	6.05	1.74	135.08	3.12	158.65	m	174.52
Stairflights; 1500mm wide; 180mm thick waist; 178mm risers; includes formwork to soffits, risers and strings										
strings 325mm wide..........	23.24	10.00	25.57	7.56	2.17	168.74	3.90	198.20	m	218.02
string 325mm wide; junction with wall..........	23.24	10.00	25.57	7.56	2.17	168.74	3.90	198.20	m	218.02
Stairflights; 1000mm wide; 155mm thick waist; 178mm undercut risers; includes formwork to soffits, risers and strings										
strings 300mm wide..........	18.59	10.00	20.45	6.05	1.74	135.08	3.12	158.65	m	174.52
string 300mm wide; junction with wall..........	18.59	10.00	20.45	6.05	1.74	135.08	3.12	158.65	m	174.52
Stairflights; 1500mm wide; 180mm thick waist; 178mm undercut risers; includes formwork to soffits, risers and strings										
strings 325mm wide..........	23.24	10.00	25.57	7.56	2.17	168.74	3.90	198.20	m	218.02
string 325mm wide; junction with wall..........	23.24	10.00	25.57	7.56	2.17	168.74	3.90	198.20	m	218.02
Formwork and basic finish; coating with retarding agent; Sloping top surfaces										
Top formwork sloping exceeding 15 degrees..........	5.02	15.00	5.77	1.50	0.54	34.99	0.86	41.61	m²	45.77
Formwork and basic finish; coating with retarding agent; Steps in top surfaces										
plain vertical										
height not exceeding 250mm	1.82	12.50	2.04	0.75	0.18	16.25	0.41	18.70	m	20.57
height 250 - 500mm	2.24	12.50	2.52	0.88	0.27	19.88	0.54	22.94	m	25.23
Wall kickers										
Plain										
straight..........	1.19	15.00	1.36	0.25	0.09	5.83	0.27	7.46	m	8.21
curved 2m radius..........	2.79	15.00	3.21	0.66	0.17	14.46	0.54	18.22	m	20.04
curved 10m radius..........	1.73	15.00	1.99	0.40	0.12	9.00	0.34	11.33	m	12.47
curved 20m radius..........	1.35	15.00	1.55	0.30	0.10	6.89	0.30	8.74	m	9.61
Suspended										
straight..........	1.34	15.00	1.54	0.28	0.10	6.52	0.30	8.36	m	9.19
curved 2m radius..........	3.07	15.00	3.53	0.73	0.19	16.02	0.60	20.15	m	22.17
curved 10m radius..........	1.92	15.00	2.20	0.44	0.13	9.87	0.38	12.45	m	13.70
curved 20m radius..........	1.46	15.00	1.68	0.33	0.11	7.58	0.33	9.58	m	10.54
Special finish formwork										
Extra over a basic finish for a fine formed finish										
slabs..........	-	-	-	-	0.35	4.82	-	4.82	m²	5.30
walls..........	-	-	-	-	0.35	4.82	-	4.82	m²	5.30
beams..........	-	-	-	-	0.35	4.82	-	4.82	m²	5.30
columns..........	-	-	-	-	0.35	4.82	-	4.82	m²	5.30
Soffits of horizontal work; Expamet Hy-rib permanent shuttering and reinforcement										
Reference 2611, 4.23 kg/m²; to soffits of slabs; horizontal; one rib side laps; 150mm end laps										
slab thickness 75mm; strutting and supports at 750mm centres; height to soffit										
not exceeding 3.00m	20.68	7.50	22.23	1.19	0.30	25.99	1.61	49.82	m²	54.81
3.00 - 4.50m..........	20.68	7.50	22.23	1.37	0.34	29.85	1.98	54.06	m²	59.46
slab thickness 100mm; strutting and supports at 650mm centres; height to soffit										
not exceeding 3.00m	20.68	7.50	22.23	1.21	0.30	26.36	1.82	50.40	m²	55.44
3.00 - 4.50m..........	20.68	7.50	22.23	1.39	0.35	30.35	2.44	55.03	m²	60.53
slab thickness 125mm; strutting and supports at 550mm centres; height to soffit										
not exceeding 3.00m	20.68	7.50	22.23	1.23	0.31	26.86	2.04	51.13	m²	56.24
3.00 - 4.50m..........	20.68	7.50	22.23	1.41	0.35	30.72	2.95	55.90	m²	61.49
slab thickness 150mm; strutting and supports at 450mm centres; height to soffit										
not exceeding 3.00m	20.68	7.50	22.23	1.25	0.31	27.23	2.30	51.75	m²	56.93
3.00 - 4.50m..........	20.68	7.50	22.23	1.44	0.36	31.41	3.43	57.07	m²	62.78

Labour hourly rates: (except Specialists) Craft Operatives 18.37 Labourer 13.76 Rates are national average prices. Refer to REGIONAL VARIATIONS for indicative levels of overall pricing in regions	MATERIALS			LABOUR				RATES		
	Del to Site	Waste	Material Cost	Craft Optve	Lab	Labour Cost	Sunds	Nett Rate	Unit	Gross rate (10%)
	£	%	£	Hrs	Hrs	£	£	£		£

FORMWORK (Cont'd)

Soffits of horizontal work; Expamet Hy-rib permanent shuttering and reinforcement (Cont'd)

Reference 2411, 6.35 kg/m²; to soffits of slabs; horizontal; one rib side laps; 150mm end laps

slab thickness 75mm; strutting and supports at 850mm centres; height to soffit

not exceeding 3.00m	30.67	7.50	32.97	1.22	0.31	26.68	1.61	61.25	m²	67.38
3.00 - 4.50m	30.67	7.50	32.97	1.40	0.35	30.53	1.98	65.48	m²	72.03

slab thickness 100mm; strutting and supports at 750mm centres; height to soffit

not exceeding 3.00m	30.67	7.50	32.97	1.24	0.31	27.04	1.82	61.83	m²	68.01
3.00 - 4.50m	30.67	7.50	32.97	1.43	0.36	31.22	2.44	66.64	m²	73.30

slab thickness 125mm; strutting and supports at 650mm centres; height to soffit

not exceeding 3.00m	30.67	7.50	32.97	1.25	0.32	27.37	2.04	62.37	m²	68.61
3.00 - 4.50m	30.67	7.50	32.97	1.45	0.36	31.59	2.95	67.51	m²	74.26

slab thickness 150mm; strutting and supports at 550mm centres; height to soffit

not exceeding 3.00m	30.67	7.50	32.97	1.28	0.32	27.92	2.30	63.18	m²	69.50
3.00 - 4.50m	30.67	7.50	32.97	1.47	0.37	32.10	3.43	68.50	m²	75.35

slab thickness 175mm; strutting and supports at 450mm centres; height to soffit

not exceeding 3.00m	30.67	7.50	32.97	1.31	0.33	28.61	2.64	64.21	m²	70.63
3.00 - 4.50m	30.67	7.50	32.97	1.51	0.38	32.97	4.09	70.03	m²	77.03

slab thickness 200mm; strutting and supports at 350mm centres; height to soffit

not exceeding 3.00m	30.67	7.50	32.97	1.34	0.34	29.29	2.97	65.23	m²	71.76
3.00 - 4.50m	30.67	7.50	32.97	1.54	0.39	33.66	4.77	71.39	m²	78.53

Reference 2611, 4.23 kg/m²; to soffits of arched slabs; one rib side laps

900mm span; 75mm rise; height to soffit

not exceeding 3.00m	20.68	7.50	22.23	1.31	0.33	28.61	1.18	52.02	m²	57.22

1200mm span; 75mm rise; one row strutting and supports per span; height to soffit

not exceeding 3.00m	20.68	7.50	22.23	1.31	0.33	28.61	1.61	52.44	m²	57.69
3.00 - 4.50m	20.68	7.50	22.23	1.49	0.37	32.46	1.98	56.67	m²	62.34

Reference 2411, 6.35 kg/m²; to soffits of arched slabs; one rib side laps

1500mm span; 100mm rise; two rows strutting and supports per span; height to soffit

not exceeding 3.00m	30.67	7.50	32.97	1.62	0.40	35.26	3.42	71.65	m²	78.82
3.00 - 4.50m	30.67	7.50	32.97	1.84	0.46	40.13	4.57	77.67	m²	85.44

1800mm span; 150mm rise; two rows strutting and supports per span; height to soffit

not exceeding 3.00m	30.67	7.50	32.97	1.62	0.40	35.26	3.42	71.65	m²	78.82
3.00 - 4.50m	30.67	7.50	32.97	1.84	0.46	40.13	4.57	77.67	m²	85.44

REINFORCEMENT

Mild steel bars; Reinforcement bars; BS 4449, hot rolled plain round mild steel including hooks and tying wire, and spacers and chairs which are at the discretion of the Contractor

Straight

6mm	1048.28	7.50	1126.90	51.00	5.00	1005.67	40.32	2172.89	t	2390.18
8mm	1050.00	7.50	1128.75	44.00	5.00	877.08	35.70	2041.53	t	2245.68
10mm	975.00	7.50	1048.12	36.00	5.00	730.12	28.98	1807.23	t	1987.95
12mm	883.76	7.50	950.04	31.00	5.00	638.27	24.78	1613.09	t	1774.40
16mm	885.58	7.50	952.00	24.00	5.00	509.68	20.38	1482.06	t	1630.26
20mm	874.07	7.50	939.63	20.00	5.00	436.20	11.14	1386.97	t	1525.67
25mm	863.24	7.50	927.98	20.00	5.00	436.20	11.14	1375.32	t	1512.86
32mm	909.90	7.50	978.15	20.00	5.00	436.20	11.14	1425.49	t	1568.03

Bent

6mm	1048.28	7.50	1126.90	51.00	5.00	1005.67	40.32	2172.89	t	2390.18
8mm	1050.00	7.50	1128.75	44.00	5.00	877.08	35.70	2041.53	t	2245.68
10mm	975.00	7.50	1048.12	36.00	5.00	730.12	28.98	1807.23	t	1987.95
12mm	883.76	7.50	950.04	31.00	5.00	638.27	24.78	1613.09	t	1774.40
16mm	885.58	7.50	952.00	24.00	5.00	509.68	20.38	1482.06	t	1630.26
20mm	874.07	7.50	939.63	20.00	5.00	436.20	11.14	1386.97	t	1525.67
25mm	863.24	7.50	927.98	20.00	5.00	436.20	11.14	1375.32	t	1512.86
32mm	909.90	7.50	978.15	20.00	5.00	436.20	11.14	1425.49	t	1568.03

Links

6mm	1048.28	7.50	1126.90	66.00	5.00	1281.22	40.32	2448.44	t	2693.28
8mm	1050.00	7.50	1128.75	66.00	5.00	1281.22	35.70	2445.67	t	2690.24

High yield steel bars; Reinforcement bars; BS 4449, hot rolled deformed high yield steel including hooks and tying wire, and spacers and chairs which are at the discretion of the Contractor

Straight

6mm	1136.99	7.50	1222.26	51.00	5.00	1005.67	40.32	2268.25	t	2495.08
8mm	1157.69	7.50	1244.52	44.00	5.00	877.08	35.70	2157.30	t	2373.03
10mm	852.89	7.50	916.86	36.00	5.00	730.12	28.98	1675.96	t	1843.56
12mm	888.89	7.50	955.56	31.00	5.00	638.27	24.78	1618.61	t	1780.47
16mm	875.00	7.50	940.63	24.00	5.00	509.68	20.38	1470.69	t	1617.75
20mm	835.80	7.50	898.49	20.00	5.00	436.20	11.14	1345.83	t	1480.41
25mm	931.17	7.50	1001.01	20.00	5.00	436.20	11.14	1448.35	t	1593.19
32mm	929.95	7.50	999.70	20.00	5.00	436.20	11.14	1447.04	t	1591.74

IN-SITU CONCRETE WORKS (side margin)

Labour hourly rates: (except Specialists) Craft Operatives 18.37 Labourer 13.76 Rates are national average prices. Refer to REGIONAL VARIATIONS for indicative levels of overall pricing in regions	MATERIALS			LABOUR				RATES		
	Del to Site £	Waste %	Material Cost £	Craft Optve Hrs	Lab Hrs	Labour Cost £	Sunds £	Nett Rate £	Unit	Gross rate (10%) £
REINFORCEMENT (Cont'd)										
High yield steel bars; Reinforcement bars; BS 4449, hot rolled deformed high yield steel including hooks and tying wire, and spacers and chairs which are at the discretion of the Contractor (Cont'd)										
Bent										
6mm	1136.99	7.50	1222.26	51.00	5.00	1005.67	40.32	2268.25	t	2495.08
8mm	1157.69	7.50	1244.52	44.00	5.00	877.08	35.70	2157.30	t	2373.03
10mm	852.89	7.50	916.86	36.00	5.00	730.12	28.98	1675.96	t	1843.56
12mm	888.89	7.50	955.56	31.00	5.00	638.27	24.78	1618.61	t	1780.47
16mm	875.00	7.50	940.63	24.00	5.00	509.68	20.38	1470.69	t	1617.75
20mm	835.80	7.50	898.49	20.00	5.00	436.20	11.14	1345.83	t	1480.41
25mm	931.17	7.50	1001.01	20.00	5.00	436.20	11.14	1448.35	t	1593.19
32mm	929.95	7.50	999.70	20.00	5.00	436.20	11.14	1447.04	t	1591.74
Links										
6mm	1136.99	7.50	1222.26	66.00	5.00	1281.22	40.32	2543.80	t	2798.18
8mm	1157.69	7.50	1244.52	66.00	5.00	1281.22	35.70	2561.44	t	2817.58
Mild steel bars - Take delivery, cut bend and fix; Take delivery, cut, bend and fix reinforcing rods; including hooks and tying wire, and spacers and chairs which are at the discretion of the Contractor										
Straight										
6mm	-	-	-	55.00	5.00	1079.15	40.32	1119.47	t	1231.42
8mm	-	-	-	48.00	5.00	950.56	35.70	986.26	t	1084.89
10mm	-	-	-	40.00	5.00	803.60	28.98	832.58	t	915.84
12mm	-	-	-	34.00	5.00	693.38	24.78	718.16	t	789.98
16mm	-	-	-	26.00	5.00	546.42	20.38	566.80	t	623.48
20mm	-	-	-	22.00	5.00	472.94	11.14	484.08	t	532.49
25mm	-	-	-	22.00	5.00	472.94	11.14	484.08	t	532.49
32mm	-	-	-	22.00	5.00	472.94	11.14	484.08	t	532.49
Bent										
6mm	-	-	-	75.00	5.00	1446.55	40.32	1486.87	t	1635.56
8mm	-	-	-	66.00	5.00	1281.22	35.70	1316.92	t	1448.61
10mm	-	-	-	55.00	5.00	1079.15	28.98	1108.13	t	1218.94
12mm	-	-	-	47.00	5.00	932.19	24.78	956.97	t	1052.67
16mm	-	-	-	37.00	5.00	748.49	20.38	768.87	t	845.76
20mm	-	-	-	31.00	5.00	638.27	11.14	649.41	t	714.35
25mm	-	-	-	29.00	5.00	601.53	11.14	612.67	t	673.94
32mm	-	-	-	29.00	5.00	601.53	11.14	612.67	t	673.94
Links										
6mm	-	-	-	99.00	5.00	1887.43	40.32	1927.75	t	2120.53
8mm	-	-	-	99.00	5.00	1887.43	35.70	1923.13	t	2115.44
Mild steel bars - Take delivery and fix; Take delivery and fix reinforcing rods supplied cut to length and bent; including tying wire, and spacers and chairs which are at the discretion of the Contractor										
Straight										
6mm	-	-	-	51.00	5.00	1005.67	40.32	1045.99	t	1150.59
8mm	-	-	-	44.00	5.00	877.08	35.70	912.78	t	1004.06
10mm	-	-	-	36.00	5.00	730.12	28.98	759.10	t	835.01
12mm	-	-	-	31.00	5.00	638.27	24.78	663.05	t	729.36
16mm	-	-	-	24.00	5.00	509.68	20.38	530.06	t	583.07
20mm	-	-	-	20.00	5.00	436.20	11.14	447.34	t	492.07
25mm	-	-	-	20.00	5.00	436.20	11.14	447.34	t	492.07
32mm	-	-	-	20.00	5.00	436.20	11.14	447.34	t	492.07
Bent										
6mm	-	-	-	51.00	5.00	1005.67	40.32	1045.99	t	1150.59
8mm	-	-	-	44.00	5.00	877.08	35.70	912.78	t	1004.06
10mm	-	-	-	36.00	5.00	730.12	28.98	759.10	t	835.01
12mm	-	-	-	31.00	5.00	638.27	24.78	663.05	t	729.36
16mm	-	-	-	24.00	5.00	509.68	20.38	530.06	t	583.07
20mm	-	-	-	20.00	5.00	436.20	11.14	447.34	t	492.07
25mm	-	-	-	20.00	5.00	436.20	11.14	447.34	t	492.07
Links										
6mm	-	-	-	66.00	5.00	1281.22	40.32	1321.54	t	1453.69
8mm	-	-	-	66.00	5.00	1281.22	35.70	1316.92	t	1448.61
High yield stainless steel bars; Reinforcement bars; high yield stainless steel BS 6744, Type 2 (minimum yield stress 460 N/mm²); including hooks and tying wire, and spacers and chairs which are at the discretion of the Contractor										
Straight										
8mm	2592.31	7.50	2786.73	83.00	10.00	1662.31	138.60	4587.64	t	5046.41
10mm	2620.14	7.50	2816.65	69.00	10.00	1405.13	100.00	4321.78	t	4753.96
12mm	2660.61	7.50	2860.15	60.00	10.00	1239.80	94.60	4194.55	t	4614.01
16mm	2474.07	7.50	2659.63	50.00	10.00	1056.10	81.40	3797.13	t	4176.84
20mm	2467.07	7.50	2652.10	41.00	10.00	890.77	60.50	3603.37	t	3963.71
25mm	2613.28	7.50	2809.28	33.00	10.00	743.81	44.00	3597.09	t	3956.80
Bent										
8mm	2592.31	7.50	2786.73	83.00	10.00	1662.31	138.60	4587.64	t	5046.41
10mm	2620.14	7.50	2816.65	69.00	10.00	1405.13	100.00	4321.78	t	4753.96
12mm	2660.61	7.50	2860.15	60.00	10.00	1239.80	94.60	4194.55	t	4614.01
16mm	2474.07	7.50	2659.63	50.00	10.00	1056.10	81.40	3797.13	t	4176.84
20mm	2467.07	7.50	2652.10	41.00	10.00	890.77	60.50	3603.37	t	3963.71
25mm	2613.28	7.50	2809.28	33.00	10.00	743.81	44.00	3597.09	t	3956.80

Labour hourly rates: (except Specialists) Craft Operatives 18.37 Labourer 13.76 Rates are national average prices. Refer to REGIONAL VARIATIONS for indicative levels of overall pricing in regions	MATERIALS			LABOUR				RATES		
	Del to Site	Waste	Material Cost	Craft Optve	Lab	Labour Cost	Sunds	Nett Rate		Gross rate (10%)
	£	%	£	Hrs	Hrs	£	£	£	Unit	£
REINFORCEMENT (Cont'd)										
High yield stainless steel bars; Reinforcement bars; high yield stainless steel BS 6744, Type 2 (minimum yield stress 460 N/mm²); including hooks and tying wire, and spacers and chairs which are at the discretion of the Contractor (Cont'd)										
Links										
6mm	2630.14	7.50	2827.40	130.00	10.00	2525.70	138.60	5491.70	t	6040.87
8mm	2592.31	7.50	2786.73	115.00	10.00	2250.15	100.00	5136.88	t	5650.57
Mesh; Reinforcement fabric; BS 4483, hard drawn plain round steel; welded; including laps, tying wire, all cutting and bending, and spacers and chairs which are at the discretion of the Contractor										
Reference A98; 1.54 kg/m²; 200mm side laps; 200mm end laps										
generally	1.23	17.50	1.45	0.07	0.01	1.42	0.08	2.95	m²	3.25
strips in one width										
750mm wide	1.23	17.50	1.45	0.11	0.01	2.16	0.08	3.69	m²	4.06
900mm wide	1.23	17.50	1.45	0.10	0.01	1.97	0.08	3.50	m²	3.86
1050mm wide	1.23	17.50	1.45	0.09	0.01	1.79	0.08	3.32	m²	3.65
1200mm wide	1.23	17.50	1.45	0.08	0.01	1.61	0.08	3.14	m²	3.45
Reference A142; 2.22kg/m²; 200mm side laps; 200mm end laps										
generally	1.78	17.50	2.09	0.08	0.02	1.74	0.12	3.95	m²	4.35
strips in one width										
750mm wide	1.78	17.50	2.09	0.12	0.02	2.48	0.12	4.69	m²	5.16
900mm wide	1.78	17.50	2.09	0.11	0.02	2.30	0.12	4.51	m²	4.96
1050mm wide	1.78	17.50	2.09	0.10	0.02	2.11	0.12	4.32	m²	4.75
1200mm wide	1.78	17.50	2.09	0.10	0.02	2.11	0.12	4.32	m²	4.75
Reference A193; 3.02kg/m²; 200mm side laps; 200mm end laps										
generally	2.42	17.50	2.84	0.08	0.02	1.74	0.16	4.75	m²	5.22
strips in one width										
750mm wide	2.42	17.50	2.84	0.12	0.02	2.48	0.16	5.48	m²	6.03
900mm wide	2.42	17.50	2.84	0.11	0.02	2.30	0.16	5.30	m²	5.83
1050mm wide	2.42	17.50	2.84	0.10	0.02	2.11	0.16	5.12	m²	5.63
1200mm wide	2.42	17.50	2.84	0.10	0.02	2.11	0.16	5.12	m²	5.63
Reference A252; 3.95kg/m²; 200mm side laps; 200mm end laps										
generally	3.17	17.50	3.72	0.09	0.02	1.93	0.20	5.85	m²	6.43
strips in one width										
750mm wide	3.17	17.50	3.72	0.14	0.02	2.85	0.20	6.77	m²	7.44
900mm wide	3.17	17.50	3.72	0.13	0.02	2.66	0.20	6.58	m²	7.24
1050mm wide	3.17	17.50	3.72	0.12	0.02	2.48	0.20	6.40	m²	7.04
1200mm wide	3.17	17.50	3.72	0.11	0.02	2.30	0.20	6.22	m²	6.84
Reference A393; 6.16 kg/m²; 200mm side laps; 200mm end laps										
generally	4.94	17.50	5.80	0.10	0.02	2.11	0.30	8.21	m²	9.03
strips in one width										
750mm wide	4.94	17.50	5.80	0.15	0.02	3.03	0.30	9.13	m²	10.04
900mm wide	4.94	17.50	5.80	0.14	0.02	2.85	0.30	8.95	m²	9.84
1050mm wide	4.94	17.50	5.80	0.13	0.02	2.66	0.30	8.76	m²	9.64
1200mm wide	4.94	17.50	5.80	0.12	0.02	2.48	0.30	8.58	m²	9.44
Reference B1131; 10.90 kg/m²; 100mm side laps; 200mm end laps										
generally	11.28	17.50	13.26	0.14	0.03	2.98	0.54	16.78	m²	18.46
strips in one width										
750mm wide	11.28	17.50	13.26	0.21	0.03	4.27	0.54	18.07	m²	19.87
900mm wide	11.28	17.50	13.26	0.20	0.03	4.09	0.54	17.88	m²	19.67
1050mm wide	11.28	17.50	13.26	0.18	0.03	3.72	0.54	17.52	m²	19.27
1200mm wide	11.28	17.50	13.26	0.17	0.03	3.54	0.54	17.33	m²	19.07
Reference B196; 3.05 kg/m²; 100mm side laps; 200mm end laps										
generally	2.37	17.50	2.79	0.08	0.02	1.74	0.30	4.84	m²	5.32
strips in one width										
750mm wide	2.37	17.50	2.79	0.12	0.02	2.48	0.30	5.57	m²	6.13
900mm wide	2.37	17.50	2.79	0.11	0.02	2.30	0.30	5.39	m²	5.93
1050mm wide	2.37	17.50	2.79	0.10	0.02	2.11	0.30	5.20	m²	5.72
1200mm wide	2.37	17.50	2.79	0.10	0.02	2.11	0.30	5.20	m²	5.72
Reference B283; 3.73 kg/m²; 100mm side laps; 200mm end laps										
generally	2.99	17.50	3.51	0.08	0.02	1.74	0.18	5.44	m²	5.98
strips in one width										
750mm wide	2.99	17.50	3.51	0.12	0.02	2.48	0.18	6.17	m²	6.79
900mm wide	2.99	17.50	3.51	0.11	0.02	2.30	0.18	5.99	m²	6.59
1050mm wide	2.99	17.50	3.51	0.10	0.02	2.11	0.18	5.80	m²	6.38
1200mm wide	2.99	17.50	3.51	0.10	0.02	2.11	0.18	5.80	m²	6.38
Reference B385; 4.53 kg/m²; 100mm side laps; 200mm end laps										
generally	5.38	17.50	6.32	0.09	0.02	1.93	0.22	8.47	m²	9.31
strips in one width										
750mm wide	5.38	17.50	6.32	0.14	0.02	2.85	0.22	9.39	m²	10.32
900mm wide	5.38	17.50	6.32	0.13	0.02	2.66	0.22	9.20	m²	10.12
1050mm wide	5.38	17.50	6.32	0.12	0.02	2.48	0.22	9.02	m²	9.92
1200mm wide	5.38	17.50	6.32	0.11	0.02	2.30	0.22	8.84	m²	9.72
Reference B503; 5.93 kg/m²; 100mm side laps; 200mm end laps										
generally	6.66	17.50	7.83	0.10	0.02	2.11	0.30	10.24	m²	11.27
strips in one width										
750mm wide	6.66	17.50	7.83	0.15	0.02	3.03	0.30	11.16	m²	12.28
900mm wide	6.66	17.50	7.83	0.14	0.02	2.85	0.30	10.98	m²	12.08
1050mm wide	6.66	17.50	7.83	0.13	0.02	2.66	0.30	10.79	m²	11.87
1200mm wide	6.66	17.50	7.83	0.12	0.02	2.48	0.30	10.61	m²	11.67
Reference B785; 8.14 kg/m²; 100mm side laps; 200mm end laps										
generally	8.43	17.50	9.90	0.11	0.03	2.43	0.40	12.73	m²	14.01

IN-SITU CONCRETE WORKS

Labour hourly rates: (except Specialists) Craft Operatives 18.37 Labourer 13.76 Rates are national average prices. Refer to REGIONAL VARIATIONS for indicative levels of overall pricing in regions	MATERIALS			LABOUR				RATES		
	Del to Site	Waste	Material Cost	Craft Optve	Lab	Labour Cost	Sunds	Nett Rate	Unit	Gross rate (10%)
	£	%	£	Hrs	Hrs	£	£	£		£
REINFORCEMENT (Cont'd)										
Mesh; Reinforcement fabric; BS 4483, hard drawn plain round steel; welded; including laps, tying wire, all cutting and bending, and spacers and chairs which are at the discretion of the Contractor (Cont'd)										
Reference B785; 8.14 kg/m²; 100mm side laps; 200mm end laps (Cont'd)										
strips in one width										
750mm wide	8.43	17.50	9.90	0.16	0.03	3.35	0.40	13.65	m²	15.02
900mm wide	8.43	17.50	9.90	0.15	0.03	3.17	0.40	13.47	m²	14.82
1050mm wide	8.43	17.50	9.90	0.14	0.03	2.98	0.40	13.28	m²	14.61
1200mm wide	8.43	17.50	9.90	0.13	0.03	2.80	0.40	13.10	m²	14.41
Reference C283; 2.61 kg/m²; 100mm side laps; 400mm end laps										
generally	2.14	17.50	2.51	0.08	0.02	1.74	0.14	4.39	m²	4.83
strips in one width										
750mm wide	2.14	17.50	2.51	0.12	0.02	2.48	0.14	5.13	m²	5.64
900mm wide	2.14	17.50	2.51	0.11	0.02	2.30	0.14	4.95	m²	5.44
1050mm wide	2.14	17.50	2.51	0.10	0.02	2.11	0.14	4.76	m²	5.24
1200mm wide	2.14	17.50	2.51	0.10	0.02	2.11	0.14	4.76	m²	5.24
Reference C385; 3.41 kg/m²; 100mm side laps; 400mm end laps										
generally	2.73	17.50	3.21	0.08	0.02	1.74	0.18	5.14	m²	5.65
strips in one width										
750mm wide	2.73	17.50	3.21	0.12	0.02	2.48	0.18	5.87	m²	6.46
900mm wide	2.73	17.50	3.21	0.11	0.02	2.30	0.18	5.69	m²	6.26
1050mm wide	2.73	17.50	3.21	0.10	0.02	2.11	0.18	5.50	m²	6.05
1200mm wide	2.73	17.50	3.21	0.10	0.02	2.11	0.18	5.50	m²	6.05
Reference C503; 4.34 kg/m²; 100mm side laps; 400mm end laps										
generally	3.48	17.50	4.09	0.09	0.02	1.93	0.22	6.24	m²	6.86
strips in one width										
750mm wide	3.48	17.50	4.09	0.14	0.02	2.85	0.22	7.15	m²	7.87
900mm wide	3.48	17.50	4.09	0.13	0.02	2.66	0.22	6.97	m²	7.67
1050mm wide	3.48	17.50	4.09	0.12	0.02	2.48	0.22	6.79	m²	7.47
1200mm wide	3.48	17.50	4.09	0.11	0.02	2.30	0.22	6.60	m²	7.26
Reference C636; 5.55 kg/m²; 100mm side laps; 400mm end laps										
generally	4.45	17.50	5.23	0.10	0.02	2.11	0.28	7.62	m²	8.38
strips in one width										
750mm wide	4.45	17.50	5.23	0.15	0.02	3.03	0.28	8.54	m²	9.39
900mm wide	4.45	17.50	5.23	0.14	0.02	2.85	0.28	8.35	m²	9.19
1050mm wide	4.45	17.50	5.23	0.13	0.02	2.66	0.28	8.17	m²	8.99
1200mm wide	4.45	17.50	5.23	0.12	0.02	2.48	0.28	7.99	m²	8.78
Reference C785; 6.72 kg/m²; 100mm side laps; 400mm end laps										
generally	5.39	17.50	6.33	0.10	0.02	2.11	0.34	8.78	m²	9.66
strips in one width										
750mm wide	5.39	17.50	6.33	0.15	0.02	3.03	0.34	9.70	m²	10.67
900mm wide	5.39	17.50	6.33	0.14	0.02	2.85	0.34	9.51	m²	10.47
1050mm wide	5.39	17.50	6.33	0.13	0.02	2.66	0.34	9.33	m²	10.26
1200mm wide	5.39	17.50	6.33	0.12	0.02	2.48	0.34	9.15	m²	10.06
Reference D49; 0.77 kg/m²; 100mm side laps; 100mm end laps										
bent	2.09	17.50	2.46	0.04	0.01	0.87	0.12	3.45	m²	3.80
Reference D98; 1.54 kg/m²; 200mm side laps; 200mm end laps										
bent	2.08	17.50	2.44	0.04	0.01	0.87	0.12	3.43	m²	3.78
Expamet flattened security mesh expanded steel reinforcement (uncoated); including laps										
Reference HD1; 150mm side laps; 150mm end laps										
generally	10.06	17.50	11.82	0.12	0.02	2.48	0.40	14.70	m²	16.17
JOINTS IN CONCRETE										
Formed joints										
Incorporating 10mm thick Korkpak; Grace Construction Products Limited - Serviciced; formwork; reinforcement laid continuously across joint										
in concrete, depth not exceeding 150mm; horizontal	5.05	5.00	5.30	-	0.08	1.10	7.03	13.44	m	14.78
in concrete, depth 150 - 300mm; horizontal	10.10	5.00	10.60	-	0.09	1.24	12.31	24.16	m	26.57
in concrete, depth 300 - 450mm; horizontal	15.15	5.00	15.90	-	0.10	1.38	17.91	35.19	m	38.71
Incorporating 13mm thick Korkpak; Grace Construction Products Limited - Serviciced; formwork; reinforcement laid continuously across joint										
in concrete, depth not exceeding 150mm; horizontal	5.89	5.00	6.18	-	0.08	1.10	7.03	14.32	m	15.75
in concrete, depth 150 - 300mm; horizontal	11.78	5.00	12.37	-	0.09	1.24	12.31	25.92	m	28.51
in concrete, depth 300 - 450mm; horizontal	17.67	5.00	18.55	-	0.10	1.38	17.91	37.84	m	41.62
Incorporating 19mm thick Korkpak; Grace Construction Products Limited - Serviciced; formwork; reinforcement laid continuously across joint										
in concrete, depth not exceeding 150mm; horizontal	6.77	5.00	7.11	-	0.08	1.10	7.03	15.24	m	16.77
in concrete, depth 150 - 300mm; horizontal	13.54	5.00	14.21	-	0.09	1.24	12.31	27.77	m	30.54
in concrete, depth 300 - 450mm; horizontal	20.30	5.00	21.32	-	0.10	1.38	17.91	40.61	m	44.67
Incorporating 195mm wide "Serviseal 195"; external face type p.v.c waterstop, Grace Construction Products Limited - Serviciced; heat welded joints; formwork; reinforcement laid continuously across joint										
in concrete, depth not exceeding 150mm; horizontal	10.11	10.00	11.13	-	0.44	6.05	7.03	24.21	m	26.64
in concrete, depth 150 - 300mm; horizontal	10.11	10.00	11.13	-	0.44	6.05	12.31	29.49	m	32.44
in concrete, depth 300 - 450mm; horizontal	10.11	10.00	11.13	-	0.44	6.05	17.91	35.09	m	38.60

Labour hourly rates: (except Specialists) Craft Operatives 18.37 Labourer 13.76 Rates are national average prices. Refer to REGIONAL VARIATIONS for indicative levels of overall pricing in regions	MATERIALS			LABOUR				RATES		
	Del to Site	Waste	Material Cost	Craft Optve	Lab	Labour Cost	Sunds	Nett Rate		Gross rate (10%)
	£	%	£	Hrs	Hrs	£	£	£	Unit	£
JOINTS IN CONCRETE (Cont'd)										
Formed joints (Cont'd)										
Incorporating 195mm wide "Serviseal 195"; external face type p.v.c waterstop, Grace Construction Products Limited - Servicised; heat welded joints; formwork; reinforcement laid continuously across joint (Cont'd)										
vertical L piece	23.45	10.00	25.79	-	0.22	3.03	-	28.82	nr	31.70
flat L piece	14.07	10.00	15.48	-	0.22	3.03	-	18.50	nr	20.35
flat T piece	23.22	10.00	25.55	-	0.28	3.85	-	29.40	nr	32.34
flat X piece	35.65	10.00	39.22	-	0.33	4.54	-	43.76	nr	48.14
in concrete, width not exceeding 150mm; vertical	10.11	10.00	11.13	-	0.55	7.57	7.03	25.73	m	28.30
in concrete, width 150 - 300mm; vertical	10.11	10.00	11.13	-	0.55	7.57	12.31	31.01	m	34.11
in concrete, width 300 - 450mm; vertical	10.11	10.00	11.13	-	0.55	7.57	17.91	36.60	m	40.26
flat L piece	14.07	10.00	15.48	-	0.22	3.03	-	18.50	nr	20.35
flat T piece	23.22	10.00	25.55	-	0.28	3.85	-	29.40	nr	32.34
flat X piece	35.65	10.00	39.22	-	0.33	4.54	-	43.76	nr	48.14
Incorporating 240mm wide "Serviseal 240"; heavy duty section external face type p.v.c. waterstop, Grace Construction Products Limited - Servicised; heat welded joints; formwork; reinforcement laid continuously across joint										
in concrete, depth not exceeding 150mm; horizontal	12.88	10.00	14.17	-	0.50	6.88	7.03	28.09	m	30.89
in concrete, depth 150 - 300mm; horizontal	12.88	10.00	14.17	-	0.50	6.88	12.31	33.37	m	36.70
in concrete, depth 300 - 450mm; horizontal	12.88	10.00	14.17	-	0.50	6.88	17.91	38.96	m	42.86
vertical L piece	25.54	10.00	28.09	-	0.22	3.03	-	31.12	nr	34.23
flat L piece	16.39	10.00	18.02	-	0.22	3.03	-	21.05	nr	23.16
flat T piece	26.84	10.00	29.52	-	0.28	3.85	-	33.38	nr	36.71
flat X piece	38.92	10.00	42.81	-	0.33	4.54	-	47.35	nr	52.09
in concrete, width not exceeding 150mm; vertical	12.88	10.00	14.17	-	0.61	8.39	7.03	29.60	m	32.56
in concrete, width 150 - 300mm; vertical	12.88	10.00	14.17	-	0.61	8.39	12.31	34.88	m	38.37
in concrete, width 300 - 450mm; vertical	12.88	10.00	14.17	-	0.61	8.39	17.91	40.47	m	44.52
flat L piece	16.39	10.00	18.02	-	0.22	3.03	-	21.05	nr	23.16
flat T piece	26.84	10.00	29.52	-	0.28	3.85	-	33.38	nr	36.71
flat X piece	38.92	10.00	42.81	-	0.33	4.54	-	47.35	nr	52.09
Incorporating 320mm wide "Serviseal K 320"; external face type p.v.c. kicker joint waterstop, Grace Construction Products Limited - Servicised; heat welded joints; formwork; reinforcement laid continuously across joint										
in concrete, depth not exceeding 150mm; vertical	17.46	10.00	19.21	-	0.66	9.08	7.03	35.32	m	38.85
in concrete, depth 150 - 300mm; vertical	17.46	10.00	19.21	-	0.66	9.08	12.31	40.60	m	44.66
in concrete, depth 300 - 450mm; vertical	17.46	10.00	19.21	-	0.66	9.08	17.91	46.20	m	50.82
vertical L piece	30.00	10.00	33.00	-	0.28	3.85	-	36.86	nr	40.54
flat L piece	35.26	10.00	38.78	-	0.28	3.85	-	42.64	nr	46.90
flat T piece	50.57	10.00	55.63	-	0.33	4.54	-	60.17	nr	66.18
flat X piece	63.11	10.00	69.43	-	0.39	5.37	-	74.79	nr	82.27
Incorporating 210mm wide "PVC Edgetie 210"; centre bulb type internally placed p.v.c. waterstop, Grace Construction Products Limited - Servicised; heat welded joints; formwork; reinforcement laid continuously across joint										
in concrete, depth not exceeding 150mm; horizontal	8.90	10.00	9.80	-	0.44	6.05	7.03	22.88	m	25.17
in concrete, depth 150 - 300mm; horizontal	8.90	10.00	9.80	-	0.44	6.05	12.31	28.16	m	30.98
in concrete, depth 300 - 450mm; horizontal	8.90	10.00	9.80	-	0.44	6.05	17.91	33.76	m	37.14
flat L piece	17.29	10.00	19.02	-	0.22	3.03	-	22.05	nr	24.25
flat T piece	22.92	10.00	25.21	-	0.28	3.85	-	29.06	nr	31.97
flat X piece	34.64	10.00	38.10	-	0.33	4.54	-	42.64	nr	46.90
in concrete, width not exceeding 150mm; vertical	8.90	10.00	9.80	-	0.55	7.57	7.03	24.40	m	26.84
in concrete, width 150 - 300mm; vertical	8.90	10.00	9.80	-	0.55	7.57	12.31	29.68	m	32.65
in concrete, width 300 - 450mm; vertical	8.90	10.00	9.80	-	0.55	7.57	17.91	35.27	m	38.80
vertical L piece	22.15	10.00	24.36	-	0.22	3.03	-	27.39	nr	30.13
flat L piece	17.29	10.00	19.02	-	0.22	3.03	-	22.05	nr	24.25
vertical T piece	34.86	10.00	38.35	-	0.28	3.85	-	42.20	nr	46.42
flat T piece	22.92	10.00	25.21	-	0.28	3.85	-	29.06	nr	31.97
flat X piece	34.64	10.00	38.10	-	0.33	4.54	-	42.64	nr	46.90
Incorporating 260mm wide "PVC Edgetie 260"; centre bulb type internally placed p.v.c. waterstop, Grace Construction Products Limited - Servicised; heat welded joints; formwork; reinforcement laid continuously across joint										
in concrete, depth not exceeding 150mm; horizontal	10.68	10.00	11.75	-	0.50	6.88	7.03	25.66	m	28.23
in concrete, depth 150 - 300mm; horizontal	10.68	10.00	11.75	-	0.50	6.88	12.31	30.94	m	34.04
in concrete, depth 300 - 450mm; horizontal	10.68	10.00	11.75	-	0.50	6.88	17.91	36.54	m	40.19
flat L piece	21.70	10.00	23.87	-	0.22	3.03	-	26.89	nr	29.58
flat T piece	29.89	10.00	32.88	-	0.28	3.85	-	36.73	nr	40.41
flat X piece	41.98	10.00	46.18	-	0.35	4.82	-	51.00	nr	56.10
in concrete, width not exceeding 150mm; vertical	10.68	10.00	11.75	-	0.61	8.39	7.03	27.18	m	29.89
in concrete, width 150 - 300mm; vertical	10.68	10.00	11.75	-	0.61	8.39	12.31	32.46	m	35.70
in concrete, width 300 - 450mm; vertical	10.68	10.00	11.75	-	0.61	8.39	17.91	38.05	m	41.86
vertical L piece	30.51	10.00	33.56	-	0.22	3.03	-	36.59	nr	40.25
flat L piece	21.70	10.00	23.87	-	0.22	3.03	-	26.89	nr	29.58
vertical T piece	41.64	10.00	45.81	-	0.28	3.85	-	49.66	nr	54.63
flat T piece	29.89	10.00	32.88	-	0.28	3.85	-	36.73	nr	40.41
flat X piece	41.98	10.00	46.18	-	0.33	4.54	-	50.72	nr	55.79
Incorporating 10 x 150mm wide PVC "Servitite"; bulb type internally placed waterstop; Grace Construction Products Limited - Servicised; heat welded joints fixed with "Secura" clips; formwork; reinforcement laid continuously across joint										
in concrete, depth not exceeding 150mm; horizontal	13.79	10.00	15.17	-	0.50	6.88	7.03	29.08	m	31.99
in concrete, depth 150 - 300mm; horizontal	13.79	10.00	15.17	-	0.50	6.88	12.31	34.36	m	37.80
in concrete, depth 300 - 450mm; horizontal	13.79	10.00	15.17	-	0.50	6.88	17.91	39.96	m	43.95
flat L piece	22.04	10.00	24.24	-	0.22	3.03	-	27.27	nr	29.99
flat T piece	31.75	10.00	34.93	-	0.28	3.85	-	38.78	nr	42.66

IN-SITU CONCRETE WORKS

IN-SITU CONCRETE WORKS

Labour hourly rates: (except Specialists) Craft Operatives 18.37 Labourer 13.76 Rates are national average prices. Refer to REGIONAL VARIATIONS for indicative levels of overall pricing in regions	MATERIALS			LABOUR				RATES		
	Del to Site	Waste	Material Cost	Craft Optve	Lab	Labour Cost	Sunds	Nett Rate	Unit	Gross rate (10%)
	£	%	£	Hrs	Hrs	£	£	£		£
JOINTS IN CONCRETE (Cont'd)										
Formed joints (Cont'd)										
Incorporating 10 x 150mm wide PVC "Servitite"; bulb type internally placed waterstop; Grace Construction Products Limited - Servicised; heat welded joints fixed with "Secura" clips; formwork; reinforcement laid continuously across joint (Cont'd)										
flat X piece	38.93	10.00	42.82	-	0.35	4.82	-	47.64	nr	52.40
in concrete, width not exceeding 150mm; vertical	13.79	10.00	15.17	-	0.61	8.39	7.03	30.59	m	33.65
in concrete, width 150 - 300mm; vertical	13.79	10.00	15.17	-	0.61	8.39	12.31	35.87	m	39.46
in concrete, width 300 - 450mm; vertical	13.79	10.00	15.17	-	0.61	8.39	17.91	41.47	m	45.62
vertical L piece	21.92	10.00	24.12	-	0.22	3.03	-	27.14	nr	29.86
flat L piece	22.04	10.00	24.24	-	0.22	3.03	-	27.27	nr	29.99
vertical T piece	36.84	10.00	40.52	-	0.28	3.85	-	44.38	nr	48.81
flat T piece	31.75	10.00	34.93	-	0.28	3.85	-	38.78	nr	42.66
flat X piece	38.93	10.00	42.82	-	0.33	4.54	-	47.36	nr	52.10
Incorporating 10 x 230mm wide PVC "Servitite"; bulb type internally placed waterstop; Grace Construction Products Limited - Servicised; heat welded joints fixed with "Secura" clips; formwork; reinforcement laid continuously across joint										
in concrete, depth not exceeding 150mm; horizontal	20.23	10.00	22.25	-	0.50	6.88	7.03	36.17	m	39.78
in concrete, depth 150 - 300mm; horizontal	20.23	10.00	22.25	-	0.50	6.88	12.31	41.45	m	45.59
in concrete, depth 300 - 450mm; horizontal	20.23	10.00	22.25	-	0.50	6.88	17.91	47.04	m	51.75
flat L piece	26.10	10.00	28.71	-	0.22	3.03	-	31.74	nr	34.92
flat T piece	37.86	10.00	41.64	-	0.28	3.85	-	45.50	nr	50.05
flat X piece	47.41	10.00	52.15	-	0.35	4.82	-	56.96	nr	62.66
in concrete, width not exceeding 150mm; vertical	20.23	10.00	22.25	-	0.61	8.39	7.03	37.68	m	41.45
in concrete, width 150 - 300mm; vertical	20.23	10.00	22.25	-	0.61	8.39	12.31	42.96	m	47.26
in concrete, width 300 - 450mm; vertical	20.23	10.00	22.25	-	0.61	8.39	17.91	48.55	m	53.41
vertical L piece	32.04	10.00	35.24	-	0.22	3.03	-	38.27	nr	42.10
flat L piece	26.10	10.00	28.71	-	0.22	3.03	-	31.74	nr	34.92
vertical T piece	38.93	10.00	42.82	-	0.28	3.85	-	46.68	nr	51.34
flat T piece	37.86	10.00	41.64	-	0.28	3.85	-	45.50	nr	50.05
flat X piece	47.41	10.00	52.15	-	0.33	4.54	-	56.69	nr	62.36
Incorporating 10 x 305mm wide PVC "Servitite"; bulb type internally placed waterstop; Grace Construction Products Limited - Servicised; heat welded joints fixed with "Secura" clips; formwork; reinforcement laid continuously across joint										
in concrete, depth not exceeding 150mm; horizontal	33.39	10.00	36.73	-	0.50	6.88	7.03	50.65	m	55.71
in concrete, depth 150 - 300mm; horizontal	33.39	10.00	36.73	-	0.50	6.88	12.31	55.93	m	61.52
in concrete, depth 300 - 450mm; horizontal	33.39	10.00	36.73	-	0.50	6.88	17.91	61.52	m	67.67
flat L piece	42.83	10.00	47.11	-	0.22	3.03	-	50.14	nr	55.15
flat T piece	60.18	10.00	66.19	-	0.28	3.85	-	70.05	nr	77.05
flat X piece	81.99	10.00	90.18	-	0.35	4.82	-	95.00	nr	104.50
in concrete, width not exceeding 150mm; vertical	33.39	10.00	36.73	-	0.61	8.39	7.03	52.16	m	57.38
in concrete, width 150 - 300mm; vertical	33.39	10.00	36.73	-	0.61	8.39	12.31	57.44	m	63.19
in concrete, width 300 - 450mm; vertical	33.39	10.00	36.73	-	0.61	8.39	17.91	63.04	m	69.34
vertical L piece	47.12	10.00	51.84	-	0.22	3.03	-	54.86	nr	60.35
flat L piece	42.83	10.00	47.11	-	0.22	3.03	-	50.14	nr	55.15
vertical T piece	73.45	10.00	80.80	-	0.28	3.85	-	84.65	nr	93.12
flat T piece	60.18	10.00	66.19	-	0.28	3.85	-	70.05	nr	77.05
flat X piece	81.99	10.00	90.18	-	0.33	4.54	-	94.73	nr	104.20
Incorporating 445mm wide Reference 2411, 6.34 kg/m² Expamet Hy-rib permanent shuttering and reinforcement; 150mm end laps; temporary supports; formwork laid continuously across joint										
in concrete, depth not exceeding 150mm; horizontal	4.60	7.50	4.95	0.25	0.07	5.56	7.57	18.08	m	19.88
in concrete, depth 150 - 300mm; horizontal	9.20	7.50	9.89	0.25	0.07	5.56	13.24	28.69	m	31.56
in concrete, width not exceeding 150mm; vertical	4.60	7.50	4.95	0.39	0.10	8.54	7.57	21.06	m	23.17
in concrete, width 150 - 300mm; vertical	9.20	7.50	9.89	0.39	0.10	8.54	13.24	31.68	m	34.84
Incorporating 445mm wide Reference 2611, 4.86 kg/m² Expamet Hy-rib permanent shuttering and reinforcement; 150mm end laps; temporary supports; formwork laid continuously across joint										
in concrete, depth not exceeding 150mm; horizontal	3.10	7.50	3.33	0.25	0.07	5.56	7.57	16.47	m	18.11
in concrete, depth 150 - 300mm; horizontal	6.20	7.50	6.67	0.25	0.07	5.56	13.24	25.47	m	28.02
in concrete, width not exceeding 150mm; vertical	3.10	7.50	3.33	0.39	0.10	8.54	7.57	19.45	m	21.39
in concrete, width 150 - 300mm; vertical	6.20	7.50	6.67	0.39	0.10	8.54	13.24	28.45	m	31.30
Incorporating 445mm wide Reference 2811, 3.39 kg/m² Expamet Hy-rib permanent shuttering and reinforcement; 150mm end laps; temporary supports; formwork laid continuously across joint										
in concrete, depth not exceeding 150mm; horizontal	2.10	-	2.10	0.25	0.07	5.56	7.57	15.23	m	16.75
in concrete, depth 150 - 300mm; horizontal	4.20	-	4.20	0.25	0.07	5.56	13.24	23.00	m	25.30
in concrete, width not exceeding 150mm; vertical	2.10	-	2.10	0.39	0.10	8.54	7.57	18.22	m	20.04
in concrete, width 150 - 300mm; vertical	4.20	-	4.20	0.39	0.10	8.54	13.24	25.99	m	28.58
Sealant to joint; Grace Construction Products Limited - Servicised. Servijoint ONE DW (one-part sealant); including preparation, cleaning and priming										
10 x 25mm; horizontal	5.46	5.00	5.73	-	0.15	2.06	-	7.80	m	8.57
13 x 25mm; horizontal	6.92	5.00	7.26	-	0.20	2.75	-	10.01	m	11.02
19 x 25mm; horizontal	10.03	5.00	10.54	-	0.25	3.44	-	13.98	m	15.37
Sealant to joint; Grace Construction Products Limited - Servicised. "Paraseal" pouring grade 2 part grey polysulphide sealant; including preparation, cleaners, primers and sealant										
10 x 25mm; horizontal	3.55	5.00	3.73	-	0.11	1.51	-	5.24	m	5.76
13 x 25mm; horizontal	4.62	5.00	4.85	-	0.12	1.65	-	6.50	m	7.15
19 x 25mm; horizontal	6.72	5.00	7.06	-	0.13	1.79	-	8.85	m	9.73

Labour hourly rates: (except Specialists) Craft Operatives 18.37 Labourer 13.76 Rates are national average prices. Refer to REGIONAL VARIATIONS for indicative levels of overall pricing in regions	MATERIALS			LABOUR				RATES		
	Del to Site	Waste	Material Cost	Craft Optve	Lab	Labour Cost	Sunds	Nett Rate		Gross rate (10%)
	£	%	£	Hrs	Hrs	£	£	£	Unit	£
ACCESSORIES CAST IN TO IN-SITU CONCRETE										
Cast in accessories										
Galvanized steel dowels										
12mm diameter x 150mm long	0.77	10.00	0.85	-	0.09	1.24	-	2.08	nr	2.29
12mm diameter x 300mm long	1.17	10.00	1.29	-	0.08	1.10	-	2.39	nr	2.62
12mm diameter x 600mm long	1.75	10.00	1.93	-	0.12	1.65	-	3.58	nr	3.94
Galvanized expanded metal tie										
300 x 50mm; bending and temporarily fixing to formwork	0.94	10.00	1.04	-	0.13	1.79	-	2.83	nr	3.11
Galvanized steel dovetailed masonry slots with 3mm thick twisted tie to suit 50mm cavity										
100mm long; temporarily fixing to formwork	0.95	10.00	1.05	0.11	-	2.02	-	3.07	nr	3.38
Stainless steel channel with anchors at 250mm centres										
28 x 15mm; temporarily fixing to formwork	14.02	5.00	14.72	0.28	-	5.14	-	19.87	m	21.85
28 x 15 x 150mm long; temporarily fixing to formwork..................	2.36	5.00	2.48	0.12	-	2.20	-	4.68	nr	5.15
M10 bolt 50mm long with `T' head nut and washers	4.84	5.00	5.08	0.05	-	0.92	-	6.00	nr	6.60
fishtailed tie to suit 50mm cavity....................	1.95	5.00	2.05	0.06	-	1.10	-	3.15	nr	3.46
Stainless steel angle drilled at 450mm centres										
70 x 90 x 5mm	36.00	5.00	37.80	0.80	-	14.70	-	52.50	m	57.75
Copper dovetailed masonry slots with 3mm thick twisted tie to suit 50mm cavity										
100mm long; temporarily fixing to formwork	10.15	10.00	11.16	0.11	-	2.02		13.18	nr	14.50
Bolt boxes; The Expanded Metal Co. Ltd										
reference 220 for use with poured concrete; 75mm diameter										
150mm long....................	1.48	10.00	1.63	0.28	-	5.14	0.86	7.63	nr	8.39
225mm long....................	1.76	10.00	1.94	0.33	-	6.06	0.86	8.85	nr	9.74
300mm long....................	1.98	10.00	2.18	0.39	-	7.16	0.86	10.20	nr	11.22
reference 220 for use with poured concrete; 100mm diameter										
375mm long....................	2.86	10.00	3.15	0.44	-	8.08	0.86	12.08	nr	13.29
450mm long....................	3.24	10.00	3.56	0.50	-	9.19	0.86	13.60	nr	14.96
600mm long....................	3.58	10.00	3.94	0.55	-	10.10	0.86	14.90	nr	16.39
reference 220 for use with vibrated concrete; wrap with single layer of thin polythene sheet; 75mm diameter										
150mm long....................	1.56	10.00	1.72	0.28	-	5.14	1.03	7.89	nr	8.68
225mm long....................	1.85	10.00	2.04	0.33	-	6.06	1.03	9.13	nr	10.05
300mm long....................	2.18	10.00	2.40	0.39	-	7.16	1.03	10.60	nr	11.66
reference 220 for use with vibrated concrete; wrap with single layer of thin polythene sheet; 100mm diameter										
375mm long....................	3.04	10.00	3.34	0.44	-	8.08	1.03	12.46	nr	13.71
450mm long....................	3.43	10.00	3.77	0.50	-	9.19	1.03	13.99	nr	15.39
600mm long....................	3.77	10.00	4.15	0.55	-	10.10	1.03	15.29	nr	16.81
Steel rag bolts										
M 10 x 100mm long....................	0.08	5.00	0.08	0.10	-	1.84	-	1.92	nr	2.11
M 10 x 160mm long....................	0.17	5.00	0.17	0.12	-	2.20	-	2.38	nr	2.62
M 12 x 100mm long....................	0.17	5.00	0.18	0.10	-	1.84	-	2.01	nr	2.21
M 12 x 160mm long....................	0.43	5.00	0.45	0.12	-	2.20	-	2.65	nr	2.92
M 12 x 200mm long....................	0.47	5.00	0.50	0.14	-	2.57	-	3.07	nr	3.38
M 16 x 120mm long....................	0.51	5.00	0.53	0.10	-	1.84	-	2.37	nr	2.61
M 16 x 160mm long....................	0.65	5.00	0.68	0.12	-	2.20	-	2.89	nr	3.18
M 16 x 200mm long....................	0.70	5.00	0.73	0.14	-	2.57	-	3.30	nr	3.63
M 16 x 300mm long....................	2.28	5.00	2.39	0.16	-	2.94	-	5.33	nr	5.86
M 20 x 450mm long....................	9.53	5.00	10.01	0.20	-	3.67	-	13.68	nr	15.05
M 24 x 450mm long....................	13.40	5.00	14.07	0.25	-	4.59	-	18.67	nr	20.53
IN-SITU CONCRETE SUNDRIES										
Grouting; Cement and sand (1; 3)										
Grouting.										
stanchion bases....................	3.31	10.00	3.64	-	0.45	6.19	1.58	11.41	nr	12.55

This page left blank intentionally

Labour hourly rates: (except Specialists) Craft Operatives 18.37 Labourer 13.76 Rates are national average prices. Refer to REGIONAL VARIATIONS for indicative levels of overall pricing in regions	MATERIALS			LABOUR				RATES		
	Del to Site	Waste	Material Cost	Craft Optve	Lab	Labour Cost	Sunds	Nett Rate		Gross rate (10%)
	£	%	£	Hrs	Hrs	£	£	£	Unit	£
PRECAST/COMPOSITE CONCRETE DECKING & FLOORING										
Composite concrete work; Prestressed concrete beams and 100mm building blocks; hoist bed and grout										
155mm thick floors										
beams at 510mm centres	27.11	2.50	27.79	-	1.15	15.82	0.68	44.29	m²	48.72
beams at 510mm and 285mm centres	34.67	2.50	35.54	-	1.05	14.45	0.68	50.66	m²	55.73
double beams at 624mm centres	42.23	2.50	43.29	-	1.00	13.76	0.68	57.72	m²	63.49
beams at 285mm centres	42.23	2.50	43.29	-	1.15	15.82	0.68	59.78	m²	65.76
double beams at 399mm centres	69.12	2.50	70.84	-	1.05	14.45	0.68	85.97	m²	94.56
beams at 114mm centres with concrete between..........................	86.03	2.50	88.18	-	1.00	13.76	0.68	102.61	m²	112.88
225mm thick floors										
beams at 530mm centres	30.73	2.50	31.49	-	1.15	15.82	0.68	47.99	m²	52.79
beams at 305mm centres	48.31	2.50	49.52	-	1.05	14.45	0.68	64.64	m²	71.11
beams at 140mm centres with concrete between..........................	100.73	2.50	103.25	-	1.00	13.76	0.68	117.68	m²	129.45
Passive fall arrest provision										
Add to the foregoing rates for providing safety decking, airbags or netting as appropriate to the works										
total area less than 100 m²......................	7.60	2.50	7.79	-	-	-	-	7.79	m²	8.57
total area over 100m²	7.54	2.50	7.73	-	-	-	-	7.73	m²	8.50

This page left blank intentionally

Labour hourly rates: (except Specialists) Craft Operatives 18.37 Labourer 13.76 Rates are national average prices. Refer to REGIONAL VARIATIONS for indicative levels of overall pricing in regions	MATERIALS			LABOUR				RATES		
	Del to Site	Waste	Material Cost	Craft Optve	Lab	Labour Cost	Sunds	Nett Rate		Gross rate (10%)
	£	%	£	Hrs	Hrs	£	£	£	Unit	£
PRECAST CONCRETE SLABS										
Precast concrete goods; Precast concrete hollowcore floors and hoist bed and grout (Hanson Concrete Products Ltd)										
150 mm thick planks 750 wide										
span between supports 7.6m; UDL not exceeding 2KN/m²	58.34	2.50	59.80	-	-	-	-	59.80	m²	65.78
span between supports 6.6m; UDL not exceeding 4KN/m²	58.34	2.50	59.80	-	-	-	-	59.80	m²	65.78
span between supports 5.9m; UDL not exceeding 6KN/m²	58.34	2.50	59.80	-	-	-	-	59.80	m²	65.78
span between supports 5.4m; UDL not exceeding 8KN/m²	58.34	2.50	59.80	-	-	-	-	59.80	m²	65.78
span between supports 5.0m; UDL not exceeding 10KN/m²	58.34	2.50	59.80	-	-	-	-	59.80	m²	65.78
200 mm thick planks 750 wide										
span between supports 9.3m; UDL not exceeding 2KN/m²	63.86	2.50	65.46	-	-	-	-	65.46	m²	72.00
span between supports 8.1m; UDL not exceeding 4KN/m²	63.86	2.50	65.46	-	-	-	-	65.46	m²	72.00
span between supports 7.3m; UDL not exceeding 6KN/m²	63.86	2.50	65.46	-	-	-	-	65.46	m²	72.00
span between supports 6.7m; UDL not exceeding 8KN/m²	63.86	2.50	65.46	-	-	-	-	65.46	m²	72.00
span between supports 6.2m; UDL not exceeding 10KN/m²	63.86	2.50	65.46	-	-	-	-	65.46	m²	72.00
250 mm thick planks 750 wide										
span between supports 10.6m; UDL not exceeding 2KN/m²	69.68	2.50	71.42	-	-	-	-	71.42	m²	78.56
span between supports 9.3m; UDL not exceeding 4KN/m²	69.68	2.50	71.42	-	-	-	-	71.42	m²	78.56
span between supports 8.5m; UDL not exceeding 6KN/m²	69.68	2.50	71.42	-	-	-	-	71.42	m²	78.56
span between supports 7.8m; UDL not exceeding 8KN/m²	69.68	2.50	71.42	-	-	-	-	71.42	m²	78.56
span between supports 7.2m; UDL not exceeding 10KN/m²	69.68	2.50	71.42	-	-	-	-	71.42	m²	78.56
300 mm thick planks 750 wide										
span between supports 11.5m; UDL not exceeding 2KN/m²	75.46	2.50	77.35	-	-	-	-	77.35	m²	85.08
span between supports 10.3m; UDL not exceeding 4KN/m²	75.46	2.50	77.35	-	-	-	-	77.35	m²	85.08
span between supports 9.4m; UDL not exceeding 6KN/m²	75.46	2.50	77.35	-	-	-	-	77.35	m²	85.08
span between supports 8.7m; UDL not exceeding 8KN/m²	75.46	2.50	77.35	-	-	-	-	77.35	m²	85.08
span between supports 8.1m; UDL not exceeding 10KN/m²	75.46	2.50	77.35	-	-	-	-	77.35	m²	85.08
Passive fall arrest provision										
Add to the foregoing rates for providing safety decking, airbags or netting as appropriate to the works										
total area less than 100 m²	7.60	2.50	7.79	-	-	-	-	7.79	m²	8.57
total area over 100m²	7.54	2.50	7.73	-	-	-	-	7.73	m²	8.50

This page left blank intentionally

Labour hourly rates: (except Specialists) Craft Operatives 18.37 Labourer 13.76 Rates are national average prices. Refer to REGIONAL VARIATIONS for indicative levels of overall pricing in regions	MATERIALS			LABOUR				RATES		
	Del to Site	Waste	Material Cost	Craft Optve	Lab	Labour Cost	Sunds	Nett Rate		Gross rate (10%)
	£	%	£	Hrs	Hrs	£	£	£	Unit	£
BRICK/BLOCK WALLING										
Walls; brickwork; Common bricks, BS EN 772, Category M, 215 x 102.5 x 65mm, compressive strength 20.5 N/mm²; in cement-lime mortar (1:2:9)										
Walls										
102mm thick; stretcher bond	18.64	5.00	19.58	1.45	1.10	41.77	3.67	65.02	m²	71.53
215mm thick; English bond	37.60	5.00	39.48	2.35	1.85	68.63	8.58	116.69	m²	128.36
327mm thick; English bond	56.56	5.00	59.39	2.85	2.30	84.00	13.48	156.88	m²	172.57
Skins of hollow walls										
102mm thick; stretcher bond	18.64	5.00	19.58	1.45	1.10	41.77	3.67	65.02	m²	71.53
215mm thick; English bond	37.60	5.00	39.48	2.35	1.85	68.63	8.58	116.69	m²	128.36
Walls; building against concrete (ties measured separately); vertical										
102mm thick; stretcher bond	18.64	5.00	19.58	1.60	1.20	45.90	3.67	69.16	m²	76.07
Walls; building against old brickwork; tie new to old with 200mm stainless steel housing type 4 wall ties - 6/m²; vertical										
102mm thick; stretcher bond	19.07	10.00	20.05	1.60	1.20	45.90	3.67	69.63	m²	76.59
215mm thick; English bond	38.03	10.00	39.96	2.65	2.05	76.89	8.58	125.43	m²	137.97
Walls; bonding to stonework; including extra material; vertical										
102mm thick; stretcher bond	19.70	5.00	20.68	1.60	1.20	45.90	3.67	70.26	m²	77.29
215mm thick; English bond	39.71	5.00	41.70	2.65	2.05	76.89	8.58	127.17	m²	139.88
327mm thick; English bond	58.67	5.00	61.61	3.15	2.50	92.27	13.48	167.36	m²	184.09
440mm thick; English bond	77.32	5.00	81.18	3.70	3.00	109.25	18.39	208.82	m²	229.70
Walls; bonding to old brickwork; cutting pockets; including extra material; vertical										
102mm thick; stretcher bond	19.70	5.00	20.68	3.65	2.55	102.14	3.67	126.49	m²	139.14
215mm thick; English bond	39.71	5.00	41.70	4.60	3.30	129.91	8.58	180.19	m²	198.21
Walls; curved 2m radius; including extra material; vertical										
102mm thick; stretcher bond	18.64	5.00	19.58	2.85	2.05	80.56	3.67	103.81	m²	114.20
215mm thick; English bond	37.60	5.00	39.48	5.20	3.75	147.12	8.58	195.19	m²	214.71
327mm thick; English bond	56.56	5.00	59.39	7.10	5.10	200.60	13.48	273.48	m²	300.83
Walls; curved 6m radius; including extra material; vertical										
102mm thick; stretcher bond	18.64	5.00	19.58	2.15	1.55	60.82	3.67	84.07	m²	92.48
215mm thick; English bond	37.60	5.00	39.48	3.75	2.80	107.42	8.58	155.48	m²	171.03
327mm thick; English bond	56.56	5.00	59.39	4.95	3.70	141.84	13.48	214.72	m²	236.19
Skins of hollow walls; curved 2m radius; including extra material; vertical										
102mm thick; stretcher bond	18.64	5.00	19.58	2.85	2.05	80.56	3.67	103.81	m²	114.20
215mm thick; English bond	37.60	5.00	39.48	5.20	3.75	147.12	8.58	195.19	m²	214.71
Skins of hollow walls; curved 6m radius; including extra material; vertical										
102mm thick; stretcher bond	18.64	5.00	19.58	2.15	1.55	60.82	3.67	84.07	m²	92.48
215mm thick; English bond	37.60	5.00	39.48	3.75	2.80	107.42	8.58	155.48	m²	171.03
Isolated piers										
215mm thick; English bond	37.60	5.00	39.48	4.70	3.40	133.12	8.58	181.19	m²	199.31
327mm thick; English bond	56.56	5.00	59.39	5.70	4.20	162.50	13.48	235.38	m²	258.92
440mm thick; English bond	75.21	5.00	78.97	6.60	4.95	189.35	18.39	286.71	m²	315.38
Chimney stacks										
440mm thick; English bond	75.21	5.00	78.97	6.60	5.45	196.23	18.39	293.59	m²	322.95
890mm thick; English bond	150.42	5.00	157.94	10.40	8.80	312.14	36.78	506.85	m²	557.54
Projections										
215mm wide x 112mm projection	4.21	5.00	4.42	0.30	0.23	8.68	0.98	14.07	m	15.48
215mm wide x 215mm projection	8.42	5.00	8.85	0.50	0.39	14.55	1.97	25.36	m	27.90
327mm wide x 112mm projection	6.32	5.00	6.64	0.45	0.35	13.08	1.47	21.19	m	23.31
327mm wide x 215mm projection	12.64	5.00	13.27	0.75	0.59	21.90	2.94	38.11	m	41.92
Projections; horizontal										
215mm wide x 112mm projection	4.21	5.00	4.42	0.30	0.23	8.68	0.98	14.07	m	15.48
215mm wide x 215mm projection	8.42	5.00	8.85	0.50	0.39	14.55	1.97	25.36	m	27.90
327mm wide x 112mm projection	6.32	5.00	6.64	0.45	0.35	13.08	1.47	21.19	m	23.31
327mm wide x 215mm projection	12.64	5.00	13.27	0.75	0.59	21.90	2.94	38.11	m	41.92
Projections; bonding to old brickwork; cutting pockets; including extra material; vertical										
215mm wide x 112mm projection	7.02	5.00	7.37	0.93	0.65	26.03	1.63	35.03	m	38.54
215mm wide x 215mm projection	11.23	5.00	11.79	1.13	0.81	31.90	2.61	46.31	m	50.94
327mm wide x 112mm projection	10.53	5.00	11.06	1.40	0.98	39.20	2.44	52.71	m	57.98
327mm wide x 215mm projection	15.45	5.00	16.22	1.70	1.22	48.02	3.60	67.83	m	74.62
Projections; bonding to old brickwork; cutting pockets; including extra material; horizontal										
215mm wide x 112mm projection	7.02	5.00	7.37	0.93	0.65	26.03	1.63	35.03	m	38.54
215mm wide x 215mm projection	11.23	5.00	11.79	1.13	0.81	31.90	2.61	46.31	m	50.94
327mm wide x 112mm projection	10.53	5.00	11.06	1.40	0.98	39.20	2.44	52.71	m	57.98

Labour hourly rates: (except Specialists) Craft Operatives 18.37 Labourer 13.76 Rates are national average prices. Refer to REGIONAL VARIATIONS for indicative levels of overall pricing in regions	MATERIALS			LABOUR				RATES		
	Del to Site	Waste	Material Cost	Craft Optve	Lab	Labour Cost	Sunds	Nett Rate	Unit	Gross rate (10%)
	£	%	£	Hrs	Hrs	£	£	£		£
BRICK/BLOCK WALLING (Cont'd)										
Walls; brickwork; Common bricks, BS EN 772, Category M, 215 x 102.5 x 65mm, compressive strength 20.5 N/mm²; in cement-lime mortar (1:2:9) (Cont'd)										
Projections; bonding to old brickwork; cutting pockets; including extra material; horizontal (Cont'd)										
327mm wide x 215mm projection	15.45	5.00	16.22	1.70	1.22	48.02	3.60	67.83	m	74.62
Arches including centering; flat										
flat										
112mm high on face; 215mm thick; width of exposed soffit 215mm; bricks-on-edge	4.21	5.00	4.42	1.50	1.00	41.32	0.98	46.71	m	51.38
215mm high on face; 112mm thick; width of exposed soffit 112mm; bricks-on-end	4.21	5.00	4.42	1.45	1.00	40.40	0.98	45.79	m	50.37
Arches including centering; semi-circular										
112mm high on face; 215mm thick; width of exposed soffit 215mm; bricks-on-edge	6.61	5.00	6.94	2.15	1.35	58.07	1.54	66.56	m	73.22
215mm high on face; 215mm thick; width of exposed soffit 215mm; bricks-on-edge	13.23	5.00	13.89	2.95	1.95	81.02	3.08	97.99	m	107.79
Closing cavities										
50mm wide with brickwork 102mm thick	2.05	5.00	2.16	0.40	0.27	11.06	0.48	13.70	m	15.07
75mm wide with brickwork 102mm thick	2.05	5.00	2.16	0.40	0.27	11.06	0.48	13.70	m	15.07
Closing cavities; horizontal										
50mm wide with slates	2.64	5.00	2.77	0.66	0.44	18.18	0.48	21.43	m	23.57
75mm wide with slates	2.64	5.00	2.77	0.66	0.44	18.18	0.48	21.43	m	23.57
50mm wide with brickwork 102mm thick	2.05	5.00	2.16	0.40	0.27	11.06	0.48	13.70	m	15.07
75mm wide with brickwork 102mm thick	2.05	5.00	2.16	0.40	0.27	11.06	0.48	13.70	m	15.07
Bonding ends to existing common brickwork; cutting pockets; extra material										
walls; bonding every third course										
102mm thick	1.40	5.00	1.47	0.37	0.25	10.24	0.33	12.04	m	13.24
215mm thick	2.81	5.00	2.95	0.66	0.44	18.18	0.66	21.79	m	23.96
327mm thick	4.21	5.00	4.42	0.95	0.63	26.12	0.98	31.52	m	34.67
Walls; brickwork; Second hard stock bricks, BS EN 772, Category M, 215 x 102.5 x 65mm; in cement-lime mortar (1:2:9)										
Walls										
102mm thick; stretcher bond	46.91	5.00	49.25	1.45	1.10	41.77	3.67	94.70	m²	104.17
215mm thick; English bond	94.61	5.00	99.34	2.35	1.85	68.63	8.58	176.54	m²	194.19
327mm thick; English bond	142.31	5.00	149.42	2.85	2.30	84.00	13.48	246.91	m²	271.60
Skins of hollow walls										
102mm thick; stretcher bond	46.91	5.00	49.25	1.45	1.10	41.77	3.67	94.70	m²	104.17
215mm thick; English bond	94.61	5.00	99.34	2.35	1.85	68.63	8.58	176.54	m²	194.19
Walls; building against concrete (ties measured separately); vertical										
102mm thick; stretcher bond	46.91	5.00	49.25	1.60	1.20	45.90	3.67	98.83	m²	108.71
Walls; building against old brickwork; tie new to old with 200mm stainless steel housing type 4 wall ties - 6/m²; vertical										
102mm thick; stretcher bond	47.33	10.00	49.72	1.60	1.20	45.90	3.67	99.30	m²	109.23
215mm thick; English bond	95.03	10.00	99.81	2.65	2.05	76.89	8.58	185.28	m²	203.80
Walls; bonding to stonework; including extra material; vertical										
102mm thick; stretcher bond	49.55	5.00	52.03	1.60	1.20	45.90	3.88	101.82	m²	112.00
215mm thick; English bond	99.91	5.00	104.90	2.65	2.05	76.89	9.07	190.87	m²	209.95
327mm thick; English bond	147.61	5.00	154.99	3.15	2.50	92.27	13.98	261.23	m²	287.36
440mm thick; English bond	194.51	5.00	204.24	3.70	3.00	109.25	18.90	332.39	m²	365.63
Walls; bonding to old brickwork; cutting pockets; including extra material; vertical										
102mm thick; stretcher bond	49.55	5.00	52.03	3.65	2.55	102.14	3.88	158.05	m²	173.86
215mm thick; English bond	99.91	5.00	104.90	4.60	3.30	129.91	9.07	243.89	m²	268.28
Walls; curved 2m radius; including extra material; vertical										
102mm thick; stretcher bond	46.91	5.00	49.25	2.85	2.05	80.56	3.67	133.49	m²	146.84
215mm thick; English bond	94.61	5.00	99.34	5.20	3.75	147.12	8.58	255.04	m²	280.54
327mm thick; English bond	142.31	5.00	149.42	7.10	5.10	200.60	13.48	363.51	m²	399.86
Walls; curved 6m radius; including extra material; vertical										
102mm thick; stretcher bond	46.91	5.00	49.25	2.15	1.55	60.82	3.67	113.75	m²	125.12
215mm thick; English bond	94.61	5.00	99.34	3.75	2.80	107.42	8.58	215.33	m²	236.86
327mm thick; English bond	142.31	5.00	149.42	4.95	3.70	141.84	13.48	304.75	m²	335.22
Skins of hollow walls; curved 2m radius; including extra material; vertical										
102mm thick; stretcher bond	46.91	5.00	49.25	2.85	2.05	80.56	3.67	133.49	m²	146.84
215mm thick; English bond	94.61	5.00	99.34	5.20	3.75	147.12	8.58	255.04	m²	280.54
Skins of hollow walls; curved 6m radius; including extra material; vertical										
102mm thick; stretcher bond	46.91	5.00	49.25	2.15	1.55	60.82	3.67	113.75	m²	125.12
215mm thick; English bond	94.61	5.00	99.34	3.75	2.80	107.42	8.58	215.33	m²	236.86
Isolated piers										
215mm thick; English bond	94.61	5.00	99.34	4.70	3.40	133.12	8.58	241.04	m²	265.14
327mm thick; English bond	142.31	5.00	149.42	5.70	4.20	162.50	13.48	325.41	m²	357.95
440mm thick; English bond	189.21	5.00	198.67	6.60	4.95	189.35	18.39	406.41	m²	447.06
Chimney stacks										
440mm thick; English bond	189.21	5.00	198.67	6.60	5.45	196.23	18.39	413.29	m²	454.62
890mm thick; English bond	378.42	5.00	397.34	10.40	8.80	312.14	36.78	746.26	m²	820.88

Labour hourly rates: (except Specialists) Craft Operatives 18.37 Labourer 13.76 Rates are national average prices. Refer to REGIONAL VARIATIONS for indicative levels of overall pricing in regions	MATERIALS			LABOUR				RATES		
	Del to Site	Waste	Material Cost	Craft Optve	Lab	Labour Cost	Sunds	Nett Rate	Unit	Gross rate (10%)
	£	%	£	Hrs	Hrs	£	£	£		£
BRICK/BLOCK WALLING (Cont'd)										
Walls; brickwork; Second hard stock bricks, BS EN 772, Category M, 215 x 102.5 x 65mm; in cement-lime mortar (1:2:9) (Cont'd)										
Projections										
215mm wide x 112mm projection	10.60	5.00	11.13	0.30	0.23	8.68	0.98	20.78	m	22.86
215mm wide x 215mm projection	21.19	5.00	22.25	0.50	0.39	14.55	1.97	38.77	m	42.65
327mm wide x 112mm projection	15.90	5.00	16.69	0.45	0.35	13.08	1.47	31.25	m	34.37
327mm wide x 215mm projection	31.80	5.00	33.39	0.75	0.59	21.90	2.94	58.23	m	64.05
Projections; horizontal										
215mm wide x 112mm projection	10.60	5.00	11.13	0.30	0.23	8.68	0.98	20.78	m	22.86
215mm wide x 215mm projection	21.19	5.00	22.25	0.50	0.39	14.55	1.97	38.77	m	42.65
327mm wide x 112mm projection	15.90	5.00	16.69	0.45	0.35	13.08	1.47	31.25	m	34.37
327mm wide x 215mm projection	31.80	5.00	33.39	0.75	0.59	21.90	2.94	58.23	m	64.05
Projections; bonding to old brickwork; cutting pockets; including extra material; vertical										
215mm wide x 112mm projection	17.66	5.00	18.54	0.93	0.65	26.03	1.63	46.20	m	50.82
215mm wide x 215mm projection	28.25	5.00	29.67	1.13	0.81	31.90	2.61	64.18	m	70.60
327mm wide x 112mm projection	26.50	5.00	27.82	1.40	0.98	39.20	2.44	69.47	m	76.42
327mm wide x 215mm projection	38.86	5.00	40.80	1.70	1.22	48.02	3.60	92.42	m	101.66
Projections; bonding to old brickwork; cutting pockets; including extra material; horizontal										
215mm wide x 112mm projection	17.66	5.00	18.54	0.93	0.65	26.03	1.63	46.20	m	50.82
215mm wide x 215mm projection	28.25	5.00	29.67	1.13	0.81	31.90	2.61	64.18	m	70.60
327mm wide x 112mm projection	26.50	5.00	27.82	1.40	0.98	39.20	2.44	69.47	m	76.42
327mm wide x 215mm projection	38.86	5.00	40.80	1.70	1.22	48.02	3.60	92.42	m	101.66
Arches including centering; flat										
112mm high on face; 215mm thick; width of exposed soffit 215mm; bricks-on-edge	10.60	5.00	11.13	1.50	1.00	41.32	0.98	53.42	m	58.76
215mm high on face; 112mm thick; width of exposed soffit 112mm; bricks-on-end	10.60	5.00	11.13	1.45	1.00	40.40	0.98	52.50	m	57.75
Arches including centering; semi-circular										
112mm high on face; 215mm thick; width of exposed soffit 215mm; bricks-on-edge	16.64	5.00	17.47	2.15	1.35	58.07	1.54	77.09	m	84.80
215mm high on face; 215mm thick; width of exposed soffit 215mm; bricks-on-edge	33.29	5.00	34.95	2.95	1.95	81.02	3.08	119.05	m	130.95
Closing cavities										
50mm wide with brickwork 102mm thick	5.17	5.00	5.43	0.40	0.27	11.06	0.48	16.97	m	18.67
75mm wide with brickwork 102mm thick	5.17	5.00	5.43	0.40	0.27	11.06	0.48	16.97	m	18.67
Closing cavities; horizontal										
50mm wide with slates	2.64	5.00	2.77	0.66	0.44	18.18	0.48	21.43	m	23.57
75mm wide with slates	2.64	5.00	2.77	0.66	0.44	18.18	0.48	21.43	m	23.57
50mm wide with brickwork 102mm thick	5.17	5.00	5.43	0.40	0.27	11.06	0.48	16.97	m	18.67
75mm wide with brickwork 102mm thick	5.17	5.00	5.43	0.40	0.27	11.06	0.48	16.97	m	18.67
Bonding ends to existing common brickwork; cutting pockets; extra material walls; bonding every third course										
102mm thick	3.53	5.00	3.71	0.37	0.25	10.24	0.33	14.27	m	15.70
215mm thick	7.06	5.00	7.41	0.66	0.44	18.18	0.66	26.25	m	28.88
327mm thick	10.60	5.00	11.13	0.95	0.63	26.12	0.98	38.22	m	42.04
Walls; brickwork; Engineering bricks, BS EN 772, Category F, 215 x 102.5 x 65mm, Class B; in cement-lime mortar (1:2:9)										
Walls										
102mm thick; stretcher bond	19.06	5.00	20.01	1.60	1.20	45.90	3.67	69.59	m²	76.55
215mm thick; English bond	38.44	5.00	40.36	2.60	2.05	75.97	8.58	124.91	m²	137.40
327mm thick; English bond	57.82	5.00	60.71	3.15	2.55	92.95	13.48	167.15	m²	183.86
Skins of hollow walls										
102mm thick; stretcher bond	19.06	5.00	20.01	1.60	1.20	45.90	3.67	69.59	m²	76.55
215mm thick; English bond	38.44	5.00	40.36	2.60	2.05	75.97	8.58	124.91	m²	137.40
Walls; building against concrete (ties measured separately); vertical										
102mm thick; stretcher bond	19.06	5.00	20.01	1.75	1.30	50.04	3.67	73.72	m²	81.09
Walls; building against old brickwork; tie new to old with 200mm stainless steel housing type 4 wall ties - 6/m²; vertical										
102mm thick; stretcher bond	19.49	10.00	20.48	1.75	1.30	50.04	3.67	74.19	m²	81.61
215mm thick; English bond	38.87	10.00	40.83	2.90	2.25	84.23	8.58	133.64	m²	147.01
Walls; bonding to stonework; including extra material; vertical										
102mm thick; stretcher bond	20.13	5.00	21.14	1.75	1.30	50.04	3.88	75.06	m²	82.57
215mm thick; English bond	40.59	5.00	42.62	2.90	2.25	84.23	9.07	135.93	m²	149.52
327mm thick; English bond	59.97	5.00	62.97	3.45	2.75	101.22	13.98	178.17	m²	195.98
440mm thick; English bond	79.03	5.00	82.98	4.05	3.30	119.81	18.90	221.69	m²	243.85
Walls; bonding to old brickwork; cutting pockets; including extra material; vertical										
102mm thick; stretcher bond	20.13	5.00	21.14	4.00	2.80	112.01	3.88	137.03	m²	150.74
215mm thick; English bond	40.59	5.00	42.62	5.05	3.65	142.99	9.07	194.69	m²	214.16
Walls; curved 2m radius; including extra material; vertical										
102mm thick; stretcher bond	19.06	5.00	20.01	3.15	2.25	88.83	3.67	112.51	m²	123.76
215mm thick; English bond	38.44	5.00	40.36	5.70	4.10	161.12	8.58	210.06	m²	231.07
327mm thick; English bond	57.82	5.00	60.71	7.80	5.60	220.34	13.48	294.53	m²	323.99

MASONRY

Labour hourly rates: (except Specialists) Craft Operatives 18.37 Labourer 13.76 Rates are national average prices. Refer to REGIONAL VARIATIONS for indicative levels of overall pricing in regions	MATERIALS			LABOUR				RATES		
	Del to Site	Waste	Material Cost	Craft Optve	Lab	Labour Cost	Sunds	Nett Rate	Unit	Gross rate (10%)
	£	%	£	Hrs	Hrs	£	£	£		£
BRICK/BLOCK WALLING (Cont'd)										
Walls; brickwork; Engineering bricks, BS EN 772, Category F, 215 x 102.5 x 65mm, Class B; in cement-lime mortar (1:2:9) (Cont'd)										
Walls; curved 6m radius; including extra material; vertical										
102mm thick; stretcher bond	19.06	5.00	20.01	2.35	1.70	66.56	3.67	90.25	m²	99.27
215mm thick; English bond	38.44	5.00	40.36	4.10	3.10	117.97	8.58	166.91	m²	183.60
327mm thick; English bond	57.82	5.00	60.71	5.45	4.05	155.84	13.48	230.04	m²	253.04
Skins of hollow walls; curved 2m radius; including extra material; vertical										
102mm thick; stretcher bond	19.06	5.00	20.01	3.15	2.25	88.83	3.67	112.51	m²	123.76
215mm thick; English bond	38.44	5.00	40.36	5.70	4.10	161.12	8.58	210.06	m²	231.07
Skins of hollow walls; curved 6m radius; including extra material; vertical										
102mm thick; stretcher bond	19.06	5.00	20.01	2.35	1.70	66.56	3.67	90.25	m²	99.27
215mm thick; English bond	38.44	5.00	40.36	4.10	3.10	117.97	8.58	166.91	m²	183.60
Isolated piers										
215mm thick; English bond	38.44	5.00	40.36	5.15	3.75	146.21	8.58	195.14	m²	214.66
327mm thick; English bond	57.82	5.00	60.71	6.25	4.60	178.11	13.48	252.30	m²	277.53
440mm thick; English bond	76.87	5.00	80.72	7.25	5.45	208.17	18.39	307.28	m²	338.01
Chimney stacks										
440mm thick; English bond	76.87	5.00	80.72	7.25	5.45	208.17	18.39	307.28	m²	338.01
890mm thick; English bond	153.75	5.00	161.44	11.45	8.80	331.42	36.78	529.64	m²	582.60
Projections										
215mm wide x 112mm projection	4.31	5.00	4.52	0.33	0.25	9.50	0.98	15.00	m	16.50
215mm wide x 215mm projection	8.61	5.00	9.04	0.55	0.42	15.88	1.97	26.89	m	29.58
327mm wide x 112mm projection	6.46	5.00	6.78	0.50	0.38	14.41	1.47	22.67	m	24.93
327mm wide x 215mm projection	12.92	5.00	13.57	0.83	0.63	23.92	2.94	40.42	m	44.46
Projections; horizontal										
215mm wide x 112mm projection	4.31	5.00	4.52	0.33	0.25	9.50	0.98	15.00	m	16.50
215mm wide x 215mm projection	8.61	5.00	9.04	0.55	0.42	15.88	1.97	26.89	m	29.58
327mm wide x 112mm projection	6.46	5.00	6.78	0.50	0.38	14.41	1.47	22.67	m	24.93
327mm wide x 215mm projection	12.92	5.00	13.57	0.83	0.63	23.92	2.94	40.42	m	44.46
Projections; bonding to old brickwork; cutting pockets; including extra material; vertical										
215mm wide x 112mm projection	7.17	5.00	7.53	1.02	0.72	28.64	1.63	37.81	m	41.59
215mm wide x 215mm projection	11.48	5.00	12.05	1.24	0.89	35.03	2.61	49.69	m	54.66
327mm wide x 112mm projection	10.77	5.00	11.30	1.54	1.08	43.15	2.44	56.90	m	62.59
327mm wide x 215mm projection	15.79	5.00	16.58	1.87	1.34	52.79	3.60	72.97	m	80.26
Projections; bonding to old brickwork; cutting pockets; including extra material; horizontal										
215mm wide x 112mm projection	7.17	5.00	7.53	1.02	0.72	28.64	1.63	37.81	m	41.59
215mm wide x 215mm projection	11.48	5.00	12.05	1.24	0.89	35.03	2.61	49.69	m	54.66
327mm wide x 112mm projection	10.77	5.00	11.30	1.54	1.08	43.15	2.44	56.90	m	62.59
327mm wide x 215mm projection	15.79	5.00	16.58	1.87	1.34	52.79	3.60	72.97	m	80.26
Arches including centering; flat										
112mm high on face; 215mm thick; width of exposed soffit 215mm; bricks-on-edge	4.31	5.00	4.52	1.60	1.05	43.84	0.98	49.34	m	54.27
215mm high on face; 112mm thick; width of exposed soffit 112mm; bricks-on-end	4.31	5.00	4.52	1.55	1.10	43.61	0.98	49.11	m	54.02
Arches including centering; semi-circular										
112mm high on face; 215mm thick; width of exposed soffit 215mm; bricks-on-edge	6.76	5.00	7.10	2.25	1.40	60.60	1.54	69.24	m	76.16
215mm high on face; 215mm thick; width of exposed soffit 215mm; bricks-on-edge	13.52	5.00	14.20	3.10	2.05	85.16	3.08	102.43	m	112.67
Closing cavities										
50mm wide with brickwork 102mm thick	2.10	5.00	2.20	0.44	0.30	12.21	0.48	14.90	m	16.38
75mm wide with brickwork 102mm thick	2.10	5.00	2.20	0.44	0.30	12.21	0.48	14.90	m	16.38
Closing cavities; horizontal										
50mm wide with slates	2.64	5.00	2.77	0.66	0.44	18.18	0.48	21.43	m	23.57
75mm wide with slates	2.64	5.00	2.77	0.66	0.44	18.18	0.48	21.43	m	23.57
50mm wide with brickwork 102mm thick	2.10	5.00	2.20	0.44	0.30	12.21	0.48	14.90	m	16.38
75mm wide with brickwork 102mm thick	2.10	5.00	2.20	0.44	0.30	12.21	0.48	14.90	m	16.38
Bonding ends to existing common brickwork; cutting pockets; extra material										
walls; bonding every third course										
102mm thick	1.43	5.00	1.51	0.41	0.28	11.38	0.33	13.22	m	14.54
215mm thick	2.87	5.00	3.01	0.73	0.48	20.01	0.66	23.69	m	26.06
327mm thick	4.31	5.00	4.52	1.05	0.69	28.78	0.98	34.28	m	37.71
Walls; brickwork; Engineering bricks, BS EN 772, Category F, 215 x 102.5 x 65mm, Class A; in cement-lime mortar (1:2:9)										
Walls										
102mm thick; stretcher bond	44.25	5.00	46.46	1.60	1.20	45.90	3.67	96.04	m²	105.65
215mm thick; English bond	89.25	5.00	93.71	2.60	2.05	75.97	8.58	178.26	m²	196.09
327mm thick; English bond	134.25	5.00	140.96	3.15	2.55	92.95	13.48	247.40	m²	272.14
Skins of hollow walls										
102mm thick; stretcher bond	44.25	5.00	46.46	1.60	1.20	45.90	3.67	96.04	m²	105.65
215mm thick; English bond	89.25	5.00	93.71	2.60	2.05	75.97	8.58	178.26	m²	196.09
Walls; building against concrete (ties measured separately); vertical										
102mm thick; stretcher bond	44.25	5.00	46.46	1.75	1.30	50.04	3.67	100.17	m²	110.19

Labour hourly rates: (except Specialists) Craft Operatives 18.37 Labourer 13.76 Rates are national average prices. Refer to REGIONAL VARIATIONS for indicative levels of overall pricing in regions	MATERIALS			LABOUR				RATES		
	Del to Site	Waste	Material Cost	Craft Optve	Lab	Labour Cost	Sunds	Nett Rate	Unit	Gross rate (10%)
	£	%	£	Hrs	Hrs	£	£	£		£
BRICK/BLOCK WALLING (Cont'd)										
Walls; brickwork; Engineering bricks, BS EN 772, Category F, 215 x 102.5 x 65mm, Class A; in cement-lime mortar (1:2:9) (Cont'd)										
Walls; building against old brickwork; tie new to old with 200mm stainless steel housing type 4 wall ties - 6/m²; vertical										
102mm thick; stretcher bond	44.68	10.00	46.93	1.75	1.30	50.04	3.67	100.65	m²	110.71
215mm thick; English bond	89.68	10.00	94.18	2.90	2.25	84.23	8.58	187.00	m²	205.70
Walls; bonding to stonework; including extra material; vertical										
102mm thick; stretcher bond	46.75	5.00	49.08	1.75	1.30	50.04	3.88	103.01	m²	113.31
215mm thick; English bond	94.25	5.00	98.97	2.90	2.25	84.23	9.07	192.27	m²	211.50
327mm thick; English bond	139.25	5.00	146.22	3.45	2.75	101.22	13.98	261.41	m²	287.55
440mm thick; English bond	183.50	5.00	192.68	4.05	3.30	119.81	18.90	331.38	m²	364.52
Walls; bonding to old brickwork; cutting pockets; including extra material; vertical										
102mm thick; stretcher bond	46.75	5.00	49.08	4.00	2.80	112.01	3.88	164.98	m²	181.48
215mm thick; English bond	94.25	5.00	98.97	5.05	3.65	142.99	9.07	251.03	m²	276.14
Walls; curved 2m radius; including extra material; vertical										
102mm thick; stretcher bond	44.25	5.00	46.46	3.15	2.25	88.83	3.67	138.96	m²	152.86
215mm thick; English bond	89.25	5.00	93.71	5.70	4.10	161.12	8.58	263.42	m²	289.76
327mm thick; English bond	134.25	5.00	140.96	7.80	5.60	220.34	13.48	374.79	m²	412.27
Walls; curved 6m radius; including extra material; vertical										
102mm thick; stretcher bond	44.25	5.00	46.46	2.35	1.70	66.56	3.67	116.70	m²	128.37
215mm thick; English bond	89.25	5.00	93.71	4.10	3.10	117.97	8.58	220.27	m²	242.29
327mm thick; English bond	134.25	5.00	140.96	5.45	4.05	155.84	13.48	310.29	m²	341.32
Skins of hollow walls; curved 2m radius; including extra material; vertical										
102mm thick; stretcher bond	44.25	5.00	46.46	3.15	2.25	88.83	3.67	138.96	m²	152.86
215mm thick; English bond	89.25	5.00	93.71	5.70	4.10	161.12	8.58	263.42	m²	289.76
327mm thick; English bond	134.25	5.00	140.96	7.80	5.60	220.34	13.48	374.79	m²	412.27
Skins of hollow walls; curved 6m radius; including extra material; vertical										
102mm thick; stretcher bond	44.25	5.00	46.46	2.35	1.70	66.56	3.67	116.70	m²	128.37
215mm thick; English bond	89.25	5.00	93.71	4.10	3.10	117.97	8.58	220.27	m²	242.29
Isolated piers										
215mm thick; English bond	89.25	5.00	93.71	5.15	3.75	146.21	8.58	248.50	m²	273.35
327mm thick; English bond	134.25	5.00	140.96	6.25	4.60	178.11	13.48	332.56	m²	365.81
440mm thick; English bond	178.50	5.00	187.42	7.25	5.45	208.17	18.39	413.99	m²	455.39
Chimney stacks										
440mm thick; English bond	178.50	5.00	187.42	7.25	5.45	208.17	18.39	413.99	m²	455.39
890mm thick; English bond	357.00	5.00	374.85	11.45	8.80	331.42	36.78	743.05	m²	817.36
Projections										
215mm wide x 112mm projection	10.00	5.00	10.50	0.33	0.25	9.50	0.98	20.97	m	23.07
215mm wide x 215mm projection	20.00	5.00	20.99	0.55	0.42	15.88	1.97	38.84	m	42.73
327mm wide x 112mm projection	15.00	5.00	15.75	0.50	0.38	14.41	1.47	31.63	m	34.80
327mm wide x 215mm projection	30.00	5.00	31.50	0.83	0.63	23.92	2.94	58.36	m	64.19
Projections; horizontal										
215mm wide x 112mm projection	10.00	5.00	10.50	0.33	0.25	9.50	0.98	20.97	m	23.07
215mm wide x 215mm projection	20.00	5.00	20.99	0.55	0.42	15.88	1.97	38.84	m	42.73
327mm wide x 112mm projection	15.00	5.00	15.75	0.50	0.38	14.41	1.47	31.63	m	34.80
327mm wide x 215mm projection	30.00	5.00	31.50	0.83	0.63	23.92	2.94	58.36	m	64.19
Projections; bonding to old brickwork; cutting pockets; including extra material; vertical										
215mm wide x 112mm projection	16.66	5.00	17.49	1.02	0.72	28.64	1.63	47.77	m	52.55
215mm wide x 215mm projection	26.66	5.00	27.99	1.24	0.89	35.03	2.61	65.62	m	72.19
327mm wide x 112mm projection	25.00	5.00	26.25	1.54	1.08	43.15	2.44	71.84	m	79.03
327mm wide x 215mm projection	36.66	5.00	38.49	1.87	1.34	52.79	3.60	94.88	m	104.37
Projections; bonding to old brickwork; cutting pockets; including extra material; horizontal										
215mm wide x 112mm projection	16.66	5.00	17.49	1.02	0.72	28.64	1.63	47.77	m	52.55
215mm wide x 215mm projection	26.66	5.00	27.99	1.24	0.89	35.03	2.61	65.62	m	72.19
327mm wide x 112mm projection	25.00	5.00	26.25	1.54	1.08	43.15	2.44	71.84	m	79.03
327mm wide x 215mm projection	36.66	5.00	38.49	1.87	1.34	52.79	3.60	94.88	m	104.37
Arches including centering; flat										
112mm high on face; 215mm thick; width of exposed soffit 215mm; bricks-on-edge	10.00	5.00	10.50	1.60	1.05	43.84	0.98	55.31	m	60.84
215mm high on face; 112mm thick; width of exposed soffit 112mm; bricks-on-end	10.00	5.00	10.50	1.55	1.10	43.61	0.98	55.08	m	60.59
Arches including centering; semi-circular										
112mm high on face; 215mm thick; width of exposed soffit 215mm; bricks-on-edge	15.70	5.00	16.48	2.25	1.40	60.60	1.54	78.62	m	86.49
215mm high on face; 215mm thick; width of exposed soffit 215mm; bricks-on-edge	31.40	5.00	32.97	3.10	2.05	85.16	3.08	121.20	m	133.32
Closing cavities										
50mm wide with brickwork 102mm thick	4.88	5.00	5.12	0.44	0.30	12.21	0.48	17.81	m	19.59
75mm wide with brickwork 102mm thick	4.88	5.00	5.12	0.44	0.30	12.21	0.48	17.81	m	19.59
Closing cavities; horizontal										
50mm wide with slates	2.64	5.00	2.77	0.66	0.44	18.18	0.48	21.43	m	23.57
75mm wide with slates	2.64	5.00	2.77	0.66	0.44	18.18	0.48	21.43	m	23.57
50mm wide with brickwork 102mm thick	4.88	5.00	5.12	0.44	0.30	12.21	0.48	17.81	m	19.59
75mm wide with brickwork 102mm thick	4.88	5.00	5.12	0.44	0.30	12.21	0.48	17.81	m	19.59

MASONRY

Labour hourly rates: (except Specialists) Craft Operatives 18.37 Labourer 13.76 Rates are national average prices. Refer to REGIONAL VARIATIONS for indicative levels of overall pricing in regions	MATERIALS			LABOUR				RATES		
	Del to Site	Waste	Material Cost	Craft Optve	Lab	Labour Cost	Sunds	Nett Rate	Unit	Gross rate (10%)
	£	%	£	Hrs	Hrs	£	£	£		£
BRICK/BLOCK WALLING (Cont'd)										
Walls; brickwork; Engineering bricks, BS EN 772, Category F, 215 x 102.5 x 65mm, Class A; in cement-lime mortar (1:2:9) (Cont'd)										
Bonding ends to existing common brickwork; cutting pockets; extra material										
walls; bonding every third course										
102mm thick	3.33	5.00	3.50	0.41	0.28	11.38	0.33	15.21	m	16.73
215mm thick	6.66	5.00	6.99	0.73	0.48	20.01	0.66	27.67	m	30.43
327mm thick	10.00	5.00	10.50	1.05	0.69	28.78	0.98	40.26	m	44.28
For other mortar mixes, ADD or DEDUCT as follows:										
For each half brick thickness										
1:1:6 cement-lime mortar............	-	-	-	-	-	-	0.54	0.54	m²	0.59
1:4 cement mortar............	-	-	-	-	-	-	0.17	0.17	m²	0.18
1:3 cement mortar............	-	-	-	-	-	-	0.22	0.22	m²	0.25
For using Sulphate Resisting cement in lieu of Portland cement, ADD as follows										
For each half brick thickness										
1:2:9 cement-lime mortar............	-	-	-	-	-	-	0.15	0.15	m²	0.17
1:1:6 cement-lime mortar............	-	-	-	-	-	-	0.21	0.21	m²	0.23
1:4 cement mortar............	-	-	-	-	-	-	0.34	0.34	m²	0.38
1:3 cement mortar............	-	-	-	-	-	-	0.44	0.44	m²	0.48
Walls; brickwork; Staffordshire Blue wirecut bricks, BS EN 772, Category F, 215 x 102.5 x 65mm; in cement mortar (1:3)										
Walls										
215mm thick; English bond	92.34	5.00	96.96	2.60	2.05	75.97	9.10	182.04	m²	200.24
Walls; brickwork; Staffordshire Blue pressed bricks, BS EN 772, Category F, 215 x 102.5 x 65mm; in cement mortar (1:3)										
Walls										
215mm thick; English bond	84.29	5.00	88.51	2.60	2.05	75.97	9.10	173.58	m²	190.94
Walls; brickwork; Composite walling of Staffordshire Blue bricks, BS EN 772, Category F, 215 x 102.5 x 65mm; in cement mortar (1:3); wirecut bricks backing; pressed bricks facing; weather struck pointing as work proceeds										
Walls										
English bond; facework one side............	85.55	5.00	89.82	2.90	2.25	84.23	9.10	183.16	m²	201.48
Flemish bond; facework one side............	86.22	5.00	90.53	2.90	2.25	84.23	9.10	183.87	m²	202.26
Walls; brickwork; Common bricks, BS EN 772, Category M, 215 x 102.5 x 65mm, compressive strength 20.5 N/mm²; in cement-lime mortar (1:2:9); flush smooth pointing as work proceeds										
Walls										
stretcher bond; facework one side										
102mm thick	18.64	5.00	19.58	1.75	1.30	50.04	3.67	73.29	m²	80.62
stretcher bond; facework both sides										
102mm thick	18.64	5.00	19.58	2.05	1.50	58.30	3.67	81.55	m²	89.70
English bond; facework one side										
215mm thick	37.60	5.00	39.48	2.65	2.05	76.89	8.58	124.95	m²	137.45
English bond; facework both sides										
215mm thick	37.60	5.00	39.48	2.95	2.25	85.15	8.58	133.22	m²	146.54
Skins of hollow walls										
stretcher bond; facework one side										
102mm thick	18.64	5.00	19.58	1.75	1.30	50.04	3.67	73.29	m²	80.62
Walls; brickwork; Engineering bricks, BS EN 772, Category F, 215 x 102.5 x 65mm, Class B; in cement-lime mortar (1:2:9); flush smooth pointing as work proceeds										
Walls										
stretcher bond; facework one side										
102mm thick	19.06	5.00	20.01	1.90	1.40	54.17	3.67	77.85	m²	85.64
stretcher bond; facework both sides										
102mm thick	19.06	5.00	20.01	2.20	1.60	62.43	3.67	86.11	m²	94.73
English bond; facework one side										
215mm thick	38.44	5.00	40.36	2.90	2.25	84.23	8.58	133.17	m²	146.49
English bond; facework both sides										
215mm thick	38.44	5.00	40.36	3.20	2.45	92.50	8.58	141.43	m²	155.58
Skins of hollow walls										
stretcher bond; facework one side										
102mm thick	19.06	5.00	20.01	1.90	1.40	54.17	3.67	77.85	m²	85.64
Walls; brickwork; Engineering bricks, BS EN 772, Category F, 215 x 102.5 x 65mm, Class A; in cement-lime mortar (1:2:9); flush smooth pointing as work proceeds										
Walls										
stretcher bond; facework one side										
102mm thick	44.25	5.00	46.46	1.90	1.40	54.17	3.67	104.30	m²	114.73
stretcher bond; facework both sides										
102mm thick	44.25	5.00	46.46	2.20	1.60	62.43	3.67	112.57	m²	123.82
English bond; facework one side										
215mm thick	89.25	5.00	93.71	2.90	2.25	84.23	8.58	186.53	m²	205.18

Labour hourly rates: (except Specialists) Craft Operatives 18.37 Labourer 13.76 Rates are national average prices. Refer to REGIONAL VARIATIONS for indicative levels of overall pricing in regions	MATERIALS			LABOUR				RATES		
	Del to Site	Waste	Material Cost	Craft Optve	Lab	Labour Cost	Sunds	Nett Rate		Gross rate (10%)
	£	%	£	Hrs	Hrs	£	£	£	Unit	£
BRICK/BLOCK WALLING (Cont'd)										
Walls; brickwork; Engineering bricks, BS EN 772, Category F, 215 x 102.5 x 65mm, Class A; in cement-lime mortar (1:2:9); flush smooth pointing as work proceeds (Cont'd)										

Walls (Cont'd)
  English bond; facework both sides
    215mm thick
	89.25	5.00	93.71	3.20	2.45	92.50	8.58	194.79	m²	214.27

Skins of hollow walls
  stretcher bond; facework one side
    102mm thick
	44.25	5.00	46.46	1.90	1.40	54.17	3.67	104.30	m²	114.73

Walls; brickwork; Facing bricks, second hard stocks, BS EN 772, Category M, 215 x 102.5 x 65mm; in cement-lime mortar (1:2:9); flush smooth pointing as work proceeds										

Walls
  stretcher bond; facework one side
    102mm thick
	46.91	5.00	49.25	1.75	1.30	50.04	3.67	102.96	m²	113.26

stretcher bond; facework both sides 102mm thick	46.91	5.00	49.25	2.05	1.50	58.30	3.67	111.22	m²	122.35
English bond; facework one side 215mm thick	94.61	5.00	99.34	2.65	2.05	76.89	8.58	184.80	m²	203.28
English bond; facework both sides 215mm thick	94.61	5.00	99.34	2.95	2.25	85.15	8.58	193.07	m²	212.37

Skins of hollow walls
  stretcher bond; facework one side
    102mm thick
	46.91	5.00	49.25	1.75	1.30	50.04	3.67	102.96	m²	113.26

Walls; brickwork; Facing bricks p.c. £200.00 per 1000, 215 x 102.5 x 65mm; in cement-lime mortar (1:2:9); flush smooth pointing as work proceeds										

Walls
  stretcher bond; facework one side
    102mm thick
	11.80	5.00	12.39	1.75	1.30	50.04	3.67	66.10	m²	72.71

stretcher bond; facework both sides 102mm thick	11.80	5.00	12.39	2.05	1.50	58.30	3.67	74.36	m²	81.80
English bond; facework both sides 215mm thick	23.80	5.00	24.99	2.65	2.05	76.89	8.58	110.46	m²	121.50
Flemish bond; facework both sides 215mm thick	23.80	5.00	24.99	2.95	2.25	85.15	8.58	118.72	m²	130.59

Skins of hollow walls
  stretcher bond; facework one side
    102mm thick
	11.80	5.00	12.39	1.75	1.30	50.04	3.67	66.10	m²	72.71

extra; special bricks; vertical angles; squint; BS 4729, type AN.1.1. (p.c. £200.00 per 100)	26.66	5.00	27.99	0.25	0.15	6.66	-	34.65	m	38.11
extra; special bricks; intersections; birdsmouth; BS 4729, type AN.4.1. (p.c. £200.00 per 100)	26.66	5.00	27.99	0.25	0.15	6.66	-	34.65	m	38.11

Walls; building overhand
  stretcher bond; facework one side
    102mm thick
	11.80	5.00	12.39	2.15	1.55	60.82	3.67	76.89	m²	84.58

stretcher bond; facework both sides 102mm thick	11.80	5.00	12.39	2.45	1.75	69.09	3.67	85.15	m²	93.67
English bond; facework both sides 215mm thick	23.80	5.00	24.99	3.05	2.30	87.68	8.58	121.25	m²	133.37
Flemish bond; facework both sides 215mm thick	23.80	5.00	24.99	3.35	2.50	95.94	8.58	129.51	m²	142.46

Skins of hollow walls; building overhand
  stretcher bond; facework one side
    102mm thick
	11.80	5.00	12.39	2.15	1.55	60.82	3.67	76.89	m²	84.58

Arches including centering; flat 112mm high on face; 215mm thick; width of exposed soffit 112mm; bricks-on-edge	2.67	5.00	2.80	1.05	0.80	30.30	0.98	34.07	m	37.48
112mm high on face; 215mm thick; width of exposed soffit 215mm; bricks-on-edge	2.67	5.00	2.80	1.50	1.00	41.32	0.98	45.09	m	49.60
215mm high on face; 112mm thick; width of exposed soffit 112mm; bricks-on-end	2.67	5.00	2.80	1.45	1.00	40.40	0.98	44.17	m	48.59
Arches including centering; segmental 215mm high on face; 215mm thick; width of exposed soffit 112mm; bricks-on-edge	5.33	5.00	5.60	2.40	1.65	66.79	1.97	74.36	m	81.79
215mm high on face; 215mm thick; width of exposed soffit 215mm; bricks-on-edge	5.33	5.00	5.60	2.95	1.95	81.02	1.97	88.59	m	97.45
Arches including centering; semi-circular 215mm high on face; 215mm thick; width of exposed soffit 112mm; bricks-on-edge	5.33	5.00	5.60	2.40	1.65	66.79	1.97	74.36	m	81.79
215mm high on face; 215mm thick; width of exposed soffit 215mm; bricks-on-edge	5.33	5.00	5.60	2.95	1.95	81.02	1.97	88.59	m	97.45
Facework copings; bricks-on-edge; pointing top and each side 215 x 102.5mm; horizontal	2.67	5.00	2.80	0.70	0.50	19.74	0.98	23.51	m	25.86
215 x 102.5mm; with oversailing course and cement fillets both sides; horizontal	2.67	5.00	2.80	1.10	0.75	30.53	0.98	34.30	m	37.73
327 x 102.5mm; horizontal	4.00	5.00	4.20	1.05	0.75	29.61	1.47	35.28	m	38.81
327 x 102.5mm; with oversailing course and cement fillets both sides; horizontal	4.00	5.00	4.20	1.50	1.05	42.00	1.47	47.67	m	52.44
extra; galvanized iron coping cramp at angle or end	2.70	5.00	2.84	0.04	0.03	1.15	-	3.98	nr	4.38

MASONRY

Labour hourly rates: (except Specialists) Craft Operatives 18.37 Labourer 13.76 Rates are national average prices. Refer to REGIONAL VARIATIONS for indicative levels of overall pricing in regions	MATERIALS			LABOUR				RATES		
	Del to Site	Waste	Material Cost	Craft Optve	Lab	Labour Cost	Sunds	Nett Rate	Unit	Gross rate (10%)
	£	%	£	Hrs	Hrs	£	£	£		£
BRICK/BLOCK WALLING (Cont'd)										
Walls; brickwork; Facing bricks p.c. £250.00 per 1000, 215 x 102.5 x 65mm; in cement-lime mortar (1:2:9); flush smooth pointing as work proceeds										
Walls										
stretcher bond; facework one side										
102mm thick	14.75	5.00	15.49	1.75	1.30	50.04	3.67	69.20	m²	76.12
stretcher bond; facework both sides										
102mm thick	14.75	5.00	15.49	2.05	1.50	58.30	3.67	77.46	m²	85.21
English bond; facework both sides										
215mm thick	29.75	5.00	31.24	2.65	2.05	76.89	8.58	116.71	m²	128.38
Flemish bond; facework both sides										
215mm thick	29.75	5.00	31.24	2.95	2.25	85.15	8.58	124.97	m²	137.47
Skins of hollow walls										
stretcher bond; facework one side										
102mm thick	14.75	5.00	15.49	1.75	1.30	50.04	3.67	69.20	m²	76.12
extra; special bricks; vertical angles; squint; BS 4729, type AN.1.1. (p.c. £250.00 per 100)...............	33.33	5.00	34.99	0.25	0.15	6.66	-	41.65	m	45.81
extra; special bricks; intersections; birdsmouth; BS 4729, type AN.4.1. (p.c. £250.00 per 100)...............	33.33	5.00	34.99	0.25	0.15	6.66	-	41.65	m	45.81
Walls; building overhand										
stretcher bond; facework one side										
102mm thick	14.75	5.00	15.49	2.15	1.55	60.82	3.67	79.99	m²	87.98
stretcher bond; facework both sides										
102mm thick	14.75	5.00	15.49	2.45	1.75	69.09	3.67	88.25	m²	97.07
English bond; facework both sides										
215mm thick	29.75	5.00	31.24	3.05	2.30	87.68	8.58	127.49	m²	140.24
Flemish bond; facework both sides										
215mm thick	29.75	5.00	31.24	3.35	2.50	95.94	8.58	135.76	m²	149.33
Skins of hollow walls										
stretcher bond; facework one side										
102mm thick	14.75	5.00	15.49	2.15	1.55	60.82	3.67	79.99	m²	87.98
Arches including centering; flat										
112mm high on face; 215mm thick; width of exposed soffit 112mm; bricks-on-edge	3.33	5.00	3.50	1.05	0.80	30.30	0.98	34.77	m	38.25
112mm high on face; 215mm thick; width of exposed soffit 215mm; bricks-on-edge	3.33	5.00	3.50	1.50	1.00	41.32	0.98	45.79	m	50.37
215mm high on face; 112mm thick; width of exposed soffit 112mm; bricks-on-end	3.33	5.00	3.50	1.45	1.00	40.40	0.98	44.87	m	49.36
Arches including centering; segmental										
215mm high on face; 215mm thick; width of exposed soffit 112mm; bricks-on-edge	6.66	5.00	7.00	2.40	1.65	66.79	1.97	75.76	m	83.33
215mm high on face; 215mm thick; width of exposed soffit 215mm; bricks-on-edge	6.66	5.00	7.00	2.95	1.95	81.02	1.97	89.99	m	98.99
Arches including centering; semi-circular										
215mm high on face; 215mm thick; width of exposed soffit 112mm; bricks-on-edge	6.66	5.00	7.00	2.40	1.65	66.79	1.97	75.76	m	83.33
215mm high on face; 215mm thick; width of exposed soffit 215mm; bricks-on-edge	6.66	5.00	7.00	2.95	1.95	81.02	1.97	89.99	m	98.99
Facework copings; bricks-on-edge; pointing top and each side										
215 x 102.5mm; horizontal	3.33	5.00	3.50	0.70	0.50	19.74	0.98	24.21	m	26.63
215 x 102.5mm; with oversailing course and cement fillets both sides; horizontal	3.33	5.00	3.50	1.10	0.75	30.53	0.98	35.00	m	38.50
327 x 102.5mm; horizontal	5.00	5.00	5.25	1.05	0.75	29.61	1.47	36.33	m	39.96
327 x 102.5mm; with oversailing course and cement fillets both sides; horizontal	5.00	5.00	5.25	1.50	1.05	42.00	1.47	48.72	m	53.60
extra; galvanized iron coping cramp at angle or end	2.70	5.00	2.84	0.04	0.03	1.15	-	3.98	nr	4.38
Walls; brickwork; Facing bricks p.c. £300.00 per 1000, 215 x 102.5 x 65mm; in cement-lime mortar (1:2:9); flush smooth pointing as work proceeds										
Walls										
stretcher bond; facework one side										
102mm thick	17.70	5.00	18.58	1.75	1.30	50.04	3.67	72.30	m²	79.53
stretcher bond; facework both sides										
102mm thick	17.70	5.00	18.58	2.05	1.50	58.30	3.67	80.56	m²	88.61
English bond; facework both sides										
215mm thick	35.70	5.00	37.49	2.65	2.05	76.89	8.58	122.95	m²	135.25
Flemish bond; facework both sides										
215mm thick	35.70	5.00	37.49	2.95	2.25	85.15	8.58	131.22	m²	144.34
Skins of hollow walls										
stretcher bond; facework one side										
102mm thick	17.70	5.00	18.58	1.75	1.30	50.04	3.67	72.30	m²	79.53
extra; special bricks; vertical angles; squint; BS 4729, type AN.1.1. (p.c. £300.00 per 100)...............	39.99	5.00	41.99	0.25	0.15	6.66	-	48.65	m	53.51
extra; special bricks; intersections; birdsmouth; BS 4729, type AN.4.1. (p.c. £300.00 per 100)...............	39.99	5.00	41.99	0.25	0.15	6.66	-	48.65	m	53.51
Walls; building overhand										
stretcher bond; facework one side										
102mm thick	17.70	5.00	18.58	2.15	1.55	60.82	3.67	83.08	m²	91.39
stretcher bond; facework both sides										
102mm thick	17.70	5.00	18.58	2.45	1.75	69.09	3.67	91.35	m²	100.48
English bond; facework both sides										
215mm thick	35.70	5.00	37.49	3.05	2.30	87.68	8.58	133.74	m²	147.12
Flemish bond; facework both sides										
215mm thick	35.70	5.00	37.49	3.35	2.50	95.94	8.58	142.00	m²	156.20

Labour hourly rates: (except Specialists) Craft Operatives 18.37 Labourer 13.76 Rates are national average prices. Refer to REGIONAL VARIATIONS for indicative levels of overall pricing in regions	MATERIALS			LABOUR				RATES		
	Del to Site £	Waste %	Material Cost £	Craft Optve Hrs	Lab Hrs	Labour Cost £	Sunds £	Nett Rate £	Unit	Gross rate (10%) £
BRICK/BLOCK WALLING (Cont'd)										
Walls; brickwork; Facing bricks p.c. £300.00 per 1000, 215 x 102.5 x 65mm; in cement-lime mortar (1:2:9); flush smooth pointing as work proceeds (Cont'd)										
Skins of hollow walls										
stretcher bond; facework one side										
102mm thick	17.70	5.00	18.58	2.15	1.55	60.82	3.67	83.08	m²	91.39
Arches including centering; flat										
112mm high on face; 215mm thick; width of exposed soffit										
112mm; bricks-on-edge	4.00	5.00	4.20	1.05	0.80	30.30	0.98	35.47	m	39.02
112mm high on face; 215mm thick; width of exposed soffit										
215mm; bricks-on-edge	4.00	5.00	4.20	1.50	1.00	41.32	0.98	46.49	m	51.14
215mm high on face; 112mm thick; width of exposed soffit										
112mm; bricks-on-end	4.00	5.00	4.20	1.45	1.00	40.40	0.98	45.57	m	50.13
Arches including centering; segmental										
215mm high on face; 215mm thick; width of exposed soffit										
112mm; bricks-on-edge	8.00	5.00	8.40	2.40	1.65	66.79	1.97	77.15	m	84.87
215mm high on face; 215mm thick; width of exposed soffit										
215mm; bricks-on-edge	8.00	5.00	8.40	2.95	1.95	81.02	1.97	91.39	m	100.53
Arches including centering; semi-circular										
215mm high on face; 215mm thick; width of exposed soffit										
112mm; bricks-on-edge	8.00	5.00	8.40	2.40	1.65	66.79	1.97	77.15	m	84.87
215mm high on face; 215mm thick; width of exposed soffit										
215mm; bricks-on-edge	8.00	5.00	8.40	2.95	1.95	81.02	1.97	91.39	m	100.53
Facework copings; bricks-on-edge; pointing top and each side										
215 x 102.5mm; horizontal	4.00	5.00	4.20	0.70	0.50	19.74	0.98	24.91	m	27.40
215 x 102.5mm; with oversailing course and cement fillets both sides; horizontal	4.00	5.00	4.20	1.10	0.75	30.53	0.98	35.70	m	39.27
327 x 102.5mm; horizontal	6.00	5.00	6.30	1.05	0.75	29.61	1.47	37.38	m	41.12
327 x 102.5mm; with oversailing course and cement fillets both sides; horizontal	6.00	5.00	6.30	1.50	1.05	42.00	1.47	49.77	m	54.75
extra; galvanized iron coping cramp at angle or end	2.70	5.00	2.84	0.04	0.03	1.15	-	3.98	nr	4.38
Walls; brickwork; Facing bricks p.c. £350.00 per 1000, 215 x 102.5 x 65mm; in cement-lime mortar (1:2:9); flush smooth pointing as work proceeds										
Walls										
stretcher bond; facework one side										
102mm thick	20.65	5.00	21.68	1.75	1.30	50.04	3.67	75.39	m²	82.93
stretcher bond; facework both sides										
102mm thick	20.65	5.00	21.68	2.05	1.50	58.30	3.67	83.66	m²	92.02
English bond; facework both sides										
215mm thick	41.65	5.00	43.73	2.65	2.05	76.89	8.58	129.20	m²	142.12
Flemish bond; facework both sides										
215mm thick	41.65	5.00	43.73	2.95	2.25	85.15	8.58	137.46	m²	151.21
Skins of hollow walls										
stretcher bond; facework one side										
102mm thick	20.65	5.00	21.68	1.75	1.30	50.04	3.67	75.39	m²	82.93
extra; special bricks; vertical angles; squint; BS 4729, type AN.1.1. (p.c. £350.00 per 100)	46.65	5.00	48.99	0.25	0.15	6.66	-	55.64	m	61.21
extra; special bricks; intersections; birdsmouth; BS 4729, type AN.4.1. (p.c. £350.00 per 100)	46.65	5.00	48.99	0.25	0.15	6.66	-	55.64	m	61.21
Walls; building overhand										
stretcher bond; facework one side										
102mm thick	20.65	5.00	21.68	2.15	1.55	60.82	3.67	86.18	m²	94.80
stretcher bond; facework both sides										
102mm thick	20.65	5.00	21.68	2.45	1.75	69.09	3.67	94.44	m²	103.89
English bond; facework both sides										
215mm thick	41.65	5.00	43.73	3.05	2.30	87.68	8.58	139.99	m²	153.99
Flemish bond; facework both sides										
215mm thick	41.65	5.00	43.73	3.35	2.50	95.94	8.58	148.25	m²	163.08
Arches including centering; flat										
112mm high on face; 215mm thick; width of exposed soffit										
112mm; bricks-on-edge	4.67	5.00	4.90	1.05	0.80	30.30	0.98	36.17	m	39.79
112mm high on face; 215mm thick; width of exposed soffit										
215mm; bricks-on-edge	4.67	5.00	4.90	1.50	1.00	41.32	0.98	47.19	m	51.91
215mm high on face; 112mm thick; width of exposed soffit										
112mm; bricks-on-end	4.67	5.00	4.90	1.45	1.00	40.40	0.98	46.27	m	50.90
Arches including centering; segmental										
215mm high on face; 215mm thick; width of exposed soffit										
112mm; bricks-on-edge	9.33	5.00	9.80	2.40	1.65	66.79	1.97	78.55	m	86.41
215mm high on face; 215mm thick; width of exposed soffit										
215mm; bricks-on-edge	9.33	5.00	9.80	2.95	1.95	81.02	1.97	92.79	m	102.06
Arches including centering; semi-circular										
215mm high on face; 215mm thick; width of exposed soffit										
112mm; bricks-on-edge	9.33	5.00	9.80	2.40	1.65	66.79	1.97	78.55	m	86.41
215mm high on face; 215mm thick; width of exposed soffit										
215mm; bricks-on-edge	9.33	5.00	9.80	2.95	1.95	81.02	1.97	92.79	m	102.06
Facework copings; bricks-on-edge; pointing top and each side										
215 x 102.5mm; horizontal	4.67	5.00	4.90	0.70	0.50	19.74	0.98	25.61	m	28.17
215 x 102.5mm; with oversailing course and cement fillets both sides; horizontal	4.67	5.00	4.90	1.10	0.75	30.53	0.98	36.40	m	40.04
327 x 102.5mm; horizontal	7.00	5.00	7.35	1.05	0.75	29.61	1.47	38.43	m	42.27
327 x 102.5mm; with oversailing course and cement fillets both sides; horizontal	7.00	5.00	7.35	1.50	1.05	42.00	1.47	50.82	m	55.91

MASONRY

	MATERIALS			LABOUR				RATES		
Labour hourly rates: (except Specialists) Craft Operatives 18.37 Labourer 13.76 Rates are national average prices. Refer to REGIONAL VARIATIONS for indicative levels of overall pricing in regions	Del to Site	Waste	Material Cost	Craft Optve	Lab	Labour Cost	Sunds	Nett Rate	Unit	Gross rate (10%)
	£	%	£	Hrs	Hrs	£	£	£		£
BRICK/BLOCK WALLING (Cont'd)										
Walls; brickwork; Facing bricks p.c. £350.00 per 1000, 215 x 102.5 x 65mm; in cement-lime mortar (1:2:9); flush smooth pointing as work proceeds (Cont'd)										
Facework copings; bricks-on-edge; pointing top and each side (Cont'd)										
extra; galvanized iron coping cramp at angle or end	2.70	5.00	2.84	0.04	0.03	1.15	-	3.98	nr	4.38
Facing bricks p.c. £400.00 per 1000, 215 x 102.5 x 65mm; in cement-lime mortar (1:2:9); flush smooth pointing as work proceeds										
Walls										
stretcher bond; facework one side										
102mm thick	23.60	5.00	24.78	1.75	1.30	50.04	3.67	78.49	m²	86.34
stretcher bond; facework both sides										
102mm thick	23.60	5.00	24.78	2.05	1.50	58.30	3.67	86.75	m²	95.43
English bond; facework both sides										
215mm thick	47.60	5.00	49.98	2.65	2.05	76.89	8.58	135.45	m²	148.99
Flemish bond; facework both sides										
215mm thick	47.60	5.00	49.98	2.95	2.25	85.15	8.58	143.71	m²	158.08
Skins of hollow walls										
stretcher bond; facework one side										
102mm thick	23.60	5.00	24.78	1.75	1.30	50.04	3.67	78.49	m²	86.34
extra; special bricks; vertical angles; squint; BS 4729, type AN.1.1. (p.c. £400.00 per 100)	53.32	5.00	55.99	0.25	0.15	6.66	-	62.64	m	68.91
extra; special bricks; intersections; birdsmouth; BS 4729, type AN.4.1. (p.c. £400.00 per 100)	53.32	5.00	55.99	0.25	0.15	6.66	-	62.64	m	68.91
Walls; building overhand										
stretcher bond; facework one side										
102mm thick	23.60	5.00	24.78	2.15	1.55	60.82	3.67	89.28	m²	98.21
stretcher bond; facework both sides										
102mm thick	23.60	5.00	24.78	2.45	1.75	69.09	3.67	97.54	m²	107.30
English bond; facework both sides										
215mm thick	47.60	5.00	49.98	3.05	2.30	87.68	8.58	146.24	m²	160.86
Flemish bond; facework both sides										
215mm thick	47.60	5.00	49.98	3.35	2.50	95.94	8.58	154.50	m²	169.95
Skins of hollow walls; building overhand										
stretcher bond; facework one side										
102mm thick	23.60	5.00	24.78	2.15	1.55	60.82	3.67	89.28	m²	98.21
Arches including centering; flat										
112mm high on face; 215mm thick; width of exposed soffit 112mm; bricks-on-edge	5.33	5.00	5.60	1.05	0.80	30.30	0.98	36.87	m	40.56
112mm high on face; 215mm thick; width of exposed soffit 215mm; bricks-on-edge	5.33	5.00	5.60	1.50	1.00	41.32	0.98	47.89	m	52.68
215mm high on face; 112mm thick; width of exposed soffit 112mm; bricks-on-end	5.33	5.00	5.60	1.45	1.00	40.40	0.98	46.97	m	51.67
Arches including centering; segmental										
215mm high on face; 215mm thick; width of exposed soffit 112mm; bricks-on-edge	10.66	5.00	11.20	2.40	1.65	66.79	1.97	79.95	m	87.95
215mm high on face; 215mm thick; width of exposed soffit 215mm; bricks-on-edge	10.66	5.00	11.20	2.95	1.95	81.02	1.97	94.19	m	103.60
Arches including centering; semi-circular										
215mm high on face; 215mm thick; width of exposed soffit 112mm; bricks-on-edge	10.66	5.00	11.20	2.40	1.65	66.79	1.97	79.95	m	87.95
215mm high on face; 215mm thick; width of exposed soffit 215mm; bricks-on-edge	10.66	5.00	11.20	2.95	1.95	81.02	1.97	94.19	m	103.60
Facework copings; bricks-on-edge; pointing top and each side										
215 x 102.5mm; horizontal	5.33	5.00	5.60	0.70	0.50	19.74	0.98	26.31	m	28.94
215 x 102.5mm; with oversailing course and cement fillets both sides; horizontal	5.33	5.00	5.60	1.10	0.75	30.53	0.98	37.10	m	40.81
327 x 102.5mm; horizontal	8.00	5.00	8.40	1.05	0.75	29.61	1.47	39.48	m	43.43
327 x 102.5mm; with oversailing course and cement fillets both sides; horizontal	8.00	5.00	8.40	1.50	1.05	42.00	1.47	51.87	m	57.06
extra; galvanized iron coping cramp at angle or end	2.70	5.00	2.84	0.04	0.03	1.15	-	3.98	nr	4.38
Walls; brickwork; Composite walling of bricks 215 x 102.5x 65mm; in cement-lime mortar (1:2:9); common bricks BS EN 772 Category M backing, compressive strength 20.5 N/mm²; facing bricks p.c. £200.00 per 1000; flush smooth pointing as work proceeds										
Walls										
English bond; facework one side										
215mm thick	26.96	5.00	28.31	2.65	2.05	76.89	8.58	113.78	m²	125.16
327mm thick	45.61	5.00	47.89	3.15	2.50	92.27	13.48	153.64	m²	169.00
Flemish bond; facework one side										
215mm thick	28.12	5.00	29.53	2.65	2.05	76.89	8.58	115.00	m²	126.50
327mm thick	46.77	5.00	49.11	3.15	2.50	92.27	13.48	154.86	m²	170.34
Walls; building overhand										
English bond; facework one side										
215mm thick	26.96	5.00	28.31	3.05	2.35	88.36	8.58	125.26	m²	137.78
327mm thick	45.61	5.00	47.89	3.55	2.80	103.74	13.48	165.12	m²	181.63
Flemish bond; facework one side										
215mm thick	28.12	5.00	29.53	3.05	2.35	88.36	8.58	126.47	m²	139.12
327mm thick	46.77	5.00	49.11	3.55	2.80	103.74	13.48	166.33	m²	182.97

Labour hourly rates: (except Specialists) Craft Operatives 18.37 Labourer 13.76 Rates are national average prices. Refer to REGIONAL VARIATIONS for indicative levels of overall pricing in regions	MATERIALS			LABOUR				RATES		
	Del to Site	Waste	Material Cost	Craft Optve	Lab	Labour Cost	Sunds	Nett Rate		Gross rate (10%)
	£	%	£	Hrs	Hrs	£	£	£	Unit	£
BRICK/BLOCK WALLING (Cont'd)										
Walls; brickwork; Composite walling of bricks 215 x 102.5 x 65mm; in cement-lime mortar (1:2:9); common bricks BS EN 772 Category M backing, compressive strength 20.5 N/mm²; facing bricks p.c. £250.00 per 1000; flush smooth pointing as work proceeds										
Walls										
English bond; facework one side										
215mm thick	31.41	5.00	32.98	2.65	2.05	76.89	8.58	118.45	m²	130.30
327mm thick	50.06	5.00	52.56	3.15	2.50	92.27	13.48	158.31	m²	174.14
Flemish bond; facework one side										
215mm thick	32.07	5.00	33.68	2.65	2.05	76.89	8.58	119.15	m²	131.06
327mm thick	50.72	5.00	53.25	3.15	2.50	92.27	13.48	159.00	m²	174.91
Walls; building overhand										
English bond; facework one side										
215mm thick	31.41	5.00	32.98	3.05	2.35	88.36	8.58	129.93	m²	142.92
327mm thick	50.06	5.00	52.56	3.55	2.80	103.74	13.48	169.79	m²	186.77
Flemish bond; facework one side										
215mm thick	32.07	5.00	33.68	3.05	2.35	88.36	8.58	130.62	m²	143.68
327mm thick	50.72	5.00	53.25	3.55	2.80	103.74	13.48	170.48	m²	187.53
Walls; brickwork; Composite walling of bricks 215 x 102.5 x 65mm; in cement-lime mortar (1:2:9); common bricks BS EN 772 Category M backing, compressive strength 20.5 N/mm²; facing bricks p.c. £300.00 per 1000; flush smooth pointing as work proceeds										
Walls										
English bond; facework one side										
215mm thick	35.86	5.00	37.66	2.65	2.05	76.89	8.58	123.13	m²	135.44
327mm thick	54.51	5.00	57.23	3.15	2.50	92.27	13.48	162.98	m²	179.28
Flemish bond; facework one side										
215mm thick	36.02	5.00	37.83	2.65	2.05	76.89	8.58	123.29	m²	135.62
327mm thick	54.67	5.00	57.40	3.15	2.50	92.27	13.48	163.15	m²	179.47
Walls; building overhand										
English bond; facework one side										
215mm thick	35.86	5.00	37.66	3.05	2.35	88.36	8.58	134.60	m²	148.06
327mm thick	54.51	5.00	57.23	3.55	2.80	103.74	13.48	174.46	m²	191.91
Flemish bond; facework one side										
215mm thick	36.02	5.00	37.83	3.05	2.35	88.36	8.58	134.77	m²	148.25
327mm thick	54.67	5.00	57.40	3.55	2.80	103.74	13.48	174.63	m²	192.09
Walls; brickwork; Composite walling of bricks 215 x 102.5 x 65mm; in cement-lime mortar (1:2:9); common bricks BS EN 772 Category M backing, compressive strength 20.5 N/mm²; facing bricks p.c. £350.00 per 1000; flush smooth pointing as work proceeds										
Walls										
English bond; facework one side										
215mm thick	40.31	5.00	42.33	2.65	2.05	76.89	8.58	127.80	m²	140.58
327mm thick	58.96	5.00	61.91	3.15	2.50	92.27	13.48	167.66	m²	184.42
Flemish bond; facework one side										
215mm thick	39.97	5.00	41.97	2.65	2.05	76.89	8.58	127.44	m²	140.19
327mm thick	58.62	5.00	61.55	3.15	2.50	92.27	13.48	167.30	m²	184.03
Walls; building overhand										
English bond; facework one side										
215mm thick	40.31	5.00	42.33	3.05	2.35	88.36	8.58	139.27	m²	153.20
327mm thick	58.96	5.00	61.91	3.55	2.80	103.74	13.48	179.13	m²	197.05
Flemish bond; facework one side										
215mm thick	39.97	5.00	41.97	3.05	2.35	88.36	8.58	138.92	m²	152.81
327mm thick	58.62	5.00	61.55	3.55	2.80	103.74	13.48	178.78	m²	196.65
Composite walling of bricks 215 x 102.5x 65mm; in cement-lime mortar (1:2:9); common bricks BS EN 772 Category M backing, compressive strength 20.5 N/mm²; facing bricks p.c. £400.00 per 1000; flush smooth pointing as work proceeds										
Walls										
English bond; facework one side										
215mm thick	44.76	5.00	47.00	2.65	2.05	76.89	8.58	132.47	m²	145.72
327mm thick	63.41	5.00	66.58	3.15	2.50	92.27	13.48	172.33	m²	189.56
Flemish bond; facework one side										
215mm thick	43.92	5.00	46.12	2.65	2.05	76.89	8.58	131.59	m²	144.75
327mm thick	62.57	5.00	65.70	3.15	2.50	92.27	13.48	171.45	m²	188.59
Walls; building overhand										
English bond; facework one side										
215mm thick	44.76	5.00	47.00	3.05	2.35	88.36	8.58	143.95	m²	158.34
327mm thick	63.41	5.00	66.58	3.55	2.80	103.74	13.48	183.81	m²	202.19
Flemish bond; facework one side										
215mm thick	43.92	5.00	46.12	3.05	2.35	88.36	8.58	143.06	m²	157.37
327mm thick	62.57	5.00	65.70	3.55	2.80	103.74	13.48	182.92	m²	201.22
For each £10.00 difference in cost of 1000 facing bricks, add or deduct the following										
Wall; in facing bricks										
102mm thick; English bond; facework one side.............	0.90	5.00	0.94	-	-	-	-	0.94	m²	1.04
215mm thick; facework both sides.........................	1.20	5.00	1.26	-	-	-	-	1.26	m²	1.39
Composite wall; common brick backing; facing bricks facework										
215mm thick; English bond; facework one side.............	0.90	5.00	0.94	-	-	-	-	0.94	m²	1.04

MASONRY

Labour hourly rates: (except Specialists) Craft Operatives 18.37 Labourer 13.76 Rates are national average prices. Refer to REGIONAL VARIATIONS for indicative levels of overall pricing in regions	MATERIALS			LABOUR				RATES		
	Del to Site	Waste	Material Cost	Craft Optve	Lab	Labour Cost	Sunds	Nett Rate		Gross rate (10%)
	£	%	£	Hrs	Hrs	£	£	£	Unit	£
BRICK/BLOCK WALLING (Cont'd)										
For each £10.00 difference in cost of 1000 facing bricks, add or deduct the following (Cont'd)										
Composite wall; common brick backing; facing bricks facework (Cont'd)										
215mm thick; Flemish bond; facework one side	0.80	5.00	0.84	-	-	-	-	0.84	m²	0.92
Extra over flush smooth pointing for the following types of pointing										
Pointing; as work proceeds										
ironing in joints	-	-	-	0.03	0.02	0.83	-	0.83	m²	0.91
weathered pointing	-	-	-	0.03	0.02	0.83	-	0.83	m²	0.91
recessed pointing	-	-	-	0.04	0.03	1.15	-	1.15	m²	1.26
Raking out joints and pointing; on completion										
flush smooth pointing with cement-lime mortar (1:2:9)	-	-	-	0.75	0.50	20.66	0.24	20.90	m²	22.99
flush smooth pointing with coloured cement-lime mortar (1:2:9)	-	-	-	0.75	0.50	20.66	0.33	20.99	m²	23.09
weathered pointing with cement-lime mortar (1:2:9)	-	-	-	0.83	0.55	22.82	0.24	23.06	m²	25.36
weathered pointing with coloured cement-lime mortar (1:2:9)	-	-	-	0.83	0.55	22.82	0.33	23.15	m²	25.46
Extra over cement-lime mortar (1:2:9) for bedding and pointing for using coloured cement-lime mortar (1:2:9) for the following										
Wall										
102mm thick; pointing both sides	-	-	-	-	-	-	1.25	1.25	m²	1.37
215mm thick; pointing both sides	-	-	-	-	-	-	2.89	2.89	m²	3.18
Walls; Blockwork; Aerated concrete blocks, Thermalite, 440 x 215mm, Shield blocks (4.0 N/mm²); in cement-lime mortar (1:1:6)										
Walls										
75mm thick; stretcher bond	10.70	5.00	11.23	0.60	0.50	17.90	1.12	30.26	m²	33.29
90mm thick; stretcher bond	11.00	5.00	11.55	0.80	0.65	23.64	1.41	36.60	m²	40.26
100mm thick; stretcher bond	12.44	5.00	13.06	0.80	0.65	23.64	1.41	38.11	m²	41.92
140mm thick; stretcher bond	17.42	5.00	18.29	0.90	0.75	26.85	2.12	47.26	m²	51.98
190mm thick; stretcher bond	24.00	5.00	25.20	0.95	0.80	28.46	2.81	56.46	m²	62.11
Skins of hollow walls										
75mm thick; stretcher bond	10.70	5.00	11.23	0.60	0.50	17.90	1.12	30.26	m²	33.29
90mm thick; stretcher bond	11.00	5.00	11.55	0.80	0.65	23.64	1.41	36.60	m²	40.26
100mm thick; stretcher bond	12.44	5.00	13.06	0.80	0.65	23.64	1.41	38.11	m²	41.92
140mm thick; stretcher bond	17.42	5.00	18.29	0.90	0.75	26.85	2.12	47.26	m²	51.98
190mm thick; stretcher bond	24.00	5.00	25.20	0.95	0.80	28.46	2.81	56.46	m²	62.11
Closing cavities; vertical										
50mm wide with blockwork 100mm thick	3.11	5.00	3.27	0.20	0.10	5.05	0.30	8.62	m	9.48
75mm wide with blockwork 100mm thick	3.11	5.00	3.27	0.20	0.10	5.05	0.30	8.62	m	9.48
Closing cavities; horizontal										
50mm wide with blockwork 100mm thick	3.11	5.00	3.27	0.20	0.10	5.05	0.30	8.62	m	9.48
75mm wide with blockwork 100mm thick	3.11	5.00	3.27	0.20	0.10	5.05	0.30	8.62	m	9.48
Bonding ends to common brickwork; forming pockets; extra material										
walls; bonding every third course										
75mm	2.67	5.00	2.81	0.15	0.10	4.13	-	6.94	m	7.63
90mm	2.75	5.00	2.89	0.25	0.20	7.34	-	10.23	m	11.26
100mm	3.11	5.00	3.27	0.25	0.20	7.34	-	10.61	m	11.67
140mm	4.36	5.00	4.57	0.35	0.25	9.87	-	14.44	m	15.89
190mm	6.00	5.00	6.30	0.40	0.30	11.48	-	17.78	m	19.55
Bonding ends to existing common brickwork; cutting pockets; extra material										
walls; bonding every third course										
75mm	2.67	5.00	2.81	0.30	0.20	8.26	-	11.07	m	12.18
90mm	2.75	5.00	2.89	0.50	0.40	14.69	-	17.58	m	19.33
100mm	3.11	5.00	3.27	0.50	0.40	14.69	-	17.95	m	19.75
140mm	4.36	5.00	4.57	0.70	0.50	19.74	-	24.31	m	26.74
190mm	6.00	5.00	6.30	0.80	0.60	22.95	-	29.25	m	32.18
Bonding ends to existing concrete blockwork; cutting pockets; extra material										
walls; bonding every third course										
75mm	2.67	5.00	2.81	0.25	0.20	7.34	-	10.15	m	11.17
90mm	2.75	5.00	2.89	0.35	0.30	10.56	-	13.45	m	14.79
100mm	3.11	5.00	3.27	0.40	0.30	11.48	-	14.74	m	16.22
140mm	4.36	5.00	4.57	0.55	0.40	15.61	-	20.18	m	22.20
190mm	6.00	5.00	6.30	0.65	0.50	18.82	-	25.12	m	27.63
Walls; Blockwork; Aerated concrete blocks, Thermalite, 440 x 215mm, Turbo blocks (2.8 N/mm²); in cement-lime mortar (1:1:6)										
Walls										
100mm thick; stretcher bond	14.89	5.00	15.63	0.80	0.65	23.64	1.41	40.68	m²	44.75
115mm thick; stretcher bond	15.10	5.00	15.85	0.80	0.65	23.64	1.41	40.90	m²	45.00
125mm thick; stretcher bond	16.78	5.00	17.61	0.85	0.70	25.25	2.12	44.98	m²	49.47
130mm thick; stretcher bond	17.36	5.00	18.23	0.85	0.70	25.25	2.12	45.59	m²	50.15
150mm thick; stretcher bond	20.62	5.00	21.65	0.90	0.75	26.85	2.12	50.62	m²	55.68
190mm thick; stretcher bond	26.79	5.00	28.13	0.95	0.80	28.46	2.81	59.39	m²	65.33
200mm thick; stretcher bond	29.77	5.00	31.26	1.00	0.85	30.07	2.81	64.13	m²	70.54
215mm thick; stretcher bond	32.03	5.00	33.63	1.00	0.85	30.07	3.51	67.21	m²	73.93

Labour hourly rates: (except Specialists) Craft Operatives 18.37 Labourer 13.76 Rates are national average prices. Refer to REGIONAL VARIATIONS for indicative levels of overall pricing in regions	MATERIALS			LABOUR				RATES		
	Del to Site £	Waste %	Material Cost £	Craft Optve Hrs	Lab Hrs	Labour Cost £	Sunds £	Nett Rate £	Unit	Gross rate (10%) £
BRICK/BLOCK WALLING (Cont'd)										
Walls; Blockwork; Aerated concrete blocks, Thermalite, 440 x 215mm, Turbo blocks (2.8 N/mm²); in cement-lime mortar (1:1:6) (Cont'd)										
Skins of hollow walls										
100mm thick; stretcher bond	14.89	5.00	15.63	0.80	0.65	23.64	1.41	40.68	m²	44.75
115mm thick; stretcher bond	15.10	5.00	15.85	0.80	0.65	23.64	1.41	40.90	m²	45.00
125mm thick; stretcher bond	16.78	5.00	17.61	0.85	0.70	25.25	2.12	44.98	m²	49.47
130mm thick; stretcher bond	17.36	5.00	18.23	0.85	0.70	25.25	2.12	45.59	m²	50.15
150mm thick; stretcher bond	20.62	5.00	21.65	0.90	0.75	26.85	2.12	50.62	m²	55.68
190mm thick; stretcher bond	26.79	5.00	28.13	0.95	0.80	28.46	2.81	59.39	m²	65.33
200mm thick; stretcher bond	29.77	5.00	31.26	1.00	0.85	30.07	2.81	64.13	m²	70.54
215mm thick; stretcher bond	32.03	5.00	33.63	1.00	0.85	30.07	3.51	67.21	m²	73.93
Closing cavities										
50mm wide with blockwork 100mm thick	3.72	5.00	3.91	0.20	0.10	5.05	0.30	9.26	m	10.18
75mm wide with blockwork 100mm thick	3.72	5.00	3.91	0.20	0.10	5.05	0.30	9.26	m	10.18
Closing cavities; horizontal										
50mm wide with blockwork 100mm thick	3.72	5.00	3.91	0.20	0.10	5.05	0.30	9.26	m	10.18
75mm wide with blockwork 100mm thick	3.72	5.00	3.91	0.20	0.10	5.05	0.30	9.26	m	10.18
Walls; Blockwork; Aerated concrete blocks, Thermalite, 440 x 215mm, Turbo blocks (2.8 N/mm²); in thin joint mortar										
Walls										
100mm thick; stretcher bond	18.56	5.00	19.49	0.64	0.52	18.91	0.70	39.10	m²	43.02
115mm thick; stretcher bond	19.32	5.00	20.29	0.64	0.52	18.91	0.76	39.96	m²	43.95
125mm thick; stretcher bond	21.36	5.00	22.43	0.68	0.56	20.20	1.06	43.69	m²	48.06
130mm thick; stretcher bond	22.13	5.00	23.24	0.68	0.56	20.20	1.06	44.49	m²	48.94
150mm thick; stretcher bond	26.13	5.00	27.43	0.72	0.60	21.48	1.06	49.97	m²	54.97
190mm thick; stretcher bond	33.76	5.00	35.45	0.76	0.64	22.77	1.40	59.62	m²	65.59
200mm thick; stretcher bond	37.11	5.00	38.96	0.80	0.68	24.05	1.40	64.42	m²	70.86
215mm thick; stretcher bond	39.92	5.00	41.92	0.80	0.68	24.05	1.75	67.72	m²	74.50
Skins of hollow walls										
100mm thick; stretcher bond	19.39	5.00	20.36	0.80	0.65	23.64	1.05	45.05	m²	49.55
125mm thick; stretcher bond	21.36	5.00	22.43	0.68	0.56	20.20	1.06	43.69	m²	48.06
130mm thick; stretcher bond	22.13	5.00	23.24	0.68	0.56	20.20	1.06	44.49	m²	48.94
150mm thick; stretcher bond	26.13	5.00	27.43	0.72	0.60	21.48	1.06	49.97	m²	54.97
190mm thick; stretcher bond	33.76	5.00	35.45	0.76	0.64	22.77	1.40	59.62	m²	65.59
200mm thick; stretcher bond	37.11	5.00	38.96	0.80	0.68	24.05	1.40	64.42	m²	70.86
215mm thick; stretcher bond	39.92	5.00	41.92	0.80	0.68	24.05	1.75	67.72	m²	74.50
Closing cavities										
50mm wide with blockwork 100mm thick	4.02	5.00	4.22	0.20	0.10	5.05	0.15	9.42	m	10.37
75mm wide with blockwork 100mm thick	4.02	5.00	4.22	0.20	0.10	5.05	0.15	9.42	m	10.37
Closing cavities; horizontal										
50mm wide with blockwork 100mm thick	4.02	5.00	4.22	0.20	0.10	5.05	0.15	9.42	m	10.37
75mm wide with blockwork 100mm thick	4.02	5.00	4.22	0.20	0.10	5.05	0.15	9.42	m	10.37
Walls; Blockwork; Aerated concrete blocks, Thermalite, (Large Format) Turbo blocks (2.8 N/mm²); in thin joint mortar										
Walls										
100mm thick; stretcher bond	15.77	5.00	16.55	0.80	0.65	23.64	0.70	40.90	m²	44.99
115mm thick; stretcher bond	17.90	5.00	18.80	0.52	0.42	15.33	0.76	34.89	m²	38.37
125mm thick; stretcher bond	19.45	5.00	20.42	0.55	0.45	16.30	1.06	37.77	m²	41.55
130mm thick; stretcher bond	20.09	5.00	21.10	0.55	0.45	16.30	1.13	38.53	m²	42.38
150mm thick; stretcher bond	22.93	5.00	24.08	0.58	0.49	17.40	1.06	42.53	m²	46.78
190mm thick; stretcher bond	28.73	5.00	30.17	0.62	0.52	18.54	1.40	50.11	m²	55.13
200mm thick; stretcher bond	31.67	5.00	33.26	0.65	0.55	19.51	1.40	54.17	m²	59.58
215mm thick; stretcher bond	33.36	5.00	35.03	0.65	0.55	19.51	1.75	56.29	m²	61.92
Skins of hollow walls										
100mm thick; stretcher bond	15.77	5.00	16.55	0.80	0.65	23.64	0.70	40.90	m²	44.99
115mm thick; stretcher bond	17.90	5.00	18.80	0.52	0.42	15.33	0.76	34.89	m²	38.37
125mm thick; stretcher bond	19.45	5.00	20.42	0.55	0.45	16.30	1.06	37.77	m²	41.55
130mm thick; stretcher bond	20.09	5.00	21.10	0.55	0.45	16.30	1.13	38.53	m²	42.38
150mm thick; stretcher bond	22.93	5.00	24.08	0.58	0.49	17.40	1.06	42.53	m²	46.78
190mm thick; stretcher bond	28.73	5.00	30.17	0.62	0.52	18.54	1.40	50.11	m²	55.13
200mm thick; stretcher bond	31.67	5.00	33.26	0.65	0.55	19.51	1.40	54.17	m²	59.58
215mm thick; stretcher bond	33.36	5.00	35.03	0.65	0.55	19.51	1.75	56.29	m²	61.92
Closing cavities										
50mm wide with blockwork 100mm thick	3.55	5.00	3.73	0.20	0.10	5.05	0.15	8.93	m	9.83
75mm wide with blockwork 100mm thick	3.55	5.00	3.73	0.20	0.10	5.05	0.15	8.93	m	9.83
Closing cavities; horizontal										
50mm wide with blockwork 100mm thick	3.55	5.00	3.73	0.20	0.10	5.05	0.15	8.93	m	9.83
75mm wide with blockwork 100mm thick	3.55	5.00	3.73	0.20	0.10	5.05	0.15	8.93	m	9.83
Walls; Blockwork; Aerated concrete blocks, Thermalite, 440 x 215mm, Party wall blocks (4.0 N/mm²); in cement-lime mortar (1:1:6)										
Walls										
215mm thick; stretcher bond	27.10	5.00	28.45	1.00	0.85	30.07	3.51	62.03	m²	68.23
Walls; Blockwork; Aerated concrete blocks, Thermalite, 440 x 215mm, Hi-Strength 7 blocks (7.0 N/mm²); in cement-lime mortar (1:1:6)										
Walls										
100mm thick; stretcher bond	18.33	5.00	19.25	0.80	0.65	23.64	1.41	44.30	m²	48.73
140mm thick; stretcher bond	25.66	5.00	26.94	0.90	0.75	26.85	2.12	55.91	m²	61.50

MASONRY

Labour hourly rates: (except Specialists) Craft Operatives 18.37 Labourer 13.76 Rates are national average prices. Refer to REGIONAL VARIATIONS for indicative levels of overall pricing in regions	MATERIALS			LABOUR				RATES		
	Del to Site	Waste	Material Cost	Craft Optve	Lab	Labour Cost	Sunds	Nett Rate	Unit	Gross rate (10%)
	£	%	£	Hrs	Hrs	£	£	£		£

BRICK/BLOCK WALLING (Cont'd)

Walls; Blockwork; Aerated concrete blocks, Thermalite, 440 x 215mm, Hi-Strength 7 blocks (7.0 N/mm²); in cement-lime mortar (1:1:6) (Cont'd)

	Del to Site	Waste	Material Cost	Craft Optve	Lab	Labour Cost	Sunds	Nett Rate	Unit	Gross rate (10%)
Walls (Cont'd)										
150mm thick; stretcher bond	26.85	5.00	28.19	0.90	0.75	26.85	2.12	57.16	m²	62.88
190mm thick; stretcher bond	30.03	5.00	31.53	0.95	0.80	28.46	2.81	62.80	m²	69.08
200mm thick; stretcher bond	36.04	5.00	37.84	1.00	0.85	30.07	2.81	70.71	m²	77.78
215mm thick; stretcher bond	39.42	5.00	41.39	1.00	0.85	30.07	3.51	74.97	m²	82.46
Skins of hollow walls										
100mm thick; stretcher bond	18.33	5.00	19.25	0.80	0.65	23.64	1.41	44.30	m²	48.73
140mm thick; stretcher bond	25.66	5.00	26.94	0.90	0.75	26.85	2.12	55.91	m²	61.50
150mm thick; stretcher bond	26.85	5.00	28.19	0.90	0.75	26.85	2.12	57.16	m²	62.88
190mm thick; stretcher bond	30.03	5.00	31.53	0.95	0.80	28.46	2.81	62.80	m²	69.08
200mm thick; stretcher bond	36.04	5.00	37.84	1.00	0.85	30.07	2.81	70.71	m²	77.78
215mm thick; stretcher bond	39.42	5.00	41.39	1.00	0.85	30.07	3.51	74.97	m²	82.46
Closing cavities										
50mm wide with blockwork 100mm thick	4.58	5.00	4.81	0.20	0.10	5.05	0.30	10.16	m	11.18
75mm wide with blockwork 100mm thick	4.58	5.00	4.81	0.20	0.10	5.05	0.30	10.16	m	11.18
Closing cavities; horizontal										
50mm wide with blockwork 100mm thick	4.58	5.00	4.81	0.20	0.10	5.05	0.30	10.16	m	11.18
75mm wide with blockwork 100mm thick	4.58	5.00	4.81	0.20	0.10	5.05	0.30	10.16	m	11.18

Walls; Blockwork; Aerated concrete blocks, Thermalite, 440 x 215mm, Trenchblocks (4.0 N/mm²); in cement mortar (1:4)

	Del to Site	Waste	Material Cost	Craft Optve	Lab	Labour Cost	Sunds	Nett Rate	Unit	Gross rate (10%)
Walls										
255mm thick; stretcher bond	30.90	5.00	32.44	1.30	1.05	38.33	2.92	73.70	m²	81.07
275mm thick; stretcher bond	32.43	5.00	34.05	1.45	1.15	42.46	3.51	80.02	m²	88.02
305mm thick; stretcher bond	35.38	5.00	37.15	1.65	1.25	47.51	3.51	88.17	m²	96.99
355mm thick; stretcher bond	36.00	5.00	37.80	1.95	1.50	56.46	4.08	98.34	m²	108.18

Walls; Blockwork; Aerated concrete blocks, Tarmac Toplite foundation blocks; 440 x 215mm, (3.5 N/mm²); in cement mortar (1:4)

	Del to Site	Waste	Material Cost	Craft Optve	Lab	Labour Cost	Sunds	Nett Rate	Unit	Gross rate (10%)
Walls										
260mm thick; stretcher bond	25.85	5.00	27.14	1.40	0.95	38.79	3.51	69.44	m²	76.39
275mm thick; stretcher bond	27.50	5.00	28.87	1.60	1.05	43.84	3.51	76.22	m²	83.85
300mm thick; stretcher bond	34.38	5.00	36.10	1.70	1.10	46.37	3.51	85.97	m²	94.57

Walls; Blockwork; Tarmac Topblock, medium density concrete block, Hemelite, 440 x 215mm, solid Standard blocks (3.5 N/mm²); in cement-lime mortar (1:1:6)

	Del to Site	Waste	Material Cost	Craft Optve	Lab	Labour Cost	Sunds	Nett Rate	Unit	Gross rate (10%)
Walls										
75mm thick; stretcher bond	7.30	5.00	7.66	0.75	0.65	22.72	1.12	31.51	m²	34.66
90mm thick; stretcher bond	8.14	5.00	8.54	0.90	0.75	26.85	1.41	36.81	m²	40.49
100mm thick; stretcher bond	7.95	5.00	8.35	0.90	0.75	26.85	1.41	36.61	m²	40.27
140mm thick; stretcher bond	11.23	5.00	11.80	1.00	0.85	30.07	2.12	43.98	m²	48.38
190mm thick; stretcher bond	17.85	5.00	18.74	1.05	0.90	31.67	2.81	53.22	m²	58.54
215mm thick; stretcher bond	20.26	5.00	21.28	1.10	0.95	33.28	3.51	58.07	m²	63.87
Skins of hollow walls										
75mm thick; stretcher bond	7.30	5.00	7.66	0.75	0.65	22.72	1.12	31.51	m²	34.66
90mm thick; stretcher bond	8.14	5.00	8.54	0.90	0.75	26.85	1.41	36.81	m²	40.49
100mm thick; stretcher bond	7.95	5.00	8.35	0.90	0.75	26.85	1.41	36.61	m²	40.27
140mm thick; stretcher bond	11.23	5.00	11.80	1.00	0.85	30.07	2.12	43.98	m²	48.38
190mm thick; stretcher bond	17.85	5.00	18.74	1.05	0.90	31.67	2.81	53.22	m²	58.54
215mm thick; stretcher bond	20.26	5.00	21.28	1.10	0.95	33.28	3.51	58.07	m²	63.87
Closing cavities										
50mm wide with blockwork 100mm thick	1.99	5.00	2.09	0.20	0.10	5.05	0.30	7.44	m	8.18
75mm wide with blockwork 100mm thick	1.99	5.00	2.09	0.20	0.10	5.05	0.30	7.44	m	8.18
Closing cavities; horizontal										
50mm wide with blockwork 100mm thick	1.99	5.00	2.09	0.20	0.10	5.05	0.30	7.44	m	8.18
75mm wide with blockwork 100mm thick	1.99	5.00	2.09	0.20	0.10	5.05	0.30	7.44	m	8.18
Bonding ends to common brickwork; forming pockets; extra material										
walls; bonding every third course										
75mm	1.82	5.00	1.92	0.25	0.20	7.34	-	9.26	m	10.19
90mm	2.03	5.00	2.14	0.30	0.25	8.95	-	11.09	m	12.20
100mm	1.99	5.00	2.09	0.30	0.25	8.95	-	11.04	m	12.14
140mm	2.81	5.00	2.95	0.40	0.30	11.48	-	14.43	m	15.87
190mm	4.46	5.00	4.69	0.45	0.35	13.08	-	17.77	m	19.54
215mm	5.07	5.00	5.32	0.55	0.40	15.61	-	20.93	m	23.02
Bonding ends to existing common brickwork; cutting pockets; extra material										
walls; bonding every third course										
75mm	1.82	5.00	1.92	0.50	0.40	14.69	-	16.60	m	18.27
90mm	2.03	5.00	2.14	0.60	0.50	17.90	-	20.04	m	22.04
100mm	1.99	5.00	2.09	0.60	0.50	17.90	-	19.99	m	21.99
140mm	2.81	5.00	2.95	0.80	0.60	22.95	-	25.90	m	28.49
190mm	4.46	5.00	4.69	0.90	0.70	26.17	-	30.85	m	33.94
215mm	5.07	5.00	5.32	1.10	0.80	31.22	-	36.53	m	40.19
Bonding ends to existing concrete blockwork; cutting pockets; extra material										
walls; bonding every third course										
75mm	1.82	5.00	1.92	0.40	0.30	11.48	-	13.39	m	14.73
90mm	2.03	5.00	2.14	0.45	0.35	13.08	-	15.22	m	16.74

Labour hourly rates: (except Specialists) Craft Operatives 18.37 Labourer 13.76 Rates are national average prices. Refer to REGIONAL VARIATIONS for indicative levels of overall pricing in regions	MATERIALS			LABOUR				RATES		
	Del to Site	Waste	Material Cost	Craft Optve	Lab	Labour Cost	Sunds	Nett Rate		Gross rate (10%)
	£	%	£	Hrs	Hrs	£	£	£	Unit	£
BRICK/BLOCK WALLING (Cont'd)										
Walls; Blockwork; Tarmac Topblock, medium density concrete block, Hemelite, 440 x 215mm, solid Standard blocks (3.5 N/mm²); in cement-lime mortar (1:1:6) (Cont'd)										
Bonding ends to existing concrete blockwork; cutting pockets; extra material (Cont'd)										
walls; bonding every third course (Cont'd)										
100mm	1.99	5.00	2.09	0.45	0.35	13.08	-	15.17	m	16.69
140mm	2.81	5.00	2.95	0.60	0.45	17.21	-	20.16	m	22.18
190mm	4.46	5.00	4.69	0.70	0.55	20.43	-	25.11	m	27.62
215mm	5.07	5.00	5.32	0.85	0.60	23.87	-	29.19	m	32.11
Walls; Blockwork; Tarmac Topblock, medium density concrete block, Hemelite, 440 x 215mm, solid Standard blocks (7.0 N/mm²); in cement-lime mortar (1:1:6)										
Walls										
90mm thick; stretcher bond	9.24	5.00	9.70	0.90	0.75	26.85	1.41	37.97	m²	41.76
100mm thick; stretcher bond	8.10	5.00	8.51	0.90	0.75	26.85	1.41	36.77	m²	40.44
140mm thick; stretcher bond	11.75	5.00	12.34	1.00	0.85	30.07	2.12	44.52	m²	48.97
190mm thick; stretcher bond	20.21	5.00	21.22	1.05	0.90	31.67	2.81	55.70	m²	61.27
215mm thick; stretcher bond	22.87	5.00	24.01	1.10	0.95	33.28	3.51	60.80	m²	66.88
Skins of hollow walls										
90mm thick; stretcher bond	9.24	5.00	9.70	0.90	0.75	26.85	1.41	37.97	m²	41.76
100mm thick; stretcher bond	8.10	5.00	8.51	0.90	0.75	26.85	1.41	36.77	m²	40.44
140mm thick; stretcher bond	11.75	5.00	12.34	1.00	0.85	30.07	2.12	44.52	m²	48.97
190mm thick; stretcher bond	20.21	5.00	21.22	1.05	0.90	31.67	2.81	55.70	m²	61.27
215mm thick; stretcher bond	22.87	5.00	24.01	1.10	0.95	33.28	3.51	60.80	m²	66.88
Closing cavities										
50mm wide with blockwork 100mm thick	2.03	5.00	2.13	0.20	0.10	5.05	0.30	7.48	m	8.22
75mm wide with blockwork 100mm thick	2.03	5.00	2.13	0.20	0.10	5.05	0.30	7.48	m	8.22
Closing cavities; horizontal										
50mm wide with blockwork 100mm thick	2.03	5.00	2.13	0.20	0.10	5.05	0.30	7.48	m	8.22
75mm wide with blockwork 100mm thick	2.03	5.00	2.13	0.20	0.10	5.05	0.30	7.48	m	8.22
Walls; Blockwork; Tarmac Topblock, fair face concrete blocks, Lignacite, 440 x 215mm, solid Standard blocks (7.0 N/mm²); in cement-lime mortar (1:1:6)										
Walls										
100mm thick; stretcher bond	15.00	5.00	15.75	0.95	0.80	28.46	1.41	45.62	m²	50.18
140mm thick; stretcher bond	19.80	5.00	20.79	1.10	0.90	32.59	2.12	55.50	m²	61.05
190mm thick; stretcher bond	28.50	5.00	29.92	1.15	0.95	34.20	2.81	66.93	m²	73.62
Skins of hollow walls										
100mm thick; stretcher bond	15.00	5.00	15.75	0.95	0.80	28.46	1.41	45.62	m²	50.18
140mm thick; stretcher bond	19.80	5.00	20.79	1.10	0.90	32.59	2.12	55.50	m²	61.05
190mm thick; stretcher bond	28.50	5.00	29.92	1.15	0.95	34.20	2.81	66.93	m²	73.62
Closing cavities										
50mm wide with blockwork 100mm thick	3.75	5.00	3.94	0.20	0.10	5.05	0.30	9.29	m	10.22
75mm wide with blockwork 100mm thick	3.75	5.00	3.94	0.20	0.10	5.05	0.30	9.29	m	10.22
Closing cavities; horizontal										
50mm wide with blockwork 100mm thick	3.75	5.00	3.94	0.20	0.10	5.05	0.30	9.29	m	10.22
75mm wide with blockwork 100mm thick	3.75	5.00	3.94	0.20	0.10	5.05	0.30	9.29	m	10.22
Walls; Blockwork; Tarmac Topblock, dense concrete blocks, Topcrete, 440 x 215mm, solid Standard blocks (7.0 N/mm²); in cement-lime mortar (1:1:6)										
Walls										
90mm thick; stretcher bond	10.32	5.00	10.84	1.15	0.95	34.20	1.41	46.45	m²	51.09
100mm thick; stretcher bond	11.23	5.00	11.79	1.20	1.00	35.80	1.41	49.01	m²	53.91
140mm thick; stretcher bond	13.76	5.00	14.44	1.30	1.10	39.02	2.12	55.57	m²	61.13
190mm thick; stretcher bond	20.64	5.00	21.68	1.40	1.15	41.54	2.81	66.02	m²	72.62
215mm thick; stretcher bond	23.24	5.00	24.40	1.50	1.20	44.07	3.51	71.98	m²	79.17
Skins of hollow walls										
90mm thick; stretcher bond	10.32	5.00	10.84	1.15	0.95	34.20	1.41	46.45	m²	51.09
100mm thick; stretcher bond	11.23	5.00	11.79	1.20	1.00	35.80	1.41	49.01	m²	53.91
140mm thick; stretcher bond	13.76	5.00	14.44	1.30	1.10	39.02	2.12	55.57	m²	61.13
190mm thick; stretcher bond	20.64	5.00	21.68	1.40	1.15	41.54	2.81	66.02	m²	72.62
215mm thick; stretcher bond	23.24	5.00	24.40	1.50	1.20	44.07	3.51	71.98	m²	79.17
Closing cavities										
50mm wide with blockwork 100mm thick	2.81	5.00	2.95	0.20	0.10	5.05	0.30	8.30	m	9.13
75mm wide with blockwork 100mm thick	2.81	5.00	2.95	0.20	0.10	5.05	0.30	8.30	m	9.13
Closing cavities; horizontal										
50mm wide with blockwork 100mm thick	2.81	5.00	2.95	0.20	0.10	5.05	0.30	8.30	m	9.13
75mm wide with blockwork 100mm thick	2.81	5.00	2.95	0.20	0.10	5.05	0.30	8.30	m	9.13
Walls; Blockwork; Tarmac Topblock, dense concrete blocks, Topcrete, 440 x 215mm, cellular Standard blocks (7.0 N/mm²); in cement-lime mortar (1:1:6)										
Walls										
100mm thick; stretcher bond	7.95	5.00	8.35	1.20	1.00	35.80	1.41	45.56	m²	50.12
140mm thick; stretcher bond	11.85	5.00	12.44	1.30	1.10	39.02	2.12	53.57	m²	58.93
Skins of hollow walls										
100mm thick; stretcher bond	7.95	5.00	8.35	1.20	1.00	35.80	1.41	45.56	m²	50.12
140mm thick; stretcher bond	11.85	5.00	12.44	1.30	1.10	39.02	2.12	53.57	m²	58.93

MASONRY *(side margin)*

Labour hourly rates: (except Specialists) Craft Operatives 18.37 Labourer 13.76 Rates are national average prices. Refer to REGIONAL VARIATIONS for indicative levels of overall pricing in regions	MATERIALS			LABOUR				RATES		
	Del to Site	Waste	Material Cost	Craft Optve	Lab	Labour Cost	Sunds	Nett Rate	Unit	Gross rate (10%)
	£	%	£	Hrs	Hrs	£	£	£		£
BRICK/BLOCK WALLING (Cont'd)										
Walls; Blockwork; Tarmac Topblock, dense concrete blocks, Topcrete, 440 x 215mm hollow standard blocks (7.0 N/mm²); in cement-lime mortar (1:1:6)										
Walls										
215mm thick; stretcher bond	15.35	5.00	16.12	1.50	0.81	38.70	3.51	58.33	m²	64.16
Walls; Blockwork; Aerated concrete blocks, Thermalite, 440 x 215mm, Smooth Face blocks (4.0 N/mm²); in cement-lime mortar (1:1:6); flush smooth pointing as work proceeds										
Walls										
stretcher bond; facework one side										
100mm thick	13.86	5.00	14.55	1.00	0.80	29.38	1.41	45.34	m²	49.88
150mm thick	28.88	5.00	30.32	1.10	0.90	32.59	2.25	65.16	m²	71.68
200mm thick	38.17	5.00	40.08	1.15	0.95	34.20	2.94	77.22	m²	84.94
215mm thick	31.18	5.00	32.74	1.20	1.00	35.80	3.51	72.06	m²	79.26
stretcher bond; facework both sides										
100mm thick	13.86	5.00	14.55	1.20	0.95	35.12	1.41	51.08	m²	56.19
150mm thick	28.88	5.00	30.32	1.30	1.05	38.33	2.25	70.90	m²	77.99
200mm thick	38.17	5.00	40.08	1.35	1.10	39.94	2.94	82.96	m²	91.25
215mm thick	31.18	5.00	32.74	1.40	1.15	41.54	3.51	77.80	m²	85.58
Skins of hollow walls										
stretcher bond; facework one side										
100mm thick	13.86	5.00	14.55	1.00	0.80	29.38	1.41	45.34	m²	49.88
150mm thick	28.88	5.00	30.32	1.10	0.90	32.59	2.25	65.16	m²	71.68
200mm thick	38.17	5.00	40.08	1.15	0.95	34.20	2.94	77.22	m²	84.94
215mm thick	31.18	5.00	32.74	1.20	1.00	35.80	3.51	72.06	m²	79.26
Walls; Blockwork; Tarmac Topblock, fair face concrete blocks, Lignacite, 440 x 215mm, solid Standard blocks (7.0 N/mm²); in cement-lime mortar (1:1:6); flush smooth pointing as work proceeds										
Walls										
stretcher bond; facework one side										
100mm thick	15.00	5.00	15.75	1.15	0.95	34.20	1.41	51.36	m²	56.49
140mm thick	19.80	5.00	20.79	1.30	1.05	38.33	2.12	61.23	m²	67.36
190mm thick	28.50	5.00	29.92	1.35	1.10	39.94	2.81	72.67	m²	79.93
stretcher bond; facework both sides										
100mm thick	15.00	5.00	15.75	1.35	1.10	39.94	1.41	57.10	m²	62.81
140mm thick	19.80	5.00	20.79	1.50	1.20	44.07	2.12	66.97	m²	73.67
190mm thick	28.50	5.00	29.92	1.55	1.25	45.67	2.81	78.40	m²	86.24
Skins of hollow walls										
stretcher bond; facework one side										
100mm thick	15.00	5.00	15.75	1.15	0.95	34.20	1.41	51.36	m²	56.49
140mm thick	19.80	5.00	20.79	1.30	1.05	38.33	2.12	61.23	m²	67.36
190mm thick	28.50	5.00	29.92	1.35	1.10	39.94	2.81	72.67	m²	79.93
Flue linings; Firebricks, 215 x 102.5 x 65mm; in fire cement										
bonding to surrounding brickwork with headers -4/m²										
112mm thick; stretcher bond	106.20	5.00	111.51	1.90	1.40	54.17	10.05	175.73	m²	193.30
built clear of main brickwork but with one header in each course set projecting to contact main work										
112mm thick; stretcher bond	106.20	5.00	111.51	1.90	1.40	54.17	10.05	175.73	m²	193.30
112mm thick; stretcher bond; in segmental top to flue	106.20	5.00	111.51	2.50	1.80	70.69	10.05	192.25	m²	211.48
Clay flue linings										
Clay flue linings, BS EN 1457; rebated joints; jointed in cement mortar (1:3)										
150mm diameter, class A1	37.98	5.00	39.88	0.45	0.35	13.08	0.44	53.39	m	58.73
terminal Type 6F	47.51	5.00	49.89	0.45	0.35	13.08	0.22	63.19	nr	69.51
185mm diameter, class A1	45.54	5.00	47.82	0.50	0.40	14.69	0.57	63.07	m	69.38
terminal Type 6F	30.14	5.00	31.64	0.50	0.40	14.69	0.25	46.59	nr	51.24
225mm diameter, class A1	69.67	5.00	73.15	0.55	0.45	16.30	0.62	90.06	m	99.07
terminal Type 6F	40.85	5.00	42.89	0.55	0.45	16.30	0.33	59.51	nr	65.46
185 x 185mm, class A1	50.54	5.00	53.06	0.55	0.45	16.30	0.57	69.93	m	76.92
terminal Type 4D	34.86	5.00	36.60	0.55	0.45	16.30	0.25	53.15	nr	58.47
Extra over walls for perimeters and abutments; Precast concrete copings BS 5642 Part 2; bedding in cement-lime mortar (1:1:6)										
Copings; figure 1; horizontal										
75 x 200mm; splayed; rebated joints	25.62	2.50	26.26	0.25	0.13	6.38	0.24	32.88	m	36.17
extra; stopped ends	2.57	-	2.57	-	-	-	-	2.57	nr	2.82
extra; internal angles	9.81	-	9.81	-	-	-	-	9.81	nr	10.79
100 x 300mm; splayed; rebated joints	39.75	2.50	40.74	0.50	0.25	12.62	0.38	53.74	m	59.12
extra; stopped ends	3.50	-	3.50	-	-	-	-	3.50	nr	3.85
extra; internal angles	19.57	-	19.57	-	-	-	-	19.57	nr	21.53
75 x 200mm; saddleback; rebated joints	24.47	2.50	25.08	0.25	0.13	6.38	0.24	31.70	m	34.87
extra; hipped ends	1.11	-	1.11	-	-	-	-	1.11	nr	1.22
extra; internal angles	8.39	-	8.39	-	-	-	-	8.39	nr	9.23
100 x 300mm; saddleback; rebated joints	32.06	2.50	32.87	0.50	0.25	12.62	0.38	45.87	m	50.45
extra; hipped ends	2.65	-	2.65	-	-	-	-	2.65	nr	2.92
extra; internal angles	11.21	-	11.21	-	-	-	-	11.21	nr	12.33

Labour hourly rates: (except Specialists) Craft Operatives 18.37 Labourer 13.76 Rates are national average prices. Refer to REGIONAL VARIATIONS for indicative levels of overall pricing in regions	MATERIALS			LABOUR				RATES		
	Del to Site	Waste	Material Cost	Craft Optve	Lab	Labour Cost	Sunds	Nett Rate	Unit	Gross rate (10%)
	£	%	£	Hrs	Hrs	£	£	£		£
BRICK/BLOCK WALLING (Cont'd)										
Extra over walls for perimeters and abutments; Restraint channels										
Ancon 25/14 restraint system Channel; fixing to steelwork with M8 screws complete with plate washer at 450mm centres; incorporating ties at 450mm centres										
100mm long SD25 ties	6.89	5.00	7.24	0.50	0.25	12.62	0.94	20.81	m	22.89
125mm long SD25 ties	7.05	5.00	7.41	0.50	0.25	12.62	1.11	21.14	m	23.26
150mm long SD25 ties	7.21	5.00	7.57	0.50	0.25	12.62	1.26	21.46	m	23.60
Extra over walls for perimeters and abutments; Wall profiles										
Stainless steel Simpson 'Strong Tie' wall extension profiles; plugging and screwing to brickwork and building ties in to joints of new walls										
60 - 250mm wall thickness; Ref. C2K	3.20	5.00	3.36	0.30	-	5.51	-	8.87	m	9.76
Extra over walls for opening perimeters; Slate and tile sills										
Clay plain roofing tiles, red, BS EN 1304, machine made, 265 x 165 x 13mm; in cement-lime mortar										
one course 150mm wide; set weathering	1.68	5.00	1.76	0.45	0.30	12.39	0.33	14.49	m	15.94
two courses 150mm wide; set weathering	3.36	5.00	3.53	0.75	0.50	20.66	0.64	24.83	m	27.31
Extra over walls for opening perimeters; Precast concrete sills; BS 5642 Part 1; bedding in cement lime mortar (1:1:6)										
Sills; figure 2 or figure 4										
50 x 150 x 300mm splayed and grooved	13.08	2.50	13.41	0.15	0.15	4.82	0.17	18.39	nr	20.23
50 x 150 x 400mm splayed and grooved	17.21	2.50	17.64	0.20	0.20	6.43	0.19	24.26	nr	26.69
50 x 150 x 700mm splayed and grooved	22.18	2.50	22.73	0.30	0.30	9.64	0.28	32.65	nr	35.92
50 x 150 x 1300mm splayed and grooved	39.11	2.50	40.09	0.50	0.50	16.07	0.42	56.58	nr	62.23
extra for stooled end	6.85	-	6.85	-	-	-	-	6.85	nr	7.53
75 x 150 x 300mm splayed and grooved	14.37	2.50	14.72	0.17	0.17	5.46	0.17	20.35	nr	22.39
75 x 150 x 400mm splayed and grooved	18.50	2.50	18.96	0.22	0.22	7.07	0.19	26.22	nr	28.84
75 x 150 x 700mm splayed and grooved	28.11	2.50	28.82	0.35	0.35	11.25	0.28	40.35	nr	44.38
75 x 150 x 1300mm splayed and grooved	41.91	2.50	42.96	0.55	0.55	17.67	0.42	61.05	nr	67.16
extra for stooled end	8.89	-	8.89	-	-	-	-	8.89	nr	9.78
100 x 150 x 300mm splayed and grooved	12.98	2.50	13.30	0.20	0.20	6.43	0.17	19.90	nr	21.88
100 x 150 x 400mm splayed and grooved	19.80	2.50	20.30	0.25	0.25	8.03	0.19	28.53	nr	31.38
100 x 150 x 700mm splayed and grooved	30.19	2.50	30.95	0.40	0.40	12.85	0.28	44.08	nr	48.49
100 x 150 x 1300mm splayed and grooved	56.01	2.50	57.41	0.60	0.60	19.28	0.42	77.11	nr	84.82
extra for stooled end	11.58	-	11.58	-	-	-	-	11.58	nr	12.74
Extra over walls for opening perimeters; Cavity Closers										
Fire rated cavity closers by Cavity Trays Ltd built into masonry										
Cavi 120 type V; Cavicloser vertical	8.16	5.00	8.57	0.33	0.16	8.26	-	16.83	m	18.51
Cavi 120 type V; Cavicloser horizontal	8.16	5.00	8.57	0.33	0.16	8.26	-	16.83	m	18.51
Fire rated Cavicheck fire integrity stop by Cavity Trays Ltd built into masonry										
Cavi 240 type CFIS; Cavicheck vertical	3.07	5.00	3.22	0.33	0.16	8.26	-	11.48	m	12.63
Cavi 240 type CFIS; Cavicheck horizontal	3.07	5.00	3.22	0.33	0.16	8.26	-	11.48	m	12.63
Fire rated party wall integrity barrier by Cavity Trays Ltd built into masonry										
Cavi 240 type PWIB; Party wall integrity barrier vertical	8.21	5.00	8.62	0.33	0.16	8.26	-	16.88	m	18.57
Cavi 240 type PWIB; Party wall integrity barrier horizontal	8.21	5.00	8.62	0.33	0.16	8.26	-	16.88	m	18.57
Thermabate insulated cavity closers; fixing to timber with nails										
Thermabate 50; vertical	4.68	5.00	4.91	0.33	0.16	8.26	-	13.18	m	14.49
Thermabate 75; vertical	4.82	5.00	5.06	0.33	0.16	8.26	-	13.32	m	14.65
Thermabate 100; vertical	5.06	5.00	5.32	0.33	0.16	8.26	-	13.58	m	14.94
Thermabate 50; horizontal	4.68	5.00	4.91	0.33	0.16	8.26	-	13.18	m	14.49
Thermabate 75; horizontal	4.82	5.00	5.06	0.33	0.16	8.26	-	13.32	m	14.65
Thermabate 100; horizontal	5.06	5.00	5.32	0.33	0.16	8.26	-	13.58	m	14.94
Thermabate insulated cavity closers; fixing to masonry with PVC-U ties at 225mm centres										
Thermabate 50; vertical	4.68	5.00	4.91	0.33	0.16	8.26	-	13.18	m	14.49
Thermabate 75; vertical	4.82	5.00	5.06	0.33	0.16	8.26	-	13.32	m	14.65
Thermabate 100; vertical	5.06	5.00	5.32	0.33	0.16	8.26	-	13.58	m	14.94
Thermabate 50; horizontal	4.68	5.00	4.91	0.33	0.16	8.26	-	13.18	m	14.49
Thermabate 75; horizontal	4.82	5.00	5.06	0.33	0.16	8.26	-	13.32	m	14.65
Thermabate 100; horizontal	5.06	5.00	5.32	0.33	0.16	8.26	-	13.58	m	14.94
Extra over walls for opening perimeters; Lintels; concrete										
Precast concrete lintels; BS8500, designed mix C25, 20mm aggregate minimum cement content 360 kg/m³; vibrated; reinforcement bars BS 4449 hot rolled plain round mild steel; bedding in cement-lime mortar (1:1:6)										
75 x 150 x 1200mm; reinforced with 1.20 kg of 12mm bars	11.49	2.50	11.78	0.18	0.09	4.55	0.17	16.49	nr	18.14
75 x 150 x 1650mm; reinforced with 1.60 kg of 12mm bars	16.50	2.50	16.91	0.25	0.13	6.38	0.21	23.50	nr	25.85
102 x 150 x 1200mm; reinforced with 1.20 kg of 12mm bars	19.20	2.50	19.68	0.30	0.15	7.58	0.18	27.43	nr	30.18
102 x 150 x 1650mm; reinforced with 1.60 kg of 12mm bars	26.37	2.50	27.03	0.41	0.20	10.28	0.21	37.52	nr	41.28
215 x 150 x 1200mm; reinforced with 2.40 kg of 12mm bars	52.62	2.50	53.94	0.60	0.30	15.15	0.27	69.36	nr	76.29
215 x 150 x 1650mm; reinforced with 3.20 kg of 12mm bars	72.36	2.50	74.17	0.83	0.41	20.89	0.36	95.42	nr	104.96
215 x 225 x 1200mm; reinforced with 2.40 kg of 12mm bars	74.56	2.50	76.42	0.78	0.39	19.70	0.28	96.40	nr	106.04
215 x 225 x 1650mm; reinforced with 3.20 kg of 12mm bars	86.00	2.50	88.15	1.07	0.53	26.95	0.38	115.47	nr	127.02
327 x 150 x 1200mm; reinforced with 3.60 kg of 12mm bars	75.70	2.50	77.59	0.78	0.39	19.70	0.28	97.57	nr	107.33
327 x 150 x 1650mm; reinforced with 4.80 kg of 12mm bars	86.50	2.50	88.66	1.07	0.54	27.09	0.38	116.12	nr	127.74
327 x 225 x 1200mm; reinforced with 3.60 kg of 12mm bars	91.00	2.50	93.28	1.20	0.60	30.30	0.34	123.92	nr	136.31
327 x 225 x 1650mm; reinforced with 4.80 kg of 12mm bars	105.00	2.50	107.62	1.65	0.83	41.73	0.44	149.79	nr	164.77

MASONRY

MASONRY

Labour hourly rates: (except Specialists) Craft Operatives 18.37 Labourer 13.76 Rates are national average prices. Refer to REGIONAL VARIATIONS for indicative levels of overall pricing in regions	MATERIALS			LABOUR				RATES		
	Del to Site £	Waste %	Material Cost £	Craft Optve Hrs	Lab Hrs	Labour Cost £	Sunds £	Nett Rate £	Unit	Gross rate (10%) £

BRICK/BLOCK WALLING (Cont'd)

Extra over walls for opening perimeters; Lintels; concrete (Cont'd)

Precast concrete lintels; BS8500, designed mix C25, 20mm aggregate minimum cement content 360 kg/m³; vibrated; reinforcement bars BS 4449 hot rolled plain round mild steel; bedding in cement-lime mortar (1:1:6) (Cont'd)

	Del to Site £	Waste %	Material Cost £	Craft Optve Hrs	Lab Hrs	Labour Cost £	Sunds £	Nett Rate £	Unit	Gross rate £
75 x 150 x 1200mm; reinforced with 1.20 kg of 12mm bars; fair finish two faces	13.00	2.50	13.32	0.18	0.09	4.55	0.17	18.03	nr	19.84
75 x 150 x 1650mm; reinforced with 1.60 kg of 12mm bars; fair finish two faces	17.50	2.50	17.94	0.25	0.13	6.38	0.21	24.53	nr	26.98
102 x 150 x 1200mm; reinforced with 1.20 kg of 12mm bars; fair finish two faces	20.20	2.50	20.70	0.30	0.15	7.58	0.17	28.44	nr	31.29
102 x 150 x 1650mm; reinforced with 1.60 kg of 12mm bars; fair finish two faces	27.37	2.50	28.05	0.41	0.20	10.28	0.21	38.55	nr	42.40
215 x 150 x 1200mm; reinforced with 2.40 kg of 12mm bars; fair finish two faces	53.62	2.50	54.96	0.60	0.30	15.15	0.27	70.38	nr	77.42
215 x 150 x 1650mm; reinforced with 3.20 kg of 12mm bars; fair finish two faces	73.36	2.50	75.19	0.83	0.41	20.89	0.36	96.44	nr	106.09
215 x 225 x 1200mm; reinforced with 2.40 kg of 12mm bars; fair finish two faces	71.56	2.50	73.35	0.78	0.39	19.70	0.28	93.33	nr	102.66
215 x 225 x 1650mm; reinforced with 3.20 kg of 12mm bars; fair finish two faces	87.00	2.50	89.17	1.07	0.53	26.95	0.38	116.50	nr	128.15
327 x 150 x 1200mm; reinforced with 3.60 kg of 12mm bars; fair finish two faces	76.70	2.50	78.62	0.78	0.39	19.70	0.28	98.60	nr	108.46
327 x 150 x 1650mm; reinforced with 4.80 kg of 12mm bars; fair finish two faces	87.50	2.50	89.69	1.07	0.54	27.09	0.38	117.15	nr	128.86
327 x 225 x 1200mm; reinforced with 3.60 kg of 12mm bars; fair finish two faces	92.00	2.50	94.30	1.20	0.60	30.30	0.33	124.93	nr	137.42
327 x 225 x 1650mm; reinforced with 4.80 kg of 12mm bars; fair finish two faces	106.00	2.50	108.65	1.65	0.83	41.73	0.50	150.88	nr	165.96

Boot lintels

	Del to Site £	Waste %	Material Cost £	Craft Optve Hrs	Lab Hrs	Labour Cost £	Sunds £	Nett Rate £	Unit	Gross rate £
215 x 150 x 1200mm; reinforced with 3.60 kg of 12mm bars and 0.51 kg of 6mm links	54.00	2.50	55.35	0.60	0.30	15.15	0.25	70.75	nr	77.83
215 x 150 x 1800mm; reinforced with 5.19 kg of 12mm bars and 0.77 kg of 6mm links	80.00	2.50	82.00	0.90	0.45	22.73	0.38	105.10	nr	115.61
215 x 225 x 1200mm; reinforced with 3.60 kg of 12mm bars and 0.62 kg of 6mm links	80.00	2.50	82.00	0.78	0.39	19.70	0.27	101.96	nr	112.16
215 x 225 x 1800mm; reinforced with 5.19 kg of 12mm bars and 0.93 kg of 6mm links	120.00	2.50	123.00	1.18	0.59	29.80	0.39	153.18	nr	168.50
252 x 150 x 1200mm; reinforced with 3.60 kg of 12mm bars and 0.56 kg of 6mm links	62.00	2.50	63.55	0.72	0.36	18.18	0.33	82.06	nr	90.27
252 x 150 x 1800mm; reinforced with 5.19 kg of 12mm bars and 0.83 kg of 6mm links	94.00	2.50	96.35	1.08	0.54	27.27	0.45	124.07	nr	136.48
252 x 225 x 1200mm; reinforced with 3.60 kg of 12mm bars and 0.67 kg of 6mm links	94.00	2.50	96.35	0.90	0.45	22.73	0.33	119.40	nr	131.35
252 x 225 x 1800mm; reinforced with 5.19 kg of 12mm bars and 1.00 kg of 6mm links	140.00	2.50	143.50	1.35	0.67	34.02	0.47	177.98	nr	195.78

Precast prestressed concrete lintels; Stressline Ltd; Roughcast; bedding in cement-lime mortar (1:1:6)

	Del to Site £	Waste %	Material Cost £	Craft Optve Hrs	Lab Hrs	Labour Cost £	Sunds £	Nett Rate £	Unit	Gross rate £
100 x 70mm										
900mm long	4.75	2.50	4.87	0.24	0.12	6.06	0.36	11.29	nr	12.42
1050mm long	5.25	2.50	5.38	0.27	0.14	6.89	0.39	12.66	nr	13.92
1200mm long	6.35	2.50	6.51	0.31	0.16	7.90	0.47	14.87	nr	16.36
1500mm long	7.70	2.50	7.89	0.38	0.19	9.60	0.56	18.04	nr	19.85
1800mm long	8.99	2.50	9.21	0.44	0.22	11.11	0.62	20.94	nr	23.03
2100mm long	10.55	2.50	10.81	0.50	0.25	12.62	0.70	24.14	nr	26.56
2400mm long	12.15	2.50	12.45	0.55	0.28	13.96	0.81	27.22	nr	29.94
2700mm long	13.70	2.50	14.04	0.60	0.30	15.15	0.87	30.06	nr	33.07
3000mm long	15.10	2.50	15.48	0.66	0.33	16.67	0.98	33.12	nr	36.43
150 x 70mm										
900mm long	6.30	2.50	6.46	0.25	0.13	6.38	0.36	13.20	nr	14.52
1050mm long	7.35	2.50	7.53	0.28	0.14	7.07	0.42	15.02	nr	16.53
1200mm long	8.40	2.50	8.61	0.32	0.16	8.08	0.47	17.16	nr	18.87
1500mm long	10.05	2.50	10.30	0.39	0.20	9.92	0.57	20.79	nr	22.87
1800mm long	12.15	2.50	12.45	0.45	0.23	11.43	0.63	24.52	nr	26.97
2100mm long	13.99	2.50	14.34	0.51	0.26	12.95	0.75	28.04	nr	30.84
2400mm long	16.15	2.50	16.55	0.57	0.29	14.46	0.81	31.83	nr	35.01
2700mm long	17.95	2.50	18.40	0.62	0.31	15.66	0.86	34.91	nr	38.40
3000mm long	20.05	2.50	20.55	0.68	0.34	17.17	1.00	38.73	nr	42.60
225 x 70mm										
900mm long	11.55	2.50	11.84	0.26	0.13	6.57	0.36	18.76	nr	20.64
1050mm long	13.20	2.50	13.53	0.29	0.15	7.39	0.42	21.34	nr	23.48
1200mm long	15.05	2.50	15.43	0.33	0.17	8.40	0.47	24.29	nr	26.72
1500mm long	18.95	2.50	19.42	0.40	0.20	10.10	0.57	30.09	nr	33.10
1800mm long	22.45	2.50	23.01	0.46	0.23	11.62	0.63	35.26	nr	38.78
2100mm long	26.35	2.50	27.01	0.52	0.26	13.13	0.75	40.89	nr	44.98
2400mm long	30.10	2.50	30.85	0.58	0.29	14.65	0.86	46.35	nr	50.99
2700mm long	33.75	2.50	34.59	0.63	0.32	15.98	0.96	51.53	nr	56.68
3000mm long	37.50	2.50	38.44	0.69	0.35	17.49	1.00	56.93	nr	62.63
255 x 70mm										
900mm long	18.03	2.50	18.48	0.26	0.13	6.57	0.41	25.45	nr	28.00
1050mm long	20.83	2.50	21.35	0.30	0.15	7.58	0.44	29.36	nr	32.30
1200mm long	23.82	2.50	24.41	0.33	0.17	8.40	0.50	33.31	nr	36.64
1500mm long	29.76	2.50	30.51	0.40	0.20	10.10	0.57	41.18	nr	45.30
1800mm long	35.71	2.50	36.60	0.47	0.24	11.94	0.68	49.22	nr	54.14
2100mm long	41.66	2.50	42.70	0.53	0.27	13.45	0.78	56.93	nr	62.62
2400mm long	47.64	2.50	48.83	0.60	0.30	15.15	0.84	64.82	nr	71.30
2700mm long	53.58	2.50	54.92	0.64	0.32	16.16	0.96	72.04	nr	79.25
3000mm long	59.53	2.50	61.02	0.70	0.35	17.68	1.00	79.70	nr	87.67

Labour hourly rates: (except Specialists) Craft Operatives 18.37 Labourer 13.76 Rates are national average prices. Refer to REGIONAL VARIATIONS for indicative levels of overall pricing in regions	MATERIALS			LABOUR				RATES		
	Del to Site £	Waste %	Material Cost £	Craft Optve Hrs	Lab Hrs	Labour Cost £	Sunds £	Nett Rate £	Unit	Gross rate (10%) £
BRICK/BLOCK WALLING (Cont'd)										
Extra over walls for opening perimeters; Lintels; concrete (Cont'd)										
Precast prestressed concrete lintels; Stressline Ltd; Roughcast; bedding in cement-lime mortar (1:1:6) (Cont'd)										
100 x 145mm										
900mm long	9.49	2.50	9.72	0.26	0.13	6.57	0.36	16.65	nr	18.31
1050mm long	11.93	2.50	12.22	0.29	0.15	7.39	0.42	20.03	nr	22.04
1200mm long	11.93	2.50	12.22	0.33	0.17	8.40	0.50	21.12	nr	23.23
1500mm long	14.84	2.50	15.21	0.40	0.20	10.10	0.57	25.88	nr	28.47
1800mm long	17.97	2.50	18.42	0.46	0.23	11.62	0.68	30.71	nr	33.78
2100mm long	21.47	2.50	22.00	0.52	0.26	13.13	0.75	35.88	nr	39.47
2400mm long	25.97	2.50	26.62	0.58	0.29	14.65	0.86	42.12	nr	46.33
2700mm long	29.15	2.50	29.88	0.63	0.32	15.98	0.96	46.82	nr	51.50
3000mm long	32.33	2.50	33.14	0.69	0.35	17.49	1.00	51.63	nr	56.80
Extra over walls for opening perimeters; Lintels; steel										
Galvanized steel lintels; SUPERGALV (BIRTLEY) lintels reference CB 50; bedding in cement-lime mortar (1:1:6)										
125mm deep x 750mm long	16.74	2.50	17.16	0.28	0.28	9.00	-	26.15	nr	28.77
125mm deep x 1200mm long	25.95	2.50	26.60	0.38	0.38	12.21	-	38.81	nr	42.69
125mm deep x 1350mm long	30.88	2.50	31.65	0.41	0.41	13.17	-	44.83	nr	49.31
125mm deep x 1500mm long	34.06	2.50	34.91	0.44	0.44	14.14	-	49.05	nr	53.95
125mm deep x 1650mm long	38.53	2.50	39.49	0.47	0.47	15.10	-	54.59	nr	60.05
165mm deep x 1800mm long	42.02	2.50	43.07	0.50	0.50	16.07	-	59.14	nr	65.05
165mm deep x 1950mm long	45.76	2.50	46.90	0.53	0.53	17.03	-	63.93	nr	70.33
165mm deep x 2100mm long	48.26	2.50	49.47	0.56	0.56	17.99	-	67.46	nr	74.21
165mm deep x 2250mm long	55.40	2.50	56.78	0.60	0.60	19.28	-	76.06	nr	83.67
165mm deep x 2400mm long	58.50	2.50	59.96	0.63	0.63	20.24	-	80.20	nr	88.22
215mm deep x 2550mm long	66.65	2.50	68.32	0.67	0.67	21.53	-	89.84	nr	98.83
215mm deep x 2850mm long	90.94	2.50	93.21	0.71	0.71	22.81	-	116.03	nr	127.63
215mm deep x 3000mm long	95.68	2.50	98.07	0.75	0.75	24.10	-	122.17	nr	134.39
215mm deep x 3300mm long	111.51	2.50	114.30	0.81	0.81	26.03	-	140.32	nr	154.36
215mm deep x 3600mm long	126.40	2.50	129.56	0.88	0.88	28.27	-	157.83	nr	173.62
215mm deep x 3900mm long	153.09	2.50	156.92	0.94	0.94	30.20	-	187.12	nr	205.83
Galvanized steel lintels; SUPERGALV (BIRTLEY) lintels reference CB 50 H.D.; bedding in cement-lime mortar (1:1:6)										
165mm deep x 750mm long	20.78	2.50	21.30	0.28	0.28	9.00	-	30.30	nr	33.33
165mm deep x 1050mm long	26.50	2.50	27.16	0.34	0.34	10.92	-	38.09	nr	41.90
165mm deep x 1200mm long	30.18	2.50	30.93	0.38	0.38	12.21	-	43.14	nr	47.46
165mm deep x 1350mm long	36.70	2.50	37.62	0.41	0.41	13.17	-	50.79	nr	55.87
165mm deep x 1500mm long	40.45	2.50	41.46	0.44	0.44	14.14	-	55.60	nr	61.16
215mm deep x 1650mm long	48.40	2.50	49.61	0.47	0.47	15.10	-	64.71	nr	71.18
215mm deep x 1800mm long	52.84	2.50	54.16	0.50	0.50	16.07	-	70.23	nr	77.25
215mm deep x 2100mm long	62.33	2.50	63.89	0.56	0.56	17.99	-	81.88	nr	90.07
215mm deep x 2250mm long	78.37	2.50	80.33	0.60	0.60	19.28	-	99.61	nr	109.57
215mm deep x 2400mm long	91.95	2.50	94.25	0.63	0.63	20.24	-	114.49	nr	125.94
215mm deep x 2550mm long	99.90	2.50	102.40	0.67	0.67	21.53	-	123.92	nr	136.32
215mm deep x 3000mm long	117.97	2.50	120.92	0.78	0.78	25.06	-	145.98	nr	160.58
Galvanized steel lintels; SUPERGALV (BIRTLEY) lintels reference CB 70; bedding in cement-lime mortar (1:1:6)										
120mm deep x 750mm long	16.22	2.50	16.63	0.29	0.29	9.32	-	25.94	nr	28.54
120mm deep x 1200mm long	25.77	2.50	26.41	0.40	0.40	12.85	-	39.27	nr	43.19
120mm deep x 1350mm long	30.27	2.50	31.03	0.43	0.43	13.82	-	44.84	nr	49.33
120mm deep x 1500mm long	33.63	2.50	34.47	0.46	0.46	14.78	-	49.25	nr	54.18
120mm deep x 1650mm long	38.09	2.50	39.04	0.49	0.49	15.74	-	54.79	nr	60.26
160mm deep x 1800mm long	41.36	2.50	42.39	0.53	0.53	17.03	-	59.42	nr	65.37
160mm deep x 1950mm long	45.35	2.50	46.48	0.56	0.56	17.99	-	64.48	nr	70.92
160mm deep x 2100mm long	48.25	2.50	49.46	0.59	0.59	18.96	-	68.41	nr	75.25
215mm deep x 2250mm long	55.45	2.50	56.84	0.63	0.63	20.24	-	77.08	nr	84.79
215mm deep x 2400mm long	59.12	2.50	60.60	0.66	0.66	21.21	-	81.80	nr	89.98
215mm deep x 2550mm long	67.96	2.50	69.66	0.70	0.70	22.49	-	92.15	nr	101.36
215mm deep x 2850mm long	89.78	2.50	92.02	0.75	0.75	24.10	-	116.12	nr	127.73
215mm deep x 3000mm long	96.44	2.50	98.85	0.79	0.79	25.38	-	124.23	nr	136.66
215mm deep x 3300mm long	111.52	2.50	114.31	0.85	0.85	27.31	-	141.62	nr	155.78
215mm deep x 3600mm long	122.58	2.50	125.64	0.92	0.92	29.56	-	155.20	nr	170.72
215mm deep x 3900mm long	149.67	2.50	153.41	0.99	0.99	31.81	-	185.22	nr	203.74
Galvanized steel lintels; SUPERGALV (BIRTLEY) lintels reference CB 70 H.D.; bedding in cement-lime mortar (1:1:6)										
160mm deep x 750mm long	19.78	2.50	20.27	0.29	0.29	9.32	-	29.59	nr	32.55
160mm deep x 1050mm long	26.46	2.50	27.12	0.36	0.36	11.57	-	38.69	nr	42.56
160mm deep x 1200mm long	29.65	2.50	30.39	0.40	0.40	12.85	-	43.24	nr	47.57
160mm deep x 1350mm long	35.88	2.50	36.78	0.43	0.43	13.82	-	50.59	nr	55.65
160mm deep x 1500mm long	39.73	2.50	40.72	0.46	0.46	14.78	-	55.50	nr	61.05
215mm deep x 1650mm long	52.16	2.50	53.46	0.49	0.49	15.74	-	69.21	nr	76.13
215mm deep x 1800mm long	55.51	2.50	56.90	0.53	0.53	17.03	-	73.93	nr	81.32
215mm deep x 2100mm long	63.83	2.50	65.43	0.59	0.59	18.96	-	84.38	nr	92.82
215mm deep x 2250mm long	78.73	2.50	80.70	0.63	0.63	20.24	-	100.94	nr	111.03
215mm deep x 2400mm long	90.33	2.50	92.59	0.66	0.66	21.21	-	113.79	nr	125.17
215mm deep x 2550mm long	99.45	2.50	101.94	0.70	0.70	22.49	-	124.43	nr	136.87
215mm deep x 3300mm long	115.90	2.50	118.80	0.82	0.82	26.35	-	145.14	nr	159.66
Galvanized steel lintels; SUPERGALV (BIRTLEY) lintels reference CB 50/130; bedding in cement-lime mortar (1:1:6)										
120mm deep x 750mm long	17.12	2.50	17.55	0.31	0.31	9.96	-	27.51	nr	30.26
120mm deep x 1200mm long	27.21	2.50	27.89	0.42	0.42	13.49	-	41.38	nr	45.52
120mm deep x 1350mm long	31.18	2.50	31.96	0.45	0.45	14.46	-	46.42	nr	51.06
120mm deep x 1500mm long	34.48	2.50	35.34	0.48	0.48	15.42	-	50.76	nr	55.84
120mm deep x 1650mm long	40.12	2.50	41.12	0.52	0.52	16.71	-	57.83	nr	63.61
165mm deep x 1800mm long	43.21	2.50	44.29	0.55	0.55	17.67	-	61.96	nr	68.16

MASONRY

Labour hourly rates: (except Specialists) Craft Operatives 18.37 Labourer 13.76 Rates are national average prices. Refer to REGIONAL VARIATIONS for indicative levels of overall pricing in regions	MATERIALS			LABOUR				RATES		
	Del to Site	Waste	Material Cost	Craft Optve	Lab	Labour Cost	Sunds	Nett Rate	Unit	Gross rate (10%)
	£	%	£	Hrs	Hrs	£	£	£		£

BRICK/BLOCK WALLING (Cont'd)

Extra over walls for opening perimeters; Lintels; steel (Cont'd)

Galvanized steel lintels; SUPERGALV (BIRTLEY) lintels reference CB 50/130; bedding in cement-lime mortar (1:1:6) (Cont'd)

	Del to Site	Waste	Material Cost	Craft Optve	Lab	Labour Cost	Sunds	Nett Rate	Unit	Gross rate
165mm deep x 1950mm long	47.68	2.50	48.87	0.58	0.58	18.64	-	67.51	nr	74.26
165mm deep x 2100mm long	51.51	2.50	52.80	0.62	0.62	19.92	-	72.72	nr	79.99
165mm deep x 2250mm long	59.03	2.50	60.51	0.66	0.66	21.21	-	81.71	nr	89.88
165mm deep x 2400mm long	62.98	2.50	64.55	0.69	0.69	22.17	-	86.72	nr	95.40
215mm deep x 2550mm long	71.37	2.50	73.15	0.74	0.74	23.78	-	96.93	nr	106.62
215mm deep x 2850mm long	100.58	2.50	103.09	0.78	0.78	25.06	-	128.16	nr	140.97
215mm deep x 3000mm long	103.97	2.50	106.57	0.83	0.83	26.67	-	133.24	nr	146.56
215mm deep x 3300mm long	116.11	2.50	119.01	0.89	0.89	28.60	-	147.61	nr	162.37
215mm deep x 3600mm long	127.86	2.50	131.06	0.97	0.97	31.17	-	162.22	nr	178.44
215mm deep x 3900mm long	187.38	2.50	192.06	1.03	1.03	33.09	-	225.16	nr	247.67

Galvanized steel lintels; SUPERGALV (BIRTLEY) lintels reference CB 50/130 H.D.; bedding in cement-lime mortar (1:1:6)

160mm deep x 750mm long	30.71	2.50	31.48	0.31	0.31	9.96	-	41.44	nr	45.58
160mm deep x 1200mm long	49.48	2.50	50.72	0.42	0.42	13.49	-	64.21	nr	70.63
160mm deep x 1350mm long	55.26	2.50	56.64	0.45	0.45	14.46	-	71.10	nr	78.21
160mm deep x 1500mm long	61.40	2.50	62.94	0.48	0.48	15.42	-	78.36	nr	86.19
215mm deep x 1650mm long	65.81	2.50	67.46	0.52	0.52	16.71	-	84.16	nr	92.58
215mm deep x 1800mm long	70.89	2.50	72.66	0.55	0.55	17.67	-	90.33	nr	99.37
215mm deep x 1950mm long	76.76	2.50	78.68	0.58	0.58	18.64	-	97.31	nr	107.05
215mm deep x 2250mm long	97.41	2.50	99.85	0.66	0.66	21.21	-	121.05	nr	133.16
215mm deep x 2400mm long	107.30	2.50	109.98	0.69	0.69	22.17	-	132.15	nr	145.37
215mm deep x 2550mm long	119.57	2.50	122.56	0.74	0.74	23.78	-	146.34	nr	160.97
215mm deep x 2700mm long	133.68	2.50	137.02	0.80	0.80	25.70	-	162.73	nr	179.00
215mm deep x 3300mm long	179.52	2.50	184.01	0.86	0.86	27.63	-	211.64	nr	232.80

Galvanized steel lintels; SUPERGALV (BIRTLEY) lintels reference CB 70/130; bedding in cement-lime mortar (1:1:6)

115mm deep x 750mm long	21.40	2.50	21.94	0.31	0.31	9.96	-	31.90	nr	35.09
115mm deep x 1200mm long	33.02	2.50	33.85	0.42	0.42	13.49	-	47.34	nr	52.07
115mm deep x 1350mm long	38.00	2.50	38.95	0.45	0.45	14.46	-	53.41	nr	58.75
115mm deep x 1500mm long	41.70	2.50	42.74	0.48	0.48	15.42	-	58.16	nr	63.98
115mm deep x 1650mm long	51.42	2.50	52.71	0.52	0.52	16.71	-	69.41	nr	76.35
155mm deep x 1800mm long	57.36	2.50	58.79	0.55	0.55	17.67	-	76.47	nr	84.11
155mm deep x 1950mm long	62.14	2.50	63.69	0.58	0.58	18.64	-	82.33	nr	90.56
155mm deep x 2100mm long	66.05	2.50	67.70	0.62	0.62	19.92	-	87.62	nr	96.38
155mm deep x 2250mm long	89.35	2.50	91.58	0.66	0.66	21.21	-	112.79	nr	124.07
155mm deep x 2400mm long	95.30	2.50	97.68	0.69	0.69	22.17	-	119.85	nr	131.84
210mm deep x 2550mm long	105.64	2.50	108.28	0.74	0.74	23.78	-	132.06	nr	145.26
210mm deep x 2850mm long	140.87	2.50	144.39	0.78	0.78	25.06	-	169.45	nr	186.40
210mm deep x 3000mm long	157.30	2.50	161.23	0.83	0.83	26.67	-	187.90	nr	206.69
210mm deep x 3300mm long	173.02	2.50	177.35	0.89	0.89	28.60	-	205.94	nr	226.54

Galvanized steel lintels; SUPERGALV (BIRTLEY) lintels reference HD50; bedding in cement-lime mortar (1:1:6)

225mm deep x 3900mm long	182.52	2.50	187.08	1.02	1.02	32.77	-	219.86	nr	241.84
225mm deep x 4200mm long	204.45	2.50	209.56	1.05	1.05	33.74	-	243.30	nr	267.63
225mm deep x 4500mm long	223.70	2.50	229.29	1.09	1.09	35.02	-	264.31	nr	290.75
225mm deep x 4800mm long	237.91	2.50	243.86	1.12	1.12	35.99	-	279.84	nr	307.83

Galvanized steel lintels; SUPERGALV (BIRTLEY) lintels reference HD 70; bedding in cement-lime mortar (1:1:6)

225mm deep x 3000mm long	132.17	2.50	135.47	0.85	0.85	27.31	-	162.78	nr	179.06
225mm deep x 3300mm long	144.46	2.50	148.07	0.89	0.89	28.60	-	176.67	nr	194.33
225mm deep x 3600mm long	154.09	2.50	157.94	0.92	0.92	29.56	-	187.50	nr	206.25

Galvanized steel lintels; SUPERGALV (BIRTLEY) lintels reference SB100; bedding in cement-lime mortar (1:1:6)

75mm deep x 750mm long	11.75	2.50	12.04	0.45	0.45	14.46	-	26.50	nr	29.15
75mm deep x 900mm long	13.72	2.50	14.06	0.48	0.48	15.42	-	29.49	nr	32.43
75mm deep x 1050mm long	16.26	2.50	16.67	0.51	0.51	16.39	-	33.05	nr	36.36
75mm deep x 1200mm long	18.06	2.50	18.51	0.54	0.54	17.35	-	35.86	nr	39.45
75mm deep x 1500mm long	22.63	2.50	23.20	0.60	0.60	19.28	-	42.47	nr	46.72
140mm deep x 1800mm long	32.33	2.50	33.14	0.65	0.65	20.88	-	54.02	nr	59.43
140mm deep x 1950mm long	36.78	2.50	37.70	0.69	0.69	22.17	-	59.87	nr	65.86
140mm deep x 2100mm long	39.06	2.50	40.04	0.73	0.73	23.45	-	63.49	nr	69.84
140mm deep x 2250mm long	42.02	2.50	43.07	0.77	0.77	24.74	-	67.81	nr	74.59
140mm deep x 2400mm long	45.28	2.50	46.41	0.82	0.82	26.35	-	72.76	nr	80.03
140mm deep x 2550mm long	48.85	2.50	50.07	0.87	0.87	27.95	-	78.02	nr	85.83
140mm deep x 2700mm long	51.93	2.50	53.23	0.92	0.92	29.56	-	82.79	nr	91.07
215mm deep x 2850mm long	79.06	2.50	81.04	0.93	0.93	29.88	-	110.92	nr	122.01
215mm deep x 3000mm long	83.80	2.50	85.90	0.95	0.95	30.52	-	116.42	nr	128.06

Galvanized steel lintels; SUPERGALV (BIRTLEY) lintels reference SB140; bedding in cement-lime mortar (1:1:6)

140mm deep x 750mm long	14.09	2.50	14.44	0.45	0.45	14.46	-	28.90	nr	31.79
140mm deep x 900mm long	16.84	2.50	17.26	0.48	0.48	15.42	-	32.68	nr	35.95
140mm deep x 1050mm long	19.86	2.50	20.36	0.51	0.51	16.39	-	36.74	nr	40.42
140mm deep x 1200mm long	22.71	2.50	23.28	0.54	0.54	17.35	-	40.63	nr	44.69
140mm deep x 1350mm long	25.76	2.50	26.40	0.57	0.57	18.31	-	44.72	nr	49.19
140mm deep x 1500mm long	29.13	2.50	29.86	0.60	0.60	19.28	-	49.14	nr	54.05
140mm deep x 1650mm long	34.98	2.50	35.85	0.62	0.62	19.92	-	55.78	nr	61.35
140mm deep x 1800mm long	38.03	2.50	38.98	0.65	0.65	20.88	-	59.87	nr	65.85
140mm deep x 1950mm long	41.68	2.50	42.72	0.69	0.69	22.17	-	64.89	nr	71.38
140mm deep x 2100mm long	46.85	2.50	48.02	0.73	0.73	23.45	-	71.48	nr	78.62
140mm deep x 2250mm long	51.67	2.50	52.96	0.77	0.77	24.74	-	77.70	nr	85.47
140mm deep x 2400mm long	54.25	2.50	55.61	0.82	0.82	26.35	-	81.95	nr	90.15
140mm deep x 2550mm long	57.49	2.50	58.93	0.87	0.87	27.95	-	86.88	nr	95.57
140mm deep x 2700mm long	60.87	2.50	62.39	0.92	0.92	29.56	-	91.95	nr	101.15
215mm deep x 2850mm long	88.23	2.50	90.44	0.95	0.95	30.52	-	120.96	nr	133.06

Labour hourly rates: (except Specialists) Craft Operatives 18.37 Labourer 13.76 Rates are national average prices. Refer to REGIONAL VARIATIONS for indicative levels of overall pricing in regions	MATERIALS			LABOUR				RATES		
	Del to Site £	Waste %	Material Cost £	Craft Optve Hrs	Lab Hrs	Labour Cost £	Sunds £	Nett Rate £	Unit	Gross rate (10%) £

BRICK/BLOCK WALLING (Cont'd)

Extra over walls for opening perimeters; Lintels; steel (Cont'd)

Galvanized steel lintels; SUPERGALV (BIRTLEY) lintels reference SB140; bedding in cement-lime mortar (1:1:6) (Cont'd)

	Del to Site	Waste	Material Cost	Craft Optve	Lab	Labour Cost	Sunds	Nett Rate	Unit	Gross rate
215mm deep x 3000mm long	91.86	2.50	94.16	0.97	0.97	31.17	-	125.32	nr	137.85

Galvanized steel lintels; SUPERGALV (BIRTLEY) lintels reference SBL200; bedding in cement-lime mortar (1:1:6)

	Del to Site	Waste	Material Cost	Craft Optve	Lab	Labour Cost	Sunds	Nett Rate	Unit	Gross rate
142mm deep x 750mm long	19.22	2.50	19.70	0.45	0.45	14.46	-	34.16	nr	37.57
142mm deep x 900mm long	23.77	2.50	24.36	0.48	0.48	15.42	-	39.79	nr	43.77
142mm deep x 1050mm long	26.77	2.50	27.44	0.51	0.51	16.39	-	43.83	nr	48.21
142mm deep x 1200mm long	31.17	2.50	31.95	0.54	0.54	17.35	-	49.30	nr	54.23
142mm deep x 1350mm long	34.19	2.50	35.04	0.60	0.60	19.28	-	54.32	nr	59.76
142mm deep x 1500mm long	39.81	2.50	40.81	0.65	0.65	20.88	-	61.69	nr	67.86
142mm deep x 1650mm long	43.74	2.50	44.83	0.67	0.67	21.53	-	66.36	nr	73.00
142mm deep x 1800mm long	49.27	2.50	50.50	0.72	0.72	23.13	-	73.64	nr	81.00
142mm deep x 1950mm long	52.19	2.50	53.49	0.76	0.76	24.42	-	77.91	nr	85.70
142mm deep x 2100mm long	57.67	2.50	59.11	0.80	0.77	25.29	-	84.40	nr	92.84
142mm deep x 2250mm long	64.06	2.50	65.66	0.77	0.80	25.15	-	90.81	nr	99.90
142mm deep x 2400mm long	69.47	2.50	71.21	0.82	0.82	26.35	-	97.55	nr	107.31
142mm deep x 2550mm long	75.20	2.50	77.08	0.87	0.87	27.95	-	105.03	nr	115.54
142mm deep x 2700mm long	81.14	2.50	83.17	0.92	0.92	29.56	-	112.73	nr	124.00
218mm deep x 2850mm long	107.01	2.50	109.69	0.93	0.93	29.88	-	139.57	nr	153.52
218mm deep x 3000mm long	113.50	2.50	116.34	0.95	0.95	30.52	-	146.86	nr	161.55

Galvanized steel lintels; SUPERGALV (BIRTLEY) internal door lintels reference INT100; bedding in cement-lime mortar (1:1:6)

	Del to Site	Waste	Material Cost	Craft Optve	Lab	Labour Cost	Sunds	Nett Rate	Unit	Gross rate
100mm wide, 900mm long	4.06	2.50	4.16	0.33	0.33	10.60	-	14.76	nr	16.24
100mm wide, 1050mm long	4.66	2.50	4.78	0.36	0.36	11.57	-	16.34	nr	17.98
100mm wide, 1200mm long	5.16	2.50	5.29	0.38	0.38	12.21	-	17.50	nr	19.25

Special purpose blocks or stones; Precast concrete padstones; BS 8500, designed mix C25, 20mm aggregate, minimum cement content 360 kg/m³; vibrated; reinforced at contractor's discretion; bedding in cement-lime mortar (1:1:6)

Padstones

	Del to Site	Waste	Material Cost	Craft Optve	Lab	Labour Cost	Sunds	Nett Rate	Unit	Gross rate
215 x 215 x 75mm	5.45	2.50	5.59	0.06	0.03	1.52	0.17	7.27	nr	8.00
215 x 215 x 150mm	9.79	2.50	10.04	0.08	0.04	2.02	0.17	12.22	nr	13.45
327 x 215 x 150mm	14.69	2.50	15.06	0.11	0.05	2.71	0.18	17.95	nr	19.74
327 x 327 x 150mm	22.48	2.50	23.04	0.17	0.08	4.22	0.19	27.46	nr	30.21
440 x 215 x 150mm	17.75	2.50	18.20	0.17	0.08	4.22	0.21	22.63	nr	24.89
440 x 327 x 150mm	30.05	2.50	30.80	0.20	0.10	5.05	0.22	36.08	nr	39.69
440 x 440 x 150mm	40.07	2.50	41.07	0.20	0.10	5.05	0.24	46.36	nr	51.00

Forming cavity

	Del to Site	Waste	Material Cost	Craft Optve	Lab	Labour Cost	Sunds	Nett Rate	Unit	Gross rate
Cavity with 200mm stainless steel housing type 4 wall ties, -4/m² built in width of cavity 50mm	0.29	10.00	0.31	0.10	0.07	2.80	-	3.12	m²	3.43
Cavity with 200mm stainless steel housing type 4 wall ties -4/m² built in; 50mm fibreglass resin bonded slab cavity insulation width of cavity 50mm	2.33	10.00	2.57	0.35	0.25	9.87	1.62	14.06	m²	15.46
Cavity with 203mm stainless steel vertical twisted wall ties -4/m² built in; 50mm fibreglass resin bonded slab cavity insulation width of cavity 50mm	4.05	10.00	4.46	0.35	0.25	9.87	1.62	15.95	m²	17.54
Cavity with 200mm stainless steel housing type 4 wall ties -4/m² built in; 75mm fibreglass resin bonded slab cavity insulation width of cavity 75mm	2.71	10.00	2.98	0.40	0.25	10.79	1.62	15.39	m²	16.93
Cavity with stainless steel vertical twisted wall ties -4/m² built in; 75mm fibreglass resin bonded slab cavity insulation width of cavity 75mm	4.43	10.00	4.87	0.40	0.25	10.79	1.62	17.28	m²	19.01
Cavity with 205mm Staifix TJ2 stainless steel thin joint wall ties - 4/m² built in; 25mm Celotex CW4025 cavity insulation boards width of cavity 50mm	6.05	10.00	6.65	0.35	0.35	11.25	1.62	19.52	m²	21.47
Cavity with 230mm Staifix TJ2 stainless steel thin joint wall ties - 4/m² built in; 40mm Celotex CW4040 cavity insulation boards width of cavity 75mm	7.09	10.00	7.80	0.40	0.35	12.16	1.62	21.58	m²	23.74
Cavity with 255mm Staifix TJ2 stainless steel thin joint wall ties - 4/m² built in; 50mm Celotex CW4050 cavity insulation boards width of cavity 80mm	8.22	10.00	9.04	0.40	0.35	12.16	1.62	22.82	m²	25.10
Cavity with 255mm Staifix TJ2 stainless steel thin joint wall ties - 4/m² built in; 75mm Celotex CW4075 cavity insulation boards width of cavity 100mm	10.81	10.00	11.90	0.45	0.40	13.77	1.62	27.29	m²	30.01
Cavity with 255mm Staifix TJ2 stainless steel thin joint wall ties - 4/m² built in; 100mm Celotex CW4100 cavity insulation boards width of cavity 100mm	12.38	10.00	13.62	0.45	0.40	13.77	1.62	29.01	m²	31.91

Damp proof courses

BS 6515 Polythene; 100mm laps; bedding in cement mortar (1:3); no allowance made for laps

	Del to Site	Waste	Material Cost	Craft Optve	Lab	Labour Cost	Sunds	Nett Rate	Unit	Gross rate
100mm wide; vertical	0.08	5.00	0.08	0.10	-	1.84	-	1.92	m	2.11
225mm wide; vertical	0.18	5.00	0.19	0.15	-	2.76	-	2.94	m	3.24
width exceeding 300mm; vertical	0.79	5.00	0.83	0.45	-	8.27	-	9.09	m2	10.00
100mm wide; horizontal	0.08	5.00	0.08	0.06	-	1.10	-	1.18	m	1.30
225mm wide; horizontal	0.18	5.00	0.19	0.09	-	1.65	-	1.84	m	2.02
width exceeding 300mm; horizontal	0.79	5.00	0.83	0.30	-	5.51	-	6.34	m2	6.97

Labour hourly rates: (except Specialists) Craft Operatives 18.37 Labourer 13.76 Rates are national average prices. Refer to REGIONAL VARIATIONS for indicative levels of overall pricing in regions	MATERIALS			LABOUR				RATES		
	Del to Site	Waste	Material Cost	Craft Optve	Lab	Labour Cost	Sunds	Nett Rate	Unit	Gross rate (10%)
	£	%	£	Hrs	Hrs	£	£	£		£
BRICK/BLOCK WALLING (Cont'd)										
Damp proof courses (Cont'd)										
BS 6515 Polythene; 100mm laps; bedding in cement mortar (1:3); no allowance made for laps (Cont'd)										
cavity trays; 275mm wide; horizontal	0.24	5.00	0.25	0.17	-	3.12	-	3.37	m	3.71
cavity trays; width exceeding 300mm; horizontal	0.79	5.00	0.83	0.50	-	9.19	-	10.01	m²	11.01
Hyload, pitch polymer; 100mm laps; bedding in cement lime mortar (1:1:6); no allowance made for laps										
100mm wide; vertical	0.48	5.00	0.51	0.10	-	1.84	-	2.34	m	2.58
225mm wide; vertical	1.09	5.00	1.14	0.15	-	2.76	-	3.90	m	4.29
width exceeding 300mm; vertical	4.83	5.00	5.07	0.45	-	8.27	-	13.34	m²	14.67
100mm wide; horizontal	0.48	5.00	0.51	0.06	-	1.10	-	1.61	m	1.77
225mm wide; horizontal	1.09	5.00	1.14	0.09	-	1.65	-	2.79	m	3.07
width exceeding 300mm; horizontal	4.83	5.00	5.07	0.30	-	5.51	-	10.58	m²	11.64
Synthaprufe bituminous latex emulsion; two coats brushed on; blinded with sand										
225mm wide	1.18	5.00	1.23	-	0.15	2.06	0.10	3.40	m	3.74
width exceeding 300mm; vertical	3.91	5.00	4.10	-	0.30	4.13	0.21	8.44	m²	9.29
Synthaprufe bituminous latex emulsion; three coats brushed on; blinded with sand										
225mm wide	1.62	5.00	1.70	-	0.20	2.75	0.10	4.55	m	5.01
width exceeding 300mm; vertical	5.88	5.00	6.17	-	0.42	5.78	0.24	12.19	m²	13.41
Bituminous emulsion; two coats brushed on										
225mm wide	0.48	5.00	0.50	-	0.12	1.65	0.10	2.26	m	2.48
width exceeding 300mm; vertical	1.58	5.00	1.66	-	0.26	3.58	0.21	5.45	m²	6.00
Bituminous emulsion; three coats brushed on										
225mm wide	0.88	5.00	0.92	-	0.18	2.48	0.10	3.51	m	3.86
width exceeding 300mm; vertical	3.06	5.00	3.21	-	0.36	4.95	0.24	8.40	m²	9.24
One course slates in cement mortar (1:3)										
112mm wide; vertical	1.37	15.00	1.57	0.20	0.14	5.60	-	7.17	m	7.89
225mm wide; vertical	2.73	15.00	3.14	0.40	0.27	11.06	-	14.20	m	15.62
width exceeding 300mm; vertical	9.36	15.00	10.76	1.20	0.80	33.05	-	43.82	m²	48.20
112mm wide; horizontal	1.37	15.00	1.57	0.12	0.10	3.67	-	5.24	m	5.77
225mm wide; horizontal	2.73	15.00	3.14	0.25	0.20	7.34	-	10.48	m	11.53
width exceeding 300mm; horizontal	9.36	15.00	10.76	0.70	0.50	19.74	-	30.50	m²	33.55
Two courses slates in cement mortar (1:3)										
112mm wide; vertical	2.73	15.00	3.14	0.33	0.22	9.09	-	12.23	m	13.45
225mm wide; vertical	5.46	15.00	6.28	0.66	0.42	17.90	-	24.18	m	26.60
width exceeding 300mm; vertical	18.72	15.00	21.53	2.00	1.35	55.32	-	76.84	m²	84.53
112mm wide; horizontal	2.73	15.00	3.14	0.20	0.13	5.46	-	8.60	m	9.46
225mm wide; horizontal	5.46	15.00	6.28	0.40	0.26	10.93	-	17.20	m	18.93
width exceeding 300mm; horizontal	18.72	15.00	21.53	1.20	0.80	33.05	-	54.58	m²	60.04
Cavity trays										
Type G Cavitray by Cavity Trays Ltd in										
half brick skin of cavity wall	9.08	5.00	9.53	0.20	-	3.67	-	13.21	m	14.53
external angle	7.94	5.00	8.34	0.10	-	1.84	-	10.17	nr	11.19
internal angle	7.94	5.00	8.34	0.10	-	1.84	-	10.17	nr	11.19
Type X Cavitray by Cavity Trays Ltd to suit 40 degree pitched roof complete with attached code 4 lead flashing and dress over tiles in										
half brick skin of cavity wall	49.64	5.00	52.12	0.20	-	3.67	-	55.80	m	61.38
ridge tray	11.98	5.00	12.58	0.10	-	1.84	-	14.42	nr	15.86
catchment tray; long	8.87	5.00	9.31	0.10	-	1.84	-	11.15	nr	12.27
corner catchment angle tray	15.70	5.00	16.48	0.10	-	1.84	-	18.32	nr	20.15
Type W Cavity weep/ventilator by Cavity Trays Ltd in										
half brick skin of cavity wall	0.59	5.00	0.62	0.10	-	1.84	-	2.45	nr	2.70
extension duct	0.94	5.00	0.99	0.10	-	1.84	-	2.82	nr	3.11
Joint reinforcement										
Expamet grade 304/S15 Exmet reinforcement, stainless steel; 150mm laps										
65mm wide	5.07	5.00	5.32	0.04	0.02	1.01	0.05	6.37	m	7.01
115mm wide	7.07	5.00	7.43	0.05	0.03	1.33	0.05	8.80	m	9.68
175mm wide	11.35	5.00	11.92	0.06	0.03	1.52	0.06	13.50	m	14.85
225mm wide	15.48	5.00	16.25	0.07	0.04	1.84	0.08	18.16	m	19.98
Brickforce, galvanised steel; 150mm laps										
60mm wide ref GBF30W60	1.50	5.00	1.57	0.04	0.02	1.01	0.05	2.63	m	2.89
60mm wide ref GBF35W60	2.03	5.00	2.14	0.04	0.02	1.01	0.05	3.19	m	3.51
60mm wide ref GBF40W60	1.30	5.00	1.36	0.04	0.02	1.01	0.05	2.42	m	2.66
60mm wide ref GBF45W60	3.56	5.00	3.74	0.04	0.02	1.01	0.05	4.79	m	5.27
60mm wide ref GBF50W60	1.92	5.00	2.01	0.04	0.02	1.01	0.05	3.07	m	3.37
160mm wide ref GBF30W160	4.18	5.00	4.39	0.05	0.03	1.33	0.05	5.77	m	6.35
160mm wide ref GBF35W160	5.68	5.00	5.97	0.05	0.03	1.33	0.05	7.34	m	8.08
160mm wide ref GBF40W160	6.83	5.00	7.17	0.05	0.03	1.33	0.05	8.54	m	9.40
160mm wide ref GBF45W160	8.90	5.00	9.35	0.05	0.03	1.33	0.05	10.73	m	11.80
160mm wide ref GBF50W160	11.38	5.00	11.95	0.05	0.03	1.33	0.05	13.33	m	14.66
Brickforce, stainless steel; 150mm laps										
60mm wide ref SBF30W60	4.50	5.00	4.72	0.04	0.02	1.01	0.05	5.77	m	6.35
60mm wide ref SBF35W60	6.11	5.00	6.41	0.04	0.02	1.01	0.05	7.47	m	8.22
60mm wide ref SBF40W60	3.24	5.00	3.40	0.04	0.02	1.01	0.05	4.46	m	4.91
60mm wide ref SBF45W60	9.89	5.00	10.38	0.04	0.02	1.01	0.05	11.43	m	12.58
60mm wide ref SBF50W60	4.84	5.00	5.09	0.04	0.02	1.01	0.05	6.14	m	6.76
160mm wide ref SBF30W160	5.10	5.00	5.35	0.05	0.03	1.33	0.05	6.73	m	7.40
160mm wide ref SBF35W160	7.51	5.00	7.89	0.05	0.03	1.33	0.05	9.26	m	10.19

Labour hourly rates: (except Specialists) Craft Operatives 18.37 Labourer 13.76 Rates are national average prices. Refer to REGIONAL VARIATIONS for indicative levels of overall pricing in regions	MATERIALS			LABOUR				RATES		
	Del to Site	Waste	Material Cost	Craft Optve	Lab	Labour Cost	Sunds	Nett Rate	Unit	Gross rate (10%)
	£	%	£	Hrs	Hrs	£	£	£		£
BRICK/BLOCK WALLING (Cont'd)										
Joint reinforcement (Cont'd)										
Brickforce, stainless steel; 150mm laps (Cont'd)										
160mm wide ref SBF40W160	9.14	5.00	9.59	0.05	0.03	1.33	0.05	10.97	m	12.07
160mm wide ref SBF45W160	11.20	5.00	11.76	0.05	0.03	1.33	0.05	13.13	m	14.45
160mm wide ref SBF50W160	16.92	5.00	17.76	0.05	0.03	1.33	0.05	19.14	m	21.05
Pointing; Pointing in flashings										
Cement mortar (1:3)										
horizontal	-	-	-	0.40	-	7.35	0.33	7.68	m	8.45
horizontal; in old wall	-	-	-	0.45	-	8.27	0.33	8.60	m	9.46
stepped	-	-	-	0.65	-	11.94	0.52	12.47	m	13.71
stepped; in old wall	-	-	-	0.70	-	12.86	0.52	13.38	m	14.72
Joints										
Expansion joints in facing brickwork 15mm wide; vertical; filling with Servicised Ltd, Aerofill 1 filler, Vertiseal compound pointing one side; including preparation, cleaners, primers and sealers										
102mm thick wall	5.41	7.50	5.82	0.09	0.18	4.13	-	9.95	m	10.94
215mm thick wall	6.94	7.50	7.46	0.10	0.20	4.59	-	12.05	m	13.26
Expansion joints in blockwork 15mm wide; vertical; filling with Servicised Ltd, Aerofill 1 filler, Vertiseal compound pointing one side; including preparation, cleaners, primers and sealers										
100mm thick wall	5.41	7.50	5.82	0.09	0.18	4.13	-	9.95	m	10.94
Expansion joints in glass blockwork 10mm wide; vertical; in filling with compressible material, polysulphide sealant both sides including preparation, cleaners, primers and sealers										
80mm thick wall	6.75	7.50	7.26	0.09	0.18	4.13	-	11.39	m	12.53
Wedging and pinning										
Two courses slates in cement mortar (1:3)										
width of wall 215mm	4.68	15.00	5.38	0.50	0.25	12.62	0.19	18.20	m	20.02
Creasing										
Nibless creasing tiles, red, machine made, 265 x 165 x 10mm; in cement-lime mortar (1:1:6)										
one course 253mm wide	4.66	5.00	4.90	0.50	0.35	14.00	0.34	19.24	m	21.17
one course 365mm wide	7.56	5.00	7.94	0.70	0.55	20.43	0.50	28.86	m	31.75
two courses 253mm wide	9.32	5.00	9.79	0.85	0.55	23.18	0.70	33.68	m	37.05
two courses 365mm wide	15.12	5.00	15.88	1.15	0.85	32.82	0.99	49.69	m	54.66
Proprietary and individual spot items; Chimney pots										
Clay chimney pots; set and flaunched in cement mortar (1:3)										
tapered roll top; 600mm high	41.29	5.00	43.35	0.75	0.63	22.45	0.57	66.37	nr	73.01
tapered roll top; 750mm high	55.48	5.00	58.25	0.88	0.75	26.49	0.72	85.46	nr	94.01
tapered roll top; 900mm high	74.48	5.00	78.20	1.00	0.93	31.17	0.86	110.23	nr	121.25
Proprietary and individual spot items; Air bricks										
Clay air bricks, BS 493, square hole pattern; opening with slate lintel over										
common brick wall 102mm thick; opening size										
215 x 65mm	3.08	-	3.08	0.20	0.10	5.05	0.86	8.98	nr	9.88
215 x 140mm	4.57	-	4.57	0.25	0.13	6.38	0.90	11.85	nr	13.04
215 x 215mm	11.47	-	11.47	0.30	0.15	7.58	0.99	20.03	nr	22.04
common brick wall 215mm thick; opening size										
215 x 65mm	3.08	-	3.08	0.30	0.15	7.58	1.62	12.28	nr	13.50
215 x 140mm	4.57	-	4.57	0.38	0.19	9.60	1.70	15.86	nr	17.45
215 x 215mm	11.47	-	11.47	0.45	0.23	11.43	1.82	24.72	nr	27.19
common brick wall 215mm thick; facework one side; opening size										
215 x 65mm	3.08	-	3.08	0.40	0.20	10.10	1.62	14.80	nr	16.28
215 x 140mm	4.57	-	4.57	0.50	0.25	12.62	1.70	18.89	nr	20.78
215 x 215mm	11.47	-	11.47	0.60	0.30	15.15	1.82	28.44	nr	31.28
common brick wall 327mm thick; facework one side; opening size										
215 x 65mm	3.08	-	3.08	0.60	0.30	15.15	2.61	20.84	nr	22.92
215 x 140mm	4.57	-	4.57	0.70	0.35	17.68	2.74	24.99	nr	27.49
215 x 215mm	11.47	-	11.47	0.80	0.40	20.20	2.84	34.51	nr	37.96
common brick wall 440mm thick; facework one side; opening size										
215 x 65mm	3.08	-	3.08	0.80	0.40	20.20	3.48	26.76	nr	29.44
215 x 140mm	4.57	-	4.57	0.90	0.45	22.73	3.60	30.90	nr	33.98
215 x 215mm	11.47	-	11.47	1.00	0.50	25.25	3.72	40.44	nr	44.48
cavity wall 252mm thick with 102mm facing brick outer skin, 100mm block inner skin and 50mm cavity; sealing cavity with slates in cement mortar (1:3); rendering all round with cement mortar (1:3); opening size										
215 x 65mm	3.08	-	3.08	0.60	0.30	15.15	3.13	21.37	nr	23.50
215 x 140mm	4.57	-	4.57	0.70	0.35	17.68	3.86	26.10	nr	28.71
215 x 215mm	11.47	-	11.47	0.80	0.40	20.20	4.51	36.19	nr	39.80
Proprietary and individual spot items; Fires and fire parts										
Solid one piece or two piece firebacks; BS 1251; bed and joint in fire cement; concrete filling at back										
fire size 400mm	62.80	2.50	64.37	1.00	0.67	27.59	1.62	93.58	nr	102.94
fire size 450mm	75.60	2.50	77.49	1.10	0.73	30.25	1.80	109.54	nr	120.50

MASONRY

Labour hourly rates: (except Specialists) Craft Operatives 18.37 Labourer 13.76 Rates are national average prices. Refer to REGIONAL VARIATIONS for indicative levels of overall pricing in regions	MATERIALS			LABOUR				RATES		
	Del to Site	Waste	Material Cost	Craft Optve	Lab	Labour Cost	Sunds	Nett Rate	Unit	Gross rate (10%)
	£	%	£	Hrs	Hrs	£	£	£		£
BRICK/BLOCK WALLING (Cont'd)										
Proprietary and individual spot items; Fires and fire parts (Cont'd)										
Frets and stools; black vitreous enamelled; place in position										
fire size 400mm	38.84	-	38.84	0.33	0.22	9.09	-	47.93	nr	52.72
fire size 450mm	48.82	-	48.82	0.36	0.24	9.92	-	58.73	nr	64.61
Frets and stools; lustre finish; place in position										
fire size 400mm	54.04	-	54.04	0.33	0.22	9.09	-	63.13	nr	69.44
fire size 450mm	58.96	-	58.96	0.36	0.24	9.92	-	68.87	nr	75.76
Threefold brick sets										
fire size 400mm	21.82	2.50	22.36	1.00	0.67	27.59	1.62	51.57	nr	56.73
fire size 450mm	22.98	2.50	23.56	1.00	0.67	27.59	1.80	52.95	nr	58.24
Proprietary and individual spot items; Gas flue blocks										
Dunbrik concrete gas flue blocks; bedding and jointing in flue joint silicone sealant										
recess block reference 1RM, 460 x 140 x 222mm	8.02	5.00	8.42	0.30	-	5.51	1.80	15.73	nr	17.31
deep and wide recess block reference MS, 554 x 272 x 222mm	10.44	5.00	10.96	0.35	-	6.43	1.95	19.34	nr	21.28
gather block reference 1GM, 460 x 140 x 222mm	12.80	5.00	13.44	0.45	-	8.27	1.80	23.51	nr	25.86
deep and wide gather block reference MG, 554 x 272 x 222mm	19.53	5.00	20.51	0.50	-	9.19	1.95	31.64	nr	34.81
Dense backup block 100mm reference MD100, 525 x 100 x 215	5.99	5.00	6.29	0.20	-	3.67	0.90	10.86	nr	11.95
straight flue block reference 2M, 320 x 140 x 225mm	5.47	5.00	5.74	0.25	-	4.59	1.73	12.06	nr	13.27
straight flue block reference 2M/150, 320 x 140 x 150mm	4.13	5.00	4.34	0.25	-	4.59	1.73	10.65	nr	11.72
straight flue block reference 2M/75, 320 x 140 x 75mm	4.13	5.00	4.34	0.30	-	5.51	1.73	11.57	nr	12.73
95mm lateral offset block reference 3ME, 340 x 140 x 225mm	6.15	5.00	6.46	0.37	-	6.80	1.80	15.05	nr	16.56
125mm backset block reference 2BM, 245 x 265 x 260mm	19.11	5.00	20.07	0.35	-	6.43	1.80	28.30	nr	31.12
Top exit transfer block reference 4M, 265 x 200 x 160mm	12.28	5.00	12.89	0.35	-	6.43	1.73	21.05	nr	23.15
Side exit transfer block reference 5M, 340 x 140 x 225mm	17.98	5.00	18.88	0.40	-	7.35	1.80	28.03	nr	30.83
Proprietary and individual spot items; Arch bars										
Steel flat arch bar										
30 x 6mm	5.70	-	5.70	0.70	-	12.86	-	18.56	m	20.41
50 x 6mm	9.54	-	9.54	0.70	-	12.86	-	22.40	m	24.64
Steel angle arch bar										
50 x 50 x 6mm	14.78	-	14.78	0.70	-	12.86	-	27.64	m	30.41
75 x 50 x 6mm	18.69	-	18.69	0.80	-	14.70	-	33.38	m	36.72
Proprietary and individual spot items; Building in										
Building in metal windows; building in lugs; bedding in cement mortar (1:3), pointing with Secomastic standard mastic one side										
200mm high; lugs to brick jambs; plugging and screwing to brick sill and concrete head										
500mm wide	-	-	-	0.53	0.22	12.76	1.39	14.16	nr	15.57
600mm wide	-	-	-	0.58	0.24	13.96	1.74	15.70	nr	17.27
900mm wide	-	-	-	0.74	0.29	17.58	2.38	19.97	nr	21.97
1200mm wide	-	-	-	0.90	0.34	21.21	3.08	24.29	nr	26.72
1500mm wide	-	-	-	1.06	0.39	24.84	3.72	28.56	nr	31.41
1800mm wide	-	-	-	1.22	0.44	28.47	4.36	32.83	nr	36.11
500mm high; lugs to brick jambs; plugging and screwing to brick sill and concrete head										
500mm wide	-	-	-	0.64	0.25	15.20	2.16	17.36	nr	19.09
600mm wide	-	-	-	0.69	0.27	16.39	2.37	18.76	nr	20.64
900mm wide	-	-	-	0.85	0.32	20.02	3.02	23.03	nr	25.34
1200mm wide	-	-	-	1.02	0.37	23.83	3.67	27.50	nr	30.25
1500mm wide	-	-	-	1.18	0.43	27.59	4.36	31.96	nr	35.15
1800mm wide	-	-	-	1.34	0.48	31.22	4.97	36.19	nr	39.80
700mm high; lugs to brick jambs; plugging and screwing to brick sill and concrete head										
500mm wide	-	-	-	0.75	0.28	17.63	2.58	20.21	nr	22.23
600mm wide	-	-	-	0.80	0.30	18.82	2.81	21.63	nr	23.79
900mm wide	-	-	-	0.97	0.35	22.63	3.48	26.11	nr	28.73
1200mm wide	-	-	-	1.13	0.41	26.40	4.12	30.52	nr	33.58
1500mm wide	-	-	-	1.30	0.46	30.21	4.77	34.98	nr	38.48
1800mm wide	-	-	-	1.46	0.52	33.98	5.40	39.38	nr	43.31
900mm high; lugs to brick jambs; plugging and screwing to brick sill and concrete head										
500mm wide	-	-	-	0.85	0.30	19.74	2.98	22.73	nr	25.00
600mm wide	-	-	-	0.91	0.32	21.12	3.15	24.27	nr	26.70
900mm wide	-	-	-	1.08	0.38	25.07	3.87	28.94	nr	31.83
1200mm wide	-	-	-	1.25	0.44	29.02	4.51	33.53	nr	36.89
1500mm wide	-	-	-	1.42	0.50	32.97	5.18	38.14	nr	41.95
1800mm wide	-	-	-	1.58	0.55	36.59	5.79	42.38	nr	46.62
1100mm high; lugs to brick jambs; plugging and screwing to brick sill and concrete head										
500mm wide	-	-	-	0.96	0.33	22.18	3.46	25.64	nr	28.21
600mm wide	-	-	-	1.02	0.35	23.55	3.62	27.17	nr	29.89
900mm wide	-	-	-	1.19	0.41	27.50	4.29	31.79	nr	34.97
1200mm wide	-	-	-	1.36	0.47	31.45	4.95	36.40	nr	40.04
1500mm wide	-	-	-	1.54	0.53	35.58	5.61	41.19	nr	45.31
1800mm wide	-	-	-	1.71	0.59	39.53	6.27	45.80	nr	50.38
1300mm high; lugs to brick jambs; plugging and screwing to brick sill and concrete head										
500mm wide	-	-	-	1.07	0.36	24.61	3.83	28.43	nr	31.28
600mm wide	-	-	-	1.12	0.38	25.80	4.05	29.85	nr	32.84
900mm wide	-	-	-	1.30	0.44	29.94	4.71	34.65	nr	38.11
1200mm wide	-	-	-	1.48	0.50	34.07	5.37	39.44	nr	43.38
1500mm wide	-	-	-	1.65	0.57	38.15	6.05	44.20	nr	48.62
1800mm wide	-	-	-	1.83	0.63	42.29	6.69	48.98	nr	53.87

	Del to Site £	Waste %	Material Cost £	Craft Optve Hrs	Lab Hrs	Labour Cost £	Sunds £	Nett Rate £	Unit	Gross rate (10%) £

Labour hourly rates: (except Specialists)
Craft Operatives 18.37   Labourer 13.76
Rates are national average prices.
Refer to REGIONAL VARIATIONS for indicative levels of overall pricing in regions

BRICK/BLOCK WALLING (Cont'd)

Proprietary and individual spot items; Building in (Cont'd)

Building in metal windows; building in lugs; bedding in cement mortar (1:3), pointing with Secomastic standard mastic one side (Cont'd)

1500mm high; lugs to brick jambs; plugging and screwing to brick sill and concrete head

Description	Del to Site £	Waste %	Material Cost £	Craft Optve Hrs	Lab Hrs	Labour Cost £	Sunds £	Nett Rate £	Unit	Gross rate (10%) £
500mm wide	-	-	-	1.18	0.39	27.04	4.28	31.32	nr	34.45
600mm wide	-	-	-	1.23	0.41	28.24	4.45	32.69	nr	35.96
900mm wide	-	-	-	1.42	0.47	32.55	5.13	37.68	nr	41.45
1200mm wide	-	-	-	1.59	0.54	36.64	5.78	42.41	nr	46.66
1500mm wide	-	-	-	1.77	0.60	40.77	6.41	47.18	nr	51.89
1800mm wide	-	-	-	1.95	0.67	45.04	7.11	52.15	nr	57.37
2100mm high; lugs to brick jambs; plugging and screwing to brick sill and concrete head										
500mm wide	-	-	-	1.39	0.44	31.59	5.52	37.11	nr	40.82
600mm wide	-	-	-	1.45	0.46	32.97	5.74	38.71	nr	42.58
900mm wide	-	-	-	1.64	0.53	37.42	6.39	43.81	nr	48.19
1200mm wide	-	-	-	1.82	0.60	41.69	7.11	48.80	nr	53.68
1500mm wide	-	-	-	2.01	0.67	46.14	7.71	53.85	nr	59.24
1800mm wide	-	-	-	2.19	0.74	50.41	8.40	58.81	nr	64.69

Building in factory glazed metal windows; screwing with galvanized screws; bedding in cement mortar (1:3), pointing with Secomastic standard mastic one side

300mm high; plugging and screwing lugs to brick jambs and sill and concrete head

Description	Del to Site £	Waste %	Material Cost £	Craft Optve Hrs	Lab Hrs	Labour Cost £	Sunds £	Nett Rate £	Unit	Gross rate (10%) £
600mm wide	-	-	-	0.70	0.30	16.99	1.93	18.92	nr	20.81
900mm wide	-	-	-	0.84	0.34	20.11	2.61	22.72	nr	24.99
1200mm wide	-	-	-	0.98	0.38	23.23	3.26	26.49	nr	29.14
1500mm wide	-	-	-	1.12	0.41	26.22	3.91	30.13	nr	33.14
1800mm wide	-	-	-	1.26	0.45	29.34	4.55	33.88	nr	37.27
2400mm wide	-	-	-	1.54	0.53	35.58	5.88	41.46	nr	45.61
3000mm wide	-	-	-	1.82	0.60	41.69	7.17	48.86	nr	53.75
700mm high; plugging and screwing lugs to brick jambs and sill and concrete head										
600mm wide	-	-	-	0.96	0.40	23.14	2.81	25.94	nr	28.54
900mm wide	-	-	-	1.17	0.48	28.10	3.48	31.58	nr	34.74
1200mm wide	-	-	-	1.37	0.57	33.01	4.17	37.18	nr	40.90
1500mm wide	-	-	-	1.58	0.64	37.83	4.77	42.60	nr	46.86
1800mm wide	-	-	-	1.79	0.73	42.93	5.40	48.33	nr	53.16
2400mm wide	-	-	-	2.21	0.90	52.98	6.74	59.72	nr	65.69
3000mm wide	-	-	-	2.62	1.06	62.72	8.07	70.79	nr	77.86
900mm high; plugging and screwing lugs to brick jambs and sill and concrete head										
600mm wide	-	-	-	1.09	0.45	26.22	3.22	29.44	nr	32.38
900mm wide	-	-	-	1.33	0.55	32.00	3.87	35.87	nr	39.46
1200mm wide	-	-	-	1.57	0.66	37.92	4.51	42.44	nr	46.68
1500mm wide	-	-	-	1.81	0.76	43.71	5.18	48.88	nr	53.77
1800mm wide	-	-	-	2.06	0.87	49.81	5.79	55.60	nr	61.16
2400mm wide	-	-	-	2.54	1.08	61.52	7.14	68.66	nr	75.53
3000mm wide	-	-	-	3.02	1.29	73.23	8.46	81.69	nr	89.86
1100mm high; plugging and screwing lugs to brick jambs and sill and concrete head										
600mm wide	-	-	-	1.22	0.50	29.29	3.63	32.92	nr	36.21
900mm wide	-	-	-	1.49	0.62	35.90	4.29	40.19	nr	44.21
1200mm wide	-	-	-	1.77	0.75	42.83	4.93	47.77	nr	52.55
1500mm wide	-	-	-	2.04	0.88	49.58	5.61	55.19	nr	60.71
1800mm wide	-	-	-	2.32	1.01	56.52	6.27	62.79	nr	69.06
2400mm wide	-	-	-	2.87	1.27	70.20	7.56	77.76	nr	85.53
3000mm wide	-	-	-	3.42	1.52	83.74	8.90	92.64	nr	101.90
1300mm high; plugging and screwing lugs to brick jambs and sill and concrete head										
600mm wide	-	-	-	1.34	0.54	32.05	4.29	36.34	nr	39.97
900mm wide	-	-	-	1.66	0.70	40.13	4.71	44.84	nr	49.32
1200mm wide	-	-	-	1.96	0.85	47.70	5.39	53.09	nr	58.39
1500mm wide	-	-	-	2.28	0.99	55.51	6.05	61.55	nr	67.71
1800mm wide	-	-	-	2.59	1.15	63.40	6.69	70.09	nr	77.10
2400mm wide	-	-	-	3.21	1.45	78.92	7.95	86.87	nr	95.56
3000mm wide	-	-	-	3.83	1.75	94.44	9.26	103.69	nr	114.06
1500mm high; plugging and screwing lugs to brick jambs and sill and concrete head										
600mm wide	-	-	-	1.47	0.59	35.12	4.45	39.58	nr	43.54
900mm wide	-	-	-	1.82	0.77	44.03	5.13	49.16	nr	54.07
1200mm wide	-	-	-	2.16	0.94	52.61	5.78	58.39	nr	64.23
1500mm wide	-	-	-	2.51	1.11	61.38	6.47	67.85	nr	74.63
1800mm wide	-	-	-	2.85	1.29	70.10	7.11	77.21	nr	84.94
2400mm wide	-	-	-	3.54	1.64	87.60	8.41	96.01	nr	105.61
3000mm wide	-	-	-	4.23	1.98	104.95	9.70	114.65	nr	126.12
2100mm high; plugging and screwing lugs to brick jambs and sill and concrete head										
600mm wide	-	-	-	1.86	0.74	44.35	5.74	50.10	nr	55.11
900mm wide	-	-	-	2.31	0.98	55.92	6.39	62.31	nr	68.54
1200mm wide	-	-	-	2.75	1.22	67.30	7.09	74.40	nr	81.84
1500mm wide	-	-	-	3.20	1.46	78.87	7.71	86.58	nr	95.24
1800mm wide	-	-	-	3.65	1.71	90.58	8.35	98.94	nr	108.83
2400mm wide	-	-	-	4.54	2.19	113.53	9.68	123.21	nr	135.53
3000mm wide	-	-	-	5.43	2.67	136.49	10.97	147.45	nr	162.20

Building in wood windows; building in lugs; bedding in cement mortar (1:3), pointing with Secomastic standard mastic one side

768mm high; lugs to brick jambs

Description	Del to Site £	Waste %	Material Cost £	Craft Optve Hrs	Lab Hrs	Labour Cost £	Sunds £	Nett Rate £	Unit	Gross rate (10%) £
438mm wide	-	-	-	1.43	0.09	27.51	2.61	30.12	nr	33.13

MASONRY

Labour hourly rates: (except Specialists) Craft Operatives 18.37 Labourer 13.76 Rates are national average prices. Refer to REGIONAL VARIATIONS for indicative levels of overall pricing in regions	MATERIALS			LABOUR				RATES		
	Del to Site	Waste	Material Cost	Craft Optve	Lab	Labour Cost	Sunds	Nett Rate	Unit	Gross rate (10%)
	£	%	£	Hrs	Hrs	£	£	£		£
BRICK/BLOCK WALLING (Cont'd)										
Proprietary and individual spot items; Building in (Cont'd)										
Building in wood windows; building in lugs; bedding in cement mortar (1:3), pointing with Secomastic standard mastic one side (Cont'd)										
768mm high; lugs to brick jambs (Cont'd)										
641mm wide	-	-	-	1.64	0.09	31.37	3.08	34.44	nr	37.88
1225mm wide	-	-	-	2.24	0.10	42.52	4.38	46.90	nr	51.60
1809mm wide	-	-	-	2.86	0.11	54.05	5.72	59.77	nr	65.74
2394mm wide	-	-	-	3.46	0.12	65.21	7.05	72.26	nr	79.49
920mm high; lugs to brick jambs										
438mm wide	-	-	-	1.61	0.09	30.81	2.91	33.72	nr	37.10
641mm wide	-	-	-	1.82	0.10	34.81	3.35	38.15	nr	41.97
1225mm wide	-	-	-	2.45	0.11	46.52	4.71	51.23	nr	56.35
1809mm wide	-	-	-	3.07	0.13	58.18	6.05	64.23	nr	70.65
2394mm wide	-	-	-	3.70	0.14	69.90	7.35	77.25	nr	84.97
1073mm high; lugs to brick jambs										
438mm wide	-	-	-	1.80	0.10	34.44	3.26	37.70	nr	41.47
641mm wide	-	-	-	2.01	0.11	38.44	3.72	42.16	nr	46.37
1225mm wide	-	-	-	2.66	0.12	50.52	5.05	55.57	nr	61.13
1809mm wide	-	-	-	3.30	0.14	62.55	6.34	68.89	nr	75.78
2394mm wide	-	-	-	3.95	0.16	74.76	7.70	82.46	nr	90.70
1225mm high; lugs to brick jambs										
438mm wide	-	-	-	1.98	0.10	37.75	3.60	41.35	nr	45.48
641mm wide	-	-	-	2.20	0.11	41.93	4.02	45.95	nr	50.54
1225mm wide	-	-	-	2.87	0.13	54.51	5.37	59.88	nr	65.87
1809mm wide	-	-	-	3.52	0.16	66.86	6.63	73.49	nr	80.84
2394mm wide	-	-	-	4.19	0.18	79.45	7.98	87.43	nr	96.17
1378mm high; lugs to brick jambs										
438mm wide	-	-	-	2.16	0.10	41.06	3.90	44.96	nr	49.45
641mm wide	-	-	-	2.39	0.11	45.42	4.36	49.78	nr	54.76
1225mm wide	-	-	-	3.07	0.14	58.32	5.70	64.02	nr	70.42
1809mm wide	-	-	-	3.75	0.17	71.23	7.01	78.23	nr	86.05
2394mm wide	-	-	-	4.43	0.20	84.13	8.34	92.47	nr	101.72
Building in fireplace interior with slabbed tile surround and loose hearth tiles for 400mm opening; bed and joint interior in fire cement and fill with concrete at back; plug and screw on surround lugs; bed hearth tiles in cement mortar (1:3) and point in										
with firebrick back	-	-	-	5.00	3.50	140.01	6.63	146.64	nr	161.30
with back boiler unit and self contained flue	-	-	-	6.00	4.00	165.26	7.98	173.24	nr	190.56
GLASS BLOCK WALLING										
Walls; Glass blockwork; Hollow glass blocks, white, cross ribbed, in cement mortar (1:3); continuous joints; flat recessed pointing as work proceeds										
Screens or panels; facework both sides										
190 x 190mm blocks; vertical; 80mm thick	47.25	10.00	51.98	3.00	3.00	96.39	74.65	223.02	m²	245.32
240 x 240mm blocks; vertical; 80mm thick	143.84	10.00	158.22	2.50	2.50	80.33	60.09	298.64	m²	328.50
NATURAL STONE RUBBLE WALLING AND DRESSINGS										
Walls; Natural stone; Stone rubble work; random stones of Yorkshire limestone; bedding and jointing in lime mortar (1:3); uncoursed										
Walls; tapering both sides; including extra material										
500mm thick	63.00	15.00	72.45	3.75	3.75	120.49	12.38	205.31	m²	225.84
600mm thick	70.00	15.00	80.50	4.50	4.50	144.59	14.85	239.94	m²	263.93
Walls; Stone rubble work; squared rubble face stones of Yorkshire limestone; bedding and jointing in lime mortar (1:3); irregular coursed; courses average 150mm high										
Walls; vertical; face stones 100 - 150mm on bed; bonding to brickwork; including extra material; scappled or axed face; weather struck pointing as work proceeds										
150mm thick; faced one side	53.00	15.00	60.95	3.00	3.00	96.39	4.12	161.46	m²	177.61
Walls; vertical; face stones 100 - 150mm on bed; bonding to brickwork; including extra material; hammer dressed face; weather struck pointing as work proceeds										
150mm thick; faced one side	53.00	15.00	60.95	3.30	3.30	106.03	4.12	171.10	m²	188.21
Walls; vertical; face stones 100 - 150mm on bed; bonding to brickwork; including extra material; rock worked face; pointing with a parallel joint as work proceeds										
150mm thick; faced one side	57.00	15.00	65.55	3.60	3.60	115.67	4.12	185.34	m²	203.88
Walls; Stone rubble work; squared rubble face stones of Yorkshire limestone; bedding and jointing in lime mortar (1:3); regular coursed; courses average 150mm high										
Walls; vertical; face stones 100 - 150mm on bed; bonding to brickwork; including extra material; scappled or axed face; weather struck pointing as work proceeds										
150mm thick; faced one side	70.00	15.00	80.50	3.20	3.20	102.82	4.12	187.44	m²	206.19
Walls; vertical; face stones 100 - 150mm on bed; bonding to brickwork; including extra material; hammer dressed face; weather struck pointing as work proceeds										
150mm thick; faced one side	70.00	15.00	80.50	3.50	3.50	112.46	4.12	197.08	m²	216.79

Labour hourly rates: (except Specialists) Craft Operatives 18.37 Labourer 13.76 Rates are national average prices. Refer to REGIONAL VARIATIONS for indicative levels of overall pricing in regions	MATERIALS			LABOUR				RATES		
	Del to Site	Waste	Material Cost	Craft Optve	Lab	Labour Cost	Sunds	Nett Rate		Gross rate (10%)
	£	%	£	Hrs	Hrs	£	£	£	Unit	£
NATURAL STONE RUBBLE WALLING AND DRESSINGS (Cont'd)										
Walls; Stone rubble work; squared rubble face stones of Yorkshire limestone; bedding and jointing in lime mortar (1:3); regular coursed; courses average 150mm high (Cont'd)										
Walls; vertical; face stones 100 - 150mm on bed; bonding to brickwork; including extra material; rock worked face; pointing with a parallel joint as work proceeds										
150mm thick; faced one side............	72.00	15.00	82.80	3.80	3.80	122.09	4.12	209.02	m²	229.92
Walls; Stone rubble work; random rubble backing and squared rubble face stones of Yorkshire limestone; bedding and jointing in lime mortar (1:3); irregular coursed; courses average 150mm high										
Walls; vertical; face stones 100 - 150mm on bed; scappled or axed face; weather struck pointing as work proceeds										
350mm thick; faced one side..............	81.00	15.00	93.15	4.00	4.00	128.52	8.62	230.30	m²	253.32
500mm thick; faced one side..............	102.00	15.00	117.30	4.75	4.75	152.62	12.38	282.29	m²	310.52
Walls; vertical; face stones 100 - 150mm on bed; hammer dressed face; weather struck pointing as work proceeds										
350mm thick; faced one side..............	81.00	15.00	93.15	4.30	4.30	138.16	8.62	239.93	m²	263.93
500mm thick; faced one side..............	102.00	15.00	117.30	5.05	5.05	162.26	12.38	291.93	m²	321.12
Walls; vertical; face stones 100 - 150mm on bed; rock worked face; pointing with a parallel joint as work proceeds										
350mm thick; faced one side..............	81.00	15.00	93.15	4.60	4.60	147.80	8.62	249.57	m²	274.53
500mm thick; faced one side..............	102.00	15.00	117.30	5.35	5.35	171.90	12.38	301.57	m²	331.73
Grooves										
12 x 25mm..............	-	-	-	0.25	-	4.59	-	4.59	m	5.05
12 x 38mm..............	-	-	-	0.30	-	5.51	-	5.51	m	6.06
Arches; relieving										
225mm high on face, 180mm wide on soffit	119.00	5.00	124.95	1.50	0.75	37.88	0.98	163.80	m	180.18
225mm high on face, 250mm wide on soffit	163.00	5.00	171.15	2.00	1.00	50.50	1.35	223.00	m	245.30
Special purpose blocks or stones; Quoin stones scappled or axed face; attached; faced two adjacent faces										
250 x 200 x 350mm	32.00	5.00	33.60	0.50	0.25	12.62	0.42	46.64	nr	51.31
250 x 200 x 500mm	47.00	5.00	49.35	0.80	0.40	20.20	0.60	70.15	nr	77.17
250 x 250 x 350mm	42.00	5.00	44.10	0.65	0.35	16.76	0.52	61.38	nr	67.52
250 x 250 x 500mm	59.00	5.00	61.95	0.90	0.45	22.73	0.75	85.42	nr	93.97
380 x 200 x 350mm	50.00	5.00	52.50	0.80	0.40	20.20	0.68	73.38	nr	80.71
380 x 200 x 500mm	70.00	5.00	73.50	1.00	0.50	25.25	0.90	99.65	nr	109.61
rock worked face; attached; faced two adjacent faces										
250 x 200 x 350mm	32.00	5.00	33.60	0.55	0.25	13.54	0.42	47.56	nr	52.32
250 x 200 x 500mm	43.00	5.00	45.15	0.85	0.40	21.12	0.60	66.87	nr	73.56
250 x 250 x 350mm	37.00	5.00	38.85	0.70	0.35	17.68	0.52	57.05	nr	62.76
250 x 250 x 500mm	53.00	5.00	55.65	0.95	0.45	23.64	0.75	80.04	nr	88.05
380 x 200 x 350mm	44.00	5.00	46.20	0.85	0.40	21.12	0.68	67.99	nr	74.79
380 x 200 x 500mm	64.00	5.00	67.20	1.10	0.50	27.09	0.90	95.19	nr	104.71
Stone rubble work; random rubble backing and squared rubble face stones of Yorkshire limestone; bedding and jointing in lime mortar (1:3); regular coursed; courses average 150mm high										
Walls; vertical; face stones 100 - 150mm on bed; scappled or axed face; weather struck pointing as work proceeds										
350mm thick; faced one side..............	102.00	15.00	117.30	4.20	4.20	134.95	8.62	260.87	m²	286.96
Walls; vertical; face stones 125 - 200mm on bed; scappled or axed face; weather struck pointing as work proceeds										
500mm thick; faced one side..............	125.00	15.00	143.75	4.95	4.95	159.04	12.38	315.17	m²	346.69
Walls; vertical; face stones 100 - 150mm on bed; hammer dressed face; weather struck pointing as work proceeds										
350mm thick; faced one side..............	102.00	15.00	117.30	4.50	4.50	144.59	8.62	270.51	m²	297.56
500mm thick; faced one side..............	125.00	15.00	143.75	5.25	5.25	168.68	12.38	324.81	m²	357.29
Walls; vertical; face stones 100 - 150mm on bed; rock worked face; pointing with a parallel joint as work proceeds										
350mm thick; faced one side..............	102.00	15.00	117.30	4.80	4.80	154.22	8.62	280.15	m²	308.16
500mm thick; faced one side..............	125.00	15.00	143.75	5.55	5.55	178.32	12.38	334.45	m²	367.89
NATURAL STONE ASHLAR WALLING AND DRESSINGS										
Walls; Natural stone										
Note stonework is work for a specialist; prices being obtained for specific projects. Some firms that specialise in this class of work are included within the list at the end of this section										
Natural stonework dressings; natural Dorset limestone; Portland Whitbed; bedding and jointing in mason's mortar (1:3:12); flush smooth pointing as work proceeds; slurrying with weak lime mortar and cleaning down on completion										
Walls; vertical; building against brickwork; BS EN 845 Fig 1 specification 3.5 wall ties - 4/m² built in										
50mm thick; plain and rubbed one side	175.00	2.50	179.38	4.00	2.60	109.26	8.25	296.88	m²	326.57
75mm thick; plain and rubbed one side	238.00	2.50	243.95	4.50	3.15	126.01	9.00	378.96	m²	416.85
100mm thick; plain and rubbed one side	315.00	2.50	322.88	5.00	3.70	142.76	9.75	475.39	m²	522.93

MASONRY

Labour hourly rates: (except Specialists) Craft Operatives 18.37 Labourer 13.76 Rates are national average prices. Refer to REGIONAL VARIATIONS for indicative levels of overall pricing in regions	MATERIALS			LABOUR				RATES		
	Del to Site	Waste	Material Cost	Craft Optve	Lab	Labour Cost	Sunds	Nett Rate	Unit	Gross rate (10%)
	£	%	£	Hrs	Hrs	£	£	£		£
NATURAL STONE ASHLAR WALLING AND DRESSINGS (Cont'd)										
Natural stonework; natural Dorset limestone, Portland Whitbed; bedding and jointing in cement-lime mortar (1:2:9); flush smooth pointing as work proceeds; Bands										
Band courses; moulded; horizontal										
225 x 125mm	175.00	2.50	179.38	1.60	1.15	45.22	4.50	229.09	m	252.00
250 x 150mm	211.00	2.50	216.27	1.80	1.35	51.64	5.25	273.17	m	300.48
300 x 150mm	249.00	2.50	255.22	2.00	1.55	58.07	6.15	319.44	m	351.39
Natural stonework; natural Dorset limestone, Portland Whitbed; bedding and jointing in cement-lime mortar (1:2:9); flush smooth pointing as work proceeds; Extra over walls for perimeters and abutments										
Copings; horizontal										
plain and rubbed faces -2; weathered and rubbed faces -1; throats -2; cramped joints with stainless steel cramps										
300 x 50mm	121.00	2.50	124.02	0.75	0.50	20.66	6.30	150.98	m	166.08
300 x 75mm	142.00	2.50	145.55	0.90	0.70	26.17	7.05	178.76	m	196.64
375 x 100mm	196.00	2.50	200.90	1.35	1.00	38.56	8.48	247.93	m	272.73
Natural stonework; natural Dorset limestone, Portland Whitbed; bedding and jointing in cement-lime mortar (1:2:9); flush smooth pointing as work proceeds; Extra over walls for opening perimeters										
Lintels										
plain and rubbed faces -3; splayed and rubbed faces -1										
200 x 100mm	75.00	2.50	76.88	1.25	0.85	34.66	1.88	113.41	m	124.75
225 x 125mm	96.00	2.50	98.40	1.50	1.10	42.69	2.40	143.49	m	157.84
Sills										
plain and rubbed faces -3; sunk weathered and rubbed faces -1; grooves -1; throats -1										
200 x 75mm	101.00	2.50	103.52	1.00	0.65	27.31	2.85	133.69	m	147.06
250 x 75mm	121.00	2.50	124.02	1.20	0.80	33.05	3.30	160.38	m	176.41
300 x 75mm	140.00	2.50	143.50	1.40	1.00	39.48	3.67	186.65	m	205.32
Jamb stones; attached										
plain and rubbed faces -3; splayed and rubbed faces -1; rebates -1; grooves -1										
175 x 75mm	77.00	2.50	78.92	1.00	0.65	27.31	2.47	108.71	m	119.59
200 x 100mm	85.00	2.50	87.12	1.25	1.00	36.72	3.22	127.07	m	139.78
Natural stonework, natural Yorkshire sandstone, Bolton Wood; bedding and jointing in cement-lime mortar (1:2:9); flush smooth pointing as work proceeds; Extra over walls for perimeters and abutments										
Copings; horizontal										
plain and rubbed faces -2; weathered and rubbed faces -2; throats -2; cramped joints with stainless steel cramps										
300 x 75mm	80.00	2.50	82.00	0.90	0.70	26.17	4.50	112.67	m	123.93
300 x 100mm	94.00	2.50	96.35	1.10	0.80	31.22	4.95	132.51	m	145.77
Kerbs; horizontal										
plain and sawn faces - 4										
150 x 150mm	50.00	2.50	51.25	0.85	0.65	24.56	1.65	77.46	m	85.20
225 x 150mm	74.00	2.50	75.85	1.20	1.00	35.80	2.47	114.13	m	125.54
Cover stones										
75 x 300; rough edges -2; plain and sawn face -1	42.00	2.50	43.05	0.85	0.65	24.56	3.30	70.91	m	78.00
Templates										
150 x 300; rough edges -2; plain and sawn face -1	94.00	2.50	96.35	1.50	1.20	44.07	4.12	144.54	m	159.00
Steps; plain										
plain and rubbed top and front										
225 x 75mm	38.00	2.50	38.95	0.75	0.55	21.35	3.30	63.60	m	69.96
225 x 150mm	74.00	2.50	75.85	1.30	1.00	37.64	4.12	117.62	m	129.38
300 x 150mm	99.00	2.50	101.47	1.60	1.30	47.28	4.95	153.71	m	169.08
Natural stonework, natural Yorkshire sandstone, Bolton Wood; bedding and jointing in cement-lime mortar (1:2:9); flush smooth pointing as work proceeds; Extra over walls for opening perimeters										
Sills										
plain and rubbed faces -3; sunk weathered and rubbed faces -1; throats -1										
175 x 100mm	73.00	2.50	74.82	1.25	0.85	34.66	1.88	111.36	m	122.49
250 x 125mm	113.00	2.50	115.82	1.40	1.00	39.48	2.47	157.78	m	173.56
Natural stonework, natural Yorkshire sandstone, Bolton Wood; bedding and jointing in cement-lime mortar (1:2:9); flush smooth pointing as work proceeds; Special purpose blocks or stones										
Landings										
75 x 900 x 900mm; sawn edges -4; plain and rubbed face -1	134.00	2.50	137.35	2.00	1.75	60.82	4.95	203.12	nr	223.43

Labour hourly rates: (except Specialists) Craft Operatives 18.37 Labourer 13.76 Rates are national average prices. Refer to REGIONAL VARIATIONS for indicative levels of overall pricing in regions	MATERIALS			LABOUR				RATES		
	Del to Site £	Waste %	Material Cost £	Craft Optve Hrs	Lab Hrs	Labour Cost £	Sunds £	Nett Rate £	Unit	Gross rate (10%) £
ARTIFICIAL/CAST STONE WALLING AND DRESSINGS										
Walls; Cast stone; Reconstructed stone blocks, Marshalls Mono Ltd. Cromwell coursed random length Pitched Face buff walling blocks; in cement-lime mortar (1:1:6); flat recessed pointing as work proceeds										
Walls										
90mm thick; alternate courses 102mm and 140mm high blocks; stretcher bond; facework one side	43.89	5.00	46.08	1.90	1.40	54.17	2.12	102.37	m²	112.60
90mm thick; one course 102mm high blocks and two courses 140mm high blocks; stretcher bond; facework one side	43.89	5.00	46.08	1.90	1.40	54.17	2.12	102.37	m²	112.60
Walls; Cast stone; Cast stonework dressings; simulated Dorset limestone; Portland Whitbed; bedding and jointing in cement-lime mortar (1:2:9); flush smooth pointing as work proceeds										
Walls; vertical; building against brickwork; BS EN 845 Fig 1 specification 3.5 wall ties -4/m² built in										
100mm thick; plain and rubbed one side	95.55	5.00	100.33	4.00	2.50	107.88	8.25	216.46	m²	238.10
Fair raking cutting										
100 thick	19.11	5.00	20.07	1.00	0.50	25.25	-	45.32	m	49.85
Bands; Band courses; plain; horizontal										
225 x 125mm	37.54	5.00	39.41	1.40	1.00	39.48	2.03	80.92	m	89.01
extra; external return	25.99	5.00	27.29	-	-	-	-	27.29	nr	30.02
250 x 150mm	49.09	5.00	51.54	1.50	1.10	42.69	2.47	96.71	m	106.38
extra; external return	25.99	5.00	27.29	-	-	-	-	27.29	nr	30.02
300 x 150mm	54.86	5.00	57.61	1.60	1.20	45.90	2.85	106.36	m	117.00
extra; external return	25.99	5.00	27.29	-	-	-	-	27.29	nr	30.02
Extra over walls for perimeters and abutments										
Copings; horizontal										
plain and rubbed faces -2; weathered and rubbed faces -2; throats -2; cramped joints with stainless steel cramps										
300 x 50mm	28.88	5.00	30.32	0.50	0.30	13.31	3.75	47.38	m	52.12
extra; internal angles	25.99	5.00	27.29	-	-	-	-	27.29	nr	30.02
extra; external angles	25.99	5.00	27.29	-	-	-	-	27.29	nr	30.02
300 x 75mm	37.54	5.00	39.41	0.60	0.35	15.84	4.12	59.38	m	65.32
extra; internal angles	25.99	5.00	27.29	-	-	-	-	27.29	nr	30.02
extra; external angles	25.99	5.00	27.29	-	-	-	-	27.29	nr	30.02
375 x 100mm	62.08	5.00	65.19	0.75	0.50	20.66	4.95	90.79	m	99.87
extra; internal angles	25.99	5.00	27.29	-	-	-	-	27.29	nr	30.02
extra; external angles	25.99	5.00	27.29	-	-	-	-	27.29	nr	30.02
Extra over walls for opening perimeters; Cast stonework; simulated Dorset limestone; Portland Whitbed; bedding and jointing in cement-lime mortar (1:2:9); flush smooth pointing as work proceeds										
Lintels										
plain and rubbed faces -3; splayed and rubbed faces -1										
200 x 100mm	36.22	5.00	38.04	1.00	0.60	26.63	1.20	65.86	m	72.45
225 x 125mm	56.31	5.00	59.12	1.25	0.75	33.28	1.65	94.05	m	103.46
Sills										
plain and rubbed faces -3; sunk weathered and rubbed faces -1; grooves -1; throats -1										
200 x 75mm	34.65	5.00	36.38	0.90	0.50	23.41	1.65	61.45	m	67.59
extra; stoolings	8.66	5.00	9.10	-	-	-	-	9.10	nr	10.01
250 x 75mm	37.54	5.00	39.41	1.00	0.60	26.63	2.03	68.07	m	74.87
extra; stoolings	8.66	5.00	9.10	-	-	-	-	9.10	nr	10.01
300 x 75mm	44.76	5.00	46.99	1.25	0.75	33.28	2.47	82.75	m	91.03
extra; stoolings	8.66	5.00	9.10	-	-	-	-	9.10	nr	10.01
Jamb stones; attached										
plain and rubbed faces -3; splayed and rubbed faces -1; rebates -1; grooves -1										
175 x 75mm	37.54	5.00	39.41	0.90	0.50	23.41	1.27	64.10	m	70.51
200 x 100mm	47.64	5.00	50.03	1.00	0.60	26.63	2.47	79.13	m	87.04
Steps; plain										
300 x 150mm; plain and rubbed tread and riser	53.42	5.00	56.09	1.25	0.90	35.35	3.30	94.74	m	104.21
Steps; spandril										
250mm wide tread; 180mm high riser; plain and rubbed tread and riser; carborundum finish to tread	73.63	5.00	77.31	2.00	1.25	53.94	2.47	133.73	m	147.10
Special purpose blocks or stones										
Landings										
150 x 900 x 900mm; plain and rubbed top surface	131.38	5.00	137.95	1.50	1.00	41.32	3.30	182.57	nr	200.82

This page left blank intentionally

Labour hourly rates: (except Specialists) Craft Operatives 18.37 Labourer 13.76 Rates are national average prices. Refer to REGIONAL VARIATIONS for indicative levels of overall pricing in regions	MATERIALS			LABOUR				RATES		
	Del to Site £	Waste %	Material Cost £	Craft Optve Hrs	Lab Hrs	Labour Cost £	Sunds £	Nett Rate £	Unit	Gross rate (10%) £
STRUCTURAL STEELWORK										
Framed members, framing and fabrication										
Prices generally for fabricated steelwork										
Note prices can vary considerably dependent upon the character of the work; the following are average prices for work fabricated and delivered to site										
Framing, fabrication; including shop and site black bolts, nuts and washers for structural framing to structural framing connections										
Weldable steel, BS EN 10025 : 1993 Grade S275JR (formerly BS 7668 Grade 43B), hot rolled sections BS 4-1 (Euronorm 54); welded fabrication in accordance with BS EN 1994-1										
Columns; length over 1.00 but not exceeding 9.00m										
weight less than 25 Kg/m	810.00	-	810.00	-	-	-	-	810.00	t	891.00
weight 25-50 Kg/m	810.00	-	810.00	-	-	-	-	810.00	t	891.00
weight 50-100 Kg/m	810.00	-	810.00	-	-	-	-	810.00	t	891.00
weight exceeding 100 Kg/m	810.00	-	810.00	-	-	-	-	810.00	t	891.00
Beams; length over 1.00 but not exceeding 9.00m										
weight less than 25 Kg/m	810.00		810.00	-	-	-	-	810.00	t	891.00
weight 25-50 Kg/m	810.00	-	810.00	-	-	-	-	810.00	t	891.00
weight 50-100 Kg/m	810.00	-	810.00	-	-	-	-	810.00	t	891.00
weight exceeding 100 Kg/m	810.00	-	810.00	-	-	-	-	810.00	t	891.00
Beams, castellated; length over 1.00 but not exceeding 9.00m										
weight less than 25 Kg/m	1350.00	-	1350.00	-	-	-	-	1350.00	t	1485.00
weight 25-50 Kg/m	1350.00	-	1350.00	-	-	-	-	1350.00	t	1485.00
weight 50-100 Kg/m	1350.00	-	1350.00	-	-	-	-	1350.00	t	1485.00
weight exceeding 100 Kg/m	1350.00	-	1350.00	-	-	-	-	1350.00	t	1485.00
Beams, curved; length over 1.00 but not exceeding 9.00m										
weight less than 25 Kg/m	1600.00	-	1600.00	-	-	-	-	1600.00	t	1760.00
weight 25-50 Kg/m	1600.00	-	1600.00	-	-	-	-	1600.00	t	1760.00
weight 50-100 Kg/m	1600.00	-	1600.00	-	-	-	-	1600.00	t	1760.00
weight exceeding 100 Kg/m	1600.00	-	1600.00	-	-	-	-	1600.00	t	1760.00
Bracings, tubular; length over 1.00 but not exceeding 9.00m										
weight less than 25 Kg/m	1150.00	-	1150.00	-	-	-	-	1150.00	t	1265.00
weight 25-50 Kg/m	1150.00	-	1150.00	-	-	-	-	1150.00	t	1265.00
weight 50-100 Kg/m	1150.00	-	1150.00	-	-	-	-	1150.00	t	1265.00
weight exceeding 100 Kg/m	1150.00	-	1150.00	-	-	-	-	1150.00	t	1265.00
Purlins and cladding rails; length over 1.00 but not exceeding 9.00m										
weight less than 25 Kg/m	1115.00	-	1115.00	-	-	-	-	1115.00	t	1226.50
weight 25-50 Kg/m	1115.00	-	1115.00	-	-	-	-	1115.00	t	1226.50
weight 50-100 Kg/m	1115.00	-	1115.00	-	-	-	-	1115.00	t	1226.50
weight exceeding 100 Kg/m	1115.00	-	1115.00	-	-	-	-	1115.00	t	1226.50
Trusses; length over 1.00 but not exceeding 9.00m										
portal frames	1045.00	-	1045.00	-	-	-	-	1045.00	t	1149.50
trusses 12m to 18m span	1620.00	-	1620.00	-	-	-	-	1620.00	t	1782.00
trusses 6m to 12m span	1675.00	-	1675.00	-	-	-	-	1675.00	t	1842.50
trusses up to 6m span	1780.00	-	1780.00	-	-	-	-	1780.00	t	1958.00
Weldable steel BS EN 10210 : 1994 Grade S275JOH (formerly BS 7668 Grade 43C), hot rolled sections BS EN 10210-2 (Euronorm 57); welded fabrication in accordance with BS EN 1994-1										
column, square hollow section; length over 1.00 but not exceeding 9.00m										
weight less than 25 Kg/m	1135.00	-	1135.00	-	-	-	-	1135.00	t	1248.50
weight 25-50 Kg/m	1135.00	-	1135.00	-	-	-	-	1135.00	t	1248.50
weight 50-100 Kg/m	1135.00	-	1135.00	-	-	-	-	1135.00	t	1248.50
weight exceeding 100 Kg/m	1135.00	-	1135.00	-	-	-	-	1135.00	t	1248.50
column, rectangular hollow section; length over 1.00 but not exceeding 9.00m										
weight less than 25 Kg/m	1150.00	-	1150.00	-	-	-	-	1150.00	t	1265.00
weight 25-50 Kg/m	1150.00	-	1150.00	-	-	-	-	1150.00	t	1265.00
weight 50-100 Kg/m	1150.00	-	1150.00	-	-	-	-	1150.00	t	1265.00
weight exceeding 100 Kg/m	1150.00	-	1150.00	-	-	-	-	1150.00	t	1265.00
Framed members, permanent erection on site										
Note the following prices for erection include for the site to be reasonably clear, ease of access and the erection carried out during normal working hours										
Permanent erection of fabricated steelwork on site with bolted connections; weldable steel, BS EN 10025 : 1993 Grade S275JR (formerly BS 7668 Grade 43B), hot rolled sections BS 4-1 (Euronorm 54); welded fabrication in accordance with BS 5950-1										
framing	-	-	-	23.00	-	422.51	-	422.51	t	464.76
Columns; length over 1.00 but not exceeding 9.00m										
weight less than 25 Kg/m	-	-	-	23.00	-	422.51	-	422.51	t	464.76

STRUCTURAL METALWORK

Labour hourly rates: (except Specialists) Craft Operatives 18.37 Labourer 13.76 Rates are national average prices. Refer to REGIONAL VARIATIONS for indicative levels of overall pricing in regions	MATERIALS			LABOUR				RATES		
	Del to Site £	Waste %	Material Cost £	Craft Optve Hrs	Lab Hrs	Labour Cost £	Sunds £	Nett Rate £	Unit	Gross rate (10%) £
STRUCTURAL STEELWORK (Cont'd)										
Framed members, permanent erection on site (Cont'd)										
Permanent erection of fabricated steelwork on site with bolted connections; weldable steel, BS EN 10025 : 1993 Grade S275JR (formerly BS 7668 Grade 43B), hot rolled sections BS 4-1 (Euronorm 54); welded fabrication in accordance with BS 5950-1 (Cont'd)										
Columns; length over 1.00 but not exceeding 9.00m (Cont'd)										
weight 25-50 Kg/m	-	-	-	23.00	-	422.51	-	422.51	t	464.76
weight 50-100 Kg/m	-	-	-	23.00	-	422.51	-	422.51	t	464.76
weight exceeding 100 Kg/m	-	-	-	23.00	-	422.51	-	422.51	t	464.76
Beams; length over 1.00 but not exceeding 9.00m										
weight less than 25 Kg/m	-	-	-	23.00	-	422.51	-	422.51	t	464.76
weight 25-50 Kg/m	-	-	-	23.00	-	422.51	-	422.51	t	464.76
weight 50-100 Kg/m	-	-	-	23.00	-	422.51	-	422.51	t	464.76
weight exceeding 100 Kg/m	-	-	-	23.00	-	422.51	-	422.51	t	464.76
Beams, castellated; length over 1.00 but not exceeding 9.00m										
weight less than 25 Kg/m	-	-	-	23.00	-	422.51	-	422.51	t	464.76
weight 25-50 Kg/m	-	-	-	23.00	-	422.51	-	422.51	t	464.76
weight 50-100 Kg/m	-	-	-	23.00	-	422.51	-	422.51	t	464.76
weight exceeding 100 Kg/m	-	-	-	23.00	-	422.51	-	422.51	t	464.76
Beams, curved; length over 1.00 but not exceeding 9.00m										
weight less than 25 Kg/m	-	-	-	23.00	-	422.51	-	422.51	t	464.76
weight 25-50 Kg/m	-	-	-	23.00	-	422.51	-	422.51	t	464.76
weight 50-100 Kg/m	-	-	-	23.00	-	422.51	-	422.51	t	464.76
weight exceeding 100 Kg/m	-	-	-	23.00	-	422.51	-	422.51	t	464.76
Bracings, tubular; length over 1.00 but not exceeding 9.00m										
weight less than 25 Kg/m	-	-	-	23.00	-	422.51	-	422.51	t	464.76
weight 25-50 Kg/m	-	-	-	23.00	-	422.51	-	422.51	t	464.76
weight 50-100 Kg/m	-	-	-	23.00	-	422.51	-	422.51	t	464.76
weight exceeding 100 Kg/m	-	-	-	23.00	-	422.51	-	422.51	t	464.76
Purlins and cladding rails; length over 1.00 but not exceeding 9.00m										
weight less than 25 Kg/m	-	-	-	23.00	-	422.51	-	422.51	t	464.76
weight 25-50 Kg/m	-	-	-	23.00	-	422.51	-	422.51	t	464.76
weight 50-100 Kg/m	-	-	-	23.00	-	422.51	-	422.51	t	464.76
weight exceeding 100 Kg/m	-	-	-	23.00	-	422.51	-	422.51	t	464.76
Trusses; length over 1.00 but not exceeding 9.00m										
portal frames	-	-	-	23.00	-	422.51	-	422.51	t	464.76
trusses 12m to 18m span	-	-	-	23.00	-	422.51	-	422.51	t	464.76
trusses 6m to 12m span	-	-	-	23.00	-	422.51	-	422.51	t	464.76
trusses up to 6m span	-	-	-	23.00	-	422.51	-	422.51	t	464.76
Weldable steel BS EN 10210 : 1994 Grade S275JOH (formerly BS 7668 Grade 43C), hot rolled sections BS EN 10210-2 (Euronorm 57); welded fabrication in accordance with BS 5950-1										
column, square hollow section; length over 1.00 but not exceeding 9.00m										
weight less than 25 Kg/m	-	-	-	23.00	-	422.51	-	422.51	t	464.76
weight 25-50 Kg/m	-	-	-	23.00	-	422.51	-	422.51	t	464.76
weight 50-100 Kg/m	-	-	-	23.00	-	422.51	-	422.51	t	464.76
weight exceeding 100 Kg/m	-	-	-	23.00	-	422.51	-	422.51	t	464.76
column, rectangular hollow section; length over 1.00 but not exceeding 9.00m										
weight less than 25 Kg/m	-	-	-	23.00	-	422.51	-	422.51	t	464.76
weight 25-50 Kg/m	-	-	-	23.00	-	422.51	-	422.51	t	464.76
weight 50-100 Kg/m	-	-	-	23.00	-	422.51	-	422.51	t	464.76
weight exceeding 100 Kg/m	-	-	-	23.00	-	422.51	-	422.51	t	464.76
Isolated structural members, fabrication										
Weldable steel, BS EN 10149 : 1996 Grade S275, hot rolled sections; various dimensions										
plain member; beam										
weight less than 25 Kg/m	1020.00	-	1020.00	-	-	-	-	1020.00	t	1122.00
weight 25-50 Kg/m	1020.00	-	1020.00	-	-	-	-	1020.00	t	1122.00
weight 50-100 Kg/m	1020.00	-	1020.00	-	-	-	-	1020.00	t	1122.00
weight exceeding 100 Kg/m	1020.00	-	1020.00	-	-	-	-	1020.00	t	1122.00
Steel Metsec short span lattice joists primed at works; treated timber inserts in top and bottom chords; full depth end seatings for bolting to supporting structure										
200mm deep, 8.2 kg/m	27.19	2.50	27.87	-	-	-	-	27.87	m	30.66
250mm deep, 8.5 kg/m	27.19	2.50	27.87	-	-	-	-	27.87	m	30.66
300mm deep, 10.7 kg/m	29.91	2.50	30.66	-	-	-	-	30.66	m	33.72
350mm deep, 11.6 kg/m	32.63	2.50	33.45	-	-	-	-	33.45	m	36.79
350mm deep, 12.8 kg/m	46.23	2.50	47.38	-	-	-	-	47.38	m	52.12
Steel Metsec short span lattice joists primed at works; treated timber inserts in top and bottom chords; full depth end seatings for bolting to supporting structure										
200mm deep, 8.2 kg/m	27.19	2.50	27.87	-	-	-	-	27.87	m	30.66
250mm deep, 8.5 kg/m	27.19	2.50	27.87	-	-	-	-	27.87	m	30.66
300mm deep, 10.7 kg/m	29.91	2.50	30.66	-	-	-	-	30.66	m	33.72
350mm deep, 11.6 kg/m	32.63	2.50	33.45	-	-	-	-	33.45	m	36.79
350mm deep, 12.8 kg/m	46.23	2.50	47.38	-	-	-	-	47.38	m	52.12
Steel Metsec intermediate span lattice joists primed at works; treated timber inserts in top and bottom chords; full depth end seatings for bolting to supporting structure										
450mm deep, 12.2 kg/m	48.95	2.50	50.17	-	-	-	-	50.17	m	55.19
500mm deep, 15.8 kg/m	51.66	2.50	52.96	-	-	-	-	52.96	m	58.25
550mm deep, 19.4 kg/m	54.38	2.50	55.74	-	-	-	-	55.74	m	61.32
600mm deep, 22.5 kg/m	76.14	2.50	78.04	-	-	-	-	78.04	m	85.85

Labour hourly rates: (except Specialists) Craft Operatives 18.37 Labourer 13.76 Rates are national average prices. Refer to REGIONAL VARIATIONS for indicative levels of overall pricing in regions	MATERIALS			LABOUR				RATES		
	Del to Site	Waste	Material Cost	Craft Optve	Lab	Labour Cost	Sunds	Nett Rate	Unit	Gross rate (10%)
	£	%	£	Hrs	Hrs	£	£	£		£

STRUCTURAL STEELWORK (Cont'd)

Isolated structural members, fabrication (Cont'd)

Steel Metsec intermediate span lattice joists primed at works; treated timber inserts in top and bottom chords; full depth end seatings for bolting to supporting structure (Cont'd)										
650mm deep, 29.7 kg/m	78.86	2.50	80.83	-	-	-	-	80.83	m	88.91
Steel Metsec intermediate span lattice joists primed at works; treated timber inserts in top and bottom chords; full depth end seatings for bolting to supporting structure										
450mm deep, 12.2 kg/m	48.95	2.50	50.17	-	-	-	-	50.17	m	55.19
500mm deep, 15.8 kg/m	51.66	2.50	52.96	-	-	-	-	52.96	m	58.25
550mm deep, 19.4 kg/m	54.38	2.50	55.74	-	-	-	-	55.74	m	61.32
600mm deep, 22.5 kg/m	76.14	2.50	78.04	-	-	-	-	78.04	m	85.85
650mm deep, 29.7 kg/m	78.86	2.50	80.83	-	-	-	-	80.83	m	88.91
Steel Metsec long span lattice joists primed at works; treated timber inserts in top and bottom chords; full depth end seatings for bolting to supporting structure										
700mm deep, 39.2 kg/m	100.61	2.50	103.13	-	-	-	-	103.13	m	113.44
800mm deep, 44.1 kg/m	103.33	2.50	105.91	-	-	-	-	105.91	m	116.50
900mm deep, 45.3 kg/m	130.52	2.50	133.78	-	-	-	-	133.78	m	147.16
1000mm deep, 46.1 kg/m	133.24	2.50	136.57	-	-	-	-	136.57	m	150.23
1500mm deep, 54.2 kg/m	141.40	2.50	144.93	-	-	-	-	144.93	m	159.43
Steel Metsec long span lattice joists primed at works; treated timber inserts in top and bottom chords; full depth end seatings for bolting to supporting structure										
700mm deep, 39.2 kg/m	100.61	2.50	103.13	-	-	-	-	103.13	m	113.44
800mm deep, 44.1 kg/m	103.33	2.50	105.91	-	-	-	-	105.91	m	116.50
900mm deep, 45.3 kg/m	130.52	2.50	133.78	-	-	-	-	133.78	m	147.16
1000mm deep, 46.1 kg/m	133.24	2.50	136.57	-	-	-	-	136.57	m	150.23
1500mm deep, 54.2 kg/m	141.40	2.50	144.93	-	-	-	-	144.93	m	159.43

Isolated structural members, permanent erection on site

Weldable steel, BS EN 10149 : 1996 Grade S275, hot rolled sections; various dimensions										
plain member; beam										
weight less than 25 Kg/m	-	-	-	44.12	-	810.48	-	810.48	t	891.53
weight 25-50 Kg/m	-	-	-	41.18	-	756.48	-	756.48	t	832.12
weight 50-100 Kg/m	-	-	-	41.18	-	756.48	-	756.48	t	832.12
weight exceeding 100 Kg/m	-	-	-	36.75	-	675.10	-	675.10	t	742.61
Steel Metsec short span lattice joists primed at works; treated timber inserts in top and bottom chords; full depth end seatings for bolting to supporting structure; hoist and fix 3.00m above ground level										
Reference MB22 - 220mm deep, 10 kg/m	-	-	-	0.66	0.08	13.23	-	13.23	m	14.55
Reference MB27 - 270mm deep, 10 kg/m	-	-	-	0.68	0.09	13.73	-	13.73	m	15.10
Reference MB30 - 300mm deep, 11 kg/m	-	-	-	0.75	0.09	15.02	-	15.02	m	16.52
Reference MB35 - 350mm deep, 12 kg/m	-	-	-	0.81	0.10	16.26	-	16.26	m	17.88
Reference MD35 - 350mm deep, 17 kg/m	-	-	-	0.90	0.11	18.05	-	18.05	m	19.85
Steel Metsec short span lattice joists primed at works; treated timber inserts in top and bottom chords; full depth end seatings for bolting to supporting structure; hoist and fix 6.00m above ground level										
Reference MB22 - 220mm deep, 10 kg/m	-	-	-	0.72	0.09	14.46	-	14.46	m	15.91
Reference MB27 - 270mm deep, 10 kg/m	-	-	-	0.75	0.09	15.02	-	15.02	m	16.52
Reference MB30 - 300mm deep, 11 kg/m	-	-	-	0.82	0.10	16.44	-	16.44	m	18.08
Reference MB35 - 350mm deep, 12 kg/m	-	-	-	0.89	0.11	17.86	-	17.86	m	19.65
Reference MD35 - 350mm deep, 17 kg/m	-	-	-	0.98	0.12	19.65	-	19.65	m	21.62
Steel Metsec intermediate span lattice joists primed at works; treated timber inserts in top and bottom chords; full depth end seatings for bolting to supporting structure; hoist and fix 3.00m above ground level										
Reference MD45 - 450mm deep, 18 kg/m	-	-	-	0.85	0.11	17.13	-	17.13	m	18.84
Reference MD50 - 500mm deep, 19 kg/m	-	-	-	1.03	0.13	20.71	-	20.71	m	22.78
Reference MD55 - 550mm deep, 20 kg/m	-	-	-	1.26	0.16	25.35	-	25.35	m	27.88
Reference MG60W - 600mm deep, 28 kg/m	-	-	-	1.35	0.17	27.14	-	27.14	m	29.85
Reference MG65W - 650mm deep, 29 kg/m	-	-	-	1.78	0.22	35.73	-	35.73	m	39.30
Steel Metsec intermediate span lattice joists primed at works; treated timber inserts in top and bottom chords; full depth end seatings for bolting to supporting structure; hoist and fix 6.00m above ground level										
Reference MD45 - 450mm deep, 18 kg/m	-	-	-	0.94	0.12	18.92	-	18.92	m	20.81
Reference MD50 - 500mm deep, 19 kg/m	-	-	-	1.13	0.14	22.68	-	22.68	m	24.95
Reference MD55 - 550mm deep, 20 kg/m	-	-	-	1.39	0.17	27.87	-	27.87	m	30.66
Reference MG60W - 600mm deep, 28 kg/m	-	-	-	1.49	0.19	29.99	-	29.99	m	32.98
Reference MG65W - 650mm deep, 29 kg/m	-	-	-	1.96	0.25	39.45	-	39.45	m	43.39
Steel Metsec long span lattice joists primed at works; treated timber inserts in top and bottom chords; full depth end seatings for bolting to supporting structure; hoist and fix 3.00m above ground level										
Reference MJ70W - 700mm deep, 39 kg/m	-	-	-	1.96	0.25	39.45	-	39.45	m	43.39
Reference MJ80W - 800mm deep, 38 kg/m	-	-	-	2.21	0.28	44.45	-	44.45	m	48.90
Reference ML90W - 900mm deep, 48 kg/m	-	-	-	2.27	0.28	45.55	-	45.55	m	50.11
Reference ML100W - 1000mm deep, 48 kg/m	-	-	-	2.31	0.29	46.43	-	46.43	m	51.07
Reference MML150W - 1500mm deep, 52 kg/m	-	-	-	2.40	0.30	48.22	-	48.22	m	53.04

STRUCTURAL METALWORK

Labour hourly rates: (except Specialists) Craft Operatives 18.37 Labourer 13.76 Rates are national average prices. Refer to REGIONAL VARIATIONS for indicative levels of overall pricing in regions	MATERIALS			LABOUR				RATES		
	Del to Site	Waste	Material Cost	Craft Optve	Lab	Labour Cost	Sunds	Nett Rate	Unit	Gross rate (10%)
	£	%	£	Hrs	Hrs	£	£	£		£

STRUCTURAL STEELWORK (Cont'd)

Isolated structural members, permanent erection on site (Cont'd)

Steel Metsec long span lattice joists primed at works; treated timber inserts in top and bottom chords; full depth end seatings for bolting to supporting structure; hoist and fix 6.00m above ground level

	Del to Site	Waste	Material Cost	Craft Optve	Lab	Labour Cost	Sunds	Nett Rate	Unit	Gross rate
Reference MJ70W - 700mm deep, 39 kg/m	-	-	-	2.16	0.27	43.39	-	43.39	m	47.73
Reference MJ80W - 800mm deep, 38 kg/m	-	-	-	2.43	0.30	48.77	-	48.77	m	53.64
Reference ML90W - 900mm deep, 48 kg/m	-	-	-	2.50	0.31	50.19	-	50.19	m	55.21
Reference ML100W - 1000mm deep, 48 kg/m	-	-	-	2.54	0.32	51.06	-	51.06	m	56.17
Reference MML150W - 1500mm deep, 52 kg/m	-	-	-	2.64	0.33	53.04	-	53.04	m	58.34

Allowance for fittings

To framed members

	Del to Site	Waste	Material Cost	Craft Optve	Lab	Labour Cost	Sunds	Nett Rate	Unit	Gross rate
Components for jointing members including brackets, supports and the like	810.00	-	810.00	23.00	-	422.51	-	1232.51	t	1355.76

To isolated members

	Del to Site	Waste	Material Cost	Craft Optve	Lab	Labour Cost	Sunds	Nett Rate	Unit	Gross rate
Components for jointing members including brackets, supports and the like	1020.00	-	1020.00	44.12	-	810.48	-	1830.48	t	2013.53

Cold rolled purlins, cladding rails and the like; Zed purlins and cladding rails; Metsec, galvanized steel

Purlins and cladding rails; sleeved system (section only); fixing to cleats on frame members at 6000mm centres; hoist and fix 3.00m above ground level; fixing with bolts

	Del to Site	Waste	Material Cost	Craft Optve	Lab	Labour Cost	Sunds	Nett Rate	Unit	Gross rate
reference 14214; 3.03 kg/m	27.01	2.50	27.68	0.20	0.03	4.09	-	31.77	m	34.95
reference 14216; 3.47 kg/m	30.49	2.50	31.25	0.20	0.03	4.09	-	35.34	m	38.87
reference 17214; 3.66 kg/m	31.94	2.50	32.74	0.25	0.03	5.01	-	37.74	m	41.52
reference 17216; 4.11 kg/m	36.31	2.50	37.22	0.25	0.03	5.01	-	42.22	m	46.44
reference 20216; 4.49 kg/m	39.63	2.50	40.63	0.30	0.04	6.06	-	46.69	m	51.36
reference 20218; 5.03 kg/m	44.71	2.50	45.83	0.30	0.04	6.06	-	51.89	m	57.08
reference 20220; 5.57 kg/m	49.29	2.50	50.52	0.30	0.04	6.06	-	56.58	m	62.24
reference 23218; 5.73 kg/m	50.74	2.50	52.01	0.33	0.04	6.61	-	58.62	m	64.48
reference 23220; 6.34 kg/m	56.27	2.50	57.68	0.33	0.04	6.61	-	64.29	m	70.72

Purlins and cladding rails; sleeved system (section only); fixing to cleats on frame members at 6000mm centres; hoist and fix 6.00m above ground level; fixing with bolts

	Del to Site	Waste	Material Cost	Craft Optve	Lab	Labour Cost	Sunds	Nett Rate	Unit	Gross rate
reference 14214; 3.03 kg/m	27.01	2.50	27.68	0.22	0.03	4.45	-	32.14	m	35.35
reference 14216; 3.47 kg/m	30.49	2.50	31.25	0.22	0.03	4.45	-	35.70	m	39.27
reference 17214; 3.66 kg/m	31.94	2.50	32.74	0.28	0.04	5.69	-	38.43	m	42.28
reference 17216; 4.11 kg/m	36.31	2.50	37.22	0.28	0.04	5.69	-	42.91	m	47.20
reference 20216; 4.49 kg/m	39.63	2.50	40.63	0.33	0.04	6.61	-	47.24	m	51.96
reference 20218; 5.03 kg/m	44.71	2.50	45.83	0.33	0.04	6.61	-	52.44	m	57.69
reference 20220; 5.57 kg/m	49.29	2.50	50.52	0.33	0.04	6.61	-	57.13	m	62.84
reference 23218; 5.73 kg/m	50.74	2.50	52.01	0.36	0.05	7.30	-	59.31	m	65.24
reference 23220; 6.34 kg/m	56.27	2.50	57.68	0.36	0.05	7.30	-	64.98	m	71.47

Purlins and cladding rails; Side rail sleeved system (section only); fixing to cleats on stanchions at 6000mm centres; hoist and fix 3.00m above ground level; fixing with bolts

	Del to Site	Waste	Material Cost	Craft Optve	Lab	Labour Cost	Sunds	Nett Rate	Unit	Gross rate
reference 14214; 3.03 kg/m	27.01	2.50	27.68	0.20	0.03	4.09	-	31.77	m	34.95
reference 14216; 3.47 kg/m	30.49	2.50	31.25	0.20	0.03	4.09	-	35.34	m	38.87
reference 17215; 3.85 kg/m	31.94	2.50	32.74	0.25	0.03	5.01	-	37.74	m	41.52
reference 17216; 4.11 kg/m	36.31	2.50	37.22	0.25	0.03	5.01	-	42.22	m	46.44
reference 20216; 4.49 kg/m	39.63	2.50	40.63	0.30	0.04	6.06	-	46.69	m	51.36
reference 20218; 5.03 kg/m	44.71	2.50	45.83	0.30	0.04	6.06	-	51.89	m	57.08
reference 20220; 5.57 kg/m	49.29	2.50	50.52	0.30	0.04	6.06	-	56.58	m	62.24
reference 23218; 5.73 kg/m	50.74	2.50	52.01	0.33	0.04	6.61	-	58.62	m	64.48
reference 23223; 7.26 kg/m	56.27	2.50	57.68	0.33	0.04	6.61	-	64.29	m	70.72

Purlins and cladding rails; Side rail sleeved system (section only); fixing to cleats on stanchions at 6000mm centres; hoist and fix 6.00m above ground level; fixing with bolts

	Del to Site	Waste	Material Cost	Craft Optve	Lab	Labour Cost	Sunds	Nett Rate	Unit	Gross rate
reference 14214; 3.03 kg/m	27.01	2.50	27.68	0.22	0.03	4.45	-	32.14	m	35.35
reference 14216; 3.47 kg/m	30.49	2.50	31.25	0.22	0.03	4.45	-	35.70	m	39.27
reference 17215; 3.85 kg/m	31.94	2.50	32.74	0.28	0.04	5.69	-	38.43	m	42.28
reference 17216; 4.11 kg/m	36.31	2.50	37.22	0.28	0.04	5.69	-	42.91	m	47.20
reference 20216; 4.49 kg/m	39.63	2.50	40.63	0.33	0.04	6.61	-	47.24	m	51.96
reference 20218; 5.03 kg/m	44.71	2.50	45.83	0.33	0.04	6.61	-	52.44	m	57.69
reference 20220; 5.57 kg/m	49.29	2.50	50.52	0.33	0.04	6.61	-	57.13	m	62.84
reference 23218; 5.73 kg/m	50.74	2.50	52.01	0.36	0.05	7.30	-	59.31	m	65.24
reference 23223; 7.26 kg/m	56.27	2.50	57.68	0.36	0.05	7.30	-	64.98	m	71.47

Sag rods; Round-lok anti-sag rods; to purlins

	Del to Site	Waste	Material Cost	Craft Optve	Lab	Labour Cost	Sunds	Nett Rate	Unit	Gross rate
purlins at 1150mm centres	9.51	2.50	9.74	0.25	0.03	5.01	-	14.75	nr	16.22
purlins at 1350mm centres	11.03	2.50	11.31	0.25	0.03	5.01	-	16.31	nr	17.94
purlins at 1550mm centres	12.46	2.50	12.77	0.33	0.04	6.61	-	19.39	nr	21.33
purlins at 1700mm centres	14.05	2.50	14.40	0.33	0.04	6.61	-	21.01	nr	23.11
purlins at 1950mm centres	14.47	2.50	14.83	0.33	0.04	6.61	-	21.45	nr	23.59

Other; Black metal purlin cleats; weld on

	Del to Site	Waste	Material Cost	Craft Optve	Lab	Labour Cost	Sunds	Nett Rate	Unit	Gross rate
Reference 142 for 142mm deep purlins	9.43	2.50	9.67	-	-	-	-	9.67	nr	10.64
Reference 172 for 172mm deep purlins	11.14	2.50	11.42	-	-	-	-	11.42	nr	12.57
Reference 202 for 202mm deep purlins	11.72	2.50	12.01	-	-	-	-	12.01	nr	13.22
Reference 232 for 232mm deep purlins	12.91	2.50	13.23	-	-	-	-	13.23	nr	14.55
Reference 262 for 262mm deep purlins	14.12	2.50	14.47	-	-	-	-	14.47	nr	15.92

Other; Black metal side rail cleats; weld on

	Del to Site	Waste	Material Cost	Craft Optve	Lab	Labour Cost	Sunds	Nett Rate	Unit	Gross rate
Reference 142 for 142mm wide rails	9.43	2.50	9.67	-	-	-	-	9.67	nr	10.64
Reference 172 for 172mm wide rails	11.14	2.50	11.42	-	-	-	-	11.42	nr	12.57

Labour hourly rates: (except Specialists) Craft Operatives 18.37 Labourer 13.76 Rates are national average prices. Refer to REGIONAL VARIATIONS for indicative levels of overall pricing in regions	MATERIALS			LABOUR				RATES		
	Del to Site	Waste	Material Cost	Craft Optve	Lab	Labour Cost	Sunds	Nett Rate	Unit	Gross rate (10%)
	£	%	£	Hrs	Hrs	£	£	£		£
STRUCTURAL STEELWORK (Cont'd)										
Cold rolled purlins, cladding rails and the like; Zed purlins and cladding rails; Metsec, galvanized steel (Cont'd)										
Other; Black metal side rail cleats; weld on (Cont'd)										
Reference 202 for 202mm wide rails	11.72	2.50	12.01	-	-	-	-	12.01	nr	13.22
Reference 232 for 232mm wide rails	12.91	2.50	13.23	-	-	-	-	13.23	nr	14.55
Other; Galvanized side rail supports; 122-262 series; weld on										
Galvanized side rail supports; 122-262 series; weld on										
rail 1000mm long	27.76	2.50	28.45	0.35	0.04	6.98	-	35.43	nr	38.98
rail 1400mm long	30.28	2.50	31.04	0.35	0.04	6.98	-	38.02	nr	41.82
rail 1600mm long	33.59	2.50	34.43	0.45	0.06	9.09	-	43.52	nr	47.87
rail 1800mm long	40.45	2.50	41.46	0.45	0.06	9.09	-	50.55	nr	55.61
Other; Diagonal tie wire ropes (assembled with end brackets); bolt on										
1700mm long	54.15	2.50	55.50	0.30	0.04	6.06	-	61.56	nr	67.72
2200mm long	60.58	2.50	62.10	0.30	0.04	6.06	-	68.16	nr	74.98
2600mm long	67.02	2.50	68.70	0.40	0.05	8.04	-	76.73	nr	84.41
3600mm long	82.75	2.50	84.82	0.40	0.05	8.04	-	92.85	nr	102.14
Metal decking; Galvanized steel troughed decking; 0.7mm thick metal, 35mm overall depth; natural soffit and top surface; 150mm end laps and one corrugation side laps										
Decking; fixed to steel rails at 1200mm centres with self tapping screws; drilling holes										
sloping; 10 degrees pitch	-	-	Specialist	-	-	Specialist	-	49.20	m²	54.12
Extra over decking for										
raking cutting	-	-	Specialist	-	-	Specialist	-	8.31	m	9.14
holes 50mm diameter; formed on site	-	-	Specialist	-	-	Specialist	-	22.97	nr	25.26
Metal decking; Galvanized steel troughed decking; 0.7mm thick metal, 48mm overall depth; natural soffit and top surface; 150mm end laps and one corrugation side laps										
Decking; fixed to steel rails at 1500mm centres with self tapping screws; drilling holes										
sloping; 10 degrees pitch	-	-	Specialist	-	-	Specialist	-	49.20	m²	54.12
Extra over decking for										
raking cutting	-	-	Specialist	-	-	Specialist	-	8.31	m	9.14
holes 50mm diameter; formed on site	-	-	Specialist	-	-	Specialist	-	22.97	nr	25.26
Metal decking; Galvanized steel troughed decking; 0.7mm thick metal, 63mm overall depth; natural soffit and top surface; 150mm end laps and one corrugation side laps										
Decking; fixed to steel rails at 2000mm centres with self tapping screws; drilling holes										
sloping; 10 degrees pitch	-	-	Specialist	-	-	Specialist	-	52.53	m²	57.78
Extra over decking for										
raking cutting	-	-	Specialist	-	-	Specialist	-	8.31	m	9.14
holes 50mm diameter; formed on site	-	-	Specialist	-	-	Specialist	-	22.97	nr	25.26
Metal decking; Galvanized steel troughed decking; 0.7mm thick metal, 100mm overall depth; natural soffit and top surface; 150mm end laps and one corrugation side laps										
Decking; fixed to steel rails at 3500mm centres with self tapping screws; drilling holes										
sloping; 10 degrees pitch	-	-	Specialist	-	-	Specialist	-	55.56	m²	61.12
Extra over decking for										
raking cutting	-	-	Specialist	-	-	Specialist	-	8.31	m	9.14
holes 50mm diameter; formed on site	-	-	Specialist	-	-	Specialist	-	22.97	nr	25.26
Metal decking; Galvanized steel troughed decking; 0.9mm thick metal, 35mm overall depth; natural soffit and top surface; 150mm end laps and one corrugation side laps										
Decking; fixed to steel rails at 1500mm centres with self tapping screws; drilling holes										
sloping; 10 degrees pitch	-	-	Specialist	-	-	Specialist	-	55.33	m²	60.87
Extra over decking for										
raking cutting	-	-	Specialist	-	-	Specialist	-	8.31	m	9.14
holes 50mm diameter; formed on site	-	-	Specialist	-	-	Specialist	-	22.97	nr	25.26
Metal decking; Galvanized steel troughed decking; 0.9mm thick metal, 48mm overall depth; natural soffit and top surface; 150mm end laps and one corrugation side laps										
Decking; fixing to steel rails at 2000mm centres with self tapping screws; drilling holes										
sloping; 10 degrees pitch	-	-	Specialist	-	-	Specialist	-	55.38	m²	60.92
Extra over decking for										
raking cutting	-	-	Specialist	-	-	Specialist	-	8.31	m	9.14
holes 50mm diameter; formed on site	-	-	Specialist	-	-	Specialist	-	22.97	nr	25.26

Labour hourly rates: (except Specialists) Craft Operatives 18.37 Labourer 13.76 Rates are national average prices. Refer to REGIONAL VARIATIONS for indicative levels of overall pricing in regions	MATERIALS			LABOUR				RATES		
	Del to Site	Waste	Material Cost	Craft Optve	Lab	Labour Cost	Sunds	Nett Rate	Unit	Gross rate (10%)
	£	%	£	Hrs	Hrs	£	£	£		£
STRUCTURAL STEELWORK (Cont'd)										
Metal decking; Galvanized steel troughed decking; 0.9mm thick metal, 63mm overall depth; natural soffit and top surface; 150mm end laps and one corrugation side laps										
Decking; fixing to steel rails at 2500mm centres with self tapping screws; drilling holes										
sloping; 10 degrees pitch................	-	-	Specialist	-	-	Specialist	-	59.25	m²	65.18
Extra over decking for										
raking cutting....................	-	-	Specialist	-	-	Specialist	-	8.31	m	9.14
holes 50mm diameter; formed on site	-	-	Specialist	-	-	Specialist	-	22.97	nr	25.26
Metal decking; Galvanized steel troughed decking; 0.9mm thick metal, 100mm overall depth; natural soffit and top surface; 150mm end laps and one corrugation side laps										
Decking; fixing to steel rails at 4000mm centres with self tapping screws; drilling holes										
sloping; 10 degrees pitch................	-	-	Specialist	-	-	Specialist	-	62.12	m²	68.33
Extra over decking for										
raking cutting....................	-	-	Specialist	-	-	Specialist	-	8.31	m	9.14
holes 50mm diameter; formed on site	-	-	Specialist	-	-	Specialist	-	22.97	nr	25.26
Metal decking; Galvanized steel troughed decking; 1.2mm thick metal, 35mm overall depth; natural soffit and top surface; 150mm end laps and one corrugation side laps										
Decking; fixing to steel rails at 2000mm centres with self tapping screws; drilling holes										
sloping; 10 degrees pitch................	-	-	Specialist	-	-	Specialist	-	62.29	m²	68.52
Extra over decking for										
raking cutting....................	-	-	Specialist	-	-	Specialist	-	8.31	m	9.14
holes 50mm diameter; formed on site	-	-	Specialist	-	-	Specialist	-	22.97	nr	25.26
Metal decking; Galvanized steel troughed decking; 1.2mm thick metal, 48mm overall depth; natural soffit and top surface; 150mm end laps and one corrugation side laps										
Decking; fixing to steel rails at 2500mm centres with self tapping screws; drilling holes										
sloping; 10 degrees pitch................	-	-	Specialist	-	-	Specialist	-	62.29	m²	68.52
Extra over decking for										
raking cutting....................	-	-	Specialist	-	-	Specialist	-	8.31	m	9.14
holes 50mm diameter; formed on site	-	-	Specialist	-	-	Specialist	-	22.97	nr	25.26
Metal decking; Galvanized steel troughed decking; 1.2mm thick metal, 63mm overall depth; natural soffit and top surface; 150mm end laps and one corrugation side laps										
Decking; fixing to steel rails at 3000mm centres with self tapping screws; drilling holes										
sloping; 10 degrees pitch................	-	-	Specialist	-	-	Specialist	-	65.82	m²	72.40
Extra over decking for										
raking cutting....................	-	-	Specialist	-	-	Specialist	-	8.31	m	9.14
holes 50mm diameter; formed on site	-	-	Specialist	-	-	Specialist	-	22.97	nr	25.26
Metal decking; Galvanized steel troughed decking; 1.2mm thick metal, 100mm overall depth; natural soffit and top surface; 150mm end laps and one corrugation side laps										
Decking; fixing to steel rails at 4500mm centres with self tapping screws; drilling holes										
sloping; 10 degrees pitch................	-	-	Specialist	-	-	Specialist	-	71.29	m²	78.42
Extra over decking for										
raking cutting....................	-	-	Specialist	-	-	Specialist	-	8.31	m	9.14
holes 50mm diameter; formed on site	-	-	Specialist	-	-	Specialist	-	22.97	nr	25.26
Holding down bolts or assemblies; mild steel; supply only										
rag or indented bolts; M 10 with nuts and washers										
100mm long................	1.48	5.00	1.55	-	-	-	-	1.55	nr	1.71
160mm long................	1.54	5.00	1.62	-	-	-	-	1.62	nr	1.78
rag or indented bolts; M 12 with nuts and washers										
100mm long................	1.60	5.00	1.68	-	-	-	-	1.68	nr	1.85
160mm long................	1.85	5.00	1.94	-	-	-	-	1.94	nr	2.14
200mm long................	2.08	5.00	2.18	-	-	-	-	2.18	nr	2.40
rag or indented bolts; M 16 with nuts and washers										
120mm long................	2.41	5.00	2.53	-	-	-	-	2.53	nr	2.78
160mm long................	2.76	5.00	2.90	-	-	-	-	2.90	nr	3.19
200mm long................	3.22	5.00	3.38	-	-	-	-	3.38	nr	3.72
holding down bolt assembly; M 20 bolt with 100 x 100 x 10mm plate washer tack welded to head; with nuts and washers										
300mm long................	5.55	5.00	5.83	-	-	-	-	5.83	nr	6.41
350mm long................	6.33	5.00	6.65	-	-	-	-	6.65	nr	7.31
450mm long................	9.64	5.00	10.12	-	-	-	-	10.12	nr	11.13
Special bolts; Black bolts; BS 4190 grade 4.6; galvanized										
M 8; with nuts and washers										
50mm long................	0.16	5.00	0.17	0.07	-	1.29	-	1.45	nr	1.60
60mm long................	0.18	5.00	0.19	0.07	-	1.29	-	1.48	nr	1.62
70mm long................	0.20	5.00	0.21	0.08	-	1.47	-	1.68	nr	1.84
80mm long................	0.22	5.00	0.23	0.08	-	1.47	-	1.70	nr	1.87

Labour hourly rates: (except Specialists) Craft Operatives 18.37 Labourer 13.76 Rates are national average prices. Refer to REGIONAL VARIATIONS for indicative levels of overall pricing in regions	MATERIALS			LABOUR				RATES		
	Del to Site	Waste	Material Cost	Craft Optve	Lab	Labour Cost	Sunds	Nett Rate	Unit	Gross rate (10%)
	£	%	£	Hrs	Hrs	£	£	£		£
STRUCTURAL STEELWORK (Cont'd)										
Special bolts; Black bolts; BS 4190 grade 4.6; galvanized (Cont'd)										
M 10; with nuts and washers										
40mm long	0.27	5.00	0.28	0.07	-	1.29	-	1.57	nr	1.72
50mm long	0.30	5.00	0.32	0.07	-	1.29	-	1.60	nr	1.76
60mm long	0.34	5.00	0.35	0.07	-	1.29	-	1.64	nr	1.80
70mm long	0.37	5.00	0.39	0.08	-	1.47	-	1.86	nr	2.05
80mm long	0.40	5.00	0.42	0.08	-	1.47	-	1.89	nr	2.08
90mm long	0.44	5.00	0.46	0.08	-	1.47	-	1.93	nr	2.13
100mm long	0.48	5.00	0.50	0.08	-	1.47	-	1.97	nr	2.17
120mm long	0.56	5.00	0.59	0.09	-	1.65	-	2.24	nr	2.46
140mm long	0.64	5.00	0.67	0.10	-	1.84	-	2.51	nr	2.76
150mm long	0.68	5.00	0.72	0.12	-	2.20	-	2.92	nr	3.21
M 12; with nuts and washers										
40mm long	0.38	5.00	0.40	0.08	-	1.47	-	1.87	nr	2.05
50mm long	0.43	5.00	0.45	0.08	-	1.47	-	1.92	nr	2.11
60mm long	0.48	5.00	0.50	0.08	-	1.47	-	1.97	nr	2.17
70mm long	0.53	5.00	0.56	0.08	-	1.47	-	2.03	nr	2.23
80mm long	0.58	5.00	0.61	0.08	-	1.47	-	2.08	nr	2.29
90mm long	0.63	5.00	0.66	0.08	-	1.47	-	2.13	nr	2.35
100mm long	0.68	5.00	0.72	0.09	-	1.65	-	2.37	nr	2.61
120mm long	0.79	5.00	0.82	0.09	-	1.65	-	2.48	nr	2.73
140mm long	0.89	5.00	0.93	0.12	-	2.20	-	3.14	nr	3.45
150mm long	0.94	5.00	0.98	0.12	-	2.20	-	3.19	nr	3.51
160mm long	0.99	5.00	1.04	0.12	-	2.20	-	3.24	nr	3.57
180mm long	1.09	5.00	1.14	0.13	-	2.39	-	3.53	nr	3.89
200mm long	1.19	5.00	1.25	0.14	-	2.57	-	3.82	nr	4.20
220mm long	2.02	5.00	2.13	0.15	-	2.76	-	4.88	nr	5.37
240mm long	2.31	5.00	2.43	0.16	-	2.94	-	5.37	nr	5.90
260mm long	2.65	5.00	2.78	0.17	-	3.12	-	5.90	nr	6.49
300mm long	3.50	5.00	3.67	0.20	-	3.67	-	7.35	nr	8.08
M 16; with nuts and washers										
50mm long	0.82	5.00	0.86	0.08	-	1.47	-	2.33	nr	2.56
60mm long	0.87	5.00	0.91	0.08	-	1.47	-	2.38	nr	2.62
70mm long	0.99	5.00	1.04	0.08	-	1.47	-	2.51	nr	2.76
80mm long	1.07	5.00	1.13	0.09	-	1.65	-	2.78	nr	3.06
90mm long	1.20	5.00	1.25	0.09	-	1.65	-	2.91	nr	3.20
100mm long	1.20	5.00	1.26	0.10	-	1.84	-	3.09	nr	3.40
120mm long	1.43	5.00	1.50	0.10	-	1.84	-	3.34	nr	3.68
140mm long	1.55	5.00	1.63	0.12	-	2.20	-	3.83	nr	4.22
150mm long	1.64	5.00	1.72	0.13	-	2.39	-	4.11	nr	4.52
160mm long	1.82	5.00	1.91	0.14	-	2.57	-	4.48	nr	4.93
180mm long	2.00	5.00	2.10	0.14	-	2.57	-	4.67	nr	5.13
200mm long	2.21	5.00	2.32	0.15	-	2.76	-	5.07	nr	5.58
220mm long	2.69	5.00	2.83	0.15	-	2.76	-	5.58	nr	6.14
M 20; with nuts and washers										
60mm long	1.46	5.00	1.54	0.08	-	1.47	-	3.00	nr	3.31
70mm long	1.63	5.00	1.71	0.08	-	1.47	-	3.18	nr	3.49
80mm long	1.95	5.00	2.05	0.09	-	1.65	-	3.70	nr	4.07
90mm long	2.09	5.00	2.20	0.10	-	1.84	-	4.04	nr	4.44
100mm long	2.02	5.00	2.13	0.10	-	1.84	-	3.96	nr	4.36
140mm long	2.35	5.00	2.47	0.13	-	2.39	-	4.85	nr	5.34
150mm long	2.69	5.00	2.83	0.14	-	2.57	-	5.40	nr	5.94
180mm long	3.27	5.00	3.43	0.15	-	2.76	-	6.19	nr	6.81
200mm long	3.70	5.00	3.88	0.16	-	2.94	-	6.82	nr	7.50
220mm long	4.31	5.00	4.53	0.17	-	3.12	-	7.65	nr	8.42
240mm long	4.99	5.00	5.24	0.18	-	3.31	-	8.55	nr	9.40
260mm long	5.40	5.00	5.67	0.20	-	3.67	-	9.35	nr	10.28
300mm long	6.13	5.00	6.44	0.23	-	4.23	-	10.66	nr	11.73
M 24; with nuts and washers										
70mm long	2.48	5.00	2.60	0.10	-	1.84	-	4.44	nr	4.88
80mm long	3.00	5.00	3.15	0.10	-	1.84	-	4.99	nr	5.49
90mm long	3.22	5.00	3.38	0.12	-	2.20	-	5.58	nr	6.14
100mm long	3.16	5.00	3.32	0.12	-	2.20	-	5.53	nr	6.08
130mm long	3.66	5.00	3.84	0.13	-	2.39	-	6.23	nr	6.85
150mm long	4.30	5.00	4.51	0.15	-	2.76	-	7.27	nr	7.99
180mm long	4.93	5.00	5.17	0.15	-	2.76	-	7.93	nr	8.72
200mm long	5.30	5.00	5.57	0.16	-	2.94	-	8.51	nr	9.36
M 30; with nuts and washers										
110mm long	5.58	5.00	5.85	0.17	-	3.12	-	8.98	nr	9.87
140mm long	5.91	5.00	6.21	0.18	-	3.31	-	9.51	nr	10.46
200mm long	7.54	5.00	7.91	0.21	-	3.86	-	11.77	nr	12.95
Special bolts; High strength friction grip bolts; BS 14399 Part 1 - general grade										
M 16; with nuts and washers										
50mm long	1.20	5.00	1.26	0.09	-	1.65	-	2.91	nr	3.20
60mm long	1.44	5.00	1.51	0.09	-	1.65	-	3.17	nr	3.48
75mm long	1.80	5.00	1.89	0.10	-	1.84	-	3.73	nr	4.10
80mm long	1.92	5.00	2.02	0.10	-	1.84	-	3.85	nr	4.24
90mm long	2.16	5.00	2.27	0.12	-	2.20	-	4.47	nr	4.92
M 20; with nuts and washers										
60mm long	2.12	5.00	2.23	0.09	-	1.65	-	3.88	nr	4.27
70mm long	2.26	5.00	2.37	0.10	-	1.84	-	4.21	nr	4.63
80mm long	2.35	5.00	2.47	0.10	-	1.84	-	4.30	nr	4.73
90mm long	2.47	5.00	2.59	0.12	-	2.20	-	4.80	nr	5.28
M 24; with nuts and washers										
65mm long	3.12	5.00	3.28	0.13	-	2.39	-	5.66	nr	6.23
80mm long	3.84	5.00	4.03	0.13	-	2.39	-	6.42	nr	7.06
90mm long	4.32	5.00	4.54	0.14	-	2.57	-	7.11	nr	7.82
100mm long	4.80	5.00	5.04	0.14	-	2.57	-	7.61	nr	8.37
120mm long	5.28	5.00	5.54	0.16	-	2.94	-	8.48	nr	9.33

Labour hourly rates: (except Specialists) Craft Operatives 18.37 Labourer 13.76 Rates are national average prices. Refer to REGIONAL VARIATIONS for indicative levels of overall pricing in regions	MATERIALS			LABOUR				RATES		
	Del to Site	Waste	Material Cost	Craft Optve	Lab	Labour Cost	Sunds	Nett Rate	Unit	Gross rate (10%)
	£	%	£	Hrs	Hrs	£	£	£		£
STRUCTURAL STEELWORK (Cont'd)										
Special bolts; High strength friction grip bolts; BS 14399 Part 1 - general grade (Cont'd)										
M 24; with nuts and washers (Cont'd)										
140mm long	6.24	5.00	6.55	0.17	-	3.12	-	9.67	nr	10.64
150mm long	8.16	5.00	8.57	0.18	-	3.31	-	11.87	nr	13.06
Surface treatment; Off site at works										
Note notwithstanding the requirement of NRM to measure painting on structural steelwork in m², painting off site has been given in tonnes of structural steelwork in accordance with normal steelwork contractors practice										
Blast cleaning										
surfaces of steelwork	-	-	Specialist	-	-	Specialist	-	115.50	t	127.05
One coat micaceous oxide primer, 75 microns										
surfaces of steelwork	-	-	Specialist	-	-	Specialist	-	129.00	t	141.90
Two coats micaceous oxide primer, 150 microns										
surfaces of steelwork	-	-	Specialist	-	-	Specialist	-	240.00	t	264.00
STRUCTURAL ALUMINIUM WORK										
Metal decking; Aluminium troughed decking; 0.9mm thick metal 35mm overall depth; natural soffit and top surface; 150mm end laps and one corrugation side laps										
Decking; fixing to steel rails at 900mm centres with self tapping screws; drilling holes										
sloping; 10 degrees pitch	-	-	Specialist	-	-	Specialist	-	62.21	m²	68.43
Extra over decking for										
raking cutting	-	-	Specialist	-	-	Specialist	-	8.31	m	9.14
holes 50mm diameter; formed on site	-	-	Specialist	-	-	Specialist	-	22.97	nr	25.26
Metal decking; Aluminium troughed decking; 0.9mm thick metal, 48mm overall depth; natural soffit and top surface; 150mm end laps and one corrugation side laps										
Decking; fixing to steel rails at 1200mm centres with self tapping screws; drilling holes										
sloping; 10 degrees pitch	-	-	Specialist	-	-	Specialist	-	62.21	m²	68.43
Extra over decking for										
raking cutting	-	-	Specialist	-	-	Specialist	-	8.31	m	9.14
holes 50mm diameter; formed on site	-	-	Specialist	-	-	Specialist	-	22.97	nr	25.26
Metal decking; Aluminium troughed decking; 0.9mm thick metal, 63mm overall depth; natural soffit and top surface; 150mm end laps and one corrugation side laps										
Decking; fixing to steel rails at 1500mm centres with self tapping screws; drilling holes										
sloping; 10 degrees pitch	-	-	Specialist	-	-	Specialist	-	66.68	m²	73.34
Extra over decking for										
raking cutting	-	-	Specialist	-	-	Specialist	-	8.31	m	9.14
holes 50mm diameter; formed on site	-	-	Specialist	-	-	Specialist	-	22.97	nr	25.26
Metal decking; Aluminium troughed decking; 0.9mm thick metal, 100mm overall depth; natural soffit and top surface; 150mm end laps and one corrugation side laps										
Decking; fixing to steel rails at 2900mm centres with self tapping screws; drilling holes										
sloping; 10 degrees pitch	-	-	Specialist	-	-	Specialist	-	71.44	m²	78.59
Extra over decking for										
raking cutting	-	-	Specialist	-	-	Specialist	-	8.31	m	9.14
holes 50mm diameter; formed on site	-	-	Specialist	-	-	Specialist	-	22.97	nr	25.26
Metal decking; Aluminium troughed decking; 1.2mm thick metal, 35mm overall depth; natural soffit and top surface; 150mm end laps and one corrugation side laps										
Decking; fixing to steel rails at 1200mm centres with self tapping screws; drilling holes										
sloping; 10 degrees pitch	-	-	Specialist	-	-	Specialist	-	68.82	m²	75.70
Extra over decking for										
raking cutting	-	-	Specialist	-	-	Specialist	-	8.31	m	9.14
holes 50mm diameter; formed on site	-	-	Specialist	-	-	Specialist	-	22.97	nr	25.26
Metal decking; Aluminium troughed decking; 1.2mm thick metal, 48mm overall depth; natural soffit and top surface; 150mm end laps and one corrugation side laps										
Decking; fixing to steel rails at 1700mm centres with self tapping screws; drilling holes										
sloping; 10 degrees pitch	-	-	Specialist	-	-	Specialist	-	68.82	m²	75.70
Extra over decking for										
raking cutting	-	-	Specialist	-	-	Specialist	-	8.31	m	9.14
holes 50mm diameter; formed on site	-	-	Specialist	-	-	Specialist	-	22.97	nr	25.26

Labour hourly rates: (except Specialists) Craft Operatives 18.37 Labourer 13.76 Rates are national average prices. Refer to REGIONAL VARIATIONS for indicative levels of overall pricing in regions	MATERIALS			LABOUR				RATES		
	Del to Site £	Waste %	Material Cost £	Craft Optve Hrs	Lab Hrs	Labour Cost £	Sunds £	Nett Rate £	Unit	Gross rate (10%) £
STRUCTURAL ALUMINIUM WORK (Cont'd)										
Metal decking; Aluminium troughed decking; 1.2mm thick metal, 63mm overall depth; natural soffit and top surface; 150mm end laps and one corrugation side laps										
Decking; fixing to steel rails at 2000mm centres with self tapping screws; drilling holes										
sloping; 10 degrees pitch..........	-	-	Specialist	-	-	Specialist	-	74.06	m²	81.46
Extra over decking for										
raking cutting	-	-	Specialist	-	-	Specialist	-	8.31	m	9.14
holes 50mm diameter; formed on site	-	-	Specialist	-	-	Specialist	-	22.97	nr	25.26
Metal decking; Aluminium troughed decking; 1.2mm thick metal, 100mm overall depth; natural soffit and top surface; 150mm end laps and one corrugation side laps										
Decking; fixing to steel rails at 3000mm centres with self tapping screws; drilling holes										
sloping; 10 degrees pitch..........	-	-	Specialist	-	-	Specialist	-	80.11	m²	88.13
Extra over decking for										
raking cutting	-	-	Specialist	-	-	Specialist	-	8.31	m	9.14
holes 50mm diameter; formed on site	-	-	Specialist	-	-	Specialist	-	22.97	nr	25.26

This page left blank intentionally

Labour hourly rates: (except Specialists) Craft Operatives 18.37 Labourer 13.76. Rates are national average prices. Refer to REGIONAL VARIATIONS for indicative levels of overall pricing in regions	MATERIALS			LABOUR				RATES		
	Del to Site £	Waste %	Material Cost £	Craft Optve Hrs	Lab Hrs	Labour Cost £	Sunds £	Nett Rate £	Unit	Gross rate (10%) £

TIMBER FRAMING

Engineered or prefabricated members/items

Roof trusses; Trussed Rafters to PD6693; Crendon Timber Engineering; softwood, sawn, impregnated; plated joints

22.5 degree Standard duo pitch; 450mm overhangs; fixing with clips (included elsewhere); span over wall plates

5000mm	24.18	2.50	24.79	1.10	0.14	22.13	8.25	55.17	nr	60.69
6000mm	28.69	2.50	29.41	1.20	0.15	24.11	9.00	62.52	nr	68.77
7000mm	32.08	2.50	32.88	1.30	0.16	26.08	9.75	68.71	nr	75.58
8000mm	39.97	2.50	40.97	1.40	0.17	28.06	10.50	79.53	nr	87.48
9000mm	50.91	2.50	52.18	1.50	0.19	30.17	11.25	93.60	nr	102.96
10000mm	61.06	2.50	62.58	1.60	0.20	32.14	12.00	106.73	nr	117.40

35 degree Standard duo pitch; 450mm overhangs; fixing with clips (included elsewhere); span over wall plates

5000mm	26.25	2.50	26.91	1.10	0.14	22.13	8.25	57.29	nr	63.02
6000mm	30.76	2.50	31.53	1.20	0.15	24.11	9.00	64.64	nr	71.10
7000mm	34.14	2.50	35.00	1.30	0.16	26.08	9.75	70.83	nr	77.91
8000mm	48.67	2.50	49.88	1.40	0.17	28.06	10.50	88.44	nr	97.28
9000mm	54.30	2.50	55.66	1.50	0.19	30.17	11.25	97.08	nr	106.79

45 degree Standard duo pitch; 450mm overhangs; fixing with clips (included elsewhere); span over wall plates

5000mm	29.24	2.50	29.97	1.10	0.14	22.13	8.25	60.36	nr	66.39
6000mm	38.26	2.50	39.22	1.20	0.15	24.11	9.00	72.33	nr	79.56
7000mm	48.41	2.50	49.62	1.30	0.16	26.08	9.75	85.45	nr	94.00
8000mm	69.88	2.50	71.63	1.40	0.17	28.06	10.50	110.19	nr	121.21
9000mm	80.03	2.50	82.03	1.50	0.19	30.17	11.25	123.45	nr	135.80

22.5 degree Bobtail duo pitch; 450mm overhangs; fixing with clips (included elsewhere); span over wall plates

4000mm	27.29	2.50	27.97	1.00	0.13	20.16	7.50	55.63	nr	61.19
5000mm	32.92	2.50	33.75	1.10	0.14	22.13	8.25	64.13	nr	70.54
6000mm	40.82	2.50	41.84	1.20	0.15	24.11	9.00	74.95	nr	82.44
7000mm	49.84	2.50	51.08	1.30	0.16	26.08	9.75	86.92	nr	95.61
8000mm	57.73	2.50	59.17	1.40	0.17	28.06	10.50	97.73	nr	107.50
9000mm	65.62	2.50	67.26	1.50	0.19	30.17	11.25	108.68	nr	119.55

35 degree Bobtail duo pitch; 450mm overhangs; fixing with clips (included elsewhere); span over wall plates

4000mm	27.58	2.50	28.27	1.00	0.13	20.16	7.50	55.93	nr	61.52
5000mm	36.60	2.50	37.51	1.10	0.14	22.13	8.25	67.90	nr	74.69
6000mm	42.24	2.50	43.29	1.20	0.15	24.11	9.00	76.40	nr	84.04
7000mm	49.00	2.50	50.23	1.30	0.16	26.08	9.75	86.06	nr	94.66
8000mm	59.15	2.50	60.63	1.40	0.17	28.06	10.50	99.18	nr	109.10
9000mm	67.04	2.50	68.72	1.50	0.19	30.17	11.25	110.14	nr	121.15

45 degree Bobtail duo pitch; 450mm overhangs; fixing with clips (included elsewhere); span over wall plates

4000mm	24.59	2.50	25.20	1.00	0.13	20.16	7.50	52.86	nr	58.15
5000mm	35.86	2.50	36.76	1.10	0.14	22.13	8.25	67.14	nr	73.86
6000mm	47.14	2.50	48.32	1.20	0.15	24.11	9.00	81.42	nr	89.57
7000mm	60.67	2.50	62.19	1.30	0.16	26.08	9.75	98.02	nr	107.82
8000mm	80.96	2.50	82.99	1.40	0.17	28.06	10.50	121.54	nr	133.70
9000mm	86.60	2.50	88.77	1.50	0.19	30.17	11.25	130.19	nr	143.20

22.5 degree Monopitch; 450mm overhangs; fixing with clips (included elsewhere); span over wall plates

2000mm	15.68	2.50	16.07	0.80	0.10	16.07	6.00	38.14	nr	41.95
3000mm	18.49	2.50	18.96	0.90	0.11	18.05	6.75	43.75	nr	48.13
4000mm	24.13	2.50	24.74	1.00	0.13	20.16	7.50	52.39	nr	57.63
5000mm	29.77	2.50	30.51	1.10	0.14	22.13	8.25	60.90	nr	66.99
6000mm	38.79	2.50	39.76	1.20	0.15	24.11	9.00	72.87	nr	80.15

35 degree Monopitch; 450mm overhangs; fixing with clips (included elsewhere); span over wall plates

2000mm	19.49	2.50	19.97	0.80	0.10	16.07	6.00	42.04	nr	46.25
3000mm	25.12	2.50	25.75	0.90	0.11	18.05	6.75	50.55	nr	55.60
4000mm	29.63	2.50	30.37	1.00	0.13	20.16	7.50	58.03	nr	63.84
5000mm	36.40	2.50	37.31	1.10	0.14	22.13	8.25	67.69	nr	74.46
6000mm	46.55	2.50	47.71	1.20	0.15	24.11	9.00	80.82	nr	88.90

45 degree Monopitch; 450mm overhangs; fixing with clips (included elsewhere); span over wall plates

2000mm	15.81	2.50	16.20	0.80	0.10	16.07	6.00	38.28	nr	42.10
3000mm	23.70	2.50	24.29	0.90	0.11	18.05	6.75	49.09	nr	54.00
4000mm	30.47	2.50	31.23	1.00	0.13	20.16	7.50	58.89	nr	64.78

22.5 degree duo pitch Girder Truss; fixing with clips (included elsewhere); span over wall plates

5000mm	91.27	2.50	93.56	1.10	0.14	22.13	8.25	123.94	nr	136.33
6000mm	103.68	2.50	106.27	1.20	0.15	24.11	9.00	139.38	nr	153.31
7000mm	114.95	2.50	117.82	1.30	0.16	26.08	9.75	153.66	nr	169.02
8000mm	137.50	2.50	140.94	1.40	0.17	28.06	10.50	179.50	nr	197.45
9000mm	148.78	2.50	152.50	1.50	0.19	30.17	11.25	193.91	nr	213.31
10000mm	193.88	2.50	198.72	1.60	0.20	32.14	12.00	242.87	nr	267.15

35 degree duo pitch Girder Truss; fixing with clips (included elsewhere); span over wall plates

5000mm	91.58	2.50	93.87	1.10	0.14	22.13	8.25	124.25	nr	136.67
6000mm	102.85	2.50	105.42	1.20	0.15	24.11	9.00	138.53	nr	152.38

CARPENTRY

Labour hourly rates: (except Specialists) Craft Operatives 18.37 Labourer 13.76 Rates are national average prices. Refer to REGIONAL VARIATIONS for indicative levels of overall pricing in regions	MATERIALS			LABOUR				RATES		
	Del to Site	Waste	Material Cost	Craft Optve	Lab	Labour Cost	Sunds	Nett Rate		Gross rate (10%)
	£	%	£	Hrs	Hrs	£	£	£	Unit	£

TIMBER FRAMING (Cont'd)

Engineered or prefabricated members/items (Cont'd)

Roof trusses; Trussed Rafters to PD6693; Crendon Timber Engineering; softwood, sawn, impregnated; plated joints (Cont'd)

35 degree duo pitch Girder Truss; fixing with clips (included elsewhere); span over wall plates (Cont'd)

	Del to Site	Waste	Material Cost	Craft Optve	Lab	Labour Cost	Sunds	Nett Rate	Unit	Gross rate (10%)
7000mm	116.38	2.50	119.29	1.30	0.16	26.08	9.75	155.12	nr	170.64
8000mm	138.93	2.50	142.40	1.40	0.17	28.06	10.50	180.96	nr	199.06
9000mm	159.23	2.50	163.21	1.50	0.19	30.17	11.25	204.63	nr	225.09
10000mm	193.05	2.50	197.88	1.60	0.20	32.14	12.00	242.02	nr	266.22

45 degree duo pitch Girder Truss; fixing with clips (included elsewhere); span over wall plates

	Del to Site	Waste	Material Cost	Craft Optve	Lab	Labour Cost	Sunds	Nett Rate	Unit	Gross rate (10%)
5000mm	73.40	2.50	75.24	1.10	0.14	22.13	8.25	105.62	nr	116.18
6000mm	107.23	2.50	109.91	1.20	0.15	24.11	9.00	143.02	nr	157.32
7000mm	138.80	2.50	142.27	1.30	0.16	26.08	9.75	178.10	nr	195.91
8000mm	163.60	2.50	167.69	1.40	0.17	28.06	10.50	206.25	nr	226.87
9000mm	194.04	2.50	198.90	1.50	0.19	30.17	11.25	240.31	nr	264.35

22.5 degree pitch gable ladder; 450mm overhang; span over wall plate

	Del to Site	Waste	Material Cost	Craft Optve	Lab	Labour Cost	Sunds	Nett Rate	Unit	Gross rate (10%)
5000mm	33.27	2.50	34.10	1.10	0.14	22.13	8.25	64.48	nr	70.93
6000mm	35.52	2.50	36.41	1.20	0.15	24.11	9.00	69.52	nr	76.47
7000mm	38.90	2.50	39.88	1.30	0.16	26.08	9.75	75.71	nr	83.28
8000mm	44.54	2.50	45.65	1.40	0.17	28.06	10.50	84.21	nr	92.63
9000mm	46.80	2.50	47.97	1.50	0.19	30.17	11.25	89.38	nr	98.32
10000mm	50.59	2.50	51.86	1.60	0.20	32.14	12.00	96.00	nr	105.60

35 degree pitch gable ladder; 450mm overhang; span over wall plate

	Del to Site	Waste	Material Cost	Craft Optve	Lab	Labour Cost	Sunds	Nett Rate	Unit	Gross rate (10%)
5000mm	29.31	2.50	30.04	1.10	0.14	22.13	8.25	60.42	nr	66.47
6000mm	39.46	2.50	40.44	1.20	0.15	24.11	9.00	73.55	nr	80.90
7000mm	49.29	2.50	50.52	1.30	0.16	26.08	9.75	86.35	nr	94.99
8000mm	60.56	2.50	62.08	1.40	0.17	28.06	10.50	100.63	nr	110.70
9000mm	69.58	2.50	71.32	1.50	0.19	30.17	11.25	112.74	nr	124.01
10000mm	78.60	2.50	80.57	1.60	0.20	32.14	12.00	124.71	nr	137.18

45 degree pitch gable ladder; 450mm overhang; span over wall plate

	Del to Site	Waste	Material Cost	Craft Optve	Lab	Labour Cost	Sunds	Nett Rate	Unit	Gross rate (10%)
5000mm	43.94	2.50	45.04	1.10	0.14	22.13	8.25	75.42	nr	82.96
6000mm	54.09	2.50	55.44	1.20	0.15	24.11	9.00	88.55	nr	97.40
7000mm	63.11	2.50	64.68	1.30	0.16	26.08	9.75	100.52	nr	110.57
8000mm	70.06	2.50	71.81	1.40	0.17	28.06	10.50	110.37	nr	121.40
9000mm	79.08	2.50	81.06	1.50	0.19	30.17	11.25	122.48	nr	134.72
10000mm	86.97	2.50	89.15	1.60	0.20	32.14	12.00	133.29	nr	146.62

TIMBER FIRST FIXINGS

Stress grading softwood

General Structural (GS) grade
included in rates

Special Structural (SS) grade
add 10% to materials prices

Flame proofing treatment to softwood

For timbers treated with proofing process to Class 1
add 100% to materials prices

Primary or structural timbers; Softwood, sawn, BS 4978, GS grade; carcassing

Rafters and associated roof timbers

	Del to Site	Waste	Material Cost	Craft Optve	Lab	Labour Cost	Sunds	Nett Rate	Unit	Gross rate (10%)
25 x 75mm	0.38	15.00	0.44	0.14	0.02	2.85	0.03	3.31	m	3.65
25 x 125mm	0.86	15.00	0.99	0.16	0.02	3.21	0.05	4.25	m	4.67
32 x 175mm	2.00	15.00	2.30	0.22	0.03	4.45	0.06	6.82	m	7.50
32 x 225mm	2.57	15.00	2.96	0.25	0.03	5.01	0.06	8.03	m	8.83
38 x 75mm	0.82	15.00	0.94	0.15	0.02	3.03	0.03	4.00	m	4.40
38 x 100mm	1.10	15.00	1.26	0.18	0.02	3.58	0.05	4.89	m	5.38
50 x 75mm	0.93	15.00	1.07	0.18	0.02	3.58	0.05	4.70	m	5.17
50 x 100mm	1.22	15.00	1.40	0.20	0.03	4.09	0.05	5.53	m	6.09
50 x 125mm	1.53	15.00	1.76	0.23	0.03	4.64	0.06	6.46	m	7.10
50 x 150mm	1.84	15.00	2.12	0.26	0.03	5.19	0.06	7.37	m	8.10
50 x 225mm	2.66	15.00	3.06	0.36	0.05	7.30	0.09	10.45	m	11.50
50 x 250mm	3.88	15.00	4.46	0.40	0.05	8.04	0.09	12.59	m	13.85
75 x 100mm	2.02	15.00	2.32	0.26	0.03	5.19	0.06	7.57	m	8.33
75 x 150mm	2.93	15.00	3.37	0.35	0.05	7.12	0.09	10.58	m	11.63
100 x 150mm	4.32	15.00	4.97	0.45	0.06	9.09	0.10	14.17	m	15.58
100 x 225mm	6.93	15.00	7.97	0.60	0.08	12.12	0.15	20.24	m	22.27

Wall plates

	Del to Site	Waste	Material Cost	Craft Optve	Lab	Labour Cost	Sunds	Nett Rate	Unit	Gross rate (10%)
38 x 75mm	0.82	15.00	0.94	0.18	0.02	3.58	0.05	4.57	m	5.03
38 x 100mm	1.10	15.00	1.26	0.21	0.03	4.27	0.05	5.58	m	6.14
50 x 75mm	0.93	15.00	1.07	0.21	0.03	4.27	0.05	5.39	m	5.92
50 x 100mm	1.22	15.00	1.40	0.24	0.03	4.82	0.06	6.28	m	6.91
75 x 150mm	2.02	15.00	2.32	0.40	0.05	8.04	0.09	10.45	m	11.49
38 x 75mm; fixing by bolting	0.82	15.00	0.94	0.22	0.03	4.45	-	5.40	m	5.94
38 x 100mm; fixing by bolting	1.10	15.00	1.26	0.25	0.03	5.01	-	6.27	m	6.90
50 x 75mm; fixing by bolting	0.93	15.00	1.07	0.25	0.03	5.01	-	6.07	m	6.68
50 x 100mm; fixing by bolting	1.22	15.00	1.40	0.29	0.04	5.88	-	7.28	m	8.01
75 x 150mm; fixing by bolting	2.02	15.00	2.32	0.48	0.06	9.64	-	11.97	m	13.16

Roof joists; flat

	Del to Site	Waste	Material Cost	Craft Optve	Lab	Labour Cost	Sunds	Nett Rate	Unit	Gross rate (10%)
50 x 175mm	2.07	15.00	2.38	0.24	0.03	4.82	0.06	7.26	m	7.99
50 x 200mm	2.36	15.00	2.71	0.27	0.04	5.51	0.06	8.28	m	9.11
50 x 225mm	2.66	15.00	3.06	0.29	0.04	5.88	0.08	9.01	m	9.91
50 x 250mm	3.88	15.00	4.46	0.32	0.04	6.43	0.08	10.97	m	12.06

Labour hourly rates: (except Specialists) Craft Operatives 18.37 Labourer 13.76 Rates are national average prices. Refer to REGIONAL VARIATIONS for indicative levels of overall pricing in regions	MATERIALS			LABOUR				RATES		
	Del to Site £	Waste %	Material Cost £	Craft Optve Hrs	Lab Hrs	Labour Cost £	Sunds £	Nett Rate £	Unit	Gross rate (10%) £
TIMBER FIRST FIXINGS (Cont'd)										
Primary or structural timbers; Softwood, sawn, BS 4978, GS grade; carcassing (Cont'd)										
Roof joists; flat (Cont'd)										
75 x 175mm	3.30	15.00	3.79	0.33	0.04	6.61	0.08	10.48	m	11.53
75 x 200mm	3.77	15.00	4.34	0.35	0.05	7.12	0.09	11.54	m	12.70
75 x 225mm	4.24	15.00	4.88	0.40	0.05	8.04	0.09	13.00	m	14.30
75 x 250mm	6.19	15.00	7.12	0.45	0.06	9.09	0.10	16.32	m	17.95
Floor joists										
38 x 75mm	0.82	15.00	0.94	0.12	0.01	2.34	0.03	3.32	m	3.65
38 x 100mm	1.10	15.00	1.26	0.14	0.02	2.85	0.03	4.14	m	4.56
50 x 75mm	0.93	15.00	1.07	0.14	0.02	2.85	0.03	3.95	m	4.34
50 x 100mm	1.22	15.00	1.40	0.16	0.02	3.21	0.05	4.66	m	5.13
50 x 150mm	1.84	15.00	2.12	0.20	0.03	4.02	0.05	6.19	m	6.81
50 x 175mm	2.07	15.00	2.38	0.24	0.03	4.82	0.06	7.26	m	7.99
50 x 200mm	2.36	15.00	2.71	0.27	0.04	5.51	0.06	8.28	m	9.11
50 x 225mm	2.66	15.00	3.06	0.29	0.04	5.88	0.08	9.01	m	9.91
50 x 250mm	3.88	15.00	4.46	0.32	0.04	6.43	0.08	10.97	m	12.06
75 x 150mm	2.93	15.00	3.37	0.29	0.04	5.88	0.08	9.32	m	10.25
75 x 175mm	3.30	15.00	3.79	0.33	0.04	6.61	0.08	10.48	m	11.53
75 x 200mm	3.77	15.00	4.34	0.35	0.05	7.12	0.09	11.54	m	12.70
75 x 225mm	4.24	15.00	4.88	0.40	0.05	8.04	0.09	13.00	m	14.30
75 x 250mm	6.19	15.00	7.12	0.45	0.06	9.09	0.10	16.32	m	17.95
Partition and wall members										
38 x 75mm	0.82	15.00	0.94	0.18	0.02	3.58	0.05	4.57	m	5.03
38 x 100mm	1.10	15.00	1.26	0.21	0.03	4.27	0.05	5.58	m	6.14
50 x 75mm	0.93	15.00	1.07	0.21	0.03	4.27	0.05	5.39	m	5.92
50 x 100mm	1.22	15.00	1.40	0.24	0.03	4.82	0.06	6.28	m	6.91
75 x 100mm	2.02	15.00	2.32	0.30	0.04	6.06	0.08	8.46	m	9.31
38 x 75mm; fixing to masonry	0.82	15.00	0.94	0.36	0.04	7.16	0.09	8.20	m	9.02
38 x 100mm; fixing to masonry	1.10	15.00	1.26	0.39	0.05	7.85	0.09	9.21	m	10.13
50 x 75mm; fixing to masonry	0.93	15.00	1.07	0.39	0.05	7.85	0.09	9.01	m	9.91
50 x 100mm; fixing to masonry	1.22	15.00	1.40	0.43	0.05	8.59	0.10	10.10	m	11.10
75 x 100mm; fixing to masonry	2.02	15.00	2.32	0.50	0.06	10.01	0.12	12.45	m	13.70
Strutting; herringbone; depth of joist 175mm										
50 x 50mm	1.15	15.00	1.32	0.65	0.08	13.04	0.17	14.53	m	15.98
Strutting; herringbone; depth of joist 200mm										
50 x 50mm	1.34	15.00	1.55	0.65	0.08	13.04	0.17	14.75	m	16.23
Strutting; herringbone; depth of joist 225mm										
50 x 50mm	1.54	15.00	1.77	0.65	0.08	13.04	0.17	14.97	m	16.47
Strutting; herringbone; depth of joist 250mm										
50 x 50mm	1.73	15.00	1.99	0.65	0.08	13.04	0.17	15.19	m	16.71
Strutting; block; depth of joist 100mm										
50 x 100mm	1.22	15.00	1.40	0.65	0.08	13.04	0.17	14.61	m	16.07
Strutting; block; depth of joist 150mm										
50 x 150mm	1.84	15.00	2.12	0.65	0.08	13.04	0.17	15.32	m	16.85
Strutting; block; depth of joist 175mm										
50 x 175mm	2.07	15.00	2.38	0.70	0.09	14.10	0.17	16.64	m	18.31
Strutting; block; depth of joist 200mm										
50 x 200mm	2.36	15.00	2.71	0.70	0.09	14.10	0.17	16.98	m	18.67
Strutting; block; depth of joist 225mm										
50 x 225mm	2.66	15.00	3.06	0.75	0.09	15.02	0.18	18.25	m	20.08
Strutting; block; depth of joist 250mm										
50 x 250mm	3.88	15.00	4.46	0.75	0.09	15.02	0.18	19.66	m	21.62
Noggins to joists										
50 x 50mm	0.64	15.00	0.74	0.28	0.04	5.69	0.08	6.51	m	7.16
50 x 75mm	0.93	15.00	1.07	0.33	0.04	6.61	0.08	7.76	m	8.53
wrought surfaces										
plain; 50mm wide	-	-	-	0.15	-	2.76	-	2.76	m	3.03
plain; 75mm wide	-	-	-	0.20	-	3.67	-	3.67	m	4.04
plain; 100mm wide	-	-	-	0.25	-	4.59	-	4.59	m	5.05
plain; 150mm wide	-	-	-	0.30	-	5.51	-	5.51	m	6.06
plain; 200mm wide	-	-	-	0.35	-	6.43	-	6.43	m	7.07
Primary or structural timbers; Softwood, sawn, BS 4978, GS grade; carcassing; impregnated										
Rafters and associated roof timbers										
25 x 75mm	0.75	15.00	0.86	0.14	0.02	2.85	0.03	3.74	m	4.11
25 x 125mm	1.23	15.00	1.42	0.16	0.02	3.21	0.05	4.68	m	5.14
32 x 175mm	2.21	15.00	2.54	0.22	0.03	4.45	0.06	7.06	m	7.76
32 x 225mm	2.84	15.00	3.26	0.25	0.03	5.01	0.06	8.33	m	9.16
38 x 75mm	1.12	15.00	1.29	0.15	0.02	3.03	0.03	4.35	m	4.79
38 x 100mm	1.50	15.00	1.72	0.18	0.02	3.58	0.05	5.35	m	5.88
50 x 75mm	1.17	15.00	1.34	0.18	0.02	3.58	0.05	4.97	m	5.47
50 x 100mm	1.42	15.00	1.63	0.20	0.03	4.09	0.05	5.76	m	6.34
50 x 125mm	1.62	15.00	1.87	0.23	0.03	4.64	0.06	6.57	m	7.22
50 x 150mm	2.00	15.00	2.30	0.26	0.03	5.19	0.06	7.55	m	8.30
50 x 225mm	2.83	15.00	3.26	0.36	0.05	7.30	0.09	10.65	m	11.71
50 x 250mm	3.16	15.00	3.63	0.40	0.05	8.04	0.09	11.76	m	12.94
75 x 100mm	1.75	15.00	2.01	0.26	0.03	5.19	0.06	7.26	m	7.99
75 x 150mm	2.93	15.00	3.37	0.35	0.05	7.12	0.09	10.58	m	11.63
100 x 150mm	5.08	15.00	5.84	0.45	0.06	9.09	0.10	15.04	m	16.54
100 x 225mm	7.62	15.00	8.76	0.60	0.08	12.12	0.15	21.04	m	23.14

Labour hourly rates: (except Specialists) Craft Operatives 18.37 Labourer 13.76 Rates are national average prices. Refer to REGIONAL VARIATIONS for indicative levels of overall pricing in regions	MATERIALS			LABOUR				RATES		
	Del to Site	Waste	Material Cost	Craft Optve	Lab	Labour Cost	Sunds	Nett Rate	Unit	Gross rate (10%)
	£	%	£	Hrs	Hrs	£	£	£		£
TIMBER FIRST FIXINGS (Cont'd)										
Primary or structural timbers; Softwood, sawn, BS 4978, GS grade; carcassing; impregnated (Cont'd)										
Wall plates										
38 x 75mm	1.12	15.00	1.29	0.18	0.02	3.58	0.05	4.92	m	5.41
38 x 100mm	1.50	15.00	1.72	0.21	0.03	4.27	0.05	6.04	m	6.64
50 x 75mm	1.17	15.00	1.34	0.21	0.03	4.27	0.05	5.66	m	6.22
50 x 100mm	1.42	15.00	1.63	0.24	0.03	4.82	0.06	6.51	m	7.16
75 x 150mm	1.75	15.00	2.01	0.40	0.05	8.04	0.09	10.14	m	11.15
38 x 75mm; fixing by bolting	1.12	15.00	1.29	0.22	0.03	4.45	-	5.74	m	6.32
38 x 100mm; fixing by bolting	1.50	15.00	1.72	0.25	0.03	5.01	-	6.73	m	7.40
50 x 75mm; fixing by bolting	1.17	15.00	1.34	0.25	0.03	5.01	-	6.35	m	6.98
50 x 100mm; fixing by bolting	1.42	15.00	1.63	0.29	0.04	5.88	-	7.51	m	8.26
75 x 150mm; fixing by bolting	1.75	15.00	2.01	0.48	0.06	9.64	-	11.66	m	12.82
Roof joists; flat										
50 x 100mm	1.42	15.00	1.63	0.20	0.03	4.02	0.06	5.71	m	6.28
50 x 175mm	2.25	15.00	2.59	0.24	0.03	4.82	0.06	7.47	m	8.22
50 x 200mm	0.52	15.00	0.60	0.27	0.04	5.51	0.06	6.17	m	6.79
50 x 225mm	2.83	15.00	3.26	0.29	0.04	5.88	0.08	9.21	m	10.13
50 x 250mm	3.16	15.00	3.63	0.32	0.04	6.43	0.08	10.14	m	11.15
75 x 175mm	3.30	15.00	3.79	0.33	0.04	6.61	0.08	10.48	m	11.53
75 x 200mm	3.77	15.00	4.34	0.35	0.05	7.12	0.09	11.54	m	12.70
75 x 225mm	4.24	15.00	4.88	0.40	0.05	8.04	0.09	13.00	m	14.30
75 x 250mm	7.38	15.00	8.49	0.45	0.06	9.09	0.10	17.69	m	19.45
Floor joists										
38 x 75mm	1.12	15.00	1.29	0.12	0.01	2.34	0.03	3.66	m	4.03
38 x 100mm	1.50	15.00	1.72	0.14	0.02	2.85	0.03	4.60	m	5.06
50 x 75mm	1.17	15.00	1.34	0.14	0.02	2.85	0.03	4.22	m	4.64
50 x 100mm	1.42	15.00	1.63	0.16	0.02	3.21	0.05	4.89	m	5.38
50 x 150mm	2.00	15.00	2.30	0.20	0.03	4.02	0.05	6.37	m	7.01
50 x 175mm	2.25	15.00	2.59	0.24	0.03	4.82	0.06	7.47	m	8.22
50 x 200mm	0.52	15.00	0.60	0.27	0.04	5.51	0.06	6.17	m	6.79
50 x 225mm	2.83	15.00	3.26	0.29	0.04	5.88	0.08	9.21	m	10.13
50 x 250mm	3.16	15.00	3.63	0.32	0.04	6.43	0.08	10.14	m	11.15
75 x 150mm	2.93	15.00	3.37	0.29	0.04	5.88	0.08	9.32	m	10.25
75 x 175mm	3.30	15.00	3.79	0.33	0.04	6.61	0.08	10.48	m	11.53
75 x 200mm	3.77	15.00	4.34	0.35	0.05	7.12	0.09	11.54	m	12.70
75 x 225mm	4.24	15.00	4.88	0.40	0.05	8.04	0.09	13.00	m	14.30
75 x 250mm	7.38	15.00	8.49	0.45	0.06	9.09	0.10	17.69	m	19.45
Partition and wall members										
38 x 75mm	1.12	15.00	1.29	0.18	0.02	3.58	0.05	4.92	m	5.41
38 x 100mm	1.50	15.00	1.72	0.21	0.03	4.27	0.05	6.04	m	6.64
50 x 75mm	1.17	15.00	1.34	0.21	0.03	4.27	0.05	5.66	m	6.22
50 x 100mm	1.42	15.00	1.63	0.24	0.03	4.82	0.06	6.51	m	7.16
75 x 100mm	1.75	15.00	2.01	0.30	0.04	6.06	0.08	8.15	m	8.96
38 x 75mm; fixing to masonry	1.12	15.00	1.29	0.36	0.04	7.16	0.09	8.54	m	9.40
38 x 100mm; fixing to masonry	1.50	15.00	1.72	0.39	0.05	7.85	0.09	9.66	m	10.63
50 x 75mm; fixing to masonry	1.17	15.00	1.34	0.39	0.05	7.85	0.09	9.28	m	10.21
50 x 100mm; fixing to masonry	1.42	15.00	1.63	0.43	0.05	8.59	0.10	10.32	m	11.35
75 x 100mm; fixing to masonry	1.75	15.00	2.01	0.50	0.06	10.01	0.12	12.14	m	13.36
Strutting; herringbone; depth of joist 175mm										
50 x 50mm	1.50	15.00	1.72	0.65	0.08	13.04	0.17	14.93	m	16.42
Strutting; herringbone; depth of joist 200mm										
50 x 50mm	1.75	15.00	2.01	0.65	0.08	13.04	0.17	15.22	m	16.74
Strutting; herringbone; depth of joist 225mm										
50 x 50mm	2.00	15.00	2.30	0.65	0.08	13.04	0.17	15.51	m	17.06
Strutting; herringbone; depth of joist 250mm										
50 x 50mm	2.25	15.00	2.59	0.65	0.08	13.04	0.17	15.79	m	17.37
Strutting; block; depth of joist 100mm										
50 x 100mm	1.42	15.00	1.63	0.65	0.08	13.04	0.17	14.84	m	16.32
Strutting; block; depth of joist 150mm										
50 x 150mm	2.00	15.00	2.30	0.65	0.08	13.04	0.17	15.51	m	17.06
Strutting; block; depth of joist 175mm										
50 x 175mm	2.25	15.00	2.59	0.70	0.09	14.10	0.17	16.85	m	18.53
Strutting; block; depth of joist 200mm										
50 x 200mm	0.52	15.00	0.60	0.70	0.09	14.10	0.17	14.86	m	16.35
Strutting; block; depth of joist 225mm										
50 x 225mm	2.83	15.00	3.26	0.75	0.09	15.02	0.18	18.45	m	20.30
Strutting; block; depth of joist 250mm										
50 x 250mm	3.16	15.00	3.63	0.75	0.09	15.02	0.18	18.83	m	20.71
Noggins to joists										
50 x 50mm	0.83	15.00	0.96	0.28	0.04	5.69	0.08	6.73	m	7.40
50 x 75mm	1.17	15.00	1.34	0.33	0.04	6.61	0.08	8.03	m	8.83
wrought surfaces										
plain; 50mm wide	-	-	-	0.15	-	2.76	-	2.76	m	3.03
plain; 75mm wide	-	-	-	0.20	-	3.67	-	3.67	m	4.04
plain; 100mm wide	-	-	-	0.25	-	4.59	-	4.59	m	5.05
plain; 150mm wide	-	-	-	0.30	-	5.51	-	5.51	m	6.06
plain; 200mm wide	-	-	-	0.35	-	6.43	-	6.43	m	7.07

Labour hourly rates: (except Specialists) Craft Operatives 18.37 Labourer 13.76 Rates are national average prices. Refer to REGIONAL VARIATIONS for indicative levels of overall pricing in regions	MATERIALS			LABOUR				RATES		
	Del to Site	Waste	Material Cost	Craft Optve	Lab	Labour Cost	Sunds	Nett Rate	Unit	Gross rate (10%)
	£	%	£	Hrs	Hrs	£	£	£		£

TIMBER FIRST FIXINGS (Cont'd)

Primary or structural timbers; Oak; sawn carcassing

Rafters and associated roof timbers

25 x 75mm	4.73	10.00	5.20	0.25	0.03	5.01	0.06	10.27	m	11.30
25 x 125mm	4.73	10.00	5.20	0.28	0.04	5.69	0.08	10.97	m	12.07
32 x 175mm	11.04	10.00	12.14	0.39	0.05	7.85	0.09	20.09	m	22.09
32 x 225mm	14.77	10.00	16.25	0.44	0.06	8.91	0.10	25.26	m	27.79
38 x 75mm	7.63	10.00	8.39	0.26	0.03	5.19	0.06	13.64	m	15.01
38 x 100mm	7.63	10.00	8.39	0.32	0.04	6.43	0.08	14.90	m	16.39
50 x 75mm	12.70	10.00	13.97	0.32	0.04	6.43	0.08	20.47	m	22.52
50 x 100mm	12.70	10.00	13.97	0.35	0.04	6.98	0.09	21.04	m	23.14
50 x 125mm	15.87	10.00	17.46	0.40	0.05	8.04	0.09	25.58	m	28.14
50 x 150mm	19.05	10.00	20.95	0.46	0.06	9.28	0.12	30.35	m	33.39
50 x 225mm	31.08	10.00	34.19	0.63	0.08	12.67	0.15	47.01	m	51.71
50 x 250mm	34.54	10.00	37.99	0.70	0.09	14.10	0.17	52.26	m	57.48
75 x 100mm	14.40	10.00	15.84	0.46	0.06	9.28	0.12	25.24	m	27.76
75 x 150mm	21.60	10.00	23.76	0.61	0.08	12.31	0.15	36.22	m	39.84

Wall plates

38 x 75mm	7.63	10.00	8.39	0.32	0.04	6.43	0.08	14.90	m	16.39
38 x 100mm	7.63	10.00	8.39	0.37	0.05	7.48	0.09	15.97	m	17.56
50 x 75mm	12.70	10.00	13.97	0.37	0.05	7.48	0.09	21.54	m	23.70
50 x 100mm	12.70	10.00	13.97	0.42	0.05	8.40	0.10	22.48	m	24.73
75 x 150mm	14.40	10.00	15.84	0.52	0.06	10.38	0.12	26.34	m	28.97
38 x 75mm; fixing by bolting	7.63	10.00	8.39	0.38	0.05	7.67	-	16.06	m	17.67
38 x 100mm; fixing by bolting	7.63	10.00	8.39	0.44	0.05	8.77	-	17.16	m	18.88
50 x 75mm; fixing by bolting	12.70	10.00	13.97	0.44	0.06	8.91	-	22.88	m	25.17
50 x 100mm; fixing by bolting	12.70	10.00	13.97	0.50	0.06	10.01	-	23.98	m	26.38
75 x 150mm; fixing by bolting	14.40	10.00	15.84	0.62	0.07	12.35	-	28.19	m	31.01

Roof joists; flat

50 x 175mm	22.22	10.00	24.44	0.42	0.05	8.40	0.10	32.95	m	36.25
50 x 200mm	27.63	10.00	30.39	0.47	0.06	9.46	0.12	39.97	m	43.97
50 x 225mm	31.08	10.00	34.19	0.51	0.06	10.19	0.12	44.50	m	48.95
50 x 250mm	34.54	10.00	37.99	0.56	0.07	11.25	0.14	49.38	m	54.32
75 x 175mm	28.80	10.00	31.68	0.58	0.07	11.62	0.14	43.43	m	47.78
75 x 200mm	28.80	10.00	31.68	0.61	0.08	12.31	0.15	44.14	m	48.55
75 x 225mm	32.40	10.00	35.64	0.70	0.09	14.10	0.17	49.90	m	54.89
75 x 250mm	36.00	10.00	39.60	0.79	0.10	15.89	0.19	55.68	m	61.25
100 x 150mm	21.60	10.00	23.76	0.61	0.08	12.31	0.15	36.22	m	39.84
100 x 225mm	32.40	10.00	35.64	0.90	0.13	18.32	0.25	54.22	m	59.64

Floor joists

38 x 75mm	7.63	10.00	8.39	0.21	0.03	4.27	0.05	12.71	m	13.98
38 x 100mm	7.63	10.00	8.39	0.24	0.03	4.82	0.06	13.27	m	14.60
50 x 75mm	12.70	10.00	13.97	0.24	0.03	4.82	0.06	18.85	m	20.74
50 x 100mm	12.70	10.00	13.97	0.28	0.04	5.69	0.08	19.74	m	21.71
50 x 175mm	22.22	10.00	24.44	0.42	0.05	8.40	0.10	32.95	m	36.25
50 x 200mm	27.63	10.00	30.39	0.47	0.06	9.46	0.12	39.97	m	43.97
50 x 225mm	31.08	10.00	34.19	0.51	0.06	10.19	0.12	44.50	m	48.95
50 x 250mm	34.54	10.00	37.99	0.56	0.07	11.25	0.14	49.38	m	54.32
75 x 150mm	21.60	10.00	23.76	0.51	0.06	10.19	0.12	34.07	m	37.48
75 x 175mm	28.80	10.00	31.68	0.58	0.07	11.62	0.14	43.43	m	47.78
75 x 200mm	28.80	10.00	31.68	0.61	0.08	12.31	0.15	44.14	m	48.55
75 x 225mm	32.40	10.00	35.64	0.70	0.09	14.10	0.17	49.90	m	54.89
75 x 250mm	36.00	10.00	39.60	0.79	0.10	15.89	0.19	55.68	m	61.25

Partition and wall members

38 x 75mm	7.63	10.00	8.39	0.32	0.04	6.43	0.08	14.90	m	16.39
38 x 100mm	7.63	10.00	8.39	0.37	0.05	7.48	0.09	15.97	m	17.56
50 x 75mm	12.70	10.00	13.97	0.37	0.05	7.48	0.09	21.54	m	23.70
50 x 100mm	12.70	10.00	13.97	0.42	0.05	8.40	0.10	22.48	m	24.73
75 x 100mm	14.40	10.00	15.84	0.52	0.06	10.38	0.12	26.34	m	28.97
38 x 75mm; fixing to masonry	7.63	10.00	8.39	0.53	0.07	10.70	0.14	19.23	m	21.15
38 x 100mm; fixing to masonry	7.63	10.00	8.39	0.59	0.07	11.80	0.15	20.34	m	22.38
50 x 75mm; fixing to masonry	12.70	10.00	13.97	0.59	0.07	11.80	0.15	25.92	m	28.51
50 x 100mm; fixing to masonry	12.70	10.00	13.97	0.65	0.08	13.04	0.17	27.18	m	29.89
75 x 100mm; fixing to masonry	14.40	10.00	15.84	0.78	0.10	15.70	0.19	31.74	m	34.91

Strutting; herringbone; depth of joist 175mm

50 x 50mm	22.86	10.00	25.15	1.10	0.14	22.13	0.27	47.55	m	52.30

Strutting; herringbone; depth of joist 200mm

50 x 50mm	26.67	10.00	29.34	1.10	0.14	22.13	0.27	51.74	m	56.91

Strutting; herringbone; depth of joist 225mm

50 x 50mm	30.48	10.00	33.53	1.10	0.14	22.13	0.27	55.93	m	61.52

Strutting; herringbone; depth of joist 250mm

50 x 50mm	34.29	10.00	37.72	1.10	0.14	22.13	0.27	60.12	m	66.13

Strutting; block; depth of joist 150mm

50 x 150mm	19.05	10.00	20.95	1.10	0.14	22.13	0.27	43.36	m	47.69

Strutting; block; depth of joist 175mm

50 x 175mm	22.22	10.00	24.44	1.20	0.15	24.11	0.30	48.85	m	53.74

Strutting; block; depth of joist 200mm

50 x 200mm	27.63	10.00	30.39	1.20	0.15	24.11	0.30	54.80	m	60.28

Strutting; block; depth of joist 225mm

50 x 225mm	31.08	10.00	34.19	1.30	0.16	26.08	0.31	60.59	m	66.64

Strutting; block; depth of joist 250mm

50 x 250mm	34.54	10.00	37.99	1.30	0.16	26.08	0.31	64.39	m	70.83

Labour hourly rates: (except Specialists) Craft Operatives 18.37 Labourer 13.76 Rates are national average prices. Refer to REGIONAL VARIATIONS for indicative levels of overall pricing in regions	MATERIALS			LABOUR				RATES		
	Del to Site	Waste	Material Cost	Craft Optve	Lab	Labour Cost	Sunds	Nett Rate	Unit	Gross rate (10%)
	£	%	£	Hrs	Hrs	£	£	£		£
TIMBER FIRST FIXINGS (Cont'd)										
Primary or structural timbers; Oak; sawn carcassing (Cont'd)										
Noggins to joists										
50 x 50mm	12.70	10.00	13.97	0.50	0.06	10.01	0.12	24.10	m	26.51
50 x 75mm	12.70	10.00	13.97	0.60	0.08	12.12	0.15	26.24	m	28.87
wrought surfaces										
plain; 50mm wide	-	-	-	0.30	-	5.51	-	5.51	m	6.06
plain; 75mm wide	-	-	-	0.40	-	7.35	-	7.35	m	8.08
plain; 100mm wide	-	-	-	0.50	-	9.19	-	9.19	m	10.10
plain; 150mm wide	-	-	-	0.60	-	11.02	-	11.02	m	12.12
plain; 200mm wide	-	-	-	0.70	-	12.86	-	12.86	m	14.14
Beams; laminated beams; softwood, wrought, GS grade										
Glued laminated beams										
65 x 150 x 4000mm	78.71	2.50	80.68	0.60	0.07	11.99	5.70	98.37	nr	108.20
65 x 175 x 4000mm	91.69	2.50	93.98	0.80	0.10	16.07	6.38	116.43	nr	128.07
65 x 200 x 4000mm	104.62	2.50	107.24	0.90	0.11	18.05	7.12	132.41	nr	145.65
65 x 225 x 4000mm	118.19	2.50	121.14	1.10	0.14	22.13	8.85	152.13	nr	167.34
65 x 250 x 4000mm	131.76	2.50	135.06	1.20	0.15	24.11	9.60	168.77	nr	185.64
65 x 250 x 6000mm	197.54	2.50	202.48	1.30	0.16	26.08	10.43	238.99	nr	262.89
65 x 275 x 4000mm	144.62	2.50	148.24	1.30	0.16	26.08	10.43	184.74	nr	203.22
65 x 275 x 6000mm	219.75	2.50	225.24	1.40	0.18	28.19	11.18	264.61	nr	291.07
65 x 300 x 4000mm	155.93	2.50	159.83	1.40	0.18	28.19	11.18	199.20	nr	219.12
65 x 300 x 6000mm	235.26	2.50	241.14	1.50	0.19	30.17	12.07	283.39	nr	311.73
65 x 325 x 4000mm	173.21	2.50	177.54	1.60	0.20	32.14	12.82	222.51	nr	244.76
65 x 325 x 6000mm	254.69	2.50	261.06	1.70	0.21	34.12	13.65	308.83	nr	339.67
65 x 325 x 8000mm	340.69	2.50	349.20	1.80	0.22	36.09	14.40	399.70	nr	439.67
90 x 150 x 4000mm	99.82	2.50	102.31	0.90	0.11	18.05	7.12	127.48	nr	140.23
90 x 175 x 4000mm	117.63	2.50	120.57	1.10	0.14	22.13	8.85	151.55	nr	166.71
90 x 200 x 4000mm	133.82	2.50	137.16	1.20	0.15	24.11	9.60	170.87	nr	187.96
90 x 225 x 4000mm	149.45	2.50	153.19	1.30	0.16	26.08	10.43	189.69	nr	208.66
90 x 250 x 4000mm	168.76	2.50	172.98	1.40	0.18	28.19	11.18	212.35	nr	233.59
90 x 250 x 6000mm	249.31	2.50	255.55	1.50	0.19	30.17	12.07	297.79	nr	327.57
90 x 275 x 4000mm	184.53	2.50	189.15	1.50	0.19	30.17	12.07	231.39	nr	254.53
90 x 275 x 6000mm	277.91	2.50	284.85	1.60	0.20	32.14	12.82	329.82	nr	362.80
90 x 300 x 4000mm	199.64	2.50	204.63	1.70	0.21	34.12	13.65	252.40	nr	277.64
90 x 300 x 6000mm	304.33	2.50	311.93	1.80	0.22	36.09	14.40	362.43	nr	398.67
90 x 325 x 4000mm	220.67	2.50	226.19	1.80	0.22	36.09	14.40	276.69	nr	304.35
90 x 325 x 6000mm	328.06	2.50	336.26	1.90	0.24	38.21	15.30	389.77	nr	428.75
90 x 325 x 8000mm	434.37	2.50	445.23	2.00	0.25	40.18	15.98	501.38	nr	551.52
90 x 350 x 4000mm	233.11	2.50	238.94	1.90	0.24	38.21	15.30	292.45	nr	321.69
90 x 350 x 6000mm	353.98	2.50	362.83	2.00	0.25	40.18	15.98	418.99	nr	460.89
90 x 350 x 8000mm	472.69	2.50	484.51	2.10	0.26	42.15	16.80	543.47	nr	597.81
90 x 375 x 4000mm	249.31	2.50	255.55	2.00	0.25	40.18	15.98	311.70	nr	342.87
90 x 375 x 6000mm	375.55	2.50	384.94	2.10	0.26	42.15	16.80	443.89	nr	488.28
90 x 375 x 8000mm	503.46	2.50	516.04	2.20	0.28	44.27	17.55	577.86	nr	635.65
90 x 375 x 10000mm	629.19	2.50	644.92	2.30	0.29	46.24	18.38	709.54	nr	780.49
90 x 400 x 4000mm	270.88	2.50	277.65	2.20	0.28	44.27	17.55	339.47	nr	373.41
90 x 400 x 6000mm	401.47	2.50	411.51	2.30	0.29	46.24	18.38	476.12	nr	523.74
90 x 400 x 8000mm	536.37	2.50	549.78	2.40	0.30	48.22	19.20	617.19	nr	678.91
90 x 400 x 10000mm	669.64	2.50	686.38	2.50	0.31	50.19	20.02	756.60	nr	832.26
90 x 425 x 6000mm	448.96	2.50	460.18	2.40	0.30	48.22	19.20	527.60	nr	580.36
90 x 425 x 8000mm	565.51	2.50	579.64	2.50	0.31	50.19	20.02	649.86	nr	714.85
90 x 425 x 10000mm	704.73	2.50	722.35	2.60	0.33	52.30	20.77	795.42	nr	874.97
90 x 425 x 12000mm	846.08	2.50	867.24	2.70	0.34	54.28	21.60	943.11	nr	1037.42
90 x 450 x 6000mm	453.27	2.50	464.60	2.50	0.31	50.19	20.02	534.81	nr	588.29
90 x 450 x 8000mm	603.29	2.50	618.37	2.60	0.33	52.30	20.77	691.45	nr	760.59
90 x 450 x 10000mm	752.20	2.50	771.01	2.70	0.34	54.28	21.60	846.88	nr	931.57
90 x 450 x 12000mm	904.38	2.50	926.99	2.80	0.35	56.25	22.42	1005.67	nr	1106.24
115 x 250 x 4000mm	213.14	2.50	218.47	1.70	0.21	34.12	13.65	266.24	nr	292.86
115 x 250 x 6000mm	315.24	2.50	323.12	1.80	0.22	36.09	14.40	373.62	nr	410.98
115 x 275 x 4000mm	231.49	2.50	237.27	1.80	0.22	36.09	14.40	287.77	nr	316.54
115 x 275 x 6000mm	353.98	2.50	362.83	1.90	0.24	38.21	15.30	416.34	nr	457.97
115 x 300 x 4000mm	256.32	2.50	262.72	1.90	0.24	38.21	15.30	316.23	nr	347.85
115 x 300 x 6000mm	384.74	2.50	394.35	2.00	0.25	40.18	15.98	450.51	nr	495.56
115 x 325 x 4000mm	277.91	2.50	284.85	2.00	0.25	40.18	15.98	341.01	nr	375.11
115 x 325 x 6000mm	417.66	2.50	428.10	2.10	0.26	42.15	16.80	487.05	nr	535.76
115 x 325 x 8000mm	557.91	2.50	571.86	2.20	0.28	44.27	17.55	633.68	nr	697.04
115 x 350 x 4000mm	303.78	2.50	311.37	2.20	0.28	44.27	17.55	373.19	nr	410.51
115 x 350 x 6000mm	453.27	2.50	464.60	2.30	0.29	46.24	18.38	529.21	nr	582.13
115 x 350 x 8000mm	603.29	2.50	618.37	2.40	0.30	48.22	19.20	685.78	nr	754.36
115 x 375 x 4000mm	325.38	2.50	333.51	2.30	0.29	46.24	18.38	398.13	nr	437.94
115 x 375 x 6000mm	484.56	2.50	496.68	2.40	0.30	48.22	19.20	564.09	nr	620.50
115 x 375 x 8000mm	649.34	2.50	665.58	2.50	0.31	50.19	20.02	735.79	nr	809.37
115 x 400 x 4000mm	344.27	2.50	352.88	2.40	0.30	48.22	19.20	420.30	nr	462.33
115 x 400 x 6000mm	518.02	2.50	530.97	2.50	0.31	50.19	20.02	601.19	nr	661.31
115 x 400 x 8000mm	690.69	2.50	707.95	2.60	0.33	52.30	20.77	781.03	nr	859.13
115 x 400 x 10000mm	864.44	2.50	886.05	2.70	0.34	54.28	21.60	961.93	nr	1058.13
115 x 425 x 6000mm	536.37	2.50	549.78	2.60	0.33	52.30	20.77	622.86	nr	685.14
115 x 425 x 8000mm	717.11	2.50	735.03	2.70	0.34	54.28	21.60	810.91	nr	892.00
115 x 425 x 10000mm	892.50	2.50	914.81	2.80	0.35	56.25	22.42	993.49	nr	1092.84
115 x 425 x 12000mm	1068.42	2.50	1095.13	2.90	0.36	58.23	23.25	1176.61	nr	1294.27
115 x 450 x 6000mm	567.66	2.50	581.85	2.80	0.35	56.25	22.42	660.52	nr	726.58
115 x 450 x 8000mm	757.60	2.50	776.54	2.90	0.36	58.23	23.25	858.02	nr	943.82
115 x 450 x 10000mm	949.70	2.50	973.44	3.00	0.38	60.34	24.00	1057.78	nr	1163.56
115 x 450 x 12000mm	1139.64	2.50	1168.13	3.10	0.39	62.31	24.83	1255.27	nr	1380.80
115 x 475 x 6000mm	603.29	2.50	618.37	2.90	0.36	58.23	23.25	699.84	nr	769.83
115 x 475 x 8000mm	805.07	2.50	825.19	3.00	0.38	60.34	24.00	909.53	nr	1000.48
115 x 475 x 10000mm	1007.10	2.50	1032.28	3.10	0.39	62.31	24.83	1119.42	nr	1231.36
115 x 475 x 12000mm	1203.32	2.50	1233.40	3.20	0.40	64.29	25.65	1323.34	nr	1455.67
115 x 475 x 14000mm	1402.97	2.50	1438.05	3.30	0.41	66.26	26.40	1530.71	nr	1683.78

Labour hourly rates: (except Specialists) Craft Operatives 18.37 Labourer 13.76 Rates are national average prices. Refer to REGIONAL VARIATIONS for indicative levels of overall pricing in regions	MATERIALS			LABOUR				RATES		
	Del to Site	Waste	Material Cost	Craft Optve	Lab	Labour Cost	Sunds	Nett Rate		Gross rate (10%)
	£	%	£	Hrs	Hrs	£	£	£	Unit	£

TIMBER FIRST FIXINGS (Cont'd)

Beams; laminated beams; softwood, wrought, GS grade (Cont'd)

Glued laminated beams (Cont'd)

	Del to Site £	Waste %	Material Cost £	Craft Optve Hrs	Lab Hrs	Labour Cost £	Sunds £	Nett Rate £	Unit	Gross rate £
115 x 500 x 6000mm	636.73	2.50	652.65	3.00	0.38	60.34	24.00	736.99	nr	810.69
115 x 500 x 8000mm	845.01	2.50	866.14	3.10	0.39	62.31	24.83	953.28	nr	1048.61
115 x 500 x 10000mm	1061.40	2.50	1087.94	3.20	0.40	64.29	25.65	1177.88	nr	1295.67
115 x 500 x 12000mm	1272.32	2.50	1304.13	3.30	0.41	66.26	26.40	1396.80	nr	1536.47
115 x 500 x 14000mm	1481.76	2.50	1518.80	3.40	0.42	68.24	27.15	1614.19	nr	1775.61
140 x 350 x 4000mm	365.31	2.50	374.44	2.40	0.30	48.22	19.20	441.86	nr	486.04
140 x 350 x 6000mm	553.09	2.50	566.92	2.50	0.31	50.19	20.02	637.14	nr	700.85
140 x 350 x 8000mm	736.01	2.50	754.41	2.60	0.33	52.30	20.77	827.49	nr	910.24
140 x 375 x 4000mm	389.03	2.50	398.76	2.50	0.31	50.19	20.02	468.97	nr	515.87
140 x 375 x 6000mm	610.30	2.50	625.56	2.60	0.33	52.30	20.77	698.63	nr	768.50
140 x 375 x 8000mm	781.34	2.50	800.87	2.70	0.34	54.28	21.60	876.75	nr	964.43
140 x 375 x 10000mm	975.61	2.50	1000.00	2.80	0.35	56.25	22.42	1078.67	nr	1186.54
140 x 400 x 6000mm	617.29	2.50	632.72	2.80	0.35	56.25	22.42	711.40	nr	782.54
140 x 400 x 8000mm	828.83	2.50	849.55	2.90	0.36	58.23	23.25	931.02	nr	1024.13
140 x 400 x 10000mm	1032.80	2.50	1058.62	3.00	0.38	60.34	24.00	1142.96	nr	1257.26
140 x 425 x 6000mm	666.95	2.50	683.62	2.90	0.36	58.23	23.25	765.10	nr	841.61
140 x 425 x 8000mm	892.50	2.50	914.81	3.00	0.38	60.34	24.00	999.15	nr	1099.07
140 x 425 x 10000mm	1115.88	2.50	1143.78	3.10	0.39	62.31	24.83	1230.92	nr	1354.01
140 x 425 x 12000mm	1339.30	2.50	1372.78	3.20	0.40	64.29	25.65	1462.72	nr	1608.99
140 x 450 x 6000mm	702.56	2.50	720.12	3.00	0.38	60.34	24.00	804.46	nr	884.90
140 x 450 x 8000mm	937.84	2.50	961.29	3.10	0.39	62.31	24.83	1048.42	nr	1153.27
140 x 450 x 10000mm	1169.86	2.50	1199.11	3.20	0.40	64.29	25.65	1289.05	nr	1417.95
140 x 450 x 12000mm	1402.97	2.50	1438.05	3.30	0.41	66.26	26.40	1530.71	nr	1683.78
140 x 475 x 6000mm	735.93	2.50	754.33	3.10	0.39	62.31	24.83	841.47	nr	925.62
140 x 475 x 8000mm	980.45	2.50	1004.96	3.20	0.40	64.29	25.65	1094.90	nr	1204.39
140 x 475 x 10000mm	1227.60	2.50	1258.29	3.30	0.41	66.26	26.40	1350.95	nr	1486.05
140 x 475 x 12000mm	1474.21	2.50	1511.06	3.40	0.42	68.24	27.15	1606.45	nr	1767.10
140 x 475 x 14000mm	1711.62	2.50	1754.41	3.50	0.44	70.35	27.98	1852.73	nr	2038.01
165 x 425 x 6000mm	785.66	2.50	805.30	3.10	0.39	62.31	24.83	892.44	nr	981.68
165 x 425 x 8000mm	1048.98	2.50	1075.20	3.20	0.40	64.29	25.65	1165.14	nr	1281.65
165 x 425 x 10000mm	1310.98	2.50	1343.76	3.30	0.41	66.26	26.40	1436.42	nr	1580.06
165 x 425 x 12000mm	1570.49	2.50	1609.76	3.40	0.42	68.24	27.15	1705.14	nr	1875.66
165 x 450 x 6000mm	819.13	2.50	839.61	3.20	0.40	64.29	25.65	929.55	nr	1022.50
165 x 450 x 8000mm	1094.33	2.50	1121.69	3.30	0.41	66.26	26.40	1214.35	nr	1335.78
165 x 450 x 10000mm	1367.36	2.50	1401.54	3.40	0.42	68.24	27.15	1496.93	nr	1646.62
165 x 450 x 12000mm	1642.55	2.50	1683.62	3.50	0.44	70.35	27.98	1781.94	nr	1960.14
165 x 475 x 6000mm	868.78	2.50	890.49	3.40	0.42	68.24	27.15	985.88	nr	1084.47
165 x 475 x 8000mm	1160.28	2.50	1189.29	3.50	0.44	70.35	27.98	1287.61	nr	1416.37
165 x 475 x 10000mm	1448.29	2.50	1484.50	3.60	0.45	72.32	28.80	1585.62	nr	1744.18
165 x 475 x 12000mm	1742.38	2.50	1785.94	3.70	0.46	74.30	29.62	1889.87	nr	2078.85
165 x 475 x 14000mm	2030.00	2.50	2080.75	3.80	0.48	76.41	30.38	2187.53	nr	2406.29
190 x 475 x 6000mm	1006.37	2.50	1031.53	3.60	0.45	72.32	28.80	1132.65	nr	1245.92
190 x 475 x 8000mm	1338.74	2.50	1372.21	3.70	0.46	74.30	29.62	1476.13	nr	1623.75
190 x 475 x 10000mm	1673.85	2.50	1715.70	3.80	0.48	76.41	30.38	1822.48	nr	2004.73
190 x 475 x 12000mm	2008.42	2.50	2058.63	3.90	0.49	78.39	31.20	2168.21	nr	2385.04
190 x 475 x 14000mm	2342.97	2.50	2401.55	4.00	0.50	80.36	32.03	2513.93	nr	2765.33

Softwood, sawn; supports

Butt jointed supports

	Del to Site £	Waste %	Material Cost £	Craft Optve Hrs	Lab Hrs	Labour Cost £	Sunds £	Nett Rate £	Unit	Gross rate £
width exceeding 600mm; 19 x 38mm at 300mm centres; fixing to masonry	0.73	15.00	0.84	1.00	0.13	20.16	0.24	21.24	m²	23.37
width exceeding 600mm; 25 x 50mm at 300mm centres; fixing to masonry	1.13	15.00	1.30	1.10	0.14	22.13	0.27	23.71	m²	26.08

Framed supports

	Del to Site £	Waste %	Material Cost £	Craft Optve Hrs	Lab Hrs	Labour Cost £	Sunds £	Nett Rate £	Unit	Gross rate £
width exceeding 600mm; 38 x 50mm members at 400mm centres one way; 38 x 50mm subsidiary members at 400mm centres one way	2.60	15.00	2.99	1.00	0.13	20.16	0.24	23.39	m²	25.73
width exceeding 600mm; 38 x 50mm members at 500mm centres one way; 38 x 50mm subsidiary members at 400mm centres one way	2.34	15.00	2.69	0.80	0.10	16.07	0.19	18.96	m²	20.85
width exceeding 600mm; 50 x 50mm members at 400mm centres one way; 50 x 50mm subsidiary members at 400mm centres one way	3.20	15.00	3.68	1.10	0.14	22.13	0.27	26.08	m²	28.69
width exceeding 600mm; 50 x 50mm members at 500mm centres one way; 50 x 50mm subsidiary members at 400mm centres one way	2.88	15.00	3.31	0.90	0.11	18.05	0.22	21.58	m²	23.74
width exceeding 600mm; 19 x 38mm members at 300mm centres one way; 19 x 38mm subsidiary members at 300mm centres one way; fixing to masonry	1.47	15.00	1.68	2.25	0.28	45.19	0.54	47.41	m²	52.15
width exceeding 600mm; 25 x 50mm members at 300mm centres one way; 25 x 50mm subsidiary members at 300mm centres one way; fixing to masonry	2.26	15.00	2.60	2.35	0.29	47.16	0.57	50.33	m²	55.37
width 300mm; 38 x 50mm members longitudinally; 38 x 50mm subsidiary members at 400mm centres laterally	2.60	15.00	2.99	0.40	0.05	8.04	0.09	11.12	m	12.23
width 300mm; 50 x 50mm members longitudinally; 50 x 50mm subsidiary members at 400mm centres laterally	3.20	15.00	3.68	0.45	0.06	9.09	0.10	12.88	m	14.16

Individual supports

	Del to Site £	Waste %	Material Cost £	Craft Optve Hrs	Lab Hrs	Labour Cost £	Sunds £	Nett Rate £	Unit	Gross rate £
6 x 38mm	0.09	10.00	0.10	0.08	0.01	1.61	0.01	1.72	m	1.89
12 x 38mm	0.17	10.00	0.18	0.08	0.01	1.61	0.01	1.80	m	1.98
16 x 22mm	0.13	15.00	0.15	0.08	0.01	1.61	0.01	1.77	m	1.95
16 x 38mm	0.22	15.00	0.25	0.08	0.01	1.61	0.01	1.88	m	2.06
16 x 50mm	0.29	15.00	0.33	0.09	0.01	1.79	0.01	2.13	m	2.35
16 x 75mm	0.43	15.00	0.49	0.10	0.01	1.97	0.03	2.50	m	2.75
19 x 38mm	0.22	15.00	0.25	0.09	0.01	1.79	0.01	2.06	m	2.26
19 x 50mm	0.34	15.00	0.39	0.10	0.01	1.97	0.03	2.40	m	2.64
19 x 75mm	0.51	15.00	0.59	0.10	0.01	1.97	0.03	2.59	m	2.85

CARPENTRY

Labour hourly rates: (except Specialists) Craft Operatives 18.37 Labourer 13.76 Rates are national average prices. Refer to REGIONAL VARIATIONS for indicative levels of overall pricing in regions	MATERIALS			LABOUR				RATES		
	Del to Site	Waste	Material Cost	Craft Optve	Lab	Labour Cost	Sunds	Nett Rate	Unit	Gross rate (10%)
	£	%	£	Hrs	Hrs	£	£	£		£

TIMBER FIRST FIXINGS (Cont'd)

Softwood, sawn; supports (Cont'd)

Individual supports (Cont'd)

	Del to Site £	Waste %	Material Cost £	Craft Optve Hrs	Lab Hrs	Labour Cost £	Sunds £	Nett Rate £	Unit	Gross rate (10%) £
19 x 100mm	0.65	15.00	0.75	0.11	0.01	2.16	0.03	2.94	m	3.23
19 x 150mm	1.02	15.00	1.18	0.12	0.02	2.48	0.03	3.69	m	4.05
25 x 50mm	0.34	15.00	0.39	0.10	0.01	1.97	0.03	2.40	m	2.64
25 x 100mm	0.51	15.00	0.59	0.12	0.02	2.48	0.03	3.10	m	3.41
25 x 150mm	1.03	15.00	1.18	0.18	0.03	3.72	0.05	4.95	m	5.44
25 x 175mm	1.20	15.00	1.38	0.21	0.03	4.27	0.06	5.71	m	6.28
25 x 200mm	1.37	15.00	1.58	0.19	0.02	3.77	0.05	5.39	m	5.92
38 x 50mm	0.52	15.00	0.60	0.11	0.01	2.16	0.03	2.79	m	3.06
50 x 50mm	0.64	15.00	0.74	0.12	0.02	2.48	0.03	3.25	m	3.57
50 x 150mm	1.84	15.00	2.12	0.17	0.02	3.40	0.05	5.56	m	6.12
50 x 300mm	4.44	15.00	5.11	0.23	0.03	4.64	0.06	9.80	m	10.78
75 x 75mm	1.54	15.00	1.77	0.17	0.02	3.40	0.05	5.21	m	5.74
75 x 100mm	2.02	15.00	2.32	0.18	0.02	3.58	0.05	5.95	m	6.54
75 x 150mm	2.93	15.00	3.37	0.21	0.03	4.27	0.06	7.70	m	8.47
100 x 100mm	3.30	15.00	3.79	0.19	0.02	3.77	0.05	7.61	m	8.37
100 x 150mm	4.32	15.00	4.97	0.23	0.03	4.64	0.06	9.67	m	10.63
100 x 200mm	6.16	15.00	7.08	0.26	0.04	5.33	0.08	12.49	m	13.73
100 x 300mm	9.45	15.00	10.87	0.40	0.05	8.04	0.12	19.02	m	20.93
150 x 150mm	7.88	15.00	9.06	0.30	0.04	6.06	0.09	15.21	m	16.73
6 x 38mm; fixing to masonry	0.09	10.00	0.10	0.23	0.03	4.64	0.06	4.80	m	5.28
12 x 38mm; fixing to masonry	0.17	10.00	0.18	0.23	0.03	4.64	0.06	4.88	m	5.37
19 x 38mm; fixing to masonry	0.22	10.00	0.25	0.24	0.03	4.82	0.06	5.13	m	5.65
25 x 50mm; fixing to masonry	0.34	15.00	0.39	0.25	0.03	5.01	0.06	5.46	m	6.00
38 x 50mm; fixing to masonry	0.52	15.00	0.60	0.27	0.03	5.37	0.06	6.03	m	6.63
50 x 50mm; fixing to masonry	0.64	15.00	0.74	0.28	0.04	5.69	0.08	6.51	m	7.16

Softwood, sawn, impregnated; supports

Butt jointed supports

	Del to Site £	Waste %	Material Cost £	Craft Optve Hrs	Lab Hrs	Labour Cost £	Sunds £	Nett Rate £	Unit	Gross rate (10%) £
width exceeding 600mm; 50mm wide x 25mm average depth at 450mm centres	2.04	10.00	2.25	0.30	0.04	6.06	0.08	8.38	m²	9.22
width exceeding 600mm; 50mm wide x 25mm average depth at 600mm centres	1.53	10.00	1.68	0.23	0.03	4.64	0.06	6.38	m²	7.02
width exceeding 600mm; 50mm wide x 50mm average depth at 450mm centres	2.78	10.00	3.05	0.36	0.05	7.30	0.09	10.44	m²	11.49
width exceeding 600mm; 50mm wide x 50mm average depth at 600mm centres	2.08	10.00	2.28	0.27	0.03	5.37	0.06	7.72	m²	8.49
width exceeding 600mm; 50mm wide x 63mm average depth at 450mm centres	4.11	10.00	4.52	0.40	0.05	8.04	0.09	12.64	m²	13.91
width exceeding 600mm; 50mm wide x 63mm average depth at 600mm centres	3.07	10.00	3.38	0.30	0.04	6.06	0.08	9.51	m²	10.47
width exceeding 600mm; 50mm wide x 75mm average depth at 450mm centres	5.22	10.00	5.74	0.43	0.05	8.59	0.10	14.43	m²	15.87
width exceeding 600mm; 50mm wide x 75mm average depth at 600mm centres	3.90	10.00	4.29	0.33	0.04	6.61	0.08	10.98	m²	12.08
width exceeding 600mm; 19 x 38mm at 300mm centres; fixing to masonry	0.95	15.00	1.10	1.00	0.13	20.16	0.24	21.49	m²	23.64
width exceeding 600mm; 25 x 50mm at 300mm centres; fixing to masonry	1.20	15.00	1.38	1.10	0.14	22.13	0.27	23.78	m²	26.16

Framed supports

	Del to Site £	Waste %	Material Cost £	Craft Optve Hrs	Lab Hrs	Labour Cost £	Sunds £	Nett Rate £	Unit	Gross rate (10%) £
width exceeding 600mm; 38 x 50mm members at 400mm centres one way; 38 x 50mm subsidiary members at 400mm centres one way	3.74	15.00	4.30	1.00	0.13	20.16	0.24	24.70	m²	27.17
width exceeding 600mm; 38 x 50mm members at 500mm centres one way; 38 x 50mm subsidiary members at 400mm centres one way	3.37	15.00	3.87	0.80	0.10	16.07	0.19	20.14	m²	22.15
width exceeding 600mm; 50 x 50mm members at 400mm centres one way; 50 x 50mm subsidiary members at 400mm centres one way	4.17	15.00	4.79	1.10	0.14	22.13	0.27	27.20	m²	29.91
width exceeding 600mm; 50 x 50mm members at 500mm centres one way; 50 x 50mm subsidiary members at 400mm centres one way	3.75	15.00	4.31	0.90	0.11	18.05	0.22	22.58	m²	24.84
width exceeding 600mm; 19 x 38mm members at 300mm centres one way; 19 x 38mm subsidiary members at 300mm centres one way; fixing to masonry	1.90	15.00	2.19	2.25	0.28	45.19	0.54	47.92	m²	52.71
width exceeding 600mm; 25 x 50mm members at 300mm centres one way; 25 x 50mm subsidiary members at 300mm centres one way; fixing to masonry	2.40	15.00	2.76	2.35	0.29	47.16	0.57	50.49	m²	55.54
width 300mm; 38 x 50mm members longitudinally; 38 x 50mm subsidiary members at 400mm centres laterally	3.74	15.00	4.30	0.40	0.05	8.04	0.09	12.43	m	13.67
width 300mm; 50 x 50mm members longitudinally; 50 x 50mm subsidiary members at 400mm centres laterally	4.17	15.00	4.79	0.45	0.06	9.09	0.10	13.99	m	15.39

Individual supports

	Del to Site £	Waste %	Material Cost £	Craft Optve Hrs	Lab Hrs	Labour Cost £	Sunds £	Nett Rate £	Unit	Gross rate (10%) £
16 x 16mm	0.11	15.00	0.13	0.08	0.01	1.61	0.01	1.75	m	1.92
16 x 38mm	0.24	15.00	0.28	0.08	0.01	1.61	0.01	1.90	m	2.09
16 x 50mm	0.32	15.00	0.37	0.09	0.01	1.79	0.01	2.17	m	2.39
16 x 75mm	0.47	15.00	0.54	0.10	0.01	1.97	0.03	2.55	m	2.80
19 x 38mm	0.29	15.00	0.33	0.09	0.01	1.79	0.01	2.13	m	2.35
19 x 50mm	0.37	15.00	0.43	0.10	0.01	1.97	0.03	2.43	m	2.68
19 x 75mm	0.56	15.00	0.65	0.10	0.01	1.97	0.03	2.65	m	2.91
19 x 100mm	0.75	15.00	0.86	0.11	0.01	2.16	0.03	3.05	m	3.35
19 x 150mm	1.12	15.00	1.29	0.12	0.02	2.48	0.03	3.80	m	4.18
25 x 50mm	0.36	15.00	0.41	0.10	0.01	1.97	0.03	2.42	m	2.66
25 x 100mm	0.99	15.00	1.14	0.12	0.02	2.48	0.03	3.65	m	4.01
25 x 150mm	1.20	15.00	1.38	0.18	0.03	3.72	0.05	5.14	m	5.66
25 x 175mm	1.72	15.00	1.97	0.21	0.03	4.27	0.06	6.30	m	6.93
25 x 200mm	1.97	15.00	2.26	0.19	0.02	3.77	0.05	6.07	m	6.68
38 x 50mm	0.75	15.00	0.86	0.11	0.01	2.16	0.03	3.05	m	3.35

CARPENTRY

Labour hourly rates: (except Specialists) Craft Operatives 18.37 Labourer 13.76 Rates are national average prices. Refer to REGIONAL VARIATIONS for indicative levels of overall pricing in regions	MATERIALS			LABOUR				RATES		
	Del to Site	Waste	Material Cost	Craft Optve	Lab	Labour Cost	Sunds	Nett Rate	Unit	Gross rate (10%)
	£	%	£	Hrs	Hrs	£	£	£		£
TIMBER FIRST FIXINGS (Cont'd)										
Softwood, sawn, impregnated; supports (Cont'd)										
Individual supports (Cont'd)										
50 x 50mm...............	0.83	15.00	0.96	0.12	0.02	2.48	0.03	3.47	m	3.81
50 x 150mm...............	2.00	15.00	2.30	0.17	0.02	3.40	0.05	5.74	m	6.32
50 x 300mm...............	5.91	15.00	6.79	0.23	0.03	4.64	0.06	11.49	m	12.64
75 x 75mm...............	1.54	15.00	1.77	0.17	0.02	3.40	0.05	5.21	m	5.74
75 x 100mm...............	1.75	15.00	2.01	0.18	0.02	3.58	0.05	5.64	m	6.20
75 x 150mm...............	2.93	15.00	3.37	0.21	0.03	4.27	0.06	7.70	m	8.47
100 x 100mm...............	3.30	15.00	3.79	0.19	0.02	3.77	0.05	7.61	m	8.37
100 x 150mm...............	5.08	15.00	5.84	0.23	0.03	4.64	0.06	10.54	m	11.59
100 x 200mm...............	6.78	15.00	7.80	0.26	0.04	5.33	0.08	13.20	m	14.52
100 x 300mm...............	10.40	15.00	11.96	0.40	0.05	8.04	0.12	20.12	m	22.13
150 x 150mm...............	7.53	15.00	8.66	0.30	0.04	6.06	0.09	14.81	m	16.29
16 x 38mm; fixing to masonry...............	0.24	15.00	0.28	0.23	0.03	4.64	0.06	4.98	m	5.47
19 x 38mm; fixing to masonry...............	0.29	15.00	0.33	0.24	0.03	4.82	0.06	5.21	m	5.73
25 x 50mm; fixing to masonry...............	0.36	15.00	0.41	0.25	0.03	5.01	0.06	5.48	m	6.03
38 x 50mm; fixing to masonry...............	0.75	15.00	0.86	0.27	0.03	5.37	0.06	6.29	m	6.92
50 x 50mm; fixing to masonry...............	0.83	15.00	0.96	0.28	0.04	5.69	0.08	6.73	m	7.40
25 x 25mm; 1 labours...............	0.29	10.00	0.32	0.11	0.01	2.16	0.03	2.51	m	2.76
25 x 25mm; 2 labours...............	0.35	10.00	0.38	0.11	0.01	2.16	0.03	2.57	m	2.83
25 x 38mm; 1 labours...............	0.38	10.00	0.42	0.11	0.01	2.16	0.03	2.61	m	2.87
25 x 38mm; 2 labours...............	0.48	10.00	0.53	0.11	0.01	2.16	0.03	2.72	m	2.99
triangular; extreme dimensions										
25 x 25mm...............	0.29	10.00	0.32	0.11	0.01	2.16	0.03	2.51	m	2.76
38 x 75mm...............	0.66	10.00	0.73	0.13	0.02	2.66	0.03	3.42	m	3.76
50 x 75mm...............	0.66	10.00	0.73	0.15	0.02	3.03	0.03	3.79	m	4.17
50 x 100mm...............	0.81	10.00	0.89	0.17	0.02	3.40	0.05	4.33	m	4.77
75 x 100mm...............	1.62	10.00	1.78	0.19	0.02	3.77	0.05	5.59	m	6.15
rounded roll for lead										
50 x 50mm...............	1.80	10.00	1.98	0.22	0.03	4.45	0.06	6.49	m	7.14
50 x 75mm...............	2.80	10.00	3.08	0.25	0.03	5.01	0.06	8.15	m	8.96
rounded roll for lead; birdsmouthed on to ridge or hip										
50 x 50mm...............	1.80	10.00	1.98	0.38	0.05	7.67	0.09	9.74	m	10.71
50 x 75mm...............	2.80	10.00	3.08	0.41	0.05	8.22	0.10	11.40	m	12.55
rounded roll for zinc										
32 x 44mm...............	1.64	10.00	1.80	0.16	0.02	3.21	0.05	5.06	m	5.57
Softwood, wrought; supports										
Butt jointed supports										
width exceeding 600mm; 19 x 38mm at 600mm centres; fixing to masonry...............	1.00	10.00	1.10	1.00	0.13	20.16	0.24	21.50	m²	23.65
width exceeding 600mm; 25 x 50mm at 600mm centres; fixing to masonry...............	1.60	10.00	1.76	1.10	0.14	22.13	0.27	24.16	m²	26.58
width exceeding 600mm; 19 x 38mm at 600mm centres; fixing to masonry with screws...............	1.07	10.00	1.17	1.40	0.16	27.92	0.08	29.17	m²	32.09
width exceeding 600mm; 25 x 50mm at 600mm centres; fixing to masonry with screws...............	1.68	10.00	1.84	1.50	0.21	30.44	0.08	32.36	m²	35.60
width exceeding 600mm; 19 x 38mm at 600mm centres; fixing to masonry with screws...............	1.12	10.00	1.23	1.80	0.20	35.82	0.15	37.20	m²	40.92
width exceeding 600mm; 25 x 50mm at 600mm centres; plugged and screwed to masonry...............	1.73	10.00	1.90	1.90	0.21	37.79	0.18	39.87	m²	43.86
Framed supports										
width exceeding 600mm; 38 x 50mm members at 400mm centres one way; 38 x 50mm subsidiary members at 400mm centres one way...............	3.65	10.00	4.02	1.00	0.13	20.16	0.24	24.41	m²	26.86
width exceeding 600mm; 38 x 50mm members at 500mm centres one way; 38 x 50mm subsidiary members at 400mm centres one way...............	3.29	10.00	3.61	0.80	0.10	16.07	0.19	19.88	m²	21.87
width exceeding 600mm; 50 x 50mm members at 400mm centres one way; 50 x 50mm subsidiary members at 400mm centres one way...............	4.80	10.00	5.28	1.10	0.14	22.13	0.27	27.68	m²	30.45
width exceeding 600mm; 50 x 50mm members at 500mm centres one way; 50 x 50mm subsidiary members at 400mm centres one way...............	4.32	10.00	4.75	0.90	0.11	18.05	0.22	23.02	m²	25.33
width exceeding 600mm; 19 x 38mm members at 600mm centres one way; 19 x 38mm subsidiary members at 600mm centres one way; fixing to masonry...............	2.00	10.00	2.20	2.25	0.28	45.19	0.54	47.92	m²	52.72
width exceeding 600mm; 25 x 50mm members at 600mm centres one way; 25 x 50mm subsidiary members at 600mm centres one way; fixing to masonry...............	3.20	10.00	3.52	2.35	0.29	47.16	0.57	51.25	m²	56.37
width not exceeding 600mm; 38 x 50mm members longitudinally; 38 x 50mm subsidiary members at 400mm centres laterally...............	3.65	10.00	4.02	0.40	0.05	8.04	0.09	12.14	m	13.36
width not exceeding 600mm; 50 x 50mm members longitudinally; 50 x 50mm subsidiary members at 400mm centres laterally...............	4.80	10.00	5.28	0.45	0.06	9.09	0.10	14.48	m	15.92
Individual supports										
6 x 38mm...............	0.20	10.00	0.22	0.08	0.01	1.61	0.01	1.84	m	2.03
12 x 38mm...............	0.22	10.00	0.24	0.08	0.01	1.61	0.01	1.86	m	2.05
19 x 38mm...............	0.30	10.00	0.33	0.09	0.01	1.79	0.01	2.14	m	2.35
25 x 50mm...............	0.48	10.00	0.53	0.10	0.01	1.97	0.03	2.53	m	2.79
38 x 50mm...............	0.73	10.00	0.80	0.11	0.01	2.16	0.03	2.99	m	3.29
50 x 50mm...............	0.96	10.00	1.06	0.12	0.02	2.48	0.03	3.57	m	3.92
50 x 75mm...............	1.44	10.00	1.58	0.14	0.02	2.85	0.03	4.46	m	4.91
50 x 100mm...............	1.85	10.00	2.04	0.15	0.02	3.03	0.05	5.11	m	5.62
50 x 150mm...............	2.77	10.00	3.05	0.17	0.02	3.40	0.05	6.49	m	7.14
50 x 200mm...............	3.74	10.00	4.11	0.20	0.02	3.95	0.05	8.11	m	8.92
50 x 300mm...............	5.70	10.00	6.27	0.23	0.03	4.64	0.06	10.97	m	12.06
75 x 75mm...............	2.16	10.00	2.38	0.17	0.02	3.40	0.05	5.82	m	6.40
75 x 100mm...............	2.77	10.00	3.05	0.18	0.02	3.58	0.05	6.67	m	7.34
75 x 150mm...............	4.15	10.00	4.57	0.21	0.03	4.27	0.06	8.90	m	9.79

Labour hourly rates: (except Specialists) Craft Operatives 18.37 Labourer 13.76 Rates are national average prices. Refer to REGIONAL VARIATIONS for indicative levels of overall pricing in regions	MATERIALS			LABOUR				RATES		
	Del to Site	Waste	Material Cost	Craft Optve	Lab	Labour Cost	Sunds	Nett Rate	Unit	Gross rate (10%)
	£	%	£	Hrs	Hrs	£	£	£		£
TIMBER FIRST FIXINGS (Cont'd)										
Softwood, wrought; supports (Cont'd)										
Individual supports (Cont'd)										
100 x 100mm..	4.55	10.00	5.01	0.19	0.02	3.77	0.05	8.82	m	9.70
100 x 150mm..	6.60	10.00	7.26	0.23	0.03	4.64	0.06	11.96	m	13.15
150 x 150mm..	8.80	10.00	9.68	0.30	0.04	6.06	0.09	15.83	m	17.41
6 x 38mm; fixing to masonry............................	0.20	10.00	0.22	0.23	0.03	4.64	0.06	4.92	m	5.41
12 x 38mm; fixing to masonry..........................	0.22	10.00	0.24	0.23	0.03	4.64	0.06	4.94	m	5.43
19 x 38mm; fixing to masonry..........................	0.30	10.00	0.33	0.24	0.03	4.82	0.06	5.21	m	5.73
25 x 50mm; fixing to masonry..........................	0.48	10.00	0.53	0.25	0.03	5.01	0.06	5.59	m	6.15
38 x 50mm; fixing to masonry..........................	0.73	10.00	0.80	0.27	0.03	5.37	0.06	6.24	m	6.86
50 x 50mm; fixing to masonry..........................	0.96	10.00	1.06	0.28	0.04	5.69	0.08	6.83	m	7.51
6 x 38mm, fixing with screws..........................	0.22	10.00	0.24	0.23	0.01	4.36	0.01	4.62	m	5.08
12 x 38mm, fixing with screws........................	0.24	10.00	0.27	0.23	0.01	4.36	0.01	4.64	m	5.11
19 x 38mm, fixing with screws........................	0.32	10.00	0.35	0.20	0.01	3.81	0.01	4.18	m	4.59
25 x 50mm, fixing with screws........................	0.50	10.00	0.55	0.20	0.01	3.81	0.03	4.39	m	4.83
38 x 50mm, fixing with screws........................	0.77	10.00	0.84	0.21	0.01	4.00	0.03	4.87	m	5.35
50 x 50mm, fixing with screws........................	1.02	10.00	1.12	0.27	0.02	5.24	0.03	6.38	m	7.02
50 x 75mm, fixing with screws........................	1.50	10.00	1.64	0.28	0.02	5.42	0.03	7.09	m	7.80
50 x 100mm, fixing with screws......................	1.91	10.00	2.09	0.30	0.02	5.79	0.03	7.91	m	8.70
50 x 150mm, fixing with screws......................	2.85	10.00	3.14	0.40	0.02	7.62	0.05	10.80	m	11.88
50 x 200mm, fixing with screws......................	3.82	10.00	4.20	0.42	0.02	7.99	0.05	12.24	m	13.46
50 x 300mm, fixing with screws......................	5.81	10.00	6.39	0.53	0.03	10.15	0.06	16.60	m	18.26
6 x 38mm; plug and screwing to masonry	0.24	10.00	0.26	0.38	0.03	7.39	0.06	7.71	m	8.48
12 x 38mm; plug and screwing to masonry	0.26	10.00	0.29	0.38	0.03	7.39	0.06	7.74	m	8.51
19 x 38mm; plug and screwing to masonry	0.33	10.00	0.36	0.29	0.03	5.74	0.06	6.16	m	6.78
25 x 50mm; plug and screwing to masonry	0.51	10.00	0.56	0.30	0.03	5.92	0.06	6.55	m	7.20
38 x 50mm; plug and screwing to masonry	0.84	10.00	0.92	0.39	0.03	7.58	0.06	8.56	m	9.41
50 x 50mm; plug and screwing to masonry	1.09	10.00	1.19	0.42	0.04	8.27	0.08	9.53	m	10.49
50 x 75mm; plug and screwing to masonry	1.57	10.00	1.72	0.44	0.04	8.63	0.08	10.43	m	11.47
50 x 100mm; plug and screwing to masonry ...	1.98	10.00	2.17	0.46	0.04	9.00	0.08	11.25	m	12.37
50 x 150mm; plug and screwing to masonry ...	2.97	10.00	3.25	0.79	0.04	15.06	0.90	19.22	m	21.14
50 x 200mm; plug and screwing to masonry ...	3.94	10.00	4.32	0.81	0.04	15.43	0.10	19.86	m	21.84
TIMBER, METAL AND PLASTIC BOARDING, SHEETING, DECKING, CASINGS AND LININGS										
Softwood, sawn; fixing to timber - boarded flooring										
Boarding to floors, square edges; 19mm thick, 75mm wide boards width exceeding 600mm................	4.93	7.50	5.30	0.60	0.08	12.12	0.15	17.58	m²	19.33
Boarding to floors, square edges; 25mm thick, 125mm wide boards width exceeding 600mm................	6.32	7.50	6.79	0.55	0.07	11.07	0.14	18.00	m²	19.80
Boarding to floors, square edges; 32mm thick, 150mm wide boards width exceeding 600mm................	9.60	7.50	10.32	0.55	0.07	11.07	0.14	21.52	m²	23.67
Softwood, wrought; fixing to timber - boarded flooring										
Boarding to floors, square edges; 19mm thick, 75mm wide boards width exceeding 600mm................	7.47	7.50	8.03	0.70	0.09	14.10	0.17	22.29	m²	24.52
Boarding to floors, square edges; 25mm thick, 125mm wide boards width exceeding 600mm................	9.17	7.50	9.85	0.65	0.08	13.04	0.17	23.06	m²	25.37
Boarding to floors, square edges; 32mm thick, 150mm wide boards width exceeding 600mm................	8.95	7.50	9.63	0.65	0.08	13.04	0.17	22.83	m²	25.12
Boarding to floors, tongued and grooved joints; 19mm thick, 75mm wide boards width exceeding 600mm................	7.60	7.50	8.17	0.80	0.10	16.07	0.19	24.44	m²	26.88
Boarding to floors, tongued and grooved joints; 25mm thick, 125mm wide boards width exceeding 600mm................	8.56	7.50	9.20	0.75	0.10	15.15	0.18	24.54	m²	26.99
Boarding to floors, tongued and grooved joints; 32mm thick, 150mm wide boards width exceeding 600mm................	15.33	7.50	16.48	0.75	0.10	15.15	0.18	31.81	m²	34.99
Abutments										
19mm thick softwood boarding										
raking cutting..	0.75	7.50	0.80	0.06	-	1.10	-	1.90	m	2.10
curved cutting...	1.12	7.50	1.20	0.09	-	1.65	-	2.86	m	3.14
25mm thick softwood boarding										
raking cutting..	0.92	7.50	0.99	0.06	-	1.10	-	2.09	m	2.30
curved cutting...	1.38	7.50	1.48	0.09	-	1.65	-	3.13	m	3.44
32mm thick softwood boarding										
raking cutting..	0.90	7.50	0.96	0.06	-	1.10	-	2.06	m	2.27
curved cutting...	1.34	7.50	1.44	0.09	-	1.65	-	3.10	m	3.41
Surface treatment of existing wood flooring; Sanding and sealing existing wood flooring										
Machine sanding width exceeding 600mm................	-	-	-	0.50	-	9.19	1.00	10.19	m²	11.20
Prepare, one priming coat and one finish coat of seal width exceeding 600mm................	-	-	-	0.16	-	2.94	0.50	3.44	m²	3.78

Labour hourly rates: (except Specialists) Craft Operatives 18.37 Labourer 13.76 Rates are national average prices. Refer to REGIONAL VARIATIONS for indicative levels of overall pricing in regions	MATERIALS			LABOUR				RATES		
	Del to Site	Waste	Material Cost	Craft Optve	Lab	Labour Cost	Sunds	Nett Rate		Gross rate (10%)
	£	%	£	Hrs	Hrs	£	£	£	Unit	£
TIMBER, METAL AND PLASTIC BOARDING, SHEETING, DECKING, CASINGS AND LININGS (Cont'd)										
Softwood, sawn, impregnated; fixing to timber; roof boarding										
Boarding to roofs, butt joints; 19mm thick, 75mm wide boards; external										
width exceeding 600mm	6.27	7.50	6.74	0.60	0.07	11.99	0.15	18.87	m²	20.76
width exceeding 600mm; sloping	6.27	7.50	6.74	0.70	0.09	14.10	0.17	21.00	m²	23.10
Boarding to roofs, butt joints; 25mm thick, 125mm wide boards; external										
width exceeding 600mm	7.12	7.50	7.65	0.55	0.07	11.07	0.14	18.86	m²	20.74
width exceeding 600mm; sloping	7.12	7.50	7.65	0.65	0.08	13.04	0.17	20.86	m²	22.95
Softwood, wrought, impregnated; fixing to timber; roof boarding										
Boarding to roofs, tongued and grooved joints; 19mm thick, 75mm wide boards external										
width exceeding 600mm	10.50	7.50	11.29	0.80	0.10	16.07	0.19	27.55	m²	30.31
extra; cross rebated and rounded drip; 50mm wide	0.46	10.00	0.51	0.20	0.03	4.09	0.05	4.64	m	5.10
extra; dovetailed cesspool 225 x 225 x 150mm	2.10	10.00	2.31	1.40	0.17	28.06	0.34	30.71	nr	33.78
extra; dovetailed cesspool 300 x 300 x 150mm	3.15	10.00	3.47	1.60	0.20	32.14	0.39	36.00	nr	39.60
width exceeding 600mm; sloping	10.50	7.50	11.29	0.90	0.11	18.05	0.22	29.56	m²	32.52
Boarding to roofs, tongued and grooved joints; 25mm thick, 125mm wide boards external										
width exceeding 600mm	11.70	7.50	12.58	0.75	0.09	15.02	0.18	27.77	m²	30.55
extra; cross rebated and rounded drip; 50mm wide	0.46	10.00	0.51	0.20	0.03	4.09	0.05	4.64	m	5.10
extra; dovetailed cesspool 225 x 225 x 150mm	2.34	10.00	2.57	1.50	0.19	30.17	0.36	33.10	nr	36.41
extra; dovetailed cesspool 300 x 300 x 150mm	3.51	10.00	3.86	1.70	0.21	34.12	0.42	38.40	nr	42.24
width exceeding 600mm; sloping	11.70	7.50	12.58	0.85	0.11	17.13	0.21	29.92	m²	32.91
Abutments										
19mm thick softwood boarding										
raking cutting	1.25	7.50	1.35	0.35	-	6.43	-	7.78	m	8.55
curved cutting	1.88	7.50	2.02	0.55	-	10.10	-	12.12	m	13.34
rebates	-	-	-	0.18	-	3.31	-	3.31	m	3.64
grooves	-	-	-	0.18	-	3.31	-	3.31	m	3.64
chamfers	-	-	-	0.13	-	2.39	-	2.39	m	2.63
25mm thick softwood boarding										
raking cutting	1.28	7.50	1.38	0.45	-	8.27	-	9.64	m	10.61
curved cutting	1.92	7.50	2.07	0.65	-	11.94	-	14.01	m	15.41
rebates	-	-	-	0.18	-	3.31	-	3.31	m	3.64
grooves	-	-	-	0.18	-	3.31	-	3.31	m	3.64
chamfers	-	-	-	0.13	-	2.39	-	2.39	m	2.63
Douglas Fir plywood, unsanded select sheathing quality, WBP bonded; tongued and grooved joints long edges										
15mm thick sheeting to floors										
width exceeding 600mm	3.34	7.50	3.59	0.29	0.04	5.88	0.15	9.62	m²	10.58
18mm thick sheeting to floors										
width exceeding 600mm	4.96	7.50	5.34	0.30	0.04	6.06	0.15	11.55	m²	12.70
Birch faced plywood, BB quality, WBP bonded; tongued and grooved joints all edges										
12mm thick sheeting to floors										
width exceeding 600mm	6.53	7.50	7.02	0.27	0.04	5.51	0.15	12.68	m²	13.95
15mm thick sheeting to floors										
width exceeding 600mm	7.42	7.50	7.98	0.29	0.04	5.88	0.15	14.01	m²	15.41
18mm thick sheeting to floors										
width exceeding 600mm	8.52	7.50	9.15	0.30	0.04	6.06	0.15	15.37	m²	16.90
Wood particle board (chipboard), BS EN 312, type P4 load bearing flooring, square edges; butt joints										
18mm thick sheeting to floors										
width exceeding 600mm	4.86	7.50	5.22	0.30	0.04	6.06	0.15	11.43	m²	12.58
22mm thick sheeting to floors										
width exceeding 600mm	6.93	7.50	7.45	0.30	0.04	6.06	0.15	13.66	m²	15.03
Wood particle board (chipboard), BS EN 312, type P5 load bearing moisture resistant flooring, square edges; butt joints										
18mm thick sheeting to floors										
width exceeding 600mm	5.90	7.50	6.35	0.30	0.04	6.06	0.15	12.56	m²	13.81
22mm thick sheeting to floors										
width exceeding 600mm	7.60	7.50	8.17	0.30	0.04	6.06	0.15	14.39	m²	15.82
Wood particle board (chipboard), BS EN 312, type P6 heavyduty load bearing flooring, square edges; butt joints										
18mm thick sheeting to floors										
width exceeding 600mm	5.83	7.50	6.27	0.30	0.04	6.06	0.15	12.48	m²	13.73
22mm thick sheeting to floors										
width exceeding 600mm	6.80	7.50	7.31	0.30	0.04	6.06	0.15	13.52	m²	14.87
38mm thick sheeting to floors										
width exceeding 600mm	18.57	7.50	19.96	0.40	0.10	8.72	0.22	28.91	m²	31.80

Labour hourly rates: (except Specialists) Craft Operatives 18.37 Labourer 13.76 Rates are national average prices. Refer to REGIONAL VARIATIONS for indicative levels of overall pricing in regions	MATERIALS			LABOUR				RATES		
	Del to Site	Waste	Material Cost	Craft Optve	Lab	Labour Cost	Sunds	Nett Rate	Unit	Gross rate (10%)
	£	%	£	Hrs	Hrs	£	£	£		£
TIMBER, METAL AND PLASTIC BOARDING, SHEETING, DECKING, CASINGS AND LININGS (Cont'd)										
Wood particle board (chipboard), BS EN 312, type P4 load bearing flooring; tongued and grooved joints										
18mm thick sheeting to floors width exceeding 600mm..................	4.97	7.50	5.34	0.30	0.04	6.06	0.15	11.55	m²	12.71
22mm thick sheeting to floors width exceeding 600mm..................	6.25	7.50	6.72	0.30	0.04	6.06	0.15	12.93	m²	14.22
Wood particle board (chipboard), BS EN 312, type P5 load bearing moisture resistant flooring; tongued and grooved joints										
18mm thick sheeting to floors width exceeding 600mm..................	4.08	7.50	4.39	0.30	0.04	6.06	0.15	10.60	m²	11.66
22mm thick sheeting to floors width exceeding 600mm..................	5.90	7.50	6.35	0.30	0.04	6.06	0.15	12.56	m²	13.81
Abutments										
12mm thick plywood linings raking cutting..................	0.65	7.50	0.70	0.03	-	0.55	-	1.25	m	1.38
curved cutting..................	0.98	7.50	1.05	0.05	-	0.92	-	1.97	m	2.17
15mm thick plywood linings raking cutting..................	0.74	7.50	0.80	0.04	-	0.73	-	1.53	m	1.69
curved cutting..................	1.11	7.50	1.20	0.06	-	1.10	-	2.30	m	2.53
18mm thick plywood linings raking cutting..................	0.85	7.50	0.92	0.04	-	0.73	-	1.65	m	1.82
curved cutting..................	1.28	7.50	1.37	0.06	-	1.10	-	2.48	m	2.72
18mm thick wood particle board (chipboard) linings raking cutting..................	0.49	7.50	0.52	0.03	-	0.55	-	1.07	m	1.18
curved cutting..................	0.73	7.50	0.78	0.05	-	0.92	-	1.70	m	1.87
22mm thick wood particle board (particle board (chipboard)) linings raking cutting..................	0.69	7.50	0.75	0.04	-	0.73	-	1.48	m	1.63
curved cutting..................	1.04	7.50	1.12	0.06	-	1.10	-	2.22	m	2.44
Floating insulated floors; Wood particle board (chipboard), BS EN 312, type P4 load bearing flooring; glued tongued and grooved joints										
18mm thick sheeting laid loose over insulation width exceeding 600mm..................	4.97	7.50	5.34	0.30	0.04	6.06	0.15	11.55	m²	12.71
22mm thick sheeting to floors width exceeding 600mm..................	6.25	7.50	6.72	0.30	0.04	6.06	0.15	12.93	m²	14.22
Floating insulated floors; Wood particle board (chipboard), BS EN 312, type P5 load bearing moisture resistant flooring; glued tongued and grooved joints										
18mm thick sheeting laid loose over insulation width exceeding 600mm..................	4.08	7.50	4.39	0.30	0.04	6.06	0.15	10.60	m²	11.66
22mm thick sheeting to floors width exceeding 600mm..................	5.90	7.50	6.35	0.30	0.04	6.06	0.15	12.56	m²	13.81
Sound insulated floors; Foam backed softwood battens covered with tongued and grooved particle board (chipboard) or plywood panels										
19mm particle board (chipboard) panels to level sub-floors width exceeding 600mm; battens at 450mm centres; elevation 50mm..................	8.46	10.00	9.13	0.27	0.03	5.37	0.39	14.89	m²	16.38
75mm..................	12.63	10.00	13.51	0.27	0.03	5.37	0.42	19.30	m²	21.23
width exceeding 600mm; battens at 600mm centres; elevation 50mm..................	7.58	10.00	8.20	0.27	0.03	5.37	0.33	13.91	m²	15.30
75mm..................	10.70	10.00	11.48	0.27	0.03	5.37	0.34	17.20	m²	18.92
18mm particle board (chipboard) panels with improved moisture resistance to level sub-floors width exceeding 600mm; battens at 450mm centres; elevation 50mm..................	9.39	10.00	10.15	0.27	0.03	5.37	0.39	15.91	m²	17.50
75mm..................	13.56	10.00	14.53	0.27	0.03	5.37	0.42	20.32	m²	22.36
width exceeding 600mm; battens at 600mm centres; elevation 50mm..................	8.51	10.00	9.23	0.27	0.03	5.37	0.33	14.93	m²	16.42
75mm..................	11.63	10.00	12.50	0.27	0.03	5.37	0.34	18.22	m²	20.04
18mm plywood panels to level sub-floors width exceeding 600mm; battens at 450mm centres; elevation 50mm..................	8.45	10.00	9.12	0.27	0.03	5.37	0.39	14.88	m²	16.37
75mm..................	12.62	10.00	13.50	0.27	0.03	5.37	0.42	19.29	m²	21.22
Fibreboards and hardboards; Standard hardboard, BS EN 622; butt joints										
3.2mm thick linings to walls width exceeding 600mm..................	1.61	5.00	1.69	0.33	0.04	6.61	0.50	8.80	m²	9.67
300mm wide..................	0.48	10.00	0.53	0.20	0.01	3.81	0.30	4.64	m	5.11
4.8mm thick linings to walls width exceeding 600mm..................	3.55	5.00	3.72	0.34	0.04	6.80	0.51	11.03	m²	12.13
300mm wide..................	1.06	10.00	1.17	0.20	0.01	3.81	0.30	5.28	m	5.81

Labour hourly rates: (except Specialists) Craft Operatives 18.37 Labourer 13.76. Rates are national average prices. Refer to REGIONAL VARIATIONS for indicative levels of overall pricing in regions	MATERIALS			LABOUR				RATES		
	Del to Site £	Waste %	Material Cost £	Craft Optve Hrs	Lab Hrs	Labour Cost £	Sunds £	Nett Rate £	Unit	Gross rate (10%) £
TIMBER, METAL AND PLASTIC BOARDING, SHEETING, DECKING, CASINGS AND LININGS (Cont'd)										
Fibreboards and hardboards; Standard hardboard, BS EN 622; butt joints (Cont'd)										
6.4mm thick linings to walls										
width exceeding 600mm	4.53	5.00	4.76	0.35	0.04	6.98	0.52	12.26	m²	13.49
300mm wide	1.36	10.00	1.49	0.21	0.01	4.00	0.31	5.81	m	6.39
3.2mm thick sheeting to floors										
width exceeding 600mm	1.61	5.00	1.69	0.17	0.02	3.40	0.25	5.34	m²	5.87
3.2mm thick linings to ceilings										
width exceeding 600mm	1.61	5.00	1.69	0.41	0.05	8.22	0.62	10.52	m²	11.57
300mm wide	0.48	10.00	0.53	0.25	0.02	4.87	0.38	5.77	m	6.35
4.8mm thick linings to ceilings										
width exceeding 600mm	3.55	5.00	3.72	0.43	0.05	8.59	0.64	12.96	m²	14.25
300mm wide	1.06	10.00	1.17	0.26	0.02	5.05	0.39	6.61	m	7.27
6.4mm thick linings to ceilings										
width exceeding 600mm	4.53	5.00	4.76	0.44	0.05	8.77	0.66	14.19	m²	15.61
300mm wide	1.36	10.00	1.49	0.26	0.02	5.05	0.39	6.94	m	7.63
Fibreboards and hardboards; Tempered hardboard, BS EN 622; butt joints										
3.2mm thick linings to walls										
width exceeding 600mm	4.01	5.00	4.21	0.33	0.04	6.61	0.50	11.32	m²	12.45
300mm wide	1.20	10.00	1.32	0.20	0.01	3.81	0.30	5.43	m	5.98
4.8mm thick linings to walls										
width exceeding 600mm	4.32	5.00	4.53	0.34	0.04	6.80	0.51	11.84	m²	13.02
300mm wide	1.29	10.00	1.42	0.20	0.01	3.81	0.30	5.54	m	6.09
6.4mm thick linings to walls										
width exceeding 600mm	4.54	5.00	4.77	0.35	0.04	6.98	0.52	12.28	m²	13.50
300mm wide	1.36	10.00	1.50	0.21	0.01	4.00	0.31	5.81	m	6.39
3.2mm thick linings to ceilings										
width exceeding 600mm	4.01	5.00	4.21	0.41	0.05	8.22	0.62	13.05	m²	14.35
300mm wide	1.20	10.00	1.32	0.25	0.02	4.87	0.38	6.57	m	7.22
4.8mm thick linings to ceilings										
width exceeding 600mm	4.32	5.00	4.53	0.43	0.05	8.59	0.64	13.76	m²	15.14
300mm wide	1.29	10.00	1.42	0.26	0.02	5.05	0.39	6.87	m	7.55
6.4mm thick linings to ceilings										
width exceeding 600mm	4.54	5.00	4.77	0.44	0.05	8.77	0.66	14.20	m²	15.62
300mm wide	1.36	10.00	1.50	0.26	0.02	5.05	0.39	6.94	m	7.64
Fibreboards and hardboards; Flame retardant hardboard, BS EN 622 tested to BS 476 Part 7 Class 1; butt joints										
3.2mm thick linings to walls										
width exceeding 600mm	4.47	5.00	4.70	0.33	0.04	6.61	0.50	11.80	m²	12.98
300mm wide	1.34	10.00	1.48	0.20	0.01	3.81	0.30	5.59	m	6.15
4.8mm thick linings to walls										
width exceeding 600mm	8.94	5.00	9.39	0.34	0.04	6.80	0.51	16.70	m²	18.37
300mm wide	2.68	10.00	2.95	0.20	0.01	3.81	0.30	7.06	m	7.77
6.4mm thick linings to walls										
width exceeding 600mm	9.78	5.00	10.26	0.35	0.04	6.98	0.52	17.77	m²	19.55
300mm wide	2.93	10.00	3.23	0.21	0.01	4.00	0.31	7.54	m	8.29
3.2mm thick linings to ceilings										
width exceeding 600mm	4.47	5.00	4.70	0.41	0.05	8.22	0.62	13.53	m²	14.88
300mm wide	1.34	10.00	1.48	0.25	0.02	4.87	0.38	6.72	m	7.39
4.8mm thick linings to ceilings										
width exceeding 600mm	8.94	5.00	9.39	0.43	0.05	8.59	0.64	18.62	m²	20.49
300mm wide	2.68	10.00	2.95	0.26	0.02	5.05	0.39	8.39	m	9.23
6.4mm thick linings to ceilings										
width exceeding 600mm	9.78	5.00	10.26	0.44	0.06	8.91	0.66	19.83	m²	21.82
300mm wide	2.93	10.00	3.23	0.26	0.02	5.05	0.39	8.67	m	9.53
Fibreboards and hardboards; Medium density fibreboard, BS EN 622, type MDF; butt joints										
9mm thick linings to walls										
width exceeding 600mm	3.86	5.00	4.05	0.35	0.04	6.98	0.52	11.56	m²	12.71
300mm wide	1.16	10.00	1.27	0.21	0.01	4.00	0.31	5.58	m	6.14
12mm thick linings to walls										
width exceeding 600mm	4.45	5.00	4.67	0.38	0.05	7.67	0.57	12.91	m²	14.20
300mm wide	1.33	10.00	1.47	0.23	0.02	4.50	0.34	6.31	m	6.94
18mm thick linings to walls										
width exceeding 600mm	4.78	5.00	5.02	0.41	0.05	8.22	0.62	13.86	m²	15.24
300mm wide	1.43	10.00	1.58	0.24	0.02	4.68	0.36	6.62	m	7.28
25mm thick linings to walls										
width exceeding 600mm	8.04	5.00	8.44	0.44	0.06	8.91	0.66	18.01	m²	19.81
300mm wide	2.41	10.00	2.65	0.26	0.02	5.05	0.39	8.09	m	8.90
9mm thick linings to ceilings										
width exceeding 600mm	3.86	5.00	4.05	0.44	0.06	8.91	0.66	13.62	m²	14.98
300mm wide	1.16	10.00	1.27	0.26	0.02	5.05	0.39	6.71	m	7.39

CARPENTRY

	MATERIALS			LABOUR				RATES		
Labour hourly rates: (except Specialists) Craft Operatives 18.37 Labourer 13.76 Rates are national average prices. Refer to REGIONAL VARIATIONS for indicative levels of overall pricing in regions	Del to Site	Waste	Material Cost	Craft Optve	Lab	Labour Cost	Sunds	Nett Rate	Unit	Gross rate (10%)
	£	%	£	Hrs	Hrs	£	£	£		£

TIMBER, METAL AND PLASTIC BOARDING, SHEETING, DECKING, CASINGS AND LININGS (Cont'd)

Fibreboards and hardboards; Medium density fibreboard, BS EN 622, type MDF; butt joints (Cont'd)

	Del to Site	Waste	Material Cost	Craft Optve	Lab	Labour Cost	Sunds	Nett Rate	Unit	Gross rate
12mm thick linings to ceilings										
width exceeding 600mm	4.45	5.00	4.67	0.48	0.06	9.64	0.72	15.03	m²	16.54
300mm wide	1.33	10.00	1.47	0.29	0.02	5.60	0.44	7.50	m	8.26
18mm thick linings to ceilings										
width exceeding 600mm	4.78	5.00	5.02	0.52	0.07	10.52	0.78	16.32	m²	17.95
300mm wide	1.43	10.00	1.58	0.31	0.02	5.97	0.47	8.01	m	8.81
25mm thick linings to ceilings										
width exceeding 600mm	8.04	5.00	8.44	0.56	0.07	11.25	0.84	20.53	m²	22.58
300mm wide	2.41	10.00	2.65	0.34	0.02	6.52	0.51	9.68	m	10.65

Fibreboards and hardboards; Insulating softboard, BS EN 622; butt joints

	Del to Site	Waste	Material Cost	Craft Optve	Lab	Labour Cost	Sunds	Nett Rate	Unit	Gross rate
13mm thick linings to walls										
width exceeding 600mm	4.06	5.00	4.26	0.38	0.05	7.67	0.57	12.50	m²	13.75
300mm wide	1.32	10.00	1.45	0.23	0.02	4.50	0.34	6.30	m	6.93
13mm thick linings to ceilings										
width exceeding 600mm	4.06	5.00	4.26	0.48	0.06	9.64	0.72	14.63	m²	16.09
300mm wide	1.32	10.00	1.45	0.29	0.02	5.60	0.44	7.49	m	8.24

Roof boarding; Plywood BS EN 636, II/III grade, WBP bonded; butt joints

	Del to Site	Waste	Material Cost	Craft Optve	Lab	Labour Cost	Sunds	Nett Rate	Unit	Gross rate
18mm thick sheeting to roofs; external										
width exceeding 600mm	7.47	10.00	8.21	0.30	0.04	6.06	0.08	14.35	m²	15.78
extra; cesspool 225 x 225 x 150mm	1.49	10.00	1.64	1.40	0.18	28.19	0.34	30.18	nr	33.20
extra; cesspool 300 x 300 x 150mm	2.05	10.00	2.26	1.50	0.19	30.17	0.36	32.79	nr	36.07
width exceeding 600mm; sloping	7.47	10.00	8.21	0.40	0.05	8.04	0.09	16.34	m²	17.97
Abutments										
raking cutting	0.75	10.00	0.82	0.30	-	5.51	-	6.33	m	6.97
curved cutting	1.12	10.00	1.23	0.45	-	8.27	-	9.50	m	10.45
rebates	-	-	-	0.18	-	3.31	-	3.31	m	3.64
grooves	-	-	-	0.18	-	3.31	-	3.31	m	3.64
chamfers	-	-	-	0.13	-	2.39	-	2.39	m	2.63
24mm thick sheeting to roofs; external										
width exceeding 600mm	9.82	10.00	10.80	0.35	0.04	6.98	0.08	17.85	m²	19.64
extra; cesspool 225 x 225 x 150mm	1.96	10.00	2.16	1.50	0.19	30.17	0.36	32.69	nr	35.96
extra; cesspool 300 x 300 x 150mm	2.70	10.00	2.97	1.60	0.20	32.14	0.39	35.50	nr	39.05
width exceeding 600mm; sloping	9.82	10.00	10.80	0.45	0.06	9.09	0.10	19.99	m²	21.99
Abutments										
raking cutting	0.98	10.00	1.08	0.40	-	7.35	-	8.43	m	9.27
curved cutting	1.47	10.00	1.62	0.60	-	11.02	-	12.64	m	13.91
rebates	-	-	-	0.18	-	3.31	-	3.31	m	3.64
grooves	-	-	-	0.18	-	3.31	-	3.31	m	3.64
chamfers	-	-	-	0.13	-	2.39	-	2.39	m	2.63

Roof boarding; Wood particle board (chipboard), BS EN 312, type 1 standard; butt joints

	Del to Site	Waste	Material Cost	Craft Optve	Lab	Labour Cost	Sunds	Nett Rate	Unit	Gross rate
12mm thick sheeting to roofs; external										
width exceeding 600mm	5.10	10.00	5.61	0.25	0.03	5.01	0.06	10.67	m²	11.74
extra; cesspool 225 x 225 x 150mm	1.02	10.00	1.12	1.30	0.16	26.08	0.31	27.52	nr	30.27
extra; cesspool 300 x 300 x 150mm	1.40	10.00	1.54	1.40	0.18	28.19	0.34	30.08	nr	33.09
width exceeding 600mm; sloping	5.10	10.00	5.61	0.35	0.04	6.98	0.09	12.68	m²	13.94
Abutments										
raking cutting	0.51	10.00	0.56	0.25	-	4.59	-	5.15	m	5.67
curved cutting	0.76	10.00	0.84	0.35	-	6.43	-	7.27	m	8.00
rebates	-	-	-	0.18	-	3.31	-	3.31	m	3.64
grooves	-	-	-	0.18	-	3.31	-	3.31	m	3.64
chamfers	-	-	-	0.13	-	2.39	-	2.39	m	2.63
18mm thick sheeting to roofs; external										
width exceeding 600mm	6.58	10.00	7.23	0.30	0.04	6.06	0.08	13.37	m²	14.71
extra; cesspool 225 x 225 x 150mm	1.32	10.00	1.45	1.40	0.18	28.19	0.34	29.99	nr	32.99
extra; cesspool 300 x 300 x 150mm	1.81	10.00	1.99	1.50	0.19	30.17	0.36	32.52	nr	35.77
width exceeding 600mm; sloping	6.58	10.00	7.23	0.40	0.05	8.04	0.09	15.36	m²	16.90
Abutments										
raking cutting	0.66	10.00	0.72	0.30	-	5.51	-	6.23	m	6.86
curved cutting	0.99	10.00	1.09	0.45	-	8.27	-	9.35	m	10.29
rebates	-	-	-	0.18	-	3.31	-	3.31	m	3.64
grooves	-	-	-	0.18	-	3.31	-	3.31	m	3.64
chamfers	-	-	-	0.13	-	2.39	-	2.39	m	2.63

Softwood, wrought, impregnated; gutter boarding

	Del to Site	Waste	Material Cost	Craft Optve	Lab	Labour Cost	Sunds	Nett Rate	Unit	Gross rate
Gutter boards including sides, tongued and grooved joints										
width exceeding 600mm; 19mm thick	8.26	10.00	9.09	2.00	0.25	40.18	0.48	49.75	m²	54.73
width exceeding 600mm; 25mm thick	9.51	10.00	10.47	2.25	0.28	45.19	0.54	56.19	m²	61.81
width not exceeding 600mm; 19mm thick										
150mm wide	1.24	10.00	1.36	0.30	0.04	6.06	0.08	7.50	m	8.25
225mm wide	1.86	10.00	2.05	0.40	0.05	8.04	0.09	10.17	m	11.19
width not exceeding 600mm; 25mm thick										
150mm wide	1.43	10.00	1.57	0.33	0.04	6.61	0.08	8.26	m	9.08
225mm wide	2.14	10.00	2.35	0.44	0.05	8.77	0.10	11.23	m	12.35
cesspool 225 x 225 x 150mm deep	1.76	10.00	1.94	1.50	0.19	30.17	0.36	32.47	nr	35.71

Labour hourly rates: (except Specialists) Craft Operatives 18.37 Labourer 13.76 Rates are national average prices. Refer to REGIONAL VARIATIONS for indicative levels of overall pricing in regions	MATERIALS			LABOUR				RATES		
	Del to Site £	Waste %	Material Cost £	Craft Optve Hrs	Lab Hrs	Labour Cost £	Sunds £	Nett Rate £	Unit	Gross rate (10%) £
TIMBER, METAL AND PLASTIC BOARDING, SHEETING, DECKING, CASINGS AND LININGS (Cont'd)										
Softwood, wrought, impregnated; gutter boarding (Cont'd)										
Chimney gutter boards, butt joints										
width not exceeding 600mm; 25mm thick x 100mm average wide.	0.95	10.00	1.05	0.50	0.06	10.01	0.12	11.18	m	12.29
gusset end	0.52	10.00	0.57	0.35	0.04	6.98	0.09	7.64	nr	8.41
width not exceeding 600mm; 25mm thick x 175mm average wide.	1.66	10.00	1.83	0.65	0.08	13.04	0.17	15.04	m	16.54
gusset end	0.72	10.00	0.79	0.40	0.05	8.04	0.09	8.92	nr	9.81
Lier boards; tongued and grooved joints										
width exceeding 600mm; 19mm thick	8.26	10.00	9.09	0.90	0.11	18.05	0.22	27.36	m²	30.10
width exceeding 600mm; 25mm thick	9.51	10.00	10.47	1.00	0.12	20.02	0.24	30.73	m²	33.80
Valley sole boards; butt joints										
100mm wide	0.95	10.00	1.05	0.45	0.06	9.09	0.10	10.24	m	11.27
150mm wide	1.43	10.00	1.57	0.50	0.06	10.01	0.12	11.70	m	12.87
Plywood BS EN 636, II/III grade, WBP bonded, butt joints; gutter boarding										
Gutter boards including sides, butt joints										
width exceeding 600mm; 18mm thick	8.42	10.00	9.26	1.50	0.19	30.17	0.36	39.79	m²	43.77
width exceeding 600mm; 25mm thick	12.07	10.00	13.27	1.65	0.21	33.20	0.41	46.88	m²	51.57
width not exceeding 600mm; 18mm thick										
150mm wide	1.26	10.00	1.39	0.20	0.03	4.09	0.05	5.52	m	6.07
225mm wide	1.89	10.00	2.08	0.22	0.03	4.45	0.06	6.60	m	7.26
width not exceeding 600mm; 25mm thick										
150mm wide	1.81	10.00	1.99	0.23	0.03	4.64	0.06	6.69	m	7.36
225mm wide	2.72	10.00	2.99	0.25	0.03	5.01	0.06	8.05	m	8.86
cesspool 225 x 225 x 150mm deep	2.23	10.00	2.46	1.50	0.19	30.17	0.36	32.99	nr	36.28
Softwood, wrought, impregnated; eaves and verge boarding										
Fascia and barge boards										
width not exceeding 600mm; 25mm thick										
150mm wide	1.43	10.00	1.57	0.22	0.03	4.45	0.06	6.08	m	6.69
200mm wide	1.90	10.00	2.09	0.23	0.03	4.64	0.06	6.79	m	7.47
Fascia and barge boards, tongued, grooved and veed joints										
width not exceeding 600mm; 25mm thick										
225mm wide	2.14	10.00	2.35	0.30	0.04	6.06	0.08	8.49	m	9.34
250mm wide	2.38	10.00	2.62	0.32	0.04	6.43	0.08	9.12	m	10.03
Eaves or verge soffit boards										
width exceeding 600mm; 25mm thick	9.51	10.00	10.47	1.00	0.12	20.02	0.24	30.73	m²	33.80
width not exceeding 600mm; 25mm thick										
225mm wide	2.14	10.00	2.35	0.30	0.04	6.06	0.08	8.49	m	9.34
Plywood BS EN 636, II/III grade, WBP bonded, butt joints; eaves and verge boarding										
Fascia and barge boards										
width not exceeding 600mm; 18mm thick										
150mm wide	1.26	10.00	1.39	0.22	0.03	4.45	0.06	5.90	m	6.49
225mm wide	1.89	10.00	2.08	0.23	0.03	4.64	0.06	6.78	m	7.46
250mm wide	2.10	10.00	2.32	0.24	0.03	4.82	0.06	7.20	m	7.92
Eaves or verge soffit boards										
width exceeding 600mm; 18mm thick	8.42	10.00	9.26	0.75	0.09	15.02	0.18	24.46	m²	26.90
width not exceeding 600mm; 18mm thick										
225mm wide	1.89	10.00	2.08	0.25	0.03	5.01	0.06	7.15	m	7.86
Fascia and bargeboards; PVC										
Fascia; 17mm thick; mahogany woodgrain square edge with groove and return leg; fixing to timber with Polytop screws										
175mm wide	6.41	5.00	6.73	0.20	0.03	4.09	0.72	11.54	m	12.69
225mm wide	7.74	5.00	8.13	0.20	0.03	4.09	0.72	12.93	m	14.23
375mm wide	13.20	5.00	13.86	0.25	0.04	5.07	1.44	20.37	m	22.41
Fascia; 22mm thick; white solid bullnose with soffit groove; fixing to timber with Polytop screws										
150mm wide	7.00	5.00	7.35	0.20	0.03	4.09	0.72	12.16	m	13.37
200mm wide	9.58	5.00	10.06	0.20	0.03	4.09	0.72	14.87	m	16.35
250mm wide	11.82	5.00	12.41	0.20	0.03	4.09	1.44	17.94	m	19.73
Fascia; 9mm thick; white square edge board; fixing to timber with Polytop screws										
100mm wide	1.25	5.00	1.31	0.30	0.04	6.06	0.72	8.09	m	8.90
150mm wide	1.78	5.00	1.87	0.30	0.04	6.06	0.72	8.65	m	9.52
225mm wide	2.65	5.00	2.78	0.30	0.04	6.06	0.72	9.56	m	10.52
Multi purpose 9mm thick; white pre-vented soffit board; fixing to timber with Polytop screws										
150mm wide	2.57	5.00	2.70	0.40	0.06	8.17	0.72	11.59	m	12.75
225mm wide	3.76	5.00	3.95	0.40	0.06	8.17	1.44	13.56	m	14.92
300mm wide	5.00	5.00	5.25	0.60	0.08	12.12	2.16	19.53	m	21.48
Multi purpose 9mm thick soffit board; fixing to timber with Polytop screws										
100mm wide	1.25	5.00	1.31	0.40	0.06	8.17	0.72	10.21	m	11.23
175mm wide	2.01	5.00	2.11	0.40	0.06	8.17	0.72	11.00	m	12.10
200mm wide	2.39	5.00	2.51	0.40	0.06	8.17	1.44	12.12	m	13.34
300mm wide	3.69	5.00	3.87	0.60	0.08	12.12	2.16	18.16	m	19.97

CARPENTRY

Labour hourly rates: (except Specialists) Craft Operatives 18.37 Labourer 13.76 Rates are national average prices. Refer to REGIONAL VARIATIONS for indicative levels of overall pricing in regions	MATERIALS			LABOUR				RATES		
	Del to Site	Waste	Material Cost	Craft Optve	Lab	Labour Cost	Sunds	Nett Rate		Gross rate (10%)
	£	%	£	Hrs	Hrs	£	£	£	Unit	£

METAL AND PLASTIC ACCESSORIES

Metal fixings, fastenings and fittings

Steel; tie rods and straps

tie rods and straps										
Tie rods										
19mm diameter; threaded.............................	8.77	2.50	8.99	0.25	0.03	5.01	-	13.99	m	15.39
extra; for nut and washer.............................	0.53	2.50	0.54	0.20	0.03	4.09	-	4.63	nr	5.09
25mm diameter; threaded.............................	19.52	2.50	20.00	0.30	0.04	6.06	-	26.07	m	28.67
extra; for nut and washer.............................	1.00	2.50	1.02	0.30	0.04	6.06	-	7.09	nr	7.80
Straps; 3 x 38; holes -4; fixing with bolts (bolts included elsewhere)										
500mm long...	0.49	2.50	0.50	0.13	0.02	2.66	-	3.17	nr	3.48
750mm long...	0.61	2.50	0.62	0.19	0.02	3.77	-	4.39	nr	4.82
1000mm long...	0.84	2.50	0.86	0.25	0.03	5.01	-	5.86	nr	6.45
Straps; 6 x 50; holes -4; fixing with bolts (bolts included elsewhere)										
500mm long...	1.04	2.50	1.07	0.16	0.02	3.21	-	4.29	nr	4.71
750mm long...	1.45	2.50	1.49	0.23	0.03	4.64	-	6.13	nr	6.74
1000mm long...	1.69	2.50	1.73	0.30	0.04	6.06	-	7.79	nr	8.57
Steel; galvanized; straps										
Straps and clips; Expamet Bat; light duty strapping; fixing with nails										
600mm long...	0.81	5.00	0.85	0.15	0.02	3.03	0.03	3.91	nr	4.30
800mm long...	1.22	5.00	1.28	0.17	0.02	3.40	0.05	4.72	nr	5.20
1000mm long...	0.07	5.00	0.07	0.18	0.02	3.58	0.05	3.70	nr	4.07
1200mm long...	1.68	5.00	1.76	0.19	0.02	3.77	0.05	5.57	nr	6.13
1600mm long...	2.86	5.00	3.00	0.20	0.02	3.95	0.05	7.00	nr	7.70
Straps and clips; Expamet Bat; light duty strapping; twists -1; fixing with nails										
600mm long...	0.81	5.00	0.85	0.15	0.02	3.03	0.03	3.91	nr	4.30
800mm long...	1.22	5.00	1.28	0.17	0.02	3.40	0.05	4.72	nr	5.20
1000mm long...	1.35	5.00	1.42	0.18	0.02	3.58	0.05	5.04	nr	5.55
1200mm long...	1.68	5.00	1.76	0.19	0.02	3.77	0.05	5.57	nr	6.13
1600mm long...	2.86	5.00	3.00	0.20	0.02	3.95	0.05	7.00	nr	7.70
Straps and clips; Expamet Bat; heavy duty strapping; bends -1; nailing one end (building in included elsewhere)										
600mm long...	1.50	5.00	1.57	0.08	0.02	1.74	0.03	3.35	nr	3.68
800mm long...	1.91	5.00	2.01	0.08	0.02	1.74	0.03	3.78	nr	4.16
1000mm long...	2.55	5.00	2.68	0.09	0.02	1.93	0.03	4.64	nr	5.10
1200mm long...	2.94	5.00	3.09	0.09	0.02	1.93	0.03	5.05	nr	5.55
1600mm long...	3.84	5.00	4.03	0.10	0.02	2.11	0.03	6.17	nr	6.79
Straps and clips; Expamet Bat; heavy duty strapping; bends -1; twists -1; nailing one end (building in included elsewhere)										
600mm long...	1.50	5.00	1.57	0.08	0.02	1.74	0.03	3.35	nr	3.68
800mm long...	1.91	5.00	2.01	0.08	0.02	1.74	0.03	3.78	nr	4.16
1000mm long...	2.55	5.00	2.68	0.09	0.02	1.93	0.03	4.64	nr	5.10
1200mm long...	2.94	5.00	3.09	0.09	0.02	1.93	0.03	5.05	nr	5.55
1600mm long...	3.84	5.00	4.03	0.10	0.02	2.11	0.03	6.17	nr	6.79
truss clips; fixing with nails										
for 38mm thick members	0.30	5.00	0.31	0.20	0.02	3.95	0.05	4.31	nr	4.74
for 50mm thick members	0.30	5.00	0.31	0.20	0.02	3.95	0.05	4.31	nr	4.74
Steel; galvanised; shoes										
Post shoe fixing with bolts to concrete										
to suit 100 x 100 post...............................	4.12	5.00	4.33	0.40	0.04	7.90	1.05	13.28	nr	14.61
to suit 150 x 150 post...............................	4.73	5.00	4.97	0.44	0.05	8.77	0.10	13.84	nr	15.23
Bolts; steel; black, BS 4190 grade 4.6										
M 8; with nuts and washers										
30mm long...	0.10	5.00	0.10	0.07	0.01	1.42	-	1.53	nr	1.68
40mm long...	0.10	5.00	0.10	0.08	0.01	1.61	-	1.71	nr	1.88
50mm long...	0.10	5.00	0.11	0.08	0.01	1.61	-	1.72	nr	1.89
60mm long...	0.12	5.00	0.13	0.08	0.01	1.61	-	1.73	nr	1.91
70mm long...	0.13	5.00	0.14	0.10	0.01	1.97	-	2.11	nr	2.32
80mm long...	0.14	5.00	0.15	0.10	0.01	1.97	-	2.12	nr	2.33
90mm long...	0.15	5.00	0.16	0.10	0.01	1.97	-	2.13	nr	2.35
100mm long...	0.19	5.00	0.20	0.10	0.01	1.97	-	2.18	nr	2.39
120mm long...	0.19	5.00	0.20	0.10	0.01	1.97	-	2.18	nr	2.39
M 10; with nuts and washers										
40mm long...	0.16	5.00	0.17	0.08	0.01	1.61	-	1.78	nr	1.96
50mm long...	0.19	5.00	0.20	0.08	0.01	1.61	-	1.81	nr	1.99
60mm long...	0.21	5.00	0.22	0.08	0.01	1.61	-	1.82	nr	2.01
70mm long...	0.22	5.00	0.24	0.08	0.01	1.61	-	1.84	nr	2.03
80mm long...	0.23	5.00	0.24	0.10	0.01	1.97	-	2.21	nr	2.43
90mm long...	0.25	5.00	0.26	0.10	0.01	1.97	-	2.23	nr	2.46
100mm long...	0.31	5.00	0.33	0.10	0.01	1.97	-	2.31	nr	2.54
120mm long...	0.37	5.00	0.39	0.10	0.01	1.97	-	2.36	nr	2.60
140mm long...	0.43	5.00	0.45	0.11	0.02	2.30	-	2.75	nr	3.02
150mm long...	0.34	5.00	0.36	0.12	0.02	2.48	-	2.83	nr	3.12
M 12; with nuts and washers										
40mm long...	0.25	5.00	0.26	0.10	0.01	1.97	-	2.24	nr	2.46
50mm long...	0.27	5.00	0.28	0.10	0.01	1.97	-	2.26	nr	2.48
60mm long...	0.32	5.00	0.33	0.10	0.01	1.97	-	2.31	nr	2.54
70mm long...	0.31	5.00	0.32	0.10	0.01	1.97	-	2.30	nr	2.53
80mm long...	0.38	5.00	0.40	0.10	0.01	1.97	-	2.37	nr	2.61
90mm long...	0.37	5.00	0.39	0.10	0.01	1.97	-	2.36	nr	2.60
100mm long...	0.39	5.00	0.41	0.11	0.01	2.16	-	2.56	nr	2.82
120mm long...	0.44	5.00	0.46	0.13	0.02	2.66	-	3.12	nr	3.43
140mm long...	0.47	5.00	0.50	0.12	0.02	2.48	-	2.98	nr	3.27
150mm long...	0.48	5.00	0.50	0.13	0.02	2.66	-	3.17	nr	3.48
160mm long...	0.72	5.00	0.76	0.13	0.02	2.66	-	3.42	nr	3.76

Labour hourly rates: (except Specialists) Craft Operatives 18.37 Labourer 13.76 Rates are national average prices. Refer to REGIONAL VARIATIONS for indicative levels of overall pricing in regions	MATERIALS			LABOUR				RATES		
	Del to Site	Waste	Material Cost	Craft Optve	Lab	Labour Cost	Sunds	Nett Rate		Gross rate (10%)
	£	%	£	Hrs	Hrs	£	£	£	Unit	£
METAL AND PLASTIC ACCESSORIES (Cont'd)										
Metal fixings, fastenings and fittings (Cont'd)										
Bolts; steel; black, BS 4190 grade 4.6 (Cont'd)										
M 12; with nuts and washers (Cont'd)										
180mm long	0.60	5.00	0.63	0.14	0.02	2.85	-	3.47	nr	3.82
200mm long	0.72	5.00	0.76	0.17	0.02	3.40	-	4.15	nr	4.57
220mm long	0.68	5.00	0.71	0.17	0.02	3.40	-	4.11	nr	4.52
240mm long	1.02	5.00	1.07	0.17	0.02	3.40	-	4.47	nr	4.92
250mm long	0.96	5.00	1.01	0.18	0.02	3.58	-	4.59	nr	5.05
M 16; with nuts and washers										
50mm long	0.52	5.00	0.54	0.10	0.01	1.97	-	2.52	nr	2.77
60mm long	0.57	5.00	0.59	0.10	0.01	1.97	-	2.57	nr	2.83
70mm long	0.61	5.00	0.64	0.10	0.01	1.97	-	2.61	nr	2.88
80mm long	0.74	5.00	0.77	0.10	0.01	1.97	-	2.75	nr	3.02
90mm long	0.84	5.00	0.88	0.10	0.01	1.97	-	2.85	nr	3.14
100mm long	0.84	5.00	0.88	0.11	0.01	2.16	-	3.04	nr	3.34
120mm long	0.86	5.00	0.91	0.13	0.02	2.66	-	3.57	nr	3.93
140mm long	1.00	5.00	1.05	0.12	0.02	2.48	-	3.53	nr	3.88
150mm long	1.04	5.00	1.09	0.13	0.02	2.66	-	3.75	nr	4.13
160mm long	1.11	5.00	1.17	0.14	0.02	2.85	-	4.02	nr	4.42
180mm long	1.24	5.00	1.31	0.16	0.02	3.21	-	4.52	nr	4.97
200mm long	1.26	5.00	1.33	0.17	0.02	3.40	-	4.72	nr	5.20
220mm long	1.28	5.00	1.34	0.17	0.02	3.40	-	4.74	nr	5.21
300mm long	2.14	-	2.14	0.18	0.02	3.58	-	5.72	nr	6.30
M 20; with nuts and washers										
60mm long	0.81	5.00	0.85	0.10	0.01	1.97	-	2.82	nr	3.11
70mm long	0.89	5.00	0.93	0.10	0.01	1.97	-	2.91	nr	3.20
80mm long	1.18	5.00	1.24	0.10	0.01	1.97	-	3.22	nr	3.54
90mm long	1.18	5.00	1.24	0.11	0.01	2.16	-	3.40	nr	3.74
100mm long	1.18	5.00	1.24	0.12	0.02	2.48	-	3.72	nr	4.09
120mm long	1.49	5.00	1.57	0.14	0.02	2.85	-	4.42	nr	4.86
140mm long	1.66	5.00	1.74	0.13	0.02	2.66	-	4.40	nr	4.84
150mm long	1.42	5.00	1.49	0.14	0.02	2.85	-	4.34	nr	4.78
160mm long	1.67	5.00	1.75	0.17	0.02	3.40	-	5.15	nr	5.67
180mm long	1.88	5.00	1.97	0.17	0.02	3.40	-	5.37	nr	5.91
200mm long	2.50	5.00	2.62	0.16	0.02	3.21	-	5.84	nr	6.42
220mm long	2.55	5.00	2.68	0.17	0.02	3.40	-	6.08	nr	6.69
240mm long	3.75	5.00	3.94	0.18	0.02	3.58	-	7.52	nr	8.28
260mm long	4.34	5.00	4.56	0.20	0.02	3.95	-	8.51	nr	9.36
280mm long	4.58	5.00	4.81	0.21	0.02	4.13	-	8.94	nr	9.83
300mm long	4.93	5.00	5.17	0.23	0.02	4.50	-	9.67	nr	10.64
M 24; with nuts and washers										
70mm long	1.69	5.00	1.78	0.10	0.01	1.97	-	3.75	nr	4.13
80mm long	1.75	5.00	1.84	0.11	0.01	2.16	-	4.00	nr	4.40
90mm long	2.09	5.00	2.19	0.12	0.01	2.34	-	4.53	nr	4.98
100mm long	2.16	5.00	2.27	0.12	0.02	2.48	-	4.75	nr	5.23
120mm long	2.34	5.00	2.46	0.12	0.02	2.48	-	4.94	nr	5.43
140mm long	2.84	5.00	2.98	0.13	0.02	2.66	-	5.64	nr	6.21
150mm long	2.87	5.00	3.01	0.15	0.02	3.03	-	6.04	nr	6.65
160mm long	2.89	5.00	3.04	0.15	0.02	3.03	-	6.07	nr	6.67
180mm long	2.94	5.00	3.09	0.15	0.02	3.03	-	6.12	nr	6.73
200mm long	3.17	5.00	3.33	0.16	0.02	3.21	-	6.55	nr	7.20
220mm long	3.53	5.00	3.70	0.17	0.02	3.40	-	7.10	nr	7.81
240mm long	3.78	5.00	3.97	0.18	0.02	3.58	-	7.55	nr	8.31
260mm long	4.44	5.00	4.66	0.21	0.02	4.13	-	8.79	nr	9.67
280mm long	4.66	5.00	4.90	0.23	0.02	4.50	-	9.40	nr	10.34
300mm long	4.94	5.00	5.18	0.25	0.02	4.87	-	10.05	nr	11.06
M 12; with nuts and 38 x 38 x 3mm square plate washers										
100mm long	0.65	5.00	0.68	0.11	0.01	2.16	-	2.84	nr	3.12
120mm long	0.77	5.00	0.81	0.13	0.02	2.66	-	3.47	nr	3.82
160mm long	1.72	5.00	1.81	0.13	0.02	2.66	-	4.47	nr	4.92
180mm long	1.79	5.00	1.88	0.14	0.02	2.85	-	4.73	nr	5.20
200mm long	1.87	5.00	1.96	0.17	0.02	3.40	-	5.36	nr	5.90
M 12; with nuts and 50 x 50 x 3mm square plate washers										
100mm long	0.70	5.00	0.73	0.11	0.01	2.16	-	2.89	nr	3.18
120mm long	0.82	5.00	0.86	0.13	0.02	2.66	-	3.52	nr	3.88
160mm long	1.77	5.00	1.86	0.13	0.02	2.66	-	4.52	nr	4.97
180mm long	1.84	5.00	1.93	0.14	0.02	2.85	-	4.78	nr	5.26
200mm long	1.92	5.00	2.02	0.17	0.02	3.40	-	5.41	nr	5.96
Bolts; zinc plated, BS 4933 Grade 4.6, cup head, square neck										
M 6; with nuts and washers										
25mm long	0.10	5.00	0.10	0.08	0.01	1.61	-	1.71	nr	1.88
50mm long	0.11	5.00	0.12	0.08	0.01	1.61	-	1.72	nr	1.89
75mm long	0.16	5.00	0.17	0.10	0.01	1.97	-	2.14	nr	2.36
100mm long	0.22	5.00	0.23	0.10	0.01	1.97	-	2.21	nr	2.43
150mm long	0.31	5.00	0.33	0.11	0.01	2.16	-	2.48	nr	2.73
M 8; with nuts and washers										
25mm long	0.16	5.00	0.17	0.08	0.01	1.61	-	1.78	nr	1.95
50mm long	0.17	5.00	0.18	0.08	0.01	1.61	-	1.79	nr	1.96
75mm long	0.22	5.00	0.23	0.10	0.01	1.97	-	2.21	nr	2.43
100mm long	0.32	5.00	0.34	0.10	0.01	1.97	-	2.31	nr	2.54
150mm long	0.46	5.00	0.48	0.11	0.01	2.16	-	2.64	nr	2.91
M 10; with nuts and washers										
50mm long	0.27	5.00	0.28	0.08	0.01	1.61	-	1.89	nr	2.08
75mm long	0.31	5.00	0.33	0.10	0.01	1.97	-	2.30	nr	2.53
100mm long	0.46	5.00	0.48	0.10	0.01	1.97	-	2.46	nr	2.70
150mm long	0.66	5.00	0.69	0.12	0.02	2.48	-	3.17	nr	3.49
M 12; with nuts and washers										
50mm long	0.39	5.00	0.41	0.10	0.01	1.97	-	2.38	nr	2.62
75mm long	0.52	5.00	0.55	0.10	0.01	1.97	-	2.52	nr	2.77
100mm long	0.63	5.00	0.66	0.11	0.01	2.16	-	2.82	nr	3.10
150mm long	0.88	5.00	0.92	0.13	0.02	2.66	-	3.59	nr	3.95

CARPENTRY

CARPENTRY

	MATERIALS			LABOUR				RATES		
Labour hourly rates: (except Specialists) Craft Operatives 18.37 Labourer 13.76 Rates are national average prices. Refer to REGIONAL VARIATIONS for indicative levels of overall pricing in regions	Del to Site	Waste	Material Cost	Craft Optve	Lab	Labour Cost	Sunds	Nett Rate	Unit	Gross rate (10%)
	£	%	£	Hrs	Hrs	£	£	£		£
METAL AND PLASTIC ACCESSORIES (Cont'd)										
Metal fixings, fastenings and fittings (Cont'd)										
Bolts; steel; galvanized; BS 4190 grade 4.6, hexagon head										
M 12; with nuts and washers										
100mm long	0.95	5.00	1.00	0.11	0.01	2.16	-	3.16	nr	3.47
120mm long	1.18	5.00	1.24	0.13	0.02	2.66	-	3.90	nr	4.29
160mm long	2.46	5.00	2.58	0.13	0.02	2.66	-	5.25	nr	5.77
180mm long	2.59	5.00	2.72	0.14	0.02	2.85	-	5.57	nr	6.12
200mm long	2.70	5.00	2.84	0.17	0.02	3.40	-	6.23	nr	6.86
M 16; with nuts and washers										
100mm long	1.59	5.00	1.67	0.11	0.01	2.16	-	3.83	nr	4.21
120mm long	2.09	5.00	2.19	0.13	0.02	2.66	-	4.86	nr	5.34
160mm long	3.30	5.00	3.46	0.14	0.02	2.85	-	6.31	nr	6.94
180mm long	3.50	5.00	3.67	0.16	0.02	3.21	-	6.89	nr	7.58
200mm long	3.73	5.00	3.92	0.17	0.02	3.40	-	7.31	nr	8.05
M 20; with nuts and washers										
100mm long	2.91	5.00	3.06	0.12	0.02	2.48	-	5.54	nr	6.09
120mm long	3.91	5.00	4.11	0.11	0.01	2.16	-	6.26	nr	6.89
160mm long	4.94	5.00	5.19	0.17	0.02	3.40	-	8.59	nr	9.44
180mm long	5.12	5.00	5.38	0.17	0.02	3.40	-	8.77	nr	9.65
200mm long	5.46	5.00	5.73	0.17	0.02	3.40	-	9.13	nr	10.04
M 12; with nuts and 38 x 38 x 3mm square plate washers										
100mm long	1.05	5.00	1.10	0.11	0.01	2.16	-	3.26	nr	3.59
120mm long	1.26	5.00	1.32	0.13	0.02	2.66	-	3.99	nr	4.38
160mm long	2.55	5.00	2.68	0.13	0.02	2.66	-	5.34	nr	5.87
180mm long	2.68	5.00	2.81	0.14	0.02	2.85	-	5.66	nr	6.23
200mm long	2.80	5.00	2.94	0.17	0.02	3.40	-	6.34	nr	6.97
M 12; with nuts and 50 x 50 x 3mm square plate washers										
100mm long	1.11	5.00	1.17	0.11	0.01	2.16	-	3.32	nr	3.66
120mm long	1.31	5.00	1.38	0.13	0.02	2.66	-	4.04	nr	4.44
160mm long	2.62	5.00	2.75	0.13	0.02	2.66	-	5.41	nr	5.96
180mm long	2.76	5.00	2.90	0.14	0.02	2.85	-	5.74	nr	6.32
200mm long	2.95	5.00	3.10	0.17	0.02	3.40	-	6.50	nr	7.15
Bolts; expanding; bolt projecting Rawlbolts; drilling masonry with nuts and washers; reference										
44 - 505 (M6 10P)	0.71	5.00	0.74	0.24	0.03	4.82	-	5.56	nr	6.12
44 - 510 (M6 25P)	0.82	5.00	0.86	0.24	0.03	4.82	-	5.68	nr	6.25
44 - 515 (M6 60P)	0.82	5.00	0.87	0.24	0.03	4.82	-	5.69	nr	6.26
44 - 555 (M8 10P)	0.93	5.00	0.97	0.26	0.03	5.19	-	6.16	nr	6.78
44 - 560 (M8 25P)	0.96	5.00	1.01	0.26	0.03	5.19	-	6.20	nr	6.82
44 - 565 (M8 60P)	1.04	5.00	1.09	0.26	0.03	5.19	-	6.28	nr	6.91
44 - 565 (M8 60P)	1.10	5.00	1.15	0.30	0.04	6.06	-	7.22	nr	7.94
44 - 610 (M10 30P)	1.15	5.00	1.20	0.30	0.04	6.06	-	7.27	nr	7.99
44 - 615 (M10 60P)	1.12	5.00	1.17	0.30	0.04	6.06	-	7.23	nr	7.96
44 - 655 (M12 15P)	1.79	5.00	1.87	0.32	0.04	6.43	-	8.30	nr	9.13
44 - 660 (M12 30P)	1.87	5.00	1.96	0.32	0.04	6.43	-	8.39	nr	9.23
44 - 665 (M12 75P)	2.32	5.00	2.43	0.32	0.04	6.43	-	8.86	nr	9.75
44 - 705 (M16 15P)	4.19	5.00	4.40	0.36	0.05	7.30	-	11.70	nr	12.87
44 - 710 (M16 35P)	4.29	5.00	4.51	0.36	0.05	7.30	-	11.81	nr	12.99
44 - 715 (M16 75P)	4.43	5.00	4.65	0.36	0.05	7.30	-	11.95	nr	13.15
44 - 755 (M20 15P)	7.18	5.00	7.54	0.40	0.05	8.04	-	15.57	nr	17.13
44 - 760 (M20 30P)	7.62	5.00	8.01	0.40	0.05	8.04	-	16.04	nr	17.65
44 - 765 (M20 100P)	8.00	5.00	8.40	0.40	0.05	8.04	-	16.44	nr	18.08
Bolts; expanding; Sleeve Anchor (Rawlok); bolt projecting type; drilling masonry with nuts and washers; reference										
69 - 506 M5; 56 long; (Max fixture thickness 25mm)	0.11	5.00	0.11	0.19	0.02	3.77	-	3.88	nr	4.27
69 - 508; M6; 40 long; (Max fixture thickness 10mm)	0.15	5.00	0.16	0.22	0.03	4.45	-	4.61	nr	5.07
69 - 510; M6; 65 long; (Max fixture thickness 35mm)	0.17	5.00	0.18	0.22	0.03	4.45	-	4.64	nr	5.10
69 - 514; M8; 50 long; (Max fixture thickness 10mm)	0.22	5.00	0.23	0.24	0.03	4.82	-	5.06	nr	5.56
69 - 516; M8; 75 long; (Max fixture thickness 36mm)	0.30	5.00	0.31	0.24	0.03	4.82	-	5.14	nr	5.65
69 - 518; M8; 95 long; (Max fixture thickness 55mm)	0.37	5.00	0.39	0.24	0.03	4.82	-	5.21	nr	5.73
69 - 520; M10; 60 long; (Max fixture thickness 10mm)	0.38	5.00	0.40	0.26	0.03	5.19	-	5.59	nr	6.15
69 - 524; M10; 100 long; (Max fixture thickness 50mm)	0.52	5.00	0.55	0.26	0.03	5.19	-	5.74	nr	6.31
69 - 525; M10; 130 long; (Max fixture thickness 80mm)	0.67	5.00	0.70	0.26	0.03	5.19	-	5.89	nr	6.48
69 - 526 M12; 60 long; (Max fixture thickness 10mm)	1.29	5.00	1.35	0.30	0.04	6.06	-	7.41	nr	8.15
69 - 528; M12; 110 long; (Max fixture thickness 55mm)	0.94	5.00	0.98	0.30	0.04	6.06	-	7.04	nr	7.75
69 - 530; M12; 145 long; (Max fixture thickness 85mm)	1.22	5.00	1.28	0.30	0.04	6.06	-	7.34	nr	8.07
Bolts; expanding; loose bolt Rawlbolts; drilling masonry with washers; reference										
44 - 015 (M6 10L)	0.76	5.00	0.79	0.24	0.03	4.82	-	5.62	nr	6.18
44 - 020 (M6 25L)	0.76	5.00	0.79	0.24	0.03	4.82	-	5.62	nr	6.18
44 - 025 (M6 40L)	0.89	5.00	0.94	0.24	0.03	4.82	-	5.76	nr	6.34
44 - 055 (M8 10L)	0.89	5.00	0.94	0.26	0.03	5.19	-	6.13	nr	6.74
44 - 060 (M8 25L)	0.97	5.00	1.02	0.26	0.03	5.19	-	6.20	nr	6.82
44 - 065 (M8 40L)	1.03	5.00	1.08	0.26	0.03	5.19	-	6.27	nr	6.90
44 - 105 (M10 10L)	1.04	5.00	1.09	0.30	0.04	6.06	-	7.15	nr	7.87
44 - 110 (M10 25L)	1.15	5.00	1.21	0.30	0.04	6.06	-	7.27	nr	8.00
44 - 115 (M10 50L)	1.17	5.00	1.23	0.30	0.04	6.06	-	7.29	nr	8.02
44 - 120 (M10 75L)	1.27	5.00	1.33	0.30	0.04	6.06	-	7.40	nr	8.14
44 - 155 (M12 10L)	2.01	5.00	2.11	0.32	0.04	6.43	-	8.54	nr	9.39
44 - 160 (M12 25L)	2.08	5.00	2.19	0.32	0.04	6.43	-	8.62	nr	9.48
44 - 165 (M12 40L)	2.06	5.00	2.16	0.32	0.04	6.43	-	8.59	nr	9.45
44 - 170 (M12 60L)	2.61	5.00	2.74	0.32	0.04	6.43	-	9.17	nr	10.09
44 - 205 (M16 15L)	4.16	5.00	4.37	0.36	0.05	7.30	-	11.67	nr	12.84
44 - 210 (M16 30L)	4.39	5.00	4.61	0.36	0.05	7.30	-	11.91	nr	13.10
44 - 215 (M16 60L)	5.13	5.00	5.39	0.36	0.05	7.30	-	12.69	nr	13.96
44 - 255 (M20 60L)	12.74	5.00	13.38	0.40	0.05	8.04	-	21.42	nr	23.56
44 - 260 (M20 100L)	12.98	5.00	13.63	0.40	0.05	8.04	-	21.67	nr	23.84
44 - 310 (M24 150L)	15.97	5.00	16.76	0.45	0.05	8.95	-	25.72	nr	28.29

152

Labour hourly rates: (except Specialists) Craft Operatives 18.37 Labourer 13.76 Rates are national average prices. Refer to REGIONAL VARIATIONS for indicative levels of overall pricing in regions	MATERIALS			LABOUR				RATES		
	Del to Site	Waste	Material Cost	Craft Optve	Lab	Labour Cost	Sunds	Nett Rate		Gross rate (10%)
	£	%	£	Hrs	Hrs	£	£	£	Unit	£
METAL AND PLASTIC ACCESSORIES (Cont'd)										
Metal fixings, fastenings and fittings (Cont'd)										
Bolts; safety anchors										
Bolts; safety anchors; Spit Triga high performance safety anchors; bolt head version; drilling masonry										
Spit Triga high performance safety anchors; bolt head version; drilling masonry										
with washers, reference										
M10; 105mm long, ref 050689 (Max fixture thickness 20mm)	2.74	5.00	2.87	0.30	0.03	5.92	-	8.80	nr	9.68
M12; 120mm long, ref 050697 (Max fixture thickness 25mm)	4.60	5.00	4.83	0.32	0.03	6.29	-	11.12	nr	12.24
M16; 145mm long, ref 050705 (Max fixture thickness 25mm)	10.44	5.00	10.97	0.36	0.04	7.16	-	18.13	nr	19.94
M20; 170mm long, ref 050711 (Max fixture thickness 25mm)	14.35	5.00	15.07	0.40	0.04	7.90	-	22.97	nr	25.26
Bolts; safety anchors; Spit Fix high performance through bolt BZP anchors; drilling masonry										
with nuts and washers, reference										
M6; 55mm long; Ref 050520 (Max fixture thickness 20mm)	0.20	5.00	0.21	0.22	0.03	4.45	-	4.66	nr	5.13
M6; 85mm long; Ref 050530 (Max fixture thickness 50mm)	0.21	5.00	0.22	0.22	0.03	4.45	-	4.68	nr	5.15
M8; 55mm long; Ref 050535 (Max fixture thickness 5mm)	0.21	5.00	0.23	0.24	0.03	4.82	-	5.05	nr	5.55
M8; 90mm long; Ref 056420 (Max fixture thickness 40mm)	0.25	5.00	0.26	0.24	0.03	4.82	-	5.08	nr	5.59
M8; 130mm long; Ref 056430 (Max fixture thickness 80mm)	0.30	5.00	0.32	0.24	0.03	4.82	-	5.14	nr	5.65
M10; 75mm long; Ref 056530 (Max fixture thickness 15mm)	0.31	5.00	0.33	0.26	0.03	5.19	-	5.52	nr	6.07
M10; 96mm long; Ref 056540 (Max fixture thickness 36mm)	0.35	5.00	0.36	0.26	0.03	5.19	-	5.55	nr	6.11
M12; 100mm long; Ref 055335 (Max fixture thickness 25mm)	0.47	5.00	0.49	0.30	0.03	5.92	-	6.42	nr	7.06
M12; 140mm long; Ref 056590 (Max fixture thickness 65mm)	0.57	5.00	0.60	0.30	0.03	5.92	-	6.52	nr	7.17
M12; 180mm long; Ref 056650 (Max fixture thickness 105mm) .	0.73	5.00	0.76	0.30	0.03	5.92	-	6.69	nr	7.35
M12; 220mm long; Ref 056660 (Max fixture thickness 145mm) ..	1.35	5.00	1.42	0.30	0.03	5.92	-	7.34	nr	8.07
M16; 125mm long; Ref 056700 (Max fixture thickness 30mm)	0.96	5.00	1.01	0.32	0.03	6.29	-	7.30	nr	8.03
M16; 170mm long; Ref 056710 (Max fixture thickness 75mm)	1.32	5.00	1.38	0.32	0.03	6.29	-	7.67	nr	8.44
M20; 160mm long; Ref 056730 (Max fixture thickness 50mm)	2.74	5.00	2.88	0.36	0.04	7.16	-	10.04	nr	11.05
M20; 215mm long; Ref 056740 (Max fixture thickness 105mm) ..	4.13	5.00	4.33	0.36	0.04	7.16	-	11.50	nr	12.65
Bolts; expanding; steel; plated; hook Rawlbolts; drilling masonry										
with nuts and washers; reference										
44 - 401 (M6H)...	1.00	5.00	1.05	0.24	0.03	4.82	-	5.88	nr	6.46
44 - 406 (M8H)...	1.24	5.00	1.30	0.26	0.03	5.19	-	6.49	nr	7.14
44 - 411 (M10H)..	1.75	5.00	1.84	0.30	0.04	6.06	-	7.90	nr	8.69
44 - 416 (M12H)..	2.98	5.00	3.12	0.32	0.04	6.43	-	9.55	nr	10.51
Bolts; expanding; steel; plated; eye Rawlbolts; drilling masonry										
with nuts and washers; reference										
44 - 432 (M6)..	1.00	5.00	1.05	0.24	0.03	4.82	-	5.88	nr	6.46
44 - 437 (M8)..	1.19	5.00	1.25	0.26	0.03	5.19	-	6.44	nr	7.08
44 - 442 (M10)..	1.75	5.00	1.84	0.30	0.04	6.06	-	7.90	nr	8.69
44 - 447 (M12)..	2.83	5.00	2.97	0.32	0.04	6.43	-	9.40	nr	10.34
Bolts; expanding; bolt projecting Rawl Throughbolts; drilling masonry										
with nuts and washers; reference										
RXPT06065/5 (M6 x 65mm)..	0.24	5.00	0.25	0.22	0.03	4.45	-	4.70	nr	5.17
RXPT06085/25 (M6 x 85mm)..	0.24	5.00	0.25	0.22	0.03	4.45	-	4.70	nr	5.17
RXPT08065/15 (M8 x 65mm)..	0.27	5.00	0.28	0.24	0.03	4.82	-	5.11	nr	5.62
RXPT08085/20 (M8 x 85mm)..	0.37	5.00	0.38	0.24	0.03	4.82	-	5.21	nr	5.73
RXPT08095/30 (M8 x 95mm)..	0.31	5.00	0.33	0.24	0.03	4.82	-	5.15	nr	5.67
RXPT08115/50 (M8 x 115mm)......................................	0.33	5.00	0.34	0.24	0.03	4.82	-	5.16	nr	5.68
RXPT10065/5 (M10 x 65mm)..	0.36	5.00	0.37	0.26	0.03	5.19	-	5.56	nr	6.12
RXPT10080/10 (M10 x 80mm)......................................	0.37	5.00	0.39	0.26	0.03	5.19	-	5.58	nr	6.14
RXPT10095/25 (M10 x 95mm)......................................	0.41	5.00	0.43	0.26	0.03	5.19	-	5.62	nr	6.18
RXPT10140/70 (M10 x 140mm)....................................	0.47	5.00	0.50	0.26	0.03	5.19	-	5.69	nr	6.26
RXPT12080/5 (M12 x 80mm)..	0.42	5.00	0.45	0.30	0.04	6.06	-	6.51	nr	7.16
RXPT12100/5 (M12 x 100mm)......................................	0.44	5.00	0.47	0.30	0.04	6.06	-	6.53	nr	7.18
RXPT12125/30 (M12 x 125mm)....................................	0.58	5.00	0.60	0.30	0.04	6.06	-	6.67	nr	7.33
RXPT12140/45 (M12 x 140mm)....................................	0.61	5.00	0.64	0.30	0.04	6.06	-	6.70	nr	7.37
RXPT12150/55 (M12 x 150mm)....................................	0.64	5.00	0.67	0.30	0.04	6.06	-	6.73	nr	7.41
RXPT16105/10 (M16 x 105mm)....................................	0.92	5.00	0.96	0.32	0.04	6.43	-	7.39	nr	8.13
RXPT16150/30 (M16 x 150mm)....................................	1.14	5.00	1.19	0.32	0.04	6.43	-	7.62	nr	8.38
RXPT16180/60 (M16 x 180mm)....................................	1.19	5.00	1.24	0.32	0.04	6.43	-	7.67	nr	8.44
RXPT16220/100 (M16 x 220mm)..................................	2.19	5.00	2.30	0.32	0.04	6.43	-	8.73	nr	9.60
RXPT20125/5 (M20 x 125mm)......................................	1.83	5.00	1.92	0.36	0.05	7.30	-	9.22	nr	10.14
RXPT20160/20 (M20 x 160mm)....................................	2.13	5.00	2.24	0.36	0.05	7.30	-	9.54	nr	10.50
RXPT20200/60 (M20 x 200mm)....................................	3.58	5.00	3.76	0.36	0.05	7.30	-	11.06	nr	12.16
RXPT20300/160 (M20 x 300mm)..................................	4.64	5.00	4.88	0.36	0.05	7.30	-	12.18	nr	13.39
RXPT24180/20 (M24 x 180mm)....................................	5.24	5.00	5.50	0.40	0.05	8.04	-	13.54	nr	14.89
RXPT24260/100 (M24 x 260mm)..................................	6.49	5.00	6.81	0.40	0.05	8.04	-	14.85	nr	16.33
Bolts; expanding; Stainless steel grade 316; bolt projecting Rawl Throughbolts; drilling masonry										
with nuts and washers; reference										
RXPTA406050/10 (M6 x 50mm)....................................	0.61	5.00	0.64	0.22	0.03	4.45	-	5.09	nr	5.60
RXPTA408075/10 (M8 x 75mm)....................................	0.77	5.00	0.81	0.24	0.03	4.82	-	5.64	nr	6.20
RXPTA408095/30 (M8 x 95mm)....................................	1.01	5.00	1.06	0.24	0.03	4.82	-	5.88	nr	6.47
RXPTA410080/10 (M10 x 80mm)..................................	1.37	5.00	1.44	0.26	0.03	5.19	-	6.63	nr	7.29
RXPTA410115/45 (M10 x 115mm)................................	1.60	5.00	1.68	0.26	0.03	5.19	-	6.87	nr	7.56
RXPTA410130/60 (M10 x 130mm)................................	1.77	5.00	1.86	0.26	0.03	5.19	-	7.05	nr	7.76
RXPTA412100/5 (M12 x 100mm)..................................	2.17	5.00	2.28	0.30	0.04	6.06	-	8.34	nr	9.17
RXPTA412125/30 (M12 x 125mm)................................	2.38	5.00	2.50	0.30	0.04	6.06	-	8.56	nr	9.42
RXPTA412150/55 (M12 x 150mm)................................	2.59	5.00	2.71	0.30	0.04	6.06	-	8.78	nr	9.65
RXPTA416180/60 (M16 x 180mm)................................	5.80	5.00	6.09	0.32	0.04	6.43	-	12.52	nr	13.77
RXPTA416220/100 (M16 x 220mm)..............................	12.62	5.00	13.25	0.32	0.04	6.43	-	19.68	nr	21.65
RXPTA420160/20 (M20 x 160mm)................................	9.19	5.00	9.65	0.36	0.05	7.30	-	16.95	nr	18.65
RXPTA420300/160 (M20 x 300mm)..............................	24.51	5.00	25.74	0.36	0.05	7.30	-	33.04	nr	36.35
RXPTA424260/100 (M24 x 260mm)..............................	52.11	5.00	54.71	0.40	0.05	8.04	-	62.75	nr	69.02

Labour hourly rates: (except Specialists) Craft Operatives 18.37 Labourer 13.76 Rates are national average prices. Refer to REGIONAL VARIATIONS for indicative levels of overall pricing in regions	MATERIALS			LABOUR				RATES		
	Del to Site £	Waste %	Material Cost £	Craft Optve Hrs	Lab Hrs	Labour Cost £	Sunds £	Nett Rate £	Unit	Gross rate (10%) £

METAL AND PLASTIC ACCESSORIES (Cont'd)

Metal fixings, fastenings and fittings (Cont'd)

Chemical anchors
  Chemical anchors; Kemfix capsules and standard studs; drilling masonry

capsule reference 60-428; stud reference 60-708; with nuts and washers (M8 x 110mm)	0.92	5.00	0.96	0.29	0.04	5.88	-	6.84	nr	7.52
capsule reference 60-430; stud reference 60-710; with nuts and washers (M10 x 130mm)	1.06	5.00	1.11	0.32	0.04	6.43	-	7.54	nr	8.29
capsule reference 60-432; stud reference 60-712; with nuts and washers (M12 x 160mm)	1.38	5.00	1.45	0.36	0.05	7.30	-	8.76	nr	9.63

Chemical anchors; Kemfix capsules and stainless steel studs; drilling masonry

capsule reference 60-428; stud reference R-STUDS-08110-A4; with nuts and washers (M8 x 110mm)	2.26	5.00	2.37	0.29	0.04	5.88	-	8.25	nr	9.07
capsule reference 60-430; stud reference R-STUDS-10130-A4; with nuts and washers (M10 x 130mm)	3.17	5.00	3.33	0.32	0.04	6.43	-	9.75	nr	10.73
capsule reference 60-432; stud reference R-STUDS-12160-A4; with nuts and washers (M12 x 160mm)	4.90	5.00	5.14	0.36	0.05	7.30	-	12.44	nr	13.69
capsule reference 60-436; stud reference R-STUDS-16190-A4; with nuts and washers (M16 x 190mm)	9.37	5.00	9.83	0.40	0.05	8.04	-	17.87	nr	19.66
capsule reference 60-442; stud reference R-STUDS-20260-A4; with nuts and washers (M20 x 260mm)	18.73	5.00	19.67	0.42	0.05	8.40	-	28.07	nr	30.88
capsule reference 60-446; stud reference R-STUDS-24300-A4; with nuts and washers (M24 x 295mm)	30.10	5.00	31.60	0.46	0.06	9.28	-	40.88	nr	44.97

Chemical anchors; Kemfix capsules and standard internal threaded sockets; drilling masonry

capsule reference 60-428; socket reference 60-623 (M8 x 90mm)	1.30	5.00	1.36	0.36	0.05	7.30	-	8.66	nr	9.53
capsule reference 60-430; socket reference 60-626 (M10 x 100mm)	1.89	5.00	1.98	0.40	0.05	8.04	-	10.02	nr	11.02
capsule reference 60-432; socket reference 60-628 (M12 x 80mm)	2.09	5.00	2.19	0.42	0.05	8.40	-	10.60	nr	11.66
capsule reference 60-436; socket reference 60-630 (M16 x 95mm)	3.31	5.00	3.47	0.46	0.06	9.28	-	12.75	nr	14.02

Chemcial anchors; Kemfix capsules and stainless steel internal threaded sockets; drilling masonry

capsule reference 60-428; socket reference 60-988 (M8 x 90mm)	2.95	5.00	3.09	0.36	0.05	7.30	-	10.40	nr	11.43
capsule reference 60-430; socket reference 60-992 (M10 x 100mm)	3.29	5.00	3.45	0.40	0.05	8.04	-	11.49	nr	12.63
capsule reference 60-432; socket reference 60-993 (M12 x 100mm)	4.15	5.00	4.35	0.42	0.05	8.40	-	12.76	nr	14.03
capsule reference 60-436; socket reference 60-995 (M16 x 125mm)	7.83	5.00	8.22	0.46	0.06	9.28	-	17.50	nr	19.25

Chemical anchors in low density material; Kemfix capsules, mesh sleeves and standard studs; drilling masonry

capsule reference 60-428; sleeve reference 60-805; stud reference 60-708; with nuts and washers (M8 x 110mm)	2.29	5.00	2.40	0.25	0.03	5.01	-	7.40	nr	8.15
capsule reference 60-430; sleeve reference 60-809; stud reference 60-710; with nuts and washers (M10 x 130mm)	2.60	5.00	2.73	0.32	0.04	6.43	-	9.16	nr	10.07
capsule reference 60-432; sleeve reference 60-115; stud reference 60-712; with nuts and washers (M12 x 160mm)	2.98	5.00	3.13	0.36	0.05	7.30	-	10.43	nr	11.47
capsule reference 60-436; sleeve reference 60-117; stud reference 60-716; with nuts and washers (M16 x 190mm)	3.30	5.00	3.47	0.42	0.05	8.40	-	11.87	nr	13.06

Chemical anchors in low density material; Kemfix capsules, mesh sleeves and stainless steel studs; drilling masonry

capsule reference 60-428; sleeve reference 60-805; stud reference RSTUDS-08110-A4; with nuts and washers (M8 x 110mm)	3.63	5.00	3.81	0.29	0.04	5.88	-	9.69	nr	10.65
capsule reference 60-430; sleeve reference 60-809; stud reference RSTUDS-10130-A4; with nuts and washers (M10 x 130mm)	4.71	5.00	4.94	0.32	0.04	6.43	-	11.37	nr	12.51
capsule reference 60-432; sleeve reference 60-115; stud reference RSTUDS-12160-A4; with nuts and washers (M12 x 160mm)	6.49	5.00	6.82	0.36	0.05	7.30	-	14.12	nr	15.53
capsule reference 60-436; sleeve reference 60-117; stud reference RSTUDS-16190-A4; with nuts and washers (M16 x 190mm)	10.97	5.00	11.52	0.42	0.05	8.40	-	19.92	nr	21.91

Chemical anchors in low density material; Kemfix capsules, mesh sleeves and standard internal threaded sockets; drilling masonry

capsule reference 60-428; sleeve reference 60-805; socket reference 60-623 (M8 x 90mm)	2.67	5.00	2.80	0.29	0.04	5.88	-	8.68	nr	9.55
capsule reference 60-430; sleeve reference 60-809; socket reference 60-626 (M10 x 100mm)	3.43	5.00	3.60	0.32	0.04	6.43	-	10.03	nr	11.03
capsule reference 60-432; sleeve reference 60-113; socket reference 60-674 (M12 x 100mm)	3.68	5.00	3.87	0.36	0.05	7.30	-	11.17	nr	12.28
capsule reference 60-436; sleeve reference 60-117; socket reference 60-676 (M16 x 125mm)	4.91	5.00	5.16	0.40	0.05	8.04	-	13.19	nr	14.51

Chemical anchors in low density material; Kemfix capsules, mesh sleeves and stainless steel internal threaded sockets; drilling masonry

capsule reference 60-428; sleeve reference 60-805; socket reference 60-988 (M8 x 90mm)	4.32	5.00	4.53	0.29	0.04	5.88	-	10.41	nr	11.45
capsule reference 60-430; sleeve reference 60-809; socket reference 60-992 (M10 x 100mm)	4.82	5.00	5.06	0.32	0.04	6.43	-	11.49	nr	12.64
capsule reference 60-436; sleeve reference 60-117; socket reference 60-995 (M16 x 125mm)	9.44	5.00	9.91	0.40	0.05	8.04	-	17.95	nr	19.74

Spit Maxima high performance chemical anchors; capsules and zinc coated steel studs; drilling masonry

capsule reference M8 stud reference SM8; with nuts and washers	1.10	5.00	1.15	0.28	0.03	5.56	-	6.71	nr	7.38

Labour hourly rates: (except Specialists) Craft Operatives 18.37 Labourer 13.76 Rates are national average prices. Refer to REGIONAL VARIATIONS for indicative levels of overall pricing in regions	MATERIALS			LABOUR				RATES		
	Del to Site	Waste	Material Cost	Craft Optve	Lab	Labour Cost	Sunds	Nett Rate	Unit	Gross rate (10%)
	£	%	£	Hrs	Hrs	£	£	£		£

METAL AND PLASTIC ACCESSORIES (Cont'd)

Metal fixings, fastenings and fittings (Cont'd)

Chemical anchors (Cont'd)

Spit Maxima high performance chemical anchors; capsules and zinc coated steel studs; drilling masonry (Cont'd)

	Del to Site £	Waste %	Material Cost £	Craft Optve Hrs	Lab Hrs	Labour Cost £	Sunds £	Nett Rate £	Unit	Gross rate £
capsule reference M10; stud reference SM10; with nuts and washers	1.33	5.00	1.39	0.31	0.04	6.25	-	7.64	nr	8.40
capsule reference M12; stud reference SM12; with nuts and washers	1.68	5.00	1.77	0.35	0.03	6.84	-	8.61	nr	9.47
capsule reference M16; stud reference SM16; with nuts and washers	2.54	5.00	2.66	0.38	0.03	7.39	-	10.06	nr	11.06
capsule reference M20; stud reference SM20; with nuts and washers	5.25	5.00	5.52	0.40	0.04	7.90	-	13.41	nr	14.76
capsule reference M24; stud reference SM24; with nuts and washers	9.61	5.00	10.09	0.44	0.04	8.63	-	18.72	nr	20.59
Spit Maxima high performance chemical anchors; capsules and stainless steel studs Grade 316 (A4); drilling masonry										
capsule reference M8 stud reference SM8i; with nuts and washers	2.30	5.00	2.42	0.28	0.03	5.56	-	7.97	nr	8.77
capsule reference M10; stud reference SM10i; with nuts and washers	3.42	5.00	3.60	0.31	0.03	6.11	-	9.70	nr	10.67
capsule reference M12; stud reference SM12i; with nuts and washers	5.47	5.00	5.74	0.35	0.03	6.84	-	12.58	nr	13.84
capsule reference M16; stud reference SM16i; with nuts and washers	9.27	5.00	9.73	0.38	0.03	7.39	-	17.12	nr	18.84
capsule reference M20; stud reference SM20i; with nuts and washers	19.49	5.00	20.47	0.40	0.04	7.90	-	28.36	nr	31.20
capsule reference M24; stud reference SM24i; with nuts and washers	40.59	5.00	42.62	0.44	0.04	8.63	-	51.25	nr	56.38
Holding down bolts or assemblies; mild steel										
rag or indented bolts; M 10 with nuts and washers										
100mm long	1.48	5.00	1.55	0.10	-	1.84	-	3.39	nr	3.73
160mm long	1.54	5.00	1.62	0.11	-	2.02	-	3.64	nr	4.00
rag or indented bolts; M 12 with nuts and washers										
100mm long	1.60	5.00	1.68	0.10	-	1.84	-	3.52	nr	3.87
160mm long	1.85	5.00	1.94	0.13	-	2.39	-	4.33	nr	4.76
200mm long	2.08	5.00	2.18	0.14	-	2.57	-	4.76	nr	5.23
rag or indented bolts; M 16 with nuts and washers										
120mm long	2.41	5.00	2.53	0.11	-	2.02	-	4.55	nr	5.01
160mm long	2.76	5.00	2.90	0.14	-	2.57	-	5.47	nr	6.02
200mm long	3.22	5.00	3.38	0.16	-	2.94	-	6.32	nr	6.95
holding down bolt assembly; M 20 bolt with 100 x 100 x 10mm plate washer tack welded to head; with nuts and washers										
300mm long	5.55	5.00	5.83	0.28	-	5.14	-	10.97	nr	12.07
350mm long	6.33	5.00	6.65	0.32	-	5.88	-	12.52	nr	13.78
450mm long	9.64	5.00	10.12	0.55	-	10.10	-	20.22	nr	22.24
High strength friction grip bolts; BS 14399 Part 1 - general grade										
M 16; with nuts and washers										
50mm long	1.20	5.00	1.26	0.09	-	1.65	-	2.91	nr	3.20
60mm long	1.44	5.00	1.51	0.09	-	1.65	-	3.17	nr	3.48
75mm long	1.80	5.00	1.89	0.10	-	1.84	-	3.73	nr	4.10
80mm long	1.92	5.00	2.02	0.10	-	1.84	-	3.85	nr	4.24
90mm long	2.16	5.00	2.27	0.12	-	2.20	-	4.47	nr	4.92
M 20; with nuts and washers										
60mm long	2.12	5.00	2.23	0.09	-	1.65	-	3.88	nr	4.27
70mm long	2.26	5.00	2.37	0.10	-	1.84	-	4.21	nr	4.63
80mm long	2.35	5.00	2.47	0.10	-	1.84	-	4.30	nr	4.73
90mm long	2.47	5.00	2.59	0.12	-	2.20	-	4.80	nr	5.28
M 24; with nuts and washers										
65mm long	3.12	5.00	3.28	0.13	-	2.39	-	5.66	nr	6.23
80mm long	3.84	5.00	4.03	0.13	-	2.39	-	6.42	nr	7.06
90mm long	4.32	5.00	4.54	0.14	-	2.57	-	7.11	nr	7.82
100mm long	4.80	5.00	5.04	0.14	-	2.57	-	7.61	nr	8.37
110mm long	5.28	5.00	5.54	0.16	-	2.94	-	8.48	nr	9.33
130mm long	6.24	5.00	6.55	0.17	-	3.12	-	9.67	nr	10.64
150mm long	8.16	5.00	8.57	0.18	-	3.31	-	11.87	nr	13.06
M 30; with nuts and washers										
80mm long	3.84	5.00	4.03	0.15	-	2.76	-	6.79	nr	7.47
100mm long	6.60	5.00	6.93	0.17	-	3.12	-	10.05	nr	11.06
Steel; galvanized; joist hangers										
Joist hangers; BAT Building Products Ltd; (building in where required included elsewhere)										
SPH type S, for 50 x 100mm joist	1.54	2.50	1.58	-	-	-	-	1.58	nr	1.74
SPH type S, for 50 x 125mm joist	1.60	2.50	1.64	-	-	-	-	1.64	nr	1.80
SPH type S, for 50 x 150mm joist	1.26	2.50	1.29	-	-	-	-	1.29	nr	1.42
SPH type S, for 50 x 175mm joist	1.34	2.50	1.37	-	-	-	-	1.37	nr	1.51
SPH type S, for 50 x 200mm joist	1.47	2.50	1.51	-	-	-	-	1.51	nr	1.66
SPH type S, for 50 x 225mm joist	1.58	2.50	1.62	-	-	-	-	1.62	nr	1.78
SPH type S, for 50 x 250mm joist	2.52	2.50	2.58	-	-	-	-	2.58	nr	2.84
SPH type S, for 75 x 100mm joist	2.35	2.50	2.41	-	-	-	-	2.41	nr	2.65
SPH type S, for 75 x 125mm joist	2.40	2.50	2.46	-	-	-	-	2.46	nr	2.71
SPH type S, for 75 x 150mm joist	2.75	2.50	2.82	-	-	-	-	2.82	nr	3.10
SPH type S, for 75 x 175mm joist	2.58	2.50	2.64	-	-	-	-	2.64	nr	2.91
SPH type S, for 75 x 200mm joist	2.75	2.50	2.82	-	-	-	-	2.82	nr	3.10
SPH type S, for 75 x 225mm joist	2.94	2.50	3.01	-	-	-	-	3.01	nr	3.31
SPH type S, for 75 x 250mm joist	3.13	2.50	3.21	-	-	-	-	3.21	nr	3.53
SPH type R, for 50 x 100mm joist	4.37	2.50	4.48	-	-	-	-	4.48	nr	4.93
SPH type R, for 50 x 125mm joist	4.54	2.50	4.65	-	-	-	-	4.65	nr	5.12
SPH type R, for 50 x 150mm joist	4.35	2.50	4.46	-	-	-	-	4.46	nr	4.90
SPH type R, for 50 x 175mm joist	4.19	2.50	4.29	-	-	-	-	4.29	nr	4.72

Labour hourly rates: (except Specialists) Craft Operatives 18.37 Labourer 13.76 Rates are national average prices. Refer to REGIONAL VARIATIONS for indicative levels of overall pricing in regions	MATERIALS			LABOUR				RATES		
	Del to Site	Waste	Material Cost	Craft Optve	Lab	Labour Cost	Sunds	Nett Rate	Unit	Gross rate (10%)
	£	%	£	Hrs	Hrs	£	£	£		£
METAL AND PLASTIC ACCESSORIES (Cont'd)										
Metal fixings, fastenings and fittings (Cont'd)										
Steel; galvanized; joist hangers (Cont'd)										
Joist hangers; BAT Building Products Ltd; (building in where required included elsewhere) (Cont'd)										
SPH type R, for 50 x 200mm joist	5.50	2.50	5.64	-	-	-	-	5.64	nr	6.20
SPH type R, for 50 x 225mm joist	6.00	2.50	6.15	-	-	-	-	6.15	nr	6.76
SPH type R, for 50 x 250mm joist	7.15	2.50	7.33	-	-	-	-	7.33	nr	8.06
SPH type R, for 63 x 100mm joist	6.45	2.50	6.61	-	-	-	-	6.61	nr	7.27
SPH type R, for 63 x 125mm joist	6.60	2.50	6.76	-	-	-	-	6.76	nr	7.44
SPH type R, for 63 x 150mm joist	6.90	2.50	7.07	-	-	-	-	7.07	nr	7.78
SPH type R, for 63 x 175mm joist	7.10	2.50	7.28	-	-	-	-	7.28	nr	8.01
SPH type R, for 63 x 200mm joist	7.25	2.50	7.43	-	-	-	-	7.43	nr	8.17
SPH type R, for 63 x 225mm joist	6.13	2.50	6.28	-	-	-	-	6.28	nr	6.91
SPH type R, for 63 x 250mm joist	8.25	2.50	8.46	-	-	-	-	8.46	nr	9.30
SPH type ST for 50 x 100mm joist	8.66	2.50	8.88	-	-	-	-	8.88	nr	9.76
SPH type ST for 50 x 125mm joist	8.72	2.50	8.94	-	-	-	-	8.94	nr	9.83
SPH type ST for 50 x 150mm joist	8.22	2.50	8.43	-	-	-	-	8.43	nr	9.27
SPH type ST for 50 x 175mm joist	8.62	2.50	8.84	-	-	-	-	8.84	nr	9.72
SPH type ST for 50 x 200mm joist	8.98	2.50	9.20	-	-	-	-	9.20	nr	10.12
SPH type ST for 50 x 225mm joist	8.75	2.50	8.97	-	-	-	-	8.97	nr	9.87
SPH type ST for 50 x 250mm joist	12.93	2.50	13.25	-	-	-	-	13.25	nr	14.58
Speedy Minor type for the following size joists; fixing with nails										
38 x 100mm	0.39	2.50	0.40	0.15	0.02	3.03	0.03	3.46	nr	3.81
50 x 100mm	0.35	2.50	0.36	0.15	0.02	3.03	0.03	3.42	nr	3.76
Speedy short leg type for the following size joists fixing with nails										
38 x 175mm	1.22	2.50	1.25	0.15	0.02	3.03	0.03	4.31	nr	4.74
50 x 175mm	1.08	2.50	1.11	0.15	0.02	3.03	0.03	4.17	nr	4.58
Speedy Standard Leg type for the following size joists; fixing with nails										
38 x 100mm	0.60	2.50	0.62	0.15	0.02	3.03	0.03	3.68	nr	4.04
50 x 175mm	0.60	2.50	0.62	0.15	0.02	3.03	0.03	3.68	nr	4.04
63 x 225mm	0.63	2.50	0.65	0.15	0.02	3.03	0.03	3.71	nr	4.08
75 x 225mm	0.63	2.50	0.65	0.17	0.02	3.40	0.05	4.09	nr	4.50
100 x 225mm	0.66	2.50	0.68	0.17	0.02	3.40	0.05	4.12	nr	4.53
Steel; galvanized; joist struts										
Herringbone joist struts; BAT Building Products Ltd; to suit joists at the following centres; fixing with nails										
400mm	0.43	5.00	0.45	0.25	0.03	5.01	0.06	5.52	m	6.07
450mm	0.49	5.00	0.51	0.25	0.03	5.01	0.06	5.58	m	6.14
600mm	0.55	5.00	0.58	0.20	0.03	4.09	0.06	4.72	m	5.20
Steel; galvanized; truss plates and framing anchors										
Truss plates; BAT Building Products Ltd; fixing with nails										
51 x 114mm	0.25	5.00	0.26	0.35	0.04	6.98	0.09	7.33	nr	8.07
76 x 254mm	0.65	5.00	0.68	1.00	0.13	20.16	0.24	21.08	nr	23.19
114 x 152mm	0.72	5.00	0.76	1.00	0.13	20.16	0.24	21.15	nr	23.27
114 x 254mm	1.30	5.00	1.36	1.50	0.19	30.17	0.36	31.89	nr	35.08
152 x 152mm	0.85	5.00	0.89	1.60	0.20	32.14	0.39	33.43	nr	36.77
Framing anchors; BAT Building Products Ltd.										
type A; fixing with nails	0.65	5.00	0.68	0.20	0.02	3.95	0.05	4.68	nr	5.14
type B; fixing with nails	0.65	5.00	0.68	0.20	0.02	3.95	0.05	4.68	nr	5.14
type C; fixing with nails	0.47	5.00	0.49	0.20	0.02	3.95	0.05	4.49	nr	4.94
Steel timber connectors; BS EN 912										
split ring connectors table 1										
63mm diameter	2.70	5.00	2.84	0.08	0.01	1.61	-	4.44	nr	4.89
100 mm diameter	13.30	5.00	13.96	0.08	0.01	1.61	-	15.57	nr	17.13
shear plate connectors, table 2										
67mm diameter	2.70	5.00	2.84	0.08	0.01	1.61	-	4.44	nr	4.89
single sided round toothed-plate connectors, table 4										
38mm diameter	0.38	5.00	0.40	0.08	0.01	1.61	-	2.01	nr	2.21
51mm diameter	0.31	5.00	0.33	0.08	0.01	1.61	-	1.93	nr	2.13
64mm diameter	0.52	5.00	0.55	0.08	0.01	1.61	-	2.15	nr	2.37
76mm diameter	0.62	5.00	0.65	0.08	0.01	1.61	-	2.26	nr	2.48
double sided round toothed-plate connector, table 4										
38mm diameter	0.42	5.00	0.44	0.08	0.01	1.61	-	2.05	nr	2.25
51mm diameter	0.35	5.00	0.37	0.08	0.01	1.61	-	1.97	nr	2.17
64mm diameter	0.62	5.00	0.65	0.08	0.01	1.61	-	2.26	nr	2.48
76mm diameter	0.66	5.00	0.69	0.08	0.01	1.61	-	2.30	nr	2.53
Cast iron connectors; BS EN 912										
shear plate connectors, table 2										
102mm diameter	13.04	5.00	13.69	0.12	0.01	2.34	-	16.03	nr	17.64

Labour hourly rates: (except Specialists) Craft Operatives 18.37 Labourer 13.76 Rates are national average prices. Refer to REGIONAL VARIATIONS for indicative levels of overall pricing in regions	MATERIALS			LABOUR				RATES		
	Del to Site	Waste	Material Cost	Craft Optve	Lab	Labour Cost	Sunds	Nett Rate		Gross rate (10%)
	£	%	£	Hrs	Hrs	£	£	£	Unit	£
BITUMINOUS FELTS										
Roofing; felt BS EN 13707, comprising 1 layer perfored underlay, 1 layer sanded underlay, 1 layer mineral surface top sheet										
Roof coverings; exceeding 500mm wide; layers of felt -3; bonding with hot bitumen compound to timber base										
pitch 7.5 degrees from horizontal	14.32	15.00	16.79	0.46	0.06	9.28	-	26.06	m²	28.67
pitch 40 degrees from horizontal	14.32	15.00	16.79	0.76	0.09	15.20	-	31.99	m²	35.18
pitch 75 degrees from horizontal	14.32	15.00	16.79	0.92	0.12	18.55	-	35.34	m²	38.87
Roof coverings; exceeding 500mm wide; layers of felt -3; bonding with hot bitumen compound to cement and sand or concrete base										
pitch 7.5 degrees from horizontal	14.32	15.00	16.79	0.46	0.06	9.28	-	26.06	m²	28.67
pitch 40 degrees from horizontal	14.32	15.00	16.79	0.76	0.09	15.20	-	31.99	m²	35.18
pitch 75 degrees from horizontal	14.32	15.00	16.79	0.92	0.12	18.55	-	35.34	m²	38.87
Eaves trim; aluminium; silver anodized butt joints over matching sleeve pieces 200mm long; fixing to timber base with aluminium alloy screws; bedding in mastic										
63.5mm wide x 44.4mm face depth	3.09	5.00	3.24	0.40	-	7.35	2.25	12.84	m	14.12
63.5mm wide x 76.2mm face depth	3.71	5.00	3.89	0.40	-	7.35	2.37	13.61	m	14.97
104.8mm wide x 44.4mm face depth	4.40	5.00	4.62	0.40	-	7.35	2.37	14.34	m	15.78
104.8mm wide x 76.2mm face depth	31.42	5.00	33.00	0.50	-	9.19	2.65	44.84	m	49.32
Eaves trim; glass fibre reinforced butt joints; fixing to timber base with stainless steel screws; bedding in mastic	29.23	5.00	30.69	0.35	-	6.43	2.65	39.77	m	43.75
Upstands; skirtings; layers of felt - 3; bonding with hot bitumen compound to brickwork base										
girth O - 200mm	2.86	15.00	3.36	0.28	0.03	5.56	-	8.91	m	9.80
girth 200 - 400mm	5.73	15.00	6.71	0.46	0.06	9.28	-	15.99	m	17.59
Collars around pipes; standard and the like; layers of felt -1; bonding with hot bitumen compound to metal base										
50mm diameter x 150mm long; hole in 3 layer covering	1.18	15.00	1.36	0.17	0.02	3.40	-	4.76	nr	5.23
100mm diameter x 150mm long; hole in 3 layer covering	1.54	15.00	1.78	0.22	0.03	4.45	-	6.24	nr	6.86
Roof ventilators										
plastic; setting in position	4.99	2.50	5.11	0.85	-	15.61	0.82	21.55	nr	23.71
aluminium; setting in position	8.14	2.50	8.34	0.85	-	15.61	0.82	24.78	nr	27.26
Roofing; BS EN 13707, all layers type 3B -1.8 kg/m², 75mm laps; fully bonding layers with hot bitumen bonding compound										
Roof coverings; exceeding 500mm wide; layers of felt -2; bonding with hot bitumen compound to timber base										
pitch 7.5 degrees from horizontal	6.65	15.00	7.85	0.38	0.05	7.67	-	15.52	m²	17.07
pitch 40 degrees from horizontal	6.65	15.00	7.85	0.63	0.08	12.67	-	20.53	m²	22.58
pitch 75 degrees from horizontal	6.65	15.00	7.85	0.76	0.09	15.20	-	23.05	m²	25.36
Roof coverings; exceeding 500mm wide; layers of felt -2; bonding with hot bitumen compound to cement and sand or concrete base										
pitch 7.5 degrees from horizontal	6.65	15.00	7.85	0.38	0.05	7.67	-	15.52	m²	17.07
pitch 40 degrees from horizontal	6.65	15.00	7.85	0.63	0.08	12.67	-	20.53	m²	22.58
pitch 75 degrees from horizontal	6.65	15.00	7.85	0.76	0.09	15.20	-	23.05	m²	25.36
Roof coverings; exceeding 500mm wide; layers of felt -3; bonding with hot bitumen compound to timber base										
pitch 7.5 degrees from horizontal	9.97	15.00	11.78	0.55	0.07	11.07	-	22.84	m²	25.13
pitch 40 degrees from horizontal	9.97	15.00	11.78	0.92	0.12	18.55	-	30.33	m²	33.36
pitch 75 degrees from horizontal	9.97	15.00	11.78	1.10	0.14	22.13	-	33.91	m²	37.30
Roof coverings; exceeding 500mm wide; layers of felt -3; bonding with hot bitumen compound to cement and sand or concrete base										
pitch 7.5 degrees from horizontal	9.97	15.00	11.78	0.55	0.07	11.07	-	22.84	m²	25.13
pitch 40 degrees from horizontal	9.97	15.00	11.78	0.92	0.12	18.55	-	30.33	m²	33.36
pitch 75 degrees from horizontal	9.97	15.00	11.78	1.10	0.14	22.13	-	33.91	m²	37.30
Roofing; felt BS EN 13707, bottom and intermediate layers type 3B -1.8 kg/m², top layer green mineral surface 3.8 kg/m², 75mm laps; fully bonding layers with hot bitumen bonding compound										
Roof coverings; exceeding 500mm wide; layers of felt -2; bonding with hot bitumen compound to timber base										
pitch 7.5 degrees from horizontal	7.77	15.00	9.14	0.38	0.05	7.67	-	16.81	m²	18.49
pitch 40 degrees from horizontal	7.77	15.00	9.14	0.63	0.08	12.67	-	21.81	m²	24.00
pitch 75 degrees from horizontal	7.77	15.00	9.14	0.76	0.09	15.20	-	24.34	m²	26.77
Roof coverings; exceeding 500mm wide; layers of felt -2; bonding with hot bitumen compound to cement and sand or concrete base										
pitch 7.5 degrees from horizontal	7.77	15.00	9.14	0.38	0.05	7.67	-	16.81	m²	18.49
pitch 40 degrees from horizontal	7.77	15.00	9.14	0.63	0.08	12.67	-	21.81	m²	24.00

157

Labour hourly rates: (except Specialists) Craft Operatives 18.37 Labourer 13.76 Rates are national average prices. Refer to REGIONAL VARIATIONS for indicative levels of overall pricing in regions	MATERIALS			LABOUR				RATES		
	Del to Site	Waste	Material Cost	Craft Optve	Lab	Labour Cost	Sunds	Nett Rate	Unit	Gross rate (10%)
	£	%	£	Hrs	Hrs	£	£	£		£
BITUMINOUS FELTS (Cont'd)										
Roofing; felt BS EN 13707, bottom and intermediate layers type 3B -1.8 kg/m², top layer green mineral surface 3.8 kg/m², 75mm laps; fully bonding layers with hot bitumen bonding compound (Cont'd)										
Roof coverings; exceeding 500mm wide; layers of felt -2; bonding with hot bitumen compound to cement and sand or concrete base (Cont'd)										
pitch 75 degrees from horizontal	7.77	15.00	9.14	0.76	0.09	15.20	-	24.34	m²	26.77
Roof coverings; exceeding 500mm wide; layers of felt -3; bonding with hot bitumen compound to timber base										
pitch 7.5 degrees from horizontal	11.09	15.00	13.07	0.55	0.07	11.07	-	24.13	m²	26.55
pitch 40 degrees from horizontal	11.09	15.00	13.07	0.92	0.12	18.55	-	31.62	m²	34.78
pitch 75 degrees from horizontal	11.09	15.00	13.07	1.10	0.14	22.13	-	35.20	m²	38.72
Roof coverings; exceeding 500mm wide; layers of felt -3; bonding with hot bitumen compound to cement and sand or concrete base										
pitch 7.5 degrees from horizontal	11.09	15.00	13.07	0.55	0.07	11.07	-	24.13	m²	26.55
pitch 40 degrees from horizontal	11.09	15.00	13.07	0.92	0.12	18.55	-	31.62	m²	34.78
pitch 75 degrees from horizontal	11.09	15.00	13.07	1.10	0.14	22.13	-	35.20	m²	38.72
Upstands; skirtings; layers of felt - 3; bonding with hot bitumen compound to brickwork base										
girth O - 200mm	2.22	15.00	2.61	0.28	0.03	5.56	-	8.17	m	8.99
girth 200 - 400mm	4.44	15.00	5.23	0.46	0.06	9.28	-	14.50	m	15.95
Collars around pipes; standard and the like; layers of felt -1; bonding with hot bitumen compound to metal base										
50mm diameter x 150mm long; hole in 3 layer covering	0.75	15.00	0.86	0.17	0.02	3.40	-	4.26	nr	4.69
100mm diameter x 150mm long; hole in 3 layer covering	1.02	15.00	1.19	0.22	0.03	4.45	-	5.64	nr	6.21
Roofing; sheet, IKO/Ruberoid Building Products hot bonded system, bottom layer perforated slate underlay (formerly type 3G); intermediate layer glass fibre underlay (formerly lype 3B); top layer glass fibre SBS mineral surfaced cap sheet, 50mm side laps, 75mm end laps; partially bonding bottom layer, fully bonding other layers in hot bonding bitumen										
Roof coverings; exceeding 500mm wide; layers of sheet -3; bonding with hot bitumen compound to timber base										
pitch not exceeding 5.0 degrees from horizontal	17.54	15.00	20.63	0.55	0.28	13.96	-	34.59	m²	38.04
Roof coverings; exceeding 500mm wide; layers of sheet -3; bonding with hot bitumen compound to cement and sand or concrete base										
pitch not exceeding 5.0 degrees from horizontal	17.54	15.00	20.63	0.55	0.28	13.96	-	34.59	m²	38.04
Roof coverings; exceeding 500mm wide; layers of sheet -3; bonding with hot bitumen compound to metal base										
pitch not exceeding 5.0 degrees from horizontal	17.54	15.00	20.63	0.55	0.28	13.96	-	34.59	m²	38.04
Upstands; skirtings; layers of sheet -2; bonding with hot bitumen compound to brickwork base										
girth O - 200mm	2.51	15.00	2.95	0.28	0.05	5.83	-	8.78	m	9.66
girth 200 - 400mm	5.02	15.00	5.90	0.46	0.06	9.28	-	15.17	m	16.69
Roofing; sheet, IKO/Ruberoid Building Products hot bonded system, bottom layer perforated slate underlay (formerly type 3G); intermediate layer glass fibre underlay (formerly lype 3B); top layer glass fibre SBS mineral surfaced cap sheet, 50mm side laps, 75mm end laps; partially bonding bottom layer, fully bonding other layers in hot bonding bitumen; 25mm Kingspan Thermaroof TR21 (PIR) Board on a vapour control layer										
Roof coverings; exceeding 500mm wide; layers of sheet -3; bonding with hot bitumen compound to timber base										
pitch not exceeding 5.0 degrees from horizontal	24.94	15.00	29.13	1.12	0.36	25.53	1.66	56.33	m²	61.96
Roof coverings; exceeding 500mm wide; layers of sheet -3; bonding with hot bitumen compound to cement and sand or concrete base										
pitch not exceeding 5.0 degrees from horizontal	24.94	15.00	29.13	1.18	0.36	26.63	1.66	57.43	m²	63.17
Roof coverings; exceeding 500mm wide; layers of sheet -3; bonding with hot bitumen compound to metal base										
pitch not exceeding 5.0 degrees from horizontal	24.94	15.00	29.13	1.16	0.36	26.26	1.66	57.06	m²	62.77
Roofing; sheet, IKO/Ruberoid Building Products hot bonded system, bottom layer perforated slate underlay (formerly type 3G); intermediate layer glass fibre underlay (formerly lype 3B); top layer glass fibre SBS mineral surfaced cap sheet, 50mm side laps, 75mm end laps; partially bonding bottom layer, fully bonding other layers in hot bonding bitumen; 50mm Kingspan Thermaboard TR21 (PIR) Board										
Roof coverings; exceeding 500mm wide; layers of sheet -3; bonding with hot bitumen compound to timber base										
pitch not exceeding 5.0 degrees from horizontal	28.00	15.00	32.66	1.20	0.36	27.00	1.66	61.32	m²	67.45
Roof coverings; exceeding 500mm wide; layers of sheet -3; bonding with hot bitumen compound to cement and sand or concrete base										
pitch not exceeding 5.0 degrees from horizontal	28.00	15.00	32.66	1.26	0.36	28.10	1.66	62.42	m²	68.67
Roof coverings; exceeding 500mm wide; layers of sheet -3; bonding with hot bitumen compound to metal base										
pitch not exceeding 5.0 degrees from horizontal	28.00	15.00	32.66	1.24	0.36	27.73	1.66	62.06	m²	68.26

Labour hourly rates: (except Specialists) Craft Operatives 18.37 Labourer 13.76 Rates are national average prices. Refer to REGIONAL VARIATIONS for indicative levels of overall pricing in regions	MATERIALS			LABOUR				RATES		
	Del to Site £	Waste %	Material Cost £	Craft Optve Hrs	Lab Hrs	Labour Cost £	Sunds £	Nett Rate £	Unit	Gross rate (10%) £
BITUMINOUS FELTS (Cont'd)										
Roofing; sheet, IKO/Ruberoid Building Products hot bonded system, bottom layer perforated slate underlay (formerly type 3G); intermediate layer glass fibre underlay (formerly lype 3B); top layer glass fibre SBS mineral surfaced cap sheet, 50mm side laps, 75mm end laps; partially bonding bottom layer, fully bonding other layers in hot bonding bitumen; 100mm Kingspan Thermaboard TR21 (PIR) Board										
Roof coverings; exceeding 500mm wide; layers of sheet -3; bonding with hot bitumen compound to timber base pitch not exceeding 5.0 degrees from horizontal	37.79	15.00	41.06	1.34	0.40	30.12	1.66	72.84	m²	80.13
Roof coverings; exceeding 500mm wide; layers of sheet -3; bonding with hot bitumen compound to cement and sand or concrete base pitch not exceeding 5.0 degrees from horizontal	37.79	15.00	41.06	1.36	0.40	30.49	1.66	73.21	m²	80.53
Roof coverings; exceeding 500mm wide; layers of sheet -3; bonding with hot bitumen compound to metal base pitch not exceeding 5.0 degrees from horizontal	37.79	15.00	41.06	1.34	0.40	30.12	1.66	72.84	m²	80.13
Roofing; sheet, IKO/Ruberoid Building Products Torch-on system, bottom layer Universal T-O Underlay; top layer Permatorch T-O slate surfaced cap sheet, 75mm side and end laps; fully bonding both layers (by torching)										
Roof coverings; exceeding 500mm wide; layers of sheet -3; bonding to timber base pitch not exceeding 5.0 degrees from horizontal	8.29	15.00	9.53	0.55	0.28	13.96	4.30	27.80	m²	30.58
Roof coverings; exceeding 500mm wide; layers of sheet -3; bonding to cement and sand or concrete base pitch not exceeding 5.0 degrees from horizontal	8.29	15.00	9.53	0.55	0.28	13.96	4.30	27.80	m²	30.58
Roof coverings; exceeding 500mm wide; layers of sheet -3; bonding to metal base pitch not exceeding 5.0 degrees from horizontal	8.29	15.00	9.53	0.59	0.28	14.69	4.30	28.53	m²	31.38

Upstands; skirtings; layers of sheet -2; bonding with hot bitumen compound to brickwork base
    girth 0 - 200mm ....
    girth 200 - 400mm ....
	1.66
	3.32
		15.00
		15.00
			1.91
			3.81
				0.28
				0.46
					0.05
					0.06
						5.83
						9.28
							0.87
							1.74
								8.61
								14.83
									m
									m
										9.47
										16.31

Roofing; sheet, IKO/Ruberoid Building Products Torch-on system, bottom layer Universal T-O Underlay; top layer Permatorch T-O slate surfaced cap sheet, 75mm side and end laps; fully bonding both layers (by torching), 25mm Kingspan Thermaroof TR21 (PIR Board) insulation with bitumen primed surface										
Roof coverings; exceeding 500mm wide; layers of sheet -3; bonding to timber base pitch not exceeding 5.0 degrees from horizontal	15.68	15.00	18.04	1.12	0.36	25.53	4.30	47.87	m²	52.66
Roof coverings; exceeding 500mm wide; layers of sheet -3; bonding to cement and sand or concrete base pitch not exceeding 5.0 degrees from horizontal	15.68	15.00	18.04	1.12	0.36	25.53	4.30	47.87	m²	52.66
Roof coverings; exceeding 500mm wide; layers of sheet -3; bonding to metal base pitch not exceeding 5.0 degrees from horizontal	15.68	15.00	18.04	1.17	0.36	26.45	4.30	48.79	m²	53.67
Roofing; sheet, IKO/Ruberoid Building Products Torch-on system, bottom layer Universal T-O Underlay; top layer Permatorch T-O slate surfaced cap sheet, 75mm side and end laps; fully bonding both layers (by torching), 50mm Kingspan Thermaroof TR21 (PIR Board) insulation with bitumen primed surface; Superbar vapour control layer										
Roof coverings; exceeding 500mm wide; layers of sheet -3; bonding to timber base pitch not exceeding 5.0 degrees from horizontal	18.75	15.00	21.56	1.20	0.36	27.00	4.30	52.86	m²	58.15
Roofing; sheet, Ruberoid Building Products Torch-on system, bottom layer Universal T-O Underlay; top layer Permatorch T-O slate surfaced cap sheet, 75mm side and end laps; fully bonding both layers (by torching), 100mm Kingspan Thermaroof TR21 (PIR Board) insulation with bitumen primed surface; Superbar vapour control layer										
Roof coverings; exceeding 500mm wide; layers of sheet -3; bonding to timber base pitch not exceeding 5.0 degrees from horizontal	28.53	15.00	29.96	1.20	0.36	27.00	4.30	61.27	m²	67.39
PLASTIC SHEETS										
Sarnafil polymeric waterproofing membrane ref: S327-12EL; Sarnabar mechanically fastened system; 85mm thick; Sarnatherm S CFC & HCFC free (0.25 U-value) rigid urethane insulation board mechanically fastened; Sarnavap 1000E vapour control layer loose laid all laps sealed										
Roof Coverings; exceeding 500mm wide Pitch not exceeding 5 degrees; to metal decking or the like..........	-	-	Specialist	-	-	Specialist	-	84.60	m2	93.06

159

Labour hourly rates: (except Specialists) Craft Operatives 18.37 Labourer 13.76 Rates are national average prices. Refer to REGIONAL VARIATIONS for indicative levels of overall pricing in regions	MATERIALS			LABOUR				RATES		
	Del to Site £	Waste %	Material Cost £	Craft Optve Hrs	Lab Hrs	Labour Cost £	Sunds £	Nett Rate £	Unit	Gross rate (10%) £
PLASTIC SHEETS (Cont'd)										
Sarnafil polymeric waterproofing membrane ref: G410-12ELF fleece backed membrane; fully adhered system; 90mm thick Sarnatherm G CFC & HCFC free (0.25 U-value) insulation board bedded in hot bitumen; BS EN 13707 type 5U felt vapour control layer bonded in hot bitumen; prime concrete with spirit bitumen priming solution										
Roof Coverings; exceeding 500mm wide										
Pitch not exceeding 5 degrees; to concrete base or the like	-	-	Specialist	-	-	Specialist	-	108.17	m2	118.98
Eaves Detail; Sarnametal drip edge to gutter; standard Sarnafil detail 1.3										
not exceeding 200 girth	-	-	Specialist	-	-	Specialist	-	32.00	m	35.19
Upstands; skirting to brickwork with galvanised steel counter flashing to top edge; standard Sarnafil detail 2.3										
not exceeding 200 girth	-	-	Specialist	-	-	Specialist	-	47.85	m	52.64
200 - 400 girth	-	-	Specialist	-	-	Specialist	-	51.04	m	56.15
400 - 600 girth	-	-	Specialist	-	-	Specialist	-	57.85	m	63.64
Skirting/Upstands; Skirting to brickwork with Sarnametal Raglet to chase; standard Sarnafil detail 2.8										
not exceeding 200 girth	-	-	Specialist	-	-	Specialist	-	66.63	m	73.29
200 - 400 girth	-	-	Specialist	-	-	Specialist	-	69.82	m	76.81
400 - 600 girth	-	-	Specialist	-	-	Specialist	-	76.67	m	84.33
Coverings to Kerb; Parapet Flashing; Sarnatrim 50 deep on face 100 fixing arm; standard Sarnafil detail 1.1										
not exceeding 200 girth	-	-	Specialist	-	-	Specialist	-	58.69	m	64.56
200 - 400 girth	-	-	Specialist	-	-	Specialist	-	61.89	m	68.08
400 - 600 girth	-	-	Specialist	-	-	Specialist	-	68.71	m	75.59
Outlets and Dishing to Gullies										
fix Sarnadrain pvc rainwater outlet; 110 diameter; weld membrane to same; fit plastic leafguard	-	-	-	-	-	-	1.66	177.66	no	195.43
Collars around Pipe Standards and The Like										
50 diameter x 150 high	-	-	Specialist	-	-	Specialist	-	55.17	no	60.69
100 diameter x 150 high	-	-	Specialist	-	-	Specialist	-	55.17	no	60.69

This section continues on the next page

Labour hourly rates: (except Specialists) Craft Operatives 23.99 Labourer 13.76 Rates are national average prices. Refer to REGIONAL VARIATIONS for indicative levels of overall pricing in regions	MATERIALS			LABOUR				RATES		
	Del to Site	Waste	Material Cost	Craft Optve	Lab	Labour Cost	Sunds	Nett Rate	Unit	Gross rate (10%)
	£	%	£	Hrs	Hrs	£	£	£		£
SHEET METALS; LEAD										
Sheet lead										
Technical data										
Code 3, 1.32mm thick, 14.97 kg/m², colour code green										
Code 4, 1.80mm thick, 20.41 kg/m², colour code blue										
Code 5, 2.24mm thick, 25.40 kg/m², colour code red										
Code 6, 2.65mm thick, 30.05 kg/m², colour code black										
Code 7, 3.15mm thick, 35.72 kg/m², colour code white										
Code 8, 3.55mm thick, 40.26 kg/m², colour code orange										
Milled lead sheet, BS EN 12588										
Nr 3 roof coverings; exceeding 500mm wide; fixing to timber with milled lead cleats and galvanized screws; sloping										
pitch 7.5 degrees from horizontal	26.31	5.00	27.62	4.40	-	105.56	0.38	133.55	m²	146.91
pitch 40 degrees from horizontal	26.31	5.00	27.62	5.05	-	121.15	0.38	149.15	m²	164.06
pitch 75 degrees from horizontal	26.31	5.00	27.62	5.70	-	136.74	0.38	164.74	m²	181.21
Nr 4 roof coverings; exceeding 500mm wide; fixing to timber with milled lead cleats and galvanized screws; sloping										
pitch 7.5 degrees from horizontal	35.75	5.00	37.53	5.05	-	121.15	0.45	159.13	m²	175.05
pitch 40 degrees from horizontal	35.75	5.00	37.53	5.70	-	136.74	0.45	174.73	m²	192.20
pitch 75 degrees from horizontal	35.75	5.00	37.53	6.35	-	152.34	0.45	190.32	m²	209.35
Nr 5 roof coverings; exceeding 500mm wide; fixing to timber with milled lead cleats and galvanized screws; sloping										
pitch 7.5 degrees from horizontal	44.61	5.00	46.84	5.70	-	136.74	0.54	184.12	m²	202.53
pitch 40 degrees from horizontal	44.61	5.00	46.84	6.35	-	152.34	0.54	199.71	m²	219.69
pitch 75 degrees from horizontal	44.61	5.00	46.84	6.95	-	166.73	0.54	214.11	m²	235.52
Extra										
oil patination to surfaces	0.07	10.00	0.08	0.23	-	5.52	-	5.60	m²	6.16
Nr 3 wall coverings; exceeding 500mm wide; fixing to timber with milled lead cleats and galvanized screws										
vertical	26.31	5.00	27.62	8.00	-	191.92	0.38	219.92	m²	241.91
Nr 4 wall coverings; exceeding 500mm wide; fixing to timber with milled lead cleats and galvanized screws										
vertical	35.75	5.00	37.53	8.65	-	207.51	0.45	245.50	m²	270.05
Nr 5 wall coverings; exceeding 500mm wide; fixing to timber with milled lead cleats and galvanized screws										
vertical	44.61	5.00	46.84	9.25	-	221.91	0.54	269.29	m²	296.21
Extra for forming										
Nr 3 edges										
welted	-	-	-	0.20	-	4.80	-	4.80	m	5.28
beaded	-	-	-	0.20	-	4.80	-	4.80	m	5.28
Nr 4 edges										
welted	-	-	-	0.20	-	4.80	-	4.80	m	5.28
beaded	-	-	-	0.20	-	4.80	-	4.80	m	5.28
Nr 5 edges										
welted	-	-	-	0.20	-	4.80	-	4.80	m	5.28
beaded	-	-	-	0.20	-	4.80	-	4.80	m	5.28
Nr 3 seams										
leadburned	1.50	-	1.50	0.23	-	5.52	-	7.02	m	7.72
Nr 4 seams										
leadburned	2.00	-	2.00	0.23	-	5.52	-	7.52	m	8.27
Nr 5 seams										
leadburned	2.50	-	2.50	0.23	-	5.52	-	8.02	m	8.82
Nr 3 dressings										
corrugated roofing; fibre cement; down corrugations	-	-	-	0.29	-	6.96	-	6.96	m	7.65
corrugated roofing; fibre cement; across corrugations	-	-	-	0.29	-	6.96	-	6.96	m	7.65
glass and glazing bars; timber	-	-	-	0.29	-	6.96	-	6.96	m	7.65
Nr 4 dressings										
corrugated roofing; fibre cement; down corrugations	-	-	-	0.29	-	6.96	-	6.96	m	7.65
corrugated roofing; fibre cement; across corrugations	-	-	-	0.29	-	6.96	-	6.96	m	7.65
glass and glazing bars; timber	-	-	-	0.29	-	6.96	-	6.96	m	7.65
Nr 5 dressings										
corrugated roofing; fibre cement; down corrugations	-	-	-	0.29	-	6.96	-	6.96	m	7.65
corrugated roofing; fibre cement; across corrugations	-	-	-	0.29	-	6.96	-	6.96	m	7.65
glass and glazing bars; timber	-	-	-	0.29	-	6.96	-	6.96	m	7.65
Flashings; horizontal										
Nr 3; 150mm lapped joints; fixing to masonry with milled lead clips and lead wedges										
150mm girth	4.09	5.00	4.30	0.46	-	11.04	0.38	15.71	m	17.28
240mm girth	6.52	5.00	6.85	0.74	-	17.75	0.38	24.98	m	27.47
300mm girth	8.19	5.00	8.60	0.92	-	22.07	0.38	31.05	m	34.15
Nr 4; 150mm lapped joints; fixing to masonry with milled lead clips and lead wedges										
150mm girth	5.61	5.00	5.89	0.50	-	11.99	0.45	18.34	m	20.17
240mm girth	8.80	5.00	9.24	0.82	-	19.67	0.45	29.36	m	32.29
300mm girth	11.22	5.00	11.78	1.02	-	24.47	0.45	36.70	m	40.37
Nr 5; 150mm lapped joints; fixing to masonry with milled lead clips and lead wedges										
150mm girth	6.98	5.00	7.33	0.58	-	13.91	0.54	21.78	m	23.96

SHEET ROOF COVERINGS

Labour hourly rates: (except Specialists) Craft Operatives 23.99 Labourer 13.76 Rates are national average prices. Refer to REGIONAL VARIATIONS for indicative levels of overall pricing in regions	MATERIALS			LABOUR				RATES		
	Del to Site £	Waste %	Material Cost £	Craft Optve Hrs	Lab Hrs	Labour Cost £	Sunds £	Nett Rate £	Unit	Gross rate (10%) £
SHEET METALS; LEAD (Cont'd)										
Milled lead sheet, BS EN 12588 (Cont'd)										
Flashings; horizontal (Cont'd)										
Nr 5; 150mm lapped joints; fixing to masonry with milled lead clips and lead wedges (Cont'd)										
240mm girth	11.07	5.00	11.63	0.92	-	22.07	0.54	34.24	m	37.66
300mm girth	13.95	5.00	14.65	1.15	-	27.59	0.54	42.78	m	47.06
Nr 3; 150mm lapped joints; fixing to timber with copper nails										
150mm girth	4.09	5.00	4.30	0.46	-	11.04	0.44	15.77	m	17.35
240mm girth	6.52	5.00	6.85	0.74	-	17.75	0.44	25.04	m	27.54
300mm girth	8.19	5.00	8.60	0.92	-	22.07	0.44	31.11	m	34.22
Nr 4; 150mm lapped joints; fixing to timber with copper nails										
150mm girth	5.61	5.00	5.89	0.50	-	11.99	0.44	18.32	m	20.15
240mm girth	8.80	5.00	9.24	0.82	-	19.67	0.44	29.34	m	32.28
300mm girth	11.22	5.00	11.78	1.02	-	24.47	0.44	36.69	m	40.36
Nr 5; 150mm lapped joints; fixing to timber with copper nails										
150mm girth	6.98	5.00	7.33	0.58	-	13.91	0.44	21.67	m	23.84
240mm girth	11.07	5.00	11.63	0.92	-	22.07	0.44	34.13	m	37.54
300mm girth	13.95	5.00	14.65	1.15	-	27.59	0.44	42.67	m	46.94
Flashings; stepped										
Nr 3; 150mm lapped joints; fixing to masonry with milled lead clips and lead wedges										
180mm girth	4.85	5.00	5.10	0.74	-	17.75	0.52	23.37	m	25.71
240mm girth	6.52	5.00	6.85	0.98	-	23.51	0.52	30.88	m	33.97
300mm girth	8.19	5.00	8.60	1.24	-	29.75	0.52	38.87	m	42.76
Nr 4; 150mm lapped joints; fixing to masonry with milled lead clips and lead wedges										
180mm girth	6.67	5.00	7.01	0.82	-	19.67	0.69	27.37	m	30.11
240mm girth	8.80	5.00	9.24	1.10	-	26.39	0.69	36.32	m	39.95
300mm girth	11.22	5.00	11.78	1.36	-	32.63	0.69	45.10	m	49.61
Nr 5; 150mm lapped joints; fixing to masonry with milled lead clips and lead wedges										
180mm girth	8.34	5.00	8.76	0.92	-	22.07	0.82	31.65	m	34.82
240mm girth	11.07	5.00	11.63	1.22	-	29.27	0.82	41.72	m	45.89
300mm girth	13.95	5.00	14.65	1.52	-	36.46	0.82	51.94	m	57.13
Aprons; horizontal										
Nr 3; 150mm lapped joints; fixing to masonry with milled lead clips and lead wedges										
150mm girth	4.09	5.00	4.30	0.46	-	11.04	0.38	15.71	m	17.28
240mm girth	6.52	5.00	6.85	0.74	-	17.75	0.38	24.98	m	27.47
300mm girth	8.19	5.00	8.60	0.92	-	22.07	0.38	31.05	m	34.15
450mm girth	12.28	5.00	12.90	1.38	-	33.11	0.52	46.53	m	51.18
Nr 4; 150mm lapped joints; fixing to masonry with milled lead clips and lead wedges										
150mm girth	5.61	5.00	5.89	0.50	-	11.99	0.45	18.34	m	20.17
240mm girth	8.80	5.00	9.24	0.82	-	19.67	0.45	29.36	m	32.29
300mm girth	11.22	5.00	11.78	1.02	-	24.47	0.45	36.70	m	40.37
450mm girth	16.68	5.00	17.52	1.52	-	36.46	0.69	54.67	m	60.14
Nr 5; 150mm lapped joints; fixing to masonry with milled lead clips and lead wedges										
150mm girth	6.98	5.00	7.33	0.58	-	13.91	0.54	21.78	m	23.96
240mm girth	11.07	5.00	11.63	0.92	-	22.07	0.54	34.24	m	37.66
300mm girth	13.95	5.00	14.65	1.15	-	27.59	0.54	42.78	m	47.06
450mm girth	20.93	5.00	21.98	1.72	-	41.26	0.82	64.06	m	70.47
Hips; sloping; dressing over slating and tiling										
Nr 3; 150mm lapped joints; fixing to timber with milled lead clips										
240mm girth	6.52	5.00	6.85	0.74	-	17.75	0.38	24.98	m	27.47
300mm girth	8.19	5.00	8.60	0.92	-	22.07	0.38	31.05	m	34.15
450mm girth	12.28	5.00	12.90	1.38	-	33.11	0.52	46.53	m	51.18
Nr 4; 150mm lapped joints; fixing to timber with milled lead clips										
240mm girth	8.80	5.00	9.24	0.82	-	19.67	0.45	29.36	m	32.29
300mm girth	11.22	5.00	11.78	1.02	-	24.47	0.45	36.70	m	40.37
450mm girth	16.68	5.00	17.52	1.52	-	36.46	0.69	54.67	m	60.14
Nr 5; 150mm lapped joints; fixing to timber with milled lead clips										
240mm girth	11.07	5.00	11.63	0.92	-	22.07	0.54	34.24	m	37.66
300mm girth	13.95	5.00	14.65	1.15	-	27.59	0.54	42.78	m	47.06
450mm girth	20.93	5.00	21.98	1.72	-	41.26	0.82	64.06	m	70.47
Kerbs; horizontal										
Nr 3; 150mm lapped joints; fixing to timber with copper nails										
240mm girth	6.52	5.00	6.85	0.74	-	17.75	0.38	24.98	m	27.47
300mm girth	8.19	5.00	8.60	0.92	-	22.07	0.38	31.05	m	34.15
450mm girth	12.28	5.00	12.90	1.38	-	33.11	0.52	46.53	m	51.18
Nr 4; 150mm lapped joints; fixing to timber with copper nails										
240mm girth	8.80	5.00	9.24	0.82	-	19.67	0.45	29.36	m	32.29
300mm girth	11.22	5.00	11.78	1.02	-	24.47	0.45	36.70	m	40.37
450mm girth	16.68	5.00	17.52	1.52	-	36.46	0.69	54.67	m	60.14
Nr 5; 150mm lapped joints; fixing to timber with copper nails										
240mm girth	11.07	5.00	11.63	0.92	-	22.07	0.54	34.24	m	37.66
300mm girth	13.95	5.00	14.65	1.15	-	27.59	0.54	42.78	m	47.06
450mm girth	20.93	5.00	21.98	1.72	-	41.26	0.82	64.06	m	70.47

Labour hourly rates: (except Specialists) Craft Operatives 23.99 Labourer 13.76 Rates are national average prices. Refer to REGIONAL VARIATIONS for indicative levels of overall pricing in regions	MATERIALS			LABOUR				RATES		
	Del to Site	Waste	Material Cost	Craft Optve	Lab	Labour Cost	Sunds	Nett Rate	Unit	Gross rate (10%)
	£	%	£	Hrs	Hrs	£	£	£		£
SHEET METALS; LEAD (Cont'd)										
Milled lead sheet, BS EN 12588 (Cont'd)										
Ridges; horizontal; dressing over slating and tiling										
Nr 3; 150mm lapped joints; fixing to timber with milled lead clips										
240mm girth	6.52	5.00	6.85	0.74	-	17.75	0.38	24.98	m	27.47
300mm girth	8.19	5.00	8.60	0.92	-	22.07	0.38	31.05	m	34.15
450mm girth	12.28	5.00	12.90	1.38	-	33.11	0.52	46.53	m	51.18
Nr 4; 150mm lapped joints; fixing to timber with milled lead clips										
240mm girth	8.80	5.00	9.24	0.82	-	19.67	0.45	29.36	m	32.29
300mm girth	11.22	5.00	11.78	1.02	-	24.47	0.45	36.70	m	40.37
450mm girth	16.68	5.00	17.52	1.52	-	36.46	0.69	54.67	m	60.14
Nr 5; 150mm lapped joints; fixing to timber with milled lead clips										
240mm girth	11.07	5.00	11.63	0.92	-	22.07	0.54	34.24	m	37.66
300mm girth	13.95	5.00	14.65	1.15	-	27.59	0.54	42.78	m	47.06
450mm girth	20.93	5.00	21.98	1.72	-	41.26	0.82	64.06	m	70.47
Cavity gutters										
Nr 3; 150mm lapped joints; bedding in cement mortar (1:3)										
225mm girth	6.22	5.00	6.53	0.74	-	17.75	0.69	24.97	m	27.47
360mm girth	9.86	5.00	10.35	1.10	-	26.39	1.03	37.78	m	41.55
Valleys; sloping										
Nr 3; dressing over tilting fillets -1; 150mm lapped joints; fixing to timber with copper nails										
240mm girth	6.52	5.00	6.85	0.74	-	17.75	0.38	24.98	m	27.47
300mm girth	8.19	5.00	8.60	0.92	-	22.07	0.38	31.05	m	34.15
450mm girth	12.28	5.00	12.90	1.38	-	33.11	0.52	46.53	m	51.18
Nr 4; dressing over tilting fillets -1; 150mm lapped joints; fixing to timber with copper nails										
240mm girth	8.80	5.00	9.24	0.82	-	19.67	0.45	29.36	m	32.29
300mm girth	11.22	5.00	11.78	1.02	-	24.47	0.45	36.70	m	40.37
450mm girth	16.68	5.00	17.52	1.52	-	36.46	0.69	54.67	m	60.14
Nr 5; dressing over tilting fillets -1; 150mm lapped joints; fixing to timber with copper nails										
240mm girth	11.07	5.00	11.63	0.92	-	22.07	0.54	34.24	m	37.66
300mm girth	13.95	5.00	14.65	1.15	-	27.59	0.54	42.78	m	47.06
450mm girth	20.93	5.00	21.98	1.72	-	41.26	0.82	64.06	m	70.47
Spot items										
Nr 3 collars around pipes, standards and the like; 150mm long; soldered joints to metal covering										
50mm diameter	22.50	5.00	23.62	1.00	-	23.99	-	47.61	nr	52.38
100mm diameter	22.50	5.00	23.62	1.15	-	27.59	-	51.21	nr	56.33
Nr 4 collars around pipes, standards and the like; 150mm long; soldered joints to metal covering										
50mm diameter	24.29	5.00	25.50	1.15	-	27.59	-	53.09	nr	58.40
100mm diameter	24.29	5.00	25.50	1.15	-	27.59	-	53.09	nr	58.40
Nr 5 collars around pipes, standards and the like; 150mm long; soldered joints to metal covering										
50mm diameter	30.50	5.00	32.02	1.15	-	27.59	-	59.61	nr	65.57
100mm diameter	30.50	5.00	32.02	1.15	-	27.59	-	59.61	nr	65.57
Nr 3 dots										
cast lead	1.80	5.00	1.89	0.77	-	18.47	-	20.36	nr	22.40
soldered	17.50	5.00	18.38	0.86	-	20.63	-	39.01	nr	42.91
Nr 4 dots										
cast lead	2.40	5.00	2.52	0.77	-	18.47	-	20.99	nr	23.09
soldered	17.50	5.00	18.38	0.86	-	20.63	-	39.01	nr	42.91
Nr 5 dots										
cast lead	3.30	5.00	3.46	0.77	-	18.47	-	21.94	nr	24.13
soldered	17.50	5.00	18.38	0.86	-	20.63	-	39.01	nr	42.91
Fittings										
Nr 3 soakers and slates; handed to others for fixing										
180 x 180mm	0.97	5.00	1.02	0.23	-	5.52	-	6.54	nr	7.19
180 x 300mm	1.62	5.00	1.71	0.23	-	5.52	-	7.22	nr	7.95
450 x 450mm	6.09	5.00	6.39	0.23	-	5.52	-	11.91	nr	13.10
Nr 4 soakers and slates; handed to others for fixing										
180 x 180mm	1.34	5.00	1.41	0.23	-	5.52	-	6.92	nr	7.62
180 x 300mm	2.23	5.00	2.34	0.23	-	5.52	-	7.86	nr	8.65
450 x 450mm	8.53	5.00	8.95	0.23	-	5.52	-	14.47	nr	15.92
Nr 5 soakers and slates; handed to others for fixing										
180 x 180mm	1.71	5.00	1.79	0.23	-	5.52	-	7.31	nr	8.04
180 x 300mm	2.84	5.00	2.98	0.23	-	5.52	-	8.50	nr	9.35
450 x 450mm	10.51	5.00	11.03	0.23	-	5.52	-	16.55	nr	18.20
SHEET METALS; ALUMINIUM										
Aluminium sheet, BS EN 485 grade S1BO, commercial purity										
0.60mm thick roof coverings; exceeding 500mm wide; fixing to timber with aluminium cleats and aluminium alloy screws; sloping										
pitch 7.5 degrees from horizontal	15.81	5.00	16.60	5.05	-	121.15	0.21	137.96	m²	151.76
pitch 40 degrees from horizontal	15.81	5.00	16.60	5.70	-	136.74	0.21	153.56	m²	168.91
pitch 75 degrees from horizontal	15.81	5.00	16.60	6.35	-	152.34	0.21	169.15	m²	186.06

Labour hourly rates: (except Specialists) Craft Operatives 23.99 Labourer 13.76 Rates are national average prices. Refer to REGIONAL VARIATIONS for indicative levels of overall pricing in regions	MATERIALS			LABOUR				RATES		
	Del to Site	Waste	Material Cost	Craft Optve	Lab	Labour Cost	Sunds	Nett Rate	Unit	Gross rate (10%)
	£	%	£	Hrs	Hrs	£	£	£		£
SHEET METALS; ALUMINIUM (Cont'd)										
Aluminium sheet, BS EN 485 grade S1BO, commercial purity (Cont'd)										
0.60mm thick edges										
welted	-	-	-	0.20	-	4.80	-	4.80	m	5.28
beaded	-	-	-	0.20	-	4.80	-	4.80	m	5.28
Flashings; horizontal 0.60mm thick; 150mm lapped joints; fixing to masonry with aluminium clips and wedges										
150mm girth	2.06	5.00	2.17	0.51	-	12.23	0.21	14.61	m	16.07
240mm girth	3.30	5.00	3.46	0.82	-	19.67	0.21	23.35	m	25.68
300mm girth	4.12	5.00	4.33	1.02	-	24.47	0.21	29.01	m	31.91
Flashings; stepped 0.60mm thick; 150mm lapped joints; fixing to masonry with aluminium clips and wedges										
180mm girth	2.47	5.00	2.60	0.82	-	19.67	0.31	22.59	m	24.84
240mm girth	3.30	5.00	3.46	1.09	-	26.15	0.31	29.93	m	32.92
300mm girth	4.12	5.00	4.33	1.37	-	32.87	0.31	37.51	m	41.26
Aprons; horizontal 0.60mm thick; 150mm lapped joints; fixing to masonry with aluminium clips and wedges										
150mm girth	2.06	5.00	2.17	0.51	-	12.23	0.21	14.61	m	16.07
180mm girth	2.47	5.00	2.60	0.82	-	19.67	0.21	22.48	m	24.73
240mm girth	3.30	5.00	3.46	1.02	-	24.47	0.21	28.14	m	30.96
450mm girth	6.19	5.00	6.50	1.53	-	36.70	0.31	43.52	m	47.87
Hips; sloping 0.60mm thick; 150mm lapped joints; fixing to timber with aluminium clips										
240mm girth	3.30	5.00	3.46	0.82	-	19.67	0.21	23.35	m	25.68
300mm girth	4.12	5.00	4.33	1.02	-	24.47	0.21	29.01	m	31.91
450mm girth	6.19	5.00	6.50	1.53	-	36.70	0.31	43.52	m	47.87
Kerbs; horizontal 0.60mm thick; 150mm lapped joints; fixing to timber with aluminium clips										
240mm girth	3.30	5.00	3.46	0.82	-	19.67	0.21	23.35	m	25.68
300mm girth	4.12	5.00	4.33	1.02	-	24.47	0.21	29.01	m	31.91
450mm girth	6.19	5.00	6.50	1.53	-	36.70	0.31	43.52	m	47.87
Ridges; horizontal 0.60mm thick; 150mm lapped joints; fixing to timber with aluminium clips										
240mm girth	3.30	5.00	3.46	0.82	-	19.67	0.21	23.35	m	25.68
300mm girth	4.12	5.00	4.33	1.02	-	24.47	0.21	29.01	m	31.91
450mm girth	6.19	5.00	6.50	1.53	-	36.70	0.31	43.52	m	47.87
Valleys; sloping 0.60mm thick; 150mm lapped joints; fixing to timber with aluminium clips										
240mm girth	3.30	5.00	3.46	0.82	-	19.67	0.21	23.35	m	25.68
300mm girth	4.12	5.00	4.33	1.02	-	24.47	0.21	29.01	m	31.91
450mm girth	6.19	5.00	6.50	1.53	-	36.70	0.31	43.52	m	47.87
0.60mm thick soakers and slates handed to others for fixing										
180 x 180mm	0.45	5.00	0.47	0.23	-	5.52	-	5.99	nr	6.58
180 x 300mm	1.24	5.00	1.30	0.23	-	5.52	-	6.82	nr	7.50
SHEET METALS; COPPER										
Copper sheet; BS EN 1172										
0.60mm thick roof coverings; exceeding 500mm wide; fixing to timber with copper cleats and copper nails; sloping										
pitch 7.5 degrees from horizontal	78.26	5.00	82.18	5.05	-	121.15	0.62	203.94	m²	224.34
pitch 40 degrees from horizontal	78.26	5.00	82.18	5.70	-	136.74	0.62	219.54	m²	241.49
pitch 75 degrees from horizontal	78.26	5.00	82.18	6.35	-	152.34	0.62	235.13	m²	258.64
0.70mm thick roof coverings; exceeding 500mm wide; fixing to timber with copper cleats and copper nails; sloping										
pitch 7.5 degrees from horizontal	86.25	5.00	90.56	5.70	-	136.74	0.78	228.09	m²	250.89
pitch 40 degrees from horizontal	86.25	5.00	90.56	6.35	-	152.34	0.78	243.68	m²	268.05
pitch 75 degrees from horizontal	86.25	5.00	90.56	6.95	-	166.73	0.78	258.07	m²	283.88
Extra for forming										
0.60mm thick edges										
welted	-	-	-	0.17	-	4.08	-	4.08	m	4.49
beaded	-	-	-	0.17	-	4.08	-	4.08	m	4.49
0.70mm thick edges										
welted	-	-	-	0.17	-	4.08	-	4.08	m	4.49
beaded	-	-	-	0.17	-	4.08	-	4.08	m	4.49
Flashings; horizontal 0.70mm thick; 150mm lapped joints; fixing to masonry with copper clips and wedges										
150mm girth	11.25	5.00	11.81	0.51	-	12.23	0.62	24.66	m	27.13

Labour hourly rates: (except Specialists) Craft Operatives 23.99 Labourer 13.76 Rates are national average prices. Refer to REGIONAL VARIATIONS for indicative levels of overall pricing in regions	MATERIALS			LABOUR				RATES		
	Del to Site	Waste	Material Cost	Craft Optve	Lab	Labour Cost	Sunds	Nett Rate	Unit	Gross rate (10%)
	£	%	£	Hrs	Hrs	£	£	£		£
SHEET METALS; COPPER (Cont'd)										
Copper sheet; BS EN 1172 (Cont'd)										
Flashings; horizontal (Cont'd) 0.70mm thick; 150mm lapped joints; fixing to masonry with copper clips and wedges (Cont'd)										
240mm girth	18.00	5.00	18.90	0.82	-	19.67	0.62	39.19	m	43.11
300mm girth	22.50	5.00	23.62	1.02	-	24.47	0.62	48.71	m	53.58
Flashings; stepped 0.70mm thick; 150mm lapped joints; fixing to masonry with copper clips and wedges										
180mm girth	13.50	5.00	14.17	0.82	-	19.67	0.93	34.78	m	38.25
240mm girth	18.00	5.00	18.90	1.09	-	26.15	0.93	45.98	m	50.58
300mm girth	22.50	5.00	23.62	1.37	-	32.87	0.93	57.42	m	63.16
Aprons; horizontal 0.70mm thick; 150mm lapped joints; fixing to timber with copper clips										
150mm girth	11.25	5.00	11.81	0.51	-	12.23	0.62	24.66	m	27.13
240mm girth	18.00	5.00	18.90	0.82	-	19.67	0.62	39.19	m	43.11
300mm girth	22.50	5.00	23.62	1.02	-	24.47	0.62	48.71	m	53.58
450mm girth	33.75	5.00	35.44	1.53	-	36.70	0.93	73.07	m	80.38
Hips; sloping 0.70mm thick; 150mm lapped joints; fixing to timber with copper clips										
240mm girth	18.00	5.00	18.90	0.71	-	17.03	0.54	36.47	m	40.12
300mm girth	22.50	5.00	23.62	1.02	-	24.47	0.62	48.71	m	53.58
450mm girth	33.75	5.00	35.44	1.53	-	36.70	0.93	73.07	m	80.38
Kerbs; horizontal 0.70mm thick; 150mm lapped joints; fixing to timber with copper clips										
240mm girth	18.00	5.00	18.90	0.82	-	19.67	0.62	39.19	m	43.11
300mm girth	22.50	5.00	23.62	1.02	-	24.47	0.62	48.71	m	53.58
450mm girth	33.75	5.00	35.44	1.53	-	36.70	0.93	73.07	m	80.38
Ridges; horizontal 0.70mm thick; 150mm lapped joints; fixing to timber with copper clips										
240mm girth	18.00	5.00	18.90	0.82	-	19.67	0.62	39.19	m	43.11
300mm girth	22.50	5.00	23.62	1.02	-	24.47	0.62	48.71	m	53.58
450mm girth	33.75	5.00	35.44	1.53	-	36.70	0.93	73.07	m	80.38
Valleys; sloping 0.70mm thick; 150mm lapped joints; fixing to timber with copper clips										
240mm girth	18.00	5.00	18.90	0.82	-	19.67	0.62	39.19	m	43.11
300mm girth	22.50	5.00	23.62	1.02	-	24.47	0.62	48.71	m	53.58
450mm girth	33.75	5.00	35.44	1.53	-	36.70	0.93	73.07	m	80.38
SHEET METALS; ZINC										
Zinc alloy sheet; BS EN 988										
0.8mm thick roof coverings; exceeding 500mm wide; roll cap; fixing to timber with zinc clips; sloping										
pitch 7.5 degrees from horizontal	37.88	5.00	39.77	5.70	-	136.74	0.30	176.82	m²	194.50
pitch 40 degrees from horizontal	37.88	5.00	39.77	6.35	-	152.34	0.30	192.41	m²	211.65
pitch 75 degrees from horizontal	37.88	5.00	39.77	6.95	-	166.73	0.30	206.80	m²	227.48
Extra for forming 0.65mm thick edges										
welted	-	-	-	0.20	-	4.80	-	4.80	m	5.28
beaded	-	-	-	0.20	-	4.80	-	4.80	m	5.28
0.8mm thick edges										
welted	-	-	-	0.20	-	4.80	-	4.80	m	5.28
beaded	-	-	-	0.20	-	4.80	-	4.80	m	5.28
Flashings; horizontal 0.8mm thick; 150mm lapped joints; fixing to masonry with zinc clips and wedges										
150mm girth	5.68	5.00	5.97	0.51	-	12.23	0.22	18.43	m	20.27
240mm girth	9.09	5.00	9.55	0.82	-	19.67	0.22	29.44	m	32.39
300mm girth	11.36	5.00	11.93	1.02	-	24.47	0.22	36.63	m	40.29
Flashings; stepped 0.8mm thick; 150mm lapped joints; fixing to masonry with zinc clips and wedges										
180mm girth	6.82	5.00	7.16	0.71	-	17.03	0.30	24.49	m	26.94
240mm girth	9.09	5.00	9.55	1.09	-	26.15	0.34	36.04	m	39.64
300mm girth	11.36	5.00	11.93	1.37	-	32.87	0.34	45.14	m	49.66
Aprons; horizontal 0.8mm thick; 150mm lapped joints; fixing to masonry with zinc clips and wedges										
150mm girth	5.68	5.00	5.97	0.92	-	22.07	0.47	28.50	m	31.35
240mm girth	9.09	5.00	9.55	0.82	-	19.67	0.22	29.44	m	32.39
300mm girth	11.36	5.00	11.93	1.02	-	24.47	0.22	36.63	m	40.29
450mm girth	17.05	5.00	17.90	1.53	-	36.70	0.34	54.95	m	60.44

Labour hourly rates: (except Specialists) Craft Operatives 23.99 Labourer 13.76 Rates are national average prices. Refer to REGIONAL VARIATIONS for indicative levels of overall pricing in regions	MATERIALS			LABOUR				RATES		
	Del to Site	Waste	Material Cost	Craft Optve	Lab	Labour Cost	Sunds	Nett Rate	Unit	Gross rate (10%)
	£	%	£	Hrs	Hrs	£	£	£		£
SHEET METALS; ZINC (Cont'd)										
Zinc alloy sheet; BS EN 988 (Cont'd)										
Hips; sloping										
0.8mm thick; 150mm lapped joints; fixing to timber with zinc clips										
240mm girth	9.09	5.00	9.55	0.82	-	19.67	0.22	29.44	m	32.39
300mm girth	11.36	5.00	11.93	1.02	-	24.47	0.22	36.63	m	40.29
450mm girth	17.05	5.00	17.90	1.53	-	36.70	0.34	54.95	m	60.44
Kerbs; horizontal										
0.8mm thick; 150mm lapped joints; fixing to timber with zinc clips										
240mm girth	9.09	5.00	9.55	0.82	-	19.67	0.22	29.44	m	32.39
300mm girth	11.36	5.00	11.93	1.02	-	24.47	0.22	36.63	m	40.29
450mm girth	17.05	5.00	17.90	1.53	-	36.70	0.34	54.95	m	60.44
Ridges; horizontal										
0.8mm thick; 150mm lapped joints; fixing to timber with zinc clips										
240mm girth	9.09	5.00	9.55	0.82	-	19.67	0.22	29.44	m	32.39
300mm girth	11.36	5.00	11.93	1.02	-	24.47	0.22	36.63	m	40.29
450mm girth	17.05	5.00	17.90	1.53	-	36.70	0.34	54.95	m	60.44
Valleys; sloping										
0.8mm thick; 150mm lapped joints; fixing to timber with zinc clips										
240mm girth	9.09	5.00	9.55	0.82	-	19.67	0.22	29.44	m	32.39
300mm girth	11.36	5.00	11.93	1.02	-	24.47	0.22	36.63	m	40.29
450mm girth	17.05	5.00	17.90	1.53	-	36.70	0.34	54.95	m	60.44
Fittings										
0.8mm thick soakers and slates -; handed to others for fixing										
180 x 180mm	1.23	5.00	1.29	0.23	-	5.52	-	6.81	nr	7.49
180 x 300mm	2.05	5.00	2.15	0.23	-	5.52	-	7.67	nr	8.43
450 x 450mm	7.67	5.00	8.05	0.23	-	5.52	-	13.57	nr	14.93
0.8mm thick collars around pipes, standards and the like; 150mm long; soldered joints to metal covering										
50mm diameter	11.50	5.00	12.07	1.00	-	23.99	-	36.06	nr	39.67
100mm diameter	13.25	5.00	13.91	1.00	-	23.99	-	37.90	nr	41.69

Labour hourly rates: (except Specialists) Craft Operatives 18.37 Labourer 13.76 Rates are national average prices. Refer to REGIONAL VARIATIONS for indicative levels of overall pricing in regions	MATERIALS			LABOUR				RATES		
	Del to Site £	Waste %	Material Cost £	Craft Optve Hrs	Lab Hrs	Labour Cost £	Sunds £	Nett Rate £	Unit	Gross rate (10%) £
PLAIN TILING										
Quantities required for 1m² of tiling										

Plain tiles to 65mm lap
  tiles, 60nr
  battens, 10m										

Plain tiles to 85mm lap
  tiles, 70nr
  battens, 11m										

Roof coverings; clayware Machine made plain tiles BS EN 1304, red, 265 x 165mm; fixing every fourth course with two galvanized nails per tile to 65mm lap; 38 x 19mm pressure impregnated softwood battens fixed with galvanized nails; twin-ply underlay 145gsm with polypropylene; 150mm laps; fixing with galvanized steel clout nails										
Pitched 50 degrees from horizontal										
generally	23.39	5.00	24.74	0.72	0.36	18.18	2.34	45.26	m²	49.78
holes	-	-	-	0.58	0.29	14.65	-	14.65	nr	16.11
Boundary work										
abutments; square	2.10	5.00	2.20	0.28	0.14	7.07	-	9.28	m	10.20
abutments; raking	3.65	5.00	3.83	0.42	0.21	10.61	-	14.44	m	15.88
double course at top edges with clips and nailing each tile with two aluminium nails	2.10	5.00	2.20	0.34	0.17	8.59	5.72	16.51	m	18.16
double course at eaves; Machine made eaves tile	2.10	5.00	2.20	0.35	0.18	8.91	0.82	11.94	m	13.13
Machine made ridge tiles, half round; in 300mm effective lengths; butt jointed; bedding and pointing in coloured cement-lime mortar (1:1:6)	14.69	5.00	15.42	0.35	0.18	8.91	0.63	24.96	m	27.45
verges; bed and point in coloured cement-lime mortar (1:1:6)	2.10	5.00	2.20	0.40	0.20	10.10	0.63	12.94	m	14.23
verges; single extra undercloak course plain tiles; bed and point in coloured cement-lime mortar (1:1:6)	4.14	5.00	4.35	0.58	0.29	14.65	0.63	19.62	m	21.58
Machine made valley tiles; angular	43.31	5.00	45.48	0.55	0.28	13.89	0.63	59.99	m	65.99
Machine made hip tiles, half round; in 300mm effective lengths; butt jointed; bedding and pointing in coloured cement-lime mortar (1:1:6)	14.69	5.00	15.42	0.35	0.18	8.91	0.63	24.96	m	27.45
Machine made hip tiles; arris; bedding and pointing in cement mortar	42.88	5.00	45.03	0.55	0.28	13.89	0.63	59.55	m	65.50
Machine made hip tiles; bonnet pattern; bedding and pointing in coloured cement-lime mortar (1:1:6)	31.31	5.00	32.88	0.55	0.28	13.89	0.63	47.39	m	52.13
Roof coverings; clayware Machine made plain tiles BS EN 1304, red, 265 x 165mm; fixing every fourth course with two galvanized nails per tile to 65mm lap; 50 x 19mm pressure impregnated softwood battens fixed with galvanized nails; twin-ply underlay 145gsm with polypropylene; 150mm laps; fixing with galvanized steel clout nails										
Pitched 50 degrees from horizontal										
generally	24.59	5.00	26.06	0.70	0.35	17.68	2.34	46.07	m²	50.68
Roof coverings; clayware Machine made plain tiles BS EN 1304, red, 265 x 165mm; fixing every course with two galvanized nails per tile to 65mm lap; 38 x 19mm pressure impregnated softwood battens fixed with galvanized nails; twin-ply underlay 145gsm with polypropylene; 150mm laps; fixing with galvanized steel clout nails										
Pitched 50 degrees from horizontal										
generally	23.39	5.00	24.74	0.85	0.43	21.53	4.50	50.77	m²	55.85
Roof coverings; clayware Machine made plain tiles BS EN 1304, red, 265 x 165mm; fixing every fourth course with two copper nails per tile to 65mm lap; 38 x19mm pressure impregnated softwood battens fixed with galvanized nails; twin-ply underlay 145gsm with polypropylene; 150mm laps; fixing with galvanized steel clout nails										
Pitched 50 degrees from horizontal										
generally	23.39	5.00	24.74	0.70	0.35	17.68	3.60	46.01	m²	50.61
Roof coverings; clayware Machine made plain tiles BS EN 1304, red, 265 x 165mm; fixing every course with two copper nails per tile to 65mm lap; 38 x 19mm pressure impregnated softwood battens fixed with galvanized nails; twin-ply underlay 145gsm with polypropylene; 150mm laps; fixing with galvanized steel clout nails										
Pitched 50 degrees from horizontal										
generally	23.39	5.00	24.74	0.85	0.43	21.53	8.25	54.52	m²	59.97

Labour hourly rates: (except Specialists) Craft Operatives 18.37 Labourer 13.76 Rates are national average prices. Refer to REGIONAL VARIATIONS for indicative levels of overall pricing in regions	MATERIALS			LABOUR				RATES		
	Del to Site £	Waste %	Material Cost £	Craft Optve Hrs	Lab Hrs	Labour Cost £	Sunds £	Nett Rate £	Unit	Gross rate (10%) £
PLAIN TILING (Cont'd)										
Roof coverings; clayware Machine made plain tiles BS EN 1304, red, 265 x 165mm; fixing every fourth course with two galvanized nails per tile to 85mm lap; 38 x 19mm pressure impregnated softwood battens fixed with galvanized nails; twin-ply underlay 145gsm with polypropylene; 150mm laps; fixing with galvanized steel clout nails										
Pitched 40 degrees from horizontal										
generally..........	27.03	5.00	28.57	0.80	0.40	20.20	2.34	51.11	m²	56.22
Wall coverings; clayware Machine made plain tiles BS EN 1304, red, 265 x 165mm; fixing every course with two galvanized nails per tile to 35mm lap; 38 x 19mm pressure impregnated softwood battens fixed with galvanized nails; twin-ply underlay 145gsm with polypropylene; 150mm laps; fixing with galvanized steel clout nails										
Vertical										
generally..........	20.36	5.00	21.54	0.85	0.43	21.53	3.42	46.49	m²	51.14
holes..........	-	-	-	0.87	0.44	21.97	-	21.97	nr	24.16
Boundary work										
abutments; square..........	2.10	5.00	2.20	0.42	0.21	10.61	-	12.81	m	14.09
abutments; raking..........	3.65	5.00	3.83	0.63	0.31	15.91	-	19.74	m	21.71
double course at eaves; Machine made eaves tile..........	2.10	5.00	2.20	0.52	0.27	13.36	0.82	16.39	m	18.03
verges; bed and point in coloured cement-lime mortar (1:1:6)..........	2.10	5.00	2.20	0.60	0.30	15.15	0.63	17.99	m	19.78
verges; single extra undercloak course plain tiles; bed and point in coloured cement-lime mortar (1:1:6)..........	4.14	5.00	4.35	0.87	0.44	21.97	0.63	26.94	m	29.64
vertical angle tiles; machine made..........	38.37	10.00	42.20	0.87	0.44	21.97	0.63	64.80	m	71.28
Roof coverings; clayware Hand made plain tiles BS EN 1304, red, 265 x 165mm; fixing every fourth course with two galvanized nails per tile to 65mm lap; 38 x 19mm pressure impregnated softwood battens fixed with galvanized nails; twin-ply underlay 145gsm with polypropylene; 150mm laps; fixing with galvanized steel clout nails										
Pitched 50 degrees from horizontal										
generally..........	53.99	5.00	56.87	0.72	0.36	18.18	2.34	77.39	m²	85.13
holes..........	-	-	-	0.58	0.29	14.65	-	14.65	nr	16.11
Boundary work										
abutments; square..........	5.10	5.00	5.36	0.28	0.14	7.07	-	12.43	m	13.67
abutments; raking..........	8.50	5.00	8.93	0.42	0.21	10.61	-	19.53	m	21.48
double course at top edges with clips and nailing each tile with two aluminium nails..........	5.10	5.00	5.36	0.34	0.17	8.59	5.72	19.66	m	21.62
double course at eaves; Hand made eaves tile..........	5.10	5.00	5.36	0.35	0.18	8.91	0.82	15.09	m	16.59
Hand made ridge tiles, half round; in 300mm effective lengths; butt jointed; bedding and pointing in coloured cement-lime mortar (1:1:6)..........	16.48	5.00	17.31	0.35	0.18	8.91	0.63	26.84	m	29.53
verges; bed and point in coloured cement-lime mortar (1:1:6)..........	5.10	5.00	5.36	0.40	0.20	10.10	0.63	16.09	m	17.69
verges; single extra undercloak course plain tiles; bed and point in coloured cement-lime mortar (1:1:6)..........	10.20	5.00	10.71	0.58	0.29	14.65	0.63	25.99	m	28.58
Hand made valley tiles; angular; bedding and pointing in cement mortar..........	39.76	5.00	41.75	0.55	0.28	13.89	0.63	56.27	m	61.89
Hand made hip tiles, half round; in 300mm effective lengths; butt jointed; bedding and pointing in coloured cement-lime mortar (1:1:6)..........	16.48	5.00	17.31	0.35	0.18	8.91	0.63	26.84	m	29.53
Hand made hip tiles; angular..........	39.76	5.00	41.75	0.55	0.28	13.89	0.63	56.27	m	61.89
Hand made hip tiles; bonnet pattern; bedding and pointing in coloured cement-lime mortar (1:1:6)..........	39.76	5.00	41.75	0.55	0.28	13.89	0.63	56.27	m	61.89
Roof coverings; clayware Hand made plain tiles BS EN 1304, red, 265 x 165mm; fixing every fourth course with two galvanized nails per tile to 85mm lap; 38 x 19mm pressure impregnated softwood battens fixed with galvanized nails; twin-ply underlay 145gsm with polypropylene; 150mm laps; fixing with galvanized steel clout nails										
Pitched 40 degrees from horizontal										
generally..........	62.73	5.00	66.06	0.80	0.40	20.20	2.30	88.55	m²	97.41
Fittings										
Hip irons										
galvanized mild steel; 32 x 3 x 380mm girth; scrolled; fixing with galvanized steel screws to timber..........	0.80	5.00	0.84	0.12	0.06	3.03	-	3.87	nr	4.26
Lead soakers										
fixing only..........	-	-	-	0.35	0.18	8.91	-	8.91	nr	9.80
Wall coverings; clayware hand made plain tiles BS EN 1304, red, 265 x 165mm; fixing every course with two galvanized nails per tile to 35mm lap; 38 x 19mm pressure impregnated softwood battens fixed with galvanized nails; twin-ply underlay 145gsm with polypropylene; 150mm laps; fixing with galvanized steel clout nails										
Vertical										
generally..........	20.36	5.00	21.54	0.85	0.43	21.53	3.42	46.49	m²	51.14
holes..........	-	-	-	0.87	0.44	21.97	-	21.97	nr	24.16
Boundary work										
abutments; square..........	5.10	5.00	5.36	0.42	0.21	10.61	-	15.96	m	17.56
abutments; raking..........	8.93	5.00	9.37	0.63	0.31	15.91	-	25.28	m	27.81
double course at eaves; Hand made eaves tile..........	5.10	5.00	5.36	0.52	0.27	13.36	0.82	19.54	m	21.49
verges; bed and point in coloured cement-lime mortar (1:1:6)..........	5.10	5.00	5.36	0.60	0.30	15.15	0.63	21.14	m	23.25

Labour hourly rates: (except Specialists) Craft Operatives 18.37 Labourer 13.76 Rates are national average prices. Refer to REGIONAL VARIATIONS for indicative levels of overall pricing in regions	MATERIALS			LABOUR				RATES		
	Del to Site £	Waste %	Material Cost £	Craft Optve Hrs	Lab Hrs	Labour Cost £	Sunds £	Nett Rate £	Unit	Gross rate (10%) £
PLAIN TILING (Cont'd)										
Wall coverings; clayware hand made plain tiles BS EN 1304, red, 265 x 165mm; fixing every course with two galvanized nails per tile to 35mm lap; 38 x 19mm pressure impregnated softwood battens fixed with galvanized nails; twin-ply underlay 145gsm with polypropylene; 150mm laps; fixing with galvanized steel clout nails (Cont'd)										
Vertical (Cont'd)										
Boundary work (Cont'd)										
verges; single extra undercloak course plain tiles; bed and point in coloured cement-lime mortar (1:1:6)	10.20	5.00	10.71	0.87	0.44	21.97	0.63	33.31	m	36.64
vertical angle tiles; hand made	49.59	10.00	54.55	0.87	0.44	21.97	0.63	77.15	m	84.86
Roof coverings; Redland Plain granular faced tiles, 268 x 165mm; fixing every fifth course with two aluminium nails to each tile to 65mm lap; 38 x 19mm pressure impregnated softwood battens fixed with galvanized nails; twin-ply underlay 145gsm with polypropylene; 150mm laps; fixing with galvanized steel clout nails										
Pitched 40 degrees from horizontal										
generally	23.39	5.00	24.74	0.70	0.35	17.68	1.73	44.14	m²	48.55
holes	-	-	-	0.58	0.29	14.65	-	14.65	nr	16.11
Boundary work										
abutments; square	2.04	5.00	2.14	0.28	0.14	7.07	-	9.21	m	10.13
abutments; raking	3.57	5.00	3.75	0.42	0.21	10.61	-	14.35	m	15.79
double course at top edges with clips and nailing each tile with two aluminium nails	2.06	5.00	2.16	0.34	0.17	8.59	5.72	16.46	m	18.11
double course at eaves and nailing each tile with two aluminium nails	2.04	5.00	2.14	0.35	0.18	8.91	0.76	11.81	m	12.99
ridge or hip tiles, Plain angle; butt jointed; bedding and pointing in tinted cement mortar (1:3)	10.09	5.00	10.59	0.35	0.18	8.91	0.63	20.13	m	22.14
ridge or hip tiles, half round; butt jointed; bedding and pointing in tinted cement mortar (1:3)	6.53	5.00	6.86	0.35	0.18	8.91	0.63	16.40	m	18.04
verges; Redland Dry Verge system with plain tiles and tile-and-a-half tiles in alternate courses, clips and aluminium nails	14.01	5.00	14.71	0.28	0.14	7.07	1.62	23.40	m	25.74
verges; plain tile undercloak; tiles and undercloak bedded and pointed in tinted cement mortar (1:3) and with plain tiles and tile-and-a-half tiles in alternate courses and aluminium nails	4.08	5.00	4.28	0.28	0.14	7.07	0.68	12.03	m	13.23
valley tiles, Plain valley tiles; bedding and pointing in cement mortar	26.27	5.00	27.58	0.55	0.28	13.89	0.63	42.10	m	46.31
hip tiles; angular; bedding and pointing in cement mortar	29.89	5.00	31.39	0.55	0.28	13.89	0.63	45.90	m	50.49
hip tiles; bonnet; bedding and pointing in coloured cement-lime mortar (1:1:6)	29.89	5.00	31.39	0.55	0.28	13.89	0.63	45.90	m	50.49
Roof coverings; Redland Plain granular faced tiles, 268 x 165mm; fixing every fifth course with two aluminium nails to each tile to 65mm lap; 50 x 25mm pressure impregnated softwood battens fixed with galvanized nails; twin-ply underlay 145gsm with polypropylene; 150mm laps; fixing with galvanized steel clout nails										
Pitched 40 degrees from horizontal										
generally	23.39	5.00	24.74	0.70	0.35	17.68	1.73	44.14	m²	48.55
Boundary work										
abutments; square	2.04	5.00	2.14	0.28	0.15	7.21	-	9.35	m	10.28
abutments; raking	3.57	5.00	3.75	0.43	0.21	10.79	-	14.54	m	15.99
abutments; curved to 3000mm radius	5.44	5.00	5.71	0.56	0.28	14.14	-	19.85	m	21.84
Roof coverings; Redland Plain granular faced tiles, 268 x 165mm; fixing each tile with two aluminium nails to 35mm lap; 38 x 19mm pressure impregnated softwood battens fixed with galvanized nails; twin-ply underlay 145gsm with polypropylene; 150mm laps; fixing with galvanized steel clout nails										
Vertical										
generally	20.36	5.00	21.54	0.83	0.42	21.03	3.42	45.98	m²	50.58
holes	-	-	-	0.87	0.44	21.97	-	21.97	nr	24.16
Boundary work										
abutments; square	2.04	5.00	2.14	0.42	0.21	10.61	-	12.75	m	14.02
abutments; raking	3.57	5.00	3.75	0.63	0.31	15.91	-	19.66	m	21.62
double course at eaves; Machine made eaves tile	2.04	5.00	2.14	0.52	0.27	13.36	0.82	16.33	m	17.96
verges; bed and point in coloured cement-lime mortar (1:1:6)	2.04	5.00	2.14	0.60	0.30	15.15	0.63	17.92	m	19.71
verges; single extra undercloak course plain tiles; bed and point in coloured cement-lime mortar (1:1:6)	4.08	5.00	4.28	0.87	0.44	21.97	0.63	26.88	m	29.57
vertical angle tiles	27.39	5.00	28.76	0.87	0.44	21.97	0.63	51.36	m	56.49
Roof coverings; Plain through coloured tiles, 268 x 165mm; fixing every fifth course with two aluminium nails to each tile to 65mm lap; 38 x 19mm pressure impregnated softwood battens fixed with galvanized nails; twin-ply underlay 145gsm with polypropylene; 150mm laps; fixing with galvanized steel clout nails										
Pitched 40 degrees from horizontal										
generally	23.39	5.00	24.74	0.70	0.35	17.68	1.73	44.14	m²	48.55
Boundary work										
double course at top edges with clips and nailing each tile with two aluminium nails	2.04	5.00	2.14	0.34	0.17	8.59	5.72	16.44	m	18.09
double course at eaves and nailing each tile with two aluminium nails	2.04	5.00	2.14	0.34	0.17	8.59	0.76	11.49	m	12.64
ridge or hip tiles, Plain angle; butt jointed; bedding and pointing in tinted cement mortar (1:3)	10.09	5.00	10.59	0.35	0.18	8.91	0.63	20.13	m	22.14

TILE AND SLATE ROOF AND WALL COVERINGS

Labour hourly rates: (except Specialists) Craft Operatives 18.37 Labourer 13.76 Rates are national average prices. Refer to REGIONAL VARIATIONS for indicative levels of overall pricing in regions	MATERIALS			LABOUR				RATES		
	Del to Site £	Waste %	Material Cost £	Craft Optve Hrs	Lab Hrs	Labour Cost £	Sunds £	Nett Rate £	Unit	Gross rate (10%) £
PLAIN TILING (Cont'd)										
Roof coverings; Plain through coloured tiles, 268 x 165mm; fixing every fifth course with two aluminium nails to each tile to 65mm lap; 38 x 19mm pressure impregnated softwood battens fixed with galvanized nails; twin-ply underlay 145gsm with polypropylene; 150mm laps; fixing with galvanized steel clout nails (Cont'd)										
Pitched 40 degrees from horizontal (Cont'd)										
Boundary work (Cont'd)										
ridge or hip tiles, half round; butt jointed; bedding and pointing in tinted cement mortar (1:3)	6.53	5.00	6.85	0.35	0.18	8.91	0.63	16.39	m	18.03
verges; Redland Dry Verge system with plain tiles and tile-and-a-half tiles in alternate courses, clips and aluminium nails	14.01	5.00	14.71	0.28	0.14	7.07	1.62	23.40	m	25.74
verges; plain tile undercloak; tiles and undercloak bedded and pointed in tinted cement mortar (1:3) and with plain tiles and tile-and-a-half tiles in alternate courses and aluminium nails	4.08	5.00	4.28	0.28	0.14	7.07	0.68	12.03	m	13.23
valley tiles, Plain valley tiles; bedding and pointing in cement mortar	26.27	5.00	27.58	0.55	0.28	13.89	0.63	42.10	m	46.31
Roof coverings; Redland Heathland plain granular faced tiles, 268 x 165mm; fixing every fifth course with two aluminium nails to each tile to 65mm lap; 38 x 19mm pressure impregnated softwood battens fixed with galvanized nails; twin-ply underlay 145gsm with polypropylene; 150mm laps; fixing with galvanized steel clout nails										
Pitched 40 degrees from horizontal										
generally	26.39	5.00	27.89	0.70	0.35	17.68	1.73	47.29	m²	52.02
Boundary work										
double course at top edges with clips and nailing each tile with two aluminium nails	2.34	5.00	2.46	0.34	0.17	8.59	5.72	16.76	m	18.43
double course at eaves and nailing each tile with two aluminium nails	2.34	5.00	2.46	0.34	0.17	8.59	0.76	11.81	m	12.99
ridge or hip tiles, Redland Plain angle; butt jointed; bedding and pointing in tinted cement mortar (1:3)	10.09	5.00	10.59	0.35	0.18	8.91	0.63	20.13	m	22.14
ridge or hip tiles, Redland half round; butt jointed; bedding and pointing in tinted cement mortar (1:3)	6.53	5.00	6.85	0.35	0.18	8.91	0.63	16.39	m	18.03
verges; Redland Dry Verge system with plain tiles and tile-and-a-half tiles in alternate courses, clips and aluminium nails	13.14	5.00	13.80	0.28	0.14	7.07	1.62	22.49	m	24.74
verges; plain tile undercloak; tiles and undercloak bedded and pointed in tinted cement mortar (1:3) and with plain tiles and tile-and-a-half tiles in alternate courses and aluminium nails	2.34	5.00	2.46	0.28	0.14	7.07	0.68	10.20	m	11.22
valley tiles, Redland Plain valley tiles	14.02	5.00	14.72	0.35	0.18	8.91	0.63	24.26	m	26.68
Roof coverings; Redland Heathland plain through coloured tiles, 268 x 165mm; fixing every fifth course with two aluminium nails to each tile to 65mm lap; 38 x 19mm pressure impregnated softwood battens fixed with galvanized nails; twin-ply underlay 145gsm with polypropylene 150mm laps; fixing with galvanized steel clout nails										
Pitched 40 degrees from horizontal										
generally	26.39	5.00	27.89	0.70	0.35	17.68	1.73	47.29	m²	52.02
Boundary work										
double course at top edges with clips and nailing each tile with two aluminium nails	3.18	5.00	3.34	0.34	0.17	8.59	5.72	17.64	m	19.40
double course at eaves and nailing each tile with two aluminium nails	3.18	5.00	3.34	0.34	0.17	8.59	0.76	12.69	m	13.96
ridge or hip tiles, Plain angle; butt jointed; bedding and pointing in tinted cement mortar (1:3)	10.09	5.00	10.59	0.35	0.18	8.91	0.63	20.13	m	22.14
ridge or hip tiles, half round; butt jointed; bedding and pointing in tinted cement mortar (1:3)	14.69	5.00	15.42	0.35	0.18	8.91	0.63	24.96	m	27.45
verges; Dry Verge system with plain tiles and tile-and-a-half tiles in alternate courses, clips and aluminium nails	15.15	5.00	15.91	0.28	0.14	7.07	1.62	24.60	m	27.06
verges; plain tile undercloak; tiles and undercloak bedded and pointed in tinted cement mortar (1:3) and with plain tiles and tile-and-a-half tiles in alternate courses and aluminium nails	5.52	5.00	5.80	0.28	0.14	7.07	0.68	13.54	m	14.90
valley tiles, Plain valley tiles; bedding and pointing in cement mortar	29.89	5.00	31.39	0.55	0.28	13.89	0.63	45.90	m	50.49
hip tiles; arris; bedding and pointing in cement mortar	29.89	5.00	31.39	0.55	0.28	13.89	0.63	45.90	m	50.49
hip tiles; bonnet; bedding and pointing in coloured cement-lime mortar (1:1:6)	29.89	5.00	31.39	0.55	0.28	13.89	0.63	45.90	m	50.49
Fittings										
Ventilator tiles										
Ridge Vent ridge ventilation terminal 450mm long for										
half round ridge	58.50	5.00	61.42	1.45	0.73	36.68	0.99	99.10	nr	109.01
universal angle ridge	58.50	5.00	61.42	1.45	0.73	36.68	0.99	99.10	nr	109.01
Thru Vent ventilator tiles for										
Stonewold slates	15.24	5.00	16.00	0.50	0.25	12.62	0.33	28.96	nr	31.85
Regent tiles	15.24	5.00	16.00	0.50	0.25	12.62	0.33	28.96	nr	31.85
Grovebury double pantiles	15.24	5.00	16.00	0.50	0.25	12.62	0.33	28.96	nr	31.85
Fenland pantiles	15.24	5.00	16.00	0.50	0.25	12.62	0.33	28.96	nr	31.85
Redland 49 tiles	15.24	5.00	16.00	0.50	0.25	12.62	0.33	28.96	nr	31.85
Renown tiles	15.24	5.00	16.00	0.50	0.25	12.62	0.33	28.96	nr	31.85
Redland 50 double roman tiles	15.24	5.00	16.00	0.50	0.25	12.62	0.33	28.96	nr	31.85
Redland plain tiles	12.68	5.00	13.31	0.56	0.28	14.14	0.38	27.83	nr	30.61
Redland Downland plain tiles	12.68	5.00	13.31	0.56	0.28	14.14	0.38	27.83	nr	30.61
Eaves ventilation										
Insulation Interupter and roof space ventilator										
Felt support eaves ventilator tray fixing with aluminium nails	1.70	5.00	1.78	0.25	0.12	6.31	0.05	8.14	m	8.96

Labour hourly rates: (except Specialists) Craft Operatives 18.37 Labourer 13.76 Rates are national average prices. Refer to REGIONAL VARIATIONS for indicative levels of overall pricing in regions	MATERIALS			LABOUR				RATES		
	Del to Site	Waste	Material Cost	Craft Optve	Lab	Labour Cost	Sunds	Nett Rate	Unit	Gross rate (10%)
	£	%	£	Hrs	Hrs	£	£	£		£
PLAIN TILING (Cont'd)										
Fittings (Cont'd)										
Eaves ventilation (Cont'd)										
Over fascia ventilator										
10mm	2.30	-	2.30	0.30	0.15	7.58	0.05	9.92	m	10.91
25mm	6.75	-	6.75	0.30	0.15	7.58	0.05	14.37	m	15.81
Gas terminals										
Gas Flue ridge terminal, 450mm long with sealing gasket and fixing brackets										
half round ridge	98.28	5.00	103.19	1.68	0.84	42.42	0.99	146.60	nr	161.26
universal angle ridge	98.28	5.00	103.19	1.68	0.84	42.42	0.99	146.60	nr	161.26
Gas Flue ridge terminal, 450mm long with sealing gasket and fixing brackets with adaptor for										
half round ridge	118.63	5.00	124.56	1.68	0.84	42.42	0.99	167.97	nr	184.77
universal angle ridge	118.63	5.00	124.56	1.68	2.84	69.94	0.99	195.49	nr	215.04
extra for extension adaptor and gasket	118.63	5.00	124.56	0.35	0.17	8.77	0.38	133.71	nr	147.08
Hip irons										
galvanized mild steel; 32 x 3 x 380mm girth; scrolled; fixing with galvanized steel screws to timber	0.80	5.00	0.84	0.11	0.06	2.85	-	3.69	nr	4.05
Lead soakers										
fixing only	-	-	-	0.34	0.17	8.59	-	8.59	nr	9.44
INTERLOCKING TILING										
Roof coverings; Redland Stonewold through coloured slates, 430 x 380mm; fixing every fourth course with galvanized nails to 75mm lap; 38 x 19mm pressure impregnated softwood battens fixed with galvanized nails; twin-ply underlay 145gsm with polypropylene; 150mm laps; fixing with galvanized steel clout nails										
Pitched 40 degrees from horizontal										
generally	36.84	5.00	38.78	0.48	0.24	12.12	1.22	52.11	m²	57.33
Boundary work										
abutments; square	3.55	5.00	3.73	0.12	0.06	3.03	-	6.76	m	7.44
abutments; raking	5.33	5.00	5.60	0.17	0.09	4.36	-	9.96	m	10.95
abutments; curved to 3000mm radius	7.11	5.00	7.46	0.23	0.12	5.88	-	13.34	m	14.67
supplementary fixing eaves course with one clip per slate	0.60	5.00	0.63	0.03	0.02	0.83	-	1.46	m	1.60
ridge or hip tiles, Redland third round; butt jointed; bedding and pointing in tinted cement mortar (1:3)	6.76	5.00	7.10	0.23	0.12	5.88	0.63	13.61	m	14.97
ridge or hip tiles, Redland half round; butt jointed; bedding and pointing in tinted cement mortar (1:3)	6.76	5.00	7.10	0.23	0.12	5.88	0.63	13.61	m	14.97
ridge tiles, Redland universal angle; butt jointed; bedding and pointing in tinted cement mortar (1:3)	6.76	5.00	7.10	0.23	0.12	5.88	0.63	13.61	m	14.97
ridge tiles, Redland universal half round type Monopitch; butt jointed; fixing with aluminium nails and bedding and pointing in tinted cement mortar (1:3)	23.50	5.00	24.67	0.23	0.12	5.88	0.63	31.18	m	34.30
ridge tiles, Redland Dry Ridge system with half round ridge tiles; fixing with stainless steel batten straps and ring shanked fixing nails with neoprene washers and sleeves; polypropylene ridge seals and uPVC profile filler units	15.71	5.00	16.50	0.45	0.23	11.43	-	27.93	m	30.72
ridge tiles, Redland Dry Ridge system with universal angle ridge tiles; fixing with stainless steel batten straps and ring shanked fixing nails with neoprene washers and sleeves; polypropylene ridge seals and uPVC profile filler units	15.71	5.00	16.50	0.45	0.23	11.43	-	27.93	m	30.72
ridge tiles, Redland Dry Vent Ridge system with half round ridge tiles; fixing with stainless steel batten straps and ring shanked nails with neoprene washers and sleeves; polypropylene ridge seals, PVC air flow control units and uPVC ventilated profile filler units	14.71	5.00	15.44	0.45	0.23	11.43	-	26.87	m	29.56
ridge tiles, Redland Dry Vent Ridge system with Universal angle ridge tiles; fixing with stainless steel batten straps and ring shanked nails with neoprene washers and sleeves; polypropylene ridge seals, PVC air flow control units and uPVC ventilated profile filler units	14.71	5.00	15.44	0.45	0.23	11.43	-	26.87	m	29.56
verges; Redland Dry Verge system with half slates, full slates and clips	26.50	5.00	27.82	0.28	0.14	7.07	2.88	37.78	m	41.55
verges; Redland Dry Verge system with half slates, verge slates and clips	27.80	5.00	29.19	0.28	0.14	7.07	2.88	39.14	m	43.05
verges; 150 x 6mm fibre cement undercloak, slates and undercloak bedded and pointed in tinted cement mortar (1:3) and with half slates, full slates and clips	19.30	5.00	20.26	0.23	0.12	5.88	3.40	29.55	m	32.50
verges; 150 x 6mm fibre cement undercloak, slates and undercloak bedded and pointed in tinted cement mortar (1:3) and with half slates, verge slates and clips	14.00	5.00	14.70	0.23	0.12	5.88	3.40	23.98	m	26.38
valley tiles, Redland Universal valley troughs; laid with 100mm laps	33.00	5.00	34.65	0.56	0.28	14.14	0.63	49.42	m	54.36
Roof coverings; Redland Stonewold through coloured slates, 430 x 380mm; fixing every fourth course with galvanized nails to 75mm lap; 38 x 25mm pressure impregnated softwood battens fixed with galvanized nails; twin-ply underlay 145gsm with polypropylene; 150mm laps; fixing with galvanized steel clout nails										
Pitched 40 degrees from horizontal										
generally	36.96	5.00	38.91	0.47	0.24	11.94	1.22	52.06	m²	57.27

TILE AND SLATE ROOF AND WALL COVERINGS

Labour hourly rates: (except Specialists) Craft Operatives 18.37 Labourer 13.76 Rates are national average prices. Refer to REGIONAL VARIATIONS for indicative levels of overall pricing in regions	MATERIALS			LABOUR				RATES		
	Del to Site	Waste	Material Cost	Craft Optve	Lab	Labour Cost	Sunds	Nett Rate	Unit	Gross rate (10%)
	£	%	£	Hrs	Hrs	£	£	£		£
INTERLOCKING TILING (Cont'd)										
Roof coverings; Redland Stonewold through coloured slates, 430 x 380mm; fixing each course with galvanized nails to 75mm; 38 x 25mm pressure impregnated softwood battens fixed with galvanized nails; twin-ply underlay 145gsm with polypropylene; 150mm laps; fixing with galvanized steel clout nails										
Pitched 40 degrees from horizontal										
generally	36.96	5.00	38.91	0.52	0.26	13.13	3.51	55.55	m²	61.11
Roof coverings; Redland Stonewold through coloured slates, 430 x 380mm; fixing every fourth course with galvanized nails to 75mm lap; 38 x 25mm pressure impregnated softwood battens fixed with galvanized nails; underlay, BS EN 13707 type 1F aluminium foil surfaced reinforced bitumen felt; 150mm laps; fixing with galvanized steel clout nails										
Pitched 40 degrees from horizontal										
generally	36.96	5.00	38.91	0.48	0.24	12.12	1.22	52.25	m²	57.47
Roof coverings; Redland Stonewold through coloured slates, 430 x 380mm; fixing every fourth course with galvanized nails to 75mm lap; 38 x 25mm pressure impregnated softwood battens fixed with galvanized nails; underlay, BS EN 13707 type 1F reinforced bitumen felt with 50mm glass fibre insulation bonded on; 150mm laps; fixing with galvanized steel clout nails										
Pitched 40 degrees from horizontal										
generally	36.96	5.00	38.91	0.58	0.29	14.65	1.22	54.77	m²	60.25
Roof coverings; Redland Regent granular faced tiles, 418 x 332mm; fixing every fourth course with galvanized nails to 75mm lap; 38 x 25mm pressure impregnated softwood battens fixed with galvanized nails; twin-ply underlay 145gsm with polypropylene; 150mm laps; fixing with galvanized steel clout nails										
Pitched 40 degrees from horizontal										
generally	9.48	5.00	10.05	0.47	0.24	11.94	1.37	23.36	m²	25.69
Boundary work										
reform eaves filler unit and eaves clip to each eaves tile	0.60	5.00	0.63	0.07	0.04	1.84	1.07	3.53	m	3.88
ridge or hip tiles, Redland third round; butt jointed; bedding and pointing in tinted cement mortar (1:3); dentil slips in pan of each tile set in bedding	6.47	5.00	6.79	0.34	0.17	8.59	2.44	17.82	m	19.60
ridge or hip tiles, Redland half round; butt jointed; bedding and pointing in tinted cement mortar (1:3); dentil slips in pan of each tile set in bedding	6.47	5.00	6.79	0.34	0.17	8.59	2.44	17.82	m	19.60
ridge tiles, Redland universal half round type Monopitch; butt jointed; fixing with aluminium nails and bedding and pointing in tinted cement mortar (1:3); dentil slips in pan of each tile set in bedding	23.50	5.00	24.67	0.34	0.17	8.59	2.44	35.71	m	39.28
verges; cloaked verge tiles and aluminium nails	13.50	5.00	14.17	0.17	0.09	4.36	0.10	18.64	m	20.51
verges; half tiles, cloaked verge tiles and aluminium nails	15.16	5.00	15.92	0.23	0.12	5.88	0.18	21.97	m	24.17
verges; 150 x 6mm fibre cement undercloak; tiles and undercloak bedded and pointed in tinted cement mortar (1:3) and with standard tiles and clips	3.49	5.00	3.66	0.23	0.12	5.88	3.40	12.95	m	14.24
verges; 150 x 6mm fibre cement undercloak; tiles and undercloak bedded and pointed in tinted cement mortar (1:3) and with verge tiles and clips	3.49	5.00	3.66	0.23	0.12	5.88	3.40	12.95	m	14.24
verges; 150 x 6mm fibre cement undercloak; tiles and undercloak bedded and pointed in tinted cement mortar (1:3) and with half tiles, standard tiles and clips	16.99	5.00	17.84	0.23	0.12	5.88	3.40	27.12	m	29.83
verges; 150 x 6mm fibre cement undercloak; tiles and undercloak bedded and pointed in tinted cement mortar (1:3) and with half tiles, verge tiles and clips	16.99	5.00	17.84	0.23	0.12	5.88	3.40	27.12	m	29.83
valley tiles, Redland universal valley troughs; laid with 100mm laps	33.00	5.00	34.65	0.34	0.17	8.59	0.63	43.86	m	48.25
Roof coverings; Redland Regent granular faced tiles, 418 x 332mm; fixing every fourth course with aluminium nails to 75mm lap; 38 x 25mm pressure impregnated softwood battens fixed with galvanized nails; twin-ply underlay 145gsm with polypropylene; 150mm laps; fixing with galvanized steel clout nails										
Pitched 40 degrees from horizontal										
generally	9.48	5.00	10.05	0.47	0.24	11.94	1.61	23.60	m²	25.96
Roof coverings; Redland Regent granular faced tiles, 418 x 332mm; fixing every fourth course with galvanized nails to 100mm lap; 38 x 25mm pressure impregnated softwood battens fixed with galvanized nails; twin-ply underlay 145gsm with polypropylene; 150mm laps; fixing with galvanized steel clout nails										
Pitched 40 degrees from horizontal										
generally	10.14	5.00	10.75	0.47	0.24	11.94	0.98	23.66	m²	26.03
Boundary work										
abutments; square	0.87	5.00	0.92	0.11	0.06	2.85	-	3.76	m	4.14
abutments; raking	1.30	5.00	1.37	0.17	0.09	4.36	-	5.73	m	6.30
abutments; curved to 3000mm radius	1.74	5.00	1.83	0.23	0.12	5.88	-	7.71	m	8.48

	MATERIALS			LABOUR				RATES		
Labour hourly rates: (except Specialists) Craft Operatives 18.37 Labourer 13.76 Rates are national average prices. Refer to REGIONAL VARIATIONS for indicative levels of overall pricing in regions	Del to Site £	Waste %	Material Cost £	Craft Optve Hrs	Lab Hrs	Labour Cost £	Sunds £	Nett Rate £	Unit	Gross rate (10%) £

INTERLOCKING TILING (Cont'd)

Roof coverings; Redland Regent through coloured tiles, 418 x 332mm; fixing every fourth course with galvanized nails to 75mm lap; 38 x 25mm pressure impregnated softwood battens fixed with galvanized nails; twin-ply underlay 145gsm with polypropylene; 150mm laps; fixing with galvanized steel clout nails

Description	Del to Site £	Waste %	Material Cost £	Craft Optve Hrs	Lab Hrs	Labour Cost £	Sunds £	Nett Rate £	Unit	Gross rate (10%) £
Pitched 40 degrees from horizontal										
generally	9.48	5.00	10.05	0.47	0.24	11.94	1.37	23.36	m²	25.69
Boundary work										
reform eaves filler unit and eaves clip to each eaves tile	0.60	5.00	0.63	0.07	0.04	1.84	1.07	3.53	m	3.88
verges; cloaked verge tiles and aluminium nails	13.50	5.00	14.17	0.17	0.09	4.36	0.10	18.64	m	20.51
verges; 150 x 6mm fibre cement undercloak; tiles and undercloak bedded and pointed in tinted cement mortar (1:3) and with standard tiles and clips	3.49	5.00	3.66	0.23	0.12	5.88	3.40	12.95	m	14.24
verges; 150 x 6mm fibre cement undercloak; tiles and undercloak bedded and pointed in tinted cement mortar (1:3) and with verge tiles and clips	3.49	5.00	3.66	0.23	0.12	5.88	3.40	12.95	m	14.24
verges; 150 x 6mm fibre cement undercloak; tiles and undercloak bedded and pointed in tinted cement mortar (1:3) and with half tiles, standard tiles and clips	16.99	5.00	17.84	0.23	0.12	5.88	3.40	27.12	m	29.83
verges; 150 x 6mm fibre cement undercloak; tiles and undercloak bedded and pointed in tinted cement mortar (1:3) and with half tiles, verge tiles and clips	16.99	5.00	17.84	0.23	0.12	5.88	3.40	27.12	m	29.83
valley tiles, Redland universal valley troughs; laid with 100mm laps	33.00	5.00	34.65	0.34	0.17	8.59	0.63	43.86	m	48.25

Roof coverings; Redland Regent through coloured tiles, 418 x 332mm; fixing every fourth course with galvanized nails to 100mm lap; 38 x 25mm pressure impregnated softwood battens fixed with galvanized nails; twin-ply underlay 145gsm with polypropylene; 150mm laps; fixing with galvanized steel clout nails

Description	Del to Site £	Waste %	Material Cost £	Craft Optve Hrs	Lab Hrs	Labour Cost £	Sunds £	Nett Rate £	Unit	Gross rate (10%) £
Pitched 40 degrees from horizontal										
generally	10.14	5.00	10.75	0.47	0.24	11.94	1.37	24.05	m²	26.46
Boundary work										
abutments; square	0.87	5.00	0.92	0.11	0.06	2.85	-	3.76	m	4.14
abutments; raking	1.30	5.00	1.37	0.17	0.09	4.36	-	5.73	m	6.30
abutments; curved to 3000mm radius	1.74	5.00	1.83	0.23	0.12	5.88	-	7.71	m	8.48

Roof coverings; Redland Grovebury granular faced double pantiles, 418 x 332mm; fixing every fourth course with galvanized nails to 75mm lap; 38 x 22mm pressure impregnated softwood battens fixed with galvanized nails; twin-ply underlay 145gsm with polypropylene; 150mm laps; fixing with galvanized steel clout nails

Description	Del to Site £	Waste %	Material Cost £	Craft Optve Hrs	Lab Hrs	Labour Cost £	Sunds £	Nett Rate £	Unit	Gross rate (10%) £
Pitched 40 degrees from horizontal										
generally	9.48	5.00	10.05	0.47	0.24	11.94	1.37	23.36	m²	25.69
Boundary work										
reform eaves filler unit and eaves clip to each eaves tile	0.60	5.00	0.63	0.07	0.04	1.84	1.07	3.53	m	3.88
ridge or hip tiles, Redland third round; butt jointed; bedding and pointing in tinted cement mortar (1:3)	6.47	5.00	6.79	0.34	0.17	8.59	3.40	18.78	m	20.66
ridge or hip tiles, Redland half round; butt jointed; bedding and pointing in tinted cement mortar (1:3)	6.47	5.00	6.79	0.34	0.17	8.59	3.40	18.78	m	20.66
verges; cloaked verge tiles and aluminium nails	13.50	5.00	14.17	0.17	0.09	4.36	0.10	18.64	m	20.51
verges; half tiles, cloaked verge tiles and aluminium nails	15.16	5.00	15.92	0.23	0.12	5.88	0.17	21.96	m	24.16
verges; 150 x 6mm fibre cement undercloak; tiles and undercloak bedded and pointed in tinted cement mortar (1:3) and with standard tiles and clips	3.49	5.00	3.66	0.23	0.12	5.88	3.40	12.95	m	14.24
verges; 150 x 6mm fibre cement undercloak; tiles and undercloak bedded and pointed in tinted cement mortar (1:3) and with treble roll verge tiles and clips	3.49	5.00	3.66	0.23	0.12	5.88	3.40	12.95	m	14.24
valley tiles, Redland Universal valley troughs; laid with 100mm laps	33.00	5.00	34.65	0.34	0.17	8.59	0.63	43.86	m	48.25

Roof coverings; Redland Grovebury through coloured double pantiles, 418 x 332mm; fixing every fourth course with galvanized nails to 75mm lap; 38 x 22mm pressure impregnated softwood battens fixed with galvanized nails; twin-ply underlay 145gsm with polypropylene; 150mm laps; fixing with galvanized steel clout nails

Description	Del to Site £	Waste %	Material Cost £	Craft Optve Hrs	Lab Hrs	Labour Cost £	Sunds £	Nett Rate £	Unit	Gross rate (10%) £
Pitched 40 degrees from horizontal										
generally	9.48	5.00	10.05	0.47	0.24	11.94	1.37	23.36	m²	25.69
Boundary work										
reform eaves filler unit and eaves clip to each eaves tile	0.60	5.00	0.63	0.07	0.04	1.84	1.07	3.53	m	3.88
ridge or hip tiles, Redland third round; butt jointed; bedding and pointing in tinted cement mortar (1:3)	6.47	5.00	6.79	0.34	0.17	8.59	0.63	16.01	m	17.61
ridge or hip tiles, Redland half round; butt jointed; bedding and pointing in tinted cement mortar (1:3)	6.47	5.00	6.79	0.34	0.17	8.59	0.63	16.01	m	17.61
verges; cloaked verge tiles and aluminium nails	13.50	5.00	14.17	0.17	0.09	4.36	0.10	18.64	m	20.51
verges; half tiles, cloaked verge tiles and aluminium nails	15.16	5.00	15.92	0.23	0.12	5.88	0.17	21.96	m	24.16
verges; 150 x 6mm fibre cement undercloak; tiles and undercloak bedded and pointed in tinted cement mortar (1:3) and with standard tiles and clips	3.49	5.00	3.66	0.23	0.12	5.88	3.40	12.95	m	14.24
verges; 150 x 6mm fibre cement undercloak; tiles and undercloak bedded and pointed in tinted cement mortar (1:3) and with treble roll verge tiles and clips	3.49	5.00	3.66	0.23	0.12	5.88	3.40	12.95	m	14.24
valley tiles, Redland Universal valley troughs; laid with 100mm laps	33.00	5.00	34.65	0.34	0.17	8.59	0.63	43.86	m	48.25

TILE AND SLATE ROOF AND WALL COVERINGS

Labour hourly rates: (except Specialists) Craft Operatives 18.37 Labourer 13.76 Rates are national average prices. Refer to REGIONAL VARIATIONS for indicative levels of overall pricing in regions	MATERIALS			LABOUR				RATES		
	Del to Site £	Waste %	Material Cost £	Craft Optve Hrs	Lab Hrs	Labour Cost £	Sunds £	Nett Rate £	Unit	Gross rate (10%) £
INTERLOCKING TILING (Cont'd)										
Roof coverings; Redland Fenland through coloured pantiles, 381 x 227mm; fixing every fourth course with galvanized nails to 75mm lap; 38 x 25mm pressure impregnated softwood battens fixed with galvanized nails; twin-ply underlay 145gsm with polypropylene; 150mm laps; fixing with galvanized steel clout nails										
Pitched 40 degrees from horizontal generally	16.38	5.00	17.31	0.61	0.31	15.47	1.98	34.76	m²	38.24
Boundary work										
abutments; square	1.47	5.00	1.54	0.11	0.06	2.85	-	4.39	m	4.83
abutments; raking	2.20	5.00	2.31	0.17	0.09	4.36	-	6.67	m	7.33
abutments; curved to 3000mm radius	2.93	5.00	3.08	0.23	0.12	5.88	-	8.96	m	9.85
reform eaves filler unit and aluminium nails to each eaves tile	0.60	5.00	0.63	0.07	0.04	1.84	1.07	3.53	m	3.88
ridge or hip tiles, Redland third round; butt jointed; bedding and pointing in tinted cement mortar (1:3)	6.76	5.00	7.10	0.34	0.17	8.59	0.63	16.32	m	17.95
ridge or hip tiles, Redland half round; butt jointed; bedding and pointing in tinted cement mortar (1:3)	6.76	5.00	7.10	0.34	0.17	8.59	0.63	16.32	m	17.95
verges; plain tile undercloak; tiles and undercloak bedded and pointed in tinted cement mortar (1:3) and with standard tiles and aluminium nails	2.70	5.00	2.83	0.11	0.06	2.85	0.68	6.36	m	6.99
verges; plain tile undercloak; tiles and undercloak bedded and pointed in tinted cement mortar (1:3) and with standard tiles and clips	2.70	5.00	2.83	0.11	0.06	2.85	5.18	10.86	m	11.94
verges; 150 x 6mm fibre cement undercloak; tiles and undercloak bedded and pointed in tinted cement mortar (1:3) and with standard tiles and aluminium nails	3.70	5.00	3.88	0.17	0.09	4.36	0.72	8.97	m	9.86
verges; 150 x 6mm fibre cement undercloak; tiles and undercloak bedded and pointed in tinted cement mortar (1:3) and with standard tiles and clips	3.70	5.00	3.88	0.17	0.09	4.36	5.18	13.42	m	14.76
valley tiles, Redland Universal valley troughs; laid with 100mm laps	33.00	5.00	34.65	0.34	0.17	8.59	0.63	43.86	m	48.25
Roof coverings; Redland Fenland through coloured pantiles, 381 x 227mm; fixing every fourth course with aluminium nails to 75mm lap; 38 x 25mm pressure impregnated softwood battens fixed with galvanized nails; twin-ply underlay 145gsm with polypropylene; 150mm laps; fixing with galvanized steel clout nails										
Pitched 40 degrees from horizontal generally	16.38	5.00	17.31	0.61	0.31	15.47	1.98	34.76	m²	38.24
Roof coverings; Redland Fenland through coloured pantiles, 381 x 227mm; fixing every fourth course with galvanized nails to 100mm lap; 38 x 25mm pressure impregnated softwood battens fixed with galvanized nails; twin-ply underlay 145gsm with polypropylene; 150mm laps; fixing with galvanized steel clout nails										
Pitched 40 degrees from horizontal generally	17.73	5.00	18.73	0.66	0.33	16.67	2.12	37.51	m²	41.26
Roof coverings; Redland 49 granular faced tiles, 381 x 227mm; fixing every fourth course with galvanized nails to 75mm lap; 38 x 25mm pressure impregnated softwood battens fixed with galvanized nails; twin-ply underlay 145gsm with polypropylene; 150mm laps; fixing with galvanized steel clout nails										
Pitched 40 degrees from horizontal generally	11.05	5.00	11.70	0.61	0.31	15.47	1.98	29.15	m²	32.07
Boundary work										
supplementary fixing eaves course with one aluminium nail per tile	-	-	-	0.02	0.01	0.50	0.05	0.55	m	0.61
ridge or hip tiles, Redland third round; butt jointed; bedding and pointing in tinted cement mortar (1:3)	6.76	5.00	7.10	0.34	0.17	8.59	0.63	16.32	m	17.95
ridge or hip tiles, Redland half round; butt jointed; bedding and pointing in tinted cement mortar (1:3)	6.76	5.00	7.10	0.34	0.17	8.59	0.63	16.32	m	17.95
verges; Redland Dry Verge system with standard tiles and clips	15.27	5.00	16.03	0.28	0.14	7.07	2.88	25.98	m	28.58
verges; Redland Dry Verge system with verge tiles and clips	15.27	5.00	16.03	0.28	0.14	7.07	2.88	25.98	m	28.58
verges; plain tile undercloak; tiles and undercloak bedded and pointed in tinted cement mortar (1:3) and with standard tiles and aluminium nails	1.77	5.00	1.86	0.11	0.06	2.85	0.72	5.42	m	5.97
verges; plain tile undercloak; tiles and undercloak bedded and pointed in tinted cement mortar (1:3) and with verge tiles and aluminium nails	1.77	5.00	1.86	0.11	0.06	2.85	0.72	5.42	m	5.97
verges; 150 x 6mm fibre cement undercloak; tiles and undercloak bedded and pointed in tinted cement mortar (1:3) and with standard tiles and aluminium nails	2.77	5.00	2.91	0.17	0.09	4.36	0.72	7.99	m	8.79
verges; 150 x 6mm fibre cement undercloak; tiles and undercloak bedded and pointed in tinted cement mortar (1:3) and with verge tiles and aluminium nails	2.77	5.00	2.91	0.17	0.09	4.36	0.72	7.99	m	8.79
valley tiles, Redland Universal valley troughs; laid with 100mm laps	33.00	5.00	34.65	0.34	0.17	8.59	0.63	43.86	m	48.25
Roof coverings; Redland 49 granular faced tiles, 381 x 227mm; fixing every fourth course with galvanized nails to 100mm lap; 38 x 25mm pressure impregnated softwood battens fixed with galvanized nails; twin-ply underlay 145gsm with polypropylene; 150mm laps; fixing with galvanized steel clout nails										
Pitched 40 degrees from horizontal generally	11.93	5.00	12.63	0.66	0.33	16.67	2.12	31.41	m²	34.55

Labour hourly rates: (except Specialists) Craft Operatives 18.37 Labourer 13.76 Rates are national average prices. Refer to REGIONAL VARIATIONS for indicative levels of overall pricing in regions	MATERIALS			LABOUR				RATES		
	Del to Site £	Waste %	Material Cost £	Craft Optve Hrs	Lab Hrs	Labour Cost £	Sunds £	Nett Rate £	Unit	Gross rate (10%) £
INTERLOCKING TILING (Cont'd)										
Roof coverings; Redland 49 granular faced tiles, 381 x 227mm; fixing every fourth course with galvanized nails to 100mm lap; 38 x 25mm pressure impregnated softwood battens fixed with galvanized nails; twin-ply underlay 145gsm with polypropylene; 150mm laps; fixing with galvanized steel clout nails (Cont'd)										
Pitched 40 degrees from horizontal (Cont'd)										
Boundary work										
abutments; square	1.05	5.00	1.10	0.11	0.06	2.85	-	3.95	m	4.34
abutments; raking	1.58	5.00	1.65	0.17	0.09	4.36	-	6.02	m	6.62
abutments; curved to 3000mm radius	1.66	5.00	1.75	0.23	0.12	5.88	-	7.62	m	8.39
Roof coverings; Redland 49 through coloured tiles, 381 x 227mm; fixing every fourth course with galvanized nails to 75mm lap; 38 x 25mm pressure impregnated softwood battens fixed with galvanized nails; twin-ply underlay 145gsm with polypropylene; 150mm laps; fixing with galvanized steel clout nails										
Pitched 40 degrees from horizontal										
generally	11.05	5.00	11.70	0.61	0.31	15.47	1.98	29.15	m²	32.07
Boundary work										
abutments; square	0.96	5.00	1.01	0.11	0.06	2.85	-	3.86	m	4.24
abutments; raking	1.45	5.00	1.52	0.17	0.09	4.36	-	5.88	m	6.47
abutments; curved to 3000mm radius	1.95	5.00	2.04	0.23	0.12	5.88	-	7.92	m	8.71
supplementary fixing eaves course with one aluminium nail per tile	-	-	-	0.02	0.01	0.50	0.05	0.55	m	0.61
ridge or hip tiles, Redland third round; butt jointed; bedding and pointing in tinted cement mortar (1:3)	6.76	5.00	7.10	0.34	0.17	8.59	0.63	16.32	m	17.95
ridge or hip tiles, Redland half round; butt jointed; bedding and pointing in tinted cement mortar (1:3)	6.76	5.00	7.10	0.34	0.17	8.59	0.63	16.32	m	17.95
verges; Redland Dry Verge system with standard tiles and clips.	15.27	5.00	16.03	0.28	0.14	7.07	2.88	25.98	m	28.58
verges; Redland Dry Verge system with verge tiles and clips	15.27	5.00	16.03	0.28	0.14	7.07	2.88	25.98	m	28.58
verges; plain tile undercloak; tiles and undercloak bedded and pointed in tinted cement mortar (1:3) and with standard tiles and aluminium nails	1.77	5.00	1.86	0.11	0.06	2.85	0.72	5.42	m	5.97
verges; plain tile undercloak; tiles and undercloak bedded and pointed in tinted cement mortar (1:3) and with verge tiles and aluminium nails	1.77	5.00	1.86	0.11	0.06	2.85	0.72	5.42	m	5.97
verges; 150 x 6mm fibre cement undercloak; tiles and undercloak bedded and pointed in tinted cement mortar (1:3) and with standard tiles and aluminium nails	2.77	5.00	2.91	0.17	0.09	4.36	0.72	7.99	m	8.79
verges; 150 x 6mm fibre cement undercloak; tiles and undercloak bedded and pointed in tinted cement mortar (1:3) and with verge tiles and aluminium nails	2.77	5.00	2.91	0.17	0.09	4.36	0.72	7.99	m	8.79
valley tiles, Redland Universal valley troughs; laid with 100mm laps	33.00	5.00	34.65	0.34	0.17	8.59	0.63	43.86	m	48.25
Roof coverings; Redland 49 through coloured tiles, 381 x 227mm; fixing every fourth course with galvanized nails to 100mm lap; 38 x 25mm pressure impregnated softwood battens fixed with galvanized nails; twin-ply underlay 145gsm with polypropylene; 150mm laps; fixing with galvanized steel clout nails										
Pitched 40 degrees from horizontal										
generally	11.93	5.00	12.63	0.66	0.33	16.67	2.12	31.41	m²	34.55
Roof coverings; Redland Renown granular faced tiles, 418 x 330mm; fixing every fourth course with galvanized nails to 75mm lap; 38 x 25mm pressure impregnated softwood battens fixed with galvanized nails; twin-ply underlay 145gsm with polypropylene; 150mm laps; fixing with galvanized steel clout nails										
Pitched 40 degrees from horizontal										
generally	9.48	5.00	10.05	0.47	0.24	11.94	0.86	22.85	m²	25.13
Boundary work										
abutments; square	0.81	5.00	0.85	0.11	0.06	2.85	-	3.69	m	4.06
abutments; raking	1.21	5.00	1.27	0.17	0.09	4.36	-	5.63	m	6.20
abutments; curved to 3000mm radius	1.61	5.00	1.69	0.23	0.12	5.88	-	7.57	m	8.32
reform eaves filler unit and aluminium nails to each eaves tile	0.60	5.00	0.63	0.07	0.04	1.84	1.07	3.53	m	3.88
ridge or hip tiles, Redland third round; butt jointed; bedding and pointing in tinted cement mortar (1:3)	6.76	5.00	7.10	0.34	0.17	8.59	0.63	16.32	m	17.95
ridge or hip tiles, Redland half round; butt jointed; bedding and pointing in tinted cement mortar (1:3)	6.76	5.00	7.10	0.34	0.17	8.59	0.63	16.32	m	17.95
verges; cloaked verge tiles and aluminium nails	2.74	5.00	2.88	0.17	0.09	4.36	0.17	7.40	m	8.14
verges; half tiles, cloaked verge tiles and aluminium nails	15.99	5.00	16.79	0.17	0.09	4.36	0.72	21.87	m	24.06
verges; plain tile undercloak; tiles and undercloak bedded and pointed in tinted cement mortar (1:3) and with standard tiles and aluminium nails	2.49	5.00	2.61	0.11	0.06	2.85	0.72	6.18	m	6.80
verges; plain tile undercloak; tiles and undercloak bedded and pointed in tinted cement mortar (1:3) and with verge tiles and aluminium nails	2.49	5.00	2.61	0.11	0.06	2.85	0.72	6.18	m	6.80
verges; 150 x 6mm fibre cement undercloak; tiles and undercloak bedded and pointed in tinted cement mortar (1:3) and with standard tiles and aluminium nails	3.49	5.00	3.66	0.17	0.09	4.36	0.72	8.75	m	9.62
verges; 150 x 6mm fibre cement undercloak; tiles and undercloak bedded and pointed in tinted cement mortar (1:3) and with verge tiles and aluminium nails	3.49	5.00	3.66	0.17	0.09	4.36	0.72	8.75	m	9.62
valley tiles, Redland Universal valley troughs; laid with 100mm laps	25.30	5.00	26.56	0.34	0.17	8.59	0.63	35.78	m	39.36

Labour hourly rates: (except Specialists) Craft Operatives 18.37 Labourer 13.76 Rates are national average prices. Refer to REGIONAL VARIATIONS for indicative levels of overall pricing in regions	MATERIALS			LABOUR				RATES		
	Del to Site £	Waste %	Material Cost £	Craft Optve Hrs	Lab Hrs	Labour Cost £	Sunds £	Nett Rate £	Unit	Gross rate (10%) £
INTERLOCKING TILING (Cont'd)										
Roof coverings; Redland 50 granular faced double roman tiles, 418 x 330mm; fixing every fourth course with galvanized nails to 75mm lap; 38 x 25mm pressure impregnated softwood battens fixed with galvanized nails; twin-ply underlay 145gsm with polypropylene; 150mm laps; fixing with galvanized steel clout nails										
Pitched 40 degrees from horizontal										
generally....................	9.67	5.00	10.26	0.47	0.24	11.94	1.03	23.23	m²	25.55
Boundary work										
reform eaves filler unit and aluminium nails to each eaves tile	0.60	5.00	0.63	0.07	0.04	1.84	1.07	3.53	m	3.88
ridge or hip tiles, Redland third round; butt jointed; bedding and pointing in tinted cement mortar (1:3)	6.76	5.00	7.10	0.34	0.17	8.59	0.63	16.32	m	17.95
ridge or hip tiles, Redland half round; butt jointed; bedding and pointing in tinted cement mortar (1:3)	6.76	5.00	7.10	0.34	0.17	8.59	0.63	16.32	m	17.95
verges; cloaked verge tiles and aluminium nails....................	16.05	5.00	16.85	0.17	0.09	4.36	0.10	21.32	m	23.45
verges; half tiles, cloaked verge tiles and aluminium nails..........	15.20	5.00	15.96	0.23	0.12	5.88	0.10	21.94	m	24.14
verges; plain tile undercloak; tiles and undercloak bedded and pointed in tinted cement mortar (1:3) and with standard tiles and aluminium nails	2.55	5.00	2.68	0.11	0.06	2.85	0.72	6.24	m	6.87
verges; plain tile undercloak; tiles and undercloak bedded and pointed in tinted cement mortar (1:3) and with treble roll verge tiles and aluminium nails............	2.55	5.00	2.68	0.11	0.06	2.85	0.72	6.24	m	6.87
verges; 150 x 6mm fibre cement undercloak; tiles and undercloak bedded and pointed in tinted cement mortar (1:3) and with standard tiles and aluminium nails....................	3.55	5.00	3.73	0.17	0.09	4.36	0.72	8.81	m	9.69
verges; 150 x 6mm fibre cement undercloak; tiles and undercloak bedded and pointed in tinted cement mortar (1:3) and with treble roll verge tiles and aluminium nails....................	3.55	5.00	3.73	0.17	0.09	4.36	0.72	8.81	m	9.69
valley tiles, Redland Universal valley troughs; laid with 100mm laps....................	33.00	5.00	34.65	0.34	0.17	8.59	0.63	43.86	m	48.25
Roof coverings; Redland 50 through coloured double roman tiles, 418 x 330mm; fixing every fourth course with galvanized nails to 75mm lap; 38 x 25mm pressure impregnated softwood battens fixed with galvanized nails; twin-ply underlay 145gsm with polypropylene; 150mm laps; fixing with galvanized steel clout nails										
Pitched 40 degrees from horizontal										
generally....................	9.67	5.00	10.26	0.47	0.24	11.94	1.03	23.23	m²	25.55
Boundary work										
reform eaves filler unit and aluminium nails to each eaves tile	0.60	5.00	0.63	0.07	0.04	1.84	1.07	3.53	m	3.88
ridge or hip tiles, Redland third round; butt jointed; bedding and pointing in tinted cement mortar (1:3)	6.76	5.00	7.10	0.34	0.17	8.59	0.63	16.32	m	17.95
ridge or hip tiles, Redland half round; butt jointed; bedding and pointing in tinted cement mortar (1:3)	6.76	5.00	7.10	0.34	0.17	8.59	0.63	16.32	m	17.95
verges; cloaked verge tiles and aluminium nails....................	16.05	5.00	16.85	0.17	0.09	4.36	0.10	21.32	m	23.45
verges; half tiles, cloaked verge tiles and aluminium nails..........	16.05	5.00	16.85	0.17	0.09	4.36	0.10	21.32	m	23.45
verges; plain tile undercloak; tiles and undercloak bedded and pointed in tinted cement mortar (1:3) and with standard tiles and aluminium nails	2.55	5.00	2.68	0.11	0.06	2.85	0.72	6.24	m	6.87
verges; plain tile undercloak; tiles and undercloak bedded and pointed in tinted cement mortar (1:3) and with treble roll verge tiles and aluminium nails............	2.55	5.00	2.68	0.11	0.06	2.85	0.72	6.24	m	6.87
verges; 150 x 6mm fibre cement undercloak; tiles and undercloak bedded and pointed in tinted cement mortar (1:3) and with standard tiles and aluminium nails....................	3.55	5.00	3.73	0.17	0.09	4.36	0.72	8.81	m	9.69
verges; 150 x 6mm fibre cement undercloak; tiles and undercloak bedded and pointed in tinted cement mortar (1:3) and with treble roll verge tiles and aluminium nails....................	3.55	5.00	3.73	0.17	0.09	4.36	0.72	8.81	m	9.69
valley tiles, Redland Universal valley troughs; laid with 100mm laps....................	33.00	5.00	34.65	0.34	0.17	8.59	0.63	43.86	m	48.25
Fittings										
Ventilator tiles										
Ridge Vent ridge ventilation terminal 450mm long for										
half round ridge....................	73.50	5.00	77.17	1.45	0.73	36.68	0.99	114.85	nr	126.33
universal angle ridge....................	73.50	5.00	77.17	1.45	0.73	36.68	0.99	114.85	nr	126.33
Thruvent tiles for										
Stonewold slates....................	23.53	5.00	24.71	0.56	0.28	14.14	0.38	39.22	nr	43.14
Regent tiles....................	23.53	5.00	24.71	0.56	0.28	14.14	0.38	39.22	nr	43.14
Grovebury double pantiles....................	23.53	5.00	24.71	0.56	0.28	14.14	0.38	39.22	nr	43.14
Fenland pantiles	23.53	5.00	24.71	0.56	0.28	14.14	0.38	39.22	nr	43.14
Redland 49 tiles....................	23.53	5.00	24.71	0.56	0.28	14.14	0.38	39.22	nr	43.14
Renown tiles....................	23.53	5.00	24.71	0.56	0.28	14.14	0.38	39.22	nr	43.14
Redland 50 double roman tiles	23.53	5.00	24.71	0.56	0.28	14.14	0.38	39.22	nr	43.14
Red line ventilation tile complete with underlay seal and fixing clips for										
Stonewold slates....................	23.50	5.00	24.67	1.12	0.56	28.28	0.38	53.33	nr	58.66
Regent tiles....................	23.50	5.00	24.67	1.12	0.56	28.28	0.38	53.33	nr	58.66
Grovebury double pantiles....................	23.50	5.00	24.67	1.12	0.56	28.28	0.38	53.33	nr	58.66
Renown tiles....................	23.50	5.00	24.67	1.12	0.56	28.28	0.38	53.33	nr	58.66
Redland 50 double roman tiles	23.50	5.00	24.67	1.12	0.56	28.28	0.38	53.33	nr	58.66
Eaves ventilation										
Insulation Interupter and roof space ventilator										
Felt support eaves ventilator tray fixing with aluminium nails.......	1.70	5.00	1.78	0.25	0.12	6.31	0.05	8.14	m	8.96
Over fascia ventilator										
10mm....................	2.30	-	2.30	0.30	0.15	7.58	0.05	9.92	m	10.91
25mm....................	6.75	-	6.75	0.30	0.15	7.58	0.05	14.37	m	15.81

Labour hourly rates: (except Specialists) Craft Operatives 18.37 Labourer 13.76 Rates are national average prices. Refer to REGIONAL VARIATIONS for indicative levels of overall pricing in regions	MATERIALS			LABOUR				RATES		
	Del to Site	Waste	Material Cost	Craft Optve	Lab	Labour Cost	Sunds	Nett Rate	Unit	Gross rate (10%)
	£	%	£	Hrs	Hrs	£	£	£		£
INTERLOCKING TILING (Cont'd)										
Fittings (Cont'd)										
Gas terminals										
Gas Flue ridge terminal, 450mm long with sealing gasket and fixing brackets										
half round ridge..	131.11	5.00	137.67	1.69	0.84	42.60	0.99	181.26	nr	199.39
extra for 150mm extension adaptor	18.00	5.00	18.90	0.34	0.17	8.59	0.38	27.86	nr	30.65
Hip irons										
galvanized mild steel; 32 x 3 x 380mm girth; scrolled; fixing with galvanized steel screws to timber....................................	0.80	5.00	0.84	0.11	0.06	2.85	-	3.69	nr	4.05
Lead soakers										
fixing only	-	-	-	0.34	0.17	8.59	-	8.59	nr	9.44
FIBRE CEMENT SLATING										
Quantities required for 1m² of slating										
Slates to 70mm lap										
400 x 240mm, 25.3nr										
Slates to 102mm lap										
600 x 300mm, 13nr										
500 x 250mm, 19.5nr										
400 x 200mm, 32nr										
Roof coverings; asbestos-free cement slates, 600 x 300mm; centre fixing with copper nails and copper disc rivets to 102mm lap; 38 x 19mm pressure impregnated softwood battens fixed with galvanized nails; twin-ply underlay 145gsm with polypropylene; 150mm laps; fixing with galvanized steel clout nails										
Pitched 30 degrees from horizontal										
generally..	10.52	5.00	11.15	0.60	0.30	15.15	2.40	28.70	m²	31.57
holes...	-	-	-	0.58	0.29	14.65	-	14.65	nr	16.11
Boundary work										
abutments; square....................................	0.90	5.00	0.94	0.23	0.12	5.88	-	6.82	m	7.50
abutments; raking....................................	1.35	5.00	1.41	0.28	0.14	7.07	-	8.48	m	9.33
abutments; curved to 3000mm radius............	1.79	5.00	1.88	0.35	0.18	8.91	-	10.79	m	11.87
double course at eaves.............................	2.41	5.00	2.54	0.17	0.09	4.36	1.10	7.99	m	8.79
verges; slate undercloak and point in cement mortar (1:3)	1.38	5.00	1.45	0.09	0.05	2.34	0.90	4.69	m	5.16
ridges or hips; asbestos free cement; fixing with nails	13.39	5.00	14.06	0.23	0.12	5.88	0.63	20.57	m	22.63
Roof coverings; asbestos-free cement slates, 600 x 300mm; centre fixing with copper nails and copper disc rivets to 102mm lap; 50 x 25mm pressure impregnated softwood battens fixed with galvanized nails; twin-ply underlay 145gsm with polypropylene; 150mm laps; fixing with galvanized steel clout nails										
Pitched 30 degrees from horizontal										
generally..	11.00	5.00	11.68	0.64	0.32	16.16	2.40	30.24	m²	33.26
Roof coverings; asbestos-free cement slates, 600 x 300mm; centre fixing with copper nails and copper disc rivets to 102mm lap (close boarding on rafters included elsewhere); twin-ply underlay 145gsm with polypropylene; 150mm laps; fixing with galvanized steel clout nails										
Pitched 30 degrees from horizontal										
generally..	9.56	5.00	10.10	0.48	0.24	12.12	1.75	23.97	m²	26.37
Roof coverings; asbestos-free cement slates, 500 x 250mm; centre fixing with copper nails and copper disc rivets to 102mm lap; 38 x 19mm pressure impregnated softwood battens fixed with galvanized nails; twin-ply underlay 145gsm with polypropylene; 150mm laps; fixing with galvanized steel clout nails										
Pitched 30 degrees from horizontal										
generally..	15.05	5.00	15.92	0.67	0.34	16.99	3.27	36.18	m²	39.79
holes...	-	-	-	0.58	0.29	14.65	-	14.65	nr	16.11
Boundary work										
abutments; square....................................	1.33	5.00	1.39	0.23	0.12	5.88	-	7.27	m	8.00
abutments; raking....................................	1.99	5.00	2.09	0.35	0.18	8.91	-	11.00	m	12.10
abutments; curved to 3000mm radius............	2.65	5.00	2.78	0.46	0.23	11.62	-	14.40	m	15.84
double course at eaves.............................	2.72	5.00	2.86	0.21	0.11	5.37	1.17	9.40	m	10.34
ridges or hips; asbestos free cement; fixing with nails	13.39	5.00	14.06	0.23	0.12	5.88	0.63	20.57	m	22.63
verges; slate undercloak and point in cement mortar (1:3)	1.70	5.00	1.78	0.12	0.06	3.03	0.96	5.78	m	6.35
Roof coverings; asbestos-free cement slates, 500 x 250mm; centre fixing with copper nails and copper disc rivets to 102mm lap; 50 x 25mm pressure impregnated softwood battens fixed with galvanized nails; twin-ply underlay 145gsm with polypropylene; 150mm laps; fixing with galvanized steel clout nails										
Pitched 30 degrees from horizontal										
generally..	15.65	5.00	16.58	0.72	0.36	18.18	3.27	38.03	m²	41.83

Labour hourly rates: (except Specialists) Craft Operatives 18.37 Labourer 13.76 Rates are national average prices. Refer to REGIONAL VARIATIONS for indicative levels of overall pricing in regions	MATERIALS			LABOUR				RATES		
	Del to Site	Waste	Material Cost	Craft Optve	Lab	Labour Cost	Sunds	Nett Rate	Unit	Gross rate (10%)
	£	%	£	Hrs	Hrs	£	£	£		£
FIBRE CEMENT SLATING (Cont'd)										
Roof coverings; asbestos-free cement slates, 400 x 200mm; centre fixing with copper nails and copper disc rivets to 102mm lap; 38 x 19mm pressure impregnated softwood battens fixed with galvanized nails; twin-ply underlay 145gsm with polypropylene; 150mm laps; fixing with galvanized steel clout nails										
Pitched 30 degrees from horizontal										
generally..	24.75	5.00	26.12	0.71	0.36	18.00	4.97	49.08	m²	53.99
holes..	-	-	-	0.58	0.29	14.65	-	14.65	nr	16.11
Boundary work										
abutments; square........................	2.27	5.00	2.39	0.29	0.15	7.39	-	9.78	m	10.75
abutments; raking........................	3.41	5.00	3.58	0.44	0.22	11.11	-	14.69	m	16.16
abutments; curved to 3000mm radius..........	4.54	5.00	4.77	0.58	0.29	14.65	-	19.42	m	21.36
double course at eaves....................	3.55	5.00	3.73	0.23	0.97	17.57	1.30	22.60	m	24.87
ridges or hips; asbestos free cement; fixing with nails........	13.39	5.00	14.06	0.23	0.12	5.88	0.63	20.57	m	22.63
verges; slate undercloak and point in cement mortar (1:3).......	2.13	5.00	2.24	0.14	0.07	3.54	1.02	6.79	m	7.47
Roof coverings; asbestos-free cement slates, 400 x 200mm; centre fixing with copper nails and copper disc rivets to 102mm lap; 50 x 25mm pressure impregnated softwood battens fixed with galvanized nails; twin-ply underlay 145gsm with polypropylene; 150mm laps; fixing with galvanized steel clout nails										
Pitched 30 degrees from horizontal										
generally..	25.47	5.00	26.91	0.80	0.40	20.20	4.97	52.07	m²	57.28
Fittings										
Hip irons										
galvanized mild steel; 32 x 3 x 380mm girth; scrolled; fixing with galvanized steel screws to timber...........	0.80	5.00	0.84	0.11	0.06	2.85	-	3.69	nr	4.05
Lead soakers										
fixing only	-	-	-	0.34	0.17	8.59	-	8.59	nr	9.44
NATURAL SLATING										
Quantities required for 1m² of slating										
Slates to 75mm lap 600 x 300mm, 12.65nr 500 x 250mm, 18.9nr 400 x 200mm, 31nr										
Roof coverings; blue/grey slates, 600 x 300mm, 6.5mm thick; fixing with slate nails to 75mm lap; 38 x 19mm pressure impregnated softwood battens fixed with galvanized nails; twin-ply underlay 145gsm with polypropylene; 150mm laps; fixing with galvanized steel clout nails										
Pitched 30 degrees from horizontal										
generally..	107.81	5.00	113.31	0.58	0.29	14.65	0.84	128.79	m²	141.67
holes..	-	-	-	0.58	0.29	14.65	-	14.65	nr	16.11
Boundary work										
abutments; square........................	10.33	5.00	10.85	0.17	0.09	4.36	-	15.21	m	16.73
abutments; raking........................	15.54	5.00	16.32	0.26	0.13	6.57	-	22.88	m	25.17
double course at eaves....................	29.40	5.00	30.87	0.16	0.08	4.04	0.75	35.66	m	39.23
verges; slate undercloak and point in cement mortar (1:3).......	16.80	5.00	17.64	0.09	0.05	2.34	0.69	20.67	m	22.74
Roof coverings; blue/grey slates, 600 x 300mm, 6.5mm thick; fixing with aluminium nails to 75mm lap; 38 x 19mm pressure impregnated softwood battens fixed with galvanized nails; twin-ply underlay 145gsm with polypropylene; 150mm laps; fixing with galvanized steel clout nails										
Pitched 30 degrees from horizontal										
generally..	107.81	5.00	113.31	0.58	0.29	14.65	1.66	129.62	m²	142.58
Roof coverings; blue/grey slates, 600 x 300mm, 6.5mm thick; fixing with copper nails to 75mm laps; 38 x 19mm pressure impregnated softwood battens fixed with galvanized nails; twin-ply underlay 145gsm with polypropylene; 150mm laps; fixing with galvanized steel clout nails										
Pitched 30 degrees from horizontal										
generally..	107.81	5.00	113.31	0.58	0.29	14.65	2.07	130.02	m²	143.02
Roof coverings; blue/grey slates, 600 x 300mm, 6.5mm thick; fixing with slate nails to 75mm lap; 50 x 25mm pressure impregnated softwood battens fixed with galvanized nails; twin-ply underlay 145gsm with polypropylene; 150mm laps; fixing with galvanized steel clout nails										
Pitched 30 degrees from horizontal										
generally..	108.29	5.00	113.83	0.62	0.31	15.66	0.84	130.33	m²	143.36

Labour hourly rates: (except Specialists) Craft Operatives 18.37 Labourer 13.76 Rates are national average prices. Refer to REGIONAL VARIATIONS for indicative levels of overall pricing in regions	MATERIALS			LABOUR				RATES		
	Del to Site	Waste	Material Cost	Craft Optve	Lab	Labour Cost	Sunds	Nett Rate	Unit	Gross rate (10%)
	£	%	£	Hrs	Hrs	£	£	£		£
NATURAL SLATING (Cont'd)										
Roof coverings; blue/grey slates, 600 x 300mm, 6.5mm thick; fixing with slate nails to 75mm lap; 50 x 25mm pressure impregnated softwood counterbattens at 1067mm centres and 38 x 19mm pressure impregnated softwood battens fixed with galvanized nails; twin-ply underlay 145gsm with polypropylene; 150mm laps; fixing with galvanized steel clout nails										
Pitched 30 degrees from horizontal										
generally............	108.29	5.00	113.83	0.62	0.31	15.66	1.71	131.20	m²	144.32
Roof coverings; blue/grey slates, 500 x 250mm, 6.5mm thick; fixing with slate nails to 75mm lap; 38 x 19mm pressure impregnated softwood battens fixed with galvanized nails; twin-ply underlay 145gsm with polypropylene; 150mm laps; fixing with galvanized steel clout nails										
Pitched 30 degrees from horizontal										
generally............	71.72	5.00	75.42	0.67	0.34	16.99	1.25	93.66	m²	103.02
holes............	-	-	-	0.58	0.29	14.65	-	14.65	nr	16.11
Boundary work										
abutments; square............	6.73	5.00	7.07	0.23	0.12	5.88	-	12.95	m	14.24
abutments; raking............	10.10	5.00	10.61	0.35	0.18	8.91	-	19.51	m	21.46
double course at eaves............	14.80	5.00	15.54	0.20	0.10	5.05	1.20	21.79	m	23.97
verges; slate undercloak and point in cement mortar (1:3)	9.25	5.00	9.71	0.12	0.06	3.03	0.70	13.45	m	14.79
Roof coverings; blue/grey slates, 500 x 250mm, 6.5mm thick; fixing with aluminium nails to 75mm lap; 38 x 19mm pressure impregnated softwood battens fixed with galvanized nails; twin-ply underlay 145gsm with polypropylene; 150mm laps; fixing with galvanized steel clout nails										
Pitched 30 degrees from horizontal										
generally............	71.72	5.00	75.42	0.67	0.34	16.99	2.47	94.89	m²	104.37
Roof coverings; blue/grey slates, 500 x 250mm, 6.5mm thick; fixing with copper nails to 75mm lap; 38 x 19mm pressure impregnated softwood battens fixed with galvanized nails; twin-ply underlay 145gsm with polypropylene; 150mm laps; fixing with galvanized steel clout nails										
Pitched 30 degrees from horizontal										
generally............	71.72	5.00	75.42	0.67	0.34	16.99	3.09	95.50	m²	105.05
Roof coverings; blue/grey slates, 500 x 250mm, 6.5mm thick; fixing with slate nails to 75mm lap; 50 x 25mm pressure impregnated softwood battens fixed with galvanized nails; twin-ply underlay 145gsm with polypropylene; 150mm laps; fixing with galvanized steel clout nails										
Pitched 30 degrees from horizontal										
generally............	72.32	5.00	76.08	0.72	0.36	18.18	1.25	95.51	m²	105.06
Roof coverings; blue/grey slates, 400 x 200mm, 6.5mm thick; fixing with slate nails to 75mm lap; 38 x 19mm pressure impregnated softwood battens fixed with galvanized nails; twin-ply underlay 145gsm with polypropylene; 150mm laps; fixing with galvanized steel clout nails										
Pitched 30 degrees from horizontal										
generally............	53.18	5.00	55.97	0.80	0.40	20.20	2.07	78.24	m²	86.06
holes............	-	-	-	0.58	0.29	14.65	-	14.65	nr	16.11
Boundary work										
abutments; square............	4.95	5.00	5.20	0.29	0.15	7.39	-	12.59	m	13.85
abutments; raking	7.42	5.00	7.80	0.44	0.22	11.11	-	18.91	m	20.80
double course at eaves............	9.90	5.00	10.39	0.23	0.12	5.88	0.62	16.89	m	18.57
verges; slate undercloak and point in cement mortar (1:3)	4.95	5.00	5.20	0.14	0.07	3.54	0.72	9.45	m	10.40
Roof coverings; blue/grey slates, 400 x 200mm, 6.5mm thick; fixing with aluminium nails to 75mm lap; 38 x 19mm pressure impregnated softwood battens fixed with galvanized nails; twin-ply underlay 145gsm with polypropylene; 150mm laps; fixing with galvanized steel clout nails										
Pitched 30 degrees from horizontal										
generally............	53.18	5.00	55.97	0.80	0.40	20.20	4.09	80.26	m²	88.29
Roof coverings; blue/grey slates, 400 x200mm, 6.5mm thick; fixing with copper nails to 75mmlap; 38 x 19mm pressure impregnated softwood battens fixed with galvanized nails; twin-ply underlay 145gsm with polypropylene; 150mm laps; fixing with galvanized steel clout nails										
Pitched 30 degrees from horizontal										
generally............	53.18	5.00	55.97	0.80	0.40	20.20	5.11	81.28	m²	89.41
Roof coverings; blue/grey slates, 400 x 200mm, 6.5mm thick; fixing with slate nails to 75mm lap; 50 x 25mm pressure impregnated softwood battens fixed with galvanized nails; twin-ply underlay 145gsm with polypropylene; 150mm laps; fixing with galvanized steel clout nails										
Pitched 30 degrees from horizontal										
generally............	53.90	5.00	56.76	0.87	0.44	22.04	2.07	80.87	m²	88.95

Labour hourly rates: (except Specialists) Craft Operatives 18.37 Labourer 13.76 Rates are national average prices. Refer to REGIONAL VARIATIONS for indicative levels of overall pricing in regions	MATERIALS			LABOUR				RATES		
	Del to Site £	Waste %	Material Cost £	Craft Optve Hrs	Lab Hrs	Labour Cost £	Sunds £	Nett Rate £	Unit	Gross rate (10%) £
NATURAL SLATING (Cont'd)										
Roof coverings; green slates, best random fixed in diminishing courses with alloy nails to 75mm lap; 38 x 19mm pressure impregnated softwood battens fixed with galvanized nails; twin-ply underlay 145gsm with polypropylene; 150mm laps; fixing with galvanized steel clout nails										
Pitched 30 degrees from horizontal										
generally..	135.18	5.00	142.07	0.86	0.43	21.72	3.05	166.83	m²	183.52
Boundary work										
abutments; square....................................	13.32	5.00	13.98	0.23	0.12	5.88	-	19.86	m	21.84
abutments; raking.....................................	19.97	5.00	20.97	0.35	0.18	8.91	-	29.88	m	32.87
double course at eaves.............................	29.05	5.00	30.50	0.23	0.12	5.88	1.51	37.90	m	41.69
verges; slate undercloak and point in cement mortar (1:3)	19.37	5.00	20.34	0.23	0.12	5.88	0.93	27.14	m	29.86
hips, valleys and angles; close mitred	21.79	5.00	22.88	0.40	0.20	10.10	0.98	33.95	m	37.35
Fittings										
Hip irons										
galvanized mild steel; 32 x 3 x 380mm girth; scrolled; fixing with galvanized steel screws to timber....................	0.80	5.00	0.84	0.11	0.06	2.85	-	3.69	nr	4.05
Lead soakers										
fixing only	-	-	-	0.34	0.17	8.59	-	8.59	nr	9.44
NATURAL OR ARTIFICIAL STONE SLATING										
Roof coverings; Redland Cambrian interlocking riven textured slates, 300 x 336mm; fixing every slate with two stainless steel nails and one stainless steel clip to 50mm laps; 38 x 19mm pressure impregnated softwood battens fixed with galvanized nails; twin-ply underlay 145gsm with polypropylene; 150mm laps; fixing with galvanized steel clout nails										
Pitched 40 degrees from horizontal										
generally..	33.54	5.00	35.32	0.54	0.27	13.64	2.94	51.90	m²	57.09
Boundary work										
abutments; square....................................	3.19	5.00	3.35	0.11	0.06	2.85	-	6.20	m	6.82
abutments; raking.....................................	4.80	5.00	5.04	0.17	0.09	4.36	-	9.40	m	10.34
abutments; curved to 3000mm radius..........	6.38	5.00	6.70	0.23	0.12	5.88	-	12.58	m	13.84
supplementary fixing at eaves with stainless steel eaves clip to each slate	0.52	5.00	0.55	0.05	0.03	1.33	-	1.88	m	2.07
ridge or hip tiles, Redland third round; butt jointed; bedding and pointing in tinted cement mortar (1:3)	6.76	5.00	7.10	0.23	0.12	5.88	0.63	13.61	m	14.97
ridge or hip tiles, Redland half round; butt jointed; bedding and pointing in tinted cement mortar (1:3)	6.76	5.00	7.10	0.23	0.12	5.88	0.63	13.61	m	14.97
ridge or hip tiles, Redland universal angle; butt jointed; bedding and pointing in tinted cement mortar (1:3)	15.03	5.00	15.78	0.23	0.12	5.88	0.63	22.28	m	24.51
ridge tiles, Redland Dry Ridge system with half round ridge tiles; fixing with stainless steel batten straps and ring shanked fixing nails with neoprene washers and sleeves; polypropylene ridge seals and uPVC profile filler units	19.44	5.00	20.42	0.45	0.23	11.43	-	31.85	m	35.03
ridge tiles, Redland Dry Ridge system with universal angle ridge tiles; fixing with stainless steel batten straps and ring shanked fixing nails with neoprene washers and sleeves; polypropylene ridge seals and uPVC profile filler units	19.81	5.00	20.81	0.45	0.23	11.43	-	32.24	m	35.46
ridge tiles, Redland Dry Vent Ridge system with half round ridge tiles; fixing with stainless steel batten straps and ring shanked nails with neoprene washers and sleeves; polypropylene ridge seals, PVC air flow control units and uPVC ventilated profile filler units...................	19.44	5.00	20.42	0.45	0.23	11.43	-	31.85	m	35.03
ridge tiles, Redland Dry Vent Ridge system with Universal angle ridge tiles; fixing with stainless steel batten straps and ring shanked nails with neoprene washers and sleeves; polypropylene ridge seals, PVC air flow control units and uPVC ventilated profile filler units...................	19.81	5.00	20.81	0.45	0.23	11.43	-	32.24	m	35.46
Monoridge filler units in conjunction with top tile, bed tile in nonsetting mastic sealant and screw on through filler unit with screws, washers and caps...............	43.03	5.00	45.18	0.20	0.12	5.33	-	50.51	m	55.56
verges; 150 x 6mm fibre cement undercloak; slates and undercloak bedded and pointed in tinted cement mortar (1:3) with full slates and slate and a half slates in alternate courses and stainless steel verge clips	8.64	5.00	9.07	0.23	0.12	5.88	3.40	18.35	m	20.19
verges; 150 x 6mm fibre cement undercloak; slates and undercloak bedded and pointed in tinted cement mortar (1:3) with verge slates and slate and a half verge slates in alternate courses and stainless steel verge clips...................	8.64	5.00	9.07	0.23	0.12	5.88	3.40	18.35	m	20.19
valley tiles, Redland Universal valley troughs; laid with 100mm laps	42.90	5.00	45.04	0.50	0.28	13.04	0.63	58.71	m	64.58
Roof coverings; Redland Cambrian interlocking riven textured slates, 300 x 336mm; fixing every slate with two stainless steel nails and one stainless steel clip to 90mm lap; 38 x 25mm pressure impregnated softwood battens fixed with galvanized nails; twin-ply underlay 145gsm with polypropylene; 150mm laps; fixing with galvanized steel clout nails										
Pitched 40 degrees from horizontal										
generally..	39.95	5.00	42.07	0.62	0.31	15.66	3.80	61.52	m²	67.67
Boundary work										
abutments; square....................................	3.82	5.00	4.01	0.11	0.06	2.85	-	6.85	m	7.54
abutments; raking	5.71	5.00	6.00	0.17	0.09	4.36	-	10.36	m	11.39
abutments; curved to 3000mm radius..........	7.63	5.00	8.01	0.23	0.12	5.88	-	13.89	m	15.28

Labour hourly rates: (except Specialists) Craft Operatives 18.37 Labourer 13.76 Rates are national average prices. Refer to REGIONAL VARIATIONS for indicative levels of overall pricing in regions	MATERIALS			LABOUR				RATES		
	Del to Site £	Waste %	Material Cost £	Craft Optve Hrs	Lab Hrs	Labour Cost £	Sunds £	Nett Rate £	Unit	Gross rate (10%) £
NATURAL OR ARTIFICIAL STONE SLATING (Cont'd)										
Fittings										
Ventilator tiles										
Ridge Vent ridge ventilation terminal 450mm long for										
half round ridge	52.50	5.00	55.12	1.45	0.73	36.68	-	91.81	nr	100.99
universal angle ridge	52.50	5.00	55.12	1.45	0.73	36.68	-	91.81	nr	100.99
Redland Cambrian Thruvent interlocking slate complete with weather cap, underlay seal and fixing clips										
slate	66.99	5.00	70.34	0.56	0.28	14.14	-	84.48	nr	92.93
Gas terminals										
Gas Flue ridge terminal, 450mm long with sealing gasket and fixing brackets										
half round ridge	131.11	5.00	137.67	1.50	0.75	37.88	-	175.54	nr	193.09
Gas Flue ridge terminal Mark III, 450mm long with sealing gasket and fixing brackets with 150mm R type adaptor for										
half round ridge	131.11	5.00	137.67	1.50	0.75	37.88	48.09	223.63	nr	245.99
Hip irons										
galvanized mild steel; 32 x 3 x 380mm girth; scrolled; fixing with galvanized steel screws to timber	0.80	5.00	0.84	0.11	0.06	2.85	-	3.69	nr	4.05
Lead soakers										
fixing only	-	-	-	0.34	0.17	8.59	-	8.59	nr	9.44
Fittings										
Timloc uPVC Mark 3 eaves ventilators; reference 1124 fixing with nails to timber at 400mm centres										
330mm wide	5.35	5.00	5.62	0.28	0.14	7.07	0.45	13.14	m	14.45
Timloc uPVC Mark 3 eaves ventilators; reference 1123 fixing with nails to timber at 600mm centres										
330mm wide	1.70	5.00	1.78	0.22	0.11	5.56	0.45	7.79	m	8.57
Timlock uPVC soffit ventilators; reference 1137, 10mm airflow; fixing with screws to timber										
type C	1.70	5.00	1.79	0.17	0.06	3.95	0.18	5.92	m	6.51
Timlock uPVC Mark 2 eaves ventilators; reference 1122; fitting between trusses at 600mm centres										
300mm girth	2.58	5.00	2.71	0.22	0.11	5.56	0.45	8.72	m	9.59
Timlock polypropylene over-fascia ventilators; reference 3011; fixing with screws to timber										
to top of fascia	3.93	5.00	4.13	0.22	0.11	5.56	0.27	9.96	m	10.95
TIMBER OR BITUMINOUS FELT SHINGLES										
Roof coverings; Western Red Cedar Shingles; fixing every course with silicon bronze annular ring nails to 25mm lap; 38 x 19mm pressure impregnated softwood battens fixed with galvanized nails; twin-ply underlay 145gsm with polypropylene; 150mm laps; fixing with galvanized steel clout nails										
Pitched 45 degrees from horizontal										
generally	27.33	5.00	28.85	0.55	0.55	17.67	-	46.53	m²	51.18
Boundary work										
abutments; square	10.83	5.00	11.37	0.60	0.60	19.28	-	30.65	m	33.71
abutments; raking	10.83	5.00	11.37	1.00	1.00	32.13	-	43.50	m	47.85
abutments; curved to 3000mm radius	10.83	5.00	11.37	1.50	1.50	48.19	-	59.56	m	65.52
eaves course	8.12	5.00	8.53	0.40	0.40	12.85	-	21.38	m	23.52
verges	10.83	5.00	11.37	0.60	0.60	19.28	-	30.65	m	33.71
top cutting to ridge or hip per single side	10.83	5.00	11.37	0.60	0.60	19.28	-	30.65	m	33.71
cutting to valleys and gutters	10.83	5.00	11.37	1.00	1.00	32.13	-	43.50	m	47.85

This page left blank intentionally

Labour hourly rates: (except Specialists) Craft Operatives 18.37 Labourer 13.76 Rates are national average prices. Refer to REGIONAL VARIATIONS for indicative levels of overall pricing in regions	MATERIALS			LABOUR				RATES		
	Del to Site	Waste	Material Cost	Craft Optve	Lab	Labour Cost	Sunds	Nett Rate	Unit	Gross rate (10%)
	£	%	£	Hrs	Hrs	£	£	£		£

MASTIC ASPHALT ROOFING

Roofing; BS 6925 (limestone aggregate); sand rubbing; BS EN 13707 type 4A sheathing felt isolating membrane, butt joints

	Del to Site	Waste	Material Cost	Craft Optve	Lab	Labour Cost	Sunds	Nett Rate	Unit	Gross rate
Roofing 20mm thick; coats of asphalt -2; to concrete base; pitch not exceeding 6 degrees from horizontal										
width exceeding 500mm	-	-	Specialist	-	-	Specialist	-	44.19	m²	48.61
extra; solid water check roll	-	-	Specialist	-	-	Specialist	-	24.72	m	27.19
Roofing 20mm thick; coats of asphalt -2; to concrete base; pitch 30 degrees from horizontal										
width exceeding 500mm	-	-	Specialist	-	-	Specialist	-	70.83	m²	77.91
Roofing 20mm thick; coats of asphalt -2; to concrete base; pitch 45 degrees from horizontal										
width exceeding 500mm	-	-	Specialist	-	-	Specialist	-	92.34	m²	101.57
Paint two coats of Solar reflective roof paint on surfaces of asphalt										
width exceeding 500mm	-	-	Specialist	-	-	Specialist	-	11.84	m²	13.02
Skirtings 13mm thick; coats of asphalt -2; to brickwork base										
girth not exceeding 150mm	-	-	Specialist	-	-	Specialist	-	29.10	m	32.01
Coverings to kerbs 20mm thick; coats of asphalt -2; to concrete base										
girth 600mm	-	-	Specialist	-	-	Specialist	-	60.21	m	66.23
Jointing new roofing to existing										
20mm thick	-	-	Specialist	-	-	Specialist	-	37.10	m	40.80
Edge trim; aluminium; silver anodized priming with bituminous primer; butt joints with internal jointing sleeves; fixing with aluminium alloy screws to timber; working two coat asphalt into grooves -1										
63.5mm wide x 44.4mm face depth	3.09	5.00	3.24	0.36	-	6.61	1.88	11.73	m	12.90
63.5mm wide x 76.2mm face depth	3.71	5.00	3.89	0.36	-	6.61	1.97	12.47	m	13.72
104.8mm wide x 44.4mm face depth	4.40	5.00	4.62	0.36	-	6.61	1.97	13.20	m	14.52
104.8mm wide x 76.2mm face depth	31.42	5.00	33.00	0.44	-	8.08	2.23	43.31	m	47.65
76.2mm fixing arm at 10 degrees x 38.1mm face depth	9.00	5.00	9.45	0.36	-	6.61	1.97	18.03	m	19.83
Edge trim; glass fibre reinforced butt joints; fixing with stainless steel screws to timber; working two coat asphalt into grooves -1	29.23	5.00	30.69	0.30	-	5.51	2.23	38.43	m	42.28
Roof Ventilators										
plastic; setting in position	4.99	2.50	5.11	0.75	-	13.78	0.74	19.63	nr	21.59
aluminium; setting in position	8.14	2.50	8.34	0.75	-	13.78	0.74	22.86	nr	25.14

Roofing; BS 6925 (limestone aggregate); covering with 13mm white spar chippings in hot bitumen; BS EN 13707 type 4A sheathing felt isolating membrane; butt joints

	Del to Site	Waste	Material Cost	Craft Optve	Lab	Labour Cost	Sunds	Nett Rate	Unit	Gross rate
Roofing 20mm thick; coats of asphalt -2; to concrete base; pitch not exceeding 6 degrees from horizontal										
width not exceeding 500mm	-	-	Specialist	-	-	Specialist	-	47.43	m²	52.17
Roofing 20mm thick; coats of asphalt -2; to concrete base; pitch 30 degrees from horizontal										
width not exceeding 500mm	-	-	Specialist	-	-	Specialist	-	71.08	m²	78.19

ASPHALT TANKING OR DAMP PROOFING

Tanking and damp proofing BS 6925 (limestone aggregate)

	Del to Site	Waste	Material Cost	Craft Optve	Lab	Labour Cost	Sunds	Nett Rate	Unit	Gross rate
Tanking and damp proofing 13mm thick; coats of asphalt -1; to concrete base; horizontal										
work subsequently covered										
width not exceeding 500mm	-	-	Specialist	-	-	Specialist	-	17.95	m	19.75
width exceeding 500mm	-	-	Specialist	-	-	Specialist	-	28.25	m²	31.07
Tanking and damp proofing 20mm thick; coats of asphalt -2; to concrete base; horizontal										
work subsequently covered										
width not exceeding 500mm	-	-	Specialist	-	-	Specialist	-	22.58	m	24.83
width exceeding 500mm	-	-	Specialist	-	-	Specialist	-	41.06	m²	45.16
Tanking and damp proofing 30mm thick; coats of asphalt -3; to concrete base; horizontal										
work subsequently covered										
width not exceeding 500mm	-	-	Specialist	-	-	Specialist	-	32.25	m	35.48
width exceeding 500mm	-	-	Specialist	-	-	Specialist	-	58.58	m²	64.43
Tanking and damp proofing 20mm thick; coats of asphalt -2; to concrete base; vertical										
work subsequently covered										
width not exceeding 500mm	-	-	Specialist	-	-	Specialist	-	51.51	m	56.66
width exceeding 500mm	-	-	Specialist	-	-	Specialist	-	93.66	m²	103.03

WATERPROOFING

Labour hourly rates: (except Specialists) Craft Operatives 18.37 Labourer 13.76 Rates are national average prices. Refer to REGIONAL VARIATIONS for indicative levels of overall pricing in regions	MATERIALS			LABOUR				RATES		
	Del to Site £	Waste %	Material Cost £	Craft Optve Hrs	Lab Hrs	Labour Cost £	Sunds £	Nett Rate £	Unit	Gross rate (10%) £
ASPHALT TANKING OR DAMP PROOFING (Cont'd)										
Tanking and damp proofing BS 6925 (limestone aggregate) (Cont'd)										
Tanking and damp proofing 20mm thick; coats of asphalt -3; to concrete base; vertical										
work subsequently covered										
width not exceeding 500mm	-	-	Specialist	-	-	Specialist	-	71.22	m	78.34
width exceeding 500mm	-	-	Specialist	-	-	Specialist	-	129.51	m²	142.46
Internal angle fillets to concrete base; priming base with bitumen two coats; work subsequently covered	-	-	Specialist	-	-	Specialist	-	13.52	m	14.87
APPLIED LIQUID TANKING OR DAMP PROOFING										
Synthaprufe bituminous emulsion; blinding with sand										
Coverings, coats -2; to concrete base; horizontal										
width not exceeding 500mm	2.59	10.00	2.84	-	0.13	1.79	0.10	4.74	m	5.21
width exceeding 500mm	5.17	10.00	5.69	-	0.20	2.75	0.19	8.64	m²	9.50
Coverings, coats -3; to concrete base; horizontal										
width not exceeding 500mm	3.86	10.00	4.25	-	0.18	2.48	0.10	6.83	m	7.51
width exceeding 500mm	7.72	10.00	8.49	-	0.28	3.85	0.19	12.54	m²	13.79
Coverings; coats -2; to concrete base; vertical										
width not exceeding 500mm	2.59	10.00	2.84	-	0.20	2.75	0.10	5.70	m	6.27
width exceeding 500mm	5.17	10.00	5.69	-	0.30	4.13	0.19	10.01	m²	11.01
Coverings; coats -3; to concrete base; vertical										
width not exceeding 500mm	3.86	10.00	4.25	-	0.28	3.85	0.10	8.20	m	9.02
width exceeding 500mm	7.72	10.00	8.49	-	0.42	5.78	0.19	14.47	m²	15.91
Bituminous emulsion										
Coverings, coats -2; to concrete base; horizontal										
width not exceeding 500mm	1.00	10.00	1.10	-	0.12	1.72	0.10	2.93	m	3.22
width exceeding 500mm	2.01	10.00	2.21	-	0.20	2.75	0.21	5.17	m²	5.69
Coverings, coats -3; to concrete base; horizontal										
width not exceeding 500mm	1.41	10.00	1.55	-	0.18	2.48	0.10	4.13	m	4.54
width exceeding 500mm	2.81	10.00	3.09	-	0.28	3.85	0.21	7.16	m²	7.87
Coverings; coats -2; to concrete base; vertical										
width not exceeding 500mm	1.00	10.00	1.10	-	0.19	2.61	0.10	3.82	m	4.21
width exceeding 500mm	2.01	10.00	2.21	-	0.30	4.13	0.21	6.55	m²	7.20
Coverings; coats -3; to concrete base; vertical										
width not exceeding 500mm	1.41	10.00	1.55	-	0.27	3.72	0.10	5.37	m	5.90
width exceeding 500mm	2.81	10.00	3.09	-	0.42	5.78	0.21	9.08	m²	9.99
R.I.W. liquid asphaltic composition										
Coverings, coats -2; to concrete base; horizontal										
width not exceeding 500mm	6.95	10.00	7.65	-	0.15	2.06	0.10	9.82	m	10.80
width exceeding 500mm	13.91	10.00	15.30	-	0.23	3.16	0.21	18.67	m²	20.54
Coverings, coats -3; to concrete base; horizontal										
width not exceeding 500mm	10.51	10.00	11.56	-	0.21	2.89	0.10	14.56	m	16.01
width exceeding 500mm	20.94	10.00	23.03	-	0.32	4.40	0.21	27.65	m²	30.41
Coverings; coats -2; to concrete base; vertical										
width not exceeding 500mm	6.95	10.00	7.65	-	0.23	3.16	0.10	10.92	m	12.01
width exceeding 500mm	13.91	10.00	15.30	-	0.35	4.82	0.21	20.32	m²	22.36
Coverings; coats -3; to concrete base; vertical										
width not exceeding 500mm	10.51	10.00	11.56	-	0.32	4.40	0.10	16.07	m	17.68
width exceeding 500mm	20.94	10.00	23.03	-	0.48	6.60	0.21	29.85	m²	32.83
FLEXIBLE SHEET TANKING OR DAMP PROOFING										
Polythene sheeting; 100mm welted laps										
Tanking and damp proofing; 125mu										
horizontal; on concrete base	0.28	20.00	0.33	-	0.06	0.83	-	1.16	m²	1.27
vertical; on concrete base	0.28	20.00	0.33	-	0.09	1.24	-	1.57	m²	1.73
Tanking and damp proofing; 250mu										
horizontal; on concrete base	0.28	20.00	0.34	-	0.09	1.24	-	1.58	m²	1.74
vertical; on concrete base	0.28	20.00	0.34	-	0.14	1.93	-	2.27	m²	2.50
Tanking and damp proofing; 300mu										
horizontal; on concrete base	0.34	20.00	0.41	-	0.11	1.51	-	1.92	m²	2.11
vertical; on concrete base	0.34	20.00	0.41	-	0.17	2.34	-	2.75	m²	3.02
Bituthene 500X self adhesive damp proof membrane; 25mm lapped joints										
Tanking and damp proofing										
horizontal; on concrete base	8.89	20.00	10.67	-	0.18	2.48	-	13.15	m²	14.46
vertical; on concrete base	8.89	20.00	10.67	-	0.27	3.72	-	14.39	m²	15.83
Bituthene 500X self adhesive damp proof membrane; 75mm lapped joints										
Tanking and damp proofing										
horizontal; on concrete base	9.74	20.00	11.69	-	0.23	3.16	-	14.85	m²	16.34
vertical; on concrete base; priming with Servicised B primer	9.74	20.00	11.69	-	0.35	4.82	0.39	16.89	m²	18.58

Labour hourly rates: (except Specialists) Craft Operatives 18.37 Labourer 13.76 Rates are national average prices. Refer to REGIONAL VARIATIONS for indicative levels of overall pricing in regions	MATERIALS			LABOUR				RATES		
	Del to Site £	Waste %	Material Cost £	Craft Optve Hrs	Lab Hrs	Labour Cost £	Sunds £	Nett Rate £	Unit	Gross rate (10%) £
FLEXIBLE SHEET TANKING OR DAMP PROOFING (Cont'd)										
British Sisalkraft Ltd 728 damp proof membrane; 150mm laps sealed with tape										
Tanking and damp proofing horizontal; on concrete base	1.32	20.00	1.58	-	0.08	1.10	0.17	2.85	m²	3.13
Visqueen 1200 super damp proof membrane; 150mm laps sealed with tape and mastic										
Tanking and damp proofing horizontal; on concrete base	0.52	20.00	0.63	-	0.08	1.10	0.17	1.89	m²	2.08

This page left blank intentionally

Labour hourly rates: (except Specialists) Craft Operatives 18.37 Labourer 13.76 Rates are national average prices. Refer to REGIONAL VARIATIONS for indicative levels of overall pricing in regions	MATERIALS			LABOUR				RATES		
	Del to Site	Waste	Material Cost	Craft Optve	Lab	Labour Cost	Sunds	Nett Rate	Unit	Gross rate (10%)
	£	%	£	Hrs	Hrs	£	£	£		£
METAL FRAMED SYSTEMS TO WALLS AND CEILINGS; **UNDER PURLIN/INSIDE RAIL PANEL LININGS**										
Gypliner metal grid fixing system; square edge sheeting of **one layer 12.5mm thick Gyproc wallboard, BS EN 520, butt** **joints; screwing to mild steel grid.**										
12.5mm thick linings to walls										
width exceeding 300mm....................................	6.29	5.00	6.64	0.80	0.40	20.20	0.40	27.24	m²	29.97
width not exceeding 300mm............................	7.50	5.00	7.88	0.48	0.12	10.47	0.40	18.74	m	20.62
12.5mm thick linings to ceilings										
width exceeding 300mm....................................	6.21	5.00	6.56	1.00	0.50	25.25	0.40	32.21	m²	35.43
width not exceeding 300mm............................	7.50	5.00	7.88	0.60	0.15	13.09	0.40	21.36	m	23.50
METAL FRAMED SYSTEMS TO WALLS AND CEILINGS										
Metal stud partitions; proprietary partitions; Gyproc Gypwall **metal stud; tapered edge sheeting of one layer 12.5mm thick** **Gyproc wallboard for direct decoration, BS EN 520, butt joints;** **applying continuous beads of Gyproc acoustical sealant to** **perimeters of construction both sides, fixing with Pozidriv** **head screws to steel frame comprising 50mm head and floor** **channels, 48mm vertical studs at 600mm centres; butt joints** **filled with joint filler tape and joint finish, spot filling; fixing to** **masonry with cartridge fired nails**										
Height 2100 - 2400mm, 75mm thick										
boarded both sides....................................	19.77	5.00	20.75	2.50	1.25	63.13	-	83.88	m	92.27
Metal stud partitions; proprietary partitions; Gyproc Gypwall **metal stud; tapered edge sheeting of two layers 12.5mm thick** **Gyproc wallboard for direct decoration, BS EN 520, butt joints;** **applying continuous beads of Gyproc acoustical sealant to** **perimeters of construction both sides, fixing with Pozidriv** **head screws to steel frame comprising 50mm head and floor** **channels, 48mm vertical studs at 600mm centres; butt joints** **filled with joint filler tape and joint finish, spot filling; fixing to** **masonry with cartridge fired nails**										
Height 2100 - 2400mm, 100mm thick										
boarded both sides....................................	28.00	5.00	29.40	2.80	1.40	70.70	-	100.10	m	110.11
Height 2400 - 2700mm, 100mm thick										
boarded both sides....................................	30.37	5.00	31.88	3.15	1.58	79.61	-	111.49	m	122.64
Height 2700 - 3000mm, 100mm thick										
boarded both sides....................................	32.73	5.00	34.37	3.50	1.75	88.38	-	122.74	m	135.02
Angles to partitions; fixing end studs together through partition; extending sheeting to one side across end of partition; joints filled with joint filler tape and joint finish Plain										
75mm thick partitions...............................	0.70	10.00	0.74	0.10	-	1.84	-	2.57	m	2.83
100mm thick partitions.............................	0.73	10.00	0.78	0.10	-	1.84	-	2.61	m	2.88

This page left blank intentionally

Labour hourly rates: (except Specialists) Craft Operatives 18.37 Labourer 13.76 Rates are national average prices. Refer to REGIONAL VARIATIONS for indicative levels of overall pricing in regions	MATERIALS			LABOUR				RATES		
	Del to Site £	Waste %	Material Cost £	Craft Optve Hrs	Lab Hrs	Labour Cost £	Sunds £	Nett Rate £	Unit	Gross rate (10%) £
PATENT GLAZING, CURTAIN WALLING & RAINSCREEN CLADDING										
Patent glazing with aluminium alloy bars 2000mm long with aluminium wings and seatings for glass, spaced at approximately 600mm centres, glazed with BS 952 Georgian wired cast, 7mm thick										
Walls; vertical surfaces; exceeding 600 wide										
single tier	153.36	5.00	161.03	1.70	1.70	54.62	3.87	219.52	m²	241.47
multi-tier	162.81	5.00	170.95	1.95	1.95	62.65	3.87	237.48	m²	261.22
Roof areas; exceeding 600 wide										
single tier	138.84	5.00	145.78	1.68	1.68	53.98	3.87	203.63	m²	224.00
multi-tier	148.19	5.00	155.60	1.95	1.95	62.65	3.87	222.12	m²	244.33
Patent glazing with aluminium alloy bars 2000mm long with aluminium wings and seatings for glass, spaced at approximately 600mm centres, glazed with BS 952 Georgian wired polished, 6mm thick										
Walls; vertical surfaces; exceeding 600 wide										
single tier	222.76	5.00	233.90	1.90	1.90	61.05	3.87	298.81	m²	328.69
multi-tier	236.25	5.00	248.06	2.20	2.20	70.69	3.87	322.62	m²	354.88
Roof areas; exceeding 600 wide										
single tier	208.20	5.00	218.61	1.90	1.90	61.05	3.87	283.53	m²	311.88
multi-tier	222.97	5.00	234.12	2.20	2.20	70.69	3.87	308.67	m²	339.54
Patent glazing with aluminium alloy bars 3000mm long with aluminium wings and seatings for glass, spaced at approximately 600mm centres, glazed with BS 952 Georgian wired cast, 7mm thick										
Walls; vertical surfaces; exceeding 600 wide										
single tier	176.29	5.00	185.11	1.70	1.70	54.62	3.87	243.60	m²	267.96
multi-tier	187.47	5.00	196.84	1.95	1.95	62.65	3.87	263.36	m²	289.70
Roof areas; exceeding 600 wide										
single tier	163.04	5.00	171.20	1.70	1.70	54.62	3.87	229.69	m²	252.66
multi-tier	172.84	5.00	181.48	1.95	1.95	62.65	3.87	248.01	m²	272.81
Patent glazing with aluminium alloy bars 3000mm long with aluminium wings and seatings for glass, spaced at approximately 600mm centres, glazed with BS 952 Georgian wired polished, 6mm thick										
Walls; vertical surfaces; exceeding 600 wide										
single tier	245.07	5.00	257.32	1.90	1.90	61.05	3.87	322.24	m²	354.46
multi-tier	156.87	5.00	164.71	2.20	2.20	70.69	3.87	239.27	m²	263.20
Roof areas; exceeding 600 wide										
single tier	230.92	5.00	242.46	1.90	1.90	61.05	3.87	307.38	m²	338.12
multi-tier	247.38	5.00	259.75	2.20	2.20	70.69	3.87	334.30	m²	367.74
Extra over patent glazing in roof areas with aluminium alloy bars 2000mm long with aluminium wings and seatings for glass, spaced at approximately 600mm centres, glazed with BS 952 Georgian wired polished, 6mm thick										
Opening lights including opening gear										
600 x 600mm	279.71	5.00	293.69	5.00	5.00	160.65	-	454.34	nr	499.78
600 x 900mm	337.46	5.00	354.33	6.30	6.30	202.42	-	556.75	nr	612.43
Extra over patent glazing in roof areas with aluminium alloy bars 2000mm long with aluminium wings and seatings for glass, spaced at approximately 600mm centres, glazed with BS 952 Georgian wired cast, 7mm thick										
Opening lights including opening gear										
600 x 600mm	280.24	5.00	294.26	4.60	4.60	147.80	-	442.06	nr	486.26
600 x 900mm	337.47	5.00	354.34	5.75	5.75	184.75	-	539.09	nr	593.00
Extra over patent glazing in vertical surfaces with aluminium alloy bars 3000mm long with aluminium wings and seatings for glass, spaced at approximately 600mm centres, glazed with BS 952 Georgian wired cast, 7mm thick										
Opening lights including opening gear										
600 x 600mm	403.39	5.00	423.56	2.30	2.30	73.90	-	497.46	nr	547.20
600 x 900mm	488.12	5.00	512.53	3.50	3.50	112.46	-	624.99	nr	687.48

CLADDING AND COVERING

CLADDING AND COVERING

Labour hourly rates: (except Specialists) Craft Operatives 18.37 Labourer 13.76 Rates are national average prices. Refer to REGIONAL VARIATIONS for indicative levels of overall pricing in regions	MATERIALS			LABOUR				RATES		
	Del to Site	Waste	Material Cost	Craft Optve	Lab	Labour Cost	Sunds	Nett Rate	Unit	Gross rate (10%)
	£	%	£	Hrs	Hrs	£	£	£		£
PATENT GLAZING, CURTAIN WALLING & RAINSCREEN CLADDING (Cont'd)										
Extra over patent glazing to vertical surfaces with aluminium alloy bars 3000mm long with aluminium wings and seatings for glass, spaced at approximately 600mm centres, glazed with BS 952 Georgian wired polished, 6mm thick										
Opening lights including opening gear										
600 x 600mm	403.35	5.00	423.51	2.55	2.55	81.93	-	505.45	nr	555.99
600 x 900mm	488.04	5.00	512.44	3.75	3.75	120.49	-	632.93	nr	696.22
WEATHERBOARDING										
Softwood, wrought, impregnated										
Boarding to walls, shiplapped joints; 19mm thick, 150mm wide boards										
width exceeding 600mm; fixing to timber	17.78	10.00	19.56	0.85	0.10	16.99	0.21	36.76	m²	40.44
Boarding to walls, shiplapped joints; 25mm thick, 150mm wide boards										
width exceeding 600mm; fixing to timber	24.17	10.00	26.58	0.90	0.11	18.05	0.22	44.85	m²	49.34
Western Red Cedar										
Boarding to walls, shiplapped joints; 19mm thick, 150mm wide boards										
width exceeding 600mm; fixing to timber	36.00	10.00	39.60	0.90	0.12	18.18	0.22	58.01	m²	63.81
Boarding to walls, shiplapped joints; 25mm thick, 150mm wide boards										
width exceeding 600mm; fixing to timber	38.72	10.00	42.59	0.95	0.12	19.10	0.22	61.92	m²	68.11
Abutments										
19mm thick softwood boarding										
raking cutting	1.78	10.00	1.96	0.07	-	1.29	-	3.24	m	3.57
curved cutting	2.67	10.00	2.93	0.10	-	1.84	-	4.77	m	5.25
25mm thick softwood boarding										
raking cutting	2.42	10.00	2.66	0.08	-	1.47	-	4.13	m	4.54
curved cutting	3.62	10.00	3.99	0.12	-	2.20	-	6.19	m	6.81
19mm thick Western Red Cedar boarding										
raking cutting	3.60	10.00	3.96	0.09	-	1.65	-	5.61	m	6.17
curved cutting	5.40	10.00	5.94	0.13	-	2.39	-	8.33	m	9.16
25mm thick Western Red Cedar boarding										
raking cutting	3.87	10.00	4.26	0.10	-	1.84	-	6.10	m	6.71
curved cutting	5.81	10.00	6.39	0.15	-	2.76	-	9.14	m	10.06
PROFILED SHEET CLADDING OR ROOFING; FIBRE CEMENT										
Roof coverings; corrugated reinforced cement Profile 3 sheeting, standard grey colour; lapped one and a half corrugations at sides and 150mm at ends										
Coverings; fixing to timber joists at 900mm general spacing with galvanized mild steel drive screws and washers; drilling holes										
width exceeding 600mm; pitch 30 degrees from horizontal	12.32	15.00	14.17	0.28	0.28	9.00	0.72	23.88	m²	26.27
Coverings; fixing to steel purlins at 900mm general spacing with galvanized hook bolts and washers; drilling holes										
width exceeding 600mm; pitch 30 degrees from horizontal	12.32	15.00	14.17	0.38	0.38	12.21	1.47	27.85	m²	30.63
Extra for										
Holes										
for pipes, standards or the like	-	-	-	0.58	-	10.65	-	10.65	nr	11.72
Boundary work										
Ridges										
two piece plain angular adjustable ridge tiles	23.32	10.00	25.66	0.23	0.23	7.39	-	33.05	m	36.35
Bottom edges; Aprons/sills										
apron flashings	23.06	10.00	25.37	0.15	0.15	4.82	-	30.19	m	33.21
Eaves										
eaves filler pieces	1.68	10.00	1.84	0.12	0.12	3.86	-	5.70	m	6.27
Verge barge boards										
standard barge boards	13.07	10.00	14.38	0.12	0.12	3.86	-	18.24	m	20.06
Finials										
standard ridge cap finials	34.60	10.00	38.06	0.17	0.17	5.46	-	43.52	nr	47.87
Cutting										
raking	2.46	15.00	2.83	0.28	0.28	9.00	-	11.83	m	13.01
Roof coverings; corrugated reinforced cement Profile 3 sheeting, standard coloured; lapped one and a half corrugations at sides and 150mm at ends										
Coverings; fixing to timber joists at 900mm general spacing with galvanized mild steel drive screws and washers; drilling holes										
width exceeding 600mm; pitch 30 degrees from horizontal	15.83	15.00	18.21	0.28	0.28	9.00	0.96	28.16	m²	30.98
Coverings; fixing to steel purlins at 900mm general spacing with galvanized hook bolts and washers; drilling holes										
width exceeding 600mm; pitch 30 degrees from horizontal	15.83	15.00	18.21	0.38	0.38	12.21	1.89	32.31	m²	35.54
Extra for										
Holes										
for pipes, standards or the like	-	-	-	0.58	-	10.65	-	10.65	nr	11.72

Labour hourly rates: (except Specialists) Craft Operatives 18.37 Labourer 13.76 Rates are national average prices. Refer to REGIONAL VARIATIONS for indicative levels of overall pricing in regions	MATERIALS			LABOUR				RATES		
	Del to Site £	Waste %	Material Cost £	Craft Optve Hrs	Lab Hrs	Labour Cost £	Sunds £	Nett Rate £	Unit	Gross rate (10%) £
PROFILED SHEET CLADDING OR ROOFING; FIBRE CEMENT (Cont'd)										
Roof coverings; corrugated reinforced cement Profile 3 sheeting, standard coloured; lapped one and a half corrugations at sides and 150mm at ends (Cont'd)										
Boundary work										
Ridges										
two piece plain angular adjustable ridge tiles	39.50	10.00	43.45	0.23	0.23	7.39	-	50.84	m	55.92
Bottom edges; Aprons/sills										
apron flashings	19.60	10.00	21.56	0.15	0.15	4.82	-	26.38	m	29.02
Eaves										
eaves filler pieces	1.68	10.00	1.84	0.12	0.12	3.86	-	5.70	m	6.27
Barge boards										
standard barge boards	18.76	10.00	20.64	0.12	0.12	3.86	-	24.49	m	26.94
Finials										
standard ridge cap finials	39.50	10.00	43.45	0.17	0.17	5.46	-	48.91	nr	53.80
Cutting										
raking	3.17	15.00	3.64	0.28	0.28	9.00	-	12.64	m	13.90
Roof coverings; corrugated reinforced cement Profile 6 sheeting, standard grey colour; lapped half a corrugation at sides and 150mm at ends										
Coverings; fixing to timber joists at 900mm general spacing with galvanized mild steel drive screws and washers; drilling holes; exceeding 600 wide										
pitch 30 degrees from horizontal	9.53	15.00	10.96	0.28	0.28	9.00	0.78	20.74	m²	22.81
Coverings; fixing to steel purlins at 900mm general spacing with galvanized hook bolts and washers; drilling holes; exceeding 600 wide										
pitch 30 degrees from horizontal	9.53	15.00	10.96	0.38	0.38	12.21	1.53	24.70	m²	27.17
Extra for										
Holes										
for pipes, standards or the like	-	-	-	0.58	-	10.65	-	10.65	nr	11.72
Boundary work										
Ridges										
two piece plain angular adjustable ridge tiles	21.98	10.00	24.18	0.23	0.23	7.39	-	31.57	m	34.72
Bottom edges; Aprons/sills										
apron flashings	12.20	10.00	13.42	0.15	0.15	4.82	-	18.24	m	20.06
Eaves										
eaves filler pieces	2.07	10.00	2.28	0.12	0.12	3.86	-	6.13	m	6.75
Barge boards										
standard barge boards	13.07	10.00	14.38	0.12	0.12	3.86	-	18.24	m	20.06
Finials										
standard ridge cap finials	14.04	10.00	15.44	0.17	0.17	5.46	-	20.91	nr	23.00
Cutting										
raking	1.91	15.00	2.19	0.28	0.28	9.00	-	11.19	m	12.31
Roof coverings; corrugated reinforced cement Profile 6 sheeting, standard coloured; lapped half a corrugation at sides and 150mm at ends										
Coverings; fixing to timber joists at 900mm general spacing with galvanized mild steel drive screws and washers; drilling holes; exceeding 600 wide										
pitch 30 degrees from horizontal	10.97	15.00	12.61	0.28	0.28	9.00	0.94	22.55	m²	24.81
Coverings; fixing to steel purlins at 900mm general spacing with galvanized hook bolts and washers; drilling holes; exceeding 600 wide										
pitch 30 degrees from horizontal	10.97	15.00	12.61	0.38	0.38	12.21	1.89	26.71	m²	29.38
Extra for										
Holes										
for pipes, standards or the like	-	-	-	0.58	-	10.65	-	10.65	nr	11.72
Boundary work										
Ridges										
two piece plain angular adjustable ridge tiles	34.98	10.00	38.48	0.23	0.23	7.39	-	45.87	m	50.45
Bottom edges; Aprons/sills										
apron flashings	19.89	10.00	21.88	0.15	0.15	4.82	-	26.70	m	29.37
Eaves										
eaves filler pieces	2.07	10.00	2.28	0.12	0.12	3.86	-	6.13	m	6.75
Barge boards										
standard barge boards	11.10	10.00	12.21	0.12	0.12	3.86	-	16.06	m	17.67
Finials										
standard ridge cap finials	44.97	10.00	49.47	0.17	0.17	5.46	-	54.93	nr	60.42
Cutting										
raking	2.19	15.00	2.52	0.28	0.28	9.00	-	11.52	m	12.67
PROFILED SHEET CLADDING OR ROOFING; METAL										
Wall cladding; PVC colour coated both sides galvanized steel profiled sheeting 0.50mm thick and with sheets secured at seams and laps										
Coverings; fixing to steel rails at 900mm general spacing with galvanized hook bolts and washers; drilling holes; exceeding 600 wide										
pitch 90 degrees from horizontal	5.85	5.00	6.14	0.40	0.40	12.85	1.77	20.76	m²	22.84
Angles; internal										
vertical corner flashings	5.67	5.00	5.95	0.17	0.17	5.46	-	11.42	m	12.56

191

Labour hourly rates: (except Specialists) Craft Operatives 18.37 Labourer 13.76 Rates are national average prices. Refer to REGIONAL VARIATIONS for indicative levels of overall pricing in regions	MATERIALS			LABOUR				RATES		
	Del to Site	Waste	Material Cost	Craft Optve	Lab	Labour Cost	Sunds	Nett Rate	Unit	Gross rate (10%)
	£	%	£	Hrs	Hrs	£	£	£		£
PROFILED SHEET CLADDING OR ROOFING; METAL (Cont'd)										
Wall cladding; PVC colour coated both sides galvanized steel profiled sheeting 0.50mm thick and with sheets secured at seams and laps (Cont'd)										
Angles; external vertical corner flashings	5.67	5.00	5.95	0.17	0.17	5.46	-	11.42	m	12.56
Filler blocks PVC	1.60	5.00	1.68	0.12	0.12	3.86	-	5.54	m	6.09
Cutting raking	1.17	5.00	1.23	0.28	0.28	9.00	-	10.22	m	11.25
Roof coverings; PVC colour coated both sides galvanized steel profiled sheeting 0.70mm thick and with sheets secured at seams and laps										
Coverings; fixing to steel purlins at 900mm general spacing with galvanized hook bolts and washers; drilling holes; exceeding 600 wide pitch 30 degrees from horizontal	5.85	5.00	6.14	0.35	0.35	11.25	1.83	19.22	m²	21.14

Boundary work
  Ridges
    ridge cappings
	6.17	5.00	6.47	0.17	0.17	5.46	-	11.94	m	13.13

Flashings gable flashings	5.67	5.00	5.95	0.17	0.17	5.46	-	11.41	m	12.55
eaves flashings	4.50	5.00	4.72	0.17	0.17	5.46	-	10.19	m	11.21
Filler blocks PVC	1.60	5.00	1.68	0.12	0.12	3.86	-	5.54	m	6.09
Cutting raking	1.17	5.00	1.23	0.28	0.28	9.00	-	10.22	m	11.25
PROFILED SHEET CLADDING OR ROOFING; PLASTICS										
Roof coverings; standard corrugated glass fibre reinforced translucent sheeting; lapped one and a half corrugations at sides and 150mm at ends										
Coverings; fixing to timber joists at 900mm general spacing with galvanized mild steel drive screws and washers; drilling holes; exceeding 600 wide pitch 30 degrees from horizontal	7.46	5.00	7.83	0.28	0.28	9.00	0.50	17.32	m²	19.06
Coverings; fixing to steel purlins at 900mm general spacing with galvanized hook bolts and washers; drilling holes; exceeding 600 wide pitch 30 degrees from horizontal	7.46	5.00	7.83	0.35	0.35	11.25	0.62	19.69	m²	21.66
Cutting raking	1.49	5.00	1.57	0.28	0.28	9.00	-	10.56	m	11.62
Roof coverings; fire resisting corrugated glass fibre reinforced translucent sheeting; lapped one and a half corrugations at sides and 150mm at ends										
Coverings; fixing to timber joists at 900mm general spacing with galvanized mild steel drive screws and washers; drilling holes; exceeding 600 wide pitch 30 degrees from horizontal	20.39	5.00	21.41	0.28	0.28	9.00	0.50	30.90	m²	33.99
Coverings; fixing to steel purlins at 900mm general spacing with galvanized hook bolts and washers; drilling holes; exceeding 600 wide pitch 30 degrees from horizontal	20.39	5.00	21.41	0.35	0.35	11.25	0.80	33.45	m²	36.79
Cutting raking	4.08	5.00	4.28	0.28	0.28	9.00	-	13.28	m	14.61
Roof coverings; standard vinyl corrugated sheeting; lapped one and a half corrugations at sides and 150mm at ends										
Coverings; fixing to timber joists at 900mm general spacing with galvanized mild steel drive screws and washers; drilling holes; exceeding 600 wide pitch 30 degrees from horizontal	9.23	5.00	9.69	0.28	0.28	9.00	0.80	19.48	m²	21.43
Coverings; fixing to steel purlins at 900mm general spacing with galvanized hook bolts and washers; drilling holes; exceeding 600 wide pitch 30 degrees from horizontal	9.23	5.00	9.69	0.35	0.35	11.25	0.80	21.73	m²	23.90
Cutting raking	1.85	5.00	1.94	0.28	0.28	9.00	-	10.93	m	12.03
Wall cladding; Swish Products high impact rigid uPVC profiled sections; colour white; secured with starter sections and clips										

Coverings with shiplap profile code C002 giving 150mm cover; fixing to timber; exceeding 600 wide
  vertical cladding
    sections applied horizontally
	59.53	5.00	62.51	0.42	0.42	13.49	2.95	78.96	m²	86.85

sections applied vertically	59.53	5.00	62.51	0.42	0.42	13.49	2.95	78.96	m²	86.85

Coverings with open V profile code C003 giving 150mm cover; fixing to timber; exceeding 600 wide
  vertical cladding
    sections applied horizontally
	59.53	5.00	62.51	0.45	0.45	14.46	4.22	81.18	m²	89.30

Labour hourly rates: (except Specialists) Craft Operatives 18.37 Labourer 13.76 Rates are national average prices. Refer to REGIONAL VARIATIONS for indicative levels of overall pricing in regions	MATERIALS			LABOUR				RATES		
	Del to Site £	Waste %	Material Cost £	Craft Optve Hrs	Lab Hrs	Labour Cost £	Sunds £	Nett Rate £	Unit	Gross rate (10%) £
PROFILED SHEET CLADDING OR ROOFING; PLASTICS (Cont'd)										
Wall cladding; Swish Products high impact rigid uPVC profiled sections; colour white; secured with starter sections and clips (Cont'd)										
Coverings with open V profile code C003 giving 150mm cover; fixing to timber; exceeding 600 wide (Cont'd) vertical cladding (Cont'd) sections applied vertically	59.53	5.00	62.51	0.45	0.45	14.46	4.22	81.18	m²	89.30
Coverings with open V profile code C269 giving 100mm cover; fixing to timber; exceeding 600 wide										
exceeding 600 wide vertical cladding sections applied horizontally	67.24	5.00	70.60	0.38	0.38	12.21	2.95	85.77	m²	94.34
sections applied vertically	67.24	5.00	70.60	0.38	0.38	12.21	2.95	85.77	m²	94.34
Vertical angles section C030 for vertically applied section	6.80	5.00	7.14	0.11	0.11	3.53	0.24	10.91	m	12.01
PANEL OR SLAB CLADDING OR ROOFING										
Blue/grey slate facings; natural riven finish; bedding, jointing and pointing in gauged mortar (1:2:9)										
450 x 600 x 20mm units to walls on brickwork or blockwork base plain, width exceeding 600mm	102.08	2.50	104.64	2.00	2.00	64.26	0.76	169.66	m²	186.63
450 x 600 x 30mm units to walls on brickwork or blockwork base plain, width exceeding 600mm	134.17	2.50	137.52	2.60	2.60	83.54	0.86	221.92	m²	244.11
450 x 600 x 40mm units to walls on brickwork or blockwork base plain, width exceeding 600mm	147.45	2.50	151.14	2.90	2.90	93.18	1.18	245.50	m²	270.05
450 x 600 x 50mm units to walls on brickwork or blockwork base plain, width exceeding 600mm	181.99	2.50	186.54	3.45	3.45	110.85	1.46	298.84	m²	328.72
Blue/grey slate facings; fine rubbed finish, one face; bedding, jointing and pointing in gauged mortar (1:2:9)										
750 x 1200 x 20mm units to walls on brickwork or blockwork base plain, width exceeding 600mm	252.90	2.50	259.23	2.30	2.30	73.90	2.20	335.33	m²	368.86
extra; rubbed square edges	20.29	-	20.29	-	-	-	-	20.29	m	22.31
extra; half rounded edges	62.16	-	62.16	-	-	-	-	62.16	m	68.38
extra; full rounded edges	62.16	-	62.16	-	-	-	-	62.16	m	68.38
extra; rebated joints	29.38	-	29.38	-	-	-	-	29.38	m	32.32
750 x 1200 x 30mm units to walls on brickwork or blockwork base plain, width exceeding 600mm	302.28	2.50	309.84	2.60	2.60	83.54	2.34	395.72	m²	435.29
extra; rubbed square edges	25.59	-	25.59	-	-	-	-	25.59	m	28.15
extra; half rounded edges	70.99	-	70.99	-	-	-	-	70.99	m	78.09
extra; full rounded edges	70.99	-	70.99	-	-	-	-	70.99	m	78.09
extra; rebated joints	34.91	-	34.91	-	-	-	-	34.91	m	38.40
750 x 1200 x 40mm units to walls on brickwork or blockwork base plain, width exceeding 600mm	309.84	2.50	317.59	2.90	2.90	93.18	2.64	413.41	m²	454.75
extra; rubbed square edges	25.59	-	25.59	-	-	-	-	25.59	m	28.15
extra; half rounded edges	70.99	-	70.99	-	-	-	-	70.99	m	78.09
extra; full rounded edges	70.99	-	70.99	-	-	-	-	70.99	m	78.09
extra; rebated joints	34.91	-	34.91	-	-	-	-	34.91	m	38.40
750 x 1200 x 50mm units to walls on brickwork or blockwork base plain, width exceeding 600mm	328.86	2.50	337.08	3.45	3.45	110.85	2.92	450.85	m²	495.94
extra; rubbed square edges	25.59	-	25.59	-	-	-	-	25.59	m	28.15
extra; half rounded edges	70.99	-	70.99	-	-	-	-	70.99	m	78.09
extra; full rounded edges	70.99	-	70.99	-	-	-	-	70.99	m	78.09
extra; rebated joints	34.15	-	34.15	-	-	-	-	34.15	m	37.56
Blue/grey slate facings; fine rubbed finish, both faces; bedding, jointing and pointing in gauged mortar (1:2:9)										
750 x 1200 x 20mm units to walls on brickwork or blockwork base plain, width exceeding 600mm	296.15	2.50	303.56	2.90	2.90	93.18	3.66	400.39	m²	440.43
750 x 1200 x 30mm units to walls on brickwork or blockwork base plain, width exceeding 600mm	345.71	2.50	354.36	3.15	3.15	101.21	3.86	459.42	m²	505.36
750 x 1200 x 40mm units to walls on brickwork or blockwork base plain, width exceeding 600mm	355.07	2.50	363.94	3.45	3.45	110.85	4.16	478.95	m²	526.84
750 x 1200 x 50mm units to walls on brickwork or blockwork base plain, width exceeding 600mm	376.08	2.50	385.48	4.00	4.00	128.52	4.35	518.35	m²	570.19
Green slate; bedding, jointing and pointing in gauged mortar (1:2:9)										
30mm thick units to sills on brickwork or blockwork base width 125mm; weathered, throated and grooved	74.48	2.50	76.34	0.65	0.65	20.88	0.96	98.18	m	108.00
width 190mm; weathered, throated and grooved	98.54	2.50	101.01	0.70	0.70	22.49	1.34	124.83	m	137.32
40mm thick units to sills on brickwork or blockwork base width 125mm	87.73	2.50	89.92	0.70	0.70	22.49	0.96	113.37	m	124.71
extra; stooling for jambs	27.61	-	27.61	-	-	-	-	27.61	nr	30.38
extra; notching for jambs and mullions	27.61	-	27.61	-	-	-	-	27.61	nr	30.38
width 190mm	133.35	2.50	136.68	0.85	0.85	27.31	1.25	165.24	m	181.76
extra; stooling for jambs	27.61	-	27.61	-	-	-	-	27.61	nr	30.38
extra; notching for jambs and mullions	27.61	-	27.61	-	-	-	-	27.61	nr	30.38

CLADDING AND COVERING

Labour hourly rates: (except Specialists) Craft Operatives 18.37 Labourer 13.76 Rates are national average prices. Refer to REGIONAL VARIATIONS for indicative levels of overall pricing in regions	MATERIALS			LABOUR				RATES		
	Del to Site	Waste	Material Cost	Craft Optve	Lab	Labour Cost	Sunds	Nett Rate	Unit	Gross rate (10%)
	£	%	£	Hrs	Hrs	£	£	£		£
PANEL OR SLAB CLADDING OR ROOFING (Cont'd)										
Green slate; bedding, jointing and pointing in gauged mortar (1:2:9) (Cont'd)										
50mm thick units to sills on brickwork or blockwork base										
width 125mm	108.70	2.50	111.41	0.85	0.85	27.31	0.96	139.68	m	153.65
extra; stooling for jambs	27.55	-	27.55	-	-	-	-	27.55	nr	30.31
extra; notching for jambs and mullions	27.55	-	27.55	-	-	-	-	27.55	nr	30.31
width 190mm	165.22	2.50	169.35	1.20	1.20	38.56	1.25	209.15	m	230.06
extra; stooling for jambs	27.55	-	27.55	-	-	-	-	27.55	nr	30.31
extra; notching for jambs and mullions	27.55	-	27.55	-	-	-	-	27.55	nr	30.31
30mm units to window boards on brickwork or blockwork base										
width 230mm	119.29	2.50	122.27	0.85	0.85	27.31	1.58	151.16	m	166.27
25mm units to combined sills and window boards on brickwork or blockwork base										
width 360mm	156.52	2.50	160.44	1.15	1.15	36.95	2.20	199.59	m	219.55

Labour hourly rates: (except Specialists) Craft Operatives 18.37 Labourer 13.76 Rates are national average prices. Refer to REGIONAL VARIATIONS for indicative levels of overall pricing in regions	MATERIALS			LABOUR				RATES		
	Del to Site	Waste	Material Cost	Craft Optve	Lab	Labour Cost	Sunds	Nett Rate	Unit	Gross rate (10%)
	£	%	£	Hrs	Hrs	£	£	£		£
UNFRAMED ISOLATED TRIMS, SKIRTINGS AND SUNDRY ITEMS										
Softwood, wrought - skirtings, architraves, picture rails and cover fillets										
Skirtings, picture rails, architraves and the like; (finished sizes)										
19 x 100mm	1.32	10.00	1.45	0.25	0.03	5.01	0.06	6.52	m	7.17
19 x 150mm	2.91	10.00	3.20	0.28	0.04	5.69	0.08	8.97	m	9.87
25 x 100mm	1.65	10.00	1.82	0.27	0.03	5.37	0.06	7.25	m	7.97
25 x 150mm	2.79	10.00	3.07	0.31	0.04	6.25	0.08	9.39	m	10.33
25 x 50mm chamfered and rounded	1.15	10.00	1.26	0.25	0.03	5.01	0.06	6.33	m	6.96
25 x 63mm chamfered and rounded	1.15	10.00	1.26	0.28	0.04	5.69	0.08	7.03	m	7.74
25 x 75mm chamfered and rounded	1.41	10.00	1.55	0.27	0.03	5.37	0.06	6.98	m	7.68
32 x 100mm chamfered and rounded	3.50	10.00	3.85	0.31	0.04	6.25	0.08	10.17	m	11.19
15 x 50mm; mouldings -1	1.20	10.00	1.32	0.21	0.03	4.27	0.06	5.65	m	6.22
15 x 125mm; mouldings -1	1.65	10.00	1.82	0.25	0.03	5.01	0.06	6.88	m	7.57
19 x 50mm; mouldings -1	1.20	10.00	1.32	0.22	0.03	4.45	0.06	5.83	m	6.42
19 x 125mm; mouldings -1	1.65	10.00	1.82	0.27	0.03	5.37	0.06	7.25	m	7.97
25 x 50mm; mouldings -1	1.98	10.00	2.18	0.22	0.03	4.45	0.06	6.69	m	7.36
25 x 63mm; mouldings -1	1.99	10.00	2.19	0.24	0.03	4.82	0.06	7.07	m	7.78
25 x 75mm; mouldings -1	2.29	10.00	2.52	0.25	0.03	5.01	0.06	7.58	m	8.34
32 x 100mm; mouldings -1	2.52	10.00	2.77	0.30	0.04	6.06	0.08	8.91	m	9.80
38 x 75mm; mouldings -1	3.65	10.00	4.02	0.28	0.04	5.69	0.08	9.78	m	10.76
Cover fillets, stops, trims, beads, nosings and the like; (finished sizes)										
12 x 50mm; mouldings -1	0.69	10.00	0.76	0.21	0.03	4.27	0.06	5.09	m	5.60
19 x 19mm; mouldings -1	0.69	10.00	0.76	0.19	0.02	3.77	0.05	4.57	m	5.03
19 x 50mm; mouldings -1	0.69	10.00	0.76	0.22	0.03	4.45	0.06	5.27	m	5.80
25 x 25mm; mouldings -1	0.99	10.00	1.09	0.21	0.03	4.27	0.06	5.42	m	5.96
25 x 38mm; mouldings -1	1.15	10.00	1.26	0.22	0.03	4.45	0.06	5.78	m	6.36
19 x 19mm; chamfers -1	0.69	10.00	0.76	0.19	0.02	3.77	0.05	4.57	m	5.03
25 x 25mm; chamfers -1	0.99	10.00	1.09	0.21	0.03	4.27	0.06	5.42	m	5.96
25 x 38mm; chamfers -1	1.15	10.00	1.26	0.22	0.03	4.45	0.06	5.78	m	6.36
Oak, wrought - skirtings, architraves, picture rails and cover fillets										
Skirtings, picture rails, architraves and the like; (finished sizes)										
19 x 100mm	4.57	10.00	5.03	0.42	0.05	8.40	0.15	13.59	m	14.94
19 x 150mm	6.93	10.00	7.62	0.48	0.06	9.64	0.15	17.41	m	19.15
25 x 100mm	6.22	10.00	6.84	0.46	0.06	9.28	0.15	16.26	m	17.89
25 x 150mm	9.35	10.00	10.29	0.48	0.06	9.64	0.15	20.08	m	22.09
extra; ends	1.00	10.00	1.10	0.05	0.01	1.06	0.01	2.17	nr	2.39
extra; angles	2.00	-	2.00	0.10	0.01	1.97	0.01	3.99	nr	4.39
extra; mitres	2.00	-	2.00	0.10	0.01	1.97	0.03	4.00	nr	4.41
extra; intersections	2.00	-	2.00	0.10	0.01	1.97	0.03	4.00	nr	4.41
25 x 50mm; splays -1	4.17	10.00	4.58	0.37	0.05	7.48	0.15	12.22	m	13.44
25 x 63mm; splays -1	4.96	10.00	5.45	0.41	0.05	8.22	0.15	13.82	m	15.21
25 x 75mm; splays -1	5.42	10.00	5.96	0.42	0.05	8.40	0.15	14.51	m	15.96
32 x 100mm; splays -1	7.92	10.00	8.71	0.48	0.06	9.64	0.15	18.50	m	20.35
38 x 75mm; splays -1	8.48	10.00	9.32	0.48	0.06	9.64	0.15	19.12	m	21.03
extra; ends	1.33	-	1.33	0.05	0.01	1.06	0.01	2.40	nr	2.64
extra; angles	2.26	-	2.26	0.10	0.01	1.97	0.03	4.26	nr	4.69
extra; mitres	2.26	-	2.26	0.10	0.01	1.97	0.03	4.26	nr	4.69
extra; intersections	2.26	-	2.26	0.10	0.01	1.97	0.03	4.26	nr	4.69
15 x 50mm; mouldings -1	2.92	10.00	3.21	0.36	0.05	7.30	0.15	10.66	m	11.73
15 x 125mm; mouldings -1	6.48	10.00	7.13	0.42	0.05	8.40	0.15	15.69	m	17.25
19 x 50mm; mouldings -1	4.08	10.00	4.49	0.37	0.05	7.48	0.15	12.13	m	13.34
19 x 125mm; mouldings -1	6.48	10.00	7.13	0.46	0.06	9.28	0.15	16.56	m	18.21
25 x 50mm; mouldings -1	4.17	10.00	4.58	0.37	0.05	7.48	0.15	12.22	m	13.44
25 x 63mm; mouldings -1	4.96	10.00	5.45	0.41	0.05	8.22	0.15	13.82	m	15.21
25 x 75mm; mouldings -1	5.42	10.00	5.96	0.42	0.05	8.40	0.15	14.51	m	15.96
32 x 100mm; mouldings -1	7.92	10.00	8.71	0.48	0.06	9.64	0.15	18.50	m	20.35
38 x 75mm; mouldings -1	8.46	10.00	9.30	0.48	0.06	9.64	0.15	19.10	m	21.01
extra; ends	1.33	-	1.33	0.05	0.01	1.06	0.01	2.40	nr	2.64
extra; angles	2.26	-	2.26	0.10	0.01	1.97	0.03	4.26	nr	4.69
extra; mitres	2.26	-	2.26	0.10	0.01	1.97	0.03	4.26	nr	4.69
extra; intersections	2.26	-	2.26	0.10	0.01	1.97	0.03	4.26	nr	4.69
15 x 125mm; chamfers -1	6.50	10.00	7.15	0.42	0.05	8.40	0.15	15.70	m	17.27
19 x 125mm; chamfers -1	6.50	10.00	7.15	0.46	0.06	9.28	0.15	16.58	m	18.23
25 x 75mm; chamfers -2	5.00	10.00	5.50	0.42	0.05	8.40	0.15	14.05	m	15.46
25 x 100mm; chamfers -2	8.60	10.00	9.46	0.44	0.06	8.91	0.15	18.52	m	20.37
Ash, wrought - skirtings, architraves, picture rails and cover fillets										
Cover fillets, stops, trims, beads, nosings and the like; (finished sizes)										
12 x 50mm; mouldings -1	3.21	10.00	3.53	0.36	0.05	7.30	0.12	10.95	m	12.05
19 x 19mm; mouldings -1	1.64	10.00	1.81	0.32	0.04	6.43	0.12	8.35	m	9.19
19 x 50mm; mouldings -1	3.83	10.00	4.22	0.37	0.05	7.48	0.12	11.82	m	13.00
25 x 25mm; mouldings -1	2.50	10.00	2.75	0.36	0.05	7.30	0.12	10.17	m	11.19

GENERAL JOINERY (side margin)

Labour hourly rates: (except Specialists) Craft Operatives 18.37 Labourer 13.76 Rates are national average prices. Refer to REGIONAL VARIATIONS for indicative levels of overall pricing in regions	MATERIALS			LABOUR				RATES		
	Del to Site	Waste	Material Cost	Craft Optve	Lab	Labour Cost	Sunds	Nett Rate	Unit	Gross rate (10%)
	£	%	£	Hrs	Hrs	£	£	£		£
UNFRAMED ISOLATED TRIMS, SKIRTINGS AND SUNDRY ITEMS (Cont'd)										
Ash, wrought - skirtings, architraves, picture rails and cover fillets (Cont'd)										
Cover fillets, stops, trims, beads, nosings and the like; (finished sizes) (Cont'd)										
25 x 38mm; mouldings -1	3.33	10.00	3.67	0.37	0.05	7.48	0.12	11.27	m	12.40
19 x 19mm; chamfers -1	1.64	10.00	1.81	0.32	0.04	6.43	0.12	8.35	m	9.19
25 x 25mm; chamfers -1	2.50	10.00	2.75	0.36	0.05	7.30	0.12	10.17	m	11.19
25 x 38mm; chamfers -1	3.33	10.00	3.67	0.37	0.05	7.48	0.12	11.27	m	12.40
Medium density fibreboard, skirtings, architraves, picture rails and cover fillets										
Skirtings, picture rails, architraves and the like; (finished sizes)										
18 x 44mm; chamfered	0.89	10.00	0.98	0.28	0.04	5.69	0.09	6.76	m	7.44
18 x 94mm; chamfered	1.85	10.00	2.04	0.29	0.04	5.88	0.09	8.00	m	8.81
18 x 144mm; chamfered	2.64	10.00	2.90	0.31	0.04	6.25	0.09	9.24	m	10.16
18 x 44mm; moulded	0.89	10.00	0.98	0.29	0.04	5.88	0.09	6.95	m	7.64
18 x 68mm; moulded	1.13	10.00	1.24	0.29	0.04	5.88	0.09	7.21	m	7.93
18 x 94mm; moulded	1.85	10.00	2.04	0.30	0.04	6.06	0.09	8.19	m	9.01
18 x 144mm; moulded	2.64	10.00	2.90	0.32	0.04	6.43	0.09	9.42	m	10.36
18 x 219mm; moulded	4.43	10.00	4.87	0.35	0.04	6.98	0.09	11.94	m	13.13
Cover fillets, stops, trims, beads, nosings and the like; (finished sizes)										
18 x 58mm Dado rail	1.72	10.00	1.89	0.33	0.02	6.34	0.09	8.32	m	9.15
20 x 69mm Dado rail	2.40	10.00	2.64	0.35	0.02	6.70	0.09	9.43	m	10.38
Softwood; wrought cover fillets and mouldings										
Cover strips, mouldings and the like; (finished sizes)										
12 x 32mm; mouldings -1	0.69	10.00	0.76	0.19	0.02	3.77	0.05	4.57	m	5.03
12 x 50mm; mouldings -1	0.69	10.00	0.76	0.21	0.03	4.27	0.06	5.09	m	5.60
19 x 19mm; mouldings -1	0.69	10.00	0.76	0.19	0.02	3.77	0.05	4.57	m	5.03
19 x 50mm; mouldings -1	0.69	10.00	0.76	0.22	0.03	4.45	0.06	5.27	m	5.80
25 x 25mm; mouldings -1	0.99	10.00	1.09	0.21	0.03	4.27	0.06	5.42	m	5.96
25 x 38mm; mouldings -1	1.15	10.00	1.26	0.22	0.03	4.45	0.06	5.78	m	6.36
Oak; cover fillets and mouldings										
Cover strips, mouldings and the like; (finished sizes)										
12 x 34mm; mouldings -1	2.26	10.00	2.48	0.32	0.04	6.43	0.10	9.02	m	9.92
12 x 50mm; mouldings -1	2.75	10.00	3.03	0.36	0.05	7.30	0.12	10.45	m	11.49
19 x 19mm; mouldings -1	1.76	10.00	1.93	0.32	0.04	6.43	0.12	8.48	m	9.33
19 x 50mm; mouldings -1	3.33	10.00	3.67	0.37	0.05	7.48	0.12	11.27	m	12.40
25 x 38mm; mouldings -1	3.50	10.00	3.85	0.37	0.05	7.48	0.12	11.45	m	12.60
Maple; cover fillets and mouldings										
Cover strips, mouldings and the like; (finished sizes)										
12 x 34mm; mouldings -1	2.73	10.00	3.01	0.32	0.04	6.43	0.10	9.54	m	10.49
12 x 50mm; mouldings -1	3.45	10.00	3.80	0.36	0.05	7.30	0.12	11.22	m	12.34
19 x 19mm; mouldings -1	2.23	10.00	2.46	0.32	0.04	6.43	0.12	9.01	m	9.91
19 x 50mm; mouldings -1	5.00	10.00	5.50	0.37	0.05	7.48	0.12	13.10	m	14.42
25 x 38mm; mouldings -1	4.17	10.00	4.58	0.37	0.05	7.48	0.12	12.19	m	13.41
Softwood, wrought - cappings										
Cover fillets, stops, trims, beads, nosings and the like; (finished sizes)										
25 x 50mm; level; rebates -1; mouldings -1	1.98	10.00	2.18	0.22	0.03	4.45	0.06	6.69	m	7.36
50 x 75mm; level; rebates -1; mouldings -1	3.44	10.00	3.78	0.31	0.04	6.25	0.08	10.10	m	11.11
50 x 100mm; level; rebates -1; mouldings -1	6.40	10.00	7.04	0.34	0.04	6.80	0.08	13.91	m	15.30
25 x 50mm; ramped; rebates -1; mouldings -1	1.77	10.00	1.95	0.33	0.04	6.61	0.08	8.63	m	9.50
50 x 75mm; ramped; rebates -1; mouldings -1	3.44	10.00	3.78	0.46	0.06	9.28	0.12	13.18	m	14.50
50 x 100mm; ramped; rebates -1; mouldings -1	6.40	10.00	7.04	0.51	0.06	10.19	0.12	17.35	m	19.09
Oak; wrought - cappings										
Cover fillets, stops, trims, beads, nosings and the like; (finished sizes)										
25 x 50mm; level; rebates -1; mouldings -1	5.75	10.00	6.33	0.37	0.05	7.48	0.09	13.90	m	15.29
50 x 75mm; level; rebates -1; mouldings -1	13.77	10.00	15.15	0.53	0.07	10.70	0.14	25.99	m	28.59
extra; ends	2.50	-	2.50	0.08	0.01	1.61	0.01	4.12	nr	4.53
extra; angles	3.75	-	3.75	0.13	0.02	2.66	0.03	6.44	nr	7.09
extra; mitres	3.75	-	3.75	0.13	0.02	2.66	0.03	6.44	nr	7.09
extra; rounded corners not exceeding 300mm girth	3.75	-	3.75	0.53	0.07	10.70	0.14	14.58	nr	16.04
50 x 100mm; level; rebates -1; mouldings -1	17.54	10.00	19.29	0.58	0.07	11.62	0.14	31.04	nr	34.15
extra; ends	3.12	-	3.12	0.09	0.01	1.79	0.01	4.93	nr	5.42
extra; angles	5.00	-	5.00	0.14	0.02	2.85	0.03	7.88	nr	8.66
extra; mitres	5.00	-	5.00	0.14	0.02	2.85	0.03	7.88	nr	8.66
extra; rounded corners not exceeding 300mm girth	5.00	-	5.00	0.58	0.07	11.62	0.14	16.75	nr	18.43
25 x 50mm; ramped; rebates -1; mouldings -1	4.60	10.00	5.06	0.55	0.07	11.07	0.14	16.26	m	17.89
50 x 75mm; ramped; rebates -1; mouldings -1	13.77	10.00	15.15	0.80	0.10	16.07	0.19	31.42	m	34.56
extra; ends	19.69	-	19.69	0.12	0.02	2.48	0.03	22.20	nr	24.42
extra; angles	3.75	-	3.75	0.20	0.03	4.09	0.05	7.88	nr	8.67
extra; mitres	3.75	-	3.75	0.20	0.03	4.09	0.05	7.88	nr	8.67
extra; rounded corners not exceeding 300mm girth	3.75	-	3.75	0.80	0.10	16.07	0.19	20.02	nr	22.02
50 x 100mm; ramped; rebates -1; mouldings -1	17.54	10.00	19.29	0.90	0.11	18.05	0.22	37.56	m	41.32
extra; ends	3.12	-	3.12	0.13	0.02	2.66	0.03	5.82	nr	6.40
extra; angles	5.00	-	5.00	0.22	0.03	4.45	0.06	9.51	nr	10.47
extra; mitres	5.00	-	5.00	0.22	0.03	4.45	0.06	9.51	nr	10.47
extra; rounded corners not exceeding 300mm girth	5.00	-	5.00	0.90	0.11	18.05	0.22	23.27	nr	25.60

Labour hourly rates: (except Specialists) Craft Operatives 18.37 Labourer 13.76 Rates are national average prices. Refer to REGIONAL VARIATIONS for indicative levels of overall pricing in regions	MATERIALS			LABOUR				RATES		
	Del to Site	Waste	Material Cost	Craft Optve	Lab	Labour Cost	Sunds	Nett Rate		Gross rate (10%)
	£	%	£	Hrs	Hrs	£	£	£	Unit	£

UNFRAMED ISOLATED TRIMS, SKIRTINGS AND SUNDRY ITEMS (Cont'd)

Oak; wrought - cappings

Cover fillets, stops, trims, beads, nosings and the like; (finished sizes)

25 x 50mm; level; rebates -1; mouldings -1	5.25	10.00	5.78	0.33	0.04	6.61	0.08	12.46	m	13.71
50 x 75mm; level; rebates -1; mouldings -1	10.73	10.00	11.80	0.46	0.06	9.28	0.12	21.20	m	23.32
extra; ends	1.52	-	1.52	0.07	0.01	1.42	0.01	2.96	nr	3.25
extra; angles	2.57	-	2.57	0.12	0.02	2.48	0.03	5.08	nr	5.59
extra; mitres	2.57	-	2.57	0.12	0.02	2.48	0.03	5.08	nr	5.59
extra; rounded corners not exceeding 300mm girth	2.57	-	2.57	0.46	0.06	9.28	0.12	11.97	nr	13.16
50 x 100mm; level; rebates -1; mouldings -1	13.15	10.00	14.47	0.50	0.06	10.01	0.12	24.60	m	27.06
extra; ends	1.91	-	1.91	0.08	0.01	1.61	0.01	3.53	nr	3.89
extra; angles	3.16	-	3.16	0.13	0.02	2.66	0.03	5.85	nr	6.44
extra; mitres	3.16	-	3.16	0.13	0.02	2.66	0.03	5.85	nr	6.44
extra; rounded corners not exceeding 300mm girth	3.16	-	3.16	0.50	0.06	10.01	0.12	13.29	nr	14.62
25 x 50mm; ramped; rebates -1; mouldings -1	4.30	10.00	4.73	0.50	0.06	10.01	0.12	14.86	m	16.35
50 x 75mm; ramped; rebates -1; mouldings -1	10.73	10.00	11.80	0.70	0.09	14.10	0.17	26.07	m	28.67
extra; ends	1.52	10.00	1.67	0.10	0.01	1.97	0.03	3.68	nr	4.04
extra; angles	2.57	-	2.57	0.17	0.02	3.40	0.05	6.01	nr	6.61
extra; mitres	2.57	-	2.57	0.17	0.02	3.40	0.05	6.01	nr	6.61
extra; rounded corners not exceeding 300mm girth	2.57	-	2.57	0.70	0.09	14.10	0.17	16.83	nr	18.52
50 x 100mm; ramped; rebates -1; mouldings -1	13.15	10.00	14.47	0.75	0.09	15.02	0.18	29.66	m	32.63
extra; ends	1.91	-	1.91	0.11	0.01	2.16	0.03	4.10	nr	4.51
extra; angles	3.16	-	3.16	0.19	0.02	3.77	0.05	6.97	nr	7.67
extra; mitres	3.16	-	3.16	0.19	0.02	3.77	0.05	6.97	nr	7.67
extra; rounded corners not exceeding 300mm girth	3.16	-	3.16	0.75	0.09	15.02	0.18	18.36	nr	20.19

Softwood, wrought - isolated shelves and worktops

Cover fillets, stops, trims, beads, nosings and the like; (finished sizes)

19 x 12mm	0.46	10.00	0.51	0.18	0.02	3.58	0.05	4.13	m	4.55
19 x 16mm	0.55	10.00	0.61	0.19	0.02	3.77	0.05	4.42	m	4.86
19 x 18mm	0.64	10.00	0.70	0.19	0.02	3.77	0.05	4.51	m	4.97
19 x 22mm	0.71	10.00	0.78	0.20	0.03	4.09	0.05	4.91	m	5.40
19 x 25mm	0.80	10.00	0.88	0.20	0.03	4.09	0.05	5.01	m	5.51

Isolated shelves and worktops

25 x 150mm	2.79	10.00	3.07	0.20	0.03	4.09	0.05	7.20	m	7.92
25 x 225mm	4.38	10.00	4.82	0.25	0.03	5.01	0.06	9.88	m	10.87
25 x 300mm	6.30	10.00	6.93	0.30	0.04	6.06	0.08	13.07	m	14.37
32 x 150mm	3.96	10.00	4.36	0.22	0.03	4.45	0.06	8.87	m	9.76
32 x 225mm	6.57	10.00	7.23	0.27	0.03	5.37	0.06	12.66	m	13.93
32 x 300mm	7.90	10.00	8.69	0.33	0.04	6.61	0.08	15.38	m	16.92
25 x 450mm; cross tongued	12.10	2.50	12.40	0.35	0.04	6.98	0.09	19.47	m	21.42
25 x 600mm; cross tongued	16.10	2.50	16.50	0.45	0.06	9.09	0.10	25.70	m	28.27
32 x 450mm; cross tongued	15.50	2.50	15.89	0.40	0.06	8.17	0.10	24.17	m	26.58
32 x 600mm; cross tongued	20.60	2.50	21.11	0.50	0.06	10.01	0.12	31.25	m	34.37
38 x 450mm; cross tongued	18.40	2.50	18.86	0.45	0.06	9.09	0.10	28.06	m	30.86
38 x 600mm; cross tongued	24.80	2.50	25.42	0.55	0.07	11.07	0.14	36.62	m	40.28
25 x 450mm overall; 25 x 50mm slats spaced 25mm apart	5.88	10.00	6.47	1.00	0.13	20.16	0.24	26.87	m	29.55
25 x 450mm overall; 25 x 50mm slats spaced 32mm apart	5.88	10.00	6.47	0.90	0.11	18.05	0.22	24.74	m	27.21
25 x 600mm overall; 25 x 50mm slats spaced 25mm apart	7.56	10.00	8.32	1.25	0.16	25.16	0.30	33.78	m	37.16
25 x 600mm overall; 25 x 50mm slats spaced 32mm apart	7.56	10.00	8.32	1.15	0.14	23.05	0.28	31.65	m	34.82
25 x 900mm overall; 25 x 50mm slats spaced 25mm apart	10.92	10.00	12.01	1.90	0.24	38.21	0.47	50.68	m	55.75
25 x 900mm overall; 25 x 50mm slats spaced 32mm apart	10.08	10.00	11.09	1.80	0.23	36.23	0.44	47.75	m	52.53

Oak, wrought - isolated shelves and worktops

Cover fillets, stops, trims, beads, nosings and the like; (finished sizes)

19 x 12mm	1.25	10.00	1.38	0.31	0.04	6.25	0.08	7.70	m	8.46
19 x 16mm	1.50	10.00	1.65	0.33	0.04	6.61	0.08	8.34	m	9.17
19 x 18mm	1.84	10.00	2.02	0.33	0.04	6.61	0.08	8.71	m	9.58
19 x 22mm	1.88	10.00	2.07	0.35	0.04	6.98	0.09	9.14	m	10.05
19 x 25mm	1.91	10.00	2.10	0.35	0.04	6.98	0.09	9.17	m	10.09

Isolated shelves and worktops

25 x 150mm	9.43	10.00	10.37	0.35	0.04	6.98	0.09	17.44	m	19.19
25 x 225mm	14.15	10.00	15.56	0.44	0.06	8.91	0.10	24.58	m	27.04
25 x 300mm	23.50	10.00	25.85	0.53	0.07	10.70	0.14	36.68	m	40.35
32 x 150mm	14.35	10.00	15.79	0.38	0.05	7.67	0.09	23.54	m	25.90
32 x 225mm	21.53	10.00	23.68	0.47	0.06	9.46	0.12	33.26	m	36.59
32 x 300mm	30.00	10.00	33.00	0.56	0.07	11.25	0.14	44.39	m	48.82
25 x 450mm; cross tongued	39.00	2.50	39.97	0.60	0.08	12.12	0.15	52.25	m	57.47
25 x 600mm; cross tongued	52.00	2.50	53.30	0.80	0.10	16.07	0.19	69.57	m	76.52
32 x 450mm; cross tongued	49.50	2.50	50.74	0.70	0.09	14.10	0.17	65.00	m	71.50
32 x 600mm; cross tongued	66.00	2.50	67.65	0.90	0.11	18.05	0.22	85.92	m	94.51
38 x 450mm; cross tongued	78.40	2.50	80.36	0.80	0.10	16.07	0.19	96.63	m	106.29
38 x 600mm; cross tongued	104.50	2.50	107.11	1.00	0.13	20.16	0.24	127.51	m	140.26

Plywood BS EN 636, II/III grade, INT bonded; butt joints - isolated shelves and worktops

Cover fillets, stops, trims, beads, nosings and the like

50 x 6.5mm	0.14	20.00	0.17	0.16	0.02	3.21	0.08	3.45	m	3.80
50 x 12mm	0.26	20.00	0.31	0.16	0.02	3.21	0.08	3.60	m	3.96
50 x 15mm	0.60	20.00	0.72	0.17	0.02	3.40	0.08	4.19	m	4.61
50 x 19mm	0.36	20.00	0.44	0.18	0.02	3.58	0.08	4.09	m	4.50
100 x 6.5mm	0.28	20.00	0.33	0.16	0.02	3.21	0.08	3.62	m	3.98
100 x 12mm	0.51	20.00	0.62	0.17	0.02	3.40	0.08	4.09	m	4.50
100 x 15mm	1.20	20.00	1.44	0.18	0.02	3.58	0.08	5.10	m	5.61
100 x 19mm	0.73	20.00	0.87	0.19	0.02	3.77	0.08	4.71	m	5.18

	MATERIALS			LABOUR				RATES		
Labour hourly rates: (except Specialists) Craft Operatives 18.37 Labourer 13.76 Rates are national average prices. Refer to REGIONAL VARIATIONS for indicative levels of overall pricing in regions	Del to Site	Waste	Material Cost	Craft Optve	Lab	Labour Cost	Sunds	Nett Rate	Unit	Gross rate (10%)
	£	%	£	Hrs	Hrs	£	£	£		£

UNFRAMED ISOLATED TRIMS, SKIRTINGS AND SUNDRY ITEMS (Cont'd)

Plywood BS EN 636, II/III grade, INT bonded; butt joints - isolated shelves and worktops (Cont'd)

Isolated shelves and worktops

	Del to Site	Waste	Material Cost	Craft Optve	Lab	Labour Cost	Sunds	Nett Rate	Unit	Gross rate
6.5 x 150mm	0.41	20.00	0.50	0.18	0.02	3.58	0.08	4.15	m	4.57
6.5 x 225mm	1.16	20.00	1.39	0.23	0.02	4.50	0.08	5.96	m	6.56
6.5 x 300mm	3.60	20.00	4.32	0.28	0.04	5.69	0.08	10.09	m	11.10
6.5 x 450mm	3.26	20.00	3.92	0.33	0.04	6.61	0.08	10.61	m	11.67
6.5 x 600mm	1.66	20.00	1.99	0.38	0.05	7.67	0.08	9.73	m	10.70
12 x 150mm	0.77	20.00	0.93	0.19	0.02	3.77	0.08	4.77	m	5.24
12 x 225mm	2.70	20.00	3.24	0.24	0.03	4.82	0.08	8.14	m	8.95
12 x 300mm	2.18	20.00	2.61	0.29	0.04	5.88	0.08	8.56	m	9.42
12 x 450mm	1.24	20.00	1.49	0.34	0.04	6.80	0.08	8.36	m	9.20
12 x 600mm	3.09	20.00	3.70	0.39	0.05	7.85	0.08	11.63	m	12.79
15 x 150mm	1.80	20.00	2.16	0.23	0.03	4.64	0.08	6.88	m	7.56
15 x 225mm	1.63	20.00	1.96	0.25	0.03	5.01	0.08	7.04	m	7.74
15 x 300mm	0.83	20.00	0.99	0.30	0.04	6.06	0.08	7.13	m	7.84
15 x 450mm	2.31	20.00	2.78	0.35	0.04	6.98	0.08	9.83	m	10.82
15 x 600mm	7.21	20.00	8.65	0.40	0.05	8.04	0.08	16.76	m	18.44
19 x 150mm	1.09	20.00	1.31	0.20	0.03	4.09	0.08	5.47	m	6.01
19 x 225mm	0.62	20.00	0.74	0.25	0.03	5.01	0.08	5.83	m	6.41
19 x 300mm	1.54	20.00	1.85	0.30	0.04	6.06	0.08	7.99	m	8.79
19 x 450mm	5.41	20.00	6.49	0.35	0.04	6.98	0.08	13.54	m	14.90
19 x 600mm	4.35	20.00	5.22	0.40	0.05	8.04	0.08	13.33	m	14.67

Blockboard, 2/2 grade, INT bonded, butt joints - isolated shelves and worktops

Isolated shelves and worktops

	Del to Site	Waste	Material Cost	Craft Optve	Lab	Labour Cost	Sunds	Nett Rate	Unit	Gross rate
18 x 150mm	1.86	20.00	2.23	0.23	0.03	4.64	0.14	7.00	m	7.70
18 x 225mm	2.79	20.00	3.34	0.29	0.04	5.88	0.14	9.36	m	10.29
18 x 300mm	3.71	20.00	4.46	0.34	0.04	6.80	0.14	11.39	m	12.53
18 x 450mm	5.57	20.00	6.69	0.40	0.05	8.04	0.14	14.86	m	16.34
18 x 600mm	7.43	20.00	8.92	0.46	0.06	9.28	0.14	18.33	m	20.16

Wood chipboard BS EN 312, type I standard; butt joints - isolated shelves and worktops

Isolated shelves and worktops

	Del to Site	Waste	Material Cost	Craft Optve	Lab	Labour Cost	Sunds	Nett Rate	Unit	Gross rate
12 x 150mm	0.53	20.00	0.63	0.19	0.02	3.77	0.08	4.47	m	4.92
12 x 225mm	0.79	20.00	0.95	0.24	0.03	4.82	0.08	5.84	m	6.43
12 x 300mm	1.05	20.00	1.26	0.29	0.04	5.88	0.08	7.21	m	7.94
12 x 450mm	1.58	20.00	1.89	0.34	0.04	6.80	0.08	8.76	m	9.64
12 x 600mm	2.10	20.00	2.52	0.39	0.05	7.85	0.08	10.45	m	11.49
18 x 150mm	0.73	20.00	0.87	0.20	0.03	4.09	0.08	5.03	m	5.54
18 x 225mm	1.09	20.00	1.31	0.25	0.03	5.01	0.08	6.39	m	7.03
18 x 300mm	1.45	20.00	1.74	0.30	0.04	6.06	0.08	7.88	m	8.67
18 x 450mm	2.18	20.00	2.61	0.35	0.04	6.98	0.08	9.67	m	10.64
18 x 600mm	2.91	20.00	3.49	0.40	0.05	8.04	0.08	11.60	m	12.76
25 x 150mm	1.04	20.00	1.24	0.20	0.03	4.09	0.08	5.41	m	5.95
25 x 225mm	1.55	20.00	1.87	0.25	0.03	5.01	0.08	6.95	m	7.64
25 x 300mm	2.07	20.00	2.49	0.30	0.04	6.06	0.08	8.62	m	9.49
25 x 450mm	3.11	20.00	3.73	0.35	0.04	6.98	0.08	10.79	m	11.86
25 x 600mm	4.15	20.00	4.97	0.40	0.05	8.04	0.08	13.09	m	14.39

Blockboard, 2/2 grade, INT bonded, faced with 1.5mm laminated plastic sheet, BS EN 438, classified HGS, with 1.2mm laminated plastic sheet balance veneer; butt joints

Isolated shelves and worktops

	Del to Site	Waste	Material Cost	Craft Optve	Lab	Labour Cost	Sunds	Nett Rate	Unit	Gross rate
18 x 150mm	1.86	20.00	2.23	0.40	0.05	8.04	0.14	10.40	m	11.44
18 x 225mm	2.79	20.00	3.34	0.50	0.06	10.01	0.14	13.49	m	14.84
18 x 300mm	3.71	20.00	4.46	0.60	0.08	12.12	0.14	16.72	m	18.39
18 x 450mm	5.57	20.00	6.69	0.70	0.09	14.10	0.14	20.92	m	23.01
18 x 600mm	7.43	20.00	8.92	0.80	0.10	16.07	0.14	25.12	m	27.63

Softwood, wrought - window boards

Cover fillets, stops, trims, beads, nosings and the like; (finished sizes)

mouldings -1; tongued on

	Del to Site	Waste	Material Cost	Craft Optve	Lab	Labour Cost	Sunds	Nett Rate	Unit	Gross rate
25 x 50mm	1.98	10.00	2.18	0.25	0.03	5.01	0.06	7.24	m	7.97
25 x 75mm	2.29	10.00	2.52	0.26	0.03	5.19	0.06	7.77	m	8.54
32 x 50mm	2.57	10.00	2.83	0.28	0.04	5.69	0.08	8.60	m	9.46
32 x 75mm	3.94	10.00	4.33	0.29	0.04	5.88	0.08	10.29	m	11.32

rounded edges -1; tongued on

	Del to Site	Waste	Material Cost	Craft Optve	Lab	Labour Cost	Sunds	Nett Rate	Unit	Gross rate
25 x 50mm	1.98	10.00	2.18	0.25	0.03	5.01	0.06	7.24	m	7.97
25 x 75mm	2.29	10.00	2.52	0.26	0.03	5.19	0.06	7.77	m	8.54
32 x 50mm	2.57	10.00	2.83	0.28	0.04	5.69	0.08	8.60	m	9.46
32 x 75mm	3.94	10.00	4.33	0.29	0.04	5.88	0.08	10.29	m	11.32

Window boards

	Del to Site	Waste	Material Cost	Craft Optve	Lab	Labour Cost	Sunds	Nett Rate	Unit	Gross rate
25 x 150mm; mouldings -1	2.82	10.00	3.10	0.30	0.04	6.06	0.08	9.24	m	10.16
32 x 150mm; mouldings -1	5.31	10.00	5.84	0.33	0.04	6.61	0.08	12.53	m	13.78
25 x 150mm; rounded edges -1	2.82	10.00	3.10	0.30	0.04	6.06	0.08	9.24	m	10.16
32 x 150mm; rounded edges -1	5.31	10.00	5.84	0.33	0.04	6.61	0.08	12.53	m	13.78

Afrormosia, wrought - window boards

Cover fillets, stops, trims, beads, nosings and the like; (finished sizes)

mouldings -1; tongued on

	Del to Site	Waste	Material Cost	Craft Optve	Lab	Labour Cost	Sunds	Nett Rate	Unit	Gross rate
25 x 50mm	6.01	10.00	6.61	0.38	0.05	7.67	0.09	14.37	m	15.81
25 x 75mm	9.34	10.00	10.27	0.40	0.05	8.04	0.09	18.40	m	20.24

GENERAL JOINERY

Labour hourly rates: (except Specialists) Craft Operatives 18.37 Labourer 13.76 Rates are national average prices. Refer to REGIONAL VARIATIONS for indicative levels of overall pricing in regions	MATERIALS			LABOUR				RATES		
	Del to Site £	Waste %	Material Cost £	Craft Optve Hrs	Lab Hrs	Labour Cost £	Sunds £	Nett Rate £	Unit	Gross rate (10%) £
UNFRAMED ISOLATED TRIMS, SKIRTINGS AND SUNDRY ITEMS (Cont'd)										
Afrormosia, wrought - window boards (Cont'd)										
Cover fillets, stops, trims, beads, nosings and the like; (finished sizes) (Cont'd)										
mouldings -1; tongued on (Cont'd)										
32 x 50mm	7.22	10.00	7.95	0.42	0.05	8.40	0.10	16.46	m	18.10
32 x 75mm	11.07	10.00	12.18	0.44	0.06	8.91	0.10	21.20	m	23.32
rounded edges -1; tongued on										
25 x 50mm	6.01	10.00	6.61	0.38	0.05	7.67	0.09	14.37	m	15.81
25 x 75mm	9.34	10.00	10.27	0.40	0.05	8.04	0.09	18.40	m	20.24
32 x 50mm	7.22	10.00	7.95	0.42	0.05	8.40	0.10	16.46	m	18.10
32 x 75mm	11.07	10.00	12.18	0.44	0.06	8.91	0.10	21.20	m	23.32
Window boards										
25 x 150mm; mouldings -1	20.25	10.00	22.27	0.52	0.07	10.52	0.14	32.93	m	36.22
32 x 150mm; mouldings -1	24.06	10.00	26.47	0.58	0.07	11.62	0.14	38.22	m	42.04
25 x 150mm; rounded edges -1	20.25	10.00	22.27	0.52	0.07	10.52	0.14	32.93	m	36.22
32 x 150mm; rounded edges -1	24.06	10.00	26.47	0.58	0.07	11.62	0.14	38.22	m	42.04
Sapele; wrought - window boards										
Cover fillets, stops, trims, beads, nosings and the like; (finished sizes)										
mouldings -1; tongued on										
25 x 50mm	6.00	10.00	6.60	0.37	0.05	7.48	0.09	14.17	m	15.59
25 x 75mm	9.13	10.00	10.05	0.39	0.05	7.85	0.09	17.99	m	19.79
32 x 50mm	7.00	10.00	7.70	0.41	0.05	8.22	0.10	16.02	m	17.62
32 x 75mm	10.91	10.00	12.00	0.43	0.05	8.59	0.10	20.69	m	22.76
rounded edges -1; tongued on										
25 x 50mm	6.00	10.00	6.60	0.37	0.05	7.48	0.09	14.17	m	15.59
25 x 75mm	9.13	10.00	10.05	0.39	0.05	7.85	0.09	17.99	m	19.79
32 x 50mm	7.00	10.00	7.70	0.41	0.05	8.22	0.10	16.02	m	17.62
32 x 75mm	10.91	10.00	12.00	0.43	0.05	8.59	0.10	20.69	m	22.76
Window boards										
25 x 150mm; mouldings -1	13.16	10.00	14.48	0.45	0.06	9.09	0.10	23.68	m	26.05
32 x 150mm; mouldings -1	15.60	10.00	17.16	0.50	0.06	10.01	0.12	27.29	m	30.02
25 x 150mm; rounded edges -1	13.16	10.00	14.48	0.45	0.06	9.09	0.10	23.68	m	26.05
32 x 150mm; rounded edges -1	15.60	10.00	17.16	0.50	0.06	10.01	0.12	27.29	m	30.02
Moisture resistant medium density fibreboard - MDF, BS EN 622										
Window boards										
25 x 225mm; nosed and tongued	4.60	10.00	5.06	0.30	0.04	6.06	0.08	11.20	m	12.32
25 x 250mm; nosed and tongued	4.95	10.00	5.44	0.32	0.04	6.43	0.08	11.95	m	13.14
Duct fronts; plywood BS EN 636, II/III grade, MR bonded; butt joints; fixing to timber with screws, countersinking										
12mm thick linings to walls										
width exceeding 300mm	16.36	20.00	19.63	1.00	0.13	20.16	0.24	40.03	m²	44.03
width exceeding 300mm; fixing to timber with brass screws and cups	16.36	20.00	19.63	1.50	0.19	30.17	0.87	50.67	m²	55.74
Blockboard, 2/2 grade, MR bonded; butt joints										
18mm thick linings to walls										
width exceeding 300mm	8.57	20.00	10.29	1.20	0.15	24.11	0.30	34.70	m²	38.17
width exceeding 300mm; fixing to timber with brass screws and cups	8.57	20.00	10.29	1.75	0.22	35.17	0.93	46.39	m²	51.03
Extra over plywood BS EN 636, II/III grade, MR bonded; butt joints; 12mm thick linings to walls; width exceeding 300mm										
Access panels										
300 x 600mm	-	-	-	0.50	0.05	9.87	0.12	9.99	nr	10.99
600 x 600mm	-	-	-	0.65	0.07	12.90	0.15	13.05	nr	14.36
600 x 900mm	-	-	-	0.80	0.08	15.80	0.19	15.99	nr	17.59
Extra over blockboard, 2/2 grade, MR bonded, butt joints; 18mm thick linings to walls; width exceeding 300mm										
Access panels										
300 x 600mm	-	-	-	0.60	0.07	11.99	0.15	12.14	nr	13.35
600 x 600mm	-	-	-	0.80	0.08	15.80	0.19	15.99	nr	17.59
600 x 900mm	-	-	-	1.10	0.11	21.72	0.27	21.99	nr	24.19
Pipe casings; standard hardboard, BS EN 622, type SHA; butt joints										
3mm thick linings to pipe ducts and casings										
width exceeding 300mm	4.04	20.00	4.85	0.80	0.10	16.07	0.19	21.12	m²	23.23
width not exceeding 300mm	1.21	20.00	1.45	0.25	0.03	5.01	0.06	6.52	m	7.17
Plywood BS EN 636, II/III grade, MR bonded; butt joints; fixing to timber with screws, countersinking										
12mm thick linings to pipe ducts and casings										
width exceeding 300mm	16.36	20.00	19.63	1.50	0.19	30.17	0.36	50.16	m²	55.18
width not exceeding 300mm	4.91	20.00	5.89	0.50	0.06	10.01	0.12	16.02	m	17.62

Labour hourly rates: (except Specialists) Craft Operatives 18.37 Labourer 13.76 Rates are national average prices. Refer to REGIONAL VARIATIONS for indicative levels of overall pricing in regions	MATERIALS			LABOUR				RATES		
	Del to Site	Waste	Material Cost	Craft Optve	Lab	Labour Cost	Sunds	Nett Rate	Unit	Gross rate (10%)
	£	%	£	Hrs	Hrs	£	£	£		£
UNFRAMED ISOLATED TRIMS, SKIRTINGS AND SUNDRY ITEMS (Cont'd)										
Pendock Profiles Ltd preformed plywood casings; white melamine finish; butt joints; fixing to timber with polytop white screws, countersinking										
5mm thick casings; horizontal										
reference TK45110, 45 x 110mm	11.26	20.00	13.52	0.40	0.05	8.04	0.30	21.85	m	24.04
extra; internal corner	9.24	2.50	9.47	0.25	0.03	5.01	0.30	14.78	nr	16.25
extra; external corner	12.79	2.50	13.11	0.25	0.03	5.01	0.30	18.42	nr	20.26
extra; stop end	4.02	2.50	4.12	0.15	0.02	3.03	0.15	7.30	nr	8.03
reference TK45150, 45 x 150mm	14.00	20.00	16.80	0.42	0.05	8.40	0.30	25.50	m	28.05
extra; internal corner	10.76	2.50	11.03	0.25	0.03	5.01	0.30	16.33	nr	17.97
extra; external corner	14.43	2.50	14.79	0.25	0.03	5.01	0.30	20.10	nr	22.11
extra; stop end	5.86	2.50	6.01	0.15	0.02	3.03	0.15	9.19	nr	10.11
reference TK45190, 45 x 190mm	19.29	20.00	23.15	0.44	0.05	8.77	0.30	32.22	m	35.44
extra; internal corner	11.09	2.50	11.37	0.25	0.03	5.01	0.30	16.67	nr	18.34
extra; external corner	14.86	2.50	15.23	0.25	0.03	5.01	0.30	20.54	nr	22.59
extra; stop end	7.43	2.50	7.62	0.15	0.02	3.03	0.15	10.80	nr	11.88
5mm thick casings; vertical										
reference TK45110, 45 x 110mm	11.26	20.00	13.52	0.40	0.05	8.04	0.30	21.85	m	24.04
reference TK45150, 45 x 150mm	14.00	20.00	16.80	0.42	0.05	8.40	0.30	25.50	m	28.05
5mm thick casings; horizontal										
reference MX 1010, 100 x 100mm	15.85	20.00	19.02	0.42	0.05	8.40	0.30	27.73	m	30.50
extra; stop end	6.53	2.50	6.70	0.15	0.02	3.03	0.15	9.88	nr	10.87
8mm thick casings; horizontal										
reference MX 1515, 150 x 150mm	27.70	20.00	33.24	0.45	0.06	9.09	0.30	42.63	m	46.90
extra; internal corner	33.57	2.50	34.41	0.25	0.03	5.01	0.30	39.71	nr	43.69
extra; external corner	40.21	2.50	41.22	0.25	0.03	5.01	0.30	46.52	nr	51.17
extra; stop end	7.06	2.50	7.24	0.15	0.02	3.03	0.15	10.42	nr	11.46
reference MX 1520, 200 x 150mm	37.31	20.00	44.77	0.46	0.06	9.28	0.30	54.35	m	59.79
extra; stop end	7.06	2.50	7.24	0.15	0.02	3.03	0.15	10.42	nr	11.46
reference MX 2020, 200 x 200mm	41.02	20.00	49.23	0.47	0.06	9.46	0.30	58.99	m	64.89
extra; internal corner	47.28	2.50	48.46	0.25	0.03	5.01	0.30	53.77	nr	59.14
extra; external corner	54.58	2.50	55.94	0.25	0.03	5.01	0.30	61.25	nr	67.37
extra; stop end	7.06	2.50	7.24	0.15	0.02	3.03	0.15	10.42	nr	11.46
reference MX 2030, 200 x 300mm	40.74	20.00	48.88	0.48	0.06	9.64	0.30	58.83	m	64.71
extra; stop end	8.36	2.50	8.57	0.15	0.02	3.03	0.15	11.75	nr	12.92
reference MX 3030, 300 x 300mm	46.90	20.00	56.28	0.50	0.06	10.01	0.30	66.59	m	73.25
extra; internal corner	52.77	2.50	54.09	0.25	0.03	5.01	0.30	59.39	nr	65.33
extra; external corner	58.84	2.50	60.31	0.25	0.03	5.01	0.30	65.62	nr	72.18
extra; stop end	11.92	2.50	12.22	0.15	0.02	3.03	0.15	15.40	nr	16.94
Laminated plastic sheet, BS EN 438 classification HGS										
Cover fillets, stops, trims, beads, nosings and the like; fixing with adhesive										
1.2 x 16mm	0.44	20.00	0.52	0.20	0.03	4.09	0.09	4.70	m	5.17
1.2 x 18mm	0.49	20.00	0.59	0.21	0.03	4.27	0.10	4.97	m	5.46
1.2 x 22mm	0.60	20.00	0.72	0.23	0.03	4.64	0.12	5.48	m	6.03
1.2 x 25mm	0.68	20.00	0.82	0.25	0.03	5.01	0.14	5.96	m	6.56
1.2 x 75mm	2.05	20.00	2.46	0.30	0.04	6.06	0.42	8.94	m	9.83
Isolated shelves and worktops; fixing with adhesive										
1.2 x 150mm	4.10	20.00	4.92	0.35	0.04	6.98	0.84	12.74	m	14.01
1.2 x 225mm	6.15	20.00	7.37	0.40	0.05	8.04	1.26	16.67	m	18.34
1.2 x 300mm	8.19	20.00	9.83	0.45	0.06	9.09	1.68	20.60	m	22.66
1.2 x 450mm	12.29	20.00	14.75	0.53	0.07	10.70	2.54	27.98	m	30.78
1.2 x 600mm	16.39	20.00	19.66	0.60	0.08	12.12	3.38	35.16	m	38.68
FLOOR, WALL AND CEILING BOARDING, SHEETING, PANELLING, LININGS AND CASINGS										
Strip flooring; hardwood, wrought, selected for transparent finish; sanded, two coats sealer finish										
Boarding to floors, tongued and grooved joints; 22mm thick, 75mm wide boards; fixing on and with 25 x 50mm impregnated softwood battens to masonry										
width exceeding 300mm; secret fixing to timber	43.00	7.50	46.19	1.25	0.15	25.03	0.30	71.52	m²	78.67
Boarding to floors, tongued and grooved joints; 12mm thick, 75mm wide boards; overlay										
width exceeding 300mm; secret fixing to timber	47.09	7.50	50.62	1.00	0.13	20.16	0.24	71.02	m²	78.12
Semi-sprung floors; foam backed softwood battens covered with hardwood tongued and grooved strip flooring										
22mm Standard Beech strip flooring to level sub-floors width exceeding 600mm; battens at approximately 400mm centres; elevation										
75mm	-	-	Specialist	-	-	Specialist	-	73.65	m²	81.01
22mm "Sylva Squash" Beech strip flooring to level sub-floors width exceeding 600mm; battens at approximately 430mm centres; elevation										
75mm	-	-	Specialist	-	-	Specialist	-	78.75	m²	86.62
20mm Prime Maple strip flooring to level sub-floors width exceeding 600mm; battens at approximately 600mm centres; elevation										
75mm	-	-	Specialist	-	-	Specialist	-	59.92	m²	65.92

Labour hourly rates: (except Specialists) Craft Operatives 18.37 Labourer 13.76 Rates are national average prices. Refer to REGIONAL VARIATIONS for indicative levels of overall pricing in regions	MATERIALS			LABOUR				RATES		
	Del to Site £	Waste %	Material Cost £	Craft Optve Hrs	Lab Hrs	Labour Cost £	Sunds £	Nett Rate £	Unit	Gross rate (10%) £
FLOOR, WALL AND CEILING BOARDING, SHEETING, PANELLING, LININGS AND CASINGS (Cont'd)										
Semi-sprung floors; foam backed softwood battens covered with hardwood tongued and grooved strip flooring (Cont'd)										
20mm First Grade Maple strip flooring to level sub-floors width exceeding 600mm; battens at approximately 600mm centres; elevation 75mm..........	-	-	Specialist	-	-	Specialist	-	65.06	m²	71.56
Softwood, wrought; fixing to timber; wall boarding										
Boarding to walls, tongued, grooved and veed joints; 19mm thick, 100mm wide boards width exceeding 600mm..........	7.67	10.00	8.43	1.00	0.13	20.16	0.24	28.83	m²	31.72
Boarding to walls, tongued, grooved and veed joints; 19mm thick, 150mm wide boards width exceeding 600mm..........	7.53	10.00	8.29	0.70	0.09	14.10	0.17	22.55	m²	24.80
Boarding to walls, tongued, grooved and veed joints; 25mm thick, 100mm wide boards width exceeding 600mm..........	9.89	10.00	10.88	1.05	0.13	21.08	0.25	32.21	m²	35.43
Boarding to walls, tongued, grooved and veed joints; 25mm thick, 150mm wide boards width exceeding 600mm..........	10.50	10.00	11.55	0.75	0.09	15.02	0.18	26.75	m²	29.42
Boarding to ceilings, tongued, grooved and veed joints; 19mm thick, 100mm wide boards width exceeding 600mm..........	7.67	10.00	8.43	1.50	0.19	30.17	0.36	38.96	m²	42.86
Boarding to ceilings, tongued, grooved and veed joints; 19mm thick, 150mm wide boards width exceeding 600mm..........	7.53	10.00	8.29	1.00	0.13	20.16	0.24	28.69	m²	31.55
Boarding to ceilings, tongued, grooved and veed joints; 25mm thick, 100mm wide boards width exceeding 600mm..........	9.89	10.00	10.88	1.55	0.19	31.09	0.38	42.34	m²	46.57
Boarding to ceilings, tongued, grooved and veed joints; 25mm thick, 150mm wide boards width exceeding 600mm..........	10.50	10.00	11.55	1.05	0.13	21.08	0.25	32.88	m²	36.17
Western Red Cedar, wrought; fixing to timber; wall boarding										
Boarding to walls, tongued, grooved and veed joints; 19mm thick, 100mm wide boards width exceeding 600mm..........	36.00	10.00	39.60	1.05	0.13	21.08	0.25	60.93	m²	67.03
Boarding to walls, tongued, grooved and veed joints; 19mm thick, 150mm wide boards width exceeding 600mm..........	33.00	10.00	36.30	0.75	0.10	15.15	0.18	51.63	m²	56.80
Boarding to walls, tongued, grooved and veed joints; 25mm thick, 100mm wide boards width exceeding 600mm..........	40.70	10.00	44.77	1.10	0.14	22.13	0.27	67.17	m²	73.89
Boarding to walls, tongued, grooved and veed joints; 25mm thick, 150mm wide boards width exceeding 600mm..........	34.64	10.00	38.11	0.80	0.10	16.07	0.19	54.37	m²	59.81
Knotty Pine, wrought, selected for transparent finish; fixing to timber; wall boarding										
Boarding to walls, tongued, grooved and veed joints; 19mm thick, 100mm wide boards width exceeding 600mm..........	30.56	10.00	33.61	1.30	0.16	26.08	0.31	60.01	m²	66.01
Boarding to walls, tongued, grooved and veed joints; 19mm thick, 150mm wide boards width exceeding 600mm..........	33.57	10.00	36.93	0.90	0.11	18.05	0.22	55.20	m²	60.72
Boarding to walls, tongued, grooved and veed joints; 25mm thick, 100mm wide boards width exceeding 600mm..........	42.00	10.00	46.20	1.35	0.17	27.14	0.33	73.67	m²	81.04
Boarding to walls, tongued, grooved and veed joints; 25mm thick, 150mm wide boards width exceeding 600mm..........	37.86	10.00	41.64	0.95	0.12	19.10	0.22	60.97	m²	67.07
Boarding to ceilings, tongued, grooved and veed joints; 19mm thick, 100mm wide boards width exceeding 600mm..........	30.56	10.00	33.61	1.85	0.23	37.15	0.45	71.21	m²	78.33
Boarding to ceilings, tongued, grooved and veed joints; 19mm thick, 150mm wide boards width exceeding 600mm..........	33.57	10.00	36.93	1.35	0.17	27.14	0.33	64.40	m²	70.84
Boarding to ceilings, tongued, grooved and veed joints; 25mm thick, 100mm wide boards width exceeding 600mm..........	42.00	10.00	46.20	1.90	0.24	38.21	0.47	84.87	m²	93.36
Boarding to ceilings, tongued, grooved and veed joints; 25mm thick, 150mm wide boards width exceeding 600mm..........	37.86	10.00	41.64	1.40	0.18	28.19	0.34	70.18	m²	77.20

GENERAL JOINERY *(side)*

Labour hourly rates: (except Specialists) Craft Operatives 18.37 Labourer 13.76 Rates are national average prices. Refer to REGIONAL VARIATIONS for indicative levels of overall pricing in regions	MATERIALS			LABOUR				RATES		
	Del to Site	Waste	Material Cost	Craft Optve	Lab	Labour Cost	Sunds	Nett Rate	Unit	Gross rate (10%)
	£	%	£	Hrs	Hrs	£	£	£		£
FLOOR, WALL AND CEILING BOARDING, SHEETING, PANELLING, LININGS AND CASINGS (Cont'd)										
Sapele, wrought, selected for transparent finish; fixing to timber; wall boarding										
Boarding to walls, tongued, grooved and veed joints; 19mm thick, 100mm wide boards										
width exceeding 600mm	80.35	10.00	88.38	1.45	0.18	29.11	0.36	117.86	m²	129.64
Boarding to walls, tongued, grooved and veed joints; 19mm thick, 150mm wide boards										
width exceeding 600mm	78.43	10.00	86.27	1.00	0.12	20.02	0.24	106.53	m²	117.19
Boarding to walls, tongued, grooved and veed joints; 25mm thick, 100mm wide boards										
width exceeding 600mm	119.00	10.00	130.90	1.50	0.19	30.17	0.36	161.43	m²	177.57
Boarding to walls, tongued, grooved and veed joints; 25mm thick, 150mm wide boards										
width exceeding 600mm	114.35	10.00	125.79	1.05	0.13	21.08	0.25	147.12	m²	161.83
Boarding to ceilings, tongued, grooved and veed joints; 19mm thick, 100mm wide boards										
width exceeding 600mm	80.35	10.00	88.38	2.00	0.25	40.18	0.50	129.06	m²	141.96
Boarding to ceilings, tongued, grooved and veed joints; 19mm thick, 150mm wide boards										
width exceeding 600mm	78.43	10.00	86.27	1.50	0.19	30.17	0.36	116.80	m²	128.48
Boarding to ceilings, tongued, grooved and veed joints; 25mm thick, 100mm wide boards										
width exceeding 600mm	119.00	10.00	130.90	2.05	0.26	41.24	0.50	172.63	m²	189.89
Boarding to ceilings, tongued, grooved and veed joints; 25mm thick, 150mm wide boards										
width exceeding 600mm	114.35	10.00	125.79	1.55	0.19	31.09	0.38	157.25	m²	172.98
Abutments										
19mm thick softwood boarding										
raking cutting	0.77	10.00	0.84	0.06	-	1.10	-	1.95	m	2.14
curved cutting	1.15	10.00	1.26	0.09	-	1.65	-	2.92	m	3.21
25mm thick softwood boarding										
raking cutting	0.99	10.00	1.09	0.07	-	1.29	-	2.37	m	2.61
curved cutting	1.48	10.00	1.63	0.10	-	1.84	-	3.47	m	3.82
19mm thick Western Red Cedar boarding										
raking cutting	3.60	10.00	3.96	0.07	-	1.29	-	5.25	m	5.77
curved cutting	5.40	10.00	5.94	0.10	-	1.84	-	7.78	m	8.55
25mm thick Western Red Cedar boarding										
raking cutting	4.07	10.00	4.48	0.08	-	1.47	-	5.95	m	6.54
curved cutting	6.11	10.00	6.72	0.11	-	2.02	-	8.74	m	9.61
19 mm thick hardwood boarding										
raking cutting	8.03	10.00	8.84	0.14	-	2.57	-	11.41	m	12.55
curved cutting	12.05	10.00	13.26	0.21	-	3.86	-	17.12	m	18.83
25 mm thick hardwood boarding										
raking cutting	11.90	10.00	13.09	0.16	-	2.94	-	16.03	m	17.63
curved cutting	17.85	10.00	19.64	0.24	-	4.41	-	24.04	m	26.45
Finished angles										
External; 19mm thick softwood boarding										
tongued and mitred	-	-	-	0.67	-	12.31	-	12.31	m	13.54
External; 25mm thick softwood boarding										
tongued and mitred	-	-	-	0.70	-	12.86	-	12.86	m	14.14
External; 19mm thick hardwood boarding										
tongued and mitred	-	-	-	0.95	-	17.45	-	17.45	m	19.20
External; 25mm thick hardwood boarding										
tongued and mitred	-	-	-	1.00	-	18.37	-	18.37	m	20.21
Solid wood panelling; softwood, wrot - wall panelling										
19mm thick panelled linings to walls; square framed; including grounds										
width exceeding 600mm	19.40	2.50	19.88	1.60	0.20	32.14	0.41	52.43	m²	57.68
width exceeding 600mm; fixing to masonry with screws	19.40	2.50	19.88	2.60	0.32	52.17	0.66	72.71	m²	79.98
25mm thick panelled linings to walls; square framed; including grounds										
width exceeding 600mm	25.95	2.50	26.60	1.70	0.21	34.12	0.44	61.15	m²	67.27
width exceeding 600mm; fixing to masonry with screws	25.95	2.50	26.60	2.70	0.34	54.28	0.70	81.58	m²	89.74
19mm thick panelled linings to walls; square framed; obstructed by integral services; including grounds										
width exceeding 600mm	19.40	2.50	19.88	2.10	0.26	42.15	0.54	62.58	m²	68.84
width exceeding 600mm; fixing to masonry with screws	19.40	2.50	19.88	3.10	0.39	62.31	0.80	82.99	m²	91.29
25mm thick panelled linings to walls; square framed; obstructed by integral services; including grounds										
width exceeding 600mm	25.95	2.50	26.60	2.20	0.27	44.13	0.57	71.30	m²	78.43
width exceeding 600mm; fixing to masonry with screws	25.95	2.50	26.60	3.20	0.40	64.29	0.82	91.71	m²	100.88
19mm thick panelled linings to walls; moulded; including grounds										
width exceeding 600mm	15.67	2.50	16.06	1.70	0.21	34.12	0.44	50.61	m²	55.67

Labour hourly rates: (except Specialists) Craft Operatives 18.37 Labourer 13.76 Rates are national average prices. Refer to REGIONAL VARIATIONS for indicative levels of overall pricing in regions	MATERIALS			LABOUR				RATES		
	Del to Site £	Waste %	Material Cost £	Craft Optve Hrs	Lab Hrs	Labour Cost £	Sunds £	Nett Rate £	Unit	Gross rate (10%) £
FLOOR, WALL AND CEILING BOARDING, SHEETING, PANELLING, LININGS AND CASINGS (Cont'd)										
Solid wood panelling; softwood, wrot - wall panelling (Cont'd)										
19mm thick panelled linings to walls; moulded; including grounds (Cont'd)										
width exceeding 600mm; fixing to masonry with screws	15.67	2.50	16.06	2.70	0.34	54.28	0.70	71.04	m²	78.14
25mm thick panelled linings to walls; moulded; including grounds										
width exceeding 600mm	20.67	2.50	21.18	1.80	0.22	36.09	0.45	57.73	m²	63.50
width exceeding 600mm; fixing to masonry with screws	20.67	2.50	21.18	2.80	0.35	56.25	0.72	78.16	m²	85.97
19mm thick panelled linings to walls; moulded; obstructed by integral services; including grounds										
width exceeding 600mm	15.67	2.50	16.06	2.20	0.27	44.13	0.57	60.76	m²	66.83
width exceeding 600mm; fixing to masonry with screws	15.67	2.50	16.06	3.20	0.40	64.29	0.82	81.17	m²	89.29
25mm thick panelled linings to walls; moulded; obstructed by integral services; including grounds										
width exceeding 600mm	20.67	2.50	21.18	2.30	0.29	46.24	0.58	68.01	m²	74.81
width exceeding 600mm; fixing to masonry with screws	20.67	2.50	21.18	3.30	0.41	66.26	0.86	88.30	m²	97.13
Solid wood panelling; Oak, wrot, selected for transparent finish - wall panelling										
19mm thick panelled linings to walls; square framed; including grounds										
width exceeding 600mm	106.62	2.50	109.29	2.25	0.28	45.19	0.58	155.06	m²	170.57
width exceeding 600mm; fixing to masonry with screws	106.62	2.50	109.29	3.25	0.41	65.34	0.84	175.47	m²	193.02
25mm thick panelled linings to walls; square framed; including grounds										
width exceeding 600mm	140.30	2.50	143.80	2.35	0.29	47.16	0.60	191.56	m²	210.72
width exceeding 600mm; fixing to masonry with screws	140.30	2.50	143.80	3.35	0.42	67.32	0.87	211.99	m²	233.19
19mm thick panelled linings to walls; square framed; obstructed by integral services; including grounds										
width exceeding 600mm	106.62	2.50	109.29	2.75	0.34	55.20	0.70	165.19	m²	181.71
width exceeding 600mm; fixing to masonry with screws	106.62	2.50	109.29	3.75	0.47	75.35	0.96	185.61	m²	204.17
25mm thick panelled linings to walls; square framed; obstructed by integral services; including grounds										
width exceeding 600mm	140.30	2.50	143.80	2.85	0.36	57.31	0.72	201.83	m²	222.01
width exceeding 600mm; fixing to masonry with screws	140.30	2.50	143.80	3.85	0.48	77.33	0.99	222.12	m²	244.34
19mm thick panelled linings to walls; moulded; including grounds										
width exceeding 600mm	117.29	2.50	120.22	2.40	0.30	48.22	0.62	169.05	m²	185.96
width exceeding 600mm; fixing to masonry with screws	117.29	2.50	120.22	3.40	0.42	68.24	0.87	189.33	m²	208.26
25mm thick panelled linings to walls; moulded; including grounds										
width exceeding 600mm	154.33	2.50	158.18	2.50	0.31	50.19	0.64	209.02	m²	229.92
width exceeding 600mm; fixing to masonry with screws	154.33	2.50	158.18	3.50	0.44	70.35	0.90	229.43	m²	252.38
19mm thick panelled linings to walls; moulded; obstructed by integral services; including grounds										
width exceeding 600mm	117.29	2.50	120.22	2.90	0.36	58.23	0.74	179.18	m²	197.10
width exceeding 600mm; fixing to masonry with screws	117.29	2.50	120.22	3.90	0.49	78.39	0.99	199.60	m²	219.55
25mm thick panelled linings to walls; moulded; obstructed by integral services; including grounds										
width exceeding 600mm	154.33	2.50	158.18	3.00	0.37	60.20	0.76	219.15	m²	241.07
width exceeding 600mm; fixing to masonry with screws	154.33	2.50	158.18	4.00	0.50	80.36	1.02	239.56	m²	263.52
Solid wood panelling; Sapele, wrot, selected for transparent finish - wall panelling										
19mm thick panelled linings to walls; square framed; including grounds										
width exceeding 600mm	80.14	2.50	82.14	2.05	0.26	41.24	0.52	123.90	m²	136.29
width exceeding 600mm; fixing to masonry with screws	80.14	2.50	82.14	3.05	0.38	61.26	0.80	144.19	m²	158.61
25mm thick panelled linings to walls; square framed; including grounds										
width exceeding 600mm	105.45	2.50	108.08	2.15	0.27	43.21	0.56	151.85	m²	167.03
width exceeding 600mm; fixing to masonry with screws	105.45	2.50	108.08	3.15	0.39	63.23	0.81	172.12	m²	189.34
19mm thick panelled linings to walls; square framed; obstructed by integral services; including grounds										
width exceeding 600mm	80.14	2.50	82.14	2.55	0.32	51.25	0.66	134.05	m²	147.45
width exceeding 600mm; fixing to masonry with screws	80.14	2.50	82.14	3.55	0.44	71.27	0.92	154.33	m²	169.76
25mm thick panelled linings to walls; square framed; obstructed by integral services; including grounds										
width exceeding 600mm	105.45	2.50	108.08	2.65	0.33	53.22	0.68	161.98	m²	178.18
width exceeding 600mm; fixing to masonry with screws	105.45	2.50	108.08	3.65	0.46	73.38	0.93	182.39	m²	200.63
19mm thick panelled linings to walls; moulded; including grounds										
width exceeding 600mm	88.15	2.50	90.36	2.15	0.27	43.21	0.56	134.12	m²	147.53
width exceeding 600mm; fixing to masonry with screws	88.15	2.50	90.36	3.15	0.39	63.23	0.81	154.40	m²	169.84
25mm thick panelled linings to walls; moulded; including grounds										
width exceeding 600mm	115.99	2.50	118.89	2.30	0.29	46.24	0.58	165.72	m²	182.29
width exceeding 600mm; fixing to masonry with screws	115.99	2.50	118.89	3.30	0.41	66.26	0.86	186.01	m²	204.61
19mm thick panelled linings to walls; moulded; obstructed by integral services; including grounds										
width exceeding 600mm	88.15	2.50	90.36	2.65	0.33	53.22	0.68	144.25	m²	158.68
width exceeding 600mm; fixing to masonry with screws	88.15	2.50	90.36	3.65	0.46	73.38	0.93	164.67	m²	181.13

Labour hourly rates: (except Specialists) Craft Operatives 18.37 Labourer 13.76 Rates are national average prices. Refer to REGIONAL VARIATIONS for indicative levels of overall pricing in regions	MATERIALS			LABOUR				RATES		
	Del to Site	Waste	Material Cost	Craft Optve	Lab	Labour Cost	Sunds	Nett Rate	Unit	Gross rate (10%)
	£	%	£	Hrs	Hrs	£	£	£		£
FLOOR, WALL AND CEILING BOARDING, SHEETING, PANELLING, LININGS AND CASINGS (Cont'd)										
Solid wood panelling; Sapele, wrot, selected for transparent finish - wall panelling (Cont'd)										
25mm thick panelled linings to walls; moulded; obstructed by integral services; including grounds										
width exceeding 600mm..............	115.99	2.50	118.89	2.80	0.35	56.25	0.72	175.86	m²	193.45
width exceeding 600mm; fixing to masonry with screws	115.99	2.50	118.89	3.80	0.47	76.27	0.98	196.14	m²	215.75
Veneered plywood panelling; pre-finished, decorative veneers; butt joints - wall panelling. Note. prices are for panelling with Afromosia, Ash, Beech, Cherry, Elm, Oak, Knotted Pine, Sapele or Teak faced veneers										
4mm thick linings to walls										
width exceeding 600mm..............	13.67	15.00	15.72	0.80	0.10	16.07	0.19	31.99	m²	35.19
width exceeding 600mm; fixing to timber with adhesive.................	13.67	15.00	15.72	0.95	0.12	19.10	3.00	37.82	m²	41.61
width exceeding 600mm; fixing to plaster with adhesive.............	13.67	15.00	15.72	0.90	0.11	18.05	3.00	36.77	m²	40.44
Veneered plywood panelling; flame retardant, pre-finished, decorative, veneers not matched; random V-grooves on face; butt joints - wall panelling. Note. prices are for panelling with Afromosia, Ash, Beech, Cherry, Elm, Oak, Knotted Pine, Sapele or Teak faced veneers										
4mm thick linings to walls										
width exceeding 600mm..............	36.69	15.00	42.19	0.80	0.10	16.07	0.19	58.46	m²	64.31
width exceeding 600mm; fixing to timber with adhesive.................	36.69	15.00	42.19	0.95	0.12	19.10	3.00	64.30	m²	70.73
width exceeding 600mm; fixing to plaster with adhesive.............	36.69	15.00	42.19	0.90	0.11	18.05	3.00	63.24	m²	69.56
Veneered blockboard panelling; decorative veneers; butt joints - wall panelling; Oak faced one side										
18mm thick linings to walls fixed to timber										
width exceeding 600mm..............	23.32	10.00	25.65	0.80	0.10	16.07	0.19	41.92	m²	46.11
Veneered medium density fibre board panelling; decorative veneers; butt joints - wall panelling; Pine faced one side										
6mm thick linings to walls fixed to timber										
width exceeding 600mm..............	10.25	10.00	11.28	0.80	0.10	16.07	0.19	27.54	m²	30.30
19mm thick linings to walls fixed to timber										
width exceeding 600mm..............	13.17	10.00	14.49	1.00	0.20	21.12	0.30	35.91	m²	39.50
Veneered medium density fibre board panelling; butt joints - wall panelling; Oak faced one side										
6mm thick linings to walls fixed to timber										
width exceeding 600mm..............	9.90	10.00	10.89	0.80	0.10	16.07	0.19	27.16	m²	29.87
13mm thick linings to walls fixed to timber										
width exceeding 600mm..............	11.58	10.00	12.73	0.90	0.15	18.60	0.24	31.57	m²	34.73
19mm thick linings to walls fixed to timber										
width exceeding 600mm..............	14.26	10.00	15.69	0.80	0.10	16.07	0.19	31.95	m²	35.15
Veneered medium density fibre board panelling; butt joints - wall panelling; Maple faced one side										
19mm thick linings to walls fixed to timber										
width exceeding 600mm..............	18.07	10.00	19.88	0.80	0.10	16.07	0.19	36.14	m²	39.76
IRONMONGERY										
Fix only ironmongery (prime cost sum for supply included elsewhere)										
To softwood										
butt hinges; 50mm..............	-	-	-	0.17	-	3.12	-	3.12	nr	3.44
butt hinges; 100mm..............	-	-	-	0.17	-	3.12	-	3.12	nr	3.44
rising hinges; 100mm	-	-	-	0.30	-	5.51	-	5.51	nr	6.06
tee hinges; 150mm..............	-	-	-	0.45	-	8.27	-	8.27	nr	9.09
tee hinges; 300mm..............	-	-	-	0.55	-	10.10	-	10.10	nr	11.11
tee hinges; 450mm..............	-	-	-	0.65	-	11.94	-	11.94	nr	13.13
hook and band hinges; 300mm..............	-	-	-	0.90	-	16.53	-	16.53	nr	18.19
hook and band hinges; 450mm..............	-	-	-	1.10	-	20.21	-	20.21	nr	22.23
hook and band hinges; 900mm..............	-	-	-	1.65	-	30.31	-	30.31	nr	33.34
collinge hinges; 600mm..............	-	-	-	1.00	-	18.37	-	18.37	nr	20.21
collinge hinges; 750mm..............	-	-	-	1.25	-	22.96	-	22.96	nr	25.26
collinge hinges; 900mm..............	-	-	-	1.55	-	28.47	-	28.47	nr	31.32
single action floor springs and top centres	-	-	-	2.75	-	50.52	-	50.52	nr	55.57
double action floor springs and top centres	-	-	-	3.30	-	60.62	-	60.62	nr	66.68
coil springs..............	-	-	-	0.30	-	5.51	-	5.51	nr	6.06
overhead door closers	-	-	-	1.65	-	30.31	-	30.31	nr	33.34
concealed overhead door closers..............	-	-	-	2.75	-	50.52	-	50.52	nr	55.57
Perko door closers..............	-	-	-	1.65	-	30.31	-	30.31	nr	33.34
door selectors..............	-	-	-	2.20	0.28	44.27	-	44.27	nr	48.69
cabin hooks..............	-	-	-	0.25	-	4.59	-	4.59	nr	5.05
fanlight catches	-	-	-	0.40	-	7.35	-	7.35	nr	8.08
roller catches..............	-	-	-	0.40	-	7.35	-	7.35	nr	8.08
casement fasteners	-	-	-	0.40	-	7.35	-	7.35	nr	8.08
sash fasteners..............	-	-	-	1.10	-	20.21	-	20.21	nr	22.23
sash screws..............	-	-	-	0.55	-	10.10	-	10.10	nr	11.11
mortice latches..............	-	-	-	0.70	Hrs	12.86	-	12.86	nr	14.14
night latches	-	-	-	1.10	-	20.21	-	20.21	nr	22.23
Norfolk latches..............	-	-	-	0.90	-	16.53	-	16.53	nr	18.19

Labour hourly rates: (except Specialists) Craft Operatives 18.37 Labourer 13.76 Rates are national average prices. Refer to REGIONAL VARIATIONS for indicative levels of overall pricing in regions	MATERIALS			LABOUR				RATES		
	Del to Site	Waste	Material Cost	Craft Optve	Lab	Labour Cost	Sunds	Nett Rate		Gross rate (10%)
	£	%	£	Hrs	Hrs	£	£	£	Unit	£
IRONMONGERY (Cont'd)										
Fix only ironmongery (prime cost sum for supply included elsewhere) (Cont'd)										
To softwood (Cont'd)										
rim latches	-	-	-	0.95	-	17.45	-	17.45	nr	19.20
Suffolk latches	-	-	-	0.90	-	16.53	-	16.53	nr	18.19
budget locks	-	-	-	0.75	-	13.78	-	13.78	nr	15.16
cylinder locks	-	-	-	1.10	-	20.21	-	20.21	nr	22.23
dead locks	-	-	-	0.90	-	16.53	-	16.53	nr	18.19
mortice locks	-	-	-	1.10	-	20.21	-	20.21	nr	22.23
rim locks	-	-	-	0.90	-	16.53	-	16.53	nr	18.19
automatic coin collecting locks	-	-	-	2.20	-	40.41	-	40.41	nr	44.46
casement stays	-	-	-	0.30	-	5.51	-	5.51	nr	6.06
quadrant stays	-	-	-	0.30	-	5.51	-	5.51	nr	6.06
barrel bolts; 150mm	-	-	-	0.35	-	6.43	-	6.43	nr	7.07
barrel bolts; 250mm	-	-	-	0.45	-	8.27	-	8.27	nr	9.09
door bolts; 150mm	-	-	-	0.35	-	6.43	-	6.43	nr	7.07
door bolts; 250mm	-	-	-	0.45	-	8.27	-	8.27	nr	9.09
monkey tail bolts; 300mm	-	-	-	0.40	-	7.35	-	7.35	nr	8.08
monkey tail bolts; 450mm	-	-	-	0.45	-	8.27	-	8.27	nr	9.09
monkey tail bolts; 600mm	-	-	-	0.55	-	10.10	-	10.10	nr	11.11
flush bolts; 200mm	-	-	-	0.75	-	13.78	-	13.78	nr	15.16
flush bolts; 450mm	-	-	-	1.10	-	20.21	-	20.21	nr	22.23
indicating bolts	-	-	-	1.10	-	20.21	-	20.21	nr	22.23
panic bolts; to single door	-	-	-	1.65	-	30.31	-	30.31	nr	33.34
panic bolts; to double doors	-	-	-	2.20	-	40.41	-	40.41	nr	44.46
knobs	-	-	-	0.25	-	4.59	-	4.59	nr	5.05
lever handles	-	-	-	0.25	-	4.59	-	4.59	nr	5.05
sash lifts	-	-	-	0.20	-	3.67	-	3.67	nr	4.04
pull handles; 150mm	-	-	-	0.20	-	3.67	-	3.67	nr	4.04
pull handles; 225mm	-	-	-	0.25	-	4.59	-	4.59	nr	5.05
pull handles; 300mm	-	-	-	0.30	-	5.51	-	5.51	nr	6.06
back plates	-	-	-	0.30	-	5.51	-	5.51	nr	6.06
escutcheon plates	-	-	-	0.25	-	4.59	-	4.59	nr	5.05
kicking plates	-	-	-	0.55	-	10.10	-	10.10	nr	11.11
letter plates	-	-	-	1.65	-	30.31	-	30.31	nr	33.34
push plates; 225mm	-	-	-	0.25	-	4.59	-	4.59	nr	5.05
push plates; 300mm	-	-	-	0.30	-	5.51	-	5.51	nr	6.06
shelf brackets	-	-	-	0.30	-	5.51	-	5.51	nr	6.06
sash cleats	-	-	-	0.25	-	4.59	-	4.59	nr	5.05
To softwood and brickwork										
cabin hooks	-	-	-	0.30	-	5.51	-	5.51	nr	6.06
To hardwood or the like										
butt hinges; 50mm	-	-	-	0.25	-	4.59	-	4.59	nr	5.05
butt hinges; 100mm	-	-	-	0.25	-	4.59	-	4.59	nr	5.05
rising hinges; 100mm	-	-	-	0.40	-	7.35	-	7.35	nr	8.08
tee hinges; 150mm	-	-	-	0.60	-	11.02	-	11.02	nr	12.12
tee hinges; 300mm	-	-	-	0.75	-	13.78	-	13.78	nr	15.16
tee hinges; 450mm	-	-	-	0.90	-	16.53	-	16.53	nr	18.19
hook and band hinges; 300mm	-	-	-	1.20	-	22.04	-	22.04	nr	24.25
hook and band hinges; 450mm	-	-	-	1.55	-	28.47	-	28.47	nr	31.32
hook and band hinges; 900mm	-	-	-	2.20	-	40.41	-	40.41	nr	44.46
collinge hinges; 600mm	-	-	-	1.45	-	26.64	-	26.64	nr	29.30
collinge hinges; 750mm	-	-	-	1.85	-	33.98	-	33.98	nr	37.38
collinge hinges; 900mm	-	-	-	2.25	-	41.33	-	41.33	nr	45.47
single action floor springs and top centres	-	-	-	4.10	-	75.32	-	75.32	nr	82.85
double action floor springs and top centres	-	-	-	4.90	-	90.01	-	90.01	nr	99.01
coil springs	-	-	-	0.45	-	8.27	-	8.27	nr	9.09
overhead door closers	-	-	-	2.45	-	45.01	-	45.01	nr	49.51
concealed overhead door closers	-	-	-	4.10	-	75.32	-	75.32	nr	82.85
Perko door closers	-	-	-	2.45	-	45.01	-	45.01	nr	49.51
door selectors	-	-	-	3.30	0.40	66.12	-	66.12	nr	72.74
cabin hooks	-	-	-	0.35	-	6.43	-	6.43	nr	7.07
fanlight catches	-	-	-	0.55	-	10.10	-	10.10	nr	11.11
roller catches	-	-	-	0.55	-	10.10	-	10.10	nr	11.11
casement fasteners	-	-	-	0.55	-	10.10	-	10.10	nr	11.11
sash fasteners	-	-	-	1.65	-	30.31	-	30.31	nr	33.34
sash screws	-	-	-	0.80	-	14.70	-	14.70	nr	16.17
mortice latches	-	-	-	1.10	-	20.21	-	20.21	nr	22.23
night latches	-	-	-	1.65	-	30.31	-	30.31	nr	33.34
Norfolk latches	-	-	-	1.30	-	23.88	-	23.88	nr	26.27
rim latches	-	-	-	0.80	-	14.70	-	14.70	nr	16.17
Suffolk latches	-	-	-	1.30	-	23.88	-	23.88	nr	26.27
budget locks	-	-	-	1.10	-	20.21	-	20.21	nr	22.23
cylinder locks	-	-	-	1.65	-	30.31	-	30.31	nr	33.34
dead locks	-	-	-	1.30	-	23.88	-	23.88	nr	26.27
mortice locks	-	-	-	1.65	-	30.31	-	30.31	nr	33.34
rim locks	-	-	-	1.30	-	23.88	-	23.88	nr	26.27
automatic coin collecting locks	-	-	-	3.30	-	60.62	-	60.62	nr	66.68
casement stays	-	-	-	0.40	-	7.35	-	7.35	nr	8.08
quadrant stays	-	-	-	0.40	-	7.35	-	7.35	nr	8.08
barrel bolts; 150mm	-	-	-	0.45	-	8.27	-	8.27	nr	9.09
barrel bolts; 250mm	-	-	-	0.60	-	11.02	-	11.02	nr	12.12
door bolts; 150mm	-	-	-	0.45	-	8.27	-	8.27	nr	9.09
door bolts; 250mm	-	-	-	0.60	-	11.02	-	11.02	nr	12.12
monkey tail bolts; 300mm	-	-	-	0.55	-	10.10	-	10.10	nr	11.11
monkey tail bolts; 450mm	-	-	-	0.65	-	11.94	-	11.94	nr	13.13
monkey tail bolts; 600mm	-	-	-	0.75	-	13.78	-	13.78	nr	15.16
flush bolts; 200mm	-	-	-	1.10	-	20.21	-	20.21	nr	22.23
flush bolts; 450mm	-	-	-	1.65	-	30.31	-	30.31	nr	33.34
indicating bolts	-	-	-	1.65	-	30.31	-	30.31	nr	33.34

Labour hourly rates: (except Specialists) Craft Operatives 18.37 Labourer 13.76 Rates are national average prices. Refer to REGIONAL VARIATIONS for indicative levels of overall pricing in regions	MATERIALS			LABOUR				RATES		
	Del to Site	Waste	Material Cost	Craft Optve	Lab	Labour Cost	Sunds	Nett Rate	Unit	Gross rate (10%)
	£	%	£	Hrs	Hrs	£	£	£		£
IRONMONGERY (Cont'd)										
Fix only ironmongery (prime cost sum for supply included elsewhere) (Cont'd)										
To hardwood or the like (Cont'd)										
panic bolts; to single door	-	-	-	2.45	-	45.01	-	45.01	nr	49.51
panic bolts; to double doors	-	-	-	3.30	-	60.62	-	60.62	nr	66.68
knobs	-	-	-	0.35	-	6.43	-	6.43	nr	7.07
lever handles	-	-	-	0.35	-	6.43	-	6.43	nr	7.07
sash lifts	-	-	-	0.35	-	6.43	-	6.43	nr	7.07
pull handles; 150mm	-	-	-	0.30	-	5.51	-	5.51	nr	6.06
pull handles; 225mm	-	-	-	0.35	-	6.43	-	6.43	nr	7.07
pull handles; 300mm	-	-	-	0.40	-	7.35	-	7.35	nr	8.08
back plates	-	-	-	0.40	-	7.35	-	7.35	nr	8.08
escutcheon plates	-	-	-	0.35	-	6.43	-	6.43	nr	7.07
kicking plates	-	-	-	0.80	-	14.70	-	14.70	nr	16.17
letter plates	-	-	-	2.45	-	45.01	-	45.01	nr	49.51
push plates; 225mm	-	-	-	0.35	-	6.43	-	6.43	nr	7.07
push plates; 300mm	-	-	-	0.40	-	7.35	-	7.35	nr	8.08
shelf brackets	-	-	-	0.45	-	8.27	-	8.27	nr	9.09
sash cleats	-	-	-	0.35	-	6.43	-	6.43	nr	7.07
To hardwood and brickwork										
cabin hooks	-	-	-	0.40	-	7.35	-	7.35	nr	8.08
Water bars; steel; galvanized										
Water bars; to concrete										
flat section; setting in groove in mastic										
25 x 3 x 900mm long	4.70	5.00	4.94	0.75	0.09	15.02	0.68	20.63	nr	22.69
40 x 3 x 900mm long	5.23	5.00	5.49	0.75	0.09	15.02	0.82	21.34	nr	23.47
40 x 6 x 900mm long	8.66	5.00	9.10	0.75	0.09	15.02	0.82	24.94	nr	27.43
50 x 6 x 900mm long	10.88	5.00	11.42	0.75	0.09	15.02	0.82	27.26	nr	29.99
Water bars; to hardwood										
flat section; setting in groove in mastic										
25 x 3 x 900mm long	4.70	5.00	4.94	0.65	0.08	13.04	0.68	18.65	nr	20.52
40 x 3 x 900mm long	5.23	5.00	5.49	0.65	0.08	13.04	0.82	19.36	nr	21.30
40 x 6 x 900mm long	8.66	5.00	9.10	0.65	0.08	13.04	0.82	22.96	nr	25.26
50 x 6 x 900mm long	10.88	5.00	11.42	0.65	0.08	13.04	0.82	25.29	nr	27.81
Sliding door gear, P.C Henderson Ltd.										
To softwood										
interior straight sliding door gear sets for commercial and domestic doors; Senator single door set comprising track, hangers, end stops and bottom guide; for doors 20 - 35mm thick, maximum weight 25kg maximum 900mm wide	32.88	2.50	33.70	1.65	-	30.31	-	64.01	nr	70.41
extra for pelmet 1850mm long	35.30	2.50	36.18	0.70	-	12.86	-	49.04	nr	53.95
extra for pelmet end cap	2.76	2.50	2.83	0.11	-	2.02	-	4.85	nr	5.33
extra for Doorseal deflector guide	2.98	2.50	3.05	0.40	-	7.35	-	10.40	nr	11.44
interior straight sliding door gear sets for commercial and domestic doors; Phantom single door set comprising top assembly, hangers, adjustable nylon guide and door stops; for doors 30 - 50mm thick, maximum weight 45kg, 610 - 915mm wide	55.45	2.50	56.84	1.95	-	35.82	-	92.66	nr	101.92
extra for pelmet 1850mm long	35.30	2.50	36.18	0.70	-	12.86	-	49.04	nr	53.95
extra for pelmet end cap	2.76	2.50	2.83	0.11	-	2.02	-	4.85	nr	5.33
extra for door guide	2.98	2.50	3.05	0.40	-	7.35	-	10.40	nr	11.44
interior straight sliding door gear sets for commercial and domestic doors; Marathon 55 Junior Nr J2 single door set comprising top assembly, hangers, end stops, inverted guide channel and nylon guide; for doors 32 - 50mm thick, maximum weight, 55kg, 400 - 750mm wide; forming groove for guide channel	47.79	2.50	48.98	2.65	-	48.68	-	97.67	nr	107.43
interior straight sliding door gear sets for commercial and domestic doors; Marathon 55 Junior Nr J3 single door set comprising top assembly, hangers, end stops, inverted guide channel and nylon guide; for doors 32 - 50mm thick, maximum weight, 55kg, 750 - 900mm wide; forming groove for guide channel	50.45	2.50	51.71	2.85	-	52.35	-	104.07	nr	114.47
interior straight sliding door gear sets for commercial and domestic doors; Marathon 55 Junior Nr J4 single door set comprising top assembly, hangers, end stops, inverted guide channel and nylon guide; for doors 32 - 50mm thick, maximum weight, 55kg, 900 - 1050mm wide; forming groove for guide channel	54.59	2.50	55.95	3.10	-	56.95	-	112.90	nr	124.19
interior straight sliding door gear sets for commercial and domestic doors; Marathon 55 Junior Nr J5 single door set comprising top assembly, hangers, end stops, inverted guide channel and nylon guide; for doors 32 - 50mm thick, maximum weight, 55kg, 1050 - 1200mm wide; forming groove for guide channel	61.72	2.50	63.26	3.30	-	60.62	-	123.88	nr	136.27
interior straight sliding door gear sets for commercial and domestic doors; Marathon 55 Junior Nr J6 single door set comprising top assembly, hangers, end stops, inverted guide channel and nylon guide; for doors 32 - 50mm thick, maximum weight, 55kg, 1200 - 1500mm wide; forming groove for guide channel	102.39	2.50	104.95	3.50	-	64.30	-	169.24	nr	186.17
extra for pelmet 1850mm long	35.30	2.50	36.18	0.70	-	12.86	-	49.04	nr	53.95
extra for pelmet end cap	2.76	2.50	2.83	0.11	-	2.02	-	4.85	nr	5.33

Labour hourly rates: (except Specialists) Craft Operatives 18.37 Labourer 13.76 Rates are national average prices. Refer to REGIONAL VARIATIONS for indicative levels of overall pricing in regions	MATERIALS			LABOUR				RATES		
	Del to Site £	Waste %	Material Cost £	Craft Optve Hrs	Lab Hrs	Labour Cost £	Sunds £	Nett Rate £	Unit	Gross rate (10%) £
IRONMONGERY (Cont'd)										
Sliding door gear, P.C Henderson Ltd. (Cont'd)										
To softwood (Cont'd)										
interior straight sliding door gear sets for commercial and domestic doors; Marathon 90 Senior Nr S3 single door set comprising top assembly, hangers, end stops, inverted guide channel and nylon guide; for doors 32 - 50mm thick maximum weight 90kg, 750 - 900mm wide; forming groove for guide channel...............	67.68	2.50	69.37	3.10	-	56.95	-	126.32	nr	138.95
interior straight sliding door gear sets for commercial and domestic doors; Marathon 90 Senior Nr S4 single door set comprising top assembly, hangers, end stops, inverted guide channel and nylon guide; for doors 32 - 50mm thick maximum weight 90kg, 900 - 1050mm wide; forming groove for guide channel...............	75.15	2.50	77.03	3.30	-	60.62	-	137.65	nr	151.41
interior straight sliding door gear sets for commercial and domestic doors; Marathon 90 Senior Nr S5 single door set comprising top assembly, hangers, end stops, inverted guide channel and nylon guide; for doors 32 - 50mm thick maximum weight 90kg, 1050 - 1200mm wide; forming groove for guide channel...............	82.92	2.50	84.99	3.50	-	64.30	-	149.29	nr	164.22
interior straight sliding door gear sets for commercial and domestic doors; Marathon 90 Senior Nr S6 single door set comprising top assembly, hangers, end stops, inverted guide channel and nylon guide; for doors 32 - 50mm thick maximum weight 90kg, 1200 - 1500mm wide; forming groove for guide channel...............	132.92	2.50	136.24	3.65	-	67.05	-	203.29	nr	223.62
extra for pelmet 1850mm long...............	35.30	2.50	36.18	0.70	-	12.86	-	49.04	nr	53.95
extra for pelmet end cap...............	2.76	2.50	2.83	0.11	-	2.02	-	4.85	nr	5.33
interior straight sliding door gear sets for wardrobe and cupboard doors; Single Top Nr ST12 single door set comprising track, hangers, guides and safety stop; for door 16 - 35mm thick, maximum weight 25kg, maximum 900mm wide; one door to 600mm wide opening	19.46	2.50	19.95	1.40	-	25.72	-	45.66	nr	50.23
interior straight sliding door gear sets for wardrobe and cupboard doors; Single Top Nr ST15 single door set comprising track, hangers, guides and safety stop; for door 16 - 35mm thick, maximum weight 25kg, maximum 900mm wide; one door to 750mm wide opening	21.90	2.50	22.45	1.50	-	27.56	-	50.00	nr	55.00
interior straight sliding door gear sets for wardrobe and cupboard doors; Single Top Nr ST18 single door set comprising track, hangers, guides and safety stop; for door 16 - 35mm thick, maximum weight 25kg, maximum 900mm wide; one door to 900mm wide opening	24.79	2.50	25.41	1.60	-	29.39	-	54.80	nr	60.28
interior straight sliding door gear sets for wardrobe and cupboard doors; Double Top Nr W12 bi-passing door set comprising double track section, hangers, guides and safety stop; for doors 16 - 35mm thick, maximum weight 25kg, maximum 900mm wide; two doors in 1200mm wide opening	33.37	2.50	34.20	2.20	-	40.41	-	74.62	nr	82.08
interior straight sliding door gear sets for wardrobe and cupboard doors; Double Top Nr W15 bi-passing door set comprising double track section, hangers, guides and safety stop; for doors 16 - 35mm thick, maximum weight 25kg, maximum 900mm wide; two doors in 1500mm wide opening	37.19	2.50	38.12	2.30	-	42.25	-	80.37	nr	88.41
interior straight sliding door gear sets for wardrobe and cupboard doors; Double Top Nr W18 bi-passing door set comprising double track section, hangers, guides and safety stop; for doors 16 - 35mm thick, maximum weight 25kg, maximum 900mm wide; two doors in 1800mm wide opening	41.24	2.50	42.27	2.40	-	44.09	-	86.36	nr	94.99
interior straight sliding door gear sets for wardrobe and cupboard doors; Double Top Nr W24 bi-passing door set comprising double track section, hangers, guides and safety stop; for doors 16 - 35mm thick, maximum weight 25kg, maximum 900mm wide; three doors in 2400mm wide opening...............	54.11	2.50	55.46	2.65	-	48.68	-	104.14	nr	114.56
interior straight sliding door gear sets for wardrobe and cupboard doors; Bifold Nr B10-2 folding door set comprising top guide track, top and bottom pivots; top guide and hinges; for doors 20 - 35mm thick; maximum weight 14kg each leaf, maximum 530mm wide; two doors in 1000mm wide opening...............	30.68	2.50	31.45	3.30	-	60.62	-	92.07	nr	101.27
interior straight sliding door gear sets for wardrobes and cupboard doors; Bifold Nr B15-4 folding door set comprising top guide track, top and bottom pivots, top guide and hinges; for doors 20 - 35mm thick, maximum weight 14kg each leaf, maximum 530mm wide; four doors in 1500mm wide opening with aligner...............	52.02	2.50	53.32	6.60	-	121.24	-	174.56	nr	192.02
interior straight sliding door gear sets for wardrobes and cupboard doors; Bifold Nr B20-4 folding door set comprising top guide track, top and bottom pivots, top guide and hinges; for doors 20 - 35mm thick, maximum weight 14kg each leaf, maximum 530mm wide; four doors in 2000mm wide opening with aligner...............	56.98	2.50	58.40	7.15	-	131.35	-	189.75	nr	208.73
interior straight sliding door gear sets for built in cupboard doors; Slipper Nr SS4 double passing door set comprising two top tracks, sliders, safety stop, flush pulls and guides; for doors 16 - 30mm thick, maximum weight 9kg, maximum 900mm wide; two doors in 1200mm wide opening...............	23.54	2.50	24.13	2.05	-	37.66	-	61.79	nr	67.97
interior straight sliding door gear sets for built in cupboard doors; Slipper Nr SS5 double passing door set comprising two top tracks, sliders, safety stop, flush pulls and guides; for doors 16 - 30mm thick, maximum weight 9kg, maximum 900mm wide; two doors in 1500mm wide opening...............	27.68	2.50	28.37	2.20	-	40.41	-	68.79	nr	75.66

Labour hourly rates: (except Specialists) Craft Operatives 18.37 Labourer 13.76 Rates are national average prices. Refer to REGIONAL VARIATIONS for indicative levels of overall pricing in regions	MATERIALS			LABOUR				RATES		
	Del to Site	Waste	Material Cost	Craft Optve	Lab	Labour Cost	Sunds	Nett Rate	Unit	Gross rate (10%)
	£	%	£	Hrs	Hrs	£	£	£		£

IRONMONGERY (Cont'd)

Sliding door gear, P.C Henderson Ltd. (Cont'd)

To softwood (Cont'd)

	Del to Site £	Waste %	Material Cost £	Craft Optve Hrs	Lab Hrs	Labour Cost £	Sunds £	Nett Rate £	Unit	Gross rate (10%) £
interior straight sliding door gear sets for built in cupboard doors; Slipper Nr SS6 double passing door set comprising two top tracks, sliders, safety stop, flush pulls and guides; for doors 16 - 30mm thick, maximum weight 9kg, maximum 900mm wide; two doors in 1800mm wide opening	30.90	2.50	31.67	2.35	-	43.17	-	74.84	nr	82.33
interior straight sliding door gear sets for cupboards, book cases and cabinet work; Loretto Nr D4 bi-passing door set comprising two top guide channels, two bottom rails, nylon guides and bottom rollers; for doors 20 - 45mm thick; maximum weight 23kg, maximum 900mm wide; two doors in 1200mm wide opening; forming grooves for top guide channels and bottom rails; forming sinkings for bottom rollers	39.86	2.50	40.86	2.75	-	50.52	-	91.37	nr	100.51
interior straight sliding door gear sets for cupboards, book cases and cabinet work; Loretto Nr D5 bi-passing door set comprising two top guide channels, two bottom rails, nylon guides and bottom rollers; for doors 20 - 45mm thick; maximum weight 23kg, maximum 900mm wide; two doors in 1500mm wide opening; forming grooves for top guide channels and bottom rails; forming sinkings for bottom rollers	44.34	2.50	45.45	2.85	-	52.35	-	97.80	nr	107.58
interior straight sliding door gear sets for cupboards, book cases and cabinet work; Loretto Nr D6 bi-passing door set comprising two top guide channels, two bottom rails, nylon guides and bottom rollers; for doors 20 - 45mm thick; maximum weight 23kg, maximum 900mm wide; two doors in 1800mm wide opening; forming grooves for top guide channels and bottom rails; forming sinkings for bottom rollers	50.14	2.50	51.39	3.00	-	55.11	-	106.50	nr	117.15
extra for mortice fixed bottom rollers (pair)	9.93	2.50	10.18	0.40	-	7.35	-	17.53	nr	19.28
extra for flush pull	5.02	2.50	5.15	0.30	-	5.51	-	10.66	nr	11.72
interior straight room divider gear sets; Husky 50; Nr H50/15 steel track sliding door gear; 1500mm track length	43.60	2.50	44.69	6.40	-	117.57	-	162.26	nr	178.48
interior straight sliding door gear sets for glass panels; Zenith Nr Z12 double passing door set comprising double top guide, bottom rail, glass rail with rubber glazing strip, end caps and bottom rollers; for panels 6mm thick, maximum weight 16kg or 1m² per panel; two panels nr 1200mm wide opening	60.50	2.50	62.01	1.65	-	30.31	-	92.32	nr	101.56
interior straight sliding door gear sets for glass panels; Zenith Nr Z15 double passing door set comprising double top guide, bottom rail, glass rail with rubber glazing strip, end caps and bottom rollers; for panels 6mm thick, maximum weight 16kg or 1m² per panel; two panels nr 1500mm wide opening	69.70	2.50	71.44	1.80	-	33.07	-	104.51	nr	114.96
interior straight sliding door gear sets for glass panels; Zenith Nr Z18 double passing door set comprising double top guide, bottom rail, glass rail with rubber glazing strip, end caps and bottom rollers; for panels 6mm thick, maximum weight 16kg or 1m² per panel; two panels nr 1800mm wide opening	78.02	2.50	79.97	2.00	-	36.74	-	116.71	nr	128.38
extra for finger pull	4.06	2.50	4.16	0.45	-	8.27	-	12.43	nr	13.67
extra for cylinder lock	19.67	2.50	20.16	1.10	-	20.21	-	40.37	nr	44.41
extra for dust seal	2.67	2.50	2.74	0.35	-	6.43	-	9.17	m	10.08

Hinges; standard quality

To softwood

	Del to Site £	Waste %	Material Cost £	Craft Optve Hrs	Lab Hrs	Labour Cost £	Sunds £	Nett Rate £	Unit	Gross rate (10%) £
backflap hinges; steel										
25mm	0.20	2.50	0.21	0.15	-	2.76	-	2.96	nr	3.26
38mm	0.24	2.50	0.24	0.17	-	3.12	-	3.37	nr	3.70
50mm	0.34	2.50	0.35	0.17	-	3.12	-	3.47	nr	3.82
63mm	0.64	2.50	0.66	0.17	-	3.12	-	3.78	nr	4.16
75mm	1.00	2.50	1.02	0.17	-	3.12	-	4.15	nr	4.56
butt hinges; steel; light medium pattern										
38mm	0.07	2.50	0.07	0.17	-	3.12	-	3.20	nr	3.52
50mm	0.09	2.50	0.09	0.17	-	3.12	-	3.22	nr	3.54
63mm	0.10	2.50	0.11	0.17	-	3.12	-	3.23	nr	3.55
75mm	0.12	2.50	0.12	0.17	-	3.12	-	3.24	nr	3.57
100mm	0.28	2.50	0.29	0.17	-	3.12	-	3.41	nr	3.75
butt hinges; steel; strong pattern										
75mm	0.32	2.50	0.33	0.17	-	3.12	-	3.45	nr	3.80
100mm	0.47	2.50	0.49	0.17	-	3.12	-	3.61	nr	3.97
butt hinges; steel; electro brass finish; washered; No.1838										
75mm	0.60	-	0.60	0.17	-	3.12	-	3.72	nr	4.10
butt hinges; cast iron; light										
50mm	0.53	2.50	0.55	0.17	-	3.12	-	3.67	nr	4.04
63mm	1.37	2.50	1.40	0.17	-	3.12	-	4.52	nr	4.98
75mm	8.62	2.50	8.84	0.17	-	3.12	-	11.96	nr	13.16
100mm	11.96	2.50	12.26	0.17	-	3.12	-	15.38	nr	16.92
butt hinges; brass, brass pin										
38mm	0.57	2.50	0.58	0.17	-	3.12	-	3.70	nr	4.07
50mm	0.73	2.50	0.75	0.17	-	3.12	-	3.87	nr	4.26
63mm	0.85	2.50	0.88	0.17	-	3.12	-	4.00	nr	4.40
75mm	1.19	2.50	1.22	0.17	-	3.12	-	4.34	nr	4.78
100mm	3.00	2.50	3.07	0.17	-	3.12	-	6.20	nr	6.82
butt hinges; brass, steel washers, steel pin										
75mm	1.34	2.50	1.37	0.17	-	3.12	-	4.49	nr	4.94
100mm	2.87	2.50	2.94	0.17	-	3.12	-	6.06	nr	6.67
butt hinges; brass, B.M.A., steel washers, steel pin										
75mm	4.10	2.50	4.20	0.17	-	3.12	-	7.33	nr	8.06
100mm	5.30	2.50	5.43	0.17	-	3.12	-	8.55	nr	9.41
butt hinges; brass, chromium plated, steel washers, steel pin										
75mm	3.93	2.50	4.03	0.17	-	3.12	-	7.15	nr	7.87
100mm	4.59	2.50	4.71	0.17	-	3.12	-	7.83	nr	8.61

Labour hourly rates: (except Specialists) Craft Operatives 18.37 Labourer 13.76 Rates are national average prices. Refer to REGIONAL VARIATIONS for indicative levels of overall pricing in regions	MATERIALS			LABOUR				RATES		
	Del to Site £	Waste %	Material Cost £	Craft Optve Hrs	Lab Hrs	Labour Cost £	Sunds £	Nett Rate £	Unit	Gross rate (10%) £
IRONMONGERY (Cont'd)										
Hinges; standard quality (Cont'd)										
To softwood (Cont'd)										
butt hinges; aluminium, stainless steel pin; nylon washers										
75mm	5.37	2.50	5.51	0.17	-	3.12	-	8.63	nr	9.49
100mm	6.24	2.50	6.40	0.17	-	3.12	-	9.52	nr	10.48
rising hinges; steel										
75mm	0.86	2.50	0.88	0.30	-	5.51	-	6.39	nr	7.03
100mm	2.48	2.50	2.54	0.30	-	5.51	-	8.05	nr	8.86
spring hinges; steel, lacquered; single action										
100mm	5.83	2.50	5.98	0.57	-	10.47	-	16.45	nr	18.09
125mm	9.52	2.50	9.76	0.62	-	11.39	-	21.15	nr	23.26
150mm	20.00	2.50	20.50	0.67	-	12.31	-	32.81	nr	36.09
spring hinges; steel, lacquered; double action										
75mm	6.92	2.50	7.09	0.67	-	12.31	-	19.40	nr	21.34
100mm	8.06	2.50	8.26	0.67	-	12.31	-	20.57	nr	22.63
125mm	17.49	2.50	17.93	0.77	-	14.14	-	32.07	nr	35.28
150mm	20.25	2.50	20.76	0.77	-	14.14	-	34.90	nr	38.39
tee hinges; japanned; light										
150mm	0.52	2.50	0.53	0.45	-	8.27	-	8.80	nr	9.68
230mm	0.73	2.50	0.75	0.50	-	9.19	-	9.93	nr	10.93
300mm	1.07	2.50	1.09	0.55	-	10.10	-	11.20	nr	12.32
375mm	2.07	2.50	2.12	0.60	-	11.02	-	13.14	nr	14.45
450mm	2.73	2.50	2.80	0.65	-	11.94	-	14.74	nr	16.21
tee hinges; galvanised; heavy										
300mm	2.48	2.50	2.54	0.60	-	11.02	-	13.56	nr	14.92
375mm	3.73	2.50	3.82	0.65	-	11.94	-	15.76	nr	17.34
450mm	5.40	2.50	5.53	0.70	-	12.86	-	18.39	nr	20.23
600mm	7.90	2.50	8.09	0.75	-	13.78	-	21.87	nr	24.06
hook and band hinges; on plate; heavy; black										
300mm	3.73	2.50	3.82	0.90	-	16.53	-	20.36	nr	22.39
450mm	6.65	2.50	6.81	1.10	-	20.21	-	27.02	nr	29.72
610mm	9.15	2.50	9.37	1.30	-	23.88	-	33.26	nr	36.58
914mm	17.90	2.50	18.34	1.65	-	30.31	-	48.65	nr	53.52
collinge hinges; cup for wood; best quality										
610mm	126.10	2.50	129.25	1.00	-	18.37	-	147.62	nr	162.38
762mm	161.85	2.50	165.90	1.25	-	22.96	-	188.86	nr	207.74
914mm	189.15	2.50	193.88	1.55	-	28.47	-	222.35	nr	244.59
parliament hinges; steel; 100mm	1.77	2.50	1.82	0.30	-	5.51	-	7.33	nr	8.06
cellar flap hinges; wrought, welded, plain joint; 450mm	5.19	2.50	5.32	0.55	-	10.10	-	15.42	nr	16.96
To hardwood or the like										
backflap hinges; steel										
25mm	0.20	2.50	0.21	0.25	-	4.59	-	4.80	nr	5.28
38mm	0.24	2.50	0.24	0.25	-	4.59	-	4.84	nr	5.32
50mm	0.34	2.50	0.35	0.25	-	4.59	-	4.94	nr	5.44
63mm	0.64	2.50	0.66	0.25	-	4.59	-	5.25	nr	5.78
75mm	1.00	2.50	1.02	0.25	-	4.59	-	5.62	nr	6.18
butt hinges; steel; light medium pattern										
38mm	0.07	2.50	0.07	0.25	-	4.59	-	4.67	nr	5.13
50mm	0.09	2.50	0.09	0.25	-	4.59	-	4.69	nr	5.16
63mm	0.10	2.50	0.11	0.25	-	4.59	-	4.70	nr	5.17
75mm	0.12	2.50	0.12	0.25	-	4.59	-	4.71	nr	5.18
100mm	0.28	2.50	0.29	0.25	-	4.59	-	4.88	nr	5.37
butt hinges; steel; strong pattern										
75mm	0.32	2.50	0.33	0.25	-	4.59	-	4.92	nr	5.41
100mm	0.47	2.50	0.49	0.25	-	4.59	-	5.08	nr	5.59
butt hinges; cast iron; light										
50mm	0.53	2.50	0.55	0.25	-	4.59	-	5.14	nr	5.65
63mm	1.37	2.50	1.40	0.25	-	4.59	-	5.99	nr	6.59
75mm	8.62	2.50	8.84	0.25	-	4.59	-	13.43	nr	14.78
100mm	11.96	2.50	12.26	0.25	-	4.59	-	16.85	nr	18.53
butt hinges; brass, brass pin										
38mm	0.57	2.50	0.58	0.25	-	4.59	-	5.17	nr	5.69
50mm	0.73	2.50	0.75	0.25	-	4.59	-	5.34	nr	5.87
63mm	0.85	2.50	0.88	0.25	-	4.59	-	5.47	nr	6.01
75mm	1.19	2.50	1.22	0.25	-	4.59	-	5.81	nr	6.40
100mm	3.00	2.50	3.07	0.25	-	4.59	-	7.67	nr	8.43
butt hinges; brass, steel washers, steel pin										
75mm	1.34	2.50	1.37	0.25	-	4.59	-	5.96	nr	6.56
100mm	2.87	2.50	2.94	0.25	-	4.59	-	7.53	nr	8.28
butt hinges; brass, B.M.A., steel washers, steel pin										
75mm	4.10	2.50	4.20	0.25	-	4.59	-	8.80	nr	9.67
100mm	5.30	2.50	5.43	0.25	-	4.59	-	10.02	nr	11.02
butt hinges; brass, chromium plated, steel washers, steel pin										
75mm	3.93	2.50	4.03	0.25	-	4.59	-	8.62	nr	9.49
100mm	4.59	2.50	4.71	0.25	-	4.59	-	9.30	nr	10.23
butt hinges; aluminium, stainless steel pin; nylon washers										
75mm	5.37	2.50	5.51	0.25	-	4.59	-	10.10	nr	11.11
100mm	6.24	2.50	6.40	0.25	-	4.59	-	10.99	nr	12.09
rising hinges; steel										
75mm	0.86	2.50	0.88	0.40	-	7.35	-	8.23	nr	9.05
100mm	2.48	2.50	2.54	0.40	-	7.35	-	9.89	nr	10.88
spring hinges; steel, lacquered; single action										
100mm	5.83	2.50	5.98	0.85	-	15.61	-	21.59	nr	23.75
125mm	9.52	2.50	9.76	0.90	-	16.53	-	26.29	nr	28.92
150mm	20.00	2.50	20.50	0.95	-	17.45	-	37.95	nr	41.75
spring hinges; steel, lacquered; double action										
75mm	6.92	2.50	7.09	0.90	-	16.53	-	23.63	nr	25.99
100mm	8.06	2.50	8.26	0.95	-	17.45	-	25.71	nr	28.28
125mm	17.49	2.50	17.93	1.05	-	19.29	-	37.22	nr	40.94
150mm	20.25	2.50	20.76	1.10	-	20.21	-	40.96	nr	45.06

Labour hourly rates: (except Specialists) Craft Operatives 18.37 Labourer 13.76 Rates are national average prices. Refer to REGIONAL VARIATIONS for indicative levels of overall pricing in regions	MATERIALS			LABOUR				RATES		
	Del to Site	Waste	Material Cost	Craft Optve	Lab	Labour Cost	Sunds	Nett Rate	Unit	Gross rate (10%)
	£	%	£	Hrs	Hrs	£	£	£		£
IRONMONGERY (Cont'd)										
Hinges; standard quality (Cont'd)										
To hardwood or the like (Cont'd)										
tee hinges; japanned; light										
150mm	0.52	2.50	0.53	0.60	-	11.02	-	11.56	nr	12.71
230mm	0.73	2.50	0.75	0.65	-	11.94	-	12.69	nr	13.96
300mm	1.07	2.50	1.09	0.75	-	13.78	-	14.87	nr	16.36
375mm	2.07	2.50	2.12	0.80	-	14.70	-	16.81	nr	18.50
450mm	2.73	2.50	2.80	0.90	-	16.53	-	19.33	nr	21.26
tee hinges; galvanised; heavy										
300mm	2.48	2.50	2.54	0.80	-	14.70	-	17.24	nr	18.96
375mm	3.73	2.50	3.82	0.85	-	15.61	-	19.44	nr	21.38
450mm	5.40	2.50	5.53	0.90	-	16.53	-	22.06	nr	24.27
600mm	7.90	2.50	8.09	1.00	-	18.37	-	26.46	nr	29.11
hook and band hinges; on plate; heavy										
300mm	3.73	2.50	3.82	1.20	-	22.04	-	25.87	nr	28.45
450mm	6.65	2.50	6.81	1.55	-	28.47	-	35.29	nr	38.81
610mm	9.15	2.50	9.37	1.90	-	34.90	-	44.28	nr	48.71
914mm	17.90	2.50	18.34	2.20	-	40.41	-	58.76	nr	64.63
collinge hinges; cup for wood; best quality										
610mm	126.10	2.50	129.25	1.45	-	26.64	-	155.89	nr	171.48
762mm	161.85	2.50	165.90	1.85	-	33.98	-	199.88	nr	219.87
914mm	189.15	2.50	193.88	2.25	-	41.33	-	235.21	nr	258.73
parliament hinges; steel; 100mm	1.77	2.50	1.82	0.40	-	7.35	-	9.16	nr	10.08
cellar flap hinges; wrought, welded, plain joint; 450mm	5.19	2.50	5.32	0.85	-	15.61	-	20.93	nr	23.02
Floor springs; standard quality										
To softwood										
single action floor springs and top centres; stainless steel cover	175.50	2.50	179.89	2.75	-	50.52	-	230.40	nr	253.45
single action floor springs and top centres; stainless steel cover; hold open	175.50	2.50	179.89	2.75	-	50.52	-	230.40	nr	253.45
double action floor springs and top centres; stainless steel cover	163.50	2.50	167.59	3.30	-	60.62	-	228.21	nr	251.03
double action floor springs and top centres; stainless steel cover; hold open	163.50	2.50	167.59	3.30	-	60.62	-	228.21	nr	251.03
To hardwood or the like										
single action floor springs and top centres; stainless steel cover	175.50	2.50	179.89	4.10	-	75.32	-	255.20	nr	280.72
single action floor springs and top centres; stainless steel cover; hold open	175.50	2.50	179.89	4.10	-	75.32	-	255.20	nr	280.72
double action floor springs and top centres; stainless steel cover	163.50	2.50	167.59	4.90	-	90.01	-	257.60	nr	283.36
double action floor springs and top centres; stainless steel cover; hold open	163.50	2.50	167.59	4.90	-	90.01	-	257.60	nr	283.36
Door closers; standard quality										
To softwood										
coil door springs; japanned	1.60	2.50	1.64	0.30	-	5.51	-	7.15	nr	7.87
overhead door closers; liquid check and spring; "Briton" 2003 series; silver; light doors	38.50	2.50	39.46	1.65	-	30.31	-	69.77	nr	76.75
overhead door closers; liquid check and spring; "Briton" 2003V series; silver; medium doors	44.50	2.50	45.61	1.65	-	30.31	-	75.92	nr	83.52
overhead door closers; liquid check and spring; "Briton" 2004E series; silver; heavy doors	79.50	2.50	81.49	1.65	-	30.31	-	111.80	nr	122.98
concealed overhead door closers; liquid check and spring; medium doors	109.00	2.50	111.72	2.75	-	50.52	-	162.24	nr	178.47
Perko door closers; brass plate	10.50	2.50	10.76	1.65	-	30.31	-	41.07	nr	45.18
To hardwood or the like										
coil door springs; japanned	1.60	2.50	1.64	0.45	-	8.27	-	9.91	nr	10.90
overhead door closers; liquid check and spring; "Briton" 2003 series; silver; light doors	38.50	2.50	39.46	2.45	-	45.01	-	84.47	nr	92.92
overhead door closers; liquid check and spring; "Briton" 2003V series; silver; medium doors	44.50	2.50	45.61	2.45	-	45.01	-	90.62	nr	99.68
overhead door closers; liquid check and spring; "Briton" 2004E series; silver; heavy doors	79.50	2.50	81.49	2.45	-	45.01	-	126.49	nr	139.14
concealed overhead door closers; liquid check and spring; medium doors	109.00	2.50	111.72	4.10	-	75.32	-	187.04	nr	205.75
Perko door closers; brass plate	10.50	2.50	10.76	2.45	-	45.01	-	55.77	nr	61.35
Door selectors; standard quality										
To softwood or the like										
Roller arm door selector; polished nickel finish	27.50	2.50	28.19	2.20	0.28	44.27	-	72.45	nr	79.70
To hardwood or the like										
Roller arm door selector; polished nickel finish	27.50	2.50	28.19	3.30	0.41	66.26	-	94.45	nr	103.90
Locks and latches; standard quality										
To softwood										
magnetic catches; 6lb pull	0.12	2.50	0.12	0.25	-	4.59	-	4.71	nr	5.18
magnetic catches; 9lb pull	0.26	2.50	0.26	0.25	-	4.59	-	4.86	nr	5.34
magnetic catches; 13lb pull	0.19	2.50	0.20	0.25	-	4.59	-	4.79	nr	5.27
roller catches; mortice; nylon; 18mm	0.94	2.50	0.97	0.40	-	7.35	-	8.31	nr	9.14
roller catches; mortice; nylon; 27mm	1.79	2.50	1.84	0.40	-	7.35	-	9.18	nr	10.10
roller catches; surface, adjustable; nylon; 16mm	0.19	2.50	0.20	0.30	-	5.51	-	5.71	nr	6.28
roller catches; mortice, adjustable; satin chrome plated; 25 x 22 x 10mm	3.08	2.50	3.16	0.45	-	8.27	-	11.43	nr	12.57
roller catches; double ball; brass; 43mm	0.88	2.50	0.91	0.35	-	6.43	-	7.33	nr	8.07
mortice latches; stamped steel case; 75mm	0.65	2.50	0.67	0.70	-	12.86	-	13.53	nr	14.88
mortice latches; locking; stamped steel case; 75mm	3.35	2.50	3.43	0.70	-	12.86	-	16.29	nr	17.92
cylinder mortice latches; "Union" key operated outside, knob inside, bolt held by slide; B.M.A.; for end set to suit 13mm rebate	41.15	2.50	42.18	0.80	-	14.70	-	56.87	nr	62.56

Labour hourly rates: (except Specialists) Craft Operatives 18.37 Labourer 13.76 Rates are national average prices. Refer to REGIONAL VARIATIONS for indicative levels of overall pricing in regions	MATERIALS			LABOUR				RATES		
	Del to Site	Waste	Material Cost	Craft Optve	Lab	Labour Cost	Sunds	Nett Rate		Gross rate (10%)
	£	%	£	Hrs	Hrs	£	£	£	Unit	£

IRONMONGERY (Cont'd)

Locks and latches; standard quality (Cont'd)

To softwood (Cont'd)

cylinder mortice latches; "Union" key operated outside, knob inside, bolt held by slide; chromium plated; for end set to suit 13mm rebate	41.15	2.50	42.18	0.80	-	14.70	-	56.87	nr	62.56
cylinder rim night latches; "Legge" 707; silver finish	27.91	2.50	28.61	1.10	-	20.21	-	48.81	nr	53.70
cylinder rim night latches; "Union" 1022; silver case; chromium plated cylinder	25.54	2.50	26.18	1.10	-	20.21	-	46.39	nr	51.02
cylinder rim night latches; "Yale" 88; grey case; chromium plated cylinder	27.91	2.50	28.61	1.10	-	20.21	-	48.81	nr	53.70
Suffolk latches; japanned; medium; size Nr 2	1.54	2.50	1.58	0.90	-	16.53	-	18.11	nr	19.92
Suffolk latches; japanned; medium; size Nr 3	3.85	2.50	3.95	0.90	-	16.53	-	20.48	nr	22.53
Suffolk latches; japanned; heavy; size Nr 4	2.29	2.50	2.35	0.90	-	16.53	-	18.88	nr	20.77
Suffolk latches; galvanized; size Nr 4	2.29	2.50	2.35	0.90	-	16.53	-	18.88	nr	20.77
cupboard locks; 1 lever; brass; 63mm	9.40	2.50	9.63	0.80	-	14.70	-	24.33	nr	26.76
cupboard locks; 2 lever; brass; 50mm	15.04	2.50	15.42	0.80	-	14.70	-	30.11	nr	33.12
cupboard locks; 2 lever; brass; 63mm	14.17	2.50	14.52	0.80	-	14.70	-	29.22	nr	32.14
cupboard locks; 4 lever; brass; 63mm	19.50	2.50	19.99	0.80	-	14.70	-	34.68	nr	38.15
rim dead lock; japanned case; 76mm	5.91	2.50	6.05	0.90	-	16.53	-	22.59	nr	24.85
rim dead lock; japanned case; 100mm	7.22	2.50	7.40	0.90	-	16.53	-	23.93	nr	26.32
rim dead lock; japanned case; 140mm	8.05	2.50	8.25	0.90	-	16.53	-	24.79	nr	27.26
mortice dead locks; japanned case; 75mm	3.85	2.50	3.95	0.90	-	16.53	-	20.48	nr	22.53
mortice locks; three levers	3.85	2.50	3.95	1.10	-	20.21	-	24.15	nr	26.57
mortice locks; five levers	13.80	2.50	14.14	1.10	-	20.21	-	34.35	nr	37.79
rim locks; japanned case; 150 x 100mm	7.89	2.50	8.09	0.90	-	16.53	-	24.62	nr	27.08
rim locks; japanned case; 150 x 75mm	8.88	2.50	9.10	0.90	-	16.53	-	25.64	nr	28.20
rim locks; japanned case; strong pattern; 150 x 100mm	21.52	2.50	22.06	0.90	-	16.53	-	38.59	nr	42.45
padlocks, "Squire", anti-pilfer bolt, warded spring; hardened steel; 20mm; LP6	3.32	2.50	3.40	-	-	-	-	3.40	nr	3.74
padlocks, "Squire", anti-pilfer bolt, warded spring; hardened steel; 30mm; LP8	4.13	2.50	4.23	-	-	-	-	4.23	nr	4.66
padlocks, "Squire", anti-pilfer bolt, warded spring; hardened steel; 38mm; LP9	5.71	2.50	5.85	-	-	-	-	5.85	nr	6.44
padlocks, "Squire", anti-pilfer bolt, 4 pin tumblers; brass; 40mm; CSL1	12.29	2.50	12.60	-	-	-	-	12.60	nr	13.86
padlocks, "Squire", 4 levers; galvanised case; 220	6.10	2.50	6.25	-	-	-	-	6.25	nr	6.88
padlocks, "Squire", 4 levers; brass case; 25mm; LN2	6.08	2.50	6.23	-	-	-	-	6.23	nr	6.86
padlocks; "Squire Stronghold" pattern; 50mm; SS50S	34.15	2.50	35.00	-	-	-	-	35.00	nr	38.50
hasps and staples; japanned wire; light; 75mm	0.71	2.50	0.73	0.35	-	6.43	-	7.16	nr	7.87
hasps and staples; japanned wire; light; 100mm	0.83	2.50	0.85	0.35	-	6.43	-	7.28	nr	8.01
hasps and staples; galvanized safety; 75mm	1.88	2.50	1.93	0.35	-	6.43	-	8.36	nr	9.20
hasps and staples; galvanized medium; 200mm	5.71	2.50	5.85	0.35	-	6.43	-	12.28	nr	13.51
hasps and staples; galvanized heavy; 200mm	6.63	2.50	6.80	0.35	-	6.43	-	13.23	nr	14.55
locking bars; japanned; 200mm	8.58	2.50	8.80	0.50	-	9.19	-	17.98	nr	19.78
locking bars; japanned; 300mm	18.58	2.50	19.05	0.50	-	9.19	-	28.23	nr	31.06
locking bars; japanned; 400mm	38.50	2.50	39.46	0.50	-	9.19	-	48.65	nr	53.51

To hardwood or the like

magnetic catches; 6lb pull	0.12	2.50	0.12	0.35	-	6.43	-	6.55	nr	7.20
magnetic catches; 9lb pull	0.26	2.50	0.26	0.35	-	6.43	-	6.69	nr	7.36
magnetic catches; 13lb pull	0.19	2.50	0.20	0.35	-	6.43	-	6.63	nr	7.29
roller catches; mortice; nylon; 18mm	0.94	2.50	0.97	0.55	-	10.10	-	11.07	nr	12.18
roller catches; mortice; nylon; 27mm	1.79	2.50	1.84	0.55	-	10.10	-	11.94	nr	13.13
roller catches; surface, adjustable; nylon; 16mm	0.19	2.50	0.20	0.45	-	8.27	-	8.46	nr	9.31
roller catches; mortice, adjustable; satin chrome plated; 25 x 22 x 10mm	3.08	2.50	3.16	0.65	-	11.94	-	15.10	nr	16.61
roller catches; double ball; brass; 43mm	0.88	2.50	0.91	0.50	-	9.19	-	10.09	nr	11.10
mortice latches; stamped steel case; 75mm	0.65	2.50	0.67	1.10	-	20.21	-	20.87	nr	22.96
mortice latches; locking; stamped steel case; 75mm	3.35	2.50	3.43	1.10	-	20.21	-	23.64	nr	26.00
cylinder mortice latches; "Union" key operated outside, knob inside, bolt held by slide; B.M.A.; for end set to suit 13mm rebate	41.15	2.50	42.18	1.20	-	22.04	-	64.22	nr	70.65
cylinder mortice latches; "Union" key operated outside, knob inside, bolt held by slide; chromium plated; for end set to suit 13mm rebate	41.15	2.50	42.18	1.20	-	22.04	-	64.22	nr	70.65
cylinder rim night latches; "Legge" 707; silver finish	27.91	2.50	28.61	1.65	-	30.31	-	58.92	nr	64.81
cylinder rim night latches; "Union" 1022; silver case; chromium plated cylinder	25.54	2.50	26.18	1.65	-	30.31	-	56.49	nr	62.14
cylinder rim night latches; "Yale" 88; grey case; chromium plated cylinder	27.91	2.50	28.61	1.65	-	30.31	-	58.92	nr	64.81
Suffolk latches; japanned; medium; size Nr 2	1.54	2.50	1.58	1.30	-	23.88	-	25.46	nr	28.01
Suffolk latches; japanned; medium; size Nr 3	3.85	2.50	3.95	1.30	-	23.88	-	27.83	nr	30.61
Suffolk latches; japanned; heavy; size Nr 4	2.29	2.50	2.35	1.30	-	23.88	-	26.23	nr	28.85
Suffolk latches; galvanized; size Nr 4	2.29	2.50	2.35	1.30	-	23.88	-	26.23	nr	28.85
cupboard locks; 1 lever; brass; 63mm	9.40	2.50	9.63	1.20	-	22.04	-	31.68	nr	34.85
cupboard locks; 2 lever; brass; 50mm	15.04	2.50	15.42	1.20	-	22.04	-	37.46	nr	41.21
cupboard locks; 2 lever; brass; 63mm	14.17	2.50	14.52	1.20	-	22.04	-	36.57	nr	40.22
cupboard locks; 4 lever; brass; 63mm	19.50	2.50	19.99	1.20	-	22.04	-	42.03	nr	46.23
rim dead lock; japanned case; 76mm	5.91	2.50	6.05	1.30	-	23.88	-	29.94	nr	32.93
rim dead lock; japanned case; 100mm	7.22	2.50	7.40	1.30	-	23.88	-	31.28	nr	34.41
rim dead lock; japanned case; 140mm	8.05	2.50	8.25	1.30	-	23.88	-	32.13	nr	35.35
mortice dead locks; japanned case; 75mm	3.85	2.50	3.95	1.30	-	23.88	-	27.83	nr	30.61
mortice locks; three levers	3.85	2.50	3.95	1.65	-	30.31	-	34.26	nr	37.68
mortice locks; five levers	13.80	2.50	14.14	1.65	-	30.31	-	44.46	nr	48.90
rim locks; japanned case; 150 x 100mm	7.89	2.50	8.09	1.30	-	23.88	-	31.97	nr	35.17
rim locks; japanned case; 150 x 75mm	8.88	2.50	9.10	1.30	-	23.88	-	32.98	nr	36.28
rim locks; japanned case; strong pattern; 150 x 100mm	21.52	2.50	22.06	1.30	-	23.88	-	45.94	nr	50.53
padlocks, "Squire", anti-pilfer bolt, warded spring; hardened steel; 20mm; LP6	3.32	2.50	3.40	-	-	-	-	3.40	nr	3.74
padlocks, "Squire", anti-pilfer bolt, warded spring; hardened steel; 30mm; LP8	4.13	2.50	4.23	-	-	-	-	4.23	nr	4.66

GENERAL JOINERY

Labour hourly rates: (except Specialists) Craft Operatives 18.37 Labourer 13.76 Rates are national average prices. Refer to REGIONAL VARIATIONS for indicative levels of overall pricing in regions	MATERIALS			LABOUR				RATES		
	Del to Site	Waste	Material Cost	Craft Optve	Lab	Labour Cost	Sunds	Nett Rate	Unit	Gross rate (10%)
	£	%	£	Hrs	Hrs	£	£	£		£
IRONMONGERY (Cont'd)										
Locks and latches; standard quality (Cont'd)										
To hardwood or the like (Cont'd)										
padlocks, "Squire", anti-pilfer bolt, warded spring; hardened steel; 38mm; LP9	5.71	2.50	5.85	-	-	-	-	5.85	nr	6.44
padlocks; "Squire", anti-pilfer bolt, 4 pin tumblers; brass; 40mm; CSL1	12.29	2.50	12.60	-	-	-	-	12.60	nr	13.86
padlocks; "Squire", 4 levers; galvanised case; 220	6.10	2.50	6.25	-	-	-	-	6.25	nr	6.88
padlocks; "Squire", 4 levers; brass case; 25mm; LN1	6.08	2.50	6.23	-	-	-	-	6.23	nr	6.86
padlocks; "Squire Stronghold" pattern; 50mm; SS50S	34.15	2.50	35.00	-	-	-	-	35.00	nr	38.50
hasps and staples; japanned wire; light; 75mm	0.71	2.50	0.73	0.50	-	9.19	-	9.91	nr	10.90
hasps and staples; japanned wire; light; 100mm	0.83	2.50	0.85	0.55	-	10.10	-	10.96	nr	12.05
hasps and staples; galvanised safety; 75mm	1.88	2.50	1.93	0.55	-	10.10	-	12.03	nr	13.24
hasps and staples; galvanized medium; 200mm	5.71	2.50	5.85	0.55	-	10.10	-	15.95	nr	17.55
hasps and staples; galvanized heavy; 200mm	6.63	2.50	6.80	0.55	-	10.10	-	16.90	nr	18.59
locking bars; japanned; 200mm	8.58	2.50	8.80	0.65	-	11.94	-	20.74	nr	22.81
locking bars; japanned; 300mm	18.58	2.50	19.05	0.65	-	11.94	-	30.99	nr	34.09
locking bars; japanned; 400mm	38.50	2.50	39.46	0.65	-	11.94	-	51.40	nr	56.54
Bolts; standard quality										
To softwood										
barrel bolts; japanned, steel barrel; medium										
100mm	0.42	2.50	0.43	0.30	-	5.51	-	5.94	nr	6.54
150mm	0.62	2.50	0.64	0.35	-	6.43	-	7.07	nr	7.77
150mm	1.03	2.50	1.06	0.35	-	6.43	-	7.49	nr	8.24
barrel bolts; japanned, steel barrel; heavy										
100mm	3.85	2.50	3.95	0.30	-	5.51	-	9.46	nr	10.40
150mm	4.62	2.50	4.74	0.35	-	6.43	-	11.16	nr	12.28
200mm	4.49	2.50	4.60	0.40	-	7.35	-	11.95	nr	13.15
250mm	6.82	2.50	6.99	0.45	-	8.27	-	15.26	nr	16.78
300mm	8.14	2.50	8.34	0.50	-	9.19	-	17.53	nr	19.28
barrel bolts; extruded brass, round brass shoot; 25mm wide										
75mm	1.37	2.50	1.40	0.30	-	5.51	-	6.92	nr	7.61
100mm	1.73	2.50	1.77	0.30	-	5.51	-	7.28	nr	8.01
150mm	2.28	2.50	2.34	0.35	-	6.43	-	8.77	nr	9.64
barrel bolts; extruded brass; B.M.A., round brass shoot; 25mm wide										
75mm	11.47	2.50	11.75	0.30	-	5.51	-	17.26	nr	18.99
100mm	12.67	2.50	12.98	0.30	-	5.51	-	18.49	nr	20.34
150mm	14.63	2.50	15.00	0.35	-	6.43	-	21.43	nr	23.57
barrel bolts; extruded brass, chromium plated, round brass shoot; 25mm wide										
75mm	1.74	2.50	1.78	0.30	-	5.51	-	7.29	nr	8.02
100mm	2.18	2.50	2.23	0.30	-	5.51	-	7.75	nr	8.52
150mm	2.93	2.50	3.00	0.35	-	6.43	-	9.43	nr	10.38
barrel bolts; extruded aluminium, S.A.A., round aluminium shoot; 25mm wide										
75mm	0.42	2.50	0.43	0.30	-	5.51	-	5.94	nr	6.54
100mm	0.51	2.50	0.52	0.30	-	5.51	-	6.03	nr	6.64
150mm	0.78	2.50	0.80	0.35	-	6.43	-	7.23	nr	7.95
monkey tail bolts										
japanned; 300mm	2.97	2.50	3.05	0.40	-	7.35	-	10.40	nr	11.44
japanned; 450mm	8.29	2.50	8.50	0.45	-	8.27	-	16.77	nr	18.44
japanned; 610mm	12.46	2.50	12.77	0.55	-	10.10	-	22.87	nr	25.16
flush bolts; brass										
100mm	4.90	2.50	5.02	0.55	-	10.10	-	15.13	nr	16.64
150mm	9.13	2.50	9.36	0.65	-	11.94	-	21.30	nr	23.43
200mm	11.63	2.50	11.92	0.75	-	13.78	-	25.70	nr	28.27
flush bolts; S.C.P.										
100mm	5.79	2.50	5.94	0.55	-	10.10	-	16.04	nr	17.64
150mm	9.13	2.50	9.36	0.65	-	11.94	-	21.30	nr	23.43
200mm	11.63	2.50	11.92	0.75	-	13.78	-	25.70	nr	28.27
lever action flush bolts; B.M.A.										
150mm	31.20	2.50	31.98	0.65	-	11.94	-	43.92	nr	48.31
200 x 25mm	33.70	2.50	34.54	0.75	-	13.78	-	48.32	nr	53.15
lever action flush bolts; S.A.A.										
150mm	9.72	2.50	9.96	0.65	-	11.94	-	21.90	nr	24.09
200mm	10.59	2.50	10.85	0.75	-	13.78	-	24.63	nr	27.10
necked bolts; extruded brass, round brass shoot; 25mm wide										
75mm	1.52	2.50	1.56	0.30	-	5.51	-	7.07	nr	7.78
100mm	1.85	2.50	1.90	0.30	-	5.51	-	7.41	nr	8.15
150mm	2.57	2.50	2.63	0.35	-	6.43	-	9.06	nr	9.97
necked bolts; extruded aluminium, S.A.A., round aluminium shoot; 25mm wide										
75mm	0.44	2.50	0.45	0.30	-	5.51	-	5.96	nr	6.56
100mm	0.56	2.50	0.57	0.30	-	5.51	-	6.09	nr	6.69
150mm	0.87	2.50	0.89	0.35	-	6.43	-	7.32	nr	8.05
indicating bolts; S.A.A.	2.52	2.50	2.58	1.10	-	20.21	-	22.79	nr	25.07
indicating bolts; B.M.A.	36.75	2.50	37.67	1.10	-	20.21	-	57.88	nr	63.66
panic bolts; to single door; iron, bronzed or silver	44.95	2.50	46.07	1.65	-	30.31	-	76.38	nr	84.02
panic bolts; to single door; aluminium box, steel shoots and cross rail, anodised silver	56.55	2.50	57.96	1.65	-	30.31	-	88.27	nr	97.10
panic bolts; to double doors; iron bronzed or silver	83.57	2.50	85.66	2.20	-	40.41	-	126.07	nr	138.68
panic bolts; to double doors; aluminium box, steel shoots and cross rail, anodised silver	100.05	2.50	102.55	2.20	-	40.41	-	142.97	nr	157.26
padlock bolts; galvanized; heavy										
150mm	2.43	2.50	2.49	0.35	-	6.43	-	8.92	nr	9.82
200mm	3.27	2.50	3.35	0.35	-	6.43	-	9.78	nr	10.76
250mm	3.84	2.50	3.94	0.50	-	9.19	-	13.12	nr	14.43
375mm	5.43	2.50	5.57	0.50	-	9.19	-	14.75	Unit	16.23

Labour hourly rates: (except Specialists) Craft Operatives 18.37 Labourer 13.76 Rates are national average prices. Refer to REGIONAL VARIATIONS for indicative levels of overall pricing in regions	MATERIALS			LABOUR				RATES		
	Del to Site	Waste	Material Cost	Craft Optve	Lab	Labour Cost	Sunds	Nett Rate	Unit	Gross rate (10%)
	£	%	£	Hrs	Hrs	£	£	£		£
IRONMONGERY (Cont'd)										
Bolts; standard quality (Cont'd)										
To hardwood or the like										
barrel bolts; japanned, steel barrel; medium										
100mm	0.42	2.50	0.43	0.40	-	7.35	-	7.78	nr	8.56
150mm	0.62	2.50	0.64	0.45	-	8.27	-	8.90	nr	9.79
150mm	1.03	2.50	1.06	0.45	-	8.27	-	9.33	nr	10.26
barrel bolts; japanned, steel barrel; heavy										
100mm	3.85	2.50	3.95	0.40	-	7.35	-	11.29	nr	12.42
150mm	4.62	2.50	4.74	0.45	-	8.27	-	13.00	nr	14.30
200mm	4.49	2.50	4.60	0.55	-	10.10	-	14.71	nr	16.18
250mm	6.82	2.50	6.99	0.60	-	11.02	-	18.01	nr	19.81
300mm	8.14	2.50	8.34	0.70	-	12.86	-	21.20	nr	23.32
barrel bolts; extruded brass, round brass shoot; 25mm wide										
75mm	1.37	2.50	1.40	0.40	-	7.35	-	8.75	nr	9.63
100mm	1.73	2.50	1.77	0.40	-	7.35	-	9.12	nr	10.03
150mm	2.28	2.50	2.34	0.45	-	8.27	-	10.60	nr	11.66
barrel bolts; extruded brass; B.M.A., round brass shoot; 25mm wide										
75mm	11.47	2.50	11.75	0.40	-	7.35	-	19.10	nr	21.01
100mm	12.67	2.50	12.98	0.40	-	7.35	-	20.33	nr	22.36
150mm	14.63	2.50	15.00	0.45	-	8.27	-	23.27	nr	25.59
barrel bolts; extruded brass, chromium plated, round brass shoot; 25mm wide										
75mm	1.74	2.50	1.78	0.40	-	7.35	-	9.13	nr	10.04
100mm	2.18	2.50	2.23	0.40	-	7.35	-	9.58	nr	10.54
150mm	2.93	2.50	3.00	0.45	-	8.27	-	11.27	nr	12.40
barrel bolts; extruded aluminium, S.A.A., round aluminium shoot; 25mm wide										
75mm	0.42	2.50	0.43	0.40	-	7.35	-	7.78	nr	8.56
100mm	0.51	2.50	0.52	0.40	-	7.35	-	7.87	nr	8.66
150mm	0.78	2.50	0.80	0.45	-	8.27	-	9.07	nr	9.97
monkey tail bolts										
japanned; 300mm	2.97	2.50	3.05	0.55	-	10.10	-	13.15	nr	14.47
japanned; 450mm	8.29	2.50	8.50	0.65	-	11.94	-	20.44	nr	22.48
japanned; 610mm	12.46	2.50	12.77	0.75	-	13.78	-	26.55	nr	29.20
flush bolts; brass										
100mm	4.90	2.50	5.02	0.90	-	16.53	-	21.56	nr	23.71
150mm	9.13	2.50	9.36	1.00	-	18.37	-	27.73	nr	30.50
200mm	11.63	2.50	11.92	1.10	-	20.21	-	32.13	nr	35.34
flush bolts; S.C.P.										
100mm	5.79	2.50	5.94	0.90	-	16.53	-	22.47	nr	24.72
150mm	9.13	2.50	9.36	1.00	-	18.37	-	27.73	nr	30.50
200mm	11.63	2.50	11.92	1.10	-	20.21	-	32.13	nr	35.34
lever action flush bolts; B.M.A.										
150mm	31.20	2.50	31.98	1.00	-	18.37	-	50.35	nr	55.39
200 x 25mm	33.70	2.50	34.54	1.10	-	20.21	-	54.75	nr	60.22
lever action flush bolts; S.A.A.										
150mm	9.72	2.50	9.96	1.00	-	18.37	-	28.33	nr	31.17
200mm	10.59	2.50	10.85	1.10	-	20.21	-	31.06	nr	34.17
necked bolts; extruded brass, round brass shoot; 25mm wide										
75mm	1.52	2.50	1.56	0.40	-	7.35	-	8.91	nr	9.80
100mm	1.85	2.50	1.90	0.40	-	7.35	-	9.24	nr	10.17
150mm	2.57	2.50	2.63	0.45	-	8.27	-	10.90	nr	11.99
necked bolts; extruded aluminium, S.A.A., round aluminium shoot; 25mm wide										
75mm	0.44	2.50	0.45	0.40	-	7.35	-	7.80	nr	8.58
100mm	0.56	2.50	0.57	0.40	-	7.35	-	7.92	nr	8.71
150mm	0.87	2.50	0.89	0.45	-	8.27	-	9.16	nr	10.07
indicating bolts										
S.A.A.	2.52	2.50	2.58	1.65	-	30.31	-	32.89	nr	36.18
B.M.A.	36.75	2.50	37.67	1.65	-	30.31	-	67.98	nr	74.78
panic bolts										
to single door; iron, bronzed or silver	44.95	2.50	46.07	2.45	-	45.01	-	91.08	nr	100.19
to single door; aluminium box, steel shoots and cross rail, anodised silver	56.55	2.50	57.96	2.45	-	45.01	-	102.97	nr	113.27
to double doors; iron bronzed or silver	83.57	2.50	85.66	3.30	-	60.62	-	146.28	nr	160.91
to double doors; aluminium box, steel shoots and cross rail, anodised silver	100.05	2.50	102.55	3.30	-	60.62	-	163.17	nr	179.49
padlock bolts; galvanized; heavy										
150mm	2.43	2.50	2.49	0.50	-	9.19	-	11.68	nr	12.85
200mm	3.27	2.50	3.35	0.50	-	9.19	-	12.53	nr	13.79
250mm	3.84	2.50	3.94	0.75	-	13.78	-	17.72	nr	19.49
375mm	5.43	2.50	5.57	0.75	-	13.78	-	19.35	nr	21.28
Door handles; standard quality										
To softwood										
lever handles; spring action; B.M.A., best quality; 41 x 150mm	35.09	2.50	35.97	0.25	-	4.59	-	40.56	nr	44.62
lever handles; spring action; chromium plated, housing quality; 41 x 150mm	10.40	2.50	10.66	0.25	-	4.59	-	15.25	nr	16.78
lever handles; spring action; chromium plated, best quality; 41 x 150mm	13.66	2.50	14.00	0.25	-	4.59	-	18.59	nr	20.45
lever handles; spring action; S.A.A. housing quality; 41 x 150mm	12.06	2.50	12.36	0.25	-	4.59	-	16.95	nr	18.65
lever handles; spring action; S.A.A., best quality; 41 x 150mm	14.85	2.50	15.22	0.25	-	4.59	-	19.81	nr	21.80
lever handles; spring action; plastic/nylon, housing quality; 41 x 150mm	29.89	2.50	30.64	0.25	-	4.59	-	35.23	nr	38.75
pull handles; B.M.A.; 150mm	12.81	2.50	13.14	0.20	-	3.67	-	16.81	nr	18.49
pull handles; B.M.A.; 225mm	24.01	2.50	24.61	0.25	-	4.59	-	29.21	nr	32.13
pull handles; B.M.A.; 300mm	27.03	2.50	27.70	0.30	-	5.51	-	33.21	nr	36.54
pull handles; S.A.A.; 150mm	1.95	2.50	2.00	0.20	-	3.67	-	5.67	nr	6.24
pull handles; S.A.A.; 225mm	5.63	2.50	5.77	0.25	-	4.59	-	10.36	nr	11.40
pull handles; S.A.A.; 300mm	6.22	2.50	6.38	0.30	-	5.51	-	11.89	nr	13.08

GENERAL JOINERY

Labour hourly rates: (except Specialists) Craft Operatives 18.37 Labourer 13.76 Rates are national average prices. Refer to REGIONAL VARIATIONS for indicative levels of overall pricing in regions	MATERIALS			LABOUR				RATES		
	Del to Site	Waste	Material Cost	Craft Optve	Lab	Labour Cost	Sunds	Nett Rate		Gross rate (10%)
	£	%	£	Hrs	Hrs	£	£	£	Unit	£
IRONMONGERY (Cont'd)										
Door handles; standard quality (Cont'd)										
To softwood (Cont'd)										
pull handles on 50 x 250mm plate, lettered; B.M.A.; 225mm	55.44	2.50	56.83	0.45	-	8.27	-	65.09	nr	71.60
pull handles on 50 x 250mm plate, lettered; S.A.A.; 225mm	38.83	2.50	39.80	0.45	-	8.27	-	48.07	nr	52.87
To hardwood or the like										
lever handles; spring action; B.M.A., best quality; 41 x 150mm	35.09	2.50	35.97	0.35	-	6.43	-	42.40	nr	46.64
lever handles; spring action; chromium plated, housing quality; 41 x 150mm	10.40	2.50	10.66	0.35	-	6.43	-	17.09	nr	18.80
lever handles; spring action; chromium plated, best quality; 41 x 150mm	13.66	2.50	14.00	0.35	-	6.43	-	20.43	nr	22.47
lever handles; spring action; S.A.A. housing quality; 41 x 150mm	12.06	2.50	12.36	0.35	-	6.43	-	18.79	nr	20.67
lever handles; spring action; S.A.A., best quality; 41 x 150mm	14.85	2.50	15.22	0.35	-	6.43	-	21.65	nr	23.82
lever handles; spring action; plastic/nylon, housing quality; 41 x 150mm	29.89	2.50	30.64	0.35	-	6.43	-	37.07	nr	40.78
pull handles; B.M.A.; 150mm	12.81	2.50	13.14	0.30	-	5.51	-	18.65	nr	20.51
pull handles; B.M.A.; 225mm	24.01	2.50	24.61	0.35	-	6.43	-	31.04	nr	34.15
pull handles; B.M.A.; 300mm	27.03	2.50	27.70	0.40	-	7.35	-	35.05	nr	38.56
pull handles; S.A.A.; 150mm	1.95	2.50	2.00	0.30	-	5.51	-	7.51	nr	8.26
pull handles; S.A.A.; 225mm	5.63	2.50	5.77	0.35	-	6.43	-	12.20	nr	13.42
pull handles; S.A.A.; 300mm	6.22	2.50	6.38	0.40	-	7.35	-	13.72	nr	15.10
pull handles on 50 x 250mm plate, lettered; B.M.A.; 225mm	55.44	2.50	56.83	0.65	-	11.94	-	68.77	nr	75.64
pull handles on 50 x 250mm plate, lettered; S.A.A.; 225mm	38.83	2.50	39.80	0.65	-	11.94	-	51.74	nr	56.92
Door furniture; standard quality										
To softwood										
knobs; real B.M.A.; surface fixing	9.99	2.50	10.24	0.25	-	4.59	-	14.83	nr	16.31
knobs; S.A.A.; surface fixing	5.82	2.50	5.96	0.25	-	4.59	-	10.56	nr	11.61
knobs; real B.M.A.; secret fixing	16.14	2.50	16.54	0.35	-	6.43	-	22.97	nr	25.27
knobs; S.A.A.; secret fixing	8.39	2.50	8.60	0.35	-	6.43	-	15.03	nr	16.54
kicking plate; SAA; 750 x 150	3.70	2.50	3.79	0.55	-	10.10	-	13.90	nr	15.29
kicking plate; SAA; 840 x 150	4.45	2.50	4.56	0.55	-	10.10	-	14.66	nr	16.13
push plate; SAA; 300 x 75	2.00	2.50	2.05	0.30	-	5.51	-	7.56	nr	8.32
sign plate; SAA; black engraved 300 x 75	4.24	2.50	4.34	0.30	-	5.51	-	9.85	nr	10.84
To hardwood or the like										
knobs; real B.M.A.; surface fixing	9.99	2.50	10.24	0.35	-	6.43	-	16.67	nr	18.33
knobs; S.A.A.; surface fixing	5.82	2.50	5.96	0.35	-	6.43	-	12.39	nr	13.63
knobs; real B.M.A.; secret fixing	16.14	2.50	16.54	0.50	-	9.19	-	25.73	nr	28.30
knobs; S.A.A.; secret fixing	8.39	2.50	8.60	0.50	-	9.19	-	17.79	nr	19.57
kicking plate; SAA; 750 x 150	3.70	2.50	3.79	0.80	-	14.70	-	18.49	nr	20.34
kicking plate; SAA; 840 x 150	4.45	2.50	4.56	0.80	-	14.70	-	19.26	nr	21.18
push plate; SAA; 300 x 75	2.00	2.50	2.05	0.40	-	7.35	-	9.40	nr	10.34
sign plate; SAA; black engraved 300 x 75	4.24	2.50	4.34	0.40	-	7.35	-	11.69	nr	12.86
Window furniture; standard quality										
To softwood										
fanlight catches; brass	4.00	2.50	4.10	0.40	-	7.35	-	11.45	nr	12.59
fanlight catches; B.M.A.	7.70	2.50	7.89	0.40	-	7.35	-	15.24	nr	16.76
fanlight catches; chromium plated	4.40	2.50	4.51	0.40	-	7.35	-	11.86	nr	13.04
casement fasteners; wedge plate; black malleable iron	5.83	2.50	5.98	0.40	-	7.35	-	13.32	nr	14.66
casement fasteners; wedge plate; B.M.A.	3.10	2.50	3.18	0.40	-	7.35	-	10.53	nr	11.58
casement fasteners; wedge plate; chromium plated	4.06	2.50	4.16	0.40	-	7.35	-	11.51	nr	12.66
casement fasteners; wedge plate; S.A.A.	2.31	2.50	2.37	0.40	-	7.35	-	9.72	nr	10.69
sash fasteners; brass; 70mm	3.40	2.50	3.49	1.10	-	20.21	-	23.69	nr	26.06
sash fasteners; B.M.A.; 70mm	7.15	2.50	7.33	1.10	-	20.21	-	27.54	nr	30.29
sash fasteners; chromium plated; 70mm	3.03	2.50	3.10	1.10	-	20.21	-	23.31	nr	25.64
casement stays; two pins; grey malleable iron										
200mm	5.00	2.50	5.12	0.40	-	7.35	-	12.47	nr	13.72
250mm	5.22	2.50	5.36	0.40	-	7.35	-	12.70	nr	13.97
300mm	5.42	2.50	5.55	0.40	-	7.35	-	12.90	nr	14.19
casement stays; two pins; B.M.A.										
200mm	2.55	2.50	2.61	0.40	-	7.35	-	9.96	nr	10.96
250mm	2.85	2.50	2.92	0.40	-	7.35	-	10.27	nr	11.30
300mm	3.30	2.50	3.38	0.40	-	7.35	-	10.73	nr	11.80
casement stays; two pins; chromium plated										
200mm	3.00	2.50	3.07	0.40	-	7.35	-	10.42	nr	11.47
250mm	3.30	2.50	3.38	0.40	-	7.35	-	10.73	nr	11.80
300mm	3.75	2.50	3.84	0.40	-	7.35	-	11.19	nr	12.31
casement stays; two pins; S.A.A.										
200mm	2.50	2.50	2.56	0.40	-	7.35	-	9.91	nr	10.90
250mm	2.73	2.50	2.80	0.40	-	7.35	-	10.14	nr	11.16
300mm	2.90	2.50	2.98	0.40	-	7.35	-	10.32	nr	11.36
sash lifts										
polished brass; 50mm	0.81	2.50	0.83	0.25	-	4.59	-	5.42	nr	5.96
B.M.A.; 50mm	1.60	2.50	1.63	0.25	-	4.59	-	6.23	nr	6.85
chromium plated; 50mm	1.52	2.50	1.55	0.25	-	4.59	-	6.15	nr	6.76
flush lifts										
brass; 75mm	0.95	2.50	0.97	0.55	-	10.10	-	11.08	nr	12.18
B.M.A.; 75mm	4.71	2.50	4.83	0.55	-	10.10	-	14.93	nr	16.42
chromium plated; 75mm	1.15	2.50	1.18	0.55	-	10.10	-	11.28	nr	12.41
sash cleats										
polished brass; 76mm	3.84	2.50	3.93	0.25	-	4.59	-	8.53	nr	9.38
B.M.A.; 76mm	4.12	2.50	4.23	0.25	-	4.59	-	8.82	nr	9.70
chromium plated; 76mm	3.84	2.50	3.93	0.25	-	4.59	-	8.53	nr	9.38
sash pulleys										
frame and wheel	1.60	2.50	1.64	0.55	-	10.10	-	11.74	nr	12.92
To hardwood or the like										
fanlight catches; brass	4.00	2.50	4.10	0.55	-	10.10	-	14.20	nr	15.62
fanlight catches; B.M.A.	7.70	2.50	7.89	0.55	-	10.10	-	18.00	nr	19.80

Labour hourly rates: (except Specialists) Craft Operatives 18.37 Labourer 13.76 Rates are national average prices. Refer to REGIONAL VARIATIONS for indicative levels of overall pricing in regions	MATERIALS			LABOUR				RATES		
	Del to Site	Waste	Material Cost	Craft Optve	Lab	Labour Cost	Sunds	Nett Rate	Unit	Gross rate (10%)
	£	%	£	Hrs	Hrs	£	£	£		£
IRONMONGERY (Cont'd)										
Window furniture; standard quality (Cont'd)										
To hardwood or the like (Cont'd)										
fanlight catches; chromium plated	4.40	2.50	4.51	0.55	-	10.10	-	14.61	nr	16.07
casement fasteners; wedge plate; black malleable iron	5.83	2.50	5.98	0.55	-	10.10	-	16.08	nr	17.69
casement fasteners; wedge plate; B.M.A.	3.10	2.50	3.18	0.55	-	10.10	-	13.28	nr	14.61
casement fasteners; wedge plate; chromium plated	4.06	2.50	4.16	0.55	-	10.10	-	14.26	nr	15.69
casement fasteners; wedge plate; S.A.A.	2.31	2.50	2.37	0.55	-	10.10	-	12.47	nr	13.72
sash fasteners; brass; 70mm	3.40	2.50	3.49	1.65	-	30.31	-	33.80	nr	37.18
sash fasteners; B.M.A.; 70mm	7.15	2.50	7.33	1.65	-	30.31	-	37.64	nr	41.40
sash fasteners; chromium plated; 70mm	3.03	2.50	3.10	1.65	-	30.31	-	33.41	nr	36.75
casement stays; two pins; grey malleable iron										
200mm	5.00	2.50	5.12	0.55	-	10.10	-	15.23	nr	16.75
250mm	5.22	2.50	5.36	0.55	-	10.10	-	15.46	nr	17.01
300mm	5.42	2.50	5.55	0.55	-	10.10	-	15.66	nr	17.22
casement stays; two pins; B.M.A.										
200mm	2.55	2.50	2.61	0.55	-	10.10	-	12.72	nr	13.99
250mm	2.85	2.50	2.92	0.55	-	10.10	-	13.02	nr	14.33
300mm	3.30	2.50	3.38	0.55	-	10.10	-	13.49	nr	14.83
casement stays; two pins; chromium plated										
200mm	3.00	2.50	3.07	0.55	-	10.10	-	13.18	nr	14.50
250mm	3.30	2.50	3.38	0.55	-	10.10	-	13.49	nr	14.83
300mm	3.75	2.50	3.84	0.55	-	10.10	-	13.95	nr	15.34
casement stays; two pins; S.A.A.										
200mm	2.50	2.50	2.56	0.55	-	10.10	-	12.66	nr	13.93
250mm	2.73	2.50	2.80	0.55	-	10.10	-	12.90	nr	14.19
300mm	2.90	2.50	2.98	0.55	-	10.10	-	13.08	nr	14.39
sash lifts										
polished brass; 50mm	0.81	2.50	0.83	0.35	-	6.43	-	7.26	nr	7.98
B.M.A.; 50mm	1.60	2.50	1.63	0.35	-	6.43	-	8.06	nr	8.87
chromium plated; 50mm	1.52	2.50	1.55	0.35	-	6.43	-	7.98	nr	8.78
flush lifts										
brass; 75mm	0.95	2.50	0.97	0.80	-	14.70	-	15.67	nr	17.24
B.M.A.; 75mm	4.71	2.50	4.83	0.80	-	14.70	-	19.52	nr	21.47
chromium plated; 75mm	1.15	2.50	1.18	0.80	-	14.70	-	15.87	nr	17.46
sash cleats										
polished brass; 76mm	3.84	2.50	3.93	0.35	-	6.43	-	10.36	nr	11.40
B.M.A.; 76mm	4.12	2.50	4.23	0.35	-	6.43	-	10.66	nr	11.72
chromium plated; 76mm	3.84	2.50	3.93	0.35	-	6.43	-	10.36	nr	11.40
sash pulleys										
frame and wheel	1.60	2.50	1.64	0.80	-	14.70	-	16.34	nr	17.97
Window furniture; trickle vents - to softwood or the like										
Frame vent; PVCu white										
265mm long	2.99	2.50	3.06	0.22	0.03	4.45	0.25	7.77	nr	8.55
366mm long	2.99	2.50	3.06	0.33	0.04	6.61	0.30	9.98	nr	10.97
Canopy grille; PVCu white										
364mm long	2.49	2.50	2.55	0.22	0.03	4.45	0.12	7.13	nr	7.84
Canopy grille; aluminium with epoxy paint finish										
412mm long	3.17	2.50	3.25	0.33	0.03	6.47	0.18	9.90	nr	10.89
Window furniture; trickle vents - to hardwood or the like										
Frame vent; PVCu white										
265mm long	2.99	2.50	3.06	0.33	0.05	6.75	0.25	10.07	nr	11.08
366mm long	2.99	2.50	3.06	0.50	0.07	10.15	0.30	13.51	nr	14.86
Canopy grille; PVCu white										
364mm long	2.49	2.50	2.55	0.33	0.05	6.75	0.12	9.42	nr	10.36
reference TV CG4; 308 x 15mm slot size; fixing with screws and clip on covers	3.17	2.50	3.25	0.50	0.07	10.15	0.18	13.58	nr	14.94
Canopy grille; aluminium with epoxy paint finish										
412mm long	3.17	2.50	3.25	0.50	0.07	10.15	0.18	13.58	nr	14.94
Letter plates; standard quality										
To softwood										
letter plates; plain; polished brass; 356mm wide	24.11	2.50	24.71	1.65	-	30.31	-	55.02	nr	60.52
letter plates; plain; anodised silver; 350mm wide	19.62	2.50	20.11	1.65	-	30.31	-	50.42	nr	55.46
letter plates; gravity flap; polished brass; 280mm wide	18.94	2.50	19.42	1.65	-	30.31	-	49.73	nr	54.70
letter plates; gravity flap; chromium plated; 280mm wide	19.82	2.50	20.31	1.65	-	30.31	-	50.62	nr	55.68
letter plates; gravity flap, satin nickel; 280mm wide	21.54	2.50	22.08	1.75	-	32.15	-	54.23	nr	59.65
postal knockers; polished brass; frame 254 x 79mm; opening 185 x 43m	15.50	2.50	15.89	1.75	-	32.15	-	48.03	nr	52.84
To hardwood or the like										
letter plates; plain; polished brass; 356mm wide	24.11	2.50	24.71	2.50	-	45.93	-	70.64	nr	77.70
letter plates; plain; anodised silver; 350mm wide	19.62	2.50	20.11	2.50	-	45.93	-	66.03	nr	72.64
letter plates; gravity flap; polished brass; 280mm wide	18.94	2.50	19.42	2.50	-	45.93	-	65.34	nr	71.87
letter plates; gravity flap; chromium plated; 280mm wide	19.82	2.50	20.31	2.50	-	45.93	-	66.24	nr	72.86
letter plates; gravity flap, satin nickel; 280mm wide	21.54	2.50	22.08	2.65	-	48.68	-	70.76	nr	77.84
postal knockers; polished brass; frame 254 x 79mm; opening 185 x 43m	15.50	2.50	15.89	2.65	-	48.68	-	64.57	nr	71.02
Security locks; standard quality										
To hardwood or the like										
window catches; locking; polished brass	5.41	2.50	5.54	1.20	-	22.04	-	27.59	nr	30.35
dual screws; sanded brass	1.88	2.50	1.93	1.00	-	18.37	-	20.30	nr	22.33
window stops; locking; brass	2.07	2.50	2.12	1.00	-	18.37	-	20.49	nr	22.54
mortice latches; locking	3.90	2.50	4.00	3.50	-	64.30	-	68.29	nr	75.12
double cylinder automatic deadlatches; high security; Brasslux	65.40	2.50	67.04	3.50	-	64.30	-	131.33	nr	144.46

GENERAL JOINERY

Labour hourly rates: (except Specialists) Craft Operatives 18.37 Labourer 13.76 Rates are national average prices. Refer to REGIONAL VARIATIONS for indicative levels of overall pricing in regions	MATERIALS			LABOUR				RATES		
	Del to Site	Waste	Material Cost	Craft Optve	Lab	Labour Cost	Sunds	Nett Rate	Unit	Gross rate (10%)
	£	%	£	Hrs	Hrs	£	£	£		£
IRONMONGERY (Cont'd)										
Security locks; standard quality (Cont'd)										
To hardwood or the like (Cont'd)										
double cylinder automatic deadlatches; standard security; chromium plated	33.20	2.50	34.03	3.50	-	64.30	-	98.33	nr	108.16
mortice deadlocks; Union 3G115PB	31.63	2.50	32.42	3.50	-	64.30	-	96.72	nr	106.39
mortice locks; two bolt (sash lock) - 5 lever high security	12.75	2.50	13.07	3.50	-	64.30	-	77.36	nr	85.10
metal window locks; brass	6.84	2.50	7.01	1.50	-	27.56	-	34.57	nr	38.02
security door chains; steel chain; brass	2.03	2.50	2.08	0.60	-	11.02	-	13.10	nr	14.41
security door chains; steel chain; chromium plated	2.49	2.50	2.55	0.60	-	11.02	-	13.57	nr	14.93
security mortice bolts; loose keys; satin stainless steel; 64mm long	3.85	2.50	3.95	1.20	-	22.04	-	25.99	nr	28.59
security mortice bolts; loose keys; satin stainless steel; 152mm long	5.50	2.50	5.64	1.30	-	23.88	-	29.52	nr	32.47
security hinge bolts; Satin Chrome	5.61	2.50	5.75	1.80	-	33.07	-	38.82	nr	42.70
door viewers; chromium plated	3.29	2.50	3.37	1.50	-	27.56	-	30.93	nr	34.02
Shelf brackets; standard quality										
To softwood										
shelf brackets; grey finished										
100 x 75mm	0.25	2.50	0.26	0.25	-	4.59	-	4.85	nr	5.33
150 x 125mm	0.25	2.50	0.26	0.25	-	4.59	-	4.85	nr	5.33
225 x 175mm	0.37	2.50	0.38	0.25	-	4.59	-	4.97	nr	5.47
300 x 250mm	0.49	2.50	0.50	0.25	-	4.59	-	5.09	nr	5.60
350 x 300mm	0.64	2.50	0.65	0.25	-	4.59	-	5.24	nr	5.77
Cabin hooks; standard quality										
To softwood										
cabin hooks and eyes; black japanned										
100mm	0.32	2.50	0.33	0.25	-	4.59	-	4.92	nr	5.41
150mm	0.43	2.50	0.44	0.25	-	4.59	-	5.04	nr	5.54
200mm	0.50	2.50	0.51	0.25	-	4.59	-	5.11	nr	5.62
250mm	1.68	2.50	1.73	0.25	-	4.59	-	6.32	nr	6.95
cabin hooks and eyes; polished brass										
100mm	1.11	2.50	1.14	0.25	-	4.59	-	5.73	nr	6.30
200mm	1.29	2.50	1.32	0.25	-	4.59	-	5.92	nr	6.51
To hardwood or the like										
cabin hooks and eyes; black japanned										
100mm	0.32	2.50	0.33	0.35	-	6.43	-	6.76	nr	7.44
150mm	0.43	2.50	0.44	0.35	-	6.43	-	6.87	nr	7.56
200mm	0.50	2.50	0.51	0.35	-	6.43	-	6.94	nr	7.64
250mm	1.68	2.50	1.73	0.35	-	6.43	-	8.15	nr	8.97
cabin hooks and eyes; polished brass										
100mm	1.11	2.50	1.14	0.35	-	6.43	-	7.57	nr	8.32
200mm	1.29	2.50	1.32	0.35	-	6.43	-	7.75	nr	8.53
To softwood and brickwork										
cabin hooks and eyes; black japanned										
100mm	0.32	2.50	0.33	0.30	-	5.51	-	5.84	nr	6.42
150mm	0.43	2.50	0.44	0.30	-	5.51	-	5.96	nr	6.55
200mm	0.50	2.50	0.51	0.30	-	5.51	-	6.02	nr	6.63
250mm	1.68	2.50	1.73	0.30	-	5.51	-	7.24	nr	7.96
cabin hooks and eyes; polished brass										
100mm	1.11	2.50	1.14	0.30	-	5.51	-	6.65	nr	7.31
200mm	1.29	2.50	1.32	0.30	-	5.51	-	6.83	nr	7.52
To hardwood and brickwork										
cabin hooks and eyes; black japanned										
100mm	0.32	2.50	0.33	0.40	-	7.35	-	7.68	nr	8.45
150mm	0.43	2.50	0.44	0.40	-	7.35	-	7.79	nr	8.57
200mm	0.50	2.50	0.51	0.40	-	7.35	-	7.86	nr	8.65
250mm	1.68	2.50	1.73	0.40	-	7.35	-	9.07	nr	9.98
cabin hooks and eyes; polished brass										
100mm	1.11	2.50	1.14	0.40	-	7.35	-	8.49	nr	9.33
200mm	1.29	2.50	1.32	0.40	-	7.35	-	8.67	nr	9.54
Draught seals and strips; standard quality										
To softwood										
draught excluders; plastic foam, self-adhesive										
900mm long	0.35	10.00	0.38	0.15	-	2.76	-	3.14	nr	3.45
2000mm long	0.78	10.00	0.86	0.30	-	5.51	-	6.37	nr	7.00
draught excluders; aluminium section, rubber tubing										
900mm long	0.41	10.00	0.45	0.17	0.02	3.40	-	3.85	nr	4.24
2000mm long	0.92	10.00	1.01	0.35	0.04	6.98	-	7.99	nr	8.79
draught excluders; aluminium section, vinyl seal										
900mm long	2.86	10.00	3.14	0.17	0.02	3.40	-	6.54	nr	7.20
2000mm long	6.35	10.00	6.98	0.35	0.04	6.98	-	13.96	nr	15.36
draught excluders; plastic moulding, nylon brush										
900mm long	4.17	10.00	4.58	0.17	0.02	3.40	-	7.98	nr	8.78
draught excluders; aluminium base, flexible arch										
900mm long	10.83	10.00	11.92	0.25	0.03	5.01	-	16.92	nr	18.61
draught excluders; aluminium threshold, flexible arch										
900mm long	10.83	10.00	11.92	0.30	0.04	6.06	-	17.98	nr	19.78
Draught seals and strips; Sealmaster Ltd										
Threshold seals										
reference BDA; fixing to masonry with screws	22.00	2.50	22.55	0.45	0.06	9.09	0.33	31.97	m	35.16
reference BDB; fixing to masonry with screws	28.20	2.50	28.91	0.45	0.06	9.09	0.45	38.45	m	42.30
reference WTSHED (weather board); fixing to timber with screws	50.21	2.50	51.46	0.70	0.10	14.24	0.52	66.22	m	72.85
surface mounted stop strip; fixing to timber with screws	20.80	2.50	21.32	0.33	0.05	6.75	0.33	28.40	m	31.24

Labour hourly rates: (except Specialists) Craft Operatives 18.37 Labourer 13.76 Rates are national average prices. Refer to REGIONAL VARIATIONS for indicative levels of overall pricing in regions	MATERIALS			LABOUR				RATES		
	Del to Site	Waste	Material Cost	Craft Optve	Lab	Labour Cost	Sunds	Nett Rate	Unit	Gross rate (10%)
	£	%	£	Hrs	Hrs	£	£	£		£
IRONMONGERY (Cont'd)										
Drawer pulls; standard quality										
To softwood										
drawer pulls; brass; 100mm	3.04	2.50	3.12	0.20	-	3.67	-	6.79	nr	7.47
drawer pulls; chromium plated; 100mm.............	3.04	2.50	3.12	0.20	-	3.67	-	6.79	nr	7.47
To hardwood or the like										
drawer pulls; brass; 100mm	3.04	2.50	3.12	0.30	-	5.51	-	8.63	nr	9.49
drawer pulls; chromium plated; 100mm.............	3.04	2.50	3.12	0.30	-	5.51	-	8.63	nr	9.49
Hooks; standard quality										
To softwood										
cup hooks; polished brass; 25mm	0.08	2.50	0.08	0.07	-	1.29	-	1.37	nr	1.51
hat and coat hooks; brass	2.02	2.50	2.07	0.30	-	5.51	-	7.58	nr	8.34
hat and coat hooks; chromium plated...............	2.02	2.50	2.07	0.30	-	5.51	-	7.58	nr	8.34
hat and coat hooks; S.A.A.	0.50	2.50	0.51	0.30	-	5.51	-	6.02	nr	6.63
To hardwood or the like										
cup hooks; polished brass; 25mm	0.08	2.50	0.08	0.10	-	1.84	-	1.92	nr	2.11
hat and coat hooks; brass	2.02	2.50	2.07	0.45	-	8.27	-	10.34	nr	11.37
hat and coat hooks; chromium plated...............	2.02	2.50	2.07	0.45	-	8.27	-	10.34	nr	11.37
hat and coat hooks; S.A.A.	0.50	2.50	0.51	0.45	-	8.27	-	8.78	nr	9.66

GENERAL JOINERY

This page left blank intentionally

Labour hourly rates: (except Specialists) Craft Operatives 18.37 Labourer 13.76 Rates are national average prices. Refer to REGIONAL VARIATIONS for indicative levels of overall pricing in regions	MATERIALS			LABOUR				RATES		
	Del to Site	Waste	Material Cost	Craft Optve	Lab	Labour Cost	Sunds	Nett Rate		Gross rate (10%)
	£	%	£	Hrs	Hrs	£	£	£	Unit	£

TIMBER WINDOWS AND SHOP FRONTS

Casements in softwood, fully finished in white by manufacturer; Bedding in cement mortar (1:3); pointing with Silicone standard mastic one side

Magnet Trade; softwood Statesman Professional; factory double glazed; trickle vents; 158mm cill: casement windows without bars; softwood sub-sills; hinges; fasteners

	Del to Site	Waste	Material Cost	Craft Optve	Lab	Labour Cost	Sunds	Nett Rate	Unit	Gross rate
488 x 900mm overall; N09V	302.26	2.50	309.99	2.11	0.77	49.26	2.43	361.68	nr	397.85
631 x 750mm overall; 107V	312.70	2.50	320.68	2.10	0.77	49.21	2.43	372.32	nr	409.55
631 x 900mm overall; 109V	318.22	2.50	326.36	2.31	0.85	54.17	2.43	382.96	nr	421.26
631 x 1050mm overall; 110V	328.50	2.50	336.91	2.47	0.90	57.79	2.43	397.14	nr	436.85
631 x 1200mm overall; 112v	338.76	2.50	347.44	2.63	0.95	61.42	2.43	411.30	nr	452.43
915 x 900mm overall; 2N09W	392.48	2.50	402.51	2.63	0.95	61.31	2.43	466.25	nr	512.87
915 x 1050mm overall; 2N10W	405.66	2.50	416.03	2.84	1.02	66.13	2.43	484.60	nr	533.06
915 x 1200mm overall; 2N12W	418.83	2.50	429.55	3.00	1.07	69.76	2.43	501.74	nr	551.92
1200 x 900mm overall; 209W	450.58	2.50	462.10	2.99	1.07	69.65	2.43	534.18	nr	587.60
1200 x 1050mm overall; 210CV	544.90	2.50	558.79	3.20	1.15	74.61	2.43	635.83	nr	699.41
1200 x 1050mm overall; 210W	471.06	2.50	483.10	3.20	1.15	74.61	2.43	560.14	nr	616.15
1200 x 1200mm overall; 212CV	567.75	2.50	582.23	3.36	1.20	78.24	2.43	662.90	nr	729.19
1200 x 1200mm overall; 212W	490.70	2.50	503.26	3.36	1.20	78.24	2.43	583.92	nr	642.31
1200 x 1500mm overall; 215W	531.20	2.50	544.78	3.52	1.25	81.86	2.43	629.07	nr	691.98
1342 x 1050mm overall; 3N10CC	602.47	2.50	617.84	3.53	1.25	82.05	2.43	702.32	nr	772.55
1769 x 1050mm overall; 310CVC	758.50	2.50	777.80	4.03	1.45	93.95	2.43	874.17	nr	961.59
1769 x 1200mm overall; 312CVC	793.91	2.50	814.12	4.19	1.50	97.57	2.43	914.12	nr	1005.53

Sash windows in softwood, wrought; fully finished in white by manufacturer; Bedding in cement mortar (1:3); pointing with Silicone standard mastic one side

Magnet Victorian Statesman Professional; factory double glazed; trickle vents; 158mm cill: non bar mock sash softwood windows; glazing beads; fasteners

	Del to Site	Waste	Material Cost	Craft Optve	Lab	Labour Cost	Sunds	Nett Rate	Unit	Gross rate
631 x 900mm overall; VBS109B	434.02	2.50	445.06	2.81	1.10	66.79	2.43	514.28	nr	565.71
631 x 1050mm overall; VBS110B	444.56	2.50	455.88	3.12	1.22	74.14	2.43	532.45	nr	585.69
631 x 1200mm overall; VBS112B	466.58	2.50	478.47	3.38	1.32	80.29	2.43	561.19	nr	617.31
631 x 1350mm overall; VBS113B	479.26	2.50	491.48	3.69	1.45	87.77	2.43	581.68	nr	639.85
1200 x 900mm overall; VBS209B	582.95	2.50	597.78	3.74	1.45	88.66	2.43	688.86	nr	757.75
1200 x 1050mm overall; VBS210B	607.27	2.50	622.72	3.95	1.52	93.48	2.43	718.63	nr	790.49
1200 x 1200mm overall; VBS212B	643.07	2.50	659.43	4.26	1.65	100.96	2.43	762.82	nr	839.10
1200 x 1200mm overall; VBN212BMOE	1478.77	2.50	1516.03	4.26	1.65	100.96	2.43	1619.42	nr	1781.36
1200 x 1350mm overall; VBS213B	667.45	2.50	684.45	4.62	1.80	109.64	2.43	796.51	nr	876.17
1769 x 1050mm overall; VBS310BB	667.94	2.50	684.98	4.83	1.85	114.18	2.43	801.59	nr	881.75
1769 x 1200mm overall; VBS312BB	990.79	2.50	1015.91	5.29	2.05	125.35	2.43	1143.69	nr	1258.06
1769 x 1350mm overall; VBS313BB	1029.40	2.50	1055.51	5.55	2.15	131.50	2.43	1189.44	nr	1308.38

Windows in softwood, base coat stained by manufacturer; Bedding in cement mortar (1:3); pointing with Silicone standard mastic one side

Jeld-Wen Ltd. Softwood Sovereign windows plain casement non bar type side hung; glazing beads; weatherstripping, hinges; fasteners

	Del to Site	Waste	Material Cost	Craft Optve	Lab	Labour Cost	Sunds	Nett Rate	Unit	Gross rate
630 x 750mm overall; W107C	169.72	2.50	174.12	2.45	0.95	58.12	2.43	234.67	nr	258.14
630 x 900mm overall; W109C	176.59	2.50	181.19	2.81	1.10	66.79	2.43	250.41	nr	275.46
630 x 1050mm overall; W110C	183.90	2.50	188.70	3.12	1.22	74.19	2.43	265.32	nr	291.85
630 x 1200mm overall; W112C	191.94	2.50	196.96	3.38	1.32	80.29	2.43	279.68	nr	307.65
630 x 1350mm overall; W113C	206.13	2.50	211.52	3.69	1.45	87.77	2.43	301.73	nr	331.90
1200 x 750mm overall; W207C	231.40	2.50	237.42	3.38	1.30	79.98	2.43	319.83	nr	351.81
1200 x 900mm overall; W209C	240.65	2.50	246.91	3.74	1.45	88.66	2.43	338.00	nr	371.80
1200 x 1050mm overall; W210C	250.31	2.50	256.83	3.95	1.52	93.48	2.43	352.74	nr	388.01
1200 x 1200mm overall; W212C	260.69	2.50	267.50	4.26	1.65	100.96	2.43	370.89	nr	407.98
1200 x 1350mm overall; W213C	277.24	2.50	284.48	4.57	1.77	108.31	2.43	395.21	nr	434.73
1770 x 750mm overall; W307CC	346.08	2.50	355.03	4.38	1.65	103.16	2.43	460.63	nr	506.69
1770 x 900mm overall; W309CC	360.62	2.50	369.95	4.62	1.77	109.19	2.43	481.57	nr	529.72
1770 x 1050mm overall; W310CC	375.99	2.50	385.73	4.83	1.85	114.15	2.43	502.30	nr	552.54
1770 x 1200mm overall; W312CC	392.85	2.50	403.02	5.19	2.00	122.82	2.43	528.27	nr	581.10
1770 x 1350mm overall; W313CC	421.99	2.50	432.91	5.55	2.15	131.50	2.43	566.84	nr	623.53
2339 x 900mm overall; W409CMC	422.87	2.50	433.83	5.60	2.15	132.38	2.43	568.64	nr	625.51
2339 x 1050mm overall; W410CMC	215.59	2.50	221.39	5.96	2.30	141.06	2.43	364.88	nr	401.37
2339 x 1200mm overall; W412CMC	459.80	2.50	471.72	6.23	2.42	147.67	2.43	621.82	nr	684.00
2339 x 1350mm overall; W413CMC	491.30	2.50	504.03	6.58	2.55	155.89	2.43	662.35	nr	728.58

Jeld-Wen Ltd. Softwood Alpha Sovereign windows casement with vents top hung type; glazing beads; weatherstripping, hinges; fasteners

	Del to Site	Waste	Material Cost	Craft Optve	Lab	Labour Cost	Sunds	Nett Rate	Unit	Gross rate
630 x 750mm overall; W107V	179.23	2.50	183.88	2.35	0.90	55.59	2.43	241.90	nr	266.09
630 x 900mm overall; W109V	183.17	2.50	187.93	2.51	0.95	59.22	2.43	249.58	nr	274.54
630 x 1050mm overall; W110V	187.11	2.50	191.99	2.87	1.10	67.89	2.43	262.31	nr	288.55
630 x 1200mm overall; W112V	191.05	2.50	196.04	3.18	1.22	75.24	2.43	273.71	nr	301.09
630 x 1350mm overall; W113V	194.99	2.50	200.10	3.44	1.32	81.39	2.43	283.92	nr	312.31
915 x 900mm overall; W2N09C	220.10	2.50	225.82	3.18	1.22	75.13	2.43	303.38	nr	333.72
915 x 1350mm overall; W2N13C	231.92	2.50	237.99	3.91	1.50	92.39	2.43	332.81	nr	366.09

Labour hourly rates: (except Specialists) Craft Operatives 18.37 Labourer 13.76 Rates are national average prices. Refer to REGIONAL VARIATIONS for indicative levels of overall pricing in regions	MATERIALS			LABOUR				RATES		
	Del to Site £	Waste %	Material Cost £	Craft Optve Hrs	Lab Hrs	Labour Cost £	Sunds £	Nett Rate £	Unit	Gross rate (10%) £

TIMBER WINDOWS AND SHOP FRONTS (Cont'd)

Windows in softwood, base coat stained by manufacturer; Bedding in cement mortar (1:3); pointing with Silicone standard mastic one side (Cont'd)

Jeld-Wen Ltd. Softwood Alpha Sovereign windows casement with vents top hung type; glazing beads; weatherstripping, hinges; fasteners (Cont'd)

	Del to Site £	Waste %	Material Cost £	Craft Optve Hrs	Lab Hrs	Labour Cost £	Sunds £	Nett Rate £	Unit	Gross rate (10%) £
915 x 750mm overall; W2N07CV	269.58	2.50	276.52	2.82	1.07	66.45	2.43	345.40	nr	379.94
915 x 900mm overall; W2N09CV	281.05	2.50	288.29	3.08	1.17	72.61	2.43	363.32	nr	399.66
915 x 1050mm overall; W2N10CV	290.70	2.50	298.20	3.39	1.30	80.09	2.43	380.72	nr	418.79
915 x 1200mm overall; W2N12CV	301.10	2.50	308.88	3.65	1.40	86.24	2.43	397.55	nr	437.30
1200 x 900mm overall; W209T	300.49	2.50	308.25	3.44	1.30	81.08	2.43	391.76	nr	430.94
1200 x 1050mm overall; W210T	310.17	2.50	318.19	3.80	1.45	89.76	2.43	410.38	nr	451.42
1200 x 1200mm overall; W212T	319.41	2.50	327.68	4.01	1.52	94.58	2.43	424.69	nr	467.16
1200 x 1350mm overall; W213T	329.07	2.50	337.60	4.32	1.65	102.06	2.43	442.09	nr	486.30
1200 x 1500mm overall; W215T	339.46	2.50	348.27	4.28	1.60	100.64	2.43	451.34	nr	496.47
1770 x 1050mm overall; W310CVC	441.98	2.50	453.36	4.43	1.65	104.05	2.43	559.84	nr	615.82
1770 x 1200mm overall; W312CVC	458.83	2.50	470.66	4.74	1.77	111.39	2.43	584.48	nr	642.93
1770 x 1350mm overall; W313CVC	487.98	2.50	500.55	4.95	1.85	116.35	2.43	619.33	nr	681.26
1770 x 1500mm overall; W315CVC	504.68	2.50	517.69	5.31	2.00	125.03	2.43	645.15	nr	709.66

Windows in hardwood, base coat stained by manufacturer; Bedding in cement mortar (1:3); pointing with Silicone standard mastic one side

Jeld-Wen Ltd. Oak Stormsure windows; plain side hung casement; factory double glazed; weatherstripping, hinges; fasteners; factory finished stain by manufacturer

	Del to Site £	Waste %	Material Cost £	Craft Optve Hrs	Lab Hrs	Labour Cost £	Sunds £	Nett Rate £	Unit	Gross rate (10%) £
630 x 750mm overall; OLEW107C	610.64	2.50	626.07	3.02	1.22	72.26	2.43	700.77	nr	770.84
630 x 900mm overall; OLEW109C	650.01	2.50	666.44	3.26	1.32	78.09	2.43	746.96	nr	821.65
630 x 1050mm overall; OLEW110C	690.02	2.50	707.47	3.22	1.32	77.31	2.43	787.22	nr	865.94
630 x 1200mm overall; OLEW112CH	731.20	2.50	749.70	3.48	1.37	82.82	2.43	834.94	nr	918.44
630 x 1350mm overall; OLEW113C	782.02	2.50	801.81	3.54	1.37	83.92	2.43	888.16	nr	976.97
1200 x 750mm overall; OLEW207C	901.20	2.50	923.96	3.38	1.30	79.98	2.43	1006.37	nr	1107.01
1200 x 900mm overall; OLEW209C	957.34	2.50	981.52	3.64	1.40	86.13	2.43	1070.08	nr	1177.09
1200 x 1050mm overall; OLEW210C	1014.13	2.50	1039.75	3.95	1.52	93.48	2.43	1135.66	nr	1249.22
1200 x 1200mm overall; OLEW212C	1072.10	2.50	1099.19	4.30	1.65	101.62	2.43	1203.24	nr	1323.56
1200 x 1350mm overall; OLEW213C	1139.70	2.50	1168.49	4.57	1.77	108.31	2.43	1279.23	nr	1407.15
1770 x 750mm overall; OLEW307CC	1325.77	2.50	1359.21	4.31	1.65	101.84	2.43	1463.48	nr	1609.83
1770 x 900mm overall; OLEW309CC	1407.11	2.50	1442.61	4.62	1.77	109.19	2.43	1554.23	nr	1709.65
1770 x 1050mm overall; OLEW310CC	1489.77	2.50	1527.35	4.83	1.85	114.15	2.43	1643.93	nr	1808.32
1770 x 1200mm overall; OLEW312CC	1574.76	2.50	1614.49	5.19	2.00	122.82	2.43	1739.74	nr	1913.71
1770 x 1350mm overall; OLEW313CC	1679.02	2.50	1721.37	5.55	2.15	131.50	2.43	1855.30	nr	2040.83
2339 x 900mm overall; OLEW409CMC	1758.25	2.50	1802.59	5.80	2.25	137.43	2.43	1942.45	nr	2136.70
2339 x 1050mm overall; OLEW410CMC	1863.82	2.50	1910.82	5.96	2.30	141.06	2.43	2054.31	nr	2259.75
2339 x 1200mm overall; OLEW412CMC	1971.73	2.50	2021.44	6.27	2.42	148.41	2.43	2172.28	nr	2389.51
2339 x 1350mm overall; OLEW413CMC	2098.89	2.50	2151.80	6.53	2.52	154.56	2.43	2308.79	nr	2539.67

Jeld-Wen Ltd. Oak Stormsure windows; plain side hung casement with vents; factory double glazed; weatherstripping, hinges; fasteners; factory finished stain by manufacturer

	Del to Site £	Waste %	Material Cost £	Craft Optve Hrs	Lab Hrs	Labour Cost £	Sunds £	Nett Rate £	Unit	Gross rate (10%) £
630 x 750mm overall; OLEW107V	660.23	2.50	676.90	2.50	0.95	59.03	2.43	738.37	nr	812.20
630 x 900mm overall; OLEW109V	678.94	2.50	696.09	3.26	1.08	74.78	2.43	773.31	nr	850.64
630 x 1050mm overall; OLEW110V	703.70	2.50	721.49	3.12	1.22	74.14	2.43	798.06	nr	877.86
630 x 1200mm overall; OLEW112V	728.45	2.50	746.88	3.33	1.30	79.10	2.43	828.41	nr	911.25
630 x 1350mm overall; OLEW113V	753.22	2.50	772.29	3.79	1.50	90.30	2.43	865.02	nr	951.52
915 x 900mm overall; OLEW2N09W	860.43	2.50	882.16	2.53	1.40	65.67	2.43	950.25	nr	1045.28
915 x 1350mm overall; OLEW2N13W	945.86	2.50	969.78	4.06	1.57	96.11	2.43	1068.32	nr	1175.16
915 x 750mm overall; OLEW2N07CV	1013.73	2.50	1039.07	3.12	1.22	74.03	2.43	1115.73	nr	1227.30
915 x 900mm overall; OLEW2N09CV	1056.89	2.50	1083.53	3.33	1.30	78.99	2.43	1164.95	nr	1281.44
915 x 1050mm overall; OLEW2N10CV	1102.28	2.50	1130.07	3.64	1.42	86.33	2.43	1218.83	nr	1340.71
915 x 1200mm overall; OLEW2N12CV	1156.51	2.50	1185.67	3.80	1.47	89.96	2.43	1278.06	nr	1405.87
1200 x 900mm overall; OLEW209T	1150.98	2.50	1180.01	3.79	1.47	89.85	2.43	1272.28	nr	1399.51
1200 x 1050mm overall; OLEW210T	1201.94	2.50	1232.26	3.95	1.52	93.48	2.43	1328.17	nr	1460.98
1200 x 1200mm overall; OLEW212T	1258.09	2.50	1289.83	4.26	1.65	100.96	2.43	1393.22	nr	1532.54
1200 x 1350mm overall; OLEW213T	1314.88	2.50	1348.06	4.62	1.80	109.64	2.43	1460.12	nr	1606.14
1200 x 1500mm overall; OLEW215T	1410.49	2.50	1446.08	4.68	1.80	110.74	2.43	1559.25	nr	1715.17
1770 x 1050mm overall; OLEW310CVC	1752.73	2.50	1796.88	4.93	1.90	116.67	2.43	1915.99	nr	2107.58
1770 x 1200mm overall; OLEW312CVC	1843.86	2.50	1890.31	5.29	2.05	125.35	2.43	2018.09	nr	2219.90
1770 x 1350mm overall; OLEW313CVC	1954.28	2.50	2003.51	5.55	2.15	131.50	2.43	2137.44	nr	2351.18
1770 x 1500mm overall; OLEW315CVC	2100.52	2.50	2153.42	5.81	2.25	137.65	2.43	2293.50	nr	2522.85

Jeld-Wen Ltd. Oak Stormsure windows - horizontal glazing bar type; factory glazed 24mm low E glass; weatherstripping, hinges; fasteners; base coat stain by manufacturer

	Del to Site £	Waste %	Material Cost £	Craft Optve Hrs	Lab Hrs	Labour Cost £	Sunds £	Nett Rate £	Unit	Gross rate (10%) £
630 x 900mm overall; OLEWH109CSDL	709.60	2.50	727.52	2.81	1.10	66.79	2.43	796.74	nr	876.42
630 x 1050mm overall; OLEWH110CSDL	753.95	2.50	773.00	3.12	1.22	74.14	2.43	849.57	nr	934.53
630 x 1200mm overall; OLEWH112CSDL	799.64	2.50	819.68	3.38	1.32	80.29	2.43	902.41	nr	992.65
630 x 1350mm overall; OLEWH113CSDL	854.64	2.50	876.25	3.69	1.45	87.77	2.43	966.45	nr	1063.09
1200 x 900mm overall; OLEWH209CSDL	1086.32	2.50	1113.73	3.74	1.45	88.66	2.43	1204.82	nr	1325.30
1200 x 1050mm overall; OLEWH210CSDL	1152.62	2.50	1181.70	3.95	1.52	93.48	2.43	1277.61	nr	1405.37
1200 x 1200mm overall; OLEWH212CSDL	1220.00	2.50	1250.88	4.26	1.65	100.96	2.43	1354.27	nr	1489.70
1200 x 1350mm overall; OLEWH213CSDL	1297.20	2.50	1329.93	4.62	1.80	109.64	2.43	1442.00	nr	1586.20
1770 x 1050mm overall; OLEWH310CCSDL	1692.09	2.50	1734.73	4.83	1.85	114.15	2.43	1851.31	nr	2036.44
1770 x 1200mm overall; OLEWH312CCSDL	1790.93	2.50	1836.06	5.29	2.05	125.35	2.43	1963.83	nr	2160.22
1770 x 1350mm overall; OLEWH313CCSDL	1909.02	2.50	1957.12	5.55	2.15	131.50	2.43	2091.05	nr	2300.16
2339 x 1050mm overall; OLEWH410CMCSDL	2140.61	2.50	2194.53	5.86	2.30	139.22	2.43	2336.18	nr	2569.80
2339 x 1200mm overall; OLEWH412CMCSDL	2267.51	2.50	2324.62	6.27	2.42	148.41	2.43	2475.46	nr	2723.00
2339 x 1350mm overall; OLEWH413CMCSDL	2413.65	2.50	2474.43	6.63	2.57	157.08	2.43	2633.95	nr	2897.34

Labour hourly rates: (except Specialists) Craft Operatives 18.37 Labourer 13.76 Rates are national average prices. Refer to REGIONAL VARIATIONS for indicative levels of overall pricing in regions	MATERIALS			LABOUR				RATES		
	Del to Site	Waste	Material Cost	Craft Optve	Lab	Labour Cost	Sunds	Nett Rate	Unit	Gross rate (10%)
	£	%	£	Hrs	Hrs	£	£	£		£
TIMBER WINDOWS AND SHOP FRONTS (Cont'd)										
Purpose made windows										
Note										
Notwithstanding the requirements of the NRM purpose made windows are shown here in square metres and excludes bedding and pointing										
Purpose made windows in softwood, wrought										
38mm moulded casements or fanlights										
in one pane	85.87	2.50	88.01	1.50	0.19	30.17	0.36	118.54	m²	130.40
divided into panes 0.10 - 0.50m²	163.72	2.50	167.81	1.50	0.19	30.17	0.36	198.34	m²	218.18
divided into panes not exceeding 0.10m²	241.71	2.50	247.75	1.50	0.19	30.17	0.36	278.28	m²	306.11
50mm moulded casements or fanlights										
in one pane	159.70	2.50	163.69	1.60	0.20	32.14	0.39	196.22	m²	215.84
divided into panes 0.10 - 0.50m²	180.71	2.50	185.23	1.60	0.20	32.14	0.39	217.76	m²	239.54
divided into panes not exceeding 0.10m²	201.19	2.50	206.22	1.60	0.20	32.14	0.39	238.76	m²	262.63
38mm moulded casements with semi-circular heads (measured square)										
in one pane	277.36	2.50	284.30	1.65	0.21	33.20	0.41	317.90	m²	349.69
divided into panes 0.10 - 0.50m²	298.38	2.50	305.84	1.65	0.21	33.20	0.41	339.44	m²	373.39
divided into panes not exceeding 0.10m²	319.39	2.50	327.37	1.65	0.21	33.20	0.41	360.98	m²	397.08
50mm moulded casements with semi-circular heads (measured square)										
in one pane	319.39	2.50	327.37	1.75	0.22	35.17	0.42	362.97	m²	399.27
divided into panes 0.10 - 0.50m²	339.88	2.50	348.37	1.75	0.22	35.17	0.42	383.97	m²	422.37
divided into panes not exceeding 0.10m²	360.89	2.50	369.91	1.75	0.22	35.17	0.42	405.51	m²	446.06
38mm bullseye casements										
457mm diameter in one pane	208.02	2.50	213.22	1.00	0.12	20.02	0.24	233.49	nr	256.83
762mm diameter in one pane	277.36	2.50	284.30	1.50	0.19	30.17	0.36	314.83	nr	346.31
50mm bullseye casements										
457mm diameter in one pane	222.21	2.50	227.76	1.10	0.14	22.13	0.27	250.17	nr	275.18
762mm diameter in one pane	298.38	2.50	305.84	1.65	0.21	33.20	0.41	339.44	nr	373.39
Labours										
check throated edge	0.74	2.50	0.75	-	-	-	-	0.75	m	0.83
rebated and splayed bottom rail	0.74	2.50	0.75	-	-	-	-	0.75	m	0.83
rebated and beaded meeting stile	1.00	2.50	1.02	-	-	-	-	1.02	m	1.13
fitting and hanging casement or fanlight on butts (included elsewhere)										
38mm	-	-	-	0.67	0.08	13.41	-	13.41	nr	14.75
50mm	-	-	-	0.74	0.09	14.83	-	14.83	nr	16.32
fitting and hanging casement or fanlight on sash centres (included elsewhere)										
38mm	-	-	-	1.50	0.19	30.17	-	30.17	nr	33.19
50mm	-	-	-	1.65	0.21	33.20	-	33.20	nr	36.52
Purpose made windows in Afrormosia, wrought										
38mm moulded casements or fanlights										
in one pane	137.93	2.50	141.37	2.25	0.28	45.19	0.56	187.11	m²	205.83
divided into panes 0.10 - 0.50m²	265.90	2.50	272.55	2.25	0.28	45.19	0.56	318.29	m²	350.12
divided into panes not exceeding 0.10m²	362.16	2.50	371.21	2.25	0.28	45.19	0.56	416.96	m²	458.65
50mm moulded casements or fanlights										
in one pane	247.42	2.50	253.61	2.40	0.30	48.22	0.58	302.41	m²	332.65
divided into panes 0.10 - 0.50m²	279.99	2.50	286.99	2.40	0.30	48.22	0.58	335.79	m²	369.37
divided into panes not exceeding 0.10m²	312.04	2.50	319.84	2.40	0.30	48.22	0.58	368.64	m²	405.50
38mm moulded casements with semi-circular heads (measured square)										
in one pane	430.23	2.50	440.99	2.45	0.31	49.27	0.60	490.86	m²	539.94
divided into panes 0.10 - 0.50m²	462.80	2.50	474.37	2.45	0.31	49.27	0.60	524.24	m²	576.67
divided into panes not exceeding 0.10m²	494.84	2.50	507.22	2.45	0.31	49.27	0.60	557.09	m²	612.80
50mm moulded casements with semi-circular heads (measured square)										
in one pane	494.84	2.50	507.22	2.60	0.32	52.17	0.63	560.01	m²	616.01
divided into panes 0.10 - 0.50m²	527.41	2.50	540.60	2.60	0.32	52.17	0.63	593.39	m²	652.73
divided into panes not exceeding 0.10m²	559.46	2.50	573.44	2.60	0.32	52.17	0.63	626.24	m²	688.86
38mm bullseye casements										
457mm diameter in one pane	323.07	2.50	331.14	1.50	0.19	30.17	0.36	361.67	nr	397.84
762mm diameter in one pane	430.23	2.50	440.99	2.25	0.28	45.19	0.56	486.73	nr	535.40
50mm bullseye casements										
457mm diameter in one pane	344.61	2.50	353.22	1.65	0.21	33.20	0.41	386.83	nr	425.51
762mm diameter in one pane	462.80	2.50	474.37	2.45	0.31	49.27	0.60	524.24	nr	576.67
Labours										
check throated edge	1.16	2.50	1.18	-	-	-	-	1.18	m	1.30
rebated and splayed bottom rail	1.16	2.50	1.18	-	-	-	-	1.18	m	1.30
rebated and beaded meeting stile	1.58	2.50	1.62	-	-	-	-	1.62	m	1.78
fitting and hanging casement or fanlight on butts (included elsewhere)										
38mm	-	-	-	1.00	0.12	20.02	-	20.02	nr	22.02
50mm	-	-	-	1.10	0.14	22.13	-	22.13	nr	24.35
fitting and hanging casement or fanlight on sash centres (included elsewhere)										
38mm	-	-	-	2.25	0.28	45.19	-	45.19	nr	49.70
50mm	-	-	-	2.45	0.31	49.27	-	49.27	nr	54.20

Labour hourly rates: (except Specialists) Craft Operatives 18.37 Labourer 13.76 Rates are national average prices. Refer to REGIONAL VARIATIONS for indicative levels of overall pricing in regions	MATERIALS			LABOUR				RATES		
	Del to Site	Waste	Material Cost	Craft Optve	Lab	Labour Cost	Sunds	Nett Rate	Unit	Gross rate (10%)
	£	%	£	Hrs	Hrs	£	£	£		£
TIMBER WINDOWS AND SHOP FRONTS (Cont'd)										
Purpose made windows in European Oak, wrought										
38mm moulded casements or fanlights										
in one pane....................	199.98	2.50	204.98	3.00	0.37	60.20	0.74	265.91	m²	292.50
divided into panes 0.10 - 0.50m².............	385.34	2.50	394.97	3.00	0.37	60.20	0.74	455.91	m²	501.50
divided into panes not exceeding 0.10m²...........	505.41	2.50	518.05	3.00	0.37	60.20	0.74	578.99	m²	636.88
50mm moulded casements or fanlights										
in one pane....................	358.26	2.50	367.22	3.20	0.40	64.29	0.78	432.29	m²	475.52
divided into panes 0.10 - 0.50m².............	405.07	2.50	415.20	3.20	0.40	64.29	0.78	480.26	m²	528.29
divided into panes not exceeding 0.10m²...........	451.87	2.50	463.17	3.20	0.40	64.29	0.78	528.24	m²	581.06
38mm moulded casements with semi-circular heads (measured square)										
in one pane....................	622.92	2.50	638.49	3.30	0.41	66.26	0.81	705.56	m²	776.12
divided into panes 0.10 - 0.50m².............	669.72	2.50	686.46	3.30	0.41	66.26	0.81	753.54	m²	828.89
divided into panes not exceeding 0.10m²...........	716.53	2.50	734.44	3.30	0.41	66.26	0.81	801.51	m²	881.66
50mm moulded casements with semi-circular heads (measured square)										
in one pane....................	716.53	2.50	734.44	3.50	0.44	70.35	0.86	805.64	m²	886.21
divided into panes 0.10 - 0.50m².............	763.33	2.50	782.41	3.50	0.44	70.35	0.86	853.62	m²	938.98
divided into panes not exceeding 0.10m²...........	810.14	2.50	830.39	3.50	0.44	70.35	0.86	901.59	m²	991.75
38mm bullseye casements										
457mm diameter in one pane..........	467.48	2.50	479.16	2.00	0.25	40.18	0.50	519.84	nr	571.82
762mm diameter in one pane..........	622.92	2.50	638.49	3.00	0.37	60.20	0.74	699.42	nr	769.37
50mm bullseye casements										
457mm diameter in one pane..........	498.68	2.50	511.15	2.20	0.27	44.13	0.54	555.82	nr	611.40
762mm diameter in one pane..........	669.72	2.50	686.46	3.30	0.41	66.26	0.81	753.54	nr	828.89
Labours										
check throated edge..........	1.62	2.50	1.66	-	-	-	-	1.66	m	1.82
rebated and splayed bottom rail..........	1.62	2.50	1.66	-	-	-	-	1.66	m	1.82
rebated and beaded meeting stile..........	2.31	2.50	2.37	-	-	-	-	2.37	m	2.61
fitting and hanging casement or fanlight on butts (included elsewhere)										
38mm..........	-	-	-	1.30	0.16	26.08	-	26.08	nr	28.69
50mm..........	-	-	-	1.45	0.18	29.11	-	29.11	nr	32.02
fitting and hanging casement or fanlight on sash centres (included elsewhere)										
38mm..........	-	-	-	3.00	0.37	60.20	-	60.20	nr	66.22
50mm..........	-	-	-	3.30	0.41	66.26	-	66.26	nr	72.89
Sash windows in softwood, wrought, preservative treated; pre-finished white by manufacturer; Bedding in cement mortar (1:3); pointing with Silicone standard mastic one side										
Magnet Trade Craftsman softwood sliding sash windows; factory double glazed; trickle vents; 168mm cill; weather stripping; pre-tensioned balances; chrome fittings; fully finished in white										
635 x 1050mm overall; CDH 0610..........	642.27	2.50	658.55	4.57	1.90	110.06	2.33	770.94	nr	848.03
635 x 1350mm overall; CDH 0613..........	678.93	2.50	696.01	5.04	2.07	121.03	2.33	819.53	nr	901.48
860 x 1350mm overall; CDH 0813..........	777.64	2.50	797.34	5.01	2.07	120.48	2.33	920.14	nr	1012.16
1085 x 1050mm overall; CDH 1010..........	784.54	2.50	804.41	5.01	2.07	120.48	2.33	927.21	nr	1019.93
1085 x 1350mm overall; CDH 1013..........	853.31	2.50	874.94	5.48	2.25	131.59	2.33	1008.85	nr	1109.74
1085 x 1650mm overall; CDH 1016; toughened glass..........	934.74	5.00	935.21	5.90	2.40	141.37	2.33	1078.90	nr	1186.79
Magnet Trade; factory glazed; Georgian; weatherstripping; pre-tensioned balances; chrome fittings; fully finished in white; fixing to masonry with galvanized steel cramps -4 nr, 25 x 3 x 150mm girth, flat section, holes -2										
860 x 1050mm overall; CDH0810GEO..........	840.65	2.50	861.90	3.49	1.30	81.96	2.33	946.19	nr	1040.81
860 x 1350m overall; CDH0813GEO..........	915.69	2.50	938.85	3.89	1.50	92.06	2.33	1033.23	nr	1136.56
1085 x 1350m overall; CDH1013GEO..........	1025.13	2.50	1051.05	4.38	1.70	103.82	2.33	1157.19	nr	1272.91
Sash windows; purpose made										
Note										
Notwithstanding the requirements of the NRM purpose made windows are shown here in square metres and excludes bedding and pointing										
Sash windows; purpose made; mainly in softwood, wrought										
Cased frames; 25mm inside and outside linings; 32mm pulley stiles and head; 10mm beads, back linings, etc; 76mm oak sunk weathered and throated sills; 38mm moulded sashes										
in one pane..........	337.67	2.50	346.11	2.50	0.31	50.19	0.62	396.92	m²	436.61
divided into panes 0.50 - 1.00m²..........	283.49	2.50	290.58	2.50	0.31	50.19	0.62	341.39	m²	375.53
divided into panes 0.10 - 0.50m²..........	310.46	2.50	318.23	2.50	0.31	50.19	0.62	369.03	m²	405.93
divided into panes not exceeding 0.10m²..........	351.23	2.50	360.01	2.50	0.31	50.19	0.62	410.82	m²	451.90
Cased frames; 25mm inside and outside linings; 32mm pulley stiles and head; 10mm beads, back linings, etc; 76mm oak sunk weathered and throated sills; 50mm moulded sashes										
in one pane..........	311.09	2.50	318.87	2.75	0.34	55.20	0.68	374.74	m²	412.21
divided into panes 0.50 - 1.00m²..........	323.64	2.50	331.73	2.75	0.34	55.20	0.68	387.60	m²	426.36
divided into panes 0.10 - 0.50m²..........	351.23	2.50	360.01	2.75	0.34	55.20	0.68	415.88	m²	457.47
divided into panes not exceeding 0.10m²..........	391.37	2.50	401.16	2.75	0.34	55.20	0.68	457.03	m²	502.73
extra; windows in three lights with boxed mullions..........	48.00	2.50	49.20	0.75	0.09	15.02	0.18	64.40	nr	70.84
extra; windows with moulded horns..........	54.00	2.50	55.35	-	-	-		55.35	nr	60.88
extra; deep bottom rails and draught beads..........	6.50	2.50	6.66	-	-	-		6.66	m	7.33

Labour hourly rates: (except Specialists) Craft Operatives 18.37 Labourer 13.76 Rates are national average prices. Refer to REGIONAL VARIATIONS for indicative levels of overall pricing in regions	MATERIALS			LABOUR				RATES		
	Del to Site £	Waste %	Material Cost £	Craft Optve Hrs	Lab Hrs	Labour Cost £	Sunds £	Nett Rate £	Unit	Gross rate (10%) £
TIMBER WINDOWS AND SHOP FRONTS (Cont'd)										
Sash windows; purpose made; mainly in softwood, wrought (Cont'd)										
Labours										
throats...............................	0.70	2.50	0.72	-	-	-	-	0.72	m	0.79
grooves for jamb linings...................	0.82	2.50	0.84	-	-	-	-	0.84	m	0.92
fitting and hanging sashes in double hung windows; providing brass faced iron pulleys, best flax cords and iron weights										
38mm thick weighing 14 lbs per sash	13.39	2.50	13.72	1.40	0.17	28.06	-	41.78	nr	45.96
50mm thick weighing 20 lbs per sash	17.35	2.50	17.78	1.40	0.17	28.06	-	45.84	nr	50.42
Note										
for small frames and sashes, i.e. 1.25m² and under, add 20% to the foregoing prices										
Sash windows; purpose made; in Afrormosia, wrought										
Cased frames; 25mm inside and outside linings; 32mm pulley stiles and head; 10mm beads, back linings, etc; 76mm oak sunk weathered and throated sills; 38mm moulded sashes										
in one pane.............................	2256.80	2.50	2313.22	4.40	0.55	88.40	1.08	2402.70	m²	2642.97
divided into panes 0.50 - 1.00m²	2780.38	2.50	2849.89	4.40	0.55	88.40	1.08	2939.36	m²	3233.30
divided into panes 0.10 - 0.50m²	3425.43	2.50	3511.06	4.40	0.55	88.40	1.08	3600.54	m²	3960.59
divided into panes not exceeding 0.10m²	4220.12	2.50	4325.63	4.40	0.55	88.40	1.08	4415.10	m²	4856.61
Cased frames; 25mm inside and outside linings; 32mm pulley stiles and head; 10mm beads, back linings, etc; 76mm oak sunk weathered and throated sills; 50mm moulded sashes										
in one pane.............................	3033.14	2.50	3108.97	4.80	0.60	96.43	1.17	3206.57	m²	3527.23
divided into panes 0.50 - 1.00m²	3736.83	2.50	3830.25	4.80	0.60	96.43	1.17	3927.85	m²	4320.64
divided into panes 0.10 - 0.50m²	4603.77	2.50	4718.87	4.80	0.60	96.43	1.17	4816.47	m²	5298.11
divided into panes not exceeding 0.10m²	5671.85	2.50	5813.64	4.80	0.60	96.43	1.17	5911.24	m²	6502.37
extra; windows in three lights with boxed mullions	99.68	2.50	102.17	1.30	0.16	26.08	0.31	128.57	nr	141.43
extra; windows with moulded horns............	68.32	2.50	70.03	-	-	-	-	70.03	nr	77.03
extra; deep bottom rails and draught beads	13.44	2.50	13.78	-	-	-	-	13.78	m	15.15
Labours										
throats...............................	2.24	2.50	2.30	-	-	-	-	2.30	m	2.53
grooves for jamb linings...................	2.24	2.50	2.30	-	-	-	-	2.30	m	2.53
fitting and hanging sashes in double hung windows; providing brass faced iron pulleys, best flax cords and iron weights										
38mm thick weighing 14 lbs per sash	13.39	2.50	13.72	2.50	0.31	50.19	-	63.91	nr	70.30
50mm thick weighing 20 lbs per sash	17.35	2.50	17.78	2.50	0.31	50.19	-	67.97	nr	74.77
Note										
for small frames and sashes, i.e. 1.25m² and under, add 20% to the foregoing prices										
Sash windows; purpose made; in European Oak, wrought										
Cased frames; 25mm inside and outside linings; 32mm pulley stiles and head; 10mm beads, back linings, etc; 76mm oak sunk weathered and throated sills; 38mm moulded sashes										
in one pane.............................	711.20	2.50	728.98	5.00	0.62	100.38	1.22	830.58	m²	913.63
divided into panes 0.50 - 1.00m²	876.20	2.50	898.10	5.00	0.62	100.38	1.22	999.70	m²	1099.67
divided into panes 0.10 - 0.50m²	1079.48	2.50	1106.46	5.00	0.62	100.38	1.22	1208.06	m²	1328.87
divided into panes not exceeding 0.10m²	1329.91	2.50	1363.16	5.00	0.62	100.38	1.22	1464.76	m²	1611.23
Cased frames; 25mm inside and outside linings; 32mm pulley stiles and head; 10mm beads, back linings, etc; 76mm oak sunk weathered and throated sills; 50mm moulded sashes										
in one pane.............................	955.85	2.50	979.75	5.50	0.69	110.53	1.35	1091.63	m²	1200.79
divided into panes 0.50 - 1.00m²	1177.61	2.50	1207.05	5.50	0.69	110.53	1.35	1318.93	m²	1450.82
divided into panes 0.10 - 0.50m²	1450.82	2.50	1487.09	5.50	0.69	110.53	1.35	1598.97	m²	1758.86
divided into panes not exceeding 0.10m²	1787.41	2.50	1832.09	5.50	0.69	110.53	1.35	1943.97	m²	2138.37
extra; windows in three lights with boxed mullions	184.80	2.50	189.42	1.55	0.19	31.09	0.38	220.88	nr	242.97
extra; windows with moulded horns............	60.48	2.50	61.99	-	-	-	-	61.99	nr	68.19
extra; deep bottom rails and draught beads	19.60	2.50	20.09	-	-	-	-	20.09	m	22.10
Labours										
throats...............................	2.24	2.50	2.30	-	-	-	-	2.30	m	2.53
grooves for jamb linings...................	2.24	2.50	2.30	-	-	-	-	2.30	m	2.53
fitting and hanging sashes in double hung windows; providing brass faced iron pulleys, best flax cords and iron weights										
38mm thick weighing 14 lbs per sash	13.39	2.50	13.72	2.80	0.35	56.25	-	69.97	nr	76.97
50mm thick weighing 20 lbs per sash	17.35	2.50	17.78	2.80	0.35	56.25	-	74.03	nr	81.44
Note										
for small frames and sashes, i.e. 1.25m² and under, add 20% to the foregoing prices										
Note										
Bedding in cement mortar (1:3); pointing with Silicone standard mastic one side' is included with standard windows as noted above; adjustment will need to be made for two part polysuphide pointing and bedding and pointing to purpose made frames										
Bedding and pointing frames										
Bedding in cement mortar (1:3); pointing with Silicone standard mastic one side										
wood frames............................	1.03	5.00	1.11	0.20	-	3.67	-	4.79	m	5.27
Bedding in cement mortar (1:3); pointing with coloured two part polysulphide mastic one side										
wood frames............................	2.55	5.00	2.79	0.20	-	3.67	-	6.46	m	7.11

WINDOWS, SCREENS AND LIGHTS

Labour hourly rates: (except Specialists) Craft Operatives 18.37 Labourer 13.76 Rates are national average prices. Refer to REGIONAL VARIATIONS for indicative levels of overall pricing in regions	MATERIALS			LABOUR				RATES		
	Del to Site £	Waste %	Material Cost £	Craft Optve Hrs	Lab Hrs	Labour Cost £	Sunds £	Nett Rate £	Unit	Gross rate (10%) £
TIMBER ROOFLIGHTS, SKYLIGHTS AND LANTERNLIGHTS										
Skylights in softwood, wrought										
Chamfered, straight bar										
38mm	106.27	2.50	108.93	1.50	0.19	30.17	0.36	139.46	m²	153.40
50mm	123.00	2.50	126.07	1.75	0.22	35.17	0.42	161.67	m²	177.84
63mm	129.56	2.50	132.80	2.00	0.25	40.18	0.50	173.47	m²	190.82
Moulded, straight bar										
38mm	111.93	2.50	114.73	1.50	0.19	30.17	0.36	145.26	m²	159.78
50mm	128.64	2.50	131.85	1.75	0.22	35.17	0.42	167.45	m²	184.19
63mm	145.55	2.50	149.19	2.00	0.25	40.18	0.50	189.86	m²	208.85
Skylights in Oak, wrought										
Chamfered, straight bar										
38mm	296.23	2.50	303.63	2.30	0.29	46.24	0.56	350.43	m²	385.47
50mm	343.38	2.50	351.96	2.70	0.34	54.28	0.66	406.90	m²	447.59
63mm	389.50	2.50	399.24	3.10	0.39	62.31	0.75	462.30	m²	508.53
Moulded, straight bar										
38mm	312.62	2.50	320.44	2.30	0.29	46.24	0.56	367.24	m²	403.96
50mm	358.75	2.50	367.72	2.70	0.34	54.28	0.66	422.66	m²	464.92
63mm	405.90	2.50	416.05	3.10	0.39	62.31	0.75	479.11	m²	527.02
Skylight kerbs in softwood, wrought										
Kerbs; dovetailed at angles										
38 x 225mm....................	14.17	2.50	14.52	0.31	0.04	6.25	0.08	20.84	m	22.92
50 x 225mm....................	16.97	2.50	17.40	0.35	0.04	6.98	0.08	24.45	m	26.90
38 x 225mm; chamfers -1 nr....	14.85	2.50	15.22	0.31	0.04	6.25	0.08	21.54	m	23.70
50 x 225mm; chamfers -1 nr....	17.68	2.50	18.12	0.35	0.04	6.98	0.08	25.18	m	27.70
Kerbs; in two thicknesses to circular skylights										
38 x 225mm....................	42.41	2.50	43.47	0.47	0.06	9.46	0.12	53.05	m	58.36
50 x 225mm....................	50.84	2.50	52.11	0.53	0.07	10.70	0.12	62.93	m	69.22
38 x 225mm; chamfers -1 nr....	43.87	2.50	44.97	0.47	0.06	9.46	0.12	54.55	m	60.00
50 x 225mm; chamfers -1 nr....	52.33	2.50	53.63	0.53	0.07	10.70	0.12	64.45	m	70.90
Skylight kerbs in Oak, wrought										
Kerbs; dovetailed at angles										
38 x 225mm....................	64.61	2.50	66.23	0.62	0.08	12.49	0.15	78.87	m	86.76
50 x 225mm....................	77.54	2.50	79.47	0.69	0.08	13.78	0.17	93.42	m	102.76
38 x 225mm; chamfers -1 nr....	67.84	2.50	69.53	0.62	0.08	12.49	0.15	82.18	m	90.39
50 x 225mm; chamfers -1 nr....	80.75	2.50	82.77	0.69	0.08	13.78	0.17	96.71	m	106.38
Kerbs; in two thicknesses to circular skylights										
38 x 225mm....................	193.84	2.50	198.69	0.94	0.12	18.92	0.22	217.83	m	239.61
50 x 225mm....................	232.19	2.50	237.99	1.06	0.13	21.26	0.25	259.51	m	285.46
38 x 225mm; chamfers -1 nr....	200.67	2.50	205.69	0.94	0.12	18.92	0.22	224.83	m	247.31
50 x 225mm; chamfers -1 nr....	238.49	2.50	244.45	1.06	0.13	21.26	0.25	265.97	m	292.57
Roof windows in Nordic red pine, wrought, treated, including flashings for slate up to 8mm thick or tiles up to 45mm in profile										
Velux roof windows GGL/GPL 3066 Pine Range, factory glazed clear 'Extra Low Energy 66' triple glazed sealed unit; laminated inner; enhanced strength outer pane; fixing to timber with screws; dressing flashings										
550 x 780mm overall, CK02 GGL Centre Pivot	317.54	2.50	325.48	4.50	2.25	113.63	-	439.10	nr	483.01
550 x 980mm overall, CK04 GGL Centre Pivot	327.27	2.50	335.45	5.25	2.62	132.49	-	467.95	nr	514.74
660 x 1180mm overall, FK06 GGL Centre Pivot	422.68	2.50	433.25	5.25	2.62	132.49	-	565.74	nr	622.31
780 x 980mm overall, MK04 GGL Centre Pivot....................	347.09	2.50	355.77	5.00	2.50	126.25	-	482.02	nr	530.22
780 x 1180mm overall, MK06 GGL Centre Pivot....................	373.30	2.50	382.63	5.00	2.50	126.25	-	508.88	nr	559.77
780 x 1400mm overall, MK08 GGL Centre Pivot....................	400.35	2.50	410.36	5.50	2.75	138.88	-	549.23	nr	604.16
940 x 1600mm overall, PK10 GGL Centre Pivot....................	462.05	2.50	473.60	5.50	2.75	138.88	-	612.48	nr	673.72
1140 x 1180mm overall, SK06 GGL Centre Pivot	452.14	2.50	463.44	6.00	3.00	151.50	-	614.94	nr	676.44
1340 x 980mm overall, UK04 GGL Centre Pivot	456.44	2.50	467.85	6.00	3.00	151.50	-	619.35	nr	681.29
550 x 980mm overall, CK04 GPL Top Hung	392.48	2.50	402.29	5.75	2.87	145.12	-	547.41	nr	602.15
660 x 1180mm overall, FK06 GPL Top Hung....................	436.02	2.50	446.92	5.75	2.87	145.12	-	592.04	nr	651.24
780 x 980mm overall, MK04 GPL Top Hung....................	447.24	2.50	458.42	5.75	2.87	145.12	-	603.54	nr	663.89
780 x 1180mm overall, MK06 GPL Top Hung....................	453.35	2.50	464.68	6.50	3.25	164.12	-	628.81	nr	691.69
780 x 1400mm overall, MK08 GPL Top Hung....................	487.80	2.50	499.99	6.50	3.25	164.12	-	664.12	nr	730.53
1140 x 1180mm overall, SK06 GPL Top Hung....................	558.86	2.50	572.83	6.50	3.25	164.12	-	736.96	nr	810.65
Velux roof windows GGL/GPL 3070 Pine Range, factory glazed clear 'Protec-star' double glazed sealed unit; 6.4mm laminated inner; toughened outer pane; fixing to timber with screws; dressing flashings										
550 x 780mm overall, CK02 GGL Centre Pivot	214.16	2.50	219.51	4.50	2.25	113.63	-	333.14	nr	366.45
550 x 980mm overall, CK04 GGL Centre Pivot	224.15	2.50	229.75	5.25	2.62	132.49	-	362.25	nr	398.47
660 x 1180mm overall, FK06 GGL Centre Pivot	259.15	2.50	265.63	5.25	2.62	132.49	-	398.12	nr	437.93
780 x 980mm overall, MK04 GGL Centre Pivot....................	244.99	2.50	251.11	5.00	2.50	126.25	-	377.36	nr	415.10
780 x 1180mm overall, MK06 GGL Centre Pivot....................	272.49	2.50	279.30	5.00	2.50	126.25	-	405.55	nr	446.11
780 x 1400mm overall, MK08 GGL Centre Pivot....................	299.98	2.50	307.48	5.50	2.75	138.88	-	446.35	nr	490.99
940 x 1600mm overall, PK10 GGL Centre Pivot....................	364.99	2.50	374.11	5.50	2.75	138.88	-	512.99	nr	564.29
1140 x 1180mm overall, SK06 GGL Centre Pivot	354.16	2.50	363.01	6.00	3.00	151.50	-	514.51	nr	565.97
1340 x 980mm overall, UK04 GGL Centre Pivot	357.82	2.50	366.77	6.00	3.00	151.50	-	518.27	nr	570.09
550 x 980mm overall, CK04 GPL Top Hung	292.49	2.50	299.80	5.75	2.87	145.12	-	444.92	nr	489.41
660 x 1180mm overall, FK06 GPL Top Hung....................	338.32	2.50	346.78	5.75	2.87	145.12	-	491.90	nr	541.09
780 x 980mm overall, MK04 GPL Top Hung....................	319.15	2.50	327.13	5.75	2.87	145.12	-	472.25	nr	519.47
780 x 1180mm overall, MK06 GPL Top Hung....................	355.82	2.50	364.72	6.50	3.25	164.12	-	528.84	nr	581.72
780 x 1400mm overall, MK08 GPL Top Hung....................	391.65	2.50	401.44	6.50	3.25	164.12	-	565.57	nr	622.12
1140 x 1180mm overall, SK06 GPL Top Hung....................	465.82	2.50	477.47	6.50	3.25	164.12	-	641.59	nr	705.75

Labour hourly rates: (except Specialists) Craft Operatives 18.37 Labourer 13.76 Rates are national average prices. Refer to REGIONAL VARIATIONS for indicative levels of overall pricing in regions	MATERIALS			LABOUR				RATES		
	Del to Site	Waste	Material Cost	Craft Optve	Lab	Labour Cost	Sunds	Nett Rate	Unit	Gross rate (10%)
	£	%	£	Hrs	Hrs	£	£	£		£

TIMBER ROOFLIGHTS, SKYLIGHTS AND LANTERNLIGHTS (Cont'd)

Roof windows in Nordic red pine, wrought, treated, including flashings for slate up to 8mm thick or tiles up to 45mm in profile (Cont'd)

Velux roof windows GGL 3060 Pine Range, factory glazed clear '5-star' double glazed sealed unit; 6.4mm laminated inner; toughened outer pane coated; fixing to timber with screws; dressing flashings

	Del	Waste	Mat	Craft	Lab	Lab Cost	Sunds	Nett	Unit	Gross
550 x 780mm overall, CK02 GGL Centre Pivot	260.82	2.50	267.34	4.50	2.25	113.63	-	380.97	nr	419.06
550 x 980mm overall, CK04 GGL Centre Pivot	270.82	2.50	277.59	5.25	2.62	132.49	-	410.08	nr	451.09
660 x 1180mm overall, FK06 GGL Centre Pivot	305.82	2.50	313.47	5.25	2.62	132.49	-	445.96	nr	490.56
780 x 980mm overall, MK04 GGL Centre Pivot	291.65	2.50	298.94	5.00	2.50	126.25	-	425.19	nr	467.71
780 x 1180mm overall, MK06 GGL Centre Pivot	319.15	2.50	327.13	5.00	2.50	126.25	-	453.38	nr	498.72
780 x 1400mm overall, MK08 GGL Centre Pivot	346.65	2.50	355.32	5.50	2.75	138.88	-	494.19	nr	543.61
940 x 1600mm overall, PK10 GGL Centre Pivot	410.82	2.50	421.09	5.50	2.75	138.88	-	559.97	nr	615.96
1140 x 1180mm overall, SK06 GGL Centre Pivot	400.82	2.50	410.84	6.00	3.00	151.50	-	562.34	nr	618.57
1340 x 980mm overall, UK04 GGL Centre Pivot	404.48	2.50	414.59	6.00	3.00	151.50	-	566.09	nr	622.70

Velux roof windows GGL 307021U Pine Electric Range, factory glazed clear 'Protect-star' double glazed sealed unit; 6.4mm laminated inner; toughened outer pane; fixing to timber with screws; dressing flashings

550 x 780mm overall, CK02 GGL Centre Pivot	428.71	2.50	439.43	4.60	2.30	116.15	-	555.58	nr	611.14
550 x 980mm overall, CK04 GGL Centre Pivot	438.43	2.50	449.39	5.35	2.67	135.02	-	584.41	nr	642.85
660 x 1180mm overall, FK06 GGL Centre Pivot	471.60	2.50	483.39	5.35	2.67	135.02	-	618.41	nr	680.25
780 x 980mm overall, MK04 GGL Centre Pivot	458.26	2.50	469.72	5.35	2.55	133.37	-	603.08	nr	663.39
780 x 1180mm overall, MK06 GGL Centre Pivot	484.47	2.50	496.58	5.10	2.55	128.78	-	625.36	nr	687.89
780 x 1400mm overall, MK08 GGL Centre Pivot	511.52	2.50	524.31	5.60	2.80	141.40	-	665.71	nr	732.28
940 x 1600mm overall, PK10 GGL Centre Pivot	573.22	2.50	587.55	5.60	2.80	141.40	-	728.95	nr	801.85
1140 x 1180mm overall, SK06 GGL Centre Pivot	563.31	2.50	577.39	6.10	3.05	154.03	-	731.42	nr	804.56
1340 x 980mm overall, UK04 GGL Centre Pivot	567.61	2.50	581.80	6.10	3.05	154.03	-	735.83	nr	809.41

Roof windows in pine with white polyurethane finish, including flashings for slate up to 8mm thick or tiles up to 45mm in profile

Velux roof windows GGU/GPU 0066 White Polyurethane Range, factory glazed clear 'Extra Low Energy 66' triple glazed sealed unit; laminated inner; enhanced strength outer pane; fixing to timber with screws; dressing flashings

550 x 780mm overall, CK02 GGU Centre Pivot	353.12	2.50	361.95	4.50	2.25	113.63	-	475.57	nr	523.13
550 x 980mm overall, CK04 GGU Centre Pivot	364.32	2.50	373.43	5.25	2.62	132.49	-	505.92	nr	556.51
660 x 1180mm overall, FK06 GGU Centre Pivot	403.41	2.50	413.50	5.25	2.62	132.49	-	545.99	nr	600.59
780 x 980mm overall, MK04 GGU Centre Pivot	388.59	2.50	398.30	5.00	2.50	126.25	-	524.55	nr	577.01
780 x 1180mm overall, MK06 GGU Centre Pivot	420.74	2.50	431.26	5.00	2.50	126.25	-	557.51	nr	613.26
780 x 1400mm overall, MK08 GGU Centre Pivot	450.75	2.50	462.02	5.50	2.75	138.88	-	600.89	nr	660.98
1140 x 1180mm overall, SK06 GGU Centre Pivot	514.39	2.50	527.25	6.00	3.00	151.50	-	678.75	nr	746.62
660 x 1180mm overall, FK06 GPU Top Hung	480.49	2.50	492.50	5.75	2.87	145.12	-	637.62	nr	701.38
780 x 1180mm overall, MK06 GPU Top Hung	500.78	2.50	513.30	6.50	3.25	164.12	-	677.42	nr	745.17
780 x 1400mm overall, MK08 GPU Top Hung	539.68	2.50	553.17	6.50	3.25	164.12	-	717.30	nr	789.03
1140 x 1180mm overall, SK06 GPU Top Hung	619.63	2.50	635.12	6.50	3.25	164.12	-	799.25	nr	879.17

Velux roof windows GGU 007030 White Polyurethane Range, Solar powered via solar technology with no requirement for an external power supply, factory glazed clear 'Protec-star' double glazed sealed unit; 6.4mm laminated inner; toughened outer pane; fixing to timber with screws; dressing flashings

550 x 780mm overall, CK02 GGU Centre Pivot	489.48	2.50	501.72	4.50	2.25	113.63	-	615.34	nr	676.88
550 x 980mm overall, CK04 GGU Centre Pivot	500.69	2.50	513.21	5.25	2.62	132.49	-	645.70	nr	710.27
660 x 1180mm overall, FK06 GGU Centre Pivot	539.78	2.50	553.27	5.25	2.62	132.49	-	685.77	nr	754.34
780 x 980mm overall, MK04 GGU Centre Pivot	524.96	2.50	538.08	5.00	2.50	126.25	-	664.33	nr	730.77
780 x 1180mm overall, MK06 GGU Centre Pivot	557.10	2.50	571.03	5.00	2.50	126.25	-	697.28	nr	767.01
780 x 1400mm overall, MK08 GGU Centre Pivot	587.11	2.50	601.79	5.50	2.75	138.88	-	740.66	nr	814.73
1140 x 1180mm overall, SK06 GGU Centre Pivot	650.76	2.50	667.03	6.00	3.00	151.50	-	818.53	nr	900.38

Velux roof windows GGU/GPU 0034 White Polyurethane Range, factory glazed obscured double glazed sealed unit; obscured inner; toughened outer pane coated; fixing to timber with screws; dressing flashings

550 x 780mm overall, CK02 GGU Centre Pivot	271.59	2.50	278.38	4.50	2.25	113.63	-	392.00	nr	431.21
550 x 980mm overall, CK04 GGU Centre Pivot	282.80	2.50	289.87	5.25	2.62	132.49	-	422.36	nr	464.60
660 x 1180mm overall, FK06 GGU Centre Pivot	321.89	2.50	329.94	5.25	2.62	132.49	-	462.43	nr	508.67
780 x 980mm overall, MK04 GGU Centre Pivot	307.07	2.50	314.75	5.00	2.50	126.25	-	441.00	nr	485.10
780 x 1180mm overall, MK06 GGU Centre Pivot	339.21	2.50	347.69	5.00	2.50	126.25	-	473.94	nr	521.33
780 x 1400mm overall, MK08 GGU Centre Pivot	369.22	2.50	378.45	5.50	2.75	138.88	-	517.33	nr	569.06
1140 x 1180mm overall, SK06 GGU Centre Pivot	432.87	2.50	443.69	6.00	3.00	151.50	-	595.19	nr	654.71
660 x 1180mm overall, FK06 GPU Top Hung	398.97	2.50	408.94	5.72	2.87	144.57	-	553.51	nr	608.86
780 x 1180mm overall, MK06 GPU Top Hung	419.25	2.50	429.73	6.50	3.25	164.12	-	593.86	nr	653.24
780 x 1400mm overall, MK08 GPU Top Hung	458.16	2.50	469.61	6.50	3.25	164.12	-	633.74	nr	697.11
1140 x 1180mm overall, SK06 GPU Top Hung	538.11	2.50	551.56	6.50	3.25	164.12	-	715.69	nr	787.26

Velux roof windows GGU 007021U Integra White Polyurethane Electric Range, factory glazed clear 'Protect-star' double glazed sealed unit; 6.4mm laminated inner; toughened outer pane; fixing to timber with screws; dressing flashings

550 x 780mm overall, CK02 GGU Centre Pivot	464.28	2.50	475.89	4.60	2.30	116.15	-	592.04	nr	651.24
550 x 980mm overall, CK04 GGU Centre Pivot	475.49	2.50	487.38	5.35	2.67	135.02	-	622.40	nr	684.64
660 x 1180mm overall, FK06 GGU Centre Pivot	514.58	2.50	527.44	5.35	2.67	135.02	-	662.46	nr	728.71
780 x 980mm overall, MK04 GGU Centre Pivot	499.76	2.50	512.25	5.10	2.55	128.78	-	641.03	nr	705.13
780 x 1180mm overall, MK06 GGU Centre Pivot	531.90	2.50	545.20	5.10	2.55	128.78	-	673.97	nr	741.37
780 x 1400mm overall, MK08 GGU Centre Pivot	561.91	2.50	575.96	5.60	2.80	141.40	-	717.36	nr	789.09
1140 x 1180mm overall, SK06 GGU Centre Pivot	625.56	2.50	641.20	6.10	3.05	154.03	-	795.22	nr	874.75

WINDOWS, SCREENS AND LIGHTS

Labour hourly rates: (except Specialists) Craft Operatives 18.37 Labourer 13.76 Rates are national average prices. Refer to REGIONAL VARIATIONS for indicative levels of overall pricing in regions	MATERIALS			LABOUR				RATES		
	Del to Site	Waste	Material Cost	Craft Optve	Lab	Labour Cost	Sunds	Nett Rate	Unit	Gross rate (10%)
	£	%	£	Hrs	Hrs	£	£	£		£
METAL WINDOWS AND SHOP FRONTS										
Windows in galvanized steel; Bedding in cement mortar (1:3); pointing with Silicone standard mastic one side										
Crittall Windows Ltd. Duralife Homelight; weatherstripping; fixing to masonry with lugs										
fixed lights										
508 x 292mm, NG5	35.50	2.50	36.48	0.93	0.25	20.52	-	57.01	nr	62.71
508 x 923mm, NC5	49.36	2.50	50.77	1.55	0.35	33.33	-	84.09	nr	92.50
508 x 1218mm, ND5	56.54	2.50	58.16	1.92	0.41	40.91	-	99.07	nr	108.98
997 x 628mm, NE13	64.62	2.50	66.43	1.77	0.40	38.02	-	104.44	nr	114.89
997 x 923mm, NC13	75.43	2.50	77.55	2.01	0.44	42.94	-	120.49	nr	132.54
997 x 1218mm, ND13	83.70	2.50	86.05	2.39	0.51	50.85	-	136.90	nr	150.59
1486 x 628mm, NE14	77.69	2.50	79.88	2.35	0.53	50.39	-	130.27	nr	143.30
1486 x 923mm, NC14	87.39	2.50	89.87	2.59	0.58	55.63	-	145.50	nr	160.05
1486 x 1218mm, ND14	98.85	2.50	101.65	2.98	0.66	63.86	-	165.51	nr	182.06
top hung lights										
508 x 292mm, NG1	115.24	2.50	118.22	0.93	0.25	20.52	-	138.74	nr	152.62
508 x 457mm, NH1	125.82	2.50	129.08	1.13	0.25	24.12	-	153.21	nr	168.53
508 x 628mm, NE1	135.56	2.50	139.08	1.31	0.29	28.13	-	167.21	nr	183.93
997 x 628mm, NE13E	190.25	2.50	195.20	1.77	0.32	36.92	-	232.12	nr	255.33
997 x 923mm, NC13C	205.11	2.50	210.47	2.01	0.40	42.39	-	252.86	nr	278.15
bottom hung lights										
508 x 628mm, NL1	176.01	2.50	180.54	1.31	0.32	28.54	-	209.08	nr	229.99
side hung lights										
508 x 628mm, NES1	147.40	2.50	151.22	1.31	0.32	28.54	-	179.77	nr	197.74
508 x 923mm, NC1	158.82	2.50	162.96	1.55	0.35	33.33	-	196.29	nr	215.92
508 x 1067mm, NC01	168.47	2.50	172.87	1.73	0.38	37.01	-	209.88	nr	230.87
508 x 1218mm, ND1	179.42	2.50	184.11	1.92	0.41	40.91	-	225.02	nr	247.52
mixed lights										
279 x 923mm, NC6F	121.97	2.50	125.16	1.46	0.35	31.65	-	156.82	nr	172.50
508 x 923mm, NC5F	140.85	2.50	144.54	1.55	0.35	33.33	-	177.87	nr	195.66
508 x 1067mm, NC05F	144.43	2.50	148.23	1.73	0.38	37.01	-	185.24	nr	203.76
997 x 292mm, NG2	142.28	2.50	145.99	1.37	0.33	29.63	-	175.62	nr	193.18
997 x 457mm, NH2	155.62	2.50	159.69	1.56	0.37	33.79	-	193.47	nr	212.82
997 x 628mm, NE2	169.16	2.50	173.58	1.77	0.40	38.02	-	211.60	nr	232.76
997 x 628mm, NES2	200.09	2.50	205.29	1.77	0.40	38.02	-	243.31	nr	267.64
997 x 923mm, NC2	218.82	2.50	224.52	2.01	0.44	42.94	-	267.46	nr	294.21
997 x 923mm, NC2F	289.16	2.50	296.62	2.01	0.44	42.94	-	339.56	nr	373.52
997 x 1067mm, NC02	237.93	2.50	244.13	2.20	0.47	46.81	-	290.93	nr	320.03
997 x 1067mm, NC02F	315.28	2.50	323.41	2.20	0.47	46.81	-	370.22	nr	407.24
997 x 1218mm, ND2	253.53	2.50	260.13	2.39	0.51	50.85	-	310.98	nr	342.08
997 x 1218mm, ND2F	319.03	2.50	327.27	2.39	0.51	50.85	-	378.12	nr	415.93
997 x 1513mm, NDV2FSB	367.59	2.50	377.08	2.63	0.58	56.37	-	433.45	nr	476.79
1486 x 628mm, NE3	213.22	2.50	218.80	2.35	0.51	50.11	-	268.91	nr	295.81
1486 x 923mm, NC4	360.28	2.50	369.58	2.59	0.58	55.63	-	425.21	nr	467.73
1486 x 923mm, NC4F	439.91	2.50	451.19	2.59	0.58	55.63	-	506.82	nr	557.51
1486 x 1067mm, NC04	387.96	2.50	397.96	2.65	0.58	56.68	-	454.64	nr	500.11
1486 x 1067mm, NC04F	465.00	2.50	476.93	2.79	0.61	59.66	-	536.59	nr	590.25
1486 x 1218mm, ND4	413.90	2.50	424.57	2.98	0.66	63.86	-	488.43	nr	537.27
1486 x 1218mm, ND4F	491.02	2.50	503.62	2.98	0.66	63.86	-	567.48	nr	624.23
1486 x 1513mm, NDV4FSB	567.49	2.50	582.03	3.24	0.69	69.01	-	651.04	nr	716.15
1994 x 923mm, NC11F	477.72	2.50	490.01	3.18	0.69	67.86	-	557.86	nr	613.65
1994 x 1218mm, ND11F	517.11	2.50	530.42	3.59	0.81	77.00	-	607.43	nr	668.17
1994 x 1513mm, NDV11FSB	603.87	2.50	619.38	3.87	0.86	82.98	-	702.36	nr	772.60
Extra for										
mullions										
292mm high	19.70	2.50	20.20	0.20	0.10	5.05	-	25.25	nr	27.77
457mm high	27.56	2.50	28.25	0.30	0.15	7.58	-	35.82	nr	39.40
628mm high	35.59	2.50	36.48	0.40	0.20	10.10	-	46.58	nr	51.24
923mm high	49.48	2.50	50.72	0.50	0.25	12.62	-	63.34	nr	69.68
1067mm high	57.30	2.50	58.73	0.60	0.30	15.15	-	73.88	nr	81.27
1218mm high	64.48	2.50	66.09	0.80	0.40	20.20	-	86.29	nr	94.92
1513mm high	78.35	2.50	80.31	1.00	0.50	25.25	-	105.56	nr	116.12
transoms										
279mm wide	20.44	2.50	20.95	0.20	0.10	5.05	-	26.00	nr	28.60
508mm wide	31.50	2.50	32.28	0.30	0.15	7.58	-	39.86	nr	43.84
997mm wide	82.13	2.50	84.19	0.60	0.30	15.15	-	99.34	nr	109.27
1143mm wide	91.23	2.50	93.51	1.00	0.50	25.25	-	118.76	nr	130.63
Controlair Ventilators, permanent, mill finish										
508mm wide	52.61	2.50	53.92	-	-	-	-	53.92	nr	59.31
997mm wide	70.65	2.50	72.42	-	-	-	-	72.42	nr	79.66
1486mm wide	159.15	2.50	163.13	-	-	-	-	163.13	nr	179.45
Controlair Ventilators, permanent, mill finish, flyscreen										
508mm wide	62.53	2.50	64.09	-	-	-	-	64.09	nr	70.50
997mm wide	89.08	2.50	91.31	-	-	-	-	91.31	nr	100.44
1486mm wide	184.22	2.50	188.83	-	-	-	-	188.83	nr	207.71
Controlair Ventilators, adjustable										
508mm wide 1-LT	62.70	2.50	64.27	-	-	-	-	64.27	nr	70.70
997mm wide 2-LT	79.28	2.50	81.26	-	-	-	-	81.26	nr	89.39
Controlair Ventilators, adjustable, flyscreen										
508mm wide 1-LT	72.24	2.50	74.04	-	-	-	-	74.04	nr	81.45
997mm wide 2-LT	97.71	2.50	100.15	-	-	-	-	100.15	nr	110.17
locks with Parkes locking handles for										
side hung lights	65.06	2.50	66.68	-	-	-	-	66.68	nr	73.35
horizontally pivoted lights	65.06	2.50	66.68	-	-	-	-	66.68	nr	73.35

Labour hourly rates: (except Specialists) Craft Operatives 18.37 Labourer 13.76 Rates are national average prices. Refer to REGIONAL VARIATIONS for indicative levels of overall pricing in regions	MATERIALS			LABOUR				RATES		
	Del to Site	Waste	Material Cost	Craft Optve	Lab	Labour Cost	Sunds	Nett Rate	Unit	Gross rate (10%)
	£	%	£	Hrs	Hrs	£	£	£		£
METAL WINDOWS AND SHOP FRONTS (Cont'd)										
Windows in galvanized steel; polyester powder coated; white matt; Bedding in cement mortar (1:3); pointing with Silicone standard mastic one side										
Crittall Windows Ltd. Duralife Homelight; weatherstripping; fixing to masonry with lugs										
fixed lights										
508 x 292mm, NG5	41.38	2.50	42.51	0.93	0.25	20.52	-	63.04	nr	69.34
508 x 923mm, NC5	57.39	2.50	58.99	1.55	0.35	33.33	-	92.32	nr	101.55
508 x 1218mm, ND5	65.73	2.50	67.58	1.92	0.41	40.91	-	108.49	nr	119.34
997 x 628mm, NE13	75.27	2.50	77.35	1.77	0.40	38.02	-	115.37	nr	126.90
997 x 923mm, NC13	88.15	2.50	90.58	2.01	0.44	42.94	-	133.52	nr	146.88
997 x 1218mm, ND13	102.17	2.50	104.99	2.39	0.51	50.85	-	155.84	nr	171.42
1486 x 628mm, NE14	90.67	2.50	93.19	2.35	0.53	50.39	-	143.58	nr	157.94
1486 x 923mm, NC14	101.68	2.50	104.51	2.59	0.58	55.63	-	160.14	nr	176.15
1486 x 1218mm, ND14	115.02	2.50	118.22	2.98	0.66	63.86	-	182.08	nr	200.29
top hung lights										
508 x 292mm, NG1	128.19	2.50	131.49	0.93	0.25	20.52	-	152.01	nr	167.21
508 x 457mm, NH1	143.48	2.50	147.18	1.13	0.25	24.12	-	171.31	nr	188.44
508 x 628mm, NE1	154.16	2.50	158.15	1.31	0.29	28.13	-	186.28	nr	204.91
997 x 628mm, NE13E	213.95	2.50	219.49	1.77	0.32	36.92	-	256.41	nr	282.05
997 x 923mm, NC13C	232.50	2.50	238.54	2.01	0.40	42.39	-	280.94	nr	309.03
bottom hung lights										
508 x 628mm, NL1	217.69	2.50	223.27	1.31	0.32	28.54	-	251.81	nr	276.99
side hung lights										
508 x 628mm, NES1	165.52	2.50	169.80	1.31	0.32	28.54	-	198.34	nr	218.17
508 x 923mm, NC1	178.47	2.50	183.10	1.55	0.35	33.33	-	216.42	nr	238.07
508 x 1067mm, NC01	189.30	2.50	194.22	1.73	0.38	37.01	-	231.23	nr	254.35
508 x 1218mm, ND1	201.88	2.50	207.14	1.92	0.41	40.91	-	248.05	nr	272.85
mixed lights										
279 x 923mm, NC6F	139.47	2.50	143.11	1.46	0.35	31.65	-	174.76	nr	192.24
508 x 923mm, NC5F	157.93	2.50	162.05	1.55	0.35	33.33	-	195.38	nr	214.91
508 x 1067mm, NC05F	162.07	2.50	166.31	1.73	0.38	37.01	-	203.32	nr	223.65
997 x 292mm, NG2	167.23	2.50	171.56	1.37	0.33	29.63	-	201.20	nr	221.32
997 x 457mm, NH2	176.27	2.50	180.85	1.56	0.37	33.79	-	214.64	nr	236.10
997 x 628mm, NE2	194.30	2.50	199.36	1.77	0.40	38.02	-	237.37	nr	261.11
997 x 628mm, NES2	225.78	2.50	231.62	1.77	0.40	38.02	-	269.63	nr	296.60
997 x 923mm, NC2	246.55	2.50	252.95	2.01	0.44	42.94	-	295.89	nr	325.48
997 x 923mm, NC2F	322.88	2.50	331.18	2.01	0.44	42.94	-	374.12	nr	411.53
997 x 1067mm, NC02	266.78	2.50	273.70	2.20	0.47	46.81	-	320.51	nr	352.56
997 x 1067mm, NC02F	352.29	2.50	361.35	2.20	0.47	46.81	-	408.16	nr	448.97
997 x 1218mm, ND2	288.26	2.50	295.73	2.39	0.51	50.85	-	346.58	nr	381.24
997 x 1218mm, ND2F	356.41	2.50	365.58	2.39	0.51	50.85	-	416.43	nr	458.07
997 x 1513mm, NDV2FSB	413.27	2.50	423.90	2.63	0.58	56.37	-	480.27	nr	528.30
1486 x 628mm, NE3	252.55	2.50	259.12	2.35	0.51	50.11	-	309.23	nr	340.15
1486 x 923mm, NC4	402.22	2.50	412.56	2.59	0.58	55.63	-	468.19	nr	515.01
1486 x 923mm, NC4F	490.61	2.50	503.16	2.59	0.58	55.63	-	558.79	nr	614.67
1486 x 1067mm, NC04	453.58	2.50	465.22	2.79	0.58	59.25	-	524.47	nr	576.92
1486 x 1067mm, NC04F	524.13	2.50	537.54	2.79	0.61	59.66	-	597.20	nr	656.92
1486 x 1218mm, ND4	489.05	2.50	501.60	2.98	0.66	63.86	-	565.46	nr	622.01
1486 x 1218mm, ND4F	552.62	2.50	566.76	2.98	0.66	63.86	-	630.62	nr	693.68
1486 x 1513mm, NDV4FSB	672.28	2.50	689.44	3.24	0.69	69.01	-	758.46	nr	834.30
1994 x 923mm, NC11F	537.16	2.50	550.93	3.18	0.69	67.86	-	618.79	nr	680.67
1994 x 1218mm, ND11F	583.50	2.50	598.47	3.60	0.81	77.19	-	675.66	nr	743.22
1994 x 1513mm, NDV11FSB	678.40	2.50	695.78	3.87	0.86	82.98	-	778.76	nr	856.63
Extra for										
mullions										
292mm high	24.01	2.50	24.61	0.20	0.10	5.05	-	29.66	nr	32.62
457mm high	31.88	2.50	32.68	0.30	0.15	7.58	-	40.25	nr	44.28
628mm high	40.20	2.50	41.20	0.40	0.20	10.10	-	51.30	nr	56.43
923mm high	54.23	2.50	55.58	0.50	0.25	12.62	-	68.21	nr	75.03
1067mm high	62.35	2.50	63.91	0.60	0.30	15.15	-	79.06	nr	86.97
1218mm high	69.51	2.50	71.25	0.80	0.40	20.20	-	91.45	nr	100.60
1513mm high	83.66	2.50	85.75	1.00	0.50	25.25	-	111.00	nr	122.10
transoms										
279mm wide	25.49	2.50	26.13	0.20	0.10	5.05	-	31.18	nr	34.30
508mm wide	36.90	2.50	37.82	0.30	0.15	7.58	-	45.40	nr	49.94
997mm wide	100.56	2.50	103.08	0.60	0.30	15.15	-	118.23	nr	130.05
1143mm wide	110.14	2.50	112.89	1.00	0.50	25.25	-	138.14	nr	151.95
Controlair Ventilators, permanent, mill finish										
508mm wide	52.61	2.50	53.92	-	-	-	-	53.92	nr	59.31
997mm wide	70.65	2.50	72.42	-	-	-	-	72.42	nr	79.66
1486mm wide	159.15	2.50	163.13	-	-	-	-	163.13	nr	179.45
Controlair Ventilators, permanent, mill finish, flyscreen										
508mm wide	62.53	2.50	64.09	-	-	-	-	64.09	nr	70.50
997mm wide	89.08	2.50	91.31	-	-	-	-	91.31	nr	100.44
1486mm wide	184.22	2.50	188.83	-	-	-	-	188.83	nr	207.71
Controlair Ventilators, adjustable										
508mm wide 1-LT	62.70	2.50	64.27	-	-	-	-	64.27	nr	70.70
997mm wide 2-LT	79.28	2.50	81.26	-	-	-	-	81.26	nr	89.39
Controlair Ventilators, adjustable, flyscreen										
508mm wide 1-LT	72.24	2.50	74.04	-	-	-	-	74.04	nr	81.45
997mm wide 2-LT	97.71	2.50	100.15	-	-	-	-	100.15	nr	110.17
locks with Parkes locking handles for										
side hung lights	65.06	2.50	66.68	-	-	-	-	66.68	nr	73.35
horizontally pivoted lights	65.06	2.50	66.68	-	-	-	-	66.68	nr	73.35

Labour hourly rates: (except Specialists) Craft Operatives 18.37 Labourer 13.76 Rates are national average prices. Refer to REGIONAL VARIATIONS for indicative levels of overall pricing in regions	MATERIALS			LABOUR				RATES		
	Del to Site	Waste	Material Cost	Craft Optve	Lab	Labour Cost	Sunds	Nett Rate		Gross rate (10%)
	£	%	£	Hrs	Hrs	£	£	£	Unit	£
METAL WINDOWS AND SHOP FRONTS (Cont'd)										
Windows in aluminium; polyester powder finish; white matt; Bedding in cement mortar (1:3); pointing with Silicone standard mastic one side										
Casement range; factory Low E double glazed; fixing to masonry with lugs										
fixed lights										
600 x 400mm	98.61	2.50	101.19	1.21	0.35	27.04	-	128.24	nr	141.06
600 x 800mm	120.73	2.50	123.92	1.81	0.46	39.58	-	163.49	nr	179.84
600 x 1000mm	135.29	2.50	138.86	2.04	0.58	45.46	-	184.32	nr	202.75
600 x 1200mm	145.70	2.50	149.56	2.41	0.68	53.63	-	203.19	nr	223.51
600 x 1600mm	173.53	2.50	178.13	2.57	0.68	56.57	-	234.69	nr	258.16
800 x 400mm	107.47	2.50	110.30	1.45	0.39	32.00	-	142.30	nr	156.53
800 x 800mm	135.29	2.50	138.86	2.17	0.63	48.53	-	187.40	nr	206.14
800 x 1000mm	148.55	2.50	152.48	2.43	0.71	54.41	-	206.89	nr	227.58
800 x 1200mm	164.61	2.50	168.97	2.89	0.89	65.34	-	234.30	nr	257.73
800 x 1600mm	195.19	2.50	200.35	3.05	0.89	68.27	-	268.63	nr	295.49
1200 x 400mm	129.69	2.50	133.12	1.77	0.44	38.57	-	171.69	nr	188.86
1200 x 800mm	153.11	2.50	157.18	2.61	0.76	58.40	-	215.58	nr	237.14
1200 x 1000mm	182.03	2.50	186.84	2.92	0.86	65.47	-	252.31	nr	277.54
1200 x 1200mm	199.44	2.50	204.71	3.44	1.91	89.47	-	294.18	nr	323.60
1200 x 1600mm	234.36	2.50	240.55	3.60	1.91	92.41	-	332.97	nr	366.26
1400 x 800mm	177.78	2.50	182.48	2.96	1.71	77.90	-	260.39	nr	286.43
1400 x 1000mm	198.04	2.50	203.27	3.31	1.85	86.26	-	289.54	nr	318.49
1400 x 1400mm	235.76	2.50	241.99	4.01	2.12	102.83	-	344.82	nr	379.31
1400 x 1600mm	281.17	2.50	288.56	4.09	2.12	104.30	-	392.86	nr	432.15
top hung casement										
600 x 800mm	269.08	2.50	275.97	1.81	0.52	40.40	-	316.38	nr	348.02
600 x 1000mm	291.99	2.50	299.48	2.04	0.58	45.46	-	344.94	nr	379.43
600 x 1200mm	310.30	2.50	318.28	2.41	0.68	53.63	-	371.90	nr	409.09
800 x 800mm	291.99	2.50	299.48	2.17	0.63	48.53	-	348.01	nr	382.82
800 x 1000mm	314.85	2.50	322.94	2.43	0.71	54.41	-	377.35	nr	415.08
800 x 1200mm	336.21	2.50	344.86	2.89	0.89	65.34	-	410.19	nr	451.21
top hung fanlights										
600 x400mm	266.86	2.50	273.65	1.21	0.35	27.04	-	300.69	nr	330.76
800 x 600mm	297.58	2.50	305.19	1.53	0.39	33.47	-	338.66	nr	372.53
1200 x 600mm	323.80	2.50	332.11	1.85	0.44	40.04	-	372.15	nr	409.37
Casement range; factory Low E double glazed, one pane obscure glass; fixing to masonry with lugs										
fixed lights										
600 x 400mm	98.56	2.50	101.14	1.21	0.35	27.04	-	128.18	nr	141.00
600 x 800mm	122.13	2.50	125.35	1.81	0.46	39.58	-	164.93	nr	181.42
600 x 1000mm	135.34	2.50	138.92	2.04	0.58	45.46	-	184.37	nr	202.81
600 x 1200mm	148.50	2.50	152.43	1.69	0.68	40.40	-	192.83	nr	212.12
600 x 1600mm	177.73	2.50	182.43	2.57	0.68	56.57	-	239.00	nr	262.90
800 x 400mm	107.57	2.50	110.40	1.45	0.39	32.00	-	142.40	nr	156.64
800 x 800mm	136.74	2.50	140.35	2.17	0.63	48.53	-	188.88	nr	207.77
800 x 1000mm	151.30	2.50	155.30	2.43	0.71	54.41	-	209.71	nr	230.68
1200 x 400mm	130.99	2.50	134.46	1.77	0.44	38.57	-	173.03	nr	190.33
1200 x 800mm	167.36	2.50	171.79	2.61	0.76	58.40	-	230.19	nr	253.21
top hung casement										
600 x 800mm	270.65	2.50	277.58	1.81	0.52	40.40	-	317.99	nr	349.79
600 x 1000mm	293.49	2.50	301.02	2.04	0.58	45.46	-	346.48	nr	381.12
600 x 1200mm	313.36	2.50	321.41	2.41	0.68	53.63	-	375.04	nr	412.54
800 x 800mm	212.51	2.50	218.02	2.17	0.63	48.53	-	266.55	nr	293.20
800 x 1000mm	316.36	2.50	324.49	2.43	0.71	54.41	-	378.90	nr	416.79
800 x 1200mm	337.71	2.50	346.40	2.89	0.89	65.34	-	411.73	nr	452.91
top hung fanlights										
600 x400mm	269.83	2.50	276.69	0.81	0.35	19.70	-	296.39	nr	326.03
800 x 600mm	300.58	2.50	308.26	1.53	0.39	33.47	-	341.73	nr	375.91
1200 x 600mm	331.32	2.50	339.82	1.85	0.44	40.04	-	379.86	nr	417.85
Sliding windows in aluminium; polyester powder finish; white matt; Bedding in cement mortar (1:3); pointing with Silicone standard mastic one side										
Factory double glazed, 24mm - 4/16/4 units with clear 'Low E' glass; fixing to masonry with lugs										
vertical sliders										
600 x 800mm	438.58	2.50	449.71	1.81	0.52	40.40	-	490.12	nr	539.13
600 x 1000mm	514.09	2.50	527.13	2.04	0.58	45.46	-	572.59	nr	629.85
600 x 1200mm	601.95	2.50	617.22	2.26	0.62	50.05	-	667.26	nr	733.99
800 x 800mm	743.59	2.50	762.37	2.17	0.63	48.53	-	810.90	nr	891.99
800 x 1000mm	866.90	2.50	888.79	2.43	0.71	54.41	-	943.20	nr	1037.52
800 x 1200mm	969.31	2.50	993.79	2.71	0.81	60.93	-	1054.71	nr	1160.19
1200 x 800mm	1550.11	2.50	1589.11	2.84	0.86	64.00	-	1653.11	nr	1818.42
1200 x 1000mm	1649.53	2.50	1691.03	3.13	0.98	70.98	-	1762.01	nr	1938.21
1200 x 1200mm	1789.94	2.50	1834.97	3.44	1.08	78.05	-	1913.03	nr	2104.33
1400 x 800mm	1961.53	2.50	2010.83	3.50	1.14	79.98	-	2090.81	nr	2299.89
1400 x 1000mm	2089.94	2.50	2142.47	3.85	1.28	88.34	-	2230.81	nr	2453.89
1400 x 1200mm	2186.35	2.50	2241.32	3.93	1.11	87.47	-	2328.79	nr	2561.66
horizontal sliders										
1200 x 800mm	411.11	2.50	421.63	2.61	0.76	58.40	-	480.03	nr	528.04
1200 x 1000mm	433.41	2.50	444.50	2.92	0.86	65.47	-	509.98	nr	560.98
1200 x 1400mm	488.85	2.50	501.38	3.29	0.98	73.92	-	575.30	nr	632.83
1400 x 800mm	432.03	2.50	443.09	2.96	0.87	66.35	-	509.44	nr	560.38
1400 x 1000mm	460.94	2.50	472.75	3.31	1.01	74.70	-	547.45	nr	602.19
1400 x 1400mm	511.26	2.50	524.38	3.74	1.14	84.39	-	608.77	nr	669.64
1400 x 1600mm	537.49	2.50	551.29	4.09	1.28	92.75	-	644.03	nr	708.44
1600 x 800mm	458.64	2.50	470.39	3.33	1.00	74.93	-	545.32	nr	599.85
1600 x 1000mm	487.50	2.50	500.00	3.71	1.16	84.11	-	584.11	nr	642.52
1600 x 1600mm	567.98	2.50	582.56	4.56	1.48	104.13	-	686.70	nr	755.37

Labour hourly rates: (except Specialists) Craft Operatives 18.37 Labourer 13.76 Rates are national average prices. Refer to REGIONAL VARIATIONS for indicative levels of overall pricing in regions	MATERIALS			LABOUR				RATES		
	Del to Site	Waste	Material Cost	Craft Optve	Lab	Labour Cost	Sunds	Nett Rate	Unit	Gross rate (10%)
	£	%	£	Hrs	Hrs	£	£	£		£
METAL WINDOWS AND SHOP FRONTS (Cont'd)										
Sliding windows in aluminium; polyester powder finish; white matt; Bedding in cement mortar (1:3); pointing with Silicone standard mastic one side (Cont'd)										
Window sills in galvanized pressed steel BS 6510; fixing to metal for opening										
508mm wide; AWA	30.00	2.50	30.75	0.50	0.25	12.62	-	43.38	nr	47.71
508mm wide; AWB	33.20	2.50	34.03	0.50	0.25	12.62	-	46.65	nr	51.32
508mm wide; PS	34.40	2.50	35.26	0.50	0.25	12.62	-	47.88	nr	52.67
508mm wide; RPS	34.40	2.50	35.26	0.50	0.25	12.62	-	47.88	nr	52.67
628mm wide; AWA	33.33	2.50	34.17	0.60	0.30	15.15	-	49.32	nr	54.25
628mm wide; AWB	36.23	2.50	37.14	0.60	0.30	15.15	-	52.29	nr	57.52
628mm wide; PS	40.90	2.50	41.92	0.60	0.30	15.15	-	57.07	nr	62.78
628mm wide; RPS	40.90	2.50	41.92	0.60	0.30	15.15	-	57.07	nr	62.78
997mm wide; AWA	46.84	2.50	48.01	0.70	0.35	17.68	-	65.68	nr	72.25
997mm wide; AWB	52.82	2.50	54.14	0.70	0.35	17.68	-	71.82	nr	79.00
997mm wide; PS	47.26	2.50	48.44	0.70	0.35	17.68	-	66.11	nr	72.73
997mm wide; RPS	48.72	2.50	49.94	0.70	0.35	17.68	-	67.61	nr	74.37
1237mm wide; AWA	52.83	2.50	54.15	0.80	0.40	20.20	-	74.35	nr	81.79
1237mm wide; AWB	54.54	2.50	55.91	0.80	0.40	20.20	-	76.11	nr	83.72
1237mm wide; PS	49.29	2.50	50.52	0.80	0.40	20.20	-	70.72	nr	77.79
1237mm wide; RPS	51.10	2.50	52.38	0.80	0.40	20.20	-	72.58	nr	79.83
1486mm wide; AWA	59.09	2.50	60.56	0.90	0.45	22.73	-	83.29	nr	91.62
1486mm wide; AWB	60.96	2.50	62.48	0.90	0.45	22.73	-	85.20	nr	93.72
1486mm wide; PS	54.61	2.50	55.98	0.90	0.45	22.73	-	78.70	nr	86.57
1486mm wide; RPS	56.76	2.50	58.18	0.90	0.45	22.73	-	80.91	nr	89.00
1846mm wide; AWA	63.28	2.50	64.86	1.00	0.50	25.25	-	90.11	nr	99.12
1846mm wide; AWB	65.71	2.50	67.36	1.00	0.50	25.25	-	92.61	nr	101.87
1846mm wide; PS	57.67	2.50	59.11	1.00	0.50	25.25	-	84.36	nr	92.80
1846mm wide; RPS	60.33	2.50	61.84	1.00	0.50	25.25	-	87.09	nr	95.80
METAL ROOFLIGHTS, SKYLIGHTS AND LANTERNLIGHTS										
Rooflights										
Pyramid rooflight; mainly in aluminium; powder coat finish; non-ventilating base frame; factory double glazed with toughened clear glass										
fixing to masonry with screws; internal size										
600 x 600mm	725.00	2.50	743.12	2.00	1.00	50.50	-	793.62	nr	872.99
600 x 900mm	916.00	2.50	938.90	2.30	1.15	58.08	-	996.97	nr	1096.67
900 x 900mm	916.00	2.50	938.90	2.30	1.15	58.08	-	996.97	nr	1096.67
900 x 1200mm	1254.00	-	1254.00	2.50	1.25	63.13	-	1317.12	nr	1448.84
1200 x 1200mm	1254.00	-	1254.00	2.80	1.40	70.70	-	1324.70	nr	1457.17
Fixed flat rooflight; mainly in aluminium; powder coat finish; non-ventilating base frame; factory triple glazed with toughened clear glass										
fixing to masonry with screws; internal size										
600 x 600mm	667.00	2.50	683.67	2.00	1.00	50.50	-	734.17	nr	807.59
600 x 900mm	667.00	2.50	683.67	2.30	1.15	58.08	-	741.75	nr	815.92
900 x 900mm	667.00	2.50	683.67	2.30	1.15	58.08	-	741.75	nr	815.92
900 x 1200mm	858.00	-	858.00	2.50	1.25	63.13	-	921.12	nr	1013.24
1200 x 1200mm	1208.00	-	1208.00	2.80	1.40	70.70	-	1278.70	nr	1406.57
METAL SCREENS, BORROWED LIGHTS AND FRAMES										
Screens in mild steel										
Screens; 6 x 50 x 50mm angle framing to perimeter; 75 x 75 x 3mm mesh infill; welded connections										
2000 x 2000mm overall; fixing to masonry with screws	417.17	5.00	438.03	6.00	3.00	151.50	-	589.53	nr	648.48
3000 x 2000mm overall; fixing to masonry with screws	516.07	5.00	541.87	8.00	4.00	202.00	-	743.87	nr	818.26
Screens; 6 x 50 x 50mm angle framing to perimeter; 75 x 75 x 5mm mesh infill; welded connections										
2000 x 2000mm overall; fixing to masonry with screws	432.38	5.00	454.00	6.00	3.00	151.50	-	605.50	nr	666.05
Screens; 6 x 50 x 75mm angle framing to perimeter; 75 x 75 x 3mm mesh infill; welded connections										
3000 x 2000mm overall; fixing to masonry with screws	635.26	5.00	667.02	8.00	4.00	202.00	-	869.02	nr	955.93
Screens; 6 x 50 x 50mm angle framing to perimeter; 6 x 51 x 102mm tee mullion; 75 x 75 x 3mm mesh infill; welded connections 2000 x 4000mm overall; mullions -1 nr; fixing to masonry with screws	1088.00	5.00	1142.40	8.00	4.00	202.00	-	1344.40	nr	1478.84
Screens; 38.1 x 38.1 x 3.2mm hollow section framing to perimeter; 75 x 75 x 3mm mesh infill; welded connections										
2000 x 2000mm overall; fixing to masonry with screws	558.00	5.00	585.90	6.00	3.00	151.50	-	737.40	nr	811.14
Screens; 38.1 x 38.1 x 3.2mm hollow section framing to perimeter; 75 x 75 x 5mm mesh infill; welded connections										
2000 x 2000mm overall; fixing to masonry with screws	600.00	5.00	630.00	6.00	3.00	151.50	-	781.50	nr	859.65
Screens; 76.2 x 38.1 x 4mm hollow section framing to perimeter; 75 x 75 x 3mm mesh infill; welded connections										
2000 x 2000mm overall; fixing to masonry with screws	610.00	5.00	640.50	6.00	3.00	151.50	-	792.00	nr	871.20
Grilles in mild steel										
Grilles; 13 x 51mm flat bar framing to perimeter; 13mm diameter vertical infill bars at 150mm centres; welded connections										
1000 x 1000mm overall; fixing to masonry with screws	425.00	5.00	446.25	6.00	3.00	151.50	-	597.75	nr	657.53
2000 x 1000mm overall; fixing to masonry with screws	571.00	5.00	599.55	8.00	4.00	202.00	-	801.55	nr	881.71

Labour hourly rates: (except Specialists) Craft Operatives 18.37 Labourer 13.76 Rates are national average prices. Refer to REGIONAL VARIATIONS for indicative levels of overall pricing in regions	MATERIALS			LABOUR				RATES		
	Del to Site	Waste	Material Cost	Craft Optve	Lab	Labour Cost	Sunds	Nett Rate	Unit	Gross rate (10%)
	£	%	£	Hrs	Hrs	£	£	£		£
METAL SCREENS, BORROWED LIGHTS AND FRAMES (Cont'd)										
Grilles in mild steel (Cont'd)										
Grilles; 13 x 51mm flat bar framing to perimeter; 13mm diameter vertical infill bars at 300mm centres; welded connections										
1000 x 1000mm overall; fixing to masonry with screws..........	406.00	5.00	426.30	4.00	2.00	101.00	-	527.30	nr	580.03
Grilles; 13 x 51mm flat bar framing to perimeter; 18mm diameter vertical infill bars at 150mm centres; welded connections										
1000 x 1000mm overall; fixing to masonry with screws..........	412.10	5.00	432.70	4.00	2.00	101.00	-	533.70	nr	587.08
Glazing frames in mild steel										
Glazing frames; 13 x 38 x 3mm angle framing to perimeter; welded connections; 13 x 13mm glazing beads, fixed with screws										
500 x 500mm overall; fixing to timber with screws..........	251.10	5.00	263.65	3.00	1.50	75.75	-	339.41	nr	373.35
Glazing frames; 15 x 21 x 3mm angle framing to perimeter; welded connections; 13 x 13mm glazing beads, fixed with screws										
500 x 1000mm overall; fixing to timber with screws..........	406.00	5.00	426.30	3.00	1.50	75.75	-	502.05	nr	552.26
Glazing frames; 18 x 25 x 3mm angle framing to perimeter; welded connections; 9 x 25mm glazing beads, fixed with screws										
1000 x 1000mm overall; fixing to timber with screws..........	488.20	5.00	512.61	4.00	2.00	101.00	-	613.61	nr	674.97
Glazing frames; 18 x 25 x 3mm angle framing to perimeter; welded connections; 13 x 13mm glazing beads, fixed with screws										
1000 x 1000mm overall; fixing to timber with screws..........	526.25	5.00	552.56	4.00	2.00	101.00	-	653.56	nr	718.92
Glazing frames; 18 x 38 x 3mm angle framing to perimeter; welded connections; 9 x 25mm glazing beads, fixed with screws										
1000 x 1000mm overall; fixing to timber with screws..........	655.55	5.00	688.33	3.00	1.50	75.75	-	764.08	nr	840.49
Glazing frames; 18 x 25 x 3mm angle framing to perimeter; welded connections; 15 x 15 x 3mm channel glazing beads; fixed with screws										
1000 x 1000mm overall; fixing to timber with screws..........	536.50	5.00	563.32	3.00	1.50	75.75	-	639.07	nr	702.98
Glazing frames; 18 x 38 x 3mm angle framing to perimeter; welded connections; 15 x 15 x 3mm channel glazing beads; fixed with screws										
1000 x 1000mm overall; fixing to timber with screws..........	655.55	5.00	688.33	3.00	1.50	75.75	-	764.08	nr	840.49
Glazing frames in aluminium										
Glazing frames; 13 x 38 x 3mm angle framing to perimeter; welded connections; 13 x 13mm glazing beads, fixed with screws										
500 x 500mm overall; fixing to timber with screws..........	424.50	5.00	445.72	3.00	1.50	75.75	-	521.47	nr	573.62
Glazing frames; 15 x 21 x 3mm angle framing to perimeter; welded connections; 13 x 13mm glazing beads, fixed with screws										
500 x 1000mm overall; fixing to timber with screws..........	434.20	5.00	455.91	3.00	1.50	75.75	-	531.66	nr	584.83
Glazing frames; 18 x 25 x 3mm angle framing to perimeter; welded connections; 9 x 25mm glazing beads, fixed with screws										
1000 x 1000mm overall; fixing to timber with screws..........	459.20	5.00	482.16	4.00	2.00	101.00	-	583.16	nr	641.48
Glazing frames; 18 x 25 x 3mm angle framing to perimeter; welded connections; 13 x 13mm glazing beads, fixed with screws										
1000 x 1000mm overall; fixing to timber with screws..........	469.70	5.00	493.18	4.00	2.00	101.00	-	594.18	nr	653.60
Glazing frames; 18 x 38 x 3mm angle framing to perimeter; welded connections; 9 x 25mm glazing beads, fixed with screws										
1000 x 1000mm overall; fixing to timber with screws..........	532.00	5.00	558.60	4.00	2.00	101.00	-	659.60	nr	725.56
Glazing frames; 18 x 25 x 3mm angle framing to perimeter; welded connections; 15 x 15 x 3mm channel glazing beads; fixed with screws										
1000 x 1000mm overall; fixing to timber with screws..........	479.50	5.00	503.47	4.00	2.00	101.00	-	604.47	nr	664.92
Glazing frames; 18 x 38 x 3mm angle framing to perimeter; welded connections; 15 x 15 x 3mm channel glazing beads; fixed with screws										
1000 x 1000mm overall; fixing to timber with screws..........	546.50	5.00	573.82	4.00	2.00	101.00	-	674.82	nr	742.31
Note										
Bedding in cement mortar (1:3); pointing with Silicone standard mastic one side' is included with standard windows as noted above; adjustment will need to be made for two part polysuphide pointing and bedding and pointing to purpose made frames, grilles, screens and the like										
Bedding and pointing frames										
Pointing with Silicone										
standard metal frames one side	1.13	10.00	1.25	0.14	-	2.57	-	3.82	m	4.20
Bedding in cement mortar (1:3); pointing with Silicone standard mastic one side										
metal frames..	1.26	5.00	1.36	0.20	-	3.67	-	5.04	m	5.54

Labour hourly rates: (except Specialists) Craft Operatives 18.37 Labourer 13.76 Rates are national average prices. Refer to REGIONAL VARIATIONS for indicative levels of overall pricing in regions	MATERIALS			LABOUR				RATES		
	Del to Site	Waste	Material Cost	Craft Optve	Lab	Labour Cost	Sunds	Nett Rate	Unit	Gross rate (10%)
	£	%	£	Hrs	Hrs	£	£	£		£

PLASTIC WINDOWS AND SHOP FRONTS

Windows in uPVC; white; Bedding in cement mortar (1:3); pointing with Silicone standard mastic one side

Windows; factory glazed 24mm double glazed units; hinges; fastenings

fixed light; fixing to masonry with cleats and screws; overall size										
600 x 600mm	61.20	2.50	62.88	3.08	1.30	74.47	2.33	139.67	nr	153.63
600 x 900mm	66.22	2.50	68.05	3.85	1.62	93.02	2.33	163.39	nr	179.73
600 x 1050mm	68.73	2.50	70.64	4.21	1.77	101.69	2.33	174.66	nr	192.13
600 x 1200mm	71.24	2.50	73.23	4.62	1.95	111.70	2.33	187.26	nr	205.98
600 x 1500mm toughened glass	115.77	2.50	118.92	5.39	2.27	130.25	2.33	251.49	nr	276.64
750 x 600mm	64.28	2.50	66.05	3.49	1.47	84.34	2.33	152.71	nr	167.98
750 x 900mm	70.39	2.50	72.35	4.26	1.80	103.02	2.33	177.70	nr	195.47
750 x 1050mm	74.00	2.50	76.07	4.62	1.95	111.70	2.33	190.09	nr	209.10
750 x 1200mm	76.51	2.50	78.66	4.98	2.10	120.38	2.33	201.36	nr	221.50
750 x 1500mm; toughened glass	132.03	2.50	135.60	5.75	2.42	138.93	2.33	276.85	nr	304.53
900 x 600mm	67.34	2.50	69.21	3.85	1.62	93.02	2.33	164.55	nr	181.00
900 x 900mm	74.56	2.50	76.64	4.62	1.95	111.70	2.33	190.67	nr	209.73
900 x 1050mm	78.17	2.50	80.36	4.88	2.05	117.85	2.33	200.54	nr	220.59
900 x 1200mm	81.78	2.50	84.07	5.39	2.27	130.25	2.33	216.65	nr	238.31
900 x 1500mm toughened glass	147.18	2.50	151.15	6.11	2.57	147.60	2.33	301.07	nr	331.18
1200 x 600mm	76.83	2.50	78.96	4.37	1.82	105.32	2.33	186.61	nr	205.27
1200 x 900mm	82.91	2.50	85.24	5.14	2.15	124.01	2.33	211.57	nr	232.72
1200 x 1050mm	86.52	2.50	88.95	5.50	2.30	132.68	2.33	223.96	nr	246.36
1200 x 1200mm	91.23	2.50	93.80	5.96	2.50	143.89	2.33	240.01	nr	264.01
1200 x 1500mm toughened glass	178.59	2.50	183.37	6.78	2.85	163.76	2.33	349.46	nr	384.41
fixed light with fixed light; fixing to masonry with cleats and screws; overall size										
600 x 1350mm	102.65	2.50	105.47	5.04	2.10	121.48	2.33	229.28	nr	252.20
600 x 1500mm toughened glass	136.30	2.50	139.96	5.39	2.27	130.25	2.33	272.54	nr	299.79
600 x 1800mm toughened glass	169.40	2.50	173.92	5.81	2.42	140.03	2.33	316.27	nr	347.90
600 x 2100mm toughened glass	183.41	2.50	188.32	6.23	2.57	149.81	2.33	340.45	nr	374.50
750 x 1350mm	107.62	2.50	110.56	5.39	2.27	130.25	2.33	243.14	nr	267.45
750 x 1500mm toughened glass	159.53	2.50	163.79	5.75	2.42	138.93	2.33	305.04	nr	335.54
750 x 1800mm toughened glass	187.63	2.50	192.62	6.17	2.57	148.71	2.33	343.66	nr	378.02
750 x 2100mm toughened glass	204.72	2.50	210.18	6.64	2.75	159.82	2.33	372.32	nr	409.55
900 x 1350mm	111.79	2.50	114.85	5.75	2.42	138.93	2.33	256.10	nr	281.71
900 x 1500mm toughened glass	174.68	2.50	179.33	6.11	2.57	147.60	2.33	329.26	nr	362.19
900 x 1800mm toughened glass	205.85	2.50	211.31	6.48	2.70	156.19	2.33	369.83	nr	406.81
900 x 2100mm toughened glass	224.92	2.50	230.90	7.00	2.90	168.49	2.33	401.72	nr	441.89
1200 x 1350mm	123.44	2.50	126.83	6.42	2.70	155.09	2.33	284.24	nr	312.66
1200 x 1500mm toughened glass	207.19	2.50	212.69	6.78	2.85	163.76	2.33	378.78	nr	416.66
1200 x 1800mm toughened glass	243.41	2.50	249.86	7.25	3.02	174.74	2.33	426.92	nr	469.61
1200 x 2100mm toughened glass	268.64	2.50	275.75	7.67	3.17	184.52	2.33	462.59	nr	508.85
overall, tilt/turn; fixing to masonry with cleats and screws; overall size										
600 x 650mm	120.22	2.50	123.37	3.08	1.30	74.47	2.33	200.16	nr	220.18
600 x 900mm	135.85	2.50	139.43	3.85	1.62	93.02	2.33	234.77	nr	258.25
600 x 1050mm	141.94	2.50	145.68	4.21	1.80	102.11	2.33	250.11	nr	275.12
600 x 1200mm	145.70	2.50	149.56	4.62	1.95	111.70	2.33	263.59	nr	289.95
750 x 650mm	123.40	2.50	126.65	3.49	1.47	84.34	2.33	213.31	nr	234.64
750 x 900mm	141.34	2.50	145.08	4.26	1.80	103.02	2.33	250.42	nr	275.47
750 x 1050mm	148.59	2.50	152.52	4.62	1.95	111.70	2.33	266.54	nr	293.20
750 x 1200mm	152.35	2.50	156.39	4.98	2.10	120.38	2.33	279.10	nr	307.01
750 x 1350mm	169.99	2.50	174.49	5.39	2.27	130.25	2.33	307.06	nr	337.77
750 x 1500mm toughened glass	224.32	2.50	230.20	5.75	2.42	138.93	2.33	371.45	nr	408.59
900 x 900mm	156.07	2.50	160.19	4.62	1.95	111.70	2.33	274.21	nr	301.63
900 x 1050mm	164.46	2.50	168.81	4.88	2.05	117.85	2.33	288.98	nr	317.88
900 x 1200mm	168.24	2.50	172.69	5.39	2.27	130.25	2.33	305.27	nr	335.79
900 x 1350mm	187.03	2.50	191.97	5.75	2.42	138.93	2.33	333.22	nr	366.55
900 x 1500mm toughened glass	251.24	2.50	257.80	6.11	2.57	147.60	2.33	407.73	nr	448.51
tilt/turn sash with fixed side light; fixing to masonry with cleats and screws; overall size										
900 x 1350mm	211.28	2.50	216.83	5.75	2.42	138.93	2.33	358.08	nr	393.89
900 x 1500mm toughened glass	276.65	2.50	283.85	6.11	2.57	147.60	2.33	433.78	nr	477.16
900 x 2100mm toughened glass	328.16	2.50	336.72	7.00	2.90	168.49	2.33	507.54	nr	558.29
1200 x 1350mm	222.27	2.50	228.13	6.42	2.70	155.09	2.33	385.54	nr	424.10
1200 x 1500mm toughened glass	308.55	2.50	316.59	6.78	2.85	163.76	2.33	482.68	nr	530.95
1200 x 2100mm toughened glass	371.43	2.50	381.11	7.67	3.17	184.52	2.33	567.95	nr	624.75
1500 x 900mm	209.44	5.00	209.87	6.18	3.09	156.05	2.33	368.24	nr	405.06
1500 x 1200mm	239.18	5.00	239.67	7.05	3.53	178.01	2.33	420.01	nr	462.01
1500 x 1350mm	248.61	5.00	249.13	7.30	3.65	184.33	2.33	435.78	nr	479.35
1500 x 1500mm toughened glass	354.77	5.00	355.32	7.92	3.96	199.98	2.33	557.63	nr	613.39
1800 x 900mm	231.78	2.50	237.90	6.38	2.65	153.66	2.33	393.89	nr	433.27
1800 x 1050mm	243.64	2.50	250.07	6.79	2.82	163.54	2.33	415.93	nr	457.52
1800 x 1200mm	263.58	2.50	270.53	7.25	3.02	174.74	2.33	447.59	nr	492.35
1800 x 1350mm	273.13	2.50	280.33	6.50	3.25	164.12	2.33	446.78	nr	491.46
1800 x 1500mm toughened glass	400.08	2.50	410.48	8.12	3.40	195.95	2.33	608.75	nr	669.63
tilt/turn sash with centre fixed light; fixing to masonry with cleats and screws; overall size										
2400 x 1200mm	373.34	2.50	383.10	8.74	3.65	210.78	2.33	596.21	nr	655.83
2400 x 1350mm	388.66	2.50	398.83	9.20	3.85	221.98	2.33	623.13	nr	685.45
2400 x 1500mm toughened glass	558.61	2.50	573.04	9.56	4.00	230.66	2.33	806.02	nr	886.62

Windows; factory glazed 40mm triple glazed units; hinges; fastenings

fixed light; fixing to masonry with cleats and screws; overall size										
600 x 600mm	82.48	2.50	84.68	3.08	1.30	74.47	2.33	161.48	nr	177.62
600 x 1200mm	136.42	2.50	140.05	4.62	1.95	111.70	2.33	254.07	nr	279.48
600 x 1500mm toughened glass	182.87	2.50	187.69	5.39	2.27	130.25	2.33	320.27	nr	352.29
750 x 600mm	102.11	2.50	104.82	3.49	1.47	84.34	2.33	191.48	nr	210.63
750 x 1500mm toughened glass	246.23	2.50	252.65	5.75	2.42	138.93	2.33	393.90	nr	433.30

Labour hourly rates: (except Specialists) Craft Operatives 18.37 Labourer 13.76 Rates are national average prices. Refer to REGIONAL VARIATIONS for indicative levels of overall pricing in regions	MATERIALS			LABOUR				RATES		
	Del to Site	Waste	Material Cost	Craft Optve	Lab	Labour Cost	Sunds	Nett Rate	Unit	Gross rate (10%)
	£	%	£	Hrs	Hrs	£	£	£		£
PLASTIC WINDOWS AND SHOP FRONTS (Cont'd)										
Windows in uPVC; white; Bedding in cement mortar (1:3); pointing with Silicone standard mastic one side (Cont'd)										
Windows; factory glazed 40mm triple glazed units; hinges; fastenings (Cont'd)										
fixed light; fixing to masonry with cleats and screws; overall size (Cont'd)										
900 x 600mm	112.08	2.50	115.06	3.85	1.62	93.02	2.33	210.40	nr	231.44
900 x 1500mm toughened glass	278.06	2.50	285.30	6.11	2.57	147.60	2.33	435.22	nr	478.75
1200 x 600mm	140.62	2.50	144.35	4.37	1.82	105.32	2.33	252.00	nr	277.20
1200 x 1200mm	234.94	2.50	241.10	5.96	2.50	143.89	2.33	387.31	nr	426.04
1200 x 1500mm toughened glass	342.20	2.50	351.08	6.78	2.85	163.76	2.33	517.17	nr	568.89
fixed light with opening top light; fixing to masonry with cleats and screws; overall size										
600 x 900mm	163.53	2.50	167.79	3.85	1.62	93.02	2.33	263.13	nr	289.45
600 x 1400mm	206.55	2.50	211.96	5.00	2.15	121.43	2.33	335.72	nr	369.29
600 x 1500mm toughened glass	256.42	2.50	263.08	5.39	2.27	130.25	2.33	395.66	nr	435.22
750 x 900mm	189.66	2.50	194.60	4.26	1.80	103.02	2.33	299.95	nr	329.95
750 x 1500mm toughened glass	308.73	2.50	316.72	5.75	2.27	136.86	2.33	455.90	nr	501.49
900 x 900mm	202.89	2.50	208.18	4.62	1.95	111.70	2.33	322.21	nr	354.43
900 x 1500mm toughened glass	340.56	2.50	349.36	6.11	2.57	147.60	2.33	499.29	nr	549.22
1200 x 900mm	253.80	2.50	260.40	5.14	2.15	124.01	2.33	386.73	nr	425.40
1200 x 1500mm toughened glass	414.15	2.50	424.83	6.78	2.85	163.76	2.33	590.92	nr	650.01
Note										
Bedding in cement mortar (1:3); pointing with Silicone standard mastic one side' is included with standard windows as noted above; adjustment will need to be made for two part polysuphide pointing and bedding and pointing to purpose made frames										
Bedding and pointing frames										
Pointing with Silicone standard mastic										
plastics frames one side	2.04	10.00	2.24	0.14	-	2.57	-	4.81	m	5.30
Bedding in cement mortar (1:3); pointing with Silicone standard mastic one side										
plastics frames	2.73	5.00	2.98	0.20	-	3.67	-	6.66	m	7.32
PLASTIC ROOFLIGHTS, SKYLIGHTS AND LANTERNLIGHTS										
Roof domelights; polycarbonate; doubleglazed										
Rooflights, 'Stardome' from the National Domelight Company; double skin UV protected polycarbonate domes with 200mm high insulated base frame										
un-ventilated upstands; fixing to masonry with screws; roof opening size										
600 x 600mm	267.93	2.50	274.63	1.80	0.90	45.45	-	320.08	nr	352.09
600 x 900mm	358.00	2.50	366.95	2.00	1.00	50.50	-	417.45	nr	459.19
900 x 900mm	485.00	2.50	497.12	2.00	1.00	50.50	-	547.62	nr	602.39
900 x 1200mm	532.81	2.50	546.13	2.20	1.10	55.55	-	601.68	nr	661.85
1200 x 1200mm	574.00	2.50	588.35	2.50	1.25	63.13	-	651.47	nr	716.62
louvre ventilated upstands; fixing to masonry with screws; roof opening size										
600 x 600mm	375.00	2.50	384.38	1.80	0.90	45.45	-	429.83	nr	472.81
600 x 900mm	461.00	2.50	472.52	2.00	1.00	50.50	-	523.03	nr	575.33
900 x 900mm	598.36	2.50	613.32	2.00	1.00	50.50	-	663.82	nr	730.20
900 x 1200mm	660.00	2.50	676.50	2.20	1.10	55.55	-	732.05	nr	805.26
1200 x 1200mm	780.00	2.50	799.50	2.50	1.25	63.13	-	862.62	nr	948.89
hinged ventilated upstands; winding rod; fixing to masonry with screws; roof opening size										
600 x 600mm	397.00	2.50	406.92	1.80	0.90	45.45	-	452.38	nr	497.61
600 x 900mm	492.00	2.50	504.30	2.00	1.00	50.50	-	554.80	nr	610.28
900 x 900mm	698.00	2.50	715.45	2.00	1.00	50.50	-	765.95	nr	842.54
900 x 1200mm	728.00	2.50	746.20	2.20	1.10	55.55	-	801.75	nr	881.92
1200 x 1200mm	812.00	2.50	832.30	2.50	1.25	63.13	-	895.42	nr	984.97

Labour hourly rates: (except Specialists) Craft Operatives 18.37 Labourer 13.76 Rates are national average prices. Refer to REGIONAL VARIATIONS for indicative levels of overall pricing in regions	MATERIALS			LABOUR				RATES		
	Del to Site £	Waste %	Material Cost £	Craft Optve Hrs	Lab Hrs	Labour Cost £	Sunds £	Nett Rate £	Unit	Gross rate (10%) £
TIMBER DOORS AND FRAMES										
Door sets mainly in softwood, wrought										
Door sets; 28mm thick jambs, head and transome with 12mm thick stop to suit 100mm thick wall; 14mm thick hardwood threshold; honeycomb core plywood faced flush door lipped two long edges; 6mm plywood transome panel fixed with pinned beads; 65mm snap in hinges; 57mm backset mortice latch; fixing frame to masonry with screws										
526 x 2040 x 40mm flush door -1 nr; basic dimensions										
600 x 2100mm	136.09	2.50	139.49	1.25	0.16	25.16	-	164.66	nr	181.12
600 x 2400mm	166.40	2.50	170.56	1.35	0.17	27.14	-	197.70	nr	217.47
626 x 2040 x 40mm flush door -1 nr; basic dimensions										
700 x 2100mm	136.09	2.50	139.49	1.25	0.16	25.16	-	164.66	nr	181.12
700 x 2400mm	166.40	2.50	170.56	1.35	0.17	27.14	-	197.70	nr	217.47
726 x 2040 x 40mm flush door -1 nr; basic dimensions										
800 x 2100mm	136.09	2.50	139.49	1.25	0.16	25.16	-	164.66	nr	181.12
800 x 2400mm	166.40	2.50	170.56	1.35	0.17	27.14	-	197.70	nr	217.47
826 x 2040 x 40mm flush door -1 nr; basic dimensions										
900 x 2100mm	142.01	2.50	145.56	1.25	0.16	25.16	-	170.73	nr	187.80
900 x 2400mm	169.95	2.50	174.19	1.35	0.17	27.14	-	201.33	nr	221.47
Doors in softwood, wrought										
Internal panelled doors; SA; Magnet Trade; pre-glazed										
762 x 1981 x 34mm (f sizes)	116.61	2.50	119.53	1.20	0.16	24.25	-	143.77	nr	158.15
Doors in softwood, wrought, preservative treated										
External ledged and braced doors; L & B Magnet Trade										
686 x 1981 x 44mm (f sizes)	51.20	2.50	52.48	1.10	0.14	22.13	-	74.61	nr	82.08
762 x 1981 x 44mm (f sizes)	51.20	2.50	52.48	1.30	0.16	26.08	-	78.56	nr	86.42
838 x 1981 x 44mm (f sizes)	51.20	2.50	52.48	1.50	0.19	30.17	-	82.65	nr	90.91
External framed, ledged and braced doors; YX; Magnet Trade										
686 x 1981 x 44mm (f sizes)	68.12	2.50	69.82	1.10	0.14	22.13	-	91.96	nr	101.15
762 x 1981 x 44mm (f sizes)	68.12	2.50	69.82	1.30	0.16	26.08	-	95.91	nr	105.50
813 x 2032 x 44mm (f sizes)	68.12	2.50	69.82	1.50	0.19	30.17	-	99.99	nr	109.99
838 x 1981 x 44mm (f sizes)	68.12	2.50	69.82	1.50	0.19	30.17	-	99.99	nr	109.99
External panelled doors; KXT CDS; Magnet Trade										
762 x 1981 x 44mm (f sizes)	116.86	2.50	119.78	1.30	0.16	26.08	-	145.86	nr	160.45
External panelled doors; 2XG CDS; Magnet Trade										
762 x 1981 x 44mm (f sizes)	89.13	2.50	91.36	1.30	0.16	26.08	-	117.45	nr	129.19
813 x 2032 x 44mm (f sizes)	89.13	2.50	91.36	1.50	0.19	30.17	-	121.53	nr	133.69
838 x 1981 x 44mm (f sizes)	89.13	2.50	91.36	1.50	0.19	30.17	-	121.53	nr	133.69
External panelled doors; 2XGG CDS; Magnet Trade										
762 x 1981 x 44mm (f sizes)	84.65	2.50	86.77	1.30	0.16	26.08	-	112.85	nr	124.13
838 x 1981 x 44mm (f sizes)	84.65	2.50	86.77	1.50	0.19	30.17	-	116.94	nr	128.63
Garage doors; MFL, pair, side hung; Magnet Trade										
1981 x 2134 x 44mm (f sizes) overall	243.28	2.50	249.36	3.50	0.44	70.35	-	319.71	nr	351.68
2134 x 2134 x 44mm (f sizes) overall	250.11	2.50	256.36	3.65	0.46	73.38	-	329.74	nr	362.71
Garage doors; 301, pair, side hung; Magnet Trade										
1981 x 2134 x 44mm (f sizes) overall	317.37	2.50	325.30	3.50	0.44	70.35	-	395.65	nr	435.22
2134 x 2134 x 44mm (f sizes) overall	340.22	2.50	348.72	3.65	0.46	73.38	-	422.10	nr	464.31
Doors in Hardwood (solid, laminated or veneered), wrought										
External panelled doors; Carolina M & T; pre-glazed obscure toughened glass; Magnet Trade										
813 x 2032 x 44mm (f sizes)	210.54	2.50	215.80	2.30	0.29	46.24	-	262.04	nr	288.24
838 x 1981 x 44mm (f sizes)	208.20	2.50	213.40	2.30	0.29	46.24	-	259.64	nr	285.61
External panelled doors; Alicante M & T; pre-glazed obscure toughened glass; Magnet Trade										
813 x 2032 x 44mm (f sizes)	309.86	2.50	317.60	2.30	0.29	46.24	-	363.84	nr	400.23
838 x 1981 x 44mm (f sizes)	307.52	2.50	315.20	2.30	0.29	46.24	-	361.44	nr	397.59
External panelled doors; Kendall M & T; pre-glazed; Magnet Trade										
762 x 1981 x 44mm (f sizes)	260.26	2.50	266.77	2.10	0.26	42.15	-	308.92	nr	339.81
838 x 1981 x 44mm (f sizes)	262.67	2.50	269.23	2.30	0.29	46.24	-	315.47	nr	347.02
External panelled doors; Manilla M & T; Magnet Trade										
813 x 2032 x 44mm (f sizes)	157.82	2.50	161.77	2.30	0.29	46.24	-	208.01	nr	228.81
838 x 1981 x 44mm (f sizes)	133.51	2.50	136.85	2.30	0.29	46.24	-	183.09	nr	201.40
External panelled doors; M & T Richmond; Magnet Trade										
838 x 1981 x 44mm (f sizes)	139.68	2.50	143.18	2.30	0.29	46.24	-	189.42	nr	208.36
External panelled doors; M & T Richmond, pre-glazed with clear glass; Magnet Trade										
762 x 1981 x 44mm (f sizes)	204.16	2.50	209.27	2.10	0.26	42.15	-	251.42	nr	276.57
838 x 1981 x 44mm (f sizes)	206.57	2.50	211.73	2.30	0.29	46.24	-	257.98	nr	283.77

DOORS, SHUTTERS AND HATCHES

Labour hourly rates: (except Specialists) Craft Operatives 18.37 Labourer 13.76 Rates are national average prices. Refer to REGIONAL VARIATIONS for indicative levels of overall pricing in regions	MATERIALS			LABOUR				RATES		
	Del to Site	Waste	Material Cost	Craft Optve	Lab	Labour Cost	Sunds	Nett Rate		Gross rate (10%)
	£	%	£	Hrs	Hrs	£	£	£	Unit	£
TIMBER DOORS AND FRAMES (Cont'd)										
Doors in Hardwood (solid, laminated or veneered), wrought (Cont'd)										
External panelled doors; Steeton M & T; pre-glazed with clear glass; Magnet Trade										
838 x 1981 x 44mm (f sizes)	193.31	2.50	198.14	2.30	0.29	46.24	-	244.38	nr	268.82
External panelled doors; Geneva Oak; Magnet Trade										
838 x 1981 x 44mm (f sizes)	243.43	2.50	249.51	2.30	0.29	46.24	-	295.75	nr	325.33
External panelled doors; Chesham Oak, preglazed with part obscure glass; Magnet Trade										
838 x 1981 x 44mm (f sizes)	290.72	2.50	297.99	2.30	0.29	46.24	-	344.23	nr	378.65
Internal panelled doors; Louis Oak Bolection Mouldings										
762 x 1981 x 34mm (f sizes)	132.66	2.50	135.98	2.10	0.26	42.15	-	178.14	nr	195.95
Internal panelled doors; Louis Oak Bolection Mouldings; pre-glazed bevelled glass										
762 x 1981 x 34mm (f sizes)	144.62	2.50	148.24	2.10	0.26	42.15	-	190.40	nr	209.43
Internal panelled doors; Mexicana Oak										
762 x 1981 x 34mm (f sizes)	116.54	2.50	119.46	2.10	0.26	42.15	-	161.61	nr	177.77
Internal panelled doors; Mexicana Oak FD30 fire door										
762 x 1981 x 44mm (f sizes)	168.68	2.50	172.89	2.10	0.26	42.15	-	215.05	nr	236.55
Internal panelled doors; Pattern 10; Magnet Trade										
762 x 1981 x 34mm (f sizes)	80.01	2.50	82.02	2.10	0.26	42.15	-	124.17	nr	136.59
Doors in hardboard, embossed										
Internal panelled doors; 6 panel; Magnet Trade										
610 x 1981 x 34mm (f sizes)	23.61	2.50	24.20	1.60	0.20	32.14	-	56.34	nr	61.98
686 x 1981 x 34mm (f sizes)	23.61	2.50	24.20	1.70	0.21	34.12	-	58.32	nr	64.15
762 x 1981 x 34mm (f sizes)	23.61	2.50	24.20	1.90	0.24	38.21	-	62.40	nr	68.64
826 x 2040 x 34mm (f sizes)	23.61	2.50	24.20	1.90	0.24	38.21	-	62.40	nr	68.64
Internal panelled doors; 4 Panel Grained; Magnet Trade										
686 x 1981 x 34mm (f sizes)	20.50	2.50	21.01	1.70	0.21	34.12	-	55.13	nr	60.65
762 x 1981 x 34mm (f sizes)	20.50	2.50	21.01	1.90	0.24	38.21	-	59.22	nr	65.14
838 x 1981 x 34mm (f sizes)	20.50	2.50	21.01	1.90	0.24	38.21	-	59.22	nr	65.14
Flush doors										
Internal; Magnaseal; Magnet Trade										
610 x 1981 x 34mm	22.16	2.50	22.71	1.80	0.23	36.23	-	58.94	nr	64.84
686 x 1981 x 34mm	22.16	2.50	22.71	1.70	0.21	34.12	-	56.83	nr	62.51
762 x 1981 x 34mm	22.16	2.50	22.71	1.90	0.24	38.21	-	60.92	nr	67.01
838 x 1981 x 34mm	22.16	2.50	22.71	2.20	0.28	44.27	-	66.98	nr	73.68
Internal; veneered paintgrade facings; Magnet Trade										
381 x 1981 x 34mm	21.96	2.50	22.51	1.10	0.14	22.13	-	44.65	nr	49.11
457 x 1981 x 34mm	21.96	2.50	22.51	1.20	0.15	24.11	-	46.62	nr	51.28
533 x 1981 x 34mm	21.96	2.50	22.51	1.30	0.16	26.08	-	48.60	nr	53.45
610 x 1981 x 34mm	21.96	2.50	22.51	1.40	0.18	28.19	-	50.71	nr	55.78
686 x 1981 x 34mm	21.96	2.50	22.51	1.50	0.19	30.17	-	52.68	nr	57.95
711 x 1981 x 34mm	21.96	2.50	22.51	1.60	0.20	32.14	-	54.66	nr	60.12
762 x 1981 x 34mm	21.96	2.50	22.51	1.70	0.21	34.12	-	56.63	nr	62.29
838 x 1981 x 34mm	25.35	2.50	25.98	1.90	0.24	38.21	-	64.19	nr	70.61
Internal; Sapele; Magnet Trade										
610 x 1981 x 34mm	43.55	2.50	44.64	1.60	0.20	32.14	-	76.78	nr	84.46
686 x 1981 x 34mm	43.55	2.50	44.64	1.70	0.21	34.12	-	78.76	nr	86.63
762 x 1981 x 34mm	43.55	2.50	44.64	2.10	0.26	42.15	-	86.79	nr	95.47
838 x 1981 x 34mm	43.55	2.50	44.64	2.10	0.26	42.15		86.79	nr	95.47
External; FD30, veneered paintgrade facings; Magnet Trade										
762 x 1981 x 44mm	62.15	2.50	63.70	1.80	0.23	36.23	-	99.93	nr	109.92
838 x 1981 x 44mm	62.15	2.50	63.70	2.00	0.25	40.18	-	103.88	nr	114.27
External; FD30, veneered paintgrade facings; Magnet Trade										
762 x 1981 x 44mm; glazing aperture 457 x 457mm	137.57	2.50	141.01	1.80	0.23	36.23	-	177.24	nr	194.97
838 x 1981 x 44mm; glazing aperture 457 x 457mm	144.36	2.50	147.97	2.00	0.25	40.18	-	188.15	nr	206.96
Internal; fire resisting; 6 Panel smooth; FD30; Magnet Trade										
838 x 1981 x 44mm	67.28	2.50	68.96	2.20	0.28	44.27	-	113.23	nr	124.55
Internal; fire resisting; veneered paintgrade facings; FD30; Magnet Trade										
686 x 1981 x 44mm	61.60	2.50	63.14	1.70	0.21	34.12	-	97.26	nr	106.98
762 x 1981 x 44mm	61.60	2.50	63.14	1.90	0.24	38.21	-	101.35	nr	111.48
813 x 2032 x 44mm	61.60	2.50	63.14	2.20	0.28	44.27	-	107.41	nr	118.15
838 x 1981 x 44mm	61.60	2.50	63.14	2.20	0.28	44.27	-	107.41	nr	118.15
Internal; fire resisting; veneered paintgrade facings; GO FD30; Magnet Trade										
762 x 1981 x 44mm; pre-glazed clear wired glass aperture 450 x 450mm	133.00	2.50	136.33	1.90	0.24	38.21	-	174.53	nr	191.99
813 x 2032 x 44mm; pre-glazed clear wired glass aperture 450 x 450mm	133.00	2.50	136.33	2.20	0.29	44.40	-	180.73	nr	198.81
838 x 1981 x 44mm; pre-glazed clear wired glass aperture 450 x 450mm	133.00	2.50	136.33	2.20	0.29	44.40	-	180.73	nr	198.81
External; fire resisting; veneered paintgrade facings; FD30; Magnet Trade										
762 x 1981 x 44mm	62.15	2.50	63.70	1.90	0.24	38.21	-	101.91	nr	112.10
838 x 1981 x 44mm	62.15	2.50	63.70	2.30	0.29	46.24	-	109.94	nr	120.94

Labour hourly rates: (except Specialists) Craft Operatives 18.37 Labourer 13.76 Rates are national average prices. Refer to REGIONAL VARIATIONS for indicative levels of overall pricing in regions	MATERIALS			LABOUR				RATES		
	Del to Site	Waste	Material Cost	Craft Optve	Lab	Labour Cost	Sunds	Nett Rate	Unit	Gross rate (10%)
	£	%	£	Hrs	Hrs	£	£	£		£
TIMBER DOORS AND FRAMES (Cont'd)										
Flush doors (Cont'd)										
Internal; fire resisting; Sapele facings; FD30; Magnet Trade										
762 x 1981 x 44mm	71.50	2.50	73.29	2.20	0.29	44.40	-	117.69	nr	129.46
838 x 1981 x 44mm	71.50	2.50	73.29	2.40	0.30	48.22	-	121.50	nr	133.65
Panelled doors in softwood, wrought										
38mm square framed (or chamfered or moulded one or both sides)										
two panel	151.36	2.50	155.14	0.85	0.11	17.13	-	172.27	m²	189.50
four panel	168.24	2.50	172.45	0.85	0.11	17.13	-	189.58	m²	208.53
six panel	178.93	2.50	183.41	0.85	0.11	17.13	-	200.54	m²	220.59
add if upper panels open moulded in small squares for glass	72.00	2.50	73.80	-	-	-	-	73.80	m²	81.18
50mm square framed (or chamfered or moulded one or both sides)										
two panel	199.16	2.50	204.13	1.00	0.11	19.88	-	224.02	m²	246.42
four panel	221.37	2.50	226.91	1.00	0.11	19.88	-	246.79	m²	271.47
six panel	235.44	2.50	241.33	1.00	0.11	19.88	-	261.21	m²	287.33
add if upper panels open moulded in small squares for glass	94.73	2.50	97.10	-	-	-	-	97.10	m²	106.81
Panelled doors in Afromosia, wrought										
38mm square framed (or chamfered or moulded one or both sides)										
two panel	287.54	2.50	294.73	1.50	0.19	30.17	-	324.89	m²	357.38
four panel	306.74	2.50	314.41	1.50	0.19	30.17	-	344.58	m²	379.03
six panel	325.91	2.50	334.06	1.50	0.19	30.17	-	364.23	m²	400.65
add if upper panels open moulded in small squares for glass	54.64	2.50	56.00	-	-	-	-	56.00	m²	61.60
50mm square framed (or chamfered or moulded one or both sides)										
two panel	316.32	2.50	324.23	1.75	0.22	35.17	-	359.40	m²	395.34
four panel	335.48	2.50	343.87	1.75	0.22	35.17	-	379.04	m²	416.95
six panel	354.66	2.50	363.52	1.75	0.22	35.17	-	398.70	m²	438.57
add if upper panels open moulded in small squares for glass	54.64	2.50	56.00	-	-	-	-	56.00	m²	61.60
Panelled doors in Sapele, wrought										
38mm square framed (or chamfered or moulded one or both sides)										
two panel	251.36	2.50	257.65	1.25	0.16	25.16	-	282.81	m²	311.09
four panel	271.01	2.50	277.78	1.25	0.16	25.16	-	302.95	m²	333.24
six panel	290.65	2.50	297.92	1.25	0.16	25.16	-	323.08	m²	355.39
add if upper panels open moulded in small squares for glass	47.13	2.50	48.31	-	-	-	-	48.31	m²	53.14
50mm square framed (or chamfered or moulded one or both sides)										
two panel	271.01	2.50	277.78	1.50	0.19	30.17	-	307.95	m²	338.75
four panel	290.65	2.50	297.92	1.50	0.19	30.17	-	328.08	m²	360.89
six panel	310.29	2.50	318.05	1.50	0.19	30.17	-	348.22	m²	383.04
add if upper panels open moulded in small squares for glass	47.13	2.50	48.31	-	-	-	-	48.31	m²	53.14
Garage doors in softwood, wrought										
Side hung; framed, tongued and grooved boarded										
2134 x 1981 x 44mm overall, pair	342.61	2.50	351.17	4.10	0.51	82.33	-	433.51	nr	476.86
2134 x 2134 x 44mm overall, pair	356.32	2.50	365.23	4.25	0.53	85.37	-	450.59	nr	495.65
Up and over; framed, tongued and grooved boarded; up and over door gear										
2134 x 1981 x 44mm	479.65	2.50	491.64	8.00	1.00	160.72	-	652.36	nr	717.60
Garage doors in Western Red Cedar, wrought										
Side hung; framed, tongued and grooved boarded										
2134 x 1981 x 44mm overall, pair	480.85	2.50	492.87	4.50	0.56	90.37	-	583.24	nr	641.56
2134 x 2134 x 44mm overall, pair	493.83	2.50	506.18	4.75	0.59	95.38	-	601.56	nr	661.71
Up and over; framed, tongued and grooved boarded; up and over door gear										
2134 x 1981 x 44mm	610.81	2.50	626.08	8.40	1.05	168.76	-	794.83	nr	874.32
Garage doors in Oak, wrought										
Side hung; framed, tongued and grooved boarded										
2134 x 1981 x 44mm overall, pair	1285.84	2.50	1317.99	7.15	0.90	143.73	-	1461.72	nr	1607.89
2134 x 2134 x 44mm overall, pair	1357.28	2.50	1391.22	7.40	0.95	149.01	-	1540.23	nr	1694.25
Up and over; framed, tongued and grooved boarded; up and over door gear										
2134 x 1981 x 44mm	1500.16	2.50	1537.66	12.00	1.50	241.08	-	1778.74	nr	1956.62
Folding patio doors white aluminium outside; Oak inside										
Double glazed with 24mm Low E units, Magnet Trade, doors and frames fully finished, fixing frame to masonry with screws										
2090 x 1790mm overall	2824.35	2.50	2894.96	11.00	1.50	222.71	-	3117.67	nr	3429.43
2090 x 2390mm overall	3163.41	2.50	3242.49	12.00	1.50	241.08	-	3483.57	nr	3831.93
2090 X 3590mm overall	4689.16	2.50	4806.39	12.50	2.00	257.15	-	5063.53	nr	5569.89
Door frames in softwood, wrought										
Note										
Notwithstanding the requirements of the NRM frames shown in metres exclude bedding and pointing										
38 x 75mm jambs	3.85	2.50	3.95	0.18	0.02	3.58	0.05	7.57	m	8.33
38 x 100mm jambs	4.80	2.50	4.92	0.18	0.02	3.58	0.05	8.55	m	9.40
50 x 75mm jambs	4.75	2.50	4.87	0.20	0.03	4.09	0.05	9.00	m	9.90
50 x 100mm jambs	6.00	2.50	6.15	0.20	0.03	4.09	0.05	10.28	m	11.31
50 x 125mm jambs	7.25	2.50	7.43	0.20	0.03	4.09	0.05	11.56	m	12.72
63 x 100mm jambs	7.30	2.50	7.48	0.20	0.03	4.09	0.05	11.61	m	12.78
75 x 100mm jambs	8.50	2.50	8.71	0.22	0.03	4.45	0.06	13.23	m	14.55

DOORS, SHUTTERS AND HATCHES

Labour hourly rates: (except Specialists) Craft Operatives 18.37 Labourer 13.76 Rates are national average prices. Refer to REGIONAL VARIATIONS for indicative levels of overall pricing in regions	MATERIALS			LABOUR				RATES		
	Del to Site	Waste	Material Cost	Craft Optve	Lab	Labour Cost	Sunds	Nett Rate		Gross rate (10%)
	£	%	£	Hrs	Hrs	£	£	£	Unit	£

TIMBER DOORS AND FRAMES (Cont'd)

Note (Cont'd)

Notwithstanding the requirements of the NRM frames shown in metres exclude bedding and pointing (Cont'd)

	Del to Site	Waste	Material Cost	Craft Optve	Lab	Labour Cost	Sunds	Nett Rate	Unit	Gross rate (10%)
75 x 113mm jambs	9.48	2.50	9.72	0.22	0.03	4.45	0.06	14.23	m	15.65
100 x 100mm jambs	11.00	2.50	11.27	0.25	0.03	5.01	0.06	16.34	m	17.97
100 x 113mm jambs	12.30	2.50	12.61	0.27	0.03	5.37	0.06	18.04	m	19.84
100 x 125mm jambs	13.50	2.50	13.84	0.27	0.03	5.37	0.06	19.27	m	21.20
113 x 113mm jambs	13.77	2.50	14.11	0.27	0.03	5.37	0.06	19.55	m	21.50
113 x 125mm jambs	15.13	2.50	15.51	0.30	0.04	6.06	0.08	21.64	m	23.81
113 x 150mm jambs	17.95	2.50	18.40	0.30	0.04	6.06	0.08	24.54	m	26.99
38 x 75mm jambs; labours -1	3.85	2.50	3.95	0.18	0.02	3.58	0.05	7.57	m	8.33
38 x 75mm jambs; labours -2	4.85	2.50	4.97	0.18	0.02	3.58	0.05	8.60	m	9.46
38 x 75mm jambs; labours -3	5.85	2.50	6.00	0.18	0.02	3.58	0.05	9.62	m	10.59
38 x 75mm jambs; labours -4	6.85	2.50	7.02	0.18	0.02	3.58	0.05	10.65	m	11.71
38 x 75mm heads	3.85	2.50	3.95	0.18	0.02	3.58	0.05	7.57	m	8.33
38 x 100mm heads	4.80	2.50	4.92	0.18	0.02	3.58	0.05	8.55	m	9.40
50 x 75mm heads	4.75	2.50	4.87	0.20	0.03	4.09	0.05	9.00	m	9.90
50 x 100mm heads	6.00	2.50	6.15	0.20	0.03	4.09	0.05	10.28	m	11.31
50 x 125mm heads	7.25	2.50	7.43	0.20	0.03	4.09	0.05	11.56	m	12.72
63 x 100mm heads	7.30	2.50	7.48	0.20	0.03	4.09	0.05	11.61	m	12.78
75 x 100mm heads	8.50	2.50	8.71	0.22	0.03	4.45	0.06	13.23	m	14.55
75 x 113mm heads	9.48	2.50	9.72	0.22	0.03	4.45	0.06	14.23	m	15.65
100 x 100mm heads	11.00	2.50	11.27	0.25	0.03	5.01	0.06	16.34	m	17.97
100 x 113mm heads	12.30	2.50	12.61	0.27	0.03	5.37	0.06	18.04	m	19.84
100 x 125mm heads	13.50	2.50	13.84	0.27	0.03	5.37	0.06	19.27	m	21.20
113 x 113mm heads	13.77	2.50	14.11	0.27	0.03	5.37	0.06	19.55	m	21.50
113 x 125mm heads	15.13	2.50	15.51	0.30	0.04	6.06	0.08	21.64	m	23.81
113 x 150mm heads	17.95	2.50	18.40	0.30	0.04	6.06	0.08	24.54	m	26.99
38 x 75mm heads; labours -1	3.85	2.50	3.95	0.18	0.02	3.58	0.05	7.57	m	8.33
38 x 75mm heads; labours -2	4.85	2.50	4.97	0.18	0.02	3.58	0.05	8.60	m	9.46
38 x 75mm heads; labours -3	5.85	2.50	6.00	0.18	0.02	3.58	0.05	9.62	m	10.59
38 x 75mm heads; labours -4	6.85	2.50	7.02	0.18	0.02	3.58	0.05	10.65	m	11.71
63 x 125mm sills	8.88	2.50	9.10	0.23	0.03	4.64	0.06	13.80	m	15.18
75 x 125mm sills	10.38	2.50	10.64	0.24	0.03	4.82	0.06	15.52	m	17.07
75 x 150mm sills	12.25	2.50	12.56	0.25	0.03	5.01	0.06	17.62	m	19.38
63 x 125mm sills; labours -1	8.88	2.50	9.10	0.23	0.03	4.64	0.06	13.80	m	15.18
63 x 125mm sills; labours -2	9.88	2.50	10.13	0.23	0.03	4.64	0.06	14.82	m	16.31
63 x 125mm sills; labours -3	10.88	2.50	11.15	0.23	0.03	4.64	0.06	15.85	m	17.43
63 x 125mm sills; labours -4	11.88	2.50	12.18	0.23	0.03	4.64	0.06	16.87	m	18.56
63 x 100mm mullions	8.30	2.50	8.51	0.05	0.01	1.06	-	9.56	m	10.52
75 x 100mm mullions	9.50	2.50	9.74	0.05	0.01	1.06	-	10.79	m	11.87
75 x 113mm mullions	10.48	2.50	10.74	0.05	0.01	1.06	-	11.80	m	12.98
100 x 100mm mullions	12.00	2.50	12.30	0.05	0.01	1.06	-	13.36	m	14.69
100 x 125mm mullions	13.30	2.50	13.63	0.05	0.01	1.06	-	14.69	m	16.16
63 x 100mm mullions; labours -1	8.30	2.50	8.51	0.05	0.01	1.06	-	9.56	m	10.52
63 x 100mm mullions; labours -2	8.30	2.50	8.51	0.05	0.01	1.06	-	9.56	m	10.52
63 x 100mm mullions; labours -3	9.30	2.50	9.53	0.05	0.01	1.06	-	10.59	m	11.65
63 x 100mm mullions; labours -4	10.30	2.50	10.56	0.05	0.01	1.06	-	11.61	m	12.77
63 x 100mm transoms	8.30	2.50	8.51	0.05	0.01	1.06	-	9.56	m	10.52
75 x 100mm transoms	9.50	2.50	9.74	0.05	0.01	1.06	-	10.79	m	11.87
75 x 113mm transoms	10.48	2.50	10.74	0.05	0.01	1.06	-	11.80	m	12.98
100 x 100mm transoms	12.00	2.50	12.30	0.05	0.01	1.06	-	13.36	m	14.69
100 x 125mm transoms	13.30	2.50	13.63	0.05	0.01	1.06	-	14.69	m	16.16
63 x 100mm transoms; labours -1	8.30	2.50	8.51	0.05	0.01	1.06	-	9.56	m	10.52
63 x 100mm transoms; labours -2	8.30	2.50	8.51	0.05	0.01	1.06	-	9.56	m	10.52
63 x 100mm transoms; labours -3	9.30	2.50	9.53	0.05	0.01	1.06	-	10.59	m	11.65
63 x 100mm transoms; labours -4	10.30	2.50	10.56	0.05	0.01	1.06	-	11.61	m	12.77

Internal door frame sets; supplied unassembled
38 x 63mm jambs and head; stops (supplied loose); assembling

	Del to Site	Waste	Material Cost	Craft Optve	Lab	Labour Cost	Sunds	Nett Rate	Unit	Gross rate (10%)
for 686 x 1981mm doors	16.60	2.50	17.01	1.00	0.13	20.16	0.24	37.41	nr	41.16
for 762 x 1981mm doors	16.60	2.50	17.01	1.00	0.13	20.16	0.24	37.41	nr	41.16

38 x 75mm jambs and head; stops (supplied loose); assembling

	Del to Site	Waste	Material Cost	Craft Optve	Lab	Labour Cost	Sunds	Nett Rate	Unit	Gross rate (10%)
for 686 x 1981mm doors	17.90	2.50	18.35	1.00	0.13	20.16	0.24	38.75	nr	42.62
for 762 x 1981mm doors	17.90	2.50	18.35	1.00	0.13	20.16	0.24	38.75	nr	42.62

50 x 100mm jambs and head; stops (supplied loose); assembling

	Del to Site	Waste	Material Cost	Craft Optve	Lab	Labour Cost	Sunds	Nett Rate	Unit	Gross rate (10%)
for 686 x 1981mm doors	24.50	2.50	25.11	1.00	0.13	20.16	0.24	45.51	nr	50.06
for 762 x 1981mm doors	24.50	2.50	25.11	1.00	0.13	20.16	0.24	45.51	nr	50.06

50 x 113mm jambs and head; stops (supplied loose); assembling

	Del to Site	Waste	Material Cost	Craft Optve	Lab	Labour Cost	Sunds	Nett Rate	Unit	Gross rate (10%)
for 686 x 1981mm doors	27.00	2.50	27.67	1.00	0.13	20.16	0.24	48.07	nr	52.88
for 762 x 1981mm doors	27.00	2.50	27.67	1.00	0.13	20.16	0.24	48.07	nr	52.88

External door frame sets; one coat external primer before delivery to site; Bedding in cement mortar (1:3); pointing with Silicone standard mastic one side
63 x 75mm jambs and head; rebates -1

	Del to Site	Waste	Material Cost	Craft Optve	Lab	Labour Cost	Sunds	Nett Rate	Unit	Gross rate (10%)
for 762 x 1981 x 50mm doors	25.86	2.50	26.79	1.95	0.13	37.52	0.24	64.55	nr	71.00
for 838 x 1981 x 50mm doors	25.94	2.50	26.87	1.96	0.13	37.79	0.24	64.91	nr	71.40

63 x 88mm jambs and head; rebates -1

	Del to Site	Waste	Material Cost	Craft Optve	Lab	Labour Cost	Sunds	Nett Rate	Unit	Gross rate (10%)
for 762 x 1981 x 50mm doors	28.36	2.50	29.35	1.95	0.13	37.52	0.24	67.11	nr	73.82
for 838 x 1981 x 50mm doors	28.44	2.50	29.43	1.96	0.13	37.79	0.24	67.47	nr	74.22

63 x 75mm jambs and head, rebates -1 nr; hardwood sill

	Del to Site	Waste	Material Cost	Craft Optve	Lab	Labour Cost	Sunds	Nett Rate	Unit	Gross rate (10%)
for 762 x 1981 x 50mm doors	26.14	2.50	27.12	2.35	0.16	45.32	0.30	72.74	nr	80.01
for 838 x 1981 x 50mm doors	56.30	2.50	58.04	2.38	0.16	45.89	0.30	104.23	nr	114.65

63 x 88mm jambs and head, rebates -1 nr; hardwood sill

	Del to Site	Waste	Material Cost	Craft Optve	Lab	Labour Cost	Sunds	Nett Rate	Unit	Gross rate (10%)
for 762 x 1981 x 50mm doors	57.64	2.50	59.41	2.35	0.16	45.32	0.30	105.03	nr	115.53
for 838 x 1981 x 50mm doors	57.80	2.50	59.58	2.38	0.16	45.89	0.30	105.77	nr	116.34
for 1168 x 1981 x 50mm doors	58.48	2.50	60.32	2.76	0.19	53.32	0.36	113.99	nr	125.39

Labour hourly rates: (except Specialists) Craft Operatives 18.37 Labourer 13.76 Rates are national average prices. Refer to REGIONAL VARIATIONS for indicative levels of overall pricing in regions	MATERIALS			LABOUR				RATES		
	Del to Site £	Waste %	Material Cost £	Craft Optve Hrs	Lab Hrs	Labour Cost £	Sunds £	Nett Rate £	Unit	Gross rate (10%) £

TIMBER DOORS AND FRAMES (Cont'd)

Note (Cont'd)

External garage door frame sets; one coat external primer before delivery to site; supplied unassembled; Bedding in cement mortar (1:3); pointing with Silicone standard mastic one side
75 x 100mm jambs and head; assembling

	Del to Site £	Waste %	Material Cost £	Craft Optve Hrs	Lab Hrs	Labour Cost £	Sunds £	Nett Rate £	Unit	Gross rate (10%) £
for 2134 x 1981mm side hung doors	67.27	2.50	69.32	3.47	0.28	67.58	0.56	137.45	nr	151.19
for 2134 x 2134mm side hung doors	67.59	2.50	69.66	3.53	0.28	68.70	0.56	138.91	nr	152.80
75 x 75mm jambs and head; assembling										
for 2134 x 1981mm up and over doors	55.77	2.50	57.53	3.47	0.28	67.58	0.56	125.66	nr	138.23

Door frames in Afromosia, wrought, selected for transparent finish

Note

Notwithstanding the requirements of the NRM frames shown in metres exclude bedding and pointing

	Del to Site £	Waste %	Material Cost £	Craft Optve Hrs	Lab Hrs	Labour Cost £	Sunds £	Nett Rate £	Unit	Gross rate (10%) £
38 x 75mm jambs	15.89	2.50	16.29	0.31	0.04	6.25	0.08	22.61	m	24.87
38 x 100mm jambs	18.31	2.50	18.76	0.31	0.04	6.25	0.08	25.08	m	27.59
50 x 75mm jambs	18.31	2.50	18.76	0.35	0.04	6.98	0.09	25.83	m	28.42
50 x 100mm jambs	21.50	2.50	22.04	0.35	0.04	6.98	0.09	29.11	m	32.02
50 x 125mm jambs	23.84	2.50	24.44	0.35	0.04	6.98	0.09	31.51	m	34.66
63 x 100mm jambs	23.84	2.50	24.44	0.35	0.04	6.98	0.09	31.51	m	34.66
75 x 100mm jambs	30.61	2.50	31.38	0.38	0.05	7.67	0.09	39.14	m	43.05
75 x 113mm jambs	34.96	2.50	35.83	0.38	0.05	7.67	0.09	43.59	m	47.95
100 x 100mm jambs	38.97	2.50	39.95	0.44	0.05	8.77	0.10	48.82	m	53.70
100 x 113mm jambs	44.51	2.50	45.62	0.47	0.06	9.46	0.12	55.20	m	60.72
100 x 125mm jambs	50.10	2.50	51.36	0.47	0.06	9.46	0.12	60.94	m	67.03
113 x 113mm jambs	50.10	2.50	51.36	0.47	0.06	9.46	0.12	60.94	m	67.03
113 x 125mm jambs	55.61	2.50	57.00	0.52	0.07	10.52	0.14	67.65	m	74.42
113 x 150mm jambs	61.24	2.50	62.77	0.52	0.07	10.52	0.14	73.42	m	80.76
38 x 75mm jambs; labours -1	16.14	2.50	16.54	0.31	0.04	6.25	0.08	22.86	m	25.15
38 x 75mm jambs; labours -2	16.31	2.50	16.71	0.31	0.04	6.25	0.08	23.03	m	25.34
38 x 75mm jambs; labours -3	16.56	2.50	16.98	0.31	0.04	6.25	0.08	23.30	m	25.63
38 x 75mm jambs; labours -4	16.82	2.50	17.24	0.31	0.04	6.25	0.08	23.56	m	25.92
38 x 75mm heads	15.89	2.50	16.29	0.31	0.04	6.25	0.08	22.61	m	24.87
38 x 100mm heads	18.31	2.50	18.76	0.31	0.04	6.25	0.08	25.08	m	27.59
50 x 75mm heads	18.31	2.50	18.76	0.31	0.04	6.25	0.08	25.08	m	27.59
50 x 100mm heads	21.50	2.50	22.04	0.35	0.04	6.98	0.09	29.11	m	32.02
50 x 125mm heads	23.84	2.50	24.44	0.35	0.04	6.98	0.09	31.51	m	34.66
63 x 100mm heads	23.84	2.50	24.44	0.35	0.04	6.98	0.09	31.51	m	34.66
75 x 100mm heads	30.61	2.50	31.38	0.38	0.05	7.67	0.09	39.14	m	43.05
75 x 113mm heads	34.96	2.50	35.83	0.38	0.05	7.67	0.09	43.59	m	47.95
100 x 100mm heads	38.97	2.50	39.95	0.44	0.05	8.77	0.10	48.82	m	53.70
100 x 113mm heads	44.51	2.50	45.62	0.47	0.06	9.46	0.12	55.20	m	60.72
100 x 125mm heads	50.10	2.50	51.36	0.47	0.06	9.46	0.12	60.94	m	67.03
113 x 113mm heads	50.10	2.50	51.36	0.47	0.06	9.46	0.12	60.94	m	67.03
113 x 125mm heads	55.61	2.50	57.00	0.52	0.07	10.52	0.14	67.65	m	74.42
113 x 150mm heads	61.55	2.50	63.08	0.52	0.07	10.52	0.14	73.74	m	81.11
38 x 75mm heads; labours -1	16.14	2.50	16.54	0.31	0.04	6.25	0.08	22.86	m	25.15
38 x 75mm heads; labours -2	16.31	2.50	16.71	0.31	0.04	6.25	0.08	23.03	m	25.34
38 x 75mm heads; labours -3	16.56	2.50	16.98	0.31	0.04	6.25	0.08	23.30	m	25.63
38 x 75mm heads; labours -4	16.82	2.50	17.24	0.31	0.04	6.25	0.08	23.56	m	25.92
63 x 125mm sills	36.13	2.50	37.04	0.40	0.05	8.04	0.09	45.16	m	49.68
75 x 125mm sills	41.32	2.50	42.35	0.44	0.06	8.91	0.10	51.37	m	56.50
75 x 150mm sills	46.52	2.50	47.68	0.45	0.06	9.09	0.10	56.88	m	62.57
63 x 125mm sills; labours -1	36.56	2.50	37.47	0.40	0.05	8.04	0.09	45.60	m	50.16
63 x 125mm sills; labours -2	36.73	2.50	37.64	0.40	0.05	8.04	0.09	45.77	m	50.35
63 x 125mm sills; labours -3	36.88	2.50	37.80	0.40	0.05	8.04	0.09	45.93	m	50.52
63 x 125mm sills; labours -4	37.06	2.50	37.99	0.40	0.05	8.04	0.09	46.11	m	50.73
63 x 100mm mullions	23.84	2.50	24.44	0.05	0.01	1.06	-	25.49	m	28.04
75 x 100mm mullions	30.61	2.50	31.38	0.05	0.01	1.06	-	32.43	m	35.68
75 x 113mm mullions	34.96	2.50	35.83	0.05	0.01	1.06	-	36.89	m	40.58
100 x 100mm mullions	38.97	2.50	39.95	0.05	0.01	1.06	-	41.00	m	45.10
100 x 125mm mullions	50.10	2.50	51.36	0.05	0.01	1.06	-	52.41	m	57.65
63 x 100mm mullions; labours -1	24.08	2.50	24.69	0.05	0.01	1.06	-	25.74	m	28.32
63 x 100mm mullions; labours -2	24.25	2.50	24.86	0.05	0.01	1.06	-	25.91	m	28.51
63 x 100mm mullions; labours -3	24.42	2.50	25.03	0.05	0.01	1.06	-	26.09	m	28.69
63 x 100mm mullions; labours -4	24.60	2.50	25.22	0.05	0.01	1.06	-	26.27	m	28.90
63 x 100mm transoms	23.84	2.50	24.44	0.05	0.01	1.06	-	25.49	m	28.04
75 x 100mm transoms	30.61	2.50	31.38	0.05	0.01	1.06	-	32.43	m	35.68
75 x 113mm transoms	34.96	2.50	35.83	0.05	0.01	1.06	-	36.89	m	40.58
100 x 100mm transoms	38.97	2.50	39.95	0.05	0.01	1.06	-	41.00	m	45.10
100 x 125mm transoms	50.12	2.50	51.37	0.05	0.01	1.06	-	52.43	m	57.67
63 x 100mm transoms; labours -1	24.08	2.50	24.69	0.05	0.01	1.06	-	25.74	m	28.32
63 x 100mm transoms; labours -2	24.25	2.50	24.86	0.05	0.01	1.06	-	25.91	m	28.51
63 x 100mm transoms; labours -3	24.42	2.50	25.03	0.05	0.01	1.06	-	26.09	m	28.69
63 x 100mm transoms; labours -4	24.60	2.50	25.22	0.05	0.01	1.06	-	26.27	m	28.90

Door frames in Sapele, wrought, selected for transparent finish

Note

Notwithstanding the requirements of the NRM frames shown in metres exclude bedding and pointing

	Del to Site £	Waste %	Material Cost £	Craft Optve Hrs	Lab Hrs	Labour Cost £	Sunds £	Nett Rate £	Unit	Gross rate (10%) £
38 x 75mm jambs	11.03	2.50	11.31	0.27	0.03	5.37	0.06	16.74	m	18.41
38 x 100mm jambs	14.71	2.50	15.07	0.27	0.03	5.37	0.06	20.51	m	22.56
50 x 75mm jambs	14.51	2.50	14.88	0.30	0.04	6.06	0.08	21.01	m	23.11
50 x 100mm jambs	19.35	2.50	19.83	0.30	0.04	6.06	0.08	25.97	m	28.57
50 x 125mm jambs	24.19	2.50	24.79	0.30	0.04	6.06	0.08	30.93	m	34.02
63 x 100mm jambs	24.38	2.50	24.99	0.30	0.04	6.06	0.08	31.13	m	34.24
75 x 100mm jambs	29.02	2.50	29.75	0.33	0.04	6.61	0.08	36.44	m	40.08

DOORS, SHUTTERS AND HATCHES

Labour hourly rates: (except Specialists) Craft Operatives 18.37 Labourer 13.76 Rates are national average prices. Refer to REGIONAL VARIATIONS for indicative levels of overall pricing in regions	MATERIALS			LABOUR				RATES		
	Del to Site	Waste	Material Cost	Craft Optve	Lab	Labour Cost	Sunds	Nett Rate	Unit	Gross rate (10%)
	£	%	£	Hrs	Hrs	£	£	£		£

TIMBER DOORS AND FRAMES (Cont'd)

Note (Cont'd)

Notwithstanding the requirements of the NRM frames shown in metres exclude bedding and pointing (Cont'd)

	Del to Site £	Waste %	Material Cost £	Craft Optve Hrs	Lab Hrs	Labour Cost £	Sunds £	Nett Rate £	Unit	Gross rate (10%) £
75 x 113mm jambs	32.80	2.50	33.62	0.33	0.04	6.61	0.08	40.31	m	44.34
100 x 100mm jambs	38.70	2.50	39.67	0.37	0.05	7.48	0.09	47.24	m	51.97
100 x 113mm jambs	43.73	2.50	44.82	0.40	0.05	8.04	0.10	52.97	m	58.26
100 x 125mm jambs	48.38	2.50	49.58	0.40	0.05	8.04	0.10	57.73	m	63.50
113 x 113mm jambs	49.42	2.50	50.66	0.40	0.05	8.04	0.10	58.80	m	64.68
113 x 125mm jambs	54.67	2.50	56.04	0.45	0.06	9.09	0.10	65.23	m	71.76
113 x 150mm jambs	65.60	2.50	67.24	0.45	0.06	9.09	0.10	76.43	m	84.08
38 x 75mm jambs; labours -1	12.32	2.50	12.63	0.27	0.03	5.37	0.06	18.06	m	19.87
38 x 75mm jambs; labours -2	13.61	2.50	13.95	0.27	0.03	5.37	0.06	19.38	m	21.32
38 x 75mm jambs; labours -3	14.90	2.50	15.27	0.27	0.03	5.37	0.06	20.70	m	22.78
38 x 75mm jambs; labours -4	16.19	2.50	16.59	0.27	0.03	5.37	0.06	22.03	m	24.23
38 x 75mm heads	11.03	2.50	11.31	0.27	0.03	5.37	0.06	16.74	m	18.41
38 x 100mm heads	14.71	2.50	15.07	0.27	0.03	5.37	0.06	20.51	m	22.56
50 x 75mm heads	14.51	2.50	14.88	0.30	0.04	6.06	0.08	21.01	m	23.11
50 x 100mm heads	19.35	2.50	19.83	0.30	0.04	6.06	0.08	25.97	m	28.57
50 x 125mm heads	24.19	2.50	24.79	0.30	0.04	6.06	0.08	30.93	m	34.02
63 x 100mm heads	24.38	2.50	24.99	0.30	0.04	6.06	0.08	31.13	m	34.24
75 x 100mm heads	29.02	2.50	29.75	0.33	0.04	6.61	0.08	36.44	m	40.08
75 x 113mm heads	32.80	2.50	33.62	0.33	0.04	6.61	0.08	40.31	m	44.34
100 x 100mm heads	38.70	2.50	39.67	0.37	0.05	7.48	0.09	47.24	m	51.97
100 x 113mm heads	43.73	2.50	44.82	0.40	0.05	8.04	0.10	52.97	m	58.26
100 x 125mm heads	48.38	2.50	49.58	0.40	0.05	8.04	0.10	57.73	m	63.50
113 x 113mm heads	49.42	2.50	50.66	0.40	0.05	8.04	0.10	58.80	m	64.68
113 x 125mm heads	54.67	2.50	56.04	0.45	0.06	9.09	0.10	65.23	m	71.76
113 x 150mm heads	65.60	2.50	67.24	0.45	0.06	9.09	0.10	76.43	m	84.08
38 x 75mm heads; labours -1	12.32	2.50	12.63	0.27	0.03	5.37	0.06	18.06	m	19.87
38 x 75mm heads; labours -2	13.61	2.50	13.95	0.27	0.03	5.37	0.06	19.38	m	21.32
38 x 75mm heads; labours -3	14.90	2.50	15.27	0.27	0.03	5.37	0.06	20.70	m	22.78
38 x 75mm heads; labours -4	16.19	2.50	16.59	0.27	0.03	5.37	0.06	22.03	m	24.23
63 x 125mm sills	31.77	2.50	32.57	0.35	0.04	6.98	0.09	39.64	m	43.60
75 x 125mm sills	37.58	2.50	38.52	0.37	0.05	7.48	0.09	46.09	m	50.70
75 x 150mm sills	44.83	2.50	45.95	0.37	0.05	7.48	0.09	53.52	m	58.88
63 x 125mm sills; labours -1	31.77	2.50	32.57	0.35	0.04	6.98	0.09	39.64	m	43.60
63 x 125mm sills; labours -2	33.06	2.50	33.89	0.35	0.04	6.98	0.09	40.96	m	45.06
63 x 125mm sills; labours -3	34.35	2.50	35.21	0.35	0.04	6.98	0.09	42.28	m	46.51
63 x 125mm sills; labours -4	35.64	2.50	36.53	0.35	0.04	6.98	0.09	43.60	m	47.96
63 x 100mm mullions	26.96	2.50	27.64	0.05	0.01	1.06	-	28.69	m	31.56
75 x 100mm mullions	31.60	2.50	32.40	0.05	0.01	1.06	-	33.45	m	36.80
75 x 113mm mullions	35.38	2.50	36.27	0.05	0.01	1.06	-	37.33	m	41.06
100 x 100mm mullions	41.28	2.50	42.31	0.05	0.01	1.06	-	43.37	m	47.70
100 x 125mm mullions	50.96	2.50	52.23	0.05	0.01	1.06	-	53.28	m	58.61
63 x 100mm mullions; labours -1	26.96	2.50	27.64	0.05	0.01	1.06	-	28.69	m	31.56
63 x 100mm mullions; labours -2	26.96	2.50	27.64	0.05	0.01	1.06	-	28.69	m	31.56
63 x 100mm mullions; labours -3	28.25	2.50	28.96	0.05	0.01	1.06	-	30.01	m	33.01
63 x 100mm mullions; labours -4	29.54	2.50	30.28	0.05	0.01	1.06	-	31.34	m	34.47
63 x 100mm transoms	26.96	2.50	27.64	0.05	0.01	1.06	-	28.69	m	31.56
75 x 100mm transoms	31.60	2.50	32.40	0.05	0.01	1.06	-	33.45	m	36.80
75 x 113mm transoms	35.38	2.50	36.27	0.05	0.01	1.06	-	37.33	m	41.06
100 x 100mm transoms	41.28	2.50	42.31	0.05	0.01	1.06	-	43.37	m	47.70
100 x 125mm transoms	50.96	2.50	52.23	0.05	0.01	1.06	-	53.28	m	58.61
63 x 100mm transoms; labours -1	26.96	2.50	27.64	0.05	0.01	1.06	-	28.69	m	31.56
63 x 100mm transoms; labours -2	26.96	2.50	27.64	0.05	0.01	1.06	-	28.69	m	31.56
63 x 100mm transoms; labours -3	28.25	2.50	28.96	0.05	0.01	1.06	-	30.01	m	33.01
63 x 100mm transoms; labours -4	29.54	2.50	30.28	0.05	0.01	1.06	-	31.34	m	34.47

Door linings in softwood, wrought

linings; fixing to masonry with screws

	Del to Site £	Waste %	Material Cost £	Craft Optve Hrs	Lab Hrs	Labour Cost £	Sunds £	Nett Rate £	Unit	Gross rate (10%) £
32 x 113mm	2.65	2.50	2.71	0.40	0.05	8.04	0.10	10.85	m	11.94
32 x 150mm	3.46	2.50	3.54	0.40	0.05	8.04	0.10	11.69	m	12.85
32 x 225mm	5.19	2.50	5.32	0.60	0.07	11.99	0.15	17.45	m	19.20
32 x 330mm	7.61	2.50	7.80	0.60	0.07	11.99	0.15	19.93	m	21.93
38 x 113mm	3.09	2.50	3.17	0.40	0.05	8.04	0.10	11.31	m	12.44
38 x 150mm	4.11	2.50	4.21	0.40	0.05	8.04	0.10	12.35	m	13.59
38 x 225mm	6.16	2.50	6.31	0.60	0.07	11.99	0.15	18.45	m	20.29
38 x 330mm	9.03	2.50	9.26	0.60	0.07	11.99	0.15	21.40	m	23.54

linings

	Del to Site £	Waste %	Material Cost £	Craft Optve Hrs	Lab Hrs	Labour Cost £	Sunds £	Nett Rate £	Unit	Gross rate (10%) £
32 x 113mm; labours -1	3.70	2.50	3.79	0.20	0.02	3.95	0.06	7.80	m	8.58
32 x 113mm; labours -2	4.22	2.50	4.33	0.20	0.02	3.95	0.06	8.34	m	9.17
32 x 113mm; labours -3	4.75	2.50	4.86	0.20	0.02	3.95	0.06	8.87	m	9.76
32 x 113mm; labours -4	5.27	2.50	5.40	0.20	0.02	3.95	0.06	9.41	m	10.35
38 x 113mm; labours -1	4.14	2.50	4.25	0.20	0.02	3.95	0.06	8.26	m	9.08
38 x 113mm; labours -2	4.67	2.50	4.79	0.20	0.02	3.95	0.06	8.79	m	9.67
38 x 113mm; labours -3	5.19	2.50	5.32	0.20	0.02	3.95	0.06	9.33	m	10.27
38 x 113mm; labours -4	5.72	2.50	5.86	0.20	0.02	3.95	0.06	9.87	m	10.86

Internal door lining sets; supplied unassembled; fixing to masonry with screws

	Del to Site £	Waste %	Material Cost £	Craft Optve Hrs	Lab Hrs	Labour Cost £	Sunds £	Nett Rate £	Unit	Gross rate (10%) £
32 x 115mm rebated linings; assembling										
for 610 x 1981mm doors	14.75	2.50	15.12	2.00	0.25	40.18	0.48	55.78	nr	61.36
for 686 x 1981mm doors	14.75	2.50	15.12	2.00	0.25	40.18	0.48	55.78	nr	61.36
for 762 x 1981mm doors	14.75	2.50	15.12	2.00	0.25	40.18	0.48	55.78	nr	61.36
32 x 140mm rebated linings; assembling										
for 610 x 1981mm doors	16.65	2.50	17.07	2.00	0.25	40.18	0.48	57.73	nr	63.50
for 686 x 1981mm doors	16.65	2.50	17.07	2.00	0.25	40.18	0.48	57.73	nr	63.50
for 762 x 1981mm doors	16.65	2.50	17.07	2.00	0.25	40.18	0.48	57.73	nr	63.50

Door linings in Oak, wrought, selected for transparent finish

linings; fixing to masonry with screws

	Del to Site £	Waste %	Material Cost £	Craft Optve Hrs	Lab Hrs	Labour Cost £	Sunds £	Nett Rate £	Unit	Gross rate (10%) £
32 x 113mm	12.00	2.50	12.30	0.60	0.08	12.12	0.15	24.57	m	27.03

Labour hourly rates: (except Specialists) Craft Operatives 18.37 Labourer 13.76 Rates are national average prices. Refer to REGIONAL VARIATIONS for indicative levels of overall pricing in regions	MATERIALS			LABOUR				RATES		
	Del to Site £	Waste %	Material Cost £	Craft Optve Hrs	Lab Hrs	Labour Cost £	Sunds £	Nett Rate £	Unit	Gross rate (10%) £

TIMBER DOORS AND FRAMES (Cont'd)

Door linings in Oak, wrought, selected for transparent finish (Cont'd)

linings; fixing to masonry with screws (Cont'd)

32 x 150mm	13.31	2.50	13.64	0.60	0.08	12.12	0.15	25.92	m	28.51
32 x 225mm	23.76	2.50	24.35	0.90	0.11	18.05	0.22	42.63	m	46.89
32 x 330mm	23.73	2.50	24.32	0.90	0.11	18.05	0.22	42.59	m	46.85
38 x 113mm	16.65	2.50	17.07	0.60	0.08	12.12	0.15	29.34	m	32.27
38 x 150mm	20.19	2.50	20.69	0.60	0.08	12.12	0.15	32.97	m	36.26
38 x 225mm	27.10	2.50	27.78	0.90	0.11	18.05	0.22	46.05	m	50.65
38 x 330mm	43.11	2.50	44.19	0.90	0.11	18.05	0.22	62.46	m	68.71

linings

32 x 113mm; labours -1	13.42	2.50	13.76	0.30	0.04	6.06	0.09	19.91	m	21.90
32 x 113mm; labours -2	13.56	2.50	13.90	0.30	0.04	6.06	0.09	20.05	m	22.06
32 x 113mm; labours -3	13.68	2.50	14.02	0.30	0.04	6.06	0.09	20.17	m	22.19
32 x 113mm; labours -4	13.82	2.50	14.17	0.30	0.04	6.06	0.09	20.32	m	22.35
38 x 113mm; labours -1	15.62	2.50	16.01	0.30	0.04	6.06	0.09	22.16	m	24.38
38 x 113mm; labours -2	15.76	2.50	16.15	0.30	0.04	6.06	0.09	22.31	m	24.54
38 x 113mm; labours -3	15.89	2.50	16.29	0.30	0.04	6.06	0.09	22.44	m	24.68
38 x 113mm; labours -4	16.02	2.50	16.42	0.30	0.04	6.06	0.09	22.57	m	24.83

Door linings in Sapele, wrought, selected for transparent finish

linings; fixing to masonry with screws

32 x 113mm	17.57	2.50	18.01	0.55	0.07	11.07	0.14	29.21	m	32.13
32 x 150mm	21.09	2.50	21.62	0.55	0.07	11.07	0.14	32.82	m	36.10
32 x 225mm	31.63	2.50	32.42	0.80	0.08	15.80	0.21	48.43	m	53.27
32 x 330mm	49.20	2.50	50.43	0.80	0.08	15.80	0.21	66.44	m	73.08
38 x 113mm	20.58	2.50	21.09	0.55	0.07	11.07	0.14	32.29	m	35.52
38 x 150mm	24.59	2.50	25.20	0.55	0.07	11.07	0.14	36.40	m	40.04
38 x 225mm	36.13	2.50	37.04	0.80	0.08	15.80	0.21	53.04	m	58.35
38 x 330mm	58.22	2.50	59.67	0.80	0.08	15.80	0.21	75.68	m	83.25

linings

32 x 113mm; labours -1	17.88	2.50	18.33	0.27	0.03	5.37	0.06	23.76	m	26.13
32 x 113mm; labours -2	18.07	2.50	18.52	0.27	0.03	5.37	0.06	23.96	m	26.35
32 x 113mm; labours -3	18.27	2.50	18.72	0.27	0.03	5.37	0.06	24.16	m	26.57
32 x 113mm; labours -4	18.47	2.50	18.93	0.27	0.03	5.37	0.06	24.37	m	26.80
38 x 113mm; labours -1	20.89	2.50	21.41	0.27	0.03	5.37	0.06	26.84	m	29.52
38 x 113mm; labours -2	21.09	2.50	21.62	0.27	0.03	5.37	0.06	27.05	m	29.76
38 x 113mm; labours -3	21.28	2.50	21.82	0.27	0.03	5.37	0.06	27.25	m	29.97
38 x 113mm; labours -4	21.49	2.50	22.03	0.27	0.03	5.37	0.06	27.46	m	30.21

Roller shutters in wood

Pole and hook operation

2185 x 2135mm	1570.00	-	1570.00	12.00	12.00	385.56	37.50	1993.06	nr	2192.37

TIMBER SHUTTERS AND HATCHES

Trap doors in softwood, wrought

19mm matchboarding on 25 x 75mm ledges

457 x 610mm	28.92	2.50	29.64	0.18	0.02	3.58	-	33.22	nr	36.54
610 x 610mm	35.76	2.50	36.66	0.19	0.02	3.77	-	40.42	nr	44.46
762 x 610mm	44.10	2.50	45.20	0.20	0.02	3.95	-	49.15	nr	54.07

Trap doors in B.C. Pine, wrought

19mm matchboarding on 25 x 75mm ledges

457 x 610mm	35.82	2.50	36.71	0.18	0.02	3.58	-	40.29	nr	44.32
610 x 610mm	50.63	2.50	51.90	0.19	0.02	3.77	-	55.66	nr	61.23
762 x 610mm	64.72	2.50	66.34	0.20	0.02	3.95	-	70.29	nr	77.32

METAL DOORS AND FRAMES

Doors and sidelights in galvanized steel; door sets

Crittall Windows Ltd. Duralife Homelight; weatherstripping; fixing to masonry with lugs; Bedding in cement mortar (1:3); pointing with Silicone standard mastic one side

761 x 2056mm doors NA15	1138.18	2.50	1166.97	3.93	1.40	91.40	-	1258.37	nr	1384.21
997 x 2056mm doors NA2	1710.73	2.50	1753.86	4.22	1.50	98.18	-	1852.04	nr	2037.25
1143 x 2056mm doors NA25	1743.09	2.50	1787.04	4.48	1.60	104.31	-	1891.36	nr	2080.49
279 x 2056mm sidelights, NA6	274.30	2.50	281.44	2.13	0.60	47.46	-	328.90	nr	361.79
508 x 2056mm sidelights, NA5	332.82	2.50	341.45	2.53	0.75	56.72	-	398.17	nr	437.99
997 x 2056mm sidelights, NA13F	629.44	2.50	645.54	3.02	0.90	67.88	-	713.42	nr	784.76

Doors and sidelights in galvanized steel; polyester powder coated; white matt; door sets

Crittall Windows Ltd. Duralife Homelight; weatherstripping; fixing to masonry with lugs; Bedding in cement mortar (1:3); pointing with Silicone standard mastic one side

761 x 2056mm doors NA15	1474.38	2.50	1511.57	3.60	1.40	85.34	-	1596.92	nr	1756.61
997 x 2056mm doors NA2	2215.82	2.50	2271.58	4.22	1.50	98.18	-	2369.76	nr	2606.74
1143 x 2056mm doors NA25	2258.32	2.50	2315.16	4.48	1.60	104.31	-	2419.47	nr	2661.42
279 x 2056mm sidelights, NA6	349.05	2.50	358.06	2.13	0.60	47.46	-	405.52	nr	446.07
508 x 2056mm sidelights, NA5	423.68	2.50	434.58	2.53	0.75	56.72	-	491.30	nr	540.43
997 x 2056mm sidelights, NA13F	786.73	2.50	806.76	3.02	0.90	67.88	-	874.64	nr	962.10

Labour hourly rates: (except Specialists) Craft Operatives 18.37 Labourer 13.76 Rates are national average prices. Refer to REGIONAL VARIATIONS for indicative levels of overall pricing in regions	MATERIALS			LABOUR				RATES		
	Del to Site	Waste	Material Cost	Craft Optve	Lab	Labour Cost	Sunds	Nett Rate	Unit	Gross rate (10%)
	£	%	£	Hrs	Hrs	£	£	£		£
METAL DOORS AND FRAMES (Cont'd)										
Garage doors; Cardale Gemini in galvanized steel; Plastisol finish by manufacturer										
Overhead garage doors; tensioning device Tracked spring counterbalanced										
fixing to timber with screws; for opening size										
2134 x 1981mm	561.67	2.50	575.71	4.00	4.00	128.52	-	704.23	nr	774.65
2134 x 2134mm	604.17	2.50	619.27	4.00	4.00	128.52	-	747.79	nr	822.57
Overhead garage doors; counterbalanced by springs										
4267 x 1981mm	1010.00	2.50	1035.25	6.00	6.00	192.78	-	1228.03	nr	1350.83
4267 x 2134mm	1025.83	2.50	1051.48	6.00	6.00	192.78	-	1244.26	nr	1368.69
METAL SHUTTERS AND HATCHES										
Roller shutters in steel, galvanized										
Note roller shutters are always purpose made to order and the following prices are indicative only. Firm quotations should always be obtained, bedding and pointing is excluded.										
Crank handle operation 4572 x 3658mm overall; fixing to masonry with screws	1910.65	2.50	1958.42	16.00	16.00	514.08	-	2472.50	nr	2719.75
Endless hand chain operation 5486 x 4267mm overall; fixing to masonry with screws	2001.29	2.50	2051.32	20.00	20.00	642.60	-	2693.92	nr	2963.31
Electric motor operation 6096 x 6096mm overall; fixing to masonry with screws	3730.66	2.50	3823.93	30.00	30.00	963.90	-	4787.83	nr	5266.61
Roller shutters in aluminium										
Endless hand chain operation 3553 x 3048mm overall; fixing to masonry with screws	1593.00	2.50	1632.82	20.00	20.00	642.60	-	2275.42	nr	2502.97
Note										
Bedding in cement mortar (1:3); pointing with Silicone standard mastic one side' is included with standard door sets as noted above; adjustment will need to be made for two part polysuphide pointing and bedding and pointing to purpose made frames etc										
Bedding and pointing frames										
Pointing with Silicone standard mastic metal frames one side	1.36	10.00	1.49	0.12	0.06	3.03	-	4.52	m	4.98
Bedding in cement mortar (1:3); pointing with Silicone standard mastic one side metal frames one side	1.48	5.00	1.61	0.20	0.10	5.05	-	6.66	m	7.33
PLASTIC DOORS & FRAMES										
Flexible doors mainly in plastics and rubber										
Standard doors; galvanised steel frame; top and bottom plates with pivoting arrangements and spring unit; Duo panel 7mm clear PVC/rubber panel doors										
fixing to masonry; for opening size										
1800 x 2100mm	932.72	2.50	956.03	4.00	4.00	128.52	-	1084.55	nr	1193.01
2440 x 2400mm	1444.97	2.50	1481.09	5.00	5.00	160.65	-	1641.74	nr	1805.92
Standard doors; galvanised steel frame; top and bottom plates with pivoting arrangements and spring unit; Duo panel 7mm clear PVC/black PVC panel doors										
fixing to masonry; for opening size										
1800 x 2100mm	821.23	2.50	841.76	4.00	4.00	128.52	-	970.28	nr	1067.30
2440 x 2400mm	1272.24	2.50	1304.05	5.00	5.00	160.65	-	1464.70	nr	1611.17
Standard doors; galvanised steel frame; top and bottom plates with pivoting arrangements and spring unit; 7mm clear PVC doors										
fixing to masonry; for opening size										
1800 x 2100mm	801.74	2.50	821.78	3.85	3.85	123.70	-	945.48	nr	1040.03
2440 x 2400mm	1086.80	2.50	1113.97	4.40	4.40	141.37	-	1255.34	nr	1380.88
Note										
Bedding in cement mortar (1:3); pointing with Silicone standard mastic one side' is included with external door sets a snoted above; adjustment will need to be made for two part polysuphide pointing and bedding and pointing to purpose made external frames										
Bedding and pointing frames										
Bedding in cement mortar (1:3); pointing one side wood frames	0.35	5.00	0.37	0.14	-	2.57	-	2.94	m	3.23
Bedding in cement mortar (1:3); pointing each side wood frames	0.47	5.00	0.49	0.20	-	3.67	-	4.16	m	4.58
Bedding in cement mortar (1:3); pointing with Silicone standard mastic one side wood frames one side	10.92	5.00	11.84	0.20	-	3.67	-	15.52	m	17.07

Labour hourly rates: (except Specialists) Craft Operatives 18.37 Labourer 13.76 Rates are national average prices. Refer to REGIONAL VARIATIONS for indicative levels of overall pricing in regions	MATERIALS			LABOUR				RATES		
	Del to Site	Waste	Material Cost	Craft Optve	Lab	Labour Cost	Sunds	Nett Rate	Unit	Gross rate (10%)
	£	%	£	Hrs	Hrs	£	£	£		£
PLASTIC DOORS & FRAMES (Cont'd)										
Flexible strip curtains mainly in pvc										
Strip curtains; 200mm wide x 2.0mm thick clear pvc strip curtains, suspended from stainless steel hook-on plates and track; fixing to masonry										
double overlap; for opening size										
1800 x 2100mm	127.01	2.50	130.18	2.00	2.00	64.26	-	194.44	nr	213.89
2100 x 2440mm	172.17	2.50	176.47	2.30	2.30	73.90	-	250.37	nr	275.41
single overlap; for opening size										
1800 x 2100mm	104.98	2.50	107.60	1.80	1.80	57.83	-	165.44	nr	181.98
2100 x 2440mm	142.31	2.50	145.86	2.10	2.10	67.47	-	213.34	nr	234.67
Strip curtains; 200mm wide x 2.0mm thick clear pvc strip curtains, with sliding door kit; fixing to masonry										
double overlap; for opening size										
1800 x 2100mm	413.66	2.50	424.00	2.20	2.20	70.69	-	494.69	nr	544.15
2100 x 2440mm	505.86	2.50	518.50	2.53	2.53	81.29	-	599.79	nr	659.77
single overlap; for opening size										
1800 x 2100mm	391.63	2.50	401.42	2.00	2.00	64.26	-	465.68	nr	512.25
2100 x 2440mm	476.00	2.50	487.90	2.33	2.33	74.86	-	562.76	nr	619.04
Strip curtains; 300mm wide x 2.5mm thick clear pvc strip curtains, suspended from stainless steel hook-on plates and track; fixing to masonry										
double overlap; for opening size										
2440 x 2740mm	266.75	2.50	273.42	2.64	2.64	84.82	-	358.24	nr	394.07
2740 x 2900mm	317.05	2.50	324.97	2.94	2.94	94.46	-	419.44	nr	461.38
single overlap; for opening size										
2440 x 2740mm	202.45	2.50	207.51	2.44	2.44	78.40	-	285.91	nr	314.50
2740 x 2900mm	240.62	2.50	246.63	2.74	2.74	88.04	-	334.67	nr	368.14
Strip curtains; 300mm wide x 3.0mm thick clear pvc strip curtains, with sliding door kit; fixing to masonry										
double overlap; for opening size										
2440 x 2740mm	647.48	2.50	663.67	2.90	2.90	93.18	-	756.85	nr	832.53
2740 x 2900mm	730.80	2.50	749.07	3.23	3.23	103.78	-	852.85	nr	938.13
single overlap; for opening size										
2440 x 2740mm	583.18	2.50	597.76	2.68	2.68	86.11	-	683.87	nr	752.26
2740 x 2900mm	655.16	2.50	671.54	3.01	3.01	96.71	-	768.25	nr	845.07
Patio doors in uPVC; white										
Double glazed clear Planibel 'A' toughened glass, one opening leaf, fixing frame to masonry with screws; Bedding in cement mortar (1:3); pointing with Silicone standard mastic one side										
1500 x 2020mm overall, including cill and vent	527.32	2.50	540.93	13.43	1.50	267.39	-	808.32	nr	889.15
1790 x 2090mm overall, including cill and vent	558.41	2.50	572.83	13.55	1.50	269.59	-	842.42	nr	926.66
2390 x 2090mm overall, including cill and vent	613.20	2.50	629.06	13.79	1.50	274.00	-	903.06	nr	993.37
French doors in uPVC; white										
Double glazed clear Planibel 'A' toughened glass, pair opening doors; Bedding in cement mortar (1:3); pointing with Silicone standard mastic one side										
1490 x 2090mm overall, including cill and vent	471.09	2.50	483.29	13.43	1.50	267.39	-	750.68	nr	825.75
1790 x 2090mm overall, including cill and vent	496.04	2.50	508.90	13.55	1.50	269.59	-	778.49	nr	856.34
2090 x 2090mm overall, including cill, pair of 300 side panels and vent	609.70	2.50	625.44	15.67	2.00	315.41	-	940.86	nr	1034.94
2400 x 2090mm overall; including cill, pair of 300 side panels and vent	719.63	2.50	738.15	15.79	2.00	317.62	-	1055.77	nr	1161.35
PLASTIC SHUTTERS AND HATCHES										
Loft hatch door and frame: insulated white plastic; hinged; lock										
Sets										
686 x 856 overall fitting	30.83	2.50	31.60	1.00	0.15	20.43	-	52.03	nr	57.23
GLASS DOORS										
Glass doors; glass, BS 952, toughened clear, polished plate; fittings finish BMA, satin chrome or polished chrome; prices include the provision of floor springs but exclude handles										
12mm thick										
750mm wide x 2150mm high	1609.00	-	1609.00	9.00	9.00	289.17	-	1898.17	nr	2087.99
762mm wide x 2134mm high	1519.00	-	1519.00	9.00	9.00	289.17	-	1808.17	nr	1988.99
800mm wide x 2150mm high	1664.00	-	1664.00	9.00	9.00	289.17	-	1953.17	nr	2148.49
838mm wide x 2134mm high	1597.00	-	1597.00	9.00	9.00	289.17	-	1886.17	nr	2074.79
850mm wide x 2150mm high	1712.00	-	1712.00	9.00	9.00	289.17	-	2001.17	nr	2201.29
900mm wide x 2150mm high	1769.00	-	1769.00	9.00	9.00	289.17	-	2058.17	nr	2263.99
914mm wide x 2134mm high	1662.00	-	1662.00	9.00	9.00	289.17	-	1951.17	nr	2146.29
950mm wide x 2150mm high	1879.00	-	1879.00	10.00	10.00	321.30	-	2200.30	nr	2420.33
1000mm wide x 2150mm high	1994.00	-	1994.00	10.00	10.00	321.30	-	2315.30	nr	2546.83
1100mm wide x 2150mm high	2225.00	-	2225.00	11.00	11.00	353.43	-	2578.43	nr	2836.27
1200mm wide x 2150mm high	2329.00	-	2329.00	11.00	11.00	353.43	-	2682.43	nr	2950.67
ASSOCIATED IRONMONGERY										
Intumescent strips and smoke seals										
white self adhesive										
10 x 4mm intumescent strip, half hour application; setting into groove in timber frame or door	1.43	10.00	1.58	0.15	-	2.76	-	4.33	m	4.77

Labour hourly rates: (except Specialists) Craft Operatives 18.37 Labourer 13.76 Rates are national average prices. Refer to REGIONAL VARIATIONS for indicative levels of overall pricing in regions	MATERIALS			LABOUR				RATES		
	Del to Site £	Waste %	Material Cost £	Craft Optve Hrs	Lab Hrs	Labour Cost £	Sunds £	Nett Rate £	Unit	Gross rate (10%) £
ASSOCIATED IRONMONGERY (Cont'd)										
Intumescent strips and smoke seals (Cont'd)										
white self adhesive (Cont'd)										
20 x 4mm intumescent strip, one hour application; setting into groove in timber frame or door	1.73	10.00	1.90	0.15	-	2.76	-	4.66	m	5.12
15 x 4mm intumescent strip with integral cold smoke seal, half hour application; setting into groove in timber frame or door.........	1.62	10.00	1.78	0.15	-	2.76	-	4.54	m	4.99
20 x 4mm intumescent strip with integral cold smoke seal, one hour application; setting into groove in timber frame or door..........	1.78	10.00	1.95	0.15	-	2.76	-	4.71	m	5.18
12 x 3mm intumescent strip, half hour application; fixing to both sides of glass behind glazing beads..............	0.71	10.00	0.78	0.30	-	5.51	-	6.29	m	6.92
Lorient Polyproducts Ltd; System 36 Plus										
15 x 12mm glazing channel to suit 9-11mm glass; fitting over the edge of the pane..............	5.83	10.00	6.41	0.15	-	2.76	-	9.17	m	10.09
Lorient Polyproducts Ltd; System 90 Plus										
27 x 27mm glazing channel reference LG2727; fitting over the edge of the pane..............	8.93	10.00	9.83	0.15	-	2.76	-	12.58	m	13.84
intumescent lining reference LX4402 to suit 44mm thick doors; fixing to timber with adhesive	2.22	10.00	2.44	0.30	-	5.51	-	7.95	m	8.75
intumescent lining reference LX5402 to suit 54mm thick doors; fixing to timber with adhesive	2.78	10.00	3.06	0.30	-	5.51	-	8.57	m	9.43
Mann McGowan Fabrications Ltd										
Pyroglaze 30 half hour application; fixing to both sides of glass behind glazing beads..............	0.29	10.00	0.31	0.35	-	6.43	-	6.74	m	7.42
Pyroglaze 60 one hour application; fixing to both sides of glass behind glazing beads..............	1.42	10.00	1.57	0.35	-	6.43	-	8.00	m	8.80

Labour hourly rates: (except Specialists) Craft Operatives 18.37 Labourer 13.76 Rates are national average prices. Refer to REGIONAL VARIATIONS for indicative levels of overall pricing in regions	MATERIALS			LABOUR				RATES		
	Del to Site	Waste	Material Cost	Craft Optve	Lab	Labour Cost	Sunds	Nett Rate	Unit	Gross rate (10%)
	£	%	£	Hrs	Hrs	£	£	£		£
TIMBER STAIRCASES										
Stairs in softwood, wrought										
Straight flight staircase and balustrade; 25mm treads; 19mm risers, glued; wedged and blocked; 25mm wall string; 38mm outer string; 75 x 75mm newels; 25 x 25mm balusters; 38 x 75mm hardwood handrail										
864mm wide x 2600mm rise overall; balustrade to one side; fixing to masonry with screws	970.30	2.50	994.56	15.00	1.90	301.69	3.60	1299.86	nr	1429.84
864mm wide x 2600mm rise overall with 3 nr winders at bottom; balustrade to one side; fixing to masonry with screws	1164.40	2.50	1193.51	20.00	2.50	401.80	4.88	1600.18	nr	1760.20
864mm wide x 2600mm rise overall with 3 nr winders at top; balustrade to one side; fixing to masonry with screws	1164.40	2.50	1193.51	20.00	2.50	401.80	4.88	1600.18	nr	1760.20
Balustrades in softwood, wrought										
Isolated balustrades; 25 x 25mm balusters at 150mm centres, housed construction (handrail included elsewhere)										
914mm high; fixing to timber with screws	48.54	2.50	49.75	0.80	0.10	16.07	0.19	66.02	m	72.62
Isolated balustrades; 38 x 38mm balusters at 150mm centres, housed construction (handrail included elsewhere)										
914mm high; fixing to timber with screws	71.14	2.50	72.92	1.15	0.14	23.05	0.28	96.26	m	105.88
Isolated balustrades; 50 x 50mm balusters at 150mm centres, housed construction (handrail included elsewhere)										
914mm high; fixing to timber with screws	93.46	2.50	95.80	1.30	0.16	26.08	0.31	122.19	m	134.41
Isolated balustrades; 25 x 25mm balusters at 150mm centres, 50mm mopstick handrail, housed construction										
914mm high; fixing to timber with screws	81.35	2.50	83.38	1.20	0.15	24.11	0.30	107.79	m	118.57
extra; ramps	49.93	2.50	51.18	1.10	0.14	22.13	0.27	73.58	nr	80.94
extra; wreaths	99.85	2.50	102.35	2.15	0.27	43.21	0.52	146.08	nr	160.69
extra; bends	49.93	2.50	51.18	0.70	0.09	14.10	0.17	65.44	nr	71.98
Isolated balustrades; 38 x 38mm balusters at 150mm centres, 50mm mopstick handrail, housed construction										
914mm high; fixing to timber with screws	103.54	2.50	106.13	1.55	0.19	31.09	0.38	137.59	m	151.35
extra; ramps	49.93	2.50	51.18	1.10	0.14	22.13	0.27	73.58	nr	80.94
extra; wreaths	99.85	2.50	102.35	2.15	0.27	43.21	0.52	146.08	nr	160.69
extra; bends	49.93	2.50	51.18	0.70	0.09	14.10	0.17	65.44	nr	71.98
Isolated balustrades; 50 x 50mm balusters at 150mm centres, 50mm mopstick handrail, housed construction										
914mm high; fixing to timber with screws	125.72	2.50	128.86	1.75	0.22	35.17	0.42	164.46	m	180.90
extra; ramps	49.93	2.50	51.18	1.10	0.14	22.13	0.27	73.58	nr	80.94
extra; wreaths	99.84	2.50	102.34	2.15	0.27	43.21	0.52	146.07	nr	160.68
extra; bends	49.93	2.50	51.18	0.70	0.09	14.10	0.17	65.44	nr	71.98
Isolated balustrades; 25 x 25mm balusters at 150mm centres, 75 x 100mm moulded handrail; housed construction										
914mm high; fixing to timber with screws	99.84	2.50	102.34	1.25	0.16	25.16	0.30	127.80	m	140.58
extra; ramps	73.95	2.50	75.80	2.40	0.30	48.22	0.58	124.60	nr	137.06
extra; wreaths	147.91	2.50	151.61	4.80	0.60	96.43	1.17	249.21	nr	274.13
extra; bends	73.95	2.50	75.80	1.45	0.18	29.11	0.36	105.27	nr	115.80
Isolated balustrades; 38 x 38mm balusters at 150mm centres, 75 x 100mm moulded handrail; housed construction										
914mm high; fixing to timber with screws	15.26	2.50	15.64	1.60	0.20	32.14	0.39	48.18	m	52.99
extra; ramps	74.00	2.50	75.85	2.40	0.30	48.22	0.58	124.65	nr	137.12
extra; wreaths	147.90	2.50	151.60	4.80	0.60	96.43	1.17	249.20	nr	274.12
extra; bends	73.95	2.50	75.80	1.45	0.18	29.11	0.36	105.27	nr	115.80
Isolated balustrades; 50 x 50mm balusters at 150mm centres, 75 x 100mm moulded handrail; housed construction										
914mm high; fixing to timber with screws	144.20	2.50	147.80	1.80	0.23	36.23	0.44	184.47	m	202.92
extra; ramps	74.00	2.50	75.85	2.40	0.30	48.22	0.58	124.65	nr	137.12
extra; wreaths	147.90	2.50	151.60	4.80	0.60	96.43	1.17	249.20	nr	274.12
extra; bends	73.95	2.50	75.80	1.45	0.18	29.11	0.36	105.27	nr	115.80
Balustrades in European Oak, wrought, selected for transparent finish										
Isolated balustrades; 25 x 25mm balusters at 150mm centres, 50mm mopstick handrail, housed construction										
914mm high; fixing to timber with screws	162.70	2.50	166.77	2.05	0.26	41.24	0.50	208.50	m	229.35
extra; ramps	99.90	2.50	102.40	1.80	0.23	36.23	0.44	139.06	nr	152.97
extra; wreaths	199.70	2.50	204.69	3.60	0.45	72.32	0.88	277.90	nr	305.69
extra; bends	99.90	2.50	102.40	1.30	0.16	26.08	0.31	128.80	nr	141.67
Isolated balustrades; 38 x 38mm balusters at 150mm centres, 50mm mopstick handrail, housed construction										
914mm high; fixing to timber with screws	43.62	2.50	44.71	2.60	0.26	51.34	0.62	96.67	m	106.33
extra; ramps	99.90	2.50	102.40	1.80	0.23	36.23	0.44	139.06	nr	152.97
extra; wreaths	199.70	2.50	204.69	3.60	0.45	72.32	0.88	277.90	nr	305.69
extra; bends	99.90	2.50	102.40	1.30	0.16	26.08	0.31	128.80	nr	141.67

Labour hourly rates: (except Specialists) Craft Operatives 18.37 Labourer 13.76 Rates are national average prices. Refer to REGIONAL VARIATIONS for indicative levels of overall pricing in regions	MATERIALS			LABOUR				RATES		
	Del to Site	Waste	Material Cost	Craft Optve	Lab	Labour Cost	Sunds	Nett Rate		Gross rate (10%)
	£	%	£	Hrs	Hrs	£	£	£	Unit	£
TIMBER STAIRCASES (Cont'd)										
Balustrades in European Oak, wrought, selected for transparent finish (Cont'd)										
Isolated balustrades; 50 x 50mm balusters at 150mm centres, 50mm mopstick handrail, housed construction										
914mm high; fixing to timber with screws	251.46	2.50	257.75	3.00	0.38	60.34	0.74	318.82	m	350.70
extra; ramps	99.86	2.50	102.36	1.80	0.23	36.23	0.44	139.02	nr	152.92
extra; wreaths	199.68	2.50	204.67	3.60	0.45	72.32	0.88	277.88	nr	305.67
extra; bends	99.86	2.50	102.36	1.30	0.16	26.08	0.31	128.75	nr	141.63
Isolated balustrades; 25 x 25mm balusters at 150mm centres, 75 x 100mm moulded handrail; housed construction										
914mm high; fixing to timber with screws	199.68	2.50	204.67	2.15	0.27	43.21	0.52	248.41	m	273.25
extra; ramps	147.91	2.50	151.61	3.70	0.46	74.30	0.90	226.81	nr	249.49
extra; wreaths	295.83	2.50	303.23	7.45	0.93	149.65	1.82	454.69	nr	500.16
extra; bends	147.91	2.50	151.61	2.60	0.33	52.30	0.63	204.54	nr	224.99
Isolated balustrades; 38 x 38mm balusters at 150mm centres, 75 x 100mm moulded handrail; housed construction										
914mm high; fixing to timber with screws	244.06	2.50	250.16	2.70	0.34	54.28	0.66	305.10	m	335.61
extra; ramps	147.91	2.50	151.61	3.70	0.46	74.30	0.90	226.81	nr	249.49
extra; wreaths	295.83	2.50	303.23	7.45	0.93	149.65	1.82	454.69	nr	500.16
extra; bends	147.91	2.50	151.61	2.60	0.33	52.30	0.63	204.54	nr	224.99
Isolated balustrades; 50 x 50mm balusters at 150mm centres, 75 x 100mm moulded handrail; housed construction										
914mm high; fixing to timber with screws	288.44	2.50	295.65	3.10	0.39	62.31	0.75	358.71	m	394.59
extra; ramps	147.91	2.50	151.61	3.70	0.46	74.30	0.90	226.81	nr	249.49
extra; wreaths	295.83	2.50	303.23	7.45	0.93	149.65	1.82	454.69	nr	500.16
extra; bends	147.91	2.50	151.61	2.60	0.33	52.30	0.63	204.54	nr	224.99
Handrails in softwood, wrought										
Associated handrails										
50mm mopstick; fixing through metal backgrounds with screws	27.74	5.00	29.13	0.40	0.05	8.04	0.09	37.25	m	40.98
extra; ramps	49.93	5.00	52.43	1.20	0.15	24.11	0.30	76.83	nr	84.52
extra; wreaths	99.85	5.00	104.84	2.40	0.30	48.22	0.58	153.64	nr	169.01
extra; bends	49.93	5.00	52.43	1.20	0.15	24.11	0.30	76.83	nr	84.52
75 x 100mm; moulded; fixing through metal backgrounds with screws	46.23	5.00	48.54	0.50	0.06	10.01	0.12	58.67	m	64.54
extra; ramps	73.95	5.00	77.65	1.45	0.18	29.11	0.36	107.12	nr	117.83
extra; wreaths	147.92	5.00	155.32	2.90	0.36	58.23	0.70	214.25	nr	235.67
extra; bends	73.95	5.00	77.65	1.45	0.18	29.11	0.36	107.12	nr	117.83
Handrails in African Mahogany, wrought, selected for transparent finish										
Associated handrails										
50mm mopstick; fixing through metal backgrounds with screws	31.52	5.00	33.10	0.60	0.08	12.12	0.15	45.37	m	49.91
extra; ramps	56.74	5.00	59.58	1.80	0.23	36.23	0.44	96.24	nr	105.87
extra; wreaths	113.46	5.00	119.13	3.60	0.45	72.32	0.88	192.34	nr	211.58
extra; bends	56.74	5.00	59.58	1.80	0.23	36.23	0.44	96.24	nr	105.87
75 x 100mm; moulded; fixing through metal backgrounds with screws	52.53	5.00	55.16	0.70	0.09	14.10	0.17	69.42	m	76.36
extra; ramps	84.05	5.00	88.25	2.15	0.27	43.21	0.52	131.99	nr	145.19
extra; wreaths	168.08	5.00	176.48	4.30	0.54	86.42	1.05	263.96	nr	290.35
extra; bends	84.05	5.00	88.25	2.15	0.27	43.21	0.52	131.99	nr	145.19
Handrails in European Oak, wrought, selected for transparent finish										
Associated handrails										
50mm mopstick; fixing through metal backgrounds with screws	55.49	5.00	58.26	0.60	0.08	12.12	0.15	70.54	m	77.59
extra; ramps	99.86	5.00	104.85	1.80	0.23	36.23	0.44	141.52	nr	155.67
extra; wreaths	199.68	5.00	209.66	3.60	0.45	72.32	0.88	282.87	nr	311.16
extra; bends	99.86	5.00	104.85	1.80	0.23	36.23	0.44	141.52	nr	155.67
75 x 100mm; moulded; fixing through metal backgrounds with screws	92.47	5.00	97.09	0.70	0.09	14.10	0.17	111.36	m	122.49
extra; ramps	147.92	5.00	155.32	2.15	0.27	43.21	0.52	199.05	nr	218.96
extra; wreaths	295.83	5.00	310.62	4.30	0.54	86.42	1.05	398.09	nr	437.90
extra; bends	147.91	5.00	155.31	2.15	0.26	43.07	0.52	198.90	nr	218.79
METAL STAIRCASES										
Staircases in steel										
Straight flight staircases; 180 x 10mm flat stringers, shaped ends to top and bottom; 6 x 250mm on plain raised pattern plate treads, with 50 x 50 x 6mm shelf angles and 40 x 40 x 6mm stiffening bars, bolted to stringers; welded, cleated and bolted connections										
770mm wide x 3000mm going x 2600mm rise overall; fixing to masonry with 4 nr Rawlbolts	-	-	Specialist	-	-	Specialist	-	2722.50	nr	2994.75
920mm wide x 3000mm going x 2600mm rise overall; fixing to masonry with 4 nr Rawlbolts	-	-	Specialist	-	-	Specialist	-	2880.00	nr	3168.00
Straight flight staircases, 180 x 10mm flat stringers, shaped ends to top and bottom; 6 x 250mm tray treads, with three 6mm diameter reinforcing bars welded to inside, and 50 x 50 x 6mm shelf angles, bolted to stringers; welded, cleated and bolted connections										
770mm wide x 3000mm going x 2600mm rise overall; fixing to masonry with 4 nr Rawlbolts	-	-	Specialist	-	-	Specialist	-	2580.00	nr	2838.00
920mm wide x 3000mm going x 2600mm rise overall; fixing to masonry with 4 nr Rawlbolts	-	-	Specialist	-	-	Specialist	-	2670.00	nr	2937.00

Labour hourly rates: (except Specialists) Craft Operatives 18.37 Labourer 13.76 Rates are national average prices. Refer to REGIONAL VARIATIONS for indicative levels of overall pricing in regions	MATERIALS			LABOUR				RATES		
	Del to Site	Waste	Material Cost	Craft Optve	Lab	Labour Cost	Sunds	Nett Rate	Unit	Gross rate (10%)
	£	%	£	Hrs	Hrs	£	£	£		£
METAL STAIRCASES (Cont'd)										
Staircases in steel (Cont'd)										
Straight flight staircases; 178 x 76mm channel stringers, shaped ends to top and bottom; 6 x 250mm on plain raised pattern plate treads, with 50 x 50 x 6mm shelf angles and 40 x 40 x 6mm stiffening bars, bolted to stringers; welded, cleated and bolted connections										
770mm wide x 3000mm going x 2600mm rise overall; fixing to masonry with 4 nr Rawlbolts..........	-	-	Specialist	-	-	Specialist	-	2947.50	nr	3242.25
920mm wide x 3000mm going x 2600mm rise overall; fixing to masonry with 4 nr Rawlbolts..........	-	-	Specialist	-	-	Specialist	-	3360.00	nr	3696.00
Straight flight staircases and balustrades; 180 x 10mm flat stringers, shaped ends to top and bottom; 6 x 250mm on plain raised pattern plate treads with 50 x 50 x 6mm shelf angles and 40 x 40 x 6mm stiffening bars, bolted to stringers; 915mm high balustrade to both sides consisting of 25mm diameter solid bar handrail and 32mm diameter solid bar standards at 250mm centres with base plate welded on and bolted to face of stringer, and ball type joints at intersections; welded, cleated and bolted connections										
770mm wide x 3000mm going x 2600mm rise overall; fixing to masonry with 4 nr Rawlbolts..........	-	-	Specialist	-	-	Specialist	-	3960.00	nr	4356.00
Straight flight staircases and balustrades; 180 x 10mm flat stringers, shaped ends to top and bottom; 6 x 250mm on plain raised pattern plate treads with 50 x 50 x 6mm shelf angles and 40 x 40 x 6mm stiffening bars, bolted to stringers; 915mm high balustrade to both sides consisting of 25mm diameter solid bar handrail and 32mm diameter solid bar standards at 250mm centres with base plate welded on and bolted to top of stringer, and ball type joints at intersections; welded, cleated and bolted connections										
770mm wide x 3000mm going x 2600mm rise overall; fixing to masonry with 4 nr Rawlbolts..........	-	-	Specialist	-	-	Specialist	-	3960.00	nr	4356.00
Straight flight staircases and balustrades; 180 x 10mm flat stringers, shaped ends to top and bottom; 6 x 250mm on plain raised pattern plate treads with 50 x 50 x 6mm shelf angles and 40 x 40 x 6mm stiffening bars, bolted to stringers; 915mm high balustrade to one side consisting of 25mm diameter solid bar handrail and 32mm diameter solid bar standards at 250mm centres with base plate welded on and bolted to face of stringer, and ball type joints at intersections; welded, cleated and bolted connections										
770mm wide x 3000mm going x 2600mm rise overall; fixing to masonry with 4 nr Rawlbolts..........	-	-	Specialist	-	-	Specialist	-	3690.00	nr	4059.00
Straight flight staircases and balustrades; 180 x 10mm flat stringers, shaped ends to top and bottom; 6 x 250mm on plain raised pattern plate treads with 50 x 50 x 6mm shelf angles and 40 x 40 x 6mm stiffening bars, bolted to stringers; 915mm high balustrade to one side consisting of 25mm diameter solid bar handrail and 32mm diameter solid bar standards at 250mm centres with base plate welded on and bolted to top of stringer, and ball type joints at intersections; welded, cleated and bolted connections										
770mm wide x 3000mm going x 2600mm rise overall; fixing to masonry with 4 nr Rawlbolts..........	-	-	Specialist	-	-	Specialist	-	3690.00	nr	4059.00
Straight flight staircases and balustrades; 180 x 10mm flat stringers, shaped ends to top and bottom; 6 x 250mm on plain raised pattern plate treads with 50 x 50 x 6mm shelf angles and 40 x 40 x 6mm stiffening bars, bolted to stringers; 1070mm high balustrade to both sides consisting of 25mm diameter solid bar handrail and 32mm diameter solid bar standards at 250mm centres with base plate welded on and bolted to top of stringer, and ball type joints at intersections; welded, cleated and bolted connections										
770mm wide x 3000mm going x 2600mm rise overall; fixing to masonry with 4 nr Rawlbolts..........	-	-	Specialist	-	-	Specialist	-	3997.50	nr	4397.25
Straight flight staircases and balustrades; 180 x 10mm flat stringers, shaped ends to top and bottom; 6 x 250mm on plain raised pattern plate treads with 50 x 50 x 6mm shelf angles and 40 x 40 x 6mm stiffening bars, bolted to stringers; 1070mm high balustrade to one side consisting of 25mm diameter solid bar handrail and 32mm diameter solid bar standards at 250mm centres with base plate welded on and bolted to top of stringer, and ball type joints at intersections; welded, cleated and bolted connections										
770mm wide x 3000mm going x 2600mm rise overall; fixing to masonry with 4 nr Rawlbolts..........	-	-	Specialist	-	-	Specialist	-	3727.50	nr	4100.25
Straight flight staircases and balustrades; 180 x 10mm flat stringers, shaped ends to top and bottom; 6 x 250mm on plain raised pattern plate treads with 50 x 50 x 6mm shelf angles and 40 x 40 x 6mm stiffening bars, bolted to stringers; 915mm high balustrade to one side consisting of 25mm diameter solid bar handrail and intermediate rail, and 32mm diameter solid bar standards at 250mm centres with base plate welded on and bolted to top of stringer, and ball type joints at intersections										
770mm wide x 3000mm going x 2600mm rise overall; fixing to masonry with 4 nr Rawlbolts..........	-	-	Specialist	-	-	Specialist	-	3697.50	nr	4067.25

Labour hourly rates: (except Specialists) Craft Operatives 18.37 Labourer 13.76 Rates are national average prices. Refer to REGIONAL VARIATIONS for indicative levels of overall pricing in regions	MATERIALS			LABOUR				RATES		
	Del to Site	Waste	Material Cost	Craft Optve	Lab	Labour Cost	Sunds	Nett Rate	Unit	Gross rate (10%)
	£	%	£	Hrs	Hrs	£	£	£		£
METAL STAIRCASES (Cont'd)										
Staircases in steel (Cont'd)										
Straight flight staircases and balustrades; 180 x 10mm flat stringers, shaped ends to top and bottom; 6 x 250mm on plain raised pattern plate treads with 50 x 50 x 6mm shelf angles and 40 x 40 x 6mm stiffening bars, bolted to stringers; 1070mm high balustrade to one side consisting of 25mm diameter solid bar handrail and intermediate rail, and 32mm diameter solid bar standards at 250mm centres with base plate welded on and bolted to top of stringer, and ball type joints at intersections										
770mm wide x 3000mm going x 2600mm rise overall; fixing to masonry with 4 nr Rawlbolts..........	-	-	Specialist	-	-	Specialist	-	3727.50	nr	4100.25
Quarter landing staircases, in two flights; 180 x 10mm flat stringers, shaped ends to top and bottom; 6 x 250mm on plain raised pattern plate treads, with 50 x 50 x 6mm shelf angles and 40 x 40 x 6mm stiffening bars, bolted to stringers; 6mm on plain raised pattern plate landing, welded on; 100 x 8mm flat kicking plates welded on; welded, cleated and bolted connections										
770mm wide x 2000mm going first flight excluding landing x 1000mm going second flight excluding landing x 2600mm rise overall; 770 x 770mm landing overall; fixing to masonry with 4 nr Rawlbolts..........	-	-	Specialist	-	-	Specialist	-	5955.00	nr	6550.50
920mm wide x 2000mm going first flight excluding landing x 1000mm going second flight excluding landing x 2600mm rise overall; 920 x 920mm landing overall; fixing to masonry with 4 nr Rawlbolts..........	-	-	Specialist	-	-	Specialist	-	6255.00	nr	6880.50
Half landing staircases, in two flights; 180 x 10mm flat stringers, shaped ends to top and bottom; 6 x 250mm on plain raised pattern plate treads, with 50 x 50 x 6mm shelf angles and 40 x 40 x 6mm stiffening bars, bolted to stringers; 6mm on plain raised pattern plate landing, welded on; 100 x 8mm flat kicking plates welded on; welded, cleated and bolted connections										
770mm wide x 2000mm going first flight excluding landing x 1000mm going second flight excluding landing x 2600mm rise overall; 770 x 1640mm landing overall; fixing to masonry with 4 nr Rawlbolts..........	-	-	Specialist	-	-	Specialist	-	6675.00	nr	7342.50
920mm wide x 2000mm going first flight excluding landing x 1000mm going second flight excluding landing x 2600mm rise overall; 920 x 1940mm landing overall; fixing to masonry with 4 nr Rawlbolts..........	-	-	Specialist	-	-	Specialist	-	7575.00	nr	8332.50
Balustrades in steel										
Isolated balustrades; 6 x 38mm flat core rail; 13mm diameter balusters at 250mm centres; welded fabrication ground to smooth finish; casting into mortices in concrete; wedging in position; temporary wedges										
838mm high; level	189.00	2.50	193.72	1.00	0.50	25.25	-	218.97	m	240.87
extra; ramps	32.50	2.50	33.31	1.00	0.50	25.25	-	58.56	nr	64.42
extra; wreaths	54.10	2.50	55.45	2.00	1.00	50.50	-	105.95	nr	116.55
extra; bends	29.50	2.50	30.24	1.00	0.50	25.25	-	55.49	nr	61.04
838mm high; raking	196.30	2.50	201.21	1.25	0.65	31.91	-	233.11	m	256.43
914mm high; level	208.60	2.50	213.82	1.00	0.50	25.25	-	239.07	m	262.97
extra; ramps	32.40	2.50	33.21	1.00	0.50	25.25	-	58.46	nr	64.31
extra; wreaths	54.15	2.50	55.50	2.00	1.00	50.50	-	106.00	nr	116.60
extra; bends	29.55	2.50	30.29	1.00	0.50	25.25	-	55.54	nr	61.09
914mm high; raking	220.60	2.50	226.12	1.25	0.65	31.91	-	258.02	m	283.82
Isolated balustrades; 6 x 38mm flat core rail; 13 x 13mm balusters at 250mm centres; welded fabrication ground to smooth finish; casting into mortices in concrete; wedging in position; temporary wedges										
838mm high; level	198.47	2.50	203.43	1.00	0.50	25.25	-	228.68	m	251.55
838mm high; raking	205.20	2.50	210.33	1.00	0.50	25.25	-	235.58	m	259.14
914mm high; level	215.60	2.50	220.99	1.00	0.50	25.25	-	246.24	m	270.86
914mm high; raking	230.00	2.50	235.75	1.00	0.50	25.25	-	261.00	m	287.10
Isolated balustrades; 13 x 51mm rounded handrail; 13mm diameter balusters at 250mm centres; welded fabrication ground to smooth finish; casting into mortices in concrete; wedging in position; temporary wedges										
838mm high; level	229.80	2.50	235.54	1.00	0.50	25.25	-	260.80	m	286.87
extra; ramps	35.70	2.50	36.59	1.00	0.50	25.25	-	61.84	nr	68.03
extra; wreaths	59.40	2.50	60.88	2.00	1.00	50.50	-	111.38	nr	122.52
extra; bends	32.20	2.50	33.01	1.00	0.50	25.25	-	58.26	nr	64.08
838mm high; raking	234.00	2.50	239.85	1.00	0.50	25.25	-	265.10	m	291.61
914mm high; level	244.00	2.50	250.10	1.00	0.50	25.25	-	275.35	m	302.88
extra; ramps	35.70	2.50	36.59	1.00	0.50	25.25	-	61.84	nr	68.03
extra; wreaths	59.40	2.50	60.88	2.00	1.00	50.50	-	111.38	nr	122.52
extra; bends	32.20	2.50	33.01	1.00	0.50	25.25	-	58.26	nr	64.08
914mm high; raking	254.10	2.50	260.45	1.00	0.50	25.25	-	285.70	m	314.27
Isolated balustrades; 6 x 38mm flat core rail; 10 x 51mm flat bottom rail; 13mm diameter infill balusters at 250mm centres; 25 x 25mm standards at 3000mm centres; welded fabrication ground to smooth finish; casting into mortices in concrete; wedging in position; temporary wedges										
838mm high; level	254.00	2.50	260.35	1.00	0.50	25.25	-	285.60	m	314.16
extra; ramps	61.10	2.50	62.63	1.00	0.50	25.25	-	87.88	nr	96.67
extra; wreaths	76.30	2.50	78.21	2.00	1.00	50.50	-	128.71	nr	141.58
extra; bends	54.10	2.50	55.45	1.00	0.50	25.25	-	80.70	nr	88.77
838mm high; raking	262.70	2.50	269.27	1.00	0.50	25.25	-	294.52	m	323.97

	MATERIALS			LABOUR				RATES		
Labour hourly rates: (except Specialists) Craft Operatives 18.37 Labourer 13.76 Rates are national average prices. Refer to REGIONAL VARIATIONS for indicative levels of overall pricing in regions	Del to Site	Waste	Material Cost	Craft Optve	Lab	Labour Cost	Sunds	Nett Rate		Gross rate (10%)
	£	%	£	Hrs	Hrs	£	£	£	Unit	£

METAL STAIRCASES (Cont'd)

Balustrades in steel (Cont'd)

Isolated balustrades; 6 x 38mm flat core rail; 10 x 51mm flat bottom rail; 13mm diameter infill balusters at 250mm centres; 25 x 25mm standards at 3000mm centres; welded fabrication ground to smooth finish; casting into mortices in concrete; wedging in position; temporary wedges (Cont'd)

	Del to Site	Waste	Material Cost	Craft Optve	Lab	Labour Cost	Sunds	Nett Rate	Unit	Gross rate (10%)
914mm high; level	279.00	2.50	285.98	1.00	0.50	25.25	-	311.23	m	342.35
extra; ramps	61.10	2.50	62.63	1.00	0.50	25.25	-	87.88	nr	96.67
extra; wreaths	76.30	2.50	78.21	2.00	1.00	50.50	-	128.71	nr	141.58
extra; bends	54.10	2.50	55.45	1.00	0.50	25.25	-	80.70	nr	88.77

Balustrades in steel tubing, BS 1387, medium grade; galvanized after fabrication

Isolated balustrades; 40mm diameter handrails; 40mm diameter standards at 1500mm centres; welded fabrication ground to smooth finish; casting into mortices in concrete; wedging in position; temporary wedges

	Del to Site	Waste	Material Cost	Craft Optve	Lab	Labour Cost	Sunds	Nett Rate	Unit	Gross rate (10%)
838mm high; level	118.70	2.50	121.67	0.75	0.40	19.28	-	140.95	m	155.04
838mm high; raking	127.20	2.50	130.38	0.75	0.40	19.28	-	149.66	m	164.63
914mm high; level	127.10	2.50	130.28	0.75	0.40	19.28	-	149.56	m	164.51
914mm high; raking	135.60	2.50	138.99	0.75	0.40	19.28	-	158.27	m	174.10

Isolated balustrades; 40mm diameter handrails; 40mm diameter standards at 1500mm centres, with 6 x 75mm diameter fixing plates welded on to end, holes -3 nr; welded fabrication ground to smooth finish; fixing to masonry with Rawlbolts

	Del to Site	Waste	Material Cost	Craft Optve	Lab	Labour Cost	Sunds	Nett Rate	Unit	Gross rate (10%)
838mm high; level	135.50	2.50	138.89	1.00	0.50	25.25	-	164.14	m	180.55
838mm high; raking	144.20	2.50	147.80	1.00	0.50	25.25	-	173.05	m	190.36
914mm high; level	161.35	2.50	165.38	1.00	0.50	25.25	-	190.63	m	209.70
914mm high; raking	178.11	2.50	182.56	1.00	0.50	25.25	-	207.81	m	228.59

Spiral staircases in steel

Domestic spiral staircases powder coated grey before delivery to site; comprising tread modules complete with centre column, tread support, metal tread, tread baluster, handrail section, PVC handrail cover, landing platform and with attachment brackets for fixing to floor and newel extending from centre column; assembling and bolting together

	Del to Site	Waste	Material Cost	Craft Optve	Lab	Labour Cost	Sunds	Nett Rate	Unit	Gross rate (10%)
1600mm diameter; for floor to floor height units between 2400mm and 2640mm; fixing base and ground plates to timber	1208.90	2.50	1239.12	30.00	15.00	757.50	-	1996.62	nr	2196.28
extra; additional tread modules, 180 - 220mm per rise	104.50	2.50	107.11	1.50	0.75	37.88	-	144.99	nr	159.49
extra; riser bars; set of 11nr	154.00	2.50	157.85	8.80	4.40	222.20	-	380.05	nr	418.06
extra; riser bars; single	13.29	2.50	13.62	0.80	0.40	20.20	-	33.82	nr	37.20
stairwell balustrading to match staircase including PVC handrail cover	109.94	2.50	112.69	1.75	0.90	44.53	-	157.23	m	172.95

Exterior spiral staircases; galvanised; comprising profiled perforated treads complete with centre core, spacers, baluster bar and section of handrail, tread riser bars and standard landing with balustrade one side, bars at 115mm centres; assembling and bolting together

	Del to Site	Waste	Material Cost	Craft Optve	Lab	Labour Cost	Sunds	Nett Rate	Unit	Gross rate (10%)
1600mm diameter; ground to first floor assembly, with base and ground plates; for floor to floor heights between 2520mm and 2860mm; fixing base plates and landing to masonry with expanding bolts	1754.50	2.50	1798.36	40.00	20.00	1010.00	-	2808.36	nr	3089.20
extra; single tread modules maximum 220mm per rise	115.50	2.50	118.39	1.50	0.75	37.88	-	156.26	nr	171.89
extra; two tread modules maximum 220mm per rise with post extension	290.51	2.50	297.77	3.00	1.50	75.75	-	373.52	nr	410.88
extra; three tread modules maximum 220mm per rise with post extension	373.07	2.50	382.39	4.00	2.00	101.00	-	483.39	nr	531.73

Ladders in steel

Vertical ladders; 65 x 10mm flat stringers; 65 x 10 x 250mm girth stringer brackets, bent once, bolted to stringers; 20mm diameter solid bar rungs welded to stringers; fixing to masonry with expanding bolts

	Del to Site	Waste	Material Cost	Craft Optve	Lab	Labour Cost	Sunds	Nett Rate	Unit	Gross rate (10%)
395mm wide x 3000mm long overall; brackets -6 nr; rungs -7 nr	433.50	2.50	444.34	10.00	10.00	321.30	-	765.64	nr	842.20
395mm wide x 5000mm long overall; brackets -8 nr; rungs -14 nr	624.58	2.50	640.19	12.00	12.00	385.56	-	1025.75	nr	1128.33
470mm wide x 3000mm long overall; brackets -6 nr; rungs -7 nr	528.49	2.50	541.70	10.00	10.00	321.30	-	863.00	nr	949.30
470mm wide x 5000mm long overall; brackets -8 nr; rungs -14 nr	926.58	2.50	949.74	12.00	12.00	385.56	-	1335.30	nr	1468.83
395mm wide x 3000mm long overall, stringers rising and returning 900mm above top fixing points; brackets -6 nr; rungs -7 nr	528.49	2.50	541.70	10.00	10.00	321.30	-	863.00	nr	949.30
395mm wide x 5000mm long overall, stringers rising and returning 900mm above top fixing points; brackets -8 nr; rungs -14 nr	926.58	2.50	949.74	12.00	12.00	385.56	-	1335.30	nr	1468.83
470mm wide x 3000mm long overall, stringers rising and returning 900mm above top fixing points; brackets -6 nr; rungs -7 nr	781.00	2.50	800.52	10.00	10.00	321.30	-	1121.82	nr	1234.01
470mm wide x 5000mm long overall, stringers rising and returning 900mm above top fixing points; brackets -8 nr; rungs -14 nr	1125.62	2.50	1153.76	12.00	12.00	385.56	-	1539.32	nr	1693.25

Vertical ladders; 65 x 10mm flat stringers; 65 x 10 x 250mm girth stringer brackets, bent once, bolted to stringers; 20mm diameter solid bar rungs welded to stringers; 65 x 10mm flat back hoops 900mm diameter, welded to stringers

	Del to Site	Waste	Material Cost	Craft Optve	Lab	Labour Cost	Sunds	Nett Rate	Unit	Gross rate (10%)
395mm wide x 3000mm long overall; brackets -6 nr; rungs -7 nr; hoops -7 nr	988.35	2.50	1013.06	15.00	15.00	481.95	-	1495.01	nr	1644.51
395mm wide x 5000mm long overall; brackets -8 nr; rungs -14 nr; hoops -14 nr	1236.55	2.50	1267.46	20.00	20.00	642.60	-	1910.06	nr	2101.07

Labour hourly rates: (except Specialists) Craft Operatives 18.37 Labourer 13.76 Rates are national average prices. Refer to REGIONAL VARIATIONS for indicative levels of overall pricing in regions	MATERIALS			LABOUR				RATES		
	Del to Site £	Waste %	Material Cost £	Craft Optve Hrs	Lab Hrs	Labour Cost £	Sunds £	Nett Rate £	Unit	Gross rate (10%) £
METAL STAIRCASES (Cont'd)										
Ladders in steel (Cont'd)										
Vertical ladders; 65 x 10mm flat stringers; 65 x 10 x 250mm girth stringer brackets, bent once, bolted to stringers; 20mm diameter solid bar rungs welded to stringers; 65 x 10mm flat back hoops 900mm diameter, welded to stringers (Cont'd)										
470mm wide x 3000mm long overall; brackets -6 nr; rungs -7 nr; hoops -7 nr	1091.78	2.50	1119.07	15.00	15.00	481.95	-	1601.02	nr	1761.13
470mm wide x 5000mm long overall; brackets -8 nr; rungs -14 nr; hoops -14 nr	1395.22	2.50	1430.10	20.00	20.00	642.60	-	2072.70	nr	2279.97
395mm wide x 3000mm long overall, stringers rising and returning 900mm above top fixing points; brackets -6 nr; rungs -7 nr; hoops -7 nr	1077.58	2.50	1104.52	15.00	15.00	481.95	-	1586.47	nr	1745.12
395mm wide x 5000mm long overall, stringers rising and returning 900mm above top fixing points; brackets -8 nr; rungs -14 nr; hoops -14 nr	1338.39	2.50	1371.85	20.00	20.00	642.60	-	2014.45	nr	2215.89
470mm wide x 3000mm long overall, stringers rising and returning 900mm above top fixing points; brackets -6 nr; rungs -7 nr; hoops -7 nr	1221.71	2.50	1252.25	15.00	15.00	481.95	-	1734.20	nr	1907.62
470mm wide x 5000mm long overall, stringers rising and returning 900mm above top fixing points; brackets -8 nr; rungs -14 nr; hoops -14 nr	1592.34	2.50	1632.15	20.00	20.00	642.60	-	2274.75	nr	2502.22
Ship type ladders; 75 x 10mm flat stringers, shaped and cleated ends to tops and bottoms; 65 x 10 x 350mm girth stringer brackets, bent once, bolted to stringers; 8 x 75mm on plain raised pattern plate treads with 50 x 50 x 6mm shelf angles, bolted to stringers; 25mm diameter solid bar handrails to both sides, with flattened ends bolted to stringers; 32mm diameter x 175mm long solid bar standards with three way ball type joint at intersection with handrail, and flattened ends bolted to stringers; fixing to masonry with expanding bolts										
420mm wide x 3000mm long overall, handrails rising and returning 900mm above top of stringers; brackets -4 nr; treads -12 nr; standards -6 nr	892.26	2.50	914.57	10.00	10.00	321.30	-	1235.87	nr	1359.45
420mm wide x 5000mm long overall, handrails rising and returning 900mm above top of stringers; brackets -4 nr; treads -21 nr; standards -10 nr	1262.89	2.50	1294.46	12.00	12.00	385.56	-	1680.02	nr	1848.02
470mm wide x 3000mm long overall, handrails rising and returning 900mm above top of stringers; brackets -4 nr; treads -12 nr; standards -6 nr	1036.39	2.50	1062.30	10.00	10.00	321.30	-	1383.60	nr	1521.96
470mm wide x 5000mm long overall, handrails rising and returning 900mm above top of stringers; brackets -4 nr; treads -21 nr; standards -10 nr	1400.16	2.50	1435.16	12.00	12.00	385.56	-	1820.72	nr	2002.80
Ship type ladders; 75 x 10mm flat stringers, shaped and cleated ends to tops and bottoms; 65 x 10 x 350mm girth stringer brackets, bent once, bolted to stringers; 8 x 75mm on plain raised pattern plate treads with 50 x 50 x 6mm shelf angles, bolted to stringers; 25mm diameter solid bar handrails to both sides, with flattened ends bolted to stringers; 32mm diameter x 330mm long solid bar standards with three way ball type joint at intersection with handrail, and flattened ends bolted to stringers; fixing to masonry with expanding bolts										
420mm wide x 3000mm long overall, handrails rising and returning 900mm above top of stringers; brackets -4 nr; treads -12 nr; standards -6 nr	892.26	2.50	914.57	10.00	10.00	321.30	-	1235.87	nr	1359.45
420mm wide x 5000mm long overall, handrails rising and returning 900mm above top of stringers; brackets -4 nr; treads -21 nr; standards -10 nr	1283.48	2.50	1315.57	12.00	12.00	385.56	-	1701.13	nr	1871.24
470mm wide x 3000mm long overall, handrails rising and returning 900mm above top of stringers; brackets -4 nr; treads -12 nr; standards -6 nr	1111.90	2.50	1139.70	10.00	10.00	321.30	-	1461.00	nr	1607.10
470mm wide x 5000mm long overall, handrails rising and returning 900mm above top of stringers; brackets -4 nr; treads -21 nr; standards -10 nr	1420.76	2.50	1456.28	12.00	12.00	385.56	-	1841.84	nr	2026.02

Labour hourly rates: (except Specialists) Craft Operatives 18.37 Labourer 13.76 Rates are national average prices. Refer to REGIONAL VARIATIONS for indicative levels of overall pricing in regions	MATERIALS			LABOUR				RATES		
	Del to Site	Waste	Material Cost	Craft Optve	Lab	Labour Cost	Sunds	Nett Rate	Unit	Gross rate (10%)
	£	%	£	Hrs	Hrs	£	£	£		£
GENERAL METALWORK										
Wire mesh; galvanized wire netting, BS EN 10223-2; Butt joints										
13mm mesh x Nr 22 gauge; across members at 450mm centres										
soffit	1.00	10.00	1.10	-	0.14	1.93	0.05	3.07	m²	3.37
vertical	1.00	10.00	1.10	-	0.10	1.38	0.05	2.52	m²	2.77
25mm mesh x Nr 19 gauge; across members at 450mm centres										
soffit	1.85	10.00	2.04	-	0.14	1.93	0.05	4.01	m²	4.41
vertical	1.85	10.00	2.04	-	0.10	1.38	0.05	3.46	m²	3.80
38mm mesh x Nr 19 gauge; across members at 450mm centres										
soffit	1.33	10.00	1.46	-	0.14	1.93	0.05	3.43	m²	3.78
vertical	1.33	10.00	1.46	-	0.10	1.38	0.05	2.88	m²	3.17
50mm mesh x Nr 19 gauge; across members at 450mm centres										
soffit	0.46	10.00	0.51	-	0.14	1.93	0.05	2.48	m²	2.73
vertical	0.46	10.00	0.51	-	0.10	1.38	0.05	1.93	m²	2.12

METALWORK

This page left blank intentionally

Labour hourly rates: (except Specialists) Craft Operatives 18.37 Labourer 13.76 Rates are national average prices. Refer to REGIONAL VARIATIONS for indicative levels of overall pricing in regions	MATERIALS			LABOUR				RATES		
	Del to Site	Waste	Material Cost	Craft Optve	Lab	Labour Cost	Sunds	Nett Rate	Unit	Gross rate (10%)
	£	%	£	Hrs	Hrs	£	£	£		£
GLASS										
Glass; BS 952, clear float										
3mm thick to wood rebates with putty										
not exceeding 0.15m²	38.65	5.00	40.58	0.96	0.10	19.01	0.48	60.07	m²	66.08
0.15 - 4.00m²	38.65	5.00	40.58	0.48	0.05	9.51	0.24	50.33	m²	55.36
4mm thick to wood rebates with putty										
not exceeding 0.15m²	38.99	5.00	40.94	0.96	0.10	19.01	0.48	60.43	m²	66.47
0.15 - 4.00m²	38.99	5.00	40.94	0.48	0.05	9.51	0.24	50.69	m²	55.75
5mm thick to wood rebates with putty										
not exceeding 0.15m²	56.72	5.00	59.56	1.08	0.10	21.22	0.48	81.25	m²	89.38
0.15 - 4.00m²	56.72	5.00	59.56	0.54	0.05	10.61	0.24	70.40	m²	77.44
6mm thick to wood rebates with putty										
not exceeding 0.15m²	57.05	5.00	59.90	1.08	0.10	21.22	0.48	81.60	m²	89.76
0.15 - 4.00m²	57.05	5.00	59.90	0.54	0.05	10.61	0.24	70.75	m²	77.83
3mm thick to wood rebates with bradded wood beads (included elsewhere) and putty										
not exceeding 0.15m²	38.65	5.00	40.58	1.20	0.10	23.42	0.48	64.48	m²	70.93
0.15 - 4.00m²	38.65	5.00	40.58	0.60	0.05	11.71	0.24	52.53	m²	57.79
4mm thick to wood rebates with bradded wood beads (included elsewhere) and putty										
not exceeding 0.15m²	38.99	5.00	40.94	1.20	0.10	23.42	0.48	64.84	m²	71.32
0.15 - 4.00m²	38.99	5.00	40.94	0.60	0.05	11.71	0.24	52.89	m²	58.18
5mm thick to wood rebates with bradded wood beads (included elsewhere) and putty										
not exceeding 0.15m²	56.72	5.00	59.56	1.32	0.10	25.62	0.48	85.66	m²	94.23
0.15 - 4.00m²	56.72	5.00	59.56	0.66	0.05	12.81	0.24	72.61	m²	79.87
6mm thick to wood rebates with bradded wood beads (included elsewhere) and putty										
not exceeding 0.15m²	57.05	5.00	59.90	1.32	0.10	25.62	0.48	86.01	m²	94.61
0.15 - 4.00m²	57.05	5.00	59.90	0.66	0.05	12.81	0.24	72.95	m²	80.25
3mm thick to wood rebates with screwed wood beads and glazing strip (included elsewhere)										
not exceeding 0.15m²	38.65	5.00	40.58	1.68	0.10	32.24	0.48	73.30	m²	80.63
0.15 - 4.00m²	38.65	5.00	40.58	0.84	0.05	16.12	0.24	56.94	m²	62.64
4mm thick to wood rebates with screwed wood beads and glazing strip (included elsewhere)										
not exceeding 0.15m²	38.99	5.00	40.94	1.68	0.10	32.24	0.48	73.66	m²	81.02
0.15 - 4.00m²	38.99	5.00	40.94	0.84	0.05	16.12	0.24	57.30	m²	63.03
5mm thick to wood rebates with screwed wood beads and glazing strip (included elsewhere)										
not exceeding 0.15m²	56.72	5.00	59.56	1.80	0.10	34.44	0.48	94.48	m²	103.93
0.15 - 4.00m²	56.72	5.00	59.56	0.90	0.05	17.22	0.24	77.02	m²	84.72
6mm thick to wood rebates with screwed wood beads and glazing strip (included elsewhere)										
not exceeding 0.15m²	57.05	5.00	59.90	1.80	0.10	34.44	0.48	94.82	m²	104.31
0.15 - 4.00m²	57.05	5.00	59.90	0.90	0.05	17.22	0.24	77.36	m²	85.10
3mm thick to metal rebates with metal casement glazing compound										
not exceeding 0.15m²	38.65	5.00	40.58	1.08	0.10	21.22	0.56	62.35	m²	68.59
0.15 - 4.00m²	38.65	5.00	40.58	0.54	0.05	10.61	0.27	51.46	m²	56.61
4mm thick to metal rebates with metal casement glazing compound										
not exceeding 0.15m²	38.99	5.00	40.94	1.08	0.10	21.22	0.56	62.71	m²	68.98
0.15 - 4.00m²	38.99	5.00	40.94	0.54	0.05	10.61	0.27	51.82	m²	57.00
5mm thick to metal rebates with metal casement glazing compound										
not exceeding 0.15m²	56.72	5.00	59.56	1.20	0.10	23.42	0.56	83.53	m²	91.88
0.15 - 4.00m²	56.72	5.00	59.56	0.60	0.05	11.71	0.27	71.54	m²	78.69
6mm thick to metal rebates with metal casement glazing compound										
not exceeding 0.15m²	57.05	5.00	59.90	1.20	0.10	23.42	0.56	83.88	m²	92.27
0.15 - 4.00m²	57.05	5.00	59.90	0.60	0.05	11.71	0.27	71.88	m²	79.07
3mm thick to metal rebates with clipped metal beads and gaskets (included elsewhere)										
not exceeding 0.15m²	38.65	5.00	40.58	1.44	0.10	27.83	0.48	68.89	m²	75.78
0.15 - 4.00m²	38.65	5.00	40.58	0.72	0.05	13.91	0.24	54.74	m²	60.21
4mm thick to metal rebates with clipped metal beads and gaskets (included elsewhere)										
not exceeding 0.15m²	38.99	5.00	40.94	1.44	0.10	27.83	0.48	69.25	m²	76.17
0.15 - 4.00m²	38.99	5.00	40.94	0.72	0.05	13.91	0.24	55.09	m²	60.60

GLAZING

Labour hourly rates: (except Specialists) Craft Operatives 18.37 Labourer 13.76 Rates are national average prices. Refer to REGIONAL VARIATIONS for indicative levels of overall pricing in regions	MATERIALS			LABOUR				RATES		
	Del to Site £	Waste %	Material Cost £	Craft Optve Hrs	Lab Hrs	Labour Cost £	Sunds £	Nett Rate £	Unit	Gross rate (10%) £
GLASS (Cont'd)										
Glass; BS 952, clear float (Cont'd)										
5mm thick to metal rebates with clipped metal beads and gaskets (included elsewhere)										
not exceeding 0.15m²	56.72	5.00	59.56	1.56	0.10	30.03	0.48	90.07	m²	99.08
0.15 - 4.00m²	56.72	5.00	59.56	0.78	0.05	15.02	0.24	74.81	m²	82.29
6mm thick to metal rebates with clipped metal beads and gaskets (included elsewhere)										
not exceeding 0.15m²	57.05	5.00	59.90	1.56	0.10	30.03	0.48	90.42	m²	99.46
0.15 - 4.00m²	57.05	5.00	59.90	0.78	0.05	15.02	0.24	75.16	m²	82.68
Glass; BS 952, rough cast										
6mm thick to wood rebates with putty										
not exceeding 0.15m²	49.46	5.00	51.93	1.08	0.10	21.22	0.48	73.63	m²	80.99
0.15 - 4.00m²	49.46	5.00	51.93	0.54	0.05	10.61	0.24	62.78	m²	69.06
6mm thick to wood rebates with bradded wood beads (included elsewhere) and putty										
not exceeding 0.15m²	49.46	5.00	51.93	1.32	0.10	25.62	0.48	78.04	m²	85.84
0.15 - 4.00m²	49.46	5.00	51.93	0.66	0.05	12.81	0.24	64.99	m²	71.48
6mm thick to wood rebates with screwed wood beads and glazing strip (included elsewhere)										
not exceeding 0.15m²	49.46	5.00	51.93	1.80	0.10	34.44	0.48	86.85	m²	95.54
0.15 - 4.00m²	49.46	5.00	51.93	0.90	0.05	17.22	0.24	69.39	m²	76.33
6mm thick to metal rebates with metal casement glazing compound										
not exceeding 0.15m²	49.46	5.00	51.93	1.20	0.10	23.42	0.56	75.91	m²	83.50
0.15 - 4.00m²	49.46	5.00	51.93	0.60	0.05	11.71	0.27	63.91	m²	70.30
6mm thick to metal rebates with clipped metal beads and gaskets (included elsewhere)										
not exceeding 0.15m²	49.46	5.00	51.93	1.56	0.10	30.03	0.56	82.52	m²	90.77
0.15 - 4.00m²	49.46	5.00	51.93	0.78	0.05	15.02	0.27	67.22	m²	73.94
Glass; BS 952, Pyroshield clear										
6mm thick to wood rebates with screwed beads and intumescent glazing strip (included elsewhere)										
not exceeding 0.15m²	91.07	5.00	95.62	2.16	0.10	41.06	-	136.68	m²	150.35
not exceeding 0.15m²; aligning panes with adjacent panes	91.07	8.50	98.81	2.40	0.10	45.46	-	144.27	m²	158.70
0.15 - 4.00m²	91.07	5.00	95.62	1.08	0.05	20.53	-	116.15	m²	127.77
0.15 - 4.00m²; aligning panes with adjacent panes	91.07	8.50	98.81	1.20	0.05	22.73	-	121.54	m²	133.70
6mm thick to wood rebates with screwed beads and Pyroglazing strip (included elsewhere)										
not exceeding 0.15m²	91.07	5.00	95.62	2.16	0.10	41.06	-	136.68	m²	150.35
not exceeding 0.15m²; aligning panes with adjacent panes	91.07	8.50	98.81	2.40	0.10	45.46	-	144.27	m²	158.70
0.15 - 4.00m²	91.07	5.00	95.62	1.08	0.05	20.53	-	116.15	m²	127.77
0.15 - 4.00m²; aligning panes with adjacent panes	91.07	8.50	98.81	1.20	0.05	22.73	-	121.54	m²	133.70
6mm thick to wood rebates with screwed beads and Lorient System 90 (included elsewhere)										
not exceeding 0.15m²	91.07	5.00	95.62	2.40	0.10	45.46	-	141.09	m²	155.20
not exceeding 0.15m²; aligning panes with adjacent panes	91.07	8.50	98.81	2.64	0.10	49.87	-	148.68	m²	163.55
0.15 - 4.00m²	91.07	5.00	95.62	1.20	0.05	22.73	-	118.36	m²	130.19
0.15 - 4.00m²; aligning panes with adjacent panes	91.07	8.50	98.81	1.32	0.05	24.94	-	123.75	m²	136.12
6mm thick to metal rebates with screwed metal beads and mild steel flat strips (included elsewhere)										
not exceeding 0.15m²	91.07	5.00	95.62	2.16	0.10	41.06	-	136.68	m²	150.35
not exceeding 0.15m²; aligning panes with adjacent panes	91.07	8.50	98.81	2.40	0.10	45.46	-	144.27	m²	158.70
0.15 - 4.00m²	91.07	5.00	95.62	1.08	0.05	20.53	-	116.15	m²	127.77
0.15 - 4.00m²; aligning panes with adjacent panes	91.07	8.50	98.81	1.20	0.05	22.73	-	121.54	m²	133.70
Glass; BS 952, Pyroshield safety clear										
6mm thick to wood rebates with screwed beads and intumescent glazing strip (included elsewhere)										
not exceeding 0.15m²	106.87	5.00	112.21	2.16	0.10	41.06	-	153.27	m²	168.60
not exceeding 0.15m²; aligning panes with adjacent panes	106.87	8.50	115.95	2.40	0.10	45.46	-	161.42	m²	177.56
0.15 - 4.00m²	106.87	5.00	112.21	1.08	0.05	20.53	-	132.74	m²	146.02
0.15 - 4.00m²; aligning panes with adjacent panes	106.87	8.50	115.95	1.20	0.05	22.73	-	138.69	m²	152.55
6mm thick to wood rebates with screwed beads and Pyroglazing strip (included elsewhere)										
not exceeding 0.15m²	106.87	5.00	112.21	2.16	0.10	41.06	-	153.27	m²	168.60
not exceeding 0.15m²; aligning panes with adjacent panes	106.87	8.50	115.95	2.40	0.10	45.46	-	161.42	m²	177.56
0.15 - 4.00m²	106.87	5.00	112.21	1.08	0.05	20.53	-	132.74	m²	146.02
0.15 - 4.00m²; aligning panes with adjacent panes	106.87	8.50	115.95	1.20	0.05	22.73	-	138.69	m²	152.55
6mm thick to wood rebates with screwed beads and Lorient System 90 (included elsewhere)										
not exceeding 0.15m²	106.87	5.00	112.21	2.40	0.10	45.46	-	157.68	m²	173.45
not exceeding 0.15m²; aligning panes with adjacent panes	106.87	8.50	115.95	2.64	0.10	49.87	-	165.83	m²	182.41
0.15 - 4.00m²	106.87	5.00	112.21	1.20	0.05	22.73	-	134.95	m²	148.44
0.15 - 4.00m²; aligning panes with adjacent panes	106.87	8.50	115.95	1.32	0.05	24.94	-	140.89	m²	154.98
6mm thick to metal rebates with screwed metal beads and mild steel flat strips (included elsewhere)										
not exceeding 0.15m²	106.87	5.00	112.21	2.16	0.10	41.06	-	153.27	m²	168.60
not exceeding 0.15m²; aligning panes with adjacent panes	106.87	8.50	115.95	2.40	0.10	45.46	-	161.42	m²	177.56
0.15 - 4.00m²	106.87	5.00	112.21	1.08	0.05	20.53	-	132.74	m²	146.02
0.15 - 4.00m²; aligning panes with adjacent panes	106.87	8.50	115.95	1.20	0.05	22.73	-	138.69	m²	152.55

Labour hourly rates: (except Specialists) Craft Operatives 18.37 Labourer 13.76 Rates are national average prices. Refer to REGIONAL VARIATIONS for indicative levels of overall pricing in regions	MATERIALS			LABOUR				RATES		
	Del to Site £	Waste %	Material Cost £	Craft Optve Hrs	Lab Hrs	Labour Cost £	Sunds £	Nett Rate £	Unit	Gross rate (10%) £
GLASS (Cont'd)										
Glass; BS 952, Pyroshield texture										
7mm thick to wood rebates with screwed beads and intumescent glazing strip (included elsewhere)										
not exceeding 0.15m²	43.57	5.00	45.75	2.16	0.10	41.06	-	86.80	m²	95.48
not exceeding 0.15m²; aligning panes with adjacent panes	43.57	8.50	47.27	2.40	0.10	45.46	-	92.74	m²	102.01
0.15 - 4.00m²	43.57	5.00	45.75	1.08	0.05	20.53	-	66.28	m²	72.90
0.15 - 4.00m²; aligning panes with adjacent panes	43.57	8.50	47.27	1.20	0.05	22.73	-	70.01	m²	77.01
7mm thick to wood rebates with screwed beads and Pyroglazing strip (included elsewhere)										
not exceeding 0.15m²	43.57	5.00	45.75	2.16	0.10	41.06	-	86.80	m²	95.48
not exceeding 0.15m²; aligning panes with adjacent panes	43.57	8.50	47.27	2.40	0.10	45.46	-	92.74	m²	102.01
0.15 - 4.00m²	43.57	5.00	45.75	1.08	0.05	20.53	-	66.28	m²	72.90
0.15 - 4.00m²; aligning panes with adjacent panes	43.57	8.50	47.27	1.20	0.05	22.73	-	70.01	m²	77.01
7mm thick to wood rebates with screwed beads and Lorient System 90 (included elsewhere)										
not exceeding 0.15m²	43.57	5.00	45.75	2.40	0.10	45.46	-	91.21	m²	100.33
not exceeding 0.15m²; aligning panes with adjacent panes	43.57	8.50	47.27	2.64	0.10	49.87	-	97.15	m²	106.86
0.15 - 4.00m²	43.57	5.00	45.75	1.20	0.05	22.73	-	68.48	m²	75.33
0.15 - 4.00m²; aligning panes with adjacent panes	43.57	8.50	47.27	1.32	0.05	24.94	-	72.21	m²	79.43
7mm thick to metal rebates with screwed metal beads and mild steel flat strips (included elsewhere)										
not exceeding 0.15m²	43.57	5.00	45.75	2.16	0.10	41.06	-	86.80	m²	95.48
not exceeding 0.15m²; aligning panes with adjacent panes	43.57	8.50	47.27	2.40	0.10	45.46	-	92.74	m²	102.01
0.15 - 4.00m²	43.57	5.00	45.75	1.08	0.05	20.53	-	66.28	m²	72.90
0.15 - 4.00m²; aligning panes with adjacent panes	43.57	8.50	47.27	1.20	0.05	22.73	-	70.01	m²	77.01
Glass; BS 952, Pyroshield safety texture										
7mm thick to wood rebates with screwed beads and intumescent glazing strip (included elsewhere)										
not exceeding 0.15m²	55.97	5.00	58.77	2.16	0.10	41.06	-	99.82	m²	109.81
not exceeding 0.15m²; aligning panes with adjacent panes	55.97	8.50	60.73	2.40	0.10	45.46	-	106.19	m²	116.81
0.15 - 4.00m²	55.97	5.00	58.77	1.08	0.05	20.53	-	79.30	m²	87.23
0.15 - 4.00m²; aligning panes with adjacent panes	55.97	8.50	60.73	1.20	0.05	22.73	-	83.46	m²	91.81
7mm thick to wood rebates with screwed beads and Pyroglazing strip (included elsewhere)										
not exceeding 0.15m²	55.97	5.00	58.77	2.16	0.10	41.06	-	99.82	m²	109.81
not exceeding 0.15m²; aligning panes with adjacent panes	55.97	8.50	60.73	2.40	0.10	45.46	-	106.19	m²	116.81
0.15 - 4.00m²	55.97	5.00	58.77	1.08	0.05	20.53	-	79.30	m²	87.23
0.15 - 4.00m²; aligning panes with adjacent panes	55.97	8.50	60.73	1.20	0.05	22.73	-	83.46	m²	91.81
7mm thick to wood rebates with screwed beads and Lorient System 90 (included elsewhere)										
not exceeding 0.15m²	55.97	5.00	58.77	2.40	0.10	45.46	-	104.23	m²	114.66
not exceeding 0.15m²; aligning panes with adjacent panes	55.97	8.50	60.73	2.64	0.10	49.87	-	110.60	m²	121.66
0.15 - 4.00m²	55.97	5.00	58.77	1.20	0.05	22.73	-	81.50	m²	89.65
0.15 - 4.00m²; aligning panes with adjacent panes	55.97	8.50	60.73	1.32	0.05	24.94	-	85.66	m²	94.23
7mm thick to metal rebates with screwed metal beads and mild steel flat strips (included elsewhere)										
not exceeding 0.15m²	55.97	5.00	58.77	2.16	0.10	41.06	-	99.82	m²	109.81
not exceeding 0.15m²; aligning panes with adjacent panes	55.97	8.50	60.73	2.40	0.10	45.46	-	106.19	m²	116.81
0.15 - 4.00m²	55.97	5.00	58.77	1.08	0.05	20.53	-	79.30	m²	87.23
0.15 - 4.00m²; aligning panes with adjacent panes	55.97	8.50	60.73	1.20	0.05	22.73	-	83.46	m²	91.81
Glass; BS 952, white patterned										
4mm thick to wood rebates with putty										
not exceeding 0.15m²	36.26	5.00	38.07	0.96	0.10	19.01	0.48	57.56	m²	63.32
0.15 - 4.00m²	36.26	5.00	38.07	0.48	0.05	9.51	0.24	47.82	m²	52.60
6mm thick to wood rebates with putty										
not exceeding 0.15m²	65.60	5.00	68.88	1.08	0.10	21.22	0.48	90.58	m²	99.63
0.15 - 4.00m²	65.60	5.00	68.88	0.54	0.05	10.61	0.24	79.73	m²	87.70
4mm thick to wood rebates with bradded wood beads (included elsewhere) and putty										
not exceeding 0.15m²	36.26	5.00	38.07	1.20	0.10	23.42	0.48	61.97	m²	68.17
0.15 - 4.00m²	36.26	5.00	38.07	0.60	0.05	11.71	0.24	50.02	m²	55.03
6mm thick to wood rebates with bradded wood beads (included elsewhere) and putty										
not exceeding 0.15m²	65.60	5.00	68.88	1.32	0.10	25.62	0.48	94.98	m²	104.48
0.15 - 4.00m²	65.60	5.00	68.88	0.66	0.05	12.81	0.24	81.93	m²	90.13
4mm thick to wood rebates with screwed wood beads and glazing strip (included elsewhere)										
not exceeding 0.15m²	36.26	5.00	38.07	1.68	0.10	32.24	0.48	70.79	m²	77.87
0.15 - 4.00m²	36.26	5.00	38.07	0.84	0.05	16.12	0.24	54.43	m²	59.87
6mm thick to wood rebates with screwed wood beads and glazing strip (included elsewhere)										
not exceeding 0.15m²	65.60	5.00	68.88	1.80	0.10	34.44	0.48	103.80	m²	114.18
0.15 - 4.00m²	65.60	5.00	68.88	0.90	0.05	17.22	0.24	86.34	m²	94.98
4mm thick to metal rebates with metal casement glazing compound										
not exceeding 0.15m²	36.26	5.00	38.07	1.08	0.10	21.22	0.56	59.84	m²	65.83
0.15 - 4.00m²	36.26	5.00	38.07	0.54	0.05	10.61	0.27	48.95	m²	53.85

GLAZING

	MATERIALS			LABOUR				RATES		
Labour hourly rates: (except Specialists) Craft Operatives 18.37 Labourer 13.76 Rates are national average prices. Refer to REGIONAL VARIATIONS for indicative levels of overall pricing in regions	Del to Site £	Waste %	Material Cost £	Craft Optve Hrs	Lab Hrs	Labour Cost £	Sunds £	Nett Rate £	Unit	Gross rate (10%) £
GLASS (Cont'd)										
Glass; BS 952, white patterned (Cont'd)										
6mm thick to metal rebates with metal casement glazing compound										
not exceeding 0.15m²	65.60	5.00	68.88	1.20	0.10	23.42	0.56	92.85	m²	102.14
0.15 - 4.00m²	65.60	5.00	68.88	0.60	0.05	11.71	0.27	80.86	m²	88.95
4mm thick to metal rebates with clipped metal beads and gaskets (included elsewhere)										
not exceeding 0.15m²	36.26	5.00	38.07	1.44	0.10	27.83	0.48	66.38	m²	73.02
0.15 - 4.00m²	36.26	5.00	38.07	0.72	0.05	13.91	0.24	52.23	m²	57.45
6mm thick to metal rebates with clipped metal beads and gaskets (included elsewhere)										
not exceeding 0.15m²	65.60	5.00	68.88	1.56	0.10	30.03	0.48	99.39	m²	109.33
0.15 - 4.00m²	65.60	5.00	68.88	0.78	0.05	15.02	0.24	84.14	m²	92.55
Glass; BS 952, tinted patterned										
4mm thick to wood rebates with putty										
not exceeding 0.15m²	53.66	5.00	56.34	0.96	0.10	19.01	0.48	75.83	m²	83.42
0.15 - 4.00m²	53.66	5.00	56.34	0.48	0.05	9.51	0.24	66.09	m²	72.70
6mm thick to wood rebates with putty										
not exceeding 0.15m²	71.50	5.00	75.07	1.08	0.10	21.22	0.48	96.77	m²	106.45
0.15 - 4.00m²	71.50	5.00	75.07	0.54	0.05	10.61	0.24	85.92	m²	94.52
4mm thick to wood rebates with bradded wood beads (included elsewhere) and putty										
not exceeding 0.15m²	53.66	5.00	56.34	1.20	0.10	23.42	0.48	80.24	m²	88.27
0.15 - 4.00m²	53.66	5.00	56.34	0.60	0.05	11.71	0.24	68.29	m²	75.12
6mm thick to wood rebates with bradded wood beads (included elsewhere) and putty										
not exceeding 0.15m²	71.50	5.00	75.07	1.32	0.10	25.62	0.48	101.18	m²	111.30
0.15 - 4.00m²	71.50	5.00	75.07	0.66	0.05	12.81	0.24	88.13	m²	96.94
4mm thick to wood rebates with screwed wood beads and glazing strip (included elsewhere)										
not exceeding 0.15m²	53.66	5.00	56.34	1.68	0.10	32.24	0.48	89.06	m²	97.97
0.15 - 4.00m²	53.66	5.00	56.34	0.84	0.05	16.12	0.24	72.70	m²	79.97
6mm thick to wood rebates with screwed wood beads and glazing strip (included elsewhere)										
not exceeding 0.15m²	71.50	5.00	75.07	1.80	0.10	34.44	0.48	110.00	m²	121.00
0.15 - 4.00m²	71.50	5.00	75.07	0.90	0.05	17.22	0.24	92.54	m²	101.79
4mm thick to metal rebates with metal casement glazing compound										
not exceeding 0.15m²	53.66	5.00	56.34	1.08	0.10	21.22	0.54	78.10	m²	85.91
0.15 - 4.00m²	53.66	5.00	56.34	0.54	0.05	10.61	0.27	67.22	m²	73.94
6mm thick to metal rebates with metal casement glazing compound										
not exceeding 0.15m²	71.50	5.00	75.07	1.20	0.10	23.42	0.54	99.03	m²	108.94
0.15 - 4.00m²	71.50	5.00	75.07	0.60	0.05	11.71	0.27	87.06	m²	95.76
4mm thick to metal rebates with clipped metal beads and gaskets (included elsewhere)										
not exceeding 0.15m²	53.66	5.00	56.34	1.44	0.10	27.83	0.48	84.65	m²	93.12
0.15 - 4.00m²	53.66	5.00	56.34	0.72	0.05	13.91	0.24	70.50	m²	77.55
6mm thick to metal rebates with clipped metal beads and gaskets (included elsewhere)										
not exceeding 0.15m²	71.50	5.00	75.07	1.56	0.10	30.03	0.48	105.59	m²	116.15
0.15 - 4.00m²	71.50	5.00	75.07	0.78	0.05	15.02	0.24	90.33	m²	99.36

Patterns available

4mm Bronze tint
  Autumn; Cotswold; Everglade; Sycamore

4mm white
  Arctic

6mm white
  Deep Flemish

4 and 6mm white
  Autumn; Cotswold; Driftwood; Everglade; Flemish; Linkon; Mayflower; Reeded; Stippolyte

Glass; BS 952, antisun float; grey										
4mm thick to wood rebates with putty										
not exceeding 2400 x 1200mm	74.60	5.00	78.33	0.72	0.12	14.88	0.34	93.55	m²	102.91
6mm thick to wood rebates with putty										
not exceeding 5950 x 3150mm	107.60	5.00	112.98	0.78	0.12	15.98	0.34	129.30	m²	142.24
10mm thick to wood rebates with putty										
not exceeding 5950 x 3150mm	192.38	5.00	202.00	1.80	0.12	34.72	0.44	237.15	m²	260.87
12mm thick to wood rebates with putty										
not exceeding 5950 x 3150mm	266.54	5.00	279.87	2.40	0.12	45.74	0.54	326.15	m²	358.76
4mm thick to wood rebates with bradded wood beads (included elsewhere) and putty										
not exceeding 2400 x 1200mm	74.60	5.00	78.33	0.84	0.12	17.08	0.34	95.76	m²	105.33

Labour hourly rates: (except Specialists) Craft Operatives 18.37 Labourer 13.76 Rates are national average prices. Refer to REGIONAL VARIATIONS for indicative levels of overall pricing in regions	MATERIALS			LABOUR				RATES		
	Del to Site £	Waste %	Material Cost £	Craft Optve Hrs	Lab Hrs	Labour Cost £	Sunds £	Nett Rate £	Unit	Gross rate (10%) £
GLASS (Cont'd)										
Glass; BS 952, antisun float; grey (Cont'd)										
6mm thick to wood rebates with bradded wood beads (included elsewhere) and putty not exceeding 5950 x 3150mm	107.60	5.00	112.98	0.90	0.12	18.18	0.34	131.51	m²	144.66
10mm thick to wood rebates with bradded wood beads (included elsewhere) and putty not exceeding 5950 x 3150mm	192.38	5.00	202.00	1.92	0.12	36.92	0.44	239.36	m²	263.29
12mm thick to wood rebates with bradded wood beads (included elsewhere) and putty not exceeding 5950 x 3150mm	266.54	5.00	279.87	2.52	0.12	47.94	0.54	328.35	m²	361.19
4mm thick to wood rebates with screwed wood beads and glazing strip (included elsewhere) not exceeding 2400 x 1200mm	74.60	5.00	78.33	1.08	0.12	21.49	-	99.82	m²	109.80
6mm thick to wood rebates with screwed wood beads and glazing strip (included elsewhere) not exceeding 5950 x 3150mm	107.60	5.00	112.98	1.14	0.12	22.59	-	135.57	m²	149.13
10mm thick to wood rebates with screwed wood beads and glazing strip (included elsewhere) not exceeding 5950 x 3150mm	192.38	5.00	202.00	2.04	0.12	39.13	-	241.12	m²	265.24
12mm thick to wood rebates with screwed wood beads and glazing strip (included elsewhere) not exceeding 5950 x 3150mm	266.54	5.00	279.87	2.64	0.12	50.15	-	330.02	m²	363.02
Glass; BS 952, antisun float; bronze										
4mm thick to wood rebates with putty not exceeding 2400 x 1200mm	74.60	5.00	78.33	0.72	0.12	14.88	0.34	93.55	m²	102.91
6mm thick to wood rebates with putty not exceeding 5950 x 3150mm	107.60	5.00	112.98	0.78	0.12	15.98	0.34	129.30	m²	142.24
10mm thick to wood rebates with putty not exceeding 5950 x 3150mm	192.38	5.00	202.00	1.80	0.12	34.72	0.44	237.15	m²	260.87
12mm thick to wood rebates with putty not exceeding 5950 x 3150mm	265.55	5.00	278.83	2.40	0.12	45.74	0.54	325.11	m²	357.62
4mm thick to wood rebates with bradded wood beads (included elsewhere) and putty not exceeding 2400 x 1200mm	74.60	5.00	78.33	0.84	0.12	17.08	0.34	95.76	m²	105.33
6mm thick to wood rebates with bradded wood beads (included elsewhere) and putty not exceeding 5950 x 3150mm	107.60	5.00	112.98	0.90	0.12	18.18	0.34	131.51	m²	144.66
10mm thick to wood rebates with bradded wood beads (included elsewhere) and putty not exceeding 5950 x 3150mm	192.38	5.00	202.00	1.92	0.12	36.92	0.44	239.36	m²	263.29
12mm thick to wood rebates with bradded wood beads (included elsewhere) and putty not exceeding 5950 x 3150mm	265.55	5.00	278.83	2.52	0.12	47.94	0.54	327.31	m²	360.04
4mm thick to wood rebates with screwed wood beads and glazing strip (included elsewhere) not exceeding 2400 x 1200mm	74.60	5.00	78.33	1.08	0.12	21.49	-	99.82	m²	109.80
6mm thick to wood rebates with screwed wood beads and glazing strip (included elsewhere) not exceeding 5950 x 3150mm	107.60	5.00	112.98	1.14	0.12	22.59	-	135.57	m²	149.13
10mm thick to wood rebates with screwed wood beads and glazing strip (included elsewhere) not exceeding 5950 x 3150mm	192.38	5.00	202.00	2.04	0.12	39.13	-	241.12	m²	265.24
12mm thick to wood rebates with screwed wood beads and glazing strip (included elsewhere) not exceeding 5950 x 3150mm	265.55	5.00	278.83	2.64	0.12	50.15	-	328.98	m²	361.87
Glass; BS 952, antisun float; green										
6mm thick to wood rebates with putty not exceeding 3150 x 2050mm	137.93	5.00	144.83	0.78	0.12	15.98	0.34	161.15	m²	177.27
6mm thick to wood rebates with bradded wood beads (included elsewhere) and putty not exceeding 3150 x 2050mm	137.93	5.00	144.83	0.90	0.12	18.18	0.34	163.36	m²	179.69
6mm thick to wood rebates with screwed wood beads and glazing strip (included elsewhere) not exceeding 3150 x 2050mm	137.93	5.00	144.83	1.14	0.12	22.59	-	167.42	m²	184.16
Glass; BS 952, clear float										
10mm thick to wood rebates with screwed wood beads and glazing strip (included elsewhere) not exceeding 5950 x 3150mm	114.83	5.00	120.57	2.70	0.12	51.25	-	171.82	m²	189.00
12mm thick to wood rebates with screwed wood beads and glazing strip (incleded elsewhere) not exceeding 5950 x 3150mm	146.36	5.00	153.68	3.30	0.12	62.27	-	215.95	m²	237.55

GLAZING

Labour hourly rates: (except Specialists) Craft Operatives 18.37 Labourer 13.76 Rates are national average prices. Refer to REGIONAL VARIATIONS for indicative levels of overall pricing in regions	MATERIALS			LABOUR				RATES		
	Del to Site	Waste	Material Cost	Craft Optve	Lab	Labour Cost	Sunds	Nett Rate	Unit	Gross rate (10%)
	£	%	£	Hrs	Hrs	£	£	£		£
GLASS (Cont'd)										
Glass; BS 952, clear float (Cont'd)										
15mm thick to wood rebates with screwed wood beads and glazing strip (included elsewhere) not exceeding 2950 x 2000mm	221.86	5.00	232.95	4.20	0.12	78.81	-	311.76	m²	342.93
19mm thick to wood rebates with screwed wood beads and glazing strip (included elsewhere) not exceeding 2950 x 2000mm	314.26	5.00	329.97	4.80	0.12	89.83	-	419.80	m²	461.78
25mm thick to wood rebates with screwed wood beads and glazing strip (included elsewhere) not exceeding 2950 x 2000mm	498.87	5.00	523.81	5.40	0.12	100.85	-	624.66	m²	687.13
Glass; BS 952, toughened clear float										
4mm thick to metal rebates with screwed metal beads and gaskets (included elsewhere) not exceeding 2400 x 1300mm	59.62	5.00	62.60	0.72	0.12	14.88	-	77.48	m²	85.23
5mm thick to metal rebates with screwed metal beads and gaskets (included elsewhere) not exceeding 2500 x 1520mm	89.13	5.00	93.59	0.72	0.12	14.88	-	108.46	m²	119.31
6mm thick to metal rebates with screwed metal beads and gaskets (included elsewhere) not exceeding 2500 x 1520mm	89.47	5.00	93.94	0.78	0.12	15.98	-	109.92	m²	120.92
10mm thick to metal rebates with screwed metal beads and gaskets (included elsewhere) not exceeding 2500 x 1520mm	142.07	5.00	149.17	1.80	0.12	34.72	-	183.89	m²	202.28
12mm thick to metal rebates with screwed metal beads and gaskets (included elsewhere) not exceeding 2500 x 1520mm	259.75	5.00	272.74	2.40	0.12	45.74	-	318.48	m²	350.32
Glass; BS 952, toughened white patterned										
4mm thick to metal rebates with screwed metal beads and gaskets (included elsewhere) not exceeding 2100 x 1300mm	120.14	5.00	126.15	0.72	0.12	14.88	-	141.02	m²	155.13
6mm thick to metal rebates with screwed metal beads and gaskets (included elsewhere) not exceeding 2100 x 1300mm	125.43	5.00	131.70	0.78	0.12	15.98	-	147.68	m²	162.45
Glass; BS 952, toughened tinted patterned										
4mm thick to metal rebates with screwed metal beads and gaskets (included elsewhere) not exceeding 2100 x 1300mm	120.15	5.00	126.16	0.72	0.12	14.88	-	141.04	m²	155.14
6mm thick to metal rebates with screwed metal beads and gaskets (included elsewhere) not exceeding 2100 x 1300mm	125.43	5.00	131.70	0.78	0.12	15.98	-	147.68	m²	162.45
Patterns available										
4mm tinted Everglade										
4mm white Reeded										
4mm tinted Autumn; Cotswold										

4 and 6mm white
    Autumn; Cotswold; Driftwood; Everglade; Flemish; Linkon;
    Mayflower; Stippolyte; Sycamore										

Glass; BS 952, toughened antisun float; grey										
4mm thick to metal rebates with screwed metal beads and gaskets (included elsewhere) not exceeding 2100 x 1250mm	109.37	5.00	114.84	0.72	0.12	14.88	-	129.72	m²	142.69
6mm thick to metal rebates with screwed metal beads and gaskets (included elsewhere) not exceeding 2500 x 1520mm	114.02	5.00	119.72	0.78	0.12	15.98	-	135.70	m²	149.27
10mm thick to metal rebates with screwed metal beads and gaskets (included elsewhere) not exceeding 2500 x 1520mm	225.67	5.00	236.95	1.80	0.12	34.72	-	271.67	m²	298.84
12mm thick to metal rebates with screwed metal beads and gaskets (included elsewhere) not exceeding 2500 x 1520mm	296.95	5.00	311.80	2.40	0.12	45.74	-	357.54	m²	393.29
Glass; BS 952, toughened antisun float; bronze										
4mm thick to metal rebates with screwed metal beads and gaskets (included elsewhere) not exceeding 2100 x 1250mm	109.28	5.00	114.74	0.72	0.12	14.88	-	129.62	m²	142.58
6mm thick to metal rebates with screwed metal beads and gaskets (included elsewhere) not exceeding 2500 x 1520mm	114.02	5.00	119.72	0.78	0.12	15.98	-	135.70	m²	149.27

Labour hourly rates: (except Specialists) Craft Operatives 18.37 Labourer 13.76 Rates are national average prices. Refer to REGIONAL VARIATIONS for indicative levels of overall pricing in regions	MATERIALS			LABOUR				RATES		
	Del to Site £	Waste %	Material Cost £	Craft Optve Hrs	Lab Hrs	Labour Cost £	Sunds £	Nett Rate £	Unit	Gross rate (10%) £
GLASS (Cont'd)										
Glass; BS 952, toughened antisun float; bronze (Cont'd)										
10mm thick to metal rebates with screwed metal beads and gaskets (included elsewhere)										
not exceeding 2500 x 1520mm	227.18	5.00	238.54	1.80	0.12	34.72	-	273.26	m²	300.58
12mm thick to metal rebates with screwed metal beads and gaskets (included elsewhere)										
not exceeding 2500 x 1520mm	296.95	5.00	311.80	2.40	0.12	45.74	-	357.54	m²	393.29
Glass; BS 952, laminated safety, clear float										
4.4mm thick to metal rebates with screwed metal beads and gaskets (included elsewhere)										
not exceeding 2100 x 1200mm	71.65	5.00	75.23	1.92	0.12	36.92	-	112.15	m²	123.37
6.4mm thick to metal rebates with screwed metal beads and gaskets (included elsewhere)										
not exceeding 3210 x 2000mm	60.83	5.00	63.87	2.40	0.12	45.74	-	109.61	m²	120.57
Glass; BS 952, laminated anti-bandit, clear float										
7.5mm thick to metal rebates with screwed metal beads and gaskets (included elsewhere)										
not exceeding 3210 x 2000mm	145.82	5.00	153.11	3.00	0.12	56.76	-	209.87	m²	230.86
9.5mm thick to metal rebates with screwed metal beads and gaskets (included elsewhere)										
not exceeding 3600 x 2500mm	153.01	5.00	160.66	3.30	0.12	62.27	-	222.93	m²	245.23
11.5mm thick to metal rebates with screwed metal beads and gaskets (included elsewhere)										
not exceeding 3180 x 2000mm	178.68	5.00	187.61	3.90	0.12	73.29	-	260.91	m²	287.00
not exceeding 4500 x 2500mm	214.62	5.00	225.35	3.90	0.12	73.29	-	298.65	m²	328.51
Sealed glazed units; factory made double glazed hermetically sealed units										
Two panes BS 952, clear float 4mm thick; to metal rebates with screwed metal beads and gaskets (included elsewhere)										
521mm wide x 421mm high	18.39	5.00	19.31	1.15	-	21.13	-	40.43	nr	44.48
521mm wide x 621mm high	27.13	5.00	28.49	1.45	-	26.64	-	55.12	nr	60.64
740mm wide x 740mm high	45.90	5.00	48.19	2.00	-	36.74	-	84.93	nr	93.43
848mm wide x 848mm high	60.28	5.00	63.29	2.30	-	42.25	-	105.54	nr	116.10
1048mm wide x 1048mm high	92.08	5.00	96.68	2.90	-	53.27	-	149.96	nr	164.95
1148mm wide x 1248mm high	120.10	5.00	126.10	3.45	-	63.38	-	189.48	nr	208.43
Two panes BS 952, clear float 5 or 6mm thick; to metal rebates with screwed metal beads and gaskets (included elsewhere)										
521mm wide x 421mm high	27.08	5.00	28.43	1.25	-	22.96	-	51.40	nr	56.54
521mm wide x 621mm high	39.95	5.00	41.95	1.60	-	29.39	-	71.34	nr	78.47
740mm wide x 740mm high	67.63	5.00	71.01	2.20	-	40.41	-	111.43	nr	122.57
848mm wide x 848mm high	88.80	5.00	93.24	2.50	-	45.93	-	139.16	nr	153.08
1048mm wide x 1048mm high	109.74	5.00	115.23	3.20	-	58.78	-	174.01	nr	191.41
1148mm wide x 1248mm high	162.75	5.00	170.89	3.80	-	69.81	-	240.69	nr	264.76
Inner pane BS 952, clear float 4mm thick; outer pane BS 952, white patterned 4mm thick; to metal rebates with screwed metal beads and gaskets (included elsewhere)										
521mm wide x 421mm high	22.44	5.00	23.56	1.15	-	21.13	-	44.69	nr	49.16
521mm wide x 621mm high	32.94	5.00	34.59	1.45	-	26.64	-	61.22	nr	67.35
740mm wide x 740mm high	55.76	5.00	58.55	2.00	-	36.74	-	95.29	nr	104.82
848mm wide x 848mm high	73.21	5.00	76.87	2.30	-	42.25	-	119.12	nr	131.03
1048mm wide x 1048mm high	90.48	5.00	95.00	2.90	-	53.27	-	148.28	nr	163.10
1148mm wide x 1248mm high	145.87	5.00	153.16	3.45	-	63.38	-	216.54	nr	238.19
Inner pane BS 952, clear float 4mm thick; outer pane BS 952, white patterned 6mm thick; to metal rebates with screwed metal beads and gaskets (included elsewhere)										
521mm wide x 421mm high	26.75	5.00	28.09	1.25	-	22.96	-	51.05	nr	56.16
521mm wide x 621mm high	39.44	5.00	41.41	1.60	-	29.39	-	70.80	nr	77.88
740mm wide x 740mm high	66.75	5.00	70.09	2.20	-	40.41	-	110.50	nr	121.55
848mm wide x 848mm high	87.66	5.00	92.04	2.50	-	45.93	-	137.97	nr	151.76
1048mm wide x 1048mm high	108.33	5.00	113.75	3.20	-	58.78	-	172.53	nr	189.78
1148mm wide x 1248mm high	174.65	5.00	183.38	3.80	-	69.81	-	253.19	nr	278.51
Drilling										
Hole through sheet or float glass; not exceeding 6mm thick										
6 - 15mm diameter	4.19	-	4.19	-	-	-	-	4.19	nr	4.61
16 - 38mm diameter	5.92	-	5.92	-	-	-	-	5.92	nr	6.51
exceeding 38mm diameter	11.94	-	11.94	-	-	-	-	11.94	nr	13.13
Hole through sheet or float glass; not exceeding 10mm thick										
6 - 15mm diameter	5.42	-	5.42	-	-	-	-	5.42	nr	5.97
16 - 38mm diameter	8.36	-	8.36	-	-	-	-	8.36	nr	9.20
exceeding 38mm diameter	14.50	-	14.50	-	-	-	-	14.50	nr	15.95
Hole through sheet or float glass; not exceeding 12mm thick										
6 - 15mm diameter	6.72	-	6.72	-	-	-	-	6.72	nr	7.39
16 - 38mm diameter	9.50	-	9.50	-	-	-	-	9.50	nr	10.45
exceeding 38mm diameter	17.20	-	17.20	-	-	-	-	17.20	nr	18.92
Hole through sheet or float glass; not exceeding 19mm thick										
6 - 15mm diameter	8.40	-	8.40	-	-	-	-	8.40	nr	9.24
16 - 38mm diameter	11.94	-	11.94	-	-	-	-	11.94	nr	13.13
exceeding 38mm diameter	21.23	-	21.23	-	-	-	-	21.23	nr	23.35

GLAZING

GLAZING

	MATERIALS			LABOUR				RATES		
Labour hourly rates: (except Specialists) Craft Operatives 18.37 Labourer 13.76 Rates are national average prices. Refer to REGIONAL VARIATIONS for indicative levels of overall pricing in regions	Del to Site £	Waste %	Material Cost £	Craft Optve Hrs	Lab Hrs	Labour Cost £	Sunds £	Nett Rate £	Unit	Gross rate (10%) £

GLASS (Cont'd)

Drilling (Cont'd)

Hole through sheet or float glass; not exceeding 25mm thick

6 - 15mm diameter	10.56	-	10.56	-	-	-	-	10.56	nr	11.62
16 - 38mm diameter	14.96	-	14.96	-	-	-	-	14.96	nr	16.46
exceeding 38mm diameter	26.38	-	26.38	-	-	-	-	26.38	nr	29.01

For wired and laminated glass
add 50%

For countersunk holes
add 33 1/3%

Bedding edges of panes

Wash leather strips

to edges of 3mm thick glass or the like	1.16	5.00	1.22	0.06	-	1.10	-	2.32	m	2.55
to edges of 6mm thick glass or the like	1.40	5.00	1.47	0.07	-	1.29	-	2.76	m	3.03

Butyl rubber glazing strips

to edges of 3mm thick glass or the like	1.08	5.00	1.13	0.06	-	1.10	-	2.24	m	2.46
to edges of 6mm thick glass or the like	1.30	5.00	1.36	0.07	-	1.29	-	2.65	m	2.92

POLYCARBONATE

UVA stabilised polycarbonate sheet latex paper masked both sides

3mm thick to metal rebates with screwed metal beads and gaskets (included elsewhere) standard grade	31.02	5.00	32.58	1.20	0.12	23.70	-	56.27	m²	61.90
4mm thick to metal rebates with screwed metal beads and gaskets (included elsewhere) standard grade	29.27	5.00	30.73	1.20	0.12	23.70	-	54.43	m²	59.87
5mm thick to metal rebates with screwed metal beads and gaskets (included elsewhere) standard grade	48.81	5.00	51.25	1.20	0.12	23.70	-	74.95	m²	82.44
6mm thick to metal rebates with screwed metal beads and gaskets (included elsewhere) standard grade	58.62	5.00	61.55	1.38	0.12	27.00	-	88.55	m²	97.41
8mm thick to metal rebates with screwed metal beads and gaskets (included elsewhere) standard grade	91.33	5.00	95.90	1.62	0.12	31.41	-	127.31	m²	140.04
9.5mm thick to metal rebates with screwed metal beads and gaskets (included elsewhere) standard grade	123.14	5.00	129.30	1.80	0.12	34.72	-	164.01	m²	180.41
3mm thick to metal rebates with screwed metal beads and gaskets (included elsewhere) abrasion resistant hard coated grade	71.37	5.00	74.94	1.20	0.12	23.70	-	98.63	m²	108.49
4mm thick to metal rebates with screwed metal beads and gaskets (included elsewhere) abrasion resistant hard coated grade	92.39	5.00	97.01	1.20	0.12	23.70	-	120.70	m²	132.77
5mm thick to metal rebates with screwed metal beads and gaskets (included elsewhere) abrasion resistant hard coated grade	115.48	5.00	121.25	1.20	0.12	23.70	-	144.95	m²	159.44
6mm thick to metal rebates with screwed metal beads and gaskets (included elsewhere) abrasion resistant hard coated grade	140.13	5.00	147.14	1.38	0.12	27.00	-	174.14	m²	191.56
8mm thick to metal rebates with screwed metal beads and gaskets (included elsewhere) abrasion resistant hard coated grade	184.76	5.00	194.00	1.62	0.12	31.41	-	225.41	m²	247.95
9.5mm thick to metal rebates with screwed metal beads and gaskets (included elsewhere) abrasion resistant hard coated grade	225.99	5.00	237.29	1.80	0.12	34.72	-	272.00	m²	299.20

GLASS REINFORCED PLASTIC

GRP vandal resistant glazing; Georgian, clear, smooth both faces or crinkle one face

3mm thick to wood rebates with screwed wood beads (included elsewhere) and putty

general purpose grade	36.99	5.00	38.84	1.15	0.12	22.78	0.22	61.84	m²	68.03
fire retardant grade, class 2	42.55	5.00	44.68	1.15	0.12	22.78	0.22	67.68	m²	74.45
fire retardant grade, class 0	57.35	5.00	60.21	1.15	0.12	22.78	0.22	83.22	m²	91.54

4mm thick to wood rebates with screwed wood beads (included elsewhere) and putty

general purpose grade	41.21	5.00	43.27	1.15	0.12	22.78	0.22	66.27	m²	72.90
fire retardant grade, class 2	47.40	5.00	49.77	1.15	0.12	22.78	0.22	72.77	m²	80.05
fire retardant grade, class 0	63.88	5.00	67.08	1.15	0.12	22.78	0.22	90.08	m²	99.09

6mm thick to wood rebates with screwed wood beads (included elsewhere) and putty

general purpose grade	58.12	5.00	61.03	1.30	0.12	25.53	0.22	86.79	m²	95.47
fire retardant grade, class 2	60.77	5.00	63.81	1.30	0.12	25.53	0.22	89.57	m²	98.53
fire retardant grade, class 0	81.92	5.00	86.01	1.30	0.12	25.53	0.22	111.77	m²	122.95

Labour hourly rates: (except Specialists) Craft Operatives 18.37 Labourer 13.76 Rates are national average prices. Refer to REGIONAL VARIATIONS for indicative levels of overall pricing in regions	MATERIALS			LABOUR				RATES		
	Del to Site	Waste	Material Cost	Craft Optve	Lab	Labour Cost	Sunds	Nett Rate	Unit	Gross rate (10%)
	£	%	£	Hrs	Hrs	£	£	£		£
GLASS REINFORCED PLASTIC (Cont'd)										
GRP vandal resistant glazing; Georgian, clear, smooth both faces or crinkle one face (Cont'd)										
3mm thick to metal rebates with screwed metal beads and gaskets (included elsewhere)										
general purpose grade	36.99	5.00	38.84	1.15	0.12	22.78	-	61.62	m²	67.78
fire retardant grade, class 2	42.55	5.00	44.68	1.15	0.12	22.78	-	67.45	m²	74.20
fire retardant grade, class 0	57.35	5.00	60.21	1.15	0.12	22.78	-	82.99	m²	91.29
4mm thick to metal rebates with screwed metal beads and gaskets (included elsewhere)										
general purpose grade	41.21	5.00	43.27	1.15	0.12	22.78	-	66.05	m²	72.65
fire retardant grade, class 2	47.40	5.00	49.77	1.15	0.12	22.78	-	72.55	m²	79.80
fire retardant grade, class 0	63.88	5.00	67.08	1.15	0.12	22.78	-	89.85	m²	98.84
6mm thick to metal rebates with screwed metal beads and gaskets (included elsewhere)										
general purpose grade	58.12	5.00	61.03	1.30	0.12	25.53	-	86.56	m²	95.22
fire retardant grade, class 2	60.77	5.00	63.81	1.30	0.12	25.53	-	89.34	m²	98.28
fire retardant grade, class 0	81.92	5.00	86.01	1.30	0.12	25.53	-	111.55	m²	122.70
GRP Vandal resistant glazing; plain, opaque colours or clear, smooth both faces or crinkle one face										
3mm thick to wood rebates with screwed wood beads (included elsewhere) and putty										
general purpose grade	35.74	5.00	37.53	1.15	0.12	22.78	0.22	60.53	m²	66.58
fire retardant grade, class 2	41.10	5.00	43.15	1.15	0.12	22.78	0.22	66.15	m²	72.77
fire retardant grade, class 0	55.39	5.00	58.16	1.15	0.12	22.78	0.22	81.16	m²	89.28
4mm thick to wood rebates with screwed wood beads (included elsewhere) and putty										
general purpose grade	40.38	5.00	42.39	1.15	0.12	22.78	0.22	65.40	m²	71.94
fire retardant grade, class 2	46.42	5.00	48.74	1.15	0.12	22.78	0.22	71.74	m²	78.92
fire retardant grade, class 0	62.57	5.00	65.70	1.15	0.12	22.78	0.22	88.70	m²	97.57
6mm thick to wood rebates with screwed wood beads (included elsewhere) and putty										
general purpose grade	59.03	5.00	61.98	1.30	0.12	25.53	0.22	87.74	m²	96.51
fire retardant grade, class 2	61.71	5.00	64.79	1.30	0.12	25.53	0.22	90.55	m²	99.61
fire retardant grade, class 0	83.17	5.00	87.33	1.30	0.12	25.53	0.22	113.09	m²	124.39
3mm thick to metal rebates with screwed metal beads and gaskets (included elsewhere)										
general purpose grade	35.74	5.00	37.53	1.15	0.12	22.78	-	60.30	m²	66.34
fire retardant grade, class 2	41.10	5.00	43.15	1.15	0.12	22.78	-	65.93	m²	72.52
fire retardant grade, class 0	55.39	5.00	58.16	1.15	0.12	22.78	-	80.94	m²	89.03
4mm thick to metal rebates with screwed metal beads and gaskets (included elsewhere)										
general purpose grade	40.38	5.00	42.39	1.15	0.12	22.78	-	65.17	m²	71.69
fire retardant grade, class 2	46.42	5.00	48.74	1.15	0.12	22.78	-	71.52	m²	78.67
fire retardant grade, class 0	62.57	5.00	65.70	1.15	0.12	22.78	-	88.48	m²	97.32
6mm thick to metal rebates with screwed metal beads and gaskets (included elsewhere)										
general purpose grade	59.03	5.00	61.98	1.30	0.12	25.53	-	87.51	m²	96.27
fire retardant grade, class 2	61.71	5.00	64.79	1.30	0.12	25.53	-	90.33	m²	99.36
fire retardant grade, class 0	83.17	5.00	87.33	1.30	0.12	25.53	-	112.86	m²	124.15
GRP Vandal resistant glazing; diamond, clear, smooth both faces or crinkle one face										
3mm thick to wood rebates with screwed wood beads (included elsewhere) and putty										
general purpose grade	36.99	5.00	38.84	1.15	0.12	22.78	0.22	61.84	m²	68.03
fire retardant grade, class 2	42.55	5.00	44.68	1.15	0.12	22.78	0.22	67.68	m²	74.45
fire retardant grade, class 0	57.35	5.00	60.21	1.15	0.12	22.78	0.22	83.22	m²	91.54
4mm thick to wood rebates with screwed wood beads (included elsewhere) and putty										
general purpose grade	42.32	5.00	44.44	1.15	0.12	22.78	0.22	67.44	m²	74.18
fire retardant grade, class 2	48.67	5.00	51.10	1.15	0.12	22.78	0.22	74.10	m²	81.51
fire retardant grade, class 0	65.59	5.00	68.87	1.15	0.12	22.78	0.22	91.88	m²	101.06
6mm thick to wood rebates with screwed wood beads (included elsewhere) and putty										
general purpose grade	58.57	5.00	61.50	1.30	0.12	25.53	0.22	87.26	m²	95.98
fire retardant grade, class 2	67.37	5.00	70.73	1.30	0.12	25.53	0.22	96.49	m²	106.14
fire retardant grade, class 0	90.78	5.00	95.32	1.30	0.12	25.53	0.22	121.08	m²	133.19
3mm thick to metal rebates with screwed metal beads and gaskets (included elsewhere)										
general purpose grade	36.99	5.00	38.84	1.15	0.12	22.78	-	61.62	m²	67.78
fire retardant grade, class 2	42.55	5.00	44.68	1.15	0.12	22.78	-	67.45	m²	74.20
fire retardant grade, class 0	57.35	5.00	60.21	1.15	0.12	22.78	-	82.99	m²	91.29
4mm thick to metal rebates with screwed metal beads and gaskets (included elsewhere)										
general purpose grade	42.32	5.00	44.44	1.15	0.12	22.78	-	67.21	m²	73.93
fire retardant grade, class 2	48.67	5.00	51.10	1.15	0.12	22.78	-	73.88	m²	81.26
fire retardant grade, class 0	65.59	5.00	68.87	1.15	0.12	22.78	-	91.65	m²	100.82
6mm thick to metal rebates with screwed metal beads and gaskets (included elsewhere)										
general purpose grade	58.57	5.00	61.50	1.30	0.12	25.53	-	87.03	m²	95.73
fire retardant grade, class 2	67.37	5.00	70.73	1.30	0.12	25.53	-	96.27	m²	105.89

Labour hourly rates: (except Specialists) Craft Operatives 18.37 Labourer 13.76 Rates are national average prices. Refer to REGIONAL VARIATIONS for indicative levels of overall pricing in regions	MATERIALS			LABOUR				RATES		
	Del to Site	Waste	Material Cost	Craft Optve	Lab	Labour Cost	Sunds	Nett Rate		Gross rate (10%)
	£	%	£	Hrs	Hrs	£	£	£	Unit	£
GLASS REINFORCED PLASTIC (Cont'd)										
GRP Vandal resistant glazing; diamond, clear, smooth both faces or crinkle one face (Cont'd)										
6mm thick to metal rebates with screwed metal beads and gaskets (included elsewhere) (Cont'd)										
fire retardant grade, class 0	90.78	5.00	95.32	1.30	0.12	25.53	-	120.86	m²	132.94
POLYESTER WINDOW FILMS										
Polyester window films; Durable Berkeley Company Ltd.										
3M Scotchshield Safety films; applying to glass										
type SH4CLL, optically clear	-	-	Specialist	-	-	Specialist	-	43.38	m²	47.72
type SH4S1L, combination solar/safety..................	-	-	Specialist	-	-	Specialist	-	52.05	m²	57.26
3M Scotchtint Solar Control films; applying to glass										
type P18, silver..	-	-	Specialist	-	-	Specialist	-	47.72	m²	52.49
type RE15S1X, external	-	-	Specialist	-	-	Specialist	-	60.72	m²	66.79
type RE35NEARL, neutral..................................	-	-	Specialist	-	-	Specialist	-	47.72	m²	52.49
3M Scotchtint Plus All Seasons insulating films; applying to glass										
type LE20 S1AR (silver)...................................	-	-	Specialist	-	-	Specialist	-	60.72	m²	66.79
type LE35 AMARL (bronze).................................	-	-	Specialist	-	-	Specialist	-	60.72	m²	66.79
type LE50 AMARL (bronze).................................	-	-	Specialist	-	-	Specialist	-	60.72	m²	66.79
MIRRORS										
BS 952, clear float, SG, silvered and protected with copper backing										
6mm thick; fixing to masonry with brass screws, chromium plated dome covers, rubber sleeves and washers										
holes 6mm diameter -4; edges polished										
254 x 400mm..	42.28	5.00	44.39	0.50	-	9.19	0.62	54.19	nr	59.61
300 x 460mm..	44.01	5.00	46.21	0.50	-	9.19	0.62	56.01	nr	61.61
360 x 500mm..	46.30	5.00	48.61	0.50	-	9.19	0.62	58.41	nr	64.26
460 x 560mm..	53.91	5.00	56.60	0.65	-	11.94	0.62	69.16	nr	76.08
460 x 600mm..	55.43	5.00	58.20	0.65	-	11.94	0.62	70.76	nr	77.84
460 x 900mm..	66.82	5.00	70.16	0.70	-	12.86	0.62	83.63	nr	91.99
500 x 680mm..	69.63	5.00	73.11	0.65	-	11.94	0.62	85.67	nr	94.24
600 x 900mm..	76.68	5.00	80.51	0.75	-	13.78	0.62	94.90	nr	104.39
holes 6mm diameter -4; edges bevelled										
254 x 400mm..	47.10	5.00	49.45	0.50	-	9.19	0.62	59.25	nr	65.18
300 x 460mm..	49.09	5.00	51.55	0.50	-	9.19	0.62	61.35	nr	67.48
360 x 500mm..	51.69	5.00	54.28	0.50	-	9.19	0.62	64.08	nr	70.48
460 x 560mm..	60.76	5.00	63.80	0.65	-	11.94	0.62	76.35	nr	83.99
460 x 600mm..	63.61	5.00	66.79	0.65	-	11.94	0.62	79.34	nr	87.28
460 x 900mm..	79.76	5.00	83.75	0.70	-	12.86	0.62	97.22	nr	106.94
500 x 680mm..	71.86	5.00	75.45	0.65	-	11.94	0.62	88.01	nr	96.81
600 x 900mm..	90.93	5.00	95.48	0.75	-	13.78	0.62	109.87	nr	120.86
HACKING OUT EXISTING GLASS										
Hacking out existing glass; preparing for re-glazing										
Float glass										
wood rebates...	-	-	-	0.44	0.44	14.14	0.25	14.39	m	15.83
wood rebates and screwed wood beads; storing beads for re-use.	-	-	-	0.52	0.52	16.71	0.25	16.96	m	18.66
metal rebates..	-	-	-	0.46	0.46	14.78	0.25	15.03	m	16.54
metal rebates and screwed metal beads; storing beads for re-use	-	-	-	0.55	0.55	17.67	0.25	17.93	m	19.72
Float glass behind guard bars in position										
wood rebates and screwed wood beads; storing beads for re-use.	-	-	-	0.85	0.85	27.31	0.25	27.57	m	30.32
metal rebates and screwed metal beads; storing beads for re-use	-	-	-	0.90	0.90	28.92	0.25	29.17	m	32.09

Labour hourly rates: (except Specialists) Craft Operatives 18.37 Labourer 13.76 Rates are national average prices. Refer to REGIONAL VARIATIONS for indicative levels of overall pricing in regions	MATERIALS			LABOUR				RATES		
	Del to Site £	Waste %	Material Cost £	Craft Optve Hrs	Lab Hrs	Labour Cost £	Sunds £	Nett Rate £	Unit	Gross rate (10%) £
IN-SITU, TILED, BLOCK, MOSAIC, SHEET, APPLIED LIQUID OR PLANTED FINISHES										
Screeds, beds and toppings; Mortar, cement and sand (1:3) - screeds										
19mm work to floors on concrete base; one coat; screeded										
level and to falls only not exceeding 15 degrees from horizontal ...	2.49	5.00	2.61	0.24	0.15	6.47	-	9.08	m²	9.99
to falls and crossfalls and to slopes not exceeding 15 degrees from horizontal	2.49	5.00	2.61	0.34	0.15	8.31	-	10.92	m²	12.01
25mm work to floors on concrete base; one coat; screeded										
level and to falls only not exceeding 15 degrees from horizontal ...	3.21	5.00	3.36	0.26	0.16	6.98	-	10.34	m²	11.38
to falls and crossfalls and to slopes not exceeding 15 degrees from horizontal	3.21	5.00	3.36	0.36	0.16	8.81	-	12.18	m²	13.40
32mm work to floors on concrete base; one coat; screeded										
level and to falls only not exceeding 15 degrees from horizontal ...	4.05	5.00	4.25	0.29	0.18	7.80	-	12.05	m²	13.26
to falls and crossfalls and to slopes not exceeding 15 degrees from horizontal	4.05	5.00	4.25	0.39	0.18	9.64	-	13.89	m²	15.28
38mm work to floors on concrete base; one coat; screeded										
level and to falls only not exceeding 15 degrees from horizontal ...	4.78	5.00	5.00	0.31	0.19	8.31	-	13.31	m²	14.64
to falls and crossfalls and to slopes not exceeding 15 degrees from horizontal	4.78	5.00	5.00	0.41	0.19	10.15	-	15.15	m²	16.66
50mm work to floors on concrete base; one coat; screeded										
level and to falls only not exceeding 15 degrees from horizontal ...	6.22	5.00	6.52	0.36	0.21	9.50	-	16.02	m²	17.62
to falls and crossfalls and to slopes not exceeding 15 degrees from horizontal	6.22	5.00	6.52	0.46	0.21	11.34	-	17.86	m²	19.64
19mm work to floors on concrete base; one coat; troweled										
level and to falls only not exceeding 15 degrees from horizontal ...	2.49	5.00	2.61	0.36	0.15	8.68	-	11.29	m²	12.41
to falls and crossfalls and to slopes not exceeding 15 degrees from horizontal	2.49	5.00	2.61	0.46	0.15	10.51	-	13.12	m²	14.43
25mm work to floors on concrete base; one coat; troweled										
level and to falls only not exceeding 15 degrees from horizontal ...	3.21	5.00	3.36	0.38	0.16	9.18	-	12.55	m²	13.80
to falls and crossfalls and to slopes not exceeding 15 degrees from horizontal	3.21	5.00	3.36	0.48	0.16	11.02	-	14.38	m²	15.82
32mm work to floors on concrete base; one coat; troweled										
level and to falls only not exceeding 15 degrees from horizontal ...	4.05	5.00	4.25	0.41	0.18	10.01	-	14.26	m²	15.68
to falls and crossfalls and to slopes not exceeding 15 degrees from horizontal	4.05	5.00	4.25	0.51	0.18	11.85	-	16.09	m²	17.70
38mm work to floors on concrete base; one coat; troweled										
level and to falls only not exceeding 15 degrees from horizontal ...	4.78	5.00	5.00	0.43	0.19	10.51	-	15.52	m²	17.07
to falls and crossfalls and to slopes not exceeding 15 degrees from horizontal	4.78	5.00	5.00	0.53	0.19	12.35	-	17.35	m²	19.09
50mm work to floors on concrete base; one coat; troweled										
level and to falls only not exceeding 15 degrees from horizontal ...	6.22	5.00	6.52	0.48	0.21	11.71	-	18.22	m²	20.05
to falls and crossfalls and to slopes not exceeding 15 degrees from horizontal	6.22	5.00	6.52	0.58	0.21	13.54	-	20.06	m²	22.07
19mm work to floors on concrete base; one coat; floated										
level and to falls only not exceeding 15 degrees from horizontal ...	2.49	5.00	2.61	0.34	0.15	8.31	-	10.92	m²	12.01
to falls and crossfalls and to slopes not exceeding 15 degrees from horizontal	2.49	5.00	2.61	0.44	0.15	10.15	-	12.75	m²	14.03
25mm work to floors on concrete base; one coat; floated										
level and to falls only not exceeding 15 degrees from horizontal ...	3.21	5.00	3.36	0.36	0.16	8.81	-	12.18	m²	13.40
to falls and crossfalls and to slopes not exceeding 15 degrees from horizontal	3.21	5.00	3.36	0.46	0.16	10.65	-	14.02	m²	15.42
32mm work to floors on concrete base; one coat; floated										
level and to falls only not exceeding 15 degrees from horizontal ...	4.05	5.00	4.25	0.39	0.18	9.64	-	13.89	m²	15.28
to falls and crossfalls and to slopes not exceeding 15 degrees from horizontal	4.05	5.00	4.25	0.49	0.18	11.48	-	15.72	m²	17.30
38mm work to floors on concrete base; one coat; floated										
level and to falls only not exceeding 15 degrees from horizontal ...	4.78	5.00	5.00	0.41	0.19	10.15	-	15.15	m²	16.66
to falls and crossfalls and to slopes not exceeding 15 degrees from horizontal	4.78	5.00	5.00	0.51	0.19	11.98	-	16.99	m²	18.68
50mm work to floors on concrete base; one coat; floated										
level and to falls only not exceeding 15 degrees from horizontal ...	6.22	5.00	6.52	0.46	0.21	11.34	-	17.86	m²	19.64
to falls and crossfalls and to slopes not exceeding 15 degrees from horizontal	6.22	5.00	6.52	0.56	0.21	13.18	-	19.69	m²	21.66
13mm work to walls on brickwork or blockwork base; one coat; screeded										
width exceeding 600mm..	1.56	5.00	1.64	0.40	0.20	10.10	-	11.74	m²	12.91
width; 300mm...	0.48	10.00	0.53	0.24	0.06	5.23	-	5.76	m	6.34

FLOOR, WALL, CEILING AND ROOF FINISHINGS

Labour hourly rates: (except Specialists) Craft Operatives 18.37 Labourer 13.76 Rates are national average prices. Refer to REGIONAL VARIATIONS for indicative levels of overall pricing in regions	MATERIALS			LABOUR				RATES		
	Del to Site	Waste	Material Cost	Craft Optve	Lab	Labour Cost	Sunds	Nett Rate	Unit	Gross rate (10%)
	£	%	£	Hrs	Hrs	£	£	£		£
IN-SITU, TILED, BLOCK, MOSAIC, SHEET, APPLIED LIQUID OR PLANTED FINISHES (Cont'd)										
Screeds, beds and toppings; Mortar, cement and sand (1:3) - screeds (Cont'd)										
13mm work to walls on brickwork or blockwork base; one coat; troweled										
width exceeding 600mm	1.56	5.00	1.64	0.52	0.20	12.30	-	13.94	m²	15.34
width: 300mm	0.48	10.00	0.53	0.31	0.06	6.52	-	7.05	m	7.75
13mm work to walls on brickwork or blockwork base; one coat; floated										
width exceeding 600mm	1.56	5.00	1.64	0.50	0.20	11.94	-	13.58	m²	14.93
width: 300mm	0.48	10.00	0.53	0.30	0.06	6.34	-	6.86	m	7.55
Screeds, beds and toppings; Mortar, cement and sand (1:3) - paving										
25mm work to floors on concrete base; one coat; troweled										
level and to falls only not exceeding 15 degrees from horizontal	3.21	5.00	3.36	0.38	0.16	9.18	-	12.55	m²	13.80
to falls and crossfalls and to slopes not exceeding 15 degrees from horizontal	3.21	5.00	3.36	0.48	0.16	11.02	-	14.38	m²	15.82
to slopes exceeding 15 degrees from horizontal	3.21	5.00	3.36	0.53	0.16	11.94	-	15.30	m²	16.83
32mm work to floors on concrete base; one coat; troweled										
level and to falls only not exceeding 15 degrees from horizontal	4.05	5.00	4.25	0.41	0.18	10.01	-	14.26	m²	15.68
to falls and crossfalls and to slopes not exceeding 15 degrees from horizontal	4.05	5.00	4.25	0.51	0.18	11.85	-	16.09	m²	17.70
to slopes exceeding 15 degrees from horizontal	4.05	5.00	4.25	0.56	0.18	12.76	-	17.01	m²	18.71
38mm work to floors on concrete base; one coat; troweled										
level and to falls only not exceeding 15 degrees from horizontal	4.78	5.00	5.00	0.43	0.19	10.51	-	15.52	m²	17.07
to falls and crossfalls and to slopes not exceeding 15 degrees from horizontal	4.78	5.00	5.00	0.53	0.19	12.35	-	17.35	m²	19.09
to slopes exceeding 15 degrees from horizontal	4.78	5.00	5.00	0.58	0.19	13.27	-	18.27	m²	20.10
50mm work to floors on concrete base; one coat; troweled										
level and to falls only not exceeding 15 degrees from horizontal	6.22	5.00	6.52	0.48	0.21	11.71	-	18.22	m²	20.05
to falls and crossfalls and to slopes not exceeding 15 degrees from horizontal	6.22	5.00	6.52	0.58	0.21	13.54	-	20.06	m²	22.07
to slopes exceeding 15 degrees from horizontal	6.22	5.00	6.52	0.63	0.21	14.46	-	20.98	m²	23.08
19mm work to treads on concrete base; screeded										
width 275mm	1.01	10.00	1.11	0.15	0.04	3.31	-	4.42	m	4.86
25mm work to treads on concrete base; screeded										
width 275mm	1.25	10.00	1.37	0.16	0.04	3.49	-	4.86	m	5.35
32mm work to treads on concrete base; screeded										
width 275mm	1.61	10.00	1.77	0.18	0.05	3.99	-	5.76	m	6.34
38mm work to treads on concrete base; screeded										
width 275mm	1.97	10.00	2.17	0.20	0.05	4.36	-	6.53	m	7.18
19mm work to treads on concrete base; floated										
width 275mm	1.01	10.00	1.11	0.20	0.04	4.22	-	5.33	m	5.87
25mm work to treads on concrete base; floated										
width 275mm	1.25	10.00	1.37	0.21	0.04	4.41	-	5.78	m	6.36
32mm work to treads on concrete base; floated										
width 275mm	1.61	10.00	1.77	0.23	0.05	4.91	-	6.68	m	7.35
38mm work to treads on concrete base; floated										
width 275mm	1.97	10.00	2.17	0.25	0.05	5.28	-	7.45	m	8.19
19mm work to treads on concrete base; troweled										
width 275mm	1.01	10.00	1.11	0.25	0.04	5.14	-	6.25	m	6.88
25mm work to treads on concrete base; troweled										
width 275mm	1.25	10.00	1.37	0.26	0.04	5.33	-	6.70	m	7.37
32mm work to treads on concrete base; troweled										
width 275mm	1.61	10.00	1.77	0.28	0.05	5.83	-	7.60	m	8.36
38mm work to treads on concrete base; troweled										
width 275mm	1.97	10.00	2.17	0.30	0.05	6.20	-	8.37	m	9.20
19mm work to plain risers on concrete base; keyed										
height 175mm	0.60	10.00	0.66	0.30	0.03	5.92	-	6.58	m	7.24
25mm work to plain risers on concrete base; keyed										
height 175mm	0.84	10.00	0.92	0.35	0.03	6.84	-	7.77	m	8.54
32mm work to plain risers on concrete base; keyed										
height 175mm	0.96	10.00	1.06	0.40	0.03	7.76	-	8.82	m	9.70
19mm work to plain risers on concrete base; troweled										
height 175mm	0.60	10.00	0.66	0.50	0.03	9.60	-	10.26	m	11.28
25mm work to plain risers on concrete base; troweled										
height 175mm	0.84	10.00	0.92	0.55	0.03	10.52	-	11.44	m	12.58
32mm work to plain risers on concrete base; troweled										
height 175mm	0.96	10.00	1.06	0.60	0.03	11.43	-	12.49	m	13.74
19mm work to undercut risers on concrete base; keyed										
height 175mm	0.60	10.00	0.66	0.35	0.03	6.84	-	7.50	m	8.25
25mm work to undercut risers on concrete base; keyed										
height 175mm	0.84	10.00	0.92	0.40	0.03	7.76	-	8.69	m	9.55

Labour hourly rates: (except Specialists) Craft Operatives 18.37 Labourer 13.76 Rates are national average prices. Refer to REGIONAL VARIATIONS for indicative levels of overall pricing in regions	MATERIALS			LABOUR				RATES		
	Del to Site £	Waste %	Material Cost £	Craft Optve Hrs	Lab Hrs	Labour Cost £	Sunds £	Nett Rate £	Unit	Gross rate (10%) £
IN-SITU, TILED, BLOCK, MOSAIC, SHEET, APPLIED LIQUID OR PLANTED FINISHES (Cont'd)										
Screeds, beds and toppings; Mortar, cement and sand (1:3) - paving (Cont'd)										
32mm work to undercut risers on concrete base; keyed height 175mm	0.96	10.00	1.06	0.45	0.03	8.68	-	9.74	m	10.71
19mm work to undercut risers on concrete base; troweled height 175mm	0.60	10.00	0.66	0.55	0.03	10.52	-	11.18	m	12.29
25mm work to undercut risers on concrete base; troweled height 175mm	0.84	10.00	0.92	0.60	0.03	11.43	-	12.36	m	13.60
32mm work to undercut risers on concrete base; troweled height 175mm	0.96	10.00	1.06	0.65	0.03	12.35	-	13.41	m	14.75
19mm work to skirtings on brickwork or blockwork base										
height 100mm	0.36	10.00	0.40	0.70	0.05	13.55	-	13.94	m	15.34
height 150mm	0.48	10.00	0.53	0.80	0.05	15.38	-	15.91	m	17.50
25mm work to skirtings on brickwork or blockwork base										
height 100mm	0.48	10.00	0.53	0.75	0.05	14.47	-	14.99	m	16.49
height 150mm	0.72	10.00	0.79	0.85	0.05	16.30	-	17.09	m	18.80
If paving oil proofed, add										
25mm work	0.28	10.00	0.30	-	0.03	0.41	-	0.72	m²	0.79
32mm work	0.35	10.00	0.39	-	0.03	0.41	-	0.80	m²	0.88
38mm work	0.42	10.00	0.46	-	0.04	0.55	-	1.01	m²	1.11
50mm work	0.55	10.00	0.61	-	0.05	0.69	-	1.29	m²	1.42
Screeds, beds and toppings; Granolithic screed; steel troweled										
25mm work to floors on concrete base; one coat										
level and to falls only not exceeding 15 degrees from horizontal	4.54	5.00	4.75	0.42	0.18	10.19	-	14.94	m²	16.44
to falls and crossfalls and to slopes not exceeding 15 degrees from horizontal	4.54	5.00	4.75	0.52	0.18	12.03	-	16.78	m²	18.46
to slopes exceeding 15 degrees from horizontal	4.54	5.00	4.75	0.57	0.18	12.95	-	17.70	m²	19.47
32mm work to floors on concrete base; one coat										
level and to falls only not exceeding 15 degrees from horizontal	5.70	5.00	5.96	0.45	0.20	11.02	-	16.98	m²	18.68
to falls and crossfalls and to slopes not exceeding 15 degrees from horizontal	5.70	5.00	5.96	0.55	0.20	12.86	-	18.82	m²	20.70
to slopes exceeding 15 degrees from horizontal	5.70	5.00	5.96	0.61	0.20	13.96	-	19.92	m²	21.91
38mm work to floors on concrete base; one coat										
level and to falls only not exceeding 15 degrees from horizontal	6.69	5.00	7.00	0.48	0.21	11.71	-	18.71	m²	20.58
to falls and crossfalls and to slopes not exceeding 15 degrees from horizontal	6.69	5.00	7.00	0.58	0.21	13.54	-	20.54	m²	22.60
to slopes exceeding 15 degrees from horizontal	6.69	5.00	7.00	0.63	0.21	14.46	-	21.46	m²	23.61
50mm work to floors on concrete base; one coat										
level and to falls only not exceeding 15 degrees from horizontal	8.66	5.00	9.08	0.54	0.24	13.22	-	22.30	m²	24.53
to falls and crossfalls and to slopes not exceeding 15 degrees from horizontal	8.66	5.00	9.08	0.64	0.24	15.06	-	24.13	m²	26.55
to slopes exceeding 15 degrees from horizontal	8.66	5.00	9.08	0.69	0.24	15.98	-	25.05	m²	27.56
25mm work to treads on concrete base										
width 275mm	1.75	10.00	1.92	0.25	0.05	5.28	-	7.20	m	7.92
32mm work to treads on concrete base										
width 275mm	2.25	10.00	2.46	0.30	0.06	6.34	-	8.80	m	9.68
38mm work to treads on concrete base										
width 275mm	2.74	10.00	3.01	0.35	0.06	7.26	-	10.26	m	11.29
50mm work to treads on concrete base										
width 275mm	3.57	10.00	3.91	0.40	0.07	8.31	-	12.22	m	13.45
13mm work to plain risers on concrete base										
height 175mm	0.49	10.00	0.54	0.50	0.05	9.87	-	10.42	m	11.46
19mm work to plain risers on concrete base										
height 175mm	0.82	10.00	0.91	0.55	0.05	10.79	-	11.70	m	12.87
25mm work to plain risers on concrete base										
height 175mm	1.15	10.00	1.27	0.60	0.05	11.71	-	12.98	m	14.28
32mm work to plain risers on concrete base										
height 175mm	1.32	10.00	1.45	0.65	0.05	12.63	-	14.08	m	15.49
13mm work to undercut risers on concrete base										
height 175mm	0.49	10.00	0.54	0.55	0.05	10.79	-	11.34	m	12.47
19mm work to undercut risers on concrete base										
height 175mm	0.82	10.00	0.91	0.60	0.05	11.71	-	12.62	m	13.88
25mm work to undercut risers on concrete base										
height 175mm	1.15	10.00	1.27	0.65	0.05	12.63	-	13.90	m	15.29
32mm work to undercut risers on concrete base										
height 175mm	1.32	10.00	1.45	0.70	0.05	13.55	-	15.00	m	16.50
13mm work to strings on concrete base										
height 300mm	0.99	10.00	1.09	0.70	0.05	13.55	-	14.63	m	16.10
19mm work to strings on concrete base										
height 300mm	1.48	10.00	1.63	0.75	0.05	14.47	-	16.10	m	17.71
25mm work to strings on concrete base										
height 300mm	1.81	10.00	1.99	0.80	0.05	15.38	-	17.38	m	19.12

FLOOR, WALL, CEILING AND ROOF FINISHINGS *(vertical left margin)*

Labour hourly rates: (except Specialists) Craft Operatives 18.37 Labourer 13.76 Rates are national average prices. Refer to REGIONAL VARIATIONS for indicative levels of overall pricing in regions	MATERIALS			LABOUR				RATES		
	Del to Site £	Waste %	Material Cost £	Craft Optve Hrs	Lab Hrs	Labour Cost £	Sunds £	Nett Rate £	Unit	Gross rate (10%) £
IN-SITU, TILED, BLOCK, MOSAIC, SHEET, APPLIED LIQUID OR PLANTED FINISHES (Cont'd)										
Screeds, beds and toppings; Granolithic screed; steel troweled (Cont'd)										
32mm work to strings on concrete base height 300mm..........	2.31	10.00	2.54	0.85	0.06	16.44	-	18.98	m	20.88
13mm work to strings on brickwork or blockwork base height 275mm	0.82	10.00	0.91	0.65	0.05	12.63	-	13.53	m	14.89
19mm work to strings on brickwork or blockwork base height 275mm	1.32	10.00	1.45	0.70	0.05	13.55	-	15.00	m	16.50
25mm work to strings on brickwork or blockwork base height 275mm	1.65	10.00	1.81	0.75	0.05	14.47	-	16.28	m	17.91
32mm work to strings on brickwork or blockwork base height 275mm	2.14	10.00	2.36	0.80	0.06	15.52	-	17.88	m	19.67
13mm work to aprons on concrete base height 150mm	0.49	10.00	0.54	0.50	0.05	9.87	-	10.42	m	11.46
19mm work to aprons on concrete base height 150mm	0.66	10.00	0.72	0.55	0.05	10.79	-	11.52	m	12.67
25mm work to aprons on concrete base height 150mm	0.99	10.00	1.09	0.60	0.05	11.71	-	12.80	m	14.08
32mm work to aprons on concrete base height 150mm	1.15	10.00	1.27	0.65	0.05	12.63	-	13.90	m	15.29
13mm work to skirtings on concrete base height 150mm	0.49	10.00	0.54	0.75	0.05	14.47	-	15.01	m	16.51
19mm work to skirtings on concrete base height 150mm	0.66	10.00	0.72	0.80	0.05	15.38	-	16.11	m	17.72
25mm work to skirtings on concrete base height 150mm	0.99	10.00	1.09	0.85	0.05	16.30	-	17.39	m	19.13
32mm work to skirtings on concrete base height 150mm	1.15	10.00	1.27	0.90	0.05	17.22	-	18.49	m	20.34
25mm linings to channels on concrete base										
150mm girth on face; to falls	1.04	10.00	1.14	0.20	0.05	4.36	-	5.50	m	6.05
225mm girth on face; to falls	1.42	10.00	1.56	0.25	0.05	5.28	-	6.84	m	7.52
32mm linings to channels on concrete base										
150mm girth on face; to falls	1.21	10.00	1.32	0.23	0.05	4.91	-	6.23	m	6.86
225mm girth on face; to falls	1.92	10.00	2.10	0.28	0.05	5.83	-	7.93	m	8.72
38mm linings to channels on concrete base										
150mm girth on face; to falls	1.54	10.00	1.68	0.25	0.05	5.28	-	6.96	m	7.66
225mm girth on face; to falls	2.25	10.00	2.46	0.30	0.05	6.20	-	8.66	m	9.53
50mm linings to channels on concrete base										
150mm girth on face; to falls	1.87	10.00	2.05	0.30	0.05	6.20	-	8.25	m	9.07
225mm girth on face; to falls	2.91	10.00	3.19	0.35	0.05	7.12	-	10.30	m	11.34
Rounded angles and intersections										
10 - 100mm radius	-	-	-	0.18	-	3.31	-	3.31	m	3.64
If paving tinted, add according to tint										
25mm work	2.56	10.00	2.82	-	0.03	0.41	-	3.23	m²	3.55
32mm work	3.28	10.00	3.61	-	0.03	0.41	-	4.02	m²	4.42
38mm work	3.89	10.00	4.28	-	0.04	0.55	-	4.83	m²	5.32
50mm work	5.12	10.00	5.64	-	0.05	0.69	-	6.32	m²	6.96
If carborundum troweled into surface										
add	3.00	10.00	3.30	0.08	0.04	2.02	-	5.32	m²	5.85
Screeds, beds and toppings; Surface hardeners on paving										
Proprietary surface hardener (Epoxy Resin Based)										
two coats	16.00	10.00	17.60	-	0.20	2.75	-	20.35	m²	22.38
three coats	21.92	10.00	24.11	-	0.25	3.44	-	27.55	m²	30.30
Polyurethane floor sealer										
two coats	1.86	10.00	2.05	-	0.20	2.75	-	4.80	m²	5.28
three coats	2.40	10.00	2.64	-	0.25	3.44	-	6.08	m²	6.69
Nitoflor Lithurin concrete surface dressing										
one coat	0.79	10.00	0.87	-	0.17	2.34	-	3.21	m²	3.53
two coats	1.58	10.00	1.73	-	0.24	3.30	-	5.04	m²	5.54
If on old floors										
add for cleaning and degreasing	0.67	5.00	0.71	-	0.35	4.82	-	5.52	m²	6.07
Screeds, beds and toppings; Vermiculite screed consisting of cement and Vermiculite aggregate finished with 20mm cement and sand (1:4) screeded bed										
45mm work to roofs on concrete base; two coats to falls and crossfalls and to slopes not exceeding 15 degrees from horizontal	6.01	5.00	6.30	0.45	0.48	14.87	-	21.17	m²	23.29
60mm work to roofs on concrete base; two coats to falls and crossfalls and to slopes not exceeding 15 degrees from horizontal	8.29	5.00	8.69	0.48	0.51	15.84	-	24.52	m²	26.98

Labour hourly rates: (except Specialists) Craft Operatives 18.37 Labourer 13.76 Rates are national average prices. Refer to REGIONAL VARIATIONS for indicative levels of overall pricing in regions	MATERIALS			LABOUR				RATES		
	Del to Site £	Waste %	Material Cost £	Craft Optve Hrs	Lab Hrs	Labour Cost £	Sunds £	Nett Rate £	Unit	Gross rate (10%) £
IN-SITU, TILED, BLOCK, MOSAIC, SHEET, APPLIED LIQUID OR PLANTED FINISHES (Cont'd)										
Screeds, beds and toppings; Vermiculite screed consisting of cement and Vermiculite aggregate finished with 20mm cement and sand (1:4) screeded bed (Cont'd)										
70mm work to roofs on concrete base; two coats to falls and crossfalls and to slopes not exceeding 15 degrees from horizontal	9.80	5.00	10.28	0.51	0.54	16.80	-	27.08	m²	29.79
80mm work to roofs on concrete base; two coats to falls and crossfalls and to slopes not exceeding 15 degrees from horizontal	11.31	5.00	11.87	0.53	0.56	17.44	-	29.31	m²	32.24
Screeds, beds and toppings; Lightweight concrete screed consisting of cement and lightweight aggregate, medium grade, 10 - 5 gauge, 800 kg/m³ (1:10) finished with 15mm cement and sand (1:4) troweled bed										
50mm work to roofs on concrete base; two coats to falls and crossfalls and to slopes not exceeding 15 degrees from horizontal	7.47	5.00	7.83	0.60	0.51	18.04	-	25.87	m²	28.46
75mm work to roofs on concrete base; two coats to falls and crossfalls and to slopes not exceeding 15 degrees from horizontal	11.58	5.00	12.14	0.65	0.56	19.65	-	31.79	m²	34.97
100mm work to roofs on concrete base; two coats to falls and crossfalls and to slopes not exceeding 15 degrees from horizontal	15.68	5.00	16.45	0.70	0.61	21.25	-	37.71	m²	41.48
Screeds, beds and toppings; Division strips										
Aluminium dividing strips; setting in bed 5 x 25mm; flat section	4.27	5.00	4.48	0.15	-	2.76	-	7.24	m	7.96
Brass dividing strips; setting in bed 5 x 25mm; flat section	26.90	5.00	28.24	0.15	-	2.76	-	31.00	m	34.10
Screeds, beds and toppings; Screed reinforcement										
Reinforcement; galvanized wire netting, BS EN 10223-2, 13mm mesh, 22 gauge wire; 150mm laps; placing in position floors	2.24	5.00	2.36	0.05	-	0.92	-	3.27	m²	3.60
Reinforcement; galvanized wire netting, BS EN 10223-2, 25mm mesh, 20 gauge wire; 150mm laps; placing in position floors	1.39	5.00	1.46	0.05	-	0.92	-	2.38	m²	2.61
Reinforcement; galvanized wire netting, BS EN 10223-2, 38mm mesh, 19 gauge wire; 150mm laps; placing in position floors	1.05	5.00	1.10	0.05	-	0.92	-	2.02	m²	2.22
Reinforcement; galvanized wire netting, BS EN 10223-2, 50mm mesh, 19 gauge wire; 150mm laps; placing in position floors	0.81	5.00	0.85	0.05	-	0.92	-	1.77	m²	1.94
Screeds, beds and toppings; Mastic asphalt flooring/floor underlays										
Paving; BS 6925 (limestone aggregate), brown, smooth floated finish										
Flooring and underlay 15mm thick; coats of asphalt - 1; to concrete base; flat										
width not exceeding 150mm	-	-	Specialist	-	-	Specialist	-	39.38	m²	43.31
width 150 - 225mm	-	-	Specialist	-	-	Specialist	-	33.00	m²	36.30
width 225 - 300mm	-	-	Specialist	-	-	Specialist	-	26.10	m²	28.71
width 300 - 600mm	-	-	Specialist	-	-	Specialist	-	21.75	m²	23.93
width exceeding 600mm	-	-	Specialist	-	-	Specialist	-	19.73	m²	21.70
Skirtings 15mm thick; coats of asphalt - 1; no underlay; to brickwork base										
girth not exceeding 150mm	-	-	Specialist	-	-	Specialist	-	12.68	m	13.94
Paving; BS 6925 (limestone aggregate), red, smooth floated finish Flooring and underlay 15mm thick; coats of asphalt - 1; to concrete base; flat										
width not exceeding 150mm	-	-	Specialist	-	-	Specialist	-	39.75	m²	43.73
width 150 - 225mm	-	-	Specialist	-	-	Specialist	-	35.55	m²	39.10
width 225 - 300mm	-	-	Specialist	-	-	Specialist	-	28.42	m²	31.27
width 300 - 600mm	-	-	Specialist	-	-	Specialist	-	21.98	m²	24.17
width exceeding 600mm	-	-	Specialist	-	-	Specialist	-	21.08	m²	23.18
Skirtings 15mm thick; coats of asphalt - 1; no underlay; to brickwork base										
girth not exceeding 150mm	-	-	Specialist	-	-	Specialist	-	14.10	m	15.51
Screeds, beds and toppings; Trowelled bitumen/resin/rubber-latex flooring										
One coat levelling screeds										
3mm work to floors on concrete base; one coat level and to falls only not exceeding 15 degrees from horizontal	2.62	5.00	2.75	0.40	0.05	8.04	-	10.78	m²	11.86
3mm work to floors on existing timber boarded base; one coat level and to falls only not exceeding 15 degrees from horizontal	2.62	5.00	2.75	0.50	0.06	10.01	-	12.76	m²	14.03

FLOOR, WALL, CEILING AND ROOF FINISHINGS (side margin)

Labour hourly rates: (except Specialists) Craft Operatives 18.37 Labourer 13.76 Rates are national average prices. Refer to REGIONAL VARIATIONS for indicative levels of overall pricing in regions	MATERIALS			LABOUR				RATES		
	Del to Site £	Waste %	Material Cost £	Craft Optve Hrs	Lab Hrs	Labour Cost £	Sunds £	Nett Rate £	Unit	Gross rate (10%) £
IN-SITU, TILED, BLOCK, MOSAIC, SHEET, APPLIED LIQUID OR PLANTED FINISHES (Cont'd)										
Screeds, beds and toppings; Plasterboard; baseboard underlay										
Baseboarding; Gyproc Handiboard; 5mm joints, filled with plaster and scrimmed; fixing with Galvanized nails, 30mm long at 400mm centres										
9.5mm work to walls on timber base										
width exceeding 600mm	2.28	5.00	2.39	0.29	0.15	7.39	-	9.78	m²	10.76
width; 300mm	0.71	10.00	0.78	0.17	0.05	3.81	-	4.59	m	5.05
9.5mm work to isolated columns on timber base										
width exceeding 600mm	2.28	5.00	2.39	0.44	0.15	10.15	-	12.54	m²	13.79
width; 300mm	0.75	10.00	0.82	0.26	0.05	5.46	-	6.28	m	6.91
9.5mm work to ceilings on timber base										
width exceeding 600mm	2.28	5.00	2.39	0.35	0.15	8.49	-	10.88	m²	11.97
width; 300mm	0.75	10.00	0.82	0.21	0.05	4.55	-	5.36	m	5.90
9.5mm work to isolated beams on timber base										
width exceeding 600mm	2.28	5.00	2.39	0.44	0.15	10.15	-	12.54	m²	13.79
width; 300mm	0.75	10.00	0.82	0.26	0.05	5.46	-	6.28	m	6.91
Baseboarding; Gyproc Handiboard; 5mm joints, filled with plaster and scrimmed; fixing with adhesive dabs										
9.5mm work to walls on masonry base										
width exceeding 600mm	2.31	5.00	2.42	0.29	0.15	7.39	-	9.81	m²	10.79
width; 300mm	0.73	10.00	0.80	0.17	0.05	3.81	-	4.61	m	5.07
Baseboarding; Gyproc Handiboard; 3mm joints, filled with plaster; fixing with Galvanized nails, 30mm long at 400mm centres										
12.5mm work to walls on timber base										
width exceeding 600mm	3.53	5.00	3.70	0.35	0.18	8.91	-	12.60	m²	13.87
width; 300mm	1.09	10.00	1.19	0.21	0.05	4.55	-	5.74	m	6.31
12.5mm work to isolated columns on timber base										
width exceeding 600mm	3.53	5.00	3.70	0.53	0.18	12.21	-	15.91	m²	17.50
width; 300mm	1.09	10.00	1.19	0.32	0.05	6.57	-	7.76	m	8.53
12.5mm work to ceilings on timber base										
width exceeding 600mm	3.48	5.00	3.65	0.42	0.18	10.19	-	13.84	m²	15.23
width; 300mm	1.09	10.00	1.19	0.25	0.05	5.28	-	6.47	m	7.12
12.5mm work to isolated beams on timber base										
width exceeding 600mm	3.53	5.00	3.70	0.53	0.18	12.21	-	15.91	m²	17.50
width; 300mm	1.09	10.00	1.19	0.32	0.05	6.57	-	7.76	m	8.53
Baseboarding; Gyproc plank, square edge; 5mm joints, filled with plaster and scrimmed										
19mm work to walls on timber base										
width exceeding 600mm	4.78	5.00	5.02	0.47	0.24	11.94	-	16.95	m²	18.65
width; 300mm	1.48	10.00	1.62	0.28	0.07	6.11	-	7.72	m	8.50
19mm work to isolated columns on timber base										
width exceeding 600mm	4.78	5.00	5.02	0.71	0.24	16.35	-	21.36	m²	23.50
width; 300mm	1.48	10.00	1.62	0.43	0.07	8.86	-	10.48	m	11.53
19mm work to ceilings on timber base										
width exceeding 600mm	4.78	5.00	5.02	0.56	0.24	13.59	-	18.61	m²	20.47
width; 300mm	1.48	10.00	1.62	0.34	0.07	7.21	-	8.83	m	9.71
19mm work to isolated beams on timber base										
width exceeding 600mm	4.78	5.00	5.02	0.71	0.24	16.35	-	21.36	m²	23.50
width; 300mm	1.48	10.00	1.62	0.43	0.07	8.86	-	10.48	m	11.53
Baseboarding; Gyproc square edge wallboard; 3mm joints filled with plaster and scrimmed; fixing with Galvanized nails, 30mm long at 400mm centres										
9.5mm work to walls on timber base										
width exceeding 600mm	1.52	5.00	1.60	0.29	0.15	7.39	-	8.99	m²	9.89
width; 300mm	0.49	10.00	0.53	0.17	0.05	3.81	-	4.34	m	4.78
12.5mm work to walls on timber base										
width exceeding 600mm	1.60	5.00	1.68	0.35	0.18	8.91	-	10.59	m²	11.64
width; 300mm	0.51	10.00	0.56	0.21	0.05	4.55	-	5.10	m	5.61
15mm work to walls on timber base										
width exceeding 600mm	2.19	5.00	2.29	0.41	0.21	10.42	-	12.71	m²	13.99
width; 300mm	0.69	10.00	0.75	0.25	0.06	5.42	-	6.17	m	6.78
9.5mm work to isolated columns on timber base										
width exceeding 600mm	1.52	5.00	1.60	0.44	0.15	10.15	-	11.74	m²	12.92
width; 300mm	0.49	10.00	0.53	0.26	0.05	5.46	-	5.99	m	6.59
12.5mm work to isolated columns on timber base										
width exceeding 600mm	1.60	5.00	1.68	0.53	0.18	12.21	-	13.89	m²	15.28
width; 300mm	0.51	10.00	0.56	0.32	0.05	6.57	-	7.12	m	7.84
15mm work to isolated columns on timber base										
width exceeding 600mm	2.19	5.00	2.29	0.62	0.21	14.28	-	16.57	m²	18.23
width; 300mm	0.69	10.00	0.75	0.37	0.06	7.62	-	8.37	m	9.21
9.5mm work to ceilings on timber base										
width exceeding 600mm	1.52	5.00	1.60	0.35	0.15	8.49	-	10.09	m²	11.10
width; 300mm	0.49	10.00	0.53	0.21	0.05	4.55	-	5.08	m	5.58
12.5mm work to ceilings on timber base										
width exceeding 600mm	1.60	5.00	1.68	0.42	0.18	10.19	-	11.87	m²	13.06
width; 300mm	0.51	10.00	0.56	0.25	0.05	5.28	-	5.84	m	6.42
15mm work to ceilings on timber base										
width exceeding 600mm	2.19	5.00	2.29	0.49	0.25	12.44	-	14.73	m²	16.21
width; 300mm	0.69	10.00	0.75	0.29	0.08	6.43	-	7.18	m	7.90
9.5mm work to isolated beams on timber base										
width exceeding 600mm	1.52	5.00	1.60	0.44	0.15	10.15	-	11.74	m²	12.92
width; 300mm	0.49	10.00	0.53	0.26	0.05	5.46	-	5.99	m	6.59
12.5mm work to isolated beams on timber base										
width exceeding 600mm	1.60	5.00	1.68	0.53	0.18	12.21	-	13.89	m²	15.28
width; 300mm	0.51	10.00	0.56	0.32	0.05	6.57	-	7.12	m	7.84

Labour hourly rates: (except Specialists) Craft Operatives 18.37 Labourer 13.76 Rates are national average prices. Refer to REGIONAL VARIATIONS for indicative levels of overall pricing in regions	MATERIALS			LABOUR				RATES		
	Del to Site	Waste	Material Cost	Craft Optve	Lab	Labour Cost	Sunds	Nett Rate		Gross rate (10%)
	£	%	£	Hrs	Hrs	£	£	£	Unit	£

IN-SITU, TILED, BLOCK, MOSAIC, SHEET, APPLIED LIQUID OR PLANTED FINISHES (Cont'd)

Screeds, beds and toppings; Plasterboard; baseboard underlay (Cont'd)

Baseboarding; Gyproc square edge wallboard; 3mm joints filled with plaster and scrimmed; fixing with Galvanized nails, 30mm long at 400mm centres (Cont'd)

	Del to Site £	Waste %	Material Cost £	Craft Optve Hrs	Lab Hrs	Labour Cost £	Sunds £	Nett Rate £	Unit	Gross rate (10%) £
15mm work to isolated beams on timber base										
width exceeding 600mm	2.19	5.00	2.29	0.62	0.21	14.28	-	16.57	m²	18.23
width; 300mm	0.69	10.00	0.75	0.37	0.06	7.62	-	8.37	m	9.21

Baseboarding; Gyproc square edge vapourcheck wallboard; 3mm joints filled with plaster and scrimmed; fixing with Galvanized nails, 30mm long at 400mm centres

	Del to Site £	Waste %	Material Cost £	Craft Optve Hrs	Lab Hrs	Labour Cost £	Sunds £	Nett Rate £	Unit	Gross rate (10%) £
9.5mm work to walls on timber base										
width exceeding 600mm	3.07	5.00	3.22	0.29	0.15	7.39	-	10.61	m²	11.67
width; 300mm	0.95	10.00	1.04	0.17	0.05	3.81	-	4.85	m	5.34
12.5mm work to walls on timber base										
width exceeding 600mm	3.03	5.00	3.17	0.35	0.18	8.91	-	12.08	m²	13.29
width; 300mm	0.94	10.00	1.03	0.21	0.05	4.55	-	5.57	m	6.13
15mm work to walls on timber base										
width exceeding 600mm	4.03	5.00	4.23	0.41	0.21	10.42	-	14.65	m²	16.12
width; 300mm	1.24	10.00	1.36	0.25	0.06	5.42	-	6.78	m	7.45
9.5mm work to isolated columns on timber base										
width exceeding 600mm	3.07	5.00	3.22	0.44	0.15	10.15	-	13.37	m²	14.71
width; 300mm	0.95	10.00	1.04	0.26	0.05	5.46	-	6.51	m	7.16
12.5mm work to isolated columns on timber base										
width exceeding 600mm	3.03	5.00	3.17	0.53	0.18	12.21	-	15.39	m²	16.93
width; 300mm	0.94	10.00	1.03	0.32	0.05	6.57	-	7.59	m	8.35
15mm work to isolated columns on timber base										
width exceeding 600mm	4.03	5.00	4.23	0.62	0.21	14.28	-	18.51	m²	20.36
width; 300mm	1.24	10.00	1.36	0.37	0.06	7.62	-	8.98	m	9.88
9.5mm work to ceilings on timber base										
width exceeding 600mm	3.07	5.00	3.22	0.35	0.15	8.49	-	11.72	m²	12.89
width; 300mm	0.95	10.00	1.04	0.21	0.05	4.55	-	5.59	m	6.15
12.5mm work to ceilings on timber base										
width exceeding 600mm	3.03	5.00	3.17	0.42	0.18	10.19	-	13.37	m²	14.70
width; 300mm	0.94	10.00	1.03	0.25	0.05	5.28	-	6.31	m	6.94
15mm work to ceilings on timber base										
width exceeding 600mm	4.03	5.00	4.23	0.49	0.25	12.44	-	16.67	m²	18.34
width; 300mm	1.24	10.00	1.36	0.29	0.08	6.43	-	7.79	m	8.57
9.5mm work to isolated beams on timber base										
width exceeding 600mm	3.07	5.00	3.22	0.44	0.15	10.15	-	13.37	m²	14.71
width; 300mm	0.95	10.00	1.04	0.26	0.05	5.46	-	6.51	m	7.16
12.5mm work to isolated beams on timber base										
width exceeding 600mm	3.03	5.00	3.17	0.53	0.18	12.21	-	15.39	m²	16.93
width; 300mm	0.94	10.00	1.03	0.32	0.05	6.57	-	7.59	m	8.35
15mm work to isolated beams on timber base										
width exceeding 600mm	4.03	5.00	4.23	0.62	0.21	14.28	-	18.51	m²	20.36
width; 300mm	1.24	10.00	1.36	0.37	0.06	7.62	-	8.98	m	9.88

Baseboarding; two layers 9.5mm Gyproc square edge wallboard; second layer with 3mm joints filled with plaster and scrimmed; fixing with Galvanized nails, 30mm long at 400mm centres

	Del to Site £	Waste %	Material Cost £	Craft Optve Hrs	Lab Hrs	Labour Cost £	Sunds £	Nett Rate £	Unit	Gross rate (10%) £
19mm work to walls on timber base										
width exceeding 600mm	3.07	5.00	3.21	0.60	0.30	15.15	-	18.36	m²	20.20
width; 300mm	0.99	10.00	1.08	0.36	0.09	7.85	-	8.93	m	9.82
19mm work to isolated columns on timber base										
width exceeding 600mm	3.07	5.00	3.21	0.80	0.30	18.82	-	22.04	m²	24.24
width; 300mm	0.99	10.00	1.08	0.48	0.09	10.06	-	11.13	m	12.25
19mm work to ceilings on timber base										
width exceeding 600mm	3.07	5.00	3.21	0.72	0.30	17.35	-	20.57	m²	22.62
width; 300mm	0.99	10.00	1.08	0.43	0.09	9.14	-	10.21	m	11.23
19mm work to isolated beams on timber base										
width exceeding 600mm	3.07	5.00	3.21	0.80	0.30	18.82	-	22.04	m²	24.24
width; 300mm	0.99	10.00	1.08	0.48	0.09	10.06	-	11.13	m	12.25

Baseboarding; one layer 9.5mm and one layer 12.5mm Gyproc square edge wallboard; second layer with 3mm joints filled with plaster and scrimmed; fixing with Galvanized nails, 30mm long at 400mm centres

	Del to Site £	Waste %	Material Cost £	Craft Optve Hrs	Lab Hrs	Labour Cost £	Sunds £	Nett Rate £	Unit	Gross rate (10%) £
22mm work to walls on timber base										
width exceeding 600mm	3.15	5.00	3.30	0.65	0.33	16.48	-	19.78	m²	21.75
width; 300mm	1.01	10.00	1.10	0.39	0.10	8.54	-	9.64	m	10.61
22mm work to isolated columns on timber base										
width exceeding 600mm	3.15	5.00	3.30	0.98	0.33	22.54	-	25.84	m²	28.42
width; 300mm	1.01	10.00	1.10	0.59	0.10	12.21	-	13.32	m	14.65
22mm work to ceilings on timber base										
width exceeding 600mm	3.15	5.00	3.30	0.78	0.33	18.87	-	22.17	m²	24.38
width; 300mm	1.01	10.00	1.10	0.47	0.10	10.01	-	11.11	m	12.22
22mm work to isolated beams on timber base										
width exceeding 600mm	3.15	5.00	3.30	0.98	0.33	22.54	-	25.84	m²	28.42
width; 300mm	1.01	10.00	1.10	0.59	0.10	12.21	-	13.32	m	14.65

Baseboarding; two layers 12.5mm Gyproc square edge wallboard; second layer with 3mm joints filled with plaster and scrimmed; fixing with Galvanized nails, 30mm long at 400mm centres

	Del to Site £	Waste %	Material Cost £	Craft Optve Hrs	Lab Hrs	Labour Cost £	Sunds £	Nett Rate £	Unit	Gross rate (10%) £
25mm work to walls on timber base										
width exceeding 600mm	3.23	5.00	3.38	0.70	0.35	17.68	-	21.05	m²	23.16
width; 300mm	1.04	10.00	1.13	0.42	0.11	9.23	-	10.36	m	11.39
25mm work to isolated columns on timber base										
width exceeding 600mm	3.23	5.00	3.38	1.05	0.35	24.10	-	27.48	m²	30.23
width; 300mm	1.04	10.00	1.13	0.64	0.11	13.27	-	14.40	m	15.84

	MATERIALS			LABOUR				RATES		
Labour hourly rates: (except Specialists) Craft Operatives 18.37 Labourer 13.76 Rates are national average prices. Refer to REGIONAL VARIATIONS for indicative levels of overall pricing in regions	Del to Site £	Waste %	Material Cost £	Craft Optve Hrs	Lab Hrs	Labour Cost £	Sunds £	Nett Rate £	Unit	Gross rate (10%) £
IN-SITU, TILED, BLOCK, MOSAIC, SHEET, APPLIED LIQUID OR PLANTED FINISHES (Cont'd)										
Screeds, beds and toppings; Plasterboard; baseboard underlay (Cont'd)										
Baseboarding; two layers 12.5mm Gyproc square edge wallboard; second layer with 3mm joints filled with plaster and scrimmed; fixing with Galvanized nails, 30mm long at 400mm centres (Cont'd)										
25mm work to ceilings on timber base										
width exceeding 600mm	3.23	5.00	3.38	0.84	0.35	20.25	-	23.63	m²	25.99
width; 300mm	1.04	10.00	1.13	0.50	0.11	10.70	-	11.83	m	13.01
25mm work to isolated beams on timber base										
width exceeding 600mm	3.23	5.00	3.38	1.05	0.35	24.10	-	27.48	m²	30.23
width; 300mm	1.04	10.00	1.13	0.64	0.11	13.27	-	14.40	m	15.84
Baseboarding; two layers 15mm Gyproc square edge wallboard; second layer with 3mm joints filled with plaster and scrimmed; fixing with Galvanized nails, 30mm long at 400mm centres										
30mm work to walls on timber base										
width exceeding 600mm	4.39	5.00	4.60	0.75	0.38	19.01	-	23.61	m²	25.97
width; 300mm	1.39	10.00	1.51	0.45	0.11	9.78	-	11.29	m	12.42
30mm work to isolated columns on timber base										
width exceeding 600mm	4.39	5.00	4.60	1.13	0.38	25.99	-	30.59	m²	33.65
width; 300mm	1.39	10.00	1.51	0.68	0.11	14.01	-	15.52	m	17.07
30mm work to ceilings on timber base										
width exceeding 600mm	4.39	5.00	4.60	0.90	0.38	21.76	-	26.37	m²	29.00
width; 300mm	1.39	10.00	1.51	0.54	0.11	11.43	-	12.95	m	14.24
30mm work to isolated beams on timber base										
width exceeding 600mm	4.39	5.00	4.60	1.13	0.38	25.99	-	30.59	m²	33.65
width; 300mm	1.39	10.00	1.51	0.68	0.11	14.01	-	15.52	m	17.07
Baseboarding; two layers; first layer 9.5mm Gyproc vapourcheck wallboard, second layer 9.5mm Gyproc wallboard both layers with square edge boards; second layer with 3mm joints filled with plaster and scrimmed; fixing with Galvanized nails, 30mm long at 400mm centres										
19mm work to walls on timber base										
width exceeding 600mm	4.62	5.00	4.84	0.60	0.30	15.15	-	19.99	m²	21.99
width; 300mm	1.45	10.00	1.59	0.36	0.09	7.85	-	9.44	m	10.38
19mm work to isolated columns on timber base										
width exceeding 600mm	4.62	5.00	4.84	0.80	0.30	18.82	-	23.66	m²	26.03
width; 300mm	1.45	10.00	1.59	0.48	0.09	10.06	-	11.64	m	12.81
19mm work to ceilings on timber base										
width exceeding 600mm	4.62	5.00	4.84	0.72	0.30	17.35	-	22.19	m²	24.41
width; 300mm	1.45	10.00	1.59	0.43	0.09	9.14	-	10.72	m	11.80
19mm work to isolated beams on timber base										
width exceeding 600mm	4.62	5.00	4.84	0.80	0.30	18.82	-	23.66	m²	26.03
width; 300mm	1.45	10.00	1.59	0.48	0.09	10.06	-	11.64	m	12.81
Baseboarding; two layers; first layer 12.5mm Gyproc vapourcheck wallboard, second layer 9.5mm Gyproc wallboard, both layers with square edge boards; second layer with 3mm joints filled with plaster and scrimmed; fixing with Galvanized nails, 30mm long at 400mm centres										
22mm work to walls on timber base										
width exceeding 600mm	4.57	5.00	4.79	0.65	0.33	16.48	-	21.27	m²	23.40
width; 300mm	1.44	10.00	1.57	0.39	0.10	8.54	-	10.11	m	11.12
22mm work to isolated columns on timber base										
width exceeding 600mm	4.57	5.00	4.79	0.98	0.33	22.54	-	27.33	m²	30.07
width; 300mm	1.44	10.00	1.57	0.59	0.10	12.21	-	13.79	m	15.16
22mm work to ceilings on timber base										
width exceeding 600mm	4.57	5.00	4.79	0.78	0.33	18.87	-	23.66	m²	26.03
width; 300mm	1.44	10.00	1.57	0.47	0.10	10.01	-	11.58	m	12.74
22mm work to isolated beams on timber base										
width exceeding 600mm	4.57	5.00	4.79	0.98	0.33	22.54	-	27.33	m²	30.07
width; 300mm	1.44	10.00	1.57	0.59	0.10	12.21	-	13.79	m	15.16
Baseboarding; two layers; first layer 12.5mm Gyproc vapourcheck wallboard, second layer 12.5mm Gyproc wallboard, both layers with square edge boards; second layer with 3mm joints filled with plaster and scrimmed; fixing with Galvanized nails, 30mm long at 400mm centres										
25mm work to walls on timber base										
width exceeding 600mm	4.65	5.00	4.87	0.70	0.35	17.68	-	22.55	m²	24.80
width; 300mm	1.46	10.00	1.60	0.42	0.11	9.23	-	10.83	m	11.91
25mm work to isolated columns on timber base										
width exceeding 600mm	4.65	5.00	4.87	1.05	0.35	24.10	-	28.98	m²	31.88
width; 300mm	1.46	10.00	1.60	0.64	0.11	13.27	-	14.87	m	16.36
25mm work to ceilings on timber base										
width exceeding 600mm	4.65	5.00	4.87	0.84	0.35	20.25	-	25.12	m²	27.63
width; 300mm	1.46	10.00	1.60	0.50	0.11	10.70	-	12.30	m	13.53
25mm work to isolated beams on timber base										
width exceeding 600mm	4.65	5.00	4.87	1.05	0.35	24.10	-	28.98	m²	31.88
width; 300mm	1.46	10.00	1.60	0.64	0.11	13.27	-	14.87	m	16.36
Baseboarding; two layers; first layer 15mm Gyproc vapourcheck wallboard, second layer 15mm Gyproc wallboard, both layers with square edge boards; second layer with 3mm joints filled with plaster and scrimmed; fixing with Galvanized nails, 30mm long at 400mm centres										
30mm work to walls on timber base										
width exceeding 600mm	6.24	5.00	6.54	0.75	0.38	19.01	-	25.55	m²	28.11
width; 300mm	1.94	10.00	2.12	0.45	0.11	9.78	-	11.90	m	13.09
30mm work to isolated columns on timber base										
width exceeding 600mm	6.24	5.00	6.54	1.13	0.38	25.99		32.53	m²	35.78

Labour hourly rates: (except Specialists) Craft Operatives 18.37 Labourer 13.76 Rates are national average prices. Refer to REGIONAL VARIATIONS for indicative levels of overall pricing in regions	MATERIALS			LABOUR				RATES		
	Del to Site	Waste	Material Cost	Craft Optve	Lab	Labour Cost	Sunds	Nett Rate		Gross rate (10%)
	£	%	£	Hrs	Hrs	£	£	£	Unit	£
IN-SITU, TILED, BLOCK, MOSAIC, SHEET, APPLIED LIQUID OR PLANTED FINISHES (Cont'd)										
Screeds, beds and toppings; Plasterboard; baseboard underlay (Cont'd)										
Baseboarding; two layers; first layer 15mm Gyproc vapourcheck wallboard, second layer 15mm Gyproc wallboard, both layers with square edge boards; second layer with 3mm joints filled with plaster and scrimmed; fixing with Galvanized nails, 30mm long at 400mm centres (Cont'd)										
30mm work to isolated columns on timber base (Cont'd)										
width; 300mm	1.94	10.00	2.12	0.68	0.11	14.01	-	16.13	m	17.74
30mm work to ceilings on timber base										
width exceeding 600mm	6.24	5.00	6.54	0.90	0.38	21.76	-	28.31	m²	31.14
width; 300mm	1.94	10.00	2.12	0.54	0.11	11.43	-	13.56	m	14.91
30mm work to isolated beams on timber base										
width exceeding 600mm	6.24	5.00	6.54	1.13	0.38	25.99	-	32.53	m²	35.78
width; 300mm	1.94	10.00	2.12	0.68	0.11	14.01	-	16.13	m	17.74
Screeds, beds and toppings; damp wall treatment: Newlath underlay; fixing with masonry nails.										
Work to walls on brickwork or blockwork base										
width exceeding 600mm	7.46	5.00	7.80	0.40	0.20	10.10	-	17.90	m²	19.69
width; 300mm	2.24	5.00	2.34	0.24	0.06	5.23	-	7.58	m	8.33
Finish to floors; clay floor quarries, BS EN ISO 10545, terracotta; 3mm joints, symmetrical layout; bedding in 10mm cement mortar (1:3); pointing in cement mortar (1:3); on cement and sand base										
150 x 150 x 12.5mm units to floors level or to falls only not exceeding 15 degrees from horizontal										
plain	22.39	5.00	23.51	0.80	0.50	21.58	-	45.08	m²	49.59
150 x 150 x 20mm units to floors level or to falls only not exceeding 15 degrees from horizontal										
plain	38.30	5.00	40.21	1.00	0.60	26.63	-	66.84	m²	73.52
225 x 225 x 25mm units to floors level or to falls only not exceeding 15 degrees from horizontal										
plain	32.75	5.00	34.38	0.70	0.45	19.05	-	53.43	m²	58.78
150 x 150 x 12.5mm units to floors to falls and crossfalls and to slopes not exceeding 15 degrees from horizontal										
plain	22.39	5.00	23.51	0.88	0.50	23.05	-	46.55	m²	51.21
150 x 150 x 20mm units to floors to falls and crossfalls and to slopes not exceeding 15 degrees from horizontal										
plain	38.30	5.00	40.21	1.10	0.60	28.46	-	68.67	m²	75.54
225 x 225 x 25mm units to floors to falls and crossfalls and to slopes not exceeding 15 degrees from horizontal										
plain	32.75	5.00	34.38	0.75	0.45	19.97	-	54.35	m²	59.79
Extra for pointing with tinted mortar										
150 x 150 x 12.5mm tiles	1.46	10.00	1.60	0.10	-	1.84	-	3.44	m²	3.78
150 x 150 x 20mm tiles	1.46	10.00	1.60	0.10	-	1.84	-	3.44	m²	3.78
225 x 225 x 25mm tiles	1.46	10.00	1.60	0.07	-	1.29	-	2.89	m²	3.18
150 x 150 x 12.5mm units to skirtings on brickwork or blockwork base										
height 150mm; square top edge	3.71	10.00	4.06	0.30	0.08	6.61	-	10.67	m	11.74
height 150mm; rounded top edge	13.18	10.00	14.48	0.30	0.08	6.61	-	21.10	m	23.20
height 150mm; rounded top edge and cove at bottom	17.43	10.00	19.15	0.30	0.08	6.61	-	25.77	m	28.34
150 x 150 x 20mm units to skirtings on brickwork or blockwork base										
height 150mm; square top edge	5.25	10.00	5.76	0.38	0.09	8.22	-	13.98	m	15.37
height 150mm; rounded top edge	21.35	10.00	23.47	0.38	0.09	8.22	-	31.69	m	34.85
Finish to floors; clay floor quarries, BS EN ISO 10545, blended; 3mm joints symmetrical layout; bedding in 10mm cement mortar (1:3); pointing in cement mortar (1:3); on cement and sand base										
150 x 150 x 12.5mm units to floors level or to falls only not exceeding 15 degrees from horizontal										
plain	22.39	5.00	23.51	0.80	0.50	21.58	-	45.08	m²	49.59
150 x 150 x 20mm units to floors level or to falls only not exceeding 15 degrees from horizontal										
plain	38.30	5.00	40.21	1.00	0.60	26.63	-	66.84	m²	73.52
225 x 225 x 25mm units to floors level or to falls only not exceeding 15 degrees from horizontal										
plain	20.75	5.00	21.78	0.70	0.45	19.05	-	40.83	m²	44.92
150 x 150 x 12.5mm units to floors to falls and crossfalls and to slopes not exceeding 15 degrees from horizontal										
plain	22.39	5.00	23.51	0.88	0.50	23.05	-	46.55	m²	51.21
150 x 150 x 20mm units to floors to falls and crossfalls and to slopes not exceeding 15 degrees from horizontal										
plain	38.30	5.00	40.21	1.10	0.60	28.46	-	68.67	m²	75.54
225 x 225 x 25mm units to floors to falls and crossfalls and to slopes not exceeding 15 degrees from horizontal										
plain	20.75	5.00	21.78	0.75	0.45	19.97	-	41.75	m²	45.93

FLOOR, WALL, CEILING AND ROOF FINISHINGS *(vertical side text)*

	MATERIALS			LABOUR				RATES		
Labour hourly rates: (except Specialists) Craft Operatives 18.37 Labourer 13.76 Rates are national average prices. Refer to REGIONAL VARIATIONS for indicative levels of overall pricing in regions	Del to Site	Waste	Material Cost	Craft Optve	Lab	Labour Cost	Sunds	Nett Rate	Unit	Gross rate (10%)
	£	%	£	Hrs	Hrs	£	£	£		£

IN-SITU, TILED, BLOCK, MOSAIC, SHEET, APPLIED LIQUID OR PLANTED FINISHES (Cont'd)

Finish to floors; clay floor quarries, BS EN ISO 10545, blended; 3mm joints symmetrical layout; bedding in 10mm cement mortar (1:3); pointing in cement mortar (1:3); on cement and sand base (Cont'd)

	Del to Site £	Waste %	Material Cost £	Craft Optve Hrs	Lab Hrs	Labour Cost £	Sunds £	Nett Rate £	Unit	Gross rate (10%) £
Extra for pointing with tinted mortar										
150 x 150 x 12.5mm tiles	1.46	10.00	1.60	0.10	-	1.84	-	3.44	m²	3.78
150 x 150 x 20mm tiles	1.46	10.00	1.60	0.10	-	1.84	-	3.44	m²	3.78
225 x 225 x 25mm tiles	1.46	10.00	1.60	0.07	-	1.29	-	2.89	m²	3.18
150 x 150 x 12.5mm units to skirtings on brickwork or blockwork base										
height 150mm; square top edge	3.71	10.00	4.06	0.30	0.08	6.61	-	10.67	m	11.74
height 150mm; rounded top edge	13.18	10.00	14.48	0.30	0.08	6.61	-	21.10	m	23.20
height 150mm; rounded top edge and cove at bottom	24.85	10.00	27.32	0.30	0.08	6.61	-	33.93	m	37.32
150 x 150 x 20mm units to skirtings on brickwork or blockwork base										
height 150mm; square top edge	5.74	10.00	6.30	0.38	0.09	8.22	-	14.51	m	15.97
height 150mm; rounded top edge	21.35	10.00	23.47	0.38	0.09	8.22	-	31.69	m	34.85

Finish to floors; clay floor quarries, BS EN ISO 10545, dark; 3mm joints, symmetrical layout; bedding in 10mm cement mortar (1:3); pointing in cement mortar (1:3); on cement and sand base

	Del to Site £	Waste %	Material Cost £	Craft Optve Hrs	Lab Hrs	Labour Cost £	Sunds £	Nett Rate £	Unit	Gross rate (10%) £
150 x 150 x 12.5mm units to floors level or to falls only not exceeding 15degrees from horizontal										
plain	38.39	5.00	40.31	0.80	0.50	21.58	-	61.89	m²	68.08
150 x 150 x 12.5mm units to floors to falls and crossfalls and to slopes not exceeding 15 degrees from horizontal										
plain	38.39	5.00	40.31	0.88	0.50	23.05	-	63.36	m²	69.70
150 x 150 x 12.5mm units to skirtings on brickwork or blockwork base										
height 150mm; square top edge	6.32	10.00	6.93	0.30	0.08	6.61	-	13.54	m	14.90
height 150mm; rounded top edge	16.39	10.00	18.01	0.30	0.08	6.61	-	24.62	m	27.09
height 150mm; rounded top edge and cove at bottom	30.97	10.00	34.05	0.30	0.08	6.61	-	40.67	m	44.73

Finish to floors; ceramic floor tiles, BS EN ISO 10545, fully vitrified, red; 3mm joints, symmetrical layout; bedding in 10mm cement mortar (1:3); pointing in cement mortar (1:3); on cement and sand base

	Del to Site £	Waste %	Material Cost £	Craft Optve Hrs	Lab Hrs	Labour Cost £	Sunds £	Nett Rate £	Unit	Gross rate (10%) £
100 x 100 x 9mm units to floors level or to falls only not exceeding 15 degrees from horizontal										
plain	23.19	5.00	24.35	1.60	0.90	41.78	-	66.13	m²	72.74
150 x 150 x 9mm units to floors level or to falls only not exceeding 15 degrees from horizontal										
plain	24.86	5.00	26.10	0.90	0.55	24.10	-	50.20	m²	55.22
200 x 200 x 12mm units to floors level or to falls only not exceeding 15 degrees from horizontal										
plain	25.80	5.00	27.09	0.60	0.40	16.53	-	43.62	m²	47.98
100 x 100 x 9mm units to floors to falls and crossfalls and to slopes not exceeding 15 degrees from horizontal										
plain	23.19	5.00	24.35	1.76	0.90	44.72	-	69.07	m²	75.97
150 x 150 x 9mm units to floors to falls and crossfalls and to slopes not exceeding 15 degrees from horizontal										
plain	24.86	5.00	26.10	0.99	0.55	25.75	-	51.86	m²	57.04
200 x 200 x 12mm units to floors to falls and crossfalls and to slopes not exceeding 15 degrees from horizontal										
plain	25.80	5.00	27.09	0.66	0.40	17.63	-	44.72	m²	49.19
Extra for pointing with tinted mortar										
100 x 100 x 9mm tiles	1.50	10.00	1.65	0.15	-	2.76	-	4.41	m²	4.85
152 x 152 x 12mm tiles	1.50	10.00	1.65	0.10	-	1.84	-	3.49	m²	3.84
200 x 200 x 12mm tiles	1.50	10.00	1.65	0.08	-	1.47	-	3.12	m²	3.43
100 x 100 x 9mm units to skirtings on brickwork or blockwork base										
height 100mm; square top edge	2.38	10.00	2.61	0.36	0.09	7.85	-	10.46	m	11.50
height 100mm; rounded top edge	11.86	10.00	13.04	0.36	0.09	7.85	-	20.89	m	22.98
152 x 152 x 12mm units to skirtings on brickwork or blockwork base										
height 152mm; square top edge	4.95	10.00	5.42	0.33	0.08	7.16	-	12.59	m	13.85
height 152mm; rounded top edge	19.78	10.00	21.74	0.33	0.08	7.16	-	28.90	m	31.79
height 115mm; rounded top edge and cove at bottom	27.59	10.00	30.33	0.33	0.08	7.16	-	37.49	m	41.24
200 x 200 x 12mm units to skirtings on brickwork or blockwork base										
height 200mm; square top edge	5.40	10.00	5.91	0.30	0.08	6.61	-	12.52	m	13.77

Finish to floors; ceramic floor tiles, BS EN ISO 10545, fully vitrified, cream; 3mm joints, symmetrical layout; bedding in 10mm cement mortar (1:3); pointing in cement mortar (1:3); on cement and sand base

	Del to Site £	Waste %	Material Cost £	Craft Optve Hrs	Lab Hrs	Labour Cost £	Sunds £	Nett Rate £	Unit	Gross rate (10%) £
100 x 100 x 9mm units to floors level or to falls only not exceeding 15 degrees from horizontal										
plain	27.74	5.00	29.13	1.60	0.90	41.78	-	70.91	m²	78.00

Labour hourly rates: (except Specialists) Craft Operatives 18.37 Labourer 13.76 Rates are national average prices. Refer to REGIONAL VARIATIONS for indicative levels of overall pricing in regions	MATERIALS			LABOUR				RATES		
	Del to Site £	Waste %	Material Cost £	Craft Optve Hrs	Lab Hrs	Labour Cost £	Sunds £	Nett Rate £	Unit	Gross rate (10%) £
IN-SITU, TILED, BLOCK, MOSAIC, SHEET, APPLIED LIQUID OR PLANTED FINISHES (Cont'd)										
Finish to floors; ceramic floor tiles, BS EN ISO 10545, fully vitrified, cream; 3mm joints, symmetrical layout; bedding in 10mm cement mortar (1:3); pointing in cement mortar (1:3); on cement and sand base (Cont'd)										
150 x 150 x 9mm units to floors level or to falls only not exceeding 15 degrees from horizontal										
plain....................	27.74	5.00	29.13	0.90	0.55	24.10	-	53.23	m²	58.55
200 x 200 x 12mm units to floors level or to falls only not exceeding 15 degrees from horizontal										
plain....................	28.80	5.00	30.24	0.60	0.40	16.53	-	46.77	m²	51.44
100 x 100 x 9mm units to floors to falls and crossfalls and to slopes not exceeding 15 degrees from horizontal										
plain....................	27.74	5.00	29.13	1.76	0.90	44.72	-	73.84	m²	81.23
150 x 150 x 9mm units to floors to falls and crossfalls and to slopes not exceeding 15 degrees from horizontal										
plain....................	27.74	5.00	29.13	0.99	0.55	25.75	-	54.88	m²	60.37
200 x 200 x 12mm units to floors to falls and crossfalls and to slopes not exceeding 15 degrees from horizontal										
plain....................	28.80	5.00	30.24	0.66	0.40	17.63	-	47.87	m²	52.66
Extra for pointing with tinted mortar										
100 x 100 x 9mm tiles....................	1.50	10.00	1.65	0.15	-	2.76	-	4.41	m²	4.85
152 x 152 x 12mm tiles....................	1.50	10.00	1.65	0.10	-	1.84	-	3.49	m²	3.84
200 x 200 x 12mm tiles....................	1.50	10.00	1.65	0.08	-	1.47	-	3.12	m²	3.43
100 x 100 x 9mm units to skirtings on brickwork or blockwork base										
height 100mm; square top edge....................	2.83	10.00	3.11	0.36	0.09	7.85	-	10.96	m	12.05
height 100mm; rounded top edge....................	15.60	10.00	17.15	0.36	0.09	7.85	-	25.00	m	27.50
152 x 152 x 12mm units to skirtings on brickwork or blockwork base										
height 152mm; square top edge....................	5.24	10.00	5.75	0.33	0.08	7.16	-	12.91	m	14.20
height 152mm; rounded top edge....................	27.38	10.00	30.10	0.33	0.08	7.16	-	37.26	m	40.99
height 152mm; rounded top edge and cove at bottom	27.59	10.00	30.33	0.33	0.08	7.16	-	37.49	m	41.24
200 x 200 x 12mm units to skirtings on brickwork or blockwork base										
height 200mm; square top edge....................	6.10	10.00	6.68	0.30	0.08	6.61	-	13.29	m	14.62
Finish to floors; ceramic floor tiles, BS EN ISO 10545, fully vitrified, black; 3mm joints, symmetrical layout; bedding in 10mm cement mortar (1:3); pointing in cement mortar (1:3); on cement and sand base										
100 x 100 x 9mm units to floors level or to falls only not exceeding 15 degrees from horizontal										
plain....................	30.64	5.00	32.17	1.60	0.90	41.78	-	73.95	m²	81.35
152 x 152 x 12mm units to floors level or to falls only not exceeding 15 degrees from horizontal										
plain....................	30.64	5.00	32.17	0.90	0.55	24.10	-	56.28	m²	61.90
200 x 200 x 12mm units to floors level or to falls only not exceeding 15 degrees from horizontal										
plain....................	29.80	5.00	31.29	0.60	0.40	16.53	-	47.82	m²	52.60
100 x 100 x 9mm units to floors to falls and crossfalls and to slopes not exceeding 15 degrees from horizontal										
plain....................	30.64	5.00	32.17	1.76	0.90	44.72	-	76.89	m²	84.58
152 x 152 x 12mm units to floors to falls and crossfalls and to slopes not exceeding 15 degrees from horizontal										
plain....................	30.64	5.00	32.17	0.99	0.55	25.75	-	57.93	m²	63.72
200 x 200 x 12mm units to floors to falls and crossfalls and to slopes not exceeding 15 degrees from horizontal										
plain....................	29.80	5.00	31.29	0.66	0.40	17.63	-	48.92	m²	53.81
Extra for pointing with tinted mortar										
100 x 100 x 9mm tiles....................	1.50	10.00	1.65	0.15	-	2.76	-	4.41	m²	4.85
152 x 152 x 12mm tiles....................	1.50	10.00	1.65	0.10	-	1.84	-	3.49	m²	3.84
200 x 200 x 12mm tiles....................	1.50	10.00	1.65	0.08	-	1.47	-	3.12	m²	3.43
100 x 100 x 9mm units to skirtings on brickwork or blockwork base										
height 100mm; square top edge....................	2.94	10.00	3.22	0.36	0.09	7.85	-	11.07	m	12.18
height 100mm; rounded top edge....................	17.22	10.00	18.93	0.36	0.09	7.85	-	26.78	m	29.46
152 x 152 x 12mm units to skirtings on brickwork or blockwork base										
height 152mm; square top edge....................	5.54	10.00	6.07	0.33	0.08	7.16	-	13.24	m	14.56
height 152mm; rounded top edge....................	28.78	10.00	31.64	0.33	0.08	7.16	-	38.81	m	42.69
height 152mm; rounded top edge and cove at bottom	28.15	10.00	30.95	0.33	0.08	7.16	-	38.11	m	41.92
200 x 200 x 12mm units to skirtings on brickwork or blockwork base										
height 200mm; square top edge....................	6.70	10.00	7.34	0.30	0.08	6.61	-	13.95	m	15.35

Labour hourly rates: (except Specialists) Craft Operatives 18.37 Labourer 13.76 Rates are national average prices. Refer to REGIONAL VARIATIONS for indicative levels of overall pricing in regions	MATERIALS			LABOUR				RATES		
	Del to Site £	Waste %	Material Cost £	Craft Optve Hrs	Lab Hrs	Labour Cost £	Sunds £	Nett Rate £	Unit	Gross rate (10%) £

IN-SITU, TILED, BLOCK, MOSAIC, SHEET, APPLIED LIQUID OR PLANTED FINISHES (Cont'd)

Finish to floors; terrazzo tiles, BS EN 13748, aggregate size random, ground, grouted and polished, standard colour range; 3mm joints, symmetrical layout; bedding in 40mm cement mortar (1:3); grouting with white cement; in-situ margins

	Del to Site £	Waste %	Material Cost £	Craft Optve Hrs	Lab Hrs	Labour Cost £	Sunds £	Nett Rate £	Unit	Gross rate £
300 x 300 x 28mm units to floors on concrete base; level or to falls only not exceeding 15 degrees from horizontal plain	44.87	5.00	47.33	1.10	0.65	29.15	-	76.48	m²	84.13
305 x 305 x 28mm units to treads on concrete base width 292mm; rounded nosing	41.15	5.00	43.31	0.75	0.19	16.39	-	59.70	m	65.67
305 x 305 x 28mm units to plain risers on concrete base height 165mm	7.53	5.00	7.96	0.50	0.11	10.70	-	18.66	m	20.53

Finish to floors and walls; terrazzo, white cement and white marble chippings (2:5); polished; on cement and sand base

	Del to Site £	Waste %	Material Cost £	Craft Optve Hrs	Lab Hrs	Labour Cost £	Sunds £	Nett Rate £	Unit	Gross rate £
16mm work to floors; one coat; floated level and to falls only not exceeding 15 degrees from horizontal; laid in bays, average size 610 x 610mm	18.85	5.00	19.79	2.25	1.13	56.88	-	76.67	m²	84.34
6mm work to walls width exceeding 600mm	7.07	5.00	7.42	3.60	1.80	90.90	-	98.32	m²	108.15
width 300mm	3.53	10.00	3.89	2.16	0.54	47.11	-	51.00	m	56.10
16mm work to treads; one coat width 279mm	8.25	5.00	8.66	3.00	0.32	59.51	-	68.17	m	74.99
extra; two line carborundum non-slip inlay	21.00	5.00	22.05	0.25	-	4.59	-	26.64	m	29.31
16mm work to undercut risers; one coat height 178mm	4.71	5.00	4.95	2.00	0.20	39.49	-	44.44	m	48.88
6mm work to strings; one coat height 150mm	1.18	5.00	1.24	1.80	0.27	36.78	-	38.02	m	41.82
height 200mm	2.36	5.00	2.47	2.50	0.36	50.88	-	53.35	m	58.69
6mm work to skirtings height 75mm	1.18	5.00	1.24	1.25	0.14	24.89	-	26.13	m	28.74
height 75mm; curved to 3000mm radius	1.18	5.00	1.24	1.67	0.14	32.60	-	33.84	m	37.23
16mm work to skirtings on brickwork or blockwork base height 150mm	4.71	5.00	4.95	1.60	0.27	33.11	-	38.06	m	41.86
height 150mm; curved to 3000mm radius	4.71	5.00	4.95	2.13	0.27	42.84	-	47.79	m	52.57
Rounded angles and intersections 10 - 100mm radius	-	-	-	0.18	-	3.31	-	3.31	m	3.64
Coves 25mm girth	-	-	-	0.40	-	7.35	-	7.35	m	8.08
40mm girth	-	-	-	0.60	-	11.02	-	11.02	m	12.12

Finish to floors; dividing strips

	Del to Site £	Waste %	Material Cost £	Craft Optve Hrs	Lab Hrs	Labour Cost £	Sunds £	Nett Rate £	Unit	Gross rate £
Plastic dividing strips; setting in bed and finishing 6 x 16mm; flat section	2.52	-	2.52	0.25	-	4.59	-	7.11	m	7.82

Finish to floors; Maple wood blocks, tongued and grooved joints; symmetrical herringbone pattern layout, two block plain borders; fixing with adhesive; sanding, one coat sealer

	Del to Site £	Waste %	Material Cost £	Craft Optve Hrs	Lab Hrs	Labour Cost £	Sunds £	Nett Rate £	Unit	Gross rate £
70 x 230 x 20mm units to floors on cement and sand base; level or to falls only not exceeding 15 degrees from horizontal plain	40.53	5.00	42.55	1.50	1.25	44.76	-	87.31	m²	96.04

Finish to floors; Merbau wood blocks, tongued and grooved joints; symmetrical herringbone pattern layout, two block plain borders; fixing with adhesive; sanding, one coat sealer

	Del to Site £	Waste %	Material Cost £	Craft Optve Hrs	Lab Hrs	Labour Cost £	Sunds £	Nett Rate £	Unit	Gross rate £
70 x 230 x 20mm units to floors on cement and sand base; level or to falls only not exceeding 15 degrees from horizontal plain	30.68	5.00	32.21	1.50	1.25	44.76	-	76.96	m²	84.66

Finish to floors; Oak wood blocks, tongued and grooved joints; symmetrical herringbone pattern layout, two block plain borders; fixing with adhesive; sanding, one coat sealer

	Del to Site £	Waste %	Material Cost £	Craft Optve Hrs	Lab Hrs	Labour Cost £	Sunds £	Nett Rate £	Unit	Gross rate £
70 x 230 x 20mm units to floors on cement and sand base; level or to falls only not exceeding 15 degrees from horizontal plain	34.26	5.00	35.97	1.50	1.25	44.76	-	80.73	m²	88.80

Finish to floors; rubber floor tiles; butt joints, symmetrical layout; fixing with adhesive; on cement and sand base

	Del to Site £	Waste %	Material Cost £	Craft Optve Hrs	Lab Hrs	Labour Cost £	Sunds £	Nett Rate £	Unit	Gross rate £
610 x 610 x 2.5mm units to floors level or to falls only not exceeding 15 degrees from horizontal width exceeding 600mm	30.61	5.00	32.28	0.35	0.23	9.59	-	41.87	m²	46.06
width 300mm	9.58	10.00	10.54	0.21	0.07	4.82	-	15.36	m	16.89
500 x 500 x 2.6mm units to floors level or to falls only not exceeding 15 degrees from horizontal width exceeding 600mm	29.63	5.00	31.24	0.35	0.23	9.59	-	40.84	m²	44.92
width 300mm	9.28	10.00	10.21	0.21	0.07	4.82	-	15.03	m	16.54

Labour hourly rates: (except Specialists) Craft Operatives 18.37 Labourer 13.76 Rates are national average prices. Refer to REGIONAL VARIATIONS for indicative levels of overall pricing in regions	MATERIALS			LABOUR				RATES		
	Del to Site £	Waste %	Material Cost £	Craft Optve Hrs	Lab Hrs	Labour Cost £	Sunds £	Nett Rate £	Unit	Gross rate (10%) £
IN-SITU, TILED, BLOCK, MOSAIC, SHEET, APPLIED LIQUID OR PLANTED FINISHES (Cont'd)										
Finish to floors; smooth finish rubber matting; butt joints; fixing with adhesive; on cement and sand base										
3mm work to floors level or to falls only not exceeding 15 degrees from horizontal										
width exceeding 600mm	8.81	5.00	9.38	0.20	0.15	5.74	-	15.12	m²	16.63
width 300mm	4.15	10.00	4.57	0.12	0.05	2.89	-	7.46	m	8.20
4mm work to floors level or to falls only not exceeding 15 degrees from horizontal										
width exceeding 600mm	10.48	5.00	11.14	0.20	0.15	5.74	-	16.88	m²	18.56
width 300mm	3.86	10.00	4.25	0.12	0.05	2.89	-	7.14	m	7.86
6mm work to floors level or to falls only not exceeding 15 degrees from horizontal										
width exceeding 600mm	13.63	5.00	14.44	0.20	0.15	5.74	-	20.18	m²	22.20
width 300mm	4.61	10.00	5.07	0.12	0.05	2.89	-	7.97	m	8.76
Finish to floors; PVC floor tiles, BS EN 654; butt joints, symmetrical layout; fixing with adhesive; two coats sealer; on cement and sand base										
300 x 300 x 2mm units to floors level or to falls only not exceeding 15 degrees from horizontal										
width exceeding 600mm	8.57	5.00	9.08	0.45	0.23	11.43	-	20.51	m²	22.56
width 300mm	2.81	10.00	3.10	0.27	0.07	5.92	-	9.02	m	9.92
300 x 300 x 3.2mm units to floors level or to falls only not exceeding 15 degrees from horizontal										
width exceeding 600mm	12.42	5.00	13.12	0.45	0.23	11.43	-	24.55	m²	27.01
width 300mm	3.97	10.00	4.37	0.27	0.07	5.92	-	10.29	m	11.32
Finish to floors; fully flexible PVC heavy duty floor tiles, BS EN 649 type A; butt joints, symmetrical layout; fixing with adhesive; two coats sealer; on cement and sand base										
610 x 610 x 2mm units to floors level or to falls only not exceeding 15 degrees from horizontal										
width exceeding 600mm	8.57	5.00	9.08	0.45	0.23	11.43	-	20.51	m²	22.56
width 300mm	2.81	10.00	3.10	0.27	0.07	5.92	-	9.02	m	9.92
610 x 610 x 2.5mm units to floors level or to falls only not exceeding 15 degrees from horizontal										
width exceeding 600mm	10.57	5.00	11.18	0.45	0.23	11.43	-	22.61	m²	24.87
width 300mm	3.41	10.00	3.76	0.27	0.07	5.92	-	9.68	m	10.65
Finish to floors; fully flexible PVC heavy duty sheet, BS EN 649 Type A; butt joints; welded; fixing with adhesive; two coats sealer; on cement and sand base										
2mm work to floors level or to falls only not exceeding 15 degrees from horizontal										
width exceeding 600mm	9.57	5.00	10.13	0.30	0.15	7.58	-	17.70	m²	19.47
width 300mm	3.11	10.00	3.43	0.18	0.05	3.99	-	7.42	m	8.16
3.2mm work to floors level or to falls only not exceeding 15 degrees from horizontal										
width exceeding 600mm	11.57	5.00	12.23	0.30	0.15	7.58	-	19.80	m²	21.78
width 300mm	3.71	10.00	4.09	0.18	0.05	3.99	-	8.08	m	8.89
Finish to floors; PVC coved skirting; fixing with adhesive										
Skirtings on plaster base										
height 100mm - Set in	2.09	5.00	2.21	0.20	0.01	3.81	-	6.02	m	6.62
height 100mm - Sit on	2.09	5.00	2.21	0.20	0.02	3.95	-	6.16	m	6.77
Finish to floors; cork tiles, BS 8203; butt joints; symmetrical layout; fixing with adhesive; two coats seal; on cement and sand base										
300 x 300 x 3.2mm units to floors level or to falls only not exceeding 15 degrees from horizontal										
width exceeding 600mm	21.72	5.00	23.15	0.75	0.23	16.94	-	40.09	m²	44.10
width 300mm	6.83	10.00	7.51	0.45	0.07	9.23	-	16.74	m	18.42
300 x 300 x 4.75mm units to floors level or to falls only not exceeding 15 degrees from horizontal										
width exceeding 600mm	24.68	5.00	26.25	0.75	0.23	16.94	-	43.20	m²	47.52
width 300mm	7.73	10.00	8.51	0.45	0.07	9.23	-	17.74	m	19.51
Finish to floors; linoleum tiles, BS EN 548 & 670 butt joints, symmetrical layout; fixing with adhesive; on cement and sand base										
330 x 330 x 2mm units to floors level or to falls only not exceeding 15 degrees from horizontal										
width exceeding 600mm	16.27	5.00	17.15	0.35	0.23	9.59	-	26.74	m²	29.42
width 300mm	5.08	5.00	5.36	0.21	0.07	4.82	-	10.18	m	11.20
500 x 500 x 2.5mm units to floors level or to falls only not exceeding 15 degrees from horizontal										
width exceeding 600mm	18.27	5.00	19.25	0.35	0.23	9.59	-	28.84	m²	31.73
width 300mm	5.68	5.00	5.99	0.21	0.07	4.82	-	10.81	m	11.89

FLOOR, WALL, CEILING AND ROOF FINISHINGS *(side margin)*

Labour hourly rates: (except Specialists) Craft Operatives 18.37 Labourer 13.76 Rates are national average prices. Refer to REGIONAL VARIATIONS for indicative levels of overall pricing in regions	MATERIALS			LABOUR				RATES		
	Del to Site £	Waste %	Material Cost £	Craft Optve Hrs	Lab Hrs	Labour Cost £	Sunds £	Nett Rate £	Unit	Gross rate (10%) £
IN-SITU, TILED, BLOCK, MOSAIC, SHEET, APPLIED LIQUID OR PLANTED FINISHES (Cont'd)										
Finish to floors; linoleum sheet, BS EN 548 & 670 butt joints; laying loose; on cement and sand base										
2.5mm work to floors level or to falls only not exceeding 15 degrees from horizontal										
width exceeding 600mm....................	14.95	5.00	15.70	0.15	0.13	4.54	-	20.24	m²	22.27
width 300mm....................	4.49	10.00	4.93	0.09	0.04	2.20	-	7.14	m	7.85
3.2mm work to floors level or to falls only not exceeding 15 degrees from horizontal										
width exceeding 600mm....................	18.50	5.00	19.42	0.15	0.13	4.54	-	23.97	m²	26.37
width 300mm....................	5.55	10.00	6.11	0.09	0.04	2.20	-	8.31	m	9.14
Finish to floors; carpet tiles; heavy ribbed, polypropylene/nylon fibre bonded pc £16.00/m²; butt joints, symetrical layout; fixing with adhesive; on cement and sand base										
500 x 500 units to floors; level or to falls only not exceeding 15 degrees from horizontal										
heavy contract grade width exceeding 600mm	17.10	5.00	17.97	0.35	0.23	9.59	-	27.56	m²	30.32
Finish to floors; carpet tiles; looped pile, polypropylene pc £15.00/m²; butt joints, symetrical layout; fixing with adhesive; on cement and sand base										
500 x 500 units to floors; level or to falls only not exceeding 15 degrees from horizontal										
general contract grade width exceeding 600mm	15.10	5.00	15.87	0.35	0.23	9.59	-	25.46	m²	28.01
Finish to floors; carpet tiles; hard twist cut pile, polypropylene pc £19.00/m²; butt joints, symetrical layout; fixing with adhesive; on cement and sand base										
500 x 500 units to floors; level or to falls only not exceeding 15 degrees from horizontal										
heavy contract grade width exceeding 600mm	19.10	5.00	20.07	0.35	0.23	9.59	-	29.66	m²	32.63
Finish to floors; fitted carpeting; fibre bonded polypropylene/nylon pc £11.00/m²; fixing with adhesive; on cement and sand base										
Work to floors; level or to falls only not exceeding 15 degrees from horizontal										
heavy contract grade width exceeding 600mm	11.30	5.00	11.88	0.30	0.20	8.26	-	20.14	m²	22.16
Finish to floors; fitted carpeting; ribbed fibre bonded polypropylene/nylonpc £12.00/m²; fixing with adhesive; on cement and sand base										
Work to floors; level or to falls only not exceeding 15 degrees from horizontal										
heavy contract grade width exceeding 600mm	12.30	5.00	12.93	0.30	0.20	8.26	-	21.19	m²	23.31
Finish to floors; underlay to carpeting; Tredaire Brio 8mm; fixing with tacks; on timber base										
Work to floors; level or to falls only not exceeding 15 degrees from horizontal										
width exceeding 300mm....................	1.90	5.00	1.99	0.15	0.08	3.86	0.22	6.07	m²	6.68
Finish to floors; fitted carpeting; Tufted wool/nylon, pc £25.00 m²; fixing with tackless grippers; on timber base										
Work to floors; level or to falls only not exceeding 15 degrees from horizontal										
width exceeding 300mm....................	25.00	5.00	26.25	0.30	0.20	8.26	0.22	34.74	m²	38.21
Finish to floors; fitted carpeting; Axminster wool/nylon, pc £30.00m²; fixing with tackless grippers; on timber base										
Work to floors; level or to falls only not exceeding 15 degrees from horizontal										
width exceeding 300mm....................	30.00	5.00	31.50	0.30	0.20	8.26	0.22	39.99	m²	43.99
Finish to floors; fitted carpeting; Wilton wool/nylon, pc £35.00 m²; fixing with tackless grippers; on timber base										
Work to floors; level or to falls only not exceeding 15 degrees from horizontal										
width exceeding 300mm....................	35.00	5.00	36.75	0.30	0.20	8.26	0.22	45.24	m²	49.76
Raised access floors; Microfloor "Bonded 600" light grade full access system; 600 x 600mm high density particle board panels, BS EN 312 and DIN 68761; 100 x 100mm precast lightweight concrete pedestals at 600mm centres fixed to sub-floor with epoxy resin adhesive										
Thickness of panel 30mm finished floor height										
50mm....................	-	-	Specialist	-	-	Specialist	-	28.86	m²	31.75
75mm....................	-	-	Specialist	-	-	Specialist	-	29.07	m²	31.98

Labour hourly rates: (except Specialists) Craft Operatives 18.37 Labourer 13.76 Rates are national average prices. Refer to REGIONAL VARIATIONS for indicative levels of overall pricing in regions	MATERIALS			LABOUR				RATES		
	Del to Site	Waste	Material Cost	Craft Optve	Lab	Labour Cost	Sunds	Nett Rate		Gross rate (10%)
	£	%	£	Hrs	Hrs	£	£	£	Unit	£
IN-SITU, TILED, BLOCK, MOSAIC, SHEET, APPLIED LIQUID OR PLANTED FINISHES (Cont'd)										
Raised access floors; Microfloor "Bonded 600" light grade full access system; 600 x 600mm high density particle board panels, BS EN 312 and DIN 68761; 100 x 100mm precast lightweight concrete pedestals at 600mm centres fixed to sub-floor with epoxy resin adhesive (Cont'd)										
Thickness of panel 30mm (Cont'd) finished floor height (Cont'd)										
100mm	-	-	Specialist	-	-	Specialist	-	29.43	m²	32.37
125mm	-	-	Specialist	-	-	Specialist	-	29.75	m²	32.72
150mm	-	-	Specialist	-	-	Specialist	-	30.34	m²	33.38
175mm	-	-	Specialist	-	-	Specialist	-	30.83	m²	33.91
200mm	-	-	Specialist	-	-	Specialist	-	31.09	m²	34.20
Raised access floors; Microfloor "Bonded 600" medium grade full access system; 600 x 600mm high density particle board panels, BS EN 312 and DIN 68761; 100 x 100mm precast lightweight concrete pedestals at 600mm centres fixed to sub-floor with epoxy resin adhesive										
Thickness of panel 38mm finished floor height										
50mm	-	-	Specialist	-	-	Specialist	-	32.82	m²	36.10
75mm	-	-	Specialist	-	-	Specialist	-	32.82	m²	36.10
100mm	-	-	Specialist	-	-	Specialist	-	33.45	m²	36.80
125mm	-	-	Specialist	-	-	Specialist	-	33.74	m²	37.11
150mm	-	-	Specialist	-	-	Specialist	-	34.83	m²	38.31
175mm	-	-	Specialist	-	-	Specialist	-	35.12	m²	38.63
200mm	-	-	Specialist	-	-	Specialist	-	35.38	m²	38.92
Raised access floors; Microfloor "Bonded 600" office loadings grade full access system; 600 x 600mm high density particle board panels, BS EN 634 and DIN 68761; 100 x 100mm precast lightweight concrete pedestals at 600mm centres fixed to sub-floor with epoxy resin adhesive										
Thickness of panel 30mm finished floor height										
50mm	-	-	Specialist	-	-	Specialist	-	22.47	m²	24.72
75mm	-	-	Specialist	-	-	Specialist	-	22.60	m²	24.87
100mm	-	-	Specialist	-	-	Specialist	-	23.10	m²	25.41
125mm	-	-	Specialist	-	-	Specialist	-	23.39	m²	25.72
150mm	-	-	Specialist	-	-	Specialist	-	23.95	m²	26.35
175mm	-	-	Specialist	-	-	Specialist	-	24.50	m²	26.94
200mm	-	-	Specialist	-	-	Specialist	-	24.75	m²	27.23
Finish to Walls, isolated columns, ceilings and isolated beams; Portland cement work										
Plain face, first and finishing coats cement and sand (1:3), total 13mm thick; wood floated										
13mm work to walls on brickwork or blockwork base										
width exceeding 600mm	1.56	5.00	1.64	0.60	0.30	15.15	-	16.79	m²	18.47
width exceeding 600mm; dubbing average 6mm thick	2.28	5.00	2.40	0.84	0.44	21.49	-	23.88	m²	26.27
width exceeding 600mm; dubbing average 12mm thick	3.00	5.00	3.15	0.88	0.58	24.15	-	27.30	m²	30.03
width exceeding 600mm; dubbing average 19mm thick	3.84	5.00	4.03	0.92	0.74	27.08	-	31.12	m²	34.23
width exceeding 600mm; curved to 3000mm radius	1.56	5.00	1.64	0.80	0.30	18.82	-	20.46	m²	22.51
width 300mm	0.48	10.00	0.53	0.36	0.09	7.85	-	8.38	m	9.22
width 300mm; dubbing average 6mm thick	0.72	10.00	0.79	0.53	0.17	12.08	-	12.87	m	14.15
width 300mm; dubbing average 12mm thick	0.96	10.00	1.06	0.53	0.17	12.08	-	13.13	m	14.45
width 300mm; dubbing average 19mm thick	1.20	10.00	1.32	0.55	0.22	13.13	-	14.45	m	15.90
width 300mm; curved to 3000mm radius	0.48	10.00	0.53	0.48	0.09	10.06	-	10.58	m	11.64
Work forming flush skirting; 13mm thick										
height 150mm	0.24	50.00	0.36	0.60	0.05	11.71	-	12.07	m	13.28
height 225mm	0.36	50.00	0.54	0.65	0.07	12.90	-	13.44	m	14.79
height 150mm; curved to 3000mm radius	0.24	50.00	0.36	0.80	0.05	15.38	-	15.74	m	17.32
height 225mm; curved to 3000mm radius	0.36	50.00	0.54	0.86	0.07	16.76	-	17.30	m	19.03
Work forming projecting skirting; 13mm projection										
height 150mm	0.24	50.00	0.36	0.90	0.05	17.22	-	17.58	m	19.34
height 225mm	0.36	50.00	0.54	0.98	0.07	18.97	-	19.51	m	21.46
height 150mm; curved to 3000mm radius	0.24	50.00	0.36	1.20	0.05	22.73	-	23.09	m	25.40
height 225mm; curved to 3000mm radius	0.36	50.00	0.54	1.29	0.07	24.66	-	25.20	m	27.72
Rounded angles										
radius 10 - 100mm	-	-	-	0.30	-	5.51	-	5.51	m	6.06
If plain face waterproofed										
add	0.06	10.00	0.07	-	0.02	0.28	-	0.35	m²	0.38
Finish to Walls, isolated columns, ceilings and isolated beams; Rough cast										
Render and dry dash; first and finishing coats cement and sand (1:3), total 15mm thick; wood floated; dry dash of pea shingle; external										
15mm work to walls on brickwork or blockwork base										
width exceeding 600mm	2.03	5.00	2.15	0.85	0.43	21.53	-	23.68	m²	26.04
extra; spatterdash coat	1.15	10.00	1.27	0.23	0.12	5.88	-	7.15	m²	7.86
width exceeding 600mm; curved to 3000mm radius	2.03	5.00	2.15	1.15	0.43	27.04	-	29.19	m²	32.11
width 300mm	0.46	10.00	0.51	0.51	0.13	11.16	-	11.66	m	12.83
extra; spatterdash coat	0.03	10.00	0.04	0.14	0.04	3.12	-	3.16	m	3.47
width 300mm; curved to 3000mm radius	0.46	10.00	0.51	0.69	0.13	14.46	-	14.97	m	16.47

Labour hourly rates: (except Specialists) Craft Operatives 18.37 Labourer 13.76 Rates are national average prices. Refer to REGIONAL VARIATIONS for indicative levels of overall pricing in regions	MATERIALS			LABOUR				RATES		
	Del to Site	Waste	Material Cost	Craft Optve	Lab	Labour Cost	Sunds	Nett Rate	Unit	Gross rate (10%)
	£	%	£	Hrs	Hrs	£	£	£		£

IN-SITU, TILED, BLOCK, MOSAIC, SHEET, APPLIED LIQUID OR PLANTED FINISHES (Cont'd)

Finish to Walls, isolated columns, ceilings and isolated beams; Rough cast (Cont'd)

Render and wet dash; first and finishing coats cement and sand (1:3), total 15mm thick; wood floated; wet dash of crushed stone or shingle and cement slurry

	Del to Site	Waste	Material Cost	Craft Optve	Lab	Labour Cost	Sunds	Nett Rate	Unit	Gross rate
15mm work to walls on brickwork or blockwork base										
width exceeding 600mm	3.06	5.00	3.28	0.95	0.48	24.06	-	27.33	m²	30.06
extra; spatterdash coat	0.69	10.00	0.75	0.28	0.14	7.07	-	7.82	m²	8.61
width exceeding 600mm; curved to 3000mm radius	3.06	5.00	3.28	1.24	0.48	29.38	-	32.66	m²	35.92
width 300mm	1.05	10.00	1.15	0.57	0.14	12.40	-	13.55	m	14.90
extra; spatterdash coat	0.33	10.00	0.36	0.17	0.04	3.67	-	4.03	m	4.44
width 300mm; curved to 3000mm radius	1.86	10.00	2.04	0.74	0.14	15.52	-	17.56	m	19.32

Finish to Walls, isolated columns, ceilings and isolated beams; Tyrolean finish

Render and Tyrolean finish; first and finishing coats cement and sand (1:3), total 15mm thick; wood floated; Tyrolean finish of "Cullamix" mixture applied by machine

	Del to Site	Waste	Material Cost	Craft Optve	Lab	Labour Cost	Sunds	Nett Rate	Unit	Gross rate
15mm work to walls on brickwork or blockwork base										
width exceeding 600mm	3.24	5.00	3.40	1.05	0.30	23.42	-	28.03	m²	30.84
width 300mm	1.08	10.00	1.19	0.63	0.09	12.81	-	14.73	m	16.20

Finish to Walls, isolated columns, ceilings and isolated beams; Hardwall plastering

Plaster, BS 1191, Part 1, Class B; finishing coat of board finish, 3mm thick; steel troweled

	Del to Site	Waste	Material Cost	Craft Optve	Lab	Labour Cost	Sunds	Nett Rate	Unit	Gross rate
3mm work to walls on concrete or plasterboard base										
width exceeding 600mm	0.78	5.00	0.82	0.25	0.13	6.38	0.02	7.22	m²	7.95
width exceeding 600mm; curved to 3000mm radius	0.78	5.00	0.82	0.33	0.13	7.85	0.02	8.69	m²	9.56
width 300mm	0.39	10.00	0.43	0.15	0.04	3.31	0.01	3.74	m	4.12
width 300mm; curved to 3000mm radius	0.39	10.00	0.43	0.20	0.04	4.22	0.01	4.66	m	5.13
3mm work to isolated columns on concrete or plasterboard base										
width exceeding 600mm	0.78	5.00	0.82	0.38	0.13	8.77	0.02	9.61	m²	10.57
width 300mm	0.39	10.00	0.43	0.23	0.04	4.78	0.01	5.21	m	5.73
3mm work to ceilings on concrete or plasterboard base										
width exceeding 600mm	0.78	5.00	0.82	0.30	0.13	7.30	0.02	8.14	m²	8.96
width exceeding 600mm; 3.50 - 5.00m above floor	0.78	5.00	0.82	0.33	0.13	7.85	0.02	8.69	m²	9.56
width 300mm	0.39	10.00	0.43	0.18	0.04	3.86	0.01	4.29	m	4.72
width 300mm; 3.50 - 5.00m above floor	0.39	10.00	0.43	0.20	0.04	4.22	0.01	4.66	m	5.13
3mm work to isolated beams on concrete or plasterboard base										
width exceeding 600mm	0.78	5.00	0.82	0.38	0.13	8.77	0.02	9.61	m²	10.57
width 300mm	0.39	10.00	0.43	0.23	0.04	4.78	0.01	5.21	m	5.73

Plaster; first coat of hardwall, 11mm thick; finishing coat of multi-finish 2mm thick; steel troweled

	Del to Site	Waste	Material Cost	Craft Optve	Lab	Labour Cost	Sunds	Nett Rate	Unit	Gross rate
13mm work to walls on brickwork or blockwork base										
width exceeding 600mm	3.11	5.00	3.27	0.45	0.23	11.43	0.06	14.76	m²	16.24
width exceeding 600mm; curved to 3000mm radius	3.11	5.00	3.27	0.60	0.23	14.19	0.06	17.52	m²	19.27
width 300mm	1.41	10.00	1.55	0.27	0.07	5.92	0.02	7.49	m	8.24
width 300mm; curved to 3000mm radius	1.41	10.00	1.55	0.36	0.07	7.58	0.02	9.15	m	10.06
13mm work to isolated columns on concrete base										
width exceeding 600mm	3.11	5.00	3.27	0.68	0.23	15.66	0.06	18.99	m²	20.88
width 300mm	1.41	10.00	1.55	0.41	0.07	8.49	0.02	10.07	m	11.07
13mm work to ceilings on concrete base										
width exceeding 600mm	3.11	5.00	3.27	0.54	0.23	13.08	0.06	16.41	m²	18.06
width exceeding 600mm; 3.50 - 5.00m above floor	3.11	5.00	3.27	0.59	0.23	14.00	0.06	17.33	m²	19.07
width 300mm	1.41	10.00	1.55	0.32	0.07	6.84	0.02	8.41	m	9.25
width 300mm; 3.50 - 5.00m above floor	1.41	10.00	1.55	0.35	0.07	7.39	0.02	8.96	m	9.86
13mm work to isolated beams on concrete base										
width exceeding 600mm	3.11	5.00	3.27	0.68	0.23	15.66	0.06	18.99	m²	20.88
width 300mm	1.41	10.00	1.55	0.41	0.07	8.49	0.02	10.07	m	11.07
Rounded angles										
radius 10 - 100mm	-	-	-	0.24	-	4.41	-	4.41	m	4.85

Plaster; Thistle Universal one coat plaster, 13mm thick; steel troweled; Note: the thickness is from the face of the metal lathing

	Del to Site	Waste	Material Cost	Craft Optve	Lab	Labour Cost	Sunds	Nett Rate	Unit	Gross rate
13mm work to walls on metal lathing base										
width exceeding 600mm	5.81	5.00	6.10	0.35	0.18	8.91	0.08	15.08	m²	16.59
width exceeding 600mm; curved to 3000mm radius	5.81	5.00	6.10	0.47	0.18	11.11	0.08	17.28	m²	19.01
width 300mm	2.42	10.00	2.66	0.21	0.05	4.55	0.02	7.23	m	7.95
width 300mm; curved to 3000mm radius	2.42	10.00	2.66	0.28	0.05	5.83	0.02	8.52	m	9.37
13mm work to isolated columns on metal lathing base										
width exceeding 600mm	5.81	5.00	6.10	0.53	0.18	12.21	0.08	18.39	m²	20.22
width 300mm	2.42	10.00	2.66	0.32	0.05	6.57	0.02	9.25	m	10.18
13mm work to ceilings on metal lathing base										
width exceeding 600mm	5.81	5.00	6.10	0.42	0.18	10.19	0.08	16.37	m²	18.00
width exceeding 600mm; 3.50 - 5.00m above floor	5.81	5.00	6.10	0.46	0.18	10.93	0.08	17.10	m²	18.81
width 300mm	2.42	10.00	2.66	0.25	0.05	5.28	0.02	7.97	m	8.76
width 300mm; 3.50 - 5.00m above floor	2.42	10.00	2.66	0.28	0.05	5.83	0.02	8.52	m	9.37
13mm work to isolated beams on metal lathing base										
width exceeding 600mm	5.81	5.00	6.10	0.53	0.18	12.21	0.08	18.39	m²	20.22
width 300mm	2.42	10.00	2.66	0.32	0.05	6.57	0.02	9.25	m	10.18
Rounded angles										
radius 10 - 100mm	-	-	-	0.18	-	3.31	-	3.31	m	3.64

Labour hourly rates: (except Specialists) Craft Operatives 18.37 Labourer 13.76 Rates are national average prices. Refer to REGIONAL VARIATIONS for indicative levels of overall pricing in regions	MATERIALS			LABOUR				RATES		
	Del to Site	Waste	Material Cost	Craft Optve	Lab	Labour Cost	Sunds	Nett Rate	Unit	Gross rate (10%)
	£	%	£	Hrs	Hrs	£	£	£		£

IN-SITU, TILED, BLOCK, MOSAIC, SHEET, APPLIED LIQUID OR PLANTED FINISHES (Cont'd)

Finish to Walls, isolated columns, ceilings and isolated beams; lightweight plastering

Plaster, Thistle; pre-mixed; floating coat of browning, 11mm thick; finishing coat of multi-finish, 2mm thick; steel troweled
13mm work to walls on brickwork or blockwork base

	Del to Site £	Waste %	Material Cost £	Craft Optve Hrs	Lab Hrs	Labour Cost £	Sunds £	Nett Rate £	Unit	Gross rate (10%) £
width exceeding 600mm	2.32	5.00	2.44	0.40	0.20	10.10	0.05	12.59	m²	13.84
width exceeding 600mm; curved to 3000mm radius	2.32	5.00	2.44	0.53	0.20	12.49	0.05	14.97	m²	16.47
width exceeding 600mm; dubbing average 6mm thick	3.41	5.00	3.58	0.63	0.38	16.80	0.07	20.45	m²	22.50
width exceeding 600mm; dubbing average 6mm thick; curved to 3000mm radius	3.41	5.00	3.58	0.84	0.38	20.66	0.07	24.31	m²	26.74
width exceeding 600mm; dubbing average 12mm thick	4.22	5.00	4.43	0.63	0.38	16.80	0.10	21.33	m²	23.46
width exceeding 600mm; dubbing average 12mm thick; curved to 3000mm radius	4.22	5.00	4.43	0.84	0.38	20.66	0.10	25.19	m²	27.71
width exceeding 600mm; dubbing average 19mm thick	5.31	5.00	5.57	0.65	0.49	18.68	0.10	24.35	m²	26.79
width exceeding 600mm; dubbing average 19mm thick; curved to 3000mm radius	5.58	5.00	5.86	0.87	0.49	22.72	0.10	28.68	m²	31.55
width 300mm	1.03	5.00	1.09	0.24	0.06	5.23	0.02	6.34	m	6.97
width 300mm; curved to 3000mm radius	1.03	5.00	1.09	0.32	0.06	6.70	0.02	7.81	m	8.59
width 300mm; dubbing average 6mm thick	1.57	5.00	1.66	0.36	0.09	7.85	0.02	9.53	m	10.49
width 300mm; dubbing average 6mm thick; curved to 3000mm radius	1.57	5.00	1.66	0.48	0.09	10.06	0.02	11.74	m	12.91
width 300mm; dubbing average 12mm thick	1.84	5.00	1.95	0.38	0.09	8.22	0.03	10.19	m	11.21
width 300mm; dubbing average 12mm thick; curved to 3000mm radius	1.84	5.00	1.95	0.51	0.09	10.61	0.03	12.58	m	13.84
width 300mm; dubbing average 19mm thick	2.39	5.00	2.52	0.39	0.15	9.23	0.03	11.77	m	12.95
width 300mm; dubbing average 19mm thick; curved to 3000mm radius	2.39	5.00	2.52	0.52	0.15	11.62	0.03	14.16	m	15.58
Rounded angles radius 10 - 100mm	-	-	-	0.24	-	4.41	-	4.41	m	4.85

Plaster, Thistle; pre-mixed; floating coat of bonding 8mm thick; finishing coat of multi-finish, 2mm thick; steel troweled
10mm work to walls on concrete or plasterboard base

	Del to Site £	Waste %	Material Cost £	Craft Optve Hrs	Lab Hrs	Labour Cost £	Sunds £	Nett Rate £	Unit	Gross rate (10%) £
width exceeding 600mm	2.59	5.00	2.72	0.40	0.20	10.10	0.05	12.87	m²	14.16
width exceeding 600mm; curved to 3000mm radius	2.59	5.00	2.72	0.53	0.20	12.49	0.05	15.26	m²	16.78
width 300mm	1.14	10.00	1.26	0.24	0.06	5.23	0.02	6.51	m	7.16
width 300mm; curved to 3000mm radius	1.14	10.00	1.26	0.32	0.06	6.70	0.02	7.98	m	8.78
10mm work to isolated columns on concrete or plasterboard base										
width exceeding 600mm	2.59	5.00	2.72	0.60	0.20	13.77	0.05	16.54	m²	18.20
width 300mm	1.14	10.00	1.26	0.36	0.06	7.44	0.02	8.71	m	9.59
10mm work to ceilings on concrete or plasterboard base										
width exceeding 600mm	2.59	5.00	2.72	0.48	0.20	11.57	0.05	14.34	m²	15.77
width exceeding 600mm; 3.50 - 5.00m above floor	2.59	5.00	2.72	0.52	0.20	12.30	0.05	15.08	m²	16.58
width 300mm	1.14	10.00	1.26	0.29	0.06	6.15	0.02	7.43	m	8.17
width 300mm; 3.50 - 5.00m above floor	1.14	10.00	1.26	0.31	0.06	6.52	0.02	7.80	m	8.57
10mm work to isolated beams on concrete or plasterboard base										
width exceeding 600mm	2.59	5.00	2.72	0.60	0.20	13.77	0.05	16.54	m²	18.20
width 300mm	1.14	10.00	1.26	0.36	0.06	7.44	0.02	8.71	m	9.59
Rounded angles radius 10 - 100mm	-	-	-	0.24	-	4.41	-	4.41	m	4.85

Plaster, Thistle; pre-mixed; floating coat of bonding 11mm thick; finishing coat of multi-finish, 2mm thick; steel troweled
13mm work to ceilings on precast concrete beam and infill blocks base

	Del to Site £	Waste %	Material Cost £	Craft Optve Hrs	Lab Hrs	Labour Cost £	Sunds £	Nett Rate £	Unit	Gross rate (10%) £
width exceeding 600mm	3.52	5.00	3.70	0.48	0.20	11.57	0.08	15.34	m²	16.88
width 300mm	1.77	10.00	1.94	0.29	0.06	6.15	0.02	8.12	m	8.93

Plaster, Thistle; tough coat, 11mm thick; finish 2mm thick steel troweled; Note: the thickness is from the face of the metal lathing
13mm work to walls on metal lathing base

	Del to Site £	Waste %	Material Cost £	Craft Optve Hrs	Lab Hrs	Labour Cost £	Sunds £	Nett Rate £	Unit	Gross rate (10%) £
width exceeding 600mm	2.97	5.00	3.12	0.55	0.28	13.96	0.05	17.12	m²	18.84
width exceeding 600mm; curved to 3000mm radius	2.97	5.00	3.12	0.73	0.28	17.26	0.05	20.43	m²	22.47
width 300mm	1.31	10.00	1.44	0.33	0.08	7.16	0.02	8.62	m	9.48
width 300mm; curved to 3000mm radius	1.31	10.00	1.44	0.44	0.08	9.18	0.02	10.64	m	11.70
13mm work to isolated columns on metal lathing base										
width exceeding 600mm	2.97	5.00	3.12	0.83	0.28	19.10	0.05	22.27	m²	24.49
width 300mm	1.31	10.00	1.44	0.50	0.08	10.29	0.02	11.74	m	12.91
13mm work to ceilings on metal lathing base										
width exceeding 600mm	2.97	5.00	3.12	0.66	0.28	15.98	0.05	19.14	m²	21.06
width exceeding 600mm; 3.50 - 5.00m above floor	2.97	5.00	3.12	0.72	0.28	17.08	0.05	20.25	m²	22.27
width 300mm	1.31	10.00	1.44	0.40	0.08	8.45	0.02	9.90	m	10.89
width 300mm; 3.50 - 5.00m above floor	1.31	10.00	1.44	0.43	0.08	9.00	0.02	10.45	m	11.50
13mm work to isolated beams on metal lathing base										
width exceeding 600mm	2.97	5.00	3.12	0.83	0.28	19.10	0.05	22.27	m²	24.49
width 300mm	1.31	10.00	1.44	0.50	0.08	10.29	0.02	11.74	m	12.91
Rounded angles radius 10 - 100mm	-	-	-	0.30	-	5.51	-	5.51	m	6.06

Finish to Walls, isolated columns, ceilings and isolated beams; applied liquid bonding fluid

Prepare and apply bonding fluid to receive plaster or cement rendering
Work to walls on existing cement and sand base

	Del to Site £	Waste %	Material Cost £	Craft Optve Hrs	Lab Hrs	Labour Cost £	Sunds £	Nett Rate £	Unit	Gross rate (10%) £
width exceeding 300mm	0.43	10.00	0.48	0.20	-	3.67	-	4.15	m²	4.57
Work to walls on existing glazed tiling base										
width exceeding 300mm	0.33	10.00	0.36	0.15	-	2.76	-	3.11	m²	3.43
Work to walls on existing painted surface										
width exceeding 300mm	0.39	10.00	0.43	0.18	-	3.31	-	3.74	m²	4.11
Work to walls on existing concrete base										
width exceeding 300mm	0.43	10.00	0.48	0.20	-	3.67	-	4.15	m²	4.57

Labour hourly rates: (except Specialists) Craft Operatives 18.37 Labourer 13.76 Rates are national average prices. Refer to REGIONAL VARIATIONS for indicative levels of overall pricing in regions	MATERIALS			LABOUR				RATES		
	Del to Site £	Waste %	Material Cost £	Craft Optve Hrs	Lab Hrs	Labour Cost £	Sunds £	Nett Rate £	Unit	Gross rate (10%) £
IN-SITU, TILED, BLOCK, MOSAIC, SHEET, APPLIED LIQUID OR PLANTED FINISHES (Cont'd)										
Finish to Walls, isolated columns, ceilings and isolated beams; applied liquid bonding fluid (Cont'd)										
Prepare and apply bonding fluid to receive plaster or cement rendering (Cont'd)										
Work to ceilings on existing cement and sand base										
width exceeding 300mm	0.43	10.00	0.48	0.24	-	4.41	-	4.89	m²	5.38
Work to ceilings on existing painted surface										
width exceeding 300mm	0.39	10.00	0.43	0.22	-	4.04	-	4.47	m²	4.92
Work to ceilings on existing concrete base										
width exceeding 300mm	0.43	10.00	0.48	0.24	-	4.41	-	4.89	m²	5.38
Finish to Walls, isolated columns, ceilings and isolated beams; dry lining plasterboard										
Linings; tapered edge sheeting of one layer 9.5mm thick Gyproc wall board for direct decoration, BS EN 520, butt joints; fixing with galvanized nails to timber base; butt joints filled with joint filler tape and joint finish, spot filling										
Walls										
over 600mm wide	1.55	5.00	1.63	0.40	0.20	10.10	-	11.73	m²	12.90
Columns; 4nr faces										
total girth 600 - 1200mm	1.94	10.00	2.12	0.88	0.21	19.06	-	21.18	m	23.30
total girth 1200 - 1800mm	2.86	10.00	3.13	1.31	0.32	28.47	-	31.60	m	34.76
Ceilings										
generally	1.56	5.00	1.64	0.40	0.18	9.82	-	11.46	m²	12.61
Beams; isolated; 3nr faces										
total girth 600 - 1200mm	1.89	10.00	2.07	0.88	0.21	19.06	-	21.13	m	23.24
Reveals and soffits of openings and recesses										
width 300mm	0.51	10.00	0.56	0.21	0.05	4.55	-	5.10	m	5.61
Finish to Walls, isolated columns, ceilings and isolated beams; linings; tapered edge sheeting of one layer 12.5mm thick Gyproc wallboard for direct decoration, BS EN 520 butt joints; fixing with galvanized nails to timber base; butt joints filled with joint filler tape and joint finish, spot filling										
Walls										
over 600mm wide	1.63	5.00	1.71	0.47	0.24	11.94	-	13.65	m²	15.01
Columns; 4nr faces										
total girth 600 - 1200mm	2.04	10.00	2.23	1.02	0.24	22.04	-	24.27	m	26.70
total girth 1200 - 1800mm	3.01	10.00	3.29	1.53	0.36	33.06	-	36.35	m	39.99
Ceilings										
generally	1.64	5.00	1.72	0.46	0.20	11.20	-	12.92	m²	14.21
Beams; isolated; 3nr faces										
total girth 600 - 1200mm	1.99	10.00	2.18	1.02	0.24	22.04	-	24.22	m	26.64
Reveals and soffits of openings and recesses										
width 300mm	0.54	10.00	0.58	0.21	0.06	4.68	-	5.27	m	5.79
Finish to Walls, isolated columns, ceilings and isolated beams; linings; tapered edge sheeting of one layer 15mm thick Gyproc wall board for direct decoration, BS EN 520, butt joints; fixing with galvanized nails to timber base; butt joints filled with joint filler tape and joint finish, spot filling										
Walls										
over 600mm wide	2.21	5.00	2.32	0.54	0.28	13.77	-	16.10	m²	17.70
Columns; 4nr faces										
total girth 600 - 1200mm	2.74	10.00	3.00	1.16	0.27	25.02	-	28.02	m	30.83
total girth 1200 - 1800mm	4.06	10.00	4.45	1.74	0.41	37.61	-	42.05	m	46.26
Ceilings										
generally	2.22	5.00	2.33	0.52	0.22	12.58	-	14.91	m²	16.40
Beams; isolated; 3nr faces										
total girth 600 - 1200mm	2.69	10.00	2.95	1.16	0.27	25.02	-	27.97	m	30.77
Reveals and soffits of openings and recesses										
width 300mm	0.71	10.00	0.78	0.22	0.06	4.87	-	5.64	m	6.21
Finish to Walls, isolated columns, ceilings and isolated beams										
Finish to Walls, isolated columns, ceilings and isolated beams; linings; tapered edge sheeting of one layer 12.5mm thick Gyproc vapourcheck wallboard for direct decoration, BS EN 520, butt joints; fixing with galvanized nails to timber base; butt joints filled with joint filler tape and joint finish, spot filling										
Walls										
over 600mm wide	3.05	5.00	3.20	0.47	0.24	11.94	-	15.14	m²	16.66
Columns; 4nr faces										
total girth 600 - 1200mm	3.75	10.00	4.11	1.02	0.24	22.04	-	26.15	m	28.76
total girth 1200 - 1800mm	5.57	10.00	6.11	1.53	0.36	33.06	-	39.17	m	43.09
Ceilings										
generally	3.06	5.00	3.21	0.46	0.20	11.20	-	14.42	m²	15.86
Beams; isolated; 3nr faces										
total girth 600 - 1200mm	3.70	10.00	4.05	1.02	0.24	22.04	-	26.09	m	28.70
Reveals and soffits of openings and recesses										
width 300mm	0.96	10.00	1.05	0.21	0.06	4.68	-	5.74	m	6.31
Finish to Walls, isolated columns, ceilings and isolated beams; linings; tapered edge sheeting of one layer 15mm thick Gyproc vapourcheck wallboard for direct decoration, BS EN 520, butt joints; fixing with galvanized nails to timber base; butt joints filled with joint filler tape and joint finish, spot filling										
Walls										
over 600mm wide	4.06	5.00	4.26	0.54	0.28	13.77	-	18.03	m²	19.84

Labour hourly rates: (except Specialists) Craft Operatives 18.37 Labourer 13.76 Rates are national average prices. Refer to REGIONAL VARIATIONS for indicative levels of overall pricing in regions	MATERIALS			LABOUR				RATES		
	Del to Site £	Waste %	Material Cost £	Craft Optve Hrs	Lab Hrs	Labour Cost £	Sunds £	Nett Rate £	Unit	Gross rate (10%) £
IN-SITU, TILED, BLOCK, MOSAIC, SHEET, APPLIED LIQUID OR PLANTED FINISHES (Cont'd)										
Finish to Walls, isolated columns, ceilings and isolated beams; linings; tapered edge sheeting of one layer 15mm thick Gyproc vapourcheck wallboard for direct decoration, BS EN 520, butt joints; fixing with galvanized nails to timber base; butt joints filled with joint filler tape and joint finish, spot filling (Cont'd)										
Columns; 4nr faces										
total girth 600 - 1200mm	4.95	10.00	5.44	1.16	0.27	25.02	-	30.46	m	33.51
total girth 1200 - 1800mm	7.38	10.00	8.10	1.74	0.41	37.61	-	45.71	m	50.28
Ceilings										
generally	4.07	5.00	4.27	0.52	0.22	12.58	-	16.85	m²	18.54
Beams; isolated; 3nr faces										
total girth 600 - 1200mm	4.90	10.00	5.38	1.16	0.27	25.02	-	30.41	m	33.45
Reveals and soffits of openings and recesses										
width 300mm	1.26	10.00	1.39	0.22	0.06	4.87	-	6.25	m	6.88
Finish to Walls, isolated columns, ceilings and isolated beams; linings; tapered edge sheeting of one layer 19mm thick Gyproc plank for direct decoration, BS EN 520, butt joints; fixing with galvanized nails to timber base; butt joints filled with joint filler tape and joint finish, spot filling										
Walls										
over 600mm wide	4.42	5.00	4.65	0.60	0.30	15.15	-	19.80	m²	21.77
Columns; 4nr faces										
total girth 600 - 1200mm	5.36	10.00	5.88	1.50	0.30	31.68	-	37.57	m	41.32
total girth 1200 - 1800mm	7.99	10.00	8.77	2.25	0.45	47.52	-	56.30	m	61.93
Ceilings										
generally	4.43	5.00	4.65	0.60	0.25	14.46	-	19.11	m²	21.03
Beams; isolated; 3nr faces										
total girth 600 - 1200mm	5.31	10.00	5.83	1.50	0.30	31.68	-	37.51	m	41.26
Reveals and soffits of openings and recesses										
width 300mm	1.37	10.00	1.50	0.30	0.08	6.61	-	8.11	m	8.92
Finish to Walls, isolated columns, ceilings and isolated beams; linings; two layers of Gypsum wallboard, first layer square edge sheeting 9.5mm thick, second layer tapered edge sheeting 9.5mm thick for direct decoration, BS EN 520, butt joints; fixing with galvanized nails to timber base; butt joints of second layer filled with joint filler tape and joint finish, spot filling										
Walls										
over 600mm wide	3.09	5.00	3.25	0.70	0.35	17.68	-	20.92	m²	23.02
Columns; 4nr faces										
total girth 600 - 1200mm	3.80	10.00	4.16	1.62	0.36	34.71	-	38.87	m	42.76
total girth 1200 - 1800mm	5.64	10.00	6.18	2.43	0.54	52.07	-	58.25	m	64.07
Ceilings										
generally	3.10	5.00	3.26	0.70	0.30	16.99	-	20.25	m²	22.27
Beams; isolated; 3nr faces										
total girth 600 - 1200mm	2.05	10.00	2.24	1.62	0.36	34.71	-	36.95	m	40.64
Reveals and soffits of openings and recesses										
width 300mm	0.98	10.00	1.07	0.33	0.09	7.30	-	8.37	m	9.20
Finish to Walls, isolated columns, ceilings and isolated beams; linings; two layers of Gypsum wallboard, first layer square edge sheeting 9.5mm thick, second layer tapered edge sheeting 12.5mm thick for direct decoration, BS EN 520, butt joints; fixing with galvanized nails to timber base; butt joints of second layer filled with joint filler tape and joint finish, spot filling										
Walls										
over 600mm wide	3.17	5.00	3.33	0.84	0.40	20.93	-	24.27	m²	26.69
Columns; 4nr faces										
total girth 600 - 1200mm	3.89	10.00	4.18	1.78	0.39	38.07	-	42.24	m	46.46
total girth 1200 - 1800mm	5.79	10.00	6.21	2.66	0.59	56.98	-	63.19	m	69.51
Ceilings										
generally	3.18	5.00	3.34	0.76	0.33	18.50	-	21.84	m²	24.03
Beams; isolated; 3nr faces										
total girth 600 - 1200mm	3.84	10.00	4.12	1.78	0.39	38.07	-	42.19	m	46.41
Reveals and soffits of openings and recesses										
width 300mm	1.00	10.00	1.07	0.36	0.10	7.99	-	9.06	m	9.97
Finish to Walls, isolated columns, ceilings and isolated beams; linings; two layers of Gypsum wallboard, first layer square edge sheeting 12.5mm thick, second layer tapered edge sheeting 12.5mm thick for direct decoration, BS EN 520, butt joints; fixing with galvanized nails to timber base; butt joints of second layer filled with joint filler tape and joint finish, spot filling										
Walls										
over 600mm wide	3.25	5.00	3.42	0.82	0.43	20.98	-	24.40	m²	26.84
Columns; 4nr faces										
total girth 600 - 1200mm	3.99	10.00	4.37	1.92	0.42	41.05	-	45.41	m	49.96
total girth 1200 - 1800mm	5.93	10.00	6.50	2.88	0.63	61.57	-	68.07	m	74.88
Ceilings										
generally	3.26	5.00	3.43	0.82	0.35	19.88	-	23.31	m²	25.64
Beams; isolated; 3nr faces										
total girth 600 - 1200mm	3.94	10.00	4.31	1.92	0.42	41.05	-	45.36	m	49.90
Reveals and soffits of openings and recesses										
width 300mm	1.02	10.00	1.12	0.39	0.11	8.68	-	9.80	m	10.78

Labour hourly rates: (except Specialists) Craft Operatives 18.37 Labourer 13.76 Rates are national average prices. Refer to REGIONAL VARIATIONS for indicative levels of overall pricing in regions	MATERIALS			LABOUR				RATES		
	Del to Site	Waste	Material Cost	Craft Optve	Lab	Labour Cost	Sunds	Nett Rate	Unit	Gross rate (10%)
	£	%	£	Hrs	Hrs	£	£	£		£
IN-SITU, TILED, BLOCK, MOSAIC, SHEET, APPLIED LIQUID OR PLANTED FINISHES (Cont'd)										
Finish to Walls, isolated columns, ceilings and isolated beams; linings; two layers of Gypsum wallboard, first layer square edge sheeting 15mm thick, second layer tapered edge sheeting 15mm thick for direct decoration, BS EN 520, butt joints; fixing with galvanized nails to timber base; butt joints of second layer filled with joint filler tape and joint finish, spot filling										
Walls										
over 600mm wide	4.42	5.00	4.64	0.90	0.46	22.86	-	27.50	m²	30.25
Columns; 4nr faces										
total girth 600 - 1200mm..............	5.39	10.00	5.91	1.68	0.45	37.05	-	42.96	m	47.25
total girth 1200 - 1800mm.............	8.03	10.00	8.81	2.52	0.68	55.65	-	64.45	m	70.90
Ceilings										
generally	4.43	5.00	4.65	0.86	0.38	21.03	-	25.68	m²	28.25
Beams; isolated; 3nr faces										
total girth 600 - 1200mm..............	5.34	10.00	5.85	1.68	0.45	37.05	-	42.91	m	47.20
Reveals and soffits of openings and recesses										
width 300mm	1.37	10.00	1.50	0.45	0.11	9.78	-	11.28	m	12.41
Finish to Walls, isolated columns, ceilings and isolated beams; linings; two layers of Gypsum wallboard, first layer vapourcheck square edge sheeting 12.5mm thick, second layer tapered edge sheeting 12.5mm thick for direct decoration, BS EN 520, butt joints; fixing with galvanized nails to timber base; butt joints of second layer filled with joint filler tape and joint finish, spot filling										
Walls										
over 600mm wide	4.68	5.00	4.91	0.85	0.43	21.53	-	26.44	m²	29.09
Columns; 4nr faces										
total girth 600 - 1200mm..............	5.69	10.00	6.07	1.92	0.42	41.05	-	47.12	m	51.83
total girth 1200 - 1800mm.............	8.49	10.00	9.05	2.88	0.63	61.57	-	70.63	m	77.69
Ceilings										
generally	4.69	5.00	4.92	0.82	0.35	19.88	-	24.80	m²	27.28
Beams; isolated; 3nr faces										
total girth 600 - 1200mm..............	5.64	10.00	6.02	1.92	0.42	41.05	-	47.07	m	51.77
Reveals and soffits of openings and recesses										
width 300mm	1.45	10.00	1.54	0.39	0.11	8.68	-	10.22	m	11.24
Finish to Walls, isolated columns, ceilings and isolated beams; linings; two layers of Gypsum wallboard, first layer vapourcheck square edge sheeting 15mm thick, second layer tapered edge sheeting 15mm thick for direct decoration, BS EN 520, butt joints; fixing with galvanized nails to timber base; butt joints of second layer filled with joint filler tape and joint finish, spot filling										
Walls										
over 600mm wide	6.27	5.00	6.58	0.90	0.46	22.86	-	29.44	m²	32.39
Columns; 4nr faces										
total girth 600 - 1200mm..............	7.60	10.00	8.11	1.68	0.45	37.05	-	45.16	m	49.68
total girth 1200 - 1800mm.............	11.35	10.00	12.11	2.52	0.68	55.65	-	67.76	m	74.53
Ceilings										
generally	6.28	5.00	6.59	0.86	0.38	21.03	-	27.62	m²	30.38
Beams; isolated; 3nr faces										
total girth 600 - 1200mm..............	7.55	10.00	8.06	1.68	0.45	37.05	-	45.11	m	49.62
Reveals and soffits of openings and recesses										
width 300mm	1.93	10.00	2.05	0.45	0.11	9.78	-	11.83	m	13.02
Finish to Walls, isolated columns, ceilings and isolated beams; linings; tapered edge sheeting of one layer 12.5mm thick Gyproc wallboard for direct decoration, BS EN 520 butt joints; fixing with Drywall screws to and including Gypframe MF suspended ceiling system to timber joists; butt joints filled with joint filler tape and joint finish, spot filling										
Ceilings										
generally....................	6.67	5.00	7.01	0.60	0.26	14.60	-	21.61	m²	23.77
Finish to Walls, isolated columns, ceilings and isolated beams; linings; tapered edge sheeting of one layer 15mm thick Gyproc wall board for direct decoration, BS EN 520, butt joints; fixing with Drywall screws to and including Gypframe MF suspended ceiling system to timber joists; butt joints filled with joint filler tape and joint finish, spot filling										
Ceilings										
generally....................	7.35	5.00	7.71	0.66	0.28	15.98	-	23.69	m²	26.06
Finish to Walls, isolated columns, ceilings and isolated beams; linings; tapered edge sheeting of one layer 12.5mm thick Gyproc vapourcheck wallboard for direct decoration, BS EN 520, butt joints; fixing with Drywall screws to and including Gypframe MF suspended ceiling system to timber joists; butt joints filled with joint filler tape and joint finish, spot filling										
Ceilings										
generally....................	8.19	5.00	8.60	0.60	0.26	14.60	-	23.19	m²	25.51

Labour hourly rates: (except Specialists) Craft Operatives 18.37 Labourer 13.76 Rates are national average prices. Refer to REGIONAL VARIATIONS for indicative levels of overall pricing in regions	MATERIALS			LABOUR				RATES		
	Del to Site £	Waste %	Material Cost £	Craft Optve Hrs	Lab Hrs	Labour Cost £	Sunds £	Nett Rate £	Unit	Gross rate (10%) £
IN-SITU, TILED, BLOCK, MOSAIC, SHEET, APPLIED LIQUID OR PLANTED FINISHES (Cont'd)										
Finish to Walls, isolated columns, ceilings and isolated beams; linings; tapered edge sheeting of one layer 15mm thick Gyproc vapourcheck wallboard for direct decoration, BS EN 520, butt joints; fixing with Drywall screws to and including Gypframe MF suspended ceiling system to timber joists; butt joints filled with joint filler tape and joint finish, spot filling										
Ceilings generally	9.19	5.00	9.65	0.66	0.28	15.98	-	25.63	m²	28.19
Finish to Walls, isolated columns, ceilings and isolated beams; linings; tapered edge sheeting of one layer 19mm thick Gyproc plank for direct decoration, BS EN 520, butt joints; fixing with Drywall screws to and including Gypframe MF suspended ceiling system to timber joists; butt joints filled with joint filler tape and joint finish, spot filling										
Ceilings generally	9.56	5.00	10.03	0.74	0.31	17.86	-	27.89	m²	30.68
Finish to Walls, isolated columns, ceilings and isolated beams; linings; two layers of Gypsum wallboard, first layer square edge sheeting 12.5mm thick, second layer tapered edge sheeting 12.5mm thick for direct decoration, BS EN 520, butt joints; fixing with Drywall screws to and including Gypframe MF suspended ceiling system to timber joists; butt joints of second layer filled with joint filler tape and joint finish, spot filling										
Ceilings generally	8.39	5.00	8.81	0.96	0.41	23.28	-	32.08	m²	35.29
Finish to Walls, isolated columns, ceilings and isolated beams; linings; two layers of Gypsum wallboard, first layer square edge sheeting 15mm thick, second layer tapered edge sheeting 15mm thick for direct decoration, BS EN 520, butt joints; fixing with Drywall screws to and including Gypframe MF suspended ceiling system to timber joists; butt joints of second layer filled with joint filler tape and joint finish, spot filling										
Ceilings generally	9.55	5.00	10.03	1.00	0.44	24.42	-	34.46	m²	37.90
Finish to Walls, isolated columns, ceilings and isolated beams; linings; two layers of Gypsum wallboard, first layer vapourcheck square edge sheeting 12.5mm thick, second layer tapered edge sheeting 12.5mm thick for direct decoration, BS EN 520, butt joints; fixing with Drywall screws to and including Gypframe MF suspended ceiling system to timber joists; butt joints of second layer filled with joint filler tape and joint finish, spot filling										
Ceilings generally	9.81	5.00	10.30	0.96	0.41	23.28	-	33.58	m²	36.94
Finish to Walls, isolated columns, ceilings and isolated beams; linings; two layers of Gypsum wallboard, first layer vapourcheck square edge sheeting 15mm thick, second layer tapered edge sheeting 15mm thick for direct decoration, BS EN 520, butt joints; fixing with Drywall screws to and including Gypframe MF suspended ceiling system to timber joists; butt joints of second layer filled with joint filler tape and joint finish, spot filling										
Ceilings generally	11.40	5.00	11.97	1.00	0.44	24.42	-	36.39	m²	40.03
Finish to Walls, isolated columns, ceilings and isolated beams; extra for 50mm acoustic insulation laid over plasterboard on metal framing to include acoustic hangers and mastic seal to perimeters										
Ceilings generally	7.04	5.00	7.39	0.10	0.05	2.53	-	9.92	m²	10.91
Finish to Walls, isolated columns, ceilings and isolated beams; linings; tapered edged sheeting of one layer 22mm thick Gyproc Thermaline BASIC board vapour check grade for direct decoration, butt joints; fixing with galvanized nails to timber base; butt joints filled with joint filler tape and joint finish, spot filling										
Walls over 600mm wide	4.14	5.00	4.35	0.72	0.38	18.46	-	22.80	m²	25.08
Ceilings generally	4.15	5.00	4.36	0.72	0.30	17.35	-	21.71	m²	23.88
Reveals and soffits of openings and recesses width 300mm	1.29	10.00	1.41	0.36	0.09	7.85	-	9.26	m	10.19
Finish to Walls, isolated columns, ceilings and isolated beams; linings; tapered edged sheeting of one layer 30mm thick Gyproc thermaline BASIC board vapour check grade for direct decoration, butt joints; fixing with galvanized nails to timber base; butt joints filled with joint filler tape and joint finish, spot filling										
Walls over 600mm wide	4.60	5.00	4.83	0.85	0.43	21.53	-	26.36	m²	29.00
Ceilings generally	4.61	5.00	4.84	0.84	0.36	20.38	-	25.22	m²	27.75

Labour hourly rates: (except Specialists) Craft Operatives 18.37 Labourer 13.76 Rates are national average prices. Refer to REGIONAL VARIATIONS for indicative levels of overall pricing in regions	MATERIALS			LABOUR				RATES		
	Del to Site £	Waste %	Material Cost £	Craft Optve Hrs	Lab Hrs	Labour Cost £	Sunds £	Nett Rate £	Unit	Gross rate (10%) £
IN-SITU, TILED, BLOCK, MOSAIC, SHEET, APPLIED LIQUID OR PLANTED FINISHES (Cont'd)										
Finish to Walls, isolated columns, ceilings and isolated beams; linings; tapered edged sheeting of one layer 30mm thick Gyproc thermaline BASIC board vapour check grade for direct decoration, butt joints; fixing with galvanized nails to timber base; butt joints filled with joint filler tape and joint finish, spot filling (Cont'd)										
Reveals and soffits of openings and recesses width 300mm	1.43	10.00	1.56	0.42	0.11	9.23	-	10.79	m	11.87
Finish to Walls, isolated columns, ceilings and isolated beams; linings; tapered edged sheeting of one layer 40mm thick Gyproc Thermaline BASIC board vapour check grade for direct decoration, butt joints; fixing with galvanized nails to timber base; butt joints filled with joint filler tape and joint finish, spot filling										
Walls										
over 600mm wide	5.20	5.00	5.46	0.96	0.49	24.38	-	29.84	m²	32.82
Ceilings										
generally	5.21	5.00	5.47	0.96	0.42	23.41	-	28.89	m²	31.77
Reveals and soffits of openings and recesses width 300mm	1.61	10.00	1.76	0.48	0.12	10.47	-	12.23	m	13.45
Finish to Walls, isolated columns, ceilings and isolated beams; linings; tapered edged sheeting of one layer 50mm thick Gyproc Thermaline SUPER board vapour check grade for direct decoration, butt joints; fixing with galvanized nails to timber base; butt joints filled with joint filler tape and joint finish, spot filling										
Walls										
over 600mm wide	12.50	5.00	13.12	1.10	0.56	27.91	-	41.04	m²	45.14
Ceilings										
generally	12.51	5.00	13.13	1.08	0.48	26.44	-	39.58	m²	43.54
Reveals and soffits of openings and recesses width 300mm	3.80	10.00	4.17	0.54	0.14	11.85	-	16.02	m	17.62
Finish to walls; ceramic tiles, BS 6431, glazed white; 2mm joints, symmetrical layout; bedding in 10mm cement mortar (1:3); pointing with neat white cement; on brickwork or blockwork base										
108 x 108 x 6.5mm units to walls										
plain, width exceeding 300mm	18.20	5.00	19.11	1.30	0.75	34.20	-	53.31	m²	58.64
plain, width 300mm	5.99	10.00	6.57	0.78	0.47	20.80	-	27.36	m	30.10
152 x 152 x 5.5mm units to walls										
plain, width exceeding 300mm	12.74	5.00	13.38	1.00	0.60	26.63	-	40.00	m²	44.00
plain, width 300mm	3.93	10.00	4.30	0.60	0.36	15.98	-	20.28	m	22.30
Finish to walls; ceramic tile cills, BS 6431, glazed white; 2mm joints, symmetrical layout; bedding in 10mm cement mortar (1:3); pointing with neat white cement; on brickwork or blockwork base - cills										
108 x 108 x 6.5mm units to cills on brickwork or blockwork base										
width 150mm; rounded angle	6.28	5.00	6.60	0.39	0.11	8.68	-	15.27	m	16.80
width 225mm; rounded angle	8.27	5.00	8.69	0.54	0.17	12.26	-	20.95	m	23.04
width 300mm; rounded angle	9.45	5.00	9.93	0.69	0.23	15.84	-	25.77	m	28.35
152 x 152 x 5.5mm units to cills on brickwork or blockwork base										
width 150mm; rounded angle	7.28	5.00	7.65	0.33	0.09	7.30	-	14.95	m	16.44
width 225mm; rounded angle	8.28	5.00	8.70	0.44	0.14	10.01	-	18.71	m	20.58
width 300mm; rounded angle	9.40	5.00	9.87	0.55	0.18	12.58	-	22.45	m	24.70
Finish to walls; ceramic tiles, BS 6431, glazed white; 2mm joints; symmetrical layout; fixing with thin bed adhesive; pointing with neat white cement; on plaster base										
108 x 108 x 6.5mm units to walls										
plain, width exceeding 300mm	20.46	5.00	21.66	1.30	0.75	34.20	-	55.86	m²	61.45
plain, width 300mm	7.19	10.00	7.90	0.78	0.47	20.80	-	28.70	m	31.57
152 x 152 x 5.5mm units to walls										
plain, width exceeding 300mm	15.00	5.00	15.93	1.00	0.60	26.63	-	42.56	m²	46.81
plain, width 300mm	5.13	10.00	5.64	0.60	0.36	15.98	-	21.62	m	23.78
Finish to walls; ceramic tiles, BS 6431, glazed white; 2mm joints, symmetrical layout; fixing with thick bed adhesive; pointing with neat white cement; on cement and sand base										
108 x 108 x 6.5mm units to walls										
plain, width exceeding 300mm	22.20	5.00	23.58	1.30	0.75	34.20	-	57.78	m²	63.56
plain, width 300mm	7.97	10.00	8.76	0.78	0.47	20.80	-	29.56	m	32.51
152 x 152 x 5.5mm units to walls										
plain, width exceeding 300mm	16.75	5.00	17.85	1.00	0.60	26.63	-	44.47	m²	48.92
plain, width 300mm	5.91	10.00	6.50	0.60	0.36	15.98	-	22.47	m	24.72
Finish to walls; ceramic tiles, BS 6431, glazed light colour; 2mm joints symmetrical layout; bedding in 10mm cement mortar (1:3); pointing with neat white cement; on cement and sand base										
108 x 108 x 6.5mm units to walls										
plain, width exceeding 300mm	18.20	5.00	19.11	1.30	0.75	34.20	-	53.31	m²	58.64
plain, width 300mm	5.87	10.00	6.44	0.78	0.47	20.80	-	27.24	m	29.96

Labour hourly rates: (except Specialists) Craft Operatives 18.37 Labourer 13.76 Rates are national average prices. Refer to REGIONAL VARIATIONS for indicative levels of overall pricing in regions	MATERIALS			LABOUR				RATES		
	Del to Site £	Waste %	Material Cost £	Craft Optve Hrs	Lab Hrs	Labour Cost £	Sunds £	Nett Rate £	Unit	Gross rate (10%) £
IN-SITU, TILED, BLOCK, MOSAIC, SHEET, APPLIED LIQUID OR PLANTED FINISHES (Cont'd)										
Finish to walls; ceramic tiles, BS 6431, glazed light colour; 2mm joints symmetrical layout; bedding in 10mm cement mortar (1:3); pointing with neat white cement; on cement and sand base (Cont'd)										
152 x 152 x 5.5mm units to walls										
plain, width exceeding 300mm	18.14	5.00	19.05	1.00	0.60	26.63	-	45.67	m²	50.24
plain, width 300mm	5.48	10.00	6.01	0.60	0.36	15.98	-	21.99	m	24.19
Finish to walls; ceramic tiles, BS 6431, glazed light colour; 2mm joints, symmetrical layout; fixing with thin bed adhesive; pointing with neat white cement; on plaster base										
108 x 108 x 6.5mm units to walls										
plain, width exceeding 300mm	20.46	5.00	21.66	1.30	0.75	34.20	-	55.86	m²	61.45
plain, width 300mm	7.07	10.00	7.78	0.78	0.47	20.80	-	28.57	m	31.43
152 x 152 x 5.5mm units to walls										
plain, width exceeding 300mm	20.40	5.00	21.60	1.00	0.60	26.63	-	48.23	m²	53.05
plain, width 300mm	6.68	10.00	7.35	0.60	0.36	15.98	-	23.33	m	25.66
Finish to walls; ceramic tiles, 6431, glazed light colour; 2mm joints, symmetrical layout; fixing with thick bed adhesive; pointing with neat white cement; on cement and sand base										
108 x 108 x 6.5mm units to walls										
plain, width exceeding 300mm	22.20	5.00	23.58	1.30	0.75	34.20	-	57.78	m²	63.56
plain, width 300mm	7.85	10.00	8.64	0.78	0.47	20.80	-	29.43	m	32.37
152 x 152 x 5.5mm units to walls										
plain, width exceeding 300mm	22.15	5.00	23.52	1.00	0.60	26.63	-	50.14	m²	55.16
plain, width 300mm	7.46	10.00	8.21	0.60	0.36	15.98	-	24.18	m	26.60
Finish to walls; ceramic tiles,6431, glazed dark colour; 2mm joints, symmetrical layout; bedding in 10mm cement mortar (1:3); pointing with neat white cement; on brickwork or blockwork base										
108 x 108 x 6.5mm units to walls										
plain, width exceeding 300mm	24.35	5.00	25.57	1.30	0.75	34.20	-	59.77	m²	65.75
plain, width 300mm	8.02	10.00	8.80	0.78	0.47	20.80	-	29.59	m	32.55
152 x 152 x 5.5mm units to walls										
plain, width exceeding 300mm	19.29	5.00	20.26	1.00	0.60	26.63	-	46.88	m²	51.57
plain, width 300mm	5.94	10.00	6.52	0.60	0.36	15.98	-	22.50	m	24.75
Finish to walls; ceramic tiles,6431, glazed dark colour; 2mm joints, symmetrical layout; fixing with thin bed adhesive; pointing with neat white cement; on plaster base										
108 x 108 x 6.5mm units to walls										
plain, width exceeding 300mm	26.61	5.00	28.12	1.30	0.75	34.20	-	62.32	m²	68.55
plain, width 300mm	9.22	10.00	10.14	0.78	0.47	20.80	-	30.93	m	34.03
152 x 152 x 5.5mm units to walls										
plain, width exceeding 300mm	21.55	5.00	22.81	1.00	0.60	26.63	-	49.43	m²	54.38
plain, width 300mm	7.14	10.00	7.86	0.60	0.36	15.98	-	23.83	m	26.22
Finish to walls; ceramic tiles,6431, glazed dark colour; 2mm joints, symmetrical layout; fixing with thick bed adhesive; pointing with neat white cement; on cement and sand base										
108 x 108 x 6.5mm units to walls										
plain, width exceeding 300mm	28.35	5.00	30.04	1.30	0.75	34.20	-	64.24	m²	70.66
plain, width 300mm	10.00	10.00	11.00	0.78	0.47	20.80	-	31.79	m	34.97
152 x 152 x 5.5mm units to walls										
plain, width exceeding 300mm	23.30	5.00	24.73	1.00	0.60	26.63	-	51.35	m²	56.49
plain, width 300mm	7.92	10.00	8.72	0.60	0.36	15.98	-	24.69	m	27.16
Finish to walls; extra over ceramic tiles, glazed white; 2mm joints, symmetrical layout; bedding or fixing in any material to general surfaces on any base										
Special tiles										
rounded edge	6.94	5.00	7.28	-	-	-	-	7.28	m	8.01
external angle	13.87	5.00	14.56	-	-	-	-	14.56	m	16.02
internal angle to skirting	-	-	-	0.07	-	1.29	-	1.29	nr	1.41
external angle to skirting	-	-	-	0.07	-	1.29	-	1.29	nr	1.41
internal angle	-	-	-	0.10	-	1.84	-	1.84	m	2.02
Finish to walls; extra over ceramic tiles, glazed light colour; 2mm joints, symmetrical layout; bedding or fixing in any material to general surfaces on any base										
Special tiles										
rounded edge	7.05	5.00	7.40	-	-	-	-	7.40	m	8.14
external angle	13.87	5.00	14.56	-	-	-	-	14.56	m	16.02
internal angle to skirting	-	-	-	0.07	-	1.29	-	1.29	nr	1.41
external angle to skirting	-	-	-	0.07	-	1.29	-	1.29	nr	1.41
internal angle	-	-	-	0.10	-	1.84	-	1.84	m	2.02

Labour hourly rates: (except Specialists) Craft Operatives 18.37 Labourer 13.76 Rates are national average prices. Refer to REGIONAL VARIATIONS for indicative levels of overall pricing in regions	MATERIALS			LABOUR				RATES		
	Del to Site	Waste	Material Cost	Craft Optve	Lab	Labour Cost	Sunds	Nett Rate	Unit	Gross rate (10%)
	£	%	£	Hrs	Hrs	£	£	£		£
IN-SITU, TILED, BLOCK, MOSAIC, SHEET, APPLIED LIQUID OR PLANTED FINISHES (Cont'd)										
Finish to walls; extra over ceramic tiles, glazed dark colour; 2mm joints, symmetrical layout; bedding or fixing in any material to general surfaces on any base										
Special tiles										
rounded edge	8.28	5.00	8.69	-	-	-	-	8.69	m	9.56
external angle bead	11.49	5.00	12.07	0.22	-	4.04	-	16.11	m	17.72
internal angle to skirting	-	-	-	0.07	-	1.29	-	1.29	nr	1.41
external angle to skirting	-	-	-	0.07	-	1.29	-	1.29	nr	1.41
internal angle	-	-	-	0.10	-	1.84	-	1.84	m	2.02
Finish to walls; clay quarry tile cills, BS EN ISO 10545, terracotta; 3mm joints, symmetrical layout; bedding in 10mm cement mortar (1:3); pointing in cement mortar (1:3); on cement and sand base - cills										
152 x 152 x 12.5mm units to cills on brickwork or blockwork base										
width 152mm; rounded angle	13.19	10.00	14.49	0.30	0.08	6.61	-	21.11	m	23.22
width 225mm; rounded angle	15.11	5.00	16.51	0.40	0.12	9.00	-	25.51	m	28.06
width 300mm; rounded angle	17.03	5.00	18.53	0.51	0.17	11.71	-	30.24	m	33.26
Coves; plasterboard										
Gyproc cove; fixing with adhesive										
90mm girth	1.41	10.00	1.55	0.18	0.09	4.55	-	6.09	m	6.70
extra; ends	-	-	-	0.06	-	1.10	-	1.10	nr	1.21
extra; internal angles	-	-	-	0.30	-	5.51	-	5.51	nr	6.06
extra; external angles	-	-	-	0.30	-	5.51	-	5.51	nr	6.06
127mm girth	1.40	10.00	1.54	0.18	0.09	4.55	-	6.08	m	6.69
extra; ends	-	-	-	0.06	-	1.10	-	1.10	nr	1.21
extra; internal angles	-	-	-	0.30	-	5.51	-	5.51	nr	6.06
extra; external angles	-	-	-	0.30	-	5.51	-	5.51	nr	6.06
Gyproc cove; fixing with nails to timber										
90mm girth	1.24	10.00	1.36	0.12	0.06	3.03	0.06	4.45	m	4.90
extra; ends	-	-	-	0.06	-	1.10	-	1.10	nr	1.21
extra; internal angles	-	-	-	0.30	-	5.51	-	5.51	nr	6.06
extra; external angles	-	-	-	0.30	-	5.51	-	5.51	nr	6.06
127mm girth	1.23	10.00	1.35	0.12	0.06	3.03	0.06	4.44	m	4.89
extra; ends	-	-	-	0.06	-	1.10	-	1.10	nr	1.21
extra; internal angles	-	-	-	0.30	-	5.51	-	5.51	nr	6.06
extra; external angles	-	-	-	0.30	-	5.51	-	5.51	nr	6.06
Finish to walls; movement joints										
6mm x 6mm silicon sealant gun applied	0.31	5.00	0.33	0.15	0.08	3.86	-	4.18	m	4.60
Beads and stops; for plaster and renders										
galvanized steel angle beads										
fixing to brickwork or blockwork with masonry nails										
Standard	0.33	10.00	0.37	0.10	-	1.84	-	2.20	m	2.42
3mm for Thin Coat	0.32	10.00	0.35	0.10	-	1.84	-	2.18	m	2.40
6mm for Thin Coat	0.32	10.00	0.35	0.10	-	1.84	-	2.18	m	2.40
galvanized steel stop beads										
fixing to brickwork or blockwork with masonry nails										
3mm for Thin Coat	0.52	10.00	0.58	0.10	-	1.84	-	2.41	m	2.65
6mm for Thin Coat	0.72	10.00	0.79	0.10	-	1.84	-	2.63	m	2.89
10mm	0.74	10.00	0.81	0.10	-	1.84	-	2.65	m	2.92
13mm	0.74	10.00	0.81	0.10	-	1.84	-	2.65	m	2.92
16mm	1.00	10.00	1.10	0.10	-	1.84	-	2.94	m	3.23
19mm	1.00	10.00	1.10	0.10	-	1.84	-	2.94	m	3.23
galvanized steel plasterboard edging beads										
fixing to timber with galvanized nails, 30mm long at 400mm centres										
10mm	6.80	10.00	6.98	0.10	-	1.84	-	8.82	m	9.70
13mm	6.80	10.00	6.98	0.10	-	1.84	-	8.82	m	9.70
Beads and stops; external; for plaster and renders										
Stainless steel stop beads										
fixing to brickwork or blockwork with masonry nails										
16-20mm	0.83	10.00	0.91	0.10	-	1.84	-	2.75	m	3.03
Stainless steel angle beads										
fixing to brickwork or blockwork with masonry nails										
16-20mm	2.91	10.00	3.20	0.10	-	1.84	-	5.04	m	5.54
Nosings										
Heavy duty aluminium alloy stair nosings with anti-slip inserts; 46mm wide with single line insert; fixing to timber with screws										
11.5mm drop	11.86	5.00	12.45	0.40	-	7.35	0.08	19.88	m	21.87
22mm drop	13.79	5.00	14.48	0.40	-	7.35	0.08	21.91	m	24.10
25mm drop	15.67	5.00	16.45	0.40	-	7.35	0.08	23.88	m	26.27
32mm drop	16.78	5.00	17.62	0.40	-	7.35	0.08	25.05	m	27.55
38mm drop	18.31	5.00	19.22	0.40	-	7.35	0.08	26.65	m	29.32
46mm drop	22.73	5.00	23.87	0.40	-	7.35	0.08	31.29	m	34.42
Heavy duty aluminium alloy stair nosings with anti-slip inserts; 80mm wide with two line inserts; fixing to timber with screws										
22mm drop	19.75	5.00	20.73	0.55	-	10.10	0.12	30.96	m	34.05
25mm drop	21.00	5.00	22.05	0.55	-	10.10	0.12	32.27	m	35.50
32mm drop	26.69	5.00	28.03	0.55	-	10.10	0.12	38.25	m	42.07
51mm drop	27.65	5.00	29.03	0.55	-	10.10	0.12	39.25	m	43.18

Labour hourly rates: (except Specialists) Craft Operatives 18.37 Labourer 13.76 Rates are national average prices. Refer to REGIONAL VARIATIONS for indicative levels of overall pricing in regions	MATERIALS			LABOUR				RATES		
	Del to Site £	Waste %	Material Cost £	Craft Optve Hrs	Lab Hrs	Labour Cost £	Sunds £	Nett Rate £	Unit	Gross rate (10%) £
IN-SITU, TILED, BLOCK, MOSAIC, SHEET, APPLIED LIQUID OR PLANTED FINISHES (Cont'd)										
Nosings (Cont'd)										
Heavy duty aluminium alloy stair nosings with anti-slip inserts; 80mm wide with two line inserts; fixing to timber with screws (Cont'd)										
63mm drop	30.13	5.00	31.64	0.55	-	10.10	0.12	41.86	m	46.05
Heavy duty aluminium alloy stair nosings with anti-slip inserts; 46mm wide with single line insert; fixing to masonry with screws										
11.5mm drop	11.86	5.00	12.45	0.65	-	11.94	0.12	24.51	m	26.96
22mm drop	13.79	5.00	14.48	0.65	-	11.94	0.12	26.54	m	29.19
25mm drop	15.67	5.00	16.45	0.65	-	11.94	0.12	28.51	m	31.37
32mm drop	16.78	5.00	17.62	0.65	-	11.94	0.12	29.68	m	32.65
38mm drop	18.31	5.00	19.22	0.65	-	11.94	0.12	31.28	m	34.41
46mm drop	22.73	5.00	23.87	0.65	-	11.94	0.12	35.93	m	39.52
Heavy duty aluminium alloy stair nosings with anti-slip inserts; 80mm wide with two line inserts; fixing to masonry with screws										
22mm drop	19.75	5.00	20.73	0.80	-	14.70	0.18	35.61	m	39.17
25mm drop	21.00	5.00	22.05	0.80	-	14.70	0.18	36.93	m	40.62
32mm drop	26.69	5.00	28.03	0.80	-	14.70	0.18	42.90	m	47.19
51mm drop	27.65	5.00	29.03	0.80	-	14.70	0.18	43.90	m	48.29
63mm drop	30.13	5.00	31.64	0.80	-	14.70	0.18	46.51	m	51.16
Aluminium stair nosings with anti-slip inserts; 46mm wide with single line insert; fixing to timber with screws										
22mm drop	10.93	5.00	11.48	0.40	-	7.35	0.08	18.90	m	20.79
33mm drop	12.80	5.00	13.44	0.40	-	7.35	0.08	20.87	m	22.95
Aluminium stair nosings with anti-slip inserts; 66mm wide with single line insert; fixing to timber with screws										
22mm drop	13.79	5.00	14.48	0.45	-	8.27	0.12	22.86	m	25.15
25mm drop	12.90	5.00	13.54	0.45	-	8.27	0.12	21.93	m	24.12
32mm drop	16.78	5.00	17.62	0.45	-	8.27	0.12	26.00	m	28.60
Aluminium stair nosings with anti-slip inserts; 80mm wide with two line inserts; fixing to timber with screws										
22mm drop	15.82	5.00	16.62	0.55	-	10.10	0.12	26.84	m	29.52
28mm drop	17.87	5.00	18.76	0.55	-	10.10	0.12	28.99	m	31.89
51mm drop	22.65	5.00	23.78	0.55	-	10.10	0.12	34.01	m	37.41
Aluminium alloy stair nosings with anti-slip inserts; 46mm wide with single line insert; fixing to masonry with screws										
22mm drop	10.93	5.00	11.48	0.65	-	11.94	0.12	23.54	m	25.89
32mm drop	12.80	5.00	13.44	0.65	-	11.94	0.12	25.50	m	28.05
Aluminium alloy stair nosings with anti-slip inserts; 66mm wide with single line insert; fixing to masonry with screws										
22mm drop	13.79	5.00	14.48	0.70	-	12.86	0.18	27.52	m	30.27
25mm drop	12.90	5.00	13.54	0.70	-	12.86	0.18	26.58	m	29.24
32mm drop	16.78	5.00	17.62	0.70	-	12.86	0.18	30.66	m	33.72
Aluminium alloy stair nosings with anti-slip inserts; 80mm wide with two line inserts; fixing to masonry with screws										
22mm drop	15.82	5.00	16.62	0.80	-	14.70	0.18	31.49	m	34.64
28mm drop	17.87	5.00	18.76	0.80	-	14.70	0.18	33.64	m	37.00
51mm drop	22.65	5.00	23.78	0.80	-	14.70	0.18	38.66	m	42.52
Aluminium alloy carpet nosings with anti-slip inserts; with single line insert; fixing to timber with screws										
65mm wide x 24mm drop	22.44	5.00	23.56	0.45	-	8.27	0.12	31.94	m	35.14
116mm wide overall with carpet gripper x 42mm drop	32.05	5.00	33.65	0.50	-	9.19	0.18	43.02	m	47.32
Aluminium alloy carpet nosings with anti-slip inserts; with two line inserts; fixing to timber with screws										
80mm wide x 31mm drop	25.65	5.00	26.93	0.55	-	10.10	0.12	37.15	m	40.87
Aluminium alloy carpet nosings with anti-slip inserts; with single line insert; fixing to masonry with screws										
65mm wide x 24mm drop	22.44	5.00	23.56	0.70	-	12.86	0.18	36.60	m	40.26
116mm wide overall with carpet gripper x 42mm drop	32.05	5.00	33.65	0.75	-	13.78	0.24	47.67	m	52.44
Aluminium alloy carpet nosings with anti-slip inserts; with two line inserts; fixing to masonry with screws										
80mm wide x 31mm drop	25.65	5.00	26.93	0.80	-	14.70	0.18	41.80	m	45.98
Metal mesh lathing; anti-crack strips										
Lathing; BB galvanized expanded metal lath, 9mm mesh X 0.500mm thick x 0.89kg/m²; butt joints fixing with nails										
100mm wide to walls; one edge to timber; one edge to brickwork or blockwork	0.44	5.00	0.46	0.10	-	1.84	0.05	2.35	m	2.58
100mm wide to walls; one edge to timber; one edge to concrete	0.44	5.00	0.46	0.12	-	2.20	0.06	2.72	m	3.00
100mm wide to walls; one edge to brickwork or blockwork; one edge to concrete	0.44	5.00	0.46	0.14	-	2.57	0.07	3.10	m	3.41
Lathing; BB galvanized expanded metal lath, 9mm mesh x 0.725mm thick x 1.11 kg/m²; butt joints; fixing with nails										
100mm wide to walls; one edge to timber; one edge to brickwork or blockwork	0.67	5.00	0.70	0.10	-	1.84	0.05	2.59	m	2.85
100mm wide to walls; one edge to timber; one edge to concrete	0.67	5.00	0.70	0.12	-	2.20	0.06	2.97	m	3.26
100mm wide to walls; one edge to brickwork or blockwork; one edge to concrete	0.67	5.00	0.70	0.14	-	2.57	0.07	3.35	m	3.68

Labour hourly rates: (except Specialists) Craft Operatives 18.37 Labourer 13.76 Rates are national average prices. Refer to REGIONAL VARIATIONS for indicative levels of overall pricing in regions	MATERIALS			LABOUR				RATES		
	Del to Site	Waste	Material Cost	Craft Optve	Lab	Labour Cost	Sunds	Nett Rate	Unit	Gross rate (10%)
	£	%	£	Hrs	Hrs	£	£	£		£
IN-SITU, TILED, BLOCK, MOSAIC, SHEET, APPLIED LIQUID OR PLANTED FINISHES (Cont'd)										
Metal mesh lathing; BB galvanized expanded metal lath, 9mm mesh x 0.500mm thick x 0.89 kg/m²; butt joints; fixing with galvanized staples for plastered coatings										
Work to walls										
width exceeding 600mm	4.38	5.00	4.60	0.15	0.08	3.86	0.08	8.53	m²	9.38
width 300mm	1.53	10.00	1.69	0.09	0.02	1.93	0.05	3.66	m	4.02
Work to isolated columns										
width exceeding 600mm	4.38	5.00	4.60	0.23	0.08	5.33	0.12	10.04	m²	11.04
width 300mm	1.53	10.00	1.69	0.14	0.02	2.85	0.07	4.60	m	5.06
Work to ceilings										
width exceeding 600mm	4.38	5.00	4.60	0.18	0.08	4.41	0.09	9.09	m²	10.00
width 300mm	1.53	10.00	1.69	0.11	0.02	2.30	0.05	4.04	m	4.44
Work to isolated beams										
width exceeding 600mm	4.38	5.00	4.60	0.23	0.08	5.33	0.12	10.04	m²	11.04
width 300mm	1.53	10.00	1.69	0.14	0.02	2.85	0.07	4.60	m	5.06
Metal mesh lathing; BB galvanized expanded metal lath 9mm mesh x 0.725mm thick x 1.11 kg/m²; butt joints; fixing with galvanized staples for plastered coatings										
Work to walls										
width exceeding 600mm	6.70	5.00	7.03	0.15	0.08	3.86	0.07	10.96	m²	12.05
width 300mm	2.34	10.00	2.58	0.09	0.02	1.93	0.07	4.58	m	5.03
Work to isolated columns										
width exceeding 600mm	6.70	5.00	7.03	0.23	0.08	5.33	0.07	12.43	m²	13.67
width 300mm	2.34	10.00	2.58	0.14	0.02	2.85	0.07	5.50	m	6.04
Work to ceilings										
width exceeding 600mm	6.70	5.00	7.03	0.18	0.08	4.41	0.07	11.51	m²	12.66
width 300mm	2.34	10.00	2.58	0.11	0.02	2.30	0.07	4.94	m	5.44
Work to isolated beams										
width exceeding 600mm	6.70	5.00	7.03	0.23	0.08	5.33	0.07	12.43	m²	13.67
width 300mm	2.34	10.00	2.58	0.14	0.02	2.85	0.07	5.50	m	6.04
Metal mesh lathing; galvanized expanded metal lath 9mm mesh x 1.61 kg/m²; butt joints; fixing with galvanized staples										
Work to walls										
width exceeding 600mm	8.70	5.00	9.14	0.15	0.08	3.86	0.07	13.06	m²	14.37
width 300mm	3.05	10.00	3.35	0.09	0.02	1.93	0.07	5.35	m	5.88
Work to isolated columns										
width exceeding 600mm	8.70	5.00	9.14	0.23	0.08	5.33	0.07	14.53	m²	15.99
width 300mm	3.05	10.00	3.35	0.14	0.02	2.85	0.07	6.27	m	6.89
Work to ceilings										
width exceeding 600mm	8.70	5.00	9.14	0.18	0.08	4.41	0.07	13.62	m²	14.98
width 300mm	3.05	10.00	3.35	0.11	0.02	2.30	0.07	5.72	m	6.29
Work to isolated beams										
width exceeding 600mm	8.70	5.00	9.14	0.23	0.08	5.33	0.07	14.53	m²	15.99
width 300mm	3.05	10.00	3.35	0.14	0.02	2.85	0.07	6.27	m	6.89
Metal mesh lathing; Expamet galvanized Rib-Lath, 0.300mm thick x 1.16 kg/m²; butt joints; fixing with galvanized staples										
Work to walls										
width exceeding 600mm	6.50	5.00	6.82	0.15	0.08	3.86	0.07	10.75	m²	11.83
width 300mm	2.52	10.00	2.77	0.09	0.02	1.93	0.07	4.77	m	5.25
Work to isolated columns										
width exceeding 600mm	6.50	5.00	6.82	0.23	0.08	5.33	0.07	12.22	m²	13.44
width 300mm	2.52	10.00	2.77	0.14	0.02	2.85	0.07	5.69	m	6.26
Work to ceilings										
width exceeding 600mm	6.50	5.00	6.82	0.18	0.08	4.41	0.07	11.30	m²	12.43
width 300mm	2.52	10.00	2.77	0.11	0.02	2.30	0.07	5.14	m	5.65
Work to isolated beams										
width exceeding 600mm	6.50	5.00	6.82	0.23	0.08	5.33	0.07	12.22	m²	13.44
width 300mm	2.52	10.00	2.77	0.14	0.02	2.85	0.07	5.69	m	6.26
Metal mesh lathing; Expamet galvanized Rib-Lath, 0.500mm thick x 1.86 kg/m²; butt joints; fixing with galvanized staples										
Work to walls										
width exceeding 600mm	8.20	5.00	8.61	0.15	0.08	3.86	0.07	12.54	m²	13.79
width 300mm	2.87	10.00	3.16	0.09	0.02	1.93	0.07	5.16	m	5.67
Work to isolated columns										
width exceeding 600mm	8.20	5.00	8.61	0.23	0.08	5.33	0.07	14.01	m²	15.41
width 300mm	2.87	10.00	3.16	0.14	0.02	2.85	0.07	6.07	m	6.68
Work to ceilings										
width exceeding 600mm	8.20	5.00	8.61	0.18	0.08	4.41	0.07	13.09	m²	14.40
width 300mm	2.87	10.00	3.16	0.11	0.02	2.30	0.07	5.52	m	6.08
Work to isolated beams										
width exceeding 600mm	8.20	5.00	8.61	0.23	0.08	5.33	0.07	14.01	m²	15.41
width 300mm	2.87	10.00	3.16	0.14	0.02	2.85	0.07	6.07	m	6.68

Labour hourly rates: (except Specialists) Craft Operatives 18.37 Labourer 13.76 Rates are national average prices. Refer to REGIONAL VARIATIONS for indicative levels of overall pricing in regions	MATERIALS			LABOUR				RATES		
	Del to Site £	Waste %	Material Cost £	Craft Optve Hrs	Lab Hrs	Labour Cost £	Sunds £	Nett Rate £	Unit	Gross rate (10%) £
IN-SITU, TILED, BLOCK, MOSAIC, SHEET, APPLIED LIQUID OR PLANTED FINISHES (Cont'd)										
Metal mesh lathing; Expamet stainless steel Rib-Lath, 1.48 kg/m²; butt joints; fixing with stainless steel staples										
Work to walls										
width exceeding 600mm	17.64	5.00	18.52	0.15	0.08	3.86	0.07	22.45	m²	24.69
width 300mm	6.41	10.00	7.05	0.09	0.02	1.93	0.07	9.05	m	9.95
Work to isolated columns										
width exceeding 600mm	17.64	5.00	18.52	0.23	0.08	5.33	0.07	23.92	m²	26.31
width 300mm	6.41	10.00	7.05	0.14	0.02	2.85	0.07	9.97	m	10.96
Work to ceilings										
width exceeding 600mm	17.64	5.00	18.52	0.18	0.08	4.41	0.07	23.00	m²	25.30
width 300mm	6.41	10.00	7.05	0.11	0.02	2.30	0.07	9.41	m	10.36
Work to isolated beams										
width exceeding 600mm	17.64	5.00	18.52	0.23	0.08	5.33	0.07	23.92	m²	26.31
width 300mm	6.41	10.00	7.05	0.14	0.02	2.85	0.07	9.97	m	10.96
Metal mesh lathing; stainless steel expanded metal lath 1.11kg/m² for render; butt joints; fixing with galvanized staples										
Work to walls										
width exceeding 600mm	17.64	5.00	18.52	0.15	0.08	3.86	0.07	22.45	m²	24.69
width 300mm	6.17	10.00	6.79	0.09	0.02	1.93	0.07	8.79	m	9.67
Work to isolated columns										
width exceeding 600mm	17.64	5.00	18.52	0.23	0.08	5.33	0.07	23.92	m²	26.31
width 300mm	6.17	10.00	6.79	0.14	0.02	2.85	0.07	9.71	m	10.68
Work to ceilings										
width exceeding 600mm	17.64	5.00	18.52	0.18	0.08	4.41	0.07	23.00	m²	25.30
width 300mm	6.17	10.00	6.79	0.11	0.02	2.30	0.07	9.16	m	10.07
Work to isolated beams										
width exceeding 600mm	17.64	5.00	18.52	0.23	0.08	5.33	0.07	23.92	m²	26.31
width 300mm	6.17	10.00	6.79	0.14	0.02	2.85	0.07	9.71	m	10.68
Metal mesh lathing; galvanized Red-rib lath, 0.500mm thick x 1.91kg/m², butt joints; fixing with galvanized staples										
Work to walls										
width exceeding 600mm	9.57	5.00	10.05	0.15	0.08	3.86	0.07	13.97	m²	15.37
width 300mm	3.35	10.00	3.68	0.09	0.02	1.93	0.07	5.68	m	6.25
Work to isolated columns										
width exceeding 600mm	9.57	5.00	10.05	0.23	0.08	5.33	0.07	15.44	m²	16.99
width 300mm	3.35	10.00	3.68	0.14	0.02	2.85	0.07	6.60	m	7.26
Work to ceilings										
width exceeding 600mm	9.57	5.00	10.05	0.18	0.08	4.41	0.07	14.53	m²	15.98
width 300mm	3.35	10.00	3.68	0.11	0.02	2.30	0.07	6.05	m	6.66
Work to isolated beams										
width exceeding 600mm	9.57	5.00	10.05	0.23	0.08	5.33	0.07	15.44	m²	16.99
width 300mm	3.35	10.00	3.68	0.14	0.02	2.85	0.07	6.60	m	7.26
Metal mesh lathing; extra cost of fixing lathing to brickwork, blockwork or concrete with cartridge fired nails in lieu of to timber with staples										
Work to walls										
width exceeding 600mm	-	-	-	0.10	-	1.84	0.15	1.99	m²	2.19
width 300mm	-	-	-	0.06	-	1.10	0.09	1.19	m	1.31
Metal mesh lathing; extra cost of fixing lathing to steel with tying wire in lieu of to timber with staples										
Work to walls										
width exceeding 600mm	-	-	-	0.15	-	2.76	0.22	2.98	m²	3.28
width 300mm	-	-	-	0.09	-	1.65	0.14	1.79	m	1.97
Work to ceilings										
width exceeding 600mm	-	-	-	0.15	-	2.76	0.22	2.98	m²	3.28
width 300mm	-	-	-	0.09	-	1.65	0.14	1.79	m	1.97
Work to isolated beams										
width exceeding 600mm	-	-	-	0.15	-	2.76	0.22	2.98	m²	3.28
width 300mm	-	-	-	0.09	-	1.65	0.14	1.79	m	1.97
Work to isolated columns										
width exceeding 600mm	-	-	-	0.15	-	2.76	0.22	2.98	m²	3.28
width 300mm	-	-	-	0.09	-	1.65	0.14	1.79	m	1.97
Metal mesh lathing; Expamet galvanized arch formers and fix to brickwork or blockwork with galvanized nails, 30mm long at 400mm centres										
Arch corners										
372mm radius	12.65	5.00	13.28	0.33	0.17	8.40	1.38	23.06	nr	25.37
452mm radius	15.15	5.00	15.91	0.33	0.17	8.40	1.69	25.99	nr	28.59
602mm radius	21.40	5.00	22.47	0.33	0.17	8.40	2.10	32.97	nr	36.27
752mm radius	32.00	5.00	33.60	0.33	0.17	8.40	2.76	44.76	nr	49.24
Semi-circular arches										
372mm radius	25.30	5.00	26.56	0.67	0.34	16.99	2.75	46.30	nr	50.93
397mm radius	26.50	5.00	27.82	0.67	0.34	16.99	2.83	47.64	nr	52.40
412mm radius	27.00	5.00	28.35	0.67	0.34	16.99	2.92	48.26	nr	53.08

Labour hourly rates: (except Specialists) Craft Operatives 18.37 Labourer 13.76 Rates are national average prices. Refer to REGIONAL VARIATIONS for indicative levels of overall pricing in regions	MATERIALS			LABOUR				RATES		
	Del to Site	Waste	Material Cost	Craft Optve	Lab	Labour Cost	Sunds	Nett Rate	Unit	Gross rate (10%)
	£	%	£	Hrs	Hrs	£	£	£		£
IN-SITU, TILED, BLOCK, MOSAIC, SHEET, APPLIED LIQUID OR PLANTED FINISHES (Cont'd)										
Metal mesh lathing; Expamet galvanized arch formers and fix to brickwork or blockwork with galvanized nails, 30mm long at 400mm centres (Cont'd)										
Semi-circular arches (Cont'd)										
452mm radius	30.30	5.00	31.81	0.67	0.34	16.99	3.38	52.18	nr	57.40
602mm radius	42.80	5.00	44.94	0.67	0.34	16.99	4.26	66.18	nr	72.80
752mm radius	64.00	5.00	67.20	0.67	0.34	16.99	5.86	90.05	nr	99.05
Elliptical arches										
1220mm wide x 340mm rise	31.50	5.00	33.07	0.67	0.34	16.99	5.30	55.36	nr	60.90
1370mm wide x 360mm rise	39.50	5.00	41.47	0.67	0.34	16.99	5.40	63.86	nr	70.25
1520mm wide x 380mm rise	40.00	5.00	42.00	0.67	0.34	16.99	5.95	64.94	nr	71.44
1830mm wide x 410mm rise	44.50	5.00	46.72	1.00	0.50	25.25	6.33	78.31	nr	86.14
2130mm wide x 430mm rise	49.50	5.00	51.97	1.00	0.50	25.25	6.82	84.04	nr	92.44
2440mm wide x 440mm rise	57.80	5.00	60.69	1.00	0.50	25.25	6.95	92.89	nr	102.18
3050mm wide x 520mm rise	66.10	5.00	69.40	1.00	0.50	25.25	7.24	101.89	nr	112.08
Spandrel arches										
760mm wide x 180mm radius x 220mm rise	30.25	5.00	31.76	0.67	0.34	16.99	3.58	52.33	nr	57.56
910mm wide x 180mm radius x 240mm rise	32.60	5.00	34.23	0.67	0.34	16.99	3.79	55.01	nr	60.51
1220mm wide x 230mm radius x 290mm rise	40.75	5.00	42.79	0.67	0.34	16.99	4.88	64.65	nr	71.12
1520mm wide x 230mm radius x 330mm rise	41.55	5.00	43.63	0.67	0.34	16.99	5.43	66.04	nr	72.64
1830mm wide x 230mm radius x 360mm rise	50.50	5.00	53.02	0.67	0.34	16.99	5.49	75.50	nr	83.05
2130mm wide x 230mm radius x 370mm rise	56.50	5.00	59.32	0.67	0.34	16.99	6.50	82.81	nr	91.09
2440mm wide x 230mm radius x 390mm rise	67.30	5.00	70.67	1.33	0.67	33.65	6.90	111.22	nr	122.34
3050mm wide x 230mm radius x 440mm rise	72.00	5.00	75.60	1.33	0.67	33.65	7.29	116.54	nr	128.19
Bulls-eyes										
222mm radius	21.28	5.00	22.35	0.17	0.09	4.36	3.44	30.15	nr	33.17
Soffit strips										
155mm wide	3.01	5.00	3.17	0.09	0.05	2.34	0.35	5.86	m	6.44
Make-up pieces										
600mm long	10.33	5.00	10.85	0.17	0.09	4.36	0.84	16.05	nr	17.65
Metal mesh lathing; Expamet galvanized circular window formers and fix to brickwork or blockwork with galvanized nails, 30mm long at 400mm centres										
Circular windows										
594mm diameter	19.90	5.00	20.89	1.33	0.67	33.65	3.00	57.55	nr	63.30
Accessories; Plaster ventilators										
Fibrous plaster ventilator; fixing in plastered wall										
Plain										
229 x 79mm	2.73	5.00	2.87	0.15	-	2.76	-	5.62	nr	6.18
229 x 152mm	3.33	5.00	3.50	0.18	-	3.31	-	6.80	nr	7.48
229 x 229mm	4.00	5.00	4.20	0.20	-	3.67	-	7.87	nr	8.66
Flyproof										
229 x 79mm	2.73	5.00	2.87	0.20	-	3.67	-	6.54	nr	7.19
229 x 152mm	3.33	5.00	3.50	0.23	-	4.23	-	7.72	nr	8.49
229 x 229mm	4.00	5.00	4.20	0.25	-	4.59	-	8.79	nr	9.67

Labour hourly rates: (except Specialists) Craft Operatives 18.37 Labourer 13.76 Rates are national average prices. Refer to REGIONAL VARIATIONS for indicative levels of overall pricing in regions	MATERIALS			LABOUR				RATES		
	Del to Site	Waste	Material Cost	Craft Optve	Lab	Labour Cost	Sunds	Nett Rate		Gross rate (10%)
	£	%	£	Hrs	Hrs	£	£	£	Unit	£
PAINTING & CLEAR FINISHINGS; EMULSION PAINTING										
Mist coat, one full coat emulsion paint										
Concrete general surfaces girth exceeding 300mm	0.21	10.00	0.23	0.15	-	2.76	0.08	3.06	m²	3.37
Concrete general surfaces 3.50 - 5.00m above floor girth exceeding 300mm	0.21	10.00	0.23	0.17	-	3.12	0.09	3.44	m²	3.78
Plaster general surfaces girth exceeding 300mm isolated surfaces, girth not exceeding 300mm	0.17 0.50	10.00 10.00	0.18 0.55	0.13 0.07	- -	2.39 1.29	0.06 0.04	2.64 1.87	m² m	2.90 2.06
Plaster general surfaces 3.50 - 5.00m above floor girth exceeding 300mm	0.17	10.00	0.18	0.15	-	2.76	0.08	3.01	m²	3.31
Plasterboard general surfaces girth exceeding 300mm	0.17	10.00	0.18	0.13	-	2.39	0.06	2.64	m²	2.90
Plasterboard general surfaces 3.50 - 5.00m above floor girth exceeding 300mm	0.17	10.00	0.18	0.15	-	2.76	0.08	3.01	m²	3.31
Brickwork general surfaces girth exceeding 300mm	0.21	10.00	0.23	0.17	-	3.12	0.09	3.44	m²	3.78
Paper covered general surfaces girth exceeding 300mm	0.17	10.00	0.18	0.14	-	2.57	0.07	2.82	m²	3.11
Paper covered general surfaces 3.50 - 5.00m above floor girth exceeding 300mm	0.17	10.00	0.18	0.16	-	2.94	0.08	3.20	m²	3.52
Mist coat, two full coats emulsion paint										
Concrete general surfaces girth exceeding 300mm	0.35	10.00	0.38	0.21	-	3.86	0.10	4.35	m²	4.78
Concrete general surfaces 3.50 - 5.00m above floor girth exceeding 300mm	0.35	10.00	0.38	0.24	-	4.41	0.12	4.91	m²	5.40
Plaster general surfaces girth exceeding 300mm isolated surfaces, girth not exceeding 300mm	0.28 0.08	10.00 10.00	0.31 0.09	0.19 0.10	- -	3.49 1.84	0.09 0.05	3.90 1.98	m² m	4.29 2.18
Plaster general surfaces 3.50 - 5.00m above floor girth exceeding 300mm	0.28	10.00	0.31	0.22	-	4.04	0.11	4.46	m²	4.91
Plasterboard general surfaces girth exceeding 300mm	0.28	10.00	0.31	0.19	-	3.49	0.09	3.90	m²	4.29
Plasterboard general surfaces 3.50 - 5.00m above floor girth exceeding 300mm	0.28	10.00	0.31	0.22	-	4.04	0.11	4.46	m²	4.91
Brickwork general surfaces girth exceeding 300mm	0.35	10.00	0.38	0.24	-	4.41	0.12	4.91	m²	5.40
Paper covered general surfaces girth exceeding 300mm	0.28	10.00	0.31	0.20	-	3.67	0.10	4.08	m²	4.49
Paper covered general surfaces 3.50 - 5.00m above floor girth exceeding 300mm	0.28	10.00	0.31	0.30	-	5.51	0.15	5.97	m²	6.57
PAINTING & CLEAR FINISHINGS; CEMENT PAINTING										
One coat waterproof cement paint; external work										
Cement rendered general surfaces girth exceeding 300mm isolated surfaces, girth not exceeding 300mm	0.25 0.25	10.00 10.00	0.27 0.27	0.11 0.04	- -	2.02 0.73	0.05 0.02	2.35 1.03	m² m	2.59 1.13
Concrete general surfaces girth exceeding 300mm	0.25	10.00	0.27	0.11	-	2.02	0.05	2.35	m²	2.59
Brickwork general surfaces girth exceeding 300mm	0.29	10.00	0.32	0.13	-	2.39	0.06	2.77	m²	3.05
Rough cast general surfaces girth exceeding 300mm	0.33	10.00	0.36	0.17	-	3.12	0.09	3.57	m²	3.93
Two coats waterproof cement paint; external work										
Cement rendered general surfaces girth exceeding 300mm isolated surfaces, girth not exceeding 300mm	0.66 0.20	10.00 10.00	0.72 0.22	0.22 0.08	- -	4.04 1.47	0.11 0.04	4.88 1.73	m² m	5.36 1.90
Concrete general surfaces girth exceeding 300mm	0.50	10.00	0.55	0.22	-	4.04	0.11	4.70	m²	5.17
Brickwork general surfaces girth exceeding 300mm	0.58	10.00	0.64	0.24	-	4.41	0.12	5.17	m²	5.68

DECORATION

DECORATION

Labour hourly rates: (except Specialists) Craft Operatives 18.37 Labourer 13.76 Rates are national average prices. Refer to REGIONAL VARIATIONS for indicative levels of overall pricing in regions	MATERIALS			LABOUR				RATES		
	Del to Site	Waste	Material Cost	Craft Optve	Lab	Labour Cost	Sunds	Nett Rate	Unit	Gross rate (10%)
	£	%	£	Hrs	Hrs	£	£	£		£
PAINTING & CLEAR FINISHINGS; CEMENT PAINTING (Cont'd)										
Two coats waterproof cement paint; external work (Cont'd)										
Rough cast general surfaces girth exceeding 300mm	0.66	10.00	0.72	0.33	-	6.06	0.17	6.95	m²	7.65
One coat sealer, one coat waterproof cement paint; external work										
Cement rendered general surfaces girth exceeding 300mm .... isolated surfaces, girth not exceeding 300mm	1.90 0.55	10.00 10.00	2.09 0.60	0.18 0.06	- -	3.31 1.10	0.09 0.03	5.49 1.74	m² m	6.04 1.91
Concrete general surfaces girth exceeding 300mm	1.82	10.00	2.01	0.18	-	3.31	0.09	5.40	m²	5.94
Brickwork general surfaces girth exceeding 300mm	1.86	10.00	2.05	0.21	-	3.86	0.10	6.01	m²	6.61
Rough cast general surfaces girth exceeding 300mm	1.90	10.00	2.09	0.27	-	4.96	0.14	7.19	m²	7.91
Two coats sealer, one coat waterproof cement paint; external work										
Cement rendered general surfaces girth exceeding 300mm .... isolated surfaces, girth not exceeding 300mm	2.99 0.89	10.00 10.00	3.29 0.98	0.34 0.11	- -	6.25 2.02	0.17 0.05	9.70 3.05	m² m	10.67 3.36
Concrete general surfaces girth exceeding 300mm	2.57	10.00	2.83	0.34	-	6.25	0.17	9.24	m²	10.17
Brickwork general surfaces girth exceeding 300mm	2.95	10.00	3.24	0.39	-	7.16	0.19	10.60	m²	11.66
Rough cast general surfaces girth exceeding 300mm	2.99	10.00	3.29	0.51	-	9.37	0.25	12.91	m²	14.20
One coat textured masonry paint; external work										
Cement rendered general surfaces girth exceeding 300mm .... isolated surfaces, girth not exceeding 300mm	0.43 0.17	10.00 10.00	0.48 0.19	0.11 0.04	- -	2.02 0.73	0.05 0.02	2.55 0.95	m² m	2.81 1.04
Concrete general surfaces girth exceeding 300mm	0.61	10.00	0.67	0.11	-	2.02	0.05	2.74	m²	3.02
Brickwork general surfaces girth exceeding 300mm	1.08	10.00	1.19	0.13	-	2.39	0.06	3.64	m²	4.01
Rough cast general surfaces girth exceeding 300mm	1.43	10.00	1.57	0.17	-	3.12	0.09	4.78	m²	5.26
Two coats textured masonry paint; external work										
Cement rendered general surfaces girth exceeding 300mm .... isolated surfaces, girth not exceeding 300mm	1.21 0.35	10.00 10.00	1.33 0.38	0.22 0.08	- -	4.04 1.47	0.11 0.04	5.48 1.89	m² m	6.03 2.08
Concrete general surfaces girth exceeding 300mm	1.21	10.00	1.33	0.22	-	4.04	0.11	5.48	m²	6.03
Brickwork general surfaces girth exceeding 300mm	2.16	10.00	2.38	0.24	-	4.41	0.12	6.91	m²	7.60
Rough cast general surfaces girth exceeding 300mm	2.85	10.00	3.14	0.33	-	6.06	0.17	9.37	m²	10.30
One coat stabilizing solution, one coat textured masonry paint; external work										
Cement rendered general surfaces girth exceeding 300mm .... isolated surfaces, girth not exceeding 300mm	0.91 0.27	10.00 10.00	1.01 0.29	0.18 0.06	- -	3.31 1.10	0.09 0.03	4.40 1.42	m² m	4.84 1.57
Concrete general surfaces girth exceeding 300mm	0.91	10.00	1.01	0.18	-	3.31	0.09	4.40	m²	4.84
Brickwork general surfaces girth exceeding 300mm	1.39	10.00	1.53	0.21	-	3.86	0.10	5.49	m²	6.04
Rough cast general surfaces girth exceeding 300mm	1.74	10.00	1.91	0.27	-	4.96	0.14	7.00	m²	7.70
One coat stabilizing solution, two coats textured masonry paint; external work										
Cement rendered general surfaces girth exceeding 300mm .... isolated surfaces, girth not exceeding 300mm	1.52 0.44	10.00 10.00	1.67 0.48	0.34 0.11	- -	6.25 2.02	0.17 0.05	8.09 2.56	m² m	8.90 2.81
Concrete general surfaces girth exceeding 300mm	1.52	10.00	1.67	0.34	-	6.25	0.17	8.09	m²	8.90
Brickwork general surfaces girth exceeding 300mm	2.47	10.00	2.72	0.39	-	7.16	0.19	10.08	m²	11.08
Rough cast general surfaces girth exceeding 300mm	3.16	10.00	3.48	0.51	-	9.37	0.25	13.10	m²	14.41

Labour hourly rates: (except Specialists) Craft Operatives 18.37 Labourer 13.76 Rates are national average prices. Refer to REGIONAL VARIATIONS for indicative levels of overall pricing in regions	MATERIALS			LABOUR				RATES		
	Del to Site £	Waste %	Material Cost £	Craft Optve Hrs	Lab Hrs	Labour Cost £	Sunds £	Nett Rate £	Unit	Gross rate (10%) £

PAINTING & CLEAR FINISHINGS; PRESERVATIVE TREATMENT

One coat creosote BS 144; external work (sawn timber)

Wood general surfaces

girth exceeding 300mm	0.43	10.00	0.48	0.10	-	1.84	0.05	2.37	m²	2.60
isolated surfaces, girth not exceeding 300mm	0.13	10.00	0.14	0.04	-	0.73	0.02	0.90	m	0.99

Two coats creosote BS 144; external work (sawn timber)

Wood general surfaces

girth exceeding 300mm	0.88	10.00	0.97	0.20	-	3.67	0.10	4.74	m²	5.22
isolated surfaces, girth not exceeding 300mm	0.26	10.00	0.29	0.07	-	1.29	0.04	1.61	m	1.77

One coat wood preservative, internal work (sawn timber)

Wood general surfaces

girth exceeding 300mm	0.31	10.00	0.34	0.11	-	2.02	0.05	2.41	m²	2.65
isolated surfaces, girth not exceeding 300mm	0.10	10.00	0.11	0.04	-	0.73	0.02	0.86	m	0.95

One coat wood preservative, internal work (wrought timber)

Wood general surfaces

girth exceeding 300mm	0.26	10.00	0.28	0.09	-	1.65	0.05	1.98	m²	2.18
isolated surfaces, girth not exceeding 300mm	0.08	10.00	0.08	0.03	-	0.55	0.01	0.65	m	0.72

One coat wood preservative, external work (sawn timber)

Wood general surfaces

girth exceeding 300mm	0.31	10.00	0.34	0.10	-	1.84	0.05	2.22	m²	2.45
isolated surfaces, girth not exceeding 300mm	0.10	10.00	0.11	0.04	-	0.73	0.02	0.86	m	0.95

One coat wood preservative, external work (wrought timber)

Wood general surfaces

girth exceeding 300mm	0.26	10.00	0.28	0.08	-	1.47	0.04	1.79	m²	1.97
isolated surfaces, girth not exceeding 300mm	0.08	10.00	0.08	0.02	-	0.37	0.01	0.46	m	0.51

Two coats wood preservative, internal work (sawn timber)

Wood general surfaces

girth exceeding 300mm	0.61	10.00	0.67	0.22	-	4.04	0.11	4.83	m²	5.31
isolated surfaces, girth not exceeding 300mm	0.19	10.00	0.21	0.08	-	1.47	0.04	1.72	m	1.89

Two coats wood preservative, internal work (wrought timber)

Wood general surfaces

girth exceeding 300mm	0.51	10.00	0.56	0.18	-	3.31	0.09	3.96	m²	4.35
isolated surfaces, girth not exceeding 300mm	0.15	10.00	0.17	0.05	-	0.92	0.03	1.11	m	1.22

Two coats wood preservative, external work (sawn timber)

Wood general surfaces

girth exceeding 300mm	0.61	10.00	0.67	0.20	-	3.67	0.10	4.45	m²	4.89
isolated surfaces, girth not exceeding 300mm	0.19	10.00	0.21	0.07	-	1.29	0.04	1.53	m	1.68

Two coats wood preservative, external work (wrought timber)

Wood general surfaces

girth exceeding 300mm	0.51	10.00	0.56	0.16	-	2.94	0.08	3.58	m²	3.94
isolated surfaces, girth not exceeding 300mm	0.15	10.00	0.17	0.05	-	0.92	0.03	1.11	m	1.22

Two coats Sadolins High Performance Extra Durable Varnish

Wood general surfaces

girth exceeding 300mm	3.22	10.00	3.54	0.30	-	5.51	0.15	9.20	m²	10.12
isolated surfaces, girth not exceeding 300mm	0.96	10.00	1.06	0.11	-	2.02	0.05	3.14	m	3.45

Wood glazed doors

girth exceeding 300mm; panes, area not exceeding 0.10m²	1.84	10.00	2.02	0.98	-	18.00	0.49	20.51	m²	22.57
girth exceeding 300mm; panes, area 0.10 - 0.50m²	1.30	10.00	1.43	0.71	-	13.04	0.35	14.83	m²	16.31
girth exceeding 300mm; panes, area 0.50 - 1.00m²	1.15	10.00	1.26	0.60	-	11.02	0.30	12.59	m²	13.84
girth exceeding 300mm; panes, area exceeding 1.00m²	1.15	10.00	1.26	0.52	-	9.55	0.26	11.08	m²	12.18

Wood partially glazed doors

girth exceeding 300mm; panes, area 0.50 - 1.00m²	1.76	10.00	1.94	0.60	-	11.02	0.30	13.26	m²	14.59
girth exceeding 300mm; panes, area exceeding 1.00m²	1.76	10.00	1.94	0.58	-	10.65	0.29	12.88	m²	14.17

Wood windows and screens

girth exceeding 300mm; panes, area not exceeding 0.10m²	1.84	10.00	2.02	1.09	-	20.02	0.55	22.59	m²	24.85
girth exceeding 300mm; panes, area 0.10 - 0.50m²	1.30	10.00	1.43	0.78	-	14.33	0.39	16.15	m²	17.77
girth exceeding 300mm; panes, area 0.50 - 1.00m²	1.15	10.00	1.26	0.66	-	12.12	0.34	13.73	m²	15.10
girth exceeding 300mm; panes, area exceeding 1.00m²	1.15	10.00	1.26	0.58	-	10.65	0.29	12.21	m²	13.43

One coat Sadolins Classic; two coats Sadolins Extra Woodstain; external work

Wood general surfaces

girth exceeding 300mm	2.70	10.00	2.98	0.55	-	10.10	0.28	13.35	m²	14.69
isolated surfaces, girth not exceeding 300mm	0.80	10.00	0.88	0.19	-	3.49	0.09	4.46	m	4.91

Wood glazed doors

girth exceeding 300mm; panes, area not exceeding 0.10m²	1.48	10.00	1.63	0.86	-	15.80	0.43	17.86	m²	19.64
girth exceeding 300mm; panes, area 0.10 - 0.50m²	1.06	10.00	1.17	0.62	-	11.39	0.31	12.87	m²	14.16
girth exceeding 300mm; panes, area 0.50 - 1.00m²	0.96	10.00	1.05	0.52	-	9.55	0.26	10.87	m²	11.95
girth exceeding 300mm; panes, area exceeding 1.00m²	0.96	10.00	1.05	0.46	-	8.45	0.23	9.73	m²	10.71

Wood partially glazed doors

girth exceeding 300mm; panes, area 0.50 - 1.00m²	1.48	10.00	1.63	0.52	-	9.55	0.26	11.44	m²	12.59
girth exceeding 300mm; panes, area exceeding 1.00m²	1.48	10.00	1.63	0.50	-	9.19	0.25	11.06	m²	12.17

Labour hourly rates: (except Specialists) Craft Operatives 18.37 Labourer 13.76 Rates are national average prices. Refer to REGIONAL VARIATIONS for indicative levels of overall pricing in regions	MATERIALS			LABOUR				RATES		
	Del to Site	Waste	Material Cost	Craft Optve	Lab	Labour Cost	Sunds	Nett Rate	Unit	Gross rate (10%)
	£	%	£	Hrs	Hrs	£	£	£		£
PAINTING & CLEAR FINISHINGS; PRESERVATIVE TREATMENT (Cont'd)										
One coat Sadolins Classic; two coats Sadolins Extra Woodstain; external work (Cont'd)										
Wood windows and screens										
girth exceeding 300mm; panes, area not exceeding 0.10m²	1.48	10.00	1.63	0.98	-	18.00	0.49	20.12	m²	22.13
girth exceeding 300mm; panes, area 0.10 - 0.50m²	1.06	10.00	1.17	0.72	-	13.23	0.36	14.76	m²	16.23
girth exceeding 300mm; panes, area 0.50 - 1.00m²	0.96	10.00	1.05	0.61	-	11.21	0.31	12.56	m²	13.82
girth exceeding 300mm; panes, area exceeding 1.00m²	0.96	10.00	1.05	0.54	-	9.92	0.27	11.24	m²	12.37
Wood railings fences and gates; open type										
girth exceeding 300mm	2.07	10.00	2.28	0.55	-	10.10	0.28	12.65	m²	13.92
isolated surfaces, girth not exceeding 300mm	0.80	10.00	0.88	0.19	-	3.49	0.09	4.46	m	4.91
Wood railings fences and gates; close type										
girth exceeding 300mm	2.70	10.00	2.98	0.55	-	10.10	0.28	13.35	m²	14.69
PAINTING & CLEAR FINISHINGS; OIL PAINTING WALLS AND CEILINGS										
One coat primer, one undercoat, one coat full gloss finish										
Concrete general surfaces										
girth exceeding 300mm	1.35	10.00	1.48	0.33	-	6.06	0.17	7.71	m²	8.48
Concrete general surfaces 3.50 - 5.00m above floor										
girth exceeding 300mm	1.35	10.00	1.48	0.36	-	6.61	0.18	8.28	m²	9.11
Brickwork general surfaces										
girth exceeding 300mm	1.57	10.00	1.73	0.37	-	6.80	0.19	8.71	m²	9.58
Plasterboard general surfaces										
girth exceeding 300mm	1.24	10.00	1.36	0.30	-	5.51	0.15	7.02	m²	7.72
Plasterboard general surfaces 3.50 - 5.00m above floor										
girth exceeding 300mm	1.24	10.00	1.36	0.33	-	6.06	0.17	7.59	m²	8.35
Plaster general surfaces										
girth exceeding 300mm	1.24	10.00	1.36	0.30	-	5.51	0.15	7.02	m²	7.72
isolated surfaces, girth not exceeding 300mm	2.28	10.00	2.51	0.13	-	2.39	0.06	4.96	m	5.46
Plaster general surfaces 3.50 - 5.00m above floor										
girth exceeding 300mm	1.24	10.00	1.36	0.33	-	6.06	0.17	7.59	m²	8.35
One coat primer, two undercoats, one coat full gloss finish										
Concrete general surfaces										
girth exceeding 300mm	1.63	10.00	1.79	0.44	-	8.08	0.22	10.09	m²	11.10
Concrete general surfaces 3.50 - 5.00m above floor										
girth exceeding 300mm	1.63	10.00	1.79	0.48	-	8.82	0.24	10.85	m²	11.93
Brickwork general surfaces										
girth exceeding 300mm	1.97	10.00	2.17	0.50	-	9.19	0.25	11.61	m²	12.77
Plasterboard general surfaces										
girth exceeding 300mm	1.68	10.00	1.85	0.40	-	7.35	0.20	9.40	m²	10.34
Plasterboard general surfaces 3.50 - 5.00m above floor										
girth exceeding 300mm	1.51	10.00	1.67	0.44	-	8.08	0.22	9.97	m²	10.96
Plaster general surfaces										
girth exceeding 300mm	1.51	10.00	1.67	0.40	-	7.35	0.20	9.21	m²	10.13
isolated surfaces, girth not exceeding 300mm	0.47	10.00	0.51	0.18	-	3.31	0.09	3.91	m	4.30
Plaster general surfaces 3.50 - 5.00m above floor										
girth exceeding 300mm	1.51	10.00	1.67	0.44	-	8.08	0.22	9.97	m²	10.96
One coat primer, one undercoat, one coat eggshell finish										
Concrete general surfaces										
girth exceeding 300mm	1.38	10.00	1.52	0.33	-	6.06	0.17	7.75	m²	8.52
Concrete general surfaces 3.50 - 5.00m above floor										
girth exceeding 300mm	1.38	10.00	1.52	0.36	-	6.61	0.18	8.32	m²	9.15
Brickwork general surfaces										
girth exceeding 300mm	1.72	10.00	1.89	0.37	-	6.80	0.19	8.87	m²	9.76
Plasterboard general surfaces										
girth exceeding 300mm	1.27	10.00	1.40	0.30	-	5.51	0.15	7.06	m²	7.76
Plasterboard general surfaces 3.50 - 5.00m above floor										
girth exceeding 300mm	1.27	10.00	1.40	0.33	-	6.06	0.17	7.62	m²	8.38
Plaster general surfaces										
girth exceeding 300mm	1.27	10.00	1.40	0.30	-	5.51	0.15	7.06	m²	7.76
isolated surfaces, girth not exceeding 300mm	2.29	10.00	2.52	0.13	-	2.39	0.06	4.97	m	5.47
Plaster general surfaces 3.50 - 5.00m above floor										
girth exceeding 300mm	1.27	10.00	1.40	0.33	-	6.06	0.17	7.62	m²	8.38
One coat primer, two undercoats, one coat eggshell finish										
Concrete general surfaces										
girth exceeding 300mm	1.66	10.00	1.83	0.44	-	8.08	0.22	10.13	m²	11.14
Concrete general surfaces 3.50 - 5.00m above floor										
girth exceeding 300mm	1.66	10.00	1.83	0.48	-	8.82	0.24	10.89	m²	11.97
Brickwork general surfaces										
girth exceeding 300mm	2.02	10.00	2.22	0.50	-	9.19	0.25	11.65	m²	12.82

Labour hourly rates: (except Specialists) Craft Operatives 18.37 Labourer 13.76 Rates are national average prices. Refer to REGIONAL VARIATIONS for indicative levels of overall pricing in regions	MATERIALS			LABOUR				RATES		
	Del to Site £	Waste %	Material Cost £	Craft Optve Hrs	Lab Hrs	Labour Cost £	Sunds £	Nett Rate £	Unit	Gross rate (10%) £
PAINTING & CLEAR FINISHINGS; OIL PAINTING WALLS AND CEILINGS (Cont'd)										
One coat primer, two undercoats, one coat eggshell finish (Cont'd)										
Plasterboard general surfaces girth exceeding 300mm	1.55	10.00	1.70	0.40	-	7.35	0.20	9.25	m²	10.17
Plasterboard general surfaces 3.50 - 5.00m above floor girth exceeding 300mm	1.55	10.00	1.70	0.44	-	8.08	0.22	10.00	m²	11.00
Plaster general surfaces girth exceeding 300mm .... isolated surfaces, girth not exceeding 300mm	1.55 0.48	10.00 10.00	1.70 0.53	0.40 0.18	- -	7.35 3.31	0.20 0.09	9.25 3.92	m² m	10.17 4.31
Plaster general surfaces 3.50 - 5.00m above floor girth exceeding 300mm	1.55	10.00	1.70	0.44	-	8.08	0.04	9.82	m²	10.80
PAINTING & CLEAR FINISHINGS; SPRAY PAINTING										
Spray one coat primer, one undercoat, one coat full gloss finish										
Concrete general surfaces girth exceeding 300mm	1.46	10.00	1.61	0.26	-	4.78	0.13	6.52	m²	7.17
Brickwork general surfaces girth exceeding 300mm	1.78	10.00	1.96	0.30	-	5.51	0.15	7.62	m²	8.38
Plaster general surfaces girth exceeding 300mm	1.35	10.00	1.48	0.24	-	4.41	0.12	6.01	m²	6.61
Spray one coat primer, two undercoats, one coat full gloss finish										
Concrete general surfaces girth exceeding 300mm	1.78	10.00	1.96	0.31	-	5.69	0.16	7.81	m²	8.59
Brickwork general surfaces girth exceeding 300mm	2.21	10.00	2.43	0.35	-	6.43	0.17	9.03	m²	9.94
Plaster general surfaces girth exceeding 300mm	1.67	10.00	1.84	0.28	-	5.14	0.14	7.12	m²	7.83
Spray one coat primer, one basecoat, one coat multicolour finish										
Concrete general surfaces girth exceeding 300mm	2.82	10.00	3.10	0.39	-	7.16	0.19	10.46	m²	11.51
Brickwork general surfaces girth exceeding 300mm	3.40	10.00	3.74	0.41	-	7.53	0.20	11.47	m²	12.62
Plaster general surfaces girth exceeding 300mm	2.56	10.00	2.82	0.37	-	6.80	0.19	9.80	m²	10.78
PAINTING & CLEAR FINISHINGS; ACRYLATED RUBBER PAINTING										
One coat primer, two coats acrylated rubber paint										
Concrete general surfaces girth exceeding 300mm	7.02	10.00	7.73	0.33	-	6.06	0.17	13.95	m²	15.35
Brickwork general surfaces girth exceeding 300mm	8.38	10.00	9.22	0.35	-	6.43	0.17	15.82	m²	17.40
Plasterboard general surfaces girth exceeding 300mm	6.30	10.00	6.93	0.32	-	5.88	0.16	12.97	m²	14.27
Plaster general surfaces girth exceeding 300mm .... isolated surfaces, girth not exceeding 300mm	6.30 1.84	10.00 10.00	6.93 2.02	0.32 0.14	- -	5.88 2.57	0.16 0.07	12.97 4.66	m² m	14.27 5.13
PAINTING & CLEAR FINISHINGS; PLASTIC FINISH										
Textured plastic coating - Stippled finish										
Concrete general surfaces girth exceeding 300mm	0.43	10.00	0.48	0.25	-	4.59	0.12	5.20	m²	5.71
Concrete general surfaces 3.50 - 5.00m above floor girth exceeding 300mm	0.43	10.00	0.48	0.28	-	5.14	0.14	5.76	m²	6.34
Brickwork general surfaces girth exceeding 300mm	0.52	10.00	0.57	0.30	-	5.51	0.15	6.23	m²	6.85
Plasterboard general surfaces girth exceeding 300mm	0.39	10.00	0.43	0.20	-	3.67	0.10	4.20	m²	4.62
Plasterboard general surfaces 3.50 - 5.00m above floor girth exceeding 300mm	0.39	10.00	0.43	0.22	-	4.04	0.11	4.58	m²	5.04
Plaster general surfaces girth exceeding 300mm	0.39	10.00	0.43	0.20	-	3.67	0.10	4.20	m²	4.62
Plaster general surfaces 3.50 - 5.00m above floor girth exceeding 300mm	0.39	10.00	0.43	0.22	-	4.04	0.11	4.58	m²	5.04
Textured plastic coating - Combed Finish										
Concrete general surfaces girth exceeding 300mm	0.52	10.00	0.57	0.30	-	5.51	0.15	6.23	m²	6.85

DECORATION

Labour hourly rates: (except Specialists) Craft Operatives 18.37 Labourer 13.76 Rates are national average prices. Refer to REGIONAL VARIATIONS for indicative levels of overall pricing in regions	MATERIALS			LABOUR				RATES		
	Del to Site	Waste	Material Cost	Craft Optve	Lab	Labour Cost	Sunds	Nett Rate	Unit	Gross rate (10%)
	£	%	£	Hrs	Hrs	£	£	£		£
PAINTING & CLEAR FINISHINGS; PLASTIC FINISH (Cont'd)										
Textured plastic coating - Combed Finish (Cont'd)										
Concrete general surfaces 3.50 - 5.00m above floor										
girth exceeding 300mm	0.52	10.00	0.57	0.33	-	6.06	0.17	6.80	m²	7.48
Brickwork general surfaces										
girth exceeding 300mm	0.62	10.00	0.68	0.35	-	6.43	0.17	7.29	m²	8.02
Plasterboard general surfaces										
girth exceeding 300mm	0.47	10.00	0.51	0.25	-	4.59	0.12	5.23	m²	5.75
Plasterboard general surfaces 3.50 - 5.00m above floor										
girth exceeding 300mm	0.47	10.00	0.51	0.27	-	4.96	0.14	5.61	m²	6.17
Plaster general surfaces										
girth exceeding 300mm	0.47	10.00	0.51	0.25	-	4.59	0.12	5.23	m²	5.75
Plaster general surfaces 3.50 - 5.00m above floor										
girth exceeding 300mm	0.47	10.00	0.51	0.27	-	4.96	0.14	5.61	m²	6.17
PAINTING & CLEAR FINISHINGS; OIL PAINTING METALWORK										
One undercoat, one coat full gloss finish on ready primed metal surfaces										
Iron or steel structural work										
girth exceeding 300mm	0.55	10.00	0.61	0.31	-	5.69	0.16	6.46	m²	7.11
isolated surfaces, girth not exceeding 300mm	0.17	10.00	0.19	0.13	-	2.39	0.06	2.64	m	2.90
Iron or steel structural members of roof trusses, lattice girders, purlins and the like										
girth exceeding 300mm	0.55	10.00	0.61	0.40	-	7.35	0.20	8.16	m²	8.97
isolated surfaces, girth not exceeding 300mm	0.17	10.00	0.19	0.13	-	2.39	0.06	2.64	m	2.90
Two undercoats, one coat full gloss finish on ready primed metal surfaces										
Iron or steel structural work										
girth exceeding 300mm	0.83	10.00	0.91	0.46	-	8.45	0.23	9.60	m²	10.55
isolated surfaces, girth not exceeding 300mm	0.26	10.00	0.28	0.15	-	2.76	0.08	3.11	m	3.42
Iron or steel structural members of roof trusses, lattice girders, purlins and the like										
girth exceeding 300mm	0.83	10.00	0.91	0.59	-	10.84	0.29	12.05	m²	13.25
isolated surfaces, girth not exceeding 300mm	0.26	10.00	0.28	0.20	-	3.67	0.10	4.06	m	4.46
One coat primer, one undercoat, one coat full gloss finish on metal surfaces										
Iron or steel general surfaces										
girth exceeding 300mm	2.26	10.00	2.48	0.43	-	7.90	0.22	10.60	m²	11.66
isolated surfaces, girth not exceeding 300mm	0.71	10.00	0.78	0.14	-	2.57	0.07	3.42	m	3.77
Galvanized glazed doors, windows or screens										
girth exceeding 300mm; panes, area not exceeding 0.10m²	1.23	10.00	1.35	0.79	-	14.51	0.40	16.26	m²	17.88
girth exceeding 300mm; panes, area 0.10 - 0.50m²	0.91	10.00	1.00	0.56	-	10.29	0.28	11.57	m²	12.72
girth exceeding 300mm; panes, area 0.50 - 1.00m²	0.79	10.00	0.87	0.47	-	8.63	0.23	9.74	m²	10.71
girth exceeding 300mm; panes, area exceeding 1.00m²	0.79	10.00	0.87	0.43	-	7.90	0.22	8.98	m²	9.88
Iron or steel structural work										
girth exceeding 300mm	2.26	10.00	2.48	0.46	-	8.45	0.23	11.16	m²	12.28
isolated surfaces, girth not exceeding 300mm	0.71	10.00	0.78	0.20	-	3.67	0.10	4.56	m	5.01
Iron or steel structural members of roof trusses, lattice girders, purlins and the like										
girth exceeding 300mm	2.26	10.00	2.48	0.59	-	10.84	0.29	13.61	m²	14.98
isolated surfaces, girth not exceeding 300mm	0.71	10.00	0.78	0.20	-	3.67	0.10	4.56	m	5.01
Iron or steel services										
girth exceeding 300mm	2.26	10.00	2.48	0.59	-	10.84	0.29	13.61	m²	14.98
isolated surfaces, girth not exceeding 300mm	0.71	10.00	0.78	0.20	-	3.67	0.10	4.56	m	5.01
isolated areas not exceeding 0.50m² irrespective of girth	1.15	10.00	1.26	0.59	-	10.84	0.29	12.40	nr	13.64
Copper services										
girth exceeding 300mm	2.26	10.00	2.48	0.59	-	10.84	0.29	13.61	m²	14.98
isolated surfaces, girth not exceeding 300mm	0.71	10.00	0.78	0.20	-	3.67	0.10	4.56	m	5.01
Galvanized services										
girth exceeding 300mm	2.26	10.00	2.48	0.59	-	10.84	0.29	13.61	m²	14.98
isolated surfaces, girth not exceeding 300mm	0.71	10.00	0.78	0.20	-	3.67	0.10	4.56	m	5.01
One coat primer, one undercoat, one coat full gloss finish on metal surfaces; external work										
Iron or steel general surfaces										
girth exceeding 300mm	2.26	10.00	2.48	0.46	-	8.45	0.23	11.16	m²	12.28
isolated surfaces, girth not exceeding 300mm	0.71	10.00	0.78	0.18	-	3.31	0.09	4.18	m	4.60
Galvanized glazed doors, windows or screens										
girth exceeding 300mm; panes, area not exceeding 0.10m²	1.23	10.00	1.35	0.83	-	15.25	0.41	17.01	m²	18.71
girth exceeding 300mm; panes, area 0.10 - 0.50m²	0.91	10.00	1.00	0.59	-	10.84	0.29	12.13	m²	13.35
girth exceeding 300mm; panes, area 0.50 - 1.00m²	0.79	10.00	0.87	0.51	-	9.37	0.25	10.49	m²	11.54
girth exceeding 300mm; panes, area exceeding 1.00m²	0.79	10.00	0.87	0.46	-	8.45	0.23	9.55	m²	10.50
Iron or steel structural work										
girth exceeding 300mm	2.26	10.00	2.48	0.46	-	8.45	0.23	11.16	m²	12.28
isolated surfaces, girth not exceeding 300mm	0.71	10.00	0.78	0.20	-	3.67	0.10	4.56	m	5.01

Labour hourly rates: (except Specialists) Craft Operatives 18.37 Labourer 13.76 Rates are national average prices. Refer to REGIONAL VARIATIONS for indicative levels of overall pricing in regions	MATERIALS			LABOUR				RATES		
	Del to Site £	Waste %	Material Cost £	Craft Optve Hrs	Lab Hrs	Labour Cost £	Sunds £	Nett Rate £	Unit	Gross rate (10%) £
PAINTING & CLEAR FINISHINGS; OIL PAINTING METALWORK (Cont'd)										
One coat primer, one undercoat, one coat full gloss finish on metal surfaces; external work (Cont'd)										
Iron or steel structural members of roof trusses, lattice girders, purlins and the like										
girth exceeding 300mm	2.26	10.00	2.48	0.59	-	10.84	0.29	13.61	m²	14.98
isolated surfaces, girth not exceeding 300mm	0.71	10.00	0.78	0.20	-	3.67	0.10	4.56	m	5.01
Iron or steel railings, fences and gates; plain open type										
girth exceeding 300mm	1.74	10.00	1.92	0.43	-	7.90	0.22	10.03	m²	11.03
isolated surfaces, girth not exceeding 300mm	0.71	10.00	0.78	0.14	-	2.57	0.07	3.42	m	3.77
Iron or steel railings, fences and gates; close type										
girth exceeding 300mm	2.26	10.00	2.48	0.36	-	6.61	0.18	9.27	m²	10.20
Iron or steel railings, fences and gates; ornamental type										
girth exceeding 300mm	1.74	10.00	1.92	0.73	-	13.41	0.37	15.69	m²	17.26
Iron or steel eaves gutters										
girth exceeding 300mm	2.26	10.00	2.48	0.50	-	9.19	0.25	11.92	m²	13.11
isolated surfaces, girth not exceeding 300mm	0.71	10.00	0.78	0.17	-	3.12	0.09	3.99	m	4.39
Galvanized eaves gutters										
girth exceeding 300mm	2.26	10.00	2.48	0.50	-	9.19	0.25	11.92	m²	13.11
isolated surfaces, girth not exceeding 300mm	0.71	10.00	0.78	0.17	-	3.12	0.09	3.99	m	4.39
Iron or steel services										
girth exceeding 300mm	2.26	10.00	2.48	0.59	-	10.84	0.29	13.61	m²	14.98
isolated surfaces, girth not exceeding 300mm	0.71	10.00	0.78	0.20	-	3.67	0.10	4.56	m	5.01
isolated areas not exceeding 0.50m² irrespective of girth	1.15	10.00	1.26	0.59	-	10.84	0.29	12.40	nr	13.64
Copper services										
girth exceeding 300mm	2.26	10.00	2.48	0.59	-	10.84	0.29	13.61	m²	14.98
isolated surfaces, girth not exceeding 300mm	0.71	10.00	0.78	0.20	-	3.67	0.10	4.56	m	5.01
Galvanized services										
girth exceeding 300mm	2.26	10.00	2.48	0.59	-	10.84	0.29	13.61	m²	14.98
isolated surfaces, girth not exceeding 300mm	0.71	10.00	0.78	0.20	-	3.67	0.10	4.56	m	5.01
One coat primer, two undercoats, one coat full gloss finish on metal surfaces										
Iron or steel general surfaces										
girth exceeding 300mm	2.53	10.00	2.79	0.57	-	10.47	0.28	13.54	m²	14.90
isolated surfaces, girth not exceeding 300mm	0.80	10.00	0.88	0.19	-	3.49	0.09	4.46	m	4.91
Galvanized glazed doors, windows or screens										
girth exceeding 300mm; panes, area not exceeding 0.10m²	1.38	10.00	1.51	1.06	-	19.47	0.53	21.52	m²	23.67
girth exceeding 300mm; panes, area 0.10 - 0.50m²	1.02	10.00	1.12	0.75	-	13.78	0.38	15.27	m²	16.80
girth exceeding 300mm; panes, area 0.50 - 1.00m²	0.87	10.00	0.96	0.63	-	11.57	0.31	12.85	m²	14.14
girth exceeding 300mm; panes, area exceeding 1.00m²	0.87	10.00	0.96	0.57	-	10.47	0.28	11.72	m²	12.89
Iron or steel structural work										
girth exceeding 300mm	2.53	10.00	2.79	0.62	-	11.39	0.31	14.49	m²	15.93
isolated surfaces, girth not exceeding 300mm	0.80	10.00	0.88	0.24	-	4.41	0.12	5.41	m	5.95
Iron or steel structural members of roof trusses, lattice girders, purlins and the like										
girth exceeding 300mm	2.53	10.00	2.79	0.79	-	14.51	0.40	17.69	m²	19.46
isolated surfaces, girth not exceeding 300mm	0.80	10.00	0.88	0.24	-	4.41	0.12	5.41	m	5.95
Iron or steel services										
girth exceeding 300mm	2.53	10.00	2.79	0.69	-	12.68	0.34	15.81	m²	17.39
isolated surfaces, girth not exceeding 300mm	0.80	10.00	0.88	0.23	-	4.23	0.12	5.22	m	5.74
isolated areas not exceeding 0.50m² irrespective of girth	1.28	10.00	1.40	0.69	-	12.68	0.34	14.43	nr	15.87
Copper services										
girth exceeding 300mm	2.53	10.00	2.79	0.69	-	12.68	0.34	15.81	m²	17.39
isolated surfaces, girth not exceeding 300mm	0.80	10.00	0.88	0.23	-	4.23	0.12	5.22	m	5.74
Galvanized services										
girth exceeding 300mm	2.53	10.00	2.79	0.69	-	12.68	0.34	15.81	m²	17.39
isolated surfaces, girth not exceeding 300mm	0.80	10.00	0.88	0.23	-	4.23	0.12	5.22	m	5.74
One coat primer, two undercoats, one coat full gloss finish on metal surfaces; external work										
Iron or steel general surfaces										
girth exceeding 300mm	2.53	10.00	2.79	0.62	-	11.39	0.31	14.49	m²	15.93
isolated surfaces, girth not exceeding 300mm	0.80	10.00	0.88	0.21	-	3.86	0.10	4.84	m	5.32
Galvanized glazed doors, windows or screens										
girth exceeding 300mm; panes, area not exceeding 0.10m²	1.38	10.00	1.51	1.10	-	20.21	0.55	22.27	m²	24.50
girth exceeding 300mm; panes, area 0.10 - 0.50m²	1.02	10.00	1.12	0.79	-	14.51	0.40	16.02	m²	17.63
girth exceeding 300mm; panes, area 0.50 - 1.00m²	0.87	10.00	0.96	0.67	-	12.31	0.34	13.60	m²	14.97
girth exceeding 300mm; panes, area exceeding 1.00m²	0.87	10.00	0.96	0.62	-	11.39	0.31	12.66	m²	13.93
Iron or steel structural work										
girth exceeding 300mm	2.53	10.00	2.79	0.62	-	11.39	0.31	14.49	m²	15.93
isolated surfaces, girth not exceeding 300mm	0.80	10.00	0.88	0.24	-	4.41	0.12	5.41	m	5.95
Iron or steel structural members of roof trusses, lattice girders, purlins and the like										
girth exceeding 300mm	2.53	10.00	2.79	0.79	-	14.51	0.40	17.69	m²	19.46
isolated surfaces, girth not exceeding 300mm	0.80	10.00	0.88	0.24	-	4.41	0.12	5.41	m	5.95
Iron or steel railings, fences and gates; plain open type										
girth exceeding 300mm	1.95	10.00	2.15	0.57	-	10.47	0.28	12.91	m²	14.20

DECORATION

Labour hourly rates: (except Specialists) Craft Operatives 18.37 Labourer 13.76 Rates are national average prices. Refer to REGIONAL VARIATIONS for indicative levels of overall pricing in regions	MATERIALS			LABOUR				RATES		
	Del to Site	Waste	Material Cost	Craft Optve	Lab	Labour Cost	Sunds	Nett Rate		Gross rate (10%)
	£	%	£	Hrs	Hrs	£	£	£	Unit	£
PAINTING & CLEAR FINISHINGS; OIL PAINTING METALWORK (Cont'd)										
One coat primer, two undercoats, one coat full gloss finish on metal surfaces; external work (Cont'd)										
Iron or steel railings, fences and gates; plain open type (Cont'd)										
isolated surfaces, girth not exceeding 300mm	0.80	10.00	0.88	0.19	-	3.49	0.09	4.46	m	4.91
Iron or steel railings, fences and gates; close type										
girth exceeding 300mm	2.53	10.00	2.79	0.48	-	8.82	0.24	11.84	m²	13.03
Iron or steel railings, fences and gates; ornamental type										
girth exceeding 300mm	1.95	10.00	2.15	0.97	-	17.82	0.49	20.45	m²	22.50
Iron or steel eaves gutters										
girth exceeding 300mm	2.53	10.00	2.79	0.66	-	12.12	0.33	15.24	m²	16.76
isolated surfaces, girth not exceeding 300mm	0.80	10.00	0.88	0.22	-	4.04	0.11	5.03	m	5.53
Galvanized eaves gutters										
girth exceeding 300mm	2.53	10.00	2.79	0.66	-	12.12	0.33	15.24	m²	16.76
isolated surfaces, girth not exceeding 300mm	0.80	10.00	0.88	0.22	-	4.04	0.11	5.03	m	5.53
Iron or steel services										
girth exceeding 300mm	2.53	10.00	2.79	0.69	-	12.68	0.34	15.81	m²	17.39
isolated surfaces, girth not exceeding 300mm	0.80	10.00	0.88	0.23	-	4.23	0.12	5.22	m	5.74
isolated areas not exceeding 0.50m² irrespective of girth..............	1.30	10.00	1.43	0.69	-	12.68	0.34	14.45	nr	15.89
Copper services										
girth exceeding 300mm	2.53	10.00	2.79	0.69	-	12.68	0.34	15.81	m²	17.39
isolated surfaces, girth not exceeding 300mm	0.80	10.00	0.88	0.23	-	4.23	0.12	5.22	m	5.74
Galvanized services										
girth exceeding 300mm	2.53	10.00	2.79	0.69	-	12.68	0.34	15.81	m²	17.39
isolated surfaces, girth not exceeding 300mm	0.80	10.00	0.88	0.23	-	4.23	0.12	5.22	m	5.74
One coat primer, one undercoat, two coats full gloss finish on metal surfaces										
Iron or steel general surfaces										
girth exceeding 300mm	2.53	10.00	2.79	0.57	-	10.47	0.28	13.54	m²	14.90
isolated surfaces, girth not exceeding 300mm	0.80	10.00	0.88	0.19	-	3.49	0.09	4.46	m	4.91
Galvanized glazed doors, windows or screens										
girth exceeding 300mm; panes, area not exceeding 0.10m²...........	1.38	10.00	1.51	1.06	-	19.47	0.53	21.52	m²	23.67
girth exceeding 300mm; panes, area 0.10 - 0.50m²	1.02	10.00	1.12	0.75	-	13.78	0.38	15.27	m²	16.80
girth exceeding 300mm; panes, area 0.50 - 1.00m²	0.87	10.00	0.96	0.63	-	11.57	0.31	12.85	m²	14.14
girth exceeding 300mm; panes, area exceeding 1.00m²................	0.87	10.00	0.96	0.57	-	10.47	0.28	11.72	m²	12.89
Iron or steel structural work										
girth exceeding 300mm	2.53	10.00	2.79	0.62	-	11.39	0.31	14.49	m²	15.93
isolated surfaces, girth not exceeding 300mm	0.80	10.00	0.88	0.24	-	4.41	0.12	5.41	m	5.95
Iron or steel structural members of roof trusses, lattice girders, purlins and the like										
girth exceeding 300mm	2.53	10.00	2.79	0.79	-	14.51	0.40	17.69	m²	19.46
isolated surfaces, girth not exceeding 300mm	0.80	10.00	0.88	0.24	-	4.41	0.12	5.41	m	5.95
Iron or steel services										
girth exceeding 300mm	2.53	10.00	2.79	0.69	-	12.68	0.34	15.81	m²	17.39
isolated surfaces, girth not exceeding 300mm	0.80	10.00	0.88	0.23	-	4.23	0.12	5.22	m	5.74
isolated areas not exceeding 0.50m² irrespective of girth..............	1.30	10.00	1.43	0.69	-	12.68	0.34	14.45	nr	15.89
Copper services										
girth exceeding 300mm	2.53	10.00	2.79	0.69	-	12.68	0.34	15.81	m²	17.39
isolated surfaces, girth not exceeding 300mm	0.80	10.00	0.88	0.23	-	4.23	0.12	5.22	m	5.74
Galvanized services										
girth exceeding 300mm	2.53	10.00	2.79	0.69	-	12.68	0.34	15.81	m²	17.39
isolated surfaces, girth not exceeding 300mm	0.80	10.00	0.88	0.23	-	4.23	0.12	5.22	m	5.74
One coat primer, one undercoat, two coats full gloss finish on metal surfaces; external work										
Iron or steel general surfaces										
girth exceeding 300mm	2.53	10.00	2.79	0.62	-	11.39	0.31	14.49	m²	15.93
isolated surfaces, girth not exceeding 300mm	0.80	10.00	0.88	0.21	-	3.86	0.10	4.84	m	5.32
Galvanized glazed doors, windows or screens										
girth exceeding 300mm; panes, area not exceeding 0.10m²...........	1.38	10.00	1.51	1.10	-	20.21	0.55	22.27	m²	24.50
girth exceeding 300mm; panes, area 0.10 - 0.50m²	1.02	10.00	1.12	0.79	-	14.51	0.40	16.02	m²	17.63
girth exceeding 300mm; panes, area 0.50 - 1.00m²	0.87	10.00	0.96	0.67	-	12.31	0.34	13.60	m²	14.97
girth exceeding 300mm; panes, area exceeding 1.00m²................	0.87	10.00	0.96	0.62	-	11.39	0.31	12.66	m²	13.93
Iron or steel structural work										
girth exceeding 300mm	2.53	10.00	2.79	0.62	-	11.39	0.31	14.49	m²	15.93
isolated surfaces, girth not exceeding 300mm	0.80	10.00	0.88	0.62	-	11.39	0.31	12.58	m²	13.83
Iron or steel structural members of roof trusses, lattice girders, purlins and the like										
girth exceeding 300mm	2.53	10.00	2.79	0.62	-	11.39	0.31	14.49	m²	15.93
isolated surfaces, girth not exceeding 300mm	0.80	10.00	0.88	0.24	-	4.41	0.12	5.41	m	5.95
Iron or steel railings, fences and gates; plain open type										
girth exceeding 300mm	1.95	10.00	2.15	0.57	-	10.47	0.28	12.91	m²	14.20
isolated surfaces, girth not exceeding 300mm	0.80	10.00	0.88	0.19	-	3.49	0.09	4.46	m	4.91
Iron or steel railings, fences and gates; close type										
girth exceeding 300mm	2.53	10.00	2.79	0.48	-	8.82	0.24	11.84	m²	13.03
Iron or steel railings, fences and gates; ornamental type										
girth exceeding 300mm	1.95	10.00	2.15	0.97	-	17.82	0.49	20.45	m²	22.50

Labour hourly rates: (except Specialists) Craft Operatives 18.37 Labourer 13.76 Rates are national average prices. Refer to REGIONAL VARIATIONS for indicative levels of overall pricing in regions	MATERIALS			LABOUR				RATES		
	Del to Site	Waste	Material Cost	Craft Optve	Lab	Labour Cost	Sunds	Nett Rate	Unit	Gross rate (10%)
	£	%	£	Hrs	Hrs	£	£	£		£
PAINTING & CLEAR FINISHINGS; OIL PAINTING METALWORK (Cont'd)										
One coat primer, one undercoat, two coats full gloss finish on metal surfaces; external work (Cont'd)										
Iron or steel eaves gutters										
girth exceeding 300mm	2.53	10.00	2.79	0.66	-	12.12	0.33	15.24	m²	16.76
isolated surfaces, girth not exceeding 300mm	0.80	10.00	0.88	0.22	-	4.04	0.11	5.03	m	5.53
Galvanized eaves gutters										
girth exceeding 300mm	2.53	10.00	2.79	0.66	-	12.12	0.33	15.24	m²	16.76
isolated surfaces, girth not exceeding 300mm	0.80	10.00	0.88	0.22	-	4.04	0.11	5.03	m	5.53
Iron or steel services										
girth exceeding 300mm	2.53	10.00	2.79	0.69	-	12.68	0.34	15.81	m²	17.39
isolated surfaces, girth not exceeding 300mm	0.80	10.00	0.88	0.23	-	4.23	0.12	5.22	m	5.74
isolated areas not exceeding 0.50m² irrespective of girth	1.30	10.00	1.43	0.69	-	12.68	0.34	14.45	nr	15.89
Copper services										
girth exceeding 300mm	2.53	10.00	2.79	0.69	-	12.68	0.34	15.81	m²	17.39
isolated surfaces, girth not exceeding 300mm	0.80	10.00	0.88	0.23	-	4.23	0.12	5.22	m	5.74
Galvanized services										
girth exceeding 300mm	2.53	10.00	2.79	0.69	-	12.68	0.34	15.81	m²	17.39
isolated surfaces, girth not exceeding 300mm	0.80	10.00	0.88	0.23	-	4.23	0.12	5.22	m	5.74
PAINTING & CLEAR FINISHINGS; METALLIC PAINTING										
One coat aluminium or silver metallic paint										
Iron or steel radiators; panel type										
girth exceeding 300mm	0.79	10.00	0.87	0.15	-	2.76	0.08	3.70	m²	4.07
Iron or steel radiators; column type										
girth exceeding 300mm	0.79	10.00	0.87	0.22	-	4.04	0.11	5.02	m²	5.53
Iron or steel services										
isolated surfaces, girth not exceeding 300mm	0.24	10.00	0.27	0.06	-	1.10	0.03	1.40	m	1.54
Copper services										
isolated surfaces, girth not exceeding 300mm	0.24	10.00	0.27	0.06	-	1.10	0.03	1.40	m	1.54
Galvanized services										
isolated surfaces, girth not exceeding 300mm	0.24	10.00	0.27	0.06	-	1.10	0.03	1.40	m	1.54
Two coats aluminium or silver metallic paint										
Iron or steel radiators; panel type										
girth exceeding 300mm	1.59	10.00	1.75	0.31	-	5.69	0.16	7.59	m²	8.35
Iron or steel radiators; column type										
girth exceeding 300mm	1.59	10.00	1.75	0.44	-	8.08	0.22	10.05	m²	11.05
Iron or steel services										
isolated surfaces, girth not exceeding 300mm	0.49	10.00	0.54	0.11	-	2.02	0.05	2.61	m	2.87
Copper services										
isolated surfaces, girth not exceeding 300mm	0.49	10.00	0.54	0.11	-	2.02	0.05	2.61	m	2.87
Galvanized services										
isolated surfaces, girth not exceeding 300mm	0.49	10.00	0.54	0.11	-	2.02	0.05	2.61	m	2.87
One coat gold or bronze metallic paint										
Iron or steel radiators; panel type										
girth exceeding 300mm	1.04	10.00	1.14	0.15	-	2.76	0.08	3.97	m²	4.37
Iron or steel radiators; column type										
girth exceeding 300mm	1.04	10.00	1.14	0.22	-	4.04	0.11	5.29	m²	5.82
Iron or steel services										
isolated surfaces, girth not exceeding 300mm	0.31	10.00	0.34	0.06	-	1.10	0.03	1.47	m	1.61
Copper services										
isolated surfaces, girth not exceeding 300mm	0.31	10.00	0.34	0.06	-	1.10	0.03	1.47	m	1.61
Galvanized services										
isolated surfaces, girth not exceeding 300mm	0.31	10.00	0.34	0.06	-	1.10	0.03	1.47	m	1.61
Two coats gold or bronze metallic paint										
Iron or steel radiators; panel type										
girth exceeding 300mm	2.07	10.00	2.28	0.31	-	5.69	0.16	8.13	m²	8.95
Iron or steel radiators; column type										
girth exceeding 300mm	2.07	10.00	2.28	0.44	-	8.08	0.22	10.58	m²	11.64
Iron or steel services										
isolated surfaces, girth not exceeding 300mm	0.61	10.00	0.67	0.11	-	2.02	0.05	2.75	m	3.02
Copper services										
isolated surfaces, girth not exceeding 300mm	0.61	10.00	0.67	0.11	-	2.02	0.05	2.75	m	3.02
Galvanized services										
isolated surfaces, girth not exceeding 300mm	0.61	10.00	0.67	0.11	-	2.02	0.05	2.75	m	3.02
PAINTING & CLEAR FINISHINGS; BITUMINOUS PAINT										
One coat black bitumen paint; external work										
Iron or steel general surfaces										
girth exceeding 300mm	0.26	10.00	0.29	0.14	-	2.57	0.07	2.93	m²	3.23

Labour hourly rates: (except Specialists) Craft Operatives 18.37 Labourer 13.76 Rates are national average prices. Refer to REGIONAL VARIATIONS for indicative levels of overall pricing in regions	MATERIALS			LABOUR				RATES		
	Del to Site £	Waste %	Material Cost £	Craft Optve Hrs	Lab Hrs	Labour Cost £	Sunds £	Nett Rate £	Unit	Gross rate (10%) £
PAINTING & CLEAR FINISHINGS; BITUMINOUS PAINT (Cont'd)										
One coat black bitumen paint; external work (Cont'd)										
Iron or steel eaves gutters										
girth exceeding 300mm	0.26	10.00	0.29	0.15	-	2.76	0.08	3.12	m²	3.43
Iron or steel services										
girth exceeding 300mm	0.26	10.00	0.29	0.17	-	3.12	0.09	3.50	m²	3.85
isolated surfaces, girth not exceeding 300mm	0.08	10.00	0.09	0.06	-	1.10	0.03	1.22	m	1.35
Two coats black bitumen paint; external work										
Iron or steel general surfaces										
girth exceeding 300mm	0.53	10.00	0.58	0.29	-	5.33	0.14	6.05	m²	6.66
Iron or steel eaves gutters										
girth exceeding 300mm	0.53	10.00	0.58	0.31	-	5.69	0.16	6.43	m²	7.07
isolated surfaces, girth not exceeding 300mm	0.16	10.00	0.17	0.10	-	1.84	0.05	2.06	m	2.26
Iron or steel services										
girth exceeding 300mm	0.53	10.00	0.58	0.33	-	6.06	0.17	6.81	m²	7.49
isolated surfaces, girth not exceeding 300mm	0.16	10.00	0.17	0.10	-	1.84	0.05	2.06	m	2.26
PAINTING & CLEAR FINISHINGS; OIL PAINTING WOODWORK										
One coat primer; carried out on site before fixing members										
Wood general surfaces										
girth exceeding 300mm	0.30	10.00	0.33	0.18	-	3.31	0.09	3.73	m²	4.10
isolated surfaces, girth not exceeding 300mm	0.09	10.00	0.10	0.07	-	1.29	0.04	1.42	m	1.56
One undercoat, one coat full gloss finish on ready primed wood surfaces										
Wood general surfaces										
girth exceeding 300mm	0.55	10.00	0.61	0.35	-	6.43	0.17	7.21	m²	7.94
isolated surfaces, girth not exceeding 300mm	0.17	10.00	0.19	0.12	-	2.20	0.06	2.45	m	2.70
Wood glazed doors										
girth exceeding 300mm; panes, area not exceeding 0.10m²	0.30	10.00	0.33	0.59	-	10.84	0.29	11.46	m²	12.61
girth exceeding 300mm; panes, area 0.10 - 0.50m²	0.21	10.00	0.23	0.42	-	7.72	0.21	8.16	m²	8.98
girth exceeding 300mm; panes, area 0.50 - 1.00m²	0.21	10.00	0.23	0.35	-	6.43	0.17	6.84	m²	7.52
girth exceeding 300mm; panes, area exceeding 1.00m²	0.21	10.00	0.23	0.32	-	5.88	0.16	6.27	m²	6.90
Wood partially glazed doors										
girth exceeding 300mm; panes, area not exceeding 0.10m²	0.43	10.00	0.47	0.47	-	8.63	0.23	9.34	m²	10.27
girth exceeding 300mm; panes, area 0.10 - 0.50m²	0.34	10.00	0.38	0.39	-	7.16	0.19	7.73	m²	8.51
girth exceeding 300mm; panes, area 0.50 - 1.00m²	0.30	10.00	0.33	0.35	-	6.43	0.17	6.93	m²	7.63
girth exceeding 300mm; panes, area exceeding 1.00m²	0.30	10.00	0.33	0.34	-	6.25	0.17	6.74	m²	7.42
Wood windows and screens										
girth exceeding 300mm; panes, area not exceeding 0.10m²	0.43	10.00	0.47	0.65	-	11.94	0.32	12.73	m²	14.01
girth exceeding 300mm; panes, area 0.10 - 0.50m²	0.34	10.00	0.38	0.46	-	8.45	0.23	9.06	m²	9.96
girth exceeding 300mm; panes, area 0.50 - 1.00m²	0.30	10.00	0.33	0.39	-	7.16	0.19	7.69	m²	8.46
girth exceeding 300mm; panes, area exceeding 1.00m²	0.30	10.00	0.33	0.35	-	6.43	0.17	6.93	m²	7.63
One undercoat, one coat full gloss finish on ready primed wood surfaces; external work										
Wood general surfaces										
girth exceeding 300mm	0.55	10.00	0.61	0.37	-	6.80	0.19	7.59	m²	8.35
isolated surfaces, girth not exceeding 300mm	0.17	10.00	0.19	0.13	-	2.39	0.06	2.64	m	2.90
Wood glazed doors										
girth exceeding 300mm; panes, area not exceeding 0.10m²	0.30	10.00	0.33	0.59	-	10.84	0.29	11.46	m²	12.61
girth exceeding 300mm; panes, area 0.10 - 0.50m²	0.21	10.00	0.23	0.42	-	7.72	0.21	8.16	m²	8.98
girth exceeding 300mm; panes, area 0.50 - 1.00m²	0.21	10.00	0.23	0.35	-	6.43	0.17	6.84	m²	7.52
girth exceeding 300mm; panes, area exceeding 1.00m²	0.21	10.00	0.23	0.32	-	5.88	0.16	6.27	m²	6.90
Wood partially glazed doors										
girth exceeding 300mm; panes, area not exceeding 0.10m²	0.43	10.00	0.47	0.47	-	8.63	0.23	9.34	m²	10.27
girth exceeding 300mm; panes, area 0.10 - 0.50m²	0.34	10.00	0.38	0.39	-	7.16	0.19	7.73	m²	8.51
girth exceeding 300mm; panes, area 0.50 - 1.00m²	0.30	10.00	0.33	0.35	-	6.43	0.17	6.93	m²	7.63
girth exceeding 300mm; panes, area exceeding 1.00m²	0.30	10.00	0.33	0.34	-	6.25	0.17	6.74	m²	7.42
Wood windows and screens										
girth exceeding 300mm; panes, area not exceeding 0.10m²	0.43	10.00	0.47	0.67	-	12.31	0.34	13.11	m²	14.42
girth exceeding 300mm; panes, area 0.10 - 0.50m²	0.34	10.00	0.38	0.48	-	8.82	0.24	9.43	m²	10.38
girth exceeding 300mm; panes, area 0.50 - 1.00m²	0.30	10.00	0.33	0.41	-	7.53	0.20	8.07	m²	8.87
girth exceeding 300mm; panes, area exceeding 1.00m²	0.30	10.00	0.33	0.37	-	6.80	0.19	7.31	m²	8.04
Wood railings fences and gates; open type										
girth exceeding 300mm	0.43	10.00	0.47	0.28	-	5.14	0.14	5.75	m²	6.33
isolated surfaces, girth not exceeding 300mm	0.17	10.00	0.19	0.09	-	1.65	0.05	1.89	m	2.07
Wood railings fences and gates; close type										
girth exceeding 300mm	0.55	10.00	0.61	0.24	-	4.41	0.12	5.14	m²	5.65
Two undercoats, one coat full gloss finish on ready primed wood surfaces										
Wood general surfaces										
girth exceeding 300mm	0.83	10.00	0.91	0.51	-	9.37	0.25	10.54	m²	11.59
isolated surfaces, girth not exceeding 300mm	0.26	10.00	0.28	0.17	-	3.12	0.09	3.49	m	3.84
Wood glazed doors										
girth exceeding 300mm; panes, area not exceeding 0.10m²	0.45	10.00	0.49	0.84	-	15.43	0.42	16.34	m²	17.98
girth exceeding 300mm; panes, area 0.10 - 0.50m²	0.32	10.00	0.35	0.61	-	11.21	0.31	11.86	m²	13.05
girth exceeding 300mm; panes, area 0.50 - 1.00m²	0.32	10.00	0.35	0.51	-	9.37	0.25	9.98	m²	10.97

Labour hourly rates: (except Specialists) Craft Operatives 18.37 Labourer 13.76 Rates are national average prices. Refer to REGIONAL VARIATIONS for indicative levels of overall pricing in regions	MATERIALS			LABOUR				RATES		
	Del to Site	Waste	Material Cost	Craft Optve	Lab	Labour Cost	Sunds	Nett Rate	Unit	Gross rate (10%)
	£	%	£	Hrs	Hrs	£	£	£		£
PAINTING & CLEAR FINISHINGS; OIL PAINTING WOODWORK (Cont'd)										
Two undercoats, one coat full gloss finish on ready primed wood surfaces (Cont'd)										
Wood glazed doors (Cont'd)										
girth exceeding 300mm; panes, area exceeding 1.00m²	0.32	10.00	0.35	0.45	-	8.27	0.22	8.84	m²	9.73
Wood partially glazed doors										
girth exceeding 300mm; panes, area not exceeding 0.10m²	0.64	10.00	0.70	0.67	-	12.31	0.34	13.35	m²	14.68
girth exceeding 300mm; panes, area 0.10 - 0.50m²	0.51	10.00	0.56	0.56	-	10.29	0.28	11.13	m²	12.24
girth exceeding 300mm; panes, area 0.50 - 1.00m²	0.45	10.00	0.49	0.51	-	9.37	0.25	10.12	m²	11.13
girth exceeding 300mm; panes, area exceeding 1.00m²	0.45	10.00	0.49	0.48	-	8.82	0.24	9.55	m²	10.51
Wood windows and screens										
girth exceeding 300mm; panes, area not exceeding 0.10m²	0.64	10.00	0.70	0.92	-	16.90	0.46	18.06	m²	19.87
girth exceeding 300mm; panes, area 0.10 - 0.50m²	0.51	10.00	0.56	0.67	-	12.31	0.34	13.21	m²	14.53
girth exceeding 300mm; panes, area 0.50 - 1.00m²	0.45	10.00	0.49	0.56	-	10.29	0.28	11.06	m²	12.17
girth exceeding 300mm; panes, area exceeding 1.00m²	0.45	10.00	0.49	0.50	-	9.19	0.25	9.93	m²	10.92
Two undercoats, one coat full gloss finish on ready primed wood surfaces; external work										
Wood general surfaces										
girth exceeding 300mm	0.83	10.00	0.91	0.54	-	9.92	0.27	11.10	m²	12.22
isolated surfaces, girth not exceeding 300mm	0.26	10.00	0.28	0.19	-	3.49	0.09	3.87	m	4.25
Wood glazed doors										
girth exceeding 300mm; panes, area not exceeding 0.10m²	0.45	10.00	0.49	0.84	-	15.43	0.42	16.34	m²	17.98
girth exceeding 300mm; panes, area 0.10 - 0.50m²	0.32	10.00	0.35	0.61	-	11.21	0.31	11.86	m²	13.05
girth exceeding 300mm; panes, area 0.50 - 1.00m²	0.32	10.00	0.35	0.51	-	9.37	0.25	9.98	m²	10.97
girth exceeding 300mm; panes, area exceeding 1.00m²	0.32	10.00	0.35	0.45	-	8.27	0.22	8.84	m²	9.73
Wood partially glazed doors										
girth exceeding 300mm; panes, area not exceeding 0.10m²	0.64	10.00	0.70	0.67	-	12.31	0.34	13.35	m²	14.68
girth exceeding 300mm; panes, area 0.10 - 0.50m²	0.51	10.00	0.56	0.56	-	10.29	0.28	11.13	m²	12.24
girth exceeding 300mm; panes, area 0.50 - 1.00m²	0.45	10.00	0.49	0.51	-	9.37	0.25	10.12	m²	11.13
girth exceeding 300mm; panes, area exceeding 1.00m²	0.45	10.00	0.49	0.48	-	8.82	0.24	9.55	m²	10.51
Wood windows and screens										
girth exceeding 300mm; panes, area not exceeding 0.10m²	0.64	10.00	0.70	0.96	-	17.64	0.48	18.82	m²	20.70
girth exceeding 300mm; panes, area 0.10 - 0.50m²	0.51	10.00	0.56	0.70	-	12.86	0.35	13.77	m²	15.15
girth exceeding 300mm; panes, area 0.50 - 1.00m²	0.45	10.00	0.49	0.59	-	10.84	0.29	11.63	m²	12.79
girth exceeding 300mm; panes, area exceeding 1.00m²	0.45	10.00	0.49	0.53	-	9.74	0.26	10.49	m²	11.54
Wood railings fences and gates; open type										
girth exceeding 300mm	0.64	10.00	0.70	0.42	-	7.72	0.21	8.63	m²	9.49
isolated surfaces, girth not exceeding 300mm	0.26	10.00	0.28	0.14	-	2.57	0.07	2.92	m	3.22
Wood railings fences and gates; close type										
girth exceeding 300mm	0.83	10.00	0.91	0.36	-	6.61	0.18	7.71	m²	8.48
One coat primer, one undercoat, one coat full gloss finish on wood surfaces										
Wood general surfaces										
girth exceeding 300mm	0.86	10.00	0.94	0.51	-	9.37	0.25	10.57	m²	11.62
isolated surfaces, girth not exceeding 300mm	0.26	10.00	0.28	0.17	-	3.12	0.09	3.49	m	3.84
Wood glazed doors										
girth exceeding 300mm; panes, area not exceeding 0.10m²	0.47	10.00	0.52	0.84	-	15.43	0.42	16.37	m²	18.01
girth exceeding 300mm; panes, area 0.10 - 0.50m²	0.34	10.00	0.38	0.61	-	11.21	0.31	11.89	m²	13.08
girth exceeding 300mm; panes, area 0.50 - 1.00m²	0.32	10.00	0.35	0.51	-	9.37	0.25	9.98	m²	10.97
girth exceeding 300mm; panes, area exceeding 1.00m²	0.32	10.00	0.35	0.45	-	8.27	0.22	8.85	m²	9.73
Wood partially glazed doors										
girth exceeding 300mm; panes, area not exceeding 0.10m²	0.66	10.00	0.73	0.67	-	12.31	0.34	13.37	m²	14.71
girth exceeding 300mm; panes, area 0.10 - 0.50m²	0.51	10.00	0.57	0.56	-	10.29	0.28	11.13	m²	12.25
girth exceeding 300mm; panes, area 0.50 - 1.00m²	0.47	10.00	0.52	0.51	-	9.37	0.25	10.14	m²	11.16
girth exceeding 300mm; panes, area exceeding 1.00m²	0.47	10.00	0.52	0.48	-	8.82	0.24	9.58	m²	10.53
Wood windows and screens										
girth exceeding 300mm; panes, area not exceeding 0.10m²	0.66	10.00	0.73	0.92	-	16.90	0.46	18.09	m²	19.90
girth exceeding 300mm; panes, area 0.10 - 0.50m²	0.51	10.00	0.57	0.67	-	12.31	0.34	13.21	m²	14.53
girth exceeding 300mm; panes, area 0.50 - 1.00m²	0.47	10.00	0.52	0.56	-	10.29	0.28	11.09	m²	12.19
girth exceeding 300mm; panes, area exceeding 1.00m²	0.47	10.00	0.52	0.50	-	9.19	0.25	9.95	m²	10.95
One coat primer, one undercoat, one coat full gloss finish on wood surfaces; external work										
Wood general surfaces										
girth exceeding 300mm	0.86	10.00	0.94	0.54	-	9.92	0.27	11.13	m²	12.25
isolated surfaces, girth not exceeding 300mm	0.26	10.00	0.28	0.19	-	3.49	0.09	3.87	m	4.25
Wood glazed doors										
girth exceeding 300mm; panes, area not exceeding 0.10m²	0.47	10.00	0.52	0.84	-	15.43	0.42	16.37	m²	18.01
girth exceeding 300mm; panes, area 0.10 - 0.50m²	0.34	10.00	0.38	0.61	-	11.21	0.31	11.89	m²	13.08
girth exceeding 300mm; panes, area 0.50 - 1.00m²	0.32	10.00	0.35	0.51	-	9.37	0.04	9.76	m²	10.74
girth exceeding 300mm; panes, area exceeding 1.00m²	0.32	10.00	0.35	0.45	-	8.27	0.22	8.85	m²	9.73
Wood partially glazed doors										
girth exceeding 300mm; panes, area not exceeding 0.10m²	0.66	10.00	0.73	0.67	-	12.31	0.34	13.37	m²	14.71
girth exceeding 300mm; panes, area 0.10 - 0.50m²	0.51	10.00	0.57	0.56	-	10.29	0.28	11.13	m²	12.25
girth exceeding 300mm; panes, area 0.50 - 1.00m²	0.47	10.00	0.52	0.51	-	9.37	0.25	10.14	m²	11.16
girth exceeding 300mm; panes, area exceeding 1.00m²	0.47	10.00	0.52	0.48	-	8.82	0.24	9.58	m²	10.53
Wood windows and screens										
girth exceeding 300mm; panes, area not exceeding 0.10m²	0.66	10.00	0.73	0.96	-	17.64	0.48	18.85	m²	20.73
girth exceeding 300mm; panes, area 0.10 - 0.50m²	0.51	10.00	0.57	0.70	-	12.86	0.35	13.77	m²	15.15

DECORATION

Labour hourly rates: (except Specialists) Craft Operatives 18.37 Labourer 13.76 Rates are national average prices. Refer to REGIONAL VARIATIONS for indicative levels of overall pricing in regions	MATERIALS			LABOUR				RATES		
	Del to Site	Waste	Material Cost	Craft Optve	Lab	Labour Cost	Sunds	Nett Rate	Unit	Gross rate (10%)
	£	%	£	Hrs	Hrs	£	£	£		£
PAINTING & CLEAR FINISHINGS; OIL PAINTING WOODWORK (Cont'd)										
One coat primer, one undercoat, one coat full gloss finish on wood surfaces; external work (Cont'd)										
Wood windows and screens (Cont'd)										
girth exceeding 300mm; panes, area 0.50 - 1.00m²	0.47	10.00	0.52	0.59	-	10.84	0.29	11.65	m²	12.82
girth exceeding 300mm; panes, area exceeding 1.00m²	0.47	10.00	0.52	0.53	-	9.74	0.26	10.52	m²	11.57
Wood railings fences and gates; open type										
girth exceeding 300mm	0.66	10.00	0.73	0.42	-	7.72	0.21	8.66	m²	9.52
isolated surfaces, girth not exceeding 300mm	0.26	10.00	0.28	0.14	-	2.57	0.07	2.92	m	3.22
Wood railings fences and gates; close type										
girth exceeding 300mm	0.86	10.00	0.94	0.36	-	6.61	0.18	7.74	m²	8.51
One coat primer, two undercoats, one coat full gloss finish on wood surfaces										
Wood general surfaces										
girth exceeding 300mm	1.13	10.00	1.25	0.66	-	12.12	0.33	13.70	m²	15.07
isolated surfaces, girth not exceeding 300mm	0.34	10.00	0.38	0.22	-	4.04	0.11	4.53	m	4.98
Wood glazed doors										
girth exceeding 300mm; panes, area not exceeding 0.10m²	0.62	10.00	0.68	1.10	-	20.21	0.55	21.44	m²	23.58
girth exceeding 300mm; panes, area 0.10 - 0.50m²	0.45	10.00	0.49	0.79	-	14.51	0.40	15.40	m²	16.94
girth exceeding 300mm; panes, area 0.50 - 1.00m²	0.43	10.00	0.47	0.66	-	12.12	0.33	12.93	m²	14.22
girth exceeding 300mm; panes, area exceeding 1.00m²	0.43	10.00	0.47	0.59	-	10.84	0.29	11.60	m²	12.76
Wood partially glazed doors										
girth exceeding 300mm; panes, area not exceeding 0.10m²	0.88	10.00	0.97	0.88	-	16.17	0.44	17.57	m²	19.33
girth exceeding 300mm; panes, area 0.10 - 0.50m²	0.68	10.00	0.75	0.73	-	13.41	0.37	14.53	m²	15.98
girth exceeding 300mm; panes, area 0.50 - 1.00m²	0.62	10.00	0.68	0.66	-	12.12	0.33	13.14	m²	14.45
girth exceeding 300mm; panes, area exceeding 1.00m²	0.62	10.00	0.68	0.63	-	11.57	0.31	12.57	m²	13.83
Wood windows and screens										
girth exceeding 300mm; panes, area not exceeding 0.10m²	0.88	10.00	0.97	1.21	-	22.23	0.61	23.80	m²	26.18
girth exceeding 300mm; panes, area 0.10 - 0.50m²	0.68	10.00	0.75	0.87	-	15.98	0.44	17.17	m²	18.89
girth exceeding 300mm; panes, area 0.50 - 1.00m²	0.62	10.00	0.68	0.73	-	13.41	0.37	14.46	m²	15.90
girth exceeding 300mm; panes, area exceeding 1.00m²	0.62	10.00	0.68	0.65	-	11.94	0.32	12.95	m²	14.24
One coat primer, two undercoats, one coat full gloss finish on wood surfaces; external work										
Wood general surfaces										
girth exceeding 300mm	1.13	10.00	1.25	0.70	-	12.86	0.35	14.46	m²	15.90
isolated surfaces, girth not exceeding 300mm	0.34	10.00	0.38	0.24	-	4.41	0.12	4.91	m²	5.40
Wood glazed doors										
girth exceeding 300mm; panes, area not exceeding 0.10m²	0.62	10.00	0.68	1.10	-	20.21	0.55	21.44	m²	23.58
girth exceeding 300mm; panes, area 0.10 - 0.50m²	0.45	10.00	0.49	0.79	-	14.51	0.40	15.40	m²	16.94
girth exceeding 300mm; panes, area 0.50 - 1.00m²	0.43	10.00	0.47	0.66	-	12.12	0.33	12.93	m²	14.22
girth exceeding 300mm; panes, area exceeding 1.00m²	0.43	10.00	0.47	0.59	-	10.84	0.29	11.60	m²	12.76
Wood partially glazed doors										
girth exceeding 300mm; panes, area not exceeding 0.10m²	0.88	10.00	0.97	0.88	-	16.17	0.44	17.57	m²	19.33
girth exceeding 300mm; panes, area 0.10 - 0.50m²	0.68	10.00	0.75	0.73	-	13.41	0.37	14.53	m²	15.98
girth exceeding 300mm; panes, area 0.50 - 1.00m²	0.62	10.00	0.68	0.66	-	12.12	0.33	13.14	m²	14.45
girth exceeding 300mm; panes, area exceeding 1.00m²	0.62	10.00	0.68	0.63	-	11.57	0.31	12.57	m²	13.83
Wood windows and screens										
girth exceeding 300mm; panes, area not exceeding 0.10m²	0.88	10.00	0.97	1.25	-	22.96	0.62	24.55	m²	27.01
girth exceeding 300mm; panes, area 0.10 - 0.50m²	0.68	10.00	0.75	0.91	-	16.72	0.46	17.93	m²	19.72
girth exceeding 300mm; panes, area 0.50 - 1.00m²	0.62	10.00	0.68	0.77	-	14.14	0.38	15.21	m²	16.73
girth exceeding 300mm; panes, area exceeding 1.00m²	0.62	10.00	0.68	0.69	-	12.68	0.34	13.70	m²	15.07
Wood railings fences and gates; open type										
girth exceeding 300mm	0.88	10.00	0.97	0.55	-	10.10	0.28	11.34	m²	12.48
isolated surfaces, girth not exceeding 300mm	0.34	10.00	0.38	0.19	-	3.49	0.09	3.96	m	4.36
Wood railings fences and gates; close type										
girth exceeding 300mm	1.13	10.00	1.25	0.48	-	8.82	0.24	10.31	m²	11.34
One coat primer, one undercoat, two coats full gloss finish on wood surfaces										
Wood general surfaces										
girth exceeding 300mm	1.13	10.00	1.25	0.66	-	12.12	0.33	13.70	m²	15.07
isolated surfaces, girth not exceeding 300mm	0.34	10.00	0.38	0.22	-	4.04	0.11	4.53	m	4.98
Wood glazed doors										
girth exceeding 300mm; panes, area not exceeding 0.10m²	0.62	10.00	0.68	1.10	-	20.21	0.55	21.44	m²	23.58
girth exceeding 300mm; panes, area 0.10 - 0.50m²	0.45	10.00	0.49	0.79	-	14.51	0.40	15.40	m²	16.94
girth exceeding 300mm; panes, area 0.50 - 1.00m²	0.43	10.00	0.47	0.66	-	12.12	0.33	12.93	m²	14.22
girth exceeding 300mm; panes, area exceeding 1.00m²	0.43	10.00	0.47	0.59	-	10.84	0.29	11.60	m²	12.76
Wood partially glazed doors										
girth exceeding 300mm; panes, area not exceeding 0.10m²	0.88	10.00	0.97	0.88	-	16.17	0.44	17.57	m²	19.33
girth exceeding 300mm; panes, area 0.10 - 0.50m²	0.68	10.00	0.75	0.73	-	13.41	0.37	14.53	m²	15.98
girth exceeding 300mm; panes, area 0.50 - 1.00m²	0.62	10.00	0.68	0.66	-	12.12	0.33	13.14	m²	14.45
girth exceeding 300mm; panes, area exceeding 1.00m²	0.62	10.00	0.68	0.63	-	11.57	0.31	12.57	m²	13.83
Wood windows and screens										
girth exceeding 300mm; panes, area not exceeding 0.10m²	0.88	10.00	0.97	1.21	-	22.23	0.61	23.80	m²	26.18
girth exceeding 300mm; panes, area 0.10 - 0.50m²	0.68	10.00	0.75	0.87	-	15.98	0.44	17.17	m²	18.89
girth exceeding 300mm; panes, area 0.50 - 1.00m²	0.62	10.00	0.68	0.73	-	13.41	0.37	14.46	m²	15.90
girth exceeding 300mm; panes, area exceeding 1.00m²	0.62	10.00	0.68	0.65	-	11.94	0.32	12.95	m²	14.24

Labour hourly rates: (except Specialists) Craft Operatives 18.37 Labourer 13.76 Rates are national average prices. Refer to REGIONAL VARIATIONS for indicative levels of overall pricing in regions	MATERIALS			LABOUR				RATES		
	Del to Site	Waste	Material Cost	Craft Optve	Lab	Labour Cost	Sunds	Nett Rate		Gross rate (10%)
	£	%	£	Hrs	Hrs	£	£	£	Unit	£
PAINTING & CLEAR FINISHINGS; OIL PAINTING WOODWORK (Cont'd)										
One coat primer, one undercoat, two coats full gloss finish wood surfaces; external work										
Wood general surfaces										
girth exceeding 300mm	1.13	10.00	1.25	0.70	-	12.86	0.35	14.46	m²	15.90
isolated surfaces, girth not exceeding 300mm	0.34	10.00	0.38	0.24	-	4.41	0.12	4.91	m	5.40
Wood glazed doors										
girth exceeding 300mm; panes, area not exceeding 0.10m²........	0.62	10.00	0.68	1.10	-	20.21	0.55	21.44	m²	23.58
girth exceeding 300mm; panes, area 0.10 - 0.50m²...............	0.45	10.00	0.49	0.79	-	14.51	0.40	15.40	m²	16.94
girth exceeding 300mm; panes, area 0.50 - 1.00m²...............	0.43	10.00	0.47	0.66	-	12.12	0.33	12.93	m²	14.22
girth exceeding 300mm; panes, area exceeding 1.00m²........	0.43	10.00	0.47	0.59	-	10.84	0.29	11.60	m²	12.76
Wood partially glazed doors										
girth exceeding 300mm; panes, area not exceeding 0.10m²........	0.88	10.00	0.97	0.88	-	16.17	0.44	17.57	m²	19.33
girth exceeding 300mm; panes, area 0.10 - 0.50m²...............	0.68	10.00	0.75	0.73	-	13.41	0.37	14.53	m²	15.98
girth exceeding 300mm; panes, area 0.50 - 1.00m²...............	0.62	10.00	0.68	0.66	-	12.12	0.33	13.14	m²	14.45
girth exceeding 300mm; panes, area exceeding 1.00m²........	0.62	10.00	0.68	0.63	-	11.57	0.31	12.57	m²	13.83
Wood windows and screens										
girth exceeding 300mm; panes, area not exceeding 0.10m²........	0.88	10.00	0.97	1.25	-	22.96	0.62	24.55	m²	27.01
girth exceeding 300mm; panes, area 0.10 - 0.50m²...............	0.68	10.00	0.75	0.91	-	16.72	0.46	17.93	m²	19.72
girth exceeding 300mm; panes, area 0.50 - 1.00m²...............	0.62	10.00	0.68	0.77	-	14.14	0.38	15.21	m²	16.73
girth exceeding 300mm; panes, area exceeding 1.00m²........	0.62	10.00	0.68	0.69	-	12.68	0.34	13.70	m²	15.07
Wood railings fences and gates; open type										
girth exceeding 300mm	0.88	10.00	0.97	0.55	-	10.10	0.28	11.34	m²	12.48
isolated surfaces, girth not exceeding 300mm	0.34	10.00	0.38	0.19	-	3.49	0.09	3.96	m	4.36
Wood railings fences and gates; close type										
girth exceeding 300mm	1.13	10.00	1.25	0.48	-	8.82	0.24	10.31	m²	11.34
PAINTING & CLEAR FINISHINGS; POLYURETHANE LACQUER										
Two coats polyurethane lacquer										
Wood general surfaces										
girth exceeding 300mm	0.88	10.00	0.97	0.44	-	8.08	0.22	9.27	m²	10.20
isolated surfaces, girth not exceeding 300mm	0.26	10.00	0.29	0.14	-	2.57	0.07	2.93	m	3.23
Wood glazed doors										
girth exceeding 300mm; panes, area not exceeding 0.10m²........	0.50	10.00	0.55	0.74	-	13.59	0.37	14.51	m²	15.96
girth exceeding 300mm; panes, area 0.10 - 0.50m²...............	0.35	10.00	0.39	0.53	-	9.74	0.26	10.39	m²	11.43
girth exceeding 300mm; panes, area 0.50 - 1.00m²...............	0.32	10.00	0.36	0.44	-	8.08	0.22	8.66	m²	9.52
girth exceeding 300mm; panes, area exceeding 1.00m²........	0.32	10.00	0.36	0.40	-	7.35	0.20	7.90	m²	8.69
Wood partially glazed doors										
girth exceeding 300mm; panes, area not exceeding 0.10m²........	0.70	10.00	0.78	0.59	-	10.84	0.29	11.91	m²	13.10
girth exceeding 300mm; panes, area 0.10 - 0.50m²...............	0.53	10.00	0.58	0.48	-	8.82	0.24	9.64	m²	10.60
girth exceeding 300mm; panes, area 0.50 - 1.00m²...............	0.50	10.00	0.55	0.44	-	8.08	0.22	8.85	m²	9.74
girth exceeding 300mm; panes, area exceeding 1.00m²........	0.50	10.00	0.55	0.42	-	7.72	0.21	8.47	m²	9.32
Wood windows and screens										
girth exceeding 300mm; panes, area not exceeding 0.10m²........	0.70	10.00	0.78	0.81	-	14.88	0.41	16.06	m²	17.67
girth exceeding 300mm; panes, area 0.10 - 0.50m²...............	0.53	10.00	0.58	0.58	-	10.65	0.29	11.53	m²	12.68
girth exceeding 300mm; panes, area 0.50 - 1.00m²...............	0.50	10.00	0.55	0.48	-	8.82	0.24	9.61	m²	10.57
girth exceeding 300mm; panes, area exceeding 1.00m²........	0.50	10.00	0.55	0.44	-	8.08	0.22	8.85	m²	9.74
Two coats polyurethane lacquer; external work										
Wood general surfaces										
girth exceeding 300mm	0.88	10.00	0.97	0.46	-	8.45	0.23	9.65	m²	10.61
isolated surfaces, girth not exceeding 300mm	0.26	10.00	0.29	0.15	-	2.76	0.08	3.12	m	3.43
Wood glazed doors										
girth exceeding 300mm; panes, area not exceeding 0.10m²........	0.50	10.00	0.55	0.74	-	13.59	0.37	14.51	m²	15.96
girth exceeding 300mm; panes, area 0.10 - 0.50m²...............	0.35	10.00	0.39	0.53	-	9.74	0.26	10.39	m²	11.43
girth exceeding 300mm; panes, area 0.50 - 1.00m²...............	0.32	10.00	0.36	0.44	-	8.08	0.22	8.66	m²	9.52
girth exceeding 300mm; panes, area exceeding 1.00m²........	0.32	10.00	0.36	0.40	-	7.35	0.20	7.90	m²	8.69
Wood partially glazed doors										
girth exceeding 300mm; panes, area not exceeding 0.10m²........	0.70	10.00	0.78	0.59	-	10.84	0.29	11.91	m²	13.10
girth exceeding 300mm; panes, area 0.10 - 0.50m²...............	0.53	10.00	0.58	0.48	-	8.82	0.24	9.64	m²	10.60
girth exceeding 300mm; panes, area 0.50 - 1.00m²...............	0.50	10.00	0.55	0.44	-	8.08	0.22	8.85	m²	9.74
girth exceeding 300mm; panes, area exceeding 1.00m²........	0.50	10.00	0.55	0.42	-	7.72	0.21	8.47	m²	9.32
Wood windows and screens										
girth exceeding 300mm; panes, area not exceeding 0.10m²........	0.70	10.00	0.78	0.84	-	15.43	0.42	16.63	m²	18.29
girth exceeding 300mm; panes, area 0.10 - 0.50m²...............	0.53	10.00	0.58	0.61	-	11.21	0.31	12.09	m²	13.30
girth exceeding 300mm; panes, area 0.50 - 1.00m²...............	0.50	10.00	0.55	0.51	-	9.37	0.25	10.17	m²	11.19
girth exceeding 300mm; panes, area exceeding 1.00m²........	0.50	10.00	0.55	0.46	-	8.45	0.23	9.23	m²	10.15
Three coats polyurethane lacquer										
Wood general surfaces										
girth exceeding 300mm	1.29	10.00	1.42	0.52	-	9.55	0.26	11.23	m²	12.36
isolated surfaces, girth not exceeding 300mm	0.38	10.00	0.42	0.18	-	3.31	0.09	3.82	m	4.20
Wood glazed doors										
girth exceeding 300mm; panes, area not exceeding 0.10m²........	0.73	10.00	0.81	0.86	-	15.80	0.43	17.04	m²	18.74
girth exceeding 300mm; panes, area 0.10 - 0.50m²...............	0.53	10.00	0.58	0.62	-	11.39	0.31	12.28	m²	13.51
girth exceeding 300mm; panes, area 0.50 - 1.00m²...............	0.47	10.00	0.52	0.52	-	9.55	0.26	10.33	m²	11.36
girth exceeding 300mm; panes, area exceeding 1.00m²............	0.47	10.00	0.52	0.46	-	8.45	0.23	9.20	m²	10.12
Wood partially glazed doors										
girth exceeding 300mm; panes, area not exceeding 0.10m²........	1.03	10.00	1.13	0.69	-	12.68	0.34	14.15	m²	15.57
girth exceeding 300mm; panes, area 0.10 - 0.50m²...............	0.76	10.00	0.84	0.63	-	11.57	0.31	12.73	m²	14.00

DECORATION

Labour hourly rates: (except Specialists) Craft Operatives 18.37 Labourer 13.76 Rates are national average prices. Refer to REGIONAL VARIATIONS for indicative levels of overall pricing in regions	MATERIALS			LABOUR				RATES		
	Del to Site	Waste	Material Cost	Craft Optve	Lab	Labour Cost	Sunds	Nett Rate	Unit	Gross rate (10%)
	£	%	£	Hrs	Hrs	£	£	£		£
PAINTING & CLEAR FINISHINGS; POLYURETHANE LACQUER (Cont'd)										
Three coats polyurethane lacquer (Cont'd)										
Wood partially glazed doors (Cont'd)										
girth exceeding 300mm; panes, area 0.50 - 1.00m²	0.70	10.00	0.78	0.52	-	9.55	0.26	10.59	m²	11.65
girth exceeding 300mm; panes, area exceeding 1.00m²	0.70	10.00	0.78	0.50	-	9.19	0.25	10.21	m²	11.23
Wood windows and screens										
girth exceeding 300mm; panes, area not exceeding 0.10m²	0.73	10.00	0.81	0.95	-	17.45	0.47	18.73	m²	20.61
girth exceeding 300mm; panes, area 0.10 - 0.50m²	0.53	10.00	0.58	0.68	-	12.49	0.34	13.41	m²	14.75
girth exceeding 300mm; panes, area 0.50 - 1.00m²	0.47	10.00	0.52	0.57	-	10.47	0.28	11.27	m²	12.40
girth exceeding 300mm; panes, area exceeding 1.00m²	0.47	10.00	0.52	0.51	-	9.37	0.25	10.14	m²	11.15
Three coats polyurethane lacquer; external work										
Wood general surfaces										
girth exceeding 300mm	1.29	10.00	1.42	0.55	-	10.10	0.28	11.80	m²	12.98
isolated surfaces, girth not exceeding 300mm	0.38	10.00	0.42	0.19	-	3.49	0.09	4.01	m	4.41
Wood glazed doors										
girth exceeding 300mm; panes, area not exceeding 0.10m²	0.73	10.00	0.81	0.86	-	15.80	0.43	17.04	m²	18.74
girth exceeding 300mm; panes, area 0.10 - 0.50m²	0.53	10.00	0.58	0.62	-	11.39	0.31	12.28	m²	13.51
girth exceeding 300mm; panes, area 0.50 - 1.00m²	0.47	10.00	0.52	0.52	-	9.55	0.26	10.33	m²	11.36
girth exceeding 300mm; panes, area exceeding 1.00m²	0.47	10.00	0.52	0.46	-	8.45	0.23	9.20	m²	10.12
Wood partially glazed doors										
girth exceeding 300mm; panes, area not exceeding 0.10m²	1.03	10.00	1.13	0.69	-	12.68	0.34	14.15	m²	15.57
girth exceeding 300mm; panes, area 0.10 - 0.50m²	0.76	10.00	0.84	0.63	-	11.57	0.31	12.73	m²	14.00
girth exceeding 300mm; panes, area 0.50 - 1.00m²	0.70	10.00	0.78	0.52	-	9.55	0.26	10.59	m²	11.65
girth exceeding 300mm; panes, area exceeding 1.00m²	0.70	10.00	0.78	0.50	-	9.19	0.25	10.21	m²	11.23
Wood windows and screens										
girth exceeding 300mm; panes, area not exceeding 0.10m²	0.73	10.00	0.81	0.98	-	18.00	0.49	19.30	m²	21.23
girth exceeding 300mm; panes, area 0.10 - 0.50m²	0.53	10.00	0.58	0.72	-	13.23	0.36	14.17	m²	15.58
girth exceeding 300mm; panes, area 0.50 - 1.00m²	0.47	10.00	0.52	0.61	-	11.21	0.31	12.03	m²	13.23
girth exceeding 300mm; panes, area exceeding 1.00m²	0.47	10.00	0.52	0.54	-	9.92	0.27	10.71	m²	11.78
PAINTING & CLEAR FINISHINGS; FIRE RETARDANT PAINTS AND VARNISHES										
Two coats fire retardant paint										
Wood general surfaces										
girth exceeding 300mm	6.57	10.00	7.22	0.40	-	7.35	0.20	14.77	m²	16.25
isolated surfaces, girth not exceeding 300mm	1.99	10.00	2.19	0.13	-	2.39	0.06	4.64	m	5.11
Two coats fire retardant varnish, one overcoat varnish										
Wood general surfaces										
girth exceeding 300mm	9.53	10.00	10.49	0.54	-	9.92	0.27	20.68	m²	22.75
isolated surfaces, girth not exceeding 300mm	2.90	10.00	3.19	0.18	-	3.31	0.09	6.59	m	7.25
PAINTING & CLEAR FINISHINGS; OILING HARDWOOD										
Two coats raw linseed oil										
Wood general surfaces										
girth exceeding 300mm	1.12	10.00	1.23	0.44	-	8.08	0.22	9.53	m²	10.48
isolated surfaces, girth not exceeding 300mm	0.33	10.00	0.37	0.15	-	2.76	0.08	3.20	m	3.52
PAINTING & CLEAR FINISHINGS; FRENCH AND WAX POLISHING										
Stain; two coats white polish										
Wood general surfaces										
girth exceeding 300mm	4.03	10.00	4.44	0.56	-	10.29	0.28	15.01	m²	16.51
isolated surfaces, girth not exceeding 300mm	1.27	10.00	1.40	0.19	-	3.49	0.09	4.98	m	5.48
Open grain French polish										
Wood general surfaces										
girth exceeding 300mm	5.13	10.00	5.65	2.64	-	48.50	1.32	55.47	m²	61.01
isolated surfaces, girth not exceeding 300mm	1.62	10.00	1.78	0.88	-	16.17	0.44	18.39	m	20.23
Stain; body in; fully French polish										
Wood general surfaces										
girth exceeding 300mm	7.33	10.00	8.07	3.96	-	72.75	1.98	82.79	m²	91.07
isolated surfaces, girth not exceeding 300mm	2.31	10.00	2.55	1.32	-	24.25	0.66	27.45	m	30.20
PAINTING & CLEAR FINISHINGS; ROAD MARKINGS										
One coat solvent based road marking paint; external work										
Concrete general surfaces										
isolated surfaces, 50mm wide	0.07	5.00	0.08	0.10	-	1.84	0.05	1.96	m	2.16
isolated surfaces, 100mm wide	0.15	5.00	0.15	0.12	-	2.20	0.06	2.42	m	2.66
Two coats solvent based road marking paint; external work										
Concrete general surfaces										
isolated surfaces, 50mm wide	0.15	5.00	0.15	0.15	-	2.76	0.08	2.99	m	3.28
isolated surfaces, 100mm wide	0.30	5.00	0.31	0.18	-	3.31	0.09	3.71	m	4.08

Labour hourly rates: (except Specialists) Craft Operatives 18.37 Labourer 13.76 Rates are national average prices. Refer to REGIONAL VARIATIONS for indicative levels of overall pricing in regions	MATERIALS			LABOUR				RATES		
	Del to Site £	Waste %	Material Cost £	Craft Optve Hrs	Lab Hrs	Labour Cost £	Sunds £	Nett Rate £	Unit	Gross rate (10%) £
PAINTING & CLEAR FINISHINGS; ROAD MARKINGS (Cont'd)										
Prime and apply non-reflective self adhesive road marking tape; external work										
Concrete general surfaces										
isolated surfaces, 50mm wide	1.38	5.00	1.45	0.10	-	1.84	0.05	3.33	m	3.67
isolated surfaces, 100mm wide	1.80	5.00	1.89	0.13	-	2.39	0.06	4.35	m	4.78
Prime and apply reflective self adhesive road marking tape; external work										
Concrete general surfaces										
isolated surfaces, 50mm wide	1.58	5.00	1.66	0.10	-	1.84	0.05	3.54	m	3.90
isolated surfaces, 100mm wide	2.00	5.00	2.10	0.13	-	2.39	0.06	4.56	m	5.01
PAINTING & CLEAR FINISHINGS; REDECORATIONS: EMULSION PAINTING										
Generally										
Note: the following rates include for the cost of all preparatory work, e.g. washing down, etc.										
Two coats emulsion paint, existing emulsion painted surfaces										
Concrete general surfaces										
girth exceeding 300mm	0.28	10.00	0.31	0.32	-	5.88	0.16	6.35	m²	6.98
Plaster general surfaces										
girth exceeding 300mm	0.22	10.00	0.25	0.28	-	5.14	0.14	5.53	m²	6.08
Plasterboard general surfaces										
girth exceeding 300mm	0.22	10.00	0.25	0.29	-	5.33	0.14	5.72	m²	6.29
Brickwork general surfaces										
girth exceeding 300mm	0.28	10.00	0.31	0.32	-	5.88	0.16	6.35	m²	6.98
Paper covered general surfaces										
girth exceeding 300mm	0.22	10.00	0.25	0.30	-	5.51	0.15	5.91	m²	6.50
Two full coats emulsion paint, existing washable distempered surfaces										
Concrete general surfaces										
girth exceeding 300mm	0.28	10.00	0.31	0.33	-	6.06	0.17	6.54	m²	7.19
Plaster general surfaces										
girth exceeding 300mm	0.22	10.00	0.25	0.29	-	5.33	0.14	5.72	m²	6.29
Plasterboard general surfaces										
girth exceeding 300mm	0.22	10.00	0.25	0.30	-	5.51	0.15	5.91	m²	6.50
Brickwork general surfaces										
girth exceeding 300mm	0.28	10.00	0.31	0.33	-	6.06	0.17	6.54	m²	7.19
Paper covered general surfaces										
girth exceeding 300mm	0.22	10.00	0.25	0.30	-	5.51	0.15	5.91	m²	6.50
Two full coats emulsion paint, existing textured plastic coating surfaces										
Concrete general surfaces										
girth exceeding 300mm	0.28	10.00	0.31	0.36	-	6.61	0.18	7.10	m²	7.81
Plaster general surfaces										
girth exceeding 300mm	0.22	10.00	0.25	0.32	-	5.88	0.16	6.29	m²	6.91
Plasterboard general surfaces										
girth exceeding 300mm	0.22	10.00	0.25	0.33	-	6.06	0.17	6.47	m²	7.12
Brickwork general surfaces										
girth exceeding 300mm	0.29	10.00	0.32	0.36	-	6.61	0.18	7.11	m²	7.82
Paper covered general surfaces										
girth exceeding 300mm	0.22	10.00	0.25	0.33	-	6.06	0.17	6.47	m²	7.12
Mist coat, two full coats emulsion paint, existing non-washable distempered surfaces										
Concrete general surfaces										
girth exceeding 300mm	0.35	10.00	0.38	0.38	-	6.98	0.19	7.55	m²	8.31
Plaster general surfaces										
girth exceeding 300mm	0.28	10.00	0.31	0.34	-	6.25	0.17	6.73	m²	7.40
Plasterboard general surfaces										
girth exceeding 300mm	0.28	10.00	0.31	0.35	-	6.43	0.17	6.92	m²	7.61
Brickwork general surfaces										
girth exceeding 300mm	0.35	10.00	0.38	0.38	-	6.98	0.19	7.55	m²	8.31
Paper covered general surfaces										
girth exceeding 300mm	0.28	10.00	0.31	0.35	-	6.43	0.17	6.92	m²	7.61
Two full coats emulsion paint, existing gloss painted surfaces										
Concrete general surfaces										
girth exceeding 300mm	0.28	10.00	0.31	0.33	-	6.06	0.17	6.54	m²	7.19
Plaster general surfaces										
girth exceeding 300mm	0.22	10.00	0.25	0.29	-	5.33	0.14	5.72	m²	6.29

303

DECORATION

Labour hourly rates: (except Specialists) Craft Operatives 18.37 Labourer 13.76 Rates are national average prices. Refer to REGIONAL VARIATIONS for indicative levels of overall pricing in regions	MATERIALS			LABOUR				RATES		
	Del to Site	Waste	Material Cost	Craft Optve	Lab	Labour Cost	Sunds	Nett Rate	Unit	Gross rate (10%)
	£	%	£	Hrs	Hrs	£	£	£		£
PAINTING & CLEAR FINISHINGS; REDECORATIONS: EMULSION PAINTING (Cont'd)										
Two full coats emulsion paint, existing gloss painted surfaces (Cont'd)										
Plasterboard general surfaces girth exceeding 300mm	0.22	10.00	0.25	0.30	-	5.51	0.15	5.91	m²	6.50
Brickwork general surfaces girth exceeding 300mm	0.29	10.00	0.32	0.33	-	6.06	0.17	6.55	m²	7.20
Paper covered general surfaces girth exceeding 300mm	0.22	10.00	0.25	0.30	-	5.51	0.15	5.91	m²	6.50
PAINTING & CLEAR FINISHINGS; REDECORATIONS: CEMENT PAINTING										
Generally										
Note: the following rates include for the cost of all preparatory work, e.g. washing down, etc.										
One coat sealer, one coat waterproof cement paint; external work										
Cement rendered general surfaces girth exceeding 300mm	1.90	10.00	2.09	0.36	-	6.61	0.18	8.89	m²	9.77
Concrete general surfaces girth exceeding 300mm	1.82	10.00	2.01	0.43	-	7.90	0.22	10.12	m²	11.13
Brickwork general surfaces girth exceeding 300mm	1.86	10.00	2.05	0.43	-	7.90	0.22	10.16	m²	11.18
Rough cast general surfaces girth exceeding 300mm	1.90	10.00	2.09	0.52	-	9.55	0.26	11.91	m²	13.10
Two coats sealer, one coat waterproof cement paint; external work										
Cement rendered general surfaces girth exceeding 300mm	2.99	10.00	3.29	0.47	-	8.63	0.23	12.16	m²	13.37
Concrete general surfaces girth exceeding 300mm	2.57	10.00	2.83	0.58	-	10.65	0.29	13.77	m²	15.15
Brickwork general surfaces girth exceeding 300mm	2.95	10.00	3.24	0.58	-	10.65	0.29	14.19	m²	15.61
Rough cast general surfaces girth exceeding 300mm	2.99	10.00	3.29	0.70	-	12.86	0.35	16.50	m²	18.15
One coat waterproof cement paint existing cement painted surfaces; external work										
Cement rendered general surfaces girth exceeding 300mm	0.25	10.00	0.27	0.27	-	4.96	0.14	5.37	m²	5.91
Concrete general surfaces girth exceeding 300mm	0.25	10.00	0.27	0.34	-	6.25	0.17	6.69	m²	7.36
Brickwork general surfaces girth exceeding 300mm	0.29	10.00	0.32	0.34	-	6.25	0.17	6.73	m²	7.41
Rough cast general surfaces girth exceeding 300mm	0.33	10.00	0.36	0.44	-	8.08	0.22	8.67	m²	9.53
Two coats waterproof cement paint existing cement painted surfaces; external work										
Cement rendered general surfaces girth exceeding 300mm	0.66	10.00	0.72	0.38	-	6.98	0.19	7.90	m²	8.68
Concrete general surfaces girth exceeding 300mm	0.50	10.00	0.55	0.49	-	9.00	0.25	9.80	m²	10.77
Brickwork general surfaces girth exceeding 300mm	0.58	10.00	0.64	0.49	-	9.00	0.25	9.88	m²	10.87
Rough cast general surfaces girth exceeding 300mm	0.66	10.00	0.72	0.60	-	11.02	0.30	12.05	m²	13.25
One coat stabilizing solution, one coat textured masonry paint, existing undecorated surfaces; external work										
Cement rendered general surfaces girth exceeding 300mm	0.74	10.00	0.81	0.35	-	6.43	0.17	7.42	m²	8.16
Concrete general surfaces girth exceeding 300mm	0.91	10.00	1.01	0.42	-	7.72	0.21	8.93	m²	9.82
Brickwork general surfaces girth exceeding 300mm	1.39	10.00	1.53	0.42	-	7.72	0.21	9.45	m²	10.40
Rough cast general surfaces girth exceeding 300mm	1.74	10.00	1.91	0.52	-	9.55	0.26	11.72	m²	12.89
One coat stabilizing solution, two coats textured masonry paint, existing undecorated surfaces; external work										
Cement rendered general surfaces girth exceeding 300mm	1.52	10.00	1.67	0.47	-	8.63	0.23	10.54	m²	11.59

Labour hourly rates: (except Specialists) Craft Operatives 18.37 Labourer 13.76 Rates are national average prices. Refer to REGIONAL VARIATIONS for indicative levels of overall pricing in regions	MATERIALS			LABOUR				RATES		
	Del to Site £	Waste %	Material Cost £	Craft Optve Hrs	Lab Hrs	Labour Cost £	Sunds £	Nett Rate £	Unit	Gross rate (10%) £
PAINTING & CLEAR FINISHINGS; REDECORATIONS: CEMENT PAINTING (Cont'd)										
One coat stabilizing solution, two coats textured masonry paint, existing undecorated surfaces; external work (Cont'd)										
Concrete general surfaces girth exceeding 300mm	1.52	10.00	1.67	0.57	-	10.47	0.28	12.43	m²	13.67
Brickwork general surfaces girth exceeding 300mm	2.47	10.00	2.72	0.57	-	10.47	0.28	13.47	m²	14.82
Rough cast general surfaces girth exceeding 300mm	3.16	10.00	3.48	0.70	-	12.86	0.35	16.69	m²	18.36
One coat textured masonry paint, existing cement painted surfaces; external work										
Cement rendered general surfaces girth exceeding 300mm	0.43	10.00	0.48	0.26	-	4.78	0.13	5.38	m²	5.92
Concrete general surfaces girth exceeding 300mm	0.61	10.00	0.67	0.33	-	6.06	0.17	6.89	m²	7.58
Brickwork general surfaces girth exceeding 300mm	1.08	10.00	1.19	0.33	-	6.06	0.17	7.42	m²	8.16
Rough cast general surfaces girth exceeding 300mm	1.43	10.00	1.57	0.43	-	7.90	0.22	9.68	m²	10.65
Two coats textured masonry paint, existing cement painted surfaces; external work										
Cement rendered general surfaces girth exceeding 300mm	1.21	10.00	1.33	0.39	-	7.16	0.19	8.69	m²	9.56
Concrete general surfaces girth exceeding 300mm	1.21	10.00	1.33	0.48	-	8.82	0.24	10.39	m²	11.43
Brickwork general surfaces girth exceeding 300mm	2.16	10.00	2.38	0.48	-	8.82	0.24	11.44	m²	12.58
Rough cast general surfaces girth exceeding 300mm	2.85	10.00	3.14	0.62	-	11.39	0.31	14.84	m²	16.32
PAINTING & CLEAR FINISHINGS; REDECORATIONS: OIL PAINTING WALLS AND CEILINGS										
Generally										
Note: the following rates include for the cost of all preparatory work, e.g. washing down, etc.										
One undercoat, one coat full gloss finish, existing gloss painted surfaces										
Concrete general surfaces girth exceeding 300mm	0.58	10.00	0.63	0.38	-	6.98	0.19	7.80	m²	8.58
Brickwork general surfaces girth exceeding 300mm	0.62	10.00	0.68	0.39	-	7.16	0.19	8.04	m²	8.84
Plasterboard general surfaces girth exceeding 300mm	0.53	10.00	0.59	0.35	-	6.43	0.17	7.19	m²	7.91
Plaster general surfaces girth exceeding 300mm	0.53	10.00	0.59	0.34	-	6.25	0.17	7.00	m²	7.70
One undercoat, one coat eggshell finish, existing gloss painted surfaces										
Concrete general surfaces girth exceeding 300mm	0.61	10.00	0.67	0.38	-	6.98	0.19	7.84	m²	8.63
Brickwork general surfaces girth exceeding 300mm	0.77	10.00	0.84	0.39	-	7.16	0.19	8.20	m²	9.02
Plasterboard general surfaces girth exceeding 300mm	0.57	10.00	0.62	0.35	-	6.43	0.17	7.23	m²	7.95
Plaster general surfaces girth exceeding 300mm	0.57	10.00	0.62	0.34	-	6.25	0.17	7.04	m²	7.74
Two undercoats, one coat full gloss finish, existing gloss painted surfaces										
Concrete general surfaces girth exceeding 300mm	0.85	10.00	0.94	0.48	-	8.82	0.24	10.00	m²	11.00
Brickwork general surfaces girth exceeding 300mm	1.02	10.00	1.13	0.48	-	8.82	0.24	10.18	m²	11.20
Plasterboard general surfaces girth exceeding 300mm	0.81	10.00	0.89	0.45	-	8.27	0.22	9.38	m²	10.32
Plaster general surfaces girth exceeding 300mm	0.81	10.00	0.89	0.44	-	8.08	0.22	9.19	m²	10.11
Two undercoats, one coat eggshell finish, existing gloss painted surfaces										
Concrete general surfaces girth exceeding 300mm	0.89	10.00	0.98	0.48	-	8.82	0.24	10.03	m²	11.04

DECORATION

Labour hourly rates: (except Specialists) Craft Operatives 18.37 Labourer 13.76 Rates are national average prices. Refer to REGIONAL VARIATIONS for indicative levels of overall pricing in regions	MATERIALS			LABOUR				RATES		
	Del to Site £	Waste %	Material Cost £	Craft Optve Hrs	Lab Hrs	Labour Cost £	Sunds £	Nett Rate £	Unit	Gross rate (10%) £
PAINTING & CLEAR FINISHINGS; REDECORATIONS: OIL PAINTING WALLS AND CEILINGS (Cont'd)										
Two undercoats, one coat eggshell finish, existing gloss painted surfaces (Cont'd)										
Brickwork general surfaces girth exceeding 300mm	1.07	10.00	1.17	0.48	-	8.82	0.24	10.23	m²	11.25
Plasterboard general surfaces girth exceeding 300mm	0.84	10.00	0.93	0.45	-	8.27	0.22	9.42	m²	10.36
Plaster general surfaces girth exceeding 300mm	0.84	10.00	0.93	0.44	-	8.08	0.22	9.23	m²	10.15
One coat primer, one undercoat, one coat full gloss finish, existing washable distempered or emulsion painted surfaces										
Concrete general surfaces girth exceeding 300mm	2.28	10.00	2.50	0.48	-	8.82	0.24	11.56	m²	12.72
Brickwork general surfaces girth exceeding 300mm	2.71	10.00	2.98	0.48	-	8.82	0.24	12.03	m²	13.24
Plasterboard general surfaces girth exceeding 300mm	2.08	10.00	2.29	0.45	-	8.27	0.22	10.78	m²	11.86
Plaster general surfaces girth exceeding 300mm	2.08	10.00	2.29	0.44	-	8.08	0.22	10.59	m²	11.65
One coat primer, one undercoat, one coat eggshell finish, existing washable distempered or emulsion painted surfaces										
Concrete general surfaces girth exceeding 300mm	2.31	10.00	2.54	0.48	-	8.82	0.24	11.60	m²	12.76
Brickwork general surfaces girth exceeding 300mm	2.86	10.00	3.14	0.48	-	8.82	0.24	12.20	m²	13.42
Plasterboard general surfaces girth exceeding 300mm	2.11	10.00	2.32	0.45	-	8.27	0.22	10.81	m²	11.90
Plaster general surfaces girth exceeding 300mm	2.11	10.00	2.32	0.44	-	8.08	0.22	10.63	m²	11.69
One coat primer, two undercoats, one coat full gloss finish, existing washable distempered or emulsion painted surfaces										
Concrete general surfaces girth exceeding 300mm	2.55	10.00	2.81	0.58	-	10.65	0.29	13.75	m²	15.13
Brickwork general surfaces girth exceeding 300mm	3.11	10.00	3.42	0.59	-	10.84	0.29	14.56	m²	16.01
Plasterboard general surfaces girth exceeding 300mm	2.36	10.00	2.59	0.55	-	10.10	0.28	12.97	m²	14.27
Plaster general surfaces girth exceeding 300mm	2.36	10.00	2.59	0.54	-	9.92	0.27	12.78	m²	14.06
One coat primer, two undercoats, one coat eggshell finish, existing washable distempered or emulsion painted surfaces										
Concrete general surfaces girth exceeding 300mm	2.59	10.00	2.85	0.58	-	10.65	0.29	13.79	m²	15.17
Brickwork general surfaces girth exceeding 300mm	3.15	10.00	3.47	0.59	-	10.84	0.29	14.60	m²	16.06
Plasterboard general surfaces girth exceeding 300mm	2.39	10.00	2.63	0.55	-	10.10	0.28	13.01	m²	14.31
Plaster general surfaces girth exceeding 300mm	2.39	10.00	2.63	0.54	-	9.92	0.27	12.82	m²	14.10
One coat primer, one undercoat, one coat full gloss finish, existing non-washable distempered surfaces										
Concrete general surfaces girth exceeding 300mm	2.28	10.00	2.50	0.48	-	8.82	0.24	11.56	m²	12.72
Brickwork general surfaces girth exceeding 300mm	2.71	10.00	2.98	0.48	-	8.82	0.24	12.03	m²	13.24
Plasterboard general surfaces girth exceeding 300mm	2.08	10.00	2.29	0.45	-	8.27	0.22	10.78	m²	11.86
Plaster general surfaces girth exceeding 300mm	2.08	10.00	2.29	0.44	-	8.08	0.22	10.59	m²	11.65
One coat primer, one undercoat, one coat eggshell finish, existing non-washable distempered surfaces										
Concrete general surfaces girth exceeding 300mm	2.31	10.00	2.54	0.48	-	8.82	0.24	11.60	m²	12.76
Brickwork general surfaces girth exceeding 300mm	2.86	10.00	3.14	0.48	-	8.82	0.24	12.20	m²	13.42
Plasterboard general surfaces girth exceeding 300mm	2.11	10.00	2.32	0.45	-	8.27	0.22	10.81	m²	11.90
Plaster general surfaces girth exceeding 300mm	2.11	10.00	2.32	0.44	-	8.08	0.22	10.63	m²	11.69

Labour hourly rates: (except Specialists) Craft Operatives 18.37 Labourer 13.76 Rates are national average prices. Refer to REGIONAL VARIATIONS for indicative levels of overall pricing in regions	MATERIALS			LABOUR				RATES		
	Del to Site	Waste	Material Cost	Craft Optve	Lab	Labour Cost	Sunds	Nett Rate	Unit	Gross rate (10%)
	£	%	£	Hrs	Hrs	£	£	£		£
PAINTING & CLEAR FINISHINGS; REDECORATIONS: OIL PAINTING WALLS AND CEILINGS (Cont'd)										
One coat primer, two undercoats, one coat full gloss finish, existing non-washable distempered surfaces										
Concrete general surfaces girth exceeding 300mm	2.55	10.00	2.81	0.58	-	10.65	0.29	13.75	m²	15.13
Brickwork general surfaces girth exceeding 300mm	3.11	10.00	3.42	0.59	-	10.84	0.29	14.56	m²	16.01
Plasterboard general surfaces girth exceeding 300mm	2.36	10.00	2.59	0.55	-	10.10	0.28	12.97	m²	14.27
Plaster general surfaces girth exceeding 300mm	2.36	10.00	2.59	0.54	-	9.92	0.27	12.78	m²	14.06
One coat primer, two undercoats, one coat eggshell finish, existing non-washable distempered surfaces										
Concrete general surfaces girth exceeding 300mm	2.59	10.00	2.85	0.58	-	10.65	0.29	13.79	m²	15.17
Brickwork general surfaces girth exceeding 300mm	3.15	10.00	3.47	0.59	-	10.84	0.29	14.60	m²	16.06
Plasterboard general surfaces girth exceeding 300mm	2.39	10.00	2.63	0.55	-	10.10	0.28	13.01	m²	14.31
Plaster general surfaces girth exceeding 300mm	2.39	10.00	2.63	0.54	-	9.92	0.27	12.82	m²	14.10
PAINTING & CLEAR FINISHINGS; REDECORATIONS: PLASTIC FINISH										
Generally										
Note: the following rates include for the cost of all preparatory work, e.g. washing down, etc.										
Textured plastic coating - stippled finish, existing washable distempered or emulsion painted surfaces										
Concrete general surfaces girth exceeding 300mm	0.43	10.00	0.48	0.40	-	7.35	0.20	8.03	m²	8.83
Brickwork general surfaces girth exceeding 300mm	0.52	10.00	0.57	0.45	-	8.27	0.22	9.06	m²	9.97
Plasterboard general surfaces girth exceeding 300mm	0.39	10.00	0.43	0.34	-	6.25	0.17	6.84	m²	7.53
Plaster general surfaces girth exceeding 300mm	0.39	10.00	0.43	0.34	-	6.25	0.17	6.84	m²	7.53
Textured plastic coating - combed finish, existing washable distempered or emulsion painted surfaces										
Concrete general surfaces girth exceeding 300mm	0.52	10.00	0.57	0.45	-	8.27	0.22	9.06	m²	9.97
Brickwork general surfaces girth exceeding 300mm	0.62	10.00	0.68	0.50	-	9.19	0.25	10.12	m²	11.13
Plasterboard general surfaces girth exceeding 300mm	0.47	10.00	0.51	0.40	-	7.35	0.20	8.06	m²	8.87
Plaster general surfaces girth exceeding 300mm	0.47	10.00	0.51	0.40	-	7.35	0.20	8.06	m²	8.87
Textured plastic coating - stippled finish, existing non-washable distempered surfaces										
Concrete general surfaces girth exceeding 300mm	0.43	10.00	0.48	0.45	-	8.27	0.22	8.97	m²	9.87
Brickwork general surfaces girth exceeding 300mm	0.52	10.00	0.57	0.50	-	9.19	0.25	10.01	m²	11.01
Plasterboard general surfaces girth exceeding 300mm	0.39	10.00	0.43	0.40	-	7.35	0.20	7.98	m²	8.77
Plaster general surfaces girth exceeding 300mm	0.39	10.00	0.43	0.40	-	7.35	0.20	7.98	m²	8.77
Textured plastic coating - combed finish, existing non-washable distempered surfaces										
Concrete general surfaces girth exceeding 300mm	0.52	10.00	0.57	0.50	-	9.19	0.25	10.01	m²	11.01
Brickwork general surfaces girth exceeding 300mm	0.62	10.00	0.68	0.56	-	10.29	0.28	11.25	m²	12.38
Plasterboard general surfaces girth exceeding 300mm	0.47	10.00	0.51	0.45	-	8.27	0.22	9.00	m²	9.91
Plaster general surfaces girth exceeding 300mm	0.47	10.00	0.51	0.45	-	8.27	0.22	9.00	m²	9.91

DECORATION

Labour hourly rates: (except Specialists) Craft Operatives 18.37 Labourer 13.76 Rates are national average prices. Refer to REGIONAL VARIATIONS for indicative levels of overall pricing in regions	MATERIALS			LABOUR				RATES		
	Del to Site	Waste	Material Cost	Craft Optve	Lab	Labour Cost	Sunds	Nett Rate	Unit	Gross rate (10%)
	£	%	£	Hrs	Hrs	£	£	£		£

PAINTING & CLEAR FINISHINGS; REDECORATIONS: CLEAN OUT GUTTERS

Generally

Note: the following rates include for the cost of all preparatory work, e.g. washing down, etc.

Clean out gutters prior to repainting, staunch joints with red lead, bituminous compound or mastic

	Del to Site	Waste	Material Cost	Craft Optve	Lab	Labour Cost	Sunds	Nett Rate	Unit	Gross rate
Iron or steel eaves gutters generally	-	-	-	0.06	-	1.10	0.09	1.19	m	1.31

PAINTING & CLEAR FINISHINGS; REDECORATIONS: OIL PAINTING METALWORK

Generally

Note: the following rates include for the cost of all preparatory work, e.g. washing down, etc.

One undercoat, one coat full gloss finish, existing gloss painted metal surfaces

	Del to Site	Waste	Material Cost	Craft Optve	Lab	Labour Cost	Sunds	Nett Rate	Unit	Gross rate
Iron or steel general surfaces										
girth exceeding 300mm	0.55	10.00	0.61	0.47	-	8.63	0.23	9.48	m²	10.43
isolated surfaces, girth not exceeding 300mm	0.17	10.00	0.19	0.16	-	2.94	0.08	3.21	m	3.53
Galvanized glazed doors, windows or screens										
girth exceeding 300mm; panes, area not exceeding 0.10m²	0.30	10.00	0.33	0.88	-	16.17	0.44	16.93	m²	18.63
girth exceeding 300mm; panes, area 0.10 - 0.50m²	0.21	10.00	0.23	0.64	-	11.76	0.32	12.31	m²	13.54
girth exceeding 300mm; panes, area 0.50 - 1.00m²	0.17	10.00	0.19	0.53	-	9.74	0.26	10.19	m²	11.21
girth exceeding 300mm; panes, area exceeding 1.00m²	0.17	10.00	0.19	0.46	-	8.45	0.23	8.87	m²	9.75
Iron or steel structural work										
girth exceeding 300mm	0.55	10.00	0.61	0.52	-	9.55	0.26	10.42	m²	11.46
isolated surfaces, girth not exceeding 300mm	0.17	10.00	0.19	0.21	-	3.86	0.10	4.15	m	4.57
Iron or steel structural members of roof trusses, lattice girders, purlins and the like										
girth exceeding 300mm	0.55	10.00	0.61	0.65	-	11.94	0.32	12.88	m²	14.16
isolated surfaces, girth not exceeding 300mm	0.17	10.00	0.19	0.22	-	4.04	0.11	4.34	m	4.77
Iron or steel services										
girth exceeding 300mm	0.55	10.00	0.61	0.65	-	11.94	0.32	12.88	m²	14.16
isolated surfaces, girth not exceeding 300mm	0.17	10.00	0.19	0.22	-	4.04	0.11	4.34	m	4.77
isolated areas not exceeding 0.50m² irrespective of girth	0.30	10.00	0.33	0.65	-	11.94	0.32	12.59	nr	13.85
Copper services										
girth exceeding 300mm	0.55	10.00	0.61	0.65	-	11.94	0.32	12.88	m²	14.16
isolated surfaces, girth not exceeding 300mm	0.17	10.00	0.19	0.22	-	4.04	0.11	4.34	m	4.77
Galvanized services										
girth exceeding 300mm	0.55	10.00	0.61	0.65	-	11.94	0.32	12.88	m²	14.16
isolated surfaces, girth not exceeding 300mm	0.17	10.00	0.19	0.22	-	4.04	0.11	4.34	m	4.77

One undercoat, one coat full gloss finish, existing gloss painted metal surfaces; external work

	Del to Site	Waste	Material Cost	Craft Optve	Lab	Labour Cost	Sunds	Nett Rate	Unit	Gross rate
Iron or steel general surfaces										
girth exceeding 300mm	0.55	10.00	0.61	0.50	-	9.19	0.25	10.04	m²	11.05
isolated surfaces, girth not exceeding 300mm	0.17	10.00	0.19	0.17	-	3.12	0.09	3.40	m	3.74
Galvanized glazed doors, windows or screens										
girth exceeding 300mm; panes, area not exceeding 0.10m²	0.30	10.00	0.33	0.90	-	16.53	0.45	17.31	m²	19.04
girth exceeding 300mm; panes, area 0.10 - 0.50m²	0.21	10.00	0.23	0.67	-	12.31	0.34	12.88	m²	14.17
girth exceeding 300mm; panes, area 0.50 - 1.00m²	0.17	10.00	0.19	0.55	-	10.10	0.28	10.57	m²	11.62
girth exceeding 300mm; panes, area exceeding 1.00m²	0.17	10.00	0.19	0.49	-	9.00	0.25	9.43	m²	10.38
Iron or steel structural work										
girth exceeding 300mm	0.55	10.00	0.61	0.52	-	9.55	0.26	10.42	m²	11.46
isolated surfaces, girth not exceeding 300mm	0.17	10.00	0.19	0.21	-	3.86	0.10	4.15	m	4.57
Iron or steel structural members of roof trusses, lattice girders, purlins and the like										
girth exceeding 300mm	0.55	10.00	0.61	0.65	-	11.94	0.32	12.88	m²	14.16
isolated surfaces, girth not exceeding 300mm	0.17	10.00	0.19	0.22	-	4.04	0.11	4.34	m	4.77
Iron or steel railings, fences and gates; plain open type										
girth exceeding 300mm	0.43	10.00	0.47	0.49	-	9.00	0.25	9.72	m²	10.69
isolated surfaces, girth not exceeding 300mm	0.17	10.00	0.19	0.16	-	2.94	0.08	3.21	m	3.53
Iron or steel railings, fences and gates; close type										
girth exceeding 300mm	0.55	10.00	0.61	0.45	-	8.27	0.22	9.10	m²	10.01
Iron or steel railings, fences and gates; ornamental type										
girth exceeding 300mm	0.43	10.00	0.47	0.90	-	16.53	0.45	17.45	m²	19.20
Iron or steel eaves gutters										
girth exceeding 300mm	0.55	10.00	0.61	0.53	-	9.74	0.26	10.61	m²	11.67
isolated surfaces, girth not exceeding 300mm	0.17	10.00	0.19	0.17	-	3.12	0.09	3.40	m	3.74
Galvanized eaves gutters										
girth exceeding 300mm	0.55	10.00	0.61	0.53	-	9.74	0.26	10.61	m²	11.67
isolated surfaces, girth not exceeding 300mm	0.17	10.00	0.19	0.17	-	3.12	0.09	3.40	m	3.74
Iron or steel services										
girth exceeding 300mm	0.55	10.00	0.61	0.65	-	11.94	0.32	12.88	m²	14.16
isolated surfaces, girth not exceeding 300mm	0.17	10.00	0.19	0.22	-	4.04	0.11	4.34	m	4.77
isolated areas not exceeding 0.50m² irrespective of girth	0.30	10.00	0.33	0.65	-	11.94	0.32	12.59	nr	13.85

Labour hourly rates: (except Specialists) Craft Operatives 18.37 Labourer 13.76 Rates are national average prices. Refer to REGIONAL VARIATIONS for indicative levels of overall pricing in regions	MATERIALS			LABOUR				RATES		
	Del to Site	Waste	Material Cost	Craft Optve	Lab	Labour Cost	Sunds	Nett Rate		Gross rate (10%)
	£	%	£	Hrs	Hrs	£	£	£	Unit	£

PAINTING & CLEAR FINISHINGS; REDECORATIONS: OIL PAINTING METALWORK (Cont'd)

One undercoat, one coat full gloss finish, existing gloss painted metal surfaces; external work (Cont'd)

Copper services										
girth exceeding 300mm	0.55	10.00	0.61	0.65	-	11.94	0.32	12.88	m²	14.16
isolated surfaces, girth not exceeding 300mm	0.17	10.00	0.19	0.22	-	4.04	0.11	4.34	m	4.77
Galvanized services										
girth exceeding 300mm	0.55	10.00	0.61	0.65	-	11.94	0.32	12.88	m²	14.16
isolated surfaces, girth not exceeding 300mm	0.17	10.00	0.19	0.22	-	4.04	0.11	4.34	m	4.77

Two coats full gloss finish, existing gloss painted metal surfaces

Iron or steel general surfaces										
girth exceeding 300mm	0.55	10.00	0.61	0.47	-	8.63	0.23	9.48	m²	10.43
isolated surfaces, girth not exceeding 300mm	0.17	10.00	0.19	0.16	-	2.94	0.08	3.21	m	3.53
Galvanized glazed doors, windows or screens										
girth exceeding 300mm; panes, area not exceeding 0.10m²	0.30	10.00	0.33	0.88	-	16.17	0.44	16.93	m²	18.63
girth exceeding 300mm; panes, area 0.10 - 0.50m²	0.21	10.00	0.23	0.64	-	11.76	0.32	12.31	m²	13.54
girth exceeding 300mm; panes, area 0.50 - 1.00m²	0.17	10.00	0.19	0.53	-	9.74	0.26	10.19	m²	11.21
girth exceeding 300mm; panes, area exceeding 1.00m²	0.17	10.00	0.19	0.46	-	8.45	0.23	8.87	m²	9.75
Iron or steel structural work										
girth exceeding 300mm	0.55	10.00	0.61	0.52	-	9.55	0.26	10.42	m²	11.46
isolated surfaces, girth not exceeding 300mm	0.17	10.00	0.19	0.21	-	3.86	0.10	4.15	m	4.57
Iron or steel structural members of roof trusses, lattice girders, purlins and the like										
girth exceeding 300mm	0.55	10.00	0.61	0.65	-	11.94	0.32	12.88	m²	14.16
isolated surfaces, girth not exceeding 300mm	0.17	10.00	0.19	0.22	-	4.04	0.11	4.34	m	4.77
Iron or steel services										
girth exceeding 300mm	0.55	10.00	0.61	0.65	-	11.94	0.32	12.88	m²	14.16
isolated surfaces, girth not exceeding 300mm	0.17	10.00	0.19	0.22	-	4.04	0.11	4.34	m	4.77
isolated areas not exceeding 0.50m² irrespective of girth	0.28	10.00	0.30	0.65	-	11.94	0.32	12.57	nr	13.83
Copper services										
girth exceeding 300mm	0.55	10.00	0.61	0.65	-	11.94	0.32	12.88	m²	14.16
isolated surfaces, girth not exceeding 300mm	0.17	10.00	0.19	0.22	-	4.04	0.11	4.34	m	4.77
Galvanized services										
girth exceeding 300mm	0.55	10.00	0.61	0.65	-	11.94	0.32	12.88	m²	14.16
isolated surfaces, girth not exceeding 300mm	0.17	10.00	0.19	0.22	-	4.04	0.11	4.34	m	4.77

Two coats full gloss finish, existing gloss painted metal surfaces; external work

Iron or steel general surfaces										
girth exceeding 300mm	0.55	10.00	0.61	0.50	-	9.19	0.25	10.04	m²	11.05
isolated surfaces, girth not exceeding 300mm	0.17	10.00	0.19	0.17	-	3.12	0.09	3.40	m	3.74
Galvanized glazed doors, windows or screens										
girth exceeding 300mm; panes, area not exceeding 0.10m²	0.30	10.00	0.33	0.90	-	16.53	0.45	17.31	m²	19.04
girth exceeding 300mm; panes, area 0.10 - 0.50m²	0.21	10.00	0.23	0.67	-	12.31	0.34	12.88	m²	14.17
girth exceeding 300mm; panes, area 0.50 - 1.00m²	0.17	10.00	0.19	0.55	-	10.10	0.28	10.57	m	11.62
girth exceeding 300mm; panes, area exceeding 1.00m²	0.17	10.00	0.19	0.49	-	9.00	0.25	9.43	m²	10.38
Iron or steel structural work										
girth exceeding 300mm	0.55	10.00	0.61	0.52	-	9.55	0.26	10.42	m²	11.46
isolated surfaces, girth not exceeding 300mm	0.17	10.00	0.19	0.21	-	3.86	0.10	4.15	m	4.57
Iron or steel structural members of roof trusses, lattice girders, purlins and the like										
girth exceeding 300mm	0.55	10.00	0.61	0.65	-	11.94	0.32	12.88	m²	14.16
isolated surfaces, girth not exceeding 300mm	0.17	10.00	0.19	0.22	-	4.04	0.11	4.34	m	4.77
Iron or steel railings, fences and gates; plain open type										
girth exceeding 300mm	0.43	10.00	0.47	0.49	-	9.00	0.25	9.72	m²	10.69
isolated surfaces, girth not exceeding 300mm	0.17	10.00	0.19	0.16	-	2.94	0.08	3.21	m	3.53
Iron or steel railings, fences and gates; close type										
girth exceeding 300mm	0.55	10.00	0.61	0.45	-	8.27	0.22	9.10	m²	10.01
Iron or steel railings, fences and gates; ornamental type										
girth exceeding 300mm	0.43	10.00	0.47	0.82	-	15.06	0.41	15.94	m²	17.54
Iron or steel eaves gutters										
girth exceeding 300mm	0.55	10.00	0.61	0.53	-	9.74	0.26	10.61	m²	11.67
isolated surfaces, girth not exceeding 300mm	0.17	10.00	0.19	0.17	-	3.12	0.09	3.40	m	3.74
Galvanized eaves gutters										
girth exceeding 300mm	0.55	10.00	0.61	0.53	-	9.74	0.26	10.61	m²	11.67
isolated surfaces, girth not exceeding 300mm	0.17	10.00	0.19	0.17	-	3.12	0.09	3.40	m	3.74
Iron or steel services										
girth exceeding 300mm	0.55	10.00	0.61	0.65	-	11.94	0.32	12.88	m²	14.16
isolated surfaces, girth not exceeding 300mm	0.17	10.00	0.19	0.22	-	4.04	0.11	4.34	m	4.77
isolated areas not exceeding 0.50m² irrespective of girth	0.28	10.00	0.30	0.65	-	11.94	0.32	12.57	nr	13.83
Copper services										
girth exceeding 300mm	0.55	10.00	0.61	0.65	-	11.94	0.32	12.88	m²	14.16
isolated surfaces, girth not exceeding 300mm	0.17	10.00	0.19	0.22	-	4.04	0.11	4.34	m	4.77
Galvanized services										
girth exceeding 300mm	0.55	10.00	0.61	0.65	-	11.94	0.32	12.88	m²	14.16
isolated surfaces, girth not exceeding 300mm	0.17	10.00	0.19	0.22	-	4.04	0.11	4.34	m	4.77

DECORATION

Labour hourly rates: (except Specialists) Craft Operatives 18.37 Labourer 13.76 Rates are national average prices. Refer to REGIONAL VARIATIONS for indicative levels of overall pricing in regions	MATERIALS			LABOUR				RATES		
	Del to Site £	Waste %	Material Cost £	Craft Optve Hrs	Lab Hrs	Labour Cost £	Sunds £	Nett Rate £	Unit	Gross rate (10%) £
PAINTING & CLEAR FINISHINGS; REDECORATIONS: OIL PAINTING METALWORK (Cont'd)										
Two undercoats, one coat full gloss finish, existing gloss painted metal surfaces										
Iron or steel general surfaces										
girth exceeding 300mm	0.83	10.00	0.91	0.62	-	11.39	0.31	12.61	m²	13.88
isolated surfaces, girth not exceeding 300mm	0.26	10.00	0.28	0.21	-	3.86	0.10	4.24	m	4.67
Galvanized glazed doors, windows or screens										
girth exceeding 300mm; panes, area not exceeding 0.10m²	0.45	10.00	0.49	1.15	-	21.13	0.57	22.19	m²	24.41
girth exceeding 300mm; panes, area 0.10 - 0.50m²	0.32	10.00	0.35	0.83	-	15.25	0.41	16.01	m²	17.62
girth exceeding 300mm; panes, area 0.50 - 1.00m²	0.26	10.00	0.28	0.69	-	12.68	0.34	13.30	m²	14.63
girth exceeding 300mm; panes, area exceeding 1.00m²	0.26	10.00	0.28	0.61	-	11.21	0.31	11.79	m²	12.97
Iron or steel structural work										
girth exceeding 300mm	0.83	10.00	0.91	0.67	-	12.31	0.34	13.56	m²	14.91
isolated surfaces, girth not exceeding 300mm	0.26	10.00	0.28	0.27	-	4.96	0.14	5.38	m	5.91
Iron or steel structural members of roof trusses, lattice girders, purlins and the like										
girth exceeding 300mm	0.83	10.00	0.91	0.84	-	15.43	0.42	16.77	m²	18.44
isolated surfaces, girth not exceeding 300mm	0.26	10.00	0.28	0.28	-	5.14	0.14	5.57	m	6.12
Iron or steel services										
girth exceeding 300mm	0.83	10.00	0.91	0.84	-	15.43	0.42	16.77	m²	18.44
isolated surfaces, girth not exceeding 300mm	0.26	10.00	0.28	0.28	-	5.14	0.14	5.57	m	6.12
isolated areas not exceeding 0.50m² irrespective of girth	0.43	10.00	0.47	0.84	-	15.43	0.42	16.32	nr	17.95
Copper services										
girth exceeding 300mm	0.83	10.00	0.91	0.84	-	15.43	0.42	16.77	m²	18.44
isolated surfaces, girth not exceeding 300mm	0.26	10.00	0.28	0.28	-	5.14	0.14	5.57	m	6.12
Galvanized services										
girth exceeding 300mm	0.83	10.00	0.91	0.84	-	15.43	0.42	16.77	m²	18.44
isolated surfaces, girth not exceeding 300mm	0.26	10.00	0.28	0.28	-	5.14	0.14	5.57	m	6.12
Two undercoats, one coat full gloss finish, existing gloss painted metal surfaces; external work										
Iron or steel general surfaces										
girth exceeding 300mm	0.83	10.00	0.91	0.65	-	11.94	0.32	13.18	m²	14.50
isolated surfaces, girth not exceeding 300mm	0.26	10.00	0.28	0.22	-	4.04	0.11	4.43	m	4.88
Galvanized glazed doors, windows or screens										
girth exceeding 300mm; panes, area not exceeding 0.10m²	0.45	10.00	0.49	1.18	-	21.68	0.59	22.76	m²	25.04
girth exceeding 300mm; panes, area 0.10 - 0.50m²	0.32	10.00	0.35	0.86	-	15.80	0.43	16.58	m²	18.24
girth exceeding 300mm; panes, area 0.50 - 1.00m²	0.26	10.00	0.28	0.72	-	13.23	0.36	13.87	m²	15.25
girth exceeding 300mm; panes, area exceeding 1.00m²	0.26	10.00	0.28	0.64	-	11.76	0.32	12.36	m²	13.59
Iron or steel structural work										
girth exceeding 300mm	0.83	10.00	0.91	0.67	-	12.31	0.34	13.56	m²	14.91
isolated surfaces, girth not exceeding 300mm	0.26	10.00	0.28	0.27	-	4.96	0.14	5.38	m	5.91
Iron or steel structural members of roof trusses, lattice girders, purlins and the like										
girth exceeding 300mm	0.83	10.00	0.91	0.84	-	15.43	0.42	16.77	m²	18.44
isolated surfaces, girth not exceeding 300mm	0.26	10.00	0.28	0.28	-	5.14	0.14	5.57	m	6.12
Iron or steel railings, fences and gates; plain open type										
girth exceeding 300mm	0.64	10.00	0.70	0.63	-	11.57	0.31	12.59	m²	13.85
isolated surfaces, girth not exceeding 300mm	0.26	10.00	0.28	0.21	-	3.86	0.10	4.24	m	4.67
Iron or steel railings, fences and gates; close type										
girth exceeding 300mm	0.83	10.00	0.91	0.57	-	10.47	0.28	11.67	m²	12.84
Iron or steel railings, fences and gates; ornamental type										
girth exceeding 300mm	0.64	10.00	0.70	1.06	-	19.47	0.53	20.71	m²	22.78
Iron or steel eaves gutters										
girth exceeding 300mm	0.83	10.00	0.91	0.68	-	12.49	0.34	13.75	m²	15.12
isolated surfaces, girth not exceeding 300mm	0.26	10.00	0.28	0.23	-	4.23	0.12	4.62	m	5.08
Galvanized eaves gutters										
girth exceeding 300mm	0.83	10.00	0.91	0.68	-	12.49	0.34	13.75	m²	15.12
isolated surfaces, girth not exceeding 300mm	0.26	10.00	0.28	0.23	-	4.23	0.12	4.62	m	5.08
Iron or steel services										
girth exceeding 300mm	0.83	10.00	0.91	0.84	-	15.43	0.42	16.77	m²	18.44
isolated surfaces, girth not exceeding 300mm	0.26	10.00	0.28	0.28	-	5.14	0.14	5.57	m	6.12
isolated areas not exceeding 0.50m² irrespective of girth	0.43	10.00	0.47	0.84	-	15.43	0.42	16.32	nr	17.95
Copper services										
girth exceeding 300mm	0.83	10.00	0.91	0.84	-	15.43	0.42	16.77	m²	18.44
isolated surfaces, girth not exceeding 300mm	0.26	10.00	0.28	0.28	-	5.14	0.14	5.57	m	6.12
Galvanized services										
girth exceeding 300mm	0.83	10.00	0.91	0.84	-	15.43	0.42	16.77	m²	18.44
isolated surfaces, girth not exceeding 300mm	0.26	10.00	0.28	0.28	-	5.14	0.14	5.57	m	6.12
One undercoat, two coats full gloss finish, existing gloss painted metal surfaces										
Iron or steel general surfaces										
girth exceeding 300mm	0.83	10.00	0.91	0.62	-	11.39	0.31	12.61	m²	13.88
isolated surfaces, girth not exceeding 300mm	0.26	10.00	0.28	0.21	-	3.86	0.10	4.24	m	4.67
Galvanized glazed doors, windows or screens										
girth exceeding 300mm; panes, area not exceeding 0.10m²	0.45	10.00	0.49	1.16	-	21.31	0.58	22.38	m²	24.62
girth exceeding 300mm; panes, area 0.10 - 0.50m²	0.32	10.00	0.35	0.83	-	15.25	0.41	16.01	m²	17.62

Labour hourly rates: (except Specialists) Craft Operatives 18.37 Labourer 13.76 Rates are national average prices. Refer to REGIONAL VARIATIONS for indicative levels of overall pricing in regions	MATERIALS			LABOUR				RATES		
	Del to Site £	Waste %	Material Cost £	Craft Optve Hrs	Lab Hrs	Labour Cost £	Sunds £	Nett Rate £	Unit	Gross rate (10%) £
PAINTING & CLEAR FINISHINGS; REDECORATIONS: OIL PAINTING METALWORK (Cont'd)										
One undercoat, two coats full gloss finish, existing gloss painted metal surfaces (Cont'd)										
Galvanized glazed doors, windows or screens (Cont'd)										
girth exceeding 300mm; panes, area 0.50 - 1.00m²	0.26	10.00	0.28	0.69	-	12.68	0.34	13.30	m²	14.63
girth exceeding 300mm; panes, area exceeding 1.00m²	0.26	10.00	0.28	0.61	-	11.21	0.31	11.79	m²	12.97
Iron or steel structural work										
girth exceeding 300mm	0.83	10.00	0.91	0.67	-	12.31	0.34	13.56	m²	14.91
isolated surfaces, girth not exceeding 300mm	0.26	10.00	0.28	0.27	-	4.96	0.14	5.38	m	5.91
Iron or steel structural members of roof trusses, lattice girders, purlins and the like										
girth exceeding 300mm	0.83	10.00	0.91	0.84	-	15.43	0.42	16.77	m²	18.44
isolated surfaces, girth not exceeding 300mm	0.26	10.00	0.28	0.28	-	5.14	0.14	5.57	m	6.12
Iron or steel services										
girth exceeding 300mm	0.83	10.00	0.91	0.84	-	15.43	0.42	16.77	m²	18.44
isolated surfaces, girth not exceeding 300mm	0.26	10.00	0.28	0.28	-	5.14	0.14	5.57	m	6.12
isolated areas not exceeding 0.50m² irrespective of girth	0.45	10.00	0.49	0.84	-	15.43	0.42	16.34	nr	17.98
Copper services										
girth exceeding 300mm	0.83	10.00	0.91	0.84	-	15.43	0.42	16.77	m²	18.44
isolated surfaces, girth not exceeding 300mm	0.26	10.00	0.28	0.28	-	5.14	0.14	5.57	m	6.12
Galvanized services										
girth exceeding 300mm	0.83	10.00	0.91	0.84	-	15.43	0.42	16.77	m²	18.44
isolated surfaces, girth not exceeding 300mm	0.26	10.00	0.28	0.28	-	5.14	0.14	5.57	m	6.12
One undercoat, two coats full gloss finish, existing gloss painted metal surfaces; external work										
Iron or steel general surfaces										
girth exceeding 300mm	0.83	10.00	0.91	0.65	-	11.94	0.32	13.18	m²	14.50
isolated surfaces, girth not exceeding 300mm	0.26	10.00	0.28	0.22	-	4.04	0.11	4.43	m	4.88
Galvanized glazed doors, windows or screens										
girth exceeding 300mm; panes, area not exceeding 0.10m²	0.45	10.00	0.49	1.18	-	21.68	0.59	22.76	m²	25.04
girth exceeding 300mm; panes, area 0.10 - 0.50m²	0.32	10.00	0.35	0.86	-	15.80	0.43	16.58	m²	18.24
girth exceeding 300mm; panes, area 0.50 - 1.00m²	0.26	10.00	0.28	0.72	-	13.23	0.36	13.87	m²	15.25
girth exceeding 300mm; panes, area exceeding 1.00m²	0.26	10.00	0.28	0.64	-	11.76	0.32	12.36	m²	13.59
Iron or steel structural work										
girth exceeding 300mm	0.83	10.00	0.91	0.67	-	12.31	0.34	13.56	m²	14.91
isolated surfaces, girth not exceeding 300mm	0.26	10.00	0.28	0.27	-	4.96	0.14	5.38	m	5.91
Iron or steel structural members of roof trusses, lattice girders, purlins and the like										
girth exceeding 300mm	0.83	10.00	0.91	0.84	-	15.43	0.42	16.77	m²	18.44
isolated surfaces, girth not exceeding 300mm	0.26	10.00	0.28	0.28	-	5.14	0.14	5.57	m	6.12
Iron or steel railings, fences and gates; plain open type										
girth exceeding 300mm	0.64	10.00	0.70	0.64	-	11.76	0.32	12.78	m²	14.06
isolated surfaces, girth not exceeding 300mm	0.26	10.00	0.28	0.21	-	3.86	0.10	4.24	m	4.67
Iron or steel railings, fences and gates; close type										
girth exceeding 300mm	0.83	10.00	0.91	0.57	-	10.47	0.28	11.67	m²	12.84
Iron or steel railings, fences and gates; ornamental type										
girth exceeding 300mm	0.64	10.00	0.70	1.06	-	19.47	0.53	20.71	m²	22.78
Iron or steel eaves gutters										
girth exceeding 300mm	0.83	10.00	0.91	0.68	-	12.49	0.34	13.75	m²	15.12
isolated surfaces, girth not exceeding 300mm	0.26	10.00	0.28	0.23	-	4.23	0.12	4.62	m	5.08
Galvanized eaves gutters										
girth exceeding 300mm	0.83	10.00	0.91	0.68	-	12.49	0.34	13.75	m²	15.12
isolated surfaces, girth not exceeding 300mm	0.26	10.00	0.28	0.23	-	4.23	0.12	4.62	m	5.08
Iron or steel services										
girth exceeding 300mm	0.83	10.00	0.91	0.84	-	15.43	0.42	16.77	m²	18.44
isolated surfaces, girth not exceeding 300mm	0.26	10.00	0.28	0.28	-	5.14	0.14	5.57	m	6.12
isolated areas not exceeding 0.50m² irrespective of girth	0.45	10.00	0.49	0.84	-	15.43	0.42	16.34	nr	17.98
Copper services										
girth exceeding 300mm	0.83	10.00	0.91	0.84	-	15.43	0.42	16.77	m²	18.44
isolated surfaces, girth not exceeding 300mm	0.26	10.00	0.28	0.28	-	5.14	0.14	5.57	m	6.12
Galvanized services										
girth exceeding 300mm	0.83	10.00	0.91	0.84	-	15.43	0.42	16.77	m²	18.44
isolated surfaces, girth not exceeding 300mm	0.26	10.00	0.28	0.28	-	5.14	0.14	5.57	m	6.12
PAINTING & CLEAR FINISHINGS; REDECORATIONS: OIL PAINTING WOODWORK										
Generally										
Note: the following rates include for the cost of all preparatory work, e.g. washing down, etc.										
One undercoat, one coat full gloss finish, existing gloss painted wood surfaces										
Wood general surfaces										
girth exceeding 300mm	0.55	10.00	0.61	0.45	-	8.27	0.22	9.10	m²	10.01
isolated surfaces, girth not exceeding 300mm	0.17	10.00	0.19	0.15	-	2.76	0.08	3.02	m	3.32

DECORATION

Labour hourly rates: (except Specialists) Craft Operatives 18.37 Labourer 13.76 Rates are national average prices. Refer to REGIONAL VARIATIONS for indicative levels of overall pricing in regions	MATERIALS			LABOUR				RATES		
	Del to Site £	Waste %	Material Cost £	Craft Optve Hrs	Lab Hrs	Labour Cost £	Sunds £	Nett Rate £	Unit	Gross rate (10%) £
PAINTING & CLEAR FINISHINGS; REDECORATIONS: OIL PAINTING WOODWORK (Cont'd)										
One undercoat, one coat full gloss finish, existing gloss painted wood surfaces (Cont'd)										
Wood glazed doors										
girth exceeding 300mm; panes, area not exceeding 0.10m²	0.30	10.00	0.33	0.79	-	14.51	0.40	15.24	m²	16.76
girth exceeding 300mm; panes, area 0.10 - 0.50m²	0.21	10.00	0.23	0.57	-	10.47	0.28	10.99	m²	12.09
girth exceeding 300mm; panes, area 0.50 - 1.00m²	0.21	10.00	0.23	0.48	-	8.82	0.24	9.29	m²	10.22
girth exceeding 300mm; panes, area exceeding 1.00m²	0.21	10.00	0.23	0.41	-	7.53	0.20	7.97	m²	8.77
Wood partially glazed doors										
girth exceeding 300mm; panes, area not exceeding 0.10m²	0.43	10.00	0.47	0.67	-	12.31	0.34	13.11	m²	14.42
girth exceeding 300mm; panes, area 0.10 - 0.50m²	0.34	10.00	0.38	0.54	-	9.92	0.27	10.57	m²	11.62
girth exceeding 300mm; panes, area 0.50 - 1.00m²	0.30	10.00	0.33	0.48	-	8.82	0.24	9.39	m²	10.32
girth exceeding 300mm; panes, area exceeding 1.00m²	0.30	10.00	0.33	0.43	-	7.90	0.22	8.44	m²	9.29
Wood windows and screens										
girth exceeding 300mm; panes, area not exceeding 0.10m²	0.43	10.00	0.47	0.85	-	15.61	0.43	16.51	m²	18.16
girth exceeding 300mm; panes, area 0.10 - 0.50m²	0.34	10.00	0.38	0.61	-	11.21	0.31	11.89	m²	13.07
girth exceeding 300mm; panes, area 0.50 - 1.00m²	0.30	10.00	0.33	0.52	-	9.55	0.26	10.14	m²	11.15
girth exceeding 300mm; panes, area exceeding 1.00m²	0.30	10.00	0.33	0.44	-	8.08	0.22	8.63	m²	9.49
One undercoat, one coat full gloss finish, existing gloss painted wood surfaces; external work										
Wood general surfaces										
girth exceeding 300mm	0.55	10.00	0.61	0.48	-	8.82	0.24	9.67	m²	10.63
isolated surfaces, girth not exceeding 300mm	0.17	10.00	0.19	0.16	-	2.94	0.08	3.21	m	3.53
Wood glazed doors										
girth exceeding 300mm; panes, area not exceeding 0.10m²	0.30	10.00	0.33	0.81	-	14.88	0.41	15.61	m²	17.17
girth exceeding 300mm; panes, area 0.10 - 0.50m²	0.21	10.00	0.23	0.59	-	10.84	0.29	11.37	m²	12.50
girth exceeding 300mm; panes, area 0.50 - 1.00m²	0.21	10.00	0.23	0.49	-	9.00	0.25	9.48	m²	10.43
girth exceeding 300mm; panes, area exceeding 1.00m²	0.21	10.00	0.23	0.42	-	7.72	0.21	8.16	m²	8.98
Wood partially glazed doors										
girth exceeding 300mm; panes, area not exceeding 0.10m²	0.43	10.00	0.47	0.69	-	12.68	0.34	13.49	m²	14.84
girth exceeding 300mm; panes, area 0.10 - 0.50m²	0.34	10.00	0.38	0.56	-	10.29	0.28	10.94	m²	12.04
girth exceeding 300mm; panes, area 0.50 - 1.00m²	0.30	10.00	0.33	0.49	-	9.00	0.25	9.57	m²	10.53
girth exceeding 300mm; panes, area exceeding 1.00m²	0.30	10.00	0.33	0.44	-	8.08	0.22	8.63	m²	9.49
Wood windows and screens										
girth exceeding 300mm; panes, area not exceeding 0.10m²	0.43	10.00	0.47	0.89	-	16.35	0.44	17.26	m²	18.99
girth exceeding 300mm; panes, area 0.10 - 0.50m²	0.34	10.00	0.38	0.65	-	11.94	0.32	12.64	m²	13.90
girth exceeding 300mm; panes, area 0.50 - 1.00m²	0.30	10.00	0.33	0.55	-	10.10	0.28	10.71	m²	11.78
girth exceeding 300mm; panes, area exceeding 1.00m²	0.30	10.00	0.33	0.47	-	8.63	0.23	9.20	m²	10.12
Wood railings fences and gates; open type										
girth exceeding 300mm	0.43	10.00	0.47	0.41	-	7.53	0.20	8.21	m²	9.03
isolated surfaces, girth not exceeding 300mm	0.17	10.00	0.19	0.13	-	2.39	0.06	2.64	m	2.90
Wood railings fences and gates; close type										
girth exceeding 300mm	0.55	10.00	0.61	0.37	-	6.80	0.19	7.59	m²	8.35
Two coats full gloss finish, existing gloss painted wood surfaces										
Wood general surfaces										
girth exceeding 300mm	0.55	10.00	0.61	0.45	-	8.27	0.22	9.10	m²	10.01
isolated surfaces, girth not exceeding 300mm	0.17	10.00	0.19	0.15	-	2.76	0.08	3.02	m	3.32
Wood glazed doors										
girth exceeding 300mm; panes, area not exceeding 0.10m²	0.30	10.00	0.33	0.79	-	14.51	0.40	15.24	m²	16.76
girth exceeding 300mm; panes, area 0.10 - 0.50m²	0.21	10.00	0.23	0.57	-	10.47	0.28	10.99	m²	12.09
girth exceeding 300mm; panes, area 0.50 - 1.00m²	0.21	10.00	0.23	0.48	-	8.82	0.24	9.29	m²	10.22
girth exceeding 300mm; panes, area exceeding 1.00m²	0.21	10.00	0.23	0.41	-	7.53	0.20	7.97	m²	8.77
Wood partially glazed doors										
girth exceeding 300mm; panes, area not exceeding 0.10m²	0.43	10.00	0.47	0.67	-	12.31	0.34	13.11	m²	14.42
girth exceeding 300mm; panes, area 0.10 - 0.50m²	0.34	10.00	0.38	0.54	-	9.92	0.27	10.57	m²	11.62
girth exceeding 300mm; panes, area 0.50 - 1.00m²	0.30	10.00	0.33	0.48	-	8.82	0.24	9.39	m²	10.32
girth exceeding 300mm; panes, area exceeding 1.00m²	0.30	10.00	0.33	0.43	-	7.90	0.22	8.44	m²	9.29
Wood windows and screens										
girth exceeding 300mm; panes, area not exceeding 0.10m²	0.43	10.00	0.47	0.85	-	15.61	0.43	16.51	m²	18.16
girth exceeding 300mm; panes, area 0.10 - 0.50m²	0.34	10.00	0.38	0.61	-	11.21	0.31	11.89	m²	13.07
girth exceeding 300mm; panes, area 0.50 - 1.00m²	0.30	10.00	0.33	0.52	-	9.55	0.26	10.14	m²	11.15
girth exceeding 300mm; panes, area exceeding 1.00m²	0.30	10.00	0.33	0.44	-	8.08	0.22	8.63	m²	9.49
Two coats full gloss finish, existing gloss painted wood surfaces; external work										
Wood general surfaces										
girth exceeding 300mm	0.55	10.00	0.61	0.48	-	8.82	0.24	9.67	m²	10.63
isolated surfaces, girth not exceeding 300mm	0.17	10.00	0.19	0.16	-	2.94	0.08	3.21	m	3.53
Wood glazed doors										
girth exceeding 300mm; panes, area not exceeding 0.10m²	0.30	10.00	0.33	0.81	-	14.88	0.41	15.61	m²	17.17
girth exceeding 300mm; panes, area 0.10 - 0.50m²	0.21	10.00	0.23	0.59	-	10.84	0.29	11.37	m²	12.50
girth exceeding 300mm; panes, area 0.50 - 1.00m²	0.21	10.00	0.23	0.49	-	9.00	0.25	9.48	m²	10.43
girth exceeding 300mm; panes, area exceeding 1.00m²	0.21	10.00	0.23	0.42	-	7.72	0.21	8.16	m²	8.98
Wood partially glazed doors										
girth exceeding 300mm; panes, area not exceeding 0.10m²	0.43	10.00	0.47	0.69	-	12.68	0.34	13.49	m²	14.84
girth exceeding 300mm; panes, area 0.10 - 0.50m²	0.34	10.00	0.38	0.56	-	10.29	0.28	10.94	m²	12.04
girth exceeding 300mm; panes, area 0.50 - 1.00m²	0.30	10.00	0.33	0.49	-	9.00	0.25	9.57	m²	10.53
girth exceeding 300mm; panes, area exceeding 1.00m²	0.30	10.00	0.33	0.44	-	8.08	0.22	8.63	m²	9.49

Labour hourly rates: (except Specialists) Craft Operatives 18.37 Labourer 13.76 Rates are national average prices. Refer to REGIONAL VARIATIONS for indicative levels of overall pricing in regions	MATERIALS			LABOUR				RATES		
	Del to Site £	Waste %	Material Cost £	Craft Optve Hrs	Lab Hrs	Labour Cost £	Sunds £	Nett Rate £	Unit	Gross rate (10%) £

PAINTING & CLEAR FINISHINGS; REDECORATIONS: OIL PAINTING WOODWORK (Cont'd)

Two coats full gloss finish, existing gloss painted wood surfaces; external work (Cont'd)

Wood windows and screens

girth exceeding 300mm; panes, area not exceeding 0.10m²	0.43	10.00	0.47	0.89	-	16.35	0.44	17.26	m²	18.99
girth exceeding 300mm; panes, area 0.10 - 0.50m²	0.34	10.00	0.38	0.65	-	11.94	0.32	12.64	m²	13.90
girth exceeding 300mm; panes, area 0.50 - 1.00m²	0.30	10.00	0.33	0.55	-	10.10	0.28	10.71	m²	11.78
girth exceeding 300mm; panes, area exceeding 1.00m²	0.30	10.00	0.33	0.47	-	8.63	0.23	9.20	m²	10.12

Wood railings fences and gates; open type

girth exceeding 300mm	0.43	10.00	0.47	0.41	-	7.53	0.20	8.21	m²	9.03
isolated surfaces, girth not exceeding 300mm	0.17	10.00	0.19	0.13	-	2.39	0.06	2.64	m	2.90

Wood railings fences and gates; close type

girth exceeding 300mm	0.55	10.00	0.61	0.37	-	6.80	0.19	7.59	m²	8.35

Two undercoats, one coat full gloss finish, existing gloss painted wood surfaces

Wood general surfaces

girth exceeding 300mm	0.83	10.00	0.91	0.61	-	11.21	0.31	12.43	m²	13.67
isolated surfaces, girth not exceeding 300mm	0.26	10.00	0.28	0.20	-	3.67	0.10	4.06	m	4.46

Wood glazed doors

girth exceeding 300mm; panes, area not exceeding 0.10m²	0.45	10.00	0.49	1.04	-	19.10	0.52	20.12	m²	22.13
girth exceeding 300mm; panes, area 0.10 - 0.50m²	0.32	10.00	0.35	0.76	-	13.96	0.38	14.69	m²	16.16
girth exceeding 300mm; panes, area 0.50 - 1.00m²	0.32	10.00	0.35	0.64	-	11.76	0.32	12.43	m²	13.67
girth exceeding 300mm; panes, area exceeding 1.00m²	0.32	10.00	0.35	0.54	-	9.92	0.27	10.54	m²	11.60

Wood partially glazed doors

girth exceeding 300mm; panes, area not exceeding 0.10m²	0.64	10.00	0.70	0.87	-	15.98	0.44	17.12	m²	18.83
girth exceeding 300mm; panes, area 0.10 - 0.50m²	0.51	10.00	0.56	0.71	-	13.04	0.35	13.96	m²	15.36
girth exceeding 300mm; panes, area 0.50 - 1.00m²	0.45	10.00	0.49	0.64	-	11.76	0.32	12.57	m²	13.83
girth exceeding 300mm; panes, area exceeding 1.00m²	0.45	10.00	0.49	0.57	-	10.47	0.28	11.25	m²	12.37

Wood windows and screens

girth exceeding 300mm; panes, area not exceeding 0.10m²	0.64	10.00	0.70	1.12	-	20.57	0.56	21.84	m²	24.02
girth exceeding 300mm; panes, area 0.10 - 0.50m²	0.51	10.00	0.56	0.82	-	15.06	0.41	16.04	m²	17.64
girth exceeding 300mm; panes, area 0.50 - 1.00m²	0.45	10.00	0.49	0.69	-	12.68	0.34	13.51	m²	14.86
girth exceeding 300mm; panes, area exceeding 1.00m²	0.45	10.00	0.49	0.59	-	10.84	0.29	11.63	m²	12.79

Two undercoats, one coat full gloss finish, existing gloss painted wood surfaces; external work

Wood general surfaces

girth exceeding 300mm	0.83	10.00	0.91	0.65	-	11.94	0.32	13.18	m²	14.50
isolated surfaces, girth not exceeding 300mm	0.26	10.00	0.28	0.22	-	4.04	0.11	4.43	m	4.88

Wood glazed doors

girth exceeding 300mm; panes, area not exceeding 0.10m²	0.45	10.00	0.49	1.26	-	23.15	0.63	24.27	m²	26.70
girth exceeding 300mm; panes, area 0.10 - 0.50m²	0.32	10.00	0.35	0.78	-	14.33	0.39	15.07	m²	16.58
girth exceeding 300mm; panes, area 0.50 - 1.00m²	0.32	10.00	0.35	0.65	-	11.94	0.32	12.62	m²	13.88
girth exceeding 300mm; panes, area exceeding 1.00m²	0.32	10.00	0.35	0.55	-	10.10	0.28	10.73	m²	11.80

Wood partially glazed doors

girth exceeding 300mm; panes, area not exceeding 0.10m²	0.64	10.00	0.70	0.89	-	16.35	0.44	17.50	m²	19.25
girth exceeding 300mm; panes, area 0.10 - 0.50m²	0.51	10.00	0.56	0.73	-	13.41	0.37	14.34	m²	15.77
girth exceeding 300mm; panes, area 0.50 - 1.00m²	0.45	10.00	0.49	0.65	-	11.94	0.32	12.76	m²	14.03
girth exceeding 300mm; panes, area exceeding 1.00m²	0.45	10.00	0.49	0.58	-	10.65	0.29	11.44	m²	12.58

Wood windows and screens

girth exceeding 300mm; panes, area not exceeding 0.10m²	0.64	10.00	0.70	1.18	-	21.68	0.59	22.97	m²	25.27
girth exceeding 300mm; panes, area 0.10 - 0.50m²	0.51	10.00	0.56	0.87	-	15.98	0.44	16.98	m²	18.68
girth exceeding 300mm; panes, area 0.50 - 1.00m²	0.45	10.00	0.49	0.73	-	13.41	0.37	14.27	m²	15.69
girth exceeding 300mm; panes, area exceeding 1.00m²	0.45	10.00	0.49	0.63	-	11.57	0.31	12.38	m²	13.62

Wood railings fences and gates; open type

girth exceeding 300mm	0.64	10.00	0.70	0.55	-	10.10	0.28	11.08	m²	12.19
isolated surfaces, girth not exceeding 300mm	0.26	10.00	0.28	0.18	-	3.31	0.09	3.68	m	4.05

Wood railings fences and gates; close type

girth exceeding 300mm	0.83	10.00	0.91	0.49	-	9.00	0.25	10.16	m²	11.18

One undercoat, two coats full gloss finish, existing gloss painted wood surfaces

Wood general surfaces

girth exceeding 300mm	0.83	10.00	0.91	0.61	-	11.21	0.31	12.43	m²	13.67
isolated surfaces, girth not exceeding 300mm	0.26	10.00	0.28	0.20	-	3.67	0.10	4.06	m	4.46

Wood glazed doors

girth exceeding 300mm; panes, area not exceeding 0.10m²	0.45	10.00	0.49	1.04	-	19.10	0.52	20.12	m²	22.13
girth exceeding 300mm; panes, area 0.10 - 0.50m²	0.32	10.00	0.35	0.76	-	13.96	0.38	14.69	m²	16.16
girth exceeding 300mm; panes, area 0.50 - 1.00m²	0.32	10.00	0.35	0.64	-	11.76	0.32	12.43	m²	13.67
girth exceeding 300mm; panes, area exceeding 1.00m²	0.32	10.00	0.35	0.54	-	9.92	0.27	10.54	m²	11.60

Wood partially glazed doors

girth exceeding 300mm; panes, area not exceeding 0.10m²	0.64	10.00	0.70	0.87	-	15.98	0.44	17.12	m²	18.83
girth exceeding 300mm; panes, area 0.10 - 0.50m²	0.51	10.00	0.56	0.71	-	13.04	0.35	13.96	m²	15.36
girth exceeding 300mm; panes, area 0.50 - 1.00m²	0.45	10.00	0.49	0.64	-	11.76	0.32	12.57	m²	13.83
girth exceeding 300mm; panes, area exceeding 1.00m²	0.45	10.00	0.49	0.57	-	10.47	0.28	11.25	m²	12.37

Wood windows and screens

girth exceeding 300mm; panes, area not exceeding 0.10m²	0.64	10.00	0.70	1.12	-	20.57	0.56	21.84	m²	24.02
girth exceeding 300mm; panes, area 0.10 - 0.50m²	0.51	10.00	0.56	0.82	-	15.06	0.41	16.04	m²	17.64
girth exceeding 300mm; panes, area 0.50 - 1.00m²	0.45	10.00	0.49	0.69	-	12.68	0.34	13.51	m²	14.86
girth exceeding 300mm; panes, area exceeding 1.00m²	0.45	10.00	0.49	0.59	-	10.84	0.29	11.63	m²	12.79

DECORATION

Labour hourly rates: (except Specialists) Craft Operatives 18.37 Labourer 13.76 Rates are national average prices. Refer to REGIONAL VARIATIONS for indicative levels of overall pricing in regions	MATERIALS			LABOUR				RATES		
	Del to Site	Waste	Material Cost	Craft Optve	Lab	Labour Cost	Sunds	Nett Rate		Gross rate (10%)
	£	%	£	Hrs	Hrs	£	£	£	Unit	£
PAINTING & CLEAR FINISHINGS; REDECORATIONS: OIL PAINTING WOODWORK (Cont'd)										
One undercoat, two coats full gloss finish, existing gloss painted wood surfaces; external work										
Wood general surfaces										
girth exceeding 300mm	0.83	10.00	0.91	0.65	-	11.94	0.32	13.18	m²	14.50
isolated surfaces, girth not exceeding 300mm	0.26	10.00	0.28	0.22	-	4.04	0.11	4.43	m	4.88
Wood glazed doors										
girth exceeding 300mm; panes, area not exceeding 0.10m².........	0.45	10.00	0.49	1.06	-	19.47	0.53	20.49	m²	22.54
girth exceeding 300mm; panes, area 0.10 - 0.50m²..................	0.32	10.00	0.35	0.78	-	14.33	0.39	15.07	m²	16.58
girth exceeding 300mm; panes, area 0.50 - 1.00m²..................	0.32	10.00	0.35	0.65	-	11.94	0.32	12.62	m²	13.88
girth exceeding 300mm; panes, area exceeding 1.00m².............	0.32	10.00	0.35	0.55	-	10.10	0.28	10.73	m²	11.80
Wood partially glazed doors										
girth exceeding 300mm; panes, area not exceeding 0.10m².........	0.64	10.00	0.70	0.89	-	16.35	0.44	17.50	m²	19.25
girth exceeding 300mm; panes, area 0.10 - 0.50m²..................	0.51	10.00	0.56	0.73	-	13.41	0.37	14.34	m²	15.77
girth exceeding 300mm; panes, area 0.50 - 1.00m²..................	0.45	10.00	0.49	0.65	-	11.94	0.32	12.76	m²	14.03
girth exceeding 300mm; panes, area exceeding 1.00m².............	0.45	10.00	0.49	0.58	-	10.65	0.29	11.44	m²	12.58
Wood windows and screens										
girth exceeding 300mm; panes, area not exceeding 0.10m².........	0.64	10.00	0.70	1.18	-	21.68	0.59	22.97	m²	25.27
girth exceeding 300mm; panes, area 0.10 - 0.50m²..................	0.51	10.00	0.56	0.87	-	15.98	0.44	16.98	m²	18.68
girth exceeding 300mm; panes, area 0.50 - 1.00m²..................	0.45	10.00	0.49	0.73	-	13.41	0.37	14.27	m²	15.69
girth exceeding 300mm; panes, area exceeding 1.00m².............	0.45	10.00	0.49	0.63	-	11.57	0.31	12.38	m²	13.62
Wood railings fences and gates; open type										
girth exceeding 300mm	0.64	10.00	0.70	0.55	-	10.10	0.28	11.08	m²	12.19
isolated surfaces, girth not exceeding 300mm	0.26	10.00	0.28	0.18	-	3.31	0.09	3.68	m	4.05
Wood railings fences and gates; close type										
girth exceeding 300mm	0.83	10.00	0.91	0.49	-	9.00	0.25	10.16	m²	11.18
Burn off, one coat primer, one undercoat, one coat full gloss finish, existing painted wood surfaces										
Wood general surfaces										
girth exceeding 300mm	0.86	10.00	0.94	0.90	-	16.53	0.90	18.38	m²	20.21
isolated surfaces, girth not exceeding 300mm	0.26	10.00	0.28	0.26	-	4.78	0.26	5.32	m	5.85
Wood glazed doors										
girth exceeding 300mm; panes, area not exceeding 0.10m².........	0.47	10.00	0.52	1.50	-	27.56	1.50	29.57	m²	32.53
girth exceeding 300mm; panes, area 0.10 - 0.50m²..................	0.34	10.00	0.38	1.08	-	19.84	1.08	21.30	m²	23.43
girth exceeding 300mm; panes, area 0.50 - 1.00m²..................	0.32	10.00	0.35	0.90	-	16.53	0.90	17.79	m²	19.57
girth exceeding 300mm; panes, area exceeding 1.00m².............	0.32	10.00	0.35	0.80	-	14.70	0.80	15.85	m²	17.43
Wood partially glazed doors										
girth exceeding 300mm; panes, area not exceeding 0.10m².........	0.66	10.00	0.73	1.20	-	22.04	1.20	23.97	m²	26.37
girth exceeding 300mm; panes, area 0.10 - 0.50m²..................	0.51	10.00	0.57	0.99	-	18.19	0.99	19.74	m²	21.72
girth exceeding 300mm; panes, area 0.50 - 1.00m²..................	0.47	10.00	0.52	0.90	-	16.53	0.90	17.95	m²	19.75
girth exceeding 300mm; panes, area exceeding 1.00m².............	0.47	10.00	0.52	0.86	-	15.80	0.86	17.18	m²	18.89
Wood windows and screens										
girth exceeding 300mm; panes, area not exceeding 0.10m².........	0.66	10.00	0.73	1.65	-	30.31	1.65	32.69	m²	35.96
girth exceeding 300mm; panes, area 0.10 - 0.50m²..................	0.51	10.00	0.57	1.18	-	21.68	1.18	23.42	m²	25.76
girth exceeding 300mm; panes, area 0.50 - 1.00m²..................	0.47	10.00	0.52	0.99	-	18.19	0.99	19.70	m²	21.66
girth exceeding 300mm; panes, area exceeding 1.00m².............	0.47	10.00	0.52	0.88	-	16.17	0.88	17.56	m²	19.32
Burn off, one coat primer, one undercoat, one coat full gloss finish, existing painted wood surfaces; external work										
Wood general surfaces										
girth exceeding 300mm	0.86	10.00	0.94	0.95	-	17.45	0.95	19.34	m²	21.28
isolated surfaces, girth not exceeding 300mm	0.26	10.00	0.28	0.29	-	5.33	0.29	5.90	m	6.49
Wood glazed doors										
girth exceeding 300mm; panes, area not exceeding 0.10m².........	0.47	10.00	0.52	1.50	-	27.56	1.50	29.57	m²	32.53
girth exceeding 300mm; panes, area 0.10 - 0.50m²..................	0.34	10.00	0.38	1.08	-	19.84	1.08	21.30	m²	23.43
girth exceeding 300mm; panes, area 0.50 - 1.00m²..................	0.32	10.00	0.35	0.90	-	16.53	0.90	17.79	m²	19.57
girth exceeding 300mm; panes, area exceeding 1.00m².............	0.32	10.00	0.35	0.80	-	14.70	0.80	15.85	m²	17.43
Wood partially glazed doors										
girth exceeding 300mm; panes, area not exceeding 0.10m².........	0.66	10.00	0.73	1.20	-	22.04	1.20	23.97	m²	26.37
girth exceeding 300mm; panes, area 0.10 - 0.50m²..................	0.51	10.00	0.57	0.99	-	18.19	0.99	19.74	m²	21.72
girth exceeding 300mm; panes, area 0.50 - 1.00m²..................	0.47	10.00	0.52	0.90	-	16.53	0.90	17.95	m²	19.75
girth exceeding 300mm; panes, area exceeding 1.00m².............	0.47	10.00	0.52	0.86	-	15.80	0.86	17.18	m²	18.89
Wood windows and screens										
girth exceeding 300mm; panes, area not exceeding 0.10m².........	0.66	10.00	0.73	1.69	-	31.05	1.69	33.47	m²	36.81
girth exceeding 300mm; panes, area 0.10 - 0.50m²..................	0.51	10.00	0.57	1.22	-	22.41	1.22	24.20	m²	26.62
girth exceeding 300mm; panes, area 0.50 - 1.00m²..................	0.47	10.00	0.52	1.03	-	18.92	1.03	20.47	m²	22.52
girth exceeding 300mm; panes, area exceeding 1.00m².............	0.47	10.00	0.52	0.92	-	16.90	0.92	18.34	m²	20.17
Wood railings fences and gates; open type										
girth exceeding 300mm	0.66	10.00	0.73	0.81	-	14.88	0.81	16.42	m²	18.06
isolated surfaces, girth not exceeding 300mm	0.26	10.00	0.28	0.24	-	4.41	0.24	4.93	m	5.42
Wood railings fences and gates; close type										
girth exceeding 300mm	0.86	10.00	0.94	0.76	-	13.96	0.76	15.66	m²	17.23
Burn off, one coat primer, two undercoats, one coat full gloss finish, existing painted wood surfaces										
Wood general surfaces										
girth exceeding 300mm	1.13	10.00	1.25	1.06	-	19.47	1.06	21.78	m²	23.96
isolated surfaces, girth not exceeding 300mm	0.34	10.00	0.38	0.32	-	5.88	0.32	6.58	Unit	7.23

Labour hourly rates: (except Specialists) Craft Operatives 18.37 Labourer 13.76 Rates are national average prices. Refer to REGIONAL VARIATIONS for indicative levels of overall pricing in regions	MATERIALS			LABOUR				RATES		
	Del to Site	Waste	Material Cost	Craft Optve	Lab	Labour Cost	Sunds	Nett Rate		Gross rate (10%)
	£	%	£	Hrs	Hrs	£	£	£	Unit	£

PAINTING & CLEAR FINISHINGS; REDECORATIONS: OIL PAINTING WOODWORK (Cont'd)

Burn off, one coat primer, two undercoats, one coat full gloss finish, existing painted wood surfaces (Cont'd)

	Del to Site £	Waste %	Material Cost £	Craft Optve Hrs	Lab Hrs	Labour Cost £	Sunds £	Nett Rate £	Unit	Gross rate (10%) £
Wood glazed doors										
girth exceeding 300mm; panes, area not exceeding 0.10m²	0.62	10.00	0.68	1.76	-	32.33	1.76	34.77	m²	38.25
girth exceeding 300mm; panes, area 0.10 - 0.50m²	0.45	10.00	0.49	1.27	-	23.33	1.27	25.09	m²	27.60
girth exceeding 300mm; panes, area 0.50 - 1.00m²	0.43	10.00	0.47	1.06	-	19.47	1.06	21.00	m²	23.10
girth exceeding 300mm; panes, area exceeding 1.00m²	0.43	10.00	0.47	0.95	-	17.45	0.95	18.87	m²	20.76
Wood partially glazed doors										
girth exceeding 300mm; panes, area not exceeding 0.10m²	0.88	10.00	0.97	1.41	-	25.90	1.41	28.28	m²	31.10
girth exceeding 300mm; panes, area 0.10 - 0.50m²	0.68	10.00	0.75	1.17	-	21.49	1.17	23.42	m²	25.76
girth exceeding 300mm; panes, area 0.50 - 1.00m²	0.62	10.00	0.68	1.06	-	19.47	1.06	21.22	m²	23.34
girth exceeding 300mm; panes, area exceeding 1.00m²	0.62	10.00	0.68	1.00	-	18.37	1.00	20.05	m²	22.06
Wood windows and screens										
girth exceeding 300mm; panes, area not exceeding 0.10m²	0.88	10.00	0.97	1.94	-	35.64	1.94	38.54	m²	42.40
girth exceeding 300mm; panes, area 0.10 - 0.50m²	0.68	10.00	0.75	1.40	-	25.72	1.40	27.87	m²	30.66
girth exceeding 300mm; panes, area 0.50 - 1.00m²	0.62	10.00	0.68	1.17	-	21.49	1.17	23.35	m²	25.68
girth exceeding 300mm; panes, area exceeding 1.00m²	0.62	10.00	0.68	1.03	-	18.92	1.03	20.63	m²	22.70

Burn off, one coat primer, two undercoats, one coat full gloss finish, existing painted wood surfaces; external work

	Del to Site £	Waste %	Material Cost £	Craft Optve Hrs	Lab Hrs	Labour Cost £	Sunds £	Nett Rate £	Unit	Gross rate (10%) £
Wood general surfaces										
girth exceeding 300mm	1.13	10.00	1.25	1.11	-	20.39	1.11	22.75	m²	25.02
isolated surfaces, girth not exceeding 300mm	0.34	10.00	0.38	0.35	-	6.43	0.35	7.16	m	7.87
Wood glazed doors										
girth exceeding 300mm; panes, area not exceeding 0.10m²	0.62	10.00	0.68	1.76	-	32.33	1.76	34.77	m²	38.25
girth exceeding 300mm; panes, area 0.10 - 0.50m²	0.45	10.00	0.49	1.27	-	23.33	1.27	25.09	m²	27.60
girth exceeding 300mm; panes, area 0.50 - 1.00m²	0.43	10.00	0.47	1.06	-	19.47	1.06	21.00	m²	23.10
girth exceeding 300mm; panes, area exceeding 1.00m²	0.43	10.00	0.47	0.95	-	17.45	0.95	18.87	m²	20.76
Wood partially glazed doors										
girth exceeding 300mm; panes, area not exceeding 0.10m²	0.88	10.00	0.97	1.41	-	25.90	1.41	28.28	m²	31.10
girth exceeding 300mm; panes, area 0.10 - 0.50m²	0.68	10.00	0.75	1.17	-	21.49	1.17	23.42	m²	25.76
girth exceeding 300mm; panes, area 0.50 - 1.00m²	0.62	10.00	0.68	1.06	-	19.47	1.06	21.22	m²	23.34
girth exceeding 300mm; panes, area exceeding 1.00m²	0.62	10.00	0.68	1.00	-	18.37	1.00	20.05	m²	22.06
Wood windows and screens										
girth exceeding 300mm; panes, area not exceeding 0.10m²	0.88	10.00	0.97	1.99	-	36.56	1.99	39.51	m²	43.46
girth exceeding 300mm; panes, area 0.10 - 0.50m²	0.68	10.00	0.75	1.45	-	26.64	1.45	28.84	m²	31.72
girth exceeding 300mm; panes, area 0.50 - 1.00m²	0.62	10.00	0.68	1.22	-	22.41	1.22	24.31	m²	26.75
girth exceeding 300mm; panes, area exceeding 1.00m²	0.62	10.00	0.68	1.09	-	20.02	1.09	21.80	m²	23.98
Wood railings fences and gates; open type										
girth exceeding 300mm	0.88	10.00	0.97	0.95	-	17.45	0.95	19.37	m²	21.30
isolated surfaces, girth not exceeding 300mm	0.34	10.00	0.38	0.29	-	5.33	0.29	5.99	m	6.59
Wood railings fences and gates; close type										
girth exceeding 300mm	1.13	10.00	1.25	0.88	-	16.17	0.88	18.29	m²	20.12

Burn off, one coat primer, one undercoat, two coats full gloss finish, existing painted wood surfaces

	Del to Site £	Waste %	Material Cost £	Craft Optve Hrs	Lab Hrs	Labour Cost £	Sunds £	Nett Rate £	Unit	Gross rate (10%) £
Wood general surfaces										
girth exceeding 300mm	1.13	10.00	1.25	1.06	-	19.47	1.06	21.78	m²	23.96
isolated surfaces, girth not exceeding 300mm	0.34	10.00	0.38	0.32	-	5.88	0.32	6.58	m	7.23
Wood glazed doors										
girth exceeding 300mm; panes, area not exceeding 0.10m²	0.62	10.00	0.68	1.76	-	32.33	1.76	34.77	m²	38.25
girth exceeding 300mm; panes, area 0.10 - 0.50m²	0.45	10.00	0.49	1.27	-	23.33	1.27	25.09	m²	27.60
girth exceeding 300mm; panes, area 0.50 - 1.00m²	0.43	10.00	0.47	1.06	-	19.47	1.06	21.00	m²	23.10
girth exceeding 300mm; panes, area exceeding 1.00m²	0.43	10.00	0.47	0.95	-	17.45	0.95	18.87	m²	20.76
Wood partially glazed doors										
girth exceeding 300mm; panes, area not exceeding 0.10m²	0.88	10.00	0.97	1.41	-	25.90	1.41	28.28	m²	31.10
girth exceeding 300mm; panes, area 0.10 - 0.50m²	0.68	10.00	0.75	1.17	-	21.49	1.17	23.42	m²	25.76
girth exceeding 300mm; panes, area 0.50 - 1.00m²	0.62	10.00	0.68	1.06	-	19.47	1.06	21.22	m²	23.34
girth exceeding 300mm; panes, area exceeding 1.00m²	0.62	10.00	0.68	1.00	-	18.37	1.00	20.05	m²	22.06
Wood windows and screens										
girth exceeding 300mm; panes, area not exceeding 0.10m²	0.88	10.00	0.97	1.94	-	35.64	1.94	38.54	m²	42.40
girth exceeding 300mm; panes, area 0.10 - 0.50m²	0.68	10.00	0.75	1.40	-	25.72	1.40	27.87	m²	30.66
girth exceeding 300mm; panes, area 0.50 - 1.00m²	0.62	10.00	0.68	1.17	-	21.49	1.17	23.35	m²	25.68
girth exceeding 300mm; panes, area exceeding 1.00m²	0.62	10.00	0.68	1.03	-	18.92	1.03	20.63	m²	22.70

Burn off, one coat primer, one undercoat, two coats full gloss finish, existing painted wood surfaces; external work

	Del to Site £	Waste %	Material Cost £	Craft Optve Hrs	Lab Hrs	Labour Cost £	Sunds £	Nett Rate £	Unit	Gross rate (10%) £
Wood general surfaces										
girth exceeding 300mm	1.13	10.00	1.25	1.11	-	20.39	1.11	22.75	m²	25.02
isolated surfaces, girth not exceeding 300mm	0.34	10.00	0.38	0.35	-	6.43	0.35	7.16	m	7.87
Wood glazed doors										
girth exceeding 300mm; panes, area not exceeding 0.10m²	0.62	10.00	0.68	1.76	-	32.33	1.76	34.77	m²	38.25
girth exceeding 300mm; panes, area 0.10 - 0.50m²	0.45	10.00	0.49	1.27	-	23.33	1.27	25.09	m²	27.60
girth exceeding 300mm; panes, area 0.50 - 1.00m²	0.43	10.00	0.47	1.06	-	19.47	1.06	21.00	m²	23.10
girth exceeding 300mm; panes, area exceeding 1.00m²	0.43	10.00	0.47	0.95	-	17.45	0.95	18.87	m²	20.76
Wood partially glazed doors										
girth exceeding 300mm; panes, area not exceeding 0.10m²	0.88	10.00	0.97	1.41	-	25.90	1.41	28.28	m²	31.10
girth exceeding 300mm; panes, area 0.10 - 0.50m²	0.68	10.00	0.75	1.17	-	21.49	1.17	23.42	m²	25.76
girth exceeding 300mm; panes, area 0.50 - 1.00m²	0.62	10.00	0.68	1.06	-	19.47	1.06	21.22	m²	23.34
girth exceeding 300mm; panes, area exceeding 1.00m²	0.62	10.00	0.68	1.00	-	18.37	1.00	20.05	m²	22.06

DECORATION *(side margin)*

Labour hourly rates: (except Specialists) Craft Operatives 18.37 Labourer 13.76 Rates are national average prices. Refer to REGIONAL VARIATIONS for indicative levels of overall pricing in regions	MATERIALS			LABOUR				RATES		
	Del to Site £	Waste %	Material Cost £	Craft Optve Hrs	Lab Hrs	Labour Cost £	Sunds £	Nett Rate £	Unit	Gross rate (10%) £
PAINTING & CLEAR FINISHINGS; REDECORATIONS: OIL PAINTING WOODWORK (Cont'd)										
Burn off, one coat primer, one undercoat, two coats full gloss finish, existing painted wood surfaces; external work (Cont'd)										
Wood windows and screens										
girth exceeding 300mm; panes, area not exceeding 0.10m²	0.88	10.00	0.97	1.99	-	36.56	1.99	39.51	m²	43.46
girth exceeding 300mm; panes, area 0.10 - 0.50m²	0.68	10.00	0.75	1.45	-	26.64	1.45	28.84	m²	31.72
girth exceeding 300mm; panes, area 0.50 - 1.00m²	0.62	10.00	0.68	1.22	-	22.41	1.22	24.31	m²	26.75
girth exceeding 300mm; panes, area exceeding 1.00m²	0.62	10.00	0.68	1.09	-	20.02	1.09	21.80	m²	23.98
Wood railings fences and gates; open type										
girth exceeding 300mm	0.88	10.00	0.97	0.95	-	17.45	0.95	19.37	m²	21.30
isolated surfaces, girth not exceeding 300mm	0.34	10.00	0.38	0.29	-	5.33	0.29	5.99	m	6.59
Wood railings fences and gates; close type										
girth exceeding 300mm	1.13	10.00	1.25	0.88	-	16.17	0.88	18.29	m²	20.12
PAINTING & CLEAR FINISHINGS; REDECORATIONS: POLYURETHANE LACQUER										
Generally										
Note: the following rates include for the cost of all preparatory work, e.g. washing down, etc.										
Two coats polyurethane lacquer, existing lacquered surfaces										
Wood general surfaces										
girth exceeding 300mm	0.88	10.00	0.97	0.54	-	9.92	0.27	11.16	m²	12.28
isolated surfaces, girth not exceeding 300mm	0.26	10.00	0.29	0.18	-	3.31	0.09	3.69	m	4.06
Wood glazed doors										
girth exceeding 300mm; panes, area not exceeding 0.10m²	0.50	10.00	0.55	0.92	-	16.90	0.46	17.91	m²	19.70
girth exceeding 300mm; panes, area 0.10 - 0.50m²	0.35	10.00	0.39	0.67	-	12.31	0.34	13.03	m²	14.33
girth exceeding 300mm; panes, area 0.50 - 1.00m²	0.32	10.00	0.36	0.55	-	10.10	0.28	10.73	m²	11.81
girth exceeding 300mm; panes, area exceeding 1.00m²	0.32	10.00	0.36	0.48	-	8.82	0.24	9.41	m²	10.35
Wood partially glazed doors										
girth exceeding 300mm; panes, area not exceeding 0.10m²	0.70	10.00	0.78	0.77	-	14.14	0.38	15.31	m²	16.84
girth exceeding 300mm; panes, area 0.10 - 0.50m²	0.53	10.00	0.58	0.62	-	11.39	0.31	12.28	m²	13.51
girth exceeding 300mm; panes, area 0.50 - 1.00m²	0.50	10.00	0.55	0.55	-	10.10	0.28	10.93	m²	12.02
girth exceeding 300mm; panes, area exceeding 1.00m²	0.50	10.00	0.55	0.50	-	9.19	0.25	9.98	m²	10.98
Wood windows and screens										
girth exceeding 300mm; panes, area not exceeding 0.10m²	0.70	10.00	0.78	0.99	-	18.19	0.50	19.46	m²	21.40
girth exceeding 300mm; panes, area 0.10 - 0.50m²	0.53	10.00	0.58	0.65	-	11.94	0.32	12.85	m²	14.13
girth exceeding 300mm; panes, area 0.50 - 1.00m²	0.50	10.00	0.55	0.59	-	10.84	0.29	11.68	m²	12.85
girth exceeding 300mm; panes, area exceeding 1.00m²	0.50	10.00	0.55	0.52	-	9.55	0.26	10.36	m²	11.40
Two coats polyurethane lacquer, existing lacquered surfaces; external work										
Wood general surfaces										
girth exceeding 300mm	0.88	10.00	0.97	0.57	-	10.47	0.28	11.73	m²	12.90
isolated surfaces, girth not exceeding 300mm	0.26	10.00	0.29	0.19	-	3.49	0.09	3.88	m	4.26
Wood glazed doors										
girth exceeding 300mm; panes, area not exceeding 0.10m²	0.50	10.00	0.55	0.94	-	17.27	0.47	18.29	m²	20.12
girth exceeding 300mm; panes, area 0.10 - 0.50m²	0.35	10.00	0.39	0.68	-	12.49	0.34	13.22	m²	14.54
girth exceeding 300mm; panes, area 0.50 - 1.00m²	0.32	10.00	0.36	0.56	-	10.29	0.28	10.92	m²	12.01
girth exceeding 300mm; panes, area exceeding 1.00m²	0.32	10.00	0.36	0.49	-	9.00	0.25	9.60	m²	10.56
Wood partially glazed doors										
girth exceeding 300mm; panes, area not exceeding 0.10m²	0.70	10.00	0.78	0.79	-	14.51	0.40	15.68	m²	17.25
girth exceeding 300mm; panes, area 0.10 - 0.50m²	0.53	10.00	0.58	0.63	-	11.57	0.31	12.47	m²	13.72
girth exceeding 300mm; panes, area 0.50 - 1.00m²	0.50	10.00	0.55	0.56	-	10.29	0.28	11.12	m²	12.23
girth exceeding 300mm; panes, area exceeding 1.00m²	0.50	10.00	0.55	0.51	-	9.37	0.25	10.17	m²	11.19
Wood windows and screens										
girth exceeding 300mm; panes, area not exceeding 0.10m²	0.70	10.00	0.78	1.04	-	19.10	0.52	20.40	m²	22.44
girth exceeding 300mm; panes, area 0.10 - 0.50m²	0.53	10.00	0.58	0.76	-	13.96	0.38	14.92	m²	16.42
girth exceeding 300mm; panes, area 0.50 - 1.00m²	0.50	10.00	0.55	0.63	-	11.57	0.31	12.44	m²	13.68
girth exceeding 300mm; panes, area exceeding 1.00m²	0.50	10.00	0.55	0.55	-	10.10	0.28	10.93	m²	12.02
Three coats polyurethane lacquer, existing lacquered surfaces										
Wood general surfaces										
girth exceeding 300mm	1.29	10.00	1.42	0.62	-	11.39	0.31	13.12	m²	14.43
isolated surfaces, girth not exceeding 300mm	0.38	10.00	0.42	0.21	-	3.86	0.10	4.38	m	4.82
Wood glazed doors										
girth exceeding 300mm; panes, area not exceeding 0.10m²	0.73	10.00	0.81	1.04	-	19.10	0.52	20.43	m²	22.48
girth exceeding 300mm; panes, area 0.10 - 0.50m²	0.53	10.00	0.58	0.76	-	13.96	0.38	14.92	m²	16.42
girth exceeding 300mm; panes, area 0.50 - 1.00m²	0.47	10.00	0.52	0.63	-	11.57	0.31	12.41	m²	13.65
girth exceeding 300mm; panes, area exceeding 1.00m²	0.47	10.00	0.52	0.54	-	9.92	0.27	10.71	m²	11.78
Wood partially glazed doors										
girth exceeding 300mm; panes, area not exceeding 0.10m²	1.03	10.00	1.13	0.87	-	15.98	0.44	17.55	m²	19.30
girth exceeding 300mm; panes, area 0.10 - 0.50m²	0.76	10.00	0.84	0.77	-	14.14	0.38	15.37	m²	16.91
girth exceeding 300mm; panes, area 0.50 - 1.00m²	0.70	10.00	0.78	0.63	-	11.57	0.31	12.66	m²	13.93
girth exceeding 300mm; panes, area exceeding 1.00m²	0.70	10.00	0.78	0.58	-	10.65	0.29	11.72	m²	12.89
Wood windows and screens										
girth exceeding 300mm; panes, area not exceeding 0.10m²	0.73	10.00	0.81	1.13	-	20.76	0.56	22.13	m²	24.34
girth exceeding 300mm; panes, area 0.10 - 0.50m²	0.53	10.00	0.58	0.82	-	15.06	0.41	16.06	m²	17.66
girth exceeding 300mm; panes, area 0.50 - 1.00m²	0.47	10.00	0.52	0.68	-	12.49	0.34	13.35	m²	14.68
girth exceeding 300mm; panes, area exceeding 1.00m²	0.47	10.00	0.52	0.59	-	10.84	0.29	11.65	m²	12.82

Labour hourly rates: (except Specialists) Craft Operatives 18.37 Labourer 13.76 Rates are national average prices. Refer to REGIONAL VARIATIONS for indicative levels of overall pricing in regions	MATERIALS			LABOUR				RATES		
	Del to Site £	Waste %	Material Cost £	Craft Optve Hrs	Lab Hrs	Labour Cost £	Sunds £	Nett Rate £	Unit	Gross rate (10%) £
PAINTING & CLEAR FINISHINGS; REDECORATIONS: POLYURETHANE LACQUER (Cont'd)										
Three coats polyurethane lacquer, existing lacquered surfaces; external work										
Wood general surfaces										
girth exceeding 300mm	1.29	10.00	1.42	0.66	-	12.12	0.33	13.88	m²	15.26
isolated surfaces, girth not exceeding 300mm	0.38	10.00	0.42	0.22	-	4.04	0.11	4.57	m	5.03
Wood glazed doors										
girth exceeding 300mm; panes, area not exceeding 0.10m²	0.73	10.00	0.81	1.06	-	19.47	0.53	20.81	m²	22.89
girth exceeding 300mm; panes, area 0.10 - 0.50m²	0.53	10.00	0.58	0.77	-	14.14	0.38	15.11	m²	16.62
girth exceeding 300mm; panes, area 0.50 - 1.00m²	0.47	10.00	0.52	0.64	-	11.76	0.32	12.59	m²	13.85
girth exceeding 300mm; panes, area exceeding 1.00m²	0.47	10.00	0.52	0.54	-	9.92	0.27	10.71	m²	11.78
Wood partially glazed doors										
girth exceeding 300mm; panes, area not exceeding 0.10m²	1.03	10.00	1.13	0.89	-	16.35	0.44	17.93	m²	19.72
girth exceeding 300mm; panes, area 0.10 - 0.50m²	0.76	10.00	0.84	0.78	-	14.33	0.39	15.56	m²	17.11
girth exceeding 300mm; panes, area 0.50 - 1.00m²	0.70	10.00	0.78	0.64	-	11.76	0.32	12.85	m²	14.14
girth exceeding 300mm; panes, area exceeding 1.00m²	0.70	10.00	0.78	0.59	-	10.84	0.29	11.91	m²	13.10
Wood windows and screens										
girth exceeding 300mm; panes, area not exceeding 0.10m²	0.73	10.00	0.81	1.18	-	21.68	0.59	23.07	m²	25.38
girth exceeding 300mm; panes, area 0.10 - 0.50m²	0.53	10.00	0.58	0.87	-	15.98	0.44	17.00	m²	18.70
girth exceeding 300mm; panes, area 0.50 - 1.00m²	0.47	10.00	0.52	0.73	-	13.41	0.37	14.29	m²	15.72
girth exceeding 300mm; panes, area exceeding 1.00m²	0.47	10.00	0.52	0.63	-	11.57	0.31	12.41	m²	13.65
PAINTING & CLEAR FINISHINGS; REDECORATIONS: FRENCH AND WAX POLISHING										
Generally										
Note: the following rates include for the cost of all preparatory work, e.g. washing down, etc.										
Stain; two coats white polish										
Wood general surfaces										
girth exceeding 300mm	4.03	10.00	4.43	0.40	-	7.35	0.20	11.98	m²	13.18
isolated surfaces, girth not exceeding 300mm	1.27	10.00	1.40	0.14	-	2.57	0.07	4.04	m	4.44
Strip old polish, oil, two coats wax polish, existing polished surfaces										
Wood general surfaces										
girth exceeding 300mm	4.03	10.00	4.43	1.23	-	22.60	0.62	27.64	m²	30.40
isolated surfaces, girth not exceeding 300mm	1.27	10.00	1.40	0.37	-	6.80	0.19	8.38	m	9.22
Strip old polish, oil, stain, two coats wax polish, existing polished surfaces										
Wood general surfaces										
girth exceeding 300mm	4.03	10.00	4.43	1.35	-	24.80	0.68	29.90	m²	32.89
isolated surfaces, girth not exceeding 300mm	1.27	10.00	1.40	0.42	-	7.72	0.21	9.32	m	10.25
Open grain French polish existing polished surfaces										
Wood general surfaces										
girth exceeding 300mm	5.13	10.00	5.65	2.64	-	48.50	1.32	55.47	m²	61.01
isolated surfaces, girth not exceeding 300mm	1.62	10.00	1.78	0.88	-	16.17	0.44	18.39	m	20.23
Fully French polish existing polished surfaces										
Wood general surfaces										
girth exceeding 300mm	7.33	10.00	8.07	1.54	-	28.29	0.77	37.13	m²	40.84
isolated surfaces, girth not exceeding 300mm	2.31	10.00	2.55	0.52	-	9.55	0.26	12.36	m	13.59
Strip old polish, oil, open grain French polish, existing polished surfaces										
Wood general surfaces										
girth exceeding 300mm	5.13	10.00	5.65	3.43	-	63.01	1.72	70.37	m²	77.41
isolated surfaces, girth not exceeding 300mm	1.62	10.00	1.78	1.14	-	20.94	0.57	23.29	m	25.62
Strip old polish, oil, stain, open grain French polish, existing polished surfaces										
Wood general surfaces										
girth exceeding 300mm	5.13	10.00	5.65	3.55	-	65.21	1.77	72.64	m²	79.90
isolated surfaces, girth not exceeding 300mm	1.62	10.00	1.78	1.19	-	21.86	0.60	24.24	m	26.66
Strip old polish, oil, body in, fully French polish existing polished surfaces										
Wood general surfaces										
girth exceeding 300mm	7.33	10.00	8.07	4.63	-	85.05	2.32	95.44	m²	104.98
isolated surfaces, girth not exceeding 300mm	2.31	10.00	2.55	1.54	-	28.29	0.77	31.61	m	34.77
Strip old polish, oil, stain, body in, fully French polish, existing polished surfaces										
Wood general surfaces										
girth exceeding 300mm	7.33	10.00	8.07	4.74	-	87.07	2.37	97.51	m²	107.26
isolated surfaces, girth not exceeding 300mm	2.31	10.00	2.55	1.58	-	29.02	0.79	32.36	m	35.60

DECORATION

DECORATION

Labour hourly rates: (except Specialists) Craft Operatives 18.37 Labourer 13.76 Rates are national average prices. Refer to REGIONAL VARIATIONS for indicative levels of overall pricing in regions	MATERIALS			LABOUR				RATES		
	Del to Site £	Waste %	Material Cost £	Craft Optve Hrs	Lab Hrs	Labour Cost £	Sunds £	Nett Rate £	Unit	Gross rate (10%) £
PAINTING & CLEAR FINISHINGS; REDECORATIONS: WATER REPELLENT										
One coat solvent based water repellent, existing surfaces										
Cement rendered general surfaces girth exceeding 300mm	0.15	5.00	0.15	0.12	-	2.20	0.06	2.42	m²	2.66
Stone general surfaces girth exceeding 300mm	0.18	5.00	0.19	0.13	-	2.39	0.06	2.64	m²	2.90
Brickwork general surfaces girth exceeding 300mm	0.18	5.00	0.19	0.13	-	2.39	0.06	2.64	m²	2.90
PAINTING & CLEAR FINISHINGS; REDECORATIONS: REMOVAL OF MOULD GROWTH										
Apply fungicide to existing decorated and infected surfaces										
Cement rendered general surfaces girth exceeding 300mm	0.12	5.00	0.13	0.22	-	4.04	0.11	4.28	m²	4.70
Concrete general surfaces girth exceeding 300mm	0.15	5.00	0.15	0.22	-	4.04	0.11	4.30	m²	4.74
Plaster general surfaces girth exceeding 300mm	0.11	5.00	0.11	0.22	-	4.04	0.11	4.26	m²	4.69
Brickwork general surfaces girth exceeding 300mm	0.15	5.00	0.15	0.24	-	4.41	0.12	4.68	m²	5.15
Apply fungicide to existing decorated and infected surfaces; external work										
Cement rendered general surfaces girth exceeding 300mm	0.12	5.00	0.13	0.24	-	4.41	0.12	4.65	m²	5.12
Concrete general surfaces girth exceeding 300mm	0.15	5.00	0.15	0.24	-	4.41	0.12	4.68	m²	5.15
Plaster general surfaces girth exceeding 300mm	0.11	5.00	0.11	0.24	-	4.41	0.12	4.64	m²	5.10
Brickwork general surfaces girth exceeding 300mm	0.15	5.00	0.15	0.26	-	4.78	0.13	5.06	m²	5.57
DECORATIVE PAPERS OR FABRICS										
Lining paper (prime cost sum for supply and allowance for waste included elsewhere); sizing; applying adhesive; hanging; butt joints										
Plaster walls and columns exceeding 0.50m²	0.05	-	0.05	0.13	-	2.39	0.06	2.51	m²	2.76
Plaster ceilings and beams exceeding 0.50m²	0.05	-	0.05	0.14	-	2.57	0.07	2.70	m²	2.97
Pulp paper (prime cost sum for supply and allowance for waste included elsewhere); sizing; applying adhesive; hanging; butt joints										
Plaster walls and columns exceeding 0.50m²	0.05	-	0.05	0.31	-	5.69	0.16	5.90	m²	6.50
Plaster ceilings and beams exceeding 0.50m²	0.05	-	0.05	0.34	-	6.25	0.17	6.47	m²	7.12
Washable paper (prime cost sum for supply and allowance for waste included elsewhere); sizing; applying adhesive; hanging; butt joints										
Plaster walls and columns exceeding 0.50m²	0.07	-	0.07	0.24	-	4.41	0.12	4.60	m²	5.06
Plaster ceilings and beams exceeding 0.50m²	0.07	-	0.07	0.26	-	4.78	0.13	4.98	m²	5.48
Vinyl coated paper (prime cost sum for supply and allowance for waste included elsewhere); sizing; applying adhesive; hanging; butt joints										
Plaster walls and columns exceeding 0.50m²	0.07	-	0.07	0.24	-	4.41	0.12	4.60	m²	5.06
Plaster ceilings and beams exceeding 0.50m²	0.07	-	0.07	0.26	-	4.78	0.13	4.98	m²	5.48
Embossed or textured paper (prime cost sum for supply and allowance for waste included elsewhere); sizing; applying adhesive; hanging; butt joints										
Plaster walls and columns exceeding 0.50m²	0.07	-	0.07	0.24	-	4.41	0.12	4.60	m²	5.06
Plaster ceilings and beams exceeding 0.50m²	0.15	-	0.15	0.26	-	4.78	0.13	5.05	m²	5.56

Labour hourly rates: (except Specialists) Craft Operatives 18.37 Labourer 13.76 Rates are national average prices. Refer to REGIONAL VARIATIONS for indicative levels of overall pricing in regions	MATERIALS			LABOUR				RATES		
	Del to Site £	Waste %	Material Cost £	Craft Optve Hrs	Lab Hrs	Labour Cost £	Sunds £	Nett Rate £	Unit	Gross rate (10%) £
DECORATIVE PAPERS OR FABRICS (Cont'd)										
Woodchip paper (prime cost sum for supply and allowance for waste included elsewhere); sizing; applying adhesive; hanging; butt joints										
Plaster walls and columns exceeding 0.50m²................................	0.05	-	0.05	0.24	-	4.41	0.12	4.58	m²	5.04
Plaster ceilings and beams exceeding 0.50m²................................	0.05	-	0.05	0.26	-	4.78	0.13	4.96	m²	5.46
Paper border strips (prime cost sum for supply and allowance for waste included elsewhere); sizing; applying adhesive; hanging; butt joints										
Border strips										
25mm wide to papered walls and columns......................................	0.01	-	0.01	0.09	-	1.65	0.05	1.70	m	1.88
75mm wide to papered walls and columns......................................	0.02	-	0.02	0.10	-	1.84	0.05	1.91	m	2.10
Lining paper; 1000g; sizing; applying adhesive; hanging; butt joints										
Plaster walls and columns exceeding 0.50m²................................	0.39	12.00	0.43	0.13	-	2.39	0.06	2.89	m²	3.18
Plaster ceilings and beams exceeding 0.50m²................................	0.39	12.00	0.43	0.14	-	2.57	0.07	3.08	m²	3.38
Pulp paper pc £4.00 roll; sizing; applying adhesive; hanging; butt joints										
Plaster walls and columns exceeding 0.50m²................................	0.86	30.00	1.09	0.24	-	4.41	0.12	5.62	m²	6.19
Plaster ceilings and beams exceeding 0.50m²................................	0.86	25.00	1.05	0.26	-	4.78	0.13	5.96	m²	6.56
Pulp paper (24" drop pattern match) pc £6.00 roll; sizing; applying adhesive; hanging; butt joints										
Plaster walls and columns exceeding 0.50m²................................	1.25	40.00	1.74	0.31	-	5.69	0.16	7.58	m²	8.34
Plaster ceilings and beams exceeding 0.50m²................................	1.25	35.00	1.68	0.33	-	6.06	0.17	7.90	m²	8.69
Washable paper pc £6.00 roll; sizing; applying adhesive; hanging; butt joints										
Plaster walls and columns exceeding 0.50m²................................	1.27	25.00	1.57	0.24	-	4.41	0.12	6.10	m²	6.71
Plaster ceilings and beams exceeding 0.50m²................................	1.27	20.00	1.51	0.26	-	4.78	0.13	6.42	m²	7.06
Vinyl coated paper pc £6.50 roll; sizing; applying adhesive; hanging; butt joints										
Plaster walls and columns exceeding 0.50m²................................	1.37	25.00	1.70	0.24	-	4.41	0.12	6.23	m²	6.85
Plaster ceilings and beams exceeding 0.50m²................................	1.37	20.00	1.63	0.26	-	4.78	0.13	6.54	m²	7.19
Vinyl coated paper pc £7.00 roll; sizing; applying adhesive; hanging; butt joints										
Plaster walls and columns exceeding 0.50m²................................	1.47	25.00	1.82	0.24	-	4.41	0.12	6.35	m²	6.99
Plaster ceilings and beams exceeding 0.50m²................................	1.47	20.00	1.75	0.26	-	4.78	0.13	6.66	m²	7.32
Embossed paper pc £6.00 roll; sizing; applying adhesive; hanging; butt joints										
Plaster walls and columns exceeding 0.50m²................................	1.35	25.00	1.65	0.24	-	4.41	0.12	6.18	m²	6.79
Plaster ceilings and beams exceeding 0.50m²................................	1.35	20.00	1.59	0.26	-	4.78	0.13	6.49	m²	7.14
Textured paper pc £6.50 roll; sizing; applying adhesive; hanging; butt joints										
Plaster walls and columns exceeding 0.50m²................................	1.41	25.00	1.74	0.24	-	4.41	0.11	6.25	m²	6.88
Plaster ceilings and beams exceeding 0.50m²................................	1.41	20.00	1.67	0.26	-	4.78	0.12	6.57	m²	7.22
Woodchip paper pc £1.50 roll; sizing; applying adhesive; hanging; butt joints										
Plaster walls and columns exceeding 0.50m²................................	0.35	25.00	0.43	0.24	-	4.41	0.11	4.95	m²	5.44
Plaster ceilings and beams exceeding 0.50m²................................	0.35	20.00	0.42	0.26	-	4.78	0.12	5.31	m²	5.84

DECORATION

Labour hourly rates: (except Specialists) Craft Operatives 18.37 Labourer 13.76 Rates are national average prices. Refer to REGIONAL VARIATIONS for indicative levels of overall pricing in regions	MATERIALS			LABOUR				RATES		
	Del to Site £	Waste %	Material Cost £	Craft Optve Hrs	Lab Hrs	Labour Cost £	Sunds £	Nett Rate £	Unit	Gross rate (10%) £
DECORATIVE PAPERS OR FABRICS (Cont'd)										
Paper border strips pc £0.50m; applying adhesive; hanging; butt joints										
Border strips 25mm wide to papered walls and columns	0.51	10.00	0.56	0.09	-	1.65	0.05	2.25	m	2.48
Paper border strips pc £1.00m; applying adhesive; hanging; butt joints										
Border strips 75mm wide to papered walls and columns	1.02	10.00	1.12	0.10	-	1.84	0.05	3.01	m	3.31
Hessian wall covering (prime cost sum for supply and allowance for waste included elsewhere); sizing; applying adhesive; hanging; butt joints										
Plaster walls and columns exceeding 0.50m²	0.22	-	0.22	0.44	-	8.08	0.22	8.52	m²	9.38
Textile hessian paper backed wall covering (prime cost sum for supply and allowance for waste included elsewhere); sizing; applying adhesive; hanging; butt joints										
Plaster walls and columns exceeding 0.50m²	0.22	-	0.22	0.36	-	6.61	0.18	7.01	m²	7.71
Hessian wall covering pc £4.00m²; sizing; applying adhesive; hanging; butt joints										
Plaster walls and columns exceeding 0.50m²	4.72	30.00	6.07	0.24	-	4.41	0.12	10.60	m²	11.66
Textile hessian paper backed wall covering pc £8.00m²; sizing; applying adhesive; hanging; butt joints										
Plaster walls and columns exceeding 0.50m²	8.22	25.00	10.22	0.36	-	6.61	0.18	17.01	m²	18.71
DECORATIVE PAPERS OR FABRICS; REDECORATIONS										
Stripping existing paper; lining paper (prime cost sum for supply and allowance for waste included elsewhere); sizing; applying adhesive hanging; butt joints										
Hard building board walls and columns exceeding 0.50m²	0.05	-	0.05	0.32	-	5.88	0.29	6.22	m²	6.85
Plaster walls and columns exceeding 0.50m²	0.05	-	0.05	0.28	-	5.14	0.28	5.48	m²	6.03
Hard building board ceilings and beams exceeding 0.50m²	0.05	-	0.05	0.35	-	6.43	0.35	6.83	m²	7.52
Plaster ceilings and beams exceeding 0.50m²	0.05	-	0.05	0.30	-	5.51	0.30	5.87	m²	6.45
Stripping existing paper; lining paper and cross lining (prime cost sum for supply and allowance for waste included elsewhere); sizing; applying adhesive; hanging; butt joints										
Hard building board walls and columns exceeding 0.50m²	0.11	-	0.11	0.45	-	8.27	0.45	8.83	m²	9.71
Plaster walls and columns exceeding 0.50m²	0.11	-	0.11	0.41	-	7.53	0.41	8.05	m²	8.86
Hard building board ceilings and beams exceeding 0.50m²	0.11	-	0.11	0.50	-	9.19	0.50	9.80	m²	10.77
Plaster ceilings and beams exceeding 0.50m²	0.11	-	0.11	0.44	-	8.08	0.44	8.63	m²	9.50
Stripping existing paper; pulp paper (prime cost sum for supply and allowance for waste included elsewhere); applying adhesive; hanging; butt joints										
Hard building board walls and columns exceeding 0.50m²	0.05	-	0.05	0.43	-	7.90	0.43	8.38	m²	9.22
Plaster walls and columns exceeding 0.50m²	0.05	-	0.05	0.39	-	7.16	0.39	7.61	m²	8.37
Hard building board ceilings and beams exceeding 0.50m²	0.05	-	0.05	0.47	-	8.63	0.47	9.16	m²	10.07
Plaster ceilings and beams exceeding 0.50m²	0.05	-	0.05	0.42	-	7.72	0.42	8.19	m²	9.01
Stripping existing washable or vinyl coated paper; washable paper (prime cost sum for supply and allowance for waste included elsewhere); applying adhesive; hanging; butt joints										
Hard building board walls and columns exceeding 0.50m²	0.07	-	0.07	0.44	-	8.08	0.44	8.60	m²	9.45
Plaster walls and columns exceeding 0.50m²	0.07	-	0.07	0.40	-	7.35	0.40	7.82	m²	8.60
Hard building board ceilings and beams exceeding 0.50m²	0.07	-	0.07	0.48	-	8.82	0.48	9.37	m²	10.31

Labour hourly rates: (except Specialists) Craft Operatives 18.37 Labourer 13.76 Rates are national average prices. Refer to REGIONAL VARIATIONS for indicative levels of overall pricing in regions	MATERIALS			LABOUR				RATES		
	Del to Site £	Waste %	Material Cost £	Craft Optve Hrs	Lab Hrs	Labour Cost £	Sunds £	Nett Rate £	Unit	Gross rate (10%) £
DECORATIVE PAPERS OR FABRICS; REDECORATIONS (Cont'd)										
Stripping existing washable or vinyl coated paper; washable paper (prime cost sum for supply and allowance for waste included elsewhere); applying adhesive; hanging; butt joints (Cont'd)										
Plaster ceilings and beams exceeding 0.50m².............	0.07	-	0.07	0.43	-	7.90	0.43	8.40	m²	9.24
Stripping existing washable or vinyl coated paper; vinyl coated paper (prime cost sum for supply and allowance for waste included elsewhere); sizing; applying adhesive; hanging; butt joints										
Hard building board walls and columns exceeding 0.50m².............	0.07	-	0.07	0.44	-	8.08	0.44	8.60	m²	9.45
Plaster walls and columns exceeding 0.50m².............	0.07	-	0.07	0.40	-	7.35	0.40	7.82	m²	8.60
Hard building board ceilings and beams exceeding 0.50m².............	0.07	-	0.07	0.48	-	8.82	0.48	9.37	m²	10.31
Plaster ceilings and beams exceeding 0.50m².............	0.07	-	0.07	0.43	-	7.90	0.43	8.40	m²	9.24
Stripping existing paper; embossed or textured paper; (prime cost sum for supply and allowance for waste included elsewhere); sizing; applying adhesive; hanging; butt joints										
Hard building board walls and columns exceeding 0.50m².............	0.15	-	0.15	0.43	-	7.90	0.43	8.48	m²	9.32
Plaster walls and columns exceeding 0.50m².............	0.15	-	0.15	0.39	-	7.16	0.39	7.70	m²	8.47
Hard building board ceilings and beams exceeding 0.50m².............	0.15	-	0.15	0.47	-	8.63	0.47	9.25	m²	10.18
Plaster ceilings and beams exceeding 0.50m².............	0.15	-	0.15	0.42	-	7.72	0.42	8.28	m²	9.11
Stripping existing paper; woodchip paper (prime cost sum for supply and allowance for waste included elsewhere); sizing; applying adhesive; hanging; butt joints										
Hard building board walls and columns exceeding 0.50m².............	0.05	-	0.05	0.43	-	7.90	0.43	8.38	m²	9.22
Plaster walls and columns exceeding 0.50m².............	0.05	-	0.05	0.39	-	7.16	0.39	7.61	m²	8.37
Hard building board ceilings and beams exceeding 0.50m².............	0.05	-	0.05	0.47	-	8.63	0.47	9.16	m²	10.07
Plaster ceilings and beams exceeding 0.50m².............	0.05	-	0.05	0.42	-	7.72	0.42	8.19	m²	9.01
Stripping existing varnished or painted paper; woodchip paper (prime cost sum for supply and allowance for waste included elsewhere); sizing; applying adhesive; hanging; butt joints										
Hard building board walls and columns exceeding 0.50m².............	0.05	-	0.05	0.57	-	10.47	0.57	11.10	m²	12.21
Plaster walls and columns exceeding 0.50m².............	0.05	-	0.05	0.53	-	9.74	0.53	10.32	m²	11.35
Hard building board ceilings and beams exceeding 0.50m².............	0.05	-	0.05	0.63	-	11.57	0.63	12.26	m²	13.48
Plaster ceilings and beams exceeding 0.50m².............	0.05	-	0.05	0.58	-	10.65	0.58	11.29	m²	12.42
Stripping existing paper; lining paper pc £1.25 roll; sizing; applying adhesive; hanging; butt joints										
Hard building board walls and columns exceeding 0.50m².............	0.39	12.00	0.43	0.32	-	5.88	0.59	6.90	m²	7.59
Plaster walls and columns exceeding 0.50m².............	0.39	12.00	0.43	0.28	-	5.14	0.28	5.86	m²	6.44
Hard building board ceilings and beams exceeding 0.50m².............	0.39	12.00	0.43	0.35	-	6.43	0.35	7.21	m²	7.93
Plaster ceilings and beams exceeding 0.50m².............	0.39	12.00	0.43	0.30	-	5.51	0.30	6.24	m²	6.87
Stripping existing paper; lining paper pc £1.25 roll and cross lining pc £1.25 roll; sizing; applying adhesive; hanging; butt joints										
Hard building board walls and columns exceeding 0.50m².............	0.79	12.00	0.87	0.45	-	8.27	0.45	9.58	m²	10.54
Plaster walls and columns exceeding 0.50m².............	0.79	12.00	0.87	0.41	-	7.53	0.37	8.77	m²	9.65

DECORATION

	MATERIALS			LABOUR				RATES		
Labour hourly rates: (except Specialists) Craft Operatives 18.37 Labourer 13.76 Rates are national average prices. Refer to REGIONAL VARIATIONS for indicative levels of overall pricing in regions	Del to Site £	Waste %	Material Cost £	Craft Optve Hrs	Lab Hrs	Labour Cost £	Sunds £	Nett Rate £	Unit	Gross rate (10%) £
DECORATIVE PAPERS OR FABRICS; REDECORATIONS (Cont'd)										
Stripping existing paper; lining paper pc £1.25 roll and cross lining pc £1.25 roll; sizing; applying adhesive; hanging; butt joints (Cont'd)										
Hard building board ceilings and beams exceeding 0.50m²	0.79	12.00	0.87	0.50	-	9.19	0.50	10.55	m²	11.61
Plaster ceilings and beams exceeding 0.50m²	0.79	12.00	0.87	0.44	-	8.08	0.44	9.39	m²	10.33
Stripping existing paper; pulp paper pc £4.00 roll; sizing; applying adhesive; hanging; butt joints										
Hard building board walls and columns exceeding 0.50m²	0.86	30.00	1.09	0.43	-	7.90	0.43	9.42	m²	10.37
Plaster walls and columns exceeding 0.50m²	0.86	30.00	1.09	0.39	-	7.16	0.39	8.65	m²	9.51
Hard building board ceilings and beams exceeding 0.50m²	0.86	25.00	1.05	0.47	-	8.63	0.47	10.16	m²	11.17
Plaster ceilings and beams exceeding 0.50m²	0.86	25.00	1.05	0.42	-	7.72	0.42	9.19	m²	10.11
Stripping existing paper; pulp paper with 24" drop pattern; pc £6.00 roll; sizing; applying adhesive; hanging; butt joints										
Hard building board walls and columns exceeding 0.50m²	1.25	40.00	1.74	0.50	-	9.19	0.50	11.42	m²	12.56
Plaster walls and columns exceeding 0.50m²	1.25	40.00	1.74	0.45	-	8.27	0.45	10.45	m²	11.50
Hard building board ceilings and beams exceeding 0.50m²	1.25	35.00	1.68	0.54	-	9.92	0.54	12.13	m²	13.35
Plaster ceilings and beams exceeding 0.50m²	1.25	35.00	1.68	0.48	-	8.82	0.48	10.97	m²	12.07
Stripping existing washable or vinyl coated paper; washable paper pc £6.00 roll; sizing; applying adhesive; hanging; butt joints										
Hard building board walls and columns exceeding 0.50m²	1.27	25.00	1.57	0.44	-	8.08	0.44	10.10	m²	11.10
Plaster walls and columns exceeding 0.50m²	1.27	25.00	1.57	0.40	-	7.35	0.40	9.32	m²	10.25
Hard building board ceilings and beams exceeding 0.50m²	1.27	20.00	1.51	0.48	-	8.82	0.48	10.81	m²	11.89
Plaster ceilings and beams exceeding 0.50m²	1.27	20.00	1.51	0.43	-	7.90	0.43	9.84	m²	10.83
Stripping existing washable or vinyl coated paper; vinyl coated paper pc £6.50 roll; sizing; applying adhesive; hanging; butt joints										
Hard building board walls and columns exceeding 0.50m²	1.37	25.00	1.70	0.44	-	8.08	0.44	10.22	m²	11.24
Plaster walls and columns exceeding 0.50m²	1.37	25.00	1.70	0.40	-	7.35	0.40	9.45	m²	10.39
Hard building board ceilings and beams exceeding 0.50m²	1.37	20.00	1.63	0.48	-	8.82	0.48	10.93	m²	12.02
Plaster ceilings and beams exceeding 0.50m²	1.37	20.00	1.63	0.43	-	7.90	0.43	9.96	m²	10.96
Stripping existing washable or vinyl coated paper; vinyl coated paper pc £8.00 roll; sizing; applying adhesive; hanging; butt joints										
Hard building board walls and columns exceeding 0.50m²	1.67	25.00	2.07	0.44	-	8.08	0.44	10.60	m²	11.65
Plaster walls and columns exceeding 0.50m²	1.67	25.00	2.07	0.40	-	7.35	0.40	9.82	m²	10.80
Hard building board ceilings and beams exceeding 0.50m²	1.67	20.00	1.99	0.48	-	8.82	0.48	11.29	m²	12.42
Plaster ceilings and beams exceeding 0.50m²	1.67	20.00	1.99	0.43	-	7.90	0.43	10.32	m²	11.35
Stripping existing paper; embossed paper pc £6.00 roll; sizing; applying adhesive; hanging; butt joints										
Hard building board walls and columns exceeding 0.50m²	1.35	25.00	1.65	0.43	-	7.90	0.43	9.98	m²	10.97
Plaster walls and columns exceeding 0.50m²	1.35	25.00	1.65	0.39	-	7.16	0.39	9.20	m²	10.12
Hard building board ceilings and beams exceeding 0.50m²	1.35	20.00	1.59	0.47	-	8.63	0.47	10.69	m²	11.76

Labour hourly rates: (except Specialists) Craft Operatives 18.37 Labourer 13.76 Rates are national average prices. Refer to REGIONAL VARIATIONS for indicative levels of overall pricing in regions	MATERIALS			LABOUR				RATES		
	Del to Site	Waste	Material Cost	Craft Optve	Lab	Labour Cost	Sunds	Nett Rate	Unit	Gross rate (10%)
	£	%	£	Hrs	Hrs	£	£	£		£
DECORATIVE PAPERS OR FABRICS; REDECORATIONS (Cont'd)										
Stripping existing paper; embossed paper pc £6.00 roll; sizing; applying adhesive; hanging; butt joints (Cont'd)										
Plaster ceilings and beams exceeding 0.50m²..	1.35	20.00	1.59	0.42	-	7.72	0.42	9.72	m²	10.70
Stripping existing paper; textured paper pc £6.50 roll; sizing; applying adhesive; hanging; butt joints										
Hard building board walls and columns exceeding 0.50m²..	1.41	25.00	1.74	0.43	-	7.90	0.43	10.06	m²	11.07
Plaster walls and columns exceeding 0.50m²..	1.41	25.00	1.74	0.39	-	7.16	0.39	9.29	m²	10.22
Hard building board ceilings and beams exceeding 0.50m²..	1.41	20.00	1.67	0.47	-	8.63	0.47	10.77	m²	11.85
Plaster ceilings and beams exceeding 0.50m²..	1.41	20.00	1.67	0.42	-	7.72	0.42	9.81	m²	10.79
Stripping existing paper; woodchip paper pc £1.50 roll; sizing; applying adhesive; hanging; butt joints										
Hard building board walls and columns exceeding 0.50m²..	0.35	25.00	0.43	0.43	-	7.90	0.43	8.76	m²	9.64
Plaster walls and columns exceeding 0.50m²..	0.35	25.00	0.43	0.39	-	7.16	0.39	7.98	m²	8.78
Hard building board ceilings and beams exceeding 0.50m²..	0.35	20.00	0.42	0.47	-	8.63	0.47	9.52	m²	10.47
Plaster ceilings and beams exceeding 0.50m²..	0.35	20.00	0.42	0.42	-	7.72	0.42	8.55	m²	9.41
Washing down old distempered or painted surfaces; lining paper pc £1.25 roll; sizing; applying adhesive; hanging; butt joints										
Hard building board walls and columns exceeding 0.50m²..	0.39	12.00	0.43	0.22	-	4.04	0.11	4.59	m²	5.04
Plaster walls and columns exceeding 0.50m²..	0.39	12.00	0.43	0.22	-	4.04	0.11	4.59	m²	5.04
Hard building board ceilings and beams exceeding 0.50m²..	0.39	12.00	0.43	0.23	-	4.23	0.12	4.77	m²	5.25
Plaster ceilings and beams exceeding 0.50m²..	0.39	12.00	0.43	0.23	-	4.23	0.12	4.77	m²	5.25
Lining papered walls and columns exceeding 0.50m²..	0.39	12.00	0.43	0.22	-	4.04	0.11	4.59	m²	5.04
Lining papered ceilings and beams exceeding 0.50m²..	0.39	12.00	0.43	0.23	-	4.23	0.12	4.77	m²	5.25
Washing down old distempered or painted surfaces; lining paper pc £1.25 roll; and cross lining pc £1.25 roll; sizing; applying adhesive; hanging; butt joints										
Hard building board walls and columns exceeding 0.50m²..	0.79	12.00	0.87	0.35	-	6.43	0.17	7.47	m²	8.22
Plaster walls and columns exceeding 0.50m²..	0.79	12.00	0.87	0.35	-	6.43	0.17	7.47	m²	8.22
Hard building board ceilings and beams exceeding 0.50m²..	0.79	12.00	0.87	0.37	-	6.80	0.19	7.85	m²	8.63
Plaster ceilings and beams exceeding 0.50m²..	0.79	12.00	0.87	0.37	-	6.80	0.19	7.85	m²	8.63
Washing down old distempered or painted surfaces; pulp paper pc £4.00 roll; sizing; applying adhesive; hanging; butt joints										
Hard building board walls and columns exceeding 0.50m²..	0.86	30.00	1.09	0.33	-	6.06	0.17	7.32	m²	8.05
Plaster walls and columns exceeding 0.50m²..	0.86	30.00	1.09	0.33	-	6.06	0.17	7.32	m²	8.05
Hard building board ceilings and beams exceeding 0.50m²..	0.86	25.00	1.05	0.35	-	6.43	0.17	7.66	m²	8.43
Plaster ceilings and beams exceeding 0.50m²..	0.86	25.00	1.05	0.35	-	6.43	0.17	7.66	m²	8.43
Washing down old distempered or painted surfaces; pulp paper with 24" drop pattern; pc £4.00 roll; sizing; applying adhesive; hanging; butt joints										
Hard building board walls and columns exceeding 0.50m²..	1.25	40.00	1.74	0.40	-	7.35	0.20	9.28	m²	10.21
Plaster walls and columns exceeding 0.50m²..	1.25	40.00	1.74	0.40	-	7.35	0.20	9.28	m²	10.21

DECORATION

Labour hourly rates: (except Specialists) Craft Operatives 18.37 Labourer 13.76 Rates are national average prices. Refer to REGIONAL VARIATIONS for indicative levels of overall pricing in regions	MATERIALS			LABOUR				RATES		
	Del to Site £	Waste %	Material Cost £	Craft Optve Hrs	Lab Hrs	Labour Cost £	Sunds £	Nett Rate £	Unit	Gross rate (10%) £
DECORATIVE PAPERS OR FABRICS; REDECORATIONS (Cont'd)										
Washing down old distempered or painted surfaces; pulp paper with 24" drop pattern; pc £4.00 roll; sizing; applying adhesive; hanging; butt joints (Cont'd)										
Hard building board ceilings and beams exceeding 0.50m²	1.25	35.00	1.68	0.42	-	7.72	0.21	9.60	m²	10.56
Plaster ceilings and beams exceeding 0.50m²	1.25	35.00	1.68	0.42	-	7.72	0.21	9.60	m²	10.56
Washing down old distempered or painted surfaces; washable paper pc £6.00 roll; sizing; applying adhesive; hanging; butt joints										
Hard building board walls and columns exceeding 0.50m²	1.27	25.00	1.57	0.33	-	6.06	0.17	7.80	m²	8.58
Plaster walls and columns exceeding 0.50m²	1.27	25.00	1.57	0.33	-	6.06	0.17	7.80	m²	8.58
Hard building board ceilings and beams exceeding 0.50m²	1.27	20.00	1.51	0.35	-	6.43	0.17	8.12	m²	8.93
Plaster ceilings and beams exceeding 0.50m²	1.27	20.00	1.51	0.35	-	6.43	0.16	8.10	m²	8.91
Washing down old distempered or painted surfaces; vinyl coated paper pc £6.50 roll; sizing; applying adhesive; hanging; butt joints										
Hard building board walls and columns exceeding 0.50m²	1.37	25.00	1.70	0.33	-	6.06	0.17	7.92	m²	8.72
Plaster walls and columns exceeding 0.50m²	1.37	25.00	1.70	0.33	-	6.06	0.17	7.92	m²	8.72
Hard building board ceilings and beams exceeding 0.50m²	1.37	20.00	1.63	0.35	-	6.43	0.17	8.24	m²	9.06
Plaster ceilings and beams exceeding 0.50m²	1.37	20.00	1.63	0.35	-	6.43	0.17	8.24	m²	9.06
Washing down old distempered or painted surfaces; vinyl coated paper pc £7.00 roll; sizing; applying adhesive; hanging; butt joints										
Hard building board walls and columns exceeding 0.50m²	1.47	25.00	1.82	0.33	-	6.06	0.17	8.05	m²	8.85
Plaster walls and columns exceeding 0.50m²	1.47	25.00	1.82	0.33	-	6.06	0.17	8.05	m²	8.85
Hard building board ceilings and beams exceeding 0.50m²	1.47	20.00	1.75	0.35	-	6.43	0.17	8.36	m²	9.19
Plaster ceilings and beams exceeding 0.50m²	1.47	20.00	1.75	0.35	-	6.43	0.17	8.36	m²	9.19
Washing down old distempered or painted surfaces; embossed paper pc £6.00 roll; sizing; applying adhesive; hanging; butt joints										
Hard building board walls and columns exceeding 0.50m²	1.35	25.00	1.65	0.33	-	6.06	0.17	7.87	m²	8.66
Plaster walls and columns exceeding 0.50m²	1.35	25.00	1.65	0.33	-	6.06	0.17	7.87	m²	8.66
Hard building board ceilings and beams exceeding 0.50m²	1.35	20.00	1.59	0.35	-	6.43	0.17	8.19	m²	9.01
Plaster ceilings and beams exceeding 0.50m²	1.35	20.00	1.59	0.35	-	6.43	0.17	8.19	m²	9.01
Washing down old distempered or painted surfaces; textured paper pc £6.50 roll; sizing; applying adhesive; hanging; butt joints										
Hard building board walls and columns exceeding 0.50m²	1.41	25.00	1.74	0.33	-	6.06	0.17	7.96	m²	8.76
Plaster walls and columns exceeding 0.50m²	1.41	25.00	1.74	0.33	-	6.06	0.17	7.96	m²	8.76
Hard building board ceilings and beams exceeding 0.50m²	1.41	20.00	1.67	0.35	-	6.43	0.17	8.27	m²	9.10
Plaster ceilings and beams exceeding 0.50m²	1.41	20.00	1.67	0.35	-	6.43	0.17	8.27	m²	9.10
Washing down old distempered or painted surfaces; woodchip paper pc £1.50 roll; sizing; applying adhesive; hanging; butt joints										
Hard building board walls and columns exceeding 0.50m²	0.35	25.00	0.43	0.33	-	6.06	0.17	6.66	m²	7.32
Plaster walls and columns exceeding 0.50m²	0.35	25.00	0.43	0.33	-	6.06	0.17	6.66	m²	7.32

Labour hourly rates: (except Specialists) Craft Operatives 18.37 Labourer 13.76 Rates are national average prices. Refer to REGIONAL VARIATIONS for indicative levels of overall pricing in regions	MATERIALS			LABOUR				RATES		
	Del to Site	Waste	Material Cost	Craft Optve	Lab	Labour Cost	Sunds	Nett Rate	Unit	Gross rate (10%)
	£	%	£	Hrs	Hrs	£	£	£		£
DECORATIVE PAPERS OR FABRICS; REDECORATIONS (Cont'd)										
Washing down old distempered or painted surfaces; woodchip paper pc £1.50 roll; sizing; applying adhesive; hanging; butt joints (Cont'd)										
Hard building board ceilings and beams exceeding 0.50m²	0.35	20.00	0.42	0.35	-	6.43	0.17	7.02	m²	7.72
Plaster ceilings and beams exceeding 0.50m²	0.35	20.00	0.42	0.35	-	6.43	0.17	7.02	m²	7.72
Stripping existing paper; hessian wall covering (prime cost sum for supply and allowance for waste included elsewhere); sizing; applying adhesive; hanging; butt joints										
Hard building board walls and columns exceeding 0.50m²	0.22	-	0.22	0.63	-	11.57	0.63	12.42	m²	13.67
Plaster walls and columns exceeding 0.50m²	0.22	-	0.22	0.58	-	10.65	0.58	11.45	m²	12.60
Stripping existing paper; textile hessian paper backed wall covering (prime cost sum for supply and allowance for waste included elsewhere); sizing; applying adhesive; hanging; butt joints										
Hard building board walls and columns exceeding 0.50m²	0.22	-	0.22	0.55	-	10.10	0.55	10.87	m²	11.96
Plaster walls and columns exceeding 0.50m²	0.22	-	0.22	0.51	-	9.37	0.51	10.10	m²	11.11
Stripping existing hessian; hessian surfaced wall paper (prime cost sum for supply and allowance for waste included elsewhere); sizing; applying adhesive; hanging; butt joints										
Hard building board walls and columns exceeding 0.50m²	0.22	-	0.22	1.03	-	18.92	1.03	20.17	m²	22.19
Plaster walls and columns exceeding 0.50m²	0.22	-	0.22	0.99	-	18.19	0.99	19.40	m²	21.34
Stripping existing hessian; textile hessian paper backed wall covering (prime cost sum for supply and allowance for waste included elsewhere); sizing; applying adhesive; hanging; butt joints										
Hard building board walls and columns exceeding 0.50m²	0.22	-	0.22	0.96	-	17.64	0.96	18.82	m²	20.70
Plaster walls and columns exceeding 0.50m²	0.22	-	0.22	0.91	-	16.72	0.91	17.85	m²	19.63
Stripping existing hessian, paper backed; hessian surfaced wall paper (prime cost sum for supply and allowance for waste included elsewhere); sizing; applying adhesive; hanging; butt joints										
Hard building board walls and columns exceeding 0.50m²	0.22	-	0.22	0.63	-	11.57	0.58	12.37	m²	13.61
Plaster walls and columns exceeding 0.50m²	0.22	-	0.22	0.58	-	10.65	0.58	11.45	m²	12.60
Stripping existing hessian, paper backed; textile hessian paper backed wall covering (prime cost sum for supply and allowance for waste included elsewhere); sizing; applying adhesive; hanging; butt joints										
Hard building board walls and columns exceeding 0.50m²	0.22	-	0.22	0.55	-	10.10	0.55	10.87	m²	11.96
Plaster walls and columns exceeding 0.50m²	0.22	-	0.22	0.51	-	9.37	0.51	10.10	m²	11.11
Stripping existing paper; hessian surfaced wall paper pc £4.50m²; sizing; applying adhesive; hanging; butt joints										
Hard building board walls and columns exceeding 0.50m²	4.72	30.00	6.07	0.63	-	11.57	0.63	18.27	m²	20.10
Plaster walls and columns exceeding 0.50m²	4.72	30.00	6.07	0.58	-	10.65	0.58	17.30	m²	19.04
Stripping existing paper; textile hessian paper backed wall covering pc £7.50m²; sizing; applying adhesive; hanging; butt joints										
Hard building board walls and columns exceeding 0.50m²	4.72	30.00	6.07	0.55	-	10.10	0.55	16.72	m²	18.40
Plaster walls and columns exceeding 0.50m²	4.72	30.00	6.07	0.51	-	9.37	0.51	15.95	m²	17.54
Stripping existing hessian; hessian surfaced wall paper pc £4.50m²; sizing; applying adhesive; hanging; butt joints										
Hard building board walls and columns exceeding 0.50m²	4.72	30.00	6.07	1.03	-	18.92	1.03	26.02	m²	28.62

Labour hourly rates: (except Specialists) Craft Operatives 18.37 Labourer 13.76 Rates are national average prices. Refer to REGIONAL VARIATIONS for indicative levels of overall pricing in regions	MATERIALS			LABOUR				RATES		
	Del to Site £	Waste %	Material Cost £	Craft Optve Hrs	Lab Hrs	Labour Cost £	Sunds £	Nett Rate £	Unit	Gross rate (10%) £
DECORATIVE PAPERS OR FABRICS; REDECORATIONS (Cont'd)										
Stripping existing hessian; hessian surfaced wall paper pc £4.50m²; sizing; applying adhesive; hanging; butt joints (Cont'd)										
Plaster walls and columns exceeding 0.50m²........................	4.72	30.00	6.07	0.99	-	18.19	0.99	25.25	m²	27.77
Stripping existing hessian; textile hessian paper backed wall covering pc £8.00m²; sizing; applying adhesive; hanging; butt joints										
Hard building board walls and columns exceeding 0.50m²........................	4.72	30.00	6.07	0.96	-	17.64	0.96	24.67	m²	27.13
Plaster walls and columns exceeding 0.50m²........................	4.72	30.00	6.07	0.91	-	16.72	0.91	23.70	m²	26.07
Stripping existing hessian, paper backed; hessian surfaced wall paper pc £4.50m²; sizing; applying adhesive; hanging; butt joints										
Hard building board walls and columns exceeding 0.50m²........................	4.72	30.00	6.07	0.63	-	11.57	0.63	18.27	m²	20.10
Plaster walls and columns exceeding 0.50m²........................	4.72	30.00	6.07	0.58	-	10.65	0.53	17.25	m²	18.98
Stripping existing hessian, paper backed; textile hessian paper backed wall covering pc £8.00m²; sizing; applying adhesive; hanging; butt joints										
Hard building board walls and columns exceeding 0.50m²........................	8.22	25.00	10.22	0.55	-	10.10	0.50	20.82	m²	22.91
Plaster walls and columns exceeding 0.50m²........................	0.92	25.00	1.12	0.51	-	9.37	-	10.49	m²	11.54
Expanded polystyrene sheet 2mm thick; sizing; applying adhesive; hanging; butt joints										
Plaster walls and columns exceeding 0.50m²........................	0.60	25.00	0.75	0.36	-	6.61	0.29	7.66	m²	8.42
Plaster ceilings and beams exceeding 0.50m²........................	0.60	25.00	0.75	0.43	-	7.90	0.29	8.94	m²	9.84

Labour hourly rates: (except Specialists) Craft Operatives 18.37 Labourer 13.76 Rates are national average prices. Refer to REGIONAL VARIATIONS for indicative levels of overall pricing in regions	MATERIALS			LABOUR				RATES		
	Del to Site	Waste	Material Cost	Craft Optve	Lab	Labour Cost	Sunds	Nett Rate		Gross rate (10%)
	£	%	£	Hrs	Hrs	£	£	£	Unit	£
DEMOUNTABLE SUSPENDED CEILINGS										
Suspended ceilings; 1200 x 600mm bevelled, grooved and rebated asbestos-free fire resisting tiles; laying in position in metal suspension system of main channel members on wire or rod hangers										
Depth of suspension 150 - 500mm										
9mm thick linings; fixing hangers to masonry...............	17.95	5.00	18.85	0.50	0.25	12.62	6.53	38.00	m²	41.80
9mm thick linings not exceeding 300mm wide; fixing hangers to masonry..................	5.39	10.00	5.92	0.30	0.08	6.54	6.53	18.99	m	20.89
Suspended ceilings; 600 x 600mm bevelled, grooved and rebated asbestos-free fire resisting tiles; laying in position in metal suspension system of main channel members on wire or rod hangers										
Depth of suspension 150 - 500mm										
9mm thick linings; fixing hangers to masonry...............	17.60	5.00	18.48	0.60	0.30	15.15	6.53	40.15	m²	44.17
9mm thick linings not exceeding 300mm wide; fixing hangers to masonry..................	5.28	10.00	5.81	0.36	0.09	7.85	6.53	20.18	m	22.20
Suspended ceilings; 1200 x 600mm bevelled, grooved and rebated plain mineral fibre tiles; laying in position in metal suspension system of main channel members on wire or rod hangers										
Depth of suspension 150 - 500mm										
15.8mm thick linings; fixing hangers to masonry...............	10.90	5.00	11.44	0.50	0.25	12.62	6.53	30.60	m²	33.65
15.8mm thick linings not exceeding 300mm wide; fixing hangers to masonry..................	3.27	10.00	3.60	0.30	0.08	6.54	3.90	14.04	m	15.44
Suspended ceilings; 600 x 600mm bevelled, grooved and rebated plain mineral fibre tiles; laying in position in metal suspension system of main channel members on wire or rod hangers										
Depth of suspension 150 - 500mm										
15.8mm thick linings; fixing hangers to masonry...............	10.60	5.00	11.13	0.60	0.30	15.15	3.97	30.26	m²	33.28
15.8mm thick linings not exceeding 300mm wide; fixing hangers to masonry..................	3.18	10.00	3.50	0.36	0.09	7.85	3.90	15.25	m	16.77
Suspended ceilings; 1200 x 600mm bevelled, grooved and rebated textured mineral fibre tiles; laying in position in metal suspension system of main channel members on wire or rod hangers										
Depth of suspension 150 - 500mm										
15.8mm thick linings; fixing hangers to masonry...............	13.65	5.00	14.33	0.50	0.25	12.62	6.53	33.48	m²	36.83
15.8mm thick linings not exceeding 300mm wide; fixing hangers to masonry..................	4.09	10.00	4.50	0.30	0.08	6.54	3.90	14.95	m	16.44
Suspended ceilings; 600 x 600mm bevelled, grooved and rebated regular drilled mineral fibre tiles; laying in position in metal suspension system of main channel members on wire or rod hangers										
Depth of suspension 150 - 500mm										
15.8mm thick linings; fixing hangers to masonry...............	13.65	5.00	14.33	0.60	0.30	15.15	3.97	33.46	m²	36.80
15.8mm thick linings not exceeding 300mm wide; fixing hangers to masonry..................	4.09	10.00	4.50	0.36	0.09	7.85	2.40	14.76	m	16.23
Suspended ceilings; 1200 x 600mm bevelled, grooved and rebated patterned tiles p.c. £15.00/m²; laying in position in metal suspension system of main channel members on wire or rod hangers										
Depth of suspension 150 - 500mm										
15.8mm thick linings; fixing hangers to masonry...............	16.50	5.00	17.32	0.50	0.25	12.62	6.53	36.47	m²	40.12
15.8mm thick linings not exceeding 300mm wide; fixing hangers to masonry..................	4.95	10.00	5.45	0.30	0.08	6.54	3.90	15.89	m	17.48
Suspended ceilings; 600 x 600mm bevelled, grooved and rebated patterned tiles p.c. £15.00/m²; laying in position in metal suspension system of main channel members on wire or rod hangers										
Depth of suspension 150 - 500mm										
15.8mm thick linings; fixing hangers to masonry...............	16.50	5.00	17.32	0.60	0.30	15.15	3.97	36.45	m²	40.10
15.8mm thick linings not exceeding 300mm wide; fixing hangers to masonry..................	4.95	10.00	5.45	0.36	0.09	7.85	2.40	15.70	m	17.27
Suspended ceiling; British Gypsum Ltd, 1200 x 600mm Vinyl faced tiles; laid into grid system										
Depth of suspension 150-500mm										
10mm thick linings; fixing hangers to masonry...............	6.05	5.00	6.35	0.60	0.11	12.54	6.53	25.41	m²	27.95

327

Labour hourly rates: (except Specialists) Craft Operatives 18.37 Labourer 13.76 Rates are national average prices. Refer to REGIONAL VARIATIONS for indicative levels of overall pricing in regions	MATERIALS			LABOUR				RATES		
	Del to Site	Waste	Material Cost	Craft Optve	Lab	Labour Cost	Sunds	Nett Rate	Unit	Gross rate (10%)
	£	%	£	Hrs	Hrs	£	£	£		£
DEMOUNTABLE SUSPENDED CEILINGS (Cont'd)										
Suspended ceiling; British Gypsum Ltd, 1200 x 600mm Vinyl faced tiles; laid into grid system (Cont'd)										
Depth of suspension 150-500mm (Cont'd) 10mm thick linings not exceeding 300mm wide; fixing hangers to masonry............	1.82	10.00	2.00	0.36	0.09	7.85	6.53	16.37	m	18.01
Edge trims										
Plain angle section; white stove enamelled aluminium; fixing to masonry with screws at 450mm centres.........................	1.15	5.00	1.21	0.40	0.20	10.10	-	11.31	m	12.44

Labour hourly rates: (except Specialists) Craft Operatives 18.37 Labourer 13.76 Rates are national average prices. Refer to REGIONAL VARIATIONS for indicative levels of overall pricing in regions	MATERIALS			LABOUR				RATES		
	Del to Site £	Waste %	Material Cost £	Craft Optve Hrs	Lab Hrs	Labour Cost £	Sunds £	Nett Rate £	Unit	Gross rate (10%) £
BOARD, SHEET, QUILT, SPRAYED, LOOSE FILL OR FOAMED INSULATION AND FIRE PROTECTION INSTALLATIONS										
Boards; Masterboard Class O fire resisting boards; butt joints										
6mm thick linings to walls; plain areas; vertical										
width exceeding 300mm	12.72	5.00	13.36	0.45	0.06	9.09	0.68	23.13	m²	25.44
width not exceeding 300mm	3.82	10.00	4.20	0.27	0.02	5.24	0.41	9.84	m	10.82
9mm thick linings to walls; plain areas; vertical										
width exceeding 300mm	25.01	5.00	26.26	0.49	0.06	9.83	0.74	36.83	m²	40.51
width not exceeding 300mm	7.50	10.00	8.25	0.29	0.02	5.60	0.44	14.29	m	15.72
12mm thick linings to walls; plain areas; vertical										
width exceeding 300mm	32.89	5.00	34.53	0.53	0.07	10.70	0.80	46.02	m²	50.63
width not exceeding 300mm	9.87	10.00	10.85	0.32	0.02	6.15	0.48	17.49	m	19.23
6mm thick linings to ceilings; plain areas; soffit										
width exceeding 300mm	12.72	5.00	13.36	0.57	0.07	11.43	0.86	25.65	m²	28.22
width not exceeding 300mm	3.82	10.00	4.20	0.34	0.02	6.52	0.51	11.23	m	12.35
9mm thick linings to ceilings; plain areas; soffit										
width exceeding 300mm	25.01	5.00	26.26	0.61	0.08	12.31	0.92	39.49	m²	43.43
width not exceeding 300mm	7.50	10.00	8.25	0.37	0.02	7.07	0.56	15.88	m	17.47
12mm thick linings to ceilings; plain areas; soffit										
width exceeding 300mm	32.89	5.00	34.53	0.67	0.08	13.41	1.00	48.94	m²	53.84
width not exceeding 300mm	9.87	10.00	10.85	0.40	0.02	7.62	0.60	19.08	m	20.98
Boards; Supalux fire resisting boards; sanded; butt joints; fixing to timber with self tapping screws										
6mm thick linings to walls; plain areas; vertical										
width exceeding 300mm	21.33	5.00	22.40	0.45	0.06	9.09	0.68	32.17	m²	35.38
width not exceeding 300mm	6.40	10.00	7.04	0.27	0.02	5.24	0.41	12.68	m	13.95
9mm thick linings to walls; plain areas; vertical										
width exceeding 300mm	28.34	5.00	29.76	0.49	0.06	9.83	0.74	40.32	m²	44.35
width not exceeding 300mm	8.50	10.00	9.35	0.29	0.02	5.60	0.44	15.39	m	16.93
12mm thick linings to walls; plain areas; vertical										
width exceeding 300mm	37.95	5.00	39.85	0.53	0.07	10.70	0.80	51.34	m²	56.48
width not exceeding 300mm	11.39	10.00	12.52	0.32	0.02	6.15	0.48	19.16	m	21.07
15mm thick linings to walls; plain areas; vertical										
width exceeding 300mm	52.35	5.00	54.97	0.53	0.07	10.70	0.80	66.46	m²	73.11
width not exceeding 300mm	15.70	10.00	17.28	0.32	0.02	6.15	0.48	23.91	m	26.30
6mm thick linings to ceilings; plain areas; soffit										
width exceeding 300mm	21.33	5.00	22.40	0.57	0.07	11.43	0.86	34.69	m²	38.16
width not exceeding 300mm	6.40	10.00	7.04	0.34	0.02	6.52	0.51	14.07	m	15.48
9mm thick linings to ceilings; plain areas; soffit										
width exceeding 300mm	28.34	5.00	29.76	0.61	0.08	12.31	0.92	42.98	m²	47.28
width not exceeding 300mm	8.50	10.00	9.35	0.37	0.02	7.07	0.56	16.98	m	18.68
12mm thick linings to ceilings; plain areas; soffit										
width exceeding 300mm	37.95	5.00	39.85	0.67	0.08	13.41	1.00	54.26	m²	59.69
width not exceeding 300mm	11.39	10.00	12.52	0.40	0.02	7.62	0.60	20.75	m	22.82
15mm thick linings to ceilings; plain areas; soffit										
width exceeding 300mm	52.35	5.00	54.97	0.67	0.08	13.41	1.00	69.38	m²	76.32
width not exceeding 300mm	15.70	10.00	17.28	0.40	0.02	7.62	0.60	25.50	m	28.05
6mm thick casings to isolated beams or the like										
total girth not exceeding 600mm	21.33	10.00	23.47	1.13	0.14	22.68	1.70	47.84	m²	52.63
total girth 600 - 1200mm	21.33	5.00	22.40	0.68	0.09	13.73	1.02	37.15	m²	40.86
9mm thick casings to isolated beams or the like										
total girth not exceeding 600mm	28.34	10.00	31.17	1.21	0.15	24.29	1.82	57.28	m²	63.01
total girth 600 - 1200mm	28.34	5.00	29.76	0.74	0.09	14.83	1.11	45.70	m²	50.27
12mm thick casings to isolated beams or the like										
total girth not exceeding 600mm	37.95	10.00	41.75	1.33	0.17	26.77	2.00	70.51	m²	77.57
total girth 600 - 1200mm	37.95	5.00	39.85	0.80	0.10	16.07	1.20	57.12	m²	62.83
15mm thick casings to isolated beams or the like										
total girth not exceeding 600mm	52.35	10.00	57.58	1.33	0.17	26.77	2.00	86.35	m²	94.99
total girth 600 - 1200mm	52.35	5.00	54.97	0.80	0.10	16.07	1.20	72.24	m²	79.46
6mm thick casings to isolated columns or the like										
total girth 600 - 1200mm	21.33	10.00	23.47	1.13	0.14	22.68	1.70	47.84	m²	52.63
total girth 1200 - 1800mm	21.33	5.00	22.40	0.68	0.09	13.73	1.02	37.15	m²	40.86
9mm thick casings to isolated columns or the like										
total girth 600 - 1200mm	28.34	10.00	31.17	1.21	0.15	24.29	1.82	57.28	m²	63.01
total girth 1200 - 1800mm	28.34	5.00	29.76	0.74	0.09	14.83	1.11	45.70	m²	50.27
12mm thick casings to isolated columns or the like										
total girth 600 - 1200mm	37.95	10.00	41.75	1.33	0.17	26.77	2.00	70.51	m²	77.57

Labour hourly rates: (except Specialists) Craft Operatives 18.37 Labourer 13.76 Rates are national average prices. Refer to REGIONAL VARIATIONS for indicative levels of overall pricing in regions	MATERIALS			LABOUR				RATES		
	Del to Site	Waste	Material Cost	Craft Optve	Lab	Labour Cost	Sunds	Nett Rate		Gross rate (10%)
	£	%	£	Hrs	Hrs	£	£	£	Unit	£
BOARD, SHEET, QUILT, SPRAYED, LOOSE FILL OR FOAMED INSULATION AND FIRE PROTECTION INSTALLATIONS (Cont'd)										
Boards; Supalux fire resisting boards; sanded; butt joints; fixing to timber with self tapping screws (Cont'd)										
12mm thick casings to isolated columns or the like (Cont'd) total girth 1200 - 1800mm......................................	37.95	5.00	39.85	0.80	0.10	16.07	1.20	57.12	m²	62.83
15mm thick casings to isolated columns or the like total girth 600 - 1200mm...................................... total girth 1200 - 1800mm......................................	52.35 52.35	10.00 5.00	57.58 54.97	1.33 0.80	0.17 0.10	26.77 16.07	2.00 1.20	86.35 72.24	m² m²	94.99 79.46
9mm thick linings to pipe ducts and casings width exceeding 300mm...................................... width not exceeding 300mm......................................	94.00 28.20	20.00 20.00	112.80 33.84	1.50 0.50	0.19 0.06	30.17 10.01	0.36 0.12	143.33 43.97	m² m	157.66 48.37
Boards; Vermiculux fire resisting boards; butt joints; fixing to steel with self tapping screws, countersinking										
20mm thick casings to isolated beams or the like total girth not exceeding 600mm...................................... total girth 600 - 1200mm......................................	27.47 27.47	10.00 5.00	30.21 28.84	1.50 0.90	0.19 0.11	30.17 18.05	2.25 1.35	62.63 48.24	m² m²	68.90 53.06
30mm thick casings to isolated beams or the like total girth not exceeding 600mm...................................... total girth 600 - 1200mm......................................	43.76 43.76	10.00 5.00	48.14 45.95	1.65 1.00	0.21 0.13	33.20 20.16	2.47 1.50	83.81 67.61	m² m²	92.20 74.37
40mm thick casings to isolated beams or the like total girth not exceeding 600mm...................................... total girth 600 - 1200mm......................................	67.99 67.99	10.00 5.00	74.79 71.39	1.80 1.08	0.23 0.14	36.23 21.77	2.70 1.62	113.72 94.78	m² m²	125.10 104.26
50mm thick casings to isolated beams or the like total girth not exceeding 600mm...................................... total girth 600 - 1200mm......................................	90.51 90.51	10.00 5.00	99.56 95.04	1.95 1.17	0.24 0.15	39.12 23.56	2.92 1.75	141.61 120.35	m² m²	155.78 132.39
60mm thick casings to isolated beams or the like total girth not exceeding 600mm...................................... total girth 600 - 1200mm......................................	107.92 107.92	10.00 5.00	118.71 113.31	2.10 1.26	0.26 0.16	42.15 25.35	3.15 1.89	164.02 140.55	m² m²	180.42 154.61
20mm thick casings to isolated columns or the like total girth 600 - 1200mm...................................... total girth 1200 - 1800mm......................................	27.47 27.47	10.00 5.00	30.21 28.84	1.50 0.90	0.19 0.11	30.17 18.05	2.25 1.35	62.63 48.24	m² m²	68.90 53.06
30mm thick casings to isolated columns or the like total girth 600 - 1200mm...................................... total girth 1200 - 1800mm......................................	43.76 43.76	10.00 5.00	48.14 45.95	1.65 1.00	0.21 0.13	33.20 20.16	2.47 1.50	83.81 67.61	m² m²	92.20 74.37
40mm thick casings to isolated columns or the like total girth 600 - 1200mm...................................... total girth 1200 - 1800mm......................................	67.99 67.99	10.00 5.00	74.79 71.39	1.80 1.08	0.23 0.14	36.23 21.77	2.70 1.62	113.72 94.78	m² m²	125.10 104.26
50mm thick casings to isolated columns or the like total girth 600 - 1200mm...................................... total girth 1200 - 1800mm......................................	90.51 90.51	10.00 5.00	99.56 95.04	1.95 1.17	0.24 0.15	39.12 23.56	2.92 1.75	141.61 120.35	m² m²	155.78 132.39
60mm thick casings to isolated columns or the like total girth 600 - 1200mm...................................... total girth 1200 - 1800mm......................................	107.92 107.92	10.00 5.00	118.71 113.31	2.10 1.26	0.26 0.16	42.15 25.35	3.15 1.89	164.02 140.55	m² m²	180.42 154.61
50 x 25mm x 20g mild steel fixing angle fixing to masonry .... fixing to steel	2.60 2.60	5.00 5.00	2.73 2.73	0.40 0.40	0.05 0.05	8.04 8.04	0.60 0.60	11.37 11.37	m m	12.50 12.50
Boards; Vicuclad fire resisting boards; butt joints; fixing with screws; countersinking										
18mm thick casings to isolated beams or the like; including noggins total girth not exceeding 600mm...................................... total girth 600 - 1200mm......................................	29.71 29.71	10.00 5.00	32.50 31.20	1.42 0.85	0.18 0.11	28.56 17.13	2.13 1.27	63.19 49.60	m² m²	69.51 54.56
25mm thick casings to isolated beams or the like; including noggins total girth not exceeding 600mm...................................... total girth 600 - 1200mm......................................	32.33 32.33	10.00 5.00	35.35 33.94	1.58 0.95	0.20 0.12	31.78 19.10	2.37 1.42	69.50 54.47	m² m²	76.45 59.92
35mm thick casings to isolated beams or the like; including noggins total girth not exceeding 600mm...................................... total girth 600 - 1200mm......................................	37.72 38.21	10.00 5.00	41.25 40.12	1.65 1.00	0.21 0.13	33.20 20.16	2.47 1.50	76.93 61.78	m² m²	84.62 67.96
18mm thick casings to isolated columns or the like total girth 600 - 1200mm...................................... total girth 1200 - 1800mm......................................	27.26 27.26	10.00 5.00	29.92 28.62	1.42 0.85	0.18 0.11	28.56 17.13	2.13 1.27	60.62 47.03	m² m²	66.68 51.73
25mm thick casings to isolated columns or the like total girth 600 - 1200mm...................................... total girth 1200 - 1800mm......................................	29.88 29.88	10.00 5.00	32.78 31.37	1.58 0.95	0.20 0.12	31.78 19.10	2.37 1.42	66.93 51.90	m² m²	73.62 57.09
35mm thick casings to isolated columns or the like total girth 600 - 1200mm...................................... total girth 1200 - 1800mm......................................	35.27 35.27	10.00 5.00	38.68 37.04	1.65 1.00	0.21 0.13	33.20 20.16	2.47 1.50	74.36 58.70	m² m²	81.79 64.57
Sheets; Waterproof reinforced building paper; grade A1F; 150mm lapped joints										
Across members at 450mm centres vertical...	1.12	5.00	1.18	0.17	0.02	3.40	0.03	4.60	m²	5.06

Labour hourly rates: (except Specialists) Craft Operatives 18.37 Labourer 13.76 Rates are national average prices. Refer to REGIONAL VARIATIONS for indicative levels of overall pricing in regions	MATERIALS			LABOUR				RATES		
	Del to Site £	Waste %	Material Cost £	Craft Optve Hrs	Lab Hrs	Labour Cost £	Sunds £	Nett Rate £	Unit	Gross rate (10%) £
BOARD, SHEET, QUILT, SPRAYED, LOOSE FILL OR FOAMED INSULATION AND FIRE PROTECTION INSTALLATIONS (Cont'd)										
Sheets; Waterproof reinforced building paper; reflection (thermal) grade, single sided; 150mm lapped joints										
Across members at 450mm centres vertical............	1.76	5.00	1.85	0.17	0.02	3.40	0.03	5.28	m²	5.80
Sheets; Waterproof reinforced building paper; reflection (thermal) grade, double sided; 150mm lapped joints										
Across members at 450mm centres vertical............	2.75	5.00	2.89	0.17	0.02	3.40	0.03	6.32	m²	6.95
Sheets; British Sisalkraft Ltd Insulex 714 vapour control layer; 150mm laps sealed with tape										
Across members at 450mm centres horizontal; fixing to timber with stainless steel staples..........	2.77	10.00	3.05	0.10	0.01	1.97	0.09	5.11	m²	5.63
Plain areas horizontal; laid loose........	2.77	10.00	3.05	0.06	0.01	1.24	-	4.29	m²	4.72
Sheets; Mineral wool insulation quilt; butt joints										
100mm thick; across members at 450mm centres horizontal; laid loose........	1.63	15.00	1.87	0.12	0.01	2.34	-	4.21	m²	4.64
Sheets; Glass fibre insulation quilt; butt joints										
100mm thick; across members at 450mm centres horizontal; laid loose........	1.63	15.00	1.87	0.09	0.01	1.79	-	3.66	m²	4.03
200mm thick; across members at 450mm centres horizontal; laid loose........	3.25	15.00	3.74	0.09	0.01	1.79	-	5.53	m²	6.08
100mm thick; between members at 450mm centres horizontal; laid loose........	1.63	15.00	1.87	0.15	0.02	3.03	-	4.90	m²	5.39
200mm thick; between members at 450mm centres horizontal; laid loose........	3.25	15.00	3.74	0.15	0.02	3.03	-	6.77	m²	7.44
Sheets; Sheeps wool insulation quilt; butt joints										
200mm thick; across members at 450mm centres horizontal; laid loose........	16.22	10.00	17.84	0.09	0.01	1.79	-	19.63	m²	21.59
200mm thick; between members at 450mm centres horizontal; laid loose........	16.22	10.00	17.84	0.15	0.02	3.03	-	20.87	m²	22.96
Sheets; Knauf 'Earthwool Loftroll 40'; butt joints										
100mm thick; between members at 450mm centres horizontal; laid loose........	2.33	10.00	2.56	0.10	0.01	1.97	-	4.54	m²	4.99
150mm thick; between members at 450mm centres horizontal; laid loose........	3.42	10.00	3.76	0.10	0.01	1.97	-	5.74	m²	6.31
200mm thick; between members at 450mm centres horizontal; laid loose........	4.75	10.00	5.22	0.20	0.03	4.09	-	9.31	m²	10.24
100mm thick; between members at 450mm centres; 100mm thick across members at 450mm centres horizontal; laid loose........	4.66	10.00	5.13	0.36	0.05	7.30	-	12.43	m²	13.67
150mm thick; between members at 450mm centres; 100mm thick across members at 450mm centres horizontal; laid loose........	5.75	10.00	6.33	0.28	0.05	5.83	-	12.16	m²	13.37
Sheets; Rockwool Rollbatts; insulation quilt; butt joints										
100mm thick; between members at 450mm centres horizontal; laid loose........	4.43	10.00	4.87	0.10	0.01	1.97	-	6.84	m²	7.53
150mm thick; between members at 450mm centres horizontal; laid loose........	6.09	10.00	6.70	0.15	0.02	3.03	-	9.73	m²	10.70
100mm thick; between members at 450mm centres; 100mm thick across members at 450mm centres horizontal; laid loose........	8.85	10.00	9.74	0.20	0.03	4.09	-	13.83	m²	15.21
Sheets; Glass fibre medium density insulation board; butt joints										
50mm thick; plain areas horizontal; bedding in bitumen........	2.13	10.00	2.34	0.20	0.02	3.95	2.25	8.54	m²	9.40
75mm thick; plain areas horizontal; bedding in bitumen........	2.38	10.00	2.62	0.22	0.03	4.45	2.25	9.32	m²	10.25
100mm thick; plain areas horizontal; bedding in bitumen........	3.68	10.00	4.05	0.25	0.03	5.01	2.25	11.30	m²	12.43
50mm thick; across members at 450mm centres vertical........	2.13	10.00	2.34	0.20	0.02	3.95	0.05	6.34	m²	6.97
75mm thick; across members at 450mm centres vertical........	2.38	10.00	2.62	0.22	0.03	4.45	0.05	7.11	m²	7.83
100mm thick; across members at 450mm centres vertical........	3.68	10.00	4.05	0.25	0.03	5.01	0.05	9.10	m²	10.01

Labour hourly rates: (except Specialists) Craft Operatives 18.37 Labourer 13.76 Rates are national average prices. Refer to REGIONAL VARIATIONS for indicative levels of overall pricing in regions	MATERIALS			LABOUR				RATES		
	Del to Site £	Waste %	Material Cost £	Craft Optve Hrs	Lab Hrs	Labour Cost £	Sunds £	Nett Rate £	Unit	Gross rate (10%) £
BOARD, SHEET, QUILT, SPRAYED, LOOSE FILL OR FOAMED INSULATION AND FIRE PROTECTION INSTALLATIONS (Cont'd)										
Sheets; Dow Construction Products; Styrofoam Floormate 300A; butt joints										
50mm thick; plain areas										
horizontal; laid loose....................	6.65	10.00	7.31	0.10	0.01	1.97	-	9.29	m²	10.21
100mm thick; plain areas										
horizontal; laid loose....................	13.29	10.00	14.62	0.15	0.02	3.03	-	17.65	m²	19.42
150mm thick; plain areas										
horizontal; laid loose....................	20.44	10.00	22.49	0.20	0.03	4.09	-	26.58	m²	29.23
Sheets; Sempatap latex foam sheeting; butt joints										
10mm thick; plain areas										
soffit; fixing with adhesive....................	16.24	15.00	18.68	1.25	1.25	40.16	3.60	62.44	m²	68.68
vertical; fixing with adhesive	16.24	15.00	18.68	1.00	1.00	32.13	3.60	54.41	m²	59.85
Sheets; Sempafloor SBR latex foam sheeting with coated non woven polyester surface; butt joints										
4.5mm thick; plain areas										
horizontal; fixing with adhesive....................	19.00	15.00	21.85	0.75	0.75	24.10	3.08	49.02	m²	53.92
Sheets; Expanded polystyrene sheeting; butt joints										
13mm thick; plain areas										
horizontal; laid loose....................	1.33	5.00	1.40	0.09	0.05	2.34	-	3.74	m²	4.11
vertical; fixing with adhesive	1.33	5.00	1.40	0.27	0.14	6.89	1.65	9.93	m²	10.93
19mm thick; plain areas										
horizontal; laid loose....................	1.33	5.00	1.40	0.10	0.05	2.53	-	3.92	m²	4.31
vertical; fixing with adhesive	1.33	5.00	1.40	0.28	1.06	19.73	1.65	22.78	m²	25.05
25mm thick; plain areas										
horizontal; laid loose....................	1.33	5.00	1.40	0.12	0.06	3.03	-	4.43	m²	4.87
vertical; fixing with adhesive	1.33	5.00	1.40	0.30	0.15	7.58	1.65	10.63	m²	11.69
50mm thick; plain areas										
horizontal; laid loose....................	2.67	5.00	2.80	0.14	0.07	3.54	-	6.34	m²	6.97
vertical; fixing with adhesive	2.67	5.00	2.80	0.32	0.16	8.08	1.65	12.53	m²	13.79
75mm thick; plain areas										
horizontal; laid loose....................	4.00	5.00	4.20	0.18	0.09	4.55	-	8.75	m²	9.62
vertical; fixing with adhesive	4.00	5.00	4.20	0.36	0.18	9.09	1.65	14.94	m²	16.44
Sheets; Expanded polystyrene sheeting, non-flammable; butt joints										
13mm thick; plain areas										
horizontal; laid loose....................	1.80	5.00	1.89	0.09	0.05	2.34	-	4.24	m²	4.66
vertical; fixing with adhesive	1.80	5.00	1.89	0.27	0.14	6.89	1.65	10.43	m²	11.47
19mm thick; plain areas										
horizontal; laid loose....................	2.14	5.00	2.25	0.10	0.05	2.53	-	4.78	m²	5.25
vertical; fixing with adhesive	2.14	5.00	2.25	0.28	0.14	7.07	1.65	10.97	m²	12.07
25mm thick; plain areas										
horizontal; laid loose....................	1.67	5.00	1.76	0.12	0.06	3.03	-	4.79	m²	5.26
vertical; fixing with adhesive	1.67	5.00	1.76	0.30	0.15	7.58	1.65	10.98	m²	12.08
50mm thick; plain areas										
horizontal; laid loose....................	3.35	5.00	3.51	0.14	0.07	3.54	-	7.05	m²	7.75
vertical; fixing with adhesive	3.35	5.00	3.51	0.32	0.16	8.08	1.65	13.24	m²	14.57
75mm thick; plain areas										
horizontal; laid loose....................	5.85	5.00	6.14	0.18	0.09	4.55	-	10.69	m²	11.76
vertical; fixing with adhesive	5.85	5.00	6.14	0.36	0.18	9.09	1.65	16.88	m²	18.57

Labour hourly rates: (except Specialists) Craft Operatives 18.37 Labourer 13.76 Rates are national average prices. Refer to REGIONAL VARIATIONS for indicative levels of overall pricing in regions	MATERIALS			LABOUR				RATES		
	Del to Site	Waste	Material Cost	Craft Optve	Lab	Labour Cost	Sunds	Nett Rate	Unit	Gross rate (10%)
	£	%	£	Hrs	Hrs	£	£	£		£
GENERAL FIXTURES, FURNISHINGS AND EQUIPMENT; CURTAIN TRACKS										
Fixing only curtain track										
Metal or plastic track with fittings										
fixing with screws to softwood	-	-	-	0.50	-	9.19	-	9.19	m	10.10
fixing with screws to hardwood	-	-	-	0.70	-	12.86	-	12.86	m	14.14
GENERAL FIXTURES, FURNISHINGS AND EQUIPMENT; BLINDS										
Internal blinds										
Venetian blinds; stove enamelled aluminium alloy slats 25mm wide; plain colours										
1200mm drop; fixing to timber with screws										
1000mm wide	27.88	2.50	28.57	1.00	0.13	20.16	0.24	48.97	nr	53.87
2000mm wide	46.52	2.50	47.68	1.50	0.19	30.17	0.36	78.21	nr	86.03
3000mm wide	60.91	2.50	62.43	2.00	0.25	40.18	0.50	103.11	nr	113.42
Venetian blinds; stove enamelled aluminium alloy slats 50mm wide; plain colours										
1200mm drop; fixing to timber with screws										
1000mm wide	29.42	2.50	30.15	1.15	0.14	23.05	0.28	53.49	nr	58.84
2000mm wide	44.12	2.50	45.23	1.70	0.21	34.12	0.42	79.77	nr	87.74
3000mm wide	56.88	2.50	58.30	2.25	0.28	45.19	0.56	104.04	nr	114.44
Venetian blinds; stove enamelled aluminium alloy slats 25mm wide; plain colours with wand control										
1200mm drop; fixing to timber with screws										
1000mm wide	51.06	2.50	52.33	1.00	0.13	20.16	0.24	72.73	nr	80.01
2000mm wide	76.53	2.50	78.45	1.50	0.19	30.17	0.36	108.98	nr	119.87
3000mm wide	98.69	2.50	101.16	2.00	0.25	40.18	0.50	141.83	nr	156.02
Venetian blinds; stove enamelled aluminium alloy slats 50mm wide; plain colours with wand control										
1200mm drop; fixing to timber with screws										
1000mm wide	51.06	2.50	52.33	1.15	0.14	23.05	0.28	75.67	nr	83.24
2000mm wide	76.53	2.50	78.45	1.70	0.21	34.12	0.42	112.99	nr	124.28
3000mm wide	98.69	2.50	101.16	2.25	0.28	45.19	0.56	146.90	nr	161.59
Venetian blinds; stove enamelled aluminium alloy slats 25mm wide; plain colours with channels for dimout										
1200mm drop; fixing to timber with screws										
1000mm wide	53.41	2.50	54.75	2.00	0.25	40.18	0.50	95.42	nr	104.97
2000mm wide	87.88	2.50	90.08	2.50	0.31	50.19	0.62	140.88	nr	154.97
3000mm wide	118.04	2.50	120.99	3.00	0.38	60.34	0.74	182.07	nr	200.27
Venetian blinds; stove enamelled aluminium alloy slats 50mm wide; plain colours with channels for dimout										
1200mm drop; fixing to timber with screws										
1000mm wide	100.15	2.50	102.65	2.25	0.28	45.19	0.56	148.39	nr	163.23
2000mm wide	123.09	2.50	126.17	2.80	0.35	56.25	0.69	183.11	nr	201.42
3000mm wide	145.20	2.50	148.83	3.35	0.42	67.32	0.82	216.97	nr	238.67
Roller blinds; sprung ratchet action; flame retardent material										
1200mm drop; fixing to timber with screws										
1000mm wide	35.39	2.50	36.28	1.00	0.13	20.16	0.24	56.68	nr	62.34
2000mm wide	54.36	2.50	55.72	1.50	0.19	30.17	0.36	86.25	nr	94.87
3000mm wide	66.99	2.50	68.67	2.00	0.25	40.18	0.50	109.34	nr	120.28
Roller blinds; sprung ratchet action; patterned polyester type material										
1200mm drop; fixing to timber with screws										
1000mm wide	38.75	2.50	39.72	1.00	0.13	20.16	0.24	60.12	nr	66.13
2000mm wide	67.78	2.50	69.48	1.50	0.19	30.17	0.36	100.01	nr	110.01
3000mm wide	102.29	2.50	104.85	2.00	0.25	40.18	0.50	145.52	nr	160.07
Roller blinds; side chain operation; blackout material										
1200mm drop; fixing to timber with screws										
1000mm wide	35.39	2.50	36.28	1.15	0.14	23.05	0.28	59.61	nr	65.57
2000mm wide	60.34	2.50	61.85	1.70	0.21	34.12	0.42	96.38	nr	106.02
3000mm wide	87.09	2.50	89.27	2.25	0.28	45.19	0.56	135.01	nr	148.51
Roller blinds; 100% blackout; natural anodised box and channels										
1200mm drop; fixing to timber with screws										
1000mm wide	180.00	2.50	184.50	2.00	0.25	40.18	0.50	225.18	nr	247.69
2000mm wide	467.05	2.50	478.73	2.75	0.34	55.20	0.68	534.60	nr	588.06
3000mm wide	676.26	2.50	693.17	3.50	0.44	70.35	0.86	764.37	nr	840.81
Vertical louvre blinds; 89mm wide louvres in standard material										
1200mm drop; fixing to timber with screws										
1000mm wide	28.55	2.50	29.26	0.90	0.11	18.05	0.22	47.54	nr	52.29
2000mm wide	44.18	2.50	45.29	1.35	0.17	27.14	0.33	72.76	nr	80.03
3000mm wide	59.14	2.50	60.62	1.80	0.23	36.23	0.44	97.29	nr	107.01

FURNITURE, FITTINGS AND EQUIPMENT

Labour hourly rates: (except Specialists) Craft Operatives 18.37 Labourer 13.76 Rates are national average prices. Refer to REGIONAL VARIATIONS for indicative levels of overall pricing in regions	MATERIALS			LABOUR				RATES		
	Del to Site	Waste	Material Cost	Craft Optve	Lab	Labour Cost	Sunds	Nett Rate	Unit	Gross rate (10%)
	£	%	£	Hrs	Hrs	£	£	£		£
GENERAL FIXTURES, FURNISHINGS AND EQUIPMENT; BLINDS (Cont'd)										
Internal blinds (Cont'd)										
Vertical louvre blinds; 127mm wide louvres in standard material 1200mm drop; fixing to timber with screws										
1000mm wide	28.55	2.50	29.26	0.95	0.12	19.10	0.22	48.59	nr	53.45
2000mm wide	44.18	2.50	45.29	1.40	0.18	28.19	0.34	73.83	nr	81.21
3000mm wide	59.14	2.50	60.62	1.85	0.23	37.15	0.45	98.22	nr	108.04
External blinds; manually operated										
Venetian blinds; stove enamelled aluminium slats 80mm wide; natural anodised side guides; excluding boxing 1200mm drop; fixing to timber with screws										
1000mm wide	-	-	Specialist	-	-	Specialist	-	373.74	nr	411.12
2000mm wide	-	-	Specialist	-	-	Specialist	-	556.25	nr	611.87
3000mm wide	-	-	Specialist	-	-	Specialist	-	749.21	nr	824.13
Queensland awnings; acrylic material; natural anodised arms and boxing 1000mm projection										
2000mm long	-	-	Specialist	-	-	Specialist	-	876.09	nr	963.70
3000mm long	-	-	Specialist	-	-	Specialist	-	1313.33	nr	1444.66
4000mm long	-	-	Specialist	-	-	Specialist	-	1752.18	nr	1927.40
Rollscreen vertical drop roller blinds; mesh material; natural anodised side guides and boxing 1200mm drop; fixing to timber with screws										
1000mm wide	-	-	Specialist	-	-	Specialist	-	364.99	nr	401.49
2000mm wide	-	-	Specialist	-	-	Specialist	-	547.57	nr	602.32
3000mm wide	-	-	Specialist	-	-	Specialist	-	729.76	nr	802.74
Quandrant canopy; acrylic material; natural anodised frames 1000mm projection										
2000mm long	-	-	Specialist	-	-	Specialist	-	483.24	nr	531.56
3000mm long	-	-	Specialist	-	-	Specialist	-	738.77	nr	812.64
4000mm long	-	-	Specialist	-	-	Specialist	-	1004.72	nr	1105.20
Foldaway awning; acrylic material; natural anodised arms and front rail 2000mm projection										
2000mm long	-	-	Specialist	-	-	Specialist	-	783.97	nr	862.37
3000mm long	-	-	Specialist	-	-	Specialist	-	1150.73	nr	1265.80
4000mm long	-	-	Specialist	-	-	Specialist	-	1496.67	nr	1646.34
External blinds; electrically operated										
Venetian blinds; stove enamelled aluminium slats 80mm wide; natural anodised side guides; excluding boxing 1200mm drop; fixing to timber with screws										
1000mm wide	-	-	Specialist	-	-	Specialist	-	858.70	nr	944.57
2000mm wide	-	-	Specialist	-	-	Specialist	-	1049.93	nr	1154.92
3000mm wide	-	-	Specialist	-	-	Specialist	-	1230.71	nr	1353.78
Queensland awnings; acrylic material; natural anodised arms and boxing 1000mm projection										
2000mm long	-	-	Specialist	-	-	Specialist	-	1004.72	nr	1105.20
3000mm long	-	-	Specialist	-	-	Specialist	-	1434.08	nr	1577.48
4000mm long	-	-	Specialist	-	-	Specialist	-	1825.19	nr	2007.71
Rollscreen vertical drop roller blinds; mesh material; natural anodised side guides and boxing 1200mm drop; fixing to timber with screws										
1000mm wide	-	-	Specialist	-	-	Specialist	-	695.31	nr	764.84
2000mm wide	-	-	Specialist	-	-	Specialist	-	876.09	nr	963.70
3000mm wide	-	-	Specialist	-	-	Specialist	-	1058.61	nr	1164.47
Quandrant canopy; acrylic material; natural anodised frames 1000mm projection										
2000mm long	-	-	Specialist	-	-	Specialist	-	629.26	nr	692.18
3000mm long	-	-	Specialist	-	-	Specialist	-	876.09	nr	963.70
4000mm long	-	-	Specialist	-	-	Specialist	-	1168.12	nr	1284.93
Foldaway awning; acrylic material; natural anodised arms and front rail 2000mm projection										
2000mm long	-	-	Specialist	-	-	Specialist	-	1149.01	nr	1263.91
3000mm long	-	-	Specialist	-	-	Specialist	-	1505.35	nr	1655.89
4000mm long	-	-	Specialist	-	-	Specialist	-	1912.11	nr	2103.32
GENERAL FIXTURES, FURNISHINGS AND EQUIPMENT; STORAGE SYSTEMS										
Boltless Office Shelving; light grey steel boltless storage systems; steel shelves; Welco.; units 900mm wide x 1850mm high; assembling										
Open shelving with six shelves; placing in position										
300mm deep code 958-064, starter bay	161.00	2.50	165.02	1.00	0.50	25.25	-	190.27	nr	209.30
300mm deep code 958-065, extension bay	128.00	2.50	131.20	1.00	0.50	25.25	-	156.45	nr	172.10
300mm deep code 958-066, back panel set	53.30	2.50	54.63	0.50	0.25	12.62	-	67.26	nr	73.98
300mm deep code 958-068, decorative steel end panel	53.75	2.50	55.09	0.50	0.25	12.62	-	67.72	nr	74.49
extra; additional steel shelf code 958-067	15.73	2.50	16.12	0.20	0.10	5.05	-	21.17	Unit	23.28

Labour hourly rates: (except Specialists) Craft Operatives 18.37 Labourer 13.76 Rates are national average prices. Refer to REGIONAL VARIATIONS for indicative levels of overall pricing in regions	MATERIALS			LABOUR				RATES		
	Del to Site	Waste	Material Cost	Craft Optve	Lab	Labour Cost	Sunds	Nett Rate	Unit	Gross rate (10%)
	£	%	£	Hrs	Hrs	£	£	£		£
GENERAL FIXTURES, FURNISHINGS AND EQUIPMENT; SHELVING SYSTEMS										
Wide access boltless office shelving systems; upright frames connected with shelf beams; chipboard shelves; Welco.; assembling										
Open shelving with three shelves; placing in position										
1220mm wide x 2000mm high x 600mm deep code 133-100	178.00	2.50	182.45	1.30	0.65	32.83	-	215.27	nr	236.80
1220mm wide x 2000mm high x 900mm deep code 133-103	193.00	2.50	197.82	1.40	0.70	35.35	-	233.18	nr	256.49
1220mm wide x 2000mm high x 1200mm deep code 133-106	209.00	2.50	214.22	1.50	0.75	37.88	-	252.10	nr	277.31
2440mm wide x 2000mm high x 600mm deep code 133-102	249.00	2.50	255.22	1.40	0.70	35.35	-	290.57	nr	319.63
2440mm wide x 2000mm high x 900mm deep code 133-105	270.00	2.50	276.75	1.50	0.75	37.88	-	314.62	nr	346.09
2440mm wide x 2000mm high x 1200mm deep code 133-108	304.00	2.50	311.60	1.60	0.80	40.40	-	352.00	nr	387.20
2440mm wide x 2500mm high x 600mm deep code 133-111	269.00	2.50	275.73	1.50	0.75	37.88	-	313.60	nr	344.96
2440mm wide x 2500mm high x 900mm deep code 133-114	300.00	2.50	307.50	1.60	0.80	40.40	-	347.90	nr	382.69
2440mm wide x 2500mm high x 1200mm deep code 133-117	330.00	2.50	338.25	1.70	0.85	42.93	-	381.18	nr	419.29
GENERAL FIXTURES, FURNISHINGS AND EQUIPMENT; SHELVING SUPPORT SYSTEMS										
Twin slot shelf supports in steel with white enamelled finish										
Wall uprights										
430mm long; fixing to masonry with screws	0.94	2.50	0.96	0.25	-	4.59	-	5.56	nr	6.11
1000mm long; fixing to masonry with screws	1.92	2.50	1.97	0.45	-	8.27	-	10.23	nr	11.26
1600mm long; fixing to masonry with screws	2.83	2.50	2.90	0.55	-	10.10	-	13.00	nr	14.30
2400mm long; fixing to masonry with screws	4.29	2.50	4.40	0.70	-	12.86	-	17.26	nr	18.98
Straight brackets										
120mm long	0.54	2.50	0.55	0.05	-	0.92	-	1.47	nr	1.62
270mm long	0.88	2.50	0.90	0.05	-	0.92	-	1.82	nr	2.00
370mm long	1.15	2.50	1.18	0.07	-	1.29	-	2.46	nr	2.71
470mm long	1.92	2.50	1.97	0.07	-	1.29	-	3.25	nr	3.58
GENERAL FIXTURES, FURNISHINGS AND EQUIPMENT; MAT RIMS										
Matwells in Aluminium										
Mat frames; welded fabrication 34 x 26 x 6mm angle section; angles mitred; plain lugs -4, welded on; mat space										
610 x 457mm	39.55	2.50	40.54	1.00	1.00	32.13	-	72.67	nr	79.94
762 x 457mm	45.33	2.50	46.46	1.25	1.25	40.16	-	86.63	nr	95.29
914 x 610mm	56.74	2.50	58.16	1.50	1.50	48.19	-	106.35	nr	116.99
Matwells in polished brass										
Mat frames; brazed fabrication 38 x 38 x 6mm angle section; angles mitred; plain lugs -4, welded on; mat space										
610 x 457mm	183.55	2.50	188.14	1.25	1.25	40.16	-	228.30	nr	251.13
762 x 457mm	216.92	2.50	222.34	1.50	1.50	48.19	-	270.54	nr	297.59
914 x 610mm	283.66	2.50	290.75	1.75	1.75	56.23	-	346.98	nr	381.68
GENERAL FIXTURES, FURNISHINGS AND EQUIPMENT; CLOTHES LOCKERS										
Wet area lockers; Welco; laminate doors; with cam locks										
1 compartment; placing in position										
300 x 300 x 1800mm; code 935-009	228.00	2.50	233.70	-	0.60	8.26	-	241.96	nr	266.15
300 x 450 x 1800mm; code 935-010	228.00	2.50	233.70	-	0.65	8.94	-	242.64	nr	266.91
2 compartment; placing in position										
300 x 300 x 1800mm; code 935-011	262.00	2.50	268.55	-	0.60	8.26	-	276.81	nr	304.49
300 x 450 x 1800mm; code 935-012	237.00	2.50	242.92	-	0.65	8.94	-	251.87	nr	277.06
3 compartment; placing in position										
300 x 300 x 1800mm; code 935-013	292.00	2.50	299.30	-	0.60	8.26	-	307.56	nr	338.31
300 x 450 x 1800mm; code 935-014	266.00	2.50	272.65	-	0.65	8.94	-	281.59	nr	309.75
4 compartment; placing in position										
300 x 300 x 1800mm; code 935-015	340.00	2.50	348.50	-	0.60	8.26	-	356.76	nr	392.43
300 x 450 x 1800mm; code 935-016	295.00	-	295.00	-	0.65	8.94	-	303.94	nr	334.34
GENERAL FIXTURES, FURNISHINGS AND EQUIPMENT; CLOAK ROOM EQUIPMENT										
Static square tube double sided coat racks; Welco; hardwood seats; in assembled units										
Racks 1500mm long x 1675mm high x 600mm deep; placing in position; 40mm square tube										
10 hooks and angle framed steel mesh single shoe tray, reference 456-102 and 456-109; placing in position	619.00	2.50	634.47	-	0.60	8.26	-	642.73	nr	707.00
10 hooks, reference 456-102; placing in position	543.00	2.50	556.58	-	0.60	8.26	-	564.83	nr	621.31
10 hooks and 5 baskets reference 456-102 and 456-106; placing in position	633.00	2.50	648.83	-	0.60	8.26	-	657.08	nr	722.79
10 hooks and 10 baskets reference 456-102 and 456-107; placing in position	657.00	2.50	673.42	-	0.60	8.26	-	681.68	nr	749.85
Static square tube single sided coat racks; Welco; mobile; in assembled units										
Racks 1500mm long x 1825mm high x 600mm deep; placing in position; 40mm square tube										
15 hangers and top tray, reference 456-103	463.08	2.50	474.66	-	0.60	8.26	-	482.91	nr	531.20

FURNITURE, FITTINGS AND EQUIPMENT

Labour hourly rates: (except Specialists) Craft Operatives 18.37 Labourer 13.76 Rates are national average prices. Refer to REGIONAL VARIATIONS for indicative levels of overall pricing in regions	MATERIALS			LABOUR				RATES		
	Del to Site	Waste	Material Cost	Craft Optve	Lab	Labour Cost	Sunds	Nett Rate	Unit	Gross rate (10%)
	£	%	£	Hrs	Hrs	£	£	£		£
GENERAL FIXTURES, FURNISHINGS AND EQUIPMENT; **CLOAK ROOM EQUIPMENT (Cont'd)**										
Free-standing bench seats; Welco; steel square tube framing; **hardwood seats; placing in position**										
Bench seat with shelf 1500mm long x 300mm deep and 450mm high; placing in position; 40mm square tube										
single sided 300mm deep, reference 787-009	187.00	2.50	191.68	-	0.70	9.63	-	201.31	nr	221.44
double sided 600mm deep reference 787-010	247.00	2.50	253.17	-	0.70	9.63	-	262.81	nr	289.09
Wall rack Welco; hardwood										
1200mm long; fixing to masonry										
6 hooks, reference 787-020	44.80	2.50	45.92	-	0.68	9.41	-	55.33	nr	60.87
1800mm long; fixing to masonry										
12 hooks, reference 787-021	54.85	2.50	56.22	-	1.03	14.12	-	70.34	nr	77.37
GENERAL FIXTURES, FURNISHINGS AND EQUIPMENT; HAT **AND COAT RAILS**										
Hat and coat rails										
Rails in softwood, wrought										
25 x 75mm; chamfers -2	2.14	10.00	2.35	0.29	0.04	5.88	0.08	8.31	m	9.14
25 x 100mm; chamfers -2	2.84	10.00	3.12	0.29	0.04	5.88	0.08	9.08	m	9.98
25 x 75mm; chamfers -2; fixing to masonry with screws	2.14	10.00	2.35	0.57	0.07	11.43	0.14	13.92	m	15.32
25 x 100mm; chamfers -2; fixing to masonry with screws	2.84	10.00	3.12	0.57	0.07	11.43	0.14	14.69	m	16.16
Rails in Afromosia, wrought, selected for transparent finish										
25 x 75mm; chamfers -2	6.96	10.00	7.65	0.57	0.07	11.43	0.14	19.22	m	21.14
25 x 100mm; chamfers -2	9.30	10.00	10.23	0.57	0.07	11.43	0.14	21.80	m	23.98
25 x 75mm; chamfers -2; fixing to masonry with screws, countersinking and pellating	6.96	10.00	7.65	1.15	0.14	23.05	0.28	30.99	m	34.09
25 x 100mm; chamfers -2; fixing to masonry with screws, countersinking and pellating	9.30	10.00	10.23	1.15	0.14	23.05	0.28	33.57	m	36.92
Hat and coat hooks										
B.M.A. finish	2.95	2.50	3.02	0.40	-	7.35	-	10.37	nr	11.41
chromium plated	4.96	2.50	5.08	0.40	-	7.35	-	12.43	nr	13.68
SAA finish	1.20	2.50	1.23	0.40	-	7.35	-	8.58	nr	9.44
Welco; steel; coloured finish; reference 456-122	3.78	2.50	3.87	0.40	-	7.35	-	11.22	nr	12.34
KITCHEN FITTINGS										
Standard melamine finish on chipboard units, with backs - self **assembly**										
Wall units										
fixing to masonry with screws										
400 x 300 x 600mm	57.00	2.50	58.42	1.75	0.22	35.17	-	93.60	nr	102.96
400 x 300 x 900mm	97.00	2.50	99.42	1.85	0.23	37.15	-	136.57	nr	150.23
500 x 300 x 600mm	60.00	2.50	61.50	1.90	0.25	38.34	-	99.84	nr	109.83
500 x 300 x 900mm	102.00	2.50	104.55	1.95	0.25	39.26	-	143.81	nr	158.19
600 x 300 x 600mm	65.00	2.50	66.62	1.95	0.25	39.26	-	105.89	nr	116.48
600 x 300 x 900mm	106.00	2.50	108.65	2.05	0.26	41.24	-	149.89	nr	164.87
1000 x 300 x 600mm	67.00	2.50	68.67	2.35	0.30	47.30	-	115.97	nr	127.57
1000 x 300 x 900mm	172.00	2.50	176.30	2.50	0.31	50.19	-	226.49	nr	249.14
1200 x 300 x 600mm	153.00	2.50	156.82	2.50	0.31	50.19	-	207.02	nr	227.72
1200 x 300 x 900mm	175.00	2.50	179.38	2.70	0.35	54.42	-	233.79	nr	257.17
Floor units on plinths and with plastic faced worktops; without drawers										
fixing to masonry with screws										
400 x 600 x 900mm	142.00	2.50	145.55	2.00	0.25	40.18	-	185.73	nr	204.30
500 x 600 x 900mm	156.00	2.50	159.90	2.20	0.28	44.27	-	204.17	nr	224.58
600 x 600 x 900mm	167.00	2.50	171.18	2.40	0.30	48.22	-	219.39	nr	241.33
1000 x 600 x 900mm	279.00	2.50	285.98	3.20	0.40	64.29	-	350.26	nr	385.29
1200 x 600 x 900mm	303.00	2.50	310.57	3.50	0.44	70.35	-	380.92	nr	419.02
Floor units on plinths and with plastic faced worktops; with one drawer										
fixing to masonry with screws										
500 x 600 x 900mm	64.00	2.50	65.60	2.20	0.28	44.27	-	109.87	nr	120.85
600 x 600 x 900mm	66.00	2.50	67.65	2.40	0.30	48.22	-	115.87	nr	127.45
1000 x 600 x 900mm	68.00	2.50	69.70	3.20	0.40	64.29	-	133.99	nr	147.39
1200 x 600 x 900mm	362.00	2.50	371.05	3.50	0.44	70.35	-	441.40	nr	485.54
Floor units on plinths and with plastic faced worktops; with four drawers										
fixing to masonry with screws										
500 x 600 x 900mm	272.00	2.50	278.80	2.20	0.28	44.27	-	323.07	nr	355.37
600 x 600 x 900mm	287.00	2.50	294.17	2.40	0.30	48.22	-	342.39	nr	376.63
Sink units on plinths										
1200 x 600 x 900mm, with one drawer; fixing to masonry with screws	257.00	2.50	263.42	3.50	0.44	70.35	-	333.77	nr	367.15
1200 x 600 x 900mm, without drawer; fixing to masonry with screws	196.00	2.50	200.90	3.50	0.44	70.35	-	271.25	nr	298.37
1500 x 600 x 900mm, without drawer; fixing to masonry with screws	308.00	2.50	315.70	3.75	0.47	75.35	-	391.05	nr	430.16
Store cupboards on plinths										
fixing to masonry with screws										
500 x 600 x 1950mm without shelves	227.00	2.50	232.67	3.60	0.45	72.32	-	305.00	nr	335.50
500 x 600 x 1950mm with shelves	242.00	2.50	248.05	3.60	0.45	72.32	-	320.37	nr	352.41
600 x 600 x 1950mm without shelves	247.00	2.50	253.17	4.00	0.50	80.36	-	333.54	nr	366.89
600 x 600 x 1950mm with shelves	260.00	2.50	266.50	4.00	0.50	80.36	-	346.86	nr	381.55

Labour hourly rates: (except Specialists) Craft Operatives 18.37 Labourer 13.76 Rates are national average prices. Refer to REGIONAL VARIATIONS for indicative levels of overall pricing in regions	MATERIALS			LABOUR				RATES		
	Del to Site £	Waste %	Material Cost £	Craft Optve Hrs	Lab Hrs	Labour Cost £	Sunds £	Nett Rate £	Unit	Gross rate (10%) £
KITCHEN FITTINGS (Cont'd)										
Laminate plastic faced units complete with ironmongery and with backs - self assembly										
Wall units										
400 x 300 x 600mm, with one door; fixing to masonry with screws	150.00	2.50	153.75	1.75	0.22	35.17	-	188.92	nr	207.82
500 x 300 x 600mm, with one door; fixing to masonry with screws	161.00	2.50	165.02	1.85	0.23	37.15	-	202.17	nr	222.39
1000 x 300 x 600mm, with two doors; fixing to masonry with screws	283.00	2.50	290.07	2.35	0.30	47.30	-	337.37	nr	371.11
1200 x 300 x 600mm, with two doors; fixing to masonry with screws	307.00	2.50	314.67	2.50	0.31	50.19	-	364.87	nr	401.35
Floor units on plinths and with worktops										
500 x 600 x 900mm, with one drawer and one cupboard; fixing to masonry with screws	390.00	2.50	399.75	2.20	0.28	44.27	-	444.02	nr	488.42
500 x 600 x 900mm, with four drawers; fixing to masonry with screws	544.00	2.50	557.60	2.40	0.30	48.22	-	605.82	nr	666.40
1000 x 600 x 900mm, with two drawers and two cupboards; fixing to masonry with screws	671.00	2.50	687.77	3.20	0.40	64.29	-	752.06	nr	827.27
1500 x 600 x 900mm, with three drawers and three cupboards; fixing to masonry with screws	867.00	2.50	888.67	3.85	0.48	77.33	-	966.00	nr	1062.60
Sink units on plinths										
1000 x 600 x 900mm, with one drawer and two cupboards, fixing to masonry with screws	515.00	2.50	527.88	3.65	0.47	73.52	-	601.39	nr	661.53
1500 x 600 x 900mm, with two drawers and three cupboards; fixing to masonry with screws	770.00	2.50	789.25	3.75	0.50	75.77	-	865.02	nr	951.52
Tall units on plinths										
500 x 600 x 1950mm, broom cupboard; fixing to masonry with screws	473.00	2.50	484.82	3.75	0.50	75.77	-	560.59	nr	616.65
600 x 600 x 1950mm, larder unit; fixing to masonry with screws	494.00	2.50	506.35	3.75	0.50	75.77	-	582.12	nr	640.33
Hardwood veneered units complete with ironmongery but without backs - self assembly										
Wardrobe units										
fixing to masonry with screws										
1000 x 600 x 2175mm	548.00	2.50	561.70	4.75	0.60	95.51	-	657.21	nr	722.93
1500 x 600 x 2175mm	688.00	2.50	705.20	4.75	0.60	95.51	-	800.71	nr	880.78
Pre-Assembled kitchen units, white melamine finish; Module 600 range; Magnet PLC										
Drawerline units; 500 deep; fixing to masonry with screws										
floor units										
300mm F36D	96.67	2.50	99.09	1.25	0.20	25.71	-	124.80	nr	137.28
500mm F56D	113.33	2.50	116.16	1.25	0.23	26.13	-	142.29	nr	156.52
600mm F66D	125.00	2.50	128.12	1.25	0.23	26.13	-	154.25	nr	169.68
1000mm F106D	170.00	2.50	174.25	1.75	0.33	36.69	-	210.94	nr	232.03
hob floor units										
1000mm H106D	119.17	2.50	122.15	1.75	0.30	36.28	-	158.42	nr	174.27
sink units										
1000mm S106D	192.92	2.50	197.74	1.75	0.30	36.28	-	234.02	nr	257.42
Tall units; fixing to masonry with screws										
floor units										
500mm L56	90.83	2.50	93.10	2.50	0.35	50.74	-	143.84	nr	158.23
Drawer units; fixing to masonry with screws										
four drawer floor units										
500mm F56	238.33	2.50	244.29	1.25	0.30	27.09	-	271.38	nr	298.52
Wall units; fixing to masonry with screws										
standard units 720mm high										
300mm W37	56.67	2.50	58.09	1.25	0.22	25.99	-	84.08	nr	92.48
500mm W57	68.33	2.50	70.04	1.25	0.23	26.13	-	96.17	nr	105.78
600mm W67	73.75	2.50	75.59	1.25	0.30	27.09	-	102.68	nr	112.95
1000mm W107	113.33	2.50	116.16	2.20	0.38	45.64	-	161.81	nr	177.99
standard corner units 720mm high										
600mm CW67	136.25	2.50	139.66	2.20	0.38	45.64	-	185.30	nr	203.83
Worktops; fixing with screws										
round front edge										
38mm worktops, round front edge; premium range	47.64	25.00	59.55	1.50	0.19	30.17	-	89.72	m	98.69
Plinth										
Plin 27	15.77	25.00	19.71	0.25	0.03	5.01	-	24.72	m	27.19
Cornice										
Minster 500	15.59	25.00	19.48	0.25	0.03	5.01	-	24.49	m	26.94
Pelmet										
Minster 500	15.59	25.00	19.48	0.25	0.03	5.01	-	24.49	m	26.94
Worktop trim										
coloured	15.60	25.00	19.50	0.25	0.03	5.01	-	24.51	m	26.96
Ancillaries										
Worktops; fixing with screws										
round front edge										
38mm worktops, round front edge; premium range	47.64	25.00	59.55	1.50	0.19	30.17	-	89.72	m	98.69
Plinth										
Lunar night upstand	15.77	25.00	19.71	0.25	0.03	5.01	-	24.72	m	27.19
Cornice										
trim 27	15.59	25.00	19.48	0.25	0.03	5.01	-	24.49	m	26.94

FURNITURE, FITTINGS AND EQUIPMENT

Labour hourly rates: (except Specialists) Craft Operatives 18.37 Labourer 13.76 Rates are national average prices. Refer to REGIONAL VARIATIONS for indicative levels of overall pricing in regions	MATERIALS			LABOUR				RATES		
	Del to Site	Waste	Material Cost	Craft Optve	Lab	Labour Cost	Sunds	Nett Rate	Unit	Gross rate (10%)
	£	%	£	Hrs	Hrs	£	£	£		£
KITCHEN FITTINGS (Cont'd)										
Ancillaries (Cont'd)										
Pelmet trim27	15.59	25.00	19.48	0.25	0.03	5.01	-	24.49	m	26.94
Worktop trim colourfil	15.60	25.00	19.50	0.25	0.03	5.01	-	24.51	m	26.96

This section continues
on the next page

Labour hourly rates: (except Specialists) Craft Operatives 23.99 Labourer 13.76 Rates are national average prices. Refer to REGIONAL VARIATIONS for indicative levels of overall pricing in regions	MATERIALS			LABOUR				RATES		
	Del to Site	Waste	Material Cost	Craft Optve	Lab	Labour Cost	Sunds	Nett Rate		Gross rate (10%)
	£	%	£	Hrs	Hrs	£	£	£	Unit	£
SANITARY APPLIANCES AND FITTINGS										
Attendance on sanitary fittings										
Note: Sanitary fittings are usually included as p.c. items or provisional sums, the following are the allowances for attendance by the main contractor i.e. unloading, storing and distributing fittings for fixing by the plumber and returning empty cases and packings										
Sinks, fireclay										
610 x 457 x 254mm	-	-	-	-	1.00	13.76	0.81	14.57	nr	16.03
762 x 508 x 254mm	-	-	-	-	1.25	17.20	1.02	18.22	nr	20.04
add if including tubular stands	-	-	-	-	0.75	10.32	0.68	10.99	nr	12.09
Wash basins, 559 x 406mm; earthenware or fireclay										
single	-	-	-	-	0.80	11.01	0.68	11.68	nr	12.85
range of 4 with cover overlaps between basins	-	-	-	-	2.50	34.40	1.83	36.23	nr	39.85
Combined sinks and drainers; stainless steel										
1067 x 533mm	-	-	-	-	1.40	19.26	1.20	20.46	nr	22.51
1600 x 610mm	-	-	-	-	1.70	23.39	1.42	24.82	nr	27.30
add if including tubular stands	-	-	-	-	0.75	10.32	0.68	10.99	nr	12.09
Combined sinks and drainers; porcelain enamel										
1067 x 533mm	-	-	-	-	1.40	19.26	1.20	20.46	nr	22.51
1600 x 610mm	-	-	-	-	1.70	23.39	1.42	24.82	nr	27.30
Baths excluding panels; pressed steel										
694 x 1688 x 570mm overall	-	-	-	-	2.50	34.40	1.77	36.17	nr	39.79
W.C. suites; china										
complete with WWP, flush pipe etc.	-	-	-	-	1.40	19.26	1.20	20.46	nr	22.51
Block pattern urinals; with glazed ends, back and channel in one piece with separate tread including flushing cistern etc.										
single stall	-	-	-	-	2.75	37.84	2.03	39.87	nr	43.85
range of four with loose overlaps	-	-	-	-	6.00	82.56	4.72	87.29	nr	96.01
Wall urinals, bowl type, with flushing cistern, etc.										
single	-	-	-	-	1.40	19.26	1.20	20.46	nr	22.51
range of 3 with two divisions	-	-	-	-	3.30	45.41	2.43	47.84	nr	52.62
Slab urinals with one return end, flushing cistern, sparge pipe, etc.										
1219mm long x 1067mm high	-	-	-	-	2.75	37.84	2.70	40.54	nr	44.59
1829mm long x 1067mm high	-	-	-	-	3.75	51.60	4.05	55.65	nr	61.22
Fix only appliances (prime cost sum for supply included elsewhere)										
Sink units, stainless steel; combined overflow and waste outlet, plug and chain; pair pillar taps fixing on base unit with metal clips										
1000 x 600mm	-	-	-	3.00	-	71.97	1.20	73.17	nr	80.49
1600 x 600mm	-	-	-	3.50	-	83.96	1.20	85.17	nr	93.68
Sinks; white glazed fireclay; waste outlet, plug, chain and stay; cantilever brackets; pair bib taps fixing brackets to masonry with screws; sealing at back with white sealant										
610 x 457 x 254mm	-	-	-	4.00	-	95.96	1.95	97.91	nr	107.70
762 x 508 x 254mm	-	-	-	4.00	-	95.96	1.95	97.91	nr	107.70
Sinks; white glazed fireclay; waste outlet, plug, chain and stay; legs and screw-to-wall bearers; pair bib taps fixing legs to masonry with screws; sealing at back with mastic sealant										
610 x 457 x 254mm	-	-	-	4.50	-	107.96	1.95	109.90	nr	120.90
762 x 508 x 254mm	-	-	-	4.50	-	107.96	1.95	109.90	nr	120.90
Wash basins; vitreous china; waste outlet, plug, chain and stay; screw to wall brackets; pair pillar taps										
560 x 406mm; fixing brackets to masonry with screws; sealing at back with mastic sealant	-	-	-	3.00	-	71.97	1.95	73.92	nr	81.31
range of four, each 559 x 406mm with overlap strips; fixing brackets to masonry with screws; sealing overlap strips and at back with mastic sealant	-	-	-	12.75	-	305.87	7.43	313.30	nr	344.63
Wash basins; vitreous china; waste outlet, plug, chain and stay; legs and screw-to-wall bearers										
560 x 406mm; fixing legs to masonry with screws; sealing at back with mastic sealant	-	-	-	3.50	-	83.96	1.95	85.92	nr	94.51
range of 4, each 559 x 406mm with overlap strips; fixing legs to masonry with screws; sealing overlap strips and at back with mastic sealant	-	-	-	13.75	-	329.86	7.43	337.29	nr	371.02
Baths; enamelled pressed steel; combined overflow and waste outlet, plug and chain; metal cradles; pair pillar taps										
700 x 1700 x 570mm overall; sealing at walls with mastic sealant	-	-	-	5.50	-	131.94	4.35	136.29	nr	149.92

Labour hourly rates: (except Specialists) Craft Operatives 23.99 Labourer 13.76 Rates are national average prices. Refer to REGIONAL VARIATIONS for indicative levels of overall pricing in regions	MATERIALS			LABOUR				RATES		
	Del to Site	Waste	Material Cost	Craft Optve	Lab	Labour Cost	Sunds	Nett Rate	Unit	Gross rate (10%)
	£	%	£	Hrs	Hrs	£	£	£		£
SANITARY APPLIANCES AND FITTINGS (Cont'd)										
Fix only appliances (prime cost sum for supply included elsewhere) (Cont'd)										
High level W.C. suites; vitreous china pan; plastics cistern with valveless fittings, ball valve, chain and pull handle; flush pipe; plastics seat and cover										
fixing pan and cistern brackets to masonry with screws; bedding pan in mastic; jointing pan to drain with cement mortar (1:2) and gaskin joint	-	-	-	4.50	-	107.96	2.62	110.58	nr	121.64
fixing pan and cistern brackets to masonry with screws; bedding pan in mastic; jointing pan to soil pipe with Multikwik connector	-	-	-	4.50	-	107.96	2.62	110.58	nr	121.64
Low level W.C. suites; vitreous china pan; plastics cistern with valveless fittings and ball valve; flush pipe; plastics seat and cover										
fixing pan and cistern brackets to masonry with screws; bedding pan in mastic; jointing pan to drain with cement mortar (1:2) and gaskin joint	-	-	-	4.50	-	107.96	2.62	110.58	nr	121.64
fixing pan and cistern brackets to masonry with screws; bedding pan in mastic; jointing pan to soil pipe with Multikwik connector	-	-	-	4.50	-	107.96	2.62	110.58	nr	121.64
Wall urinals; vitreous china bowl; automatic cistern; spreads and flush pipe; waste outlet										
single bowl; fixing bowl, cistern and pipe brackets to masonry with screws..................	-	-	-	4.50	-	107.96	2.03	109.98	nr	120.98
range of three bowls; fixing bowls, divisions, cistern and pipe brackets to masonry with screws..................	-	-	-	11.50	-	275.89	4.12	280.01	nr	308.01
Single stall urinals; white glazed fireclay in one piece; white glazed fireclay tread; plastics automatic cistern; spreader and flush pipe; waste outlet and domed grating										
bedding stall and tread in cement mortar (1:4) and jointing tread with waterproof jointing compound; fixing cistern and pipe brackets to masonry with screws..................	-	-	-	7.00	-	167.93	4.95	172.88	nr	190.17
Slab urinals; range of four; white glazed fireclay back, ends, divisions, channel and tread; automatic cistern; flush and sparge pipes; waste outlet and domed grating										
bedding back, ends, channel and tread in cement mortar (1:4) and jointing with waterproof jointing compound; fixing divisions, cistern and pipe brackets to masonry with screws..................	-	-	-	19.00	-	455.81	15.75	471.56	nr	518.72
Showers; white glazed fireclay tray 760 x 760 x 180mm; waste outlet; recessed valve and spray head										
bedding tray in cement mortar (1:4); fixing valve head to masonry with screws; sealing at walls with mastic sealant..................	-	-	-	4.75	-	113.95	6.38	120.33	nr	132.36
Shower curtains with rails, hooks, supports and fittings										
straight, 914mm long x 1829mm drop; fixing supports to masonry with screws..................	-	-	-	1.00	-	23.99	0.50	24.49	nr	26.93
angled, 1676mm girth x 1829mm drop; fixing supports to masonry with screws	-	-	-	1.50	-	35.99	0.68	36.66	nr	40.33
Supply and fix appliances - sinks										
Note: the following prices include joints to copper services and wastes but do not include traps										
Sink units; single bowl with single drainer and back ledge; stainless steel; BS EN 13310 Type A; 38mm waste, plug and chain to BS EN 274 Part 1 with combined overflow; pair 13mm pillar taps										
fixing on base unit with metal clips										
1000 x 600mm..................	102.50	2.50	105.06	3.00	-	71.97	1.20	178.23	nr	196.06
1200 x 600mm..................	113.75	2.50	116.59	3.50	-	83.96	1.20	201.76	nr	221.93
Sink units; single bowl with double drainer and back ledge; stainless steel; BS EN 13310 Type B; 38mm waste, plug and chain to BS EN 274 Part 1 with combined overflow; pair 13mm pillar taps										
fixing on base unit with metal clips										
1500 x 600mm..................	130.00	2.50	133.25	3.00	-	71.97	1.58	206.79	nr	227.47
Sinks; white glazed fireclay; BS 1206 with weir overflow; 38mm slotted waste, chain, stay and plug to BS EN 274 Part 1; aluminium alloy cantilever brackets										
screwing chain stay; fixing brackets to masonry with screws; sealing at back with mastic sealant										
455 x 380 x 205mm	137.87	2.50	141.32	3.50	-	83.96	1.95	227.23	nr	249.95
610 x 455 x 255mm	208.71	2.50	213.93	3.50	-	83.96	1.95	299.84	nr	329.83
760 x 455 x 255mm	319.50	2.50	327.49	3.50	-	83.96	1.95	413.40	nr	454.74
Sinks; white glazed fireclay; BS 1206 with wier overflow; 38mm slotted waste, chain, stay and plug to BS EN 274 Part 1; painted legs and screw-to-wall bearers										
screwing chain stay; fixing legs to masonry with screws; sealing at back with mastic sealant										
455 x 380 x 205mm	163.00	2.50	167.07	3.50	-	83.96	1.95	252.99	nr	278.29
610 x 455 x 255mm	214.00	2.50	219.35	3.50	-	83.96	1.95	305.26	nr	335.79
760 x 455 x 255mm	471.00	2.50	482.77	3.50	-	83.96	1.95	568.69	nr	625.56

Labour hourly rates: (except Specialists) Craft Operatives 23.99 Labourer 13.76 Rates are national average prices. Refer to REGIONAL VARIATIONS for indicative levels of overall pricing in regions	MATERIALS			LABOUR				RATES		
	Del to Site	Waste	Material Cost	Craft Optve	Lab	Labour Cost	Sunds	Nett Rate		Gross rate (10%)
	£	%	£	Hrs	Hrs	£	£	£	Unit	£

SANITARY APPLIANCES AND FITTINGS (Cont'd)

Supply and fix appliances - lavatory basins

Wash basins; white vitreous china; BS 5506 Part 3; 32mm slotted waste, chain, stay and plug to BS EN 274 Part 1; painted cantilever brackets; pair 13 pillar taps
  sealing at back with mastic sealant
    560 x 406mm....................
	71.00	2.50	72.78	3.50	-	83.96	1.95	158.69	nr	174.56

Wash basins; white vitreous china; BS 5506 Part 3; 32mm slotted waste, chain, stay and plug to BS EN 274 Part 1; painted legs and screw-to-wall bearers; pair 13mm taps
  fixing legs to masonry with screws; sealing at back with mastic sealant
    560 x 406mm....................
	81.00	2.50	83.03	4.10	-	98.36	1.95	183.33	nr	201.67

Wash basins; white vitreous china; BS 5506 Part 3; 32mm slotted waste, chain, stay and plug to BS EN 274 Part 1; painted towel rail brackets; pair 13mm pillar taps
  fixing brackets to masonry with screws; sealing at back with mastic sealant
    560 x 406mm....................
	63.50	2.50	65.09	4.00	-	95.96	1.95	163.00	nr	179.30

Wash basins, angle type; white vitreous china; BS 5506 Part 3; 32mm slotted waste, chain, stay and plug to BS EN 274 Part 1; concealed brackets; pair 13 pillar taps
  fixing brackets to masonry with screws; sealing at back with mastic sealant
    457 x 431mm....................
	91.50	2.50	93.79	4.00	-	95.96	1.95	191.70	nr	210.87

Wash basins with pedestal; white vitreous china; BS 5506 Part 3; 32mm slotted waste, chain, stay and plug to BS EN 274 Part 1; pair 13mm pillar taps
  fixing basin to masonry with screws; sealing at back with mastic sealant
    560 x 406mm....................
	89.25	2.50	91.48	4.50	-	107.96	1.95	201.39	nr	221.52

Supply and fix appliances - baths

Baths; white enamelled pressed steel; BS EN 232; 32mm overflow with front grid and 38mm waste, chain and plug to BS EN 274 Part 1 with combined overflow; metal cradles with adjustable feet; pair 19mm pillar taps
  sealing at walls with mastic sealant
    700 x 1700 x 570mm overall...................
	119.00	2.50	121.97	5.50	-	131.94	4.35	258.27	nr	284.10

724 x 1500 x 570mm overall	109.00	2.50	111.72	5.50	-	131.94	4.35	248.02	nr	272.82
Baths; white acrylic; BS EN 198; 32mm overflow with front grid and 38mm waste, chain and plug to BS EN 274 Part 1 with combined overflow; metal cradles with adjustable feet and wall fixing brackets; pair 19mm pillar taps 700 x 1700 x 570mm overall; fixing brackets to masonry with screws; sealing at walls with mastic sealant...................	144.00	2.50	147.60	5.00	-	119.95	4.35	271.90	nr	299.09

Supply and fix appliances - W.C. pans

White vitreous china pans to BS EN 997 pan with horizontal outlet; fixing pan to timber with screws; bedding pan in mastic; jointing pan to soil pipe with Multikwik connector...................	41.00	2.50	42.02	2.50	-	59.97	1.05	103.05	nr	113.36
pan with horizontal outlet; fixing pan to masonry with screws; bedding pan in mastic; jointing pan to soil pipe with Multikwik connector...................	41.00	2.50	42.02	3.00	-	71.97	1.50	115.49	nr	127.04
Plastics W.C. seats; BS 1254, type 1 black; fixing to pan...................	11.66	2.50	11.95	0.30	-	7.20	-	19.15	nr	21.06
Plastics W.C. seats and covers; BS 1254, type 1 black; fixing to pan...................	14.23	2.50	14.59	0.30	-	7.20	-	21.78	nr	23.96
coloured; fixing to pan	25.40	2.50	26.03	0.30	-	7.20	-	33.23	nr	36.56
white; fixing to pan...................	20.17	2.50	20.67	0.30	-	7.20	-	27.87	nr	30.66

Supply and fix appliances - cisterns

9 litre black plastics cistern with valveless fittings, chain and pull handle to BS 1125; 13mm piston type high pressure ball valve to BS 1212 Part 1 with 127mm plastics float to BS 2456 fixing cistern brackets to masonry with screws...................	76.05	2.50	77.95	1.00	-	23.99	0.50	102.44	nr	112.68
Plastics flush pipe to BS 1125; adjustable for back wall fixing...................	3.55	2.50	3.64	0.75	-	17.99	0.50	22.13	nr	24.34
for side wall fixing...................	3.95	2.50	4.05	0.75	-	17.99	0.50	22.54	nr	24.79

Supply and fix appliances - W.C. suites

High level W.C. suites; white vitreous china pan to BS EN 997; 9 litre black plastics cistern with valveless fittings, chain and pull handle and plastics flush pipe to BS 1125; 13mm piston type high pressure ball valve to BS 1212 Part 1 with 127mm plastics float to BS 2456; black plastics seat and cover to BS 1254 type 1 pan with S or P trap conversion bend to BS 5627; fixing pan to timber with screws; fixing cistern brackets to masonry with screws; bedding pan in mastic; jointing pan to soil pipe with Multikwik connector...................	115.50	2.50	118.39	4.00	-	95.96	2.62	216.97	nr	238.67

341

FURNITURE, FITTINGS AND EQUIPMENT

Labour hourly rates: (except Specialists) Craft Operatives 23.99 Labourer 13.76 Rates are national average prices. Refer to REGIONAL VARIATIONS for indicative levels of overall pricing in regions	MATERIALS			LABOUR				RATES		
	Del to Site	Waste	Material Cost	Craft Optve	Lab	Labour Cost	Sunds	Nett Rate	Unit	Gross rate (10%)
	£	%	£	Hrs	Hrs	£	£	£		£
SANITARY APPLIANCES AND FITTINGS (Cont'd)										
Supply and fix appliances - W.C. suites (Cont'd)										
High level W.C. suites; white vitreous china pan to BS EN 997; 9 litre black plastics cistern with valveless fittings, chain and pull handle and plastics flush pipe to BS 1125; 13mm piston type high pressure ball valve to BS 1212 Part 1 with 127mm plastics float to BS 2456; black plastics seat and cover to BS 1254 type 1 (Cont'd)										
pan with S or P trap conversion bend to BS 5627; fixing pan and cistern brackets to masonry with screws; bedding pan in mastic; jointing pan to soil pipe with Multikwik connector	115.50	2.50	118.39	4.50	-	107.96	2.62	228.97	nr	251.86
Low level W.C. suites; white vitreous china pan to BS EN 997; 9 litre white vitreous china cistern with valveless fittings and plastics flush bend to BS 1125; 13mm piston type high pressure ball valve to BS 1212 Part 1 with 127mm plastics float to BS 2456; white plastics seat and cover to BS 1254 type 1										
pan with S or P trap conversion bend to BS 5627; fixing pan to timber with screws; fixing cistern brackets to masonry with screws; bedding pan in mastic; jointing pan to soil pipe with Multikwik connector	116.25	2.50	119.16	4.00	-	95.96	2.62	217.74	nr	239.52
pan with S or P trap conversion bend to BS 5627; fixing pan and cistern brackets to masonry with screws; bedding pan in mastic; jointing pan to soil pipe with Multikwik connector	116.25	2.50	119.16	4.50	-	107.96	2.62	229.74	nr	252.71
Supply and fix appliances - urinals										
Wall urinals; white vitreous china bowl to BS 5520; 4.5 litre white vitreous china automatic cistern to BS 1876; spreader and stainless steel flush pipe; 32mm plastics waste outlet										
fixing bowl, cistern and pipe brackets to masonry with screws.......	207.00	2.50	212.17	4.50	-	107.96	2.03	322.15	nr	354.37
Wall urinals; range of 2; white vitreous china bowls to BS 5520; white vitreous china divisions; 9 litre white plastics automatic cistern to BS 1876; stainless steel flush pipes and spreaders; 32mm plastics waste outlets										
fixing bowls, divisions, cistern and pipe brackets to masonry with screws	399.00	2.50	408.97	9.00	-	215.91	3.00	627.88	nr	690.67
Wall urinals; range of 3; white vitreous china bowls to BS 5520; white vitreous china divisions; 14 litre white plastics automatic cistern to BS 1876; stainless steel flush pipes and spreaders; 32mm plastics waste outlets										
fixing bowls, divisions, cistern and pipe brackets to masonry with screws	526.00	2.50	539.15	13.50	-	323.86	3.97	866.99	nr	953.69
1200mm long stainless steel wall hung trough urinal with plastic; white; automatic cistern to BS 1876; spreader and stainless steel flush pipe; waste outlet and domed grating										
fixing urinal cistern and pipe brackets to masonry with screws	409.00	2.50	419.22	5.00	-	119.95	4.95	544.12	nr	598.54
1800mm long stainless steel wall hung trough urinal with plastic; white; automatic cistern to BS 1876; spreader and stainless steel flush pipe; waste outlet and domed grating										
fixing urinal cistern and pipe brackets to masonry with screws	479.00	2.50	490.97	7.00	-	167.93	8.55	667.46	nr	734.20
2400mm long stainless steel wall hung trough urinal with plastic; white; automatic cistern to BS 1876; spreader and stainless steel flush pipe; waste outlet and domed grating										
fixing urinal cistern and pipe brackets to masonry with screws	565.00	2.50	579.12	10.00	-	239.90	12.00	831.03	nr	914.13
Supply and fix appliances - showers										
Showers; white glazed Armastone tray, 760 x 760 x 180mm; 38mm waste outlet; surface mechanical valve										
bedding tray in cement mortar (1:4); fixing valve and head to masonry with screws; sealing at walls with mastic sealant.............	270.00	2.50	276.75	4.75	-	113.95	6.38	397.08	nr	436.79
Showers; white glazed Armastone tray, 760 x 760 x 180mm; 38mm waste outlet; recessed thermostatic valve and swivel spray head										
bedding tray in cement mortar (1:4); fixing valve and head to masonry with screws; sealing at walls with mastic sealant.............	385.00	2.50	394.62	4.75	-	113.95	6.38	514.95	nr	566.45
Showers; white moulded acrylic tray with removable front and side panels, 760 x 760 x 260mm with adjustable metal cradle; 38mm waste outlet; surface fixing mechanical valve, flexible tube hand spray and slide bar										
fixing cradle, valve and slide bar to masonry with screws; sealing at walls with mastic sealant	315.00	2.50	322.88	4.75	-	113.95	4.28	441.10	nr	485.21
Showers; white moulded acrylic tray with removable front and side panels, 760 x 760 x 260mm with adjustable metal cradle; 38mm waste outlet; surface fixing thermostatic valve, flexible tube hand spray and slide bar										
fixing cradle, valve and slide bar to masonry with screws; sealing at walls with mastic sealant	430.00	2.50	440.75	4.75	-	113.95	4.28	558.98	nr	614.88
Shower curtains; nylon; anodised aluminium rail, glider hooks, end and suspension fittings										
straight, 914mm long x 1829mm drop	41.50	2.50	42.54	1.20	-	28.79	0.60	71.93	nr	79.12
angled, 1676mm girth x 1829mm drop...................................	58.00	2.50	59.45	1.50	-	35.99	0.75	96.18	nr	105.80

Labour hourly rates: (except Specialists) Craft Operatives 23.99 Labourer 13.76 Rates are national average prices. Refer to REGIONAL VARIATIONS for indicative levels of overall pricing in regions	MATERIALS			LABOUR				RATES		
	Del to Site £	Waste %	Material Cost £	Craft Optve Hrs	Lab Hrs	Labour Cost £	Sunds £	Nett Rate £	Unit	Gross rate (10%) £
SANITARY APPLIANCES AND FITTINGS (Cont'd)										
Supply and fix appliances - showers (Cont'd)										
Shower curtains; heavy duty plastic; anodised aluminium rail, glider hooks, end and suspension fitting; fixing supports to masonry with screws										
straight, 914mm long x 1829mm drop	46.50	2.50	47.66	1.20	-	28.79	0.60	77.05	nr	84.76
angled, 1676mm girth x 1829mm drop	63.25	2.50	64.83	1.50	-	35.99	0.75	101.57	nr	111.72

This section continues
on the next page

FURNITURE, FITTINGS AND EQUIPMENT

Labour hourly rates: (except Specialists) Craft Operatives 18.37 Labourer 13.76 Rates are national average prices. Refer to REGIONAL VARIATIONS for indicative levels of overall pricing in regions	MATERIALS			LABOUR				RATES		
	Del to Site £	Waste %	Material Cost £	Craft Optve Hrs	Lab Hrs	Labour Cost £	Sunds £	Nett Rate £	Unit	Gross rate (10%) £
SANITARY APPLIANCES AND FITTINGS (Cont'd)										
Supply and fix bath panels										
Hardboard with bearers										
Front panel; bearers fixed to masonry	42.82	2.50	44.16	1.20	0.15	24.11	0.31	68.58	nr	75.44
End panel; bearers fixed to masonry	26.21	2.50	27.03	0.70	0.08	13.96	0.17	41.15	nr	45.27
Bath angle strip	5.25	2.50	5.38	0.12	0.02	2.48	-	7.86	nr	8.65
Bath Panel Plastic with plinth										
Front panel and plinth fixed to masonry	32.15	2.50	32.95	0.70	0.08	13.96	0.17	47.08	nr	51.79
End panel and plinth fixed to masonry	26.25	2.50	26.91	0.70	0.08	13.96	0.10	40.97	nr	45.07
NOTICES AND SIGNS										
Signwriting in gloss paint; one coat										
Letters or numerals, Helvetica medium style, on painted or varnished surfaces										
50mm high	0.02	10.00	0.02	0.14	-	2.57	0.00	2.60	nr	2.86
extra; shading	0.02	10.00	0.02	0.08	-	1.47	0.00	1.50	nr	1.65
extra; outline	0.02	10.00	0.02	0.12	-	2.20	0.00	2.23	nr	2.46
100mm high	0.02	10.00	0.02	0.28	-	5.14	0.01	5.18	nr	5.69
extra; shading	0.02	10.00	0.02	0.16	-	2.94	0.00	2.97	nr	3.26
extra; outline	0.02	10.00	0.02	0.24	-	4.41	0.01	4.44	nr	4.89
150mm high	0.04	10.00	0.05	0.42	-	7.72	0.01	7.78	nr	8.56
extra; shading	0.02	10.00	0.02	0.24	-	4.41	0.01	4.44	nr	4.89
extra; outline	0.02	10.00	0.02	0.36	-	6.61	0.01	6.65	nr	7.32
200mm high	0.04	10.00	0.05	0.56	-	10.29	0.02	10.35	nr	11.39
extra; shading	0.02	10.00	0.02	0.32	-	5.88	0.01	5.91	nr	6.50
extra; outline	0.02	10.00	0.02	0.48	-	8.82	0.02	8.86	nr	9.75
300mm high	0.06	10.00	0.07	0.84	-	15.43	0.03	15.53	nr	17.08
extra; shading	0.04	10.00	0.05	0.48	-	8.82	0.02	8.88	nr	9.77
extra; outline	0.04	10.00	0.05	0.72	-	13.23	0.03	13.30	nr	14.63
stops	-	-	-	0.04	-	0.73	-	0.73	nr	0.81
Signwriting in gloss paint; two coats										
Letters or numerals, Helvetica medium style, on painted or varnished surfaces										
50mm high	0.02	10.00	0.02	0.25	-	4.59	0.01	4.63	nr	5.09
100mm high	0.04	10.00	0.05	0.50	-	9.19	0.02	9.25	nr	10.18
150mm high	0.04	10.00	0.05	0.75	-	13.78	0.03	13.85	nr	15.23
200mm high	0.06	10.00	0.07	1.00	-	18.37	0.04	18.48	nr	20.32
250mm high	0.06	10.00	0.07	1.25	-	22.96	0.05	23.08	nr	25.39
300mm high	0.09	10.00	0.09	1.50	-	27.56	0.05	27.70	nr	30.47
stops	-	-	-	0.07	-	1.29	0.00	1.29	nr	1.42

Labour hourly rates: (except Specialists) Craft Operatives 23.99 Labourer 13.76 Rates are national average prices. Refer to REGIONAL VARIATIONS for indicative levels of overall pricing in regions	MATERIALS			LABOUR				RATES		
	Del to Site £	Waste %	Material Cost £	Craft Optve Hrs	Lab Hrs	Labour Cost £	Sunds £	Nett Rate £	Unit	Gross rate (10%) £
RAINWATER INSTALLATIONS; PIPEWORK										
Cast iron pipes and fittings, BS 460 Type A sockets; dry joints										
Pipes; straight										
75mm; in or on supports (included elsewhere)....................	24.18	5.00	25.39	0.67	-	16.07	-	41.46	m	45.61
extra; shoes............................	19.47	2.00	19.86	0.84	-	20.15	-	40.01	nr	44.01
extra; bends............................	18.04	2.00	18.40	0.67	-	16.07	-	34.47	nr	37.92
extra; offset bends 75mm projection...........	22.81	2.00	23.27	0.67	-	16.07	-	39.34	nr	43.27
extra; offset bends 150mm projection..........	22.81	2.00	23.27	0.67	-	16.07	-	39.34	nr	43.27
extra; offset bends 230mm projection..........	26.29	2.00	26.82	0.67	-	16.07	-	42.89	nr	47.18
extra; offset bends 305mm projection..........	26.29	2.00	26.82	0.67	-	16.07	-	42.89	nr	47.18
extra; branches........................	31.47	2.00	32.10	0.67	-	16.07	-	48.17	nr	52.99
100mm; in or on supports (included elsewhere).......	34.42	5.00	36.14	0.80	-	19.19	-	55.33	m	60.87
extra; shoes............................	28.25	2.00	28.81	1.00	-	23.99	-	52.81	nr	58.09
extra; bends............................	24.27	2.00	24.76	0.80	-	19.19	-	43.95	nr	48.34
extra; offset bends 75mm projection...........	42.87	2.00	43.73	0.80	-	19.19	-	62.92	nr	69.21
extra; offset bends 150mm projection..........	42.87	2.00	43.73	0.80	-	19.19	-	62.92	nr	69.21
extra; offset bends 230mm projection..........	51.61	2.00	52.64	0.80	-	19.19	-	71.83	nr	79.02
extra; offset bends 305mm projection..........	51.61	2.00	52.64	0.80	-	19.19	-	71.83	nr	79.02
extra; branches........................	37.42	2.00	38.17	0.80	-	19.19	-	57.36	nr	63.10
Cast iron pipes and fittings, BS 460 Type A sockets; ears cast on; dry joints										
Pipes; straight										
63mm; ears fixing to masonry with galvanized pipe nails and distance pieces...................	30.58	5.00	32.12	0.57	-	13.67	-	45.79	m	50.37
extra; shoes............................	24.90	2.00	25.40	0.71	-	17.03	-	42.43	nr	46.68
extra; bends............................	16.09	2.00	16.41	0.57	-	13.67	-	30.09	nr	33.10
extra; offset bends 75mm projection...........	23.54	2.00	24.01	0.57	-	13.67	-	37.69	nr	41.46
extra; offset bends 150mm projection..........	23.54	2.00	24.01	0.57	-	13.67	-	37.69	nr	41.46
extra; offset bends 230mm projection..........	26.81	2.00	27.35	0.57	-	13.67	-	41.02	nr	45.12
extra; offset bends 305mm projection..........	26.81	2.00	27.35	0.57	-	13.67	-	41.02	nr	45.12
extra; branches........................	28.91	2.00	29.49	0.57	-	13.67	-	43.16	nr	47.48
75mm; ears fixing to masonry with galvanized pipe nails and distance pieces...................	30.58	5.00	32.12	0.67	-	16.07	-	48.19	m	53.01
extra; shoes............................	24.90	2.00	25.40	0.84	-	20.15	-	45.55	nr	50.11
extra; bends............................	19.06	2.00	19.44	0.67	-	16.07	-	35.52	nr	39.07
extra; offset bends 75mm projection...........	23.54	2.00	24.01	0.67	-	16.07	-	40.09	nr	44.10
extra; offset bends 150mm projection..........	23.54	2.00	24.01	0.67	-	16.07	-	40.09	nr	44.10
extra; offset bends 230mm projection..........	26.81	2.00	27.35	0.67	-	16.07	-	43.42	nr	47.76
extra; offset bends 305mm projection..........	26.81	2.00	27.35	0.67	-	16.07	-	43.42	nr	47.76
extra; branches........................	31.66	2.00	32.30	0.67	-	16.07	-	48.37	nr	53.21
100mm; ears fixing to masonry with galvanized pipe nails and distance pieces...................	40.70	5.00	42.75	0.80	-	19.19	-	61.94	m	68.13
extra; shoes............................	31.66	2.00	32.30	1.00	-	23.99	-	56.29	nr	61.91
extra; bends............................	25.90	2.00	26.42	0.80	-	19.19	-	45.61	nr	50.17
extra; offset bends 75mm projection...........	42.35	2.00	43.19	0.80	-	19.19	-	62.38	nr	68.62
extra; offset bends 150mm projection..........	42.35	2.00	43.19	0.80	-	19.19	-	62.38	nr	68.62
extra; offset bends 230mm projection..........	50.54	2.00	51.55	0.80	-	19.19	-	70.75	nr	77.82
extra; offset bends 305mm projection..........	50.54	2.00	51.55	0.80	-	19.19	-	70.75	nr	77.82
extra; branches........................	37.24	2.00	37.99	0.80	-	19.19	-	57.18	nr	62.90
Pipe supports										
Steel holderbats										
for 75mm pipes; fixing to masonry with galvanized screws...........	7.55	2.00	7.70	0.17	-	4.08	-	11.78	nr	12.96
for 100mm pipes; fixing to masonry with galvanized screws..........	8.30	2.00	8.47	0.17	-	4.08	-	12.55	nr	13.80
Cast iron rectangular pipes and fittings; ears cast on; dry joints										
Pipes; straight										
100 x 75mm; ears fixing to masonry with galvanized pipe nails and distance pieces....................	86.22	5.00	90.54	1.00	-	23.99	-	114.53	m	125.98
extra; shoes (front)	76.09	2.00	77.61	1.00	-	23.99	-	101.60	nr	111.76
extra; bends (front).....................	72.00	2.00	73.44	0.75	-	17.99	-	91.43	nr	100.58
extra; offset bends (front) 150 projection........	89.47	2.00	91.26	0.75	-	17.99	-	109.25	nr	120.18
extra; offset bends (front) 300 projection........	125.78	2.00	128.30	0.75	-	17.99	-	146.29	nr	160.92
Cast aluminium pipes and fittings Section B, ears cast on; dry joints										
Pipes; straight										
63mm; ears fixing to masonry with galvanized screws................	12.42	5.00	13.04	0.50	-	11.99	-	25.03	m	27.54
extra; shoes............................	7.92	2.00	8.08	0.63	-	15.11	-	23.19	nr	25.51
extra; bends............................	9.06	2.00	9.24	0.50	-	11.99	-	21.24	nr	23.36
extra; offset bends 75mm projection...........	17.43	2.00	17.78	0.50	-	11.99	-	29.77	nr	32.75
extra; offset bends 150mm projection..........	21.61	2.00	22.04	0.50	-	11.99	-	34.04	nr	37.44
extra; offset bends 225mm projection..........	24.49	2.00	24.98	0.50	-	11.99	-	36.97	nr	40.67
extra; offset bends 300mm projection..........	27.42	2.00	27.97	0.50	-	11.99	-	39.96	nr	43.96
75mm; ears fixing to masonry with galvanized pipe nails............	14.50	5.00	15.22	0.67	-	16.07	-	31.29	m	34.42
extra; shoes............................	10.98	2.00	11.20	0.84	-	20.15	-	31.35	nr	34.49
extra; bends............................	11.87	2.00	12.11	0.67	-	16.07	-	28.18	nr	31.00

Labour hourly rates: (except Specialists) Craft Operatives 23.99 Labourer 13.76 Rates are national average prices. Refer to REGIONAL VARIATIONS for indicative levels of overall pricing in regions	MATERIALS			LABOUR				RATES		
	Del to Site £	Waste %	Material Cost £	Craft Optve Hrs	Lab Hrs	Labour Cost £	Sunds £	Nett Rate £	Unit	Gross rate (10%) £
RAINWATER INSTALLATIONS; PIPEWORK (Cont'd)										
Cast aluminium pipes and fittings Section B, ears cast on; dry joints (Cont'd)										
Pipes; straight (Cont'd)										
extra; offset bends 75mm projection	21.12	2.00	21.54	0.67	-	16.07	-	37.62	nr	41.38
extra; offset bends 150mm projection	24.25	2.00	24.73	0.67	-	16.07	-	40.80	nr	44.88
extra; offset bends 225mm projection	27.04	2.00	27.58	0.67	-	16.07	-	43.66	nr	48.02
extra; offset bends 300mm projection	30.03	2.00	30.63	0.67	-	16.07	-	46.70	nr	51.37
100mm; ears fixing to masonry with galvanized pipe nails	23.58	5.00	24.76	0.80	-	19.19	-	43.95	m	48.35
extra; shoes	13.10	2.00	13.36	1.00	-	23.99	-	37.35	nr	41.08
extra; bends	17.36	2.00	17.71	0.80	-	19.19	-	36.90	nr	40.59
extra; offset bends 75mm projection	23.85	2.00	24.32	0.80	-	19.19	-	43.52	nr	47.87
extra; offset bends 150mm projection	27.71	2.00	28.27	0.80	-	19.19	-	47.46	nr	52.20
extra; offset bends 225mm projection	31.35	2.00	31.98	0.80	-	19.19	-	51.17	nr	56.28
extra; offset bends 300mm projection	35.00	2.00	35.71	0.80	-	19.19	-	54.90	nr	60.39
PVC-U pipes and fittings; push fit joints; pipework and supports self coloured										
Pipes; straight										
68mm; in standard holderbats fixing to masonry with galvanized screws	3.96	5.00	4.14	0.40	-	9.60	-	13.73	m	15.10
extra; shoes	2.21	2.00	2.25	0.31	-	7.44	-	9.69	nr	10.66
extra; bends	2.69	2.00	2.74	0.25	-	6.00	-	8.74	nr	9.62
extra; offset bends (spigot)	3.21	2.00	3.27	0.25	-	6.00	-	9.27	nr	10.20
extra; offset bends (socket)	3.15	2.00	3.21	0.25	-	6.00	-	9.21	nr	10.13
extra; angled branches	7.42	2.00	7.57	0.25	-	6.00	-	13.57	nr	14.92
RAINWATER INSTALLATIONS; PIPEWORK ANCILLARIES										
Aluminium pipework ancillaries										
Rainwater heads; square type										
250 x 180 x 180mm; 63mm outlet spigot; dry joint to pipe; fixing to masonry with galvanized screws	20.21	1.00	20.42	0.50	-	11.99	-	32.42	nr	35.66
250 x 180 x 180mm; 75mm outlet spigot; dry joint to pipe; fixing to masonry with galvanized screws	21.17	1.00	21.39	0.67	-	16.07	-	37.46	nr	41.20
250 x 180 x 180mm; 100mm outlet spigot; dry joint to pipe; fixing to masonry with galvanized screws	34.80	1.00	35.15	0.80	-	19.19	-	54.34	nr	59.78
Cast iron pipework ancillaries										
Rainwater heads; BS 460 hopper type (flat pattern)										
210 x 160 x 185mm; 65mm outlet spigot; dry joint to pipe; fixing to masonry with galvanized screws	42.24	1.00	42.67	0.50	-	11.99	-	54.66	nr	60.13
210 x 160 x 185mm; 75mm outlet spigot; dry joint to pipe; fixing to masonry with galvanized screws	42.24	1.00	42.67	0.67	-	16.07	-	58.74	nr	64.61
250 x 180 x 175mm; 100mm outlet spigot; dry joint to pipe; fixing to masonry with galvanized screws	69.56	1.00	70.26	0.80	-	19.19	-	89.45	nr	98.40
Rainwater heads; BS 460 square type (flat pattern)										
250 x 180 x 175mm; 65mm outlet spigot; dry joint to pipe; fixing to masonry with galvanized screws	61.55	1.00	62.17	0.50	-	11.99	-	74.16	nr	81.58
250 x 180 x 175mm; 75mm outlet spigot; dry joint to pipe; fixing to masonry with galvanized screws	3.39	1.00	3.42	0.67	-	16.07	-	19.50	nr	21.45
250 x 180 x 175mm; 100mm outlet spigot; dry joint to pipe; fixing to masonry with galvanized screws	61.55	1.00	62.17	0.80	-	19.19	-	81.36	nr	89.50
Roof outlets; luting flange for asphalt										
flat grating; 100mm outlet spigot; coupling joint to cast iron pipe	105.96	1.00	107.02	1.00	-	23.99	-	131.01	nr	144.11
domical grating; 100mm outlet spigot; coupling joint to cast iron pipe	118.85	1.00	120.04	1.00	-	23.99	-	144.03	nr	158.43
Plastics pipework ancillaries self coloured black										
Rainwater heads; square type										
280 x 155 x 230mm; 68mm outlet spigot; push fit joint to plastics pipe; fixing to masonry with galvanized screws	13.71	1.00	13.86	0.50	-	11.99	-	25.85	nr	28.44
252 x 195 x 210mm; 110mm outlet spigot; push fit joint to plastics pipe; fixing to masonry with galvanized screws	27.01	1.00	27.29	0.80	-	19.19	-	46.48	nr	51.13
Plastics pipework ancillaries, self coloured grey										
Roof outlets; luting flange for asphalt										
domical grating; 75mm outlet spigot; push fit joint to plastics pipe	33.03	1.00	33.36	0.37	-	8.88	-	42.24	nr	46.46
domical grating; 100mm outlet spigot; push fit joint to plastics pipe	32.90	1.00	33.23	0.37	-	8.88	-	42.11	nr	46.32
RAINWATER INSTALLATIONS; GUTTERS										
Cast iron half round gutters and fittings, BS 460; bolted and mastic joints										
Gutters; straight										
100mm; in standard fascia brackets fixing to timber with galvanized screws	15.25	5.00	15.95	0.50	-	11.99	-	27.95	m	30.74
extra; stopped ends	3.79	2.00	3.87	0.50	-	11.99	-	15.86	nr	17.45
extra; running outlets	11.31	2.00	11.54	0.50	-	11.99	-	23.53	nr	25.88
extra; stopped ends with outlet	15.10	2.00	15.40	0.50	-	11.99	-	27.40	nr	30.14
extra; angles	11.41	2.00	11.64	0.50	-	11.99	-	23.63	nr	26.00
114mm; in standard fascia brackets fixing to timber with galvanized screws	15.88	5.00	16.61	0.50	-	11.99	-	28.60	m	31.46
extra; stopped ends	4.98	2.00	5.08	0.50	-	11.99	-	17.07	nr	18.78
extra; running outlets	12.16	2.00	12.40	0.50	-	11.99	-	24.40	nr	26.84
extra; stopped ends with outlet	17.14	2.00	17.48	0.50	-	11.99	-	29.48	nr	32.43
extra; angles	11.73	2.00	11.96	0.50	-	11.99	-	23.96	nr	26.36

Labour hourly rates: (except Specialists) Craft Operatives 23.99 Labourer 13.76 Rates are national average prices. Refer to REGIONAL VARIATIONS for indicative levels of overall pricing in regions	MATERIALS			LABOUR				RATES		
	Del to Site £	Waste %	Material Cost £	Craft Optve Hrs	Lab Hrs	Labour Cost £	Sunds £	Nett Rate £	Unit	Gross rate (10%) £
RAINWATER INSTALLATIONS; GUTTERS (Cont'd)										
Cast iron half round gutters and fittings, BS 460; bolted and mastic joints (Cont'd)										
Gutters; straight (Cont'd)										
125mm; in standard fascia brackets fixing to timber with galvanized screws	18.65	5.00	19.50	0.57	-	13.67	-	33.18	m	36.49
extra; stopped ends	5.25	2.00	5.36	0.57	-	13.67	-	19.03	nr	20.93
extra; running outlets	14.50	2.00	14.79	0.57	-	13.67	-	28.46	nr	31.31
extra; stopped ends with outlet	19.75	2.00	20.15	0.57	-	13.67	-	33.82	nr	37.20
extra; angles	13.81	2.00	14.09	0.57	-	13.67	-	27.76	nr	30.54
150mm; in standard fascia brackets fixing to timber with galvanized screws	30.18	5.00	31.60	0.75	-	17.99	-	49.59	m	54.55
extra; stopped ends	11.62	2.00	11.85	0.75	-	17.99	-	29.84	nr	32.83
extra; running outlets	20.84	2.00	21.26	0.75	-	17.99	-	39.25	nr	43.17
extra; stopped ends with outlet	32.46	2.00	33.11	0.75	-	17.99	-	51.10	nr	56.21
extra; angles	20.84	2.00	21.26	0.75	-	17.99	-	39.25	nr	43.17
Cast iron OG gutters and fittings, BS 460; bolted and mastic joints										
Gutters; straight										
100mm; fixing to timber with galvanized screws	14.67	5.00	15.40	0.50	-	11.99	-	27.39	m	30.13
extra; stopped ends	3.08	2.00	3.14	0.50	-	11.99	-	15.14	nr	16.65
extra; running outlets	10.54	2.00	10.75	0.50	-	11.99	-	22.75	nr	25.02
extra; angles	10.54	2.00	10.75	0.50	-	11.99	-	22.75	nr	25.02
114mm; fixing to timber with galvanized screws	15.55	5.00	16.33	0.50	-	11.99	-	28.33	m	31.16
extra; stopped ends	4.12	2.00	4.20	0.50	-	11.99	-	16.20	nr	17.82
extra; running outlets	11.47	2.00	11.70	0.50	-	11.99	-	23.69	nr	26.06
extra; angles	11.47	2.00	11.70	0.50	-	11.99	-	23.69	nr	26.06
125mm; fixing to timber with galvanized screws	17.10	5.00	17.96	0.57	-	13.67	-	31.63	m	34.80
extra; stopped ends	4.12	2.00	4.20	0.57	-	13.67	-	17.88	nr	19.66
extra; running outlets	12.51	2.00	12.76	0.57	-	13.67	-	26.44	nr	29.08
extra; angles	12.51	2.00	12.76	0.57	-	13.67	-	26.44	nr	29.08
Cast iron moulded (stock sections) gutters and fittings; bolted and mastic joints										
Gutters; straight										
100 x 75mm; in standard fascia brackets fixing to timber with galvanized screws	27.56	5.00	28.88	0.75	-	17.99	-	46.87	m	51.56
extra; stopped ends	9.80	2.00	10.00	0.75	-	17.99	-	27.99	nr	30.79
extra; running outlets	19.26	2.00	19.65	0.75	-	17.99	-	37.64	nr	41.40
extra; angles	25.71	2.00	26.22	0.75	-	17.99	-	44.21	nr	48.64
extra; clips	10.76	2.00	10.98	0.75	-	17.99	-	28.97	nr	31.87
125 x 100mm; in standard fascia brackets fixing to timber with galvanized screws	39.18	5.00	41.07	1.00	-	23.99	-	65.06	m	71.56
extra; stopped ends	12.78	2.00	13.04	1.00	-	23.99	-	37.03	nr	40.73
extra; running outlets	36.96	2.00	37.70	1.00	-	23.99	-	61.69	nr	67.86
extra; angles	36.96	2.00	37.70	1.00	-	23.99	-	61.69	nr	67.86
extra; clips	12.78	2.00	13.04	1.00	-	23.99	-	37.03	nr	40.73
Cast iron box gutters and fittings; bolted and mastic joints										
Gutters; straight										
100 x 75mm; in or on brackets	42.96	5.00	45.01	0.75	-	17.99	-	63.00	m	69.30
extra; stopped ends	5.62	2.00	5.74	0.75	-	17.99	-	23.73	nr	26.10
extra; running outlets	19.40	2.00	19.78	0.75	-	17.99	-	37.78	nr	41.55
extra; angles	26.26	2.00	26.79	0.75	-	17.99	-	44.78	nr	49.26
extra; clips	7.30	2.00	7.45	0.75	-	17.99	-	25.44	nr	27.98
Cast aluminium half round gutters and fittings, Section A; bolted and mastic joints										
Gutters; straight										
102mm; in standard fascia brackets fixing to timber with galvanized screws	13.61	5.00	14.26	0.50	-	11.99	-	26.25	m	28.88
extra; stopped ends	3.79	2.00	3.87	0.50	-	11.99	-	15.86	nr	17.45
extra; running outlets	8.63	2.00	8.80	0.50	-	11.99	-	20.80	nr	22.88
extra; angles	7.93	2.00	8.09	0.50	-	11.99	-	20.08	nr	22.09
127mm; in standard fascia brackets fixing to timber with galvanized screws	18.43	5.00	19.29	0.57	-	13.67	-	32.96	m	36.26
extra; stopped ends	4.98	2.00	5.08	0.57	-	13.67	-	18.75	nr	20.63
extra; running outlets	10.43	2.00	10.64	0.57	-	13.67	-	24.31	nr	26.74
extra; angles	10.15	2.00	10.35	0.57	-	13.67	-	24.03	nr	26.43
PVC-U half round gutters and fittings; push fit connector joints; gutterwork and supports self coloured grey										
Gutters; straight										
114mm; in standard fascia brackets fixing to timber with galvanized screws	2.62	2.00	2.67	0.33	-	7.92	-	10.59	m	11.65
extra; stopped ends	2.13	2.00	2.17	0.33	-	7.92	-	10.09	nr	11.10
extra; running outlets	3.27	2.00	3.34	0.33	-	7.92	-	11.25	nr	12.38
extra; angles	4.55	2.00	4.64	0.33	-	7.92	-	12.56	nr	13.81
FOUL DRAINAGE INSTALLATIONS; PIPEWORK										
Cast iron soil and waste pipes and fittings, BS 416; bolted and synthetic rubber gasket couplings										
Pipes; straight										
75mm; in or on supports (included elsewhere)	87.79	5.00	92.05	0.67	-	16.07	-	108.12	m	118.93
extra; pipes with inspection door bolted and sealed	82.73	2.00	84.39	0.67	-	16.07	-	100.46	nr	110.50
extra; short radius bends	49.52	2.00	50.51	0.67	-	16.07	-	66.58	nr	73.24
extra; short radius bends with inspection door bolted and sealed	75.32	2.00	76.83	0.67	-	16.07	-	92.90	nr	102.20

Labour hourly rates: (except Specialists) Craft Operatives 23.99 Labourer 13.76 Rates are national average prices. Refer to REGIONAL VARIATIONS for indicative levels of overall pricing in regions	MATERIALS			LABOUR				RATES		
	Del to Site £	Waste %	Material Cost £	Craft Optve Hrs	Lab Hrs	Labour Cost £	Sunds £	Nett Rate £	Unit	Gross rate (10%) £

FOUL DRAINAGE INSTALLATIONS; PIPEWORK (Cont'd)

Cast iron soil and waste pipes and fittings, BS 416; bolted and synthetic rubber gasket couplings (Cont'd)

Pipes; straight (Cont'd)										
extra; offset bends 150mm projection	62.69	2.00	63.94	0.67	-	16.07	-	80.01	nr	88.01
extra; single angled branches	76.46	2.00	77.99	1.00	-	23.99	-	101.98	nr	112.18
extra; single angled branches with inspection door bolted and sealed	136.18	2.00	138.90	1.00	-	23.99	-	162.89	nr	179.18
extra; double angled branches	128.32	2.00	130.88	1.33	-	31.91	-	162.79	nr	179.07
boss pipes with one threaded boss	87.49	2.00	89.24	0.67	-	16.07	-	105.32	nr	115.85
100mm; in or on supports (included elsewhere)	42.39	5.00	44.27	0.80	-	19.19	-	63.46	m	69.81
extra; pipes with inspection door bolted and sealed	106.96	2.00	109.10	0.80	-	19.19	-	128.29	nr	141.12
extra; short radius bends	81.19	2.00	82.81	0.80	-	19.19	-	102.01	nr	112.21
extra; short radius bends with inspection door bolted and sealed	115.78	2.00	118.10	0.80	-	19.19	-	137.29	nr	151.02
extra; offset bends 150mm projection	97.80	2.00	99.76	0.80	-	19.19	-	118.95	nr	130.85
extra; offset bends 300mm projection	113.74	2.00	116.01	0.80	-	19.19	-	135.21	nr	148.73
extra; single angled branches	135.21	2.00	137.91	1.20	-	28.79	-	166.70	nr	183.37
extra; single angled branches with inspection door bolted and sealed	156.99	2.00	160.13	1.20	-	28.79	-	188.91	nr	207.81
extra; double angled branches	159.39	2.00	162.58	1.60	-	38.38	-	200.97	nr	221.06
extra; double angled branches with inspection door bolted and sealed	219.00	2.00	223.38	1.60	-	38.38	-	261.76	nr	287.94
extra; single anti-syphon branches, angled branch	138.17	2.00	140.94	1.20	-	28.79	-	169.72	nr	186.70
extra; straight wc connectors	64.15	2.00	65.43	0.80	-	19.19	-	84.62	nr	93.08
extra; roof connectors	50.11	2.00	51.11	0.80	-	19.19	-	70.30	nr	77.33
boss pipes with one boss	106.06	2.00	108.18	0.80	-	19.19	-	127.37	nr	140.11
boss pipes with two bosses	122.78	2.00	125.24	0.80	-	19.19	-	144.43	nr	158.87
50mm; in standard wall fixing bracket (fixing included elsewhere)	79.83	5.00	83.70	0.55	-	13.19	-	96.90	m	106.59
extra; short radius bends	45.98	2.00	46.90	0.55	-	13.19	-	60.10	nr	66.11
extra; short radius bends with inspection door bolted and sealed	96.35	2.00	98.27	0.55	-	13.19	-	111.47	nr	122.62
extra; single angled branches	59.31	2.00	60.49	0.82	-	19.67	-	80.16	nr	88.18
extra; single angled branches with inspection door bolted and sealed	132.39	2.00	135.04	0.82	-	19.67	-	154.71	nr	170.18

Cast iron lightweight soil and waste pipes and fittings

Pipes; straight										
50mm; in or on supports (included elsewhere)	22.25	5.00	23.28	0.55	-	13.19	-	36.48	m	40.13
extra; short pipe with access door	55.96	2.00	57.08	0.55	-	13.19	-	70.27	nr	77.30
extra; bends	31.35	2.00	31.98	0.55	-	13.19	-	45.17	nr	49.69
extra; offset bend 130mm projection	40.56	2.00	41.37	0.55	-	13.19	-	54.56	nr	60.02
extra; single angled branch	48.28	2.00	49.24	0.82	-	19.67	-	68.91	nr	75.81
extra; plug	14.17	2.00	14.45	0.27	-	6.48	-	20.93	nr	23.02
extra; universal plug with one inlet	22.08	2.00	22.52	0.14	-	3.36	-	25.88	nr	28.47
extra; step coupling for connection to BS 416 pipe	15.36	2.00	15.67	0.40	-	9.60	-	25.27	nr	27.79
75mm; in or on supports (included elsewhere)	70.01	5.00	73.42	0.67	-	16.07	-	89.49	m	98.44
extra; short pipe with access door	60.25	2.00	61.45	0.67	-	16.07	-	77.53	nr	85.28
extra; bends	35.14	2.00	35.85	0.67	-	16.07	-	51.92	nr	57.11
extra; offset bend 130mm projection	55.80	2.00	56.91	0.67	-	16.07	-	72.99	nr	80.29
extra; single angled branch	52.23	2.00	53.28	1.00	-	23.99	-	77.27	nr	84.99
extra; diminishing piece 75/50	36.79	2.00	37.52	0.67	-	16.07	-	53.60	nr	58.96
extra; plug	15.36	2.00	15.67	0.33	-	7.92	-	23.59	nr	25.95
extra; step coupling for connection to BS 416 pipe	15.36	2.00	15.67	0.50	-	11.99	-	27.67	nr	30.43
100mm; in or on supports (included elsewhere)	83.49	5.00	87.55	0.80	-	19.19	-	106.74	m	117.41
extra; short pipe with access door	68.07	2.00	69.43	0.80	-	19.19	-	88.62	nr	97.48
extra; bends	43.73	2.00	44.61	0.80	-	19.19	-	63.80	nr	70.18
extra; long radius bend	73.24	2.00	74.71	0.80	-	19.19	-	93.90	nr	103.29
extra; offset bend 75mm projection	65.87	2.00	67.19	0.80	-	19.19	-	86.38	nr	95.02
extra; offset bend 130mm projection	74.12	2.00	75.60	0.80	-	19.19	-	94.79	nr	104.27
extra; single angled branch	67.79	2.00	69.15	1.20	-	28.79	-	97.94	nr	107.73
extra; double angled branch	100.88	2.00	102.90	1.60	-	38.38	-	141.28	nr	155.41
extra; diminishing piece 100/50	46.62	2.00	47.55	0.80	-	19.19	-	66.75	nr	73.42
extra; diminishing piece 100/75	47.51	2.00	48.46	0.80	-	19.19	-	67.65	nr	74.41
extra; plug	19.42	2.00	19.81	0.40	-	9.60	-	29.41	nr	32.35
extra; universal plug with three inlets	25.15	2.00	25.65	0.40	-	9.60	-	35.25	nr	38.77
extra; step coupling for connection to BS 416 pipe	19.81	2.00	20.21	0.60	-	14.39	-	34.60	nr	38.07
extra; traditional joint connector	41.01	2.00	41.83	0.80	-	19.19	-	61.02	nr	67.12
extra; roof connector (asphalt)	54.62	2.00	55.72	0.80	-	19.19	-	74.91	nr	82.40

Pipe supports

galvanized steel vertical bracket										
for 50mm pipes; for fixing bolt(fixing included elsewhere)	9.56	2.00	9.75	-	-	-	-	9.75	nr	10.73
for 75mm pipes; for fixing bolt(fixing included elsewhere)	10.48	2.00	10.69	-	-	-	-	10.69	nr	11.76
for 100mm pipes; for fixing bolt(fixing included elsewhere)	10.74	2.00	10.95	-	-	-	-	10.95	nr	12.05

MuPVC pipes and fittings, BS EN 1329-1 Section Three; solvent welded joints

Pipes; straight										
32mm; in standard plastics pipe brackets fixing to timber with screws	2.33	10.00	2.51	0.33	-	7.92	-	10.43	m	11.47
extra; straight expansion couplings	1.44	2.00	1.47	0.25	-	6.00	-	7.47	nr	8.21
extra; connections to plastics pipe socket; socket adaptor; solvent welded joint	1.33	2.00	1.36	0.25	-	6.00	-	7.35	nr	8.09
extra; connections to plastics pipe boss; boss adaptor; solvent welded joint	1.33	2.00	1.36	0.25	-	6.00	-	7.35	nr	8.09
extra; fittings; one end	1.33	2.00	1.36	0.13	-	3.12	-	4.48	nr	4.92
extra; fittings; two ends	1.00	2.00	1.02	0.25	-	6.00	-	7.02	nr	7.72
extra; fittings; three ends	1.10	2.00	1.12	0.25	-	6.00	-	7.12	nr	7.83
40mm; in standard plastics pipe brackets fixing to timber with screws	2.72	10.00	2.94	0.33	-	7.92	-	10.86	m	11.94
extra; straight expansion couplings	1.44	2.00	1.47	0.25	-	6.00	-	7.47	nr	8.21

Labour hourly rates: (except Specialists) Craft Operatives 23.99 Labourer 13.76 Rates are national average prices. Refer to REGIONAL VARIATIONS for indicative levels of overall pricing in regions	MATERIALS			LABOUR				RATES		
	Del to Site £	Waste %	Material Cost £	Craft Optve Hrs	Lab Hrs	Labour Cost £	Sunds £	Nett Rate £	Unit	Gross rate (10%) £
FOUL DRAINAGE INSTALLATIONS; PIPEWORK (Cont'd)										
MuPVC pipes and fittings, BS EN 1329-1 Section Three; solvent welded joints (Cont'd)										
Pipes; straight (Cont'd)										
extra; connections to plastics pipe socket; socket adaptor; solvent welded joint	1.13	2.00	1.15	0.25	-	6.00	-	7.15	nr	7.87
extra; connections to plastics pipe boss; boss adaptor; solvent welded joint	1.33	2.00	1.36	0.25	-	6.00	-	7.35	nr	8.09
extra; fittings; one end	1.33	2.00	1.36	0.13	-	3.12	-	4.48	nr	4.92
extra; fittings; two ends	1.00	2.00	1.02	0.25	-	6.00	-	7.02	nr	7.72
extra; fittings; three ends	1.10	2.00	1.12	0.25	-	6.00	-	7.12	nr	7.83
50mm; in standard plastics pipe brackets fixing to timber with screws	4.24	10.00	4.56	0.40	-	9.60	-	14.16	m	15.57
extra; straight expansion couplings	2.21	2.00	2.25	0.33	-	7.92	-	10.17	nr	11.19
extra; connections to plastics pipe socket; socket adaptor; solvent welded joint	1.98	2.00	2.02	0.33	-	7.92	-	9.94	nr	10.93
extra; connections to plastics pipe boss; boss adaptor; solvent welded joint	2.47	2.00	2.52	0.33	-	7.92	-	10.44	nr	11.48
extra; fittings; one end	2.00	2.00	2.04	0.17	-	4.08	-	6.12	nr	6.73
extra; fittings; two ends	1.89	2.00	1.93	0.33	-	7.92	-	9.84	nr	10.83
extra; fittings; three ends	1.89	2.00	1.93	0.33	-	7.92	-	9.84	nr	10.83
32mm; in standard plastics pipe brackets fixing to masonry with screws	2.36	10.00	2.54	0.39	-	9.36	-	11.90	m	13.09
40mm; in standard plastics pipe brackets fixing to masonry with screws	2.75	10.00	2.97	0.39	-	9.36	-	12.33	m	13.56
50mm; in standard plastics pipe brackets fixing to masonry with screws	4.26	10.00	4.59	0.46	-	11.04	-	15.63	m	17.19
Polypropylene pipes and fittings; butyl 'O' ring joints										
Pipes; straight										
32mm; in standard plastics pipe brackets fixing to timber with screws	1.38	10.00	1.48	0.33	-	7.92	-	9.39	m	10.33
extra; connections to plastics pipe socket; socket adaptor; solvent welded joint	1.91	2.00	1.95	0.25	-	6.00	-	7.95	nr	8.74
extra; connections to plastics pipe boss; boss adaptor; solvent welded joint	1.91	2.00	1.95	0.25	-	6.00	-	7.95	nr	8.74
extra; fittings; one end	1.68	2.00	1.71	0.13	-	3.12	-	4.83	nr	5.32
extra; fittings; two ends	0.95	2.00	0.97	0.25	-	6.00	-	6.97	nr	7.66
extra; fittings; three ends	1.51	2.00	1.54	0.25	-	6.00	-	7.54	nr	8.29
40mm; in standard plastics pipe brackets fixing to timber with screws	1.68	10.00	1.80	0.33	-	7.92	-	9.71	m	10.68
extra; connections to plastics pipe socket; socket adaptor; solvent welded joint	1.91	2.00	1.95	0.33	-	7.92	-	9.86	nr	10.85
extra; connections to plastics pipe boss; boss adaptor; solvent welded joint	1.91	2.00	1.95	0.33	-	7.92	-	9.86	nr	10.85
extra; fittings; one end	1.92	2.00	1.96	0.17	-	4.08	-	6.04	nr	6.64
extra; fittings; two ends	0.97	2.00	0.99	0.33	-	7.92	-	8.91	nr	9.80
extra; fittings; three ends	1.55	2.00	1.58	0.33	-	7.92	-	9.50	nr	10.45
32mm; in standard plastics pipe brackets fixing to masonry with screws	1.41	10.00	1.51	0.39	-	9.36	-	10.86	m	11.95
40mm; in standard plastics pipe brackets fixing to masonry with screws	1.71	10.00	1.83	0.39	-	9.36	-	11.18	m	12.30
PVC-U pipes and fittings, BS 4514; rubber ring joints; pipework self coloured grey										
Pipes; straight										
82mm; in or on supports (included elsewhere)	12.35	5.00	12.90	0.44	-	10.56	-	23.45	m	25.80
extra; pipes with inspection door	24.08	2.00	24.56	0.33	-	7.92	-	32.48	nr	35.73
extra; short radius bends	13.51	2.00	13.78	0.33	-	7.92	-	21.70	nr	23.87
extra; single angled branches	19.85	2.00	20.25	0.33	-	7.92	-	28.16	nr	30.98
pipes with one boss socket	13.71	2.00	13.98	0.33	-	7.92	-	21.90	nr	24.09
110mm; in or on supports (included elsewhere)	10.73	5.00	11.20	0.50	-	11.99	-	23.20	m	25.52
extra; pipes with inspection door	22.55	2.00	23.00	0.50	-	11.99	-	35.00	nr	38.50
extra; short radius bends	14.10	2.00	14.38	0.50	-	11.99	-	26.38	nr	29.01
extra; single angled branches	20.12	2.00	20.52	0.50	-	11.99	-	32.52	nr	35.77
extra; straight wc connectors	6.09	2.00	6.21	0.50	-	11.99	-	18.21	nr	20.03
pipes with one boss socket	11.30	2.00	11.53	0.50	-	11.99	-	23.52	nr	25.87
Air admittance valve	50.15	2.00	51.15	0.50	-	11.99	-	63.15	nr	69.46
160mm; in or on supports (included elsewhere)	19.79	5.00	20.68	0.75	-	17.99	-	38.67	m	42.54
extra; pipes with inspection door	37.46	2.00	38.21	0.75	-	17.99	-	56.20	nr	61.82
extra; short radius bends	30.54	2.00	31.15	0.75	-	17.99	-	49.14	nr	54.06
extra; single angled branches	50.29	2.00	51.30	0.75	-	17.99	-	69.29	nr	76.22
pipes with one boss socket	15.38	2.00	15.69	0.75	-	17.99	-	33.68	nr	37.05
PVC-U pipes and fittings, BS 4514; solvent weld joints; pipework self coloured grey										
Pipes; straight										
110mm; in or on supports (included elsewhere)	10.73	5.00	11.20	0.53	-	12.71	-	23.92	m	26.31
extra; pipes with inspection door	22.55	2.00	23.00	0.53	-	12.71	-	35.72	nr	39.29
extra; short radius bends	14.10	2.00	14.38	0.53	-	12.71	-	27.10	nr	29.81
extra; single angled branches	20.12	2.00	20.52	0.53	-	12.71	-	33.24	nr	36.56
extra; straight wc connectors	6.09	2.00	6.21	0.53	-	12.71	-	18.93	nr	20.82
pipes with one boss socket	11.30	2.00	11.53	0.53	-	12.71	-	24.24	nr	26.66
Air admittance valve	46.00	2.00	46.92	0.53	-	12.71	-	59.63	nr	65.60
160mm; in or on supports (included elsewhere)	19.79	5.00	20.68	0.80	-	19.19	-	39.87	m	43.86
extra; pipes with inspection door	37.46	2.00	38.21	0.80	-	19.19	-	57.40	nr	63.14
extra; short radius bends	30.54	2.00	31.15	0.80	-	19.19	-	50.34	nr	55.38
extra; single angled branches	50.29	2.00	51.30	0.80	-	19.19	Sunds	70.49	nr	77.54
pipes with one boss socket	15.38	2.00	15.69	0.80	-	19.19	-	34.88	nr	38.37

Labour hourly rates: (except Specialists) Craft Operatives 23.99 Labourer 13.76 Rates are national average prices. Refer to REGIONAL VARIATIONS for indicative levels of overall pricing in regions	MATERIALS			LABOUR				RATES		
	Del to Site £	Waste %	Material Cost £	Craft Optve Hrs	Lab Hrs	Labour Cost £	Sunds £	Nett Rate £	Unit	Gross rate (10%) £
FOUL DRAINAGE INSTALLATIONS; PIPEWORK (Cont'd)										
Plastics pipework ancillaries, self coloured grey										
Weathering aprons										
fitted to 82mm pipes	4.93	2.00	5.03	0.33	-	7.92	-	12.95	nr	14.24
fitted to 110mm pipes	3.71	2.00	3.78	0.50	-	11.99	-	15.78	nr	17.36
Pipe supports for PVC-U soil and vent pipes, rubber ring joints; self coloured grey										
Plastics coated metal holderbats										
for 82mm pipes; for building in (fixing included elsewhere)	4.80	2.00	4.90	-	-	-	-	4.90	nr	5.39
for 110mm pipes; for building in (fixing included elsewhere)	5.98	2.00	6.10	-	-	-	-	6.10	nr	6.71
for 82mm pipes; fixing to timber with galvanized screws	3.87	2.00	3.95	0.17	-	4.08	-	8.03	nr	8.83
for 110mm pipes; fixing to timber with galvanized screws	5.69	2.00	5.81	0.17	-	4.08	-	9.89	nr	10.87
for 160mm pipes; fixing to timber with galvanized screws	9.10	2.00	9.29	0.17	-	4.08	-	13.36	nr	14.70
Pipe supports; solvent weld joints										
Plastic brackets										
for 110mm pipes; for building in (fixing included elsewhere)	7.29	2.00	7.44	-	-	-	-	7.44	nr	8.18
for 110mm pipes; fixing to timber with galvanized screws	3.54	2.00	3.61	0.25	-	6.00	-	9.61	nr	10.57
FOUL DRAINAGE INSTALLATIONS; PIPEWORK ANCILLARIES										
Balloon gratings										
Plastic balloon gratings										
fitted to 50mm pipes	1.78	3.00	1.83	0.10	-	2.40	-	4.23	nr	4.66
fitted to 82mm pipes	2.68	3.00	2.76	0.10	-	2.40	-	5.16	nr	5.68
fitted to 110mm pipes	2.72	3.00	2.80	0.10	-	2.40	-	5.20	nr	5.72
Plastics pipework ancillaries self coloured white										
Traps; BS EN 274										
P, two piece, 75mm seal, inlet with coupling nut, outlet with seal ring socket										
32mm outlet...............................	2.85	2.00	2.91	0.33	-	7.92	-	10.82	nr	11.91
40mm outlet...............................	3.15	2.00	3.21	0.33	-	7.92	-	11.13	nr	12.24
S, two piece, 75mm seal, inlet with coupling nut, outlet with seal ring socket										
32mm outlet...............................	3.31	2.00	3.38	0.33	-	7.92	-	11.29	nr	12.42
40mm outlet...............................	3.70	2.00	3.77	0.33	-	7.92	-	11.69	nr	12.86
P, two piece, 75mm seal, with flexible polypropylene pipe for overflow connection, inlet with coupling nut, outlet with seal ring socket										
40mm outlet...............................	4.19	2.00	4.27	0.50	-	11.99	-	16.27	nr	17.90
Copper pipework ancillaries										
Traps; Section Two (Solid Drawn)										
P, two piece, 75mm seal, inlet with coupling nut, compression outlet										
35mm outlet...............................	71.49	2.00	72.92	0.33	-	7.92	-	80.84	nr	88.92
42mm outlet...............................	84.29	2.00	85.98	0.33	-	7.92	-	93.89	nr	103.28
54mm outlet...............................	302.31	2.00	308.36	0.70	-	16.79	-	325.15	nr	357.66
S, two piece, 75mm seal, inlet with coupling nut, compression outlet										
35mm outlet...............................	75.75	2.00	77.26	0.33	-	7.92	-	85.18	nr	93.70
42mm outlet...............................	91.20	2.00	93.02	0.33	-	7.92	-	100.94	nr	111.03
54mm outlet...............................	326.80	2.00	333.34	0.70	-	16.79	-	350.13	nr	385.14
bath, two piece, 75mm seal, overflow connection, cleaning eye and plug, inlet with coupling nut, compression outlet										
42mm outlet...............................	117.68	2.00	120.03	0.50	-	11.99	-	132.03	nr	145.23

Labour hourly rates: (except Specialists) Craft Operatives 23.99 Labourer 13.76 Rates are national average prices. Refer to REGIONAL VARIATIONS for indicative levels of overall pricing in regions	PLANT AND TRANSPORT			LABOUR				RATES		
	Plant Cost	Trans Cost	P and T Cost	Craft Optve	Lab	Labour Cost	Sunds	Nett Rate	Unit	Gross rate (10%)
	£	£	£	Hrs	Hrs	£	£	£		£

DRAINAGE

Note

Notwithstanding the requirements of NRM2 excavation, disposal, beds, benching and surrounds have not been included with the items for drain runs and manholes and prices for specific drain runs and manholes etc should be calculated from the details below.

STORM WATER AND FOUL DRAIN SYSTEMS; DRAIN RUNS

Excavating trenches by machine to receive pipes not exceeding 200mm nominal size; disposing of surplus excavated material by removing from site

Excavations commencing from natural ground level
  average depth

	Plant Cost	Trans Cost	P and T Cost	Craft Optve	Lab	Labour Cost	Sunds	Nett Rate	Unit	Gross rate
500mm	2.44	6.22	8.67	-	0.23	3.16	1.02	12.85	m	14.14
750mm	3.95	6.22	10.17	-	0.34	4.68	1.53	16.38	m	18.02
1000mm	5.32	6.22	11.54	-	0.45	6.19	2.06	19.79	m	21.76
1250mm	6.82	6.22	13.04	-	0.61	8.39	2.57	24.00	m	26.40
1500mm	8.19	6.22	14.41	-	0.73	10.04	3.08	27.53	m	30.28
1750mm	9.69	6.22	15.91	-	0.86	11.83	3.59	31.33	m	34.46
2000mm	11.01	6.22	17.23	-	0.98	13.48	4.09	34.81	m	38.29
2250mm	14.04	6.22	20.26	-	1.69	23.25	4.62	48.13	m	52.95
2500mm	15.57	6.22	21.80	-	1.88	25.87	5.13	52.80	m	58.08
2750mm	17.25	6.22	23.47	-	2.06	28.35	5.64	57.46	m	63.20
3000mm	18.78	6.22	25.01	-	2.25	30.96	6.15	62.12	m	68.33
3250mm	20.46	6.22	26.68	-	2.44	33.57	6.66	66.91	m	73.61
3500mm	21.95	6.22	28.17	-	2.63	36.19	7.17	71.53	m	78.68
3750mm	23.62	6.22	29.84	-	2.82	38.80	7.70	76.34	m	83.97
4000mm	25.20	6.22	31.43	-	3.00	41.28	8.20	80.91	m	89.00
4250mm	26.83	6.22	33.05	-	4.25	58.48	8.72	100.25	m	110.27
4500mm	28.37	6.22	34.59	-	4.50	61.92	9.23	105.73	m	116.31
4750mm	30.04	6.22	36.26	-	4.75	65.36	9.73	111.36	m	122.49
5000mm	31.53	6.22	37.75	-	6.87	94.53	11.28	143.56	m	157.92
5250mm	33.20	6.22	39.42	-	6.56	90.27	10.77	140.46	m	154.50
5500mm	34.74	6.22	40.96	-	6.87	94.53	11.28	146.77	m	161.45
5750mm	36.41	6.22	42.63	-	7.18	98.80	11.79	153.22	m	168.54
6000mm	38.37	6.22	44.60	-	7.50	103.20	12.30	160.10	m	176.11

Excavating trenches by machine to receive pipes 225mm nominal size; disposing of surplus excavated material by removing from site

Excavations commencing from natural ground level
  average depth

	Plant Cost	Trans Cost	P and T Cost	Craft Optve	Lab	Labour Cost	Sunds	Nett Rate	Unit	Gross rate
500mm	2.63	6.57	9.20	-	0.24	3.30	1.02	13.52	m	14.87
750mm	4.13	6.57	10.70	-	0.36	4.95	1.53	17.18	m	18.90
1000mm	5.63	6.57	12.20	-	0.47	6.47	2.06	20.72	m	22.79
1250mm	7.11	6.57	13.68	-	0.64	8.81	2.57	25.06	m	27.56
1500mm	8.68	6.57	15.25	-	0.77	10.60	3.08	28.92	m	31.82
1750mm	10.19	6.57	16.76	-	0.90	12.38	3.59	32.72	m	36.00
2000mm	11.67	6.57	18.24	-	1.03	14.17	4.09	36.51	m	40.16
2250mm	14.78	6.57	21.35	-	1.77	24.36	4.62	50.33	m	55.36
2500mm	16.54	6.57	23.11	-	1.97	27.11	5.13	55.34	m	60.88
2750mm	18.21	6.57	24.78	-	2.16	29.72	5.64	60.14	m	66.15
3000mm	19.88	6.57	26.45	-	2.36	32.47	6.15	65.07	m	71.58
3250mm	21.55	6.57	28.12	-	2.56	35.23	6.66	70.01	m	77.01
3500mm	23.31	6.57	29.88	-	2.76	37.98	7.17	75.02	m	82.53
3750mm	24.98	6.57	31.55	-	2.96	40.73	7.70	79.97	m	87.97
4000mm	26.51	6.57	33.08	-	3.15	43.34	8.20	84.62	m	93.09
4250mm	28.32	6.57	34.89	-	4.46	61.37	8.72	104.97	m	115.47
4500mm	30.12	6.57	36.69	-	4.70	64.67	9.23	110.59	m	121.65
4750mm	31.75	6.57	38.31	-	4.98	68.52	9.73	116.57	m	128.23
5000mm	33.42	6.57	39.99	-	6.56	90.27	11.28	141.53	m	155.68
5250mm	35.09	6.57	41.66	-	6.89	94.81	10.77	147.23	m	161.96
5500mm	36.89	6.57	43.46	-	7.20	99.07	11.28	153.81	m	169.19
5750mm	38.51	6.57	45.08	-	7.54	103.75	11.79	160.62	m	176.69
6000mm	40.18	6.57	46.75	-	7.85	108.02	12.30	167.07	m	183.78

Excavating trenches by machine to receive pipes 300mm nominal size; disposing of surplus excavated material by removing from site

Excavations commencing from natural ground level
  average depth

	Plant Cost	Trans Cost	P and T Cost	Craft Optve	Lab	Labour Cost	Sunds	Nett Rate	Unit	Gross rate
500mm	2.90	7.26	10.16	-	0.25	3.44	1.02	14.62	m	16.08
750mm	4.63	7.26	11.89	-	0.37	5.09	1.53	18.51	m	20.36
1000mm	6.31	7.26	13.57	-	0.50	6.88	2.06	22.51	m	24.76
1250mm	8.04	7.26	15.31	-	0.67	9.22	2.57	27.09	m	29.80
1500mm	9.78	7.26	17.04	-	0.80	11.01	3.08	31.12	m	34.23
1750mm	11.46	7.26	18.72	-	0.95	13.07	3.59	35.38	m	38.92

Labour hourly rates: (except Specialists) Craft Operatives 23.99 Labourer 13.76 Rates are national average prices. Refer to REGIONAL VARIATIONS for indicative levels of overall pricing in regions	PLANT AND TRANSPORT			LABOUR				RATES		
	Plant Cost £	Trans Cost £	P and T Cost £	Craft Optve Hrs	Lab Hrs	Labour Cost £	Sunds £	Nett Rate £	Unit	Gross rate (10%) £

STORM WATER AND FOUL DRAIN SYSTEMS; DRAIN RUNS (Cont'd)

Excavating trenches by machine to receive pipes 300mm nominal size; disposing of surplus excavated material by removing from site (Cont'd)

Excavations commencing from natural ground level (Cont'd) average depth (Cont'd)

	Plant Cost £	Trans Cost £	P and T Cost £	Craft Optve Hrs	Lab Hrs	Labour Cost £	Sunds £	Nett Rate £	Unit	Gross rate (10%) £
2000mm	13.19	7.26	20.45	-	1.10	15.14	4.09	39.68	m	43.65
2250mm	16.71	7.26	23.97	-	1.86	25.59	4.62	54.18	m	59.60
2500mm	18.64	7.26	25.90	-	2.07	28.48	5.13	59.52	m	65.47
2750mm	20.58	7.26	27.84	-	2.25	30.96	5.64	64.44	m	70.88
3000mm	22.47	7.26	29.73	-	2.48	34.12	6.15	70.00	m	77.00
3250mm	24.40	7.26	31.66	-	2.68	36.88	6.66	75.20	m	82.72
3500mm	26.29	7.26	33.55	-	2.90	39.90	7.17	80.62	m	88.69
3750mm	28.22	7.26	35.48	-	3.10	42.66	7.70	85.84	m	94.42
4000mm	30.16	7.26	37.42	-	3.30	45.41	8.20	91.03	m	100.14
4250mm	32.05	7.26	39.31	-	4.68	64.40	8.72	112.42	m	123.66
4500mm	33.98	7.26	41.24	-	4.95	68.11	9.23	118.58	m	130.44
4750mm	35.87	7.26	43.13	-	5.23	71.96	9.73	124.83	m	137.31
5000mm	37.81	7.26	45.07	-	6.88	94.67	11.28	151.02	m	166.12
5250mm	39.74	7.26	47.00	-	7.22	99.35	10.77	157.12	m	172.83
5500mm	41.63	7.26	48.89	-	7.55	103.89	11.28	164.06	m	180.46
5750mm	43.56	7.26	50.82	-	7.90	108.70	11.79	171.32	m	188.45
6000mm	45.50	7.26	52.76	-	8.25	113.52	12.30	178.58	m	196.44

Excavating trenches by machine to receive pipes 400mm nominal size; disposing of surplus excavated material by removing from site

Excavations commencing from natural ground level average depth

	Plant Cost £	Trans Cost £	P and T Cost £	Craft Optve Hrs	Lab Hrs	Labour Cost £	Sunds £	Nett Rate £	Unit	Gross rate (10%) £
2250mm	19.51	8.64	28.15	-	2.03	27.93	4.62	60.71	m	66.78
2500mm	21.66	8.64	30.31	-	2.26	31.10	5.13	66.53	m	73.19
2750mm	23.95	8.64	32.59	-	2.47	33.99	5.64	72.22	m	79.44
3000mm	26.15	8.64	34.79	-	2.70	37.15	6.15	78.09	m	85.90
3250mm	28.43	8.64	37.07	-	2.92	40.18	6.66	83.91	m	92.31
3500mm	30.63	8.64	39.27	-	3.16	43.48	7.17	89.93	m	98.92
3750mm	32.91	8.64	41.56	-	3.38	46.51	7.70	95.76	m	105.34
4000mm	35.11	8.64	43.76	-	3.60	49.54	8.20	101.50	m	111.65
4250mm	37.40	8.64	46.04	-	5.10	70.18	8.72	124.93	m	137.43
4500mm	39.55	8.64	48.20	-	5.40	74.30	9.23	131.72	m	144.90
4750mm	28.60	8.64	37.24	-	5.70	78.43	10.15	125.83	m	138.41
5000mm	44.03	8.64	52.68	-	7.50	103.20	11.28	167.16	m	183.88
5250mm	46.32	8.64	54.96	-	7.85	108.02	10.77	173.75	m	191.12
5500mm	48.52	8.64	57.16	-	8.25	113.52	11.28	181.96	m	200.16
5750mm	50.80	8.64	59.45	-	8.60	118.34	11.79	189.57	m	208.53
6000mm	35.97	8.64	44.61	-	9.00	123.84	12.84	181.29	m	199.42

Excavating trenches by machine to receive pipes 450mm nominal size; disposing of surplus excavated material by removing from site

Excavations commencing from natural ground level average depth

	Plant Cost £	Trans Cost £	P and T Cost £	Craft Optve Hrs	Lab Hrs	Labour Cost £	Sunds £	Nett Rate £	Unit	Gross rate (10%) £
2250mm	20.87	8.99	29.86	-	2.11	29.03	4.62	63.51	m	69.86
2500mm	23.29	8.99	32.28	-	2.35	32.34	5.13	69.74	m	76.72
2750mm	25.57	8.99	34.56	-	2.58	35.50	5.64	75.70	m	83.27
3000mm	28.03	8.99	37.02	-	2.81	38.67	6.15	81.84	m	90.02
3250mm	28.84	8.99	37.83	-	3.05	41.97	6.66	86.46	m	95.10
3500mm	32.87	8.99	41.86	-	3.29	45.27	7.17	94.30	m	103.73
3750mm	35.15	8.99	44.14	-	3.52	48.44	7.70	100.27	m	110.30
4000mm	37.62	8.99	46.61	-	3.75	51.60	8.20	106.41	m	117.05
4250mm	40.03	8.99	49.02	-	5.30	72.93	8.72	130.67	m	143.73
4500mm	42.45	8.99	51.44	-	5.63	77.47	9.23	138.13	m	151.95
4750mm	44.73	8.99	53.72	-	5.95	81.87	9.73	145.33	m	159.86
5000mm	47.20	8.99	56.19	-	7.80	107.33	11.28	174.80	m	192.27
5250mm	49.61	8.99	58.60	-	8.20	112.83	10.77	182.21	m	200.43
5500mm	52.03	8.99	61.02	-	8.60	118.34	11.28	190.64	m	209.70
5750mm	54.31	8.99	63.30	-	8.95	123.15	11.79	198.25	m	218.07
6000mm	56.78	8.99	65.77	-	9.35	128.66	12.30	206.72	m	227.40

Excavating trenches by hand to receive pipes not exceeding 200mm nominal size; disposing of surplus excavated material on site

Excavations commencing from natural ground level average depth

	Plant Cost £	Trans Cost £	P and T Cost £	Craft Optve Hrs	Lab Hrs	Labour Cost £	Sunds £	Nett Rate £	Unit	Gross rate (10%) £
500mm	-	-	-	-	1.15	15.82	1.02	16.84	m	18.53
750mm	-	-	-	-	2.05	28.21	1.53	29.74	m	32.71
1000mm	-	-	-	-	2.95	40.59	2.06	42.65	m	46.91
1250mm	-	-	-	-	3.85	52.98	2.57	55.54	m	61.10
1500mm	-	-	-	-	4.75	65.36	3.08	68.43	m	75.28
1750mm	-	-	-	-	5.65	77.74	3.59	81.33	m	89.46
2000mm	-	-	-	-	6.50	89.44	4.09	93.54	m	102.89
2250mm	-	-	-	-	7.65	105.26	4.62	109.88	m	120.87
2500mm	-	-	-	-	8.80	121.09	5.13	126.22	m	138.84
2750mm	-	-	-	-	9.95	136.91	5.64	142.55	m	156.81
3000mm	-	-	-	-	11.10	152.74	6.15	158.89	m	174.77
3250mm	-	-	-	-	12.25	168.56	6.66	175.22	m	192.74
3500mm	-	-	-	-	13.45	185.07	7.17	192.24	m	211.47
3750mm	-	-	-	-	14.60	200.90	7.70	208.59	m	229.45
4000mm	-	-	-	-	15.80	217.41	8.20	225.61	m	248.17

DRAINAGE BELOW GROUND

Labour hourly rates: (except Specialists) Craft Operatives 23.99 Labourer 13.76 Rates are national average prices. Refer to REGIONAL VARIATIONS for indicative levels of overall pricing in regions	PLANT AND TRANSPORT			LABOUR				RATES		
	Plant Cost	Trans Cost	P and T Cost	Craft Optve	Lab	Labour Cost	Sunds	Nett Rate	Unit	Gross rate (10%)
	£	£	£	Hrs	Hrs	£	£	£		£

STORM WATER AND FOUL DRAIN SYSTEMS; DRAIN RUNS (Cont'd)

Excavating trenches by hand to receive pipes 225mm nominal size; disposing of surplus excavated material on site

Excavations commencing from natural ground level
average depth

500mm	-	-	-	-	1.28	17.61	1.02	18.63	m	20.50
750mm	-	-	-	-	2.23	30.68	1.53	32.21	m	35.44
1000mm	-	-	-	-	3.18	43.76	2.06	45.81	m	50.39
1250mm	-	-	-	-	4.13	56.83	2.57	59.39	m	65.33
1500mm	-	-	-	-	5.08	69.90	3.08	72.98	m	80.27
1750mm	-	-	-	-	6.03	82.97	3.59	86.56	m	95.21
2000mm	-	-	-	-	6.95	95.63	4.09	99.73	m	109.70
2250mm	-	-	-	-	8.23	113.24	4.62	117.86	m	129.65
2500mm	-	-	-	-	9.45	130.03	5.13	135.16	m	148.68
2750mm	-	-	-	-	10.70	147.23	5.64	152.87	m	168.16
3000mm	-	-	-	-	11.95	164.43	6.15	170.58	m	187.64
3250mm	-	-	-	-	13.20	181.63	6.66	188.29	m	207.12
3500mm	-	-	-	-	14.45	198.83	7.17	206.00	m	226.60
3750mm	-	-	-	-	15.70	216.03	7.70	223.73	m	246.10
4000mm	-	-	-	-	17.00	233.92	8.20	242.12	m	266.34

Excavating trenches by hand to receive pipes 300mm nominal size; disposing of surplus material on site

Excavations commencing from natural ground level
average depth

500mm	-	-	-	-	1.40	19.26	1.02	20.28	m	22.31
750mm	-	-	-	-	2.40	33.02	1.53	34.55	m	38.01
1000mm	-	-	-	-	3.40	46.78	2.06	48.84	m	53.72
1250mm	-	-	-	-	4.40	60.54	2.57	63.11	m	69.42
1500mm	-	-	-	-	5.40	74.30	3.08	77.38	m	85.12
1750mm	-	-	-	-	6.40	88.06	3.59	91.65	m	100.81
2000mm	-	-	-	-	7.40	101.82	4.09	105.92	m	116.51
2250mm	-	-	-	-	8.80	121.09	4.62	125.71	m	138.28
2500mm	-	-	-	-	10.10	138.98	5.13	144.11	m	158.52
2750mm	-	-	-	-	11.45	157.55	5.64	163.19	m	179.51
3000mm	-	-	-	-	12.75	175.44	6.15	181.59	m	199.75
3250mm	-	-	-	-	14.10	194.02	6.66	200.68	m	220.74
3500mm	-	-	-	-	15.45	212.59	7.17	219.76	m	241.74
3750mm	-	-	-	-	16.80	231.17	7.70	238.86	m	262.75
4000mm	-	-	-	-	18.15	249.74	8.20	257.95	m	283.74

This section continues
on the next page

DRAINAGE BELOW GROUND *(vertical side text)*

	MATERIALS			LABOUR				RATES		
Labour hourly rates: (except Specialists) Craft Operatives 18.37 Labourer 13.76 Rates are national average prices. Refer to REGIONAL VARIATIONS for indicative levels of overall pricing in regions	Del to Site	Waste	Material Cost	Craft Optve	Lab	Labour Cost	Sunds	Nett Rate	Unit	Gross rate (10%)
	£	%	£	Hrs	Hrs	£	£	£		£

STORM WATER AND FOUL DRAIN SYSTEMS; DRAIN RUNS (Cont'd)

Beds, benchings and covers; Granular material, 10mm nominal size pea shingle, to be obtained off site

Beds

400 x 150mm	2.78	10.00	3.06	-	0.15	2.06	-	5.12	m	5.63
450 x 150mm	3.10	10.00	3.41	-	0.16	2.20	-	5.62	m	6.18
525 x 150mm	3.61	10.00	3.98	-	0.17	2.34	-	6.31	m	6.95
600 x 150mm	4.17	10.00	4.59	-	0.22	3.03	-	7.61	m	8.38
700 x 150mm	4.86	10.00	5.35	-	0.24	3.30	-	8.65	m	9.52
750 x 150mm	5.24	10.00	5.76	-	0.27	3.72	-	9.47	m	10.42

Beds and surrounds

400 x 150mm bed; 150mm thick surround to 100mm internal diameter pipes -1	7.41	10.00	8.15	-	0.46	6.33	-	14.48	m	15.93
450 x 150mm bed; 150mm thick surround to 150mm internal diameter pipes -1	9.36	10.00	10.29	-	0.52	7.16	-	17.45	m	19.20
525 x 150mm bed; 150mm thick surround to 225mm internal diameter pipes -1	12.74	10.00	14.02	-	0.68	9.36	-	23.37	m	25.71
600 x 150mm bed; 150mm thick surround to 300mm internal diameter pipes -1	16.68	10.00	18.35	-	0.84	11.56	-	29.91	m	32.90
700 x 150mm bed; 150mm thick surround to 400mm internal diameter pipes -1	22.70	10.00	24.97	-	1.00	13.76	-	38.73	m	42.61
750 x 150mm bed; 150mm thick surround to 450mm internal diameter pipes -1	25.95	10.00	28.54	-	1.16	15.96	-	44.50	m	48.95

Beds and filling to half pipe depth

400 x 150mm overall; to 100mm internal diameter pipes - 1	6.49	10.00	7.14	-	0.12	1.65	-	8.79	m	9.66
450 x 225mm overall; to 150mm internal diameter pipes - 1	7.78	10.00	8.56	-	0.20	2.75	-	11.31	m	12.45
525 x 263mm overall; to 225mm internal diameter pipes - 1	11.86	10.00	13.05	-	0.28	3.85	-	16.90	m	18.59
600 x 300mm overall; to 300mm internal diameter pipes - 1	16.68	10.00	18.35	-	0.36	4.95	-	23.30	m	25.63
700 x 350mm overall; to 400mm internal diameter pipes -1	24.32	10.00	26.76	-	0.49	6.74	-	33.50	m	36.85
750 x 375mm overall; to 450mm internal diameter pipes -1	28.63	10.00	31.50	-	0.56	7.71	-	39.20	m	43.12

Beds, benchings and covers; sand; to be obtained off site

Beds

400 x 50mm	0.46	5.00	0.48	-	0.04	0.55	-	1.03	m	1.13
400 x 100mm	0.91	5.00	0.96	-	0.08	1.10	-	2.06	m	2.26
450 x 50mm	0.50	5.00	0.53	-	0.05	0.69	-	1.21	m	1.34
450 x 100mm	1.03	5.00	1.08	-	0.09	1.24	-	2.32	m	2.55
525 x 50mm	0.59	5.00	0.62	-	0.06	0.83	-	1.45	m	1.59
525 x 100mm	1.19	5.00	1.24	-	0.11	1.51	-	2.76	m	3.03
600 x 50mm	0.68	5.00	0.72	-	0.06	0.83	-	1.54	m	1.70
600 x 100mm	1.37	5.00	1.44	-	0.12	1.65	-	3.09	m	3.40
700 x 50mm	0.80	5.00	0.84	-	0.07	0.96	-	1.80	m	1.98
700 x 100mm	1.60	5.00	1.68	-	0.14	1.93	-	3.60	m	3.96
750 x 50mm	0.84	5.00	0.89	-	0.08	1.10	-	1.99	m	2.19
750 x 100mm	1.71	5.00	1.80	-	0.15	2.06	-	3.86	m	4.25

Beds, benchings and covers; plain in-situ concrete; BS 8500, ordinary prescribed mix, ST3, 20mm aggregate

Beds

400 x 100mm	3.92	5.00	4.12	-	0.18	2.48	2.94	9.53	m	10.49
400 x 150mm	5.88	5.00	6.17	-	0.27	3.72	4.41	14.30	m	15.73
450 x 100mm	4.41	5.00	4.63	-	0.20	2.75	2.94	10.32	m	11.35
450 x 150mm	6.57	5.00	6.89	-	0.30	4.13	4.41	15.43	m	16.97
525 x 100mm	5.10	5.00	5.35	-	0.24	3.30	2.94	11.59	m	12.75
525 x 150mm	7.64	5.00	8.03	-	0.35	4.82	4.41	17.25	m	18.98
600 x 100mm	5.88	5.00	6.17	-	0.27	3.72	2.94	12.83	m	14.11
600 x 150mm	8.82	5.00	9.26	-	0.38	5.23	4.41	18.90	m	20.79
700 x 100mm	6.86	5.00	7.20	-	0.32	4.40	2.94	14.55	m	16.00
700 x 150mm	10.29	5.00	10.80	-	0.47	6.47	4.41	21.68	m	23.85
750 x 100mm	7.35	5.00	7.72	-	0.34	4.68	2.94	15.34	m	16.87
750 x 150mm	10.97	5.00	11.52	-	0.51	7.02	4.41	22.95	m	25.25

Beds and surrounds

400 x 100mm bed; 150mm thick surround to 100mm internal diameter pipes -1	13.72	5.00	14.40	-	0.72	9.91	5.76	30.07	m	33.08
400 x 150mm bed; 150mm thick surround to 100mm internal diameter pipes -1	15.68	5.00	16.46	-	0.81	11.15	6.47	34.07	m	37.48
450 x 100mm bed; 150mm thick surround to 150mm internal diameter pipes -1	17.64	5.00	18.52	-	0.91	12.52	6.47	37.51	m	41.26
450 x 150mm bed; 150mm thick surround to 150mm internal diameter pipes -1	19.79	5.00	20.78	-	1.01	13.90	7.18	41.87	m	46.05
525 x 100mm bed; 150mm thick surround to 225mm internal diameter pipes -1	24.40	5.00	25.62	-	1.24	17.06	7.18	49.87	m	54.85
525 x 150mm bed; 150mm thick surround to 225mm internal diameter pipes -1	26.95	5.00	28.29	-	1.36	18.71	8.27	55.27	m	60.80
600 x 100mm bed; 150mm thick surround to 300mm internal diameter pipes -1	32.34	5.00	33.95	-	1.62	22.29	8.64	64.88	m	71.37
600 x 150mm bed; 150mm thick surround to 300mm internal diameter pipes -1	35.28	5.00	37.04	-	1.76	24.22	9.35	70.60	m	77.66
700 x 100mm bed; 150mm thick surround to 400mm internal diameter pipes -1	44.59	5.00	46.81	-	2.21	30.41	10.06	87.29	m	96.02
700 x 150mm bed; 150mm thick surround to 400mm internal diameter pipes -1	48.02	5.00	50.42	-	2.36	32.47	10.80	93.69	m	103.06
750 x 100mm bed; 150mm thick surround to 450mm internal diameter pipes -1	51.44	5.00	54.02	-	2.53	34.81	10.80	99.63	m	109.59
750 x 150mm bed; 150mm thick surround to 450mm internal diameter pipes -1	55.07	5.00	57.82	-	2.70	37.15	11.51	106.48	m	117.13

Labour hourly rates: (except Specialists) Craft Operatives 18.37 Labourer 13.76 Rates are national average prices. Refer to REGIONAL VARIATIONS for indicative levels of overall pricing in regions	MATERIALS			LABOUR				RATES		
	Del to Site	Waste	Material Cost	Craft Optve	Lab	Labour Cost	Sunds	Nett Rate	Unit	Gross rate (10%)
	£	%	£	Hrs	Hrs	£	£	£		£
STORM WATER AND FOUL DRAIN SYSTEMS; DRAIN RUNS (Cont'd)										
Beds, benchings and covers; plain in-situ concrete; BS 8500, ordinary prescribed mix, ST3, 20mm aggregate (Cont'd)										
Beds and haunchings										
400 x 150mm; to 100mm internal diameter pipes - 1	7.55	5.00	7.92	-	0.40	5.50	4.32	17.75	m	19.52
450 x 150mm; to 150mm internal diameter pipes - 1	9.31	5.00	9.77	-	0.46	6.33	4.32	20.42	m	22.47
525 x 150mm; to 225mm internal diameter pipes - 1	12.05	5.00	12.66	-	0.62	8.53	4.32	25.51	m	28.06
600 x 150mm; to 300mm internal diameter pipes - 1	15.09	5.00	15.84	-	0.81	11.15	4.32	31.31	m	34.44
700 x 150mm; to 400mm internal diameter pipes - 1	19.50	5.00	20.48	-	1.10	15.14	4.32	39.93	m	43.92
750 x 150mm; to 450mm internal diameter pipes - 1	21.85	5.00	22.94	-	1.35	18.58	4.32	45.84	m	50.42
Vertical casings										
400 x 400mm to 100mm internal diameter pipes - 1	33.74	5.00	34.53	-	0.80	11.01	-	45.53	m	50.09
450 x 450mm to 150mm internal diameter pipes - 1	40.12	5.00	41.11	-	1.01	13.90	-	55.00	m	60.50
525 x 525mm to 225mm internal diameter pipes - 1	50.88	5.00	52.23	-	1.38	18.99	-	71.22	m	78.34
600 x 600mm to 300mm internal diameter pipes - 1	62.37	5.00	64.14	-	1.80	24.77	-	88.90	m	97.79
700 x 700mm to 400mm internal diameter pipes - 1	79.63	5.00	82.03	-	2.45	33.71	-	115.74	m	127.31
750 x 750mm to 450mm internal diameter pipes - 1	88.94	5.00	91.69	-	2.81	38.67	-	130.36	m	143.40
Drains; vitrified clay pipes and fittings, normal; flexible mechanical joints										
Pipework in trenches										
375mm	135.14	5.00	141.90	0.55	0.18	12.58	-	154.48	m	169.93
extra; bends	578.56	5.00	607.49	0.39	-	7.16	-	614.65	nr	676.12
450mm	265.39	5.00	278.66	0.70	0.20	15.61	-	294.27	nr	323.70
extra; bends	986.55	5.00	1035.88	0.70	-	12.86	-	1048.74	nr	1153.61
Pipework in trenches; vertical										
375mm	135.14	5.00	141.90	0.60	0.18	13.50	-	155.40	m	170.94
450mm	265.39	5.00	278.66	0.77	0.20	16.90	-	295.56	m	325.11
Drains; vitrified clay pipes and fittings; sleeve joints, push-fit polypropylene standard ring flexible couplings; Hepworth Supersleeve										
Pipework in trenches										
100mm	10.08	5.00	10.59	0.16	0.06	3.76	-	14.35	m	15.79
extra; bends	9.31	5.00	9.78	0.08	-	1.47	-	11.25	nr	12.37
extra; branches	20.10	5.00	21.10	0.16	-	2.94	-	24.04	nr	26.45
150mm	24.75	5.00	25.99	0.18	0.08	4.41	-	30.39	m	33.43
extra; bends	24.36	5.00	25.58	0.09	-	1.65	-	27.23	nr	29.95
extra; branches	36.28	5.00	38.09	0.18	-	3.31	-	41.40	nr	45.54
Pipework in trenches; vertical										
100mm	10.08	5.00	10.59	0.18	0.06	4.13	-	14.72	m	16.19
150mm	24.75	5.00	25.99	0.20	0.08	4.77	-	30.76	m	33.84
Drains; vitrified clay pipes and fittings; sleeve joints, push-fit polypropylene neoprene ring flexible couplings; Hepworth Supersleeve										
Pipework in trenches										
100mm	12.52	5.00	13.14	0.16	0.06	3.76	-	16.91	m	18.60
150mm	27.84	5.00	29.23	0.18	0.08	4.41	-	33.64	m	37.00
Pipework in trenches; vertical										
100mm	12.52	5.00	13.14	0.18	0.06	4.13	-	17.28	m	19.00
150mm	27.84	5.00	29.23	0.20	0.08	4.77	-	34.00	m	37.40
Drains; vitrified clay pipes and fittings, BS EN 295, normal; cement mortar (1:3)										
Pipework in trenches										
100mm	9.17	5.00	9.64	0.33	0.03	6.47	-	16.12	m	17.73
extra; double collars	12.30	5.00	12.92	0.10	-	1.84	-	14.76	nr	16.24
extra; bends	8.75	5.00	9.20	0.20	-	3.67	-	12.87	nr	14.16
extra; rest bends	15.44	5.00	16.22	0.20	-	3.67	-	19.89	nr	21.88
extra; branches	17.46	5.00	18.35	0.22	-	4.04	-	22.39	nr	24.63
150mm	12.75	5.00	13.40	0.44	0.04	8.63	-	22.03	m	24.23
extra; double collars	20.56	5.00	21.60	0.15	-	2.76	-	24.36	nr	26.79
extra; bends	14.77	5.00	15.52	0.27	-	4.96	-	20.48	nr	22.53
extra; rest bends	25.65	5.00	26.95	0.27	-	4.96	-	31.91	nr	35.10
extra; branches	29.84	5.00	31.35	0.28	-	5.14	-	36.49	nr	40.14
225mm	69.65	5.00	73.16	0.55	0.07	11.07	-	84.22	m	92.65
extra; double collars	21.21	5.00	22.30	0.23	-	4.23	-	26.52	nr	29.17
300mm	102.38	5.00	107.52	0.80	0.10	16.07	-	123.59	m	135.95
Pipework in trenches; vertical										
100mm	9.17	5.00	9.64	0.50	0.03	9.60	-	19.24	m	21.16
150mm	12.75	5.00	13.40	0.48	0.04	9.37	-	22.76	m	25.04
225mm	69.65	5.00	73.16	0.60	0.07	11.99	-	85.14	m	93.66
300mm	102.38	5.00	107.52	0.88	0.10	17.54	-	125.06	m	137.57
Drains; Wavin; PVC-U pipes and fittings, BS EN 13598-1; ring seal joints										
Pipework in trenches										
110mm, ref. 4D.076	3.92	5.00	4.12	0.14	0.02	2.85	-	6.96	m	7.66
extra; connections to cast iron and clay sockets, ref. 4D.107	12.37	5.00	12.99	0.08	-	1.47	-	14.46	nr	15.90
extra; 45 degree bends, ref. 4D.163	5.95	5.00	6.25	0.14	0.02	2.85	-	9.09	nr	10.00
extra; branches, ref. 4D.210	10.55	5.00	11.08	0.14	0.02	2.85	-	13.92	nr	15.32
extra; connections cast iron and clay spigot, ref. 4D.128	12.37	5.00	12.99	0.08	-	1.47	-	14.46	nr	15.90
160mm, ref. 6D.076	8.46	5.00	8.88	0.20	0.03	4.09	-	12.97	m	14.26
extra; connections to cast iron and clay sockets, ref. 6D.107	24.84	5.00	26.08	0.10	-	1.84	-	27.92	nr	30.71

Labour hourly rates: (except Specialists) Craft Operatives 18.37 Labourer 13.76 Rates are national average prices. Refer to REGIONAL VARIATIONS for indicative levels of overall pricing in regions	MATERIALS			LABOUR				RATES		
	Del to Site £	Waste %	Material Cost £	Craft Optve Hrs	Lab Hrs	Labour Cost £	Sunds £	Nett Rate £	Unit	Gross rate (10%) £

STORM WATER AND FOUL DRAIN SYSTEMS; DRAIN RUNS (Cont'd)

Drains; Wavin; PVC-U pipes and fittings, BS EN 13598-1; ring seal joints (Cont'd)

	Del to Site £	Waste %	Material Cost £	Craft Optve Hrs	Lab Hrs	Labour Cost £	Sunds £	Nett Rate £	Unit	Gross rate (10%) £
Pipework in trenches (Cont'd)										
extra; 45 degree bends, ref. 6D.163	23.97	5.00	25.17	0.20	0.03	4.09	-	29.26	nr	32.18
extra; branches, ref. 6D.210	48.39	5.00	50.81	0.20	0.03	4.09	-	54.90	nr	60.39
extra; connections to cast iron and clay spigot, ref. 6D.128	26.00	5.00	27.30	0.10	-	1.84	-	29.14	nr	32.05
Pipework in trenches; vertical										
110mm, ref. 4D.076	3.92	5.00	4.12	0.28	0.02	5.42	-	9.54	m	10.49
160mm, ref. 6D.076	8.46	5.00	8.88	0.24	0.03	4.82	-	13.70	m	15.07

Drains; PVC-U solid wall concentric external rib-reinforced Wavin "Ultra-Rib" pipes and fittings with sealing rings to joints

	Del to Site £	Waste %	Material Cost £	Craft Optve Hrs	Lab Hrs	Labour Cost £	Sunds £	Nett Rate £	Unit	Gross rate (10%) £
Pipework in trenches										
150mm 6UR 046	26.38	5.00	27.70	0.20	0.02	3.95	-	31.65	m	34.81
extra; bends 6UR 563	35.77	5.00	37.55	0.20	-	3.67	-	41.23	nr	45.35
extra; branches 6UR 213	86.48	5.00	90.81	0.20	-	3.67	-	94.48	nr	103.93
extra; connections to clayware pipe ends 6UR 129	70.56	5.00	74.09	0.10	-	1.84	-	75.93	nr	83.52
225mm 9UR 046	56.80	5.00	59.64	0.30	0.03	5.92	-	65.57	m	72.12
extra; bends 9UR 563	150.44	5.00	157.96	0.30	-	5.51	-	163.47	nr	179.82
extra; branches 9UR 213	215.89	5.00	226.69	0.30	-	5.51	-	232.20	nr	255.42
extra; connections to clayware pipe ends 9UR 109	181.89	5.00	190.98	0.12	-	2.20	-	193.19	nr	212.51
300mm 12UR 043	92.04	5.00	96.64	0.38	0.05	7.67	-	104.31	m	114.74
extra; bends 12UR 563	273.20	5.00	286.86	0.38	-	6.98	-	293.84	nr	323.23
extra; branches 12UR 213	693.14	5.00	727.80	0.38	-	6.98	-	734.78	nr	808.26
extra; connections to clayware pipe ends 12UR 112	478.44	5.00	502.36	0.15	-	2.76	-	505.11	nr	555.63
Pipework in trenches; vertical										
150mm	26.38	5.00	27.70	0.40	0.02	7.62	-	35.32	m	38.86
225mm	56.80	5.00	59.64	0.38	0.03	7.39	-	67.04	m	73.74
300mm	92.04	5.00	96.64	0.48	0.05	9.51	-	106.15	m	116.76

Drains; unreinforced concrete pipes and fittings BS 5911, Class H, flexible joints

	Del to Site £	Waste %	Material Cost £	Craft Optve Hrs	Lab Hrs	Labour Cost £	Sunds £	Nett Rate £	Unit	Gross rate (10%) £
Pipework in trenches										
300mm	20.94	5.00	21.99	0.59	0.24	14.14	-	47.52	m	52.27
extra; bends	206.01	5.00	216.32	0.30	0.12	7.16	-	229.17	nr	252.09
extra; 100mm branches	121.75	5.00	127.83	0.24	0.06	5.23	-	135.91	nr	149.51
extra; 150mm branches	121.70	5.00	127.79	0.40	0.11	8.86	-	141.87	nr	156.06
extra; 225mm branches	246.32	5.00	258.63	0.56	0.16	12.49	-	278.71	nr	306.59
extra; 300mm branches	282.15	5.00	296.25	0.72	0.22	16.25	-	322.95	nr	355.24
375mm	25.97	5.00	27.27	0.72	0.30	17.35	-	58.86	m	64.74
extra; bends	272.61	5.00	286.24	0.36	0.15	8.68	-	302.04	nr	332.24
extra; 100mm branches	136.86	5.00	143.70	0.24	0.06	5.23	-	151.78	nr	166.96
extra; 150mm branches	136.86	5.00	143.70	0.40	0.11	8.86	-	157.78	nr	173.56
extra; 225mm branches	272.61	5.00	286.24	0.56	0.16	12.49	-	306.33	nr	336.96
extra; 300mm branches	310.22	5.00	325.73	0.72	0.22	16.25	-	352.42	nr	387.68
extra; 375mm branches	347.81	5.00	365.20	0.88	0.27	19.88	-	397.89	nr	437.68
450mm	32.72	5.00	34.36	0.85	0.37	20.71	-	72.62	m	79.88
extra; bends	314.91	5.00	330.66	0.43	0.18	10.38	-	349.57	nr	384.53
extra; 100mm branches	155.11	5.00	162.86	0.24	0.06	5.23	-	170.95	nr	188.04
extra; 150mm branches	155.11	5.00	162.86	0.40	0.11	8.86	-	176.95	nr	194.64
extra; 225mm branches	291.41	5.00	305.98	0.56	0.16	12.49	-	326.06	nr	358.67
extra; 300mm branches	329.01	5.00	345.46	0.72	0.22	16.25	-	372.16	nr	409.37
extra; 375mm branches	366.62	5.00	384.95	0.88	0.27	19.88	-	417.64	nr	459.41
extra; 450mm branches	432.42	5.00	454.04	1.04	0.33	23.65	-	493.34	nr	542.68
525mm	40.17	5.00	42.18	0.98	0.43	23.92	-	86.50	m	95.15
extra; bends	398.68	5.00	418.61	0.49	0.22	12.03	-	441.08	nr	485.19
extra; 100mm branches	171.98	5.00	180.58	0.24	0.06	5.23	-	188.66	nr	207.53
extra; 150mm branches	171.98	5.00	180.58	0.40	0.11	8.86	-	194.66	nr	214.13
extra; 225mm branches	315.00	5.00	330.75	0.56	0.16	12.49	-	350.83	nr	385.91
extra; 300mm branches	354.38	5.00	372.09	0.72	0.22	16.25	-	398.79	nr	438.67
extra; 375mm branches	393.75	5.00	413.44	0.88	0.27	19.88	-	446.13	nr	490.75
extra; 450mm branches	462.65	5.00	485.79	1.04	0.33	23.65	-	525.09	nr	577.60
600mm	50.45	5.00	52.97	1.10	0.50	27.09	-	103.78	m	114.16
extra; bends	465.32	5.00	488.58	0.55	0.25	13.54	-	513.99	nr	565.39
extra; 100mm branches	173.36	5.00	182.03	0.24	0.06	5.23	-	190.11	nr	209.12
extra; 150mm branches	173.36	5.00	182.03	0.40	0.11	8.86	-	196.11	nr	215.72
extra; 225mm branches	315.86	5.00	331.65	0.56	0.16	12.49	-	351.73	nr	386.90
extra; 300mm branches	353.46	5.00	371.13	0.72	0.22	16.25	-	397.83	nr	437.61
extra; 375mm branches	391.05	5.00	410.61	0.88	0.27	19.88	-	443.30	nr	487.63
extra; 450mm branches	456.86	5.00	479.71	1.04	0.33	23.65	-	519.01	nr	570.91

Drains; reinforced concrete pipes and fittings, BS 5911, Class H, flexible joints

	Del to Site £	Waste %	Material Cost £	Craft Optve Hrs	Lab Hrs	Labour Cost £	Sunds £	Nett Rate £	Unit	Gross rate (10%) £
Pipework in trenches										
450mm	60.29	5.00	63.30	0.85	0.37	20.71	-	101.57	m	111.72
extra; bends	602.97	5.00	633.12	0.43	0.18	10.38	-	652.04	nr	717.24
extra; 100mm branches	168.27	5.00	176.68	0.24	0.06	5.23	-	184.76	nr	203.24
extra; 150mm branches	172.08	5.00	180.68	0.40	0.11	8.86	-	194.76	nr	214.24
extra; 225mm branches	177.86	5.00	186.75	0.56	0.16	12.49	-	206.83	nr	227.52
extra; 300mm branches	216.35	5.00	227.17	0.72	0.22	16.25	-	253.87	nr	279.25
extra; 375mm branches	245.23	5.00	257.50	0.88	0.27	19.88	-	290.19	nr	319.21
extra; 450mm branches	283.74	5.00	297.93	1.04	0.33	23.65	-	337.23	nr	370.95
525mm	73.70	5.00	77.38	0.98	0.43	23.92	-	121.71	m	133.88
extra; bends	736.97	5.00	773.82	0.49	0.22	12.03	-	796.29	nr	875.92
extra; 100mm branches	189.58	5.00	199.06	0.24	0.06	5.23	-	207.14	nr	227.86
extra; 150mm branches	193.44	5.00	203.11	0.40	0.11	8.86	-	217.19	nr	238.91

Labour hourly rates: (except Specialists) Craft Operatives 18.37 Labourer 13.76 Rates are national average prices. Refer to REGIONAL VARIATIONS for indicative levels of overall pricing in regions	MATERIALS			LABOUR				RATES		
	Del to Site	Waste	Material Cost	Craft Optve	Lab	Labour Cost	Sunds	Nett Rate	Unit	Gross rate (10%)
	£	%	£	Hrs	Hrs	£	£	£		£
STORM WATER AND FOUL DRAIN SYSTEMS; DRAIN RUNS (Cont'd)										
Drains; reinforced concrete pipes and fittings, BS 5911, Class H, flexible joints (Cont'd)										
Pipework in trenches (Cont'd)										
extra; 225mm branches........................	199.21	5.00	209.17	0.56	0.16	12.49	-	229.25	nr	252.18
extra; 300mm branches........................	228.09	5.00	239.49	0.72	0.22	16.25	-	266.19	nr	292.80
extra; 375mm branches........................	266.59	5.00	279.92	0.88	0.27	19.88	-	312.62	nr	343.88
extra; 450mm branches........................	305.10	5.00	320.36	1.04	0.33	23.65	-	359.66	nr	395.63
600mm	80.39	5.00	84.41	1.10	0.50	27.09	-	135.22	m	148.74
extra; bends..	803.96	5.00	844.15	0.55	0.25	13.54	-	869.56	nr	956.52
extra; 100mm branches........................	171.87	5.00	180.47	0.24	0.06	5.23	-	188.55	nr	207.40
extra; 150mm branches........................	214.24	5.00	224.95	0.40	0.11	8.86	-	239.03	nr	262.93
extra; 225mm branches........................	220.01	5.00	231.01	0.56	0.16	12.49	-	251.09	nr	276.20
extra; 300mm branches........................	248.88	5.00	261.33	0.72	0.22	16.25	-	288.02	nr	316.82
extra; 375mm branches........................	287.39	5.00	301.76	0.88	0.27	19.88	-	334.45	nr	367.90
extra; 450mm branches........................	325.90	5.00	342.19	1.04	0.33	23.65	-	381.49	nr	419.64

This section continues
on the next page

DRAINAGE BELOW GROUND

Labour hourly rates: (except Specialists) Craft Operatives 23.99 Labourer 13.76 Rates are national average prices. Refer to REGIONAL VARIATIONS for indicative levels of overall pricing in regions	MATERIALS			LABOUR				RATES		
	Del to Site £	Waste %	Material Cost £	Craft Optve Hrs	Lab Hrs	Labour Cost £	Sunds £	Nett Rate £	Unit	Gross rate (10%) £
STORM WATER AND FOUL DRAIN SYSTEMS; DRAIN RUNS (Cont'd)										
Drains; cast iron pipes and cast iron fittings BS 437, coated; bolted cast iron coupling and synthetic rubber gasket joints										
Pipework in trenches										
100mm	41.00	5.00	42.88	1.00	-	23.99	-	66.87	m	73.56
extra; connectors, large socket for clayware	34.95	2.00	35.65	1.00	-	23.99	-	59.64	nr	65.60
extra; bends	59.43	2.00	60.62	1.00	-	23.99	-	84.61	nr	93.07
extra; branches	91.88	2.00	93.72	1.50	-	35.99	-	129.70	nr	142.67
extra; branches with access	127.75	2.00	130.31	1.50	-	35.99	-	166.29	nr	182.92
150mm	81.99	5.00	85.73	1.50	-	35.99	-	121.72	m	133.89
extra; connectors, large socket for clayware	61.16	2.00	62.38	1.50	-	35.99	-	98.37	nr	108.21
extra; diminishing pieces, reducing to 100mm	111.08	2.00	113.30	1.50	-	35.99	-	149.29	nr	164.22
extra; bends	115.43	2.00	117.74	1.50	-	35.99	-	153.72	nr	169.10
extra; branches	201.24	2.00	205.26	2.25	-	53.98	-	259.24	nr	285.17
extra; branches with access	268.03	2.00	273.39	2.25	-	53.98	-	327.37	nr	360.10
in runs not exceeding 3m										
100mm	52.92	5.00	55.03	1.50	-	35.99	-	91.01	m	100.12
150mm	106.32	5.00	110.55	2.25	-	53.98	-	164.52	m	180.98
Pipework in trenches; vertical										
100mm	52.92	5.00	55.03	1.20	-	28.79	-	83.82	m	92.20
150mm	106.32	5.00	110.55	1.80	-	43.18	-	153.73	m	169.10
Drains; stainless steel pipes and fittings; ring seal push fit joints (AISI 316)										
Pipework in trenches										
75mm	26.79	5.00	28.13	0.60	-	14.39	-	42.52	m	46.78
extra; bends	21.09	2.00	21.51	0.60	-	14.39	-	35.90	nr	39.49
extra; branches	22.48	2.00	22.93	0.90	-	21.59	-	44.52	nr	48.97
110mm	37.50	5.00	39.38	0.75	-	17.99	-	57.37	m	63.11
extra; diminishing pieces, reducing to 75mm	19.82	2.00	20.21	0.75	-	17.99	-	38.21	nr	42.03
extra; bends	29.58	2.00	30.17	0.75	-	17.99	-	48.17	nr	52.98
extra; branches	31.40	2.00	32.03	1.12	-	26.87	-	58.90	nr	64.79
in runs not exceeding 3m										
75mm	26.79	5.00	28.13	0.90	-	21.59	-	49.72	m	54.69
110mm	37.50	5.00	39.38	1.12	-	26.87	-	66.25	m	72.87
Pipework in trenches; vertical										
75mm	26.79	5.00	28.13	0.75	-	17.99	-	46.12	m	50.73
110mm	37.50	5.00	39.38	0.94	-	22.55	-	61.93	m	68.12

This section continues
on the next page

Labour hourly rates: (except Specialists) Craft Operatives 18.37 Labourer 13.76 Rates are national average prices. Refer to REGIONAL VARIATIONS for indicative levels of overall pricing in regions	MATERIALS			LABOUR				RATES		
	Del to Site £	Waste %	Material Cost £	Craft Optve Hrs	Lab Hrs	Labour Cost £	Sunds £	Nett Rate £	Unit	Gross rate (10%) £
STORM WATER AND FOUL DRAIN SYSTEMS; ACCESSORIES										
Vitrified clay accessories										
Gullies; Supersleve; joint to pipe; bedding and surrounding in concrete to BS 8500, ordinary prescribed mix ST3, 20mm aggregate; Hepworth Building Products										
100mm outlet, reference RG1/1, trapped, round; 255mm diameter top	73.40	5.00	77.07	0.60	0.08	12.12	-	89.19	nr	98.11
100mm outlet, reference SG2/1, trapped, square; 150 x 150 top	82.94	5.00	87.08	0.60	0.08	12.12	-	99.21	nr	109.13
100mm outlet, reference SG1/1, trapped, reversible, round; 100mm trap; hopper reference SH1, 100mm outlet, 150 x 150mm rebated top	123.53	5.00	129.70	1.00	0.10	19.75	-	149.45	nr	164.39
100mm outlet, reference SG1/1, trapped, reversible, round; 100mm trap; hopper reference SH2, 100mm outlet, 100mm horizontal inlet, 150 x 150mm rebated top	125.79	5.00	132.08	1.15	0.10	22.50	-	154.58	nr	170.04
100mm outlet, reference SG1/1, trapped, reversible, round; 100mm trap; raising pieces - 1, reference RRP2/2, 100mm diameter, 225mm high; hopper reference SH1, 100mm outlet, 150 x 150mm rebated top	200.10	5.00	210.10	1.40	0.15	27.78	-	237.89	nr	261.68
100mm outlet, reference SG1/1, trapped, reversible, round; 100mm trap; raising pieces - 1, reference RRP2/2, 100mm diameter, 225mm high; hopper reference SH2, 100mm outlet, 100mm horizontal inlet, 150 x 150mm rebated top	202.37	5.00	212.48	1.45	0.16	28.84	-	241.32	nr	265.45
100mm outlet, reference RGP5, trapped, round; 225mm diameter x 600mm deep; perforated galvanized steel bucket reference IBP3	244.83	5.00	257.08	2.50	0.25	49.37	-	306.44	nr	337.09
100mm outlet, reference RGR1, trapped, round; 300mm diameter x 600mm deep	232.94	5.00	244.59	2.50	0.33	50.47	-	295.06	nr	324.56
150mm outlet, reference RGR2, trapped, round; 300mm diameter x 600mm deep	241.02	5.00	253.07	2.50	0.33	50.47	-	303.54	nr	333.89
150mm outlet, reference RGR3, trapped, round; 400mm diameter x 750mm deep	267.46	5.00	280.83	4.00	0.45	79.67	-	360.50	nr	396.55
150mm outlet, reference RGR4, trapped, round; 450mm diameter x 900mm deep	350.28	5.00	367.79	5.50	0.50	107.92	-	475.71	nr	523.28
100mm outlet, reference RGU1, trapped; internal size 600 x 450 x 600mm deep; perforated tray and galvanized cover and frame	1181.94	5.00	1241.04	10.00	1.50	204.34	-	1445.38	nr	1589.91
Rainwater shoes; Supersleve; joint to pipe; bedding and surrounding in concrete to BS 8500, ordinary prescribed mix ST3, 20mm aggregate										
100mm outlet, reference RRW/S3/1 trapless, round; 100mm vertical inlet, 250 x 150mm rectangular access opening	77.27	5.00	81.14	0.50	0.05	9.87	-	91.01	nr	100.11
Unreinforced concrete accessories										
Gullies; BS 5911; joint to pipe; bedding and surrounding in concrete to BS 8500 ordinary prescribed mix ST3, 20mm aggregate										
150mm outlet, trapped, round; 375mm diameter x 750mm deep internally, stopper	80.47	5.00	84.49	3.30	0.50	67.50	-	151.99	nr	167.19
150mm outlet, trapped, round; 375mm diameter x 900mm deep internally, stopper	99.19	5.00	104.15	3.50	0.50	71.18	-	175.33	nr	192.86
150mm outlet, trapped, round; 450mm diameter x 750mm deep internally, stopper	82.54	5.00	86.66	3.50	0.60	72.55	-	159.21	nr	175.14
150mm outlet, trapped, round; 450mm diameter x 900mm deep internally, stopper	102.90	5.00	108.04	3.70	0.60	76.23	-	184.26	nr	202.69
150mm outlet, trapped, round; 450mm diameter x 1050mm deep internally, stopper	117.45	5.00	123.33	4.00	0.60	81.74	-	205.06	nr	225.57
150mm outlet, trapped, round; 450mm diameter x 1200mm deep internally, stopper	145.59	5.00	152.87	4.50	0.60	90.92	-	243.79	nr	268.17
PVC-U accessories										
Gullies; Osma; joint to pipe; bedding and surrounding in concrete to BS 8500, ordinary prescribed mix ST3, 20mm aggregate										
110mm outlet trap, base reference 4D.500, outlet bend with access reference 4D.569, inlet raising piece 300mm long, plain hopper reference 4D.503	103.50	5.00	108.67	1.00	0.20	21.12	-	129.79	nr	142.77
110mm outlet bottle gulley, reference 4D.900 trapped, round; 200mm diameter top with grating reference 4D.919	80.48	5.00	84.50	1.50	0.25	31.00	-	115.50	nr	127.05
Sealed rodding eyes; Osma; joint to pipe; surrounding in concrete to BS 8500; ordinary prescribed mix ST3, 20mm aggregate										
110mm outlet, reference 4D.360, with square cover	60.90	5.00	63.94	1.80	0.45	39.26	-	103.20	nr	113.52
Cast iron accessories, painted black										
Gratings; Hepworth Building Products										
reference IG6C; 140mm diameter	6.68	-	6.68	-	0.03	0.41	-	7.09	nr	7.80
reference IG7C; 197mm diameter	6.15	-	6.15	-	0.03	0.41	-	6.56	nr	7.22
reference IG8C; 284mm diameter	14.90	-	14.90	-	0.03	0.41	-	15.31	nr	16.84
reference IG2C; 150 x 150mm	4.05	-	4.05	-	0.03	0.41	-	4.46	nr	4.91
reference IG3C; 225 x 225mm	14.65	-	14.65	-	0.03	0.41	-	15.06	nr	16.57
reference IG4C; 300 x 300mm	31.75	-	31.75	-	0.03	0.41	-	32.16	nr	35.38
Gratings and frames; bedding frames in cement mortar (1:3)										
225mm diameter	75.77	15.00	87.14	1.70	-	31.23	-	118.37	nr	130.20
150 x 150mm with lock and key	18.63	15.00	21.43	1.25	-	22.96	-	44.39	nr	48.83
230 x 230mm with lock and key	26.00	15.00	29.90	1.75	-	32.15	-	62.05	nr	68.25
300 x 300mm with lock and key	98.93	15.00	113.77	1.90	-	34.90	-	148.68	nr	163.54

DRAINAGE BELOW GROUND (side margin)

Labour hourly rates: (except Specialists) Craft Operatives 18.37 Labourer 13.76 Rates are national average prices. Refer to REGIONAL VARIATIONS for indicative levels of overall pricing in regions	MATERIALS			LABOUR				RATES		
	Del to Site £	Waste %	Material Cost £	Craft Optve Hrs	Lab Hrs	Labour Cost £	Sunds £	Nett Rate £	Unit	Gross rate (10%) £
STORM WATER AND FOUL DRAIN SYSTEMS; ACCESSORIES (Cont'd)										
Cast iron accessories, painted black (Cont'd)										
Sealing plates and frames; bedding frames in cement mortar (1:3)										
140mm diameter	29.78	15.00	34.25	1.00	-	18.37	-	52.62	nr	57.88
197mm diameter	33.53	15.00	38.56	1.25	-	22.96	-	61.53	nr	67.68
273mm diameter	53.52	15.00	61.54	1.70	-	31.23	-	92.77	nr	102.05
150 x 150mm	28.77	15.00	33.09	1.25	-	22.96	-	56.05	nr	61.66
225 x 225mm	52.19	15.00	60.02	1.75	-	32.15	-	92.17	nr	101.38
Alloy accessories										
Gratings										
140mm diameter	2.75	15.00	3.16	-	0.03	0.41	-	3.58	nr	3.93
197mm diameter	5.25	15.00	6.04	-	0.03	0.41	-	6.45	nr	7.10
284mm diameter	8.36	15.00	9.61	-	0.03	0.41	-	10.03	nr	11.03
150 x 150mm	5.00	15.00	5.75	-	0.03	0.41	-	6.16	nr	6.78
225 x 225mm	6.68	15.00	7.68	-	0.03	0.41	-	8.09	nr	8.90
300 x 300mm	9.20	15.00	10.58	-	0.03	0.41	-	10.99	nr	12.09
Gratings and frames; bedding frames in cement mortar (1:3)										
193mm diameter	41.65	15.00	47.90	1.25	-	22.96	-	70.86	nr	77.94
150 x 150mm with lock and key	12.79	15.00	14.71	1.25	-	22.96	-	37.67	nr	41.44
225 x 225mm with lock and key	13.63	15.00	15.68	1.75	-	32.15	-	47.83	nr	52.61
Sealing plates and frames; bedding frames in cement mortar (1:3)										
263mm diameter	9.43	15.00	10.85	1.70	-	31.23	-	42.08	nr	46.28
150 x 150mm	10.27	15.00	11.81	1.25	-	22.96	-	34.78	nr	38.25
Cast iron accessories, painted black										
Gratings and frames; Stanton PLC.; bedding frames in cement mortar (1:3)										
reference HY811 Watergate; 370 x 305mm	123.21	15.00	141.69	1.00	-	18.37	-	160.06	nr	176.07
reference HY812; 370 x 430mm	130.27	15.00	149.82	2.50	-	45.93	-	195.74	nr	215.31
reference HY815; 510 x 360mm	121.77	15.00	140.03	3.33	-	61.17	-	201.20	nr	221.32
reference HP894 Waterflow; 325 x 312mm	151.39	15.00	174.09	4.25	-	78.07	-	252.17	nr	277.38
reference HP895; 325 x 437mm	295.05	15.00	339.30	2.75	-	50.52	-	389.82	nr	428.80
Precast polyester concrete ACO Drain surface drainage systems; ACO Technologies plc; bedding and haunching in concrete to BS 8500 ordinary prescribed mix ST5, 20mm aggregate										
Multidrain 100D Channel system with resin composite non rusting lockable grating										
constant depth or complete with 0.6% fall	45.80	5.00	48.09	1.50	1.95	54.39	-	102.48	m	112.72
extra; Multidrain M100D end cap	6.06	5.00	6.36	-	0.20	2.75	-	9.11	nr	10.03
ParkDrain one piece channel system with integral grating										
constant depth or complete with 0.6% fall	133.34	5.00	140.01	1.70	2.10	60.12	-	200.13	m	220.15
extra; end cap	6.06	5.00	6.36	-	0.20	2.75	-	9.11	nr	10.03
Raindrain recycled polymer concrete channel and galvanised steel grating										
invert depth 115mm	9.68	5.00	10.16	1.40	1.65	48.42	-	58.59	m	64.44
extra; end cap	3.42	5.00	3.59	-	0.20	2.75	-	6.34	nr	6.98
extra; rodding access unit	51.77	5.00	54.36	-	0.50	6.88	-	61.24	nr	67.36
extra; universal Gully assembly complete with bucket	602.78	5.00	632.92	1.00	1.00	32.13	-	665.05	nr	731.55
KerbDrain Class D400 combined kerb drainage system 305mm x 150mm x 500mm	61.82	5.00	64.91	1.20	1.45	42.00	-	106.91	nr	117.60
extra; rodding access	231.66	5.00	243.24	-	0.50	6.88	-	250.12	nr	275.14
extra; Class D400 Gully	476.91	5.00	500.76	1.00	1.00	32.13	-	532.89	nr	586.17

This section continues
on the next page

Labour hourly rates: (except Specialists) Craft Operatives 18.37 Labourer 13.76 Rates are national average prices. Refer to REGIONAL VARIATIONS for indicative levels of overall pricing in regions	PLANT AND TRANSPORT			LABOUR				RATES		
	Plant Cost £	Trans Cost £	P and T Cost £	Craft Optve Hrs	Lab Hrs	Labour Cost £	Sunds £	Nett Rate £	Unit	Gross rate (10%) £
STORM WATER AND FOUL DRAIN SYSTEMS; MANHOLES, INSPECTION CHAMBERS AND SOAKAWAYS										
Excavating - by machine										
Excavating pits commencing from natural ground level; maximum depth not exceeding										
0.25m....................	7.12	-	7.12	-	0.84	11.56	-	18.68	m³	20.54
1.00m....................	9.49	-	9.49	-	0.88	12.11	-	21.60	m³	23.76
2.00m....................	10.44	-	10.44	-	1.00	13.76	-	24.20	m³	26.62
4.00m....................	11.39	-	11.39	-	1.27	17.48	-	28.86	m³	31.75
Working space allowance to excavation										
pits....................	9.49	-	9.49	-	0.60	8.26	-	17.75	m²	19.52
Compacting										
bottoms of excavations....................	2.09	-	2.09	-	-	-	-	2.09	m²	2.30
Excavated material depositing on site in temporary spoil heaps where directed										
25m....................	3.80	-	3.80	-	-	-	-	3.80	m³	4.18
50m....................	4.75	-	4.75	-	-	-	-	4.75	m³	5.22
100m....................	5.69	-	5.69	-	-	-	-	5.69	m³	6.26
Excavating - by hand										
Excavating pits commencing from natural ground level; maximum depth not exceeding										
0.25m....................	-	-	-	-	3.35	46.10	-	46.10	m³	50.71
1.00m....................	-	-	-	-	3.50	48.16	-	48.16	m³	52.98
2.00m....................	-	-	-	-	4.00	55.04	-	55.04	m³	60.54
4.00m....................	-	-	-	-	5.10	70.18	-	70.18	m³	77.19
Working space allowance to excavation										
pits....................	-	-	-	-	3.12	42.93	-	42.93	m²	47.22
Compacting										
bottoms of excavations....................	-	-	-	-	0.12	1.65	-	1.65	m²	1.82
Excavated material depositing on site in temporary spoil heaps where directed										
25m....................	-	-	-	-	1.25	17.20	-	17.20	m³	18.92
50m....................	-	-	-	-	1.50	20.64	-	20.64	m³	22.70
100m....................	-	-	-	-	2.00	27.52	-	27.52	m³	30.27

This section continues
on the next page

DRAINAGE BELOW GROUND *(side heading)*

	MATERIALS			LABOUR				RATES		
Labour hourly rates: (except Specialists) Craft Operatives 18.37 Labourer 13.76 Rates are national average prices. Refer to REGIONAL VARIATIONS for indicative levels of overall pricing in regions	Del to Site £	Waste %	Material Cost £	Craft Optve Hrs	Lab Hrs	Labour Cost £	Sunds £	Nett Rate £	Unit	Gross rate (10%) £

STORM WATER AND FOUL DRAIN SYSTEMS; MANHOLES, INSPECTION CHAMBERS AND SOAKAWAYS (Cont'd)

Earthwork support

	Del to Site £	Waste %	Material Cost £	Craft Optve Hrs	Lab Hrs	Labour Cost £	Sunds £	Nett Rate £	Unit	Gross rate (10%) £
Earthwork support distance between opposing faces not exceeding 2.00m; maximum depth not exceeding										
1.00m	0.81	-	0.81	-	0.20	2.75	-	3.57	m²	3.92
2.00m	1.09	-	1.09	-	0.25	3.44	-	4.53	m²	4.98
4.00m	1.36	-	1.36	-	0.30	4.13	-	5.49	m²	6.03
distance between opposing faces 2.00 - 4.00m; maximum not exceeding										
1.00m	1.63	-	1.63	-	0.21	2.89	-	4.52	m²	4.97
2.00m	1.90	-	1.90	-	0.26	3.58	-	5.48	m²	6.03
4.00m	2.17	-	2.17	-	0.32	4.40	-	6.58	m²	7.23
Earthwork support; in unstable ground distance between opposing faces not exceeding 2.00m; maximum depth not exceeding										
1.00m	2.99	-	2.99	-	0.33	4.54	-	7.53	m²	8.28
2.00m	3.26	-	3.26	-	0.42	5.78	-	9.04	m²	9.94
4.00m	3.53	-	3.53	-	0.50	6.88	-	10.41	m²	11.45
distance between opposing faces 2.00 - 4.00m; maximum depth not exceeding										
1.00m	3.53	-	3.53	-	0.35	4.82	-	8.35	m²	9.18
2.00m	3.80	-	3.80	-	0.43	5.92	-	9.72	m²	10.69
4.00m	4.07	-	4.07	-	0.53	7.29	-	11.37	m²	12.50
Excavated material arising from excavations										
Filling to excavation average thickness exceeding 0.25m	-	-	-	-	1.33	18.30	-	18.30	m³	20.13
Excavated material obtained from on site spoil heaps 25m distant										
Filling to excavation average thickness not exceeding 0.25m	-	-	-	-	1.75	24.08	-	24.08	m³	26.49
average thickness exceeding 0.25m	-	-	-	-	1.50	20.64	-	20.64	m³	22.70
Excavated material obtained from on site spoil heaps 50m distant										
Filling to excavation average thickness not exceeding 0.25m	-	-	-	-	2.00	27.52	-	27.52	m³	30.27
average thickness exceeding 0.25m	-	-	-	-	1.85	25.46	-	25.46	m³	28.00
Excavated material obtained from on site spoil heaps 100m distant										
Filling to excavation average thickness not exceeding 0.25m	-	-	-	-	2.50	34.40	-	34.40	m³	37.84
average thickness exceeding 0.25m	-	-	-	-	2.25	30.96	-	30.96	m³	34.06
Hard, dry broken brick or stone, 100 - 75mm gauge, to be obtained off site										
Filling to excavation average thickness exceeding 0.25m	25.06	25.00	31.32	-	1.25	17.20	-	48.52	m³	53.37
Plain in-situ concrete; BS 8500, ordinary prescribed mix ST3, 20mm aggregate										
Beds; poured on or against earth or unblinded hardcore thickness 150 - 450mm	97.99	5.00	102.89	-	4.00	55.04	-	157.93	m³	173.72
thickness not exceeding 150mm	97.99	5.00	102.89	-	4.30	59.17	-	162.06	m³	178.26
Benching in bottoms; rendering with 13mm cement and sand (1:2) before final set										
600 x 450 x average 225mm; trowelling	6.90	5.00	7.34	-	0.54	7.43	-	14.77	nr	16.25
675 x 675 x average 225mm; trowelling	11.64	5.00	12.38	-	0.92	12.66	-	25.03	nr	27.54
800 x 675 x average 225mm; trowelling	13.81	5.00	14.69	-	1.10	15.14	-	29.82	nr	32.80
800 x 800 x average 300mm; trowelling	22.69	5.00	24.13	-	1.73	23.80	-	47.94	nr	52.73
1025 x 800 x average 300mm; trowelling	28.36	5.00	30.17	-	2.21	30.41	-	60.58	nr	66.63
Reinforced in-situ concrete; BS 8500, designed mix ST4, 20mm aggregate, minimum cement content 240 kg/m³; vibrated										
Slabs thickness 150 - 450mm	99.73	5.00	104.72	-	3.50	48.16	-	152.88	m³	168.17
thickness not exceeding 150mm	99.73	5.00	104.72	-	4.00	55.04	-	159.76	m³	175.74
Formwork and basic finish										
Edges of suspended slabs; plain vertical height not exceeding 250mm	0.88	12.50	0.98	0.75	0.15	15.84	-	16.83	m	18.51
Soffits of slabs; horizontal slab thickness not exceeding 200mm; height to soffit not exceeding 1.50m	11.29	10.00	12.42	2.20	0.42	46.19	-	58.61	m²	64.47
slab thickness not exceeding 200mm; height to soffit 1.50 - 3.00m	11.29	10.00	12.42	2.00	0.42	42.52	-	54.94	m²	60.43

Labour hourly rates: (except Specialists) Craft Operatives 18.37　Labourer 13.76 Rates are national average prices. Refer to REGIONAL VARIATIONS for indicative levels of overall pricing in regions	MATERIALS			LABOUR				RATES		
	Del to Site	Waste	Material Cost	Craft Optve	Lab	Labour Cost	Sunds	Nett Rate		Gross rate (10%)
	£	%	£	Hrs	Hrs	£	£	£	Unit	£
STORM WATER AND FOUL DRAIN SYSTEMS; MANHOLES, INSPECTION CHAMBERS AND SOAKAWAYS (Cont'd)										
Formwork and basic finish (Cont'd)										
Holes; rectangular										
girth 1.00 - 2.00m; depth not exceeding 250mm	1.75	12.50	1.97	0.50	0.05	9.87	-	11.84	nr	13.03
girth 2.00 - 3.00m; depth not exceeding 250mm	2.62	12.50	2.95	0.80	0.10	16.07	-	19.03	nr	20.93
Reinforcement fabric; BS 4483; hard drawn plain round steel, welded; including laps, tying wire, all cutting and bending and spacers and chairs which are at the discretion of the Contractor										
Reference A142, 2.22kg/m²; 200mm side laps; 200mm end laps generally..............	1.78	15.00	2.05	0.25	0.02	4.87	-	6.91	m²	7.60
Common bricks, BS EN 772, Category M, 215 x 102.5 x 65mm, compressive strength 20.5 N/mm²; in cement mortar (1:3)										
Walls; vertical										
102mm thick; stretcher bond	23.93	5.00	25.60	1.33	1.45	44.38	-	69.98	m²	76.98
215mm thick; English bond	47.55	5.00	50.86	2.30	2.54	77.20	-	128.06	m²	140.87
Milton Hall Second hard stock bricks, BS EN 772, Category M, 215 x 102.5 x 65mm; in cement mortar (1:3)										
Walls; vertical										
102mm thick; stretcher bond	53.15	5.00	56.28	1.33	1.45	44.38	-	100.66	m²	110.73
215mm thick; English bond	105.51	5.00	111.72	2.30	2.54	77.20	-	188.92	m²	207.81
Engineering bricks, Category F, BS EN 772, 215 x 102.5 x 65mm, class B; in cement mortar (1:3)										
Walls; vertical										
102mm thick; stretcher bond	24.36	5.00	26.04	1.33	1.45	44.38	-	70.43	m²	77.47
215mm thick; English bond	48.40	5.00	51.75	2.30	2.54	77.20	-	128.95	m²	141.85
Common bricks, BS EN 772, Category M, 215 x 102.5 x 65mm, compressive strength 20.5 N/mm²; in cement mortar (1:3); flush smooth pointing as work proceeds										
Walls; vertical										
102mm thick; stretcher bond; facework one side..........................	23.93	5.00	25.60	1.53	1.65	50.81	-	76.41	m²	84.05
215mm thick; English bond; facework one side..........................	47.55	5.00	50.86	2.50	2.74	83.63	-	134.49	m²	147.94
Milton Hall Second hard stock bricks, BS EN 772, Category M, 215 x 102.5 x 65mm; in cement mortar (1:3); flush smooth pointing as work proceeds										
Walls; vertical										
102mm thick; stretcher bond; facework one side..........................	53.15	5.00	56.28	1.53	1.65	50.81	-	107.09	m²	117.79
215mm thick; English bond; facework one side..........................	105.51	5.00	111.72	2.50	2.74	83.63	-	195.34	m²	214.88
Engineering bricks, BS EN 772, Category F, 215 x 102.5 x 65mm, class B; in cement mortar (1:3); flush smooth pointing as work proceeds										
Walls; vertical										
102mm thick; stretcher bond; facework one side......................	24.36	5.00	26.04	1.53	1.65	50.81	-	76.85	m²	84.54
215mm thick; English bond; facework one side......................	48.40	5.00	51.75	2.50	2.74	83.63	-	135.38	m²	148.91
Accessories/sundry items for brick/block/stone walling										
Building in ends of pipes; 100mm diameter										
making good fair face one side										
102mm brickwork..............	0.12	15.00	0.13	0.08	0.08	2.57	-	2.70	nr	2.97
215mm brickwork..............	0.23	15.00	0.27	0.13	0.13	4.18	-	4.44	nr	4.89
Building in ends of pipes; 150mm diameter										
making good fair face one side										
102mm brickwork..............	0.12	15.00	0.13	0.13	0.13	4.18	-	4.31	nr	4.74
215mm brickwork..............	0.23	15.00	0.27	0.20	0.20	6.43	-	6.69	nr	7.36
Building in ends of pipes; 225mm diameter										
making good fair face one side										
102mm brickwork..............	0.23	15.00	0.27	0.20	0.20	6.43	-	6.69	nr	7.36
215mm brickwork..............	0.47	15.00	0.54	0.25	0.25	8.03	-	8.57	nr	9.42
Building in ends of pipes; 300mm diameter										
making good fair face one side										
102mm brickwork..............	0.35	15.00	0.40	0.25	0.25	8.03	-	8.43	nr	9.28
215mm brickwork..............	0.70	15.00	0.80	0.33	0.33	10.60	-	11.41	nr	12.55
Mortar, cement and sand (1:3)										
13mm work to walls on brickwork base; one coat; trowelled width exceeding 300mm..............	1.51	15.00	1.74	0.40	0.40	12.85	-	14.59	m²	16.05
13mm work to floors on concrete base; one coat; trowelled level and to falls only not exceeding 15 degrees from horizontal ...	1.51	15.00	1.74	0.60	0.60	19.28	-	21.02	m²	23.12

	MATERIALS			LABOUR				RATES		
Labour hourly rates: (except Specialists) Craft Operatives 18.37 Labourer 13.76 Rates are national average prices. Refer to REGIONAL VARIATIONS for indicative levels of overall pricing in regions	Del to Site £	Waste %	Material Cost £	Craft Optve Hrs	Lab Hrs	Labour Cost £	Sunds £	Nett Rate £	Unit	Gross rate (10%) £

STORM WATER AND FOUL DRAIN SYSTEMS; MANHOLES, INSPECTION CHAMBERS AND SOAKAWAYS (Cont'd)

Channels in bottoms; vitrified clay, BS 65, normal, glazed; cement mortar (1:3) joints; bedding in cement mortar (1:3)

	Del to Site £	Waste %	Material Cost £	Craft Optve Hrs	Lab Hrs	Labour Cost £	Sunds £	Nett Rate £	Unit	Gross rate (10%) £
Half section										
straight; 600mm effective length										
100mm	9.92	5.00	10.45	0.30	0.03	5.92	-	16.37	nr	18.01
150mm	14.30	5.00	15.06	0.40	0.04	7.90	-	22.96	nr	25.25
straight; 1000mm effective length										
100mm	15.19	5.00	16.02	0.50	0.05	9.87	-	25.89	nr	28.48
150mm	21.05	5.00	22.20	0.67	0.07	13.27	-	35.47	nr	39.02
225mm	50.72	5.00	53.38	0.84	0.08	16.53	-	69.91	nr	76.90
300mm	103.56	5.00	108.87	1.25	0.15	25.03	-	133.90	nr	147.29
curved; 500mm effective length										
100mm	14.55	5.00	15.31	0.30	0.03	5.92	-	21.24	nr	23.36
150mm	20.65	5.00	21.72	0.40	0.25	10.79	-	32.51	nr	35.76
225mm	79.75	5.00	83.81	0.50	0.05	9.87	-	93.68	nr	103.05
300mm	135.32	5.00	142.18	0.75	0.09	15.02	-	157.20	nr	172.92
curved; 900mm effective length										
100mm	14.90	5.00	15.71	0.60	0.06	11.85	-	27.56	nr	30.32
150mm	21.11	5.00	22.26	0.80	0.08	15.80	-	38.06	nr	41.86
225mm	80.21	5.00	84.34	1.00	0.10	19.75	-	104.09	nr	114.50
300mm	135.79	5.00	142.72	1.50	0.18	30.03	-	172.75	nr	190.02
Three quarter section										
branch bends										
100mm	39.66	5.00	42.34	0.33	0.03	6.47	-	48.81	nr	53.69
150mm	58.95	5.00	62.83	0.50	0.05	9.87	-	72.71	nr	79.98
225mm	210.34	5.00	220.98	1.00	0.10	19.75	-	240.72	nr	264.80

Channels in bottoms; P.V.C.; 'O' ring joints; bedding in cement mortar (1:3)

	Del to Site £	Waste %	Material Cost £	Craft Optve Hrs	Lab Hrs	Labour Cost £	Sunds £	Nett Rate £	Unit	Gross rate (10%) £
For 610mm clear opening										
straight										
110mm	37.97	5.00	40.22	0.40	0.03	7.76	-	47.98	nr	52.78
160mm	52.38	5.00	55.46	0.50	0.04	9.74	-	65.20	nr	71.72
short bend 3/4 section										
110mm	25.32	5.00	26.94	0.20	0.01	3.81	-	30.75	nr	33.82
160mm	42.15	5.00	44.72	0.30	0.02	5.79	-	50.51	nr	55.56
long radius bend										
110mm	53.95	5.00	57.34	0.33	0.02	6.34	-	63.68	nr	70.05
160mm	98.28	5.00	104.13	0.50	0.04	9.74	-	113.87	nr	125.25

Precast concrete; standard or stock pattern units; BS 5911; jointing and pointing in cement mortar (1:3)

	Del to Site £	Waste %	Material Cost £	Craft Optve Hrs	Lab Hrs	Labour Cost £	Sunds £	Nett Rate £	Unit	Gross rate (10%) £
Chamber or shaft sections										
900mm internal diameter										
250mm high	32.48	5.00	34.12	0.58	1.16	26.62	-	60.73	nr	66.80
500mm high	51.90	5.00	54.50	0.67	1.34	30.75	-	85.25	nr	93.77
750mm high	50.54	5.00	53.08	0.87	1.74	39.92	-	93.01	nr	102.31
1000mm high	67.36	5.00	70.74	0.93	1.86	42.68	-	113.42	nr	124.76
1050mm internal diameter										
250mm high	49.33	5.00	51.82	0.81	1.62	37.17	-	88.99	nr	97.89
500mm high	82.86	5.00	87.03	0.87	1.74	39.92	-	126.95	nr	139.65
750mm high	97.46	5.00	102.36	0.93	1.86	42.68	-	145.03	nr	159.54
1000mm high	115.29	5.00	121.08	0.99	1.98	45.43	-	166.51	nr	183.16
1200mm internal diameter										
250mm high	59.85	5.00	62.87	0.99	1.98	45.43	-	108.30	nr	119.13
500mm high	99.68	5.00	104.69	1.04	2.08	47.73	-	152.41	nr	167.65
750mm high	115.49	5.00	121.29	1.22	2.44	55.99	-	177.27	nr	195.00
1000mm high	153.92	5.00	161.64	1.28	2.56	58.74	-	220.38	nr	242.41
1500mm internal diameter										
500mm high	99.68	5.00	104.69	1.04	2.08	47.73	-	152.41	nr	167.65
750mm high	115.49	5.00	121.29	1.22	2.44	55.99	-	177.27	nr	195.00
1000mm high	153.92	5.00	161.64	1.28	2.56	58.74	-	220.38	nr	242.41
1800mm internal diameter										
500mm high	99.68	5.00	104.69	1.04	2.08	47.73	-	152.41	nr	167.65
750mm high	115.49	5.00	121.29	1.22	2.44	55.99	-	177.27	nr	195.00
1000mm high	153.92	5.00	161.64	1.28	2.56	58.74	-	220.38	nr	242.41
2100mm internal diameter										
500mm high	99.68	5.00	104.69	1.04	2.08	47.73	-	152.41	nr	167.65
750mm high	115.49	5.00	121.29	1.22	2.44	55.99	-	177.27	nr	195.00
1000mm high	153.92	5.00	161.64	1.28	2.56	58.74	-	220.38	nr	242.41
2400mm internal diameter										
750mm high	115.49	5.00	121.29	1.22	2.44	55.99	-	177.27	nr	195.00
1000mm high	153.92	5.00	161.64	1.28	2.56	58.74	-	220.38	nr	242.41
2700mm internal diameter										
750mm high	115.49	5.00	121.29	1.22	2.44	55.99	-	177.27	nr	195.00
1000mm high	153.92	5.00	161.64	1.28	2.56	58.74	-	220.38	nr	242.41
3000mm internal diameter										
750mm high	115.49	5.00	121.29	1.22	2.44	55.99	-	177.27	nr	195.00
1000mm high	153.92	5.00	161.64	1.28	2.56	58.74	-	220.38	nr	242.41

DRAINAGE BELOW GROUND

Labour hourly rates: (except Specialists) Craft Operatives 18.37 Labourer 13.76 Rates are national average prices. Refer to REGIONAL VARIATIONS for indicative levels of overall pricing in regions	MATERIALS			LABOUR				RATES		
	Del to Site £	Waste %	Material Cost £	Craft Optve Hrs	Lab Hrs	Labour Cost £	Sunds £	Nett Rate £	Unit	Gross rate (10%) £
STORM WATER AND FOUL DRAIN SYSTEMS; MANHOLES, INSPECTION CHAMBERS AND SOAKAWAYS (Cont'd)										
Precast concrete; standard or stock pattern units; BS 5911; jointing and pointing in cement mortar (1:3) (Cont'd)										
Cutting holes										
cutting holes for pipes; 100mm diameter; making good; pointing...	38.23	-	38.23	-	0.08	1.10	-	39.33	nr	43.27
cutting holes for pipes; 150mm diameter; making good; pointing...	38.35	-	38.35	-	0.10	1.38	-	39.73	nr	43.70
cutting holes for pipes; 225mm diameter; making good; pointing...	65.58	-	65.58	-	0.12	1.65	-	67.23	nr	73.96
cutting holes for pipes; 300mm diameter; making good; pointing...	75.82	-	75.82	-	0.15	2.06	-	77.88	nr	85.67
Step irons										
galvanized malleable cast iron step irons BS 1247, cast in	6.00	5.00	6.30	-	0.15	2.06	-	8.36	nr	9.20
Cover slabs, heavy duty										
for chamber or shaft sections 900mm internal diameter, 125mm thick; access opening 600mm x 600mm......................................	64.47	5.00	67.72	1.50	2.25	58.52	-	126.24	nr	138.86
for chamber or shaft sections 1050mm internal diameter, 125mm thick; access opening 600mm x 600mm..................................	87.00	5.00	91.37	2.00	3.00	78.02	-	169.39	nr	186.33
for chamber or shaft sections 1200mm internal diameter, 125mm thick; access opening 600mm x 600mm..................................	89.17	5.00	93.65	2.50	3.75	97.53	-	191.17	nr	210.29
for chamber or shaft sections 1350mm internal diameter, 125mm thick; access opening 600mm x 600mm..................................	128.72	5.00	135.17	3.00	4.50	117.03	-	252.20	nr	277.42
for chamber or shaft sections 1500mm internal diameter, 150mm thick; access opening 600mm x 600mm..................................	148.14	5.00	155.59	3.50	5.25	136.54	-	292.12	nr	321.33
for chamber or shaft sections 1800mm internal diameter, 150mm thick access opening 600mm x 600mm..................................	262.90	5.00	276.08	4.00	6.00	156.04	-	432.12	nr	475.34
for chamber or shaft sections 2100mm internal diameter, 150mm thick access opening 600mm x 600mm..................................	599.92	5.00	629.98	4.50	6.75	175.55	-	805.52	nr	886.07
for chamber or shaft sections 2400mm internal diameter, 150mm thick access opening 600mm x 600mm..................................	1044.76	5.00	1097.06	5.00	7.50	195.05	-	1292.11	nr	1421.32
for chamber or shaft sections 2700mm internal diameter, 150mm thick access opening 600mm x 600mm..................................	1352.19	5.00	1419.87	5.50	8.25	214.56	-	1634.42	nr	1797.86
for chamber or shaft sections 3000mm internal diameter, 150mm thick access opening 600mm x 600mm..................................	1591.58	5.00	1671.23	6.00	9.00	234.06	-	1905.29	nr	2095.82
Precast concrete inspection chambers; BS 5911; jointing and pointing in cement mortar (1:3)										
Rectangular inspection chambers, internal size 450 x 600mm										
100mm high..	9.42	5.00	9.91	0.50	0.35	14.00	-	23.92	nr	26.31
150mm high..	11.51	5.00	12.11	0.65	0.45	18.13	-	30.24	nr	33.27
250mm high..	14.30	5.00	15.04	0.75	0.53	21.07	-	36.11	nr	39.72
extra; galvanized malleable cast iron step irons BS 1247, built in ..	6.00	5.00	6.30	-	-	-	-	6.30	nr	6.93
Rectangular inspection chambers, internal size 760 x 610mm										
150mm high..	16.82	5.00	17.70	1.05	0.75	29.61	-	47.30	nr	52.03
225mm high..	19.88	5.00	20.91	1.21	0.88	34.34	-	55.24	nr	60.77
extra; galvanized malleable cast iron step irons BS 1247, built in ..	6.00	5.00	6.30	-	-	-	-	6.30	nr	6.93
Rectangular inspection chambers, internal size 990 x 610mm										
150mm high..	18.93	5.00	19.92	1.05	0.75	29.61	-	49.52	nr	54.48
Rectangular inspection chambers, internal size 1200 x 750mm										
150mm high..	26.09	5.00	27.44	1.20	0.85	33.74	-	61.18	nr	67.29
200mm high..	32.07	5.00	33.72	1.38	1.00	39.11	-	72.83	nr	80.11
Base units										
450 x 610mm internal size; 360mm deep..................................	45.29	5.00	47.59	1.05	0.75	29.61	-	77.20	nr	84.92
760 x 610mm internal size; 360mm deep..................................	54.86	5.00	57.65	1.05	0.75	29.61	-	87.25	nr	95.98
990 x 610mm internal size; 360mm deep..................................	73.03	5.00	76.74	1.05	0.75	29.61	-	106.35	nr	116.99
Cover slab with integral cover										
450 x 600mm internal size..	31.91	5.00	33.54	1.05	0.75	29.61	-	63.15	nr	69.46
760 x 610mm internal size..	51.22	5.00	53.82	1.05	0.75	29.61	-	83.43	nr	91.77
990 x 610mm internal size..	73.91	5.00	77.67	1.05	0.75	29.61	-	107.27	nr	118.00
Osma; Shallow inspection chamber ref. 4D.960 Polypropylene with preformed benching. Light duty single sealed cover and frame ref. 4D.961 set in position, 250mm diameter; for 100mm diameter pipes, depth to invert										
Chamber or shaft sections and benching										
600mm	114.28	5.00	119.99	2.00	0.10	38.12	-	158.11	nr	173.92
Osma; universal inspection chamber; Polypropylene; 450mm diameter										
preformed chamber base benching, ref. 4D.922; for 100mm pipes..	105.36	5.00	110.63	1.10	0.10	21.58	-	132.21	nr	145.43
chamber shaft 185mm effective length, ref. 4D925	19.43	5.00	20.40	-	0.50	6.88	-	27.28	nr	30.01
cast iron cover and frame; single seal light duty to suit, ref. 4D.942..	91.52	5.00	96.10	0.50	-	9.19	-	105.28	nr	115.81
Polypropylene sealed cover and frame, ref. 4D.969......................	48.27	5.00	50.68	0.50	-	9.19	-	59.87	nr	65.86

Labour hourly rates: (except Specialists) Craft Operatives 23.99 Labourer 13.76 Rates are national average prices. Refer to REGIONAL VARIATIONS for indicative levels of overall pricing in regions	MATERIALS			LABOUR				RATES		
	Del to Site £	Waste %	Material Cost £	Craft Optve Hrs	Lab Hrs	Labour Cost £	Sunds £	Nett Rate £	Unit	Gross rate (10%) £
STORM WATER AND FOUL DRAIN SYSTEMS; MANHOLES, INSPECTION CHAMBERS AND SOAKAWAYS (Cont'd)										
Coated cast iron bolted inspection chambers										
100mm diameter straight chamber; coupling joints; bedding in cement mortar (1:3)										
with one branch each side..	261.98	2.00	267.23	3.25	-	77.97	-	345.20	nr	379.72
with two branches each side	793.45	2.00	809.34	4.75	-	113.95	-	923.29	nr	1015.62
150 x 100mm diameter straight chamber; coupling joints; bedding in cement mortar (1:3)										
with one branch one side..	337.70	2.00	344.47	3.25	-	77.97	-	422.43	nr	464.68
150mm diameter straight chamber; coupling joints; bedding in cement mortar (1:3)										
with one branch each side..	428.65	2.00	437.23	4.50	-	107.96	-	545.19	nr	599.71
with two branches each side	1322.34	2.00	1348.81	6.50	-	155.93	-	1504.74	nr	1655.22

This section continues
on the next page

Labour hourly rates: (except Specialists) Craft Operatives 18.37 Labourer 13.76 Rates are national average prices. Refer to REGIONAL VARIATIONS for indicative levels of overall pricing in regions	MATERIALS			LABOUR				RATES		
	Del to Site	Waste	Material Cost	Craft Optve	Lab	Labour Cost	Sunds	Nett Rate		Gross rate (10%)
	£	%	£	Hrs	Hrs	£	£	£	Unit	£
STORM WATER AND FOUL DRAIN SYSTEMS; OTHER TANKS AND PITS										
Fibreglass settlement tanks; including cover and frame										
Settlement tank; fibreglass; set in position										
1000mm deep to invert of inlet, standard grade										
2700 litres capacity	423.00	-	423.00	0.90	1.80	41.30	-	464.30	nr	510.73
3750 litres capacity	541.00	-	541.00	1.00	2.00	45.89	-	586.89	nr	645.58
4500 litres capacity	623.00	-	623.00	1.10	2.20	50.48	-	673.48	nr	740.83
7500 litres capacity	1247.00	-	1247.00	1.20	2.40	55.07	-	1302.07	nr	1432.27
9000 litres capacity	1334.00	-	1334.00	1.30	2.60	59.66	-	1393.66	nr	1533.02
1500mm deep to invert of inlet, heavy grade										
2700 litres capacity	900.00	-	900.00	0.90	1.80	41.30	-	941.30	nr	1035.43
3750 litres capacity	992.00	-	992.00	1.00	2.00	45.89	-	1037.89	nr	1141.68
4500 litres capacity	1081.00	-	1081.00	1.10	2.20	50.48	-	1131.48	nr	1244.63
7500 litres capacity	1455.00	-	1455.00	1.20	2.40	55.07	-	1510.07	nr	1661.07
9000 litres capacity	1711.00	-	1711.00	1.30	2.60	59.66	-	1770.66	nr	1947.72
STORM WATER AND FOUL DRAIN SYSTEMS; SUNDRIES										
Manhole step irons										
Step irons; BS 1247 malleable cast iron; galvanized; building in to joints										
figure 1 general purpose pattern; 110mm tail	5.05	-	5.05	0.25	-	4.59	-	9.64	nr	10.61
figure 1 general purpose pattern; 225mm tail	5.73	-	5.73	0.25	-	4.59	-	10.32	nr	11.35
Vitrified clay intercepting traps										
Intercepting traps; BS 65; joint to pipe; building into side of manhole; bedding and surrounding in concrete to BS 8500 ordinary prescribed mix C15P, 20mm aggregate										
cleaning arm and stopper										
100mm outlet, ref. R1 1/1	190.25	5.00	199.76	1.25	1.00	36.72	-	236.48	nr	260.13
150mm outlet, ref. R1 1/2	273.68	5.00	287.36	1.75	1.50	52.79	-	340.15	nr	374.17
225mm outlet, ref. R1 1/3	804.76	5.00	845.00	2.25	2.00	68.85	-	913.85	nr	1005.24
Aluminium; fresh air inlets										
Air inlet valve, mica flap; set in cement mortar (1:3)										
100mm diameter	31.07	5.00	32.63	1.15	-	21.13	-	53.76	nr	59.13
STORM WATER AND FOUL DRAIN SYSTEMS; COVERS AND FRAMES										
Manhole covers - A15 light duty										
Access covers; ductile iron; BS EN 124; coated; bedding frame in cement mortar (1:3); bedding cover in grease and sand										
clear opening 450mm dia	14.57	2.50	15.22	0.65	0.65	20.88	-	36.11	nr	39.72
clear opening 450 x 450mm	16.69	2.50	17.40	0.65	0.65	20.88	-	38.28	nr	42.11
clear opening 600 x 450mm	35.77	2.50	36.95	0.75	0.75	24.10	-	61.05	nr	67.16
clear opening 600 x 600mm	61.13	2.50	62.95	1.25	1.25	40.16	-	103.11	nr	113.42
recessed; clear opening 600 x 450mm	31.84	2.50	32.93	1.15	1.15	36.95	-	69.88	nr	76.86
recessed; clear opening 600 x 600mm	45.06	2.50	46.63	1.45	1.45	46.59	-	93.21	nr	102.54
recessed; clear opening 750 x 600mm	117.39	2.50	120.76	1.60	1.60	51.41	-	172.17	nr	189.39
double seal; clear opening 450 x 450mm	51.63	2.50	53.21	1.30	1.30	41.77	-	94.98	nr	104.48
double seal; clear opening 600 x 450mm	52.43	2.50	54.03	1.45	1.45	46.59	-	100.62	nr	110.68
double seal; clear opening 600 x 600mm	59.94	2.50	61.73	1.60	1.60	51.41	-	113.14	nr	124.45
lifting keys	1.65	2.50	1.69	-	-		-	1.69	nr	1.86
Manhole covers - B125 medium duty										
Access covers; Stanton PLC; BS EN 124; coated; bedding frame in cement mortar (1:3); bedding cover in grease and sand										
600mm diameter	80.13	2.50	82.42	1.65	1.65	53.01	-	135.44	nr	148.98
600 x 450mm	42.00	2.50	43.34	1.43	1.43	45.95	-	89.29	nr	98.21
600 x 600mm	51.87	2.50	53.61	1.80	1.80	57.83	-	111.44	nr	122.58
lifting keys	5.46	2.50	5.60	-	-	-	-	5.60	nr	6.16
Manhole covers - D400 heavy duty										
Access covers; BS EN 124; coated; bedding frame in cement mortar (1:3); bedding cover in grease and sand										
600mm diameter	108.93	2.50	111.94	1.89	1.89	60.73	-	172.67	nr	189.94
600 x 600mm	90.25	2.50	92.95	1.89	1.89	60.73	-	153.67	nr	169.04
750 x 750mm	176.54	2.50	181.82	2.25	2.25	72.29	-	254.11	nr	279.53
lifting keys	6.50	2.50	6.66	-	-	-	-	6.66	nr	7.33
STORM WATER AND FOUL DRAIN SYSTEMS; WORK TO EXISTING DRAINS										
Work to disused drains										
Fly ash filling to disused drain										
100mm diameter	0.66	-	0.66	0.40	-	7.35	-	8.01	m	8.81
150mm diameter	1.21	-	1.21	0.60	-	11.02	-	12.23	m	13.46
225mm diameter	2.60	-	2.60	0.90	-	16.53	-	19.13	m	21.04
Sealing end of disused drain with plain in-situ concrete										
100mm diameter	1.04	5.00	1.09	0.20	-	3.67	-	4.76	nr	5.24
150mm diameter	1.96	5.00	2.06	0.30	-	5.51	-	7.57	nr	8.33

Labour hourly rates: (except Specialists) Craft Operatives 18.37 Labourer 13.76 Rates are national average prices. Refer to REGIONAL VARIATIONS for indicative levels of overall pricing in regions	MATERIALS			LABOUR				RATES		
	Del to Site £	Waste %	Material Cost £	Craft Optve Hrs	Lab Hrs	Labour Cost £	Sunds £	Nett Rate £	Unit	Gross rate (10%) £
STORM WATER AND FOUL DRAIN SYSTEMS; WORK TO EXISTING DRAINS (Cont'd)										
Work to disused drains (Cont'd)										
Sealing end of disused drain with plain in-situ concrete (Cont'd)										
225mm diameter......................................	3.14	5.00	3.29	0.40	-	7.35	-	10.64	nr	11.70
LAND DRAINAGE										
Drains; clayware pipes, BS 1196; butt joints										
Pipework in trenches										
75mm	7.80	5.00	8.19	-	0.22	3.03	-	11.22	m	12.34
100mm	11.07	5.00	11.62	-	0.24	3.30	-	14.92	m	16.41
150mm	25.50	5.00	26.77	-	0.30	4.13	-	30.90	m	33.99
Drains; perforated plastic twinwall land drain pipes										
Pipework in trenches										
80mm	0.88	5.00	0.92	-	0.20	2.75	-	3.68	m	4.04
100mm	1.33	5.00	1.39	-	0.22	3.03	-	4.42	m	4.86
160mm	3.30	5.00	3.46	-	0.25	3.44	-	6.90	m	7.60
Drains; Concrete; integral jointed pipes, BS EN1916										
Pipework in trenches										
300mm	22.36	5.00	23.48	-	0.42	5.78	-	29.26	m	32.19
375mm	27.73	5.00	29.12	-	0.48	6.60	-	35.72	m	39.29
450mm	32.69	5.00	34.33	-	0.54	7.43	-	41.76	m	45.93
525mm	42.90	5.00	45.04	-	0.60	8.26	-	53.30	m	58.63
600mm	53.88	5.00	56.57	-	0.66	9.08	-	65.65	m	72.22
Drains; clayware perforated pipes, butt joints; 150mm hardcore to sides and top										
Pipework in trenches										
100mm	17.91	5.00	19.18	-	0.42	5.78	-	24.96	m	27.45
150mm	35.49	5.00	37.72	-	0.54	7.43	-	45.15	m	49.67
225mm	63.74	5.00	67.57	-	0.60	8.26	-	75.82	m	83.40

Labour hourly rates: (except Specialists) Craft Operatives 18.37 Labourer 13.76 Rates are national average prices. Refer to REGIONAL VARIATIONS for indicative levels of overall pricing in regions	MATERIALS			LABOUR				RATES		
	Del to Site	Waste	Material Cost	Craft Optve	Lab	Labour Cost	Sunds	Nett Rate		Gross rate (10%)
	£	%	£	Hrs	Hrs	£	£	£	Unit	£
SITE WORKS										
Note										
Notwithstanding the requirements of NRM2 excavation and disposal has not been included with the items and should be added as appropriate and as detailed below.										
ROAD AND PATH PAVINGS; GRANULAR SUB-BASES										
Hard, dry, broken brick or stone to be obtained off site; wheeling not exceeding 25m - by machine										
Filling to make up levels; depositing in layers 150mm maximum thickness										
average thickness not exceeding 0.25m	25.06	25.00	31.32	-	-	-	4.39	35.72	m³	39.29
average thickness exceeding 0.25m	25.06	25.00	31.32	-	-	-	3.64	34.97	m³	38.46
Surface packing to filling										
to vertical or battered faces	-	-	-	-	-	-	0.75	0.75	m²	0.83
Compacting with 680 kg vibratory roller										
filling; blinding with sand, ashes or similar fine material	0.34	33.00	0.45	-	0.05	0.69	0.27	1.41	m²	1.55
Compacting with 6 - 8 tonnes smooth wheeled roller										
filling; blinding with sand, ashes or similar fine material	0.34	33.00	0.45	-	0.05	0.69	0.52	1.67	m²	1.83
Hard, dry, broken brick or stone to be obtained off site; wheeling not exceeding 50m - by machine										
Filling to make up levels; depositing in layers 150mm maximum thickness										
average thickness not exceeding 0.25m	25.06	25.00	31.32	-	-	-	5.11	36.43	m³	40.08
average thickness exceeding 0.25m	25.06	25.00	31.32	-	-	-	4.39	35.72	m³	39.29
Surface packing to filling										
to vertical or battered faces	-	-	-	-	-	-	0.75	0.75	m²	0.83
Compacting with 680 kg vibratory roller										
filling; blinding with sand, ashes or similar fine material	0.34	33.00	0.45	-	0.05	0.69	0.27	1.41	m²	1.55
Compacting with 6 - 8 tonnes smooth wheeled roller										
filling; blinding with sand, ashes or similar fine material	0.34	33.00	0.45	-	0.05	0.69	0.48	1.62	m²	1.79
MOT Type 1 to be obtained off site; wheeling not exceeding 25m - by machine										
Filling to make up levels; depositing in layers 150mm maximum thickness										
average thickness not exceeding 0.25m	34.56	25.00	43.20	-	-	-	5.11	48.32	m³	53.15
average thickness exceeding 0.25m	34.56	25.00	43.20	-	-	-	4.24	47.44	m³	52.19
Surface packing to filling										
to vertical or battered faces	-	-	-	-	-	-	0.75	0.75	m²	0.83
Compacting with 680 kg vibratory roller										
filling; blinding with sand, ashes or similar fine material	0.34	33.00	0.45	-	0.05	0.69	0.27	1.41	m²	1.55
Compacting with 6 - 8 tonnes smooth wheeled roller										
filling; blinding with sand, ashes or similar fine material	0.34	33.00	0.45	-	0.05	0.69	0.52	1.67	m²	1.83
MOT Type 1 to be obtained off site; wheeling not exceeding 50m - by machine										
Filling to make up levels; depositing in layers 150mm maximum thickness										
average thickness not exceeding 0.25m	34.56	25.00	43.20	-	-	-	5.11	48.32	m³	53.15
average thickness exceeding 0.25m	34.56	25.00	43.20	-	-	-	4.39	47.60	m³	52.35
Surface packing to filling										
to vertical or battered faces	-	-	-	-	-	-	0.75	0.75	m²	0.83
Compacting with 680 kg vibratory roller										
filling; blinding with sand, ashes or similar fine material	0.34	33.00	0.45	-	0.05	0.69	0.27	1.41	m²	1.55
Compacting with 6 - 8 tonnes smooth wheeled roller										
filling; blinding with sand, ashes or similar fine material	0.34	33.00	0.45	-	0.05	0.69	0.52	1.67	m²	1.83
MOT Type 2 to be obtained off site; wheeling not exceeding 25m - by machine										
Filling to make up levels; depositing in layers 150mm maximum thickness										
average thickness not exceeding 0.25m	34.56	25.00	43.20	-	-	-	4.39	47.60	m³	52.35
average thickness exceeding 0.25m	34.56	25.00	43.20	-	-	-	3.64	46.85	m³	51.53
Surface packing to filling										
to vertical or battered faces	-	-	-	-	-	-	0.75	0.75	m²	0.83

	MATERIALS			LABOUR				RATES		
Labour hourly rates: (except Specialists) Craft Operatives 18.37 Labourer 13.76 Rates are national average prices. Refer to REGIONAL VARIATIONS for indicative levels of overall pricing in regions	Del to Site £	Waste %	Material Cost £	Craft Optve Hrs	Lab Hrs	Labour Cost £	Sunds £	Nett Rate £	Unit	Gross rate (10%) £

ROAD AND PATH PAVINGS; GRANULAR SUB-BASES (Cont'd)

MOT Type 2 to be obtained off site; wheeling not exceeding 25m - by machine (Cont'd)

Compacting with 680 kg vibratory roller filling; blinding with sand, ashes or similar fine material	0.34	33.00	0.45	-	0.05	0.69	0.27	1.41	m²	1.55
Compacting with 6 - 8 tonnes smooth wheeled roller filling; blinding with sand, ashes or similar fine material	0.34	33.00	0.45	-	0.05	0.69	0.52	1.67	m²	1.83

MOT Type 2 to be obtained off site; wheeling not exceeding 50m - by machine

Filling to make up levels; depositing in layers 150mm maximum thickness										
average thickness not exceeding 0.25m	34.56	25.00	43.20	-	-	-	5.11	48.32	m³	53.15
average thickness exceeding 0.25m	34.56	25.00	43.20	-	-	-	4.39	47.60	m³	52.35
Surface packing to filling to vertical or battered faces	-	-	-	-	-	-	0.75	0.75	m²	0.83
Compacting with 680 kg vibratory roller filling; blinding with sand, ashes or similar fine material	0.34	33.00	0.45	-	0.05	0.69	0.27	1.41	m²	1.55
Compacting with 6 - 8 tonnes smooth wheeled roller filling; blinding with sand, ashes or similar fine material	0.34	33.00	0.45	-	0.05	0.69	0.52	1.67	m²	1.83

Sand to be obtained off site; wheeling not exceeding 25m - by machine

Filling to make up levels; depositing in layers 150mm maximum thickness										
average thickness not exceeding 0.25m	22.80	33.00	30.32	-	-	-	4.65	34.97	m³	38.47
Compacting with 680 kg vibratory roller filling	-	-	-	-	-	-	0.27	0.27	m²	0.30
Compacting with 6 - 8 tonnes smooth wheeled roller filling	-	-	-	-	-	-	0.52	0.52	m²	0.58

Sand to be obtained off site; wheeling not exceeding 50m - by machine

Filling to make up levels; depositing in layers 150mm maximum thickness										
average thickness not exceeding 0.25m	22.80	33.00	30.32	-	-	-	6.28	36.61	m³	40.27
Compacting with 680 kg vibratory roller filling	-	-	-	-	-	-	0.27	0.27	m²	0.30
Compacting with 6 - 8 tonnes smooth wheeled roller filling	-	-	-	-	-	-	0.52	0.52	m²	0.58

Hoggin to be obtained off site; wheeling not exceeding 25m - by machine

Filling to make up levels; depositing in layers 150mm maximum thickness										
average thickness not exceeding 0.25m	29.76	33.00	39.58	-	-	-	5.10	44.68	m³	49.15
average thickness exceeding 0.25m	29.76	33.00	39.58	-	-	-	4.24	43.83	m³	48.21
Compacting with 680 kg vibratory roller filling	-	-	-	-	-	-	0.27	0.27	m²	0.30
Compacting with 6 - 8 tonnes smooth wheeled roller filling	-	-	-	-	-	-	0.52	0.52	m²	0.58

Hoggin to be obtained off site; wheeling not exceeding 50m - by machine

Filling to make up levels; depositing in layers 150mm maximum thickness										
average thickness not exceeding 0.25m	29.76	33.00	39.58	-	-	-	4.39	43.98	m³	48.37
average thickness exceeding 0.25m	29.76	33.00	39.58	-	-	-	3.64	43.23	m³	47.55
Compacting with 680 kg vibratory roller filling	-	-	-	-	-	-	0.27	0.27	m²	0.30
Compacting with 6 - 8 tonnes smooth wheeled roller filling	-	-	-	-	-	-	0.52	0.52	m²	0.58

Hard, dry, broken brick or stone to be obtained off site; wheeling not exceeding 25m - by hand

Filling to make up levels; depositing in layers 150mm maximum thickness										
average thickness not exceeding 0.25m	25.06	25.00	31.32	-	1.50	20.64	-	51.96	m³	57.16
average thickness exceeding 0.25m	25.06	25.00	31.32	-	1.25	17.20	-	48.52	m³	53.37
Compacting with 680 kg vibratory roller filling; blinding with sand, ashes or similar fine material	0.34	33.00	0.45	-	0.05	0.69	0.27	1.41	m²	1.55
Compacting with 6 - 8 tonnes smooth wheeled roller filling; blinding with sand, ashes or similar fine material	0.34	33.00	0.45	-	0.05	0.69	0.52	1.67	m²	1.83

Labour hourly rates: (except Specialists) Craft Operatives 18.37 Labourer 13.76 Rates are national average prices. Refer to REGIONAL VARIATIONS for indicative levels of overall pricing in regions	MATERIALS			LABOUR				RATES		
	Del to Site £	Waste %	Material Cost £	Craft Optve Hrs	Lab Hrs	Labour Cost £	Sunds £	Nett Rate £	Unit	Gross rate (10%) £
ROAD AND PATH PAVINGS; GRANULAR SUB-BASES (Cont'd)										
Hard, dry, broken brick or stone to be obtained off site; wheeling not exceeding 50m - by hand										
Filling to make up levels; depositing in layers 150mm maximum thickness										
average thickness not exceeding 0.25m	25.06	25.00	31.32	-	1.75	24.08	-	55.40	m³	60.94
average thickness exceeding 0.25m	25.06	25.00	31.32	-	1.50	20.64	-	51.96	m³	57.16
Compacting with 680 kg vibratory roller										
filling; blinding with sand, ashes or similar fine material	0.34	33.00	0.45	-	0.05	0.69	0.27	1.41	m²	1.55
Compacting with 6 - 8 tonnes smooth wheeled roller										
filling; blinding with sand, ashes or similar fine material	0.34	33.00	0.45	-	0.05	0.69	0.52	1.67	m²	1.83
MOT Type 1 to be obtained off site; wheeling not exceeding 25m - by hand										
Filling to make up levels; depositing in layers 150mm maximum thickness										
average thickness not exceeding 0.25m	34.56	25.00	43.20	-	1.50	20.64	-	63.84	m³	70.22
average thickness exceeding 0.25m	34.56	25.00	43.20	-	1.25	17.20	-	60.40	m³	66.44
Compacting with 680 kg vibratory roller										
filling; blinding with sand, ashes or similar fine material	0.34	33.00	0.45	-	0.05	0.69	0.27	1.41	m²	1.55
Compacting with 6 - 8 tonnes smooth wheeled roller										
filling; blinding with sand, ashes or similar fine material	0.34	33.00	0.45	-	0.05	0.69	0.52	1.67	m²	1.83
MOT Type 1 to be obtained off site; wheeling not exceeding 50m - by hand										
Filling to make up levels; depositing in layers 150mm maximum thickness										
average thickness not exceeding 0.25m	34.56	25.00	43.20	-	1.75	24.08	-	67.28	m³	74.01
average thickness exceeding 0.25m	34.56	25.00	43.20	-	1.50	20.64	-	63.84	m³	70.22
Compacting with 680 kg vibratory roller										
filling; blinding with sand, ashes or similar fine material	0.34	33.00	0.45	-	0.05	0.69	0.27	1.41	m²	1.55
Compacting with 6 - 8 tonnes smooth wheeled roller										
filling; blinding with sand, ashes or similar fine material	0.34	33.00	0.45	-	0.05	0.69	0.52	1.67	m²	1.83
MOT Type 2 to be obtained off site; wheeling not exceeding 25m - by hand										
Filling to make up levels; depositing in layers 150mm maximum thickness										
average thickness not exceeding 0.25m	34.56	25.00	43.20	-	1.50	20.64	-	63.84	m³	70.22
average thickness exceeding 0.25m	34.56	25.00	43.20	-	1.25	17.20	-	60.40	m³	66.44
Compacting with 680 kg vibratory roller										
filling; blinding with sand, ashes or similar fine material	0.34	33.00	0.45	-	0.05	0.69	0.27	1.41	m²	1.55
Compacting with 6 - 8 tonnes smooth wheeled roller										
filling; blinding with sand, ashes or similar fine material	0.34	33.00	0.45	-	0.05	0.69	0.52	1.67	m²	1.83
MOT Type 2 to be obtained off site; wheeling not exceeding 50m - by hand										
Filling to make up levels; depositing in layers 150mm maximum thickness										
average thickness not exceeding 0.25m	34.56	25.00	43.20	-	1.75	24.08	-	67.28	m³	74.01
average thickness exceeding 0.25m	34.56	25.00	43.20	-	1.50	20.64	-	63.84	m³	70.22
Compacting with 680 kg vibratory roller										
filling; blinding with sand, ashes or similar fine material	0.34	33.00	0.45	-	0.05	0.69	0.27	1.41	m²	1.55
Compacting with 6 - 8 tonnes smooth wheeled roller										
filling; blinding with sand, ashes or similar fine material	0.34	33.00	0.45	-	0.05	0.69	0.52	1.67	m²	1.83
Sand to be obtained off site; wheeling not exceeding 25m - by hand										
Filling to make up levels; depositing in layers 150mm maximum thickness										
average thickness not exceeding 0.25m	22.80	33.00	30.32	-	1.60	22.02	-	52.34	m³	57.57
Compacting with 680 kg vibratory roller										
filling	-	-	-	-	-	-	0.27	0.27	m²	0.30
Compacting with 6 - 8 tonnes smooth wheeled roller										
filling	-	-	-	-	-	-	0.52	0.52	m²	0.58
Sand to be obtained off site; wheeling not exceeding 50m - by hand										
Filling to make up levels; depositing in layers 150mm maximum thickness										
average thickness not exceeding 0.25m	22.80	33.00	30.32	-	1.85	25.46	-	55.78	m³	61.36
Compacting with 680 kg vibratory roller										
filling	-	-	-	-	-	-	0.27	0.27	m²	0.30
Compacting with 6 - 8 tonnes smooth wheeled roller										
filling	-	-	-	-	-	-	0.52	0.52	m²	0.58

Labour hourly rates: (except Specialists) Craft Operatives 18.37 Labourer 13.76 Rates are national average prices. Refer to REGIONAL VARIATIONS for indicative levels of overall pricing in regions	MATERIALS			LABOUR				RATES		
	Del to Site	Waste	Material Cost	Craft Optve	Lab	Labour Cost	Sunds	Nett Rate	Unit	Gross rate (10%)
	£	%	£	Hrs	Hrs	£	£	£		£
ROAD AND PATH PAVINGS; GRANULAR SUB-BASES (Cont'd)										
Hoggin to be obtained off site; wheeling not exceeding 25m - by hand										
Filling to make up levels; depositing in layers 150mm maximum thickness										
average thickness not exceeding 0.25m	29.76	33.00	39.58	-	1.50	20.64	-	60.22	m³	66.24
average thickness exceeding 0.25m	29.76	33.00	39.58	-	1.25	17.20	-	56.78	m³	62.46
Compacting with 680 kg vibratory roller										
filling	-	-	-	-	-	-	0.27	0.27	m²	0.30
Compacting with 6 - 8 tonnes smooth wheeled roller										
filling	-	-	-	-	-	-	0.52	0.52	m²	0.58
Hoggin to be obtained off site; wheeling not exceeding 50m - by hand										
Filling to make up levels; depositing in layers 150mm maximum thickness										
average thickness not exceeding 0.25m	29.76	33.00	39.58	-	1.75	24.08	-	63.66	m³	70.03
average thickness exceeding 0.25m	29.76	33.00	39.58	-	1.50	20.64	-	60.22	m³	66.24
Compacting with 680 kg vibratory roller										
filling	-	-	-	-	-	-	0.27	0.27	m²	0.30
Compacting with 6 - 8 tonnes smooth wheeled roller										
filling	-	-	-	-	-	-	0.52	0.52	m²	0.58
ROAD AND PATH PAVINGS; IN-SITU CONCRETE										
Plain in-situ concrete; mix 1:8, all in aggregate										
Beds; poured on or against earth or unblinded hardcore										
thickness not exceeding 150mm	130.10	7.50	139.86	-	4.68	64.40	2.70	206.96	m³	227.65
Plain in-situ concrete; mix 1:6; all in aggregate										
Beds; poured on or against earth or unblinded hardcore										
thickness 150 - 450mm	137.06	7.50	147.34	-	4.40	60.54	2.70	210.58	m³	231.64
thickness not exceeding 150mm	137.06	7.50	147.34	-	4.68	64.40	2.70	214.43	m³	235.88
Plain in-situ concrete; BS 8500, ordinary prescribed mix ST4, 20mm aggregate										
Beds										
thickness not exceeding 150mm	99.73	5.00	104.72	-	4.68	64.40	2.70	171.82	m³	189.00
Beds; poured on or against earth or unblinded hardcore										
thickness not exceeding 150mm	99.73	5.00	104.72	-	4.68	64.40	2.70	171.82	m³	189.00
Reinforced in-situ concrete; BS 8500, designed mix C20, 20mm aggregate, minimum cement content 240 kg/m³; vibrated										
Beds										
thickness 150 - 450mm	96.25	5.00	101.06	-	4.68	64.40	0.75	166.21	m³	182.83
thickness not exceeding 150mm	96.25	5.00	101.06	-	4.95	68.11	0.75	169.92	m³	186.91
Formwork and basic finish										
Edges of beds										
height not exceeding 250mm	1.16	12.50	1.30	0.58	0.12	12.31	0.36	13.97	m	15.37
height 250 - 500mm	2.04	12.50	2.29	1.06	0.21	22.36	0.68	25.33	m	27.86
Formwork steel forms										
Edges of beds										
height not exceeding 250mm	3.22	15.00	3.71	0.40	0.08	8.45	0.36	12.51	m	13.77
height 250 - 500mm	5.56	15.00	6.39	0.70	0.14	14.79	0.68	21.85	m	24.03
Reinforcement fabric; BS 4483, hard drawn plain round steel; welded; including laps, tying wire, all cutting and bending, and spacers and chairs which are at the discretion of the Contractor										
Reference C283, 2.61 kg/m²; 100mm side laps; 400mm end laps										
generally	2.14	17.50	2.51	0.06	0.02	1.38	0.10	3.99	m²	4.39
Reference C385, 3.41 kg/m²; 100mm side laps; 400mm end laps										
generally	2.73	17.50	3.21	0.06	0.02	1.38	0.14	4.72	m²	5.20
Reference C503, 4.34 kg/m²; 100mm side laps; 400mm end laps										
generally	3.48	17.50	4.09	0.07	0.02	1.56	0.17	5.81	m²	6.39
Reference C636, 5.55 kg/m²; 100mm side laps; 400mm end laps										
generally	4.45	17.50	5.23	0.07	0.02	1.56	0.21	7.00	m²	7.70
Reference C785, 6.72 kg/m²; 100mm side laps; 400mm end laps										
generally	5.39	17.50	6.33	0.08	0.02	1.74	0.25	8.33	m²	9.16
Formed joints										
Sealant to joint; Grace Construction Products. Serviced joints, Paraseal (two-part sealant) including preparation, cleaners and primers										
10 x 25mm	3.55	5.00	3.73	-	0.11	1.51	-	5.24	m	5.76
12.5 x 25mm	4.62	5.00	4.85	-	0.12	1.65	-	6.50	m	7.15

Labour hourly rates: (except Specialists) Craft Operatives 18.37 Labourer 13.76 Rates are national average prices. Refer to REGIONAL VARIATIONS for indicative levels of overall pricing in regions	MATERIALS			LABOUR				RATES		
	Del to Site	Waste	Material Cost	Craft Optve	Lab	Labour Cost	Sunds	Nett Rate	Unit	Gross rate (10%)
	£	%	£	Hrs	Hrs	£	£	£		£
ROAD AND PATH PAVINGS; IN-SITU CONCRETE (Cont'd)										
Worked finish on in-situ concrete										
Tamping by mechanical means										
level surfaces	-	-	-	-	0.16	2.20	0.44	2.64	m²	2.90
sloping surfaces...	-	-	-	-	0.33	4.54	0.57	5.11	m²	5.62
surfaces to falls...	-	-	-	-	0.44	6.05	0.57	6.62	m²	7.29
Trowelling										
level surfaces	-	-	-	-	0.36	4.95	-	4.95	m²	5.45
sloping surfaces...	-	-	-	-	0.44	6.05	-	6.05	m²	6.66
surfaces to falls...	-	-	-	-	0.55	7.57	-	7.57	m²	8.32
Rolling with an indenting roller										
level surfaces	-	-	-	-	0.11	1.51	0.42	1.93	m²	2.13
sloping surfaces...	-	-	-	-	0.17	2.34	0.42	2.76	m²	3.04
surfaces to falls...	-	-	-	-	0.33	4.54	0.54	5.08	m²	5.59
Accessories cast into in-situ concrete										
Mild steel dowels; half coated with bitumen										
16mm diameter x 400mm long......................	0.45	10.00	0.50	-	0.06	0.83	0.27	1.59	nr	1.75
20mm diameter x 500mm long; with plastic compression cap........	0.86	10.00	0.95	-	0.08	1.10	0.27	2.32	nr	2.55
ROAD AND PATH PAVINGS; COATED MACADAM/ASPHALT										
Coated macadam, BS 594987; base course 28mm nominal size aggregate, 50mm thick; wearing course 6mm nominal size aggregate, 20mm thick; rolled with 3 - 4 tonne roller; external										
70mm roads on concrete base; to falls and crossfalls, and slopes not exceeding 15 degrees from horizontal										
generally..............	-	-	Specialist	-	-	Specialist	-	20.73	m²	22.80
70mm pavings on concrete base; to falls and crossfalls, and slopes not exceeding 15 degrees from horizontal										
generally..............	-	-	Specialist	-	-	Specialist	-	23.06	m²	25.36
Coated macadam, BS 594987; base course 28mm nominal size aggregate, 50mm thick; wearing course 6mm nominal size aggregate, 20mm thick; rolled with 3 - 4 tonne roller; dressing surface with coated grit brushed on and lightly rolled; external										
70mm roads on concrete base; to falls and crossfalls, and slopes not exceeding 15 degrees from horizontal										
generally..............	-	-	Specialist	-	-	Specialist	-	25.11	m²	27.62
70mm pavings on concrete base; to falls and crossfalls, and slopes not exceeding 15 degrees from horizontal										
generally..............	-	-	Specialist	-	-	Specialist	-	27.38	m²	30.11
Coated macadam, BS 594987; base course 40mm nominal size aggregate, 65mm thick; wearing course 10mm nominal size aggregate, 25mm thick; rolled with 6 - 8 tonne roller; external										
90mm roads on concrete base; to falls and crossfalls, and slopes not exceeding 15 degrees from horizontal										
generally..............	-	-	Specialist	-	-	Specialist	-	25.11	m²	27.62
90mm pavings on concrete base; to falls and crossfalls, and slopes not exceeding 15 degrees from horizontal										
generally..............	-	-	Specialist	-	-	Specialist	-	27.54	m²	30.29
Coated macadam, BS 594987; base course 40mm nominal size aggregate, 65mm thick; wearing course 10mm nominal size aggregate, 25mm thick; rolled with 6 - 8 tonne roller; dressing surface with coated grit brushed on and lightly rolled; external										
90mm roads on concrete base; to falls and crossfalls, and slopes not exceeding 15 degrees from horizontal										
generally..............	-	-	Specialist	-	-	Specialist	-	29.79	m²	32.77
90mm pavings on concrete base; to falls and crossfalls, and slopes not exceeding 15 degrees from horizontal										
generally..............	-	-	Specialist	-	-	Specialist	-	30.83	m²	33.91
Fine cold asphalt, BS 594987; single course 6mm nominal size aggregate; rolled with 3 - 4 tonne roller; external										
12mm roads on concrete base; to falls and crossfalls, and slopes not exceeding 15 degrees from horizontal										
generally..............	-	-	Specialist	-	-	Specialist	-	9.45	m²	10.40
19mm roads on concrete base; to falls and crossfalls, and slopes not exceeding 15 degrees from horizontal										
generally..............	-	-	Specialist	-	-	Specialist	-	10.95	m²	12.05
25mm roads on concrete base; to falls and crossfalls, and slopes not exceeding 15 degrees from horizontal										
generally..............	-	-	Specialist	Craft -	-	Specialist	-	13.01	m²	14.31
12mm pavings on concrete base; to falls and crossfalls, and slopes not exceeding 15 degrees from horizontal										
generally..............	-	-	Specialist	-	Hrs -	Specialist	-	10.02	m²	11.02

Labour hourly rates: (except Specialists) Craft Operatives 18.37 Labourer 13.76 Rates are national average prices. Refer to REGIONAL VARIATIONS for indicative levels of overall pricing in regions	MATERIALS			LABOUR				RATES		
	Del to Site £	Waste %	Material Cost £	Craft Optve Hrs	Lab Hrs	Labour Cost £	Sunds £	Nett Rate £	Unit	Gross rate (10%) £
ROAD AND PATH PAVINGS; COATED MACADAM/ASPHALT (Cont'd)										
Fine cold asphalt, BS 594987; single course 6mm nominal size aggregate; rolled with 3 - 4 tonne roller; external (Cont'd)										
19mm pavings on concrete base; to falls and crossfalls, and slopes not exceeding 15 degrees from horizontal										
generally..........	-	-	Specialist	-	-	Specialist	-	12.57	m²	13.83
25mm pavings on concrete base; to falls and crossfalls, and slopes not exceeding 15 degrees from horizontal										
generally..........	-	-	Specialist	-	-	Specialist	-	14.39	m²	15.82
ROAD AND PATH PAVINGS; GRAVEL/HOGGIN/WOODCHIP										
Gravel paving; first layer MOT type 1 75mm thick; intermediate layer coarse gravel 75mm thick; wearing layer blinding gravel 38mm thick; well water and roll										
188mm roads on blinded hardcore base; to falls and crossfalls and slopes not exceeding 15 degrees from horizontal										
generally..........	11.36	10.00	12.88	-	0.44	6.05	-	19.77	m²	21.75
Gravel paving; single layer blinding gravel; well water and roll										
50mm pavings on blinded hardcore base; to falls and crossfalls and slopes not exceeding 15 degrees from horizontal										
generally..........	3.11	10.00	3.42	-	0.15	2.06	-	5.69	m²	6.26
Gravel paving; first layer sand 50mm thick; wearing layer blinding gravel 38mm thick; well water and roll										
88mm pavings on blinded hardcore base; to falls and crossfalls and slopes not exceeding 15 degrees from horizontal										
generally..........	4.77	10.00	5.24	-	0.26	3.58	-	9.03	m²	9.93
extra; "Colas" and 10mm shingle dressing; one coat	9.70	5.00	10.19	-	0.17	2.34	-	12.53	m²	13.78
extra; "Colas" and 10mm shingle dressing; two coats..................	12.94	5.00	13.59	-	0.27	3.72	-	17.30	m²	19.03

This section continues
on the next page

SITE WORKS

Labour hourly rates: (except Specialists) Craft Operatives 18.37 Labourer 13.76. Rates are national average prices. Refer to REGIONAL VARIATIONS for indicative levels of overall pricing in regions	PLANT AND TRANSPORT			LABOUR				RATES		
	Plant Cost	Trans Cost	P and T Cost	Craft Optve	Lab	Labour Cost	Sunds	Nett Rate	Unit	Gross rate (10%)
	£	£	£	Hrs	Hrs	£	£	£		£

ROAD AND PATH PAVINGS; KERBS/EDGINGS/CHANNELS/PAVING ACCESSORIES

Excavating - by machine

	Plant Cost £	Trans Cost £	P and T Cost £	Craft Optve Hrs	Lab Hrs	Labour Cost £	Sunds £	Nett Rate £	Unit	Gross rate (10%) £
Trenches width not exceeding 0.30m										
maximum depth not exceeding 0.25m	8.27	-	8.27	-	1.71	23.53	-	31.79	m³	34.97
maximum depth not exceeding 1.00m	8.60	-	8.60	-	1.80	24.77	-	33.37	m³	36.71
Extra over any types of excavating irrespective of depth										
excavating below ground water level	7.29	-	7.29	-	-	-	-	7.29	m³	8.02
excavating in running silt, running sand or liquid mud	12.52	-	12.52	-	-	-	-	12.52	m³	13.77
Breaking out existing materials; extra over any types of excavating irrespective of depth										
hard rock	20.42	-	20.42	-	-	-	-	20.42	m³	22.46
concrete	16.96	-	16.96	-	-	-	-	16.96	m³	18.65
reinforced concrete	25.47	-	25.47	-	-	-	-	25.47	m³	28.02
brickwork, blockwork or stonework	11.32	-	11.32	-	-	-	-	11.32	m³	12.45
100mm diameter drain and concrete bed under	1.59	-	1.59	-	-	-	-	1.59	m	1.75
Breaking out existing hard pavings; extra over any types of excavating irrespective of depth										
concrete 150mm thick	2.67	-	2.67	-	-	-	-	2.67	m²	2.94
reinforced concrete 200mm thick	5.36	-	5.36	-	-	-	-	5.36	m²	5.90
brickwork, blockwork or stonework 100mm thick	1.26	-	1.26	-	-	-	-	1.26	m²	1.39
coated macadam or asphalt 75mm thick	0.45	-	0.45	-	-	-	-	0.45	m²	0.50
Disposal of excavated material										
depositing on site in temporary spoil heaps where directed										
25m	-	3.66	3.66	-	0.12	1.65	-	5.31	m³	5.84
100m	-	4.17	4.17	-	0.12	1.65	-	5.82	m³	6.41
800m	-	6.42	6.42	-	0.12	1.65	-	8.07	m³	8.87
1600m	-	7.26	7.26	-	0.12	1.65	-	8.91	m³	9.80
extra for each additional 1600m	-	1.13	1.13	-	-	-	-	1.13	m³	1.24
removing from site to tip a) Inert	-	34.58	34.58	-	-	-	-	34.58	m³	38.04
removing from site to tip b) Active	-	71.46	71.46	-	-	-	-	71.46	m³	78.61
removing from site to tip c) Contaminated (Guide price - always seek a quotation for specialist disposal costs.)	-	207.47	207.47	-	-	-	-	207.47	m³	228.22
Disposal of preserved top soil										
depositing on site in temporary spoil heaps where directed										
25m	-	3.66	3.66	-	0.12	1.65	-	5.31	m³	5.84
50m	-	3.82	3.82	-	0.12	1.65	-	5.47	m³	6.02
100m	-	4.17	4.17	-	0.12	1.65	-	5.82	m³	6.41

Surface treatments

	Plant Cost £	Trans Cost £	P and T Cost £	Craft Optve Hrs	Lab Hrs	Labour Cost £	Sunds £	Nett Rate £	Unit	Gross rate (10%) £
Compacting										
bottoms of excavations	0.41	-	0.41	-	-	-	-	0.41	m²	0.45

Excavating - by hand

	Plant Cost £	Trans Cost £	P and T Cost £	Craft Optve Hrs	Lab Hrs	Labour Cost £	Sunds £	Nett Rate £	Unit	Gross rate (10%) £
Trenches width not exceeding 0.30m										
maximum depth not exceeding 0.25m	-	-	-	-	4.25	58.48	-	58.48	m³	64.33
maximum depth not exceeding 1.00m	-	-	-	-	4.50	61.92	-	61.92	m³	68.11
Extra over any types of excavating irrespective of depth										
excavating below ground water level	-	-	-	-	1.50	20.64	-	20.64	m³	22.70
excavating in running silt, running sand or liquid mud	-	-	-	-	4.50	61.92	-	61.92	m³	68.11
Breaking out existing materials; extra over any types of excavating irrespective of depth										
hard rock	-	-	-	-	10.00	137.60	-	137.60	m³	151.36
concrete	-	-	-	-	7.00	96.32	-	96.32	m³	105.95
reinforced concrete	-	-	-	-	12.00	165.12	-	165.12	m³	181.63
brickwork, blockwork or stonework	-	-	-	-	4.00	55.04	-	55.04	m³	60.54
100mm diameter drain and concrete bed under	-	-	-	-	1.00	13.76	-	13.76	m	15.14
Breaking out existing hard pavings; extra over any types of excavating irrespective of depth										
concrete 150mm thick	-	-	-	-	1.50	20.64	-	20.64	m²	22.70
reinforced concrete 200mm thick	-	-	-	-	2.00	27.52	-	27.52	m²	30.27
brickwork, blockwork or stonework 100mm thick	-	-	-	-	0.70	9.63	-	9.63	m²	10.60
coated macadam or asphalt 75mm thick	-	-	-	-	0.25	3.44	-	3.44	m²	3.78
Disposal of excavated material										
depositing on site in temporary spoil heaps where directed										
25m	-	-	-	-	1.25	17.20	-	17.20	m³	18.92
50m	-	-	-	-	1.50	20.64	-	20.64	m³	22.70
100m	-	-	-	-	2.00	27.52	-	27.52	m³	30.27
extra for removing beyond 100m and not exceeding 400m	-	-	-	-	1.55	21.33	3.27	24.60	m³	27.06
extra for removing beyond 100m and not exceeding 800m	-	-	-	-	1.55	21.33	4.26	25.59	m³	28.15
depositing on site in permanent spoil heaps average 25m distant	-	-	-	-	1.25	17.20	-	17.20	m³	18.92
removing from site to tip										
a) Inert	-	-	-	-	1.35	18.58	22.97	41.55	m³	45.70
b) Active	-	-	-	-	1.35	18.58	194.64	213.22	m³	234.54
c) Contaminated (guide price - always seek a quote for specialist disposal cost	-	-	-	-	1.35	18.58	247.40	265.98	m³	292.57

SITE WORKS

Labour hourly rates: (except Specialists) Craft Operatives 18.37 Labourer 13.76 Rates are national average prices. Refer to REGIONAL VARIATIONS for indicative levels of overall pricing in regions	PLANT AND TRANSPORT			LABOUR				RATES		
	Plant Cost	Trans Cost	P and T Cost	Craft Optve	Lab	Labour Cost	Sunds	Nett Rate	Unit	Gross rate (10%)
	£	£	£	Hrs	Hrs	£	£	£		£
ROAD AND PATH PAVINGS; **KERBS/EDGINGS/CHANNELS/PAVING ACCESSORIES** **(Cont'd)**										
Surface treatments										
Compacting bottoms of excavations...	-	-	-	-	0.12	1.65	-	1.65	m²	1.82

This section continues
on the next page

SITE WORKS

	Plant Cost	Trans Cost	P and T Cost	Craft Optve	Lab	Labour Cost	Sunds	Nett Rate	Unit	Gross rate (10%)
ROAD AND PATH PAVINGS; **KERBS/EDGINGS/CHANNELS/PAVING ACCESSORIES** **(Cont'd)**	£	£	£	Hrs	Hrs					£

Labour hourly rates: (except Specialists) Craft Operatives 18.37 Labourer 13.76 Rates are national average prices. Refer to REGIONAL VARIATIONS for indicative levels of overall pricing in regions	MATERIALS			LABOUR				RATES		
	Del to Site	Waste	Material Cost	Craft Optve	Lab	Labour Cost	Sunds	Nett Rate	Unit	Gross rate (10%)
	£	%	£	Hrs	Hrs	£	£	£		£
ROAD AND PATH PAVINGS; KERBS/EDGINGS/CHANNELS/PAVING ACCESSORIES (Cont'd)										
Earthwork support										
Earthwork support maximum depth not exceeding 1.00m; distance between opposing faces not exceeding 2.00m....................	0.81	-	0.81	-	0.20	2.75	0.03	3.60	m²	3.96
Earthwork support; curved Maximum depth not exceeding 1.00m; distance between opposing faces not exceeding 2.00m....................	1.09	-	1.09	-	0.30	4.13	0.06	5.27	m²	5.80
Earthwork support; unstable ground maximum depth not exceeding 1.00m; distance between opposing faces not exceeding 2.00m....................	1.46	-	1.46	-	0.36	4.95	0.03	6.44	m²	7.09
Earthwork support; next to roadways maximum depth not exceeding 1.00m; distance between opposing faces not exceeding 2.00m....................	1.60	-	1.60	-	0.40	5.50	0.06	7.17	m²	7.88
Precast concrete; standard or stock pattern units; BS EN 1340 Part 1; bedding, jointing and pointing in cement mortar (1:3); on plain in-situ concrete foundation; BS 8500 ordinary prescribed mix ST4, 20mm aggregate										
Kerbs; Figs. 1, 2, 6 and 7; concrete foundation and haunching; formwork										
125 x 255mm kerb; 400 x 200mm foundation	16.60	5.00	17.65	-	1.09	15.00	-	32.65	m	35.92
125 x 255mm kerb; 400 x 200mm foundation; curved 10.00m radius	26.01	5.00	27.53	-	1.16	15.96	-	43.49	m	47.84
150 x 305mm kerb; 400 x 200mm foundation	21.25	5.00	22.53	-	1.16	15.96	-	38.49	m	42.34
150 x 305mm kerb; 400 x 200mm foundation; curved 10.00m radius..........	30.20	5.00	31.93	-	1.26	17.34	-	49.27	m	54.20
Edgings; Figs. 10, 11, 12 and 13; concrete foundation and haunching; formwork										
50 x 150mm edging; 300 x 150mm foundation.....................	9.87	5.00	10.49	-	0.78	10.73	-	21.22	m	23.34
50 x 200mm edging; 300 x 150mm foundation.............................	10.66	5.00	11.31	-	0.90	12.38	-	23.70	m	26.07
50 x 250mm edging; 300 x 150mm foundation.............................	12.17	5.00	12.90	-	1.02	14.04	-	26.94	m	29.63
Channels; Fig. 8; concrete foundation and haunching; formwork										
255 x 125mm channel; 450 x 150mm foundation.....................	16.21	5.00	17.19	-	0.86	11.83	-	29.03	m	31.93
255 x 125mm channel; 450 x 150mm foundation; curved 10.00 radius..........	23.01	5.00	24.33	-	0.93	12.80	-	37.13	m	40.84
Quadrants; Fig. 14; concrete foundation and haunching; formwork										
305 x 305 x 150mm quadrant; 500 x 500 x 150mm foundation......	19.31	5.00	20.37	-	0.61	8.39	-	28.77	nr	31.65
305 x 305 x 255mm quadrant; 500 x 500 x 150mm foundation......	18.32	5.00	19.34	-	0.69	9.49	-	28.83	nr	31.71
455 x 455 x 150mm quadrant; 650 x 650 x 150mm foundation......	22.41	5.00	23.67	-	0.85	11.70	-	35.37	nr	38.91
455 x 455 x 255mm quadrant; 650 x 650 x 150mm foundation......	23.40	5.00	24.71	-	0.94	12.93	-	37.65	nr	41.41
Granite; standard units; BS EN 1343; bedding jointing and pointing in cement mortar (1:3); on plain in-situ concrete foundation; BS 8500 ordinary prescribed mix ST4, 20mm aggregate										
Edge kerb; concrete foundation and haunching; formwork										
150 x 300mm; 300 x 200mm foundation.....................	33.12	2.50	34.35	0.55	0.72	20.01	-	54.36	m	59.79
150 x 300mm; 300 x 200mm foundation; curved external radius exceeding 1000mm...............	39.76	2.50	41.27	0.88	0.72	26.07	-	67.35	m	74.08
200 x 300mm; 350 x 200mm foundation	54.76	2.50	56.58	0.66	0.80	23.13	-	79.71	m	87.68
200 x 300mm; 350 x 200mm foundation; curved external radius exceeding 1000mm...............	70.32	2.50	72.65	0.99	0.80	29.19	-	101.84	m	112.03
Flat kerb; concrete foundation and haunching; formwork										
300 x 150mm; 450 x 200mm foundation	36.04	2.50	37.49	0.55	0.80	21.11	-	58.60	m	64.46
300 x 150mm; 450 x 200mm foundation; curved external radius exceeding 1000mm...............	62.83	2.50	65.06	0.88	0.80	27.17	-	92.24	m	101.46
300 x 200mm; 450 x 200mm foundation	56.71	2.50	58.67	0.66	0.83	23.55	-	82.22	m	90.44
300 x 200mm; 450 x 200mm foundation; curved external radius exceeding 1000mm...............	84.83	2.50	87.61	0.99	0.83	29.61	-	117.22	m	128.94
ROAD AND PATH PAVINGS; INTERLOCKING BRICK AND BLOCKS, SLABS, BRICKS, BLOCKS, SETTS AND COBBLES										
York stone paving; 13mm thick bedding, jointing and pointing in cement mortar (1:3)										
Paving on blinded hardcore base										
50mm thick; to falls and crossfalls and to slopes not exceeding 15 degrees from horizontal................	95.80	10.00	100.57	0.30	0.30	9.64	1.56	111.77	m²	122.94
75mm thick; to falls and crossfalls and to slopes not exceeding 15 degrees from horizontal................	150.16	10.00	157.68	0.36	0.36	11.57	1.65	170.89	m²	187.98
extra; rubbed top surface.......................	18.80	5.00	19.73	-	-	-	-	19.73	m²	21.71
Treads on concrete base										
50mm thick; 300mm wide................	92.44	5.00	97.06	0.36	0.36	11.57	0.45	109.08	m	119.99

SITE WORKS

377

SITE WORKS

Labour hourly rates: (except Specialists) Craft Operatives 18.37 Labourer 13.76 Rates are national average prices. Refer to REGIONAL VARIATIONS for indicative levels of overall pricing in regions	MATERIALS			LABOUR				RATES		
	Del to Site £	Waste %	Material Cost £	Craft Optve Hrs	Lab Hrs	Labour Cost £	Sunds £	Nett Rate £	Unit	Gross rate (10%) £
ROAD AND PATH PAVINGS; INTERLOCKING BRICK AND BLOCKS, SLABS, BRICKS, BLOCKS, SETTS AND COBBLES (Cont'd)										
Precast concrete flags, BS EN 1339 Part 1, natural finish; 6mm joints, symmetrical layout; bedding in 13mm cement mortar (1:3); jointing and pointing with lime and sand (1:2)										
600 x 450 x 50mm units to pavings on sand, granular or blinded hardcore base										
50mm thick; to falls and crossfalls and to slopes not exceeding 15 degrees from horizontal..................	16.29	5.00	17.01	0.30	0.30	9.64	1.59	28.24	m²	31.07
600 x 600 x 50mm units to pavings on sand, granular or blinded hardcore base										
50mm thick; to falls and crossfalls and to slopes not exceeding 15 degrees from horizontal..................	13.62	5.00	14.21	0.28	0.28	9.00	1.50	24.71	m²	27.18
600 x 750 x 50mm units to pavings on sand, granular or blinded hardcore base										
50mm thick; to falls and crossfalls and to slopes not exceeding 15 degrees from horizontal..................	13.50	5.00	14.05	0.25	0.50	11.47	1.61	27.13	m²	29.84
600 x 900 x 50mm units to pavings on sand, granular or blinded hardcore base										
50mm thick; to falls and crossfalls and to slopes not exceeding 15 degrees from horizontal..................	12.15	5.00	12.64	0.23	0.46	10.55	1.62	24.82	m²	27.30
600 x 450 x 63mm units to pavings on sand, granular or blinded hardcore base										
63mm thick; to falls and crossfalls and to slopes not exceeding 15 degrees from horizontal..................	27.05	5.00	28.28	0.30	0.30	9.64	1.59	39.51	m²	43.46
600 x 600 x 63mm units to pavings on sand, granular or blinded hardcore base										
63mm thick; to falls and crossfalls and to slopes not exceeding 15 degrees from horizontal..................	22.65	5.00	23.67	0.28	0.28	9.00	1.50	34.17	m²	37.58
600 x 750 x 63mm units to pavings on sand, granular or blinded hardcore base										
63mm thick; to falls and crossfalls and to slopes not exceeding 15 degrees from horizontal..................	22.62	5.00	23.63	0.25	0.50	11.47	1.51	36.62	m²	40.28
600 x 900 x 63mm units to pavings on sand, granular or blinded hardcore base										
63mm thick; to falls and crossfalls and to slopes not exceeding 15 degrees from horizontal..................	22.60	5.00	23.62	0.23	0.46	10.55	1.53	35.70	m²	39.27
Precast concrete flags, BS EN 1339 Part 1, natural finish; 6mm joints, symmetrical layout; bedding in 13mm lime mortar (1:3); jointing and pointing with lime and sand (1:2)										
600 x 450 x 50mm units to pavings on sand, granular or blinded hardcore base										
50mm thick; to falls and crossfalls and to slopes not exceeding 15 degrees from horizontal..................	16.48	5.00	17.20	0.30	0.30	9.64	1.59	28.43	m²	31.27
600 x 600 x 50mm units to pavings on sand, granular or blinded hardcore base										
50mm thick; to falls and crossfalls and to slopes not exceeding 15 degrees from horizontal..................	13.80	5.00	14.39	0.28	0.28	9.00	1.50	24.89	m²	27.38
600 x 750 x 50mm units to pavings on sand, granular or blinded hardcore base										
50mm thick; to falls and crossfalls and to slopes not exceeding 15 degrees from horizontal..................	13.21	5.00	13.77	0.25	0.50	11.47	1.51	26.76	m²	29.43
600 x 900 x 50mm units to pavings on sand, granular or blinded hardcore base										
50mm thick; to falls and crossfalls and to slopes not exceeding 15 degrees from horizontal..................	11.87	5.00	12.36	0.23	0.46	10.55	1.59	24.50	m²	26.96
600 x 450 x 63mm units to pavings on sand, granular or blinded hardcore base										
63mm thick; to falls and crossfalls and to slopes not exceeding 15 degrees from horizontal..................	26.76	5.00	28.00	0.30	0.30	9.64	1.59	39.23	m²	43.15
600 x 600 x 63mm units to pavings on sand, granular or blinded hardcore base										
63mm thick; to falls and crossfalls and to slopes not exceeding 15 degrees from horizontal..................	22.37	5.00	23.39	0.28	0.28	9.00	1.50	33.88	m²	37.27
600 x 750 x 63mm units to pavings on sand, granular or blinded hardcore base										
63mm thick; to falls and crossfalls and to slopes not exceeding 15 degrees from horizontal..................	22.34	5.00	23.35	0.25	0.50	11.47	1.51	36.34	m²	39.97
600 x 900 x 63mm units to pavings on sand, granular or blinded hardcore base										
63mm thick; to falls and crossfalls and to slopes not exceeding 15 degrees from horizontal..................	22.32	5.00	23.34	0.23	0.46	10.55	1.53	35.42	m²	38.96

Labour hourly rates: (except Specialists) Craft Operatives 18.37 Labourer 13.76 Rates are national average prices. Refer to REGIONAL VARIATIONS for indicative levels of overall pricing in regions	MATERIALS			LABOUR				RATES		
	Del to Site £	Waste %	Material Cost £	Craft Optve Hrs	Lab Hrs	Labour Cost £	Sunds £	Nett Rate £	Unit	Gross rate (10%) £
ROAD AND PATH PAVINGS; INTERLOCKING BRICK AND BLOCKS, SLABS, BRICKS, BLOCKS, SETTS AND COBBLES (Cont'd)										
Precast concrete flags, BS EN 1339 Part 1, natural finish; 6mm joints, symmetrical layout; bedding in 25mm sand; jointing and pointing with lime and sand (1:2)										
600 x 450 x 50mm units to pavings on sand, granular or blinded hardcore base										
50mm thick; to falls and crossfalls and to slopes not exceeding 15 degrees from horizontal	14.79	5.00	15.52	0.30	0.30	9.64	1.59	26.75	m²	29.42
600 x 600 x 50mm units to pavings on sand, granular or blinded hardcore base										
50mm thick; to falls and crossfalls and to slopes not exceeding 15 degrees from horizontal	12.42	5.00	13.02	0.28	0.28	9.00	1.50	23.52	m²	25.87
600 x 750 x 50mm units to pavings on sand, granular or blinded hardcore base										
50mm thick; to falls and crossfalls and to slopes not exceeding 15 degrees from horizontal	11.53	5.00	12.09	0.25	0.50	11.47	1.51	25.08	m²	27.58
600 x 900 x 50mm units to pavings on sand, granular or blinded hardcore base										
50mm thick; to falls and crossfalls and to slopes not exceeding 15 degrees from horizontal	10.19	5.00	10.68	0.23	0.46	10.55	1.53	22.76	m²	25.04
Keyblok precast concrete paving blocks; Marshalls Mono Ltd Driveline 50; in standard units; bedding on sand; covering with sand, compacting with plate vibrator, sweeping off surplus										
200 x 100mm units to pavings on 50mm sand base; natural colour symmetrical half bond layout; level and to falls only										
50mm thick	30.99	10.00	34.09	-	1.35	18.58	2.13	54.80	m²	60.28
laid in straight herringbone pattern; level and to falls only										
50mm thick	30.99	10.00	34.09	-	1.50	20.64	2.13	56.86	m²	62.55
Brick paving on concrete base; 13mm thick bedding, jointing and pointing in cement mortar (1:3)										
Paving to falls and crossfalls and to slopes not exceeding 15 degrees from horizontal										
25mm thick Brick paviors P.C. £300.00 per 1000	18.00	5.00	18.90	2.75	1.55	71.85	1.56	92.31	m²	101.54
65mm thick Facing bricks P.C. £210.00 per 1000	12.60	5.00	13.23	2.75	1.55	71.85	1.56	86.64	m²	95.30
50mm thick Staffordshire blue chequered paviors P.C. £500.00 per 1000	30.00	5.00	31.50	2.75	1.55	71.85	1.56	104.91	m²	115.40
Granite sett paving on concrete base; 13mm thick bedding and grouting in cement mortar (1:3)										
Paving to falls and crossfalls and to slopes not exceeding 15 degrees from horizontal										
125mm thick; new	79.08	2.50	80.95	1.80	2.20	63.34	3.46	147.75	m²	162.53
150mm thick; new	94.66	2.50	96.91	2.00	2.35	69.08	3.96	169.94	m²	186.94
100mm thick; reclaimed	73.49	2.50	75.24	1.65	2.00	57.83	2.97	136.04	m²	149.65
150mm thick; reclaimed	85.28	2.50	87.31	1.80	2.20	63.34	3.46	154.11	m²	169.52
200mm thick; reclaimed	95.66	2.50	97.93	2.00	2.35	69.08	3.96	170.97	m²	188.06
Cobble paving set in concrete bed (measured separately) tight butted, dry grouted with cement and sand (1:3) watered and brushed										
Paving level and to falls only										
plain	20.40	2.50	20.91	1.65	1.65	53.01	3.83	77.75	m²	85.52
set to pattern	20.40	2.50	20.91	2.00	2.00	64.26	3.83	88.99	m²	97.89
Crazy paving on blinded hardcore base, bedding in 38mm thick sand; pointing in cement mortar (1:4)										
Paving to falls and crossfalls and to slopes not exceeding 15 degrees from horizontal										
50mm thick precast concrete flag	12.50	2.50	12.81	0.40	0.40	12.85	3.22	28.89	m²	31.78
38-50mm thick York stone	34.00	2.50	34.85	0.60	0.60	19.28	4.65	58.78	m²	64.66
38mm thick Westmorland green slate	165.00	2.50	169.12	0.60	0.60	19.28	3.22	191.63	m²	210.79
Grass concrete paving; voids filled with vegetable soil and sown with grass seed										
Precast concrete perforated slabs										
120mm thick units on 25mm thick sand bed; level and to falls only	50.46	2.50	52.00	0.32	0.32	10.28	-	62.29	m²	68.51
HARD LANDSCAPING; SITE FURNITURE										
Plain ordinary portland cement precast concrete bollards; Broxap Ltd; setting in ground, excavating hole, removing surplus spoil, filling with concrete mix ST4; working around base										
200mm diameter										
600mm high above ground level; Smooth Grey; BX03 3707	69.00	-	69.00	0.50	0.50	16.07	4.65	89.71	nr	98.69
225mm diameter tapering										
900mm high above ground level; Smooth Grey; BX03 3706	65.00	-	65.00	0.90	0.90	28.92	4.65	98.57	nr	108.42

Labour hourly rates: (except Specialists) Craft Operatives 18.37 Labourer 13.76 Rates are national average prices. Refer to REGIONAL VARIATIONS for indicative levels of overall pricing in regions	MATERIALS			LABOUR				RATES		
	Del to Site £	Waste %	Material Cost £	Craft Optve Hrs	Lab Hrs	Labour Cost £	Sunds £	Nett Rate £	Unit	Gross rate (10%) £
HARD LANDSCAPING; SITE FURNITURE (Cont'd)										
Cast Iron litter bin with galvanized liner; Broxap Ltd; setting in position										
Round; heavy duty										
90 litres capacity; Medium; BX 2319	599.00	-	599.00	1.00	1.00	32.13	0.48	631.61	nr	694.77
150 litres capacity; Large; BX 2306..	649.00	-	649.00	1.00	1.00	32.13	0.57	681.70	nr	749.87
Benches and seats; Steel framed with recycled plastic slats; Broxap Ltd; fixed or free-standing										
Bench; placing in position										
1800mm long seat; five plastic slats; BX 17 4014	282.00	-	282.00	0.50	0.50	16.07	-	298.07	nr	327.87
1800mm long bench; three plastic slats; BX 17 4009....................	229.00	-	229.00	0.50	0.50	16.07	-	245.07	nr	269.57
Steel cycle rack, Broxap Ltd; galvanized steel construction; free standing										
Single sided										
Fishpond rack for 6 cycles; BXMW/FIS	331.00	-	331.00	1.00	0.25	21.81	-	352.81	nr	388.09
Double sided										
Fishpond rack for 11 cycles; BXMW/FIS	423.00	-	423.00	1.50	0.40	33.06	-	456.06	nr	501.66
Steel front wheel cycle supports; Broxap Ltd; galvanized steel construction, pillar mounted; bolting base plate to concrete										
Stands at 800mm centres; BXMW/GS										
5 stands...	195.00	-	195.00	4.00	1.00	87.24	-	282.24	nr	310.46
6 stands...	244.00	-	244.00	6.00	1.50	130.86	-	374.86	nr	412.35
Galvanized Steel shelters, Broxap Ltd; box section framing; 4mm 'ClearView' roofing; assembling; bolting to concrete										
Horizontal loading, single sided; Lightwood; BXMW/LIG										
3600mm long...	1383.00	-	1383.00	8.00	2.00	174.48	-	1557.48	nr	1713.23
5000mm long...	1627.00	-	1627.00	8.00	2.00	174.48	-	1801.48	nr	1981.63

Labour hourly rates: (except Specialists) Craft Operatives 18.37 Labourer 13.76 Rates are national average prices. Refer to REGIONAL VARIATIONS for indicative levels of overall pricing in regions	MATERIALS			LABOUR				RATES		
	Del to Site	Waste	Material Cost	Craft Optve	Lab	Labour Cost	Sunds	Nett Rate		Gross rate (10%)
	£	%	£	Hrs	Hrs	£	£	£	Unit	£
POST AND WIRE FENCING										
Fencing; strained wire; BS 1722-2; concrete posts and struts; 4.00mm diameter galvanized mild steel line wires; galvanized steel fittings and accessories; backfilling around posts in concrete mix ST4										
900mm high fencing; type SC90; wires -3; posts with rounded tops										
posts at 2743mm centres; 600mm into ground	7.10	2.50	7.21	-	0.75	10.32	-	17.53	m	19.28
extra; end posts with struts -1..	39.47	2.50	40.33	-	1.00	13.76	-	54.09	nr	59.50
extra; angle posts with struts -2...	60.57	2.50	61.90	-	1.50	20.64	-	82.54	nr	90.79
1050mm high fencing; type SC105A; wires-5; posts rounded tops										
posts at 2743mm centres; 600mm into ground	8.46	2.50	8.61	-	0.80	11.01	-	19.62	m	21.58
extra; end posts with struts -1..	46.44	2.50	47.48	-	1.10	15.14	-	62.61	nr	68.88
extra; angle posts with struts -2...	68.09	2.50	69.61	-	1.65	22.70	-	92.31	nr	101.54
CHAIN LINK FENCING										
Fencing; chain link; BS 1722-1; concrete posts; galvanized mesh, line and tying wires; galvanized steel fittings and accessories; backfilling around posts in concrete mix ST4										
900mm high fencing; type GLC 90; posts with rounded tops										
posts at 2743mm centres; 600mm into ground	10.57	5.00	10.98	-	1.00	13.76	-	24.74	m	27.21
extra; end posts with struts -1..	49.59	5.00	51.83	-	1.20	16.51	-	68.34	nr	75.17
extra; angle posts with struts -2...	69.21	5.00	72.30	-	1.60	22.02	-	94.32	nr	103.75
1200mm high fencing; type GLC 120; posts with rounded tops										
posts at 2743mm centres; 600mm into ground	12.92	5.00	13.44	-	1.10	15.14	-	28.58	m	31.43
extra; end posts with struts -1..	57.36	5.00	59.98	-	1.40	19.26	-	79.24	nr	87.17
extra; angle posts with struts -2...	84.30	5.00	88.15	-	1.80	24.77	-	112.91	nr	124.21
1500mm high fencing; posts with rounded tops										
posts at 2743mm centres; 600mm into ground	16.51	5.00	17.21	-	1.20	16.51	-	33.72	m	37.10
extra; end posts with struts -1..	72.16	5.00	75.52	-	1.55	21.33	-	96.85	nr	106.53
extra; angle posts with struts -2...	104.18	5.00	109.02	-	2.00	27.52	-	136.54	nr	150.20
1800mm high fencing; posts with rounded tops										
posts at 2743mm centres; 600mm into ground	17.78	5.00	18.55	-	1.45	19.95	-	38.50	m	42.35
extra; end posts with struts -1..	75.96	5.00	79.51	-	1.65	22.70	-	102.21	nr	112.44
extra; angle posts with struts -2...	110.90	5.00	116.08	-	2.15	29.58	-	145.66	nr	160.23
Fencing; chain link; BS 1722-1; concrete posts; plastics coated Grade A mesh, line and tying wire; galvanized steel fittings and accessories; backfilling around posts in concrete mix ST4										
900mm high fencing; type PLC 90A; posts with rounded tops										
posts at 2743mm centres; 600mm into ground	10.54	5.00	10.94	-	1.00	13.76	-	24.70	m	27.17
extra; end posts with struts -1..	49.59	5.00	51.83	-	1.20	16.51	-	68.34	nr	75.17
extra; angle posts with struts -2...	69.21	5.00	72.30	-	1.60	22.02	-	94.32	nr	103.75
1200mm high fencing; type PLC 120; posts with rounded tops										
posts at 2743mm centres; 600mm into ground	12.29	5.00	12.78	-	1.10	15.14	-	27.92	m	30.71
extra; end posts with struts -1..	57.36	5.00	59.98	-	1.40	19.26	-	79.24	nr	87.17
extra; angle posts with struts -2...	84.30	5.00	88.15	-	1.80	24.77	-	112.91	nr	124.21
1400mm high fencing; type PLC 140A; posts with rounded tops										
posts at 2743mm centres; 600mm into ground	15.59	5.00	16.25	-	1.20	16.51	-	32.76	m	36.04
extra; end posts with struts -1..	72.16	5.00	75.52	-	1.55	21.33	-	96.85	nr	106.53
extra; angle posts with struts -2...	104.18	5.00	109.02	-	2.00	27.52	-	136.54	nr	150.20
1800mm high fencing; type PLC 180; posts with rounded tops										
posts at 2743mm centres; 600mm into ground	16.80	5.00	17.52	-	1.45	19.95	-	37.47	m	41.22
extra; end posts with struts -1..	75.96	5.00	79.51	-	1.65	22.70	-	102.21	nr	112.44
extra; angle posts with struts -2...	110.90	5.00	116.08	-	2.15	29.58	-	145.66	nr	160.23
Fencing; chain link; BS 1722-1; rolled steel angle posts; galvanized mesh, line and tying wires; galvanized steel fittings and accessories; backfilling around posts in concrete mix ST4										
900mm high fencing; type GLS 90; posts with rounded tops										
posts at 2743mm centres; 600mm into ground	10.44	2.50	10.63	-	0.65	8.94	-	19.57	m	21.53
extra; end posts with struts -1..	38.65	2.50	39.47	-	1.00	13.76	-	53.23	nr	58.55
extra; angle posts with struts -2...	50.76	2.50	51.81	-	1.60	22.02	-	73.83	nr	81.21
1200mm high fencing; type GLS 120; posts with rounded tops										
posts at 2743mm centres; 600mm into ground	12.22	2.50	12.45	-	0.75	10.32	-	22.77	m	25.05
extra; end posts with struts -1..	46.09	2.50	47.09	-	1.10	15.14	-	62.23	nr	68.45
extra; angle posts with struts -2...	66.61	2.50	68.06	-	1.75	24.08	-	92.14	nr	101.35
1400mm high fencing; type GLS 140A; posts with rounded tops										
posts at 2743mm centres; 600mm into ground	14.07	2.50	14.34	-	0.80	11.01	-	25.35	m	27.89
extra; end posts with struts -1..	51.26	2.50	52.40	-	1.50	20.64	-	73.04	nr	80.34
extra; angle posts with struts -2...	71.88	2.50	73.45	-	1.90	26.14	-	99.60	nr	109.56

FENCING

Labour hourly rates: (except Specialists) Craft Operatives 18.37 Labourer 13.76 Rates are national average prices. Refer to REGIONAL VARIATIONS for indicative levels of overall pricing in regions	MATERIALS			LABOUR				RATES		
	Del to Site £	Waste %	Material Cost £	Craft Optve Hrs	Lab Hrs	Labour Cost £	Sunds £	Nett Rate £	Unit	Gross rate (10%) £
CHAIN LINK FENCING (Cont'd)										
Fencing; chain link; BS 1722-1; rolled steel angle posts; galvanized mesh, line and tying wires; galvanized steel fittings and accessories; backfilling around posts in concrete mix ST4 (Cont'd)										
1800mm high fencing; type GLS 180; posts with rounded tops										
posts at 2743mm centres; 600mm into ground	15.51	2.50	15.82	-	1.00	13.76	-	29.58	m	32.54
extra; end posts with struts -1	59.82	2.50	61.17	-	1.60	22.02	-	83.19	nr	91.50
extra; angle posts with struts -2	83.32	2.50	85.18	-	2.10	28.90	-	114.08	nr	125.49
Fencing; chain link; BS 1722 Part 1; rolled steel angle posts; plastics coated Grade A mesh, line and tying wires; galvanized steel fittings and accessories; backfilling around posts in concrete mix ST4										
900mm high fencing; type PLS 90A; posts with rounded tops										
posts at 2743mm centres; 600mm into ground	10.63	5.00	11.01	-	0.65	8.94	-	19.95	m	21.95
extra; end posts with struts -1	38.65	5.00	40.28	-	1.00	13.76	-	54.04	nr	59.45
extra; angle posts with struts -2	50.76	5.00	52.86	-	1.60	22.02	-	74.87	nr	82.36
1200mm high fencing; type PLS 120; posts with rounded tops										
posts at 2743mm centres; 600mm into ground	11.81	5.00	12.25	-	0.75	10.32	-	22.57	m	24.83
extra; end posts with struts -1	46.09	5.00	48.10	-	1.10	15.14	-	63.23	nr	69.56
extra; angle posts with struts -2	66.61	5.00	69.50	-	1.75	24.08	-	93.58	nr	102.94
1400mm high fencing; type PLS 140A; posts with rounded tops										
posts at 2743mm centres; 600mm into ground	13.37	5.00	13.89	-	0.80	11.01	-	24.90	m	27.38
extra; end posts with struts -1	51.26	5.00	53.53	-	1.50	20.64	-	74.17	nr	81.59
extra; angle posts with struts -2	71.88	5.00	75.03	-	1.90	26.14	-	101.17	nr	111.29
1800mm high fencing; type PLS 180; posts with rounded tops										
posts at 2743mm centres; 600mm into ground	14.90	5.00	15.50	-	1.00	13.76	-	29.26	m	32.18
extra; end posts with struts -1	59.82	5.00	62.52	-	1.60	22.02	-	84.53	nr	92.99
extra; angle posts with struts -2	83.32	5.00	87.04	-	2.10	28.90	-	115.94	nr	127.53
CHESTNUT FENCING										
Fencing; cleft chestnut pale; BS 1722 Part 4; sweet chestnut posts and struts; galvanized accessories; backfilling around posts in concrete mix ST4										
900mm high fencing; type CW90										
posts at 2000mm centres; 600mm into ground	21.12	5.00	21.72	-	0.25	3.44	-	25.16	m	27.68
extra; end posts with struts - 1	41.01	5.00	42.16	-	0.30	4.13	-	46.29	nr	50.92
extra; angle posts with struts - 2	52.76	5.00	54.05	-	0.40	5.50	-	59.55	nr	65.51
1200mm high fencing; type CW 120										
posts at 2000mm centres; 600mm into ground	25.64	5.00	26.47	-	0.17	2.34	-	28.81	m	31.70
extra; end posts with struts - 1	42.08	5.00	43.29	-	0.45	6.19	-	49.48	nr	54.43
extra; angle posts with struts - 2	54.90	5.00	56.30	-	0.60	8.26	-	64.55	nr	71.01
1500mm high fencing; type CW 150										
posts at 2000mm centres; 600mm into ground	30.18	5.00	31.23	-	0.22	3.03	-	34.26	m	37.69
extra; end posts with struts - 1	45.78	5.00	47.17	-	0.50	6.88	-	54.05	nr	59.46
extra; angle posts with struts - 2	60.59	5.00	62.26	-	0.65	8.94	-	71.21	nr	78.33
BOARDED FENCING										
Fencing; close boarded; BS 1722 Part 5; sawn softwood posts, rails, pales, gravel boards and centre stumps, pressure impregnated with preservative; backfilling around posts in concrete mix ST4										
1200mm high fencing; type BW 120										
posts at 3000mm centres; 600mm into ground	33.36	5.00	34.44	1.25	-	22.96	-	57.40	m	63.14
1500mm high fencing; type BW 150										
posts at 3000mm centres; 600mm into ground	39.41	5.00	40.79	1.50	-	27.56	-	68.34	m	75.18
1800mm high fencing; type BW 180A										
posts at 3000mm centres; 750mm into ground	43.46	5.00	45.04	1.67	-	30.68	-	75.72	m	83.29
Fencing; wooden palisade; BS 1722-5; softwood posts, rails, pales and stumps; pressure impregnated with preservative; backfilling around posts in concrete mix ST4										
1050mm high fencing; type WPW 105; 75 x 19mm rectangular pales with pointed tops										
posts at 3000mm centres; 600mm into ground	30.44	5.00	31.37	1.00	-	18.37	-	49.74	m	54.72
1200mm high fencing; type WPW 120; 75 x 19mm rectangular pales with pointed tops										
posts at 3000mm centres; 600mm into ground	31.70	5.00	32.69	1.25	-	22.96	-	55.65	m	61.22
STEEL RAILINGS										
Ornamental steel railings; Cannock Gates (UK) Ltd.; primed at works; fixing in brick openings										
Marlborough Railings										
460mm high for 920mm gap width	39.60	1.50	40.19	0.75	0.38	19.01	0.31	59.52	nr	65.47
460mm high for 1830mm gap width	64.68	1.50	65.65	1.00	0.50	25.25	0.31	91.22	nr	100.34
460mm high for 2740mm gap width	91.74	1.50	93.12	1.50	0.75	37.88	0.31	131.31	nr	144.44
610mm high for 920mm gap width	45.54	1.50	46.22	0.75	0.38	19.01	0.31	65.54	nr	72.10
610mm high for 1830mm gap width	72.60	1.50	73.69	1.00	0.50	25.25	0.31	99.25	nr	109.18
610mm high for 2740mm gap width	97.68	1.50	99.15	1.50	0.75	37.88	0.31	137.34	nr	151.07
920mm high for 920mm gap width	52.80	1.50	53.59	1.50	0.75	37.88	0.31	91.78	nr	100.96
920mm high for 1830mm gap width	76.56	1.50	77.71	0.75	0.38	19.01	0.31	97.03	nr	106.73

FENCING

Labour hourly rates: (except Specialists) Craft Operatives 18.37 Labourer 13.76 Rates are national average prices. Refer to REGIONAL VARIATIONS for indicative levels of overall pricing in regions	MATERIALS			LABOUR				RATES		
	Del to Site £	Waste %	Material Cost £	Craft Optve Hrs	Lab Hrs	Labour Cost £	Sunds £	Nett Rate £	Unit	Gross rate (10%) £

STEEL RAILINGS (Cont'd)

Ornamental steel railings; Cannock Gates (UK) Ltd.; primed at works; fixing in brick openings (Cont'd)

Marlborough Railings (Cont'd)

	Del to Site	Waste	Material Cost	Craft Optve	Lab	Labour Cost	Sunds	Nett Rate	Unit	Gross rate
920mm high for 2740mm gap width	102.96	1.50	104.50	1.50	0.75	37.88	0.39	142.77	nr	157.05
Royal Railings										
460mm high for 920mm gap width	73.26	1.50	74.36	0.75	0.38	19.01	0.31	93.68	nr	103.05
460mm high for 1830mm gap width	102.96	1.50	104.50	1.00	0.50	25.25	0.31	130.07	nr	143.08
460mm high for 2740mm gap width	132.00	1.50	133.98	1.50	0.75	37.88	0.31	172.17	nr	189.39
610mm high for 920mm gap width	82.50	1.50	83.74	0.75	0.38	19.01	0.31	103.06	nr	113.36
610mm high for 1830mm gap width	112.20	1.50	113.88	1.00	0.50	25.25	0.31	139.45	nr	153.39
610mm high for 2740mm gap width	143.88	1.50	146.04	1.50	0.75	37.88	0.31	184.23	nr	202.65
920mm high for 920mm gap width	93.72	1.50	95.13	1.50	0.75	37.88	0.31	133.32	nr	146.65
920mm high for 1830mm gap width	123.42	1.50	125.27	0.75	0.38	19.01	0.31	144.59	nr	159.05
920mm high for 2740mm gap width	153.78	1.50	156.09	1.50	0.75	37.88	0.39	194.35	nr	213.79

GUARD RAILS

Aluminium guard rail of tubing to BS 1387 medium grade with Kee Klamp fittings

	Del to Site	Waste	Material Cost	Craft Optve	Lab	Labour Cost	Sunds	Nett Rate	Unit	Gross rate
Rail or standard										
42mm diameter	19.65	5.00	20.63	0.25	-	4.59	0.45	25.67	m	28.24
extra; flanged end	5.49	5.00	5.76	-	-	-	-	5.76	nr	6.34
extra; bend No.15-7	6.06	5.00	6.36	0.15	-	2.76	-	9.12	nr	10.03
extra; three-way intersection No.20-7	9.09	5.00	9.54	0.15	-	2.76	-	12.30	nr	13.53
extra; three-way intersection No.25-7	8.62	5.00	9.05	0.15	-	2.76	-	11.81	nr	12.99
extra; four-way intersection No.21-7	7.01	5.00	7.36	0.15	-	2.76	-	10.12	nr	11.13
extra; four-way intersection No.26-7	6.62	5.00	6.95	0.15	-	2.76	-	9.71	nr	10.68
extra; five-way intersection No.35-7	9.95	5.00	10.45	0.15	-	2.76	-	13.20	nr	14.52
extra; five-way intersection No.40-7	14.90	5.00	15.64	0.15	-	2.76	-	18.40	nr	20.24
extra; floor plate No.61-7	8.03	5.00	8.43	0.15	-	2.76	-	11.19	nr	12.31
Infill panel fixed with clips										
50 x 50mm welded mesh	32.50	5.00	34.12	0.80	-	14.70	0.30	49.12	m²	54.03

PRECAST CONCRETE POSTS

Precast concrete fence posts; BS 1722 excavating holes, backfilling around posts in concrete mix C20, disposing of surplus materials, earthwork support

	Del to Site	Waste	Material Cost	Craft Optve	Lab	Labour Cost	Sunds	Nett Rate	Unit	Gross rate
Fence posts for three wires, housing pattern										
100 x 100mm tapering intermediate post 1570mm long; 600mm into ground	12.95	2.50	13.18	0.25	0.25	8.03	-	21.21	nr	23.34
100 x 100mm square strainer post (end or intermediate) 1570mm long; 600mm into ground	14.43	2.50	14.70	0.25	0.25	8.03	-	22.73	nr	25.00
Clothes line post										
125 x 125mm tapering, 2670mm long	36.52	2.50	37.34	0.33	0.33	10.60	-	47.94	nr	52.74
Close boarded fence post										
94 x 100mm, 2745mm long	25.57	2.50	26.12	0.40	0.40	12.85	-	38.97	nr	42.87
Slotted fence post										
94 x 100mm, 2625mm long	15.65	2.50	15.95	0.40	0.40	12.85	-	28.80	nr	31.68
Chain link fence post for 1800mm high fencing										
125 x 75mm tapering intermediate post, 2620mm long	19.45	2.50	19.84	0.33	0.33	10.60	-	30.45	nr	33.49
125 x 125mm end post, 1 strut, 2620mm long	47.22	2.50	48.22	0.66	0.66	21.21	-	69.42	nr	76.37
125 x 125mm angle post, 2 strut, 2620mm long	65.22	2.50	66.58	0.99	0.99	31.81	-	98.38	nr	108.22
125 x 125mm gate post, 2620mm long	29.22	2.50	29.86	0.33	0.33	10.60	-	40.46	nr	44.51
150 x 150mm gate post, 2620mm long	38.65	2.50	39.52	0.33	0.33	10.60	-	50.13	nr	55.14

GATES

Gates; impregnated wrought softwood; featheredge pales; including ring latch and heavy hinges

	Del to Site	Waste	Material Cost	Craft Optve	Lab	Labour Cost	Sunds	Nett Rate	Unit	Gross rate
Gates (posts included elsewhere)										
900 x 1150mm	48.35	2.50	49.56	0.50	0.50	16.07	0.25	65.88	nr	72.47
900 x 1450mm	50.54	2.50	51.80	0.50	0.50	16.07	0.25	68.12	nr	74.94

Ornamental steel gates; Cannock Gates (UK) Ltd.; primed at works; fixing in brick openings

	Del to Site	Waste	Material Cost	Craft Optve	Lab	Labour Cost	Sunds	Nett Rate	Unit	Gross rate
Clifton, 920mm high										
single gate, for 760 mm wide opening	43.56	1.50	44.21	1.50	0.75	37.88	0.41	82.49	nr	90.74
single gate, for 840 mm wide opening	46.20	1.50	46.89	1.50	0.75	37.88	0.41	85.17	nr	93.69
single gate, for 920 mm wide opening	48.67	1.50	49.41	1.50	0.75	37.88	0.41	87.69	nr	96.45
single gate, for 1000 mm wide opening	51.15	1.50	51.92	1.50	0.75	37.88	0.41	90.20	nr	99.22
single gate, for 1070 mm wide opening	53.62	1.50	54.43	1.50	0.75	37.88	0.41	92.71	nr	101.98
double gate, for 2300 mm wide opening	89.93	1.50	91.27	2.25	1.12	56.81	0.74	148.82	nr	163.70
double gate, for 2600 mm wide opening	96.53	1.50	97.97	2.25	1.12	56.81	0.74	155.52	nr	171.07
double gate, for 2740 mm wide opening	99.82	1.50	101.32	2.25	1.12	56.81	0.74	158.87	nr	174.76
double gate, for 3050 mm wide opening	106.43	1.50	108.02	2.25	1.12	56.81	0.74	165.57	nr	182.13
double gate, for 3660 mm wide opening	112.20	1.50	113.88	2.25	1.12	56.81	0.74	171.43	nr	188.57
Clifton, 1220mm high										
single gate, for 760 mm wide opening	51.97	1.50	52.75	1.50	0.75	37.88	0.74	91.36	nr	100.50
single gate, for 1070 mm wide opening	61.88	1.50	62.80	2.25	1.12	56.81	0.74	120.35	nr	132.39
Clifton, 1830mm high										
single gate, for 760 mm wide opening	63.53	1.50	64.48	1.50	0.75	37.88	0.41	102.76	nr	113.03
single gate, for 1070 mm wide opening	76.72	1.50	77.88	1.50	0.75	37.88	0.41	116.16	nr	127.77
double gate, for 2300 mm wide opening	169.12	1.50	171.66	2.25	2.25	72.29	0.74	244.69	nr	269.16

Labour hourly rates: (except Specialists) Craft Operatives 18.37 Labourer 13.76 Rates are national average prices. Refer to REGIONAL VARIATIONS for indicative levels of overall pricing in regions	MATERIALS			LABOUR				RATES		
	Del to Site £	Waste %	Material Cost £	Craft Optve Hrs	Lab Hrs	Labour Cost £	Sunds £	Nett Rate £	Unit	Gross rate (10%) £
GATES (Cont'd)										
Ornamental steel gates; Cannock Gates (UK) Ltd.; primed at works; fixing in brick openings (Cont'd)										
Clifton, 1830mm high (Cont'd)										
double gate, for 2740 mm wide opening	184.80	1.50	187.57	2.25	2.25	72.29	0.74	260.60	nr	286.66
double gate, for 3050 mm wide opening	195.52	1.50	198.46	2.25	2.25	72.29	0.74	271.49	nr	298.63
Winchester, 920mm high										
double gate, for 2300 mm wide opening	93.06	1.50	94.46	2.25	1.12	56.81	0.74	152.00	nr	167.20
double gate, for 2740 mm wide opening	99.82	1.50	101.32	2.25	1.12	56.81	0.74	158.87	nr	174.76
double gate, for 3050 mm wide opening	106.43	1.50	108.02	2.25	1.12	56.81	0.74	165.57	nr	182.13
double gate, for 3660 mm wide opening	113.03	1.50	114.72	2.25	1.12	56.81	0.74	172.27	nr	189.49
Winchester, 1220mm high										
single gate, for 760 mm wide opening	52.14	1.50	52.92	1.50	0.75	37.88	0.41	91.20	nr	100.32
single gate, for 1070 mm wide opening	62.04	1.50	62.97	1.50	0.75	37.88	0.41	101.25	nr	111.38
Winchester, 1830mm high										
single gate, for 760 mm wide opening	63.53	1.50	64.48	1.50	0.75	37.88	0.41	102.76	nr	113.03
single gate, for 1070 mm wide opening	76.72	1.50	77.88	1.50	0.75	37.88	0.41	116.16	nr	127.77
double gate, for 2300 mm wide opening	169.12	1.50	171.66	2.25	2.25	72.29	0.74	244.69	nr	269.16
double gate, for 2740 mm wide opening	183.98	1.50	186.73	2.25	2.25	72.29	0.74	259.76	nr	285.74
double gate, for 3050 mm wide opening	195.52	1.50	198.46	2.25	2.25	72.29	0.74	271.49	nr	298.63
Westminster, 920mm high										
single gate, for 760 mm wide opening	57.75	1.50	58.62	1.50	0.75	37.88	0.41	96.90	nr	106.59
double gate, for 2300 mm wide opening	107.25	1.50	108.86	2.25	1.12	56.81	0.74	166.41	nr	183.05
double gate, for 2600 mm wide opening	118.80	1.50	120.58	2.25	1.12	56.81	0.74	178.13	nr	195.94
double gate, for 2900 mm wide opening	130.35	1.50	132.31	2.25	1.12	56.81	0.74	189.85	nr	208.84
double gate, for 3200 mm wide opening	141.90	1.50	144.03	2.25	1.12	56.81	0.74	201.58	nr	221.73
double gate, for 3500 mm wide opening	153.45	1.50	155.75	2.25	1.12	56.81	0.74	213.30	nr	234.63
double gate, for 3660 mm wide opening	159.23	1.50	161.61	2.25	1.12	56.81	0.74	219.16	nr	241.08
Westminster, 1830mm high										
single gate, for 760 mm wide opening	82.50	1.50	83.74	1.50	0.75	37.88	0.41	122.02	nr	134.22
single gate, for 1000 mm wide opening	92.40	1.50	93.79	1.50	0.75	37.88	0.41	132.07	nr	145.27
single gate, for 1220 mm wide opening	99.00	1.50	100.49	1.50	0.75	37.88	0.41	138.76	nr	152.64
Marlborough, 920mm high										
single gate, for 760 mm wide opening	57.75	1.50	58.62	1.50	0.75	37.88	0.41	96.90	nr	106.59
double gate, for 2300 mm wide opening	107.25	1.50	108.86	2.25	1.12	56.81	0.74	166.41	nr	183.05
double gate, for 3660 mm wide opening	159.23	1.50	161.61	2.25	1.12	56.81	0.74	219.16	nr	241.08
folding (4 piece) gate, for 2300 mm wide opening	173.25	1.50	175.85	3.00	1.50	75.75	0.75	252.35	nr	277.58
folding (4 piece) gate, for 3660 mm wide opening	232.65	1.50	236.14	3.00	1.50	75.75	0.75	312.64	nr	343.90
Marlborough, 1220mm high										
single gate, for 760 mm wide opening	59.40	1.50	60.29	1.50	0.75	37.88	0.41	98.57	nr	108.43
double gate, for 2300 mm wide opening	127.88	1.50	129.79	2.25	1.12	56.81	0.74	187.34	nr	206.07
double gate, for 3050 mm wide opening	189.75	1.50	192.60	2.25	1.12	56.81	0.74	250.14	nr	275.16
Marlborough, 1830mm high										
single gate, for 760 mm wide opening	82.50	1.50	83.74	1.50	0.75	37.88	0.41	122.02	nr	134.22
single gate, for 1000 mm wide opening	92.40	1.50	93.79	1.50	0.75	37.88	0.41	132.07	nr	145.27
double gate, for 2300 mm wide opening	321.75	1.50	326.58	3.00	3.00	96.39	0.90	423.87	nr	466.25
double gate, for 3660 mm wide opening	561.00	1.50	569.41	3.30	3.00	101.90	1.50	672.82	nr	740.10
Royal Talisman, 1220mm high										
single gate, for 920 mm wide opening	206.25	1.50	209.34	1.75	0.88	44.19	0.60	254.13	nr	279.54
single gate, for 1070 mm wide opening	231.00	1.50	234.46	1.75	0.88	44.19	0.60	279.25	nr	307.18
double gate, for 2440 mm wide opening	521.40	1.50	529.22	3.00	1.50	75.75	1.20	606.17	nr	666.79
double gate, for 3200 mm wide opening	636.90	1.50	646.45	3.00	1.50	75.75	1.20	723.40	nr	795.74
double gate, for 4270 mm wide opening	749.10	1.50	760.34	3.60	1.80	90.90	1.20	852.44	nr	937.68
Royal Talisman, 1830mm high										
single gate, for 920 mm wide opening	276.38	1.50	280.52	1.75	0.88	44.19	0.60	325.31	nr	357.84
single gate, for 1070 mm wide opening	309.38	1.50	314.02	1.75	0.88	44.19	0.60	358.80	nr	394.68
arched double gate, for 2440 mm wide opening	808.50	1.50	820.63	3.00	3.00	96.39	1.20	918.22	nr	1010.04
arched double gate, for 3200 mm wide opening	911.62	1.50	925.30	3.00	3.00	96.39	1.20	1022.89	nr	1125.18
arched double gate, for 4270 mm wide opening	1014.75	1.50	1029.97	3.50	3.50	112.46	1.20	1143.63	nr	1257.99
arched double gate, for 4880 mm wide opening	1056.00	1.50	1071.84	3.50	3.50	112.46	1.20	1185.50	nr	1304.04
double gate, for 2440 mm wide opening	808.50	1.50	820.63	3.00	3.00	96.39	1.20	918.22	nr	1010.04
double gate, for 4880 mm wide opening	1056.00	1.50	1071.84	3.50	3.50	112.46	1.20	1185.50	nr	1304.04
Extra cost of fixing gates to and including pair of steel gate posts; backfilling around posts in concrete mix ST4										
for single or double gates 920 high; posts section 'C', size 50 x 50mm	42.34	1.50	43.03	0.33	0.17	8.40	-	51.78	nr	56.96
for single or double gates 1220 high; posts section 'C', size 50 x 50mm	47.29	1.50	48.06	0.33	0.17	8.40	-	56.81	nr	62.49
for single or double gates 1830 high; posts section 'C', size 50 x 50mm	60.71	1.50	61.69	0.33	0.17	8.40	-	70.51	nr	77.56
for single or double gates 1220 high; posts section 'B', size 75 x 75mm	109.99	1.50	111.70	0.33	0.17	8.40	-	120.45	nr	132.49
for single or double gates 1830 high; posts section 'B', size 75 x 75m	154.76	1.50	157.15	0.33	0.17	8.40	-	165.97	nr	182.56
for single or double gates 1220 high; posts section 'A', size 100 x 100mm	174.56	1.50	177.25	0.33	0.17	8.40	-	186.06	nr	204.67
for single or double gates 1830 high; posts section 'A', size 100 x 100mm	216.14	1.50	219.47	0.33	0.17	8.40	-	228.39	nr	251.23
Gates in chain link fencing										
Gate with 40 x 40 x 5mm painted angle iron framing, braces and rails, infilled with 50mm x 10 1/2 G mesh										
1200 x 1800mm; galvanized mesh	261.54	1.50	265.46	6.00	6.00	192.78	-	458.24	nr	504.07
1200 x 1800mm; plastic coated mesh	261.54	1.50	265.46	6.00	6.00	192.78	-	458.24	nr	504.07

FENCING (side margin)

Labour hourly rates: (except Specialists) Craft Operatives 18.37 Labourer 13.76 Rates are national average prices. Refer to REGIONAL VARIATIONS for indicative levels of overall pricing in regions	MATERIALS			LABOUR				RATES		
	Del to Site	Waste	Material Cost	Craft Optve	Lab	Labour Cost	Sunds	Nett Rate	Unit	Gross rate (10%)
	£	%	£	Hrs	Hrs	£	£	£		£
GATES (Cont'd)										
Gates in chain link fencing (Cont'd)										
Gate post of 80 x 80 x 6mm painted angle iron; back-filling around posts in concrete mix C30 to suit 1800mm high gate	61.90	1.50	62.75	1.50	1.50	48.19	-	110.95	nr	122.04

This page left blank intentionally

Labour hourly rates: (except Specialists) Craft Operatives 18.37 Labourer 13.76 Rates are national average prices. Refer to REGIONAL VARIATIONS for indicative levels of overall pricing in regions	MATERIALS			LABOUR				RATES		
	Del to Site	Waste	Material Cost	Craft Optve	Lab	Labour Cost	Sunds	Nett Rate	Unit	Gross rate (10%)
	£	%	£	Hrs	Hrs	£	£	£		£
CULTIVATING										
Cultivating tilth										
Surfaces of natural ground digging over one spit deep; removing debris; weeding.................	-	-	-	-	0.25	3.44	-	3.44	m²	3.78
Surfaces of filling digging over one spit deep	-	-	-	-	0.30	4.13	-	4.13	m²	4.54
SURFACE APPLICATIONS										
Weedkillers										
Selective weedkiller and feed, 0.03 kg/m²; applying by spreader general surfaces...........................	0.05	5.00	0.05	-	0.05	0.69	-	0.74	m²	0.82
Fertilizer										
Bone meal, 0.06 kg/m²; raking in general surfaces...............................	0.03	5.00	0.04	-	0.05	0.69	-	0.72	m²	0.80
Pre-seeding fertilizer at the rate of 0.04 kg/m² general surfaces...............................	0.13	5.00	0.13	-	0.05	0.69	-	0.82	m²	0.90
General grass fertilizer at the rate of 0.03 kg/m² general surfaces...............................	0.03	5.00	0.03	-	0.05	0.69	-	0.71	m²	0.79
SEEDING										
Grass seed										
Grass seed, 0.035 kg/m², raking in; rolling; maintaining for 12 months after laying general surfaces...............................	0.13	5.00	0.13	-	0.20	2.75	-	2.88	m²	3.17
Stone pick, roll and cut grass with a rotary cutter and remove arisings general surfaces...............................	-	-	-	-	0.10	1.38	-	1.38	m²	1.51
TURFING										
Turf										
Take turf from stack, wheel not exceeding 100m, lay, roll and water, maintaining for 12 months after laying general surfaces...............................	-	-	-	-	0.50	6.88	-	6.88	m²	7.57
Rolawn® Medallion turf; cultivated; lay, roll and water; maintaining for 12 months after laying general surfaces...............................	3.20	5.00	3.36	-	0.50	6.88	-	10.24	m²	11.26
TREES, SHRUBS, HEDGE PLANT, PLANTS, BULBS, CORMS AND TUBERS										
Planting trees, shrubs and hedge plants										
Note; Not withstanding the requirements of NRM2 Tree stakes are measured separately as noted below, all other staking is deemed included.										
Excavate or form pit, hole or trench, dig over ground in bottom, spread and pack around roots with finely broken soil, refill with top soil with one third by volume of farmyard manure incorporated, water in, remove surplus excavated material and provide labelling										
small tree......................	0.77	-	0.77	-	0.85	11.70	0.58	13.05	nr	14.36
medium tree..................	2.61	-	2.61	-	1.65	22.70	0.84	26.15	nr	28.77
large tree.....................	7.81	-	7.81	-	3.30	45.41	1.08	54.30	nr	59.73
shrub	0.53	-	0.53	-	0.55	7.57	0.24	8.34	nr	9.17
hedge plant..................	0.53	-	0.53	-	0.28	3.85	0.24	4.62	nr	5.09
60mm diameter treated softwood tree stake, pointed and driven into ground and with two PVC tree ties secured around tree and nailed to stake										
2100mm long..................	3.89	-	3.89	-	0.33	4.54	2.07	10.50	nr	11.55
2400mm long..................	4.37	-	4.37	-	0.33	4.54	2.07	10.98	nr	12.08
75mm diameter treated softwood tree stake, pointed and driven into ground and with two PVC tree ties secured around tree and nailed to stake										
3000mm long..................	7.50	-	7.50	-	0.35	4.82	2.17	14.49	nr	15.94
BULBS, CORMS AND TUBERS										
Planting herbaceous plants, bulbs, corms and tubers										
Provide planting bed of screened and conditioned Rolawn® topsoil										
150mm thick..................	19.57	5.00	20.54	-	0.75	10.32	-	30.86	m²	33.95
225mm thick..................	29.35	5.00	30.82	-	1.12	15.41	-	46.23	m²	50.85

SOFT LANDSCAPING

Labour hourly rates: (except Specialists) Craft Operatives 18.37 Labourer 13.76 Rates are national average prices. Refer to REGIONAL VARIATIONS for indicative levels of overall pricing in regions	MATERIALS			LABOUR				RATES		
	Del to Site	Waste	Material Cost	Craft Optve	Lab	Labour Cost	Sunds	Nett Rate	Unit	Gross rate (10%)
	£	%	£	Hrs	Hrs	£	£	£		£
BULBS, CORMS AND TUBERS (Cont'd)										
Planting herbaceous plants, bulbs, corms and tubers (Cont'd)										
Form hole and plant										
herbaceous plants	0.75	5.00	0.78	-	0.20	2.75	-	3.53	nr	3.89
bulbs, corms and tubers	0.25	5.00	0.26	-	0.10	1.38	-	1.64	nr	1.80
Mulching after planting										
25mm bark; Rolawn®	2.53	-	2.53	-	0.10	1.38	-	3.91	m²	4.30
75mm farmyard manure	5.00	-	5.00	-	0.11	1.51	-	6.51	m²	7.16
PLANT CONTAINERS										
Precast concrete plant containers; Broxap Ltd; setting in position										
Plant container (soil filling included elsewhere)										
1400 x 700 x 600mm high; BX25 6532....................	1160.00	-	1160.00	0.60	0.60	19.28	0.45	1179.73	nr	1297.70
1500 x 1300 x 580mm high; BX25 6530....................	1420.00	-	1420.00	0.80	0.80	25.70	0.45	1446.15	nr	1590.77
Recycled plastic planter; Broxap Ltd; setting in position										
Plant container (soil filling included elsewhere)										
1200mm hexagonal x 622mm high; BX71 8003-L....................	335.00	-	335.00	0.50	0.50	16.07	0.48	351.55	nr	386.70
900 x 900 x 900mm high; BX71 8012-L	323.00	-	323.00	0.50	0.50	16.07	0.48	339.55	nr	373.50

Labour hourly rates: (except Specialists) Craft Operatives 23.99 Labourer 13.76 Rates are national average prices. Refer to REGIONAL VARIATIONS for indicative levels of overall pricing in regions	MATERIALS			LABOUR				RATES		
	Del to Site £	Waste %	Material Cost £	Craft Optve Hrs	Lab Hrs	Labour Cost £	Sunds £	Nett Rate £	Unit	Gross rate (10%) £
PRIMARY EQUIPMENT										
Gas fired boilers										
Gas fired condensing combination boilers for central heating and hot water; automatically controlled by thermostat with electrical control, gas governor and flame failure device										
wall mounted boiler; fanned flue; approximate output rating										
15 KW	638.42	1.00	644.80	7.50	-	179.93	-	824.73	nr	907.20
18 KW	682.85	1.00	689.68	7.50	-	179.93	-	869.60	nr	956.56
24 KW	701.08	1.00	708.09	8.50	-	203.91	-	912.01	nr	1003.21
Gas fired condensing system boilers for central heating and hot water; automatically controlled by thermostat with electrical control, gas governor and flame failure device										
floor standing boiler; fanned flue; output rating:										
30 KW	1221.62	1.00	1233.84	8.00	-	191.92	-	1425.76	nr	1568.33
40 KW	1465.00	1.00	1479.65	8.50	-	203.91	-	1683.57	nr	1851.92
wall mounted boiler; fanned flue; output rating:										
12 KW	710.28	1.00	717.38	7.50	-	179.93	-	897.31	nr	987.04
15 KW	746.22	1.00	753.68	7.50	-	179.93	-	933.61	nr	1026.97
18 KW	782.15	1.00	789.97	7.50	-	179.93	-	969.90	nr	1066.89
24 KW	834.65	1.00	843.00	7.50	-	179.93	-	1022.92	nr	1125.21
30 KW	902.82	1.00	911.85	8.00	-	191.92	-	1103.77	nr	1214.15
35 KW	1005.09	1.00	1015.14	8.50	-	203.91	-	1219.06	nr	1340.96
Oil fired boilers										
Oil fired boilers for central heating and hot water; automatically controlled by thermostat, with electrical control box										
wall mounted boiler; conventional flue; approximate output rating										
12/18 KW	1385.48	1.00	1399.33	6.50	-	155.93	-	1555.27	nr	1710.80
free standing boiler; conventional flue; approximate output rating										
25 KW	1079.63	1.00	1090.43	6.00	-	143.94	-	1234.37	nr	1357.80
32 KW	1147.19	1.00	1158.66	6.00	-	143.94	-	1302.60	nr	1432.86
46 KW	1680.00	1.00	1696.80	7.50	-	179.93	-	1876.72	nr	2064.40
Coal fired boilers										
Solid fuel boilers for central heating and hot water; thermostatically controlled; with tools										
free standing boiler, gravity feed; approximate output rating										
13.2 KW	2159.10	1.00	2180.69	6.00	-	143.94	-	2324.63	nr	2557.09
17.6 KW	2514.60	1.00	2539.75	6.00	-	143.94	-	2683.69	nr	2952.05
23.5 KW	2815.20	1.00	2843.35	7.50	-	179.93	-	3023.28	nr	3325.60
GRP water storage tanks										
GRP Sarena mfg Ltd; pre insulated water storage tanks with cover										
reference SC10; 18 litres	113.00	2.00	115.26	1.50	-	35.99	-	151.24	nr	166.37
reference SC20; 55 litres	167.00	2.00	170.34	1.50	-	35.99	-	206.32	nr	226.96
reference SC30; 86 litres	196.00	2.00	199.92	1.75	-	41.98	-	241.90	nr	266.09
reference SC40; 114 litres	230.00	2.00	234.60	1.75	-	41.98	-	276.58	nr	304.24
reference SC50; 159 litres	263.00	2.00	268.26	2.00	-	47.98	-	316.24	nr	347.86
reference SC60; 191 litres	285.00	2.00	290.70	2.00	-	47.98	-	338.68	nr	372.55
reference Domestic; 227 litres	304.00	2.00	310.08	2.15	-	51.58	-	361.66	nr	397.82
reference SC70; 227 litres	321.00	2.00	327.42	2.50	-	59.97	-	387.39	nr	426.13
reference SC80; 264 litres	371.00	2.00	378.42	2.75	-	65.97	-	444.39	nr	488.83
reference SC100/1; 327 litres	444.00	2.00	452.88	3.00	-	71.97	-	524.85	nr	577.34
reference SC100/2; 336 litres	488.00	2.00	497.76	3.00	-	71.97	-	569.73	nr	626.70
reference SC125; 423 litres	543.00	2.00	553.86	4.00	-	95.96	-	649.82	nr	714.80
reference SC150; 491 litres	639.00	2.00	651.78	4.00	-	95.96	-	747.74	nr	822.51
reference SC200; 709 litres	820.00	2.00	836.40	4.00	-	95.96	-	932.36	nr	1025.60
reference SC250; 900 litres	935.00	2.00	953.70	6.00	-	143.94	-	1097.64	nr	1207.40
reference SC350; 1250 litres	1104.00	2.00	1126.08	7.00	-	167.93	-	1294.01	nr	1423.41
reference SC500; 1727 litres	1350.00	2.00	1377.00	8.00	-	191.92	-	1568.92	nr	1725.81
reference SC600; 2137 litres	1592.00	2.00	1623.84	9.00	-	215.91	-	1839.75	nr	2023.72
reference SC600H; 2350 litres	1939.00	2.00	1977.78	9.50	-	227.90	-	2205.68	nr	2426.25
reference SC1000; 3364 litres	2444.00	2.00	2492.88	12.00	-	287.88	-	2780.76	nr	3058.84
drilled holes for 13mm pipes	-	-	-	0.20	-	4.80	-	4.80	nr	5.28
drilled holes for 19mm pipes	-	-	-	0.20	-	4.80	-	4.80	nr	5.28
drilled holes for 25mm pipes	-	-	-	0.25	-	6.00	-	6.00	nr	6.60
drilled holed for 32mm pipes	-	-	-	0.25	-	6.00	-	6.00	nr	6.60
drilled holes for 38mm pipes	-	-	-	0.33	-	7.92	-	7.92	nr	8.71
drilled holes for 51mm pipes	-	-	-	0.33	-	7.92	-	7.92	nr	8.71
Plastics water storage cisterns										
Rectangular cold water storage cisterns; plastics; BS 4213 with sealed lid and byelaw 30 kit										
18 litres	22.98	2.00	23.44	1.25	-	29.99	-	53.43	nr	58.77
114 litres	67.85	2.00	69.21	1.75	-	41.98	-	111.19	nr	122.31
190 litres	107.25	2.00	109.39	2.00	-	47.98	-	157.38	nr	173.11
227 litres	111.63	2.00	113.86	2.50	-	59.97	-	173.84	nr	191.22
drilled holes for 13mm pipes	-	-	-	0.13	-	3.12	-	3.12	nr	3.43

Labour hourly rates: (except Specialists) Craft Operatives 23.99 Labourer 13.76 Rates are national average prices. Refer to REGIONAL VARIATIONS for indicative levels of overall pricing in regions	MATERIALS			LABOUR				RATES		
	Del to Site	Waste	Material Cost	Craft Optve	Lab	Labour Cost	Sunds	Nett Rate		Gross rate (10%)
	£	%	£	Hrs	Hrs	£	£	£	Unit	£

PRIMARY EQUIPMENT (Cont'd)

Plastics water storage cisterns (Cont'd)

Rectangular cold water storage cisterns; plastics; BS 4213 with sealed lid and byelaw 30 kit (Cont'd)

drilled holes for 19mm pipes....................................	-	-	-	0.13	-	3.12	-	3.12	nr	3.43
drilled holes for 25mm pipes....................................	-	-	-	0.17	-	4.08	-	4.08	nr	4.49
drilled holed for 32mm pipes....................................	-	-	-	0.17	-	4.08	-	4.08	nr	4.49
drilled holes for 38mm pipes....................................	-	-	-	0.22	-	5.28	-	5.28	nr	5.81
drilled holes for 51mm pipes....................................	-	-	-	0.22	-	5.28	-	5.28	nr	5.81

Copper direct hot water cylinders; pre-insulated

Direct hot water cylinders; copper cylinder, BS 1566 Grade 3

reference 3; 116 litres; four bosses	160.00	2.00	163.20	2.00	-	47.98	-	211.18	nr	232.30
reference 7; 120 litres; four bosses	150.51	2.00	153.52	2.00	-	47.98	-	201.50	nr	221.65
reference 8; 144 litres; four bosses	166.37	2.00	169.70	2.00	-	47.98	-	217.68	nr	239.45
reference 9; 166 litres; four bosses	230.75	2.00	235.36	2.00	-	47.98	-	283.35	nr	311.68

Copper indirect hot water cylinders; pre-insulated

Double feed indirect hot water cylinders; copper cylinder, BS 1566 Part 1 Grade 3

reference 3; 114 litres; four bosses	163.44	2.00	166.71	2.00	-	47.98	-	214.69	nr	236.16
reference 7; 117 litres; four bosses	163.45	2.00	166.72	2.00	-	47.98	-	214.70	nr	236.17
reference 8; 140 litres; four bosses	180.79	2.00	184.41	2.00	-	47.98	-	232.39	nr	255.62
reference 9; 162 litres; four bosses	245.00	2.00	249.90	2.00	-	47.98	-	297.88	nr	327.67

Single feed indirect hot water cylinders; copper cylinder, BS 1566 Part 2 Grade 3

reference 7; 108 litres; four bosses	476.53	2.00	486.06	2.00	-	47.98	-	534.04	nr	587.44
reference 8; 130 litres; four bosses	503.83	2.00	513.91	2.00	-	47.98	-	561.89	nr	618.08

Copper combination hot water storage units; pre-insulated

Combination direct hot water storage units; copper unit, BS 3198 with lid

450mm diameter x 1050mm high; 115 litres hot water; 45 litres cold water	317.25	2.00	323.60	3.00	-	71.97	-	395.57	nr	435.12
450mm diameter x 1200mm high; 115 litres hot water; 115 litres cold water	351.50	2.00	358.53	3.00	-	71.97	-	430.50	nr	473.55

Combination double feed indirect hot water storage units; copper unit, BS 3198 with lid

450mm diameter x 1050mm high; 115 litres hot water; 45 litres cold water	264.03	2.00	269.31	3.00	-	71.97	-	341.28	nr	375.41
450mm diameter x 1200mm high; 115 litres hot water; 115 litres cold water	297.27	2.00	303.22	3.00	-	71.97	-	375.19	nr	412.70

Combination single feed indirect hot water storage units; copper unit, BS 3198 with lid

450mm diameter x 1200mm high; 115 litres hot water; 115 litres cold water	515.00	2.00	525.30	3.00	-	71.97	-	597.27	nr	657.00

Plastic single skin fuel oil storage tanks

Oil storage tanks; Plastic single skin tank complete with watchman probe, transmitter and long life battery with gate valve and filter.

1300 litre, 1700L x 1050W x 1150H	372.00	-	372.00	3.50	-	83.96	-	455.96	nr	501.56
1800 litre, 2055L x 1130W x 1145H	491.00	-	491.00	3.75	-	89.96	-	580.96	nr	639.06
2500 litre, 2250L x 1265W x 1320H	598.50	-	598.50	4.00	-	95.96	-	694.46	nr	763.91

Oil storeage tanks; Plastic Bunded tank complete with watchman probe, transmitter and long life battery, Bundman leak indicator. Bund drain off point with gate valve and filter.

1300 litre, 1900L x 1220W x 1330H	986.00	-	986.00	3.50	-	83.96	-	1069.96	nr	1176.96
1800 litre, 2530L x 1430W x 1430H	1042.75	-	1042.75	3.75	-	89.96	-	1132.71	nr	1245.98
2500 litre, 2360L x 1280W x 1380H	1145.25	-	1145.25	4.00	-	95.96	-	1241.21	nr	1365.33

Accelerator pumps

Variable head accelerator pumps for forced central heating; small bore indirect systems
BSP connections; with valves

20mm....................................	89.38	2.00	91.17	1.00	-	23.99	-	115.16	nr	126.67
25mm....................................	90.56	2.00	92.37	1.00	-	23.99	-	116.36	nr	128.00

TERMINAL EQUIPMENT AND FITTINGS

Radiators

Pressed steel radiators single panel with convector; air cock, plain plug
450mm high; fixing brackets to masonry with screws

600mm long................................	23.22	1.00	23.45	3.75	-	89.96	-	113.41	nr	124.76
900mm long................................	29.85	1.00	30.15	3.75	-	89.96	-	120.11	nr	132.12
1200mm long................................	35.85	1.00	36.21	3.75	-	89.96	-	126.17	nr	138.79
1600mm long................................	53.25	1.00	53.78	3.75	-	89.96	-	143.74	nr	158.12
2400mm long................................	79.88	1.00	80.67	3.75	-	89.96	-	170.64	nr	187.70

600mm high; fixing brackets to masonry with screws

600mm long................................	24.07	1.00	24.31	3.75	-	89.96	-	114.27	nr	125.70
900mm long................................	35.14	1.00	35.49	3.75	-	89.96	-	125.45	nr	138.00
1200mm long................................	40.85	1.00	41.26	3.75	-	89.96	-	131.22	nr	144.34
1600mm long................................	58.64	1.00	59.23	3.75	-	89.96	-	149.19	nr	164.11
2400mm long................................	87.96	1.00	88.84	3.75	-	89.96	-	178.80	nr	196.68

700mm high; fixing brackets to masonry with screws

600mm long................................	31.67	1.00	31.99	3.75	-	89.96	-	121.95	nr	134.14

Labour hourly rates: (except Specialists) Craft Operatives 23.99 Labourer 13.76 Rates are national average prices. Refer to REGIONAL VARIATIONS for indicative levels of overall pricing in regions	MATERIALS			LABOUR				RATES		
	Del to Site	Waste	Material Cost	Craft Optve	Lab	Labour Cost	Sunds	Nett Rate	Unit	Gross rate (10%)
	£	%	£	Hrs	Hrs	£	£	£		£
TERMINAL EQUIPMENT AND FITTINGS (Cont'd)										
Radiators (Cont'd)										
Pressed steel radiators single panel with convector; air cock, plain plug (Cont'd)										
700mm high; fixing brackets to masonry with screws (Cont'd)										
900mm long	54.37	1.00	54.91	3.75	-	89.96	-	144.88	nr	159.36
1200mm long	97.50	1.00	98.47	3.75	-	89.96	-	188.44	nr	207.28
1600mm long	130.02	1.00	131.32	3.75	-	89.96	-	221.28	nr	243.41
2400mm long	195.03	1.00	196.98	3.75	-	89.96	-	286.94	nr	315.64
Pressed steel radiators; double panel with single convector; air cock, plain plug										
450mm high; fixing brackets to masonry with screws										
600mm long	27.33	1.00	27.60	3.75	-	89.96	-	117.57	nr	129.32
900mm long	39.91	1.00	40.31	3.75	-	89.96	-	130.27	nr	143.30
1200mm long	46.38	1.00	46.84	3.75	-	89.96	-	136.81	nr	150.49
1600mm long	73.67	1.00	74.41	3.75	-	89.96	-	164.37	nr	180.81
2400mm long	110.50	1.00	111.61	3.75	-	89.96	-	201.57	nr	221.73
600mm high; fixing brackets to masonry with screws										
600mm long	34.18	1.00	34.52	3.75	-	89.96	-	124.48	nr	136.93
900mm long	47.15	1.00	47.62	3.75	-	89.96	-	137.58	nr	151.34
1200mm long	83.83	1.00	84.67	3.75	-	89.96	-	174.63	nr	192.09
1600mm long	193.58	1.00	195.52	3.75	-	89.96	-	285.48	nr	314.03
2400mm long	258.11	1.00	260.69	3.75	-	89.96	-	350.65	nr	385.72
700mm high; fixing brackets to masonry with screws										
600mm long	54.04	1.00	54.58	3.75	-	89.96	-	144.54	nr	159.00
900mm long	109.03	1.00	110.12	3.75	-	89.96	-	200.08	nr	220.09
1200mm long	145.36	1.00	146.81	3.75	-	89.96	-	236.78	nr	260.45
1600mm long	218.05	1.00	220.23	3.75	-	89.96	-	310.19	nr	341.21
2400mm long	290.74	1.00	293.65	3.75	-	89.96	-	383.61	nr	421.97
Brass valves for radiators; self colour										
Valves for radiators, BS 2767; wheel head										
inlet for copper; angle pattern										
15mm	7.46	2.00	7.61	0.30	-	7.20	-	14.81	nr	16.29
22mm	10.15	2.00	10.35	0.40	-	9.60	-	19.95	nr	21.94
Valves for radiators, BS 2767; lock shield										
inlet for copper; angle pattern										
15mm	7.46	2.00	7.61	0.30	-	7.20	-	14.81	nr	16.29
22mm	10.15	2.00	10.35	0.40	-	9.60	-	19.95	nr	21.94
Brass valves for radiators; chromium plated										
Valves for radiators, BS 2767; wheel head										
inlet for copper; angle pattern										
15mm	9.00	2.00	9.18	0.30	-	7.20	-	16.38	nr	18.01
22mm	11.77	2.00	12.01	0.40	-	9.60	-	21.60	nr	23.76
inlet for copper; straight pattern										
15mm	10.45	2.00	10.66	0.30	-	7.20	-	17.86	nr	19.64
22mm	11.77	2.00	12.01	0.40	-	9.60	-	21.60	nr	23.76
Valves for radiators, BS 2767; lock shield										
inlet for copper; angle pattern										
15mm	9.00	2.00	9.18	0.30	-	7.20	-	16.38	nr	18.01
22mm	11.77	2.00	12.01	0.40	-	9.60	-	21.60	nr	23.76
inlet for copper; straight pattern										
15mm	9.00	2.00	9.18	0.30	-	7.20	-	16.38	nr	18.01
22mm	11.77	2.00	12.01	0.40	-	9.60	-	21.60	nr	23.76
Brass thermostatic valves for radiators; chromium plated										
Thermostatic valves for radiators, one piece; ABS plastics head										
inlet for copper; angle pattern										
15mm	13.30	2.00	13.57	0.30	-	7.20	-	20.76	nr	22.84
inlet for copper; straight pattern										
15mm	13.30	2.00	13.57	0.30	-	7.20	-	20.76	nr	22.84
inlet for iron; angle pattern										
19mm	26.86	2.00	27.40	0.40	-	9.60	-	36.99	nr	40.69

This section continues
on the next page

391

Labour hourly rates: (except Specialists) Craft Operatives 28.34 Labourer 13.76 Rates are national average prices. Refer to REGIONAL VARIATIONS for indicative levels of overall pricing in regions	MATERIALS			LABOUR				RATES		
	Del to Site	Waste	Material Cost	Craft Optve	Lab	Labour Cost	Sunds	Nett Rate	Unit	Gross rate (10%)
	£	%	£	Hrs	Hrs	£	£	£		£
TERMINAL EQUIPMENT AND FITTINGS (Cont'd)										
Extract fans										
Extract fans including shutter, window mounted										
Note: the following prices exclude cutting holes in glass										
102mm diameter, with incorporated pull cord switch	19.05	2.50	19.53	1.25	–	35.42	–	54.95	nr	60.45
102mm diameter - low voltage, with incorporated pull cord switch.	52.10	2.50	53.40	1.25	–	35.42	–	88.83	nr	97.71
257mm diameter, with ceiling mounted switch	152.52	2.50	156.34	1.50	–	42.51	–	198.85	nr	218.73
324mm diameter, with ceiling mounted switch	242.52	2.50	248.59	1.50	–	42.51	–	291.10	nr	320.21
Extract fans including shutter, wall mounted with wall duct										
Note: the following prices exclude forming hole and making good										
102mm diameter, with incorporated pull cord switch	19.05	2.50	19.53	1.50	–	42.51	–	62.04	nr	68.24
102mm diameter - low voltage, with incorporated pull cord switch.	52.10	2.50	53.40	1.50	–	42.51	–	95.91	nr	105.50
257mm diameter, with ceiling mounted switch	173.62	2.50	177.97	2.00	–	56.68	–	234.65	nr	258.11
324mm diameter, with ceiling mounted switch	266.52	2.50	273.19	2.00	–	56.68	–	329.87	nr	362.85
Extract fans including shutter, built in wall; controlled by a PIR switch unit										
290 x 321 mm..	171.75	2.50	176.04	2.50	0.08	71.95	–	247.99	nr	272.79
409 x 433 mm..	301.75	2.50	309.29	2.70	0.08	77.62	–	386.91	nr	425.60
473 x 476 mm..	346.75	2.50	355.42	2.80	0.08	80.45	–	435.87	nr	479.46

This section continues
on the next page

MECHANICAL SERVICES

Labour hourly rates: (except Specialists) Craft Operatives 23.99 Labourer 13.76 Rates are national average prices. Refer to REGIONAL VARIATIONS for indicative levels of overall pricing in regions	MATERIALS			LABOUR				RATES		
	Del to Site £	Waste %	Material Cost £	Craft Optve Hrs	Lab Hrs	Labour Cost £	Sunds £	Nett Rate £	Unit	Gross rate (10%) £
PIPEWORK										
Steel pipes and fittings; screwed and PTFE tape joints; pipework black or red primer										
Pipes, straight in malleable iron pipe brackets fixing to timber with screws										
15mm: pipe and fittings										
15mm pipe	4.64	5.00	4.86	0.50	-	11.99	-	16.86	m	18.54
extra; made bends	-	-	-	0.17	-	4.08	-	4.08	nr	4.49
extra; connections to copper pipe ends; compression joint	1.18	2.00	1.20	0.10	-	2.40	-	3.60	nr	3.96
extra; fittings; one end	0.73	2.00	0.74	0.13	-	3.12	-	3.86	nr	4.24
extra; fittings; two ends	0.74	2.00	0.76	0.25	-	6.00	-	6.76	nr	7.43
extra; fittings; three ends	1.02	2.00	1.04	0.25	-	6.00	-	7.04	nr	7.74
20mm; pipe and fittings										
20mm pipe	5.73	5.00	6.00	0.57	-	13.67	-	19.68	m	21.65
extra; made bends	-	-	-	0.25	-	6.00	-	6.00	nr	6.60
extra; connections to copper pipe ends; compression joint	1.85	2.00	1.89	0.17	-	4.08	-	5.97	nr	6.56
extra; fittings; one end	0.83	2.00	0.85	0.17	-	4.08	-	4.93	nr	5.42
extra; fittings; two ends	1.02	2.00	1.04	0.33	-	7.92	-	8.96	nr	9.85
extra; fittings; three ends	1.48	2.00	1.51	0.33	-	7.92	-	9.43	nr	10.37
25mm; pipe and fittings										
25mm pipe	7.33	5.00	7.69	0.67	-	16.07	-	23.76	m	26.13
extra; made bends	-	-	-	0.33	-	7.92	-	7.92	nr	8.71
extra; connections to copper pipe ends; compression joint	3.59	2.00	3.66	0.25	-	6.00	-	9.66	nr	10.62
extra; fittings; one end	1.04	2.00	1.06	0.25	-	6.00	-	7.06	nr	7.76
extra; fittings; two ends	1.57	2.00	1.60	0.50	-	11.99	-	13.59	nr	14.95
extra; fittings; three ends	2.13	2.00	2.17	0.50	-	11.99	-	14.16	nr	15.58
32mm; pipe and fittings										
32mm pipe	9.59	5.00	10.05	0.80	-	19.19	-	29.24	m	32.16
extra; made bends	-	-	-	0.50	-	11.99	-	11.99	nr	13.19
extra; connections to copper pipe ends; compression joint	6.57	2.00	6.70	0.33	-	7.92	-	14.61	nr	16.08
extra; fittings; one end	1.70	2.00	1.73	0.33	-	7.92	-	9.65	nr	10.61
extra; fittings; two ends	2.69	2.00	2.75	0.67	-	16.07	-	18.82	nr	20.70
extra; fittings; three ends	3.72	2.00	3.80	0.67	-	16.07	-	19.87	nr	21.86
40mm; pipe and fittings										
40mm pipe	12.68	5.00	13.28	1.00	-	23.99	-	37.27	m	41.00
extra; made bends	-	-	-	0.67	-	16.07	-	16.07	nr	17.68
extra; connections to copper pipe ends; compression joint	9.35	2.00	9.54	0.50	-	11.99	-	21.53	nr	23.68
extra; fittings; one end	1.46	2.00	1.49	0.42	-	10.08	-	11.57	nr	12.72
extra; fittings; two ends	2.72	2.00	2.78	0.83	-	19.91	-	22.69	nr	24.96
extra; fittings; three ends	2.93	2.00	2.99	0.83	-	19.91	-	22.90	nr	25.19
50mm; pipe and fittings										
50mm pipe	14.99	5.00	15.71	1.33	-	31.91	-	47.62	m	52.38
extra; made bends	-	-	-	1.00	-	23.99	-	23.99	nr	26.39
extra; connections to copper pipe ends; compression joint	13.68	2.00	13.95	0.67	-	16.07	-	30.03	nr	33.03
extra; fittings; one end	2.14	2.00	2.18	0.50	-	11.99	-	14.17	nr	15.59
extra; fittings; two ends	4.35	2.00	4.44	1.00	-	23.99	-	28.43	nr	31.27
extra; fittings; three ends	4.75	2.00	4.85	1.00	-	23.99	-	28.84	nr	31.72
Pipes straight in malleable iron pipe brackets fixing to masonry with screws										
15mm	4.64	5.00	4.87	0.53	-	12.71	-	17.58	m	19.34
20mm	5.73	5.00	6.01	0.60	-	14.39	-	20.40	m	22.44
25mm	7.34	5.00	7.69	0.70	-	16.79	-	24.48	m	26.93
32mm	9.60	5.00	10.05	0.83	-	19.91	-	29.96	m	32.96
40mm	12.70	5.00	13.30	1.03	-	24.71	-	38.01	m	41.81
50mm	14.99	5.00	15.72	1.36	-	32.63	-	48.35	m	53.18
Pipes straight in malleable iron single rings, 165mm long screwed both ends steel rods and malleable iron backplates fixing to timber with screws										
15mm	5.16	5.00	5.41	0.60	-	14.39	-	19.80	m	21.78
20mm	6.30	5.00	6.60	0.67	-	16.07	-	22.68	m	24.94
25mm	7.80	5.00	8.18	0.77	-	18.47	-	26.65	m	29.32
32mm	9.90	5.00	10.37	0.90	-	21.59	-	31.96	m	35.16
40mm	12.96	5.00	13.57	1.10	-	26.39	-	39.96	m	43.96
50mm	15.20	5.00	15.94	1.43	-	34.31	-	50.25	m	55.27
Steel pipes and fittings; screwed and PTFE tape joints; galvanized										
Pipes, straight in malleable iron pipe brackets fixing to timber with screws										
15mm; pipe and fittings										
15mm pipe	5.03	5.00	5.28	0.50	-	11.99	-	17.28	m	19.00
extra; made bends	-	-	-	0.17	-	4.08	-	4.08	nr	4.49
extra; connections to copper pipe ends; compression joint	2.71	2.00	2.76	0.10	-	2.40	-	5.16	nr	5.68
extra; fittings; one end	0.56	2.00	0.57	0.13	-	3.12	-	3.69	nr	4.05
extra; fittings; two ends	0.78	2.00	0.79	0.25	-	6.00	-	6.79	nr	7.47
extra; fittings; three ends	1.10	2.00	1.12	0.25	-	6.00	-	7.12	nr	7.83
20mm; pipe and fittings										
20mm pipe	5.62	5.00	5.90	0.57	-	13.67	-	19.58	m	21.53
extra; made bends	-	-	-	0.25	-	6.00	-	6.00	nr	6.60
extra; connections to copper pipe ends; compression joint	3.02	2.00	3.08	0.17	-	4.08	-	7.16	nr	7.88
extra; fittings; one end	0.66	2.00	0.67	0.17	-	4.08	-	4.75	nr	5.23

Labour hourly rates: (except Specialists) Craft Operatives 23.99 Labourer 13.76 Rates are national average prices. Refer to REGIONAL VARIATIONS for indicative levels of overall pricing in regions	MATERIALS			LABOUR				RATES		
	Del to Site	Waste	Material Cost	Craft Optve	Lab	Labour Cost	Sunds	Nett Rate	Unit	Gross rate (10%)
	£	%	£	Hrs	Hrs	£	£	£		£

PIPEWORK (Cont'd)

Steel pipes and fittings; screwed and PTFE tape joints; galvanized (Cont'd)

Pipes, straight in malleable iron pipe brackets fixing to timber with screws (Cont'd)

	Del to Site £	Waste %	Material Cost £	Craft Optve Hrs	Lab Hrs	Labour Cost £	Sunds £	Nett Rate £	Unit	Gross rate £
20mm; pipe and fittings (Cont'd)										
extra; fittings; two ends	1.01	2.00	1.03	0.33	-	7.92	-	8.94	nr	9.84
extra; fittings; three ends	1.26	2.00	1.29	0.33	-	7.92	-	9.20	nr	10.12
25mm; pipe and fittings										
25mm pipe	8.37	5.00	8.78	0.67	-	16.07	-	24.86	m	27.34
extra; made bends	-	-	-	0.33	-	7.92	-	7.92	nr	8.71
extra; connections to copper pipe ends; compression joint	4.55	2.00	4.64	0.25	-	6.00	-	10.63	nr	11.70
extra; fittings; one end	0.82	2.00	0.84	0.25	-	6.00	-	6.83	nr	7.52
extra; fittings; two ends	1.45	2.00	1.48	0.50	-	11.99	-	13.47	nr	14.82
extra; fittings; three ends	1.86	2.00	1.90	0.50	-	11.99	-	13.89	nr	15.28
32mm; pipe and fittings										
32mm pipe	9.40	5.00	9.87	0.80	-	19.19	-	29.06	m	31.96
extra; made bends	-	-	-	0.50	-	11.99	-	11.99	nr	13.19
extra; connections to copper pipe ends; compression joint	5.34	2.00	5.45	0.33	-	7.92	-	13.37	nr	14.70
extra; fittings; one end	1.45	2.00	1.48	0.33	-	7.92	-	9.39	nr	10.33
extra; fittings; two ends	2.40	2.00	2.45	0.67	-	16.07	-	18.53	nr	20.38
extra; fittings; three ends	3.10	2.00	3.16	0.67	-	16.07	-	19.23	nr	21.16
40mm; pipe and fittings										
40mm pipe	10.94	5.00	11.49	1.00	-	23.99	-	35.48	m	39.03
extra; made bends	-	-	-	0.67	-	16.07	-	16.07	nr	17.68
extra; connections to copper pipe ends; compression joint	8.36	2.00	8.53	0.50	-	11.99	-	20.52	nr	22.57
extra; fittings; one end	1.67	2.00	1.70	0.42	-	10.08	-	11.78	nr	12.96
extra; fittings; two ends	3.75	2.00	3.82	0.83	-	19.91	-	23.74	nr	26.11
extra; fittings; three ends	4.19	2.00	4.27	0.83	-	19.91	-	24.18	nr	26.60
50mm; pipe and fittings										
50mm pipe	15.76	5.00	16.54	1.33	-	31.91	-	48.45	m	53.30
extra; made bends	-	-	-	1.00	-	23.99	-	23.99	nr	26.39
extra; connections to copper pipe ends; compression joint	9.54	2.00	9.74	0.67	-	16.07	-	25.81	nr	28.39
extra; fittings; one end	2.46	2.00	2.51	0.50	-	11.99	-	14.50	nr	15.95
extra; fittings; two ends	4.71	2.00	4.81	1.00	-	23.99	-	28.80	nr	31.68
extra; fittings; three ends	6.39	2.00	6.52	1.00	-	23.99	-	30.51	nr	33.56
Pipes straight in malleable iron pipe brackets fixing to masonry with screws										
15mm	5.27	5.00	5.53	0.53	-	12.71	-	18.24	m	20.07
20mm	5.93	5.00	6.21	0.60	-	14.39	-	20.61	m	22.67
25mm	8.76	5.00	9.19	0.70	-	16.79	-	25.98	m	28.58
32mm	9.93	5.00	10.41	0.83	-	19.91	-	30.32	m	33.35
40mm	11.62	5.00	12.18	1.03	-	24.71	-	36.89	m	40.58
50mm	16.77	5.00	17.58	1.36	-	32.63	-	50.20	m	55.22
Pipes straight in malleable iron single rings, 165mm long screwed both ends steel tubes and malleable iron backplates fixing to timber with screws										
15mm	6.10	5.00	6.40	0.60	-	14.39	-	20.80	m	22.88
20mm	7.06	5.00	7.40	0.67	-	16.07	-	23.47	m	25.82
25mm	9.84	5.00	10.32	0.77	-	18.47	-	28.80	m	31.67
32mm	10.98	5.00	11.52	0.90	-	21.59	-	33.11	m	36.42
40mm	12.70	5.00	13.32	1.10	-	26.39	-	39.71	m	43.68
50mm	18.04	5.00	18.92	1.43	-	34.31	-	53.22	m	58.54

Stainless steel pipes, BS EN 10312 Part 2; C.P. copper fittings, capillary, BS EN 1254 Part 2

Pipes, straight in C.P. pipe brackets fixing to timber with screws

	Del to Site £	Waste %	Material Cost £	Craft Optve Hrs	Lab Hrs	Labour Cost £	Sunds £	Nett Rate £	Unit	Gross rate £
15mm; pipe and fittings										
15mm pipe	12.50	5.00	13.11	0.40	-	9.60	-	22.71	m	24.98
extra; fittings; one end	10.12	3.00	10.42	0.06	-	1.44	-	11.86	nr	13.05
extra; fittings; two ends	2.66	3.00	2.74	0.12	-	2.88	-	5.62	nr	6.18
extra; fittings; three ends	9.67	3.00	9.96	0.12	-	2.88	-	12.84	nr	14.12
22mm; pipe and fittings										
22mm pipe	14.36	5.00	15.06	0.43	-	10.32	-	25.38	m	27.92
extra; fittings; one end	12.74	3.00	13.12	0.10	-	2.40	-	15.52	nr	17.07
extra; fittings; two ends	2.98	3.00	3.07	0.20	-	4.80	-	7.87	nr	8.65
extra; fittings; three ends	12.06	3.00	12.42	0.20	-	4.80	-	17.22	nr	18.94
28mm; pipe and fittings										
28mm pipe	16.62	5.00	17.44	0.48	-	11.52	-	28.96	m	31.85
extra; fittings; one end	15.20	3.00	15.66	0.15	-	3.60	-	19.25	nr	21.18
extra; fittings; two ends	3.25	3.00	3.35	0.30	-	7.20	-	10.54	nr	11.60
extra; fittings; three ends	14.56	3.00	15.00	0.30	-	7.20	-	22.19	nr	24.41
35mm; pipe and fittings										
35mm pipe	23.77	5.00	24.91	0.60	-	14.39	-	39.31	m	43.24
extra; fittings; one end	19.37	3.00	19.95	0.20	-	4.80	-	24.75	nr	27.22
extra; fittings; two ends	19.00	3.00	19.57	0.40	-	9.60	-	29.17	nr	32.08
extra; fittings; three ends	26.53	3.00	27.33	0.40	-	9.60	-	36.92	nr	40.61
42mm; pipe and fittings										
42mm pipe	25.98	5.00	27.23	0.80	-	19.19	-	46.42	m	51.06
extra; fittings; one end	25.12	3.00	25.87	0.30	-	7.20	-	33.07	nr	36.38
extra; fittings; two ends	27.23	3.00	28.05	0.60	-	14.39	-	42.44	nr	46.68
extra; fittings; three ends	32.70	3.00	33.68	0.60	-	14.39	-	48.08	nr	52.88

Labour hourly rates: (except Specialists) Craft Operatives 23.99 Labourer 13.76 Rates are national average prices. Refer to REGIONAL VARIATIONS for indicative levels of overall pricing in regions	MATERIALS			LABOUR				RATES		
	Del to Site	Waste	Material Cost	Craft Optve	Lab	Labour Cost	Sunds	Nett Rate		Gross rate (10%)
	£	%	£	Hrs	Hrs	£	£	£	Unit	£
PIPEWORK (Cont'd)										
Stainless steel pipes, BS EN 10312 Part 2; C.P. copper fittings, capillary, BS EN 1254 Part 2 (Cont'd)										
Pipes straight in C.P. pipe brackets fixing to masonry with screws										
15mm	12.51	5.00	13.12	0.45	-	10.80	-	23.91	m	26.30
22mm	14.36	5.00	15.07	0.48	-	11.52	-	26.59	m	29.24
28mm	16.63	5.00	17.45	0.53	-	12.71	-	30.16	m	33.18
35mm	23.78	5.00	24.92	0.65	-	15.59	-	40.51	m	44.56
42mm	25.99	5.00	27.23	0.85	-	20.39	-	47.63	m	52.39
Stainless steel pipes, BS EN 10312 Part 2; C.P. copper alloy fittings, compression, BS EN 1254 Part 2, Type A										
Pipes, straight in C.P. pipe brackets fixing to timber with screws										
15mm; pipe and fittings										
15mm pipe	13.45	5.00	14.08	0.40	-	9.60	-	23.68	m	26.04
extra; fittings; one end	10.06	3.00	10.36	0.06	-	1.44	-	11.80	nr	12.98
extra; fittings; two ends	14.58	3.00	15.02	0.12	-	2.88	-	17.90	nr	19.69
extra; fittings; three ends	20.54	3.00	21.16	0.12	-	2.88	-	24.03	nr	26.44
22mm; pipe and fittings										
22mm pipe	17.50	5.00	18.31	0.43	-	10.32	-	28.62	m	31.48
extra; fittings; one end	16.30	3.00	16.79	0.10	-	2.40	-	19.19	nr	21.11
extra; fittings; two ends	28.97	3.00	29.84	0.20	-	4.80	-	34.64	nr	38.10
extra; fittings; three ends	42.43	3.00	43.70	0.20	-	4.80	-	48.50	nr	53.35
28mm; pipe and fittings										
28mm pipe	20.87	5.00	21.81	0.48	-	11.52	-	33.32	m	36.66
extra; fittings; one end	26.32	3.00	27.11	0.15	-	3.60	-	30.71	nr	33.78
extra; fittings; two ends	39.52	3.00	40.71	0.30	-	7.20	-	47.90	nr	52.69
extra; fittings; three ends	58.03	3.00	59.77	0.30	-	7.20	-	66.97	nr	73.66
35mm; pipe and fittings										
35mm pipe	35.75	5.00	37.25	0.60	-	14.39	-	51.64	m	56.80
extra; fittings; one end	33.04	3.00	34.03	0.20	-	4.80	-	38.83	nr	42.71
extra; fittings; two ends	94.96	3.00	97.81	0.40	-	9.60	-	107.40	nr	118.15
extra; fittings; three ends	149.23	3.00	153.71	0.40	-	9.60	-	163.30	nr	179.63
42mm; pipe and fittings										
42mm pipe	39.99	5.00	41.65	0.80	-	19.19	-	60.84	m	66.93
extra; fittings; one end	49.73	3.00	51.22	0.30	-	7.20	-	58.42	nr	64.26
extra; fittings; two ends	148.30	3.00	152.75	0.60	-	14.39	-	167.14	nr	183.86
extra; fittings; three ends	205.44	3.00	211.60	0.60	-	14.39	-	226.00	nr	248.60
Pipes straight in C.P. pipe brackets fixing to masonry with screws										
15mm	13.45	5.00	14.09	0.45	-	10.80	-	24.88	m	27.37
22mm	17.51	5.00	18.31	0.48	-	11.52	-	29.83	m	32.81
28mm	20.87	5.00	21.82	0.53	-	12.71	-	34.53	m	37.98
35mm	35.75	5.00	37.25	0.65	-	15.59	-	52.85	m	58.13
42mm	39.99	5.00	41.66	0.85	-	20.39	-	62.05	m	68.25
Copper pipes, BS EN 1057 Type X; copper fittings, capillary, BS EN 1254 Part 2										
Pipes, straight in two piece copper spacing clips to timber with screws										
15mm; pipe and fitings										
15mm pipe	1.82	5.00	1.91	0.33	-	7.92	-	9.83	m	10.81
extra; made bends	-	-	-	0.08	-	1.92	-	1.92	nr	2.11
extra; connections to iron pipe ends; screwed joint	1.20	3.00	1.24	0.10	-	2.40	-	3.63	nr	4.00
extra; fittings; one end	0.13	3.00	0.13	0.05	-	1.20	-	1.33	nr	1.47
extra; fittings; two ends	0.13	3.00	0.13	0.10	-	2.40	-	2.53	nr	2.79
extra; fittings; three ends	0.37	3.00	0.38	0.10	-	2.40	-	2.78	nr	3.06
22mm; pipe and fittings										
22mm pipe	3.50	5.00	3.67	0.36	-	8.64	-	12.31	m	13.54
extra; made bends	-	-	-	0.13	-	3.12	-	3.12	nr	3.43
extra; connections to iron pipe ends; screwed joint	1.89	3.00	1.95	0.17	-	4.08	-	6.03	nr	6.63
extra; fittings; one end	0.31	3.00	0.32	0.09	-	2.16	-	2.48	nr	2.73
extra; fittings; two ends	0.35	3.00	0.36	0.17	-	4.08	-	4.44	nr	4.88
extra; fittings; three ends	0.98	3.00	1.01	0.17	-	4.08	-	5.09	nr	5.60
28mm; pipe and fittings										
28mm pipe	4.70	5.00	4.93	0.40	-	9.60	-	14.53	m	15.98
extra; made bends	-	-	-	0.17	-	4.08	-	4.08	nr	4.49
extra; connections to iron pipe ends; screwed joint	22.15	3.00	22.81	0.25	-	6.00	-	28.81	nr	31.69
extra; fittings; one end	0.67	3.00	0.69	0.13	-	3.12	-	3.81	nr	4.19
extra; fittings; two ends	0.72	3.00	0.74	0.25	-	6.00	-	6.74	nr	7.41
extra; fittings; three ends	2.53	3.00	2.61	0.25	-	6.00	-	8.60	nr	9.46
35mm; pipe and fittings										
35mm pipe	9.92	5.00	10.41	0.50	-	11.99	-	22.40	m	24.64
extra; made bends	-	-	-	0.25	-	6.00	-	6.00	nr	6.60
extra; connections to iron pipe ends; screwed joint	43.17	3.00	44.47	0.33	-	7.92	-	52.38	nr	57.62
extra; fittings; one end	1.50	3.00	1.54	0.17	-	4.08	-	5.62	nr	6.19
extra; fittings; two ends	1.68	3.00	1.73	0.33	-	7.92	-	9.65	nr	10.61
extra; fittings; three ends	4.31	3.00	4.44	0.33	-	7.92	-	12.36	nr	13.59
42mm; pipe and fittings										
42mm pipe	11.92	5.00	12.51	0.67	-	16.07	-	28.59	m	31.45
extra; made bends	-	-	-	0.33	-	7.92	-	7.92	nr	8.71
extra; connections to iron pipe ends; screwed joint	46.35	3.00	47.74	0.50	-	11.99	-	59.74	nr	65.71

Labour hourly rates: (except Specialists) Craft Operatives 23.99 Labourer 13.76 Rates are national average prices. Refer to REGIONAL VARIATIONS for indicative levels of overall pricing in regions	MATERIALS			LABOUR				RATES		
	Del to Site	Waste	Material Cost	Craft Optve	Lab	Labour Cost	Sunds	Nett Rate	Unit	Gross rate (10%)
	£	%	£	Hrs	Hrs	£	£	£		£

PIPEWORK (Cont'd)

Copper pipes, BS EN 1057 Type X; copper fittings, capillary, BS EN 1254 Part 2 (Cont'd)

Pipes, straight in two piece copper spacing clips to timber with screws (Cont'd)

42mm; pipe and fittings (Cont'd)										
extra; fittings; one end	1.93	3.00	1.99	0.25	-	6.00	-	7.99	nr	8.78
extra; fittings; two ends	2.02	3.00	2.08	0.50	-	11.99	-	14.08	nr	15.48
extra; fittings; three ends	6.67	3.00	6.87	0.50	-	11.99	-	18.87	nr	20.75
54mm; pipe and fittings										
54mm pipe	15.75	5.00	16.53	1.00	-	23.99	-	40.52	m	44.57
extra; made bends	-	-	-	0.50	-	11.99	-	11.99	nr	13.19
extra; connections to iron pipe ends; screwed joint	93.01	3.00	95.80	0.67	-	16.07	-	111.87	nr	123.06
extra; fittings; one end	3.33	3.00	3.43	0.33	-	7.92	-	11.35	nr	12.48
extra; fittings; two ends	3.53	3.00	3.64	0.67	-	16.07	-	19.71	nr	21.68
extra; fittings; three ends	10.62	3.00	10.94	0.67	-	16.07	-	27.01	nr	29.71

Pipes straight in two piece copper spacing clips fixing to masonry with screws

15mm	1.83	5.00	1.92	0.36	-	8.64	-	10.55	m	11.61
22mm	3.50	5.00	3.68	0.39	-	9.36	-	13.03	m	14.34
28mm	4.71	5.00	4.94	0.43	-	10.32	-	15.26	m	16.78
35mm	9.92	5.00	10.41	0.53	-	12.71	-	23.13	m	25.44
42mm	11.93	5.00	12.52	0.70	-	16.79	-	29.31	m	32.24
54mm	15.76	5.00	16.53	1.03	-	24.71	-	41.24	m	45.37

Pipes straight in pressed brass pipe brackets fixing to timber with screws

15mm	2.00	5.00	2.10	0.33	-	7.92	-	10.01	m	11.01
22mm	3.76	5.00	3.94	0.36	-	8.64	-	12.58	m	13.84
28mm	5.20	5.00	5.46	0.40	-	9.60	-	15.06	m	16.56
35mm	10.32	5.00	10.83	0.50	-	11.99	-	22.83	m	25.11
42mm	12.41	5.00	13.03	0.67	-	16.07	-	29.10	m	32.01
54mm	16.40	5.00	17.21	1.00	-	23.99	-	41.20	m	45.32

Pipes straight in cast brass pipe brackets fixing to timber with screws

15mm	2.01	5.00	2.11	0.33	-	7.92	-	10.02	m	11.03
22mm	3.84	5.00	4.03	0.36	-	8.64	-	12.67	m	13.94
28mm	5.16	5.00	5.41	0.40	-	9.60	-	15.01	m	16.51
35mm	10.71	5.00	11.24	0.50	-	11.99	-	23.24	m	25.56
42mm	12.94	5.00	13.59	0.67	-	16.07	-	29.66	m	32.62
54mm	17.09	5.00	17.93	1.00	-	23.99	-	41.92	m	46.11

Pipes straight In cast brass single rings and back plates fixing to timber with screws

15mm	2.37	5.00	2.48	0.36	-	8.64	-	11.12	m	12.23
22mm	4.09	5.00	4.30	0.39	-	9.36	-	13.65	m	15.02
28mm	5.37	5.00	5.64	0.43	-	10.32	-	15.95	m	17.55
35mm	10.70	5.00	11.23	0.53	-	12.71	-	23.94	m	26.34
42mm	12.65	5.00	13.28	0.70	-	16.79	-	30.07	m	33.08
54mm	16.75	5.00	17.58	1.03	-	24.71	-	42.29	m	46.52
67mm	22.58	5.00	23.69	1.50	-	35.99	-	59.67	m	65.64
76mm	35.55	5.00	37.30	2.00	-	47.98	-	85.28	m	93.81
108mm	51.24	5.00	53.73	3.00	-	71.97	-	125.70	m	138.27

Copper pipes, BS EN 1057 Type X; copper alloy fittings, compression, BS EN 1254 Part 2, Type A

Pipes, straight in C.P. pipe brackets fixing to timber with screws

15mm; pipe and fittings										
15mm pipe	2.16	5.00	2.26	0.33	-	7.92	-	10.18	m	11.19
extra; made bends	-	-	-	0.08	-	1.92	-	1.92	nr	2.11
extra; connections to iron pipe ends; screwed joint	1.19	3.00	1.23	0.10	-	2.40	-	3.62	nr	3.99
extra; fittings; one end	0.94	3.00	0.97	0.05	-	1.20	-	2.17	nr	2.38
extra; fittings; two ends	1.09	3.00	1.12	0.10	-	2.40	-	3.52	nr	3.87
extra; fittings; three ends	1.97	3.00	2.03	0.10	-	2.40	-	4.43	nr	4.87
22mm; pipe and fittings										
22mm pipe	4.20	5.00	4.40	0.36	-	8.64	-	13.03	m	14.34
extra; made bends	-	-	-	0.13	-	3.12	-	3.12	nr	3.43
extra; connections to iron pipe ends; screwed joint	2.15	3.00	2.21	0.17	-	4.08	-	6.29	nr	6.92
extra; fittings; one end	1.27	3.00	1.31	0.09	-	2.16	-	3.47	nr	3.81
extra; fittings; two ends	2.32	3.00	2.39	0.17	-	4.08	-	6.47	nr	7.11
extra; fittings; three ends	4.52	3.00	4.66	0.17	-	4.08	-	8.73	nr	9.61
28mm; pipe and fittings										
28mm pipe	5.60	5.00	5.86	0.40	-	9.60	-	15.45	m	17.00
extra; made bends				0.17	-	4.08	-	4.08	nr	4.49
extra; connections to iron pipe ends; screwed joint	3.40	3.00	3.50	0.25	-	6.00	-	9.50	nr	10.45
extra; fittings; one end	2.17	3.00	2.24	0.13	-	3.12	-	5.35	nr	5.89
extra; fittings; two ends	3.08	3.00	3.17	0.25	-	6.00	-	9.17	nr	10.09
extra; fittings; three ends	7.03	3.00	7.24	0.25	-	6.00	-	13.24	nr	14.56
35mm; pipe and fittings										
35mm pipe	13.57	5.00	14.17	0.50	-	11.99	-	26.17	m	28.78
extra; made bends				0.25	-	6.00	-	6.00	nr	6.60
extra; connections to iron pipe ends; screwed joint	10.48	3.00	10.79	0.33	-	7.92	-	18.71	nr	20.58
extra; fittings; one end	10.24	3.00	10.55	0.17	-	4.08	-	14.63	nr	16.09

Labour hourly rates: (except Specialists) Craft Operatives 23.99 Labourer 13.76 Rates are national average prices. Refer to REGIONAL VARIATIONS for indicative levels of overall pricing in regions	MATERIALS			LABOUR				RATES		
	Del to Site	Waste	Material Cost	Craft Optve	Lab	Labour Cost	Sunds	Nett Rate		Gross rate (10%)
	£	%	£	Hrs	Hrs	£	£	£	Unit	£

PIPEWORK (Cont'd)

Copper pipes, BS EN 1057 Type X; copper alloy fittings, compression, BS EN 1254 Part 2, Type A (Cont'd)

Pipes, straight in C.P. pipe brackets fixing to timber with screws (Cont'd)

	Del to Site £	Waste %	Material Cost £	Craft Optve Hrs	Lab Hrs	Labour Cost £	Sunds £	Nett Rate £	Unit	Gross rate (10%) £
35mm; pipe and fittings (Cont'd)										
extra; fittings; two ends	11.94	3.00	12.30	0.33	-	7.92	-	20.21	nr	22.24
extra; fittings; three ends	21.69	3.00	22.34	0.33	-	7.92	-	30.26	nr	33.28
42mm; pipe and fittings										
42mm pipe	16.60	5.00	17.33	0.67	-	16.07	-	33.41	m	36.75
extra; made bends	-	-	-	0.33	-	7.92	-	7.92	nr	8.71
extra; connections to iron pipe ends; screwed joint	13.21	3.00	13.61	0.50	-	11.99	-	25.60	nr	28.16
extra; fittings; one end	16.11	3.00	16.59	0.25	-	6.00	-	22.59	nr	24.85
extra; fittings; two ends	15.22	3.00	15.68	0.50	-	11.99	-	27.67	nr	30.44
extra; fittings; three ends	34.97	3.00	36.02	0.50	-	11.99	-	48.01	nr	52.82
54mm; pipe and fittings										
54mm pipe	22.60	5.00	23.59	1.00	-	23.99	-	47.58	m	52.33
extra; made bends	-	-	-	0.50	-	11.99	-	11.99	nr	13.19
extra; connections to iron pipe ends; screwed joint	20.97	3.00	21.60	0.67	-	16.07	-	37.67	nr	41.44
extra; fittings; one end	23.28	3.00	23.98	0.33	-	7.92	-	31.90	nr	35.08
extra; fittings; two ends	22.59	3.00	23.27	0.67	-	16.07	-	39.34	nr	43.28
extra; fittings; three ends	53.82	3.00	55.43	0.67	-	16.07	-	71.51	nr	78.66
Pipes; straight; in two piece copper spacing clips fixing to masonry with screws										
15mm	2.17	5.00	2.27	0.36	-	8.64	-	10.90	m	11.99
22mm	4.21	5.00	4.40	0.39	-	9.36	-	13.76	m	15.14
28mm	5.60	5.00	5.86	0.43	-	10.32	-	16.18	m	17.80
35mm	13.58	5.00	14.18	0.53	-	12.71	-	26.89	m	29.58
42mm	16.61	5.00	17.34	0.70	-	16.79	-	34.13	m	37.55
54mm	22.61	5.00	23.59	1.03	-	24.71	-	48.30	m	53.13
Pipes straight in pressed brass pipe brackets fixing to timber with screws										
15mm	2.33	5.00	2.44	0.33	-	7.92	-	10.36	m	11.40
22mm	4.46	5.00	4.67	0.36	-	8.64	-	13.31	m	14.64
28mm	6.10	5.00	6.38	0.40	-	9.60	-	15.98	m	17.57
35mm	13.98	5.00	14.60	0.50	-	11.99	-	26.59	m	29.25
42mm	17.09	5.00	17.85	0.67	-	16.07	-	33.92	m	37.31
54mm	23.26	5.00	24.27	1.00	-	23.99	-	48.26	m	53.09
Pipes straight in cast brass pipe brackets fixing to timber with screws										
15mm	2.34	5.00	2.45	0.33	-	7.92	-	10.37	m	11.41
22mm	4.55	5.00	4.76	0.36	-	8.64	-	13.40	m	14.74
28mm	6.05	5.00	6.33	0.40	-	9.60	-	15.93	m	17.52
35mm	14.37	5.00	15.01	0.50	-	11.99	-	27.00	m	29.70
42mm	17.62	5.00	18.40	0.67	-	16.07	-	34.48	m	37.93
54mm	23.94	5.00	24.99	1.00	-	23.99	-	48.98	m	53.88
Pipes straight in cast brass single rings and back plates fixing to timber with screws										
15mm	2.70	5.00	2.83	0.33	-	7.92	-	10.75	m	11.82
22mm	4.80	5.00	5.02	0.36	-	8.64	-	13.66	m	15.03
28mm	6.27	5.00	6.56	0.40	-	9.60	-	16.16	m	17.77
35mm	14.35	5.00	14.99	0.50	-	11.99	-	26.99	m	29.69
42mm	17.33	5.00	18.10	0.67	-	16.07	-	34.17	m	37.59
54mm	23.61	5.00	24.64	1.00	-	23.99	-	48.63	m	53.49
66.7mm	74.95	5.00	77.63	1.50	-	35.99	-	113.61	m	124.97
76.1mm	114.75	5.00	118.88	2.00	-	47.98	-	166.86	m	183.54
108mm	178.52	5.00	184.82	3.00	-	71.97	-	256.79	m	282.47

Copper pipes, BS EN 1057 Type X; copper alloy fittings, compression, BS EN 1254 Part 2 Type B

Pipes, straight in two piece copper spacing clips to timber with screws

	Del to Site £	Waste %	Material Cost £	Craft Optve Hrs	Lab Hrs	Labour Cost £	Sunds £	Nett Rate £	Unit	Gross rate (10%) £
15mm; pipe and fittings										
15mm pipe	2.96	5.00	3.09	0.33	-	7.92	-	11.01	m	12.11
extra; made bends	-	-	-	0.08	-	1.92	-	1.92	nr	2.11
extra; connections to iron pipe ends; screwed joint	2.96	3.00	3.05	0.13	-	3.12	-	6.17	nr	6.78
extra; fittings; one end	4.57	3.00	4.71	0.07	-	1.68	-	6.39	nr	7.03
extra; fittings; two ends	3.53	3.00	3.64	0.13	-	3.12	-	6.75	nr	7.43
extra; fittings; three ends	6.04	3.00	6.22	0.13	-	3.12	-	9.34	nr	10.27
22mm; pipe and fittings										
22mm pipe	5.27	5.00	5.49	0.36	-	8.64	-	14.13	m	15.54
extra; made bends	-	-	-	0.13	-	3.12	-	3.12	nr	3.43
extra; connections to iron pipe ends; screwed joint	5.24	3.00	5.40	0.21	-	5.04	-	10.44	nr	11.48
extra; fittings; one end	5.44	3.00	5.60	0.11	-	2.64	-	8.24	nr	9.07
extra; fittings; two ends	5.54	3.00	5.71	0.21	-	5.04	-	10.74	nr	11.82
extra; fittings; three ends	9.29	3.00	9.57	0.21	-	5.04	-	14.61	nr	16.07
28mm; pipe and fittings										
28mm pipe	10.10	5.00	10.50	0.40	-	9.60	-	20.09	m	22.10
extra; made bends	-	-	-	0.17	-	4.08	-	4.08	nr	4.49
extra; connections to iron pipe ends; screwed joint	9.90	3.00	10.20	0.31	-	7.44	-	17.63	nr	19.40
extra; fittings; one end	18.03	3.00	18.57	0.16	-	3.84	-	22.41	nr	24.65

MECHANICAL SERVICES (side margin)

Labour hourly rates: (except Specialists) Craft Operatives 23.99 Labourer 13.76. Rates are national average prices. Refer to REGIONAL VARIATIONS for indicative levels of overall pricing in regions	MATERIALS			LABOUR				RATES		
	Del to Site £	Waste %	Material Cost £	Craft Optve Hrs	Lab Hrs	Labour Cost £	Sunds £	Nett Rate £	Unit	Gross rate (10%) £

PIPEWORK (Cont'd)

Copper pipes, BS EN 1057 Type X; copper alloy fittings, compression, BS EN 1254 Part 2 Type B (Cont'd)

Pipes, straight in two piece copper spacing clips to timber with screws (Cont'd)

	Del to Site £	Waste %	Material Cost £	Craft Optve Hrs	Lab Hrs	Labour Cost £	Sunds £	Nett Rate £	Unit	Gross rate (10%) £
28mm; pipe and fittings (Cont'd)										
extra; fittings; two ends	15.08	3.00	15.53	0.31	-	7.44	-	22.97	nr	25.27
extra; fittings; three ends	24.05	3.00	24.77	0.31	-	7.44	-	32.21	nr	35.43
35mm; pipe and fittings										
35mm pipe	18.28	5.00	19.02	0.50	-	11.99	-	31.02	m	34.12
extra; made bends	-	-	-	0.25	-	6.00	-	6.00	nr	6.60
extra; connections to iron pipe ends; screwed joint	32.52	3.00	33.49	0.40	-	9.60	-	43.09	nr	47.40
extra; fittings; one end	19.88	3.00	20.48	0.20	-	4.80	-	25.27	nr	27.80
extra; fittings; two ends	30.76	3.00	31.68	0.40	-	9.60	-	41.28	nr	45.41
extra; fittings; three ends	44.03	3.00	45.35	0.40	-	9.60	-	54.95	nr	60.44
42mm; pipe and fittings										
42mm pipe	23.35	5.00	24.28	0.67	-	16.07	-	40.35	m	44.39
extra; made bends	-	-	-	0.33	-	7.92	-	7.92	nr	8.71
extra; connections to iron pipe ends; screwed joint	48.78	3.00	50.25	0.63	-	15.11	-	65.36	nr	71.90
extra; fittings; one end	33.12	3.00	34.11	0.32	-	7.68	-	41.79	nr	45.97
extra; fittings; two ends	33.75	3.00	34.76	0.63	-	15.11	-	49.88	nr	54.86
extra; fittings; three ends	68.06	3.00	70.10	0.63	-	15.11	-	85.22	nr	93.74
54mm; pipe and fittings										
54mm pipe	32.57	5.00	33.85	1.00	-	23.99	-	57.84	m	63.62
extra; made bends	-	-	-	0.50	-	11.99	-	11.99	nr	13.19
extra; connections to iron pipe ends; screwed joint	72.02	3.00	74.18	0.83	-	19.91	-	94.09	nr	103.50
extra; fittings; one end	76.24	3.00	78.53	0.42	-	10.08	-	88.60	nr	97.47
extra; fittings; two ends	72.05	3.00	74.21	0.83	-	19.91	-	94.12	nr	103.54
extra; fittings; three ends	113.82	3.00	117.23	0.83	-	19.91	-	137.15	nr	150.86

Pipes straight in two piece copper spacing clips fixing to masonry with screws

	Del to Site £	Waste %	Material Cost £	Craft Optve Hrs	Lab Hrs	Labour Cost £	Sunds £	Nett Rate £	Unit	Gross rate (10%) £
15mm	2.97	5.00	3.10	0.36	-	8.64	-	11.73	m	12.91
22mm	5.27	5.00	5.50	0.39	-	9.36	-	14.86	m	16.34
28mm	10.11	5.00	10.50	0.43	-	10.32	-	20.82	m	22.90
35mm	18.28	5.00	19.03	0.53	-	12.71	-	31.74	m	34.91
42mm	23.35	5.00	24.28	0.70	-	16.79	-	41.08	m	45.18
54mm	32.57	5.00	33.85	1.03	-	24.71	-	58.56	m	64.42

Pipes straight in pressed brass pipe brackets fixing to timber with screws

	Del to Site £	Waste %	Material Cost £	Craft Optve Hrs	Lab Hrs	Labour Cost £	Sunds £	Nett Rate £	Unit	Gross rate (10%) £
15mm	3.14	5.00	3.27	0.36	-	8.64	-	11.91	m	13.10
22mm	5.53	5.00	5.77	0.39	-	9.36	-	15.12	m	16.63
28mm	10.60	5.00	11.02	0.43	-	10.32	-	21.34	m	23.47
35mm	18.68	5.00	19.45	0.53	-	12.71	-	32.16	m	35.38
42mm	23.84	5.00	24.79	0.70	-	16.79	-	41.59	m	45.74
54mm	33.22	5.00	34.53	1.03	-	24.71	-	59.24	m	65.17

Pipes straight in cast brass pipe brackets fixing to timber with screws

	Del to Site £	Waste %	Material Cost £	Craft Optve Hrs	Lab Hrs	Labour Cost £	Sunds £	Nett Rate £	Unit	Gross rate (10%) £
15mm	3.15	5.00	3.28	0.36	-	8.64	-	11.92	m	13.11
22mm	5.61	5.00	5.86	0.39	-	9.36	-	15.21	m	16.73
28mm	10.56	5.00	10.97	0.43	-	10.32	-	21.29	m	23.42
35mm	19.07	5.00	19.85	0.53	-	12.71	-	32.57	m	35.83
42mm	24.37	5.00	25.35	0.70	-	16.79	-	42.14	m	46.36
54mm	33.91	5.00	35.25	1.03	-	24.71	-	59.96	m	65.96

Pipes straight in cast brass single rings and back plates fixing to timber with screws

	Del to Site £	Waste %	Material Cost £	Craft Optve Hrs	Lab Hrs	Labour Cost £	Sunds £	Nett Rate £	Unit	Gross rate (10%) £
15mm	3.51	5.00	3.66	0.36	-	8.64		12.30	m	13.53
22mm	5.86	5.00	6.12	0.39	-	9.36		15.47	m	17.02
28mm	10.77	5.00	11.20	0.43	-	10.32		21.52	m	23.67
35mm	19.06	5.00	19.84	0.53	-	12.71		32.55	m	35.81
42mm	24.07	5.00	25.04	0.70	-	16.79		41.83	m	46.02
54mm	33.57	5.00	34.90	1.03	-	24.71		59.61	m	65.57

Copper pipes, BS EN 1057 Type Y; copper fittings, capillary, BS EN 1254 Part 2

Pipes; straight; in trenches

	Del to Site £	Waste %	Material Cost £	Craft Optve Hrs	Lab Hrs	Labour Cost £	Sunds £	Nett Rate £	Unit	Gross rate (10%) £
15mm; pipe and fittings										
15mm pipe	3.51	5.00	3.68	0.20	-	4.80	-	8.48	m	9.33
extra; made bends	-	-	-	0.10	-	2.40	-	2.40	nr	2.64
extra; connections to iron pipe ends; screwed joint	1.20	3.00	1.24	0.16	-	3.84	-	5.07	nr	5.58
extra; fittings; one end	0.13	3.00	0.13	0.08	-	1.92	-	2.05	nr	2.26
extra; fittings; two ends	0.13	3.00	0.13	0.16	-	3.84	-	3.97	nr	4.37
extra; fittings; three ends	0.37	3.00	0.38	0.16	-	3.84	-	4.22	nr	4.64
22mm; pipe and fittings										
22mm pipe	9.21	5.00	9.67	0.22	-	5.28	-	14.94	m	16.44
extra; made bends	-	-	-	0.16	-	3.84	-	3.84	nr	4.22
extra; connections to iron pipe ends; screwed joint	1.89	3.00	1.95	0.25	-	6.00	-	7.94	nr	8.74
extra; fittings; one end	0.31	3.00	0.32	0.13	-	3.12	-	3.44	nr	3.78
extra; fittings; two ends	0.35	3.00	0.36	0.25	-	6.00	-	6.36	nr	6.99
extra; fittings; three ends	0.98	3.00	1.01	0.25	-	6.00	-	7.01	nr	7.71

Labour hourly rates: (except Specialists) Craft Operatives 23.99 Labourer 13.76 Rates are national average prices. Refer to REGIONAL VARIATIONS for indicative levels of overall pricing in regions	MATERIALS			LABOUR				RATES		
	Del to Site	Waste	Material Cost	Craft Optve	Lab	Labour Cost	Sunds	Nett Rate	Unit	Gross rate (10%)
	£	%	£	Hrs	Hrs	£	£	£		£
PIPEWORK (Cont'd)										
Copper pipes, BS EN 1057 Type Y; non-dezincifiable fittings; compression, BS EN 1254 Part 2, Type B										
Pipes; straight; in trenches										
15mm; pipe and fittings										
15mm pipe	3.60	5.00	3.78	0.20	-	4.80	-	8.58	m	9.44
extra; made bends	-	-	-	0.10	-	2.40	-	2.40	nr	2.64
extra; connections to iron pipe ends; screwed joint	1.60	3.00	1.65	0.16	-	3.84	-	5.49	nr	6.04
extra; fittings; one end	0.95	3.00	0.98	0.08	-	1.92	-	2.90	nr	3.19
extra; fittings; two ends	1.59	3.00	1.64	0.16	-	3.84	-	5.48	nr	6.02
extra; fittings; three ends	2.51	3.00	2.59	0.16	-	3.84	-	6.42	nr	7.07
22mm; pipe and fittings										
22mm pipe	9.36	5.00	9.83	0.22	-	5.28	-	15.11	m	16.62
extra; made bends	-	-	-	0.16	-	3.84	-	3.84	nr	4.22
extra; connections to iron pipe ends; screwed joint	2.25	3.00	2.32	0.25	-	6.00	-	8.31	nr	9.15
extra; fittings; one end	1.62	3.00	1.67	0.13	-	3.12	-	4.79	nr	5.27
extra; fittings; two ends	2.71	3.00	2.79	0.25	-	6.00	-	8.79	nr	9.67
extra; fittings; three ends	4.58	3.00	4.72	0.25	-	6.00	-	10.71	nr	11.79
28mm; pipe and fittings										
28mm pipe	8.48	5.00	8.90	0.24	-	5.76	-	14.66	m	16.12
extra; made bends	-	-	-	0.20	-	4.80	-	4.80	nr	5.28
extra; connections to iron pipe ends; screwed joint	3.47	3.00	3.57	0.37	-	8.88	-	12.45	nr	13.70
extra; fittings; one end	4.46	3.00	4.59	0.19	-	4.56	-	9.15	nr	10.07
extra; fittings; two ends	4.14	3.00	4.26	0.37	-	8.88	-	13.14	nr	14.45
extra; fittings; three ends	8.83	3.00	9.09	0.37	-	8.88	-	17.97	nr	19.77
42mm; pipe and fittings										
42mm pipe	16.71	5.00	17.52	0.35	-	8.40	-	25.92	m	28.51
extra; made bends	-	-	-	0.40	-	9.60	-	9.60	nr	10.56
extra; connections to iron pipe ends; screwed joint	7.41	3.00	7.63	0.76	-	18.23	-	25.86	nr	28.45
extra; fittings; one end	21.82	3.00	22.47	0.38	-	9.12	-	31.59	nr	34.75
extra; fittings; two ends	20.15	3.00	20.75	0.76	-	18.23	-	38.99	nr	42.89
extra; fittings; three ends	21.04	3.00	21.67	0.76	-	18.23	-	39.90	nr	43.89
Polybutylene pipes; Wavin Hepworth Hep2O flexible plumbing system BS 7291 Class H; polybutylene demountable fittings										
Pipes, flexible in pipe clips to timber with screws										
15mm; pipe and fittings										
15mm pipe	2.08	10.00	2.23	0.37	-	8.88	-	11.11	m	12.22
extra; connector to copper	2.93	2.00	2.99	0.10	-	2.40	-	5.39	nr	5.93
extra; fittings; one end	1.60	2.00	1.63	0.05	-	1.20	-	2.83	nr	3.11
extra; fittings; two ends	1.35	2.00	1.38	0.10	-	2.40	-	3.78	nr	4.15
extra; fittings; three ends	2.32	2.00	2.37	0.10	-	2.40	-	4.77	nr	5.24
22mm; pipe and fittings										
22mm pipe	2.88	10.00	3.13	0.43	-	10.32	-	13.44	m	14.79
extra; connector to copper	3.46	2.00	3.53	0.17	-	4.08	-	7.61	nr	8.37
extra; fittings; one end	2.48	2.00	2.53	0.09	-	2.16	-	4.69	nr	5.16
extra; fittings; two ends	1.89	2.00	1.93	0.17	-	4.08	-	6.01	nr	6.61
Refer; fittings; three ends	3.15	2.00	3.21	0.17	-	4.08	-	7.29	nr	8.02
28mm; pipe and fittings										
28mm pipe	4.37	10.00	4.75	0.50	-	11.99	-	16.74	m	18.42
extra; fittings; one end	3.97	2.00	4.05	0.17	-	4.08	-	8.13	nr	8.94
extra; fittings; two ends	5.15	2.00	5.25	0.33	-	7.92	-	13.17	nr	14.49
extra; fittings; three ends	8.81	2.00	8.99	0.33	-	7.92	-	16.90	nr	18.59
Pipes flexible in pipe clips fixing to masonry with screws										
15mm	2.11	10.00	2.27	0.33	-	7.92	-	10.19	m	11.20
22mm	2.89	10.00	3.14	0.37	-	8.88	-	12.02	m	13.22
28mm	4.38	10.00	4.76	0.50	-	11.99	-	16.75	m	18.43
Polythene pipes, BS EN 12201 Blue; MDPE fittings, compression										
Pipes; straight; in trenches										
20mm; pipe and fittings										
20mm pipe	0.58	10.00	0.63	0.17	-	4.08	-	4.71	m	5.18
extra; connections to copper pipe ends; compression joint	4.31	3.00	4.44	0.16	-	3.84	-	8.28	nr	9.11
extra; fittings; one end	4.14	3.00	4.26	0.08	-	1.92	-	6.18	nr	6.80
extra; fittings; two ends	3.14	3.00	3.23	0.16	-	3.84	-	7.07	nr	7.78
extra; fittings; three ends	6.07	3.00	6.25	0.16	-	3.84	-	10.09	nr	11.10
25mm; pipe and fittings										
25mm pipe	0.73	10.00	0.79	0.20	-	4.80	-	5.59	m	6.15
extra; connections to copper pipe ends; compression joint	6.55	3.00	6.75	0.25	-	6.00	-	12.74	nr	14.02
extra; fittings; one end	1.91	3.00	1.97	0.13	-	3.12	-	5.09	nr	5.59
extra; fittings; two ends	3.51	3.00	3.62	0.25	-	6.00	-	9.61	nr	10.57
extra; fittings; three ends	9.39	3.00	9.67	0.25	-	6.00	-	15.67	nr	17.24
32mm; pipe and fittings										
32mm pipe	1.28	10.00	1.39	0.24	-	5.76	-	7.15	m	7.86
extra; connections to copper pipe ends; compression joint	10.79	3.00	11.11	0.37	-	8.88	-	19.99	nr	21.99
extra; fittings; one end	6.81	3.00	7.01	0.19	-	4.56	-	11.57	nr	12.73
extra; fittings; two ends	8.06	3.00	8.30	0.37	-	8.88	-	17.18	nr	18.90
extra; fittings; three ends	13.17	3.00	13.57	0.37	-	8.88	-	22.44	nr	24.69

Labour hourly rates: (except Specialists) Craft Operatives 23.99 Labourer 13.76 Rates are national average prices. Refer to REGIONAL VARIATIONS for indicative levels of overall pricing in regions	MATERIALS			LABOUR				RATES		
	Del to Site £	Waste %	Material Cost £	Craft Optve Hrs	Lab Hrs	Labour Cost £	Sunds £	Nett Rate £	Unit	Gross rate (10%) £

PIPEWORK (Cont'd)

Polythene pipes, BS EN 12201 Blue; MDPE fittings, compression (Cont'd)

Pipes; straight; in trenches (Cont'd)
50mm; pipe and fittings

50mm pipe	3.08	10.00	3.34	0.40	-	9.60	-	12.93	m	14.23
extra; fittings; two ends	19.38	3.00	19.96	0.60	-	14.39	-	34.36	nr	37.79
extra; fittings; three ends	29.50	3.00	30.38	0.60	-	14.39	-	44.78	nr	49.26

63mm; pipe and fittings

63mm pipe	6.47	10.00	6.92	0.54	-	12.95	-	19.87	m	21.86
extra; fittings; two ends	28.44	3.00	29.29	1.00	-	23.99	-	53.28	nr	58.61
extra; fittings; three ends	44.02	3.00	45.34	1.00	-	23.99	-	69.33	nr	76.26

PVC-U pipes, BS 3505 Class E; fittings, solvent welded joints

Pipes, straight in standard plastics pipe brackets fixing to timber with screws
3/8"; pipe and fittings

3/8" pipe	3.73	5.00	3.84	0.33	-	7.92	-	11.75	m	12.93
extra; connections to iron pipe ends; screwed joint	1.19	2.00	1.21	0.10	-	2.40	-	3.61	nr	3.97
extra; fittings; one end	0.75	2.00	0.76	0.05	-	1.20	-	1.96	nr	2.16
extra; fittings; two ends	1.14	2.00	1.17	0.10	-	2.40	-	3.57	nr	3.92
extra; fittings; three ends	1.25	2.00	1.28	0.10	-	2.40	-	3.68	nr	4.05

1/2"; pipe and fittings

1/2" pipe	4.32	5.00	4.44	0.37	-	8.88	-	13.31	m	14.64
extra; connections to iron pipe ends; screwed joint	1.10	2.00	1.12	0.17	-	4.08	-	5.20	nr	5.72
extra; fittings; one end	0.46	2.00	0.47	0.09	-	2.16	-	2.63	nr	2.89
extra; fittings; two ends	0.49	2.00	0.50	0.17	-	4.08	-	4.58	nr	5.04
extra; fittings; three ends	0.74	2.00	0.75	0.17	-	4.08	-	4.83	nr	5.32

3/4"; pipe and fittings

3/4" pipe	3.55	5.00	3.66	0.43	-	10.32	-	13.97	m	15.37
extra; connections to iron pipe ends; screwed joint	1.27	2.00	1.30	0.25	-	6.00	-	7.29	nr	8.02
extra; fittings; one end	0.54	2.00	0.55	0.13	-	3.12	-	3.67	nr	4.04
extra; fittings; two ends	0.54	2.00	0.55	0.25	-	6.00	-	6.55	nr	7.20
extra; fittings; three ends	0.94	2.00	0.96	0.25	-	6.00	-	6.96	nr	7.65

1"; pipe and fittings

1" pipe	3.04	5.00	3.15	0.50	-	11.99	-	15.14	m	16.66
extra; connections to iron pipe ends; screwed joint	1.55	2.00	1.58	0.33	-	7.92	-	9.50	nr	10.45
extra; fittings; one end	0.62	2.00	0.63	0.17	-	4.08	-	4.71	nr	5.18
extra; fittings; two ends	0.63	2.00	0.64	0.33	-	7.92	-	8.56	nr	9.42
extra; fittings; three ends	1.39	2.00	1.42	0.33	-	7.92	-	9.33	nr	10.27

1 1/4"; pipe and fittings

1 1/4" pipe	3.86	5.00	4.01	0.60	-	14.39	-	18.40	m	20.24
extra; connections to iron pipe ends; screwed joint	2.03	2.00	2.07	0.50	-	11.99	-	14.07	nr	15.47
extra; fittings; one end	0.95	2.00	0.97	0.25	-	6.00	-	6.97	nr	7.66
extra; fittings; two ends	1.11	2.00	1.13	0.50	-	11.99	-	13.13	nr	14.44
extra; fittings; three ends	2.00	2.00	2.04	0.50	-	11.99	-	14.03	nr	15.44

1 1/2"; pipe and fittings

1 1/2" pipe	5.06	5.00	5.24	0.75	-	17.99	-	23.23	m	25.56
extra; connections to iron pipe ends; screwed joint	2.62	2.00	2.67	0.67	-	16.07	-	18.75	nr	20.62
extra; fittings; one end	1.58	2.00	1.61	0.33	-	7.92	-	9.53	nr	10.48
extra; fittings; two ends	1.31	2.00	1.34	0.67	-	16.07	-	17.41	nr	19.15
extra; fittings; three ends	2.86	2.00	2.92	0.67	-	16.07	-	18.99	nr	20.89

Pipes straight In standard plastics pipe brackets fixing to masonry with screws

3/8"	3.78	5.00	3.89	0.39	-	9.36	-	13.25	m	14.57
1/2"	4.37	5.00	4.49	0.43	-	10.32	-	14.81	m	16.29
3/4"	3.58	5.00	3.69	0.49	-	11.76	-	15.44	m	16.98
1"	2.93	5.00	3.03	0.56	-	13.43	-	16.46	m	18.11
1 1/4"	3.88	5.00	4.02	0.66	-	15.83	-	19.86	m	21.84
1 1/2"	5.12	5.00	5.31	0.81	-	19.43	-	24.74	m	27.21

Pipes; straight; in trenches

3/8"	0.84	5.00	0.88	0.17	-	4.08	-	4.96	m	5.46
1/2"	0.83	5.00	0.87	0.18	-	4.32	-	5.19	m	5.71
3/4"	1.16	5.00	1.21	0.20	-	4.80	-	6.01	m	6.61
1"	1.48	5.00	1.55	0.25	-	6.00	-	7.55	m	8.30
1 1/4"	2.33	5.00	2.44	0.29	-	6.96	-	9.40	m	10.34
1 1/2"	3.01	5.00	3.15	0.33	-	7.92	-	11.07	m	12.18

PVC-U pipes and fittings; solvent welded joints; pipework self coloured white

Pipes, straight in standard plastics pipe brackets fixing to timber with screws
21mm; pipe and fittings

21mm pipe	1.14	5.00	1.18	0.25	-	6.00	-	7.17	m	7.89
extra; fittings; two ends	0.83	2.00	0.85	0.20	-	4.80	-	5.64	nr	6.21
extra; fittings; three ends	0.83	2.00	0.85	0.20	-	4.80	-	5.64	nr	6.21
extra; fittings; tank connector	0.86	2.00	0.88	0.20	-	4.80	-	5.68	nr	6.24

Labour hourly rates: (except Specialists) Craft Operatives 23.99 Labourer 13.76 Rates are national average prices. Refer to REGIONAL VARIATIONS for indicative levels of overall pricing in regions	MATERIALS			LABOUR				RATES		
	Del to Site	Waste	Material Cost	Craft Optve	Lab	Labour Cost	Sunds	Nett Rate	Unit	Gross rate (10%)
	£	%	£	Hrs	Hrs	£	£	£		£
PIPEWORK (Cont'd)										
Ductile iron pipes and fittings, Tyton socketed flexible joints										
Pipes; straight; in trenches										
80mm; pipe and fittings										
80mm pipe	29.88	5.00	31.38	0.80	-	19.19	-	50.57	m	55.63
extra; bends, 90 degree	62.45	2.00	63.70	0.80	-	19.19	-	82.89	nr	91.18
extra; duckfoot bends, 90 degree	142.35	2.00	145.20	0.80	-	19.19	-	164.39	nr	180.83
extra; tees	91.82	2.00	93.66	1.20	-	28.79	-	122.44	nr	134.69
extra; hydrant tees	125.39	2.00	127.90	1.20	-	28.79	-	156.69	nr	172.35
extra; branches, 45 degree	290.00	2.00	295.80	1.20	-	28.79	-	324.59	nr	357.05
extra; flanged sockets	49.31	2.00	50.30	0.80	-	19.19	-	69.49	nr	76.44
extra; flanged spigots	46.11	2.00	47.03	0.80	-	19.19	-	66.22	nr	72.85
100mm; pipe and fittings										
100mm pipe	29.88	5.00	31.38	1.00	-	23.99	-	55.37	m	60.90
extra; bends, 90 degree	65.75	2.00	67.07	1.00	-	23.99	-	91.06	nr	100.16
extra; duckfoot bends, 90 degree	154.25	2.00	157.33	1.00	-	23.99	-	181.32	nr	199.46
extra; tees	97.83	2.00	99.79	1.50	-	35.99	-	135.77	nr	149.35
extra; hydrant tees	92.49	2.00	94.34	1.50	-	35.99	-	130.32	nr	143.36
extra; branches, 45 degree	43.20	2.00	44.06	1.50	-	35.99	-	80.05	nr	88.05
extra; flanged sockets	53.34	2.00	54.41	1.00	-	23.99	-	78.40	nr	86.24
extra; flanged spigots	48.69	2.00	49.66	1.00	-	23.99	-	73.65	nr	81.02
150mm; pipe and fittings										
150mm pipe	36.14	5.00	37.95	1.50	-	35.99	-	73.93	m	81.33
extra; bends, 90 degree	134.16	2.00	136.84	1.50	-	35.99	-	172.83	nr	190.11
extra; duckfoot bends, 90 degree	337.15	2.00	343.89	1.50	-	35.99	-	379.88	nr	417.87
extra; tees	182.20	2.00	185.84	2.25	-	53.98	-	239.82	nr	263.80
extra; hydrant tees	167.91	2.00	171.27	2.25	-	53.98	-	225.25	nr	247.77
extra; branches, 45 degree	474.86	2.00	484.36	2.25	-	53.98	-	538.33	nr	592.17
extra; flanged sockets	81.27	2.00	82.90	1.50	-	35.99	-	118.88	nr	130.77
extra; flanged spigots	56.47	2.00	57.60	1.50	-	35.99	-	93.58	nr	102.94
Gas flue pipes comprising galvanised steel outer skin and aluminium inner skin with air space between, BS 715; socketed joints										
Pipes; straight										
100mm	24.80	5.00	26.04	0.80	-	19.19	-	45.23	m	49.76
extra; connections to appliance	5.82	2.00	5.94	0.60	-	14.39	-	20.33	nr	22.36
extra; adjustable pipes	17.72	2.00	18.07	0.60	-	14.39	-	32.47	nr	35.72
extra; terminals	18.34	2.00	18.71	0.60	-	14.39	-	33.10	nr	36.41
extra; elbows 0- 90 degrees	14.60	2.00	14.89	0.60	-	14.39	-	29.29	nr	32.21
extra; tees	30.88	2.00	31.50	0.60	-	14.39	-	45.89	nr	50.48
125mm	28.44	5.00	29.87	0.96	-	23.03	-	52.90	m	58.19
extra; connections to appliance	6.94	2.00	7.08	0.80	-	19.19	-	26.27	nr	28.90
extra; adjustable pipes	21.88	2.00	22.32	0.80	-	19.19	-	41.51	nr	45.66
extra; terminals	38.94	2.00	39.72	0.80	-	19.19	-	58.91	nr	64.80
extra; elbows 0- 90 degrees	18.30	2.00	18.67	0.80	-	19.19	-	37.86	nr	41.64
extra; tees	41.36	2.00	42.19	0.80	-	19.19	-	61.38	nr	67.52
150mm	29.27	5.00	30.73	1.20	-	28.79	-	59.52	m	65.47
extra; connections to appliance	7.84	2.00	8.00	1.20	-	28.79	-	36.78	nr	40.46
extra; adjustable pipes	23.66	2.00	24.13	1.20	-	28.79	-	52.92	nr	58.21
extra; terminals	23.54	2.00	24.01	1.20	-	28.79	-	52.80	nr	58.08
extra; elbows 0- 90 degrees	19.74	2.00	20.13	1.20	-	28.79	-	48.92	nr	53.82
extra; tees	50.38	2.00	51.39	1.20	-	28.79	-	80.18	nr	88.19
Flue supports; galvanised steel wall bands fixing to masonry with screws										
for 100mm pipes	7.78	2.00	7.94	0.20	-	4.80	-	12.73	nr	14.01
for 125mm pipes	8.76	2.00	8.94	0.20	-	4.80	-	13.73	nr	15.11
for 150mm pipes	10.20	2.00	10.40	0.20	-	4.80	-	15.20	nr	16.72
Fire stop spacers										
for 100mm pipes	3.26	2.00	3.33	0.30	-	7.20	-	10.52	nr	11.57
for 125mm pipes	3.50	2.00	3.57	0.40	-	9.60	-	13.17	nr	14.48
for 150mm pipes	3.54	2.00	3.61	0.60	-	14.39	-	18.00	nr	19.81
PIPE ANCILLARIES										
Brass stopcocks										
Stopcocks; crutch head; DZR; joints to polythene BS EN 12201 each end										
15mm	6.49	2.00	6.62	0.33	-	7.92	-	14.54	nr	15.99
20mm	8.00	2.00	8.16	0.33	-	7.92	-	16.08	nr	17.68
Stopcocks, crutch head; compression joints to copper (Type A) each end										
15mm	2.69	2.00	2.74	0.25	-	6.00	-	8.74	nr	9.62
22mm	6.54	2.00	6.67	0.33	-	7.92	-	14.59	nr	16.05
28mm	13.84	2.00	14.12	0.50	-	11.99	-	26.11	nr	28.72
Stopcocks, crutch head; DZR; compression joints to copper (Type A) each end										
15mm	5.05	2.00	5.15	0.25	-	6.00	-	11.15	nr	12.26
22mm	8.07	2.00	8.23	0.33	-	7.92	-	16.15	nr	17.76

Labour hourly rates: (except Specialists) Craft Operatives 23.99 Labourer 13.76 Rates are national average prices. Refer to REGIONAL VARIATIONS for indicative levels of overall pricing in regions	MATERIALS			LABOUR				RATES		
	Del to Site	Waste	Material Cost	Craft Optve	Lab	Labour Cost	Sunds	Nett Rate		Gross rate (10%)
	£	%	£	Hrs	Hrs	£	£	£	Unit	£
PIPE ANCILLARIES (Cont'd)										
Brass stopcocks (Cont'd)										
Stopcocks, crutch head; DZR; compression joints to copper (Type A) (Cont'd)										
each end (Cont'd)										
28mm	14.15	2.00	14.43	0.50	-	11.99	-	26.43	nr	29.07
35mm	40.20	2.00	41.00	0.67	-	16.07	-	57.08	nr	62.79
42mm	49.36	2.00	50.35	0.75	-	17.99	-	68.34	nr	75.17
54mm	89.18	2.00	90.96	1.00	-	23.99	-	114.95	nr	126.45
Polybutylene stopcocks										
Stopcocks; fitted with Hep2O ends										
each end										
15mm	9.23	2.00	9.42	0.25	-	6.00	-	15.42	nr	16.96
22mm	11.35	2.00	11.58	0.33	-	7.92	-	19.49	nr	21.44
Copper alloy ball valves										
Lever ball valves; screwed and PTFE joints										
each end threaded internally										
15mm	3.97	2.00	4.05	0.25	-	6.00	-	10.05	nr	11.05
22mm	5.87	2.00	5.99	0.33	-	7.92	-	13.90	nr	15.29
28mm	9.12	2.00	9.30	0.50	-	11.99	-	21.30	nr	23.43
35mm	60.86	2.00	62.08	0.67	-	16.07	-	78.15	nr	85.97
42mm	89.14	2.00	90.92	0.75	-	17.99	-	108.92	nr	119.81
54mm	111.96	2.00	114.20	1.00	-	23.99	-	138.19	nr	152.01
Copper alloy gate valves										
Gate valves; BS EN 12288 series B; compression joints to copper (Type A)										
each end										
15mm	2.98	2.00	3.04	0.25	-	6.00	-	9.04	nr	9.94
22mm	4.18	2.00	4.26	0.33	-	7.92	-	12.18	nr	13.40
28mm	6.70	2.00	6.83	0.50	-	11.99	-	18.83	nr	20.71
35mm	28.50	2.00	29.07	0.67	-	16.07	-	45.14	nr	49.66
42mm	36.04	2.00	36.76	0.75	-	17.99	-	54.75	nr	60.23
54mm	60.50	2.00	61.71	1.00	-	23.99	-	85.70	nr	94.27
Gate valves; DZR; fitted with Hep2O ends										
each end										
15mm	9.43	2.00	9.62	0.25	-	6.00	-	15.62	nr	17.18
22mm	13.40	2.00	13.67	0.33	-	7.92	-	21.58	nr	23.74
Copper alloy body float valves										
Float operated valves for low pressure; BS 1212; copper float inlet threaded externally; fixing to steel										
13mm; part 1	11.36	2.00	11.59	0.33	-	7.92	-	19.50	nr	21.45
13mm; part 2	11.36	2.00	11.59	0.33	-	7.92	-	19.50	nr	21.45
Float operated valves for low pressure; BS 1212; plastics float inlet threaded externally; fixing to steel										
13mm; part 1	5.20	2.00	5.30	0.33	-	7.92	-	13.22	nr	14.54
13mm; part 2	5.20	2.00	5.30	0.33	-	7.92	-	13.22	nr	14.54
Float operated valves for high pressure; BS 1212; copper float inlet threaded externally; fixing to steel										
13mm; part 1	12.88	2.00	13.14	0.33	-	7.92	-	21.05	nr	23.16
13mm; part 2	10.86	2.00	11.08	0.33	-	7.92	-	18.99	nr	20.89
19mm; part 1	18.74	2.00	19.11	0.50	-	11.99	-	31.11	nr	34.22
25mm; part 1	43.19	2.00	44.05	1.00	-	23.99	-	68.04	nr	74.85
Float operated valves for high pressure; BS 1212; plastics float inlet threaded externally; fixing to steel										
13mm; part 1	6.72	2.00	6.85	0.33	-	7.92	-	14.77	nr	16.25
13mm; part 2	4.70	2.00	4.79	0.33	-	7.92	-	12.71	nr	13.98
19mm; part 1	10.06	2.00	10.26	0.50	-	11.99	-	22.26	nr	24.48
25mm; part 1	25.37	2.00	25.88	1.00	-	23.99	-	49.87	nr	54.85
Brass draining taps										
Drain cocks; square head Type 2										
13mm	1.14	2.00	1.16	0.10	-	2.40	-	3.56	nr	3.92
19mm	4.00	2.00	4.08	0.13	-	3.12	-	7.20	nr	7.92
INSULATION AND FIRE PROTECTION										
Thermal insulation; foil faced glass fibre preformed lagging, butt joints in the running length; secured with metal bands										
19mm thick insulation to copper pipework										
around one pipe										
15mm	2.42	3.00	2.49	0.20	-	4.80	-	7.29	m	8.02
22mm	2.56	3.00	2.64	0.22	-	5.28	-	7.91	m	8.71
28mm	2.71	3.00	2.79	0.25	-	6.00	-	8.79	m	9.67
35mm	2.91	3.00	3.00	0.30	-	7.20	-	10.19	m	11.21
42mm	3.31	3.00	3.41	0.33	-	7.92	-	11.33	m	12.46
54mm	3.77	3.00	3.88	0.40	-	9.60	-	13.48	m	14.83

Labour hourly rates: (except Specialists) Craft Operatives 23.99 Labourer 13.76 Rates are national average prices. Refer to REGIONAL VARIATIONS for indicative levels of overall pricing in regions	MATERIALS			LABOUR				RATES		
	Del to Site £	Waste %	Material Cost £	Craft Optve Hrs	Lab Hrs	Labour Cost £	Sunds £	Nett Rate £	Unit	Gross rate (10%) £
INSULATION AND FIRE PROTECTION (Cont'd)										
Thermal insulation; foil faced glass fibre preformed lagging, butt joints in the running length; secured with metal bands (Cont'd)										
25mm thick insulation to copper pipework										
around one pipe										
15mm............	2.71	3.00	2.79	0.20	-	4.80	-	7.59	m	8.35
22mm............	2.89	3.00	2.98	0.22	-	5.28	-	8.25	m	9.08
28mm............	3.12	3.00	3.21	0.25	-	6.00	-	9.21	m	10.13
35mm............	3.51	3.00	3.62	0.30	-	7.20	-	10.81	m	11.89
42mm............	3.84	3.00	3.96	0.33	-	7.92	-	11.87	m	13.06
54mm............	4.57	3.00	4.71	0.40	-	9.60	-	14.30	m	15.73
Thermal insulation; foamed polyurethane preformed lagging, butt joints in the running length; secured with adhesive bands										
13mm thick insulation to copper pipework										
around one pipe										
15mm............	1.45	3.00	1.49	0.17	-	4.08	-	5.57	m	6.13
22mm............	1.76	3.00	1.81	0.18	-	4.32	-	6.13	m	6.74
28mm............	2.07	3.00	2.13	0.20	-	4.80	-	6.93	m	7.62
35mm............	2.23	3.00	2.30	0.22	-	5.28	-	7.57	m	8.33
42mm............	2.61	3.00	2.69	0.25	-	6.00	-	8.69	m	9.55
54mm............	3.21	3.00	3.31	0.33	-	7.92	-	11.22	m	12.35
INSULATION AND FIRE PROTECTION TO EQUIPMENT										
Thermal insulation; glass fibre filled insulating jacket in strips, white pvc covering both sides; secured with metal straps and wire holder ring at top										
75mm thick insulation to equipment										
sides and tops of cylinders; overall size										
400mm diameter x 1050mm high	13.00	5.00	13.65	0.75	-	17.99	-	31.64	nr	34.81
450mm diameter x 900mm high	11.38	5.00	11.95	0.75	-	17.99	-	29.94	nr	32.94
450mm diameter x 1050mm high	12.75	5.00	13.39	0.75	-	17.99	-	31.38	nr	34.52
450mm diameter x 1200mm high	15.00	5.00	15.75	0.75	-	17.99	-	33.74	nr	37.12
HEATING/COOLING/REFRIGERATION SYSTEMS; LOW TEMPERATURE HOT WATER HEATING (SMALL SCALE)										
Central heating installations - indicative prices										
Note										
the following are indicative prices for installation in two storey three bedroomed dwellings with a floor area of approximately 85m²										
Installation; boiler, copper piping, pressed steel radiators, heated towel rail; providing hot water to sink, bath, lavatory basin; complete with all necessary pumps, controls etc										
solid fuel fired	-	-	Specialist	-	-	Specialist	-	5587.50	it	6146.25
gas fired............	-	-	Specialist	-	-	Specialist	-	5250.00	it	5775.00
oil fired, including oil storage tank	-	-	Specialist	-	-	Specialist	-	6450.00	it	7095.00

This page left blank intentionally

Labour hourly rates: (except Specialists) Craft Operatives 28.34 Labourer 13.76 Rates are national average prices. Refer to REGIONAL VARIATIONS for indicative levels of overall pricing in regions	MATERIALS			LABOUR				RATES		
	Del to Site	Waste	Material Cost	Craft Optve	Lab	Labour Cost	Sunds	Nett Rate		Gross rate (10%)
	£	%	£	Hrs	Hrs	£	£	£	Unit	£
PRIMARY EQUIPMENT										
Three phase Switchgear and Distribution boards										
Switch fuse, mild steel enclosure, fixed to masonry with screws; plugging										
32 amp; three pole and solid neutral, including fuses....................	83.20	2.50	85.28	2.20	-	62.35	0.75	148.38	nr	163.22
Distribution boards, TP & N, IP41 protection, fitted with 125A incomer switch, fixed to masonry with screws; plugging										
8 way times three phase board..........................	157.00	2.50	160.93	5.95	-	168.62	0.75	330.30	nr	363.33
16 ways times three phase board..........................	238.45	2.50	244.41	9.70	-	274.90	0.75	520.06	nr	572.07
Accessories for Distribution boards										
Type B single pole MCB 10KA	4.40	2.50	4.51	0.14	-	3.97	-	8.48	nr	9.33
Type B three pole MCB 10KA	18.50	2.50	18.96	0.15	-	4.25	-	23.21	nr	25.53
Type C single pole MCB 10KA	4.40	2.50	4.51	0.14	-	3.97	-	8.48	nr	9.33
Type C three pole MCB 10KA	18.50	2.50	18.96	0.15	-	4.25	-	23.21	nr	25.53
TERMINAL EQUIPMENT AND FITTINGS										
Accessories; white plastics, with boxes, fixing to masonry with screws; plugging										
Flush plate switches										
10 amp; one gang; one way; single pole	1.22	2.50	1.25	0.42	-	11.90	0.38	13.52	nr	14.88
10 amp; two gang; one way; single pole.............................	2.41	2.50	2.47	0.43	-	12.19	0.38	15.03	nr	16.53
Surface plate switches										
10 amp; one gang; one way; single pole	4.82	2.50	4.94	0.42	-	11.90	0.38	17.21	nr	18.94
10 amp; two gang; one way; single pole.............................	2.86	2.50	2.93	0.43	-	12.19	0.38	15.49	nr	17.04
Ceiling switches										
6 amp; one gang; one way; single pole	2.53	2.50	2.59	0.43	-	12.19	0.38	15.15	nr	16.66
Flush switched socket outlets										
13 amp; one gang.............................	0.68	2.50	0.69	0.43	-	12.19	0.38	13.25	nr	14.58
13 amp; two gang.............................	3.23	2.50	3.31	0.43	-	12.19	0.38	15.88	nr	17.46
Surface switched socket outlets										
13 amp; one gang.............................	1.16	2.50	1.19	0.43	-	12.19	0.38	13.75	nr	15.12
13 amp; two gang.............................	4.32	2.50	4.43	0.43	-	12.19	0.38	16.99	nr	18.69
Flush switched socket outlets with RCD (residual current device) protected at 30 mAmp										
13 amp; two gang.............................	45.64	2.50	46.78	0.43	-	12.19	0.38	59.34	nr	65.28
Flush fused connection units										
13 amp; one gang; switched; flexible outlet.............................	4.05	2.50	4.15	0.50	-	14.17	0.38	18.70	nr	20.57
13 amp; one gang; switched; flexible outlet; pilot lamp...............	6.03	2.50	6.18	0.50	-	14.17	0.38	20.72	nr	22.79
Accessories, metalclad, with boxes, fixing to masonry with screws; plugging										
Surface plate switches										
5 amp; one gang; two way; single pole..	2.46	2.50	2.52	0.42	-	11.90	0.38	14.80	nr	16.28
5 amp; two gang; two way; single pole	2.45	2.50	2.51	0.43	-	12.19	0.38	15.07	nr	16.58
Surface switched socket outlets										
13 amp; one gang.............................	2.66	2.50	2.72	0.43	-	12.19	0.38	15.29	nr	16.81
13 amp; two gang.............................	3.96	2.50	4.06	0.43	Hrs	12.19	0.38	16.62	nr	18.28
Weatherproof accessories, with boxes, fixing to masonry with screws; plugging										
Surface plate switches										
20 amp; one gang; two way; single pole.............................	2.91	2.50	2.98	0.43	-	12.19	0.38	15.54	nr	17.10
Fluorescent lamp fittings, mains voltage operations; switch start; including lamps										
Batten type; single tube										
1200mm; 36W	12.70	2.50	13.02	0.75	-	21.26	0.75	35.02	nr	38.52
1500mm; 58W	12.50	2.50	12.81	1.00	-	28.34	0.75	41.90	nr	46.09
1800mm; 70W	15.95	2.50	16.35	1.00	-	28.34	0.75	45.44	nr	49.98
Batten type; twin tube										
1200mm; 36W	22.80	2.50	23.37	0.75	-	21.26	0.75	45.38	nr	49.91
1500mm; 58W	23.40	2.50	23.98	1.00	-	28.34	0.75	53.07	nr	58.38
1800mm; 70W	26.90	2.50	27.57	1.00	-	28.34	0.75	56.66	nr	62.33
Metal trough reflector; single tube										
1200mm; 36W	20.95	2.50	21.47	1.00	-	28.34	0.75	50.56	nr	55.62
1500mm; 58W	22.40	2.50	22.96	1.20	-	34.01	0.75	57.72	nr	63.49
1800mm; 70W	27.45	2.50	28.14	1.20	-	34.01	0.75	62.89	nr	69.18
Metal trough reflector; twin tube										
1200mm; 36W	31.05	2.50	31.83	1.00	-	28.34	0.75	60.92	nr	67.01
1500mm; 58W	33.30	2.50	34.13	1.20	-	34.01	0.75	68.89	nr	75.78

Labour hourly rates: (except Specialists) Craft Operatives 28.34 Labourer 13.76 Rates are national average prices. Refer to REGIONAL VARIATIONS for indicative levels of overall pricing in regions	MATERIALS			LABOUR				RATES		
	Del to Site	Waste	Material Cost	Craft Optve	Lab	Labour Cost	Sunds	Nett Rate	Unit	Gross rate (10%)
	£	%	£	Hrs	Hrs	£	£	£		£
TERMINAL EQUIPMENT AND FITTINGS (Cont'd)										
Fluorescent lamp fittings, mains voltage operations; switch start; including lamps (Cont'd)										
Metal trough reflector; twin tube (Cont'd)										
1800mm; 70W	38.40	2.50	39.36	1.20	-	34.01	0.75	74.12	nr	81.53
Plastics diffused type; single tube										
1200mm; 36W	20.70	2.50	21.22	1.00	-	28.34	0.75	50.31	nr	55.34
1500mm; 58W	21.50	2.50	22.04	1.20	-	34.01	0.75	56.80	nr	62.48
1800mm; 70W	25.45	2.50	26.09	1.20	-	34.01	0.75	60.84	nr	66.93
Plastics diffused type; twin tube										
1200mm; 36W	33.30	2.50	34.13	1.00	-	28.34	0.75	63.22	nr	69.54
1500mm; 58W	36.40	2.50	37.31	1.20	-	34.01	0.75	72.07	nr	79.27
1800mm; 70W	40.90	2.50	41.92	1.20	-	34.01	0.75	76.68	nr	84.35
Lighting fittings complete with lamps										
Bulkhead; polycarbonate										
16W Round with energy saving lamp	9.95	2.50	10.20	0.65	-	18.42	0.38	28.99	nr	31.89
Ceramic Wall Uplighter										
20W compact fluorescent lamp	17.50	2.50	17.94	0.75	-	21.26	0.38	39.57	nr	43.52
Ceiling sphere										
150mm, 42W halogen lamp	7.60	2.50	7.79	0.65	-	18.42	0.38	26.59	nr	29.24
180mm, 70W halogen lamp	8.80	2.50	9.02	0.65	-	18.42	0.38	27.82	nr	30.60
Commercial Down light; recessed										
HF Electronic PL Downlight 2 x 18w	31.40	2.50	32.18	0.65	-	18.42	0.38	50.98	nr	56.08
Lighting track; low Voltage										
1m track, transformer and 3 x 50W halogen lamps	57.27	2.50	58.70	1.20	-	34.01	0.75	93.46	nr	102.81
Plug-in fitting; 50W halogen lamp	10.49	2.50	10.75	0.20	-	5.67	0.38	16.80	nr	18.47
Fan heaters										
3kw unit fan heaters										
for commercial application	124.17	2.50	127.27	1.50	-	42.51	0.75	170.53	nr	187.58
Thermostatic switch units										
with concealed setting	23.99	2.50	24.59	1.00	-	28.34	0.38	53.30	nr	58.64
Tubular heaters										
Less than 150w per metre										
305mm long	11.50	2.50	11.79	1.00	-	28.34	1.12	41.25	nr	45.38
610mm long	12.99	2.50	13.31	1.00	-	28.34	1.12	42.78	nr	47.06
915mm long	15.00	2.50	15.38	1.00	-	28.34	1.50	45.22	nr	49.74
1220mm long	18.25	2.50	18.71	1.25	-	35.42	1.50	55.63	nr	61.19
1830mm long	26.75	2.50	27.42	1.50	-	42.51	2.25	72.18	nr	79.40
Tubular Heater Interlinking Kit										
To stack an additional tubular heater	5.75	2.50	5.89	0.40	-	11.34	0.75	17.98	nr	19.78
Night storage heaters; installed complete in new building including wiring										
Plastics insulated and sheathed cabled										
1.7KW	307.47	2.50	315.51	6.50	-	184.21	-	499.72	nr	549.70
2.55KW	379.74	2.50	389.60	7.00	-	198.38	-	587.98	nr	646.78
3.4KW	461.10	2.50	472.98	7.50	-	212.55	-	685.53	nr	754.08
Mineral insulated copper sheathed cables										
1.7KW	375.97	2.50	390.75	10.00	-	283.40	-	674.15	nr	741.56
2.55KW	448.22	2.50	464.80	10.50	-	297.57	-	762.37	nr	838.61
3.4KW	529.56	2.50	548.16	11.00	-	311.74	-	859.90	nr	945.89
Bell equipment										
Transformers										
4-8-12V	6.40	2.50	6.56	0.40	-	11.34	0.36	18.26	nr	20.08
Bells for transformer operation										
chime 2-note	13.70	2.50	14.04	0.50	-	14.17	0.36	28.57	nr	31.43
76mm domestic type	8.95	2.50	9.17	0.50	-	14.17	0.36	23.70	nr	26.07
180mm round bell type	32.75	2.50	33.57	0.40	-	11.34	0.36	45.26	nr	49.79
Bell pushes										
domestic	2.08	2.50	2.13	0.25	-	7.09	-	9.22	nr	10.14
industrial/weatherproof	5.45	2.50	5.59	0.30	-	8.50	-	14.09	nr	15.50
Immersion heaters and thermostats										
Note: the following prices exclude Plumber's work										
2 or 3KW immersion heaters; without flanges										
305mm; non-withdrawable elements	10.20	2.50	10.45	0.65	-	18.42	6.97	35.85	nr	39.44
457mm; non-withdrawable elements	10.90	2.50	11.17	0.80	-	22.67	6.97	40.82	nr	44.90
686mm; non-withdrawable elements	11.90	2.50	12.20	0.95	-	26.92	6.97	46.10	nr	50.71
Thermostats										
immersion heater	5.60	2.50	5.74	0.50	-	14.17	0.38	20.28	nr	22.31
16 amp a.c. for air heating; without switch	24.45	2.50	25.06	0.40	-	11.34	0.38	36.77	nr	40.45
Water heaters										
Note: the following prices exclude Plumber's work										

ELECTRICAL SERVICES

Labour hourly rates: (except Specialists) Craft Operatives 28.34 Labourer 13.76 Rates are national average prices. Refer to REGIONAL VARIATIONS for indicative levels of overall pricing in regions	MATERIALS			LABOUR				RATES		
	Del to Site	Waste	Material Cost	Craft Optve	Lab	Labour Cost	Sunds	Nett Rate	Unit	Gross rate (10%)
	£	%	£	Hrs	Hrs	£	£	£		£
TERMINAL EQUIPMENT AND FITTINGS (Cont'd)										
Water heaters (Cont'd)										
Storage type units; oversink, single outlet										
5 liter, 2kW	85.99	-	85.99	1.50	-	42.51	7.72	136.22	nr	149.85
7 litre, 3kW	97.00	-	97.00	1.50	-	42.51	7.72	147.24	nr	161.96
10 litre, 2kW	124.50	-	124.50	1.50	-	42.51	7.72	174.74	nr	192.21
Storage type units; unvented; multi-point										
30 litre, 3kW	330.75	-	330.75	4.00	-	113.36	11.97	456.08	nr	501.69
50 litre, 3kW	342.75	-	342.75	4.50	-	127.53	11.97	482.25	nr	530.48
80 litre, 3kW	358.50	-	358.50	5.25	-	148.79	15.45	522.73	nr	575.01
100 litre, 3kW	391.50	-	391.50	6.00	-	170.04	15.45	576.99	nr	634.69
Shower units										
Note: the following prices exclude Plumber's work										
Instantaneous shower units complete with fittings	175.00	2.50	179.38	4.00	-	113.36	11.60	304.33	nr	334.76
Street lighting columns and set in ground; excavate hole, remove surplus spoil, filling with concrete mix ST4, working around base										
Galvanised steel columns; nominal mounting height										
4.00m										
70w son post top sphere lantern; Part night photo cell	398.84	5.00	399.40	4.75	9.50	265.33	31.93	696.67	nr	766.34
28w LED lantern; Single arm mounted; Full night photo cell	301.10	5.00	301.66	4.75	9.50	265.33	31.93	598.93	nr	658.82
5.00m										
36w PLL lantern; Single arm mounted: Part night photo cell	365.42	5.00	365.98	4.75	9.50	265.33	49.68	680.99	nr	749.09
2 off 36w PLL lanterns; Double arm mounted: Part night photo cell	530.45	5.00	531.01	4.75	9.50	265.33	49.68	846.02	nr	930.63
6.00m										
50w (5500 lumens) LED lantern; Single arm mounted: Part night photo cell	402.61	5.00	403.29	4.75	9.50	265.33	108.24	776.86	nr	854.55
2 off 50w (5500 lumens) LED lanterns; Double arm mounted: Part night photo cell	576.11	5.00	576.79	4.75	9.50	265.33	108.24	950.36	nr	1045.40
8.00m										
80w (8800 lumens) LED lantern; Single arm mounted: Full night photo cell	571.35	5.00	572.16	4.75	9.50	265.33	108.24	945.73	nr	1040.30
2 off 80w (8800 lumens) LED lanterns; Double arm mounted: Full night photo cell	780.85	5.00	781.66	4.75	9.50	265.33	108.24	1155.23	nr	1270.75
10.00m										
100w (11000 lumens) LED lantern; Single arm mounted: Full night photo cell	687.95	5.00	688.94	4.75	9.50	265.33	108.24	1062.52	nr	1168.77
2 off 100w (11000 lumens) LED lanterns; Double arm mounted: Full night photo cell	697.45	5.00	698.44	4.75	9.50	265.33	108.24	1072.02	nr	1179.22
CABLE CONTAINMENT										
Steel trunking										
Straight lighting trunking; PVC lid										
50 x 50mm	3.39	5.00	3.56	0.40	-	11.34	0.75	15.65	m	17.21
extra; end cap	0.88	-	0.88	0.08	-	2.27	0.38	3.52	nr	3.87
extra; internal angle	6.38	-	6.38	0.25	-	7.09	0.38	13.84	nr	15.22
extra; external angle	6.38	-	6.38	0.25	-	7.09	0.38	13.84	nr	15.22
extra; tee	6.98	-	6.98	0.25	-	7.09	0.38	14.44	nr	15.88
extra; flat bend	6.38	-	6.38	0.40	-	11.34	0.38	18.09	nr	19.90
extra; trunking suspension fixed to 10mm studding	1.60	-	1.60	0.25	-	7.09	0.75	9.43	nr	10.38
Straight dado trunking, two compartment; steel lid										
150 x 40mm	17.79	5.00	18.68	0.45	-	12.75	0.75	32.18	m	35.40
extra; end cap	4.49	-	4.49	0.08	-	2.27	0.38	7.13	nr	7.85
extra; internal angle	16.77	-	16.77	0.35	-	9.92	0.38	27.06	nr	29.77
extra; external angle	16.77	-	16.77	0.35	-	9.92	0.38	27.06	nr	29.77
extra; single socket plate	8.50	-	8.50	0.16	-	4.53	0.38	13.41	nr	14.75
extra; twin socket plate	8.50	-	8.50	0.16	-	4.53	0.38	13.41	nr	14.75
Straight dado trunking, three compartment; steel lid										
200 x 40mm	24.58	5.00	25.81	0.50	-	14.17	1.12	41.11	m	45.22
extra; end cap	17.05	-	17.05	0.08	-	2.27	0.38	19.69	nr	21.66
extra; internal angle	26.55	-	26.55	0.40	-	11.34	0.38	38.26	nr	42.09
extra; external angle	26.55	-	26.55	0.40	-	11.34	0.38	38.26	nr	42.09
extra; single socket plate	13.78	-	13.78	0.16	-	4.53	0.38	18.69	nr	20.56
extra; twin socket plate	13.78	-	13.78	0.16	-	4.53	0.38	18.69	nr	20.56
Straight skirting trunking; three compartment; steel lid										
200 x 40mm	22.63	5.00	23.77	0.50	-	14.17	1.12	39.06	m	42.97
extra; internal bend	24.65	-	24.65	0.40	-	11.34	0.38	36.36	nr	40.00
extra; external bend	24.65	-	24.65	0.40	-	11.34	0.38	36.36	nr	40.00
extra; single socket plate	9.98	-	9.98	0.16	-	4.53	0.38	14.89	nr	16.38
extra; twin socket plate	9.98	-	9.98	0.16	-	4.53	0.38	14.89	nr	16.38
Straight underfloor trunking; two compartments										
150 x 25mm	9.48	5.00	9.95	0.70	-	19.84	-	29.79	m	32.77
extra; floor junction boxes; four way; adjustable frame and trap	53.72	-	53.72	1.85	-	52.43	-	106.15	nr	116.77
extra; horizontal bend	27.50	-	27.50	0.50	-	14.17	-	41.67	nr	45.84
PVC trunking; white										
Straight mini-trunking; clip on lid										
16 x 16mm	0.70	5.00	0.73	0.20	-	5.67	0.75	7.15	m	7.87
extra; end cap	0.48	-	0.48	0.04	-	1.13	0.19	1.81	nr	1.99
extra; internal bend	0.45	-	0.45	0.08	-	2.27	0.19	2.91	nr	3.20

Labour hourly rates: (except Specialists) Craft Operatives 28.34 Labourer 13.76 Rates are national average prices. Refer to REGIONAL VARIATIONS for indicative levels of overall pricing in regions	MATERIALS			LABOUR				RATES		
	Del to Site	Waste	Material Cost	Craft Optve	Lab	Labour Cost	Sunds	Nett Rate	Unit	Gross rate (10%)
	£	%	£	Hrs	Hrs	£	£	£		£
CABLE CONTAINMENT (Cont'd)										
PVC trunking; white (Cont'd)										
Straight mini-trunking; clip on lid (Cont'd)	0.45	-	0.45	0.08	-	2.27	0.19	2.91	nr	3.20
extra; external bend	0.45	-	0.45	0.08	-	2.27	0.19	2.91	nr	3.20
extra; flat bend	0.83	-	0.83	0.17	-	4.82	0.19	5.84	nr	6.43
extra; equal tee	0.73	5.00	0.77	0.20	-	5.67	0.19	6.63	m	7.30
25 x 16mm	0.47	-	0.47	0.04	-	1.13	0.19	1.80	nr	1.98
extra; end cap	0.50	-	0.50	0.08	-	2.27	0.19	2.96	nr	3.26
extra; internal bend	0.45	-	0.45	0.08	-	2.27	0.19	2.91	nr	3.20
extra; external bend	0.50	-	0.50	0.08	-	2.27	0.19	2.96	nr	3.26
extra; flat bend	0.91	-	0.91	0.17	-	4.82	0.19	5.92	nr	6.52
extra; equal tee	1.32	5.00	1.38	0.25	-	7.09	0.19	8.66	m	9.53
40 x 16mm	0.50	-	0.50	0.04	-	1.13	0.19	1.83	nr	2.01
extra; end cap	0.57	-	0.57	0.08	-	2.27	0.19	3.03	nr	3.34
extra; internal bend	0.52	-	0.52	0.08	-	2.27	0.19	2.98	nr	3.28
extra; external bend	0.57	-	0.57	0.08	-	2.27	0.19	3.03	nr	3.34
extra; flat bend	0.82	-	0.82	0.17	-	4.82	0.19	5.83	nr	6.42
extra; equal tee										
Straight dado trunking, three compartment; clip on lid	13.31	5.00	13.98	0.40	-	11.34	0.75	26.06	m	28.67
145 x 50mm	1.79	-	1.79	0.08	-	2.27	0.38	4.43	nr	4.88
extra; end caps	4.79	-	4.79	0.30	-	8.50	0.38	13.67	nr	15.03
extra; internal bend; adjustable	3.99	-	3.99	0.30	-	8.50	0.38	12.87	nr	14.15
extra; external bend; adjustable	38.93	-	38.93	0.40	-	11.34	0.38	50.64	nr	55.71
extra; tee	29.92	-	29.92	0.30	-	8.50	0.38	38.80	nr	42.68
extra; flat angle	1.79	-	1.79	0.20	-	5.67	0.38	7.83	nr	8.62
extra; straight coupler	19.98	5.00	20.98	0.45	-	12.75	0.75	34.48	m	37.93
180 x 57mm curved profile	3.19	-	3.19	0.08	-	2.27	0.38	5.83	nr	6.42
extra; end caps	5.99	-	5.99	0.30	-	8.50	0.38	14.87	nr	16.35
extra; internal bend	5.79	-	5.79	0.30	-	8.50	0.38	14.67	nr	16.13
extra; external bend	3.79	-	3.79	0.20	-	5.67	0.38	9.83	nr	10.82
extra; straight coupler										
Straight skirting trunking, three compartment; clip on lid	8.32	5.00	8.73	0.35	-	9.92	0.75	19.40	m	21.34
170 x 50mm	1.16	-	1.16	0.06	-	1.70	0.38	3.24	nr	3.56
extra; end cap	2.99	-	2.99	0.25	-	7.09	0.38	10.45	nr	11.50
extra; internal corner	2.88	-	2.88	0.25	-	7.09	0.38	10.34	nr	11.37
extra; external corner	1.00	-	1.00	0.16	-	4.53	0.38	5.91	nr	6.50
extra; straight coupler	1.35	-	1.35	0.25	-	7.09	0.38	8.81	nr	9.69
extra; socket adaptor, box one gang	1.44	-	1.44	0.25	-	7.09	0.38	8.90	nr	9.79
extra; socket adaptor box, two gang										
PVC heavy guage conduit and fittings; push fit joints; spacer bar saddles at 600mm centres										
Conduits; straight										
20mm diameter; plugging to masonry; to surfaces	0.26	10.00	0.29	0.17	-	4.82	0.63	5.74	m	6.31
extra; small circular terminal boxes	0.68	2.50	0.70	0.10	-	2.83	0.41	3.94	nr	4.33
extra; small circular angle boxes	0.69	2.50	0.71	0.11	-	3.12	0.41	4.23	nr	4.65
extra; small circular three way boxes	0.69	2.50	0.71	0.13	-	3.68	0.41	4.80	nr	5.28
extra; small circular through way boxes	0.69	2.50	0.71	0.11	-	3.12	0.41	4.23	nr	4.65
25mm diameter; plugging to masonry; to surfaces	0.39	10.00	0.43	0.18	-	5.10	0.63	6.16	m	6.77
extra; small circular terminal boxes	0.82	2.50	0.85	0.10	-	2.83	0.41	4.08	nr	4.49
extra; small circular angle boxes	0.82	2.50	0.85	0.11	-	3.12	0.41	4.37	nr	4.80
extra; small circular three way boxes	0.82	2.50	0.85	0.13	-	3.68	0.41	4.93	nr	5.43
extra; small circular through way boxes	0.82	2.50	0.85	0.11	-	3.12	0.41	4.37	nr	4.80
20mm diameter; plugging to masonry; in chases	0.26	10.00	0.29	0.15	-	4.25	0.63	5.17	m	5.69
extra; small circular terminal boxes	0.68	2.50	0.70	0.10	-	2.83	0.41	3.94	nr	4.33
extra; small circular angle boxes	0.69	2.50	0.71	0.11	-	3.12	0.41	4.23	nr	4.65
extra; small circular three way boxes	0.69	2.50	0.71	0.13	-	3.68	0.41	4.80	nr	5.28
extra; small circular through way boxes	0.69	2.50	0.71	0.11	-	3.12	0.41	4.23	nr	4.65
25mm diameter; plugging to masonry; in chases	0.39	10.00	0.43	0.16	-	4.53	0.41	5.37	m	5.90
extra; small circular terminal boxes	0.82	2.50	0.85	0.10	-	2.83	0.41	4.08	nr	4.49
extra; small circular angle boxes	0.82	2.50	0.85	0.11	-	3.12	0.41	4.37	nr	4.80
extra; small circular three way boxes	0.82	2.50	0.85	0.13	-	3.68	0.41	4.93	nr	5.43
extra; small circular through way boxes	0.82	2.50	0.85	0.11	-	3.12	0.41	4.37	nr	4.80
Oval conduits; straight										
20mm (nominal size); plugging to masonry; in chases	0.26	10.00	0.29	0.15	-	4.25	0.63	5.17	m	5.69
25mm (nominal size); plugging to masonry in chases	0.43	10.00	0.47	0.16	-	4.53	0.63	5.63	m	6.20
20mm (nominal size); fixing to timber	0.26	10.00	0.29	0.13	-	3.68	0.31	4.29	m	4.72
25mm (nominal size); fixing to timber	0.43	10.00	0.47	0.15	-	4.25	0.31	5.04	m	5.54
Steel welded heavy gauge conduits and fittings, screwed joints; spacer bar saddles at 1000mm centres; enamelled black inside and outside by manufacturer										
Conduits; straight										
20mm diameter; plugging to masonry; to surfaces	2.12	10.00	2.32	0.33	-	9.35	0.41	12.08	m	13.28
extra; small circular terminal boxes	1.79	2.50	1.83	0.13	-	3.68	0.41	5.92	nr	6.52
extra; small circular angle boxes	2.00	2.50	2.05	0.16	-	4.53	0.41	6.99	nr	7.69
extra; small circular three way boxes	2.21	2.50	2.27	0.20	-	5.67	0.41	8.34	nr	9.17
extra; small circular through way boxes	2.00	2.50	2.05	0.16	-	4.53	0.41	6.99	nr	7.69
25mm diameter; plugging to masonry; to surfaces	2.82	10.00	3.09	0.42	-	11.90	0.41	15.40	m	16.94
extra; small circular terminal boxes	2.70	2.50	2.77	0.13	-	3.68	0.41	6.86	nr	7.54
extra; small circular angle boxes	2.97	2.50	3.04	0.16	-	4.53	0.41	7.98	nr	8.78
extra; small circular three way boxes	3.24	2.50	3.32	0.20	-	5.67	0.41	9.39	nr	10.33
extra; small circular through way boxes	2.97	2.50	3.04	0.16	-	4.53	0.41	7.98	nr	8.78
32mm diameter; plugging to masonry; to surfaces	4.44	10.00	4.88	0.50	-	14.17	0.41	19.45	m	21.40

ELECTRICAL SERVICES

Labour hourly rates: (except Specialists) Craft Operatives 28.34 Labourer 13.76 Rates are national average prices. Refer to REGIONAL VARIATIONS for indicative levels of overall pricing in regions	MATERIALS			LABOUR				RATES		
	Del to Site	Waste	Material Cost	Craft Optve	Lab	Labour Cost	Sunds	Nett Rate		Gross rate (10%)
	£	%	£	Hrs	Hrs	£	£	£	Unit	£

CABLE CONTAINMENT (Cont'd)

Steel welded heavy gauge conduits and fittings, screwed joints; spacer bar saddles at 1000mm centres; galvanized inside and outside by manufacturer

Conduits; straight

20mm diameter; plugging to masonry; to surfaces	1.34	10.00	1.47	0.33	-	9.35	0.41	11.22	m	12.35
extra; small circular terminal boxes	1.80	2.50	1.84	0.13	-	3.68	0.41	5.93	nr	6.52
extra; small circular angle boxes	1.88	2.50	1.93	0.16	-	4.53	0.41	6.87	nr	7.56
extra; small circular three way boxes	2.09	2.50	2.14	0.20	-	5.67	0.41	8.22	nr	9.04
extra; small circular through way boxes	1.88	2.50	1.93	0.16	-	4.53	0.41	6.87	nr	7.56
25mm diameter; plugging to masonry; to surfaces	1.82	10.00	1.99	0.42	-	11.90	0.38	14.27	m	15.70
extra; small circular terminal boxes	2.71	2.50	2.78	0.13	-	3.68	0.41	6.86	nr	7.55
extra; small circular angle boxes	2.02	2.50	2.07	0.16	-	4.53	0.41	7.01	nr	7.71
extra; small circular three way boxes	2.56	2.50	2.62	0.20	-	5.67	0.41	8.69	nr	9.56
extra; small circular through way boxes	2.29	2.50	2.34	0.16	-	4.53	0.41	7.28	nr	8.01
32mm diameter; plugging to masonry; to surfaces	5.05	10.00	5.55	0.50	-	14.17	0.38	20.09	m	22.10

galvanized standard cable tray and fittings

Light duty; straight

50mm tray	2.17	5.00	2.28	0.25	-	7.09	0.14	9.50	m	10.45
extra; flat bend	5.45	-	5.45	0.20	-	5.67	0.38	11.49	nr	12.64
extra; tee	5.81	-	5.81	0.25	-	7.09	0.38	13.27	nr	14.60
extra; cross piece	6.54	-	6.54	0.40	-	11.34	0.75	18.63	nr	20.49
extra; riser bend	3.62	-	3.62	0.20	-	5.67	0.38	9.66	nr	10.63
100mm tray	2.91	5.00	3.06	0.33	-	9.35	0.14	12.54	m	13.80
extra; flat bend	5.60	-	5.60	0.23	-	6.52	0.38	12.49	nr	13.74
extra; tee	7.87	-	7.87	0.30	-	8.50	0.38	16.75	nr	18.42
extra; cross piece	9.37	-	9.37	0.41	-	11.62	0.75	21.74	nr	23.91
extra; riser bend	3.67	-	3.67	0.23	-	6.52	0.38	10.56	nr	11.62
extra; reducer 100 to 50mm	6.62	-	6.62	0.23	-	6.52	0.38	13.51	nr	14.86
150mm tray	3.28	5.00	3.44	0.42	-	11.90	0.14	15.48	m	17.03
extra; flat bend	6.01	-	6.01	0.25	-	7.09	0.38	13.47	nr	14.82
extra; tee	9.15	-	9.15	0.33	-	9.35	0.38	18.88	nr	20.76
extra; cross piece	12.34	-	12.34	0.42	-	11.90	0.75	24.99	nr	27.49
extra; riser bend	4.08	-	4.08	0.25	-	7.09	0.38	11.54	nr	12.69
extra; reducer 150 to 100mm	7.32	-	7.32	0.25	-	7.09	0.38	14.78	nr	16.26
300mm tray	6.50	5.00	6.82	0.50	-	14.17	0.14	21.13	m	23.24
extra; flat bend	11.02	-	11.02	0.33	-	9.35	0.38	20.75	nr	22.82
extra; tee	16.74	-	16.74	0.50	-	14.17	0.38	31.28	nr	34.41
extra; cross piece	22.33	-	22.33	0.58	-	16.44	0.75	39.52	nr	43.47
extra; riser bend	8.40	-	8.40	0.33	-	9.35	0.38	18.13	nr	19.94
extra; reducer 300 to 150mm	8.88	-	8.88	0.33	-	9.35	0.38	18.61	nr	20.47
450mm tray	18.03	5.00	18.93	0.58	-	16.44	0.14	35.50	m	39.05
extra; flat bend	24.96	-	24.96	0.50	-	14.17	0.38	39.50	nr	43.46
extra; tee	30.41	-	30.41	0.75	-	21.26	0.38	52.04	nr	57.24
extra; cross piece	38.08	-	38.08	1.00	-	28.34	0.75	67.17	nr	73.89
extra; riser bend	11.00	-	11.00	0.50	-	14.17	0.38	25.55	nr	28.10
extra; reducer 450 to 300mm	13.11	-	13.11	0.50	-	14.17	0.38	27.65	nr	30.42

Medium duty; straight

75mm tray	2.09	5.00	2.20	0.26	-	7.37	0.14	9.70	m	10.67
extra; flat bend	5.62	-	5.62	0.21	-	5.95	0.38	11.95	nr	13.14
extra; tee	7.56	-	7.56	0.26	-	7.37	0.38	15.30	nr	16.83
extra; cross piece	10.57	-	10.57	0.42	-	11.90	0.75	23.22	nr	25.55
extra; riser bend	3.35	-	3.35	0.21	-	5.95	0.38	9.68	nr	10.64
100mm tray	2.35	5.00	2.46	0.34	-	9.64	0.14	12.23	m	13.46
extra; flat bend	5.75	-	5.75	0.24	-	6.80	0.38	12.93	nr	14.22
extra; tee	7.51	-	7.51	0.31	-	8.79	0.38	16.67	nr	18.34
extra; cross piece	10.51	-	10.51	0.43	-	12.19	0.75	23.45	nr	25.79
extra; riser bend	3.67	-	3.67	0.24	-	6.80	0.38	10.85	nr	11.93
extra; reducer 100 to 75mm	5.94	-	5.94	0.24	-	6.80	0.38	13.12	nr	14.43
150mm tray	3.24	5.00	3.40	0.44	-	12.47	0.14	16.00	m	17.60
extra; flat bend	6.16	-	6.16	0.26	-	7.37	0.38	13.90	nr	15.29
extra; tee	8.74	-	8.74	0.34	-	9.64	0.38	18.75	nr	20.63
extra; cross piece	12.23	-	12.23	0.44	-	12.47	0.75	25.45	nr	27.99
extra; riser bend	4.08	-	4.08	0.26	-	7.37	0.38	11.82	nr	13.01
extra; reducer 150 to 100mm	6.20	-	6.20	0.26	-	7.37	0.38	13.94	nr	15.34
300mm tray	7.52	5.00	7.90	0.52	-	14.74	0.14	22.77	m	25.04
extra; flat bend	11.29	-	11.29	0.34	-	9.64	0.38	21.30	nr	23.43
extra; tee	13.70	-	13.70	0.52	-	14.74	0.38	28.81	nr	31.69
extra; cross piece	19.15	-	19.15	0.61	-	17.29	0.75	37.19	nr	40.91
extra; riser bend	8.40	-	8.40	0.34	-	9.64	0.38	18.41	nr	20.25
extra; reducer 300 to 150mm	9.15	-	9.15	0.34	-	9.64	0.38	19.16	nr	21.08
450mm tray	9.59	5.00	10.07	0.61	-	17.29	0.14	27.49	m	30.24
extra; flat bend	19.61	-	19.61	0.52	-	14.74	0.38	34.72	nr	38.19
extra; tee	22.44	-	22.44	0.78	-	22.11	0.38	44.92	nr	49.41
extra; cross piece	31.40	-	31.40	1.05	-	29.76	0.75	61.91	nr	68.10
extra; riser bend	11.00	-	11.00	0.52	-	14.74	0.38	26.11	nr	28.72
extra; reducer 450 to 300mm	9.47	-	9.47	0.52	-	14.74	0.38	24.58	nr	27.04

Heavy duty; straight

75mm tray	2.47	5.00	2.59	0.27	-	7.65	0.14	10.38	m	11.42
extra; flat bend	6.27	-	6.27	0.22	-	6.23	0.38	12.88	nr	14.17
extra; tee	7.68	-	7.68	0.27	-	7.65	0.38	15.71	nr	17.28
extra; cross piece	9.60	-	9.60	0.44	-	12.47	0.75	22.82	nr	25.10
extra; riser bend	5.44	-	5.44	0.22	-	6.23	0.38	12.05	nr	13.25
100mm tray	2.90	5.00	3.04	0.36	-	10.20	0.14	13.38	m	14.72
extra; flat bend	6.87	-	6.87	0.25	-	7.09	0.38	14.33	nr	15.76
extra; tee	8.60	-	8.60	0.33	-	9.35	0.38	18.33	nr	20.16
extra; cross piece	9.96	-	9.96	0.45	-	12.75	0.75	23.46	nr	25.81
extra; riser bend	5.82	-	5.82	0.25	-	7.09	0.38	13.28	nr	14.61
extra; reducer 100 to 75mm	6.70	-	6.70	0.25	-	7.09	0.38	14.16	nr	15.58

	MATERIALS			LABOUR				RATES		
Labour hourly rates: (except Specialists) Craft Operatives 28.34 Labourer 13.76 Rates are national average prices. Refer to REGIONAL VARIATIONS for indicative levels of overall pricing in regions	Del to Site £	Waste %	Material Cost £	Craft Optve Hrs	Lab Hrs	Labour Cost £	Sunds £	Nett Rate £	Unit	Gross rate (10%) £
CABLE CONTAINMENT (Cont'd)										
galvanized standard cable tray and fittings (Cont'd)										
Heavy duty; straight (Cont'd)										
150mm tray	3.41	5.00	3.58	0.46	-	13.04	0.14	16.76	m	18.43
extra; flat bend	8.01	-	8.01	0.27	-	7.65	0.38	16.04	nr	17.64
extra; tee	10.04	-	10.04	0.36	-	10.20	0.38	20.62	nr	22.68
extra; cross piece	12.54	-	12.54	0.46	-	13.04	0.75	26.33	nr	28.96
extra; riser bend	6.72	-	6.72	0.27	-	7.65	0.38	14.75	nr	16.22
extra; reducer 150 to 100mm	7.00	-	7.00	0.27	-	7.65	0.38	15.03	nr	16.53
300mm tray	6.93	5.00	7.27	0.55	-	15.59	0.14	22.99	m	25.29
extra; flat bend	13.71	-	13.71	0.36	-	10.20	0.38	24.29	nr	26.72
extra; tee	19.91	-	19.91	0.55	-	15.59	0.38	35.87	nr	39.46
extra; cross piece	24.89	-	24.89	0.64	-	18.14	0.75	43.78	nr	48.16
extra; riser bend	13.02	-	13.02	0.36	-	10.20	0.38	23.60	nr	25.96
extra; reducer 300 to 150mm	9.13	-	9.13	0.36	-	10.20	0.38	19.71	nr	21.68
450mm tray	14.21	5.00	14.92	0.64	-	18.14	0.14	33.20	m	36.52
extra; flat bend	18.86	-	18.86	0.55	-	15.59	0.38	34.82	nr	38.30
extra; tee	33.67	-	33.67	0.83	-	23.52	0.38	57.57	nr	63.32
extra; cross piece	43.83	-	43.83	1.10	-	31.17	0.38	75.38	nr	82.92
extra; riser bend	17.12	-	17.12	0.55	-	15.59	0.38	33.08	nr	36.39
extra; reducer 450 to 300mm	10.70	-	10.70	0.55	-	15.59	0.38	26.66	nr	29.33
Supports for cable tray										
Cantilever arms; mild steel; hot dip galvanized; fixing to masonry										
100mm wide	3.05	2.50	3.13	0.33	-	9.35	0.38	12.85	nr	14.14
300mm wide	7.90	2.50	8.10	0.33	-	9.35	0.38	17.82	nr	19.61
450mm wide	9.91	2.50	10.16	0.33	-	9.35	0.38	19.88	nr	21.87
Stand-off bracket; mild steel; hot dip galvanized; fixing to masonry										
75mm wide	1.66	2.50	1.70	0.33	-	9.35	0.38	11.43	nr	12.57
100mm wide	2.00	2.50	2.05	0.33	-	9.35	0.38	11.78	nr	12.95
150mm wide	2.05	2.50	2.10	0.33	-	9.35	0.38	11.83	nr	13.01
300mm wide	3.58	2.50	3.67	0.42	-	11.90	0.38	15.95	nr	17.54
450mm wide	3.93	2.50	4.03	0.42	-	11.90	0.38	16.31	nr	17.94
CABLES										
PVC insulated cables; single core; reference 6491X; stranded copper conductors; to BS 6004										
Drawn into conduits or ducts or laid or drawn into trunking										
1.5mm²	0.00	15.00	0.00	0.03	-	0.85	-	0.85	m	0.94
2.5mm²	0.00	15.00	0.00	0.03	-	0.85	-	0.85	m	0.94
4.0mm²	0.00	15.00	0.00	0.04	-	1.13	-	1.14	m	1.25
6.0mm²	0.01	15.00	0.01	0.05	-	1.42	-	1.42	m	1.57
10.0mm²	0.01	15.00	0.01	0.05	-	1.42	-	1.43	m	1.57
16.0mm²	0.01	15.00	0.02	0.08	-	2.13	-	2.14	m	2.36
PVC double insulated cables; single core; reference 6181Y; stranded copper conductors; to BS 6004										
Fixed to timber with clips										
10.0mm²	1.28	15.00	1.47	0.08	-	2.27	-	3.74	m	4.11
16.0mm²	1.88	15.00	2.16	0.10	-	2.83	-	5.00	m	5.50
25.0mm²	3.00	15.00	3.45	0.15	-	4.25	-	7.70	m	8.47
35.0mm²	5.38	15.00	6.19	0.20	-	5.67	-	11.85	m	13.04
PVC insulated and PVC sheathed cables; multicore; copper conductors and bare earth continuity conductor 6242Y; to BS 6004										
Drawn into conduits or ducts or laid or drawn into trunking										
1.0mm²; twin with bare earth	0.00	15.00	0.00	0.05	-	1.42	-	1.42	m	1.56
1.5mm²; twin with bare earth	0.00	15.00	0.00	0.05	-	1.42	-	1.42	m	1.56
2.5mm²; twin with bare earth	0.00	15.00	0.00	0.05	-	1.42	-	1.42	m	1.56
4.0mm²; twin with bare earth	0.01	15.00	0.01	0.06	-	1.70	-	1.71	m	1.88
6.0mm²; twin with bare earth	0.08	15.00	0.09	0.07	-	1.98	-	2.08	m	2.28
Fixed to timber with clips										
1.0mm²; twin with bare earth	0.00	15.00	0.00	0.07	-	1.98	0.09	2.08	m	2.28
1.5mm²; twin with bare earth	0.00	15.00	0.00	0.07	-	1.98	0.09	2.08	m	2.28
2.5mm²; twin with bare earth	0.00	15.00	0.00	0.07	-	1.98	0.10	2.09	m	2.30
4.0mm²; twin with bare earth	0.01	15.00	0.01	0.08	-	2.27	0.12	2.39	m	2.63
6.0mm²; twin with bare earth	0.08	15.00	0.09	0.09	-	2.55	0.12	2.76	m	3.04
PVC insulated SWA armoured and PVC sheathed cables; multicore; copper conductors										
Laid and laced on cable tray										
16mm²; three core	4.42	5.00	4.64	0.17	-	4.82	0.17	9.62	m	10.59
Fixed to masonry with screwed clips at average 300mm centres; plugging										
16mm²; three core	4.42	5.00	4.64	0.25	-	7.09	1.41	13.14	m	14.45
Mineral insulated copper sheathed cables; PVC outer sheath; copper conductors; 600V light duty to BS EN 60702-1										
Fixed to masonry with screwed clips at average 300mm centres; plugging										
1.5mm²; two core	3.78	10.00	4.16	0.17	-	4.82	1.41	10.39	m	11.43
extra; termination including gland, seal and shroud	1.86	2.50	1.90	0.42	-	11.90	-	13.81	nr	15.19
2.5mm²; two core	4.78	10.00	5.26	0.17	-	4.82	1.41	11.49	m	12.64
extra; termination including gland, seal and shroud	1.86	2.50	1.90	0.42	-	11.90	-	13.81	nr	15.19
4.0mm²; two core	7.19	10.00	7.91	0.20	-	5.67	1.41	14.99	m	16.49
extra; termination including gland, seal and shroud	1.91	2.50	1.96	0.50	-	14.17	-	16.13	nr	17.74

Labour hourly rates: (except Specialists) Craft Operatives 28.34 Labourer 13.76 Rates are national average prices. Refer to REGIONAL VARIATIONS for indicative levels of overall pricing in regions	MATERIALS			LABOUR				RATES		
	Del to Site £	Waste %	Material Cost £	Craft Optve Hrs	Lab Hrs	Labour Cost £	Sunds £	Nett Rate £	Unit	Gross rate (10%) £
FINAL CIRCUITS; DOMESTIC LIGHTING AND POWER (SMALL SCALE)										
Electric wiring										
Note the following approximate prices of various types of installations are dependant on the number and disposition of points; lamps and fittings together with cutting and making good are excluded										
Electric wiring in new building										
Insulated Consumer unit, fixed to masonry with screws; plugging										
2 usable ways 60A DP main Switch	23.05	2.50	23.63	1.10	-	31.17	0.38	55.18	nr	60.69
6 usable ways 100A DP main Switch	23.00	2.50	23.57	2.00	-	56.68	0.38	80.63	nr	88.69
10 usable ways 100A DP main Switch	24.50	2.50	25.11	2.60	-	73.68	0.38	99.17	nr	109.09
14 usable ways 100A DP main Switch	26.00	2.50	26.65	3.50	-	99.19	0.38	126.21	nr	138.84
19 usable ways 100A DP main Switch	59.40	2.50	60.89	5.00	-	141.70	0.38	202.96	nr	223.26
2 usable ways 63A, 30mA RCD main Switch	72.60	2.50	74.42	1.15	-	32.59	0.38	107.38	nr	118.12
6 usable ways 80A, 30mA RCD main Switch	70.90	2.50	72.67	2.05	-	58.10	0.38	131.14	nr	144.26
10 usable ways 80A, 30mA RCD main Switch	72.40	2.50	74.21	2.65	-	75.10	0.38	149.69	nr	164.65
14 usable ways 80A, 30mA RCD main Switch	73.90	2.50	75.75	3.55	-	100.61	0.38	176.73	nr	194.40
19 usable ways 100A, 30mA RCD main Switch	125.20	2.50	128.33	5.05	-	143.12	0.38	271.82	nr	299.00
Weatherproof Insulated Consumer unit, fixed to masonry with screws; plugging										
2 usable ways 60A DP main Switch	30.50	2.50	31.26	1.40	-	39.68	0.38	71.31	nr	78.44
6 usable ways 100A DP main Switch	32.90	2.50	33.72	2.35	-	66.60	0.38	100.70	nr	110.77
2 usable ways 63A 30mA RCD main Switch	80.05	2.50	82.05	1.35	-	38.26	0.38	120.69	nr	132.75
6 usable ways 80A 30mA RCD main Switch	80.80	2.50	82.82	2.40	-	68.02	0.38	151.21	nr	166.33
Domestic Metal consumer units and accessories, fixed to masonry with screws; plugging										
2 usable ways 60A DP main Switch	24.65	2.50	25.27	1.10	-	31.17	0.38	56.82	nr	62.50
6 usable ways 100A DP main Switch	38.00	2.50	38.95	2.00	-	56.68	0.38	96.01	nr	105.61
10 usable ways 100A DP main Switch	43.00	2.50	44.08	2.60	-	73.68	0.38	118.13	nr	129.95
14 usable ways 100A DP main Switch	45.00	2.50	46.12	3.50	-	99.19	0.38	145.69	nr	160.26
19 usable ways 100A DP main Switch	59.40	2.50	60.89	5.00	-	141.70	0.38	202.96	nr	223.26
Domestic Split Insulated consumer unit and accessories, fixed to masonry with screws; plugging										
8 useable ways with 100A main switch and 80A 30mA RCD	43.00	2.50	44.08	2.30	-	65.18	0.38	109.63	nr	120.60
17th Edition Insulated consumer unit and accessories, fixed to masonry with screws; plugging										
10 useable ways with 100A main switch and 80A 30mA RCD and 63A 30mA RCD	61.00	2.50	62.52	2.60	-	73.68	0.38	136.58	nr	150.24
15 useable ways with 100A main switch and 80A 30mA RCD and 63A 30mA RCD	67.00	2.50	68.67	3.80	-	107.69	0.38	176.74	nr	194.42
Miniature Circuit Breakers fitted in above consumer units										
3A to 50A single pole MCB	2.90	2.50	2.97	0.13	-	3.68	-	6.66	nr	7.32
6A to 50A, 30mA single pole RCBO	24.95	2.50	25.57	0.14	-	3.83	-	29.40	nr	32.34
Installation with PVC insulated and sheathed cables										
lighting points	7.58	2.50	7.82	3.25	-	92.11	-	99.92	nr	109.91
socket outlets; 5A	9.30	2.50	9.57	4.40	-	124.70	-	134.27	nr	147.70
socket outlets; 13A ring main	5.98	2.50	6.16	3.70	-	104.86	-	111.01	nr	122.12
socket outlets; 13A radial circuit	6.89	2.50	7.10	3.85	-	109.11	-	116.20	nr	127.83
cooker points; 45A	14.33	2.50	14.88	6.30	-	178.54	-	193.42	nr	212.76
immersion heater points	9.11	2.50	9.40	4.60	-	130.36	-	139.77	nr	153.74
shaver sockets (transformer)	25.59	2.50	26.32	1.82	-	51.58	-	77.90	nr	85.69
Installation with black enamel heavy gauge conduit with PVC cables										
lighting points	34.89	2.50	36.97	4.65	-	131.78	-	168.75	nr	185.62
socket outlets; 5A	34.01	2.50	36.07	4.70	-	133.20	-	169.26	nr	186.19
socket outlets; 13A ring main	27.23	2.50	28.81	4.25	-	120.44	-	149.26	nr	164.18
socket outlets; 13A radial circuit	31.67	2.50	33.67	6.00	-	170.04	-	203.71	nr	224.08
cooker points; 45A	51.14	2.50	54.44	9.65	-	273.48	-	327.92	nr	360.72
immersion heater points	39.87	2.50	42.38	7.15	-	202.63	-	245.01	nr	269.51
shaver sockets (transformer)	53.38	2.50	55.92	2.53	-	71.70	-	127.62	nr	140.38
Installation with black enamel heavy gauge conduit with coaxial cable										
TV/FM sockets	50.69	2.50	53.54	5.75	-	162.96	-	216.50	nr	238.15
Installation with black enamel heavy gauge conduit with 2 pair cable										
telephone points	47.55	2.50	50.28	5.45	-	154.45	-	204.73	nr	225.21
Electric wiring in extension to existing building										
Installation with PVC insulated and sheathed cables										
lighting points	7.58	2.50	7.82	3.25	-	92.11	-	99.92	nr	109.91
socket outlets; 5A	9.30	2.50	9.57	4.40	-	124.70	-	134.27	nr	147.70
socket outlets; 13A ring main	5.98	2.50	6.16	3.70	-	104.86	-	111.01	nr	122.12
socket outlets; 13A radial circuit	6.89	2.50	7.10	3.85	-	109.11	-	116.20	nr	127.83
cooker points; 45A	14.33	2.50	14.88	6.30	-	178.54	-	193.42	nr	212.76
immersion heater points	9.11	2.50	9.40	4.60	-	130.36	-	139.77	nr	153.74
shaver sockets (transformer)	25.59	2.50	26.32	1.82	-	51.58	-	77.90	nr	85.69
Installation with mineral insulated copper sheathed cables										
lighting points	25.15	2.50	26.63	6.20	-	175.71	-	202.34	nr	222.57
socket outlets; 5A	48.56	2.50	52.64	4.85	-	137.45	-	190.09	nr	209.10
socket outlets; 13A ring main	35.73	2.50	38.78	4.15	-	117.61	-	156.39	nr	172.03
socket outlets; 13A radial circuit	46.21	2.50	50.24	5.55	-	157.29	-	207.52	nr	228.28
immersion heater points	59.10	2.50	64.17	6.60	-	187.04	-	251.21	nr	276.33
shaver sockets (transformer)	75.20	2.50	80.66	3.30	-	93.52	-	174.19	nr	191.60

ELECTRICAL SERVICES

Labour hourly rates: (except Specialists) Craft Operatives 28.34 Labourer 13.76 Rates are national average prices. Refer to REGIONAL VARIATIONS for indicative levels of overall pricing in regions	MATERIALS			LABOUR				RATES		
	Del to Site	Waste	Material Cost	Craft Optve	Lab	Labour Cost	Sunds	Nett Rate	Unit	Gross rate (10%)
	£	%	£	Hrs	Hrs	£	£	£		£
FINAL CIRCUITS; DOMESTIC LIGHTING AND POWER (SMALL SCALE) (Cont'd)										
Electric wiring in extension to existing building (Cont'd)										
Installation with black enamel heavy gauge conduit with PVC cables										
lighting points	34.89	2.50	36.97	4.65	-	131.78	-	168.75	nr	185.62
socket outlets; 5A	34.01	2.50	36.07	4.70	-	133.20	-	169.26	nr	186.19
socket outlets; 13A ring main	27.23	2.50	28.81	4.25	-	120.44	-	149.26	nr	164.18
socket outlets; 13A radial circuit	31.67	2.50	33.67	6.00	-	170.04	-	203.71	nr	224.08
cooker points; 45A	51.14	2.50	54.44	9.65	-	273.48	-	327.92	nr	360.72
immersion heater points	39.87	2.50	42.38	7.15	-	202.63	-	245.01	nr	269.51
shaver sockets (transformer)...............	53.38	2.50	55.92	2.53	-	71.70	-	127.62	nr	140.38
Installation with black enamel heavy gauge conduit with coaxial cable										
TV/FM sockets....................................	50.69	2.50	53.54	5.75	-	162.96	-	216.50	nr	238.15
Installation with black enamel heavy gauge conduit with 2 pair cable										
telephone points	47.55	2.50	50.28	5.45	-	154.45	-	204.73	nr	225.21
Electric wiring in extending existing installation in existing building										
Installation with PVC insulated and sheathed cables										
lighting points	7.58	2.50	7.82	3.75	-	106.28	-	114.09	nr	125.50
socket outlets; 5A	9.30	2.50	9.57	5.05	-	143.12	-	152.69	nr	167.96
socket outlets; 13A ring main	5.98	2.50	6.16	4.25	-	120.44	-	126.60	nr	139.26
socket outlets; 13A radial circuit	6.89	2.50	7.10	4.45	-	126.11	-	133.21	nr	146.53
cooker points; 45A	14.33	2.50	14.88	7.25	-	205.46	-	220.34	nr	242.37
immersion heater points	9.11	2.50	9.40	5.30	-	150.20	-	159.60	nr	175.56
shaver sockets (transformer)...............	25.59	2.50	26.32	2.09	-	59.23	-	85.55	nr	94.11
Installation with mineral insulated copper sheathed cables										
lighting points	25.15	2.50	26.63	7.15	-	202.63	-	229.26	nr	252.19
socket outlets; 5A	48.56	2.50	52.64	5.60	-	158.70	-	211.35	nr	232.48
socket outlets; 13A ring main	35.73	2.50	38.78	4.80	-	136.03	-	174.81	nr	192.29
socket outlets; 13A radial circuit	46.21	2.50	50.24	6.40	-	181.38	-	231.61	nr	254.77
immersion heater points	59.10	2.50	64.17	7.60	-	215.38	-	279.55	nr	307.51
shaver sockets (transformer)...............	75.20	2.50	80.66	3.80	-	107.69	-	188.36	nr	207.19
Installation with black enamel heavy gauge conduit with PVC cables										
lighting points	34.89	2.50	36.97	5.35	-	151.62	-	188.58	nr	207.44
socket outlets; 5A	34.01	2.50	36.07	5.40	-	153.04	-	189.10	nr	208.01
socket outlets; 13A ring main	27.23	2.50	28.81	4.90	-	138.87	-	167.68	nr	184.45
socket outlets; 13A radial circuit	31.67	2.50	33.67	6.90	-	195.55	-	229.21	nr	252.13
cooker points; 45A	51.14	2.50	54.44	11.10	-	314.57	-	369.02	nr	405.92
immersion heater points	39.87	2.50	42.38	8.25	-	233.81	-	276.18	nr	303.80
shaver sockets (transformer)...............	53.38	2.50	55.92	2.92	-	82.75	-	138.67	nr	152.54
Installation with black enamel heavy gauge conduit with coaxial cable										
TV/FM sockets....................................	50.69	2.50	53.54	6.60	-	187.04	-	240.59	nr	264.64
Installation with black enamel heavy gauge conduit with 2 pair cable										
telephone points	47.55	2.50	50.28	6.25	-	177.12	-	227.41	nr	250.15
TESTING										
Testing; existing installations										
Point										
from	-	-	-	2.25	-	63.76	-	63.76	nr	70.14
to	-	-	-	3.85	-	109.11	-	109.11	nr	120.02
Complete installation; three bedroom house										
from	-	-	-	11.00	-	311.74	-	311.74	nr	342.91
to	-	-	-	17.50	-	495.95	-	495.95	nr	545.55
FIRE AND LIGHTNING PROTECTION										
Fire detection and alarm; Terminal equipment and fittings										
Fittings and equipment										
fire alarm 2 zone panel with batteries and charger....................	87.00	2.50	89.17	3.30	-	93.52	0.75	183.45	nr	201.79
fire alarm 4 zone panel with batteries and charger....................	109.50	2.50	112.24	4.40	-	124.70	0.75	237.68	nr	261.45
fire alarm 8 zone panel with batteries and charger....................	178.00	2.50	182.45	6.60	-	187.04	0.75	370.24	nr	407.27
Ionisation smoke detector....................	6.80	2.50	6.97	0.30	-	8.50	0.75	16.22	nr	17.84
Optical smoke detector.......................	12.95	2.50	13.27	0.30	-	8.50	0.75	22.53	nr	24.78
Smoke detector mounting base with sounder.........................	12.45	2.50	12.76	0.75	-	21.26	0.75	34.77	nr	38.24
Manual surface mounted call point........	6.45	2.50	6.61	0.75	-	21.26	0.75	28.62	nr	31.48
Emergency lighting										
General purpose 8 watt non maintained....	12.65	2.50	12.97	1.00	-	28.34	0.75	42.06	nr	46.26
General purpose 8 watt maintained........	18.50	2.50	18.96	1.00	-	28.34	0.75	48.05	nr	52.86
General purpose 8 watt mains...............	12.65	2.50	12.97	1.00	-	28.34	0.75	42.06	nr	46.26
Lightning protection; conductors										
Note: the following are indicative average prices										
Air termination rods										
610mm long; in clips, fixing to masonry........................	-	%	Specialist	-	-	Specialist	-	62.25	nr	68.47

ELECTRICAL SERVICES

Labour hourly rates: (except Specialists) Craft Operatives 28.34 Labourer 13.76 Rates are national average prices. Refer to REGIONAL VARIATIONS for indicative levels of overall pricing in regions	MATERIALS			LABOUR				RATES		
	Del to Site £	Waste %	Material Cost £	Craft Optve Hrs	Lab Hrs	Labour Cost £	Sunds £	Nett Rate £	Unit	Gross rate (10%) £
FIRE AND LIGHTNING PROTECTION (Cont'd)										
Lightning protection; conductors (Cont'd)										
Air termination tape										
fixing to masonry	-	-	Specialist	-	-	Specialist	-	15.00	m	16.50
Copper tape down conductors										
19 x 3mm; in clips, fixing to masonry..............	-	-	Specialist	-	-	Specialist	-	22.58	m	24.83
25 x 3mm; in clips, fixing to masonry..............	-	-	Specialist	-	-	Specialist	-	24.15	m	26.57
Test clamps										
with securing bolts	-	-	Specialist	-	-	Specialist	-	34.65	nr	38.12
Copper earth rods; tape connectors										
2438 x 16mm; driving into ground	-	-	Specialist	-	-	Specialist	-	220.50	nr	242.55

This page left blank intentionally

Labour hourly rates: (except Specialists) Craft Operatives 18.37 Labourer 13.76 Rates are national average prices. Refer to REGIONAL VARIATIONS for indicative levels of overall pricing in regions	MATERIALS			LABOUR				RATES		
	Del to Site	Waste	Material Cost	Craft Optve	Lab	Labour Cost	Sunds	Nett Rate		Gross rate (10%)
	£	%	£	Hrs	Hrs	£	£	£	Unit	£
LIFTS										
Generally										
Note: the following are average indicative prices										
Light passenger lifts; standard range										
Electro Hydraulic drive; 630 kg, 8 person, 0.63m/s, 3 stop										
basic; primed car and entrances	-	-	Specialist	-	-	Specialist	-	23250.00	nr	25575.00
median; laminate walls, standard carpet floor, painted entrances	-	-	Specialist	-	-	Specialist	-	27300.00	nr	30030.00
ACVF drive; 1000 kg, 13 person, 0.63m/s, 4 stop										
basic; primed car and entrances	-	-	Specialist	-	-	Specialist	-	51900.00	nr	57090.00
median; laminate walls, standard carpet floor, painted entrances	-	-	Specialist	-	-	Specialist	-	53850.00	nr	59235.00
Extras on the above lifts										
800/900mm landing doors in cellulose paint finish	-	-	Specialist	-	-	Specialist	-	225.00	nr	247.50
800/900mm landing doors in brushed stainless steel	-	-	Specialist	-	-	Specialist	-	630.00	nr	693.00
800/900mm landing doors in brushed brass	-	-	Specialist	-	-	Specialist	-	1815.00	nr	1996.50
landing position indicators	-	-	Specialist	-	-	Specialist	-	355.50	nr	391.05
landing direction arrows and gongs	-	-	Specialist	-	-	Specialist	-	210.00	nr	231.00
car preference key switch	-	-	Specialist	-	-	Specialist	-	180.00	nr	198.00
car entrance safety (Progard L)	-	-	Specialist	-	-	Specialist	-	600.00	nr	660.00
fluorescent emergency light	-	-	Specialist	-	-	Specialist	-	558.00	nr	613.80
half height mirror to rear wall	-	-	Specialist	-	-	Specialist	-	630.00	nr	693.00
full height mirror to rear wall	-	-	Specialist	-	-	Specialist	-	915.00	nr	1006.50
vandal resistant buttons; per car/entrance	-	-	Specialist	-	-	Specialist	-	204.00	nr	224.40
car bumper rail	-	-	Specialist	-	-	Specialist	-	1230.00	nr	1353.00
trailing cable for Warden Call System	-	-	Specialist	-	-	Specialist	-	523.50	nr	575.85
car telephone	-	-	Specialist	-	-	Specialist	-	411.00	nr	452.10
General purpose passenger lifts										
ACVF drive; 630 kg, 8 person, 1.0m/s, 4 stop median; laminate walls, standard carpet floor, painted entrances	-	-	Specialist	-	-	Specialist	-	53400.00	nr	58740.00
Variable speed a/c drive; 1000 kg, 13 person, 1.6m/s, 5 stop median; laminate walls, standard carpet floor, painted entrances	-	-	Specialist	-	-	Specialist	-	57300.00	nr	63030.00
Variable speed a/c drive; 1600 kg, 21 person, 1.6m/s, 6 stop median; laminate walls, standard carpet floor, painted entrances	-	-	Specialist	-	-	Specialist	-	138000.00	nr	151800.00
If a high quality architectural finish is required, e.g. veneered panelling walls, marble floors, special ceiling, etc, add to the above prices										
from	-	-	Specialist	-	-	Specialist	-	22500.00	nr	24750.00
to	-	-	Specialist	-	-	Specialist	-	63000.00	nr	69300.00
Bed/Passenger lifts										
Electro Hydraulic drive; 1800 kg, 24 person, 0.63m/s, 3 stop standard specification			Specialist	-	-	Specialist	-	72300.00	nr	79530.00
Variable speed a/c drive; 2000 kg, 26 person, 1.6m/s, 4 stop standard specification			Specialist	-	-	Specialist	-	131250.00	nr	144375.00
Goods lifts (direct coupled)										
Electro Hydraulic drive; 1500 kg, 0.4m/s, 3 stop stainless steel car lining with chequer plate floor and galvanised shutters			Specialist	-	-	Specialist	-	57300.00	nr	63030.00
Electro Hydraulic drive; 2000 kg, 0.4m/s, 3 stop stainless steel car lining with chequer plate floor and galvanised shutters	-	-	Specialist	-	-	Specialist	-	59100.00	nr	65010.00
Service hoists										
Single speed a/c drive; 50 kg, 0.63m/s, 2 stop standard specification	-	-	Specialist	-	-	Specialist	-	8100.00	nr	8910.00
Single speed a/c drive; 250 kg, 0.4m/s, 2 stop standard specification	-	-	Specialist	-	-	Specialist	-	10275.00	nr	11302.50
Extra on the above hoists two hour fire resisting shutters; per landing	-	-	Specialist	-	-	Specialist	-	340.50	nr	374.55
ESCALATORS										
Escalators										
30 degree inclination; 3.00m vertical rise department store specification	-	-	Specialist	-	-	Specialist	-	87000.00	nr	95700.00

TRANSPORTATION

BUILDERS WORK IN CONNECTION WITH INSTALLATIONS

Labour hourly rates: (except Specialists) Craft Operatives 18.37 Labourer 13.76 Rates are national average prices. Refer to REGIONAL VARIATIONS for indicative levels of overall pricing in regions	MATERIALS			LABOUR				RATES		
	Del to Site	Waste	Material Cost	Craft Optve	Lab	Labour Cost	Sunds	Nett Rate		Gross rate (10%)
	£	%	£	Hrs	Hrs	£	£	£	Unit	£
WORK FOR SERVICES INSTALLATIONS IN NEW BUILDINGS; **MECHANICAL INSTALLATIONS**										
Cutting holes for services										
For ducts through concrete; making good										
rectangular ducts not exceeding 1.00m girth										
150mm thick	-	-	-	1.60	-	29.39	0.76	30.16	nr	33.17
200mm thick	-	-	-	1.90	-	34.90	0.92	35.82	nr	39.40
300mm thick	-	-	-	2.20	-	40.41	1.07	41.48	nr	45.63
rectangular ducts 1.00-2.00m girth										
150mm thick	-	-	-	3.20	-	58.78	1.70	60.48	nr	66.53
200mm thick	-	-	-	3.80	-	69.81	1.85	71.65	nr	78.82
300mm thick	-	-	-	4.40	-	80.83	2.14	82.97	nr	91.27
rectangular ducts 2.00-3.00m girth										
150mm thick	-	-	-	4.00	-	73.48	1.95	75.43	nr	82.97
200mm thick	-	-	-	4.75	-	87.26	2.31	89.57	nr	98.52
300mm thick	-	-	-	5.50	-	101.04	2.64	103.68	nr	114.04
For pipes through concrete; making good										
pipes not exceeding 55mm nominal size										
150mm thick	-	-	-	0.90	-	16.53	0.48	17.01	nr	18.71
200mm thick	-	-	-	1.15	-	21.13	0.62	21.74	nr	23.91
300mm thick	-	-	-	1.45	-	26.64	0.76	27.40	nr	30.14
pipes 55 - 110mm nominal size										
150mm thick	-	-	-	1.12	-	20.57	0.60	21.17	nr	23.29
200mm thick	-	-	-	1.36	-	24.98	0.70	25.69	nr	28.26
300mm thick	-	-	-	1.80	-	33.07	0.96	34.03	nr	37.43
pipes exceeding 110mm nominal size										
150mm thick	-	-	-	1.35	-	24.80	0.68	25.47	nr	28.02
200mm thick	-	-	-	1.72	-	31.60	0.86	32.45	nr	35.70
300mm thick	-	-	-	2.18	-	40.05	1.11	41.16	nr	45.27
For ducts through reinforced concrete; making good										
rectangular ducts not exceeding 1.00m girth										
150mm thick	-	-	-	2.40	-	44.09	1.11	45.20	nr	49.72
200mm thick	-	-	-	2.85	-	52.35	1.30	53.66	nr	59.03
300mm thick	-	-	-	3.30	-	60.62	1.53	62.15	nr	68.37
rectangular ducts 1.00-2.00m girth										
150mm thick	-	-	-	4.80	-	88.18	2.25	90.43	nr	99.47
200mm thick	-	-	-	5.70	-	104.71	2.65	107.36	nr	118.10
300mm thick	-	-	-	6.60	-	121.24	3.09	124.33	nr	136.77
rectangular ducts 2.00-3.00m girth										
150mm thick	-	-	-	6.00	-	110.22	2.79	113.01	nr	124.31
200mm thick	-	-	-	7.00	-	128.59	3.32	131.90	nr	145.10
300mm thick	-	-	-	8.25	-	151.55	3.78	155.33	nr	170.87
For pipes through reinforced concrete; making good										
pipes not exceeding 55mm nominal size										
150mm thick	-	-	-	1.35	-	24.80	0.69	25.49	nr	28.04
200mm thick	-	-	-	1.75	-	32.15	0.90	33.05	nr	36.35
300mm thick	-	-	-	2.20	-	40.41	1.12	41.54	nr	45.69
pipes 55 - 110mm nominal size										
150mm thick	-	-	-	1.70	-	31.23	0.87	32.10	nr	35.31
200mm thick	-	-	-	2.05	-	37.66	1.05	38.71	nr	42.58
300mm thick	-	-	-	2.70	-	49.60	1.38	50.98	nr	56.08
pipes exceeding 110mm nominal size										
150mm thick	-	-	-	2.00	-	36.74	1.02	37.76	nr	41.54
200mm thick	-	-	-	2.60	-	47.76	1.32	49.08	nr	53.99
300mm thick	-	-	-	3.20	-	58.78	1.63	60.42	nr	66.46
For ducts through brickwork; making good										
rectangular ducts not exceeding 1.00m girth										
102.5mm thick	-	-	-	1.20	-	22.04	0.52	22.57	nr	24.83
215mm thick	-	-	-	1.40	-	25.72	0.63	26.35	nr	28.98
327.5mm thick	-	-	-	1.65	-	30.31	0.75	31.06	nr	34.17
rectangular ducts 1.00-2.00m girth										
102.5mm thick	-	-	-	2.40	-	44.09	1.05	45.14	nr	49.65
215mm thick	-	-	-	2.80	-	51.44	1.26	52.70	nr	57.97
327.5mm thick	-	-	-	3.30	-	60.62	1.50	62.12	nr	68.33
rectangular ducts 2.00-3.00m girth										
102.5mm thick	-	-	-	3.00	-	55.11	1.35	56.46	nr	62.11
215mm thick	-	-	-	3.50	-	64.30	1.58	65.87	nr	72.46
327.5mm thick	-	-	-	4.10	-	75.32	1.88	77.19	nr	84.91
For pipes through brickwork; making good										
pipes not exceeding 55mm nominal size										
102.5mm thick	-	-	-	0.40	-	7.35	0.21	7.56	nr	8.31
215mm thick	-	-	-	0.66	-	12.12	0.33	12.45	nr	13.70
327.5mm thick	-	-	-	1.00	-	18.37	0.51	18.88	nr	20.77
pipes 55 - 110mm nominal size										
102.5mm thick	-	-	-	0.53	-	9.74	0.27	10.01	nr	11.01
215mm thick	-	-	-	0.88	-	16.17	0.45	16.62	nr	18.28
327.5mm thick	-	-	-	1.33	-	24.43	0.68	25.11	nr	27.62

BUILDERS WORK IN CONNECTION WITH INSTALLATIONS

Labour hourly rates: (except Specialists) Craft Operatives 18.37 Labourer 13.76 Rates are national average prices. Refer to REGIONAL VARIATIONS for indicative levels of overall pricing in regions	MATERIALS			LABOUR				RATES		
	Del to Site	Waste	Material Cost	Craft Optve	Lab	Labour Cost	Sunds	Nett Rate	Unit	Gross rate (10%)
	£	%	£	Hrs	Hrs	£	£	£		£
WORK FOR SERVICES INSTALLATIONS IN NEW BUILDINGS; MECHANICAL INSTALLATIONS (Cont'd)										
Cutting holes for services (Cont'd)										
For pipes through brickwork; making good (Cont'd)										
pipes exceeding 110mm nominal size										
102.5mm thick	-	-	-	0.67	-	12.31	0.34	12.65	nr	13.92
215mm thick	-	-	-	1.10	-	20.21	0.56	20.76	nr	22.84
327.5mm thick	-	-	-	1.67	-	30.68	0.84	31.52	nr	34.67
For ducts through brickwork; making good fair face or facings one side										
rectangular ducts not exceeding 1.00m girth										
102.5mm thick	-	-	-	1.45	-	26.64	0.82	27.46	nr	30.21
215mm thick	-	-	-	1.65	-	30.31	0.98	31.29	nr	34.41
327.5mm thick	-	-	-	1.90	-	34.90	1.05	35.95	nr	39.55
rectangular ducts 1.00-2.00m girth										
102.5mm thick	-	-	-	2.70	-	49.60	1.65	51.25	nr	56.37
215mm thick	-	-	-	3.10	-	56.95	1.95	58.90	nr	64.79
327.5mm thick	-	-	-	3.60	-	66.13	2.10	68.23	nr	75.06
rectangular ducts 2.00-3.00m girth										
102.5mm thick	-	-	-	3.40	-	62.46	2.10	64.56	nr	71.01
215mm thick	-	-	-	3.90	-	71.64	2.47	74.12	nr	81.53
327.5mm thick	-	-	-	4.50	-	82.67	2.62	85.29	nr	93.82
For pipes through brickwork; making good fair face or facings one side										
pipes not exceeding 55mm nominal size										
102.5mm thick	-	-	-	0.50	-	9.19	0.27	9.46	nr	10.40
215mm thick	-	-	-	0.76	-	13.96	0.39	14.35	nr	15.79
327.5mm thick	-	-	-	1.10	-	20.21	0.56	20.76	nr	22.84
pipes 55 - 110mm nominal size										
102.5mm thick	-	-	-	0.68	-	12.49	0.33	12.82	nr	14.10
215mm thick	-	-	-	1.03	-	18.92	0.50	19.42	nr	21.36
327.5mm thick	-	-	-	1.48	-	27.19	0.74	27.92	nr	30.71
pipes exceeding 110mm nominal size										
102.5mm thick	-	-	-	0.87	-	15.98	0.41	16.39	nr	18.03
215mm thick	-	-	-	1.30	-	23.88	0.60	24.48	nr	26.93
327.5mm thick	-	-	-	1.87	-	34.35	0.88	35.24	nr	38.76
For ducts through blockwork; making good										
rectangular ducts not exceeding 1.00m girth										
100mm thick	-	-	-	0.90	-	16.53	0.52	17.06	nr	18.76
140mm thick	-	-	-	1.00	-	18.37	0.63	19.00	nr	20.90
190mm thick	-	-	-	1.10	-	20.21	0.75	20.96	nr	23.05
rectangular ducts 1.00-2.00m girth										
100mm thick	-	-	-	1.80	-	33.07	1.05	34.12	nr	37.53
140mm thick	-	-	-	1.95	-	35.82	1.26	37.08	nr	40.79
190mm thick	-	-	-	2.10	-	38.58	1.50	40.08	nr	44.08
rectangular ducts 2.00-3.00m girth										
100mm thick	-	-	-	2.25	-	41.33	1.35	42.68	nr	46.95
140mm thick	-	-	-	2.45	-	45.01	1.58	46.58	nr	51.24
190mm thick	-	-	-	2.65	-	48.68	1.88	50.56	nr	55.61
For pipes through blockwork; making good										
pipes not exceeding 55mm nominal size										
100mm thick	-	-	-	0.30	-	5.51	0.19	5.71	nr	6.28
140mm thick	-	-	-	0.40	-	7.35	0.31	7.66	nr	8.43
190mm thick	-	-	-	0.50	-	9.19	0.48	9.66	nr	10.63
pipes 55 - 110mm nominal size										
100mm thick	-	-	-	0.40	-	7.35	0.25	7.60	nr	8.36
140mm thick	-	-	-	0.53	-	9.74	0.42	10.16	nr	11.17
190mm thick	-	-	-	0.66	-	12.12	0.64	12.77	nr	14.05
pipes exceeding 110mm nominal size										
100mm thick	-	-	-	0.50	-	9.19	0.33	9.52	nr	10.47
140mm thick	-	-	-	0.68	-	12.49	0.52	13.02	nr	14.32
190mm thick	-	-	-	0.83	-	15.25	0.81	16.06	nr	17.66
For ducts through existing brickwork with plaster finish; making good each side										
rectangular ducts not exceeding 1.00m girth										
102.5mm thick	-	-	-	1.95	-	35.82	0.90	36.72	nr	40.39
215mm thick	-	-	-	2.25	-	41.33	1.12	42.46	nr	46.70
327.5mm thick	-	-	-	2.55	-	46.84	1.27	48.12	nr	52.93
rectangular ducts 1.00-2.00m girth										
102.5mm thick	-	-	-	3.60	-	66.13	1.80	67.93	nr	74.73
215mm thick	-	-	-	4.20	-	77.15	2.25	79.40	nr	87.34
327.5mm thick	-	-	-	4.80	-	88.18	2.55	90.73	nr	99.80
rectangular ducts 2.00-3.00m girth										
102.5mm thick	-	-	-	4.50	-	82.67	2.70	85.37	nr	93.90
215mm thick	-	-	-	5.25	-	96.44	3.38	99.82	nr	109.80
327.5mm thick	-	-	-	6.05	-	111.14	3.83	114.96	nr	126.46
For ducts through existing brickwork with plaster finish; making good one side										
rectangular ducts not exceeding 1.00m girth										
102.5mm thick	-	-	-	1.75	-	32.15	0.81	32.96	nr	36.25
215mm thick	-	-	-	2.05	-	37.66	1.02	38.68	nr	42.55
327.5mm thick	-	-	-	2.30	-	42.25	1.14	43.39	nr	47.73
rectangular ducts 1.00-2.00m girth										
102.5mm thick	-	-	-	3.25	-	59.70	1.62	61.32	nr	67.45
215mm thick	-	-	-	3.80	-	69.81	2.03	71.83	nr	79.01
327.5mm thick	-	-	-	4.35	-	79.91	2.33	82.23	nr	90.46
rectangular ducts 2.00-3.00m girth										
102.5mm thick	-	-	-	4.05	-	74.40	2.40	76.80	nr	84.48

Labour hourly rates: (except Specialists) Craft Operatives 18.37 Labourer 13.76 Rates are national average prices. Refer to REGIONAL VARIATIONS for indicative levels of overall pricing in regions	MATERIALS			LABOUR				RATES		
	Del to Site	Waste	Material Cost	Craft Optve	Lab	Labour Cost	Sunds	Nett Rate		Gross rate (10%)
	£	%	£	Hrs	Hrs	£	£	£	Unit	£
WORK FOR SERVICES INSTALLATIONS IN NEW BUILDINGS; MECHANICAL INSTALLATIONS (Cont'd)										
Cutting holes for services (Cont'd)										
For ducts through existing brickwork with plaster finish; making good one side (Cont'd)										
rectangular ducts 2.00-3.00m girth (Cont'd)										
215mm thick	-	-	-	4.75	-	87.26	3.00	90.26	nr	99.28
327.5mm thick	-	-	-	5.45	-	100.12	3.45	103.57	nr	113.92
For pipes through existing brickwork with plaster finish; making good each side										
pipes not exceeding 55mm nominal size										
102.5mm thick	-	-	-	0.65	-	11.94	0.33	12.27	nr	13.50
215mm thick	-	-	-	1.00	-	18.37	0.47	18.84	nr	20.72
327.5mm thick	-	-	-	1.45	-	26.64	0.66	27.30	nr	30.03
pipes 55 - 110mm nominal size										
102.5mm thick	-	-	-	0.95	-	17.45	0.39	17.84	nr	19.63
215mm thick	-	-	-	1.35	-	24.80	0.60	25.40	nr	27.94
327.5mm thick	-	-	-	1.95	-	35.82	0.88	36.71	nr	40.38
pipes exceeding 110mm nominal size										
102.5mm thick	-	-	-	1.20	-	22.04	0.48	22.52	nr	24.78
215mm thick	-	-	-	1.75	-	32.15	0.72	32.87	nr	36.15
327.5mm thick	-	-	-	2.50	-	45.93	1.07	46.99	nr	51.69
For pipes through existing brickwork with plaster finish; making good one side										
pipes not exceeding 55mm nominal size										
102.5mm thick	-	-	-	0.60	-	11.02	0.27	11.29	nr	12.42
215mm thick	-	-	-	0.90	-	16.53	0.39	16.92	nr	18.62
327.5mm thick	-	-	-	1.30	-	23.88	0.56	24.44	nr	26.88
pipes 55 - 110mm nominal size										
102.5mm thick	-	-	-	0.85	-	15.61	0.33	15.94	nr	17.54
215mm thick	-	-	-	1.20	-	22.04	0.50	22.54	nr	24.79
327.5mm thick	-	-	-	1.75	-	32.15	0.74	32.88	nr	36.17
pipes exceeding 110mm nominal size										
102.5mm thick	-	-	-	1.10	-	20.21	0.41	20.61	nr	22.67
215mm thick	-	-	-	1.60	-	29.39	0.60	29.99	nr	32.99
327.5mm thick	-	-	-	2.25	-	41.33	0.81	42.14	nr	46.36
For pipes through softwood										
pipes not exceeding 55mm nominal size										
19mm thick	-	-	-	0.10	-	1.84	-	1.84	nr	2.02
50mm thick	-	-	-	0.13	-	2.39	-	2.39	nr	2.63
pipes 55 - 110mm nominal size										
19mm thick	-	-	-	0.15	-	2.76	-	2.76	nr	3.03
50mm thick	-	-	-	0.20	-	3.67	-	3.67	nr	4.04
For pipes through plywood										
pipes not exceeding 55mm nominal size										
13mm thick	-	-	-	0.10	-	1.84	-	1.84	nr	2.02
19mm thick	-	-	-	0.13	-	2.39	-	2.39	nr	2.63
pipes 55 - 110mm nominal size										
13mm thick	-	-	-	0.15	-	2.76	-	2.76	nr	3.03
19mm thick	-	-	-	0.20	-	3.67	-	3.67	nr	4.04
For pipes through existing plasterboard with skim finish; making good one side										
pipes not exceeding 55mm nominal size										
13mm thick	-	-	-	0.20	-	3.67	-	3.67	nr	4.04
pipes 55 - 110mm nominal size										
13mm thick	-	-	-	0.30	-	5.51	-	5.51	nr	6.06
Cutting chases for services										
In concrete; making good										
15mm nominal size pipes -1	-	-	-	0.30	-	5.51	-	5.51	m	6.06
22mm nominal size pipes -1	-	-	-	0.35	-	6.43	-	6.43	m	7.07
In brickwork										
15mm nominal size pipes -1	-	-	-	0.25	-	4.59	-	4.59	m	5.05
22mm nominal size pipes -1	-	-	-	0.30	-	5.51	-	5.51	m	6.06
In brickwork; making good fair face or facings										
15mm nominal size pipes -1	-	-	-	0.30	-	5.51	0.17	5.68	m	6.24
22mm nominal size pipes -1	-	-	-	0.35	-	6.43	0.25	6.68	m	7.35
In existing brickwork with plaster finish; making good										
15mm nominal size pipes -1	-	-	-	0.40	-	7.35	0.33	7.68	m	8.45
22mm nominal size pipes -1	-	-	-	0.50	-	9.19	0.51	9.70	m	10.66
In blockwork; making good										
15mm nominal size pipes -1	-	-	-	0.20	-	3.67	-	3.67	m	4.04
22mm nominal size pipes -1	-	-	-	0.25	-	4.59	-	4.59	m	5.05
Pipe and duct sleeves										
Steel pipes BS 1387 Table 2; casting into concrete; making good										
100mm long; to the following nominal size steel pipes										
15mm	0.77	5.00	0.80	0.12	0.06	3.03	-	3.83	nr	4.22
20mm	0.85	5.00	0.89	0.14	0.07	3.54	-	4.43	nr	4.87
25mm	1.04	5.00	1.09	0.15	0.08	3.86	-	4.95	nr	5.44
32mm	1.22	5.00	1.28	0.16	0.08	4.04	-	5.32	nr	5.85
40mm	1.41	5.00	1.48	0.18	0.09	4.55	-	6.02	nr	6.62
50mm	1.77	5.00	1.86	0.22	0.11	5.56	-	7.42	nr	8.16
150mm long; to the following nominal size steel pipes										
15mm	0.94	5.00	0.99	0.14	0.07	3.54	-	4.53	nr	4.98

Labour hourly rates: (except Specialists) Craft Operatives 18.37 Labourer 13.76 Rates are national average prices. Refer to REGIONAL VARIATIONS for indicative levels of overall pricing in regions	MATERIALS			LABOUR				RATES		
	Del to Site	Waste	Material Cost	Craft Optve	Lab	Labour Cost	Sunds	Nett Rate		Gross rate (10%)
	£	%	£	Hrs	Hrs	£	£	£	Unit	£

WORK FOR SERVICES INSTALLATIONS IN NEW BUILDINGS; MECHANICAL INSTALLATIONS (Cont'd)

Pipe and duct sleeves (Cont'd)

Steel pipes BS 1387 Table 2; casting into concrete; making good (Cont'd)

	Del to Site £	Waste %	Material Cost £	Craft Optve Hrs	Lab Hrs	Labour Cost £	Sunds £	Nett Rate £	Unit	Gross rate (10%) £
150mm long; to the following nominal size steel pipes (Cont'd)										
20mm	1.06	5.00	1.11	0.16	0.08	4.04	-	5.15	nr	5.67
25mm	1.32	5.00	1.39	0.17	0.09	4.36	-	5.75	nr	6.33
32mm	1.55	5.00	1.63	0.18	0.09	4.55	-	6.18	nr	6.79
40mm	1.77	5.00	1.86	0.19	0.10	4.87	-	6.73	nr	7.40
50mm	2.31	5.00	2.43	0.23	0.12	5.88	-	8.30	nr	9.13
200mm long; to the following nominal size steel pipes										
15mm	1.11	5.00	1.17	0.15	0.08	3.86	-	5.02	nr	5.53
20mm	1.25	5.00	1.31	0.17	0.09	4.36	-	5.67	nr	6.24
25mm	1.59	5.00	1.66	0.18	0.09	4.55	-	6.21	nr	6.83
32mm	1.89	5.00	1.98	0.19	0.10	4.87	-	6.85	nr	7.54
40mm	2.16	5.00	2.27	0.20	0.10	5.05	-	7.32	nr	8.05
50mm	2.85	5.00	2.99	0.24	0.12	6.06	-	9.05	nr	9.95
250mm long; to the following nominal size steel pipes										
15mm	1.29	5.00	1.36	0.15	0.08	3.86	-	5.21	nr	5.73
20mm	1.45	5.00	1.52	0.19	0.10	4.87	-	6.39	nr	7.03
25mm	1.86	5.00	1.95	0.20	0.10	5.05	-	7.00	nr	7.70
32mm	2.24	5.00	2.35	0.21	0.11	5.37	-	7.72	nr	8.49
40mm	2.55	5.00	2.68	0.22	0.11	5.56	-	8.23	nr	9.06
50mm	3.39	5.00	3.56	0.26	0.13	6.57	-	10.13	nr	11.14
100mm long; to the following nominal size copper pipes										
15mm	0.85	5.00	0.89	0.12	0.06	3.03	-	3.92	nr	4.32
22mm	0.93	5.00	0.98	0.14	0.07	3.54	-	4.52	nr	4.97
28mm	1.15	5.00	1.21	0.15	0.08	3.86	-	5.07	nr	5.58
32mm	1.34	5.00	1.41	0.16	0.08	4.04	-	5.45	nr	6.00
45mm	1.54	5.00	1.62	0.18	0.09	4.55	-	6.17	nr	6.78
54mm	1.96	5.00	2.06	0.22	0.11	5.56	-	7.62	nr	8.38
150mm long; to the following nominal size copper pipes										
15mm	1.04	5.00	1.09	0.14	0.07	3.54	-	4.63	nr	5.09
22mm	1.17	5.00	1.22	0.16	0.08	4.04	-	5.26	nr	5.79
28mm	1.46	5.00	1.53	0.17	0.09	4.36	-	5.89	nr	6.48
32mm	1.71	5.00	1.80	0.18	0.09	4.55	-	6.34	nr	6.98
45mm	1.96	5.00	2.06	0.19	0.10	4.87	-	6.93	nr	7.62
54mm	2.54	5.00	2.67	0.23	0.12	5.88	-	8.54	nr	9.40
200mm long; to the following nominal size copper pipes										
15mm	1.22	5.00	1.28	0.15	0.08	3.86	-	5.14	nr	5.65
22mm	1.38	5.00	1.44	0.17	0.09	4.36	-	5.81	nr	6.39
28mm	1.73	5.00	1.82	0.18	0.09	4.55	-	6.36	nr	7.00
32mm	2.08	5.00	2.18	0.19	0.10	4.87	-	7.05	nr	7.75
45mm	2.39	5.00	2.51	0.20	0.10	5.05	-	7.56	nr	8.32
54mm	3.14	5.00	3.30	0.24	0.12	6.06	-	9.36	nr	10.29
250mm long; to the following nominal size copper pipes										
15mm	1.37	5.00	1.43	0.17	0.09	4.36	-	5.79	nr	6.37
22mm	1.59	5.00	1.66	0.19	0.10	4.87	-	6.53	nr	7.18
28mm	2.05	5.00	2.15	0.20	0.10	5.05	-	7.20	nr	7.92
32mm	2.46	5.00	2.58	0.21	0.11	5.37	-	7.95	nr	8.75
45mm	2.79	5.00	2.93	0.22	0.11	5.56	-	8.49	nr	9.34
54mm	3.73	5.00	3.91	0.26	0.13	6.57	-	10.48	nr	11.53

Steel pipes BS 1387 Table 2; building into blockwork; bedding and pointing in cement mortar (1:3); making good

	Del to Site £	Waste %	Material Cost £	Craft Optve Hrs	Lab Hrs	Labour Cost £	Sunds £	Nett Rate £	Unit	Gross rate (10%) £
100mm long; to the following nominal size steel pipes										
15mm	0.77	5.00	0.80	0.17	0.09	4.36	-	5.17	nr	5.68
20mm	0.85	5.00	0.89	0.18	0.09	4.55	-	5.44	nr	5.98
25mm	1.04	5.00	1.09	0.19	0.10	4.87	-	5.96	nr	6.55
32mm	1.22	5.00	1.28	0.20	0.10	5.05	-	6.33	nr	6.96
40mm	1.41	5.00	1.48	0.21	0.11	5.37	-	6.85	nr	7.53
50mm	1.77	5.00	1.86	0.22	0.11	5.56	-	7.42	nr	8.16
140mm long; to the following nominal size steel pipes										
15mm	0.90	5.00	0.95	0.19	0.10	4.87	-	5.81	nr	6.40
20mm	1.02	5.00	1.07	0.20	0.10	5.05	-	6.12	nr	6.73
25mm	1.26	5.00	1.32	0.21	0.11	5.37	-	6.69	nr	7.36
32mm	1.49	5.00	1.57	0.22	0.11	5.56	-	7.12	nr	7.83
40mm	1.71	5.00	1.80	0.23	0.12	5.88	-	7.67	nr	8.44
50mm	2.20	5.00	2.32	0.24	0.12	6.06	-	8.38	nr	9.21
190mm long; to the following nominal size steel pipes										
15mm	1.08	5.00	1.14	0.21	0.11	5.37	-	6.51	nr	7.16
20mm	1.21	5.00	1.27	0.22	0.11	5.56	-	6.82	nr	7.51
25mm	1.54	5.00	1.62	0.23	0.12	5.88	-	7.50	nr	8.25
32mm	1.84	5.00	1.93	0.24	0.12	6.06	-	7.99	nr	8.79
40mm	2.08	5.00	2.18	0.25	0.13	6.38	-	8.56	nr	9.42
50mm	2.74	5.00	2.88	0.26	0.13	6.57	-	9.44	nr	10.39
215mm long; to the following nominal size steel pipes										
15mm	1.17	5.00	1.22	0.22	0.11	5.56	-	6.78	nr	7.46
20mm	1.30	5.00	1.37	0.23	0.12	5.88	-	7.24	nr	7.97
25mm	1.68	5.00	1.76	0.24	0.12	6.06	-	7.82	nr	8.61
32mm	2.01	5.00	2.11	0.25	0.13	6.38	-	8.49	nr	9.34
40mm	2.28	5.00	2.39	0.26	0.13	6.57	-	8.96	nr	9.85
50mm	3.01	5.00	3.16	0.27	0.14	6.89	-	10.05	nr	11.06
100mm long; to the following nominal size copper pipes										
15mm	0.85	5.00	0.89	0.17	0.09	4.36	-	5.25	nr	5.78
22mm	0.91	5.00	0.96	0.18	0.09	4.55	-	5.50	nr	6.05
28mm	1.15	5.00	1.21	0.19	0.10	4.87	-	6.08	nr	6.69
32mm	1.34	5.00	1.41	0.20	0.10	5.05	-	6.46	nr	7.11
45mm	1.41	5.00	1.48	0.21	0.11	5.37	-	6.85	nr	7.53
54mm	1.96	5.00	2.06	0.22	0.11	5.56	-	7.62	nr	8.38

Labour hourly rates: (except Specialists) Craft Operatives 18.37 Labourer 13.76 Rates are national average prices. Refer to REGIONAL VARIATIONS for indicative levels of overall pricing in regions	MATERIALS			LABOUR				RATES		
	Del to Site £	Waste %	Material Cost £	Craft Optve Hrs	Lab Hrs	Labour Cost £	Sunds £	Nett Rate £	Unit	Gross rate (10%) £
WORK FOR SERVICES INSTALLATIONS IN NEW BUILDINGS; MECHANICAL INSTALLATIONS (Cont'd)										
Pipe and duct sleeves (Cont'd)										
Steel pipes BS 1387 Table 2; building into blockwork; bedding and pointing in cement mortar (1:3); making good (Cont'd)										
140mm long; to the following nominal size copper pipes										
15mm	1.00	5.00	1.05	0.19	0.10	4.87	-	5.91	nr	6.51
22mm	1.11	5.00	1.17	0.20	0.10	5.05	-	6.22	nr	6.84
28mm	1.41	5.00	1.48	0.21	0.11	5.37	-	6.85	nr	7.53
32mm	1.64	5.00	1.72	0.22	0.11	5.56	-	7.27	nr	8.00
45mm	1.88	5.00	1.97	0.23	0.12	5.88	-	7.85	nr	8.63
54mm	2.44	5.00	2.56	0.24	0.12	6.06	-	8.62	nr	9.48
190mm long; to the following nominal size copper pipes										
15mm	1.20	5.00	1.26	0.21	0.11	5.37	-	6.63	nr	7.29
22mm	1.33	5.00	1.40	0.22	0.11	5.56	-	6.96	nr	7.65
28mm	1.69	5.00	1.78	0.23	0.12	5.88	-	7.65	nr	8.42
32mm	2.02	5.00	2.12	0.24	0.12	6.06	-	8.18	nr	8.99
45mm	2.29	5.00	2.40	0.25	0.13	6.38	-	8.78	nr	9.66
54mm	3.01	5.00	3.16	0.26	0.13	6.57	-	9.73	nr	10.70
215mm long; to the following nominal size copper pipes										
15mm	1.28	5.00	1.35	0.23	0.12	5.88	-	7.22	nr	7.94
22mm	1.43	5.00	1.50	0.24	0.12	6.06	-	7.56	nr	8.32
28mm	1.85	5.00	1.94	0.25	0.13	6.38	-	8.32	nr	9.15
32mm	2.19	5.00	2.30	0.26	0.13	6.57	-	8.87	nr	9.76
45mm	2.50	5.00	2.62	0.27	0.14	6.89	-	9.51	nr	10.46
54mm	3.32	5.00	3.48	0.28	0.14	7.07	-	10.55	nr	11.61
Trench covers and frames										
Duct covers in cast iron, coated; continuous covers in 610mm lengths; steel bearers										
22mm deep, for pedestrian traffic; bedding frames to concrete in cement mortar (1:3); nominal width										
150mm	191.48	-	191.48	0.50	0.50	16.07	1.02	208.57	m	229.42
225mm	215.08	-	215.08	0.55	0.55	17.67	1.02	233.77	m	257.15
300mm	227.96	-	227.96	0.65	0.65	20.88	1.02	249.87	m	274.86
375mm	269.53	-	269.53	0.70	0.70	22.49	1.02	293.04	m	322.34
450mm	304.01	-	304.01	0.75	0.75	24.10	1.02	329.13	m	362.04
Alumasc Interior Building Products Ltd., Pendock FDT floor ducting profiles; galvanized mild steel tray section with 12mm plywood cover board; fixing tray section to masonry with nails; fixing cover board with screws, countersinking										
For 50mm screeds										
reference FDT 100/50, 100mm wide	10.14	2.50	10.40	0.35	-	6.43	0.15	16.98	m	18.67
extra; stop end	2.98	2.50	3.05	0.10	-	1.84	-	4.89	nr	5.38
extra; corner	12.89	2.50	13.21	0.25	-	4.59	0.22	18.03	nr	19.83
extra; tee	14.37	2.50	14.73	0.30	-	5.51	0.22	20.47	nr	22.51
reference FDT 150/50, 150mm wide	13.41	2.50	13.74	0.40	-	7.35	0.15	21.24	m	23.36
extra; stop end	2.58	2.50	2.64	0.12	-	2.20	-	4.85	nr	5.33
extra; corner	16.00	2.50	16.40	0.30	-	5.51	0.22	22.14	nr	24.35
extra; tee	14.35	2.50	14.71	0.35	-	6.43	0.22	21.36	nr	23.50
reference FDT 200/50, 200mm wide	18.92	2.50	19.39	0.50	-	9.19	0.15	28.73	m	31.60
extra; stop end	3.58	2.50	3.67	0.15	-	2.76	-	6.43	nr	7.07
extra; corner	20.03	2.50	20.53	0.35	-	6.43	0.22	27.19	nr	29.90
extra; tee	20.03	2.50	20.53	0.40	-	7.35	0.22	28.10	nr	30.91
For 70mm screeds										
reference FDT 100/70, 100mm wide	10.96	2.50	11.23	0.35	-	6.43	0.15	17.81	m	19.59
extra; stop end	2.98	2.50	3.05	0.10	-	1.84	-	4.89	nr	5.38
extra; corner	12.89	2.50	13.21	0.25	-	4.59	0.22	18.03	nr	19.83
extra; tee	12.89	2.50	13.21	0.30	-	5.51	0.22	18.95	nr	20.84
reference FDT 150/70, 150mm wide	14.23	2.50	14.59	0.40	-	7.35	0.15	22.08	m	24.29
extra; stop end	2.98	2.50	3.05	0.12	-	2.20	-	5.26	nr	5.78
extra; corner	16.69	2.50	17.11	0.30	-	5.51	0.22	22.84	nr	25.13
extra; tee	16.69	2.50	17.11	0.35	-	6.43	0.22	23.76	nr	26.14
reference FDT 200/70, 200mm wide	17.02	2.50	17.45	0.50	-	9.19	0.15	26.78	m	29.46
extra; stop end	2.98	2.50	3.05	0.15	-	2.76	-	5.81	nr	6.39
extra; corner	16.69	2.50	17.11	0.35	-	6.43	0.22	23.76	nr	26.14
extra; tee	16.69	2.50	17.11	0.40	-	7.35	0.22	24.68	nr	27.15
Ends of supports for equipment, fittings, appliances and ancillaries										
Fix only; casting into concrete; making good										
holderbat or bracket	-	-	-	0.25	-	4.59	0.42	5.01	nr	5.51
Fix only; building into brickwork										
holderbat or bracket	-	-	-	0.10	-	1.84	-	1.84	nr	2.02
Fix only; cutting and pinning to brickwork; making good										
holderbat or bracket	-	-	-	0.12	-	2.20	0.14	2.34	nr	2.57
Fix only; cutting and pinning to brickwork; making good fair face or facings one side										
holderbat or bracket	-	-	-	0.15	-	2.76	0.19	2.95	nr	3.25
Cavity fixings for ends of supports for pipes and ducts										
Rawlnut Multi-purpose fixings, plated pan head screws; The Rawlplug Co. Ltd; fixing to soft building board, drilling M4 with 30mm long screw; product code 09 - 130 (RNT-M4x30)										
at 300mm centres	0.93	5.00	0.97	0.17	-	3.12	-	4.09	m	4.50
at 450mm centres	0.62	5.00	0.65	0.11	-	2.02	-	2.67	m	2.94
isolated	0.28	5.00	0.29	0.05	-	0.92	-	1.21	nr	1.33

BUILDERS WORK IN CONNECTION WITH INSTALLATIONS

	MATERIALS			LABOUR				RATES		
Labour hourly rates: (except Specialists) Craft Operatives 18.37 Labourer 13.76 Rates are national average prices. Refer to REGIONAL VARIATIONS for indicative levels of overall pricing in regions	Del to Site £	Waste %	Material Cost £	Craft Optve Hrs	Lab Hrs	Labour Cost £	Sunds £	Nett Rate £	Unit	Gross rate (10%) £

WORK FOR SERVICES INSTALLATIONS IN NEW BUILDINGS; MECHANICAL INSTALLATIONS (Cont'd)

Cavity fixings for ends of supports for pipes and ducts (Cont'd)

Rawlnut Multi-purpose fixings, plated pan head screws; The Rawlplug Co. Ltd; fixing to soft building board, drilling (Cont'd)

	Del to Site	Waste	Material Cost	Craft Optve	Lab	Labour Cost	Sunds	Nett Rate	Unit	Gross rate
M5 with 30mm long screw; product code 09 - 235 (RNT-M5x30)										
at 300mm centres	0.97	5.00	1.01	0.20	-	3.67	-	4.69	m	5.16
at 450mm centres	0.64	5.00	0.68	0.13	-	2.39	-	3.06	m	3.37
isolated	0.29	5.00	0.30	0.06	-	1.10	-	1.41	nr	1.55
M5 with 50mm long screw; product code 09 - 317 (RNT-M5x50)										
at 300mm centres	1.27	5.00	1.33	0.20	-	3.67	-	5.00	m	5.50
at 450mm centres	0.84	5.00	0.89	0.13	-	2.39	-	3.27	m	3.60
isolated	0.38	5.00	0.40	0.06	-	1.10	-	1.50	nr	1.65
Rawlnut Multi-purpose fixings, plated pan head screws; The Rawlplug Co. Ltd; fixing to sheet metal, drilling										
M4 with 30mm long screw; product code 09 - 130 (RNT-M4x30)										
at 300mm centres	0.93	5.00	0.97	0.33	-	6.06	-	7.03	m	7.74
at 450mm centres	0.62	5.00	0.65	0.22	-	4.04	-	4.69	m	5.16
isolated	0.28	5.00	0.29	0.10	-	1.84	-	2.13	nr	2.34
M5 with 30mm long screw; product code 09 - 235 (RNT-M5x30)										
at 300mm centres	0.97	5.00	1.01	0.40	-	7.35	-	8.36	m	9.20
at 450mm centres	0.64	5.00	0.68	0.27	-	4.96	-	5.64	m	6.20
isolated	0.29	5.00	0.30	0.12	-	2.20	-	2.51	nr	2.76
M5 with 50mm long screw; product code 09 - 317 (RNT-M5x50)										
at 300mm centres	1.27	5.00	1.33	0.40	-	7.35	-	8.68	m	9.55
at 450mm centres	0.84	5.00	0.89	0.27	-	4.96	-	5.85	m	6.43
isolated	0.38	5.00	0.40	0.12	-	2.20	-	2.60	nr	2.86
Interset high performance cavity fixings, plated pan head screws; The Rawlplug Co. Ltd; fixing to soft building board, drilling										
M4 with 40mm long screw; product code 41 - 620 (SM-04038)										
at 300mm centres	0.33	5.00	0.35	0.17	-	3.12	-	3.47	m	3.82
at 450mm centres	0.22	5.00	0.23	0.11	-	2.02	-	2.25	m	2.48
isolated	0.10	5.00	0.10	0.05	-	0.92	-	1.02	nr	1.13
M5 with 40mm long screw; product code 41 - 636 (SM-05037)										
at 300mm centres	0.40	5.00	0.42	0.20	-	3.67	-	4.09	m	4.50
at 450mm centres	0.27	5.00	0.28	0.13	-	2.39	-	2.67	m	2.93
isolated	0.12	5.00	0.13	0.06	-	1.10	-	1.23	nr	1.35
M5 with 65mm long screw; product code 41 - 652 (SM-05065)										
at 300mm centres	0.57	5.00	0.59	0.20	-	3.67	-	4.27	m	4.70
at 450mm centres	0.38	5.00	0.40	0.13	-	2.39	-	2.78	m	3.06
isolated	0.17	5.00	0.18	0.06	-	1.10	-	1.28	nr	1.41
Spring toggles plated pan head screws; The Rawlplug Co. Ltd; fixing to soft building board, drilling										
M5 with 80mm long screw; product code 94 - 439 (SPO-05080)										
at 300mm centres	0.81	5.00	0.85	0.20	-	3.67	-	4.53	m	4.98
at 450mm centres	0.54	5.00	0.57	0.13	-	2.39	-	2.96	m	3.25
isolated	0.24	5.00	0.26	0.06	-	1.10	-	1.36	nr	1.49
M6 with 60mm long screw; product code 94 - 442 (SPO-06060)										
at 300mm centres	1.09	5.00	1.15	0.20	-	3.67	-	4.82	m	5.30
at 450mm centres	0.73	5.00	0.77	0.13	-	2.39	-	3.15	m	3.47
isolated	0.33	5.00	0.34	0.06	-	1.10	-	1.45	nr	1.59
M6 with 80mm long screw; product code 94 - 464 (SPO-06080)										
at 300mm centres	1.30	5.00	1.36	0.20	-	3.67	-	5.04	m	5.54
at 450mm centres	0.87	5.00	0.91	0.13	-	2.39	-	3.30	m	3.63
isolated	0.39	5.00	0.41	0.06	-	1.10	-	1.51	nr	1.66

Casings, bearers and supports for equipment

	Del to Site	Waste	Material Cost	Craft Optve	Lab	Labour Cost	Sunds	Nett Rate	Unit	Gross rate
Softwood, sawn										
25mm boarded platforms	11.00	10.00	12.10	1.10	0.14	22.13	0.22	34.46	m²	37.90
50 x 50mm bearers	0.85	10.00	0.94	0.16	0.02	3.21	0.03	4.18	m	4.60
50 x 75mm bearers	1.25	10.00	1.38	0.19	0.02	3.77	0.03	5.17	m	5.69
50 x 100mm bearers	1.75	10.00	1.93	0.23	0.03	4.64	0.05	6.61	m	7.27
38 x 50mm bearers; framed	0.65	10.00	0.71	0.25	0.03	5.01	0.05	5.77	m	6.34
Softwood, wrought										
25mm boarded platforms; tongued and grooved	12.50	10.00	13.75	1.20	0.15	24.11	0.25	38.11	m²	41.92
19mm boarded sides; tongued and grooved	8.50	10.00	9.35	1.10	0.14	22.13	0.22	31.71	m²	34.88
extra for holes for pipes not exceeding 55mm nominal size	-	-	-	0.25	-	4.59	-	4.59	nr	5.05
19mm boarded cover; tongued and grooved; ledged; sectional	15.66	10.00	17.23	1.20	0.15	24.11	0.25	41.59	m²	45.75
Fibreboard										
13mm sides	7.95	10.00	8.74	2.10	0.26	42.15	0.44	51.33	m²	56.47
extra for holes for pipes not exceeding 55mm nominal size	-	-	-	0.20	-	3.67	-	3.67	nr	4.04
Vermiculite insulation										
packing around tank	144.18	10.00	158.60	4.80	0.60	96.43	-	255.03	m³	280.53
Slag wool insulation										
packing around tank	33.00	10.00	36.30	6.00	0.75	120.54	-	156.84	m³	172.52

WORK FOR SERVICES INSTALLATIONS IN NEW BUILDINGS; WORK FOR ELECTRICAL INSTALLATIONS

Cutting or forming holes, mortices, sinkings and chases

	Del to Site	Waste	Material Cost	Craft Optve	Lab	Labour Cost	Sunds	Nett Rate	Unit	Gross rate
Concealed steel conduits; making good										
luminaire points	-	-	-	0.70	0.35	17.68	0.52	18.20	nr	20.02
socket outlet points	-	-	-	0.60	0.30	15.15	0.45	15.60	nr	17.16
fitting outlet points	-	-	-	0.60	0.30	15.15	0.45	15.60	nr	17.16
equipment and control gear points	-	-	-	0.80	0.40	20.20	0.62	20.82	nr	22.90

Labour hourly rates: (except Specialists) Craft Operatives 18.37 Labourer 13.76 Rates are national average prices. Refer to REGIONAL VARIATIONS for indicative levels of overall pricing in regions	MATERIALS			LABOUR				RATES		
	Del to Site	Waste	Material Cost	Craft Optve	Lab	Labour Cost	Sunds	Nett Rate		Gross rate (10%)
	£	%	£	Hrs	Hrs	£	£	£	Unit	£
WORK FOR SERVICES INSTALLATIONS IN NEW BUILDINGS; **WORK FOR ELECTRICAL INSTALLATIONS (Cont'd)**										
Cutting or forming holes, mortices, sinkings and chases **(Cont'd)**										
Exposed p.v.c. conduits; making good										
luminaire points	-	-	-	0.25	0.13	6.38	0.18	6.56	nr	7.22
socket outlet points	-	-	-	0.20	0.10	5.05	0.15	5.20	nr	5.72
fitting outlet points	-	-	-	0.20	0.10	5.05	0.15	5.20	nr	5.72
equipment and control gear points	-	-	-	0.30	0.15	7.58	0.21	7.79	nr	8.56
WORK FOR SERVICES INSTALLATIONS IN EXISTING **BUILDINGS**										
Cutting on in-situ concrete										
Cutting chases										
depth not exceeding 50mm	-	-	-	-	0.19	2.61	-	2.61	m	2.88
Cutting chases; making good										
depth not exceeding 50mm	-	-	-	-	0.22	3.03	0.38	3.40	m	3.74
Cutting mortices										
50 x 50mm; depth not exceeding 100mm	-	-	-	-	0.50	6.88	-	6.88	nr	7.57
75 x 75mm; depth not exceeding 100mm	-	-	-	-	0.55	7.57	-	7.57	nr	8.32
100 x 100mm; depth 100 - 200mm	-	-	-	-	0.66	9.08	-	9.08	nr	9.99
38mm diameter; depth not exceeding 100mm	-	-	-	-	0.39	5.37	-	5.37	nr	5.90
Cutting holes										
225 x 150mm; depth not exceeding 100mm	-	-	-	-	1.32	18.16	0.51	18.67	nr	20.54
300 x 150mm; depth not exceeding 100mm	-	-	-	-	1.45	19.95	0.57	20.52	nr	22.57
300 x 300mm; depth not exceeding 100mm	-	-	-	-	1.60	22.02	0.57	22.59	nr	24.84
225 x 150mm; depth 100 - 200mm	-	-	-	-	1.89	26.01	0.72	26.73	nr	29.40
300 x 150mm; depth 100 - 200mm	-	-	-	-	2.11	29.03	0.81	29.84	nr	32.83
300 x 300mm; depth 100 - 200mm	-	-	-	-	2.32	31.92	0.81	32.73	nr	36.01
225 x 150mm; depth 200 - 300mm	-	-	-	-	2.20	30.27	0.84	31.11	nr	34.22
300 x 150mm; depth 200 - 300mm	-	-	-	-	2.43	33.44	0.92	34.35	nr	37.79
300 x 300mm; depth 200 - 300mm	-	-	-	-	2.65	36.46	0.92	37.38	nr	41.12
50mm diameter; depth not exceeding 100mm	-	-	-	-	0.85	11.70	0.30	12.00	nr	13.20
100mm diameter; depth not exceeding 100mm	-	-	-	-	0.94	12.93	0.39	13.32	nr	14.66
150mm diameter; depth not exceeding 100mm	-	-	-	-	1.03	14.17	0.39	14.56	nr	16.02
50mm diameter; depth 100 - 200mm	-	-	-	-	1.37	18.85	0.50	19.35	nr	21.28
100mm diameter; depth 100 - 200mm	-	-	-	-	1.51	20.78	0.57	21.35	nr	23.48
150mm diameter; depth 100 - 200mm	-	-	-	-	1.62	22.29	0.57	22.86	nr	25.15
50mm diameter; depth 200 - 300mm	-	-	-	-	1.63	22.43	0.64	23.07	nr	25.38
100mm diameter; depth 200 - 300mm	-	-	-	-	1.79	24.63	0.69	25.32	nr	27.85
150mm diameter; depth 200 - 300mm	-	-	-	-	1.97	27.11	0.69	27.80	nr	30.58
Grouting into mortices with cement mortar (1:1); around steel										
50 x 50mm; depth not exceeding 100mm	-	-	-	-	0.12	1.65	0.76	2.42	nr	2.66
75 x 75mm; depth not exceeding 100mm	-	-	-	-	0.13	1.79	0.76	2.55	nr	2.81
100 x 100mm; depth 100 - 200mm	-	-	-	-	0.17	2.34	0.98	3.31	nr	3.65
38mm diameter; depth not exceeding 100mm	-	-	-	-	0.09	1.24	0.56	1.79	nr	1.97
Cutting holes on reinforced in-situ concrete										
Cutting holes										
225 x 150mm; depth not exceeding 100mm	-	-	-	-	1.64	22.57	0.64	23.21	nr	25.53
300 x 150mm; depth not exceeding 100mm	-	-	-	-	1.81	24.91	0.69	25.60	nr	28.16
300 x 300mm; depth not exceeding 100mm	-	-	-	-	1.99	27.38	0.69	28.07	nr	30.88
225 x 150mm; depth 100 - 200mm	-	-	-	-	2.39	32.89	0.94	33.83	nr	37.21
300 x 150mm; depth 100 - 200mm	-	-	-	-	2.63	36.19	0.99	37.18	nr	40.90
300 x 300mm; depth 100 - 200mm	-	-	-	-	2.89	39.77	0.99	40.76	nr	44.83
225 x 150mm; depth 200 - 300mm	-	-	-	-	2.78	38.25	1.07	39.32	nr	43.25
300 x 150mm; depth 200 - 300mm	-	-	-	-	3.06	42.11	1.17	43.28	nr	47.60
300 x 300mm; depth 200 - 300mm	-	-	-	-	3.33	45.82	1.17	46.99	nr	51.69
50mm diameter; depth not exceeding 100mm	-	-	-	-	1.05	14.45	0.41	14.85	nr	16.34
100mm diameter; depth not exceeding 100mm	-	-	-	-	1.18	16.24	0.47	16.70	nr	18.37
150mm diameter; depth not exceeding 100mm	-	-	-	-	1.30	17.89	0.47	18.35	nr	20.19
50mm diameter; depth 100 - 200mm	-	-	-	-	1.72	23.67	0.66	24.33	nr	26.76
100mm diameter; depth 100 - 200mm	-	-	-	-	1.89	26.01	0.70	26.71	nr	29.38
150mm diameter; depth 100 - 200mm	-	-	-	-	2.02	27.80	0.70	28.50	nr	31.35
50mm diameter; depth 200 - 300mm	-	-	-	-	2.04	28.07	0.78	28.85	nr	31.74
100mm diameter; depth 200 - 300mm	-	-	-	-	2.24	30.82	0.87	31.69	nr	34.86
150mm diameter; depth 200 - 300mm	-	-	-	-	2.46	33.85	0.87	34.72	nr	38.19
Cutting mortices, sinking and the like for services										
In existing concrete; making good										
75 x 75 x 35mm	-	-	-	0.45	0.13	10.06	0.75	10.81	nr	11.89
150 x 75 x 35mm	-	-	-	0.60	0.15	13.09	1.12	14.21	nr	15.63
In existing brickwork										
75 x 75 x 35mm	-	-	-	0.25	0.10	5.97	0.45	6.42	nr	7.06
150 x 75 x 35mm	-	-	-	0.40	0.13	9.14	0.68	9.81	nr	10.79
In existing brickwork with plaster finish; making good										
75 x 75 x 35mm	-	-	-	0.30	0.18	7.99	0.75	8.74	nr	9.61
150 x 75 x 35mm	-	-	-	0.45	0.20	11.02	1.05	12.07	nr	13.28
In existing blockwork; making good										
75 x 75 x 35mm	-	-	-	0.25	0.08	5.69	0.50	6.19	nr	6.81
150 x 75 x 35mm	-	-	-	0.35	0.10	7.81	0.54	8.35	nr	9.18
In existing blockwork with plaster finish; making good										
75 x 75 x 35mm	-	-	-	0.40	0.13	9.14	0.68	9.81	nr	10.79
150 x 75 x 35mm	-	-	-	0.46	0.15	10.51	0.60	11.11	nr	12.23

Labour hourly rates: (except Specialists) Craft Operatives 18.37 Labourer 13.76 Rates are national average prices. Refer to REGIONAL VARIATIONS for indicative levels of overall pricing in regions	MATERIALS			LABOUR				RATES		
	Del to Site	Waste	Material Cost	Craft Optve	Lab	Labour Cost	Sunds	Nett Rate		Gross rate (10%)
	£	%	£	Hrs	Hrs	£	£	£	Unit	£
WORK FOR SERVICES INSTALLATIONS IN EXISTING BUILDINGS (Cont'd)										
Cutting chases for services										
In existing concrete; making good										
20mm nominal size conduits -1	-	-	-	0.40	0.20	10.10	0.52	10.62	m	11.69
20mm nominal size conduits -3	-	-	-	0.50	0.25	12.62	0.62	13.24	m	14.56
20mm nominal size conduits -6	-	-	-	0.60	0.30	15.15	0.70	15.86	m	17.44
In existing brickwork										
20mm nominal size conduits -1	-	-	-	0.30	0.15	7.58	0.52	8.10	m	8.91
20mm nominal size conduits -3	-	-	-	0.40	0.20	10.10	0.62	10.72	m	11.79
20mm nominal size conduits -6	-	-	-	0.50	0.25	12.62	0.70	13.33	m	14.66
In existing brickwork with plaster finish; making good										
20mm nominal size conduits -1	-	-	-	0.45	0.23	11.43	0.69	12.12	m	13.33
20mm nominal size conduits -3	-	-	-	0.55	0.28	13.96	0.78	14.74	m	16.21
20mm nominal size conduits -6	-	-	-	0.65	0.33	16.48	0.87	17.35	m	19.09
In existing blockwork										
20mm nominal size conduits -1	-	-	-	0.23	0.12	5.88	0.52	6.40	m	7.04
20mm nominal size conduits -3	-	-	-	0.30	0.15	7.58	0.62	8.19	m	9.01
20mm nominal size conduits -6	-	-	-	0.38	0.19	9.60	0.70	10.30	m	11.33
In existing blockwork with plaster finish; making good										
20mm nominal size conduits -1	-	-	-	0.38	0.19	9.60	0.69	10.28	m	11.31
20mm nominal size conduits -3	-	-	-	0.45	0.23	11.43	0.78	12.21	m	13.43
20mm nominal size conduits -6	-	-	-	0.53	0.27	13.45	0.87	14.32	m	15.75
Lifting and replacing floorboards										
For cables or conduits										
in groups 1 - 3	-	-	-	0.16	0.08	4.04	0.15	4.19	m	4.61
in groups 3 - 6	-	-	-	0.20	0.10	5.05	0.17	5.22	m	5.74
in groups exceeding 6	-	-	-	0.25	0.13	6.38	0.18	6.56	m	7.22

This section continues
on the next page

Labour hourly rates: (except Specialists) Craft Operatives 18.37 Labourer 13.76 Rates are national average prices. Refer to REGIONAL VARIATIONS for indicative levels of overall pricing in regions	PLANT AND TRANSPORT			LABOUR				RATES		
	Plant Cost	Trans Cost	P and T Cost	Craft Optve	Lab	Labour Cost	Sunds	Nett Rate	Unit	Gross rate (10%)
	£	£	£	Hrs	Hrs	£	£	£		£
WORK FOR EXTERNAL SERVICES INSTALLATIONS										
Underground service runs; excavating trenches to receive services not exceeding 200mm nominal size - by machine										
Note:										
Notwithstanding the requirements of NRM2 ducts are not included with items of excavation and the appropriate ducts must be added to to the rates as noted below										
Excavations commencing from natural ground level; compacting; backfilling with excavated material										
not exceeding 1m deep; average 750mm deep	3.08	1.04	4.12	-	0.25	3.44	3.13	10.70	m	11.77
not exceeding 1m deep; average 750mm deep; next to roadways	3.32	1.04	4.36	-	0.25	3.44	4.97	12.76	m	14.04
Excavations commencing from existing ground level; levelling and grading backfilling to receive turf										
not exceeding 1m deep; average 750mm deep	3.32	1.04	4.36	-	0.25	3.44	5.01	12.81	m	14.09
Extra over service runs										
Extra over excavating trenches irrespective of depth; breaking out existing hard pavings										
concrete 150mm thick	4.73	-	4.73	-	-	-	-	4.73	m²	5.20
reinforced concrete 200mm thick	9.79	-	9.79	-	-	-	-	9.79	m²	10.77
concrete 150mm thick; reinstating	5.74	-	5.74	-	-	-	19.59	25.33	m²	27.86
macadam paving 75mm thick	0.84	-	0.84	-	-	-	-	0.84	m²	0.93
macadam paving 75mm thick; reinstating	0.84	-	0.84	-	-	-	25.05	25.89	m²	28.48
concrete flag paving 50mm thick	-	-	-	-	0.26	3.58	-	3.58	m²	3.94
concrete flag paving 50mm thick; re-instating	-	-	-	0.50	0.76	19.64	16.45	36.10	m²	39.71
Lifting turf for preservation										
stacking on site average 50 metres distance for immediate use; watering	0.01	-	0.01	-	0.60	8.26	-	8.26	m²	9.09
Re-laying turf										
taking from stack average 50 metres distance, laying on prepared bed, watering, maintaining	0.01	-	0.01	-	0.60	8.26	-	8.26	m²	9.09
Extra over excavating trenches irrespective of depth; breaking out existing materials										
hard rock	37.13	-	37.13	-	-	-	-	37.13	m³	40.84
concrete	30.38	-	30.38	-	-	-	-	30.38	m³	33.42
reinforced concrete	45.57	-	45.57	-	-	-	-	45.57	m³	50.13
brickwork, blockwork or stonework	20.25	-	20.25	-	-	-	-	20.25	m³	22.28
Extra over excavating trenches irrespective of depth; excavating next to existing services										
electricity services -1	-	-	-	-	5.00	68.80	-	68.80	m	75.68
gas services -1	-	-	-	-	5.00	68.80	-	68.80	m	75.68
water services -1	-	-	-	-	5.00	68.80	-	68.80	m	75.68
Extra over excavating trenches irrespective of depth; excavating around existing services crossing trench										
electricity services; cables crossing -1	-	-	-	-	7.50	103.20	-	103.20	nr	113.52
gas services services crossing -1	-	-	-	-	7.50	103.20	-	103.20	nr	113.52
water services services crossing -1	-	-	-	-	7.50	103.20	-	103.20	nr	113.52
Underground service runs; excavating trenches to receive services not exceeding 200mm nominal size - by hand										
Note:										
Notwithstanding the requirements of NRM2 ducts are not included with items of excavation and the appropriate ducts must be added to to the rates as noted below										
Excavations commencing from natural ground level; compacting; backfilling with excavated material										
not exceeding 1m deep; average 750mm deep	-	-	-	-	1.20	16.51	10.19	26.70	m	29.37
not exceeding 1m deep; average 750mm deep; next to roadways	-	-	-	-	1.20	16.51	11.58	28.09	m	30.90
Excavations commencing from existing ground level; levelling and grading backfilling to receive turf										
not exceeding 1m deep; average 750mm deep	-	-	-	-	1.20	16.51	11.88	28.39	m	31.23
Extra over service runs										
Extra over excavating trenches irrespective of depth; breaking out existing hard pavings										
concrete 150mm thick	-	-	-	-	1.50	20.64	-	20.64	m²	22.70
reinforced concrete 150mm thick	-	-	-	-	3.00	41.28	-	41.28	m²	45.41
concrete 150mm thick; reinstating	-	-	-	-	3.00	41.28	19.75	61.03	m²	67.14
macadam paving 75mm thick	-	-	-	-	0.25	3.44	-	3.44	m²	3.78
macadam paving 75mm thick; reinstating	-	-	-	-	0.25	3.44	25.05	28.49	m²	31.34
concrete flag paving 50mm thick	-	-	-	-	0.26	3.58	-	3.58	m²	3.94
concrete flag paving 50mm thick; re-instating	-	-	-	-	0.76	10.46	16.69	27.15	m²	29.87
Extra over excavating trenches irrespective of depth; breaking out existing materials										
hard rock	-	-	-	-	13.35	183.70	-	183.70	m³	202.07
concrete	-	-	-	-	10.00	137.60	-	137.60	m³	151.36
reinforced concrete	-	-	-	-	15.00	206.40	-	206.40	m³	227.04
brickwork, blockwork or stonework	-	-	-	-	7.00	96.32	-	96.32	m³	105.95

Labour hourly rates: (except Specialists) Craft Operatives 18.37 Labourer 13.76 Rates are national average prices. Refer to REGIONAL VARIATIONS for indicative levels of overall pricing in regions	PLANT AND TRANSPORT			LABOUR				RATES		
	Plant Cost	Trans Cost	P and T Cost	Craft Optve	Lab	Labour Cost	Sunds	Nett Rate	Unit	Gross rate (10%)
	£	£	£	Hrs	Hrs	£	£	£		£
WORK FOR EXTERNAL SERVICES INSTALLATIONS (Cont'd)										
Extra over service runs (Cont'd)										
Extra over excavating trenches irrespective of depth; excavating next existing services										
electricity services -1	-	-	-	-	5.00	68.80	-	68.80	m	75.68
gas services -1	-	-	-	-	5.00	68.80	-	68.80	m	75.68
water services -1	-	-	-	-	5.00	68.80	-	68.80	m	75.68
Extra over excavating trenches irrespective of depth; excavating around existing services crossing trench										
electricity services; cables crossing -1...........................	-	-	-	-	7.50	103.20	-	103.20	nr	113.52
gas services services crossing -1	-	-	-	-	7.50	103.20	-	103.20	nr	113.52
water services services crossing -1	-	-	-	-	7.50	103.20	-	103.20	nr	113.52

This section continues
on the next page

Labour hourly rates: (except Specialists) Craft Operatives 18.37 Labourer 13.76 Rates are national average prices. Refer to REGIONAL VARIATIONS for indicative levels of overall pricing in regions	MATERIALS			LABOUR				RATES		
	Del to Site £	Waste %	Material Cost £	Craft Optve Hrs	Lab Hrs	Labour Cost £	Sunds £	Nett Rate £	Unit	Gross rate (10%) £
WORK FOR EXTERNAL SERVICES INSTALLATIONS (Cont'd)										
Underground ducts; vitrified clayware, BS 65 extra strength; flexible joints										
Straight										
100mm nominal size single way duct; laid in position in trench......	15.96	5.00	16.76	0.16	0.02	3.21	-	19.97	m	21.97
100mm nominal size bonded to form 2 way duct; laid in position in trench................	31.92	5.00	33.52	0.25	0.03	5.01	-	38.52	m	42.37
100mm nominal size bonded to form 4 way duct; laid in position in trench................	63.84	5.00	67.03	0.35	0.04	6.98	-	74.01	m	81.41
100mm nominal size bonded to form 6 way duct; laid in position in trench................	95.76	5.00	100.55	0.45	0.06	9.09	-	109.64	m	120.60
extra; providing and laying in position nylon draw wire................	0.16	10.00	0.18	0.02	-	0.37	-	0.54	m	0.60
Cast iron surface boxes and covers, coated to BS EN 10300; bedding in cement mortar (1:3)										
Surface boxes, BS 5834 Part 2 marked S.V. hinged lid; 150 x 150mm overall top size, 75mm deep................	20.81	15.00	20.99	0.75	0.38	19.01	0.30	40.30	nr	44.32
Surface boxes, BS 750, marked 'FIRE HYDRANT' heavy grade; minimum clear opening 230 x 380mm, minimum depth 125mm................	43.28	15.00	43.63	1.25	0.63	31.63	1.49	76.74	nr	84.42
Surface boxes, BS 5834 Part 2 marked 'W' or WATER medium grade; double triangular cover; minimum clear opening 300 x 300mm, minimum depth 100mm	52.43	15.00	52.78	1.50	0.75	37.88	1.47	92.12	nr	101.34
Stop cock pits, valve chambers and the like										
Stop cock pits; half brick thick sides of common bricks, BS EN 772, Category M, 215 x 102.5 x 65mm, compressive strength 20.5 N/mm², in cement mortar (1:3); 100mm thick base and top of plain in-situ concrete, BS 8500, ordinary prescribed mix ST4, 20mm aggregate, formwork; including all excavation, backfilling, disposal of surplus excavated material, earthwork support and compaction of ground										
600mm deep in clear; (surface boxes included elsewhere)										
internal size 225 x 225mm	188.97	5.00	190.49	1.50	1.50	48.19	0.08	243.94	nr	268.34
internal size 338 x 338mm	248.79	5.00	252.20	2.25	2.25	72.29	0.12	332.56	nr	365.82
aggregate, formwork										
750mm deep in clear; (surface boxes included elsewhere)										
150mm diameter................	205.69	5.00	207.00	1.15	1.15	36.95	0.09	249.23	nr	274.15
Stop cock keys										
tee	4.17	-	4.17	-	-	-	-	4.17	nr	4.59

This page left blank intentionally

This section gives basic prices of materials delivered to site. Prices are exclusive of Value Added Tax.

Excavation and Earthwork

Filling

Broken brick or stone filling .	25.06 m³
MOT type 1 filling	34.56 m³
MOT type 2 filling	34.56 m³
Imported topsoil	20.00 m³
Sand	22.80 m³

Gravel

Coarse gravel	40.94 m³
Blinding gravel	32.37 m³

Earthwork support

Earthwork support timber	271.49 m³

Concrete Work

Materials for site mixed concrete

40mm Aggregate	41.98 m³
20mm Aggregate	48.80 m³
10mm Aggregate	48.96 m³
Portland Cement	146.80 tonne
White Portland Cement	401.60 tonne
Sharp sand	42.56 m³
White silver sand	44.72 m³

Ready mixed concrete, prescribed mix

ST1 - 40mm aggregate	94.71 m³
ST3 - 20mm aggregate	97.99 m³
ST4 - 20mm aggregate	99.73 m³
ST5 - 20mm aggregate	101.68 m³

Ready mixed concrete, design mix

C 12/15 - 15 N/mm² - 20mm	95.12 m³
C 16/20 - 20 N/mm² - 20mm	96.25 m³
C 20/25 - 25 N/mm² - 20mm	98.61 m³
C 25/30 - 30 N/mm² - 20mm	101.27 m³

Site mixed concretes, mortars, various mixes

1:6, all in aggregate	137.06 m³
1:8, all in aggregate	130.10 m³
1:12, all in aggregate	126.27 m³

Lightweight concrete

20.5 N/mm²; vibrated	131.25 m³
26.0 N/mm²; vibrated	135.45 m³
41.5 N/mm²; vibrated	143.85 m³

Mild steel rod reinforcement

32mm diameter	909.90 tonne
25mm diameter	863.24 tonne
20mm diameter	874.07 tonne
16mm diameter	885.58 tonne
12mm diameter	883.76 tonne
10mm diameter	975.00 tonne
8mm diameter	1,050.00 tonne
6mm diameter	1,048.28 tonne

High yield rod reinforcement

32mm diameter	929.95 tonne
25mm diameter	931.17 tonne
20mm diameter	835.80 tonne
16mm diameter	875.00 tonne
12mm diameter	888.89 tonne
10mm diameter	852.89 tonne
8mm diameter	1,157.69 tonne
6mm diameter	1,136.99 tonne

Stainless steel rod reinforcement

25mm diameter	2,613.28 tonne
20mm diameter	2,467.07 tonne
16mm diameter	2,474.07 tonne
12mm diameter	2,660.61 tonne
10mm diameter	2,620.14 tonne
8mm diameter	2,592.31 tonne

Fabric reinforcement, BS Ref 4483

A393	4.94 m²
A252	3.17 m²
A193	2.42 m²
A142	1.78 m²
A98	1.23 m²
B1131	11.28 m²
B785	8.43 m²
B503	6.66 m²
B385	5.38 m²
B283	2.99 m²
B196	2.37 m²
C785	5.39 m²
C636	4.45 m²
C503	3.48 m²
C385	2.73 m²
C283	2.14 m²
D98	2.08 m²
D49	2.09 m²

Formwork

Basic finish

Sides of foundations	
height exceeding 1.00m ..	3.73 m²
height not exceeding 250mm	1.16 m
Sides of ground beams and edges of beds	
height exceeding 1.00m ..	3.73 m²
height not exceeding 250mm	1.16 m
Edges of suspended slabs	
height not exceeding 250mm	1.40 m
height 250 - 500mm	2.61 m
Sides of upstands	
height not exceeding 250mm	1.66 m
height 250 - 500mm	1.94 m
Soffits of slabs	
horizontal slab thickness not exceeding 200mm	11.29 m²
horizontal slab thickness 200 - 300mm	12.42 m²
sloping not exceeding 15 degrees slab thickness not exceeding 200mm	12.44 m²
sloping exceeding 15 degrees slab thickness not exceeding 200mm	13.01 m²
Walls	
plain	9.14 m²
plain; height exceeding 3.00m above floor level ...	8.43 m²
Beams	
attached to slabs	7.89 m²
with 50mm wide chamfers	8.03 m²
isolated	8.27 m²
isolated with 50mm wide chamfers	8.43 m²
Columns	
attached to walls	5.48 m²
attached to walls, height exceeding 3.00m above floor level	5.09 m²
Wall kickers	
straight	1.09 m
curved 2m radius	2.67 m
curved 10m radius	1.62 m
Stairflights	
1000mm wide; 155mm thick waist; 178mm risers	16.85 m
1500mm wide; 180mm thick waist; 178mm risers	21.09 m
1000mm wide; 155mm thick waist; 178mm undercut risers	16.85 m
1500mm wide; 180mm thick waist; 178mm undercut risers	21.08 m

Claymaster expanded polystyrene permanent formwork

2400 x 1200 by	
75mm thick	5.19 m²
100mm thick	6.93 m²
150mm thick	10.42 m²

Damp proofing

Liquid damp proof applications

Synthaprufe bituminous emulsion	3.86 litre
Bituminous emulsion	2.01 litre
R.I.W. liquid asphaltic composition	7.90 litre

Flexible sheet damp proof membranes

Polythene sheeting, 125mu	0.28 m²
Polythene sheeting, 250mu (1000g)	0.29 m²
Polythene sheeting, 300mu (1200g)	0.34 m²
Polythene sheeting, 300mu (Fenton)	0.34 m²

Waterproof building paper to BS 1521

British Sisalkraft Ltd 728 damp proof membrane	1.32 m²
Visqueen 300mu damp proof membrane	0.52 m²

Precast concrete lintels

lintel	
75 x 150mm 1200mm long reinf with 1.20 kg of 12mm lintel bars	11.49 nr
102 x 150mm 1200mm long reinf with 1.20 kg of 12mm lintel bars	19.20 nr
215 x 150mm 1200mm long reinf with 2.40 kg of 12mm lintel bars	52.62 nr
327 x 150mm 1200mm long reinf with 3.60 kg of 12mm lintel bars	75.70 nr
boot lintel	
252 x 150mm 1200mm long reinf with 3.60 kg of 12mm boot lintel bars and 0.56 kg of 6mm boot lintel links	62.00 nr
252 x 150mm 1800mm long reinf with 5.19 kg of 12mm boot lintel bars and 0.83 kg of 6mm boot lintel links	94.00 nr

Prestressed precast lintels

100 x 65mm	
900mm long	4.75 nr
1050mm long	5.25 nr
1200mm long	6.35 nr
1500mm long	7.70 nr
1800mm long	8.99 nr
2100mm long	10.55 nr
2400mm long	12.15 nr
2700mm long	13.70 nr
3000mm long	15.10 nr
140 x 70mm	
1200mm long	8.40 nr
3000mm long	20.05 nr
225 x 70mm	
1200mm long	15.05 nr

Concrete Work (Cont'd)

Prestressed precast lintels (Cont'd)

225 x 70mm (Cont'd)

3000mm long	37.50 nr
1200mm long	23.82 nr
3000mm long	59.53 nr

Precast concrete sills

figure 2 or figure 4; splayed and grooved

50 x 150mm

300mm long	13.08 nr
700mm long	22.18 nr
1300mm long	39.11 nr
stooled end	6.85 nr

75 x 150mm

300mm long	14.37 nr
700mm long	28.11 nr
1300mm long	41.91 nr
stooled end	8.89 nr

100 x 150mm

300mm long	12.98 nr
700mm long	30.19 nr
1300mm long	56.01 nr
stooled end	11.58 nr

Precast concrete copings; figure 1

75 x 200mm

splayed; rebated joints	25.62 m
extra; stopped ends	2.57 nr

100 x 300mm

splayed; rebated joints	39.75 m
extra; stopped ends	3.50 nr

75 x 200mm

saddleback; rebated joints	24.47 m
extra; hipped ends	1.11 nr

100 x 300mm

saddleback; rebated joints	32.07 m
extra; hipped ends	2.65 nr

Precast concrete padstones

215 x 215 x 75mm	5.45 nr
215 x 215 x 150mm	9.79 nr
327 x 215 x 150mm	14.69 nr
327 x 327 x 150mm	22.48 nr

Brickwork and Blockwork

Bricks

Common bricks plain	316.00 1000
Selected regraded bricks	216.00 1000
Second hard stock bricks	795.00 1000
Engineering bricks class A	750.00 1000
class B	323.00 1000
Staffordshire blue wirecut bricks Ibstock 2221A	776.00 1000
pressed bricks	708.33 1000
splayed plinth or bullnose	2,950.00 1000

Special bricks

Claydon Red Multi	800.00 1000
Tudors	800.00 1000
Milton Buff Ridgefaced	800.00 1000
Heathers	800.00 1000
Brecken Grey	800.00 1000
Leicester Red Stock	600.00 1000
West Hoathly Medium multi-stock	968.00 1000
Himley dark brown rustic	1,143.45 1000
Tonbridge hand made multi-coloured	1,166.79 1000
Old English Russet	500.00 1000

Miscellaneous materials for brickwork and blockwork

Cement and sand mortar (1:3)	116.43 m³
Cement lime mortar (1:1:6)	121.98 m³
Hydrated Lime	325.00 tonne

Cavity wall insulation

Glass fibre slabs 50mm (Dritherm 37)	1.93 m²
Glass fibre slabs 75mm	2.30 m²

Wall Ties

Stainless steel

200mm Housing type 4	17.89 250
Vertical twisted 203 x 19 x 3mm	501.28 1000

Extension Profiles

Catnic Stronghold Extension profiles 60-250mm ref SWC	1.30 m
Expamet Wall Starter Single Flange Stainless Steel 100-115mm	2.27 m
Simpson C2K Crocodile Wall Starter Stainless Steel	3.20 m

Thermabate insulated cavity closers for fixing to masonry with PVCU ties at 225mm centres

Thermabate 50	4.68 m
Thermabate 75	4.82 m
Thermabate 100	5.06 m

Blockwork

Thermalite Shield blocks

75mm thick blocks	10.70 m²
100mm thick blocks	12.44 m²
140mm thick blocks	17.42 m²
150mm thick blocks	18.67 m²
200mm thick blocks	25.00 m²

Thermalite Smooth Face blocks

100mm thick blocks	13.86 m²
140mm thick blocks	18.71 m²
200mm thick blocks	38.17 m²

Thermalite Turbo blocks

100mm thick blocks	14.89 m²
115mm thick blocks	15.10 m²
130mm thick blocks	17.36 m²
190mm thick blocks	26.79 m²
215mm thick blocks	32.03 m²

Thermalite Party Wall blocks

215mm thick blocks	27.10 m²

Thermalite Hi-Strength blocks

100mm thick blocks	18.33 m²
140mm thick blocks	25.66 m²

Thermalite Trench blocks

255mm thick blocks	30.90 m²
275mm thick blocks	32.43 m²
305mm thick blocks	35.38 m²
355mm thick blocks	36.00 m²

Lignacite blocks solid Fair Face 7.0 N/mm²

100mm thick blocks	15.00 m²
140mm thick blocks	19.80 m²
190mm thick blocks	28.50 m²

Toplite solid Standard blocks 7.0 N/mm²

100mm thick blocks	11.23 m²
140mm thick blocks	13.76 m²
190mm thick blocks	20.64 m²
215mm thick blocks	23.24 m²

Hemelite solid Standard blocks 3.5 N/mm²

75mm thick blocks	7.30 m²
90mm thick blocks	8.14 m²
100mm thick blocks	7.95 m²
140mm thick blocks	11.24 m²
190mm thick blocks	17.85 m²
215mm thick blocks	20.27 m²

Hemelite solid Standard blocks 7.0 N/mm²

90mm thick blocks	9.24 m²
100mm thick blocks	8.10 m²
140mm thick blocks	11.75 m²
190mm thick blocks	20.21 m²
215mm thick blocks	22.87 m²

Tarmac Topcrete cellular blocks

100mm thick	7.95 m²
140mm thick	11.85 m²

Hollow glass blocks, white, cross ribbed

190 x 190 x 80mm blocks	1.89 nr
240 x 240 x 80mm blocks	8.99 nr

Damp proof course

Hyload "Original", pitch polymer	4.83 m²
BS 6515 Polythene DPC	0.79 m²
Synthaprufe bituminous latex emulsion	2.94 litre

Bituminous emulsion	1.58 litre
Slates 350 x 225mm	780.00 1000

Cavity trays and closers

Type G Cavitray without lead

900mm long cavitray	8.18 nr
220 x 332mm external angle	7.94 nr
230 x 117mm internal angle	7.94 nr

Type X Cavitray with lead, 40 degree pitch

Intermediate tray; short	5.84 nr
Ridge tray	11.98 nr
Catchment tray; long	8.87 nr
Corner catchment angle tray	15.70 nr

Type W

Cavity weep ventilator	0.53 nr
Extension duct	0.94 nr

Fire rated cavity closers by Cavity Trays Ltd

Cavi 120 type V; Cavicloser vertical	8.16 m
Cavi 120 type V; Cavicloser horizontal	8.16 m
Cavi 240 type CFIS; Cavicheck vertical	3.07 m
Cavi 240 type CFIS; Cavicheck horizontal	3.07 m
Cavi 240 type PWIB; Party wall integrity barrier vertical sleeved	8.21 m
Cavi 240 type PWIB; Party wall integrity barrier horizontal	8.21 m

Stainless steel mesh reinforcement

65mm mesh	5.07 m
115mm mesh	7.07 m
175mm mesh	11.35 m
225mm mesh	15.48 m

Brickforce stainless steel joint reinforcement

60mm wide ref SBF30W60	4.24 m
60mm wide ref SBF50W60	4.57 m
175mm wide ref SBF30W175	4.81 m
175mm wide ref SBF50W175	15.96 m

Air Bricks

Terra cotta, red or buff, square hole

225 x 75mm	3.08 nr
215 x 140mm	4.57 nr
215 x 215mm	11.47 nr
215 x 65 Cavity liner 200 long	3.24 nr
215 x 140 Cavity liner 200 long	3.63 nr
215 x 215 Cavity liner 200 long	9.84 nr

Iron, light

225 x 75mm	9.99 nr
225 x 150mm	16.66 nr
225 x 225mm	29.16 nr

Plastic

225 x 75mm	0.74 nr
225 x 150mm	1.48 nr
225 x 225mm	2.23 nr

Arch bars

30 x 6mm flat	5.70 m
50 x 6mm flat	9.54 m
50 x 50 x 6mm angle	14.79 m
75 x 50 x 6mm angle	18.69 m

Yorkshire Limestone

Random stones 500mm thick	63.00 m²
Squared rubble face stones, irregular coursed, 100 - 150mm on bed scappled or axed face	53.00 m²
Random rubble backing and squared rubble face stones, irregular coursed; 350mm thick; faced one side	81.00 m²

Brickwork and Blockwork (Cont'd)

Portland stone, plain and rubbed one side

50mm thick dressing	175.00 m²
75mm thick dressing	238.00 m²
100mm thick dressing	315.00 m²

Simulated Dorset limestone

100mm thick, plain and rubbed one side	95.55 m²
200 x 100mm lintels	36.23 m
200 x 75mm sills	34.65 m
200 x 75mm sills extra; stoolings	8.66 nr
175 x 75mm jamb stones	37.54 m
225 x 125mm band courses	37.54 m

Roofing

Profile 3 sheeting fibre cement

Standard grey colour coverings	12.32 m²
eaves filler pieces; Polyethylene	1.68 m
two piece plain angular adjustable ridge tiles	23.32 m
standard barge boards	13.07 m
apron flashings	23.06 m
standard ridge cap finials	34.60 nr

Profile 6 sheeting

Standard coloured coverings	10.97 m²
eaves filler pieces	2.07 m
two piece plain angular adjustable ridge tiles	34.98 m
standard barge boards	11.10 m
apron flashings	19.89 m
standard ridge cap finials	44.97 nr

Treated softwood roofing battens

25 x 19mm	0.24 m
32 x 19mm	0.24 m
32 x 25mm	0.24 m
38 x 19mm	0.24 m
50 x 19mm	0.36 m
50 x 25mm	0.36 m

Roofing underfelts, reinforced bitumen

Underlay; twin-ply; 145gsm with polypropylene	0.59 m²

Tiles

Plain clay tiles, 268 x 165mm

Machine made	
plain tiles	340.00 1000
eaves tiles	0.35 nr
tile and a half	0.70 nr
ridge tiles, half round; 300mm long	4.41 nr
hip tiles; arris	6.04 nr
hip tiles; bonnet	4.41 nr
Hip irons	0.80 nr

Hand made	
clay tiles	850.00 1000
eaves tiles	0.85 nr
clay tile and a half	1.70 nr
ridge tiles, half round; 300mm long	4.95 nr
hip tiles; bonnet pattern	5.60 nr
valley tiles; angular	5.60 nr
vertical angle tile	5.70 nr

Redland Plain tiles, 268 x 165mm

Concrete granular faced	
tiles	340.00 1000
tile and a half	68.00 100

Heathland	
standard	390.00 1000
tile and a half	106.00 100

Concrete granular faced half round ridge or hip tile 450mm long	2.94 nr

Slates

Fibre cement slates

600 x 300mm	690.00 1000
500 x 250mm	680.00 1000
400 x 200mm	710.00 1000
Ridges or hips	13.39 m

Welsh Blue/grey slates, 6.5mm thick

600 x 300mm	8,400.00 1000
500 x 250mm	3,700.00 1000
400 x 200mm	1,650.00 1000

Redland Cambrian interlocking riven textured slates

300 x 336mm	2,400.00 1000
slate and a half slates to match 300 x 336mm	4,800.00 1000
150 x 6mm fibre cement undercloak	1.44 m
Stainless steel eaves clip to each slate	0.13 nr
Redland third round ridge or hip	2.94 nr
Redland half round ridge or hip	2.94 nr
Redland universal angle ridge or hip	6.53 m
Redland Dry Ridge system with half round ridge tiles	19.44 m
Redland Dry Ridge system with universal angle ridge tiles	19.81 m
Monoridge filler units	43.03 m
Universal valley troughs 2.40m long	14.30 nr

Timloc uPVC ventilators

Mark 3 eaves ventilators; reference 1124	5.35 m
Soffit ventilators; reference 1137	1.70 m
Polypropylene over-fascia ventilators; reference 3011 10mm airflow	3.93 m

Tiles

Redland single lap tiles

430 x 380mm	4,333.33 1000
418 x 332mm	830.00 1000
418 x 332mm	830.00 1000
381 x 227mm	900.00 1000
381 x 227mm	590.00 1000
418 x 330mm	830.00 1000
418 x 330mm (GRANULAR)	850.00 1000

Timber shingling

Western Red Cedar Shingles; best grade; bundle (2.32m²)	54.13 nr

Felt or similar

Roofing Felt

Perforated glass fibre underlay 2mm	1.79 m²
Sanded underlay 4mm	2.19 m²
Mineral surface top sheet 4mm	4.07 m²
Sand surfaced glassfibre underlay (formerly Type 3B) -1.8 kg/m²	1.23 m²
Top sheet green mineral surface - 3.8 kg/m²	2.35 m²
Hot bitumen bonding compound	1.05 kg

Insulating boards for felt roofing

25mm Kingspan Thermaroof TR21 (PIR) Board	6.16 m²
50mm Kingspan Thermaroof TR21 (PIR) Board	9.23 m²

Lead sheet

Code 3	26.31 m²
Code 4	35.75 m²
Code 5	44.61 m²

Aluminium roofing

0.6 mm thick sheets	15.81 m²
150 mm girth flashings	2.06 m
180 mm girth flashings	2.48 m
240 mm girth flashings	3.30 m
300 mm girth flashings	4.13 m
450 mm girth flashings	6.19 m
180 x 180 mm soakers	0.45 nr
180 x 300 mm soakers	1.24 nr
360mm girth gutter linings	4.95 m
420mm girth gutter linings	5.78 m
450mm girth gutter linings	6.19 m

Copper roofing

0.55 mm thick sheets	78.26 m²
0.70 mm thick sheets	86.25 m²

Zinc roofing - 0.80mm thick

0.80 mm thick sheets	37.88 m²
150 mm girth flashings	5.68 m
180 mm girth flashings	6.82 m
240 mm girth flashings	9.09 m
300 mm girth flashings	11.36 m
450 mm girth flashings	17.05 m
180 x 180 mm soakers	1.23 nr
180 x 300 mm soakers	2.05 nr
450 x 450 mm soakers	7.67 nr

Woodwork

Sawn Softwood

General Structural Grade SW SC3

32mm	
by 125mm	1.43 m
by 150mm	1.72 m
by 175mm	2.00 m
by 200mm	2.29 m
by 225mm	2.57 m
by 250mm	2.86 m
by 300mm	3.43 m

38mm	
by 125mm	1.37 m
by 150mm	1.56 m
by 175mm	1.82 m
by 200mm	2.09 m
by 225mm	2.35 m
by 250mm	3.40 m
by 300mm	4.08 m

44mm	
by 125mm	1.53 m
by 150mm	1.84 m
by 175mm	2.07 m
by 200mm	2.36 m
by 225mm	2.66 m
by 250mm	3.88 m
by 300mm	4.44 m

50mm	
by 50mm	0.64 m
by 75mm	0.93 m
by 100mm	1.22 m
by 125mm	1.53 m
by 150mm	1.84 m
by 175mm	2.07 m
by 200mm	2.36 m
by 225mm	2.66 m
by 250mm	3.88 m
by 300mm	4.44 m

63mm	
by 63mm	1.54 m
by 75mm	1.54 m
by 100mm	2.02 m
by 125mm	2.82 m
by 150mm	2.93 m
by 175mm	3.30 m
by 200mm	3.77 m
by 225mm	4.24 m
by 250mm	6.19 m
by 300mm	7.09 m

75mm	
by 75mm	1.54 m
by 100mm	2.02 m
by 125mm	2.93 m
by 150mm	2.93 m
by 175mm	3.30 m
by 200mm	3.77 m
by 225mm	4.24 m
by 250mm	6.19 m
by 300mm	7.09 m

100mm	
by 100mm	3.30 m

150mm	
by 150mm	7.88 m
by 250mm	13.34 m
by 300mm	14.18 m

200mm	
by 200mm	14.00 m

Woodwork (Cont'd)

Sawn Softwood (Cont'd)

General Structural Grade SW SC3 (Cont'd)

250mm
by 250mm	22.35	m

300mm
by 300mm	32.18	m

General Structural Grade SW SC3 Pressure impregnated

32mm
by 125mm	1.57	m
by 150mm	1.89	m
by 175mm	2.21	m
by 200mm	2.52	m
by 225mm	2.84	m
by 250mm	3.15	m
by 300mm	3.78	m

38mm
by 125mm	1.87	m
by 150mm	2.24	m
by 175mm	2.62	m
by 200mm	2.99	m
by 225mm	3.37	m
by 250mm	3.74	m
by 300mm	4.49	m

44mm
by 125mm	2.17	m
by 150mm	2.60	m
by 175mm	3.04	m
by 200mm	3.47	m
by 225mm	3.89	m
by 250mm	4.33	m
by 300mm	5.19	m

50mm
by 50mm	0.83	m
by 63mm	1.24	m
by 75mm	1.17	m
by 100mm	1.42	m
by 125mm	1.63	m
by 150mm	2.00	m
by 175mm	2.25	m
by 200mm	0.52	m
by 225mm	2.83	m
by 250mm	3.16	m
by 300mm	5.91	m

63mm
by 63mm	1.56	m
by 75mm	1.86	m
by 100mm	2.49	m
by 125mm	3.11	m
by 150mm	3.72	m
by 175mm	4.33	m
by 200mm	4.96	m
by 225mm	5.58	m
by 250mm	6.19	m
by 300mm	7.44	m

75mm
by 75mm	1.54	m
by 100mm	1.75	m
by 125mm	2.93	m
by 150mm	2.93	m
by 175mm	3.30	m
by 200mm	3.77	m
by 225mm	4.24	m
by 250mm	7.38	m
by 300mm	8.86	m

100mm
by 100mm	3.30	m

150mm
by 150mm	7.53	m
by 225mm	13.28	m
by 250mm	14.75	m
by 300mm	15.59	m

200mm
by 200mm	13.20	m

250mm
by 250mm	24.59	m

300mm
by 300mm	35.40	m

Softwood boarding, wrot, impregnated
19mm thick	8.26	m²
25mm thick	9.51	m²

Plywood BB/CC Far Eastern H/W
18mm thick	8.42	m²
25mm thick	12.07	m²

Wrought softwood SW PAR

6mm
by 38mm	0.20	m

12mm
by 38mm	0.22	m

16mm
by 25mm	0.28	m

19mm
by 38mm	0.30	m

25mm
by 50mm	0.48	m

32mm
by 125mm	2.00	m
by 150mm	2.30	m

38mm
by 50mm	0.73	m

50mm
by 50mm	0.96	m
by 75mm	1.44	m
by 100mm	1.85	m
by 150mm	2.77	m
by 200mm	3.74	m
by 300mm	5.70	m

75mm
by 75mm	2.16	m
by 100mm	2.77	m
by 150mm	4.15	m

100mm
by 100mm	4.55	m
by 150mm	6.60	m

150mm
by 150mm	8.80	m

Pressure Impregnated Softwood Boarding
25 x 200mm	2.08	m
25 x 225mm	2.36	m

Boarded flooring

Softwood, wrought
19 x 75mm, square edge	7.47	m²
25 x 125mm, square edge	9.17	m²
32 x 150mm, square edge	8.95	m²
19 x 75mm, tongued and grooved joint	7.60	m²
25 x 125mm, tongued and grooved joint	8.56	m²

Wall and ceiling boarding

Western Red Cedar, wrought, tongued, grooved and veed joints
19mm thick, 100mm wide	36.00	m²
25mm thick, 100mm wide	40.70	m²

Knotty Pine, wrought, selected for transparent finish
19mm thick, 100mm wide	30.56	m²
25mm thick, 100mm wide	42.00	m²

Sapele, wrought, selected for transparent finish, tongued, grooved and veed joints
19mm thick, 100mm wide	80.35	m²
25mm thick, 100mm wide	119.00	m²

Timber for First and Second Fixings
Ash	1,133.26	m³
Iroko	1,370.93	m³
Idigbo	897.71	m³
American Walnut	3,038.50	m³
Beech	921.02	m³
American Oak	1,647.09	m³
Sapele	1,223.66	m³
British Columbain Pine	799.88	m³
Teak	10,311.60	m³

Blockboard

Birch faced
16mm	12.06	m²
18mm	12.06	m²
25mm	25.00	m²

Oak faced
18mm	23.32	m²

Sapele faced
18mm	29.00	m²

Beech faced
18mm	32.00	m²

Plywood

Douglas Fir T&G, unsanded select sheathing quality, WBP bonded
15mm thick	3.34	m²
18mm thick	4.96	m²

Birch faced T&G, BB quality, WBP bonded
12mm thick	6.53	m²
15mm thick	7.42	m²
18mm thick	8.52	m²

Wood particle board (chipboard) BS EN 312

P4 load bearing boards, square edge
18mm thick	4.86	m²
22mm thick	6.93	m²

P5 load bearing moisture resistant boards, square edge
18mm thick	5.90	m²
22mm thick	7.60	m²

P4 T&G load bearing boards
18mm thick P4 T&G	4.97	m²
22mm thick P4 T&G	6.25	m²
18mm thick, moisture resisting, Furniture Grade	7.20	m²
22mm thick, moisture resisting, Furniture Grade	8.15	m²

P5 T&G load bearing moisture resistant boards
18mm thick P5 T&G	4.08	m²
22mm thick P5 T&G	5.90	m²

Hardboard, BS EN 622

Standard, type SHA; butt joints
3.2mm thick linings	1.61	m²
4.8mm thick linings	3.55	m²
6.4mm thick linings	4.53	m²

Tempered, Type THE; butt joints
3.2mm thick linings	4.01	m²
4.8mm thick linings	4.32	m²
6.4mm thick linings	4.54	m²

Medium density fibreboard (MDF) BS EN 622

Type MDF; linings to walls, butt joints
9mm thick	3.86	m²
12mm thick	4.45	m²
18mm thick	4.78	m²
25mm thick	8.04	m²

Linings; butt joints; Pine veneered
6mm thick	10.25	m²
19mm thick	13.17	m²

Linings; butt joints; Oak veneered
6mm thick	9.90	m²
13mm thick	11.58	m²
19mm thick	14.26	m²

Linings; Maple veneered
19mm thick	18.07	m²

Insulating board, BS EN 622

Softboard, type SBN butt joints
13mm thick thick linings	4.06	m²

Fire resisting boards

Masterboard Class O; butt joints
6mm thick linings	12.72	m²
9mm thick linings	25.01	m²
12mm thick linings	32.89	m²

Supalux, Sanded; butt joints
6mm thick linings	21.33	m²
9mm thick linings	28.34	m²
12mm thick linings	37.95	m²
15mm thick linings	52.35	m²

Vermiculux; butt joints
20mm thick casings	27.47	m²
30mm thick casings	43.76	m²
40mm thick casings	67.99	m²
50mm thick casings	90.51	m²
60mm thick casings	107.92	m²

Vicuclad 900; butt joints
18mm thick casings	26.03	m²
25mm thick casings	28.16	m²
30mm thick casings	32.89	m²

PVC Fascia and bargeboards

Fascia; 17mm thick mahogany woodgrain square edge with groove and return leg
175mm wide	6.41	m
225mm wide	7.74	m
400mm wide	13.20	m

Woodwork (Cont'd)

PVC Fascia and bargeboards (Cont'd)

Fascia; 22mm thick white bullnose with soffit groove

150mm wide	7.00 m
200mm wide	9.58 m
250mm wide	11.82 m

Multi purpose pre vented 9mm thick soffit board

150mm wide	2.57 m
225mm wide	3.76 m
300mm wide	5.00 m

Multi purpose 9mm thick soffit board

100mm wide	1.25 m
175mm wide	2.01 m
300mm wide	3.69 m

Skirtings, picture rails, architraves and the like

Softwood wrought, (finished sizes)

19 x 100mm	1.32 m
19 x 150mm	2.91 m
25 x 100mm	1.65 m
25 x 150mm	2.79 m
splays -1	
25 x 50mm	1.15 m
25 x 63mm	1.15 m
25 x 75mm	1.41 m
32 x 100mm	3.50 m

Skirtings, picture rails, architraves and the like

Oak, (finished sizes)

19 x 95mm	4.58 m
19 x 145mm	6.93 m
25 x 100mm	6.22 m
25 x 150mm	9.35 m

Skirtings, picture rails, architraves and the like

MDF, (finished sizes)
chamfered

18 x 44mm	0.89 m
chamfered	
18 x 94mm	1.85 m
moulded	
18 x 44mm	0.89 m
moulded	
18 x 144mm	2.64 m

Cover fillets, stops, trims, beads, nosings and the like

MDF, (finished sizes)
Dado rail

18 x 58mm	1.72 m
Dado rail	
20 x 69mm	2.40 m

Cappings

Softwood, wrought
level; rebates -1; mouldings -1

25 x 50mm	1.98 m
50 x 75mm	3.44 m
50 x 100mm	6.40 m

Window boards

Softwood wrought
mouldings -1

25 x 150mm	2.82 m
32 x 150mm	5.31 m
rounded edges -1	
25 x 150mm	2.82 m
32 x 150mm	5.31 m

MDF
nosed and tongued

25 x 219mm	4.60 m
25 x 244mm	4.95 m

Windows

Magnet Trade; softwood Statesman Professional; double glazed; trickle vents; 158mm cill: casement windows without bars; hinges; fasteners; fully finished in white

630 x 900mm overall	
109V	315.07 nr
915 x 900mm overall	
2N09W	388.75 nr
915 x 1050mm overall	
2N10W	401.62 nr
915 x 1200mm overall	
2N12W	414.48 nr
1200 x 900mm overall	
209W	446.26 nr
1200 x 1050mm overall	
210CV	540.27 nr
1200 x 1200mm overall	
212W	485.76 nr
1200 x 1500mm overall; toughened glass	
215W	525.99 nr
1769 x 1200mm overall	
312CVC	787.81 nr

Jeld-Wen Ltd. Softwood Stormsure windows plain casement non bar type side hung; unglazed; glazing beads; weatherstripping, hinges; fasteners; base coat stain by manufacturer

630 x 900mm overall	
W109C	173.45 nr
1200 x 750mm overall	
W207C	227.39 nr
1200 x 900mm overall	
W209C	236.33 nr
1200 x 1050mm overall	
W210C	245.68 nr
1200 x 1200mm overall	
W212C	255.76 nr
1200 x 1350mm overall	
W213C	272.00 nr
1770 x 750mm overall	
W307CC	340.90 nr
1770 x 900mm overall	
W309CC	355.13 nr
1770 x 1050mm overall	
W310CC	370.19 nr
1770 x 1200mm overall	
W312CC	386.74 nr
2339 x 900mm overall	
W409CMC	416.21 nr
2339 x 1050mm overall	
W410CMC	208.62 nr
2339 x 1200mm overall	
W412CMC	452.52 nr
2339 x 1350mm overall	
W413CMC	483.71 nr

Purpose made windows in softwood, wrought
38mm moulded casements or fanlights divided into panes 0.10 - 0.50m² ... 163.72 m²
50mm moulded casements or fanlights divided into panes 0.10 - 0.50m² ... 180.71 m²

Purpose made windows in Afrormosia, wrought
38mm moulded casements or fanlights divided into panes 0.10 - 0.50m² ... 265.90 m²
50mm moulded casements or fanlights divided into panes 0.10 - 0.50m² ... 279.99 m²

Velux roof windows GGL/GPL 3066 Pine Range, factory glazed clear 'Extra low energy 66' triple glazed sealed unit; laminated inner; enhanced strength outer pane

550 x 980mm overall	
CK04 GGL Centre Pivot	283.11 nr
660 x 1180mm overall	
FK06 GGL Centre Pivot	373.52 nr
940 x 1600mm overall	
PK10 GGL Centre Pivot	398.72 nr

Windows uPVC white; overall, tilt/turn, factory glazed 28mm double glazed units; hinges; fastenings; 150 cill

600 x 900mm	132.77 nr
600 x 1200mm	142.00 nr
750 x 650mm	120.63 nr
750 x 900mm	137.95 nr
750 x 1200mm	148.34 nr
900 x 900mm	152.37 nr
900 x 1200mm	163.92 nr
900 x 1500mm toughened glass	246.30 nr

Doors

Doors in softwood, wrought, Magnet Trade
Internal panelled doors
SA; glazed with bevelled glass; 762 x 1981 x 34mm (f sizes) ... 116.61 nr

Doors in softwood, wrought, preservative treated, Magnet Trade
Ext. ledged and braced doors
L & B 838 x 1981 x 44mm (f sizes) ... 51.20 nr
Ext. framed, ledged and braced doors
YX; 838 x 1981 x 44mm (f sizes) ... 68.12 nr
Ext. panelled doors
2XG CDS; 838 x 1981 x 44mm (f sizes) ... 89.13 nr
2XGG CDS; 838 x 1981 x 44mm (f sizes) ... 84.65 nr
Garage doors
MFL, pair, side hung; 2134 x 2134 x 44mm (f sizes) overall ... 250.11 nr
301, pair, side hung; 2134 x 2134 x 44mm (f sizes) overall ... 340.22 nr

Doors in hardboard, embossed, Magnet Trade
Int. panelled doors
6 Panel Smooth; 610 x 1981 x 34mm (f sizes) ... 23.61 nr
6 Panel Smooth; 838 x 1981 x 34mm (f sizes) ... 23.61 nr

Flush doors, Magnet Trade
Int. Magnaseal; embossed; unlipped

610 x 1981 x 34mm	22.16 nr
762 x 1981 x 34mm	22.16 nr
838 x 1981 x 34mm	22.16 nr
Int. veneered paintgrade facings	
381 x 1981 x 34mm	21.96 nr
610 x 1981 x 34mm	21.96 nr
762 x 1981 x 34mm	21.96 nr
838 x 1981 x 34mm	25.35 nr
Int. Sapele	
610 x 1981 x 34mm	43.55 nr
838 x 1981 x 34mm	43.55 nr

Ext. veneered paintgrade facings; FD 30
838 x 1981 x 44mm ... 62.15 nr
Int. fire resisting
6 Panel Smooth; FD30; 838 x 1981 x 44mm ... 67.28 nr
veneered paintgrade facings; FD30; 838 x 1981 x 44mm ... 61.60 nr
Ext. fire resisting veneered paintgrade facings; FD30; 838 x 1981 x 44mm ... 62.15 nr
Int. fire resisting Sapele facings; FD30; 838 x 1981 x 44mm ... 71.50 nr

Patio doors in uPVC
1500 x 2020mm overall, including cill and vent ... 520.01 nr

Staircases

Handrails
in softwood, wrought

50mm mopstick	27.74 m
75 x 100mm	46.23 m

Furniture and fixings

Wall rack Welco; oak
12 hooks; 1800 long reference 787-021 ... 54.85 nr
Hat and coat hooks
Welco; steel; coloured finish reference 456-122 ... 3.78 nr

Kitchen units, standard melamine finish with backs; self assembly
wall units

500 x 300 x 600mm	60.00 nr
500 x 300 x 900mm	102.00 nr
1000 x 300 x 600mm	67.00 nr
1000 x 300 x 900mm	172.00 nr

Furniture and fixings (Cont'd)

Kitchen units, standard melamine finish with backs; self assembly (Cont'd)

floor units without drawers
500 x 600 x 900mm	156.00 nr
1000 x 600 x 900mm	279.00 nr

floor units with one drawer
500 x 600 x 900mm	64.00 nr
1000 x 600 x 900mm	68.00 nr

floor units with four drawers
600 x 600 x 900mm	287.00 nr

sink units with one drawer
1200 x 600 x 900mm	257.00 nr

store cupboards with shelves
500 x 600 x 1950mm	242.00 nr
Nails, Oval brads, 50mm	1.23 kg
Nails, Lost Head Oval brads 50mm	1.37 kg
Nails, Round wire, plain head, 50mm	1.18 kg
Nails, cut clasp, 50mm	2.70 kg
Nails, galvanized clout, 50mm	1.56 kg
25mm masonry nails	1.15 100
40mm masonry nails	1.65 100
50mm masonry nails	1.99 100
60mm masonry nails	2.42 100
75mm masonry nails	3.38 100
80mm masonry nails	3.91 100
100mm masonry nails	5.03 100
Screws 8g 3/4" pozidrive, zinc plated, countersunk	0.57 100
Screws 8g 1" pozidrive, zinc plated, countersunk	0.79 100
Screws 8g 1 1/4" pozidrive, zinc plated, countersunk	0.91 100
Screws 8g 1 1/2" pozidrive, zinc plated, countersunk	1.03 100
Screws 10g 2" pozidrive, zinc plated, countersunk	1.80 100
Screws 10g 2 1/2" pozidrive, zinc plated, countersunk	2.36 100
Screws 10g 3" pozidrive, zinc plated, countersunk	2.82 100
5 mm plastic plug	0.32 100
6 mm plastic plug	0.42 100
7 mm plastic plug	0.65 100
8 x 60mm plastic plug	3.75 100
10 x 80mm plastic plug	8.26 100

Hardware

butt hinges
38mm; steel; light medium pattern	7.08 100
63mm; steel; light medium pattern	0.10 nr
75mm; steel; light medium pattern	0.12 nr
100mm; steel; light medium pattern	0.28 nr
38mm; brass, brass pin	0.57 nr
63mm; brass, brass pin	0.85 nr
75mm; brass, brass pin	1.19 nr
100mm; brass, brass pin	3.00 nr

cupboard locks
4 lever; brass; 63mm	19.50 nr

mortice locks
three levers	3.85 nr
five levers	13.80 nr

barrel bolts
150mm; japanned, steel barrel; medium	0.62 nr
300mm; japanned, steel barrel; heavy	8.14 nr

Insulation quilts
Glass fibre 100mm thick	1.63 m²
Glass fibre 200mm thick	3.25 m²
Sheeps wool insulation 200mm thick	16.22 m²

Structural Steelwork

Universal beams (S355JR)
914 x 419 x 343	750.00 tonne
914 x 419 x 388	750.00 tonne
914 x 305 x 201	750.00 tonne
914 x 305 x 224	750.00 tonne
914 x 305 x 253	750.00 tonne
914 x 305 x 289	750.00 tonne
838 x 292 x 176	750.00 tonne
838 x 292 x 194	750.00 tonne
838 x 292 x 226	750.00 tonne
762 x 267 x 134	750.00 tonne
762 x 267 x 147	750.00 tonne
762 x 267 x 173	750.00 tonne
762 x 267 x 197	750.00 tonne
686 x 254 x 125	750.00 tonne
686 x 254 x 140	750.00 tonne
686 x 254 x 152	750.00 tonne
686 x 254 x 170	750.00 tonne
610 x 305 x 238	750.00 tonne
610 x 305 x 179	750.00 tonne
610 x 305 x 149	750.00 tonne
610 x 229 x 140	750.00 tonne
610 x 229 x 125	750.00 tonne
610 x 229 x 113	750.00 tonne
610 x 229 x 101	750.00 tonne
533 x 210 x 122	750.00 tonne
533 x 210 x 109	750.00 tonne
533 x 210 x 101	750.00 tonne
533 x 210 x 92	750.00 tonne
533 x 210 x 82	750.00 tonne
457 x 191 x 98	750.00 tonne
457 x 191 x 89	750.00 tonne
457 x 191 x 82	750.00 tonne
457 x 191 x 74	750.00 tonne
457 x 191 x 67	750.00 tonne
457 x 152 x 82	750.00 tonne
457 x 152 x 74	750.00 tonne
457 x 152 x 67	750.00 tonne
457 x 152 x 60	750.00 tonne
457 x 152 x 52	750.00 tonne
406 x 178 x 74	750.00 tonne
406 x 178 x 67	750.00 tonne
406 x 178 x 60	750.00 tonne
406 x 178 x 54	750.00 tonne
406 x 140 x 46	750.00 tonne
406 x 140 x 39	750.00 tonne
356 x 171 x 67	750.00 tonne
356 x 171 x 57	750.00 tonne
356 x 171 x 51	750.00 tonne
356 x 171 x 45	750.00 tonne
356 x 127 x 39	750.00 tonne
356 x 127 x 33	750.00 tonne
305 x 165 x 54	750.00 tonne
305 x 165 x 46	750.00 tonne
305 x 165 x 40	750.00 tonne
305 x 127 x 48	750.00 tonne
305 x 127 x 42	750.00 tonne
305 x 127 x 37	750.00 tonne
305 x 102 x 33	750.00 tonne
305 x 102 x 28	750.00 tonne
305 x 102 x 25	750.00 tonne
254 x 102 x 28	750.00 tonne
254 x 102 x 25	750.00 tonne
254 x 102 x 22	750.00 tonne

Universal Columns (S355JR)
356 x 406 x 634	750.00 tonne
356 x 406 x 551	750.00 tonne
356 x 406 x 467	750.00 tonne
356 x 406 x 393	750.00 tonne
356 x 406 x 340	750.00 tonne
356 x 406 x 287	750.00 tonne
356 x 406 x 235	750.00 tonne
356 x 368 x 202	750.00 tonne
356 x 368 x 177	750.00 tonne
356 x 368 x 153	750.00 tonne
356 x 368 x 129	750.00 tonne
305 x 305 x 283	750.00 tonne
305 x 305 x 240	750.00 tonne
305 x 305 x 198	750.00 tonne
305 x 305 x 158	750.00 tonne
305 x 305 x 137	750.00 tonne
305 x 305 x 118	735.00 tonne
305 x 305 x 97	735.00 tonne
254 x 254 x 167	735.00 tonne
254 x 254 x 132	735.00 tonne
254 x 254 x 107	735.00 tonne
254 x 254 x 89	735.00 tonne
254 x 254 x 73	735.00 tonne
203 x 203 x 86	735.00 tonne
203 x 203 x 71	735.00 tonne
203 x 203 x 60	735.00 tonne
203 x 203 x 52	735.00 tonne
203 x 203 x 46	735.00 tonne
152 x 152 x 37	735.00 tonne
152 x 152 x 30	735.00 tonne
152 x 152 x 23	735.00 tonne

Quantity extras
10 tonnes and over	0.00 tonne
Under 10 tonnes to 5 tonnes	0.00 tonne
Under 5 tonnes to 2 tonnes	10.00 tonne
Under 2 tonnes to 1 tonne	50.00 tonne

Parallel flange channels
430 x 100 x 64	990.00 tonne
380 x 100 x 54	990.00 tonne
300 x 100 x 46	900.00 tonne
300 x 90 x 41	850.00 tonne
260 x 90 x 35	850.00 tonne
260 x 75 x 28	850.00 tonne
230 x 90 x 32	850.00 tonne
230 x 75 x 26	850.00 tonne
200 x 90 x 30	800.00 tonne
180 x 90 x 26	800.00 tonne
150 x 90 x 24	800.00 tonne
Extra; Parallel flange channels, non-standard sizes	95.00 tonne

Equal and unequal angles
200 x 150 x 18	910.00 tonne
200 x 150 x 15	910.00 tonne
200 x 150 x 12	910.00 tonne
200 x 100 x 15	850.00 tonne
200 x 100 x 12	850.00 tonne
200 x 100 x 10	850.00 tonne
200 x 200 x 24	850.00 tonne
200 x 200 x 20	850.00 tonne
200 x 200 x 18	850.00 tonne
200 x 200 x 16	850.00 tonne
150 x 90 x 15	850.00 tonne
150 x 90 x 12	850.00 tonne
150 x 90 x 10	850.00 tonne
150 x 150 x 18	850.00 tonne
150 x 150 x 15	850.00 tonne
150 x 150 x 12	850.00 tonne
150 x 150 x 10	850.00 tonne
150 x 75 x 15	850.00 tonne
150 x 75 x 12	850.00 tonne
150 x 75 x 10	850.00 tonne
120 x 120 x 15	850.00 tonne
120 x 120 x 12	810.00 tonne
120 x 120 x 10	855.00 tonne
120 x 120 x 8	810.00 tonne
125 x 75 x 12	810.00 tonne
125 x 75 x 10	810.00 tonne
125 x 75 x 8	810.00 tonne
100 x 75 x 12	810.00 tonne
100 x 75 x 10	810.00 tonne
100 x 75 x 8	810.00 tonne
100 x 65 x 10	810.00 tonne
100 x 65 x 8	810.00 tonne
100 x 65 x 7	810.00 tonne
100 x 100 x 15	810.00 tonne
100 x 100 x 12	810.00 tonne
100 x 100 x 10	810.00 tonne
100 x 100 x 8	810.00 tonne
90 x 90 x 12	810.00 tonne
90 x 90 x 10	810.00 tonne
90 x 90 x 8	810.00 tonne
90 x 90 x 7	810.00 tonne
90 x 90 x 6	810.00 tonne
Extra; angles non-standard sizes	100.00 tonne

Cold rolled zed purlins and cladding rails, Metsec purlin sleeved system
ref: 14214; 3.03 kg/m	27.01 m
ref: 14216; 3.47 kg/m	30.49 m
ref: 17214; 3.66 kg/m	31.94 m
ref: 17216; 4.11 kg/m	36.31 m
ref: 20216; 4.49 kg/m	39.63 m
ref: 20218; 5.03 kg/m	44.71 m
ref: 20220; 5.57 kg/m	49.29 m
ref: 23218; 5.73 kg/m	50.74 m
ref: 23220; 6.34 kg/m	56.27 m

Structural Steelwork (Cont'd)

Cold rolled zed purlins and cladding rails, Metsec galvanized purlin cleats

ref: 142 for 142mm deep purlins	9.43 nr
ref: 172 for 172mm deep purlins	11.14 nr
ref: 202 for 202mm deep purlins	11.72 nr
ref: 232 for 232mm deep purlins	12.91 nr
ref: 262 for 262mm deep purlins	14.12 nr

Cold rolled zed purlins and cladding rails, Metsec Round Lok anti-sag rods

1150mm centres	9.51 nr
1350mm centres	11.03 nr
1550mm centres	12.46 nr
1700mm centres	14.05 nr
1950mm centres	14.47 nr

Cold rolled zed purlins and cladding rails, Metsec side rail sleeved/single span system

ref: 14214; 3.03 kg/m	27.01 m
ref: 14216; 3.47 kg/m	30.49 m
ref: 17215; 3.85 kg/m	31.94 m
ref: 17216; 4.11 kg/m	36.31 m
ref: 20216; 4.49 kg/m	39.63 m
ref: 20218; 5.03 kg/m	44.71 m
ref: 20220; 5.57 kg/m	49.29 m
ref: 23218; 5.73 kg/m	50.74 m
ref: 23223; 7.26 kg/m	56.27 m

Cold rolled zed purlins and cladding rails, Metsec side rail cleats

ref: 142 for 142mm rails	9.43 nr
ref: 172 for 172mm rails	11.14 nr
ref: 202 for 202mm rails	11.72 nr
ref: 232 for 232mm rails	12.91 nr

Cold rolled zed purlins and cladding rails, Metsec galvanized side rail suports, 122-262 series

rail 1000mm long	27.76 nr
rail 1400mm long	30.28 nr
rail 1600mm long	33.59 nr
rail 1800mm long	40.45 nr

Cold rolled zed purlins and cladding rails, Metsec diagonal tie wire supports

1700mm long	54.15 nr
2200mm long	60.58 nr
2600mm long	67.02 nr
3600mm long	82.75 nr

Steel short span lattice joists primed at works

200mm deep, 8.2 kg/m	27.19 m
250mm deep, 8.5 kg/m	27.19 m
300mm deep, 10.7 kg/m	29.91 m
350mm deep, 11.6 kg/m	32.63 m
350mm deep, 12.8 kg/m	46.23 m

Steel intermediate span lattice joists primed at works

450mm deep, 12.2 kg/m	48.95 m
500mm deep, 15.8 kg/m	51.66 m
550mm deep, 19.4 kg/m	54.38 m
600mm deep, 22.5 kg/m	76.14 m
650mm deep, 29.7 kg/m	78.86 m

Steel long span lattice joists primed at works

700mm deep, 39.2 kg/m	100.61 m
800mm deep, 44.1 kg/m	103.33 m
900mm deep, 45.3 kg/m	130.52 m
1000mm deep, 46.1 kg/m	133.24 m
1500mm deep, 54.2 kg/m	141.40 m

Metalwork

Steel; galvanised - 27.5 x 2.5mm standard strapping

600mm long	0.81 nr
800mm long	1.22 nr
1000mm long	6.75 100
1200mm long	1.68 nr
1600mm long	2.86 nr

Truss clips

for 38mm thick members	0.30 nr
for 50mm thick members	0.30 nr

Joist hangers; BAT Building Products Ltd SPH type S

for 50 x 100mm joist	1.54 nr
for 50 x 150mm joist	1.26 nr
for 50 x 200mm joist	1.47 nr
for 50 x 250mm joist	2.52 nr
for 75 x 100mm joist	2.35 nr
for 75 x 150mm joist	2.75 nr
for 75 x 200mm joist	2.75 nr
for 75 x 250mm joist	3.13 nr

SPH type R

for 50 x 100mm joist	4.37 nr
for 50 x 150mm joist	4.35 nr
for 50 x 200mm joist	5.50 nr
for 50 x 250mm joist	7.15 nr
for 63 x 100mm joist	6.45 nr
for 63 x 150mm joist	6.90 nr
for 63 x 200mm joist	7.25 nr
for 63 x 250mm joist	8.25 nr

Steel; galvanised - truss plates and framing anchors

51 x 114mm	0.25 nr
76 x 254mm	0.65 nr
114 x 152mm	0.72 nr
114 x 254mm	1.30 nr
152 x 152mm	0.85 nr

Expanding bolts; bolt projecting Rawlbolts, ref

44505 (M6 10P)	0.71 nr
44510 (M6 25P)	0.82 nr
44605 (M10 15P)	1.10 nr
44610 (M10 30P)	1.15 nr
44615 (M10 60P)	1.12 nr
44660 (M12 30P)	1.87 nr
44710 (M16 35P)	4.29 nr
44760 (M20 30P)	7.62 nr

Chemical anchors; Kemfix capsules and standard studs

capsule reference 60-428; stud reference 60-708; with nuts and washers (M8 x 110mm)	0.92 nr
capsule reference 60-432; stud reference 60-712; with nuts and washers (M12 x 160mm)	1.38 nr

Galvanized steel lintels; SUPERGALV (BIRTLEY) lintels reference CB 50

125 x 750mm	16.74 nr
125 x 1200mm	25.95 nr
125 x 1350mm	30.88 nr
125 x 1500mm	34.06 nr
125 x 1650mm	38.53 nr
165 x 1800mm	42.02 nr
165 x 1950mm	45.76 nr
165 x 2100mm	48.26 nr
165 x 2250mm	55.40 nr
165 x 2400mm	58.50 nr
215 x 2550mm	66.65 nr
215 x 2850mm	90.94 nr
215 x 3000mm	95.68 nr
215 x 3300mm	111.51 nr
215 x 3600mm	126.40 nr
215 x 3900mm	153.09 nr

Galvanized steel lintels; SUPERGALV (BIRTLEY) lintels reference CB 50 H.D

165 x 750mm	20.78 nr
165 x 1050mm	26.50 nr
165 x 1200mm	30.18 nr
165 x 1350mm	36.70 nr
165 x 1500mm	40.45 nr
215 x 1650mm	48.40 nr
215 x 1800mm	52.84 nr
215 x 2100mm	62.33 nr
215 x 2250mm	78.37 nr
215 x 2400mm	91.95 nr
215 x 2550mm	99.90 nr
215 x 3000mm	117.97 nr

Galvanized steel lintels; SUPERGALV (BIRTLEY) lintels reference CB 70

120 x 750mm	16.22 nr
120 x 1200mm	25.77 nr
120 x 1500mm	33.63 nr
160 x 1800mm	41.36 nr
215 x 2400mm	59.12 nr
215 x 3000mm	96.44 nr
215 x 3900mm	149.67 nr

Galvanized steel lintels; SUPERGALV (BIRTLEY) lintels reference CB 70 H.D

160 x 750mm	19.78 nr
160 x 1200mm	29.65 nr
160 x 1500mm	39.73 nr
215 x 1800mm	55.51 nr
215 x 2400mm	90.33 nr
215 x 3000mm	115.90 nr

Galvanized steel lintels; SUPERGALV (BIRTLEY) lintels reference CB 50/130

120 x 750mm	17.12 nr
120 x 1200mm	27.21 nr
120 x 1500mm	34.48 nr
165 x 1800mm	43.21 nr
165 x 2400mm	62.98 nr
215 x 3900mm	187.38 nr

Galvanized steel lintels; SUPERGALV (BIRTLEY) lintels reference HS 50/130

160 x 750mm	30.71 nr
160 x 1200mm	49.48 nr
160 x 1500mm	61.40 nr
215 x 1800mm	70.89 nr
215 x 2400mm	107.30 nr
215 x 3300mm	179.52 nr

Galvanized steel lintels; SUPERGALV (BIRTLEY) lintels reference CB 70/130

115 x 750mm	21.40 nr
115 x 1200mm	33.02 nr
115 x 1500mm	41.70 nr
155 x 1800mm	57.36 nr
155 x 2400mm	95.30 nr
210 x 3300mm	173.02 nr

Galvanized steel lintels; SUPERGALV (BIRTLEY) lintels reference HS 50

225 x 3900mm	182.52 nr
225 x 4200mm	204.45 nr
225 x 4500mm	223.70 nr
225 x 4800mm	237.91 nr

Galvanized steel lintels; SUPERGALV (BIRTLEY) lintels reference HS 70

225 x 3000mm	132.17 nr
225 x 3300mm	144.46 nr
225 x 3600mm	154.09 nr

Galvanized steel lintels; SUPERGALV (BIRTLEY) lintels reference SB100

75 x 750mm	11.75 nr
75 x 1200mm	18.06 nr
140 x 1800mm	32.33 nr
140 x 2400mm	45.28 nr
215 x 3000mm	83.80 nr

Galvanized steel lintels; SUPERGALV (BIRTLEY) lintels reference SB140

140 x 750mm	14.09 nr
140 x 1200mm	22.71 nr
140 x 1800mm	38.03 nr
140 x 2400mm	54.25 nr
215 x 3000mm	91.86 nr

Galvanized steel lintels; SUPERGALV (BIRTLEY) lintels reference SBL200

142 x 750mm	19.22 nr
142 x 1200mm	31.17 nr
142 x 1800mm	49.27 nr
142 x 2400mm	69.47 nr
218 x 3000mm	113.50 nr

Galvanized steel lintels; SUPERGALV (BIRTLEY) internal door lintels reference INT100

100 x 900mm	4.06 nr
100 x 1050mm	4.66 nr
100 x 1200mm	5.16 nr

Mechanical Installations

Red Primer steel pipes, BS 1387 Table 2; steel fittings BS EN 10241

15mm

pipes	3.90 m
fittings; one end	0.73 nr
fittings; two ends	0.74 nr
fittings; three ends	1.02 nr

20mm

pipes	4.90 m
fittings; one end	0.83 nr
fittings; two ends	1.02 nr
fittings; three ends	1.48 nr

Mechanical Installations (Cont'd)

Red Primer steel pipes, BS 1387 Table 2; steel fittings BS EN 10241 (Cont'd)

25mm
pipes	6.23 m
fittings; one end	1.04 nr
fittings; two ends	1.57 nr
fittings; three ends	2.13 nr

32mm
pipes	7.89 m
fittings; one end	1.70 nr
fittings; two ends	2.70 nr
fittings; three ends	3.72 nr

40mm
pipes	10.44 m
fittings; one end	1.46 nr
fittings; two ends	2.72 nr
fittings; three ends	2.93 nr

50mm
pipes	12.74 m
fittings; one end	2.14 nr
fittings; two ends	4.35 nr
fittings; three ends	4.75 nr

Black steel pipe brackets
15mm malleable iron	0.93 nr
20mm malleable iron	1.00 nr
25mm malleable iron	1.34 nr
32mm malleable iron	1.82 nr
40mm malleable iron	2.24 nr
50mm malleable iron	2.90 nr

Galvanized steel pipes, BS 1387 Table 2; steel fittings BS EN 10241

15mm
pipes	4.51 m
pipe connectors	0.71 nr
fittings; one end	0.56 nr
fittings; two ends	0.78 nr
fittings; three ends	1.10 nr

20mm
pipes	5.07 m
pipe connectors	0.91 nr
fittings; one end	0.66 nr
fittings; two ends	1.01 nr
fittings; three ends	1.26 nr

25mm
pipes	7.66 m
pipe connectors	1.18 nr
fittings; one end	0.82 nr
fittings; two ends	1.45 nr
fittings; three ends	1.86 nr

32mm
pipes	8.58 m
pipe connectors	1.59 nr
fittings; one end	1.45 nr
fittings; two ends	2.40 nr
fittings; three ends	3.10 nr

40mm
pipes	9.77 m
pipe connectors	2.02 nr
fittings; one end	1.67 nr
fittings; two ends	3.75 nr
fittings; three ends	4.19 nr

50mm
pipes	14.42 m
pipe connectors	3.05 nr
fittings; one end	2.46 nr
fittings; two ends	4.71 nr
fittings; three ends	6.39 nr

Galvanized steel pipe brackets
15mm malleable iron	0.97 m
20mm malleable iron	1.03 nr
25mm malleable iron	1.35 nr
32mm malleable iron	1.56 nr
40mm malleable iron	2.29 nr
50mm malleable iron	2.61 m

Stainless steel pipes, BS 4127 Part 2
15mm pipes	11.10 m
22mm pipes	13.25 m
28mm pipes	15.20 m
35mm pipes	19.97 m
42mm pipes	21.42 m

Copper pipes, BS 2871 Part 1 Table X
15mm pipes	1.62 m
22mm pipes	3.23 m
28mm pipes	4.30 m
35mm pipes	9.20 m
42mm pipes	11.07 m
54mm pipes	14.53 m
67mm pipes	19.34 m
76mm pipes	31.62 m
108mm pipes	43.27 m

Copper pipes, BS EN 1057 2871 Part 1 Table Y
15mm pipes	3.50 m
22mm pipes	9.18 m
28mm pipes	8.20 m
42mm pipes	15.36 m

Copper coil pipes, EN 1057 R220 (Annealed) Table Y plastic coated
15mm pipes	4.67 m
22mm pipes	8.10 m

Capillary fittings for copper pipes

15mm
fittings; one end	0.13 nr
fittings; two ends	0.13 nr
fittings; three ends	0.37 nr

22mm
fittings; one end	0.31 nr
fittings; two ends	0.35 nr
fittings; three ends	0.98 nr

28mm
fittings; one end	0.67 nr
fittings; two ends	0.72 nr
fittings; three ends	2.53 nr

35mm
fittings; one end	1.50 nr
fittings; two ends	1.68 nr
fittings; three ends	4.31 nr

42mm
fittings; one end	1.93 nr
fittings; two ends	2.02 nr
fittings; three ends	6.67 nr

54mm
fittings; one end	3.33 nr
fittings; two ends	3.53 nr
fittings; three ends	10.62 nr

Compression fittings type A for copper pipes

15mm
fittings; one end	0.94 nr
fittings; two ends	1.09 nr
fittings; three ends	1.97 nr

22mm
fittings; one end	1.27 nr
fittings; two ends	2.32 nr
fittings; three ends	4.52 nr

28mm
fittings; one end	2.17 nr
fittings; two ends	3.08 nr
fittings; three ends	7.03 nr

35mm
fittings; one end	10.24 nr
fittings; two ends	11.94 nr
fittings; three ends	21.69 nr

42mm
fittings; one end	16.11 nr
fittings; two ends	15.22 nr
fittings; three ends	34.97 nr

54mm
fittings; one end	23.28 nr
fittings; two ends	22.59 nr
fittings; three ends	53.82 nr

Wavin Hepworth Hep20 polybutylene pipes
15mm pipes	1.14 m
22mm pipes	2.24 m
28mm pipes	3.50 m

Blue polythene pipes, BS EN 12201
20mm pipes	0.45 m
25mm pipes	0.59 m
32mm pipes	0.96 m
50mm pipes	2.31 m
63mm pipes	3.63 m

PVC-U pipes, BS 3505 Class E
3/8" pipes	0.70 m
1/2" pipes	0.75 m
3/4" pipes	1.06 m
1" pipes	1.37 m
1 1/4" pipes	2.14 m
1 1/2" pipes	2.79 m

PVC-U pipes self coloured white
21.5mm pipes	0.63 m

Ductile iron pipes
80mm pipes	29.88 m
100mm pipes	29.88 m
150mm pipes	36.14 m

Brass stopcocks

Crutch head; DZR; joints to polythene
20mm	8.00 nr

Crutch head; compression joints to copper (Type A)
15mm	2.69 nr
22mm	6.54 nr
28mm	13.84 nr
15mm	5.05 nr
22mm	8.07 nr
28mm	14.15 nr
35mm	40.20 nr
42mm	49.36 nr
54mm	89.18 nr

Polybutylene stopcocks

Fitted with Hep2O ends
15mm	9.23 nr
22mm	11.35 nr

Brass gate valves

BS 5154 series B; compression joints to copper (Type A)
15mm	2.98 nr
22mm	4.18 nr
28mm	6.70 nr
35mm	28.50 nr
42mm	36.04 nr
54mm	60.50 nr

DZR; fitted with Hep2O ends
15mm	9.43 nr
22mm	13.40 nr

Lever ball valves

Screwed and PTFE joints
15mm	3.97 nr
22mm	5.87 nr
28mm	9.12 nr
35mm	60.86 nr
42mm	89.14 nr
54mm	111.96 nr

Brass ball valves

For low pressure; BS 1212

13mm; part 1; copper
float	11.36 nr

13mm; part 2; copper
float	11.36 nr

13mm; part 1; plastics
float	5.20 nr

13mm; part 2; plastics
float	5.20 nr

For high pressure; BS 1212

13mm; part 1; copper
float	12.88 nr

13mm; part 2; copper
float	10.86 nr

19mm; part 1; copper
float	18.74 nr

25mm; part 1; copper
float	43.19 nr

13mm; part 1; plastics
float	6.72 nr

13mm; part 2; plastics
float	4.70 nr

19mm; part 1; plastics
float	10.06 nr

25mm; part 1; plastics
float	25.37 nr

GRP Sarena mfg Ltd water storage tanks with cover

Pre insulated
ref. SC10; 18 litres	113.00 nr
ref. SC20; 55 litres	167.00 nr
ref. SC30; 86 litres	196.00 nr
ref. SC40; 114 litres	230.00 nr
ref. SC50; 159 litres	263.00 nr
ref. SC60; 191 litres	285.00 nr
ref. SC70; 227 litres	321.00 nr
ref. SC100/1; 327 litres	444.00 nr
ref. SC100/2; 336 litres	488.00 nr
ref. SC125; 423 litres	543.00 nr

Mechanical Installations (Cont'd)

GRP Sarena mfg Ltd water storage tanks with cover (Cont'd)

Pre insulated (Cont'd)

ref. SC150; 491 litres......	639.00 nr
ref. SC200; 709 litres......	820.00 nr
ref. SC250; 900 litres......	935.00 nr
ref. SC350; 1250 litres.....	1,104.00 nr
ref. SC500; 1727 litres.....	1,350.00 nr
ref. SC600; 2137 litres.....	1,592.00 nr

Plastics water storage cisterns

With sealed lid / Byelaw 30 thick

18 litres...........................	22.98 nr
114 litres.........................	67.85 nr
190 litres.........................	107.25 nr
227 litres.........................	111.63 nr

Copper direct hot water cylinders

Four bosses; pre-insulated

ref. 3; 116 litres...............	160.00 nr
ref. 7; 120 litres...............	150.51 nr
ref. 8; 144 litres...............	166.37 nr
ref. 9; 166 litres...............	230.75 nr

Copper indirect hot water cylinders

Single feed; four bosses; pre-insulated

ref. 7; 108 litres...............	476.53 nr
ref. 8; 130 litres...............	503.83 nr

Cast iron rainwater pipes and fittings, BS 460 Type A sockets; dry joints

75mm

pipes..............................	24.18 m
shoes..............................	19.47 nr
bends.............................	18.04 nr
offset bends 150mm projection........................	22.81 nr
offset bends 230mm projection........................	26.29 nr
offset bends 305mm projection........................	26.29 nr
angled branches.............	31.47 nr

100mm

pipes..............................	34.42 m
shoes..............................	28.25 nr
bends.............................	24.27 nr
offset bends 75mm projection........................	42.87 nr
offset bends 150mm projection........................	42.87 nr
offset bends 230mm projection........................	51.61 nr
offset bends 305mm projection........................	51.61 nr
angled branches.............	37.42 nr

Cast iron rainwater pipes and fittings, BS 460 Type A sockets; ears cast on; dry joints

65mm

pipes..............................	30.13 m
shoes..............................	24.90 nr
bends.............................	16.09 nr

75mm

pipes..............................	30.13 m
shoes..............................	24.90 nr
bends.............................	19.06 nr

100mm

pipes..............................	40.25 m
shoes..............................	31.66 nr
bends.............................	25.90 nr

PVC-U rainwater pipes and fittings, BS EN 607; push fit joints; self coloured

68mm

pipes..............................	3.17 m
pipe brackets	1.49 nr
shoes..............................	2.21 nr
bends.............................	2.69 nr

Cast iron half round gutters and fittings, BS 460

100mm

gutters	13.20 m
gutter brackets................	4.05 nr
stopped ends..................	3.79 nr
running outlets................	11.31 nr
stopped ends with outlet..	15.10 nr
angles............................	11.41 nr

115mm

gutters	13.74 m

gutter brackets................	4.22 nr
stopped ends..................	4.98 nr
running outlets................	12.16 nr
stopped ends with outlet..	17.14 nr
angles............................	11.73 nr

125mm

gutters	16.07 m
gutter brackets................	5.10 nr
stopped ends..................	5.25 nr
running outlets................	14.50 nr
stopped ends with outlet..	19.75 nr
angles............................	13.81 nr

150mm

gutters	27.39 m
gutter brackets................	5.52 nr
stopped ends..................	11.62 nr
running outlets................	20.84 nr
stopped ends with outlet..	32.46 nr
angles............................	20.84 nr

Cast iron box gutters and fittings

100 x 75mm

gutters	39.76 m
stopped ends..................	5.62 nr
running outlets................	19.40 nr
gutter brackets................	5.73 nr

Cast aluminium half round gutters and fittings, BS EN 612 Section A

100mm

gutters	12.40 m
gutter brackets................	2.15 nr
stopped ends..................	3.79 nr
running outlets................	8.63 nr
angles............................	7.93 nr

125mm

gutters	16.11 m
gutter brackets................	4.17 nr
stopped ends..................	4.98 nr
running outlets................	10.43 nr
angles............................	10.15 nr

PVC-U half round gutters and fittings, BS 4576, Part 1; push fit connector joints; self coloured grey

114mm

gutters	2.45 nr
Union.............................	2.07 m
gutter brackets................	1.08 nr
stopped ends..................	2.13 nr
running outlets................	3.27 nr
angles............................	4.55 nr

Cast iron soil and waste pipes and fittings, BS 416

75mm

pipes..............................	83.25 m
pipe couplings	13.78 nr

100mm

pipes..............................	34.33 m
pipe couplings	24.43 nr

50mm

pipes..............................	75.70 m
pipe couplings	12.51 nr

MuPVC waste pipes and fittings, BS EN 1329-1 Section Three; solvent welded joints

32mm

pipes..............................	1.69 m
pipe couplings	0.93 nr
pipe brackets	0.30 nr
fittings; three ends	1.10 nr

40mm

pipes..............................	2.04 m
pipe couplings	1.00 nr
pipe brackets	0.32 nr
fittings; three ends	1.10 nr

50mm

pipes..............................	2.97 m
pipe couplings	1.89 nr
pipe brackets	0.61 nr
fittings; three ends	1.89 nr

Polypropylene waste pipes and fittings; butyl ring joints

32mm

pipes..............................	0.82 m
pipe couplings	0.97 nr
fittings; three ends	1.51 nr

40mm

pipes..............................	1.04 m

pipe couplings................	1.08 nr
pipe brackets..................	0.25 nr
fittings; three ends	1.55 nr

PVC-U soil and vent pipes and fittings, BS 4514; rubber ring joints; self coloured grey

82mm

pipes..............................	10.21 m
pipe couplings................	6.47 nr

110mm

pipes..............................	8.54 m
pipe couplings................	6.64 nr
short radius bends...........	14.10 nr
single angled branches....	20.12 nr
straight wc connectors.....	6.09 nr
Air admittance valve push fit.............................	50.15 nr

160mm

pipes..............................	16.27 m
pipe couplings................	10.68 nr

PVC-U soil and vent pipes, solvent joints; self coloured grey

110mm

pipes..............................	9.40 m
pipe couplings................	5.95 nr
short radius bends...........	13.98 nr
single angled branches....	18.48 nr
straight wc connectors.....	13.21 nr
Air admittance valve solvent joint	46.00 nr

Pipe supports for PVC-U soil and vent pipes

for building in

for 82mm pipes.............	4.80 nr
for 110mm pipes.............	5.98 nr

for fixing to timber

for 82mm pipes.............	3.84 nr
for 110mm pipes.............	5.66 nr
for 160mm pipes.............	9.07 nr

PVC Vent Terminal

50mm pipes.....................	1.78 nr
82mm pipes.....................	2.68 nr
110mm pipes...................	2.72 nr

Plastic traps

Two piece, 75mm seal, Inlet with coupling nut, outlet with seal ring socket

32mm outlet, P	2.85 nr
40mm outlet, P	3.15 nr
32mm outlet, S	3.31 nr
40mm outlet, S	3.70 nr
40mm outlet, P, with flexible polypropylene pipe for overflow connection	4.19 nr

Copper traps

Two piece, 75mm seal, inlet with coupling nut, compression outlet

35mm outlet, P	71.49 nr
42mm outlet, P	84.29 nr
54mm outlet, P	302.31 nr
35mm outlet, S	75.75 nr
42mm outlet, S	91.20 nr
54mm outlet, S	326.80 nr
42mm outlet, bath set, including overflow connection	117.68 nr

Pressed steel radiators single panel with convector; air cock, plain plug

450 x 600mm long............	23.22 nr
450 x 900mm long............	29.85 nr
450 x 1200mm long..........	35.85 nr
450 x 1600mm long..........	53.25 nr
450 x 2400mm long..........	79.88 nr
600 x 600mm long............	24.07 nr
600 x 900mm long............	35.14 nr
600 x 1200mm long..........	40.85 nr
600 x 1600mm long..........	58.64 nr
600 x 2400mm long..........	87.96 nr
700 x 600mm long............	31.67 nr
700 x 900mm long............	54.37 nr
700 x 1200mm long..........	97.50 nr
700 x 1600mm long..........	130.02 nr
700 x 2400mm long..........	195.03 nr

Bath Panel

Hardboard front bath panel................................	39.17 nr

Mechanical Installations (Cont'd)

Bath Panel (Cont'd)

Hardboard end bath panel .	24.02 nr

Electrical Installations

Lighting columns

Galvanised steel column; nominal mounting height

3.00m	165.53 nr
4.00m	120.75 nr
5.00m	131.25 nr
5.00m mid-hinged	298.75 nr
6.00m	157.50 nr
8.00m	311.03 nr
10.0m	423.88 nr

Optional extras

Single arm mounting bracket	35.75 nr
Double arm mounting bracket	45.25 nr
Photo cell	8.50 nr
Part night photo cell	31.75 nr
Blanking plug (no photo cell)	3.49 nr

Lanterns; Amenity/Decorative

40w LED; Regency or Modern	124.96 nr
60w LED; Regency or Modern	149.96 nr
70w son-t	135.00 nr
70w white	154.56 nr
36w PLL	155.53 nr
70w son-t post top small regency style	247.50 nr
70w son post top sphere	235.20 nr
70w son-t post top stylish amenity lantern	198.88 nr

Cobra Head LED Lanterns; 150 x 75 degrees; CRI > 80; CCT 3000K, 4500K or 6000K; 50,000hr; 5yr warranty

30w (3300 lumens)	158.00 nr
40w (4400 lumens)	158.00 nr
50w (5500 lumens)	164.00 nr
60w (6600 lumens)	164.00 nr
80w (8800 lumens)	200.00 nr
100w (11000 lumens)	200.00 nr
120w (13200 lumens)	233.00 nr
150w (16500 lumens)	233.00 nr
160w (17600 lumens)	264.00 nr
200w (22000 lumens)	264.00 nr
220w (24200 lumens)	307.00 nr
240w (26400 lumens)	307.00 nr
300w (33000 lumens)	417.00 nr

Trunking

Steel lighting trunking; PVC lid

50 x 50mm

straight body	8.15 3m
straight Lid	2.03 3m
end cap	0.88 nr
internal angle	6.38 nr
external angle	6.38 nr
tee	6.98 nr
flat bend	6.38 nr
trunking suspension (for 10mm studding)	1.60 nr

Steel dado trunking, two compartment; steel lid

150 x 40mm

straight body	18.00 2m
straight lid	17.58 2m
end cap	4.49 nr
internal angle body	7.98 nr
internal angle lid	8.79 nr
external angle body	7.98 nr
external angle lid	8.79 nr
single socket plate	8.50 nr
twin socket plate	8.50 nr

Steel dado trunking, three compartment; steel lid

200 x 40mm

straight body	21.66 2m
straight lid	27.50 2m
end cap	17.05 nr
internal angle body	8.55 nr
internal angle lid	18.00 nr

external angle body	8.55 nr
external angle lid	18.00 nr
single socket plate	13.78 nr
twin socket plate	13.78 nr

Steel skirting trunking; three compartments; steel lid

200 x 40mm

straight body	21.66 2m
straight lid	23.61 2m
internal angle body	8.55 nr
internal angle lid	16.10 nr
external angle body	8.55 nr
external angle lid	16.10 nr
single socket plate	9.98 nr
twin socket plate	9.98 nr

Steel underfloor trunking; two compartments

150 x 25mm

straight	18.95 2m
floor junction boxes, four way; adjustable frame and trap	53.72 nr
horizontal bend	27.50 nr

PVC trunking; white mini-trunking; clip on lid

16 x 16mm

straight	2.10 3m
end cap	0.48 nr
internal bend	0.45 nr
external bend	0.45 nr
flat bend	0.45 nr
equal tee	0.83 nr

25 x 16mm

straight	2.20 3m
end cap	0.47 nr
internal bend	0.50 nr
external bend	0.45 nr
flat bend	0.50 nr
equal tee	0.91 nr

40 x 16

straight	3.95 3m
end cap	0.50 nr
internal bend	0.57 nr
external bend	0.52 nr
flat bend	0.57 nr
equal tee	0.82 nr

PVC trunking; white dado trunking, three compartment; clip on lid

145 x 50mm

(3m) straight	39.93 3m
end caps	1.79 nr
internal bend; adjustable	4.79 nr
external bend; adjustable	3.99 nr
fabricated flat tee	38.93 nr
fabricated flat elbow	29.92 nr
straight coupler	1.79 nr

180 x 57mm curved profile

(3m) straight	59.94 3m
end caps	3.19 nr
internal bend	5.99 nr
external bend	5.79 nr
straight coupler	3.79 nr

PVC trunking; white skirting trunking, three compartment; clip on lid

170 x 50 mm

straight	24.95 3m
end cap	1.16 nr
internal corner	2.99 nr
external corner	2.88 nr
straight coupler	1.00 nr
socket adaptor box, one gang	1.35 nr
socket adaptor box, two gang	1.44 nr

Conduit

PVC conduits; small circular boxes with lid, gasket and screws

20mm diameter

straight tube, light guage	1.40 3m
straight tube, heavy guage	0.53 2m
terminal boxes	0.68 nr
angle boxes	0.69 nr
three way boxes	0.69 nr
through way boxes	0.69 nr

25mm diameter

straight tube, heavy guage	0.78 2m
terminal boxes	0.83 nr
angle boxes	0.83 nr
three way boxes	0.83 nr
through way boxes	0.83 nr
Repalcement PVC lid for Circular box	0.25 nr
rubber gasket for PVC or steel circular box	0.20 nr
cover screws (need 2 per box)	4.00 100

PVC oval conduits

16mm (nominal size)

straight tube	0.39 2m

20mm (nominal size)

straight tube	0.53 2m

25mm (nominal size)

straight tube	1.28 3m

Steel welded black enamelled heavy gauge conduits; small circular boxes without lid, gasket or screws

20mm diameter

straight tube	5.99 3m
terminal boxes	1.20 nr
angle boxes	1.20 nr
three way boxes	1.20 nr
through way boxes	1.20 nr
inspection bends	3.16 nr
inspection elbow	1.90 nr
couplers	0.33 nr
saddle spacer bar	0.21 nr

25mm diameter

straight tube	8.00 3m
terminal boxes	2.05 nr
angle boxes	2.05 nr
three way boxes	2.05 nr
through way boxes	2.05 nr
inspection bends	3.00 nr
inspection elbow	3.25 nr
couplers	0.52 nr
saddle spacer bar	0.27 nr

32mm diameter

straight tube	16.15 3.75m
steel lid for circular box (see above for gasket and screws)	0.10 nr

Steel welded galvanised heavy gauge conduits; small circular boxes without lid, gasket or screws

20mm diameter

straight tube	4.58 3.75m
terminal boxes	1.20 nr
angle boxes	1.08 nr
three way boxes	1.08 nr
through way boxes	1.08 nr

25mm diameter

straight tube	6.25 3.75m
terminal boxes	2.05 nr
angle boxes	1.36 nr
three way boxes	1.36 nr
through way boxes	1.36 nr
straight tube	14.75 3m
steel lid for circular box (see above for gasket and screws)	1.08 10

Cable Tray

Galvanised light duty

50mm

straight tray	6.51 3m
flat bend	5.45 nr
tee	5.81 nr
cross piece	6.54 nr
riser bend	3.62 nr

100mm

straight tray	8.73 3m
flat bend	5.60 nr
tee	7.87 nr
cross piece	9.37 nr
riser bend	3.67 nr
reducer 100 to 50mm	6.62 nr

150mm

straight tray	9.83 3m
flat bend	6.01 nr

Electrical Installations (Cont'd)

Cable Tray (Cont'd)

Galvanised light duty (Cont'd)
150mm (Cont'd)

tee	9.15 nr
cross piece	12.34 nr
riser bend	4.08 nr
reducer 150 to 100mm	7.32 nr

300mm

straight tray....................	19.49 3m
flat bend....................	11.02 nr
tee	16.74 nr
cross piece	22.33 nr
riser bend	8.40 nr
reducer 300 to 150mm	8.88 nr

450mm

straight tray....................	54.09 3m
flat bend....................	24.96 nr
tee	30.41 nr
cross piece	38.08 nr
riser bend	11.00 nr
reducer 450 to 300mm	13.11 nr

Galvanised medium duty
75mm

straight tray....................	6.28 3m
flat bend....................	5.62 nr
tee	7.56 nr
cross piece	10.57 nr
riser bend	3.35 nr

100mm

straight tray....................	7.04 3m
flat bend....................	5.75 nr
tee	7.51 nr
cross piece	10.51 nr
riser bend	3.67 nr
reducer 100 to 75mm	5.94 nr

150mm

straight tray....................	9.71 3m
flat bend....................	6.16 nr
tee	8.74 nr
cross piece	12.23 nr
riser bend	4.08 nr
reducer 150 to 100mm	6.20 nr

300mm

straight tray....................	22.56 3m
flat bend....................	11.29 nr
tee	13.70 nr
cross piece	19.15 nr
riser bend	8.40 nr
reducer 300 to 150mm	9.15 nr

450mm

straight tray....................	28.76 3m
flat bend....................	19.61 nr
tee	22.44 nr
cross piece	31.40 nr
riser bend	11.00 nr
reducer 450 to 300mm	9.47 nr

Galvanised heavy duty
75mm

straight tray....................	7.41 3m
flat bend....................	6.27 nr
tee	7.68 nr
cross piece	9.60 nr
riser bend	5.44 nr

100mm

straight tray....................	8.69 3m
flat bend....................	6.87 nr
tee	8.60 nr
cross piece	9.96 nr
riser bend	5.82 nr
reducer 100 to 75mm	6.70 nr

150mm

straight tray....................	10.24 3m
flat bend....................	8.01 nr
tee	10.04 nr
cross piece	12.54 nr
riser bend	6.72 nr
reducer 150 to 100mm	7.00 nr

300mm

straight tray....................	20.78 3m
flat bend....................	13.71 nr
tee	19.91 nr
cross piece	24.89 nr
riser bend	13.02 nr
reducer 300 to 150mm	9.13 nr

450mm

straight tray....................	42.64 3m
flat bend....................	18.86 nr
tee	33.67 nr
cross piece	43.83 nr
riser bend	17.12 nr
reducer 450 to 300mm	10.70 nr

Supports for cable tray; ms; galv

100mm cantilever arms ...	3.05 nr
300mm cantilever arms ...	7.90 nr
450mm cantilever arms ...	9.91 nr
75mm stand-off bracket...	1.66 nr
100mm stand-off bracket.	2.00 nr
150mm stand-off bracket.	2.05 nr
300mm stand-off bracket.	3.58 nr
450mm stand-off bracket.	3.93 nr

Steel Basket; bright zink plated
200 x 50mm

Straight	9.34 m
Outside Bend....................	5.29 nr
Horizontal Bend..............	6.78 nr
Inside Bend	5.56 nr
CrossPiece	8.29 nr
Horizontal Tee	7.22 nr
Surface fixing bracket 200mm with 50 mm offset....................	1.93 nr
Wall mounting bracket 210mm	2.17 nr
Basket Coupler set	1.55 nr

Cable

PVC insulated cables; single core; reference 6491X; stranded copper conductors; to BS 6004

1.5mm²	17.00 100m
2.5mm²	23.00 100m
4.0mm²	40.00 100m
6.0mm²	55.00 100m
10.0mm²	89.00 100m
16.0mm²	143.00 100m

PVC double insulated cables; single core; reference 6181Y; stranded copper conductors; to BS 6004

16.0mm² meter tails..........	94.00 50m
25.0mm² meter tails..........	3.00 m
35.0mm² meter tails..........	5.38 m

PVC insulated and PVC sheathed cables; multicore; copper conductors and bare earth continuity conductor 6242Y; to BS 6004

1.00mm²; twin and earth....	17.25 100m
1.50mm²; twin and earth....	21.50 100m
2.5mm²; twin and earth......	31.00 100m
4.00mm²; twin and earth....	55.00 100m
6.00mm²; twin and earth....	798.00 100m

PVC insulated SWA armoured and LSF sheathed cables; multicore; copper conductors to BS 6724

16mm²; 3 core	4.42 m

PVC Sheathed, polyethylene insulated cables Co-axial, braid screened, copper conductor 75 ohm

(100m)	15.00 100m

Telephone cables

4 core (100m)	8.00 100m

Mineral insulated copper sheathed cables; PVC outer sheath; copper conductors; 600V light duty to BS EN 60702-1

1.0mm²; two core cable	3.41 m
1.5mm²; two core cable	3.78 m
1.5mm²; termination including gland and seal....	14.99 10
2.5mm²; two core cable	4.78 m
2.5mm²; termination including glanda and seal..	14.99 10
4.0mm²; two core cable	7.19 m
4.0mm²; termination including gland and seal....	15.52 10
Shroud for terminations	0.36 nr

FP200 gold, fire rated cable, red outer

1.5mm; 2 core + E	57.86 100m

CAT5E cable

LSZH rated....................	48.88 305m

Switchgear and Distribution boards (Three phase)

Switch fuse, mild steel enclosure, TP & N

32 Amp....................	83.20 nr

Distribution boards TP & N, IP41 protection
8 x 3 Way TP&N
Distribution Board 125A

Max Incomer....................	138.00 nr

16 x 3 Way TP&N
Distribution Board 125A

Max Incomer....................	219.45 nr

Accessories for distribution boards
125 Amp TP Switch

Disconnector	19.00 nr
Incommer connection kit....	16.40 nr
Type B, single pole MCB 10KA....................	4.40 nr
Type B, triple pole MCB 10KA....................	18.50 nr
Type C, single pole MCB 10KA....................	4.40 nr
Type C, triple pole MCB 10KA....................	18.50 nr

Accessories and fittings

White plastics, without boxes

Plate switches 10 amp; one gang; one way; single pole....................	0.93 nr
Plate switches 10 amp; two gang; two way; single pole....................	2.12 nr
Plate switches 20amp; double pole with neon	4.08 nr
Ceiling switches 6 amp; one gang; one way; single pole....................	2.53 nr
Switched socket outlets 13 amp; one gang	1.71 5
Switched socket outlets 13 amp; two gang.............	2.80 nr
Switched socket outlets with RCD (residual current device) protected at 30 mAmp 13 amp; two gang...	45.21 nr
Fused connection units 13 amp; one gang; switched; flexible outlet	3.72 nr
Fused connection units 13 amp; one gang; switched; flexible outlet; pilot lamp	5.69 nr
Batten lampholder; shockguard....................	3.34 nr
Ceiling rose and pendant...	1.45 nr
Cooker point 45A (with neon, no socket)................	7.74 nr
Shaver socket (transformer).....................	19.81 nr
TV/FM Socket....................	5.62 nr
Telephone point, slave	3.32 nr
Socket outlet 5 Amp	5.14 nr

White plastics boxes for surface fitting

1G 16mm deep....................	0.74 nr
1G 30mm deep....................	0.82 nr
2G 30mm deep....................	1.53 nr

Galvanised boxes for flush fitting
Plaster depth box, 1G

16mm deep	0.29 nr
1G 35mm deep	0.33 nr
2G 35mm deep....................	0.43 nr

Metalclad, with boxes

Surface plate switches 5 amp; one gang; two way; single pole	2.46 nr
Surface plate switches 5 amp; two gang; two way; single pole	2.45 nr
Surface switched socket outlets 13 amp; one gang..	2.66 nr
Surface switched socket outlets 13 amp; two gang ..	3.96 nr

Weatherproof accessories, with boxes

Surface plate switches 20 amp; one gang; two way; single pole	2.91 nr

Electrical Installations (Cont'd)

Accessories and fittings (Cont'd)

Bell equipment
Transformers; 4-8-12v.......	6.40 nr

Bells for transformer operation
Chime 2-note....................	13.70 nr
73mm domestic type	8.95 nr
180 mm commercial........	32.75 nr

Bell pushes
Domestic	2.08 nr
Industrial/weatherproof.....	5.45 nr

Extras
Draw wire, 20m length, reusable	5.45 20
Grommets	1.69 100

Domestic consumer units

17th Edition consumer unit
10 useable ways and 100A main switch and 80A 30mA RCD and 63A 30mA RCD	61.00 nr
15 useable ways and 100A main switch and 80A 30mA RCD and 63A 30mA RCD	67.00 nr

Accessories for consumer units
3 to 45 Amp single pole MCB	2.90 nr
6 to 50 Amp 30mA RCBO (single unit)	24.95 nr

Lighting; Accessories and fittings

Fluorescent lamp fittings, mains voltage operations; switch start

Batten type; single tube
1200mm; 36W, with tube .	12.70 nr
1500mm; 58W, with tube .	12.50 nr
1800mm; 70W, with tube .	15.95 nr

Batten type; twin tube
1200mm; 36W, with tubes	22.80 nr
1500mm; 58W, with tubes	23.40 nr
1800mm; 70W, with tubes	26.90 nr

Metal trough reflector; for single or twin tube
1200mm	8.25 nr
1500mm	9.90 nr
1800mm	11.50 nr

Plastics prismatic type difuser; for single tube
1200mm	8.00 nr
1500mm	9.00 nr
1800mm	9.50 nr

Plastics prismatic type difuser; for twin tube
1200mm	10.50 nr
1500mm	13.00 nr
1800mm	14.00 nr

Lighting fittings complete with lamps

Bulkhead; Polycarbonate
16W Round with energy saving lamp.....................	9.95 nr

Ceramic Wall Uplighter
20W compact fluorescent lamp...............	17.50 nr

Ceiling sphere
150mm, 42W halogen lamp	7.60 nr
180mm, 70W halogen lamp	8.80 nr

Commercial Down light; recessed
HF Electronic PL Downlight 2 x 18w	31.40 nr

Lighting track; low Voltage
1m track, transformer and 3 x 50W halogen lamps.............................	57.27 nr
Plug-in fitting; 50W halogen lamp	10.49 nr

Heating

Fan heaters
3kW unit fan heaters for commercial application.......	124.17 nr
Thermostatic switch units with concealed setting........	23.99 nr

Turbular heaters (including 2 way brackets, excluding wire guards), less than 150w per metre loading
305mm long, 45w.................	11.50 nr
610mm long, 80w.................	12.99 nr
915mm long, 135w...............	15.00 nr
1220mm long, 180w.............	18.25 nr
1830mm long, 270w.............	26.75 nr
Tubular heater interlinking kit	5.75 nr

Night storage heaters
1.7KW heater	276.70 nr
2.55KW heater	348.95 nr
3.4KW heater	429.95 nr

Immersion heaters and thermostats; 2 or 3kW non-withdrawable elements
Without flanges 305mm	10.20 nr
Without flanges 457mm	10.90 nr
Without flanges 686mm	11.90 nr

Thermostats
immersion heater	5.60 nr
16 amp a.c. for air heating; without switch	24.45 nr

Water heaters; Storage type units
Oversink, single outlet, 5L, 2kW..................................	85.99 nr
Oversink, single outlet, 7L, 3kW..................................	97.00 nr
Oversink, single outlet, 10L, 2kW.............................	124.50 nr
Unvented, multipoint, 30L, 3kW..................................	330.75 nr
Unvented, multipoint, 50L, 3kW..................................	342.75 nr
Unvented, multipoint, 80L, 3kW..................................	358.50 nr

Shower units; Instantaneous
Complete with fittings	175.00 nr

Extract fans, window or wall mounted; with incorporated shutter
105mm diameter; ref DX100PC	19.05 nr
257mm diameter; ref GX9..	150.00 nr
324mm diameter; ref GX12..............................	240.00 nr
Wall duct kit for GX9	21.10 nr
Wall duct kit for GX12........	24.00 nr
290 x 321mm; ref WX6	115.00 nr
409 x 433mm; ref WX9	245.00 nr
473 x 476mm; ref WX12	290.00 nr
PIR fan controller	56.75 nr

Floor Wall and Ceiling Finishes

Screeds and toppings
Cement and sand (1:3)	120.06 m³

Surface hardener for paving
Epoxy resin based sealer...	17.78 litre
Polyurethane.....................	6.00 litre
Nitoflor Lithurin	10.50 kg

Latex Flooring
One coat levelling screeds .	0.47 kg

Gypsum plasterboard to BS 1230 Part 2 for surface finishes
9.5mm handiboard	2.17 m²
12.5mm handiboard	3.42 m²
9.5mm square edge wallboard........................	1.41 m²
12.5mm square edge wallboard........................	1.49 m²
15mm square edge wallboard........................	2.08 m²

Rendering and the like
Cement and sand (1:3)	120.06 m³
Pea shingle for pebbledash	33.00 tonne
Cement for wet slurry	146.80 tonne

Plasters
Board finish	196.00 tonne
Hardwall	299.20 tonne
Multi finish	209.20 tonne
Universal one coat	484.00 tonne
Thistle browning	271.60 tonne
Thistle bonding.................	310.40 tonne
Thistle tough coat.............	364.40 tonne

Plaster beads and stops
Angle, standard..................	0.33 m
Angle, thin ct 3mm	0.32 m
Angle, thin ct 6mm	0.32 m
Stop, thin ct 3mm	0.52 m
Stop, thin ct 6mm	0.72 m
Stop, 10mm......................	0.74 m
Stop, 13mm......................	0.74 m
Stop, 16mm......................	1.00 m
Stop, 19mm......................	1.00 m
Plasterboard edging, 10mm..............................	1.87 m
Plasterboard edging, 13mm..............................	1.87 m

Bonding fluid
Bonding fluid	3.27 litre

Metal lathing

Expanded metal lath (galv)
9mm mesh x 0.89 kg/m²...	4.38 m²
9mm mesh x 1.11 kg/m²...	6.70 m²
9mm mesh x 1.61 kg/m²...	8.70 m²

Rib-Lath
0.300mm thick x 1.16 kg/m²	6.50 m²
0.500mm thick x 1.86 kg/m²	8.20 m²

Stainless steel Rib-Lath
1.48 kg/m²	17.64 m²

Expanded metal lath (stainless steel)
1.11 kg/m²	17.64 m²

Red-rib lath
0.500mm thick x 1.91 kg/m²	9.57 m²

Galvanized arch formers

Arch corners (2 pieces)
372mm radius	12.65 nr
452mm radius	15.15 nr
602mm radius	21.40 nr
752mm radius	32.00 nr

Plasterboard coving
cove 90mm girth	1.24 m
Gyproc cove 127mm girth..	1.23 m
Coving Adhesive	0.92 kg

Plaster ventilators
Plain 229 x 79mm	2.73 nr
Plain 229 x 152mm	3.33 nr
Plain 229 x 229mm	4.00 nr

Clay floor quarry tiles - terracotta - square edge
150 x 150 x 12.5mm..........	480.00 1000
150 x 150 x 20mm.............	850.00 1000
225 x 225 x 25mm.............	1,550.00 1000

Ceramic floor tiles - red - square edge
100 x 100 x 9mm..............	21.39 m²
150 x 150 x 9mm..............	23.06 m²
200 x 200 x 12mm.............	24.00 m²

Ceramic floor tiles - cream - square edge
100 x 100 x 9mm..............	25.94 m²
150 x 150 x 9mm..............	25.94 m²
200 x 200 x 12mm.............	27.00 m²

Glazed ceramic wall tiles to BS 6431
108 x 108 x 6.5mm - white .	16.85 m²
152 x 152 x 5.5mm - white .	11.45 m²
108 x 108 x 6.5mm - light colours	16.85 m²
152 x 152 x 5.5mm - light colours	16.85 m²
108 x 108 x 6.5mm - dark colours	23.00 m²
152 x 152 x 5.5mm - dark colours	18.00 m²

Fixing materials for glazed ceramic wall tiles
Cement and sand (1:3)	120.06 m³
Tile adhesive	2.30 litre
White cement	0.36 kg

Terrazzo tiles
300 x 300 x 28mm units	35.70 m²
305 x 305 x 28mm units rounded nosing	37.80 m
305 x 165 x 28mm units	10.50 m

Terrazzo
Terrazzo mix 1:2	1,178.10 m³

Floor Wall and Ceiling Finishes (Cont'd)

Wood block flooring, tongued and grooved, 70 x 230 x 20mm units

Maple wood blocks	34.60 m²
Merbau wood blocks	24.75 m²
Oak wood blocks	28.33 m²

PVC flooring

300 x 300 x 2mm tiles	6.95 m²
300 x 300 x 3.2mm tiles	10.80 m²
610 x 610 x 2mm heavy duty tiles	6.95 m²
610 x 610 x 2.5mm heavy duty tiles	8.95 m²

Linoleum flooring

330 x 330 x 2mm tiles	14.95 m²
2.5mm sheet flooring	14.95 m²
3.2mm sheet flooring	18.50 m²
Adhesive for linoleum flooring	3.96 litre

Carpet tiling

500 x 500mm tiles heavy ribbed heavy contract	16.80 m²
500 x 500mm tiles loop pile general contract	14.80 m²
500 x 500mm tiles twist cut pile heavy contract	18.80 m²

Carpet sheeting

Fibre bonded heavy contract	11.00 m²
Ribbed fibre bonded heavy contract	12.00 m²
Adhesive for carpeting	6.00 litre

Edge fixed carpeting
Tredaire underlay; Brio

8mm	1.90 m²
Tufted wool/nylon	25.00 m²
Axminster wool/nylon	30.00 m²
Wilton wool/nylon	35.00 m²

Wallpapers and wall coverings

Lining paper	1.69 roll
Pulp paper	4.00 roll
Washable	6.00 roll
Vinyl coated	6.50 roll
Embossed	6.00 roll
Textured	6.50 roll
Woodchip	1.50 roll

Gypsum plasterboard to BS 1230 Part 2 for linings and partitions

9.5mm Handiboard	2.17 m²
12.5mm Handiboard	3.42 m²
9.5mm wallboard	1.41 m²
12.5mm wallboard	1.49 m²
12.5mm vapourcheck wallboard	2.92 m²
15mm wallboard	2.08 m²
15mm vapourcheck wallboard	3.92 m²
19mm plank with square edges	4.22 m²

Gyproc Thermaline BASIC vapour check grade plasterboard

22mm	3.97 m²
30mm	4.42 m²
40mm	5.02 m²
50mm	12.32 m²
90mm	20.69 m²

Gypsum laminated partitions plasterboard components

25 x 38mm softwood battens	0.36 m

Metal stud partitions

48mm studs	0.58 m
50mm floor and ceiling channels	1.08 m

Suspended ceiling tiles, bevelled, grooved and rebated

1200 x 600 x 9mm asbestos-free fire resisting	17.95 m²
1200 x 600 x 15.8mm plain mineral fibre	10.90 m²
600 x 600 x 15.8mm plain mineral fibre	10.60 m²

Glazing

Float glass

3mm GG quality	38.65 m²
4mm GG quality	38.99 m²
5mm GG quality	56.72 m²
6mm GG quality	57.05 m²
10mm GG quality	114.83 m²
3mm obscured ground	83.62 m²
4mm obscured ground	91.75 m²

Rough cast glass

6mm	49.46 m²

Pyroshield glass

6mm clear	91.07 m²
6mm clear, aligning panes	91.07 m²
6mm safety clear	106.87 m²
6mm safety clear, aligning panes	106.87 m²
7mm texture	43.57 m²
7mm texture, aligning panes	43.57 m²
7mm safety texture	55.97 m²
7mm safety texture, aligning panes	55.97 m²

Patterned glass

4mm white	36.26 m²
6mm white	65.60 m²
4mm tinted	53.66 m²
6mm tinted	71.50 m²

Antisun float glass

4mm grey	74.60 m²
6mm grey	107.60 m²
10mm grey	192.38 m²
12mm grey	266.54 m²
4mm bronze	74.60 m²
6mm bronze	107.60 m²
10mm bronze	192.38 m²
12mm bronze	265.55 m²
6mm green	137.93 m²

Thick float glass

10mm GG quality	114.83 m²
12mm GG quality	146.36 m²
15mm GG quality	221.86 m²
19mm GG quality	314.26 m²
25mm GG quality	498.87 m²

Toughened safety glass

4mm clear float glasses	59.62 m²
5mm clear float glasses	89.13 m²
6mm clear float glasses	89.47 m²
10mm clear float glasses	142.07 m²
12mm clear float glasses	259.75 m²
4mm white	120.14 m²
6mm white	125.43 m²
4mm tinted	120.15 m²
6mm tinted	125.43 m²

Toughened safety solar control glasses

4mm, antisun grey	109.37 m²
6mm, antisun grey	114.02 m²
10mm, antisun grey	225.67 m²
12mm, antisun grey	296.95 m²
4mm, antisun bronze	109.28 m²
6mm, antisun bronze	114.02 m²
10mm, antisun bronze	227.18 m²
12mm, antisun bronze	296.95 m²

Clear laminated glass - float quality

4.4mm safety glass	71.65 m²
6.4mm safety glass	60.83 m²
8.8mm safety glass	108.57 m²
10.8mm safety glass	127.13 m²
7.5mm anti-bandit glass	145.82 m²
9.5mm anti-bandit glass	153.01 m²
11.5mm anti-bandit glass	178.68 m²
11.5mm anti-bandit glass, large panes	214.62 m²

Georgian wired glass

7mm standard	66.00 m²
7mm safety	80.00 m²
6mm polished	108.00 m²
6mm polished safety	124.00 m²

Sundries

linseed oil putty	33.20 25kg
metal casement putty	33.20 25kg
washleather strip, 3mm glass	1.16 m
washleather strip, 6mm glass	1.40 m
butyl rubber glazing strip, 3mm glass	1.08 m
butyl rubber glazing strip, 6mm glass	1.30 m

Mirrors; 6mm clear float; copper plastic safety backing

with polished edges, 6mm diameter holes

254 x 400mm	42.28 nr
460 x 900mm	66.82 nr
600 x 900mm	76.68 nr

Painting and Decorating

Timber treatments

Creosote	263.41 200litre
Wood preservative	19.17 5litre

Emulsion paint

matt brilliant white	8.31 5litre
matt colours	12.65 5litre
silk brilliant white	11.04 5litre
silk colours	13.80 5litre

Priming, paint

wood primer	21.63 5litre
primer sealer	35.18 5litre
aluminium sealer and wood primer	32.82 5litre
alkali-resisting primer	29.73 5litre
universal primer	77.34 5litre
red oxide primer	31.44 5litre
zinc phosphate primer	34.15 5litre
acrylic primer	21.63 5litre

Undercoat paint

Brilliant white undercoat	21.33 5litre
colours	31.33 5litre

Gloss paint

Brilliant white gloss	21.33 5litre
colours	31.33 5litre

Eggshell paint

Brilliant white eggshell (oil based)	23.99 5litre
colours	35.33 5litre

Plastic finish

coating compound	16.21 25kg

Waterproof cement paint

cream or white	9.98 5litre
colours	13.33 5litre
sealer	37.46 5litre

Textured masonry paint

white	43.25 10litre
colours	54.92 10litre
Stabilising solution	61.63 20litre

Multicolour painting

primer	32.45 5litre
basecoat	63.80 5litre
finish	63.80 5litre

Metallic paint

aluminium or silver	61.02 5litre
gold or bronze	61.02 5litre

Bituminous paint

Bituminous paint	12.01 5litre

Polyurethane lacquer

clear (varnish)	29.38 5litre

Linseed oil

raw	19.58 5litre
boiled	22.49 5litre

Fire retardant paints and varnishes

fire retardant paint	99.50 5litre
fire retardant varnish	129.80 5litre
overcoat varnish	44.00 5litre

French polishing

white polish	13.50 litre
button polish	17.90 litre
garnet polish	8.80 litre
methylated spirit	1.50 litre

Road marking materials

solvent based paint

chlorinated rubber	59.00 20kg
50mm wide non-reflective self adhesive tape	6.50 5m

Painting and Decorating (Cont'd)

Road marking materials (Cont'd)

100mm wide non-reflective self adhesive tape	8.25 5m
50mm wide reflective self adhesive tape	7.50 5m
100mm wide reflective self adhesive tape	9.25 5m
road marking primer	192.39 25kg

Surface treatments

silicone based water repellent	16.24 5litre
fungicide	6.63 2.5litre

Drainage

Vitrified Clay pipes and fittings

Drains; normal; flexible mechanical joints

375mm pipes	135.14 m
375mm bend	578.56 nr
450mm pipes	265.39 m
450mm bend	986.55 nr

Drains; sleeve joints, push-fit polypropylene standard ring flexible couplings; Hepworth Supersleve

100mm pipes	6.90 m
100mm pipe couplings	5.09 nr
100mm bends	9.31 nr
100mm branches	20.10 nr
150mm pipes	17.94 m
150mm pipe couplings	11.92 nr
150mm bends	24.36 nr
150mm branches	36.28 nr

Sleeve joints, push-fit polypropylene neoprene ring flexible couplings; Hepworth Supersleve/Densleve

100mm pipes	10.74 m
100mm pipe couplings	2.84 nr
150mm pipes	23.54 m
150mm pipe couplings	7.52 nr

Drains; normal; cement mortar (1:3)

100mm pipes	9.06 m
100mm double collars	12.18 nr
100mm bends	8.63 nr
100mm rest bends	15.32 nr
100mm branches	17.35 nr
150mm pipes	12.63 m
150mm double collars	20.44 nr
150mm bends	14.66 nr
150mm rest bends	25.54 nr
150mm branches	29.73 nr
225mm pipes	69.42 m
225mm double collars	20.98 nr
300mm pipes	102.14 m

PVC-U pipes and Fittings

Drains; ring seal joints

110mm pipes	8.25 3m
110mm pipe couplings	3.55 nr
110mm connections to cast iron and clay sockets	12.37 nr
110mm 45 degree bends	5.95 nr
110mm branches	10.55 nr
110mm connections cast iron and clay spigot	12.37 nr
160mm pipes	17.50 3m
160mm pipe couplings	7.95 nr
160mm connections to cast iron and clay sockets	24.84 nr
160mm 45 degree bends	23.97 nr
160mm branches	48.39 nr
160mm connections to cast iron and clay spigot	26.00 nr

Solid wall concentric external rib-reinforced Wavin "Ultra-Rib" with sealing rings to joints

150mm pipes 6UR 046 (6m long pipe)	105.42 6m
150mm pipe couplings	26.70 nr
150mm bends 6UR 563	35.77 nr
150mm branches 6UR 213	86.48 nr
150mm connections to clayware pipe ends 6UR 129	70.56 nr
225mm 9UR 046 pipes	112.80 3m
225mm pipe couplings	58.19 nr
225mm bends 9UR 563	150.44 nr
225mm branches 9UR 213	215.89 nr
225mm connections to clayware pipe ends 9UR 109	181.89 nr
300mm 12UR 043 pipes	160.05 3m
300mm pipe couplings	117.24 nr
300mm bends 12UR 563	273.20 nr
300mm branches 12UR 213	693.14 nr
300mm connections to clayware pipe ends 12UR 112	478.44 nr
Reducers 40/32mm	1.10 nr
32mm Pipe brackets - Metal	1.83 nr
40mm Pipe brackets - Metal	2.52 nr

Concrete pipes and fittings (BS 5911)

Unreinforced, Class H, flexible joints

300mm pipes	52.35 2.5m
300mm bends	206.01 nr
300mm, 100mm branches	121.75 nr
300mm, 150mm branches	121.70 nr
300mm, 225mm branches	246.32 nr
300mm, 300mm branches	282.15 nr
375mm pipes	64.93 2.5m
375mm bends	272.61 nr
375mm, 100mm branches	136.86 nr
375mm, 150mm branches	136.86 nr
375mm, 225mm branches	272.61 nr
375mm, 300mm branches	310.22 nr
375mm, 375mm branches	347.81 nr
450mm pipes	81.80 2.5m
450mm bends	314.91 nr
450mm, 100mm branches	155.11 nr
450mm, 150mm branches	155.11 nr
450mm, 225mm branches	291.41 nr
450mm, 300mm branches	329.01 nr
450mm, 375mm branches	366.62 nr
450mm, 450mm branches	432.42 nr
525mm pipes	100.43 2.5m
525mm bends	398.68 nr
525mm, 100mm branches	171.98 nr
525mm, 150mm branches	171.98 nr
525mm, 225mm branches	315.00 nr
525mm, 300mm branches	354.37 nr
525mm, 375mm branches	393.75 nr
525mm, 450mm branches	462.65 nr
600mm pipes	126.13 2.5m
600mm bends	465.32 nr
600mm, 100mm branches	173.36 nr
600mm, 150mm branches	173.36 nr
600mm, 225mm branches	315.86 nr
600mm, 300mm branches	353.46 nr
600mm, 375mm branches	391.05 nr
600mm, 450mm branches	456.86 nr

Reinforced, Class H, flexible joints

450mm pipes	60.29 m
450mm bends	602.97 nr
450mm, 100mm branches	168.27 nr
450mm, 150mm branches	172.08 nr
450mm, 225mm branches	177.86 nr
450mm, 300mm branches	216.35 nr
450mm, 375mm branches	245.23 nr
450mm, 450mm branches	283.74 nr
525mm pipes	73.70 m
525mm bends	736.97 nr
525mm, 100mm branches	189.58 nr
525mm, 150mm branches	193.44 nr
525mm, 225mm branches	199.21 nr
525mm, 300mm branches	228.09 nr
525mm, 375mm branches	266.59 nr
525mm, 450mm branches	305.10 nr
600mm pipes	80.39 m
600mm bends	803.96 nr
600mm, 100mm branches	171.87 nr
600mm, 150mm branches	214.24 nr
600mm, 225mm branches	220.01 nr
600mm, 300mm branches	248.88 nr
600mm, 375mm branches	287.39 nr
600mm, 450mm branches	325.90 nr

Metal pipes and fittings

Cast iron, BS 437, coated; bolted cast iron coupling and synthetic rubber gasket joints

100mm pipes	35.14 m
100mm pipe couplings	17.78 nr
100mm connectors, large socket for clayware	34.95 nr
100mm bends	23.87 nr
100mm branches	38.54 nr
150mm pipes	70.01 m
150mm pipe couplings	36.31 nr
150mm connectors, large socket for clayware	61.16 nr
150mm diminishing pieces, reducing to 100mm	56.99 nr
150mm bends	42.81 nr
150mm branches	92.31 nr

Stainless steel, Blucher UK Ltd; ring seal push fit joints (AISI 316)

75mm pipes	26.79 m
75mm pipe couplings	13.54 nr
75mm bends	21.09 nr
75mm branches	22.48 nr
110mm pipes	37.50 m
110mm pipe couplings	19.14 nr
110mm diminishing pieces, reducing to 75mm	19.82 nr
110mm bends	29.58 nr
110mm branches	31.40 nr

Gullies, Channels, Gratings etc

Concrete gullies; BS 5911

375mm diameter x 750mm deep internally, stopper	48.13 nr
375mm diameter x 900mm deep internally, stopper	50.20 nr
450mm diameter x 750mm deep internally, stopper	50.20 nr
450mm diameter x 900mm deep internally, stopper	53.90 nr

Cast iron accesories, painted black gratings

140mm diameter	6.68 nr
gratings 197mm diameter	6.15 nr
gratings 150 x 150mm	4.05 nr
gratings 225 x 225mm	14.65 nr

Alloy accessories gratings

140mm diameter	2.75 nr
gratings 197mm diameter	5.25 nr
gratings 150 x 150mm	5.00 nr
gratings 225 x 225mm	6.68 nr

Cast iron accesories, painted black gratings and frames, reference HY811 Watergate

370 x 305mm	122.98 nr
gratings and frames, reference HY812 370 x 430mm	130.04 nr

Channels; vitrified clay, BS 65, normal, glazed
100mm, half section straight

600mm effective length	9.57 nr
1000mm effective length	14.49 nr

150mm, half section, straight

1000mm effective length	20.12 nr

Inspection chambers and Manholes

Precast circular concrete inspection chambers; standard or stock pattern units; BS 5911

250mm high, 900mm internal diameter	32.36 nr
500mm high, 900mm internal diameter	51.78 nr
750mm high, 900mm internal diameter	50.43 nr

Drainage (Cont'd)

Inspection chambers and Manholes (Cont'd)

Precast circular concrete inspection chambers; standard or stock pattern units; BS 5911 (Cont'd)

1000mm high, 900mm internal diameter	67.24 nr
250mm high, 1050mm internal diameter	49.10 nr
500mm high, 1050mm internal diameter	82.63 nr
750mm high, 1050mm internal diameter	97.23 nr
1000mm high, 1050mm internal diameter	115.06 nr
250mm high, 1200mm internal diameter	59.62 nr
500mm high, 1200mm internal diameter	99.45 nr
750mm high, 1200mm internal diameter	115.26 nr
1000mm high, 1200mm internal diameter	153.68 nr

Galvanized malleable cast iron steps

step irons BS 1247	6.00 nr

Cover slabs, heavy duty, 125mm thick, for chamber sections

2400mm internal diameter	1,044.18 nr
2700mm internal diameter	1,351.49 nr
3000mm internal diameter	1,590.89 nr

Precast rectangular concrete inspection chambers; standard or stock pattern units; BS 5911

150mm high, internal size 1200 x 750mm	25.62 nr
200mm high, internal size 1200 x 750mm	31.61 nr

Polypropylene shallow inspection chamber with preformed benching. Light duty single sealed cover and frame ref. 4D.961

300mm diameter; for 100mm diameter pipes ref. 4D.960, depth to invert 600mm	114.28 nr

Polypropylene universal inspection chamber; 475mm diameter

preformed chamber base benching; for 100mm pipes, ref. 4D.922	105.36 nr
chamber shaft 185mm effective length, ref. 4D925	19.43 nr
cast iron cover and frame; single seal light duty to suit, ref. 4D.942	91.52 nr
Polypropylene sealed cover and frame, ref. 4D.969	48.27 nr

Cast iron inspection chambers

100mm diameter straight with one branch	261.86 nr
100mm diameter straight with two branches	793.34 nr
150 x 100mm diameter straight with one branch	337.58 nr
150mm diameter straight with one branch	428.53 nr
150mm diameter straight with two branches	1,322.23 nr

Manhole step irons

figure 1 general purpose pattern; 110mm tail	5.05 nr
figure 1 general purpose pattern; 225mm tail	5.73 nr

Manhole covers, A15 light duty, Single Seal, ductile iron

Solid top Circular 450mm dia	11.44 nr
Solid top 450mm x 450mm	13.56 nr
Solid top 600mm x 450mm	32.64 nr
Solid top 600mm x 600mm	58.00 nr
Recessed top 600mm x 450mm	28.71 nr
Recessed top 600mm x 600mm	40.77 nr
Recessed top 750mm x 600mm	113.10 nr

Manhole covers, A15 light duty, Double Seal

Solid top 450mm x 450mm	48.50 nr
Solid top 600mm x 450mm	48.50 nr
Solid top 600mm x 600mm	56.01 nr
A15 light duty, lifting keys	1.65 nr

Manhole covers, B125 medium duty

Ref. H783 Trojan single seal; 600mm diameter	77.00 nr
Ref. H781; 600 x 450mm	38.87 nr
Ref. H742; 600 x 600mm	47.58 nr
Ref. H741; 750 x 600mm	148.00 nr
B125 medium duty, lifting keys	5.46 nr

Manhole covers, cast iron, D400 heavy duty

Ref. H9330; 600mm diameter	105.00 nr
Ref. H9320; 600 x 600mm	85.16 nr
Ref. H9420; 750 x 750mm	167.95 nr
D400 heavy duty, lifting keys	6.50 nr

Clayware land drain pipes, BS 1196; butt joints

75mm	7.80 m
100mm	11.07 m
150mm	25.50 m

Plastic land drain pipe

80mm	44.00 50m
100mm	132.65 100m
160mm	165.00 50m

Fencing

Strained wire fencing - concrete posts

900mm intermediate post (1750mm long)	11.31 nr
900mm end/angle post (1800mm long)	15.92 nr
900mm strutt	9.65 nr
900mm stretcher bar	3.00 nr
1050mm intermediate post (2400 long)	14.14 nr
1050mm end/angle post	22.34 nr
1050mm strutt	9.20 nr
1050mm stretcher bar	4.00 nr
Ratchet winders	3.00 nr
2.5mm Stirrup wire	42.00 651m
4mm Galvanized line wire	42.00 254m

Chain link fencing

900mm - concrete posts - galvanized mesh

intermediate post	11.47 nr
end straining post	24.53 nr
strutt	11.60 nr
stretcher bar	2.57 nr
chain link (2.5mm wire)	89.41 25m

1200mm - concrete posts - galvanized mesh

intermediate post	14.44 nr
end straining post	27.97 nr
strutt	12.10 nr
stretcher bar	3.39 nr
chain link	120.34 25m

1500mm - concrete posts - galvanized mesh

intermediate post	20.75 nr
end straining post	37.68 nr
strutt	16.35 nr
stretcher bar	4.23 nr
chain link	152.59 25m

1800mm - concrete posts - galvanized mesh

intermediate post	21.29 nr
end straining post	38.57 nr
strutt	18.35 nr
stretcher bar	5.14 nr
chain link	175.20 25m

Sundries

Ratchet winders	3.00 nr
2.5mm Stirrup wire	42.00 651
PPC Line wire	4.25 50m
Ring fasteners	6.85 1000

900mm - concrete posts - plastic coated mesh

intermediate post	11.47 nr
end straining post	24.53 nr
strutt	11.60 nr
stretcher bar	2.57 nr
chain link (3.15 gauge)	93.97 25m

1200mm - concrete posts - plastic coated mesh

intermediate post	14.44 nr
end straining post	27.97 nr
strutt	12.10 nr
stretcher bar	3.39 nr
chain link (3.15 gauge)	110.10 25m

1400mm - concrete posts - plastic coated mesh

intermediate post	20.75 nr
end straining post	37.68 nr
strutt	16.35 nr
stretcher bar	4.23 nr
chain link (3.15 gauge)	135.09 25m

1800mm - concrete posts - plastic coated mesh

intermediate post	21.29 nr
end straining post	38.57 nr
strutt	18.35 nr
stretcher bar	5.14 nr
chain link (3.15 gauge)	160.03 25m

900mm - steel posts - galvanized mesh

intermediate post GLS (heavy)	9.94 nr
end post with strutts	32.77 nr
angle post with strutts	41.94 nr
chain link 3 x 50mm	89.41 25m

1200mm - steel posts - galvanized mesh

intermediate post GLS (heavy)	11.36 nr
end post with strutts	40.21 nr
angle post with strutts	57.79 nr
chain link	120.34 25m

1400mm - steel posts - galvanized mesh

intermediate post GLS (heavy)	12.88 nr
end post with strutts	45.38 nr
angle post with strutts	63.06 nr
chain link	152.59 25m

1800mm - steel posts - galvanized mesh

intermediate post GLS (heavy)	14.35 nr
end post with strutts	53.94 nr
angle post with strutts	74.50 nr
chain link	175.20 25m

900mm - steel posts - plastic coated mesh

intermediate post PLS (heavy)	9.94 nr
end post with strutts	32.77 nr
angle post with strutts	41.94 nr
chain link	93.97 25m

1200mm - steel posts - plastic coated mesh

intermediate post PLS (heavy)	11.36 nr
end post with strutts	40.21 nr
angle post with strutts	57.79 nr
chain link	110.10 25m

1400mm - steel posts - plastic coated mesh

intermediate post PLS (heavy)	12.88 nr
end post with strutts	45.38 nr
angle post with strutts	63.06 nr
chain link	135.09 25m

1800mm - steel posts - plastic coated mesh

intermediate post PLS (heavy)	14.35 nr
end post with strutts	53.94 nr
angle post with strutts	74.50 nr
chain link	160.03 25m

Wooden fencing

Chestnut fencing

900mm high fencing	51.55 9.5m
900mm intermediate post (1650mm long)	2.55 nr
900mm end/angle post	12.15 nr
900mm strutts	2.73 nr
1200mm high fencing	61.55 9.5m

Fencing (Cont'd)

Wooden fencing (Cont'd)

Chestnut fencing (Cont'd)

1200mm intermediate post	4.10 nr
1200mm end/angle post	12.15 nr
1200mm strutts	3.81 nr
1500mm high fencing	45.00 4.5m
1500mm intermediate post	6.12 nr
1500mm end/angle post	13.86 nr
1500mm strutts	5.79 nr

Close boarded fencing - posts 600 into ground

1200mm posts	6.68 nr
1200mm pales	0.94 nr
1500mm posts	7.74 nr
1500mm pales	1.17 nr
1800mm posts	8.59 nr
1800mm pales	1.48 nr
Rails	2.93 m
Gravel boards 150x22	1.59 m
Centre stumps	0.53 nr
50mm nails (galv)	1.74 25kg
75mm nails (galv)	1.74 25kg

Wooden palisade fencing

1050mm posts	3.92 nr
1050mm pales	1.43 nr
1200mm posts	4.51 nr
1200mm pales	1.54 nr

Steel Railings

Aluminium guard rail of tubing to BS 1387 medium grade with Kee Klamp fittings and intersections

42mm diameter	19.65 m
flanged end (type 61) wall flange	5.49 nr
bend No.15-7	6.06 nr
three-way No.20-7	9.09 nr
three-way No.25-7	8.62 nr
four-way No.21-7	7.01 nr
four-way No.26-7	6.62 nr
five-way No.35-7	9.95 nr
five-way No.40-7	14.90 nr
floor plate No.62-7	8.03 nr

Fencing extras

Precast concrete posts

Fence posts 100 x 100mm tapering intermediate post 1570mm long	9.30 nr
Fence posts 100 x 100mm square strainer post (end or intermediate) 1570mm long	10.78 nr

Gates

Ornamental steel gates; Cannock Gates (UK) Ltd

Clifton, 920mm high single gate, for 760 mm wide opening	43.56 nr
Clifton, 920mm high double gate, for 2600 mm wide opening	96.53 nr
Clifton, 1830mm high single gate, for 1070 mm wide opening	76.73 nr
Clifton, 1830mm high double gate, for 2300 mm wide opening	169.13 nr
Winchester, 1220mm high single gate, for 1070 mm wide opening	62.04 nr
Westminster, 920mm high double gate, for 2600 mm wide opening	118.80 nr
Marlborough, 920mm high single gate, for 760 mm wide opening	57.75 nr
Marlborough, 920mm high folding (4 piece) gate, for 2300 mm wide opening	173.25 nr
Marlborough, 1830mm high single gate, for 760 mm wide opening	82.50 nr
Marlborough, 1830mm high single gate, for 1000 mm wide opening	92.40 nr
Royal Talisman, 1220mm high single gate, for 920 mm wide opening	206.25 nr
Royal Talisman, 1220mm high double gate, for 4270 mm wide opening	749.10 nr
Royal Talisman, 1830mm high single gate, for 1070 mm wide opening	309.38 nr
Royal Talisman, 1830mm high arched double gate, for 4270 mm wide opening	1,014.75 nr

External works and landscaping

Precast concrete kerbs

125 x 255 x 915mm	5.06 nr
125 x 255 x 915mm; curved 6.00m radius	9.30 nr
150 x 305 x 915mm	9.31 nr
150 x 305 x 915mm; curved 10.00m radius	13.14 nr

Precast concrete edgings

50 x 150 x 915mm	2.46 m
50 x 200 x 915mm	3.18 m
50 x 250 x 915mm	4.57 m

Granite edge kerb

150 x 300mm	23.99 m
200 x 300mm	44.66 m

York stone paving

50mm thick	92.44 m²
75mm thick	145.83 m²

Precast concrete flags, BS 7263 Part 1, natural finish

600 x 450 x 50mm	3.90 nr
600 x 600 x 50mm	4.23 nr
600 x 750 x 50mm	5.03 nr
600 x 900 x 50mm	5.31 nr
600 x 450 x 63mm	6.68 nr

Brick paving blocks

25mm thick Brick paviors	300.00 1000

Granite sett paving

125mm thick; new	75.00 m²
150mm thick; new	90.00 m²

COMPOSITE PRICES FOR APPROXIMATE ESTIMATING

The purpose of this section is to provide an easy reference to costs of "composite" items of work which would usually be measured for Approximate Estimates. Rates for items, which are not included here, may be found by reference to the preceding pages of this volume. Allowance has been made in the rates for the cost of works incidental to and forming part of the described item although not described therein, i.e. the cost of formwork to edges of slabs have been allowed in the rates for slabs; the cost of finishings to reveals has been allowed in the rates for finishings to walls.

The rates are based on rates contained in the preceding Works section of the book and include the cost of materials, labour fixing or laying and allow 10% for overheads and profit: Preliminaries are not included within the rates. The rates are related to a housing or simple commercial building contract. For contracts of higher or lower values, rates should be adjusted by the percentage adjustments given in Essential Information at the beginning of this volume. Landfill tax is included for inert waste.

The following cost plans are priced as examples on how to construct an Elemental Cost Plan, for information and to demonstrate how to use the Approximate Estimating Section.

COST PLAN FOR AN OFFICE DEVELOPMENT

The Cost Plan that follows gives an example of an Approximate Estimate for an office building as the attached sketch drawings. The estimate can be amended by using the examples of Composite Prices following or the prices included within the preceding pages. Details of the specification used are indicated within the elements. The cost plan is based on a fully measured bill of quantities priced using the detailed rates within this current edition of Laxton's. Preliminaries are shown separately.

Floor area   932 m²

Description	Item	Qty		Rate	Item Total	Cost per m²	Total
1.0 SUBSTRUCTURE							
Note: Rates allow for excavation by machine							
1.1 Foundations (271 m²)							
Clear site vegetation; excavate 150 topsoil and remove spoil.	Site clearance	288	m²	1.55	446.32	0.48	
300 x 600 reinforced concrete perimeter ground beams with attached 180 x 300 toes; damp proof membrane 50 concrete blinding; jablite; 102.5 thick common and facing bricks from toe to top of ground beam; 75 cavity filled with plain concrete; cavity weeps; damp proof courses (excavation included in ground slab excavation)	Ground beam	81	m	268.86	21,777.93	23.37	
600 x 450 reinforced concrete internal ground beams, mix C30; 50 thick concrete blinding; formwork; jablite protection (excavation included in ground slab excavation)	Ground beam	34	m	221.26	7,522.91	8.07	
1900 x 1900 x 1150 lift pit; excavate; level and compact bottoms; working space; earthwork support; 50 concrete blinding; damp proof membrane; 200 reinforced concrete slab, 200 reinforced concrete walls, formwork; Bituthene tanking and protection board	Lift pit	1	nr	2,993.75	2,993.75	3.21	
1500 x 1500 x 1000 stanchion bases; excavate; level and compact bottoms; earthwork support; reinforced concrete foundations; 102.5 bricks from top of foundation to ground slab level	Stanchion base	13	nr	1,096.28	<u>14,251.62</u> <u>46,992.52</u>	<u>15.29</u> <u>50.42</u>	46,992.52
(Element unit rate = £ 173.40 m²)							
1.4 Ground floor construction (271 m²)							
Excavate to reduced levels; level and compact bottoms; 100 hardcore filling, blinded; damp proof membrane; 50 jablite insulation; 150 reinforced concrete ground slab 2 layers A252 fabric reinforcement	Ground floor slab	271	m²	124.65	<u>33,781.05</u> <u>33,781.05</u>	<u>36.25</u> <u>36.25</u>	33,781.05
(Element unit rate = £ 124.65 m²)							
2.0 SUPERSTRUCTURE							
2.1 Frame (932 m²)							
300 x 300 reinforced concrete columns; formwork two coats RIW to exterior faces	Columns	111	m	129.65	14,391.40	15.44	
29.5m girth x 1.0m high steel frame to support roof over high level louvres	Frame	1	nr	8,470.00	8,470.00	9.09	
300 x 550 reinforced concrete beams; formwork	Beams	61	m	184.88	<u>11,277.84</u> <u>34,139.24</u>	<u>12.10</u> <u>36.63</u>	34,139.24
(Element unit rate = £ 36.63 m²)							
2.2 Upper floors (614 m²)							
275 reinforced concrete floor slab; formwork	Floor slab	614	m²	158.71	97,450.45	104.56	
300 x 400/675 x 175/300 x 150 reinforced concrete attached beams; formwork	Attached beams	273	m	142.76	<u>38,972.99</u> <u>136,423.45</u>	<u>41.82</u> <u>146.38</u>	136,423.45
(Element unit rate = £ 222.19 m²)							
2.3 Roof (403 m² on plan)							
2.3.1 Roof structure							
Pitched roof trusses, wall plates; 160 thick insulation and Netlon support	Roof structure	403	m²	56.22	22,657.47	24.31	
100 thick blockwork walls in roof space	Blockwork	27	m²	49.48	1,335.97	1.43	

COST PLAN FOR AN OFFICE DEVELOPMENT (Cont'd)

Description	Item	Qty		Rate	Item Total	Cost per m²	Total
2.0 SUPERSTRUCTURE (cont'd)							
2.3 Roof (403 m² on plan) (cont'd)							
2.3.1 Roof structure (cont'd)							
Lift shaft capping internal size 1900 x 1900; 200 thick blockwork walls and 150 reinforced concrete slab	Lift shaft	1	nr	944.04	944.04	1.01	
2.3.2 Roof coverings							
Concrete interlocking roof tiles, underlay and battens	Roof finish	465	m²	49.62	23,073.30	24.76	
Ridge	Ridge	22	m	19.37	426.14	0.46	
Hip	Hip	50	m	19.93	996.36	1.07	
Valley	Valley	6	m	48.02	288.12	0.31	
Abutments to cavity walls, lead flashings and cavity tray damp proof courses	Abutments	9	m	101.15	910.33	0.98	
Eaves framing, 25 x 225 softwood fascia 950 wide soffit and ventilator; eaves tile ventilator and decoration; 164 x 127 pressed Aluminium gutters and fittings	Eaves	102	m	291.68	29,751.38	31.92	
2.3.4 Roof drainage							
100 x 100 Aluminium rainwater pipes and rainwater heads and hopper heads	Rain water pipes	58	m	95.66	5,548.04	5.95	85,931.15
(Element unit rate = £ 213.23 m²)					85,931.15	92.20	
2.4 Stairs and ramps							
Reinforced in-situ concrete staircase; blockwork support walls screed and vinyl floor finishes; plaster and paint to soffits; to wall							
3nr 1200 wide flights, 1nr half landing and 1nr quarter landing; 3100 rise; handrail one side	Stairs	1	nr	6,115.05	6,115.05	6.56	
4nr 1200 wide flights, 2nr half landings 620 rise; plastics coated steel balustrade one side and handrail one side	Stairs	1	nr	19,022.13	19,022.13	20.41	
4nr 1200 wide flights, 2nr half landings mix; 6200 rise; plastics coated steel and glass in fill panels balustrade one side and handrail one side	Stairs	1	nr	18,632.44	18,632.44	19.99	43,769.61
					43,769.61	46.96	
2.5 External walls (614 m²)							
Cavity walls, 102.5 thick facing bricks (PC £250 per 1000), 75 wide insulated cavity, 140 thick lightweight concrete blockwork	Facing brick wall	455	m²	143.44	65,265.20	70.03	
Cavity walls, 102.5 thick facing bricks (PC £250 per 1000), built against concrete beams and columns	Facing brick wall	159	m²	116.09	18,457.58	19.80	
880 x 725 x 150 cast stonework pier cappings	Stonework	9	nr	132.11	1,188.97	1.28	
780 x 1300 x 250 cast stonework spandrels	Stonework	9	nr	264.23	2,378.10	2.55	
6500 x 2500 attached main entrance canopy	Canopy	1	nr	11,528.54	11,528.54	12.37	
Milled lead sheet on impregnated softwood boarding and framing, and building paper	Lead sheet	4	m²	376.49	1,505.97	1.62	
Expansion joint in facing brickwork	Expansion joint	118	m	20.78	2,452.37	2.63	102,776.73
(Element unit rate = £ 167.39 m²)					102,776.73	110.28	
2.6 Windows and external doors (214 m²)							
2.6.1 External windows (190 m²)							
Polyester powder coated Aluminium vertical pivot double glazed windows	Windows	190	m²	714.66	135,784.89	145.69	
Polyester powder coated Aluminium fixed louvre panels	Louvres	35	m²	298.97	10,464.12	11.23	
Heads 150 x 150 x 10 stainless steel angle bolted to concrete beam, cavity damp proof course, weep vents, plaster and decoration	Heads	98	m	117.77	11,541.70	12.38	
Reveals closing cavity with facing bricks, damp proof course, plaster and decoration	Reveals	99	m	39.15	3,875.95	4.16	
Works to reveals of openings comprising closing cavity with facing bricks, damp proof course, 13 thick cement and sand render with angle beads, and ceramic tile finish	Reveals	27	m	53.42	1,442.38	1.55	
Sills closing cavity with blockwork, damp proof course, and 19 x 150 Ash veneered MDF window boards	Sills	8	m	40.81	326.50	0.35	
Sills 230 x 150 cast stonework sills and damp proof course	Sills	36	m	65.58	2,360.86	2.53	
Sills 230 x 150 cast stonework sills, damp proof course, and 19 x 150 Ash veneered MDF window boards	Sills	36	m	347.02	12,492.79	13.40	
Sills closing cavity with blockwork, damp proof course, and ceramic tile sill finish	Sills	9	m	50.77	456.96	0.49	
Cavity wall lintels	Lintels	15	m	37.40	561.00	0.60	
Surrounds to 900 diameter circular openings; stainless steel radius lintels, 102.5 x 102.5 facing brick on edge surround, 13 thick plaster, and eggshell decoration	Lintels	8	m	220.28	1,762.23	1.89	
300 x 100 cast stonework surrounds to openings	Stonework	20	m	83.65	1,672.99	1.80	

COST PLAN FOR AN OFFICE DEVELOPMENT (Cont'd)

Description	Item	Qty		Rate	Item Total	Cost per m²	Total
2.0 SUPERSTRUCTURE (cont'd)							
2.6 Windows and external doors (214 m²) (cont'd)							
2.6.2 External doors (24 m²)							
Polyester powder coated Aluminium double glazed doors and any integral screens	External doors	24	m²	672.03	16,128.71	17.31	
Cavity wall lintels	Lintels	3	m	37.40	112.20	0.12	198,983.27
(Element unit rate = £ 929.83 m²)					198,983.27	213.50	
2.7 Internal walls and partitions (481 m²)							
100 thick blockwork walls	Block walls	231	m²	42.82	9,891.13	10.61	
200 thick blockwork walls	Block walls	113	m²	75.84	8,570.22	9.20	
215 thick common brickwork walls	Brick walls	84	m²	131.28	11,027.62	11.83	
Demountable partitions	Partitions	42	m²	44.21	1,856.94	1.99	
Extra over demountable partitions for single doors	Partitions	2	nr	42.53	85.06	0.09	
Extra over demountable partitions for double doors	Partitions	1	nr	51.05	51.05	0.05	
WC Cubicles	Cubicles	9	nr	204.83	1,843.49	1.98	
Screens in softwood glazed with 6 thick GWPP glass	Screens	11	m²	419.74	4,617.11	4.95	
Duct panels in Formica Beautyboard and 50 x 75 softwood framework	Ducts	27	m²	84.47	2,280.69	2.45	
Duct panels in veneered MDF board and 50 x 75 softwood framework	Ducts	26	m²	72.89	1,895.21	2.03	42,118.52
(Element unit rate = £ 87.56 m²)					42,118.52	45.19	
2.8 Internal doors (84 m²)							
900 x 2100 half hour fire resistant single doorsets comprising doors, softwood frames, ironmongery and decoration	Doors	45	m²	330.18	14,858.16	15.94	
900 x 2100 one hour fire resistance single doorsets comprising doors, softwood frames, ironmongery and decoration	Doors	9	m²	510.76	4,596.84	4.93	
900 x 2100 half hour fire resistance single doorsets comprising doors with 2nr vision panels, softwood frames, ironmongery and decoration	Doors	11	m²	407.88	4,486.65	4.81	
1500 x 2100 half hour fire resistance double doorsets comprising Ash veneered doors with 2nr vision panels, Ash frames, ironmongery and decoration	Doors	19	m²	392.88	7,464.77	8.01	
100 wide lintels	Lintels	41	m	25.30	1,037.43	1.11	
200 wide lintels	Lintels	7	m	65.04	455.26	0.49	32,899.12
(Element unit rate = £ 391.66 m²)					32,899.12	35.30	
3.0 INTERNAL FINISHES							
3.1 Wall finishes (1,086 m²)							
13 thick plaster to blockwork walls with eggshell paint finish	Plaster work	703	m²	23.41	16,460.11	17.66	
13 thick plaster to blockwork walls with lining paper and vinyl wall paper finish	Plaster work	124	m²	24.15	2,995.18	3.21	
10 thick plaster to concrete walls with eggshell paint finish	Plaster work	75	m²	34.63	2,596.98	2.79	
13 thick cement and sand render to blockwork walls with 150 x 150 coloured ceramic tiles finish	Tiling	184	m²	55.87	10,280.08	11.03	32,332.34
(Element unit rate = £ 29.77 m²)					32,332.34	34.69	
3.2 Floor finishes (779 m²)							
50 thick sand and cement screed with 300 x 300 x 2.5 vinyl tile finish	Vinyl tiling	19	m²	44.85	852.18	0.91	
50 thick sand and cement screed with 2 thick sheet vinyl finish	Sheet vinyl	108	m²	39.95	4,314.75	4.63	
50 thick sand and cement screed with 250 x 250 x 6 carpet tile finish	Carpet tiling	587	m²	47.94	28,139.08	30.19	
50 thick sand and cement screed with carpet finish	Carpeting	17	m²	66.59	1,132.06	1.21	
50 thick sand and cement screed with 150 x 150 x 13 ceramic tile finish	Ceramic tiling	46	m²	68.12	3,133.29	3.36	
Entrance mats comprising brass Matt frame, 60 thick cement and sand screed and matting	Matting	2	m²	306.60	613.20	0.66	
150 high x 2.5 thick vinyl skirting	Skirting	45	m	6.67	300.15	0.32	
100 high x 2 thick sheet vinyl skirtings with 20 x 30 lacquered softwood cover fillet	Skirting	116	m	25.23	2,926.77	3.14	
25 x 150 Softwood skirtings with lacquer finish	Skirting	214	m	17.25	3,692.14	3.96	
150 x 150 x 13 ceramic tile skirtings on 13 thick sand and cement render	Skirting	74	m	26.11	1,931.95	2.07	47,035.57
(Element unit rate = £ 60.38 m²)					47,035.57	50.47	
3.3 Ceiling finishes (809 m²)							
Suspended ceiling systems to offices	Suspended	587	m²	39.69	23,298.03	25.00	
Suspended ceiling systems to toilets	Suspended	46	m²	50.67	2,330.94	2.50	
Suspended ceiling systems to reception area	Suspended	18	m²	50.67	912.11	0.98	

APPROXIMATE ESTIMATING

COMPOSITE PRICES FOR APPROXIMATE ESTIMATING

COST PLAN FOR AN OFFICE DEVELOPMENT (Cont'd)

Description	Item	Qty		Rate	Item Total	Cost per m²	Total
3.0 INTERNAL FINISHES (cont'd)							
3.3 Ceiling finishes (809 m²) (cont'd)							
Suspended ceiling plain edge trims and 25 x 38 painted painted softwood shadow battens	Edge trim	300	m	18.48	5,545.05	5.95	
Gyproc MF suspended ceiling system with eggshell paint finish	Suspended	127	m²	42.68	5,420.29	5.82	
12.5 Gyproc Square edge plaster board and skim on softwood supports with eggshell paint finish	Plastering	29	m²	54.12	1,569.46	1.68	
10 thick plaster to concrete with eggshell paint finish	Plastering	2	m²	88.30	176.61	0.19	
Plaster coving	Cove	234	m	8.10	1,894.36	2.03	41,146.85
(Element unit rate = £ 50.86 m²)					41,146.85	44.15	
4.0 FITTINGS AND FURNISHINGS AND EQUIPMENT							
4.1 General fittings, furnishings and equipment							
2100 x 500 x 900 reception counter in softwood	Counter	1	nr	1,635.33	1,635.33	1.75	
2700 x 500 x 900 vanity units in Formica faced chipboard	Units	6	nr	1,065.52	6,393.11	6.86	
600x 400 x 6 thick mirrors	Mirror	15	nr	86.96	1,304.40	1.40	
Vertical fabric blinds	Blinds	161	m²	80.25	12,919.53	13.86	
900 x 250 x 19 blockboard lipped shelves on spur support system	Shelves	12	nr	45.36	544.38	0.58	
1200 x 600 x 900 sink base units in melamine finished particle board	Sink base unit	3	nr	256.45	769.35	0.83	23,566.09
					23,566.09	25.29	
5.0 SERVICES							
5.1 Sanitary appliances							
1200 x 600 stainless steel sink units with pillar taps	Sink	3	nr	193.88	581.64	0.62	
White glazed fireclay cleaners sink units with pillar taps	Cleaners sink	3	nr	623.02	1,869.06	2.01	
Vitreous china vanity basins with pillar taps	Vanity basins	12	nr	279.45	3,353.40	3.60	
Vitreous china wash hand basins and pedestal with pillar taps	Wash hand basin	1	nr	218.26	218.26	0.23	
Vitreous china wash hand basin with pillar taps	Wash hand basin	1	nr	172.02	172.02	0.18	
Vitreous china disabled wash hand basin with pillar taps	Wash hand basin	1	nr	207.96	207.96	0.22	
Close coupled vitreous china WC suites	WC suite	11	nr	268.70	2,955.74	3.17	
Showers comprising shower tray, mixer valve and fittings, and corner shower cubicle	Shower	1	nr	1,164.16	1,164.16	1.25	
Vitreous china urinals with automatic cistern	Urinals	3	nr	351.10	1,053.30	1.13	
Disabled grab rails set	Grab rails	1	nr	638.05	638.05	0.68	
Toilet roll holders	Accessories	11	nr	24.65	271.15	0.29	
Paper towel dispensers	Accessories	8	nr	40.52	324.13	0.35	
Paper towel waste bins	Accessories	8	nr	61.65	493.22	0.53	
Sanitary towel disposal units	Accessories	3	nr	1,162.64	3,487.91	3.74	
Hand dryers	Accessories	8	nr	405.23	3,241.87	3.48	20,031.86
					20,031.86	21.49	
5.2 Services equipment							
Allowance for kichenette fittings		1	nr	3,388.89	3,388.89	2.15	3,388.89
5.3 Disposal installations							
Upvc waste pipes and fittings to internal drainage	Internal Drainage	37	nr	134.48	4,975.69	5.34	4,975.69
5.4 Water installations							
Hot and cold water supplies		37	nr	333.28	12,331.35	13.23	12,331.35
5.5 Heat source							
Gas boiler for radiators and hot water from indirect storage tank		1	nr	130,349.12	130,349.12	139.86	130,349.12
5.6 Space heating and air conditioning							
Low temperature hot water heating by radiators		1	nr	4,335.27	4,335.27	4.65	4,335.27
5.7 Ventilating Systems							
Extractor fan to toilet and kitchen areas		9	nr	1,066.93	9,602.34	10.30	9,602.34
5.8 Electrical installations							
Lighting & Power		1	nr	149,321.78	149,321.78	160.22	149,321.78
5.9 Gas and other fuel installations							
To boiler room		1	nr	3,392.82	3,392.82	3.64	3,392.82
5.10 Lift and conveyor installations							
8 person 3 stop hydraulic lift		1	nr	36,468.30	36,468.30	39.13	36,468.30
5.12 Communication, security and control systems							
Burglar Alarm System		1	nr	9,957.94	9,957.94	10.68	
Cabling for telephones		1	nr	13,336.55	13,336.55	14.31	

COST PLAN FOR AN OFFICE DEVELOPMENT (Cont'd)

Description	Item	Qty		Rate	Item Total	Cost per m²	Total
5.0 SERVICES (cont'd)							
5.12 Communication, security and control systems (cont'd)							
Door entry to receptions on all floors		1	nr	2,202.77	2,202.77	2.36	25,497.26
5.14 Builders' work in connection with services							
Builders work in connection with plumbing, mechanical and electrical installations		1	nr	8,547.24	8,547.24	9.17	
Builders Profit and Attendance on Services		1	nr	1,694.44	1,694.44	1.82	10,241.68
					10,241.68	10.99	
8.0 EXTERNAL WORKS							
8.1 Site preparation works							
Clearing site vegetation, filling root voids with excavated material	Site clearance	1716	m²	1.26	2,162.16	2.32	
Excavated topsoil 150 deep and preserve on site	Clear topsoil	214	m³	10.74	2,299.30	2.47	
Dispose of surplus topsoil off site	Disposal	70	m³	36.22	2,535.40	2.72	
					6,996.86	7.51	
8.2 Roads, paths and pavings							
Excavated to reduce levels 400 deep; dispose of excavated material on site; level and compact bottoms; herbicide; 250 thick granular material filling, levelled, compacted and blinded; 70 thick two coat coated macadam pavings; white road lines and markings	Pavings	719	m²	56.10	40,334.07	43.28	
Excavate to reduce levels 400 deep; dispose of excavated material off site; level and compact bottoms; herbicide; 150 thick granular material filling, levelled compacted, and blinded; 200 x 100 x 65 thick clay paviors	Pavings	40	m²	96.59	3,863.62	4.15	
Excavate to reduce levels 400 deep; dispose of excavated material off site; level and compact bottoms; herbicide; 150 thick granular material filling, levelled, compacted and blinded; 900 x 600 x 50 thick concrete paving flags	Pavings	149	m²	50.90	7,584.33	8.14	
125 x 255 precast concrete road kerbs and foundations	Kerbs	265	m	38.27	10,141.55	10.88	
50 x 150 precast concrete edgings and foundations	Kerbs	48	m	23.06	1,106.88	1.19	
					63,030.46	67.63	
8.3 Planting							
Excavate to reduce levels 400 deep; dispose of excavated material off site; prepare sub-soil; herbicide; 150 thick preserved topsoil filling; rotavate; turf	Landscaping	55	m²	36.93	2,030.89	2.18	
Excavate to reduce levels 400 deep; dispose of excavated material off site; prepare sub-soil; herbicide; 450 thick preserved topsoil filling; cultivate; plant 450 - 600 size herbaceous plants - 3/m², and 3m high trees - 1nr per 22m²; fertiliser; 25 thick peat mulch	Planting	397	m²	69.09	27,426.83	29.43	
					29,457.73	31.61	
8.4 Fencing, railings and walls							
Palisade fencing	Fencing	62	m	52.72	3,268.64	3.51	
1800 high fencing	Fencing	96	m	40.13	3,852.79	4.13	
Extra for 1000 x 2000 gate	Gate	1	nr	472.37	472.37	0.51	7,593.80
					7,593.80	8.15	
8.6 External drainage							
Excavate trench 250 - 500 deep; lay 100 diameter vitrified clay pipe and fittings bedded and surrounded in granular material	Pipes in trenches	44	m	57.04	2,509.92	2.69	
Excavate trench 500 - 750 deep; lay 100 diameter vitrified clay pipe and fittings bedded and surrounded in granular material	Pipes in trenches	59	m	61.06	3,602.35	3.87	
Excavate trench 750 - 1000 deep; lay 100 diameter vitrified clay pipe and fittings bedded and surrounded in granular material	Pipes in trenches	49	m	64.29	3,150.29	3.38	
Excavate trench 1000 - 1250 deep; lay 100 diameter vitrified clay pipe and fittings bedded and surrounded in granular material	Pipes in trenches	4	m	66.34	265.38	0.28	
Excavate trench 250 - 500 deep; lay 100 diameter vitrified clay pipe and fittings bedded and surrounded in in-situ concrete	Pipes in trenches	8	m	67.88	543.00	0.58	
Excavate trench 1000 - 1250 deep; lay 150 diameter vitrified clay pipe and fittings bedded and surrounded in in-situ concrete	Pipes in trenches	23	m	106.14	2,441.28	2.62	
Excavate trench 1500 - 1750 deep; lay 150 diameter vitrified clay pipe and fittings bedded and surrounded in in-situ concrete	Pipes in trenches	10	m	117.60	1,175.97	1.26	

COST PLAN FOR AN OFFICE DEVELOPMENT (Cont'd)

Description	Item	Qty		Rate	Item Total	Cost per m²	Total
8.0 EXTERNAL WORKS (cont'd)							
8.6 External drainage (cont'd)							
100 diameter vertical vitrified clay pipe and fittings cased in in-situ concrete	Pipes in trenches	3	m	93.60	280.81	0.30	
100 outlet vitrified clay yard gully with cast iron grating, surrounded in in-situ concrete	Gully	1	nr	169.94	169.94	0.18	
100 outlet vitrified clay road gully with brick kerb and cast iron grating, surrounded in in-situ concrete	Gully	7	nr	356.18	2,493.26	2.68	
Inspection chambers 600 x 900 x 600 deep internally comprising excavation works, in-situ concrete bases, suspended slabs and benchings, engineering brick walls and kerbs, clay channels, galvanised iron step irons and grade A access covers and frames	Insp. chamber	4	nr	973.05	3,892.20	4.18	
Inspection chambers 600 x 900 x 600 deep internally comprising excavation works, in-situ concrete bases, suspended slabs and benchings, engineering brick walls and kerbs, clay channels, galvanised iron step irons and grade B access covers and frames	Insp. chamber	3	nr	962.54	2,887.62 23,412.03	3.10 25.12	23,412.03
8.7 External services							
Excavate trench for 4nr ducts; lay 4nr 100 diameter UPVC ducts and fittings; lay warning tape; backfill trench		20	m	120.05	2,401.04	2.58	
Sewer connection charges		1	nr	8,472.22	8,472.22	9.09	
Gas main and meter charges		1	nr	6,777.77	6,777.77	7.27	
Electricity main and meter charges		1	nr	6,777.77	6,777.77	7.27	
Water main and meter charges		1	nr	5,083.33	5,083.33 29,512.13	5.45 31.67	29,512.13
10 MAIN CONTRACTOR'S PRELIMINARIES							
Site staff, accommodation, lighting, water safety health and welfare, rubbish disposal, cleaning, drying, protection, security, plant, transport, temporary works and scaffolding with a contract period of 38 weeks		item		172,869.70	111,467.89	119.60	111,467.89
				Total			**1,583,302.76**
				Cost per m²		**1,698.82**	

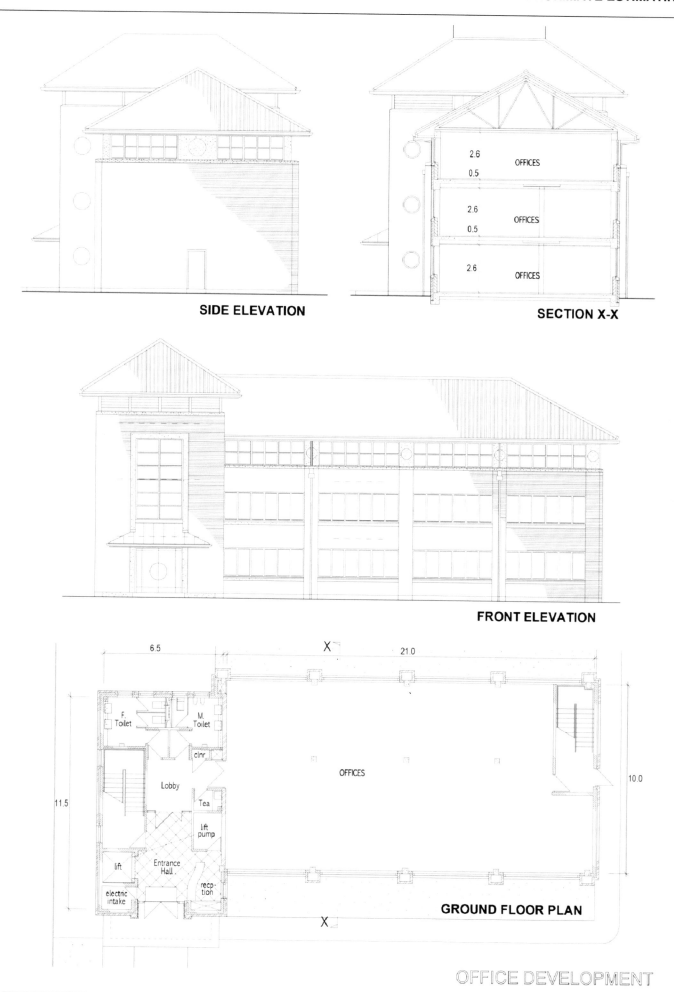

SIDE ELEVATION

SECTION X-X

FRONT ELEVATION

GROUND FLOOR PLAN

APPROXIMATE ESTIMATING

OFFICE DEVELOPMENT

COST PLAN FOR A DETACHED HOUSE

The Cost Plan that follows gives an example of an Approximate Estimate for a detached house with garage and porch as the attached sketch drawings. The estimate can be amended by using the examples of Composite Prices following or the prices included within the preceding pages. Details of the specification used are included within the elements. The cost plan is based on a fully measured bill of quantities priced using the detailed rates within this current edition of Laxton's. Preliminaries are shown separately.

Floor area

House (Including porch 5 m²)	120	m²
Garage	22	m²
Total	**142**	**m²**

Description	Item	Qty		Rate	Item Total	Cost per m²	Total
1.0 SUBSTRUCTURE							
Note: Rates allow for excavation by machine							
1.1 Foundations (86 m²)							
Clear site vegetation excavate 150 topsoil and remove spoil.	Site clearance	113	m²	7.38	834.28	5.88	
Excavate trench commencing at reduced level, average 800 deep, concrete to within 150 of ground level, construct cavity brick wall 300 high and lay pitch polymer damp proof course	Strip foundation	33	m	203.64	6,720.04	47.32	
Excavate trench commencing at reduced level, average 800mm deep, concrete to within 150 of ground level, construct 103 brick wall 300 high and lay pitch polymer damp proof course	Strip foundation	26	m	163.47	4,250.23	29.93	
Excavate trench commencing at reduced level, average 800mm deep, concrete to within 150 of ground level, construct 215 brick wall 300 high and lay pitch polymer damp proof course.	Strip foundation	2	m	353.80	707.59	4.98	12,512.14
(Element unit rate = £ 145.49 m²)					12,512.14	88.11	
1.4 Ground floor construction (86 m²)							
Excavate to reduce levels average 150 deep, lay 150mm bed of hardcore blinded, dpm and 150 concrete ground slab reinforced with Ref. A193 Fabric reinforcement.	Ground floor slab	86	m²	89.89	7,730.53	54.44	7,730.53
(Element unit rate = £ 89.89 m²)					7,730.53	54.44	
2.0 SUPERSTRUCTURE							
2.2 Upper floors (55 m²)							
175 deep floor joists with 22 chipboard finish	Softwood floor	49	m²	43.83	2,147.45	15.12	
175 deep floor joists with 22 plywood finish	Softwood floor	6	m²	58.42	350.53	2.47	2,497.98
(Element unit rate = £ 45.42 m²)					2,497.98	17.59	
2.3 Roof (110 m² on plan)							
2.3.1 Roof structure							
35 degree pitched roof with 50 x 100 rafters and ceiling joists, 100 glass fibre insulation.	Pitched roof	110	m²	79.47	8,741.99	61.56	
Gable end 275 cavity brick and blockwork facings PC £250/1000	Gable wall	29	m²	126.65	3,672.72	25.86	
Gable end 102 brick wall facings £250/1000	Gable wall	8	m²	92.35	738.80	5.20	
2.3.2 Roof coverings							
Interlocking concrete tiles, felt underlay, battens.	Roof tiling	143	m²	24.81	3,547.83	24.98	
Verge, 150 under cloak.	Verge	38	m	13.91	528.58	3.72	
Ridge, half round	Ridge	14	m	19.37	271.18	1.91	
Valley trough tiles.	Valley	2	m	48.02	96.04	0.68	
Abutments, lead flashings cavity trays.	Abutment	11	m	147.01	1,617.10	11.39	
Sheet lead cladding.	Lead roof	2	m²	609.12	1,218.24	8.58	
Eaves soffit, fascia and 100 PVC gutter	Eaves	29	m	63.38	1,838.13	12.94	
2.3.4 Roof drainage							
68 PVCU rainwater pipe.	Rainwater pipe	24	m	27.89	669.44	4.71	22,940.05
(Element unit rate = £ 208.55 m²)					22,940.05	161.55	
2.4 Stairs (1nr)							
Softwood staircase 2670 rise, balustrade one side, 910 wide two flights with half landing	Stairs	1	nr	2,008.80	2,008.80	14.15	2,008.80
2.5 External walls (152 m²)							
Hollow wall, one skin light weight blocks, one skin facings PC £250/1000, ties.	Hollow wall	127	m²	130.26	16,543.47	116.50	
Facing brick wall PC £250/1000 102 thick.	1/2B Facing wall	25	m²	90.15	2,253.65	15.87	
Facing brick wall PC £250/1000 215 thick.	1B Facing wall	3	m²	158.39	475.17	3.35	19,272.29
(Element unit rate = £ 126.79 m²)					19,272.29	135.72	
2.6 Windows and external doors (35 m²)							
2.6.1 External windows (16 m²)							
Hardwood double glazed casement windows.	HW windows	16	m²	1,266.10	20,257.68	142.66	
Galvanised steel lintel, dpc, brick soldier course.	Lintels	16	m	43.71	699.39	4.93	
Softwood window board, dpc close cavity.	Window board	15	m	10.16	152.40	1.07	
Close cavity at jamb, dpc facings to reveal.	Jamb	32	m	14.44	462.22	3.26	

COST PLAN FOR A DETACHED HOUSE (Cont'd)

Description	Item	Qty		Rate	Item Total	Cost per m²	Total
2.0 SUPERSTRUCTURE (cont'd)							
2.6 Windows and external doors (35 m²) (cont'd)							
2.6.2 External Doors (19 m²)							
Hardwood glazed doors, frames, cills, ironmongery	Hardwood doors	6	m²	767.26	4,603.57	32.42	
Hardwood patio doors.	Patio doors	9	m²	812.82	7,315.38	51.52	
Galvanised steel garage door.	Garage door	4	m²	262.36	1,049.46	7.39	
Galvanised steel lintel, dpc, brick soldier course.	Lintels	12	m	45.45	545.38	3.84	
Close cavity at jamb, dpc, facings to reveal.	Jamb	4	m	14.43	57.72	0.41	35,143.20
(Element unit rate = £ 1,004.09 m²)					35,143.20	247.49	
2.7 Internal walls and partitions (153 m²)							
Lightweight concrete block 100 thick.	Block walls	104	m²	46.50	4,835.95	34.06	
75 timber stud partition plasterboard and skim both sides.	Stud partitions	53	m²	75.10	3,980.42	28.03	8,816.36
(Element unit rate = £ 57.62 m²)					8,816.36	62.09	
2.8 Internal doors (40 m²)							
12 hardboard embossed panel door, linings ironmongery PC £20.00	Flush doors	18	m²	295.95	5,327.09	37.51	
3 softwood 2XG glazed door, linings ironmongery PC £20.00	Glazed doors	6	m²	280.72	1,684.33	11.86	
6 Plywood faced door, linings, ironmongery PC £20.00	Flush doors	12	m²	256.27	3,075.21	21.66	
1 plywood faced half hour fire check flush door, frame, ironmongery PC £60.00	Fire doors	2	m²	233.04	466.09	3.28	
Insulated white plastic ceiling hatch	Hatch	1	nr	132.38	132.38	0.93	
Precast concrete lintels.	Lintels	10	m	22.57	225.68	1.59	10,910.78
(Element unit rate = £ 272.77 m²)					10,910.78	76.84	
3.0 INTERNAL FINISHES							
3.1 Wall finishes (264 m²)							
13 light weight plaster.	Plaster work	264	m²	22.56	5,956.34	41.95	
152 x 152 wall tiling coloured PC £20.00 supply	Tiling	13	m²	43.27	562.51	3.96	6,518.85
(Element unit rate = £ 24.69 m²)					6,518.85	45.91	
3.2 Floor finishes (63 m²)							
22 Chipboard, 50 insulation.	Board flooring	63	m²	46.41	2,923.83	20.59	
3 PVC sheet flooring.	Sheet vinyl	18	m²	19.27	346.86	2.44	
25 x 125 softwood skirting.	Skirting	131	m	12.69	1,662.39	11.71	4,933.08
(Element unit rate = £ 78.30 m²)					4,933.08	34.74	
3.3 Ceiling Finishes (115 m²)							
9.5 Plasterboard and set.	Plasterboard	109	m²	19.36	2,110.24	14.86	
Plywood finish	Plywood	6	m²	59.01	354.06	2.49	2,464.30
(Element unit rate = £ 21.43 m²)					2,464.30	17.35	
4.0 FITTINGS AND FURNISHINGS AND EQUIPMENT							
4.1 General fittings, furnishings and equipment							
Kitchen units :-							
600 base unit		2	nr	240.03	480.06	3.38	
1000 base unit		6	nr	383.56	2,301.36	16.21	
500 broom cupboard		1	nr	333.55	333.55	2.35	
600 wall unit		3	nr	115.42	346.26	2.44	
1000 wall unit		2	nr	126.30	252.60	1.78	
Work tops		10	m	97.88	978.80	6.89	
Gas fire surround		1	nr	605.00	605.00	4.26	5,297.63
					5,297.63	37.31	
5.0 SERVICES							
5.1 Sanitary appliances							
Lavatory basin with pedestal.		3	nr	218.26	654.78	4.61	
Bath with front panel and plinth.		1	nr	331.51	331.51	2.33	
WC suite.		3	nr	249.44	748.32	5.27	
Stainless steel sink and drainer. (500)		1	nr	193.88	193.88	1.37	
Stainless steel sink and drainer. (600)		1	nr	219.39	219.39	1.54	
shower.		1	nr	886.43	886.43	6.24	3,034.31
					3,034.31	21.37	
5.2 Services Equipment							
Electric oven built in.		1	nr	880.00	880.00	6.20	
Gas hob built in.		1	nr	440.00	440.00	3.10	
Kitchen extract unit		1	nr	357.50	357.50	2.52	
Gas fire with balanced flue.		1	nr	440.00	440.00	3.10	2,117.50
					2,117.50	14.91	

APPROXIMATE ESTIMATING

COST PLAN FOR A DETACHED HOUSE (Cont'd)

Description	Item	Qty		Rate	Item Total	Cost per m²	Total
5.0 SERVICES (cont'd)							
5.3 Disposal Installations							
Upvc soil and vent pipe.		15	m	106.05	1,590.73	11.20	
Air admittance valve.		3	nr	69.10	207.30	1.46	1,798.03
					1,798.03	12.66	
5.4 Water Installations							
Cold water, tanks, 9 draw off points		item		1,443.54	1,443.54	10.17	
Hot water cylinder, 7 draw off points		item		1,094.39	1,094.39	7.71	2,537.93
					2,537.93	17.87	
5.5 Heat Source							
Boiler, pump, and controls.		item		1,247.04	1,247.04	8.78	1,247.04
5.6 Space heating and air conditioning							
Low temperature hot water radiator system.		item		4,950.00	4,950.00	34.86	4,950.00
5.7 Ventilating Systems							
Extract fan units.		3	nr	71.78	215.34	1.52	215.34
5.8 Electrical Installations							
Electrical Installation comprising: 16 Nr lighting points, 30 Nr outlet points, cooker points, immersion heater point and 9 Nr fittings points to include smoke alarms, bells, telephone and aerial installation.		item		5,635.77	5,635.77	39.69	5,635.77
5.9 Gas and other fuel installations							
Installation for boiler, hob and fire		item		605.00	605.00	4.26	605.00
5.14 Builders' work in connection with services							
Building work for service installation.		item		2,147.56	2,147.56	15.12	2,147.56
8.0 EXTERNAL WORKS							
8.1 Site preparation works							
Clear site vegetation excavate 150 topsoil and remove spoil.	Site clearance	479	m²	7.38	3,536.46	24.90	3,536.46
8.2 Roads, paths and pavings							
Excavate and lay 50 precast concrete flag paving on hardcore.	Paving	120	m²	97.59	11,710.61	82.47	11,710.61
8.3 Planting							
Turfed Areas	Landscaping	103	m²	53.64	5,525.39	38.91	5,525.39
8.4 Fencing, railings and walls							
Brick wall, excavation 600 deep, concrete foundation, 215 brickwork facing. PC £200/1000 damp proof course brick on edge coping 900 high	Garden wall	13	m	156.48	2,034.23	14.33	
Plastic coated chain link fencing 900 high on angle iron posts	Fencing	92	m	54.30	4,996.05	35.18	7,030.27
					7,030.27	49.51	
8.6 External drainage							
100 diameter clay pipes, excavation, 150 bed and surround, pea shingle, backfill 750 deep.		85	m	74.75	6,354.02	44.75	
Polypropylene inspection chambers; universal inspection chamber; polypropylene; 475 diameter; preformed chamber base benching; for 100mm pipe; chamber shaft, 185mm effective length; cast iron cover and frame; single seal A15 light duty to suit.		6	nr	290.24	1,741.44	12.26	8,095.46
					8,095.46	57.01	
8.7 External services							
100 diameter clay duct, excavation backfill, polythene warning tape, 500 deep.		38	m	36.27	1,378.13	9.71	
Provisional sums for External Services Gas, electric, water, telephone.		item		4,497.77	4,497.77	31.67	5,875.90
					5,875.90	41.38	
10 MAIN CONTRACTOR'S PRELIMINARIES							
Site staff, accommodation, lighting, water, safety and health and welfare, rubbish disposal cleaning, drying, protection, security, plant, transport, temporary works and scaffolding with a contract period of 28 weeks				15,644.44	14,059.46	99.01	14,059.46
				Total			**221,168.02**
				Cost per m²		**1,557.52**	

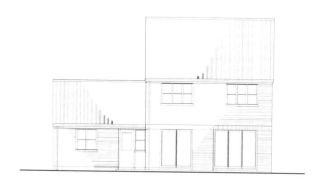

REAR ELEVATION

SIDE ELEVATION

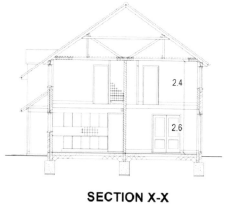

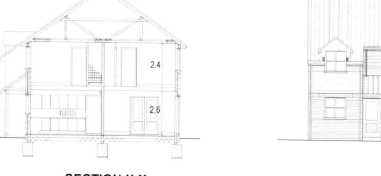

SECTION X-X

FRONT ELEVATION

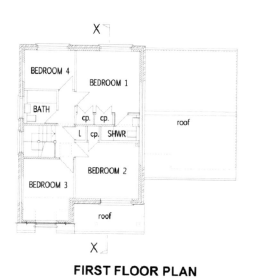

FIRST FLOOR PLAN

GROUND FLOOR PLAN

APPROXIMATE ESTIMATING

4 - BEDROOM DETACHED HOUSE

COMPOSITE PRICES

The following composite prices are based on the detailed rates herein and may be used to amend the previous example cost plans or to create new approximate budget estimates.

The composite prices can be further extended and amended by reference to the various sections.

Allowances should be made for location factors, tender values and price changes.

PRELIMINARIES £

The following preliminary items need to be added to the composite prices that follow.

A calculation can be made by reference to the preliminaries section.

Management and staff - average each
Site accommodation
Power and lighting
  add setup and remove
Heating
Water
  add setup and remove
Telephone
Safety, health and welfare
Rubbish disposal and cleaning site
Final clean
Drying out building
Security
Small plant and tools
Insurance of works
Travelling time and fares
Rates on temporary buildings
Signboards and notices
Glass breakage
Sample panels
Testing
Attendance; chainman etc (on clerk of works etc)
Setting out equipment, pegs
Provision for winter working
Defects
Sureties and bonds
Special conditions of contract
Attendance on subcontractors
Cranes - mobile tower
Cranes - static tower
Cranes - mobile crane
Hoist
Pumping
On-site transport
Transport for plant
Temporary roads
Temporary access
External access scaffold
Mobile tower
Support scaffold
Internal scaffold
Hoardings

DEMOLITIONS AND ALTERATIONS

Demolition

Demolish timber outbuilding

size 2.50 x 2.00 x 3.00 high	nr	260.68

Demolish brick lean-to and make good

size 1.50 x 1.00 x 2.00 high	nr	329.58

Demolishing parts of structures

Substructures

incl. Excavation and removal by hand

break out brickwork	m³	220.88
break out concrete	m³	269.63
break out reinforced concrete	m³	350.88

break up concrete bed

100 thick	m²	26.97
150 thick	m²	40.45
200 thick	m²	53.93

break up reinforced concrete bed

100 thick	m²	35.09
150 thick	m²	52.63
200 thick	m²	64.82

Walls

brick internal

102 thick	m²	19.76
215 thick	m²	40.72
327 thick	m²	62.37

brick external

102 thick	m²	15.03
250 thick cavity wall	m²	31.27

Floors

concrete reinforced; screed

150 thick; 50 screed	m²	67.36
200 thick; 50 screed	m²	84.28

timber boarded on joists

225 x 50 joists; floor boards	m²	15.60

Roofs

timber flat roofs

150 x 50 joists, 25 boards; felt	m²	16.42

timber pitched roof

100 x 50 rafters; joists; plain tile	m²	20.34
100 x 50 rafters; joists; interlocking tiles	m²	17.70
100 x 50 rafters; joists; slates	m²	20.22

Alterations

Includes make good finishes

Cut opening through reinforced concrete floor

150 thick x 900 x 2134 for staircase	nr	1,106.38

Cut opening through block wall

100 thick x 838 x 2032 for doorway	nr	696.63

Cut opening through brick wall

215 thick x 914 x 2082 for doorway	nr	1,023.64
250 thick cavity wall x 900 x 900 for window	nr	659.23

Cut opening through stud partition

100 thick x 838 x 2032 for doorway	nr	493.43

Fill opening in block wall

100 thick x 838 x 2032; door removed	nr	484.05

Fill opening in brick wall

102 thick x 838 x 2032; door removed	nr	591.06
215 thick x 914 x 2082; external door removed	nr	957.81
250 thick cavity wall x 914 x 2082; external door removed	nr	875.38

Fill opening in stud partition

100 thick x 838 x 2032; door removed	nr	439.46

Removal

Remove window

1486 x 923 metal	nr	27.60
1225 x 1070 wood casement; frame	nr	30.74
915 x 1525 cased frame; sashes	nr	45.85

Remove door and frame

single internal	nr	22.53
single external	nr	25.10

Remove roof coverings

3 layer felt	m²	5.29
asphalt	m²	6.62
slates	m²	14.89
tiles	m²	15.01

SUBSTRUCTURE

Note: Rates allow for excavation by machine.

APPROXIMATE ESTIMATING

COMPOSITE PRICES (Cont'd)

£

SUBSTRUCTURE (cont'd)

Strip Foundations

Excavating, 225x450 concrete, brickwork, dpc, common bricks

800 deep for 103 wall	m	90.34
1000 deep for 103 wall	m	109.42
1200 deep for 103 wall	m	141.62
800 deep for 215 wall	m	135.51
1000 deep for 215 wall	m	166.93
1200 deep for 215 wall	m	198.02
800 deep for 275 wall	m	158.39
1000 deep for 275 wall	m	198.12
1200 deep for 275 wall	m	235.46
Extra 100mm depth concrete foundation; excavation; concrete	m	11.22
Extra 100mm depth 103 brick wall in foundation; excavation; brickwork	m	9.67
Extra 100mm depth 215 brick wall in foundation; excavation; brickwork	m	15.68
Extra 100mm depth 275 brick wall in foundation; excavation; brickwork	m	17.36
Extra for four courses facing bricks P.C. £250.00 per 1000	m	1.09

Excavating, 225x600 concrete, brickwork, dpc, common bricks

800 deep for 275 wall	m	167.01
1000 deep for 275 wall	m	207.54
1200 deep for 275 wall	m	241.23
Extra 100mm depth 275 brick wall in foundation; excavation; brickwork	m	17.63
Extra for four courses facing bricks P.C. £250.00 per 1000	m	1.09

Excavating, 225x450 concrete, brickwork, dpc, Engineering bricks Class B

800 deep for 215 wall	m	141.27
1000 deep for 215 wall	m	174.25
1200 deep for 215 wall	m	206.92
800 deep for 275 wall	m	164.82
1000 deep for 275 wall	m	206.32
1200 deep for 275 wall	m	245.43
Extra 100mm depth concrete foundation; excavation; concrete	m	11.22
Extra 100mm depth 103 brick wall in	m	10.12
Extra 100mm depth 215 brick wall in foundation; excavation; brickwork	m	16.47
Extra 100mm depth 275 brick wall in foundation; excavation; brickwork	m	18.25
Extra for four courses facing bricks P.C. £250.00 per 1000	m	-0.24

Excavating, 225x450 concrete, blockwork, dpc, Concrete blocks

800 deep for 100 wall	m	75.11
1000 deep for 100 wall	m	80.90
1200 deep for 100 wall	m	93.85
Extra 100mm depth concrete foundation; excavation; concrete	m	11.22
Extra 100mm depth 100 block wall in foundation; excavation; blockwork	m	6.61
800 deep for 140 wall	m	77.06
1000 deep for 140 wall	m	92.49
1200 deep for 140 wall	m	107.92
Extra 100mm depth concrete foundation; excavation; concrete	m	11.22
Extra 100mm depth 140 block wall in foundation; excavation; blockwork	m	7.81
800 deep for 190 wall	m	86.11
1000 deep for 190 wall	m	103.87
1200 deep for 190 wall	m	121.64
Extra 100mm depth concrete foundation; excavation; concrete	m	11.22
Extra 100mm depth 190 block wall in foundation; excavation; blockwork	m	8.98

Excavating, 225 x 450 concrete, blockwork, dpc, Thermalite Trenchblocks

800 deep for 275 wall	m	104.45
1000 deep for 275 wall	m	127.08
1200 deep for 275 wall	m	149.71
Extra 100mm depth concrete foundation; excavation; concrete	m	11.22
Extra 100mm depth 190 block wall in foundation; excavation; blockwork	m	11.54

Excavating, 225x600 concrete, blockwork, dpc, Thermalite Trenchblocks

800 deep for 275 wall	m	114.13
1000 deep for 275 wall	m	137.55
1200 deep for 275 wall	m	160.65
Extra 100mm depth concrete foundation; excavation; concrete;	m	11.22
Extra 100mm depth 275 block wall in foundation; excavation; blockwork	m	11.54

Fabric reinforcement ref. A193 (no allowance for working space or formwork)

350 wide	m	2.22
600 wide	m	3.80
750 wide	m	5.35

Trench Fill Foundations

Excavating 450 wide concrete fill to 150 of ground level, brickwork, dpc, Common bricks

800 deep for 102 wall	m	94.69
1000 deep for 102 wall	m	115.08
1200 deep for 102 wall	m	135.46
800 deep for 215 wall	m	111.74
1000 deep for 215 wall	m	132.13
1200 deep for 215 wall	m	152.52
800 deep for 275 wall	m	119.99
1000 deep for 275 wall	m	140.38
1200 deep for 275 wall	m	160.77
Extra 100mm depth concrete foundation; excavation; concrete;	m	11.22
Extra for four courses facing bricks PC £250 per 1000	m	1.09

Column bases

Excavating, 450 concrete

1000 x 1000 x 1000 deep	nr	126.08
1000 x 1000 x 1200 deep	nr	142.19
1000 x 1000 x 1500 deep	nr	157.41
1200 x 1200 x 1000 deep	nr	177.14
1200 x 1200 x 1200 deep	nr	198.17
1200 x 1200 x 1500 deep	nr	218.44
1500 x 1500 x 1000 deep	nr	269.89
1500 x 1500 x 1200 deep	nr	299.34
1500 x 1500 x 1500 deep	nr	328.44
2000 x 2000 x 1000 deep	nr	467.54
2000 x 2000 x 1200 deep	nr	513.86
2000 x 2000 x 1500 deep	nr	561.02

Excavating, 750 concrete

1000 x 1000 x 1000 deep	nr	180.00
1000 x 1000 x 1200 deep	nr	196.11
1000 x 1000 x 1500 deep	nr	211.33
1000 x 1000 x 2000 deep	nr	236.70
1200 x 1200 x 1000 deep	nr	254.78
1200 x 1200 x 1200 deep	nr	275.81
1200 x 1200 x 1500 deep	nr	296.08
1200 x 1200 x 2000 deep	nr	329.87
1500 x 1500 x 1000 deep	nr	391.20
1500 x 1500 x 1200 deep	nr	420.65
1500 x 1500 x 1500 deep	nr	449.75
1500 x 1500 x 2000 deep	nr	498.26
2000 x 2000 x 1000 deep	nr	683.20
2000 x 2000 x 1200 deep	nr	729.52
2000 x 2000 x 1500 deep	nr	776.68
2000 x 2000 x 2000 deep	nr	855.28

APPROXIMATE ESTIMATING

COMPOSITE PRICES (Cont'd)

SUBSTRUCTURE (cont'd)

Column bases (cont'd)

Excavating, 750 concrete (cont'd)

		£
3000 x 3000 x 1000 deep	nr	1,509.60
3000 x 3000 x 1200 deep	nr	1,600.24
3000 x 3000 x 1500 deep	nr	1,696.05

Excavating, 1000 concrete

1000 x 1000 x 1000 deep	nr	224.93
1000 x 1000 x 1200 deep	nr	241.04
1000 x 1000 x 1500 deep	nr	256.26
1000 x 1000 x 2000 deep	nr	281.63
1200 x 1200 x 1000 deep	nr	319.48
1200 x 1200 x 1200 deep	nr	340.51
1200 x 1200 x 1500 deep	nr	360.78
1200 x 1200 x 2000 deep	nr	394.56
1500 x 1500 x 1000 deep	nr	492.29
1500 x 1500 x 1200 deep	nr	521.74
1500 x 1500 x 1500 deep	nr	550.84
1500 x 1500 x 2000 deep	nr	599.35
2000 x 2000 x 1000 deep	nr	862.92
2000 x 2000 x 1200 deep	nr	909.24
2000 x 2000 x 1500 deep	nr	956.40
2000 x 2000 x 2000 deep	nr	1,035.00
3000 x 3000 x 1000 deep	nr	1,913.97
3000 x 3000 x 1200 deep	nr	2,004.61
3000 x 3000 x 1500 deep	nr	2,100.42
3000 x 3000 x 2000 deep	nr	2,260.11

Excavating, 1500 concrete

1000 x 1000 x 1500 deep	nr	346.12
1000 x 1000 x 2000 deep	nr	371.49
1500 x 1500 x 1500 deep	nr	753.03
1500 x 1500 x 2000 deep	nr	801.53
2000 x 2000 x 1500 deep	nr	1,315.84
2000 x 2000 x 2000 deep	nr	1,394.44
3000 x 3000 x 1500 deep	nr	2,909.16
3000 x 3000 x 2000 deep	nr	3,068.85

Excavating, 450 reinforced concrete

1000 x 1000 x 1000 deep	nr	366.96
1000 x 1000 x 1200 deep	nr	398.34
1000 x 1000 x 1500 deep	nr	434.07
1000 x 1000 x 2000 deep	nr	493.62
1200 x 1200 x 1000 deep	nr	499.13
1200 x 1200 x 1200 deep	nr	536.56
1200 x 1200 x 1500 deep	nr	581.44
1200 x 1200 x 2000 deep	nr	656.24

Excavating, 750 reinforced concrete

1000 x 1000 x 1000 deep	nr	603.26
1000 x 1000 x 1200 deep	nr	633.04
1000 x 1000 x 1500 deep	nr	668.77
1000 x 1000 x 2000 deep	nr	728.32
1200 x 1200 x 1000 deep	nr	814.39
1200 x 1200 x 1200 deep	nr	851.82
1200 x 1200 x 1500 deep	nr	896.70
1200 x 1200 x 2000 deep	nr	971.50

Ground beams

Excavation, blinding, reinforced concrete

300 x 450	m	119.35
300 x 600	m	180.41
300 x 450	m	165.51
300 x 600	m	239.72

Ground slabs

Excavation, 150 hardcore, blinding, concrete

150 ground slab	m²	53.96
200 ground slab	m²	61.36
225 ground slab	m²	66.28
Extra: 50 blinding	m²	10.24
Extra: Fabric reinforcement A142	m²	5.14
Extra: Fabric reinforcement A193	m²	6.34
Slab thickening 450 wide x 150 thick	m²	13.28

Damp proof membrane

250mu Polythene sheeting	m²	1.65
2 coats Bitumen emulsion	m²	12.66
Extra: 50mm Expanded polystyrene insulation	m²	6.97
Extra: 50mm Concrete thickness	m²	7.88
Extra: 100mm Reduced level excavation	m²	3.90
Extra: 100mm hardcore filling	m²	3.66

Beam and block floors

(excludes excavation, blinding etc.)

155mm beam floor span 3.9m	m²	47.38
Extra: 150mm Reduced level excavation; 50 concrete blinding	m²	14.17

Piling

On/off site charge	nr	9,472.32
450 bored cast piles n/e 6m deep	m	55.47
610 bored cast piles n/e 6m deep	m	83.91
Cutting off tops of piles	nr	19.01

INSIDE EXISTING BUILDING

Note: rates allow for excavation by hand.

Strip foundations

Excavating, 225 concrete, brickwork, dpc Common bricks

800 deep for 103 wall	m	114.57
1000 deep for 103 wall	m	138.79
1200 deep for 103 wall	m	197.31
800 deep for 215 wall	m	161.89
1000 deep for 215 wall	m	199.54
1200 deep for 215 wall	m	246.01
Extra 100mm depth concrete foundation; excavation; concrete	m	15.51

Trench Fill Foundations

Excavating 450 wide concrete fill to 150 of ground level, brickwork, dpc Common bricks

800 deep for 102 wall	m	124.83
1000 deep for 102 wall	m	152.94
800 deep for 215 wall	m	141.88
1000 deep for 215 wall	m	148.73
Extra 100mm depth concrete foundation; excavation; concrete	m	15.51

Ground slabs

Excavation, 150 hardcore, blinding, concrete

150 ground slab	m²	81.57
200 ground slab	m²	96.14
Slab thickening 450 wide x 150 thick	m	19.67

Damp proof membrane

250 mu polythene sheeting	m²	1.65
2 coats bitumen emulsion	m²	12.66

FRAME

Reinforced concrete

200 x 200 column	m	71.43
225 x 225 column	m	80.36
300 x 300 column	m	119.08
200 x 200 attached beam	m	72.12
225 x 250 attached beam	m	90.04
300 x 450 attached beam	m	165.37
400 x 650 attached beam	m	267.18
Casing to 152 x 152 steel column	m	56.46
Casing to 203 x 203 steel column	m	73.20
Casing to 146 x 254 steel beam	m	72.67
Casing to 165 x 305 steel beam	m	85.62
Extra: wrought formwork	m²	5.17

Steel

Columns and beams	tonne	1,344.38
Roof trusses	tonne	2,295.88

APPROXIMATE ESTIMATING

COMPOSITE PRICES (Cont'd)

£

FRAME (cont'd)

Steel (cont'd)

Purlins and cladding rails	tonne	1,679.88
Small sections	tonne	891.00
Two coats primer at works	tonne	264.00
Prime, undercoat and gloss paint	m²	11.43

Glued laminated softwood beams

65x150x4000 long	nr	107.88
65x250x4000 long	nr	184.99
65x300x6000 long	nr	310.91
65x325x8000 long	nr	438.70
90x200x4000 long	nr	187.31
90x325x8000 long	nr	550.44
90x400x10000 long	nr	830.91
90x450x12000 long	nr	1,104.72
115x250x4000 long	nr	291.94
115x300x4000 long	nr	346.82
115x350x8000 long	nr	753.06
115x425x12000 long	nr	1,292.70
115x450x8000 long	nr	942.25
115x500x14000 long	nr	1,773.77
140x350x8000 long	nr	908.83
140x450x12000 long	nr	1,682.00
165x450x12000 long	nr	1,958.25
190x475x14000 long	nr	2,763.16

UPPER FLOORS

Concrete

In-situ reinforced concrete

150 thick	m²	95.34
200 thick	m²	103.50
225 thick	m²	112.55

In-situ reinforced concrete with permanent Expamet

Hy-rib formwork ref 2411 and A142 mesh reinforcement

100 thick	m²	99.24
125 thick	m²	104.28
150 thick	m²	109.43

Hollow prestressed concrete

125 thick	m²	65.78
150 thick	m²	72.00
200 thick	m²	78.56

Beam and block

200 thick	m²	47.38

Timber

Softwood joist

100 thick	m²	22.62
150 thick	m²	26.31
175 thick	m²	26.24
200 thick	m²	30.12
225 thick	m²	32.74
250 thick	m²	39.06

Floor boarding

19 square edge softwood	m²	24.14
25 square edge softwood	m²	25.02
19 tongued and grooved softwood	m²	26.45
25 tongued and grooved softwood	m²	26.58
18 tongued and grooved chipboard	m²	12.41
22 tongued and grooved chipboard	m²	14.87
15 tongued and grooved plywood, Douglas Fir unsanded	m²	14.41
18 tongued and grooved plywood, Douglas Fir unsanded	m²	14.98
15 tongued and grooved plywood, Birch Faced BB quality	m²	25.09
18 tongued and grooved plywood, Birch Faced BB quality	m²	28.78

ROOF

Roof structure

Reinforced in-situ concrete

100 thick	m²	77.73
150 thick	m²	91.03
200 thick	m²	93.72
50 lightweight screed to falls	m²	27.75
50 cement sand screed to falls	m²	17.20
50 vermiculite screed to falls	m²	23.28
50 flat glass fibre insulation board	m²	10.13
Extra for vapour barrier	m²	5.57

Softwood flat roof joists

50 x 100 joists @ 400 centres	m²	27.82
50 x 175 joists @ 400 centres	m²	29.16
50 x 200 joists @ 400 centres	m²	27.16
50 x 225 joists @ 400 centres	m²	36.14
50 x 100 joists @ 600 centres	m²	19.95
50 x 175 joists @ 600 centres	m²	22.20
50 x 200 joists @ 600 centres	m²	19.48
50 x 225 joists @ 600 centres	m²	25.50

Boarding

18 chipboard	m²	14.50
25 softwood tongued grooved	m²	14.98
18 plywood decking	m²	14.81
24 plywood decking	m²	19.45
Pitched roof with 50 x 100 rafters and ceiling joists at 400 centres	m²	45.71
Pitched roof with trussed rafters at 600 centres	m²	22.81

Roof coverings

Three layer fibre based felt	m²	28.14
Three layer high performance felt	m²	59.71
20 two coat mastic asphalt	m²	48.61
50 lightweight screed to falls	m²	27.75
50 cement sand screed to falls	m²	17.20
60 Vermiculite screed to falls	m²	26.87
50 flat glass fibre insulation board	m²	10.13
Extra for vapour barrier	m²	5.57
Profile 3 fibre cement sheeting (on slope)	m²	26.46
Profile 6 fibre cement sheeting (on slope)	m²	22.57
Colour coated galvanised steel sheeting (on slope)	m²	20.84
381 x 227 concrete interlocking tiles (on slope)	m²	38.92
268 x 165 plain concrete tiles (on slope)	m²	48.60
268 x 165 machine made clay tiles (on slope)	m²	49.51
265 x 165 hand made tiles (on slope)	m²	84.64
400 x 200 fibre cement slates (on slope)	m²	53.50
500 x 250 fibre cement slates (on slope)	m²	39.34
600 x 300 fibre cement slates (on slope)	m²	30.87
400 x 200 natural slates (on slope)	m²	85.52
500 x 250 natural slates (on slope)	m²	102.57
600 x 300 natural slates (on slope)	m²	141.28
100 Fibreglass insulation quilt	m²	6.03
100 Rockwool insulation quilt	m²	7.48
150 Rockwool insulation quilt	m²	10.62

Eaves, soffits, fascias and gutters

Eaves soffit, fascia, gutter and decoration

225 x 150, 114 pvc-u	m	49.00
225 x 150, 100 half round Cast Iron gutter	m	77.63
225 x 150, 114 half round Cast Iron gutter	m	79.69
225 x 150, 125 half round Cast Iron gutter	m	86.77
225 x 150, 150 half round Cast Iron gutter	m	116.46
225 x 150, 100 og Cast Iron gutter	m	76.23
225 x 150, 114 og Cast Iron gutter	m	78.42
225 x 150, 125 og Cast Iron gutter	m	83.80
400 x 200, 114 pvc-u	m	56.51
400 x 200, 100 half round Cast Iron gutter	m	85.14
400 x 200, 114 half round Cast Iron gutter	m	87.20
400 x 200, 125 half round Cast Iron gutter	m	94.28
400 x 200, 150 half round Cast Iron gutter	m	123.97

APPROXIMATE ESTIMATING

COMPOSITE PRICES (Cont'd)

£

ROOF (cont'd)

Eaves, soffits, fascias and gutters (cont'd)

Eaves soffit, fascia, gutter and decoration (cont'd)

400 x 200, 100 og Cast Iron gutter	m	83.74
400 x 200, 114 og Cast Iron gutter	m	85.93
400 x 200, 125 og Cast Iron gutter	m	91.31
400 x 200, 102 half round cast aluminium gutter	m	79.83
400 x 200, 127 half round cast aluminium gutter	m	90.30

Verges, hips, ridges and valleys

Verge undercloak, barge board and decoration

225 x 25	m	34.45

Hip rafter and capping

225 x 50, concrete capping tiles	m	29.32
225 x 50, clay machine made capping tiles	m	38.73
225 x 50, clay hand made capping tiles	m	40.81
225 x 50, concrete bonnet hip tiles	m	73.04
225 x 50, clay machine made bonnet hip tiles	m	63.28
225 x 50, clay hand made bonnet hip tiles	m	73.04

Ridge board and capping

175 x 32, concrete capping tile	m	25.44
175 x 32, machine made clay tile	m	34.85
175 x 32, hand made clay tile	m	36.93

Valley rafter and tile

225 x 32, concrete angular valley tile	m	54.97
225 x 32, concrete valley trough	m	32.90
225 x 32, machine made clay valley tile	m	74.65
225 x 32, hand made clay valley tile	m	70.55

Valley rafter, sole boards and leadwork

225 x 32, code 4 lead	m	116.03
225 x 32, code 5 lead	m	125.33

Rain water pipes

Rainwater pipe

68 pvc-u pipe	m	24.24
63 cast iron pipe and decoration	m	86.08
75 cast iron pipe and decoration	m	94.54
100 cast iron pipe and decoration	m	120.58
63 cast aluminium pipe	m	50.18
75 cast aluminium pipe	m	64.45
100 cast aluminium pipe	m	87.78

STAIRS

Concrete

Reinforced concrete 2670 rise, mild steel balustrade one side, straight flight

1000 wide granolithic finish	nr	3,368.71
1500 wide granolithic finish	nr	4,646.80
1000 wide pvc sheet finish non slip nosings	nr	3,976.49
1500 wide pvc sheet finish non slip nosings	nr	4,486.05
1000 wide terrazzo finish	nr	4,458.51
1500 wide terrazzo finish	nr	5,061.16

Reinforced concrete 2670 rise, mild steel balustrade one side in two flights with half landing

1000 wide granolithic finish	nr	4,009.70
1500 wide granolithic finish	nr	4,596.24
1000 wide pvc sheet finish non slip nosings	nr	4,648.88
1500 wide pvc sheet finish non slip nosings	nr	5,104.48
1000 wide terrazzo finish	nr	5,160.94
1500 wide terrazzo finish	nr	6,440.97

Timber

Softwood staircase 2670 rise, balustrade one side.

910 wide straight flight	nr	1,421.73
910 wide 1 flight with 3 winders	nr	1,749.39
910 wide 2 flight with half landing	nr	1,749.39

Steel

Mild steel staircase 2670 rise, balustrade both sides.

910 wide straight flight	nr	3,696.00
910 wide 2 flights with quarter landing	nr	6,880.50
910 wide 2 flights with half landing	nr	8,332.50

Spiral steel staircase, balustrade,

1600 diameter 2640 rise, domestic, powder coated	nr	2,588.09
1600 diameter 3000 rise, external, galvanised, decorated	nr	3,426.89

EXTERNAL WALLS

Brickwork

Common bricks

102 wall	m²	71.14
215 wall	m²	127.99
327 wall	m²	172.53

Hollow wall with stainless steel ties

275 common bricks both skins	m²	145.63
275 facings one side P.C. £200/1000	m²	145.86
275 facings one side P.C. £250/1000	m²	149.27
275 facings one side P.C. £300/1000	m²	152.68
275 facings one side P.C. £350/1000	m²	156.08

Engineering bricks Class B

102 wall	m²	75.57
215 wall	m²	135.87
327 wall	m²	182.13

Engineering bricks Class A

102 wall	m²	103.85
215 wall	m²	192.91
327 wall	m²	267.93

Facing bricks

102 wall P.C. £200 per 1000	m²	71.37
102 wall P.C. £250 per 1000	m²	74.78
102 wall P.C. £300 per 1000	m²	78.19
102 wall P.C. £350 per 1000	m²	81.59
215 wall P.C. £200 per 1000	m²	119.45
215 wall P.C. £250 per 1000	m²	126.32
215 wall P.C. £300 per 1000	m²	133.19
215 wall P.C. £350 per 1000	m²	140.07

Blockwork

Aerated concrete block 4.0n/mm²

100 wall	m²	40.47
140 wall	m²	51.27
190 wall	m²	61.35

Aerated concrete block 7.0n/mm²

100 wall	m²	47.23
140 wall	m²	59.56
190 wall	m²	66.98
215 wall	m²	81.66

Dense concrete block 7.0n/mm²

100 wall	m²	49.91
140 wall	m²	59.33
190 wall	m²	70.38
215 wall	m²	76.72

Fair face concrete block

100 wall	m²	61.35
140 wall	m²	70.03
190 wall	m²	81.75

Composite

Hollow wall in light weight blocks with stainless steel ties.

275 facings one side P.C. £200/1000	m²	115.19
275 facings one side P.C. £250/1000	m²	118.60
275 facings one side P.C. £300/1000	m²	122.01
275 facings one side P.C. £350/1000	m²	125.41

Extra for:

50 mm glass fibre slab insulation	m²	7.71

COMPOSITE PRICES (Cont'd)

£

EXTERNAL WALLS (cont'd)

Composite (cont'd)

Extra for: (cont'd)

13 mm two coat plain render painted	m²	12.60
15 mm pebbled dash render	m²	25.42

WINDOWS AND EXTERNAL DOORS

Windows

Softwood single glazed casement window

type W107C size 630 x 750	nr	406.10
type W110C size 630 x 1050	nr	463.49
type W207C size 1200 x 750	nr	578.84
type W212C size 1200 x 1200	nr	808.62
type W312CC size 1770 x 1200	nr	1,121.53
average cost	m²	301.50
purpose made in panes 0.10-0.50m²	m²	333.74

Softwood double glazed casement window

type W107C size 630 x 750	nr	485.55
type W110C size 630 x 1050	nr	563.07
type W207C size 1200 x 750	nr	684.39
type W212C size 1200 x 1200	nr	955.94
type W309CC size 1770 x 900	nr	1,258.99
type W312CC size 1770 x 1200	nr	1,325.40
average cost	m²	404.26
purpose made in panes 0.10-0.50m²	m²	498.24

Hardwood double glazed standard casement windows

type OLEW107V size 630 x 750	nr	906.06
type OLEW112V size 630 x 1200	nr	1,042.05
type OLEW2N10CV size 915 x 1050	nr	1,486.80
type OLEWH212CSDL size 1200 x 1200	nr	1,718.78
average cost	m²	553.29

Afrormosia double glazed casement windows

purpose made in panes 0.l0-0.50m²	m²	628.63

European Oak double glazed casement windows

purpose made in panes 0.10-0.50m²	m²	783.68

Softwood glazed double hung sash window

type CDH0810GEO size 860 x 1050	nr	1,200.44
type CDH0813GEO size 860 x 1350	nr	1,343.67
type CDH1013GEO size 1085 x 1350	nr	1,590.71
average cost	m²	708.80

Softwood glazed double hung sash window 38 thick; cased frames; oak cills; sash chords

size 900 x 1200	nr	821.83
size 1050 x 1200	nr	1,006.71
size 1200 x 1500	nr	1,226.55
size 1200 x 1800	nr	1,501.19
average cost	m²	644.39

Softwood glazed double hung sash window 50 thick; cased frames; oak cills; sash chords

size 1050 x 1200	nr	1,073.79
size 1200 x 1500	nr	1,318.73
size 1200 x 1800	nr	1,610.11
size 1500 x 1800	nr	1,912.97
average cost	m²	649.12

Galvanised steel windows;4mm glass

type NC6F size 279 x 923	nr	190.15
type NC05F size 508 x 1067	nr	245.12
type NE2 size 997 x 628	nr	281.00
type NC02F size 997 x 1067	nr	533.22
type ND2F size 997 x 1218	nr	512.26
type NC4F size 1486 x 923	nr	666.64
type NC04F size 1486 x 1067	nr	716.98
type ND4 size 1486 x 1218	nr	682.30
type NC11F size 1994 x 923	nr	760.96
type ND11F size 1994 x 1218	nr	863.85

average cost	m²	310.52

Aluminium double glazed; vertical sliders; cill

size 600 x 800	nr	591.03
size 600 x 1200	nr	784.71
size 800 x 800	nr	964.16
size 800 x 1200	nr	1,230.61
size 1200 x 800	nr	1,893.02
size 1200 x 1200	nr	2,177.14

Aluminium double glazed; horizontal sliders; cill

size 1200 x 800	nr	603.53
size 1200 x 1400	nr	706.51
size 1400 x 800	nr	644.87
size 1400 x 1400	nr	751.70
size 1600 x 800	nr	691.11
size 1600 x 1600	nr	842.62

Upvc windows; opening/fixed lights; factory double glazed

size 600 x 1350 window	nr	224.00
size 750 x 1350 window	nr	239.05
size 900 x 1350 window	nr	251.48
size 1200 x 1350 window	nr	278.78

Upvc windows tilt / turn factory double glazed

size 600 x 650 window	nr	188.03
size 600 x 1050 window	nr	250.61
size 750 x 1050 window	nr	265.71
size 900 x 1200 window	nr	304.32
size 900 x 1500 window toughened glass	nr	345.84

Upvc windows tilt / turn; fixed side light; factory double glazed

size 1800 x 1050 window	nr	416.09
size 1800 x 1350 window toughened glass	nr	484.66

Upvc windows tilt / turn; fixed centre light; factory double glazed

size 2400 x 1200 window	nr	598.26
size 2400 x 1350 window	nr	625.66

Doors

44 External ledged and braced door, 50 x 100 softwood frame, ironmongery P.C. £50.00 per leaf and decorate

762 x 1981 door	nr	342.81
838 x 1981 door	nr	353.22
1372 x 1981 pair doors	nr	559.77

44 External framed, ledged and braced door, 50 x 100 softwood frame, ironmongery P.C. £50.00 per leaf and decorate

762 x 1981 door	nr	361.88
838 x 1981 door	nr	372.30
1372 x 1981 pair doors	nr	597.93

44 External plywood flush door, 50 x 100 SW frame, architrave one side, ironmongery P.C. £50.00 per leaf and decorate

762 x 1981 door	nr	396.24
838 x 1981 door	nr	405.59
1372 x 1981 pair doors	nr	642.80

44 Softwood external panel door glazed panels, 50 x 100 softwood frame, architrave one side, ironmongery P.C. £50.00 decorated.

2XG size 762 x 1981	nr	489.37
2XG size 838 x 1981	nr	431.45
2XGG size 762 x 1981	nr	413.97
2XGG size 838 x 1981	nr	426.40
KXT size 762 x 1981	nr	450.29

44 Hardwood Magnet Carolina door, clear glass 50 x 100 hardwood frame architrave one side, ironmongery P.C. £50.00 decorated.

size 838 x 1981	nr	571.71
size 813 x 2032	nr	579.96

COMPOSITE PRICES (Cont'd) £

WINDOWS AND EXTERNAL DOORS (cont'd)

Doors (cont'd)

44 Hardwood Magnet Alicante door, bevelled glass 50 x 100 hardwood frame architrave one side, ironmongery P.C. £50.00 decorated.

size 838 x 1981	nr	717.10
size 813 x 2032	nr	725.33

Garage doors

Galvanised steel garage door

size 2135 x 1980	nr	1,264.51
size 4265 x 1980	nr	2,092.05

Western Red Cedar garage door

size 2135 x 1980 side hung pair	nr	1,134.71
size 2135 x 1980 up and over	nr	1,322.55

Opening in cavity wall

Galvanised steel lintel, DPC, brick

soldier course	m	104.95

Concrete boot lintel

size 252x150x1200	nr	94.68
size 252x225x1200	nr	136.34
size 252x150x1800	nr	143.10
size 252x225x1800	nr	203.27
Softwood window board, DPC, close cavity	m	25.44
Close cavity at jamb, DPC, facings to reveal	m	15.44

INTERNAL WALLS

Brickwork

Common bricks

102 wall	m²	71.14
215 wall	m²	127.99
327 wall	m²	172.53

Engineering bricks

102 wall	m²	75.57
215 wall	m²	135.87
327 wall	m²	182.13

Blockwork

Light weight concrete block 4.0 n/m²

75 wall	m²	32.12
90 wall	m²	39.63
100 wall	m²	40.47
140 wall	m²	51.27
190 wall	m²	61.35

Dense concrete block 7.0 n/m²

90 wall	m²	49.61
100 wall	m²	49.91
140 wall	m²	59.33
190 wall	m²	70.38
215 wall	m²	76.72

Hollow glass blocks

80 wall - 190 x 190 blocks	m²	245.77
80 wall - 240 x 240 blocks	m²	326.36

Partitions

Timber stud partition with plasterboard both sides

75 partition, direct decoration	m²	68.24
100 partition direct decoration	m²	72.99
75 partition with skim coat plaster and decorate	m²	83.78
100 partition with skim coat plaster and decorate	m²	88.53
Extra for 100 insulation	m²	4.57

Metal stud partition with one layer plasterboard each side

75 partition direct decoration	m²	55.72
75 partition with skim coat plaster and decorate	m²	71.26

Metal stud partition with two layers plasterboard each side

100 partition direct decoration	m²	64.05
100 partition with skim coat plaster and decorate	m²	79.59

INTERNAL DOORS

Panel doors

34 softwood panelled type SA door; glazed, 38 x 150 lining, stop, architrave both sides, ironmongery PC £20.00 per leaf; decorated

size 762 x 1981	nr	407.43

34 Magnet Victorian hardwood four panel door; 38 x 150 lining, stop, architrave both sides, ironmongery PC £20.00 per leaf; decorated

size 838 x 1981	nr	418.63
size 762 x 1981	nr	489.19

44 softwood panelled pattern 2XG door; glazed, 38 x 150 lining, stop, architrave both sides, ironmongery PC £20.00 per leaf; decorated

size 762 x 1981	nr	377.20

34 softwood panelled pattern 10 door; 4mm toughened glass to one large panel; 38 x 150 lining, stop, architrave both sides, ironmongery PC £20.00 per leaf; decorated

size 762 x 1981	nr	386.53

Softwood panelled door 38 x 150 lining, stop, architrave both sides, ironmongery PC £20.00 per leaf; decorated

38 thick x 762 x 1981 two panel	nr	532.84
38 thick x 762 x 1981 four panel	nr	560.88
38 thick x 762 x 1981 six panel	nr	578.64
50 thick x 838 x 1981 six panel	nr	677.75

Afromosia panelled door 38 x 150 swd lining, stop, architrave both sides, ironmongery PC £20.00 per leaf; decorated

38 thick x 762 x 1981 two panel	nr	771.66
38 thick x 762 x 1981 four panel	nr	803.54
38 thick x 762 x 1981 six panel	nr	835.38
50 thick x 838 x 1981 six panel	nr	891.90

Flush doors

34 flush door 38 x 150 lining, stop, architrave both sides, ironmongery PC £20.00 per leaf, decorated

762 x 1981 hardboard faced	nr	319.75
838 x 1981 hardboard faced	nr	333.21
1372 x 1981 pair hardboard faced	nr	470.37
762 x 1981 plywood faced	nr	315.15
838 x 1981 plywood faced	nr	330.30
1372 x 1981 pair plywood faced	nr	461.45
762 x 1981 Sapele faced	nr	300.57
838 x 1981 Sapele faced	nr	303.36
1372 x 1981 pair Sapele faced	nr	433.04

44 half hour fire check flush door 50 x 100 frame, architrave both sides, ironmongery PC £60.00 per leaf, decorated.

762 x 1981 plywood faced	nr	348.66
838 x 1981 plywood faced	nr	383.95
1372 x 1981 pair plywood faced	nr	602.01
762 x 1981 Sapele faced	nr	334.50
838 x 1981 Sapele faced	nr	341.38

44 half hour fire check flush door, polished wired glass panel, aperture 450 x 450mm, 50 x 100 frame, architrave both sides, ironmongery PC £60.00 per leaf, decorated.

762 x 1981 plywood faced	nr	96.42
838 x 1981 plywood faced	nr	104.39

Lintels

Precast prestressed concrete lintels

100 x 70 x 900	nr	11.97
100 x 70 x 1200	nr	15.69

COMPOSITE PRICES (Cont'd)

£

INTERNAL DOORS (cont'd)

Lintels (cont'd)

Precast prestressed concrete lintels (cont'd)

100 x 70 x 1800	nr	22.18
100 x 70 x 2400	nr	28.84
100 x 70 x 3000	nr	35.02
150 x 70 x 1200	nr	18.15
150 x 70 x 3000	nr	40.90
225 x 70 x 1200	nr	25.59
225 x 70 x 3000	nr	59.90
255 x 70 x 1200	nr	34.90
255 x 70 x 3000	nr	83.39
100 x 145 x 1800	nr	32.32
100 x 145 x 3000	nr	54.27

Precast concrete lintels

100x150x1200	nr	54.27
100x150x1650	nr	29.98
215x150x1200	nr	41.00
215x150x1650	nr	75.89

Galvanized steel lintels - Birtley

ref INT100; 100 wide x 900 long	nr	16.27
ref INT100; 100 wide x 1050 long	nr	18.02
ref INT100; 100 wide x 1200 long	nr	19.34
ref SB100; 75 deep x 1200 long	nr	38.94
ref SB100; 140 deep x 1800 long	nr	61.34
ref SB100; 215 deep x 3000 long	nr	133.04
ref SB140; 140 deep x 1200 long	nr	45.80
ref SB140; 140 deep x 1800 long	nr	68.13
ref SB140; 215 deep x 3000 long	nr	143.88
ref SBL200; 142 deep x 1200 long	nr	56.10
ref SBL200; 142 deep x 1800 long	nr	84.06
ref SBL200; 218 deep x 3000 long	nr	169.22
ref HD50; 225 deep x 3900 long	nr	251.62
ref HD50; 225 deep x 4800 long	nr	321.01
ref HD70; 225 deep x 3600 long	nr	216.13

WALL FINISHES

Plaster

Hardwall plaster in two coats

13 thick	m²	15.94
13 thick, emulsion	m²	20.31
13 thick, oil paint	m²	23.50
13 thick textured plastic coating, stipple	m²	20.47

Lightweight plaster in two coats

13 thick	m²	13.58
13 thick, emulsion	m²	17.95
13 thick, oil paint	m²	21.14
13 thick textured plastic coating, stipple	m²	18.11

Plasterboard

Plasterboard for direct decoration

9.5 thick	m²	11.55
9.5 thick, emulsion	m²	15.92
9.5 thick, oil paint	m²	19.11
9.5 thick textured plastic coating, stipple	m²	16.08
12.5 thick	m²	13.21
12.5 thick, emulsion	m²	17.58
12.5 thick, oil paint	m²	20.77
12.5 thick textured plastic coating, stipple	m²	17.74
15 thick	m²	19.30
15 thick, emulsion	m²	23.67
15 thick, oil paint	m²	26.86
15 thick textured plastic coating, stipple	m²	23.83
two layer 9.5 thick total 19mm thick	m²	20.39
two layer 9.5 + 12.5; total 22mm thick	m²	22.05
two layer 12.5; total 25mm thick	m²	23.71

Plasterboard with 3 plaster skim coat finish

9.5 thick plasterboard; skim	m²	18.66
9.5 thick plasterboard; skim, emulsion	m²	23.03
9.5 thick plasterboard; skim, oil paint	m²	26.22
9.5 thick plasterboard; skim, textured plastic coating, stipple	m²	15.42
12.5 thick plasterboard and skim	m²	19.36
12.5 thick plasterboard and skim, emulsion	m²	23.73
12.5 thick plasterboard and skim, oil paint	m²	26.92
12.5 thick plasterboard and skim, textured plastic coating; stipple	m²	23.89
15 thick plasterboard and skim	m²	23.61
15 thick plasterboard and skim, emulsion	m²	27.98
15 thick plasterboard and skim, oil paint	m²	31.17
15 thick plasterboard and skim, textured plastic coating; stipple	m²	28.14
two layer 9.5 thick total 21mm thick including skim	m²	29.97
two layer 9.5 + 12.5; total 25mm thick including skim	m²	30.47
two layer 12.5; total 28mm thick including skim	m²	30.82

Gyproc plaster cove

90 girth	m	7.69
127 girth	m	7.67

Extra

lining paper PC £1.00/roll	m²	3.11
wall paper PC £4.00/roll	m²	6.59
vinyl coated wall paper PC £7.00/roll	m²	6.87
embossed paper PC £4.50/roll	m²	11.25
wall tiling 108 x 108 x 6.5 white glazed	m²	57.71
wall tiling 152 x 152 x 5.5 white glazed	m²	43.27
wall tiling 152 x 152 x 5.5 light colour	m²	49.51

Redecorations

Stripping existing paper and lining with paper PC £1.00 per roll	m²	7.43
Two coats emulsion paint on existing	m²	6.09
One undercoat; one gloss paint on existing	m²	7.52
One undercoat; one eggshell paint on existing	m²	7.61
Textured plastic coating on existing painted plaster	m²	7.37

FLOOR FINISHES

Cement; sand bed

25 thick	m²	11.10
32 thick	m²	12.94
38 thick	m²	14.29
50 thick	m²	17.20
25 thick, quarry tile paving	m²	60.11
32 thick, quarry tile paving	m²	61.95
38 thick, quarry tile paving	m²	63.30
50 thick, quarry tile paving	m²	66.21

Cement; sand bed and flooring

25 thick, 3 pvc sheet	m²	37.80
32 thick, 3 pvc sheet	m²	39.64
38 thick, 3 pvc sheet	m²	40.99
50 thick, 3 pvc sheet	m²	43.90
25 thick, 2.5 pvc tile	m²	37.80
32 thick, 2.5 pvc tile	m²	39.64
38 thick, 2.5 pvc tile	m²	40.99
50 thick, 2.5 pvc tile	m²	43.90
25 thick, 20 oak wood block	m²	98.70
32 thick, 20 oak wood block	m²	100.54
38 thick, 20 oak wood block	m²	101.89
50 thick, 20 oak wood block	m²	104.80
25 thick, 20 maple wood block	m²	105.94
32 thick, 20 maple wood block	m²	107.78
38 thick, 20 maple wood block	m²	109.13
50 thick, 20 maple wood block	m²	112.04
25 thick, 20 Merbau wood block	m²	94.56
32 thick, 20 Merbau wood block	m²	96.40
38 thick, 20 Merbau wood block	m²	97.75
50 thick, 20 Merbau wood block	m²	100.66

COMPOSITE PRICES FOR APPROXIMATE ESTIMATING

COMPOSITE PRICES (Cont'd)

FLOOR FINISHES (cont'd)

Cement; sand bed (cont'd)

Cement; sand bed; Carpet and underlay

		£
25 thick, carpet PC £28/m²	m²	60.64
32 thick, carpet PC £28/m²	m²	62.48
38 thick, carpet PC £28/m²	m²	63.83
50 thick, carpet PC £28/m²	m²	66.74

Granolithic pavings

25 thick	m²	15.32
32 thick	m²	17.31
38 thick	m²	18.99
50 thick	m²	22.50

Granolithic skirtings

19 thick x 150 high	m	17.17
25 thick x 150 high	m	18.48
32 thick x 150 high	m	19.63

Levelling screed

Latex levelling screed; one coat

3 thick average to concrete	m²	11.70

Terrazzo

300x300x28 tiles; 40 bed; polished	m²	83.22
16 thick laid insitu in bays	m²	82.82
6 thick skirtings laid insitu 75 high	m	28.07
16 thick skirtings laid insitu 150 high	m	40.97

PVC sheet flooring

2 thick	m²	19.27
2.5 thick	m²	21.58

PVC coved skirting

100 high set in	m	6.52
100 high sit on	m	6.67

Softwood flooring

Floor boarding and insulation to concrete

22 tongued and grooved chipboard and 50 insulation board	m²	46.41

Softwood skirting decorated

25 x 50 chamfered and rounded	m	11.69
25 x 63 chamfered and rounded	m	12.44
32 x 100 chamfered and rounded	m	15.88
19 x 100 square	m	11.89
19 x 150 square	m	14.57
25 x 100 square	m	12.69
25 x 150 square	m	15.02
15 x 125 moulded	m	12.29
19 x 125 moulded	m	12.69

CEILING FINISHES

Plaster

Hardwall plaster in two coats

13 thick	m²	17.71
13 thick, emulsion	m²	21.06
13 thick, oil paint	m²	25.88
13 thick textured plastic coatings, stipple	m²	23.92

Lightweight plaster in two coats

10 thick	m²	15.47
10 thick, emulsion	m²	18.82
10 thick, oil paint	m²	23.64
10 thick textured plastic coatings, stipple	m²	21.68

Plasterboard

Plasterboard for direct decoration

9.5 thick	m²	12.62
9.5 thick, emulsion	m²	15.97
9.5 thick, oil paint	m²	20.79
9.5 thick textured plastic coatings, stipple	m²	18.83
12.5 thick	m²	15.98
12.5 thick, emulsion	m²	19.33
12.5 thick, oil paint	m²	24.15
12.5 thick textured plastic coating, stipple	m²	22.19
two layer 9.5 thick total 19mm thick	m²	22.37
two layer 9.5 + 12.5; total 22mm thick	m²	24.30
two layer 12.5; total 25mm thick	m²	26.09

Plasterboard with 3 plaster skim coat finish

9.5 thick plasterboard and skim	m²	20.84
9.5 thick plasterboard and skim, emulsion	m²	24.19
9.5 thick plasterboard and skim, oil paint	m²	29.01
9.5 thick plasterboard; skim, textured plastic coating, stipple	m²	27.05
12.5 thick plasterboard and skim	m²	21.73
12.5 thick plasterboard and skim, emulsion	m²	25.08
12.5 thick plasterboard and skim, oil paint	m²	29.90
12.5 thick plasterboard; skim, textured plastic coating, stipple	m²	27.94
two layer 9.5 thick total 21mm thick including skim	m²	33.33
two layer 9.5 + 12.5; total 25mm thick including skim	m²	34.02
two layer 12.5; total 28mm thick including skim	m²	34.57

Suspended ceilings

300 x 300 x 15.8 plain mineral fibre tile	m²	33.32
300 x 300 x 15.8 textured mineral fibre tile	m²	36.49
300 x 300 patterned tile P.C. £15.00	m²	39.78
600 x 600 x 15.8 plain mineral fibre tile	m²	32.87
600 x 600 x 15.8 drilled mineral fibre tile	m²	36.40
600 x 600 patterned tile P.C. £15.00	m²	39.69

FITTINGS AND FURNISHINGS

Mirrors

6mm bevelled edges; silvered; copper backing

250 x 400	nr	64.93
460 x 600	nr	86.96
600 x 900	nr	120.49

Venetian blinds

25mm slat; plain colours

1000 wide x 1200 drop	nr	54.85
2000 wide x 1200 drop	nr	84.86
3000 wide x 1200 drop	nr	111.69

50mm slat; plain colours

1000 wide x 1200 drop	nr	56.95
2000 wide x 1200 drop	nr	84.91
3000 wide x 1200 drop	nr	110.75

Roller blinds

self acting roller

1000 wide x 1200 drop	nr	63.00
2000 wide x 1200 drop	nr	92.75
3000 wide x 1200 drop	nr	121.35

Vertical louvre blinds

89mm wide louvres

1000 wide x 1200 drop	nr	45.50
2000 wide x 1200 drop	nr	66.91
3000 wide x 1200 drop	nr	85.22

127mm wide louvres

1000 wide x 1200 drop	nr	51.70
2000 wide x 1200 drop	nr	78.54
3000 wide x 1200 drop	nr	104.48

Kitchen units

Pre-assembled domestic kitchen units

400 floor unit	nr	203.22
500 floor unit	nr	223.39
600 floor unit	nr	240.03
1000 floor unit	nr	383.56
1200 floor unit	nr	417.12
1200 sink unit	nr	296.48
1500 sink unit	nr	428.13
500 floor one drawer unit	nr	119.66
600 floor one drawer unit	nr	126.16

COMPOSITE PRICES (Cont'd)

£

FITTINGS AND FURNISHINGS (cont'd)

Kitchen units (cont'd)

Pre-assembled domestic kitchen units (cont'd)

1000 floor one drawer unit	nr	145.66
1200 sink one drawer unit	nr	365.26
500 floor four drawer unit	nr	354.18
600 floor four drawer unit	nr	375.33
500 x 600 high wall unit	nr	108.80
600 x 600 high wall unit	nr	115.42
1000 x 600 high wall unit	nr	126.30
1200 x 600 high wall unit	nr	226.37
500 x 1950 tall store unit	nr	350.47
600 x 1950 tall store unit	nr	379.38
40 work top	m	97.88
plinth	m	27.05
cornice	m	26.80
pelmet	m	26.80

SANITARY APPLIANCES

Sanitary appliance with allowance for waste fittings, traps, overflows, taps and builder's work

Domestic building, with plastic wastes

lavatory basin	nr	188.06
lavatory basin with pedestal	nr	229.93
bath with panels to front and one end	nr	368.26
low level w.c. suite	nr	236.61
stainless steel sink and drainer	nr	205.88
shower	nr	611.43

Commercial building, with copper wastes

lavatory basin	nr	287.37
low level w.c. suite	nr	236.61
bowl type urinal	nr	454.14
stainless steel sink	nr	296.92
drinking fountain	nr	747.53
shower	nr	666.04

DISPOSAL INSTALLATIONS

Soil and vent pipe for ground and first floor

plastic wastes and SVP	nr	382.06
copper wastes and cast iron SVP	nr	1,296.56

WATER INSTALLATIONS

Hot and cold water services domestic building	nr	2,067.36
gas instantaneous sink water heater	nr	1,141.18
electric instantaneous sink water heater	nr	1,032.33
undersink water heater	nr	1,121.40
electric shower unit	nr	506.19

HEATING INSTALLATIONS

Boiler, radiators, pipework, controls and pumps

100m² domestic building, solid fuel	m²	63.50
100m² domestic building, gas	m²	58.12
100m² domestic building, oil	m²	72.45
150m² domestic building, solid fuel	m²	56.68
150m² domestic building, gas	m²	50.98
150m² domestic building, oil	m²	63.35
500m² commercial building, gas	m²	84.69
500m² commercial building, oil	m²	98.56
1000m² commercial building, gas	m²	100.47
1000m² commercial building, oil	m²	119.15
1500m² commercial building, gas	m²	116.79
1500m² commercial building, oil	m²	136.87
2000m² commercial building, gas	m²	126.89
2000m² commercial building, oil	m²	149.75
Underfloor heating with polymer pipework and insulation - excludes screed.	m²	43.92

ELECTRICAL INSTALLATIONS

Concealed installation with lighting, power, and ancillary cooker, immersion heater etc. circuit.

100m² domestic building

PVC insulated and sheathed cables	m²	70.97
MICS cables	m²	97.57
steel conduit	m²	113.68

150m² domestic building,

PVC insulated and sheathed cables	m²	63.68
MICS cables	m²	86.17
steel conduit	m²	99.47

500m² commercial building, light fittings,

steel conduit	m²	142.10

1000m² commercial building, light fittings,

steel conduit	m²	150.98

1500m² commercial building, light fittings,

steel conduit	m²	159.86

2000m² commercial building, light fittings,

steel conduit	m²	168.73

LIFT AND ESCALATOR INSTALLATIONS

Light passenger lift, laminate walls, stainless steel doors.

3 stop, 8 person	nr	34,923.90
4 stop, 13 person	nr	65,212.95

General purpose passenger lift, laminate walls, stainless steel doors.

4 stop, 8 person	nr	64,717.95
5 stop, 13 person	nr	70,092.00
6 stop, 21 person	nr	159,946.05

Service hoist

2 stop, 50 kg	nr	10,623.15

Escalator 30 degree inclination department store specification

3.00 m rise	nr	95,700.00

SITE WORK

Kerbs, edgings channels

Precast concrete kerb and foundation

125 x 255 kerb	m	45.19
150 x 305 kerb	m	53.17

Precast concrete edging and foundation

50 x 150 edging	m	31.23
50 x 200 edging	m	35.70
50 x 250 edging	m	40.85

Precast concrete channel and foundation

255 x 125 channel	m	41.49

Granite kerb on edge and foundation

150 x 300 kerb	m	72.53
200 x 300 kerb	m	98.71

Granite kerb laid flat and foundation

300 x 150 kerb	m	92.70
300 x 200 kerb	m	122.94

Pavings

Concrete; 250 excavation, 150 hardcore

150 concrete road tamped	m²	56.72
200 concrete road tamped	m²	64.32
150 concrete and 70 bitumen macadam	m²	75.79
200 concrete and 70 bitumen macadam	m²	83.39
Extra for Ref. A252 fabric reinforcement	m²	83.39
Extra for Ref. C636 fabric reinforcement	m²	7.86

Gravel; 250 excavation; 100 hardcore; 75 clinker;

75 course gravel; 38 blinding gravel	m²	34.82

Gravel; 150 excavation; 100 hardcore;

50 blinding gravel	m²	15.94

Gravel; 150 excavation; 100 hardcore; 50 clinker;

38 blinding gravel	m²	19.87

APPROXIMATE ESTIMATING

COMPOSITE PRICES (Cont'd) — £

SITE WORK (cont'd)

Pavings (cont'd)

Gravel; 150 excavation; 100 hardcore; 50 clinker; (cont'd)

38 blinding gravel with 'Colas'; Shingle dressing	m²	31.72

York Stone; 200 excavation, 150 hardcore

50 paving	m²	121.16
75 paving	m²	141.44

Precast concrete; 200 excavation, 150 hardcore

600 x 450 x 50 slabs	m²	45.16
600 x 600 x 50 slabs	m²	41.29
600 x 750 x 50 slabs	m²	43.91
600 x 900 x 50 slabs	m²	41.39
600 x 450 x 63 slabs	m²	57.58
600 x 600 x 63 slabs	m²	51.72
600 x 750 x 63 slabs	m²	54.35
600 x 900 x 63 slabs	m²	53.37
50 thick Crazy paving	m²	45.69

Precast concrete; 200 excavation, 150 hardcore

200 x 100 x 50 Blocks on 50 sand	m²	77.50

Granite Sets; 300 excavation, 100 hardcore

125 thick on 100 concrete	m²	192.83
150 thick on 100 concrete	m²	217.11

Cobbles; 200 excavation, 150 hardcore

Cobbles set in 100 concrete	m²	111.72

Westmorland Green slate; 250 excavation, 150 hardcore

38 slate on 100 concrete	m²	239.97

Seeding/Turfing

Seeding; Cultivating;digging over one spit deep; removing debris; weeding; weedkiller; Grass seed; fertilizer; maintain	m²	8.47
Seeding; 150 excavation; 150 topsoil; weedkiller; Grass seed; fertilizer; maintain	m²	17.39
Turfing; Cultivating;digging over one spit deep; removing debris; weeding; weedkiller; Turf; maintain	m²	11.88
Turfing; 150 excavation; 150 topsoil; weedkiller; Turf; maintain	m²	20.80

Walls

Brick wall, excavation 600 deep concrete foundation, 215 wall with piers at 3m centres in facings PC £200 / 1000, damp proof course.

900 high brick on edge coping	m	212.59
1800 high brick on edge coping	m	335.42
900 high 300 x 100 PCC coping	m	250.08
1800 high 300 x 100 PCC coping	m	372.91

Fences

Plastic coated chain link fencing, posts at 2743 centres.

900 high 38 x 38 angle iron posts, two line wires	m	21.63
1200 high 38 x 38 angle iron posts, two line wires	m	24.48
1400 high 38 x 38 angle iron posts, three line wires	m	27.02
1800 high 38 x 38 angle iron posts, three line wires	m	31.74
900 high concrete posts two line wires	m	26.74
1200 high concrete posts two line wires	m	30.24
1400 high concrete posts three line wires	m	35.53
1800 high concrete posts three line wires	m	40.63

Gates; chain link; posts

1200 x 1800	nr	713.63

Close boarded fencing, 100 x 100 posts at 2743 centres 25 x 100 pales, 25 x 150 gravel boards, creosote, post holes filled with concrete

1200 high two arris rails	m	59.83
1500 high three arris rails	m	69.73
1800 high three arris rails	m	76.86

Gates; timber

900 x 1150	nr	160.36
900 x 1450	nr	162.83

DRAINAGE

Drains

Vitrified clay sleeve joint drain, excavation and backfill.

100 diameter 500 deep	m	30.76
100 diameter 750 deep	m	35.15
100 diameter 1000 deep	m	39.42
100 diameter 1500 deep	m	48.94
150 diameter 500 deep	m	48.85
150 diameter 750 deep	m	53.32
150 diameter 1000 deep	m	57.67
150 diameter 1500 deep	m	67.35

Vitrified clay sleeve joint drain, excavation 150 bed and surround pea shingle and backfill.

100 diameter 500 deep	m	51.10
100 diameter 750 deep	m	55.49
100 diameter 1000 deep	m	59.76
100 diameter 1500 deep	m	69.28
150 diameter 500 deep	m	73.41
150 diameter 750 deep	m	77.88
150 diameter 1000 deep	m	82.22
150 diameter 1500 deep	m	91.90

Vitrified clay sleeve joint drain, excavation 150 bed and haunch concrete and backfill.

100 diameter 500 deep	m	51.91
100 diameter 750 deep	m	56.30
100 diameter 1000 deep	m	60.57
100 diameter 1500 deep	m	70.56
150 diameter 500 deep	m	70.19
150 diameter 750 deep	m	74.66
150 diameter 1000 deep	m	79.01
150 diameter 1500 deep	m	88.69

Vitrified clay sleeve joint drain, excavation 150 bed and surround concrete and back fill.

100 diameter 500 deep	m	67.76
100 diameter 750 deep	m	72.15
100 diameter 1000 deep	m	76.42
100 diameter 1500 deep	m	85.94
150 diameter 500 deep	m	86.83
150 diameter 750 deep	m	91.30
150 diameter 1000 deep	m	95.64
150 diameter 1500 deep	m	105.32

Upvc drain, excavation 50 mm bed sand and backfill.

110 diameter 500 deep	m	24.47
110 diameter 750 deep	m	28.86
110 diameter 1000 deep	m	33.13
110 diameter 1500 deep	m	42.65
160 diameter 500 deep	m	31.82
160 diameter 750 deep	m	36.29
160 diameter 1000 deep	m	40.64
160 diameter 1500 deep	m	50.31

Upvc drain, excavation bed, and surround pea shingle and back fill.

110 diameter 500 deep	m	32.81
110 diameter 750 deep	m	37.20
110 diameter 1000 deep	m	41.47
110 diameter 1500 deep	m	50.99
160 diameter 500 deep	m	41.36

COMPOSITE PRICES (Cont'd)

£

DRAINAGE (cont'd)

Drains (cont'd)

Upvc drain, excavation bed, and surround pea shingle and back fill. (cont'd)

160 diameter 750 deep	m	45.83
160 diameter 1000 deep	m	50.18
160 diameter 1500 deep	m	59.86

Upvc drain, excavation bed and haunch concrete and back fill.

110 diameter 500 deep	m	57.34
110 diameter 750 deep	m	61.73
110 diameter 1000 deep	m	66.00
110 diameter 1500 deep	m	75.52
160 diameter 500 deep	m	51.87
160 diameter 750 deep	m	56.34
160 diameter 1000 deep	m	60.69
160 diameter 1500 deep	m	70.37

Upvc drain, excavation bed and surround concrete and back fill.

110 diameter 500 deep	m	59.66
110 diameter 750 deep	m	64.05
110 diameter 1000 deep	m	68.32
110 diameter 1500 deep	m	77.84
160 diameter 500 deep	m	75.67
160 diameter 750 deep	m	80.14
160 diameter 1000 deep	m	84.49
160 diameter 1500 deep	m	94.17

Cast iron drain, excavation and back fill.

100 diameter 500 deep	m	87.91
100 diameter 750 deep	m	92.30
100 diameter 1000 deep	m	96.57
100 diameter 1500 deep	m	106.09
150 diameter 500 deep	m	148.33
150 diameter 750 deep	m	152.80
150 diameter 1000 deep	m	157.15
150 diameter 1500 deep	m	166.83

Cast iron drain, excavation, bed and haunch concrete and back fill.

100 diameter 500 deep	m	122.51
100 diameter 750 deep	m	126.90
100 diameter 1000 deep	m	131.17
100 diameter 1500 deep	m	140.69
150 diameter 500 deep	m	170.36
150 diameter 750 deep	m	174.83
150 diameter 1000 deep	m	179.18
150 diameter 1500 deep	m	188.85

Cast iron drain, excavation, bed and surround concrete and back fill.

100 diameter 500 deep	m	124.91
100 diameter 750 deep	m	129.30
100 diameter 1000 deep	m	133.57
100 diameter 1500 deep	m	143.09
150 diameter 500 deep	m	194.32
150 diameter 750 deep	m	198.79
150 diameter 1000 deep	m	203.14

150 diameter 1500 deep	m	212.82

Brick manhole

Excavation; 150 concrete bed; 215 class B engineering brick walls; 100 clayware main channel; concrete benching; step irons and cast iron manhole cover

600 x 450, two branches 750 deep	nr	691.43
600 x 450, two branches 1500 deep	nr	1,184.35
750 x 450, three branches 750 deep	nr	1,063.66
750 x 450, three branches 1000 deep	nr	1,261.78
750 x 450, three branches 1500 deep	nr	1,586.23
900 x 600, five branches 750 deep	nr	1,189.11
900 x 600, five branches 1000 deep	nr	1,411.05
900 x 600, five branches 1500 deep	nr	1,801.35

Concrete circular manhole

Excavation; 150 concrete bed; manhole rings; cover slab; 100 clayware main channel; concrete benching; step irons; cast iron manhole cover

900 dia., two branches 750 deep	nr	996.32
900 dia., two branches 1000 deep	nr	1,240.30
900 dia., two branches 1500 deep	nr	1,523.11
1050 dia., three branches 750 deep	nr	1,208.90
1050 dia., three branches 1000 deep	nr	1,434.49
1050 dia., three branches 1500 deep	nr	1,856.16
1200 dia., five branches 750 deep	nr	1,372.43
1200 dia., five branches 1000 deep	nr	1,609.13
1200 dia., five branches 1500 deep	nr	2,005.96

Concrete rectangular manhole

Excavation; concrete bed; manhole sections; base unit; cover slab with integral cover

600 x 450 x 500 deep	nr	370.21
600 x 450 x 750 deep	nr	482.25
600 x 450 x 1000 deep	nr	580.43

Polypropolene Inspection chamber

Excavation; Polypropolene Inspection chamber cast iron manhole cover

250 diameter, 600 deep	nr	241.49
450 diameter, 600 deep	nr	427.74
450 diameter, 1000 deep	nr	545.44

Concrete gullies

Excavation; concrete bed; gully; bricks; cast iron road grating

375 diameter x 750 deep	nr	511.60
375 diameter x 900 deep	nr	556.44
450 diameter x 750 deep	nr	532.00
450 diameter x 900 deep	nr	578.51

Vitrified clay gullies

Excavation; concrete bed; gully; grating

trapped round; 100 outlet	nr	165.47

Pvc-u gullies

Excavation; concrete bed; bottle gully; grating

200 diameter top; trapped round	nr	181.04

This page left blank intentionally

STANDARD RATES OF WAGES

WAGES

Note: Wage agreements and amendments to the following made after the end of September 2016 and notified to us prior to publication will be noted at the end of the Preliminaries section or as noted herein.

NATIONAL MINIMUM WAGE

The hourly rate for the minimum wage must apply and depends on age and whether an apprentice.
National Living Wage and the National Minimum Wage from 1st October 2016

25 and over	21 to 24	18 to 20	Under 18	Apprentice
£7.20	£6.95	£5.55	£4.00	£3.40

Apprentices are entitled to the apprentice rate if they're either aged under 19 or aged 19 or over and in the first year of their apprenticeship
Apprentices are entitled to the minimum wage for their age if they are aged 19 or over and have completed the first year of their apprenticeship

ENGLAND, SCOTLAND AND WALES

Following negotiations between the parties to the Construction Industry Joint Council the Council adopted recommendations for a one year agreement on pay and conditions,

With effect on and from Monday, 25th July 2016 the following basic rates of pay, allowances and additional payments will be applicable:

Note: The rates shown below are weekly rates. The hourly equivalents for a 39 hour week are shown in brackets.

1   ENTITLEMENT TO BASIC AND ADDITIONAL RATES -WR.1

25th July 2016	per week £	per hour £
Craft Operative	452.79	(11.61)
Skill Rate - 1	431.34	(11.06)
- 2	415.35	(10.65)
- 3	388.44	(9.96)
- 4	366.99	(9.41)
General Operative	340.47	(8.73)

Apprentice Rates – WR. 1.4.2

25th July 2016	per week £	per hour £
Year – 1	195.00	(5.00)
Year – 2	242.97	(6.23)
Year - 3 Without NVQ2	283.92	(7.28)
Year - 3 with NVQ2	361.53	(9.27)
Year - 3 with NVQ3	452.79	(11.61)
On completion of apprenticeship With NVQ2	452.79	(11.61)

2.   BONUS - WR.2

It shall be open to employers and operatives on any job to agree a bonus scheme based on measured output and productivity for any operation or operations on that particular job.

3.   STORAGE OF TOOLS - WR.12

Employers maximum liability shall be £750.00 from 30th June 2014

4.   SUBSISTENCE ALLOWANCE - WR.15

£ 36.00 per night from 1st August 2016

5   HOLIDAYS WITH PAY – WR 18.4 and 19.2

Annual Holidays – WR 18.4
A week's pay is the average of the previous 12 weeks including overtime, taxable travel allowance, bonus and regular allowances. Weeks the operative is absent due to sickness are excluded.

Public/Bank Holidays – WR 19.2
Operative's pay does not vary with work done WR 19.2.1
A week's pay is the normal weekly wage including bonus and regular allowances, but excluding overtime.
Operative's pay varies with work done WR 19.2.2
A week's pay is arrived at calculating the earnings during the normal working week averaged over the previous 12 weeks including fixed bonus and regular allowances is the normal weekly wage including bonus and regular allowances, but excluding overtime.
Weeks the operative is absent due to sickness are excluded.

6.   SICK PAY - WR.20

Effective on and from 25th July 2016 employers whose terms and conditions of employment incorporate the Working Rule Agreement of the CIJC should make a payment to operatives, who are absent from work due to sickness or injury, of £118.00 per week.

7.   BENEFIT SCHEMES WR 21

Death benefit – WR 21 1 - £32,500.00 from 1st August 2016. Such benefit to be doubled to £65,000 if death occurs either at work or travelling to and from work.

EasyBuild pension contributions – WR 21.2
The minimum employer contribution shall be £5.00 per week. Where the operative contributes between £5.01 and £10.00 per week the employer shall increase the minimum contribution to match that of the operative up to a maximum of £10.00 per week

With effect on and from Monday, 26th June 2017 the following basic rates of pay, allowances and additional payments will be applicable:

Note: The rates shown below are weekly rates. The hourly equivalents for a 39 hour week are shown in brackets.

1   ENTITLEMENT TO BASIC AND ADDITIONAL RATES -WR.1

26th June 2017	per week £	per hour £
Craft Operative	465.27	(11.93)
Skill Rate - 1	443.04	(11.36)
- 2	426.66	(10.94)
- 3	399.36	(10.24)
- 4	377.13	(9.67)
General Operative	349.83	(8.97)

Apprentice Rates – WR. 1.4.2

25th July 2016	per week £	per hour £
Year – 1	200.46	(5.14)
Year – 2	249.60	(6.40)
Year - 3 Without NVQ2	291.72	(7.48)
Year - 3 with NVQ2	371.67	(9.53)
Year - 3 with NVQ3	465.27	(11.93)
On completion of apprenticeship With NVQ2	465.27	(11.93)

GENERAL INFORMATION

WAGES (Cont'd)
ENGLAND, SCOTLAND AND WALES (Cont'd)

The following tables give the rates of wages since 2000.

			Craft Operative	Skilled Operative Rate: 1	2	3	4	General Building Operative
June	26th	2000	247.65	235.95	227.37	212.94	200.85	186.42
June	25th	2001	261.30	248.82	239.85	224.64	211.77	196.56
June	24th	2002	284.70	271.05	261.30	244.53	230.49	214.11
June	30th	2003	299.13	284.70	274.56	257.01	242.19	225.03
June	28th	2004	320.38	305.37	294.06	273.34	259.35	241.02
June	27th	2005	351.00	334.62	322.14	301.47	284.31	264.03
June	26th	2006	363.48	346.32	333.45	312.00	294.45	273.39
June	25th	2007	379.08	361.53	347.88	325.65	306.93	285.09
June	30th	2008	401.70	382.98	368.94	345.15	325.65	302.25
September	5th	2011	407.94	388.83	374.40	350.22	330.72	306.93
January	7th	2013	416.13	396.63	381.81	357.24	337.35	313.17
June	30th	2014	429.00	408.72	393.12	367.77	347.49	322.53
June	30th	2015	441.87	420.81	405.21	379.08	358.02	332.28
July	25th	2016	452.79	431.34	415.35	388.44	366.99	340.47
June	26th	2017	465.27	443.04	426.66	399.36	377.13	349.83

HOLIDAYS WITH PAY

"Template" introduced to meet the industry's Working Rule Agreements and the Working Time Regulations.
Holiday pay annual contribution calculated by multiplying number of hours holiday by hourly rate.
Employers Contributions (per week)

	Holiday Hours	Craft Operative £	General Operative £	Retirement and Death Benefit £
June 25th 2001	226	29.12	20.78	3.20 (July 2nd 2001)
June 24th 2002	226	31.73	23.86	5.40 (July 1st 2002)
June 30th 2003	226	33.34	25.08	8.40 (August 4th 2003)
June 28th 2004	226	35.73	26.86	10.90
June 27th 2005	226	39.12	29.42	10.90
June 26th 2006	226	40.51	30.46	10.90

NORTHERN IRELAND

The Joint Council for the Building and Civil Engineering Industry in Northern Ireland has agreed wage awards for the Building and Civil Engineering Industry (Northern Ireland):

Effective from	4th August 2014		3rd August 2015	
	Hourly Rate	Weekly minimum Wage	Hourly Rate	Weekly minimum Wage
New Entrant	7.09	276.51	7.23	281.97
General Construction Operative	8.33	324.87	8.50	331.50
Skilled Construction Operative	9.01	351.39	9.19	358.41
Craft	9.95	388.05	10.15	395.85
Advanced Craft	10.70	417.30	10.91	425.49
Young General Construction Operative	4.92	191.88	5.02	195.78

HOLIDAYS WITH PAY

Employers Contributions for Operatives: consolidated holiday credit

From 31st July 2000 Contributions provide holiday pay equivalent to basic wages to comply with the Working Time Regulations.

PENSION CONTRIBUTIONS

The Joint Council for the Building and Civil Engineering Industry in Northern Ireland has set the minimum level of contributions by the employer with effect from December 2013 at 3% (4% from December 2016) based on normal weekly pay (before tax) and after National Insurance contributions (NIC). The employer also makes a contribution of £1.76 per employee per week towards lump sum death in service cover for each employee.

HEATING, VENTILATING, AIR CONDITIONING, PIPING AND DOMESTIC ENGINEERING CONTRACTS

Rates agreed between the Building and Engineering Services Association (formerly HVCA) and Unite.

RATES OF WAGES

All districts of the United Kingdom
Note : Ductwork Erection Operatives, are entitled to the same rates and allowances as the parallel fitter grades shown.
wage agreement effective from 3rd October 2016

RATES AND ALLOWANCES	From 3 October 2016	From 2 October 2017	From 5 October 2015
Hourly Rate			
	£	£	£
Building Services Engineering Supervisor	20.00	20.50	16.08
Foreman	16.40	16.81	
Senior Craftsman (+ Supervisory Responsibility and 2 Units of Responsibility Allowance)	15.76	15.02	15.50
Senior Craftsman (+ Supervisory Responsibility and 1 Units of Responsibility Allowance)	15.21	14.49	14.95
Senior Craftsman (+ Supervisory Responsibility)	14.66	13.96	14.40
Senior Craftsman (+ 2 Units of Responsibility Allowance)	14.66	13.96	14.40
Senior Craftsman (+ 1 Units of Responsibility Allowance)	14.11	13.43	13.85
Senior Craftsman	13.56	12.90	13.30
Craftsman (+ 3 Units of Responsibility Allowance)	14.11	13.43	13.85
Craftsman (+ 2 Units of Responsibility Allowance)	13.56	12.90	13.30
Craftsman(+ 1 Unit of Responsibility Allowance)	13.01	12.37	12.75
Craftsman	12.46	11.84	12.20
Installer	11.26	10.72	11.04
Adult Trainee	9.50	9.04	9.31
Mate (18 and over)	9.50	9.04	9.31
Mate (16 and 17)	4.41	4.19	4.32
Modern Apprentices			
Senior	11.26	10.72	11.04
Intermediate	8.74	8.32	8.57
Junior	6.16	5.86	6.04
Responsibility Allowances (Senior Craftsmen) per hour			
Second welding skill	0.55	0.53	0.55
Gas responsibility	0.55	0.53	0.55
Supervisory responsibility	1.10	1.06	1.10
Responsibility Allowance (Craftsmen) per hour			
Second welding skill	0.55	0.53	0.55
Gas responsibility	0.55	0.53	0.55
Supervisory responsibility	0.55	0.53	0.55
DAILY ABNORMAL CONDITIONS MONEY £ per day	3.28	3.28	3.28
LODGING ALLOWANCE £ per night	To be promulgated	To be promulgated	37.25 From 28 Sept 2015

DAILY TRAVELLING ALLOWANCE
Scale 1

C = Craftsmen including installers
M & A = Mates, Apprentices, Adult Trainees.

Direct distance from centre to job in miles	From 3 Oct 2016		From 2 Oct 2017		From 5 Oct 2015	
	C	M & A	C	M & A	C	M & A
Over-Not exceeding	£	£	£	£	£	£
0-15	7.30	7.30	7.30	7.30	7.30	7.30
15-20	9.98	9.60	9.98	9.60	9.98	9.60
20-30	14.18	13.25	14.18	13.25	14.18	13.25
30-40	17.19	15.86	17.19	15.86	17.19	15.86
40-50	20.31	18.46	20.31	18.46	20.31	18.46

DAILY TRAVELLING ALLOWANCE
Scale 2 – (Outside M25)

C = Craftsmen including installers
M & A = Mates, Apprentices, Adult Trainees.

Direct distance from centre to job in miles	C	M & A	C	M & A	C	M & A
Over-Not exceeding	£	£	£	£	£	£
15-20	2.67	2.29	2.67	2.29	2.67	2.29
20-30	6.87	5.94	6.87	5.94	6.87	5.94
30-40	9.88	8.55	9.88	8.55	9.88	8.55
40-50	13.02	11.13	13.02	11.13	13.02	11.13

GENERAL INFORMATION

STANDARD RATES OF WAGES

HEATING, VENTILATING AND DOMESTIC ENGINEERING CONTRACTS (Cont'd)

WEEKLY HOLIDAY CREDIT VALUES, WELFARE CONTRIBUTIONS AND EMPLOYERS' PENSION CONTRIBUTIONS

National Agreement Grade and Allowance(s)	From 3 October 2016			From 2 October 2017			From 5 October 2015)		
	Weekly Holiday Credit £	Weekly Holiday Credit and Welfare Contribution £	Employers Pension Contribution See Table below £	Weekly Holiday Credit £	Combined Weekly Holiday Credit and Welfare Contribution £	Employers Pension Contribution See Table below £	Weekly Holiday Credit £	Combined Weekly Holiday Credit and Welfare Contribution £	Weekly Holiday Credit and Welfare Contribution plus employers Pension Contribution £
Building Services Engineering Supervisor	92.86	102.51		95.19	104.84		-	-	-
Foreman	76.15	85.80		78.06	87.71		74.73	84.38	111.52
Senior Craftsman (+ Supervisory Responsibility and 2 Units of Responsibility Allowance)	73.23	82.88		74.77	84.42		71.93	81.58	107.72
Senior Craftsman (+ Supervisory Responsibility and 1 Units of Responsibility Allowance)	70.72	80.37		72.21	81.86		69.40	79.05	104.27
Senior Craftsman (+ Supervisory Responsibility)	68.15	77.80		69.65	79.30		66.84	76.49	100.76
Senior Craftsman (+ 2 Units of Responsibility Allowance)	68.15	77.80		69.65	79.30		66.84	76.49	100.76
Senior Craftsman (+ 1 Units of Responsibility Allowance)	65.58	75.23		67.09	76.74		64.29	73.94	97.30
Senior Craftsman	63.05	72.70		64.51	74.16		61.75	71.40	93.83
Craftsman (+ 3 Units of Responsibility Allowance)	65.58	75.23		67.09	76.74		64.29	73.94	97.30
Craftsman (+ 2 Units of Responsibility Allowance)	63.05	72.70		64.51	74.16		61.75	71.40	93.83
Craftsman(+ 1 Unit of Responsibility Allowance)	60.41	70.06		61.99	71.64		59.25	68.90	90.40
Craftsman	57.89	67.54		59.40	69.05		56.69	66.34	89.63
Installer	52.32	61.97		53.58	63.23		51.34	60.99	79.62
Adult Trainee	44.14	53.79		45.25	54.90		43.27	52.92	68.64
Mate(over 18)	44.14	53.79		45.25	54.90		43.27	52.92	68.64
Mate (16 - 17)	20.48	30.13		21.00	30.65		20.07	29.72	37.01
Senior Modern Apprentice	52.32	61.97		53.58	63.23		51.34	60.99	79.62
Intermediate Modern Apprentice	40.58	50.23		41.60	51.25		39.90	49.55	64.02
Junior Modern Apprentice	28.62	38.27		29.31	38.96		28.11	37.76	47.96

WEEKLY PENSION CONTRIBUTIONS – EMPLOYERS' AND EMPLOYEE CONTRIBUTIONS

National Agreement Grade and Allowance(s)	Credit Value Category	From 5 October 2015	From 3 October 2016			From 5 June 2017		
Pension Contribution Rate – Percentage of pre-tax basic pay (hourly rate x 37½): Employer / Employee		4½% — £	5% / ¾% Employer Contribution £	Employee Contribution £	Total Contribution £	5% / 1½% Employer Contribution £	Employee Contribution £	Total Contribution £
Building Services Engineering Supervisor	y	—	37.50	5.63	43.13	37.50	11.25	48.75
Foreman	a	27.14	30.75	4.62	35.37	30.75	9.23	39.98
Senior Craftsman (+ Supervisory Responsibility and 2 Units of Responsibility Allowance)	z	26.14	29.55	4.44	33.99	29.55	8.87	38.42
Senior Craftsman (+ Supervisory Responsibility and 1 Unit of Responsibility Allowance)	b	25.22	28.52	4.28	32.80	28.52	8.56	37.08
Senior Craftsman (+ Supervisory Responsibility)	c	24.27	27.49	4.13	31.62	27.49	8.25	35.74
Senior Craftsman (+ 2 units of Responsibility Allowance)	c	24.27	27.49	4.13	31.62	27.49	8.25	35.74
Senior Craftsman (+ 1 unit of Responsibility Allowance)	d	23.36	26.46	3.97	30.43	26.46	7.94	34.40
Senior Craftsman	e	22.43	25.43	3.82	29.25	25.43	7.63	33.06
Craftsman (+ 3 units of Responsibility Allowance)	d	23.36	26.46	3.97	30.43	26.46	7.94	34.40
Craftsman (+ 2 units of Responsibility Allowance)	e	22.43	25.43	3.82	29.25	25.43	7.63	33.06
Craftsman (+ 1 unit of Responsibility Allowance)	f	21.50	24.40	3.66	28.06	24.40	7.32	31.72
Craftsman	g	20.59	23.37	3.51	26.88	23.37	7.01	30.38
Installer	h	18.63	21.12	3.17	24.29	21.12	6.34	27.46
Adult Trainee	i	15.72	17.82	2.68	20.50	17.82	5.35	23.17
Mate (18 and over)	i	15.72	17.82	2.68	20.50	17.82	5.35	23.17
Mate (aged 16 and 17)	m	7.29	8.27	1.25	9.52	8.27	2.49	10.76
Senior Modern Apprentice	h	18.63	21.12	3.17	24.29	21.12	6.34	27.46
Intermediate Modern Apprentice	j	14.47	16.39	2.46	18.85	16.39	4.92	21.31
Junior Modern Apprentice	k	10.20	11.55	1.74	13.29	11.55	3.47	15.02

GENERAL INFORMATION

HEATING, VENTILATING AND DOMESTIC ENGINEERING CONTRACTS (Cont'd)

NOTES

Welplan Ref.

y	Building Services Engineering Supervisor
a	Foreman
z	Senior Craftsman (+ Supervisory Responsibility and 2 Units of Responsibility Allowance)
b	Senior Craftsman (+ Supervisory Responsibility and 1 Unit of Responsibility Allowance)
c	Senior Craftsman (+ Supervisory Responsibility)
c	Senior Craftsman (+ 2 Units of Responsibility Allowance)
d	Senior Craftsman (+ 2 Units of Responsibility Allowance)
e	Senior Craftsman
d	Craftsmen (+ 3 Units of Responsibility Allowance)
e	Craftsman (+ 2 Units of Responsibility Allowance)
f	Craftsmen (+ 1 Units of Responsibility Allowance)
g	Craftsman
h	Installer
i	Adult Trainee
i	Mate(over 18)
m	Mate (16 - 17)
h	Senior Modern Apprentice
j	Intermediate Modern Apprentice
k	Junior Modern Apprentice

Pension contributions rates from 2nd October 2017 for employer to be 5% and Employee 1.5%, from 5th February 2018 for employer to be 5% and Employee 2.25%,

For premium rates explanations, overtime payments fuller details and general conditions the reader is advised to obtain a copy of the National Working Rule Agreement available from the Building and Engineering Services Association (B&Esa Limited) Publications Department, Old Mansions House, Eamont Bridge, Penrith, Cumbria, CA10 2BX. www.thebesa.com

STANDARD RATES OF WAGES

JOINT INDUSTRY BOARD FOR THE ELECTRICAL CONTRACTING INDUSTRY

JIB INDUSTRIAL DETERMINATION 2015/2016/2017

Terms and conditions have been agreed and a brief review issued including:-

Promulgated by the JIB National Board to be effective from the dates shown.

WAGE RATES

From **7th January 2002** two different wage rates have applied to JIB Graded operatives working on site, depending on whether the employer transports them to site or whether they provide their own transport. The two categories are:

Job Employed (Transport Provided)
Payable to an operatives who are transported to and from the job by their employer. Operatives shall also be entitled to payment for Travel Time, travelling in their own time, as detailed in the appropriate scale.

Job Employed (Own Transport)
Payable to operatives who travel by their own means to and from the job. Operatives under this category shall be entitled to payment for Travel Allowance and also when travelling in their own times to Travel Time, as detailed in the appropriate scale.

There is a separate rate to accommodate those few operatives who are employed permanently at the shop and who do not work on site. This rate is shown in the appropriate table.

WAGES 2017

From and including 5th **January 2017,** the JIB hourly rates for Job Employed operatives shall be:

(i)   National Standard Rates

Grade	Transport Provided	Own Transport	Shop Employed
Electrical / Site Technician Mechanical Technician Cable Installation Supervisor (or equivalent specialist grade)	£17.06	£17.92	£15.97
Approved Electrician Advanced Craftsperson Cable Foreman Approved Jointer (or equivalent specialist grade)	£15.08	£15.92	£13.98
Electrician Craftsperson Leading Cable Hand Jointer (or equivalent specialist grade)	£13.81	£14.68	£12.73
ECS Experienced Worker Cardholder ECS Mature Candidate Assessment Cardholder Trainee Electrician (Stage 3) Mechanical Trainee (Stage 3)	£13.14	£13.95	£12.11
Trainee Electrician (Stage 2) Mechanical Trainee (Stage 2) Senior Graded Electrical Trainee Electrical Improver	£12.42	£13.23	£11.49
Trainee Electrician (Stage 1) Mechanical Trainee (Stage 1) Adult Trainee Labourer	£10.97	£11.79	£9.83
Apprentice (Stage 4) Apprentice (Stage 3) Apprentice (Stage 2) Apprentice (Stage 1)	£10.89 £10.29 £7.11 £4.83	£11.76 £11.15 £7.96 £5.65	£10.07 £9.47 £6.54 £4.49

(ii)   London Rate

for Operatives engaged upon work falling within the terms of JIB National Working Rule 6.2 will be:

Grade	Transport Provided	Own Transport	Shop Employed
Electrical / Site Technician Mechanical Technician Cable Installation Supervisor (or equivalent specialist grade)	£19.13	£20.07	£17.88
Approved Electrician Advanced Craftsperson Cable Foreman Approved Jointer (or equivalent specialist grade)	£16.89	£17.83	£15.66
Electrician Craftsperson Leading Cable Hand Jointer (or equivalent specialist grade	£15.46	£16.45	£14.27
ECS Experienced Worker Cardholder ECS Mature Candidate Assessment Cardholder Trainee Electrician (Stage 3) Mechanical Trainee (Stage 3)	£14.69	£15.64	£13.56
Trainee Electrician (Stage 2) Mechanical Trainee (Stage 2) Senior Graded Electrical Trainee Electrical Improver	£13.91	£14.81	£13.04
Trainee Electrician (Stage 1) Mechanical Trainee (Stage 1) Adult Trainee Labourer	£12.30	£13.20	£11.02
Apprentice (Stage 4) Apprentice (Stage 3) Apprentice (Stage 2) Apprentice (Stage 1)	£12.21 £11.54 £7.98 £5.40	£13.17 £12.50 £8.89 £6.33	£11.27 £10.59 £7.32 £5.03

Further details of the wage agreement, allowances and working Rules may be obtained on written application to the Joint Industry Board for the Electrical Contracting Industry, Kingswood House, 47/51 Sidcup Hill, Sidcup, Kent DA14 6HP. Email: administration@jib.org.uk

THE NATIONAL JOINT COUNCIL FOR THE LAYING SIDE OF THE MASTIC ASPHALT INDUSTRY

For rates of wages as approved by the National Joint Council for the Laying Side of the Mastic Asphalt Industry and further information and details contact:-

Employers' Secretary, Mastic Asphalt Council Ltd, P O Box 77, Hastings. TN35 4WL
Tel: 01424 814400   www.masticasphaltcouncil.co.uk

GENERAL INFORMATION

THE JOINT INDUSTRY BOARD FOR PLUMBING MECHANICAL ENGINEERING SERVICES IN ENGLAND AND WALES

Basic hourly rates of pay inclusive of tool allowance

		From 5th January 2015 £	From 4th January 2016 £	From 2nd January 2017 £
(a)	**Operatives**			
	Technical Plumber & Gas Service Technician	15.61	16.00	16.32
	Advanced Plumber & Gas Service Engineer	14.06	14.41	14.70
	Plumber & Gas Service Fitter	12.06	12.36	12.61
(b)	**Apprentices**			
	4th year of training with NVQ level 3*	11.67	11.96	12.20
	4th year of training with NVQ level 2*..	10.57	10.83	11.05
	4th year of training.............................	9.31	9.54	9.73
	3rd year of training with NVQ level 2*..	9.19	9.42	9.61
	3rd year of training.............................	7.56	7.75	7.91
	2nd year of training.............................	6.70	6.87	7.01
	1st year of training.............................	5.84	5.99	6.11
(c)	**Adult Trainees**			
	3rd 6 months of employment	10.51	10.77	10.99
	2nd 6 months of employment	10.10	10.35	10.56
	1st 6 months of employment	9.41	9.65	9.84

- Note
- PHMES Operative covers Plumbers, Gas Service Fitters, Mechanical Pipe Fitters and Heating Fitters
- Where apprentices have achieved NVQ's, the appropriate rate is payable from the date of attainment except that it shall not be any earlier than the commencement of the promulgated year of   Training in which it applies.

WORKING HOURS AND OVERTIME

The normal working week shall be 37.5 hours
Overtime:- From 5th January 2015, 37.5 Hours at normal rates are to be worked (Monday to Friday) before overtime rates shall apply
Note: overtime hours worked Monday to Friday shall be paid at time and a half up to 8.00pm. After 8.00pm overtime shall be paid at double time.

Responsibility/Incentive Pay allowance

Employers may, In consultation with the employees concerned, enhance the basis graded rates of pay by the payment of an additional amount, as per the bands shown below, where it is agreed that their work involves extra responsibility, productivity or flexibility.

		From 5th Jan 2015
Band 1	additional hourly rate of	1p to 27p
Band 2	an additional rate of	28p to 48p
Band 3	an additional rate of	49p to 71p
Band 4	an additional rate of	72p to 93p
		From 4th Jan 2016
Band 1	additional hourly rate of	1p to 28p
Band 2	an additional rate of	29p to 49p
Band 3	an additional rate of	50p to 73p
Band 4	an additional rate of	74p to 95p
		From 2nd Jan 2017
Band 1	additional hourly rate of	1p to 29p
Band 2	an additional rate of	30p to 50p
Band 3	an additional rate of	51p to 74p
Band 4	an additional rate of	75p to 97p

This allowance forms part of an operative's basic rate of pay and shall be used to calculate premium payments. It is payable **EITHER** on a contract-by-contract basis **OR** on an annual review basis.

Explanatory Notes

Plumbing employees are **entitled to be included in the Industry Pension Scheme** – or one which provides equivalent benefits. This is **IN ADDITION to their Hourly Rates of Pay.**

From 6th April 2014 there shall be 2 basic scales:-
Existing Basic Scale   Employer @ 7.5% - Employee @ 3.75%.
Alternative Basic Scale   Employer @ 5.3% - Employee @ 2.7%.

Further information can be obtained from The Joint Industry Board for Plumbing Mechanical Engineering Services in England and Wales, Lovell House, Sandpiper Court, Phoenix Business Park, Eaton Socon St Neots, Cambridgeshire PE19 8EP Tel: 01480 476925 Fax: 01480 403081

GENERAL INFORMATION

STANDARD RATES OF WAGES

NATIONAL INSURANCES

Employers Contribution:

April 2008	12.8% of employee's earnings above the earnings threshold where exceeding £105.00 per Week
April 2009	12.8% of employee's earnings above the earnings threshold where exceeding £110.00 per week
April 2010	12.8% of employee's earnings above the earnings threshold where exceeding £110.00 per week
April 2011	13.8% of employee's earnings above the earnings threshold where exceeding £136.00 per week
April 2012	13.8% of employee's earnings above the earnings threshold where exceeding £144.00 per week
April 2013	13.8% of employee's earnings above the earnings threshold where exceeding £148.00 per week
April 2014	13.8% of employee's earnings above the earnings threshold where exceeding £153.00 per week
April 2015	13.8% of employee's earnings above the earnings threshold where exceeding £156.00 per week

CITB LEVY

The CITB levies applicable following The Industrial Training Levy (Construction Board) orders are as follows:

The Industrial Training Levy (Construction Industry Training Board) Order 2012

Date came into force	Payroll	Labour only
28th March 2012	0.5%	1.5%

Subsequent Industrial Training Levy (Construction Board) Orders keep the PAYE rate at 0.5% and the labour only subcontractor rate at 1.5%
Further information from www.citb.co.uk including details of Small Business Levy Exemption and Reduction

GENERAL INFORMATION

Mechanical Plant

The following rates are inclusive of operator as noted, fuel, consumables, reparis and maintenance, but excludes transport.

£

Excavators including Driver and Banksman
Hydraulic, full circle slew, crawler mounted with single equipment (shovel)

0.70 m³	48.80	hour
1.00 m³	50.21	hour
1.50 m³	50.12	hour

Hydraulic, offset or centre post, half circle slew, wheeled, dual purpose (back loader)

0.60 m³	45.98	hour
0.80 m³	45.98	hour
1.00 m³	47.23	hour
1.20 m³	47.45	hour

Mini excavators, full circle slew, crawler mounted with single equipment (back actor)

1.4 tonne capacity	47.45	hour
2.4 tonne capacity	49.43	hour
3.6 tonne capacity	50.08	hour

Compactor including operator
Vibrating plate

90kg	20.67	hour
120kg	20.67	hour
176kg	21.21	hour
310kg	21.76	hour

Rollers including driver
Road roller dead-weight

2.54 tonnes	26.09	hour
6.10 tonnes	27.47	hour
8.13 tonnes	30.03	hour
10.20 tonnes	31.97	hour

Vibratory pedestrian operated

Single roller 550kg	20.83	hour
Single roller 680kg	20.83	hour
Twin roller 650kg	20.83	hour
Twin roller 950kg	21.27	hour
Twin roller 1300kg	21.60	hour

Self propelled vibratory tandem roller

1.07 tonne	25.11	hour
2.03 tonne	25.61	hour
3.05 tonne	26.26	hour
6.10 tonne	27.14	hour

Pumps including attendant quarter time
Single diaphragm (excluding hoses)

51mm	7.39	hour
76mm	7.74	hour
102mm	8.71	hour

Double diaphragm (excluding hoses)

51mm	7.81	hour
76mm	8.12	hour
102mm	9.80	hour

Suction or delivery hose, flexible, including coupling, valve and strainer, per 7.62m length

51mm	0.34	hour
76mm	0.45	hour
102mm	0.79	hour

Compressors including driver quarter time
Portable, normal delivery of free air at 7kg/cm³

2.0m³/min	6.95	hour

Compressor tools with 15.3m hose (excluding operator)

Breaker including steels	2.26	hour
Light pick including steels	2.26	hour
Clay spade including blade	2.26	hour

Compressor tool sundries

Muffler	0.26	hour

Concrete Mixers Including Driver
Closed drum

0.20/0.15m³	20.05	hour
0.30/0.20m³	20.05	hour

Closed drum with swing batch weighing gear

0.40/0.30m³	23.96	hour

Mixer with batch weighing gear and power loading shovel

0.30/0.20m³	23.96	hour
0.40/0.30m³	26.81	hour

Mortar Mixers Including Driver
Mortar Mixer

0.15/0.10m³	19.02	hour

Concrete Equipment including Driver/Operator
Concrete pump, skid mounted, exclusive of piping

20/26m³	48.93	hour
30/45m³	51.53	hour
46/54m³	54.15	hour

Piping per 3.05m length

102mm diameter	0.55	hour
127mm diameter	0.60	hour
152mm diameter	0.66	hour

Vibrator poker

Petrol	17.21	hour
Diesel	17.67	hour
Electric	16.47	hour
Air	16.03	hour

Vibrator, external type, clamp on

Small	16.03	hour
Medium	17.23	hour
Large	18.00	hour

Lorries including Driver
Ordinary of the following plated gross vehicle weights

3.56 tonnes	32.27	hour
5.69 tonnes	36.88	hour
8.74 tonnes	40.34	hour
10.16 tonnes	46.10	hour

Tipper of the following plated gross vehicle weights

6.60 tonnes	39.19	hour
8.74 tonnes	41.49	hour
9.65 tonnes	46.10	hour

Vans, including driver, carrying capacity

0.4 tonnes	23.05	hour
1.0 tonnes	29.28	hour
1.25 tonnes	29.97	hour

Dumpers including Driver
Manual gravity tipping 2 wheel drive

750kg	20.42	hour
1000kg	20.42	hour
2 tonnes	20.60	hour
3 tonnes	21.25	hour

GENERAL INFORMATION

BUILDERS AND CONTRACTORS PLANT

Mechanical Plant (Cont'd)

	£	
Hydraulic tipping 4 wheel drive		
2 tonnes	21.25	hour
3 tonnes	21.25	hour
4 tonnes	22.74	hour
5 tonnes	22.74	hour

Hoists

Scaffold type hoist excluding operator
250 kg	123.56	week

Ladder type hoist excluding operator
150 kg electric	152.14	week

Goods type hoist up to 10m height including operator, excluding erecting and dismantling
Mobile type 500 kg	685.80	week
Tied-in type 750 kg	731.91	week

CHARGES FOR HIRE

The following Trade List Prices have been provided by HSS Hire Shops, to whom acknowledgement is hereby given. All hire charges are per week, day rates available on application. Guide prices only and subject to alteration without notice. All carriage to and from site charged extra.

£

Builders' Ladders

Extension Ladders (push up)
2 part, up to 6.2m	each	44.70
3 part, up to 6.0m	each	46.00
3 part, up to 9.1m	each	65.70
Ladder safety foot	each	11.20

Roof Ladders
4.6 double	each	68.40
6m double	each	77.10

Adjustable Steel Trestles, 4 boards 920mm wide (fixed leg)

	closed	extended		
no 1	0.5m	0.8m	each	8.20
no 2	0.8m	1.2m	each	8.20
no 3	1.1m	1.6m	each	8.20

Scaffold Boards
2.4m to 3.0m long wooden	5.50
4.0m long wooden	5.50

Aluminium Work Towers

Access towers (Multilevel Work Platforms) narrow width 0.85m x 1.8m (or 2.5m) and full width 1.45m x 1.8m (or 2.5m)
2.2m height	each	127.60
4.2m height	each	180.80
6.2m height	each	238.70
8.2m height	each	293.80
10.2m height	each	346.90

Stagings

Lightweight staging 600mm wide deck
2.4m	each	51.50
3.6m	each	57.60
4.8m	each	73.10
6.0m	each	75.80
6.6m	each	85.40

Scissor Lifts Electric
Narrow Aisle (2.44 x 0.81 x 1.73m) up to 10m	655.50
Slab (2.50 x 1.22 x 1.98m) up to 8m	595.80

PURCHASE PRICES

RAMSAY LADDERS

The following Trade Prices have been provided by Ramsay & Sons (Forfar) Ltd., 61, West High Street, Forfar, Angus, DD8 1BH, Tel: 01307 462255, to whom acknowledgement is hereby given. (prices exclude VAT)

Aluminium Ladders
Single (parallel sides)
Serrated round rungs fitted at 250mm centres. Ends of stiles fitted with rubber feet.

	length (m)	rungs	weight (kg)	£
SA3. 0MK	3.05	12	6	54.60
SA4. 0MK	4.05	16	8	71.50
SA5. 0ML	5.05	20	12	98.15
SA6. 0MF	6.05	24	18	145.60
SA10. 0MJ	10.05	40	40	338.65

Large D shaped rungs with twist proof joint fitted at 300mm centres. Ends of stiles fitted with rubber feet
TDS30	3.05	10	7	48.00
TDS40	4.00	13	10	61.20
TDS55	5.75	19	15	103.20

Double extending (push up type)
Serrated round rungs fitted at 250mm centres. Ends of stiles fitted with rubber feet.

	closed height (m)	extended height (m)	rungs	approx. weight (kg)	£
DE3. 0ML	3.05	5.55	24	13	136.50
DE4. 0MF	4.05	7.55	32	27	231.40
DE5. 0MF	5.05	9.30	40	33	273.00
DE6. 0MJ	6.05	11.05	47	52	467.27
DE7. 0MJ	7.05	12.75	57	63	535.60

Ropes and pulleys - Per Double Ladder   40.30

Large D shaped rungs with twist proof joint fitted at 300mm centres. Ends of stiles fitted with rubber feet
TDD20	1.85	3.03	12	9.5	66.67
TDD30	3.05	5.44	20	13.5	102.50
TDD40	4.00	7.28	26	20	144.00
TDD55	5.75	10.40	38	30	213.00

Triple Extending (push up type)
Serrated round rungs fitted at 250mm centres. Ends of stiles fitted with rubber feet.

	closed height (m)	extended height (m)	rungs	approx. weight (kg)	£
TE3. 0MF	3.05	7.30	35	32	319.80
TE4. 0MF	4.05	10.30	47	42	395.85
TE5. 0MF	5.05	13.30	59	52	484.90

Ropes and pulleys - Per triple Ladder   62.40

AFPS10	3.05	10	21.5	285.00

PURCHASE PRICES (Cont'd)

RAMSAY INDUSTRIAL CLASS 1 LADDERS: BS 2037 1984 (Cont'd)

GRP ladders
Single section

	length (m)	rungs	weight (kg)	£
AFL09	2.35	9	7	160.00
AFL12	3.05	12	9	180.00
AFL14	3.55	14	11	201.00
AFL16	4.10	16	14	217.00

Double section

AFD09	2.35	18	14	339.00
AFD12	3.05	24	17	375.00
AFD14	3.55	28	22	420.00

Aluminium heavy duty platform steps

	height to platform (m)	treads	weight (kg)	£
AFS7	1.66	7	15	149.50
AFS9	2.13	9	17	172.25
AFS11	2.60	11	19	195.00
AFS13	3.07	13	22	274.30
AFS15	3.54	15	31	330.85
AFS20	4.72	20	45	494.00
AFS24	5.66	24	54	603.20

GRP platform steps

	height to platform (m)	treads	weight (kg)	£
AFPS05	1.18	5	11	217.00
AFPS07	1.68	7	15	235.00
AFPS09	2.18	9	19	266.00

Loft Ladders

Complete with one handrail, operating pole and fixings; excludes trap door

		£
AL1	2.29m to 2.54m	206.40
AL2	2.57m to 2.82m	216.00
AL3	2.84m to 3.10m	222.40
AL4	3.12m to 3.38m	241.60
AL5	3.40m to 3.66m	280.00

Additional safety handrail per ladder   20.00

Aluminium Fixed Industrial Platform Steps (no handrails)

	height (m)	weight (kg)	
Single Sided			
DPS 01	0.240	6	86.33
DPS 02	0.475	7	108.00
DPS 03	0.710	9	130.40
DPS 04	0.945	10	153.60
DPS 06	1.415	12	198.40
DPS 08	1.880	15	228.80
Double Sided			
DPD 02	0.475	8	164.00
DPD 03	0.710	10	180.00
DPD 04	0.945	12	196.80
DPD 05	1.180	15	212.80
DPD 06	1.415	17	228.80
DPD 07	1.645	19	244.80
DPD 08	1.88	23	269.60

GENERAL INFORMATION

This page left blank intentionally

GUIDE PRICES TO BUILDING TYPES PER SQUARE METRE

The prices throughout this section are ranged over the most commonly occurring average building prices and are based upon the total floor area of all storeys measured inside external walls and over internal walls and partitions. They exclude contingencies, external works, furniture, fittings and equipment, professional fees and VAT.

Building prices are influenced significantly by location, market, local conditions, size and specification. The prices given are intended only as an indicative guide to representative building types in a U.K. location and must be used with caution, professional skill and judgement. Prices considerably outside the ranges stated can, of course, be encountered to meet specific requirements.

New Build

Industrial Buildings

	per m2 £
Agricultural Shed	455-950
Factory, light industrial for letting incoming mains only	455-1000
including lighting and power, toilets and basic office accommodation	600-1150
for owner occupation including all services and basic accommodation	550-1400
high tech: with controlled environment	800-2600
Garage/Showroom	700-1300
Workshop	650-1150
Warehouse/Store	450-1100

Administrative Buildings

Council Offices	1300-2220
Magistrates Court	1550-2740
Offices, low rise (for letting) non air conditioned	1100-2000
air conditioned	1500-2250
Offices, High rise (for owner occupation) air conditioned	1500-2500
Bank (branch)	1500-2350
Shops	600-1350
Retail Warehouse	500-1900
Shopping Mall air-conditioned	1100-2350
Police Station	1250-2050
Ambulance Station	800-1500
Fire Station	1200-1800

Health and Welfare Buildings

General Hospital	1250-2400
Private Hospital	1450-2800
Day Hospital	1300-2500
Health Centre	1000-1700
Hospice	1300-1800
Day Centre	1100-1600
Home for the Mentally Ill	1000-1500
Children's Home	1000-1700
Home for the Aged	1000-1350

Refreshment, Entertainment and Recreation Buildings

Restaurant	1300-2250
Public House	1000-2100
Kitchen with equipment	1400-2700
Theatres	1550-2350
Community Centre	1000-1600
Swimming Pool (covered)	1650-3250
Sports Hall	1000-1600
Squash Courts	850-1500
Sports Pavilion	850-1550
Golf Club House	890-1800

Religious Buildings

Church	1120-1800
Church Hall	1000-1800
Crematoria	1560-2350

Educational, Cultural and Scientific Buildings

Nursery School	1100-1800
Primary School	950-1600
Middle School	820-1600
Secondary School	880-1700
Special School	950-1850
Universities	1000-2400
College	950-1850
Training Centre	850-1600
Laboratory building	1300-2600
Computer Building	1450-3250

Libraries	1000-1650

Residential Buildings

Local Authority and Housing Association Schemes low rise flats	750-1250
4/5 person two storey houses	750-1150
Private Developments low rise flats	780-1280
4/5 person two storey houses	750-1450
Private Houses, detached	1000-2300
Sheltered Housing Bungalows	800-1100
low rise	800-1100
flats with lifts and Warden's accommodation, etc	900-1450
Student's Residence	980-1550
Hotel	1050-2350

Refurbishment

Factories	300-1200
Offices (basic refurbishment)	500-1350
Offices (high quality including air conditioning)	1250-2250
Banks	750-1450
Shops	450-1550
Hospitals	660-1550
Public Houses	620-1200
Theatres	760-1750
Churches	450-1350
Schools	450-1350
Housing & Flats	350-1950
Hotels	650-2000

GENERAL INFORMATION

This page left blank intentionally

CAPITAL ALLOWANCES

It is important to recognise that construction tax advice is a complex area because the variables that need to be considered are many and dependent on particular circumstances. This section is therefore intended only as a guide

The Act

This is the Capital Allowances Act 2001. It is amended each year in the Finance Act.

Updates from July 2013 included guidance on withdrawal of allowances on sports grounds, updates on energy saving plant and machinery and guidance on flat conversion allowances to flag the withdrawal of allowances.

Industrial Building Allowances

The allowance of 4% has been phased out.

The rate of allowance was 4% per annum on a straight line basis over 25 years as at the 2007/8 tax year the last year at which this rate could be used. In 2008/9 the rate reduced to 3% and by a further 1% per annum until the end of the 2010/11 tax year when the allowance ceased.

Hotel allowances

Industrial Building Allowances were available on capital expenditure incurred in connection with a qualifying hotel as defined in Section 279 of the Act but these have been withdrawn as above.

Agricultural Building Allowances

These were available under Part 4 of the Act and operated generally in accordance with the rules for Industrial Buildings with certain important exceptions. These have also been withdrawn.

Plant and Machinery

Sections 21-22 of the 2001 Act set out details of those assets which are specifically excluded from qualifying as plant and machinery under all circumstances. Section 23 gives a list of those items within a building which may qualify as expenditure on plant and machinery. This does not mean that the items listed automatically qualify as this will depend on the nature of the trade carried on in the building.

Case law governs the interpretation of the items listed in Section 23 and the circumstances required to confirm qualification in each separate situation. For example, the provision of a swimming pool in a hotel would be regarded as plant and machinery whereas it would not be in an office block. Relevant case law goes back as far as 1887 and there are "landmark" cases identified, on which the HM Customs and Revenue heavily rely.

Capital expenditure on solar panels, which include photovoltaic varieties, which generate electricity, and solar thermal systems, which provide hot water, from 1 April 2012 for corporation tax and 6 April 2012 for income tax all capital expenditure on solar panels is specifically designated as special rate

Rate of Allowances

On plant and machinery in the general pool the allowances are given on the basis of writing down allowances currently at 18% per annum on the reducing balance basis. This means that the full amount is never claimed but that in seventeen years nearly 95% will have been claimed.

The rate of WDA on the special rate pool of plant and machinery expenditure is 8 per cent (from April 2012).

Long Life Assets

A reduced writing down allowance of 6% per annum on the reducing balance basis is available for plant and machinery that is a long-life asset. A long-life asset is defined as plant and machinery that can reasonably be expected to have a useful economic life of at least 25 years. The depreciation policy for accounting purposes or the client's design requirements may help to determine the useful economic life.

Plant and machinery provided in connection with buildings used mainly as dwelling houses, showrooms, hotels, offices or shops or similar premises cannot be a long life asset. However assets in buildings such as cinemas, hospitals, nursing homes and factories can potentially be long life assets and will be assessed on the rules above.

From April 2012 solar panels will be treated as long-life assets

The rate of writing down allowances on long-life asset expenditure is 10% for 2011/12 and 8% from 2012/13.

Building refurbishment and alterations

Section 25 of the Act allows that where capital expenditure is incurred that is incidental to the installation of plant and machinery it can be treated as part of the expenditure on the item. This means that in an existing building, elements of structure and the like can be allowed whereas in a new building they would not be allowable. An example is a lift shaft, where inserted into an existing building would be allowable whereas a lift shaft in a new building is deemed to be structure and non allowable.

Elements in a refurbishment project may be a mixture of capital and revenue expenditure. A repair or a "like for like" replacement of part of an asset would be a revenue expense and this would enable the whole amount to be deducted from the taxable profit in the accounts covering the year of expenditure.

Green technologies

From April 2008 a payable tax credit for losses resulting from capital expenditure on certain designated "green technologies" was introduced.

CAPITAL ALLOWANCES (Cont'd)

The Enhanced Capital Allowance Scheme

Enhanced Capital Allowances (ECAs) enable a business to claim 100% first-year capital allowances on their spending on qualifying plant and machinery. There are three schemes for ECAs:

- Energy-saving plant and machinery
- Low carbon dioxide emission cars and natural gas and hydrogen refuelling infrastructure
- Water conservation plant and machinery

details of which can be found on website www.eca.gov.uk. However these products must be new or unused plant and machinery.

The allowance is not available on plant and machinery specifically excluded from qualifying for writing down allowances under the normal rules or on long-life assets.

Claims should be made on the basis of the actual cost but in the event that the price is not readily available a claim can be made as detailed in the above website.

Flats Conversion Allowance

Flat conversion allowances no longer apply after 6th April 2013.

Business Premises Renovation Allowance

Capital expenditure on the renovation of business premises in certain 'disadvantaged areas' may qualify for the Business Premises Renovation Allowance (BPRA).

The BPRA scheme took effect from 11 April 2007 and is due to end on 31 March (Income tax) / 5 April 2017 (corporation tax). It covers expenditure on the conversion or renovation of unused business premises that brings them back into business use.

You can claim a 100% initial allowance, that is, your entire qualifying capital expenditure or you can claim less than 100%, and then claim WDA of 25% of the qualifying expenditure, but the limited to the residue of the qualifying expenditure (the amount of expenditure that has not been written off).

The 100% initial allowance would accelerate the rate of allowance on plant and machinery and provide new relief on expenditure which does not currently qualify for capital allowances.

The designated disadvantaged areas are now defined as Northern Ireland and the areas specified as development areas by the Assisted Areas Order 2007.

You cannot claim BPRA on:
- the costs of acquiring the land
- extending the business premises
- developing land next to the business premises
- conversion or renovation expenditure incurred on any residential property

Excluded from the scheme are any premises that are used by businesses engaged in the following trades

- Fisheries and aquaculture
- Shipbuilding
- Synthetic fibres
- The coal industry
- The steel industry
- Farming
- The manufacture of products which imitate or substitute for milk or milk products

Research and Development Allowances

The Finance Act 2000 provided for tax relief for small and medium companies (SME) undertaking qualifying R&D activities. A 50% enhancement of qualifying expenditure can be claimed under the scheme and in some circumstances this can lead to a payable tax credit. A company is small or medium if it has fewer than 250 employees and an annual turnover not exceeding €50 million and/or a balance sheet total not exceeding €43 million. From April 2008 the enhancement will be increased to 75%.

The new rules in Finance Act 2007 will extend the support available under the SME scheme to companies with fewer than 500 employees which have an annual turnover not exceeding €100 and/or who have an annual balance sheet total not exceeding €86 million.

The Finance Act 2002 provided for tax relief for large companies and a 25% enhancement of their qualifying expenditure under this scheme. From April 2008 this will be increased to 30%.

Contaminated Land Tax Relief

Land remediation of contaminated or derelict land became available on expenditure incurred on or after 11th May 2001 and the provisions were included in the 2001 Finance Act updated from April 2009. It enables companies to claim relief on capital and revenue expenditure on qualifying land remediation expenditure. It is important to note that this is relief from corporation tax; individuals and partnerships are not eligible for this relief and that a company cannot claim where they caused the contamination.

Relief is given at the rate of 100% plus 50% for qualifying expenditure on the actual amount incurred and should be claimed within two years of the year in which the expenditure is incurred.

Summary

In order to maximise the tax relief available in claiming capital allowances, it is important to seek expert professional advice at an early date.

HOMES AND COMMUNITIES AGENCY

Affordable Housing Capital Funding Guide 2015-18

The following information is based on the Affordable Housing Capital Funding Guide 2015-18.

Introduction

The Guide contains the rules and procedures for Registered Providers (RPs) providing housing under the Affordable Homes Programme (AHP) with funding from the Homes and Communities Agency (HCA).
The main types of funding are:

- Social Housing Assistance payable under section 19(6) of the Housing and Regeneration Act 2008; and
- Financial Assistance payable under section 19(3) of the Housing and Regeneration Act 2008.

The Agency's terms and conditions for Financial Assistance (and Social Housing Assistance) provided under the 2015-18 AHP are set out in:

- The 2015-18 Affordable Homes Grant Agreement;
- The Affordable Housing Capital Funding Guide;
- The 2015-18 AHP propectus

For remaining developments funded by Social Housing Grant, the Agency's specifications are set out in a number of documents:

- the National Affordable Housing Programme 2008-11 Prospectus;
- The Affordable Housing Capital Funding Guide;
- The Funding Conditions;
- Programme Partnering Agreement;
- Grant Agreements; and
- The Social Housing Grant (Capital) General Determination 2003

New for the 2015-18 AHP

The republished Socal HomeBuy lease(s) removes the pre-emption right post 100% staircasing for prospective purchases under Social HomeBuy on shared ownership terms, whilst the deeds of variation will allow solicitors to remove the pre-emption right for existing New Build HomeBuy and Key Worker Living leaseholders.
New chapter which sets out the Agency's requirements in relation to management of a provider's approved allocation funded through the 2015-18 Affordable Homes Programme.

Scope and Format

The Guide is now organised into a number of books, each containing chapters:

GENERAL

- Introduction - General points applicable to Agency funding through the AHP;
- Programme Management AHP;
- Programme Management National AHP (NAHP);
- Procurement and Scheme Issues; including construction methods, types of scheme available, acquisition issues, inter RP transfers, planning issues and site signage;
- Management Arrangements;
- Finance – matters pertaining to the circumstances and conditions relating to grant claims..

HOUSING FOR RENT

- Affordable Rent - permanent rented housing which is new build, rehabilitation, and re-improvement, let on Affordable Rent terms;
- Social Rent - permanent rented housing which is new build, rehabilitation, and re-improvement;
- Repair - Major repairs and some rehabilitation of existing stock;

HOUSING FOR SALE

This book is divided into three sections:

(i) Shared Ownership
  including shared ownership for the elderly, rural repurchase and Protected Area repurchase
(ii) Right To Acquire
(iii)Social HomeBuy

SPECIALIST PROVISION

. This section should be read within the context of the 'General book' and the Agency's '2015-18 Affordable Homes Programme Prospectus'

- Empty Homes – there are no restrictions on the type of properties funded

Examples of empty properties could include:
  - Homes over shops in town centres.
  - Those awaiting redevelopment or sale.
  - Long term privately owned properties.
  - Non-residential properties that have been empty
  - Where appropriate properties subject to Compulsory Purchase Orders and Empty Dwelling Management orders.
  - Any other residential property which has been empty for a significant period of time, excluding existing social housing.

- Traveller Pitches – new sites including funding for new pitches on existing sites.
- Supported housing and housing for older people - includes housing for rough sleepers

Scheme types can include but are not limited to:
  - Sheltered housing
  - Shared supported houses
  - Domestic violence refuges
  - Drug / alcohol rehabilitation schemes
  - Hostels
  - Foyers
  - Extra care housing
  - Retirement villages

CARE AND SUPPORT SPECIALISED HOUSING FUND

- Sets out the requirements and procedures to be followed when developing affordable housing funded through the Department of Health's Care and Support Specialised Housing Fund.

GRANT RECOVERY

- Grant Recovery, Repayment & Recycling - including Recycled Capital Grant Fund

Affordable Housing Capital Funding Guide 2015-18 (Cont'd)

Further information and contact

The guide does not duplicate Agency requirements relating to the Offer process which are set out in the 2015-18 Affordable Homes Programme, Framework, or Regulatory Standards which are available on the Agency's website.

In many instances throughout the guide it requires grant recipients to contact their Operating Area for further guidance or approval on investment matters. Unless specified otherwise initial contact should be made with the relevant HCA lead.

The HCA lead should also be contacted with regards other queries relating to the content of the guide.

For general queries
contact the switchboard on 0300 1234 500.
For IT or IMS queries
contact the helpdesk on 01908 353 604

For a full copy of the guide and further information visit www.homesandcommunities.co.uk

HOMES AND COMMUNITIES AGENCY OFFICES
Main switchboard for all offices 0300 1234 500

Corporate Centre - Warrington

Apley House
110 Birchwood Boulevard
Birchwood
Warrington
WA3 7QH

Corporate Centre - London

Fry Building
2 Marsham Street
London
SW1P 4DF
Tel. 020 7393 2000

Bedford
Woodlands
Manton Lane
Manton Industrial Estate
Bedford
MK41 7LW

Birmingham
5 St Phillips Place
Colmore Row
Birmingham
B3 2PW

Bristol
2 Rivergate
Temple Quay
Bristol
BS1 6EH
Tel. 0300 123 4500

Cambridge
Eastbrook
Shaftesbury Road
Cambridge
CB2 8BF
Tel: 01223 374003

Chatham
Second Floor
The Observatory
Brunel
Chatham Maritime
Kent
ME4 4NT

Gateshead
St Georges House
Kingsway
Team Valley
Gateshead
NE11 0NA
Tel: 0300 123 4500

Guildford
Bridge House
1 Walnut Tree Close
Guildford
Surrey GU1 4GA

Leeds Office
1st Floor Lateral
8 City Walk
Leeds
LS11 9AT

Manchester Office
Level 1A
City Tower
Piccadilly Plaza
Manchester
M1 4BT

Nottingham
3rd Floor,
Apex Court
City Link
Nottingham
NG24 4LA

GENERAL INFORMATION

The following are extracts from the Working Rule Agreement for the Construction Industry as approved by the Construction Industry Joint Council (Revised 1st August 2016)

WR.1 ENTITLEMENT TO BASIC AND ADDITIONAL RATES OF PAY

Operatives employed to carry out work in the Building and Civil Engineering Industry are entitled to basic pay in accordance with this Working Rule (W. R. 1). Rates of pay are set out in a separate schedule, published periodically by the Council.

Classification of basic and additional pay rates of operatives:

General operative
Skilled Operative Rate:- 4
3
2
1
Craft Operative.

1.1 General Operatives

1.1.1 General Operatives employed to carry out general building and/or civil engineering work are entitled to receive the General Operatives Basic Rate of Pay.

Payment for Occasional Skilled Work

1.1.2 General Operatives, employed as such, who are required to carry out building and/or civil engineering work defined in Schedule 1, on an occasional basis, are entitled to receive the General Operative Basic Rate of Pay increased to the rate of pay specified in Schedule 1 for the hours they are engaged to carry out the defined work.

1.2 Skilled Operatives

1.2.1 Skilled Operatives engaged and employed whole time as such, who are required to carry out skilled building and/or civil engineering work defined in Schedule 1. on a continuous basis, are entitled to the Basic Rate of Pay specified in Schedule 1.

1.3 Craft operatives

Craft operatives employed to carry out craft building and/or civil engineering work are entitled to receive the Craft Operative Basic Rate of Pay.

1.4 Conditions of employment of apprentices.

1.4.1 Conditions.

An apprentice who has entered into a training service agreement is subject to the same conditions of employment as other operatives employed under the Working Rule Agreement except as provided in WR.1.4.2 to 1.4.6.

1.4.2 Wages

Rates of pay are set out in a separate schedule, published periodically by the Council. Payment under the schedule is due from the date of entry into employment as an apprentice, whether the apprentice is working on site or undergoing full-time training on an approved course, subject to the provisions of WR1.4.3. Payment under the scale is due from the beginning of the pay week during which the specified period starts.

1.4.3 Payment During Off-the-Job Training

Apprentices are entitled to be paid during normal working hours to attend approved courses off-the-job training in accordance with the requirement of their apprenticeship. payment during such attendance shall be at their normal rate of pay, but the employer may withhold payment for hours during which an apprentice, without authorisation fails to attend the course

1.4.4 Overtime

The working of overtime by apprentices under 18 years of age shall not be permitted. Where an apprentice age 18 or over is required to work overtime payment shall be in accordance with the provisions of WR.4.

1.4.5 Daily Fare and Travel Allowances

.The apprentice shall be entitled to fare and travel allowances in accordance with WR. 5.

1.4.6 Absence and Sick Pay

The employer must be notified at the earliest practical time during the first day of any absence and no later than midday. The first seven days may be covered by self certification. Thereafter absence must be covered by a certificate or certificates given by a registered medical practitioner. The apprentice shall be entitled to Statutory Sick Pay (SSP) plus Industry sick pay in accordance with WR.20 save the aggregate amount of SSP plus Industry sick pay shall not exceed a normal week's pay in accordance with WR.1.4.2.

1.4.7 Other Terms and Conditions of Engagement

The apprentice shall be subject to all other provisions and entitlements contained within the Working Rule Agreement.

Note: **Normal hourly rate**
The expression 'normal hourly rate' in this agreement means the craft, skilled or general operatives' weekly basic rate of pay as above, divided by the hours defined in WR. 3 "Working Hours". Additional payments for occasional skilled or bonus payments are not taken into account for calculating the "normal hourly rate".

WR.2 BONUS

It shall be open to employers and employees on any job to agree a bonus scheme based on measured output and productivity for any operation or operations on that particular job.

WR.3 WORKING HOURS

Working Hours
The normal working hours shall be:
Monday to Thursday   8 hours per day
Friday   7 hours per day
Total   39 hours per week
except for operatives working shifts whose working hours shall continue to be 8 hours per weekday and 40 hours per week.
The expression "normal working hours" means the number of hours prescribed above for any day (or night) when work is actually undertaken reckoned from the starting time fixed by the employer.

WR.3 WORKING HOURS (Cont'd)

3.1 Rest/Meal Breaks

3.1.1. Meal/Refreshment Breaks

At each site or job there shall be a break or breaks for rest and/or refreshment at times to be fixed by the employer. The breaks, which shall not exceed one hour per day on aggregate, shall include a meal break of not less than a half an hour.

3.1.2. Daily/Weekly Rest Breaks

Where there are objective or technical reasons concerning the organisation of work the application of the Working Time Regulations 1998 – Regulations 10(1) Daily Rest Period – 11(1) and 11(2) Weekly Rest Period is excluded.

3.2. Average Weekly Working Hours

Where there are objective or technical reasons concerning the organisation of work, average weekly working hours will be calculated by reference to a 12 month period subject to the employer complying with the general principles relating to the protection of health and safety of workers and providing equivalent compensatory rest periods or, in exceptional cases where it is not possible for objective reasons to grant such periods, ensuring appropriate protection for the operatives concerned.

WR.4 OVERTIME RATES

The employer may require overtime to be worked and the operative may not unreasonably refuse to work overtime.

Overtime will be calculated on a daily basis, but overtime premium rates will not be payable until the normal hours (39 hours-WR3) have been worked in the pay week unless the short time is authorised by the employer on compassionate or other grounds or is a certified absence due to sickness or injury.

Note: The number of hours worked in excess of normal hours will be reduced by the number of hours of unauthorised absence before the overtime premium is calculated.

Overtime shall be calculated as follows:

Monday to Friday
For the first four hours after completion of the normal working hours of the day at the rate of time and a half; thereafter at the rate of double time until starting time the following day.

Saturday
Time and a half, until completion of the first four hours, and thereafter at double time.

Sunday
At the rate of double time, until starting time on Monday morning.

When an operative is called out after completing the normal working hours of the day or night, he shall be paid at overtime rates for the additional time worked as if he had worked continuously. Any intervening period shall count to determine the rate, but shall not be paid.

Overtime shall be calculated on the normal hourly rate. Additional payments for intermittent skill or responsibility or adverse conditions and bonus shall not be included when calculating overtime payments.

In no case shall payment exceed double time.

WR.5 DAILY FARE AND TRAVEL ALLOWANCES

5.1 Extent of Payment

Operatives are entitled to a daily fare and travel allowance, measured one way from their home to the job/site. The allowances will be paid in accordance with the table published periodically by the Council. There is no entitlement to allowances under this Rule for operatives who report, at normal starting time, to a fixed establishment, such as the employers yard, shop or permanent depot. The distance travelled will be calculated by reference to WR 5.2. There is no entitlement under this Rule to allowances for distances less than 9 miles. Having due regard for health and safety an operative may be required to travel distances beyond the published scale; in which case payment for each additional kilometre should be made based on the difference between the rate for the 49th and 50th mile.

5.2 Measurement of Distance

All distances shall be measured utilising the RAC Route Planner (or similar) using the post codes of the operative's home and place of work based on the shortest distance route.

An operative's home is the address at which the operative is living while employed on the work to which travelled.

In the event that there is no post code for the operative's home and/or place of work the closest post code should be used.

5.3 Transport provided free by the employer

Where the employer provides free transport, the operative shall not he entitled to fare allowance. However, operatives who travel to the pick up point for transport provided free by the employer are entitled to fare allowance for that part of the journey, in accordance with the table.

5.4 Transfer during Working Day

An operative transferred to another place of work during work hours shall, on the day of the transfer only, be paid any fares actually incurred:

(a)   in the transfer, and

(b)   in travelling home from the place where he finishes work if this differs from the place where he reported for work at starting time, subject to deduction of half the daily fare allowance.

5.5 Emergency Work

An operative called from the home (or temporary place of residence) outside normal working hours to carry out emergency work shall be paid travelling expenses and his normal hourly rate for the time spent travelling to and from the job.

WR.6 SHIFT WORKING

6.1   Shift working means a situation in which more than one shift of not less than eight hours is worked on a job in a 24 hour period and such shifts do not overlap.

On all work which is carried out on two or more shifts in a 24 hour period the following provisions shall apply:

The first shift in the week shall be the first shift that ends after midnight on Sunday

The normal hours of a shift shall be eight hours, excluding meal breaks, notwithstanding which, the hours to be worked on any particular shift shall be established by the employer.

WR.6 SHIFT WORKING (Cont'd)

The rate payable for the normal hours of the shift shall be the operative's normal hourly rate plus, in the case of an operative completing a shift, a shift allowance of 14% of the normal rate.

An operative required to work continuously for over eight hours on a shift or shifts shall be paid at the rate of time and a half, plus a shift allowance of 14% of his normal hourly rate, for the first four hours beyond eight and thereafter at double time but such double time payment shall not be enhanced by the 14% shift allowance (i.e. the maximum rate in any circumstance shall be double the normal hourly rate).

After having worked five complete scheduled shifts in a week an operative shall on the first shift thereafter be paid at the rate of time and a half of normal rate plus 12.5% shift allowance for the first eight hours of the shift and thereafter and on any subsequent shift in that week at the rate of double time but with no shift allowance.

Where the nature of the work is such as to require an operative to remain at his workplace and remain available for work during mealtimes a shift allowance of 20% shall apply instead of the 14% or 12.5% otherwise referred to in this rule.

Where the work so requires, an operative shall be responsible for taking over from and handing over to his counterpart at commencement and on completion of duty unless otherwise instructed by his employer.

6.2 Employers and operatives may agree alternative shift working arrangements and rates of pay where, at any job or site, flexibility is essential to effect completion of the work.

Under this rule the first five complete shifts worked by an operative shall count as meeting the requirements of the Guaranteed Minimum Weekly Earnings Rule (WR. 17).

The shift allowance shall be regarded as a conditions payment and shall not be included when calculating overtime payments.

This Rule does not apply to operatives employed under the Continuous Working, Night Work, or Tunnel Work Rules.

WR.7 NIGHT WORK (see WR 29.1 for definition)

7.1 Night Work Allowance and Overtime

Where work is carried out at night by a separate gang of operatives from those working during daytime, operatives so working shall be paid at their normal hourly rate plus an allowance of 25% of the normal hourly rate.

Overtime shall be calculated on the normal hourly rate provided that in no case shall the total rate exceed double the normal hourly rate. Overtime shall therefore be paid as follows:

(a) Monday to Friday:
after completion of the normal working hours at the rate of time and a half plus the night work allowance (i.e. time and a half plus 25% of normal hourly rate) for the next four working hours and thereafter at double time.

(b) Weekends:
All hours worked on Saturday or Sunday night at double time until the start of working hours on Monday.

This rule does not apply to operatives employed on shift work, tunnel work or continuous working.

7.2 Health Assessments

Under the terms of this agreement night workers shall be provided with a free confidential health assessment before starting night work and repeated on a regular basis. The frequency of repeat assessments will vary between individuals according to factors such as type of night work, its duration and the age and health of the individual worker.

WR.8 CONTINUOUS WORKING

8.1 An operative whose normal duties are such as to require him to be available for work during mealtimes and consequently has no regular mealtime, shall be deemed a "continuous worker" and shall be responsible for taking over from and handing over to a work colleague at commencement and completion of duty unless otherwise instructed by his employer.

8.2 Continuous working payment will be calculated as follows:
(a) All times but excluding Saturday 10 p.m. to Sunday 10 p.m.
The rate of pay will be at the normal hourly rate plus 20% for the number of hours on duty on the job.

(b) Saturday 10 p.m. to Sunday 10 p.m.
The rate of pay will be at time and a half plus 20% of the normal hourly rate.
If work between 10 p.m. Saturday to 10 p.m. Sunday is not within the normal cycle of operations for the particular job, then no continuous working allowance shall be paid but the rate of payment shall be double the normal hourly rate.

The continuous working allowance shall be regarded as a conditions payment and shall not be reckoned for the purpose of calculating overtime payments.

This rule does not apply to operatives on night-work, shift work or tunnel work.

WR.9 TIDE WORK

9.1 Where work governed by tidal conditions is carried out during part only of the normal working hours, and an operative is employed on other work for the remainder of the normal working hours, the normal hourly rate shall be paid during the normal working hours a thereafter shall be in accordance with the rule on Overtime Rates.

9.2 Where work governed by tidal conditions necessitates operatives turning out for each tide and they are not employed on other work, they shall be paid a minimum for each tide of six hours pay at ordinary rates, provided they do not work more than eight hours in the two tides. Payment for hours worked beyond a total of eight on two tides shall be calculated proportionately i.e. those hours worked in excess of eight multiplied by the total hours worked and the result divided by eight, to give the number of hours to be paid in addition to the twelve paid hours (six for each tide) provided for above.

Work done on Saturday after 4 p.m. and all Sunday shall be paid at the rate of double time. Operatives shall be guaranteed eight hours at ordinary rate for time work between 4 p.m. and midnight on Saturday and 16 hours at ordinary rates for two tides work on Sunday.
The rule on Overtime Rates does not apply to this paragraph 9.2.

9.3 The table below shows how hours of overtime are calculated
Not shown here – refer to full rules

WR.10 TUNNEL WORK

The long-standing custom of the industry that tunnel work is normally carried out by day and by night is reaffirmed. Where shifts are being worked within and in connection with the driving tunnels the first period of a shift equivalent to the normal working hours specified in the Working Hours rule for that day shall be deemed to be the ordinary working day. Thereafter the next four working hours shall be paid at time and a half and thereafter at double time provided that

(a) In the case of shifts worked on Saturday the first four hours shall be paid at time and a half and thereafter at double time.

(b) In the case of shifts worked wholly on Sunday payment shall be made double time.

(c) In the case of shifts commencing on Saturday but continuing into Sunday, payment shall be made for all hours worked at double time.

(d) In case of shifts commencing on Sunday but continuing into Monday, hours worked before midnight shall be paid for at double time and thereafter four working hours calculated from midnight shall be at time and a half and thereafter at double time.

This rule does not apply to an operative employed under the Continuous Working or Shift Working Rules.

WR.11 REFUELLING, SERVICING, MAINTENANCE AND REPAIRS

Operators of mechanical plant, such as excavators, cranes, derricks, rollers, locomotives, compressors or concrete mixers and boiler attendants shall, if required, work and be paid at their normal hourly rate for half an hour before and half an hour after the working hours prescribed by the employer for preparatory or finishing work such as refuelling firing up, oiling, greasing, getting out, starting up, banking down, checking over, cleaning out and parking and securing the machine or equipment.

Refuelling, Servicing, Maintenance and Repair work carried out on Saturday and Sunday shall be paid in accordance with the Rule on Overtime Rates.

WR.12 STORAGE OF TOOLS

When practicable and reasonable on a site, job or in a workshop the employer shall provide an adequate lock-up or lock-up boxes, where the operative's tools can be securely stored. The operative shall comply with the employer's requirements as regards the storage of tools. At all times an operative shall take good care of his tools and personal property and act in a responsible manner to ensure their reasonable safety.

The employer shall accept liability for such tools up to a maximum amount specified by the Council for any loss caused by fire or theft of tools which have been properly secured by an operative in such lock-up facilities or lock-up boxes. The employer does not have liability for loss in excess of the specified amount or loss caused other than by fire or theft

WR.13 HIGHWAYS MAINTENANCE.

Where an operative is engaged in highways maintenance work the requirements of the job/contract may require that the operatives working arrangements vary from the normal provisions of this agreement. This may require, by way of example, a 12 hour working day rotating by day and night.

Such variations should be agreed in advance and the employer should advise the operative in writing of all such variations within one month.

This rule is designed to clarify how variations maybe approached.

It is open to employers and operatives to agree alternative arrangements, subject to the principles of WR.13.

13.1 Shift Working

Where an operative is required to work a shift pattern, other than five days each week, the following entitlements will need to be adjusted.

13.1.1 Statutory and Industry Sick Pay

The qualifying days and waiting days for both SSP and ISP may need to be adjusted in order not to disadvantage the operative and to take account of the shift pattern being worked.

13.1.2 Payment for Absence

The daily rate of payment for both SSP and ISP, following the first three waiting days (as per WR13.1.1) shall be based on the weekly entitlement for both SSP and ISP divided by the number of days the operative would have worked in the next seven days.

13.1.3 Annual Holiday Entitlement (including Public & Bank)

Where an operative is required to work a shift pattern, other than five days each week, the annual holiday entitlement and its accrual will need to be adjusted. Hours, shift patterns and working week need to be taken into account.

13.1.4 Call out and Standby

Where an operative is required to be available (standby) for "call out" it is open to the employer and operatives to agree appropriate working and payment arrangements.

WR.14 TRANSFER ARRANGEMENTS

14.1 At any time during the period of employment, the operative may, at the discretion of the employer be transferred from one job to another.

14.2 The employer shall have the right to transfer an operative to any site within daily travelling distance of where the operative is living. A site is within daily travelling distance if:

(a) when transport is provided free by the employer the operative can normally get from where he is living to the pickup point designated by the employer within one hour, using public transport if necessary or

(b) in any other case the operative, by using the available public transport on the most direct surface route, can normally get to the site within two hours.

14.3 Transfer to a job which requires the operative to live away from home shall be by mutual consent. The consent shall not be necessary where the operative has been in receipt of subsistence allowance in accordance with the Subsistence Allowance Rule from the same employer at any time within the preceding 12 months.

WR.15 SUBSISTENCE ALLOWANCE

When an operative is recruited on the job or site and employment commences on arrival at job or site he shall not be entitled to payment of subsistence allowance. An operative necessarily living away from the place in which he normally resides shall be entitled to a subsistence allowance of an amount specified by the Council.

Subsistence allowance shall not be paid in respect of any day on which an operative is absent from work except when that absence is due to sickness or industrial injury and he continues to live in the temporary accommodation and meets the industry sick pay requirements.

Alternatively, the employer may make suitable arrangements for a sick or injured operative to return home, the cost of which shall be met in full by the employer

An operative in receipt of subsistence allowance shall only be entitled to daily fare and travel allowances under WR.5 between his accommodation and the job if he satisfies the employer that he is living as near to the job as there is a accommodation available.

WR.16 PERIODIC LEAVE

16.1 Entitlement

When an operative is recruited on the job or site and employment commences on arrival at the job or site he shall not be entitled to the periodic leave allowances in this Rule. In other cases when an operative is recruited or a sent to a job which necessitates his living away from the place in which he normally resides, he shall he entitled to payment of his fares or conveyance in transport provided by the employer as follows:-

(a) from a convenient centre to the job at commencement

(b) to the convenient centre and back to the job at the following periodic leave intervals

   (i) for jobs up to 80 miles from the convenient centre (measured utilising the RAC Route Planner or similar on the shortest distance route), every four weeks

   (ii) for jobs over 80 miles from the convenient centre (measured utilising the RAC Route Planner or similar on the shortest distance route) at an interval fixed by mutual arrangement between the employer and operative before he goes to the job.

(c) from the job to the convenient centre at completion.

16.2 Payment of Fares and Time Travelling

Where an employer does not exercise the option to provide free transport, the obligation to pay fares may at the employer's option be discharged by the provision of a free railway or bus ticket or travel voucher or the rail fare.

Payment for the time spent travelling between the convenient centre and the job as follows:

(a) On commencement of his employment at the job, the time travelling from the convenient centre to the job, provided that an operative shall not be entitled to such payment if within one month from the date of commencement of his employment on the job he discharges himself voluntarily or is discharged for misconduct.

(b) When returning to the job (i.e. one way only) after periodic leave, provided that he returns to the job at the time specified by the employer and provided also that an operative shall not be entitled to such payment if, within one month from the date of his return to the jobs,

he discharges himself voluntarily or is discharged for misconduct.

(c) On termination of his employment on the job by his employer, the time spent travelling from the job to the convenient centre, provided that he is not discharged for misconduct.

(d) Time spent in periodic travelling is not to be reckoned as part of the normal working hours; periodic travelling time payments shall in all cases be at the operative's normal hourly rate to the nearest quarter of an hour and shall not exceed payment for eight hours per journey.

16.3 Convenient centre

The convenient centre shall be a railway station, bus station or other similar suitable place in the area in which the operative normally resides.

WR.17 GUARANTEED MINIMUM WEEKLY EARNINGS

An operative, who has been available for work for the week whether or not work has been provided by the employer shall be entitled to guaranteed minimum weekly earnings as defined in WR 1.

17.1 Loss of Guarantee

There shall be no entitlement to guaranteed minimum weekly earnings where the employer is unable to provide continuity of work due to industrial action.

17.2 Proportional Reduction

Where an operative is absent for part of normal working hours due to certified sickness or injury or for one or more days of annual or recognised public holiday the requirement for the operative to be available for work will be deemed to be met and the payment of Guaranteed Minimum Weekly Earnings will be proportionately reduced. The proportionate reduction will not apply where the employer authorises the absence on compassionate or other grounds.

17.3 Availability for Work

An operative has satisfied the requirements to remain available for work during normal working hours if he complies with the following conditions:

(a) That, unless otherwise instructed by the employer, he has presented himself for work at the starting time and location prescribed by the employer and has remained available for work during normal working hours.

(b) Carries out satisfactorily the work for which he was engaged or suitable alternative work if instructed by the employer and

(c) He complies with the instructions of the employer as to when, during normal working hours, work is to be carried out, interrupted or resumed.

17.4 Temporary Lay-off

17.4.1 Where work is temporarily stopped or is not provided by the employer the operative may be temporarily laid off. The operative shall, subject to the provisions of WR 17.4.2, be paid one fifth of his "Guaranteed Minimum Weekly Earnings" as defined in WR 17 for the day on which he is notified of the lay-off and for each of the first five days of temporary lay-off. While the stoppage of work continues and the operative is prevented from actually working, he will be required by the employer to register as available for work at the operatives local job centre.

GENERAL INFORMATION

WR.17 GUARANTEED MINIMUM WEEKLY EARNINGS (Cont'd)

17.4.2 The payment described in WR. 17.4.1 will be made, provided that in the three months prior to any lay-off there has not been a previous period or periods of lay-off in respect of which a guaranteed payment was made for 5 consecutive days or 5 days cumulative, excluding the day or days of notification of lay-off. In any such case the operative will not be entitled to a further guaranteed payment until a total of three months has elapsed from the last day of the period covered by the previous payment. Thereafter and for so long as the stoppage lasts, he shall be entitled to a further guaranteed payment of up to five days.

17.4.3 The temporary lay-off provisions may only be used when the employer has a reasonable expectation of being able to provide work within a reasonable time. In this context an example of an employer who has a reasonable expectation to be able to provide work may be where a tender has been accepted but commencement delayed, where work is temporarily stopped due to weather conditions or for some other reason outside the employer's control. Reasonable time is not legally defined, however, an employee who has been temporarily laid off for four or more consecutive weeks or six weeks cumulative in any 13 week period may claim a redundancy payment.

17.4.4 In no circumstances may the temporary lay-off rule be used where a genuine redundancy situation exists or to evade statutory obligations.

17.4.5 An operative who is temporarily laid off is entitled to payment of one fifth of his guaranteed minimum earnings for the day of notification of lay-off and for each of the first five days of the layoff subject to the limitations in WR 17.4.2

17.5 Disputes

A dispute arising under this agreement concerning guaranteed minimum payment due may, at the option of the claimant, be referred to ACAS and/or an industrial tribunal in the event of no decision by the Council.

WR.18 ANNUAL HOLIDAYS

For the current (2016) holiday year the annual holiday entitlement remains unchanged at 21 days of Industry and 8 days of Public/Bank holidays.

From January 2017 the rule becomes:-

The holiday year runs from the 1st January for each year with an annual entitlement of 22 days of industry plus 8 days of Public/Bank holidays. Total paid holiday entitlement accrues at the rate of 0.577 days per week of service in the relevant holiday year.

This is an absolute entitlement that cannot be replaced by rolling it up into basic pay, bonus or any other allowance which would result in the operative not receiving their full holiday pay when taking annual leave.

Under the provisions of the EU Working Time Directive the entitlement to paid holidays continues to accrue during employment, notwithstanding that the operative may be absent due to sickness, paternity/maternity leave etc.

18.1 **The Winter Holiday** shall be seven working days taken in conjunction with Christmas Day, Boxing Day and New Year's Day, to give a Winter Holiday of two calendar weeks. The Council shall publish the dates of each Winter Holiday. It shall be open to employers and operatives to agree that all or some of the Winter

Holiday will be taken on alternative dates. Those days of Winter Holiday that fall into the next calendar year are normally allocated to days of holiday earned in the previous year.

18.2 **Other Holidays.** The remaining 15 days of Industry Holiday may be taken at any time by agreement with the employer. An operative requesting to take paid holiday must give the employer reasonable written notice and, as a minimum, equivalent to twice the duration of holiday requested (ie. two weeks written notice to take a one week holiday) and the employer can either accept or reject the request, not later than the period equivalent to the period of holiday requested.

18.3 **Payment for Annual Holiday**
Payment for annual holiday which shall be made on the last pay-day preceding the commencement of each holiday period shall be made as follows:

18.3.1 **Calculation of pay for Annual Holiday**
A weeks' pay is the average of the previous 12 complete weeks including overtime in accordance with WR.4, taxable travel allowance in accordance with WR5.1, Bonus in accordance with WR2 and regular allowances in accordance with WR's 6, 7, 8, 9, 10, 11 & 13. Weeks during which the operative is absent due to sickness are to be excluded.

18.3.2 One day's pay is calculated by dividing a week's pay as defined by WR.18.3.1 by the contractual hours in the normal working week and multiplying by the contractual hours in the particular day.

18.4 **Leavers**

18.4.1 Operatives who leave the employment of the employer during a leave year are entitled to a compensatory payment calculated as follows:
(A÷52) x 29 - B

Where;

A is the number of complete weeks of service in the leave year.

B is the number of days leave taken by the operative in the leave year including public/bank holidays.

18.4.2 Where the number of days leave taken exceeds the operative's entitlement the employer has the right to make a deduction from payment made to the operative leaving the employment of the employer in respect of any overpayment of holiday pay. Such deduction will be calculated in accordance with WR 18.6

18.5 **Amount of the compensatory payment**
The operative is entitled to a compensatory payment for each day, or part of a day, of entitlement calculated by reference to the hourly rate of pay under WR.18.4 multiplied by the normal contractual working hours.

18.6 **General Provisions related to Annual Holiday.**
Where employment commences after the start of the leave year the operatives will be entitled to the proportion of the 30 days Annual Holiday equivalent to the proportion of the leave year calculated from the first week of employment to the last week of the leave year.

An operative has no entitlement to payment for holidays not taken during the holiday leave year or to carry forward entitlement to holiday from one holiday year to the subsequent holiday year.

WR.19 PUBLIC HOLIDAYS

19.1 The following are recognised as public holiday for the purpose of this agreement:

(a) **England and Wales**
Christmas Day, Boxing Day, New Year's Day, Good Friday, Easter Monday, the May Day, Bank Holiday, the Spring Bank Holiday and the Summer Bank Holiday shall be recognised as public holidays in England and Wales, provided that such days are generally recognised as holidays in the locality in which the work is being done.

(b) **Scotland**
Christmas Day, Boxing Day, New Year's Day, Easter Monday, the first Monday in May, the Friday immediately preceding the Annual Summer Local Trades Holiday and the Friday and Monday at the Autumn Holiday, as fixed by the competent Local Authority.

(c) **Local Variations**
Where, in any locality, any of the above public holidays is generally worked and another day is recognised instead as a general holiday, such other day shall be recognised as the alternative holiday.

(d) **Alternative Days**
When Christmas Day, Boxing Day or New Year's Day falls on a Saturday or Sunday an alternative day or days of public holiday will be promulgated. Any reference in this Rule to Christmas Day, Boxing Day or New Year's Day shall be taken to apply to the alternative day so fixed.

19.2 **Payment in respect of Public Holidays**

Payment for days of public holiday recognised under this rule shall be made by the employer to an operative in his employment at the time of each such holiday on the pay day in respect of the payweek in which such holiday occurs, except that payment for Christmas Day, Boxing Day and New Year's Day shall be made on the last pay day before the Winter Holiday.
The amount of payment for one day's pay is calculated by dividing a week's pay as defined by either by WR.19.2.1 or 19.2.2 by the contractual hours in the normal working week and multiplying by the contractual hours in the particular day.

19.2.1 and 19.2.2 - refer to full rules

19.3 **General provisions related to payment for Public Holidays**

An operative who is required to work on a public or bank holiday has the option, by arrangement with the employer, of an alternative day of holiday as soon thereafter as its mutually convenient, in which case the payment prescribed by this Rule shall be made in the respect of such alternative day instead of the public holiday. When the employment is terminated before such alternative day occurs, the operative shall receive such payment on the termination of employment.

19.4 **Payment for Work on a Public Holiday**

All hours worked on a day designated as a public holiday shall be paid for at double time.

WR.20 PAYMENT OF INDUSTRY SICK PAY

20.1 **Relationship of Industry Sick Pay with Statutory Sick Pay (SSP)**

Under existing legislation there is an entitlement to statutory sick pay. Any payment due under this rule shall be increased by an amount equivalent to any statutory sick pay that may be payable in respect of the same day of incapacity for work under the Regulations made under that Act. These are referred to elsewhere in this Rule as "SSP Regulations".

20.2 **Limit of weekly payment**

The aggregate amount of SSP plus industry Sick Pay shall not exceed a normal weeks pay in accordance with WR1.

20.3 **Qualifying Days**

For the purpose of both this Rule and the SSP Regulations, the Qualifying Days that shall generally apply in the industry are Monday to Friday in each week
While the qualifying days referred to above shall generally be the same five days as those which form the normal week of guaranteed employment under this agreement, it is accepted that there might be certain exceptions, e.g. where the particular circumstances of the workplace require continuous six or seven day working.
In these situations it is in order, where there is mutual agreement, for other days to be regarded as 'qualifying days' for the purpose of this Rule and SSP.

20.4 **Amount and Duration of Payment**

(a) An operative who during employment with an employer is absent from work on account of sickness or injury shall, subject to satisfying all the conditions set out in this Rule be paid the appropriate proportion of a weekly amount specified by the Council for each qualifying day of incapacity for work. For this purpose, the appropriate proportion due for a day shall be the weekly rate divided by the number of qualifying days specified under WR 20.3 above.

(b) During the first four continuous weeks of employment with a new employer the Operative shall be entitled to Statutory Sick Pay for absence which the employer is satisfied is due to genuine sickness or injury.

(c) After four continuous weeks of employment the Operative shall be entitled to a total of four weeks Industry sick pay in addition to SSP in respect of absence that starts after these four weeks.

(d) After 15 continuous weeks of employment the Operative shall be entitled to a total of seven weeks Industry sick pay in addition to SSP (less any sick pay received) in respect of absence that starts after these 15 weeks.

(e) After 26 continuous weeks of employment the Operative shall be entitled to a total of 10 weeks Industry sick pay in addition to SSP (less any sick pay received) in respect of absence that starts after these 26 weeks.

(f) This entitlement is based on a rolling 12 month period or single period of absence, whichever is the longer.

20.5 **Notification of Incapacity for Work**

An operative shall not be entitled to payment under this Rule unless, during the first qualifying day in the period of incapacity, his employer is notified that he is unable to work due to sickness or injury and when the incapacity for work started. Thereafter, the operative shall, at intervals not exceeding one week throughout the whole period of absence, keep the employer informed of his continuing incapacity for work. Where the employer is notified later than this rule requires, he may nevertheless make payment under the rule if satisfied that there was good cause for the delay.

GENERAL INFORMATION

WR.20 PAYMENT OF INDUSTRY SICK PAY (Cont'd)

20.6 Certification of Incapacity for Work

The whole period of absence from work shall be covered by a certificate or certificates of incapacity for work to the satisfaction of the employer. For the first seven consecutive days of sickness absence, including weekends and public holidays, a self certificate will normally suffice for this purpose. Any additional days of the same period of absence must be covered by a certificate or certificates given by a registered medical practitioner.

NOTE: For the purpose of this paragraph a self certificate means a signed statement made by the operative in a form that is approved by the employer, that he has been unable to work due to sickness/injury for the whole period specified in the statement.

20.7 Qualifying Conditions for Payment

An operative shall not be entitled to the payment prescribed in this rule unless the following conditions are satisfied:

(a) That incapacity has been notified to the employer in accordance with WR 20.5 above

(b) That the requirements of WR 20.6 above to supply certificate(s) of incapacity for work have been complied with.

(c) That the first three qualifying days (for which no payment shall be due) have elapsed each period of absence.

(d) That none of the qualifying days concerned is a day of annual or public holiday granted in accordance with the provisions of this Working Rule Agreement.

(e) That the incapacity does not arise directly or indirectly from insurrection or war, attempted suicide or self-inflicted injury, the operative's own misconduct, any gainful occupation outside working hours or participation as a professional in sports or games.

(f) That the limit of payment has not been reached.

20.8 Record of Absence

The employer shall be responsible for keeping records of absence and payments made to operatives under this Rule.

WR.21 BENEFIT SCHEMES

21.1 Accident and Death Benefit

An operative is entitled to and the employer will provide insurance cover for:

(a) accidental injury in accordance with the table below as a result of an injury sustained as a result of an accident at the place of work or an accident while travelling to or from work.

Claim Type	Cover
Qualifying claims:	
Loss of Sight in both eyes	£25,000
Loss of Sight in one eye	£10,000
Loss of Hearing in one ear	£3,000
Loss of Hearing in both ears	£10,000
Loss by amputation or the permanent loss of use of:	
An entire hand or foot (below the elbow or knee)	£25,000
An entire arm or leg (at or above the elbow or knee)	£25,000
Big Toe	£1,300
Any other toe	£600
Thumb	£4,000
Index finger (loss of at least one joint)	£4,000
Any other finger (loss of at least one joint)	£800
Total Disablement:	
12 Months	**£3,000**
24 Months	**£3,000**
Permanent	**£15,000***

*Less any payment previously made in respect of the total disablement

(b) death benefit, of £25,000.00 and provided on a 24/7 basis with the cover doubled to £50,000 if death occurs either at the place of work or travelling to or from work. Death Benefit also covers operatives who have been continuously absent from any work since the cover ceased for a period of 2 years if unemployed or 3 years if incapacitated. The death benefit payable reduces depending on the period of absence.

(c) Employers must provide cover which includes all of the above features. Both accident benefit and death benefit are available from B&CE Benefit Schemes who can be contacted on 01293 586666.

(d) The above entitlements will be the subject of periodic review.

21.2 Pension Scheme

Under the provisions of the Pensions Act 2011 the government hasl from October 2012 introduced new, statutory pensions requirements. The new provisions are to be phased in over six years. The largest employers (those employing more than 120,000) were required to comply from October 2012 whereas (those with less than 30 employees) will not be required to comply until 2017 (depending on the employers PAYE reference number). The precise date at which an employer will be required to comply with the new provisions is known as the "staging date" and the requirement will be for the employer to automatically enrol all eligible employees into a qualifying pension scheme with both the employer and employee making at least the minimum specified contributions. The level of employer and employee contributions are planned to increase in 2018 and then again in 2019.

As CIJC employers are of various sizes and will therefore have staging dates at different times it is necessary for the existing CIJC pension scheme arrangements to be maintained alongside the new arrangements for employers to introduce at their "staging date".

21.3 Existing CIJC Pension scheme Arrangements (prior to the employer's "staging date")

An employer is required to make payment on behalf of the operative of such amount or amounts as is promulgated from time to time by the relevant parties to this agreement for the purpose of providing a retirement benefit. Such benefit is an entitlement under this agreement and the employer should normally make a payment of such amount to the B&CE EasyBuild Stakeholder Pension scheme. However, if the operative and the employer shall agree in writing, payment may be made to an alternative approved pension arrangement provider. In any event the parties shall not agree to contribute less than the minimum amounts promulgated.

The current minimum employer contribution is £5.00 per week. Where the operative makes a contribution of between £5.01 and £10.00 per week the employer shall increase its contribution to match that of the operative up to a maximum of £10.00 per week. Since 30th June 2008, employers were not required to make any pension contribution for operatives who decline to make a personal pension contribution of at least £5.00 per week. Pre 30th June 2008 members who do not wish to make a contribution, continue to receive £5.00 per week employer contribution.

WR.21 BENEFIT SCHEMES (Cont'd)

21.4 **New Pension Arrangements (effective from the Employers "Staging Date")**

CIJC has nominated The People's Pension provided by B&CE as the preferred pension scheme for operatives employed under CIJC terms. An industry wide scheme is important because of the transient nature of employment and the aim to reduce the number of pension pots held by individuals. However, CIJC recognises that it is ultimately for the employer to select a qualifying pension arrangement and in the event that an employer utilises an alternative to The People's Pension the contribution levels and general arrangements in respect of pension provision shall be no less favourable than the following:

(a) Operatives within the age range 22 up to State Pension Age who earn at least the auto-enrolment earnings trigger (as defined by Department of Work and Pensions) to be auto-enrolled. Those who fall outside of this criteria will be allowed to opt-in if they wish.

(b) The existing £5.00 per week employer contribution to be maintained during Phase One (from the employers staging date until 30th September 2018) with employers making a higher level of contribution if required as a result of the operatives' level of earnings. Operatives to make a minimum £5.00 per week contribution during Phase One and a higher level of contribution if required as a result of the operatives' level of earnings.

(c) Contribution to be based on Qualifying Earnings as determined by Department of Work and Pensions.

(d) Operatives who currently receive a £5.00 per week employer contribution but make no personal contribution and decide to opt-out and therefore do not make a personal contribution of at least £5.00 per week shall continue to receive the existing £5.00 per week employer contribution.

(e) Operatives who currently neither receive an employer contribution nor make a personal contribution and decide to opt-out shall not be eligible to receive any employer contributions.

(f) Operatives who currently make a personal contribution of between £5.01 and £10.00 per week and while they continue to do so shall receive a matching employer contribution.

(g) An operative who has opted-out of the new pension arrangements may, at any time decide to opt in and providing such operative makes a minimum £5.00 per week personal contribution will receive an equal matched contribution from the employer up to a maximum of £10 per week.

(h) An operative may, at any time, decide to increase their personal weekly contribution from £5.00 up to £10.00 per week in which event such operative will receive a matching contribution from the employer.

(i) An operative may choose to make a personal contribution of any amount above £10.00 per week. In such circumstances the employer's contribution will be limited to £10.00 per week.

(j) As an objective newly employed eligible operatives should be auto-enrolled within 6 weeks of commencing employment.

(k) Under the regulations an operative may choose to opt-out at any time in the future in which case such operative shall not be entitled to receive any employer contributions.

The above arrangements are intended to be generally superior to the minimum statutory provisions. However, nothing within the above shall diminish or detract from the operative's statutory entitlements.

The above arrangements shall apply during Phase One of the new regulations. There will be subsequent phases of the new regulations and new arrangements will then apply.

WR.22 GRIEVANCE PROCEDURE

22.1 **Procedure**

Grievances are concerns, problems or complaints that operatives raise with their employers. Any issue which may give rise to or has given raise to a grievance (including issues relating to discipline) affecting the employer's workplace and operatives employed by that employer at that workplace shall be dealt with in accordance with the following procedure.

Operatives should aim to resolve most grievances informally with their line manager. This has advantages for all work places, particularly where there might be be a close working relationship between a manager and operative. It also allows for problems to be resolved quickly.

If appropriate the employer should give consideration to the use of mediation to assist in resolving the matter.

If a grievance cannot be settled informally, or the matter is considered sufficiently serious, the following procedure should be followed.

Step 1

The operative must write to the employer setting out the details of the grievance or complaint.

Step 2

The employer must investigate the allegations detailed in writing by the operative and arrange a meeting with the operative at the earliest practical opportunity.

The employer shall arrange a meeting and advise the operative of the right to be accompanied at the meeting by either a trade union representative or work colleague (WR.22.3)

Where possible, the employer should allow a companion to have a say in the date and time of the hearing. If the companion cannot attend on a proposed date, the operative can suggest an alternative time and date so long it is reasonable and it is not more than five working days after the original date.

Step 3

Following the meeting the employer shall write to the operative with a decision on their grievance and notify the operative of the right of appeal against that decision if the operative is not satisfied with it.

22.2 **Appeals**

Step 1

If the operative wishes to appeal against the employer's decision then the operative must write to the employer within five working days of the operative receiving the employer's written decision.

WR.22 GRIEVANCE PROCEDURE (Cont'd)
22.2 Appeals (Cont'd)

Step 2

The employer shall arrange a meeting at a time, date and place convenient to the operative and advise the operative of the right to be accompanied at the meeting by either a trade union representative or work colleague (WR.22.3). As far as is reasonably practicable the appeal should be with the must senior appropriate manager/director who has not previously been involved in the matter.

Step 3

Following the meeting the employer shall write to the operative with a decision on the grievance, which shall be regarded as the final stage of the grievance procedure.

22.3 The Accompanying Person (The Companion)

Accompanying an operative at a grievance hearing is a serious responsibility and the companion is entitled to a reasonable amount of paid time off to fulfil this responsibility. The time off should not only cover the hearing but also allow a reasonable amount of time to become familiar with the case and confer with the operative before the hearing. The operative must inform the employer in advance of the hearing of the identity of the proposed companion.

Companions have an important role to play in supporting the operative and should be allowed to participate as fully as possible in the hearing in order to:
- Put operative's case
- Sum up the operative's case
- Respond on the operatives behalf to any view expressed at the hearing.

The companion may confer privately with the operative either in the hearing room or outside. The companion has no right to answer questions on the operative's behalf.

22.4 Raising a Grievance

Setting out a grievance in writing is not easy – especially for those operatives whose first language is not English or have difficultly expressing themselves on paper. In these circumstances the operative should be encouraged to seek help for example from a work colleague or a trade union representative. Under the Disciplinary Discrimination Act 1995 employers are required to make reasonable adjustments which may include assisting operatives to formulate a written grievance if they are unable to do so themselves because of a disability.

22.5 Collective Grievance or Dispute

Any issue which may give rise to or has given rise to a written grievance involving more than one operative or interpretation of the Working Rule Agreement affecting the employer's workplace and operatives employed at that workplace shall be dealt with in accordance with the following procedure.

There shall be no stoppage of work, either partial or general, including a 'go slow', strike, lock out or any other kind of disruption or restriction in output or departure from normal working, in relation to any grievance unless the grievance procedure has been fully used and exhausted at all levels.

Every effort should be made by all concerned to resolve any issue at the earliest stage. To assist in the speedy resolution of a collective grievance the matter should be referred to a steward, if appointed or a full time union representative where no steward is appointed.

A written record shall be kept of meetings held and conclusions reached or decisions taken. The appropriate management or union representative should indicate at each stage of the procedure when an answer to questions arising is likely to be given, which should be as quickly as practicable.

Stage 1

If the matter then remains unresolved, and has not already been referred to a full time union representative, the steward shall report the matter to the appropriate full time union representative who shall, if he considers it appropriate, pursue any outstanding issue with the employer or his nominee after advising him in writing of the issues(s).

Stage 2

Failing resolution of the issue at stage 1, and within 28 day, or such further period as may be agreed between the parties, the full time local union representative shall report the matter up to the appropriate senior full time representative and to an appropriate representative of the employer. Such senior union representative, if there are good grounds for so doing, shall pursue the issue with the appropriate representative of the employer.

Where a collective grievance reaches this stage it would be appropriate for each party to notify the appropriate CIJC joint secretary of the grievance.

Stage 3

Failing resolution of the issue at stage 2, and within 28 days, or such further period as may be agreed between the parties, the senior union representative concerned shall, if it is decided to pursue the matter further, put the issue in writing to the employer and it is the duty of such representative and / or the employer to submit the matter, as quickly as practical, to the construction Industry Joint Council for settlement.

The decisions of the Construction Industry Joint Council shall be accepted and implemented by all concerned.

WR.23 DISCIPLINARY PROCEDURE

It is recognised that, in order to maintain good morale, the employer has the right to discipline any operative:

- who fails to perform his or her duties and responsibilities competently and in accordance with the instructions of the employer; and/or
- whose behaviour is unsatisfactory; and/or
- who fails to make appropriate use of the disputes procedure for the resolution of questions arising without recourse to strike or other industrial action.

It is equally recognised that the employer must exercise this right with fairness and care.

Cases of minor misconduct or unsatisfactory performance are usually best dealt with informally. A quiet word is often all that is required to improve an operative's conduct or performance. The informal approach may be particularly helpful in small firms, where problems can be dealt with quickly and confidentially. There will, however, be situations where matters are more serious or where an informal approach has been tried and isn't working.

If informal action doesn't bring about an improvement, or the misconduct or unsatisfactory performance is considered to be too serious to be classed as minor, the employer should provide the operative with a clear signal of their dissatisfaction by taking formal action as follows:-

GENERAL INFORMATION

WR.23 DISCIPLINARY PROCEDURE (Cont'd)

Note: The employer will not take any disciplinary action before carrying out a full investigation into the matter for which the disciplinary hearing is to be held.

If appropriate the employer should give consideration to the use of mediation to assist in resolving the matter.

Where there is cause to take disciplinary action, the employer will give the operative adequate written notice of the date, time and place of the disciplinary hearing. The notice must contain details of the complaint against the operative and advise of the right to be accompanied at the hearing or appeal stage by either a trade union representative or a work colleague. (WR.23.7). Prior to the disciplinary hearing the operative should be given copies of any documents that will be produced at the hearing.

Where possible, the employer should allow a companion to have a say in the date and time of the hearing. If the companion can't attend on a proposed date, the operative can suggest an alternative time and date so long as it is reasonable and it is not more than five working days after the original date.

Discipline shall be applied in accordance with the following procedure:

23.1 Disciplinary Action and Stages

Disciplinary action will comprise the following stages unless WR.23.4 is applicable:
(a) A written warning
(b) A final written warning
(c) Dismissal
(d) Following each of the above stages the employer will notify the operative of the decision in Writing including the right of appeal under WR.23.3.
(e) The employer shall deal with disciplinary matters without undue delay.
(f) Where an operative has been accompanied at a disciplinary or appeal hearing by a union representative the employer shall provide the representative of the union with a copy of any letter of warning or dismissal providing the operative gives express permission.

23.2 Duration of Warnings

Warnings will normally be discarded after 6 months in the case of a written warning and 12 months in the case of a final warning.

23.3 Right of Appeal

The operative shall be advised of the right to appeal at every stage of the procedure. Where the employer's organisation structure allows, the appeal should be heard by a senior manager / director of the employer who has not been involved in the disciplinary procedure. The request for an appeal must be made in writing within five working days of the date of the disciplinary decision. The employer will inform the operative of the final decision in writing.

23.4 Serious Misconduct

In exceptional circumstances and if the matter is sufficiently serious, a final written warning may be issued in the first instance.

23.5 Gross Misconduct

23.5.1 In certain circumstances the conduct may be so serious as to be referred to as gross misconduct. In such circumstances the first two stages of the disciplinary procedure, written warning, and final warning, may be omitted and the operative may be summarily dismissed without notice or pay in lieu of notice, but only after following a fair disciplinary process in line with the ACAS Code of Practice.

The employer will notify the operative of the alleged gross misconduct in writing and invite them to a disciplinary hearing advising them of their right to be accompanied at the disciplinary hearing or appeal stage by either a trade union representative or a work colleague (WR.23.7).

23.5.2 Set out below is a list, although notexhaustive, of behaviour, which will be considered by the employer to be gross misconduct:
• Being under the influence of alcohol or other stimulants or illicit drugs during working hours,

• Physical violence – actual or threatened,

• Violent, disorderly or indecent conduct.

• Deliberate damage to property,

• Theft, fraud or falsification of company records, documents or time sheets,

• Serious breach of confidence (subject to the Public Interest Disclosure legislation)

• Removal from company sites or other premises of property belonging to the company, fellow operative, client, sub-contractor, supplier or other without the approval of the employer,

• Serious breach of the employer's safety policy rules or regulations,

• Bringing the employer into serious disrepute,

• Acts of incitement to or actual acts of discrimination on grounds of sex, race, religion, belief, colour, ethnic origin, disability, age or sexual orientation.

• Serious bullying or harassment,

• Serious carelessness resulting in loss or damage – or potential loss or damage,

• Serious insubordination,

• Misuse of the employers or clients property or name.

23.5.3 Summary dismissal means termination of employment without notice or pay in lieu of notice. In circumstances where a gross misconduct is alleged to have occurred the operative will normally be suspended on full pay whilst an investigation is carried out prior to a disciplinary hearing.

23.6 Shop Stewards

Where it is proposed to take disciplinary action against a duly appointed shop steward, or other union official, then before doing so, the employer shall notify the appropriate full time official of the union concerned

23.7 The Accompanying Person (The Companion)

Accompanying an operative at a disciplinary hearing is a serious responsibility and the companion is entitled to a reasonable amount of paid time off to fulfil this responsibility. The time off should not only cover the hearing but also allow a reasonable amount of time to become familiar with the case and confer with the operative before the hearing. The operative must inform the employer in advance of the hearing of the identity of the proposed companion.

Companions have an important role to play in supporting the operative and should be allowed to participate as fully as possible the hearing in order to:

• Put the operative's case
• Sum up the operative's case

23.7 The Accompanying Person (The Companion) (Cont'd)

- Respond on the operatives behalf to any view expressed at the hearing
- The companion may confer privately with the operative, either in the hearing room or outside. The companion has no right to answer questions on the operative's behalf.
-

WR.24 TERMINATION OF EMPLOYMENT

24.1 Employer notice to Operative

The employment may be terminated at any time by mutual consent which should preferably be expressed in writing.

All outstanding wages including holiday pay are to be paid at the expiration of the period of notice and the employee advised of his entitlement to PAYE certificates or, in lieu thereof, a written statement that they will be forwarded as soon as possible.

The minimum period of notice of termination of employment that an employer shall give to an employee is:-

(a) During the first month — One day's notice

(b) After one months' continuous employment but less than two years — One week's notice

(c) After two years' continuous employment but less than twelve years — One weeks notice for each full year of employment.

(d) Twelve years' continuous employment or more — Twelve weeks' notice

24.2 Operative notice to Employer

The minimum period of notice of termination of employment that an employee shall give an employer is:

(a) During the first month - one day's notice

(b) After one month's continuous employment - one week's notice.

WR.25 TRADE UNIONS

25.1 The Employers' Organisations recognises the signatory Trade Unions within the Construction Industry Joint Council for the purposes of collective bargaining. Both parties are fully committed to the Working Rule Agreement and strongly urge employers to:

(a) recognise the trade unions who are signatories to the Agreement;

(b) ensure that all operatives are in the direct employment of the company or its sub contractors and are engaged under the terms and conditions of the Working Rule Agreement.

25.2 Deduction of Union Subscriptions

When requested by a signatory union, employers should not unreasonably refuse facilities for the deduction of union subscriptions (check-off) from the pay of union members.

25.3 Full Time Trade Union Officials

A full time official of a trade union which is party to the Agreement shall be entitled, by prior arrangement with the employer's agent or other senior representative in charge and on presenting his credentials, to visit a workplace to carry out trade union duties and to see that the Working Rule Agreement is being properly observed.

25.4 Trade Union Steward

An operative is eligible for appointment as a steward on completion of not less than four weeks continuous work in the employment of the employer. Where an operative has been properly appointed as a steward in accordance with the rules of his trade union (being a trade union signatory to the Agreement) and issued with written credentials by the trade union concerned, the trade union shall notify the employer's agent or representative of the appointment, for formal recognition by the employer of the steward. On completion of this procedure the employer will recognise the steward, unless the employer has any objection to granting recognition, in which case he shall immediately notify the trade union with a view to resolving the question.

An employer shall not be required to recognise for the purposes of representation more than one officially accredited steward for each trade or union at any one site or workplace.

25.5 Convenor Stewards

Where it is jointly agreed by the employer and the unions, having regard to the number of operatives employed by the employer at the workplace and/or the size of the workplace the recognised trade unions may appoint a convenor steward, who should normally be in the employment of the main contractor, from among the stewards and such appointment shall be confirmed in writing by the Operatives' side. On completion of this procedure the employer will recognise the Convenor Steward, unless the employer has any objection to granting recognition in which case he shall immediately notify the trade union with a view to resolving the question.

25.6 Duties and Responsibilities of Stewards and Convenor Stewards.

25.6.1 Duties of the Shop Stewards shall be:

(a) To represent the members of their union engaged on the site/factory/depot,

(b) to ensure that the Working rule is observed,

(c) to recruit members on the site/factory/depot into a signatory union,

(d) to participate in the Grievance Procedure at the appropriate stage under WR. 22,

(e) to assist in the resolution of disputes in accordance with the Working Rule Agreement.

25.6.2 Duties of the Convenor Steward, in addition to those set out in WR 25.6.1 shall be:

(a) to represent the operatives on matters concerning members of more than one union,

(b) to co-operate with Management and to assist individual Shop Stewards

(c) to ensure that disputes are resolved by negotiation in accordance with the Working Rule Agreement.

25.6.3 No Steward or Convenor Steward shall leave their place of work to conduct union business without prior permission of their immediate supervisor. Such permission should not be unreasonably withheld but should only be given where such business is urgent and relevant to the site, job or shop.

WR.25 TRADE UNIONS (Cont'd)

25.7 Steward Training

To assist them in carrying out their functions, Shop Stewards will be allowed reasonable release to attend training courses approved by their union.

25.8 Stewards Facilities

Management shall give recognised union officials and/or convenor stewards reasonable facilities for exercising their proper duties and functions. These facilities, which must not be unreasonably withheld, must not be abused. The facilities should include use of a meeting room, access to a telephone and the use of a notice board on site. If the convenor steward so requests, the employer shall provide him regularly with the names of operatives newly engaged by that contractor for work on that site, factory or depot.

25.9 Meetings

Meetings of operatives, stewards or convenor stewards may be held during working hours only with the prior consent of Management.

25.10 Blacklisting

The CIJC does not condone any form of blacklisting of any worker.

WR.26 SCAFFOLDERS

The following provisions are to be read in conjunction with the provisions of the Construction Industry Scaffolders' Record Scheme (CISRS) General Information publication. It should be noted that if a scaffolder or trainee scaffolder is not in possession of a valid CISRS card at the time of engagement, the new employer should make an application immediately. Any difficulties should be referred to the Joint Secretaries for action in accordance with the Scheme.

26.1 Scaffolders employed whole time as such come under one of the categories outlined below:

Trainee Scaffolder: an operative who can produce a current, valid CISRS Trainee card

Scaffolder: an operative who can produce a current, valid CISRS Scaffolders card or a current, valid CISRS Basic Scaffolders card.

Note: the word Basic has been dropped by the CISRS Scheme. CISRS Cards issued at this grade since June 2006 will state on both the front and reverse of the card "CISRS Scaffolder". CISRS Basic Scaffolder Cards will be recognised until the expiry date displayed on the card.

Advanced Scaffolder: an operative who can produce a current, valid CISRS Advanced Scaffolders card.

Operatives holding expired CISRS Basic or Advanced Scaffolder Cards will be paid at the Trainee rate (see WR.26.4) until their card is renewed.

CISRS have introduced a new category of card, Basic Access System Erector (BASE) for non scaffolding operatives. This will allow an operative upon completion of the appropriate CISRS course to erect some specified simple scaffolding structures using prefabricated systems scaffold, which will be restricted by height, structure and work environment. See the CISRS website www.cisrs.org.uk for further information.

Note: operatives carrying the BASE card are not fully qualified scaffolders and must not carry out any works in tube and fittings, also they must not carry out any works in prefabricated systems which exceeds the scope of their training.

26.2 No operative other than a Scaffolder or Advanced Scaffolder, as defined above, may be employed on scaffolding operations and no operative other than an Advanced Scaffolder may be employed on advanced scaffolding operations (see schedule WR 26.5 and 26.6) unless working together and under the supervision of a qualified CISRS Scaffolder or Advanced Scaffolder

26.3 The onus of proof of training and experience required under this Rule is on the operative concerned and the onus on checking the proof submitted is on the employer.

26.4 Scaffolders covered by WR 26.1 are entitled to the Skilled Operative Basic Rate of Pay 4,3,2,1 or Craft Rate as follows:

	Rate
CISRS BASE (Basic Access Systems Erector)	4
CISRS Trainee	4.
CISRS Scaffolder	Craft Rate
CISRS Advanced	Craft Rate

26.5 Approved list of scaffolding operations,

Erecting, adapting and dismantling:
Independent, putlog and birdcage scaffolds, static and mobile towers.
Beams to form gantries and openings, correctly braced.
Hoist frameworks
Protective fans.
Stack scaffolds.
Roof scaffolds
Scaffolds to form Truss out Scaffolds
Simple cantilevers
Edge protection
Proprietary systems.(see rear of card to verify product training)
Fixing sheeting/netting to scaffold framework
Interpreting simple design layout drawing for scaffolding detailed above.
Applying knowledge of Construction Regulations/Legislation/Industry best practice and guidance to operations listed above.

Note: CISRS Scaffolders / CISRS Basic Scaffolders are sufficiently qualified and deemed competent for onsite supervision for all of the operations listed above.

26.6 Advanced List

All work in list of Scaffolding operations:
Erecting, adapting or dismantling:
Suspended scaffolds
Raking or flying shores.
Other forms of designed structures, e.g. larger truss-outs, cantilevers, lifting structures, ramps, footbridges and temporary roofs.
Scaffolding or standard props (including all bracing) to form a dead shore, including adjustable bases and fork heads.
Scaffolding and proprietary systems (including levelling to within reasonable tolerances) to support formwork as laid out in engineering scaffold drawings. .(see rear of card to verify product training)
Interpreting scaffold design drawings.
Applying knowledge of Construction Regulations/Legislation/Industry best practice and guidance to operations listed above.

Note: Only CISRS Advanced Scaffolders are sufficiently qualified and deemed competent for onsite supervision for all of the operations listed above.

GENERAL INFORMATION

WR.27 HEALTH SAFETY AND WELFARE

27.1 The employers and operatives Organisation who are signatories to the Working Rule Agreement are committed to operating construction sites that provide a working environment which is both safe and free from hazards for everybody within the construction industry and for members of the public. All workers, whether operatives or management, comply with the requirement of legislation dealing with health, safety and welfare.

27.2 Trade Union Safety Representatives

Legislation provides that recognised trade unions may appoint safety representatives to represent operatives. Provision is also made for the establishment of safety committees where a formal request, in writing, is made to an employer by at least two safety representatives who have been appointed in accordance with legislation.

Trade union safety representatives are an appropriate means of consulting with those workers who are represented by a trade union, however not all workers will be represented by the appointed person and they need an alternative method of consultation.

27.3 Site Induction

Everyone working on site will go through a health and safety induction process before they are allowed to commence work on site. This induction training will concentrate on site specific health and safety factors and will be given by appropriate personnel nominated by the employer.

27.4 Consultation with the Workforce
27.5

Employers' and Operatives' Organisations wish to create an industry where everyone is valued, all views are listened to and a safe and healthy working environment is the norm. Employers are committed to worker consultation on health, safety and welfare issues. Consultation mechanisms, such as toolbox talks, notice boards and other appropriate means determined by the employer, will be made known to all workers on site and details will be included in the construction phase health and safety plan.

WR. 28 REFERENCE PERIOD

28.1 For the purpose of compliance with the Working Rule Agreement, statutory definitions, entitlements and calculations the reference period shall, subject to the requirements of WR 3.2 and WR.7.2 - "AVERAGE WEEKLY WORKING HOURS", be 12 months, in accordance with the Working time regulations 1998

WR.29 LENGTH OF NIGHT WORK WR6-10, 13

29.1 Night Worker

A night worker is defined as a worker who works at least 3 hours of his daily working time between the hours of 11.00pm and 6.00am on the majority of days worked.

29.2 Length of Night Work

The parties to this agreement recognise that working patterns occasionally arising within the construction industry require recognition in order that operations comply with the Working Time Regulations 1998. In the light of this the parties have agreed that in accordance with Regulation 23(a) of that act and any amendments or modifications thereof, the following Regulations are excluded in relation to work undertaken under the provisions of the above Working Rules:

(a) Regulation 6(1) (eight hour average limit on length of night work);

(b) Regulation 6(2) (application of average eight hour limit to night workers);

(c) Regulation 6(3) (17-week reference period); and

(d) Regulation 6(7) (eight hour absolute limit on the length of night work in the case of work involving special hazards or heavy physical or mental strain).

Whilst establishing the ability to work night shifts of longer than 8 hours, nothing in this rule places any obligation on any worker to work a night shift of longer than 8 hours.

WR.30 SUPPLEMENTARY AGREEMENTS

When it is agreed between the employer(s) and operative(s) that at any particular workplace it would be appropriate to enter into an agreement specifically for that workplace, any such agreement shall be supplementary to and not in conflict with this Working Rule Agreement. Where any dispute arises in this respect the National Working Rule Agreement takes precedence.

WR.31 DURATION OF AGREEMENT

This Agreement shall continue in force and the parties to it agree to honour its terms until the expiration of three calendar month's notice to withdraw from it, given by either the Employers' side or the Operatives' side.

SCHEDULE 1

Rates of Pay

Specified Work Establishing Entitlement to the Skilled Operative Pay Rate 4,3,2 1. or Craft Rate

Basic Rate of Pay

BAR BENDERS AND EINFORCEMENT FIXERS

Bender and fixer of Concrete Reinforcement capable of reading and understanding drawings and bending schedules and able to set out work	Craft Rate

CONCRETE

Concrete Leveller or Vibrator Operator	4
Screeder and Concrete Surface Finisher working from datum such as road-form, edge beam or wire	4
Operative required to use trowel or float (hand or powered) to produce high quality finished concrete	3

DRILLING AND BLASTING

Drills, rotary or percussive: mobile rigs, operator of	3
Operative attending drill rig	4
Shotfirer, operative in control of and responsible for explosives, including placing, connecting and detonating charges	3
Operatives attending on shotfirer including stemming	4

DRYLINERS

Operatives undergoing approved training in drylining	4
Operatives who can produce a certificate of training achievement indicating satisfactory completion of at least one unit of approved drylining training	3
Dryliners who have successfully completed their training In drylining fixing and finishing	Craft Rate

FORMWORK CARPENTERS

1st year trainee	4
2nd year trainee	3
Formwork Carpenters	Craft Rate

GANGERS AND TRADE CHARGEHANDS

(Higher Grade payments may be made at the employer's discretion)	2

GAS NETWORK OPERATIONS

Operatives who have successfully completed approved training to the standard of:

GNO Trainee	4
GNO Assistant	3
Team Leader - Services	2
Team Leader-Mains	1
Team Leader - Main and Services	1

Highways Maintenance

Lead safety fence installer: Holder of appropriate qualification in vehicle restraint systems. Team leader, erector, installer and maintenance of vehicle safety fencing.	Craft
Safety fence installer: Holder of appropriate qualification in vehicle restraint systems. Erector, installer and maintenance of vehicle safety fencing.	3
Traffic management operative: Installing, maintaining and removal of traffic management systems.	3
Trainee traffic management operative: Assist in Installing, maintaining and removal of traffic management systems.	4
Lead traffic management operative (TMF equivalent): Holder of appropriate qualification in installation, maintenance and removal of traffic management systems.	Craft

Highways Maintenance operative: Undertake routine and cyclical maintenance duties and secondary response. To include, lighting, structures and other general highways duties.	3
Highways Incident Response Operative: Holder of appropriate qualification in incident response. Routine and cyclical maintenance and incident response.	2

LINESMEN - ERECTORS

1st Grade (Skilled in all works associated with the assembly, erection, maintenance of Overhead Lines Transmission Lines on Steel Towers; concrete or Wood Poles; including all overhead lines construction elements.)	2
2nd Grade (As above but lesser degree of skill - or competent and fully skilled to carry out some of the elements of construction listed above.)	3
Linesmen-erector's mate.(Semi-skilled in works specified above and a general helper)	4

MASON PAVIORS

Operative assisting a Mason Pavior undertaking kerb laying, block and sett paving, flag laying, in natural stone and precast products	4
Operative engaged in stone pitching or dry stone walling	3

MECHANICS

Maintenance Mechanic capable of carrying out field service duties, maintenance activities and minor repairs	2
Plant Mechanic capable of carrying out major repairs and overhauls including welding work, operating turning lathe or similar machine and using electronic diagnostic equipment	1
Maintenance/Plant Mechanics' Mate on site or in depot	4
Tyre Fitter, heavy equipment tyres	2

MECHANICAL PLANT DRIVERS AND OPERATORS

Backhoe Loaders (with rear excavator bucket and front shovel and additional equipment such as blades, hydraulic hammers, and patch planers)

Backhoe, up to and including 50kW net engine power; driver of	4
Backhoe, over 50kW up to and including 100kW net engine power; driver of	3
Backhoe, over 100kW net engine power; driver of	2

Compressors and Generators

Air compressor or generators over 10 KW operator of	4

Concrete Mixers

Operative responsible for operating a concrete mixer or mortar pan up to and including 400 litres drum capacity	4
Operative responsible for operating a concrete mixer over 400 litres and up to and including 1,500 litres drum capacity	3
Operative responsible for operating a concrete mixer over 1,500 litres drum capacity	2
Operative responsible for operating a mobile self-loading and batching concrete mixer up to 2,500 litres drum capacity	2
Operative responsible for a operating a mechanical drag-shovel	4

GENERAL INFORMATION

SCHEDULE 1 (Cont'd)　　　　　Rates of Pay

Concrete Placing Equipment

Trailer mounted or static concrete pumps; self-propelled concrete placers; concrete placing booms; operator of — 3

Self-propelled Mobile Concrete Pump, with or without boom, mounted on lorry or lorry chassis; driver/operator of — 2

Mobile Cranes,

Self-propelled Mobile Crane on road wheels, rough terrain wheels or caterpillar tracks including lorry mounted:

Max. lifting capacity at min. radius, up to and including 5 Tonne; driver of — 4

Max. lifting capacity at min. radius, over 5 Tonne and up to and including 10 tonne; driver of — 3

Max. lifting capacity at min. radius, over 10 Tonne — Craft Rate

Where grabs are attached to Cranes the next higher skilled basic rate of pay applies except over 10 tonne where the rate is at the employer's discretion.

Tower Cranes (including travelling or climbing)

Up to and including 2 Tonne max. lifting capacity at min. radius; driver of — 4

Over 2 Tonne up to and including 10 Tonne max. lifting capacity at min. radius; driver of — 3

Over 10 Tonne up to and including 20 Tonne max. lifting capacity at min. radius; driver of — 2

Over 20 Tonne max. lifting capacity at min. radius; driver of — Craft Rate

Miscellaneous Cranes and Hoists

Overhead bridge crane or gantry crane up to and including 10 Tonne capacity; driver of — 3

Overhead bridge crane or gantry crane over 10 Tonne capacity up to and including 20 Tonne capacity; driver of — 2

Power driven Hoist or jib crane; driver of — 4

Slinger / Signaller appointed to attend Crane or hoist to be responsible for fastening or slinging loads and generally to direct lifting operations — 4

Dozers

Crawler dozer with standard operating weight up to and including 210 tonne ; driver of — 3

Crawler dozer with standard operating weight over 10 Tonne and up to and including 50 tonne; driver of — 2

Crawler dozer with standard operating weight over 50 Tonne; driver of — 1

Dumpers and Dump Trucks

Up to and including 10 tonne rated payload; driver of — 4

Over 10 Tonne and up to and including 20 Tonne rated payload; driver of — 3

Over 20 Tonne and up to and including 50 Tonne rated payload; driver of — 2

Over 50 Tonne and up to and including 100 Tonne rated payload; driver of — 1

Over 100 Tonne rated payload; driver of — Craft Rate

Excavators (360 degree slewing)

Excavators with standard operating weight up to and including 10 Tonne; driver of — 3

Excavator with standard operating weight over 10 Tonne and up to and including 50 Tonne; driver of — 2

Excavator with standard operating weight over 50 Tonne;; driver of — 1

Banksman appointed to attend excavator or responsible for positioning vehicles during loading of tipping — 4

Rates of Pay

Fork-Lifts Trucks and Telehandlers

Smooth or rough terrain fork lift trucks (including side loaders) and telehandlers up to and including 3 Tonne lift capacity; driver of — 3

Over 3 Tonne lift capacity; driver of. — 2

Motor Graders; driver of — 2

Motorised Scrapers; driver of — 2

Motor Vehicles (Road Licensed Vehicles) Driver and Vehicle Licensing Agency (DVLA)

Vehicles requiring a driving licence of category C1; driver of — 4

(Goods vehicle with maximum authorised mass (mam) exceeding 3.5 Tonne but not exceeding 7.5 Tonne and including such a vehicle drawing a trailer with a mam not over 750kg)

Vehicles requiring a driving licence of category C; driver of — 2

(Goods vehicle with a maximum authorised mass (mam) exceeding 3.5 Tonne and including such a vehicle drawing a trailer with mam not over 750kg)

Vehicles requiring a driving licence of category C plus E; driver of — 1

(Combination of a vehicle in category C and a trailer with maximum authorised mass over 750kg)

Power Driven Tools

Operatives using power-driven tools such as breakers, percussive drills, picks and spades, rammers and tamping machines. — 4

Power Rollers

Roller up to but not including 4 tonne operating weight; driver of — 4

Roller, over 4 Tonne operating weight and upwards; driver of — 3

Pumps

Power-driven pump(s); attendant of — 4

Shovel Loaders, (Wheeled or tracked, including skid steer)

Up to and including 2 cubic metre shovel capacity; driver of — 4

Over 2 cubic metre and up to and including 5 cubic metre shovel capacity; driver of — 3

Over 5 cubic metre shovel capacity; driver of — 2

Tractors (Wheeled or Tracked)

Tractor, when used to tow trailer and/or with mounted

Compressor up to and including 100kW rated engine power; driver of. — 4

Tractor, ditto, over 100KW up to and including 250kW rated engine power; driver of — 3

Tractor, ditto, over 250kW rated engine power; driver of — 2

Trenchers (Type wheel, chain or saw)

Trenching Machine, up to and including 50kW gross engine power; driver of — 4

Trenching Machine, over 50kW and up to and including 100kW gross engine power; driver of — 3

Trenching Machine, 100kW gross engine power; driver of — 2

Winches, Power driven winch, driver of — 4

PILING

General Skilled Piling Operative — 4

Piling Chargehand/Ganger — 3

Pile Tripod Frame Winch Driver — 3

CFA or Rotary or Driven Mobile Piling Rig Driver — 2

SCHEDULE 1 (Cont'd)   Rates of Pay

Concrete Pump Operator	3

PIPE JOINTERS

Jointers, pipes up to and including 300mm diameter	4
Jointers, pipes over 300mm diameter and up to 900mm diameter	3
Jointers, pipes over 900mm diameter	2
except in HDPE mains when experienced in butt fusion and/or electrofusion jointing operations	2

PIPELAYERS

Operative preparing the bed and laying pipes up to and including 300 mm diameter	4
Operative preparing the bed and laying pipes over 300 mm diameter and up to and including 900mm diameter.	3
Operative preparing the bed and laying pipes over 900mm diameter	2

PRE-STRESSING CONCRETE

Operative in control of and responsible for hydraulic jack and other tensioning devices engaged in post-tensioning and/or pre-tensioning concrete elements	3

RAIL

Plate layer (Not labourer in a gang)	3

ROAD SURFACING WORK (includes rolled asphalt, dense bitumen macadam and surface dressings)

Operatives employed in this category of work to be paid as follows:

Chipper	4
Gritter Operator	4
Raker	3
Paver Operator	3
Leveller on Paver	3
Road Planer Operator	3
Road Roller Driver, 4 Tonne and upwards	3
Spray Bar Operator	4

SCAFFOLDERS   See WR. 26 above

STEELWORK CONSTRUCTION

A skilled steel erector engaged in the assembly erection and fixing of steel-framed buildings and structures	1
Operative capable of and engaged in fixing simple steelwork components such as beams, girders and metal decking	3

TIMBERMAN

Timberman, installing timber supports	3
Highly skilled timberman working on complex supports using timbers of size 250mm by 125mm and above	2
Operative attending	4

TUNNELS

Operative working below ground on the construction of tunnels underground spaces or sinking shafts:

Tunnel Boring Machine operator	2
Tunnel Miner (skilled operative working at the face)	3
Tunnel Miner's assistants (operative who assists the tunnel miner).	4
Other operatives engaged in driving headings, in connection with cable and pipe laying	4
Operative driving loco	4

WELDERS

Grade 4 (Fabrication Assistant)

Welder able to tack weld using SMAW or MIG welding processes in accordance with verbal instructions and including mechanical preparation such as cutting and grinding	3

Rates of Pay

Grade 3 (Basic Skill Level)

Welder able to weld carbon and stainless steel using at least one of the following processes SMAW, GTAW, GMAW for plate-plate fillet welding in all major welding positions, including mechanical preparation and complying with fabrication drawings	2

Grade 2 (Intermediate Skill Level)

Welder able to weld carbon and stainless steel using manual SMAW, GTAW, semi-automatic MIG or MAG, and FCAW welding processes including mechanical preparation, and complying with welding procedures, specifications and fabrication drawings.	1

Grade 1 (Highest Skill Level)

Welder able to weld carbon and stainless steel using manual SMAW, GTAW, semi-automatic GMAW or MIG or MAG, and FCAW welding processes in all modes and directions in accordance with BSEN 287-1 and/or 287-2 Aluminium Fabrications including mechanical preparation and complying with welding procedures, specifications and fabrication drawings.	Craft Rate

YOUNG WORKERS

Operatives below 18 years of age will receive payment of 60% of the General Operative Basic Rate

At 18 years of age or over the payment is100% of the relevant rate.

CONSTRUCTION INDUSTRY JOINT COUNCIL

Joint Secretaries Guidance Notes on the Working Rule Agreement of the Construction Industry Joint Council.

Introduction

These Guidance Notes, whilst not forming part of the Working Rules, are intended to assist employers and operatives to understand and implement the Working Rule Agreement.

It is the intent of all parties to this Agreement that operatives employed in the building and civil engineering industry are engaged under the terms and conditions of the CIJC Working Rule Agreement.

Requests for definitions, clarification or resolution of disputes in relation to this Agreement should be addressed to the appropriate adherent body.

WR. 1 Entitlement to Basic Pay and Additional Rates of Pay.

WR. 1 sets out the entitlement to the basic rate of pay, additional payments for skilled work and occasional skilled work.

There are six basic rates of pay and rates for Apprentices under this Agreement; a General Operative rate, four rates for Skilled Operatives, a rate for a Craft Operative and rates for Apprentices.

Payment for Occasional Skilled Work

WR. 1.1.2 deals with the payment for occasional skilled work and provides that general operatives who are required to carry out work defined in Schedule 1 on an occasional basis should receive an increased rate of pay commensurate with the work they are carrying out for the period such work is undertaken. This sets out the flexibility to enable enhanced payment to be made to general operatives undertaking skilled work for a limited amount of time but should not be used where the operative is engaged whole time on skilled work.

Skilled Operatives

WR.1.2.1 sets out a permanent rate of pay for skilled operatives who are engaged whole time on the skilled activity and does not permit the operative engaged on whole time skilled work to have his pay reduced to the General Operative basic rate when occasional alternative work is undertaken.

CONSTRUCTION INDUSTRY JOINT COUNCIL (Cont'd)
Joint Secretaries Guidance Notes on the Working Rule
Agreement of the Construction Industry Joint Council.
(Cont'd)

WR.3.2 Average Weekly Working Hours.

Working Rules 3.2 and 7.2 provide, where there are objective or technical reasons, for the calculation of average weekly working hours by reference to a period of twelve months.

Whilst it is open to employers and employees to agree to work additional hours over the "normal working hours," Rule 3.2 does not give the employer the unilateral right to introduce excessive hours on a job or site. The 12 month averaging period referred to may only be applied where, for objective reasons, it is necessary to ensure completion of the work efficiently and effectively or where there are technical reasons that require additional hours to be worked.

Examples of objective and/or technical reasons which may require average weekly working hours to be calculated using a twelve month reference period are set out below. The list is not exhaustive other objective or technical reasons may apply.

Objective Reasons:
Work on infrastructure, roads, bridges, tunnels and tide work etc.
Client requirements for work to be completed within a tightly defined period, work undertaken for exhibitions, schools, retail outlets, shopfitting and banks etc.
Emergency work, glazing and structural safety etc.

Technical Reasons;
Work requiring a continuous concrete pour surfacing and coating work, tunnelling etc.

Note Any disputes regarding the validity of objective and/or technical reasons may be referred to the National Conciliation Panel of the Construction Industry Joint Council.

W.R. 4 Overtime Rates.

Where an operative who has worked overtime fails without authorisation to be available for work during normal weekly working hours he may suffer a reduction in or may not be entitled to premium payments in respect of overtime worked.

To calculate the number of hours paid at premium rate (overtime) you subtract the number of hours of unauthorised absence from the total number of hours overtime worked. This is in effect using a part of the hours of overtime worked to make the hours paid at the normal hourly rate up to 39 hours per week, the balance of overtime is paid at the appropriate premium rate.

EXAMPLES

Example 1.

An operative works three hours overtime on Monday. Tuesday and Wednesday works normal hours on, Thursday and is unavailable for work on Friday due to unauthorised absence. In these circumstances overtime premia will be calculated as follows:

	Normal Hours	Overtime Hours
Monday	8	3
Tuesday	8	3
Wednesday	8	3
Thursday	8	-
Friday	-	-
Total hours worked	(A) 32	(B) 9

Normal weekly working hours	(C)	39
Less total normal hours worked	(A)	32
Hours required to make up to 39	(D)	7
Total overtime hours	(B)	9
Less	(D)	7
Hours to be paid at premium rate	(E)	2

The operative is therefore, entitled to be paid:

(A+D)	39	Hours at "Normal Hourly Rate"
(E)	2	Hours at premium rate (time and a half)
Total	41	Hours

Example 2.

An operative works four hours overtime on a Monday, three hours on Tuesday five hours (one hour double time) on the Wednesday, no overtime on either Thursday or Friday and absents himself from work on Friday at 12 noon without authorisation and then works six hours on Saturday. This calculation would be as follows:

	Normal Hours	Overtime Hours	
Monday	8	4	
Tuesday	8	3	
Wednesday	8	5	(1 hour double time)
Thursday	8	-	
Friday	3 ½	-	
Saturday	-	6	(2 hours double time)
Total Hours worked	(A) 35½	(B) 18	

Normal weekly working hours	(C)	39
Less total normal hours worked	(A)	35 ½
Hours required to make up to 39	(D)	3 ½
Total overtime hours	(B)	18
Less	(D)	3 ½
Hours to be paid at premium rate	(E)	14 ½

The operative is, therefore, entitled to be paid:

(A+D)	39	Hours at "Normal Hourly Rate"
(E)	14 ½	Hours at premium rate (time and a half)
Total	53 ½	Hours

The entitlement to 3 hours pay at double time is extinguished in this example by the hours of unauthorised absence. If the number of hours worked at double time exceeds the number of hours of unauthorised absence the balance must be paid at the rate of double time.

Note There shall be no reduction in overtime premium payments for operatives who are absent from work with the permission of the employer or who are absent due to sickness or injury

.

N.B. **For full details of the conciliation dispute machinery please see the Constitution and Rules of the CIJC.**
Any question relating to Working Rules not covered by the notes of guidance should be referred to the to the appropriate adherent body.

GENERAL INFORMATION

PRIME COST OF DAYWORK CARRIED OUT UNDER A BUILDING CONTRACT

The Royal Institution of Chartered Surveyors and The Construction Confederation released the third edition of the Definition in June 2007.

The third edition is easy to follow and provides two options for dealing with the cost of labour; either to use the traditional Percentage Addition method or to use All-inclusive Rates.

The example calculations for a typical Standard Hourly Rate have been updated for a Craft Operative and for a General Operative, to include for a work place pension in accordance with the Pensions Act 2011,

The Definition states that it is provided 'for convenience and for use by people who choose to use it'. Provision for dayworks is generally included on all Projects and all specifiers should ensure that their documentation follows and refers to the Definition and agreement should be reached on the definition of daywork prior to ordering daywork.

As previously the percentage addition method is calculated on the labour rates current at the time the works are carried out, whilst the all-inclusive rates will be at the rates stated at the time of tender although these could be adjusted by reference to a suitable index.

The composition of Daywork includes for labour, materials and plant with incidental costs, overheads and profit being added as a percentage, except in the case of the labour all-inclusive rates where these are deemed to be included.

A summary of the requirements of the Definition follows, users should refer to the full document to calculate the applicable rate, view model documentation and examples and to understand the limitations and inclusions of the Definition:

LABOUR

The percentage addition rate provides for:
- Guaranteed minimum weekly earnings (e.g. Standard Basic Rate of Wages, Joint Board Supplement and Guaranteed Minimum Bonus Payment in the case of NJCBI rules).
- All other guaranteed minimum payments (unless included with incidental costs, overheads and profit).
- Differentials or extra payments in respect of skill, responsibility, discomfort, inconvenience or risk (excluding those in respect of supervisory responsibility).
- Payments in respect of public holidays.
- Any amounts which may become payable by the Contractor to or in respect of operatives arising from the operation of the rules referred to.
- Employer's contributions to annual holidays
- Employer's contributions to benefit schemes
- Employer's National Insurance contributions
- Any contribution, levy or tax imposed by statute, payable by the Contractor in his capacity as an employer.

Differentials or extra payments in respect of supervisory responsibility are excluded from the annual prime cost. The time of supervisory staff, principals, foremen, gangers, leading hands and similar categories, when working manually, is admissible under this Section at the appropriate rates.

The all-inclusive rate provides for:
- Items as listed above pre calculated by the tenderer inclusive of on-costs

MATERIALS
- The Prime cost of materials obtained for daywork is the invoice cost after discounts over 5%
- The Prime cost of materials supplied from stock for daywork is the current market price after discounts over 5%

PLANT
- Where hired for the daywork it is the invoice cost after discounts over 5%
- Where not hired for the daywork it is calculated in accordance with the RICS Schedule of Basic Plant Charges.
- Includes for transport, erection, dismantling and qualified operators.

CONTRACT WORK

DEFINITION OF PRIME COST OF DAYWORK CARRIED OUT UNDER A BUILDING CONTRACT

The Second Edition is reproduced by permission of the Royal Institution of Chartered Surveyors.

This Definition of Prime Cost is published by The Royal Institution of Chartered Surveyors and the National Federation of Building Trades Employers, for convenience and for use by people who choose to use it. Members of the National Federation of Building Trades Employers are not in any way debarred from defining Prime Cost and rendering their accounts for work carried out on that basis in any way they choose. Building owners are advised to reach agreement with contractors on the Definition of Prime Cost to be used prior to issuing instructions.

SECTION 1

Application

1.1 This definition provides a basis for the valuation of daywork executed under such building contracts as provide for its use (e.g. contracts embodying the Standard Forms issued by the Joint Contracts Tribunal).

1.2 It is not applicable in any other circumstances, such as jobbing or other work carried out as a separate or main contract nor in the case of daywork executed during the Defects Liability Period of contracts embodying the above mentioned Standard Forms.

CONTRACT WORK (Cont'd)
DEFINITION OF PRIME COST OF DAYWORK CARRIED OUT UNDER A BUILDING CONTRACT (Cont'd)

SECTION 2

Composition of total charges

2.1 The prime cost of daywork comprises the sum of the following costs:

(a) Labour as defined in Section 3.

(b) Materials and goods as defined in Section 4.

(c) Plant as defined in Section 5.

2.2 Incidental costs, overheads and profit as defined in Section 6, as provided in the building contract and expressed therein as percentage adjustments, are applicable to each of 2.1 (a)-(c).

SECTION 3

Labour

3.1 The standard wage rates, emoluments and expenses referred to below and the standard working hours referred to in 3.2 are those laid down for the time being in the rules or decisions of the National Joint Council for the Building Industry and the terms of the Building and Civil Engineering Annual and Public Holiday Agreements applicable to the works, or the rules or decisions or agreements of such body, other than the National Joint Council for the Building Industry, as may be applicable relating to the class of labour concerned at the time when and in the area where the daywork is executed.

3.2 Hourly base rates for labour are computed by dividing the annual prime cost of labour, based upon standard working hours and as defined in 3.4 (a)-(i), by the number of standard working hours per annum (see examples)

3.3 The hourly rates computed in accordance with 3.2 shall be applied in respect of the time spent by operatives directly engaged on daywork, including those operating mechanical plant and transport and erecting and dismantling other plant (unless otherwise expressly provided in the building contract).

3.4 The annual prime cost of labour comprises the following:

(a) Guaranteed minimum weekly earnings (e.g. Standard Basic Rate of Wages, Joint Board Supplement and Guaranteed Minimum Bonus Payment in the case of NJCBI rules).

(b) All other guaranteed minimum payments (unless included in Section 6).

(c) Differentials or extra payments in respect of skill, responsibility, discomfort, inconvenience or risk (excluding those in respect of supervisory responsibility - see 3.5).

(d) Payments in respect of public holidays.

(e) Any amounts which may become payable by the Contractor to or in respect of operatives arising from the operation of the rules referred to in 3.1 which are not provided for in 3.4 (a)-(d) or in Section 6.

(f) Employer's National Insurance contributions applicable to 3.4 (a)-(e).

(g) Employer's contributions to annual holiday credits.

(h) Employer's contributions to death benefit scheme.

(i) Any contribution, levy or tax imposed by statute, payable by the Contractor in his capacity as an employer.

3.5 Note

Differentials or extra payments in respect of supervisory responsibility are excluded from the annual prime cost (see Section 6). The time of principals, foremen, gangers, leading hands and similar categories, when working manually, is admissible under this Section at the appropriate rates for the trades concerned.

SECTION 4

Materials and Goods

4.1 The prime cost of materials and goods obtained from stockists or manufacturers is the invoice cost after deduction of all trade discounts but including cash discounts not exceeding 5 per cent and includes the cost of delivery to site

4.2 The prime cost of materials and goods supplied from the Contractor's stock is based upon the current market prices plus any appropriate handling charges.

4.3 Any Value Added Tax which is treated, or is capable of being treated, as input tax (as defined in the Finance Act, 1972) by the Contractor is excluded.

SECTION 5

Plant

5.1 The rates for plant shall be as provided in the building contract.

5.2 The costs included in this Section comprise the following:

(a) Use of mechanical plant and transport for the time employed on daywork.

(b) Use of non-mechanical plant (excluding non-mechanical hand tools) for the time employed on daywork.

5.3 Note: The use of non-mechanical hand tools and of erected scaffolding, staging, trestles or the like is excluded (see Section 6).

SECTION 6

Incidental Costs, Overheads and Profit

6.1 The percentage adjustments provided in the building contract, which are applicable to each of the totals of Sections 3, 4 and 5, comprise the following:

(a) Head Office charges

(b) Site staff including site supervision.

(c) The additional cost of overtime (other than that referred to in 6.2).

(d) Time lost due to inclement weather.

(e) The additional cost of bonuses and all other incentive payments in excess of any guaranteed minimum included in 3.4 (a).

(f) Apprentices study time.

(g) Subsistence and periodic allowances.

(h) Fares and travelling allowances.

(i) Sick pay or insurance in respect thereof.

(j) Third party and employer's liability insurance.

(k) Liability in respect of redundancy payments to employees.

(l) Employer's National Insurance contributions not included in Section 3.4

(m) Tool allowances.

CONTRACT WORK (Cont'd)
DEFINITION OF PRIME COST OF DAYWORK CARRIED OUT
UNDER A BUILDING CONTRACT (Cont'd)
SECTION 6 1(Cont'd)

(n) Use, repair and sharpening of non-mechanical hand tools.

(o) Use of erected scaffolding, staging, trestles or the like.

(p) Use of tarpaulins, protective clothing, artificial lighting, safety and welfare facilities, storage and the like that may be available on the site.

(q) Any variation to basic rates required by the Contractor in cases where the building contract provides for the use of a specified schedule of basic plant charges (to the extent that no other provision is made for such variation).

(r) All other liabilities and obligations whatsoever not specifically referred to in this Section nor chargeable under any other section.

(s) Profit.

Note: The additional cost of overtime, where specifically ordered by the Architect/Supervising Officer, shall only be chargeable in the terms of prior written agreement between the parties to the building contract.

Example of calculation of typical standard hourly base rate (as defined in Section 3) for NJCBI building craft operative and labour in Grade A areas **based upon rates ruling at 1st July 1975.**

		Rate £	Craft Operative £	Rate £	Labourer £
Guaranteed minimum weekly earnings					
Standard Basic Rate	49 wks	37.00	1813.00	31.40	1538.60
Joint Board Supplement	49 wks	5.00	245.00	4.20	205.80
Guaranteed Minimum Bonus	49 wks	4.00	196.00	3.60	176.40
			2254.00		1920.80
Employer's National Insurance Contribution at 8.5%			191.59		63.27
			2445.59		2084.07
Employer's Contributions to:					
CITB annual levy			15.00		3.00
Annual holiday credits	49 wks	2.80	137.20	2.80	137.20
Public holidays (included in guaranteed minimum weekly earnings above)			-		-
Death benefit scheme	49 wks	0.10	4.90	0.10	4.90
Annual labour cost as defined in Section 3		£	2602.69	£	2229.17

Hourly base rates as defined in Section 3, clause 3.2

Craft operative $\frac{£\ 2602.69}{1904} = £\ 1.37$

Labourer $\frac{£2229.17}{1904} = £\ 1.17$

Note: (1) Standard working hours per annum calculated as follows:

	52 weeks at 40 hours	=	2080
Less	3 weeks holiday at 40hours	= 120	
	7 days public holidays at 8 hours	= 56	176
			1904

(2) It should be noted that all labour costs incurred by the Contractor in his capacity as an employer, other than those contained in the hourly base rate, are to be taken into account under Section 6.

(3) The above example is for the convenience of users only and does not form part of the Definition; all the basic costs are subject to re-examination according to the time when and in the area where the daywork is executed.

BUILD UP OF STANDARD HOURLY BASE RATES - RATES APPLICABLE AT 25th JULY 2016

Under the JCT Standard Form, dayworks are calculated in accordance with the Definition of the Prime Cost of Dayworks carried out under a Building Contract published by the RICS and the NFBTE. The following build-up has been calculated from information provided by the Building Cost Information Service (BCIS) in liaison with the NFBTE. The example is for the calculation of the standard hourly base rates and is for convenience only and does not form part of the Definition; all basic rates are subject to re-examination according to when and where the dayworks are executed.
Note from January 2017 - The holiday year runs from the 1st January for each year with an annual entitlement of 22 days of industry plus 8 days of Public/Bank holidays

Standard working hours per annum			
52 weeks at 39 hours		2028	hours
Less - 21 days annual holiday:			
16 days at 8 hours	128		
5 days at 7 hours	35		
8 days public holiday			
7 days at 8 hours	56		
1 day at 7 hours	7	226	hours
		1802	hours

	Craft Rate £	General Operative £
Guaranteed Minimum weekly earnings		
Standard basic rate	452.79	340.47

CONTRACT WORK (Cont'd)
DEFINITION OF PRIME COST OF DAYWORK CARRIED OUT UNDER A BUILDING CONTRACT (Cont'd)
BUILD UP OF STANDARD HOURLY BASE RATES - RATES APPLICABLE AT 25th JULY 2016 (Cont'd)

Hourly Base Rate	Rate £	Craftsman £	Rate £	General Operative £
Guaranteed minimum weekly earnings 46.2. weeks at	452.79	20,921.22(£11.61)	340.47	15,731.46(£8.73)
Extra payments for skill, responsibility, discomfort, inconvenience or risk* - 1802 hours	-	-	-	-
Employer's National Insurance Contribution** - 13.8% of	12,808.70	1767.60	7618.94	1051.41
Holidays with Pay – 226 Hours	11.61	2623.86	8.73	1972.98
*** CITB Annual Levy - 0.5% of	23,545.08	117.73	17,704.44	88.52
Welfare benefit – 52 stamps at £10.90		566.80		566.80
Public holidays (included with Holidays with Pay above).		-		-
		£25,997.21		£19,411.18

Hourly Base Rate	Craftsman	General Operative
	$\frac{£25,997.21}{1802} = £14.43$	$\frac{£19,411.18}{1802} = £10.77$

Notes

* Only included in hourly base rate for operatives receiving such payments.

** National Insurance Contribution based on 13.8% above the Earnings Threshold for 52 weeks at £156.01 (=£8,112.52).

*** From 1st April 2002, the CITB levy is calculated at 0.5% of the PAYE payroll of each employee plus 1.5% of payments made under a labour only agreement. Building contractors having a combined payroll of £73,000 (2006 assessment) or less are exempt. The levy is included in the daywork rate of each working operative, the remainder being a constituent part of the overheads percentage.

**** From 29th June 1992, the Guaranteed Minimum Bonus ceased and is consolidated in the Guaranteed Minimum Earnings.

Holidays – The winter holiday is 2 calendar weeks taken in conjunction with Christmas, Boxing and New Years Day.

The Easter Holiday shall be four working days immediately following Easter Monday

The Summer Holiday shall be two calendar weeks.

DAYWORK, OVERHEADS AND PROFIT

Fixed percentage additions to cover overheads and profit no longer form part of the published Daywork schedules issued by the Royal Institution of Chartered Surveyors and Contractors are usually asked to state the percentages they require as part of their tender. These will vary from firm to firm and before adding a percentage, the list of items included in Section 6 of the above Schedule should be studied and the cost of each item assessed and the percentage overall additions worked out.

As a guide, the following percentages are extracted from recent tenders for Contract Work:
On Labour costs 80-200%
On Material costs 10-20%
On Plant costs 5-30%
Much higher percentages may occur on small projects or where the amount of daywork envisaged is low.

JOBBING WORK

DEFINITION OF PRIME COST OF BUILDING WORKS OF A JOBBING OR MAINTENANCE CHARACTER

Reproduced by permission of the Royal Institution of Chartered Surveyors.

This Definition of Prime Cost is published by the Royal Institution of Chartered Surveyors and the National Federation of Building Trades Employers for convenience and for use by people who choose to use it. Members of the National Federation of Building Trades Employers are not in any way debarred from defining prime cost and rendering their accounts for work carried out on that basis in any way they choose.

Building owners are advised to reach agreement with contractors on the Definition of Prime Cost to be used prior to issuing instructions.

SECTION 1

Application

1.1 This definition provides a basis for the valuation of work of a jobbing or maintenance character executed under such building contracts as provide for its use.

1.2 It is not applicable in any other circumstances such as daywork executed under or incidental to a building contract.

SECTION 2

Composition of Total Charges

2.1.1 The prime cost of jobbing work comprises the sum of the following costs:

(a) Labour as defined in Section 3.

(b) Materials and goods as defined in Section 4.

(c) Plant, consumable stores and services as defined in Section 5.

(d) Sub-contracts as defined in Section 6.

2.2 Incidental costs, overhead and profit as defined in Section 7 and expressed as percentage adjustments are applicable to each of 2.1 (a)-(d).

SECTION 3

Labour

3.1 Labour costs comprise all payments made to or in respect of all persons directly engaged upon the work, whether on or off the site, except those included in Section 7.

JOBBING WORK (Cont'd)
DEFINITION OF PRIME COST OF BUILDING WORKS OF A
JOBBING OR MAINTENANCE CHARACTER (Cont'd)
SECTION 3, Labour (Cont'd)

3.2 Such payments are based upon the standard wage rates, emoluments and expenses as laid down for the time being in the rules or decisions of the National Joint Council for the Building Industry and the terms of the Building and Civil Engineering Annual and Public Holiday Agreements applying to the works, or the rules or decisions or agreements of such other body as may relate to the class of labour concerned, at the time when and in the area where the work is executed, together with the Contractor's statutory obligations, including:

(a) Guaranteed minimum weekly earnings (e.g. Standard Basic Rate of Wages and Guaranteed Minimum Bonus Payment in the case of NJCBI rules).

(b) All other guaranteed minimum payments (unless included in Section 7).

(c) Payments in respect of incentive schemes or productivity agreements applicable to the works.

(d) Payments in respect of overtime normally worked; or necessitated by the particular circumstances of the work; or as otherwise agreed between parties.

(e) Differential or extra payments in respect of skill responsibility, discomfort, inconvenience or risk.

(f) Tool allowance.

(g) Subsistence and periodic allowances.

(h) Fares, travelling and lodging allowances.

(j) Employer's contributions to annual holiday credits.

(k) Employer's contributions to death benefit schemes.

(l) Any amounts which may become payable by the Contractor to or in respect of operatives arising from the operation of the rules referred to in 3.2 which are not provided for in 3.2 (a)-(k) or in Section 7.

(m) Employer's National Insurance contributions and any contributions, levy or tax imposed by statute, payable by the Contractor in his capacity as employer.

Note: Any payments normally made by the Contractor which are of a similar character to those described in 3.2 (a)-(c) but which are not within the terms of the rules and decisions referred to above are applicable subject to the prior agreement of the parties, as an alternative to 3.2(a)-(c).

3.3 The wages or salaries of supervisory staff, timekeepers, storekeepers, and the like, employed on or regularly visiting site, where the standard wage rates etc. are not applicable, are those normally paid by the Contractor together with any incidental payments of a similar character to 3.2 (c)-(k).

3.4 Where principals are working manually their time is chargeable, in respect of the trades practised, in accordance with 3.2.

SECTION 4

Materials and Goods

4.1 The prime cost of materials and goods obtained by the Contractor from stockists or manufacturers is the invoice cost after deduction of all trade discounts but including cash discounts not exceeding 5 per cent, and includes the cost of delivery to site.

4.2 The prime cost of materials and goods supplied from the Contractor's stock is based upon the current market prices plus any appropriate handling charges.

4.3 The prime cost under 4.1 and 4.2 also includes any costs of:

(a) non-returnable crates or other packaging.

(b) returning crates and other packaging less any credit obtainable.

4.4 Any Value Added Tax which is treated, or is capable of being treated, as input tax (as defined in the Finance Act, 1972 or any re-enactment thereof) by the Contractor is excluded.

SECTION 5

Plant, Consumable Stores and Services

5.1 The prime cost of plant and consumable stores as listed below is the cost at hire rates agreed between the parties or in the absence of prior agreement at rates not exceeding those normally applied in the locality at the time when the works are carried out, or on a use and waste basis where applicable:

(a) Machinery in workshops.

(b) Mechanical plant and power-operated tools.

(c) Scaffolding and scaffold boards.

(d) Non-Mechanical plant excluding hand tools.

(e) Transport including collection and disposal of rubbish.

(f) Tarpaulins and dust sheets.

(g) Temporary roadways, shoring, planking and strutting, hoarding, centering, formwork, temporary fans, partitions or the like.

(h) Fuel and consumable stores for plant and power-operated tools unless included in 5.1 (a), (b), (d) or (e) above.

(i) Fuel and equipment for drying out the works and fuel for testing mechanical services.

5.2 The prime cost also includes the net cost incurred by the Contractor of the following services, excluding any such cost included under Sections 3, 4 or 7:

(a) Charges for temporary water supply including the use of temporary plumbing and storage

(b) Charges for temporary electricity or other power and lighting including the use of temporary installations.

(c) Charges arising from work carried out by local authorities or public undertakings.

(d) Fee, royalties and similar charges.

(e) Testing of materials

(f) The use of temporary buildings including rates and telephone and including heating and lighting not charged under (b) above.

(g) The use of canteens, sanitary accommodation, protective clothing and other provision for the welfare of persons engaged in the work in accordance with the current Working Rule Agreement and any Act of Parliament, statutory instrument, rule, order, regulation

(h) The provision of safety measures necessary to comply with any Act of Parliament.

(j) Premiums or charges for any performance bonds or insurances which are required by the Building Owner and which are not referred to elsewhere in this Definition. or bye-law.

JOBBING WORK (Cont'd)
DEFINITION OF PRIME COST OF BUILDING WORKS OF A
JOBBING OR MAINTENANCE CHARACTER (Cont'd)

SECTION 6

Sub-Contracts

6.1 The prime cost of work executed by sub-contractors, whether nominated by the Building Owner or appointed by the Contractor is the amount which is due from the Contractor to the sub-contractors in accordance with the terms of the sub-contracts after deduction of all discounts except any cash discount offered by any sub-contractor to the Contractor not exceeding 2 1/2 per cent.

SECTION 7

Incidental Costs, Overheads and Profit

7.1 The percentages adjustments provided in the building contract, which are applicable to each of the totals of Sections 3-6, provide for the following:-

(a)　Head Office Charges.

(b)　Off-site staff including supervisory and other administrative staff in the Contractor's workshops and yard.

(c)　Payments in respect of public holidays.

(d)　Payments in respect of apprentices' study time.

(e)　Sick pay or insurance in respect thereof.

(f)　Third Party and employer's liability insurance.

(g)　Liability in respect of redundancy payments made to employees.

(h)　Use, repair and sharpening of non-mechanical hand tools.

(j)　Any variation to basic rates required by the Contractor in cases where the building contract provides for the use of a specified schedule of basic plant charges (to the extent that no other provision is made for such variation).

(k)　All other liabilities and obligations whatsoever not specifically referred to in this Section nor chargeable under any other section.

(l)　Profit.

SPECIMEN ACCOUNT FORMAT

If this Definition of Prime Cost is followed the Contractor's account could be in the following format:-　　　£

　　　Labour (as defined in Section 3)
　　　Add　% (see Section 7)
　　　Materials and goods (as defined in Section 4)
　　　Add　% (see Section 7)
　　　Plant, consumable stores and services (as defined in Section 5)
　　　Add　% (see Section 7)
　　　Sub-contracts (as defined in Section 6)
　　　Add　% (see Section 7)

　　　　　　　　　　　　　　　　　　　　　　　　　　　　　　　£

VAT to be added if applicable

SCHEDULE OF BASIC PLANT CHARGES

The RICS has now published the 2010 edition of the Schedule of Basic Plant Charges and which came into effect from the 1st July 2010. The Schedule includes rates for of plant covering the following sections:

- Pumps
- Concrete Equipment
- Scaffolding, Shoring, Fencing
- Testing Equipment
- Lighting Appliances and Conveyors
- Site Accommodation
- Construction Vehicles
- Temporary Services
- Excavators and Loaders
- Site Equipment
- Compaction Equipment
- Small Tools

For information the explanatory notes for use in connection with Dayworks under a Building Contract (Fifth revision - 1st May 2001) stated;-

1.　The rates in the Schedule are intended to apply solely to daywork carried out under and incidental to a Building Contract. They are NOT intended to apply to:-
(i)　Jobbing or any other work carried out as a main or separate contract; or
(ii)　Work carried out after the date of commencement of the Defects Liability Period.

2.　The rates apply only to plant and machinery already on site, whether hired or owned by the Contractor.

3.　The rates, unless otherwise stated, include the cost of fuel and power of every description, lubricating oils, grease, maintenance, sharpening of tools, replacement of spare parts, all consumable stores and for licences and insurances applicable to items of plant.

4.　The rates, unless otherwise stated, do not include the cost of drivers and attendants.

5.　The rates are base costs and may be subject to the overall adjustment for price movement, overheads and profit, quoted by the Contractor prior to the placing of the Contract.

6.　The rates should be applied to the time during which the plant is actually engaged in daywork.

7.　Whether or not plant is chargeable on daywork depends on the daywork agreement in use and the inclusion of an item of plant in this schedule does not necessarily indicate that that item is chargeable.

8.　Rates for plant not included in the Schedule or which is not on site and is specifically provided or hired for daywork shall be settled at prices which are reasonably related to the rates in the Schedule having regard to any overall adjustment quoted by the Contractor in the Conditions of Contract.

ELECTRICAL CONTRACTORS

A "Definition of Prime Cost of Daywork carried out under an Electrical Contract" agreed between the Royal Institution of Chartered Surveyors and the Electrical Contractors Associations can be obtained from the RICS, 12 Great George Street, London SW1P 3AD, Tel. 020 7222 7000

Example of calculation of typical standard hourly base rate (as defined in Section 3) for JIB Electrical Contracting Industry Approved Electrician and Electrician based upon rates applicable at 2nd January 2017

		Rate (£)	Approved Electrician	Rate (£)	Electrician
Basic Wages :	1725 hours	15.92	£27,462.00	14.68	£25,323.00
Public Holidays :	60 hours	15.92	£955.20	14.68	£880.80
Extra Payments :	Where applicable				
Sub Total :			£28,417.20		£26,203.80
National Insurance					
52 weeks Earnings threshold at £156.01 = £8,112.52					
13.80% Contribution on		£20,304.68	£2,802.05	£18,091.28	£2,496.60
Holidays with Pay :	52 stamps at £57.65		£2,997.80	£53.90	£2,802.80
Annual Labour Cost :			£34,217.05		£31,503.20
Hourly Base Rate :	Divide by 1725 hours		£19.84		£18.26

Note:

Standard working hours per annum calculated as follows	
52 weeks at 37.5 hours	1950
Less\	
hours annual holiday	165
hours public holiday	60
Standard working hours per year	1725

TERRAZZO MARBLE AND MOSAIC SPECIALISTS

Daywork rates for terrazzo, marble, mosaic and the like for:

Terrazzo Marble or Mosaic Craftsman
General Operatives
All travelling and waiting time to be paid.

Non-productive overtime or overtime premiums
Floor grinding machines (single head)
Dry sanding machines
Multi-head floor polishing machines,
Fuel, abrasive stones and discs.
Fares
Delivery and collection of materials/plant not included

Full details can be obtained from the National Federation of Terrazzo Marble & Mosaic Specialists, P.O. Box 2843, London W1A 5PG

HEATING, VENTILATING, AIR CONDITIONING, REFRIGERATION, PIPEWORK AND/OR DOMESTIC ENGINEERING CONTRACTS

A "Definition of Prime Cost of Daywork carried out under a heating, ventilating and air conditioning, refrigeration, pipework and/or domestic engineering contract" agreed between the Royal Institution of Chartered Surveyors and the Heating and Ventilating Contractors' Association can be obtained from the RICS, 12 Great George Street, London SW1P 3AD, Tel. 020 7222 7000

THE CIVIL ENGINEERING CONTRACTORS ASSOCIATION

CECA Schedules of Daywork Carried Out Incidental to Contract Work 2011

These Schedules are the Schedules referred to in the I.C.E. Conditions of Contract and have been prepared for use in connection with Dayworks carried out incidental to contract work where no other rates have been agreed. They are not intended to be applicable for Dayworks ordered to be carried out after the contract works have been substantially completed or to a contract to be carried out wholly on a daywork basis. The circumstances of such works vary so widely that the rates applicable call for special consideration and agreement between contractor and employing authority.

Contents

Introductory notes

Schedule 1.   LABOUR

Schedule 2.   MATERIALS

Schedule 3.   SUPPLEMENTARY CHARGES

Schedule 4.   PLANT

The schedules may be obtained from the ICE Bookshop at the Institution of Civil Engineers, One Great George Street, Westminster, London SW1P 3AA.

GENERAL INFORMATION

This page left blank intentionally

ROYAL INSTITUTE OF BRITISH ARCHITECTS

There is no fixed fee scale in the UK, so an architect will charge a fee taking into account the requirements of the project, his or her skills, experience, overheads, the resources needed to undertake the work, profit and competition. The RIBA issues a Fees Toolkit available for architects, to help the profession combat harmful low fee bids. The Toolkit comprises the new RIBA Fees Calculator alongside the RIBA Benchmarking service.

The following which may assist in calculating likely fees for budget purposes and are extracts from 'A guide to RIBA Standard Forms of Appointment 1999 and other Architect's Appointments', 'Standard Form of Agreement for the Appointment of an Architect (SFA/99) (Updated April 2000)', 'A Client's Guide to Engaging an Architect including guidance on fees' and 'Conditions of Engagement for the appointment of an Architect (CE/99)' are printed by permission of the Royal Institute of British Architects, 66 Portland Place, London, W1N 4AD from whom full copies can be obtained.

Note: The RIBA Agreements 2010 (2012 revision) replace the 'Standard Form of Agreement for the Appointment of an Architect (SFA/99) (Updated April 2004) and should be referred to for details.

A guide to RIBA Standard Forms of Appointment 1999 and other Architect's Appointments

The guide provides Architects with advice on the application and completion of the RIBA Forms of Appointment1999 and their various supplements, with worked examples.

Clients may also find the guide useful to understand the Architect's role and the responsibilities of the parties.

RIBA Forms of Appointment 1999

SFA/99 - Standard Form of Agreement for the appointment of an Architect

The core document from which all other forms are derived. It is suitable for use where an Architect is to provide services for a fully designed building project in a wide range of sizes or complexity and/or to provide other professional services. It is used with Articles of Agreement and includes notes on completion and an optional Services Supplement

CE/99 - Conditions of Engagement for the Appointment of an Architect

Suitable for use where an architect is to provide services for a fully designed building project and/or to provide other professional services where a letter of Appointment is preferred to the Articles of Agreement in SFA/99.It includes notes on completion, a draft Model Letter and an optional (modified) Services Supplement.

SW/99 - Small Works

Suitable for use where an architect is to provide services of a relatively straightforward nature where the cost of construction works is not expected to exceed £150,000 and use of the JCT Agreement for Minor Works is appropriate. It is used with a Letter of Appointment and includes notes on completion a draft Model Letter and an optional Schedule of Services for Small Works

SC/99 - Form of appointment as Sub-consultant

Suitable for use where a consultant wishes another consultant (sub-consultant) to perform a part of his or her responsibility but not for use where the intention is for the client to appoint consultants directly. It is used With Articles of Agreement and

includes notes on completion and a draft form of Warranty to the Client

DB1/99 - Employers Requirements

An amendment for SFA/99 and CE/99 where an architect is appointed by the employer client to prepare Employer's requirements for a design and build contract. It includes a replacement Services Supplement and notes on completion for initial appointment where a change to design and build occurs later or where 'consultant switch' or 'Novation' is contemplated

DB2/99 - Contractors proposals

An amendment for SFA/99 where an architect is appointed by the contractor client to prepare Contractor's Proposals under a design and build contract. (Use with CE/99 is not recommended.) It includes replacement Articles, Appendix and Services Supplement and notes on completion for initial appointment and for 'consultant switch' and 'Novation'.

PS/99 - Form of Appointment as Planning Supervisor

Used for the appointment as planning supervisor under the CDM Regulations 1994 of suitably qualified construction professionals. An appointment as planning supervisor is distinct from the provisions of architectural services under other RIBA Forms of Appointment It is used with Articles of Agreement and includes notes on completion. Advice on use with a Letter of Appointment is given in this Guide

PM/99 - Form of Appointment as Project Manager

Suitable for a Wide range of projects where the client wishes to appoint a Project Manager to provide a management service and/or other professional services. Does not duplicate or conflict with an architect's services under other RIBA Forms. It is used with Articles of Agreement and includes notes on completion. Not yet published, at the time of going to press.

STANDARD FORM OF AGREEMENT
For the appointment of an Architect (SFA/99)

The Standard Form of Agreement for the appointment of an Architect (SFA/99) has been issued to replace the previous Architects Appointment (1992).

The Standard Form of Agreement for the Appointment of an Architect (SFA/99) consists of:-

Articles of Agreement

Schedule One:	- Project Description
Schedule Two:	- Services
Schedule Three:	- Fees and Expenses
Schedule Four:	- Other Appointments - made under separate Agreements by the Client).
Services Supplement	- Optional
Notes for Architects	on the use and completion of SFA/99
Conditions of Engagement	

STANDARD FORM OF AGREEMENT (Cont'd)
For the appointment of an Architect (SFA/99) (Cont'd)

Conditions of Engagement - Extracts

CONDITIONS

1 General

Interpretation

1.1 The headings and notes to these conditions are for convenience only and do not affect the interpretation of the conditions.

1.2 Words denoting the masculine gender include the feminine gender and words denoting natural persons include corporations and firms and shall be construed interchangeably in that manner.

Applicable law

1.3 The law applicable to this Agreement shall be that stated in the letter of Appointment and, if not so stated, shall be the law of England and Wales.

Communications

1.4 Communications, between the Client and the Architect, including any notice or other document required under the Agreement, shall be in writing and given or served by any effective means. Communications that are not in writing shall be of no effect unless and until confirmed in writing by the sender or the other party. Communications shall take effect when received at an agreed address of the recipient. Communications sent by recorded or registered fist-class post shall be presumed to have arrived at the address to which they are posted on the second day after posting.

Public Holidays

1.5 Where under this Agreement an action is required within a specific period of days, that period shall exclude any day which is a Public Holiday.

Services Variation

1.6 In relation to the Services, either party shall advise the other upon becoming aware of:
.1 a need to vary the Services, the Project Timetable and/or the fees and/or any other part of the Agreement;
.2 any incompatibility in or between any of the Client's requirements in the Brief, any Client's instruction, the Construction Cost, the Timetable and/or the approved design; or any need to vary any part of them;
.3 any issues affecting or likely to affect the progress, quality or cost of the Project;
.4 any information or decisions required from the Client or others in connection with performance of the Services;
and the parties shall agree how to deal with the matter.

2 Obligations & authority of the Architect

Duty of care

2.1 The Architect shall in performing the Services and discharging all the obligations under this part 2 of these Conditions, exercise reasonable skill and care in conformity with the normal standards of the Architect's profession.

Architect's authority

2.2 The Architect shall act on behalf of the Client in the matters set out or necessarily implied in the Agreement.

2.3 In relation to the Services, the Architect shall obtain the authority of the Client before proceeding with the Services or initiating any Work Stage. The Architect shall confirm such authority in writing.

Appointment of Consultants or other persons

2.4 The Architect shall advise the Client on the appointment of Consultants or other persons, other than those named in Schedule 4, to design and/or carry out certain parts of the Works or to provide specialists advice if required in connection with the Project.

Appointment of Site Inspectors

2.5 The Architect shall advise the Client on the appointment of full- or part-time Site Inspectors, other than those named in Schedule 4, under separate agreements where the Architect considers that the execution of the Works warrants such appointment.

Co-operation etc

2.6 The Architect in performing the Services shall when reasonably required by any of the persons identified in Schedule 4:
.1 Co-operate with them as reasonably necessary for carrying out their services;
.2 provide them with information concerning the Services for carrying out their services;
.3 consider and where appropriate comment on their work so that they may consider making any necessary changes to their work;
.4 integrate into his work relevant information provided by them.

No alteration to Services or design

2.7 The Architect shall make no material alteration or addition to or omission from the Services or the approved design without the knowledge and consent of the Client, which consent shall be confirmed in writing by the Architect. In an emergency the Architect may make such alteration, addition or omission without the knowledge and consent of the Client but shall inform the Client without delay and subsequently confirm such action.

Visits to the Works

2.8 The Architect shall in providing the Services make such visits to the Works as the Architect at the date of the appointment reasonably expected to be necessary.

3 Obligations and authority of the Client

Client's representative

3.1 The Client shall name the person who shall exercise the powers of the Client under the Agreement and through whom all instructions shall be given.

Information, decisions, approvals and instructions

3.2 The Client shall supply, free of charge, accurate information as necessary for the proper and timely performance of the Services and to comply with CDM Regulation 11.

3.3 The Client, when requested by the Architect, shall give decisions and approvals as necessary for the proper and timely performance of the Services.

3.4 The Client shall advise the Architect of the relative priorities of the Brief, the Construction Cost and the Timetable.

STANDARD FORM OF AGREEMENT
For the appointment of an Architect (SFA/99) (Cont'd)
Conditions of Engagement (Cont'd)
Obligations & authority of the Client (Cont'd)

3.5 The Client shall have authority to issue instructions to the Architect, subject to the Architect's right of reasonable objection. Such instructions and all instructions to any Consultants or Contractors or other persons providing services in connection with the project shall be issued through the Lead Consultant.

Statutory and other consents required

3.6 The Client shall instruct the making of applications for consents under planning legislation, building acts, regulations or other statutory requirements and by freeholders and others having an interest in the Project. The Client shall pay any statutory charges and any fees, expenses and disbursements in respect of such applications.

CDM Regulations 3.7 The Client Shall, *where required by CDM Regulations:*

.1 comply with his obligations under the CDM Regulations and in any conflict between the obligations under the Regulations and this Agreement, the former shall take precedence;

.2 appoint a competent planning Supervisor;

.3 appoint a competent Principal Contractor.

Appointments and payment others

3.8 Where it is agreed Consultants, or other persons, are to be appointed, the Client shall appoint and pay them under separate agreements and shall confirm in writing to the Architect the services to be performed by such persons so appointed.

Nominations

3.9 Either the Client or the Architect may propose the appointment of such Consultants or other persons, at any time, subject to acceptance by each party.

Site Inspectors

3.10 Where it is agreed Site inspectors shall be appointed they shall be under the direction of the Lead Consultant and the Client shall appoint and pay them under separate agreements and shall confirm in writing to the Architect the services to be performed, their disciplines and the expected duration of their employment.

Responsibilities of others

3.11 The Client, in respect of any work or services in connection with the Project performed or to be performed by any person other than the Architect, shall:

.1 hold such person responsible for the competence and performance of his services and for visits to the site in connection with the work undertaken by him;

.2 ensure that such person shall co-operate with the Architect and provide to the Architect drawings and information reasonably needed for the proper and timely performance of the Services;

.3 ensure that such person shall, when requested by the Architect, consider and comment on work of the Architect in relation to their own work so that the Architect may consider making any necessary change to his work.

3.12 The client shall hold the Principle Contractor and/or other contractors appointed to undertake construction works and not the Architect responsible for their management and operational methods, for the proper carrying out and completion of the Works in compliance with the building contract and for health and safety provisions on the Site.

Legal advice

3.13 The Client shall procure such legal advice and provide such information and evidence as required for the resolution of any dispute between the Client and any parties providing services in connection with the Project.

4 Assignment and sub-letting

Assignment

4.1 Neither the Architect nor the Client shall assign the whole or any part of the Agreement without the consent of the other in writing.

Sub-letting

4.2 The Architect shall not appoint and Sub-Consultants to perform any part of the Services without the consent of the Client, which consent shall not be unreasonably withheld. The Architect shall confirm such consent in writing.

5 Payment

Fees for performance of the Services

5.1 The Fees for performance of the Services, including the anticipated Visits to the Works, shall be calculated and charged as set out in Schedule 3.
Works Stage Fees shall be:

.1 as a percentage of the Construction Costs calculated in accordance with clause 5.2.2; or

.2 as lumps sums in accordance with clause 5.3; or

.3 as time charges in accordance with clause 5.4; or

.4 other agreed method.

Percentage fees 5.2 .1 Percentage fees shall be in accordance with clause 5.2.2 or 5.3.1 as stated in Schedule 3.

.2 Where this clause 5.2.2 is stated to apply in Schedule 3 the percentage or percentages stated therein shall be applied to the Construction Costs. Until the final Construction Cost has been ascertained interim fee calculations shall be based on:

(a) before tenders have been obtained - the lowest acceptable estimate of the Construction Cost;

(b) after tenders have been obtained - the lowest acceptable tender;

(c) after the contract is let either the certified value or the anticipated final account.

The final fee shall be calculated on the ascertained gross final cost of all Works included in the Construction Cost, excluding any adjustment for loss and/or expense payable to or liquidated damages recoverable from a contractor by the Client.

Where the Client is the Contractor, the final cost shall include an allowance for the Contractor's profit and overheads.

GENERAL INFORMATION

FEES

STANDARD FORM OF AGREEMENT
For the appointment of an Architect (SFA/99) (Cont'd)
Conditions of Engagement (Cont'd)
Payment (Cont'd)

Lump sums | 5.3 A lump sum fee or fees shall be in accordance with clause 5.3.1 or 5.3.2 as stated in Schedule 3 and subject to clause 5.5. Such fee or fees shall be:
.1 calculated by applying the percentage stated in Schedule 3, to be in accordance with this clause 5.3.1, to create:
  (a) a lump sum or sums based on the Construction Cost approved by the Client at the end of Work Stage D; or
  (b) separate lump sums for each Work Stages based on the Construction Cost approved by the Client at the end of the previous stage; or
.2 a fixed lump sum or sums stated in Schedule 3 to be in accordance with this clause 5.3.2, which shall be adjusted in accordance with clause 5.6 if substantial amendments are made to the Brief and/or the Construction Cost and/or the Timetable.

Time charges | 5.4 A time based fee shall be ascertained by multiplying the time reasonably spent in the performance of the Services by the relevant hourly rate set out in Schedule 3. Time 'reasonably spent' shall include the time spent in connection with performance of the Services in travelling from and returning to the Architect's office.

Revision of lump sums and time charge and other rates | 5.5 Lump sums complying with clause 5.3.1a or 5.3.2, rates for time charges mileage and printing carried out in the Architect 's office shall be revised every 12 months in accordance with changes in the Retail Price Index*. Each 12-month period shall commence on the anniversary of the Effective Date of the Agreement, or the date of calculation of lump sums complying with clause 5.3.1a, whichever is the later.
*Retail Price Index is set out in Table 6.1 (All items) to 'Labour Market trends' published by the Office for National Statistics.

Additional Fees | 5.6 If the Architect, for reasons beyond his control is involved in extra work or incurs extra expense, for which he will not otherwise be remunerated, the Architect shall be entitled to additional fees calculated on a time basis unless otherwise agreed. Reasons for such entitlement include, but shall not be limited to:
.1 the scope of the Services or the Timetable or the period specified for any work stage is varied by the Client;
.2 the nature of the Project requires that substantial parts of the design can not be completed or must be specified provisionally or approximately before construction commences;
.3 the Architect is required to vary any item of work commenced or completed pursuant to the Agreement or to provide a new design after the Client has authorised the Architect to develop an approved design;
.4 delay or disruption by others;

.5 prolongation of any building contract(s) relating to the Project;
.6 the Architect consents to enter into any third party agreement the form or beneficiary of which had not been agreed by the Architect at the date of the Agreement;
.7 the cost of any work designed by the Architect or the cost of special equipment is excluded from the Construction Cost.
This clause 5.6 shall not apply where the extra work and/or expense to which it refers is due to a breach of the Agreement by the Architect.

5.7 Where for any reason the Architect completed provides only part of the Services specified in Schedule 2, the Architect shall be entitled to fees calculated as follows:
.1 for completed Services, as described for those Services in Schedule 3;
.2 for completed Work Stages, as apportioned for those Work Stages in Schedule 3;
.3 for services or Work Stages not completed, a fee proportionate to that described or apportioned in Schedule 3 based on the Architect's estimate of the percentage of completion.

Expenses and disbursements | 5.8 The Client shall reimburse at net cost plus the handling charge stated in Schedule 3:
.1 expenses specified in Schedule 3;
.2 expenses other than those specified and incurred with the prior authorisation of the Client;
.3 any disbursements made on the Client's behalf.

Maintain records | 5.9 If the Architect is entitled to reimbursement of time spent on Services performed on a time basis, and of expenses and disbursements, the Architect shall maintain records and shall make these available to the Client on reasonable request.

Payments | 5.10 Payments under the Agreement shall become due to the Architect on issue of the Architect's accounts. The final date for such payments by the Client shall be 30 days from the date of issue of an account.

The Architect's account shall be issued at intervals of not less than one month and shall include any additional fees, expenses or disbursements and state the basis of calculation of the amounts due.

Instalments of the fees shall be calculated on the basis of the Architect's estimate of the percentage of completion of the Work Stage or other Services or such other method specified in Schedule 3.

5.11 The Client may not withhold payment of any part of an account for a sum or sums due to the Architect under the Agreement by reason of claims or alleged claims against the Architect unless the amount to be withheld has been agreed by the Architect as due to the Client, or has been awarded in adjudication, arbitration or litigation in favour of the Client and arises out of or under the Agreement. Save as aforesaid, all rights of set-off at common law or in equity which the Client would otherwise be entitled to exercise are hereby expressly excluded.

STANDARD FORM OF AGREEMENT
For the appointment of an Architect (SFA/99) (Cont'd)
Conditions of Engagement (Cont'd)
Payment (Cont'd)

Payment notices 5.12 A written notice from the Client to the Architect:

.1 may be given within 5 days of the date of issue of an account specifying the amount the Client proposes to pay and the basis of calculation of that amount; and/or

.2 shall be given not later than 5 days before the final date for payment of any amount due to the Architect if the Client intends to withhold payment of any part of that amount stating the amount proposed to be withheld and the ground for doing so or, if there is more than one ground, each ground and the amount attributable to it.

If no such notices are given the amount due shall be the amount stated as due in the account. The Client shall not delay payment of any undisputed part of an account.

Late payment 5.13 Any sums due and remaining unpaid at the expiry of 30 days after the date of issue of an account from the Architect shall bear interest. Interest shall be payable at 8% over Bank of England base rate current at the date of issue of the account.

For the avoidance of doubt the Architect's entitlement to interest at the expiry of 30 days after the date of issue of an account shall also apply to any amount which an Adjudicator decides should be paid to the Architect.

Payment on suspension or determination 5.15 If the Client or the Architect suspends or determines performance of the Services, the Architect shall be entitled to payment of any part of the fee or other amounts due at the date of suspension or determination on issue of the Architect's account in accordance with clause 5.10.

5.16 Where the performance of the Services is suspended or determined by the Client, or suspended or determined by the Architect because of breach of the Agreement by the Client, the Architect shall be entitled to payment of all expenses and other costs necessarily incurred as a result of any suspension and any resumption or determination. On issue of the Architect's account in accordance with clause 5.10.

VAT 5.17 Fees, expenses and disbursements arising under the Agreement do not include Value Added Tax. The Client shall pay any Value Added Tax chargeable on the net value of the Architect's fees and expenses.

6 Copyright

Copyright 6.1 The Architect owns the copyright on the work produced by him in performing the Services and generally asserts the right to be identified as the author of the artistic work/work of architecture comprising the Project.

Licence 6.2 The Client shall have a licence to copy and use and allow other Consultants and contractors providing services to the Project to use and copy drawings, documents and bespoke software produced by the Architect, in performing the Services hereinafter called 'the Material', but only for purposes related to the Project on the Site or part of the Site to which the design relates.

Such purposes shall include its operation, maintenance, repairs, reinstatement, alteration, extending, promotion, leasing and/or sale, but shall exclude the reproduction of the Architect's design for any part of any extension of the Project and/or for any other project unless a licence fee in respect of any identified part of the Architect's design stated in Schedule 3.

Provided that:

.1 the Architect shall not be liable if the Material is modified other than by or with the consent of the Architect, or used for any purpose other than that for which it was prepared; or used for any unauthorised purpose;

.2 in the event of any permitted use occurring after the date of the last Service performed under the Agreement and prior to practical completion of the completion of the construction of the project, the Client shall:

(a) where the Architect has not completed Detailed Proposals (Work Stage D), obtain the Architect's consent, which consent shall not be unreasonably withheld; and/or

(b) pay to the Architect a reasonable licence fee where no licence fee is specified in Schedule 3;

.3 in the event of the Client being in default of payment of any fees or other amounts due, the Architect may suspend further use of the licence on giving 7 day's notice of the intention of doing so. Use of the licence maybe resumed on receipt of outstanding amounts.

7 Liabilities and Insurance

Limitation of warranty by Architect 7.1 Subject always to the provisions of clause 2.1, the Architect does not warrant:

.1 that the Services will be completed in accordance with the Timetable;

.2 that planning permission and other statutory approvals will be granted;

.3 the performance, work or the products of others;

.4 the solvency of any other appointed body whether or not such appointment was made on the advice of the Architect.

Time limit for action or proceeding 7.2 No action or proceedings whatsoever for any breach of this Agreement or arising out of or in connection with this Agreement whether in contract, negligence, tort or howsoever shall be commenced against the Architect after the expiry of the period stated in the Letter of Appointment from the date of the Last Services performed under the Agreement or (if earlier) practical completion of the construction of the project.

GENERAL INFORMATION

STANDARD FORM OF AGREEMENT
For the appointment of an Architect (SFA/99) (Cont'd)
Conditions of Engagement (Cont'd)
Liabilities and Insurance (Cont'd)

Architect's liability

7.3 In any action or proceedings brought against the Architect under or in connection with the Agreement whether in contract, negligence, tort or howsoever the Architect's liability for loss or damage in respect of any one occurrence or series of occurrences arising out of one event shall be limited to whichever is the lesser of the sum:
.1 stated in the Appendix; or
.2 such sum as it is just and equitable for the Architect to pay having regard to the extent of his responsibility for the loss and/or damage in question when compared with the responsibilities of contractors, sub-contractors, Consultants and other persons for that loss and/or damage. Such sums to be assessed on the basis that such persons are deemed to have provided contractual undertakings to the Client no less onerous than those of the Architect under the Agreement and had paid to the Client such sums as it would be just and equitable for them to pay having regard to the extent of their responsibility for that loss and/or damage.

Professional Indemnity Insurance

7.4 The Architect shall maintain Professional Indemnity Insurance cover in the amount stated in the Appendix for any one occurrence or series of occurrences arising out of any one event until at least the expiry of the period stated in the Appendix from the date of the Last Services, performed under the Agreement or (if earlier) practical completion of the construction of the Project provided such insurance is available at commercially reasonable rates and generally available in the insurance market to the Architect.

The Architect when requested by the Client, shall produce for inspection documentary evidence that the Professional Indemnity Insurance required under the Agreement is being maintained.

The Architect shall inform the Client if such insurance ceases to be available at commercially reasonable rates in order that the Architect and Client can discuss the best means of protecting their respective positions in respect of the project in the absence if such insurance.

Third Party Agreements

7.5 Where the Client has notified, prior to the signing of this Agreement, that he will require the Architect to enter into an agreement with a third party or third parties, the terms of which and the names or categories of other parties who will sign similar agreements are set out in an annex to this Agreement, then the Architect shall enter such agreement or agreements within a reasonable period of being requested to do so by the Client.

Rights of Third Parties

7.6 For the avoidance of doubt nothing in this Agreement shall confer or purport to confer on any third party any benefit or the right to enforce any term of this Agreement.

8 Suspension and determination

Suspension

8.1 The Client may suspend the performance of any or all of the Services by giving at least 7-days' notice to the Architect. The notice shall specify the Services affected.

8.2 The Architect may suspend performance of the Services and his obligations under the Agreement on giving at least 7-days' notice to the Client of his intention and the grounds for doing so in the event that the Client:
Is in default of any fees or other amounts due; or
Fails to comply with the requirements of the CDM Regulations.
The Architect shall resume performance of his obligations on receipt of the outstanding amounts.

8.3 If any period of suspension arising from a valid notice given under clause 8.1 or clause 8.2 exceeds 6 months the Architect shall request the Client to issue instructions. If written instructions have not been received within 30 days of the date of such request the Architect shall have the right to treat performance of any Service or his obligations affected as determined.

8.4 Any period of suspension arising from a valid notice given under clause 8.1 or clause 8.2 shall be disregarded in computing any contractual date for completion of the Services.

Determination

8.5 The Client or the Architect may by giving 14-days' notice in writing to the other determine performance of any or all of the Services and the Architect's obligations under Part 2 of these conditions stating the grounds for doing so and the Services and obligations affected.

8.6 Performance of the Services and the Architect's obligations under Part 2 of these Conditions may be determined immediately by notice from either party in the event of:
.1 insolvency of the Client or the Architect; or
.2 the Architect becoming unable to provide the Services through death or incapacity.

8.7 On determination on performance of the Services or the Architect's obligations under Part 2 of these conditions, a copy of the Material referred to in clause 6.2 shall be delivered on demand to the Client by the Architect, subject to the terms of the licence under clause 6.2 and payment of the Architect's reasonable copying charges.

8.8 Determination of the performance of the Services or the Architect's obligations shall be without prejudice to the accrued rights and remedies of either party.

9 Dispute resolution*

Negotiation or conciliation

9.1 In the event of any dispute or difference arising under the Agreement, the Client and the Architect may attempt to settle such dispute or difference by negotiation or in accordance with the RIBA Conciliation Procedure.

STANDARD FORM OF AGREEMENT
For the appointment of an Architect (SFA/99) (Cont'd)
Conditions of Engagement (Cont'd)
Dispute resolution (Cont'd)

Adjudication: England and Wales

9.2 Where the law of England and Wales is England the applicable law, any dispute or difference arising out of this Agreement may be referred to adjudication by the Client or the Architect at any time. The adjudication procedures and the Agreement for the Appointment of an Adjudicator shall be as set out in the "Model Adjudication Procedures" published by the *Construction Industry Council* current at the date of the reference. Clause 28 of the 'Model Adjudication Procedures' shall be deleted and replaced as follows: 'The adjudicator may in his discretion direct the payment of legal costs and expenses of one party by another as part of his decision. The adjudicator may determine the amount of costs to be paid or may delegate the task to an independent costs draftsman.

Adjudication: Scotland

9.3 .1 Where the law of Scotland is the applicable law, any dispute or difference touching or concerning any matter or thing arising out of this Agreement (other than with regard to the meaning or construction of this Agreement) may be referred to some independent and fit persons within 7 days of the application of the Client or the Architect and any fees which may become payable to the persons so appointed shall be within the award of that person.

.2 Any such Adjudicator appointed in terms of clause 9.3.1 hereof shall have 28 days from the date of referral within which to reach a decision on the dispute, or such longer period as is agreed between the parties after the dispute has been referred but without prejudice, to the foregoing the Adjudicator shall be permitted to extend the said period of 28 days by up to 14 days, with the consent of the party by whom the dispute was referred.

The Adjudicator shall act impartially and shall be entitled to take the initiative in ascertaining the facts and the law relating to the dispute. The decision of the Adjudicator shall be binding on both parties until the dispute is finally determined by arbitration pursuant to clause 9.5 hereof.

The Adjudicator shall not be liable for anything done or omitted in the discharge or purported discharge of his functions as Adjudicator unless the act or omission is in bad faith, and all employees or agents of the Adjudicator are similarly protected from liability subject to the same proviso.

Naming or nomination of an Adjudicator

9.4 Where no Adjudicator is named in the Agreement and the parties are unable to agree on a person to act as Adjudicator, the Adjudicator shall be a person to be nominated at the request of either party by the nominator identified in the Letter of Appointment.

Arbitration

9.5 When in accordance with the Letter of Appointment either the Client or the Architect require any dispute or difference to be referred to Arbitration the requiring party shall give notice to the other to such effect and the dispute or difference shall or referred to the arbitration and final decision of a person to be agreed between the parties or, failing agreement within 14 days of the date of the notice, the Appointor shall be the person identified in the Appendix.

Provided that where the law of England and Wales is applicable to the Agreement:

.1 the Client or the Architect may litigate any claim for a pecuniary remedy which does not exceed £5,000 or such other sum as is provided by statute pursuant to section 91 of the Arbitration Act 1996;

.2 the Client or the Architect may litigate the enforcement of any decision of an Adjudicator;

.3 where and to the extent that the claimant in any dispute which is referred to arbitration is the Architect, the arbitrator shall not have the power referred to in Section 38(3) of the Arbitration Act 1996.

Costs

9.6 The Client shall indemnify the Architect in respect of his legal and other costs in any action or proceedings, together with a reasonable sum in respect of his time spent in connection with such action or proceedings or any part thereof, if:

.1 the Architect obtains a judgement of the court or an Arbitrator's award on his favour for the recovery of fees and/or expenses under the Agreement; or

.2 the Client fails to obtain a judgement of the court or an Arbitrator's award in the Client's favour for any claim or any part of any claim against the Architect.

Architects are subject to the disciplinary sanction of the Architects Registration Board in relation to complaints of unacceptable professional conduct or serious professional incompetence.

STANDARD FORM OF AGREEMENT
For the appointment of an Architect (SFA/99) (Cont'd)

Schedule Two   **Services**

The Architect shall:

1. Perform the Services as designer, design leader, lead consultant during pre-construction and construction Work Stages.
2. Perform the Services for the Work Stages indicated below and in accordance with the Services Supplement.
3. Make visits to the Works in accordance with clause 2.8.
4. Perform any other Services identified below.

Feasibility   **A** Appraisal   **B** Strategic Brief

Pre-Construction **C** Outline Proposals   **F** Production Information
                      D Detailed Proposals   **G** Tender Documentation
                      E Final Proposals   **H** Tender Action

Construction   **J** Mobilisation   **L** After Practical Completion
                      K To Practical Completion

Other Services   *If identifying any other services the description should be sufficient to identify its scope.*

Other Activities

These activities do not form part of the services unless identified under 'Other Services'

Performance of any of these activities will attract additional fees in accordance with clause 5.6.

Activities - *These may be required in connection with:*

Sites and Buildings

o   Selection of Consultants
o   Options approval
o   Compiling, revising and editing:
    (a) strategic brief
    (b) detailed (written) brief
    (c) room data sheets
o   Selection of site and/or buildings
o   Outline planning submission
o   Environmental studies
o   Surveys, inspection or specialist investigations
o   Party wall matters
o   Two-stage tendering
o   Negotiating a price with a contractor (in-lieu tendering)
o   Use of energy in new or existing buildings
o   Value management services
o   Compiling maintenance and operational manuals
o   Specially prepared drawings of a building as built
o   Submission of plans for proposed works for approval of landlords, founders, freeholders, tenants etc
o   Applications or negotiations for statutory and other grants

Design Skills

o   Interior design services
o   Selection of furniture and fittings
o   Design of furniture and fittings
o   Landscape design services
o   Special drawings, photographs, models or technical information produced at the Client's request

Historic Buildings and Conservation

o   Detailed inspection and report
o   Historical research and archaeological records
o   Listed building consent
o   Conservation area consents
o   Grant aided works

Special Activities - These may be required in connection with:

o   Exceptional negotiations with planning or other statutory authorities
o   Revision of documents to
    (a) comply with requirements of planning or statutory authorities, or landlords, etc
    (b) comply with changes in interpretation or enactment or revisions to laws or statutory regulations
    (c) make corrections not arising from any failure of the Architect
o   Ascertainment of contractor's claims
o   Investigations and instructions relating to work not in accordance with the building contract

o   Assessment of alternative designs, materials or products proposed by the contractor
o   Dispute resolution services on behalf of the Client
o   Damage to or destruction of a building in construction or to existing buildings
    o   Determination of any contract or agreement with any other party providing services to the project
    o   Insolvency of any other party providing services to the project
    o   Valuations for mortgage or other purposes
    o   Easements or other legal agreements
    o   Investigation of building failures
o   Feed Back – post-completion evaluation

GENERAL INFORMATION

STANDARD FORM OF AGREEMENT
For the appointment of an Architect (SFA/99) (Cont'd)

Services Supplement: Design and Management.

Architect's design services

All Commissions

1.1 Receive Client's instructions

1.2 Advise Client on the need to obtain statutory approvals and of the duties of the Client under the CDM Regulations

1.3 Receive information about the site from the Client (CDM Reg 11)

1.4 Where applicable, co-operate with and pass information to the Planning Supervisor

1.5 Visit the site and carry out an initial appraisal

A Appraisal

1 Carry out studies to determine the feasibility of the Client's Requirements

2A Review with Client alternative design and construction approaches and the cost implications, *or*

2B Provide information for report on cost implications

B Strategic Brief

1 (Strategic Brief prepared by or for the Client)

C Outline Proposals

1 Receive Strategic Brief and Commence development into Project Brief

2 Prepare Outline Proposals

3A Prepare an approximation of construction cost, *or*

3B Provide information for cost planning

4 Submit to Client Outline Proposals and approximate construction cost

D Detailed Proposals

1 Complete development of Project Brief

2 Develop the Detailed Proposals from approved Outline Proposals

3A Prepare a cost estimate, *or*

3B Provide information for preparation of cost estimate

4 Consult statutory authorities

5 Submit to Client the Detailed Proposals, showing spatial arrangements, materials and appearance, and a cost estimate

6 Prepare and submit application for full planning permission

E Final Proposals

1 Develop Final Proposals from approved Detailed Proposals

2A Revise cost estimate, *or*

2B Provide information for revision of cost estimate

3 Consult statutory authorities on developed design proposals

4 Submit to Client type of construction, quality of materials, standard of workmanship and revised cost estimate

5 Advise on consequences of any subsequent changes on cost and programme

F Production Information

1 Prepare production information for tender purposes

2A Prepare schedules of rates and/or quantities and/or schedules of works for tendering purposes and revise cost estimate, *or*

2B Provide information for preparation of tender pricing documents and revision of cost estimate

3A Prepare and make submissions under building acts and/or regulations or other statutory requirements, *or*

3B Prepare and give building notice under building acts and/or regulations*
Not applicable in Scotland

3 Prepare further production information for construction purposes

G Tender Documentation

1 Prepare and collate tender documents in sufficient detail to enable a tender or tenders to be obtained

2 Where applicable pass final information to Planning Supervisor for pre-tender Health and Safety Plan

3A Prepare pre-tender cost estimate, *or*

3B Provide information for preparation of pre-tender cost estimate

H Tender Action

1 Contribute to appraisal and report on tenders/ negotiations

2 *If instructed, revise production information to meet adjustments in the tender sum*

J Mobilisation

1 Provide production information as required for the building contract and for construction

K Construction to Practical Completion

1 Make Visits to the Works in connection with the Architect's design

2 Provide further information reasonably required for construction

3 Review design information from contractors or specialists

4 Provide drawings showing the building and main lines of drainage and other information for the Health and Safety File

4 Give general advice on operation and maintenance of the building

L After Practical Completion

1 Identify defects and make final inspections

2A Settle Final Account, *or*

2B Provide information required by others for settling final account

---

If the Architect is to provide cost advice Alternative **A** applies

If a Quantity Surveyor is appointed Alternative **B** applies

---

GENERAL INFORMATION

STANDARD FORM OF AGREEMENT
For the appointment of an Architect (SFA/99) (Cont'd)
Services Supplement (Cont'd)

Architect's Management Services

As Design Leader

The Architect has authority and responsibility for:

1 directing the design process;

2 co-ordinating design of all constructional elements, including work by any Consultants, specialists or suppliers;

3 establishing the form and content of design outputs, their interfaces and a verification procedure;

5 communicating with the Client on significant design issues.

As Lead Consultant
(pre-construction)

The Architect has authority and responsibility in the pre-construction Work Stages for:

1 co-ordinating and monitoring design work;

2 communications between the Client and the Consultants; except that communications on significant design matters are dealt with as design leader;

3 advising on the need for and the scope of services by Consultants, specialists, sub-contractors or suppliers;

4 advising on methods of procuring construction;

5 receiving regular status reports from each Consultant, including the design leader;

6 developing and managing change control procedures, making or obtaining decisions necessary for time and cost control;

7 reporting to the Client as appropriate.

As Lead Consultant and
Contract Administrator

The Architect has authority and responsibility in the tender and construction Work Stages A-G for:

1 inviting and appraising a tender or tenders, including:
- considering with the Client a tenderer or a list of tenderers for the Works;
- considering with the Client appointment of a contractor, and the responsibilities of the parties and the Architect/Contract Administrator under the building contract;
- preparing the building contract and arranging for signatures;

2 administering the building contract, including:
- monitoring the progress of the Works against the Contractor's programme;
- issuing information, instructions, etc;
- preparing and certifying valuations of work carried out or completed and preparing financial reports for the Client; or
- certifying valuations of work prepared by others, and presenting financial reports prepared by others to the Client;
- collating record information including the Health and Safety File;

3 co-ordinating and monitoring the work of Consultants and Site Inspectors, if any, to the extent required for the administration of the building contract, including:
- receiving reports from such Consultants and Site Inspectors to enable decisions to be made in respect of the administration of the building contract;
- consulting any Consultant or other person whose design or specification may be affected by a Client instruction relating to the construction contract, obtaining any information required and issuing any necessary instructions to the Contractor;
- managing change control procedures and making or obtaining decisions as necessary for cost control during the construction period;
- providing information obtained during the administration of the building contract to the Consultants and Site Inspectors;

4 communications between the Client and the Consultants;

5 reporting to the Client as appropriate.

ROYAL INSTITUTE OF BRITISH ARCHITECTS

A Client's guide to Engaging an Architect including guidance on fees

The Royal Institute of British Architects Guide provides guidance for Clients on engaging an Architect, and on professional services, conditions of engagement and fees, together with information on :-

Architect's Services
Other services
Appointment
RIBA Standard Forms
Design and Build
Historic buildings
CDM Regulations

The Client – Architect Agreement
The architect
The client
Statutory requirements
Other appointments
Copyright
Liability and insurance
Suspension and determination
Dispute Resolution

Architect's Fees and Expenses
Percentage basis
Lump sums
Time basis
Expenses
Payment

Architect's Fees and Expenses

RIBA forms of appointment provide a number options for the calculation of the architect's fee, i.e.:
• a quoted percentage of the final cost of the building work;
• a fixed or calculated lump sum or sums;
• time charge; or
• an other agreed basis.

There is no 'standard' or 'recommended' basis for calculation of the fee, but the relative cost usually incurred in providing architects' normal services is indicated in the tables below. The tables recognise that the additional complexity of the work in alterations or extensions or to historic buildings may justify a higher fee than for new work. Figures 1 and 2 show indicative percentage rates in relation to construction cost for new works and works to existing buildings respectively for five classes of building (Fig 5). Figure 6 and 7 provide similar information for historic buildings.

Before fees can be agreed, client and architect should establish the requirements, the services to be provided, the approximate construction cost and the timetable. It is usually helpful to identify the proportion of the fee for each work Stage (Fig 3) and any other fees payable for additional services such as a survey.

In proposing a fee the architect will take account of the requirement that a client must be *reasonably satisfied* that designers will allocate adequate resources for the purposes of CDM Regulations.

Percentage basis

The architect's fees for normal services are expressed as a percentage of the final construction cost executed under the architect's direction. This method is best used for clearly defined building projects. Figures 1 and 2 may be a helpful point of reference in selecting the percentage(s).

Lump sums

Fixed lump sums – may be suitable where the parameters of the services, i.e. the requirements, time and cost, can be clearly defined from the outset. Then, if these are varied by more than a stated amount the lump sum itself may be varied.

Calculated lump sums – with this option the lump sums for the Work Stages are calculated by applying quoted percentages to the latest approved cost either, when the design and estimated cost have been settled at the end of Stage D or at the commencement of each stage. This method will be particularly appropriate where the project parameters cannot be pre-determined with the necessary accuracy or there are inflationary or deflationary pressures in the market place.

Time basis

This basis is best used where the scope of the work cannot be reasonably foreseen or where services do not relate to the cost of construction, for example, for:
• varied or additional services to an otherwise basic service;
• open-ended exploration work, particularly for repair and conservation work;
• feasibility studies;
• protracted planning negotiations;
• party wall services, etc.

The time spent on such services by principals and technical staff is charged at agreed rates, usually expressed as hourly rates, which may be revised at 12-monthly intervals. A range of indicative values is given in Figure 4. Time spent in travelling in conjunction with the time-charged services is chargeable, but the time of secretarial and administrative staff is usually excluded.

Expenses

In addition to the fees, expenses identified in the Agreement or Letter of Appointment for reimbursement may include:
• the cost of copies of drawings and other documents;
• the purchase of maps and printed contract documentation;
• the cost of photography;
• postage and fax charges;
• travel costs and accommodation where relevant.

Additionally, the relevant fee must accompany applications for building regulations or planning approval. If the architect agrees to make such payments on behalf of the client, these disbursements will also be chargeable, perhaps with a handling charge.

Payment

Under RIBA forms of appointment accounts for instalments of the fees and expenses will normally be issued monthly for payment within 30 days. If required, regular payments can be budgeted over a period with review at completion of say the pre-construction Work Stages. Alternatively, fees may be paid at the completion of each Work Stage.

The agreed arrangements will be set out in the Agreement or Letter of Appointment, showing the basis on which the fee is calculated and the rates that are to apply. Statutory provision apply to the payment terms.

Architects will keep and make available for inspection records of expenditure on expenses and disbursements and time spent on services carried out on a time-charged basis. Where applicable, VAT is due on architects' fees and expenses at the appropriate rate.

A Clients guide to Engaging an Architect (cont'd.)
Architect's Fees and Expenses (Cont'd)

Indicative Percentage Fee Scales – Normal Services

Figure 1
New Works

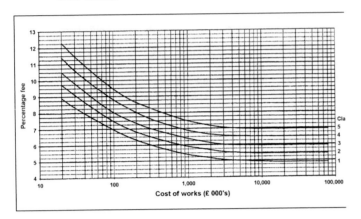

Figure 2
Works to existing buildings

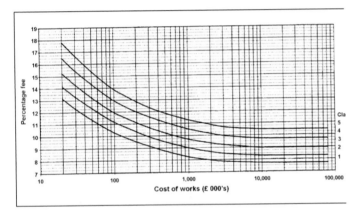

Figure 6
Historic Buildings

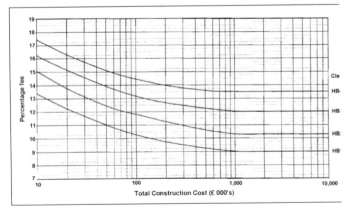

Figure 3

Proportion of fee by Work stage

Proportion of fee	%		
Work stage	C	Outline proposals	20
	D	Detailed Proposals	20
	E	Final Proposals	15
	F	Production Information	20
	G	Tender Documentation	2
	H	Tender Action	1
	J	Mobilisation	1
	K	Construction to Practical Completion	20
	L	After Practical Completion	1
			100%

Stages A and B. Appraisal and strategic brief, are normally carried out on a time-charged basis and are therefore not included above.

Figure 4

Indicative hourly rates (June 1999)

Type of work	General	Complex	Specialist
Partner/director Or equivalent	£95	£140	£180
Senior Architect	£75	£105	£140
Architect	£55	£75	£95

Rates are purely indicative and may be negotiated above or below the figures suggested to reflect the nature of the work.

Figure 7

Classification of work to Historic Buildings

Class HB1	Work of a straight forward nature where one repair element predominates (eg large areas of re-roofing).
Class HB2	Work of general repair not including a great deal; of internal alteration.
Class HB3	Work of general repair and alteration, including substantial alteration work. Repair work where the services installations are to be renewed or replaced.
Class HB4	Work including specialist conservation work, requiring an exceptional amount of detailed specification and inspection.

A Clients guide to Engaging an Architect (cont'd.)

Figure 5   Classification by building type

Type	Class 1	Class 2	Class 3	Class 4	Class 5
Industrial	• Storage sheds	• Speculative factories and warehouse • Assembly and machine workshops • Transport garages	• Purpose-built factories and warehouses		
Agricultural	• Barns and sheds	• Stables	• Animal breeding units		
Commercial	• Speculative shops • Surface car parks	• Multi-storey and underground car parks	• Supermarkets • Banks • Purpose-built shops • Office developments • Retail warehouses • Garages/ showrooms	• Department stores • Shopping centres • Food processing units • Breweries • Telecommunications and computer buildings • Restaurants • Public houses	• High risk research and production buildings • Research and development laboratories • Radio, TV and recording studios
Community		• Community halls	• Community centres • Branch libraries • Ambulance and fire stations • Bus stations • Railway stations • Airports • Police stations • Prisons • Postal Buildings • Broadcasting	• Civic centres • Churches and crematoria • Specialist libraries • Museums and art galleries • Magistrates/ County Courts	• Theatres • Opera houses • Concert halls • Cinemas • Crown Courts
Residential		• Dormitory hostels	• Estates housing and flats • Barracks • Sheltered housing • Housing for single people • Student housing	• Parsonages/ manses • Apartment blocks • Hotels • Housing for the handicapped • Housing for the frail elderly	• Houses and flats for individual clients
Education			• Primary/ nursery/ first schools	• Other schools including middle and secondary • University complexes	• University laboratories
Recreation			• Sports halls • Squash courts	• Swimming pools • Leisure complexes	• Leisure pools • Specialised complexes
Medical/Social services			• Clinics	• Health Centres • General hospitals • Nursing homes • Surgeries	• Teaching hospitals • Hospital laboratories • Dental surgeries

CONDITIONS OF ENGAGEMENT FOR THE APPOINTMENT OF AN ARCHITECT (CE/99)

The RIBA *Conditions of Engagement for the Appointment of an Architect (CE/99)* is suitable for use where an architect is to provide services for fully designed building project in a wide range of sizes or complexity and/or to provide other professional services. It is fully compliant with the requirements of the Housing Grants, Construction and Regeneration Act 1996.

The Agreement is completed by the exchange of a Letter of Appointment to which a copy of the form is attached, after completion of the Schedules.

'Notes for Architects' and a draft Model letter is included in CE/99 and may assist in identifying those matters which will affect the scope and terms of the Agreement.

It should be noted that the appointment of an architect as Planning Supervisor, where required by the CDM Regulations 1994, is separate from the provision of architectural services under CE/99 and should be made using the RIBA *Form of Appointment as Planning Supervisor (PS/99)*

A series of guides addressed directly to Clients is also published on topics associated with the appointment of an Architect, which Architects may wish to send to their Clients as appropriate and also refer to themselves. The guides can be obtained from RIBA Publications by mail order (020) 7251 0791 or from RIBA bookshops in London and the regions.

GENERAL INFORMATION

THE ROYAL INSTITUTION OF CHARTERED SURVEYORS

Consultant's appointment form

The RICS consultant's appointment forms include a full and short form of appointment, with coordinated scopes of services for:

- Building surveyors
- CDM co-ordinators
- Employer's agents
- Project managers
- Project monitors
- Quantity surveyors

Standard form

The standard form of consultant's appointment is suitable for use in relation to projects of any size or value. However, it is most likely to be used for projects where clients and consultants have experience of development projects of relatively high value. Where the services and project are more straightforward, the short form of consultant's appointment may be more appropriate.

Short form

The short form of consultant's appointment is appropriate where the project and services are more straightforward. More complex projects should use the RICS standard form of consultant's appointment.

Explanatory notes

Explanatory notes for the RICS consultant's appointment forms are also available

Appointing a Quantity Surveyor

The Scales of Professional Charges for Quantity Surveying Services have been withdrawn as has the replacement 'Appointing a Quantity Surveyor: A guide for Clients and Surveyors'. The document gives no indication of the level of fees to be anticipated although a Form of Enquiry, Schedule of Services, Fee Offer, Form of Agreement and Terms of Appointment are included. The view being that consultants are now appointed less often by reputation and recommendation and more often after a process of competition. To assist employers, quantity surveyors and those procuring professional services and where consultants are being appointed by reputation or recommendation the following will assist in assessing the reasonableness of any fee proposal.

Extracts from Appointing a Quantity Surveyor: A guide for Clients and Surveyors
Scales of Professional Charges
Example of calculation of fees for lump sum and % basis as clause 5.6 of the guide
Anticipated Percentage fees by project and value based on calculations as example.

Appointing a Quantity Surveyor

The following extracts are for information and the full document should be referred to for further advice.

Contents

RICS Guidance Notes

SECTION 1
Selection and Appointment Advice

1. Introduction
2. Notes on use and completion of documents
3. When to appoint a Quantity Surveyor
4. How to select and appoint a Quantity Surveyor
5. Specifying the quantity surveying service and determining the fee options and expense costs
6. Submission and comparison of fee offers and selection and notification of results
7. Confirmation of the Agreement between the Client and the Quantity Surveyor
8. Complaints handling and disputes resolution

SECTION 2
Form of Enquiry

1. Client
2. Project title and address
3. General project description
4. Project programme
5. Construction cost budget
6. Project tender documentation
7. Other consultants
8. Professional indemnity insurance
9. Period of limitation
10. Collateral warranties

Schedule of Services

1. Category One: General services
2. Category Two: Services particular to non-traditional methods of procurement
3. Category Three: Services not always required in Categories One and Two

Fee Offer

1. Percentage or lump sum fees
2. Time charges
3. Expenses
4. Instalment payments
5. Interest
6. Value Added Tax
7. Confirmation of fee offer

SECTION 3
Form of Agreement
Terms of Appointment

1. Introduction
2. Client's obligations
3. Assignment and subcontracting
4. Payment
5. Professional indemnity insurance
6. Copyright
7. Warranties
8. Suspension and termination
9. Consequences of suspension and termination
10. Complaints
11. Disputes
12. Arbitration
13. Liability
14. Notice

SECTION 1

Selection and Appointment Advice

This Section outlines for Clients and Surveyors when and how to select and appoint a Quantity Surveyor

1. **Introduction**

1.1 The Royal Institution of Chartered Surveyors ('the RICS') has published this Guide to assist Clients when they wish to appoint a chartered Quantity Surveyor. The contents of the Guide should also assist Quantity Surveyors when concluding an agreement with their Client.

THE ROYAL INSTITUTION OF CHARTERED SURVEYORS
Appointing a Quantity Surveyor (Cont'd)
Selection and Appointment Advice (Cont'd)

1.2   Over the last few years changes have taken place in the market place for professional services. Consultants are now appointed less often by reputation and recommendation and more often after a process of competition. Competition takes place in the private and public sectors and is supported by European Union requirements and UK Compulsory Competitive Tendering legislation.

1.3   This Guide is applicable to a range of situations - whether details of the services required are proposed to or by the Quantity Surveyor; or whether appointment is of a single Quantity Surveyor or following selection from a limited list. The Guide sets out principles to be considered and applied rather than a detailed step-by-step procedure. Details often vary from project to project depending on the type of Client and the nature of the project. A detailed procedure that attempts to cover all projects is not therefore considered appropriate.

1.4   Quantity Surveyors may be appointed directly by a Client or as part of a multidisciplinary team. The RICS recommends that the quantity surveying appointment is made directly by a Client so as to ensure that independent financial advice is made directly to that Client by the Quantity Surveyor

1.5   The Guide provides a basis for the appointment of a Quantity Surveyor so that those concerned are clear about:
- the services being requested and/or offered
- the terms and conditions of the contract
- the fee payable and the method of payment.

A Client and his Quantity Surveyor should set out clearly their requirements in their agreement so that the services to be provided and the conditions of engagement are certain. It is necessary for a Client to set out his requirements with the same clarity as when seeking tenders from contractors on construction projects. It is now compulsory for Members of the RICS to provide written notification to their Client of the terms and conditions of the appointment. This documentation is intended to form a good basis upon which this certainty can be achieved and is recommended for use between the Client and his Quantity Surveyor.

2.   Notes on use and completion of documents

2.1   **Form of Enquiry**
The Form of Enquiry sets out the details of the services a Client wishes a Quantity Surveyor to provide and should be attached to the Form of Agreement and identified as belonging to it.

2.2   **Form of Agreement and Terms of Appointment**
The Form of Agreement should be completed and signed only when the services, fees and expenses have been agreed between a Client and his Quantity Surveyor. The Form of Agreement states the names of the parties, their intentions and their arrangement.

The names and addresses of the parties should be inserted, a brief description of the project given and any site identified. If the Form of Agreement is to made as a simple contract it should be signed by both parties and the date entered. If it is to be executed as a Deed the alternative version should be used. See notes *1 and *2.

3.   **When to appoint a Quantity Surveyor**

In order that maximum benefit can be gained from his skills a Quantity Surveyor should be appointed by a Client as soon as possible in the life of a project, preferably at the inception of a scheme, so that advice can be provided on:
- the costs of the project so that a realistic budget can be set at inception and cost management can be applied throughout
- the procurement method best suited to the requirements of the Client
- the implications of the appointment of other consultants and contractors.

It is recommended that a Client and his prospective Quantity Surveyor should meet and discuss the appointment before an agreement is reached, unless the services of the Quantity Surveyor are to be restricted only to some of those from the range available and shown in the Form of Enquiry.

4.   **How to select and appoint a Quantity Surveyor**

4.1   Detailed guidance on the selection and appointment of Chartered Surveyors is given in the RICS publication A Guide to Securing the Services of a Chartered Surveyor. Chapters 2, 3 and 4 also provide useful information on preparing a brief for a Client's requirements, producing tender and contract documentation, organising competitions and evaluating offers.

4.2   Methods of selecting a Quantity Surveyor include:-

Selection based on existing knowledge
A Client may select and then appoint a Quantity Surveyor using existing knowledge of that Quantity Surveyor's performance and reputation. This knowledge may arise from a previous working relationship or be based on the recommendation of others
Selection from a panel maintained by a Client
A Client may maintain a panel of Quantity Surveyors. He will have records of their experience which will enable him to make his selection and appointment
Selection from an ad hoc list produced by a Client
If a Client is unable to make an appointment based on knowledge or reputation or by selection from a standing panel it may be more appropriate for an ad hoc list to be prepared.

4.3   Whichever of the above methods of selection is used it is important for the selection criteria to include the following:
- the financial standing of the Quantity Surveyor under consideration
- the experience, competence and reputation of each Quantity Surveyor in the area of the project/skill being considered so that selection is made using comparable standards between firms.
- the ability of each Quantity Surveyor to provide the required service at the relevant time.

There is no one correct way to use competitive tendering as a basis for appointment and no single way of complying with legislation. It is important, however, that the procedure is always fair and open. Where legislation means that competition is compulsory, reference should be made to Clause 4.8 below and any other guidance available.

THE ROYAL INSTITUTION OF CHARTERED SURVEYORS
Appointing a Quantity Surveyor (Cont'd)
Selection and Appointment Advice (Cont'd)

4.4 If competition is to be used in the selection process the following should be borne in mind:
- a Client should choose between the various methods of competition from either open, selective or restrictive lists, or single- or two-stage tenders, according to the complexity of the project which is to be the subject of competition
- the basis of selection can be on quality, competence or price, or a combination of all three. Whichever basis is chosen it must be clearly stated in the documentation inviting interest in the commission
- it is recommended that price is never regarded in isolation, as will be clear from the detailed guidance set out below.

4.5 To ensure that any competition is well organised the key ingredients required are that:
- specification of the service required is clear
- definition of the skills and competencies required to deliver the services is clear
- criteria on which offers will be evaluated will be made available with the tender documents
- any weighting to be applied to skills, competencies and price is made available with the tender documents
- each stage of the tender process is documented and the procedures for selection and invitation to tender match the requirements of the commission.

4.6 Every effort should be made to ensure that contract documentation is complete. If circumstances change to the extent that criteria are altered all tenderers should immediately be advised and given the opportunity to respond and confirm a willingness to continue.

4.7 **Private contracts**
In the case of private contracts there is no upper or lower limit to the number of private sector firms that may be invited to tender, nor formal rules as to their selection. Nevertheless, the RICS recommends that not less than 3 and not more than 6 firms are invited to submit proposals.

4.8 **Public contracts**
In the case of public sector contracts Compulsory Competitive Tendering rules and European Union rules, together with relevant Treasury regulations and DETR guidance, need to be followed. More information is available in the RICS publication A Guide to Securing the Services of a Chartered Surveyor. Chapter 2, pages 3 to 6, provides details on regulations but it is emphasised that current requirements should always be verified before seeking public sector offers.

5. **Specifying the quantity surveying service and determining the fee options and expense costs**

5.1 It is emphasised that the fee for a project can either be negotiated, sought as a sole offer or sought in competition. The following considerations may apply to any of the methods of selecting a Quantity Surveyor and determining his fee in relation to the services to be provided.

5.2 When fee quotations are to be sought from more than one Quantity Surveyor it is of paramount importance that the enquiry and/or the submissions relate precisely to the same details of service(s) required. Proper use of the Form of Enquiry should ensure that this is achieved. In particular, the services required

should be specified in detail, by using the Schedule of Services provided and/or by adding or deleting any special services not listed or required in that Schedule.

5.3 Information provided by a Client in an offer document should include, as a minimum:
- the complete scope of the project
- a full and precise description of the quantity surveying services to be provided
- the terms and conditions which will apply, preferably the Form of Agreement and Terms of Appointment contained in this Guide, together with any further requirements that will apply
- the anticipated time scales that will apply, both for the quantity surveying service and for the project
- information on the inclusion or exclusion of expense costs
- the basis of the fee upon which the offer is being invited or offered including:
  - where a fee submission is to deal separately with different components of the service, the way in which any components will be evaluated in order to give comparable totals for selection purposes; and
  - wherever a percentage is to be quoted by a Quantity Surveyor; the definition of the total construction cost to which that percentage is to be applied
- provisions for stage or instalment payments
- where the nature, size, scope, time or value of the project or of the quantity surveying service(s) are likely to vary, the method(s) of appropriately adjusting the fee.

5.4 In determining the basis of the services and the fees required the following should be considered:
- where quantity surveying services are incapable of precise definition at the time of appointment, or where they could change substantially, enough information should be contained in an agreement to enable possible variations in those services to be negotiated
- where services are likely to be executed over a particularly long period, or where the service and/or the project might be delayed or postponed, the method and timing of any increase(s) in the fees should be stated
- the basis of the fees to be offered might be one of, or a combination of:
  - a single percentage encompassing all the described services
  - separate percentages for individually defined components of services
  - a single lump sum encompassing all the described services
  - separate lump sums for individually defined components of services
  - a time charge with either hourly rates quoted or a multiplier or adjustment factor to be applied to a specified definition of hourly cost.

5.5 Where the nature or scale of the project warrants it Quantity Surveyors may be invited to:
- explain in outline their proposed method of operation for the project, and/or
- demonstrate their ability to provide the services required from their own resources, and/or
- state the extent of partner/director involvement, and/or
- name the key personnel to be allocated to the project with details of relevant experience.

THE ROYAL INSTITUTION OF CHARTERED SURVEYORS
Appointing a Quantity Surveyor (Cont'd)
Selection and Appointment Advice (Cont'd)

5.6 **Fee options**

Before offering any fee proposal a Quantity Surveyor will need to evaluate the costs of carrying out the service being offered. Options for charging fees are given below and may be used alone or in combination.

5.6.1 Percentage fees

This method of charging fees is appropriate when a construction project is reasonably straightforward and the quantity surveying service can be clearly defined but the exact amount of the total construction cost cannot be determined with much certainty

In determining a fee, judgements need to be made by the Quantity Surveyor on the size of the project, its complexity, degree of repetition, method of procurement and contract arrangements. This method may be seen as a 'broad brush' way of assessment which sometimes may be vulnerable to market forces and their influence on contractors' tenders.

Fees are expressed as a percentage of a sum, which is invariably the total construction cost, for the provision of a defined service. This sum will generally be calculated from the value of all construction contracts and other items of work carried out directly by a Client on a project.

It is advisable to have a definition of total construction cost agreed between a Client and his Quantity Surveyor Exclusions from the total construction cost should also be defined.

5.6.2 Lump sum fee

This method of charging fees is appropriate when a construction project is reasonably certain in programme time, project size and construction cost and the quantity surveying service can be clearly defined. A total fee is agreed for providing a defined service or amount of work. If appropriate, percentage-based or time-based fees may be converted to a lump sum fee once a project has become sufficiently defined.

If time, size, cost or circumstances of the project are significantly varied by more than specified amounts the lump sum fee may be varied, perhaps according to a formula contained in the Agreement. Otherwise the fee remains fixed. It should be borne in mind that a lump sum fee, whilst giving certainty to both parties at the start of a commission, may not always be appropriate unless the circumstances of the project remain significantly as stated at the time that the Agreement was signed.

5.6.3 Time charge fees

This method of charging fees is appropriate where the scope and/or the extent of the services to be provided cannot reasonably be foreseen and/or cannot reasonably be related to the cost of construction. It is often appropriate where open-ended services are required on feasibility studies, negotiation, claims consultancy etc. It is also a method that allows for additional or varied services to be provided, in addition to the provision of basic services quoted for in a fee agreement.

Fees are calculated from all the time expended on a project by partners/directors and staff and charged at rates previously agreed. The rates may be increased periodically to allow for the effects of inflation.

It is advisable to have previously agreed rates for grades of staff and partners/directors and methods of revising those rates periodically to reflect subsequent changes in salaries and costs. The inclusion of overhead allowances, generally including secretarial and administrative staff, within the rates needs careful calculation by the Quantity Surveyor Principles for the reimbursement of the time involved also need to have been agreed in advance. Sometimes this method may appear to be open ended, perhaps with little incentive for working efficiently. Periodic review of the charges is important so that progress is monitored against budgets.

5.6.4 Target cost fees

This method is appropriate where time charge fees are the basis for the Agreement but a Client wishes to have a guaranteed maximum target for fees.
Fees are recovered on a time charge basis but a 'capped' or guaranteed target cost is agreed between the parties before the work is carried out by the Quantity Surveyor If the target cost is exceeded then the Quantity Surveyor may not receive all of his costs, for instance part of his overhead allowances. If alternatively the total amount of the time charges falls below the budget cost then the Quantity Surveyor may receive an additional payment, a 'bonus' to reflect or share in the 'saving' to his Client. A formula for this should be agreed at the outset.

5.7 **Expense costs**

Expense costs are the costs to a firm such as the provision of cars, mileage-based payments to staff for travel, other travel costs, hotels, subsistence, meals, reproduction of documents, photocopying, postage and phone/fax charges, purchase costs of items bought specifically for a project, expenses in connection with appointing and engaging site-based staff, exceptional time spent by staff in travelling outside of usual hours and/or beyond usual distances etc.

In any agreement on fees it is advisable to be clear about the inclusion or exclusion of any category of expense costs within the fee charges, to identify which expenses are chargeable and to have a machinery for adjustment where necessary.

5.7.1 Recovering expense costs in fee arrangements.
Expense costs in fee bids and fee arrangements may be dealt with in a number of ways. Tender enquiries may prescribe the way required or there may be flexibility in how expense costs are included within the fee figures or shown separately from them. Three options exist for the recovery of expense costs in fee arrangements:

- **Expense costs may be included within a lump sum fee**
  Expense costs of any variety can be assessed and included within a fee and not shown separately. The fee offer thus includes a lump sum for all expense costs. This is often referred to as a 'rolled-up' offer in that the costs have not been shown separately in the bid. Although expense costs have been so included it is often necessary to consider how any adjustment to those costs may be made if changed circumstances require. For instance, if travel costs have been included in a 'rolled-up' fee and much more travel becomes necessary, it may then be required to demonstrate the basis of travel expenses that had been included in the fee offer in order to agree an adjustment

GENERAL INFORMATION

THE ROYAL INSTITUTION OF CHARTERED SURVEYORS
Appointing a Quantity Surveyor (Cont'd)
Selection and Appointment Advice (Cont'd)

- **Expense costs may be converted to a percentage of the total construction cost of the project**

  An assessment is made of the likely expense costs that may be incurred and the amount is then converted into a percentage of the total construction cost. This percentage may be added to the percentage shown as a fee offer for the quantity surveying service or it may be shown separately. The relationship between construction costs, to which a percentage fee would be applied, and the amount of expense costs that will be incurred is often tenuous and this method should only be entered into after careful consideration by a Client and his Quantity Surveyor

- **Expense costs may be paid as a separate lump sum**

  A lump sum can be calculated for expense costs and shown separately from any other lump sum, percentage-based or time charge fee for the project. Whilst this method will give a Client certainty at the outset, provision should be made for adjustment by the parties if circumstances change.

5.7.2 Constituents of expense costs and payment considerations

If expense costs are not to be included in a lump sum, percentage-based or time charge fee they may be recovered by the Quantity Surveyor on the submission of the cost records. If this method is used it is important to establish how expense costs will be calculated and reimbursed to a Quantity Surveyor by his Client. There are three methods for this:

- **Recovery of actual expense costs of disbursements incurred by a Quantity Surveyor with the authority of his Client**

  This category of cost would normally include items such as public transport costs, hotels and subsistence, and sundry expenses incurred, for instance, in obtaining financial reports on companies when considering prospective tender lists, or in establishing a site office

- **Recovery of expense costs for in-house resources provided by an organisation**

  This would include, for example, photocopying and the reproduction of tender and contract documentation. The amounts due under this category of costs are not always as clearly demonstrable as the costs referred to in the previous category. It may be helpful to set up an agreement between the parties that allows the Quantity Surveyor to recover an amount in addition to the actual material costs incurred so that some staff and overhead costs can be recovered in addition to the basic costs

- **Recovery of expense costs at market rates**

  This category is for resources provided by an organisation from its in-house facilities, for example photocopying or drawing reproduction, travel carried out by personnel using company cars etc. These costs would be recovered at what is agreed to be the market rate for the service provided.

5.8 **Implication of termination or suspension**

If a project is terminated or suspended it may become necessary to consider the implications on fees and on expenses, for instance for staff, offices, cars, rented accommodation or leased equipment taken on especially for a project. It may be that some of these costs cannot readily be avoided if termination or suspension occurs and a fee agreement should contain provisions to allow adjustments to be made, particularly to lump sum, 'rolled-up' or percentage fees.

6. **Submission and comparison of fee offers and selection and notification of results**

6.1 **Submission**

6.1.1 Clients are advised to invite fee offers only from firms of comparable capability and to make a selection taking into account value as well as the fee bid.

6.1.2 Where competitive offers have been sought, instructions on the submission of offers should be clear; giving the date, time and place for their delivery. Offers should all be opened at the same time and treated as confidential until notification of results is possible.

6.2 **Comparison and selection**

6.2.1 Competitive offers should be analysed and compared (including where there are different components of a fee evaluation, as referred to above). Comparison of offers should be on a basis which incorporates all the component parts of an offer in order to indicate the lowest offer and its relationship with other offers.

6.2.2 If two or more submissions give identical or very close results then it may be appropriate to apply a sensitivity test, namely to check the impact of possible or probable changes in the scope, size or value of the project on the fee bids.

6.2.3 In comparing offers it is advisable to weigh quality criteria against the offer price. Further detailed guidance is given in Appendix 10 of the RICS publication A Guide to Securing the Services of a Chartered Surveyor. It is usual to weigh quality of service criteria at a minimum of 60 per cent relative to price criteria at a maximum of 40 per cent.

6.3 **Notification**

6.3.1 Once a decision has been taken to appoint a Quantity Surveyor; the successful firm should be notified. Unsuccessful firms should also be notified of the decision and given information on the bids received (where appropriate this can be by the use of 'indices' which do not link the names of each firm to its offer). All notifications should be made in writing and be sent as soon as possible after the decision to appoint has been made.

THE ROYAL INSTITUTION OF CHARTERED SURVEYORS
Appointing a Quantity Surveyor (Cont'd)
Selection and Appointment Advice (Cont'd)

7. Confirmation of the Agreement between the Client and the Quantity Surveyor

7.1 The Agreement between a Client and his Quantity Surveyor should be effected by either:

- using the Form of Agreement and Terms of Appointment contained in this Guide, with cross-reference to the Form of Enquiry, Schedule of Services and Fee Offer
- using a separate form of agreement, terms and conditions, with cross reference to the services to be provided and to the fee offer
- a simple exchange of letters incorporating the information given above, making reference to the fee offer.

The 1996 Housing Grants, Construction and Regeneration Act ('the HGCR Act') introduced a right to refer any dispute to adjudication and certain provisions relating to payment. This documentation complies with the HGCR Act and includes the necessary adjudication and payment provisions. Other forms of agreement will need to comply with the HGCR Act or the adjudication and payment provisions of the Statutory Scheme for Construction Contracts will apply. An exchange of letters is unlikely to comply and the Scheme will therefore also apply to these.

8. Complaints handling and disputes resolution

8.1 Quantity Surveyors who are partners or directors in firms providing surveying services must operate an internal complaints handling procedure, which applies to disputes less than £50,000, under the RICS Bye-Laws. The RICS also sets a minimum standard of complaints handling, as laid out in its Professional Conduct – Rules of Conduct and Disciplinary Procedures. If the complaint cannot be resolved internally by the firm then the matter must go to final resolution by a third party. A reference to adjudication under the HGCR Act would be sufficient for the purposes of satisfying RICS regulations, pending any final resolution of the dispute at the end of the contract period. See also Clause 11 of the Terms of Appointment.

8.2 The HGCR Act introduced a compulsory scheme of third party neutral dispute resolution – which can occur at any time during the construction process – called adjudication. Any party may refer any dispute to an adjudicator at any time.

SECTION 2
Form of Enquiry
Schedule of Services

1. Category One: General services

The following services may be provided on any project, whatever its nature and whatever the method of procurement adopted.

☐ The following services shall, where relevant, also apply to environmental engineering service (mechanical and electrical engineering) if indicated by placing a cross in this box.

1.1 Inception and feasibility

1.1.1 ☐ Liaise with Client and other consultants to determine Client's initial requirements and subsequent development of the full brief

1.1.2 ☐ Advise on selection of other consultants if not already appointed

1.1.3 ☐ Advise on implications of proposed project and liase with other experts in developing such advice

1.1.4 ☐ Advise on feasibility of procurement options

1.1.5 ☐ Establish Client's order of priorities for quality, time and cost

1.1.6 ☐ Prepare initial budget estimate from feasibility proposals

1.1.7 ☐ Prepare overall project cost calculation and cash flow projections

1.2 Pre-contract cost control

1.2.1. ☐ Prepare and develop preliminary cost plan

1.2.2 ☐ Advise on cost of design team's proposals, including effects of site usage, shape of buildings, alternative forms of design and construction as design develops

1.2.3 ☐ Monitor cost implications during detailed design stage

1.2.4 ☐ Maintain and develop cost plan, and prepare periodic reports and updated cash flow forecasts

1.3 Tender and contractual documentation

1.3.1 ☐ Advise on tendering and contractual arrangements taking into account the Client's priorities and information available from designers

1.3.2 ☐ Advise on insurance responsibilities and liase with Client's insurance adviser

1.3.3 ☐ Advise on warranties

1.3.4 ☐ Advise on bonds for performance and other purposes

1.3.5 ☐ Prepare tender and contract documentation in conjunction with the Client and members of the design team

1.3.6 ☐ Provide copies of documentation as agreed

1.3.7 ☐ Advise on use and/or amendment of standard forms of contract or contribute to drafting of particular requirements in association with Client's legal advisers

THE ROYAL INSTITUTION OF CHARTERED SURVEYORS
Appointing a Quantity Surveyor (Cont'd)
Schedule of Services (Cont'd)
Category One: General services (Cont'd)

1.3.8 ☐ Draw up forms of contract, obtain contract drawings from members of design team and prepare and deliver to both parties contract copies of all documents

1.4 Tender selection and appraisal

1.4.1 ☐ Advise on short listing prospective tenderers

1.4.2 ☐ Investigate prospective tenderers and advise Client on financial status and experience

1.4.3 ☐ Attend interviews with tenderers

1.4.4 ☐ Arrange delivery of documents to selected tenderers

1.4.5 ☐ Check tender submissions for accuracy, level of pricing, pricing policy etc.

1.4.6 ☐ Advise on errors and qualifications and, if necessary, negotiate thereon

1.4.7 ☐ Advise on submission of programme of work and method statement

1.4.8 ☐ Prepare appropriate documentation, if required, to adjust the tender received to an acceptable contract sum

1.4.9 ☐ Review financial budget in view of tenders received and prepare revised cash flow

1.4.10 ☐ Prepare report on tenders with appropriate recommendations

1.4.11 ☐ Advise on letters of intent and issue in conjunction with Client's advisers

1.5 Interim valuations

1.5.1 ☐ Prepare recommendations for interim payments to contractors, subcontractors and suppliers in accordance with contract requirements

1.6 Post-contract cost control

1.6.1 ☐ Value designers' draft instructions for varying the project before issue

1.6.2 ☐ Prepare periodic cost reports in agreed format at specified intervals including any allocations of cost and/or copies as requested by third parties

1.7 Final account

1.7.1 ☐ Prepare the final account

1.8 Attendance at meetings

1.8.1 ☐ Attend meetings as provided for under this Agreement

1.9 Provision of printing/reproduction/copying of documents and the like

1.9.1 ☐ Provide copies of documentation as provided for under this Agreement

2. Category Two: Services particular to non-traditional methods of procurement

These services relate to particular methods of procurement and contract arrangement and should be incorporated into the Agreement as required in conjunction with services from Categories One and Three of the Schedule.

2.1 Services particular to prime cost and management or construction management contracts

2.1.1 ☐ Obtain agreement of a contractor to the amount of the approximate estimate and confirm the amount of the fee for the contract

2.1.2 ☐ Prepare recommendations for interim payments to contractor based on contractor's prime costs

2.1.3 ☐ Adjust the approximate estimate to take account of variations and price fluctuations

2.1.4 ☐ Check the final amounts due to contractors, subcontractors and suppliers

2.2 Services particular to management and construction management contracts

The terms 'management contracting' and 'construction management' mean contractual arrangements where a firm is employed for a fee to manage, organise, supervise and secure the carrying out of the work by other contractors.

2.2.1 ☐ If required, assist in drafting special forms of contract

2.2.2 ☐ Prepare tender documents for the appointment of a management contractor or construction manager

2.2.3 ☐ Attend interviews of prospective contractors or managers

2.2.4 ☐ Obtain manager's agreement to contract cost plan and confirm amount of manager's fee

2.2.5 ☐ Assist in allocation of cost plan into work packages

2.2.6 ☐ Assist in preparation of tender and contract documents

2.2.7 ☐ Price tender documents to provide an estimate comparable with tenders

2.2.8 ☐ Review cost plan as tenders are obtained and prepare revised forecasts of cash flow

THE ROYAL INSTITUTION OF CHARTERED SURVEYORS
Appointing a Quantity Surveyor (Cont'd)
Schedule of Services (Cont'd)
Category Two: Services particular to non-traditional
methods of procurement (Cont'd)

2.2.9 ☐ Prepare periodic cost reports to show effect of variations, tenders let and prime costs

2.2.10 ☐ Check the final amounts due to managers, contractors, subcontractors or works contractors and suppliers

2.3 Services particular to design and build contracts - services available to a Client

2.3.1 ☐ Draft the Client's brief, in association with the Client and his designers

2.3.2 ☐ Prepare tender documents incorporating the Client's requirements

2.3.3 ☐ Prepare contract documentation, taking into account any changes arising from the contractor's proposals

2.3.4 ☐ Prepare recommendations for interim and final payments to the contractor; including compliance with statutory requirements of the 1996 Housing Grants, Construction and Regeneration Act

2.3.5 ☐ Assist in agreement of settlement of the contractor's final account

2.4 Services particular to design and build contracts - services available to a contractor

2.4.1 ☐ Prepare bills of quantities to assist in the preparation of a contractor's tender

2.4.2 ☐ Prepare alternative cost studies to assist in determining the optimum scheme for a contractor's submission

2.4.3 ☐ Draft specifications forming the contractor's proposals

2.4.4 ☐ Assist with specialist enquiries in compiling the contractor's tender

2.4.5 ☐ Measure and price variations for submission to the Client's representative

2.4.6 ☐ Prepare applications for interim payments

2.4.7 ☐ Agree final account with Client's representative

2.5 Services particular to measured term contracts

2.5.1 ☐ Take measurements, price from agreed schedule of rates and agree totals with contractor

2.5.2 ☐ Check final amounts due to contractor(s)

3. **Category Three: Services not always required in Categories One and Two**

These services should be incorporated into the Agreement in conjunction with services from Categories One and Two of the Schedule as required. Where points are left blank, the Client and/or the Quantity Surveyor should specify their own service as required.

3.1 Bill of quantities

3.1.1 ☐ Provide bills of quantities for mechanical and engineering services

3.1.2 ☐ Price bills of quantities to provide an estimate comparable with tenders

3.2 Cost analysis

3.2.1 ☐ Prepare cost analysis based on agreed format or special requirement

3.3 Advise on financial implications as follows:

3.3.1 ☐ Cost options of developing different sites

3.3.2 ☐ Preparation of development appraisals

3.3.3 ☐ Cost implications of alternative development programmes

3.3.4 ☐ Effect of capital and revenue expenditure

3.3.5 ☐ Life cycle cost studies and estimate of annual running costs

3.3.6 ☐ Availability of grants

3.3.7 ☐ Assist in applications for grants and documentation for these

3.3.8 ☐ Evaluation of items for capital allowances, grant payments or other such matters

3.4 Advice on use of areas and provide:

3.4.1 ☐ Measurement of gross floor areas

3.4.2 ☐ Measurement of net lettable floor areas

3.5 Provide advice on contractual matters affecting the following:

3.5.1 ☐ Entitlement to liquidated and ascertained damages

3.5.2 ☐ Final assessment of VAT

3.5.3 ☐ Opinion on delays and/or disruptions and requests for extensions of time

3.5.4 ☐ Consequences of acceleration

3.5.5 ☐ Assessment of the amount of loss and expense or other such matters and if instructed carrying out negotiations with contractors to reach a settlement

GENERAL INFORMATION

THE ROYAL INSTITUTION OF CHARTERED SURVEYORS
Appointing a Quantity Surveyor (Cont'd)
Schedule of Services (Cont'd)
Category Three: Services not always required (Cont'd)

3.6   Provide value management and value engineering services as follows:

3.6.1   ☐
(state)_____

3.6.2   ☐
(state)_____

3.7   Provide risk assessment and management services as follows:

3.7.1   ☐
(state)_____

3.7.2   ☐
(state)_____

3.8   Adjudication services

3.8.1   ☐   Provide services acting as an adjudicator in construction disputes

3.8.2   ☐   Provide services in connection with advising the Client in relation to active or threatened adjudication proceedings

3.9   Provide services in connection with arbitration and/or litigationas follows:

3.9.1   ☐
(state)_____

3.9.2   ☐
(state)_____

3.10   Provide services arising from fire or other damage to buildings including preparing and negotiating claims with loss adjusters as follows:

3.10.1   ☐
(state)_____

3.10.2   ☐
(state)_____

3.11   Provide services to a contractor in connection with negotiation of claims as follows:

3.11.1   ☐
(state)_____

3.11.2   ☐
(state)_____

3.12   Project Management

Project management services are available from Quantity Surveyors and are set out in detail (together with a Form of Agreement and Guidance Notes) in the RICS publication Project Management Agreement and Conditions of Engagement.

Clients are referred to the Project Management Agreement if the service is mainly for project management. This section is intended to be used if ancillary project management is required to a mainly quantity surveying service.

Provide project management services as follows:

3.12.1   ☐
(state)_____

3.12.2   ☐
(state)_____

3.13   Provide programme co-ordination and monitoring services as follows:

3.13.1   ☐
(state)_____

3.13.2   ☐
(state)_____

3.14   Provide planning supervisor services as follows:

3.14.1   ☐
(state)_____

3.14.2   ☐
(state)_____

3.15   Provide information for use in future management and / or maintenance of the building as follows:

3.15.1   ☐
(state)_____

3.15.2   ☐
(state)_____

3.16   Any other services not listed elsewhere in the Form of Enquiry:

3.16.1   ☐
(state)_____

3.16.2   ☐
(state)_____

3.16.3   ☐
(state)_____

Other Quantity Surveying Services

Other services required or provided by quantity surveyors include:-
Planning Supervisor duties
Contractors Services – Measurement
                                        Pricing
                    Procurement advice -
                                        Subcontractors
                                        Suppliers
                    Post Contract services

Adjudication
Mediation
Arbitration
Expert Witness

FEE CALCULATION

The following is an example of the calculations necessary to establish the level of fee required for a specific project. The reader should adjust the rates as appropriate to a level based on in-house costs and the complexity, duration and scope of the project and as noted below.

Example - Quantity Surveying Fee Calculation (see clause 5.6 of Guide)

Example - Complex Project Resource Schedule

Project Value £1,500,000 (Including alteration works value £150,000)

	Hours	Rate	Cost
1. COST PLANNING			
Client Liaison	18	30.00	540.00
Attendance at meetings	18	30.00	540.00
Pre tender cost planning	60	25.00	1,500.00
			2,580.00
2. BILLS OF QUANTITIES			
Client Liaison	12	30.00	360.00
Attendance at meetings	20	30.00	600.00
Demolitions & Alterations	28	20.00	560.00
Substructures	20	20.00	400.00
Frame	12	20.00	240.00
Upper floors	10	20.00	200.00
Roof	14	20.00	280.00
Stairs	8	20.00	160.00
External walls	14	20.00	280.00
Windows & Doors	26	20.00	520.00
Internal Doors	12	20.00	240.00
Wall Finishes	12	20.00	240.00
Floor Finishes	9	20.00	180.00
Ceiling Finishes	8	20.00	160.00
Fittings & Furnishings	6	20.00	120.00
Sanitary appliances	6	20.00	120.00
Services equipment	6	20.00	120.00
Disposal installations	10	20.00	200.00
Water installations	6	20.00	120.00
Heat source	6	20.00	120.00
Space heating & air treatment	6	20.00	120.00
Ventilation installations	6	20.00	120.00
Electrical installations	18	20.00	360.00
Gas Installations	6	20.00	120.00
Lift installations	6	20.00	120.00
Protective installations	6	20.00	120.00
Communication installations	6	20.00	120.00
Special installations	6	20.00	120.00
Builders work in connection	19	20.00	380.00
External works	36	20.00	720.00
Drainage	12	20.00	240.00
External services	6	20.00	120.00
Preliminaries & Preambles	20	20.00	400.00
Pre-tender estimates	20	20.00	400.00
Tender report & Cost Analysis	12	20.00	240.00
Contract documentation	8	25.00	200.00
			9,120.00
3. POST CONTRACT			
Client liaison	20	30.00	600.00
Attendance at meetings	80	22.00	1,760.00
Valuations	66	22.00	1,452.00
Cost Reports	60	22.00	1,320.00
Pre Estimate Instructions	30	22.00	660.00
Final Account	60	22.00	1,320.00
			7,112.00
Total estimated cost			18,812.00
Add overheads		60 %	11,287.20
			30,099.20
Add profit		25 %	7,524.80
			£37,624.00

Cost Planning only

Estimated Cost			2,580.00
Add Overheads	60 %		1,548.00
			4,128.00
Add Profit	25 %		1,032.00
			5,160.00

Bills of Quantities only

Estimated Cost			9,120.00
Add Overheads	60 %		5,472.00
			14,592.00
Add Profit	25 %		3,648.00
			18,240.00

Post Contract only

Estimated Cost			7,112.00
Add Overheads	60%		4,267.20
			11,379.20
Add Profit	25%		2,844.80
			14,224.00

Total	£37,624.00

The above example excludes all disbursements – travel, reproduction of documents, forms of contracts and the like.

The rates indicated equate to:

Partner	cost rate £30.00	Charge Rate £60.00
Senior Surveyor	cost rate £25.00	Charge Rate £50.00
Taker off	cost rate £20.00	Charge Rate £40.00
Post Contract Surveyor	cost rate £22.00	Charge Rate £44.00

The hours indicated and rates used should be adjusted as appropriate to suit each individual project and the costs, overheads and profit requirements of the practice bidding for work.

Other factors that should be considered when calculating fees include timescale, workload, relationship with client and other consultants, prestige, locality of project, likely ongoing programme of work, availability of labour.

Typical Quantity Surveying fees

Typical fees calculated as above for provision of basic services

Work Value	Cost Planning	Bills of Quantities	Post Contract services
Complex Works with little repetition			
50,000	280	1,240	790
150,000	730	3,210	2,140
300,000	1,410	5,520	3,830
600,000	2,810	9,120	7,090
1,500,000	5,180	18,240	14,980
3,000,000	9,230	30,400	27,000
6,000,000	16,300	52,600	49,300
10,000,000	24,200	79,200	76,300
Less Complex work with some repetition			
50,000	280	1,130	790
150,000	730	3,040	2,140
300,000	1,400	5,010	3,830
600,000	2,800	8,000	6,800
1,500,000	5,100	14,700	13,500
3,000,000	9,200	24,700	23,600
6,000,000	16,300	43,000	41,800
10,000,000	24,200	64,600	63,500

GENERAL INFORMATION

Example - Quantity Surveying Fee Calculation (Cont'd)
Example - Complex Project Resource Schedule (Cont'd)

Typical fees calculated as above for provision of basic services

	Cost Planning	Bills of Quantities	Post Contract services

Simple works or works with a substantial amount of repetition
Work Value

Work Value	Cost Planning	Bills of Quantities	Post Contract services
50,000	230	1,010	680
150,000	560	2,700	1,800
300,000	1,130	4,500	3,260
600,000	2,020	6,860	6,080
1,500,000	3,800	12,200	12,700
3,000,000	6,800	20,300	21,800
6,000,000	11,900	34,500	38,000
10,000,000	17,300	50,600	56,900

Alteration Works – Additional Fee on value of alteration works
Work Value

Work Value	
50,000	340
150,000	1,010

Work Value

Work Value	
50,000	510
150,000	1,520

Where Provisional Sums, Prime Cost Sums, Subcontracts, Dayworks or contingencies form a substantial part of the project the above figures may require adjustment accordingly.

Mechanical and Electrical Services – Additional Fee for provision of Bills of Quantities

Value of M & E work	Bills of Quantities	Post Contract services
120,000	2,020	1,690
240,000	3,850	2,900
600,000	8,400	5,900
1,000,000	12,300	9,200
4,000,000	35,600	29,500
6,000,000	51,200	50,000

Negotiating Tenders – Additional Fee
Work Value

Work Value	
150,000	500
600,000	1,400
1,200,000	2,200
3,000,000	3,400

Decoration Works – Additional Fee on value of alteration works

---

THE ROYAL INSTITUTION OF CHARTERED SURVEYORS

The following Scales of Professional Charges printed by permission of The Royal Institution of Chartered Surveyors, 12 Great George Street, Westminster, London SW1P 3AD, are now obsolete and are reproduced for information only.

SCALE NO. 36

INCLUSIVE SCALE OF PROFESSIONAL CHARGES FOR QUANTITY SURVEYING SERVICES FOR BUILDING WORKS

(Effective from 29th July, 1988)

This Scale has been determined by the Fees Committee appointed by the Quantity Surveyors Divisional Council of The Royal Institution of Chartered Surveyors. The Scale is for guidance and is not mandatory.

1.0   GENERALLY

1.1   This scale is for use when an inclusive scale of professional charges is considered to be appropriate by mutual agreement between the employer and the quantity surveyor.

1.2   This scale does not apply to civil engineering works, housing schemes financed by local authorities and the Housing Corporation and housing improvement work for which separate scales of fees have been published.

1.3   The fees cover quantity surveying services as may be required in connection with a building project irrespective of the type of contract from initial appointment to final certification of the contractor's account such as:

(a)   Budget estimating; cost planning and advice on tendering procedures and contract arrangements.

(b)   Preparing tendering documents for main contract and specialist sub-contracts;

examining tenders received and reporting thereon or negotiating tenders and pricing with a selected contractor and/or sub-contractors.

(c)   Preparing recommendations for interim payments on account to the contractor; preparing periodic assessments of anticipated final cost and reporting thereon; measuring work and adjusting variations in accordance with the terms of the contract and preparing final account, pricing same and agreeing totals with the contractor.

(d)   Providing a reasonable number of copies of bills of quantities and other documents; normal travelling and other expenses. Additional copies of documents, abnormal travelling and other expenses (e.g. in remote areas or overseas) and the provision of checkers on site shall be charged in addition by prior arrangement with the employer.

1.4   If any of the materials used in the works are supplied by the employer or charged at a preferential rate, then the actual or estimated market value thereof shall be included in the amounts upon which fees are to be calculated.

1.5   If the quantity surveyor incurs additional costs due to exceptional delays in building operations or any other cause beyond the control of the quantity surveyor then the fees shall be adjusted by agreement between the employer and the quantity surveyor to cover the reimbursement of these additional costs.

1.6   The fees and charges are in all cases exclusive of value added tax which will be applied in accordance with legislation.

1.7   Copyright in bills of quantities and other documents prepared by the quantity surveyor is reserved to the quantity surveyor.

THE ROYAL INSTITUTION OF CHARTERED SURVEYORS (Cont'd)
SCALE NO. 36 (Cont'd)

2.0 INCLUSIVE SCALE

2.1 The fees for the services outlined in paragraph 1.3, subject to the provision of paragraph 2.2, shall be as follows:

(a) CATEGORY A. Relatively complex works and/or works with little or no repetition.

Examples:

Ambulance and fire stations; banks; cinemas; clubs; computer buildings; council offices; crematoria; fitting out of existing buildings; homes for the elderly; hospitals and nursing homes; laboratories; law courts; libraries; 'one off' houses; petrol stations; places of religious worship; police stations; public houses, licensed premises; restaurants; sheltered housing; sports pavilions; theatres; town halls; universities, polytechnics and colleges of further education (other than halls of residence and hostels); and the like.

Value of Work		Category A Fee	
	£	£	£
Up to	150,000	380 + 6.0% (Minimum Fee £3,380)	
150,000 -	300,000	9,380 + 5.0% on balance over	150,000
300,000 -	600,000	16,880 + 4.3% on balance over	300,000
600,000 -	1,500,000	29,780 + 3.4% on balance over	600,000
1,500,000 -	3,000,000	60,380 + 3.0% on balance over	1,500,000
3,000,000 -	6,000,000	105,380 + 2.8% on balance over	3,000,000
Over	6,000,000	189,380 + 2.4% on balance over	6,000,000

(b) CATEGORY B. Less complex works and/or works with some element of repetition.

Examples:

Adult education facilities; canteens; church halls; community centres; departmental stores; enclosed sports stadia and swimming baths; halls of residence; hostels; motels; offices other than those included in Categories A and C; railway stations; recreation and leisure centres; residential hotels; schools; self-contained flats and maisonettes; shops and shopping centres; supermarkets and hypermarkets; telephone exchanges; and the like.

Value of Work		Category B Fee	
	£	£	£
Up to	150,000	360 + 5.8% (Minimum Fee £3,260)	
150,000 -	300,000	9,060 + 4.7% on balance over	150,000
300,000 -	600,000	16,110 + 3.9% on balance over	300,000
600,000 -	1,500,000	27,810 + 2.8% on balance over	600,000
1,500,000 -	3,000,000	53,010 + 2.6% on balance over	1,500,000
3,000,000 -	6,000,000	92,010 + 2.4% on balance over	3,000,000
Over	6,000,000	64,010 + 2.0% on balance over	6,000,000

(c) CATEGORY C. Simple works and/or works with a substantial element of repetition.

Examples:

Factories; garages; multi-storey car parks; open-air sports stadia; structural shell offices not fitted out; warehouses; workshops; and the like.

Value of Work		Category C Fee	
	£	£	£
Up to	150,000	300 + 4.9% (Minimum Fee £2,750)	
150,000 -	300,000	7,650 + 4.1% on balance over	150,000
300,000 -	600,000	13,800 + 3.3% on balance over	300,000
600,000 -	1,500,000	23,700 + 2.5% on balance over	600,000
1,500,000 -	3,000,000	46,200 + 2.2% on balance over	1,500,000
3,000,000 -	6,000,000	79,200 + 2.0% on balance over	3,000,000
Over	6,000,000	139,200 + 1.6% on balance over	6,000,000

THE ROYAL INSTITUTION OF CHARTERED SURVEYORS (Cont'd)
SCALE NO. 36 (Cont'd)
INCLUSIVE SCALE (Cont'd)

2.1   (d) Fees shall be calculated upon the total of the final account for the whole of the work including all nominated sub-contractors' and nominated suppliers' accounts. When work normally included in a building contract is the subject of a separate contract for which the quantity surveyor has not been paid fees under any other clause hereof, the value of such work shall be included in the amount upon which fees are charged.

(e) When a contract comprises buildings which fall into more than one category, the fee shall be calculated as follows:

  (i) The amount upon which fees are chargeable shall be allocated to the categories of work applicable and the amounts so allocated expressed as percentages of the total amount upon which fees are chargeable.

(ii) Fees shall then be calculated for each category on the total amount upon which fees are chargeable.

(iii) The fee chargeable shall then be calculated by applying the percentages of work in each category to the appropriate total fee and adding the resultant amounts.

(iv) A consolidated percentage fee applicable to the total value of the work may be charged by prior agreement between the employer and the quantity surveyor. Such a percentage shall be based on this scale and on the estimated cost of the various categories of work and calculated in accordance with the principles stated above.

(f) When a project is the subject of a number of contracts then, for the purpose of calculating fees, the values of such contracts shall not be aggregated but each contract shall be taken separately and the scale of charges (paragraphs 2.1 (a) to (e)) applied as appropriate.

2.2 Air Conditioning, Heating, Ventilating and Electrical Services

(a) When the services outlined in paragraph 1.3 are provided by the quantity surveyor for the air conditioning, heating, ventilating and electrical services there shall be a fee for these services in addition to the fee calculated in accordance with paragraph 2.1 as follows:

Value of Work		Additional Fee	
	£	£	£
Up to	120,000	5.0%	
120,000 -	240,000	6,000 + 4.7% on balance over	120,000
240,000 -	480,000	11,640 + 4.0% on balance over	240,000
480,000 -	750,000	21,240 + 3.6% on balance over	480,000
750,000 -	1,000,000	30,960 + 3.0% on balance over	750,000
1,000,000 -	4,000,000	38,460 + 2.7% on balance over	1,000,000
Over	4,000,000	119,460 + 2.4% on balance over	4,000,000

(b) The value of such services, whether the subject of separate tenders or not, shall be aggregated and the total value of work so obtained used for the purpose of calculating the additional fee chargeable in accordance with paragraph (a). (Except that when more than one firm of consulting engineers is engaged on the design of these services, the separate values for which each such firm is responsible shall be aggregated and the additional fees charged shall be calculated independently on each such total value so obtained.)

(c) Fees shall be calculated upon the basis of the account for the whole of the air conditioning, heating, ventilating and electrical services for which bills of quantities and final accounts have been prepared by the quantity surveyor.

2.3 Works of Alteration

On works of alteration or repair, or on those sections of the work which are mainly works of alteration or repair, there shall be a fee of 1.0% in addition to the fee calculated in accordance with paragraphs 2.1 and 2.2.

2.4 Works of Redecoration and Associated Minor Repairs

On works of redecoration and associated minor repairs, there shall be a fee of 1.5% in addition to the fee calculated in accordance with paragraphs 2.1 and 2.2.

2.5 Generally

If the works are substantially varied at any stage or if the quantity surveyor is involved in an excessive amount of abortive work, then the fees shall be adjusted by agreement between the employer and the quantity surveyor.

THE ROYAL INSTITUTION OF CHARTERED SURVEYORS (Cont'd)
SCALE NO. 36 (Cont'd)
INCLUSIVE SCALE (Cont'd)

3.0 ADDITIONAL SERVICES

3.1 For additional services not normally necessary, such as those arising as a result of the termination of a contract before completion, liquidation, fire damage to the buildings, services in connection with arbitration, litigation and investigation of the validity of contractors' claims, services in connection with taxation matters and all similar services where the employer specifically instructs the quantity surveyor, the charges shall be in accordance with paragraph 4.0 below.

4.0 TIME CHARGES

4.1 (a) For consultancy and other services performed by a principal, a fee by arrangement according to the circumstances including the professional status and qualifications of the quantity surveyor.

(b) When a principal does work which would normally be done by a member of staff, the charge shall be calculated as paragraph 4.2 below.

4.2 (a) For services by a member of staff, the charges for which are to be based on the time involved, such charges shall be calculated on the hourly cost of the individual involved plus 145%.

4.2 (b) A member of staff shall include a principal doing work normally done by an employee (as paragraph 4.1 (b) above), technical and supporting staff, but shall exclude secretarial staff or staff engaged upon general administration.

(c) For the purpose of paragraph 4.2 (b) above, a principal's time shall be taken at the rate applicable to a senior assistant in the firm.

(d) The supervisory duties of a principal shall be deemed to be included in the addition of 145% as paragraph 4.2 (a) above and shall not be charged separately.

(e) The hourly cost to the employer shall be calculated by taking the sum of the annual cost of the member of staff of:

(i) Salary and bonus but excluding expenses;

(ii) Employer's contributions payable under any Pension and Life Assurance Schemes;

(iii) Employer's contributions made under the National Insurance Acts, the Redundancy Payments Act and any other payments made in respect of the employee by virtue of any statutory requirements; and

(iv) Any other payments or benefits made or granted by the employer in pursuance of the terms of employment of the member of staff; and dividing by 1650.

5.0 INSTALMENT PAYMENTS

5.1 In the absence of agreement to the contrary, fees shall be paid by instalments as follows:

(a) Upon acceptance by the employer of a tender for the works, one half of the fee calculated on the amount of the accepted tender.

(b) The balance by instalments at intervals to be agreed between the date of the first certificate and one month after final certification of the contractor's account.

5.2 (a) In the event of no tender being accepted, one half of the fee shall be paid within three months of completion of the tender documents. The fee shall be calculated upon the basis of the lowest original bona fide tender received. In the event of no tender being received, the fee shall be calculated upon a reasonable valuation of the works based upon the tender documents.
Note: In the foregoing context "bona fide tender" shall be deemed to mean a tender submitted in good faith without major errors of computation and not subsequently withdrawn by the tenderer.

(b) In the event of the project being abandoned at any stage other than those covered by the foregoing, the proportion of the fee payable shall be by agreement between the employer and the quantity surveyor.

SCALE NO. 37

ITEMISED SCALE OF PROFESSIONAL CHARGES
FOR QUANTITY SURVEYING SERVICES FOR
BUILDING WORKS
(29th July, 1988)

This Scale has been determined by the Fees Committee appointed by the Quantity Surveyors Divisional Council of The Royal Institution of Chartered Surveyors. The Scale is for guidance and is not mandatory.

1.0 GENERALLY

1.1 The fees are in all cases exclusive of travelling and other expenses (for which the actual disbursement is recoverable unless there is some prior arrangement for such charges) and of the cost of reproduction of bills of quantities and other documents, which are chargeable in addition at net cost.

1.2 The fees are in all cases exclusive of services in connection with the allocation of the cost of the works for purposes of calculating value added tax for which there shall be an additional fee based on the time involved (see paragraphs 19.1 and 19.2)

1.3 If any of the materials used in the works are supplied by the employer or charged at a preferential rate, then the actual or estimated market value thereof shall be included in the amounts upon which fees are to be calculated.

1.4 The fees are in all cases exclusive of preparing a specification of the materials to be used and the works to be done, but the fees for preparing bills of quantities and similar documents do include for incorporating preamble clauses describing the materials and workmanship (from instructions given by the architect and/or consulting engineer).

GENERAL INFORMATION

THE ROYAL INSTITUTION OF CHARTERED SURVEYORS (Cont'd)
SCALE NO. 37 (Cont'd)

1.5 If the quantity surveyor incurs additional costs due to exceptional delays in building operations or any other cause beyond the control of the quantity surveyor then the fees may be adjusted by agreement between the employer and the quantity surveyor to cover the reimbursement of these additional costs.

1.6 The fees and charges are in all cases exclusive of value added tax which will be applied in accordance with legislation.

1.7 Copyright in bills of quantities and other documents prepared by the quantity surveyor is reserved to the quantity surveyor.

2.0 BILLS OF QUANTITIES

2.1 Basic Scale: For preparing bills of quantities and examining tenders received and reporting thereon.

2.1 (a) CATEGORY A: Relatively complex works and/or works with little or no repetition.
Examples:
Ambulance and fire stations; banks; cinemas; clubs; computer buildings; council offices; crematoria; fitting out of existing buildings; homes for the elderly; hospitals and nursing homes; laboratories; law courts; libraries; 'one off' houses; petrol stations; places of religious worship; police stations; public houses, licensed premises; restaurants; sheltered housing; sports pavilions; theatres; town halls; universities, polytechnics and colleges of further education (other than halls of residence and hostels); and the like.

Value of Work	Category A Fee	
£	£	£
Up to 150,000	230 + 3.0% (minimum fee £1,730)	
150,000 - 300,000	4,730 + 2.3% on balance over	150,000
300,000 - 600,000	8,180 + 1.8% on balance over	300,000
600,000 - 1,500,000	13,580 + 1.5% on balance over	600,000
1,500,000 - 3,000,000	27,080 + 1.2% on balance over	1,500,000
3,000,000 - 6,000,000	45,080 + 1.1% on balance over	3,000,000
Over - 6,000,000	78,080 + 1.0% on balance over	6,000,000

2.1 (b) CATEGORY B: Less complex works and/or works with some element of repetition.
Examples:
Adult education facilities; canteens; church halls; community centres; departmental stores; enclosed sports stadia and swimming baths; halls of residence; hostels; motels; offices other than those included in Categories A and C; railway stations; recreation and leisure centres; residential hotels; schools; self-contained flats and maisonettes; shops and shopping centres; supermarkets and hypermarkets; telephone exchanges; and the like.

Value of Work	Category B Fee	
£	£	£
Up to 150,000	210 + 2.8% (minimum fee £1610)	
150,000 - 300,000	4,410 + 2.0% on balance over	150,000
300,000 - 600,000	7,410 + 1.5% on balance over	300,000
600,000 - 1,500,000	11,910 + 1.1% on balance over	600,000
1,500,000 - 3,000,000	21,810 + 1.0% on balance over	1,500,000
3,000,000 - 6,000,000	36,810 + 0.9% on balance over	3,000,000
Over 6,000,000	63,810 + 0.8% on balance over	6,000,000

(c) CATEGORY C: Simple works and/or works with a substantial element of repetition.
Examples:
Factories; garages; multi-storey car parks; open-air sports stadia; structural shell offices not fitted out; warehouses; workshops; and the like.

Value of Work	Category C Fee	
£	£	£
Up to 150,000	180 + 2.5% (minimum fee £1,430)	
150,000 - 300,000	3,930 + 1.8% on balance over	150,000
300,000 - 600,000	6,630 + 1.2% on balance over	300,000
600,000 - 1,500,000	10,230 + 0.9% on balance over	600,000
1,500,000 - 3,000,000	18,330 + 0.8% on balance over	1,500,000
3,000,000 - 6,000,000	30,330 + 0.7% on balance over	3,000,000
Over 6,000,000	51,330 + 0.6% on balance over	6,000,000

THE ROYAL INSTITUTION OF CHARTERED SURVEYORS (Cont'd)
SCALE NO. 37 (Cont'd)
BILLS OF QUANTITIES (Cont'd)

2.1 (d) The scales of fees for preparing bills of quantities (paragraphs 2.1 (a) to (c)) are overall scales based upon the inclusion of all provisional and prime cost items, subject to the provision of paragraph 2.1 (g). When work normally included in a building contract is the subject of a separate contract for which the quantity surveyor has not been paid fees under any other clause hereof, the value of such work shall be included in the amount upon which fees are charged.

(e) Fees shall be calculated upon the accepted tender for the whole of the work subject to the provisions of paragraph 2.6. In the event of no tender being accepted, fees shall be calculated upon the basis of the lowest original bona fide tender received. In the even of no such tender being received, the fees shall be calculated upon a reasonable valuation of the works based upon the original bills of quantities.

Note: In the foregoing context "bona fide tender" shall be deemed to mean a tender submitted in good faith without major errors of computation and not subsequently withdrawn by the tenderer.

(f) In calculating the amount upon which fees are charged the total of any credits and the totals of any alternative bills shall be aggregated and added to the amount described above. The value of any omission or addition forming part of an alternative bill shall not be added unless measurement or abstraction from the original dimension sheets was necessary.

(g) Where the value of the air conditioning, heating, ventilating and electrical services included in the tender documents together exceeds 25% of the amount calculated as described in paragraphs 2.1

(d) and (e), then, subject to the provisions of paragraph 2.2, no fee is chargeable on the amount by which the value of these services exceeds the said 25% in this context the term "value" excludes general contractor's profit, attendance, builder's work in connection with the services, preliminaries and any similar additions.

(h) When a contract comprises buildings which fall into more than one category, the fee shall be calculated as follows:

(i) The amount upon which fees are chargeable shall be allocated to the categories of work applicable and the amounts so allocated expressed as percentages of the total amount upon which fees are chargeable.
(ii) Fees shall then be calculated for each category on the total amount upon which fees are chargeable.
(iii) The fee chargeable shall then be calculated by applying the percentages of work in each category to the appropriate total fee and adding the resultant amounts.

(j) When a project is the subject of a number of contracts then, for the purpose of calculating fees, the values of such contracts shall not be aggregated but each contract shall be taken separately and the scale of charges (paragraphs 2.1 (a) to (h)) applied as appropriate.

(k) Where the quantity surveyor is specifically instructed to provide cost planning services the fee calculated in accordance with paragraphs 2.1 (a) to (j) shall be increased by a sum calculated in accordance with the following table and based upon the same value of work as that upon which the aforementioned fee has been calculated:

2.1 (k)   CATEGORIES A & B: (as defined in paragraphs 2.1 (a) and (b)).

Value of Work		Fee	
	£	£	£
Up to	600,000	0.7%	
600,000	- 3,000,000	4,200 + 0.4% on balance over	600,000
3,000,000	- 6,000,000	13,800 + 0.35% on balance over	3,000,000
Over	6,000,000	24,300 + 0.3% on balance over	6,000,000

CATEGORY C: (as defined in paragraph 2.1 (c)).

Value of Work		Fee	
	£	£	£
Up to	600,000	0.5%	
600,000	- 3,000,000	3,000 + 0.3% on balance over	600,000
3,000,000	- 6,000,000	10,200 + 0.25% on balance over	3,000,000
Over	6,000,000	17,700 + 0.2% on balance over	6,000,000

THE ROYAL INSTITUTION OF CHARTERED SURVEYORS (Cont'd)
SCALE NO. 37 (Cont'd)
BILLS OF QUANTITIES (Cont'd)

2.2 Air Conditioning, Heating, Ventilating and Electrical Services

(a) Where bills of quantities are prepared by the quantity surveyor for the air conditioning, heating, ventilating and electrical services there shall be a fee for these services (which shall include examining tenders received and reporting thereon), in addition to the fee calculated in accordance with paragraph 2.1 as follows:

Value of Work			Additional Fee
	£	£	£
Up to	120,000	2.5%	
120,000 -	240,000	3,000 + 2.25% on balance over	120,000
240,000 -	480,000	5,700 + 2.0% on balance over	240,000
480,000 -	750,000	10,500 + 1.75% on balance over	480,000
750,000 -	1,000,000	15,225 + 1.25% on balance over	750,000
Over	1,000,000	18,350 + 1.15% on balance over	1,000,000

(b) The values of such services, whether the subject of separate tenders or not, shall be aggregated and the total value of work so obtained used for the purpose of calculating the additional fee chargeable in accordance with paragraph (a).

(Except that when more than one firm of consulting engineers is engaged on the design of these services, the separate values for which each such firm is responsible shall be aggregated and the additional fees charged shall be calculated independently on each such total value so obtained).

(c) Fees shall be calculated upon the accepted tender for the whole of the air conditioning, heating, ventilating and electrical services for which bills of quantities have been prepared by the quantity surveyor. In the event of no tender being accepted, fees shall be calculated upon the basis of the lowest original bona fide tender received. In the event of no such tender being received, the fees shall be calculated upon a reasonable valuation of the services based upon the original bills of quantities.

Note: In the foregoing context "bona fide tender" shall be deemed to mean a tender submitted in good faith without major errors of computation and not subsequently withdrawn by the tenderer.

(d) When cost planning services are provided by the quantity surveyor for air conditioning, heating, ventilating and electrical services (or for any part of such services) there shall be an additional fee based on the time involved (see paragraphs 19.1 and 19.2). Alternatively the fee may be on a lump sum or percentage basis agreed between the employer and the quantity surveyor.

Note: The incorporation of figures for air conditioning, heating, ventilating and electrical services provided by the consulting engineer is deemed to be included in the quantity surveyor's services under paragraph 2.1.

2.3 Works of Alteration

On works of alteration or repair, or on those sections of the works which are mainly works of alteration or repair, there shall be a fee of 1.0% in addition to the fee calculated in accordance with paragraphs 2.1 and 2.2.

2.4 Works of Redecoration and Associated Minor Repairs On works of redecoration and associated minor repairs, there shall be a fee of 1.5% in addition to the fee calculated in accordance with paragraphs 2.1 and 2.2.

2.5 Bills of Quantities Prepared in Special Forms

Fees calculated in accordance with paragraphs 2.1, 2.2, 2.3 and 2.4 include for the preparation of bills of quantities on a normal trade basis. If the employer requires additional information to be provided in the bills of quantities or the bills to be prepared in an elemental, operational or similar form, then the fee may be adjusted by agreement between the employer and the quantity surveyor.

2.6 Action of Tenders

(a) When cost planning services have been provided by the quantity surveyor and a tender, when received, is reduced before acceptance and if the reductions are not necessitated by amended instructions of the employer or by the inclusion in the bills of quantities of items which the quantity surveyor has indicated could not be contained within the approved estimate, then in such a case no charge shall be made by the quantity surveyor for the preparation of bills of reductions and the fee for the preparation of the bills of quantities shall be based on the amount of the reduced tender.

(b) When cost planning services have not been provided by the quantity surveyor and if a tender, when received, is reduced before acceptance, fees are to be calculated upon the amount of the unreduced tender. When the preparation of bills of reductions is required, a fee is chargeable for preparing such bills of reductions as follows:

(i) 2.0% upon the gross amount of all omissions requiring measurement or abstraction from original dimension sheets.

(ii) 3.0% upon the gross amount of all additions requiring measurement.

(iii) 0.5% upon the gross amount of all remaining additions.

Note: The above scale for the preparation of bills of reductions applies to work in all categories.

2.7 Generally

If the works are substantially varied at any stage or if the quantity surveyor is involved in an excessive amount of abortive work, then the fees shall be adjusted by agreement between the employer and the quantity surveyor.

THE ROYAL INSTITUTION OF CHARTERED SURVEYORS (Cont'd)
SCALE NO. 37 (Cont'd)

3.0 NEGOTIATING TENDERS

3.1 (a) For negotiating and agreeing prices with a contractor:

Value of Work		Fee	
	£	£	£
Up to	150,000	0.5%	
150,000 -	600,000	750 + 0.3% on balance over	150,000
600,000 -	1,200,000	2,100 + 0.2% on balance over	600,000
Over	1,200,000	3,300 + 0.1% on balance over	1,200,000

3.1 (b) The fee shall be calculated on the total value of the works as defined in paragraphs 2.1 (d), (e), (f), (g) and (j).

(c) For negotiating and agreeing prices with a contractor for air conditioning, heating, ventilating and electrical services there shall be an additional fee as paragraph 3.1 (a) calculated on the total value of such services as defined in paragraph 2.2 (b).

4.0 CONSULTATIVE SERVICES AND PRICING BILLS OF QUANTITIES

4.1 Consultative Services

Where the quantity surveyor is appointed to prepare approximate estimates, feasibility studies or submissions for the approval of financial grants or similar services, then the fee shall be based on the time involved (see paragraphs 19.1 and 19.2) or,

alternatively, on a lump sum or percentage basis agreed between the employer and the quantity surveyor.

4.2 Pricing Bills of Quantities

(a) For pricing bills of quantities, if instructed, to provide an estimate comparable with tenders, the fee shall be one-third (33 1/3%) of the fee for negotiating and agreeing prices with a contractor, calculated in accordance with paragraphs 3.1 (a) and (b).

(b) For pricing bills of quantities, if instructed, to provide an estimate comparable with tenders for air conditioning, heating, ventilating and electrical services the fee shall be one-third (33 1/3%) of the fee calculated in accordance with paragraph 3.1 (c).

CONTRACTS BASED ON BILLS OF QUANTITIES: POST-CONTRACT SERVICES

Alternative Scales (I and II) for post-contract services are set out below to be used at the quantity surveyor's discretion by prior agreement with the employer.

ALTERNATIVE 1

5.0 OVERALL SCALE OF CHARGES FOR POST-CONTRACT SERVICES

5.1 If the quantity surveyor appointed to carry out the post-contract services did not prepare the bills of quantities then the fees in paragraphs 5.2 and 5.3 shall be increased to cover the additional services undertaken by the quantity surveyor.

5.2 Basic Scale: For taking particulars and reporting valuations for interim certificates for payments on account to the contractor, preparing periodic assessments of anticipated final cost and reporting thereon, measuring and making up bills of variations including pricing and agreeing totals with the contractor, and adjusting fluctuations in the cost of labour and materials if required by the contract.

(a) CATEGORY A: Relatively complex works and/or works with little or no repetition.

Examples:
Ambulance and fire stations; banks; cinemas; clubs; computer buildings; council offices; crematoria; fitting out of existing buildings; homes for the elderly; hospitals and nursing homes; laboratories; law courts; libraries; 'one-off' houses; petrol stations; places of religious worship; police stations; public houses, licensed premises; restaurants; sheltered housing; sports pavilions; theatres; town halls; universities, polytechnics and colleges of further education (other than halls of residence and hostels); and the like.

Value of Work		Category A Fee	
	£	£	£
Up to 150,000		150 + 2.0% (minimum fee £1,150)	
150,000 -	300,000	3,150 + 1.7% on balance over	150,000
300,000 -	600,000	5,700 + 1.6% on balance over	300,000
600,000 -	1,500,000	10,500 + 1.3% on balance over	600,000
1,500,000 -	3,000,000	22,200 + 1.2% on balance over	1,500,000
3,000,000 -	6,000,000	40,200 + 1.1% on balance over	3,000,000
Over	6,000,000	73,200 + 1.0% on balance over	6,000,000

THE ROYAL INSTITUTION OF CHARTERED SURVEYORS (Cont'd)
SCALE NO. 37 (Cont'd)
CONTRACTS BASED ON BILLS OF QUANTITIES: POST-CONTRACT SERVICES (Cont'd)
ALTERNATIVE 1 (Cont'd)
OVERALL SCALE OF CHARGES FOR POST-CONTRACT SERVICES (Cont'd)

(b) CATEGORY B: Less complex works and/or works with some element of repetition.

Examples:
Adult education facilities; canteens; church halls; community centres; departmental stores; enclosed sports stadia and swimming baths; halls of residence; hostels; motels; offices other than those included in Categories A and C; railway stations; recreation and leisure centres; residential hotels; schools; self-contained flats and maisonettes; shops and shopping centres; supermarkets and hypermarkets; telephone exchanges; and the like.

Value of Work		Category B Fee	
£		£	£
Up to 150,000		150 + 2.0% (minimum fee £1150)	
150,000	- 300,000	3,150 + 1.7% on balance over	150,000
300,000	- 600,000	5,700 + 1.5% on balance over	300,000
600,000	- 1,500,000	10,200 + 1.1% on balance over	600,000
1,500,000	- 3,000,000	20,100 + 1.0% on balance over	1,500,000
3,000,000	- 6,000,000	35,100 + 0.9% on balance over	3,000,000
Over	6,000,000	62,100 + 0.8% on balance over	6,000,000

(c) CATEGORY C: Simple works and/or works with a substantial element of repetition.

Examples:
Factories; garages; multi-storey car parks; open-air sports stadia; structural shell offices not fitted out; warehouses; workshops; and the like.

Value of Work		Category C Fee	
£		£	£
Up to 150,000		120 + 1.6% (minimum fee £920)	
150,000	- 300000	2,520 + 1.5% on balance over	150,000
300,000	- 600,000	4,770 + 1.4% on balance over	300,000
600,000	- 1,500,000	8,970 + 1.1% on balance over	600,000
1,500,000	- 3,000,000	18,870 + 0.9% on balance over	1,500,000
3,000,000	- 6,000,000	32,370 + 0.8% on balance over	3,000,000
Over	6,000,000	56,370 + 0.7% on balance over	6,000,000

(d) The scales of fees for post-contract services (paragraphs 5.2 (a) and (c) are overall scales, based upon the inclusion of all nominated sub-contractors' and nominated suppliers' accounts, subject to the provision of paragraph 5.2 (g). When work normally included in a building contract is the subject of a separate contract for which the quantity surveyor has not been paid fees under any other clause hereof, the value of such work shall be included in the amount on which fees are charged.

(e) Fees shall be calculated upon the basis of the account for the whole of the work, subject to the provisions of paragraph 5.3.

(f) In calculating the amount on which fees are charged the total of any credits is to be added to the amount described above.

(g) Where the value of air conditioning, heating, ventilating and electrical services included in the tender documents together exceeds 25% of the amount calculated as described in paragraphs 5.2 (d) and (e) above, then, subject to the provisions of paragraph 5.3, no fee is chargeable on the amount by which the value of these services exceeds the said 25%. In this context the term "value" excludes general contractor's profit, attendance, builders work in connection with the services, preliminaries and any other similar additions.

(h) When a contract comprises buildings which fall into more than one category, the fee shall be calculated as follows:

(i) The amount upon which fees are chargeable shall be allocated to the categories of work applicable and the amounts so allocated expressed as percentages of the total amount upon which fees are chargeable.

(ii) Fees shall then be calculated for each category on the total amount upon which fees are chargeable.

(iii) The fee chargeable shall then be calculated by applying the percentages of work in each category to the appropriate total fee and adding the resultant amounts.

(j) When a project is the subject of a number of contracts then, for the purposes of calculating fees, the values of such contracts shall not be aggregated but each contract shall be taken separately and the scale of charges (paragraphs 5.2 (a) to (h)) applied as appropriate.

(k) When the quantity surveyor is required to prepare valuations of materials or goods off site, an additional fee shall be charged based on the time involved (see paragraphs 19.1 and 19.2).

THE ROYAL INSTITUTION OF CHARTERED SURVEYORS (Cont'd)

SCALE NO. 37 (Cont'd)

OVERALL SCALE OF CHARGES FOR POST-CONTRACT SERVICES (Cont'd)

Air Conditioning, Heating, Ventilating and Electrical Services (Cont'd)

(l) The basic scale for post-contract services includes for a simple routine of periodically estimating final costs. When the employer specifically requests a cost monitoring service which involves the quantity surveyor in additional or abortive measurement an additional fee shall be charged based on the time involved (see paragraphs 19.1 and 19.2), or alternatively on a lump sum or percentage basis agreed between the employer and the quantity surveyor.

(m) The above overall scales of charges for post contract services assume normal conditions when the bills of quantities are based on drawings accurately depicting the building work the employer requires. If the works are materially varied to the extent that substantial remeasurement is necessary then the fee for post-contract services shall be adjusted by agreement between the employer and the quantity surveyor.

5.3 Air Conditioning, Heating, Ventilating and Electrical Services

(a) Where final accounts are prepared by the quantity surveyor for the air conditioning, heating, ventilating and electrical services there shall be a fee for these services, in addition to the fee calculated in accordance with paragraph 5.2, as follows:

Value of Work		Additional Fee	
	£	£	£
Up to	120,000	2.0%	
120,000 -	240,000	2,400 + 1.6% on balance over	120,000
240,000 -	1,000,000	4,320 + 1.25% on balance over	240,000
1,000,000 -	4,000,000	13,820 + 1.0% on balance over	1,000,000
Over	4,000,000	43,820 + 0.9% on balance over	4,000,000

(b) The values of such services, whether the subject of separate tenders or not, shall be aggregated and the total value of work so obtained used for the purpose of calculating the additional fee chargeable in accordance with paragraph (a).

(Except that when more than one firm of consulting engineers is engaged on the design of these services the separate values for which each such firm is responsible shall be aggregated and the additional fee charged shall be calculated independently on each such total value so obtained).

(c) The scope of the services to be provided by the quantity surveyor under paragraph (a) above shall be deemed to be equivalent to those described for the basic scale for post-contract services.

(d) When the quantity surveyor is required to prepare periodic valuations of materials or goods off site, an additional fee shall be charged based on the time involved (see paragraphs 19.1 and 19.2).

(e) The basic scale for post-contract services includes for a simple routine of periodically estimating final costs. When the employer specifically requests a cost monitoring service which involves the quantity surveyor in additional or abortive measurement an additional fee shall be charged based on the time involved (see paragraph 19.1 and 19.2), or alternatively on a lump sum or percentage basis agreed between the employer and the quantity surveyor.

(f) Fees shall be calculated upon the basis of the account for the whole of the air conditioning, heating, ventilating and electrical services for which final accounts have been prepared by the quantity surveyor.

Scale 37 Alternative II; - Separate Stages of Post Contract Services and Scales 38 - 47 are not reproduced here.

V. B. Johnson LLP provide full Quantity Surveying services as indicated above. Please do not hesitate to contact us at our Watford office – 01923 227236 or Wakefield office – 01924 897373 or Hull office - 01482 492191 for more specific advice on the provision of Quantity Surveying services.

GENERAL INFORMATION

FEES

BUILDING CONTROL CHARGES from 1st June 2013

Notes

Before you build, extend or convert, you or your agent must advise your local authority either by submitting Full Plans or a Building Notice. The fee payable depends on the type of work. The number of dwellings in a building and the total floor area. The following tables may be used in conjunction with the Fees Regulations as a guide to calculating the fees.

Should you submit Full Plans you will pay a plan fee at the time of submission to cover their passing or rejection.

With Full Plans submissions, for most types of work, an inspection fee covering all necessary site visits will be payable following the first inspection.

Should you submit a Building Notice, the appropriate Building Notice fee is payable at the time of submission and covers all necessary checks and site visits. The Building Notice fee is equivalent to the sum of the relevant plan fee and inspection fee.

Total estimated cost means an estimate accepted by the local authority of a reasonable cost that would be charged by a person in business to carry out the work shown or described in the application excluding VAT and any fees paid to an architect, engineer or surveyor etc. and also excluding land acquisition costs.

A supplementary charge will apply if electrical works, that are controlled under part P of the regulations, are not carried out by a competent person to self certify the work.

For further advice and to check current fees consult your local authority.

FULL PLAN AND INSPECTION CHARGES

Table 1 – New Dwellings (new build and conversions)

No. of dwellings Description	Full Plans		Building Notice	Regularisation Charge
	Plan Charge £	Inspection Fee £	£	£
Creation of 1 new dwelling.	300.00	516.00	816.00	884.00
Creation of 2 new dwellings.	300.00	679.20	979.20	1,060.08
Creation of 3 new dwellings.	300.00	875.04	1,175.04	1,272.96
Creation of 4 new dwellings.	300.00	1,070.88	1,370.88	1,485.12
Creation of 5 new dwellings.	300.00	1,266.72	1,566.72	1,697.28
Over 5 dwellings.	Subject to an individually determined charge			
Conversion of dwelling into 2 flats	300.00	352.80	652.80	707.2
Conversion of dwelling into 3 flats	300.00	420.72	750.72	813.28
Over 3 flats	Subject to an individually determined charge			
Inclusive of VAT at 20% New dwellings over 300m2 subject to an individually determined charge Regularisation Charge not subject to VAT For further information consult the Charges Scheme				

Table 2 – Domestic Extensions and Alterations

Domestic extensions, garages and non-exempt outbuildings

Description	Full Plans		Building Notice Charge £	Regularisation Charge £
	Plan Charge £	Inspection Charge £		
Extension Internal floor area under 10m^2	200.00	289.6.00	489.60	530.40
Extension over 10m^2 and under 40m^2	200.00	387.52	587.52	636.48
Extension over 40m^2 and under 100m^2	200.00	648.64	848.64	919.36
Extensions where the total floor area exceeds 100m^2	Subject to an individually determined charge			
Loft conversion under 100m^2	200.00	387.52	587.52	636.48
Loft conversion where the total floor area exceeds 100m^2	Subject to an individually determined charge			
Carport, attached garage, detached garage and outbuildings under 100m^2	200.00	159.04	359.04	388.96
Garage, conversion	200.00	159.04.00	359.04	388.96

Certain garages, carport (and conservatories) under 30m^2 may be exempt buildings
'Extension' includes attached garages and non exempt conservatories
Inclusive of VAT at 20%
Regularisation charges are not subject to VAT
For further information contact building control

BUILDING CONTROL CHARGES from 1st June 2013 (Cont'd)
FULL PLAN AND INSPECTION CHARGES (Cont'd)
Table 2 – Domestic Extensions and Alterations (Cont'd)

Domestic alterations

Description	Full Plans		Building Notice Charge £	Regularisation Charge £
	Plan Charge £	Inspection Charge £		
Replacement roof covering			195.84	212.16
Alterations (including chimney breast removal) up to £5,000 estimated cost of work	200.00	61.12	261.12	282.88
Alterations over £5,000 and up to £10,000 estimated cost of work	200.00	126.40	326.40	420.00
Alterations over £10,000 estimated cost of work	Subject to an individually determined charge			
Installation of replacement windows or external doors			195.84	200.00
Electrical works up to £5,000 estimated cost of work			277.92	414.08
Installation of solar panels			130.56	141.44
Installation of a solid fuel stove			195.84	212.16
Installation of a gas fired boiler				340.08
Installation of an oil fired boiler.				402.48
Inclusive of VAT at 20% Regularisation Charge not subject to VAT For further information consult the charges scheme				

Table 3 – Non-domestic Extensions and Alterations

Non-domestic Extensions

Description	Residential and Institutional		
	Full plans		Regularisation Charge £
	Plan Charge £	Inspection Charge £	
Extension under 10m^2	200.00	289.60	530.40
Extension over 10m^2 and under 40m^2	200.00	387.52	636.48
Extension over 40m^2 and under 100m^2	200.00	648.64	919.36

Description	Assembly and Recreational		
	Full plans		Regularisation Charge £
	Plan Charge £	Inspection Charge £	
Extension under 10m^2	200.00	289.60	530.40
Extension over 10m^2 and under 40m^2	200.00	387.52	636.48
Extension over 40m^2 and under 100m^2	200.00	648.64	919.36

Description	Industrial and Storage		
	Full plans		Regularisation Charge £
	Plan Charge £	Inspection Charge £	
Extension under 10m^2	200.00	289.60	530.40
Extension over 10m^2 and under 40m^2	200.00	387.52	636.48
Extension over 40m^2 and under 100m^2	200.00	648.64	919.36

GENERAL INFORMATION

FEES

BUILDING CONTROL CHARGES from 1st June 2011 (Cont'd)
FULL PLAN AND INSPECTION CHARGES (Cont'd)
Table 3 – Non-domestic Extensions and Alterations (Cont'd)

Non-domestic Alterations

| Description | Residential and Institutional | | |
| | Full plans | | |
	Plan Charge £	Inspection Charge £	Regularisation Charge £
Internal alterations up to £5,000 estimated cost of work	200.00	61.12	282.88
Internal alterations over £5,000 and up to £10,000 estimated cost of work	200.00	126.40	353.60
Installation of mezzanine floor up to 100m^2	200.00	159.04	388.96

| Description | Assembly and Recreational | | |
| | Full plans | | |
	Plan Charge £	Inspection Charge £	Regularisation Charge £
Internal alterations up to £5,000 estimated cost of work	200.00	61.12	282.88
Internal alterations over £5,000 and up to £10,000 estimated cost of work	200.00	126.40	353.60
Installation of mezzanine floor up to 100m^2	200.00	159.04	388.96

| Description | Industrial and Storage | | |
| | Full plans | | |
	Plan Charge £	Inspection Charge £	Regularisation Charge £
Internal alterations up to £5,000 estimated cost of work	200.00	61.12	282.88
Internal alterations over £5,000 and up to £10,000 estimated cost of work	200.00	126.40	353.60
Installation of mezzanine floor up to 100m^2	200.00	159.04	388.96

GENERAL INFORMATION

CONSTRUCTION (DESIGN AND MANAGEMENT) REGULATIONS 2015

CONSTRUCTION (DESIGN AND MANAGEMENT) REGULATIONS 2015

The Construction (Design and Management) Regulations 2015 came into force on 6th April 2015. The CDM Regulations and Amendments provide a framework in health and safety terms for the design, construction, maintenance and demolition of a structure.

If you have not followed the relevant provisions of the Regulations you may be prosecuted for breach of health and safety law.

The following is an outline of the requirements of the Regulations. No attempt has been made to cover all the detail and readers are advised to study the Regulations and Guidance in full.

CDM 2015 is divided into five parts:
* Part 1 deals with the application of CDM 2015 and definitions
* Part 2 covers the duties of clients for all construction projects. These duties apply in full for commercial clients.
* Part 3 covers the health and safety duties and roles of other dutyholders including designers, principal designers, principal contractors and contractors
* Part 4 contains general requirements for all construction sites
* Part 5 contains transitional arrangements and revocations

The regulations place duties upon clients, client's agents, principal designers, designers, principal contractors, contractors and subcontractors so that health and safety is taken into account and then co-ordinated and managed effectively throughout all stages of a construction project; from conception to execution of works on site and subsequent maintenance and repair and demolitions.

Key elements

Key elements to securing construction health and safety include:
(a) managing the risks by applying the general principles of prevention;
(b) appointing the right people and organisations at the right time;
(c) making sure everyone has the information, instruction, training and supervision they need to carry out their jobs in a way that secures health and safety;
(d) dutyholders cooperating and communicating with each other and coordinating their work
(e) consulting workers and engaging with them to promote and develop effective measures to secure health, safety and welfare.

General principles of prevention

These set out the principles dutyholders should use in their approach to identifying the measures they should take to control the risks to health and safety in a particular project
(a) avoid risks where possible;
(b) evaluate those risks that cannot be avoided
(c) put in place measures that control them at source.
(d) adapt the work to the individual and technical progress
(e) replace the dangerous by the non or less-dangerous
(f) give appropriate instructions to employees.
Due regard must be made of
* The Health and Safety at Work etc Act 1974,
* Construction (Design and Management) Regulations 2015

* The Construction (Health, Safety and Welfare) Regulations 1996
* Lifting Operations and Lifting Equipment Regulations 1998
* Health and Safety (Consultation with Employees) Regulations 1996
* Control of Substances Hazardous to Health Regulations 2002
* Electricity at Work Regulations 1989
* Regulatory Reform (Fire Safety) Order 2005
* Control of Noise at Work Regulations 2005
* The Control of Asbestos Regulations 2012,
* Management of Health and Safety at Work Regulations 1999
* The Workplace (Health, Safety and Welfare) Regulations 1992
* Control of Vibration at Work Regulations 2005,
* The Work at Height Regulations 2005,
* Provision and Use of Work Equipment Regulations 1998,
* Building Regulations 2010,
* The Occupiers' Liability Acts 1957 and 1984

Notifiable and Non-notifiable Projects

A project is notifiable that lasts more than 30 days and have more than 20 workers at any point in the project or will involve more than 500 person days, other projects are non-notifiable and do not require the appointment of a project co-ordinator. For such non-notifiable projects clients, designers and contractors will still have responsibilities and duties.
The client must submit the notice as soon as practicable before the construction phase begins, the client may request someone else do this. The notice must be displayed in the site office
The requirements of CDM 2015 also apply whether or not the project is notifiable.
The regulations do not impose duties on householders having construction work carried out on their own home however designers and contractors still have duties and domestic clients will have duties if they control the way in which construction work is carried out .

DUTIES AND DUTY HOLDERS

General duties

A designer or contractor must have the skills, knowledge, experience and organisational capability necessary to fulfil the role that they undertake in a manner that secures the health and safety of any person affected by the project.
A person who is responsible for appointing a designer or contractor to carry out work on a project must take reasonable steps to satisfy themselves that the designer or contractor fulfils the above conditions
A person with a duty or function under these Regulations must cooperate with any other person working on or in relation to a project to enable any person with a duty or function to fulfil that duty or function
A person working on a project under the control of another must report to that person anything they are aware of in relation to the project which is likely to endanger their own health or safety or that of others.
Anyone with a duty under CDM 2015 to provide health and safety information or instructions to anyone else must ensure that it is easy to understand

CONSTRUCTION (DESIGN AND MANAGEMENT) REGULATIONS 2015 (Cont'd)
DUTIES AND DUTY HOLDERS (Cont'd)

THE CLIENT
Organisations or individuals for whom a construction project is carried out.

Duties

Project Management
Make arrangements for managing a project to include:-
(a) assembling the project team – appointing designers and contractors.
(b) ensuring the roles, functions and responsibilities of the project team are clear;
(c) ensuring sufficient resources and time are allocated for each stage of the project – from concept to completion;
(d) ensuring effective mechanisms are in place for members of the project team to communicate and cooperate with each other and coordinate their activities;
(e) reasonable steps to ensure that the principal designer and principal contractor comply with their separate duties. This could take place at project progress meetings or via written updates;
(f) setting out the means to ensure that the health and safety performance of designers and contractors is maintained throughout;
(g) ensuring that workers are provided with suitable welfare facilities for the duration of construction work.

Client's Brief
Clients should prepare a clear 'client's brief' as a way of setting out the arrangements which;-
(a) sets out the main function and operational requirements of the finished project;
(b) outlines how the project is expected to be managed including its health and safety risks;
(c) sets a realistic timeframe and budget; and
(d) covers other relevant matters, such as establishing design direction and a single point of contact in the client's organisation.

Risks
Where risks involved in the work warrants, the management arrangements should also include:
(a) the expected standards of health and safety, including safe working practices,
(b) what is expected from the design team in terms of the steps they should reasonably take to ensure their designs help manage foreseeable risks during the construction phase and when maintaining and using the building once it is built;
(c) the arrangements for commissioning the new building and a well-planned handover procedure to the new user.
The principal designer should be in a position to help in making these arrangements. Clients could also draw on the advice of a competent person under the Management of Health and Safety at Work Regulations 1999

Pre-construction information
Pre-construction information is information already in the client's possession.
The client has the main duty for providing pre-construction information. This must be provided as soon as practicable to each designer and contractor

The construction phase plan
The client must ensure that a construction phase plan for the project is prepared before the construction phase begins. The plan outlines the health and safety arrangements, site rules and specific measures concerning any work involving the particular risks

The health and safety file
A health and safety file is only required for projects involving more than one contractor.
The client must ensure that the principal designer prepares a health and safety file for their project. Its purpose is to ensure that, at the end of the project, the client has information that anyone carrying out subsequent construction work on the building will need to know about in order to be able to plan and carry out the work safely and without risks to health

DOMESTIC CLIENTS
People who have construction work carried out on their own home, or the home of a family member that is not done as part of a business, whether for profit or not.

Duties
Domestic clients are within the scope of CDM 2015, but their duties as a client are normally transferred to:
■ the contractor, on a single contractor project; or;
■ the principal contractor, on a project involving more than one contractor.
However, the domestic client can choose to have a written agreement with the principal designer to carry out the client duties.

DESIGNERS
An organisation or individual, who:
(a) prepares or modifies a design for a construction project or
(b) arranges for, or instructs someone else to do so.
The term 'design' includes drawings, design details, specifications, bills of quantity and calculations prepared for the purpose of a design. Designers include architects, architectural technologists, consulting engineers, quantity surveyors, interior designers, temporary work engineers, chartered surveyors, technicians or anyone who specifies or alters a design. They can include others if they carry out design work, such as principal contractors, specialist contractors and commercial clients.

Duties
A designer must not commence work in relation to a project unless satisfied that the client is aware of the duties owed by the client under these Regulations.
When preparing or modifying a design the designer must take into account the general principles of prevention and any pre-construction information to eliminate foreseeable risks to the health or safety of any person—
(a) carrying out or liable to be affected by construction work;
(b) maintaining or cleaning a structure; or
(c) using a structure designed as a workplace
If it is not possible to eliminate these risks, the designer must, so far as is reasonably practicable—
(a) take steps to reduce or, if that is not possible, control the risks through the subsequent design process;
(b) provide information about those risks to the principal designer
(c) ensure appropriate information is included in the health and safety file.
A designer must take steps to provide, with the design, sufficient information about the design, construction or maintenance of the structure, to assist the client, other designers and contractors to comply with these Regulations.
The designer should agree with the principal designer the arrangements for sharing information to avoid omissions or duplicated effort.
Liaise with any others so that work can be coordinated to establish how different aspects of designs interact and influence health and safety. This includes temporary and permanent works designers, contractors and principal contractors so that their knowledge and experience is taken into account.

Working for domestic clients
A designer's role on a project for a domestic client is no different to the role undertaken for commercial clients.

CONSTRUCTION (DESIGN AND MANAGEMENT)
REGULATIONS 2015 (Cont'd)
DUTIES AND DUTY HOLDERS (Cont'd)

PRINCIPAL DESIGNER
Designers appointed by the client in projects involving more than one contractor. They can be an organisation or an individual with sufficient knowledge, experience and ability to carry out the role. Principal designers may have separate duties as designers
Principal designers are not a direct replacement for CDM co-ordinators. The range of duties they carry out is different to those undertaken by CDM co-ordinators

Duties

Pre-construction phase
Plan, manage, monitor and coordinate health and safety. This includes: identifying, eliminating or controlling foreseeable risks;
ensuring designers carry out their duties.
take into account the general principles of prevention and, the content of any construction phase plan and health and safety file.
ensure that all persons working in relation to the pre-construction phase cooperate with the client, the principal designer and each other.
assist the client in the provision of the pre-construction information
Prepare and provide relevant information to other dutyholders.
Provide relevant information to the principal contractor to help them plan, manage, monitor and coordinate health and safety in the construction phase.
A principal designer's role on a project for a domestic client is no different to the role undertaken for commercial clients.

Construction phase plan and health and safety file
The principal designer must assist the principal contractor in preparing the construction phase plan by providing to the principal contractor all information the principal designer holds that is relevant to the construction phase plan including;-
(a) pre-construction information obtained from the client;
(b) any information obtained from designers
During the pre-construction phase, the principal designer must prepare a health and safety file appropriate to the characteristics of the project which must contain information relating to the project which is likely to be needed during any subsequent project to ensure the health and safety of any person
If the principal designer's appointment concludes before the end of the project, the principal designer must pass the health and safety file to the principal contractor.
The health and safety file is only required for projects involving more than one contractor

Construction phase
The principal designer must ensure that the health and safety file is appropriately reviewed, updated and revised from time to time to take account of the work and any changes that have occurred
At the end of the project, the principal designer, or where there is no principal designer the principal contractor, must pass the health and safety file to the client.

PRINCIPAL CONTRACTOR
Contractors appointed by the client before the construction phase begins to coordinate the construction phase of a project where it involves more than one contractor.
Contractors must not carry out any construction work on a project unless they are satisfied that the client is aware of the duties the client has under CDM 2015.

Duties

Planning, managing and monitoring construction work
Plan, manage, monitor and coordinate health and safety in the construction phase of a project. This includes:
liaising with the client and principal designer;
preparing the construction phase plan;

organising cooperation between contractors and coordinating their work.
Ensure: suitable site inductions are provided;
steps are taken to prevent unauthorised access;
workers are consulted and engaged in securing their health and safety
welfare facilities are provided.
ensure anyone they appoint has the skills, knowledge, and experience and, where they are an organisation, the organisational capability to carry out the work in a way that secures health and safety
A systematic approach to managing should be taken to ensure workers understand:
(a) the risks and control measures on the project;
(b) who has responsibility for health and safety;
(c) that consistent standards apply throughout the project and will be checked frequently;
(d) where they can locate health and safety information which is understandable, organised and relevant to the site;
(e) that incidents will be investigated and lessons learned.
Special consideration will be required for sites that have:
(a) rights of way through them;
(b) other work areas next to them
(c) occupied houses next to them
(d) children or vulnerable people nearby.

Providing welfare facilities
The principal contractor must ensure that suitable and sufficient welfare facilities are provided and maintained throughout the construction phase. Facilities must be made available before any construction work starts and should be maintained until the end of the project.
The principal contractor should liaise with other contractors involved with the project to ensure appropriate welfare facilities are provided. Such liaison should continue for the duration of the construction phase and take account of any changes in the nature of the site.

Liaising with the principal designer
The principal contractor must liaise with the principal designer for the duration of the project. The early appointment of a principal contractor by the client will allow their construction expertise to be used from the earliest stages of designing and planning a project. They should also liaise with the principal designer throughout the construction phase on matters such as changes to the designs and the implications these changes may have for managing the health and safety risks.
Liaison should cover drawing together information the principal designer will need:
(a) to prepare the health and safety file; or
(b) that may affect the planning and management of the pre-construction phase. The pre-construction information is important for planning and managing this phase and the subsequent development of the construction phase plan
A principal contractor's role on a project for a domestic client is no different to the role undertaken for commercial clients.

Consult and engage with workers
The principal contractor must;-
(a) make and maintain arrangements which will enable the principal contractor and workers engaged in construction work to cooperate in developing, promoting and checking the effectiveness of measures to ensure the health, safety and welfare of the workers;
(b) consult those workers or their representatives in good time on matters connected with the project which may affect their health, safety or welfare,
(c) ensure that those workers or their representatives can inspect and take copies of any information which the principal contractor has, or which these Regulations require to be provided to the principal contractor, which relate to the health, safety or welfare of workers at the site, except any information
(i) the disclosure of which would be against national security;
(ii) which the principal contractor could not disclose without contravening a prohibition imposed by or under an enactment;

(iii) relating specifically to an individual, unless that individual has consented to its being disclosed;

(iv) the disclosure of which would, for reasons other than its effect on health, safety or welfare at work, cause substantial injury to the principal contractor's undertaking or, where the information was supplied to the principal contractor by another person, to the undertaking of that other person;

(v) obtained by the principal contractor for the purpose of bringing, prosecuting or defending any legal proceedings.

Construction phase plan

During the pre-construction phase, and before setting up a construction site, the principal contractor must draw up a construction phase plan or make arrangements for a construction phase plan to be drawn up.

Throughout the project the principal contractor must ensure that the construction phase plan is appropriately reviewed, updated and revised from time to time so that it continues to be sufficient to ensure that construction work is carried out, so far as is reasonably practicable, without risks to health or safety.

The construction phase plan must set out the arrangements for securing health and safety during the period construction work is carried out. These arrangements include site rules and any specific measures put in place where work involves one or more of listed risks

Health and safety file

During the project, the principal contractor must provide the principal designer with any information in the principal contractor's possession relevant to the health and safety file, for inclusion in the health and safety file.

Where the health and safety file is passed to the principal contractor, the principal contractor must ensure that the health and safety file is reviewed, updated and revised to take account of the work and any changes that have occurred.

At the end of the project where there is no principal designer the principal contractor, must pass the health and safety file to the client.

CONTRACTORS

Those who do the actual construction work and can be either an individual or a company.

Anyone who directly employs or engages construction workers or manages construction is a contractor. Contractors include sub-contractors, any individual, sole trader, self-employed worker, or business that carries out, manages or controls construction work. This also includes companies that use their own workforce to do construction work on their own premises.

The duties on contractors apply whether the workers under their control are employees, self-employed or agency workers.

Where contractors are involved in design work, including for temporary works, they also have duties as designers

Contractors must not carry out any construction work on a project unless they are satisfied that the client is aware of their duties under CDM 2015.

Duties

Plan, manage and monitor construction work under their control so that it is carried out without risks to health and safety.

For projects involving more than one contractor, coordinate their activities with others in the project team – comply with principal designer or principal contractor directions.

For single-contractor projects, prepare a construction phase plan.

Planning, managing and monitoring construction work

Contractors are required to plan, manage and monitor the construction work under their control so that it is carried out in a way that controls the risks to health and safety.

On projects involving more than one contractor, this will involve the contractor coordinating the planning, management and monitoring of their own work with that of the principal contractor, other contractors and the principal designer.

For single contractor projects, the arrangements to plan, manage and monitor the construction phase will normally be simpler.

Planning

In planning the work, the contractor must take into account the risks to those who may be affected.

On projects involving more than one contractor, each contractor must plan their own work so it is consistent with the project-wide arrangements.

On single contractor projects, the contractor is responsible for planning the construction phase and for drawing up the construction phase plan before setting up the construction site.

Managing

The arrangements for managing construction work must take into account the same issues as for principal contractors

Monitoring

The contractor should monitor their work to ensure that the health and safety precautions are appropriate, remain in place and are followed in practice.

On projects involving more than one contractor the contractor should provide the principal contractor with any relevant information that stems from their own monitoring.

Complying with directions and the construction phase plan

For projects involving more than one contractor, the contractor is required to comply with any directions to secure health and safety given to them by the principal designer or principal contractor.

Drawing up a construction phase plan

For single contractor projects, the contractor must ensure a construction phase plan is drawn up as soon as practicable before the construction site is set up.

Appointing and employing workers

Appointing workers

When a contractor employs or appoints an individual to work on a construction site, they should make enquiries to make sure the individual: has the skills, knowledge, training and experience to carry out the work they will be employed to do in a way that secures health and safety for anyone working on the site or is in the process of obtaining them.

Sole reliance should not be placed on industry certification cards or similar being presented to them as evidence that a worker has the right qualities. Nationally recognised qualifications can provide contractors with assurance that the holder has the skills, knowledge, training and experience to carry out the task(s)

Newly trained individuals need to be supervised and given the opportunity to gain positive experience of working in a range of conditions.

When appointing individuals who may be skilled but who do not have any formal qualifications, contractors may need to assess them in the working environment.

Training workers

Establish whether training is necessary for any worker.

This assessment should take account of the training required by other health and safety legislation as well as that needed to meet the requirements of CDM 2015.

Assessing training needs should be an ongoing process throughout the project.

Providing supervision

A contractor who employs or manages workers under their control must ensure that appropriate supervision is provided. The level of supervision provided will depend on the risks to health and safety involved, and the skills, knowledge, training and experience of the workers concerned.

Providing information and instructions

Contractors should provide employees and workers under their control with the information and instructions they need to carry out their work without risk to health and safety to include:

(a) suitable site induction where this has not been provided by the principal contractor.

CONSTRUCTION (DESIGN AND MANAGEMENT)
REGULATIONS 2015 (Cont'd)
DUTIES AND DUTY HOLDERS (Cont'd)
CONTRACTORS (Cont'd)
Providing information and instructions (Cont'd)

(b) the procedures to be followed in the event of serious and imminent danger to health and safety.
(ii) take account of the relevant requirements which set out provisions relating to emergency procedures, emergency routes and exits and fire detection and fire-fighting;
(c) information on the hazards on site relevant to their work, the risks associated with those hazards and the control measures put in place.

Preventing unauthorised access to the site
A contractor must not begin work on a construction site unless steps have been taken to prevent unauthorised access to the site.

Providing welfare facilities
Contractors are required to provide welfare facilities as far as is reasonably practicable. This duty only extends to the provision of welfare facilities for the contractor's own employees who are working on a construction site or anyone else working under their control. Facilities must be made available before any construction work starts and should be maintained until the end of the project.

Working for domestic clients
A contractor's role on a project for a domestic client is no different to the role undertaken for commercial clients.

WORKERS
The people who work for or under the control of contractors on a construction site.

Duties
They must: be consulted about matters which affect their health, safety and welfare;

take care of their own health and safety and others who may be affected by their actions;

report anything they see which is likely to endanger either their own or others' health and safety;

cooperate with their employer, fellow workers, contractors and other dutyholders.

PRE-CONSTRUCTION INFORMATION
Pre-construction information must include proportionate information about:
(a) the project, such as the client brief and key dates of the construction phase;
(b) the planning and management of the project such as the resources and time being allocated to each stage of the project and the arrangements to ensure there is cooperation between dutyholders and the work is coordinated;
(c) the health and safety hazards of the site, including design and construction hazards and how they will be addressed;
(d) any relevant information in an existing health and safety file.

CONSTRUCTION PHASE PLAN
During the pre-construction phase the principal contractor must draw up a construction phase plan. For single contractor projects the contractor must draw up a construction phase plan
The construction phase plan must set out the health and safety arrangements and site rules for the project during the period construction work is carried out
The principal designer must assist the principal contractor in preparing the construction phase plan to include pre-construction information obtained from the client and information obtained from designers.
Throughout the project the principal contractor must ensure that the construction phase plan is appropriately reviewed, updated and revised.
The construction phase plan must record the:
(a) health and safety arrangements for the construction phase;

(b) site rules;
(c) Any specific measures concerning listed work
Information included should be:
(a) is relevant to the project;
(b) has sufficient detail to set out the arrangements, site rules and special measures needed to manage the construction phase;
(c) proportionate to the scale and complexity of the project and the risks involved.

HEALTH AND SAFETY FILE
The health and safety file is only required for projects involving more than one contractor.
During the pre-construction phase, the principal designer must prepare a health and safety file which must contain information relating to the project which is likely to be needed during any subsequent project to ensure the health and safety of any person.
The principal designer must ensure that the health and safety file is appropriately reviewed, updated and revised from time to time to take account of the work and any changes that have occurred.
During the project, the principal contractor must provide the principal designer with any information in the principal contractor's possession relevant to the health and safety file, for inclusion in the health and safety file.
If the principal designer's appointment concludes before the end of the project, the principal designer must pass the health and safety file to the principal contractor.
Where the health and safety file is passed to the principal contractor the principal contractor must ensure that the health and safety file is appropriately reviewed, updated and revised from time to time to take account of the work and any changes that have occurred.
At the end of the project the health and safety file must be passed to the client.
The file must contain information about the current project likely to be needed to ensure health and safety during any subsequent work, such as maintenance, cleaning, refurbishment or demolition. When preparing the health and safety file, information on the following should be considered for inclusion:
(a) a brief description of the work carried out;
(b) any hazards that have not been eliminated through the design and construction processes, and how they have been addressed;
(c) key structural principles and safe working loads for floors and roofs;
(d) hazardous materials used;
(e) information regarding the removal or dismantling of installed plant and equipment;
(f) health and safety information about equipment provided for cleaning or maintaining the structure;
(g) the nature, location and markings of significant services, including underground cables; gas supply equipment; fire-fighting services etc;
(h) information and as-built drawings of the building, its plant and equipment

GENERAL REQUIREMENTS FOR ALL CONSTRUCTION SITES

Safe places of construction work
There must, so far as is reasonably practicable, be suitable and sufficient safe access and egress from every construction site or place construction work is being carried out to every other place provided for the use of any person whilst at work or have access to within a construction site;
A construction site must be, so far as is reasonably practicable, made and kept safe for, and without risks to, the health of any person at work there.
A construction site must, so far as is reasonably practicable, have sufficient working space and be arranged so that it is suitable for any person who is working or who is likely to work there, taking account of any necessary work equipment likely to be used.

CONSTRUCTION (DESIGN AND MANAGEMENT) REGULATIONS 2015 (Cont'd)
GENERAL REQUIREMENTS FOR ALL CONSTRUCTION SITES (Cont'd)

Good order and site security

Each part of a construction site must, so far as is reasonably practicable, be kept in good order and those parts in which construction work is being carried out must be kept in a reasonable state of cleanliness

Where necessary in the interests of health and safety, a construction site must, so far as is reasonably practicable, have its perimeter identified by suitable signs and be arranged so that its extent is readily identifiable; or be fenced off.

No timber or other material with projecting nails (or similar sharp object) must be used in any construction work; or be allowed to remain in any place, if the nails (or similar sharp object) may be a source of danger to any person.

Stability of structures

(1) All practicable steps must be taken, where necessary to prevent danger to any person, to ensure that any new or existing structure does not collapse if, due to the carrying out of construction work, it may become unstable; or is in a temporary state of weakness or instability.

(2) Any buttress, temporary support or temporary structure must be of such design and installed and maintained so as to withstand any foreseeable loads which may be imposed on it; and only be used for the purposes for which it was designed, and installed and is maintained.

(3) A structure must not be so loaded as to render it unsafe to any person.

Demolition or dismantling

(1) The demolition or dismantling of a structure must be planned and carried out in such a manner as to prevent danger or, where it is not practicable to prevent it, to reduce danger to as low a level as is reasonably practicable.

(2) The arrangements for carrying out such demolition or dismantling must be recorded in writing before the demolition or dismantling work begins.

Excavations

All practicable steps must be taken to prevent danger to any person including, where necessary, the provision of supports or battering

Construction work must not be carried out in an excavation where any supports or battering have been provided unless the excavation and any work equipment and materials which may affect its safety have been inspected by a competent person

Reports of inspections

Where a person who carries out an inspection is not satisfied that construction work can be carried out safely at the place inspected, that person must inform the person on whose behalf the inspection was carried out and prepare and provide the report

The Construction (Design and Management) Regulations 2015 covers the following topics.

Acknowledgement

Excerpts reproduced from HSE publications –
- further details from HSE (www.hse.gov.uk or www.hsebooks.co.uk)

V.B. Johnson LLP provides a full CDM Principal designer service. Please do not hesitate to contact any of our offices Watford (01923) 227236, Wakefield (01924) 897373 or Hull (01482 492191) for more specific advice or provision of Health and Safety Services

Construction Procurement Guide

Introduction

There are numerous issues that need to be considered when deciding upon the means of procurement of a building project, and which will determine the Form of Contract to define the obligations and responsibilities between the Parties. Procurement is the term used to enable the process of creating a contractual relationship. There are four essential elements:

1) Offer
2) Acceptance
3) Consideration (payment)
4) The intention to create a legal relationship

Factors which influence the procurement process include:

- the experience of client and the business case for the project including any particular requirements that the client and his funders, if separate, may have
- the client's desired programme for development, including the timing of the start and finish dates
- the nature, size and complexity of the project
- the state and completeness of the designs
- the degree of cost certainty, or conversely the requirement for risk aversion
- the need for competitive prices
- the apportionment of risk between client and contractor

There are eight main recognised procurement routes:

a) traditional *(bills of quantities; schedules of work; specification and drawings; and the like)*
b) target cost
c) design and build; and target cost design and build
d) design, manage and construct
e) management contracting
f) construction management
g) engineer, procure & construct (turnkey) - *not dealt with here*
h) partnering (also a process – see below)

Further specialised routes used by public authorities are PFI – the private finance initiative; and PRIME contracting. In the latter case variants of PRIME contracting have been adopted by the NHS, Defence Estates and several 'blue chip' private sector organisations. PFI and PRIME are not dealt with here. *For the HM Treasury definition of PFI see: www.hm-treasury.gov.uk/hmt/documents/public_private_ partnerships/ppp_index.cfm For a definition of PRIME contracting see: www.contracts.mod.uk/dc/pdfs/Prime_Contracting.pdf* With the exception of partnering, all of the above routes can be used with:

- a single stage process (tendered or negotiated)
- a two-stage process
- as a process to define a GMP – guaranteed maximum price

Partnering – as a route and a process

Partnering is both a procurement route and a tendering process. Although now well defined and much written about, it is still misunderstood by many professionals and contractors alike, who consider that they have been *'partnering'* with their clients long before Egan, having had longstanding and 'ongoing' relationships with clients for a considerable number of years. However, in many cases this is not partnering in the style championed in 1998 by The Egan Report, "Rethinking Construction".

Partnering seeks to create relationships between members of the whole project team including the client, design team , contractor(s) and the underlying supply chain. This creates a framework to deliver demonstrable (through KPIs) and continuing economies over time from: better design and construction; lower risk; less waste (time, material, etc); and as a result avoid disputes. Partnering workshops provide for better knowledge, communication, sharing, education and team building to create a much more purposeful relationship than merely continuing to work together with a client.! Egan's Partnering is discussed more fully later.

Single Stage – as a process

The single stage process is in many cases the conventional process of procurement. Tenders are sought based upon one of the routes set out herein and bids are considered and a contractor selected and then appointed based upon chosen criteria defined to suit the project. In most cases the 'choice criteria' involved in the contractor selection will be based upon the most economic cost, but can also be based upon programme, quality of design solution (in the case of a Design and Construct Tender) or a mixture of these criteria.

Two Stage – as a process

In the case of a two-stage process, the selection of a contractor is made by means of a first-stage tender early in the project programme before the design is well advanced. With the exception of partnering this process can be used in connection with most of the routes listed above

Normally, the first-stage is by means of a competitive tender based on limited elements of the work, for example:

- a schedule of unitary rates provisionally prepared from the design drawings as prepared at the time of the stage-one tender. *
- preliminaries costs (site staffing, plant, site setup, scaffolding, craneage, insurances, etc);
- the percentage required for overheads and profit;
- the percentages required on Dayworks;
- additions required on the sub-contract work which is to be sub-let;
- submission of CV's for proposed site and head office staff; and
- method statements - so that the contractor's understanding of the project can be assessed by the design team and the client.

* If time and design information is available, measurements might be made of selected major work elements (for example demolitions) and Bills of Approximate Quantities (or a Schedule of Works) can then be included for fuller pricing in the first-stage documents.

Contractor selection at stage-one, is made on the limited information available from the stage-one bids. If sufficient information is available further evaluation may be made on the basis of a notionally calculated contract cost generated by the Quantity Surveyor from his most up to date estimate at the time of the decision being made; together with:

- interviews and assessments of the contractor's proposed staff;
- project understanding and proposed method statements; and
- the contractor's willingness to enter into the second-stage negotiations.

At the second-stage, a Bill of Quantities/Schedule of Work or a more fully detailed Schedule of Rates or a mixture of both, is generally prepared by the Quantity Surveyor and is priced with the chosen contractor by means of 'open book' negotiation in order to calculate the contract sum.

As referred to above, a two-stage tender process can overlay any of the 'traditional' procurement routes and can bring an additional set of advantages and disadvantages to the chosen means of tendering. Set alone, two-stage procurement can provide:

- Early appointment of contractor and access to his experience of programming and buildability.
- An earlier start on site than with a single stage process, because the works can proceed, if required, before a contract sum has been negotiated and fixed.

Construction Procurement Guide (Cont'd)
Introduction (Cont'd)
Two Stage – as a process (Cont'd)

However two-stage procurement does not establish a firm contract sum at the time when the contractor is initially appointed as this is calculated later. It is therefore most appropriate for projects where an early start on site is required and where a project is reasonably well defined, in terms of the scope of work; and for which management or construction management are not considered appropriate.

Guaranteed Maximum Price (GMP) – as a process

A GMP can be tendered at outset, but more often a previously tendered lump sum contract (excluding Target type contracts) is converted to a GMP following a traditional single or two-stage tender process utilising:

- Bills of Quantity (either Firm or Approximate); or
- a Schedule of Works; or
- Specification and Drawings

Thus the definition of a GMP can occur, either before commencing the works or during the course of the works. The GMP is negotiated with the chosen contractor and his sub-contractor(s) on a basis that includes for the contractor's future design development of the scheme, albeit in many cases the initial tender methodology may not have included design.
The Contractor therefore assume a larger element of risk that with other contract relationships and takes responsibility for matters that would normally cause extensions of time and potentially additional payment(s).
There are no generally available standard Forms of Contract for a GMP, it will require a bespoke contracts or a standard contract modified by a bespoke addendum agreement.
The advantages of a GMP are that it:

- Gives cost certainty similar to that of a design and build contract whilst the client employs and retains control of the design team.
- Provides the client often with a more appropriately detailed building than might be the case if let as a design and build contract, because the client's design team retain control of detailing.
- Potentially means that the design needs to be completed to a greater degree than normal in order to let contract as a GMP, because after the GMP is agreed all variations that cause change, will attract cost and programme alterations. Therefore the design team will want to avoid such 'change'

However, it is also important to note that a GMP contract has the potential to provide 'rich' grounds for disputes; particularly in respect of whether varied work has resulted from design detailing and therefore included in the GMP, or whether it is a variation caused by client change, which is an addition to the contract sum.
Very few contractors have experience of GMP contracts and those that do have had mixed experiences. Particularly because the financial and programme risk for unforeseen circumstances is entirely borne by the contractor.
The creation of a GMP creates certainty for both parties. However, a GMP can mean that the contractor can take a heavy fall if things go wrong! Examples of GMP style contracts are those for Cardiff Arms Park and the new Wembley Stadiums.

With the exception of a GMP, the eight routes noted are served by a plethora of standard forms of contract, sub-contract, supply agreements, warranties, etc. Such forms are widely available, with versions for use in England and Wales, and Scotland, where the law relating to contract differs from English law.
The success of any procurement route and process will depend largely on the ability and experience of both the chosen designers and contractor(s).
At the time of Tender, it is therefore very important that care is exercised in the selection of tendering contractors. In all cases it is to be recommended that pre-tender interviews take place with
At the time of Tender, it is therefore very important that care is exercised in the selection of tendering contractors. In all cases it is to be recommended that pre-tender interviews take place

with the potential tenderers to confirm their suitability for the work in advance; followed by both mid-tender review meetings and post-tender interviews to assess the understanding of the project, the calibre of staff and other resourcing that each prospective contractor will provide. This is particularly important where the proposed project is complex, fast-track, or involves 'business as usual' working.
The sections following this Introduction provide an overview of some of the main procurement methods in use in the UK today. It is by not intended to cover every type of procurement route or all the issues and should be considered as a guide only.

Bills of Quantities

Basis

Often referred to as "traditional Bills of Quantities"
A lump sum tender price based upon the priced Bills of Quantities is normally tendered in competition by a pre-selected list of between three and six contractors.

Selection of a contractor is made on the basis of the tenderer pricing measured Bills of Quantities, prepared to standard rules of measurement, for example:

- SMM7 (the rules of measurement upon which Laxton's SMM is based)
- NRM – New Rules of Measurement – (the rules of measurement upon which Laxton's NRM is based)
- POMI – Principles of Measurement International
- ARM – Agreed Rules of Measurement (in the Republic of Ireland)
- CESMM – Civil Engineering Standard Method of Measurement

Bills of Quantities include definitive measurements of the major elements of the building.
Where design is not complete, 'provisional' and/or 'approximate' quantities can be included.
Additionally Provisional Sum allowances for undefined and defined works can also be included, together with provisional sum allowances for works to be carried out by 'named' specialist sub-contractors; which, if required, can be tendered separately – although this is relatively uncommon today and it is more likely that such work will be part of the main contractor's tender by means of selection from list of 'preferred sub-contractors', to then be employed as 'Domestic sub-contractors'.
This method of procurement is the oldest methodology in the UK. It is not now as popular as it was in the first 85 years of the 20th century, but is still often preferred by Public sector clients.

Advantages

- Cost certainty is generally high but depends on the degree of completeness of design at the time when the Bills of Quantity are prepared.
- Gives excellent comparison of tender bids because all tenders are based on the same measured information.
- Creates a low risk tendering environment allowing tenderers to give their most competitive price because the risk for the contractor is well understood and defined.
- This is a procurement process which is widely understood.
- Gives a good basis for measurement and valuation of variations and for the calculation of interim valuations and the eventual final account.
- Needs the design team to have prepared and developed the building design before the Bills of Quantity can be prepared and so through reduction of design risk often leads to a much greater level of programme certainty and thus the date for completion.
- Can incorporate Design and Build and Performance Specified works if required

Disadvantages

- For the level of cost certainty expected by this procurement methodology to be delivered, the design must have evolved before preparation of the Bills of Quantity is started.
- Pre-contract phase of procurement is lengthy compared to other procurement methods and so often leads to a later start on site.

Construction Procurement Guide (Cont'd)
Bills of Quantities (Cont'd)

Suitability

More appropriate for projects where time is available for design work to largely be completed so that detailed measurements can be made before the tenders are sought.
Bills of Quantity can be used on any size project but are particularly suitable for those which are extremely large and complex and where the design time required would mean a long lead-in period.

Bills of Approximate Quantities

Basis

Bills of Approximate Quantities are an alternative form of Bills of Quantities (see above) and are prepared early in the design process before a firm design is available.
The contractor is selected, normally by competition from a pre-selected list of between three and six contractors.
Contractors prepare a tender bid based on pricing a Bill of Approximate Quantities. This is essentially a traditional Bill of Quantities but with the quantities assessed from professional experience by the Quantity Surveyor rather than firmly measured, as would be the case with "Bills of Quantities".

Advantages

- Allows early appointment of a contractor and access to experience in terms of his programming and buildability skills.
- Enables an earlier start on site to be made than with Traditional Bills of Quantities.
- Can incorporate Design and Build and Performance Specified works if required

Disadvantages

Approximate Bills of Quantities do not establish a firm cost for the work at the time the contractor is appointed, thus there is less price certainty. This is because the actual cost of the works is calculated only when the design is available and detailed re-measurements have been made.

- The client proceeds to the construction stage at greater risk, despite a check being made at the tender stage by means of bids being submitted by the tendering contractors.

Suitability

More appropriate for projects for which an early start on site is required or where the design is reasonably well defined or alternatively where the work is of a repetitive nature following on from other similar projects (allowing assessments to be made of the quantity of works from previous experience), but where time is not available for full Bills of Quantities to be prepared.

Schedules of Work

Basis

A Schedule of Works document lists all of the main sections and heads of work to be undertaken by the contractor.
A lump sum tender price based upon the priced Schedule of Works and associated documents are normally tendered in competition by a pre-selected list of between three and six contractors.
The Schedule of Works must be read in conjunction with the specification and the drawings and is required to be priced by the tenderer on a lump sum basis. Normally if given, quantities of work are for guidance only and the contractor is required to make his own measurements in order to prepare his tender bid.
Often in addition to the Schedule of Works, the contractor is requested to price a 'unitary' schedule of rates relating to the principal items of material, etc to be carried out. This can assist in the valuation of variations that may be later instructed.
Provisional sum items in the Schedule of Works can be included where these can be defined. Additionally and as with Bills of Quantity, Provisional allowances for undefined and defined works can be included and if required. Also allowances for works to be carried out by 'named' specialist sub-contractors can be included and if required can then be

tendered separately – although this is relatively uncommon today and it is more likely that such work will be part of the main contractor's tender by means of his selection from list of 'preferred sub-contractors' to be employed as 'Domestic sub-contractors'.
However, it is important to note that because a Schedule of Works does not refer to a set of Measurement Rules, the documents must set out the basis (or rules) upon which Provisional Sums are included. Additionally and similarly definitions for General Attendances and Special Attendances to be included in connection with specialist sub-contractors must also be defined in the tender/contract documents.
This method of procurement became popular after the introduction of the JCT Intermediate Form of Contract in 1984. It continues to be popular today.

Advantages

- The lump sum price is firm, subject only to variations which may be instructed during the course of the contract works.
- Client risk tends to be avoided because the contractor prepares his own measurements and quantities.
- Can incorporate Design and Build and Performance Specified works if required

Disadvantages

- The design must be reasonably well advanced (like Bills of Quantity) in order that tender documentation can be prepared.
- When variations occur, the valuation of changes can be more difficult to agree with the contractor than if firm Bills of Quantities exist, because individual prices for items of work do not exist unless a 'unitary' schedule of rates has been requested as part of the Tender.
- Tenders are not as easily comparable to each other as is possible with Bills of Quantity, because the tendering contractors may interpret and price risk in the tender of document in different ways.

Suitability

More appropriate for use on small to medium sized projects. Tenderers need to carry out their own measurements and produce their own quantities.

The use of Schedules of Works is not particularly appropriate where change can be foreseen post-contract.

Specification and Drawings

Basis

A contractor is selected and appointed on the basis of a lump sum bid provided in response to a detailed Specification document, which when read in conjunction with the drawings, defines the scope of work to be carried out as the contract works.
The Specification can also include provisional allowances for unforeseen or undesigned work; and also allowances can be included for specialist works yet to be tendered. Measurements and quantities are prepared by the tenderers to enable them to price the work. Like Schedules of Work, the use and meaning of these provisional sums needs to be described and defined in the documents.
Normally, tendering is in competition with a pre-selected list of approximately five contractors.

Advantages

- The lump sum price is firm, subject only to variations which may be instructed during the course of the contract works.
- Client risk tends to be avoided because the contractor prepares his own measurements and quantities.
- Can incorporate Design and Build and Performance Specified works if required

Construction Procurement Guide (Cont'd)
Specification and Drawings (Cont'd)
Disadvantages

- The design must be well advanced in order to prepare the detailed Specification documents. This procurement route can therefore mean a later site start than with alternative procurement routes.
- Give less control of cost when variations are instructed than firm Bills of Quantities because of the lack of a defined system of measurement of the building elements.
- Tenders are not as easily comparable to each other as is possible with Bills of Quantities, because the tendering contractors may interpret and price risk in the tender document in different ways.

Suitability

Specification and Drawings are Appropriate to smaller projects such as house extensions and renovations; partly because all tenderers need to produce their own quantities and partly because of the time involved in achieving a sufficiency of the designs to enable the Specification tender documents to be prepared.
Like Schedules of Work, the use this form of procurement is not particularly appropriate where it is likely that change will be required by the client after the contract has been started.

Target Cost Contracts

Basis

A contractor is appointed, either in competition or by negotiation, on the basis of pricing of simplified Bills of Quantities or a Schedule of Works. The tender price forms the Target Cost but the contractor is paid the actual costs for executing the work, as audited by the quantity surveyor, so long as this is less than the tendered Target Cost.
The Target Cost is, in effect, a 'Guaranteed Maximum'. Any saving on the Target Cost is split, normally 50/50 (but can be varied) between the contractor and the client

Advantages

- Enables an early start to be made on site as the tender documents can be prepared and a contractor appointed before the full design is completed.
- Establishes the client's maximum financial commitment (subject to client variations). If the contractor's costs exceed the Target Cost, only the Target Cost sum is paid.
- Gives the opportunity for the client to benefit from any savings made by the contractor. It is particularly useful in cases where risks may be priced in the tender, as under this system if those risks do not materialise the saving is shared.
- Leads to a less adversarial attitude between contractor and client with both benefiting from savings made.
- Can give early warning of future financial problems as the auditing quantity surveyor has complete access to the contractor's actual costs.
- Can incorporate Design and Build and Performance Specified works if required

Disadvantages

- The contractor and design team need to fully understand the Target Cost process.
- Can prove marginally more expensive than a more traditional contract where more risk is taken by the contractor.

Suitability

More appropriate for projects where an early start is required on site, and where the client wants to establish a maximum financial commitment, together with a less adversarial relationship.

Management Contracting

Basis

The philosophy of Management Contracting differs from that of other more traditional procurement methods in that the management contractor is appointed to manage the construction of the project rather than build it.

The contractor is selected, normally in competition, early in the programme on the basis of a response to a tender enquiry document. This requires the contractor to tender a fee for pre-commencement services and construction services during and after the project as well as a lump sum or guaranteed maximum for site staffing and facilities, etc. As well as price, the required performance of the management contractor is a major factor in the selection process.
For the actual building process the works are divided into separate trade packages which are tendered by trade contractors. The selected trade-contractors are taken on by the management contractor who is responsible for managing them and in particular their ability to meet quality and programme targets.

Advantages

- Enables an early start to be made on site before design is well advanced.
- Allows flexibility for change as the works are tendered progressively. There is a reduced likelihood of claims affecting other packages in the event that major changes are made.
- Lends itself well to complex construction projects as construction can commence before design work is completed. It is important, however, that the design of each trade package is complete at the time of tendering that package.
- Leads to a less adversarial relationship between management contractor and client.
- Can incorporate Design and Build and Performance Specified works if required

Disadvantages

- Can lead to duplication of resources between trade-contractors and the management contractor and therefore higher tenders than would be the case under a more traditional route.
- As no contract sum is established, the client relies upon the Quantity Surveyor's estimate. This is, however, endorsed by the management contractor initially and subsequently firm costs are established progressively during the course of the works.
- The client accepts a greater degree of risk because he has financial responsibility for the default of sub-contractors.

Suitability

More appropriate for large, complex projects where an earlier start is required than can be made by proceeding with the full design process and tendering by a more traditional route. This is achieved by overlapping design, preparation of tender documents and construction.

Design and Build

Basis

A contractor is selected, normally in competition, to Design and Build the project. Tenders are invited on the basis of an "Employer's Requirements" document prepared normally by the client or his consultants. The Employers Requirements set out the project needs in terms specification, function and performance of the building required and if applicable will also define planning and any other restrictions.
The contractor responds with a set of "Contractor's Proposals" upon which the tender bids are assessed. Assuming the Contractors Proposals fulfil the employer's requirements, the lowest bid is often accepted, but this may not be the case because subjective consideration of the overall design and quality of the proposals may be important than cost alone.
Once appointed the contractor employs and is therefore responsible for the design team. In this connection, it should be noted that under a JCT form of Design and Build contract, the Employer, unless he/she appoints somebody else, becomes the "Employer's Agent". In the JCT contract this person is assumed to be non-technical and could be the only named party apart from the Contractor – there is normally no quantity surveyor. In most cases the client would be advised to retain the services of his own consultants including a quantity surveyor to protect his interests and act on his behalf.

Construction Procurement Guide (Cont'd)
Design and Build (Cont'd)
Basis (Cont'd)

In many cases nowadays and particularly where the project is of a complex nature or facing difficult town planning procedures, the client often appoints a design team to negotiate planning, prepare preliminary designs and specifications and a detailed design brief before seeking Design and Build proposals from contractors – this is referred to by clients as "Develop then Design and Build", or by contractors colloquially as "Design and Dump".

Develop then Design and Build ensures that the client gets a design that works for him and the risk of the final design and construction works is assumed by the contractor. In some instances and where the employer agrees or requires it, the initial design team can be novated to the contractor.

Advantages

- Establishes a fixed lump sum price, subject only to client's required changes
- Leaves responsibility with the contractor for organising and programming the design team's activities. The client is therefore not responsible for extensions of time in the event that design information is not produced on time.
- Variations (known as Changes) are normally pre-agreed and the client has the opportunity to instruct or otherwise, knowing the full consequences in terms of cost and time.
- The risk for unforeseen circumstances lies with the contractor.
- Gives single point responsibility.

Disadvantages

- Provides the client with arguably a less sophisticated building in terms of design detailing than would be the case with other forms of procurement.
- Gives less control over the work in total and of the costs of any variations required.
- Can be difficult in certain instances to precisely define the standards and quality of design required.
- Depending on the risks imposed on the contractor at tender stage can result in a higher contract value than would otherwise be the case.

Suitability

More appropriate where the client requires a firm lump sum price and where the required standards and quality can be easily defined before tenders are sought.

For more control over quality and cost, this procurement route can be used in a "Develop then Design and Build" form where the client's design team initially negotiates and obtains planning permission, prepares outline designs, specifications and a detailed brief for the Employer's Requirements, particularly where more complex buildings are required under this procurement method. In some cases the employer's design team is required to be novated to the contractor after obtaining planning permission.

Target Cost Design and Build

Basis

As with traditional Design and Build, a contractor is selected, normally in competition, to Design and Build the building. Tenders are invited on the basis of an "Employer's Requirements" which sets out the specification of the building required and defines the planning and other restrictions. The contractor is responsible for the design team, although the client should retain the services of consultants to protect his interests and act on his behalf.

The contractor's costs are audited by the Quantity Surveyor and the contractor is paid the actual cost of the work so long as it is less than the tendered Target Cost. In addition, any saving on the Target Cost Design and Build tender sum is split, normally 50/50 (but can be varied) between the contractor and the client.

Like Design and Build, a form of "Develop then Design and Build" can be adopted with Target Cost Design and Build. However because of the inherent sharing of savings the potential for such economies being made is reduced and unlikely to be as attractive to the contractor.

Advantages

- Establishes the client's maximum financial commitment (subject to client variations). If the contractor's costs exceed the Target Cost, only the Target Cost sum is paid.
- Leaves responsibility with the contractor for organising and programming the design team's activities. The client is therefore not responsible for extensions of time in the event that design information is not produced on time.
- Variations are normally pre-agreed and the client has the opportunity to instruct or otherwise knowing the consequences in terms of cost and time.
- Gives single point responsibility.
- Leads to a less adversarial attitude between contractor and client as both benefit from savings.
- Can give prior warning of future financial problems as the Quantity Surveyor has complete access to the contractor's actual costs.
- The risk for unforeseen circumstances is shared more equitably between the client and the contractor.

Disadvantages

- Provides the client with arguably a less sophisticated building in terms of design detailing then would be the case with other forms of procurement.
- Gives less control over the work in total and of the costs of any variations required.
- Can be difficult in certain instances to precisely define the standards and quality of design required.

Suitability

More appropriate for contracts where the client requires a firm lump sum price but where risks are likely to be priced by the contractor in the tender. Probably appropriate for less sophisticated buildings where the standards and quality can be easily defined.

Design, Manage and Construct

Basis

The Design, Manage and Construct route is a hybrid route between Design and Build and Management Contracting. Like Management Contracting, the Design, Manage and Construct contractor is appointed to manage the construction of the project through a series of trade-contracts. It differs, in that the contractor is also given responsibility for programming and procuring the design information, so that the design consultants are employed by the 'design and manage' contractor, rather than the client.

The Design, Manage and Construct contractor is selected, usually in competition, early in the programme on the basis of a response to a set of tender enquiry documents which requires the contractor to tender a fee for:

- pre-commencement services;
- construction services during and after the project;
- a lump sum for site staffing and facilities, etc

As well as price, the required performance of the design and manage contractor is a major factor in the selection process. For the actual building process, the works are divided into separate trade packages, for which tenders are sought from trade-contractors by means of a variety of tender routes, vut by preference on a lump sum basis. The selected trade-contractors are taken on by the design and manage contractor.

Advantages

- Gives single point responsibility.
- Enables an early start to be made on site before the design is completed.
- Creates a flexible tendering environment. Allows for ongoing change because works are tendered progressively. In turn this reduces the ongoing likelihood of claims being made by the trade-contractors, should significant changes be made.

Construction Procurement Guide (Cont'd)
Design, Manage and Construct (Cont'd)
Advantages (Cont'd)

- The 'Design, Manage and Construct' method lends itself well to complex or fast track construction because construction work is able to be started before the design is finished.
- Leads to a less confrontational relationship between the Design and Manage contractor and the client.
- Management of the design information production by the design and manage contractor has the potential to lead to fewer delays arising from lack of information.
- In most cases the Design and Manage contractor adopts overall financial responsibility for the default of sub-contractors.

Disadvantages

- Can lead to duplication of resources between trade-contractors and the Design and Manage contractor and therefore can also lead to a higher level of tenders.
- No contract sum is established. Therefore the client relies upon the Quantity Surveyor's estimate, endorsed by the Design and Manage contractor initially and then progressively firmed up as tendering progresses during the course of works.
- There is an 'arms-length' relationship between the client and the contractor's design team, which can potentially lead to lower client satisfaction with the final design.

Suitability

More appropriate for large, complex projects or fast track projects where potentially an early start is required and where the client wishes to have a single point responsibility.

Construction Management

Basis

The original philosophy of Construction Management was that the client would organise the management of the construction activities in-house. As with Management Contracting, the actual works are divided into separate trade packages that are tendered by trade contractors. However, the major difference from management contracting is that the client employs the selected trade contractors direct.

Because of the criticism of Management Contracting that has arisen over the past few years, the Construction Management approach has become more widely applied.

Not all clients have the facility to manage their own construction work in-house and it is now normal for an outside construction manager to be appointed - either an independent consultant, or a contractor. In these cases the trade contracts are still direct with the client.

Advantages

- Enables an early start to be made on site before design is well advanced.
- Allows flexibility for change as the works are tendered progressively. There is a reduced likelihood of claims affecting other packages in the event that major changes are made.
- Lends itself well to complex construction projects as construction can commence before design work is completed.
- Direct contracts between the works contractors and the client should result in stronger relationships and potentially a less adversarial situation.
- Trade Packages can include design where specialist works are required and are let on a Design and Build basis.
- The client has a greater degree of control over the works contractors.
- Can incorporate Design and Build and Performance Specified works if required

Disadvantages

- As no contract sum is established, the client relies upon the Quantity Surveyor's estimate. This is, however, endorsed by the construction manager initially and firm costs are established progressively during the course of the works.

- Organisation costs such as site accommodation, telephones, copying etc., are likely to be paid on a prime cost basis. This gives the client less cost certainty.
- The client accepts a greater degree of risk because he has financial responsibility for the default of the works contractors.
- The client takes on the burden of dealing with additional correspondence, multiple payments and possibly adjudication on disputes and extensions of time, etc., for all individual trade contractors.

Suitability

More appropriate for large, complex projects where the client is experienced in the construction process, wishes to have a "hands on" approach and where an earlier start is required than can be made by proceeding with the full design process and tendering in a more traditional way. This is achieved by overlapping design, preparation of tender documents and construction.

Measured Term Maintenance Contract

Basis

A method of contracting is used where there is much repetition of work, for example in the case of regular or planned preventive maintenance work is required, e.g. Housing, Schools, etc.

Tenders are sought on the basis of a Schedule of Rates, chosen for their applicability to the proposed work. Tenderers quote either their price for each work item, or where the Schedule of rates are a standard priced document, on the basis of a 'plus or minus' rate in each case. Additionally the tenderer will likely be asked to quote his minimum charge, or call out rate.

Contracts are generally let on a 'Framework' basis, generally for a period of between 2 and 3 years, although longer periods can be agreed and the contract often incorporates a means by which the contract can be extended for a further period of time, perhaps one or two years as an incentive for the contractor to perform. In this respect clients often define KPI's (Key Performance Indicators) as part of the contract in order to provide measurement of the performance of the contractor. After the contract commences, the Contractor measures the works carried out and raises invoices against the Schedule of Rates for each separate job. Applications for payment are based upon a monthly schedule and application for payment. Invoices are generally payable on completion of the service and monthly in arrears.

Additionally there can also be an additional annual sum per property for reactive repairs; and a further schedule of rates for renewals rather than repairs as applicable.

Payments for the reactive repairs element can be allowed to be adjusted down for any system down time or unavailability during the contract period.

Often used for repairs, maintenance, renewals and minor works by councils and other public authorities with major estate portfolios.

Advantages

- Creates a long-term integrated multi-party partnering approach between the client's estate management team and the chosen contractor, to provide better communication and response.
- Allows the development of more open pricing over the life of the term programme
- Allows proactive change and risk management, including use of a risk register.
- Creates a supply chain partnering environment whereby problem solving and alternative dispute resolution can thrive.
- Provides process for continuous improvement by allowing the measurement of improvement and related incentives which may be incorporated into the contract terms
- Certainty of cost for each work task and can be verified and easily audited.
- Can incorporate Design and Build and Performance Specified works if required

Construction Procurement Guide (Cont'd)
Measured Term Maintenance Contract (Cont'd)

Disadvantages

- Some clients believe that an open book agreement with a contractor may be a more economic way to proceed.
- Minimum charges can cause a feeling of mistrust between the client and contractor, where an apparently large number of 'small' charge jobs are carried out.
- A change in personnel may bring conflict and when things go wrong, there can be a serious breakdown between the parties

Suitability

Works well where there is a high degree of repetition in the works, for example the renewal of locks, painting, replacement of glazing, plant repairs, which can be reasonably foreseen.

Cost Reimbursable or Prime Cost

Basis

The contractor is appointed, possibly in competition, but on many occasions by negotiation, to carry out the works on the basis that he is paid his actual costs. He also receives a fixed fee or percentage for overheads and profit, which if competitive tendering is possible, can form the basis of the tender.
Some practitioners argue that this procurement method can be likened to being on a time and materials basis, or otherwise 'Daywork'; albeit with modified terms.

Advantages

- Enables a very early start to be made on site, but at the expense of cost certainty.

Disadvantages

- Overall cost control can be difficult because the contractor does not have any great incentive to work efficiently. Although interim assessments can be made, the total cost is not known until the work is largely completed.
- There is often no means to ensure that errors and bad workmanship are borne by the contractor and sub-contracted works, particularly if all works are let on a prime cost 'style' basis. It is for this reason that in many cases sub-contract packages are let on a competitive lump sum basis in order to maintain an element of cost control. Additionally it may be a term of the tender that a reduction in the costs is made to cover abortive work. Similarly, it is normal to restrict head office staff costs so that a proportion of these are included in the fixed fee; and to agree a schedule of Head office staff and site establishment costs with the tender in order to create a greater level of cost certainty.
- Risk for unforeseen circumstances is borne by the client.

Suitability

Probably a last resort method of procurement in view of the uncertainty of total cost inherent in the process. Really only suitable for projects where an immediate start on site is required, for example where fire damage or terrorist damage has occurred, or perhaps with difficult refurbishments where the scope of work cannot be established until the work progresses.

Partnering

Basis

"Effective partnering should allow formal contracts to be dispensed with"- Sir John Egan.
The partnering approach creates a trust relationship, fairness and dedication between all the parties to the contract which involves the client, the design team, the contractor(s) and sub-contractor(s) working to identified common goals.
A suite of contracts now provides the necessary framework to enable Partnering and the most popular, known as PPC 2000 is published by the Association of Consultant Architects (ACA). Partnering often allows projects to start on site many months earlier than would have been the case if a more traditional procurement method had been used.
However, whilst PPC 2000 is a sound foundation for partnering it is up to the project team to deliver the benefits. It is therefore important for the project team to be aware that "PPC 2000

cannot assist partnering teams who do not give serious thought to the best way in which its provisions are to be put into effect"
The management of risk is an important element of partnering and demands a structured approach to establish and actively maintain a project risk register.
The client can appoint a Partnering Advisor to record and document the project team relationships, the commitments made by each party and their expectations. He is an honest broker who has no affiliation to any party.
PPC 2000 is not alone as a contractual solution; other standard forms exist, for example:

- ECC (NEC 2 and 3)
- JCT 2005 suite with the JCT "Non-Binding Partnering Charter" – refer to JCT 2005 Practice Note 4 – Partnering.
- Be Collaborative Contract
- Perform21 (PSPC + Local Government Practice toolkit)
- TPC 2005 (PPC 2000 for term contracts)

Taking the PPC 2000 as an example, the documents would be as follows:

- The Project Partnering Agreement - sets out and governs the activities of all the partnering team members during the pre-construction phase and is signed as far in advance of the start on site as is possible.
- The Partnering Terms – will include The partnering terms, the project brief, the project proposal, known costs, provisional KPI's, preliminaries (site overheads) and a partnering charter. A budget at this stage is not required.
- Any Joining Agreements – this replaces sub-contract forms. Partners can join at any time.
- Any Pre Possession Agreement – the purpose of this document is to provide the ability to undertake work on or off site prior to approval to commence on site. For example: surveys and investigations, planning and building regulations approvals, Insurances and Bonds and Health and Safety plan (pre-possession only), and the like.
- The Commencement Agreement – the document includes the document includes express confirmation by all the partnering team members that to the best of their knowledge the project is ready to commence on site. It will include an 'agreed maximum price' (note not guaranteed), the project drawings, supply chain packages and associated drawings, the project timetable, KPI/Incentives, Surveys and other investigations, Planning and Building regulations, funding, H&S plan, Bond, Insurances, etc.
- The Partnering Charter – an agreed statement of the values, goals and priorities of the Partnering team members and interested parties
- The Project Brief
- The Project Proposals
- The Price Framework
- The Consultant Service Schedule
- The Consultant Payment Terms
- The KPI's – to measure and judge performance and service quality; to assess strengths and weaknesses and to act as an aid to continuous improvement; to create a financial incentive to perform well in key areas; to indicate the local social and economic impact of the construction phase of a project; and for benchmarking purposes

Advantages

- Increased client satisfaction
- Creates better value for money
- Projects benefit from the inherent consistency of the design / construction teams
- Improved predictability of out-turn costs and time - in most cases partnering projects are completed on time and within target cost and moreover handed over defect free.
- Better resource planning with a faster start on site
- Better design and whole life value
- Higher quality construction with fewer defects
- More effective procurement and administrative procedures
- Fewer disputes and more respect for people
- Reduced accidents
- Better use of resources, including Improved supply chain management

Construction Procurement Guide (Cont'd)
Partnering (Cont'd)
Advantages (Cont'd)

- Using construction work to gain social and economic benefit
- Can incorporate Design and Build and Performance Specified works if required

The following list notes some problems that have been encountered by partnering teams:

- The client must stay interested and actively involved as a partnering team member, otherwise the lack of interest and support has a tendency to encourage cynicism and apathy in team members; and a belief that ideas for innovation or early warnings will not reach the client
- Partnering team members must be committed to changing their way of working – partnering is a major departure from familiar ways of working and an uncommitted attitude is likely to undermine effective partnering and furthermore create the problems that partnering is intended to prevent.
- Failure to complete the partnering timetable – means that there is a serious risk that pre-construction activities will be delayed; particularly if members cannot rely on each other to meet deadlines and in turn relationships may suffer.
- Failure to clarify and integrate the Project Brief, Brief Proposals and the consultant services schedule – this creates a Risk of gaps in agreed activities and potential later disputes; and duplication in agreed activities wasting money and causing confusion
- KPI's and incentives not agreed early – KPI's must be firmed up not later than the date of the commencement agreement. If KPI's are neglected or targets/incentives not agreed this will reduce the motivation of the partnering team members
- Joint risk management is neglected – meaning that Risks will be transferred to constructor and priced rather than managed. This means that the opportunity to consider who is best to take the risk is missed
- Failure to appoint an independent or suitably experienced partnering adviser – This affects understanding, focus and driving forward the partnering process. A lack of an independent adviser can undermine team confidence
- Failure to read PPC 2000 – this obviously affects understanding and may lead to traditional adversarial alternatives being put into practice

Suitability

Partnering is a major departure from familiar ways of working. It is a new way of working to many of us
Professionals and constructors need to be:
- Enthusiastic
- Working to the same goal
- Co–operative and effective communicators
- Flexible
- Willing to manage the process

Partnering is about being fair and challenging about value and performance, it is not just about being friendly.
The design professionals are required be proactive and assertive during the project. A strong, determined, value/performance driven and professional stance is essential. The constructor only has guaranteed work provided value for money; quality and performance are being achieved. Underperformance must be tackled by an assertive, (not an aggressive) approach.

The NEC/ECC Contract

It is appropriate here to also mention the New Engineering Contract (referred to above), which was introduced in 1993.

Published by the Institute of Civil Engineers it was intended to respond to:
- Client dissatisfaction with existing forms - too many, too adversarial
- The inflexibility of existing contracts
- Disputes - too many, too expensive, too long to settle
- Management of contracts seen as more important than legal niceties
- UK as a member of Europe – national distinctions disappearing
- Clients looking to seek alternative procurement strategies

The Latham report of 1994 recommended that the NEC contract should become a national standard contract across the whole of the construction industry. In response, the ICE introduced various changes as recommended in the Report, creating the second edition, re-named as "The Engineering and Construction Contract" (ECC). It requires the parties to work "in a spirit of mutual trust and co-operation".
The ECC was followed by the Construction Act and as a combination then sought to reduce confrontation and to encourage all concerned to "get on with the job".

As a contract it:
- Is intended to be suitable for all the needs of the construction industry and allows a wide range of tender procurement methods
- Provides for a variety of approaches to risk allocation.
- Can be Adapted for partial design, full design or no design responsibility, and for all current contract options including target, management and cost reimbursable contracts.
- Creates a stimulus for good management

There are six main methods for seeking tendered works under a ECC partnering contract:
- A Lump sum with activity schedule
- Priced contract with Bill of Quantities (Note - the bill is used only for interim payments, not for valuation of compensation events).
- Target contract with activity schedule (lump sum quoted by tenderer as target)
- Target contract with Bill of Quantities (subject to remeasurement)
- Cost reimbursable contract
- Management contract (by involvement of a management contractor).

In addition Construction Management can also be used. It is therefore extremely flexible. Indeed it provides a contract to enable good management and a broad set of contract procurement Guide which goes well beyond the scope of this section.

Conclusion
The choice of a procurement route and tendering process is a complex matter. This section has attempted to set out the main features of the principal procurement routes and processes available.
It is intended as a broad guide to assist in understanding the main issues, advantages and disadvantages. In order to make a full assessment of what can be a set of complex issues, before entering into a contract/agreement you are strongly advised to seek professional advice.

Sustainable Construction

The terms "sustainable construction", "sustainable development" and "sustainable design" have become much used in the press in recent years and now is more and more frequently used by, seemingly, everybody. However, it is still little understood by the average person in the industry, let alone 'the street'.

Introduction

In February 2007, a UN report concluded that Scientists are now overwhelmingly confident that mankind is to blame for the warming of the global climate observed since the industrial revolution.

The Daily Telegraph reported:

'The 2,500 scientists from 130 countries who make up the Intergovernmental Panel on Climate Change said they were now more than 90 per cent certain global warming is happening because of people, as opposed to natural variation. This Report was far more confident than the IPCC's previous three reports in 2001, 1995 and 1990, which stated only that is was "likely" climate change had a human cause.

At the release of the Report in Paris Achim Steiner, the head of the UN Environment Programme, said *"Feb. 2, 2007 may be remembered as the day the question mark was removed from whether (people) are to blame for climate change,"*.

The scientists estimated that temperatures would rise by between 3.2° Fahrenheit and 7.8°F (1.8°C and 4°C) in the 21st century. This is expected to mean far higher rises in temperate latitudes and at the poles. That level of temperature rise is enough to trigger the start of the melting of land-based ice in Greenland and Antarctica, and the dying back of the Brazilian rainforest, with severe consequences for sea levels within a century.

The report records that:

* Temperatures rose 0.7 degrees in the 20th century and the 10 hottest years since records began in the 1850s have occurred since 1994.
* Between 1961 and 2003, ocean temperatures have increased at least two miles below the surface, causing seawater to expand and levels to rise by seven hundredths of an inch (1.8mm) a year.
* Changes have been found in wind patterns, prolonged droughts or floods, the thickness of the Arctic icecap and the salinity of the ocean.
* Significantly increased rainfall has been recorded in eastern parts of North and South America, northern Europe and northern and central Asia. More intense and longer droughts have been recorded over wider areas since the 1970s.

The IPCC report goes onto predict that weather extremes such as heatwaves, drought and heavy rainfall will become more frequent. Storms are likely to become more severe. Sea ice in the Arctic and Antarctic is expected to shrink drastically.

The role of the IPCC since it was founded by the UN in 1988 has been to assess on a comprehensive and open basis evidence for and against human-induced climate change. It bases its assessment mainly on peer reviewed and published scientific literature.

"We can no longer afford to ignore growing and compelling warnings from the world's leading experts."

Here in the UK, June 2007, according to the Met. Office, despite all the rain we suffered in the UK saw average monthly temperatures on average 1.5 degrees warmer than the seasonal norm.; whilst the three years preceding June 2007 were recorded as the warmest on record.

Over the last 1,000 years there were two relatively warm periods identified in the world's northern hemisphere. These were in the late 12th and 14th centuries and lasted between 10-20 years, but neither period was as extended or as hot as the period of consistent (even persistent) temperature rise, recorded over the past 50 years. At the same time the CO2 in our atmosphere is now approximately 360 ppm and still rising. These are the highest levels recorded for some 650,000 years; higher now than those recorded in polar ice just prior to the lat ice age, which have been identified as at some 260 ppm.

Published in late 2006, Sir Nicholas Stern's UK Government funded report on "Climate-Change Economics" states the world must spend 1 per cent of GDP from now on to avert disaster.

Indeed as recorded in Al Gore's book, "An Inconvenient Truth", the mean temperature and CO2 levels in the earth's atmosphere over the last 650,000 years bear a remarkable relationship to each other. Global warming now appears to be accepted as fact.

In 2007 the Construction Industry in the UK and Europe is:

a 40% Industry:
* it creates 40% of ALL emissions; and
* consumes 40% of material and energy

It is an industry that needs to become sustainable

a 30/30 Industry
* 30% of any country's GDP is construction related
* (26 million jobs in the EU)
* 30% is waste (non value transactions)

Lack of communication (interoperability) is also a key problem
This was demonstrated in a US Report in 2004, which stated:

'Inadequate interoperability in the US Capital Facilitates Industry costs approximately $15.8 billion annually, representing 2% of industry revenue.' (source ECTP)

The final report of the H M Government Low Carbon Construction Innovation and Growth Team dated Autumn 2010 stated:

The United Kingdom's commitment to reduce carbon and other greenhouse gas emissions is now a matter of legal obligation. Under the Climate Change Act 2008, emissions are targeted to fall by 26% by 2020 (by comparison with a 1990 baseline) and by no less than 80% to 2050.

Good Practice

The Breeam Good Practice Guidance: Sustainable Design and Construction August 2012 sets out planning policy and gives guidance to the achievement of sustainable development, of which sustainable design and construction is an integral part. While the guidance is not a statutory document, it nevertheless has considerable support and can be expected to be accorded appropriate weight in both plan-making and development management.

Headlines of UK Low Carbon Transition Plan as it relates to construction
Residential

Increased energy efficiency in homes to reduce emissions by 29% by 2020 (from 2008 levels)

All new homes to be zero carbon from 2016

Smart displays to be fitted to existing meters for two to three million households by 2016; and smart meters to all homes by 2020

Major retrofit programme to increase energy efficiency of existing stock

Non-Domestic Buildings

Increase in efficiency to reduce emissions by 13% by 2020 (from 2008 levels)

All new public sector buildings to be zero carbon from 2018, and private sector buildings by 2019

GENERAL INFORMATION

SUSTAINABLE CONSTRUCTION

Headlines of UK Low Carbon Transition Plan as it relates to construction (Cont'd)

Infrastructure

A more flexible, smarter grid

New nuclear power stations to provide additional 16GW of power between 2018 and 2025

Major programme of wind power and marine energy to increase electricity from renewable sources to around 30% by 2020

Programme of carbon capture from coal-fired power stations

New green communities

Infrastructure to support a more sustainable transport system, to reduce transport emissions by 14% by 2020 (from 2008 levels), including sourcing 10% of UK transport energy from sustainable renewable sources by 2020

To support this commitment, and broader issues around energy use, there is a range of tax, levy and market mechanisms in place.

Carbon, Major Climate Change and Energy Policies

Policy	Objective	Payee
Feed-in-Tariffs	Support for small-scale renewable electricity via a requirement on electricity suppliers to pay a tariff to Small-scale low-carbon generators.	Energy companies
Renewable heat incentive	Financial support to stimulate uptake of renewable heat	Paid directly by HM Treasury.
Climate Change Levy	Tax on the use of energy in industry, commerce and the public sector.	Business and public sector
Enhanced Capital Allowances	100 per cent first-year Tax relief in the form of Enhanced Capital Allowances (ECA) allow the full cost of an investment in designated energysaving plant and machinery to be written off against the taxable profits of the period in which the investment is made.	Treasury
Product Policy	Sets minimum energy efficiency standards for energy using products. Includes the Framework Directive for the Eco-design of Energy Using Products (EuP).	Product manufacturers
Building Regulations	Energy efficiency requirements (including progressive tightening of Part L) of Building Regulations.	Builders, clients
Zero-Carbon Homes (ZCH)	All new homes to be built to a zero carbon standard from 2016.	Builders, clients
Zero Carbon for New Non-Domestic	All new non-domestic to be built to zero carbon standard from 2019.	Builders, clients
Carbon Emission Reduction Target (CERT)	Obligation on energy suppliers to increase the uptake of energy efficiency and carbon reduction measures in the Household sector.	Energy companies
EPBD	Requires all EU countries to enhance their building regulations (minimum energy performance requirements), set standards for major refurbishments, and to introduce energy certification schemes for buildings (EPCs & DECs in UK). All countries are also required to have inspections of boilers and air-conditioners.	Property owners and occupiers
Smart Metering	Roll-out of smart or advanced meters to non-domestic and domestic customers not previously required to have them.	Energy companies

What is sustainable construction?

A DTI document in April 2004, entitled Sustainable Construction gave the following brief definition:

The UK strategy for more sustainable construction, Building a Better Quality of Life, suggests key themes for action by the construction industry. These are:

- design for minimum waste
- lean construction & minimise waste
- minimise energy in construction & use
- do not pollute
- preserve & enhance biodiversity
- conserve water resources
- respect people & local environment
- monitor & report, (i.e. use benchmarks)

Most of these points simply make good business sense e.g. minimising waste   Sustainability is of increasing importance to the efficient, effective & responsible operation of business.

(see www.dti.gov.uk/files/file13939.pdf)

Put concisely, sustainable construction means meeting the needs of today by creating sustainable value over the life-cycle of buildings without compromising the ability of future generations to meet their own needs.

This means not only creating the best sustainable design, but also maintaining actions in the building process itself by the use of responsible supply; and once completed sustainable operation and maintenance of buildings that meet the needs of their owners and users over the building's life span with minimal unfavourable environmental impacts whilst at the same time encouraging economic, social and cultural progress.

So sustainable construction aims to apply these principles to the construction industry by providing ways of building that:

- use less virgin material - recycle;
- use less energy;
- cause less pollution; and
- create less waste

but still provide the social and cultural benefits that construction projects have brought us throughout history.

In 2007, the BRE website states that "Sustainability is the issue of our time. In addition to the environmental and social benefits, the business benefits of operating sustainably are becoming increasingly apparent"   (www.bre.co.uk)

How do we design sustainably?

One thing is certain, in the case of knowledge based sustainability standards; there is huge confusion and a wide lack of knowledge. Equally at this time (2007), country standards in the EU are very fragmented and in each nation the standards and guidelines available tend to be used in heterogonous ways.

There is now a move to create a set of unified best practises and standards for the EU, and a number of funded initiatives have been set up:

- INNOVA
- STAND-INN
- ECTP
- ERABUILD
- EUREKA

However:

- The Lords Prayer has 56 words
- The Ten Commandments has 297 words
- The American Declaration of Independence has 300 words

How do we design sustainably? (Cont'd)

- Yet an EEC Directive on the import of caramel and caramel products required 26,911 words!

Maybe we shouldn't hold our breath....

The following are examples of some of the currently used "sustainability standards" within the EU and the UK:

- Local Building Codes and regulations – in the UK, Part L
- The bioconstruction standard
- BREEAM - for assessing and improving the environmental performance of office, retail, school and industrial buildings, and Ecohomes for housing
- Envest - for assessing environmental impacts of a building at early design stage
- EN832 – Thermal performance of buildings
- IEA Task13 – Energy efficiency guidelines
- FiSIAQ 2000 – Finnish indoor climate guideline (thermal conditions, indoor air quality, acoustic conditions)
- ISO 9223 – Corrosion of metals
- ISO 15686-2 – Service life prediction
- ISO 15686-5 – Whole life costing
- ISO/DIS 15392 - Sustainability in building construction - General principles
- LiderA – Portuguese environmental rating method
- Miljøprogram – Norwegian guidelines
- PromisE – Finnish environmental rating method
- The Strawbale construction standard
- The Zero Carbon standard

The following are also useful resources:

- Environmental profiles - a universal method of measuring the sustainability of construction products and materials
- The Green Guide to Specification
- Environmental Management Toolkits - for offices, local authorities, schools and utilities
- MaSC - to help construction companies improve business through sustainability
- SMARTWaste (website) - measuring, managing and reducing construction waste

How do we enable the sustainable construction vision?

- Start with clients (the building owners) and motivate them to see the potential for improved sustainability by creating buildings which are designed to reduce consumption by using less energy, less construction material, reduce waste and specify sustainable and recycled materials.
- Focus on "seeing is believing" – demonstrate what is possible and stop just talking about it – Do It !
- Motivate government to see the potential for efficiency and freeing up resources to create more value adding activities and improve grants and tax allowances to provide purpose behind the words. For example if you increase the thermal performance of a building by, say the use of hemp and lime blocks in the external walls and by so doing reduce or even avoid cooling of a warehouse, the capital allowances on the M&E plant is reduced or even lost, very sustainable design.
- Motivate educational institutions to see the potential for new roles and new ways of providing knowledge and training to assist in the cultural challenge of education within the industry as a whole.

The tools to assist the design and measurement processes needed to create this vision are:

- Life Cycle Assessment (LCA)
- Environmental Product Declaration (EPD)
- Life Cycle Costing (LCC)
- Energy Performance Declaration

- Environmental Impact
- Adaptability to change in use
- Reusability / Recycling
- Service Life Planning
- Social Impact
- Energy Efficiency
- The renewables toolkit – GLA published September 2004

ISO/DIS 15392 General Principles creates a framework for sustainable construction:

Its objectives are:

- improvement of the construction sector and the built environment
- reduction of adverse impacts while improving value
- stimulation of a pro-active approach
- stimulation of innovation
- decoupling of economic growth from increasing adverse impacts
- reconciliation of contradictory interests or requirements arising from short-term / long-term planning or decision-making

and the principles set out are:

- Continual Improvement
- Equity
- Global Thinking and Local Action
- Holistic Approach
- Responsibility
- Involvement of Interested Parties
- Long-Term Concern
- Precaution and Risk
- Transparency

Thus the design development process for sustainable construction should address:

- service life design
- design for required building performance
- design for healthy indoor conditions
- design for energy-efficiency
- design for low environmental risks
- design of care and maintenance
- life cycle assessment of building parts and buildings
- estimation of life cycle costs
- optimisation of the processes of maintenance
- care and maintenance
- repair and refurbishment
- reuse and recycling products or disposal of products after service life

Summary

It is not difficult to foresee major change driven by governments in the coming years. The climate Change Levy is likely to be increased substantially to enable the UK to meet its EU target of 12% of energy from renewable sources by 2010. Enhanced Capital Allowances may yet enable businesses to claim 100% of all spending on certain plant and machinery which are designated to reduce the level of energy, CO_2 emissions and demands on water resources. And in addition, continuing fiscal changes to fuel duties and The Emission Trading Scheme will in the future significantly influence the built environment. Also in the pipeline is that all buildings will have a 'building energy label' upon completion taken from its carbon emissions calculations –similar to the Home Energy Certificates now being introduced for house sales.

It can therefore be seen as an approach to design rather than a fixed set of rules and the underlying ethos includes the care in the whole development process to enable what has become known as "sustainable construction"

SUSTAINABLE CONSTRUCTION

Summary (Cont'd)

The following are given as examples of techniques and materials that can be used to enable the overall concept of sustainable construction, they are in no particular order and is not intended as an exhaustive list:

- Ground source heatpumps
- Heat recovery units in ventilation exhaust
- Geo-thermal/Ground sourced heating and cooling
- Bio-mass boilers – on their own or with Combined Heat and Power (CHP)
- Condensing boilers
- Increased efficiency in internal lighting through both control systems via daylight sensing and/or lighting systems operated by PIR sensors
- DC motor technology
- Intelligent Building Management Systems (BMS)
- Wind turbine generation
- Photo Voltaic Cells (PVC) - Solar power generation
- Solar water heating
- Sun-pipes
- Natural ventilation
- Double glazing incorporating solar / thermal control
- Rainwater harvesting

The following list of materials has largely been taken from the rematerialise project set up to define ecosmart materials, they are in no particular order and is not intended as an exhaustive list:

- Thatch
- Crushed concrete
- FSC timber
- Formaldehyde free MDF
- Intrallam – structural timber joists made from 100% recycled wood
- Hardboard
- Sundeala Boards
- Frit – trade term to describe recycled glass
- Moulded tiles made from 100% glass cullet
- T.T.U.R.A.
- Clay bricks
- Claytec – a building block made from a mix of clay, woven jute and bamboo
- Hemp and Lime blocks

The following list of materials has largely been taken from the rematerialise project set up to define ecosmart materials, they are in no particular order and is not intended as an exhaustive list (Cont'd):

- Duratext – a surfacing material for interior and external use, based upon a fibre cement compound using a proportion of recycled fibres
- Fermacell – a multipurpose tough fibreboard made from gypsum. It is highly durable and moisture resilient
- Granite – quarried locally from the UK
- Marble, quarried locally from the UK
- Slate – quarried locally from the UK
- Rastra – essentially a customizable breeze-block or lightweight concrete.
- Terrazzo using reclaimed stone and glass
- Droptec – a safety flooring manufactured from 100% recycled polyurethane foam
- H.D.P.E – from recycled plastic tapped from the post consumer waste stream.
- Jute/Thermoset composites
- Kenaf/Thermoset composites
- P.E.T. from recycled plastic tapped from the post consumer waste stream
- Natural rubber
- Recycled rubber matting and sheeting made from recycled car tyres

- Coconut matting
- Cotton
- Hemp
- Jute
- Polyester Ge0-Textile – manufactured from 100% recycled P.E.T. from the post consumer waste stream
- Polyester fleece - manufactured from 100% recycled P.E.T. (mainly soft drink bottles) from the post consumer waste stream
- Polypropylene Ge0-Textile – manufactured from 100% recycled Polypropylene from the post industrial waste stream
- Straw Ge0-Textile – manufactured from a blend of straw and a loosely woven net of biodegradable string, can be used to help avoid soil erosion on manmade slopes
- Bamboo and Bamboo plank
- Biopol – a 100% biodegradable loose fill packing made from foamed corn starch, used as a packaging material in the supply chain.
- Cork
- Flax Insulation
- Homasote – made from 100% recycled newspapers to create a board that is lightweight and cheap, can be used for pinboards
- Rye Grass Straw Board - a biodegradable building board made from agricultural rye grass straw waste
- Isobord – a biodegradable building board made from agricultural straw waste
- Strawboard - a biodegradable building board made from agricultural straw waste
- Straw bales
- Lime render
- Treeplast – an injection mouldable bio-polymer made from a blend of corn starch and natural cellulose extracted from wood waste
- Warmcel – an insulating material made from 100% recycled newspapers

Useful references:

EcoSmart Concrete - www.ecosmartconcrete.com

BREEAM – www.breeam.org

Sustainable Construction - www.sustainable-construction.org.uk

Envest – www.bre.co.uk

The Environment Agency - www.environment-agency.gov.uk

The DTI - www.dti.gov.uk/sectors/construction/

Environmental Profiles – www.bre.co.uk

The Green Guide to Construction - www.brebookshop.com

MaSC - projects.bre.co.uk/masc/

SMARTWaste - www.smartwaste.co.uk

BuildingSmart – www.buildingsmart.org.uk

Europe INNOVA and STAND-INN - www.europe-innova.org

European Construction Technology Platform - www.ectp.org

ERABUILD - www.erabuild.net

EUREKA - www.eureka.be

The London Plan - www.london.gov.uk/mayor/planning/

The "Renewables Toolkit" from the Mayor of London - mayor.london.gov.uk/mayor/environment/energy/docs/renewables_toolkit.pdf

Rematerilise - www.rematerialise.org

RSA Carbon Limited - www.rsacarbonlimited.org

Enhanced Capital Allowances - www.eca.gov.uk

V B Johnson LLP provide cost advice on sustainablility issues and prepare BREEAM assessments, Whole Life (Life Cycle) Costing, and Claims for Capital Allowances and other Tax Credits as part of our a range of specialised professional consultancy services to Clients.

AVERAGE COVERAGE OF PAINTS

The following information has been provided by the British Decorators Association, whose permission to publish is hereby acknowledged.

In this revision a range of spreading capacities is given. Figures are in square metres per litre, except for oil-bound water paint and cement-based paint which are in square metres per kilogram.

For comparative purposes figures are given for a single coat, but users are recommended to follow manufactures' recommendations as to when to use single or multicoat systems.

It is emphasised that the figures quoted in the schedule are practical figures for brush application, achieved in big scale painting work and take into account losses and wastage. They are not optima figures based upon ideal conditions of surface, nor minima figures reflecting the reverses of these conditions.

There will be instances when the figures indicated by paint manufacturers in their literature will be higher than those shown in the schedule. The Committee realise that under ideal conditions of application, and depending on such factors as the skill of the applicator, type and quality of the product, substantially better covering figures can be achieved.

The figures given below are for application by brush and to appropriate systems on each surface. They are given for guidance and are qualified to allow for variation depending on certain factors.

SCHEDULE OF AVERAGE COVERAGE OF PAINTS IN SQUARE METRES PER LITRE

Coating per litre	Finishing plaster	Wood floated rendering	Smooth concrete/ cement	Fair faced brick- work	Block work -	Roughcast pebble dash -	Hardboard	Soft fibre insulating board
Water thinned primer/undercoat								
as primer	13-15	-	-	-	-	-	10-12	7-10
as undercoat	-	-	-	-	-	-	-	10-12
Plaster primer (including building board)	9-11	8-12	9-11	7-9	5-7	2-4	8-10	7-9
Alkali resistant primer	7-11	6-8	7-11	6-8	4-6	2-4	-	-
External wall primer sealer	6-8	6-7	6-8	5-7	4-6	2-4	-	-
Undercoat	11-14	7-9	7-9	6-8	6-8	3-4	11-14	10-12
Oil based thixotropic finish	Figures should be obtained from individual manufacturers							
Eggshell/semi-gloss finish (oil based)	11-14	9-11	11-14	8-10	7-9	-	10-13	10-12
Acrylic eggshell	11-14	10-12	11-14	8-11	7-10	-	11-14	10-12
Emulsion paint								
standard	12-15	8-12	11-14	8-12	6-10	2-4	12-15	8-10
contract	10-12	7-11	10-12	7-10	5-9	2-4	10-12	7-9
Glossy emulsion.	Figures should be obtained from individual manufacturers							
Heavy textured coating	2-4	2-4	2-4	2-4	2-4	-	2-4	2-4
Masonry paint	5-7	4-6	5-7	4-6	3-5	2-4	-	-
Coating per kilogram								
Oil bound water paint	7-9	6-8	7-9	6-8	5-7	-	-	-
Cement based paint	-	4-6	6-7	3-6	3-6	2-3	-	-

Coating per litre	Fire retardent fibre insulating board	Smooth paper faced board	Hard asbestos sheet **	Structural steel work	Metal sheeting	Joinery	Smooth primed surfaces	Smooth under- coated surfaces
Woodprimer (oil based)	-	-	-	-	-	8-11	-	-
Water thinned primer/undercoat								
as primer	-	8-11	**	-	-	10-14	-	-
as undercoat.	-	10-12	-	-	-	12-15	12-15	-
Aluminium sealer +								
spirit based	-	-	-	-	-	7-9	-	-
oil based	-	-	-	-	9-13	9-13	-	-

AVERAGE COVERAGE OF PAINTS (Cont.d)
SCHEDULE OF AVERAGE COVERAGE OF PAINTS IN SQUARE METRES PER LITRE (Cont'd)

Coating per litre	Fire retardent fibre insulating board	Smooth paper faced board	Hard asbestos sheet **	Structural steel work	Metal sheeting	Joinery	Smooth primed surfaces	Smooth under-coated surfaces
Metal primer								
conventional	-	-	-	7-10	10-13	-	-	-
Specialised	Figures should be obtained from individual manufacturers							
Plaster primer (including building board)	8-10	10-12	10-12	-	-	-	-	-
Alkali resistant primer	-	-	8-10	-	-	-	-	-
External wall primer sealer	-	-	6-8	-	-	-	-	-
Undercoat	10-12	11-14	10-12	10-12	10-12	10-12	11-14	-
Gloss finish	10-12	11-14	10-12	10-12	10-12	10-12	11-14	11-14
Eggshell/semi-gloss finish (oil based)	10-12	11-14	10-12	10-12	10-12	10-12	11-14	11-14
Acrylic eggshell	10-12	11-14	-	-	11-14	11-13	11-14	-
Emulsion paint								
standard	8-10	12-15	10-12	-	-	10-12	12-15	12-15
contract	-	10-12	8-10	-	-	10-12	10-12	10-12
Heavy textured coating	2-4	2-4	2-4	2-4	2-4	2-4	2-4	2-4
Masonry paint	-	-	5-7	-	-	-	6-8	6-8
Coating per kilogram								
Oil bound water paint	-	-	7-9	-	-	-	-	-
Cement based paint	-	-	4-6	-	-	-	-	-
Glossy emulsion	Figures should be obtained from individual manufacturers							

+ Aluminium primer/sealer is normally used over "bitumen" painted surfaces.
* On some roughcast/pebbledash surfaces appreciably lower coverage may be obtained.
In many instances the coverages achieved will be affected by the suction and texture to the backing; for example the suction and texture of brickwork can vary to such an extent that coverages outside those quoted may on occasions be obtained.
It is necessary to take these factors into account when using this table.
** Owing to new legislation (COSHH) further advice regarding the encapsulation of asbestos should be sought.

APPROXIMATE WEIGHT OF SUNDRY MATERIALS
IMPERIAL

	lb per ft²	lb per ft³
Asbestos-cement sheeting		
1/4 in corrugated	3 1/4	-
1/4 in flat	2 1/4	-
Asbestos-cement slating		
Diamond	3	-
Rectangular	4	-
Asphalt 1 in thick	12	-
Ballast, river	-	120
Bituminous felt roofing	1	-
Brickwork, Commons	-	113
Cement, Portland	-	90
Cement screeding (1:3) 1/2 in	6	-
Concrete, cement		
Ballast	-	140
Brick	-	115
Clinker	-	95
Pumice	-	70
Reinforced (about 2% steel)	-	150
Cork slabs 1 in thick	1	-
Fibre boards		
1/2 in thick	3/4	-
Compressed 1/4 in thick	2/3	-
Fibrous plaster 5/8 in thick	3	-
Flooring		
1 in Magnesium Oxychloride (sawdust filler)	7 1/2	-
1 in Magnesium Oxychloride (mineral filler)	11 1/2	-
1/4 in Rubber	2 3/4	-
1 in (Nominal) softwood	2 1/4	-
1 in (Nominal) pitchpine	3	-
1 in (Nominal) hardwood	3 1/4	-
Glass, 1/4 in plate	3 1/2	-
Gravel	-	115
Lime, chalk	-	44
Partition slabs (solid)		
2 in Coke breeze or pumice	11 1/2	-
2 in Clinker	15	-
2 in Terrazzo	25	-
Partition slabs (hollow)		
2 in Coke breeze or pumice	9 1/4	-
2 in Clinker	11 1/2	-
2 in Moler	2 1/2	-
Plastering		
3/4 in Lime or gypsum	7 1/2	-
on lathing	8 3/4	-
Roof boarding, 1 in	2 1/2	-
Reconstructed stone	-	145
Sand, pit	-	90
river	-	120
Shingle	-	90
Slag wool	-	17
Slate 1 in slab	15	-
Slating, 3 in lap		
Cornish (medium)	7 1/2	-
Welsh (medium)	8 1/2	-
Westmorland (medium)	11 1/2	-
Westmorland (thin)	9	-
Terrazzo pavings 5/8 in	7	-
Tiling, roof		
Machine made 4 in gauge	13	-
Hand made 4 in gauge	14	-
Tiling floor (1/2 in)	5 3/4	-
wall (3/8 in)	4	-

GENERAL INFORMATION

APPROXIMATE WEIGHT OF SUNDRY MATERIALS IMPERIAL (Cont'd)

	lb per ft^2	lb per ft^3
Timber, seasoned		
Elm	-	39
Baltic Fir	-	35-38
Red Pine	-	40
Yellow Pine	-	28
Douglas Fir	-	33
Canadian Spruce	-	29
White Pine	-	27
Yellow Birch	-	44
Canadian Maple	-	47
Honduras Mahogany	-	34
Spanish Mahogany	-	44
English Oak	-	45
American Oak	-	47
Baltic Oak	-	46
Indian Teak	-	41
African Teak	-	60
Blackbean	-	40-47
Water	-	62 1/2
rain	-	62 1/2
sea	-	64

WEIGHTS OF VARIOUS METALS - IMPERIAL

Weight in lb per Square Foot

Inches	Thickness 1/16	1/8	1/4	3/8	1/2	5/8	3/4	7/8	1
Steel	2.55	5.1	10.2	15.3	20.4	25.5	30.6	35.7	40.8
Wrought Iron	2.50	5.0	10.0	15.0	20.0	25.0	30.0	35.0	40.0
Cast iron	2.35	4.69	9.37	14.06	18.75	23.44	28.12	32.81	37.5
Brass	2.75	5.48	10.94	16.42	21.88	27.34	32.81	38.29	43.76
Copper	2.89	5.78	11.56	17.34	23.12	28.99	36.68	40.46	46.24
Cast lead	3.70	7.39	14.78	22.17	29.56	36.95	44.34	51.73	59.12

Weights of Steel Flat Bar in lb per Foot Lineal

Width Inches	Thickness 1/8	1/4	5/16	3/8	7/16	1/2	5/8	3/4	7/8	1
1	0.43	0.85	1.06	1.28	1.49	1.70	2.13	2.55	2.98	3.40
1 1/8	0.48	0.96	1.20	1.43	1.67	1.91	2.39	2.87	3.35	3.83
1 1/4	0.53	1.06	1.33	1.59	1.86	2.13	2.66	3.19	3.72	4.25
1 3/8	0.58	1.17	1.46	1.75	2.05	2.34	2.92	3.51	4.09	4.68
1 1/2	0.64	1.28	1.59	1.91	2.23	2.55	3.19	3.83	4.46	5.10
1 5/8	0.69	1.38	1.73	2.07	2.42	2.76	3.45	4.14	4.83	5.53
1 3/4	0.74	1.49	1.86	2.23	2.60	2.98	3.72	4.46	5.21	5.95
1 7/8	0.80	1.59	1.99	2.39	2.79	3.19	3.98	4.78	5.58	6.38
2	0.85	1.70	2.13	2.55	2.98	3.40	4.25	5.10	5.95	6.80
2 1/4	0.96	1.91	2.39	2.87	3.35	3.83	4.78	5.74	6.69	7.65
2 1/2	1.06	2.13	2.66	3.19	3.72	4.25	5.31	6.38	7.44	8.50
2 3/4	1.17	2.34	2.92	3.51	4.09	4.68	5.84	7.01	8.18	9.35
3	1.28	2.55	3.19	3.83	4.46	5.10	6.38	7.65	8.93	10.20
3 1/4	1.38	2.76	3.45	4.14	4.83	5.53	6.91	8.29	9.67	11.05
3 1/2	1.49	2.98	3.72	4.46	5.21	5.95	7.44	8.93	10.41	11.90
3 3/4	1.59	3.19	3.98	4.78	5.58	6.38	7.97	9.56	11.16	12.75
4	1.70	3.40	4.25	5.10	5.95	6.80	8.50	10.20	11.90	13.60
4 1/4	1.81	3.61	4.52	5.42	6.32	7.23	9.03	10.84	12.64	14.45
4 1/2	1.91	3.83	4.78	5.74	6.69	7.65	9.56	11.48	13.39	15.30
4 3/4	2.02	4.04	5.05	6.06	7.07	8.08	10.09	12.11	14.13	16.15
5	2.13	4.25	5.31	6.38	7.44	8.50	10.63	12.75	14.88	17.00
5 1/2	2.34	4.68	5.84	7.01	8.18	9.35	11.69	14.03	16.36	18.70
6	2.55	5.10	6.38	7.56	8.93	10.20	12.75	15.30	17.85	20.40
6 1/2	2.76	5.53	6.91	8.29	9.67	11.05	13.81	16.58	19.34	22.10
7	2.98	5.95	7.44	8.93	10.41	11.90	14.88	17.85	20.83	23.80
7 1/2	3.19	6.38	7.97	9.56	11.16	12.75	15.94	19.13	22.31	25.50
8	3.40	6.80	8.50	10.20	11.90	13.60	17.00	20.40	23.80	27.20
9	3.83	7.65	9.56	11.48	13.39	15.30	19.13	22.95	26.78	30.60
10	4.25	8.50	10.63	12.75	14.88	17.00	21.25	25.50	29.75	34.00
11	4.68	9.35	11.69	14.03	16.36	18.70	23.38	28.05	32.73	37.40
12	5.10	10.20	12.75	15.30	17.85	20.40	25.50	30.60	35.70	40.80

Weights of Steel Round and Square Bars in lb per Foot Lineal

Inches	Round	Square	Inches	Round	Square	Inches	Round	Square	Inches	Round	Square
1/8	0.042	0.053	7/8	2.044	2.603	2 3/8	15.06	19.18	5	66.76	85.00
3/16	0.094	0.120	15/16	2.347	2.988	2 1/2	16.69	21.25	5 1/4	73.60	93.71
1/4	0.167	0.213	1	2.670	3.400	2 5/8	18.40	23.43	5 1/2	80.78	102.85
5/16	0.261	0.332	1 1/8	3.380	4.303	2 3/4	20.19	25.71	5 3/4	88.29	112.41
3/8	0.376	0.478	1 1/4	4.172	5.312	2 7/8	22.07	28.10	6	96.13	122.40
7/16	0.511	0.651	1 3/8	5.049	6.428	3	24.03	30.60	6 1/4	104.31	138.81
1/2	0.668	0.849	1 1/2	6.008	7.650	3 1/4	28.21	35.91	6 1/2	112.82	143.65
9/16	0.845	1.076	1 5/8	7.051	8.978	3 1/2	32.71	41.65	6 3/4	121.67	154.88
5/8	1.043	1.328	1 3/4	8.178	10.412	3 3/4	37.55	47.81	7	130.85	-
11/16	1.262	1.607	1 7/8	9.388	11.953	4	42.73	54.40	7 1/2	150.21	-
3/4	1.502	1.912	2	10.681	13.600	4 1/4	48.23	61.41	8	170.90	-
13/16	1.763	2.245	2 1/8	12.06	15.35	4 1/2	54.07	68.85	9	216.00	-
			2 1/4	13.52	17.21	4 3/4	60.25	76.71	10	267.00	-

WEIGHTS OF STEEL JOISTS TO BS 4: PART 1: 1980

Size inches	Size millimetres	Weight lb/ft	Weight kg/m	Size inches	Size millimetres	Weight lb/ft	Weight kg/m
10 x 8	254 x 203	55.0	81.85	5 x 4.5	127 x 114	18.0	26.79
10 x 4.5	254 x 114	25.0	37.20	4.5 x 4.5	114 x 114	18.0	26.79
8 x 6	203 x 152	35.0	52.09	4 x 4	102 x 102	15.5	23.07
6 x 5	152 x 127	25.0	37.20	3.5 x 3.5	89 x 89	13.0	19.35
5 x 4.5	127 x 114	20.0	29.76	3 x 3	76 x 76	8.5	12.65

WEIGHTS OF STEEL CHANNELS TO BS 4: PART 1: 1980

Size of channel in	mm	Weight lb/ft	Weight kg/m	Size of channel in	mm	Weight lb/ft	Weight kg/m
17 x 4	432 x 102	44	65.54	8 x 3.5	203 x 89	20	29.78
15 x 4	381 x 102	37	55.10	8 x 3	203 x 76	16	23.82
12 x 4	305 x 102	31	46.18	7 x 3.5	178 x 89	18	26.81
12 x 3.5	305 x 89	28	41.69	7 x 3	178 x 76	14	20.84
10 x 3.5	254 x 89	24	35.74	6 x 3.5	152 x 89	16	23.84
10 x 3	254 x 76	19	28.29	6 x 3	152 x 76	12	17.88
9 x 3.5	229 x 89	22	32.76	5 x 2.5	127 x 64	10	14.90
9 x 3	229 x 76	17.5	26.06				

WEIGHTS OF STRUCTURAL STEEL TEE BARS SPLIT FROM UNIVERSAL BEAMS

Size of Tee in	mm	Weight lb/ft	Weight kg/m	Size of Tee in	mm	Weight lb/ft	Weight kg/m
12 x 12	305 x 305	60.0	90	7 x 8	178 x 203	18.0	27
10 x 13.5	254 x 343	42.0	63	5.5 x 8	140 x 203	15.5	23
9 x 12	229 x 305	47.0	70	6.75 x 7	171 x 178	22.5	34
9 x 12	229 x 305	34.0	51	6.75 x 7	171 x 178	15.0	23
8.25 x 10.5	210 x 267	41.0	61	5 x 7	127 x 178	13.0	20
8.25 x 10.5	210 x 267	27.5	41	6.5 x 6	165 x 152	18.0	27
7 x 9	191 x 229	33.0	49	5 x 6	127 x 152	16.0	24
7 x 9	191 x 229	22.5	34	4 x 6	102 x 152	11.0	17
6 x 9	152 x 229	27.5	41	5.75 x 5	146 x 127	14.5	22
6 x 9	152 x 229	17.5	26	4 x 5	102 x 127	9.5	14
7 x 8	178 x 203	25.0	37	5.25 x 4	133 x 102	10.0	15

WEIGHTS OF STRUCTURAL STEEL TEE BARS SPLIT FROM UNIVERSAL COLUMNS

Size of Tee in	mm	Weight lb/ft	Weight kg/m	Size of Tee in	mm	Weight lb/ft	Weight kg/m
12 x 6	305 x 152	53.0	79	8 x 4	203 x 102	29.0	43
12 x 6	305 x 152	32.5	49	8 x 4	203 x 102	15.5	23
10 x 5	254 x 127	44.5	66	6 x 3	152 x 76	12.5	19
10 x 5	254 x 127	24.5	37	6 x 3	152 x 76	7.85	12

WEIGHTS OF STEEL EQUAL ANGLES TO B.S. 4848 PART 4: 1972

Size mm	Thickness mm	Weight kg/m	Size mm	Thickness mm	Weight kg/m	Size mm	Thickness mm	Weight kg/m
250 x 250	35	128.0	150 x 150	10	23.0	110 x 110	12	19.8
	32	118.0					10	16.7
	28	104.0	130 x 130	16	30.8		8	13.5
	25	93.6		12	23.5			
				10	19.8	100 x 100	15	21.9
200 x 200	24	71.1		8	16.0		12	17.8
	20	59.9					10	15.0
	18	54.2	120 x 120	15	26.6		8	12.2
	16	48.5		12	21.6			
				10	18.2	90 x 90	12	15.9
150 x 150	18	40.1		8	14.7		10	13.4
	15	33.8					8	10.9
	12	27.3	110 x 110	16	25.8		6	8.3

WEIGHTS OF STEEL UNEQUAL ANGLES TO B.S. 4848 PART 4: 1972 (except where marked *)

Size mm	Thickness mm	Weight kg/m	Size mm	Thickness mm	Weight kg/m	Size mm	Thickness mm	Weight kg/m
200 x 150	18	47.1	150 x 75	15	24.8	100 x 75	12	15.4
	15	39.6		12	20.2		10	13.0
	12	32.0		10	17.0		8	10.6
200 x 100	15	33.7	*137 x 102	9.5	17.3			
	12	27.3		7.9	14.5	100 x 65	10	12.3
	10	23.0		6.4	11.7		8	9.94
150 x 90	15	26.6	125 x 75	12	17.8		7	8.77
	12	21.6		10	15.0			
	10	18.2		8	12.2			
				6.5	9.98			

WEIGHTS OF UNIVERSAL BEAMS TO B.S. 4 PART 1: 1980

Size in	mm	Weight lb/ft	kg/m	Size in	mm	Weight lb/ft	kg/m
36 x 16.5	914 x 419	260	388	18 x 6	457 x 152	55	82
		230	343			50	74
						45	67
						40	60
						35	52
36 x 12	914 x 305	194	289	16 x 7	406 x 178	50	74
		170	253			45	67
		150	224			40	60
		135	201			36	54
33 x 11.5	838 x 292	152	226	16 x 5.5	406 x 140	31	46
		130	194			26	39
		118	176				
30 x 10.5	762 x 267	132	197	14 x 6.75	356 x 171	45	67
		116	173			38	57
		99	147			34	51
						30	45
27 x 10	686 x 254	114	170	14 x 5	356 x 127	26	39
		102	152			22	33
		94	140				
		84	125				
24 x 12	610 x 305	160	238	12 x 6.5	305 x 165	35	54
		120	179			31	46
		100	149			27	40
24 x 9	610 x 229	94	140	12 x 5	305 x 127	32	48
		84	125			28	42
		76	113			25	37
		68	101				
21 x 8.25	533 x 210	82	122	12 x 4	305 x 102	22	33
		73	109			19	28
		68	101			16.5	25
		62	92				
		55	82				
18 x 7.5	457 x 191	66	98	10 x 5.75	254 x 146	29	43
		60	89			25	37
		55	82			21	31
		50	74				
		45	67				
				10 x 4	254 x 102	19	28
						17	25
						15	22
				8 x 5.25	203 x 133	20	30
						17	25

WEIGHTS OF UNIVERSAL COLUMNS TO B.S. 4 PART 1: 1980

Size in	mm	Weight lb/ft	kg/m	Size in	mm	Weight lb/ft	kg/m
14 x 16	356 x 406	426	634	14 x 14.5	356 x 368	136	202
		370	551			119	177
		314	467			103	153
		264	393			87	129
		228	340				
		193	287				
		158	235				
12 x 12	305 x 305	109	283	10 x 10	254 x 254	112	167
		161	240			89	132
		133	198			72	107
		106	158			60	89
		92	137			49	73
		79	118				
		65	97				
8 x 8	203 x 203	58	86	6 x 6	152 x 152	25	37
		48	71			20	30
		40	60			15.7	23
		35	52				
		31	46				

BRITISH WEIGHTS AND MEASURES

Lineal Measure

2.25 in................................	= 1 nail	3 in...................................	= 1 palm
4 in....................................	= 1 hand	9 in or 4 nails.....................	= span, or 1/4 yd
12 in...................................	= 1 ft	3 ft or 4 quarters................	= 1 yd
5 quarters of yd..................	= 1 ell	6 ft or 2 yd........................	= 1 fathom
16 ft 6 in or 5.5 yd..............	= 1 rod, pole or perch	4 poles or 22 yd.................	= 1 chain
220 yd or 40 poles...............	= 1 furlong	1760 yd or 8 furlongs..........	= 1 mile = 5280 ft
7.92 in................................	= 1 link	100 links or 66 ft................	= 1 chain
10 chains............................	= 1 furlong	80 chains...........................	= 1 mile
3 miles...............................	= 1 league	2027 yd..............................	= 1 Admiralty knot
6075.5 ft.............................	= 1 nautical mile	69.04 miles........................	= 1 degree of latitude
69.16 miles.........................	= 1 degree of longitude at Equator	43.08 miles........................	= 1 degree of longitude at London
The Cheshire pole................	= 8 yd		

Square Measure

144 in^2...........................	= 1 ft^2	10 square chains.................	= 1 acre
9 ft^2..............................	= 1 yd^2	30 acres............................	= 1 yard land
100 ft^2..........................	= 1 square	100 acres...........................	= 1 hide
272.25 ft^2......................	= 1 rod, pole or perch	460 acres...........................	= 1 square mile
30.25 yd^2.......................	= 1 rod, pole or perch	1 mile long by 80 chains........	= 1 square mile
40 square rods....................	= 1 rood	7840 yd^2........................	= 1 Irish acre
4 roods or 4840 yd^2.........	= 1 acre	6084 yd^2........................	= 1 Scotch acre
43560 ft^2.......................	= 1 acre	1 mile long x 1 chain wide.......	= 8 acres
67.7264 in^2.....................	= 1 square link	10000 square links.............	= 1 square chain

Solid or Cubic Measure

1728 in^3.........................	= 1 ft^3	327 ft^3...........................	= 1 yd^3
42 ft^3 of timber................	= 1 shipping ton	108 ft^3...........................	= 1 stack of wood
165 ft^3...........................	= 1 standard of wood	128 ft^3...........................	= 1 cord of wood

Avoirdupois Weight

16 drachms........................	= 1 ounce	20 cwt................................	= 1 ton
16 ounces..........................	= 1 lb	7000 grains........................	= 1 lb avoirdupois
28 lb..................................	= 1 qr cwt	437.5 grains.......................	= 1 oz avoirdupois
112 lb................................	= 1 cwt		

Troy Weight

24 grains............................	= 1 dwt	5760 grains........................	= 1 lb troy
20 dwts..............................	= 1 ounce	480 grains..........................	= 1 oz troy
12 ounces..........................	= 1 lb		

Dry Measure

2 gal..................................	= 1 peck	6.232 gal............................	= 1 ft^3
8 gal..................................	= 1 bushel	168.264 gal.........................	= 1 yd^3
64 gal................................	= 1 quarter	4.893 gal............................	= 1 cyl ft

Liquid Measure

8.665 in^2........................	= 1 gill	4 gills.................................	= 1 pint
2 pints...............................	= 1 quart	4 quarts..............................	= 1 gal
1 gal..................................	= 277.25 in^3 or 0.16 ft^3	1 gal..................................	= 10 lb of distilled water
1 ft^3..............................	= 6.232 gal	1 cwt water........................	= 1.8 ft^3
1 bushel............................	= 1.28 ft	38 gal................................	= 1 bushel

Imperial gallons = 277.275 in^3, and 10 lb of distilled water at 62 degrees F

METRIC SYSTEM

SYMBOLS USED

Length	m = linear metre	Area	m^2 = square metre
	mm = linear millimetre		mm^2 = square millimetre
Volume	m^3 = cubic metre	Mass	kg = kilogramme
	mm^3 = cubic millimetre		g = gramme
Pressure	N/m^2 = Newton per square metre	Density	kg/m = kilogrammes per linear metre
			kg/m^3 = kilogrammes per cubic metre

CONVERSION FACTORS

To convert Imperial weights and measures into Metric or vice versa, multiply by the following conversion factors.

Length	Factor	Length	Factor
Inches to millimetres (mm)............................	25.4000	Millimetres to inches............................	0.0394
Feet to metres (m).....................................	0.3048	Metres to feet.......................................	3.281
Yards to metres (m)...................................	0.9144	Metres to yards.....................................	1.094

Area		Square metres to square feet................	10.764
Square feet to square metres (m^2)..............	0.0929	Ditto to square yards...........................	1.196
Square yards to ditto.................................	0.8361		

Volume		Cubic metres to cubic feet....................	35.315
Cubic feet to cubic metres (m^3)..................	0.0283	Ditto to cubic yards.............................	1.308
Cubic yards to ditto..................................	0.7645		

Mass		Tonnes (1000 kg) to tons......................	0.9842
Tons (2240 lbs) to tonnes........................	1.0161	Kilogrammes to cwts.............................	0.0196
Cwts to kilogrammes.................................	50.802	Kilogrammes to pounds..........................	2.205
Pounds to kilogrammes (kg)......................	0.454	Grammes to ounces..............................	0.0353
Ounces to grammes (g).............................	28.350		

Pressure		Newtons per square metre to pounds	
Pounds per square inch to Newtons per square metre (N/mm^2)...........................	6894.800	per square inch..................................	0.000145

Density		Kilogrammes per linear metre to	
Pounds per linear foot to kilogrammes per linear metre (kg/m)...........................	1.488	Pounds per linear foot...........................	0.6720
Pounds per square foot to kilogrammes per metre (k/m^2)...................................	4.882	Kilogrammes per square metre to Pounds per square foot........................	0.2048
Pounds per square yard to ditto................	0.542	Ditto to pounds per square yard.............	1.845
Pounds per cubic foot to kilogrammes per cubic metre (kg/m^3)............................	16.019	Kilogrammes per cubic metre to Pounds per cubic foot..........................	0.0624

Capacity			
Gallons to litres.......................................	4.546	Litres to gallons..................................	0.220

To convert items priced in Imperial units into Metric or vice versa, mulitply by the following conversion factors.

	£ Factor		£ Factor
Linear feet to linear metres.......................	3.281	Linear metres to linear feet.......................	0.3048
Linear yards to linear metres....................	1.094	Linear metres to linear yards...................	0.9144
Square feet to square metres....................	10.764	Square metres to square feet..................	0.0929
Square yards to square metres.................	1.196	Square metres to square yards..............	0.8361
Cubic feet to cubic metres.......................	35.315	Cubic metres to cubic feet.....................	0.028
Cubic yards to cubic metres.....................	1.308	Cubic metres to cubic yards...................	0.7645
Pounds to kilogrammes............................	2.2046	Kilogrammes to pounds........................	0.454
Cwts to kilogrammes...............................	0.0196	Kilogrammes to cwts............................	50.802
Cwts to tonnes......................................	19.684	Tonnes to cwts...................................	0.051
Tons to tonnes......................................	0.9842	Tonnes to tons...................................	1.0161
Gallons to litres.....................................	0.220	Litres to gallons.................................	4.54

GENERAL INFORMATION

CONVERSION FACTORS (Cont'd)

METRIC EQUIVALENTS

Linear inches to linear millimetres
12 in = 304.80 millimetres

Inches	0	1	2	3	4	5	6	7	8	9	10	11
0		25.40	50.80	76.20	101.60	127.00	152.40	177.80	203.20	228.60	154.00	279.40
1/32	0.79	26.19	51.59	76.99	102.39	127.79	153.19	178.59	203.99	229.39	254.79	280.19
1/16	1.59	26.99	52.39	77.79	103.19	128.59	153.99	179.39	204.79	230.19	255.59	280.99
3/32	2.38	27.78	53.18	78.58	103.98	129.38	154.78	180.18	205.58	230.98	256.38	281.78
1/8	3.18	28.58	53.98	79.38	104.78	130.18	155.58	180.98	206.38	231.78	257.18	282.58
5/32	3.97	29.37	54.77	80.17	105.57	130.97	156.37	181.77	207.17	232.57	257.97	283.37
6/16	4.76	30.16	55.56	80.96	106.36	131.76	157.16	182.56	207.96	233.36	258.76	284.16
7/32	5.56	30.96	56.36	81.76	107.16	132.56	157.96	183.36	208.76	234.16	259.56	284.96
1/4	6.35	31.75	57.15	82.55	107.95	133.35	158.75	184.15	209.55	234.95	260.35	285.75
3/8	9.53	34.93	60.33	85.73	111.13	136.53	161.93	187.33	212.73	238.13	263.53	288.93
1/2	12.70	38.10	63.50	88.90	114.30	139.70	165.10	190.50	215.90	241.30	266.70	292.10
5/8	15.88	41.28	66.68	92.08	117.48	142.88	168.28	193.68	219.08	244.48	269.88	295.28
3/4	19.05	44.45	69.85	95.25	120.65	146.05	171.45	196.85	222.25	247.65	273.05	298.45
7/8	22.23	47.63	98.43	98.43	123.83	149.23	174.63	200.03	225.43	150.83	276.23	301.63

Linear feet and inches to linear millimetres

Feet	1	2	3	4	5	6	7	8	9	10	11	12
0 in	304.8	609.6	914.4	1219.2	1524.0	1828.8	2133.6	2438.4	2743.2	3048.0	3352.8	3657.6
1 in	330.2	635.0	939.8	1244.6	1549.4	1854.2	2159.0	2463.8	2768.6	3073.4	3378.2	2683.0
2 in	355.6	660.4	965.2	1270.0	1574.8	1879.6	2184.4	2489.2	2794.0	3098.8	3403.6	3708.4
3 in	381.0	385.8	990.6	1295.4	1600.2	1905.0	2209.8	2514.6	2819.4	3124.2	3429.0	3733.8
4 in	406.4	711.2	1016.0	1320.8	1625.6	1930.4	2235.2	2540.0	2844.8	3149.6	3454.4	3759.2
5 in	431.8	736.6	1041.4	1346.4	1651.0	1955.8	2260.6	2565.4	2870.2	3175.0	3479.8	3784.6
6 in	457.2	762.0	1066.8	1371.6	1676.4	1981.2	2286.0	2590.8	2895.6	3200.4	3505.2	3810.0
7 in	482.6	787.4	1092.2	1397.0	1701.8	2006.6	2311.4	2616.2	2921.0	3225.8	3530.6	3835.4
8 in	508.0	812.8	1117.6	1422.4	1727.2	2032.0	2336.8	2641.6	2946.4	3251.2	3556.0	3860.8
9 in	533.4	838.2	1143.0	1447.8	1752.6	2057.4	2362.2	2667.0	2971.8	3276.6	3581.4	3886.2
10 in	558.8	863.6	1168.4	1473.2	1778.0	2082.8	2387.6	2692.4	2997.2	3302.0	3606.8	3911.6
11 in	584.2	889.0	1193.8	1498.6	1803.4	2108.2	2413.0	2717.8	3022.6	3327.4	3632.2	3937.0

METRIC VALUES

POWER
1000	microwatts	= 1 milliwatt
1000	milliwatts	= 1 watt
1000	watts	= 1 kilowatt
1000	kilowatts	= 1 megawatt
1000	gigawatts	= 1 terawatt

VOLUME AND CAPACITY
1000	cu millimetres	= 1 cu centimetre
1000	cu centimetres	= 1 cu decimetre
1000	cu decimetres	= 1 cu metre
1000	cu metres	= 1 cu dekametre
1000	cu dekametres	= 1 cu hectometre
1000	cu hectometre	= 1 cu kilometre
1000	microlitres	= 1 millilitres (1cc)
10	millilitres	= 1 centilitre
10	centilitres	= 1 decilitre
1000	millilitres	= 1 litre
100	centilitres	= 1 litre
1000	litres	= 1 kilolitre
1000	litres	= 1 cu metre

PRESSURE AND STRESS
1000	micropascals	= 1 millipascal
100	millipascals	= 1 microbar
1000	millipascals	= 1 pascal
100	pascals	= 1 millibar
1000	pascals	= 1 kilopascal
1000	millibars	= 1 bar
1000	kilopascals	= 1 megapascal
1000	bars	= 1 kilobar
1000	megapascals	= 1 gigapascal

VELOCITY
36	km/hour	= 1 metre/second

3600	km/hour	= 1 km/second

ELECTRICITY AND MAGNETISM
1000	picoamperes	= 1 nanoampere
1000	nanoamperes	= 1 microampere
1000	microamperes	= 1 milliampere
1000	milliamperes	= 1 ampere
1000	amperes	= 1 kiloampere
1000	millicoulombs	= 1 coulomb
1000	coulombs	= 1 kilocoulomb
1000	microvolts	= 1 millivolt
1000	millivolts	= 1 volt
1000	volts	= 1 kilovolt
1000	kilovolts	= 1 megavolt
1000	microhms	= 1 milliohm
1000	milliohms	= 1 ohm
1000	ohms	= 1 kilohms
1000	kilohms	= 1 megaohm
1000	millisiemens	= 1 siemen
1000	millihenrys	= 1 henry
1000	milliteslas	= 1 tesla

LENGTH
1000	picometres	= 1 nanometre
1000	nanometres	= 1 micrometre
1000	micrometres	= 1 millimetre
10	millimetres	= 1 centimetre
10	centimetres	= 1 decimetre
1000	millimetres	= 1 metre
100	centimetres	= 1 metre
100	metres	= 1 hectometre
1000	metres	= 1 kilometre
1000	kilometres	= 1 megametre
1852	nautical metres	= 1 international nautical mile

AREA
100	sq millimetres	= 1 sq centimetre
100	sq centimetres	= 1 sq decimetre
10000	sq centimetres	= 1 sq metre
100	sq metres	= 1 are
10	ares	= 1 dekare
10000	sq metres	= 1 hectare
100	hectares	= 1 sq kilometre

MASS
1000	nanograms	= 1 microgram
1000	micrograms	= 1 milligram
200	milligrams	= 1 metric carat
1000	milligrams	= 1 gram
25	grams	= 1 metric ounce
100	grams	= 1 hectogram
1000	grams	= 1 kilogram
100	kilograms	= 1 quintal
1000	kilograms	= 1 megagram

ENERGY (WORK AND HEAT)
10000	ergs	= 1 millijoule
1000	millijoules	= 1 joule
1000	joules	= 1 kilojoule
1000	kilojoules	= 1 megajoule
3.6	megajoules	= 1 kilowatt hour
1000	megajoules	= 1 gigajoule
1000	gigajoules	= 1 terajoule

FORCE
10	micronewtons	= 1 dyne
1000	micronewtons	= 1 millinewton
10	millinewtons	= 1 centinewton
1000	millinewtons	= 1 newton
1000	newtons	= 1 kilonewton
1000	kilonewtons	= 1 meganewton

FREQUENCY
1000	hertz	= 1 kilohertz
1000	kilohertz	= 1 megahertz
1000	megahertz	= 1 gigahertz
1000	gigahertz	= 1 terahertz
1000	terahertz	= 1 petahertz
1000	petahertz	= 1 exahertz

GENERAL INFORMATION

METRIC WEIGHTS OF MATERIALS

Material	kg/m³	Material	kg/m³
Aerated concrete	800-960	Macadam	2100
Aluminium		Mahogany	
cast	2550	African	560
rolled	2700	Honduras	540
Asphalt		Spanish	690
natural	1000	Marble	2590-2830
paving	2080	Mortar	
Ballast, loose, graded	1600	cement, set	1920-2080
Beech	700	lime, set	1600-1760
Birch		Oak	
American	640	African	960
yellow	700	American red	720
Brass		English	800-880
cast	8330	Padouk	780
rolled	8570	Paint	
Bricks (common burnt clay)		aluminium	1200
stacked	1600-1920	bituminous emulsion	1120
sand cement	1840	red lead	3120
sand lime	2080	white lead	2800
ballast	1200	zinc	2400
brickwork	1920	Pine	
Cement		American red	530
bags	1280	British Columbian	530
bulk	1281-1442	Christina	690
casks	960	Oregon	530
Concrete (cement, plain)		Pitch	650
brick aggregate	18400	Plywood	480-640
clinker	1440	Polyvinyl chloride acetate	1200-1350
stone ballast	2240	Poplar	450
Concrete (cement, reinforced)		Portland cement	
1% steel	2370	loose	1200-1360
2% steel	2420	bags	1120-1280
5% steel	2580	drums	1200
Copper		Redwood	
cast	8760	American	530
drawn or sheet	8940	Baltic	500
Cork	130-240	non-graded	430
Deal, yellow	430	Rhodesian	910
Ebony	1180-1330	Slate	
Elm		Welsh	2800
American	670	Westmorland	3000
Dutch	580	Steel	7830
English	580	Stone	
Wych	690	Ancaster	2500
Fir		Bath	2080
Douglas	530	Darley Dale	2370
Silver	480	Forest of Dean	2430
Flint	2560	Granite	2640
Foam slag	700	Hopton Wood	2530
Freestone	2243-4280	Kentish rag	2670
masonry, dressed	2400	Mansfield	2260
rubble	2240	Portland	2240
Glass		Purbeck	2700
bottle	2720	York	2240
flint, best	3080	Tarmacadam	2080
flint, heavy	5000-6000	Teak	
plate	2800	Burma, African	650
Granolithic	2240	Walnut	660
Hardcore	1920	Water	
Hoggin	1760	fresh	1001
Iroko	600	salt	1009-1201
Iron		Whitewood	460
cast	7200	Zinc	
malleable cast	7370-7500	cast	6840
wrought	7690	rolled	7190
Lead		sheets, packed	900
cast or rolled	11300		
Lime			
acetate or, bags	1280		
blue lias, ground	850		
carbonate of, barrels	1280		
chloride of, lead-lined cases	450		
grey chalk, lump	700		
grey stone, lump	800		
hydrate, bags	510		

GENERAL INFORMATION

WEIGHTS OF VARIOUS METALS

in kilogrammes per square metre

mm	in	Steel kg	Wt. iron kg	Cast iron kg	Brass kg	Copper kg	Cast Lead kg
1.59	1/16	12.45	12.21	11.47	13.43	14.11	18.07
3.18	1/8	24.90	24.41	22.90	26.76	28.22	36.08
6.35	1/4	49.80	48.82	45.75	53.41	56.44	72.16
9.53	3/8	74.70	73.24	68.65	80.17	84.66	108.24
12.70	1/2	99.60	97.65	91.55	106.83	112.88	144.32
15.88	5/8	124.50	122.06	144.44	133.49	141.54	180.41
19.05	3/4	149.40	146.47	137.29	160.19	169.32	216.49
22.23	7/8	174.30	170.88	160.19	186.95	197.54	252.57
25.40	1	199.20	195.30	183.09	213.65	225.76	288.65

WEIGHTS OF ROUND AND SQUARE STEEL BARS

mm	Round kg/m	lb/lin ft	Square kg/m	lb/lin ft	mm	Round kg/m	lb/lin ft	Square kg/m	lb/lin ft
6	0.222	0.149	0.283	0.190	20	2.466	1.657	3.139	2.110
8	0.395	0.265	0.503	0.338	25	3.854	2.590	4.905	3.296
10	0.616	0.414	0.784	0.527	32	6.313	4.243	8.035	5.400
12	0.888	0.597	1.130	0.759	40	9.864	6.629	12.554	8.437
16	1.579	1.061	2.010	1.351	50	15.413	10.358	19.617	13.183

WEIGHTS OF FLAT STEEL BARS
in kilogrammes per linear metre

Thickness in millimetres and inches

Width mm	in	3.18 (1/8 in)	6.35 (1/4 in)	7.94 (5/16 in)	9.53 (3/8 in)	11.11 (7/16 in)	12.70 (1/2 in)	15.88 (5/8 in)	19.05 (3/4 in)	22.23 (7/8 in)	25.40 (1 in)
25.40	1	0.64	1.27	1.58	1.91	2.22	2.53	3.17	3.80	4.44	5.06
28.58	1 1/8	0.71	1.43	1.79	2.13	2.49	2.84	3.56	4.27	4.99	5.70
31.75	1 1/4	0.79	1.58	1.98	2.37	2.77	3.17	3.96	4.75	5.54	6.33
34.93	1 3/8	0.86	1.74	2.17	2.60	3.05	3.48	4.35	5.22	6.09	6.97
38.10	1 1/2	0.95	1.91	2.37	2.84	3.32	3.80	4.75	5.70	6.64	7.59
41.28	1 5/8	1.03	2.05	2.58	3.08	3.60	4.11	5.13	6.16	7.19	8.23
44.45	1 3/4	1.10	2.22	2.77	3.32	3.87	4.44	5.54	6.64	7.75	8.86
50.80	2	1.27	2.53	3.17	3.80	4.44	5.06	6.33	7.59	8.86	10.12
57.15	2 1/4	1.44	2.84	3.56	4.27	4.99	5.70	7.11	8.54	9.96	11.39
63.50	2 1/2	1.58	3.17	3.96	4.75	5.54	6.33	7.90	9.50	11.07	12.68
69.85	2 3/4	1.74	3.48	4.35	5.22	6.09	6.97	8.69	10.43	12.17	13.92
76.20	3	1.91	3.80	4.75	5.70	6.64	7.59	9.50	11.39	13.29	15.18
82.55	3 1/4	2.05	4.11	5.13	6.16	7.19	8.23	10.28	12.34	14.39	16.44
88.90	3 1/2	2.22	4.44	5.54	6.64	7.75	8.86	11.07	13.29	15.49	17.71
95.25	3 3/8	2.37	4.75	5.92	7.11	8.30	9.50	11.86	14.23	16.61	18.97
101.60	4	2.53	5.06	6.33	7.59	8.86	10.12	12.65	15.18	17.71	20.24
107.95	4 1/4	2.69	5.37	6.73	8.07	9.41	10.76	13.44	16.13	18.81	21.50
114.30	4 1/2	2.84	5.70	7.11	8.54	9.96	11.39	14.23	17.08	19.93	22.77
120.65	4 3/4	3.01	6.01	7.52	9.02	10.52	12.02	15.02	18.02	21.03	24.03
127.00	5	3.17	6.33	7.90	9.50	11.07	12.65	15.82	18.97	22.14	25.30
139.70	5 1/2	3.48	6.97	8.69	10.43	12.17	13.92	17.40	20.88	24.35	27.83
152.40	6	3.80	7.59	9.50	11.39	13.29	15.18	18.97	22.77	26.56	30.36
165.10	6 1.2	4.11	8.23	10.28	12.34	14.39	16.44	20.55	24.67	28.78	32.89
177.80	7	4.44	8.86	11.07	13.29	15.49	17.71	22.14	26.56	31.00	35.42
190.50	7 1/2	4.75	9.50	11.86	14.23	16.61	18.97	23.72	28.47	33.20	37.95
203.20	8	5.06	10.12	12.65	15.18	17.71	20.24	25.30	30.36	35.42	40.48
228.60	9	5.70	11.39	14.23	17.08	19.93	22.77	28.47	34.15	39.85	45.54
254.00	10	6.73	12.65	15.82	18.97	22.14	25.30	31.62	37.95	44.27	50.50
279.40	11	6.97	13.92	17.40	20.88	24.35	27.83	34.79	41.74	48.71	55.66
304.80	12	7.59	15.18	18.97	22.77	26.56	30.36	37.95	45.54	53.13	60.72

STANDARD WIRE GAUGE (SWG)
in millimetres and inches

SWG	mm	inches	SWG	mm	inches	SWG	mm	inches
1	7.62	0.300	11	2.95	0.116	21	0.813	**0.032**
2	7.00	0.276	12	2.64	0.104	22	0.711	**0.028**
3	6.40	0.252	13	2.34	0.092	23	0.610	**0.024**
4	5.89	0.232	14	2.03	0.080	24	0.559	**0.022**
5	5.38	0.212	15	1.83	0.072	25	0.508	**0.020**
6	4.88	0.192	16	1.63	0.064	26	0.457	**0.018**
7	4.47	0.176	17	1.42	0.056	27	0.417	**0.016**
8	4.06	0.160	18	1.22	0.048	28	0.376	**0.015**
9	3.66	0.144	19	1.02	0.040	29	0.345	**0.014**
10	3.25	0.128	20	0.914	0.036	30	0.315	**0.012**

GENERAL INFORMATION

BRANDS AND TRADE NAMES

7 - Interclad (UK) Ltd
202 - Sapa Building Systems (monarch and glostal)
238 - Chalmit Lighting
242 - Crawford Hafa Ltd
325 - Sapa Building Systems (monarch and glostal)
542 - Crawford Hafa Ltd
2000 - Healthmatic Ltd
"the Mix" - Ulster Carpet Mills Ltd
(IPAF Training) - Turner Access Ltd
(PASMA Training) - Turner Access Ltd
'Airtec Stone' by Alsecco - Telling Architectural Ltd
'Airtec Stone' by Alsecco - Telling Lime Products Ltd
>B< Range : >B< ACR, >B< Oyster, >B< Press, >B< Press Carbon Steel, >B< Press Gas, >B< Press Inox, >B< Press Solar, >B< Press Tools, >B< Push - Conex Universal Limited
0.35 Loft Match - Polypipe Building Products
100 High - Hudevad Britain
1581 Silicone - Dunbrik (Yorks) Ltd
1890's (GA) - Barber Wilsons & Co Ltd
20/20 Interlocking clay plain (tile) - Sandtoft Roof Tiles Ltd
25/40/56 Series - Albany Standard Pumps
2Wire Systems - BPT Security Systems (UK) Ltd
3 Step Mat - Dycem Ltd
316 Range - James Gibbons Format Ltd
3230 Classic Fibre Bonded Carpet - Burmatex Ltd
3500 Range - EcoWater Systems Ltd
3BoxStack™ - Straight Ltd
3D Range - Twyford Bathrooms
3Dlockers - 3d Storage Systems (UK) LTD
3G Synthetic Turf - Charles Lawrence Surfaces Ltd
3M - Kenyon Paints Limited
3M Commercial Graphics - Northgate Solar Controls
4200 Sidewalk fibre bonded carpet - Burmatex Ltd
425 Range - Horstmann Controls Ltd
4250 Alarm Confirmation - ADT Fire and Security Plc
4400 Broadway fibre bonded carpet - Burmatex Ltd
500 Series - Opella Ltd
506 NCE - Otis Ltd
513 NPE - Otis Ltd
606 NCT - Otis Ltd
800 Series conventional call systems - C-TEC (Computionics) Ltd
900 Series - Weir Pumps Ltd
9235 Waterproof membrane - Laticrete International Inc. UK

A

A + K - Anders + Kern UK Ltd
A Division of Castle House Joinery - Orchard Street Furniture
A Series - Andersen/ Black Millwork
A1 Concrete Flue Liners - Dunbrik (Yorks) Ltd
A100 - Auld Valves Ltd
Aalco - Aalco Metals Ltd
Aaztec - Aaztec Cubicles
Aaztec Associates Limited - Aaztec Cubicles
Aaztec Health - Aaztec Cubicles
Abacus - Smart F & G (Shopfittings)Ltd
Abacus Lighting Ltd - Abacus Lighting Ltd
Abacus, Aeon, Alien, Arat, Atilla - Pitacs Ltd
Abanaki - Furmanite International Ltd
Abbcol - The Abbseal Group incorporating Everseal (Thermovitrine Ltd)
Abbeydale - Chiltern Invadex
Abbotsford - Brintons Ltd
ABFT - Fire Industry Association
ABG - Ingersoll-Rand European Sales Ltd
ABK - Arnull Bernard J & Co Ltd
Abloy - ASSA ABLOY Limited
Abloy Disklock Pro - Abloy Security Ltd
Abloy Exec - Abloy Security Ltd
Abloy High Profile - Abloy Security Ltd
Abloy Protec - Abloy Security Ltd
Aborslot - Jones of Oswestry Ltd
Abrobility - Gabriel & Co Ltd
Abroclamps - Gabriel & Co Ltd
Abrotube - Gabriel & Co Ltd
ABS Floors - Maxit UK
ABS Pumps - Sulzer Pumps
AbSence/ PreSense - Philips Lighting Solutions
Absolut - Schiedel Chimney Systems Ltd
Absolute - Johnson Tiles
Abtex - ABG Ltd
AC - Howden Buffalo
AC 2000 - CEM Systems Ltd
AC 2000 Airport - CEM Systems Ltd
AC 2000 Lite - CEM Systems Ltd
Academy Bedrooms - Moores Furniture Group Ltd
Acare - ArjoHuntleigh UK
ACAS - Advisory, Conciliation and Arbitration Service
Access - Anson Concise Ltd
Access - Gradus Ltd
Access Ramp - Sportsmark Group Ltd
Access Steps - Sportsmark Group Ltd
Acclaim - Chalmit Lighting
Acclaim - Dudley Thomas Ltd
Acclaim - Sidhil Care
Accoflex - DLW Flooring
Accolade - Gledhill Water Storage Ltd
Accolade - Heatrae Sadia Heating Ltd
Accommodation 2 - Checkmate Industries Ltd
Accord - Stocksigns Ltd
Accorroni - May William (Ashton) Ltd
Accuro - Draeger Safety UK Ltd
Accu-Tune - IAC Acoustics
Acczent - Tarkett Ltd
ACE - Association for Consultancy and Engineering (ACE)
ACF DuraLED - Whitecroft Lighting Limited
ACH - Acoustic Ceiling Hangers - Christie & Grey Ltd
Ackermann Spider - Ackermann Ltd
Acklaim - Ackermann Ltd
Acknowledge - Ackermann Ltd
Ackwire - Ackermann Ltd
ACO Building Drainage - ACO Technologies plc
ACO Commercial Gully - ACO Technologies plc
ACO FreeDeck - ACO Technologies plc
ACO HexDrain® - ACO Technologies plc
ACO KerbDrain - ACO Technologies plc
ACO Modular 125 - ACO Technologies plc
ACO MultiDrain™ - ACO Technologies plc
ACO RainDrain® - ACO Technologies plc
ACO StormBrixx - ACO Technologies plc
ACO Water Management - ACO Technologies plc
Acorn - Oakdale (Contracts) Ltd
Acorn, Axis - Amwell Systems Ltd
Acoustic Vents - Simon R W Ltd
Acousticel - Bicester Products Ltd
Acousticel M20AD - Sound Service (Oxford) Ltd
Acousticel R10 - Sound Service (Oxford) Ltd
Acousticurtain - Acousticabs Industrial Noise Control Ltd
Acoustifoam - Acousticabs Industrial Noise Control Ltd
Acoustilay - Sound Reduction Systems Ltd
Acoustilouvre - Acousticabs Industrial Noise Control Ltd
ACOVA - Zehnder Ltd
ACP - ACP (Concrete) Ltd
Acrylacote - Dacrylate Paints Ltd
Acrylic 2 - Dufaylite Developments Ltd
ACT - Artur Fischer (UK) Ltd
Actimatic - Wade International Ltd
Actionair - Ruskin Air Management
Activ - Pilkington Birmingham
ActiV conventional fire detectors - C-TEC (Computionics) Ltd
Activa - Full Spectrum Fluorescent Tubes - Sylvania Lighting International INC
Active 8 - Gas Measurement Instruments Ltd
Activent - Fläkt Woods Ltd
Activity Floor - British Harlequin Plc
Activ-Ox - Feedwater Ltd
Aczent - Tarkett Sommer Ltd
AD11 - Tapworks Water Softeners
Adam Equipment - Phoenix Scales Ltd
Adams Rite - ASSA ABLOY Limited
Adanced counters & Serveries - Viscount Catering Ltd
Adaptalok - Adaptaflex Ltd
Adaptaring - Adaptaflex Ltd
Adaptaseal - Adaptaflex Ltd
Adaptasteel - Adaptaflex Ltd
Adash - Cil Retail Solutions Ltd
Adcor - Grace Construction Products Ltd
Addabound - Addagrip Terraco Ltd
Addacolor - Addagrip Terraco Ltd
Addaflor - Addagrip Terraco Ltd
Addagrip1000 System - Addagrip Terraco Ltd
Addalevel - Addagrip Terraco Ltd
Addaprime - Addagrip Terraco Ltd
Addaset - Addagrip Terraco Ltd
Addastone, Addastone TP - Addagrip Terraco Ltd
Addinox - Rockwood Pigments (UK) Ltd
Adesilex Range - Mapei (UK) Ltd
Adhesin - Henkel Ltd
Adhesives Toolkit - TWI Ltd
Adjusta Kitchens - Moores Furniture Group Ltd
Admatic - Potter & Soar Ltd
Adorn - Valor
ADPRO - Xtralis
Adria Perla - Armstrong World Industries Ltd
ADS - Schueco UK
ADT - Tyco Fire and Integrated Solutions
Adva - Grace Construction Products Ltd
Advance - Nilfisk Advance Ltd
Advance - Tata Steel Europe Limited
Advance Matt Gaffa - Advance Tapes International Ltd
Advance Premier Closure Plate Tape - Advance Tapes International Ltd
Advantage - Dixon Turner Wallcoverings
Advantage - Saint-Gobain Ecophon Ltd
Advantage Respirators - MSA (Britain) Ltd
Advantra® - H.B. Fuller
Advisor & Infoboard - Arbory Group Limited
AEG - Electrolux Domestic Appliances
Aercon - Aercon Consultants Ltd
Aero - Twyford Bushboard
Aerodyne cowl - Brewer Metalcraft
Aerovent - ABB Ltd
Aerton, Argylle Radiators - Barlo Radiators Limited
AET - AET.gb Ltd
AF29 Anti-Fracture for tile and stone - Carter-Dal International
Afasil - Illbruck Sealant Systems UK Ltd
Affinity - Hille Educational Products Ltd
Affinity Disposable Respirtiors - MSA (Britain) Ltd
Afghan - Carpets of Worth Ltd
A-Frame Picnic Suite - Branson Leisure Ltd
AG132 - Light Engine Lighting Ltd
Agabekov - Crescent Lighting
Agastat - Thomas & Betts Manufacturing Services Ltd
Agglio Conglomerate Marble - Reed Harris - Reed Harris
Agglosimplex Conglomerate Marble - Marble Flooring Spec. Ltd
Agito - HAGS SMP
Agrement certificates - British Board of Agrément
Agri drain - Naylor Drainage Ltd
Aida-DENCO - DencoHappel UK Ltd
Aidalarm - Hoyles Electronic Developments Ltd
Ainsworth OSB - Wood International Agency Ltd
Ainsworth Pourform 107 - Wood International Agency Ltd
Ainsworth Pourform H00 - Wood International Agency Ltd
Air Marshall - Rapaway Energy Ltd
Air Minder - Vent-Axia
Air Miser - Rapaway Energy Ltd
Air Safety Ltd - GVS Filter Technology UK
Air Sentry - S & B Ltd
Airack - Clenaware Systems Ltd
Airbath - Astracast PLC
Aircare 2000 series - Aircare Europe Ltd
AirCore®, AirLiner® - Ryton's Building Products Ltd
Aircrete - Hanson Thermalite
Airdale - Airedale International Air Conditioning Ltd
Airdale Service - Airedale International Air Conditioning Ltd
Airdor - S & P Coil Products Ltd
AirDuct - Airflow (Nicoll Ventilators) Ltd
Airetronix - Airedale International Air Conditioning Ltd
Airfix - Flamco UK Ltd
Airguard - Aquastat Ltd
Airmaster ventilation - SAV United Kingdom Ltd
Airseal - Fillcrete Ltd
Airspeed - Warner Howard Group Ltd
Airsprung Beds Ltd - Airsprung Beds Ltd
Airstar - Rapaway Energy Ltd
Airstream - PHS Group plc
Airstrip - Simon R W Ltd
Airtec Glass Walling - Telling Architectural Ltd
Airtec Glass Walling - Telling Lime Products Ltd
Airworld - Warner Howard Group Ltd
Aitersall - Deva Tap Co Ltd
AJ - Architects Journal - EMAP
Akademy Classroom System - Portakabin Ltd
Akor - ABB Ltd
Akord - DW Windsor Lighting
Aktivated Showers - Hansgrohe
AL 50 - SG System Products Ltd
Alabastine - Polycell Products Ltd
Alag Aggregate - Lafarge Aluminates Ltd
Alarm-a-fence® - Jackson H S & Son (Fencing) Ltd
Alarmline - Linear Heat Detection - Kidde Fire Protection Ltd
AlarmSense - Apollo Fire Detectors Ltd
ALB Difutec Underfloor heating - SAV United Kingdom Ltd
Albany - Gloster Furniture Ltd
Albany Paints, Wallcoverings, Brushes, Paint rollers - Brewer C & Sons Ltd
Albany Plus - MK (MK Electric Ltd)
Albi-Grips - Albion Manufacturing
Albion - Swedecor
Albi-Slings - Albion Manufacturing
Albi-Traps - Albion Manufacturing
Alclad - Solair Ltd
Alco Controls - Emerson Climate Technologies Retail Solutions
Alcoa Extrusions - Alcoa Custom Extrudes Solutions
Alcona Range - Twyford Bathrooms
ALD - ALD Lighting Solutions
Aldwark - York Handmade Brick Co Ltd
Alfacryl - Illbruck Sealant Systems UK Ltd
Alfapower - MJ Electronics Services (International) Ltd
Alfas - Illbruck Sealant Systems UK Ltd
Alfas Bond FR - Compriband Ltd
Alfatherm - AEL
Alfed - Aluminium Federation Ltd
Alfix - Baj System Design Ltd
Aliflex - Flexible Ducting Ltd
Aligator - Alutec
Aligator ® - Marley Plumbing and Drainage
Alimaster Shelving - Bedford Shelving Ltd
Alimat - Watts Industries UK Ltd
Aline - Cil Retail Solutions Ltd
A-Line - Desking Systems Ltd
Alitherm - Smart Systems Ltd
ALKORBRIGHT - RENOLIT Cramlington Limited
ALKORDESIGN - RENOLIT Cramlington Limited
ALKORGREEN - RENOLIT Cramlington Limited
Alkorplan - Landline Ltd
ALKORPLAN - RENOLIT Cramlington Limited
ALKORPLUS - RENOLIT Cramlington Limited
ALKORSOLAR - RENOLIT Cramlington Limited
ALKORTOP - RENOLIT Cramlington Limited
Alkysil - Dacrylate Paints Ltd
All in one sealer - Sealocrete PLA Ltd
All Range - Twyford Bathrooms
Allart Deco - Allart, Frank, & Co Ltd
AllClear - Lakes Bathrooms Limited
Allegra - Hansgrohe
Allegro - Erlau AG
Allen Bradley - Electrix International Ltd
Allergi Guard - Nationwide Filter Company Ltd
Allgood Hardware - Allgood plc
Allgood Secure - Allgood plc
Allied Kingswood - Waterline Limited
alligator entrance matting - Jaymart Rubber & Plastics Ltd
Allmat - Allmat (East Surrey) Ltd
Alno - Alno (United Kingdom) Ltd
Alno 2000 - Alno (United Kingdom) Ltd
Alnova - Alno (United Kingdom) Ltd
AL-OR - Harbro Supplies Ltd
Alpha - Anson Concise Ltd
Alpha - Titan Polution Control
Alpha 24 Secure - Premdor
Alpha Cool - Airedale International Air Conditioning Ltd
Alpha Design 24 - Premdor
Alpha scan - Milliken Carpet
Alpha, Alpha Flow, Alpha Kerb, Alpha Antique - Brett Landscaping
Alphaline - ABG Ltd
Alpina - Cembrit Ltd
Alpolic - Euroclad Ltd
ALPOLIC®/fr - Booth Muirie
Altholz - LSA Projects Ltd
Althon - Althon Ltd
Althon-Lite - Althon Ltd
Altitude - Dunhams of Norwich
Alto - Ideal-Standard Ltd
Altro PU - Altro Ltd
Altroscreed Quartz - Altro Ltd
Altrosmooth - Altro Ltd
Altrotimber - Altro Ltd
ALU - Osram Ltd
Alu Combi, Alu Dano, Alu Intro, Alu Rapid, Alu Rapid Junior, Alu Rapid Super and Alu Top staging - CPS Manufacturing Co
Alu-Care - Aluline Ltd
Alucobond® - Booth Muirie
Aluglas - Clow Group Ltd
Aluglaze - Panel Systems Ltd
Aluglide - Integra Products
Alu-Lift portable aluminium gantry crane - Metreel Ltd
Aluma - Sapphire Balustrades
Alumatilt - Caldwell Hardware (UK) Ltd
Aluminex - HW Architectural Ltd
Aluminium Systems - Barlow Group
aluminium/hybrid curtain walling, windows and doors - Senior Aluminium Systems plc
Alutec - Marley Plumbing and Drainage
AluTrap - Aluline Ltd
Aluzinc - SSAB Swedish Steel Ltd
Aluzyme - Aluline Ltd
Alwitra - ICB (International Construction Bureau) Ltd
Ama Drainer - KSB Ltd
Amadeus-equipment - Amadeus
Ama-Drainer Box - KSB Ltd
Amal - Bailey Birkett Sales Division
Ama-porter - KSB Ltd
Ama-Porter ICS - KSB Ltd
Amarex - KSB Ltd
Amazon - Potter & Soar Ltd
Amazon Filters - Barr & Wray Ltd
Amber Booth - Designed for Sound Ltd
Ambi-deck - Redman Fisher Engineering
Ambi-Rad - AmbiRad Ltd
Amdega - Amdega Ltd
Amercoat - Kenyon Group Ltd
Amercoat - PPG Protective and Marine Coatings
Amerlock - Kenyon Group Ltd
Ameron - Andrews Coatings Ltd
AMF - Knauf AMF Ceilings Ltd
Amida - Genie Europe
Amie - Tunstal Healthcare (UK) Ltd
AmiLake Solutions - Amilake Southern Ltd
AMIRAN Anti-Reflective Glass - SCHOTT UK Ltd
AML Connections Ltd - Keyline
Amor metals international - Aalco Metals Ltd
Amore - Brockway Carpets Ltd
Amorim - Amorim (UK) Ltd
AMP - Tyco Fire and Integrated Solutions
Amphistoma - Adams- Hydraulics Ltd
Amtico - The Amtico Company
Ancaster - Realstone Ltd
Ancestry - Ibstock Brick Ltd
Anchorlite slate - Forticrete Roofing Products
Ancon - Ancon Building Products
AnconOptima - Ancon Building Products
Anda-Crib - PHI Group Ltd
Anders & Kern - Anders + Kern UK Ltd

A (con't)

Andersen ®, Andersen Art Glass - Andersen/ Black Millwork
Anderton interloc - Anderton Concrete Products Ltd
Andrews Air Conditioning - Andrews Sykes Hire Ltd
Andrews Heat for Hire - Andrews Sykes Hire Ltd
Andromeda, APC - Schneider Electric Ltd
Andura - Andura Coatings Ltd
Anglepoise - Anglepoise Lighting Ltd
Angletie - Redifix Ltd
Anglian - Invicta Window Films Ltd
Anglo - Quiligotti Terrazzo Tiles Limited
Ankalok Revetment - Ruthin Precast Concrete Ltd
Anki - Sparkes K & L
Annabelle Classique - Brintons Ltd
Anolight - Gooding Aluminium
Ansty - Hamworthy Heating Limited
Antel - Andrews Coatings Ltd
Anti Climb - Andura Coatings Ltd
Anti fatigue mat - Dycem Ltd
Antimicrobial Copper - Copper Development Association
Antique - Samuel Heath & Sons plc
Anti-Social Stud - Sportsmark Group Ltd
Antium - Decra Ltd
AP Acoustic Absorption Panel - Salex Acoustics Limited
Apa plywood - Wood International Agency Ltd
Apache - Ellis Patents
apachestat soft foot - Jaymart Rubber & Plastics Ltd
Aperture - BriggsAmasco
Apex Concrete Garages - Liget Compton
Apex Doors - Regency Garage Door Services Ltd
Apex heritage - Alumasc Exterior Building Products Ltd
Apex range - Dixon Turner Wallcoverings
APF International Forestry Exhibition - Confor: Promoting forestry and wood
APHC - Association of Plumbing and Heating Contractors
Apollo incinerators - Combustion Linings Ltd
Apollo Windows - HW Architectural Ltd
APP Liner - Acousticabs Industrial Noise Control Ltd
Appollo - Astracast PLC
APT Barriers - APT Controls Ltd
APT Bollards (Hydraulic) - APT Controls Ltd
Apton Partitions - Adex Interiors for industry Ltd
APW Ltd - Associated Perforators & Weavers Ltd
Aqua - Clear - Aldous & Stamp (Services) Ltd
Aqua Blue Designs - Aqua-Blue Ltd
Aqua Tackseal - Charcon Tunnels (Division of Tarmac plc)
AquaBarrier - Mat.Works Flooring Solutions
Aquachill - J & E Hall Ltd
Aquaflex - Richard Hose Ltd
Aquaflow - Armorex Ltd
Aquaflow - Formpave Ltd
Aquaflow ML - Formpave Ltd
Aquafun - Hansgrohe
Aquagrip - Viking Johnson
Aquaguard - Aquastat Ltd
Aquaheat and Aquaheat SS - Santon
Aquahib - Aquastat Ltd
Aquair, AquaTec - Johnson & Starley
Aqualine - Alumasc Exterior Building Products Ltd
Aqualine - Rangemaster
Aqualine - Santon
Aqualine Range - Amwell Systems Ltd
Aqua-Lite - Naylor Drainage Ltd
Aqualock - Respatex International Limited
Aqualock Surface DPM - Isocrete Floor Screeds Ltd
Aquamac - Schlegel UK (2006) Ltd
Aquamaster - ABB Instrumentation Ltd
Aquaminder - Setsquare
Aquamix - Watts Industries UK Ltd
Aquamixa - Aqualisa Products
Aquapanel - Knauf UK
Aquaprobe - ABB Instrumentation Ltd
Aquarelle - Tarkett Sommer Ltd
Aquarian - Aqualisa Products
Aquarius - Santon
Aquasave - HRS Hevac Ltd
Aquasave BIO - HRS Hevac Ltd
Aquasentry - Aquasentry
Aquasil - Aldous & Stamp (Services) Ltd
Aquaslab - Formpave Ltd
Aquaslot, Aquadish, Aquadrain - Jones of Oswestry Ltd
Aquasol - Conren Ltd
Aquastore - CST Industries, Inc. - UK
Aquastream - Aqualisa Products
Aquastyle - Aqualisa Products
Aquasystems - George Fischer Sales Ltd
Aquatec - Loblite Ltd
Aquatec Beluga bathlift - Bison Bede Ltd
Aquatec Elan bathlift - Bison Bede Ltd
Aquatec Fortuna bathlift - Bison Bede Ltd
Aquatek - Grace De Neef UK
Aquatique - Aqualisa Products
Aquator - Jacopa Limited
Aquatread - Tor Coatings Ltd
Aquavalve - Aqualisa Products
AquaWorks - Mat.Works Flooring Solutions
AR - The Architectural Review - EMAP
Arbo - Adshead Ratcliffe & Co Ltd
Arbocrylic - Adshead Ratcliffe & Co Ltd
Arbofoam - Adshead Ratcliffe & Co Ltd
Arbokol - Adshead Ratcliffe & Co Ltd
Arbomast - Adshead Ratcliffe & Co Ltd
Arboseal - Adshead Ratcliffe & Co Ltd
Arbosil - Adshead Ratcliffe & Co Ltd
Arbothane - Adshead Ratcliffe & Co Ltd
Arc - Illuma Lighting
ARCA - Asbestos Removal Contractors' Association
Arcade - Shackerley (Holdings) Group Ltd incorporating Designer ceramics
Arcadia Pantile - Sandtoft Roof Tiles Ltd
Arcadian - Haddonstone Ltd
Arcadian garden features - Haddonstone Ltd
Arcelik - Beko plc
Archco-Rigidon - Winn & Coales (Denso) Ltd
Archform - Helmsman
Architectural - Protim Solignum Ltd
Architectural Concrete Ltd - Marble Mosaic Co Ltd, The
Architectural coping - Alifabs (Woking) Ltd
Architectural Facing Masonry - Lignacite Ltd
Architectural Vinyls - Architectural Textiles Ltd
Architectural Wallcoverings - Architectural Textiles Ltd
Architectural Window Films - Northgate Solar Controls
Architen Landrell Associates Ltd - Architen Landrell Associates Ltd
Archmaster® - Simpson Strong-Tie®
Archtec - Cintec International Ltd
Arcitectural Textiles - Architectural Textiles Ltd
Arcus - Erlau AG
Arcus - Laidlaw Ltd
Ardenbrite - Tor Coatings Ltd
Ardesia - EBC UK Ltd
Ardex - Ardex UK Ltd
Ardex DPM - Ardex UK Ltd
Ardex Feather Finish - Ardex UK Ltd
Ardex UK Ltd - Ardex UK Ltd
Ardex-flex - Ardex UK Ltd
Arditex - Ardex UK Ltd
Arena - HAGS SMP
arena stadia seating - Metalliform Products Plc
Arena Walsall - Legrand Electric Ltd
Arenastone - Toffolo Jackson(UK) Ltd
ARES - Commercial Lighting Systems
Arezzo-25, Alto and Demi Alto - Lang+Fulton
Argelith - Ancorite Surface Protection Ltd
Argenta - Imperial Machine Co Ltd
ArGeton - Telling Architectural Ltd
ArGeton - Telling Lime Products Ltd
Argo - TSE Brownson
Argo-Braze™ - Johnson Matthey PLC - Metal Joining
Aria - Ellis J T & Co Ltd
Ariane Spiral - Loft Centre Products
Aries - Fitzgerald Lighting Ltd
Aristocast Originals LTD - Troika Contracting Ltd
Aristom - Diamond Merchants
Aritco - Gartec Ltd
Arlon - DLW Flooring
Armabuild - Elwell Buildings Ltd
Armadillo Range - Wybone Ltd
Armaflex - Davant Products Ltd
Armarr Sectional + Roller Shutters - Regency Garage Door Services Ltd
Armaseam - Alumasc Exterior Building Products Ltd
Armbond - Armstrong (Concrete Blocks), Thomas, Ltd
Armorcote - Armorex Ltd
Armordon - Attwater & Sons Ltd
Armorflex - Concrete Products (Lincoln) 1980 Ltd
Armorloe - Concrete Products (Lincoln) 1980 Ltd
Armorsol - Armorex Ltd
Armortec - Concrete Products (Lincoln) 1980 Ltd
Armourcast - Gillespie (UK) Ltd
armourclad - HMG Paints
Armourguard - HAG Shutters & Grilles Ltd
Armspan - Armfibre Ltd
Armstead - Brewer C & Sons Ltd
Armstead Trade - AkzoNobel Decorative Paints UK
Armstrong - Kitsons Insulation Products Ltd
Armstrong - Nevill Long Limited
Armstrong - New Forest Ceilings Ltd
Armstrong DLW - Armstrong Floor Products UK Ltd
Armtage - Olby H E & Co Ltd
Aro Pumps - Ingersoll-Rand European Sales Ltd
Arosa - Gloster Furniture Ltd
Arpa - Deralam Laminates Ltd
Arpax - Leigh's Paints
Arpeggio - Rawson Carpets Ltd
Arresta Rail - Unistrut Limited
Arroclip - Barlow Group
Arrogard - Barlow Group
Arrolok - Barlow Group
Arrone - HOPPE (UK) Ltd
Arrow - Walls & Ceilings International Ltd
Arstyl - NMC (UK) Ltd
Art Deco Suite - Brass Art Ltd
Artalu - Baj System Design Ltd
Artcomfort - Amorim (UK) Ltd
Arteco - British Gypsum Ltd
Artek - Astron Buildings Parker Hannifin PLC
Artemide - Quip Lighting Consultants & Suppliers Ltd
Artemis - Desking Systems Ltd
Artemis - Stocksigns Ltd
Artesit - Erlau AG
Artificial Grass - Sportsmark Group Ltd
Artikula - Anglepoise Lighting Ltd
Artile - Johnson Tiles
Artoleum - Forbo Flooring
Arunhithe - Falco UK Ltd
AS Aqua Seal - Consult Lighting Ltd
ASA - Association of Sealant Applicators Ltd (ASA)
Asbestoscheck Interactive - National Britannia Ltd
Ascari - Dimplex UK Limited
Asco - ASCO Extinguishers Co. Ltd
ASFP - Association of Specialist Fire Protection (ASFP)
Ashbury - Paragon by Heckmondwike
AshDeck™ Structural decking - Ash & Lacy Building Products Ltd
AshFab™ Specialist fabrications - Ash & Lacy Building Products Ltd
AshFix™ Fasteners - Ash & Lacy Building Products Ltd
AshFlow™ Rainwater amangement systems - Ash & Lacy Building Products Ltd
AshGrid™ Spacer support system - Ash & Lacy Building Products Ltd
AshJack™ Lightweight over roof conversion system - Ash & Lacy Building Products Ltd
Ashley - Hager Ltd
Ashmount (cigarette bins) - Glasdon UK Limited
Ashphaltic Roofing Supplies - SIG Roofing
AshScreen™ Solid, perforated and expanded screens - Ash & Lacy Building Products Ltd
AshTech™ Rainscreen cladding - Ash & Lacy Building Products Ltd
AshTray™ Structural trays - Ash & Lacy Building Products Ltd
Ashworth Speedy tester - Protimeter, GE Thermometrics (UK) Ltd
Ashworth Weyrite - Phoenix Scales Ltd
AshZip™ Standing seam roofing system - Ash & Lacy Building Products Ltd
ASIM - Xtralis
ASIS-Y - Optelma Lighting Ltd
Aspect - Dunhams of Norwich
Aspiromatic - Docherty New Vent Chimney Group
Aspirotor - Brewer Metalcraft
Aspirotor - Marflex Chimney Systems
ASR 683 - Monowa Manufacturing (UK) Ltd
ASR, Architectural System Ramps, AeroLight, Accesscentre - Enable Access
Assa Abloy - Mul-T-Lock
Asset seating - CPS Manufacturing Co
Associated Metal - Pland Stainless Ltd
Association of Fencing Contractors (ASFC) - Fencing Contractors Association
Assure - Tata Steel Colors
Assured Data Sheets - British Board of Agrément
Astera Lighting - LSA Projects Ltd
Astoria - Muraspec
Astra - Guardall Ltd
Astra Micra - Guardall Ltd
Astracast - Astracast PLC
Astralite (minor surgical lamps) - Brandon Medical
Astro - Zellweger Analytic Ltd, Sieger Division
Astron - Astron Buildings Parker Hannifin PLC
Astron MSB - Astron Buildings Parker Hannifin PLC
Astronet - Astron Buildings Parker Hannifin PLC
Astrotec - Astron Buildings Parker Hannifin PLC
ASX - Hochiki Europe (UK) Ltd
AT ™ - Pressure Coolers Ltd T/A Maestro Pressure Coolers
Atea - Kermi (UK) Ltd
Atelier Sedap - Optelma Lighting Ltd
Athletics – Non Porous Prefabricated System - Charles Lawrence Surfaces Ltd
Athletics - Porous Spray System - Charles Lawrence Surfaces Ltd
Athletics- Non Porous Sandwich System - Charles Lawrence Surfaces Ltd
Athmer - Strand Hardware Ltd
Atlantic - Signs & Labels Ltd
Atlantis - Lotus Water Garden Products Ltd
Atlas - Ellis Patents
Atlas (medical equipment pendants) - Brandon Medical
Atlas (Smart & Brown) - Bernlite Ltd
Atlas Buildings - Atlas Ward Structures Ltd
Atlas Concorde - Swedecor
Atlas Fire - Tyco Fire and Integrated Solutions
Atlas Solar - Atlas Solar
Atox - Aldous & Stamp (Services) Ltd
Atticvent - Sola Skylights
Audio System 200 - BPT Security Systems (UK) Ltd
Audiopath' Voice Alarm Systems - PEL Services Limited
Aura - Eaton Electric Limited
Aura, Aura Kerb - Brett Landscaping
Ausmark - Ramset Fasteners Ltd
Authentic Collection - Masterframe Windows Ltd
Auto M - Dantherm FiltrationLtd
Auto Pile - Skanska UK Plc
Autodor - Garador Ltd
Auto-Freway - Tuke & Bell Ltd
Autoglide - Cardale Doors Ltd
Autoglide - Silent Gliss Ltd
AutoKote - Grace Construction Products Ltd
Automat - Peek Traffic Ltd
Automax - Flowserve Flow Control
Autopa - Autopa Ltd
AutoStamp™ Franking & Weighing System - Neopost Ltd
Autostic - Minco Sampling Tech UK Ltd (Incorporating Fortafix)
Autowalk - Kone Escalators Ltd
Avalon Glass Tiles - Kirkstone
Avanta - Broag Ltd
Avanti Partitions - Logic Office Contracts
Avanza Benching System - Envopak Group Ltd
Avaya - Connaught Communications Systems Ltd
Avenue - Neptune Outdoor Furniture Ltd
Avenue Dot - Whitecroft Lighting Limited
Avipoint, Avishock - P+L Systems Ltd
avon gutter system - Dales Fabrications Ltd
Avon Industrial Doors Ltd - Avon Industrial Doors Ltd
AWS - Schueco UK
Axa - Hansgrohe
Axcess - Rackline Systems Storage Ltd
Axima - Cofely Engineering Services Ltd
Axiom - Armstrong World Industries Ltd
Axis - Dixon Turner Wallcoverings
Axis Digital - Aqualisa Products
Axjet - Howden Buffalo
Axor - Hansgrohe
Axor Citterio - Hansgrohe
Axpet - Amari Plastics Plc
Axtehal GX Fire Resistant Plastics - Plasticable Ltd
Axxo - Rivermeade Signs Ltd
Axxys - Benlowe Stairs
Azimex - Azimex Fabrications Ltd
Aztec - Scotmat Flooring Services
Aztec Boilers - TR Engineering Ltd
Aztec Europe Ltd - Flexel International Ltd

B

B G Bertha - Kair ventilation Ltd
B&ES - Building & Engineering Services Association
B&F - BF Bassett & Findley
B.P - Anderson Gibb & Wilson
Baby Ben - General Time Europe
BACS - British Association for Chemical Specialities
Bacticell - Nationwide Filter Company Ltd
Bacti-clean - Liquid Technology Ltd
Bacticoat - Nationwide Filter Company Ltd
Bacti-coat floor coating - Liquid Technology Ltd
Bacti-coat wall / ceiling coating - Liquid Technology Ltd
Bactiguard - Liquid Technology Ltd
Badu - Golden Coast Ltd
Baffles - Rockfon Ltd
Baggeridge - Baggeridge Brick PLC
Bailey - Bailey Birkett Sales Division
Bal - R J Stokes & Co Limited
Bal - Admix AD1/GT1 - Building Adhesives Ltd
Bal - Easypoxy - Building Adhesives Ltd
Bal - Floor Epoxy - Building Adhesives Ltd
Bal - Grout - Building Adhesives Ltd
Bal - Pourable Thick Bed - Building Adhesives Ltd
Bal - Rapidset - Building Adhesives Ltd
Bal - Superflex Wall & Wide Joint Grouts - Building Adhesives Ltd
Bal - Wall Green/Blue/White Star - Building Adhesives Ltd
Bal Gold Star - Building Adhesives Ltd
Bal Microseal - Building Adhesives Ltd
Bal Supergrout - Building Adhesives Ltd
Bal- Wide Joint Grout - Building Adhesives Ltd
Balley Birkett - The BSS Group
Ball-O-Star Ball Valves - Klinger Fluid Instrumentation Ltd
Ball-O-Top - Klinger Fluid Instrumentation Ltd
Balmoral - Davant Products Ltd
Balmoral - Loft Centre Products
Balmoral - Sandtoft Roof Tiles Ltd
Balmoral bunded tanks - Balmoral Tanks
Balmoral spaceformer modular buildings - Balmoral Tanks
Balmoral superfill - Balmoral Tanks
Baltic - Neptune Outdoor Furniture Ltd
Baltimore - Townscape Products Ltd
Balzaro - Girsberger London
Bamboo,Banio, Bolero, Bombe, Bosporos, Buckingham - Pitacs Ltd
Band-seal - Naylor Drainage Ltd
Barbican - Carpets of Worth Ltd
Barbican Steel Doors - Bradbury Group Ltd
Barbican® - Jackson H S & Son (Fencing) Ltd
Barclay Kellett - Albany Standard Pumps
Barcooler - Imperial Machine Co Ltd
Bardoline - Onduline Building Products Ltd
Barduct - Eaton Electric Ltd
Barent - Neptune Outdoor Furniture Ltd
Barkeller - Imperial Machine Co Ltd
Barkers engineering Ltd - Chestnut Products Limited
Barkwith engineering - Ermine Engineering Co. Ltd
Barley Twist - Benlowe Stairs
Barlo - Diamond Merchants
Barlo Radiators - Barlo Radiators Limited
Barlow metal fabrications - Barlow Group
Barnwood - Barnwood Shopfitting Ltd
Barrier Foil - Thatching Advisory Services Ltd
Barrikade - Minco Sampling Tech UK Ltd (Incorporating Fortafix)
Bartender - Imperial Machine Co Ltd
Bartoline - Bartoline Ltd
Barton & Merton - Anthony de Grey Trellises + Garden Lighting
Barwil - Barber Wilsons & Co Ltd
BASA - British Adhesives and Sealants Association
Basalton Revetment - Ruthin Precast Concrete Ltd
Baseline - Twyford Bushboard
Basepak - Flamco UK Ltd
BASF - Licensed applicators for BASF Resin Products - Ancorite Surface Protection Ltd
Basis - Flexiform Business Furniture Ltd
Bat - Allmat (East Surrey) Ltd
Bat - Expamet Building Products
Batchmatic - BSA Machine Tools Ltd
Bathscreens by Coram - Coram Showers Ltd
Batiguard - Nationwide Filter Company Ltd
Battleship - Heckmondwicke FB
Bauclad - CEP Claddings Ltd
Bauclad - Omnis
Baufix - Pump Technical Services Ltd
Baxall - Vidionics Security Systems Ltd
Bayferrox - LANXESS Inorganic Pigments Group
Bayrel - BioLab UK
BBA - British Board of Agrément
BCA - MPA Cement
BCB - B C Barton Limited
BCC - British Ceramic Confederation
BCF - British Coatings Federation
BCH - Coates LSF Ltd

B (con't)

BCIA - The Federation of Environmental Trade Association (FETA)
BCIS - Royal Institution of Chartered Surveyors (RICS)
BCIS Online - BCIS
BCMA - The Carpet Foundation
B-Comm - Kaba Ltd
BDR Thermea - Santon
Beach Glass Mosaics - Kirkstone
Beam seating, Bench seating - CPS Manufacturing Co
BEAMA - British Electrotechnical and Allied Manufacturers Association
Beamform - Rom Ltd
Beany Block - Marshalls Mono Ltd
Bearl Sandstone - Dunhouse Quarry Co Ltd
Beaufont - Benlowe Group Limited
Beaufort - BF Bassett & Findley
Beaufort - Neptune Outdoor Furniture Ltd
Beaumont Chimneys - Beaumont Ltd F E
Beauty of marble collection - Emsworth Fireplaces Ltd
Beaver - Hipkiss, H, & Co Ltd
BEBO® - Asset International Ltd
Becker Acroma - Anderson Gibb & Wilson
Beco Wallform - Beco Products Ltd
Bedale - Oakdale (Contracts) Ltd
Bee Bumps - Sportsmark Group Ltd
Beehive Grit Bin - Sportsmark Group Ltd
Beeline - Ciret Limited
Beemul - Emusol Products (LEICS) Ltd
Bekaert - Betafence Limited
Bekaert fencing - Chestnut Products Limited
Bekaert Fencing 10 yr guarantee system - SJ McBride Ltd T/A Warefence
Bekassure - Termstall Limited
Beko - Beko plc
Belair - Draeger Safety UK Ltd
Belcor - May William (Ashton) Ltd
Belgravia - S & P Coil Products Ltd
BELL - British Electric Lamps Ltd
Bell Bollard - Furnitubes International Ltd
Bell Fireplaces - Bell & Co Ltd
Bell fires - Bell & Co Ltd
Bell Twist - Brintons Ltd
Belle - Chippindale Plant Ltd
Belzona 3131 WG Membrane - Belzona PolymericLtd
Belzona 4111 Magma Quartz - Belzona PolymericLtd
Belzona 4131 Magma Screed - Belzona PolymericLtd
Belzona 4141 Magma Build - Belzona PolymericLtd
Belzona 4311 CRI Chemical Barrier - Belzona PolymericLtd
Belzona 441 Granogrip - Belzona PolymericLtd
Belzona 5122 Clear Cladding Concentrate - Belzona PolymericLtd
Belzona 5231 SG Laminate - Belzona PolymericLtd
Belzona 5811 Immersion Grade - Belzona PolymericLtd
Belzona 8111 Mouldable Wood - Belzona PolymericLtd
Bendix - Electrolux Domestic Appliances
Bendywood - The Angle Ring Company Ltd
Benga W - Jotun Henry Clark Ltd Decorative Division
Bennett's Barges - Aggregate Industries UK Ltd
BentoJect, Bentorub, Bentosteel, Bentostic - Grace De Neef UK
Bentonite Waterproofing Membrane - Cordek Ltd
Berber Twist Forte - Brockway Carpets Ltd
Bergen - Loft Centre Products
Berger Lahr - Schneider Electric Ltd
Berkeley - Carpets of Worth Ltd
Berkeley - Crittall Steel Windows Ltd
Berkeley Variations - Carpets of Worth Ltd
Berriedale - Chiltern Invadex
Berry - Dimplex UK Limited
Berry - Seagoe Technologies Ltd
Berry Beam - Berry Systems
Besaplast - Movement Joints (UK)
Beta De Luxe - Naco
Beta Naco - Naco
Beta Tower - Turner Access Ltd
Beta, Beta Flow - Brett Landscaping
Betoatlas - Grass Concrete Ltd
Betoconcept - Grass Concrete Ltd
Betokem - Maxit UK
Betonap - Grass Concrete Ltd
Betotitan - Grass Concrete Ltd
Bewater - Vidionics Security Systems Ltd
BFA - Cast Metals Federation
BFCMA - British Flue & Chimney Manufacterers Association
BFCMA - The Federation of Environmental Trade Association (FETA)
BFPSA - Fire Industry Association
BFRC - British Fenestration Rating Council
BFT gate automation and barriers - Parsons Brothers Gates
BGL - Benlowe Stairs
BICTA - Cast Metals Federation
BiDi Safe - Safetell Ltd
Bidston Door - AS Newbould Ltd
Bifjet - Howden Buffalo
Biflo - Grohe Ltd
Big Ben - General Time Europe
Big Green Chart - Sportsmark Group Ltd
Bigbond - BigHead Bonding Fasteners Ltd
Bighead® - BigHead Bonding Fasteners Ltd
Bigrings - BigHead Bonding Fasteners Ltd
Bigstrip - BigHead Bonding Fasteners Ltd
Bigwood Stokers - Hodgkinson Bennis Ltd
Bilco - Bilco UK Ltd
Bimagrip - Clare R S & Co Ltd
Binnlock Mesh Fencing - Binns Fencing Ltd
Binwall - Grass Concrete Ltd
Bioclip - Delabie UK Ltd
Biofil - Delabie UK Ltd
Biogard - Leigh's Paints
Biogaurd - Armstrong World Industries Ltd
Biogene - Corroless Corrosion Control
Bioguard - BioLab UK
Biomass Grease Traps - Progressive Product Developments Ltd
Biometric - FC Frost Ltd
Biopure - Veoila Water Technologies UKElga Ltd
Biostat - Tuke & Bell Ltd
Biotec - Titan Polution Control
Birch Interiors - Designer Radiators Direct
Birco - Marshalls Mono Ltd
Birkett - Bailey Birkett Sales Division
BIS - Walraven Ltd
Bisley - Bisley Office Equipment
Bison Beams - ACP (Concrete) Ltd
Bitubond - SCP Concrete Sealing Technology Ltd
Bituflex - SCP Concrete Sealing Technology Ltd
Bitufor - Bekaert Building Products
Bitusheet - SCP Concrete Sealing Technology Ltd
Bituthene - Grace Construction Products Ltd
Bitutint - Procter Johnson & Co Ltd
BK - Imperial Machine Co Ltd
BL Bollard - Consult Lighting Ltd
Black Beauty - Valor
Black Ebony - Manders Paint Ltd
Blackfriar - Blackfriar Paints Ltd
Blackfriar - Tor Coatings Ltd
Blackheat® - Roberts-Gordon
Blackjack Square Spiral - Loft Centre Products
Blade - Cil Retail Solutions Ltd
Blake - Tuke & Bell Ltd
Blakley - Blakley Electrics Ltd
Blanc de Bierges - Blanc de Bierges
Blast Coolers - Turnbull & Scott (Engineers) Ltd
Blaw-Knox - Ingersoll-Rand European Sales Ltd
Blaxter Sandstone - Dunhouse Quarry Co Ltd
Blazex - Britannia Fire
Blenheim - Adam Carpets Ltd
Blenheim - BF Bassett & Findley
Blenheim - Race Furniture Ltd
Blink - Shackerley (Holdings) Group Ltd incorporating Designer ceramics
Blockleys - Michelmersh Brick Holdings PLC
Bloomberg - Beko plc
Bloomsbury - Ellis J T & Co Ltd
Bloomsbury - Hamilton R & Co Ltd
BluBat - Junckers Limited
Blücher® Channel - Blücher UK Ltd
Blücher® Drain - Blücher UK Ltd
Blücher® Europipe Drainage Pipework System - Blücher UK Ltd
Blue 92 - Laticrete International Inc. UK
Blue Circle - LaFarge Cement UK
Bluesense - Ves Andover Ltd
Blueshield Pmb - Pitchmastic PmB Ltd
Bluesil - Bluestar Silicones
BM TRADA - TRADA Bookshop
BM TRADA Certification - BM TRADA
BMA - Bathroom Manufacturers Association, The
BMCA - Cast Metals Federation
BMF - Builders Merchants Federation
BMI - BCIS
BML - Booth Muirie
Bn - Wilo Samson Pumps Ltd
BN Series - NovaSpan Structures
BNL Series - NovaSpan Structures
BOB Stevensons - BOB Stevenson Ltd
Bobi Mail Boxes - JB Architectural Ltd
Boiler Buddy - Fernox
Boiler noise Silencer F2, Superconcentrate Boiler Noise Silencer F2, Boiler Noise Silencer F2 Express - Fernox
Boilermate - Gledhill Water Storage Ltd
Bold roll - Forticrete Roofing Products
Bolderaja - Boardcraft Ltd
Boldroll - Britmet Tileform Limited
Bolton Wood Yorkstone - Hard York Quarries Ltd
Bomag - Chippindale Plant Ltd
Bond It - Avocet Hardware Ltd
Bonda Contact Adhesive - Bondaglass Voss Ltd
Bonda Expanding Foam - Bondaglass Voss Ltd
Bonda G4 Damp Seal - Bondaglass Voss Ltd
Bonda PU Adhesive - Bondaglass Voss Ltd
Bonda PVA Adhesive - Bondaglass Voss Ltd
Bonda Rust Primer - Bondaglass Voss Ltd
Bonda Wood Hardener - Bondaglass Voss Ltd
Bonda Woodfill - Bondaglass Voss Ltd
Bonda Woodstop - Bondaglass Voss Ltd
Bondaflex PU - Bondaglass Voss Ltd
Bondaseal Clear - Bondaglass Voss Ltd
Bondax - Carpets of Worth Ltd
Bondel - Boardcraft Ltd
Bondite - Trade Sealants Ltd
Bondmaster - National Starch & Chemical Ltd
Bondwave - Flexible Reinforcements Ltd
Bookman - Sidhil Care
Booths - Tyco Waterworks - Samuel Booth
Bordalok - Rawson Fillings Ltd
Bosch - Arrow Supply Company Ltd
Bosch - Robert Bosch Ltd
BoSS - Youngman Group Limited
Boss White - The BSS Group
Boston Home Care, Bradshaw - Sidhil Care
Boulevard - Fitzgerald Lighting Ltd
Boulevard - Gradus Ltd
Boulevard - Marshalls Mono Ltd
Boulton & Paul - Jeld-Wen UK Ltd
Bourne London - Bourne Steel Ltd
Bourne Off Site Solutions - Bourne Steel Ltd
Bourne Parking - Bourne Steel Ltd
Bourne Special Projects - Bourne Steel Ltd
Bowater - Veka Plc
Bowls Bumper - Sportsmark Group Ltd
BowTie - Helifix
Box Ground Stone - Bath Stone - Hanson UK - Bath & Portland Stone
BPD - Zled (DPC) Co Ltd
BPF - British Plastics Federation, The
BPMA - British Pump Manufacturers Association
BPT - Vidionics Security Systems Ltd
BRA - The Federation of Environmental Trade Association (FETA)
BrackettGreen - Jacopa Limited
Bradley Washroom Products - Relcross Ltd
Bradstone - Aggregate Industries UK Ltd
Bradstone Paving - B & M Fencing Limited
Bradstone slabs - B & M Fencing Limited
Braemar suite - Armitage Shanks Ltd
Brahams - Bitmen Products Ltd
Brailliant Touch - Signs & Labels Ltd
Bramah - Bramah Security Equipment Ltd
Brand Energy & Infrastructure Services UK, Ltd - SGB, a Brand Company
Brandy Crag - Burlington Slate Ltd
Brathay Blue/Black Slate - Kirkstone
Brazilian Pinc plywood - Wood International Agency Ltd
BRC - BRC Reinforcement
BRE - Building Reserch Establishment
Breacon Radiators - Barlo Radiators Limited
Breathalyzer - Draeger Safety UK Ltd
Brecon - Sashless Window Co Ltd
Breeam - BRE
Breeza - London Fan Company Ltd
Breezair - Radiant Services Ltd
Breezax - London Fan Company Ltd
Breeze - Rycroft Ltd
Brett Landscaping and Building Products - B & M Fencing Limited
Brewer - Brewer Metalcraft
Brick & Stone - Maxit UK
Brickfill - Fillcrete Ltd
Brickforce - Masonry Reinforcement - Bekaert Building Products
Brickforce® - BRC Special Products - Bekaert Ltd
Bricklifter - Benton Co Ltd Edward
Bricklifter - Edward Benton & Co Ltd
Brickseal - Fillcrete Ltd
BrickTie - Masonry Reinforcement - Bekaert Building Products
Bricktie® - BRC Special Products - Bekaert Ltd
Bricktor - Masonry Reinforcement - Bekaert Building Products
Bricktor® - BRC Special Products - Bekaert Ltd
Bridge - Checkmate Industries Ltd
Bridge - Townscape Products Ltd
Bridge Quarry - Bridge Quarry
Bridgestone - Braithaite Engineers Ltd
Bridgman Doors - Bridgman IBC Ltd
Bridgwater - Sandtoft Roof Tiles Ltd
Brigadier - Richard Hose Ltd
Brillite - Witham Oil & Paint Ltd
Brimar Tanks - Brimar Plastics Ltd
Bristan - Olby H E & Co Ltd
Bristlex - Exitex Ltd
Bristol - Gloster Furniture Ltd
Bristol Armour - Meggitt Armour Systems
Bristol Maid - Hospital Metalcraft Ltd
Bristowes - Bitmen Products Ltd
Brit Clips - Kem Edwards Ltd
Britannia - Diamond Merchants
Britannia - Marflow Eng Ltd
Britclips® - Walraven Ltd
Britelite - Cooper Lighting and Security Ltd
Britelux - Metal Halide Lamps - Sylvania Lighting International INC
British Decorators Association - Painting & Decorating Association
British Electric Lamps Ltd - British Electric Lamps Ltd
British Glass - British Glass Manufacturers Confederation
British Gypsum - Hatmet Limited
British Gypsum - Kitsons Insulation Products Ltd
British Gypsum - Nevill Long Limited
British Monorail - Morris Material Handling Ltd
British Standards Online - British Standards Institution
Britlock - Sandtoft Roof Tiles Ltd
Briton - Allegion (UK) Ltd
Briton - Laidlaw Architectural Hardware Ltd
Briton - Laidlaw Ltd
Britool - Arrow Supply Company Ltd
Britstream - Britannia Kitchen Ventilation Ltd
BRIX - Federation of Master Builders
BRK Brands Europe Ltd - BRK Brands Europe Ltd
Broadmead - Broadmead Cast Stone
Broadrib - Heckmondwicke FB
Broadsword - TROAX UK Ltd
Broadway - Atlas Stone Products
Broadway - Race Furniture Ltd
Broag-Remeha - Broag Ltd
Brodclad - Roles Broderick Roofing Ltd
Brodeck - Roles Broderick Roofing Ltd
Brodform - Roles Broderick Roofing Ltd
Brodscreen - Roles Broderick Roofing Ltd
Brodseam - Roles Broderick Roofing Ltd
Broflame - Bollom Fire Protection Ltd
Brolac - Crown Paints Ltd
Bromgard - Feedwater Ltd
Brookes - Peek Traffic Ltd
Brooking - Clement Steel Windows
Brooking Collection - Brass Art Ltd
Brooks Roof Units - Matthews & Yates Ltd
Brosteel - Bollom Fire Protection Ltd
Broughton Controls - Geoquip Worldwide
Broughton Moor - Burlington Slate Ltd
Brown Rustic /Drag - Carlton Main Brickworks Ltd
Broxap® - Broxap Ltd
BRUCK - Optelma Lighting Ltd
Brummer - Clam Brummer Ltd
Brunel (litter bins) - Glasdon UK Limited
Brunner Mono - Anderson Gibb & Wilson
Bruynzeel - Construction Group UK Ltd
BSA - BSA Machine Tools Ltd
BSI British Standards - British Standards Institution
BSI Business Soloution - British Standards Institution
BSI Information - British Standards Institution
BSI Inspectorate - British Standards Institution
BSI Management System - British Standards Institution
BSI Product Services - British Standards Institution
BSIA - British Security Industry Association
BSSA - British Stainless Steel Association
BSWA - British Structural Waterproofing Association, The
BTA PrintMachine Eazimark™ Laserprint Management - Neopost Ltd
Bticino - Legrand Electric Ltd
Buchanan - Thomas & Betts Manufacturing Services Ltd
Buckden & Brampton - Anthony de Grey Trellises + Garden Lighting
Buckingham - Triton PLC
BUFCA - BRITISH URETHANE FOAM CONTRACTORS ASSOCIATION
Buff Rustic/ Dragwire - Carlton Main Brickworks Ltd
Buggie - Manitou (Site Lift) Ltd
Buggiescopic - Manitou (Site Lift) Ltd
Buildaid - Armorex Ltd
Builders Guild - Guild of Builders & Contractors, The
Buildex - ITW Spit
Building Bookshop - The Building Centre Group Ltd
Building Centre - The Building Centre Group Ltd
Building Cost Information Service - BCIS
Bulb-Tite Rivet - SFS Intec LTD
BulkTec - CST Industries, Inc. - UK
bullnose fascia soffit - Dales Fabrications Ltd
Bunnie Dryers - Wandsworth Elecrtrical Ltd
Bunzl - Lockhart Catering Equipment
Buroflex - Metalliform Products Plc
Bursting Stone - Burlington Slate Ltd
Bush Circulators - Pumps & Motors UK Ltd
Bushboard - Anderson C F & Son Ltd
Butler MR24 Standing seam roofing systems - CA Group Ltd - MR-24 Division
Butterley - Hanson Building Products
Button diamond - Lionweld Kennedy Ltd
Butzbach - Envirodoor Markus Ltd
Buy A Digger - www.buyadigger.com - HE Services (Plant Hire) Ltd
BVC - Bivac - Barloworld Vacuum Technology Plc
Bvent Universal Gas Vent - Rite-Vent Ltd
BWA - Bitumen Waterproofing Association
BWF - British Woodworking Federation
BWPDA - The Property Care Association
BX2000 - Horstmann Controls Ltd
Bygone Collection - Masterframe Windows Ltd
Byland - York Handmade Brick Co Ltd

C

C & R - Leofric Building Systems Ltd
C.A.T - Radiodetection Ltd
C.F.P - Westpile Ltd
C+/C plywood - Wood International Agency Ltd
C4000 Combination - Elster Metering Ltd
C6L - McMullen Facades
CA Profiles - CA Group Ltd - MR-24 Division
CABE - Design Council
Caberdek - Norbord
Caberfloor - Norbord
Caberfloor P5 - Norbord
Cable Tidy - Tower Manufacturing
Cablelink-Rapide - Ackermann Ltd
Cablemann - Ackermann Ltd
Cablofil - Legrand Electric Ltd
Cada - Kermi (UK) Ltd
Cadenza - Laird Security Hardware
Cadet seating - CPS Manufacturing Co
Cadet, Carleton - Glasdon U.K. Limited
Cadiz - Dimplex UK Limited
Cadweld® - Erico Europa (GB) Ltd
Caelum Mesh Wall System - Barton Storage Systems Ltd
Cafco - Firebarrier Services Ltd
Cafco - Promat UK Ltd
Caimi - Lesco Products Ltd
Caithness Stone - Caithness Flagstone Limited
Calder - Calder Industrial Materials Ltd
Calder, Calder Plus - Hartley & Sugden
Calderdale - Sandtoft Roof Tiles Ltd
Caldy Door - AS Newbould Ltd
Caledonian - Caledonian Modular
Caleffi - Altecnic Ltd
Calibre - BRE
Calor Autogas - Calor Gas Ltd
Calor Gas - Calor Gas Ltd
Calypso - Eltron Chromalox
Calypso Range - Evergreens Uk
Camargue - Carpets of Worth Ltd
Cambourne, Clarendon - Sidhil Care
Cambrian slates - Monier Redland Limited
Cambridge Gazebo - Anthony de Grey Trellises + Garden Lighting
Camer - Anthony de Grey Trellises + Garden Lighting
Camer Arbour seat - Anthony de Grey Trellises + Garden Lighting
Camer Rose Arch - Anthony de Grey Trellises + Garden Lighting
Camer Rose Walk - Anthony de Grey Trellises + Garden Lighting
Camfine Thermsaver - Kenilworth Daylight Centre Ltd
Camflex - Claydon Architectural Metalwork Ltd
Camjoint - Claydon Architectural Metalwork Ltd

C (con't)

Camlok - Camlok Lifting Clamps Ltd
Camouflage™ - Rockwood Pigments (UK) Ltd
Campden Pitched Stone - Atlas Stone Products
Campden Stone - Atlas Stone Products
Camtwin - Claydon Architectural Metalwork Ltd
Can-Corp - Bush Nelson PLC
Candy - British Ceramic Tile Ltd
Canexel - Vulcan Cladding Systems
Canigo' - Harbro Supplies Ltd
Cannon Consumables - Cannon Hygiene Ltd
Cannon Horticulture - Cannon Hygiene Ltd
Cannon Pest Control - Cannon Hygiene Ltd
Cannon Textile CARE - Cannon Hygiene Ltd
Canply Cofi plywood - Wood International Agency Ltd
cantech (kitchen canopies) - Mansfield Pollard & Co Ltd
Canterbury - Atlas Stone Products
Canterbury - Polybeam Ltd
Canterbury, Chaucer, Classico - Brett Landscaping
Canti-Bolt - Hi-Store Ltd
Canti-Clad - Hi-Store Ltd
Canti-Frame - Hi-Store Ltd
Canti-Guide - Hi-Store Ltd
Canti-Lec - Hi-Store Ltd
Canti-Lock - Hi-Store Ltd
Canti-Track - Hi-Store Ltd
Canti-Weld - Hi-Store Ltd
Capco - Capco Test Equipment, A division of Castlebroom Engineering Ltd
Cape - Firebarrier Services Ltd
Cape Boards - Kitsons Insulation Products Ltd
Capella - Ward Insulated Panels Ltd
Capex - Exitex Ltd
Capital & Capital clear - Arc Specialist Engineering Limited
Capital Bollard and Cycle Stand - Furnitubes International Ltd
Caplock - Ibstock Building Products Ltd
Capoplastic GP - Thermica Ltd
Cappadocia, Cat Ladder, Cengiz, Chelmsford, Clipper, Coffer, Combe, Compact4, Compact 6 - Pitacs Ltd
Cappair - Swish Building Products
Cappit - Swish Building Products
Capyt SS100 composition - Thermica Ltd
Carbital SB - Imerys Minerals Ltd
Carboline - Nullifire Ltd
Cardale - Cardale Doors Ltd
Cardale - Chiltern Invadex
Cardale - Moffett Thallon & Co Ltd
Cardale Doors - Regency Garage Door Services Ltd
Cardax - Vidionics Security Systems Ltd
Cardoc - Kent & Co (Twines) Ltd
Care LST - Hudevad Britain
Carefree Buffable - Johnson Wax Professional
Carefree Emulsion - Johnson Wax Professional
Carefree Eternum - Johnson Wax Professional
Carefree Gloss Restorer - Johnson Wax Professional
Carefree Maintainer - Johnson Wax Professional
Carefree Satin - Johnson Wax Professional
Carefree Speed Stripper - Johnson Wax Professional
Carefree Stride 1000 - Johnson Wax Professional
Carefree Stride 2000 - Johnson Wax Professional
Carefree Stride 3000 - Johnson Wax Professional
Carefree Stripper - Johnson Wax Professional
Carefree Undercoat - Johnson Wax Professional
Caremix - Altecnic Ltd
Carerscreen - Contour Showers Ltd
Caretaker III - Rapaway Energy Ltd
Cargo 2000 - Otis Ltd
Caribe - Holophane Europe Ltd
Caribe ® - Roberts-Gordon
Carina Quality Agent Assessment System - Computertel Ltd
Carina Quality Dashboard - Real Time Display - Computertel Ltd
Carina range of recording solutions - Computertel Ltd
Carlton - Triton PLC
Carlton - Tuke & Bell Ltd
Carlton Brick - Carlton Main Brickworks Ltd
Carousel - Heatrae Sadia Heating Ltd
Carr - Carr Gymnasium Equipment Ltd
Carr of Nottingham - Carr Gymnasium Equipment Ltd
Carrier air - Toshiba Carrier UK
Carron Phoenix - Carron Phoenix Ltd
CarStop - Rediweld Rubber & Plastics Ltd
Carter Coldstores - Carter Retail Equipment Ltd
Carter Parratt - Railex Systems Ltd
Carter Refridgeration Display - Carter Retail Equipment Ltd
Carter Refridgeration Services - Carter Retail Equipment Ltd
Carter-Dal - Carter-Dal International
Cascade - Calomax Ltd
Cascade - Potter & Soar Ltd
Cascade - Whitecroft Lighting Limited
Cascade Wallwash Luminaire - ITAB Prolight UK Limited
Cascamite - Wessex Resins & Adhesives Ltd
Cascamite (see Extramite) - Humbrol
Cascofil - Humbrol
Cascophen - Humbrol
Cascophen - Wessex Resins & Adhesives Ltd
Cascorez - Humbrol
Casco-Tape - Wessex Resins & Adhesives Ltd
Casella CEL - Casella Measurement Ltd
Casella ET - Casella Measurement Ltd
Casella Monitor - Casella Measurement Ltd
Cash - Bailey Birkett Sales Division
Cashflow - Masterbill Micro Systems Ltd
Casscom II - Telelarm care Ltd
Cast in situ linings - CICO Chimney Linings Ltd
CASTALDI - Optelma Lighting Ltd
Castalia cistern - Opella Ltd
Castalia Filter Clear - Opella Ltd
Castalia Tap Range - Opella Ltd
Castelco - Omeg Ltd
Castell - Castell Safety International
Castell 150 - Lok - Castell Safety International
Castelli - Haworth UK Ltd
Castle Multicem - Castle Cement Ltd
Castlemead Twist - Adam Carpets Ltd
Cataphos Paint - Sportsmark Group Ltd
Cataphos CR Paint - Sportsmark Group Ltd
Catcastle Sandstone - Dunhouse Quarry Co Ltd
Category Grid systems - SIG Interiors
Caterclad - Interclad (UK) Ltd
Caterpillar - Briggs Industrial Footwear Ltd t/a Briggs Safety Wear
Caterpillar - FG Wilson
Catnic - Catnic
Catnic - Tata Steel Europe Limited
Catnic - Wade Building Services Ltd
Catnic Classic - Catnic
Catnic Unique - Garador Ltd
Catseye - Reflecting Roadstuds Ltd
Caviarch - Cavity Trays Ltd
Cavicloser - Cavity Trays Ltd
Caviflash - Cavity Trays Ltd
Cavilintel - Cavity Trays Ltd
Caviroll - Cavity Trays Ltd
Cavitray - Cavity Trays Ltd
Cavity stop - Polypipe Building Products
Cavity Trays of Yeovil - Cavity Trays Ltd
Cavity Wall Batts - Rockwool Ltd
Cavivent - Cavity Trays Ltd
Caviweep - Cavity Trays Ltd
CBA - Concrete Block Association
Cbi - Worcester Heat Systems Ltd
CCF - Keyline
CDC - Clark Door Ltd
CDI - Worcester Heat Systems Ltd
CDX - Hochiki Europe (UK) Ltd
CE SI Wall Tiles - R J Stokes & Co Limited
CECA - Civil Engineering Contractors Association
Cedagravel - CED Ltd
Cedec footpath gravel - CED Ltd
CeeJay - Simpson Strong-Tie®
Cefndy Healthcare - Cefndy Healthcare
Cego - Laird Security Hardware
Ceillex - Salex Acoustics Limited
Celafelt - Davant Products Ltd
Celeste - Ubbink (UK) Ltd
Cellarcote - Sealocrete PLA Ltd
Cellarguard - Trade Sealants Ltd
Cellcore - Cordek Ltd
Cellcore - Jablite Ltd
Cellform - Cordek Ltd
Cellio - Armstrong World Industries Ltd
Celmac - Wirquin
Celmac Wirquin - Wirquin
Celotex - Celotex Ltd
Celsi - BFM Europe Limited
Celsius - Tata Steel Europe Limited
Celuform - Celuform Building Products
Celuform - FGF Ltd
Celutex - Celuform Building Products
Cembonit - Cembrit Ltd
Cembrit PB - Cembrit Ltd
Cementation - Skanska UK Plc
Cementitious Coating 851 - Flexcrete Technologies Limited
Cemlevel - Armorex Ltd
Cempatch - Don Construction Products Ltd
Cemprotec E. Floor - Flexcrete Technologies Limited
Cemprotec E942 - Flexcrete Technologies Limited
Cemprotec Elastic - Flexcrete Technologies Limited
Cemrok - Tarmac Building Products Ltd
Cemset - Skanska UK Plc
CemTie - Helifix
Centaur Plus - Horstmann Controls Ltd
Centaurstat - Horstmann Controls Ltd
Centerline - Crane Fluid Systems
Centra - Veolia Water Technologies UKElga Ltd
CentrAlert - Geoquip Worldwide
CentralScotland Plumbers - Ogilvie Construction Ltd
Centrel - Emergi-Lite Safety Systems
Centro - Hille Educational Products Ltd
Centurian - Laidlaw Architectural Hardware Ltd
Centurion - Forticrete Ltd
Centurion - Laidlaw Ltd
Centurion - Shackerley (Holdings) Group Ltd incorporating Designer ceramics
Centurion 12.5 - Forticrete Roofing Products
Century - Yale Security Products Ltd
Cepac - Emusol Products (LEICS) Ltd
Ceram - Lucideon
Certificates of Conformity - British Board of Agrément
Certiken - Golden Coast Ltd
Certikin - Certikin International Ltd
Cerus - Tidmarsh & Sons
Cesi - Arnull Bernard J & Co Ltd
Cetemodule DHW generator - HRS Hevac Ltd
Cetol - Akzo Noble Coatings Ltd
CFP AlarmSense 2-wire fire panels - C-TEC (Computionics) Ltd
CFP conventional fire panels - C-TEC (Computionics) Ltd
CGL - CGL Cometec Ltd
CGL Totalwall - CGL Cometec Ltd
Challenger - Abacus Lighting Ltd
Challenger - En-tout-cas Tennis Courts Ltd
Challenger - Flamco UK Ltd
Chamelneon - Oldham Lighting Ltd
Champion - Rawson Carpets Ltd
Chancery - Muraspec
Channel Plus - Horstmann Controls Ltd
Charcon - Aggregate Industries UK Ltd
Charisma - Laird Security Hardware
Charles Yorke - Symphony Group PLC, The
Charlotte Space Saver Spiral - Loft Centre Products
Charlton Gates - B & M Fencing Limited
Charnwood - Michelmersh Brick Holdings PLC
Chartered Institution of Waste Management - Chartered Institute of Wastes Management
Chartres - Formpave Ltd
Chase - Chase Equipment Ltd
Chateau - Muraspec
Checkstat - Checkmate Industries Ltd
Chelsea - Gloster Furniture Ltd
Cheltenham - Ellis J T & Co Ltd
Chemco - Brash John & Co Ltd
Chem-crete - Cross-Guard International Ltd
Chemflo - CPV Ltd
Chemset - Ramset Fasteners Ltd
Cheriton - Hamilton R & Co Ltd
Cherwell PV - Portacel
Cheshire Heritage - Maxit UK
Chesil - Hamworthy Heating Limited
Chesil - Hamworthy Heating Ltd
Chess - Flexiform Business Furniture Ltd
Chesterflex - Chesterfelt Ltd
Chestermeric - Chesterfelt Ltd
Chesterstar - Chesterfelt Ltd
Chestertorch - Chesterfelt Ltd
Chevin - Cooper Lighting and Security Ltd
Chevlol Walling - Ruthin Precast Concrete Ltd
Chichester Stoneworks Ltd - Chichester Stoneworks Limited
Childers - Kingspan Industrial Ins. Ltd
Childrey seat - Anthony de Grey Trellises + Garden Lighting
Chilstone Architectural Stonework - Chilstone
Chilstone Garden Ornaments - Chilstone
Chiltern - Macemain + Amstad Ltd
Chiltern Dynamics - BM TRADA
Chiltern International Fire - BM TRADA
Chimaster MF - Docherty New Vent Chimney Group
Chimflex LW/SB - Rite-Vent Ltd
Chimliner - Marflex Chimney Systems
Chimney Capper - Brewer Metalcraft
Chippy - Owlett - Jaton
C-Hitlite - BLV Licht-und Vakuumtecnik
Chrome bond - Howse Thomas Ltd
Chrome Plus - MK (MK Electric Ltd)
Chubb - ASSA ABLOY Limited
Chubb - Olby H E & Co Ltd
Chubb Locks - Mul-T-Lock
Chubbsafes - Gunnebo UK Limited
Churchill - BSA Machine Tools Ltd
Churchill - Gunnebo UK Limited
Churchill - Longden
Churchill - Northcot Brick Ltd
Chuted - The Safety Letterbox Company Ltd
CIArb - Chartered Institute of Arbitrators
CIBSE - Chartered Institution of Building Services Engineers (CIBSE)
CIBSE Certification - Chartered Institution of Building Services Engineers (CIBSE)
CIBSE Services - Chartered Institution of Building Services Engineers (CIBSE)
Cico - CICO Chimney Linings Ltd
Cico flex - CICO Chimney Linings Ltd
Cilplan - Cil Retail Solutions Ltd
Cimberio Valves - SAV United Kingdom Ltd
Ciment Fondu Lafarge - Lafarge Aluminates Ltd
CINCA Mosaic & Tiles - Reed Harris
Cintec - Cintec International Ltd
CIP - Construction Industry Publications
CIPEC JEP Asphaltic plug joints - Pitchmastic PmB Ltd
CIRIA - Construction Industry Research & Information Association
Cirkit - Metalliform Products Plc
CIS - Crack Inducing System - Compriband Ltd
CISA - Allegion (UK) Ltd
Cissell Dryers - Warner Howard Group Ltd
Cistermiser - Cistermiser Ltd
Citadel - SG System Products Ltd
Citadel - WF Senate
Citax - Henkel Ltd
CITB - Construction Industry Training Board
Citizen - Townscape Products Ltd
City Plumbing Supplies - Keyline
Citycabin - Wernick Group Ltd
Cityflor - Axter Ltd
Cityspace - Kone PLC
Civic - Polybeam Ltd
civic Multi - Carlton Main Brickworks Ltd
CK Station - KSB Ltd
CL - Clever - Staverton (UK) Ltd
Clairsol™ - Petrochem Carless
Clarice Range - Twyford Bathrooms
Clarity - Girsberger London
Clarity™ - H.B. Fuller
Clarke - Nilfisk Limited
Clarke "C" spray - Clarke UK Ltd
Clarus Immersive 3d Screens - Harkness Screens
Clarus XC - Harkness Screens
Class I Chimney Systems - Dunbrik (Yorks) Ltd
Classic - Airdri Ltd
Classic - Kompan Ltd
Classic - Paddock Fabrications Ltd
Classic - Rangemaster
Classic - Rivermeade Signs Ltd
Classic - Vicaima Ltd
Classic 25 - Altro Ltd
Classic 62 - Spectus Systems
Classic Bathrooms - Silverdale Bathrooms
Classic Coarse and Fine Texture - Andura Coatings Ltd
Classic Collection - Lakes Bathrooms Limited
Classic Collection - Masterframe Windows Ltd
Classic Compact Straight stairlift - Bison Bede Ltd
Classic Tile Range - Respatex International Limited
Classic Trolleys - Envopak Group Ltd
Classic, Concept - Kermi (UK) Ltd
Classic+ MkII - Airdri Ltd
Classica - Matki plc
Classicair louvres - Grille Diffuser & Louvre Co Ltd The
Classical - Saint-Gobain PAM UK
Classical express - Saint-Gobain PAM UK
Classical plus - Saint-Gobain PAM UK
Classics - Muraspec
Classidur - Blackfriar Paints Ltd
Classmate - Rawson Carpets Ltd
Claudgen - Consort Equipment Products Ltd
Claverton - Gloster Furniture Ltd
Clayboard - Dufaylite Developments Ltd
Claymaster - Cordek Ltd
Claymaster - Jablite Ltd
Claypenny - Oakdale (Contracts) Ltd
Clayslate - Sandtoft Roof Tiles Ltd
Clean Melt® - H.B. Fuller
Clean steam - Fulton Boiler Works (Great Britain) Ltd
Cleaneo Akustik (Apertura) - Knauf UK
Cleaner F3, Superconcentrate Cleaner F3, Cleaner F5 Express, Powerflushing Cleaner F5, DS40 System cleaner, System Neutraliser, HVAC Protector F3, DS-10 Dryside Cleaner - Fernox
Cleanflow - Johnson & Starley
Cleanshield ™ - Accent Hansen
Clear Erase - Tektura Plc
Clearfire 2000 - McKenzie-Martin Ltd
Clearline - Clearline Architectural
Clearstor Shelving - Moresecure Ltd
Clearview Glazed Overhead Doors - Avon Industrial Doors Ltd
Clenaglass - Clenaware Systems Ltd
Cleveland - Sashless Window Co Ltd
Clic - Signs & Labels Ltd
Clickfix coping - Alifabs (Woking) Ltd
Clients own brand if required - FG Wilson
Clifton - Formpave Ltd
Clifton - Macemain + Amstad Ltd
Clifton Landscape and design - Clifton Nurseries
Clifton little Venice - Clifton Nurseries
Climaflex - Davant Products Ltd
Climaflex - NMC (UK) Ltd
ClimEight - Benlowe Group Limited
Climmy - LTI Advanced Systems Technology Ltd
Clinicall - Tunstal Healthcare (UK) Ltd
Clino21 - Ackermann Ltd
ClinoOpt 99 - Ackermann Ltd
Clipclad Panel Management - Grant Westfield Ltd
Clip-In - PFP Electrical Products Ltd
Clipper - Calomax Ltd
Clipper - Norton Diamond Products
Clipsal - Schneider Electric Ltd
Clipstrip® - Ryton's Building Products Ltd
Clip-Top - Gradus Ltd
Clivet porducts - Clivet UK Ltd
Clorocote - Witham Oil & Paint Ltd
Club - Lappset UK Ltd
Clyde - Portacel
CME Sanitary Systems - Wirquin
cmf - Cast Metals Federation
CMIX - ITW Construction Products
CN - Construction News - EMAP
CO2 Monitor - Gas Measurement Instruments Ltd
CoachStop - Rediweld Rubber & Plastics Ltd
Coalmaster - Docherty New Vent Chimney Group
Coarse buff - Realstone Ltd
Coastline Collection - Lakes Bathrooms Limited
Cobden, Contour, Curzon - Samuel Heath & Sons plc
Cobra - Allmat (East Surrey) Ltd
Cobra - MSA (Britain) Ltd
Cocoon LST - Jaga Heating Products (UK) Ltd
Codelock - Exidor Limited
Coem - Arnull Bernard J & Co Ltd
Cofast adhesive - Combustion Linings Ltd
Cofax castables - Combustion Linings Ltd
Cofax cements - Combustion Linings Ltd
Colaquex - Emusol Products (LEICS) Ltd
Cold Cathode - Oldham Lighting Ltd
Coldblocker - Space - Ray UK
Coldseal - SCP Concrete Sealing Technology Ltd
Colflex HN - Grace De Neef UK
Colne - Portacel
Colonade - Matki plc
Colonial - Intersolar Group Ltd
Colorail - Rothley Burn Ltd
Colorcoat - Tata Steel Colors
Colorcoat - Tata Steel Europe Limited
Colorcoat Celestia - Tata Steel Colors
Colorcoat HBS200 - Tata Steel Colors
Colorduct 2 - Hotchkiss Air Supply
Colorfarm - Tata Steel Colors
Colorfects - Crown Paints Ltd
Colorlite - BLV Licht-und Vakuumtecnik
Colorpan - CEP Claddings Ltd
Colortone - Armstrong World Industries Ltd
Colortread - EMS Entrance Matting Systems Ltd
Colosseum - Metalliform Products Plc
Colour dimensions - AkzoNobel Decorative Paints UK
Colour Express - Milliken Carpet
Colour Index - Muraspec
Colour Palette - AkzoNobel Decorative Paints UK
Colour Range - Ibstock Brick Ltd
Colour Select Glazed Bricks - Ibstock Brick Ltd
Colourama and ColourTex - Spaceright Europe Limited
Colourcast - Procter Bros Limited
Colourfan - Ves Andover Ltd
Colourfast - EJOT UK Limited
Colourmark - Armorex Ltd

C (con't)

Colouroute - Clare R S & Co Ltd
Colourpave - Clare R S & Co Ltd
Colourplex™ - Rockwood Pigments (UK) Ltd
Coloursafe - Sportsmark Group Ltd
Colourseal - Movement Joints (UK)
Colourtex S - LaFarge Cement UK
Colson - Avocet Hardware Ltd
Colt Houses - W.H Colt Son & Co Ltd
Coltage - Loft Centre Products
Coltec - Cadisch MDA Ltd
Coltlite - Colt International
Columbus timers - Elkay Electrical Manufacturing Co Ltd
Column System - S & B Ltd
COMAR - H W Architectural Ltd
Combat - Range Cylinders
Combat ® - Roberts-Gordon
Combi LNHI - Babcock Wanson Uk Ltd
Combiflex - Sika Ltd
Combihatch - Howe Green Ltd
Combimate - Cistermiser Ltd
Come Back to Carpit - The Carpet Foundation
Comet - Laird Security Hardware
Cometec - CGL Cometec Ltd
Comflor - Tata Steel Europe Limited
Comfo Tube - Greenwood Airvac
Comfort Plus - Milliken Carpet
Comforto - Haworth UK Ltd
Command Elite, Command, Command Connect, Command Plus and Command Fast Link controls - Whitecroft Lighting Limited
Commander - ABB Instrumentation Ltd
Commando - Richard Hose Ltd
Commission For Architecture and the Built Environment - Design Council
Commodore Plus - Hartley & Sugden
Communicall - Tunstal Healthcare (UK) Ltd
Community - Checkmate Industries Ltd
Compac - Rycroft Ltd
Compact - Calomax Ltd
Compact - Dewhurst plc
Compact - The Safety Letterbox Company Ltd
Compact & Origin Switches - Ensto Briticent Ltd
Compact Activ - Alno (United Kingdom) Ltd
Compact grade - Bridgman IBC Ltd
Compact Line - Noral Scandinavia AB
Companion 90 - Telelarm care Ltd
Compas Sort Units - Envopak Group Ltd
COMPLi7 - Lumitron Ltd
Compriband - Compriband Ltd
Compriband - Illbruck Sealant Systems UK Ltd
Comprisil - Compriband Ltd
Compton - Bambi Air Compressors Ltd.
Conbextra - Fosroc Ltd
Concept - Ideal-Standard Ltd
Concept 2000 bath - Chiltern Invadex
Concept Suite - Armitage Shanks Ltd
Concepta - Illuma Lighting
Concertainer gabions amd mattresses - BRC Reinforcement
Concrete Cleaner - Tank Storage & Services Ltd
Concrex - Watco UK Ltd
Condensor - Poujoulat (UK) Ltd
Conductite - Laybond Products Ltd
Cone Crushers - Parker Plant Ltd
Conex Range : Conex Compression, Conex Compression Chrome Plated, Conex Compression Manifolds, Conex Valves, Conex Waste Traps - Conex Universal Limited
Confidex - Tata Steel Colors
Congress - Project Office Furniture PLC
Congrip - Conren Ltd
Conision - Potter & Soar Ltd
Connect - Metalliform Products Plc
Connect grid - Saint-Gobain Ecophon Ltd
Connexions Steel Partitioning - Bradbury Group Ltd
Conservation Rooflights - The Rooflight company
Consort - Consort Equipment Products Ltd
Consort - Marflow Eng Ltd
Consort - Vantrunk Engineering Ltd
Constable suite - Brass Art Ltd
Constant Force™ - Latchways plc
Construction Books - Professional Bookshops Limited
Construction Line - Scotfen Limited
Constructionline - Chestnut Products Limited
Constructionline - Constructionline
Constructor - Construction Group UK Ltd
Conta-clip terminals - Elkay Electrical Manufacturing Co Ltd
Contact - Girsberger London
Contact Telecoms - Designer Radiators Direct
Contain-It - George Fischer Sales Ltd
Contect Duck oil - Deb Ltd
Contemporary (Q) - Barber Wilsons & Co Ltd
Contemporary Bathrooms - Silverdale Bathrooms
Contessa - Cembrit Ltd
Contigym - Continental Sports Ltd
Contimat - Continental Sports Ltd
Continental Range - Pedley Furniture International Ltd
Continuum - Twyford Bushboard
Contitramp - Continental Sports Ltd
Contour - Airdri Ltd
Contour - Hambleside Danelaw Ltd
Contour - Levolux A.T. Limited
Contour - Valor
Contour 2 - Armitage Shanks Ltd
Contour 7000 - Redring
Contour Cisterns - Dudley Thomas Ltd
Contour EAGLE - Contour Showers Ltd
Contour ECO-DEC - Contour Showers Ltd
Contour Elite Curved stairlift - Bison Bede Ltd
Contour Factor 20 - Hambleside Danelaw Ltd
Contour FALCON - Contour Showers Ltd
Contour OTT - Hambleside Danelaw Ltd
Contour Primo - Contour Showers Ltd
Contour SNIFT - Contour Showers Ltd
Contour Stepsafe - Hambleside Danelaw Ltd
Contract - Owlett - Jaton
Contract 'E' - Eltron Chromalox
Contract Journal - Contract Journal
Control Packs - Horstmann Controls Ltd
Controlmaster - ABB Ltd
CONTURAN Display Enhancement Glass - SCHOTT UK Ltd
Conway, Carron, Compact Radiators - Barlo Radiators Limited
Cool-fit - George Fischer Sales Ltd
Coolfit - Low voltage Halogen Lamps - Sylvania Lighting International INC
Coolflow - Biddle Air Systems Ltd
Coolguard - Feedwater Ltd
Coolkote - Bonwyke Ltd
Coolkote - Invicta Window Films Ltd
COOL-PHASE® - Monodraught Ltd
Coolplex - Feedwater Ltd
CoolView (examination lamps) - Brandon Medical
coopers-uk.com - Cooper Group Ltd
Coopertiva D'imola - Arnull Bernard J & Co Ltd
Coo-Var ® - Coo-Var Ltd
Coo-Var ® - Teal & Mackrill Ltd
Cop Crag Sandstone - Dunhouse Quarry Co Ltd
Copa® - Jacopa Limited
Copar 700 - Monowa Manufacturing (UK) Ltd
CopaSac® - Jacopa Limited
Copeland Brand Products - Emerson Climate Technologies Retail Solutions
Copenhagen - Loft Centre Products
Copernit (Torch On Roofing Membrane) - Ensor Building Products Ltd
Copley Decor - Copley Decor Ltd
Copon - Andrews Coatings Ltd
Coppa Gutter - Coppa Cutta Ltd
Copper in Architecture - Copper Development Association
Copper Initiative - Copper Development Association
Copperad - The BSS Group
Copperfield - Swintex Ltd
Copper-Fin - Lochinvar Limited
Copper-Flo™ - Johnson Matthey PLC - Metal Joining
Copydex - Henkel Consumer Adhesives
CoRayVac ® - Roberts-Gordon
Corbin - Mul-T-Lock
CORGI - Corgi Technical Dervices
Corian - DuPont Corian
Corian - Du - pont - Decra Ltd
Corical Lime Paint' - Telling Lime Products Ltd
Corium - Baggeridge Brick PLC
Corkcomfort - Amorim (UK) Ltd
Corkfast - Western Cork Ltd
Corklok - Gradient Insulations (UK) Ltd
Corncockle Sandstone - Dunhouse Quarry Co Ltd
Corniche - Geberit Ltd
Cornish - Formpave Ltd
Corofil - Firebarrier Services Ltd
Corogrid - Omnis
Coroline - Ariel Plastics Ltd
Corolla - AEL
Corolux - Ariel Plastics Ltd
Corotherm - Ariel Plastics Ltd
Corpoate W20 - Crittall Steel Windows Ltd
Corporate 2000 - Crittall Steel Windows Ltd
Corporate W20 - Crittall Steel Windows Ltd
Correx - Cordek Ltd
Corridor - Saint-Gobain Ecophon Ltd
Corroban - Feedwater Ltd
Corrocure - Corroless Corrosion Control
Corrocure - Kenyon Group Ltd
Corrocure - Kenyon Paints Limited
Corrogiene - Corroless Corrosion Control
Corroguard - Corroless Corrosion Control
Corroless - Corroless Corrosion Control
corroless® - Kenyon Paints Limited
Corroshield - Corroless Corrosion Control
Corrosion Control Services Limited - Freyssinet Ltd
Corrosperse - Feedwater Ltd
Corruspan - Compriband Ltd
Corruspan - Illbruck Sealant Systems UK Ltd
Corsehill Sandstone - Dunhouse Quarry Co Ltd
Cortal - Haworth UK Ltd
Cosmofin - Beton Construction Ltd
Cost planner - Masterbill Micro Systems Ltd
Costmodelling Software - Johnson VB LLP
Cosybug - Harton Heating Appliances
Cottage Roof Tiles - Tudor Roof Tile Co Ltd
Coubro & Scrutton - PCT Group
Cougar - Catnic
Council of Industrial Design - Design Council
Countdown - Checkmate Industries Ltd
Countershield - Safetell Ltd
Country collection - Emsworth Fireplaces Ltd
County Concrete Ditch Channels - Sportsmark Group Ltd
County pantile - Sandtoft Roof Tiles Ltd
County Walling - Atlas Stone Products
Couplerbox - RFA-Tech Ltd
Courtrai - Smithbrook Building Products Ltd
Coustifoam - Sound Reduction Systems Ltd
Cove red - Realstone Ltd
Covent Garden Litter Bins - Furnitubes International Ltd
Coverlife - EBC UK Ltd
Coveseal - Armorex Ltd
Coxdome Skytube - Cox Building Products Ltd
Coxdome T.P.X - Cox Building Products Ltd
Coxdome Trade - Cox Building Products Ltd
Coxdone 2000 - Cox Building Products Ltd
Coxdone Mark 5 - Cox Building Products Ltd
Coxspan Baselock - Cox Building Products Ltd
Coxspan GlassPlank - Cox Building Products Ltd
Coxspan Modular roof glazing - Cox Building Products Ltd
CoxwindowsScape - Cox Building Products Ltd
CP Canopy - Consult Lighting Ltd
CPA - Construction Plant-hire Association
CPSA - Concrete Pipeline Systems Association
Cpv - df - CPV Ltd
CPV-Safeflo - CPV Ltd
CPV-Zurn - CPV Ltd
CrackBond - Helifix
Cractie - Redifix Ltd
Craftsman Elite - Brockway Carpets Ltd
Craftwork - Benlowe Group Limited
Cranham, Cranham Picnic Suite - Branson Leisure Ltd
Craven - Morris Material Handling Ltd
Create - Ideal-Standard Ltd
Creative Glass - Creative Glass
Creda - Applied Energy Products Ltd
Credifon® Remote Postage Meter Recrediting System - Neopost Ltd
Credo, Credo-Uno - Kermi (UK) Ltd
Crendon - Crendon Timber Engineering Ltd
Crescent - Crescent Lighting
Crescent - Geberit Ltd
Cresfinex - Exitex Ltd
Cresset - Adams- Hydraulics Ltd
Cresta - Brockway Carpets Ltd
Cresta - MSA (Britain) Ltd
Cresta Supreme - Brockway Carpets Ltd
Cresta Supreme Heathers - Brockway Carpets Ltd
Creteangle - Benton Co Ltd Edward
Creteangle - Edward Benton & Co Ltd
Cricket Gas Detectors - MSA (Britain) Ltd
Cristal - Johnson Tiles
Critic - Surge Protection Products - Erico Europa (GB) Ltd
Crittall Composite - Crittall Steel Windows Ltd
Crocodile - Simpson Strong-Tie®
Croda - Andrews Coatings Ltd
Cromleigh - Chameleon Stone Ltd
Cromleigh S.V.K - Chameleon Stone Ltd
Crompack - Cooper Lighting and Security Ltd
Crompton - ERA Products
Crompton Lighting - Cooper Lighting and Security Ltd
Crosinox - Strand Hardware Ltd
Crosland Hill Hard Yorkstone - Johnsons Wellfield Quarries Ltd
Crosland Hill Multi - Johnsons Wellfield Quarries Ltd
Cross-cote - Cross-Guard International Ltd
Crossgrip - Plastic Extruders Ltd
Crossguard - Bradbury Group Ltd
Crossvent - Howden Buffalo
Croston Plumbing and Heating - Designer Radiators Direct
Crown - Albany Standard Pumps
Crown - Brewer C & Sons Ltd
Crown Trade - Crown Paints Ltd
Crown Wool - Knauf Insulation Ltd
CRT - Range Cylinders
Cryclad - Cryotherm Insulation Ltd
Cryosil - Cryotherm Insulation Ltd
Cryostop - Cryotherm Insulation Ltd
Cryotherme - Firebarrier Services Ltd
Crystal - Sapphire Balustrades
Crystal Tanks - Water Technology Engineering Ltd
crystic - SIG Roofing
Cubase1 - Foster W H & Sons Ltd
Cubase2 - Foster W H & Sons Ltd
Cubase3 - Foster W H & Sons Ltd
Cube wall - Grant Westfield Ltd
Cubes Range - Pedley Furniture International Ltd
Cubo enclosures - Ensto Briticent Ltd
Culina - Vianen UK Ltd
Cullamix Tyrolean - Weber Building Solutions
Cullen - Allmat (East Surrey) Ltd
Cummins - Cummins Power Generation Ltd
Cummins Power Generation - Cummins Power Generation Ltd
Cupola - Hughes Safety Showers Limited
Cuprofit, Pushfit - Conex Universal Limited
Cuproright 70 - Righton Ltd
Cuproright 90 - Righton Ltd
Curaflam - Interflow UK
Curaflex - Interflow UK
Curalnn - Interflow UK
Curvetech - Stocksigns Ltd
Cushion Coil - Kleen-Tex Industries Ltd
Cutler-Hammer - Eaton Electric Ltd
Cwt y Bugail - Welsh Slate Limited
CX Window Grilles - Bradbury Group Ltd
Cyclone - Colt International
Cyclone - Saint-Gobain PAM UK

D

d Line - Allgood plc
D + H Mechantronic - Dyer Environmental Controls
D.F.D. - (Dudley Factory Doors Ltd.) - Shutter Door Repair & Maintenance Ltd
D.F.D.-(Dudley Factory Doors Ltd.) - Priory Shutter & Door Co Ltd
D50N - Movement Joints (UK)
Dac Chlor - Dacrylate Paints Ltd
Dac Crete - Dacrylate Paints Ltd
Dac flex - Dacrylate Paints Ltd
Dac Roc - Dacrylate Paints Ltd
Dac Shield - Dacrylate Paints Ltd
Dacatie - Quantum Profile Systems Ltd
Dacpol - Dacrylate Paints Ltd
Dacrylate paints - Kenyon Paints Limited
Daikin - Purified Air Ltd
Dalamatic - Donaldson Filtration (GB) Ltd
DalChoc - Dalhaus Ltd
Dales Decor - Copley Decor Ltd
Dalesauna - Dalesauna Ltd
Dalhem Panel - LSA Projects Ltd
DALI - Helvar Ltd
DalLastic - Dalhaus Ltd
DalMagnetic - Dalhaus Ltd
DalOmni - Dalhaus Ltd
DalRollazzo - Dalhaus Ltd
DalRollo - Dalhaus Ltd
DalRollo Satin - Dalhaus Ltd
DalUni - Dalhaus Ltd
DalWerk - Dalhaus Ltd
DalZone - Dalhaus Ltd
Damcor - Polypipe Building Products
DamRyt® - Ryton's Building Products Ltd
Danelaw - Hambleside Danelaw Ltd
Danesmoor - Worcester Heat Systems Ltd
Danfoss - Danfoss Ltd
Danfoss Heat Interface Units and DPCVs - SAV United Kingdom Ltd
Daniel Platt - R J Stokes & Co Limited
Danish Oil - Anglo Building Products Ltd
Darbo Particleboard - Willamette Europe Ltd
Darkroom safelights - Encapsulite International Ltd
Darksky - Cooper Lighting and Security Ltd
Daryl - Kohler UK
Datacall Aquarius - Blick Communication Systems Ltd
Datacall Gemini - Blick Communication Systems Ltd
Datacall Minder - Blick Communication Systems Ltd
Dataworks - MK (MK Electric Ltd)
Davenset - Peek Traffic Ltd
David Ball Test Sands - David Ball Group Ltd
Davis - Legrand Electric Ltd
Day-Lite - Emergi-Lite Safety Systems
Daylux - Setsquare
dBan - Interfloor Limited
DBE - Jaga Heating Products (UK) Ltd
DBI-SALA - Capital Safety Group (NE) Ltd
D-Bus - Eaton Electric Limited
DCE - Donaldson Filtration (GB) Ltd
DCE10000 - Donaldson Filtration (GB) Ltd
Dea Madore Range - Brass Art Ltd
Deaf Alerter - Deaf Alerter plc
Deanlite - Deans Blinds & Awnings (UK) Ltd
Deanmaster - Deans Blinds & Awnings (UK) Ltd
Deanox - Elementis Pigments
Deans - Deans Blinds & Awnings (UK) Ltd
Deans signs - Deans Blinds & Awnings (UK) Ltd
Deb Lime - Deb Ltd
Deb Natural - Deb Ltd
Deb Protect - Deb Ltd
Deb Restore - Deb Ltd
Deben Systems - Phoenix Scales Ltd
Deborah Services - Actavo (UK) Ltd
Debotec Torch-On - Chesterfelt Ltd
Debut - Ness Furniture Ltd
Dec - Nology - Owlett - Jaton
Decaflex - Sika Liquid Plastics Limited
Decamel - Formica Ltd
Decathlon - Hodkin Jones (Sheffield) Ltd
Decimetric - GEC Anderson Ltd
Deck drain - ABG Ltd
Deckshield - Flowcrete Plc
Decoline - Deceuninck Ltd
Decolite - The Window Film Company UK Ltd
Decomatic - Goelst
Deconyl - Plascoat Systems Ltd
Deconyl - Plastic Coatings Ltd
Decoquick - Sika Liquid Plastics Limited
Decor - Deceuninck Ltd
Décor Profile - Sarnafil Ltd
Decorail - Integra Products
Decorfix - Decorfix Ltd
Decoroc - Deceuninck Ltd
Decorscreed - Conren Ltd
Decostar - Osram Ltd
Decothane, Decothane Balcons, Decothane Ultra - Sika Liquid Plastics Limited
Decotherm - Sika Liquid Plastics Limited
Decra Classic Tile - Decra Roofing Systems
Decra Elegance Tile - Decra Roofing Systems
Decra Stratos - Decra Roofing Systems
Decra tiles - Decra Roofing Systems
Decra Vent - Decra Roofing Systems
Dedicated Micro's - Vidionics Security Systems Ltd
Dee-Organ - Signature Ltd
Deep lane Yorkstone - Hard York Quarries Ltd
Deepflow - Marley Plumbing and Drainage
Deeplas - Deceuninck Ltd
Deepstor - Construction Group UK Ltd
Defender - Gilgen Door Systems UK Ltd
Defensor - Geoquip Worldwide
Deflex - Movement Joints (UK)
Degussa - Anderson Gibb & Wilson
Deha - Halfen Ltd
Dekguard - Fosroc Ltd
Dekor - Rockfon Ltd
Dekordor ® - Vicaima Ltd
Dektite - ITW Construction Products
DEL - Golden Coast Ltd
Delabole - The Delabole Slate Co. Ltd
Delabole Slate - The Delabole Slate Co. Ltd
Delamere - Delta Balustrades
Delbraze, Delcop End Feed - Conex Universal Limited
Delchem - Opella Ltd
Delight - Glamox Electric (UK) Ltd
Delmag - Schwing Stetter (UK) Ltd
Delphis Glass Mosaics - Kirkstone
Delta - Gilgen Door Systems UK Ltd
Delta - Harbro Supplies Ltd
Delta - Sika Liquid Plastics Limited
delta gutter system - Dales Fabrications Ltd
Delta II - Capital Safety Group (NE) Ltd
DeltaBox, DeltaFoil, DeltaWing, DeltaBox Cladding and Screening and DeltaWing Cladding and Screening - Lang+Fulton

D (con't)

Deltadeck - Turner Access Ltd
Deltapox, Deneblock, Denepox, DenePur, DeneSteel - Grace De Neef UK
Deltascreen - Quartet-GBC UK Ltd
Deltos - Charles Lightfoot Ltd
DeLuxe - Hansgrohe
Deluxe Concrete Garages - Liget Compton
Demarcation Stud - Sportsmark Group Ltd
Demidekk - Jotun Henry Clark Ltd Decorative Division
Deminpac - Feedwater Ltd
Denby - Rawson Carpets Ltd
Denchem - Naylor Drainage Ltd
DencoHappel Service - DencoHappel UK Ltd
Denduct - Naylor Drainage Ltd
Denka Lifts - Denka International
Denline - Naylor Drainage Ltd
Denlok - Naylor Drainage Ltd
Denrod - Naylor Drainage Ltd
Denseal - Naylor Drainage Ltd
Densleeve - Naylor Drainage Ltd
Denso - Winn & Coales (Denso) Ltd
Densostrip - Winn & Coales (Denso) Ltd
Derablok - Deralam Laminates Ltd
Deralam - Deralam Laminates Ltd
Derawood - Deralam Laminates Ltd
Derbigum - Alumasc Exterior Building Products Ltd
Design Centre - Design Council
Design Line - Railex Systems Ltd
Design-5 - Benlowe Group Limited
Designer - Clark Door Ltd
Designer 2000 - BF Bassett & Findley
Designer Collection - Brintons Ltd
Designer Partitions - Avanti Systems
Designer Slimline Showertrays by Coram - Coram Showers Ltd
Designerlite - Lumitron Ltd
DesignLine - AudiocomPendax Ltd
Desimpel - Hanson Building Products
Desktop - Forbo Flooring
DespatchLink™ Despatch Management Software - Neopost Ltd
Desso - Armstrong Floor Products UK Ltd
Detan - Halfen Ltd
Detsafe Curtain Track - Roc Secure Ltd
Detsafe Fixings - Roc Secure Ltd
Detsafe Furniture - Roc Secure Ltd
Detsafe Glazing - Roc Secure Ltd
Developer - Checkmate Industries Ltd
Deveron Radiators - Barlo Radiators Limited
Devweld structural - Kenyon Group Ltd
Dexlux - Chalmit Lighting
DF.D- (Dudley Factory Doors Ltd.) - Neway Doors Ltd
Dialock - Hafele UK Ltd
DIAlux - Glamox Luxo Lighting Limited
Diamard - Emusol Products (LEICS) Ltd
Diamond - Cembrit Ltd
Diamond - Heckmondwicke FB
Diamond - Iles Waste Systems
Diamond - Tank Storage & Services Ltd
Diamond Access Floor - Stone Raised Access Floor - Chelsea Artisans Ltd
Diamond Décor - Impact Resistant Decorative Glass - Chelsea Artisans Ltd
Diamond Glaze - AkzoNobel Decorative Paints UK
Diamond Mirror - Impact Resistant Safety Mirror - Chelsea Artisans Ltd
Diamond Optic® - DW Windsor Lighting
Diamond Ultralite Internal/External - Chelsea Artisans Ltd
Dibond ® - Righton Ltd
Dicon Safety Products (Europe) Ltd - BRK Brands Europe Ltd
Dictator - Coburn Sliding Systems Ltd
Dielectric - Radiodetection Ltd
Diga - Kermi (UK) Ltd
Digidim - Helvar Ltd
Digifort - Tektura Plc
Digital Mercury - Hamilton R & Co Ltd
Digital Screen: Modeller, Checker, Archiver, Calculator - Harkness Screens
Digital System - BPT Security Systems (UK) Ltd
Dimensions - Anderson C F & Son Ltd
Dimple Liner - Sportsmark Group Ltd
Dimple Marker - Marking Machines International
Dimplex - Dimplex UK Limited
Diplomat - Dudley Thomas Ltd
Directflow - Johnson & Starley Ltd
Directions - Desking Systems Ltd
Direx - Universal Components Ltd
Disbocrete - PermaRock Products Ltd
Discovery - Apollo Fire Detectors Ltd
Discreet - Integra Products
Diva - Black Cat Music
Dixell Controld - Emerson Climate Technologies Retail Solutions
DK - Harbro Supplies Ltd
DM - Schiedel Chimney Systems Ltd
DMP Bonder - Isocrete Floor Screeds Ltd
DNA - Bernlite Ltd
Dobel - SSAB Swedish Steel Ltd
Dobelshield - SSAB Swedish Steel Ltd
DOCM Packs - Armitage Shanks Ltd
Doherty - Sidhil Care
DoMarKit - Safety Works & Solutions Ltd
Dome - Ves Andover Ltd
Dominator - Crane Fluid Systems
Domo - Dalesauna Ltd
DoMoGuard - Safety Works & Solutions Ltd
Domostyl - NMC (UK) Ltd
Domus - Polypipe Ventilation
Domus Radial - Polypipe Ventilation
Domus Thermal - Polypipe Ventilation
Domus Ventilation - Polypipe Ventilation Ltd
Don - Gazco Ltd
Donaldson filter components - Encon Air Systems Ltd
Doncrete Dense Concrete Blocks - Broome Bros (Doncaster) Ltd
Doncrete Paintgrade Blocks - Broome Bros (Doncaster) Ltd
Donio - Dalesauna Ltd
Donlite Building Blocks - Broome Bros (Doncaster) Ltd
Donlite Foundation Blocks - Broome Bros (Doncaster) Ltd
Donlite light weight concrete blocks - Broome Bros (Doncaster) Ltd
Donlite Paintgrade Blocks - Broome Bros (Doncaster) Ltd
Donn - Knauf AMF Ceilings Ltd
Doorfit - Doorfit Products Ltd
doorline, connectplatform - Enable Access
Doors 4 Net Floors - Contour Showers Ltd
DoorsByDesign - Bridgman IBC Ltd
Doorseal - Exidor Limited
Doosan Babcock - Doosan Power Systems Ltd
Doosan Lentjes - Doosan Power Systems Ltd
Doosan Power Systems - Doosan Power Systems Ltd
Dorchester - Hamworthy Heating Limited
Dorchester - Hamworthy Heating Ltd
Dorchester - Polybeam Ltd
Dorfab - Premdor
Dorma ED800 - Dorma Door Services Ltd
Dorma Huppe - Style Door Systems Ltd
Dormers Direct - Arbory Group Limited
Dorothea ® - Broxap Ltd
Dorothea Restorations® - Wallis Conservation Limited trading as Dorothea Restorations
Dorset collection - Emsworth Fireplaces Ltd
Dorus - Henkel Ltd
Double Roman - Russell Roof Tiles Ltd
Doublestore - Railex Systems Ltd
Doubleswing - Blair Joinery
dove gutter system - Dales Fabrications Ltd
Dovedale - Chiltern Invadex
Dovre - Gazco
Dow Corning - Geocel Ltd
DowCorning - Dow Corning Europe S.A
Downflo - Donaldson Filtration (GB) Ltd
DP5 - Kenilworth Daylight Centre Ltd
DP91 - Kenilworth Daylight Centre Ltd
DPn - Wilo Samson Pumps Ltd
Dr Martens - Briggs Industrial Footwear Ltd t/a Briggs Safety Wear
Draeger-Tubes - Draeger Safety UK Ltd
Dragon - Fulton Boiler Works (Great Britain) Ltd
Dragon HLO - Elan-Dragonair
Dragon HNG - Elan-Dragonair
Drainmaster - Pump Technical Services Ltd
Drakon and Mesa Concrete Bench Seats - Furnitubes International Ltd
Dramix - Steel Wire Fibres - Bekaert Building Products
Draper - Arrow Supply Company Ltd
Draper - Kem Edwards Ltd
Draper - Unifix Ltd
Draught Buster - Airflow (Nicoll Ventilators) Ltd
Draughtbuster - Timloc Expamet Building Products
Dravo - Johnson & Starley Ltd
Drawmet - Drawn Metal Ltd
Dreadnought - Heckmondwicke FB
Dreadnought Tiles - Hinton Perry & Davenhill Ltd
Dream Fireslide - Valor
Dreamfold - Coburn Sliding Systems Ltd
Dremel - Robert Bosch Ltd
Dricon - Arch Timber Protection
Dri-Marker - Magiboards Ltd
Dritherm - Knauf Insulation Ltd
Driva - ITW Construction Products
Driveaway Domestic Car Turntable - British Turntable Ltd
DriveRepair - Hanson UK
Drum Form - Scapa Tapes UK Ltd
Dry Fix - Repair care International
Dry Flex - Repair care International
Dry Seal - Repair care International
Dry Shake Armorshield - Armorex Ltd
Dry Shield - Repair care International
Dryangles - Atlas Stone Products
Drybase liquid-applied DPM - Safeguard Europe Ltd
DryFix - Helifix
Dryflo - Donaldson Filtration (GB) Ltd
Dryflow - Johnson & Starley
Dryseal - Hambleside Danelaw Ltd
Drystone - Atlas Stone Products
Drywall Acadamy - British Gypsum Ltd
drywall thermotek - Peter Cox Ltd
Dryzone DPC injection cream for rising damp - Safeguard Europe Ltd
DS - 3 - Fernox
DS - 40 - Fernox
DS System - Lewes Design Contracts Ltd
DSA - Drilling & Sawing Association, The
DSL - Actavo (UK) Ltd
DSMA - Door and Hardware Federation
D-stain - Milliken Carpet
DTFA and DTFN Elite - Whitecroft Lighting Limited
DTX - Guardall Ltd
Dual seal - SCP Concrete Sealing Technology Ltd
DualAir ® - Roberts-Gordon
Dual-Care - Heckmondwicke FB
Dualcase - Universal Components Ltd
Dualframe - Sapa Building Systems (monarch and glostal)
Dualheat - Redring
Dualstrip - Dufaylite Developments Ltd
Dublo - Range Cylinders
Duck Tapes - Henkel Consumer Adhesives
Ductex - Rega Ductex Ltd
Ductex - Ubbink (UK) Ltd
Ductube® - BRC Special Products - Bekaert Ltd
Duets - Forticrete Ltd
Duette® Shades - Luxaflex®
Duke - Iles Waste Systems
Dukes - Realstone Ltd
Dulux - Brewer C & Sons Ltd
Dulux - Olby H E & Co Ltd
Dulux Diamond Glaze - ICI Woodcare
Dulux Polyurethane Varnish - ICI Woodcare
Dulux Quick Dry Varnish - ICI Woodcare
Dulux Trade Ecosure - AkzoNobel Decorative Paints UK
Dulux Trade High Performance Floor Paint - AkzoNobel Decorative Paints UK
Dulux Trade Natural Wood Finishes - AkzoNobel Decorative Paints UK
Dulux Trade, Dulux Trade Gloss, Dulux Trade Emulsions - AkzoNobel Decorative Paints UK
DUNA - Reiner Fixing Devices (Ex: Hardo Fixing Devices)
Dunabout - ITW Construction Products
Dune Supreme Premier - Armstrong World Industries Ltd
Dunham Strip - Dunham Bush Ltd
Dunhouse Sandstone - Dunhouse Quarry Co Ltd
Dunseal - Dunbrik (Yorks) Ltd
Dunton - Michelmersh Brick Holdings PLC
Dunvent Ridge Vent Terminal - Dunbrik (Yorks) Ltd
Duo - Delta Balustrades
Duo-Fast - Paslode
DuoHeat - Dimplex UK Limited
Duotherm - Smart Systems Ltd
Dupar - Dewhurst plc
Duplex - Airsprung Beds Ltd
Duplex - Fillcrete Ltd
Duplex Building System - Portakabin Ltd
Duplex Coating - Jones of Oswestry Ltd
Duplo - Becker (SLIDING PARTITIONS) Ltd
DuPont Antron Supra Nylon - Scotmat Flooring Services
Dupox - Andrews Coatings Ltd
Duquesa Slate - Cembrit Ltd
Durabase - Duraflex Limited
Durable - Durable Ltd
Durabuild - Duraflex Limited
DuraDeco - Knauf UK
Duraflex Diamond Suite - Duraflex Limited
Duraflon - Plastic Coatings Ltd
Durafort - Dixon Turner Wallcoverings
Duraframe - Duraflex Limited
Duragalv - Jones of Oswestry Ltd
Duraglas - Amari Plastics Plc
Duraglas - Flexitallic Ltd
Duraglass - Duraflex Limited
Duraglass - Leigh's Paints
Duragreen - Duraflex Limited
Duraguard - Plastic Coatings Ltd
Duralay - Interfloor Limited
Duralife Polyester Powder Coat - Crittall Steel Windows Ltd
Duralite - Pryorsign
Duralock Performance Fencing - Duralock (UK) Ltd
Duralok - Duraflex Limited
Durama X - Quartet-GBC UK Ltd
DuraPine - The Outdoor Deck Company
Duraroof - Duraflex Limited
Duraroom - Duraflex Limited
Durashape - Duraflex Limited
Duraspring - Airsprung Beds Ltd
Durasteel - Promat UK Ltd
Duratec - Loblite Ltd
Duratech - Golden Coast Ltd
Duravent - Duraflex Limited
Duravit - Olby H E & Co Ltd
DuroDeck - The Outdoor Deck Company
Duropal - Anderson C F & Son Ltd
Duropal - Deralam Laminates Ltd
Durotan Contracts Limited - Durotan Ltd
Durotan Limited - Durotan Ltd
Durotan Products Limited - Durotan Ltd
Durox - Tarmac Building Products Ltd
Dušo sport shower column - Horne Engineering Ltd
Dustguard - Conren Ltd
Dustplug - Linear Ltd
Dycel Revetment - Ruthin Precast Concrete Ltd
Dycem Clean-Zone - Dycem Ltd
Dycem Protectamat - Dycem Ltd
Dycem Work - Zone - Dycem Ltd
Dyform - Navtec
DYKA - John Davidson (Pipes) Ltd
Dyke- Roof Solar - Dyke Chemicals Ltd
Dyke-Aluminium - Dyke Chemicals Ltd
Dyke-Flashing - Dyke Chemicals Ltd
Dyke-Glass - Dyke Chemicals Ltd
Dyke-Mastic - Dyke Chemicals Ltd
Dyke-Roof - Dyke Chemicals Ltd
Dyke-Roof universal primer - Dyke Chemicals Ltd
Dyke-Seal - Dyke Chemicals Ltd
Dyke-Sil - Dyke Chemicals Ltd
Dyke-Silver - Dyke Chemicals Ltd
Dyna Quell Waterstop - BRC Reinforcement
Dynabolt - Ramset Fasteners Ltd
Dynaset - Ramset Fasteners Ltd
Dyno-locks - Dyno Rod PLC
Dyno-plumbing - Dyno Rod PLC
Dyno-rod - Dyno Rod PLC
Dynotile - Rockwell Sheet Sales Ltd
Dytap Revetment - Ruthin Precast Concrete Ltd

E

E Range - James Gibbons Format Ltd
E Series - Andersen/ Black Millwork
E&T Magazine - The Institution of Engineering and Technology
E. F.S - Fulton Boiler Works (Great Britain) Ltd
E7BX - Horstmann Controls Ltd
EAC - Harbro Supplies Ltd
Earth square - Milliken Carpet
Earthstone - Wilsonart Ltd
Easby - Oakdale (Contracts) Ltd
Easi joist - Wyckham Blackwell Ltd
Easi-away - Easi-Fall International Ltd
Easicheck - Cooper Lighting and Security Ltd
Easiclamp - Viking Johnson
Easicollar - Viking Johnson
EasiCool - Airedale International Air Conditioning Ltd
Easi-Fall - Easi-Fall International Ltd
Easifiles - Railex Systems Ltd
Easi-fill - Easi-Fall International Ltd
Easi-Fit - Vicaima Ltd
Easifix - Decra Ltd
Easi-flex - Easi-Fall International Ltd
Easifoam - Vita Liquid Polymers Ltd
Easigrip - Uponor Ltd
Easi-mend - Easi-Fall International Ltd
Easipipe - Polypipe Ventilation
EasiSounds - Easi-Fall International Ltd
Easitap - Viking Johnson
Easitee - Viking Johnson
Easivent - SCS Group
Easy Change - Tapworks Water Softeners
Easy•Q - Repair care International
Easyarch - Simpson Strong-Tie®
Easybolt - Eliza Tinsley Ltd
Easyclad - Advanced Hygienic Contracting Ltd
Easyfix - Eliza Tinsley Ltd
Easy-Flo™ - Johnson Matthey PLC - Metal Joining
EasyFlue - Alpha Therm Ltd
Easy-flush - Cistermiser Ltd
Easyguard - Safety Works & Solutions Ltd
Easymark - Marking Machines International
easySwitch - Helvar Ltd
Easyway - Sapa Building Systems (monarch and glostal)
Eaton - Eaton Electric Ltd
Eauzone - Matki plc
Eazicoat - Howse Thomas Ltd
Eazimark™ OMR Coding Software - Neopost Ltd
Eazistrip - reinforcement continuity system - BRC Reinforcement
EB - Clement Steel Windows
EB Handietch - Eyre & Baxter Limited
EBC UK Ltd - EBC UK Ltd
Ebco - EBCO
Ebm-Papst - Air Control Industries Ltd
EC 2000 - ABB Ltd
EC Power CHP - SAV United Kingdom Ltd
ECB - Donaldson Filtration (GB) Ltd
ECCO 5000 - Reznor UK Ltd
Echosorba - Soundsorba Ltd
Echosorption - Bicester Products Ltd
Eclat - Coburn Sliding Systems Ltd
Eclipse - Chalmit Lighting
Eclipse - Eclipse Blind Systems Ltd
Eclipse - Fitzgerald Lighting Ltd
Eclipse - Hawker Siddley Switchgear
Eclipse - Safetell Ltd
Eclipse - Spear & Jackson
Eclipse - Veka Plc
Eclipse Premium - Tarkett Ltd
Eclipse Sliding Glazed Doors - Avanti Systems
Eclipse Spiralux - Spear & Jackson
Eclispe - Grace Construction Products Ltd
ECO - Kair ventilation Ltd
Eco - Townscape Products Ltd
Eco Collection - Solus Ceramics Ltd
Eco ICID - Schiedel Chimney Systems Ltd
ECO Lift - Phoenix Lifting Systems Ltd
Eco Parking (Grass Paver & Soil Stabilisation) - Ensor Building Products Ltd
Ecoarc - Improved Retrofit for Mercury Vapour Lamp - Sylvania Lighting International INC
EcoCaddy™ - Straight Ltd
Ecodisc - Kone PLC
Ecofibre - Gilmour Ecometal
Ecoflex - Uponor Ltd
Ecoflex/Ecoline - Axter Ltd
EcoGard - Lonsdale Metal Company
Ecoheat - IPPEC Sytsems Ltd
Ecohome - Gilmour Ecometal
Ecojoist - ITW Consutruction Products
Ecomax Reduc - Trim Acoustics
Ecomax Soundslab - Trim Acoustics
EcoMembrane - Visqueen Building Products
Ecometal - George Gilmour (Metals) Ltd
Ecometal - Gilmour Ecometal
ECOMIN - Knauf AMF Ceilings Ltd
Econet - Fläkt Woods Ltd
Econo-Air - Stokvis Industrial Boilers International Limited
Econoboost - Stokvis Industrial Boilers International Limited
Econoflame - Stokvis Industrial Boilers International Limited
Econoloc - Arvin Motion Control Ltd
Economaire, ElJan - Johnson & Starley
Economatic - Stokvis Industrial Boilers International Limited
Economix - Delabie UK Ltd
Economy 7 QTZ - Horstmann Controls Ltd
Econopak - Stokvis Industrial Boilers International Limited
Econoplate - Stokvis Industrial Boilers International Limited
Econopress - Stokvis Industrial Boilers International Limited
Econoroll - Crawford Hafa Ltd
Econospan - Gilmour Ecometal
Econostor - Stokvis Industrial Boilers International Limited
Econovent - AEL

E (con't)

Ecopack - Airedale International Air Conditioning Ltd
Ecopanel - Gilmour Ecometal
Ecophon - Nevill Long Limited
Ecophon - New Forest Ceilings Ltd
Eco-rail, Ectasy, Elixir - Pitacs Ltd
Eco-Rax - Barton Storage Systems Ltd
EcoShield, EcoForce, EcoKnight - Lochinvar Limited
EcoSort® - Straight Ltd
Ecospan - Gilmour Ecometal
Ecostream - Britannia Kitchen Ventilation Ltd
Ecotel - Airedale International Air Conditioning Ltd
Ecotherm - Prestoplan
ecotile - Ecotile Flooring Ltd
Ecotray - Gilmour Ecometal
Ecotube - Stokvis Industrial Boilers International Limited
Ecovent - Ves Andover Ltd
Ecowall - Gilmour Ecometal
Ecozip - Gilmour Ecometal
ECS - Electrotechnical Certification Scheme - Joint Industry Board for the Electrical Contracting Industry
ECS epoxy floor coating - Safeguard Europe Ltd
ECS Reset - Philips Lighting Solutions
Edale - Chiltern Invadex
Edelman Leather (Europe) - Edelman Leather
Eden - Rawson Carpets Ltd
EDENAIRE® - Eaton-Williams Group Limited
Edge - Cil Retail Solutions Ltd
Edge - MK (MK Electric Ltd)
Edge leathers - Edelman Leather
Edilkamin - Emsworth Fireplaces Ltd
Edwin clarke stairways - Ermine Engineering Co. Ltd
EEKO - Electric Elements Co, The
Eeto - Davant Products Ltd
Eetofoam - Davant Products Ltd
E-fix - Artur Fischer (UK) Ltd
Eflame collection - Emsworth Fireplaces Ltd
Egcobox® - BRC Special Products - Bekaert Ltd
Egger - Anderson C F & Son Ltd
Egger - Meyer Timber Limited
EGGER Advanced Structural Flooring - Egger (UK) Ltd
EGGER Decorative Protect K20 and + K20 - Egger (UK) Ltd
EGGER Decorative Protect Loft Panels K20 - Egger (UK) Ltd
EGGER DHF K20 - Egger (UK) Ltd
EGGER Eurospan - Egger (UK) Ltd
EGGER OSB K20 - Egger (UK) Ltd
EGGER P5 K20 - Egger (UK) Ltd
EGGER Peel Clean Xtra K20 - Egger (UK) Ltd
EGGER Protect K20 - Egger (UK) Ltd
Eiffelgres - Arnull Bernard J & Co Ltd
EIMeter - EnergyICT Ltd
Eisotech Services - Geoquip Worldwide
EJ - Zon International
ejotherm - EJOT UK Limited
Elastaseal - Tor Coatings Ltd
Elasto - Fillcrete Ltd
Elastopak, Epicol T - Grace De Neef UK
Elau AG - Schneider Electric Ltd
Elden Radiators - Barlo Radiators Limited
Eleco - Bell & Webster Concrete Ltd
Electrak - Electrak International Ltd
Electramate - Gledhill Water Storage Ltd
Electric Security Fencing Federartion (ESFF) - Fencing Contractors Association
Electrisaver - Horstmann Controls Ltd
Electrix - Electrix International Ltd
electro+ - Peter Cox Ltd
Electro-Fence - Advanced Perimeter Systems Ltd
Electrolux - Electrolux Domestic Appliances
Electrolux - Electrolux Laundry Systems
Electrolux proffessional service - Electrolux Foodservice
Electromatic - Tuke & Bell Ltd
Electronic 7 - Horstmann Controls Ltd
Electropack - Fulton Boiler Works (Great Britain) Ltd
Electrospot - Illuma Lighting
Electroway - Viscount Catering Ltd
ElectroWire - Geoquip Worldwide
Eleganza - Matki plc
Element - RB UK Ltd
Elementis - Rockwood Pigments (UK) Ltd
Elementis Pigments - Elementis Pigments
Elements Seating Range - Furnitubes International Ltd
Elequant Cost Modelling Software - Johnson VB LLP
Elequip - Peek Traffic Ltd
Elga, Elga Labwater - Veoila Water Technologies UKElga Ltd
Elica Handrail - Lewes Design Contracts Ltd
Eliminodor - Purified Air Ltd
Eliptec® - Hubbard Architectural Metalwork Ltd
Elite - Airdri Ltd
Elite - Cardale Doors Ltd
Elite - Dudley Thomas Ltd
Elite - Ellis Patents
Elite - Fitzgerald Lighting Ltd
Elite - Gradus Ltd
Elite - Nevill Long Limited
Elite - Shackerley (Holdings) Group Ltd incorporating Designer ceramics
Elite Single Point Glazing System - Avanti Systems
Elixir - Matki plc
Eliza Tinley gate fittings - B & M Fencing Limited
Eliza Tinsley - Challenge Fencing Ltd
Ellatec Lighting - Woodhouse UK Plc
Elliott Group Ltd - FineLine
Elliott Lucas - Spear & Jackson
Elliptipar - Crescent Lighting
Elster - EnergyICT Ltd
Elterwater - Burlington Slate Ltd
Embedded Retaining Walls - Westpile Ltd
emerald intelligent touchscreen terminal - CEM Systems Ltd
Emergi - Man - Emergi-Lite Safety Systems
Emergi-Lite - Thomas & Betts Manufacturing Services Ltd
Emil Ceramica - R J Stokes & Co Limited
Emir - Emmerich (Berlon) Ltd
Emit - Allgood plc
Emperor - Ellis Patents
Ems - Envirodoor Markus Ltd
Enable Access, Exitmaster - Enable Access
ENACTO™ - EnergyICT Ltd
Enamelcoat - Conren Ltd
Endura seating - CPS Manufacturing Co
ENDURO - Propex Concrete Systems
Energy Focus - Crescent Lighting
Energy Range - Twyford Bathrooms
Energy Saver F6 - Fernox
Enforcer™ - H.B. Fuller
Engineered Wall - ITW Consutruction Products
EnOcean - Helvar Ltd
Ensbury - Hamworthy Heating Limited
Ensbury - Hamworthy Heating Ltd
Ensign - Saint-Gobain PAM UK
Ensign EEZI-FIT - Saint-Gobain PAM UK
Ensor Drain (Channel Drainage Systems) - Ensor Building Products Ltd
EnsorTape (Drywall Joint Tape) - Ensor Building Products Ltd
Ensto - Ensto Briticent Ltd
Ensudisc - Ensor Building Products Ltd
Enterprise - ArjoHuntleigh UK
Enterprise - Leofric Building Systems Ltd
Enterprise coal fire - Dunsley Heat Ltd
Envirofast - Envirodoor Markus Ltd
Envirogard - Leigh's Paints
Envirograf - Firebarrier Services Ltd
Envirolite - Envirodoor Markus Ltd
Enviromantal Noise Barrier Association (ENBA) - Fencing Contractors Association
Enviromat - Linatex Ltd
Enviropave - Linatex Ltd
Enviroplast - Envirodoor Markus Ltd
Enviva - Durable Ltd
Enyiroseal - Mann McGowan Group
EOS PRO Control system - Hunter Douglas
EOS® 500 - Hunter Douglas
EP203 automatic extinguisher fire panels - C-TEC (Computionics) Ltd
Epams - Saint-Gobain PAM UK
Epcon/ Epobar - ITW Construction Products
Epidek - Leigh's Paints
Epidox - Witham Oil & Paint Ltd
Epigrip - Leigh's Paints
E-Plex - Kaba Ltd
Epocem - Sika Ltd
Epomax - ITW Construction Products
Epoxy Verti-Patch - Anglo Building Products Ltd
Epoxy Wall Shield - Anglo Building Products Ltd
Epwin Group - Windowbuild
eQ Master Air Handling Units - Fläkt Woods Ltd
eQ Prime Air Handling Units - Fläkt Woods Ltd
Equator - Marley Plumbing and Drainage
Equity Cutpile Carpet Tiles - Burmatex Ltd
ERA - ERA Products
Era - Olby H E & Co Ltd
ERA - Schlegel UK (2006) Ltd
Erdu - Hughes Safety Showers Limited
Ergonomic Kitchens - Moores Furniture Group Ltd
Erico - Kem Edwards Ltd
Erico-caddy ® - Erico Europa (GB) Ltd
Ericson - Sunrite Blinds Ltd
Eriflex®, Eriflex® Flexibar - Erico Europa (GB) Ltd
Erik Joergensen - Zon International
Eritech - Ground Rods - Erico Europa (GB) Ltd
Ermine engineering - Ermine Engineering Co. Ltd
Erosamat - ABG Ltd
Erskines - Newton John & Co Ltd
Esam - Barloworld Vacuum Technology Plc
Esavian - 1200 - Jewers Doors Ltd
Esavian - 126 - Jewers Doors Ltd
Esavian - 127 - Jewers Doors Ltd
Esavian - 128 - Jewers Doors Ltd
Escalator SafetyStrip - Kleeneze Sealtech Ltd
ESCO - Pre-finished Flooring - OSMO UK
Escofet - Woodhouse UK Plc
E-Screen - Mermet U.K
ESM - EcoWater Systems Ltd
ESM11 - EcoWater Systems Ltd
ESM15 - EcoWater Systems Ltd
ESM9 - EcoWater Systems Ltd
ESP - Hochiki Europe (UK) Ltd
Espero - Dividers Modernfold Ltd
Esplanade - Gradus Ltd
Esprit - Whitecroft Lighting Limited
ESR - Kaldewei
Essar Aquacoat SP - Smith & Rodger Ltd
Essar Aquacoat Xtra - Smith & Rodger Ltd
Essar Bar Top Lacquer - Smith & Rodger Ltd
Essar French Polish - Smith & Rodger Ltd
Essar Precatalysed Lacquers - Smith & Rodger Ltd
Essar Woodshield - Smith & Rodger Ltd
Essential - Vicaima Ltd
Essential Collection - Solus Ceramics Ltd
Essentials & EVA - Golden Coast Ltd
Essex - Railex Systems Ltd
Esthetique - Veka Plc
Etalbond - CEP Claddings Ltd
Eternal - Coburn Sliding Systems Ltd
Eternit - FGF Ltd
Eternit Fibre Cement Slates - Marley Eternit Ltd
Eternit Profiled Sheeting - Marley Eternit Ltd
Eternit-Promat - Firebarrier Services Ltd
EU 2000 - ABB Ltd
Euracom - Ackermann Ltd
Eurisol - UK Mineral Wool Association,The
Euro Lift - Phoenix Lifting Systems Ltd
Euro Shelving - Moresecure Ltd
Eurobar - Imperial Machine Co Ltd
Eurobench - Q.M.P
Eurobox - The Safety Letterbox Company Ltd
Eurocord 2000 - Rawson Carpets Ltd
Eurodek Raised Access Floors - SIG Interiors
Eurodoor 2000 - James & Bloom Ltd
Euroflex - Pitchmastic PmB Ltd
Euroflow - Johnson & Starley Ltd
Eurofold Folding Shutter - Bolton Gate Co. Ltd
Euroglas - Clow Group Ltd
Eurograde - Securikey Ltd
Euro-Guard® - Jackson H S & Son (Fencing) Ltd
Eurohose - Flexible Ducting Ltd
Euroline - Rangemaster
Euroline - S & B Ltd
Euro-meridian - Castle Care-Tech
Euronox - Wellman Robey Ltd
Europack - Fulton Boiler Works (Great Britain) Ltd
Europak - Reznor UK Ltd
European - Wellman Robey Ltd
European Fencing Industry Association - Scotfen Limited
European Technical Approvals - British Board of Agrément
EuroPipe - Blucher UK
Euroslide - Q.M.P
Eurostar - BLV Licht-und Vakuumtecnik
Eurostar Boilers - TR Engineering Ltd
Eurotec - Astron Buildings Parker Hannifin PLC
Eurotherm - Watts Industries UK Ltd
EV2008 and EV2008 No Touch - Santon
Evac+Chair - Evac+Chair International Ltd
Evacast facing masonary - Hanson Concrete Products
Evalast - Hanson Building Products UK
Evalast Blocks - Hanson Concrete Products
Evalastic - ICB (International Construction Bureau) Ltd
Evalon - ICB (International Construction Bureau) Ltd
Evamatic - KSB Ltd
Evamatic Box - KSB Ltd
Evans - Evans Vanodine International PLC
Evans Pourform - Wood International Agency Ltd
Evaporative Coolers - Radiant Services Ltd
Evelite - Amari Plastics Plc
Evenceil - Polycell Products Ltd
Evenlode - Gloster Furniture Ltd
Everbind, Everblock - Everlac (GB) Ltd
Everclad - The Abbseal Group incorporating Everseal (Thermovitrine Ltd)
Evercoat - Everlac (GB) Ltd
Everdry - Timloc Expamet Building Products
Everflex - Everlac (GB) Ltd
Everlac - Everlac (GB) Ltd
Everlast Paint grade Masonary - Hanson Concrete Products
Everlite Facing Masonary - Hanson Concrete Products
Everlite paint Grade masonary - Hanson Concrete Products
Everply - Everlac (GB) Ltd
Eversan - Everlac (GB) Ltd
Everseal - Everlac (GB) Ltd
Evertaut - Evertaut Ltd
Evertex - Everlac (GB) Ltd
Everthane - Everlac (GB) Ltd
eVision Supercars - www.evrent.co.uk - HE Services (Plant Hire) Ltd
Evo - Hyperion Wall Furniture Ltd
Evoke - Alutec
Evolution - Capital Safety Group (NE) Ltd
Evolution - Carter Retail Equipment Ltd
Evolution - Chalmit Lighting
Evolution - En-tout-cas Tennis Courts Ltd
Evolution - Mode Lighting (UK) Ltd
Evolution - T B S Fabrications
Evolution radiant strip - Dunham Bush Ltd
Evolve - Alutec
Evolve - Flexiform Business Furniture Ltd
Evo-Stik - Evode Ltd
EW - Redring
EWS® - Eaton-Williams Group Limited
EWT - Dimplex UK Limited
Excalibar - Kenrick Archibald & Sons Ltd
Excalibur - Guardall Ltd
Excalibur - TROAX UK Ltd
Excel - Axter Ltd
Excel - James Gibbons Format Ltd
Excel - Kleeneze Sealtech Ltd
Excel 5000 - Honeywell Control Systems Ltd
Excel Life Safety - Honeywell Control Systems Ltd
Excel Pro - Marlow Ropes Ltd
Excel Security Manager - Honeywell Control Systems Ltd
Excelastic - Geosynthetic Technology Ltd
Exceliner - Geosynthetic Technology Ltd
Excelsior - ADT Fire and Security Plc
Excil - LPA Group PLC
ExcluDoor - Sunray Engineering Ltd
Exclusive - Vicaima Ltd
ExecDoor - Sunray Engineering Ltd
Executive - Polybeam Ltd
Exedra - BPT Security Systems (UK) Ltd
Exhausterl blowers - BVC
Exhausto - Emsworth Fireplaces Ltd
Exide - MJ Electronics Services (International) Ltd
Exit Box - Fitzgerald Lighting Ltd
Exit:Way guidance - Hoyles Electronic Developments Ltd
Exitguard - Hoyles Electronic Developments Ltd
Exmet - Expamet Building Products
EXOS - Franke Sissons Ltd
Exos - Kaba Ltd
Expamet - Advanced Fencing Systems Ltd
Expamet - Allmat (East Surrey) Ltd
Expamet - Expamet Building Products
Expamet fencing - Chestnut Products Limited
Expandafoam - Fosroc Ltd
Expansionthene - SCP Concrete Sealing Technology Ltd
Expelex - Exitex Ltd
Exposit - Erlau AG
Expowall - Deceuninck Ltd
Express - Heatrae Sadia Heating Ltd
Express Asphalt - Aggregate Industries UK Ltd
Extaseal - SCP Concrete Sealing Technology Ltd
Extralyte Low Density Blocks - Broome Bros (Doncaster) Ltd
Extraproof - BriggsAmasco
Eye2Eye - Safetell Ltd
E-Z Rect - E-Z Rect Ltd
Ezi - lift - Didsbury Engineering Co Ltd

F

F Shield - Bollom Fire Protection Ltd
F.S.B - Hafele UK Ltd
F'Light - Glamox Electric (UK) Ltd
F2P - Artur Fischer (UK) Ltd
F2P-G - Artur Fischer (UK) Ltd
Fab&Fix - Schlegel UK (2006) Ltd
Faber - Faber Blinds UK Ltd
Faber - Hunter Douglas
Faber 1800 - Faber Blinds UK Ltd
Faber 2000 - Faber Blinds UK Ltd
Faber Maximatic - Faber Blinds UK Ltd
Faber Metalet - Faber Blinds UK Ltd
Faber Metamatic - Faber Blinds UK Ltd
Faber Midimatic - Faber Blinds UK Ltd
Faber Minimatic - Faber Blinds UK Ltd
Faber Multistop - Faber Blinds UK Ltd
Faber Rollotex - Faber Blinds UK Ltd
Faber Series One - Faber Blinds UK Ltd
Faber Softline - Faber Blinds UK Ltd
Fablok - SFS Intec LTD
Fabresa - R J Stokes & Co Limited
Fabrex - Tuke & Bell Ltd
Façade System - Paroc Panel Systems Uk Ltd
Facet - Zon International
Facette® Shades - Luxaflex®
Facom - Arrow Supply Company Ltd
Factor 4 - Interface Europe Ltd
Fagley Yorkstone - Hard York Quarries Ltd
Fairfield - Samuel Heath & Sons plc
Fairmile - Advanced Fencing Systems Ltd
Falco - Falco UK Ltd
Falcon - James Gibbons Format Ltd
Falcon - Swintex Ltd
Falcon - Water Technology Engineering Ltd
Fall Detector - Telelarm care Ltd
Fallguard vertical fall arrest ladder system - Metreel Ltd
Fanaire - Fantasia Distribution Ltd
Fantasia - Lappset UK Ltd
Fantasia ceiling fans - Fantasia Distribution Ltd
FAR Manifolds and Components - SAV United Kingdom Ltd
Farmington - Farmington Masonry LLP
Farr 3D/3D - Camfil Farr Ltd
Fascinating finishes - Kingstonian Paint Ltd
Fascinations - Carpets of Worth Ltd
Fasset - CEP Claddings Ltd
Fast Air Handling Units - Thermal Technology Sales Ltd
Fast build barriers - Bell & Webster Concrete Ltd
fast build platforms - Bell & Webster Concrete Ltd
Fast build rooms - Bell & Webster Concrete Ltd
fast build rooms - Bell & Webster Concrete Ltd
Fast build tendering - Bell & Webster Concrete Ltd
fast build walls - Bell & Webster Concrete Ltd
Fast fit - Panel & Louvre Co Ltd
Fast Gate - APT Controls Ltd
Fast rack - Legrand Electric Ltd
FastArch - EHSmith
Fastcall - Tunstal Healthcare (UK) Ltd
FastClad - EHSmith
Fastel - P4 Limited
Fastel Care - P4 Limited
Fastel SRM - P4 Limited
Fastel Wireless - P4 Limited
Fastel Within - P4 Limited
Fastelink - P4 Limited
Fastelwatch - P4 Limited
Fastfill - Flexcrete Technologies Limited
Fastie - Redifix Ltd
FastTrack™ Folder-Inserter System - Neopost Ltd
Fastwall - Fastwall Ltd
FaTec - Artur Fischer (UK) Ltd
Favourite - Valor
FCA member - Chestnut Products Limited
FCF plc - Crimeguard Protection Ltd
FEB - Strand Hardware Ltd
Fecon Filters - Vianen UK Ltd
Felix Dance Floor - Felix Design Ltd
Felix Riser Stage - Felix Design Ltd
Felix Staging - Felix Design Ltd
Feltex - Rawson Fillings Ltd
Fence Decor - Jotun Henry Clark Ltd Decorative Division
Fence Mate - Danbury Fencing Limited
Fencetone - Protim Solignum Ltd
Fencing Division - May Gurney (Technical Services) Ltd - Fencing Division
Fenlite - Hanson Building Products UK
Fenlite Blocks - Hanson Concrete Products
Fentox - Aldous & Stamp (Services) Ltd
Fep 'O' rings - Ashton Seals Ltd
Ferham - Titan Environmental Ltd
Fermacell (K10) - Fermacell
Fermacell Powerpanel H20 (H30) - Fermacell
Fermacell Powerpanel HD (H30) - Fermacell

BRANDS AND TRADE NAMES

F (con't)

Ferndale - Chiltern Invadex
Feron - Clayton Munroe
Ferramenta - Clayton Munroe
FETA - Fire Industry Association
Ffestiniog - Welsh Slate Limited
FG Contracts - Laytrad Contracts Ltd
FG Wilson - FG Wilson
Fibercast - Propex Concrete Systems
Fibercill - Fibercill
Fibermesh - Propex Concrete Systems
Fiberskirt - Fibercill
Fibertrave - Fibercill
Fibo - Lignacite (North London) Ltd
Fibo Lightweight Insulation Blocks - Stowell Concrete Ltd
Fibral - Rockfon Ltd
Fibreroll Rolling Fire Curtain - Bolton Gate Co. Ltd
Fibrespan - Armfibre Ltd
Fibrethatch - Africa Roofing Uk
Fibrocem - LaFarge Cement UK
Fibrotrace ® - Pressure Coolers Ltd T/A Maestro Pressure Coolers
Fido (dog waste bins) - Glasdon UK Limited
Fife Alarms - Fife Fire Engineers & Consultants Ltd
Fife Fire - Fife Fire Engineers & Consultants Ltd
Fiji 2 - Rangemaster
Fildek - Don Construction Products Ltd
Filestation, Filetrack, Filing Heaven - Rackline Systems Storage Ltd
Filia XP - Kermi (UK) Ltd
Fillaboard - Fillcrete Ltd
Fillcrete - Fillcrete Ltd
Filmatic - Watts Industries UK Ltd
Filon - Filon Products Ltd
Filon Citadel - Filon Products Ltd
Filon DR - Filon Products Ltd
Filon Fixsafe - Filon Products Ltd
Filon Monarch-F Barrel Vault Rooflights - Filon Products Ltd
Filon mpGRP - Filon Products Ltd
Filon Multiclad-F - Filon Products Ltd
Filon Over-roofing with Profix Spacer System - Filon Products Ltd
Filon SupaSafe - Filon Products Ltd
Filon V-Flow Valley Gutters - Filon Products Ltd
Filter Pod - Water Technology Engineering Ltd
Filterball - Marflow Eng Ltd
Filtermate - Marflow Eng Ltd
Filto-Bench - Horizon International Ltd
Filtrair - Auchard Development Co Ltd
Fimap - Barloworld Vacuum Technology Plc
Fin Cycle Stand - Furnitubes International Ltd
Find A Builder - Federation of Master Builders
Fine Uplands Twist - Adam Carpets Ltd
Fine Worcester Twist - Adam Carpets Ltd
Finebar Glazing system - Blair Joinery
Finelight, Flexiframe ®, Frenchwood ® - Andersen/ Black Millwork
Fineline - Fastwall Ltd
FineLine - FineLine
Fineline - Matki plc
Finesse - ABG Ltd
Finesse - Nevill Long Limited
Finger Protector - Safety Assured Ltd
FingerSafe - Dorma Door Services Ltd
Finned Tube - Turnbull & Scott (Engineers) Ltd
Finned Tube Radiators - Turnbull & Scott (Engineers) Ltd
Finno - Lappset UK Ltd
Finsa - Boardcraft Ltd
Finsa - Meyer Timber Limited
Finvector - Dunham Bush Ltd
Firaqua - Firwood Paints Ltd
Fire Curtain - Colt International
Fire Flexmaster - Docherty New Vent Chimney Group
Fire master - Paddock Fabrications Ltd
Fire Path - Colt International
Fire Protection Prducts - Smith & Rodger Ltd
Fire Rated - The Safety Letterbox Company Ltd
Fire Safe - Kingspan Ltd
Fire secur - Ide T & W Ltd
Fireact - Rea Metal Windows Ltd
Fireater - Tyco Safety Products
Firebeam - Photain Controls (part of GE security)
Fireblock - Dufaylite Developments Ltd
Firebrand - Hart Door Systems Ltd
Firebrand - Minco Sampling Tech UK Ltd (Incorporating Fortafix)
Firebreak - Rea Metal Windows Ltd
Firecel - Fitzpatrick Doors Ltd
Firecheck Interactive - National Britannia Ltd
Firechef - Profab Access Ltd
Firecom' Fire Telephone Systems - PEL Services Limited
Firecryl FR - Thatching Advisory Services Ltd
Fired Earth - Fired Earth
Fired Earth H2O (The Miracle of Water) - Fired Earth
Firedex - Cooper Lighting
Firedex - Cooper Lighting and Security Ltd
Firedex 3302 - Cooper Lighting
Firedex Biwire - Cooper Lighting
FireFace Plus - Sealmaster
FireFoam - Sealmaster
Fireframe - Sapa Building Systems (monarch and glostal)
FireGlaze - Sealmaster
Fireguard - TCW Services (Control) Ltd
Firelite - Valor
FireLock - Sunray Engineering Ltd
Firelok Cladding Panel - Gradient Insulations (UK) Ltd
Firelok Deck - Gradient Insulations (UK) Ltd
Firemac 2000 - McKenzie-Martin Ltd
Fireman - Clark Door Ltd
FireMaster - Cooper Group Ltd
Firemaster - Docherty New Vent Chimney Group
Firemaster - Firemaster Extinguisher Ltd
Firemaster - Premdor
Fireplan - Internal Partitions Systems
FirePlugs - Sealmaster
Firepoint - Hoyles Electronic Developments Ltd
Fireroll E240 Rolling Shutter - Bolton Gate Co. Ltd
Fireroll E30 Rolling Shutter - Bolton Gate Co. Ltd
Fireshield - Bollom Fire Protection Ltd
Fireshield - Premdor
Fireshield Partitions - Avanti Systems
Fireshield ™ - Accent Hansen
Firesleeve - Dufaylite Developments Ltd
FireSmart - Icopal Limited
Firestile - Mann McGowan Group
Firestop - Dow Corning Europe S.A
Firestop - George Fischer Sales Ltd
Firestore - Franklin Hodge Industries Ltd
Firetainer - Franklin Hodge Industries Ltd
Firetech - Knauf Insulation Ltd
Firetex - Gilgen Door Systems UK Ltd
Firetex - Leigh's Paints
Firewarn' Fire Alarm Systems - PEL Services Limited
Firewatch - TCW Services (Control) Ltd
Firglo - Firwood Paints Ltd
Firlene - Firwood Paints Ltd
Firlex - Firwood Paints Ltd
Firpavar - Firwood Paints Ltd
First Alert - BRK Brands Europe Ltd
First class - Lappset UK Ltd
First-Rol - Reflex-Rol (UK)
Firsyn - Firwood Paints Ltd
Firth Carpets - Interface Europe Ltd
Firwood - Firwood Paints Ltd
Fischer - Arrow Supply Company Ltd
Fischer - Artur Fischer (UK) Ltd
Fischer - Kem Edwards Ltd
Fissure-Clean - Above All
Fissure-Tex - Above All
Fitex - Airsprung Beds Ltd
Fitt box - PFP Electrical Products Ltd
Fitter & Poulton - PFP Electrical Products Ltd
Five 5 Star signs - Arbory Group Limited
Five Brazilian Slates - Kirkstone
Fix - Grorud Industries Ltd
Fix=r - SIG Roofing
FL Façade - Consult Lighting Ltd
Flagmaster - Zephyr The Visual Communicators
Flame Bar - Fire Protection Ltd
Flame Shield - Spaceright Europe Limited
Flamecheck - Arch Timber Protection
Flameguard - Bollom Fire Protection Ltd
Flameguard Hinged Fire Door - Bolton Gate Co. Ltd
Flametex - Airsprung Beds Ltd
Flamlac - Bollom Fire Protection Ltd
Flammtex - Vita Liquid Polymers Ltd
Flamrad - Infraglo Limited
Flanders - Forticrete Roofing Products
Flaplock - The Safety Letterbox Company Ltd
Flapmaster - Paddock Fabrications Ltd
Flashpoint infra red - Emergi-Lite Safety Systems
Flavel - Beko plc
Flavel - BFM Europe Limited
Flaviker - Arnull Bernard J & Co Ltd
Flaviker Tiles - Reed Harris
Flectoline - Rockfon Ltd
Fleet - Hamworthy Heating Limited
Fleet - Hamworthy Heating Ltd
Flex - Harbro Supplies Ltd
Flex - seal - Flex-Seal Couplings Ltd
Flexbond - Rawson Fillings Ltd
Flexcase - Universal Components Ltd
Flexcon - Flamco UK Ltd
Flexconpak - Flamco UK Ltd
FlexDrum - Flexelec(UK) Ltd
Flexfelt - Rawson Fillings Ltd
Flexfiller - Flamco UK Ltd
FlexFloor - Flexelec(UK) Ltd
Flexflyte - Flexible Ducting Ltd
Flexible Window Foil - Illbruck Sealant Systems UK Ltd
Flexi-bollard - Berry Systems
Flexibuild Sort Units - Envopak Group Ltd
Flexiburo - Flexiform Business Furniture Ltd
Flexicool - ABB Ltd
Flexideck - Safety Works & Solutions Ltd
Flexidoor - Envirodoor Markus Ltd
Flexiglas - Coburn Sliding Systems Ltd
Flexiglaze - Safetell Ltd
Flexiglide - Flexiform Business Furniture Ltd
Flexiguard - Advanced Perimeter Systems Ltd
Flexiline - Sportsmark Group Ltd
Fleximetric - Flexiform Business Furniture Ltd
Flexiphalte - BriggsAmasco
Flexi-post - Berry Systems
Flexistor - Flexiform Business Furniture Ltd
Flexithane - Tor Coatings Ltd
Flexiwall acrylic wall coating - Altro Ltd
FlexKit - Flexelec(UK) Ltd
Flexlock - Viking Johnson
FlexMat - Flexelec(UK) Ltd
Flex-net - Metway Electrical Industries Ltd
Flex-net II - Metway Electrical Industries Ltd
Flexocel - Baxenden Chemicals Ltd
Flextra® - H.B. Fuller
FlexTrace - Flexelec(UK) Ltd
Flextract 2000 - Flexible Ducting Ltd
Flexvent - Flamco UK Ltd
Flipflap - Envirodoor Markus Ltd
Floating Solar Fountain - Intersolar Group Ltd
Flocke - Mermet U.K
FloCon Commissioning Modules - SAV United Kingdom Ltd
FloDrain Channel Drainage - FloPlast Ltd
FloFit Hot and Cold Plumbing Systems - FloPlast Ltd
FloGuard - FloPlast Ltd
Flo-jo ® - Pressure Coolers Ltd T/A Maestro Pressure Coolers
Flomate - Stuart Turner Ltd
Floodline - Illuma Lighting
FloodSax® - Straight Ltd
Floor Muffle - Polypipe Building Products
Floorcote - Coo-Var Ltd
Floor-fast - Lindapter International
Flooritall - Spaceway South Ltd
Floormaker - Western Cork Ltd
Floormaster - ABB Ltd
Floormaster - Interfloor Limited
Floorshield - Springvale E P S
Floorworks - Interfloor Limited
Flopast - Allmat (East Surrey) Ltd
Florad - FloRad Heating and Cooling
Floralsilk - Floralsilk Ltd
FlorAwall - PHI Group Ltd
Florence - Whitecroft Lighting Limited
Florentine - Forticrete Ltd
Florette - Carpets of Worth Ltd
Florex - Tor Coatings Ltd
Flos/Arteluce - Quip Lighting Consultants & Suppliers Ltd
FloSaver - FloPlast Ltd
Floway - CPV Ltd
Flowcrete - Flowcrete Plc
Flowdeck - Redman Fisher Engineering
Flowforge® - Redman Fisher Engineering
Flowgrid - Redman Fisher Engineering
FlowGRiP TM - Redman Fisher Engineering
Flowguard - Pentair Thermal Management
Flowlok - Redman Fisher Engineering
Flowmax - Range Cylinders
Flowseal - Movement Joints (UK)
FlowSmart - Alpha Therm Ltd
Flowtop - Watco UK Ltd
Flowtread - Redman Fisher Engineering
Fluemaster - Docherty New Vent Chimney Group
Fluidair Economy drive system - Fluidair International Ltd
Fluidair rotary - Fluidair International Ltd
Flush - T B S Fabrications
Flush glazing - Clestra Limited
Flushplan - Internal Partitions Systems
Flushslide - Coburn Sliding Systems Ltd
Fluxite, Fernox - Fernox
FLX Floodlight - Consult Lighting Ltd
FMB - Federation of Master Builders
FMB Masterbond - Federation of Master Builders
Foamalux ® - Righton Ltd
Foamglas - Pittsburg Corning (UK) Ltd
Foamtex - Exitex Ltd
Foam-Tite - Schlegel UK (2006) Ltd
Focal Point - Brintons Ltd
Focus - James Gibbons Format Ltd
Focus - Saint-Gobain Ecophon Ltd
Focus 1000 - Noral Scandinavia AB
Focus 2000 - Noral Scandinavia AB
Fogo Montanha - Emsworth Fireplaces Ltd
Foil IP44, Foiul Single Optic and Foil Twin Optic - Whitecroft Lighting Limited
Foldaside - Coburn Sliding Systems Ltd
Foldaway - Tilley International Ltd
Folio - Girsberger London
Folyn, Fayer Radiators - Barlo Radiators Limited
Fondu - Lafarge Aluminates Ltd
Foodcheck Interactive - National Britannia Ltd
Footlights - Paragon by Heckmondwike
Footprint - Footprint Sheffield Ltd
Force Dalle - Axter Ltd
Force Line - Axter Ltd
Forceflow - Biddle Air Systems Ltd
Fordham - Astracast
Forest - Lappset UK Ltd
Forest-Saver - Earth Anchors Ltd
Forma D400 - Althon Ltd
Form-a-duct - Airflow (Nicoll Ventilators) Ltd
Format - James Gibbons Format Ltd
Format 50 - James Gibbons Format Ltd
Formica - Decra Ltd
Formica - Formica Ltd
Formica Spawall - Formica Ltd
Formica Cubewall - Formica Ltd
Formica Firewall - Formica Ltd
Formica Lifeseal - Formica Ltd
Formica Prima - Formica Ltd
Formica Unipanel - Formica Ltd
Formost - Flexiform Business Furniture Ltd
Fortic - Range Cylinders
Foscarini - Quip Lighting Consultants & Suppliers Ltd
Fospro - Hellermann Tyton
Foulmaster - Pump Technical Services Ltd
Foxtrot Underlay - Panel Agency Limited
Foxxx swimming pool kits - Fox Pool (UK) Ltd
FP McCann - Bell & Webster Concrete Ltd
fpdc - Federation of Plastering and Drywall Contractors
FPS - Federation of Piling Specialists
Framecel - Fitzpatrick Doors Ltd
Framelights - Illuma Lighting
Frametherm - Davant Products Ltd
Frameworks - Mat.Works Flooring Solutions
Frami Panel System - Mabey Hire Ltd
Franch - Zon International
Francis Pegler - Pegler Ltd
Franke Eco System - Franke UK Ltd
Franke Kitchen Systems - Franke UK Ltd
Franke Sinks - Franke UK Ltd
Franke Taps - Franke UK Ltd
Franke Triflow - Franke UK Ltd
Franki - Skanska UK Plc
FreeFlush - Cistermiser Ltd
Freespace - Project Office Furniture PLC
Freestyle - Swedecor
Freeway - Cistermiser Ltd
Freeway - Rawson Carpets Ltd
Freeway - Vantrunk Engineering Ltd
FreeWeigh™ In-Motion Electronic Weighing System - Neopost Ltd
Fresfield Lane - Michelmersh Brick Holdings PLC
Fresh Touch - Floralsilk Ltd
Freyssibar - Freyssinet Ltd
Freyssinet Makers - Freyssinet Ltd
Frico - shearflow LTD
Fridgewatch - J & E Hall Ltd
Friedland - Friedland Ltd
Frimeda - Halfen Ltd
FringeSeal - Linear Ltd
Fritztile Resilient Terrazzo - Marble Flooring Spec. Ltd
Frontrunner - Plastic Extruders Ltd
Frost - FC Frost Ltd
Frostar - Imperial Machine Co Ltd
Frostguard - Pentair Thermal Management
Froth-pak - Dow Building Solutions
FRT Exterior® - Brash John & Co Ltd
FRX (Fluidair rotary Screw) - Fluidair International Ltd
FS Screed - Conren Ltd
FSB - Allgood plc
FSB Door Furniture - JB Architectural Ltd
FST (Flexible Security Topping) - Binns Fencing Ltd
FT Gearing Landscape Services Ltd - Gearing F T Landscape Services Ltd
FTPG Olympic Anti-Stain Grout - Carter-Dal International
Fuel bank bunded storage units - Cookson and Zinn (PTL) Ltd
Fufil - Springvale E P S
Fujitsu - Purified Air Ltd
Fulcrum - Paddock Fabrications Ltd
Full-Care™ - H.B. Fuller
Fullflow - Fullflow Group Ltd
Fulton-Series J - Fulton Boiler Works (Great Britain) Ltd
Fundermax - Telling Architectural Ltd
Fundermax - Telling Lime Products Ltd
Fungishield - Glixtone Ltd
Funtheme - Design & Display Structures Ltd
Furfix Wall Profiles - Simpson Strong-Tie®
Furmanite Flange Covers - Furmanite International Ltd
Furmanite Sheet/Jointing Compound - Furmanite International Ltd
Furmanite Tube Plug - Furmanite International Ltd
Fuseal - George Fischer Sales Ltd
Fusion - Quiligotti Terrazzo Tiles Limited
Fusite - Emerson Climate Technologies Retail Solutions
Futura - T B S Fabrications
Futurewall - Optima
FW 50 - Schueco UK
FW 60 - Schueco UK
fwd products - Rivermeade Signs Ltd
Fyrebag 240 - Mann McGowan Group
Fyrex - Iles Waste Systems

G

G146 Gutter - Alifabs (Woking) Ltd
G3000 - Girsberger London
G3300 - Girsberger London
G55 Gutter - Alifabs (Woking) Ltd
Gabions - PHI Group Ltd
GAI - Guild of Architectural Ironmongers
Galaxy - Kompan Ltd
Galaxy spring operated cable reels - Metreel Ltd
Galaxy Ultra (operating theatre lamps) - Brandon Medical
Galena Spencer Ltd - Kidde Fire Protection Services Ltd
Galerie & Galerie Optimise Ranges - Twyford Bathrooms
Galleria - Brintons Ltd
Galloway - Russell Roof Tiles Ltd
Galparket - Deralam Laminates Ltd
Galtee - Meyer Timber Limited
Galtres - York Handmade Brick Co Ltd
Galvalloy - Tata Steel Colors
Gamma - Sika Liquid Plastics Limited
Gang-Nail Connectors - ITW Consutruction Products
Garaclad - Liget Compton
Garador - Garador Ltd
Garador - Regency Garage Door Services Ltd
Garbina - Unicorn Containers Ltd
Garfa - Zon International
Garment racking - Arc Specialist Engineering Limited
Garrison, Genesis, Glasdon Jubilee - Glasdon U.K. Limited
Garside Sands - Aggregate Industries UK Ltd
Gas 210 ECO - Broag Ltd
Gas 310 ECO - Broag Ltd
Gas 610 ECO - Broag Ltd
Gas zips - Power Plastics Ltd
Gascogne Limestone - Kirkstone
Gascool - Birdsall Services Ltd
Gascoseeker - Gas Measurement Instruments Ltd
Gasgard - MSA (Britain) Ltd
Gasgard gas detector - MSA (Britain) Ltd
GasSaver - Alpha Therm Ltd
Gastech - Birdsall Services Ltd
Gastyle II Ridge Vent Terminal - Dunbrik (Yorks) Ltd
Gasurveyor - Gas Measurement Instruments Ltd
Gate Automation and Access Barrier Association (GAABA) - Fencing Contractors Association
Gates - Harling Security Solutions
Gateway - Gunnebo UK Limited
Gatic Access Covers - Gatic
Gatic Drainage Gratings - Gatic
Gatic Fibrelite - Gatic
Gatic Slotdrain - Gatic
Gatic Streetwise - Gatic

G (con't)

Gatorband - Simplex Signs Limited
Gayton Door - AS Newbould Ltd
GDF Suez - Cofely Engineering Services Ltd
GEA - Garage Equipment Association Ltd
GEC Anderson - GEC Anderson Ltd
Gedina - Saint-Gobain Ecophon Ltd
Geepee - Swish Building Products
GeGe - Kaba Ltd
Gelacryl - Grace De Neef UK
Gem - Crane Fluid Systems
Gem - Tunstal Healthcare (UK) Ltd
Gemini - Forticrete Roofing Products
Gemini - Hoyles Electronic Developments Ltd
Gemini - Rangemaster
Gemini tile - Forticrete Ltd
GEN2 - Otis Ltd
General Combustion - Beverley Environmental Ltd
Genesis - Dixon Turner Wallcoverings
Genesis - Eltron Chromalox
Genesis - T B S Fabrications
Genie - Genie Europe
Genny range - Radiodetection Ltd
Genoa, and HS Grating - Lang+Fulton
Genovent - Gebhardt Kiloheat
Genpac - Sandhurst Manufacturing Co Ltd
Gent - Gent Limited
Gent Services - Gent Limited
GEO - Woodhouse UK Plc
Geocel - Geocel Ltd
GeoGard - Firestone Building Products
George Fischer - The BSS Group
Georgian - Samuel Heath & Sons plc
Georgian & Victorian Brass - Eaton Electric Limited
Geotextile - Pitchmastic PmB Ltd
Gerard - Decra Roofing Systems
Gerbergraphix - GB Sign Solutions Limited
Gerni - Nilfisk Advance Ltd
Geryon - Gilgen Door Systems UK Ltd
Geschwender - LSA Projects Ltd
Gestra - Flowserve Flow Control
GET - Schneider Electric Ltd
Gethal FSC Certiply - Wood International Agency Ltd
Getinge - ArjoHuntleigh UK
Geze - Hafele UK Ltd
GIAS - ITAB Prolight UK Limited
GIF - Interflow UK
Gif UV-C (lean) - Interflow UK
GIFAfloor - Knauf UK
Gifcrete - Don Construction Products Ltd
Gillair smoke vent products - Matthews & Yates Ltd
Gilt - Baggeridge Brick PLC
Girdlestone - Weir Pumps Ltd
GL Globe - Consult Lighting Ltd
Glade - SC Johnson Wax
Glamox - Glamox Electric (UK) Ltd
Glamox - Glamox Luxo Lighting Limited
Glamur - Jotun Henry Clark Ltd Decorative Division
Glasbord - Crane Composites
Glasflex - Flexible Ducting Ltd
Glasroc - British Gypsum Ltd
Glasroc - Nevill Long Limited
Glass Blocks - Swedecor
Glass chalkboards - Charles Lightfoot Ltd
Glass fibre Manufacturors - Aden Hynes Sculpture Studios
Glass Shower Panels by Coram - Coram Showers Ltd
Glass Technology Services - British Glass Manufacturers Confederation
Glassgard - Bonwyke Ltd
Glasswhite - Cardale Doors Ltd
Glasswood - Cardale Doors Ltd
Glasurit - HMG Paints
GlazaTherm - Lonsdale Metal Company
Glazed Bricks - Ibstock Hathernware Ltd
Glazing Accessories - Simon R W Ltd
Glen - Dimplex UK Limited
Glendyne Slate - Cembrit Ltd
Glenfield - Benlowe Group Limited
Gleno - Coverworld UK Limited
Glidebolt - Paddock Fabrications Ltd
GlideFit - Linear Ltd
Glidetrak - Rackline Systems Storage Ltd
Glidevale - Zled (DPC) Co Ltd
Global - BFM Europe Limited
Global 2 - EBC UK Ltd
Glocote - Coo-Var Ltd
Glo-dac - Dacrylate Paints Ltd
Glostal - Sapa Building Systems (monarch and glostal)
Glostal - Sapa Building Systems Ltd
Gloster - Gloster Furniture Ltd
Gloucester - Atlas Stone Products
Glover Series 2000 Extinguishers - Fife Fire Engineers & Consultants Ltd
Glow-Worm - Hepworth Heating Ltd
Gloy - Henkel Consumer Adhesives
Gluegunsdirect - Kenyon Performance Adhesives
Glulam Beams - Kingston Craftsmen Structural Timber Engineers Ltd
Glynn-Johnson Door Controls - Relcross Ltd
Go Green - ABG Ltd
Godwin Pumps - Goodwin HJ Ltd
Gold - Wilo Samson Pumps Ltd
Goldstar - Wilo Samson Pumps Ltd
Golf Range - Helmsman
Goodsmaster - Stannah Lifts Ltd
Gordon Russell - Steelcase Strafor Plc
Gourmet - Anson Concise Ltd
Goxhill Plain Tile - Sandtoft Roof Tiles Ltd
Grada - Panel & Louvre Co Ltd
Gradelux - Laidlaw Architectural Hardware Ltd
Graduat - AEL
Gradus - Gradus Ltd
Grafati Re-Gard - Andura Coatings Ltd
Grafik Eye - Tidmarsh & Sons
Grafix - Dunhams of Norwich
G-Rail 4100 - Goelst
G-Rail 5100 - Goelst
Grampian - Russell Roof Tiles Ltd
Grampian - Sashless Window Co Ltd
Granclene - Harbro Supplies Ltd
Grando UK - Buckingham Swimming Pools Ltd
Granguard Aqua - Granwood Flooring Ltd
Granirapid - Mapei (UK) Ltd
Granit - Tarkett Sommer Ltd
Granit Multisafe - Tarkett Ltd
Granit SafeT - Tarkett Ltd
Granline - Granwood Flooring Ltd
Gransprung - Granwood Flooring Ltd
Gransprung Hi-load - Granwood Flooring Ltd
Granwood - Granwood Flooring Ltd
Graphic Inlay - Kleen-Tex Industries Ltd
Grass Flow - Brett Landscaping
Grassblock - Grass Concrete Ltd
Grasscel Cellular Paving - Ruthin Precast Concrete Ltd
Grasscrete - Grass Concrete Ltd
Grasshopper - Charles Lawrence Surfaces Ltd
Grasskerb - Grass Concrete Ltd
Grasspave - ABG Ltd
Grassroad - Grass Concrete Ltd
Grate-fast - Lindapter International
Grateglow - Robinson Willey Ltd
Gravelsafe - CED Ltd
Graviner - Explosion Protection Systems - Kidde Fire Protection Ltd
Grease Bugs - Progressive Product Developments Ltd
Great-stuf - Dow Building Solutions
Greaves Portmadoc Slate - Greaves Welsh Slate Co Ltd
Green Genie - Lotus Water Garden Products Ltd
Green Help App - Institute of Specialist Surveyors and Engineers
Green2Clean - Lotus Water Garden Products Ltd
GreenScreen - Hunter Douglas
Greenstar HE - Worcester Heat Systems Ltd
greenteQ - VBH (GB) Ltd
Greenwich - Branson Leisure Ltd
Gres de Valles - R J Stokes & Co Limited
Grid Plus - MK (MK Electric Ltd)
Gridframe - Sapa Building Systems (monarch and glostal)
Grid-it - Hamilton R & Co Ltd
Gridliner - Sportsmark Group Ltd
Grid-lok - Vantrunk Engineering Ltd
Gridmaster - Mat.Works Flooring Solutions
Gridpart - Gridpart Interiors Ltd
Griffin - Eliza Tinsley Ltd
Griffin Fire - Griffin and General Fire Services Ltd
grime grabber entrance matting - Jaymart Rubber & Plastics Ltd
Grinnell Firekill - Tyco Fire and Integrated Solutions
Gripfalt SMA - Breedon Aggregates
Gripfill - Laybond Products Ltd
Gripline - Enable Access
Grippa - Tyco Water Works
Gripper handlamps - Ensto Briticent Ltd
Gripperrods - Interfloor Limited
Gripple - Kem Edwards Ltd
Griptight Plus Woodscrews - UK Fasteners Ltd
Griptop - Don Construction Products Ltd
GritKart - Barton Storage Systems Ltd
Groheart - Grohe Ltd
Grohedal - Grohe Ltd
Grohetec - Grohe Ltd
Grohetec special fittings - Grohe Ltd
Gro-Lux - Plant House /Greenhouse & Aquarium Lamps - Sylvania Lighting International INC
Ground Anchors - PHI Group Ltd
Ground Engineering - EMAP
Groundform - Cordek Ltd
Groundsman's Bible - Sportsmark Group Ltd
GRP 4 - Group Four Glassfibre Ltd
GRP System - Intergrated Polymer Systems (UK) Ltd
Grundfos - The BSS Group
GT Trolleys - Envopak Group Ltd
GTEC Acoustic Homespan - Siniat Ltd
GTEC Aqua Board - Siniat Ltd
GTEC db Board - Siniat Ltd
GTEC Drywall Fixings - Siniat Ltd
GTEC Fire Board - Siniat Ltd
GTEC Joint Compounds - Siniat Ltd
GTEC LaDura Board - Siniat Ltd
GTEC Megadeco - Siniat Ltd
GTEC Moisture Board - Siniat Ltd
GTEC Thermal Board - Siniat Ltd
Guard - Dixon Turner Wallcoverings
Guardian - Alumasc Exterior Building Products Ltd
Guardian - Intersolar Group Ltd
Guardian - The Safety Letterbox Company Ltd
Guardian LST - Jaga Heating Products (UK) Ltd
Guardmaster - Docherty New Vent Chimney Group
Guardstation - Guardall Ltd
Guardwire - Geoquip Worldwide
Guidograph - Marking Machines International
Guiting (Cotswold Stone) - Hanson UK - Bath & Portland Stone
Gulfstream - Gledhill Water Storage Ltd
Gummers - Bristan Group
Gummybins® - Straight Ltd
Gunnebo - Gunnebo UK Limited
Gun-point - Gun-Point Ltd
Gustafs - LSA Projects Ltd
GutterBrush Leaf Protection System - FloPlast Ltd
GVR Recloser - Hawker Siddley Switchgear
GVS Sectionaliser - Hawker Siddley Switchgear
GW Axial - Matthews & Yates Ltd
Gwent Weighing Systems - Phoenix Scales Ltd
GWF - Switch2 Energy Solutions Ltd
GX Geomembrane, GX Flexi, GX Barrier Membrane - Visqueen Building Products
Gypframe - British Gypsum Ltd
Gypframe - Metsec Ltd
Gypglas - Nevill Long Limited
Gyproc - British Gypsum Ltd
Gyproc - Nevill Long Limited
Gyvlon - Flowcrete Plc
Gyvlon - Isocrete Floor Screeds Ltd

H

H & R Johnson - R J Stokes & Co Limited
H & S Whelan - Stamford Products Limited
H&V News - Heating and Ventilation News - EMAP
H.S.M - Tony Team Ltd
H2Ocheck Interactive - National Britannia Ltd
H30000 - Reznor UK Ltd
H4000 Helix - Elster Metering Ltd
HA CUT AF/HA FLEX AF - Grace De Neef UK
Haddon - Tecstone - Haddonstone Ltd
Haddoncraft - Haddonstone Ltd
Haddonstone - Haddonstone Ltd
Hadrian Toilet Cubicles - Relcross Ltd
Hadron slates - Forticrete Roofing Products
Hafa - Crawford Hafa Ltd
Haffele UK Ltd - Hafele UK Ltd
HAG - HAG Shutters & Grilles Ltd
Hager - Hager Ltd
Hairfelt - Davant Products Ltd
Hajom - Olsen Doors & Windows Ltd
Halcyon - Rixonway Kitchens Ltd
Hale Hamilton Ltd - Hale Hamilton Valves Ltd
Halfen - Halfen Ltd
Hall Clips - BLM British Lead
Halla - ALD Lighting Solutions
Hallamshire - Phoenix Scales Ltd
Hallclip - SIG Roofing
Hallmark - J & E Hall Ltd
Hallmark - Marcrist International Ltd
Halls Beeline - Ciret Limited
Hallscrew - J & E Hall Ltd
Halo - Veka Plc
Halolux - Cooper Lighting
Halopar - Osram Ltd
Halstead - Diamond Merchants
Hambleton - Sashless Window Co Ltd
Hambleton - York Handmade Brick Co Ltd
Hamilton Litestat - Hamilton R & Co Ltd
Hammersmith Bollard, Cycle Stand - Furnitubes International Ltd
Hampshire collection - Emsworth Fireplaces Ltd
Handi-Access - Profab Access Ltd
Handiclamp - Viking Johnson
Handitap - Viking Johnson
Handsafe - Tank Storage & Services Ltd
Handy Angle - Link 51 (Storage Products)
Handy Tube - Link 51 (Storage Products)
Hangmann - Ackermann Ltd
Hanovia UV - Barr & Wray Ltd
Hansgrohe - Hansgrohe
Hansgrome - Olby H E & Co Ltd
Hanson Aquaflow - Visqueen Building Products
Harbex - Harris & Bailey Ltd
Harcostar® - Straight Ltd
Hard Hat - Andrews Coatings Ltd
Hardall - Hardall International Ltd
Hardrive - Westpile Ltd
Hardrow - Forticrete Ltd
Hardwood Flooring Imports - Atkinson & Kirby Ltd
Hardyhemp - Marlow Ropes Ltd
Harland Fire Extinguishers - Nu-Swift International Ltd
Harmer - Alumasc Exterior Building Products Ltd
Harmonica-in-vinyl - Becker (SLIDING PARTITIONS) Ltd
Harmonika-on-wood - Becker (SLIDING PARTITIONS) Ltd
Harmonise - Paragon by Heckmondwike
Hart - Hart Door Systems Ltd
Hart - High Sensitivity Smoke Detection - Kidde Fire Protection Ltd
Hart Door Systems - Hart Door Systems Ltd
Hart Doors - Hart Door Systems Ltd
Hartham Park Ground Stone - Bath Stone - Hanson UK - Bath & Portland Stone
Hartland - Hamilton R & Co Ltd
Harton and Harco Pak - Harton Heating Appliances
Harton Boost - Harton Heating Appliances
Harvestore - CST Industries, Inc. - UK
HAS seal - Hotchkiss Air Supply
HAScheck Interactive - National Britannia Ltd
Hasflex - Hotchkiss Air Supply
Hatcrete - Hatfield Ltd, Roy
Hathern Terra Cotta - Michelmersh Brick Holdings PLC
Hathernware - Ibstock Brick Ltd
Hathernware architectural faience - Ibstock Hathernware Ltd
Hathernware architectural terracotta - Ibstock Hathernware Ltd
Hathernware faience - Ibstock Hathernware Ltd
Hathernware terracotta - Ibstock Hathernware Ltd
Hattersley - The BSS Group
Hatz - Chippindale Plant Ltd
Hawa - Hafele UK Ltd
Hawk White - Fernox
Haworth - Haworth UK Ltd
Hazguard - Thurston Building Systems
HBS flagpoles - Harrision Flagpoles
HC Slingsby PLC - HC Slingsby PLC
HCI - Osram Ltd
HCS (HOPPE Compact System) - HOPPE (UK) Ltd
HDPE - Geberit Ltd
HE Transport Services - www.hetransportservices.co.uk - HE Services (Plant Hire) Ltd
Healthcheck Interactive - National Britannia Ltd
Healthgear - Roc Secure Ltd
Healthguard® - Earth Anchors Ltd
Heartbeat - Valor
Heatec - Loblite Ltd
Heathland - Checkmate Industries Ltd
Heathland - Monier Redland Limited
Heathland - Neptune Outdoor Furniture Ltd
Heating & Ventilating Contractors Association - Building & Engineering Services Association
Heatkeeper Homes - Walker Timber Ltd
Heatline - Ves Andover Ltd
Heatovent - Seagoe Technologies Ltd
Heat-Rad - Horizon International Ltd
Heatrae Sadia - Baxi Heating UK Ltd
Heat-Saver - Horizon International Ltd
Heatslave - Worcester Heat Systems Ltd
Heatstat T2 TMV2 thermostatic Mixing Valve - Horne Engineering Ltd
Heatwave - Jaga Heating Products (UK) Ltd
Heavy Duty Restorer - Fernox
Heavy V - Raised Floor Systems Ltd
Heavyside - Oakdale (Contracts) Ltd
Hebefix - Pump Technical Services Ltd
Hebel - Xella Aircrete Systems UK
HeiTel - Xtralis
Helagrip - Hellermann Tyton
Helatemp - Hellermann Tyton
Heldite - Heldite Ltd
HeliBar, HeliBond - Helifix
Helit - Lesco Products Ltd
Hellerman - Hellermann Tyton
Hellermann - Kem Edwards Ltd
Hellermark - Hellermann Tyton
Helmsman - Helmsman
Helsyn - Hellermann Tyton
Helvar - Bernlite Ltd
Helvin - Hellermann Tyton
Hemelite - Tarmac Building Products Ltd
Hemera external roller blind - Hallmark Blinds Ltd
Henry Nuttall - Viscount Catering Ltd
Henslowe & Fox - Commodore Kitchens Ltd
Hentage Collection - Masterframe Windows Ltd
Hep2O - Hepworth Building Products Ltd
Hep3O - Hepworth Building Products Ltd
HepBio - Hepworth Building Products Ltd
Hepflow - Hepworth Building Products Ltd
Hepseal - Hepworth Building Products Ltd
Hepsleve - Hepworth Building Products Ltd
Hepsure - Hepworth Building Products Ltd
HepvO - Hepworth Building Products Ltd
Hepworth - Sibelco UK Ltd
Heradesign - Skanda Acoustics Limited
Heraklith - Skanda Acoustics
Heraklith - Skanda Acoustics Limited
Herakustic star - Skanda Acoustics
Herakustik - Skanda Acoustics
Herakustik - Skanda Acoustics Limited
Heraldesign - Knauf AMF Ceilings Ltd
Heraperm - Isocrete Floor Screeds Ltd
Heras - Advanced Fencing Systems Ltd
Heratekta - Skanda Acoustics
Herbert - BSA Machine Tools Ltd
Herbol - HMG Paints
Hercal - Range Cylinders
Hercules - Berry Systems
Heritage - Alumasc Exterior Building Products Ltd
Heritage - Benlowe Group Limited
Heritage - K. V. Radiators
Heritage - Marshalls Mono Ltd
Heritage Bathrooms - Bristan Group
Heritage Brass - Marcus M Ltd
Heritage Range - Pedley Furniture International Ltd
Heritage Tip-Up Seating System - Race Furniture Ltd
Hermit - Glasdon U.K. Limited
Hessian 51 - Muraspec
Heswell Door - AS Newbould Ltd
Heuga - Interface Europe Ltd
HEVAC - Heating Ventilating and Air Conditioning Manufacturers Association (HEVAC)
HEVAC - The Federation of Environmental Trade Association (FETA)
Hevilifts - Hydratight Sweeney Limited
Hewetson Floors - Kingspan Access Floors Ltd
HEWI - H E W I (UK) Ltd
Heydal - Britannia Kitchen Ventilation Ltd
Heyes - PFP Electrical Products Ltd
HFB - Motorised Air Products Ltd
Hi Grip Excel Decking - CTS Bridges Ltd
Hi Grip Plus Decking - CTS Bridges Ltd
Hi Grip Standard Decking - CTS Bridges Ltd
HI Hibay - Consult Lighting Ltd
Hi Load Masonry - Simpson Strong-Tie®
Hi Spot - Mains Voltage Halogen Lamps - Sylvania Lighting International INC
Hi Watt Batteries - J.F. Poynter Ltd
HiBuild - Delta Balustrades
Hi-Cap Rainwater System - FloPlast Ltd
Hideaway - Coburn Sliding Systems Ltd
Hi-Deck - Hi-Store Ltd
Hi-Frame - Hi-Store Ltd
High Performance 2 Kitchens - Moores Furniture Group Ltd
High Performance Kitchens - Moores Furniture Group Ltd
High Performance Sports 2-part Aqueous Lacquer - Junckers Limited
high street entrance matting - Jaymart Rubber & Plastics Ltd
Highbeam Refractor - DW Windsor Lighting
Highlander - Richard Hose Ltd
Highlander and Highlander Solo stoves - Dunsley Heat Ltd
Highlux - Illuma Lighting
High-Performance - Andersen/ Black Millwork

H (con't)

Highplow - Worcester Heat Systems Ltd
Highways Agency Products - International Protective Coatings
HiJan, Hispec - Johnson & Starley
Hiline - CPV Ltd
HiLiner Concrete Garages - Liget Compton
Hill & Knowles - CrowsonFrabrics Ltd
Hill Top Sections - Hadley Industries Plc
Hilltop - Welsh Slate Limited
Hilmor - Hilmor
Hilo - MSA (Britain) Ltd
Hiltons Italian Plaster Coving - Hilton Banks Ltd
Hilume AAA - Stocksigns Ltd
Himalayan - Briggs Industrial Footwear Ltd t/a Briggs Safety Wear
Himalayan Iconic - Briggs Industrial Footwear Ltd t/a Briggs Safety Wear
HiMet - Delta Balustrades
Hinton Pavilion - Anthony de Grey Trellises + Garden Lighting
Hiper Bar - Exitex Ltd
Hiperlan - Nexans UK Ltd
Hiperlan-E - Nexans UK Ltd
Hiperlink - Nexans UK Ltd
Hipernet - Nexans UK Ltd
Hiperway - Nexans UK Ltd
Hippo - Heckmondwicke FB
Hippo - Horizon International Ltd
HIR Acoustic Absorption Panel - Salex Acoustics Limited
Hi-Slim - Oldham Lighting Ltd
Hit - BLV Licht-und Vakuumtecnik
Hitherm - Kingspan Industrial Ins. Ltd
Hit-Lite - BLV Licht-und Vakuumtecnik
Hi-Trolley - Hi-Store Ltd
Hi-way - Vantrunk Engineering Ltd
Hodge Close - Burlington Slate Ltd
Hodgkinson Bennis Stokers - Hodgkinson Bennis Ltd
Hodgkisson - TSE Brownson
Holec - Eaton Electric Ltd
Hollo-bolt - Lindapter International
Holorib - Richard Lees Steel Decking Ltd
Home 'n' dry - Johnson & Starley
Homefit - Doorfit Products Ltd
Homeguard - The Safety Letterbox Company Ltd
Homelift - Wessex Lift Co Ltd
Homelight - Crittall Steel Windows Ltd
Home-Loc 6000 - Timloc Expamet Building Products
Homelux - Homelux Nenplas
Homeqest - RB UK Ltd
Hometrim - L. B. Plastics Ltd
Homeworks - Tidmarsh & Sons
Honderson - Regency Garage Door Services Ltd
Honeywell - The BSS Group
Hoppe (UK) LTD - Moffett Thallon & Co Ltd
Hoppings decking products - B & M Fencing Limited
Hopton Wood Limcstone - Kirkstone
Horizon - Hawker Siddley Switchgear
Horizon - Hudevad Britain
Horizon - Telelarm care Ltd
Hormann - Hörmann (UK) Ltd
Hormann - Regency Garage Door Services Ltd
Horne - Horne Engineering Ltd
Horne 15 - Horne Engineering Ltd
Horne 15 4th Connection - Horne Engineering Ltd
Horne 20 - Horne Engineering Ltd
Horne ILTDU (In-line Thermal Disinfection Unit) - Horne Engineering Ltd
Hospital (HEC) - Barber Wilsons & Co Ltd
Hospital (HFC) - Barber Wilsons & Co Ltd
Hospital (NC) - Barber Wilsons & Co Ltd
Hotflo - Heatrae Sadia Heating Ltd
Hotfoil - Pentair Thermal Management
Houair Systems - British Turntable Ltd
Housesafe - Securikey Ltd
How Fire - Tyco Fire and Integrated Solutions
Howard Bird - Deva Tap Co Ltd
Hoylake Door - AS Newbould Ltd
HPA - The Federation of Environmental Trade Association (FETA)
HPI, HPL, HPS - High Pressure Series - Grosvenor Pumps Ltd
HPT Carrier - Hemsec Panel Technologies
HPT Control - Hemsec Panel Technologies
HPT Control 30 + 60 - Hemsec Panel Technologies
HPT External - Hemsec Panel Technologies
HPT Internal - Hemsec Panel Technologies
HQI - Osram Ltd
HR nails, tacks & pins - John Reynolds & Sons (Birmingham) Ltd
HRV - Kair ventilation Ltd
HS/FR Armouring Compound - Thermica Ltd
HSS Hire - Weld - HSS Hire Service Group
HSS Lift & Shift - HSS Hire Service Group
HSS One call - HSS Hire Service Group
HSS Safe & Sure - HSS Hire Service Group
HSS Workshops - HSS Hire Service Group
HSV and HSK - Lochinvar Limited
HT Range - Rycroft Ltd
HTH Door Systems - Carter Retail Equipment Ltd
HTK/TKR - EJOT UK Limited
HTV - Babcock Wanson Uk Ltd
Hufcor - Gilgen Door Systems UK Ltd
Hughes & Lancaster - Satec Service Ltd
Humber Plain Tile - Sandtoft Roof Tiles Ltd
HumidiSMART - Greenwood Airvac
Humidity Vents - Simon R W Ltd
Humidvent - Airflow Developments Ltd
Hünnebeck, a Brand company - SGB, a Brand Company
HunterDouglas - Hunter Douglas
Hush Acoustic Battens - Hush Acoustics
Hush Acoustic Flooring - Hush Acoustics
Hush Joist Strips - Hush Acoustics
Hush Panel 28 - Hush Acoustics
Hush Slab 100 - Hush Acoustics
Hush Systems - Hush Acoustics
Huwil - RB UK Ltd
HVCA - Building & Engineering Services Association
HWA's Seies 48 - H W Architectural Ltd
Hyarain - KSB Ltd
Hybox - Tata Steel Europe Limited
Hyclean - Potter & Soar Ltd
Hycor System 3000/F - Polypipe Building Products
Hyde - Hyde Brian Ltd
Hydome - Axter Ltd
Hydrakerb Barrier - APT Controls Ltd
Hydrasenta - Kohler UK
Hydroban ® - TRC (Midlands) Ltd
Hydrocol - LANXESS Inorganic Pigments Group
Hydrocol - W. Hawley & Son Ltd
Hydroduct - Grace Construction Products Ltd
Hydroferrox - W. Hawley & Son Ltd
Hydroflow - Hydropath (UK) Ltd
Hydrophilic Joint Seals - Charcon Tunnels (Division of Tarmac plc)
Hydrophilic Sealant - Charcon Tunnels (Division of Tarmac plc)
Hydrostal Pumps - Barr & Wray Ltd
HydroTec - CST Industries, Inc. - UK
Hydrotech - Alumasc Exterior Building Products Ltd
Hydrothane - Conren Ltd
Hydrotite - Charcon Tunnels (Division of Tarmac plc)
Hydrotite - Laybond Products Ltd
Hydulignum - Permali Deho Ltd
Hye Oak - Hye Oak
Hyflek - Flowcrete Plc
Hyflex - BriggsAmasco
Hyflex 10 - Hyflex Roofing
Hyflex 10 Plus - Hyflex Roofing
Hyflex 15 - Hyflex Roofing
Hyflex 15 Plus - Hyflex Roofing
Hyflex 5 - Hyflex Roofing
Hyflex Exemplar Car Park System - Hyflex Roofing
Hyflex Exemplar PMMA - Hyflex Roofing
Hyflex Exemplar PU - Hyflex Roofing
Hyflex Exemplar Balcony System - Hyflex Roofing
Hyflex Exemplar Polyester - Hyflex Roofing
Hyflex Profile - Hyflex Roofing
Hygiene - Saint-Gobain Ecophon Ltd
Hygiene Safe - Kingspan Ltd
Hygienic - Rockfon Ltd
Hygood - Tyco Safety Products
Hyload Structural Waterproofing - IKO PLC
Hyranger - Axter Ltd
Hy-rib - Expamet Building Products
Hytex - Potter & Soar Ltd
Hyweld - Potter & Soar Ltd

I

I - Worcester Heat Systems Ltd
I.C.I - Anderson Gibb & Wilson
I.C.P - Rockfon Ltd
I² Inspiration Squared - Interface Europe Ltd
IAL2 - Novoferm Europe Ltd
IAS2 - Novoferm Europe Ltd
IAT - Institute of Ashphalt Technology, the
IBC - Bridgman IBC Ltd
Ibex- Transeat - Evac+Chair International Ltd
Ibiza 2000 - Kermi (UK) Ltd
Iboflor - Checkmate Industries Ltd
Ibstock Hathernware - Ibstock Building Products Ltd
Ibstock Scottish brick - Ibstock Building Products Ltd
Ibstock Scottish Stone - Ibstock Scotish Brick
ICAM - Xtralis
ICEL - Lighting Industry Association, The
Icelert - Findlay Irvine Ltd
Icemate - CSSP (Construction Software Services Partnership)
Icepac - CSSP (Construction Software Services Partnership)
ICI - Anderson Gibb & Wilson
ICI Paints - Dulux Trade
Icon - Valor
IcoScaff - Icopal Limited
ICS - Schiedel Chimney Systems Ltd
ICS chimney system - Rite-Vent Ltd
ICS5000 chimney system - Rite-Vent Ltd
ICT - The Institute of Concrete Technology
ID Inspiration - Tarkett Ltd
Ideal - The BSS Group
Ideal - Tuke & Bell Ltd
IDEX Blagdon Pumps - Barr & Wray Ltd
iDim - Helvar Ltd
IDM - Kone PLC
IET - The Institution of Engineering and Technology
IET Electrical Standards Plus - The Institution of Engineering and Technology
IET Membership - The Institution of Engineering and Technology
IET Venues - The Institution of Engineering and Technology
IET Wiring Regulations Bookshop - Professional Bookshops Limited
IET.tv - The Institution of Engineering and Technology
IG - IG Limited
Iguana - Jaga Heating Products (UK) Ltd
IKO Single Ply - IKO PLC
Illbruk - Tremco Illbruck Ltd
Illusion - Matki plc
Image - Interface Europe Ltd
Imagine - Helvar Ltd
Imit - Altecnic Ltd
Immersion Paint Stripper - Howse Thomas Ltd
Imola - Arnull Bernard J & Co Ltd
Impact - Amwell Systems Ltd
Impact Crusher - Parker Plant Ltd
Impactafoam - Sound Reduction Systems Ltd
Impactor - Geoquip Worldwide
Impactor - Imperial Machine Co Ltd
Imperial - Miele Co Ltd
Imperial - Smart Systems Ltd
Imperial locks - Guardian Lock and Engineering Co Ltd
Impetus - Lochinvar Limited
Impex - Construction Group UK Ltd
Impressionist II - Altro Ltd
Imprint - Prismo Ltd
Impro Proximity Access Control - BPT Security Systems (UK) Ltd
Impulse - ITW Construction Products
Impulse - Mode Lighting (UK) Ltd
Index cable glands - Elkay Electrical Manufacturing Co Ltd
Index of Quality Names - The Carpet Foundation
Ind-Oil-Taper - Olley & Sons Ltd, C
Inducool - LTI Advanced Systems Technology Ltd
Indul - LTI Advanced Systems Technology Ltd
Industrial Range : Industrial Air Vent Valves, Industrial Balancing Valves, Industrial Butterfly Valves, Industrial Check Valves, Industrial Expansion Joint, Industrial Gate Valves, Industrial Y Strainer Valves - Conex Universal Limited
Inergen - ADT Fire and Security Plc
Infiltra Stop, IP - Grace De Neef UK
Infiniti - Levolux A.T. Limited
Infinity - Shackerley (Holdings) Group Ltd incorporating Designer ceramics
Infinity - Tapworks Water Softeners
Influences - Muraspec
infopanel - Stocksigns Ltd
Inforail ® - Broxap Ltd
Infrapod - Setsquare
ingenius entrance matting - Jaymart Rubber & Plastics Ltd
Ingeny - Interphone Limited
Ingersoll - Yale Security Products Ltd
Ingersoll Security - Mul-T-Lock
Ingersol-Rand - Ingersoll-Rand European Sales Ltd
Initial Electronic Security Systems Ltd - Chubb Systems Ltd
Initial Sherrock Monitoring - Initial Electronic Security Systems Ltd
Initial Sherrork Surveilance - Initial Electronic Security Systems Ltd
Initial Shorrock Fire - Initial Electronic Security Systems Ltd
Initial Shorrock Monitoring - Ingersoll-Rand European Sales Ltd
Initial Shorrock Security - Initial Electronic Security Systems Ltd
Initial Shorrock Surveillance - Ingersoll-Rand European Sales Ltd
Initial Shorrock Systems - Initial Electronic Security Systems Ltd
Initial Shorrrock - Chubb Systems Ltd
Injection Plastics - Hines & Sons, P E, Ltd
Inlay Range - Pedley Furniture International Ltd
INMP/ NDF - Babcock Wanson Uk Ltd
Innov8, Independence Versa, Inspiration - Sidhil Care
Innova - Premdor
Innovair - Biddle Air Systems Ltd
Innovation - Rawson Carpets Ltd
Innovators in Clay - Ibstock Brick Ltd
In-Oil-Board - Olley & Sons Ltd, C
Inova Interactive Seating - CPS Manufacturing Co
Inperim - Polypipe Building Products
Inpro - Gradus Ltd
insectacoat - Peter Cox Ltd
Insect-O-Cutor - P+L Systems Ltd
Inserdor - Coburn Sliding Systems Ltd
InSkew - Helifix
Inspec - The Institution of Engineering and Technology
Insta Wall - Spaceright Europe Limited
Instacoustic - InstaGroup Ltd
Instafibre - InstaGroup Ltd
Instaflex - George Fischer Sales Ltd
Install Plus - Tata Steel Europe Limited
Instant - Redring
Instant Adhesive - Loctite UK Ltd
Insta-stik - Dow Building Solutions
Insucel - Fitzpatrick Doors Ltd
Insugard - Gilgen Door Systems UK Ltd
Insuglaze - Panel Systems Ltd
INSULATH™ - Hadley Industries Plc
Insulator - Hambleside Danelaw Ltd
Insulight - Pilkington Plyglass
Insuloid - Hellermann Tyton
Insul-Quilt specialist film studio acoustic liner - Salex Acoustics Limited
Insul-Sheet - NMC (UK) Ltd
Insultube - Davant Products Ltd
Insul-Tube - NMC (UK) Ltd
Insupex - IPPEC Sytsems Ltd
InTec - Alpha Therm Ltd
Integra - Durable Ltd
Integra - Fitzgerald Lighting Ltd
Integra - Integra Products
Integra - LANXESS Inorganic Pigments Group
integra - W. Hawley & Son Ltd
Integra Monitor - BPT Security Systems (UK) Ltd
Intejan - Johnson & Starley
Inteli-Fin - Lochinvar Limited
Intellect - Cooper Lighting and Security Ltd
Intellem - Cooper Lighting and Security Ltd
Intelligent Brick Work - Ibstock Brick Ltd
Intelligent Glass - Pro Display TM Limited
Intelock - Vantrunk Engineering Ltd
Interact - Ackermann Ltd
Interbond FP thin film intumescent coating - International Protective Coatings
Intercell - Interface Europe Ltd
Intercure rapid curing epoxies - International Protective Coatings
Interface - Interface Europe Ltd
Interface Fabrics - Interface Europe Ltd
Interfine 629 iso cyonate free finishes - International Protective Coatings
Interfine Polysiloxane - International Protective Coatings
Interfloor Ltd - Interfloor Limited
Interglaze - Exitex Ltd
Interior Contracting - Parthos UK Ltd
Interline Tank linings - International Protective Coatings
Interlock - Howarth Windows & Doors Ltd
Interlock - Interclad (UK) Ltd
Interlock - Interface Europe Ltd
International Forestry Exhibition - Confor: Promoting forestry and wood
International Waterproofing Association - Bitumen Waterproofing Association
Interphone - Interphone Limited
Interplan - Internal Partitions Systems
Interplan maintanance Painting - International Protective Coatings
Interplus Surface Tolerant - International Protective Coatings
Interpretations - Paragon by Heckmondwike
Interscreen - Safetell Ltd
Interspan - Compriband Ltd
Interspan All Metal - Compriband Ltd
Interstrip - Shackerley (Holdings) Group Ltd incorporating Designer ceramics
Intertspan - Illbruck Sealant Systems UK Ltd
Interzinc 72 epoxy zinc rich - International Protective Coatings
Interzone high solid expoxies - International Protective Coatings
Intrad - Fixatrad Ltd
Intravent - Gebhardt Kiloheat
Intuclear - Bollom Fire Protection Ltd
Intufoam - Quelfire
Intulac - Bollom Fire Protection Ltd
Intumescent Compound - Sealmaster
Intusteel WB - Bollom Fire Protection Ltd
Intuwrap - Quelfire
Invadex - Chiltern Invadex
Invaflo Shower Drain - Chiltern Invadex
Invicta - Invicita Plastics Ltd
Invikta - Grosvenor Pumps Ltd
Invisidor - Biddle Air Systems Ltd
Invisirung vertical fall arrest ladder system - Metreel Ltd
IP64 Lampholders - Jerrards Plc
IP67 Ballast - Jerrards Plc
Ipe - Wilo Samson Pumps Ltd
Ipn - Wilo Samson Pumps Ltd
IPS washroom systems - Armitage Shanks Ltd
IQ - Trend Control Systems Ltd
Iq Locker - T B S Fabrications
iQ Ranges - Tarkett Ltd
Irby Door - AS Newbould Ltd
IRC - Osram Ltd
Irius - SFS Intec LTD
Iron Duke - Heckmondwicke FB
Iron-Horse - Kleen-Tex Industries Ltd
IRT - Pentair Thermal Management
IRTS - Institute of Specialist Surveyors and Engineers
iSAFE - Capital Safety Group (NE) Ltd
ISD2 - Novoferm Europe Ltd
Isis suite - Brass Art Ltd
ISO 95+ GL Polyiso roof insulation boards - Firestone Building Products
Isoconex - Conex Universal Limited
Isocrete - Flowcrete Plc
Isocrete Self Level - Isocrete Floor Screeds Ltd
Isofast - SFS Intec LTD
Isoflex - Isolated Systems Ltd
Isofoam - Baxenden Chemicals Ltd
Isogran - Isocrete Floor Screeds Ltd
Isokern - Schiedel Chimney Systems Ltd
Isokoat - Schiedel Chimney Systems Ltd
Isola Marble & Granite - Reed Harris
Isolan - Ancon Building Products
Isolgomma - Sound Service (Oxford) Ltd
Isotape - Pentair Thermal Management
Isotherm - Biddle Air Systems Ltd
Isothin - Isocrete Floor Screeds Ltd
Isoval - Clivet UK Ltd
Isowool - Nevill Long Limited
Ispotherm - Alumasc Exterior Building Products Ltd
ISSE - Institute of Specialist Surveyors and Engineers
Ista - Switch2 Energy Solutions Ltd
Italia Collection - Lakes Bathrooms Limited
Italia-80, Italia-100, Italia-80 Cladding and Screening and Italia-100 Cladding and Screening - Lang+Fulton
Italian Travertines - Kirkstone
Itconi - Lewes Design Contracts Ltd
iTrack - Whitecroft Lighting Limited
ITS (Interior Transformation Services) - ITS Ceilings & Partitions
IVE - Vitreous Enamel Association, The
Iviring & Sellers - IJK Timber Group Ltd

J

J D Interiors - Shaylor Group Plc
J R Pearson - Drawn Metal Ltd
Jabcore - Jablite Ltd
Jabfloor - Jablite Ltd
Jablite - Jablite Ltd
Jabroc - Permali Deho Ltd
Jabroof Board - Jablite Ltd

J (con't)

Jabroof Panel - Jablite Ltd
Jabroof Slimfix - Jablite Ltd
Jabsqueeze - Jablite Ltd
Jabtherm - Jablite Ltd
Jackson - Viscount Catering Ltd
Jacksons fencing® - Jackson H S & Son (Fencing) Ltd
Jacksons finefencing® - Jackson H S & Son (Fencing) Ltd
Jacuzzi® - Jacuzzi® Spa and Bath Ltd
Jakcure® - Jackson H S & Son (Fencing) Ltd
Jakob Rope Systems - MMA Architectural Systems Ltd
Jaktop® - Jackson H S & Son (Fencing) Ltd
JALI - Jali Ltd
Janitor - Deb Ltd
Janus - Johnson & Starley
Jasper - Project Office Furniture PLC
Jasper Morrison - Ideal-Standard Ltd
Jasper - Sapphire Balustrades
JB Kind Doors - JB Kind Ltd
JBBlue® - Brash John & Co Ltd
JBRED® - Brash John & Co Ltd
JCB - J C Bamford (Excavators) Ltd
JCP - Ankerit - Owlett - Jaton
Jebron - Exidor Limited
Jenny Twin - Owen Slate Services Ltd
Jenny Twin - Owens Slate Services Ltd
Jet Print - Kleen-Tex Industries Ltd
Jet stream - Vianen UK Ltd
Jetdye Digital Printing - Zephyr The Visual Communicators
Jetfloor - Hanson Concrete Products Ltd
Je-Trae - Olsen Doors & Windows Ltd
Jet-rite - Flamco UK Ltd
Jetslab - Hanson Concrete Products Ltd
Jetstream Ventilation Fan - Colt International
Jet-Thrust Fans - Fläkt Woods Ltd
Jewel - Metalliform Products Plc
J-Fix - Unifix Ltd
JFS - Fulton Boiler Works (Great Britain) Ltd
JIB - Joint Industry Board for the Electrical Contracting Industry
Jingmei Rubber - Renqiu Jingmei Rubber&Plastic Products Co., Ltd
Jizer - Deb Ltd
JM Aerofoil - Fläkt Woods Ltd
JM Landscaping - Jack Moody Ltd
JMix - Jack Moody Ltd
Jobomulch - ABG Ltd
John Carr - Jeld-Wen UK Ltd
John Watson & Carter - Johnson VB LLP
Johnson - Johnson Tiles
Johnson Controls - Johnson Controls Building Efficiency UK Ltd
Johnson Tiles - Johnson Tiles
Johnstones - Brewer C & Sons Ltd
Johnstone's Paints - Kalon Decorative Products
Johnstones Performance Coatings - Kalon Decorative Products
Johnstone's Woodworks - Kalon Decorative Products
Joltec - Ronacrete Ltd
Jomy - Baj System Design Ltd
Jones+Attwood - Jacopa Limited
Joseph Bramah - Bramah Security Equipment Ltd
Jotalakk - Jotun Henry Clark Ltd Decorative Division
Jotaplast - Jotun Henry Clark Ltd Decorative Division
Jotapro - Jotun Henry Clark Ltd Decorative Division
Jotawall - Jotun Henry Clark Ltd Decorative Division
Joy - Zon International
JPA Furniture - John Pulsford Associates Ltd
JSB - Cooper Lighting and Security Ltd
Jubilee - Clenaware Systems Ltd
Jubilee - Cooper Lighting
Jubilee - Wade International Ltd
Jubilee Gold - Howse Thomas Ltd
Juggenaut Poles - Harrision Flagpoles
Juliet/Coyne & Cicero/Sirton Radiators - Barlo Radiators Limited
Juliette Spiral - Loft Centre Products
Jumbo - Swish Building Products
Jumbotec - Swish Building Products
Jumbovent - Swish Building Products
Jung - Pump Technical Services Ltd
Jung Pumpen - Pump Technical Services Ltd
Jupiter - Fabrikat (Nottingham) Ltd
Jupiter - Guardall Ltd
Jupiter - Samuel Heath & Sons plc
Jupiter - Ubbink (UK) Ltd
Jupiter spring / motor operated cable reels - Metreel Ltd
Just Shelving - Moresecure Ltd
Jutland - Cembrit Ltd
JVC - Vidionics Security Systems Ltd
JW Green Swimming Pools Ltd - Regency Swimming Pools Ltd

K

K Bead - Kilwaughter Chemical Co Ltd
K Dash - Kilwaughter Chemical Co Ltd
K Lime - Kilwaughter Chemical Co Ltd
K Mix - Kilwaughter Chemical Co Ltd
K Post - Kilwaughter Chemical Co Ltd
K Rend - Kilwaughter Chemical Co Ltd
K Screed - Isocrete Floor Screeds Ltd
K40 - Gilgen Door Systems UK Ltd
K65 - Conex Universal Limited
K7 - Komfort Workspace PLC
Kaba - Kaba Ltd
Kaba garog - Kaba Garog
Kair - Kair ventilation Ltd
Kalahari - Architen Landrell Associates Ltd
Kalea - Wessex Lift Co Ltd
Kaloric - Kaloric Heater Co Ltd
Kalzip - Tata Steel Europe Limited
Kameleon - Colt International
Kameo - Komfort Workspace PLC
Kamstrup - Switch2 Energy Solutions Ltd
Kana - Ciret Limited
Kanaline - Flexible Ducting Ltd
Kappa - Rockwell Sheet Sales Ltd
Kappa Curvo - Rockwell Sheet Sales Ltd
Kaptan, Kare, Karnak, Kensington, Klon, Kupka - Pitacs Ltd
Kara - Neptune Outdoor Furniture Ltd
Karibia - Kompan Ltd
Kasbah Twist - Adam Carpets Ltd
K-Bond - Kenyon Group Ltd
K-Bond - Kenyon Performance Adhesives
Kederflex - Icopal Limited
Kee Anchor® - Kee Klamp Ltd
Kee Guard ® - Kee Klamp Ltd
Kee Klamp ® - Kee Klamp Ltd
Kee Koat® - Kee Klamp Ltd
Kee Line ® - Kee Klamp Ltd
Kee Lite® - Kee Klamp Ltd
Kee Mark™ - Kee Klamp Ltd
Kee nect ™ - Kee Klamp Ltd
Keep Kool - Airedale International Air Conditioning Ltd
Keim Concretal Lasur - Keim Mineral Paints Ltd
Keim Ecosil - Keim Mineral Paints Ltd
Keim Granital - Keim Mineral Paints Ltd
Keim Lotexan - Keim Mineral Paints Ltd
Keim Optil - Keim Mineral Paints Ltd
Keim Restauro - Keim Mineral Paints Ltd
Keim Royalan - Keim Mineral Paints Ltd
Keim Soldalit - Keim Mineral Paints Ltd
Keim Universal Render - Keim Mineral Paints Ltd
Keizer Venesta - IJK Timber Group Ltd
Keller Foundations - Keller Ground Engineering
Kemira - Anderson Gibb & Wilson
Kemira - Andrews Coatings Ltd
Kemlite - Crane Composites
Kemperol - Granflex (Roofing) Ltd
Kemperol 1K-PUR - Kemper System Ltd
Kemperol 2K-PUR - Kemper System Ltd
Kemperol V210 - Kemper System Ltd
Kemprotect - Granflex (Roofing) Ltd
Kempston - Hanson Building Products
Kenco - The Kenco Coffee Company
Kenco In-Cup - The Kenco Coffee Company
Kenco Singles - The Kenco Coffee Company
Kencot Linking Chair - Race Furniture Ltd
Kenngott international - Ermine Engineering Co. Ltd
Kenrick - Kenrick Archibald & Sons Ltd
KensalFlame - Gazco Ltd
Kensington - Branson Leisure Ltd
Kenstack - McKenzie-Martin Ltd
Kentish Range - Hye Oak
Kenyon floor paint - Kenyon Group Ltd
Keps - Springvale E P S
Kerabond - Mapei (UK) Ltd
Keralastic - Mapei (UK) Ltd
Keraquick - Mapei (UK) Ltd
Kerastar - Johnson Tiles
Keratint - Procter Johnson & Co Ltd
Kerbmaster - Benton Co Ltd Edward
Kerbmaster - Edward Benton & Co Ltd
Kernal - Range Cylinders
Kerridge Stone - Earl's Stone Ltd
Kerridge Stone - Kerridge Stone Ltd
Kershaw - Freeman T R
Kestrel - Owlett - Jaton
Kew - Nilfisk Limited
Keybak - Securikey Ltd
Keyblok - Marshalls Mono Ltd
Keyguard - Hoyles Electronic Developments Ltd
Keylex Digital Locks - Relcross Ltd
Keylite - Allmat (East Surrey) Ltd
Keymer - Keymer Hand Made Clay Tiles
Keymer Clayton - Keymer Hand Made Clay Tiles
Keymer Ditchling - Keymer Hand Made Clay Tiles
Keymer Inline Ventilation System - Keymer Hand Made Clay Tiles
Keymer Shire - Keymer Hand Made Clay Tiles
Keymer Traditional - Keymer Hand Made Clay Tiles
Keypac - Wernick Group Ltd
Keystone - Allmat (East Surrey) Ltd
K-Flex - Pentair Thermal Management
Khrvf100 - Kair ventilation Ltd
Khrvisu - Kair ventilation Ltd
KiddEx - Explosion Suppression agent - Kidde Fire Protection Ltd
Killaspray - Hozelock Ltd
Kimpton - Branson Leisure Ltd
Kinder - BFM Europe Limited
Kinder - Dunhams of Norwich
kinetics rubber floorings - Jaymart Rubber & Plastics Ltd
King hoists and trollys - PCT Group
King Sliding Door Gear - Lakeside Group
Kingfisher - William Hopkins & Sons Ltd
Kingfisher - Zled (DPC) Co Ltd
Kingley - Flowflex Components Ltd
Kings Fund - Sidhil Care
KingShield - Green Brook
Kingspan - FGF Ltd
Kingspan Environmental Service - Kingspan Environmental Ltd
Kingspan Klargester - Kingspan Environmental
Kingspan Water - Kingspan Environmental
Kingspan Water - Kingspan Environmental Ltd
Kingston - Gloster Furniture Ltd
Kingstonian - Kingstonian Paint Ltd
Kingswood - Waterline Limited
Kirkby - Burlington Slate Ltd
Kirkstone Volcanic Stone - Kirkstone
Kitchen Collection - Ellis J T & Co Ltd
Kitchen Confidence - Symphony Group PLC, The
Kito Bitumen Emultions - Colas Ltd
Klargester - Kingspan Environmental Ltd
Klassic - Komfort Workspace PLC
Klaxon Signals - Texecom Limited
Kleen Thru Plus - Kleen-Tex Industries Ltd
Kleenscrape - Kleen-Tex Industries Ltd
Kleen-Stat - Kleen-Tex Industries Ltd
Kleerlok - Nettlefolds Ltd
Kleertred - Exitex Ltd
Klemm - Ingersoll-Rand European Sales Ltd
Klensorb - B & D Clays & Chemicals Ltd
Klik - Hager Ltd
Klimavent - Motorised Air Products Ltd
Kliplok - Unifix Ltd
KM 3 - Komfort Workspace PLC
Knauf - Firebarrier Services Ltd
Knauf - Kitsons Insulation Products Ltd
Knauf Insulation Polyfoam - Knauf Insulation Ltd
Knight - Iles Waste Systems
Knobs Range - Pedley Furniture International Ltd
Knorr - The Kenco Coffee Company
Knotless Netting - Sportsmark Group Ltd
Kofferlite - IKM Systems Ltd
Kohlangax - BFM Europe Limited
Kohler - Kohler UK
Kolorband - Kronospan Ltd
Kolorcourt - Emusol Products (LEICS) Ltd
Kolourseal Roofing System - Kingfisher Building Products Ltd
Kombimetall - Knauf AMF Ceilings Ltd
Kombivol - Skanda Acoustics Limited
Komfire - Komfort Workspace PLC
Komfort - HEM Interiors Group Ltd
Komfort - New Forest Ceilings Ltd
Komfort CX - Komfort Workspace PLC
Kompan - Kompan Ltd
Konaflex - (Above ground drainage systems) - Servomac Limited
Kone Eco3000 - Kone PLC
Konexion - Kone PLC
Kontrakt - Pedley Furniture International Ltd
KoolDuct - Kingspan Industrial Ins. Ltd
Koolshade - Cooper Group Ltd
Kooltherm - Kingspan Industrial Ins. Ltd
Kooltherm - Kingspan Insulation Ltd
Kooltherm - Kitsons Insulation Products Ltd
Kopex - Kem Edwards Ltd
Koral - Kronospan Ltd
Korean plywood - Wood International Agency Ltd
Korifit - Adaptaflex Ltd
Koverflor - Witham Oil & Paint Ltd
Kraftex - Package Products Ltd
Krantz - Designed for Sound Ltd
Kriblok Crb Walling - Ruthin Precast Concrete Ltd
Krion - Porcelanosa Group Ltd
Kromax - Tyco Waterworks - Samuel Booth
Krono Country - Kronospan Ltd
Krono LAQ - Kronospan Ltd
Kronofix - Kronospan Ltd
Kronofloor - Kronospan Ltd
Kronoplus - Kronospan Ltd
Kronosilent - Kronospan Ltd
Kronospan - Anderson C F & Son Ltd
Kronospan - Boardcraft Ltd
Kronospan - Meyer Timber Limited
Kryptonite - Allegion (UK) Ltd
KS1000 - Kingspan Ltd
K-Seal patio range - Kingfisher Building Products Ltd
KT3 - Marlow Ropes Ltd
Kudos - Calomax Ltd
Kuterlex - Yorkshire Copper Tube
Kuterlex plus - Yorkshire Copper Tube
Kuterlon - Yorkshire Copper Tube
Kvent - Rite-Vent Ltd
Kwikastrip - BRC Reinforcement
Kwikastrip - Halfen Ltd
kwikfix fascia soffit system - Dales Fabrications Ltd
Kwikpoint - Kingfisher Building Products Ltd
Kwikroll - Gilgen Door Systems UK Ltd
Kwikstage - Mabey Hire Ltd
KX10 wood stains - Kingfisher Building Products Ltd
Kynar 500 - Arkema France
Kynar Aquatec - Arkema France

L

L2 Highflow - Armorex Ltd
La Linea - Lumitron Ltd
Label controls - Kaba Garog
Labline 2 - Norwood Partition Systems Ltd
Labren, Lanner, Linear, Lokum, Lunar, Luvre, Lstrad - Pitacs Ltd
Lacdor - Vicaima Ltd
Lachat - Zellweger Analytic Ltd, Sieger Division
Ladder Up Safety Post - Bilco UK Ltd
LadderLatch™ - Latchways plc
Laddermatic - Clow Group Ltd
Laguna - Kompan Ltd
Lakeside Group - Lakeside Group
Lakeside Lifestyle.com - Lakeside Group
Lakeside Sercurity - Lakeside Group
Lami - Lami Doors UK Ltd
Lamina - FGF Ltd
Lamina Filtertrap - Progressive Product Developments Ltd
Laminated Supplies Ltd - Halmark Panels Ltd
Lampro Systems - Crane Composites
Lamson - Barloworld Vacuum Technology Plc
Lamwood - Lilleheden Ltd
Lancashire Interiors - Designer Radiators Direct
Lancaster - BF Bassett & Findley
Landcoil - Polypipe Civils
Landflex AV - Landline Ltd
Landflex CL - Landline Ltd
Landflex EcoBarrier - Landline Ltd
Landflex GDS - Landline Ltd
Landflex HCR - Landline Ltd
Landflex LLR - Landline Ltd
Landflex VR300 - Landline Ltd
Landflex ZR - Landline Ltd
Landis & Gyr - Switch2 Energy Solutions Ltd
Landmark (Slate, Double Pantile & Double Roman) - Monier Redland Limited
Landsurveyor - Gas Measurement Instruments Ltd
Langley By Shavrin - Shavrin Levatap Co Ltd
Lanka - Swedecor
LANmark - Nexans UK Ltd
Lantana - Brockway Carpets Ltd
Lap - Polycell Products Ltd
Lapidolith - Conren Ltd
Laser - Flexiform Business Furniture Ltd
Laser Gold 20 Year & Laser Torch 15 Year - Chesterfelt Ltd
Latapoxy - Laticrete International Inc. UK
Lateral Restraint Tie (LRT.) - Redifix Ltd
Latham Clad - James Latham plc
Laticrete - Laticrete International Inc. UK
Laumans - Smithbrook Building Products Ltd
Laura Ashley - British Ceramic Tile Ltd
Lavanda - Saint-Gobain Ecophon Ltd
Lawapan ceiling and wall panels - Salex Acoustics Limited

Laxton's - V B Johnson LLP
Laxtons's Building Price Book - Johnson VB LLP
Laybond adhesives - Laybond Products Ltd
Laytrad contract furnishers - Laytrad Contracts Ltd
LB lighting - ALD Lighting Solutions
Lbofall - Leofric Building Systems Ltd
LC-35 Barrier - APT Controls Ltd
LCN Door Closers - Relcross Ltd
LCS - Kenilworth Daylight Centre Ltd
LD Barrier - APT Controls Ltd
Lead Pointing Sealant - BLM British Lead
Lead T-Pren - BLM British Lead
Lead T-Pren Plus - BLM British Lead
Lead-Cote - BLM British Lead
LeaderFlush Shapland - Moffett Thallon & Co Ltd
Leaf - Metalliform Products Plc
Leak Sealer F4, Superconcentrate Leak Sealer F4, Leak Sealer F4 Express - Fernox
Leakguard - Pentair Thermal Management
Leakmaster - Charcon Tunnels (Division of Tarmac plc)
Lee Bishop - Viscount Catering Ltd
Legge - Allegion (UK) Ltd
Legge - Laidlaw Ltd
Legno Partitions - Avanti Systems
Legrand - Legrand Electric Ltd
LEIA - Lift and Escalator Industry Association,The
Leisure - Beko plc
Leisure Sinks - Rangemaster
Lentex - Dixon Turner Wallcoverings
Lenton ® - Erico Europa (GB) Ltd
Lenton reinforcing bar couplers - BRC Reinforcement
Leromur - Grass Concrete Ltd
Les Actuels de Lucien Gau - Myddleton Hall Lighting, Division of Peerless Design Ltd
Lesco - Lesco Products Ltd
Lessman - Hyde Brian Ltd
Letterboxes @ Signs of the times - Signs Of The Times Ltd
Levelay HO - Conren Ltd
Levelmaster - Stannah Lifts Ltd
Levit - Glamox Luxo Lighting Limited
Levolux - Levolux A.T. Limited
Lewen Radiators - Barlo Radiators Limited
Lexin - Rangemaster
Leyland - Brewer C & Sons Ltd
Leyland Paints - Kalon Decorative Products
LFH - Hellermann Tyton
LHC Kitchens - Moores Furniture Group Ltd
LIA Laboratories Ltd - Lighting Industry Association, The
LIA Quality Assurance - Lighting Industry Association, The
Liberator - Allart, Frank, & Co Ltd
Liberty range of platform lifts - Wessex Lift Co Ltd
Liebig - Arrow Supply Company Ltd
Lieutenant - Neptune Outdoor Furniture Ltd
Lifeguard - ArjoHuntleigh UK
Lifestyle Curved stairlift - Bison Bede Ltd
Lifestyle Elite, Luxe, Delight, Lawn, Multi, City - Artificial Grass Ltd
Lifestyle Showers - Hansgrohe
LifeSystem - H E W I (UK) Ltd
Li-Flat - Aremco Products
Lift 'n' Lock - Airsprung Beds Ltd
Liftingclamps.com - Camlok Lifting Clamps Ltd
Lift-Slab - Hydratight Sweeney Limited
Light Coffers - Saint-Gobain Ecophon Ltd
Light Fantastic - Intersolar Group Ltd
Lighting Industry Academy - Lighting Industry Association, The
Lighting Systems UK - CU Phosco Lighting
Lightmanager - Philips Lighting Solutions
Lightmaster - Philips Lighting Solutions
Lightmatic - Optex (Europe) Ltd
Lightning NDC - Hawker Siddley Switchgear
Lightpack - Fitzgerald Lighting Ltd
Lightport - Sola Skylights
Lightrak - Electrak International Ltd
Lightseal - Illuma Lighting
Lightspan - Duplus Architectural Systems Ltd

L (con't)

Lignacite - Lignacite (North London) Ltd
Lignacite - Lignacite Ltd
Lignacrete - Lignacite (North London) Ltd
Lignacrete - Lignacite Ltd
Lignastone Architectural Dressings - Lignacite Ltd
Lilliput Nursery - Portakabin Ltd
Lime quick - Legge Thompson F & Co Ltd
Limebacker cable ramps - Lewden Electrical Industries
Limescale Remover DS-3, Superconcentrate Limescale Preventer - Fernox
Limetics/Foamglas - Telling Lime Products Ltd
Limpet BD6 self setting composition - Thermica Ltd
Limpetite Rubber Coatings - Bristol Metal Spraying & Protective Coatings Ltd
Lindapter - Kem Edwards Ltd
Lindapters - Arrow Supply Company Ltd
Lindiclip - Lindapter International
Lindum - York Handmade Brick Co Ltd
Line Lazer - Marking Machines International
Linea - Hamilton R & Co Ltd
Linea Duo - Hamilton R & Co Ltd
Linea, Minima - Amwell Systems Ltd
Linear - Rangemaster
Linect - Glamox Luxo Lighting Limited
Liner Hydratow Fast Tow Concrete Mixer - Multi Marque Production Engineering Ltd
Liner Major Concrete Mixers - Multi Marque Production Engineering Ltd
Liner Rolpanit Pan Mixers - Multi Marque Production Engineering Ltd
Liner Roughrider Dumpers - Multi Marque Production Engineering Ltd
Linergrip - Viking Johnson
Linesman - Marking Machines International
Linesman - MSA (Britain) Ltd
Linflex Fin and Narrow Filter Drain - Polypipe Civils
Linido - Amilake Southern Ltd
Link 51 - Libraco
Linosom - Tarkett Sommer Ltd
Lintie - Redifix Ltd
Linton & Grafham Planters - Anthony de Grey Trellises + Garden Lighting
Linx Range; Bollards, Railing system, Cycle stands and Litter Bins - Furnitubes International Ltd
Lion - Boardcraft Ltd
Lion Brand - Cottam Brush Ltd
LION: Lion-Farm. LION Standard, LION Softboards, LION Oil tempered - Finnish Fibreboard (UK) Ltd
Liquid 99 - British Nova Works Ltd
Liquid Trent - Portacel
Liquistore - Franklin Hodge Industries Ltd
Liquitainer - Franklin Hodge Industries Ltd
Lisbon - Swedecor
Lister-Petter - Chippindale Plant Ltd
Liteglaze - Ariel Plastics Ltd
Litelink LSC - Eaton Electric Limited
Liteminder - Setsquare
Litetile - Catnic
Litex - Lawton Tube Co Ltd, The
Little Acorns and Little Rainbows - Spaceright Europe Limited
Little crown - Kompan Ltd
Livewire Shelving - Bedford Shelving Ltd
Living Stone - Concrete Products (Lincoln) 1980 Ltd
LivLoc - Atkinson & Kirby Ltd
Livonett Wood - Tarkett Sommer Ltd
LO Lobay - Consult Lighting Ltd
Load Pro - Nico Manufacturing Ltd
Load Release System - Goelst
Loadbank - Dorman Smith Switchgear
Loadframe - Dorman Smith Switchgear
Loadlimiter - Dorman Smith Switchgear
Loadline - Dorman Smith Switchgear
Loadmaster - Cooke Brothers Ltd
Loadswitch - Dorman Smith Switchgear
Loblite - Loblite Ltd
LOC Strip - Marble Flooring Spec. Ltd
Locate - Universal Components Ltd
LocFold - Omnis
Lochrin ® - Wm. Bain Fencing Ltd
Lochrin Classic Rivetless Palisade - Wm. Bain Fencing Ltd
Lochrin® Bain - Wm. Bain Fencing Ltd
Lockhart Project services - Lockhart Catering Equipment
Lockmaster - Paddock Fabrications Ltd
Lockstone - PHI Group Ltd
Loctite - Henkel Consumer Adhesives
Loctite - Loctite UK Ltd
Log walls - PHI Group Ltd
Logic Plus - MK (MK Electric Ltd)
Logicwall Cupboard Storage - Logic Office Contracts
LokFacade - Omnis
Lokfix - Fosroc Ltd
Lokmesh - Potter & Soar Ltd
LokRoll - Omnis
London - Hanson Building Products
London Concrete - Aggregate Industries UK Ltd
Longlife - Nicholson Plastics Ltd
Longline - Nevill Long Limited
Longmore Conduit - Tyco European Tubes
Longspan - Link 51 (Storage Products)
Longspan Shelving - Moresecure Ltd
Loo-Clean - Tank Storage & Services Ltd
LookRyt® - Ryton's Building Products Ltd
Loovent - Airflow Developments Ltd
Lord's - Scotts of Thrapston Ltd
Lorient - Humphrey & Stretton PLC
Lothian - Russell Roof Tiles Ltd
Lotrak geotextiles - Don & Low Ltd
Lotrak pavelay - Don & Low Ltd
Lotus - Gazco
Lotus - Lotus Water Garden Products Ltd
Louver-Lite Ltd - Louver-Lite Ltd
Louvolite - Louver-Lite Ltd
Louvrestyle - Hallmark Blinds Ltd
Low Energy Lighting Systems - Connect Lighting Systems (UK) Ltd
Lowara Pumps - Barr & Wray Ltd
Lozaron - Schlegel UK (2006) Ltd
LP 1000 - Wessex Lift Co Ltd
LP 550 - Wessex Lift Co Ltd
LPA - LPA Group PLC
LR800 - Wessex Lift Co Ltd
LR900 - Wessex Lift Co Ltd
LS Longden - Longden
LT - Staverton (UK) Ltd
LTG - Motorised Air Products Ltd
Lucas Furniture Systems - Carleton Furniture Group Ltd
Luce Plan - Quip Lighting Consultants & Suppliers Ltd
Lumaseal - Illuma Lighting
Lumax - light shelf blind - Hallmark Blinds Ltd
Lumicom - Lighting Industry Association, The
Lumilux - Osram Ltd
Lumisty - Northgate Solar Controls
Lutron - Tidmarsh & Sons
Lux Classic - Kronospan Ltd
Lux Lok - Kronospan Ltd
Luxaclair - Pilkington Plyglass
Luxaflex ® - Luxaflex®
Luxcrete - Luxcrete Ltd
Luxe - Rangemaster
Luxflex - IKM Systems Ltd
Luxfloor - Kronospan Ltd
Luxline Plus - Triphosphor Fluorescent Tubes - Sylvania Lighting International INC
Luxo - Glamox Luxo Lighting Limited
LWC - Redring
Lynx - Compact Fluorescent Lamps - Sylvania Lighting International INC
Lysander - Brockway Carpets Ltd
Lyssand - Broxwood (scotland Ltd)
Lytag - CEMEX
Lytag - Lytag Ltd

M

M & C Energy group, Merlin Gerin, Modicon, Mita - Schneider Electric Ltd
M E Duffells - Duffells Limited
M+P Labs - Lucideon
M2 - Guardall Ltd
M2M Steel Doors - Bradbury Group Ltd
Mab Door Closers - JB Architectural Ltd
MAC - Pressure Coolers Ltd T/A Maestro Pressure Coolers
Macalloy - MacAlloy Limited
Macaw - Paragon by Heckmondwike
Macclesfield Stone - Earl's Stone Ltd
Macclex - Exitex Ltd
Macdee Wirquin - Wirquin
Machin - Amdega Ltd
Mackay Engineering - Mackay Engineering
Mackridge - McKenzie-Martin Ltd
MacMount - MacLellan Rubber Ltd
Macpherson - Akzo Noble Coatings Ltd
Macphersons - Crown Paints Ltd
Macromelt - Henkel Ltd
Macron - Tyco Safety Products
Mactac - Northgate Solar Controls
Mactie - Redifix Ltd
Macvent - MacLellan Rubber Ltd
Maddalena Water Meters - SAV United Kingdom Ltd
Madico - Invicta Window Films Ltd
Madico - Northgate Solar Controls
Madico - The Window Film Company UK Ltd
Madison - Gloster Furniture Ltd
Madonna, Marion, Maze, Meridien, Mystic - Pitacs Ltd
Maestro - Lewden Electrical Industries
Maestro - Vianen UK Ltd
Maestro ® - Pressure Coolers Ltd T/A Maestro Pressure Coolers
Maestroflo ® - Pressure Coolers Ltd T/A Maestro Pressure Coolers
Maestropac ® - Pressure Coolers Ltd T/A Maestro Pressure Coolers
Mag Master - ABB Instrumentation Ltd
Magiboards - Magiboards Ltd
MagiCAD - Glamox Luxo Lighting Limited
MagiRail - Magiboards Ltd
Magis - Laytrad Contracts Ltd
Magnapleat - Camfil Farr Ltd
Magnaseal - Magnet Ltd
Magnastar - Magnet Ltd
Magnastorm - Magnet Ltd
Magnetek - Bernlite Ltd
Magnetic accessories - Clestra Limited
Magnum Firechests - Schiedel Chimney Systems Ltd
Mahtal - James Latham plc
Mailbox - Stamford Products Limited
Mailforce - The Safety Letterbox Company Ltd
MailManager™ Mail Management Software - Neopost Ltd
Main - Myson Radiators Ltd
Majestic - Brintons Ltd
Majestic - Majestic Shower Co Ltd
Majestic - Triton PLC
Makita - Arrow Supply Company Ltd
Makrolon - Amari Plastics Plc
Makroswing - Envirodoor Markus Ltd
Malindi - Hyperion Wall Furniture Ltd
mallard (air handling units) - Mansfield Pollard & Co Ltd
Malvern Beam Mounted Seating System - Race Furniture Ltd
Mammoth hoists - PCT Group
Mamoreete - DLW Flooring
Manade - Lesco Products Ltd
Managed Garment range - Helmsman
Managers Decorating Sundries - Kalon Decorative Products
Manby - Mandor Engineering Ltd
Mancuna - Mandor Engineering Ltd
Mandor - Mandor Engineering Ltd
Mangers - Davant Products Ltd
Manhattan - Cooper Lighting
Manhattan - DW Windsor Lighting
Manhattan - Laird Security Hardware
Manhattan Furniture - Dennis & Robinson Ltd
Maniaccess - Manitou (Site Lift) Ltd
Manifestation - Optima
Manifix - Manitou (Site Lift) Ltd
Manireach - Manitou (Site Lift) Ltd
Maniscopic - Manitou (Site Lift) Ltd
Manitransit - Manitou (Site Lift) Ltd
Mann+Hummel - Vokes Air
Mannesmann - Cento Engineering Company Ltd
Manntech - Cento Engineering Company Ltd
Mansafe - Unistrut Limited
ManSafe® - Latchways plc
Manuvic jacks - PCT Group
MAOF - Schlegel UK (2006) Ltd
Mapegrip - Mapei (UK) Ltd
Mapelastic - Mapei (UK) Ltd
Mapress - Geberit Ltd
Marblex - Swish Building Products
Marcato Range - Respatex International Limited
Marchrist - Marcrist International Ltd
Mardome - Ariel Plastics Ltd
Margard - Dacrylate Paints Ltd
Marghestone Assimilated Granite - Marble Flooring Spec. Ltd
Marine - Richard Hose Ltd
Marineflex - Trade Sealants Ltd
Marineseal - Trade Sealants Ltd
Markar Hinges - Relcross Ltd
Markus - Envirodoor Markus Ltd
Marlborough - Adam Carpets Ltd
Marlborough - Ellis J T & Co Ltd
Marlborough - Gloster Furniture Ltd
Marley - DC Plastic Handrails Ltd
Marley Eternit Ltd - Marley Eternit Ltd
Marley Waterproofing - IKO PLC
Marleyrail - Brighton (Handrails), W
Marlux ® - Marlux Ltd
Marmoleum - Forbo Flooring
Marmolit - Harbro Supplies Ltd
Marples - Record Tools Ltd
Marquis - Brintons Ltd
Marrakesh - Brintons Ltd
Marryn Radiators - Barlo Radiators Limited
Marshalite - Marshalls Mono Ltd
Marstair - TEV Limited
Marston - Bailey Birkett Sales Division
Marstron - Marlow Ropes Ltd
Martin Sectional Doors - Regency Garage Door Services Ltd
Marutex Magnum stainless drilling screws - UK Fasteners Ltd
Marvac - Bailey Birkett Sales Division
Marxco - Marcrist International Ltd
MAS - Optelma Lighting Ltd
Masco - Bristan Group
Masco OneStep Installation Services - Moores Furniture Group Ltd
Masonite - Boardcraft Ltd
Masonite - Premdor
Masonite beams - Panel Agency Limited
Masons Timber Products - Mason FW & Sons Ltd
Masq - Ciret Limited
Massey Ferguson - AGCO Ltd
Mastascrew - Nettlefolds Ltd
Master - Norwood Partition Systems Ltd
Master - Securikey Ltd
Master Craftsman Forte - Brockway Carpets Ltd
Master Life, MasterMatrix - BASF plc, Construction Chemicals
Master X-Seed - BASF plc, Construction Chemicals
MasterAir, MasterCast, MasterCell - BASF plc, Construction Chemicals
Masterbill - Masterbill Micro Systems Ltd
Masterbill Elite - Masterbill Micro Systems Ltd
Masterblock - Aggregate Industries UK Ltd
Masterboard - Promat UK Ltd
Mastercraft (MC) - Barber Wilsons & Co Ltd
Masterdor - L. B. Plastics Ltd
MasterEmaco - BASF plc, Construction Chemicals
MasterFinish, MasterFlow - BASF plc, Construction Chemicals
Masterflue - Hamworthy Heating Limited
MasterGlenium, MasterKure - BASF plc, Construction Chemicals
Masterhitch - www.masterhitch.co.uk - HE Services (Plant Hire) Ltd
Masterpave - Tarmac Ltd
MasterPel, MasterPolyheed, MasterPozzolith, MasterProtect - BASF plc, Construction Chemicals
Masterpiece - Checkmate Industries Ltd
Masterplank® - Brash John & Co Ltd
MasterRheobuild, MasterRoc - BASF plc, Construction Chemicals
Masterseal - MK (MK Electric Ltd)
Masterseal - Sealmaster
MasterSeal, MasterSet - BASF plc, Construction Chemicals
Mastersolve - Tank Storage & Services Ltd
Match® - Vicaima Ltd
Matchplay - En-tout-cas Tennis Courts Ltd
Material Lab - Johnson Tiles
Mather & Platt - Tyco Fire and Integrated Solutions
Mathys - Andrews Coatings Ltd
Matki - Matki plc
Matrex - Terrapin Ltd
Matrix - Dixon Turner Wallcoverings
Matrix - Levolux A.T. Limited
Matrix - Lumitron Ltd
Matterson Cranes - PCT Group
Matthews & Yates Centrifugal Fans - Matthews & Yates Ltd
Mattseal - Armorex Ltd
Maurice Hill - BriggsAmasco
Mawrob ® - Broxap Ltd
Max - Jotun Henry Clark Ltd Decorative Division
Max - Ves Andover Ltd
Max ST - Max Appliances Ltd
Max-Econ - Horizon International Ltd
Maxi - Construction Group UK Ltd
Maxi - Isolated Systems Ltd
Maxi 60 Ceiling Systems - Sound Reduction Systems Ltd
Maxi LST - Jaga Heating Products (UK) Ltd
Maxi Sky - ArjoHuntleigh UK
Maxiboard - Sound Reduction Systems Ltd
Maxifit - Viking Johnson
Maxilift - Stannah Lifts Ltd
Maxim - J.F. Poynter Ltd
Maxim - Laird Security Hardware
Maxima - Fastwall Ltd
Maxima - Norwood Partition Systems Ltd
Maximair - McKenzie-Martin Ltd
Maximin - Max Appliances Ltd
Maximiser - Rycroft Ltd
Maximixam - O'Brian Manufacturing Ltd
Maxi-Mixer - Cardale Doors Ltd
Maxinex - Chalmit Lighting
Maxivent - Airflow Developments Ltd
Maxmatic 1000 - Max Appliances Ltd
Maxmatic 1500 - Max Appliances Ltd
Maxmatic 2000 - Max Appliances Ltd
Maxmatic 3000 - Max Appliances Ltd
Maxmatic 4000 - Max Appliances Ltd
Maxmatic 5000 - Max Appliances Ltd
Maxol 12D - Burco Maxol
Maxol M10D - Burco Maxol
Maxol M15 - Burco Maxol
Maxol Microturbo - Burco Maxol
Maxol Mirage - Burco Maxol
Maxol Montana - Burco Maxol
Maxol Morocco - Burco Maxol
Maxum R ™ - Dezurik International Ltd
Maxum V ™ - Dezurik International Ltd
Maxwell House - The Kenco Coffee Company
Maybeguard - Mabey Hire Ltd
Mayfair - Gower Furniture Ltd
Mayfair - Kaloric Heater Co Ltd
Mayfield Brol - Kingstonian Paint Ltd
MB - 1 - Fernox
MBCad - Masterbill Micro Systems Ltd
M-Bond - Isocrete Floor Screeds Ltd
M-Bond Xtra - Isocrete Floor Screeds Ltd
MBR Technology - Jacopa Limited
MC - Howden Buffalo
McPherson - Brewer C & Sons Ltd
McQuay UK Ltd - Daikin Applied (UK) Ltd
MCR Barrier - APT Controls Ltd
MCR Recycling - Jack Moody Ltd
MCS Barrier - APT Controls Ltd
mctavishramsay - Bridgman IBC Ltd
Mc-Wall - Smart Systems Ltd
MDA Scientific - Zellweger Analytic Ltd, Sieger Division
Meadrain - Mea UK Ltd
Meagard - Mea UK Ltd
Meagard S - Mea UK Ltd
Meakerb - Mea UK Ltd
Mealine - Mea UK Ltd
Meaplas - Mea UK Ltd
Mearin - Mea UK Ltd
Mecury - Fabrikat (Nottingham) Ltd
medeco - Mul-T-Lock
Medica - Veoila Water Technologies UKElga Ltd
Medical Gas - Conex Universal Limited
Medici - Forticrete Ltd
Mediclad - Interclad (UK) Ltd
Medicool (bedhead lamps) - Brandon Medical
Mediland MDF - Willamette Europe Ltd
Medilog - ArjoHuntleigh UK
Medite - Boardcraft Ltd
Medite - Meyer Timber Limited
Medite 313 MR MDF - Willamette Europe Ltd
Medite Exterior - Willamette Europe Ltd
Medite FQ - Willamette Europe Ltd
Medite FR Class 0 - Willamette Europe Ltd
Medite FR Class 1 - Willamette Europe Ltd
Medite HD - Willamette Europe Ltd
Medite MDF - Willamette Europe Ltd
Medite ZF - Willamette Europe Ltd
Medstor - Stamford Products Limited
Medway - Portacel
Medway - Potter & Soar Ltd
Meg - Abet Ltd
Megadoor - Crawford Hafa Ltd
Megaduct - Polypipe Ventilation
Megafilm, Megatac, Megaboard - Visqueen Building Products
Megafit - Viking Johnson
Megaflo - Diamond Merchants
Megaflo, MegaLife - Heatrae Sadia Heating Ltd
Megasys - Advanced Perimeter Systems Ltd
MegaTEE - Uponor Ltd
Melbury - Hamworthy Heating Limited
Melbury - Hamworthy Heating Ltd
Meleto - Panel Systems Ltd
meltaway - Uponor Ltd
MEM - Eaton Electric Ltd
Memera 2000 - Eaton Electric Limited
Memera 2000 AD - Eaton Electric Limited
Memlite BC3 - Eaton Electric Limited
Memstyle - Eaton Electric Limited
Menvier - Cooper Lighting and Security Ltd
Menvier - Cooper Safety
Menvier Security - Eaton's Security Business
Meols Door - AS Newbould Ltd

M (con't)

Mepla - Geberit Ltd
Merchant Hardware - Avocet Hardware Ltd
Mercury - Eltron Chromalox
Mercury - S & B Ltd
Mercury - Triton PLC
Meridian Garage Doors - Regency Garage Door Services Ltd
Merley - Hamworthy Heating Limited
Merley - Hamworthy Heating Ltd
Merlin - Swintex Ltd
Mermaid Panels - Mermaid Panels Ltd
Mermet - Mermet U.K
Mervene - Firwood Paints Ltd
Meshtec - Cadisch MDA Ltd
Meshtrack - Bekaert Building Products
Message Manager - Deaf Alerter plc
Mestervindu - Broxwood (scotland Ltd)
META Cubicles - Detlef Muller Ltd
Metabin - Arc Specialist Engineering Limited
Metabolt - Arc Specialist Engineering Limited
Metaclip - Arc Specialist Engineering Limited
Metagard - Leigh's Paints
Metal Halide Lamps - Sylvania Lighting International INC
Metalarc - Metal Halide Lamps - Sylvania Lighting International INC
Metalclad Plus - MK (MK Electric Ltd)
Metalliform - Metalliform Products Plc
Metalphoto - Pryorsign
Metalrax - Arc Specialist Engineering Limited
Metalset - Loctite UK Ltd
Metasys - Johnson Controls Building Efficiency UK Ltd
Metbar - Mid Essex Trading Co Ltd
Meteon - Trespa UK Ltd
Meters Direct - Switch2 Energy Solutions Ltd
Metframe - Metsec Ltd
Metiflash - Metra Non Ferrous Metals Ltd
Metizinc - Metra Non Ferrous Metals Ltd
Metlex - Triton PLC
Metpost - Challenge Fencing Ltd
Metpost - Expamet Building Products
Metpost products - B & M Fencing Limited
Metro - Harton Heating Appliances
Metro - Townscape Products Ltd
Metro Therm - Harton Heating Appliances
Metrobond - Britmet Tileform Limited
Metropole - Silent Gliss Ltd
Metropolitan - Hartley & Sugden
Metropolitan - Silent Gliss Ltd
Metrose E - Hartley & Sugden
Metrowel - Mid Essex Trading Co Ltd
Metsec - Metsec Lattice Beams Ltd
Metsec - Metsec Ltd
Metsec floor beams - Metsec Lattice Beams Ltd
Metsec lattice joists - Metsec Lattice Beams Ltd
Metsec lattice trusses - Metsec Lattice Beams Ltd
Metsec Zed purlins - Metsec Ltd
Metspan - Metsec Ltd
Metstrut - Metsec Ltd
Met-Track enclosed track system - Metreel Ltd
Metway - Metway Electrical Industries Ltd
Meva Screens - Adams- Hydraulics Ltd
Meyer - Commercial Lighting Systems
Meynell - Kohler Mira Ltd
Mezzo - Kermi (UK) Ltd
MGB - Titan Environmental Ltd
MGE Office Protection Systems - Eaton Electric Ltd
MGL Distribution - McArthur Group Limited
Michelmersh - Michelmersh Brick Holdings PLC
Micrex - Stamford Products Limited
Microbore - Wednesbury Tube
Microcard Readers - Time and Data Systems International Ltd (TDSI)
Microdek® - Metalfloor UK Ltd
Microflon - Plastic Coatings Ltd
Microfloor 600® - Workspace Technology Ltd
Microlift - Stannah Lifts Ltd
Microlock Systems - Time and Data Systems International Ltd (TDSI)
Micronex - Chalmit Lighting
MicroPac - Draeger Safety UK Ltd
MicroPoint - Paragon by Heckmondwike
Micropol - Stamford Products Limited
Microvent - Watts Industries UK Ltd
Midlift - Stannah Lifts Ltd
Miele - Miele Co Ltd
Milano - Symphony Group PLC, The
Milborne - Hamworthy Heating Limited
Millars - Bitmen Products Ltd
Millenium - Stocksigns Ltd
Millenium - Troika Contracting Ltd
Miller by Honeywell - Honeywell Safety Products
Milliken Colours - Milliken Carpet
Milltown Blend - Carlton Main Brickworks Ltd
Millwood - Project Office Furniture PLC
Milnrow Sandstone - Kerridge Stone Ltd
Milton Precast - Milton Pipes Ltd
Minerva MX - ADT Fire and Security Plc
MinGardi - Dyer Environmental Controls
Mini Canal - Jaga Heating Products (UK) Ltd
Mini Clearflow Gas Flue Blocks - Dunbrik (Yorks) Ltd
Mini Compack - Fulton Boiler Works (Great Britain) Ltd
Mini Glaze - Pearce Security Systems
Mini Locker - Helmsman
Mini Loovent - Airflow Developments Ltd
Mini Profile - Onduline Building Products Ltd
Minibore - Yorkshire Copper Tube
Mini-compacta - KSB Ltd
Miniflo Rainwater System - FloPlast Ltd
Minilift - Didsbury Engineering Co Ltd
Mini-Lynx - Compact Fluorescent Lamps - Sylvania Lighting International INC
Minima - Fastwall Ltd
Minimaster - ABB Ltd
Minipack - Ramset Fasteners Ltd
Minislate - Forticrete Ltd
Minislate - Forticrete Roofing Products
Minisoccer Goals - Sportsmark Group Ltd
Minispace - Kone PLC
Miniwarn - Draeger Safety UK Ltd
Minor - Norwood Partition Systems Ltd
Minster - Minsterstone Ltd
Minton Hollins - Johnson Tiles
Miodox - Witham Oil & Paint Ltd
Mipolam - Brighton (Handrails), W
Mipolam - DC Plastic Handrails Ltd
Mipolam EL - Gerflor Ltd
Mipolam Esprit 500 - Gerflor Ltd
Mira - Kohler Mira Ltd
Miraclean - AudiocomPendax Ltd
Miraflex - Knauf Insulation Ltd
Mirage - Dudley Thomas Ltd
Mirage - Komfort Workspace PLC
Mirage - Matki plc
Mirage - Mode Lighting (UK) Ltd
Mirage - Muraspec
Mirage - Shackerley (Holdings) Group Ltd incorporating Designer ceramics
Mirage - Whitecroft Lighting Limited
MIRANIT - Franke Sissons Ltd
MIRO - Alanod Ltd
MIROGARD Anti-Reflective Glass - SCHOTT UK Ltd
Mirrorvision - Pro Display TM Limited
Mistral - Airedale International Air Conditioning Ltd
Mistral - HEM Interiors Group Ltd
Mistral - Imperial Machine Co Ltd
Mistral - Optima
Mita - Schneider Electric Ltd
Mitix - Howse Thomas Ltd
Mitsubishi - Deralam Laminates Ltd
Mitsubishi - Purified Air Ltd
Mixcal - Altecnic Ltd
MJC - Dantherm FiltrationLtd
MJX - Dantherm FiltrationLtd
MK - MK (MK Electric Ltd)
MMS - Kone PLC
Mobile 2000 - Lappset UK Ltd
Mobile shelving - Arc Specialist Engineering Limited
Mobile VMS - Techspan Systems
Mobilier International - Haworth UK Ltd
Mobility Kerb - Brett Landscaping
Mod 4 - Leofric Building Systems Ltd
Moda Range - Twyford Bathrooms
Mode - Allgood plc
Mode - Bernlite Ltd
Moderna, Metro and Micro - Lang+Fulton
Modernfold - Dividers Modernfold Ltd
Modric - Allgood plc
MODUCEL® - Eaton-Williams Group Limited
Modul - Glamox Luxo Lighting Limited
Modulair - Biddle Air Systems Ltd
Modular - Helmsman
Modular 70 - H W Architectural Ltd
Modular Glazing - HW Architectural Ltd
Modulight - Mermet U.K
Moduline - IAC Acoustics
Modupak - Stokvis Industrial Boilers International Limited
Modus - Glasdon U.K. Limited
Moeller - Eaton Electric Ltd
Moeller Electric - Electrix International Ltd
Mogaspaan - Metra Non Ferrous Metals Ltd
Mogat - Metra Non Ferrous Metals Ltd
Molosil - Notcutt Ltd
Monaframe - Sapa Building Systems (monarch and glostal)
Monaframe - Sapa Building Systems Ltd
Monaperm 700 - Icopal Limited
Monarch - Sapa Building Systems (monarch and glostal)
Monarch - Sapa Building Systems Ltd
Monarflex - Icopal Limited
Monarsound - Icopal Limited
Mondial - Uponor Ltd
Mondo Sportflex - Altro Ltd
Mondo Sportflex Athletic Surface - Bernhards Landscapes Ltd
Mondopave - Altro Ltd
Monks Park - Bath Stone - Hanson UK - Bath & Portland Stone
Monnex - Britannia Fire
Mono - Delta Balustrades
Mono - Parthos UK Ltd
Mono Control - Tidmarsh & Sons
Monoceram - R J Stokes & Co Limited
monocoque support system - Dales Fabrications Ltd
Monoglass - Aardvark Transatlantic Ltd
Monoglass - Becker (SLIDING PARTITIONS) Ltd
Monolevel - Flexcrete Technologies Limited
Monolite - Flexcrete Technologies Limited
Monomix - Flexcrete Technologies Limited
Monoplan - Becker (SLIDING PARTITIONS) Ltd
Monopour - Flexcrete Technologies Limited
Monoprufe - Ronacrete Ltd
Monorail - Integra Products
Monoscape - Marshalls Mono Ltd
Monoshake - Don Construction Products Ltd
Monospace - Kone PLC
MonoSpace® - Kone
Monostrip - Dufaylite Developments Ltd
Monotrak, Multitrak - Rackline Systems Storage Ltd
Monowall5oo - Monowa Manufacturing (UK) Ltd
Monroe Sign Family - GB Sign Solutions Limited
Monsoon - Stuart Turner Ltd
Montage - Heckmondwicke FB
Montana - Armitage Shanks Ltd
Montanna - Libraco
Monument - Monument Tools Ltd
Monza - Laird Security Hardware
Moonlight Marker - Intersolar Group Ltd
Moorland - Cembrit Ltd
Moorwood Vulcan - Viscount Catering Ltd
Morcton Door - AS Newbould Ltd
Morion - Moravia (UK) Ltd
Morris - Morris Material Handling Ltd
Morris and Co - Sanderson
Mosaic Collection - Solus Ceramics Ltd
Mosaiq - Kompan Ltd
Motala™ - Kone
Motor Up staging - CPS Manufacturing Co
mould clear - Peter Cox Ltd
Mould Making - Aden Hynes Sculpture Studios
Mouldguard - Trade Sealants Ltd
Mouldshield - AkzoNobel Decorative Paints UK
Moventi - Zon International
Moveo - Style Door Systems Ltd
MS3 - Techspan Systems
MS4 - Techspan Systems
M-Screen - Mermet U.K
MSR Columns - Consult Lighting Ltd
MT70 IP67 waterproof fitting - Encapsulite International Ltd
MTM - Railex Systems Ltd
MTX Engineering - Multitex GRP LLP
MudularUK - Caledonian Modular
Muhlhauser - Schwing Stetter (UK) Ltd
Multi - Glamox Electric (UK) Ltd
Multibeam - Ward Insulated Panels Ltd
Multibond - Loctite UK Ltd
Multibond Gold - Laybond Products Ltd
Multibrace - Mabey Hire Ltd
Multicem - Hanson UK
Multichannel - Ward Insulated Panels Ltd
Multicryl - Laybond Products Ltd
Multideck - Ward Insulated Panels Ltd
Multi-DENCO - DencoHappel UK Ltd
Multiduct - Ward Insulated Panels Ltd
Multifix Airbrick® - Ryton's Building Products Ltd
Multiflex - Marflex Chimney Systems
Multiflex - Multikwik Ltd
MULTIFLEX - Slab formwork - PERI Ltd
Multiflush - Multikwik Ltd
Multiguard - Hoyles Electronic Developments Ltd
Multikwik - Multikwik Ltd
Multilab laboratory furniture - Grant Westfield Ltd
Multilock - Nico Manufacturing Ltd
Multipanel Shower Panels - Grant Westfield Ltd
Multi-Plate - Asset International Ltd
Multipoint - Heatrae Sadia Heating Ltd
Multipor - Xella Aircrete Systems UK
Multipot - TSE Brownson
Multirend - Thermica Ltd
Multishield ™ - Accent Hansen
Multi-system Less-abled Products - Armitage Shanks Ltd
Multitack - Laybond Products Ltd
Multitex - Multitex GRP LLP
Multi-User - Healthmatic Ltd
Multivap - Ubbink (UK) Ltd
Multivent - Ubbink (UK) Ltd
Multivent - Vent-Axia
Multiwarn - Draeger Safety UK Ltd
MUL-T-LOCK - ASSA ABLOY Limited
Muncher - Mono Pumps Ltd
Munchpump - Mono Pumps Ltd
Municiple - Fitzgerald Lighting Ltd
Muralon - Muraspec
Murek - Muraspec
Murfor - Masonry Reinforcement - Bekaert Building Products
Mutrator - Mono Pumps Ltd
Mvi - Wilo Samson Pumps Ltd
MW - Redring
MW Motorway - Consult Lighting Ltd
MWD - Waterline Limited
Myrtha - Barr & Wray Ltd
Myson - The BSS Group

N

N.T.Master & Minor - Nevill Long Limited
Nailor - Advanced Air (UK) Ltd
Nano - Delabie UK Ltd
Nanu - Girsberger London
Nappigon - Imperial Machine Co Ltd
Nappychanger - LSA Projects Ltd
Narrowline ® - Andersen/ Black Millwork
NAS - National Association of Shopfitters
NASC - National Access & Scaffolding Confederation (NASC)
Nathan Range - Claydon Architectural Metalwork Ltd
National - Kenyon Performance Adhesives
National bondmaster - Kenyon Group Ltd
National Register of Warranted Builders - Federation of Master Builders
Natre - Olsen Doors & Windows Ltd
Natte - Mermet U.K
Natural Collection - Solus Ceramics Ltd
Natural Ventilation Solutions - SE Controls
Naturdor® - Vicaima Ltd
Nature Play - HAGS SMP
Nautilus - Aldous & Stamp (Services) Ltd
Navigator - Project Office Furniture PLC
NBS - NBS
NBS - RIBA Enterprises
NBS Building - NBS
NBS Building Regulations - NBS
NBS Create - NBS
NBS Engineering Services - NBS
NBS Landscape - NBS
NBS Plus - Knauf AMF Ceilings Ltd
nce - New Civil Engineer - EMAP
nDura - Delta Balustrades
Neaco - Norton Engineering Alloys Co Ltd
Neaco Support Systems - Norton Engineering Alloys Co Ltd
Neatdek 2 - Norton Engineering Alloys Co Ltd
Neatedge - Enable Access
Neatgrille - Norton Engineering Alloys Co Ltd
Neatmat - Norton Engineering Alloys Co Ltd
NeatraZone - PHS Group plc
Neff - Diamond Merchants
Nelson - Marlow Ropes Ltd
Nemef - Mul-T-Lock
Neo - Desking Systems Ltd
Neo Pantile - Sandtoft Roof Tiles Ltd
Neopolitan - Glasdon U.K. Limited
Neopost™ Mailroom Equipment and Supplies - Neopost Ltd
Neotran - Mode Lighting (UK) Ltd
Neotronics - Zellweger Analytic Ltd, Sieger Division
Neptune - Neptune Outdoor Furniture Ltd
NET LED Lighting - NET LED Limited
Net Play - HAGS SMP
NetComms - ADT Fire and Security Plc
Network bird management - P+L Systems Ltd
Net-works - Metway Electrical Industries Ltd
New c3m/p5. Moisture Resistant Carcase - Rixonway Kitchens Ltd
New Era - Benlowe Group Limited
New Ultomat - Project Office Furniture PLC
Neway - Neway Doors Ltd
Neway - (Neway Doors Ltd.) - Shutter Door Repair & Maintenance Ltd
Neway-(Neway Doors Ltd) - Priory Shutter & Door Co Ltd
Newdawn - Newdawn & Sun Ltd
Newdome (Rooflights) - Novaglaze Limited
Newlath - Newton John & Co Ltd
Newline - Tunstal Healthcare (UK) Ltd
Newman Monmore - Tyco European Tubes
Newspack - Leofric Building Systems Ltd
Newteam - NewTeam Ltd
Newton Door - AS Newbould Ltd
Nextspeed cat 6 - Hubbell Premise Wiring
Nextstep - Otis Ltd
Nexus - Astracast PLC
Nexus - HAGS SMP
Nexus (recycling Bins) - Glasdon UK Limited
Nexxus - Chalmit Lighting
NFC Specialist Products - GVS Filter Technology UK
NFDC - National Federation of Demolition Contractors
NFRC - National Federation of Roofing Contractors Ltd., The
NHBC - National House Building Council
Niagara Rainwater system - FloPlast Ltd
Nic-cool - Nicholson Plastics Ltd
NICEIC - National Inspection Council for Electrical Installation Contracting
Nickleby - Swintex Ltd
Nico - Nico Manufacturing Ltd
Nicobond - Nicholls & Clarke Ltd
Nico-O-Grout - Nicholls & Clarke Ltd
Night Cradle - Deaf Alerter plc
Nilfisk Advance - Nilfisk Limited
Nilfisk CFM - Nilfisk Limited
Nilfisk-ALTO - Nilfisk Limited
Nilflam - Kingspan Industrial Ins. Ltd
Nimbus® - ArjoHuntleigh UK
Nite Site - Leigh's Paints
Nitobond - Fosroc Ltd
Nitofil - Fosroc Ltd
Nitoflor - Fosroc Ltd
Nitoseal - Fosroc Ltd
Nitriflex - Movement Joints (UK)
Nitromors - Henkel Consumer Adhesives
No Skidding - Kenyon Paints Limited
Noise Lock - IAC Acoustics
NoiseStopSystems - The Sound Solution
Nomafoam - NMC (UK) Ltd
Nomapack - NMC (UK) Ltd
Nomastyl - NMC (UK) Ltd
Non Com X - Arch Timber Protection
Non-slip Bench Mat - Dycem Ltd
Norbo - Booth Muirie
Norbro - Flowserve Flow Control
Nordal rainwater system - Dales Fabrications Ltd
nordic light - ITAB Prolight UK Limited
Nordica - Deceuninck Ltd
Nordpeis - Gazco
Nordyl - Dixon Turner Wallcoverings
Nori - Marshalls Mono Ltd
Norland, Norland Plus - Hartley & Sugden
Norlyn - Smart F & G (Shopfittings)Ltd
Normban - Laidlaw Architectural Hardware Ltd
Normbau - Allegion (UK) Ltd
Nor-Ray-Vac - Continuous radiant tube heating - AmbiRad Ltd
Norseman Swageless - Navtec
Norslo - Green Brook
North Star - Tapworks Water Softeners
Northern Hardwood - IJK Timber Group Ltd
NOS - Strand Hardware Ltd
Notifier Fire Detection - Fife Fire Engineers & Consultants Ltd
Nova - BPT Security Systems (UK) Ltd
Nova - Rawson Fillings Ltd
Nova - The Window Film Company UK Ltd
Nova - Zon International
Nova / Novaplex - British Nova Works Ltd
Novacare - British Nova Works Ltd
Novacryl - British Nova Works Ltd
Novaglaze - Swedecor
Novalift - British Nova Works Ltd
Novalin - Dixon Turner Wallcoverings
Novalite - British Nova Works Ltd
Novara-25 and Novara-34 - Lang+Fulton
Novaract - British Nova Works Ltd
Nova-Seal - Schlegel UK (2006) Ltd

N (con't)

Novashield - British Nova Works Ltd
Nova-Span Space - NovaSpan Structures
Novastet - British Nova Works Ltd
Novastone - Swedecor
Novatreet - British Nova Works Ltd
Novatron - British Nova Works Ltd
Novaways - British Nova Works Ltd
Novera - Regency Garage Door Services Ltd
Noviasol - Dalesauna Ltd
Novis - Samuel Heath & Sons plc
Novocon - Propex Concrete Systems
Novoferm - Envirodoor Markus Ltd
Novojet - Reznor UK Ltd
Novolit - Skanda Acoustics Limited
Novomesh - Propex Concrete Systems
NP100 Slurry - Imerys Minerals Ltd
NRG - HAGS SMP
NRWB - Federation of Master Builders
NSC4OUD1 - Tapworks Water Softeners
Nuance - Twyford Bushboard
Nuastyle taps - Armitage Shanks Ltd
Nucana - Ward's Flexible Rod Company Ltd
Nufins - Hines & Sons, P E, Ltd
Nuflex - Ward's Flexible Rod Company Ltd
Nu-klad - Kenyon Group Ltd
Nulfire - Tremco Illbruck Ltd
Nullifire - Firebarrier Services Ltd
Nullifire - Nullifire Ltd
Nutristore - CST Industries, Inc. - UK
NVS (Natural Ventilation Solutions) - SE Controls

O

Oasis - Architen Landrell Associates Ltd
Oasis - Kompan Ltd
Oasis - Water Technology Engineering Ltd
Oberflex - Abet Ltd
Oberon - Decra Roofing Systems
Oboin - Mermet U.K
Ocean - Signs & Labels Ltd
Octadoor - Righton Ltd
Octanorm - RTD Systems Ltd
OCTO 250 - Turner Access Ltd
Octoflex - Mat.Works Flooring Solutions
Odoni - Elwell Buildings Ltd
Odyssey worksurfaces - Twyford Bushboard
OEP contract - OEP Furniture Group PLC
OEP Engineering - OEP Furniture Group PLC
OEP matrix - OEP Furniture Group PLC
OfficeCool - DencoHappel UK Ltd
Ogilive Sealants - Ogilvie Construction Ltd
Ogilvie Construction Ltd - Ogilvie Construction Ltd
Oilmaster - Docherty New Vent Chimney Group
OKPOL - Sola Skylights
Old Clamp - York Handmade Brick Co Ltd
Old English pantile - Sandtoft Roof Tiles Ltd
Oldroyd Gtx geotextile cavity drainage membrane - Safeguard Europe Ltd
Oldroyd XP plaster membrane - Safeguard Europe Ltd
Oldroyd Xv cavity drainage membrane - Safeguard Europe Ltd
Oldstock Autique - Carlton Main Brickworks Ltd
Ollevibe - Olley & Sons Ltd, C
Olympia - Erlau AG
Olympian - FG Wilson
Olympic - Glasdon U.K. Limited
Olympic - Swedecor
Olympus - Carpets of Worth Ltd
OMA - Harbro Supplies Ltd
Omeg - Omeg Ltd
Omega - Anderson C F & Son Ltd
Omega - Sika Liquid Plastics Limited
Omega 4 - Cooper Lighting
Omega worksurfaces - Twyford Bushboard
Omega®, Omega Pencil Edge, Omega Flow - Brett Landscaping
Omegadeck - Turner Access Ltd
Omicron - Auchard Development Co Ltd
Omniclay - En-tout-cas Tennis Courts Ltd
Omnipex Horticulture - Omnipex Ltd
Omnitek - Grace De Neef UK
Onduline - Onduline Building Products Ltd
Ondutile - Onduline Building Products Ltd
Onduvilla - Onduline Building Products Ltd
One Step - Tank Storage & Services Ltd
Onyx - Pland Stainless Ltd
Onyx - Project Office Furniture PLC
Onyx - Sapphire Balustrades
OP 763 Waterstop - Carter-Dal International
Opal - Sapphire Balustrades
Opalux - Northgate Solar Controls
Open Options - Project Office Furniture PLC
Openlok - Unifix Ltd
Opera - AEL
Opertis - H E W I (UK) Ltd
Opiocolour Mosaics - Reed Harris
OPT539 Waterproof membrane - Carter-Dal International
Optelma - Optelma Lighting Ltd
OptiDome - CST Industries, Inc. - UK
Optiflame - Dimplex UK Limited
Optiflex - Anglepoise Lighting Ltd
Optiflex - Hughes Safety Showers Limited
Optifloat - Pilkington Birmingham
Optiguard window screens - Bradbury Group Ltd
Optilam - Pilkington Birmingham
Optima - Cardale Doors Ltd
Optima - DW Windsor Lighting
Optima - Optima
Optima - Paslode
Optima - Tarkett Sommer Ltd
Optima by Coram - Coram Showers Ltd
Optimajor Plus - Hartley & Sugden
Optimirror - Pilkington Birmingham
Option Range - Twyford Bathrooms
Options - Hands of Wycombe
Options - Sanderson
Options Range - Integra Products
Options Toilet Cubicles - Grant Westfield Ltd
Options worksurfaces - Twyford Bushboard
Optisneen - Aalco Metals Ltd
Optitherm - Pilkington Birmingham
Optitherm thermostatic tap - Horne Engineering Ltd
Optivent - ABB Ltd
Optivision - The Window Film Company UK Ltd
Optiwin - Glamox Luxo Lighting Limited
Opto - Stamford Products Limited
Opto Thermo - Aqualisa Products
Opus - Booth Muirie
OPV 2000 - Colt International
Orator - Gent Limited
Oratorio carpet tiles cut pile - Burmatex Ltd
Orbik - Bernlite Ltd
Orbis - Apollo Fire Detectors Ltd
Orbis - Delta Balustrades
Orbis - Laidlaw Architectural Hardware Ltd
Orbis - Laidlaw Ltd
Orbit - Cil Retail Solutions Ltd
Orca - Rediweld Rubber & Plastics Ltd
Orcal Infusions - Armstrong World Industries Ltd
Orchestra - Vianen UK Ltd
Orcon - Fillcrete Ltd
Original - C - Paroc Panel Systems Uk Ltd
Original - E - Paroc Panel Systems Uk Ltd
Original - I - Paroc Panel Systems Uk Ltd
Original No Butts - The No Butts Bin Company (NBB), Trading as NBB Outdoor Shelters
Origins - Baggeridge Brick PLC
Orion - Ubbink (UK) Ltd
Orsogril - Squires Metal Fabrications Ltd
Ortega - Lumitron Ltd
Ortho Mat - Interface Europe Ltd
Ortronics - Legrand Electric Ltd
OS2 - SE Controls
OSID - Xtralis
Osma - Wavin Ltd
Osma DeepLine - Wavin Ltd
Osma RoofLine - Wavin Ltd
Osma Roundline - Wavin Ltd
Osma Squareline - Wavin Ltd
Osma Stormline - Wavin Ltd
Osma SuperLine - Wavin Ltd
Osma UltraRib - Wavin Ltd
OsmaDrain - Wavin Ltd
OsmaGold - Wavin Ltd
OsmaSoil - Wavin Ltd
OsmaWeld - Wavin Ltd
OSMO - Wood Finishes - OSMO UK
Osprey - Swintex Ltd
Osram - Bernlite Ltd
Osram Dulux - Osram Ltd
OT Medipower (isolated power systems) - Brandon Medical
Otis 2000 E - Otis Ltd
Otis 2000 H - Otis Ltd
Ovalgrip - Hellermann Tyton
OWA - OWA UK Ltd
OWAconstruct - OWA UK Ltd
OWAcoustic - OWA UK Ltd
OWAcoustic Janus - OWA UK Ltd
OWAcoustic smart - OWA UK Ltd
OWAdeco - OWA UK Ltd
OWAlux - OWA UK Ltd
OWALux Clean - OWA UK Ltd
OWAspectra - OWA UK Ltd
OWAtecta - OWA UK Ltd
OWAtecta Perfora - OWA UK Ltd
Owens Corning - Kitsons Insulation Products Ltd
Oxan WP - Jotun Henry Clark Ltd Decorative Division
Oxygen - Jaga Heating Products (UK) Ltd
Oxypic - Dunsley Heat Ltd

P

P. S. Fan units - Elan-Dragonair
P.P.P - Wade International Ltd
P+L Systems Washroom - P+L Systems Ltd
P2000 - Johnson Controls Building Efficiency UK Ltd
P5 - Hudevad Britain
P50 - Britannia Fire
P5K - Hudevad Britain
PAC - Vidionics Security Systems Ltd
Pace - Caldwell Hardware (UK) Ltd
Pacemaker - Portakabin Ltd
Pacific - Deceuninck Ltd
Pacifyre® - Walraven Ltd
Pactan - Tremco Illbruck Ltd
PaintSpec Finder™ - Crown Paints Ltd
Palace - Parthos UK Ltd
Palace Switches - Wandsworth Elecrtrical Ltd
Palazzoli - Lewden Electrical Industries
Palermo and Piazza - Lang+Fulton
Pallas - Architectural Textiles Ltd
Palletstor - Moresecure Ltd
Pamec - Schwing Stetter (UK) Ltd
Panacea, Pegasus, Petit, Planal, Planet Moon, Poplar, Portofino, Porte - Pitacs Ltd
Panasonic - Connaught Communications Systems Ltd
Panasonic - Vidionics Security Systems Ltd
Panatrim - Universal Components Ltd
Panda - Horizon International Ltd
Panel Glide - Silent Gliss Ltd
Panel&Frame - Amwell Systems Ltd
Panelam® - Panel Agency Limited
Paneline - Panel Agency Limited
Panelmaster - Timeguard Ltd
Panelvent® - Panel Agency Limited
Panic Hardware - Dorma Door Services Ltd
Pan-L-Trim - Gooding Aluminium
Panorama - Muraspec
Panther - Simpson Strong-Tie®
Pantile 2000 - Britmet Tileform Limited
Paptrim Aluminium - Pitchmastic PmB Ltd
Paptrim Roof Edge Trime - Pitchmastic PmB Ltd
Parabolt - Arrow Supply Company Ltd
Parabolt - Tucker Fasteners Ltd
Parade - Iles Waste Systems
Parafon - CoolZone - Armstrong World Industries Ltd
Paragon - Macemain + Amstad Ltd
Paragon AV - Macemain + Amstad Ltd
Paraline - Marlow Ropes Ltd
Paraline - Twyford Bushboard
Paramount - Macemain + Amstad Ltd
Parat C - Draeger Safety UK Ltd
Parctile - Decra Roofing Systems
Parflu - May William (Ashton) Ltd
Pargon - Fitzgerald Lighting Ltd
Parigi - Altecnic Ltd
Parinox - Altecnic Ltd
Park Home Coatings - Everlac (GB) Ltd
Park Street Furniture - Coates LSF Ltd
Par-Ker - Porcelanosa Group Ltd
Parker Bath - ArjoHuntleigh UK
Parkiflex - Western Cork Ltd
Parkray - Hepworth Heating Ltd
Parmet - May William (Ashton) Ltd
Paroc - Knauf Insulation Ltd
Parocs Structural - Paroc Panel Systems Uk Ltd
Parri - Laytrad Contracts Ltd
Pasa XP - Kermi (UK) Ltd
Pasivent - Passivent Ltd
Paslode - ITW Construction Products
Paslode - ITW Spit
Passivent - Zled (DPC) Co Ltd
Passivent Tricklevents - Passivent Ltd
Pastoral - Alno (United Kingdom) Ltd
Patay Pumps - Pump International Ltd
Patchfast - Conren Ltd
Patented energy control - Fluidair International Ltd
Patex - Henkel Ltd
Pathfinder - Intersolar Group Ltd
Pathmaster - Tarmac Ltd
Patina - Shadbolt F R & Sons Ltd
Patination Oil - BLM British Lead
Patio - Rawson Carpets Ltd
Pavia Toscana - Kirkstone
Pavigres - R J Stokes & Co Limited
PCA - The Property Care Association
PCE - PCE Instruments UK Ltd
PDA Range of induction loop systems - C-TEC (Computionics) Ltd
Peabody Water Services - Satec Service Ltd
Peak - Chesterfelt Ltd
Peakmoor - Realstone Ltd
Peakstone - CEMEX
Pearce Homes - Pearce Construction (Barnstaple) Limited
Pearce Property Services - Pearce Construction (Barnstaple) Limited
Pearpoint - Radiodetection Ltd
Pecafil® - BRC Special Products - Bekaert Ltd
Pedalo - Erlau AG
Pegler - Olby H E & Co Ltd
Pegler - Pegler Ltd
Pelham Picnic Suite - Branson Leisure Ltd
Pelican - Amilake Southern Ltd
Pelican - Tuke & Bell Ltd
Pel-Job - Chippindale Plant Ltd
PEMKO - ASSA ABLOY Limited
Pemko Seals and Thresholds - Relcross Ltd
Pendock Casings and Enclosures - Alumasc Interior Building Products Limited
Pendock profiles - Alumasc Interior Building Products Limited
Pennine - Laidlaw Architectural Hardware Ltd
Pennine - Laidlaw Ltd
Pennine - Russell Roof Tiles Ltd
Pennine Stone - Haddonstone Ltd
Penrhyn - Welsh Slate Limited
Pensby Door - AS Newbould Ltd
Pent Concrete Garages - Liget Compton
Pent Mansard Concrete Garages - Liget Compton
People Flow™ - Kone
Pepex Pipes - Uponor Ltd
Pepperpot Stool - Hille Educational Products Ltd
Perfect Fit - Tidmarsh & Sons
Perfector - Crane Fluid Systems
PerfoLight - Enable Access
Performa - Pegler Ltd
Performing Art - Brintons Ltd
Perftec - Cadisch MDA Ltd
Pergola - Panel & Louvre Co Ltd
PERI UP - System scaffolding - PERI Ltd
Perimate DI-A - Dow Building Solutions
Perinsul - Pittsburg Corning (UK) Ltd
Period suite - Brass Art Ltd
Periscope® - Ryton's Building Products Ltd
Perko, Perkomatic, Powermatic - Samuel Heath & Sons plc
Perlux Digital 2D/3D Screens - Harkness Screens
Permaban - BRC Reinforcement
Permabond - Kenyon Group Ltd
Perma-clean, Permashield ® - Andersen/ Black Millwork
Permacrib - PHI Group Ltd
Permafoam - Isocrete Floor Screeds Ltd
Permagard - Andura Coatings Ltd
Permagard - Haynes Manufacturing (UK) Ltd
Permagold - Deva Tap Co Ltd
Permaguard - PermaRock Products Ltd
Permakote - Wessex Resins & Adhesives Ltd
Permalath - PermaRock Products Ltd
Permali - Permali Deho Ltd
Permalux - Lumitron Ltd
Permanite - IKO PLC
permaramp, swift ramp system - Enable Access
Permarend - PermaRock Products Ltd
PermaRock - PermaRock Products Ltd
Permascreed - DPC Screeding Ltd
Permaseal Roofing - PermaSeal Ltd
Permasint - Plastic Coatings Ltd
Permastrip - BLM British Lead
Permatred - Permali Deho Ltd
Permavoid - Polypipe Civils
Permawood - Permali Deho Ltd
Permoglaze - Akzo Noble Coatings Ltd
Permoglaze - Brewer C & Sons Ltd
Permoglaze - Crown Paints Ltd
Personal Surveyor - Gas Measurement Instruments Ltd
Perspectives - Gradus Ltd
Pest-Stop - P+L Systems Ltd
Petrarch - CEP Claddings Ltd
Petrel - PFP Electrical Products Ltd
Petrelux - PFP Electrical Products Ltd
Petro fast modular forecourt systems - Cookson and Zinn (PTL) Ltd
Petrochem - SSAB Swedish Steel Ltd
PetroTec - CST Industries, Inc. - UK
Pexapipe - IPPEC Sytsems Ltd
Pexatherm - IPPEC Sytsems Ltd
PFI Group - GB Sign Solutions Limited
Pfleiderer - Anderson C F & Son Ltd
Pfliederer - Deralam Laminates Ltd
PGF - Production Glassfibre
Phantom - Dudley Thomas Ltd
Pharo - Hansgrohe
Phenblox® - Walraven Ltd
Phenomenon - Paragon by Heckmondwike
Phillip Jeffries - Architectural Textiles Ltd
Phillips - Bernlite Ltd
Phlexicare - Nicholls & Clarke Ltd
Phoenix - Cooke Brothers Ltd
Phoenix - Glasdon U.K. Limited
Phoenix - Intersolar Group Ltd
Phoenix - Swintex Ltd
Phoenix - Kingfisher - Jewers Doors Ltd
Phoenix - Osprey - Jewers Doors Ltd
Phoenix - Swift - Jewers Doors Ltd
Phoenix (recycling Bins) - Glasdon UK Limited
Phoenix Fire Doors - Benlowe Group Limited
Phonic - Parthos UK Ltd
Phosco - CU Phosco Lighting
Photosorption - Bicester Products Ltd
Piazza - Erlau AG
Piccola - Draeger Safety UK Ltd
Piccolo - Stannah Lifts Ltd
PICDOR - TORMAX United Kingdom Ltd
Pickering Europe Ltd - Pickerings Ltd
Pickwick - Swintex Ltd
Picture Perf perforated metal Pictures/Logos - Graepel Perforators Ltd
Pieri - Grace Construction Products Ltd
Piggyback - Safety Assured Ltd
Pikes - Pike Signals Ltd
Piling Accessories - Cordek Ltd
Pilkington Acoustic Laminate - Pilkington UK Ltd
Pilkington Insulight - Pilkington UK Ltd
Pilkington K Glass - Pilkington UK Ltd
Pilkington 'K' Glass - Pilkington Birmingham
Pilkington Optifloat - Pilkington UK Ltd
Pilkington Optimirror Plus - Pilkington UK Ltd
Pilkington Planar Structural Glazing - Pilkington UK Ltd
Pilkington Planarclad - Pilkington UK Ltd
Pilkington Pyrodur - Pilkington UK Ltd
Pilkington Pyroshield Safety - Pilkington UK Ltd
Pilkington Pyrostop - Pilkington UK Ltd
Pilkington Satin - Pilkington UK Ltd
Pilkington Suncool - Pilkington UK Ltd
Pilkington Tiles - R J Stokes & Co Limited
Pimlico - Ellis J T & Co Ltd
Pin Panelz - Spaceright Europe Limited
Pinoleum - Tidmarsh & Sons
Pinpoint - Stocksigns Ltd
Pinsulator - BigHead Bonding Fasteners Ltd
Pioneer Mr Kitchens - Moores Furniture Group Ltd
Pipeflex - Concrete Products (Lincoln) 1980 Ltd
Pipegard - Leigh's Paints
Piper Lifeline - Tunstal Healthcare (UK) Ltd
Pirelli - Kem Edwards Ltd
Pirouette - Rackline Systems Storage Ltd
Pirthane 0 - Modern Plan Insulation Ltd
PivotSafe - Dorma Door Services Ltd
Pladek - En-tout-cas Tennis Courts Ltd
Plain - Russell Roof Tiles Ltd
Plain Tile - Decra Roofing Systems
Plan - Hudevad Britain
Planar - Ide T & W Ltd
Planar - Pilkington Birmingham
Planet - Pland Stainless Ltd
Planet Peripherals - Planet Communications Group Ltd
Planet Projects - Planet Communications Group Ltd
Planet Videoconferenceing - Planet Communications Group Ltd
Plano - Rockfon Ltd
Planrad - AEL
Plant Managers Journal & Plant Spec & Dealer Guide - Contract Journal
Planta - Erlau AG
Plantec - Cadisch MDA Ltd
Plascoat - Plastic Coatings Ltd
Plasflow - Fullflow Group Ltd
Plasgard - Lonsdale Metal Company
Plasinter - Plastic Coatings Ltd
Plassim - Brett Martin Ltd
Plastalux - Clare R S & Co Ltd
Plastex - Plastic Extruders Ltd
Plasti Drain - Hepworth Building Products Ltd
Plastic Padding - Henkel Consumer Adhesives
Plastic padding - Loctite UK Ltd
Plastilac - Trimite Ltd
Plastimetal - EBC UK Ltd

P (con't)

Plastite - Nettlefolds Ltd
Plaswood - Visqueen Building Products
Platinum Care - Heckmondwicke FB
platinum Floorshield - Springvale E P S
Platinum Fulfil - Springvale E P S
Platinum Wallshield - Springvale E P S
Platinum Warm shark - Springvale E P S
Platinum Warm Squeez - Springvale E P S
Platoon Brush - Mat.Works Flooring Solutions
Platoon Premier - Mat.Works Flooring Solutions
Playa - Ideal-Standard Ltd
Playtime - Amwell Systems Ltd
Playtime® - Jackson H S & Son (Fencing) Ltd
Playtop 3D Range - Playtop Licensing Limited
Playtop Safer Surfacing - Playtop Licensing Limited
Playtop Spheres - Playtop Licensing Limited
Playtop Walkways - Playtop Licensing Limited
Playtop with Nike Grind - Playtop Licensing Limited
Plaza (litter bins) - Glasdon UK Limited
Plaza Range - Pedley Furniture International Ltd
Plexiglas - Amari Plastics Plc
Plise Shades - Luxaflex®
PLS - Topcon GB Ltd
Plumbers bits - Traps & wc connectors - Hunter Plastics Ltd
Plumbfit - Brett Martin Ltd
Plumbfit - Davant Products Ltd
Plumbing-flexi hoses - Guest (Speedfit) Ltd, John
Plumbuild - Airflow (Nicoll Ventilators) Ltd
Plus Clips - Tower Manufacturing
Plycorapid - Laybond Products Ltd
Plyload - SCP Concrete Sealing Technology Ltd
Plylok Deck - Gradient Insulations (UK) Ltd
Plymovent Limited - Encon Air Systems Ltd
PM2000 - Gas Measurement Instruments Ltd
PmB : Waterproofing - Pitchmastic PmB Ltd
Pmb waterproofing Systmes - Pitchmastic PmB Ltd
PNC 3 Vision - Tunstal Healthcare (UK) Ltd
Podium - Barlow Group
Podium Steps - Turner Access Ltd
Polagard - Manders Paint Ltd
Polar - Komfort Workspace PLC
Polar - Rockfon Ltd
Polaris™ - Kone
Polcarb 40 S - Imerys Minerals Ltd
Polcarb 45 Slurry - Imerys Minerals Ltd
Polcarb 60 (S) - Imerys Minerals Ltd
Polcarb 60 Slurry - Imerys Minerals Ltd
Polcarb 90 - Imerys Minerals Ltd
Polcarb SB - Imerys Minerals Ltd
Policor C.C - Polypipe Building Products
Polo - Swish Building Products
Polow - Troika Contracting Ltd
Poly-Bond - Schlegel UK (2006) Ltd
Polycell - Polycell Products Ltd
Polyclens - Polycell Products Ltd
Polycote - Cross-Guard International Ltd
Polycup - Rockwell Sheet Sales Ltd
Polyduct, Polysewer, Polystorm - Polypipe Civils
Polyester - Colorpro Systems Ltd
Polyfilla - Polycell Products Ltd
Polyflex - Polyflor Ltd
Polyflor - Polyflor Ltd
Polyfoam - Knauf Insulation Ltd
PolyLay - Junckers Limited
Polymate - Nettlefolds Ltd
Polymatic - CPV Ltd
Polymer Modified Bitumen Emultions and cutbacks - Colas Ltd
Polymetron - Zellweger Analytic Ltd, Sieger Division
PolyPed - Star Uretech Ltd
Polyplumb - Polypipe Building Products Ltd
PolyPlus - Helifix
Polyrail - Brighton (Handrails), W
Polyrey - Anderson C F & Son Ltd
Polyrey - Decra Ltd
Polyripple - Polycell Products Ltd
Polyroof - PermaSeal Ltd
Polysafe - Polyflor Ltd
Polyscreed - Cross-Guard International Ltd
Polyshield - Anglo Building Products Ltd
Polyshield - Feedwater Ltd
Polysil - Modern Plan Insulation Ltd
Polystar - Rockwell Sheet Sales Ltd
Polystrippa - Polycell Products Ltd
Polystyrene - FGF Ltd
Polytan - Feedwater Ltd
Polytek - Notcutt Ltd
Polytex - Polycell Products Ltd
Polytherm - Smart Systems Ltd
Polytred - Polyflor Ltd
Polytron - Draeger Safety UK Ltd
Polyu - Rockwell Sheet Sales Ltd
PondGard - Firestone Building Products
Pool Collection - Solus Ceramics Ltd
Poolwise - BioLab UK
Pop - Tucker Fasteners Ltd
POP Rivets - Allman Fasteners Ltd
Poplar - Branson Leisure Ltd
Popstar - BLV Licht-und Vakuumtecnik
Pop-up - Ackermann Ltd
Porcupine - Concrete Products (Lincoln) 1980 Ltd
Porotherm - Wienerberger Ltd
Portadec - Hughes Safety Showers Limited
Portaflex - Hughes Safety Showers Limited
Portaheater - Hughes Safety Showers Limited
Portaro® - Vicaima Ltd
Portasilo - Portasilo Ltd
Portaspray - Bambi Air Compressors Ltd.
Portway - BFM Europe Limited
Poselco - Poselco Lighting
Posicote - Cross-Guard International Ltd
Posilok - Exitex Ltd
Posi-Strut - MiTek Industries Ltd
Posi-Web - MiTek Industries Ltd
Posners - Posners the floorstore
Possi - Pitchmastic PmB Ltd
Post + Column - Signature Ltd
Post master - Paddock Fabrications Ltd
Postfix - Hanson UK
Pottelberg - Smithbrook Building Products Ltd
Potterton - Baxi Heating UK Ltd
Pouliot - Floralsilk Ltd
Powa glide - Moresecure Ltd
Powdagalv - Howse Thomas Ltd
Powder Prime - Howse Thomas Ltd
Powdermark - Marcrist International Ltd
Power - Kenyon Performance Adhesives
Power hotmelts - Kenyon Group Ltd
Powerball - Osram Ltd
Powerbase Gas Membrane - Cordek Ltd
Powerbond - Gilmour Ecometal
PowerBreaker - Green Brook
Powercyl - Gledhill Water Storage Ltd
Power-Fin - Lochinvar Limited
Powerflow Flux - Fernox
Powerflow Solder - Fernox
Powerform 25 - Dorman Smith Switchgear
Powerform 63 - Dorman Smith Switchgear
Powerframe - Sapa Building Systems (monarch and glostal)
Powerlab 8 - S & B Ltd
Powerlink Plus - MK (MK Electric Ltd)
Powerlok - Nettlefolds Ltd
Powerman - Schneider Electric Ltd
PowerPack - Santon
PowerPlas 520 - Power Plastics Ltd
PowerPlas 541 - Power Plastics Ltd
Powerrac - Dezurik International Ltd
Powersafe - Hawker Ltd
Powerstar - Osram Ltd
Powerstock - Hamworthy Heating Limited
Powerstock - Hamworthy Heating Ltd
Powerstream - Redring
Powertrack - Legrand Electric Ltd
Powertrak - Rackline Systems Storage Ltd
Powertrak plastic drag chains & carriers - Metreel Ltd
Powervent - Watts Industries UK Ltd
Powerware - Eaton Electric Ltd
Powrmatic - The BSS Group
Pozi GP - Nettlefolds Ltd
Pozidrain - ABG Ltd
Pozidry - Vent-Axia
PPA 571 - Plascoat Systems Ltd
PPC - Brighton (Handrails), W
PPG High Performance Coatings - PPG Protective and Marine Coatings
PPG Protective & Marine Coatings - Kenyon Paints Limited
PR - Pure - Staverton (UK) Ltd
PRA - PRA Coatings Technology Centre
Precedence - Paragon by Heckmondwike
Precious Gems - Carpets of Worth Ltd
PRELAQ - SSAB Swedish Steel Ltd
Prelasti - AAC Waterproofing Ltd
Prelude - Ellis J T & Co Ltd
Premier - Dixon Turner Wallcoverings
Premier - Fitzgerald Lighting Ltd
Premier - Integra Products
Premier - Owlett - Jaton
Premier - Securikey Ltd
Premier - Xpelair Ltd
Premier by Coram - Coram Showers Ltd
Premier Fire Extinguishers - Nu-Swift International Ltd
Premier Frameless by Coram - Coram Showers Ltd
Premier hoists - PCT Group
Premier Interlink Waco UK Limited - Waco UK Limnited
Premier Plus - Heatrae Sadia Heating Ltd
Premier Service - Flexiform Business Furniture Ltd
PremierPlus and PremierPlus Solar - Santon
Premix - Delabie UK Ltd
Premseal - BRC Reinforcement
Prep - Ciret Limited
Prepakt - Westpile Ltd
Preprufe - Grace Construction Products Ltd
Presbury - Race Furniture Ltd
Prescor - Flamco UK Ltd
PreSense/ AbSence - Philips Lighting Solutions
President - Deans Blinds & Awnings (UK) Ltd
President - Parthos UK Ltd
Pressalit - Amilake Southern Ltd
Presspak - Flamco UK Ltd
Pressure Jet Marker - Marking Machines International
Prestex - Pegler Ltd
Prestige - Gent Limited
Prestige - Stocksigns Ltd
Prestige Plus - MK (MK Electric Ltd)
Prestige Range - Respatex International Limited
Presto - Girsberger London
Preston Refrigeration - Johnson Controls Building Efficiency UK Ltd
Presweb - Prestoplan
prialpas rubber floorings - Jaymart Rubber & Plastics Ltd
Price Guides Direct - Professional Bookshops Limited
Prima - Cardale Doors Ltd
Prima - Prima Security & Fencing Products
Prima - Sampson & Partners Fencing
Prima - Schiedel Chimney Systems Ltd
Prima Axal - Armstrong World Industries Ltd
Prima 2 Gas Vent - Rite-Vent Ltd
Prima Plus Single Wall System - Rite-Vent Ltd
Prima SW Single Wall System - Rite-Vent Ltd
Primacalc - Fullflow Group Ltd
Primaflow - Fullflow Group Ltd
Primary Options - Project Office Furniture PLC
Primatic - Range Cylinders
Primofit - George Fischer Sales Ltd
Primus Washers/ Ironers - Warner Howard Group Ltd
Princess suite - Brass Art Ltd
Prinmuls Lite LMP90K Polymer Emultion - Colas Ltd
Prinmuls Mac H.A.U.C. Binder - Colas Ltd
Prinmuls MP90X Polymer Emultion - Colas Ltd
Print HPL - Abet Ltd
PrintMachine™ Laserprinting Management Software - Neopost Ltd
Printsign - Rivermeade Signs Ltd
Priory - CEMEX
Priory - Priory Shutter & Door Co Ltd
Priory - (The Priory Shutter & Door Co. Ltd.) - Shutter Door Repair & Maintenance Ltd
Priory Mixture - Carlton Main Brickworks Ltd
Priory Mixtures - Carlton Main Brickworks Ltd
Priory-(The Priory Shutter & Door Co. Ltd.) - Neway Doors Ltd
Prisma - Quiligotti Terrazzo Tiles Limited
Prismafit - Johnson Tiles
Prismatics - Johnson Tiles
Pritt - Henkel Consumer Adhesives
Pro - Net Contractor - Termstall Limited
Pro - Select - Owlett - Jaton
Pro Clear - Andura Coatings Ltd
Pro Display - Pro Display TM Limited
Pro Flex - Andura Coatings Ltd
PRO II - Capital Safety Group (NE) Ltd
Pro-Balance - Crane Fluid Systems
Procast - Procter Bros Limited
PROcheck - National Britannia Ltd
Procor - Grace Construction Products Ltd
procter cast stone - Procter Cast stone & Concrete Products
procter concrete products - Procter Cast stone & Concrete Products
procter fencing products - Procter Cast stone & Concrete Products
Proctor Johnson Colour - Procter Johnson & Co Ltd
Prodigy - Oldham Lighting Ltd
Proffessional - Anson Concise Ltd
Profil - Fillcrete Ltd
Profile - Railex Systems Ltd
Profile - Samuel Heath & Sons plc
Profile - VBH (GB) Ltd
Profile 22 - Plastal (SBP Ltd)
Profile 49 - Britmet Tileform Limited
Profile, Proform - Rackline Systems Storage Ltd
Profiled Screen - Acousticabs Industrial Noise Control Ltd
Profiles - Twyford Bushboard
Pro-flite Suspended Aerofoil - ITAB Prolight UK Limited
Profoil - FloRad Heating and Cooling
Proforce - Briggs Industrial Footwear Ltd t/a Briggs Safety Wear
Proframe - Sapa Building Systems (monarch and glostal)
ProFuse - Uponor Ltd
Progef - George Fischer Sales Ltd
Programastat - Timeguard Ltd
Progress - Parthos UK Ltd
Prolift hoists and winches - PCT Group
Proline - Rangemaster
Pro-Line - Moravia (UK) Ltd
Promat - FGF Ltd
Promat - Kitsons Insulation Products Ltd
Promat, Promatect, Promo seal, Promafour, Promina - Promat UK Ltd
Promenade - Lappset UK Ltd
Promesh - Procter Bros Limited
Pronto - Girsberger London
Proofex - Fosroc Ltd
ProPile - Linear Ltd
Proplene - CPV Ltd
Propulsion - Golden Coast Ltd
ProQuest - Coutts Information Services
Proseal - Armorex Ltd
Proseal - Illuma Lighting
Proseal - Sika Ltd
Pro-Set - Wessex Resins & Adhesives Ltd
Prospex - Terrapin Ltd
Protal - Winn & Coales (Denso) Ltd
Protecsol - Gerflor Ltd
Protect and Protect OD - Tank Storage & Services Ltd
Protecta - Capital Safety Group (NE) Ltd
Protecta - Chalmit Lighting
Protecta - Unifix Ltd
Protector - Amber Doors Ltd
Protector - Securikey Ltd
Protector F1, Superconcentrate Protector F1, Protector F1 Express, Protector MB-1, Antifreeze Protector Alphi-11, HVAC Protector F1 - Fernox
Pro-Tee - Crane Fluid Systems
Protega Paints - Kenyon Paints Limited
Protim - Protim Solignum Ltd
Protim- Prevac - Protim Solignum Ltd
Protimeter Hygromaster - Protimeter, GE Thermometrics (UK) Ltd
Protimeter Mini - Protimeter, GE Thermometrics (UK) Ltd
Protimeter MMS - Protimeter, GE Thermometrics (UK) Ltd
Protimeter Surveymaster - Protimeter, GE Thermometrics (UK) Ltd
Protimeter Timbermaster - Protimeter, GE Thermometrics (UK) Ltd
Proton - Paddock Fabrications Ltd
PS Concrete Repair System - David Ball Group Ltd
PS Gel Retarder - David Ball Group Ltd
PS HDBR Adhesive - David Ball Group Ltd
PS Hi-Flow Grouting Mortar - David Ball Group Ltd
PS Liquid Retarder - David Ball Group Ltd
PS Primer Latex - David Ball Group Ltd
PS Swell Mastic - David Ball Group Ltd
PS2000 - Gas Measurement Instruments Ltd
Psicon - Geoquip Worldwide
PSM - Elster Metering Ltd
Public - Macemain + Amstad Ltd
Pudlo - David Ball Group Ltd
Pudlo CWP Waterproofer - David Ball Group Ltd
Puf - Macemain + Amstad Ltd
Pullman - Portakabin Ltd
Pulsa - ITW Construction Products
Pulsacoil - Gledhill Water Storage Ltd
pulse secure electric fencing - Chestnut Products Limited
Puma - Mode Lighting (UK) Ltd
Puma - Resdev Ltd
Pumadur - Resdev Ltd
Pumaflor - Resdev Ltd
Pumpen - Pump Technical Services Ltd
PUR - Gerflor Ltd
Purbeck Limestone - J Suttle Swanage Quarries Ltd
Pure - Isolated Systems Ltd
Pure Air - Consort Equipment Products Ltd
Purelab, Purelab Chorus - Veoila Water Technologies UKElga Ltd
Purewater - Veoila Water Technologies UKElga Ltd
Purewell - Hamworthy Heating Ltd
Purilan - Ubbink (UK) Ltd
Puriton - Uponor Ltd
Purmelt - Henkel Ltd
Purmo Radiators - Purmo-UK
Purmo Underfloor Heating - Purmo-UK
Purogene - Hertel Services
Purogene - Verna Ltd
Purton - Uponor Ltd
Push & Lock® - Ryton's Building Products Ltd
Pushfit - Tyco Water Works
PushLock™ - Latchways plc
Pushtite - Nettlefolds Ltd
PV Low Profile - Polypipe Ventilation
PVF2 - Colorpro Systems Ltd
PX Control Panel - Guardall Ltd
Pygme - Grosvenor Pumps Ltd
PYRAN S Fire-Resistant Glass - SCHOTT UK Ltd
PYRANOVA Insulated Fire-resistant Glass - SCHOTT UK Ltd
Pyratox - Aldous & Stamp (Services) Ltd
Pyrocoil - Mann McGowan Group
Pyrodur - Ide T & W Ltd
Pyrodur - Pilkington Birmingham
Pyrodur plus - Ide T & W Ltd
Pyroglaze - Mann McGowan Group
Pyrogrille - Mann McGowan Group
Pyroguard EI - Pyroguard UK Ltd
Pyroguard EW - Pyroguard UK Ltd
Pyroguard T - Pyroguard UK Ltd
Pyromas - Mann McGowan Group
Pyroshield - Pilkington Birmingham
Pyrosleeve - Mann McGowan Group
Pyrospan - Mann McGowan Group
Pyrostem - Pyroguard UK Ltd
Pyrostop - Ide T & W Ltd
Pyrostop - Pilkington Birmingham
Pyrostrip - Mann McGowan Group
Pyrovista - Mann McGowan Group
Pyxel - Grosvenor Pumps Ltd

Q

Q Doors - HAG Shutters & Grilles Ltd
Q Lean Solutions - Quantum Profile Systems Ltd
Q screed - Carter-Dal International
Qbs - Arc Specialist Engineering Limited
QD 90 - Blackfriar Paints Ltd
QHG - Motorised Air Products Ltd
Q-Lon - Schlegel UK (2006) Ltd
Q-Rail 2000 - Quartet-GBC UK Ltd
QSBIM - Masterbill Micro Systems Ltd
QSCad - Masterbill Micro Systems Ltd
QUAD - Optelma Lighting Ltd
Quadro - Franke Sissons Ltd
Quadro 4 - Rockfon Ltd
QuadroClad - Hunter Douglas
Quadro-secura - Interflow UK
Qualceram - Shires Bathrooms
Quality Mark - The Carpet Foundation
Qualube - Witham Oil & Paint Ltd
Quantec - Johnson & Starley
Quantec addressable call systems - C-TEC (Computionics) Ltd
Quantum - Calomax Ltd
Quantum - Light Engine Lighting Ltd
Quantum - Marley Plumbing and Drainage
Quantum - Quantum Profile Systems Ltd
Quantum - Rackline Systems Storage Ltd
Quantum - Valor
Quarrycast - Troika Contracting Ltd
Quarter 2 - Litex Design Ltd
Quartermaster Shelving - Bedford Shelving Ltd
Quartz - TEV Limited
Quartz Digital - Aqualisa Products
Quartzstone Quartz Tiles - Marble Flooring Spec. Ltd
Quasar - Fitzgerald Lighting Ltd
Quasilan - Baxenden Chemicals Ltd
Quatro and DemiQuattro - Lang+Fulton
Queensfil 240 - Imerys Minerals Ltd
Queensfil 25 - Imerys Minerals Ltd
Queensfil 300 - Imerys Minerals Ltd
Quelcote - Quelfire
Quelfire - Firebarrier Services Ltd
Quelfire - Quelfire
Quelfire QF1 Mortar - Quelfire
Quelfire QF2 Mortar - Quelfire
Quelfire QF4 Mortar - Quelfire
Quick Clip - Junckers Limited
Quickbridge - Mabey Hire Ltd
Quickcem - Hanson UK
Quickfit - Vantrunk Engineering Ltd

Q (con't)

Quickfit - Viking Johnson
Quickflex - Sealocrete PLA Ltd
Quickframe - IPPEC Sytsems Ltd
Quickheat - IPPEC Sytsems Ltd
Quickmast - Don Construction Products Ltd
Quickpost - Catnic
Quickscreed - Laticrete International Inc. UK
Quickslate - BLM British Lead
Quickstrip - Laybond Products Ltd
Quicktronic - Osram Ltd
Quiclean - Hansgrohe
QuietBoard - Bicester Products Ltd
Quietboard - Sound Service (Oxford) Ltd
QuietFin - Linear Ltd
QuietFloor - Bicester Products Ltd
Quietfloor - Sound Service (Oxford) Ltd
Quikaboard - QK Honeycomb Products Ltd
Quil-nazzo - Quiligotti Terrazzo Tiles Limited
Quinta - Broag Ltd
Quintesse - Matki plc
Quip - Quip Lighting Consultants & Suppliers Ltd
Quitetite - Auld Valves Ltd
QVC - Motorised Air Products Ltd
Qwood - Quantum Profile Systems Ltd
QX Control Panel - Guardall Ltd

R

R Accommodation - Rollalong Ltd
R Defence - Rollalong Ltd
R Healthcare - Rollalong Ltd
R Modular - Rollalong Ltd
R Schools - Rollalong Ltd
R Units - Santon
R.A.M - R.A.M. Perimeter Protection Ltd
R.C.C - Roofing Contractors (Cambridge) Ltd
rac - Refrigeration & Air Conditioning - EMAP
Rada - Kohler Mira Ltd
Radcrete - Radflex Contract Services Ltd
Radflex 125 - Radflex Contract Services Ltd
Radflex Expantion Joints - Radflex Contract Services Ltd
Radflex FP - Radflex Contract Services Ltd
Radflex S100 - Radflex Contract Services Ltd
Radflex S150 - Radflex Contract Services Ltd
Radflex S200 - Radflex Contract Services Ltd
Radflex WF - Radflex Contract Services Ltd
RadiAL Windows & Louvres - Midland Alloy Ltd
Radiance - Matki plc
Radipex - Uponor Ltd
Radius - Hudevad Britain
Radjoint - Radflex Contract Services Ltd
Radon Aquis Double Defence System - Proten Services Ltd
Radroof - Radflex Contract Services Ltd
Radson Radiators - Purmo-UK
Rafid - Geoquip Worldwide
Rail Master - Youngman Group Limited
Rail Track Specialists - Southern Zone - GJ Durafencing Limited
Railex - Railex Systems Ltd
Railok - Capital Safety Group (NE) Ltd
Railtrack Propducts - International Protective Coatings
Rainbox - John Davidson (Pipes) Ltd
Raincoat - Tor Coatings Ltd
Raindance - Hansgrohe
Raindance Allrounder - Hansgrohe
Raindance Puro Air - Hansgrohe
Raindance Unica - Hansgrohe
Rainsaver® - Straight Ltd
Rainspan - Booth Muirie
Rakoll® - H.B. Fuller
Rallisil - Trade Sealants Ltd
Rall-on - Wessex Lift Co Ltd
Ram Pumps - Barr & Wray Ltd
Rambolt - Ramset Fasteners Ltd
Ramesis - Oldham Lighting Ltd
Rampcentre, rollouttrackway, rampkit - Enable Access
Ramplug - Ramset Fasteners Ltd
Ramsay Access - Ramsay & Sons (Forfar) Ltd
Ramsay Ladders - Ramsay & Sons (Forfar) Ltd
Ramset - Arrow Supply Company Ltd
Ramset - ITW Construction Products
Ramset - ITW Spit
Ramset - Ramset Fasteners Ltd
Randi - Laidlaw Architectural Hardware Ltd
Randi-line - Laidlaw Ltd
Range Cooker Co - Diamond Merchants
Rangemaster - Rangemaster
Rangewood - Vulcan Cladding Systems
Rapid - Clark Door Ltd
Rapid - Fillcrete Ltd
Rapidac - Dacrylate Paints Ltd
Rapide - Project Office Furniture PLC
Rapidex® - H.B. Fuller
Rapidflow - Geberit Ltd
Rapidobad - BRC Reinforcement
Rapidplan - Wernick Group Ltd
RapidRail® - Walraven Ltd
RapidSTACK - MCL Composites Ltd
RapidStrut® - Walraven Ltd
Rapier - Kenrick Archibald & Sons Ltd
Rapier rising screen - Power Security Systems
Raumluft - Motorised Air Products Ltd
Ravenna - Paragon by Heckmondwike
Rawlplug - Arrow Supply Company Ltd
Rawlplug - Kem Edwards Ltd
Raya - Kermi (UK) Ltd
Rayovac - Pifco Ltd
Raypak - AEL
Razor - Astracast
RB Horizontal - Fulton Boiler Works (Great Britain) Ltd
RBC - Fulton Boiler Works (Great Britain) Ltd
RCC - RCC, Division of Tarmac Precast Concrete Ltd
RD Road - Consult Lighting Ltd
RD series locators and detectors - Radiodetection Ltd
Rea frame - Rea Metal Windows Ltd
Rea W20 - Rea Metal Windows Ltd
Rea W40 - Rea Metal Windows Ltd
Ready Plumbed Modules - Twyford Bushboard
Ready span - CEMEX
Readymix, Ready floor, Ready block, Ready Pave - CEMEX
Realstone Ltd - Realstone Ltd
Rebound Furniture - Roc Secure Ltd
Rebound Signmaster - Glasdon U.K. Limited
ReCooler HP - Fläkt Woods Ltd
Rectaleen - Flexible Reinforcements Ltd
Recupovent - Thermal Technology Sales Ltd
Red Box Wipes - Deb Ltd
Red Engineering Type - Carlton Main Brickworks Ltd
Red Rustic/ Dragwire - Carlton Main Brickworks Ltd
Reddicord - Reddiglaze Ltd
Reddifoam - Reddiglaze Ltd
Reddihinges - Reddiglaze Ltd
Reddilock - Reddiglaze Ltd
Reddipile - Reddiglaze Ltd
Reddiprene - Reddiglaze Ltd
Redfyre - Gazco Ltd
Redfyre Cookers - TR Engineering Ltd
Redhall - Booth Industries Ltd
Rediguard - Hoyles Electronic Developments Ltd
RediKerb - Rediweld Rubber & Plastics Ltd
Redipress - Hoyles Electronic Developments Ltd
Redi-Resin - Redifix Ltd
RedLine - Monier Redland Limited
Redring - Applied Energy Products Ltd
Redufix - Watts Industries UK Ltd
Redupress - Watts Industries UK Ltd
RedVent - Monier Redland Limited
Reemat Premium - Sika Liquid Plastics Limited
Refatex - Astron Buildings Parker Hannifin PLC
Reflections - Ideal-Standard Ltd
Reflect™ - Rockwood Pigments (UK) Ltd
RefleKto - BLV Licht-und Vakuumtecnik
Reflex - Altecnic Ltd
Reflex-Rol - Mermet U.K
Reflex-Rol - Reflex-Rol (UK)
Refresh - Britannia Kitchen Ventilation Ltd
Refresh Range - Twyford Bathrooms
Regaflex S - Rega Ductex Ltd
Regaflex W - Rega Ductex Ltd
Regency Swimming Pools - Regency Swimming Pools Ltd
Regent - Cooper Lighting
Regent (R) - Barber Wilsons & Co Ltd
Registered Specialists - The Carpet Foundation
Regulator - Checkmate Industries Ltd
Rehau - Architectural Plastics (Handrail) Ltd
Rehau - Brighton (Handrails), W
Rehau - DC Plastic Handrails Ltd
Reid Span - Reid John & Sons (Strucsteel) Ltd
Rei-Lux - Valmont Stainton
Reinforced Earth™ TerraBlock - Freyssinet Ltd
Relite - Fantasia Distribution Ltd
RELUX - ALD Lighting Solutions
Relux - Glamox Luxo Lighting Limited
Relux - ITAB Prolight UK Limited
Remington - Pifco Ltd
Remix - SIG Roofing
Rendalath® - BRC Special Products - Bekaert Ltd
Rendalite® - BRC Special Products - Bekaert Ltd
Rendcomse - Race Furniture Ltd
Renderoc - Fosroc Ltd
Renno - Lumitron Ltd
Reno - Johnson & Starley
Renovent - Ubbink (UK) Ltd
Renovisions - Interface Europe Ltd
Repair Care Mini-Profi - Repair care International
Repair Care Sander - Repair care International
Repair Care Scraper - Repair care International
Repairmaster - Tarmac Ltd
Repcoat - Don Construction Products Ltd
Repertiore - Tata Steel Colors
Reportline - National Britannia Ltd
Resdev - Resdev Ltd
Resiblock 22 - Resiblock Ltd
Resiblock ER - Resiblock Ltd
Resiblock F.R.I - Resiblock Ltd
Resiblock OR - Resiblock Ltd
Resiblock Superior - Resiblock Ltd
Resiblock Ultra - Resiblock Ltd
Resiblock Ultra Matt - Resiblock Ltd
Resiclean - Resiblock Ltd
Resicure - Benlowe Group Limited
Resiecco - Resiblock Ltd
Resistex - Leigh's Paints
ResiTie - Helifix
Resolute Furniture - Roc Secure Ltd
Resolute Products - Roc Secure Ltd
Resolute Vandal Resistant Products - Roc Secure Ltd
Resolute VRP - Roc Secure Ltd
Resotec - Richard Lees Steel Decking Ltd
Respatex - Respatex International Limited
Responseline - National Britannia Ltd
Restol - Arch Timber Protection
RESTOVER restoration glass - SCHOTT UK Ltd
Restwall - Architectural Textiles Ltd
RETINO - Lumitron Ltd
retroflect - HMG Paints
Retrofloor - Redman Fisher Engineering
RetroTie - Helifix
Retrotread - Redman Fisher Engineering
Rettig - Purmo-UK
Revelation - Spectus Systems
Revolution - Optima
Rexovent - ABB Ltd
Reynobond 33 - Amari Plastics Plc
reynobond cassette panels - Dales Fabrications Ltd
RFS - Raised Floor Systems Ltd
RFS Heavy 4x4 - Raised Floor Systems Ltd
RFS Meduim 4x4 - Raised Floor Systems Ltd
RFS Meduim V - Raised Floor Systems Ltd
RH 'U' Tube Heater - Radiant Services Ltd
RHD Double Linear Heater - Radiant Services Ltd
Rhino Flex - DLW Flooring
Rhino Quartz - DLW Flooring
Rhino Tex - DLW Flooring
Rhino Tough - Wybone Ltd
RHL Linear Heater - Radiant Services Ltd
Rhodius - Hyde Brian Ltd
Rhodorsil - Bluestar Silicones
Rhomax Engineering - Tyco Fire and Integrated Solutions
RIAS Bookshops - Royal Incorporation of Architects in Scotland
RIAS Competitions - Royal Incorporation of Architects in Scotland
RIAS Consultancy - Royal Incorporation of Architects in Scotland
RIAS CPD - Royal Incorporation of Architects in Scotland
RIAS Gallery - Royal Incorporation of Architects in Scotland
RIAS Insurance Services - Royal Incorporation of Architects in Scotland
RIAS Practice Services - Royal Incorporation of Architects in Scotland
RIBA Appointments - RIBA Enterprises
RIBA Bookshops - RIBA Enterprises
RIBA Insight - RIBA Enterprises
RIBA Journal - RIBA Enterprises
RIBA Product Selector - NBS
RIBA Publishing - RIBA Enterprises
Ribblelite - Richard Hose Ltd
Ribdeck - Richard Lees Steel Decking Ltd
Ribloc - Forticrete Ltd
Richardson - Amdega Ltd
Richter System - Knauf UK
RICS Data Products - BCIS
RIDBA - Rural Industrial Design & Building Association
Ridgicoil - Polypipe Civils
Ridgidrain Surface water Drainage System - Polypipe Civils
Ridgiduct Power HV - Polypipe Civils
Ridgiduct, Ridgigully, Rigisewer - Polypipe Civils
Ridgitreat - Polypipe Civils
Ridigistorm-XL Large Diameter Piping System - Polypipe Civils
Rigaflo - Camfil Farr Ltd
Righton PQ - Righton Ltd
Rigidal Lokrol - Rigidal Industries Ltd
Rigidal Lokroll - Rigidal Systems Ltd
Rigidal Profiled Cladding - Rigidal Systems LTD
Rigidal Thermocore - Rigidal Industries Ltd
Rigidal Thermowall - Rigidal Industries Ltd
Rigidal Trocaldek - Rigidal Industries Ltd
Rigidal Ziplok - Rigidal Industries Ltd
Rigidal Ziplok - Rigidal Systems LTD
Rigid-Cor/F - Polypipe Building Products
Rilass V - Camfil Farr Ltd
Rior - Ward's Flexible Rod Company Ltd
Ripac - CSSP (Construction Software Services Partnership)
Ripon - Ellis J T & Co Ltd
Riposeal - Shackerley (Holdings) Group Ltd incorporating Designer ceramics
Riser Showertrays by Coram - Coram Showers Ltd
Rite-Vent - Schiedel Chimney Systems Ltd
Ritz - New Haden Pumps Ltd
River Gate - Claydon Architectural Metalwork Ltd
rla - Robinson Lloyds Architecture
RMC Concrete Products - CEMEX
Roadline - Leigh's Paints
Roadmaster - Tarmac Ltd
RoadRepair - Hanson UK
Roadside - Macemain + Amstad Ltd
Robec - Tuke & Bell Ltd
Robert Lynam collection - Emsworth Fireplaces Ltd
Robertshaw - SK Environmental Ltd
Robette - Tuke & Bell Ltd
Robey - Wellman Robey Ltd
Robinson Willey - Robinson Willey Ltd
Robocal - Altecnic Ltd
Robofil - Altecnic Ltd
Robokit - Altecnic Ltd
Robolink - Altecnic Ltd
Robust Detail Socket Box - Sound Reduction Systems Ltd
Rocal - Gazco
Rockcem - Cryotherm Insulation Ltd
Rockclad - FGF Ltd
RockDelta Acoustic Barriers - Ruthin Precast Concrete Ltd
Rockfon - Nevill Long Limited
Rockfon - New Forest Ceilings Ltd
Rockliner - Cryotherm Insulation Ltd
Rockranger Primary Crushing Outfits - Parker Plant Ltd
Rocksil - Knauf Insulation Ltd
Rockwell - Rockwell Sheet Sales Ltd
Rockwood - Rockwell Sheet Sales Ltd
Rockwool - FGF Ltd
Rockwool - Firebarrier Services Ltd
Rockwool - Rockwool Ltd
ROCOM GreaseShield - Vianen UK Ltd
Rocwall Walling/Cladding - Ruthin Precast Concrete Ltd
Rodan - Franke Sissons Ltd
Rogada - Knauf AMF Ceilings Ltd
ROK - Astracast PLC
ROK® - Astracast
Rola - Bramah Security Equipment Ltd
Rolfe King - Legrand Electric Ltd
Roll-A-Door - Rolflex Doors Ltd
Rollaplay - En-tout-cas Tennis Courts Ltd
Rollatape - Advance Tapes International Ltd
Rolled Lead Sheet & Flashings - BLM British Lead
Rollercash - Safetell Ltd
Rollerseal - Kleeneze Sealtech Ltd
Roll-fix ridge - Klober Ltd
Rollgliss - Capital Safety Group (NE) Ltd
Rolls Rotary - Tuke & Bell Ltd
Rolux - Ubbink (UK) Ltd
Roma Transport - Phoenix Scales Ltd
Roma-3 and Roma-4 - Lang+Fulton
Romil - Anderson Gibb & Wilson
RonaBond - Ronacrete Ltd
RonaDeck Resin Bonded Surfacing - Ronacrete Ltd
RonaDeck Resin Bound Surfacing - Ronacrete Ltd
Ronafix - Ronacrete Ltd
RonaFloor - Ronacrete Ltd
RonaRoad - Ronacrete Ltd
RonaScreed - Ronacrete Ltd
RonaStreet - Ronacrete Ltd
Rondo - Franke Sissons Ltd
Rondo - Noral Scandinavia AB
RONDOR - TORMAX United Kingdom Ltd
Ronez - Aggregate Industries UK Ltd
Roofdek - Tata Steel Europe Limited
Roofdex - Flexcrete Technologies Limited
Roofite - Watco UK Ltd
Roofline - Celuform Building Products
Roofmate SL-A - Dow Building Solutions
Roofmax - Knauf Insulation Ltd
RoofShop - SIG Roofing
Rooftex - Conren Ltd
Room Service - Brintons Ltd
RoomSign - Rivermeade Signs Ltd
Roomstat - Horstmann Controls Ltd
Roomvent - Airflow Developments Ltd
Rootbox - Jones of Oswestry Ltd
Rootfast® - Earth Anchors Ltd
Rope Descent - Baj System Design Ltd
Rosconi Benches - Detlef Muller Ltd
Rosconi Cloakroom Systems - Detlef Muller Ltd
Rosemary Clay - Monier Redland Limited
Rosengrens - Gunnebo UK Limited
Rossetti - Symphony Group PLC, The
Rota - Ciret Limited
Rota Guard - Geoquip Worldwide
Rota spike® - Jackson H S & Son (Fencing) Ltd
Rotaflow - Gebhardt Kiloheat
Rotapak - Fluidair International Ltd
Rotasoc - Electrak International Ltd
Rotastar - Fluidair International Ltd
Rotavent - Gebhardt Kiloheat
Rotaworm - Ward's Flexible Rod Company Ltd
Rothwell - Deva Tap Co Ltd
Roto - Roto Frank Ltd
Roto - SIG Roofing
RotoLatch™ - Latchways plc
Rougeite - CED Ltd
Rough at the edges - Clayton Munroe
Rovacabin - Wernick Group Ltd
Royal Forest - Formpave Ltd
Royalux - Cooper Lighting
Royce - Morris Material Handling Ltd
Royde & Tucker HI-LOAD hinges - Bridgman IBC Ltd
RPV 2000 - Reznor UK Ltd
RS Rotary screw compressor - Fluidair International Ltd
RubberCover (EPDM single ply roofing membranes) - Ensor Building Products Ltd
RubberCover EPDM - Firestone Building Products
Rubberfuse - Intergrated Polymer Systems (UK) Ltd
RubberGard EPDM - Firestone Building Products
Rubberline - Dudley Thomas Ltd
Rubberwell - Dudley Thomas Ltd
Rubboseal - Trade Sealants Ltd
Ruberoid Building Products - IKO PLC
Rubit - Hansgrohe
Rudgwick - Baggeridge Brick PLC
Rugby - CEMEX
RUNDFLEX - Circular wall formwork - PERI Ltd
Runtal Arteplano - Zehnder Ltd
Runtal Jet - Zehnder Ltd
Runtal LST - Zehnder Ltd
Runtal RX - Zehnder Ltd
Russell Hobbs - Pifco Ltd
Rustins - Rustins Ltd
Rustique - Veka Plc
Rustiver - Mermet U.K
Rustoleum - Andrews Coatings Ltd
Rustoleum - Kenyon Group Ltd
Rutland Range - Evergreens Uk
RV Active Inra Red Beams - Geoquip Worldwide
RW Linquartz - Robinson Willey Ltd
Rylstone - GB Architectural Cladding Products Ltd
RytComb®, RytDuct®, RytHose®, Rytweep® - Ryton's Building Products Ltd
Rytons Retro Weep Vent® - Ryton's Building Products Ltd

S

S.D.R.M.-(Shutter & Door Repare & Maintenance Ltd) - Shutter Door Repair & Maintenance Ltd
S.D.R.M.-(Shutter Door Repair & Maintenance Ltd.) - Priory Shutter & Door Co Ltd
S.M.I.T.M - Telelarm care Ltd
S.V.K. Cromleigh - Chameleon Stone Ltd
S2000 CSJ - Elster Metering Ltd
S3040 Portable Reader - CEM Systems Ltd
S610 intelligent card reader - CEM Systems Ltd
Sabre - TROAX UK Ltd
Sabroe - Johnson Controls Building Efficiency UK Ltd
Sadia Refrigeration - Viscount Catering Ltd
Sadolin Classic - Sadolin UK Ltd

S (con't)

Sadolin Extra - Sadolin UK Ltd
Sadolin Fencing Wood Stain - Sadolin UK Ltd
Sadolin High Performance Varnish - Sadolin UK Ltd
Sadolin Prestige - Sadolin UK Ltd
Sadolin PV67 Heavy Duty Floor Varnish - Sadolin UK Ltd
Sadolin Quick Drying Woodstain - Sadolin UK Ltd
Sadolin Shed and Fence Protection - Sadolin UK Ltd
Sadolin Supercoat - Sadolin UK Ltd
Sadolin Superdec - Sadolin UK Ltd
Safe Call - Raidio-Based Nurse Call System - Wandsworth Elecrtrical Ltd
Safe Glaze - Tank Storage & Services Ltd
Safe guard - DLW Flooring
Safe 'n' Sound - Premdor
Safe T Epoxy - Anglo Building Products Ltd
SafeDeck - Graepel Perforators Ltd
safe-deko vinyl floorings - Jaymart Rubber & Plastics Ltd
Safegard - James & Bloom Ltd
Safegrate - Lionweld Kennedy Ltd
Safegrid - Lionweld Kennedy Ltd
Safeguard - Securikey Ltd
SafeGuard - Residant Monitoring/Patient Tagging - Wandsworth Elecrtrical Ltd
Safeguard Hydracheck - Haynes Manufacturing (UK) Ltd
Safelock - Lionweld Kennedy Ltd
SafePay - Gunnebo UK Limited
Saferail - Lionweld Kennedy Ltd
Saferstore - Barton Storage Systems Ltd
Safetread - Lionweld Kennedy Ltd
Safetred Ranges - Tarkett Ltd
Sahara - Architen Landrell Associates Ltd
Salamandre - Legrand Electric Ltd
Salex Acoustic Products - Salex Acoustics Limited
Salmen's - Fine H & Son Ltd
Salter, Salter Brecknell, Salter Abbey - Phoenix Scales Ltd
SAM - Optelma Lighting Ltd
SaMontec - Artur Fischer (UK) Ltd
Sampson and Partners Fencing - Prima Security & Fencing Products
Samson - Rockfon Ltd
Samsumo TM Safety Access Systems - Redman Fisher Engineering
Samsung - Connaught Communications Systems Ltd
Samuel Booth collection - Tyco Waterworks - Samuel Booth
Sand Drain - ABG Ltd
Sand Dressed Turf - Charles Lawrence Surfaces Ltd
Sandene - Intergrated Polymer Systems (UK) Ltd
Sanderson - Sanderson
Sandfilled Turf - Charles Lawrence Surfaces Ltd
Sandler Seating - Sandler Seating Ltd
Sandringham - Loft Centre Products
Sandringham suite - Armitage Shanks Ltd
Sandtex - Akzo Noble Coatings Ltd
Sandtex - Brewer C & Sons Ltd
Sandtex - Crown Paints Ltd
Sandtoft - Wienerberger Ltd
Sandwell - Chiltern Invadex
Sangwin Building - Sangwin Group
Sangwin Civil Engineering - Sangwin Group
Sangwin Concrete Products - Sangwin Group
Sangwin Educational Furniture - Sangwin Group
Sangwin group - Sangwin Concrete Products Ltd
Sangwin Plant Hire - Sangwin Group
Sangwin Surfacing Contractors - Sangwin Group
Sanicat - B & D Clays & Chemicals Ltd
Sani-fem - Unicorn Containers Ltd
Sanistrel - Imperial Machine Co Ltd
Sanitop - Mat.Works Flooring Solutions
Sanosol - Gerflor Ltd
Santak - Eaton Electric Ltd
Santon - Baxi Heating UK Ltd
Santric - Pland Stainless Ltd
Saphir - EJOT UK Limited
Saracen - Laird Security Hardware
Sarel, Serk Controls, Square D - Schneider Electric Ltd
Sarena Housings - Sarena Mfg Ltd
Sarena Mouldings - Sarena Mfg Ltd
Sarena Tanks - Sarena Mfg Ltd
Sarina rubber floorings - Jaymart Rubber & Plastics Ltd
Sarnafast - Sarnafil Ltd
Sarnafil - Sarnafil Ltd
Sarnalite - Sarnafil Ltd
Sarnatherm - Sarnafil Ltd
Sarnatred - Sarnafil Ltd
Sartorius - Phoenix Scales Ltd
Sasmox - Panel Agency Limited
Satalico - Satec Service Ltd
Satellite - Light Engine Lighting Ltd
Satime 5500 - Mermet U.K
Saturn - Fabrikat (Nottingham) Ltd
Saturn - S & B Ltd
Saunier Duval - Hepworth Heating Ltd
Savanna - En-tout-cas Tennis Courts Ltd
Saver CF - Draeger Safety UK Ltd
Saver PP - Draeger Safety UK Ltd
SaverDoor - Sunray Engineering Ltd
Savile Plus - Hartley & Sugden
Savolit - Skanda Acoustics
Savolit - Skanda Acoustics Limited
Savotherm - Skanda Acoustics
Savotherm - Skanda Acoustics Limited
Sawco - The Saw Centre Group
Sawcut - The Saw Centre Group
Saxon - Monier Redland Limited
SB - Wilo Samson Pumps Ltd
SBM5 Soundproofing Mat - Sound Service (Oxford) Ltd
SBP 646 - Specialist Building Products
SBS - Hawker Ltd
SC - Hudevad Britain
SCA - Sprayed Concrete Association
Scaffband - Icopal Limited
Scala - DLW Flooring
Scalpture Products - Aden Hynes Sculpture Studios
Sculpture Services - Aden Hynes Sculpture Studios
Scan Pumps - Sulzer Pumps
Scanco - McMullen Facades
Scan-Q - Zon International
Scantronic - Eaton's Security Business
Scenario - Mode Lighting (UK) Ltd
Scenestyle - Mode Lighting (UK) Ltd
Schafer Lockers - Detlef Muller Ltd
Schindler - Schindler Ltd
Schlage Locks - Relcross Ltd
Schoema - Schwing Stetter (UK) Ltd
Scholar - Ellis J T & Co Ltd
SCHOTT TERMOFROST Cold Room Doors - SCHOTT UK Ltd
Schwan - Hyde Brian Ltd
Schwarzwald - Erlau AG
Schwing - Schwing Stetter (UK) Ltd
Scimitar - TROAX UK Ltd
Scotch Buff - Realstone Ltd
Scotchshield - Durable Ltd
Scotchtint - Durable Ltd
Scottish Brick - Ibstock Brick Ltd
SCP - SCP Concrete Sealing Technology Ltd
SCP, SCP Plus - Hartley & Sugden
Screedmaster - Laybond Products Ltd
Screen Nature - Hunter Douglas
Screening Outfits - Parker Plant Ltd
Screenmaster - ABB Instrumentation Ltd
Screensorption - Bicester Products Ltd
Screwlok - Camlok Lifting Clamps Ltd
Scroll - Emerson Climate Technologies Retail Solutions
SD.RM- (Shutter Door Repair & Maintenance Ltd.) - Neway Doors Ltd
SDR - Stoves PLC
SE Ergonomic chair - Hille Educational Products Ltd
SE Stool - Hille Educational Products Ltd
Seagull - Gledhill Water Storage Ltd
Sealdeck - Laybond Products Ltd
Sealex - Klinger Fluid Instrumentation Ltd
Sealmaster - Sealmaster
Sealobond - Sealocrete PLA Ltd
Seal-O-Cork - Olley & Sons Ltd, C
Sealocrete - Olby H E & Co Ltd
Sealoflash - Sealocrete PLA Ltd
Sealofoam - Sealocrete PLA Ltd
Sealomix - Sealocrete PLA Ltd
Sealopol - Sealocrete PLA Ltd
Sealoproof - Sealocrete PLA Ltd
Sealoshield Asbestos - Sealocrete PLA Ltd
Sealotone - Sealocrete PLA Ltd
Sealtite - Movement Joints (UK)
Secar Cements - Lafarge Aluminates Ltd
Secritron Magnets - Relcross Ltd
Sector - WF Senate
Securaglide insulated roller shutters - Regency Garage Door Services Ltd
Secure-Mix - FC Frost Ltd
Securesheild ™ - Accent Hansen
Securicel - Fitzpatrick Doors Ltd
Securigard - Solaglas Laminated
Securiglass - Promat UK Ltd
Securiguard window shutter - Bolton Gate Co. Ltd
Securistore - Barton Storage Systems Ltd
securistyle - ASSA ABLOY Limited
Securitherm - Delabie UK Ltd
Security Plus® - Nico Manufacturing Ltd
SecurityLine - Briggs Industrial Footwear Ltd t/a Briggs Safety Wear
SED - The National Event For Construction - Contract Journal
Sedia - Erlau AG
Seefire - Colt International
Seepex Pumps - Barr & Wray Ltd
Segga - Metalliform Products Plc
Seip Remote Control - Regency Garage Door Services Ltd
Sela - SFS Intec LTD
Seldex - Haworth UK Ltd
Select Range - Pedley Furniture International Ltd
Selecta Shower Screens - Shires Bathrooms
Selectarail - Rothley Burn Ltd
Selectascreen - Pearce Security Systems
Self Serve - Unicorn Containers Ltd
SelfStik - Linear Ltd
Sellite - Sellite Blocks Ltd
Selux Lighting - Woodhouse UK Plc
Sembla - Allgood plc
Semiautomatic Palace - Parthos UK Ltd
Senator - Deans Blinds & Awnings (UK) Ltd
Senator - Dudley Thomas Ltd
Senator - Forticrete Roofing Products
Sensalux - Setsquare
Sensor Coil 600 - Geoquip Worldwide
Sensor flow & electronic taps - Armitage Shanks Ltd
Sensor Systems - Kingspan Environmental Ltd
Sentinal - Advanced Fencing Systems Ltd
Sentinal ATM enclosure - Pearce Security Systems
Sentinel - Fitzgerald Lighting Ltd
Sentinel - Nettlefolds Ltd
Sentinel - Shackerley (Holdings) Group Ltd incorporating Designer ceramics
Sentrilock - Kenrick Archibald & Sons Ltd
Sentry - (MK (MK Electric Ltd)
Sentry® - Jackson H S & Son (Fencing) Ltd
Sequentia - Crane Composites
SerckAudio - Flowserve Flow Control
Serenade, Softrest - Sidhil Care
Series 2100 - James Gibbons Format Ltd
Series 3000 and Series 8000 - Conex Universal Limited
Series 60 - Apollo Fire Detectors Ltd
Series 65 - Apollo Fire Detectors Ltd
Series 90 Heavy Duty Benching - Envopak Group Ltd
Series AM - Dunham Bush Ltd
Series BM - Dunham Bush Ltd
Series CM - Dunham Bush Ltd
Series E - Hille Educational Products Ltd
Series F - Dunham Bush Ltd
Series J - Fulton Boiler Works (Great Britain) Ltd
Series UH - Dunham Bush Ltd
Serina - DW Windsor Lighting
Serpo RG - Maxit UK
Serpo Rock - Maxit UK
Serristrap - Grace Construction Products Ltd
SERVERCOOL® - Eaton-Williams Group Limited
Servery - Cooper Group Ltd
ServiCare - Evac+Chair International Ltd
Servicised - Grace Construction Products Ltd
Serviflex - Uponor Ltd
Serviseal - Grace Construction Products Ltd
Servitherm - Morgan Hope Industries Ltd
Servowarm - Servowarm
SES - Shepherd Engineering Services Ltd
Sesam - Dividers Modernfold Ltd
Sesame - Profab Access Ltd
Sesame® - H.B. Fuller
Seth Thomas - General Time Europe
Setlite - Setsquare
Seventh Heaven - Interfloor Limited
severn gutter system - Dales Fabrications Ltd
Sewer Drain - Hepworth Building Products Ltd
SF30 - Sunfold Systems Ltd
SF40 - Sunfold Systems Ltd
SF50 - Sunfold Systems Ltd
SFS-Masterflash - SFS Intec LTD
SGB, a Brand company - SGB, a Brand Company
SGG Bioclean - Glassolutions Saint-Gobain Ltd
SGG Cool-Lite - Glassolutions Saint-Gobain Ltd
SGG Diamant - Glassolutions Saint-Gobain Ltd
SGG Emalit Evolution - Glassolutions Saint-Gobain Ltd
SGG Planiclear - Glassolutions Saint-Gobain Ltd
SGG Planitherm - Glassolutions Saint-Gobain Ltd
SGG Priva-Lite - Glassolutions Saint-Gobain Ltd
SGG Seralit - Glassolutions Saint-Gobain Ltd
SGG Stadip range - Glassolutions Saint-Gobain Ltd
SGG Viewclear - Glassolutions Saint-Gobain Ltd
SGT - Illbruck Sealant Systems UK Ltd
Shadacrete - W. Hawley & Son Ltd
Shadeacrete - LANXESS Inorganic Pigments Group
Shadeplay - Architen Landrell Associates Ltd
shadex shading system - Dales Fabrications Ltd
Shadflam - Shadbolt F R & Sons Ltd
Shadmaster - Shadbolt F R & Sons Ltd
Shadoglass Glass louvre - Colt International
ShaftBrace, Super Shaftbrace, Supershaft Plus - Mabey Hire Ltd
Shaftwall - Knauf UK
Shaker - Illuma Lighting
Shallovent - Thermal Technology Sales Ltd
Shapphire, Streamline - Heatrae Sadia Heating Ltd
Sharps - Inter Public Urban Systems Uk Ltd
Shavrin - Shavrin Levatap Co Ltd
Shavrin Bijoux - Shavrin Levatap Co Ltd
Shavrin Levamixa - Shavrin Levatap Co Ltd
Shavrin Levatap - Shavrin Levatap Co Ltd
Shavrin Modesta - Shavrin Levatap Co Ltd
Shaws glazed bricks - Ibstock Building Products Ltd
Shear flow - shearflow LTD
Shearail® - BRC Special Products - Bekaert Ltd
Shearfix - Ancon Building Products
Sheartech - RFA-Tech Ltd
Sheartech Grid - RFA-Tech Ltd
Sheer - Hamilton R & Co Ltd
Sheerframe - L. B. Plastics Ltd
Sheerline - L. B. Plastics Ltd
Sheerlite - L. B. Plastics Ltd
SheetGuard - Mabey Hire Ltd
Sheffield - Gilgen Door Systems UK Ltd
Shell Tixophalte - Chesterfelt Ltd
Shelter - Checkmate Industries Ltd
Shepherd - Kenrick Archibald & Sons Ltd
Sheriden - Ulster Carpet Mills Ltd
Sherrill - Vermeer United Kingdom
SHEVTEC - SE Controls
Shield - Stocksigns Ltd
Shield LST - Hudevad Britain
Shieldoor - Norwood Partition Systems Ltd
Shiluvit by Oranit - LSA Projects Ltd
Shinsei seiki - Kaba Garog
Shopline - Smart Systems Ltd
Shoresafe - Mabey Hire Ltd
Shoreseal - Shackerley (Holdings) Group Ltd incorporating Designer ceramics
Show 'N' Tell Boards - Spaceright Europe Limited
Shower Accessories by Coram - Coram Showers Ltd
Shower DEC - Contour Showers Ltd
Shower Panels - Hansgrohe
Shower Temples - Hansgrohe
Showercyl - Gledhill Water Storage Ltd
Showerforce - NewTeam Ltd
Showermate - Stuart Turner Ltd
Showermax - Range Cylinders
Showerpannel - Hansgrohe
Showerpipe - Hansgrohe
ShowerSport - ShowerSport Ltd
Shuredor - L. B. Plastics Ltd
Shutters - Harling Security Solutions
Si - Worcester Heat Systems Ltd
Sichenia - Arnull Bernard J & Co Ltd
SideShield - Linear Ltd
Sidetrak - Rackline Systems Storage Ltd
Sidewinder - Clarke UK Ltd
Sieger - Zellweger Analytic Ltd, Sieger Division
Siena and Siena Sport - Lang+Fulton
Sierra - Sierra Windows
Siesta - Erlau AG
SIG - SIG Roofing
SIG UK Exteriors - SIG Roofing
Sightline 70 - Spectus Systems
Sigma - Andrews Coatings Ltd
Sigma - Brett Landscaping
Sigma - Delta Balustrades
Sigma - Propex Concrete Systems
Sigma Coatings - PPG Protective and Marine Coatings
Sigmadeck - Turner Access Ltd
Signal - Shadbolt F R & Sons Ltd
Signe - Silent Gliss Ltd
SignEdge - Rivermeade Signs Ltd
Signet - George Fischer Sales Ltd
Signicolour - Amari Plastics Plc
Signmaster (bollards) - Glasdon UK Limited
Signscape modular systems - GB Sign Solutions Limited
SigTEL disabled refuge systems - C-TEC (Computionics) Ltd
Sika-1 - Sika Ltd
SikaBond - Sika Ltd
SikaDur - Sika Ltd
Sikaflex - Sika Ltd
SikaFloor - Sika Ltd
Sikaplan - Sika Ltd
Sikatop - Sika Ltd
Sikkens - Akzo Noble Coatings Ltd
Sikkens - Brewer C & Sons Ltd
Silavent - Polypipe Ventilation
Silbione - Bluestar Silicones
Silbralloy™ - Johnson Matthey PLC - Metal Joining
Silent Gliss - Silent Gliss Ltd
SilentMAX - Marcrist International Ltd
Silfos™ - Johnson Matthey PLC - Metal Joining
Silhouette® Shades - Luxaflex®
Silicone 4 - Dufaylite Developments Ltd
Silk Machines - Furmanite International Ltd
Silka - Xella Aircrete Systems UK
Silkalastic 625 for Metal/Asbestos Cement roofs - Sika Liquid Plastics Limited
Silktrim - James Donaldson Timber
Silver-Flo™ - Johnson Matthey PLC - Metal Joining
Silverline - Aluminium RW Supplies Ltd
Silverline - Britannia Kitchen Ventilation Ltd
Simon Fiesta - Simon R W Ltd
Simpak compactors - Randalls Fabrications Ltd
Simplex - Kaba Ltd
Simply Elegant - Hodkin Jones (Sheffield) Ltd
SimplyU - Ideal-Standard Ltd
Simpson Strong-Tie® - Simpson Strong-Tie®
Single Point Gas Alarm - Gas Measurement Instruments Ltd
Single Touch Communication Systems - BPT Security Systems (UK) Ltd
Sintamatic - Donaldson Filtration (GB) Ltd
Sintesi - Laytrad Contracts Ltd
Sirius - Valor
Sirocco - Paragon by Heckmondwike
Sirrus - Bristan Group
Sisalite - British Sisalkraft Ltd
Sisalkraft - British Sisalkraft Ltd
Sissons - Franke Sissons Ltd
SITE BOOK - British Gypsum Ltd
SiteCop - Rediweld Rubber & Plastics Ltd
Sitesealer Membrane - Cavity Trays Ltd
Sivoia - Tidmarsh & Sons
Six hundred 600 Series - Komfort Workspace PLC
SKI - Komfort Workspace PLC
Skil - Robert Bosch Ltd
SKIL - SK Environmental Ltd
Skinner for Fencing - since 1920 - WA Skinner & Co Limited
Skjold - Broxwood (scotland Ltd)
Skoda Power - Doosan Power Systems Ltd
Skum - Tyco Safety Products
Sky Tunnel - Sola Skylights
SKYDECK - Aluminum panel slab formwork - PERI Ltd
Skye, Style Moderne - Samuel Heath & Sons plc
Skyfold - Style Door Systems Ltd
Skygard - Lonsdale Metal Company
Skyline - Alumasc Exterior Building Products Ltd
Skypod - Ves Andover Ltd
Skyvane - Levolux A.T. Limited
Slab Stress - Freyssinet Ltd
Slate2000 - Britmet Tileform Limited
Slide Safe - Safety Assured Ltd
Slik Bar - Furmanite International Ltd
Slim Vent® - Ryton's Building Products Ltd
Slimdek - Tata Steel Europe Limited
Slimline - Cooke Brothers Ltd
Slimline - Dudley Thomas Ltd
Slimline - T B S Fabrications
Slimstyle - Hallmark Blinds Ltd
Slimstyle Groundfloor Treatments - HW Architectural Ltd
Slingsby - HC Slingsby PLC
Sloan & Davidson - Longbottom J & J W Ltd
Slot 20 - Air Diffusion Ltd
Slot 25 - Air Diffusion Ltd
Slotz - Winther Browne & co Ltd
Slurrystore - CST Industries, Inc. - UK
SM Range - Becker (SLIDING PARTITIONS) Ltd
SM Series - NovaSpan Structures
Smart - S & B Ltd
Smart Air Curtain - D and DWW - Vianen UK Ltd
Smart Drive 1101 - TORMAX United Kingdom Ltd
Smartank Duplux Pump Sets - Metcraft Ltd
Smartank Overfill Alarm Unit - Metcraft Ltd

S (con't)

Smartank Pump Control Pannels - Metcraft Ltd
SmartPly - Meyer Timber Limited
Smartroof - Wyckham Blackwell Ltd
SmartSash - Veka Plc
Smartscreen - Smart Systems Ltd
smartsigns - Stocksigns Ltd
SmartSystem - Vulcan Cladding Systems
Smiths - Timeguard Ltd
Smog-Eater - Horizon International Ltd
Smog-Mobile - Horizon International Ltd
Smog-Rambler - Horizon International Ltd
Smoke Ventilation Louvres - Matthews & Yates Ltd
Smokeguard - Pyroguard UK Ltd
Smokemaster curtain - Colt International
Smokescreen - The No Butts Bin Company (NBB), Trading as NBB Outdoor Shelters
SmokeSeal - Linear Ltd
SmokeStop - Cooper Group Ltd
Smokex - Poujoulat (UK) Ltd
Smometa - Photain Controls (part of GE security)
Smooth - On - Notcutt Ltd
Smooth 600 - E - Paroc Panel Systems Uk Ltd
Smoothline - Rega Ductex Ltd
Smoothline - WP Metals Ltd
Snake Way - Marshall-Tufflex Ltd
Snaptyte - Newdawn & Sun Ltd
SnickarPer - Olsen Doors & Windows Ltd
snipef - Scottish & Northern Ireland Plumbing Employers' Federation
SnoMelt - Spiral Construction Ltd
Sobrano (Natural Roofing Slates) - Ensor Building Products Ltd
Sofco ® - Broxap Ltd
Softgrip Tools - Tower Manufacturing
Softmood - Ideal-Standard Ltd
Soil nails - PHI Group Ltd
Soil Panel - PHI Group Ltd
Soilcrete - Keller Ground Engineering
Soilfrac - Keller Ground Engineering
soilmaster entrance matting - Jaymart Rubber & Plastics Ltd
Sokkia - Topcon GB Ltd
SOLA-BOOST® - Monodraught Ltd
Solair - Solair Ltd
Solaplex® - Rockwood Pigments (UK) Ltd
Solar - Rivermeade Signs Ltd
Solar Bronze - The Window Film Company UK Ltd
Solar C Brise soleil - Colt International
Solar Knight - Intersolar Group Ltd
Solar Moler - Intersolar Group Ltd
Solar Pumpkit - Intersolar Group Ltd
Solar PV - Axter Ltd
Solar PV Tiles - Monier Redland Limited
Solare Frameless Glazed Partitions - Avanti Systems
Solarfin - Colt International
Solarglaze - National Domelight Company
Solaris - Shackerley (Holdings) Group Ltd incorporating Designer ceramics
SolarSmart - Alpha Therm Ltd
Solarvent - Intersolar Group Ltd
Solarvent Turbo - Intersolar Group Ltd
Solatherm - IPPEC Sytsems Ltd
SOLA-VENT® - Monodraught Ltd
Solenium - Interface Europe Ltd
Solga - Harbro Supplies Ltd
Solidbase - Foster W H & Sons Ltd
Solignum - Protim Solignum Ltd
Solignum - Tor Coatings Ltd
Solite - Sidhil Care
Solitex Plus - Fillcrete Ltd
Solitex Wa - Fillcrete Ltd
Solitude - Cep Ceilings Ltd
Solo - Ellis J T & Co Ltd
Solo - Köttermann Ltd
Soloblock - Jones of Oswestry Ltd
Solomat - Zellweger Analytic Ltd, Sieger Division
Solos - Forticrete Roofing Products
Solosun - Ubbink (UK) Ltd
Solray - Comyn Ching & Co (Solray) Ltd
Solutions - Hands of Wycombe
Solutionz - Loblite Ltd
Solvite - Henkel Consumer Adhesives
Sombra - Saint-Gobain Ecophon Ltd
Somdrain - Geosynthetic Technology Ltd
Sommer - Tarkett Sommer Ltd
Somtube - Geosynthetic Technology Ltd
Sonacoustic seamless acoustic ceiling - Salex Acoustics Limited
Sonae - Anderson C F & Son Ltd
Sonae - Boardcraft Ltd
Sonae - Deralam Laminates Ltd
Sonafold Acoustic Folding Door - Bolton Gate Co. Ltd
Sonar - Rockfon Ltd
Sonata - Quartet-GBC UK Ltd
Sondes - Radiodetection Ltd
Sonicaid - ArjoHuntleigh UK
Sony - Vidionics Security Systems Ltd
Sorbent - B & D Clays & Chemicals Ltd
SOS - Pressure Coolers Ltd T/A Maestro Pressure Coolers
Soss Invisible Hinges - Notcutt Ltd
Sound Control Ear Muffs - MSA (Britain) Ltd
Sound Reduction Systems Ltd Acoustilay - Trim Acoustics
Sound secure - Premdor
Sound Sentinel - AET.gb Ltd
SoundBar 53 - Sound Reduction Systems Ltd
SoundBlocker - Bicester Products Ltd
SoundBlocker - Sound Reduction Systems Ltd
Soundcel - Fitzpatrick Doors Ltd
Soundmaster - Monowa Manufacturing (UK) Ltd
Soundpac - Hotchkiss Air Supply
Soundsafe - Emergi-Lite Safety Systems
Soundseal - Sound Reduction Systems Ltd
Soundshield - Knauf UK
Soundshield ™ - Accent Hansen
Soundsorba - Soundsorba Ltd
SoundStop - Sound Reduction Systems Ltd
Soundtube wall panels - Salex Acoustics Limited
Soundvision - Pro Display TM Limited
Southampton - Neptune Outdoor Furniture Ltd
Sovereign - Baggeridge Brick PLC
Sovereign - Clenaware Systems Ltd
Sovereign - Commercial Lighting Systems
Sovereign Doors - Jeld-Wen UK Ltd
Sovereign Paving - Thakeham Tiles Ltd
Sovereign Windows - Jeld-Wen UK Ltd
SP Panel - Acousticabs Industrial Noise Control Ltd
SP10 Street Scape Mortar - Carter-Dal International
SP100 - Laticrete International Inc. UK
SPAC - Brighton (Handrails), W
Space -Ray - Space - Ray UK
Spacebuild - Wernick Group Ltd
Spacemaster Bollard - Dee-Organ Ltd
Spacepac - Wernick Group Ltd
Spaceright and Space/Dividers - Spaceright Europe Limited
Spacers - Parthos UK Ltd
Spacesaver Bollard - Dee-Organ Ltd
Spacestopper Major - Dee-Organ Ltd
SPA-Clean - Above All
SPA-Coat - Above All
Spadeoak - Aggregate Industries UK Ltd
Spae - DC Plastic Handrails Ltd
Spaguard - BioLab UK
Spalex - Vita Liquid Polymers Ltd
Spalsh, Sylan - Amwell Systems Ltd
Spandex - Invicta Window Films Ltd
Spandex Modular System - GB Sign Solutions Limited
Spangard - Lonsdale Metal Company
Spania - Marshalls Mono Ltd
Spanseal - Compriband Ltd
Spanwall - McMullen Facades
Sparclean - Gerflor Ltd
Spartacus - Metalliform Products Plc
Spartan - Spartan Promenade Tiles Ltd
Sparyfilm - Promat UK Ltd
Spawall wall lining - Grant Westfield Ltd
SPC - Furmanite International Ltd
Spear & Jackson - Spear & Jackson
Spechra - T B S Fabrications
Specialist Formwork 3D - Cordek Ltd
Specifinder.com - The Building Centre Group Ltd
Specify it (online service) - IHS
Speck - Golden Coast Ltd
Specktralock non stain grout - Laticrete International Inc. UK
Spectator seating - CPS Manufacturing Co
Spector Lumenex - Tyco Fire and Integrated Solutions
Spectral Passive 3D - Harkness Screens
Spectrum - Norton Engineering Alloys Co Ltd
Spectrum Brands - Pifco Ltd
Spectrum Paint - Sanderson
Specula - Shadbolt F R & Sons Ltd
Spedec - SFS Intec LTD
Spediboil - Santon
Speed Patch - Anglo Building Products Ltd
Speed Screed - Anglo Building Products Ltd
Speed Set - Premdor
Speedal ® - Righton Ltd
Speedbolt® - Nico Manufacturing Ltd
Speedchannel cat 5E - Hubbell Premise Wiring
Speededge formwork - BRC Reinforcement
speedfab - Speedfab Midlands
Speedfit - Guest (Speedfit) Ltd, John
Speedframe - Construction Group UK Ltd
Speedlock - Construction Group UK Ltd
Speedor - Hart Door Systems Ltd
Speedplug - Ramset Fasteners Ltd
Speedrax - Arc Specialist Engineering Limited
Speedturn - BSA Machine Tools Ltd
Speedwall, Speedfit - Bridgman IBC Ltd
Speedway - Vantrunk Engineering Ltd
Spelsburg enclosures - Elkay Electrical Manufacturing Co Ltd
Spencers - Anderson Gibb & Wilson
SPF - Prima Security & Fencing Products
Sphere - Silent Gliss Ltd
Spike - SFS Intec LTD
Spikemaster - Rawson Carpets Ltd
SpillKart - Barton Storage Systems Ltd
Spiral construction - In Steel (Blacksmiths & Fabricators) Ltd
Spiral Stairs - In Steel (Blacksmiths & Fabricators) Ltd
Spiralift - Caldwell Hardware (UK) Ltd
Spiralite - Cryotherm Insulation Ltd
Spiratube - Flexible Ducting Ltd
Spiratube - HRS Heat Exchangers Ltd - HRS Hevac Ltd
Spirex - Caldwell Hardware (UK) Ltd
Spirotred - Pedley Furniture International Ltd
Spirtech - Monier Redland Limited
Spit - ITW Construction Products
Spit - ITW Spit
SPL - Stamford Products Limited
Splash (dolphin bin) - Glasdon UK Limited
Splitline - Lewden Electrical Industries
Sport M - Gerflor Ltd
Sportcel - Fitzpatrick Doors Ltd
Sporturf - En-tout-cas Tennis Courts Ltd
SPRA - Single Ply Roofing Association
Sprayed Limpet Mineral Wool-GP - Thermica Ltd
Sprayed Limpet Mineral Wool-HT - Thermica Ltd
Sprayed Limpet Mineral Wool-TI - Thermica Ltd
Sprayed Limpet Vermiculite-external - Thermica Ltd
Sprayed Limpet Vermiculite-Internal - Thermica Ltd
Spraymark - Marking Machines International
Spring steel buffer - Berry Systems
Springflex - Rawson Fillings Ltd
Springflow - Astracast
Springflow - Astracast PLC
Springlok - Rawson Fillings Ltd
Springvin - Flexible Ducting Ltd
Sprint - Paddock Fabrications Ltd
Sprint Truss Equipment - ITW Consutruction Products
Sproughton Arbour Seat - Anthony de Grey Trellises + Garden Lighting
SPS Screen - Acousticabs Industrial Noise Control Ltd
Spyder - Setsquare
Squire - Iles Waste Systems
Squire - Lochinvar Limited
Squire - Squire Henry & Sons
SRD - Stamford Products Limited
SRS acoustic floating floor - Christie & Grey Ltd
SS System - Quelfire
SSI Schaefer Noell - SSI Schaefer Ltd
SSI Schaefer Peem - SSI Schaefer Ltd
SSI 1000 - SSL Access
SSL A300 - SSL Access
SSL Melody - SSL Access
St James - Marflow Eng Ltd
STA - Solar Trade Association, The
Stabila - Hyde Brian Ltd
StackFast - EHSmith
Stadium - Flambeau
Staifix - Ancon Building Products
Stainless Steel - Swift & Sure - Arvin Motion Control Ltd
Stainton Metal - Valmont Stainton
Stairiser - Stannah Lifts Ltd
Sta-Lok Rigging - Scotia Rigging Services (Industrial Stainless Steel Rigging Specialists)
Standard Patent Glazing - Standard Patent Glazing Co Ltd
Standfast - Auld Valves Ltd
Stanford - Glasdon U.K. Limited
Stanhope - Albany Standard Pumps
Stanley - Ciret Limited
Stanley Cookers - Diamond Merchants
Stanley Hinges - Relcross Ltd
Stanton - Saint-Gobain PAM UK
Stanway High Density Stacking Chair - Race Furniture Ltd
Stanweld - Lionweld Kennedy Ltd
Stanwin - Albany Standard Pumps
Star spring operated cable reels - Metreel Ltd
Starbloc - Holden Brooke Pullen
Starcoat - Axter Ltd
Stardome - National Domelight Company
Starfix Twin - Holden Brooke Pullen
Starflex - Holden Brooke Pullen
Stargard - SG System Products Ltd
Starglaze - National Domelight Company
Starkan - Holden Brooke Pullen
Starlight - Haddonstone Ltd
Starline - Deceuninck Ltd
Starline - Holden Brooke Pullen
Starnorm - Holden Brooke Pullen
Starquest - Rawson Carpets Ltd
Starsystem - Watco UK Ltd
Startabox - RFA-Tech Ltd
Startwin - Holden Brooke Pullen
Statements of Conformity - British Board of Agrément
Statesman - Magnet Ltd
Statesman - Project Office Furniture PLC
Stayflex - Adaptaflex Ltd
Steamline - Geberit Ltd
Steden - Laidlaw Ltd
Steel Doors - Harling Security Solutions
Steel monoblock - Clestra Limited
Steelcase Strafor - Steelcase Strafor Plc
Steelcoat - Winn & Coales (Denso) Ltd
Steelite - Marlow Ropes Ltd
Steelkane - Ward's Flexible Rod Company Ltd
Steel-line - Cardale Doors Ltd
Steelplan - Elliott Ltd
Steelybin® - Straight Ltd
Steicoflex - Steico
Steicojoist - Steico
Steicolvl - Steico
Steni colour - CEP Claddings Ltd
Steni nature - CEP Claddings Ltd
Stent - Stent Foundations Ltd
Stereo and Stretto - Lang+Fulton
Steristeel - Pland
Sterling - Chalmit Lighting
Sterling - Marlow Ropes Ltd
Sterling OSB - Norbord
Sterling range of cable management systems - Marshall-Tufflex Ltd
Sterox, LP Sterox - Fernox
Sterreberg - Smithbrook Building Products Ltd
Stetter - Schwing Stetter (UK) Ltd
Stewart Film Screen - Anders + Kern UK Ltd
Stick-Lite - Encapsulite International Ltd
Sticks and Stones - Mat.Works Flooring Solutions
Stiebel Elton (OK Only) - Applied Energy Products Ltd
Stik - Iles Waste Systems
Stilpro - Althon Ltd
Stirling Range; Seats, Cycle Stands, Cycle Shelters, Bollards - Furnitubes International Ltd
Sto Rend - STO Ltd
Sto Silent - STO Ltd
Sto Therm Classic - STO Ltd
Sto Therm Eco - STO Ltd
Sto Therm Mineral - STO Ltd
Sto Therm Protect - STO Ltd
Sto Therm Robust - STO Ltd
Sto Therm Vario - STO Ltd
Sto Ventec Glass - STO Ltd
Sto Ventec R - STO Ltd
Stockmaster Shelving - Bedford Shelving Ltd
Stockrax - Link 51 (Storage Products)
Stomor Shelving - Link 51 (Storage Products)
Stone Collection - Solus Ceramics Ltd
Stone Italiana - Diespeker Marble & Terrazzo Ltd
Stone slate - Forticrete Roofing Products
Stonecor - Polypipe Building Products
Stoneraise red - Realstone Ltd
Ston-Ker - Porcelanosa Group Ltd
Stontex - Atlas Stone Products
Stopdust - Don Construction Products Ltd
Stopgap - Ball F & Co Ltd
Stoplite - Cryotherm Insulation Ltd
Storbox - Moresecure Ltd
Storemaster - Airedale International Air Conditioning Ltd
Storm - English Braids
Stormdry masonry protection cream - Safeguard Europe Ltd
Stormflo Rainwater Sytems - Hunter Plastics Ltd
Stormframe - Sapa Building Systems (monarch and glostal)
Stormgard - Corroless Corrosion Control
Stormor Euro - Link 51 (Storage Products)
Stormor XL Pallet - Link 51 (Storage Products)
Stormproof - Premdor
StormSaver Rainwater Harvesting System - FloPlast Ltd
Stortech - Rothley Burn Ltd
Stothert & Pitt - Bitmen Products Ltd
Stovax - Gazco
Stoves - Stoves PLC
Stoves Newhome - Stoves PLC
Stowlite Medium Blocks - Stowell Concrete Ltd
Stra - Maxit UK
Strada - Jaga Heating Products (UK) Ltd
Strading - SG System Products Ltd
Straightaway - Coburn Sliding Systems Ltd
Straightpoint - Phoenix Scales Ltd
Strand - DW Windsor Lighting
Strand FEB - Strand Hardware Ltd
Strate-Grip - Andura Coatings Ltd
Stratogrid - HT Martingale Ltd
Stratos - Franke Sissons Ltd
Stratotrack - HT Martingale Ltd
Stratpord Tip-Up Seating System - Race Furniture Ltd
Stratton - Hamworthy Heating Ltd
Stratus Lighting - WF Senate
Streamline - Wade International Ltd
street king entrance matting - Jaymart Rubber & Plastics Ltd
Streetcrete ™ - Broxap Ltd
Streetiron ™ - Broxap Ltd
Streetscene ® - Broxap Ltd
Stremaform® - BRC Special Products - Bekaert Ltd
Stren-Cor - Asset International Ltd
Stretch Acoustic - Acousticabs Industrial Noise Control Ltd
Strip Form - Scapa Tapes UK Ltd
Stroke - Bernhards Landscapes Ltd
Strong Water Based Lacquer - Junckers Limited
Strongbond - Chesterfelt Ltd
Strongcast concrete - Challenge Fencing Ltd
Stronghold - Catnic
Stronghold - SG System Products Ltd
Strongoat - Don Construction Products Ltd
Structural Soils (Northern) Ltd - Structural Soils Ltd
Strypit - Rustins Ltd
Stuart - Stuart Turner Ltd
Stucanet - Metal Mesh Lathing for Plaster & Render - Bekaert Building Products
Studflex vibration isolation mat - Christie & Grey Ltd
Studio - Ideal-Standard Ltd
Studio 3 - Eaton Electric Limited
Studio 5 - Eaton Electric Limited
Studio Sanderson - Sanderson
Studrail system acoustic floating floor - Christie & Grey Ltd
Sturdee - Marlow Ropes Ltd
Styccobond - Ball F & Co Ltd
Stylefold - Style Door Systems Ltd
Stylemaster - Checkmate Industries Ltd
S-type, Seren, Serhad, Serif, Sesriem, Siesta, Smyrna, Sofi, Sovran, Speira, Stanza, Stargate, Stria, Sumela,
Supra - Pitacs Ltd
Styrene - Ariel Plastics Ltd
Styroclad - Panel Systems Ltd
Styrodur - Kingspan Insulation Ltd
Styrofloor - Panel Systems Ltd
Styrofoam-A - Dow Building Solutions
Styroglaze - Panel Systems Ltd
Styroliner - Panel Systems Ltd
SubPrimo - Sound Reduction Systems Ltd
Succah - NuLite Ltd
Suffolk - Dixon Turner Wallcoverings
Suhner - Harbro Supplies Ltd
Sumo - Pump Technical Services Ltd
Sumo can crushers - Randalls Fabrications Ltd
Sumpmaster - Pump Technical Services Ltd
SUNCATCHER® - Monodraught Ltd
Suncell - CPV Ltd
Suncool - Pilkington Birmingham
Suncuva - Architen Landrell Associates Ltd
Sundeala K Grade - Sundeala Ltd
Sundeala FRB Flame Retardant Board - Sundeala Ltd
Sunflex - Olsen Doors & Windows Ltd
Sunfold - Becker (SLIDING PARTITIONS) Ltd
Sungard - Bonwyke Ltd
Sungard - Invicta Window Films Ltd
Sun-Guard: White-Out, Black-Out and Frost Matte - The Window Film Company UK Ltd
SUNPIPE® - Monodraught Ltd
Sunray - Sunray Engineering Ltd
Sunrite - Sunrite Blinds Ltd
SunScreen - Hunter Douglas
Sunscreen - Mermet U.K
Sunspot - BioLab UK

S (con't)

Sunstar - Space - Ray UK
Suntech - Deans Blinds & Awnings (UK) Ltd
Suntech systems - Deans Blinds & Awnings (UK) Ltd
Sunvizor - Levolux A.T. Limited
Supa Glide - Moresecure Ltd
Supabore - Avesta Sheffield Ltd
Supacord - Heckmondwicke FB
Supadriv - Nettlefolds Ltd
Supafil - Knauf Insulation Ltd
Supalux - FGF Ltd
Supalux - Promat UK Ltd
Supapac - Rycroft Ltd
Supascrew - Nettlefolds Ltd
Supastor - Flamco UK Ltd
Supastrike box striking plate - Guardian Lock and Engineering Co Ltd
Supatube - Avesta Sheffield Ltd
Supaxpress - Nettlefolds Ltd
Super Saphir - EJOT UK Limited
Super Seal - Hepworth Building Products Ltd
Super Top - Armorex Ltd
Superbolt - Furmanite International Ltd
Superbond - Interclad (UK) Ltd
Supercal - Range Cylinders
Supercast - Fosroc Ltd
Superchill - Heatrae Sadia Heating Ltd
Superclean coal fire - Dunsley Heat Ltd
Supercoil - Airsprung Beds Ltd
Supercoil Rolling Smoke Curtain - Bolton Gate Co. Ltd
Supercomfort - Dunham Bush Ltd
Superconcentrate Boiler Noise Silencer - Fernox
Superconcentrate Leaksealer - Fernox
Superconcentrate Limescale Preventer - Fernox
Superconcentrate Protector - Fernox
Superconcentrate Restorer - Fernox
Supercote - Witham Oil & Paint Ltd
Supercut - Unifix Ltd
Superdeck Insulated Composite System - Intergrated Polymer Systems (UK) Ltd
Superdeluxe - Premdor
Superfold 4000 high-security folding door - Bolton Gate Co. Ltd
Superfoot - Bernhards Landscapes Ltd
Superfoot plus - Bernhards Landscapes Ltd
SuperG - Saint-Gobain Ecophon Ltd
Supergalv Lintels - Birtley Group
Supergres - Arnull Bernard J & Co Ltd
Superior - Beacon Machine Tools Ltd
Superior - Hartley & Sugden
Superior Folding Shutter - Bolton Gate Co. Ltd
SuperJan - Johnson & Starley
Superline - BLV Licht-und Vakuumtecnik
Superline - Marlow Ropes Ltd
Superlintels - Jones of Oswestry Ltd
Superlite Blocks - Hanson Concrete Products
Superlite masonary - Hanson Concrete Products
Supermat - Kleen-Tex Industries Ltd
Supermix - Marlow Ropes Ltd
Supermix products - B & M Fencing Limited
Supersafe - Hawker Ltd
Supersaver gas burners - Infraglo Limited
Superseal - Kleeneze Sealtech Ltd
Superseal - Unifix Ltd
Superseal-pipe insert - Guest (Speedfit) Ltd, John
Supersleve House Drain - Hepworth Building Products Ltd
Superspan - Compriband Ltd
Superspan - Illbruck Sealant Systems UK Ltd
Superswitch - Friedland Ltd
Supertube - Polypipe Ventilation
Supervent - Airflow Developments Ltd
SuperVigilant - Auld Valves Ltd
Supervisor - Gent Limited
Supplyline - National Britannia Ltd
Supplymaster - Timeguard Ltd
SupraBlock, SupraDrain - Jones of Oswestry Ltd
Supreme Boiling Water Products - Heatrae Sadia Heating Ltd
Sure Ballistics - Specialist Building Products
Sure GRP Finishes - Specialist Building Products
Sure Rend - Specialist Building Products
Surefire - Thurston Building Systems
Surefire Adhesive - Advance Tapes International Ltd
Surefit Rainwater Sytems - Hunter Plastics Ltd
suregrip - Ecotile Flooring Ltd
Suregrip - Wincro Metal Industries Ltd
Suregrip ® - Coo-Var Ltd
Suregrip ® - Teal & Mackrill Ltd
Sureguard - Thurston Building Systems
Surell - Decra Ltd
Surell - Formica Ltd
Surelock - Laird Security Hardware
Sureloo - Thurston Building Systems
Surespace - Thurston Building Systems
Surespan - Thurston Building Systems
Suresport - Thurston Building Systems
Surface Mounted Road Blockers - APT Controls Ltd
Sussex Broadware - Focus SB Ltd
SV - Redring
S-Vap 5000E SA vapour control layer - Sika Liquid Plastics Limited
SWA - Steel Window Association
Swaledale - Oakdale (Contracts) Ltd
Swaledale - Realstone Ltd
Swan - Boardcraft Ltd
Swanglide - Integra Products
Swarfega - Deb Ltd
Swarfega Orange - Deb Ltd
Swarfega Power - Deb Ltd
Swedeglaze - Swedecor
Swedsign - Signs & Labels Ltd
Swellseal - Grace De Neef UK
Swift®bond, Swift®lock, Swift®melt, Swift®tak, Swift®therm - H.B. Fuller
Swift-and-Sure - Arvin Motion Control Ltd
Swiftplan - Wernick Group Ltd
Swiftrack - Legrand Electric Ltd
Swifts - Legrand Electric Ltd
Swiftshop - Stamford Products Limited
Swimmer - Golden Coast Ltd
Swing-Boom - Horizon International Ltd
Swish - FGF Ltd
Swisslab/therm/pan/tech/rail - Alumasc Exterior Building Products Ltd
Swisspearl - CEP Claddings Ltd
Swisspearl® - Omnis
SX Environmental - P+L Systems Ltd
Sygef - George Fischer Sales Ltd
Sygnette - Wellman Robey Ltd
Sykes Pumps - Andrews Sykes Hire Ltd
SylvaFoam - Junckers Limited
SylvaKet - Junckers Limited
SylvaRed - Junckers Limited
SylvaSport Club - Junckers Limited
SylvaSport Premium - Junckers Limited
SylvaSquash - Junckers Limited
SylvaThene - Junckers Limited
Synergetic flylights - P+L Systems Ltd
Synergy - H. B. Fuller Powder Coatings Ltd
Synergy - Legrand Electric Ltd
Synergy - Valspar Powder Coatings Ltd
Syntha Pulvin - H. B. Fuller Powder Coatings Ltd
Syntha Pulvin - Valspar Powder Coatings Ltd
Synthascreed - DPC Screeding Ltd
Synthatec - H. B. Fuller Powder Coatings Ltd
Synthatec - Valspar Powder Coatings Ltd
Synthtec Metallics - H. B. Fuller Powder Coatings Ltd
Synthtec Metallics - Valspar Powder Coatings Ltd
Syntropal - ICB (International Construction Bureau) Ltd
System 100 - BPT Security Systems (UK) Ltd
System 100 - Polypipe Ventilation
System 160, Superprop - Mabey Hire Ltd
System 2000 - NMC (UK) Ltd
System 3 - CBS (Curtain and Blind Specialists Ltd)
System 300 - BPT Security Systems (UK) Ltd
System 4 - Simplex Signs Limited
System 500 - Newton John & Co Ltd
System 6 - CBS (Curtain and Blind Specialists Ltd)
System 8 - CBS (Curtain and Blind Specialists Ltd)
System 9000 - Timloc Expamet Building Products
System Access - Saint-Gobain Ecophon Ltd
System J - Nullifire Ltd
System one - Twyford Bushboard
System S - Nullifire Ltd
System W - Nullifire Ltd
System10 - Veka Plc
System2000 range - Helmsman
Systemate - Gledhill Water Storage Ltd
SystemFit - Santon
Systemlabor - Köttermann Ltd
SystemLogic - Range Cylinders
SystemRoMedic - Amilake Southern Ltd
Syston doors - Syston Rolling Shutters Ltd
Syvac - Donaldson Filtration (GB) Ltd

T

T line counters - Platonoff & Harris Plc
T.I.P.S - Hellermann Tyton
T14 storage wall - Gifford Grant Ltd
T2 - Flexiform Business Furniture Ltd
T2000 - Reznor UK Ltd
T5 Lighting - Encapsulite International Ltd
TAC, Telemecanique, Thorsman, Tower - Schneider Electric Ltd
Tactual - Stocksigns Ltd
TakPave - Rediweld Rubber & Plastics Ltd
Talisman - Tilley International Ltd
Tambour - Barton Storage Systems Ltd
Tanalith - Arch Timber Protection
Tanalith Extra - Arch Timber Protection
Tanatone - Arch Timber Protection
Tanda - Drawn Metal Ltd
Tangent - Marflow Eng Ltd
Tango - Litex Design Ltd
Tangye Pumps - Satec Service Ltd
Taperell Taylor - Michelmersh Brick Holdings PLC
Tapes4 Builders - Advance Tapes International Ltd
Tapetex - Architectural Textiles Ltd
Tapiflex Collections - Tarkett Ltd
Taplon - ITW Construction Products
Tappex - Uponor Ltd
Taptite - Nettlefolds Ltd
Tapworks - Tapworks Water Softeners
Tapworks - Northstar - EcoWater Systems Ltd
Tapworks Domestic - AD11 - EcoWater Systems Ltd
Tapworks Domestic - AD15 - EcoWater Systems Ltd
Tapworks Domestic - NSC40 UDI - EcoWater Systems Ltd
Tapworks Domestic - Ultra 9 - EcoWater Systems Ltd
Taraflex - Gerflor Ltd
Tarasafe - Gerflor Ltd
Targha - BPT Security Systems (UK) Ltd
Tarkett - Tarkett Sommer Ltd
Tarkett Sommer - Tarkett Sommer Ltd
TAS100 - Thatch Fibreboard - Thatching Advisory Services Ltd
TAS-33 Thatch Alert - Thatching Advisory Services Ltd
Taskmaster - Doors & Hardware Ltd
Taskmaster - Taskmaster Doors Ltd
Taskmaster S series - Taskmaster Doors Ltd
Taskmaster T series - Taskmaster Doors Ltd
Taskworthy - Taskworthy Ltd
Tate Access Floors - Kingspan Access Floors Ltd
Tate Automation - Tate Fencing
Tate Floors - Hatmet Limited
TATOO - Optelma Lighting Ltd
Tauvus - Rivermeade Signs Ltd
Tayler tools - Hyde Brian Ltd
TB professional staple gun and staples for stainless steel wire - Thatching Advisory Services Ltd
T-Bar, Tardis, Tilbrook, TimeCables, Tora, Truva, Trojan, Tubo, Tudor, Twister - Pitacs Ltd
T-Box-H-Matic - Healthmatic Ltd
TBS Baffle - Acousticabs Industrial Noise Control Ltd
TBS Panels - Acousticabs Industrial Noise Control Ltd
TD - Donaldson Filtration (GB) Ltd
TD50 - Clenaware Systems Ltd
Teamac - Teal & Mackrill Ltd
TEBRAX - Tebrax Ltd
Tecbond - Kenyon Performance Adhesives
TecCast, TecLite, TecStone - Haddonstone Ltd
Tech Eco Slate - SIG Roofing
Techdek - Norton Engineering Alloys Co Ltd
Techflo - Stuart Turner Ltd
Technal - Cantifix of London Limited
Technics - Interfloor Limited
Technishield - Optima
Technistone - Haddonstone Ltd
Technomelt - Henkel Ltd
TechSpan® - Freyssinet Ltd
Techstyle - Hunter Douglas
Techtonic rooflight blinds - Hallmark Blinds Ltd
TecnoFlex - Schiedel Chimney Systems Ltd
Tecsom - Tarkett Sommer Ltd
Tectrix Fins - Hallmark Blinds Ltd
Tectum Abuse Resistant Panel - Salex Acoustics Limited
Tee Gee - Swish Building Products
Teetector Golf - Evergreens Uk
Teflon Super Protection - Scotmat Flooring Services
Tegola - Matthew Hebden
Tegometall - RB UK Ltd
Tegometall - Smart F & G (Shopfittings)Ltd
Tegula - Marshalls Mono Ltd
Tehalit - Hager Ltd
TekGrip - Star Uretech Ltd
TekSet - Star Uretech Ltd
Tektalan - Skanda Acoustics
Tektalan - Skanda Acoustics Limited
Tektura - Tektura Plc
Teleblock - Jones of Oswestry Ltd
Tele-Call - bedside TV/Telephone system - Wandsworth Elecrtical Ltd
Teleflex - Clearline Architectural
Teleguard - Youngman Group Limited
Telelight - Sandhurst Manufacturing Co Ltd
Telford Flip-Chart - Magiboards Ltd
Telling GFRC - Telling Architectural Ltd
Telling GFRC - Telling Lime Products Ltd
Tema - Optelma Lighting Ltd
Temcem - Minco Sampling Tech UK Ltd (Incorporating Fortafix)
Tem-Energy - May William (Ashton) Ltd
Tempo LST - Jaga Heating Products (UK) Ltd
Tempomatic - Delabie UK Ltd
Tempomix - Delabie UK Ltd
Tempostop - Delabie UK Ltd
Tempsoft - Delabie UK Ltd
Tempus - Zon International
Ten Brazilian Granites - Kirkstone
Ten Plus - Kompan Ltd
Tenniturf - En-tout-cas Tennis Courts Ltd
Tenon - HEM Interiors Group Ltd
Tenon Flexplus, Fire & Sound, Ovation, Vitrage, Vitrage (db), Scion - SIG Interiors
Tenon Operable Walls, Wallstore - SIG Interiors
Tenon Pace - SIG Interiors
Tenon Washrooms - SIG Interiors
Tensabarrier - Tensator Ltd
Tensaguide - Tensator Ltd
Tensar - Tensar International Ltd
Terca - Wienerberger Ltd
Terminator - Marflow Eng Ltd
Termodeck - Tarmac Building Products Ltd
Terodem - Henkel Ltd
Terokal - Henkel Ltd
Terophon - Henkel Ltd
Teroson - Beton Construction Ltd
Teroson - Henkel Ltd
Terostat - Henkel Ltd
Terra-34, Torino and Verona - Lang+Fulton
Terrabase Rustic - Addagrip Terraco Ltd
Terrabound - Addagrip Terraco Ltd
TerraClass™ - Freyssinet Ltd
Terraforce - Keller Ground Engineering
Terraglass - Diespeker Marble & Terrazzo Ltd
Terrain - Geberit Ltd
Terrain Blocker - APT Controls Ltd
Terrapin - Terrapin Ltd
Terratone - Andersen/ Black Millwork
TerraTrel™ - Freyssinet Ltd
Terrex - ABG Ltd
Terrier - Pegler Ltd
Terrings - Compriband Ltd
Terry's - The Kenco Coffee Company
Tesa - Beiersdorf (UK) Ltd
Tetra - Pifco Ltd
Texaa Panels - LSA Projects Ltd
Texitone ECO - Brett Landscaping
Texsol - Trade Sealants Ltd
Textomur - Keller Ground Engineering
Textura - Gradus Ltd
Texture - Pilkington Birmingham
TF1 Total Filter - Fernox
Thames - Portacel
Thatchbatt - Thatching Advisory Services Ltd
The Carpet Foundation - The Carpet Foundation
The Carpet Foundation Quality Mark - The Carpet Foundation
The CAST Rooflight - Clement Steel Windows
The Construction Information Service - NBS
The Design Archives - CrowsonFabrics Ltd
The Diamond System - Chelsea Artisans Ltd
The Executive Range - Steel Doors & Security Systems
The Finishing Touch - www.ubendwemend.co.uk - HE Services (Plant Hire) Ltd
The Gallery - Symphony Group PLC, The
The Invisible Lightswitch - Forbes & Lomax Ltd
The Part System - S & B Ltd
The Pyramid - The Rooflight company
The Rapid Range - Monier Redland Limited
The Resin Flooring Association - FeRFA
The Rutland Press - Royal Incorporation of Architects in Scotland
The Stationary Office - tso shop
Theben - Timeguard Ltd
Themerend - Thermica Ltd
Therapeutic Support Systems™ - ArjoHuntleigh UK
Therma - Kingspan Industrial Ins. Ltd
Thermac - McKenzie-Martin Ltd
Thermaclip - Flexible Ducting Ltd
Thermacool - Reznor UK Ltd
Thermadome - National Domelight Company
Thermafibre - Davant Products Ltd
Thermaflex - Flexible Ducting Ltd
Thermafloor - Kingspan Insulation Ltd
Thermafold Insulated Folding Door - Bolton Gate Co. Ltd
Thermaglide - Cardale Doors Ltd
Thermal Safe - Kingspan Ltd
Thermalight - National Domelight Company
Thermaline - Harton Heating Appliances
Thermaliner - Kingspan Insulation Ltd
Thermalite - Hanson Thermalite
Thermalsave Heat Re-circulators - Turnbull & Scott (Engineers) Ltd
Thermapitch - Kingspan Insulation Ltd
Thermaroof - Kingspan Insulation Ltd
Thermashield - Glixtone Ltd
Thermataper - Kingspan Insulation Ltd
Thermatex - Knauf AMF Ceilings Ltd
Thermatic - Watts Industries UK Ltd
Thermawall - Kingspan Insulation Ltd
Thermax - Birdsall Services Ltd
Thermgard - Lonsdale Metal Company
THERMIC - K. V. Radiators
THERMIC Zana-line - K. V. Radiators
Therminox - Poujoulat (UK) Ltd
Thermlock - L. B. Plastics Ltd
Thermo Guard - Kenyon Paints Limited
Thermoclear - Kenilworth Daylight Centre Ltd
Thermocor C.C - Polypipe Building Products
Thermoflue - Marflex Chimney Systems
Thermolier unit heaters - Turnbull & Scott (Engineers) Ltd
Thermoliner - The Rooflight company
Thermomax Refrigeration - Kingspan Environmental Ltd
Thermonda - Rockwell Sheet Sales Ltd
thermosave - Peter Cox Ltd
Thermoscreens - Thermoscreens Ltd
thermotek drywall - Peter Cox Ltd
Thermotone air curtains - shearflow LTD
Thioflex - Fosroc Ltd
Thirkleby - York Handmade Brick Co Ltd
Thistle - British Gypsum Ltd
Tholoflow - Fullflow Group Ltd
Thomas Dudley - Olby H E & Co Ltd
Thomas Lowe Joinery - Benlowe Stairs
Thor - Thor Hammer Co Ltd
Thorace - Thor Hammer Co Ltd
Thorex - Thor Hammer Co Ltd
Thorlite - Thor Hammer Co Ltd
Thorn Security - Tyco Fire and Integrated Solutions
Three in One 3 in 1 Mould killer - Polycell Products Ltd
Threshex - Exitex Ltd
Thrii handwash unit - Wallgate Ltd
Thunderbolt - Unifix Ltd
Thwaites - Chippindale Plant Ltd
Tibmix - Don Construction Products Ltd
TICA - Thermal Insulation Contractors Association
Tico - Tiflex Ltd
Tidy Bin - Unicorn Containers Ltd
Tiefoam - Modern Plan Insulation Ltd
Tie-Sleeve - Redifix Ltd
Tiger - Mode Lighting (UK) Ltd
Tikkurila - Andrews Coatings Ltd
Tikkurila - Valtti Specialist Coatings Ltd
Tile Association - The Tile Association
Tile-A-Door - Howe Green Ltd
Tileflex - Compriband Ltd
Tilley Lamp-Kerosene - Tilley International Ltd
Tilt Wash - Andersen/ Black Millwork
Timbacrib - Keller Ground Engineering
Timbalok Crib Walling - Ruthin Precast Concrete Ltd
Timbasol™ - Rockwood Pigments (UK) Ltd
Timb-a-Tilt - Caldwell Hardware (UK) Ltd
Timber Resin Splice - Proten Services Ltd
Timber Shade - Tidmarsh & Sons
Timbercare - Manders Paint Ltd
timberline - DLW Flooring
TimberLine - Premdor
Timbersound noise barrier - Chris Wheeler Construction Limited
Timbertie - Redifix Ltd
Timbertone - Protim Solignum Ltd
Timbridge - Panel Agency Limited
Timesaver - Saint-Gobain PAM UK
Timloc - Allmat (East Surrey) Ltd
Timonox - Crown Paints Ltd

T (con't)

Tintersol® - Rockwood Pigments (UK) Ltd
Tipform - Cordek Ltd
Titan - Fabrikat (Nottingham) Ltd
Titan - Ingersoll-Rand European Sales Ltd
Titan - Osram Ltd
Titan - PHI Group Ltd
Titan - Titan Environmental Ltd
Titan - Titan Ladders Ltd
Titan Environmental - Kingspan Environmental Ltd
Tivoli 21 carpet tiles loop pile - Burmatex Ltd
TKV - LTI Advanced Systems Technology Ltd
TL Board, TLFR Board, TD Board - Promat UK Ltd
TLC Control - Telelarm care Ltd
T-line 120 - Ves Andover Ltd
TMC Fine Texture - Andura Coatings Ltd
Toby Dustbins - Iles Waste Systems
Toesavers - Briggs Industrial Footwear Ltd t/a Briggs Safety Wear
Toledo - Townscape Products Ltd
Tomkinson - Jacopa Limited
Tonon Fan Coil Units - Thermal Technology Sales Ltd
Tony Team - Tony Team Ltd
ToolShield - Green Brook
TOP FOLDOR - TORMAX United Kingdom Ltd
Top office - OEP Furniture Group PLC
Top Star Rooflights - Klober Ltd
Top Vent - Klober Ltd
Topcem - Mapei (UK) Ltd
Topcon - Topcon GB Ltd
Topcrete - Tarmac Building Products Ltd
Topdek - Ward Insulated Panels Ltd
Topflex - Topseal Systems Ltd
Topflood - BLV Licht-und Vakuumtecnik
Topfoam - Tarmac Ltd
Topforce - Tarmac Ltd
TopForm - SFS Intec LTD
Topguard 10 Year - Chesterfelt Ltd
Toplab Base - Trespa UK Ltd
Toplab Plus - Trespa UK Ltd
Toplab Vertical - Trespa UK Ltd
Topline wall panels - Salex Acoustics Limited
Toplite - Tarmac Building Products Ltd
Topmix - Tarmac Ltd
TopProof - Tarmac Ltd
Toprail - Toprail Systems Ltd
Toproc - Tarmac Ltd
Topseal - Topseal Systems Ltd
Topspot - BLV Licht-und Vakuumtecnik
Topstore, Topdrawer, Toprax, Topshelf, Topstep - Barton Storage Systems Ltd
Topsy (litter bins) - Glasdon UK Limited
Topsy Jubilee (litter bins) - Glasdon UK Limited
Toptek - Barlow Group
Toptint - Tarmac Ltd
Torbeck - Opella Ltd
Torbeck Variflush - Opella Ltd
Torclad / Tordeck / Torlife / Torrex / Torshield / Tortread - Tor Coatings Ltd
Torcure MC - Tor Coatings Ltd
Toreador carpet tiles loop pile - Burmatex Ltd
Torit - Donaldson Filtration (GB) Ltd
Torlife WB - Tor Coatings Ltd
TORMAX - TORMAX United Kingdom Ltd
Toro - Hille Educational Products Ltd
Torprufe CRC / Torprufe EMF - Tor Coatings Ltd
Torrent - Gledhill Water Storage Ltd
Torso - Caldwell Hardware (UK) Ltd
Toscana - Marshalls Mono Ltd
Toscana - Paragon by Heckmondwike
Toshiba - Purified Air Ltd
Toshiba air conditioning - Toshiba Carrier UK
Total Care - Heckmondwicke FB
Totem - Directional Data Systems Ltd
Tough Furniture - Roc Secure Ltd
Tough-Cote™ High Build Texture - Glixtone Ltd
Tower - Pifco Ltd
Towerfill - Tower Manufacturing
Towerfire - Tower Manufacturing
Towerfoam - Tower Manufacturing
Towergrip - Tower Manufacturing
Towerpak - Tower Manufacturing
Towerseal - Tower Manufacturing
TP Plumbing & Heating - Keyline
TPC LN/B - Babcock Wanson Uk Ltd
T-Pren - Matthew Hebden
TQ - Owlett - Jaton
TR Freeman - Kershaw Mechanical Services Ltd
Trackmaster - Rawson Carpets Ltd
Trackranger Crawler Primary Crushing Outfits - Parker Plant Ltd
TRADA Technology - BM TRADA
Trade Range - Cox Building Products Ltd
Trademark - Marcrist International Ltd
Tradesman - Ibstock Brick Ltd
Traditional - Northcot Brick Ltd
Traditional Range - Respatex International Limited
Trafalgar - Light Engine Lighting Ltd
Traffic - Kompan Ltd
Trafficline - Moravia (UK) Ltd
Traffic-Line - Barton Storage Systems Ltd
TrafiCop - Rediweld Rubber & Plastics Ltd
TRAKA - ASSA ABLOY Limited
Trakelast - Tiflex Ltd
Trakway - Eve Trakway
Trammel fin drain - Don & Low Ltd
Transaction - Rackline Systems Storage Ltd
Transfastener™ - Latchways plc
Transforma - Youngman Group Limited
TransitMaster™ - Kone
Transmit - Tunstal Healthcare (UK) Ltd
Transpalace - Parthos UK Ltd
Transtor - Desking Systems Ltd
Transvario - Kone Escalators Ltd
Trashmaster - Pump Technical Services Ltd
Travel master hoists - Wessex Lift Co Ltd
TravelLED - Glamox Luxo Lighting Limited
TravelMaster™ - Kone
Travertin - Skanda Acoustics
Travertin - Skanda Acoustics Limited
Travis Perkins Trading Company - Keyline
Trav-O-lators® - Otis Ltd
Tread Safe - Anglo Building Products Ltd
TreadGUARD - Visqueen Building Products
Treadmaster - Tiflex Ltd
Treadspire - In Steel (Blacksmiths & Fabricators) Ltd
Trebitt - Jotun Henry Clark Ltd Decorative Division
Trebitt Opaque - Jotun Henry Clark Ltd Decorative Division
Tredaire - Interfloor Limited
Trellidor - Harling Security Solutions
Tremco - Tremco Illbruck Ltd
Trench Heating - Turnbull & Scott (Engineers) Ltd
Trent Bathrooms - Shires Bathrooms
trent gutter system - Dales Fabrications Ltd
Trentobond - Tremco Illbruck Ltd
Trentocrete - Tremco Illbruck Ltd
Trentodel - Tremco Illbruck Ltd
Trentoflex - Tremco Illbruck Ltd
Trentoshield - Tremco Illbruck Ltd
Trentothane - Tremco Illbruck Ltd
Trespa - Amari Plastics Plc
Trespa - Decra Ltd
Trespa - FGF Ltd
Trespa - Trespa UK Ltd
Trevi Showers - Armitage Shanks Ltd
Trevira CS - Silent Gliss Ltd
Trewor - Broxwood (scotland Ltd)
Triaqua - Multikwik Ltd
Tricity - Electrolux Domestic Appliances
Trident - Marflow Eng Ltd
Tridonic - Bernlite Ltd
Triflo Solder Ring - Conex Universal Limited
Trigon - Hamworthy Heating Ltd
Tri-Guard® - Jackson H S & Son (Fencing) Ltd
Trilax - Girsberger London
Triline - Smart Systems Ltd
Tri-lock - Grass Concrete Ltd
Trim Acoustic Cradle System - Trim Acoustics
Trim Defender Door Bars - Trim Acoustics
Trim Defender Wall - Trim Acoustics
Trimalac - Trimite Ltd
TRIMEC - ASSA ABLOY Limited
Trimite - Andrews Coatings Ltd
Trimite - Trimite Ltd
Trimtop - Swish Building Products
Triniti - Levolux A.T. Limited
Trio - Panel & Louvre Co Ltd
TRIO - Panel wall formwork - PERI Ltd
Triogen & Tylo - Golden Coast Ltd
Triple Seven - Dunbrik (Yorks) Ltd
Tri-Plugs - Tower Manufacturing
Tri-Shell - Dudley Thomas Ltd
Tritainer - Franklin Hodge Industries Ltd
Triton - Albany Standard Pumps
Triton - Triton PLC
Trixene - Baxenden Chemicals Ltd
Trizone - Airsprung Beds Ltd
Troikafix - Troika Contracting Ltd
Trojan - Bitmen Products Ltd
Trojan - Crane Fluid Systems
Trojan - Hallmark Blinds Ltd
Trojan - Shackerley (Holdings) Group Ltd incorporating Designer ceramics
Troldtekt - Skanda Acoustics
Troldtekt - Skanda Acoustics Limited
Trolleylift - Stannah Lifts Ltd
Trooper - Glasdon U.K. Limited
Tropicano - Xpelair Ltd
Trough & Waffle Moulds - Cordek Ltd
Trubert - Watts Industries UK Ltd
Trubolt - Ramset Fasteners Ltd
TruckStop - Rediweld Rubber & Plastics Ltd
Trueflue - Marflex Chimney Systems
Trugrain - Laird Security Hardware
Truline - Simpson Strong-Tie®
Trulok Prelude - Armstrong World Industries Ltd
Trumeter - Marking Machines International
Truth - Caldwell Hardware (UK) Ltd
Truth Hardware - Schlegel UK (2006) Ltd
TS60P - Ramset Fasteners Ltd
TS750P - Ramset Fasteners Ltd
T-Screen - Mermet U.K
TSV1 surface mounted shower pannel - Horne Engineering Ltd
TT - Lochinvar Limited
TT - Optelma Lighting Ltd
TT - Technology Tower - Staverton (UK) Ltd
T-T controls - T-T Pumps & T-T controls
TTA - The Tile Association
TTF - Simon R W Ltd
TTF - Timber Trade Federation
Tubarad - AEL
Tubeclamps - Tubeclamps Ltd
Tubela - Tubela Engineering Co Ltd
Tube-U-Fit Hanging Rail & Supports - Daro Factors Ltd
Tubinox - Poujoulat (UK) Ltd
Tucker - Arrow Supply Company Ltd
Tudor - Branson Leisure Ltd
Tudor Roof Tiles - Tudor Roof Tile Co Ltd
Tufanega - Deb Ltd
Tufcon 80 - Conren Ltd
Tuff Stuff - Twyford Bushboard
Tuff workbenches - Barton Storage Systems Ltd
TuffBin™ - Straight Ltd
Tufftrak - Mabey Hire Ltd
Tuftex - Vita Liquid Polymers Ltd
Tung Oil - Chestnut Products
TurboCharger - Lochinvar Limited
TurboFast - Helifix
Turbo-Flo - MSA (Britain) Ltd
Turbolite - Marcrist International Ltd
TurboSyphons - Dudley Thomas Ltd
Turkish room - Dalesauna Ltd
Turney Turbines - Pumps & Motors UK Ltd
Turnquest - H E W I (UK) Ltd
Tusc - Unistrut Limited
Tuscany - Lotus Water Garden Products Ltd
Tuscany - Muraspec
TVSS - Eaton Electric Limited
Tweeny - The Haigh Tweeny Co Ltd
Twenty-one Limestones - Kirkstone
Twiclad - Ide T & W Ltd
Twilfix - Betafence Limited
Twill - Advanced Fencing Systems Ltd
Twill Wire products - Challenge Fencing Ltd
Twilweld - United Wire Ltd
Twin & Big Twin - Helmsman
Twinarc - Two Arc Tube High Pressure Sodium Lamps - Sylvania Lighting International INC
Twindisk - BigHead Bonding Fasteners Ltd
Twinfix - Amari Plastics Plc
Twinfold Fine Twist - Brockway Carpets Ltd
Twinlok® - Walraven Ltd
Twinlux - Illuma Lighting
Twinqwik - Owlett - Jaton
TwinSash - Veka Plc
Twisco - Manitou (Site Lift) Ltd
Twist® Roller Blinds - Luxaflex®
Twosome and Super Twosome - Integra Products
TW-Tec Metal cubicles - Detlef Muller Ltd
Twyford Range - Twyford Bathrooms
Tycho - Kronospan Ltd
Tyglas - Fothergill Engineered Fabrics Ltd
Type E - Cavity Trays Ltd
Type seating - CPS Manufacturing Co
Type X - Cavity Trays Ltd
Typhoo - The Kenco Coffee Company
Tyre Tile - Mat.Works Flooring Solutions
Tyrolean Rendadash - LaFarge Cement UK
Tyton - Hellermann Tyton
Tyveks - Klober Ltd
Tyzack - Spear & Jackson

U

U. Bik - Noral Scandinavia AB
U.S.G Donn - New Forest Ceilings Ltd
U+ - Uponor Ltd
U45 Insulated Sectional Doors - Avon Industrial Doors Ltd
UbiSoil - Ubbink (UK) Ltd
Ubivent - Ubbink (UK) Ltd
UCL Ltd - Capital Safety Group (NE) Ltd
Ucrete - BASF plc, Construction Chemicals
UESA - High efficiency condensing warm air heaters - AmbiRad Ltd
Uginox Roll On - Aperam Stainless Services & Solutions UK Limited
Uginox Bright - Aperam Stainless Services & Solutions UK Limited
Uginox Leather, Uginox Linen, Uginox Lozenge - Aperam Stainless Services & Solutions UK Limited
Uginox Mat - Aperam Stainless Services & Solutions UK Limited
Uginox Patina K44 - Aperam Stainless Services & Solutions UK Limited
Uginox Sand, Uginox Square - Aperam Stainless Services & Solutions UK Limited
Uginox Top, Uginox Touch - Aperam Stainless Services & Solutions UK Limited
UL-M - Breedon Aggregates
Ulster Velvet - Ulster Carpet Mills Ltd
Ultima - Airedale International Air Conditioning Ltd
Ultima - Armstrong World Industries Ltd
Ultima - Laird Security Hardware
Ultima - Wincro Metal Industries Ltd
Ultima - For the less abled - Rixonway Kitchens Ltd
Ultimate - Owlett - Jaton
Ultimate Collection - Ball William Ltd
Ultimate flue outlet - Brewer Metalcraft
Ultimate Range - Valor
Ultra - Eaton Electric Limited
ULTRA - Lakes Bathrooms Limited
Ultra 80 - Norwood Partition Systems Ltd
Ultra 9 - Tapworks Water Softeners
Ultra Bar - Sections & Profiles Ltd
Ultra Fence - Sections & Profiles Ltd
Ultra Fortis - Uponor Ltd
Ultra Post - Sections & Profiles Ltd
Ultra Rail - Sections & Profiles Ltd
Ultra Sheet - Sections & Profiles Ltd
Ultrabond - Mapei (UK) Ltd
Ultracal - Range Cylinders
Ultracolor - Mapei (UK) Ltd
Ultra-DENCO - DencoHappel UK Ltd
Ultraflex - Dyer Environmental Controls
Ultraframe - Ultraframe PLC
Ultragard Software - Time and Data Systems International Ltd (TDSI)
Ultragard Vision-ID - Time and Data Systems International Ltd (TDSI)
Ultraglide - Integra Products
Ultragrain® - Righton Ltd
UltraGRID™ - Hadley Industries Plc
UltraLATH™ - Hadley Industries Plc
Ultralife - BLV Licht-und Vakuumtecnik
Ultralift - Caldwell Hardware (UK) Ltd
ultralight, utilityramp - Enable Access
Ultralite 500 - Ultraframe PLC
Ultralite Blocks - Hanson Concrete Products
Ultralon - Spaceright Europe Limited
Ultramastic - Mapei (UK) Ltd
UltraMEZZ™ - Hadley Industries Plc
Ultramix - Watts Industries UK Ltd
UltraPALE™ - Hadley Industries Plc
Ultraplan - Mapei (UK) Ltd
UltraPly TPO - Firestone Building Products
Ultra-Rib - Brett Martin Ltd
Ultra-Rib - Uponor Ltd
UltraSHEET™ - Hadley Industries Plc
UltraSTEEL™ - Hadley Industries Plc
Ultrastor - Norwood Partition Systems Ltd
Ultrastream - Britannia Kitchen Ventilation Ltd
UltraSTRUT™ - Hadley Industries Plc
Ultratile - Britmet Tileform Limited
Ultraturn - Righton Ltd
Unbreakable - Briggs Industrial Footwear Ltd t/a Briggs Safety Wear
Uni- Ecoloc - CEMEX
Unibond - Henkel Consumer Adhesives
Unibond No More Nails - Henkel Consumer Adhesives
Unibond No More Sealent Guns - Henkel Consumer Adhesives
Unibond Unifilla - Henkel Consumer Adhesives
Unica - Hansgrohe
Unicell - Donaldson Filtration (GB) Ltd
Unidare - Dimplex UK Limited
Unidare - Seagoe Technologies Ltd
Unidox - Witham Oil & Paint Ltd
Unieck - DLW Flooring
Unifit - NewTeam Ltd
Unifloor Flooring System - Hanson Concrete Products
Uniframe - Hanson Concrete Products
Unify - Connaught Communications Systems Ltd
Unigas - GP Burners (CIB) Ltd
Unijet - Carpark Ventilation System - SCS Group
Unikrame - Hanson Concrete Products
Unilight - Universal Components Ltd
Unilin - Meyer Timber Limited
Uniline - Rangemaster
Uniline Safety Systems - Capital Safety Group (NE) Ltd
Unilit Natural Hydraulic Lime - Telling Lime Products Ltd
Unimaster - Donaldson Filtration (GB) Ltd
Unimate - Quartet-GBC UK Ltd
UniMini - HAGS SMP
UNINEP - TORMAX United Kingdom Ltd
Union - ASSA ABLOY Limited
Union - Mul-T-Lock
Union - Olby H E & Co Ltd
Union Jack Label Fittings - Daro Factors Ltd
Uniperf - United Wire Ltd
Unipipe - Uponor Ltd
Uniplant - Witham Oil & Paint Ltd
UniPlay - HAGS SMP
Unislate - BLM British Lead
Unistrut - Kem Edwards Ltd
Unistrut - Unistrut Limited
Unit Swimming Pools - Regency Swimming Pools Ltd
Unitas - Witham Oil & Paint Ltd
Uni-Trex - Terrapin Ltd
Unity Expandable Aluminium Door & Glazing Frames - Avanti Systems
Univent smoke relief chimney - SCS Group
Universal - Matki plc
Universal Louvre - Colt International
Universal range of rainwater products - Marshall-Tufflex Ltd
Uniwall - Hanson Concrete Products
Uniweave® - Ulster Carpet Mills Ltd
Uno - Becker (SLIDING PARTITIONS) Ltd
Upat - Artur Fischer (UK) Ltd
UPE - Tyco Water Works
UPM - Meyer Timber Limited
Upton Door - AS Newbould Ltd
Uraflex - SCP Concrete Sealing Technology Ltd
Urban - Amwell Systems Ltd
Uretech - Star Uretech Ltd
Uretek - Uretek (UK) Ltd
USDA - High Efficiency warm air heaters - AmbiRad Ltd
uSee - Helvar Ltd
USVR - Watts Industries UK Ltd
Utiform - Schwing Stetter (UK) Ltd
Utility Elite Lift - Phoenix Lifting Systems Ltd
Utility Lift - Phoenix Lifting Systems Ltd
Utrain - Uponor Ltd
UV Dermagard - Bonwyke Ltd
UV Free Lighting - Encapsulite International Ltd

V

V B Johnson - **V B Johnson** LLP
V Force Vapour Control Layer - Firestone Building Products
V100 PSM - Elster Metering Ltd
V200 - Elster Metering Ltd
V210 - Elster Metering Ltd
V3 - Reznor UK Ltd
V300 Master - Elster Metering Ltd
vac (vibration and acoustic control) - Mansfield Pollard & Co Ltd
Vac Vac - Arch Timber Protection
Vacsol - Arch Timber Protection
Vacurain - John Davidson (Pipes) Ltd
Valiant - Laird Security Hardware
Valley - Iles Waste Systems
Valspar - Akzo Noble Coatings Ltd
Valtek - Flowserve Flow Control
Valtti - Valtti Specialist Coatings Ltd
Valves - Horstmann Controls Ltd
Vandal Resistant Entry Panels - BPT Security Systems (UK) Ltd
Vandalene® - Coo-Var Ltd
Vandex BB75 - Safeguard Europe Ltd - Safeguard Europe Ltd
Vandex Super - Safeguard Europe Ltd - Safeguard Europe Ltd
Vandex Unimortar - Safeguard Europe Ltd - Safeguard Europe Ltd

V (con't)

Vandgard - Anti-Climb Guards Ltd
Vanera (bedhead lamps) - Brandon Medical
Vanguard - Benlowe Group Limited
Vanguard - Forticrete Roofing Products
Vanities Range - Amwell Systems Ltd
Vanity flair - Ellis J T & Co Ltd
Vanodine - Evans Vanodine International PLC
Vanquisher - Heckmondwicke FB
Vantage Sio - Astracast
VAPAC® - Eaton-Williams Group Limited
Vapourguard - Trade Sealants Ltd
Varde - Gazco
Vari-Ceat - Ellis Patents
Varicool Spectra - LTI Advanced Systems Technology Ltd
Variflow - Grosvenor Pumps Ltd
Varihi - Emmerich (Berlon) Ltd
Vari-Level - Wade International Ltd
Varilift - Arvin Motion Control Ltd
Vario - Kompan Ltd
Vario - RTD Systems Ltd
VARIO - Wall formwork - PERI Ltd
Vario seating, Venue seating - CPS Manufacturing Co
Vari-Purpose - FC Frost Ltd
Varitone - IAC Acoustics
Varitrans - Style Door Systems Ltd
Varley TopJet - Marking Machines International
Varta - Pifco Ltd
Vasura - Erlau AG
VCD - Naco
V-CHR - K. V. Radiators
VDC - Motorised Air Products Ltd
V-DPS, V-GDS, V-AFD also V-ITL Light fittings - Vianen UK Ltd
VEA - Vitreous Enamel Association, The
VECTOR - Vantrunk Engineering Ltd
Vedette - Helmsman
Vedette Specialized range - Helmsman
Vee - Project Office Furniture PLC
VeeBee Filtration - Barr & Wray Ltd
Vega - Guardall Ltd
VEKA - ABB Ltd
VELFAC - VELFAC Ltd
Velocity - Komfort Workspace PLC
Velopa - Autopa Ltd
Velour carpet tiles fibre bonded - Burmatex Ltd
Velux - Tidmarsh & Sons
Velux - VELUX Co Ltd, The
Venetian - Symphony Group PLC, The
Veneto xf2 - Tarkett Ltd
Vent A Matic - Simon R W Ltd
VENTATEC - Knauf AMF Ceilings Ltd
Ventform - Cordek Ltd
Ventform - Jablite Ltd
Venti Seal - Trelleborg Sealing Profiles
Ventmaster AG - Docherty New Vent Chimney Group
Ventrolla - Durable Ltd
Ventrolla Perimeter Sealing System - Ventrolla Ltd
Ventrolla Sash Removal System - Ventrolla Ltd
VENTSAIR - Monodraught Ltd
Ventuno - Ideal-Standard Ltd
Ventura - McKenzie-Martin Ltd
Venture Parry - Bernlite Ltd
Venvaer - SE Controls
Verco - Verco Office Furniture Ltd
Verena - Deceuninck Ltd
Verges - Zon International
Vericure - Proten Services Ltd
Verine - BFM Europe Limited
Vermeer - Vermeer United Kingdom
Vermiculux - Promat UK Ltd
Vermont - Deceuninck Ltd
Vernacare - Verna Ltd
Vernagene - Verna Ltd
Veronastone Assimilated Limestone - Marble Flooring Spec. Ltd
Versaline - Lewden Electrical Industries
Versalite - Black Cat Music
Versalite - Righton Ltd
Versalux - Cooper Lighting
VersaTank - CST Industries, Inc. - UK
Versatemp - Clivet UK Ltd
Versiplan - Becker (SLIDING PARTITIONS) Ltd
Verti-Frame - Hi-Store Ltd
VESDA - Xtralis
Vesselpak - Flamco UK Ltd
Vestos - Arch Timber Protection
Vetcare™ - Steel Line Ltd
Vétec - Vianen UK Ltd
Vetter UK - Vetter UK
Vex - Guardall Ltd
V-Gard - MSA (Britain) Ltd
Viajoint 'D' Expansion Ioint - Pitchmastic PmB Ltd
Vianergy - Vianen UK Ltd
Vibro - Keller Ground Engineering
Victaulic - Midland Tube and Fabrications
Victaulic - Victaulic Systems
Victoriana - AEL
Victory - PCT Group
Vicuclad - Promat UK Ltd
Video System 200 - BPT Security Systems (UK) Ltd
VideoSign - Rivermeade Signs Ltd
Viewpoint - Pryorsign
Vigilant - Auld Valves Ltd
Vigilante - Blick Communication Systems Ltd
Vigilon - Gent Limited
Viking - Viking Laundry Equipment Ltd
Villa - Kompan Ltd
Village Stone - CEMEX
Vilter - Emerson Climate Technologies Retail Solutions
Vinadac - Dacrylate Paints Ltd
Vintage Collection - Masterframe Windows Ltd
Vinylcomfort - Amorim (UK) Ltd
Vinyllon - Dixon Turner Wallcoverings
Viper spike® - Jackson H S & Son (Fencing) Ltd
Viroc - FGF Ltd
Visage - Valor
Viscacid - Kenyon Group Ltd
Viscount - AEL
Viscount - Forticrete Roofing Products
Visedge - Howe Green Ltd
Vision Series - Radianr rube heaters - AmbiRad Ltd
Visions Bathrooms - Shires Bathrooms
Visit Range - Twyford Bathrooms
Visofold - Smart Systems Ltd
Visoglide - Smart Systems Ltd
Visoline - Smart Systems Ltd
Visqueen - Visqueen Building Products
Vista - Delta Balustrades
Vista - Dunhams of Norwich
Vista - Fabrikat (Nottingham) Ltd
Vista - Stocksigns Ltd
Vista modular systems - GB Sign Solutions Limited
Vista-Fix - Vista Engineering Ltd
Vistalux - Ariel Plastics Ltd
Vista-Plas - Vista Engineering Ltd
Visual Surveillance Systems - Vador Security Systems Ltd
Visualdirect - The Visual Systems Healthcare
Vitachem - Vita Liquid Polymers Ltd
Vitaplas - Vita Liquid Polymers Ltd
Vitopan Steel Clean - Kleen-Tex Industries Ltd
Vitra - Swedecor
Vitra - Vitra (UK) Ltd
Vitram - Evans Concrete Products Ltd
Vitramesh - Cadisch MDA Ltd
Vitrogres - Arnull Bernard J & Co Ltd
Vittera - Switch2 Energy Solutions Ltd
Vivak - Amari Plastics Plc
VKH - Motorised Air Products Ltd
V-Kool - Northgate Solar Controls
VLT ® - Danfoss Ltd
Vmag - Hawker Siddley Switchgear
Voidak - Dunham Bush Ltd
Vokera - Diamond Merchants
Vokes - Vokes Air
Voltan - Townscape Products Ltd
Volumeter - Gebhardt Kiloheat
Volute - Benlowe Stairs
Volvo Compact Equipment - Chippindale Plant Ltd
Von Duprin Exit Devices - Relcross Ltd
Vormatic - RB UK Ltd
Vortax - Axial - Howden Buffalo
Vor-Tec™ - Lotus Water Garden Products Ltd
Vortex - Unifix Ltd
Vortex - Water Technology Engineering Ltd
Vortibreak® - Franklin Hodge Industries Ltd
VortX - Saint-Gobain PAM UK
Vossloh Schwabe - Bernlite Ltd
VPX LN/RR - Babcock Wanson Uk Ltd
VR90 Timber Repair - Ventrolla Ltd
VSoL® - Asset International Ltd
VSS - Vador Security Systems Ltd
Vtech - Vantrunk Engineering Ltd
Vulcaboard - Vulcan Cladding Systems
Vulcalap - Vulcan Cladding Systems
Vulcalucent - Vulcan Cladding Systems
Vulcan - Ellis Patents
Vulcan - Tuke & Bell Ltd
Vulcan Tank - CST Industries, Inc. - UK
Vulcascot Floor Laid Flexible Cable Protectors - Vulcascot Cable Protectors Ltd
Vulcatuf - Vulcan Cladding Systems
Vyflex - Plastic Coatings Ltd
Vylon Plus - Tarkett Ltd
Vynarac - Hispack Limited

W

W/A formwork plywood - Wood International Agency Ltd
W/A pine doors - Wood International Agency Ltd
W20 - Clement Steel Windows
W20 Range Steel Windows - Monk Metal Windows Ltd
W40 - Clement Steel Windows
W40 Range Steel Windows - Monk Metal Windows Ltd
Wachenfeld - Spiral Construction Ltd
Wacker Silicones - Notcutt Ltd
Wade - Wade International Ltd
Waldor - Howe Green Ltd
wales and wales - Wales & Wales
Walk On Ceiling - Norwood Partition Systems Ltd
Walker Timber Frame - Walker Timber Ltd
Walksafe - McKenzie-Martin Ltd
WalkSafe® - Latchways plc
Wall panels - Saint-Gobain Ecophon Ltd
Wall Shield - Springvale E P S
Wall Tile Collection - Solus Ceramics Ltd
Wallis - Bitmen Products Ltd
Wallmate - Dow Building Solutions
Wallsorba - Soundsorba Ltd
Wallsorption - Bicester Products Ltd
Wallstor - Desking Systems Ltd
Walltalker - Tektura Plc
Walraven - Kem Edwards Ltd
Walther - Barloworld Vacuum Technology Plc
Walton Sheds - Challenge Fencing Ltd
Wamsler - Gazco
WAP - Nilfisk Limited
Warm Air Cabinet Heaters - Radiant Services Ltd
Warm Air Unit Heaters - Radiant Services Ltd
Warm Squeez - Springvale E P S
Warm Touch - H E W I (UK) Ltd
Warmafloor - Warmafloor (GB) Ltd
WarmCair - Johnson & Starley
warmerwall - Peter Cox Ltd
Warmsafe LST - Dunham Bush Ltd
Warmskark - Springvale E P S
Warren G-V - Radiodetection Ltd
Warrior - Glasdon U.K. Limited
Warrwick - Benlowe Stairs
Wascator - Electrolux Laundry Systems
Wash-Horse - Kleen-Tex Industries Ltd
Washroom Control - Cistermiser Ltd
Wasteflo - Stuart Turner Ltd
Wastematic Bench Mounted - Max Appliances Ltd
Wastematic Free Standing - Max Appliances Ltd
Wastematic Sink Mounted - Max Appliances Ltd
Watchkeepers - Rapaway Energy Ltd
Watchman - Kingspan Environmental Ltd
Waterco - Golden Coast Ltd
Watergem - Franke Sissons Ltd
Water-Horse - Kleen-Tex Industries Ltd
Waterless Urinals - Relcross Ltd
Waterseal - Sealocrete PLA Ltd
Watertight - Laticrete International Inc. UK
Watrous - FC Frost Ltd
Watts cliff - Realstone Ltd
Wauderer-Castle care-tech - Castle Care-Tech
Wavespan - Flexible Reinforcements Ltd
Wayland - Advanced Fencing Systems Ltd
Waylite 800 - Cooper Lighting
WBB Minerals - Sibelco UK Ltd
Weather Safe - Kronospan Ltd
Weather Ten - Exitex Ltd
Weatherbeaters - Owlett - Jaton
Weatherbeta - Naco
Weatherbeta 24 - Naco
Weathercor/F - Polypipe Building Products
Weatherfin - Ventrolla Ltd
WeatherFin and WeatherFin + - Linear Ltd
Weathershield - AkzoNobel Decorative Paints UK
Weathershield - The No Butts Bin Company (NBB), Trading as NBB Outdoor Shelters
WeatherTone - Vulcan Cladding Systems
Weatherwhite - Cardale Doors Ltd
Weatherwood - Cardale Doors Ltd
Weavespread - Flexible Reinforcements Ltd
Weavetop - Flexible Reinforcements Ltd
Web wall - ABG Ltd
Weber Building solutions - Weber Building Solutions
Weber cem - Weber Building Solutions
Weber certite - Weber Building Solutions
Weber cote - Weber Building Solutions
Weber fix - Weber Building Solutions
Weber Joint - Weber Building Solutions
Weber mulsifix - Weber Building Solutions
Weber Pral - Weber Building Solutions
Weber rent - Weber Building Solutions
Weber Set - Weber Building Solutions
Weber tec - Weber Building Solutions
WebRTU - EnergyICT Ltd
Wedgtec - Cadisch MDA Ltd
Weep - Airflow (Nicoll Ventilators) Ltd
Weger Air Handling Units - Thermal Technology Sales Ltd
WEHA - Harbro Supplies Ltd
Weigh-tronix - Phoenix Scales Ltd
Welcome - Enable Access
Weldmesh - Betafence Limited
Wellington - BF Bassett & Findley
Welltec - Cadisch MDA Ltd
Welmade - Q.M.P
Wenger V-Room - Black Cat Music
Wensleydale - Carpets of Worth Ltd
Wernick - Wernick Group Ltd
Wessex - Healthmatic Ltd
Wessex - Multitex GRP LLP
Wessex Doors - Regency Garage Door Services Ltd
Wessex ModuMax - Hamworthy Heating Ltd
West Pennine Gritstone - Johnsons Wellfield Quarries Ltd
West System - Wessex Resins & Adhesives Ltd
West System Epoxy - Wessex Resins & Adhesives Ltd
Westbrick - Ibstock Brick Ltd
Westclox - General Time Europe
Westdale - Chiltern Invadex
Westerland - Cembrit Ltd
Western & Drayton Fencing - Anthony de Grey Trellises + Garden Lighting
Western Cork Ltd - Westco Group Ltd
Western Isles - Scotmat Flooring Services
Westman Systems - MJ Electronics Services (International) Ltd
Westminster - Atlas Stone Products
Westminster - Branson Leisure Ltd
Westminster - Polybeam Ltd
Weston - Gloster Furniture Ltd
Westwood Ground - Bath Stone - Hanson UK - Bath & Portland Stone
WF - Westbury Filters Limited
WF1 - Westbury Filters Limited
WFB - Westbury Filters Limited
WFC - Westbury Filters Limited
WFF - Westbury Filters Limited
WFG - Westbury Filters Limited
WFH HEPA - Westbury Filters Limited
WFK - Westbury Filters Limited
WFR - Movement Joints (UK)
WFV - Westbury Filters Limited
WhatRisk.com - National Britannia Ltd
Whirlwind - Envirodoor Markus Ltd
Whispair - Xpelair Ltd
WHITE BOOK - British Gypsum Ltd
Whitehill - Whitehill Spindle Tools Ltd
Whiterock Doors - Altro Ltd
Whiterock PVC Ceiling system - Altro Ltd
Whiterock PVC Sport Cladding - Altro Ltd
Whitestar - BLV Licht-und Vakuumtecnik
WHS - Spear & Jackson
Wicanders - Amorim (UK) Ltd
Widespan - Arc Specialist Engineering Limited
Widra - Metal Wire Plaster & Render Beads - Bekaert Building Products
Wilclo - Clow Group Ltd
Wildgoose - Derby Timber Supplies
Wilka Locks & Cylinders - JB Architectural Ltd
Wilkinson's Furniture - Carleton Furniture Group Ltd
Williams Lea Group - tso shop
Williams Lea Tag - tso shop
Wilo SE - Wilo Samson Pumps Ltd
Wilo Top E - Wilo Samson Pumps Ltd
Wilo Top S - Wilo Samson Pumps Ltd
Wilo Top SD - Wilo Samson Pumps Ltd
Wilotekt-Plus - Axter Ltd
Wilson & Garden - Spaceright Europe Limited
Wilsonart - Anderson C F & Son Ltd
Wilsonart - Wilsonart Ltd
Wimbledon - Gloster Furniture Ltd
Win Drive - TORMAX United Kingdom Ltd
Winchester - BSA Machine Tools Ltd
Winchester - Sidhil Care
Wincro - Wincro Metal Industries Ltd
WINDCATCHER® - Monodraught Ltd
Windoline - Celuform Building Products
Window and Door Security Systems - Rolflex Doors Ltd
Window Care Systems - Repair care International
Window Files - Railex Systems Ltd
Windowgard - Gilgen Door Systems UK Ltd
Windowmaster Screws - UK Fasteners Ltd
Windsor - DW Windsor Lighting
Windsor Water Fittings - Deva Tap Co Ltd
Wing - Illuma Lighting
Wings - Bisley Office Equipment
Winther Browne & Co Ltd - Winther Browne & co Ltd
Wiremaster - Philips Lighting Solutions
Wiring Regulations - The Institution of Engineering and Technology
Wiring Regulations Bookshop - Professional Bookshops Limited
WirralRange - AS Newbould Ltd
Wisa-Deck - Schauman (UK) Ltd
Wispa Hoist - Chiltern Invadex
Witherley Services - Aggregate Industries UK Ltd
WMA - Wallcovering Manufacturers Association (now part of the British Coatings Federation)
WMS - Avocet Hardware Ltd
Woco - Witham Oil & Paint Ltd
Wolfin - Beton Construction Ltd
Wolfin - Pitchmastic PmB Ltd
Wonderex - Optex (Europe) Ltd
Wondertrack - Optex (Europe) Ltd
Wonderwall - Hanson Building Products
Wood Symetra - Knauf AMF Ceilings Ltd
Woodcomfort - Amorim (UK) Ltd
Wooden Lockers - Helmsman
Woodflex - Polycell Products Ltd
Woodgrip - Pedley Furniture International Ltd
Woodhead Natural York Stone - Johnsons Wellfield Quarries Ltd
Woodland - Neptune Outdoor Furniture Ltd
Woodman Twist - Brockway Carpets Ltd
Woodro - Hille Educational Products Ltd
Woods - Hamilton R & Co Ltd
Woodscape - Woodscape Ltd
Woodsorption - Bicester Products Ltd
Woodstock Accent - Adam Carpets Ltd
Woodstock Classic - Adam Carpets Ltd
Woodworks - Interfloor Limited
Woodwright - Andersen/ Black Millwork
Worcester - The BSS Group
Work-Safe-Matting - Olley & Sons Ltd, C
Workspace - Paragon by Heckmondwike
World hand dryers - Warner Howard Group Ltd
Worldspan - Clow Group Ltd
Wormald Ansul - Tyco Fire and Integrated Solutions
WP Wallpack - Consult Lighting Ltd
WPIF - Wood Panels Industries Federation
WS 7 - Redring
WSC - Wallcovering Sector Council - see - Wallcovering Manufacturers Association (now part of the British Coatings Federation)
WSG Donn Floors - Hatmet Limited
WTIF - Wall Tie Installers Federation
WW - Workwalls - Staverton (UK) Ltd
www.constructionjobsuk.com - Contract Journal
www.contractjournal.com - Contract Journal
www.electrolux.com/laundrysystems - Electrolux Laundry Systems
www.sales@firebarrier.co.uk - Firebarrier Services Ltd
Wyevale Grounds Maintenance - Western landscapes Ltd (trading as Wyevale)
Wyevale Landscapes - Western landscapes Ltd (trading as Wyevale)
Wyre Forest - Adam Carpets Ltd
Wyrem - Flexible Ducting Ltd
Wyvern Fireplaces - Wyvern Marlborough Ltd

X

X1000 - Reznor UK Ltd
XC - Harkness Screens
Xena - Marflow Eng Ltd
Xenex - CEMEX
Xenex - Gent Limited
Xenon, Xira - Samuel Heath & Sons plc
Xerra - Girsberger London
Xetex - Electrix International Ltd
XFP addressable fire panels - C-TEC (Computionics) Ltd
X-Frame Picnic Suite - Branson Leisure Ltd
XIB-IT - Spaceright Europe Limited
Xinox - Franke Sissons Ltd
XLight - Porcelanosa Group Ltd
X-Line - AudiocomPendax Ltd

X (con't)

XLnt - Hunter Douglas
Xodus - Xpelair Ltd
XP95 - Apollo Fire Detectors Ltd
XP95 Intrinsically safe - Apollo Fire Detectors Ltd
Xpelair - Applied Energy Products Ltd
Xplorer - Apollo Fire Detectors Ltd
XPS200 Plastisol - Colorpro Systems Ltd
Xtraflex - Adaptaflex Ltd
Xtraflo Rainwater System - FloPlast Ltd
Xtralift - Stannah Lifts Ltd
Xtralis - Xtralis
Xtreme - Harkness Screens
XXL Sliding Wall - Parthos UK Ltd

Y

Yale - ASSA ABLOY Limited
Yale - Olby H E & Co Ltd
Yale - Yale Security Products Ltd
YBS - Allmat (East Surrey) Ltd
Yeoman - Gazco

Yeoman Custom mouldings - Harrison Thompson & Co Ltd
Yeoman Formula One seating - Harrison Thompson & Co Ltd
Yeoman Rainguard - Harrison Thompson & Co Ltd
Yeoman Shield total surface protection - Harrison Thompson & Co Ltd
YeomanShield Total Surface Protection - Moffett Thallon & Co Ltd
Ygnette - Wellman Robey Ltd
Ygnis - Wellman Robey Ltd
Ykrlin Automobile Car Turntable - British Turntable Ltd
Yobstopper - Fitzpatrick Doors Ltd
Yoke System - S & B Ltd
York - Architectural Textiles Ltd
York - Johnson Controls Building Efficiency UK Ltd
Yorkdale - Atlas Stone Products
Yorkex - Yorkshire Copper Tube
Yorkshire - The BSS Group
Yorkshire - Yorkshire Copper Tube
Yorkshire stove - Dunsley Heat Ltd
Youngman - Youngman Group Limited

Ytong - Xella Aircrete Systems UK

Z

Zampano - Girsberger London
Zannussi - Electrolux Domestic Appliances
Zaun fencing - Chestnut Products Limited
ZD - Staverton (UK) Ltd
Zedex - Visqueen Building Products
Zeeland - Cembrit Ltd
Zehnder - Diamond Merchants
Zehnder - Greenwood Airvac
Zehnder CoMo - Zehnder Ltd
Zehnder Flatline - Zehnder Ltd
Zehnder Multicolumn - Zehnder Ltd
Zehnder Radiavecter - Zehnder Ltd
Zehnder Stratos - Zehnder Ltd
Zehnder ZBN - Zehnder Ltd
Zehnder ZIP - Zehnder Ltd
Zenith Club Class - Brintons Ltd
Zenith Range; Bollard, Seat, Litter Bin & Ash Waste Bin, Cycle Stand, Signage - Furnitubes International Ltd

Zerodec - Gillespie (UK) Ltd
Zeta - Zon International
Zeta II - Cooper Lighting
Zeta Lock - Brett Landscaping
Zetalite - Cooper Lighting and Security Ltd
ZFP addressable fire panels - C-TEC (Computionics) Ltd
Zig Zag - Zarges (UK) Ltd
Zinco - Alumasc Exterior Building Products Ltd
Zingo - HAGS SMP
Z-Led - Zled (DPC) Co Ltd
Zone: 1 - Youngman Group Limited
Zoneguard - Hoyles Electronic Developments Ltd
Zsolnay - LSA Projects Ltd
Zuccini - Legrand Electric Ltd
ZV1 - K. V. Radiators
ZV2 - K. V. Radiators
Zwick Mail Boxes - JB Architectural Ltd
Zykon - Artur Fischer (UK) Ltd
ZZ System - Quelfire

3d Storage Systems (UK) LTD - Suite 1, 30 Station Road, Ossett, Wakefield, West Yorkshire. WF5 8AD tel:(01924) 240291, fax:(01924) 261677. sales@3dlockers.co.uk, http://www.3dlockers.co.uk

3M - 3M Centre, Cain Road, Bracknell. RG12 8HT tel:(08705) 360036. commcareuk@mmm.com, www.3m.co.uk

A

A N H Refractories Europe Ltd - Dock Road South, Bromborough, Wirral, Merseyside. CH62 4SP tel:(0151) 641 5900, fax:(0151) 641 5910. sales@anheurope.co.uk, www.anhlimited.co.uk

A. F Jones (Stonemasons) - 33 Bedford Road, Reading, Berkshire. RG1 7EX tel:(0118) 957 3537, fax:(0118) 957 4334. info@afjones.co.uk, www.afjones.co.uk

AAC Waterproofing Ltd - Industrial Estate, Gaerwen, Anglesey. LL60 6HR tel:(01248) 421 955, fax:(01248) 421 052. info@prelasti.co.uk, www.prelasti.co.uk

Aalco Metals Ltd - 25 High Street, Cobham, Surrey. KT11 3DH tel:(01932) 250100, fax:(01932) 250101. marketing@amari-metals.com, www.aalco.co.uk

Aardee Security Shutters Ltd - 5 Dalsholm Ave, Dawsholm Industrial Estate, Glasgow. G20 0TS tel:(0141) 810 3444, fax:(0141) 810 3777. sales@aardee.co.uk, www.aardeesecurity.co.uk

Aardvark Transatlantic Ltd - 106 New Road, Ascot. SL5 8QH tel:(01344) 882 314, fax:(01334) 884 506. sweeks7956@aol.com, www.srindustrial.co.uk

Aztec Cubicles - Becklands Close, Bar Lane, Roecliffe, Boroughbridge, N. Yorks. YO51 9NR tel:(01423) 326400, fax:(01423) 325115. sales@aaztec.com, www.aaztec.com

Abacus Lighting Ltd - Oddicroft Lane, Sutton-in-Ashfield, Notts. NG17 5FT tel:(01623) 511111, fax:(01623) 552133. sales@abacuslighting.com, www.abacuslighting.com

Abacus Signs - Unit A6A Garth Works, Taffs Well, Cardiff, Glamorgan. CF15 7YF tel:(029) 20 811315, fax:(029) 20 813899. sales@abacussigns.co.uk, www.abacussigns.co.uk

ABB Ltd - Tower Court, Courtaulds Way, Coventry. CV6 5NX tel:(024) 76 368500, fax:(024) 76 364499. trevor.kirtley@gb.abb.com, www.abb.co.uk

ABB Instrumentation Ltd - Howard Road, Eaton Socon, St. Neots, Huntingdon. PE19 3EU tel:(01480) 475321, fax:(01480) 218361. info@gb.abb.com, www.abb.com

Abbey Nameplates Ltd - Beech Court, 127 Haslemere Road, Liphook, Hampshire. GU30 7BX tel:(0800) 092 3317, fax:(0800) 056 1362. sales@abbey.go-plus.net, www.abbeycraftsmen.co.uk

Abel Alarm Co Ltd - Detection House, 4 Vaughan Way, Leicester. LE1 4ST tel:(0116) 265 4200, fax:(0116) 251 5341. info.leicester@abelalarm.co.uk, www.abelalarm.co.uk

Abet Ltd - 70 Roding Road, London Industrial Park, London. E6 6LS tel:(020) 7473 6910, fax:(020) 7476 6935. sales@abet.ltd.uk, www.abet-ltd.co.uk

ABG Ltd - Unit E7, Meltham Mills, Meltham Mills Road, Meltham, W. Yorks. HD9 4DS tel:(01484) 852096, fax:(01484) 851562. geo@abgltd.com, www.abg-geosynthetics.com

Abloy Security Ltd - 1-3 Hatters Lane, Croxley Business Park, Watford, Herts. WD18 8QY tel:(01923) 255066, fax:(01923) 655001. sales@abloysecurity.co.uk, www.abloysecurity.co.uk

Above All - 178 High Road, Chilwell, Nottingham. NG9 5BB tel:(0115) 925 1959, fax:(0115) 943 1408. service@aboveall.co.uk, www.aboveall.co.uk

Accent Hansen - Greengate Industrial Park, Greengate, Middleton, Gt Man. M24 1SW tel:(0161) 284 4100, fax:(0161) 655 3119. sales@accenthansen.com, www.accenthansen.com

Access Industries Group Ltd - Edgeworth House, 20 High Street, Northchurch, Berkhampstead, Herts. HP4 3LS tel:(01525) 383101, fax:(01442) 878525. sales@accessequipment.net, www.accessequipment.net

Ackermann Ltd - The Arnold Centre, Paycocke Road, Basildon, Essex. SS14 3EA tel:(01268) 563252, fax:(01268) 563437. ackermann.enquiries@honeywell.com, www.ackermann.co.uk

ACO Technologies plc - ACO Business Park, Hitchin Road, Shefford, Beds. SG17 5TE tel:(01462) 816666, fax:(01462) 815895. technologies@aco.co.uk, www.aco.co.uk

Acorn Powell Ltd - 5 Brearley Court, Baird Road Waterwells Business Park, Quedgeley, Gloucester, Gloucestershire. GL2 2AF tel:(01452) 721211, fax:(01452) 721231. sales@acornpowell.co.uk, www.acornpowell.co.uk

Acousticabs Industrial Noise Control Ltd - Unit 52, Heyford Road, Pocklington Ind Estate, Pocklington, York. YO42 1NR tel:(01759) 305266, fax:(01759) 305268. info@acousticabs.com, www.acousticabs.com

ACP (Concrete) Ltd - Risehow Industrial Estate, Firmby, Maryprot, Cumbria. CA15 8PD tel:(01900) 814659, fax:(01900) 816200. sales@acp-concrete.co.uk, www.acp-concrete.co.uk

ACP Concrete Limited - Wood Lane Business Centre, Wood Lane, Uttoxeter, Staffordshire. ST14 8JR tel:(01889) 598660, fax:(01889) 568160. sales@acp-concrete.co.uk, www.acp-concrete.co.uk

Actavo (UK) Ltd - Unit C, Cedar Court Office Park, Denby Dale Road, Calder Grove, Wakefield, West Yorkshire. WF4 3QZ tel:(01924) 416000, fax:(01924) 366250. info@actavo.com, www.actavo.com

Active Carbon Filters Ltd - Unit 4, Vickers Industrial Estate, Mellishaw Lane, Morecambe, Lancs. LA3 3EN tel:(01524) 383200, fax:(01524) 389438. info@activecarbonfilters.com, www.fengroup.com

Adam Carpets Ltd - Greenhill Works, Birmingham Road, Kidderminster, Worcs. DY10 2SH tel:(01562) 829966, fax:(01562) 751741. eprescott@adamcarpets.com, www.adamcarpets.com

Adams- Hydraulics Ltd - PO Box 15, York. YO30 4TA tel:(01904) 695695, fax:(01904) 695600. sales@adamshydraulics.co.uk, www.adamshydraulics.co.uk

Adaptaflex Ltd - Station Road, Coleshill, Birmingham. B46 1HT tel:(01675) 468200, fax:(01675) 464930. sales@adaptaflex.co.uk, www.adaptaflex.com

Addagrip Terraco Ltd - Addagrip House, Bell Lane Industrial Estate, Uckfield, Sussex. TN22 1QL tel:(01825) 761333. sales@addagrip.co.uk, www.addagrip.co.uk

Aden Hynes Sculpture Studios - Unit 3F Harvey Road, Nevendon Industrial Estate, Basildon, Essex. SS13 1DA tel:(01268) 726470. aden.hynes@hotmail.com, www.sculpturestudios.co.uk

Adex Interiors for industry Ltd - 5 Avebury Court, Mark Road, Hemel Hempstead, Herts. HP2 7TA tel:(01442) 232327, fax:(01442) 262713. info@adex.co.uk, www.Adex.co.uk

Adshead Ratcliffe & Co Ltd - Derby Road, Belper, Derbyshire. DE5 1WJ tel:(01773) 826661, fax:(01773) 821215. arbo@arbo.co.uk, www.arbo.co.uk

ADT Fire and Security Plc - Security House, The Sunmmitt, Hanworth Road, Sunbury-on-Thames, Middx. TW16 5DB tel:(01932) 743333, fax:(01932) 743155. tspemealicense@tycoint.com, www.adt.co.uk

Advance Tapes International Ltd - P.O Box 122, Abbey Meadows, Leics. LE4 5RA tel:(0116) 251 0191, fax:(0116) 2652046. sales@advancetapes.com, www.advancetapes.com

Advanced Air (UK) Ltd - Burell Way, Thetford, Norfolk, Suffolk. IP24 3WB tel:(01284) 701356, fax:(01284) 701357. sales@advancedair.co.uk, www.advancedair.co.uk

Advanced Fencing Systems Ltd - Park Farm, Park Avenue, Dronfield, Derbyshire. S61 2DN tel:(0114) 2891891, fax:(0114) 2891892. office@afsfencing.co.uk, www.advancedfencingsystems.com

Advanced Hygienic Contracting Ltd - Hammerain House, Hookstone Avenue, Harrogate, N. Yorks. HG2 8ER tel:(01423) 870049, fax:(01423) 870051. advanced@cladding.co.uk, www.cladding.co.uk

Advanced Interior Solutions Ltd - The Old Registry, 20 Amersham Hill, High Wycombe, Buckinghamshire. HP13 6NZ tel:(01494) 429 884. solutions@advancedinteriors.co.uk, www.advancedinteriors.net

Advanced Perimeter Systems Ltd - 16 Cunningham Road, Springkerse Ind Est, Stirling. FK7 7TP tel:(01786) 479862, fax:(01786) 470331. admin@apsltd.net, www.apsltd.net

Advisory, Conciliation and Arbitration Service - Euston Tower, 286 Euston Road, London. NW1 3JJ tel:(08457) 474747. www.acas.org.uk

AEL - 4 Berkeley Court, Manor Park, Runcorn, Cheshire. WA7 1TQ tel:(01928) 579068, fax:(01928) 579523. sales@aelheating.com, www.aelheating.com

Aercon Consultants Ltd - Aercon Works, Alfred Road, Gravesend, Kent. DA11 7QF tel:(01268) 418822, fax:(01268) 418822

AET.gb Ltd - 201 Solent Business Centre, Millbrook Road West, Southampton, Hants. SO15 0HW tel:(08453) 700400, fax:(08453) 700401. sales@aet.gb.com, www.aet.gb.com

Africa Roofing Uk - Sunnyhills Road, Barnfields, Staffordshire. ST13 5RJ tel:(01538) 398488, fax:(01538) 398456. sales@africaroofinguk.co.uk, www.africaroofinguk.co.uk

AGC Glass UK Ltd - 6 Allerton Road, Rugby, Warwickshire. CV23 0PA tel:(01788) 535353, fax:(01788) 560853. andrea.marston@eu.agc.com, www.yourglass.com

AGCO Ltd - Abbey Park, Stoneleigh, Kenilworth. CV8 2TQ tel:(024) 76 852 164, fax:(024) 76 851 172. shop@masseyferguson.com, www.agcocorp.com

Aggregate Industries UK Ltd - Hulland Ward, Smith Hall Lane, Ashbourne, Derbyshire. DE6 3ET tel:(01335) 372222, fax:(01335) 370074. sales@aggregate.com, www.aggregate.com

Air Control Industries Ltd - Weycroft Avenue, Millwey Rise Industrial Estate, Axminister, Devon. EX13 5HU tel:(01297) 529 242, fax:(01297) 529 241. sales@aircontrolindustries.com, www.aircontrolindustries.com

Air Diffusion Ltd - Stourbridge Road, Bridgnorth, Shropshire. WV15 5BB tel:(01746) 761921, fax:(01746) 760127. sales@air-diffusion.co.uk, www.air-diffusion.co.uk

Aircare Europe Ltd - GVS Filter Technology UK, NFC House, Vickers Industrial Estate, Lancashire. LA3 3EN tel:(08707) 445588, fax:(01925) 850325. gvsuk@gvs.com, www.aircareeurope.com

Airdri Ltd - Technology House, Oakfield Industrial Est, Eynsham, Oxford. OX29 4AQ tel:(01865) 882330, fax:(01865) 881647. sales@airdri.com, www.airdri.com

Airedale International Air Conditioning Ltd - Leeds Road, Rawdon, Leeds. LS19 6JY tel:(0113) 239 1000, fax:(0113) 250 7219. enquiries@airedale.com, www.Airedale.co.uk

Airflow (Nicoll Ventilators) Ltd - Unit 3 Hamilton Business Park, Gore Road Industrial Estate, New Milton, Hants. BH25 6AX tel:(01425) 611547, fax:(01425) 638912. sales@airflow-vent.co.uk, www.airflow-vent.co.uk

Airflow Developments Ltd - Lancaster Road, Cressex Business Park, High Wycombe, Bucks. HP12 3QP tel:(01494) 525252, fax:(01494) 461073. info@airflow.co.uk, www.airflow.co.uk

Airsprung Beds Ltd - Canal Road, Trowbridge, Wilts. BA14 8RQ tel:(01225) 779101, fax:(01225) 779123. contracts@airsprungbeds.co.uk, www.airsprungbeds.co.uk

Akzo Noble Coatings Ltd - Wexham Road, Slough, Berkshire. SL5 5DS tel:(08444) 817818. sikkens.advice@akzonobel.com, www.sikkens.co.uk

AkzoNobel Decorative Paints UK - Wexham Road, Slough, Berks. SL2 5DS tel:(01753) 691690, fax:(01753) 530336. john.ashford@akzonobel.com, www.duluxtrade.co.uk

Alanod Ltd - Chippenham Drive, Kingston, Milton Keynes. MK10 0AN tel:(01908) 282044, fax:(01908) 282032. alanod@alanod.co.uk, www.alanod.co.uk

Albany Engineering Co Ltd - Church Road, Lydney, Glos. GL15 5EQ tel:(01594) 842275, fax:(01594) 842574. sales@albany-pumps.co.uk, www.albany.pumps.co.uk

Albany Standard Pumps - Richter Works, Garnett Street, Bradford, W. Yorks. BD3 9HB tel:(01274) 725351, fax:(01274) 742467. sales@albany-pumps.co.uk, www.albany-pumps.co.uk

Albion Manufacturing - The Granary, Silfield Road, Wymondham, Norfolk. NR18 9AU tel:(01953) 605983, fax:(01953) 606764. sales@albionmanufacturing.com, www.albionmanufacturing.com

Alcoa Custom Extrudes Solutions - Southam Road, Banbury, Oxon. OX16 7SN tel:(01295) 454545, fax:(01295) 454683. kuk.kawneer@alcoa.com, www.alcoa.com

ALD Lighting Solutions - Unit 6E, Southbourne Business Park, Courtlands Road, Eastbourne, East Sussex. BN22 8UY tel:(01323) 729337, fax:(01323) 732356. sales@aldlighting.com, www.aldlighting.com

Aldous & Stamp (Services) Ltd - 86-90 Avenue Road, Beckenham, Kent. BR3 4SA tel:(020) 8659 1833, fax:(020) 8676 9676. sales@aldous-stamp.co.uk, www.aldous-stamp.co.uk

Alifabs (Woking) Ltd - Kernel Court, Walnut Tree Close, Guildford, Surrey. GU1 4UD tel:(01483) 546547, fax:(01483) 546548. info@alifabs.com, www.alifabs.com

Alitex Ltd - Torbery Farm, South Harting, Petersfield. GU31 5RG tel:(01730) 826900, fax:(01730) 826901. enquiries@alitex.co.uk, www.alitex.co.uk

Allart, Frank, & Co Ltd - 15-35 Great Tindal Street, Ladywood, Birmingham. B16 8DR tel:(0121) 410 6000, fax:(0121) 410 6066. sales@allart.co.uk, www.allart.co.uk

Allaway Acoustics Ltd - 1 Queens Road, Hertford. SG14 1EN tel:(01992) 550825, fax:(01992) 554982. enquiries@allawayacoustics.co.uk, www.allawayacoustics.co.uk

Allegion (UK) Ltd - Bescot Crescent, Walsall, West Midlands. WS1 4DL tel:(01922) 707400, fax:(0191) 335 2020. info@allegion.com, www.allegion.co.uk

Allen (Concrete) Ltd - 38 Willow Lane, Micham, Surrey. CR4 4NA tel:(020) 86872222, fax:(020) 86875400. sales@allenconcrete.co.uk, www.allenconcrete.co.uk

Allen (Fencing) Ltd - Birch Walk, West Byfleet, Surrey. KT14 6EJ tel:(01932) 349607, fax:(01932) 354868. sales@allenfencing.co.uk, www.allenfencing.co.uk

Allermuir Contract Furniture Ltd - Altham Business Park, Accrington, Lancs. BB5 5YE tel:(01254) 682421, fax:(01254) 673793. sales@allermuir.com, www.allermuir.co.uk

Allgood plc - 297 Euston Road, London. NW1 3AQ tel:(020) 7387 9951, fax:(020) 7380 1232. info@allgood.co.uk, www.allgood.co.uk

Allman Fasteners Ltd - PO Box 5, Wilmslow, Cheshire. SK9 2EF tel:(01625) 537535, fax:(01625) 537635. ianallman@allman2000.fsnet.co.uk

Allmat (East Surrey) Ltd - Kenley Treatment Works, Godstone Road, Kenley, Surrey. CR8 5AE tel:(020) 8668 6666, fax:(020) 8763 2110. info@allmat.co.uk, www.allmat.co.uk

Allsebrook Pump and Electrical Services Limited - Unit 10, Vanalloys Business Park, Busgrove Lane, Stoke Row, Henley-on-Thames, Oxon. RG9 5QW tel:(01491) 680628, fax:(01491) 682318. sales-allsebrook@btconnect.com, www.allsebrookservices.co.uk

Alltype Fencing Specialists - Ye Wentes Wayes, High Road, Langdon Hills, Basildon, Essex. SS16 6HY tel:(01268) 545192. sales@alltypefencing.com, www.alltypefencing.com

Alno (United Kingdom) Ltd - Unit 1, North Downs Business Park, Dunton Green, Kent. TN13 2TL tel:(01732) 464600, fax:(01732) 462288. info@alnocontracts.co.uk, www.alnocontracts.co.uk

Alpha Mosaic & Terrazzo Ltd - Unit 2, Munro Drive, Cline Road, London. N11 2LZ tel:(020) 8368 2230, fax:(020) 8361 8720. dipesh@romamarble.co.uk

Alpha Therm Ltd - Nepicar House, London Road, Wrotham Heath, Sevenoaks, Kent. TN15 7RS tel:(01732) 783 000, fax:(0844) 871 8762. info@alpha-innovation.co.uk, www.alpha-innovation.co.uk

Altecnic Ltd - Mustang Drive, Stafford, Staffordshire. ST16 1GW tel:(01785) 218200, fax:(01785) 218201. sales@altecnic.co.uk, www.altecnic.co.uk

Althon Ltd - Vulcan Road South, Norwich. NR6 6AF tel:(01603) 488700, fax:(01603) 488598. sales@althon.co.uk, www.althon.net

Altro Ltd - Head Office, Works Road, Letchworth, Herts. SG6 1NW tel:(01462) 480480, fax:(01462) 480010. enquiries@altro.co.uk, www.altro.co.uk

Aluline Ltd - Harbour House, 1 Aldborough Street, Blyth, Northumberland. NE24 2EU tel:(01670) 544322, fax:(01670) 544340. enquiry@aluline.co.uk, www.aluline.co.uk

Alumasc Exterior Building Products Ltd - White House Works, Bold Road, Sutton, St. Helens, Merseyside. WA9 4JG tel:(01744) 648400, fax:(01744) 648401. info@alumasc-exteriors.co.uk, www.alumasc-exteriors.co.uk

Alumasc Interior Building Products Limited - Halesfield 19, Telford, Shropshire. TF7 4QT tel:(01952) 580590, fax:(01952) 587805. sales@pendock.co.uk, www.pendock.co.uk

Aluminium Federation Ltd - National Metalforming Centre, 47 Birmingham Road, West Bromwich, West Midlands. B70 6PY tel:(0121) 6016363, fax:(0870) 138 9714. alfed@alfed.org.uk, www.alfed.org.uk

Aluminium RW Supplies Ltd - Ryan House, Unit 6, Dumballs Road, Cardiff. CF10 5DF tel:(029) 20 390576, fax:(029) 20 238410. sales@arwsltd.com, www.arwsltd.com

Alutec - Unit 1 (G-H), Elms Farm Industrial Estate, Bedford. MK41 0LZ tel:(01234) 359438, fax:(01234) 357199. enquiries@marleyalutec.co.uk, www.marleyalutec.co.uk

A (con't)

Amadeus - Great Beech Barn, Kane Hythe Road, Battle, East Sussex. TN33 9QU tel:(01424) 775867. info@amadeus-equipment.co.uk, http://www.amadeus-equipment.co.uk/

Amari Plastics Plc - Holmes House, 24-30 Baker Street, Weybridge, Surrey. KT13 8AU tel:(01932) 835000, fax:(01932) 835002. wl@amariplastics.com, www.amariplastics.com

Ambec Fencing - Hall Lane, Farnworth, Bolton, Lancashire. BL4 7QF tel:(01204) 574 011, fax:(01204) 861 653. sales@ambecfencing.co.uk, http://ambecfencing.co.uk/

Amber Doors Ltd - Mason Way, Platts Common Industrial Estate, Hoyland, Barnsley, S. Yorks. S74 9TG tel:(01226) 351135, fax:(01226) 350176. enquiries@rollershutterservices.co.uk, www.amber-doors.co.uk

Amberol Ltd - The Plantation, King Street, Alfreton, Derbys. DE55 7TT tel:(01773) 830930, fax:(01773) 834191. info@amberol.co.uk, www.amberol.co.uk

AmbiRad Ltd - Fens Pool Avenue, Wallows Industrial Estate, Brierley Hill, W. Midlands. DY5 1QA tel:(01384) 489739, fax:(01384) 489707. sales@ambirad.co.uk, www.ambirad.co.uk

Amdega Ltd - Faverdale, Darlington, Co. Durham. DL3 0PW tel:(01325) 468522, fax:(01325) 489209. info@amdega.co.uk, www.amegda.co.uk

Amilake Southern Ltd - Penrose Road, Ferndown, Dorset. BH22 9FJ tel:(07857) 603031, fax:(01202) 891 716. info@amilakesouthern.co.uk, www.amilakesouthern.co.uk

AMK Fence-In Limited - Wallace Road, Parkwood Springs, Sheffield, Yorks. S3 9SR tel:(0114) 2739372, fax:(0114) 2739373. amkfencein@aol.com, www.amkfence-in.co.uk

AMO Blinds - Exhibition Centre, Leeds Road, Liversedge, W. Yorks. WF15 6JB tel:(01924) 410170, fax:(01924) 410170. showroom@amoblinds.co.uk, www.amoblinds.co.uk

Amorim (UK) Ltd - Unit 9, Horsham Court, City Business Centre, Horsham, W. Sussex. RH13 5BB tel:(01403) 750387, fax:(01403) 230124. general.auk@amorim.com, www.amorim.com

Amwell Systems Ltd - Ground Floor, Suite 2, Middlesex House, Meadway Corporate Centre, Stevenage, Hertfordshire. SG21 2EF tel:(01763) 276200, fax:(01763) 276222. contact@amwell-systems.com, http://www.amwell-systems.com

Ancon Building Products - President Way, President Park, Sheffield, S. Yorks. S4 7UR tel:(0114) 275 5224, fax:(0114) 276 8543. info@ancon.co.uk, www.ancon.co.uk

Ancorite Surface Protection Ltd - Millbuck Way, Ettiley Heath, Sandbach, Cheshire. CW11 3AB tel:(01270) 761720, fax:(01270) 761697. david.clegg@ancorite.co.uk, www.ancorite.co.uk

Anders + Kern UK Ltd - Norderstedt House, James Carter Road, Mildenhall, Suffolk. IP28 7RQ tel:(01638) 510900, fax:(01638) 510901. sales@anders-kern.co.uk, www.anders-kern.co.uk

Andersen/ Black Millwork - Andersen House, Dallow Street, Burton-on-Trent, Staffs. DE14 2PQ tel:(01283) 511122, fax:(01283) 510863. info@blackmillwork.co.uk, www.blackmillwork.co.uk

Anderson C F & Son Ltd - 228 London Road, Marks Tey, Colchester, Essex. CO6 1HD tel:(01206) 211666, fax:(01206) 212450. info@cfanderson.co.uk, www.cfanderson.co.uk

Anderson Gibb & Wilson - A Division of Charles Tennant & Co Ltd, 543 Gorgie Road, Edinburgh. E11 3AR tel:(0131) 443 4556, fax:(0131) 455 7608. mike.grace@charlestennant.co.uk

Anderton Concrete Products Ltd - Anderton Wharf, Soot Hill, Anderton, Northwich, Cheshire. CW9 6AA tel:(01606) 79436, fax:(01606) 871590. sales@andertonconcrete.co.uk, www.andertonconcrete.co.uk

Andrew Moor Associates - 14 Chamberlain Street, London. NW1 8XB tel:(020) 7586 8181, fax:(020) 7586 8484. andrew@andrewmoor.co.uk, www.andrewmoor.com

Andrews Coatings Ltd - Carver Building, Littles Lane, Wolverhampton, W. Midlands. WV1 1JY tel:(01902) 429190, fax:(01902) 426574. info@andrewscoatings.co.uk, www.antigraffiti.co.uk

Andrews Sykes Hire Ltd - Premier House, Darlington Street, Wolverhampton, W. Midlands. WV1 4JJ tel:(01902) 328700, fax:(01902) 422466. info@andrews-sykes.com, www.andrew-sykes.com

Andura Coatings Ltd - 20 Murdock Road, Bicester, Oxon. OX26 4PP tel:(01869) 240374, fax:(01869) 240375. sales@andura.co.uk, www.andura.com

Andy Thornton Ltd - Rosemount, Huddersfield Road, Elland, West Yorkshire. HX5 0EE tel:(01422) 376000, fax:(01422) 376060. marketing@andythornton.com, www.andythornton.com

Anglepoise Lighting Ltd - Unit 51, Enfield Industrial Area, Redditch, Worcs. B97 6DR tel:(01527) 63771, fax:(01527) 61232. jt@anglepoise.co.uk, www.anglepoise.co.uk

Anglia Lead Ltd - 49 Barker Street, Norwich, Norfolk. NR2 4TN tel:(01603) 626856, fax:(01603) 619171. info@alr-ltd.co.uk, www.alr-ltd.co.uk

Anglian Building Products - 59 Hurricane Way, Norwich, Norfolk. NR6 6JB tel:(01603) 428455, fax:(01603) 420500. Stephen.Preece@Angliangroup.com, www.anglianhome.co.uk

Anglo Building Products Ltd - Branksome House, Filmer Grove, Godalming, Surrey. GU7 3AB tel:(01483) 427777, fax:(01483) 428888. r.nivison@rust-oleum.eu, www.anglobuild.co.uk

Angus Fire - Thame Park Road, Thame, Oxon. OX9 3RT tel:(01844) 265000, fax:(01844) 265156. general.enquiries@angusuk.co.uk, www.angusfire.co.uk

Anixter (UK) Ltd - 1 York Road, Uxbridge, Middx. UB8 1RN tel:(01895) 818181, fax:(01895) 818182. contact.europe@anixter.com, www.anixter.co.uk

Anson Concise Ltd - 1 Eagle Close, Arnold. NG5 7FJ tel:(0115) 926 2102, fax:(0115) 967 3398. info@ansonconcise.co.uk, www.ansonconcise.co.uk

Anthony de Grey Trellises + Garden Lighting - Broadhinton Yard, 77a North Street, London. SW4 0HQ tel:(020) 7738 8866, fax:(020) 7498 9075. sales@anthonydegrey.com, www.anthonydegrey.com

Anti-Climb Guards Ltd - PO Box 51, Edenbridge, Kent. TN8 6WY tel:(01732) 865901, fax:(01732) 867567. acg@vandgard.co.uk, www.vandgard.co.uk

Antocks Lairn Ltd - Ness Furniture, 31 St. Cuthberts Way, Aycliffe Industrial Estate, Newton Aycliffe, County Durham. DL5 6XW tel:(01388) 816109. sales@antocks.co.uk, www.antocks.co.uk

APA - Youlditch Barns, Peter Tavy, Tavistock, Devon. PL19 9LY tel:(01822) 810187/8, fax:(01822) 810189. sales@alanpow.co.uk, www.alanpow.co.uk

Aperam Stainless Services & Solutions UK Limited - 9 Midland Way, Barlborough Links, Barlborough, Derbyshire. S43 4XA tel:(01246) 571660, fax:(01246) 571 661. kevin.jones@aperam.com, www.aperam.com

Apollo Fire Detectors Ltd - 36 Brookside Road, Havant, Hants. PO9 1JR tel:(023) 92492412, fax:(023) 9249 2754. enquiries@apollo-fire.co.uk, www.apollo-fire.co.uk

Applied Energy Products Ltd - Applied Energy Products Ltd, Morley Way, Peterborough, Cambs. PE2 9JJ tel:(01733) 456789, fax:(01733) 310606. Aeinfo@applied-energy.com, www.applied-energy.com

Applied Felts Ltd - Castle Bank Mills, Portobello Road, Wakefied, W. Yorks. WF1 5PS tel:(01924) 200535, fax:(01924) 366951. sales@appliedfelts.co.uk, www.appliedfelts.com

APS Masonry Ltd - Osney Mead, Oxford. OX2 0EQ tel:(01865) 254600, fax:(01865) 254935. enquiries@apsmasonry.com, www.apsmasonry.com

APT Controls Ltd - The Power House, Chantry Place, Headstone Lane, Harrow, Middx. HA3 6NY tel:(020) 8421 2411, fax:(020) 8421 3951. info@aptcontrols.co.uk, www.aptcontrols.co.uk

APW Ltd - Unit 12 Deacon Trading Estate, Earle Street, Newton-Le-Willows, Merseyside. WA12 9XD tel:(01925) 295577, fax:(01925) 295588. sales@apw.co.uk, www.apw.co.uk

Aqua-Blue Ltd - The Studio, 47 Flexford Close, Chandlers Ford, Hampshire. SO53 5RY tel:(02380) 260888, fax:(02380) 260888. aqua.blue@ntlbusiness.com, www.aquabluedesigns.net

Aquacontrol - PO Box 171, Abingdon, Oxon. OX14 4DJ tel:(01865) 407480, fax:(01865) 407480. aquacontrol@btopenworld.com

Aqualisa Products - The Flyer's Way, Westerham, Kent. TN16 1DE tel:(01959) 560000, fax:(01959) 560030. marketing@aqualisa.co.uk, www.aqualisa.co.uk

Aquasentry - Unit 2 Britannia Mills, Gelderd Road, Birstall, West Yorkshire. WF17 9QD tel:(01924) 284900, fax:(01924) 284911. info@aquasentry.co.uk, www.aquasentry.co.uk

Aquastat Ltd - Unit 11 River Road Business Park, 33 River Road, Barking, Essex. IG11 0DA tel:(020) 8591 3433, fax:(020) 8591 2113. enquiries@aquastat.co.uk, www.aquastat.co.uk

AquaTech Ltd - AGM House, London Road, Copford, Colchester. CO6 1GT tel:(01206) 215121, fax:(01206) 215131. info@aqua-techuk.co.uk, www.aquatech.co.uk

Arbory Group Limited - Holker Business Centre, Burnley Road, Colne, Lancs. BB8 8EG tel:(01254) 394417, fax:(01254) 382278. sales@trojanproducts.co.uk, www.trojanproducts.co.uk

Arc Specialist Engineering Limited - Rectory Court, Old Rectory Lane, Birmingham, W. Midlands. B48 7SX tel:(0121) 285 1560, fax:(0121) 285 1570. info@arcspecialist.co.uk, www.arcspecialist.co.uk

Arch Timber Protection - Wheldon Road, Castleford, W. Yorks. WF10 2JT tel:(01977) 71400, fax:(01977) 714001. advice@archchemicals.com, www.archtp.com

Architectural Association - 36 Bedford Square, London. WCIB 3ES tel:(0207) 8874090. development1@aaschool.ac.uk, www.aaschool.ac.uk

Architectural Plastics (Handrail) Ltd - Unit One, 2 Robert Street, Harrogate, N. Yorks. HG1 1HP tel:(01423) 561852, fax:(01423) 520728. architecturalplastics@hotmail.co.uk, www.architecturalplastics.co.uk

Architectural Textiles Ltd - 25 Moulton Road, Gazeley, Suffolk. CB8 8RA tel:(01638) 751970, fax:(01638) 750520. sales@architecturaltextiles.co.uk, www.architecturaltextiles.co.uk

Architen Landrell Associates Ltd - Station Road, Chepstow, Monmouthshire. NP16 5PF tel:(01291) 638200, fax:(01291) 621991. mail@architen.com, www.architen.com

Ardex UK Ltd - Homefield Road, Haverhill, Suffolk. CB9 8QP tel:(01440) 714939, fax:(01440) 716660. info@ardex.co.uk, www.ardex.co.uk

Aremco Products - Foxoak Street, Cradley Heath, W. Midlands. B64 5DQ tel:(01384) 568566, fax:(01384) 638919. sales@aremco-products.co.uk, www.aremco-products.co.uk

Ariel Plastics Ltd - Speedwell Ind. Est, Staveley, Derbys. S34 3JP tel:(01246) 281111, fax:(01246) 561111. info@arielplastics.com, www.arielplastics.com

ArjoHuntleigh UK - ArjoHuntleigh House, Houghton Hall Business Park, Houghton Regis, Bedfordshire. LU5 5XF tel:(01582) 745 700, fax:(01582) 745 745. sales.admin@ArjoHuntleigh.com, www.ArjoHuntleigh.co.uk

Arkema France - 429 rue d'Estienne d'Orves, 92705 Colombes Cedex. FRANCE tel:(0121) 7434811, fax:(0121) 7434811. ken.griffiths@arkema.com, www.arkema.com

Arkinstall Galvanizing Ltd - 38 Coventry Street, Birmingham, W. Midlands. B5 5NQ tel:(0121) 643 6455, fax:(0121) 643 0192. info@galvanizing.co.uk, www.galvanizing.co.uk

Armes, Williams, Ltd - Armes Trading Estate, Cronard Road, Sudbury, Suffolk. CO10 6XB tel:(01787) 372988, fax:(01787) 379383. sales@william-armes.co.uk, www.william-armes.co.uk

Armfibre Ltd - Wilstead Industrial Park, Wilstead, Bedfordshire. MK45 3PD tel:(01234) 741 444, fax:(01767) 651901. south@productionglassfibre.co.uk, www.productionglassfibre.co.uk

Armitage Shanks Ltd - Armitage, Rugeley, Staffs. WS15 4BT tel:(0870) 122 8822, fax:(01543) 413297. info@thebluebook.co.uk, www.thebluebook.co.uk

Armorex Ltd - Riverside House, Bury Road, Lavenham, Suffolk. CO10 9QD tel:(01787) 248482, fax:(01787) 248277. enquiries@uk.sika.com, www.armorex.com

Armstrong (Concrete Blocks), Thomas, Ltd - Whinfield Industrial Estate, Rowlands Gill, Tyne & Wear. NA39 1EH tel:(01207) 544214, fax:(01207) 542761. blocks@thomasarmstrong.co.uk, www.xlomasarmstrong.co.uk

Armstrong (Timber), Thomas, Ltd - Workington Road, Flimby, Maryport, Cumbria. CA15 8RY tel:(01900) 68226, fax:(01900) 870800. timber@thomasarmstrong.co.uk, www.thomasarmstrong.co.uk

Armstrong Floor Products UK Ltd - Hitching Court, Abingdon Business Park, Abingdon, Oxon. OX14 1RB tel:(01235) 554 848, fax:(01235) 553 583. sales-support@armstrong.com, www.armstrong-europe.com

Armstrong Integrated Ltd - Wenlock Way, Manchester. M12 5JL tel:(0161) 223 2223, fax:(0161) 220 9660. salesuk@armlink.com, www.armstrongpumps.com

Armstrong World Industries Ltd - Armstrong House, 38 Market Square, Uxbridge, Middlesex. UB8 1NG tel:(0800) 371849, fax:(01895) 274287. sales_support@armstrong.com, www.armstrong-ceilings.co.uk

Arnold Wragg Ltd - Unit 2, parkway, Parkway Drive, Sheffield, S. Yorks. S9 4WU tel:(0114) 2519050, fax:(0114) 2446635. sales@arnold-wragg.com, www.stag-aerospace.com

Arnull Bernard J & Co Ltd - Unit 10, Trade City, Avro Way, Brooklands, Weybridge. KT13 0YF tel:(01932) 341078, fax:(01932) 352747. bernard.arnull@easynet.co.uk, www.bernardarnull.co.uk

Arrow Supply Company Ltd - Sunbeam Road, Woburn Road Industrial Estate, Kempston, Beds. MK42 7BZ tel:(01234) 840404, fax:(01234) 840374. information@arrow-supply.co.uk, www.arrow-supply.co.uk

Artex Ltd - Artex Avenue, Newhaven, Sussex. BN9 9DD tel:(0800) 0326345. bgtechnical.enquiries@bpb.com, www.artexltd.com

Artificial Grass Ltd - Tavistock Works, Glasson Industrial Estate, Maryport, Cumbria. CA15 8NT tel:(01900) 811970, fax:(01900) 817605. sales@artificial-grass.com, www.artificial-grass.com

Artistic Plastercraft Ltd - Lyndhurst Studios, 16-18 Lyndhurst Road, Oldfield Park, Bath, Avon. BA2 3JH tel:(01225) 315404, fax:(01225) 315404. enquiries@artisticplastercraft.co.uk, www.artisticplastercraft.co.uk

Arts Council England - The Hive, 49 Lever Street, Manchester. M1 1FN tel:(0845) 300 6200. foi@artscouncil.org.uk, www.artscouncil.org.uk

Artur Fischer (UK) Ltd - Whiteley Road, Hithercroft Trading Est, Willingford, Oxon. OX10 9AT tel:(01491) 827900, fax:(01491) 827953. info@fischer.co.uk, www.fischer.co.uk

Arundel Stone Ltd - 62 Aldwick Road, Bognor Regis. PO21 2PE tel:(01243) 829151, fax:(01243) 860341. info@arundelstone.co.uk, www.arundelstone.co.uk

Arvin Motion Control Ltd - 15 New Star Road, Leicester, Leics. LE4 9JD tel:(0116) 274 3600, fax:(0116) 274 3620. tracy.ayres@camloc.com, www.camloc.com

AS Newbould Ltd - 19 Tarran Way West, Tarran Way Industrial Estate, Moreton, Wirral. CH46 4TT tel:(0151) 677 6906, fax:(0151) 678 0680. sales@newbould-joinery.co.uk, www.newbould-joinery.co.uk

Asbestos Removal Contractors' Association - Arca House, 237 Branston Road, Burton-on-Trent, Staffs. DE13 0BY tel:(01283) 566467, fax:(01283) 505770. info@arca.org.uk, www.arca.org.uk

ASCO Extinguishers Co. Ltd - Melisa House, 3 Festival Court, Brand Street, Glasgow. G51 1DR tel:(0141) 427 1144, fax:(0141) 427 6644. customer.service@asco.uk.com, www.asco.uk.com

Ascot Doors Ltd - Brittania Way Industrial Park, Union Road, Bolton, Greater Manchester. BL2 2HE tel:(01204) 545801, fax:(01204) 545800. sales@ascotdoors.co.uk

Ash & Lacy Building Products Ltd - Bromford Lane, West Bromwich, West Midlands. B70 7JJ tel:(0121) 525 1444, fax:(0121) 524 3444. info@ashandlacy.com, www.ashandlacy.com

Ashton Seals Ltd - PO Box 269, Cortonwood Drive, Cortonwood Business Park, Wombwell, S. Yorks. S73 0YP tel:(01226) 273700, fax:(01226) 756774. sales@ashton-group.co.uk, www.ashton-group.co.uk

ASSA ABLOY Limited - School Street, Willenhall, W. Midlands. WV13 3PW tel:(01902) 364500, fax:(01902) 364501. info@uniononline.co.uk, www.assaabloy.co.uk

ASSA Ltd - 75 Sumner Road, Croydon, Surrey. CR0 3LN tel:(020) 8688 5191, fax:(020) 8688 0285. sales@assa.co.uk, www.assa.co.uk

Asset International Ltd - Stephenson Street, Newport, S. Wales. NP19 4XH tel:(01633) 637505, fax:(01633) 290519. sales@multiplate.co.uk, www.assetint.co.uk

Associated Perforators & Weavers Ltd - 75, Church Street, Warrington, Cheshire. WA1 2SR tel:(01925) 632402, fax:(01925) 413810. sales@apw.co.uk, www.apw.co.uk

Association for Consultancy and Engineering (ACE) - Alliance House, 12 Caxton Street, London. SW1H 0QL tel:(020) 7222 6557, fax:(020) 7222 0750. consult@acenet.co.uk, www.acenet.co.uk

Association of Concrete Industrial Flooring Contractors (ACIFC) - 6-8 Bonhill Street, London. EC2A 4BX tel:(0844) 2499176, fax:(0844) 2499177. info@acifc.org, www.acifc.org

Association of Cost Engineers - Administrative Office, Lea House, 5 Middlewich Road, Sandbach, Cheshire. CW11 1XL tel:(01270) 764798, fax:(01270) 766180. enquiries@ACostE.org.uk, www.acoste.org.uk

Association of Interior Specialists (AIS) - Olton Bridge, 245 Warwick Road, Solihull, W. Midlands. B92 7AH tel:(0121) 707 0077, fax:(0121) 706 1949. info@ais-interiors.org.uk, www.ais-interiors.org.uk

Association of Plumbing and Heating Contractors - 12 The Pavilions, Cranmore Drive, Solihull. B90 4SB tel:(0121) 711 5030, fax:(0121) 705 7871. info@aphc.co.uk, www.aphc.co.uk

Association of Sealant Applicators Ltd (ASA) - 8 Westwood Road, Canvey Island, Essex. SS8 0ED tel:(07557) 650625, fax:(01268) 511247. arichardson.asa@hotmail.co.uk, www.associationofsealantapplicators.org

Association of Specialist Fire Protection (ASFP) - Kingsley House, Ganders Business Park, Kingsley, Bordon, Hampshire. GU35 9LU tel:(01420) 471612, fax:(01420) 471611. info@asfp.org.uk, www.asfp.org.uk

Astracast - Woodlands, Roydsdale Way, Euroway Trading Estate, Bradford, West Yorkshire. BD4 6SE tel:(01274) 654700, fax:(01274) 654176. sales@astracast.co.uk, www.astracast.com/uk/

Astracast PLC - Woodlands, Roydsdale Way, Bradford, West Yorkshire. BD4 6SE tel:(01274) 654700. sales@astracast.co.uk, www.astracast.com/uk/

Astro Lighting Ltd - 21 River Way, Mead Ind Park, Harlow, Essex. CM20 2SE tel:(01279) 427001, fax:(01279) 427002. sales@astrolighting.co.uk, www.astrolighting.co.uk

Astrofade Ltd - Kyle Road, Gateshead, Tyne & Wear. NE8 2YE tel:(0191) 420 0515, fax:(0191) 460 4185. info@astrofade.com, www.astrofade.co.uk

Astron Buildings Parker Hannifin PLC - Tachbrook Park Drive, Tachbrook Park, Warwick, Warwicks. CV34 6TU tel:(01926) 888080, fax:(01926) 885088. astron.uk@btinternet.com, www.astron-buildings.com

Atkinson & Kirby Ltd - Atkinson Road, Ormskirk, Lancs. L39 2AJ tel:(01695) 573234, fax:(01695) 586902. sales@akirby.co.uk, www.akirby.co.uk

Atlas Solar - 157 Buslingthorpe Lane, Leeds. LS7 2DQ tel:(0800) 980 8939. solar@sundwel.com, www.sundwel.com

Atlas Stone Products - Westington Quarry, Chipping Campden, Glos. GL55 6EG tel:(01386) 841104, fax:(01386) 841356. sales@atlasstone.co.uk, www.atlasstone.co.uk

A (con't)

Atlas Ward Structures Ltd - Sherburn, Malton, N. Yorks. YO17 8PZ tel:(01944) 710421, fax:(01944) 710759. enquiries@atlasward.com, www.atlasward.com

Attwater & Sons Ltd - Hopwood Street Mills, Preston, Lancs. PR1 1TH tel:(01772) 258245, fax:(01772) 203361. info@attwater.com

Auchard Development Co Ltd - Old Road, Southam, Leamington Spa, Warwicks. CV33 0HP tel:(01926) 812419, fax:(01926) 817425. sales@auchard.co.uk, www.auchard.co.uk

AudiocomPendax Ltd - 57 Suttons Park Avenue, Reading, Berkshire. RG6 1AZ tel:(0118) 966 8383, fax:(0118) 966 8895. info@aupx.com, ww.audicompendax.com

Auld Valves Ltd - Cowlairs Industrial Estate, Finlas Street, Glasgow, Scotland. G22 5DQ tel:(0141) 557 0515, fax:(0141) 558 1059. sales@auldvalves.com, www.auldvalves.com

Aura - Unit 5B, Segensworth Business Centre, Fareham, Hampshire. PO15 5RQ tel:(0845) 6522420, fax:(0845) 6522425. info@auracustom.com, www.auracanopies.com

Autobar Vending Services Limited - Apollo House, Odyssey Business Park, West End Road, Ruislip, Middlesex. HA4 6QD tel:(020) 3697 0620. enquiries@bunzlvend.com, www.autobar.co.uk

Autopa Ltd - Cottage Leap, Off Butlers Leap, Rugby, Warwicks. CV21 3XP tel:(01788) 550 556, fax:(01788) 550 265. info@autopa.co.uk, www.autopa.co.uk

Avanti Systems - Head Office, Avanti House, Albert Drive, Burgess Hill. RH15 9TN tel:(01444) 247360, fax:(01444) 247206. enquiries@avantisystems.co.uk, www.avantisystems.co.uk

Avesta Sheffield Ltd - Stelco Hardy, Blaenrhondda, Treorchy, M. Glam. CF42 5BY tel:(01443) 778621, fax:(01443) 778626. sales.bar@outokumpu.com

Avocet Hardware Ltd - Brookfoot Mills, Elland Road, Brighouse, W. Yorks. HD6 2RW tel:(01484) 711700, fax:(01484) 720124. enquiries@avocet-hardware.co.uk, www.avocet-hardware.co.uk

Avon Industrial Doors Ltd - Armstrong Way, Yate, Bristol, Avon. BS37 5NG tel:(01454) 273110, fax:(01454) 323224. sales@avondoors.co.uk, www.avondoors.co.uk

Avonside Roofing Group T/A Letchworth Roofing - Unit 2 Business Centre East, Avenue One, Letchworth Garden City, Hertfordshire. SG6 2HB tel:(01462) 755755, fax:(01462) 755750. letchworthsales@avonsidegroup.co.uk, www.letchworthroofing.co.uk

Axter Ltd - Ransomes Europark, West Road, Ipswich, Suffolk. IP3 9SX tel:(01473) 724056, fax:(01473) 723263. info@axterltd.co.uk, www.axter.co.uk

Azimex Fabrications Ltd - Cartwright House, 44 Cartwright Road, Northampton, Northants. NN2 6HF tel:(01604) 717712, fax:(01604) 791087. John@azimex.wanadoo.co.uk, www.azimex.co.uk

B

B & D Clays & Chemicals Ltd - 10 Wandle Way, Willow Lane Trading Estate, Mitcham, Surrey. CR4 4NB tel:(0844) 4774828, fax:(0208) 6485033. sales@bdclays.co.uk, www.bdclays.co.uk

B & M Fencing Limited - Reading Road, Hook, Basingstoke, Hants. RG27 9DB tel:(01256) 762739, fax:(01256) 766891. enquiries@bmfencing.co.uk, www.bmfencing.co.uk

B C Barton Limited - 1 Hainge Road, Tividale, Oldbury, West Midlands. B69 2NR tel:(0121) 557 2272, fax:(0121) 557 2276. website@bcbarton.co.uk, www.bcbarton.co.uk

B R Ainsworth (Southern) - Old Grange Farm, Bursledon, Sothampton, Hampshire. SO31 8GD tel:(01489) 606060, fax:(01489) 885258. info@ainsworth-insulation.com, www.ainsworth-insulation.com

Babcock Wanson Uk Ltd - 7 Elstree Way, Borehamwood, Herts. WD6 1SA tel:(020) 8953 7111, fax:(020) 8207 5177. info@babcock-wanson.co.uk, www.babcock-wanson.co.uk

BAC Ltd - Fanningdon Avenue, Romford, Essex. RM3 8SP tel:(01708) 724824, fax:(01708) 382326. sales@bac.uk.com, www.bacwindows.co.uk

Bacon Group - 6a Little Braxted Hall, Witham Road, Little Braxted, Witham, Essex. CM8 3EU tel:(01621) 230100. sales@bacongroup.co.uk, www.bacongroup.co.uk

Baggeridge Brick PLC - Fir Street, Sedgley, Dudley, W. Midlands. DY3 4AA tel:(01902) 880555, fax:(01902) 880432. office@wienerberger.co.uk, www.baggeridge.co.uk

Bailey Birkett Sales Division - Sharp Street, Worsley, Manchester. M28 3NA tel:(0161) 790 7741, fax:(0161) 703 8451. support@safetysystemsuk.com, www.baileybirkett.com

Bainbridge Engineering Ltd - White Cross, Cold Kirby, Thirsk, North Yorkshire. YO7 2HL tel:(01845) 597655, fax:(01845) 597528. bainbridge.engineering@btinternet.com, www.bainbridge-engineering.com

Baj System Design Ltd - Unit 45A-C, Hartlebury Trading Estate, Kidderminster. DY10 4JB tel:(01299) 250052, fax:(01299) 251153. info@jomy.be, www.jomy.co.uk

Baker Fencing Ltd - PO Box 78, Trelleck, Monmouth, Gwent. NP25 4XB tel:(01600) 860600, fax:(01600) 710827. sales@bakerfencing.co.uk, www.bakerfencing.co.uk

Ball F & Co Ltd - Churnetside Business Park, Station Road, Cheddleton, Leek, Staffs. ST13 7RS tel:(01538) 361633, fax:(01538) 361622. webmaster@f-ball.co.uk, www.f-ball.co.uk

Ball William Ltd - Ultimate House, London Road, Grays, Essex. RM20 4WB tel:(01375) 375151, fax:(01375) 393355. sales@terenceballkitchens.co.uk, www.wball.co.uk

Ballantine, Bo Ness Iron Co Ltd - Links Road, Bo'ness, Scotland. EH51 9PW tel:(01506) 822721, fax:(01506) 827326. sales@ballantinebo ness.co.uk, www.creativeironworks.co.uk

Balmoral Tanks - Balmoral Park, Loirston, Aberdeen. AB12 3GY tel:(01224) 859100, fax:(01224) 859123. tanks@balmoral.co.uk, www.balmoraltanks.com

Bambi Air Compressors Ltd. - 152 Thimble Mill Lane, Heartlands, Birmingham. B7 5HT tel:(0121) 3222299, fax:(0121) 3222297. sales@bambi-air.co.uk, www.bambi-air.co.uk

Barber Wilsons & Co Ltd - Crawley Road, Wood Green, London. N22 6AH tel:(020) 8888 3461, fax:(020) 88882041. sales@barwil.co.uk, www.barwil.co.uk

Barbour Enquiry Service - New Lodge, Drift Road, Windsor. SL4 4RQ tel:(01344) 884999, fax:(01344) 899377. enquiries@barbourehs.com, www.barbour-index.co.uk

Barker & Geary Limited - The Yard, Romsey Road, Kings Somborne, Nr Stockbridge, Hants. SO20 6PW tel:(01794) 388205, fax:(01794) 388205. info@barkerandgeary.co.uk, www.barkerandgeary.co.uk

Barlo Radiators Limited - Imperial Park, Newport, Gwent. NP10 8FS tel:(01633) 657 277, fax:(01633) 657 151. info@barlo-radiators.com, www.barlo-radiators.com

Barlow Group - 136 London Road, Sheffield, S. Yorks. S2 4NX tel:(0114) 280 3000, fax:(0114) 280 3001. info@barlowgroup.co.uk, www.barlowgroup.co.uk

Barlow Tyrie Ltd - Braintree, Essex. CM7 2RN tel:(01376) 557600, fax:(01376) 557610. sales@teak.com, www.teak.com

Barloworld Vacuum Technology Plc - Harbour Road, Gosport, Hants. PO12 1BG tel:(0870) 0106929, fax:(0870) 0106916. marketing@bvc.co.uk, www.barloworldvt.com

Barnwood Shopfitting Ltd - 203 Barnwood Road, Gloucester. GL4 3HT tel:(01452) 614124, fax:(01452) 372933. mail@barnwoodshopfitting.com, www.barnwoodshopfitting.com

Barr & Wray Ltd - 1 Buccleuch Avenue, Hillington Park, Glasgow. G52 4NR tel:(0141) 882 9991, fax:(0141) 882 3690. sales@barrandwray.com, www.barrandwray.com/

Bartoline Ltd - Barmston Close, Beverley, E. Yorks. HU17 0LG tel:(01482) 678710, fax:(01482) 872606. info@bartoline.co.uk, www.bartoline.co.uk

Barton Engineering - Rose Street, Bradley, Bilston, W.Midlands. WV14 8TS tel:(01902) 407199, fax:(01902) 495106. bartonsales@barton-engineering.co.uk, www.hublebas.co.uk

Barton Storage Systems Ltd - Barton Industrial Park, Mount Pleasant, Bilston, West Midlands. WV14 7NG tel:(01902) 499500, fax:(01902) 353098. enquiries@bartonstorage.com, www.barton-storage-systems.co.uk

BASF plc, Construction Chemicals - Earl Road, Cheadle Hulme, Cheadle, Cheshire. SK8 6QG tel:(0161) 485 6222, fax:(0161) 488 5220. info@master-builders-solutions.basf.co.uk, www.master-builders-solutions.basf.co.uk

Bassaire Ltd - Duncan Road, Park Gate, Southampton. SO31 1ZS tel:(01489) 885111, fax:(01489) 885211. sales@bassaire.co.uk, www.bassaire.co.uk

Bathroom Manufacturers Association, The - Innovation Centre 1, Keele University Science & Business Park, Newcastle-under-Lyme. ST5 5NB tel:(01782) 631619, fax:(01782) 630155. info@bathroom-association.org.uk, www.bathroom-association.org

Bathstore - PO Box 21, Boroughbridge Road, Ripon, North Yorkshire. HG4 1SL tel:(08000) 23 23 23. customerservices@wolseley.co.uk, www.bathstore.com

Bauder Limited - 70 Landseer Road, Ipswich, Suffolk. IP3 0DH tel:(01473) 257671, fax:(01473) 230761. info@bauder.co.uk, www.bauder.co.uk

Bauer Inner City Ltd - Dallam Court, Dallam Lane, Warrington, Cheshire. WA2 7LT tel:(01925) 428940, fax:(01925) 244133. Karl.Hall@bauerinnercity.co.uk, www.bauerinnercity.co.uk

Baxenden Chemicals Ltd - Paragon Works, Baxenden, Nr Accrington, Lancs. BB5 2SL tel:(01254) 872278, fax:(01284) 871 247. mail@baxchem.co.uk, www.baxchem.co.uk

Baxi Heating UK Ltd - Brooks House, Coventry Road, Warwick. CV34 4LL tel:(0844) 8711525. info@baxi.co.uk, www.baxi.co.uk

Baz-Roll Products Ltd - Porte Marsh Road, Calne, Wiltshire. SN11 9BW tel:(01249) 822222, fax:(01249) 822300. sales@bilgroup.eu, www.bilgroup.eu

BCIS - Parliament Square, London. SW1P 3AD tel:(024) 7686 8555. contact@bcis.co.uk, www.rics.org/bcis

Beacon Machine Tools Ltd - Mission Works, Purdy Road, Bilston, W. Midlands. WV4 8UB tel:(01902) 493331, fax:(01902) 493241. dheng@btclick.com, www.home.btclick.com/dheng

Beama Ltd - Westminster Tower, 3 Albert Embankment, London. SE1 7SL tel:(020) 7793 3000, fax:(020) 7793 3003. info@beama.org.uk, www.beama.org.uk

Beaumont Ltd F E - Woodlands Road, Mere, Wilts. BA12 6BT tel:(01747) 860481, fax:(01747) 861076. sales@beaumont-chimneys.co.uk, www.beaumont-chimneys.co.uk

Becker (SLIDING PARTITIONS) Ltd - Wemco House, 477 Whippendell Road, Watford, Herts. WD1 7PS tel:(01923) 236906, fax:(01923) 230149. sales@becker.uk.com, www.becker.uk.com

Beco Products Ltd - Albert Street, Brigg, North Lincolnshire. DN20 8HQ tel:(01652) 653 844. info@becowallform.co.uk, www.becowallform.co.uk

Bedford Fencing Co. Limited - Unit 8 Sargeant Turner Ind. Estate, Bromley Street , Lye, Stourbridge, W. Midlands. DY9 8HZ tel:(01384) 422688, fax:(01384) 422688. bedfordfencing@btconnect.com, www.bedfordfencing.org.uk

Bedford Shelving Ltd - Springvale Works, Elland Road, Brighouse. HD6 2RN tel:(01525) 852121, fax:(01525) 851666. sales@bedfordshelf.com, www.bedfordshelf.co.uk

Beehive Coils Ltd - Studlands Park Avenue, Newmarket, Suffolk. CB8 7AU tel:(01638) 664134, fax:(01638) 561542. info@beehivecoils.co.uk, www.beehivecoils.co.uk

Begetube UK Ltd - 8 Carsegate Road South, Inverness. IV3 8LL tel:(01463) 246600, fax:(01463) 246624. info@begetube.co.uk, www.begetube.co.uk

Beiersdorf (UK) Ltd - tesa Division, Yeomans Drive, Blakelands, Milton Keynes, Bucks. MK14 5LS tel:(01908) 211333, fax:(01908) 211555. consumer.relations.uk@beiersdorf.com, www.beiersdorf.co.uk

Bekaert Building Products - Park House Road, Low More, Bradford, West Yorkshire. BD12 0PX tel:(0114) 2427485, fax:(0114) 2427490. building.uk@bekaert.com, www.bekaert.com/building

Beko plc - Beko House, 1 Greenhill Crescent, Watford, Herts. WD18 8QU tel:(0845) 600 4904, fax:(0845) 600 4925. www.beko.co.uk

Bell & Co Ltd - Kingsthorpe Road, Northampton. NN2 6LT tel:(01604) 777500, fax:(01604) 777501. sales@abell.co.uk, www.abell.co.uk

Bell & Webster Concrete Ltd - Alma Park Road, Grantham, Lincsonshire. NG31 9SE tel:(01476) 562277, fax:(01476) 562944. bellandwebster@eleco.com, www.bellandwebster.co.uk

Belzona PolymericLtd - Claro Road, Harrogate, N. Yorks. HG1 4AY tel:(01423) 567641, fax:(01423) 505967. belzona@belzona.co.uk, www.belzona.com

Be-Modern Ltd - Western Approach, South Shields, Tyne & Wear. NE33 5QZ tel:(0191) 455 3571, fax:(0191) 456 5556. enquiries@bemodern.co.uk, www.bemodern.co.uk

Benlowe Group Limited - Benlowe Windows And Doors, Park Road, Ratby, Leicestershire. LE6 0JL tel:(0116) 2395353, fax:(0116) 2387295. info@benlowe.co.uk, www.benlowe.co.uk

Benlowe Stairs - Coppice Side Industrial Estate, Engine Lane, Brownhills, West Midlands. WS8 7ES tel:(0116) 2395353, fax:(01543) 375300. info@benlowe.co.uk, www.benlowe.co.uk

Benson Industries Ltd - Valley Mills, Valley Road, Bradford, West Yorkshire. BD1 4RU tel:(01274) 722204, fax:(01274) 394620. sales@bensonindustry.com, www.bensonindustry.com

Benton Co Ltd Edward - Creteangle Works, Brook Lane, Ferring, Worthing, W. Sussex. BN12 5LP tel:(01903) 241349, fax:(01903) 700213. sales@creteangle.com, www.creteangle.com

Be-Plas Marketing Ltd - Unit 2, Junction 8 Business Park, Ellesmere Port, Wirral. CH65 3AS tel:(0800) 413 758. sales@beplas.com, www.beplas.com

Bernhards Landscapes Ltd - Bilton Road, Rugby, Warwicks. CV22 7DT tel:(01788) 811500, fax:(01788) 816803. bernhards@btconnect.com, www.bernhards.co.uk

Bernlite Ltd - 3 Brookside, Colne Way, Watford, Herts. WD24 7QJ tel:(01923) 200160, fax:(01923) 246057. sales@bernlite.co.uk, www.bernlite.co.uk

Berry Systems - A Memberof Hill & Smith Infrastructure Group, Springvale Business and Ind. Park, Bilston, Wolverhampton, W. Midlands. WV14 0QL tel:(01902) 491100, fax:(01902) 494080. sales@berrysystems.co.uk, www.berrysystems.co.uk

Bertrams - 1 Broadland Business Park, Norwich. NR7 0WF tel:(0800) 333344, fax:(0871) 803 6709. books@bertrams.com, www.bertrams.com

Best & Lloyd Ltd - 51 Downing Street, Smethwick, Birmingham. B66 2PP tel:(0121) 565 6086, fax:(0121) 565 6087. info@bestandlloyd.com, www.bestandlloyd.co.uk

Betafence Limited - PO Box 119, Shepcote Lane, Sheffield, S. Yorks. S9 1TY tel:(0870) 120 3252, fax:(0870) 120 3242. sales.sheffield@betafence.com, www.betafence.co.uk

Beton Construction Ltd - PO Box 11, Basingstoke, Hants. RG21 8EL tel:(01256) 353146, fax:(01256) 840621. info@betonconmat.co.uk, www.betonconmat.co.uk

Bevan Funnell Group Limited - Norton Road, Newhaven, Sussex. BN9 0BZ tel:(01273) 616100, fax:(01273) 611167. sales@bevan-funnell.co.uk, www.bevan-funnell.co.uk

Beverley Environmental Ltd - Unit 49, Bolney Grange Industrial Park, Haywards Heath, West Sussex. RH17 5PB tel:(01444) 248 930. info@beverley-environmental.co.uk, www.beverley-environmental.co.uk

BF Bassett & Findley - Talbot Road North, Wellingborough, Northants. NN8 1QS tel:(01933) 224898, fax:(01933) 227731. info@bassettandfindley.co.uk, www.bassettandfindley.co.uk

BFM Europe Limited - Trentham Lakes, Stoke-on-Trent, Staffordshire. ST4 4TJ tel:(01782) 339000, fax:(01782) 339009. info@bfm-europe.com, www.bfm-europe.com

BFRC Services Ltd - 177 Bagnall Road, Basford, Nottingham, Notts. NG6 8SJ tel:(0115) 942 4200, fax:(0115) 942 4488. enquiries@bfrc.org, www.roofinguk.com

Bicester Products Ltd - 7 Crawley Mill, Dry Lane, Witney, Oxon. OX29 9TJ tel:(01993) 704810, fax:(01993) 779569. bicpro@soundservice.co.uk, www.bicpro.co.uk

Biddle Air Systems Ltd - St Mary's Road, Nuneaton, Warwicks. CV11 5AU tel:(024) 76 384233, fax:(024) 76 373621. sales@biddle-air.co.uk, www.biddle.air.co.uk

BigHead Bonding Fasteners Ltd - Unit 15-16 Elliott Road, West Howe Ind Estate, Bournemouth, Dorset. BH11 8LZ tel:(01202) 574601, fax:(01202) 578300. info@bighead.co.uk, www.bighead.co.uk

Bilco UK Ltd - Park Farm Business Centre, Fornham St Genevieve, Bury St Edmunds, Suffolk. IP28 6TS tel:(01284) 701696, fax:(01284) 702531. bilcouk@bilco.com, www.bilco.com

Bill Switchgear - Reddings Lane, Birmingham. B11 3EZ tel:(0121) 685 2080, fax:(0121) 685 2184. uksystorders@eaton.com, www.bill-switchgear.com

Billericay Fencing Ltd - Morbec Farm, Arterial Road, Wickford, Essex. SS12 9JF tel:(01268) 727712, fax:(01268) 590225. sales@billericayfencing.co.uk, www.billericay-fencing.co.uk

Billington Structures Ltd - Barnsley Road, Wombwell, Barnsley, South Yorkshire. S73 8DS tel:(01226) 340666, fax:(01266) 755947. postroom@billington-structures.co.uk, www.billington-structures.co.uk

Binder - Old Ipswich Road, Claydon, Ipswich. IP6 0AG tel:(01473) 830582, fax:(01473) 832175. info@binder.co.uk, www.binder.co.uk

Binns Fencing Ltd - Harvest House, Cranborne Road, Potters Bar, Herts. EN6 3JF tel:(01707) 855555, fax:(01707) 857565. contracts@binns-fencing.com, www.binns-fencing.com

BioLab UK - Unit P, Andoversford Industrial Estate, Andoversford, Cheltenham. GL54 4LB tel:(01242) 820115, fax:(01242) 820438. sales@biolabuk.com, www.biolabuk.com

Birdsall Services Ltd - 6 Frogmore Road, Hemel Hempstead, Herts. HP3 9RW tel:(01442) 212501, fax:(01442) 248989. sales@birdsall.co.uk, www.birdsall.co.uk

Birtley Group - Mary Avenue, Birtley, Co. Durham. DH3 1JF tel:(0191) 410 6631, fax:(0191) 410 0650. info@birtley-building.co.uk, www.birtley-building.co.uk

BIS Door Systems Ltd - 13 Hodgson Court, Hodgson Way, Wickford, Essex. SS11 8XR tel:(01268) 767566, fax:(01268) 560284. sales@bis-doors.co.uk, www.bis-doors.co.uk

Bisley Office Equipment - Queens Road, Bisley, Woking, Surrey. GU24 9BJ tel:(01483) 474577, fax:(01483) 489962. marketing@bisley.com, www.bisley.com

Bison Bede Ltd - Unit 9, No.1 Ind. Estate, Consett, Co. Durham. DH8 6ST tel:(01207) 585000, fax:(01207) 585085. sales@bisonbede.co.uk, www.bisonbede.com

Bison Manufacturing Limited - Tetron Point, William Nadin Way, Swadlincote, Derbyshire. DE11 0BB tel:(01283) 817500, fax:(01283) 220563. concrete@bison.co.uk, www.bison.co.uk

Bitmen Products Ltd - PO Box 339, Over, Cambridge, Cambs. CB4 5TU tel:(01954) 231315, fax:(01954) 231512. sales@bitmen.force9.co.uk, www.bitmen.co.uk

Bitumen Waterproofing Association - 19 Regina Crescent, Ravenshead, Nottingham. NG15 9AE tel:(01623) 430574, fax:(01623) 798098. info@bwa-europe.com, www.bwa-europe.com

B (con't)

Black Cat Music - Festival House, 4 Chapman Way, Tunbridge Wells, Kent. TN2 3EF tel:(01732) 371555, fax:(01732) 371556. sales@blackcatmusic.co.uk, www.blackcatmusic.co.uk

Blackfriar Paints Ltd - Portobello Industrial Estate, Bitley, Chester-le-Street, County Durham. DH3 2RE tel:(0191) 411 3146, fax:(0191) 492 0125. blackfriar@tor-coatings.com, www.blackfriar.co.uk

Blagg & Johnson Ltd - Newark Business Park, Brunel Drive, Newark, Notts. NG24 2EG tel:(01636) 703137, fax:(01636) 701914. info@blaggs.co.uk, www.blaggs.co.uk

Blair Joinery - 9 Baker Street, Greenock, Strathclyde. PA15 4TU tel:(01475) 721256, fax:(01475) 787364. sales@blairswindows.co.uk, www.blairsofscotland.co.uk

Blakell Europlacer Ltd - 30 Factory Road, Upton Industrial Estate, Poole, Dorset. BH16 5SL tel:(01202) 266500, fax:(01202) 266599. sales@europlacer.co.uk, www.europlacer.com

Blakley Electrics Ltd - 1 Thomas Road, Optima Park, Crayford, Kent. DA1 4GA tel:(0845) 074 0084, fax:(0845) 074 0085. sales@blakley.co.uk, www.blakley.co.uk

Blanc de Bierges - Pye Bridge Industrial Estate, Main Road, Pie Bridge, Near Alfreton, Derbyshire. DE55 4NX tel:(01733) 202566, fax:(01733) 205405. info@blancdebierges.com, http://blancdebierges.com/

Blaze Signs Ltd - 5 Patricia Way, Pysons Road, Broadstairs, Kent. CT10 2XZ tel:(01843) 601 075, fax:(01843) 867 924. info@blazesigns.co.uk, www.blaze-signs.com

Blick Communication Systems Ltd - Blick House, Bramble Road, Swindon, Wilts. SN2 8ER tel:(01793) 692401, fax:(01793) 615848. info@stanleysecuritysolutions.co.uk, www.blick.co.uk

BLM British Lead - Peartree Lane, Welwyn Garden City, Herts. AL7 3UB tel:(01707) 324595, fax:(01707) 328941. sales@britishlead.co.uk, www.britishlead.co.uk

Blount Shutters Ltd - Unit B, 734 London Road, West Thurrock, Essex. RM20 3NL tel:(08456) 860000, fax:(01708) 861272. sales@blountshutters.co.uk, www.blountshutters.co.uk

Blucher UK - Station Road Industrial Estate, Tadcaster, North Yorkshire. LS24 9SG tel:(01937) 838000. www.blucher.co.uk

Blücher UK Ltd - Station Road Industrial Estate, Tadcaster, N. Yorks. LS24 9SG tel:(01937) 838000, fax:(01937) 832454. mail@blucher.co.uk, www.blucher.co.uk

Bluestar Silicones - Wolfe Mead, Farnham Road, Bordon, Hampshire. GU35 0NH tel:(01420) 477000, fax:0483200. webmaster.silicones@bluestarsilicones.com, www.bluestarsilicones.com

BLV Licht-und Vakuumtecnik - Units 25 & 26, Rabans Close, Rabans Lane Industrial Estate, Aylesbury, Bucks. HP19 3RS tel:(01296) 399334, fax:(01296) 393422. info@blv.co.uk, www.blv.co.uk

BM TRADA - Chiltern House, Stocking Lane, Hughenden Valley, High Wycombe, Bucks. HP14 4ND tel:(01494) 569800, fax:(01494) 564895. testing@bmtrada.com, www.bmtradagroup.com/

Boardcraft Ltd - Howard Road, Eaton Socon, St Neots, Huntingdon, Cambs. PE19 3ET tel:(01480) 213266, fax:(01480) 219095. info@boardcraft.co.uk, www.boardcraft.co.uk

BOB Stevenson Ltd - Coleman Street, Derby, Derbys. DE24 8NN tel:(01332) 574112, fax:(01332) 757286. sales@bobstevenson.co.uk, www.bobstevenson.co.uk

Bodycote Metallurgical Coatings Ltd - Tyherington Business Park, Macclesfield, Cheshire. SK10 2XF tel:(01625) 505300, fax:(01625) 505320. Phil.Adams@bodycote.com, www.bodycotemetallurgicalcoatings.com

Bollom Fire Protection Ltd - Croydon Road, Elmers End, Beckenham, Kent. BR3 4BL tel:(020) 8658 2299, fax:(020) 8658 8672. enquiries@tor-coatings.com, www.bollom.com

Bolton Gate Co. Ltd - Waterloo Street, Bolton. BL1 2SP tel:(01204) 871000, fax:(01204) 871049. sales@boltongate.co.uk, www.boltongate.co.uk

Bondaglass Voss Ltd - 158-160 Ravenscroft Road, Beckenham, Kent. BR3 4TW tel:(020) 8778 0071, fax:(020) 8659 5297. bondaglass@btconnect.com

Bonwyke Ltd - Bonwyke House, 41-43 Redlands Lane, Fareham, Hants. PO14 1HL tel:(01329) 289621, fax:(01329) 822768. sales@bonwyke.co.uk, www.bonwyke.demon.co.uk

Booth Industries Ltd - PO Box 50, Nelson Street, Bolton, Lancashire. BL3 2RW tel:(01204) 366333, fax:(01204) 380888. marketing@booth-industries.co.uk, www.booth-industries.co.uk

Booth Muirie - South Caldeen Road, Coadbridge, North Lanarkshire. ML5 4EG tel:(01236) 354 500, fax:(01236) 345 515. enquiries@boothmuirie.co.uk, www.boothmuirie.co.uk

Border Concrete Products Ltd - Jedburgh Road, Kelso, Roxburghshire. TD5 8-JG tel:(01573) 224393, fax:(01573) 226360. sales@borderconcrete.co.uk, www.borderconcrete.co.uk

Bostik Ltd - Ulverscroft Road, Leicester. LE4 6BW tel:(01785) 272727, fax:(0116) 2513943. technical.service@bostik.com, www.bostik.co.uk

Bourne Steel Ltd - St Clements House, St Clements Road, Poole, Dorset. BH12 4GP tel:(01202) 746666, fax:(01202) 732002. sales@bourne-steel.co.uk, www.bournesteel.co.uk

Bovingdon Brickworks Ltd - Leyhill Road, Hemel Hempstead, Herts. HP3 0NW tel:(01442) 833176, fax:(01442) 834539. info@bovingdonbricks.co.uk, www.bovingdonbrickworks.co.uk

Bowden Fencing Ltd - Leicester Lane, Great Bowden, Market Harborough, Leics. LE16 7HA tel:(01858) 410660, fax:(01858) 433957. info@bowdenfencing.co.uk, www.bowdenfencing.co.uk

BPT Automation Ltd - Unit 16, Sovereign Park, Cleveland Way, Hemel Hempstead, Herts. HP2 7DA tel:(01442) 235355, fax:(01442) 244729. sales@bpt.co.uk, www.bptautomation.co.uk

BPT Security Systems (UK) Ltd - Unit 16, Sovereign Park, Cleveland Way, Hemel Hempstead, Herts. HP2 7DA tel:(01442) 230800, fax:(01442) 244729. sales@bpt.co.uk, www.bpt.co.uk

Bradbury Group Ltd - Dunlop Way, Queensway Enterprise Estate, Scunthorpe, N. Lincs. DN16 3RN tel:(01724) 271999, fax:(01724) 271888. sales@bradburyuk.com, www.bradburyuk.com

Braithaite Engineers Ltd - Neptune Works, Usk Way, Newport, S. Wales. NP9 2UY tel:(01633) 262141, fax:(01633) 250631. tanks@braithwaite.co.uk

Bramah Security Equipment Ltd - 7 Goodge Place, Fitzrovia, London. W1T 4SF tel:(020) 7637 8500 option 3, fax:(020) 7636 5598. lock.sales@bramah.co.uk, www.bramah.co.uk

Brandon Hire PLC - 72-75 Feeder Road, St Philips, Bristol. BS2 0TQ tel:(0117) 9413 550, fax:(0117) 9540 972. info@brandonhire.co.uk, www.brandontoolhire.co.uk

Brandon Medical - Holme Well Road, Leeds, W. Yorks. LS10 4TQ tel:(0113) 2777393, fax:(0113) 2728844. enquiries@brandon-medical.com, www.brandon-medical.com

Brannan S & Sons Ltd - Leconfield Ind Est, Cleator Moor, Cumbria. CA25 5QE tel:(01946) 816600, fax:(01946) 816625. sales@brannan.co.uk, www.brannan.co.uk

Brannan Thermometers & Gauges - Leconfield Industrial Estate, Cleator Moor, Cumbria. CA25 5QE tel:(01946) 816624, fax:(01946) 816625. sales@brannan.co.uk, www.brannan.co.uk

Branson Leisure Ltd - Fosters Croft, Foster Street, Harlow, Essex. CM17 9HS tel:(01279) 432151, fax:(01279) 432151. sales@bransonleisure.co.uk, www.bransonleisure.co.uk

Brash John & Co Ltd - The Old Shipyard, Gainsborough, Lincs. DN21 1NG tel:(01427) 613858, fax:(01427) 810218. enquiries@johnbrash.co.uk, www.johnbrash.co.uk

Brass Age Ltd - 6 Bidwell Rd, Rackheath Industrial Estate, Norwich. NR13 6PT tel:(01603) 722330, fax:(01603) 722777. sales@basystems.co.uk

Brass Art Ltd - Regent Works, Attwood Street, Lye, Stourbridge, W. Midlands. DY9 8RY tel:(01384) 894814, fax:(01384) 423824. sales@brassart.com, www.brassart.co.uk

BRC Reinforcement - Brierley Park Close, Stanton Hill, Sutton-in-Ashfield, Nottinghamshire. NG17 3FW tel:(01623) 555111, fax:(01623) 440932. sales@midlands.brc.ltd.uk, www.brc-reinforcement.co.uk

BRC Special Products - Bekaert Ltd - Park House Road, Low Moor, Bradford, West Yorkshire. BD12 0PX tel:(01142) 427 480, fax:(01142) 427 490. infobuilding@bekaert.com, www.brc-special-products.co.uk

BRE - Bucknalls Lane, Garston, Watford, Herts. WD25 9XX tel:(01923) 664000, fax:(01923) 664010. enquiries@bre.co.uk, www.bre.co.uk

Breedon Aggregates - Breedon Quarry, Breedon on the Hill, Derby. DE73 8AP tel:(01332) 862254, fax:(01332) 864320. sales@breedonaggregates.com, www.breedonaggregates.com

Brett Landscaping - Sileby Road, Barrow upon Soar, Leicestershire. LE12 8LX tel:(0845) 60 80 570, fax:(0845) 60 80 575. sales@brettpaving.co.uk, www.brettpaving.co.uk

Brett Martin Ltd - Speedwell Industrial Estate, Staveley, Chesterfield, Derbys. S43 3JP tel:(01246) 280000, fax:(01246) 280001. mail@brettmartin.com, www.brettmartin.co.uk

Brewer C & Sons Ltd - Albany House, Ashford Road, Eastbourne, Sussex. BN21 3TR tel:(01323) 411080, fax:(01323) 721435. enquiries@brewers.co.uk, www.brewers.co.uk

Brewer Metalcraft - W. Sussex. BN18 0DF tel:(01243) 539639, fax:(01243) 533184. sales@brewercowls.co.uk, www.brewercowls.co.uk

Brewer T. & Co Ltd - Old Station Yard, Springbank Road, Hither Green, London. SE13 6SS tel:(020) 8461 2471, fax:(020) 8461 4822. clapham@tbrewer.co.uk

Brick Development Association - The Building Centre, 26 Store Street, London. WC1E 7BT tel:(020) 7323 7030, fax:(020) 7580 3795. brick@brick.org.uk, www.brick.org.uk

Bridge Quarry - Bridge House, Windmill Lane, Kerridge, Macclesfield, Cheshire. SK10 5AZ tel:(01625) 572700, fax:(01625) 572700. davidtooth@tiscali.co.uk, www.bridgequarry.co.uk

Bridgman IBC Ltd - Grealtham Street, Longhill Industrial Estate (North), Hartlepool, Cleveland. TS25 1PU tel:(01429) 221111, fax:(01429) 274035. sales@bridgman-ibc.com, www.bridgman-ibc.com

Briggs Industrial Footwear Ltd t/a Briggs Safety Wear - Briggs House, 430 Thurmaston Boulevard, Leicester, Leics. LE4 9LE tel:(01162) 444700, fax:(01162) 444744. sales@briggsafetywear.co.uk, www.briggssafetywear.co.uk

BriggsAmasco - Amasco House, 101 Powke Lane, Cradley Heath, West Midlands. B64 5PX tel:(0121) 502 9600, fax:(0121) 502 9601. enquiries@briggsamasco.co.uk, www.briggsamasco.co.uk

Brighton (Handrails), W - 55 Quarry Hill, Tamworth, Staffs. B77 5BW tel:(01827) 284488, fax:(01827) 250907. wbrightonhandrails@ntlworld.com, www.wbrightonhandrails.com

Brimar Plastics Ltd - 18 Dixon Road, Brislington, Bristol. BS4 5QW tel:(0117) 971 5976, fax:(0117) 971 6839. sales@brimarplastics.co.uk, www.brimarplastics.co.uk

Brintons Ltd - PO Box 16, Exchange Street, Kidderminster, Worcs. DY10 1AG tel:(01562) 820000, fax:(01562) 515597. solutions@brintons.co.uk, www.brintons.net

brissco signs & graphics - Block 9, 25 Cater Road, Bishopsworth, Bristol, Avon. BS13 7TX tel:(0117) 311 3777, fax:(0117) 311 6777. sales@brissco.co.uk, www.brissco.com/

Bristan Group - Pooley Hall Drive, Birch Coppice Business Park, Dordon, Tamworth. B78 1SG tel:(0330) 026 6273. enquire@bristan.com, www.bristan.com

Bristol Metal Spraying & Protective Coatings Ltd - Paynes Shipyard, Coronation Road, Bristol. BS3 1RP tel:(0117) 966 2206, fax:(0117) 966 1158. sales@bmspc.co.uk, www.bmspc.co.uk

Britannia Fire - Ashwellthorpe Industrial Estate, Ashwellthorpe, Norwich, Norfolk. NR16 1ER tel:(01508) 488416, fax:(01508) 481753. sales@britannia-fire.co.uk, www.britannia-fire.co.uk

Britannia Kitchen Ventilation Ltd - 10 Highdown Road, Sydenham Industrial Estate, Leamington Spa, Warwickshire. CV31 1XT tel:(01926) 463540, fax:(01926) 463541. sales@kitchen-ventilation.co.uk, www.kitchen-ventilation.co.uk

British Adhesives and Sealants Association - 24 Laurel Close, Mepal, Ely, Cambridgeshire. CB6 2BN tel:(0330) 22 33 290, fax:(0330) 22 33 408. secretary@basaonline.org, www.basaonline.org

British Association for Chemical Specialities - Simpson House, Windsor Court, Clarence Drive, Harrowgate. HG1 2PE tel:(01423) 700249, fax:(01423) 520297. enquiries@bacsnet.org, www.bacsnet.org

British Association of Landscape Industries - Landscape House, Stoneleigh Park, National Agricultural Centre, Warwickshire. CV8 2LG tel:(024) 7669 0333, fax:(024) 7669 0077. contact@bali.org.uk, www.bali.org.uk

British Blind and Shutter Association (BBSA) - PO Box 232, Stowmarket, Suffolk. IP14 9AR tel:(01449) 780444, fax:(01449) 780444. info@bbsa.org.uk, www.bbsa.org.uk

British Board of Agrément - Bucknalls Lane, Garston, Watford, Herts. WD25 9BA tel:(01923) 665300, fax:(01923) 665301. mail@bba.star.co.uk, www.bbacerts.co.uk

British Ceramic Confederation - Federation House, Station Road, Stoke-on-Trent. ST4 2SA tel:(01782) 744631, fax:(01782) 744102. bcc@ceramfed.co.uk, www.ceramfed.co.uk

British Ceramic Tile Ltd - Heathfield, Newton Abbot, Devon. TQ12 6RF tel:(01626) 834775, fax:(01626) 834775. internalsales@britishceramictile.com, www.britishceramictile.com

British Coatings Federation - Riverbridge House, Guildford Road, Leatherhead, Surrey. KT22 9AD tel:(01372) 365989, fax:(01372) 365979. enquiry@bcf.co.uk, www.coatings.org.uk

British Constructional Steelwork Association Ltd - 4 Whitehall Court, Westminster, London. SW1A 2ES tel:(020) 7839 8566, fax:(020) 7976 1634. postroom@steelconstruction.org, www.steelconstruction.org

British Drilling Association (BDA) - Wayside, London End, Upper Boddington, Daventry, Northamptonshire. NN11 6DP tel:(01327) 264622, fax:(01327) 264623. office@britishdrillingassociation.co.uk, www.britishdrillingassociation.co.uk

British Electric Lamps Ltd - Spencer Hill Road, London. SW19 4EN tel:(01924) 893380, fax:(01924) 894320. sales@belllighting.co.uk, www.belllighting.co.uk

British Electrotechnical and Allied Manufacturers Association - Westminster Tower, 3 Albert Embankment, London. SE1 7SL tel:(020) 7793 3000, fax:(020) 7793 3003. info@beama.org.uk, www.beama.org.uk

British Fenestration Rating Council - 54 Ayres Street, London. SE1 1EU tel:(020) 7403 9200, fax:(08700) 278 493. enquiries@bfrc.org, www.bfrc.org

British Flue & Chimney Manufacterers Association - FETA, 2 Waltham Court, Milley Lane, Hare Hatch, Reading, Berks. RG10 9TH tel:(0118) 9403416, fax:(0118) 9406258. info@feta.co.uk, www.feta.co.uk/associations/bfcma

British Glass Manufacturers Confederation - 9 Churchill Road, Chapeltown, Sheffield. S35 2PY tel:(0114) 290 1850, fax:(0114) 290 1851. info@britglass.co.uk, www.britglass.org.uk

British Gypsum Ltd - Head Office, East Leake, Loughborough, Leicestershire. LE12 6HX tel:(0115) 9451000, fax:(0115) 9451901. bgtechnical.enquiries@bpb.com, www.british-gypsum.com

British Harlequin Plc - Festival House, Chapman Way, Tunbridge Wells. TN2 3EF tel:(01892) 514888, fax:(01892) 514222. enquiries@harlequinfloors.com, www.harlequinfloors.com

British Library Business Information Service - The British Library, 96 Euston Road, London. NW1 2DB tel:(020) 7412 7454, fax:(020) 7412 7453. business-information@bl.uk, www.bl.uk/bipc

British Nova Works Ltd - Neville House, Beautmont Road, Banbury, Oxon. OX16 1RB tel:(020) 8574 6531, fax:(020) 8571 7572. sales@britishnova.co.uk

British Plastics Federation, The - 5-6 Bath Place, Rivington Street, London. EC2A 3JE tel:(020) 7457 5000, fax:(020) 7457 5045. reception@bpf.co.uk, www.bpf.co.uk

British Precast Concrete Federation - The Old Rectory, Main Street, Leicester, Leics. LE3 8DG tel:(0116) 253 6161, fax:(0116) 251 4568. info@britishprecast.org, www.britishprecast.org

British Pump Manufacturers Association - The National Metalforming Centre, 47 Birmingham Road, West Bromwich. B70 6PY tel:(0121) 200 1299, fax:(0121) 200 1306. enquiry@bpma.org.uk, www.bpma.org.uk

British Refrigeration Association (BRA) - 2 Waltham Court, Milley Lane, Hare Hatch, Reading, Berkshire. RG10 9TH tel:(01189) 403416, fax:(01189) 406258. bra@feta.co.uk, www.feta.co.uk

British Security Industry Association - Kirkham House, John Comyn Drive, Worcester. WR3 7NS tel:(0845) 3893889, fax:(0845) 3890761. info@bsia.co.uk, www.bsia.co.uk

British Sisalkraft Ltd - Commissioners Road, Rochester, Strood, Kent. ME2 4ED tel:(01634) 22700, fax:(01634) 291029. ask@proctorgroup.com, www.bsk-laminating.com

British Slate Association - Channel Business Centre, Ingles Manor, Castle Hill Avenue, Folkestone, Kent. CT20 2RD tel:(01303) 856123, fax:(01303) 221095. enquiries@stone-federationgb.org.uk, www.stone-federationgb.org.uk

British Stainless Steel Association - Regus, Blades Enterprise Centre, John Street, Sheffield. S2 4SW tel:(0114) 292 2636, fax:(0114) 292 2633. admin@bssa.org.uk, www.bssa.org.uk

British Standards Institution - 389 Chiswick High Road, London. W4 4AL tel:(020) 8996 9000, fax:(020) 8996 7001. orders@bsi-global.com, www.bsi-global.com

British Structural Waterproofing Association, The - Westcott House, Catlins Lane, Pinner, Middx. HA5 2EZ tel:(020) 8866 8339, fax:(020) 8868 9971. enquiries@bswa.org.uk, www.bswa.org.uk

British Turntable Ltd - Emblem Street, Bolton. BL3 5BW tel:(01252) 319922, fax:(01252) 341872. info@movetechuk.com, www.hovair.co.uk

BRITISH URETHANE FOAM CONTRACTORS ASSOCIATION - P O Box 12, Haslemere, Surrey. GU27 3AH tel:(01428) 870150. info@bufca.co.uk, www.bufca.co.uk

British Wall Tie Association - PO Box 22, Goring, Reading. RG8 9YX tel:(01189) 842674, fax:(01189) 845396. info@bwta.org.uk, www.bwta.org.uk

British Water - 1 Queen Anne's Gate, London. SW1H 9BT tel:(020) 7957 4554, fax:(020) 7957 4565. info@britishwater.co.uk, www.britishwater.co.uk

British Woodworking Federation - The Building Centre, 26 Store Street, London. WC1E 7BT tel:(0844) 209 2610, fax:(0844) 458 6949. bwf@bwf.org.uk, www.bwf.org.uk

Britmet Tileform Limited - Spital Farm, Thorpe Mead, Banbury, Oxon. OX16 4RZ tel:(01295) 250998, fax:(01295) 271068. BritmetTileform@hotmail.com, www.britmet.co.uk

BRK Brands Europe Ltd - Gordano Gate, Portishead, Bristol, Avon. BS20 7GG tel:(01275) 845024, fax:(01275) 849255. zforman@brk.co.uk

BRK Brands Europe Ltd - P.O. Box 402, Gloucester. GL2 9YB tel:(01452) 714999, fax:(01452) 713103. info@brk.co.uk, www.brkdicon.co.uk

Broadbent - Droppingstone Farm, New Lane, Harthill, Chester. CH3 9LG tel:(01829) 782822, fax:(01829) 782820. enquiries@sbal.co.uk, www.sbal.co.uk

Broadcrown Ltd - Alliance Works, Airfield Industrial Estate, Hixon, Stafford, Staffs. ST18 0PF tel:(08794) 122200, fax:(01889) 272220. info@broadcrown.co.uk, www.broadcrown.com

B (con't)

Broadmead Cast Stone - Broadmead Works, Hart Street, Maidstone, Kent. ME16 8RE tel:(01622) 690960, fax:(01622) 765484. info@topbond.co.uk, www.kier.co.uk
Broag Ltd - Remeha House, Molly Millars Lane, Wokingham, Berks. RG41 2QP tel:(0118) 978 3434, fax:(0118) 978 6977. boilers@broag-remeha.com, www.uk.remeha.com
Brockway Carpets Ltd - Hoobrook, Kidderminster, Worcs. DY10 1XW tel:(01562) 824737, fax:(01562) 752010. sales@brockway.co.uk
Broen Valves Ltd - Unit 7 Clecton Street Business Park, Clecton Street, Tipton, W. Midlands. DY4 7TR tel:(0121) 522 4515, fax:(0121) 522 4535. broenvalves@broen.com, www.broen.co.uk
Bromag Structures Ltd - Monarch Court House, 2 Mill Lane, Benson, Wallingford, Oxon. OX10 6SA tel:(01491) 838808, fax:(01491) 834183. jo@bromag.co.uk, www.bromag.co.uk
Bromford Iron & Steel Co Ltd - Bromford Lane, West Bromwich, W. Midlands. B70 7JJ tel:(0121) 525 3110, fax:(0121) 525 4673. enquiries@bromfordsteels.co.uk, www.bromfordsteels.co.uk
Brooks Stairlifts Ltd - Hawksworth Hall, Hawksworth Lane, Guiseley, W. Yorks. LS20 8NU tel:(0800) 422 0653. www.stairlifts.co.uk
Broome Bros (Doncaster) Ltd - Lineside, Cheswold Lane, Doncaster, S. Yorks. DN5 8AR tel:(01302) 361733, fax:(01302) 328536. jeff@broomebros.co.uk
Broxap Ltd - Rowhurst Industrial Estate, Chesteron, Newcastle-under-Lyme, Staffs. ST5 6BD tel:(01782) 564411, fax:(01782) 565357. dave.challinor@broxap.com, www.broxap.com
Broxwood (scotland Ltd) - Inveralmond Way, Inveralmond Industrial Estate, Perth. PH1 3UQ tel:(01738) 444456, fax:(01738) 444452. sales@broxwood.com, www.broxwood.com
BSA Machine Tools Ltd - Mackadown Lane, Kitts Green, Birmingham. B33 0LE tel:(0121) 783 4071, fax:(0121) 784 5921. ch&ceo@bsatools.co.uk, www.bsamachinetools.co.uk
Buckingham Swimming Pools Ltd - Dalehouse Lane, Kenilworth, Warwicks. CV8 2EB tel:(01926) 852351, fax:(01926) 512387. info@buckinghampools.com, www.buckinghampools.com
Builders Merchants Federation - 1180 Elliott Court, Coventry Business Park, Herald Avenue, Coventry. CV5 6UB tel:(02476) 854980, fax:(02476) 854981. info@bmf.org.uk, www.bmf.org.uk
Building & Engineering Services Association - Esca House, 34 Palace Court, London. W2 4JG tel:(020) 7313 4900, fax:(020) 7727 9268. contact@b-es.org.uk, www.b-es.org.uk
Building Additions Ltd - Unit C1, Southgate Commerce Park, Frome, Somerset. BA11 2RY tel:(01373) 454577, fax:(01373) 454578. partitions@buildingadditions.co.uk, www.buildingadditions.co.uk
Building Adhesives Ltd - Longton Road, Trentham, Stoke-on-Trent. ST4 8JB tel:(01782) 591100, fax:(01782) 591101. info@building-adhesives.com, www.building-adhesives.com
Building Controls Industry Association (BCIA) - 92 Greenway Business Centre, Harlow Business Park, Harlow, Essex. CM10 5QE tel:(01189) 403416, fax:(01189) 406258. Karen@bcia.co.uk, www.bcia.co.uk
Building Reserch Establishment - BRE, Garston, Watford. WD25 9XX tel:(01923) 664000. enquiries@bre.co.uk, www.bre.co.uk
Bullock & Driffill Ltd - Staunton Works, Newark Road, Staunton in the Vale, Nottingham, Notts. NG13 9PF tel:(01400) 280000, fax:(01400) 280010. bullock.driffill@btopenworld.com
Bunting Magnetics Europe Limited - Northbridge Road, Berkhamptsead, Hertfordshire. HP4 1EH tel:(01442) 875081, fax:(01442) 875009. sales@buntingeurope.com, www.buntingeurope.com
Burco Maxol - Rosegrove, Burnley, Lancs. BB12 6AL tel:(0844) 8153755, fax:(0844) 8153748. sales@burco.co.uk, www.burco.co.uk
Burgess Architectural Products Ltd - PO Box 2, Brookfield Road, Hinkley, Leics. LE10 2LL tel:(01455) 618787, fax:(01455) 251061. info@burgessceilings.co.uk, www.burgessceilings.co.uk
Burlington Slate Ltd - Cavendish House, Kirkby in Furness, Cumbria. LA17 7UN tel:(01229) 889661, fax:(01229) 889466. sales@burlingtonstone.co.uk, www.burlingtonstone.co.uk
Burmatex Ltd - Victoria Mills, The Green, Ossett, W. Yorks. WF5 0AN tel:(01924) 262525, fax:(01924) 280033. info@burmatex.co.uk, www.burmatex.co.uk
Burn Fencing Limited - West End Farm, West End Lane, Balne, Goole, E. Yorks. DN14 0EH tel:(01302) 708706, fax:(01302) 707377. richard@burnfencing.co.uk, www.burnfencing.co.uk
Bush Nelson PLC - Stephenson Way, Three Bridges, Crawley, W. Sussex. RH10 1TN tel:(01293) 547361, fax:(01293) 531432. sales@bnthermic.co.uk, www.bush-nelson.com
Butterworth-Heinemann - The Boulevard, Langford Lane, Kidlington, Oxford. OX5 1GB tel:(01865) 844640, fax:(01865) 843912. directenquiries@elsevier.com, www.elsevierdirect.com
BVC - Harbour Road, Gosport, Hants. PO12 1BG tel:(023) 92584281, fax:(023) 92504648. marketing@bvc.co.uk, www.bvc.co.uk

C

C & W Fencing Ltd - The Steelworks, Bradfield Road, Wix, Manningtree, Essex. CO11 2SG tel:(01255) 871300, fax:(01255) 871309. Sales@cw-fencing.co.uk, www.cw-fencing.co.uk
C.R. Longley & Co. Ltd - Ravensthorpe Road, Thornhill Lees, Dewsbury, W. Yorks. WF12 9EF tel:(01924) 464283, fax:(01924) 459183. sales@longley.uk.com, www.longley.uk.com
CA Group Ltd - MR-24 Division - No 1 The Parade, Lodge Drive, Culcheth, Cheshire. WA3 4ES tel:(01925) 764335, fax:(01925) 763445. info@cagroup.ltd.uk, www.cagroupltd.co.uk
Cadisch MDA Ltd - Unit 1, Finchley Industrial Centre, 879 High Road, Finchley, London. N12 8QA tel:(020) 8492 7622, fax:(020) 8492 0333. info@cadisch.com, www.cadischmda.com
Caithness Flagstone Limited - 4, The Shore, Wick, Caithness. KW1 4JW tel:(01955) 605472. info@caithness-stone.co.uk, www.caithnessstone.co.uk
Calder Industrial Materials Ltd - 1 Derwent Court, Earlsway, Team Valley Trading Estate, Gateshead, Tyne & Wear. NE11 0TF tel:(0191) 482 7350, fax:(0191) 482 7351. buildingproducts@caldergroup.co.uk, www.caldergroup.co.uk
Caldwell Hardware (UK) Ltd - Herald Way, Binley Industrial Estate, Coventry, Warwicks. CV3 2RQ tel:(024) 7643 7900, fax:(024) 7643 7969. sales@caldwell.co.uk, www.caldwell.co.uk
Caledonian Modular - Carlton Works, Carlton-on-Trent, Newark, Nottinghamshire. NG23 6NT tel:(01636) 821645, fax:(01636) 821261. sales@caledonianmodular.com, www.caledonianmodular.com
Calomax Ltd - Lupton Avenue, Leeds, W. Yorks. LS9 7DD tel:(0113) 249 6681, fax:(0113) 235 0358. sales@calomax.co.uk, www.calomax.co.uk
Calor Gas Ltd - Athena Drive, Tachbrook Park, Warwick. CV34 6RL tel:(0800) 216659, fax:(0870) 4006904. enquiries@calor.co.uk, www.calor.co.uk
Calpeda Limited - Wedgwood Road Industrial Estate, Bicester, Oxon. OX26 4UL tel:(01869) 241441, fax:(01869) 240681. pumps@calpeda.co.uk, www.calpeda.co.uk
Cambridge Structures Ltd - 2 Huntingdon Street, St. Neots, Cambs. PE19 1BG tel:(01480) 477700, fax:(01480) 477766. info@cambridgestructures.com, www.cambridgestructures.com
Camfil Farr Ltd - Knowsley Park Way, Haslingden, Lancs. BB4 4RS tel:(01706) 238000, fax:(01706) 226736. dustcollectors@camfil.co.uk, www.farrapc.com
Camlok Lifting Clamps Ltd - Knutsford Way, Sealand Industrial Estate, Chester, Cheshire. CH14NZ tel:(01244) 375375, fax:(01244) 377403. sales.uk@cmco.eu, www.camlok.co.uk
Cannock Gates UK Ltd - Martindale Industrial Estate, Hawks Green, Connock, Staffs. WS11 7XT tel:(01543) 462500, fax:(01543) 506237. sales@cannockgates.co.uk, www.cannockgates.co.uk
Cannon Hygiene Ltd - Northgate, White Lund Industrial Estate, Morecambe, Lancs. LA3 3BJ tel:(0870) 444 1988, fax:(0807) 444 3938. hygiene@cannonhygiene.com, www.cannonhygiene.co.uk
Canopy Products Ltd - Paradise Works, Paradise Street, Ramsbottom. BL0 9BS tel:(01706) 822665, fax:(01706) 823333. sales@canopyproducts.co.uk, www.canopyproducts.co.uk
Cantifix of London Limited - Unit 9, Garrick Industrial Centre, Irving Way, London. NW9 6AQ tel:(020) 8203 6203, fax:(020) 8203 6454. webenquiries@cantifix.co.uk, www.cantifix.co.uk
Capco Test Equipment, A division of Castlebroom Engineering Ltd - Unit 10, Farthing Road, Ipswich, Suffolk. IP1 5AP tel:(01473) 748144, fax:(01473) 748179. sales@capco.co.uk, www.capco.co.uk
Capel Fencing Contractors Ltd - 22 Sychem Lane, Five Oak Green, Tonbridge, Kent. TNI2 6TR tel:(01892) 836036, fax:(01892) 834844. info@capelfencing.co.uk, www.capelfencing.co.uk
Capital Safety Group (NE) Ltd - 5 A Merse Road, North Moons Moat, Redditch, Worcestershire. B98 9HL tel:(01527) 548000. information@capitalsafety.com, www.capitalsafety.com
Cardale Doors Ltd - Buckingham Road Industrial Estate, Brackley, Northants. NN13 7EA tel:(01280) 703022, fax:(01280) 701138. marketing@cardale.co.uk, www.cardale.com
Carleton Furniture Group Ltd - Mill Dam Lane, Monkill, Pontefract, W. Yorks. WF8 2NS tel:(01977) 700770, fax:(01977) 708740. sales@carletonfurniture.com, www.carletonfurniture.com
Carlton Main Brickworks Ltd - Grimethorpe, Barnsley, S. Yorks. S72 7BG tel:(01226) 711521, fax:(01226) 780417. sales@carltonbrick.co.uk, www.carltonbrick.co.uk
Caroflow Ltd - Edgebarn, 11 Market Hill, Royston, SG8 9JN tel:(01763) 244446, fax:(01763) 244111. info@caro.co.uk, www.caro.co.uk
Carpets of Worth Ltd - Townshed Works, Puxton Lane, Kidederminster, Worcs. DY11 5DF tel:(01562) 745000, fax:(01562) 732827. sales@carpetsofworth.co.uk, www.carpetsofworth.co.uk
Carr Gymnasium Equipment Ltd - Ronald Street, Radford, Nottingham, Notts. NG7 3GY tel:(0115) 942 2252, fax:(0115) 942 2276. carrofnottm@btconnect.com, www.carrofnottm.co.uk
Carron Phoenix Ltd - Stenhouse Road, Carron, Falkirk, Stirlingshire. FK2 8DW tel:(01324) 638321, fax:(01324) 620978. sales@carron.com, www.carron.com
Carter Concrete Ltd - Stonehill Way, Holt Road, Cromer, Norfolk. NR27 9JW tel:(01263) 516919, fax:(01263) 512399. mail@carter-concrete.co.uk, www.carter-concrete.co.uk
Carter Environmental Engineers Ltd - Hamilton House, 2 Lawley, Midleway, Birmingham. B4 7XL tel:(0121) 250 1415, fax:(0121) 250 1400. sales@cee.co.uk, www.cee.co.uk
Carter Retail Equipment Ltd - Redhill Road, Birmingham. B25 8EY tel:(0121) 250 1000, fax:(0121) 250 1005. info@cre-ltd.co.uk, www.cre-ltd.co.uk
Carter-Dal International - New Broad Street House, New Broad Street House, London. EC2M 1NH tel:(0207) 030 3395. info@carter-dal.com, www.Carter-Dal.com
Casella Measurement Ltd - Regent House, Wolseley Road, Kempston, Bedford, Beds. MK42 7JY tel:(01234) 844100, fax:(01234) 844150. info@casellameasurement.com, www.casellameasurement.com
Cast Metals Federation - The National Metal Forming Centre, 47 Birmingham Road, West Bromwich, W. Midlands. B70 6PY tel:(0121) 601 6390, fax:(0121) 601 6391. admin@cmfed.co.uk, www.castmetalsfederation.com
Castell Safety International - Kingsbury Works, Kingsbury Road, Kingsbury, London. NW9 - 8UR tel:(020) 8200 1200, fax:(020) 8205 0055. sales@castell.co.uk, www.castell.co.uk
Castle Care-Tech - North Street, Winkfield, Windsor, Berks. SL4 4SY tel:(01344) 887788, fax:(01344) 890024. sales@castle-caretech.com, www.castle-caretech.com
Castle Cement Ltd - Park Square, 3160 Solihull Parkway, Brimingham Business Park, Birmingham, W. Midlands. B37 7YN tel:(0121) 779 7771, fax:(0121) 779 7609. enquiries@hanson.com, www.castlecement.co.uk
Caswell & Co Ltd - 6 Princewood Road, Earlstrees Ind. Estate, Corby, Northants. NN17 4AP tel:(01536) 464800, fax:(01536) 464801. sales@caswell-adhesives.co.uk, www.caswell-adhesives.co.uk
Catering Equipment Suppliers Association - 235/237 Vauxhall Bridge Road, London. SW1V 1EJ tel:(020) 7233 7724, fax:(020) 7828 0667. enquiries@cesa.org.uk, www.cesa.org.uk
Catnic - Pontypandy Industrial Estate, Caerphilly, M. Glam. CF83 3GL tel:(029) 20337900, fax:(029) 20867796. catnic.technical@tatasteel.com, www.catnic.com
Caunton Engineering Ltd - Moorgreen Industrial Park, Moorgreen, Nottingham, Notts. NG16 3QU tel:(01773) 531111, fax:(01773) 532020. sales@caunton.co.uk, www.caunton.co.uk
Cave Tab Ltd - 1 Sovereign Court, South Portway Close, Round Spinney, Northampton, Northants. NN3 8RH tel:(01604) 798500, fax:(01604) 798505. info@civica.co.uk, www.cavetab.co.uk
Cavity Trays Ltd - Administration Centre, Boundary Avenue, Yeovil, Somerset. BA22 8HU tel:(01935) 474769, fax:(01935) 428223. sales@cavitytrays.co.uk, www.cavitytrays.com
CBS (Curtain and Blind Specialists Ltd) - 1 Fellgate, Morecambe. LA3 3PE tel:(01524) 383000, fax:(01524) 383111. www.cbseurope.com, sales@cbseurope.com
CED Ltd - 728 London Road, West Thurrock, Grays, Essex. RM20 3LU tel:(01708) 867237, fax:(01708) 867230. sales@ced.ltd.uk, www.ced.ltd.uk
Cefndy Healthcare - Cefndy Road, Rhyl, Denbighshire. LL18 2HG tel:(01745) 343877, fax:(01745) 355806. cefndy.sales@denbighshire.gov.uk, www.cefndy.com
Ceiling & Lighting Ltd - 11 Leighton Avenue, Fleetwood. FY7 8BP tel:(01253) 874135, fax:(01253) 870612. info@ceiling-lighting.co.uk, www.ceiling-lighting.co.uk
Celotex Ltd - Lady Lane Industrial Estate, Hadleigh, Ipswich, Suffolk. IP7 6BA tel:(01473) 822093, fax:(01473) 820880. info@celotex.co.uk, www.celotex.co.uk
Celuform Building Products - Billet Lane, Normanby Enterprise Park, Normanby Road, Scunthorpe, North Lincolnshire. DN15 9YH tel:(08705) 920930, fax:(08700) 720930. info@celuform.co.uk, www.celuform.co.uk
CEM Systems Ltd - 195 Airport Road West, Belfast. BT3 9ED tel:(028) 90 456767, fax:(028) 90 454535. cem.sales@tycoint.com, www.cemsys.com
Cembrit Ltd - 57 Kellner Road, London. SE28 0AX tel:(020) 8301 8900, fax:(020) 8301 8901. sales@cembrit.co.uk, www.cembrit.co.uk
Cement Admixtures Association - 38A Tilehouse Green Lane, Knowle, W. Midlands. B93 9EY tel:(01564) 776362. info@admixtures.org.uk, www.admixtures.org.uk
CEMEX - CEMEX House, Coldharbour Lane, Thorpe, Egham, Surrey. TW20 8TD tel:(01932) 583600, fax:(01932) 583611. elizabeth.young@cemex.com, www.cemex.co.uk
Cento Engineering Company Ltd - Unit 6, Baddow Park, Great Baddow, Chelmsford, Essex. CM2 7SY tel:(01245) 477708, fax:(01245) 477748
Cep Ceilings Ltd - Verulam Road, Common Road Industrial Estate, Stafford. ST16 3EA tel:(01785) 223435, fax:(01785) 251309. info@cepceilings.com, www.cepceilings.com
CEP Claddings Ltd - Wainwright Close, Churchfields, Hastings, East Sussex. TN38 9PP tel:(01424) 852641, fax:(01424) 852797. claddings@cepcladdings.com, www.cepcladdings.com
Certikin International Ltd - Unit 9, Witan Park, Avenue 2, Station Lane Industrial Estate, Witney, Oxon. OX28 4FJ tel:(01993) 778855, fax:(01993) 778620. info@certikin.co.uk, www.certikin.co.uk
CFS Carpets - Arrow Valley, Claybrook Drive, Redditch, Worcestershire. B98 0FY tel:(01527) 511860, fax:(01527) 511864. sales@cfscarpets.co.uk, www.cfscarpets.co.uk
CGL Cometec Ltd - 2 Young Place, Kelvin Hill, East Kilbridge, Lanarkshire. G75 0TD tel:(01355) 235561, fax:(01355) 247189. sales@cglsystems.co.uk, www.cglsystems.co.uk
Challenge Fencing Ltd - Downside Road, Cobham, Surrey. KT11 3LY tel:(01932) 866555, fax:(01932) 866445. cobhamsales@challengefencing.com, www.challengefencing.com
Chalmit Lighting - 388 Hillington Road, Glasgow, Lanarkshire. G52 4BL tel:(0141) 882 5555, fax:(0141) 883 3704. info@chalmit.com, www.chalmit.com
Chameleon Stone Ltd - Unit 3 Longton Industrial Estate, Winterstoke Road, Weston-Super-Mare, N. Somerset. BS23 3YB tel:(01934) 616275, fax:(01934) 616279. info@chameleonstone.co.uk, www.tilesofnaturalstone.co.uk
Channel Safety Systems Ltd - 9 Petersfield Business Park, Bedford Road, Petersfield, Hants. GU32 3QA tel:(0870) 243 0931, fax:(0870) 243 0932. sales@channelsafety.co.uk, www.channelsafety.co.uk
Channel Woodcraft Ltd - Kent. tel:(01303) 850231, fax:(01303) 850734. www.channelwoodcraft.com, www.channelwoodcraft.com
Chapel Studio - Bridge Road, Hunton Bridge, Kings Langley, Herts. WD4 8RE tel:(01923) 226386, fax:(01923) 269707. customer@chapelstudio.co.uk, www.btinternet.com/~chapelstudio
Charcon Tunnels (Division of Tarmac plc) - PO Box 1, Southwell Lane, Kirby-in-Ashfield, Notts. NG17 8GQ tel:(01623) 754493, fax:(01623) 759825. buildingproducts@tarmac.co.uk, www.tarmacbuildingproducts.co.uk
Charles Lawrence Surfaces Ltd - Brunel House, Jessop Way, Newark, Notts. NG24 2ER tel:(01636) 615866, fax:(01636) 615867. sales@charleslawrencesurfaces.co.uk, www.charleslawrencesurfaces.co.uk
Charles Lightfoot Ltd - Orchard House, Heywood Road, Sale. M33 3WB tel:(0161) 973 6565. info@charleslightfoot.co.uk, www.charleslightfoot.co.uk
Charnwood Fencing Limited - Beveridge Lane, Bardon Hill, Leicester, Leics. LE67 1TB tel:(01530) 835835, fax:(01530) 814545. darren@charnwoodfencing.com, www.charnwoodfencing.com
Chartered Building Company Scheme - Englemere, Kings Ride, Ascot, Berks. SL5 7TB tel:(01344) 630 743, fax:(01344) 630771. cbcinfo@ciob.org.uk, www.cbcscheme.org.uk
Chartered Building Consultancy Scheme - Englemere, Kings Ride, Ascot, Berks. SL5 7TB tel:(01344) 630700, fax:(01344) 630777. cbcinfo@ciob.org.uk, www.cbcscheme.org.uk
Chartered Institute of Arbitrators - International Arbitration Centre, 12 Bloomsbury Square, London. WC1A 2LP tel:(020) 7421 7444, fax:(020) 7404 4023. info@ciarb.org, www.arbitrators.org
Chartered Institute of Building - 1 Arlington Square, Downshire Way, Bracknell, Berkshire. RG12 1WA tel:(01344) 630 700, fax:(01344) 306 430. info@ciob-mail.org.uk, www.ciob.org
Chartered Institute of Plumbing & Heating Engineering - 64 Station Lane, Hornchurch, Essex. RM12 6NB tel:(01708) 472791, fax:(01708) 448987. info@ciphe.org.uk, www.ciphe.org.uk
Chartered Institute of Wastes Management - 9 Saxon Court, St Peters Gardens, Northampton, Northants. NN1 1SX tel:(01604) 620426, fax:(01604) 621339. ciwm@ciwm.co.uk, www.ciwm.co.uk
Chartered Institution of Building Services Engineers (CIBSE) - Delta House, 222 Balham High Road, London. SW12 9BS tel:(020) 8675 5211, fax:(020) 8675 5449. www.cibse.org

C (con't)

Chase Equipment Ltd - Wellington House, Sangwin Road, Coseley, Bilston, W. Midlands. WV14 9EE tel:(01902) 675835, fax:(01902) 674998. sales@chaseequipment.com, www.chaseequipment.com

Checkmate Industries Ltd - Bridge House, Bridge Street, Halstead, Essex. CO9 1HT tel:(01787) 477272, fax:(01787) 476334. checkmatecarpets@btopenworld.com, www.checkmatecarpets.co.uk

Chelsea Artisans Ltd - Units 1-2A Pylon Way, Beddington Farm Road, Croydon. CR0 4XX tel:(0208) 665 0558, fax:(0208) 689 62880. info@chelsea-fusion.com, www.chelsea-fusion.com

Chesterfelt Ltd - Foxwood Way, Sheepbridge, Chesterfield, Derbys. S41 9RX tel:(01246) 268000, fax:(01246) 268001. general@chesterfelt.co.uk, www.chesterfelt.co.uk

Chestnut Products - PO Box 260, Stowmarket. IP14 9BX tel:(01473) 890118, fax:(01473) 206522. mailroom@chestnutproducts.co.uk, www.chestnutproducts.co.uk

Chestnut Products Limited - Unit 15, Gaza Trading Estate, Hildenborough, Tonbridge, Kent. TN11 8PL tel:(01732) 463 777, fax:(01732) 454 636. sales@chestnut-products.co.uk, www.chestnut-products.co.uk

Chichester Stoneworks Limited - Terminus Road, Chichester, West Sussex. PO19 8TX tel:(01243) 784225, fax:(01243) 785616. info@csworks.co.uk, www.chichesterstoneworks.co.uk

Chilstone - Victoria Park, Fordcombe Road, Langton Green, Tunbridge Wells, Kent. TN3 0RD tel:(01892) 740866, fax:(01892) 740249. office@chilstone.com, www.chilstone.com

Chiltern Invadex - Chiltern House, 6 Wedgewood Road, Bichester, Oxon. OX26 4UL tel:(01869) 246470, fax:(01869) 247214. sales@chilterninvadex.co.uk, www.chilterninvadex.co.uk

Chiorino UK Ltd - Phoenix Avenue, Green Lane Industrial Park, Featherstone, West Yorkshire. WF7 6EP tel:(01977) 691880, fax:(01977) 791 547. sales@chiorino.co.uk, www.chiorino.co.uk

Chippindale Plant Ltd - Butterbowl Works, Lower Wortley Ring Road, Leeds, W. Yorks. LS12 5AJ tel:(0113) 263 2344, fax:(0113) 279 1710. nigel.chippindale@chippindale-plant.co.uk, www.chippindale-plant.co.uk

Chris Brammall - Low Mill Business Park, Morecambe Road, Ulverston. LA12 9EE tel:(01229) 588580, fax:(01229) 588581. info@chrisbrammall.com, www.chrisbrammall.com

Chris Wheeler Construction Limited - Church Farm, Burbage, Nr Marlborough, Wilts. SN8 3AT tel:(01672) 810315, fax:(01672) 810309. cw.cw@btinternet.com, www.chriswheelerconstruction.co.uk

Christie & Grey Ltd - Morley Road, Tonbridge, Kent. TN9 1RA tel:(01732) 371100, fax:(01732) 359666. sales@christiegrey.com, www.christiegrey.com

Chromalox (UK) L td - Unit 1.22, Lombard House, 2 Purley Way, Croydon, Surrey. CR0 3JP tel:(0208) 6658900, fax:(0208) 6890571. uksales@chromalox.com, www.chromalox.co.uk

Chubb Systems Ltd - Shadsworth Road, Blackburn. BB1 2PR tel:(01254) 688583, fax:(01254) 667663. systems-sales@chubb.co.uk, www.chubbsystems.co.uk

CICO Chimney Linings Ltd - North End Wood, Hinton Road, Darsham, Suffolk. IP17 3QS tel:(01986) 784 044, fax:(01986) 784 763. cico@chimney-problems.co.uk, www.chimney-problems.co.uk

Cil Retail Solutions Ltd - Unit 8, Temple House Estate, 7 West Road, Harlow, Essex. CM20 2DU tel:(01279) 444448. info@cil.co.uk, www.cil.co.uk

Cintec International Ltd - Cintec House, 11 Gold Tops, Newport, S. Wales. NP20 4PH tel:(01633) 246614, fax:(01633) 246110. hqcintec@cintec.com, www.cintec.com

Ciret Limited - Total Logistics Building, FulFlood Road, Havant, Hampshire. PO9 5AX tel:(02392) 457450, fax:(02392) 457451. enquiries@ciret.co.uk, www.ciret.co.uk

Cistermiser Ltd - Unit 1, Woodley Park Estate, 59-69 Reading Raod, Woodley, Reading. RG5 3AN tel:(0118) 969 1611, fax:(0118) 944 1426. sales@cistermiser.co.uk, www.cistermiser.co.uk

Civic Trust - Essex Hall, 1-6 Essex Street, London. WC2R 3HU tel:(020) 7539 7900, fax:(020) 7539 7901. info@civictrust.org.uk, www.civictrust.org.uk

Civil & Marine Slag Cement Ltd - London Road, Grays, Essex. RM20 3NL tel:(01708) 864813, fax:(01708) 865907. webmail@claytonmunroe.com, www.civilandmarine.com

Civil Engineering Contractors Association - 1 Birdcage Walk, London. SW1H 9JJ tel:(0207) 340 0450. lauraellis@ceca.co.uk, www.ceca.co.uk

Clam Brummer Ltd - PO Box 413, Hatfield, Herts. AL10 1BZ tel:(01707) 274813, fax:(01707) 266846. sales@woodfillers.co.uk, www.woodfillers.co.uk

Clare R S & Co Ltd - Stanhope Street, Liverpool, Merseyside. L8 5RQ tel:(0151) 709 2902, fax:(0151) 709 0518. sales@rsclare.co.uk , www.rsclare.com

Clark Door Ltd - Unit F Central, Kingmoor Park, Carlisle, Cumbria. CA6 4SJ tel:(01228) 522321, fax:(01228) 401854. mail@clarkdoor.com, www.clarkdoor.com

Clarke Instruments Ltd - Distloc House, Old Sarum Airfield, The Portway, Salisbury, Wilts. SP4 6DZ tel:(01722) 323451, fax:(01722) 335154. sales@clarke-inst.com, www.clarke-inst.com

Clarke UK Ltd - Grange Works, Lomond Road, Coatbridge, Lanarkshire. ML5 2NN tel:(0123) 6707560, fax:(0123) 6427274. dmurray@clarkefire.com, www.clarkefire.com

Clarksteel Ltd - Station Works, Station Road, Yaxley, Peterborough. PE7 3EG tel:(01733) 240811, fax:(01733) 240201. sales@clarksteel.com, www.clarksteel.com

Clay Pipe Development Association Ltd - Copsham House, 53 Broad Street, Chesham, Bucks. HP5 3EA tel:(01494) 791456, fax:(01494) 792378. cpda@aol.com, www.cpda.co.uk

Claydon Architectural Metalwork Ltd - Units 11/12 Claydon Industrial Park, Gipping Road, Great Blakenham, Suffolk. IP6 0NL tel:(01473) 831000, fax:(01473) 832154. sales@cam-ltd.co.uk, www.cam-ltd.co.uk

Clayton Munroe - 2B Burke Road, Totness Industrial Estate, Totnes, Devon. TQ9 5XL tel:(01803) 865700, fax:(01803) 840720. sales@claytonmunroe.com, www.claytonmunroe.com

Clearline Architectural - Christopher Martin Road, Basildon, Essex. SS14 3ES tel:(01268) 522861, fax:(01268) 282994. pat.pinnock@ka-group.com, www.clearlinearchitectural.com

Clement Steel Windows - Clement House, Haslemere, Surrey. GU27 1HR tel:(01428) 647700, fax:(01428) 661369. info@clementwg.co.uk, www.clementwg.co.uk

Clenaware Systems Ltd - The Wagon Hovel, Strixton Manor Business Centre, Strixton, Wellingborough, Northamptonshire. NN29 7PA tel:(01933) 666244, fax:(01933) 665 584. info@clenaware.com, www.clenaware.com

Clestra Limited - 1st Floor, Kent House, 27-33 Upper Mulgrave Road, Cheam, Surrey. SM2 7AY tel:(020) 8773 2121, fax:(020) 8773 4793. uksales@clestra.com, www.clestra.com

Clifford Partitioning Co Ltd - Champion House, Burlington Road, New Malden. KT3 4NB tel:(0208) 7861974, fax:(0208) 7861975. cpartition@aol.com, www.cliffordpartitioning.co.uk

Clifton Nurseries - 5a Clifton Villas, Little Venice, London. W9 -2PH tel:(020) 7289 6851, fax:(020) 7286 4215. cld.admin@clifton.co.uk, www.clifton.co.uk

Climate Center - The Wolseley Center, Harrison Way, Spa Park, Royal Leamington Spa. CV31 3HH 834498. customerservices@wolseley.co.uk, www.climatecenter.co.uk

Clivet UK Ltd - 4 Kingdom Close, Segensworth East, Hants. PO15 5TJ tel:(01489) 572238, fax:(01489) 573033. l.joy@clivet-uk.co.uk, www.clivet.com

Clow Group Ltd - Diamond Ladder Factory, 562-584 Lea Bridge Road, Leyton, London. E10 7DW tel:(020) 8558 0300, fax:(020) 8558 0301. clow@ladders-direct.co.uk, www.ladders-direct.co.uk

Coates Fencing Limited - Unit 3, Barhams Close, Wylds Road, Bridgwater, Somerset. TA6 4DS tel:(01278) 423577, fax:(01278) 427760. info@coatesfencing.co.uk, www.coatesfencing.co.uk

Coates LSF Ltd - 5 Gardeners Place, West Gillibrands, Skelmersdale. WN8 9SP tel:(01695) 727011, fax:(01695) 727611. reception@coateslsf.com, www.coateslsf.com

Coblands Landscapes Ltd - South Farm Barn, South Farm Lane, Langton Green, Tunbridge Wells, Kent. TN3 9JN tel:(01892) 863535, fax:(01892) 863778. info@coblandslandscapes.co.uk, www.coblandslandscapes.com

Coburn Sliding Systems Ltd - Unit 1, Cardinal West, Cardinal Distribution Park, Cardinal Way, Godmanchester, Huntingdon, Cambridgeshire. PE29 2XN tel:(020) 8845 6680, fax:(020) 8545 6720. sales@coburn.co.uk, www.coburn.co.uk

Cofely Engineering Services Ltd - Apian House, Selinas Lane, Dagenham, Essex. RM8 1TB tel:(020) 8227 7200, fax:(020) 8227 7210. engineering.services@cofely-gdfsuez.com, www.axima-uk.com

Colas Ltd - Cakemore Road, Rowley Regis, Warley, W. Midlands. B65 0QU tel:(0121) 561 4332, fax:(0121) 559 5217. colas@colas.co.uk, www.colas.co.uk

Cold Rolled Sections Association - Robson Rhodes, Centre City Tower, 7 Hill Street, Birmingham. B5 4UU tel:(0121) 6016350, fax:(0121) 601 6373. crsa@crsauk.com, www.crsauk.com

Colebrand Ltd - Cloebrand House, 18-20 Warwick Street, London. W1R 6BE tel:(020) 7439 1000, fax:(020) 7734 3358. enquiries@colebrand.com, www.colebrand.com

Coleford Brick & Tile Co Ltd - The Royal Forest of Dean Brickworks, Hawkwell Green, Cinderford, Glos. GL14 3JJ tel:(01594) 822160, fax:(01594) 826655. sales@colefordbrick.co.uk, www.colefordbrick.co.uk

Collier W H Ltd - Church Lane, Marks Tey, Colchester, Essex. CO6 1LN tel:(01206) 210301, fax:(01206) 212540. sales@whcollier.co.uk, www.whcollier.co.uk

Collins Walker Ltd, Elec Steam & Hot Water Boilers - Unit 7a, Nottingham S & Wilford Ind. Est, Ruddington Lane, Nottingham. NG11 7EP tel:(01159) 818044, fax:(01159) 455376. enquiries@collins-walker.co.uk, www.collins/walker.co.uk

Colman Greaves - Unit 3, Rutland Street, Cockbrook, Ashton-under-Lyne, Lancs. OL6 6TX tel:(0161) 330 9316, fax:(0161) 339 5016. info@colmangreaves.com, www.colmangreaves.com

Colorpro Systems Ltd - Whitehead Estate, Docks Way, Newport, Gwent. NP20 2NW tel:(01633) 223854, fax:(01633) 220 175. info@colorgroup.co.uk, www.colorgroup.co.uk

Colt International - New Lane, Havant, Hants. PO9 2LY tel:(023) 92451111, fax:(023) 92454220. info@coltgroup.com, www.coltinfo.co.uk

Coltman Precast Concrete Ltd - London Road, Canwell, Sutton Coldfield, W. Midlands. B75 5SX tel:(01543) 480482, fax:(01543) 481587. general@coltman.co.uk, www.coltman.co.uk

Comar Architectural Aluminuim Systems - Unit 5, The Willow Centre, 17 Willow Lane, Mitcham, Surrey. CR4 4NX tel:(020) 8685 9685, fax:(020) 8646 5096. sales@parksidegrp.co.uk, www.comar-alu.co.uk

Comber Models - 17 London Lane, London. E8 3PR tel:(020) 8533 6592, fax:(020) 8533 5333. info@combermodels.com, www.combermodels.com

Combustion Linings Ltd - Jacaidem Works, Walley Street, Burslem, Stoke-on-Trent. ST6 2AH tel:(01782) 822712, fax:(01782) 823920. jeff.hurst@combustionlinings.com, www.combustionlinings.com

Commercial Lighting Systems - Units 16 &17, Chandlers Way, Swanick, Southampton. SO31 1FQ tel:(01489) 581002, fax:(01489) 576262. sales@commercial-lighting.co.uk, www.commercial-lighting.co.uk

Commodore Kitchens Ltd - Acorn House, Gumley Road, Grays, Essex. RM20 4XP tel:(01375) 382323, fax:(01375) 394955. info@commodorekitchens.co.uk, www.commodorekitchens.co.uk

Company name: Britannia Metalwork Services Ltd - Units N1-N4 Andoversford Link, Andoversford Industrial Estate, Cheltenham, Gloucestershire. GL54 4LB tel:(01242) 820 037. info@britannia.uk.com, www.britannia.uk.com

Component Developments - Halesfield 10, Telford, Shropshire. TF7 4QP tel:(01952) 588488, fax:(01952) 684395. sales@componentdevelopments.com, www.componentdevelopments.com

Composite Panel Services Ltd - CPS House, Clay Street, Hull, N. Humberside. HU8 8HA tel:(01482) 620277, fax:(01482) 587121. info@cps-hull.co.uk, www.cps-hull.co.uk

Compriband Ltd - Bentall Business Park, Glover, District II, Washington, Tyne & Wear. NE37 3JD tel:(0191) 4196860, fax:(0191) 4196861. uk.sales@tremco-illbruck.com, www.compribrand.co.uk

Computertel Ltd - CTL House, 52 Bath Street, Gravesend, Kent. DA11 0DF tel:(01474) 561111, fax:(01474) 561122. sales@computertel.co.uk, www.computertel.co.uk

Comyn Ching & Co (Solray) Ltd - Garngoch Ind. Est, Phoenix Way, Gorseinon, Swansea. SA4 9WF tel:(01792) 892211, fax:(01792) 898855. sales@solray.co.uk, www.solray.co.uk

Concept Sign & Display Ltd - Unit 320 Fort Dunlop, Fort Parkway, Birmingham. B24 9FD tel:(0121) 6930005, fax:(0121) 7473602. signs@conceptsigns.co.uk, www.conceptsigns.co.uk

Concordia Electric Wire & Cable Co. Ltd - Derwent Street, Long Eaton, Nottingham, Notts. NG10 3LP tel:(0115) 946 7400, fax:(0115) 946 1026. sales@concordiatechnologies.com, www.concordia.ltd.uk

Concrete Block Association - 60 Charles Street, Leicester. LE1 1FB tel:(0116) 222 1507, fax:(0116) 251 4568. is@britishprecast.org, www.cba-blocks.org.uk

Concrete Pipeline Systems Association - The Old Rectory, Main Street, Glenfield, Leicestershite. LE3 8DG tel:(0116) 232 5170, fax:(0116) 232 5197. email@concretepipes.co.uk, www.concretepipes.co.uk

Concrete Products (Lincoln) 1980 Ltd - Riverside Industrial Estate, Skellingthorpe Road, Saxilby, Lincoln, Lincs. LN1 2LR tel:(01522) 704158, fax:(01522) 704233. info@armortec.co.uk, www.armortec.co.uk

Concrete Repairs Ltd - Cathite House, 23a Willow Lane, Mitcham, Surrey. CR4 4TU tel:(020) 8288 4848, fax:(020) 8288 4847. jdrewett@concrete-repairs.co.uk, www.concrete.repairs.co.uk

Conex Universal Limited - Global House, 95 Vantage Point, The Pensnett Estate, Kingswinford, West Midlands. DY6 7FT tel:(0121) 557 2831 2955, fax:(0121) 520 8778. salesuk@ibpgroup.com, www.conexbanninger.com

Confor: Promoting forestry and wood - 59 George Street (3rd Floor), Edinburgh. EH2 2JG tel:(0131) 240 1410, fax:(0131) 240 1411. mail@confor.org.uk, www.confor.org.uk

Connaught Communications Systems Ltd - Systems House, Reddicap Estate, Sutton Coalfield, Birmingham. B75 7BU tel:(0121) 311 1010, fax:(0121) 311 1890. info@connaughtltd.co.uk, www.connaughtltd.co.uk

Connect Lighting Systems (UK) Ltd - 1B Wessex Gate, Portsmouth Road, Horndean, Hants. PO8 9LP tel:(023) 9257 0098, fax:(023) 9257 0097. sales@connectlighting.co.uk, www.connectlighting.co.uk

Conren Ltd - Unit 1 The Bridge Business Centre Ash Road South, Wrexham Industrial Estate, Wrexham. LL13 9UG tel:(01978) 661 991, fax:(01978) 664 664. info@conren.com, www.conren.com

Consort Equipment Products Ltd - Thornton Industrial Estate, Milford Haven, Pembs. SA73 2RT tel:(01646) 692172, fax:(01646) 695195. enquiries@consortepl.com, www.consortepl.com

Constant Air Systems Ltd - CAS House, Hillbottom Road, Sands Industrial Estate, High Wycombe, Bucks. HP12 4HJ tel:(01494) 469529, fax:(01494) 469549. admin@constantair.co.uk, www.constantair.co.uk

Construction Employers Federation - 143 Malone Road, Belfast. BT9 6SU tel:(028) 90 877143, fax:(028) 90 877155. mail@cefni.co.uk, www.cefni.co.uk

Construction Group UK Ltd - Dexion Comino Ltd, Murdock Road, Dorcan, Swindon, Wiltshire. SN3 5HY tel:(0870) 224 0220 , fax:(0870) 224 0221. enquiries@dexion.co.uk, www.dexion.co.uk

Construction Industry Publications - C/o Asendia Ltd, 2B Viking Industrial Estate, Hudson Road, Bedford. MK41 0QB tel:(0870) 078 4400, fax:(0870) 078 4401. sales@cip-books.com, www.cip-books.com

Construction Industry Research & Information Association - Griffin Court, 15 Long Lane, London. EC1A 9PN tel:(020) 7549 3300, fax:(020) 7549 3349. enquiries@ciria.org, www.ciria.org

Construction Industry Training Board - CITB, Bircham Newton, Kings Lynn, Norfolk. PE31 6RH tel:(01485) 577577, fax:(01485) 577793. publications@cskills.org, www.citb.org.uk

Construction Plant-hire Association - 27/28 Newbury Street, Barbican, London. EC1A 7HU tel:(020) 7796 3366, fax:(020) 7796 3399. enquiries@cpa.uk.net, www.cpa.uk.net

Constructionline - PO Box 6441, Basingstoke, Hants. RG21 7FN tel:(0844) 892 0312, fax:(0844) 892 0315. constructionline@capita.co.uk, www.constructionline.co.uk

Consult Lighting Ltd - 92 Grange Lane (rear Of), Barnsley, S. Yorks. S71 5QQ tel:(01226) 956820. john@consultlighting.co.uk, www.consultlighting.co.uk

Contactum Ltd - Edgware Road, London. NW2 6LF tel:(020) 8452 6366, fax:(020) 8208 3340. general@contactum.co.uk, www.contactum.co.uk

Continental Shutters Ltd - Unit 1 Heatway Industrial Estate, Manchester Way, Wantz Road, Dagenham, Essex. RH10 8PN tel:(020) 8517 8877, fax:(020) 8593 5721. sales@continentalshutters.com, www.continentalshutters.com

Continental Sports Ltd - Hill Top Road, Paddock, Huddersfield. HD1 4SD tel:(01484) 542051, fax:(01484) 539148. sales@contisports.co.uk, www.contisports.co.uk

Contour Showers Ltd - Siddorn Street, Winsford, Cheshire. CW7 2BA tel:(01606) 592586, fax:(01606) 861260. sales@contour-shower.co.uk, www.contour-showers.co.uk

Contract Flooring Association - 4C St Marys Place, The Lace Market, Nottingham. NG1 1PH tel:(0115) 941 1126, fax:(0115) 941 2238. info@cfa.org.uk, www.cfa.org.uk

Contract Journal - Reed Business Information, Quadrant House, The Quadrant, Sutton, Surrey. SM2 5AS tel:(020) 8652 4805, fax:(020) 8652 4804. contract.journale@rbi.co.uk, www.contractjournal.com

Cooke Brothers Ltd - Northgate, Aldridge, Walsall, W. Midlands. WS9 8TL tel:(01922) 740001, fax:(01922) 456227. sales@cookebrothers.co.uk, www.cookebrothers.co.uk

Cookson and Zinn (PTL) Ltd - Station Road Works, Hadleigh, Ipswich, Suffolk. IP7 5PN tel:(01473) 825200, fax:(01473) 824164. info@czltd.com, www.czltd.com

Cooper & Turner - Sheffield Road, Sheffield. S9 1RS tel:(0114) 256 0057, fax:(0114) 244 5529. sales@cooperandturner.co.uk, www.cooperandturner.co.uk

Cooper Group Ltd - Unit 18, The Tanneries, Brockhampton Lane, Havant, Hants. PO9 1JB tel:(02392) 454405, fax:(02392) 492732. info@coopersfire.com, www.coopersblinds.co.uk

Cooper Lighting - Wheatley Hall Road, Doncaster, S. Yorks. DN2 4NB tel:(01302) 303303, fax:(01302) 367155. sales@cooper-ls.com, www.jsb-electrical.com

Cooper Lighting and Security Ltd - Wheatley Hall Road, Doncaster, S. Yorks. DN2 4NB tel:(01302) 303303, fax:(01302) 367155. sales@cooper-ls.com, www.jsb-electrical.com

C (con't)

Cooper Safety - Jephson Court, Tancred Close, Royal Leamington Spa, Warwickshire. CV31 3RZ tel:(01926) 439200, fax:(01926) 439240. sales@cooper-ls.com, www.cooper-safety.com

Coo-Var Ltd - Elenshaw Works, Lockwood Street, Hull. HU2 0HN tel:(01482) 328053, fax:(01482) 219266. info@coo-var.co.uk, www.coo-var.co.uk

Copley Decor Ltd - Unit 1 Leyburn Business Park, Leyburn, N. Yorks. DL8 -5QA tel:(01969) 623410, fax:(01969) 624398. mouldings@copleydecor.co.uk, www.copleydecor.com

Coppa Cutta Ltd - 8 Bottings Industrial Est, Hillson Road, Southampton, Hants. SO30 2DY tel:(01489) 797774, fax:(01489) 796700. coppagutta@good-directions.co.uk, www.coppagutta.com

Copper Development Association - 5 Grovelands Business Centre, Boundary Way, Hemel Hempstead, Herts. HP2 7TE fax:(01442) 275716. info@copperalliance.org.uk, www.copperalliance.org.uk

Coram Showers Ltd - Stanmore Industrial Estate, Bridgnorth, Shropshire. WV15 5HP tel:(01746) 766466, fax:(01746) 764140. sales@coram.co.uk, www.coram.co.uk

Corbett & Co (Galvanising) W Ltd - New Alexandra Works, Haledane, Halesfield 1, Telford, Shropshire. TF7 4QQ tel:(01952) 412777, fax:(01952) 412888. sales@wcorbett.co.uk, www.wcorbett.co.uk

Cordek Ltd - Spring Copse Business Park, Slinfold, W. Sussex. RH13 0SZ tel:(01403) 799601, fax:(01403) 791718. info@cordek.com, www.cordek.com

Corgi Technical Dervices - Unit 8, The Park Centre, Easter Park, Benyon Road, Silchester, Berkshire. RG7 2PQ tel:(01256) 548 040, fax:(01256) 548055. Enquiries@trustcorgi.org, www.corgitechnical.com

Corroless Corrosion Control - Kelvin Way, West Bromwich, W. Midlands. B70 7JZ tel:(0121) 524 2235, fax:(0121) 553 2787. contact@corroless.com, www.corroless.com

Cottage Craft Spirals - The Barn, Gorsty Low Farm, The Wash, Chapel-en-le-Firth, High Peak. SK23 0QL tel:(01663) 750716, fax:(01663) 751093. sales@castspiralstairs.com, www.castspiralstairs.com

Cottam & Preedy Ltd - Bishopsgate Works, 68 Lower City Road, Tividale, W. Midlands. B69 2HF tel:(0121) 552 5281, fax:(0121) 552 6895. enquiries@cottamandpreedy.co.uk, www.cottamandpreedy.co.uk

Cottam Bros Ltd - Sheepfolds Industrial Estate, Sunderland, Tyne & Wear. SR5 1AZ tel:(0191) 567 1091, fax:(0191) 510 8187. sales@cottambros.com, www.cottambros.com

Cottam Brush Ltd - Unit 7, Monkton Business Park North, Hebburn. NE31 2JZ tel:(0845) 4348436, fax:(0845) 4348437. info@cottambrush.com, www.cottambrush.com

Council for Aluminium in Building - River View House, Bond's Mill, Stonehouse, Glos. GL10 3RF tel:(01453) 828851, fax:(01453) 828861. enquiries@c-a-b.org.uk, www.c-a-b.org.uk

Coutts Information Services - Avon House, Unit 9, Headlands Business Park, Salisbury Road, Ringwood, Hampshire. BH24 3PB tel:(01425) 471160, fax:(01425) 471525. publisherservicesuk@couttsinfo.com, www.couttsinfo.com

Coverad - PSS (Technologies) Group, Unit F1, Grafton Way, Basingstoke, Hampshire. RG22 6HY tel:(01256) 844 685. sales@psswww.co.uk, www.coverad.co.uk

Covers Timber & Builders Merchants - Sussex House, Quarry Lane, Chichester, West Sussex. PO19 8PE tel:(01243) 785141, fax:(01243) 531151. enquiries@covers.biz, www.covers.biz

Coverscreen UK LLP - Unit 174, 78 Marylebone High Street, London. W1U 5AP tel:(0845) 680 0409 . sales@brass-grilles.co.uk, www.brass-grilles.co.uk

Coverworld UK Limited - Mansfield Road, Bramley Vale, Chesterfield, Derbyshire. S44 5GA tel:(01246) 858222, fax:(01246) 858223. sales@coverworld.co.uk, www.coverworld.co.uk

Cowley Structural Timberwork Ltd - The Quarry, Grantham Road, Waddington, Lincoln, Lincs. LN5 9NT tel:(01522) 803800, fax:(01522) 803801. admin@cowleytimberwork.co.uk

Cox Building Products Ltd - Unit 1, Shaw Road, Bushbury, Wolverhampton, W. Midlands. WV10 9LA tel:(01902) 371800, fax:(01902) 371810. sales@coxdome.co.uk, www.coxbp.com

Cox Long Ltd - Airfield Industrial Estate, Hixon, Stafford, Staffs. ST18 0PA tel:(01889) 270166, fax:(01889) 271041. info@coxlong.com

CPS Manufacturing Co - Brunel House, Brunel Close, Harworth, Doncaster. DN11 8QA tel:(01302) 741888, fax:(01302) 741999. sales@cpsmanufacturingco.com, www.cpsmanufacturingco.com

CPV Ltd - Woodington Mill, East Wellow, Romsey, Hants. SO51 6DQ tel:(01794) 322884, fax:(01794) 322885. sales@cpv.co.uk, www.cpv.co.uk

Craft Guild of Chefs - 1 Victoria Parade, by 331 Sandycombe Road, Richmond, Surrey. TW9 3NB tel:(020) 8948 3870. enquiries@craftguildofchefs.org, www.craftguildofchefs.org

Cranborne Stone Ltd - Butts Pond Industrial Estate, Sturminster Newton, Dorset. DT10 1AZ tel:(01258) 472685, fax:(01258) 471251. sales@cranbornestone.com, www.cranbornestone.com

Crane Composites - 25 Caker Stream Road, Mill Lane Industry Estate, Alton, Hampshire. GU34 2QF tel:(01420) 593270, fax:(01420) 541124. sales@cranecomposites.com, www.cranecomposites.com

Crane Fluid Systems - Crane House, Epsilon Terrace, West Road, Ipswich, Suffolk. IP3 9FJ tel:(01473) 277 300, fax:(01473) 277 301. enquiries@cranefs.com, www.cranefs.com

Crawford Hafa Ltd - 7 Churchill Way, 35a Business Park, Chapletown, Sheffield, S. Yorks. S35 2PY tel:(0114) 257 4330, fax:(0114) 257 4399. sales.uk@crawfordsolutions.com, www.crawforduk.co.uk

Creative Glass - Design House, 20-22 Lustrum Avenue, Portrack Lane, Stockton-on-Tees. TS18 2RB tel:(01642) 603545, fax:(01642) 604667. info@creativeglass.co.uk, www.creativeglass.co.uk

Creda NOBO - Millbrook House, Grange Drive, Hedge End, Southampton, Hants. SO30 2DF tel:(0845) 601 5111, fax:(0845) 604 2369. customerservices@credaheating.co.uk, www.credaheating.co.uk

Crendon Timber Engineering Ltd - Drakes Drive, Long Crendon, Aylesbury, Bucks. HP18 9BA tel:(01844) 201020, fax:(01844) 201625. sales@crendon.co.uk, www.crendon.co.uk

Crescent Lighting - 8 Rivermead, Pipers Lane, Thatcham, Berks. RG19 4EP tel:(01635) 878888, fax:(01635) 873888. sales@crescent.co.uk, www.crescent.co.uk

Crescent of Cambridge Ltd - Edison Road, St. Ives, Cambs. PE27 3LG tel:(01480) 301522, fax:(01480) 494001. info@crescentstairs.co.uk, www.crescentstairs.co.uk

Crimeguard Protection Ltd - Units 2 & 3, The Stables, Stanmore, Beedon, Newbury. RG20 8SR tel:(01635) 281 050. info@crime-guard.co.uk, www.crime-guard.co.uk

Crittall Steel Windows Ltd - 39 Durham Street, Glasgow. G41 1BS tel:(0141) 427 4931, fax:(0141) 427 1463. hq@crittall-windows.co.uk, www.crittall-windows.co.uk

Crittall Steel Windows Ltd - Springwood Drive, Braintree, Essex. CM7 2YN tel:(01376) 324106, fax:(01376) 349662. hq@crittall-windows.co.uk, www.crittall-windows.co.uk

Cross-Guard International Ltd - Bridge House, Severn Bridge Riverside North, Bewdley, Worcs. DY12 1AB tel:(01299) 406022, fax:(01299) 406023. info@cross-guard.co.uk, www.cross-guard.co.uk

Crown Nail Co Ltd - 48 Commercial Road, Wolverhampton. WV1 3QS tel:(01902) 351806, fax:(01902) 871212. sales@crown-nail.com, www.crown-nail.com

Crown Paints Ltd - Crown House, P O Box 37, Hollins Road, Darwen, Lancashire. BB3 0BG tel:(0330) 0240310, fax:(0845) 389 9457. info@crownpaintspec.co.uk, www.crownpaintspec.co.uk

CrowsonFabrics Ltd - Crowson House, Bolton Close, Bellbrook Ind Est, Uckfield, Sussex. TN22 1QZ tel:(01202) 753277, fax:(01202) 762582. sales@crowsonfabrics.com, www.monkwell.com

Crowthorne Fencing - Englemere Sawmill, London Road, Ascot, Berks. SL5 8DG tel:(01344) 885451, fax:(01344) 893101. ascot@crowthornefencing.co.uk, www.crowthorneascot.co.uk

Cryotherm Insulation Ltd - Hirst Wood Works, Hirst Wood Road, Shipley, W. Yorks. BD18 4BU tel:(01274) 589175, fax:(01274) 593315. enquiries@cryotherm.co.uk, www.cryotherm.co.uk

Cryselco Ltd - Cryselco House, 274 Ampthill Road, Bedford. MK42 9QL tel:(01234) 273355, fax:(01234) 210867. sales@cryselco.co.uk, www.cryselco.co.uk

CSSP (Construction Software Services Partnership) - 29 London Road, Bromley, Kent. BR1 1DG tel:(020) 8460 0022, fax:(020) 8460 1196. enq@cssp.co.uk, www.cssp.co.uk

CST Industries, Inc. - UK - Cotes Park Lane, Cotes Park Industrial Estate, Alfreton, Derbyshire. DE55 4NJ tel:(01773) 835321, fax:(01773) 836578. europe@cstindustries.com, www.cstindustries.com

C-TEC (Computionics) Ltd - Challenge Way, Martland Park, Wigan, Lancs. WN5 0LD tel:(01942) 322744, fax:(01942) 829867. sales@c-tec.co.uk, www.c-tec.co.uk

CTS Bridges Ltd - Abbey Road, Shepley, Huddersfield, W. Yorks. HD8 8BX tel:(01484) 606 416, fax:(01484) 608 763. enquiries@ctsbridges.co.uk, www.ctsbridges.co.uk

CU Lighting Ltd - Great Amwell, Ware, Herts. SG12 9AT tel:(01920) 462272, fax:(01920) 461370. sales@cuphosco.co.uk, www.cuphosco.com

CU Phosco Lighting - Charles House, Lower Road, Great Amwell, Ware, Herts. SG12 9TA tel:(01920) 860600, fax:(01920) 485915. sales@cuphosco.co.uk, www.cuphosco.com

Cumberland Construction - Orchard Court, 4 Station Square, Gidea Park, Romford, Essex. RM2 6AT tel:(01708) 766 369, fax:(01708) 520231. contracts@cumberlandgroup.co.uk, www.cumberland-construction.co.uk

Cummins Power Generation Ltd - Manston Park, Columbus Avenue, Manston, Ramsgate, Kent. CT12 5BF tel:(01843) 255000, fax:(01843) 255913. cpg.uk@cummins.com, www.cumminspower.com

Cutting R C & Co - Arcadia Avenue, London. N3 2JU tel:(020) 8371 0001, fax:(020) 8371 0003. info@rccutting.co.uk, www.rccutting.co.uk

D

Dacrylate Paints Ltd - Lime Street, Kirkby in Ashfield, Nottingham, Notts. NG17 8AL tel:(01623) 753845, fax:(01623) 757151. sales@dacrylate.co.uk, www.dacrylate.co.uk

Daikin Airconditioning UK Ltd - The Heights, Brooklands, Weybridge, Surrey. KT13 0NY tel:(0845) 6419000, fax:(0845) 6419009. marketing@daikin.co.uk, www.daikin.co.uk

Daikin Applied (UK) Ltd - Formerly McQuay (UK) Ltd, Bassington Lane, Cramlington, Northumberland. NE23 8AF tel:(01670) 566159, fax:(01670) 566206. info@daikinapplied.uk, www.daikinapplied.uk

Dalair Ltd - Southern Way, Wednesbury, W. Midlands. WS10 7BU tel:(0121) 556 9944, fax:(0121) 502 3124. sales@dalair.co.uk, www.dalair.co.uk

Dales Fabrications Ltd - Crompton Road Industrial Estate, Ilkeston, Derbys. DE7 4BG tel:(0115) 930 1521, fax:(0115) 930 7625. sales@dales-eaves.co.uk, www.dales-eaves.co.uk

Dalesauna Ltd - Grombald Crag Close, St James Business Park, Knaresborough, N. Yorks. HG5 8PJ tel:(01423) 798630, fax:(01423) 798670. sales@dalesauna.co.uk, www.dalesauna.co.uk

Dalhaus Ltd - Showground Road, Bridgwater, Somerset. TA6 6AJ tel:(01278) 727727, fax:(01278) 727766. info@dalhaus.co.uk, www.dalhaus.co.uk

Dampcoursing Limited - 10-12 Dorset Road, Tottenham, London. N15 5AJ tel:(020) 8802 2233, fax:(020) 8809 1839. dampcoursingltd@btconnect.com, www.dampcoursing.com

Danbury Fencing Limited - Olivers Farm, Maldon Road, Witham, Essex. CM8 3HY tel:(01376) 502020, fax:(01376) 520500. sales@danburyfencing.com, www.danburyfencing.com

Dandy Booksellers - Opal Mews, Units 3 & 4, 31-33 Priory Park Road, London. NW6 7UP tel:(020) 7624 2993, fax:(020) 7624 5049. enquiries@dandybooksellers.com, www.dandybooksellers.com

Danfoss Ltd - Ampthill Road, Bedford, Bedfordshire. MK42 9ER tel:(01234) 364621, fax:(01234) 219705. drl_uksalesoffice@danfoss.com, www.heating.danfoss.co.uk

Danfoss Ltd - Capswood, Oxforc Street, Denham, Bucks. UB9 4LH tel:(0870) 241 7030, fax:(0870) 241 7035. drl_uksalesoffice@danfoss.com, www.danfoss.co.uk

Daniel Platt Ltd - Canal Lane, Tunstall, Stoke-on-Trent, Staffs. ST6 4NY tel:(01782) 577187, fax:(01782) 577877. sales@danielplatt.co.uk, www.danielplatt.co.uk

Dantherm FiltrationLtd - Limewood Approach, Seacroft, Leeds. LS14 1NG tel:(0113) 273 9400, fax:(0113) 265 0735. info.uk@danthermfiltration.com, www.danthermfiltration.co.uk

Dantherm Ltd - Hither Green, Clevedon, Avon. BS21 6XT tel:(01275) 876851, fax:(01275) 343086. ikf.ltd.uk@dantherm.com

Darfen Durafencing - Unit 12 Smallford Works, Smallford Lane, St. Albans, Herts. AL4 0SA tel:(01727) 828290, fax:(01727) 828299. enquiries@darfen.co.uk, www.darfen.co.uk

Darfen Durafencing, Scottish Region - Units 7-8, Rochsolloch Road Industrial Estate, Airdrie, Scotland. ML6 9BG tel:(01236) 755001, fax:(01236) 747012. scottish@darfen.co.uk, www.darfen.co.uk

Daro Factors Ltd - 80-84 Wallis Road, London. E9 5LW tel:(020) 8510 4000, fax:(020) 8510 4001. sales@daro.com, www.daro.com

Dart Valley Systems Ltd - Kemmings Close, Long Road, Paignton, Devon . TQ4 7TW tel:(01803) 592021, fax:(01803) 559016. sales@dartvalley.co.uk, www.dartvalley.co.uk

Dartford Portable Buildings Ltd - 389-397 Princes Road, Dartford, Kent. DA1 1JU tel:(01322) 229521, fax:(01322) 221948. mark@dpbl.co.uk, www.dpbl.co.uk

Davant Products Ltd - Davant House, Jugs Green Business Park, Jugs Green, Staplow, nr Ledbury, Herefordshire. HR8 1NR tel:(01531) 640880, fax:(01531) 640827. info@davant.co.uk, www.davant.co.uk

Davicon - The Wallows Industrial Estate, Dudley Road, Brierley Hill. DY5 1QA tel:(01384) 572 851, fax:(01384) 265 098. sales@davicon.com, www.davicon.com

David Ball Group Ltd - Wellington Way, Bourn Airfield, Cambridge. CB23 2TQ tel:(01954) 780687, fax:(01954) 782912. sales@davidballgroup.com, www.davidballgroup.com

DC Plastic Handrails Ltd - Unit 3 TVB Factory, Whickham Industrial Estate, Swalwell, Newcastle Upon Tyne, Tyne & Wear. NE16 3DA tel:(0191) 488 1112, fax:(0191) 488 1112. sales@dcplastics.co.uk, www.dcplastics.co.uk

De Longhi Limited - 1 Kenwood Business Park, 1-2 New Lane, Havant, Hampshire. PO9 2NH tel:(0845) 600 6845. www.delonghi.com/en-gb

Deaf Alerter plc - Enfield House, 303 Burton Road, Derby. DE23 6AG tel:(01332) 363981, fax:(01332) 293267. info@deaf-alerter.com, www.deaf-alerter.com

Deans Blinds & Awnings (UK) Ltd - Unit 4, Haslemere Industrial Estate, Ravensbury Terrace, London. SW18 4SE tel:(020) 8947 8931, fax:(020) 8947 8336. info@deansblinds.com, www.deansblinds.com

Deb Ltd - Spencer Road, Belper, Derbys. DE56 1JX tel:(01773) 596700, fax:(01773) 822548. enquiry@deb.co.uk, www.deb.co.uk

Deceuninck Ltd - Unit 2, Stanier Road, Porte Marsh Industrial Estate, Calne, Wilts. SN11 9PX tel:(01249) 816969, fax:(01249) 815234. deceuninck.ltd@deceuninck.com, www.deceuninck.com

Decorfix Ltd - Chapel Works, Chapel Lane, Lower Halstow, Sittingbourne, Kent. ME9 7AB tel:(01795) 843124, fax:(01795) 842465. enquiries@decorfix.co.uk, www.decorfix.co.uk

Decra Ltd - 34 Forest Business Park, Argall Avenue, Leyton, London. E10 7FB tel:(020) 8520 4371, fax:(020) 8521 0605. sales@decraltd.co.uk, www.decraltd.co.uk

Decra Roofing Systems - Barton Dock Road, Stretford, Manchester. M32 0YL tel:(01293) 545058, fax:(01293) 562709. technical@decra.co.uk, www.decra.co.uk/

Dee-Organ Ltd - 5 Sandyford Road, Paisley, Renfrewshire. PA3 4HP tel:(0141) 889 7000, fax:(0141) 889 7764. signs@dee-organ.co.uk

Deepdale Engineering Co Ltd - Pedmore Road, Dudley, W. Midlands. DY2 0RD tel:(01384) 480022, fax:(01384) 480489. sales@deepdale-eng.co.uk, www.deepdale-eng.co.uk

Delabie UK Ltd - Henderson House, Hithercroft Road, Wallingford, Oxon. OX10 9DG tel:(01491) 824449, fax:(01491) 825729. sales@delabie.co.uk, www.delabie.co.uk

Delta Balustrades - Millbuck Way, Sandbach, Cheshire. CW11 3JA tel:(01270) 753383, fax:(01270) 753207. info@deltabalustrades.com, www.deltabalustrades.com

Delta Fire - 8 Mission Road, Rackheath Business Estate, Norwich. NR13 6PL tel:(01603) 735000, fax:(01603) 735009. sales@deltafire.co.uk, www.deltafire.co.uk

DencoHappel UK Ltd - Dolphin House, Moreton Business Park, Moreton-on-Lugg, Hereford. HR4 8DS tel:(01432) 277277, fax:(01432) 268005. sales.enquiry@dencohappel.com, www.dencohappel.com/en-gb

Denka International - Red Roofs, Chinnor Road, Thame, Oxon. OX9 3RF tel:(01844) 216754, fax:(01844) 216141. info@denka.nl

Dennis & Robinson Ltd - Blenhiem Road, Churchill Industrial Estate, Lancing, W. Sussex. BN15 8HU tel:(01903) 524300, fax:(01903) 750679. manhattan@manhattan.co.uk, www.manhattan.co.uk

Department for Business, Innovation & Skills - 1 Victoria Street, London. SW1H 0ET tel:(020) 7215 5000, fax:(020) 7215 6740. www.bis.gov.uk

Deralam Laminates Ltd - West Coast Park, Bradley Lane, Standish Wigan, Lancs. WN6 0YR tel:(01257) 478540, fax:(01257) 478550. sales@deralam.co.uk

Derby Timber Supplies - 3 John Street, Derby, Derbys. DE1 2LU tel:(01332) 348340, fax:(01332) 385573. info@derbytimbersupplies.co.uk, www.derbytimbersupplies.co.uk

Design & Display Structures Ltd - Unit 2, 158 Yateley Street, Westminster Industrial Estate, London. SE18 5TA tel:(0844) 736 5995, fax:(0844) 736 5992. grp@design-and-display.co.uk, www.design-and-display.co.uk

Design Council - Angel Building, 407 St John Street, London. EC1V 4AB tel:(020) 7420 5200, fax:(020) 7420 5300. info@designcouncil.org.uk, www.designcouncil.org.uk

Design Windows (South West) Ltd - PO Box 165, Fishponds, Bristol, Avon. BS16 7YP tel:(0117) 9702585, fax:(0117) 9567608. sales@designwindows.co.uk, www.designwindows.co.uk

Designed for Sound Ltd - 61-67 Rectory Road, Wivenhoe, Essex. CO7 9ES tel:(01206) 827171, fax:(01206) 826936. rs@d4s.co.uk, www.d4s.co.uk

Designer Construction Ltd - The Conservatory Trade Centre, 91 Ashacre Lane, Worthing. BN13 2DE tel:(01903) 831333, fax:(01903) 830724. enquiries@livingdaylight.co.uk, www.livingdaylight.co.uk

Designer Radiators Direct - Unit 12 Old Mill Industrial Estate, Bamber Bridge, Preston. PR5 6SY tel:(01772) 367540, fax:(01772) 314516. enquiries@designerradiatorsdirect.co.uk, www.designerradiatorsdirect.co.uk

Designplan Lighting Ltd - 16 Kimpton Park Way, Kimpton Business Park, Sutton, Surrey. SM3 9QS tel:(020) 8254 2020. sales@designplan.co.uk, www.designplan.co.uk

Desking Systems Ltd - Warpsgrove Lane, Chalgrove, Oxon. OX44 7TH tel:(01865) 891444, fax:(01865) 891427. sales@desking.co.uk, www.desking.co.uk

D (con't)

Detlef Muller Ltd - 82 Chobham Road, Frimley, Camberley, Surrey. GU16 5PP tel:(01276) 61967, fax:(01276) 64711. detlef.muller@btclick.com

Deva Composites Limited - 262 Ringwood Road, Parkstone, Poole, Dorset. BH14 0RS tel:(01202) 744115, fax:(01202) 744126. info@devacomposites.com, www.devacomposites.com

Deva Tap Co Ltd - Brooklands Mill, English Street, Leigh, Lancs. WN7 3EH tel:(01942) 680177, fax:(01942) 680190. sales@uk.methven.com, www.devatap.co.uk

Devar Premier Flooring - Spiersbridge Business Park, Thornliebank, Glasgow. G46 8NL tel:(0141) 638 2203, fax:(0141) 620 0207. enquiries@devargroup.com, www.devargroup.com

Dewey Waters Ltd - Heritage Works, Winterstoke Road, Weston-Super-Mare, N. Somerset. BS24 9AN tel:(01934) 421477, fax:(01934) 421488. sales@deweywaters.co.uk, www.deweywaters.co.uk

Dewhurst plc - Inverness Road, Hounslow, Middx. TW3 3LT tel:(020) 8607 7300, fax:(020) 8572 5986. info@dewhurst.co.uk, www.dewhurst.co.uk

Dezurik International Ltd - Unit 15, Orton Way, Hayward Industrial Estate, Castle Bromich, Birmingham. B35 7BT tel:(0121) 748 6111, fax:(0121) 747 4324. sales@industrialvalve.co.uk, www.industrialvalve.co.uk

Diamond Merchants - 43 Acre Lane, Brixton, London. SW2 5TN tel:(020) 72746624, fax:(020) 7978 8370. www.diamond-merchants.co.uk

Didsbury Engineering Co Ltd - Lower Meadow Road, Brooke Park, Handforth, Wilmslow, Cheshire. SK9 3LP tel:(0161) 486 2200, fax:(0161) 486 2211. sales@didsbury.com, www.didsbury.com

Diespeker Marble & Terrazzo Ltd - 132-136 Ormside Street, Peckham, London. SE15 1TF tel:(020) 7358 0160, fax:(020) 7358 2897. sales@diespeker.co.uk, www.diespeker.co.uk

Dimplex UK Limited - Millbrook House, Grange Drive, Hedge End, Southampton. SO30 2DF tel:(0870) 077 7117, fax:(0870) 727 0109. presales@dimplex.co.uk, www.dimplex.co.uk

Dinnington Fencing Co. Limited - North Hill, Dinnington, Newcastle-upon-Tyne. NE13 7LG tel:(01661) 824046, fax:(01661) 872234. info@dinningtonfencing.co.uk, www.dinningtonfencing.co.uk

Directional Data Systems Ltd - 5 Dalsholm Avenue, Dawsholm Industrial Estate, Glasgow. G20 0TS tel:(0141) 945 4243, fax:(0141) 945 4238. sales@directionaldata.co.uk, www.directionaldata.co.uk

Dividers Modernfold Ltd - Great Gutter Lane, Willerby, Hull, E. Yorks. HU10 6BS tel:(01482) 651331, fax:(01482) 651497. sales@dividersmodernfold.com, www.esperowalls.com

Dixon Turner Wallcoverings - Brecon Works, Henfaes Lane, Welshpool, Powys. SY21 7BE tel:(0870) 606 1237, fax:(0870) 606 1239. enquiries@dixon-turner.co.uk, www.dixon-turner.co.uk

DLW Flooring - St Centurion Court, Milton Park, Addington, Oxon. OX14 4RY tel:(01642) 763224, fax:(01642) 750213. sales@dlw.co.uk, www.dlw.co.uk

Docherty New Vent Chimney Group - Unit 3, Sawmill Road, Redshute Hill Indusrial Estate, Hermitage, Newbury, Berks. RG18 9QL tel:(01635) 200145, fax:(01635) 201737. sales@docherty.co.uk

Domus Tiles Ltd - Cantebury Court, Brixton Road, Battersea, London. SW9 6TA tel:(020) 7223 5555, fax:(020) 7924 2556. service@domustiles.com, www.domustiles.co.uk

Don & Low Ltd - Newfordpark House, Glamis Road, Forfar, Angus. DD8 IFR tel:(01307) 452200, fax:(01307) 452300. lotrak@donlow.co.uk, www.donlow.co.uk

Don Construction Products Ltd - Churnetside Business Park, Station Road, Cheddleton, Leek, Staffs. ST13 7RS tel:(01538) 361799, fax:(01538) 361899. info@donconstruction.co.uk, www.donconstruction.co.uk

Donaldson Filtration (GB) Ltd - Humberstone Lane, Thumaston, Leicester, Leics. LE4 8HP tel:(0116) 269 6161, fax:(0116) 269 3028. IAF-uk@donaldson.com, www2.donaldson.com/toritdce

Door and Hardware Federation - 42 Heath Street, Tamworth, Staffs. B79 7JH tel:(01827) 52337, fax:(01827) 310827. info@dhfonline.org.uk, www.dhfonline.org.uk

Door Panels by Design - Unit 27 Moorland Mills, Moorland Industrial Est, Law Street, Cleckheaton, Bradford. BD19 3QR tel:(01274) 852488, fax:(01274) 851819. lee.harding@dpbd.co.uk, www.dpbd.co.uk

Doorfit Products Ltd - Icknield House, Heaton Street, Hockley, Birmingham, W. Midlands. B18 5BA tel:(0121) 523 4171, fax:(0121) 554 3859. enquiries@doorfit.co.uk, www.doorfit.co.uk

Doors & Hardware Ltd - Taskmaster Works, Maybrook Road, Minworth, Sutton Coldfield. B76 1AL tel:(0121) 351 5276, fax:(0121) 313 1228. sales@doors-and-hardware.com, www.doors-and-hardware.com

Doosan Power Systems Ltd - Porterfield Road, Renfrew. PA4 8DJ tel:(0141) 886 4141, fax:+44 0 141 885 3338. dps.info@doosan.com, www.doosanpowersystems.com

Dorma Door Services Ltd - Wilbury Way, Hitchen, Herts. SG4 0AB tel:(01462) 477600, fax:(01462) 477601. info@dorma-uk.co.uk, www.dorma-uk.co.uk

DORMA UK Ltd - Wilbury Way, Hitchin, Herts. SG4 0AB tel:(01462) 477600, fax:(01462) 477601. info@dorma-uk.co.uk, www.dorma-uk.co.uk

Dorman Smith Switchgear - 1 Nile Close, Nelson Court Business Centre, Ashton-on-Ribble, Preston, Lancs. PR2 2XU tel:(01772) 325389, fax:(01772) 726276. sales@dormansmith.co.uk, www.dormansmithswitchgear.com/

Dover Trussed Roof Co Ltd - Shelvin Manor, Shelvin, Canterbury, Kent. CT4 6RL tel:(01303) 844303, fax:(01303) 844342. sales@dovertruss.co.uk, www.dovertruss.co.uk

Dow Building Solutions - Diamond House, Lotus Park, Kingsbury Crescent, Staines, Middlesex. TW18 3AG tel:(020) 3139 4000, fax:(020) 8917 5413. FKLMAIL@dow.com, www.dow.com

Dow Corning Europe S.A - Parc Industriel Zone C, Rue Jules Bordet, B-7180 Seneffe, Belgium. tel:+32 64 88 80 00, fax:+32 64 88 84 01. eutech.info@dowcorning.com, www.dowcorning.com

DP Sercurity - Ryecroft Buildings, Cireon Street Green Road, Longfield, Kent. DA2 8DX tel:(01322) 278178, fax:(01322) 284315. davidpaul120@hotmail.com, www.dpsecurity.co.uk

DPC Screeding Ltd - Brunwick Industrial Estate, Brunswick Village, Newcastle-upon-Tyne. NE13 7BA tel:(0191) 236 4226, fax:(0191) 236 2242. dpcscreeding@btconnect.com

Draeger Safety UK Ltd - Blyth Riverside Business Park, Ullswater Close, Blyth, Northumberland. NE24 4RG tel:(01670) 352 891, fax:(01670) 356 266. michaela.wetter@draeger.com, http://www.draeger.com/sites/en_uk/Pages/Applications/Advisor.aspx?navID=1875

Drain Center - The Wolseley Center, Harrison Way, Spa Park, Royal Leamington Spa. CV31 3HH tel:(0870) 16 22557. customerservices@wolseley.co.uk, www.draincenter.co.uk

Draught Proofing Advisory Association - P O Box 12, Haslemere, Surrey. GU27 3AH tel:(01428) 870150. info@dpaa-association.org.uk, www.dpaa-association.org.uk

Drawn Metal Ltd - Swinnow Lane, Bramley, Leeds. LS13 4NE tel:(0113) 256 5661, fax:(0113) 239 3194. sales@drawnmetal.co.uk, www.drawmet.com

Drawn Metal Ltd - Swinnow Lane, Leeds, W. Yorks. LS13 4NE tel:(0113) 256 5661, fax:(0113) 239 3194. systems@drawnmetal.co.uk, www.dmlas.co.uk

Drilling & Sawing Association, The - Suite 5.0 North Mill, Bridge Foot, Belper, Derbyshire. DE56 1YD tel:(01773) 820000, fax:(01773) 821284. dsa@drillandsaw.org.uk, www.drillandsaw.org.uk

Drugasar Ltd - Deans Road, Swinton, Gt Man. M27 3JH tel:(0161) 793 8700, fax:(0161) 727 8057. info@drufire.co.uk, www.gasfire.info

Dry Stone Walling Association of Great Britain - Lane Farm, Crooklands, Milnthorpe, Cumbria. LA7 7NH tel:(015395) 67953. information@dswa.org.uk, www.dswa.org.uk

DSM UK Ltd - DSM House, Papermill Drive, Redditch, Worcs. B98 8QJ tel:(01527) 590590, fax:(01527) 590555. marketing.dnpe@dsm.com, www.dsm.com

Du Pont (UK) Ltd - Maylands Avenue, Hemel Hempstead, Herts. HP2 7DP tel:(01438) 734 000. ww2.dupont.com/United_Kingdom_Country/en_GB

Dudley Thomas Ltd - PO Box 28, Birmingham New Road, Dudley. DY1 4SN tel:(0121) 557 5411, fax:(0121) 557 5345. info@thomasdudley.co.uk

Dufaylite Developments Ltd - Cromwell Road, St Neots, Huntingdon, Cambs. PE19 1QW tel:(01480) 215000, fax:(01480) 405526. enquiries@dufaylite.com, www.dufaylite.com

Duffells Limited - 3 Commerce Park, 19 Commerce Way, Croydon, London. CR0 4YL tel:(020) 8662 4010, fax:(020) 8662 4039. enquiries@duffells.co.uk, http://www.duffells.com/

Dulux Trade - ICI Paints plc, Wexham Road, Slough. SL2 5DS tel:(0870) 242 1100, fax:(01753) 556 938. duluxtrade_advice@ici.com, www.duluxtrade.co.uk

Dunbrik (Yorks) Ltd - Ferry Lane, Stanley Ferry, Wakefield, W. Yorks. WF3 4LT tel:(01924) 373694, fax:(01924) 383459. Flues@dunbrik.co.uk, www.dunbrik.co.uk

Dunham Bush Ltd - European Headquaters, Downley Road, Havant, Hants. PO9 2JD tel:(02392) 477700, fax:(012392) 450396. info@dunham-bush.co.uk

Dunhams of Norwich - t/a Dunhams Washroom Systems, The Granary, School Road, Neatishead, Norwich. NR12 8BU tel:(01603) 424855, fax:(01603) 413336. info@dunhamsofnorwich.co.uk, www.dunhamsofnorwich.co.uk

Dunhouse Quarry Co Ltd - Dunhouse Quarry Works, Staindrop, Darlington. DL2 3QU tel:(0845) 3301295, fax:(01833) 660748. paul@dunhouse.co.uk, www.dunhouse.co.uk

Dunlop Adhesives - Building Adhesives, Longton Road, Trentham, Stoke-on-Trent. ST4 8JB tel:(0121) 373 8101, fax:(0121) 384 2826. info@building-adhesives.com, www.dunlop-adhesives.co.uk

Dunphy Combustion Ltd - Queensway, Rochdale, Lancs. OL11 2SL tel:(01706) 649217, fax:(01706) 655512. sharon.kuligowski@dunphy.co.uk, www.dunphy.co.uk

Dunsley Heat Ltd - Fearnough, Huddersfield Road, Holmfirth, W. Yorks. HD7 2TU tel:(01484) 682635, fax:(01484) 688428. sales@dunsleyheat.co.uk, www.dunsleyheat.co.uk

Duplus Architectural Systems Ltd - 370 Melton Road, Leicester, Leics. LE4 7SL tel:(0116) 261 0710, fax:(0116) 261 0539. sales@duplus.co.uk, www.duplus.co.uk

Duplus Architectural Systems Ltd - 370 Melton Road, Leicester. LE4 7SL tel:(0116) 2610710, fax:(0116) 2610539. sales@duplus.co.uk, www.duplus.co.uk

DuPont Corian - McD Marketing Ltd, 10 Quarry Court, Pitstone Green Business Park, Pitstone. LU7 9GW tel:(01296) 663555, fax:(01296) 663599. info@corian.co.uk, www.corian.co.uk

Dupré Minerals - Spencroft Road, Newcastle-under-Lyme, Staffs. ST5 9JE tel:(01782) 383000, fax:(01782) 383101. info@dupreminerals.com, www.dupreminerals.com

Durable Contracts Ltd - Durable House, Crabtree Manorway, Belvedere, Kent. DA17 6AB tel:(020) 8311 1211, fax:(020) 8310 7893. sales@durable-online.com, www.durable-online.com

Durable Ltd - Unit 1, 498 Reading Road, Winnersh, Reading, Berkshire. RG41 5EX tel:(0118) 989 5200, fax:(0118) 989 5209. mail@durable.co.uk, www.durable.co.uk

Duraflex Limited - Severn Drive, Tewkesbury Business Park, Tewkesbury, Glos. GL20 8SF tel:(08705) 351351, fax:(01684) 852701. info@duraflex.co.uk, www.duraflex.co.uk

Dural (UK) Ltd - Unit 40, Monckton Road Industrial Estate, Wakefield. WF2 7AL tel:(01924) 360110, fax:(01924) 360660. welcome@dural.de, www.dural.com

Duralock (UK) Ltd - 6A Enstone Business Park, Enstone, Chipping Norton. OX7 4NP tel:(01608) 678238, fax:(01608) 677170. info@duralock.com, www.duralock.com

Durey Casting Ltd - Hawley Road, Dartford, Kent. DA1 1PU tel:(01322) 272424, fax:(01322) 288073. sales@dureycastings.co.uk, www.dureycastings.co.uk

Durotan Ltd - 20 West Street, Buckingham. MK18 1HE tel:(01280) 814048, fax:(01280) 817842. sales@durotan.ltd.uk, www.durotan.ltd.uk

Duroy Fibreglass Mouldings Ltd - Mercury Yacht Harbour, Satchell Lane, Hamble, Southampton. SO31 4HQ tel:(023) 80453781, fax:(023) 80455538. duroygrp@aol.com

DW Windsor Lighting - Pindar Road, Hoddesdon, Herts. EN11 0DX tel:(01992) 474600, fax:(01992) 474601. info@dwwindsor.co.uk, www.dwwindsor.co.uk

DWA Partnership - St Johns House, 304 -310 St Albans Road, Watford, Hertfordshire. WD24 6PW tel:(01923) 227779, fax:(03332) 407363. enquiries@dwa-p.co.uk, www.dwa-p.co.uk

Dycem Ltd - Ashley Trading Estate, Bristol. BS2 9BB tel:(0117) 955 9921, fax:(0117) 954 1194. uk@dycem.com, www.dycem.com

Dyer Environmental Controls - Unit 10, Lawnhurst Trading Est, Cheadle Heath, Stockport. SK3 0SD tel:(0161) 491 4840, fax:(0161) 491 4841. enquiry@dyerenvironmental.co.uk, www.dyerenvironmental.co.uk

Dyke Chemicals Ltd - PO Box 381, Cobham. KT11 9EF tel:(01932) 866096, fax:(01932) 866097. info@dykechemicals.co.uk, www.dykechemicals.co.uk

Dyno Rod PLC - Millstream, Maidenhead Road, Windsor. SL4 5GD tel:(020) 8481 2200, fax:(020) 8481 2288. DynoOnlineBookings@britishgas.co.uk, www.dyno.com

E

E A Higginson & Co Ltd - Unit 1, Carlisle Road, London. NW9 0HD tel:(020) 8200 4848, fax:(020) 8200 8249. sales@higginson.co.uk, www.higginson.co.uk

Eagles, William, Ltd - 100 Liverpool Street, Salford, Gt Man. M5 4LP tel:(0161) 736 1661, fax:(0161) 745 7765. sales@william-eagles.co.uk, www.william-eagles.co.uk

Earl's Stone Ltd - Sycamore Quarry, Kerridge, Macclesfield, Cheshire. SK10 5AZ tel:(01782) 514353, fax:(01782) 516783. www.earlsstone.co.uk

Earth Anchors Ltd - 15 Campbell Road, Croydon, Surrey. CR0 2SQ tel:(020) 8684 9601, fax:(020) 8684 2230. info@earth-anchors.com, www.earth-anchors.com

Ease-E-Load Trolleys Ltd - Saunders House, Moor Lane, Wiltan, Birmingham. B6 7HH tel:(0121) 356 2224, fax:(0121) 344 3358. simondallow@yahoo.co.uk, www.ease-e-load.co.uk

Easi-Fall International Ltd - Unit 4, Booth Road, Sale Motorway Estate, Sale, Cheshire. M33 7JS tel:(0161) 973 0304, fax:(0161) 969 5009. sales@easifall.co.uk, www.easifall.co.uk

Eaton Electric Limited - Grimshaw Lane, Middleton, Manchester. M23 1GQ tel:(0161) 655 8900, fax:(08700) 507 525. ukresiorders@eaton.com, www.mem250.com

Eaton Electric Ltd - Grimshaw Lane, Middleton, Manchester. M24 1GQ tel:(08700) 545 333, fax:(08700) 540 333. ukcommorders@eaton.com, www.moeller.co.uk

Eaton's Security Business - Security House, Vantage Point Business Village, Mitcheldean, Glos. GL17 0SZ tel:(01594) 545400, fax:(01594) 545401. enquiries@coopersecurity.co.uk, www.coopersecurity.co.uk

Eaton-Williams Group Limited - Fircroft Way, Edenbridge, Kent. TN8 6EZ tel:(01782) 599995, fax:(01782) 599220. info@eaton-williams.com, www.eaton-williams.com

EBC UK Ltd - Kirkes Orchard, Church Street, East Markham, Newark, Notts. NG22 0QW tel:(01777) 871134, fax:(01777) 871134. rob@ebcuk.f9.co.uk, www.e-b-c-uk.com

EBCO - Nobel Road, Eley Trading Estate, Edmonton, London. N18 3DW tel:(020) 8884 4388, fax:(020) 8884 4766. wwinfo@tyco-valves.com, www.edwardbarber.demon.co.uk

Ebor Concrete Ltd - PO Box 4, Ure Bank Top, Ripon, North Yorkshire. HG4 1JE tel:(01765) 604351, fax:(01765) 690065. sales@eborconcrete.co.uk, www.eborconcrete.co.uk

Eclipse Blind Systems Ltd - Inchinnan Business Park, 10 Founain Cress, Inchinnan, Renfrew. PA4 9RE tel:(0141) 812 3322, fax:(0141) 812 5253. email@eclipseblinds.co.uk, www.eclipseblinds.co.uk

Eclipse Sprayers Ltd - 120 Beakes Road, Smethwick, W. Midlands. B67 5AB tel:(0121) 420 2494, fax:(0121) 429 1668. eclipsesales@btconnect.com, www.eclipsesprayers.co.uk

Ecolec - Sharrocks Street, Wolverhampton, West Midlands. WV1 3RP tel:(01902) 457575, fax:(01902) 457797. info@ecolec.co.uk, www.ecolec.co.uk

Ecotherm Insulation (UK) Ltd - Harvey Road, Burnt Mills Industrial Estate, Basildon, Essex. SS13 1QJ tel:(01268) 591155, fax:(01268) 597242. info@ecotherm.co.uk, www.ecotherm.co.uk

Ecotile Flooring Ltd - Unit 15 North Luton Industrial Estate, Sedgewick Road, Luton. LU4 9DT tel:(01582) 788232, fax:(020) 8929 9150. info@ecotileflooring.com, www.ecotileflooring.com

EcoWater Systems Ltd - Solar House, Mercury Park, Wooburn Green, Buckinghamshire. HP10 0HH tel:(01494) 484000, fax:(01494) 484312. info@ecowater.co.uk, www.ecowater.co.uk

Edelman Leather - 2/11 Centre Dome, Design Centre Chelsea Harbour, London. SW10 0XE tel:(0207) 351 7305, fax:(0207) 349 0515. hughk@edelmanleather.com, www.edelmanleather.com

Eden Springs (UK) Ltd - 3 Livingstone Boulevard, Blantyre, St Neots, Glasgow, Lanarkshire. G72 0BP tel:(0844) 800 3344. sales@uk.edensprings.com, www.edensprings.co.uk

Edmonds A & Co Ltd - 91 Constitution Hill, Birmingham. B19 3JY tel:(0121) 236 8351, fax:(0121) 236 4793. enquiries@edmonds.uk.com, www.edmonds.co.uk

Edward Benton & Co Ltd - Creteangle Works, Brook Lane, Ferring, Worthing, W. Sussex. BN12 5LP tel:(01903) 241349, fax:(01903) 700213. sales@creteangle.com, www.creteangle.com

EFG Office Furniture Ltd - 3 Clearwater, Lingley Mere Business Park, Lingley Green Avenue, Warrington. WA5 3UZ tel:(0845) 608 4100, fax:(0845) 604 1924. sales@efgoffice.co.uk, www.efgoffice.co.uk

EGE Carpets Ltd - Suite 1, Ground Floor, Conway House, Chorley, Lancs. PR7 1NY tel:(01257) 239000, fax:(01257) 239001. UK@egecarpet.com, www.egecarpet.com/uk

Egger (UK) Ltd - Anick Grange Road, Hexham, Northumberland. NE46 4JS tel:(0845) 602 4444, fax:(01434) 613302. building.uk@egger.com, www.egger.co.uk

Eglo UK - Unit 12 Cirrus Park, Lower Farm Road, Moulton Park, Northampton. NN3 6UR tel:(01604) 790986, fax:(01604) 670282. info-greatbritain@eglo.com, www.elgo.com

EHSmith - Westhaven House, Arleston Way, Shirley, West Midlands. B90 4LH tel:(0121) 713 7100. enquiries@ehsmith.co.uk, www.ehsmith.co.uk

EITE & Associates - 58 Carsington Cresent, Allestree, Derby. DE22 2QZ tel:(01332) 559929, fax:(01332) 559929. petereite@associates83.freeserve.co.uk

EJOT UK Limited - Hurricane Close, Sherburn Enterprise Park, Sherburn-in-Elmet, Leeds. LS25 6PB tel:(01977) 68 70 40, fax:(01977) 68 70 41. sales@ejot.co.uk, www.ejot.co.uk

COMPANY INFORMATION

E (con't)

Elan-Dragonair - Units 7-8, Dragon Industrial Centre, Fitzherbert Road, Farlington, Portsmouth. PO6 1SQ tel:(023) 92376451, fax:(023) 92370411. david@elan-dragonair.co.uk, www.elan-dragonair.co.uk

Electrak International Ltd - Unit 12, No.1 Industrial Estate, Medomsley Road, Consett, Co. Durham. DH8 6SR tel:(01207) 503400, fax:(01207) 501799. sales@electrak.co.uk, www.electrak.co.uk

Electric Center - The Wolseley Center, Harrison Way, Spa Park, Royal Leamington Spa. CV31 3HH tel:(01926) 705000, fax:(0870) 410 3933. customerservices@wolseley.co.uk, www.electric-center.co.uk

Electric Elements Co, The - Electric Elements Company Ltd, Greens Lane, Kimberley, Nottingham. NG16 2PB tel:(0115) 9459944, fax:(0115) 9384909. info@elelco.co.uk, www.elelco.co.uk

Electrical Contractors' Association (ECA) - ECA Court, 24-26 South Park, Sevenoaks, Kent. TN13 1DU tel:(020) 7313 4800. electricalcontractors@eca.co.uk, www.eca.co.uk

Electrical Review - Quadrant House, The Quadrant, Sutton. SM2 tel:(020) 8652 3492. johns@stjohnpatrick.com, www.electricalreview.co.uk

Electrical Times - Quadrant House, The Quadrant, Sutton. SM2 tel:(020) 8652 8735. rodney.jack@purplems.com, www.electricaltimes.co.uk

Electrix International Ltd - Dovecot Hill, South Church Enterprise Park, Bishop Auckland, Co. Durham. DL14 6XP tel:(01388) 774455, fax:(01388) 777359. info@electrix.co.uk, www.electrix.co.uk

Electrolux Domestic Appliances - PO Box 545, 55-57 High Street, Slough, Berks. SL1 9BG tel:(01753) 872500, fax:(01753) 872381. elsinfo@electrolux.co.uk, www.electrolux.co.uk

Electrolux Foodservice - Crystal Court, Aston Cross Business Park, Rocky Lane, Aston, Birmingham, W. Midlands. B6 5RQ tel:(0121) 220 2800, fax:(0121) 220 2801. professional@electrolux.com, www.foodservice.electrolux.com

Electrolux Laundry Systems - Unit 3A, Humphrys Road, Woodside Estate, Dunstable, Beds. LU5 4TP tel:(0870) 0604118, fax:(0870) 0604113. els.info@electrolux.co.uk, www.electrolux-wascator.co.uk

Electrosonic Ltd - Hawley Mill, Hawley Road, Dartford, Kent. DA2 7SY tel:(01322) 222211, fax:(01322) 282282. info@electrosonic-uk.com, www.electrosonic.co.uk

Elementis Pigments - Liliput Road, Bracksmills Industrial Estate, Northampton. NN4 7DT tel:(01604) 827403, fax:(01604) 827400. pigmentsinfo.eu@elementis.com, www.elementis.com

Elite Trade & Contract Kitchens Ltd - 90 Willesden Lane, Kiburn, London. NW6 7TA tel:(020) 7328 1243, fax:(020) 7328 1243. sales@elitekitchens.co.uk, www.elitekitchens.co.uk

Eliza Tinsley Ltd - Potters Lane, Wednesbury, West Midlands. WS10 0AS tel:(0121) 502 0055, fax:(0121) 502 7348. info@elizatinsley.co.uk, www.elizatinsley.co.uk

Elkay Electrical Manufacturing Co Ltd - Unit 18, Mochdre Industrial Estate, Newtown, Powys. SY16 4LF tel:(01686) 627000, fax:(01686) 628276. sales@elkay.co.uk, www.elkay.co.uk

Elliott Ltd - Manor Drive, Peterborough. PE4 7AP tel:(01733) 298700, fax:(01733) 298749. info@elliottuk.com, www.elliottuk.com

Ellis J T & Co Ltd - Kilner Bank Industrial Esatae, Silver Street, Huddersfield, W. Yorks. HD5 9BA tel:(01484) 514212, fax:(01484) 456433/533454. sales@ellisfurniture.co.uk, www.ellisfurniture.co.uk

Ellis Patents - High Street, Rillington, Malton, N. Yorks. YO17 8LA tel:(01944) 758395, fax:(01944) 758808. sales@ellispatents.co.uk, www.ellispatents.co.uk

Elster Metering Ltd - Pondwicks Road, Luton, Beds. LU1 3LJ tel:(01582) 402020, fax:(01582) 438052. water.metering@gb.elster.com, www.elstermetering.com

Elta Fans Ltd - 17 Barnes Wallis Road, Segensworth East Industrial Estate, Fareham, Hants. PO15 5ST tel:(01489) 566 500, fax:(01489) 584699. mailbox@eltafans.co.uk, www.eltafans.com

Eltron Chromalox - Eltron House, 28 Whitehorse Road, Croydon, Surrey. CR0 2JA tel:(020) 8665 8900, fax:(020) 8689 0571. uksales@chromalox.com, www.chromalox.co.uk

Elwell Buildings Ltd - Capital Works, Garratts Lane, Cradley Heath. B64 5RE tel:(0121) 561 5656, fax:(0121) 559 0505. paul.clews@elwells.co.uk, www.elwells.co.uk

Elwell Buildings Ltd - Unit 5 Excelsior Industrial Estate, Cakemore Road, Blackheath, W. Midlands. B65 0QT tel:(0121) 561 5656, fax:(0121) 559 0505. mail@elwells.co.uk, www.elwells.co.uk

EMAP - Telephone House, 69 - 77 Paul Street, London. EC2A 4NQ tel:(020) 3033 2600. telephonehouse.reception@emap.com, www.emap.com

Emergi-Lite Safety Systems - (Thomas & Betts Ltd), Bruntcliffe Lane, Morley, Leeds, W. Yorks. LS27 9LL tel:(0113) 2810600, fax:(0113) 2810601. emergi-lite_sales@tnb.com, www.emergi-lite.co.uk

Emerson Climate Technologies Retail Solutions - Tomo House, Tomo Road, Stowmarket, Suffolk. IP14 5AY tel:(01449) 672732, fax:(01449) 672736. solutions.europe@emerson.com, www.emersonclimate.com

Emerson Network Power Ltd - Fourth Avenue, Globe Park, Marlow. SL7 1YG tel:(01628) 403200. ukenquiries@emersonnetworkpower.com, www.eu.emersonnetworkpower.com

Emmerich (Berlon) Ltd - Kingsnorth Industrial Estate, Wotton Road, Ashford, Kent. TN23 6JY tel:(01233) 622684, fax:(01233) 645801. enquiries@emir.co.uk, www.emir.co.uk

EMS Entrance Matting Systems Ltd - Freiston Business Park, Priory Road, Freiston, Boston, Lincolnshire. PE22 0JZ tel:(01205) 761757, fax:(01205) 761811. info@entrance-matting.com, www.entrance-matting.com

Emsworth Fireplaces Ltd - Unit 3 Station Approach, North Street, Emsworth, Hants. PO10 7PW tel:(01243) 373 431, fax:(01243) 371023. info@emsworth.co.uk, www.emsworth.co.uk

Emusol Products (LEICS) Ltd - Unit 1, Trowel Lane, Loughborough, Leics. LE12 5RW tel:(01509) 857880, fax:(01509) 857881. info@emusolproducts.com, www.emusolproducts.com

Enable Access - Marshmoor Works, Great North Road, North Mymms, Hatfield, Herts. AL9 5SD tel:(020) 8275 0375, fax:(0208) 4490326. sales@enable-access.com, www.enable-access.com

Encapsulite International Ltd - 17 Chartwell Business Park, Chartmoor Road, Leighton Buzzard, Beds. LU7 4WG tel:(01525) 376974, fax:(01525) 850306. reply@encapsulite.co.uk, www.encapsulite.co.uk

Encon - 1 Deighton Close, Wetherby, W. Yorks. LS22 7GZ tel:(01937) 524200, fax:(01937) 524222. info@encon.co.uk, www.encon.co.uk

Encon Air Systems Ltd - 31 Quarry Park Close, Charter Gate, Moulton Park Industrial Estate, Northampton. NN3 6QB tel:(01604) 494187, fax:(01604) 645848. sales@encon-air.co.uk, www.encon-air.co.uk

EnergyICT Ltd - Tollgate Business Park, Paton Drive, Beaconside, Stafford. ST16 3EF tel:(01785) 275200, fax:(01785) 275300. info-uk@energyict.com, www.energyict.co.uk

Enfield Speciality Doors - Alexandra Road, The Ride, Enfield, Middx. EN3 7EH tel:(020) 8805 6662, fax:(020) 8443 1290. sales@enfielddoors.co.uk, www.enfielddoors.co.uk

English Architectural Glazing Ltd - Chiskwick Avenue, Mildenhall, Suffolk. IP28 7AY tel:(01638) 510000, fax:(01638) 510400. sales@eag.uk.com, www.eag.uk.com

English Braids - Spring Lane, Malvern Link, Worcestershire. WR14 1AL tel:(01684) 89222, fax:(01684) 892111. info@englishbraids.com, www.englishbraids.com

Ensor Building Products Ltd - Blackamoor Road, Blackburn, Lancs. BB1 2LQ tel:(01254) 52244, fax:(01254) 682371. ensor-enquiries@ensorbuilding.com, www.ensorbuilding.com

Ensto Briticent Ltd - Unit 6, Priory Industrial park, Airspeed Road, Christchurch, Dorset. BH23 4HD tel:(01425) 283300, fax:(01425) 280480. salesuk@ensto.com, www.ensto.com

En-tout-cas Tennis Courts Ltd - 20 Nene Valley Business Park, Oundle, Peterborough. PE8 4HN tel:(01832) 274199. info@tenniscourtsuk.co.uk, www.tenniscourtsuk.co.uk

Envair Ltd - York Avenue, Haslingden, Rossendale, Lancs. BB4 4HX tel:(01706) 228416, fax:(01706) 242205. info@envair.co.uk, www.envair.co.uk

Envirodoor Markus Ltd - Viking Close, Willerby, Hull, E. Yorks. HU10 6BS tel:(01482) 659375, fax:(01482) 655131. sales@envirodoor.com, www.envirodoor.com

Envopak Group Ltd - Edgington Way, Sidcup, Kent. DA14 5EF tel:(020) 8308 8000, fax:(020) 8300 3832. sales@envopak.co.uk, www.envopak.com

ERA Products - Straight Road, Short Heath, Willenhall, W. Midlands. WV12 5RA tel:(01922) 490000, fax:(01922) 490044. info@era-security.com, www.era-security.com

Erico Europa (GB) Ltd - 52 Milford Road, Reading, Berkshire. RG1 8LJ tel:(0808) 2344 670, fax:(0808) 2344 676. info@erico.com, www.erico.com

Erlau AG - UK Sales Office, 42 Macclesfield Road, Hazel Grove, Stockport. SK7 6BE tel:(01625) 877277, fax:(01625) 850242. erlau@zuppinger.u-net.com, www.erlau.de

Ermine Engineering Co. Ltd - Freancis House, Silver Birch Park, Great Northern Terrace, Lincoln, Lincs. LN5 8LG tel:(01522) 510977, fax:(01522) 510929. info@ermineengineering.co.uk, www.ermineengineering.co.uk

Eurobrick Systems Ltd - Unit 7, Wilverley Trading Estates, Bath Road, Brislington, Bristol. BS4 5NL tel:(0117) 971 7117, fax:(0117) 971 7217. richard@eurobrick.co.uk, www.eurobrick.co.uk

Euroclad Ltd - Wentloog Corperate Park, Wentloog Road, Cardiff. CF3 2ER tel:(029) 2079 0722, fax:(029) 2079 3149. sales@euroclad.com, www.euroclad.com

European Lead Sheet Industry Association - Bravington House, 2 Bravingtons Walk, London. N1 9AF tel:(020) 7833 8090, fax:(020) 7833 1611. davidson@ila-lead.org, www.elsia.org

Evac+Chair International Ltd - ParAid House, Weston Lane, Birmingham. B11 3RS tel:(0121) 706 6744, fax:(0121) 706 6746. info@evacchair.co.uk, www.evacchair.co.uk

Evans Concrete Products Ltd - Pye Bridge Industrial Estate, Main Road, Pie Bridge, Near Alfreton, Derbyshire. DE55 4NX tel:(01773) 529200, fax:(01773) 529217. evans@evansconcreteproducts.co.uk, www.evansconcrete.co.uk

Evans Howard Roofing Ltd - Tyburn Road, Erdington, Birmingham. B24 8NB tel:(0121) 327 1336, fax:(0121) 327 3423. info@howardevansroofingandcladding.co.uk, www.howardevansroofingandcladding.co.uk

Evans Vanodine International PLC - Brierley Road, Walton Summit Centre, Bamber Bridge, Preston. PR5 8AH tel:(01772) 322200, fax:(01772) 626000. sales@evansvanodine.co.uk, www.evansvanodine.co.uk

Eve Trakway - Bramley Vale, Chesterfield, Derbys. S44 5GA tel:(08700) 767676, fax:(08700) 737373. mail@evetrakway.co.uk, www.evetrakway.co.uk

Evergreens Uk - No. 2 Extons Units, MKT overton Ind. Est., Market Overton, Nr. Oakham, Leics. LE15 7PP tel:(01572) 768208, fax:(01572) 768261. sales@evergreensuk.com, www.evergreensuk.com

Everlac (GB) Ltd - Hawthorn House, Helions Bumpstead Road, Haverhill, Suffolk. CB9 7AA tel:(01440) 766360, fax:(01440) 768897. enquiries@everlac.co.uk, www.everlac.co.uk

Evertaut Ltd - Lions Drive, Shadsworth Business Park, Blackburn, Lancs. BB1 2QS tel:(01254) 297 880, fax:(01254) 274 859. sales@evertaut.co.uk, www.evertaut.co.uk

Evode Ltd - Common Road, Stafford. ST16 3EH tel:(01785) 257755, fax:(01785) 252337. technical.service@bostik.com, www.evode.co.uk

Exidor Limited - Progress Drive, Cannock, Staffordshire. WS11 0JE tel:(01543) 460030, fax:(01543) 573534. info@exidor.co.uk, www.exidor.co.uk

Exitex Ltd - Dundalk. Ireland tel:+353 4293 71244, fax:+353 4293 71221. info@exitex.com, www.exitex.xom

Expamet Building Products - PO Box 52, Longhill Industrial Estate (North), Hartlepool. TS25 1PR tel:(01429) 866611, fax:(01429) 866633. admin@expamet.net, www.expamet.co.uk

Expanded Piling Ltd - Cheapside Works, Waltham, Grimsby, Lincs. DN37 0JD tel:(01472) 822552, fax:(01472) 220675. info@expandedpiling.com, www.expandedpiling.com

Eyre & Baxter Limited - 229 Derbyshire Lane, Sheffield. S8 8SD tel:(0114) 250 0153, fax:(0114) 258 0856. enquiries@eyreandbaxter.co.uk, www.eryeandbaxter.co.uk

E-Z Rect Ltd - Unit 8c, Witan Park, Avenue 2, Station Lane, Witney. OX28 4FH tel:(01993) 779494, fax:(01993) 704111. mail@e-z-rect.com, www.e-z-rect.com

F

Faber Blinds UK Ltd - Kilvey Road, Brackmills, Northampton, Northants. NN4 7BQ tel:(01604) 766251, fax:(01604) 705209. contracts-uk@faber.dk, www.faberblinds.co.uk

Fabrikat (Nottingham) Ltd - Hamilton Road, Sutton-in-Ashfield, Nottingham, Notts. NG17 5LN tel:(01623) 442200, fax:(01623) 442233. sales@fabrikat.co.uk, www.fabrikat.co.uk

Fagerhult Lighting - 33-34 Dolben Street, London. SE1 0UQ tel:(020) 7403 4123, fax:(020) 7378 0906. light@fagerhult.co.uk, www.fagerhult.co.uk

Fairfield Displays & Lighting Ltd - 127 Albert Street, Fleet, Hants. GU51 3SN tel:(01252) 812211, fax:(01252) 812123. info@fairfielddisplays.co.uk, www.fairfielddisplays.co.uk

Fairport Construction Equipment - Blagden Street, Sheffield. S2 5QS tel:(01142) 767921, fax:(01142) 720965. sales@fairport.uk.com, www.fairport.uk.com

Falco UK Ltd - Unit 8, Leekbrook Way, Leekbrook, Staffs. ST13 7AP tel:(01538) 380080, fax:(01538) 386421. sales@falco.co.uk, www.falco.co.uk

Fantasia Distribution Ltd - Unit B, The Flyers Way, Westerham, Kent. TN16 1DE tel:(01959) 564440, fax:(01959) 564829. info@fantasiaceilingfans.com, www.fantasiaceilingfans.com

Farefence (NW) Ltd - Pinfold House, Pinfold Road, Worsley, Manchester. M28 5DZ tel:(0161) 799 4925, fax:(0161) 703 8542. info@farefence.co.uk, www.farefence.co.uk

Farmington Masonry LLP - Farmington Quarry, Northleach, Cheltenham, Glos. GL54 3NZ tel:(01451) 860280, fax:(01451) 860115. info@farmington.co.uk, www.farmington.co.uk

Fastcall - St John's House, 304-310 St Albans Road, Watford, Herts. WD24 6PW tel:(01923) 230843

Fastclean Blinds Ltd - 45 Townhill Road, Hamilton Road. ML63 9RH tel:(01698) 538392. gordonmichael18@yahoo.co.uk, www.fastcleanblindshamilton.co.uk

Fastwall Ltd - Units 3/10, Overfield, Thorpe Way Ind Estate, Banbury, Oxon. OX16 4XR tel:(01295) 25 25 88, fax:(01295) 25 25 84. www.fastwall.com

FC Frost Ltd - Bankside Works, Benfield Way, Braintree, Essex. CM7 3YS tel:(01376) 329 111, fax:(01376) 347 002. info@fcfrost.com, www.fcfrost.com

Federation of Building Specialist Contractors (FBSC) - Unit 9, Lakeside Industrial Estate, Stanton Harcourt, Oxford. OX29 5SL tel:(0870) 429 6355, fax:(0870) 429 6352. enquiries@fbsc.org.uk, www.fbsc.org.uk

Federation of Master Builders - David Croft House, 25 Ely Place, London. EC1N 6TD tel:(020) 7025 2900, fax:(020) 7025 2929. jaynerunacres@fmb.org.uk, www.fmb.org.uk

Federation of Piling Specialists - Forum Court, 83 Copers Cope Road, Beckenham, Kent. BR3 1NR tel:(020) 8663 0947, fax:(020) 8663 0949. fps@fps.org.uk, www.fps.org.uk

Federation of Plastering and Drywall Contractors - FPDC 4th Floor, 61 Cheapside, London. EC2V 6AX tel:(020) 7634 9480, fax:(020) 7248 3685. enquiries@fpdc.org, www.fpdc.org

Feedwater Ltd - Tarran Way, Moreton, Wirral. L46 4TP tel:(0151) 606 0808, fax:(0151) 678 5459. enquiries@feedwater.co.uk, www.feedwater.co.uk

Felix Design Ltd - Unit 15, Tiverton Way, Tiverton Business Park, Tiverton, Devon. EX16 6SR tel:(01884) 255420, fax:(01884) 242613. sales@felixstaging.co.uk, www.felixstaging.co.uk

Fencelines Ltd - Unit 16, Westbrook Road, Trafford Park, Manchester. M17 1AY tel:(0161) 848 8311, fax:(0161) 872 9643. sell@fencelines.co.uk, www.fencelines.co.uk

Fencing Contractors Association - Airport House, Purley Way, Croydon. CR0 0XZ tel:(020) 8253 4516. info@fencingcontractors.org, www.fencingcontractors.org

Fenlock-Hansen Ltd - Heworth House, William Street, Felling, Gateshead, Tyne & Wear. NE10 0JP tel:(0191) 438 3222, fax:(0191) 438 1686. sales@fendorhansen.co.uk, www.hansongroup.biz

FeRFA - The Resin Flooring Association, 16 Edward Road, Farnham, Surrey. GU9 8NP tel:(07484) 075 254, fax:(01252) 739140. lisa@ferfa.org.uk, www.ferfa.org.uk

Fermacell - Unit 2, The Courtyard, Reddicap Trading Estate, Sutton Coldfield, West Midlands. B75 7BU tel:(0870) 609 0306, fax:(0870) 240 2948. fermacell-uk@xella.com, www.fermacell.co.uk

Fernden Construction (Winchester) Ltd - Barfield Close, Winchester, Hants. SO23 9SQ tel:(01962) 866400, fax:(01962) 864139. sales@ferndenwin.co.uk, www.ferndenwin.co.uk

Fernox - Forsyth Road, Sheerwater, Woking, Surrey. GU21 5RZ tel:(01483) 793200, fax:(01483) 793201. sales@fernox.com, www.fernox.com

Ferrograph Limited - New York Way, New York Industrial Park, Newcastle-upon-Tyne. NE27 0QF tel:(0191) 280 8800, fax:(0191) 280 8810. info@ferrograph.com, www.ferrograph.com

Ferroli Ltd - Lichfield Road, Branston Industrial Estate, Burton-on-Trent. DE14 3HD tel:(0870) 728 2882, fax:(0870) 728 2883. sales@ferroli.co.uk, www.ferroli.co.uk

FG Wilson - Old Glenarm Road, Larne, Co. Antrim. BT40 1EJ tel:(028) 28261000, fax:(028) 28261111. sales@fgwilson.com, www.fgwilson.com

FGF Ltd - Shadwell House, Shadwell Street, Birmingham. B4 6LJ tel:(0121) 233 1144, fax:(0121) 212 2539. info@fgflimited.co.uk, www.fgflimited.co.uk

Fibaform Products Ltd - Unit 22, Lansil Industrial Estate, Caton Road, Lancaster. LA1 3PQ tel:(01524) 60182, fax:(01524) 389829. info@fibaform.co.uk, www.fibaform.co.uk

Fibercill - The Moorings, Hurst Business Park, Brierley Hill, W. Midlands. DY5 1UX tel:(01384) 482221, fax:(01384) 482212. mail@fibercill.com, www.fibercill.com

Fieldmount Ltd - 18B Aintree Road, Perivale, Middlesex. UB6 7LA tel:(0207) 624 8866, fax:(020) 7328 1836. info@fieldmount.co.uk, www.fieldmount.co.uk

Fife Fire Engineers & Consultants Ltd - Waverley Road, Mitchelston Industrial Estate, Kirkcaldy, Fife. KY1 3NH tel:(01592) 653661, fax:(01592) 653990. sales@ffec.co.uk, www.ffec.co.uk

Fighting Films - PO BOX 2405, Bristol. BS1 9BA tel:(0845) 408 5836, fax:(0845) 929 4540. info@fightingfilms.com, www.fightingfilms.com

Fillcrete Ltd - Maple House, 5 Over Minnis, New Ash Green, Kent. DA3 8JA tel:(01474) 872444, fax:(01474) 872426. sales@fillcrete.com, www.fillcrete.com

Filon Products Ltd - Unit 3 Ring Road, Zone 2, Burntwood Business Park, Burntwood, Staffordshire. WS7 3JQ tel:(01543) 687300, fax:(01543) 687303. sales@filon.co.uk, www.filon.co.uk

F (con't)

Findlay Irvine Ltd - Bog Road, Penicuik, Midlothian. EH26 9BU tel:(01968) 671200, fax:(01968) 671237. sales@findlayirvine.com, www.findlayirvine.com

Fine H & Son Ltd - Victoria House, 93 Manor Farm Road, Wembley, Middx. HA0 1XB tel:(020) 8997 5055, fax:(020) 8997 8410. sales@hfine.co.uk, www.hfine.co.uk

FineLine - Whitewall Road, Medway Estate, Rochester, Kent. ME2 4EW tel:(01634) 719701, fax:(01634) 716394. info@finelinewindows.co.uk, www.finelinewindows.co.uk

Finlock Gutters - Compass House, Manor Royal, Crawly, West Sussex. RH10 9PY tel:(01293) 572200, fax:(01293) 572288. finlockgutters@connaught.plc.uk, www.finlock-concretegutters.co.uk

Finnish Fibreboard (UK) Ltd - Suite 3, 4 The Limes, Ingatestone, Essex. CM4 0BE tel:(01442) 264400, fax:(01442) 266350. keith@finfib.co.uk, www.finnishfibreboard.com

Firco Ltd - Unit 3, Beckley Hill Industrial Estate, Lowwer Higham, Kent. ME3 7HX tel:(01474) 824338. info@firco.co.uk, www.firco.co.uk

Fire Escapes and Fabrications (UK) Ltd - Foldhead Mills, Newgate, Mirfield, W. Yorks. WF14 8DD tel:(01924) 498787, fax:(01924) 497424. info@fireescapes.co.uk, www.fireescapes.co.uk

Fire Industry Association - Tudor House, Kingsway Business Park, Oldfield Road, Hampton, Middlesex. TW12 2HD tel:(020) 3166 5002. info@fia.uk.com, www.fia.uk.com

Fire Protection Ltd - Millars 3, Southmill Road, Bishop's Stortford, Herts. CM23 3DH tel:(01279) 467077, fax:(01279) 466994. fire.protection@btinternet.com

Fire Security (Sprinkler Installations) Ltd - Homefield Road, Haverhill, Suffolk. CB9 8QP tel:(01440) 705815, fax:(01440) 704352. info@firesecurity.co.uk, www.firesecurity.co.uk

Firebarrier Services Ltd - The Old Bank Chambers, Brunswick Square, Torquay, Devon. TQ1 4UT tel:(01803) 291185, fax:(01803) 290026. sales@firebarrier.co.uk, www.firebarrier.co.uk

Fired Earth - Twyford Mill, Oxford Road, Adderbury, Oxon. OX17 3HP tel:(01295) 812088, fax:(01295) 810832. enquiries@firedearth.com, www.firedearth.com

Firemaster Extinguisher Ltd - Firex House, 174-176 Hither Green Lane, London. SE13 6QB tel:(020) 8852 8585, fax:(020) 8297 8020. sales@firemaster.co.uk, www.firemaster.co.uk

Firestone Building Products - Chester Road, Salterswall, Winsford, Cheshire. CW7 2NQ tel:(01606) 552026, fax:(01606) 592666. info@fbpl.co.uk, www.firestonebpe.co.uk

Firetecnics Systems Ltd - Southbank House, Black Prince Road, London. SE1 7SJ tel:(020) 7587 1493, fax:(020) 7582 3496. info@firetecnics.co.uk, www.firetecnics.com

First Stop Builders Merchants Ltd - Queens Drive, Kilmarnock, Ayrshire. KA1 3XA tel:(01563) 534818, fax:(01563) 537848. sales@firststopbm.co.uk, www.firststopbm.co.uk

Firwood Paints Ltd - Victoria Works, Oakenbottom Road, Bolton, Lancs. BL2 6DP tel:(01204) 525231, fax:(01204) 362522. sales@firwood.co.uk, www.firwood.co.uk

Fisher Scientific (UK) Ltd - Bishop Meadow Road, Loughborough, Leics. LE11 5RG tel:(01509) 231166, fax:(01509) 231893. fsuk.sales@thermofisher.com, www.fisher.co.uk

Fitchett & Woollacott Ltd - Willow Road, Lenton Lane, Nottingham. NG7 2PR tel:(0115) 993 1112, fax:(0115) 993 1151. enquiries@fitchetts.co.uk, www.fitchetts.co.uk

Fitzgerald Lighting Ltd - Normandy Way, Bodmin, Cornwall. PL31 1HH tel:(01208) 262200, fax:(01208) 262334. info@flg.co.uk, www.flg.co.uk

Fitzpatrick Doors Ltd - Milnhay Road, Langly Mill, Nottingham. NG16 4AZ tel:(01773) 530500, fax:(01773) 530040. enquiries@fitzpatrickmetaldoors.co.uk, www.fitzpatrickmetaldoors.com

Fixatrad Ltd - Unit 7, Stadium Way, Reading, Berks. RG30 6BX tel:(0118) 921 2100, fax:(0118) 921 0634. sales@intrad.com, www.intrad.com

Fläkt Woods Ltd - Axial Way, Colchester, Essex. CO4 5ZD tel:(01206) 222555, fax:(01206) 222777. info.uk@flaktwoods.com, www.flaktwoods.co.uk

Flambeau - Manston Road, Ramsgate, Kent. CT12 6HW tel:(01843) 854000, fax:(01843) 854010. eusales@flambeau.com, www.flambeau.co.uk

Flamco UK Ltd - Unit 4, St Micheals Road, Lea Green Industrial Estate, St. Helens, Merseyside. WA9 4WZ tel:(01744) 818100, fax:(01744) 830400. info@flamco.co.uk, www.flamco.co.uk

Flat Roofing Alliance (FRA) - Fields House, Gower Road, Haywards Heath, Sussex. RH16 4PL tel:(01444) 440027, fax:(01444) 415616. info@fra.org.uk, www.fra.org.uk

Fleming Buildings Ltd - 23 Auchinloch Road, Lenzie, Dunbartonshire. G66 5ET tel:(0141) 776 1181, fax:(0141) 775 1394. office@fleming-buildings.co.uk, www.fleming-buildings.co.uk

Fleming Homes Ltd - Coldstream Road, Duns, Berwickshire. TD11 3HS tel:(01361) 883785, fax:(01361) 883898. enquiries@fleminghomes.co.uk, www.fleminghomes.co.uk

Flexcrete Technologies Limited - Tomlinson Road, Leyland, Lancs. PR25 2DY tel:(0845) 260 7005, fax:(0845) 260 7006. info@flexcrete.com, www.flexcrete.com

Flexel International Ltd - Queensway Industrial Estate, Glenrothes, Fife. KY7 5QF tel:(01592) 760 928, fax:(01592) 760 929. sales@flexel.co.uk, www.flexel.co.uk

Flexelec(UK) Ltd - Unit 11, Kings Park Industrial Estate, Primrose Hill, Kings Langley, Herts. WD4 8ST tel:(01923) 274477, fax:(01923) 270264. sales@omerin.co.uk, www.flexelec.com

Flexible Ducting Ltd - Cloberfield, Milngavie, Dunbartonshire. G62 7LW tel:(0141) 956 4551, fax:(0141) 956 4847. info@flexibleducting.co.uk, www.flexibleducting.co.uk

Flexible Reinforcements Ltd - Queensway House, Queensway, Clitheroe, Lancs. BB7 1AU tel:(01200) 442266, fax:(01200) 452010. sales@flexr.co.uk, www.flexr.co.uk

Flexiform Business Furniture Ltd - The Business Furniture Centre, 1392 Leeds Road, Bradford, W. Yorks. BD3 7AE tel:(01274) 706206, fax:(01274) 660867. info@flexiform.co.uk, www.flexiform.co.uk

Flexitallic Ltd - Marsh Works, Dewsbury Rd, Cleckheaton, W. Yorks. BD19 5BT tel:(01274) 851273, fax:(01274) 851386. sales@flexitallic.com, www.flexitallic.com

Flex-Seal Couplings Ltd - Endeavour Works, Newlands Way, Valley Park, Wombwell, Barnsley, S. Yorks. S73 0UW tel:(01226) 340888, fax:(01226) 340999. sales@flexseal.co.uk, www.flexseal.co.uk

Float Glass Industries - Float Glass House, Floats Road, Roundthorn, Manchester. M23 9QA tel:(0161) 946 8000, fax:(0161) 946 8092. salesorders@floatglass.co.uk, www.floatglass.co.uk

FloPlast Ltd - Eurolink Business Park, Castle Road, Sittingbourne, Kent. ME10 3FP tel:(01795) 431731, fax:(01795) 431188. sales@floplast.co.uk, www.floplast.co.uk

FloRad Heating and Cooling - Radiant House, The Crosspath, Radlett, Hertfordshire. WD7 8HR tel:(01923) 850823, fax:(01923) 850823. info@florad.co.uk, www.florad.co.uk

Floralsilk Ltd - Meadow Drove Business Park, Meadow Drove, Bourne. PE10 0BE tel:(01778) 425 205, fax:(01778) 420 457. sales@floralsilk.co.uk, www.floralsilk.co.uk

Flowcrete Plc - Flowcrete Business Park, Booth Lane, Moston, Sandbach, Cheshire. CW11 3QF tel:(01270) 753000, fax:(01270) 753333. ukweb@flowcrete.com, www.flowcrete.co.uk

Flowflex Components Ltd - Samuel Blaser Works, Tounge Lane Industrial Estate, Buxton, Derbys. SK17 7LR tel:(01298) 77211, fax:(01298) 72362. sales@flowflex.com, www.flowflex.com

Flowserve Flow Control - Burrell Road, Haywards Heath, W. Sussex. RH16 1TL tel:(01444) 314400, fax:(01444) 314401. wvukinfo@flowserve.com, www.flowserve.com

Fluidair International Ltd - School Hill Works, Kent Street, Bolton, Gt Man. BL1 2LN tel:(01204) 559955, fax:(01204) 559966. sales@fluidair.co.uk, www.fluidair.co.uk

Fluorel Lighting Ltd - 312 Broadmead Road, Woodford Green, Essex. IG8 8PG tel:(020) 8504 9691, fax:(020) 8506 1792. info@fluorel.co.uk, www.fluorel.co.uk

Foamseal Ltd - New Street House, New Street, Petworth, W. Sussex. GU28 0AS tel:(01798) 345400, fax:(01798) 344093. mail@foamseal.co.uk, www.foamseal.co.uk

Focal Signs Ltd - 12 Wandle Way, Mitcham, Surrey. CR4 4NB tel:(020) 8687 5300, fax:(020) 8687 5301. sales@safetyshop.com, www.focalsigns.co.uk

Focus SB Ltd - Napier Road, Castleham Industrial Estate, St. Leonards-on-Sea, Sussex. TN38 9NY tel:(01424) 858060, fax:(01424) 853862. sales@focus-sb.co.uk, www.focus-sb.co.uk

Footprint Sheffield Ltd - Admiral Works, Sedgley Road, Sheffield, S. Yorks. S6 2AH tel:(01142) 327080, fax:(01142) 327089. sales@footprint-tools.co.uk, www.footprint-tools.co.uk

Forbes & Lomax Ltd - 205a St John's Hill, London. SW11 1TH tel:(020) 7738 0202, fax:(020) 7738 9224. sales@forbesandlomax.co.uk, www.forbesandlomax.co.uk

Forbo Flooring - PO Box 1, Den Road, Kirkcaldy, Fife. KY1 2SB tel:(01592) 643777, fax:(01592) 643999. info.uk@forbo.com, www.forbo-flooring.co.uk

Forest of Dean Stone Firms Ltd - Bixslade Stone Works, Parkend, Nr Lydney, Glos. GL15 4JS tel:(01594) 562304, fax:(01594) 564184. info@fodstone.co.uk, www.fodstone.co.uk

Formica Ltd - 11 Silver Fox Way, Cobalt Business Park, Newcastle Upon Tyne. NE27 0QJ tel:(0191) 2593100, fax:(0191) 2592648. formica.limited@formica.com, www.Formica.eu/uk

Formpave Ltd - Tufthorne Avenue, Coleford, Glos. GL16 8PR tel:(01594) 836999, fax:(01594) 810577. sales@formpave.co.uk, www.formpave.co.uk

Forticrete Ltd - Boss Avenue, Off Grovebury Road, Leighton Buzzard. LU7 4SD tel:(01525) 244900. info@forticrete.com, www.forticrete.com

Forticrete Roofing Products - Bridle Way, Bootle, Merseyside. L30 4UA tel:(0151) 521 3545, fax:(0151) 521 5696. info@forticrete.com, www.forticrete.co.uk

Fosroc Ltd - Drayton Manor Business Park, Coleshill Road, Tamworth, Staffordshire. B78 3TL tel:(01827) 262222, fax:(01827) 262444. enquiryuk@fosroc.com, www.fosroc.com

Foster W H & Sons Ltd - 3 Cardale Street, Rowley Regis, W. Midlands. B65 0LX tel:(0121) 561 1103, fax:(0121) 559 2620. sales@whfoster.co.uk, www.whfoster.co.uk

Fothergill Engineered Fabrics Ltd - PO Box 1, Summit, Littleborough, Lancs. OL15 0LU tel:(01706) 372414, fax:(01706) 376422. sales@fothergill.co.uk, www.fothergill.co.uk

Fountains Direct Ltd - The Office, 41 Dartnell Park, West Byfleet, Surrey. KT14 6PR tel:(01259) 722628, fax:(01259) 722732. gordon@fountains-direct.co.uk, www.fountains-direct.co.uk

Fox Pool (UK) Ltd - Mere House, Stow, Lincoln, Lincs. LN1 2BZ tel:(01427) 788662, fax:(01427) 788526. sales@grayfoxswimmingpools.co.uk, www.fox-pool.co.uk

Francis Frith Collection - Frith's Barn, Teffont, Salisbury, Wilts. SP3 5QP tel:(01722) 716376, fax:(01722) 716 881. sales@francisfrith.com, www.francisfrith.com

Franke Sissons Ltd - 14 Napier Court, Gander Lane, Barlborough Link Business Park, Barlborough, Derbyshire. S43 4PZ tel:(01246) 450 255, fax:(01246) 451 276. ws-info.uk@franke.com, www.franke.co.uk

Franke UK Ltd - West Park, Manchester Int'l Office Centre, Styal Road, Manchester, Gt Man. M22 5WB tel:(0161) 436 6280, fax:(0161) 436 2180. info.uk@franke.com, www.franke.co.uk

Franklin Hodge Industries Ltd - Jubilee Building, Faraday Raod, Westfield Trading Estate, Hereford, Herefordshire. HR4 9NS tel:(01432) 269605, fax:(01432) 277454. sales@franklinhodge.co.uk, www.franklinhodge.com

Fray Design Ltd - Ghyll Way, Airedale Business Centre, Keighley Road, Skipton. BD23 2TZ tel:(01756) 704040, fax:(01756) 704041. sales@fraydesign.co.uk, www.fraydesign.co.uk

Freeman T R - Edward Leonard House, Pembroke Avenue, Denny End Road, Waterbeach, Cambridge. CB25 9QR tel:(01223) 715810, fax:(01223) 411061. info@trfreeman.co.uk, www.trfreeman.co.uk

Freyssinet Ltd - Innovation House, Euston Way, Town Centre, Telford, Shropshire. TF3 4LT tel:(01952) 201901, fax:(01952) 201753. info@freyssinet.co.uk, www.freyssinet.co.uk

Friedland Ltd - The Arnold Centre, Parcocke Road, Basildon, Essex. SS14 3EA tel:(01268) 563000, fax:(01268) 563538. friedlandorderenquiries@honeywell.com, www.friedland.co.uk

Friends of the Earth - 26 Underwood Street, London. N1 7JQ tel:(020) 7490 1555, fax:(020) 7490 0881. info@foe.co.uk, www.foe.co.uk

Frosts Landscape Construction Ltd - Wain Close, Newport Road, Woburn Sands, Milton Keynes, Bucks. MK17 8UZ tel:(0845) 021 9001. info@frostslandscapes.com, www.frostsgroup.com

Fuchs Lubricants (UK) PLC - New Century Street, Hanley, Stoke-On-Trent. ST1 5HU tel:(08701) 200400, fax:(01782) 203775. ukwebsite@fuchs-oil.com, www.fuchslubricants.com

Fujichem Sonneborn Ltd - Jaxa Works, 91-95 Peregrine Road, Hainault, Ilford. IG6 3XH tel:(020) 8500 0251, fax:(020) 8500 3696. info@fcsonneborn.com, www.fcsonneborn.com/en/

Fullflow Group Ltd - Fullflow House, Holbrook Avenue, Holbrook, Sheffield, S. Yorks. S20 3FF tel:(0114) 247 3655, fax:(0114) 247 7805. info@uk.fullflow.com, www.fullflow.com

Fulton Boiler Works (Great Britain) Ltd - Fernhurst Road, Bristol. BS5 7FG tel:(0117) 972 3322, fax:(0117) 972 3358. uk.sales.office@fulton.com, www.fulton.com

Furmanite International Ltd - Furman House, Shap Road, Kendal, Cumbria. LA9 6RU tel:(01539) 729009, fax:(01539) 729359. uk.enquiry@furmanite.com, www.furmanite.com

Furnitubes International Ltd - 3rd Floor, Meridian House, Royal Hill, Greenwich, London. SE10 8RD tel:(020) 8378 3200, fax:(020) 8378 3250. sales@furnitubes.com, www.furnitubes.com

Furse - Wilford Road, Nottingham, Notts. NG2 1EB tel:(0115) 964 3700, fax:(0115) 986 0538. enquiry@furse.com, www.furse.com

G

G Miccoli & Sons Limited - 5 Vulcan Way, New Addington, Croydon, Surrey. CR0 9UG tel:(0208) 684 3816, fax:(020) 8689 6514. info@stonebymiccoli.co.uk, www.stonebymiccoli.co.uk

Gabriel & Co Ltd - Abro Works, 10 Hay Hall Road, Tyseley, Birmingham. B11 2AU tel:(0121) 248 3333, fax:(0121) 248 3330. contacts@gabrielco.com, www.gabrielco.com

Galvanizers Association - Wrens Court, 56 Victoria Road, Sutton Coldfield, W. Midlands. B72 1SY tel:(0121) 355 8838, fax:(0121) 355 8727. ga@hdg.org.uk, www.galvanizing.org.uk

Garador Ltd - Bunford Lane, Yeovil, Somerset. BA20 2YA tel:(01935) 443700, fax:(01935) 443744. enquiries@garador.co.uk, www.garador.co.uk

Garage Equipment Association Ltd - 2-3 Church Walk, Daventry. NN11 4BL tel:(01327) 312616, fax:(01327) 312606. info@gea.co.uk, www.gea.co.uk

Gardners Books - 1 Whittle Drive, Eastbourne, East Sussex. BN23 6QH tel:(01323) 521777, fax:(01323) 521666. sales@gardners.com, www.gardners.com

Garran Lockers Ltd - Garran House, Nantgarw Road, Caerphilly. CF83 1AQ tel:(0845) 658 8600, fax:(0845) 6588601. info@garran-lockers.co.uk, www.garran-lockers.ltd.uk

Gartec Ltd - Midshires Business Park, Smeaton Close, Aylesbury, Bucks. HP19 8HL tel:(01296) 768977, fax:(01296) 397600. sales@gartec.com, www.gartec.com

Gas Measurement Instruments Ltd - Inchinnan Business Park, Renfrew, Renfrewshire. PA4 9RG tel:(0141) 812 3211, fax:(0141) 812 7820. sales@gmiuk.com, www.gmiuk.com

Gatic - Poulton Close, Dover, Kent. CT17 0UF tel:(01304) 203545, fax:(01304) 215001. info@gaticdover.co.uk, www.gatic.com

Gazco - Osprey Road, Sowton Industrial Estate, Exeter, Devon. EX2 7JG tel:(01392) 474000, fax:(01392) 219932. info@gazco.com, www.gazco.com

Gazco Ltd - Osprey Road, Sowton Industrial Estate, Exeter, Devon. EX2 7JG tel:(01392) 216999, fax:(01392) 444148. info@gazco.com, www.gazco.com

GB Architectural Cladding Products Ltd - Spen Valley Works, Carr Street, Liversedge, West Yorkshire. WF15 6EE tel:(01924) 404045, fax:(01924) 401070. jim.gorst@artstone.co.uk, www.artstone-gbgroup.co.uk/

GB Sign Solutions Limited - Unit 5, Orion Trading Estate, Tenax Road, Trafford Park, Manchester. M17 1JT tel:(0161) 741 7270, fax:(0161) 741 7272. sales@greensigns.co.uk, www.greensigns.co.uk

GBG Fences Limited - 25 Barns Lane, Rushall, Walsall, Staffs. WS4 1HQ tel:(01922) 623207, fax:(01922) 722110. enquiries@gbgfences.co.uk

GE Lighting Ltd - Lincoln Road, Enfield, Surrey. EN1 1SB tel:(020) 8626 8500, fax:(020) 8626 8501. englishcustomercontactteam@ge.com, www.gelighting.com

Gearing F T Landscape Services Ltd - Crompton Road Depot, Stevenage, Herts. SG1 2EE tel:(01438) 369321, fax:(01483) 353039. Sales@ft-gearing.co.uk, www.ft-gearing.co.uk

Geberit Ltd - Geberit House, Acadamy Drive, Warwick, Warwickshire. CV34 6QZ tel:(0800) 077 8365, fax:(0844) 800 6604. enquiries@geberit.co.uk, www.geberit.co.uk

Gebhardt Kiloheat - Kiloheat House, Enterprise Way, Edenbridge, Kent. TN8 6HF tel:(01732) 866000, fax:(01732) 866370. sales@nicotra.co.uk, www.nicotra-gebhardt.com

GEC Anderson Ltd - Oakengrove, Shire Lane, Hastoe, Tring, Herts. HP23 6LY tel:(01442) 826999, fax:(01442) 825999. webinfo@gecanderson.co.uk, www.gecanderson.co.uk

Geemat - 1 Triangle houseq, 2 Broomhill Road, Wandsworth, London. SW18 4HX tel:(020) 8877 1441, fax:(020) 8874 8590. info@fightingfilms.com, www.fightingfilms.com

General Time Europe - 8 Heathcote Way, Warwick. CV34 6TE tel:(01926) 885400, fax:(01926) 885723. naomi@nylholdings.com, www.westclox.co.uk

Genie Europe - The Maltings, Wharf Road, Grantham, Lincs. NG31 6BH tel:(01476) 584333, fax:(01476) 584334. infoeurope@genieind.com, www.genieindustries.com

Gent Limited - 140 Waterside Road, Hamilton Industrial Park, Leicester, Leics. LE5 1TN tel:(0116) 246 2000, fax:(0116) 246 2300. gent_enquiry@gent.co.uk, www.gent.co.uk

Geocel Ltd - Western Wood Way, Langage Science Park, Plympton, Plymouth, Devon. PL7 5BG tel:(01752) 202060, fax:(01752) 202065. info@geocel.co.uk, www.geocel.co.uk

Geoquip Worldwide - 33 Stakehill Industrial Estate, Middleton, Manchester. M24 2RW tel:(0161) 655 1020, fax:(0161) 655 1021. info@geoquip.com, www.geoquip.com

Geoquip Worldwide - Units 3 & 4, Duffield Road, Little Eaton, Derbyshire. DE21 5DR tel:(01629) 824891, fax:(01629) 824896. info@geoquip.com, www.geoquip.co.uk

George Fischer Sales Ltd - Paradise Way, Coventry, W. Midlands. CV2 2ST tel:(024) 765 35535, fax:(024) 765 30450. uk.ps@georgfischer.com, ww.gfps.com

George Gilmour (Metals) Ltd - 245 Govan Road, Glasgow. G51 2SQ tel:(0141) 427 1264, fax:(0141) 427 2205. info@gilmour-ecometal.co.uk, www.gilmour-ecometal.com

Geosynthetic Technology Ltd - Nags Corner, Wiston Road, Nayland, Colchester, Essex. CO6 4LT tel:(01206) 262676, fax:(01206) 262998. sales@geosynthetic.co.uk, www.geosynthetic.co.uk

G (con't)

Gerflor Ltd - Wedgnock House, Wedgnock Lane, Warwick, Warwicks. CV34 5AP tel:(01926) 622612, fax:(01926) 401647. gerflorebayuk@gerflor.com, www.gerflor.co.uk

GEZE UK - Blenheim Way, Fradley Park, Lichfield, Staffs. WS13 8SY tel:(01543) 443000, fax:(01543) 443001. info.uk@geze.com, www.geze.co.uk

Gifford Grant Ltd - 20 Wrecclesham Hill, Farnham, Surrey. GU10 4JW tel:(01252) 816188. Enquiries@gifford-grant.com, http://www.gifford-grant.com

Gilgen Door Systems UK Ltd - Crow House, Crow Arch Lane, Ringwood, Hampshire. BH24 1PD tel:(01425) 46200, fax:(0870) 000 5299. info@gilgendoorsystems.co.uk, www.gilgendoorsystems.com

Gillespie (UK) Ltd - Alma House, 38 Crimea Road, Aldershot, Hants. GU11 1UD tel:(01252) 323311, fax:(01252) 336836. michaelt@gillespieuk.co.uk, www.gillespie.co.uk

Gilmour Ecometal - 245 Govan Road, Glasgow. G51 2SQ tel:(0141) 427 7000, fax:(0141) 427 5345. info@gilmour-ecometal.co.uk, www.gilmour-ecometal.com

Girsberger London - 140 Old Street, London. EC1V 9BJ tel:(020) 7490 3223, fax:(020) 7490 5665. info@girsberger.com, www.girsberger.com

GJ Durafencing Limited - Silverlands Park Nursery, Holloway Hill, Chertsey, Surrey. KT16 0AE tel:(01932) 568727, fax:(01932) 567799. sales@theplantationnursery.co.uk, www.gavinjones.co.uk

Glamox Electric (UK) Ltd - 5 College Street, Mews, Northampton. NN1 2QF tel:(01604) 635611, fax:(01604) 630131. info.uk@glamox.com, www.glamox.com

Glamox Luxo Lighting Limited - Unit 3, Capital Business Park, Mannor Way, Borehamwood, Hertfordshire. WD6 1GW tel:(0208) 9530540, fax:(0208) 9539580. ukoffice@glamoxluxo.com, www.glamoxluxo.co.uk

Glasdon U.K. Limited - Preston New Road, Blackpool, Lancs. FY4 4UL tel:(01253) 600410, fax:(01253) 792558. sales@glasdon-uk.co.uk, www.glasdon.com

Glasdon UK Limited - Preston New Road, Blackpool, Lancs. FY4 4UL tel:(01253) 600410, fax:(01253) 792558. sales@glasdon-uk.co.uk, www.glasdon.com

Glass and Glazing Federation - 54 Ayres Street, London. SE1 1EU tel:(020) 7939 9100, fax:(0870) 042 4266. info@ggf.org.uk, www.ggf.org.uk

Glassolutions Saint-Gobain Ltd - Herald Way, Binley, Coventry. CV3 2ZG tel:(024) 7654 7400, fax:(024) 7654 7799. enquiries@glassolutions.co.uk, www.glassolutions.co.uk

Glazzard (Dudley) Ltd - The Washington Centre, Netherton, Dudley, W. Midlands. DY2 9RE tel:(01384) 233151, fax:(01384) 250224. gdl@glazzard.co.uk, www.glazzard.co.uk

Gledhill Water Storage Ltd - Sycamore Trading Estate, Squires Gate Lane, Blackpool, Lancs. FY4 3RL tel:(01253) 474444, fax:(01253) 474445. sales@gledhill.net, www.gledhill.net

Glenigan Ltd - 41-47 Seabourne Road, Bournemouth, Dorset. BH5 2HU tel:(0800) 373771, fax:(01202) 431204. info@glenigan.com, www.glenigan.com

Glixtone Ltd - Westminster Works, Alvechurch Road, West Heath, Birmingham. B31 3PG tel:(0121) 243 1122, fax:(0121) 243 1123. info@carrscoatings.com, www.carrspaints.com

Gloster Furniture Ltd - Concorde Road, Patchway, Bristol, Avon. BS34 5TB tel:(0117) 931 5335, fax:(0117) 931 5334. uk@gloster.com, www.gloster.co.uk

Goelst - Crimple Court, Hornbeam Park, Harrogate. HG2 8PB tel:(01423) 873002, fax:(01423) 874006. info@goelstuk.com, www.goelst.nl

Gold & Wassall (Hinges) Ltd - Castle Works, Tamworth, Staffs. B79 7TH tel:(01827) 63391, fax:(01827) 310819. enquiries@goldwassallhinges.co.uk, www.goldwassallhinges.co.uk

Golden Coast Ltd - Fishleigh Road, Roundswell Commercial Park West, Barnstaple, Devon. EX31 3UA tel:(01271) 378100, fax:(01271) 371699. swimmer@goldenc.com, www.goldenc.com

Goodacres Fencing - Shoby Lodge Farm, Loughborough Road, Melton Mowbray, Leics. LE14 3PF tel:(01664) 813989, fax:(01664) 813989. kgoodacre1@aol.com

Gooding Aluminium - 1 British Wharf, Landmann Way, London. SE14 5RS tel:(020) 8692 2255, fax:(020) 8469 0031. sales@goodingalum.com, www.goodingalum.com

Goodwin HJ Ltd - Quenington, nr Cirencester, Glos. GL7 5BX tel:(01285) 750271, fax:(01285) 750352. sales@godwinpumps.co.uk, www.godwinpumps.com/

Goodwin Tanks Ltd - Pontefract Street, Derby. DE24 8JD tel:(01332) 363112, fax:(01332) 294 683. info@goodwintanks.co.uk, www.goodwintanks.co.uk

Gower Furniture Ltd - Holmfield Industrial Estate, Holmefield, Halifax, W. Yorks. HX2 9TN tel:(01422) 232200, fax:(01422) 243988. enquiries@gower-furniture.co.uk, www.gower-furniture.co.uk

GP Burners (CIB) Ltd - 2d Hargreaves Road, Groundwell Industrial Estate, Swindon, Wilts. SN25 5AZ tel:(01793) 709050, fax:(01793) 709060. info@gpburners.co.uk, www.gpburners.co.uk

Grace Construction Products Ltd - Ipswich Road, Slough, Berks. SL1 4EQ tel:(01753) 490000, fax:(01753) 490001. uksales@grace.com, www.graceconstruction.com

Grace De Neef UK - 830 Birchwood Boulevard, Birchwood, Warrington, Cheshire. CH41 3PE tel:(01925) 855 335, fax:(01925) 855 350. amanda.browne@deneef.com, www.deneef.com

Gradient Insulations (UK) Ltd - Station Road, Four Ashes, Wolverhampton, W. Midlands. WV10 7DB tel:(01902) 791888, fax:(01902) 791886. sales@gradientuk.com, www.gradientuk.com

Gradus Ltd - Park Green, Macclesfield, Cheshire. SK11 7LZ tel:(01625) 428922, fax:(01625) 433949. www.gradusworld.com

Gradwood Ltd - Landsdowne House, 85 BuxtonRoad, Stockport, Cheshire. SK2 6LR tel:(0161) 480 9629, fax:(0161) 474 7433. tonydavies@gradwood.co.uk, www.gradwood.co.uk

Graepel Perforators Ltd - Unit 5, Burtonwood Industrial Centre, Burtonwood, Warrington, Cheshire. WA5 4HX tel:(01925) 229809, fax:(01925) 228069. sales@graepel.co.uk, www.graepel.co.uk

Grando (UK) Ltd - Dalehouse Lane, Kenilworth, Warwicks. CV8 2EB tel:(01926) 854977, fax:(01926) 856772. info@grando.co.uk, www.grando.co.uk

Granflex (Roofing) Ltd - 62 Brick Kiln Lane, Parkhouse Industrial Estate, Newcastle Under Lyme. ST5 7AS tel:(01782) 202208. sales@granflexroofing.co.uk, www.granflexroofing.co.uk

Grange Fencing Limited - Halesfield 21, Telford, Shropshire. TF7 4PA tel:(01952) 587892, fax:(01952) 684461. sales@grangefen.co.uk, www.grangefen.co.uk

Grant & Livingston Ltd - Kings Road, Charfleets Industrial Estate, Canvey Island, Essex. SS8 0RA tel:(01268) 696855, fax:(01268) 697018. gandl.canvey@btconnect.com, www.grantandlivingstonltd.com

Grant Westfield Ltd - Westfield Avenue, Edinburgh, Lothian. EH11 2QH tel:(0131) 337 6262, fax:(0131) 337 2859. sales@grantwestfield.co.uk, www.grantwestfield.co.uk

Granwood Flooring Ltd - PO Box 60, Alfreton. DE55 4ZX tel:(01773) 606060, fax:(01773) 606030. sales@granwood.co.uk, www.granwood.co.uk

Grass Concrete Ltd - Duncan House, 142 Thornes Lane, Wakefield, W. Yorks. WF2 7RE tel:(01924) 379443, fax:(01924) 290289. info@grasscrete.com, www.grasscrete.com

Gravesend Fencing Limited - Lower Range Road, Denton, Gravesend, Kent. DA12 2QL tel:(01474) 326016, fax:(01474) 324562. gravesendfencing@btconnect.com, www.gravesendfencing.co.uk

Greaves Welsh Slate Co Ltd - Llechwedd Slate Mines, Blaenau Ffestiniog, Gwynedd. LL41 3NB tel:(01766) 830522, fax:(01766) 830711. info@welsh-slate.com, www.llechwedd.co.uk

Green Brook - West Road, Harlow, Essex. CM20 2BG tel:(01279) 772 772, fax:(01279) 422 007. gbe@greenbrook.co.uk, www.greenbrook.co.uk

Green Gerald - 211 Hinckley Road, Nuneaton, Warwicks. CV11 6LL tel:(02476) 325059, fax:(02476) 325059. ggarts@btinternet.com, www.geraldgreen.co.uk

Greenwood Airvac - a division Of Zehnder Group UK Ltd, Unit 4, Watchmoor Point, Camberley, Surrey. GU15 3AD tel:(01276) 605800. orders@greenwood.co.uk, www.greenwood.co.uk

Greif UK Ltd - Merseyside Works, Ellesmere Port, Cheshire. CH65 4EZ tel:(0151) 373 2000, fax:(0151) 373 2072. info.uk@greif.com, www.greif.com

Gridpart Interiors Ltd - 168 Stamford Street Central, Ashton-Under-Lyne, Lancashire. OL6 6AB tel:(0161) 883 2111. mail@gridpart.co.uk, http://www.gridpart.co.uk

Griff Chains Ltd - Quarry Road, Dudley Wood, Dudley, W. Midlands. DY2 0ED tel:(01384) 569415, fax:(01384) 410580. sales@griffchains.co.uk, www.griffchairs.co.uk

Griffin and General Fire Services Ltd - Unit F, 7 Willow Street, London. EC2A 4BH tel:(020) 7251 9379, fax:(020) 7729 5652. headoffice@griffinfire.co.uk, www.griffinfire.co.uk

Grille Diffuser & Louvre Co Ltd The - Air Diffusion Works, Woolley Bridge Road, Hollingworth, Hyde, Cheshire. SK14 7BW tel:(01457) 861538, fax:(01457) 866010. sales@grille.co.uk, www.grille.co.uk

Gripperrods Ltd - Wyrley Brook Park, Walkmill Lane, Bridgtown, Cannock, Staffs. WS11 3RX tel:(01922) 417777, fax:(01922) 419411. sales@interfloor.com, www.interfloor.com

Grohe Ltd - 1 River Road, Barking, Essex. IG11 0HD tel:(020) 8594 7292, fax:(020) 8594 8898. info-uk@grohe.com, www.grohe.co.uk

Grorud Industries Ltd - Castleside Industrial Estate, Consett, Co. Durham. DH8 8HG tel:(01207) 581485, fax:(01207) 580036. enquiries@grorud.com, www.grorud.com

Grosvenor Pumps Ltd - Trevoole, Praze, Camborne, Cornwall. TR14 0PJ tel:(01209) 831500, fax:(01209) 831939. sales@grosvenorpumps.com, www.grosvenorpumps.com

Group Four Glassfibre Ltd - Church Road Business Centre, Murston, Sittingbourne, Kent. ME10 3RS tel:(01795) 429424, fax:(01795) 476248. info@groupfourglassfibre.co.uk, www.groupfourglassfibre.co.uk

Groupco Ltd - 18 Tresham Road, Orton Southgate, Peterborough, Cambs. PE2 6SG tel:(01733) 234750, fax:(01733) 235246. sales@groupcoltd.co.uk, www.groupcoltd.co.uk

Grundfos Pumps Ltd - Groveberry Road, Leighton Buzzard, Beds. LU7 4TL tel:(01525) 850000, fax:(01525) 850011. uk-sales@grundfos.com, uk.grundfos.com

Guardall Ltd - Queen Ann Drive, Lochend Industrial Estate, Newbridge, Edinburgh, Midlothian. EH28 8PL tel:(0131) 333 2900, fax:(0131) 333 4919. sales@guardall.com, www.guardall.com

Guardian Lock and Engineering Co Ltd - Imperial Works, Wednesfield Road, Willenhall, W. Midlands. WV13 1AL tel:(01902) 635964, fax:(01902) 630675. sales@imperiallocks.co.uk, www.imperiallocks.co.u

Guardian Wire Ltd - Guardian Works, Stock Lane, Chadderton, Lancs. OL9 9EY tel:(0161) 624 6020, fax:(0161) 620 2880. sales@guardianwire.co.uk, www.guardianwire.co.uk/

Guest (Speedfit) Ltd, John - Horton Road, West Drayton, Middx. UB7 8JL tel:(01895) 449233, fax:(01895) 420321. info@johnguest.co.uk, www.johnguest.co.uk

Guild of Architectural Ironmongers - BPF House, 6 Bath Place, Rivington Street, London. EC2A 3JE tel:(020) 7033 2480, fax:(020) 7033 2486. info@gai.org.uk, www.gai.org.uk

Guild of Builders & Contractors, The - Crest House, 102-104 Church Road, Teddington, Middx. TW11 8PY tel:(020) 8977 1105, fax:(020) 8943 3151. info@buildersguild.co.uk, www.buildersguild.co.uk

Gunnebo UK Limited - Fairfax House, Pendeford Business Park, Wobaston Road, Wolverhampton, West Midlands. WV9 5HA tel:(01902) 455111, fax:(01902) 351961. info.uk@gunnebo.com, www.gunnebo.co.uk

Gun-Point Ltd - Thavies Inn House, 3-4 Holborn Circus, London. EC1N 2PL tel:(020) 7353 1759, fax:(020) 7583 7259. enquiries@gun-pointltd.com, www.gunpointlimited.co.uk

Guthrie Douglas Ltd - Unit 1 Titan Business Centre, Spartan Close, Warwick, Warwickshire. CV34 6RR tel:(01926) 452452, fax:(01926) 336417. sales@guthriedouglas.com, www.guthriedouglas.com

Guttercrest - Victoria Road, Oswestry, Shropshire. SY11 2HX tel:(01691) 663300, fax:(01691) 663311. info@guttercrest.co.uk, www.guttercrest.co.uk

GVS Filter Technology UK - NFC House, Vickers Ind. Est, Morecambe, Lancs. LA3 3EN tel:(01524) 847600, fax:(01524) 847800. gvsuk@gvs.com, ww.airsafetymedical.com

GVS Filter Technology UK - NFC House, Vickges Industrial Estate, Mellishaw Lane, Morecambe, Lancs. LA3 3EN tel:(01524) 847600, fax:(01524) 847800. gvsuk@gvs.com, www.fengroup.com

H

H E W I (UK) Ltd - Scimitar Close, Gillingham Business Park, Gillingham, Kent. ME8 0RN tel:(01634) 377688, fax:(01634) 370612. info@hewi.co.uk, www.hewi.co.uk/en/

H Pickup Mechanical & Electrical Services Ltd - Durham House, Lower Clark Street, Scarborough, N. Yorks. YO12 7PW tel:(01723) 369191, fax:(01723) 362044. pickup@hpickup.co.uk, www.hpickup.co.uk/

H W Architectural Ltd - Birds Royd Lane, Brighouse, W. Yorkshire. HD6 1NG tel:(01484) 717677, fax:(01484) 400148. hsykes@hwa.co.uk, www.hwa.co.uk

H. B. Fuller Powder Coatings Ltd - 95 Aston Church Road, Birmingham. B7 5RQ tel:(0121) 322 6900, fax:(0121) 322 6901. powdercoatings@valsparеurope.com, www.hbfuller.com

H.B. Fuller - Globe Lane Industrial Estate, Outram Road, Dunkfield, Cheshire. SK16 4XE tel:(0161) 666 0666. www.hbfuller.com

Haddonstone Ltd - The Forge House, Church Lane, East Haddon, Northampton. NN6 8DB tel:(01604) 770711, fax:(01604) 770027. info@haddonstone.co.uk, www.haddonstone.com

Hadley Industries Plc - PO Box 92, Downing Street, Smethwick, W. Midlands. B66 2PA tel:(0121) 555 1300, fax:(0121) 555 1301. sales@hadleygroup.co.uk, www.hadleygroup.co.uk

Hafele UK Ltd - Swift Valley Industrial Estate, Rugby, Warwicks. CV21 1RD tel:(01788) 542020, fax:(01788) 544440. info@hafele.co.uk, www.hafele.co.uk

HAG Shutters & Grilles Ltd - Unit 1, Oak Lane, Fishponds Trading Estate, Fishponds, Bristol. BS5 7UY tel:(0117) 9654888, fax:(0117) 9657773. info@hag.co.uk, www.hag.co.uk

Hager Ltd - Hortonwood 50, Telford, Shropshire. TF1 7FT tel:(0870) 2402400, fax:(0870) 2400400. info@hager.co.uk, www.hager.co.uk

HAGS SMP - Clockhouse Nurseries, Clockhouse Lane East, Egham, Surrey. TW20 8PG tel:(01784) 489100, fax:(01784) 431079. sales@hags-smp.co.uk, www.hags-smp.co.uk

Hale Hamilton Valves Ltd - Frays Mills Works, Cowley Road, Uxbridge, Middlesex. UB8 2AF tel:(01895) 236525, fax:(01895) 231407. _HHVWebEnquiries@circor.com, www.halehamilton.com

Halfen Ltd - A1/A2 Portland Close, Houghton Regis, Beds. LU5 5AW tel:(01582) 470300, fax:(01582) 470304. info@halfen.co.uk, www.halfen.co.uk

Hallidays UK Ltd - 3a Queen Street, Dorchester On Thames, Oxon. OX10 7HR tel:(01865) 340028. info@hallidays.com, www.hallidays.com

Hallmark Blinds Ltd - Hallmark House, 173 Caladonian Road, Islington, London. N1 0SL tel:(020) 7837 0964. info@hallmarkblinds.co.uk, www.hallmarkblinds.co.uk

Halmark Panels Ltd - Valletta House, Valletta Street, Hedon Road, Hull. HU9 5NP tel:(01482) 703222, fax:(01482) 701185. sales@hallmarkpanels.com, www.hallmarkpanels.com

Hambleside Danelaw Ltd - 2-8 Bentley Way, Royal Oak Industrial Estate, Daventry, Northants. NN11 5QH tel:(01327) 701900, fax:(01327) 701909. marketing@hambleside-danelaw.co.uk, www.hambleside-danelaw.co.uk

Hamilton Acorn Ltd - Halford Road, Attleborough, Norfolk. N17 2HZ tel:(01953) 453201, fax:(01953) 454943. info@hamilton-acorn.co.uk, www.hamilton-acorn.co.uk

Hamilton R & Co Ltd - Unit G, Quarry Industrial Estate, Mere, Wilts. BA12 6LA tel:(01747) 860088, fax:(01747) 861032. info@hamilton-litestat.com, www.hamilton-litestat.com

Hammicks Legal Bookshop - 1st Floor, The Atrium, Suites 1-5, 31-37 Church Road, Ashford, Middlesex. TW15 2UD tel:(01784) 423 321, fax:(01784) 427 959. customerservices@hammicks.com, www.hammickslegal.com

Hammonds - New Home Division - Fleming Road, Harrowbrook Industrial Estate, Hinckley, Leics. LE10 3DT tel:(01455) 251451, fax:(01455) 633981. correspondence@hammonds-uk.com, www.hammonds-uk.com

Hamworthy Heating Limited - Fleets corner, Poole, Dorset. BH17 0HH tel:(01202) 662 500, fax:(01202) 662544. sales@hamworthy-heating.com, http://hamworthy-heating.com

Hamworthy Heating Ltd - Fleets Corner, Poole, Dorset. BH17 0HH tel:(01202) 662510, fax:(01202) 662544. marketing@hamworthy-heating.com, www.hamworthy-heating.com

Hands of Wycombe - 36 Dashwood Avenue, High Wycombe, Bucks. HP12 3DX tel:(01494) 524222, fax:(01494) 526508. info@hands.co.uk, www.hands.co.uk

Hanovia Ltd - 145 Farnham Road, Slough, Berks. SL1 4XB tel:(01753) 515300, fax:(01753) 534277. sales@hanovia.co.uk, www.hanovia.co.uk

Hansgrohe - Unit D1-2, Sandown Park Trading Estate, Royal Mills, Esher, Surrey. KT10 8BL tel:(0870) 7701972, fax:(0870) 7701973. enquiries@hansgrohe.co.uk, www.hansgrohe.co.uk

Hansgrohe - Units D1 & D2, Sandown Park Trading Estate, Royal Mills, Esher, Surrey. KT10 8BL tel:(0870) 7701972, fax:(0870) 7701973. enquiries@hansgrohe.co.uk, www.hansgrohe.co.uk

Hanson Building Products - Sales Office, 222 Peterborough Road, Whittlesey, Peterborough. PE7 1PD tel:(0330) 123 1017, fax:(01733) 206040. bricks@hanson.com, www.ask-hanson.com

Hanson Building Products - Stewartby, Bedford. MK43 9LZ tel:(01773) 514011, fax:(01773) 514044. info@hansonbp.com, www.hanson.biz

Hanson Building Products - Stewartby, Bedford. MK43 9LZ tel:(08705) 258258, fax:(01234) 762040. info@hansonbrick.com, www.hansonbrick.com

Hanson Building Products UK - 222 Peterborough Road, Whittlesey, Peterborough. PE7 1PD tel:(0330) 123 1015, fax:(01733) 206170. aggregateblock.sales@hanson.biz, www.hanson.co.uk

Hanson Concrete Products - Alfreton Road, Derby. DE21 4BN tel:(01332) 364314, fax:(01332) 372208. info@lehighhanson.com, www.hanson.biz

Hanson Concrete Products - PO Box 14, Appleford Road, Sutton Courtenay, Abbingdon, Oxon. OX14 4UB tel:(01235) 848877, fax:(01235) 848767. sales@hansonconcreteproducts.com, www.hansonplc.com

Hanson Concrete Products Ltd - Hoveringham, Nottingham, Notts. NG14 7JX tel:(01636) 832000, fax:(01636) 832020. concrete@hanson.com, www.heidelbergcement.com

H (con't)

Hanson Thermalite - Hanson, Stewartby, Bedford. MK43 9LZ tel:(08705) 626 500, fax:(01234) 762040. thermalitesales@hanson.com, www.heidelbergcement.com/uk/en/hanson/products/blocks/aircrete_blocks/index.htm
Hanson UK - Hanson House, 14 Castle Hill, Maidenhead. SL6 4JJ tel:(01628) 774100. enquiries@hanson.com, www.hanson.co.uk
Hanson UK - Bath & Portland Stone - Avon Mill Lane, Keynsham, Bristol. BS31 2UG tel:(0117) 986 9631, fax:(01305) 860275. contact@bathandportlandstone.co.uk, http://bathandportlandstone.co.uk
Harbro Supplies Ltd - Morland Street, Bishop Auckland, Co. Durham. DL14 6JQ tel:(01388) 605363, fax:(01388) 603263. harbrosupplies@hotmail.com, www.habrosupplies.com
Hard York Quarries Ltd - Fagley Lane, Eccleshill, Bradford, W. Yorks. BD2 3NT tel:(01274) 637307, fax:(01274) 626146. sales@hardyorkquarries.co.uk, www.hardyorkquarries.co.uk
Hardall International Ltd - Unit 2, Fairway Works, Southfields Road, Dunstable, Beds. LU6 3EP tel:(01582) 500860, fax:(01582) 690975. sales@hardall.co.uk, www.hardall.co.uk
Harewood Products Ltd (Adboards) - Unit 1, Union Road, The Valley, Bolton. BL2 2DT tel:(01204) 395730, fax:(01204) 388018. sales@adboards.com, www.adboards.com
Hargreaves Group GB Ltd - Westleigh House, Wakefield Road, Denby Dale, Huddersfield. HD8 8QJ tel:(01484) 866634, fax:(01484) 865268. www.hargreavesgb.com
Harkness Screens - Unit A, Norton Road, Stevenage, Herts. SG1 2BB tel:(01438) 725 200, fax:(01438) 344 400. Sales@harkness-screens.com, www.harkness-screens.com
Harling Security Solutions - 235-237, Church Road, Hayes, Middx. UB3 2LG tel:(020) 85613787, fax:(020) 8848 0999. sales@harlingsecurity.com
Harris & Bailey Ltd - 50 Hastings Road, Croydon, Surrey. CR9 6BR tel:(020) 8654 3181, fax:(020) 8656 9369. mail@harris-bailey.co.uk, www.harris-bailey.co.uk
Harrision Flagpoles - Borough Road, Darlington, Co. Durham. DL1 1SW tel:(01325) 355433, fax:(01325) 461726. sales@harrisoneds.com, www.flagpoles.co.uk
Harrison Thompson & Co Ltd - Yeoman House, Whitehall Estate, Whitehall Road, Leeds. LS12 5JB tel:(0113) 279 5854, fax:(0113) 231 0406. info@yeomanshield.com, www.yeomanshield.com
Hart Door Systems Ltd - Redburn Road, Westerhope Industrial Estate, Newcastle-upon-Tyne, Tyne & Wear. NE5 1PJ tel:(0191) 214 0404, fax:(0191) 271 1611. response@speedor.com, www.speedor.com
Hartley & Sugden - Atlas Works, Gibbet Street, Halifax, W. Yorks. HX1 4DB tel:(01422) 355651, fax:(01422) 359636. sales@hartleyandsugden.co.uk, www.ormandy.biz/hsboilers
Hartley Botanic Ltd - Wellington Road, Greenfield, Oldham, Lancs. OL3 7AG tel:(01457) 873244. info@hartleybotanic.co.uk, www.hartley-botanic.co.uk
Harton Heating Appliances - Unit 6, Thustlebrook Industrial Est, Eynsham Drive, Abbey Wood. SE2 9RB tel:(020) 8310 0421, fax:(020) 8310 6785. info@hartons.globalnet.co.uk, www.hartons.co.uk
Harvey Steel Lintels Limited - Commerce Way, Whitehall Industrial Estate, Colchester, Essex. CO2 8HH tel:(01206) 792001, fax:(01206) 792022. harvey@lintels.co.uk, www.lintels.co.uk
Harviglass-Fibre Ltd - Alexandra Street, Hyde, Cheshire. SK14 1DX tel:(0161) 3682398, fax:(0161) 3681508. info@harviglass.com, www.harviglass.com
Hatfield Ltd, Roy - Fullerton Road, Rotherham, S. Yorks. S60 1DL tel:(01709) 820855, fax:(01709) 374062. info@royhatfield.com, www.royhatfield.com
Hatmet Limited - No. 3 The Courtyard, Lynton Road, Crouch End, London. N8 8SL tel:(020) 8341 0200, fax:(020) 8341 9878. trevorh@hatmet.co.uk, www.hatmet.co.uk
Hattersley - P O Box 719, Ipswitch, Suffolk. IP1 9DU tel:(01744) 458670, fax:(01744) 458671. uksales@hattersley.com, www.hattersley.com
Havering Fencing Co - 237 Chase Cross Road, Collier Row, Romford, Essex. RM5 3XS tel:(01708) 747855, fax:(01708) 721010. enquiries@haveringfencing.com, www.haveringfencing.com
Hawker Ltd - Rake Lake, Clifton Junction, Swinton, Gt Man. M27 8LR tel:(0161) 794 4611, fax:(0161) 793 6606. hawkeruk.sales@hawker.invensys.com, www.hawker.invensys.com
Hawker Siddley Switchgear - Unit 3, Blackwood Business Park, Newport Road, Blackwood, South Wales. NP12 2XH tel:(01495) 223001, fax:(01495) 225674. sales@hss-ltd.com, www.hss-ltd.com
Hawkesworth Appliance Testing - Guidance House, York Road, Thirsk. YO7 3BT tel:(01845) 524498, fax:(01845) 526884. sales@hawktest.co.uk, www.hawktest.co.uk
Hawkins Insulation Ltd - Central House, 101Central Park, Petherton Road, Hengrave, Bristol, Avon. BS14 9BZ tel:(01275) 839500, fax:(01275) 835555. enquiries@hawkinsulation.co.uk, www.hawkinsulation.co.uk
Haworth UK Ltd - Cannon Court, Brewhouse Yard, St. John Street, Clerkenwell, London. EC1V 4JQ tel:(020) 7324 1360, fax:(020) 7490 1513. info.uk@haworth.com, www.haworthuk.com/en
Hayes Cladding Systems - Brindley Road, (off Hadfield Road), Cardiff. CF11 8TL tel:(029) 2038 9954, fax:(029) 2039 0127. sales@hayescladding.co.uk, www.hayescladding.co.uk
Haynes Manufacturing (UK) Ltd - Marlowe House, Stewkins, Stourbridge, W. Midlands. DY8 4YW tel:(01384) 371416, fax:(01384) 371416. info@haynes-uk.co.uk, www.haynes-uk.co.uk
Haysom (Purbeck Stone) Ltd - Lander's Quarry, Kingston Road, Langton Matravers, Swanage, Dorset. BH19 3JP tel:(01929) 439205, fax:(01929) 439268. haysom@purbeckstone.co.uk, www.purbeckstone.co.uk
Hazard Safety Products Ltd - 55-57 Bristol Road, Edgbaston, Birmingham. B5 7TU tel:(0121) 446 4433, fax:(0121) 446 4230. sales@hazard.co.uk, www.hazard.co.uk
HC Slingsby PLC - Otley Road, Baildon, Shipley, W. Yorks. BD17 7LW tel:(0800) 294 4440, fax:(0800) 294 4442. sales@slingsby.com, www.slingsby.com
HCL Contracts Ltd - Bridge House, Commerce Road, Brentford, Middx. TW8 8LQ tel:(0800) 212867, fax:(020) 8847 9003. info@hclsafety.com, www.hclgroup.co.uk
HCL Devizes - Hopton Park, Devizes, Wiltshire. SN10 2JP tel:(01380) 732393, fax:(01380) 732701. devizes@hclsafety.com, www.hclsafety.com
HE Services (Plant Hire) Ltd - Whitewall Road, Strood, Kent. ME2 4DZ tel:(0208) 804 2000, fax:(0871) 534437. nhc@heservices.co.uk, www.heservices.co.uk
Healthmatic Ltd - 3 Wellington Square, Ayr, Ayrshire. KA7 1EN tel:(01292) 265 456, fax:(01292) 287 869. sales@healthmatic.com, www.healthmatic.com
Heat Merchants - Moydrum Road, Athlone, Westmeath, Ireland. tel:+353 90 6424000, fax:+353 90 6424067. customerservices@wolseley.co.uk, www.heatmerchants.ie
Heat Pump Association (HPA) - 2 Waltham Court, Milley Lane, Hare Hatch, Reading, Berkshire. RG10 9TH tel:(01189) 403416, fax:(01189) 406258. info@feta.co.uk, www.feta.co.uk
Heath Samuel & Sons PLC - Cobden Works, Leopold Street, Birmingham. B12 0UJ tel:(0121) 766 4200, fax:(0121) 772 3334. info@samuel-heath.com, www.samuel-heath.co.uk
Heating Ventilating and Air Conditioning Manufacturers Association (HEVAC) - 2 Waltham Court, Milley Lane, Hare Hatch, Reading, Berkshire. RG10 9TH tel:(01189) 403416, fax:(01189) 406258. info@feta.co.uk, www.feta.co.uk
HeatProfile Ltd - Unit 1, Walnut Tree Park, Walnut Tree Close, Guildford, Surrey. GU1 4TR tel:(01483) 537000, fax:(01483) 537500. sales@heatprofile.co.uk, www.heatprofile.co.uk
Heatrae Sadia Heating Ltd - Hurricane Way, Norwich. NR6 6EA tel:(01603) 420220, fax:(01603) 420149. sales@heatraesadia.com, www.heatraesadia.com
Heckmondwicke FB - PO Box 7, Wellington Mills, Liversedge, West Yorkshire. WF15 7XA tel:(01924) 406161, fax:(0800) 136769. sales@heckmondwike-fb.co.uk, www.heckmondwicke-fb.co.uk
Heldite Ltd - Heldite Centre Bristow Road, Hounslow, Middx. TW3 1UP tel:(020) 8577 9157/9257, fax:(020) 8577 9057. sales@heldite.com, www.heldite.com
Helifix - The Mille, 1000 Great West Road, Brentford, London. TW8 9DW tel:(020) 8735 5200, fax:(020) 8735 5201. sales@helifix.co.uk, www.helifix.co.uk
Hellermann Tyton - Sharston Green Buisness Park , 1 Robeson Way, Altrincham Road, Wythenshawe, Manchester. M22 4TY tel:(0161) 945 4181, fax:(0161) 954 3708. ric.kynnersley@hellermanntyton.co.uk, www.hellermanntyton.co.uk
Helmsman - Northern Way, Bury St Edmunds, Suffolk. IP32 6NH tel:(01284) 727600, fax:(01284) 727601. sales@helmsman.co.uk, www.helmsman.co.uk
Helvar Ltd - Hawley Mill, Hawley Road, Dartford, Kent. DA2 7SY tel:(01322) 617 200, fax:(01322) 617 229. uksystemsales@helvar.com, www.helvar.com
HEM Interiors Group Ltd - HEM House, Kirkstall Road, Leeds. LS4 2BT tel:(0113) 263 2222, fax:(0113) 231 0237. heminteriors@btinternet.com
Hemsec Panel Technologies - Stoney Lane, Rainhill, Prescot, Merseyside. L35 9LL tel:(0151) 426 7171, fax:(0151) 493 1331. sales@hemsec.com, www.hpt-panels.com
Henkel Consumer Adhesives - Road 5, Winsford Ind Est, Winsford, Cheshire. CW7 3QY tel:(020) 8804 3343, fax:(020) 8443 4321. technical.services@henkel.co.uk, www.henkel.com
Henkel Ltd - Wood Lane End, Hemel Hempstead, Hertfordshire. HP2 4RQ tel:(01442) 27 8000, fax:(01442) 278071. info@henkel.co.uk, www.henkel.co.uk
Hepworth Building Products Ltd - Hazlehead, Crow Edge, Sheffield, S. Yorks. S36 4HG tel:(01226) 763561, fax:(01226) 764827. customerservices@wavin.co.uk, www.hepworthbp.co.uk
Hepworth Heating Ltd - Nottingham Road, Belper, Derbys. DE56 1JT tel:(01773) 824141, fax:(01773) 820569. info@glow-worm.co.uk, www.glow-worm.co.uk
Herga Electric Ltd - Northern Way, Bury St Edmunds, Suffolk. IP32 6NN tel:(01284) 701422, fax:(01284) 753112. info@herga.com, www.herga.com
Hertel Services - Hertel (UK) Ltd, Alton House, Alton Business Park, Alton Road, Ross-on-Wye. HR9 5BP tel:(0845) 604 6729, fax:(01989) 561 189. www.hertelsolutions.com
Herz Valves UK Ltd - Progress House, Moorfield Point, Moorfield Road, Slyfield Industrial Estate, Guildford, Surrey. GU1 1RU tel:(01483) 502211, fax:(01483) 502025. sales@herzvalves.com, www.herzvalves.com
Hettich UK Ltd - Unit 200, Metroplex Business Park, Broadway, Salford, Manchester. M50 2UE tel:(0161) 872 9552, fax:(0161) 848 7605. info@uk.hettich.com, www.hettich.com
Higginson Staircases Ltd - Unit 1, Carlisle Road, London. NW9 0HD tel:(020) 8200 4848, fax:(020) 8200 8249. sales@higginson.co.uk, www.higginson.co.uk
Hill and Smith - Springvale Business & Industrial Park, Bilston, Wolverhampton. WV14 0ql tel:(01902) 499400, fax:(01902) 499419. barrier@hill-smith.co.uk, www.hill-smith.co.uk
Hillday Limited - 1 Haverscroft Industrial Estate, Attleborough, Norfolk. NR17 1YE tel:(01953) 454014, fax:(01953) 454014. hillday@btinternet.com, www.hillday.co.uk
Hille Educational Products Ltd - Unit 27, Rassau Industrial Estate, Ebbw Vale, Gwent. NP23 5SD tel:(01495) 352187, fax:(01495) 306646. info@hille.co.uk, www.hille.co.uk
Hilmor - Caxton Way, Stevenage, Herts. SG1 2DQ tel:(01438) 312466, fax:(01438) 728327. uksales@irwin.co.uk, www.hilmar.co.uk
Hilti (Great Britain) Ltd - 1 Trafford Wharf Road, Manchester. M17 1BY tel:(0800) 886100, fax:(0800) 886200. gbsales@hilti.co.uk, www.hilti.co.uk
Hilton Banks Ltd - 74 Oldfield Road, Hampton, Middx. TW12 2HR tel:(08452) 300 404, fax:(020) 8779 8294. hiltonbanks@btinternet.com
Hines & Sons, P E, Ltd - Whitbridge Lane, Stone, Staffs. ST15 8LU tel:(01785) 814921, fax:(01785) 818808. sales@hines.co.uk, www.hines.co.uk
Hinton Perry & Davenhill Ltd - Dreadnought Works, Pensnett, Brierley Hill, Staffs. DY5 4TH tel:(01384) 77405, fax:(01384) 74553. sales@dreadnought-tiles.co.uk, www.dreadnought-tiles.co.uk
Hipkiss, H, & Co Ltd - Park House, Clapgate Lane, Woodgate, Birmingham. B32 3BL tel:(0121) 421 5777, fax:(0121) 421 5333. info@hipkiss.co.uk, www.hipkiss.co.uk
Hispack Limited - 8 School Road, Downham, Billericay, Essex. CM11 1QU tel:(01268) 711499, fax:(01268) 711068. hispack@btinternet.com, www.hispack.co.uk
Hi-Store Ltd - Station Approach, Four Marks, Alton, Hants. GU34 5HN tel:(01420) 562522, fax:(01420) 564420. sales@hi-store.com, www.hi-store.com
Historic England - 1 Waterhouse Square, 138 - 142 Holborn, London. EC1N 2ST tel:(020) 7973 3700, fax:(020) 7973 3001. customers@HistoricEngland.org.uk, https://historicengland.org.uk/
Hi-Vee Ltd - Southdown, Western Road, Crowborough, Sussex. TN6 3EW tel:(01892) 662177, fax:(01892) 667225. www.hi-vee.co.uk
HMG Paints - Riverside Works, Collyhurst Road, Manchester, Gt Man. M40 7RU tel:(0161) 205 7631, fax:(0161) 205 4829. sales@hmgpaint.com, www.hmgpaint.com
HMSO Publications Office - Parliamentary Press, Mandela Way, London. SE1 tel:(020) 7394 4200. book.orders@tso.co.uk, www.legislation.gov.uk
Hoben International Ltd - Brassington Works, Manystones Lane, Matlock, Derbyshire. DE4 4HF tel:(01629) 540201, fax:(01629) 540205. sales@hobeninternational.com, www.hobeninternational.com
Hochiki Europe (UK) Ltd - Grosvenor Road, Gillingham Business Park, Gillingham, Kent. ME8 0SA tel:(01634) 260133, fax:(01634) 260132. sales@hochikieurope.com, www.hochikieurope.com
Hodgkinson Bennis Ltd - Unit 7A, Highfield Road, Little Hulton, Worsley, Manchester. M38 9SS tel:(0161) 790 4411, fax:(0161) 703 8505. enquiries@hbcombustion.com, www.hbcombustion.co.uk
Hodkin Jones (Sheffield) Ltd - Callywhite Lane, Dronfield, Sheffield. S18 2XP tel:(01246) 290890, fax:(01246) 290292. info@hodkin-jones.co.uk, www.hodkin-jones.co.uk
Holden Brooke Pullen - Wenlock Way, Manchester. M12 5JL tel:(0161) 223 2223, fax:(0161) 220 9660. sales@holdenbrookepullen.com, www.holdenbrookepullen.com
Holophane Europe Ltd - Bond Avenue, Bletchley, Milton Keynes. MK1 1JG tel:(01908) 649292, fax:(01908) 367618. info@holophane.co.uk, www.holophane.co.uk
Homelux Nenplas - Airfield Industrial, Airfield Industrial Estate, Ashbourne, Derbys. DE6 1HA tel:(01335) 347300, fax:(01335) 340333. enquiries@homeluxnenplas.com, www.homeluxnenplas.com
Honeywell Control Systems Ltd - Honeywell House, Arlington Business Park, Bracknell, Berks. RG12 1EB tel:(01344) 656000, fax:(01344) 656240. uk.infocentre@honeywell.com, www.honeywell.com/uk
Honeywell Safety Products - Edison Road, Basingstoke. RG21 6QD tel:(01256) 693200, fax:(01256) 693300. info-uk.hsp@honeywell.com, www.honeywellsafety.com
Hoover European Appliances Group - Pentrbach, Merthyr Tydfil, M. Glam. CF48 4TU tel:(01685) 721222, fax:(01685) 725694. www.hoover.co.uk
HOPPE (UK) Ltd - Gailey Park, Gravelly Way, Standeford, Wolverhampton, W. Midlands. WV10 7GW tel:(01902) 484400, fax:(01902) 484406. info.uk@hoppe.com, www.hoppe.co.uk
Horizon International Ltd - Willment Way, Avonmouth, Bristol, Avon. BS11 8DJ tel:(0117) 982 1415, fax:(0117) 982 0630. sales@horizon-int.com, www.horizon-int.com
Hörmann (UK) Ltd - Gee Road Coalville, Coalville, Leicestershire. LE67 4JW tel:(01530) 513000, fax:(01530) 513001. info@hormann.co.uk, www.hormann.co.uk
Horne Engineering Ltd - PO Box 7, Rankine Street, Johnstone, Renfrewshire. PA5 8BD tel:(01505) 321455, fax:(01505) 336287. sales@horne.co.uk, www.horne.co.uk
Horstmann Controls Ltd - South Bristol Business Park, Roman Farm Road, Bristol. BS4 1UP tel:(0117) 978 8700, fax:(0117) 978 8701. sales@horstmann.co.uk, www.horstmann.co.uk
Hospital Metalcraft Ltd - Blandford Heights, Blandford Forum, Dorset. DT11 7TG tel:(01258) 451338, fax:(01258) 455056. sales@bristolmaid.com, www.bristolmaid.com
Hotchkiss Air Supply - Heath Mill Road, Wombourne, Wolverhampton, W. Midlands. WV5 8AP tel:(01903) 895161, fax:(01903) 892045. sales@hotchkissairsupply.co.uk, www.hotchkissairsupply.co.uk
Hoval Ltd - Northgate, Newark-on-Trent, Notts. NG24 1JN tel:(01636) 72711, fax:(01636) 73532. boilersales@hoval.co.uk, www.hoval.co.uk
Howard Bros Joinery Ltd - Station Approach, Battle, Sussex. TN33 0DE tel:(01424) 773272, fax:(01424) 773836. sales@howard-bros-joinery.com, www.howard-bros-Joinery.com
Howard Evans Roofing and Cladding Ltd - C/O SPV Group, Westgate, Aldridge Walsall, West Midlands. WS9 8EX tel:(0121) 327 1336, fax:(01922) 749 515. info@howardevansroofingandcladding.co.uk, www.howardevansroofingandcladding.co.uk
Howarth Windows & Doors Ltd - The Dock, New Holland, N. Lincs. DN19 7RT tel:(01469) 530577, fax:(01469) 531559. windows&doors@howarth-timber.co.uk, www.howarth-timber.co.uk
Howden Buffalo - Old Govan Road, Renfrew, Renfrewshire. PA4 8XJ tel:(0141) 885 7500, fax:(0141) 885 7555. huk.sales@howden.com, www.howden.com
Howe Fencing & Sectional Buildings - Horse Cross, Standon Road, Standon, Nr Ware, Herts. SG11 2PU tel:(01920) 822055, fax:(01920) 822871. sales@howefencing.co.uk, www.howefencing.co.uk
Howe Green Ltd - Marsh Lane, Ware, Herts. SG12 9QQ tel:(01920) 463 230, fax:(01920) 463 231. info@howegreen.co.uk, www.howegreen.com
Howse Thomas Ltd - Cakemore Road, Rowley Regis, W. Midlands. B65 0RD tel:(0121) 559 1451, fax:(0121) 559 2722. sales@howsepaints.co.uk, www.howsepaints.co.uk
Hoyles Electronic Developments Ltd - Sandwash Close, Rainford Industrial Estate, Rainford, St. Helens, Merseyside. WA11 8LY tel:(01744) 886600, fax:(01744) 886607. info@hoyles.com, www.hoyles.com
Hozelock Ltd - Midpoint Park, Birmingham. B76 1AB tel:(0121) 313 4242. www.hozelock.com
HRS Hevac Ltd - PO Box 230, Watford, Herts. WD1 8TX tel:(01923) 232335, fax:(01923) 230266. mail@hrs.co.uk, www.hrshevac.co.uk
HSS Hire Service Group - 25 Willow Lane, Mitcham, Surrey. CR4 4TS tel:(020) 8260 3100, fax:(020) 8687 5005. hire@hss.co.uk, www.hss.co.uk
HT Martingale Ltd - Ridgeway Industrial Estate, Iver, Bucks. SL0 9HU tel:(01753) 654411, fax:(01753) 630002. www.martingalefabrications.co.uk
Hubbard Architectural Metalwork Ltd - 3 Hurricane Way, Norwich, Norfolk. NR6 6HS tel:(01603) 424817, fax:(01603) 487158. tony.hubbard@hubbardsmetalwork.co.uk, www.hubbardsmetalwork.co.uk
Hubbell Premise Wiring - Chantry Avenue, Kempston, Bedford, Beds. MK42 7RR tel:(01234) 848559, fax:(01234) 856190. sales@minitran.co.uk, www.hubbell-premise.com

H (con't)

Hudevad Britain - Hudevad House, Walton Lodge, Bridge Street, Walton-on-Thames, Surrey. KT12 1BT tel:(01932) 247835, fax:(01932) 247694. sales@hudevad.co.uk, www.hudevad.co.uk

Hufcor UK Ltd - The Maltings, Station Road, Sawbridgeworth, Hertfordshire. CM21 9JX tel:(01279) 882258. enquiries@hufcoruk.co.uk, www.hufcoruk.co.uk

Hughes Safety Showers Limited - Whitefield Road, Bredbury, Stockport, Cheshire. SK6 2SS tel:(0161) 430 6618, fax:(0161) 4307928. sales@hughes-safety-showers.co.uk, www.hughes-safety-showers.co.uk

Humbrol - Westwood, Margate, Kent. CT9 4JX tel:(01843) 233525. internetsales@hornby.com, www.humbrol.com

Humphrey & Stretton PLC - Pindar Road Industrial Estate, Hoddesdon, Herts. EN11 0EU tel:(01992) 462965, fax:(01992) 463996. enquiries@humphreystretton.com, www.humphreystretton.com/

Hunter Douglas - 8 Charter Gate, Clayfield Close, Moulton Park, Northampton. NN3 6QF tel:(01604) 766251, fax:(01604) 212863. info@hunterdouglas.co.uk, www.hunterdouglas.co.uk

Hunter Douglas Ltd (Wood Division) - Kingswick House, Kingswick Drive, Sunninghill, Berkshire. SL5 7BH tel:(01344) 292293, fax:(01344) 292214. hswood@compuserve.com

Hunter Douglas UK Ltd (Window Fashions Division) - Mersey Industrial Estate, Heaton Mersey, Stockport, Cheshire. SK4 3EQ tel:(0161) 442 9500, fax:(0161) 431 5087. russell.malley@luxaflex-sunway.co.uk, www.luxalon.com

Hunter Plastics Ltd - Nathan Way, London. SE28 0AE tel:(020) 8855 9851, fax:(020) 8317 7764. info@multikwik.com, www.hunterplastics.co.uk

Hunter Timber (Scotland) - Earls Road, Grangemouth Saw Mills, Grangemouth, Stirlingshire. FK3 8XF tel:(01324) 483294. www.finnforest.com

Huntree Fencing Ltd - Cosy Corner, Great North Road, Little Paxton, Cambs. PE19 6EH tel:(01234) 870864, fax:(01480) 471082. karen@huntreefencing.co.uk, www.huntreefencing.co.uk

Hush Acoustics - 44 Canal Street, Bootle, Liverpool. L20 8QU tel:(0151) 9332026, fax:(0151) 9441146. info@hushacoustics.co.uk, www.hushacoustics.co.uk

HW Architectural Ltd - Birds Royd Lane, Brighouse, W. Yorks. HD6 1NG tel:(01484) 717677, fax:(01484) 400148. enquiry@hwa.co.uk, www.hwa.co.uk

Hyde Brian Ltd - Stirling Road, Shirley, Solihull, W. Midlands. B90 4LZ tel:(0121) 705 7987, fax:(0121) 711 2465. sales@brianhyde.co.uk, www.brianhyde.co.uk

Hydraseeders Ltd - Coxbench, Derby, Derbys. DE21 5BH tel:(01332) 880364, fax:(01332) 883241. hydraseeders1@btconnect.com, www.hydraseeders.uk.com

Hydratight Sweeney Limited - Bently Road South, Darlaston, W. Midlands. WS10 8LQ tel:(0121) 5050 600, fax:(0121) 5050 800. enquiry@hydratight.com, www.hevilifts.com

Hydropath (UK) Ltd - Unit F, Acorn Park Ind Est, Nottingham, Notts. NG7 2TR tel:(01923) 210028, fax:(01923) 800243. sales@hydropath.com, www.hydropath.co.uk

Hydrovane Compressor Co Ltd - Claybrook Drive, Washford Industrial Estate, Redditch. B98 0DS tel:(01527) 525522, fax:(01527) 510862. sales@compair.com, www.hydrovane.co.uk

Hye Oak - Denton Wharf, Gravesend, Kent. DA12 2QB tel:(01474) 332291, fax:(01474) 564491. enquiries@hyeoak.co.uk, www.hyeoak.co.uk

Hyflex Roofing - Amasco House, 101 Powke Lane, Cradley Heath, West Midlands. B64 5PX tel:(0121) 502 9580, fax:(0121) 502 9581. enquiries@hyflex.co.uk, www.hyflex.co.uk

Hymo Ltd - 35 Cornwell Business Park, Salthouse Road, Brackmills, Northampton, Northants. NN4 7EX tel:(01604) 661601, fax:(01604) 660166. info@hymo.ltd.uk, www.hymo.com

Hyperion Wall Furniture Ltd - Business Park 7, Brook Way, Leatherhead, Surrey. KT22 7NA tel:(01932) 844783, fax:(01372) 362004. enquiries@hyperion-furniture.co.uk, http://hyperion-furniture.co.uk

Hy-Ten Ltd - 12 The Green, Richmond, Surrey. TW9 1PX tel:(020) 8940 7578, fax:(020) 8332 1757. admin@hy-ten.co.uk, www.hy-ten.co.uk

I

IAC Acoustics - IAC House, Moorside Road, Winchester, Hants. SO23 7US tel:(01962) 873 000, fax:(01962) 873111. info@iacl-uk.com, www.iac-noisecontrol.com

IBL Lighting Limited - 1 Farnham Road, Guildford, Surrey. GU2 4RG tel:(0844) 822 5210, fax:(0844) 822 5211. info@ibl.co.uk, www.ibl.co.uk

Ibstock Brick Ltd - Leicester Road, Ibstock, Leics. LE67 6HS tel:(01530) 261999, fax:(01530) 261888. marketing@ibstock.co.uk, www.ibstock.co.uk

Ibstock Building Products Ltd - Anstone Office, Kiveton Park Station, Kiveton Park, Sheffield, S. Yorks. S26 6NP tel:(01909) 771122, fax:(01909) 515281. marketing@ibstock.co.uk, www.ibstock.com/

Ibstock Hathernware Ltd - Station Works, Rempstone Road, Normanton on Soar, Loughborough, Leics. LE12 5EW tel:(01509) 842273, fax:(01509) 843629. enquiries@ibstock.co.uk, www.hathernware.co.uk

Ibstock Scotish Brick - Tannochside Works, Old Edingburgh Road, Tanockside, Uddingston, Strathclyde. G71 6HL tel:(01698) 810686, fax:(01698) 812364. enquiries@ibstock.co.uk, www.ibstock.co.uk

ICB (International Construction Bureau) Ltd - Unit 9, Elliott Road, West Howe Industrial Estate, Bournemouth, Dorset. BH11 8JX tel:(01202) 579208, fax:(01202) 581748. info@icb.uk.com, www.icb.uk.com

ICI Woodcare - ICI Paints plc, Wexham Road, Slough. SL2 5DS tel:(0870) 242 1100, fax:(01753) 556938. duluxtrade.advice@ici.com, www.duluxtrade.co.uk

Icopal Limited - Barton Dock Road, Stretford, Manchester. M32 0YL tel:(0161) 865 4444, fax:(0161) 864 2616. info.uk@icopal.com, www.icopal.co.uk

Icopal Limited - Barton Rock Rd, Stretford, Manchester. M32 0YC tel:(0161) 865 4444, fax:(0161) 864 2616. info.uk@icopal.com, www.monarflex.co.uk

Ide T & W Ltd - 5 Sovereign Close, Wapping, London. E1W 3HW tel:(020) 7790 2333, fax:(020) 7790 0201. contracting@idecon.co.uk, www.twigroup.co.uk

Ideal Building Systems Ltd - Lancaster Road, Carnaby Industrial Estate, Bridlington, E. Yorks. YO15 3QY tel:(01262) 606750, fax:(01262) 671960. sales@idealbuildingsystems.co.uk, www.idealbuildingsystems.co.uk

Ideal-Standard Ltd - The Bathroom Works, National Avenue, Kingston Upon Hull. HU5 4HS tel:(01482) 346461, fax:(01482) 445886. UKCustcare@idealstandard.com, www.ideal-standard.co.uk

IG Limited - Avondale Road, Cwmbran, Gwent. NP44 1XY tel:(01633) 486486, fax:(01633) 486465. info@igltd.co.uk, www.igltd.co.uk

IHS - Viewpoint One Willoughby Road, Bracknell, Berks. RG12 8FB tel:(01344) 426311, fax:(01344) 424971. systems@ihs.com, www.ihs.com

IJK Timber Group Ltd - 24-28 Duncrue Street, Belfast. BT3 9AR tel:(028) 9035 1224, fax:(028) 9035 1527. sales@ijktimber.co.uk, www.ijktimber.co.uk

IKM Systems Ltd - Unit7A, The Seedbed Business Centre, Vangaurd Way, Shoeburyness, Southend-on-Sea, Essex. SS3 9QX tel:(01702) 382321, fax:(01702) 382323. ikmsystems@btopenworld.com, www.ikmsystems.co.uk

IKO PLC - Appley Lane North, Appley Bridge, Wigan, Lancashire. WN6 9AB tel:(01257) 256097. sales@ikogroup.co.uk, www.ikogroup.co.uk

Iles Waste Systems - Valley Mills, Valley Road, Bradford, W. Yorks. BD1 4RU tel:(01274) 728837, fax:(01274) 734351. wastesystems@trevoriles.co.uk, www.ileswastesystems.co.uk

Illbruck Sealant Systems UK Ltd - Bentall Business Park, Glover, District 11, Washington, Tyne & Wear. NE37 3JD tel:(0191) 4190505, fax:(0191) 4192200. uk.info@tremco-illbruck.com, www.illbruck.com

Illuma Lighting - Sills Road, Willow Farm Business Park, Castle Donington, Derbyshire. DE74 2US tel:(01332) 818200, fax:(01332) 818222. info@illuma.co.uk, www.illuma.co.uk

Imerys Minerals Ltd - Westwood, Beverley, N. Humberside. HU17 8RQ tel:(01482) 881234, fax:(01482) 872301. perfmins@imerys.com, www.imerys.com

Imerys Roof Tiles - PO Box 10, Hadleigh, Suffolk. IP7 7WD tel:(0161) 928 4572, fax:(0161) 929 8513. enquiries.rooftiles@imerys.com, www.imerys-rooftiles.com

Imperial Machine Co Ltd - Harvey Road, Croxley Green, Herts. WD3 3AX tel:(01923) 718000, fax:(01923) 777273. sales@imco.co.uk, www.imco.co.uk

IMS UK - Arley Road, Saltley, Birmingham. B8 1BB tel:(0121) 326 3100, fax:(0121) 326 3105. a.uk@abraservice.com, www.abraservice.com/uk

In Steel (Blacksmiths & Fabricators) Ltd - United Downs Industrial Park, St. Day, Redruth, Cornwall. TR16 5HY tel:(01209) 822233, fax:(01209) 822596. enquiries@spiralstairs.uk.com, www.spiralstairs.uk.com

Industrial Brushware Ltd - Ibex House, 76-77 Malt Mill Lane, Halesowen, West Midlands. B62 8JJ tel:(0121) 559 3862, fax:(0121) 559 9404. sales@industrialbrushware.co.uk, www.industrialbrushware.co.uk

Infraglo Limited - Dannemora Drive, Greenland Road Industrial Park, Sheffield, South Yorks. S9 5DF tel:(0114) 249 5445, fax:(0114) 249 5066. info@infraglo.com, www.infraglo.com

Ingersoll-Rand European Sales Ltd - PO Box 2, Chorley New Road, Horwich, Bolton. BL6 6JN tel:(01204) 690690, fax:(01204) 690388. hibon@eu.irco.com, www.ingersollrand.com

Initial Electronic Security Systems Ltd - Shadsworth Road, Blackburn. BB1 2PR tel:(01254) 688688, fax:(01254) 662571. citsales@chubb.co.uk, www.chubbsystems.co.uk

Initial Fire Services - 2 City Place, Beehive Ring Road, Gatwick. RH6 0HA tel:(0800) 077 8963. pr@ri-facilities.com, www.initial.co.uk/fire-services

Insituform Technologies® Ltd - 24-27 Brunel Close, Park Farm Industrial Estate, Wellingborough, Northamptonshire. NN8 6QX tel:(01933) 678 266, fax:(01933) 678 637. uksales@insituform.com, www.insituform.co.uk

InstaGroup Ltd - Insta House, Ivanhoe Road, Hogwood Business Park, Finchampstead, Wokingham, Berks. RG40 4PZ tel:(0118) 932 8811, fax:(0118) 932 8314. info@instagroup.co.uk, www.instagroup.co.uk

Institute of Ashphalt Technology, the - PO Box 15690, Bathgate. EH48 9BT tel:(0131) 629 5370. info@instituteofasphalt.org, www.instituteofasphalt.org

Institute of Historic Building Conservation - Jubilee House, High Street, Tisbury, Wiltshire. SP3 6HA tel:(01747) 873133, fax:(01747) 871718. admin@ihbc.org.uk, www.ihbc.org.uk

Institute of Quarrying - 8a Regan Way, Chetwynd Business Park, Chilwell, Nottingham. NG9 6RZ tel:(0115) 972 9995. mail@quarrying.org, www.quarrying.org

Institute of Specialist Surveyors and Engineers - Amity House, 156 Ecclesfield Road, Chapeltown, Shefield. S35 1TE tel:(0800) 915 6363. enquiries@isse.org.uk, www.isse.co.uk

Institution of Civil Engineers - One Great George Street, London. SW1P 3AA tel:(020) 7222 7722, fax:(020) 7222 7500. communications@ice.org.uk, www.ice.org.uk

Institution of Electrical Engineers - Savoy Place, London. WC2R OBL tel:(020) 7240 1871, fax:(020) 7240 7735. postmaster@theiet.org, www.iee.org.uk

Institution of Mechanical Engineers - 1 Birdcage Walk, Westminster, London. SW1H 9JJ tel:(020) 7222 7899, fax:(020) 7222 4557. enquiries@imeche.org.uk, www.imeche.org.uk

Institution of Structural Engineers - 47-58 Bastwick Street, London. EC1V 3PS tel:(020) 7235 4535, fax:(020) 7235 4294. pr@istructe.org, www.istructe.org

Insulated Render & Cladding Association - PO Box 12, Haslemere, Surrey. GU27 3AH tel:(01428) 654011, fax:(01428) 651401. info@inca-ltd.org.uk, www.inca-ltd.org.uk

Insulated Render & Cladding Assocication (INCA) - PO Box 12, Haslemere, Surrey. GU27 3AH tel:(01428) 654011, fax:(01428) 651401. info@inca-ltd.org.uk, www.inca-ltd.org.uk

Integra Products - Sunflex NDC, Keys Park Road, Hednesford, Cannock, Staffordshire. WS12 2FR tel:(01543) 271421, fax:(01543) 279505. cus.care@integra-products.co.uk, www.integra-products.co.uk

Inter Public Urban Systems Uk Ltd - Suite 16, Horsehay House, Horsehay, Telford, Shropshire. TF4 3PY tel:(01952) 502012, fax:(01952) 502022. sales@interpublicurbansystems.co.uk, www.interpublicurbansystems.co.uk

Interclad (UK) Ltd - 173 Main Road, Biggin Hill, Kent. TN16 3JR tel:(01959) 572447, fax:(01959) 576974. sales@interclad.co.uk, www.interclad.co.uk

Interface - Interface Europe Ltd, Shelf Mills, Shelf, Halifax, West Yorkshire. HX3 7PA tel:(01274) 690690, fax:(01274) 694095. marketing@interface.com, www.interface.com

Interface Europe Ltd - Head office, Shelf Mills, Shelf, Halifax. HX3 7PA tel:(08705) 304030, fax:(01274) 698300. enquiries@eu.interfaceinc.com, www.interfaceeurope.com

Interface Europe Ltd - Shelf Mills, Shelf, Halifax, W. Yorks. HX3 7PA tel:(01274) 690690, fax:(01274) 694095. enquiries@eu.interfaceinc.com, www.interfaceeurope.com

Interfloor Limited - Edinburgh Road, Heathhall, Dumfries. DG1 1QA tel:(01387) 253111, fax:(01387) 255726. sales@interfloor.com, www.interfloor.com

Interflow UK - Leighton, Shrewsbury, Shropshire. SY5 6SQ tel:(01952) 510050, fax:(01952) 510967. info@interflow.co.uk, www.interflow.co.uk

Interframe Ltd - Aspen Way, Yalberton Industrial Estate, Paignton, Devon. TQ4 7QR tel:(01803) 666633, fax:(01803) 663030. mail.interframe@btconnect.com, www.interframe.co.uk

Intergrated Polymer Systems (UK) Ltd - Allen House, Harmby Road, Leyburn, N. Yorks. DL8 5QA tel:(01969) 625000, fax:(01969) 623000. info@ipsroofing.co.uk, www.ipsroofing.co.uk

Interlink Group - Interlink House, Commerse Way, Lancing Industrial Estate, Lancing, W. Sussex. BN15 8TA tel:(01903) 763663, fax:(01903) 762621. info@kingspanpanels.com, www.kingspanfabrications.com

Internal Partitions Systems - Interplan House, Dunmow Industrial Estate, Chelmsford Road, Great Dunmow, Essex. CM6 1HD tel:(01371) 874241, fax:(01371) 873848. contact@ips-interiors.co.uk, www.ips-interiors.co.uk

International Protective Coatings - Stoneygate Lane, Felling on Tyne, Tyne & Wear. NE10 0JY tel:(0191) 469 6111, fax:(0191) 495 0676. protectivecoatings@akzonobel.com, www.international-pc.com

Interpave - The Old Rectory, Main Street, Glenfield, Leicestershire. LE3 8DG tel:(01162) 325170, fax:(01162) 325197. info@paving.org.uk, www.paving.org.uk

Interphone Limited - Interphone House, 12-22 Herga Road, Wealdstone, Middx. HA3 5AS tel:(020) 8621 6000, fax:(020) 8621 6100. sales@interphone.co.uk, www.interphone.co.uk

Intersolar Group Ltd - Magdalen Center, The Oxford Science Park, Oxford, Oxon. OX4 4GA tel:(01865) 784670, fax:(01865) 784681. intersolar@intersolar.com, www.intersolar.com

Intime Fire and Security - Nimax House, 20 Ullswater Crescent, Coulsdon, Surrey. CR5 2HR tel:(020) 8763 8800, fax:(020) 8763 9996. lee.james@intimefireandsecurity.co.uk, www.intimefireandsecurity.co.uk

Intrad Architectural Systems - PJP plc, St. Albans Road West, Hatfield, Herts. AL10 0TF tel:(0800) 0643545, fax:(01707) 263614. sales@intrad.com

Invensys Controls Europe - Farnham Road, Slough, Berks. SL1 -4UH tel:(0845) 130 5522, fax:(01753) 611001. customer.care@invensys.com, www.invensyscontrolseurope.com

Invicta Plastics Ltd - Harborough Road, Oadby, Leics. LE2 4LB tel:(0116) 272 0555, fax:(0116) 272 0626. sales@invictagroup.co.uk, www.invictagroup.co.uk

Invicta Window Films Ltd - Invicta House, 7 Western Parade, Woodhatch, Reigate, Surrey. RH2 8AU tel:(0800) 393380, fax:(01737) 240845. invictafilms@btconnect.com, www.invicawindowfilms.com

IPPEC Sytsems Ltd - 21 Buntsford Drive, Buntsford Gate Business Park, Worcestershire. B60 3AJ tel:(01527) 579705, fax:(01527) 574109. info@ippec.co.uk, www.ippec.co.uk

Irish Shell Ld - Shell House, Beach Hill, Clonskeagh, Dublin 4. Ireland tel:+353 1 785177, fax:+353 1 767489. shell@shs-sales.ie, www.shell.com/

Isgus - 10 Springfield Business Centre, Stonehouse, Glos. GL10 3FX tel:(01453) 827373. sales@uk.isgus.com, www.isgus.co.uk

Isocrete Floor Screeds Ltd - Flowcrete Business Park, Booth Lane, Moston, Sandbach, Cheshire. CW11 3QF tel:(01270) 753753, fax:(01270) 753333. ukweb@flowcrete.com, www.flowcrete.co.uk

Isolated Systems Ltd - Adams Close, Heanor Gate Industrial Estate, Heanor, Derbys. DE75 7SW tel:(01773) 761226, fax:(01773) 760408. sales@isolatedsystems.com, www.isolatedsystems.com

ITAB Prolight UK Limited - 4, 5 & 6 Raynham Road, Bishop's Stortford, Herts. CM23 5PB tel:(01279) 757595, fax:(01279) 755599. enquiries@itabprolight.co.uk, www.itabprolight.co.uk

ITM Communications LtdLtd - 41 Alston Drive, Bradwell Abbey, Milton Keynes. MK13 9HA tel:(01908) 318844, fax:(01908) 318833. enquiries@itm.uk.com, www.itm.uk.com

ITS Ceilings & Partitions - 44 Portman Road, Reading, Berks. RG30 1EA tel:(0118) 9500225, fax:(0118) 9503267. info@itsprojects.co.uk, www.itsprojects.co.uk

ITT Flygt Ltd - Colwick, Nottingham. NG4 2AN tel:(0115) 940 0111, fax:(0115) 940 0444. flygtgb@flygt.com, www.flygt.com

ITW Automotive Finishing UK - Anchorbrook Industrial Estate, Lockside, Aldridge, Walsall. WS9 8EG tel:(01922) 423700, fax:(01922) 423705. marketing-uk@itwifeuro.com, www.itwautomotivefinishing.co.uk

ITW Construction Products - Fleming Way, Crawley, W.Sussex. RH10 9DP tel:(01293) 523372, fax:(01293) 515186. gmason@itwcp.co.uk, www.itwcp.co.uk

ITW Consutruction Products - 3rd Floor, Westmead House, Farnborough, Hampshire. GU14 7LP tel:(01252) 551960, fax:(01252) 543436. gangnail@itw-industry.com, www.itw-industry.com

ITW Spit - Fleming Way, Crawley, W. Sussex. RH10 2QR tel:(01293) 523372, fax:(01293) 515186. gmason@itwcp.co.uk, www.itwcp.co.uk

J

J & E Hall Ltd - Invicta House, Sir Thomas Longley Rd, Rochester, Kent. ME2 4DP tel:(01634) 731400, fax:(01634) 731401. helpline@jehall.co.uk, www.jehall.co.uk

J & M Fencing Services - Unit Q, Wrexham Road, Laindon, Basildon, Essex. SS15 6PX tel:(01268) 415233, fax:(01268) 417357. m.cottage@btconnect.com, www.jandmfencingservices.co.uk

J C Bamford (Excavators) Ltd - Rocester, Staffs. ST14 5JP tel:(01889) 590312, fax:(01889) 590588. enquiries@jcb.co.uk, www.jcb.com

J Pugh-Lewis Limited - Bushypark Farm, Pilsley, Chesterfield, Derbys. S45 8HW tel:(01773) 872362, fax:(01773) 874763. info@pugh-lewis.co.uk, www.pugh-lewis.co.uk

J Suttle Swanage Quarries Ltd - California Quarry, Panorama Road, Swanage, Dorset. BH19 2QS tel:(01929) 423576, fax:(01929) 427656. nick.crocker@btconnect.com; cjs@stone.uk.com, www.stone.uk.com

COMPANY INFORMATION

J (con't)

J W Gray Lightning Protection Ltd - Unit 1, Swanbridge Industrial Park, Black Croft Road, Witham, Essex. CM8 3YN tel:(01376) 503330, fax:(01376) 503337. enquiries@jwgray.com, www.jwgray.com

J. B. Corrie & Co Limited - Frenchmans Road, Petersfield, Hants. GU32 3AP tel:(01730) 237100, fax:(01730) 264915. sales@jbcorrie.co.uk, www.jbcorrie.co.uk

J.F. Poynter Ltd - Maxim Lamps, Unit E Consort Way, Victoria Industrial Estate, Burgess Hill. RH15 9TJ tel:(01444) 239223, fax:(01444) 245366. sales@maximlamps.co.uk, www.maxinlamps.co.uk

Jablite Ltd - Infinity House, Anderson Way, Belvedere, Kent. DA17 6BG tel:(020) 83209100, fax:(020) 83209110. sales@jablite.co.uk, www.jablite.co.uk

Jack Moody Ltd - Hollybush Farm, Warstone Road, Shareshill, Wolverhampton, West Midlands. WV10 7LX tel:(01922) 417648, fax:(01922) 413420. sales@jackmoodylimited.co.uk, www.jackmoodyltd.co.uk

Jackson H S & Son (Fencing) Ltd - Stowting Common, Ashford, Kent. TN25 6BN tel:(01233) 750393, fax:(01233) 750403. sales@jacksons-fencing.co.uk, www.jacksons-fencing.co.uk

Jacopa Limited - Cornwallis Road, Millard Industrial Estate, West Bromwich, West Midlands. B70 7JF tel:(0121) 511 2400, fax:(0121) 511 2401. info@jacopa.com, www.jacopa.com

Jacuzzi® Spa and Bath Ltd - Old Mill Lane, Low Road, Hunslet, Leeds. LS10 1RB tel:(0113) 2727430, fax:(0113) 2727445. retailersales@jacuzziemea.com, www.jacuzziuk.co.uk

Jaga Heating Products (UK) Ltd - Jaga House, Orchard Business Park, Bromyard Road, Ledbury, Herefordshire. HR8 1LG tel:(01531) 631533, fax:(01531) 631534. jaga@jaga.co.uk, www.jaga.co.uk

Jali Ltd - Albion Works, Church Lane, Barham, Canterbury, Kent. CT4 6QS tel:(01227) 833333, fax:(01227) 831950. sales@jali.co.uk, www.jali.co.uk

JAMAK Fabrication Europe - 52 & 53 Oakhill Industrial Estate, Devonshire Rd, Worsley, Manchester. M28 3PT tel:(01204) 794554, fax:(01204) 574521. sales@jamak.co.uk, www.jamak.co.uk

James & Bloom Ltd - Crossley Park, Crossley Road, Stockport, Cheshire. SK4 5DF tel:(0161) 432 5555, fax:(0161) 432 5312. jamesandbloom.co@btinternet.com

James Donaldson Timber - Elm Park Sawmills, Leven, Fife. KY8 4PS tel:(01333) 422 222, fax:(01333) 429 469. elmpark@donaldson-timber.co.uk, www.don-timber.co.uk

James Gibbons Format Ltd - Vulcon Road, Bilston, W. Midlands. WV14 7JG tel:(01902) 458585, fax:(01902) 351336. deptname@jgf.co.uk, www.jgf.co.uk

James Latham plc - Unit 2, Swallow Park, Finley Road, Hemel Hampstead, Herts. HP2 7QU tel:(020) 8806 3333, fax:(020) 8806 7249. marketing@lathams.co.uk, www.lathamtimber.co.uk

James Lever 1856 Ltd - Unit 26 Morris Green Business Park, Prescott Street, Bolton, Lancs. BL3 3PE tel:(01204) 61121, fax:(01204) 658154. sales@jameslever.co.uk, www.jameslever.co.uk

James Walker & Co Ltd - 1 Millennium Gate, Westmere Drive, Crewe, Cheshire. CW1 6AY tel:+44 0 1270 536000, fax:+44 0 1270 536100. csc@jameswalker.biz, www.jameswalker.biz

Jaymart Rubber & Plastics Ltd - Woodland Trading Estate, Edenvale Road, Westbury, Wilts. BA13 3QS tel:(01373) 864926, fax:(01373) 858454. mattings@jaymart.co.uk, www.jaymart.net

JB Architectural Ltd - Unit C, Rich Industrial Estate, Avis Way, Newhaven, East Sussex. BN9 0DU tel:(01273) 514961, fax:(01273) 516764. info@jbai.co.uk

JB Kind Ltd - Portal Place, Astron Business Park, Hearthcote Road, Swadlincote, Derbyshire. DE11 9DW tel:(01283) 554197, fax:(01283) 554182. info@jbkind.com, www.jbkind.com

Jeld-Wen UK Ltd - Retford Road, Woodhouse Mill, Sheffield, S. Yorks. S13 9WH tel:(0845) 122 2890, fax:(01302) 787383. marketing@jeld-wen.co.uk, www.jeld-wen.co.uk

Jeld-Wen UK Ltd - Retford Road, Woodhouse Mill, Sheffield, South Yorkshire. S13 9WH tel:08451222890, fax:01142542365. marketing@jeld-wen.co.uk, www.jeld-wen.co.uk

Jenkins Newell Dunford Ltd - Thrumpton Lane, Retford, Notts. DN22 7AN tel:(01777) 706777, fax:(01777) 708141. info@jnd.co.uk, www.jnd.co.uk

Jerrards Plc - Arcadia House, Cairo New Road, Croydon, London. CR0 1XP tel:(020) 8251 5555, fax:(020) 8251 5500. enquiries@jerrards.com, www.jerrards.com

Jewers Doors Ltd - Stratton Business Park, Normandy Lane, Biggleswade, Beds. SG18 8QB tel:(01767) 317090, fax:(01767) 312305. postroom@jewersdoors.co.uk, www.jewersdoors.co.uk

Jewson Ltd - High Street, Brasted, Westerham, Kent. TN16 1NG tel:(01959) 563856, fax:(01959) 564666. www.jewson.co.uk

JLC Pumps & Engineering Co Ltd - PO Box 225, Barton-le-Clay, Beds. MK45 4PN tel:(01582) 881946, fax:(01582) 881951. jbcatton@btconnect.com, www.jlcpumps.co.uk

John Davidson (Pipes) Ltd - Townfoot Industrial Estate, Longtown, Carlisle, Cumbria. CA6 5LY tel:(01228) 791503, fax:(01228) 792051. headoffice@jdpipes.co.uk, www.jdpipes.co.uk

John Pulsford Associates Ltd - Sphere Industrial Estate, Campfield Road, St. Albans, Herts. AL1 5HT tel:(01727) 840800, fax:(01727) 840083. info@jpa-furniture.com, www.jpa-furniture.com

John Reynolds & Sons (Birmingham) Ltd - Church Lane, West Bromwich, W. Midlands. B71 1DJ tel:(0121) 553 1287 / 2754, fax:(0121) 500 5460. www.mcarthur-group.com

John Watson & Carter

Suite 2, Studio 700, Princess Street, Hull. HU2 8BJ tel:(01482) 492191, fax:(03332) 407363. hull@johnwatsonandcarter.co.uk, www.vbjohnson.com

Johnson & Johnson PLC - Units 12-19, Guiness Road Trading Estate, Trafford Park, Manchester. M17 1SB tel:(0161) 872 7041, fax:(0161) 872 7351. sales@zutux.com, www.johnson-johnson.co.uk

Johnson & Starley - Rhosili Road, Brackmills, Northampton, Northants. NN4 7LZ tel:(01604) 762881, fax:(01604) 767408. marketing@johnsonandstarley.co.uk, www.johnsonandstarley.co.uk

Johnson & Starley Ltd - Dravo Division, Rhosili Road, Brackmills, Northampton. NN4 7LZ tel:(01604) 707022, fax:(01604) 706467. Dravo@johnsonandstarley.co.uk, www.dravo.co.uk

Johnson Controls Building Efficiency UK Ltd - 2, The Briars, Waterberry Drive, Waterlooville, Hampshire. PO7 7YH tel:(01236) 108 0001, fax:(01236) 738937. serviceinfo.uki@jci.com, www.johnsoncontrols.com

Johnson Matthey PLC - Metal Joining - York Way, Royston, Herts. SG8 5HJ tel:(01763) 253200, fax:(01763) 253168. mj@matthey.com, www.jm-metaljoining.com

Johnson Tiles - Harewood Street, Tunstall, Stoke-on-Trent. ST6 5JZ tel:(01782) 575575, fax:(01782) 524138. sales@johnson-tiles.com, www.johnson-tiles.com

Johnson VB LLP

St John's House, 304-310 St Albans Road, Watford, Herts. WD24 6PW tel:(01923) 227236, fax:(03332) 407363. watford@vbjohnson.co.uk, www.vbjohnson.co.uk

Johnson Wax Professional - Frimley Green, Camberley, Surrey. GU16 7AJ tel:(01276) 852852, fax:(01276) 852800. ask.uk@scj.com, www.scjohnson.co.uk

Johnsons Wellfield Quarries Ltd - Crosland Hill, Huddersfield, W. Yorkshire. HD4 7AB tel:(01484) 652 311, fax:(01484) 460007. sales@johnsons-wellfield.co.uk, www.myersgroup.co.uk/jwq/

Joint Council For The Building & Civil Engineering Industry (N. Ireland) - 143 Marlone Road, Belfast. BT9 6SU tel:(028) 90 877143, fax:(028) 90 877155. mail@cefni.co.uk, www.cefni.co.uk

Joint Industry Board for the Electrical Contracting Industry - Kingswood House, 47/51 Sidcup Hill, Sidcup, Kent. DA14 6HP tel:(020) 8302 0031, fax:(020) 8309 1103. administration@jib.org.uk, www.jib.org.uk

Jolec Electrical Supplies - Unit 2, The Empire Centre, Imperial Way, Watford, Herts. WD24 4YH tel:(01923) 243656, fax:(01923) 230301. www.jolec.co.uk

Jones of Oswestry Ltd - Haselfield 18, Telford, Shropshire. TF7 4JS tel:(01691) 653251, fax:(01691) 658222. sales@jonesofoswestry.com, www.jonesofoswestry.co.uk

Joseph Ash Ltd - The Alcora Building 2, Mucklow Hill, Halesowen. B62 8DG tel:(0121) 5042560, fax:(0121) 5042599. mickj@josephash.co.uk, www.josephash.co.uk

Jotun Henry Clark Ltd Decorative Division - Stather Road, Flixborough, Lincs. DN15 8RR tel:(01724) 400123, fax:(01724) 400100. deco.uk@jotun.co.uk, www.jotun.com

JS Humidifiers PLC - Artex Avenue, Rustington, Littlehampton, West Sussex. BN16 3LN tel:(01903) 850200, fax:(01903) 850345. info@jshumidifiers.com, www.jshumidifiers.com

Junckers Limited - Unit A, 1 Wheaton Road, Witham, Essex. CM8 3UJ tel:(01376) 534700, fax:(01376) 514401. brochures@junckers.co.uk, www.junckers.co.uk

K

K. V. Radiators - 6 Postle Close, Kilsby, Rugby, Warwickshire. CV23 8YG tel:(01788) 823286, fax:(01788) 823002. solutions@kvradiators.com, www.kvradiators.com

KAB Seating Ltd - Round Spinney, Northampton. NN3 8RS tel:(01604) 790500, fax:(01604) 790155. marketing@kabseating.com, www.kabseating.com

Kaba Garog - 14 Leacroft Road, Birchwood, Warrington. WA3 6GG tel:(0870) 000 0600, fax:(0870) 000 0642. sales@guthriedouglas.com, www.kaba-garog.co.uk

Kaba Ltd - Lower Moor Way, Tiverton Business Park, Tiverton, Devon. EX16 6SS tel:(0870) 000 5625, fax:(0870) 000 5397. info@kaba.co.uk, www.kaba.co.uk

KAC Alarm Company Ltd - KAC House, Thornhill Road, North Moons Moat, Redditch, Worcs. B98 9ND tel:(01527) 406655, fax:(01527) 406677. marketing@kac.co.uk, www.kac.co.uk

Kair ventilation Ltd - Unit 6, Chiltonran Industrial Estate, 203 Manton Lane, Lee, London. SE12 0TX tel:(08451) 662240, fax:(08451) 662250. info@kair.co.uk, www.kair.co.uk

Kaldewei - Unit 7, Sundial Court, Barnesbury Lane, Tollworth Rise South, Surbiton. KT5 9RN tel:(0870) 7772223, fax:(07766) 462958. sales-uk@kaldewei.com, www.kaldewei.com

Kalon Decorative Products - Huddersfield Road, Birstall, Batley, W. Yorks. WF17 9XA tel:(01924) 354000, fax:(01924) 354001. paint247.acuk@ppg.com, www.kalon.co.uk

Kaloric Heater Co Ltd - 31-33 Beethoven Street, London. W10 4LJ tel:(020) 8969 1367, fax:(020) 8968 8913. admin@kaloricheater.co.uk, www.kaloricheaters.co.uk

Kawneer UK Ltd - Astmoor Road, Astmoor Industrial Estate, Runcorn, Cheshire. WA7 1QQ tel:(01928) 502500, fax:(01928) 502501. kuk.kawneer@alcoa.com, www.kawneer.co.uk

Kaye Aluminium plc - Shaw Lane Industrial Estate, Ogden Road, Wheatley Hills, Doncaster, S. Yorks. DN2 4SG tel:(01302) 762500, fax:(01302) 360307. design@kayealu.co.uk, www.kayealuminium.co.uk

KCW Comercial Windows Ltd - 25A Shuttleworth Road, Goldington, Bedford, Beds. MK41 0HS tel:(01234) 269911, fax:(01234) 325034. sales@kcwwindows.com, www.kcwwindows.com

Kee Klamp Ltd - 1 Boulton Road, Reading, Berks. RG2 0NH tel:(0118) 931 1022, fax:(0118) 931 1146. sales@keesafety.com, www.keesafety.co.uk

Keim Mineral Paints Ltd - Santok Building, Deer Park Way, Telford. TF2 7NA tel:(01952) 231250, fax:(01952) 231251. sales@keimpaints.co.uk, www.keimpaints.co.uk

Keller Ground Engineering - Oxford Road, Ryton-on-Dunsmore, Coventry. CV8 3EG tel:(024) 76 511266, fax:(024) 76 305230. foundations@keller.co.uk, www.keller-uk.com

Kem Edwards Ltd - Longwood Business Park, Fordbridge Road, Sunbury-on-Thames, Middx. TW16 6AZ tel:(01932) 754700, fax:(01932) 754754. sales@kemedwards.co.uk, www.kemedwards.co.uk

Kemper System Ltd - Kemper House, 30 Kingsland Grange, Woolston, Warrington, Cheshire. WA1 4RW tel:(01925) 445532, fax:(01925) 575096. enquiries@kempersystem.co.uk, www.kempersystem.co.uk

Kenilworth Daylight Centre Ltd - Princes Drive Industrial Estate, Kenilworth, Warwicks. CV8 2FD tel:(01926) 511411, fax:(01926) 854155. daylightcentre@btinternet.com

Kenrick Archibald & Sons Ltd - PO Box 9, Union Street, West Bromwich, W. Midlands. B70 6BD tel:(0121) 553 2741, fax:(0121) 500 6332. sales@kenrick.prestel.co.uk

Kensal CMS - Kensal House, President Way, Luton, Beds. LU2 9NR tel:(01582) 425777, fax:(01582) 425776. sales@kensal.co.uk, www.kensal.co.uk

Kent & Co (Twines) Ltd - Hartley Trading Estate, Long Lane, Liverpool. L9 7DE tel:(0151) 525 1601, fax:(0151) 523 1410

Kenyon Group Ltd - Regent House, Regent Street, Oldham, Lancs. OL1 3TZ tel:(0161) 633 6328, fax:(0161) 627 5072. sales@kenyon-group.co.uk, www.kenyon-group.co.uk

Kenyon Paints Limited - Regent Street, Oldham. OL1 3TZ tel:(0161) 665 4470, fax:(0161) 627 5072. enquiries@kenyonpaints.co.uk, www.kenyonpaints.co.uk

Kenyon Performance Adhesives - Regent House, Regent Street, Oldham, Lancs. OL1 3TZ tel:(0161) 633 6328, fax:(0161) 627 5072. sales@kenyon-group.co.uk, www.4adhesives.com

Kermi (UK) Ltd - 7, Brunel Road, Earlstrees Industrial Estate, Corby, Northants. NN17 4JW tel:(01536) 400004, fax:(01536) 446614. info@kermi.co.uk, www.kermi.co.uk

Kerridge Stone Ltd - Endon Quarry, Windmill Lane, Kerridge, Cheshire. SK10 5AZ tel:(01782) 514353, fax:(01782) 516783. www.kerridgestone.co.uk

Kershaw Mechanical Services Ltd - Edward Leonard House, Pembroke Ave, Denny End Rd, Waterbeach, Cambridge. CB25 9QR tel:(01223) 715800, fax:(01223) 411061. sales@kershaw-grp.co.uk, www.kershawmechanical.co.uk/

Ketley Brick Co Ltd , The - Dreadnought Road, Pensnett, Brierley Hill, W. Midlands. DY5 4TH tel:(01384) 77405, fax:(01384) 74553. sales@ketley-brick.co.uk, www.ketley-brick.co.uk

Keyline - Unit S3, 8 Strathkelvin Place, Kirkintilloch, Glasgow. G66 1XH tel:(0141) 777 8979. www.keyline.co.uk

Keymer Hand Made Clay Tiles - Nye Road, Burgess Hill, W. Sussex. RH15 0LZ tel:(01444) 232931, fax:(01444) 871852. info@keymer.co.uk, www.keymer.co.uk

Kidde Fire Protection Ltd - Belvue Road, Northolt, Middx. UB5 5QW tel:(020) 8839 0700, fax:(020) 8845 4304. general.enquiries@kiddeuk.co.uk, www.kfp.co.uk

Kidde Fire Protection Services Ltd - 400 Dallows Road, Luton, Beds. LU1 1UR tel:(01582) 413694, fax:(01582) 402339. general.enquiries@kiddeuk.co.uk, www.kfp.co.uk

Kier Limited - Tempsford Hall, Sandy, Bedfordshire. SG19 2BD tel:(01767) 355000, fax:(01767) 355633. info@kier.co.uk, www.kier.co.uk

Kilwaughter Chemical Co Ltd - Kilwaughter Chemical Co Ltd, 9 Starbog Road, Larne, Co. Antrim. BT40 2JT tel:(028) 2826 0766, fax:(028) 2826 0136. sales@K-Rend.co.uk, www.K-Rend.co.uk

Kimberly-Clark Ltd - 1 Tower View, Kings Hill, West Malling, Kent. ME19 4HA tel:(01732) 594333, fax:(01732) 594338. kcpuk@kcc.com, www.kcc.com

Kinder-Janes Engineering Ltd - Porters Wood, St. Albans, Hertfordshire. AL3 6HU tel:(01727) 844441, fax:(01727) 844247. info@kinder-janes.co.uk, www.kinder-janes.co.uk

Kinetico Water Softeners - 11 The Maltings, Thorney, Peterborough, Cambs. PE6 0QF tel:(01733) 270463. enquiries@harvey.co.uk, www.kineticoltd.co.uk

Kingfisher Building Products Ltd - Cooper Lane, Bardsea, Ulverston, Cumbria. LA12 9RA tel:(01229) 869100, fax:(01229) 869101. kingfisheruk@tiscali.co.uk, www.kingfisherco.uk

Kingscourt Country Manor Bricks - Unit 26, Airways Industrial Estate, Santry, Dublin 17. Ireland tel:+353 1836 6901, fax:+353 1855 4743. sales@cmb.ie, www.cmb.ie

Kingsforth Security Fencing Limited - Mangham Way, Barbot Hall Industrial Estate, Rotherham, S. Yorks. S61 4RL tel:(01709) 378 977, fax:(01709) 838992. info@kingsforthfencing.co.uk, www.securityfencingyorkshire.co.uk/

Kingspan Ltd - Greenfield Business Park 2, Greenfield, Holywell, Flintshire. CH8 7GJ tel:(01352) 716100, fax:(01352) 710616. info@kingspanpanels.com, www.kingspanpanels.com

Kingspan Access Floors Ltd - Burma Drive, Marfleet, Hull. HU9 5SG tel:(01482) 781701, fax:(01482) 799276. info@kingspanaccessfloors.co.uk, www.kingspanaccessfloors.co.uk

Kingspan Environmental - College Road North, Aston Clinton, Aylesbury, Buckinghamshire. HP22 5EW tel:(01296) 633000, fax:(01296) 633001. online@kingspan.com, www.kingspanenviro.com

Kingspan Environmental - College Road North, Aston Clinton, Aylesbury, Bucks. HP22 5EW tel:(01296) 633000, fax:(01296) 633001. online@kingspan.com, www.kingspanenviro.com

Kingspan Environmental Ltd - 180 Gilford Road, Portadown, Co Armagh, Northern Ireland. BT63 5LF tel:02838364400, fax:02838364445. enquiry@kingspanenv.com, www.kingspanenv.com

Kingspan Industrial Ins. Ltd - PO Box 3, Charlestown, Glossop, Derbys. SK13 8LE tel:(01457) 861611, fax:(01457) 852319. info@kingspaninsulation.co.uk, www.kingspan.kooltherm.co.uk

Kingspan Insulation Ltd - Pembridge, Leominster, Herefordshire. HR6 9LA tel:(0870) 8508555, fax:(0870) 8508666. info@kingspaninsulation.co.uk, www.insulation.kingspan.com

Kingston Craftsmen Structural Timber Engineers Ltd - Cannon Street, Hull, East Yorkshire. HU2 0AD tel:(01482) 225171, fax:(01482) 217032. sales@kingston-craftsmen.co.uk, www.kingston-craftsmen.co.uk

Kingstonian Paint Ltd - Mayfield House, Sculcoates Lane, Hull, E. Yorks. HU5 1DR tel:(01482) 342216, fax:(01482) 493096. info@kpaints.co.uk, www.kpaints.co.uk

Kirkpatrick Ltd - Frederick Street, Walsall, Staffs. WS2 9NF tel:(01922) 620026, fax:(01922) 722525. enquiries@kirkpatrick.co.uk, www.kirkpatrick.co.uk

Kirkstone - Skelwith Bridge, Ambleside, Cumbria. LA22 9NN tel:(01539) 433296, fax:(01539) 434006. sales@kirkstone.com, www.kirkstone.com

K (con't)

Kitsons Insulation Products Ltd - Kitsons House, Centurion Way, Meridian Business Park, Leicester, Leics. LE3 2WH tel:(0116) 201 4499, fax:(0116) 201 4498. leicester@kitsonsthermal.co.uk, www.kitsonsthermal.co.uk

Kleeneze Sealtech Ltd - Ansteys Road, Hanham, Bristol, Avon. BS15 3SS tel:(0117) 958 2450, fax:(0117) 960 0141. sales@ksl.uk.com, www.ksltd.com

Kleen-Tex Industries Ltd - Causeway Mill, Express Trading Estsate, Stone Hill Road, Farnworth, Nr. Bolton. BL4 9TP tel:(01204) 863000, fax:(01204) 863001. sales@kleentexuk.com, www.kleentexuk.com

Klick Technology Ltd - Claverton Road, Wythenshawe, Manchester. M23 9FT tel:(0161) 998 9726, fax:(0161) 946 0419. webcontact@klicktechnology.co.uk, www.klicktechnology.co.uk

Klinger Fluid Instrumentation Ltd - Edgington Way, Sidcup, Kent. DA14 5AG tel:(020) 8300 7777, fax:(020) 8302 8145. enquiries@klingeruk.co.uk, www.klinger.co.uk

Klober Ltd - Ingleberry Road, Shepshed, Loughborough. LE12 9DE tel:(0800) 7833216, fax:(011509) 505535. support@klober.co.uk, www.klober.co.uk

Knauf AMF Ceilings Ltd - Thames House, 6 Church Street, Twickenham, Middlesex. TW1 3NJ tel:(020) 8892 3216, fax:(020) 8892 6866. sales@amfceilings.co.uk, www.amfceilings.co.uk

Knauf Drywall - PO Box 133, Sittingbourne, Kent. ME10 3HW tel:(01795) 424499, fax:(01795) 428651. info@knauf.co.uk, www.knaufdrywall.co.uk

Knauf Insulation Ltd - Hunter House Industrial Estate, Brenda Road, Hartlepool, Cleveland. TS25 2BE tel:(01429) 855100, fax:(01429) 855138. sales@knaufinsulation.com, www.knaufinsulation.co.uk

Knauf Insulation Ltd - PO Box 10, Stafford Road, St. Helens, Merseyside. WA10 3NS tel:(01270) 824024, fax:(01270) 824025. sales.uk@knaufinsulation.com, www.knaufinsulation.co.uk

Knauf UK - Kemsley Fields Business Park, Sittingbourne, Kent. ME9 8SR tel:(0800) 521 050, fax:(0800) 521 205. cservice@knauf.co.uk, www.knauf.co.uk

Knightbridge Furniture Productions Ltd - 191 Thornton Road, Bradford, W. Yorks. BD1 2JT tel:(01274) 731442, fax:(01274) 736641. sales@knightsbridge-furniture.co.uk, www.knightsbridge-furniture.co.uk

Knowles Ltd R E - Knowles Industrial Estate, Furness Vale, High Peak. SK23 7PJ tel:(01663) 744127, fax:(01663) 741 562

Knowles W T & Sons Ltd - Ash Grove Sanitary Works, Elland, W. Yorks. HX5 9JA tel:(01422) 372833, fax:(01422) 370900. sales@wtknowles.co.uk, www.wtknowles.co.uk

Knurr (UK) Ltd - Burrel Road, St. Ives, Cambs. PE27 3LE tel:(01480) 496125, fax:(01480) 496373. knuerr.uk@knuerr.com, www.kneurr.com

Kobi Ltd - Unit 19 Seax Court, Southfields Industrail Estate, Laindon, Essex. SS15 6SL tel:(01268) 416335, fax:(01268) 542148. cradles@kobi.co.uk, www.kobi.co.uk

Kohler Mira Ltd - Cromwell Road, Cheltenham, Glos. GL52 5EP tel:(01242) 221221, fax:(01242) 221925. mira_technical@mirashowers.com, www.mirashowers.com

Kohler UK - Cromwell Road, Cheltenahm, Gloucestershire. GL52 5EP tel:(0844) 571 0048, fax:(0844) 571 1001. info@kohler.co.uk, www.kohler.co.uk

Komfort Workspace PLC - Unit 1-10, Whittle Way, Crawley, W. Sussex. RH10 9RT tel:(01293) 592500, fax:(01293) 553271. general@komfort.com, www.komfort.com

Kommerling UK Ltd - Unit 27, Riverside Way, Uxbridge, Middx. UB8 2YF tel:(01895) 465600, fax:(01895) 465617. enquiries@profine-group.com

Kompan Ltd - 20 Denbigh Hall, Bletchley, Milton Keynes, Bucks. MK3 7QT tel:(01908) 642466, fax:(01908) 270137. KOMPAN.uk@KOMPAN.com, www.kompan.com

Kone - Global House Station Place, Fox Lane North, Chertsey, Surrey. KT16 9HW tel:(08451) 999 999. sales.marketinguk@kone.com, www.kone.co.uk

Kone Escalators Ltd - Worth Bridge Road, Keighley, W. Yorks. BD21 4YA tel:(01535) 662841, fax:(01535) 680498. sales.marketinguk@kone.com, www.kone.com

Kone PLC - Global House, Station Place, Fox Lane North, Chertsey, Surrey. KT16 9HW tel:(0845) 1 999 999, fax:(0870) 774 8347. sales.marketinguk@kone.com, www.kone.co.uk

Köttermann Ltd - Unit 8, The Courtyard, Furlong Road, Bourne End, Bucks. SL8 5AU tel:(01628) 532211, fax:(01628) 532233. systemlabor.uk@koettermann.com, www.kottermann.com

Kronospan Ltd - Chirk, Wrexham. LL14 5NT tel:(01691) 773361, fax:(01691) 773292. sales@kronospan.co.uk, www.kronospan.co.uk

KSB Ltd - 2 Cotton Way, Loughborough, Leics. LE11 5TF tel:(01509) 231872, fax:(01509) 215228. sales@kgnpillinger.com, www.ksbuk.com

Kush UK Ltd - 48/50 St Johns Street, London. EC1M 4DG tel:(01684) 850787, fax:(01684) 850758. info-uk@kusch.com, www.kusch.de

L

L. B. Plastics Ltd - Firs Works, Nether Heage, Belper, Derbys. DE56 2JJ tel:(01773) 852311, fax:(01773) 857080. sheerframe@lbplastics.co.uk, www.sheerframe.co.uk

Lab Systems Furniture Ltd - Rotary House, Bontoft Avenue, Hull, Yorks. HU5 4HF tel:(01482) 444650, fax:(01482) 444730. office@lab-systems.co.uk, www.labsystemsfurniture.co.uk

Labcraft Ltd - Thunderley Barns, Thaxted Road, Wimbish, Saffron Walden, Essex. CB10 2UT tel:(01799) 513434, fax:(01799) 513437. sales@labcraft.co.uk, www.labcraft.co.uk

Lafarge Aluminates Ltd - 730 London Road, Grays, Essex. RM20 3NJ tel:(01708) 863333, fax:(01708) 861033. www.lafarge.com

LaFarge Cement UK - Manor Court, Chiltern, Oxon. OX11 0RN tel:(0870) 6000203. info@lafargecement.co.uk, www.lafargecement.co.uk

Laidlaw Architectural Hardware Ltd - Pennine House, Dakota avenue, Manchester, Gt Man. M5 2PU tel:(0161) 848 1700, fax:(0161) 848 1700. technical@laidlaw.net, www.laidlaw.net

Laidlaw Ltd - 344 – 354 Grays Inn Road, London. WC1E 7BT tel:(020) 7436 0779, fax:(020) 7436 0740. info.london@laidlaw.net, www.laidlaw.net

Laird Security Hardware - Western Road, Silver End, Witham, Essex. CM8 3QB tel:(0121) 224 6000, fax:(0121) 520 1039. sales@lairdsecurity.co.uk, www.lairdsecurity.co.uk

Lakes Bathrooms Limited - Alexandra Way, Ashchurch, Tewkesbury, Gloucestershire. GL20 8NB tel:(01242) 620061, fax:(01684) 850912. info@lakesbathrooms.co.uk, www.lakesbathrooms.co.uk

Lakeside Group - Bruce Grove, Forest Fach Industrial Estate, Swansea, Swansea. SA5 4HS tel:(01792) 561117, fax:(01792) 587046. sales@lakesidesecurity.co.uk, www.lakesidesercurity.co.uk

Lami Doors UK Ltd - Station House, Stamford New Road, Altrincham, Cheshire. WA14 1EP tel:(0161) 9242217, fax:(0161) 9242218. lamidoors@lamidoors.com, www.lamidoors.com

Lampost Construction Co Ltd - Greenore, Co. Louth. Ireland tel:+353 42 93 73554, fax:+353 42 93 73378. lampost@iol.ie

Lampways Ltd - Allenby House, Knowles Lane, Wakfield Road, Bradford, W. Yorks. BD4 9AB tel:(01274) 686666, fax:(01274) 680157. info@europeanlampgroup.com, www.europeanlampgroup.com

Lancashire Fittings Ltd - The Science Village, Claro Road, Harrogate, N. Yorks. HG1 4AF tel:(01423) 522355, fax:(01423) 506111. kenidle@lancashirefittings.com, www.lancashirefittings.com

Landline Ltd - 1 Bluebridge Industrial Estate, Halstead, Essex. CO9 2EX tel:(01787) 476699, fax:(01787) 472507. sales@landline.co.uk, www.landline.co.uk

Landscape Institute - Charles Darwin House, 12 Roger Street, London. WC1N2JU tel:(020) 7685 2640. mail@landscapeinstitute.org, www.landscapeinstitute.org

Lane Roofing Contractors Ltd - Walsall House, 165-167 Walsall Road, Perry Barr, Birmingham. B42 1TX tel:(08450) 0667000, fax:(0121) 344 3782. info@laneroofing.co.uk, www.laneroofing.co.uk

Lang+Fulton - Unit 2b, Newbridge Industrial Estate, Edinburgh. EH28 8PJ tel:(0131) 441 1255, fax:(0131) 441 4161. sales@langandfulton.co.uk, www.langandfulton.co.uk

LANXESS Inorganic Pigments Group - Colour Works, Lichfield Road, Branston, Staffs. DE14 3WH tel:(01283) 714200, fax:(01283) 714210. simon.kentesber@lanxess.com, www.bayferrox.com

Lappset UK Ltd - Lappset House, Henson Way, Telford Industrial Estate, Kettering, Northants. NN16 8PX tel:(01536) 412612, fax:(01536) 521703. uk@lappset.com, www.lappset.co.uk

Latchways plc - Hopton Park, Devizes, Wilts. SN10 2JP tel:(01380) 732700, fax:(01380) 732701. info@latchways.com, www.latchways.com

Laticrete International Inc. UK - Hamilton House, Mabledon Place, London. WC1H 9BB tel:+44 0871 284 5959, fax:+44 0 208 886 2880. ukinfo@laticrete.com, www.laticrete.com

Laurier M & Sons - Unit 10, Triumph Trading Estate, Tariff Road, Tottenham, London. N17 0EB tel:(0844) 488 0642, fax:(020) 8365 9005. info@laurier.co.uk, www.laurier.co.uk

Lawton Tube Co Ltd, The - Torrington Avenue, Coventry. CV4 9AB tel:(024) 7646 6203, fax:(024) 7669 4183. sales@lawtontubes.co.uk, www.lawtontubes.co.uk

Laxton's Publishing Ltd

St John's House, 304-310 St Albans Road, Watford, Herts. WD24 6PW tel:(01923) 227236, fax:(03332) 407363. info@laxton-s.co.uk, www.laxton-s.co.uk

Laybond Products Ltd - Riverside, Saltney, Chester. CH4 8RS tel:(01244) 674774, fax:(01244) 680215. technical.service@bostik.com, www.laybond.com

Laytrad Contracts Ltd - Unit 3 Beza Court, Beza Road, Hunslet, Leeds. LS10 2BR tel:(0113) 271 2117. info@laytrad.co.uk, www.laytrad.co.uk

Leaflike - Olympic House, Collett, Southmead Park, Didcot. OX11 7WB tel:(01235) 515050. info@leaflike.co.uk

Leander Architectural - Fletcher Foundry, Hallsteads Close, Dove Holes, Buxton, Derbys. SK17 8BP tel:(01298) 814941, fax:(01298) 814970. sales@leanderarchitectural.co.uk, www.leanderarchitectural.co.uk

Leatherhead Food Research Association - Randalls Road, Leatherhead, Surrey. KT22 7RY tel:(01372) 376761, fax:(01372) 386228. help@lfra.co.uk, www.leatherheadfood.com

Ledlite Glass (Southend) Ltd - 168 London Road, Southend on Sea, Essex. SS1 1PH tel:(01702) 345893, fax:(01702) 435099. rob@ledlite.fsnet.co.uk, www.ledliteglass.co.uk

Legge Thompson F & Co Ltd - 1 Norfolk Street, Liverpool. L1 0BE tel:(0151) 709 7494, fax:(0151) 709 3774. sales@livergrease.co.uk, www.livergreaseltd.co.uk

Legrand Electric Ltd - Dial Lane, West Bromwich, W. Midlands. B70 0EB tel:(0845) 6054333, fax:(0845) 605 4334. legrand.sales@legrand.co.uk, www.legrand.co.uk

Legrand Electric Ltd - Great King Street North, Birmingham. B19 2LF tel:(0121) 515 0515, fax:(0121) 515 0516. legrand.sales@legrand.co.uk, www.legrand.co.uk

Legrand Power Centre - Brookside, Wednesbury, W. Midlands. WS10 0QF tel:(0121) 506 4506, fax:(0121) 506 4507. legrand.sales@legrand.co.uk, www.legrand.co.uk

LEIA - 33 Devonshire Street, London. W1G 6PY tel:(020) 7935 3013, fax:(020) 7935 3321. enquiries@leia.co.uk, www.leia.co.uk

Leicester Barfitting Co Ltd - West Avenue, Wigston, Leics. LE18 2FB tel:(0116) 288 4897, fax:(0116) 281 3122. sales@leicesterbarfitting.co.uk, www.leicesterbarfitting.co.uk

Leigh's Paints - Tower Works, Kestor Street, Bolton, Lancs. BL2 2AL tel:(01204) 521771, fax:(01204) 382115. enquiries@leighspaints.com, www.leighspaints.co.uk

Leofric Building Systems Ltd - Hillside House, Stratford Road, Mickleton, Glos. GL55 6SR tel:(01386) 430121, fax:(01386) 438135. sales@leofricbuildings.co.uk, www.leofricbuildings.co.uk

Lesco Products Ltd - Wincheap Industrial Estate, Canterbury, Kent. CT1 3RH tel:(01227) 763637, fax:(01227) 762239. sales@lesco.co.uk, www.lesco.co.uk

Levolux A.T. Limited - 24 Easville Close Eastern Avenue, Gloucester. GL4 3SJ tel:(01452) 500007, fax:(01452) 527496. info@levolux.co.uk, www.levolux.com

Lewden Electrical Industries - Argall Avenue, Leyton, London. E10 7QD tel:(020) 8539 0237, fax:(020) 8558 2718. sales@lewden.co.uk, www.lewden.net

Lewes Design Contracts Ltd - The Mill, Glynde, Lewes, Sussex. BN8 6SS tel:(01273) 858341, fax:(01273) 858200. info@spiralstairs.co.uk, www.spiralstairs.co.uk

Leyland Paint Company, The - Huddersfield Road, Birstall, Batley, W. Yorks. WF17 9XA tel:(01924) 354100, fax:(01924) 354001. leylandtrade.acuk@ppg.com, www.leyland-paints.co.uk

LHC - Royal House, 4th Floor, 2-4 Vine Street, Uxbridge, Middx. UB8 1QE tel:(01895) 274800. mail@lhc.gov.uk, www.lhc.gov.uk

Libraco - The Cowbarn, Filston Farm, Shoreham, Stevenoaks, Kent. TN14 5JU tel:(01959) 524074, fax:(01959) 525218. libraco@btconnect.com, www.libraco.uk.com

Library Furnishing Consultants - Grange House, Geddings Road, Hoddesdon, Hertfordshire. EN11 0NT tel:(01992) 454545, fax:(0800) 616629. enquiries@gresswell.co.uk, www.lfcdespatchline.co.uk

Lift and Escalator Industry Association,The - 33/34 devonshire Street, London. W1G 6PY tel:(020) 7935 3013, fax:(020) 7935 3321. enquiries@leia.co.uk, www.leia.co.uk

Liget Compton - Albion Drive, Lidget Lane Industrial Estate, Thurnscoe, Rotherham. S63 0BA tel:(01295) 770291, fax:(01295) 770748. sales@lidget.co.uk, www.lidget.co.uk

Light (Multiforms), H A - Woods Lane, Cradley Heath, Warley, W. Midlands. B64 7AL tel:(01384) 569283, fax:(01384) 633712. enquiry@multiforms.co.uk, www.multiforms.co.uk

Light Engine Lighting Ltd - The Old Power Station, Enterprise Court, Lakes Road, Braintree, Essex. CM7 3QS tel:(01376) 528701, fax:(01376) 528702. sales@lightengine.co.uk, www.lightengine.co.uk

Lighting Industry Association, The - Stafford Park 7, Telford, Shropshire. TF3 3BQ tel:(01952) 290905, fax:(01952) 290906. info@thelia.org.uk, www.thelia.org.uk

Lignacite (North London) Ltd - Meadgate Works, Nazeing, Waltham Abbey, Essex. EN9 2PD tel:(01992) 464661, fax:(01992) 445713. info@lignacite.co.uk, www.lignacite.co.uk

Lignacite Ltd - Norfolk House, High Street, Brandon, Suffolk. IP27 0AX tel:(01842) 810678, fax:(01842) 814602. info@lignacite.co.uk, www.lignacite.co.uk

Lilleheden Ltd - Digital Media Centre, County Way, Barnsley. S70 2JW tel:+44 1226 720 760, fax:+44 1226 720 701. mailuk@lilleheden.com, www.lilleheden.dk/uk

Linatex Ltd - Wilkinson House, Galway Road, Blackbush Business Park, Yateley, Hants. GU46 6GE tel:(01252) 743000, fax:(01252) 743030. sales.uk@weirminerals.com, www.linatex.net

Lindapter International - Lindsay House, Brackenbeck Road, Bradford, W. Yorks. BD7 2NF tel:(01274) 521444, fax:(01274) 521130. enquiries@lindapter.com, www.lindapter.com

Linear Ltd - Coatham Avenue, Newton Aycliffe Industrial Estate, Newton Aycliffe, Co. Durham. DL5 6DB tel:(01325) 310 151, fax:(01325) 307 200. sales@linear-ltd.com, www.linear-ltd.com

Link 51 (Storage Products) - Link House, Halesfield 6, Telford, Shropshire. TF7 4LN tel:(0800) 1695151, fax:(01384) 472599. sales@link51.co.uk, www.link51.co.uk

Lionweld Kennedy Ltd - Marsh Road, Middlesbrough, Cleveland. TS1 5JS tel:(01642) 245151, fax:(01642) 224710. sales@lionweldkennedy.co.uk, www.lionweldkennedy.co.uk

Liquid Technology Ltd - Unit 27, Tatton Court, Kingsland Grange, Woolston, Warrington, Cheshire. WA1 4RR tel:(01925) 850324, fax:(01925) 850325. info@liquidtechnologyltd.com, www.fengroup.com

Litex Design Ltd - 3 The Metro Centre, Dwight Road, Watford, Herts. WD18 9HG tel:(01923) 247254, fax:(01923) 226772. joshea@litex.demon.co.uk, www.litexuk.com

Liver Grease Oil & Chemical Co Ltd - 11 Norfolk Street, Liverpool. L1 0BE tel:(0151) 709 7494, fax:(0151) 709 3774. sales@livergrease.co.uk, www.livergreaseltd.co.uk

LMOB Electrical Contractors Ltd - Unit 28, Balfour Business Centre, Balfour Road, Southall, Middx. UB2 5BD tel:(020) 8574 6464, fax:(020) 8573 1189. sales@lmob.co.uk, www.lmob.co.uk

Loblite Ltd - Third Avenue, Team Valley, Gateshead, Tyne & Wear. NE11 0QQ tel:(0191) 487 8103, fax:(0191) 482 0270. info@loblite.co.uk, www.loblite.com

Lochinvar Limited - 7 Lombard Way, The MXL Centre, Banbury, Oxon. OX16 4TJ tel:(01295) 269981, fax:(01295) 271640. info@lochinvar.ltd.uk, www.lochinvar.ltd.uk

Lockhart Catering Equipment - Lockhart House, Brunel Road, Theale, Reading, Berks. RG7 4XE tel:(0118) 930 3900, fax:(0118) 930 0757. bceweb@bunzl.co.uk, www.lockhartcatering.co.uk

Loctite UK Ltd - Watchmead, Welwyn Garden City, Herts. AL7 1JB tel:(01707) 358800, fax:(01707) 358900. kieran.bowden@loctite-europe.com, www.loctite.co.uk

Loft Centre Products - Thicket Lane, Halnaker nr. Chichester, W. Sussex. PO18 OQS tel:(01243) 785246/785229, fax:(01243) 533184. sales@loftcentre.co.uk, www.loftcentreproducts.co.uk

Logic Office Contracts - Vestry Estate, Otford Road, Sevenoaks, Kent. TN14 5EL tel:(01732) 457636, fax:(01732) 740706. contracts.admin@logic-office.co.uk, www.logic-office.co.uk

London Concrete Ltd - London House, 77 Boston Manor Road, Brentford, Middlesex. TW8 9JQ tel:(0208) 380 7300, fax:(0208) 380 7301. www.aggregate.com

London Fan Company Ltd - 75-81 Stirling Road, Acton, London. W3 8DJ tel:(020) 8992 6923, fax:(020) 8992 6928. sales@londonfan.co.uk, www.londonfan.co.uk

London Shopfitters Ltd - Unit 6, Blackwater Close, Fairview Ind. Est, Marsh Way, Rainham, Essex. RM13 8UA tel:(01708) 552225, fax:(01708) 557567. sales@londonshopfitters.co.uk, www.londonshopfitters.co.uk

Longbottom J & J W Ltd - Bridge Foundry, Holmfirth, Huddersfield, W. Yorks. HD9 7AW tel:(01484) 682141, fax:(01484) 681513. www.longbottomfoundry.co.uk

Longden - 55 Parkwood Road, Sheffield, South Yorkshire. S3 8AH tel:(0114) 2706330. enquiries@longdendoors.co.uk, www.lslongden.co.uk

L (con't)

Lonsdale Metal Company - Unit 40 Millmead Industrial Centre, Mill Mead Road, London. N17 9QU tel:(020) 8801 4221, fax:(020) 8801 1287. info@lonsdalemetal.co.uk, www.lonsdalemetal.co.uk

Lord Isaac Ltd - Desborough Road, High Wycombe, Bucks. HP11 2QN tel:(01494) 459191, fax:(01494) 461376. info@isaaclord.co.uk, www.isaaclord.co.uk

Lotus Water Garden Products Ltd - Lotus House, Deer Park Industrial Estate, Knowle Lane, Fair Oak, Eastleigh, Hampshire. SO50 7DZ tel:(023) 8060 2602, fax:(023) 8060 2603. info@lotuswgp.com, www.lotuswatergardenproducts.com

Louver-Lite Ltd - Ashton Road, Hyde, Cheshire. SK14 4BG tel:(0161) 882 5000, fax:(0161) 882 5009. janet.dunn@louvolite.com, www.louvolite.com

Lowe & Fletcher Ltd - Moorcroft Drive, Wednesbury, W. Midlands. WS10 7DE tel:(0121) 5050400, fax:(0121) 5050420. Sales@lowe-and-fletcher.co.uk, www.lowe-and-fletcher.co.uk

LPA Group PLC - Tudor Works, Debden Road, Saffron Walden, Essex. CB11 4AN tel:(01779) 512800, fax:(01779) 512826. enquiries@lpa-group.com, www.lpa-group.com

LSA Projects Ltd - The Barn, White Horse Lane, Witham, Essex. CM8 2BU tel:(01376) 501199, fax:(01376) 502027. sales@lsaprojects.co.uk, www.lsaprojects.co.uk

LTI Advanced Systems Technology Ltd - 3 Kinsbourne Court, 96-100 Luton Road, Harpenden, Herts. AL5 3BL tel:(01582) 469 769, fax:(01582) 469 789. sales@lti-ast.co.uk, www.lti-ast.co.uk

Lucideon - Queens Road, Penkhull, Stoke-on-Trent, Staffordshire. ST4 7LQ tel:(01782) 764428, fax:(01782) 412331. enquiries@lucideon.com, www.lucideon.com

Lucideon - Queens Road, Penkhull, Stoke-on-Trent. ST4 7LQ tel:(01782) 764428, fax:(01782) 412331. enquiries@lucideon.com, www.lucideon.com

Lumitron Ltd - Unit 31, The Metro Centre Tolpits Lane, Watford, Herts. WD18 9UD tel:(01923) 226222, fax:(01923) 211300. sales@lumitron.co.uk, www.lumitron.co.uk

Luxaflex® - Hunter Douglas Ltd, Battersea Road, Heaton Mersey Industrial Estate, Stockport, Cheshire. SK4 3EQ tel:(0161) 442 9500, fax:(0161) 432 3200. info@luxaflex.co.uk, www.luxaflex.co.uk

Luxcrete Ltd - Unit 2, Firbank Industrial Estate, Dallow Road, Luton. LU1 1TW tel:(01582) 488767, fax:(01582) 724607. enquiries@luxcrete.co.uk, www.luxcrete.co.uk

Lytag Ltd - Second Floor (front), 75/77 Margaret Street, London. W1W 8SY tel:(0207) 4995242. sales@lytag.com, www.lytag.com

M

M & G Brickcutters (specialist works) - Hockley Works, Hooley Lane, Redhill, Surrey. RH1 6JE tel:(01737) 771171, fax:(01737) 771171. sales@mgbrickcutters.co.uk, www.mgbrickcutters.co.uk

Mabey Hire Ltd - 1 Railway Street, Scout Hill, Ravensthorpe, Dewsbury, W. Yorks. WF13 3EJ tel:(01924) 460601, fax:(01924) 457932. info@mabeyhire.co.uk, www.mabeyhire.co.uk

MacAlloy Limited - PO Box 71, Hawke Street, Sheffield. S9 2LN tel:(0114) 242 6704, fax:(0114) 243 1324. sales@macalloy.com, www.macalloy.com

Macemain + Amstad Ltd - Boyle Road, Willowbrook Industrial Estate, Corby, Northants. NN17 5XU tel:(01536) 401331, fax:(01536) 401298. sales@macemainamstad.com, www.macemainamstad.com

Mackay Engineering - Unit 2 Mackay Business Park, 120 Church End, Cherry Hinton, Cambridge, Cambs. CB1 3LB tel:(01223) 508222, fax:(01223) 510222. engineering@mackay.co.uk, www.mackay.co.uk

Mackwell Electronics Ltd - Virgo Place, Aldridge, W. Midlands. WS9 8UG tel:(01922) 58255, fax:(01922) 51263. sales@mackwell.com, www.mackwell.com

MacLellan Rubber Ltd - Neachells Lane, Wednesfield, Wolverhampton. WV11 3QG tel:(01902) 307711, fax:(0151) 946 5222. sales@maclellanrubber.com, www.maclellanrubber.com

MacMarney Refrigeration & Air Conditioning Ltd - The Old Forge, Stone Street, Crowfield, Ipswich, Suffolk. IP6 9SZ tel:(01449) 760560, fax:(01449) 760590. sales@macmarney.co.uk, www.macmarney.co.uk

Maco Door & Window Hardware (UK) Ltd - Eurolink Ind. Centre, Castle Road, Sittingbourne, Kent. ME10 3LY tel:(01795) 433900, fax:(01795) 433901. enquiry@macouk.net, www.maco-europe.com

Macpuarsa (UK) Ltd - 13 14 Charlton Business Centre, 15 The Avenue, Bromley, Kent. BR1 2BS tel:(01932) 223161, fax:(01932) 223181. hc@mpcorporacion.com, www.macpuarsa.es

MAGHansen Ltd - Unit 6a Spence Mills, Mill Lane, Bramley, Leeds, W. Yorks. LS13 3HE tel:(0113) 255 5111, fax:(0113) 255 1946. sales@maghansen.com, www.hansengroup.biz/maghansen/

Magiboards Ltd - Unit B3, Stafford Park 11, Telford, Shropshire. TF3 3AY tel:(01952) 292111, fax:(01952) 292280. sales@magiboards.com, www.magiboards.com

Magiglo Ltd - Lysander Close, Broadstairs, Kent. CT10 2YJ tel:(01843) 602863, fax:(01843) 860108. info@magiglo.co.uk, www.magiglo.co.uk

Magnet Ltd - Allington Way, Yarm Road Ind. Estate, Darlington, Co. Durham. DR1 4XT tel:(01325) 469441, fax:(01325) 468876. info@magnet.co.uk, www.magnet.co.uk

Magnum Ltd T/A Magnum Scaffolding - Yard Brook Estate, Stockwood Vale, Keynsham, Bristol, Avon. BS31 2AL tel:(0117) 986 0123, fax:(0117) 986 2123. chris@magnumltd.co.uk, www.magnumltd.co.uk

Magpie Furniture - Blundell Harling Ltd, 9 Albany Road, Granby Industrial Estate, Weymouth, Dorset. DT4 9TH tel:(01305) 206000, fax:(01305) 760598. sales@magpiefurniture.co.uk, www.magpiefurniture.co.uk

Magrini Ltd - Unit 5, Maybrook Industrial Estate, Brownhills, Walsall, W. Midlands. WS8 7DG tel:(01543) 375311, fax:(01543) 361172. sales@magrini.co.uk, www.magrini.co.uk

Majestic Shower Co Ltd - 1 North Place, Edinburgh Way, Harlow, Essex. CM20 2SL tel:(01279) 443644, fax:(01279) 635074. info@majesticshowers.com, www.majesticshowers.com

Maldon Fencing Co - Burnham Road, Latchingdon, Nr Chelmsford, Essex. CM3 6HA tel:(01621) 740415, fax:(01621) 740769. maldonfencing@btconnect.com, www.maldonfencing.com

Manders Paint Ltd - PO Box 9, Old Heath Road, Wolverhampton, W. Midlands. WV1 2XG tel:(01902) 871028, fax:(01902) 452435. richard@mandersconstruction.co.uk, www.manderspaints.co.uk

Mandor Engineering Ltd - Turner Street Works, Turner Street, Ashton-under-Lyne, Gt Man. OL6 8LU tel:(0161) 330 6837, fax:(0161) 308 3336. sales@mandor.co.uk, www.mandor.co.uk

Manitou (Site Lift) Ltd - Ebblake Industrial Estate, Verwood, Wimborne, Dorset. BH31 6BB tel:(01202) 825331, fax:(01202) 813027. www.manitou.com

Mann McGowan Group - Unit 4 Brook Trading Estate, Deadbrook Lane, Aldershot, Hants. GU12 4XB tel:(01252) 333601, fax:(01252) 322724. sales@mannmcgowan.co.uk, www.mannmcgowan.co.uk

Mansfield Brick Co Ltd - Sanhurst Avenue, Mansfield, Notts. NG18 4BE tel:(01623) 622441, fax:(01623) 420904. info@mansfield-sand.co.uk, www.mansfield-sand.co.uk

Mansfield Pollard & Co Ltd - Edward House, Parry Lane, Bradford. BD4 8TL tel:(01274) 77 40 50, fax:(01274) 77 54 24. salesteam@mansfieldpollard.co.uk, www.mansfieldpollard.co.uk

MANTAIR Ltd - Unit 13-14, Baker Close, Oakwood Business Park, Clacton-on-Sea, Essex. CO15 4TL tel:(01255) 476467, fax:(01255) 476817. info@mantair.com, www.mantair.com

Mapei (UK) Ltd - Mapei House, Steel Park Road, Halesowen, W. Midlands. B62 8HD tel:(0121) 508 6970, fax:(0121) 508 6960. info@mapei.co.uk, www.mapei.co.uk

Marble Flooring Spec. Ltd - Verona house, Filwood Road, Bristol. BS16 3RY tel:(0117) 9656565, fax:(0117) 9656573. maryford@marbleflooring.co.uk, www.marbleflooring.co.uk

Marble Mosaic Co Ltd, The - Winterstoke Road, Weston-Super-Mare, N. Somerset. BS23 3YE tel:(01934) 419941, fax:(01934) 625479. sales@marble-mosaic.co.uk, www.marble-mosaic.co.uk

Marcrist International Ltd - Marcrist House, Kirk Sandall Industrial Estate, Doncaster, S. Yorks. DN3 1QR tel:(01302) 890888, fax:(01302) 883864. info@marcrist.com, www.marcrist.co.uk

Marcus M Ltd - Unit 7 Narrowboat Way, Peartree Lane, Dudley, W. Midlands. DY2 0XW tel:(01384) 457900, fax:(01384) 457903. info@m-marcus.com, www.m-marcus.com

Mardale Pipes Plus Ltd - PO BOX 86, Davy Road, Astmoor Industrial Estate, Runcorn, Cheshire. WA7 1PX tel:(01928) 580555, fax:(01928) 591033. sales@ippgrp.com, www.mardale-pipes.com

Marflex Chimney Systems - Unit 40, Vale Business Park, Cowbridge, S. Glams. CF71 7PF tel:(01446) 775551, fax:(01446) 772468. sales@pd-edenhall.co.uk, www.pd-edenhall.co.uk

Marflow Eng Ltd - Britannia House, Austin Way, Hamstead Industrial Estate, Birmingham. B42 1DU tel:(0121) 358 1555, fax:(0121) 358 1444. sales@marflow.co.uk, www.marflow.co.uk

Margolis Office Interiors Ltd - 341 Euston Road, London. NW1 3AD tel:(020) 7387 8217, fax:(020) 7388 0625. info@margolisfurniture.co.uk, www.margolisfurniture.co.uk

Marking Machines International - Sportsmark House, 4 Clerewater Place, Thatcham, Berks. RG19 3RF tel:(01635) 867537, fax:(01635) 864588. sales@sportsmark.net, www.sportsmark.net

Marleton Cross Ltd - Unit 2 Alpha Close, Delta Drive, Tewkesbury Industrial Estate, Tewkesbury, Glos. GL20 8JF tel:(01684) 293311, fax:(01684) 293900. enquiries@mx-group.com, www.mx-group.com

Marley Eternit Ltd - Litchfield Road, Branston, Burton On Trent. DE14 3HD tel:(01283) 722588. info@marleyeternit.co.uk, www.marleyeternit.co.uk

Marley Plumbing and Drainage - Dickley Lane, Lenham, Maidstone, Kent. ME17 2DE tel:(01622) 858888, fax:(01622) 858725. marketing@marleypd.com, www.marley.co.uk

Marlow Ropes Ltd - Diplocks Way, Hailsham, Sussex. BN27 3JS tel:(01323) 847234, fax:(01323) 440093. industrial@marlowropes.com, www.marlowropes.com

Marlux Ltd - Unit 1, Ace Business Park, Mackadown Lane, Kitts Green, Birmingham. B33 0LD tel:(0121) 783 5777, fax:(0121) 783 1117. info@marlux.co.uk, www.marlux.co.uk

Marshalls Mono Ltd - Southowram, Halifax, Yorks. HX3 9SY tel:(01422) 306000, fax:(01422) 330185. customeradvice@marshalls.co.uk, www.marshalls.co.uk

Marshall-Tufflex Ltd - Churchfields Ind.Estate, Hastings, Sussex. TN38 9PU tel:(0870) 2403200, fax:(0870) 2403201. sales@marshall-tufflex.com, www.marshall-tufflex.com

Marsland & Co Ltd - Station Road, Edenbridge, Kent. TN8 6EE tel:(01732) 862501, fax:(01732) 866737. www.marsland-windows.co.uk

Martello Plastics Ltd - Unit 11, Ross Way, Shorncliffe Industrial Estate, Folkestone, Kent. CT20 3UJ tel:(01303) 256848, fax:(01303) 246301. mail@martelloplastics.co.uk, www.martelloplastics.co.uk

Martin & Co Ltd - 119 Camden Street, Birmingham. B1 3DJ tel:(0121) 233 2111, fax:(0121) 236 0488. sales@armacmartin.co.uk, www.martin.co.uk

Martin Childs Limited - 1 Green Way, Swaffham, Norfolk. PE37 7FD tel:(01760) 722 275. enquiries@martinchilds.com, www.martinchilds.com

Mason FW & Sons Ltd - Colwick Industrial Estate, Colwick, Nottingham, Notts. NG4 2EQ tel:(0115) 911 3500, fax:(0115) 911 35444. enquiries@masons-timber.co.uk, www.masons-timber.co.uk

Masonary Cleaning Services - 1A Alpines Road, Calow. S44 5AU tel:(01246) 209926, fax:(01246) 221620. maxblastuk@btinternet.com

Masson Seeley & Co Ltd - Howdale, Downham Market, Norfolk. PE38 9AL tel:(01366) 388000, fax:(01366) 385222. admin@masson-seeley.co.uk, www.masson-seeley.co.uk

Masterbill Micro Systems Ltd - 5b Cedar Court, Porters Wood, St. Albans, Herts. AL3 6PA tel:(01727) 855563, fax:(01727) 854626. sales@masterbill.com, www.masterbill.com

Masterframe Windows Ltd - 4 Crittall Road, Witham, Essex. CM8 3DR tel:(01376) 510410, fax:(01376) 510400. sales@masterframe.co.uk, www.masterframe.co.uk

Mastic Asphalt Council - P.O. Box 77, Hastings. TN35 4WL tel:(01424) 814400, fax:(01424) 814446. masphaltco@aol.com, www.masticasphaltcouncil.co.uk

Mat.Works Flooring Solutions - A Division of National Floorcoverings Ltd, Farfield Park, Manvers, Wath on Dearne. S63 5DB tel:(01709) 763839, fax:(01709) 763813. sales@mat-works.co.uk, www.mat-works.co.uk

Matki plc - Churchward Road, Yate, Bristol, S. Glos. BS37 5PL tel:(01454) 322888, fax:(01454) 315284. helpline@matki.co.uk, www.matki.co.uk

Mattersons Cranes - 45 Regent Street, Rochdale, Lancs. OL12 0HQ tel:(01706) 649321. matterson@pctgroup.co.uk, www.mattersoncranes.co.uk

Matthew Hebden - 54 Blacka Moor Road, Sheffield. S17 3GJ tel:(0114) 236 8122. sales@matthewhebden.co.uk, www.matthewhebden.co.uk

Matthews & Yates Ltd - Peartree Road, Stanway, Colchester, Essex. CO3 0LD tel:(01206) 543311, fax:(01206) 760497. info@systemair.co.uk, www.matthewyates.co.uk

Mawdsleys Ber Ltd - Barton Manor Works, Barton Manor Trading Estate, Midland Road, St. Philips, Bristol. BS2 0RL tel:(0117) 955 2481, fax:(0117) 955 2483. woodward1974@msn.com, www.mawdsleys.com

Max Appliances Ltd - Wheel Park Farm Ins Est, Wheel Lane, Hastings, Sussex. TN35 4SE tel:(01424) 854 444, fax:(01424) 853862. sales@max-appliances.co.uk, www.max-appliances.co.uk

Maxit UK - The Heath, Runcorn, Cheshire. WA7 4QX tel:(01928) 515656. sales@maxit.com, www.maxit.com

May Gurney (Technical Services) Ltd - Fencing Division - Trowse, Norwich, Norfolk. NR14 8SZ tel:(01603) 744440, fax:(01603) 747310. piling@maygurney.co.uk, www.maygurney.co.uk

May William (Ashton) Ltd - Cavendish Street, Ashton-under-Lyne, Lancs. OL6 7BR tel:(0161) 330 3838/4879, fax:(0161) 339 1097. sales@william-may.com, www.william-may.com

Mayplas Ltd - Building 1, Peel Mills, Chamberhall Street, Bury, Lancs. BL9 0JU tel:(0161) 447 8320. sales@mayplas.co.uk, www.mayplas.co.uk

McAlpine & Co Ltd - Hillington Industrial Estate, Glasgow. G52 4LF tel:(0141) 882 3213. www.mcalpineplumbing.com

McArthur Group Limited - Foundry Lane, Bristol, Avon. BS5 7UE tel:(0117) 943 0500, fax:(0117) 958 3536. bristolsales@macarthur.uk.com, www.mcarthur.uk.com

McCarthy & Stone Construction Services - 3 Queensway, New Milton, Hants. BH25 5PB tel:(01425) 638855, fax:(01425) 638343. info@homelife.co.uk, www.mccarthyandstone.co.uk

McKenzie-Martin Ltd - Eton Hill Works, Eton Hill Road, Radcliffe, Gt Man. M26 2US tel:(0161) 723 2234, fax:(0161) 725 9531. hilda@mckenziemartin.co.uk, www.mckenziemartin.co.uk

MCL Composites Ltd - New Street, Biddulph Moor, Stoke-on-Trent, Staffs. ST8 7NL tel:(01782) 375450, fax:(01782) 522652. sales@mcl-grp.co.uk, www.mcl-grp.co.uk

McMullen Facades - 66 Lurgan Road, Moira, Craigavon, Co. Armagh. BT67 0LX tel:(028) 9261 9688, fax:(028) 9261 9711. enquiries@mcmullenfacades.com, www.mcmullenfacades.com

Mea UK Ltd - Rectors Lane, Pentre, Deeside, Flintshire. CH5 2DH tel:(01244) 534455, fax:(01244) 534477. uk.technical@mea.de, www.mea.uk.com

Meggitt Armour Systems - Bestobell Ave, Slough, Berks. SL1 4UY tel:(01384) 357799, fax:(01384) 357700. advancedarmour@tencate.com, www.meggit.com

Mells Roofing Ltd - Beehive Works, Beehive Lane, Chelmsford, Essex. CM2 9JY tel:(01245) 262621, fax:(01245) 260060. info@mellsroofing.co.uk, www.mellsroofing.co.uk

Mentha & Hessal (Shopfitters) Ltd - 95a Linaker Street, Southport, Merseyside. PR8 5BU tel:(01704) 530800, fax:(01704) 500601. paulmentha@mentha-halsall.com, www.mentha-halsall.com/

Mercian Industrial Doors - Pearsall Drive, Oldbury, West Midlands. B69 2RA tel:(0121) 544 6124, fax:(0121) 552 6793. info@merciandoors.co.uk, www.merciandoors.co.uk

Mermaid Panels Ltd - DBC House, Laceby Business Park, Grimsby, Humb. DN37 7DP tel:(01472) 279940, fax:(01472) 752575. sales@mermaidpanels.com, www.mermaidpanels.com

Mermet U.K - Ryeford Hall, Ryeford, Nr Ross on Wye, Herefordshire. HR9 7PU tel:(01989) 750910, fax:(01989) 750768. sales@mermet.co.uk, www.mermet.co.uk

Merronbrook Ltd - Hazeley Bottom, Hartley Wintney, Hook, Hants. RG27 8LX tel:(01252) 844747, fax:(01252) 845304. sales@merronbrook.co.uk

Metal Technology Ltd - Steeple Road Industrial Estate, Steeple Road, Antrim. BT41 1AB tel:(028) 94487777, fax:(028) 94487878. sales@metaltechnology.com, www.metaltechnology.com

Metalcraft (Tottenham) Ltd - 6-40 Durnford Street, Seven Sisters Road, Tottenham, London. N15 5NQ tel:(020) 8802 1715, fax:(020) 8802 1258. sales@makingmetalwork.com, www.makingmetalwork.com

Metalfloor UK Ltd - Unit 706 Merlin Park, Ringtail Road, Merlin Park, Burscough, Lancashire. L40 8JY tel:(01704) 896061, fax:(01704) 897293. sales@metalfloor.co.uk, www.metalfloor.co.uk

Metalliform Products Plc - Chambers Road, Hoyland, Barnsley, S. Yorks. S74 0EZ tel:(01226) 350555, fax:(01226) 350112. sales@metalliform.co.uk, www.metalliform.co.uk

Metalline Signs Ltd - Unit 18, Barton Hill Trading Estate, Bristol, Avon. BS5 9TE tel:(01179) 555291, fax:(01179) 557518. info@metalline-signs.co.uk, www.metalline-signs.co.uk

Metalwood Fencing (Contracts) Limited - Hutton House, Soothouse Spring, Valley Road Industrial Estate, St. Albans, Herts. AL3 6PG tel:(01727) 861141/2, fax:(01727) 846018. info@metalwoodfencing.com, www.metalwoodfencing.com

Metcraft Ltd - Harwood Industrial Estate, Harwood Road, Littlehampton, Sussex. BN17 7BB tel:(01903) 714226, fax:(01903) 723206. sales@metcraft.co.uk, www.metcraft.co.uk

Metra Non Ferrous Metals Ltd - Pindar Road, Hoddesdon, Herts. EN11 0DE tel:(01992) 460455, fax:(01992) 451207. enquiries@metra-metals.co.uk, www.metra-metals.co.uk

Metreel Ltd - Cossall Industrial Estate, Coronation Road, Ilkeston, Derbys. DE7 5UA tel:(0115) 932 7010, fax:(0115) 930 6263. sales@metreel.co.uk, www.metreel.co.uk

Metsä Wood - Old Golf Course, Fishtoft Road, Boston, Lincs. PE21 0BJ tel:(0800) 00 44 44, fax:(01205) 354488. uk@metsagroup.com, www.finnforest.co.uk

Metsec Lattice Beams Ltd - Units 10 & 11 Rolls Royce Estate, Spring Road, Etingshall, Wolverhampton. WV4 6JX tel:(01902) 408011, fax:(01902) 490440. sales@metseclb.com, www.metseclatticebeams.com

M (con't)

Metsec Ltd - Broadwell Road, Oldbury, Warley, W. Midlands. B69 4HF tel:(0121) 601 6000, fax:(0121) 601 6109. metsecplc@metsec.com, www.metsec.com

Metway Electrical Industries Ltd - Barrie House, 18 North Street, Portslade, Brighton. BN41 1DG tel:(01273) 431600, fax:(01273) 439288. sales@metway.co.uk, www.metway.co.uk

Meyer Timber Limited - Meyer House, Hadleigh Park, Grindley Lane, Blythe Bridge, Stoke on Trent. ST11 9LW tel:(0845) 873 5000, fax:(0845) 873 5005. sales.stoke@meyertimber.com, www.meyertimber.com

Michelmersh Brick Holdings PLC - Freshfield Lane, Danehill, Sussex. RH17 7HH tel:(0844) 931 0022, fax:(01794) 368845. sales@mbhplc.co.uk, www.mbhplc.co.uk

Mid Essex Trading Co Ltd - Montrose Road, Dukes Park Industrial Estate, Springfield, Chelmsford, Essex. CM2 6TH tel:(01245) 469922, fax:(01245) 450755

Midland Alloy Ltd - Stafford Park 17, Telford, Shropshire. TF3 3DG tel:(01952) 290961, fax:(01952) 290441. info@midlandalloy.com, www.midlandalloy.com

Midland Tube and Fabrications - Corngreave Works, Unit 4, Corngreave Road, Cradley Heath, Warley, W. Midlands. B64 7DA tel:(01384) 566364, fax:(01384) 566365. Keithcadman@btconnect.com

Midland Wire Cordage Co Ltd - Wire Rope House, Eagle Rd, North Moons Moat, Redditch, Worcs. B98 9HF tel:(01527) 594150, fax:(01527) 64322. info@ormiston-wire.co.uk, www.midlandwirecordage.co.uk

Miele Co Ltd - Fairacres, Marcham Road, Abingdon, Oxon. OX14 1TW tel:(01235) 554455, fax:(01235) 554477. info@miele.co.uk, www.miele.co.uk

Milesahead Graphic Design/Consultant - 11 Aylsham Close, The Willows, Widnes, Cheshire. WA8 4FF tel:(0151) 422 0004, fax:(0151) 422 0004. Milesahead@talk21.com

Miles-Stone, Natural Stone Merchants - Quarry Yard, Woodside Avenue, Boyattwood Industrial Estate, Eastleigh, Hants. SO50 9ES tel:(023) 80613178. sales@miles-stoneuk.com, www.milesstone.co.uk

Miller Construction - Miller House, Pontefract Road, Normanton, W. Yorks. WF6 1RN tel:(01924) 224370, fax:(01924) 224380. mc.normanton@miller.co.uk, www.millerconstruction.co.uk/

Milliken Carpet - Beech Hill Plant, Gidlow Lane, Wigan, Lancs. WN6 8RN tel:(01942) 612783, fax:(01942) 826570. carpetenquiries@milliken.com, www.millikencarpeteurope.com

Milton Pipes Ltd - Milton Regis, Sittingbourne, Kent. ME10 2QF tel:(01795) 425191, fax:(01795) 420360. sales@miltonpipes.com, www.miltonpipes.co.uk

Minco Sampling Tech UK Ltd (Incorporating Fortafix) - First Drove, Fengate, Peterborough, Cambs. PE1 5BJ tel:(01429) 273252. sales@fortafix.com, www.fortafix.com

Minsterstone Ltd - Pondhayes Farm, Dinnington, Hinton St. George, Somerset. TA17 8SU tel:(01460) 52277, fax:(01460) 57865. varyl@minsterstone.ltd.uk, www.minsterstone.ltd.uk

MiTek Industries Ltd - MiTek House, Grazebrook Industrial Park, Pear Tree Lane, Dudley, W. Midlands. DY2 0XW tel:(01384) 451400, fax:(01384) 451411. roy.troman@mitek.co.uk, www.mitek.co.uk

MJ Electronics Services (International) Ltd - Unit 8, Axe Road, Colley Lane Industrial Estate, Bridgewater, Somerset. TA6 5LJ tel:(01278) 422882, fax:(01278) 453331. sales@mjelectronics.freeserve.co.uk

MK (MK Electric Ltd) - The Arnold Centre, Paycocke Road, Basildon, Essex. SS14 3EA tel:(01268) 563000, fax:(01268) 563563. mkorderenquiries@honeywell.com, www.mkelectric.co.uk

MMA Architectural Systems Ltd - Broadway House, Unit 5 Second Avenue, Midsomer Norton, Somerset. BA3 4BH tel:(0845) 1300135, fax:(0845) 1300136. sales@mma.gb.com, www.jakob.co.uk

Mockridge Labels & Nameplates Ltd - Cavendish Street, Ashton-under-Lyne, Lancs. OL6 7QL tel:(0161) 308 2331, fax:(0161) 343 1958. sales@mockridge.com, www.mockridge.com

Mode Lighting (UK) Ltd - The Maltings, 63 High Street, Ware, Herts. SG12 9AD tel:(01920) 462121, fax:(01920) 466881. sales@modelighting.com, www.modelighting.com

Modelscape - Adams House, Dickerage Lane, New Malden, Surrey. KT3 3SF tel:(020) 8949 9286, fax:(020) 8949 7418. ian@ianmorton.me.uk, www.modelscape.co.uk

Modern Plan Insulation Ltd - Church Street, Westhoughton, Bolton. BL5 3QW tel:(01942) 811839, fax:(01942) 812310. mpinsulations@btconnect.com, www.mpinsulations.co.uk

Moffat Ltd, E & R - Seabegs Road, Bonnybridge, Stirlingshire. FK4 2BS tel:(01324) 812272, fax:(01324) 814107. sales@ermoffat.co.uk, www.ermoffat.co.uk

Moffett Thallon & Co Ltd - 143 Northumberland Street, Belfast. BT13 2JF tel:(028) 90 322802, fax:(028) 90 241428. info@moffett-thallon.co.uk, www.moffett-thallon.co.uk

Monier Redland Limited - Sussex Manor Business Park, Gatwick Road, Crawley, West Sussex. RH10 9NZ tel:(08705) 601000, fax:(08705) 642742. sales.redland@monier.com, www.redland.co.uk

Monk Metal Windows Ltd - Hansons Bridge Road, Erdington, Birmingham, W. Midlands. B24 0QP tel:(0121) 351 4411, fax:(0121) 351 3673. bromilow.d@monkmetal.co.uk, www.monkmetalwindows.co.uk

Mono Pumps Ltd - Martin Street, Audenshaw, Manchester, Gt Man. M34 5JA tel:(0161) 339 9000, fax:(0161) 344 0727. info@mono-pumps.com, www.mono-pumps.com

Monodraught Ltd - Halifax House, Cressex Business Park, High Wycombe, Bucks. HP12 3SE tel:(01494) 897700, fax:(01494) 532465. info@monodraught.com, www.monodraught.com

Monowa Manufacturing (UK) Ltd - Unit 6 Llyn-yr-Eos, Parc Menter, Cross Hands Industrial Estate, Llanelli, Carmarthenshire. SA14 6RA tel:(01269) 845554, fax:(01269) 845515. sales@accordial.co.uk, www.monowa.com

Montrose Fasteners - Montrose House, Lancaster Road, High Wycombe, Bucks. HP12 3PY tel:(01494) 451227, fax:(01494) 436270. sales@themontrosegroup.com, www.themontrosegroup.com

Monument Tools Ltd - Restmor Way, Hackbridge Road, Hackbridge, Wallington, Surrey. SM6 7AH tel:(020) 8288 1100, fax:(020) 8288 1108. info@monument-tools.com, www.monument-tools.com

Moores Furniture Group Ltd - Thorp Arch Trading Estate, Wetherby, W. Yorks. LS23 7DD tel:(01937) 842394, fax:(01937) 845396. marketing@moores.co.uk, www.moores.co.uk

Moravia (UK) Ltd - Unit 9-10, Spring Mill Industrial Estate, Avening Road, Nailsworth, Stroud, Glos. GL6 0BS tel:(01453) 834778, fax:(01453) 839383. service@moravia.co.uk, www.moravia.co.uk

Moray Timber Ltd - 11 Perimeter Road, Pinefield Ind. Estate, Elgin, Morayshire. IV30 6AF tel:(01343) 545151, fax:(01343) 549518. MTLenquiries@hotmail.com, www.moraytimberelgin.co.uk

Moresecure Ltd - Haldane House, Halesfield 1, Telford. TF7 4EH tel:(01952) 683900, fax:(01952) 683982. sales@moresecure.co.uk, www.moresecure.co.uk

Morgan Hope Industries Ltd - Units 5 and 6 Blowick Industrial Park, Crowland Street, Southport, Merseyside. PR9 7RU tel:(01704) 512000, fax:(01704) 542632. info@morganhope.com, www.morganhope.com

Morris Material Handling Ltd - PO Box 7, North Road, Loughborough, Leics. LE11 1RL tel:(01509) 643200, fax:(01509) 610666. info@morriscranes.co.uk, www.morriscranes.co.uk

Morrison Chemicals Ltd - 331-337 Derby Road, Liverpool. L20 8LQ tel:(0151) 933 0044. enquiries@morrisonsgrp.co.uk, www.morrisonsgrp.co.uk

Mortar Industry Association - Gillingham House, 38-44 Gillingham Street, London. SW1V 1HU tel:(020) 7963 8000, fax:(020) 7963 8001. brian.james@mineralproducts.org, www.mortar.org.uk

Moseley GRP Product (Division of Moseley Rubber Co Ltd) - Hoyle Street, Mancunian Way, Manchester, Gt Man. M12 6HL tel:(0161) 273 3341, fax:(0161) 274 3743. sales@moseleyrubber.com, www.moseleyrubber.com

Moss WM & Sons (Stove Anamellers Ripon) Ltd - PO Box 4, Ure Bank, Ripon, N. Yorks. HG4 1JE tel:(01765) 604351, fax:(01765) 690065. info@moseleyrubber.com, www.moseleyrubber.com

Motorised Air Products Ltd - Unit 5a, Sopwith Crescent, Wickford Business Park, Wickford, Essex. SS11 8YU tel:(01268) 574442, fax:(01268) 574443. info@mapuk.com, www.mapuk.com

Mountford Rubber & Plastics Ltd - 44 Bracebridge Street, Aston, Birmingham. B6 4PE tel:(0121) 359 0135/6, fax:(0121) 333 3204. sales@mrp.uk.net, www.mountfordrubberandplastics.com

Movement Joints (UK) - Unit 3, 57 Thorby Avenue, March, Cambs. PE15 0AR tel:(01354) 607960, fax:(01354) 607833. info@mjuk.co.uk, www.mjuk.co.uk

MPA Cement - Riverside House, 4 Meadows Business Park, Station Approach, Blackwater, Camberly, Surrey. GU17 9AB tel:(01276) 608700, fax:(01276) 608701. mpacement@mineralproducts.org, www.mineralproducts.org

MSA (Britain) Ltd - Lochard House. Linnet Way, Strathclyde Business Park, Bellshill. ML4 3RA tel:(01698) 573757, fax:(01698) 740141. info@msabritain.co.uk, www.msabritain.co.uk/msa

Mueller Europe - Oxford Street, Bilston, W. Midlands. WV14 7DS tel:(01902) 499700, fax:(01902) 405838. sales@muellereurope.com, www.muellereurope.com

Multi Marque Production Engineering Ltd - 33 Monkton Road Industrial Estate, Wakefield, W. Yorks. WF2 7AL tel:(01924) 290231, fax:(01924) 382241. enquiries@multi-marque.co.uk, www.multi-marque.co.uk

Multi Mesh Ltd - Eurolink House, Lea Green Industrial Estate, St. Helens, Merseyside. WA9 4QU tel:(01744) 820 666, fax:(01744) 821 417. enquiries@multimesh.co.uk, www.multimesh.co.uk

Multibeton Ltd - 15 Oban Court, Wickford Business Park, Wickford, Essex. SS11 8YB tel:(01268) 561688, fax:(01268) 561690. multibeton@btconnect.com

Multikwik Ltd - Dickley Lane, Lenham, Maidstone, Kent. ME17 2DE tel:(01622) 852654, fax:(01622) 852723. info@multikwik.com, www.multikwik.com

Multitex GRP LLP - Unit 5, Dolphin Industrial Estate, Southampton Road, Salisbury, Wilts. SP1 2NB tel:(01722) 332139, fax:(01722) 338458. sales@wessexbps.co.uk, www.multitex.co.uk

Mul-T-Lock - Portobello Works, Wood Street, Willenhall, W. Midlands. WV13 3PW tel:(01902) 364 200, fax:(01902) 364 201. enquiries@mul-t-lock.co.uk, www.mul-t-lock.co.uk

Muraspec - 74-78 Wood Lane End, Hemel Hempstead, Herts. HP2 4RF tel:(08705) 117118, fax:(08705) 329020. customerservices@muraspec.com, www.muraspec.com

Muswell Manufacturing Co Ltd - Unit D1, Lower Park Road, New Southgate Industrial Estate, London. N11 1QD tel:(020) 8368 8738, fax:(020) 8368 4726. sales@muswell.co.uk, www.muswell.co.uk

Myddleton Hall Lighting, Division of Peerless Design Ltd - Unit 9, Brunswick Industrial Park, Brunswick Way, London. N11 1JL tel:(020) 8362 8500, fax:(020) 8362 8525. enquiries@peerlessdesigns.com, www.peerlessdisplay.com

Myson Radiators Ltd - Eastern Avenue, Team Valley, Gateshead, Tyne & Wear. NE11 0PG tel:(0191) 491 7530, fax:(0191) 491 7568. sales@myson.co.uk, www.myson.co.uk

N

N R Burnett Ltd - West Carr Lane, Hull, Humb. HU7 0AW tel:(01482) 838 800, fax:(01482) 838 800. kevinprowse@nrburnett.co.uk, www.nrburnett.co.uk

Naco - Stourbridge Road, Bridgenorth, Shropshire. W15 5BB tel:(01746) 761921, fax:(01746) 766450. sales@naco.co.uk, www.naco.co.uk

National Access & Scaffolding Confederation (NASC) - 4th Floor, 12 Bridewell Place, London. EC4V 6AP tel:(0207) 822 7400, fax:(0207) 822 7401. enquiries@nasc.org.uk, www.nasc.org.uk

National Association of Shopfitters - NAS House, 411 Limpsfield Road, Warlingham, Surrey. CR6 9HA tel:(01883) 624961, fax:(01883) 626841. enquiries@shopfitters.org, www.shopfitters.org

National Britannia Ltd - Britannia House, Caerphilly Business Park, Van Road, Caerphilly, M. Glam. CF83 3GG tel:(029) 20 852852, fax:(029) 20 867738. enquiries@santia-foodsafety.co.uk, www.nb-group.com

National Council of Master Thatchers Associations - Foxhill, Hillside, South Brent, Devon. TQ10 9AU tel:(07000) 781909. gandewakley@btinternet.com, www.ncmta.co.uk

National Domelight Company - Pyramid House, 52 Guildford Road, Lightwater, Surrey. GU18 5SD tel:(01276) 451555, fax:(01276) 453666. info@nationaldomes.com, www.nationaldomelightcompany.co.uk

National Federation of Builders - B & CE Building, Manor Royal, Crawley, West Sussex. RH10 9QP tel:(08450) 578 160, fax:(08450) 578 161. national@builders.org.uk, www.builders.org.uk

National Federation of Demolition Contractors - Resurgam House, Paradise, The Causway, Hemel Hempstead, Hertfordshire. HP2 4TF tel:(01442) 217144, fax:(01442) 218268. info@demolition-nfdc.com, www.demolition-nfdc.com

National Federation of Master Steeplejacks & Lightning Conductor Engineers - 4d St Mary's Place, The Lace Market, Nottingham, Notts. NG1 1PH tel:(0115) 955 8818, fax:(0115) 941 2238. info@atlas.org.uk, www.nfmslce.co.uk

National Federation of Roofing Contractors Ltd., The - Roofing House, 31 Worship Street, London. EC2A 2DY tel:(020) 7638 7663, fax:(020) 7256 2125. info@nfrc.co.uk, www.nfrc.co.uk

National Federation of Terrazzo, Marble & Mosaic Specialists - PO.Box 2843, London. W1A 5PG tel:(0845) 609 0050, fax:(0845) 607 8610. donaldslade@nftmms.org, www.nftmms.org

National House Building Council - NHBC House, Davy Avenue, Knowlhill, Milton Keynes, Bucks. MK5 8FP tel:(0844) 633 1000, fax:(0908) 747 255. cssupport@nhbc.co.uk, www.nhbc.co.uk

National Inspection Council for Electrical Installation Contracting - Warwick House, Houghton Hall Park, Houghton Regis, Dunstable. LU5 5ZX tel:(0870) 013 0382, fax:(01582) 539090. enquiries@niceic.com, www.niceic.org.uk

National Insulation Association Ltd - 3 Vimy Court, Vimy Road, Leighton Buzzard. LU7 1FG tel:(08451) 636363, fax:(Fax:) 1525 854918. info@nia-uk.org, www.nia-uk.org

National Starch & Chemical Ltd - Wexham Road, Slough, Berks. SL2 5DS tel:(01753) 533494, fax:(01753) 501241. Joyce.Corbett@nstarch.com, www.nationalstarch.com

Nationwide Filter Company Ltd - Unit 16, First Quarter Business Park, Blenheim Road, Epsom, Surrey. KT19 9QN tel:(01372) 728548, fax:(01372) 742831. info@nationwidefilters.com, www.fengroup.com

Nationwide Premixed Ltd - 22 Thorne Way, Woolsbridge Industrial Park, Three Legged Cross, Wimborne, Dorset. BH21 6SP tel:(01202) 824700, fax:(01202) 827757. sales@nationwidedirect.uk.com, www.nationwidepremixed.com

Navtec - Southmoor Lane, Havent, Hants. PO9 1JJ tel:(023) 9248 5777, fax:(023) 92485770. navnor@navtec.net, www.navtec.net

Naylor Drainage Ltd - Clough Green, Cawthorne, Nr Barnsley, S. Yorks. S75 4AD tel:(01226) 790591, fax:(01226) 790531. info@naylor.co.uk, www.naylor.co.uk

NBS - The Old Post Office, St. Nicholas Street, Newcastle-upon-Tyne. NE1 1RH tel:(0191) 244 5500, fax:(0191) 232 5714. info@thenbs.com, www.thenbs.com

NCMP Ltd - Central Way, Feltham, Middx. TW14 0XJ tel:(020) 8844 0940, fax:(020) 8751 5793. www.cmf.co.uk/ncmp.asp

Nederman UK - PO Box 503, 91 Walton Summit, Bamber Bridge, Preston. PR5 8AF tel:(01772) 334721, fax:(01772) 315273. info@nederman.co.uk, www.nederman.co.uk

Nendle Acoustic Company Ltd - Tacitus House, 153 High Street, Aldershot, Hants. GU11 1TT tel:(01252) 344222, fax:(01252) 333782. sales@nendle.co.uk, www.nendle.co.uk

Nenplas - Airfield Industrial Estate, Ashbourne, Derbys. DE6 1HA tel:(01335) 347300, fax:(01335) 340271. enquiries@homeluxnenplas.com, www.nenplas.co.uk

Neocrylic Signs Ltd - Unit 7, Mather Road, Eccles, Gt Man. M30 0WQ tel:(0161) 707 8933, fax:(0161) 707 8934. sales@manchester-signs.com, www.emlgroup.co.uk

Neopost Ltd - Neopost House, South Street, Romford, Essex. RM1 2AR tel:(01708) 746000, fax:(01708) 714050. mktg@neopost.co.uk, www.neopost.co.uk

Neptune Outdoor Furniture Ltd - Thompsons Lane, Marwell, Nr Winchester, Hants. SO21 1JH tel:(01962) 777799, fax:(01962) 777723. sales@nofl.co.uk, www.nofl.co.uk

Neslo Interiors - 10 Woodway Court, Thursby Road, Wirral International Business Park, Bromborough, Wirral, Merseyside. CH62 3PR tel:(0151) 334 9326, fax:(0151) 334 0668. steve.lamb@neslointeriors.co.uk, www.neslopartitioning.co.uk

Ness Furniture Ltd - Croxdale, Durham. DH6 5HT tel:(01388) 816109, fax:(01388) 812416. sales@nessfurniture.co.uk, www.nessfurniture.co.uk

NET LED Limited - 18-21 Evolution Business Park, Milton Road, Impington, Cambridge. CB24 9NG tel:(01223) 851505, fax:(01223) 851506. jack@netled.co.uk, www.netled.co.uk

Nettlefolds Ltd - Unit 7, Harworth Enterprise Park, Brunel Close, Harworth, Doncaster, Yorkshire. DN11 8SG tel:(01302) 759555. info@nettlefolds.com

Nevill Long Limited - Chartwell Drive, West Avenue, Wigston, Leicester, Leics. LE18 2FL tel:(0116) 2570670, fax:(0116) 2570044. info@nevilllong.co.uk, www.nevilllong.co.uk

New Forest Ceilings Ltd - 61-65 High Street, Totton, Southampton. SO40 9HL tel:(023) 80869510, fax:(023) 80862244. nfcl@btconnect.com, www.newforestceilings.co.uk

New Haden Pumps Ltd - Draycott Cross Road, Cheadle, Stoke on Trent, Staffs. ST10 2NW tel:(01538) 757900, fax:(01538) 757999. info@nhpumps.com, www.nhpumps.com

Neway Doors Ltd - 89-91 Rolfe Street, Smethwick, Warley, W. Midlands. B66 2AY tel:(0121) 558 6406, fax:(0121) 558 7140. sales@priory-group.com

Newdawn & Sun Ltd - Springfield Business Park, Arden Forest Inustrial Estate, Alcester, Warwicks. B49 6EY tel:(01789) 764444, fax:(01789) 400164. sales@livingspaceltd.co.uk, www.newdawn-sun.co.uk

Newey Ceilings - 1-4, South Uxbridge St, Burton-on-Trent, Staffs. DE14 3LD tel:(01865) 337200, fax:(01865) 337201. info@neweyceilings.co.uk, www.neweyceilings.co.uk

NewTeam Ltd - Brunel Road, Earlstrees Industrial Estate, Corby, Northants. NN17 4JW tel:(01536) 409222, fax:(01536) 400144. cjennings@bristan.com, www.newteamshowers.com

Newton & Frost Fencing Ltd - Downsview Yard, North Corner, Horam, Heathfield, East Sussex. TN21 9HJ tel:(01435) 813 535, fax:(01435) 813 687. enquiries@nfltd.co.uk, www.nfltd.co.uk

Newton John & Co Ltd - 12 Verney Road, London. SE16 3DH tel:(020) 7237 1217, fax:(020) 7252 2769. info@newton-membranes.co.uk, www.newton-membranes.co.uk

Nexans UK Ltd - Nexans House, Chesney Wold, Bleak Hall, Milton Keynes. MK6 1LA tel:(01908) 250850, fax:(01908) 250851. sales.uk@nexans.com, www.nexans.co.uk

N (con't)

NGF - MACHINED TIMBER SOLUTIONS LTD - Stoneyfield, Roch Valley Way, Rochdale, Lancashire. OL11 4PZ tel:(01706) 657167. neil@ngfmtsl.com, www.ngfmtsl.com

Nicholls & Clarke Ltd - Niclar House, Shoreditch High Street, London. E1 6PE tel:(020) 7247 5432, fax:(020) 7247 7738. info@nichollsandclarke.com, www.nichollsandclarke.com/

Nicholson Plastics Ltd - Riverside Road, Kirkfieldbank, Lanarkshire. ML11 9JS tel:(01555) 664316, fax:(01555) 663056. sales@nicholsonplastics.co.uk, www.nicholsonplastics.co.uk

Nico Manufacturing Ltd - 109 Oxford Road, Clacton on Sea, Essex. CO15 3TJ tel:(01255) 422333, fax:(01255) 432909. sales@nico.co.uk, www.nico.co.uk

Nilfisk Advance Ltd - Nilfisk House, Bowerbank Way, Gilwilly Ind Est, Penrith, Cumbria. IP33 3SR tel:(01768) 868995, fax:(01768) 864713. mail.uk@nilfisk.com, www.nilfisk.co.uk

Nilfisk Limited - Bowerbank Way, Gilwilly Industrial Estate, Penrith, Cumbria. CA11 9BN tel:(01768) 868995, fax:(01768) 864713. mail.uk@nilfisk.com, www.nilfisk.co.uk

Nimlok Ltd - Booth Drive, Park Farm, Wellingborough, Northants. NN8 6NL tel:(01933) 409409, fax:(01933) 409451. info@nimlok.co.uk, www.nimlok.co.uk

Nittan (UK) Ltd - Hipley Street, Old Woking, Surrey. GU22 9LQ tel:(01483) 769555, fax:(01483) 756686. sales@nittan.co.uk, www.nittan.co.uk

NMC (UK) Ltd - Unit B, Tafarnaubach Industrial Estate, Tredegar, S. Wales. NP22 3AA tel:(01495) 713266, fax:(01495) 713277. enquiries@nmc-uk.com, www.nmc-uk.com

Noral Scandinavia AB - Hultsfredsvägen 41, 598 40 VIMMERBY, Sweden. www.noral.se/?lang=eng

Norbord - Station Road, Cowie, Sterlingshire. FK7 7BQ tel:(01786) 812921, fax:(01786) 817143. info@norbord.com, www.norbord.com

North Herts Asphalte (Roofing) Ltd - 70 Hawthorn Hill, Letchworth Garden City, Herts. SG6 4HQ tel:(01462) 434 877, fax:(01462) 421539. nhasphalte@gmail.com, www.northherts-asphalteroofing.co.uk

Northcot Brick Ltd - Blockley, Nr Moreton in Marsh, Glos. GL56 9LH tel:(01386) 700551, fax:(01386) 700852. info@northcotbrick.co.uk, www.northcotbrick.co.uk

Northern Fencing Contractors Limited - Parkhill, Walton Road, Wetherby, W. Yorks. LS22 5DZ tel:(01937) 545155, fax:(01937) 580034. tmorgan@holroydconstruction.co.uk

Northern Joinery Ltd - Daniel Street, Whitworth, Rochdale, Lancs. OL12 8DA tel:(01706) 852345, fax:(01706) 853114. office@northernjoinery.co.uk, www.northernjoinery.co.uk

Northgate Solar Controls - PO Box 200, Barnet, Herts. EN4 9EW tel:(020) 8441 4545, fax:(020) 8441 4888. enquiries@northgateuk.com, www.northgateuk.com

Northwest Pre-cast Ltd - Holmefield Works, Garstang Road, Pilling, Preston. PR3 6AN tel:(01253) 790444, fax:(01253) 790085. nwprecast@talk21.com

Norton Diamond Products - Unit 2, Meridian West, Meridian Business Park, Leicester, Leics. LE3 2WX tel:(0116) 263 2302, fax:(0116) 282 7292. sales.nlx@saint-gobain.com, www.norton-diamond.com

Norton Engineering Alloys Co Ltd - Norton Grove Industrial Estate, Norton, Malton, N. Yorks. YO17 9HQ tel:(01653) 695721, fax:(01653) 600418. sales@neaco.co.uk, www.neaco.co.uk

Norwegian Log Buildings Ltd - 230 London Road, Reading, Berks. RG6 1AH tel:(0118) 9669236, fax:(0118) 9660456. sales@norwegianlog.co.uk, www.norwegianlog.co.uk

Norwood Partition Systems Ltd - Mallard Court, Mallard Way, Crewe Business Park, Crewe. CW1 6ZQ tel:(0870) 240 6405, fax:(0870) 240 6407. sales@norwood.co.uk, www.norwood.co.uk

Notcutt Ltd - Homeward Farm, Newark Lane, Ripley, Surrey. GU23 6DJ tel:(020) 8977 2252, fax:(020) 8977 6423. sales@notcutt.com, www.notcutt.com

Nova Garden Furniture Ltd - The Faversham Group, Graveney Road, Faversham, Kent. ME13 8UN tel:(01795) 535511, fax:(01795) 539215. sales@novagardenfurniture.co.uk, www.novagardenfurniture.co.uk

Nova Group Ltd - Norman Road, Board Heath, Altrincham, Cheshire. WA14 4EN tel:(0161) 941 5174, fax:(0161) 926 8405. sales@novagroup.co.uk, www.windows-conservatories.co.uk

Novaglaze Limited - Queens Mill Road, Lockwood, Huddersfield, W. Yorks. HD1 3PG tel:(01484) 517 010, fax:(01484) 517 050. sales@novaglaze.co.uk, www.novaglaze.co.uk

NovaSpan Structures - Millbuck Way, Springvale Industrial Estate, Sanobach, Cheshire. CW11 3HT tel:(01270) 768500, fax:(01270) 753330

Novoferm Europe Ltd - Brook Park, Epsom Avenue, Handforth Dean, Wilmslow. SK9 3RN tel:(0161) 4862700, fax:(0161) 4862701. info@novoferm.co.uk, www.novoferm.co.uk

NRG Fabrications - Harlestone Firs, Harlestone Road, Northampton, Northants. NN5 6UJ tel:(01604) 580022, fax:(01604) 580033. nrg@cemfencing.co.uk

NuAir Ltd - Western Industrial Estate, Caerphilly, M. Glam. CF83 1XH tel:(029) 20 885911, fax:(029) 20 887033. info@nuairegroup.com, www.nuaire.co.uk

NuLite Ltd - Unit 51 Hutton Close, Crownther Industrial Estate, District 3, Washington, Tyne & Wear. NE38 0AH tel:(0191) 419 1111, fax:(0191) 419 1123. sales@nulite-ltd.co.uk, www.nulite-ltd.co.uk

Nullifire Ltd - Torrington Avenue, Coventry. CV4 9TJ tel:(024) 7685 5000, fax:(024) 7646 9547. protect@nullifire.com, www.nullifire.com

Nu-Swift International Ltd - Elland, W. Yorks. HX5 9DS tel:(01422) 372852, fax:(01422) 379569. customer.service@nuswift.co.uk, www.nu-swift.co.uk

Nuttall, Henry (Viscount Catering Ltd) - Green Lane, Ecclesfield, Sheffield, S. Yorks. S35 9ZY tel:(0114) 257 0100, fax:(0114) 257 0251. fsuk.info@manitowoc.com, www.viscount-catering.co.uk

Nu-Way Ltd - Ten Acres, Berry Hill Industrial Estate, Droitwich, Worcestershire. WR9 9AQ tel:(01905) 794331, fax:(01905) 794017. info@nu-way.co.uk, www.nu-way.co.uk/

O

Oakdale (Contracts) Ltd - Walkerville Industrial Estate, Catterick Garrison, N. Yorks. DL9 4SA tel:(01748) 834184, fax:(01748) 833003. sales@oakdalecontracts.co.uk, www.oakdalecontracts.co.uk

Oakleaf Reproductions Ltd - Unit A Melbourn Mills, Chesham Street, Keighley, W. Yorks. BD21 4LG tel:(01535) 663274, fax:(01535) 661951. sales@oakleaf.co.uk, www.oakleaf.co.uk

O'Brian Manufacturing Ltd - Robian Way, Swadlincote, Derbys. DE11 9DH tel:(01283) 217588, fax:(01283) 215613

Oce (UK) Ltd - Oce House, Chatham Way, Brentwood, Essex. CM14 4DZ tel:(0870) 600 5544, fax:(0870) 600 1113. salesinformation@oce.co.uk, www.oce.com/uk

O'Connor Fencing Limited - Whitehaven Commercial Park, Moresby Parks, Whitehaven, Cumbria. CA28 8YD tel:(01946) 693983, fax:(01946) 693984. enquiries@oconorfencing.co.uk, http://oconnorfencing.co.uk

OEP Furniture Group PLC - 7-9 Cartersfield Road, Waltham Abbey, Essex. EN9 1JD tel:(01992) 767014, fax:(01992) 762968. enquiries@oepfurniture.co.uk, www.oepfurniture.co.uk

Ogilvie Construction Ltd - Pirnhall Works, 200 Glasgow Road, Whins of Milton, Stirling, Stirlingshire. FK7 8ES tel:(01786) 812273, fax:(01786) 816287. enq@ogilvie.co.uk

Ogley Bros Ltd - Allen Street, Smithfield, Sheffield. S3 7AS tel:(0114) 276 8948, fax:(0114) 275 8948

Olby H E & Co Ltd - 229-313 Lewisham High Street, London. SE13 6NW tel:(020) 8690 3401, fax:(020) 8690 1408. mark@heolby.co.uk

Oldham Lighting Ltd - Claudgen House, Eastwick Road, Leatherhead, Surrey. KT23 4DT tel:(020) 8946 5555, fax:(020) 8946 5522. sales@oldhamlighting.co.uk, www.oldhamlighting.com

Olivand Metal Windows Ltd - Chesley House, 43a Chesley Gardens, London. E6 3LN tel:(020) 8471 8111, fax:(020) 8552 7015. Info@olivand-steel-windows.co.uk, www.olivand-steel-windows.co.uk

Olley & Sons Ltd, C - Iberia House, 36, Southgate Avenue, Mildenhall, Suffolk. IP28 7AT tel:(01638) 712076, fax:(01638) 717304. olley-cork@btconnect.com, www.olleycork.co.uk

Olsen Doors & Windows Ltd - Unit 25, British Fields, Tuxford, Newark, Notts. NG22 0PQ tel:(01777) 874510, fax:(01777) 874519. enquiries@olsenuk.com, www.olsenuk.com

Omeg Ltd - Imberhorne Industrial Estate, East Grinstead, W. Sussex. RH19 1RJ tel:(01342) 410420, fax:(01342) 316253. sales@omeg.com, www.omeg.co.uk

Omnipex Ltd - Sienna Court, The Broadway, Maidenhead, Berks. SL6 1NJ tel:(0845) 6449051, fax:(0845) 6449052. info@plants4business.com, www.omnipex.com

Omnis - Unit 62, Blackpole Trading Estate Wes, Worcester. WR3 8JZ tel:(01905) 750500. enquiries@omnisexteriors.com, www.omnisexteriors.com

Onduline Building Products Ltd - Eardley House, 182-184 Campden Hill Road, Kensington, London. W8 7AS tel:(020) 7727 0533, fax:(020) 7792 1390. enquiries@onduline.net, www.onduline.co.uk

Opella Ltd - Twyford Road, Rotherwas Industrial Estate, Hereford, Herefordshire. HR2 6JR tel:(01432) 357331, fax:(01432) 264014. sales@opella.co.uk, www.opella.co.uk

Optelma Lighting Ltd - 14 Napier Court, The Science Park, Abingdon, Oxon. OX14 3NB tel:(01235) 553769, fax:(01235) 523005. sales@optelma.co.uk, www.optelma.co.uk

Optex (Europe) Ltd - Clivemont Road, Cordwallis Park, Maidenhead, Berks. SL6 7BU tel:(01628) 631000, fax:(01628) 636311. sales@optex-europe.com, www.optex-europe.com

Optima - Courtyard House, West End Road, High Wycombe, Buckinghamshire. HP11 2QB tel:(01494) 492600, fax:(01494) 492800. action@optima-group.co.uk, www.optimasystems.com

Opus 4 Ltd - Orchard House, Orchard Business Centre, North Farm Road, Tunbridge Wells, Kent. TN2 3DY tel:(01892) 515157, fax:(01892) 515417. hello@opus-4.com, www.opus-4.com

Orbik Electrics Ltd - Orbik House, Northgate Way, Aldridge, Walsall. WS9 8TX tel:(01922) 743515, fax:(01922) 743173. uksales@orbik.co.uk, www.orbik.co.uk

Orchard Street Furniture - 119 The Street, Crowmarsh & Gifford, Nr. Waringford, Oxon. OX10 8EF tel:(01491) 642123, fax:(01491) 642126. sales@orchardstreet.co.uk, www.orchardstreet.co.uk

Ordnance Survey - Explorer House, Adanac Drive, Southampton. SO16 0AS tel:(03456) 050505, fax:(03450) 990494. customerservices@ordnancesurvey.co.uk, www.ordnancesurvey.co.uk

OSF Ltd - Unit 6, The Four Ashes Industrial Estate, Station Road, Four Ashes, Wolverhampton, W. Midlands. WV10 7DB tel:(01902) 798080, fax:(01902) 794750. www.osfltd.com/

OSMO UK - Unit 24, Anglo Business Park, Smeaton Close, Aylesbury, Bucks. HP19 8UP tel:(01296) 481220, fax:(01296) 424090. info@osmouk.com, www.osmouk.com

Osram Ltd - Customer Service Centre, Neills Road, Bold Industrial Park, St. Helens. WA9 4XG tel:(01744) 812221, fax:(01744) 831900. csc@osram.co.uk, www.osram.co.uk

Otis Ltd - Chiswick Park Building 5 Ground Floor, 566 Chiswick High Road, London. W4 5YF tel:(020) 8495 7750, fax:(020) 8495 7751. contact@otis.com, www.otis.com/site/gb/

Outokumpu - Europa Link, Sheffield, S. Yorks. S9 1TZ tel:(01142) 613701, fax:(0114) 243 1277. PSCSales.Sheffield@outokumpu.com, ww.outokumpu.com

OWA UK Ltd - 10 Perth Trading Estate, Perth Avenue, Slough, Berkshire. SL1 4XX tel:(01753) 552489. sales@owa-ceilings.co.uk, www.owa-ceilings.co.uk

Owen Slate Services Ltd - 2 Tanysgrafell, Coed y Parc, Bethesda, Gwynedd. LL57 4AJ tel:(01248) 605575, fax:(01248) 605574. Sales@owens-slate.com, www.owens-slate.com

Owens Corning Automotive (uk) Ltd - Lune Ind Est, Lancaster, Lancs. LA1 5QP tel:(01524) 591700. chris.balmer@owenscorning.com, www.owenscorning.com

Owens Slate Services Ltd - 2 Tanysgrafell, Coed y Parc, Bethesda, Gwynedd. LL57 4AJ tel:(01248) 605575, fax:(01248) 605574. sales@owens-slate.com, www.owens-slate.com

Owlett - Jaton - Opal Way, Stone Business Park, Stone, Staffs. ST15 0SW tel:(01785) 811300, fax:(01785) 819699. info@owlett-jaton.com, www.owlett-jaton.com

Oxford Double Glazing Ltd - Ferry Hinksey Road, Oxford. OX2 0BY tel:(01865) 248287, fax:(01865) 251070. info@oxforddoubleglazing.com, www.oxforddoubleglazing.com

P

P+L Systems Ltd - Sterling House, Grimbald Crag Close, Knaresborough, N. Yorks. HG5 9PJ tel:(0800) 988 5359, fax:(01423) 863497. info@pandlsystems.com, www.pandlsystems.com

P4 Limited - 1 Wymans Way, Fakenham, Norfolk. NR21 8NT tel:(01328) 850 555, fax:(01328) 850 559. info@p4fastel.co.uk, www.p4fastel.co.uk

Pac International Ltd - 1 Park Gate Close, Bredbury, Stockport, Gt Man. SK6 2SZ tel:(0161) 494 1331, fax:(0161) 430 9658. customerservices@stanleysecurityproducts.com, www.pac.co.uk

Package Products Ltd - Little Green Works, Collyhurst Road, Manchester. M40 7RT tel:(0161) 205 4181, fax:(0161) 203 4678. sales@packagingproducts.co.uk, www.packagingproducts.co.uk

Paddock Fabrications Ltd - Fryer Road, Walsall, Bloxwich, W. Midlands. WS2 7NF tel:(01922) 711722, fax:(01922) 476021. sales@paddockfabrications.co.uk, www.paddockfabrications.com

Painting & Decorating Association - 32 Coton Road, Nuneaton, Warwicks. CV11 5TW tel:(024) 7635 3776, fax:(024) 7635 4513. info@paintingdecoratingassociation.co.uk, www.paintingdecoratingassociation.co.uk

PAL Extrusions - Darlaston Road, Wednesbury, W. Midlands. WS10 7TN tel:(0121) 526 4048, fax:(0121) 526 4658. sales@palextrusions.co.uk, www.palextrusions.co.uk

Palladio Stone - Station House, Station Road, Tisbury, Salisbury. SP3 6JT tel:(01747) 871546, fax:(01747) 871547. admin@palladiostone.com, www.palladiostone.com

Panasonic UK Ltd - Panasonic House, Willoughby Road, Bracknell, Berks. RG12 8FP tel:(01344) 862444, fax:(01344) 853704. system.solutions@eu.panasonic.com, www.panasonic.co.uk

Panel & Louvre Co Ltd - 1A Bridle Close, Finedon Road Industrial Estate, Wellingborough, Northants. NN8 4RN tel:(01933) 228500, fax:(01933) 228580. info@palcouk.com, www.ascwebindex.com/palco

Panel Agency Limited - Maple House, 5 Over Minnis, New Ash Green, Kent. DA3 8JA tel:(01474) 872578, fax:(01474) 872426. sales@panelagency.com, www.panelagency.com

Panel Systems Ltd - 3-9 Welland Close, Parkwood Industrial Estate, Rutland Road, Sheffield, S. Yorks. S3 9QY tel:(0114) 275 2881, fax:(0114) 278 6840. sales@panelsystems.co.uk, www.panelsystemsgroup.co.uk

Paragon by Heckmondwike - Farfield Park, Manvers, Wath-Upon-Dearne, Rotherham. S63 5DB tel:(0845) 601 0431, fax:(0800) 731 4521. sales@paragon-carpets.co.uk, www.pbyh.co.uk

Parker Plant Ltd - PO Box 146, Canon Street, Leicester. LE4 6HD tel:(0116) 266 5999, fax:(0116) 268 1254. sales@parkerplant.com, www.parkerplant.com

Parklines Building Ltd - Gala House, 3 Raglan Road, Edgbaston, Birmingham. B5 7RA tel:(0121) 446 6030, fax:(0121) 446 5991. sales@parklines.co.uk, www.parklines.co.uk

Paroc Panel Systems Uk Ltd - Stoney Lane, Rainhill, Prescot, Merseyside. L35 9LL tel:(0151) 4266555, fax:(0151) 426 6622. technical.insulation@paroc.com, www.paroc.com

Parsons Brothers Gates - 5 Prior Wharf, Harris Business Park, Bromsgrove, W. Midlands. B60 4FG tel:(01527) 576355, fax:(01527) 579419. gates@parsonsbrothers.com, www.parsonsbrothers.com

Parthos UK Ltd - 1 The Quadrant, Howarth Road, Maidenhead, Berks. SL6 1AP tel:(01628) 773 353, fax:(01628) 773 363. info@parthos.co.uk, www.parthos.co.uk

Parton Fibreglass Ltd - PFG House, Claymore, Tame Valley Industrial Estate, Tamworth, Staffs. B77 5DQ tel:(01827) 251899, fax:(01827) 261390. sales@pfg-tanks.com, www.pfg-tanks.com

Parts Centre - The Wolseley Center, Harrison Way, Spa Park, Royal Leamington Spa. CV31 3HH tel:(01282) 834400. customerservices@wolseley.co.uk, www.partscenter.co.uk

Paslode - Diamond Point, Fleming Way, Crawley, West Sussex. RH10 9DP tel:(0800) 652 9260. technical@itwcp.co.uk, www.paslode.co.uk

Passivent Ltd - 2 Brooklands Road, Sale, Cheshire, Notts. M33 3SS tel:(0161) 962 7113, fax:(0161) 905 2085. info@passivent.com, www.passivent.com

PCE Instruments UK Ltd - Units 12/13 Southpoint Business Park, Ensign Way, Southampton, Hampshire. SO31 4RF tel:(020) 8089 7035. info@industrial-needs.com, www.pce-instruments.com/english/

PCT Group - 37 Dalsetter Avenue, Glasgow. G15 8TE tel:(0141) 944 4000, fax:(0141) 944 9000. sales@pctgroup.co.uk, www.pctgroup.co.uk

Peal furniture (Durham) Ltd - Littleburn Industrial Estate, Langley Moor, Co. Durham. DH7 8HE tel:(0191) 378 0232, fax:(0191) 378 1660. sales@godfreysyrett.co.uk, www.godfreysyrett.co.uk

Pearce Construction (Barnstaple) Limited - Pearce House, Brannam Crescent, Roundswell Business Park, Barnstaple, Devon. EX31 3TD tel:(01271) 345261, fax:(01271) 852134. buildit@pearceb.co.uk, www.pearcebarnstaple.co.uk

Pearce Security Systems - Ensign House, Green Lane, Felling, Gateshead, Tyne & Wear. NE10 0QH tel:(0191) 438 1177, fax:(0191) 495 0227. sales@ensig.co.uk, www.ensig.co.uk

Pearce Signs Ltd - Unit 28, Regent Trade Park, Barnwell Lane, Gosport, Hampshire. PO13 0EQ tel:(01329) 238015. info@pearcesigns.com, www.pearcegroup.com

Pearman Fencing - Greystone Yard, Notting Hill Way, Lower Weare, Axbridge, Somerset. BS26 2JU tel:(01934) 733380, fax:(01934) 733398. pearmanfencing@btconnect.com, www.premier-pearman.co.uk

Pedley Furniture International Ltd - Shirehill Works, Saffron Walden, Essex. CB11 3AL tel:(01799) 522461, fax:(01799) 513403. sales@pedley.com, www.pedley.com

Peek Traffic Ltd - Hazlewood House, Lime Tree Way, Chineham Business Park, Basingstoke, Hants. RG24 8WZ tel:(01256) 891800, fax:(01256) 891870. www.peek-traffic.co.uk

Pegler Ltd - St Cathrines Avenue, Doncaster, S. Yorks. DN4 8DF tel:(01302) 560560, fax:(01302) 367661. uk.sales@pegleryorkshire.co.uk, www.pegler.co.uk

Pegson Ltd - Coalville, Leics. LE67 3GN tel:(01530) 510051, fax:(01530) 510041. sales@bl-pegson.com, www.powerscreen.com/en

PEL Services Limited - Belvue Business Centre, Belvue Road, Northolt, Middx. UB5 5QQ tel:(020) 8839 2100, fax:(020) 8841 1948. pel@pel.co.uk, www.pel.co.uk

P (con't)

Pembury Fencing Ltd - Unit 2 Church Farm, Collier Street, Marden, Tonbridge, Kent. TN12 9RT tel:(0870) 242 3707. sonia@pemburygroup.co.uk, www.pemburygroup.co.uk
Pentair Thermal Management - 3 Rutherford Road, Stephenson Industrial Estate, Washington, Tyne & Wear. NE37 3HX tel:(0800) 969013, fax:(0800) 968624. sales@pentairthermal.co.uk, www.pentairthermal.co.uk
Perform Construction (Essex) Ltd - 10 Warren Lane, Stanway, Colchester. CO3 0LW tel:(01206) 330901, fax:(01206) 330023. christine.burrows@performconstruction.com, www.performconstruction.com
PERI Ltd - Market Harborough Road, Clifton Upon Dunsmore, Rugby, Warwicks. CV23 0AN tel:(01788) 861600, fax:(01788) 861610. info@peri.ltd.uk, www.peri.ltd.uk
Permali Deho Ltd - Unit 4, Permali Park, Bristol Road, Gloucester, Glos. GL1 5SR tel:(01452) 411 607, fax:(01452) 411 617. sales@permalideho.co.uk, www.permalideho.co.uk
Perman Briggs Ltd - 224 Cheltenham Road, Longlevens, Gloucester. GL2 0JW tel:(01452) 524192, fax:(01452) 309879. sales@pearmanbriggs.co.uk, www.pearmanbriggs.co.uk
Permanoid Ltd - Hulme Hall Road, Manchester. M10 8HH tel:(0161) 205 6161, fax:(0161) 205 9325. sales@permanoid.co.uk, www.permanoid.co.uk
PermaRock Products Ltd - Jubilee Drive, Loughborough, Leics. LE11 5TW tel:(01509) 262924, fax:(01509) 230063. permarock@permarock.com, www.permarock.com
PermaSeal Ltd - Unit 20 Cloverlay Ind. Park, Cantebury Lane, Rainham, Kent. ME8 8GL tel:(0845) 838 6394, fax:(01634) 260701. info@permasealroofing.co.uk, www.permasealroofing.co.uk
Petal Postforming Limited - Gibbs House, Kennel Ride, Ascot, Berkshire. SL5 7NT tel:(01344) 893990, fax:(01344) 893 930. info@petal.co.uk, www.petal.co.uk
Peter Cox Ltd - Aniseed Park, Broadway Business Park, Chadderton, Manchester. OI9 9XA tel:(0800) 030 4701, fax:(020) 8642 0677. headoffice@petercox.com, www.petercox.com
Petrochem Carless - Cedar Court, Guildford Road, Leatherhead, Surrey. KT22 9RX tel:(01372) 360 000, fax:(01372) 380 400. cpatronas@h-c-s-group.com, www.petrochemcarless.com
PFC Corofil - Units 3 & 4 King George Trading Estate, Davis Road, Chessington, Surrey. KT9 1TT tel:(0208) 391 0533, fax:(0208) 391 2723. ndifato@pfc-corofil.net, www.pfc-corofil.co.uk
PFP Electrical Products Ltd - Fortnum Close, Mackadown Lane, Kitts Green, Birmingham. B33 0LB tel:(0121) 783 7161, fax:(0121) 783 5717. webenquiries@petrel-ex.co.uk, www.pfp-elec.co.uk
PHI Group Ltd - Hadley House, Bayshill Road, Cheltenham, Glos. GL50 3AW tel:(0870) 333 4120, fax:(0870) 333 4121. info@phigroup.co.uk, www.phigroup.co.uk
Philips Lighting - Philips Centre, Guildford Business Park, Guildford, Surrey. GU2 8XH tel:(08706) 010 101, fax:(01483) 575534. lighting.uk@philips.com, www.lighting.philips.com
Philips Lighting Solutions - Philips Centre, Guildford Business Park, Guildford, Surrey. GU2 8XH tel:(020) 8751 6514, fax:(020) 8890 438. lighting.uk@philips.com, www.ecs-control.co.uk
Phoenix Engineering Co Ltd, The - Phoenix Works, Combe Street, Chard, Somerset. TA20 1JE tel:(01460) 63531, fax:(01460) 67388. sales@phoenixeng.co.uk, www.phoenixeng.co.uk
Phoenix Lifting Systems Ltd - Unit 5B/C Castlegate Business Park, Old Sarum, Salisbury, Wiltshire. SP4 6QX tel:(01722) 410144, fax:(01722) 331814. sales@phoenix.co.uk, www.phoenixlifts.co.uk
Phoenix Scales Ltd - 34 Oldbury Road, West Bromwich, West Midlands. B70 9ED tel:(0845) 601 7464, fax:(0845) 602 4205. sales@phoenixscales.co.uk, www.phoenixscales.co.uk
Photain Controls (part of GE security) - 8 Newmarket Court, Chippenham Drive, Kingston, Milton Keynes, Buckinghamshire. MK10 0AQ tel:(01908) 281981
Photofabrication Limited - 14 Cromwell Road, St Neots, Cambs. PE19 2HP tel:(01480) 475831, fax:(01480) 475801. sales@photofab.co.uk, www.photofab.co.uk
PHS Group plc - Block B, Western Industrial Estate, Caerphilly. CF83 1XH tel:(02920) 851 000, fax:(02920) 863 288. Enquiries@phs.co.uk, www.phs.co.uk
Pickerings Ltd - Globe Elevator Works, PO Box 19, Stoke-on-Tees, Cleveland. TS20 2AD tel:(01642) 607161, fax:(01642) 677638. info@pickeringslifts.co.uk, www.pickerings.co.uk
Pickersgill Kaye Ltd - Pepper Road, Leeds. LS10 2PP tel:(0113) 277 5531, fax:(0113) 276 0221. enquiries@pkaye.co.uk, www.pkaye.co.uk
Pifco Ltd - Failsworth, Manchester. M35 0HS tel:(0161) 681 8321, fax:(0161) 682 1708. service@russellhobbs.com, www.saltoneurope.com
Pike Signals Ltd - Equipment Works, Alma Street, Birmingham, W. Midlands. B19 2RS tel:(0121) 359 4034, fax:(0121) 333 3167. enquiries@pikesignals.com, www.pikesignals.com
Pilkington Birmingham - Nechells Park Road, Nechells, Birmingham. B7 5NQ tel:(0121) 326 5300, fax:(0121) 328 4277. pilkington@respond.uk.com, www.Pilkington.com
Pilkington Plyglass - Cotes Park, Somercotes, Derbys. DE55 4PL tel:(01773) 520000, fax:(01773) 520052. pilkington@respond.uk.com, www.pilkington.com
Pilkington UK Ltd - Prescot Road, St. Helens, Merseyside. WA10 3TT tel:(01744) 692000, fax:(01744) 613049. pilkington@respond.uk.com, www.pilkington.co.uk
Pims Pumps Ltd - 22 Invincible Rd Ind Est, Farnborough, Hants. GU14 7QU tel:(01252) 513366, fax:(01252) 516404. sales@pimsgroup.co.uk, www.pimsgroup.co.uk
Pipe Centre - The Wolseley Center, Harrison Way, Spa Park, Royal Leamington Spa. CV31 3HH tel:(08701) 600909. customerservices@wolseley.co.uk, www.pipecenter.co.uk
Pitacs Ltd - Bradbourne Point, Bradbourne Drive, Tillbrook, Milton Keynes, Buckinghamshire. MK7 8AT tel:(01908) 271155, fax:(01908) 640017. info@pitacs.com, www.pitacs.com
Pitchmastic PmB Ltd - Panama House, 184 Attercliffe Road, Sheffield, S. Yorks. S4 7WZ tel:(0114) 270 0100, fax:(0114) 276 8782. info@pitchmasticpmb.co.uk, www.pitchmasticpmb.co.uk
Pittsburg Corning (UK) Ltd - 31-35 Kirby Street, Hatton Garden, London. EC1N 8TE tel:(020) 7492 1731, fax:(020) 7492 1730. info@foamglas.co.uk, www.foamglas.co.uk
Pland - Lower Wortley Ring Road, Leeds. LS12 6AA tel:(0113) 263 4184, fax:(0113) 231 0560. sales@plandstainless.co.uk, www.plandstainless.co.uk
Pland Stainless Ltd - Lower Wortley Ring Road, Leeds, W. Yorks. LS12 6AA tel:(0113) 263 4184, fax:(0113) 231 0560. sales@plandstainless.co.uk, www.plandstainless.co.uk
Planet Communications Group Ltd - Partland Tower, Partland Street, Manchester, Lancs. M1 3LD tel:(01753) 807777, fax:(01753) 807700. info@planet.co.uk, www.planet.co.uk
Plansee Tizit (UK) Ltd - Grappenhall, Warrington, Cheshire. WA4 3JX tel:(01925) 261161, fax:(01925) 267933. uk@plansee.com, www.plansee.com
Plascoat Systems Ltd - Trading Estate, Farnham, Surrey. GU90 0NY tel:(01252) 733777, fax:(01252) 721250. sales@plascoat.com, www.plascoat.com
Plasmor - PO Box 44, Womersley Road, Knottingley, W. Yorks. WF11 0DN tel:(01977) 673221, fax:(01977) 607071. knott@plasmor.co.uk, www.plasmor.co.uk
Plastal (SBP Ltd) - Alders Way, Paignton, Devon. TQ4 7QE tel:(01952) 205000, fax:(01952) 290956. estimating@plastal.co.uk, www.plastal.co.uk
Plastic Coatings Ltd - Woodbridge Meadow, Guildford, Surrey. GU1 1BG tel:(01483) 531155, fax:(01483) 533534. enquiries@plastic-coatings.com, www.plasticcoatings.co.uk
Plastic Extruders Ltd - Russell Gardens, Wickford, Essex. SS11 8DN tel:(01268) 735231, fax:(01268) 560027. sales@plastex.co.uk, www.plastex.co.uk
Plasticable Ltd - Unit 3 Riverwey Ind Est, Newmen Lane, Alton, Hants. GU34 2QL tel:(01252) 541385, fax:(01252) 373816. sales@plasticable.co.uk, www.plasticable.co.uk
Platform Lift Company Ltd, The - Millside House, Anton Business Park, Andover, Hampshire. SP10 2RW tel:(01256) 896000. info@platformliftco.co.uk, www.platformliftco.co.uk/
Platonoff & Harris Plc - 206 Mill Studio, Crane Mead, Ware, Herts. SG12 9PY tel:(01920) 444255, fax:(01920) 487673. info@platonoffharris.co.uk
Playtop Licensing Limited - Brunel House, Jessop Way, Newark, Notts. NG24 2ER tel:(01636) 642461, fax:(01636) 642478. sales@playtop.com, www.playtop.com
PLC Hunwick Ltd - Harrision Works, Kings Road, Halstead, Essex. CO9 1HD tel:(01787) 474547, fax:(01787) 475741. sales@lightning-crushers.co.uk, www.lightning-crushers.co.uk
Plumb Centre - The Wolseley Center, Harrison Way, Spa Park, Royal Leamington Spa. CV31 3HH tel:(0870) 16 22557. customerservices@wolseley.co.uk, www.plumbcenter.co.uk
Polybeam Ltd - Isleport Business Park, Isleport Road, Highbridge, Somerset. TA9 4JU tel:(01278) 780807, fax:(01278) 780907. info@polybeam.co.uk, www.polybeam.co.uk
Polycell Products Ltd - Wexham Road, Slough, Berks. SL2 5DS tel:(01753) 550000, fax:(01753) 578218. www.polycell.co.uk
Polyflor Ltd - PO Box 3, Radcliffe New Road, Whitefield, Gt Man. M45 7NR tel:(0161) 767 1122, fax:(0161) 767 1128. info@polyflor.com, www.polyflor.com
Polypipe Building Products - Broomhouse Lane, Edlington, Doncaster. DN12 1ES tel:(01709) 770 000, fax:(01709) 770 001. info@polypipe.com, www.polypipe.com/building-products
Polypipe Building Products Ltd - Broomhouse Lane, Edlington, Doncaster, S. Yorks. DN12 1ES tel:(01709) 770000, fax:(01709) 770001. info@polypipe.com, www.polypipe.com
Polypipe Civils - Unit 1A, Charnwood Business Park, North Road, Loughborough, Leics. LE11 1LE tel:(01509) 615100, fax:(01509) 610215. civils@polypipe.com, www.polypipe.com/civils
Polypipe Terrain Ltd - College Road, Aylesford, Maidstone, Kent. ME20 7PJ tel:(01622) 795200, fax:(01622) 716796. commercialenquiries@polypipe.com, www.polypipe.com
Polypipe Ventilation - Sandall Stones Road, Kirk Sandall Industrial Estate, Kirk Sandall, Doncaster. DN3 1QR tel:(03443) 715523, fax:(03443) 715524. vent.marketing@polypipe.com, www.polypipe.com/ventilation
Polypipe Ventilation Ltd - Sandall Stones Road, Kirk Sandall Industrial Estate, Kirk Sandall, Doncaster, South Yorkshire. DN3 1QR tel:+44 0 8443 715523, fax:+44 0 8443 715524. vent.info@polypipe.com, http://www.polypipe.com/ventilation
Porcelanosa Group Ltd - Unit 1 – 6, Otterspool Way, Watford, Hertfordshire. WD25 8HL tel:(01923) 831 867, fax:(01923) 691 600. group@porcelanosa.com/gb, www.porcelanosa.co.uk
Portacel - Winnall Valley Road, Winchester, Hants. SO23 0LL tel:(01962) 705200, fax:(01962) 866084. wtuk.water@siemens.com, www.portacel.co.uk
Portakabin Ltd - Huntington, York. YO32 9PT tel:(01904) 611655, fax:(01904) 611644. solutions@portakabin.com, www.portakabin.com
Portasilo Ltd - New Lane, Huntington, York. YO32 9PR tel:(01904) 624872, fax:(01904) 611760. bulk@portasilo.co.uk, www.portasilo.co.uk
Portia Engineering Ltd - PO Box 9, Star Lane, Ipswich. IP4 1JJ tel:(01473) 252334, fax:(01473) 233863. portia@btinternet.com
Poselco Lighting - 1 Metropolitan Park, Bristol Road, Greenford, Middlesex. UB6 8UW tel:(020) 8813 0101, fax:(020) 8813 0099. info@poselco.co.uk, www.poselco.co.uk
Posners the floorstore - 35a-37 Fairfax Road, Swiss Cottage, London. NW6 4EW tel:(020) 7625 8899, fax:(020) 7625 8866. sales@posnersfloorstore.co.uk, www.posnersfloorstore.co.uk
Potter & Soar Ltd - Beaumont Road, Banbury, Oxon. OX16 1SD tel:(01295) 253344, fax:(01295) 272132. potter.soar@btinternet.com, www.wiremesh.co.uk
Poujoulat (UK) Ltd - 105-109 Oyster Lane, Byfleet, Surrey. KT14 7HJ tel:(01932) 343934, fax:(01932) 343222. sales@poujoulat.co.uk, www.poujoulat.co.uk
Power Plastics Ltd - Station Road, Thirsk, Yorks. YO7 1PZ tel:(01845) 525503, fax:(01845) 525485. info@powerplastics.co.uk, www.powerplastics.com
PPG Protective and Marine Coatings - Trigate House - 2nd Floor, 210-222 Hagley Road West, Birmingham. B68 0NP tel:(01525) 375234. pmcsalesuk@ppg.com, www.ppgpmc.com
PRA Coatings Technology Centre - 14 Castle Mews, High Street, Hampton, Middx. TW12 2NP tel:(020) 8487 0800, fax:(020) 8487 0801. www.pra-world.com
Precast Flooring Federation - 60 Charles Street, Leicester, Leics. LE1 1FB tel:(0116) 253 6161, fax:(0116) 251 4568. info@precastfloors.info, www.precastfloors.info
Precolor Sales Ltd - Newport Road, Market Drayton, Shropshire. TF9 2AA tel:(01630) 657281, fax:(01630) 655545. enquiries@precolortankdivision.co.uk, www.precolor.co.uk
Premdor - Birthwaite Business Park, Huddersfield Road, Darton, Barnsley, South Yorkshire. S75 5JS tel:(0844) 209 0008, fax:(0844) 371 5333. enquiries@premdor.com, www.premdor.co.uk
Premdor - Gemini House, Hargreaves Road, Swindon, Wilts. SN25 5AJ tel:(0870) 990 7998, fax:(01793) 708280. ukmarketing@premdor.com, www.premdor.co.uk
PRE-MET Ltd - Studley Road, Redditch, Worcs. B98 7HJ tel:(01527) 510535, fax:(01527) 500868. sales@pre-met.com, www.lewis-spring.com
President Blinds Ltd - Unit 13 Forest Hill Buisness Centre, Clyde Vale, Forest Hill, London. SE23 3JF tel:(020) 8699 8885, fax:(020) 8699 8005. president@flyscreens-uk.co.uk, www.flyscreens-uk.co.uk
Pressalit Care - 100 Longwater Avenue, Green Park, Reading, Berkshire. RG2 6GP tel:(0844) 8806950, fax:(0844) 8806951. uk@pressalit.com, www.pressalitcare.com
Pressmain - 130 Princess Road, Manchester. M16 7BY tel:(01422) 349560, fax:(01274) 483459. info@aquatechpressmain.co.uk, www.aquatechpressmain.co.uk
Pressure Coolers Ltd T/A Maestro Pressure Coolers - 67-69 Nathan Way, London. SE28 0BQ tel:(020) 8302 4035, fax:(020) 8302 8933. sales@maestrointl.co.uk, www.pressurecoolers.co.uk
Presto Engineers Cutting Tools Ltd - Penistone Road, Sheffield. S6 2FN tel:(01742) 349361, fax:(01742) 347446. sales@presto-tools.com, www.presto-tools.com
Preston & Thomas Ltd - 1a Crabtree Close, Gravesend Road, Wrotham, Sevenoaks, Kent. TN15 7JL tel:(01372) 727424. prestonandthomas@hotmail.co.uk, www.prestonandthomas.co.uk
Prestoplan - Four Oaks Road, Walton Summit Centre, Preston. PR5 8AP tel:(01772) 627373, fax:(01772) 627575. susan.burrows@prestoplan.co.uk, www.prestoplan.co.uk
Prestressed Concrete Association - 60 Charles Street, Leicester, Leics. LE1 1FB tel:(0116) 253 6161, fax:(0116) 251 4568. pca@britishprecast.org, www.bridgebeanns.org.uk
Prima Security & Fencing Products - Redwell Wood Farm, Ridge Hill, Potters Bar, Herts. EN6 3NA tel:(01707) 663 400, fax:(01707) 661112. sales@sampsonfencing.co.uk, www.sampsonfencing.co.uk
Prima Systems (South East) Ltd - The Old Malt House, Easole Street, Nonington, Dover, Kent. CT15 4HF tel:(01304) 842888, fax:(01304) 842840. halfpennyd@primasystems.co.uk, www.primasystems.co.uk
Priory Castor & Engineering Co Ltd - Aston Hall Road, Aston, Birmingham. B6 7LA tel:(0121) 327 0832, fax:(0121) 322 2123. enquiries@priorycastor.co.uk, www.priorycastor.co.uk
Priory Shutter & Door Co Ltd - 89-91 Rolfe Street, Smethwick, Warley, W. Midlands. B66 2AY tel:(0121) 558 6406, fax:(0121) 558 7140. enquiries@sis-group.co.uk, www.sis-group.co.uk/priory-roller-shutters
Prismo Ltd - 5 Drumhead Road, Chorley North Industrial Park, Chorley, Lancs. PR6 7BX tel:(01302) 309335, fax:(013020) 309342. info@ennisprismo.com, www.prismo.co.uk
Pro Display TM Limited - Unit 5 Shortwood Business Park, Hoyland, Barnsley, South Yorkshire. S74 9LH tel:(0870) 766 8438, fax:(0870) 766 8437. sales@prodisplay.co.uk, www.prodisplay.com
Procter Cast stone & Concrete Products - Newhold, Aberford Rd, Garforth, Leeds, Yorks. LS25 2HG tel:(01132) 863 329, fax:(01132) 867376. websales@proctergarforth.co.uk, www.caststoneuk.co.uk
Procter Bros Limited - 4 Beaconsfield Court, Garfirth, Leeds, W. Yorks. LS25 1QH tel:(0113) 287 2777, fax:(0113) 287 1177. enquiries@procterfencing.co.uk, www.fencing-systems.co.uk
Procter Johnson & Co Ltd - Excelsior Works, Castle Park, Evans Street, Flint, Flintshire. CH6 5NT tel:(01352) 732157, fax:(01352) 735530. sales@pjcolours.com, www.pjcolours.com
Prodema Lignum - PO Box 1945, Maidenhead. SL6 2DD tel:(01628) 687022, fax:(01628) 687023. lignum@enox.co.uk, www.enox.co.uk
Production Glassfibre - Wilstead Industrial Park, Wilstead, Bedfordshire. MK45 3PD tel:(01234) 741 444. south@productionglassfibre.co.uk, www.productionglassfibre.co.uk
Profab Access Ltd - Unit 52 Fourways, Carlyon Road Ind Est, Atherstone, Warwicks. CV9 1LH tel:(01827) 718222, fax:(01827) 721092. www.profilex.com
Pro-Fence (Midlands) - 3 Barley Croft, Perton, Wolverhampton, W. Midlands. WV6 7XX tel:(01902) 894747, fax:(01902) 897007
Professional Bookshops Limited - Winkworth House, 4 Market Place, Devizes, Wiltshire. SN10 1HT tel:(01380) 820003, fax:(01380) 730218. customerservice@professionalbooks.co.uk, www.professionalbooks.co.uk
Profile 22 Systems Ltd - Stafford Park 6, Telford, Shropshire. TF3 3AT tel:(01952) 290910, fax:(01952) 290460. mail@profile22.co.uk, www.profile22.co.uk
Progress Work Place Solutions - Ground Floor, Taunton House, Waterside Court, Neptune Way, Medway City Estate, Rochester, Kent. ME2 4NZ tel:(01634) 290988, fax:(01634) 291028. info@progressfurnishing.co.uk, www.progressfurnishing.co.uk
Progressive Product Developments Ltd - 24 Beacon Bottom, Swanwick, South Hampton, Hants. SO31 7GQ tel:(01489) 576787, fax:(01489) 578463. sales@ppd-ltd.com, www.ppd-ltd.com
Project Aluminium Ltd - 418-420 Limpsfield Road, Warlington, Surrey. CR6 9LA tel:(01883) 624004, fax:(01883) 627201. lesley@projects.co.uk
Project Office Furniture PLC - 103 High Street, Waltham Cross, Herts. EN8 7AN tel:(01440) 705411, fax:(01440) 703376. enquiries@project.co.uk, www.project.co.uk
Promat UK Ltd - The Sterling Centre, Eastern Road, Bracknell, Berks. RG12 2TD tel:(01344) 381300, fax:(01344) 381301. salesuk@promat.co.uk, www.promat.co.uk
Propex Concrete Systems - Propex House, 9 Royal Court, Basil Close, Chesterfield, Deryshire. S41 7ST tel:(01246) 564200, fax:(01246) 564201. enquiries@propexinc.co.uk, www.fibermesh.com

P (con't)

Protecter Lamp & Lighting Co Ltd - Lansdowne Road, Eccles, Gt Man. M30 9PH tel:(0161) 789 5680, fax:(0161) 787 8257. dmather@protectorlamp.com, www.protectorlamp.com

Proten Services Ltd - Unit 10, Progress Business Park, Progress Road, Leigh-on-Sea, Essex. SS9 5PR tel:(01702) 471666, fax:(0844) 854 8796. info@protenservices.co.uk, www.protenservices.co.uk

Protim Solignum Ltd - Fieldhouse Lane, Marlow, Bucks. SL7 1LS tel:(01628) 486644, fax:(01628) 476757. info@osmose.co.uk, www.protimsolignum.com

Protimeter, GE Thermometrics (UK) Ltd - Crown Industrial Estate, Priorswood Road, Taunton, Somerset. TA2 8QY tel:(01823) 335 200, fax:(01823) 332 637. Taunton.cc@ge.com, www.protimeter.com

Pryorsign - Field View, Brinsworth Lane, Brinsworth, Rotherham, S. Yorks. S60 5DG tel:(01709) 839559, fax:(01709) 837659. phil@pryorsign.com, www.buyersguide.co.uk/document/pryor-sign/

PTS Plumbing Trade Supplies - PTS House, Eldon Way, Crick, Northamptonshire. NN6 7SL tel:(01788) 527700, fax:(01788) 527799. ptsplumbing@ptsplumbing.co.uk, www.ptsplumbing.co.uk

Pump International Ltd - Trevoole Praze, Camborne, Cornwall. TR14 0PJ tel:(01209) 831937, fax:(01209) 831939. sales@pumpinternational.com, www.pumpinternational.com

Pump Technical Services Ltd - Beco Works, Cricket Lane, off Kent House Lane, Beckenham, Kent. BR3 1LA tel:(020) 8778 4271, fax:(020) 8659 3576. sales@ptsjung.co.uk, www.pts-jung.co.uk

Pumps & Motors UK Ltd - Units 3-4, Abbey Grange Works, 52 Hertford Road, Barking, Essex. IG11 8BL tel:(020) 8507 2288, fax:(020) 8591 7757. sales@pumpsmotors.co.uk, www.pumpsmotors.co.uk

Purewell Timber Buildings Limited - Unit 4, Lea Green Farm, Christchurch Road, Lymington, Hants. SO41 0LA tel:(01202) 484422, fax:(01202) 490151. mail@purewelltimber.co.uk, www.purewelltimberbuildings.co.uk

Purified Air Ltd - Lyon House, Lyon Road, Romford, Essex. RM1 2BG tel:(01708) 755414, fax:(01708) 721488. enq@purifiedair.com, www.purifiedair.com/

Purmo-UK - Rettig Park, Drum Lane, Birtley, County Durham. DH2 1AB tel:(0191) 492 1700, fax:(0191) 492 9484. uk@purmo.co.uk, www.purmo.com/en

Pyramid Plastics UK Ltd - Unit 22 Corringham Road Industrial Estate, Corringham Road, Gainsborough, Linconshire. DN21 1QB tel:(01427) 677990, fax:(01427) 612204. website@pyramid-plastics.co.uk, www.pyramid-plastics.co.uk

Pyroguard UK Ltd - International House, Millford Lane, Haydock, Merseyside. WA11 9GA tel:(01942) 710720, fax:(01942) 710730. info@pyroguard.eu, www.pyroguard.eu

Q

Q.M.P - Timmis Road, Lye, Stourbridge, W. Midlands. DY9 7BQ tel:(01384) 899 800, fax:(01384) 899 801. sales@qmp.uk.com, www.qmp.uk.com

QK Honeycomb Products Ltd - Creeting Road, Stowmarket, Suffolk. IP14 5AS tel:(01449) 612145, fax:(01449) 677604. sales@qkhoneycomb.co.uk, www.qkhoneycomb.co.uk

Quantum Profile Systems Ltd - Salmon Fields, Royton, Oldham, Lancs. OL2 6JG tel:(0161) 627 4222, fax:(0161) 627 4333. info@quantum-ps.co.uk, www.quantumprofilesystems.com

Quarry Products Association - Gillingham House, 38-44 Gillingham Street, London. SW1W 1HU tel:(020) 7730 8194, fax:(020) 7730 4355. info@qpa.org, www.qpa.org

Quartet-GBC UK Ltd - Rutherford Road, Basingstoke, Hants. RG24 8PD tel:(01256) 842828, fax:(01256) 476001. sales@quartetmfg.co.uk

Quelfire - PO Box 35, Caspian Road, Altrincham, Cheshire. WA14 5QA tel:(0161) 928 7308, fax:(0161) 924 1340. sales@quelfire.co.uk, www.quelfire.co.uk

Quiligotti Terrazzo Tiles Limited - PO Box 4, Clifton Junction, Manchester. M27 8LJ tel:(0161) 727 9798, fax:(0161) 727 0457. sales@quiligotti.co.uk, www.quiligotti.co.uk

Quip Lighting Consultants & Suppliers Ltd - 71 Tenison Road, Cambridge, Cambs. CB1 2DG tel:(01223) 321277, fax:(01223) 321277. ygj36@dial.pipex.com

R

R J Stokes & Co Limited - Holbrook Industrial Estate, Rother Valley Way, Sheffield, South Yorkshire. S20 3RW tel:(0114) 2512680. info@stokestiles.co.uk, www.stokestiles.co.uk

R. C. Cutting & Company Ltd - 10-12 Arcadia Avenue, Finchley Central, London. N3 2JU tel:(020) 8371 0001, fax:(020) 8371 0003. info@rccutting.co.uk, www.rccutting.co.uk

R.A.M. Perimeter Protection Ltd - 179 Higher Hilgate, Harrop Street, Stockport, Cheshire. SK1 3JG tel:(0161) 477 4001, fax:(0161) 477 1007. ramperimeterprotection@btconnect.com, www.rampp.co.uk

R.M. Easdale & Co Ltd - 67 Washington Street, Glasgow. G3 8BB tel:(0141) 204 2708, fax:(0141) 204 3159. david@rmeasdale.com, www.rmeasdale.com

Race Furniture Ltd - Bourton Industrial Park, Bourton-on-the-Water, Glos. GL54 2HQ tel:(01451) 821446, fax:(01451) 821686. sales@racefurniture.com, www.racefurniture.com

Rackline Systems Storage Ltd - Oaktree Lane, Talke, Newcastle-under-Lyme, Staffs. ST7 1RX tel:(01782) 777666, fax:(01782) 777444. now@rackline.com, www.rackline.com

Radflex Contract Services Ltd - Unit 35, Wilks Avenue, Questor, Dartford, Kent. DA1 1JS tel:(01322) 276363, fax:(01322) 270606. expjoint@radflex.co.uk, www.radflex.co.uk

Radiant Services Ltd - Barrett House, 111 Millfields Road, Ettingshall, Wolverhampton, W. Midlands. WV4 6JQ tel:(01902) 494266, fax:(01902) 494153. radser@btconnect.com, www.radiantservices.co.uk

Radiodetection Ltd - Western Drive, Bristol. BS14 0AF tel:(0117) 976 7776, fax:(0117) 976 7775. rd.sales.uk@spx.com, www.spx.com/en/radiodetection

Railex Systems Ltd - The Wilson Building, 1 Curtain Road, London. EC2A 3JX tel:(020) 7377 1777, fax:(020) 7247 6181. hwilson@railex.co.uk, www.railex.co.uk

Railex Systems Ltd, Elite Division - Elite Works, Station Road, Manningtree, Essex. CO11 1DZ tel:(01206) 392171, fax:(01206) 391465. info@railex.co.uk, www.railexstorage.co.uk

Rainham Steel Co Ltd - Kathryn House, Manor Way, Rainham, Essex. RM13 8RE tel:(01708) 522311, fax:(01708) 559024. sales@rainhamsteel.co.uk, www.rainhamsteel.co.uk

Raised Floor Systems Ltd - Peak House, Works Road, Letchworth Garden City, Herts. SG6 1GB tel:(01462) 685898, fax:(01462) 685 353. rfs@raisedfloorsystems.co.uk, www.raisedfloorsystems.co.uk

Ramsay & Sons (Forfar) Ltd - 61 West High Street, Forfar, Angus. DD8 1BG tel:(01307) 462255, fax:(01307) 466956. enquiries@ramsayladders.co.uk, www.ramsayladders.co.uk

Ramset Fasteners Ltd - Diamond Point, Fleming Way, Crawley, W. Sussex. RH10 9DP tel:(01293) 523372, fax:(01293) 515186. productadvice@itwcp.co.uk, www.itwramset.co.uk

Randalls Fabrications Ltd - Hoyle Mill Road, Kinsley, Pontefract, W. Yorks. WF9 5JB tel:(01977) 615132, fax:(01977) 610059. sales@randallsfabrications.co.uk, www.randallsfabrications.co.uk

Range Cylinders - Tadman Street, Wakefield, W. Yorks. WF1 5QU tel:(01924) 376026, fax:(01924) 385015. sales@range-cylinders.co.uk, www.range-cylinders.co.uk

Rangemaster - Meadow Lane, Long Eaton, Notts. NG10 2AT tel:(0115) 946 4000, fax:(0115) 946 0374. sales@rangemaster.co.uk, www.rangemaster.co.uk

Rapaway Energy Ltd - 35 Park Avenue, Solihull, W. Midlands. B91 3EJ tel:(0121) 246 0441, fax:(0121) 246 0442. rapaway@btinternet.com

Rawson Carpets Ltd - Castle Bank Mills, Portobello Road, Wakefield, W. Yorks. WF1 5PS tel:(01924) 382860, fax:(01924) 366204. sales@rawsoncarpets.co.uk, www.rawsoncarpets.co.uk

Rawson Fillings Ltd - Castle Bank Mills, Portobello Road, Wakefield, W. Yorks. WF1 5PS tel:(01924) 373421, fax:(01924) 290334. sales@rawsonfillings.co.uk

RB Farquhar Manufacturing Ltd - Deveronside Works, Huntly, Aberdeenshire. AB54 4PS tel:(01466) 793231, fax:(01466) 793098. info@rbfarquhar.co.uk, http://rbf.digitalface.co.uk/

RB UK Ltd - Element House, Napier Road, Bedford, Beds. MK41 0QS tel:(01234) 272717, fax:(01234) 270202. customerservices@rbuk.co.uk, www.rbuk.co.uk

RCC, Division of Tarmac Precast Concrete Ltd - Barholm Road, Tallington, Stamford, Lincs. PE9 4RL tel:(01778) 344460, fax:(01778) 345949. buildingproducts@tarmac.co.uk, www.tarmacprecast.com

RDA Projects Ltd - Innovation House, Daleside Road, Nottingham. NG2 4DH tel:(0115) 911 0243, fax:(0115) 911 0246. advice@rdaprojects.co.uk, www.rdaproject.co.uk

Rea Metal Windows Ltd - 126-136 Green Lane, Liverpool. L13 7ED tel:(0151) 228 6373, fax:(0151) 254 1828. all@reametal.co.uk, www.reametal.co.uk

Realstone Ltd - Wingerworth, Chesterfield, Derbys. S42 6RG tel:(01246) 270244, fax:(01246) 220095. sales@realstone.co.uk, www.realstone.co.uk

Record Electrical - Atlantic Street, Altrincham, Cheshire. WA14 5DB tel:(0161) 928 6211, fax:(0161) 926 9750. info@reauk.com, www.record-electrical.co.uk

Record Power Ltd - Parkway Works, Sheffield. S9 3BL tel:(01742) 434370, fax:(01742) 617141. recordpower@recordtools.co.uk, www.recordpower.co.uk

Record Tools Ltd - Parkway Works, Sheffield. S9 3BL tel:(01742) 449066, fax:(01742) 434302. recordpower@recordtools.co.uk, www.recordpower.co.uk

Red Bank Maufacturing Co Ltd - Atherstone Road, Measham, Swandlincote, Derbys. DE12 7EL tel:(01530) 270333, fax:(01530) 273667. meashamsalesuk@hanson.biz, www.heidelbergcement.co

Reddiglaze Ltd - The Furlong, Droitwich, Worcs. WR9 9BG tel:(01905) 795432, fax:(01905) 795757. sales@reddiplex.com, www.reddiglaze.co.uk

Redditch Plastic Products - Pipers Road, Park Farm Industrial Estate, Redditch, Worcs. B98 0HU tel:(01527) 528024, fax:(01527) 520236. cons@proteusswitchgear.co.uk, www.proteusswitchgear.co.uk

Redifix Ltd - Queenslie Industrial Estate, 44 Coltness Lane, Glasgow. G33 4DR tel:(0141) 774 2020, fax:(0141) 774 3080. sales@redifix.co.uk, www.redifix.co.uk

Rediweld Rubber & Plastics Ltd - 6-10 Newman Lane, Alton, Hants. GU34 2QR tel:(01420) 543007, fax:(01420) 546740. info@rediweld.co.uk, www.rediweldtraffic.co.uk

Redman Fisher Engineering - Birmingham New Road, Tipton, W. Midlands. DY4 9AA tel:(01902) 880880, fax:(01902) 880446. flooring@redmanfisher.co.uk, www.redmanfisher.co.uk

Redring - Morley Way, Peterborough, Cambs. PE2 9JJ tel:(0845) 607 6448. aeinfo@applied-energy.com, www.redring.co.uk

Reed Harris - Riverside House, 27 Carnwath Road, Fulham, London. SW6 3HR tel:(020) 7736 7511, fax:(020) 7736 2988. architectural@reed-harris.co.uk, www/reedharris.co.uk

Reflecting Roadstuds Ltd - Boothtown, Halifax, W. Yorks. HX3 6TR tel:(01422) 360208, fax:(01422) 349075. info@percyshawcatseyes.com, www.percyshawcatseyes.com

Reflex-Rol (UK) - Ryeford Hall, Ryeford, Nr Ross-on-Wye, Herefordshire. HR9 7PU tel:(01989) 750704, fax:(01989) 750768. info@reflex-rol.co.uk, www.reflex-rol.co.uk

Rega Ductex Ltd - 21 Elton Way, Biggleswade, Beds. SG18 8NH tel:(01767) 600499, fax:(01767) 692451. sales@rega-uk.com, www.regaductex.co.uk

Regency Garage Door Services Ltd - Unit 8, Central Trading Estate, Signal Way, Swindon, Wilts. SN3 1PD tel:(01793) 611688, fax:(01793) 495144. info@regencygaragedoors.co.uk, www.regencygaragedoors.co.uk

Regency Swimming Pools Ltd - 11 Regis Road, Tettenhall, Wolverhampton, West Midlands. WV6 8RU tel:(01902) 757 757, fax:(01902) 422632. info@jwgswimming.co.uk, www.jwgswimming.co.uk

REHAU Ltd - Hill Court, Walford, Ross-on-Wye, Herefordshire. HR9 5QN tel:(01989) 762600, fax:(01989) 762601. enquiries@rehau.com, www.rehau.co.uk

Reid John & Sons (Strucsteel) Ltd - Strucsteel House, 264-266 Reid Street, Christchurch, Dorset. BH23 2BT tel:(01202) 483333, fax:(01202) 470103. sales@reidsteel.co.uk, www.reidsteel.com

Reiner Fixing Devices (Ex: Hardo Fixing Devices) - Hall Farm, Church Lane, North Ockendon, Upminster, Essex. RM14 3QH tel:(01708) 856601, fax:(01708) 852293. sales@reinerfixings.co.uk, www.reinerfixings.co.uk

Relcross Ltd - Hambleton Avenue, Devizes, Wilts. SN10 2RT tel:(01380) 729600, fax:(013080) 729888. sales@relcross.co.uk, www.relcross.co.uk

Remploy Ltd - Unit 2, Brocastle Avenue, Waterton Industrial Estate, Bridgend, Bridgend. CF31 3YN tel:(01656) 653982, fax:(01656) 767499. info@remploy.co.uk, www.remploy.co.uk

Remtox Silexine - 14 Spring Road, Smethwick, Warley, W. Midlands. B66 1PE tel:(0121) 525 2299, fax:(0121) 525 1740. sales@sovchem.co.uk, www.sovchem.co.uk

Renderplas Ltd - Number 2, 70-72 High Street, Bewdley, Worcs. DY12 2DJ tel:(01299) 888333, fax:(01299) 888234. info@renderplas.co.uk, www.renderplas.co.uk

RENOLIT Cramlington Limited - Station Road, Cramlington, Northumberland. NE23 8AQ tel:(01234) 272999, fax:(01234) 357313. renolit.cramlington@renolit.com, www.renolit.com/waterproofing-roofing/en/

Renqiu Jingmei Rubber&Plastic Products Co., Ltd - Industrial Zone, Maozhou Town, Renqiu City, Hebei Province. CHINA tel:+86-317-3384666, fax:+86-317-3385186. sales@jingmeirubber.com, www.jingmeirubber.com

Rentokil Property Care - Chartists Way, Morley, Leeds. LS27 9EG tel:(0800) 731 2343. www.rentokill.co.uk

Rentokil Specialist Hygiene - 2nd Floor, Riverbank Meadows Business Park, Camberley. GU17 9AB tel:(0800) 028 0839, fax:(01342) 326229. rsh-info-uk@rentokil-initial.com, www.rentokil-hygiene.co.uk

Rentokil Specialist Hygiene - 2nd Floor, Riverbank, Meadows Business Park, Camberley. GU17 9AB tel:(0800) 028 0839. www.retokil.co.uk

Repair care International - Unit 19, Darwell Park, Mica Close, Amington, Tamworth, Staffordshire. PE28 5GQ tel:(01827) 302 517. salesuk@repair-care.com, uk.repair-care.com

Resdev Ltd - Pumaflor House, Ainleys Industrial Estate, Elland, W. Yorks. HX5 9JP tel:(01422) 379131, fax:(01422) 370943. info@resdev.co.uk, www.resdev.co.uk

Resiblock Ltd - Resiblock House, Archers Fields Close, Basildon, Essex. SS13 1DW tel:(01268) 273344, fax:(01268) 273355. mail@resiblock.com, www.resiblock.com

Resin Flooring Association (FeRFA) - 16 Edward Road, Farnham, Surrey. GU9 8NP tel:(07484) 075 254, fax:(01252) 714250. lisa@ferfa.org.uk, www.ferfa.org.uk

Respatex International Limited - Water Meadow House, Chesham, Buckinghamshire. HP5 1LF tel:(01494) 771242, fax:(01494) 771292. sales@respatex.co.uk, www.respatex.co.uk

Revol Products Ltd - Samson Close, Killingworth, Newcastle-upon-Tyne. NE12 0DZ tel:(0191) 268 4555, fax:(0191) 216 0004. sales@revol.co.uk, www.revol.co.uk

Reynolds John & Son (Birmingham) Ltd - Church Lane, West Bromwich, W. Midlands. B71 1DJ tel:(0121) 553 2754/1287, fax:(0121) 500 5460. www.oaktreepackaging.co.uk

Reyven (Sportsfields) Ltd - Greenhill Farm, Tilsworth, Leighton Buzzard, Beds. LU7 9PU tel:(01525) 210714, fax:(01525) 211130

Reznor UK Ltd - Park Farm Road, Park Farm Industrial Estate, Folkestone, Kent. CT19 5DR tel:(01303) 259141, fax:(01303) 850002. sales@reznor.co.uk, www.reznor.co.uk

RFA-Tech Ltd - Eastern Avenue, Trent Valley, Lichfield, Staffordshire. WS13 6RN tel:(0870) 011 2881, fax:(0870) 011 2882. sales@rfa-tech.co.uk, www.rfa-tech.co.uk

Rheinzink UK - Wyvern House, 55-61 High Street, Frimley, Surrey. GU16 7HJ tel:(01276) 686725, fax:(01276) 64480. info@rheinzink.co.uk, www.rheinzink.co.uk

Rhodar Limited - Astra Park, Parkside Lane, Leeds. LS11 5SZ tel:(0113) 270 0775, fax:(0113) 270 4124. info@rhodar.co.uk, www.Rhodar.co.uk

RIBA Enterprises - The Old Post Office, St Nicholas Street, Newcastle Upon Tyne. NE1 1RH info@ribaenterprises.com, www.ribaenterprises.com

Richard Hose Ltd - Unit 7, Roman Way Centre, Longridge Road, Ribbleton, Preston, Lancs. PR2 5BB tel:(01772) 651550, fax:(01772) 651325. richards.fire@btinternet.com, www.richardsfire.co.uk

Richard Lees Steel Decking Ltd - Moor Farm Road West, The Airfield, Ashbourne, Derbys. DE6 1HD tel:(01335) 300999, fax:(01335) 300888. rlsd.decks@skanska.co.uk, www.rlsd.com

Richards H S Ltd - King Street, Smethwick, Warley, W. Midlands. B66 2JW tel:(0121) 558 2261, fax:(01785) 840 603. info@richards-paints.co.uk, www.richards-paints.co.uk

RICS Book Shop - 12 Great George Street (Parliament Square), London. SW1P 3AD tel:(024) 7686 8502, fax:(020) 7334 3851. mailorder@rics.org, www.rics.org/uk/shop/

Righton Ltd - Righton House, Brookvale Road, Witton, Birmingham. B6 7EY tel:(0121) 356 1141, fax:(0121) 331 1347. righton@righton.co.uk, www.righton.co.uk

Rigidal Industries Ltd - Blackpole Trading Estate West, Worcester. WR3 8ZJ tel:(01905) 750500, fax:(01905) 750555. sales@rigidal.co.uk, www.rigidal-industries.com

Rigidal Systems LTD - Unit 62, Blackpole Trading Estate West, Worcester. WR3 8ZJ tel:(01905) 750500, fax:(01905) 750555. sales@rigidal.co.uk, www.rigidal.co.uk

Rite-Vent Ltd - Crowther Estate, Washington, Tyne & Wear. NE38 0AQ tel:(0191) 416 1150, fax:(0191) 415 1263. sales@rite-vent.co.uk, www.rite-vent.co.uk

Ritherdon & Co Ltd - Lorne Street, Darwen, Lancs. BB3 1QW tel:(01254) 819100, fax:(01254) 819101. sales@ritherdon.co.uk, www.ritherdon.co.uk

Rivermeade Signs Ltd - Roslin Road, South Acton Industrial Estate, London. W3 8BW tel:(020) 8896 6900, fax:(020) 8752 1691. salesw@rivermeade.com, www.rivermeade.com

Rixonway Kitchens Ltd - Shaw Cross Busness Park, Dewsbury, W. Yorks. WF12 7RD tel:(01924) 431300, fax:(01924) 431301. info@rixonway.co.uk, www.rixonway.co.uk

Robert Bosch Ltd - PO Box 98, Broadwater Park, North Orbital Road, Denham, Uxbridge, Middx. UB9 5HJ tel:(01895) 834466, fax:(01895) 838388. jo.hudson@uk.bosch.com, www.bosch.co.uk

Roberts-Gordon - Oxford Street, Bilston, W. Midlands. WV14 7EG tel:(01902) 494425, fax:(01902) 403200. uksales@rg-inc.com, www.combat.co.uk

Robinson Lloyds Architecture - 5 Premier Cour, Boarden Close, Moulton Park Industrial Estate, Northampton. NN3 6LF tel:(01604) 671633, fax:(01604) 495282. projects@robinson-lloyd.com, www.robinson-lloyd.com

Robinson Willey Ltd - Mill Lane, Old Swan, Liverpool. L134AJ tel:(0151) 228 9111, fax:(0151) 228 6661. info@robinson-willey.com, www.robinson-willey.com

R (con't)

Roc Secure Ltd - 90 College Street, Bedford. MK42 8LU tel:(0845) 671 2155. sales@roc-secure.co.uk, www.roc-secure.co.uk

Rock & Alluvium Ltd - SBC House, Restmor Way, Wallington, Surrey. SM6 7AH tel:(020) 8255 2088, fax:(020) 82416934. Enquiries@rockal.com, www.rockal.com

Rockfon Ltd - Pencoed, Bridgend. CF35 6NY tel:(01656) 864696, fax:(01656) 864549. daniel.broyd@rockfon.co.uk, www.rockfon.co.uk

Rockwell Sheet Sales Ltd - Rockwell House, Birmingham Road, Millisons Wood, Coventry. CV5 9AZ tel:(01676) 523386, fax:(01676) 523630. info@rockwellsheet.com, www.rockwellsheet.com

Rockwood Pigments (UK) Ltd - Liverpool Road East, Kidsgrove, Stoke-on-Trent, Staffordshire. ST7 3AA tel:(01782) 794400, fax:(01782) 787338. info.uk@rpigments.com, www.rpigments.com

Rockwool Ltd - Pencoed, Bridgend, M. Glam. CF35 6NY tel:(01656) 862621, fax:(01656) 862302. info@rockwool.co.uk, www.rockwool.co.uk

Rogers Concrete Ltd - Puddletown Road, Wareham, Dorset. BH20 6AU tel:(01929) 462373, fax:(01929) 405326. Warehamsales@gardenstone.co.uk, www.rogersgardenstone.co.uk

Roles Broderick Roofing Ltd - 4 High Street, Woking, Surrey. GU24 8AA tel:(01276) 856604. info@rolesbroderick.com

Rolflex Doors Ltd - Unit 63, Third Avenue, The Pensnett Estate, Kingswinford, West Midlands. DY6 7XU tel:(01384) 401555, fax:(01384) 401556. sales@rolflex.co.uk, www.rolflex.co.uk

Rollalong Ltd - Woolsbridge Industrial Estate, Three Legged Cross, Wimborne, Dorset. BH21 6SF tel:(01202) 824541, fax:(01202) 812584. enquiries@rollalong.co.uk, www.rollalong.co.uk

Rollins & Sons (London) Ltd - Rollins House, 1 Parkway, Harlow. CM19 5QF tel:(01279) 401570, fax:(01279) 401580. sales@rollins.co.uk, www.rollins.co.uk

Rom Ltd - Eastern Avenue, Trent Valley, Lichfield, Staffs. WS13 6RN tel:(01543) 414111, fax:(01543) 421672. sales@rom.co.uk, www.rom.co.uk

Ronacrete Ltd - Ronac House, Flex Meadow, Harlow, Essex. CM19 5TD tel:(020) 8593 7621, fax:(020) 8595 6969. sales@ronacrete.co.uk, www.ronacrete.co.uk

Roofing Contractors (Cambridge) Ltd - 5 Winship Road, Milton, Cambridge, Cambs. CB24 6BQ tel:(01223) 423059. info@rcc-cambridge.co.uk, www.rcc-cambridge.co.uk

RoofRite - Unit 46, Tondu Enterprise Park, Bryn Road, Tondu, Bridgend, Mid Glamorgan. CF32 9BS tel:(01443) 227840. roofrite@roofs4u.co.uk, www.roofs4u.co.uk

RoofRite Stirling Scotland - 4 Riverside Drive, Riverside, Stirling. FK8 1LR tel:(01786) 464395. info@roofrite.co.uk, www.roofrite.co.uk

Rotafix Resins - Rotafix House, Abercraf, Swansea. SA9 1UX tel:(01639) 730481, fax:(01639) 730858. rotafixltd@aol.com, www.rotafix.co.uk

Rothley Burn Ltd - Macrome Road, Wolverhampton, W. Midlands. WV6 9HG tel:(01902) 756461, fax:(01902) 745554. sales@rothley.com, www.rothley.co.uk

Roto Frank Ltd - Swift Point, Rugby, Warwicks. CV21 1QH tel:(01788) 558600, fax:(01788) 558605. uk-sales@roto-frank.com, www.roto-frank.co.uk

Royal Incorporation of Architects in Scotland - 15 Rutland Square, Edinburgh. EH1 2BE tel:(0131) 229 7545, fax:(0131) 228 2188. info@rias.org.uk, www.rias.org.uk

Royal Institute of British Architects (RIBA) - 66 Portland Place, London. W1B 1AD tel:(020) 7580 5533, fax:(020) 7255 1541. info@riba.org, www.architecture.com

Royal Institution of Chartered Surveyors (RICS) - 12 Great George Street, Parliament Square, London. SW1P 3AD tel:(0870) 333 1600, fax:(020) 7334 3811. contactrics@rics.org, www.rics.org

Royal Town Planning Institute - 41 Botolph Lane, London. EC3R 8DL tel:(020) 7929 9494, fax:(020) 7929 9490. online@rtpi.org.uk, www.rtpi.org.uk

Royde & Tucker - Bilton Road, Cadwell Lane, Hitchen. SG4 0SB tel:(01462) 444444, fax:(01462) 444433. sales@ratman.co.uk, www.ratman.co.uk

Royston Lead Ltd - Pogmoor Works, Stocks Lane, Barnsley, S. Yorks. S75 2DS tel:(01226) 770110, fax:(01226) 730359. info@roystonlead.co.uk, www.roystonlead.co.uk

RTD Systems Ltd - Unit 10 Mole Business Park, Randalls Road, Leatherhead, Surrey. KT22 7BA tel:(020) 8545 2945, fax:(020) 8545 2955. sales@rtdsystems.co.uk, www.octanorm.co.uk

Ruabon Sales Limited - Hafod Tileries, Ruabon, Wrexham, Clwyd. LL14 6ET tel:(01978) 843484, fax:(01978) 843276. sales@ruabonsales.co.uk, www.ruabonsales.co.uk

Rubert & Co Ltd - Acru Works, Demmings Road, Cheadle, Cheshire. SK8 2PG tel:(0161) 428 6058, fax:(0161) 428 1146. info@rubert.co.uk, www.rubert.co.uk

Rundum Meir (UK) Ltd - 1 Troutbeck Road, Liverpool. L18 3LF tel:(0151) 280 6626, fax:(0151) 737 2504. info@rundum.co.uk, www.rundum.co.uk

Rural & Industrial Design and Building Association (RIDBA) - ATSS House, Station Road East, Stowmarket, Suffolk. IP124 1RQ tel:(01449) 676049, fax:(01449) 770028. secretary@ridba.org.uk, www.ridba.org.uk

Rural Industrial Design & Building Association - ATSS House, Station Road East, Stowmarket, Suffolk. IP14 1RQ tel:(01449) 676049, fax:(01449) 770028. secretary@rdba.org.uk, www.rdba.org.uk

Ruskin Air Management - Wilson Industrial Estate, South Street, Whitstable, Kent. CT5 3DU tel:(01227) 276100, fax:(01227) 264262. sales@actionair.co.uk, www.actionair.co.uk

Russell Roof Tiles Ltd - Nicolson Way, Wellington Road, Burton-on-trent, Staffordshire. DE14 2AW tel:(01283) 517070, fax:(01283) 516290. james.pendleton@russellrooftiles.co.uk, www.russellrooftiles.co.uk

Rustins Ltd - Waterloo Road, Cricklewood, London. NW2 7TX tel:(020) 8450 4666, fax:(020) 8452 2008. rustins@rustins.co.uk, www.rustins.co.uk

Ruthin Precast Concrete Ltd - 1st. Floor, 11 Paul Street, Taunton, Somerset. TA1 3PF tel:(01823) 274232, fax:(01823) 274231. enquiries@rpcltd.co.uk, www.rpcltd.co.uk

Ryall & Edwards Limited - Green Lane Sawmills, Outwood, Redhill, Surrey. RH1 5QP tel:(01342) 842288, fax:(01342) 843312. enquiries@ryall-edwards.co.uk, www.ryall-edwards.co.uk

Rycroft Ltd - Duncombe Road, Bradford, W. Yorks. BD8 9TB tel:(01274) 490911, fax:(01274) 498580. sales@rycroft.com, www.rycroft.com

Rycroft Ltd - Duncombe Road, Bradford. BD8 9TB tel:(01274) 4904911, fax:(01274) 498580. sales@rycroft.com, www.rycroft.co.uk

Ryton's Building Products Ltd - Design House, Orion Way, Kettering Business Park, Kettering, Northants. NN15 6NL tel:(01536) 511874, fax:(01536) 310455. admin@rytons.com, www.vents.co.uk

S

S & B Ltd - Labtec Street, Swinton, Gt Man. M27 8SE tel:(0161) 793 9333, fax:(0161) 728 2233. sales@splusb.co.uk, www.splusb.info

S & P Coil Products Ltd - SPC House, Evington Valley Road, Leicester. LE5 5LU tel:(0116) 2490044, fax:(0116) 2490033. spc@spcoils.co.uk, www.spcoils.co.uk

Saacke Combustion Services Ltd - Langstone Technology Park, Langstone Road, Havant, Hants. PO9 1SA tel:(023) 9233 3900, fax:(023) 9233 3901. ukservice@saacke.com, www.saacke.com

Sadolin UK Ltd - PO Box 37, Crown House, Hollins Road, Darwin, Lancashire. BB3 0BG tel:(0845) 034 1464, fax:(0845) 389 9457. info@crownpaintspec.co.uk, www.crownpaintspec.co.uk

Safeguard Europe Ltd - Redkiln Close, Horsham, West Sussex. RH13 5QL tel:(01403) 210204, fax:(01403) 217529. info@safeguardeurope.com, www.safeguardeurope.com

Safetell Ltd - Unit 46, Fawkes Avenue, Dartford, Kent. DA1 1JQ tel:(01322) 223233, fax:(01322) 277751. secure@safetell.co.uk, www.safetell.co.uk

Safety Assured Ltd - Home Farm Estate, Fen Lane, North Ockendon, Essex. RM14 3RD tel:(01708) 855 777, fax:(01708) 855 125. info@safetyassured.com, www.safetyassured.com

Safety Works & Solutions Ltd - Unit 6, Earith Business Park, Meadow Drove, Earith. Huntingdon, Cambs. PE28 3QF tel:(01487) 841 400, fax:(01487) 841 100. marco@safetyworksandsolutions.co.uk, www.safetyworksandsolutions.co.uk

Saint-Gobain Abrasives - Doxey Road, Stafford. ST16 1EA tel:(01785) 222000, fax:(01785) 213487. enquiries.sgauk@saint-gobain.com, ww.saint-gobain-abrasives.com

Saint-Gobain Ecophon Ltd - Old Brick Kiln, Ramsdell, Tadley, Hants. RG26 5PP tel:(01256) 850977, fax:(01256) 850600. info@ecophon.co.uk, www.ecophon.co.uk

Saint-Gobain Glass UK Ltd - Weeland Road, Eggborough, Goole. DN14 0FD tel:(01977) 666100, fax:(01977) 666200. glassinfo.uk@saint-gobain-glass.com, www.saint-gobain.co.uk

Saint-Gobain PAM UK - Lows Lane, Stanton-By-Dale, Ilkeston, Derbys. DE7 4QU tel:(0115) 930 5000, fax:(0115) 932 9513. sales.uk.pam@saint-gobain.com, www.saint-gobain-pam.co.uk

Salex Acoustics Limited - 8 London Road, Woolmer Green, Hertfordshire. SG3 6JS tel:(01438) 545772, fax:(08700) 941918. info@salexacoustics.com, www.salexacoustics.com

Sampson & Partners Fencing - Redwell Wood Farm, Ridgehill, Potters Bar, Herts. EN6 3NA tel:(01707) 663400, fax:(01707) 661112. sales@sampsonfencing.co.uk, www.sampsonandpartners.com

Sampson Windows Ltd - Maitland Road, Lion Barn Business Park, Needham Market, Ipswich, Suffolk. IP6 8NZ tel:(01449) 722922, fax:(01449) 722911. info@nordan.co.uk, www.sampsonwindows.co.uk

Samuel Heath & Sons plc - Cobden Works, Leopold Street, Birmingham. B12 0UJ tel:(0121) 766 4200. info@samuel-heath.com, www.samuel-heath.co.uk

Sanderson - Sanderson House, Oxford Road, Denham, Bucks. UB9 4DX tel:(01895) 830000, fax:(01895) 830031. contractsales@a-sanderson.co.uk, www.sanderson-uk.com

Sandhurst Manufacturing Co Ltd - Belchmire Lane, Gosberton, Spaling, Lincolnshire. PE11 4HG tel:(01775) 840020, fax:(01775) 843063. info@sandhurst-mfg.com, www.sandhurst-mfg.com

Sandler Seating Ltd - 1A Fountayne Road, London. N15 4QL tel:(020) 7799 3000, fax:(020) 7729 2843. sales@sandlerseating.com, www.sandlerseating.com

Sandtoft Roof Tiles Ltd - Belton Road, Sandtoft, Doncaster, S. Yorks. DN8 5SY tel:(01427) 871200, fax:(01427) 871222. info@sandtoft.co.uk, www.sandtoft.com

Sangwin Concrete Products Ltd - Kelsey Hill, Burstwick. HU12 9HU tel:(01964) 622339, fax:(01964) 624287. sales@sangwin.co.uk, www.sangwin.co.uk

Sangwin Group - Dansom Lane South, Hull. HU8 7LN tel:(01482) 329921, fax:(01482) 215353. sales@sangwin.co.uk, www.sangwin.co.uk

Sanitary Appliances Ltd - 591 London Road, Sutton, Surrey. SM3 9AG tel:(020) 8641 0310, fax:(020) 8641 6426. info@sanitaryappliances.co.uk, www.sanitaryappliances.co.uk

Santon - Hurricane Way, Norwich, Norfolk. NR6 6EA tel:(01603) 420100, fax:(01603) 420229. specifier@santon.co.uk, www.santon.co.uk

Sapa Building Systems (monarch and glostal) - Alexandra Way, Ashchurch, Tewkesbury, Glos. GL20 8NB tel:(01684) 297073, fax:(01684) 293904. info.buildingsystems.uk@sapagroup.com, www.sapagroup.com/uk/buildingsystems

Sapa Building Systems Ltd - Severn Drive, Tewkesbury, Gloucestershire. GL20 8SF tel:(01684) 853500, fax:(01684) 851850. info@sapabuildingsystems.co.uk, www.sapabuildingsystems.co.uk

Sapphire Balustrades - 11 Arkwright Road, Reading. RG2 0LU tel:(0844) 8800553, fax:(0844) 8800530. sales@sapphirebalustrades.com, www.sapphirebalustrades.com

Sarena Mfg Ltd - Vickers Business Centre, Priestly Road, Basingstoke, Hampshire. RG24 9NP tel:(01634) 370887, fax:(01634) 370915. sales@sarena.co.uk, www.sarena.co.uk

Sarnafil Ltd - Robberds Way, Bowthorpe Industrial Estate, Norwich, Norfolk. NR5 9Jf tel:(01603) 748985, fax:(01603) 743054. roofing@sarnafil.co.uk, www.sarnafill.co.uk

Sashless Window Co Ltd - Standard Way, Northallerton, N. Yorks. DL6 2XA tel:(01609) 780202, fax:(01609) 779820. alastair@sashless.com, www.sashless.com

Satec Service Ltd - Unit 5 Wokingham Commercial Centre, Molly Millars Lane, Wokingham, Berkshire. RG41 2RF tel:(01189) 649006, fax:(01189) 640074. info@satec.co.uk, www.satec.co.uk

SAV United Kingdom Ltd - Scandia House, Boundary Road, Woking, Surrey. GU21 5BX tel:(01483) 771 910, fax:(01483) 227 519. info@sav-systems.com, www.sav-systems.com

Sawyer & Fisher - Unit B, Aviary Court, 138 Miles Road, Epsom, Surrey. KT19 9AB tel:(01372) 742815, fax:(01372) 729710. qs@sawyerfisher.co.uk

SC Johnson Wax - Frimley Green, Camberley, Surrey. GU16 5AJ tel:(01276) 852000, fax:(01276) 852308. www.scjohnson.co.uk

Scandanavian Timber - 73 Queens Road, Cheltenham, Gloucestershire. GL50 2NH tel:(01242) 694352. finn@scandinaviantimber.com, www.scandinaviantimber.com

Scapa Tapes UK Ltd - Gordleton Industrial Park, Hannah Way, Lymington, Hants. SO41 8JD tel:(01590) 684400, fax:(01590) 683728. technical.services@henkel.co.uk, www.sellotape.co.uk

Schauman (UK) Ltd - Stags End House, Gaddesden Row, Hemel Hempstead, Herts. HP2 6HN tel:(01582) 794661, fax:(01582) 794661. media@upm.com, http://www.upm.com

Schiedel Chimney Systems Ltd - Units 8 & 9 Block A, Holton Road, Holton Heath Industrial Estate, Poole, Dorset. BH16 6LG tel:(01202) 861650, fax:(01202) 861632. info@schiedel.co.uk, www.isokern.co.uk

Schindler Ltd - Benwell House, Green Street, Sunbury On Thames, Middx. TW16 6QT tel:(01932) 758100, fax:(01932) 758258. info@gb.schindler.com, www.schindlerlifts.co.uk

Schlegel UK (2006) Ltd - Henlow Industrial Estate, Henlow Camp, Bedfordshire. SG16 6DS tel:(01462) 815 500, fax:(01462) 638 800. schlegeluk@schlegel.com, www.schlegel.com/eu/en/

Schmidlin (UK) Ltd - White Lion Court, Swan Street, Old Isleworth, Middx. TW7 6RN tel:(020) 8560 9944, fax:(020) 8568 7081. info@lindner-group.com, www.lindner-schmidlin.com

Schneider Electric Ltd - Stafford Park 5, Telford, Shropshire. TF3 3BL tel:(01952) 290029, fax:(01952) 290534. info@schneider-electric.co.uk, www.schneider-electric.co.uk

SCHOTT UK Ltd - Sales Office, Drummond Road, Stafford. ST16 3EL tel:(01785) 223166, fax:(01785) 223522. info.uk@schott.com, www.schott.com/uk

Schueco UK - Whitehall Avenue, Kingston, Milton Keynes. MK10 0AL tel:(01908) 282111, fax:(01908) 282124. mkinfobox@schueco.com, www.schueco.co.uk

Schwing Stetter (UK) Ltd - Unit 11, Perivale Industrial Park, Horsenden Lane South, Greenford, Middx. UB6 7RL tel:(020) 8997 1515, fax:(020) 8998 3517. info@schwing-stetter.co.uk, www.schwing-stetter.co.uk

Scotfen Limited - Unit 1 Allanshaw Estate, Hamilton. ML3 9FD tel:(0345) 618 7218. info@scotfen.co.uk, www.scotfen.co.uk

Scotia Rigging Services (Industrial Stainless Steel Rigging Specialists) - 68 Bridge Street, Linwood, Paisley, Renfrewshire. PA3 3DR tel:(01505) 321127, fax:(01505) 321333. enquiries@scotia-handling-rigging.co.uk

Scotmat Flooring Services - 10 Westerton Road, East Mains Industrial Estate, Broxburn, West Lothian. EH52 5AU tel:(01506) 859995, fax:(01506) 859996

Scott Fencing Limited - Brunswick Industrial Estate, Newcastle-upon-Tyne, Tyne & Wear. NE13 7BA tel:(0191) 236 5314, fax:(0191) 217 0193. admin@scottfencing.ltd.uk

Scottish & Northern Ireland Plumbing Employers' Federation - Bellevue House, 22 Hopetoun Street, Edinburgh. EH7 4GH tel:(0131) 556 0600, fax:(0131) 557 8409. info@snipef.org, www.snipef.org

Scottish Master Wrights and Builders Association (SMWBA) - Blairtummock Lodge, Campsie Glen, Glasgow. G66 7AR tel:(01360) 770 583, fax:(01360) 770 583. enquiry@smwba.org.uk, www.smwba.org.uk

Scotts of Thrapston Ltd - Bridge Street, Thrapston, Northants. NN14 4LR tel:(01832) 732366, fax:(01832) 733703. enquiries@scottsofthrapston.co.uk, www.scottsofthrapston.co.uk

SCP Concrete Sealing Technology Ltd - Crowbush Farm Business Park, Luton Road, Toddington, Bedfordshire. LU5 6HU tel:(01525) 872700, fax:(01525) 871019. info@scpwaterproofing.com, www.scpwaterproofing.com

Screenbase Ltd - Unit B Avondale Works, Woodland Way, Kingswood, Bristol. BS15 1QH tel:(0845) 1300998, fax:(0845) 1300998. sales@Screenbase.co.uk, www.Screenbase.co.uk

SCS Group - T2 Capital Business Park, Parkway, Cardiff. CF3 2PZ tel:(0870) 240 64 60, fax:(02921) 510 058. info@groupscs.co.uk, www.groupscs.co.uk

SE Controls - Lancaster House, Wellington Crescent, Fradley Park, Lichfield, Staffs. WS13 8RZ tel:(01543) 443060, fax:(01543) 443070. sales@secontrols.co.uk, www.secontrols.com

Seagoe Technologies Ltd - Church Road, Portadown, Co. Armagh. BT63 5HU tel:(028) 38333131, fax:(028) 38333042. enquiries@glendimplex.com, www.glendimplexireland.com

Sealmaster - Brewery Road, Pampisford, Cambridge, Cambs. CB22 3HG tel:(01223) 832851, fax:(01223) 837215. sales@sealmaster.co.uk, www.sealmaster.co.uk

Sealocrete PLA Ltd - Greenfield Lane, Rochdale. OL11 2LD tel:(01706) 352255, fax:(01706) 860880. technical.service@bostik.com, www.sealocrete.co.uk

Seco Aluminium Ltd - Crittall Road, Witham, Essex. CM8 3AW tel:(01376) 515141, fax:(01376) 500542. seco.aluminium@btinternet.com, www.secoaluminium.co.uk

Sections & Profiles Ltd - Gaitskell Way, Dartmouth Road, Smethwick, W. Midlands. B66 1BF tel:(0121) 555 1400, fax:(0121) 555 1431. ask.hadley@hadleygroup.co.uk, www.hadleygroup.co.uk

Securikey Ltd - PO Box 18, Aldershot, Hants. GU12 4SL tel:(01252) 311888, fax:(01252) 343950. enquires@securikey.co.uk, www.securikey.co.uk

Selkin Media Ltd - P O Box 10118, Melton Mobray, Leics. LE14 2WT tel:(01159) 232286, fax:(01159) 233816. customerservices@selkinmedia.co.uk, www.selkinshop.co.uk

Sellite Blocks Ltd - Old Quarry, Long Lane, Great Heck, N. Yorks. DN14 0BT tel:(01977) 661631, fax:(01977) 662155. sales@sellite.co.uk, www.sellite.co.uk

Senior Aluminium Systems plc - Eland Road, Denaby Main, Doncaster, S. Yorkshire. DN12 4HA tel:(01709) 772600, fax:(01709) 772601. enquiry@senioraluminium.co.uk, www.senioraluminium.co.uk

S (con't)

Sentra Access Engineering Ltd - Unit D13, Meltham Mills Industrial Estate, Holmfirth, Huddersfield. HD9 4DF tel:(01484) 851222, fax:(01484) 851333. sales@sentra-online.co.uk, www.sentra-online.co.uk
Servomac Limited - Meadows Drive, Ireland Industrial Estate, Staveley, Chesterfield. S43 3LH tel:(01246) 472631, fax:(01246) 470259. info@servomac.com, www.servomac.com
Servowarm - Unit 9, The Gateway Centre, Coronation Road, High Wycombe, Bucks. HP12 3SU tel:(01494) 474474, fax:(01494) 472906. newbusiness@npower.com, www.servowarm.co.uk
Setsquare - Valley Industries, Hadlow Road, Tonbridge, Kent. TN11 0AH tel:(01732) 851 888, fax:(01732) 851 853. sales@setsquare.co.uk, www.setsquare.co.uk
SFC (Midlands) Ltd - Unit 35, Hayhill Industrial Estate, Barrow Upon Soar, Loughborough, Leics. LE12 8LD tel:(01509) 816700. dansleeman@sfcmidlands.co.uk, www.sfcmidlands.co.uk
SFS Intec LTD - 153 Kirkstall Road, Leeds, W. Yorks. LS4 2AT tel:(0113) 2085 500, fax:(0133) 2085573. gb.leeds@sfsintec.biz, www.sfsintec.biz/uk
SG System Products Ltd - Unit 22, Wharfedale Road, Ipswich, Suffolk. IP1 4JP tel:(01473) 240055, fax:(01473) 461616. Sales@sgsystems.co.uk, www.handrailsuk.co.uk
SGB, a Brand Company - Building 100, Relay Point, Relay Drive, Tamworth. B77 5PA tel:(0844) 335 8860. ukinfo@beis.com, www.sgb.co.uk
Shackerley (Holdings) Group Ltd incorporating Designer ceramics - Ceramics House, 139 Wigan Road, Euxton, Chorley, Lancs. PR7 6JH tel:(01257) 273114, fax:(01257) 262386. info@shackerley.com, www.shackerley.com
Shadbolt F R & Sons Ltd - North Circular Road, South Chingford, London. E4 8PZ tel:(020) 8527 6441, fax:(020) 8523 2774. sales@shadbolt.co.uk, www.shadbolt.co.uk
Sharman Fencing Ltd - Robinswood Farm, Bere Regis, Wareham, Dorset. BH20 7JJ tel:(01929) 472181, fax:(01929) 472182. info@sharmanfencing.co.uk, www.sharmanfencing.co.uk
Shavrin Levatap Co Ltd - 32 Waterside, Kings Langley, Herts. WD4 8HH tel:(01923) 267678, fax:(01923) 265050. info@shavrinlevatap.co.uk, www.shavrinlevatap.co.uk/top menu.htm
Shaws of Darwen - Waterside, Darwen, Lancs. BB3 3NX tel:(01254) 775111, fax:(01254) 873462. jwilson@shawsofdarwen.com, www.shaws-terracotta.com
Shaylor Group Plc - Frederick James House, 52 Wharf Approach, Anchor Brook Business Park, Aldridge, West Midlands. WS9 8BX tel:(01922) 741570, fax:(01922) 745604. enquiries@shaylorgroup.com, www.shaylorgroup.com
shearflow LTD - Pharaoh House, Arnolde close, Medway City Estate, Rochester, Kent. ME2 4SP tel:(01634) 735 020, fax:(01634) 735 019. sales@shearflow.co.uk, www.shearflow.co.uk
Shell Bitumen - Shell House, Beach Hill, Clonskeagh, Dublin 4. Ireland tel:+353 1819 7030, fax:+353 1808 8250. shell-bitumen@shell.com, www.shell.com
Shellcast Security Shutters - Unit 14, Acorn Enterprise Centre, Hoo Farm Industrial Estate, Kidderminster, Worcs. DY11 7RA tel:(01562) 750700, fax:(01562) 750800. shellcast@btconnect.com, www.shellcast-shutters.co.uk
Shepherd Engineering Services Ltd - Mill Mount, York, N. Yorks. YO24 1GH tel:(01904) 629151, fax:(01904) 610175. mfaxon@ses-ltd.co.uk, www.ses-ltd.co.uk
Shires Bathrooms - Beckside Road, Bradford, W. Yorks. BD7 2JE tel:(01274) 521199, fax:(01274) 521583. info@shires-bathrooms.com, www.shires-bathrooms.co.uk
Showerlux UK Ltd - Sibree Road, Coventry, W. Midlands. CV3 4EL tel:(024) 76 639400, fax:(024) 76 305457. sales@showerlux.co.uk, www.showerlux.uk.com
ShowerSport Ltd - 1 North Place, Edinburgh Way, Harlow, Essex. CM20 2SL tel:(01279) 451450, fax:(01279) 635074
Shutter Door Repair & Maintenance Ltd - 89-91 Rolfe Street, Smethwick, Warley, W. Midlands. B66 2AY tel:(0121) 558 6406, fax:(0121) 558 7140. enquiries@sis-group.co.uk, www.sis-group.co.uk
Sibelco UK Ltd - Brookside Hall, Sandbach, Cheshire. CW11 4TF tel:(01270) 752 752, fax:(01270) 752753. info@sibelco.co.uk, www.sibelco.co.uk
Sidhil Care - Sidhil Business Park, Holmfield, Halifax, West Yorkshire. HX2 9NT tel:(01422) 363447, fax:(01422) 344270. sales@sidhil.com, www.sidhil.com
Siemens Building Technologies - Brunel House, Sir William Siemens Square, Frimley, Camberley, Surrey. GU16 8QD tel:(01784) 461616, fax:(01784) 464646. sales.sbt.uk@siemens.com, www.siemens.co.uk/buildingtechnologies
Sierra Windows - Alders Way, Yalberton Industrial Estate, Paignton, Devon. TQ4 7QE tel:(01803) 697000, fax:(01803) 697071. info@sierrawindows.co.uk, www.sierrawindows.co.uk
SIG Interiors - Hillsborough Works, Langsett Road, Sheffield. S6 2LW tel:(0114) 2318030, fax:(0114) 2318031. info@sigplc.com, www.sigplc.co.uk
SIG Roofing - Harding Way, St Ives, Cambridgeshire. PE27 3YJ tel:(01480) 466777, fax:(01480) 300269. info@sigroofing.co.uk, www.sigroofing.co.uk
Sign Industries - Gardyn, Forfar, Angus. DD8 2SQ tel:(01241) 828694, fax:(01241) 828331. info@signindustries.com, www.signindustries.com
Signature Ltd - Signature House, 51 Hainge Road, Tividale, Oldbury, West Midlands. B69 2NY tel:(0121) 557 0995, fax:(0121) 557 0995. sales@signatureltd.com, www.signatureltd.com
Signs & Labels Ltd - Douglas Bruce House, Corrie Way, Stockport, Cheshire. SK6 2RR tel:(0800) 132323, fax:(0161) 430 8514. sales@signsandlabels.co.uk, www.safetyshop.com
Signs Of The Times Ltd - Wingfield Road, Tebworth, Leighton Buzzard, Beds. LU7 9QG tel:(01525) 874185, fax:(01525) 875746. enquiries@sott.co.uk, www.sott.co.uk
Sika Liquid Plastics Limited - Sika House, Miller Street, Preston, Lancashire. PR1 1EA tel:(01772) 259781, fax:(01772) 255672. info@liquidplastics.co.uk, http://gbr.liquidplastics.sika.com
Sika Ltd - Watchmead, Welwyn Garden City, Herts. AL7 1BQ tel:(01707) 394444, fax:(01707) 329129. sales@uk.sika.com, www.sika.co.uk
Silavent Ltd - 60 High Street, Sandhurst, Berks. GU47 8DY tel:(01252) 878282, fax:(01252) 871212. silavent@btinternet.com, www.silavent.co.uk
Silent Gliss Ltd - Pyramid Business Park, Poorhole Lane, Broadstairs, Kent. CT10 2PT tel:(01843) 863571, fax:(01843) 864503. info@silentgliss.co.uk, www.silentgliss.co.uk
Silverdale Bathrooms - 293 Silverdale Road, Newcastle Under Lyme, Staffordshire. ST5 6EH tel:(01782) 717 175, fax:(01782) 717 166. sales@silverdalebathrooms.co.uk, www.silverdalebathrooms.co.uk
Simflex Grilles and Closures Ltd - 9 Woburn Street, Ampthill, Beds. NK45 2HP tel:(01525) 841100, fax:(01525) 405561. sales@simflex.co.uk, www.simflex.co.uk
Simon Moore Water Services - Unit 2, Poundbury West Industrial Estate, Dorchester. DT1 2BG tel:(01305) 251551, fax:(01305) 257107. info@simon-moore.com, www.simon-moore.com
Simon R W Ltd - System Works, Hatchmoor Industrial Estate, Torrington, Devon. EX38 7HP tel:(01805) 623721, fax:(01805) 624578. info@rwsimon.co.uk, www.rwsimon.co.uk
Simons Construction Ltd - 401 Monks Road, Lincoln, Lincs. LN3 4NU tel:(01522) 510000, fax:(01522) 521812. webmaster@simonsgroup.com, www.simonsgroup.com
Simplex Signs Limited - Unit C, Peter Road, Lancing Business Park, Lancing, W. Sussex. BN15 8TH tel:(01903) 750333, fax:(01903) 750444. sales@simplexltd.com, www.simplexltd.com
Simplylabs Limited - 4 Aerodrome Close, Loughborough, Leicestershire. LE11 5RJ tel:(01509) 611322, fax:(01509) 611416. sales@simplylabs.co.uk, www.simplylabs.co.uk
Simpson Strong-Tie® - Winchester Road, Cardinal Point, Tamworth, Staffs. B78 3HG tel:(01827) 255600, fax:(01827) 255616. info@strongtie.com, www.strongtie.com
Single Ply Roofing Association - Roofing House, 31 Worship Street, London. EC2A 2DY tel:(0845) 154 7188. enquiries@spra.co.uk, www.spra.co.uk
Siniat Ltd - Marsh Lane, Easton-in-Gordano, Bristol. BS20 0NF tel:(0800) 145 6033, fax:(01275) 377456. enquiryline@siniat.co.uk, www.siniat.co.uk
Sinotech Limited - Unit B1Sovereign Court, Market Harborough, Leics. LE16 9EG tel:(01858) 433 067, fax:(01858) 410 205. tdelday@sinotechltd.co.uk, www.coldform.co.uk
SJ McBride Ltd T/A Warefence - Unit H, Ventura Park, Broadshires Way, Carterton, Oxon. OX18 1AD tel:(01993) 845279, fax:(01993) 840551. info@warefence.co.uk, www.warefence.co.uk
SK Environmental Ltd - Unit 7, Greenhey Place, East Gillibrands, Skelmersdale, Lancs. WN8 9SB tel:(01695) 714600. sales@skenvironmental.com, www.skenvironmental.com
Skanda Acoustics - 67 Clywedog Road North, Wrexham Industrial Estate, Wrexham. LL13 9XN tel:(01978) 664255, fax:(01978) 661427. info@skanda-uk.com, www.savolit.co.uk
Skanda Acoustics Limited - 67 Clywedog Road North, Wrexham Industrial Estate, Wrexham. LL13 9XN tel:(01978) 664255, fax:(01978) 661427. sales@skanda-uk.com, www.skanda-uk.com
Skanska UK Plc - Maple Cross House, Denham Way, Maple Cross, Rickmansworth, Herts. WD3 9SW tel:(01923) 776666, fax:(01923) 423681. srw.enquiries@skanska.co.uk, www.skanska.co.uk
Slatesystem - Tythings Commercial Centre, Southgate Road, Wincanton, Somerset. BA9 9RZ tel:(01963) 31041, fax:(01963) 31042. sales@slatesystem.co.uk, www.sletesystem.co.uk
SLE Cladding Ltd - Tilstock Lane, Prees Heath, Whitchurch, Shropshire. SY13 3JP tel:(01948) 666321, fax:(01948) 665532. www.slecladding.co.uk/
Slipstop - P.O. Box 7404, Coalville. LE67 4ZS tel:(01530) 813 500, fax:(01530) 813 400. info@slipstop.co.uk, www.slipstop.co.uk
Smart F & G (Shopfittings)Ltd - The Shopfitting Centre, Tyseley Industrial Estate, Seeleys Road, Greet, Birmingham. B11 2LA tel:(0121) 772 5634, fax:(0121) 766 8995. info@smartsshopfittings.co.uk, www.smartsshopfittings.co.uk
Smart Systems Ltd - Arnolds Way, Yatton, N. Somerset. BS49 4QN tel:(01934) 876100, fax:(01934) 835169. sales@smartsystems.co.uk, www.smartsystems.co.uk
Smith & Choyce Ltd - 280 Barton Street, Gloucester, Glos. GL1 4JJ tel:(01452) 523531, fax:(01452) 310032. martin@smithandchoyce.co.uk
Smith & Rodger Ltd - 24-36 Elliot Street, Glasgow. G3 8EA tel:(0141) 248 6341, fax:(0141) 248 6475. info@smithandrodger.co.uk, www.frenchpolishes.com
Smithbrook Building Products Ltd - PO Box 2133, Shoreham-by-Sea. BN43 9BD tel:(01273) 573811, fax:(01273) 689021. info@smithbrookproducts.com, www.smithbrookproducts.com
Society for the Protection of Ancient Buildings (SPAB) - 37 Spital Square, London. E1 6DY tel:(020) 7377 1644, fax:(020) 7247 5296. info@spab.org.uk, www.spab.org.uk
Sola Skylights - Greenfield Industrial Estate, Tindale Crescent, Bishop Aukland, Co. Durhan. DL14 9TF tel:(01388) 451133, fax:(01388) 608444. info@solaskylights.com, www.solaskylights.co.uk
Solaglas Laminated - Saffron Way, Sittingbourne, Kent. ME10 2PD tel:(01795) 421534, fax:(01795) 473651. solaglas.gpd@saint-gobain-glass.com, www.solaglas.sggs.com
Solair Ltd - Pennington Close, Albion Road, West Bromwich, W. Midlands. B70 8BA tel:(0121) 525 2722, fax:(0121) 525 6786. sales@solair.co.uk, www.solair.co.uk
Solair GRP Architectural Products - Smeaton Road, Churchfields Industrial Estate, Sailsbury, Wilts. SP2 7NQ tel:(01722) 323036, fax:(01722) 337546. sales@solair.co.uk, www.solair.co.uk
Solar Trade Association, The - The National Energy Centre, Davy Avenue, Knowlhill, Milton Keynes. MK5 8NG tel:(01908) 442290, fax:(01908) 665577. enquiries@solar-trade.org.uk, www.solartradeassociation.org.uk
Solus Ceramics Ltd - Unit 1, Cole River Park, 285 Warwick Road, Birmingham. B11 2QX tel:(0121) 753 0777, fax:(0121) 753 0888. specify@solusceramics.com, www.solusseramics.com
Sommerfeld Flexboard Ltd - New Works Lane, Arleston Hill, Wellington,Telford, Shropshire. TF1 2JY tel:(07770) 822973. info@sommerfeld.co.uk, www.sommerfeld.co.uk
Sound Reduction Systems Ltd - Adam Street, Off Lever Street, Bolton. BL3 2AP tel:(01204) 380074, fax:(01204) 380957. info@soundreduction.co.uk, www.soundreduction.co.uk
Sound Service (Oxford) Ltd - 55 West End, Witney, Oxon. OX28 1NJ tel:(08707) 203093, fax:(01993) 779569. soundservice@btconnect.com, www.soundservice.co.uk
Soundsorba Ltd - Shaftesbury Street, High Wycombe, Bucks. HP11 2NA tel:(01494) 536888, fax:(01494) 536818. info@soundsorba.com, www.soundsorba.com
South Durham Structures Limited - Dovecot Hill, South Church Enterprise Park, Bishop Auckland, Co. Durham. DL14 6XR tel:(01388) 777350, fax:(01388) 775225. info@southdurhamstructures.co.uk, www.south-durham-structures.co.uk/
South east Bird Control - SVS House, Oliver Grove, London. SE25 6EJ tel:(020) 8123 0284. info@southeastbirdcontrol.co.uk, www.southeastbirdcontrol.co.uk
Sovereign Fencing Limited - Sovereign House, 261 Monton Road, Monton, Eccles, Gt Man. M30 9LF tel:(0161) 789 5479. sovereignfencing@btconnect.com
Space - Ray UK - Chapel Lane, Claydon, Ipswich, Suffolk. IP6 0JL tel:(01473) 830551, fax:(01473) 832055. info@spaceray.co.uk, www.spaceray.co.uk
Spaceright Europe Limited - 38 Tollpark Road, Wardpark East, Cumbernauld. G68 0LW tel:(01236) 823291, fax:(01236) 825356. info@spacerighteurope.com, www.spacerighteurope.com
Spaceway South Ltd - Premier House, Abbey Park Industrail Estate, Romsey, Hants. SO51 9DG tel:(01794) 835600, fax:(01794) 835601. sales@spaceway.co.uk, www.spaceway.co.uk
Sparkes K & L - The Forge, Claverdon, Nr Warwick, Warwicks. CV35 8P tel:(01926) 842545, fax:(01926) 842559. info@sparkesstoves.co.uk, www.sparkesstoves.co.uk
Spartan Promenade Tiles Ltd - Slough Lane, Ardleigh, Colchester, Essex. CO7 7RU tel:(01206) 230553, fax:(01206) 230516. david@spartantiles.com, www.spartantiles.com
Spear & Jackson - Atlas Way, Atlas North, Sheffield, S. Yorks. S4 7QQ tel:(0114) 281 4242, fax:(0114) 281 4252. sales@spear-and-jackson.com, www.spear-and-jackson.com
Spear & Jackson Interntional Ltd - Atlas Way, Atlasnorth, Sheffield. S4 -7QQ tel:(0114) 256 1133, fax:(0114) 243 1360. sales@neill-tools.co.uk, www.spear-and-jackson.com
Specialist Access Engineering and Maintenance Association (SAEMA) - Carthusian Court, 12 Carthusian Street, London. EC1M 6EZ tel:(020) 7397 8122, fax:(020) 7397 8121. enquiries@saema.org, www.saema.org
Specialist Building Products - PO Box 12422, Holland Park Avenue, Holland Park, London. W11 4GX tel:(020) 8458 8212, fax:(020) 8458 4116. specialistbuildingproducts@yahoo.co.uk
Spectra Glaze Services - Unit 3, Ensign House, Woodmansterne Lane, Carshalton, Surrey. RH2 7QL tel:(020) 8647 1545, fax:(020) 8647 8070. sales@spectraglaze.co.uk, www.spectraglaze.co.uk
Spectus Systems - Snape Road, Macclesfield, Cheshire. SK10 2NZ tel:(01625) 420400, fax:(01625) 433946. contacting@spectus.co.uk, www.spectussystems.com
Speedfab Midlands - Unit 10, Credenda Road, West Bromwich, West Midlands. B70 7JE tel:(0121) 541 1761, fax:(0121) 544 0028. Sales@speedfab.com, www.speedfab.com
Sperrin Metal Products Ltd - Cahore Road, Draperstown, Co. Derry. BT45 7AP tel:(028) 79628362, fax:(028) 79628972. dean@sperrin-metal.com, www.sperrin-metal.com
Spindlewood - 8 Edward Street, Bridgewater, Somerset. TA6 5ET tel:(01278) 453665, fax:(01278) 453666. info@spindlewoodturning.co.uk, www.spindlewoodturning.co.uk
Spiral Construction Ltd - Helston, Cornwall. TR13 0LW tel:(01326) 574497, fax:(01326) 574760. enquiries@spiral.uk.com, www.spiral.uk.com
Spontex Ltd - Berkley Business Park, Wainwright Road, Worcester. WR4 9ZS tel:(01305) 450300, fax:(01905) 450350. www.spontex.co.uk
Sports and Play Construction Association - Federation House, Stoneleigh Park, Kenilworth, Warwickshire. CV8 2RF tel:(024) 7641 6316, fax:(024) 7641 4773. info@sapca.org.uk, www.sapca.org.uk
Sportsmark Group Ltd - 18 St Georges Place, Semington, Trowbridge, Wiltshire. BA14 6GB tel:(01635) 867537 , fax:(01635) 864588. sales@sportsmark.net, www.sportsmark.net
Sportsmark Group Ltd - Sportsmark House, 4 Clerewater Place, Lower Way, Thatcham, Berkshire. RG19 3RF tel:(01635) 867537, fax:(01635) 864588. sales@sportsmark.net, www.sportsmark.net
Sprayed Concrete Association - Kingsley House, Ganders Business Park, Kingsley, Bordon, Hampshire. GU35 9LU tel:(01420) 471622, fax:(01420) 471611. admin@sca.org.uk, www.sca.org.uk
Springvale E P S - 75 Springvale Road, Ballyclare, Co. Antrim. BT39 0SS tel:(028) 93340203, fax:(028) 93341159. sales@springvale.com, www.springvale.com
SPX Flow Technology Crawley Ltd - The Beehive, Beehive Ring Road, Gatwick, W. Sussex. RH6 0PA tel:(01293) 527777, fax:(01293) 552640. apvwebsite@apv.com, www.apv.com
Squire Henry & Sons - Linchfield Road, New Invention, Willenhall, W. Midlands. WV12 5BD tel:(01922) 476711, fax:(01922) 493490. info@henry-squire.co.uk, www.squirelocks.co.uk
Squires Metal Fabrications Ltd - 6 Burgess Road, Ivyhouse Lane industrial Estate, Hastings, Sussex. TN35 4NR tel:(01424) 428794, fax:(01424) 431567. squiresmetal@tiscali.co.uk, www.squiresmetal.co.uk
SSAB Swedish Steel Ltd - Narrow Boat Way, Hurst Business Park, Brierley Hill, West Midlands. DY5 1UF tel:(01384) 74660, fax:(01384) 77575. sales@dobel.co.uk, www.dobel.co.uk
SSI Schaefer Ltd - 83-84 Livingstone Road, Walworth Industrial Estate, Andover, Hants. SP10 5QZ tel:(01264) 386600, fax:(01264) 386611. solutions@ssi-schaefer.co.uk, www.ssi-schaefer.co.uk
SSL Access - Arrol House, 9 Arroy Road, Glasgow. G40 3DQ tel:(0141) 551 0807, fax:(0141) 554 7803. sales@sslaccess.co.uk, www.sslaccess.co.uk
Staines Steel Gate Co - 20 Ruskin Road, Staines-Upon-Thames, Surrey. TW18 2PX tel:(01784) 454456, fax:(01784) 466668. www.stainesgates.co.uk
Stamford Products Limited - Bayley Street, Stalybridge, Cheshire. SK15 1QQ tel:(0161) 330 6511, fax:(0161) 330 5576. enquiry@stamford-products.co.uk, www.stamford-products.co.uk
Stancliffe Stone - Grangemill, Matloc, Derbyshire. DE4 4BW tel:(01629) 653 000, fax:(01629) 650 996. info@stancliffe.com, www.stancliffe.com
Standard Patent Glazing Co Ltd - Flagship House, Forge Lane, Dewsbury, W. Yorks. WF12 9EL tel:(01924) 461213, fax:(01924) 458083. sales@patent-glazing.com, www.patent-glazing.com

S (con't)

Stanley Security Solutions - Bramble Road, Swindon, Whiltshire. SN2 8ER tel:(01793) 412345, fax:(01793) 615848. info@stanleysecuritysolutions.co.uk, www.stanleysecuritysolutions.co.uk
Stanley Tools - Sheffield Business Park, Sheffield City Airport, Europa Link, Sheffield. S3 9PD tel:(0114) 244 8883, fax:(0114) 273 9038. www.stanleytools.com
Stannah Lifts Ltd - Anton Mill, Andover, Hants. SP10 2NX tel:(01264) 339090, fax:(01264) 337942. contact@stannah.co.uk, www.stannahlifts.co.uk
Star Uretech Ltd - Enterprise House, Hollin Bridge Street, Blackburn, Lancashire. BB2 4AY tel:(01254) 663444, fax:(01254) 681886. info@star-uretech.com, www.star-uretech.com
Staverton (UK) Ltd - Micklebring Way, Rotherham, South Yorkshire. S66 8QD tel:(0844) 225 7474. info@staverton.co.uk, www.staverton.co.uk/
Staytite Ltd - Staylite House, Halifax Road, Cressex Business Park, High Wycombe, Herts. HP12 3SN tel:(01494) 462322, fax:(01494) 464747. fasteners@staytite.com, www.staytite.com
Steel Doors & Security Systems - Haywood Trading Estate, 15 Holcroft Road, Halesowen, W. Midlands. B62 8ES tel:(0800) 783 5745, fax:(01384) 566977. sales@steeldoorssecurity.co.uk, www.steeldoorssecurity.co.uk
Steel Line Ltd - 415 Petre Street, Sheffield. S4 8LL tel:(0114) 231 7330, fax:(0114) 256 0330. enquiries@steelline.co.uk, www.steelline.co.uk
Steel Window Association - Steel Window Association, Unit 2 Temple Place, 247 The Broadway, London. SW19 1SD tel:(0844) 249 1355, fax:(0844) 249 1356. info@steel-window-association.co.uk, www.steel-window-association.co.uk
Steelcase Strafor Plc - 183 Eversholt Street, London. NW1 1BU chessie@mediahouse.co.uk, www.steelcase.com
Steel-Crete Ltd - Stuart House, Kingsbury Link, Trinity Road, Staffs. B87 2EX tel:(01827) 871140, fax:(01827) 871144. tony@steelcrete.com, www.steelcrete.com
Steelway Fensecure Ltd - Queensgate Works, Bilston Road, Wolverhampton, W. Midlands. WV2 2NJ tel:(01902) 451733, fax:(01902) 452256. sales@steelway.co.uk, www.steelway.co.uk
Steico - 1st Floor, New Barnes Mill, Cotton Mill Lane, St Albans, Hertfordshire. AL1 2HA tel:(01727) 515 120, fax:(01727) 836 392. a.moore@steico.com, www.steico.co.uk
Steinel (UK) Ltd - 25 Manasty Road, Axis Park, Orton Southgate, Peterborough. PE2 6UP tel:(01733) 366700, fax:(01733) 238270. steinel@steinel.co.uk, www.steinel.co.uk
Stent Foundations Ltd - Pavilion C2, Ashwood Park, Ashwood Way, Basingstoke, Hants. RG23 8BG tel:(01256) 400200, fax:(01256) 400201. foundations@stent.co.uk, www.stent.co.uk
Stevenson & Cheyne (1983) - Unit 7, Butlerfield Industrial Estate, Bonnyrigg, Midlothian. EH19 3JQ tel:(01875) 822822, fax:(01875) 823723. farser-s-smith@btconnect.com, www.platerolling.co.uk
Stiebel Eltron Ltd - Lyveden Road, Brackmills, Northampton, Northants. NN4 7ED tel:(01604) 766421, fax:(01604) 765283. info@stiebel-eltron.co.uk, www.Stiebel-eltron.co.uk
STO Ltd - 2 Gordon Avenue, Hillingdon Park, Glasgow, Scotland. G52 4TG tel:(0141) 892 8000, fax:(0141) 404 9001. info.uk@sto.com, www.sto.co.uk
Stocksigns Ltd - 43 Ormside Way, Redhill, Surrey. RH1 2-LG tel:(01737) 764764, fax:(01737) 763763. info@stocksigns.co.uk, www.stocksigns.co.uk
Stokvis Industrial Boilers International Limited - 96 R Walton Road, East Molesey, Surrey. KT8 0DL tel:(020) 8941 1212, fax:(020) 8941 4136. info@stokvisboilers.com, www.stokvisboilers.com
Stone Fasterners Ltd - Woolwich Road, London. SE7 8SL tel:(020) 8293 5080, fax:(020) 8293 4935. sales@stonefasteners.com, www.stonefasteners.com
Stone Federation Great Britain - Channel Business Centre, Ingles Manor, Castle Hill Avenue, Folkstone, Kent. CT20 2RD tel:(01303) 856123, fax:(01303) 221095. enquiries@stone-federationgb.org.uk, www.stone-federationgb.org.uk
Stoneguard (London) Ltd - Coombe Works, Coombe Road, Neasden, London. NW10 0EB tel:(020) 8450 9933, fax:(020) 8450 7372. sales@stoneguard.co.uk, www.stoneguard.co.uk
Stoneham PLc - Powerscroft Road, Foots Cray, Sidecup, Kent. DA14 5DZ tel:(020) 83008181, fax:(020) 8300 8183. kitchens@stoneham.plc.uk, www.stoneham-kitchens.co.uk/
Stonewest Ltd - 67 Westow Street, Crystal Palace, London. SE19 3RW tel:(020) 8684 6646, fax:(020) 8684 9323. info@stonewest.co.uk, www.stonewest.co.uk
Stoves PLC - Stoney Lane, Prescot, Liverpool, Merseyside. L35 2XW tel:(0151) 426 6551, fax:(0151) 426 3261. productsales@gdha.com, www.stoves.co.uk
Stowell Concrete Ltd - Arnolds Way, Yatton, N. Somerset. BS49 4QN tel:(01934) 834000, fax:(01934) 835474. sales@stowellconcrete.co.uk, www.stowellconcrete.co.uk
Strada London - Unit 2C Kimberley Business Park, Blackness Lane, Keston, Kent. BR2 6HL tel:(0808) 1786007, fax:(0808) 1789007. sales@strada.uk.com, www.strada.uk.com
Straight Ltd - No 1 Whitehall Riverside, Leeds. LS1 4BN tel:(0113) 245 2244, fax:(0843) 557 0011. info@straight.co.uk, www.straight.co.uk
Strand Hardware Ltd - Strand House, Premier Business Park, Long Street, Walsall, W. Midlands. WS2 9DY tel:(01922) 639111, fax:(01922) 626025. info@strandhardware.co.uk, www.strandhardware.co.uk
Streamtec Limited - 2 Kirkgate House, Baden-Powell Road, Arbroath, Angus. DD11 3LS tel:(01241) 436862, fax:(01241) 436787. info@streamtec.com, www.streamtec.com
Stretched Fabric Systems - 74 Compton Street, London. EC1V 0BN tel:(020) 7253 4608, fax:(020) 7253 5746. sales@stretchedfabricsystems.com, www.stretchedfabricsystems.com
Structural Soils Ltd - The Old School, Stillhouse Lane, Bedminster, Bristol. BS3 4EB tel:(0117) 947 1000, fax:(0117) 947 1004. admin@soils.co.uk, www.soils.co.uk
Stuart Turner Ltd - Market Place, Henley-on-Thames, Oxfordshire. RG9 2AD tel:(01491) 572655, fax:(01491) 573704. pumps@stuart-turner.co.uk, www.stuart-turner.co.uk
Stuarts Industrial Flooring Ltd - Stuart House, Trinity Road, Kingsbury Link, Tamworth. B78 2EX tel:(01827) 871140, fax:(01827) 871144. chris.henderson@stuarts-flooring.co.uk, www.stuarts-flooring.co.uk
Style Door Systems Ltd - The Old Stables, 42a Chorley New Road, Bolton, Lancashire. BL1 4NX tel:(01204) 845590, fax:(01204) 849065. north@style-partitions.co.uk, www.style-partitions.co.uk
Sulzer Pumps - 5th Floor, Astral Towers, Betts Way, London Road, Crawley, W. Sussex. RH10 9UY tel:(01293) 558 140, fax:(01293) 527 972. www.sulzer.com
Sundeala Ltd - Middle Mill, Cam, Dursley, Glos. GL11 5LQ tel:(01453) 542286, fax:(01453) 549085. sales@sundeala.co.uk, www.sundeala.co.uk
Sunflex Ltd - Keys Park Road, Hednesford, Cannock, Staffs. WS12 2FR tel:(01543) 271421, fax:(01543) 279505. info@sunflex.co.uk, www.sunflex.co.uk
Sunfold Systems Ltd - The Green House, 93 Norwich Road, Dereham, Norfolk. NR20 3AI tel:(020) 8742 8887, fax:(020) 8994 2525. info@sunfold.com, www.sunfoldsystems.co.uk
Sunray Engineering Ltd - Kingsnorth Industrial Estate, Wotton Road, Ashford, Kent. TN23 6LL tel:(01233) 639039, fax:(01233) 625137. sales@sunraydoors.co.uk, www.sunraydoors.co.uk
Sunrite Blinds Ltd - 4 Newhailes Ind Estate, Musselburgh. EH21 6SY tel:(0131) 669 2345, fax:(0131) 657 3595. info@sunrite.co.uk, www.sunrite.co.uk
Sunway UK Ltd - Mersey Industrial Estate, Heaton Mersey, Stockport, Cheshire. SK4 3EQ tel:(0161) 432 5303, fax:(0161) 431 5087. www.sunwayblinds.co.uk
Surespan Ltd - PO Box 52, Walsal, W. Midlands. WS2 7PL tel:(01922) 711185, fax:(01822) 497943. sales@surespancovers.com, www.surespancovers.com
Swedecor - First Floor, Rotterdam Road, Hull, E. Yorks. HU7 0XU tel:(01482) 329691, fax:(01482) 212988. info@swedecor.com, www.swedecor.com
Swintex Ltd - Derby Works, Manchester Road, Bury, Lancs. BL9 9NX tel:(0161) 761 4933, fax:(0161) 797 1146. alan.armitage@swintex.co.uk, www.swintex.com
Swish Building Products - Pioneer House, Lichfield Rd Industrial Estate, Mariner, Tamworth, Staffs. B79 7TF tel:(01827) 317200, fax:(01827) 317201. marketing@swishbp.co.uk, www.swishbp.co.uk
Switch2 Energy Solutions Ltd - High Mill, Cullingworth, Bradford, W. Yorks. BD13 5HA tel:(0870) 9996030, fax:(0870) 9996031. sales@switch2.com, www.switch2.com
Sylvania Lighting International INC - Otley Road, Charlestown, Shipley, W. Yorks. BD17 7SN tel:(01274) 532 552, fax:(01274) 531 672. info.uk@havells-sylvania.com, www.sylvania-lighting.com
Symphony Group PLC, The - Gelderd Lane, Leeds. LS12 6AL tel:(0113) 230 8000, fax:(0113) 230 8134. enquiries@symphony-group.co.uk, www.symphonygroup.co.uk
Syston Rolling Shutters Ltd - 33 Albert Street, Syston, Leicester, Leics. LE7 2JB tel:(0116) 2608841, fax:(0116) 2640846. sales@syston.com, www.syston.com
Szerelmey Ltd - 369 Kennington Lane, Vauxhall, London. SE11 5QY tel:(020) 7735 9995, fax:(020) 7793 9800. info@szerelmey.com, www.szerelmey.com

T

T B S Fabrications - Martens Road, Northbank Ind. Estate, Irlam, Manchester, Gt Man. M44 5AX tel:(0161) 775 7915, fax:(0161) 775 8929. webinfo@tbs-fabrications.com, www.abp-tbswashrooms.co.uk
TA Lucas Sails - Porchester House, Hospital Lane, Portchester, Hants. PO16 9QP tel:(02392) 373699, fax:(02392) 373656. info@lucas-sails.com, www.lucas-sails.com
Tank Storage & Services Ltd - Lilley Farm, Thurlow Road, Withersfield, Suffolk. CB9 7SA tel:(01440) 712614, fax:(01440) 712615. tom@tankstorage.co.uk, www.tankstorage.co.uk
Taperell Taylor Co - Meadhams Farm Brickworks, Blackwell Hall Lane, Lay Hill, Nr Chesham, Bucks. HP5 1TN tel:(01494) 794630, fax:(01494) 791883. www.taperelltaylor.co.uk
Tapworks Water Softeners - Solar House, Mercury Park, Wooburn Green, Buckinghamshire. HP10 0HH tel:(01494) 480621, fax:(01494) 484312. info@tapworks.co.uk, www.tapworks.co.uk
Tarkett Ltd - Lenham, Maidstone, Kent. ME17 2QX tel:(01622) 854000, fax:(01622) 854500. marketing@tarkett.com, www.tarkett.co.uk
Tarkett Sommer Ltd - Century House, Bridgwater Road, Worcester, Worcs. WR4 9FA tel:(01905) 342 700, fax:(01905) 342 777. uksales@tarkett.com, www.tarkett.co.uk
Tarmac Building Products Ltd - Millfields Road, Ettingshall, Wolverhampton, W. Midlands. WV4 6JP tel:(01902) 353 522, fax:(01902) 382219. enquiries@tarmac.co.uk, www.tarmacbuildingproducts.co.uk
Tarmac Ltd - Millfields Road, Ettingshall, Wolverhampton, W. Midlands. WV4 6JP tel:(01902) 353 522. enquiries@tarmac.co.uk, www.tarmac.co.uk
Taskmaster Doors Ltd - Unit 5 Gravelly Ind Park, Erdington, Birmingham. B24 8HZ tel:(0121) 351-5276, fax:(0121) 328-8793. sales@taskmaster.co, www.taskmaster.co
Taskworthy Ltd - 1 Stainiers Way, Hereford, Herefordshire. HR1 1JT tel:(01432) 376000, fax:(0432) 376001. sales@taskworthy.co.uk, www.taskworthy.co.uk
Tata Steel Colors - Shotton Works, Deeside, Flintshire. CH5 2NH tel:(01244) 892434, fax:(01244) 836134. colorcoat.connection@tatasteel.com, www.colorcoat-online.com
Tata Steel Europe Limited - PO Box 1, Brigg Road, Scunthorpe, North Linclnshire. DN16 1BP tel:(01724) 405060, fax:(01724) 404224. construction@tatasteel.com, www.tatasteelconstruction.com
Tate Fencing - Yellowcoat Sawmill, Hastings Road, Flimwell, Sussex. TN5 7PR tel:(01580) 879900, fax:(01580) 879677. sales@tate-fencing.co.uk, www.tate-fencing.co.uk
TBA Sealing Materials Ltd - PO Box 21, Rochdale, Lancs. OL12 7EQ tel:(01706) 715000, fax:(01706) 42284. info@tbaecp.co.uk, www.tbaecp.co.uk
TBS Fabrications Ltd - Martens Road, Northbank Industrial Park, Irlam, Manchester. M44 5AX tel:(0161) 775 1971, fax:(0161) 775 8929. webinfo@tbs-fabrications.com, www.abp-tbswashrooms.co.uk
TCW Services (Control) Ltd - 293 New Mill Road, Brockholes, Huddersfield, West Yorkshire. HD9 7AL tel:(01484) 662865, fax:(01484) 667574. sales@tcw-services.co.uk, www.tcw-services.co.uk
Teal & Mackrill Ltd - Ellenshaw Works, Lockwood Street, Hull, E. Yorks. HU2 0HN tel:(01482) 320194, fax:(01482) 219266. info@teamac.co.uk, www.teamac.co.uk
Tebrax Ltd - International House, Cray Avenue, Orpington, Kent. BR5 3RY tel:(01689) 897766, fax:(01689) 896789. brackets@tebrax.co.uk, www.tebrax.co.uk
Technical Control Systems Ltd - Treefield Industrial Estate, Gildersome, Leeds. LS27 7JU tel:(0113) 252 5977, fax:(0113) 238 0095. enquiries@tcspanels.co.uk, www.tcspanels.co.uk
Technical Timber Services Ltd - The Coach House, 27 Mail Road, Hursley, Nr Winchester. SO21 2JW tel:(01794) 516653, fax:(01794) 512632. sales@technicaltimber.co.uk, www.technicaltimber.co.uk
Techrete (UK) Ltd - Feldspar Close, Warren Park Way, Enderby, Leicester. LE19 4SD tel:(01162) 865 965, fax:(01162) 750 778. info@techrete.com, http://techrete.com
Techspan Systems - Griffin Lane, Aylesbury, Buckinghamshire. HP19 8BP tel:(01296) 673000, fax:(01296) 673002. enquiries@techspan.co.uk, www.techspan.co.uk
Teddington Controls Ltd - Daniels Lane, Holmbush, St Austell, Cornwall. PL25 3HG tel:(01726) 74400, fax:(01726) 67953. info@tedcon.com, www.tedcon.com
Tefcote Surface Systems - Central House, 4 Christchurch Road, Bournemouth, Dorset. BH1 3LT tel:(01202) 551212, fax:(01202) 559090. office@tefcote.co.uk, www.tefcote.co.uk
Tektura Plc - Customer Services, 34 Harbour Exchange Square, London. E14 9GE tel:(020) 7536 3300, fax:(020) 7536 3322. enquiries@tektura.com, www.tektura.com
Telelarm care Ltd - 1 Centurion Business Park, Bessemer Way, Rotherham, S. Yorks. S60 1FB tel:(01709) 389324, fax:(01709) 389344. info@telelarmcare.co.uk
Telling Architectural Ltd - 7 The Dell, Enterprise Drive, Four Ashes, Wolverhampton, West Midlands. WV10 7DF tel:(01902) 797 700, fax:(01902) 797 720. info@telling.co.uk, www.telling.co.uk
Telling Lime Products Ltd - 7 The Dell, Enterprise Drive, Four Ashes, Wolverhampton, West Midlands. WV10 7DF tel:(01902) 797 700, fax:(01902) 797 720. info@telling.co.uk, www.telling.co.uk
Tensar International Ltd - New Wellington Street, Blackburn, Lancs. BB2 4PJ tel:(01254) 262431, fax:(01254) 266868. info@tensar.co.uk, www.tensar.co.uk
Tensator Ltd - Danbury Court, Linford Wood, Milton Keynes. MK14 6TS tel:(01908) 271153, fax:(01908) 274572. info@tensator.com, www.tensator.co.uk
Terex Construction - Central Boulevard, ProLogis Park, Keresley End, Coventry. CV6 4BX tel:(02476) 339400, fax:(02476) 339500. customercare@terex.com, www.terexcompactequipment.com
Terminix Property Services - Oakhurst Drive, Cheadle Heath, Stockport, Cheshire. SK3 0XT tel:(0161) 491 3181, fax:(0161) 428 8138. headoffice@petercox.com, www.petercox.com
Termstall Limited - 50 Burman Street, Droylsden, Gt Man. M43 6TE tel:(0161) 370 2835, fax:(0161) 370 9264. enquiries@termstallltd.co.uk, www.termstallfencing.co.uk
Terrapin Ltd - Bond Avenue, Bletchley, Milton Keynes. MK1 1JJ tel:(01908) 270900, fax:(01908) 270052. info@terrapin-ltd.co.uk, www.terrapin-ltd.co.uk
Terry of Redditch Ltd - 19 Oxleaslow Rd, East Moons Moat, Redditch, Worcs. B98 0RE tel:(01527) 517100. info.uk@normagroup.com, www.normagroup.com
TEV Limited - Armytage Road, Brighouse, West Yorks. HD6 1QF tel:(01484) 405600, fax:(01484) 405620. sales@tevlimited.com, www.tevlimited.com
Tex Engineering Ltd - Unit 35, Claydon Ind Park, Gipping Road, Great Blakenham, Ipswich, Suffolk. IP6 0NL tel:(01870) 751 3977, fax:(0870) 751 3978. info@bitmen.co.uk, www.tex-engineering.co.uk
Texecom Limited - Bradwood Court, St. Crispin Way, Haslingden, Lancashire. BB4 4PW tel:(01706) 234800. pharries@texe.com, www.klaxonsignals.com
TH Kenyon & Sons Ltd - Kenyon House, 14a Hockerill Street, Bishop's Stortford, Herts. CM23 2DW tel:(01279) 661800, fax:(01279) 661807. sales@thkenyonplc.co.uk, www.thkenyonplc.co.uk
Thakeham Tiles Ltd - Rock Road, Heath Common, Storrington, W. Sussex. RH20 3AD tel:(01903) 742381, fax:(01903) 746341. support@thakeham.co.uk, www.Thakeham.co.uk
Thanet Ware Ltd - Ellington Works, Princes Road, Ramsgate, Kent. CT11 7RZ tel:(01843) 591076, fax:(01843) 586198
Thatching Advisory Services Ltd - 8-10 QueenStreet, Seaton, Devon. EX12 2NY tel:(08455) 20 40 60, fax:(01297) 624177. info@thatchingadvisoryservices.co.uk, http://thatchingadvisoryservices.co.uk
The Abbseal Group incorporating Everseal (Thermovitrine Ltd) - P O Box 7, Broadway, Hyde, Cheshire. SK14 4QW tel:(0161) 368 5711, fax:(0161) 366 8155. www.abbseal.co.uk
The Access Panel Company - The Old Waterworks, Winterton Road, Scunthorp. DN15 0BA tel:(0800) 071 2311, fax:(0870) 201 2311. info@accesspanels.co.uk, www.accesspanels.co.uk
The Amtico Company - Kingfield Road, Coventry. CV6 5AA tel:(024) 7686 1400, fax:(024) 7686 1552. samples@amtico.com, www.amtico.com
The Angle Ring Company Ltd - Bloomfield Road, Tipton, W. Midlands. DY4 9EH tel:(0121) 557 7241, fax:(0121) 522 4555. sales@anglering.com, www.anglering.com
The Association of Technical Lighting & Access Specialists - 4c St. Mary's Place, The Lace Market, Nottingham. NG1 1PH tel:(01159) 558818, fax:(01159) 412238. info@atlas.org.uk, www.atlas-1.org.uk
The Benbow Group - Bradley Mill, Newton Abbott, Devon. TQ12 1NF tel:(01626) 883400, fax:(01626) 355591. mail@benbowgroup.co.uk, www.benbowgroup.co.uk
The Book Depository Ltd - 60 Holborn Viaduct, London. EC1A 2FD help@support.bookdepository.com, www.bookdepository.com
The BSS Group - 7 Barton Close, Leicester. LE19 1SJ tel:(0116) 245 5500. reception@bssgroup.com, www.bssgroup.co.uk
The Building Centre Group Ltd - 26 Store Street, Camden, London. WC1E 7BT tel:(020) 7692 4000, fax:(020) 7580 9641. information@buildingcentre.co.uk, www.buildingcentre.co.uk
The Carpet Foundation - MCF Complex, 60 New Road, P. O. Box 1155, Kidderminster, Worcs. DY10 1AQ tel:(01562) 755568, fax:(01562) 865405. info@carpetfoundation.com, www.carpetfoundation.com

T (con't)

The Clearwater Group - Welsh Road East, Napton Holt, Warwickshire. CV47 1NA tel:(01926) 818283, fax:(01926) 818284. support@clearwatergroup.co.uk, www.clearwatergroup.co.uk

The Concrete Society - Riverside House, 4 Meadows Business Park, Station Approach, Blackwater, Camberly, Surrey. GU17 9AB tel:(01276) 607140, fax:(01276) 607141. j.luckey@concrete.org.uk, www.concrete.org.uk

The Cotswold Casement Company - Cotswold Business Village, London Road, Moreton-on-Marsh, Glos. GL56 0JQ tel:(01608) 650568, fax:(01608) 651699. info@cotswold-casements.co.uk, www.cotswold-casements.co.uk

The David Sharp Studio Ltd - 201A Nottinham Road, Summercoats, Alfreton. DE55 4JG tel:(01773) 606066, fax:(01773) 540737. info@david-sharp.co.uk, www.david-sharp.co.uk

The Delabole Slate Co. Ltd - Delabole Slate Quarry, Pengelly, Delabole, Cornwall. PL33 9AZ tel:(01840) 212242, fax:(01840) 212948. sales@delaboleslate.co.uk, www.delaboleslate.co.uk

The Federation of Environmental Trade Association (FETA) - 2 Waltham Court, Milley Lane, Hare Hatch, Reading, Berkshire. RG10 9TH tel:(01189) 403416, fax:(01189) 406258. info@feta.co.uk, www.feta.co.uk

The Flyscreen Company Ltd - Unit 4, Warren Business Park, Knockdown, Tetbury. GL8 8QY tel:(01454) 238288, fax:(01454) 238988. sales@flyscreen.com, www.flyscreen.com

The Fountain Company - Etherow House, Woolley Bridge Road, Glossop, Derbyshire. SK13 2NS tel:(01457) 866088, fax:(01457) 865588. fountain.co@btconnect.com, www.thefountaincompany.co.uk

The Haigh Tweeny Co Ltd - Haigh Industrial Estate, Alton Road, Ross-on-Wye, Herefordshire. HR9 5LA tel:(01989) 566222, fax:(01989) 767498. sales@tweeny.co.uk, www.tweeny.co.uk

The Institute of Concrete Technology - Riverside House, 4 Meadows Business Park, Station Approach, Blackwater, Camberley, Surrey. GU17 9AB tel:(01276) 607140, fax:(01276) 607141. ict@concrete.org.uk, http://ict.concrete.org.uk/

The Institution of Engineering and Technology - Michael Farady House, Six Hills Wayll, Stephenage, Hertfordshire. SG1 2AY tel:(01438) 313 311, fax:(01438) 765 526. postmaster@theiet.org, www.theiet.org.uk

The Kenco Coffee Company - St. Georges House, Bayshill Road, Cheltenham, Glos. GL50 3AE tel:(0800) 242000, fax:(01295) 264602. orders.kenco@krafteurope.com, www.kencocoffeecompany.co.uk

The Kitchen Bathroom Bedroom Specialists Association - Unit L4A Mill 3, Pleasley Vale Business Park, Mansfiled, Notts. NG19 8RL tel:(01623) 818808, fax:(01623) 818808. info@kbsa.org.uk, www.kbsa.org.uk

The No Butts Bin Company (NBB), Trading as NBB Outdoor Shelters - Unit 5, Sterte Road Industrial Estate, 145 Sterte Road, Poole, Dorset. BH15 2AF tel:(0800) 1777 052, fax:(0800) 1777 053. sales@nobutts.co.uk, www.nobutts.co.uk

The Outdoor Deck Company - Unit 6, Teddington Business Park, Station Road, Teddington. TW11 9BQ tel:(020) 8977 0820, fax:(020) 8977 0825. sales@outdoordeck.co.uk, www.outdoordeck.co.uk

The Property Care Association - 11 Ramsay Court, Kingfisher Way, Hinchingbrooke Business Park, Huntingdon, Cambs. PE29 6FY tel:(0844) 375 4301, fax:(01480) 417587. pca@property-care.org, www.property-care.org

The Rooflight company - Unit 8, Wychwood Business Centre, Milton Road, Shipton-under-Wychwood, Chipping Norton, Oxon. OX7 6XU tel:(01993) 833108, fax:(01993) 831066. info@therooflightcompany.co.uk, www.therooflightingcompany.co.uk

The Safety Letterbox Company Ltd - Unit B, Milland Industrial Estate, Milland Road, Neath, West Glamorgan. SA11 1NJ tel:(01639) 633 525, fax:(01639) 646 359. sales@safetyletterbox.com, www.safetyletterbox.com

The Saw Centre Group - 650 Eglinton Street, Glasgow. G5 9RP tel:(0141) 429 4444, fax:(0141) 429 5609. sales@thesawcentre.co.uk, www.thesawcentre.co.uk

The Silk Forest - Main Street, Bagworth, Leics. LE67 1DW tel:(01530) 231241, fax:(01530) 231240. info@thesilkforest.com, www.thesilkforest.com

The Sound Solution - Unit 2 Carr House Farm, Nun Monkton, Pool Lane, York. YO26 8EH tel:(08451) 306269, fax:(014123) 339153. info@noisestopsystems.co.uk, www.noisestopsystems.co.uk

The Stone Catalogue Ltd - Zone 1 HY-Tec Industrial Park, Peel Lane, Cheetham, Manchester. M8 8RJ tel:(0870) 084 3422, fax:(0870) 050 7269. info@garden-statues.com, www.thestonecatalogue.co.uk

The Tile Association - Forum Court, 83 Copers Cope Road, Beckenham, Kent. BR3 1NR tel:(020) 8663 0946, fax:(020) 8663 0949. info@tiles.org.uk, www.tiles.org.uk

The Visual Systems Healthcare - Penine Court, Standback Way, Station Raod Technology Park, Skelmanthorpe, Huddersfield, West Yorkshire. HD8 9GA tel:(01484) 865786, fax:(01484) 865788. sales@visualsystemsltd.co.uk, www.visualsystemshealthcare.co.uk

The Window Film Company UK Ltd - Unit 5 Power House, Higham Mead, Chesham, Buckinghamshire. HP5 2AH tel:(01494) 797800, fax:(01494) 794488. info@windowfilms.co.uk, www.windowfilms.co.uk

The Wykamol Group - Knowsley Road Industrial Estate, Haslingden, Rossendale, Lancs. BB4 4RX tel:(01706) 831223, fax:(01706) 214998. sales@lectros.com, www.wykamol.com

Thermal Insulation Contractors Association - TICA House, Allington Way, Yarm Road Business Park, Darlington, Co. Durham. DL1 4QB tel:(01325) 466704, fax:(01325) 487691. ralphbradley@tica-acad.co.uk, www.tica-acad.com/tica.htm

Thermal Technology Sales Ltd - Bridge House, Station Road, Westbury, Wilts. BA13 4HR tel:(01373) 865454, fax:(01373) 864425. sales@thermaltechnology.co.uk, www.thermaltechnology.co.uk

Thermica Ltd - Vulcan Street, Clough Road, Hull. HU6 7PS tel:(01482) 348771, fax:(01482) 441873. nestaan@nhb.be, www.thermica.co.uk

Thermoscreens Ltd - St. Mary's Road, Nuneaton, Warwickshire. CV11 5AU tel:(02476) 384646, fax:(02476) 388578. sales@thermoscreens.com, www.thermoscreens.com

Thomas & Betts Manufacturing Services Ltd - Bruntcliffe Lane, Morley, Leeds, W. Yorks. LS27 9LL tel:(0113) 281 0600, fax:(0113) 281 0601. mailbox@emergi-lite.co.uk

Thomas Sanderson - Waterberry Drive, Waterlooville, Hants. PO7 7UW tel:(02392) 232600, fax:(02392) 232700. www.thomas-sanderson.co.uk

Thor Hammer Co Ltd - Highlands Road, Shirley, Birmingham. B90 4NJ tel:(0121) 705 4695, fax:(0121) 705 4727. info@thorhammer.com, www.thorhammer.com

Thorlux Lighting - Merse Road, North Moons Moat, Redditch, Worcs. B98 9HH tel:(01527) 583200, fax:(01527) 584177. thorlux@thorlux.co.uk, www.thorlux.com

Thorn Lighting - ZG Lighting (UK Limited, Durhamgate, Spennymoor, Co Durham. DL16 6HL tel:(01388) 420042. wiebke-marie.friedewald@zumtobelgrou.com, www.thornlighting.com

Thurston Building Systems - Quarry Hill Industrial Estate, Hawkingcroft Road, Horbury, Wakefield, W. Yorks. WF4 6AJ tel:(01924) 265461, fax:(01924) 280246. sales@thurstongroup.co.uk, www.thurstongroup.co.uk

ThyssenKrupp Encasa, A Division of ThyssenKrupp Access Ltd - Unit E3 Eagle Court, Preston Farm Industrial Estate, Stockton -On-Tees, Cleveland. TS18 3TB tel:(01642) 704850. enquiries@tkencasa.co.uk, www.tkencasa.co.uk

Tidmarsh & Sons - Pleshey Lodge Farm, Pump Lane, Chelmsford, Essex. CM3 1HF tel:(01245) 237288. info@tidmarsh.co.uk, www.tidmarsh.co.uk

Tiflex Ltd - Tiflex House, Liskeard, Cornwall. PL14 4NB tel:(01579) 320808, fax:(01579) 320802. sales@tiflex.co.uk, www.tiflex.co.uk

Till & Whitehead Ltd - Eadon House, 2 Lonsdale Road, Heaton, Bolton, Lancs. BL1 4PW tel:(01204) 493000, fax:(01204) 493888. bolton@tillwite.com, www.tillwite.com

Tilley International Ltd - 30-32 High Street, Frimley, Surrey. GU16 5JD tel:(01276) 691996, fax:(01276) 27282. tilleyuk@globalnet.co.uk, www.tilleylamp.co.uk

Tillicoutry Quarries Ltd - Tulliallan Quarry, Kincardine, Fife. FK10 4DT tel:(01259) 730481, fax:(01259) 731201. sales@tillicoultryquarries.com, www.tillicoultryquarries.com

Timber Components (UK) Ltd - Wood Street, Grangemouth, Stirlingshire. FK3 8LH tel:(01324) 666222, fax:(01324) 666322. info@tcuk.co.uk, www.tcuk.co.uk

Timber Trade Federation - The Building Centre, 26 Store Street, London. WC1E 7BT tel:(020) 3205 0067, fax:(020) 7291 5379. ttf@ttf.co.uk, www.ttf.co.uk

Timberwise (UK) Ltd - 1 Drake Mews, Gadbrook Park, Cheshire. CW9 7XF tel:(01606) 33 36 36, fax:(01606) 33 46 64. hq@timberwise.co.uk, www.timberwise.co.uk

Timbmet Ltd - Kemp House, Chawley Works, Cumnor Hill, Oxford. OX2 9PH tel:(01865) 862223, fax:(01865) 864367. marketing@timbmet.com, www.timbmet.com

Time and Data Systems International Ltd (TDSI) - Sentinel House, Nuffield Road, Poole, Dorset. BH17 0RE tel:(01202) 666222, fax:(01202) 679730. sales@tdsi.co.uk, www.tdsi.co.uk

Timeguard Ltd - Apsley Way, London. NW2 7UR tel:(020) 8450 8944, fax:(020) 8452 5143. csc@timeguard.com, www.timeguard.com

Timloc Expamet Building Products - Rawcliffe Road, Goole, E. Yorks. DN14 6UQ tel:(01405) 765567, fax:(01405) 720479. sales@timloc.co.uk, www.timloc.co.uk

Titan Environmental Ltd - Barbot Hall Ind. Est, Mangham Rd, Rotherham, S. Yorks. S61 4RJ tel:(0870) 362448, fax:(0870) 538301. rotherham@titanenu.com, www.titanenu.com

Titan Ladders Ltd - 191-201 Mendip Road, Yatton, Bristol, Avon. BS19 4ET tel:(01934) 832161, fax:(01934) 876180. sales@titanladders.co.uk, www.titanladders.co.uk

Titan Polution Control - West Portway, Andover, Hants. SP10 3LF tel:(01264) 353222, fax:(01264) 353122. info@titanpc.co.uk, www.titanpc.co.uk

TLC Southern Ltd - Unit 10, Chelsea Fields Industrial Estate, 278 Western Road, Merton, London. SW19 2QA tel:(020) 8646 6866, fax:(020) 8646 6750. sales@tlc-direct.co.uk, www.tlc-direct.co.uk

Today Interiors Ltd - Unit 5, Orchard Park, Isaac Newton Way, Grantham, Lincs. NG31 9RT tel:(01476) 574401, fax:(01476) 590208. info@today-interiors.co.uk, www.todayinteriors.com

Toffolo Jackson(UK) Ltd - Burnfield Road, Thornliebank, Renfrewshire. G46 7TQ tel:(0141) 649 5601, fax:(0141) 632 9314. info@toffolo-jackson.co.uk, www.toffolojackson.co.uk

Tonbridge Fencing Limited - Court Lane Farm, Court Lane, Hadlow, Tonbridge, Kent. TN11 0DP tel:(01732) 852596, fax:(01732) 852593. geoff.wallace@tonbridgefencing.co.uk

Tony Team Ltd - Unit 5, Station Road, Bakewell, Derbys. DE45 1GE tel:(01629) 813859, fax:(01629) 814334. sales@tonyteam.co.uk, www.tonyteam.co.uk

Topcon GB Ltd - Topcon House, Kennet Side, Bone Lane, Newbury, Berks. RG14 5PX tel:(01635) 551120, fax:(01635) 551170. info@topconsokkia.co.uk, www.topcon-positioning.eu

Toprail Systems Ltd - Unit 17, Shepperton Business Park, Govett Avenue, Shepperton. TW17 8BA tel:(0870) 777 5557, fax:(0870) 777 5556. sales@toprail.com, www.toprail.com

Topseal Systems Ltd - Unit 1-5, 108 Hookstone Chase, Harrowgate. HG2 7HH tel:(01423) 886495, fax:(01423) 889550. enquiries@topseal.co.uk, www.topseal.co.uk

Tor Coatings Ltd - Portobello Industrial Estate, Birtley, Chester-le-Street, Co. Durham. DH3 2RE tel:(0191) 410 6611, fax:(0191) 492 0125. enquiries@tor-coatings.com, www.tor-coatings.com

TORMAX United Kingdom Ltd - Unit1, Shepperton Business Park, Govett Avenue, Sheperton, Middlesex. TW17 8BA tel:(01932) 238040, fax:(01932) 238055. sales@tormax.co.uk, www.tomax.co.uk

Tormo Ltd - Unit 7, Devonshire Business Park, 4 Chester Road, Borehamwood, Herts. WD6 1NA tel:(020) 8205 5533, fax:(020) 8201 3656. sales@tormo.co.uk, www.tormo.co.uk

Toshiba Carrier UK - United Technologies House, Guildford Road, Leatherhead, Surrey. KT22 9UT tel:(01372) 220240, fax:(01372) 220241. general.enquiries@toshiba-ac.com, www.toshiba-aircon.co.uk

Tower Flue Componants Ltd - Tower House, Vale Rise, Tonbridge, Kent. TN9 1TB tel:(01732) 351555, fax:(01732) 354445. tjt@tfc.org.uk, www.Tfc.ukco.com

Tower Manufacturing - Navigation Road, Worcester, Worcs. WR5 3DE tel:(01905) 763012, fax:(01905) 763610. clips@towerman.co.uk, www.towerman.co.uk

Town & Country Landscapes Ltd - Burnsall House, 6 Birk Dale, Bexhill-On-Sea, Sussex. TN39 3TR tel:(01424) 773834, fax:(01424) 774633. nick@tandc85.wanadoo.co.uk, www.town-and-country-landscapes.co.uk

Town and Country Planning Association - 17 Carlton House Terrace, London. SW1Y 5AS tel:(020) 7930 8903, fax:(020) 7930 3280. tcpa@tcpa.org.uk, www.tcpa.org.uk

Townscape Products Ltd - Fulwood Road South, Sutton in Ashfield, Notts. NG17 2JZ tel:(01623) 513355, fax:(01623) 440267. sales@townscape-products.co.uk, www.townscapeproducts.co.uk

TR Engineering Ltd - Thorncliffe, Chapeltown, Sheffield, S. Yorks. S35 2PZ tel:(0114) 257 2300, fax:(0114) 257 1419. info@trianco.co.uk, www.trianco.co.uk

TRADA Bookshop - Publications Team, BM TRADA, Stocking Lane, Hughenden Valley, High Wycombe. HP14 4ND tel:(01494) 569602. publications@bmtrada.com, http://bookshop.bmtrada.com

TRADA Technology Ltd - Stocking Lane, Hughenden Valley, High Wycombe. HP14 4ND tel:(01494) 569642, fax:(01494) 565487. information@trada.co.uk, www.tradatechnology.co.uk

Trade Sealants Ltd - 34 Aston Road, Waterlooville, Hampshire. PO7 7XQ tel:(02392) 251321, fax:(023) 92264307. marinemastics@gmail.com, www.marinemastics.com

Travis Perkins (Southern) - 149 Harrow Road, London. W2 6NA tel:(020) 7262 6602, fax:(020) 7724 7485. paul.lonsdale@travisperkins.co.uk, www.tradamate.co.uk

TRC (Midlands) Ltd - 1 Mount Pleasant Street, West Bromwich, W. Midlands. B70 7DL tel:(0121) 5006181, fax:(0121) 5005075. info@totalroofcontrol.co.uk, www.totalroofcontrol.co.uk

Trelleborg Ltd - Trelleborg Building Systems UK Ltd, Maybrook Rd, Castle Vale Ind. Estate, Minworth, Sutton Coldfield, W. Midlands. B76 1AX tel:(0121) 352 3800, fax:(0121) 352 3899. tssuk@trelleborg.com, www.trelleborg.com

Trelleborg Sealing Profiles - International House, Staffort Park 11, Telford, Shropshire. TF3 3AY tel:(01952) 236 017, fax:(01952) 236 012. john.connor@trelleborg.com, www.trelleborg.com/en/Sealing-Profiles

Tremco Illbruck Ltd - Coupland Road, Hindley Green, Wigan. WN2 4HT tel:(01942) 251400, fax:(01942) 251410. uk.info@tremco-illbruck.com, www.tremco-illbruck.co.uk

Trend Control Systems Ltd - Albery House, Springfield Road, Horsham, West Sussex. RH12 2PQ tel:(01403) 211888, fax:(01403) 241608. marketing@trendcontrols.com, www.trendcontrols.com

Trespa UK Ltd - 35 Calthorpe Road, Edgbaston, Birmingham. B15 1TS tel:(0808) 234 02 68, fax:(0808) 234 13 77. info.uk@trespa.com, www.trespa.com

Trico V.E. Ltd (Signage Division) - 76 Windmill Hill, Colley Gate, Halesowen, West Midlands. B63 2BZ tel:(01384) 569555, fax:(01384) 565777. info@trico-ve.co.uk, www.trico-ve.co.uk

Trim Acoustics - Unit 4, Leaside Industrial Estate, Stockingswater Lane, Enfield, London. EN3 7PH tel:(020) 8443 0099, fax:(020) 8443 1919. sales@trimacoustics.co.uk, www.trimacoustics.co.uk

Trimite Ltd - Arundel Road, Uxbridge, Middx. UB8 2SD tel:(01895) 951234, fax:(01895) 256789. info@trimite.com, www.trimite.com

Trimite Scotland Ltd - 38 Welbeck Road, Darnley Industrial Estate, Glasgow. G53 7RG tel:(0141) 881 9595, fax:(0141) 881 9333. sales@tslpaints.com, www.tslpaints.com

Trimplex Safety Tread - Trdent Works, Mulberry Way, Belvedere, Kent. DA17 6AN tel:(020) 8311 2101, fax:(020) 8312 1400. safetytread@btconnect.com, www.safetytread.co.uk

Triton PLC - Shepperton Park, Caldwell Road, Nuneaton, Warwicks. CV11 4NR tel:(024) 76 344441, fax:(024) 76 324424. serviceenquiries@tritonshowers.co.uk, www.tritonshowers.co.uk

TROAX UK Ltd - Enterprise House, Murdock Road, Dorcan, Swindon, Wilts. SN3 5HY tel:(01793) 520000, fax:(01793) 618784. info@troax.com, www.troax.com

Troika Contracting Ltd - 850 Herries Road, Sheffield, S. Yorks. S6 1QW tel:(01142) 327000, fax:(01142) 344885. sales@troikaam.co.uk, www.plasterware.com

Trox (UK) Ltd - Caxton Way, Thetford, Norfolk. IP24 3SQ tel:(01842) 754545, fax:(01842) 763051. trox@troxuk.co.uk, www.troxuk.co.uk

TSE Brownson - Sacksville Street, Skipton, N. Yorks. BD23 2PR tel:(01756) 797744, fax:(01756) 796644. info@rewardtrolleys.com, www.rewardmanufacturing.com

tso shop - TSO Customer Services, PO Box 29, Norwich. NR3 1GN tel:(0333) 200 2425. customer.services@tso.co.uk, www.tsoshop.co.uk

T-T Pumps & T-T controls - Onneley Works, Newcastle Road, Woore. CW3 9RU tel:(01630) 647200, fax:(01630) 642100. response@ttpumps.com, www.ttpumps.com

Tubeclamps Ltd - PO Box 41, Petford Street, Cradley Heath, Warley, W. Midlands. B64 6EJ tel:(01384) 565241, fax:(01384) 410490. sales@tubeclamps.co.uk, www.tubeclamps.co.uk

Tubela Engineering Co Ltd - 11 A/B Hoblongs Ind Estate, Chelmsford Road, Gt Dunmow, Essex. CM6 1JA tel:(01371) 859100, fax:(01371) 859101. tubebending@tubela.com, www.tubela.com

Tubs & Tiles - Moydrum Road, Athlone, Westmeath, Ireland. tel:+353 90 6424000, fax:+353 90 6424050. customerservices@wolseley.co.uk, www.tubstiles.ie

Tucker Fasteners Ltd - Emhart Fastening Teknologies, Walsall Road, Birmingham, W. Midlands. B42 1BP tel:(0121) 356 4811, fax:(0121) 356 1598. www.emhart.eu

Tudor Roof Tile Co Ltd - Denge Marsh Road, Lydd, Kent. TN29 9JH tel:(01797) 320202, fax:(01797) 320700. info@tudorrooftiles.co.uk, www.tudorrooftiles.co.uk

Tuke & Bell Ltd - Lombard house, No 1 Cross Keys, Lichfield, Staffs. WS13 6DN tel:(01543) 414161, fax:(01543) 250462. sales@tukeandbell.co.uk, www.tukeandbell.co.uk

Tunstal Healthcare (UK) Ltd - Whitley Lodge, Whitley Bridge, N. Yorks. DN14 0HR tel:(01977) 661234, fax:(01977) 661993. sales@tunstall.co.uk, www.tunstallgroup.com

Turnbull & Scott (Engineers) Ltd - Unit 1a, Burnfoot Industrial Estate, Hawick, Scottish Borders. TD9 8SL tel:(01450) 372 053, fax:(01450) 377 800. info@turnbull-scott.co.uk, www.turnbull-scott.co.uk

Turner Access Ltd - 65 Craigton Road, Glasgow. G51 3EQ tel:(0141) 309 5555, fax:(0141) 309 5436. enquiries@turner-access.co.uk, www.turner-access.co.uk

Tuscan Foundry Products Ltd - Cowfold, W. Sussex. RH13 8AZ tel:(0800) 174093, fax:(0845) 3450215. info@tuscanfoundry.co.uk, www.tuscanfoundry.co.uk

T (con't)

TWI - Abington Hall, Abington, Cambridge. CB21 6AL tel:(01223) 891162, fax:(01223) 892588. twi@twi.co.uk, www.twi.co.uk
TWI Ltd - Granta Park, Great Abingdon, Cambridge. CB1 6AL tel:(01223) 891162. adhesivestoolkit@twi.co.uk, www.adhesivestoolkit.com
Twyford Bathrooms - Lawton Road, Alsager, Stoke-on-Trent. ST7 2DF tel:(01270) 879777, fax:(01270) 873864. twyford.sales@twyfordbathrooms.com, www.twyfordbathrooms.com
Twyford Bushboard - 9-29 Rixon Road, Wellingborough, Northants. NN8 4BA tel:(01933) 232200, fax:(01933) 232283. sales@bushboard.co.uk, www.twyfordbushboard.co.uk
Tyco European Tubes - PO Box 13, Popes Lane, Oldbury, Warley. B69 4PF tel:(0121) 5435700, fax:(0121) 543 5750. www.tycotube.co.uk
Tyco Fire and Integrated Solutions - Tyco Park, Grimshaw Lane, Newton Heath, Manchester. M40 2WL tel:(0161) 455 4400. tfis.marketing.uk@tycoint.com, www.tycofis.co.uk
Tyco Safety Products - Burlingham House, Hewett Road, Gapton Hall Industrial Estate, Great Yarmouth, Norfolk. NR31 0MN tel:(01493) 417600, fax:(01493) 417700. macron-info@tycoint.com, www.macron-safety.com
Tyco Water Works - Nobel Road, Eley Trading Estate, Edmonton, London. N18 3DW tel:(020) 8884 4388, fax:(020) 8884 4766. enquiries@talis-group.com, www.tycowaterworks.com
Tyco Waterworks - Samuel Booth - Samuel Booth Works, Warstock Road, Kings Heath, Birmingham. B14 4RT tel:(0121) 772 2717, fax:(0121) 766 6962. tfis.marketing.uk@tycoint.com, www.tycowaterworks.com/sbooth

U

Ubbink (UK) Ltd - Borough Road, Brackley, Northants. NN13 7TB tel:(01280) 700211, fax:(01280) 705332. info@ubbink.co.uk, www.ubbink.co.uk
UK Fasteners Ltd - C1 Liddington Trading Estate, Leckhampton Road, Cheltenham, Glos. GL53 0DL tel:(01242) 577077, fax:(01242) 577078. sales@ukfasteners.co.uk, www.ukfasteners.co.uk
UK Lift Co Ltd - Westminster Wks, Sandown Road, Watford, Herts. WD24 7UB tel:(01923) 656200. info@uk-lift.co.uk, www.uk-lift.co.uk
UK Mineral Wool Association,The - PO Box 35084, London. NW1 4XE tel:(020) 7935 8532, fax:(020) 7935 8532. info@eurisol.com, www.eurisol.com
Ulster Carpet Mills Ltd - Castleisland Factory, Craigavon, Co. Armagh. BT62 1EE tel:(028) 38334433, fax:(028) 38333142. marketing@ulstercarpets.com, www.ulstercarpets.com
Ultraframe PLC - Salthill Road, Clitheroe, Lancs. BB7 1PE tel:(0500) 822340, fax:(01200) 425455. info@ultraframe.co.uk, www.ultraframe.co.uk/ultra
Unicorn Containers Ltd - 5 Ferguson Drive, Knockmore Hill Industrial Park, Lisburn, Antrim. BT28 2EX tel:(02892) 667264, fax:(02892) 625616. info@unicorn-containers.com, www.unicorn-containers.com
Unifix Ltd - Hexstone, Opal Way, Stone Business Park, Stone, Staffordshire. ST15 0SW tel:(0800) 808 7172, fax:(0800) 731 3579. sales@unifix.co.uk, www.unifix.co.uk
UNION Locks and Hardware - School Street, Willenhall, W. Midlands. WV13 3PW info@uniononline.co.uk, www.uniononline.co.uk
Unistrut Limited - Unistrut House, Delta Point Greets Green Road, West Bromwich. B70 9PL tel:(0121) 580 6300, fax:(0121) 580 6370. www.unistrut.co.uk
United Flexibles (A Division of Senior Flex) - Abercanaid, Merthyr Tydfil, M. Glam. CF48 1UX tel:(01685) 385641, fax:(01685) 389683. sales@amnitec.co.uk, www.amnitec.co.uk
United Wire Ltd - Granton Park Avenue, Edinburgh. EH5 1HT tel:(0131) 552 6241, fax:(0131) 552 8462. info@unitedwire.com, www.unitedwire.com
Universal Air Products Ltd - Unit 5 & 6 Ardley Works, London Road, Billericay, Essex. CM12 9HP tel:(01277) 634637, fax:(01277) 632655. sales@universalair.net, www.universalair.net
Universal Components Ltd - Universal House, Pennywell Road, Bristol, Avon. BS5 0ER tel:(0117) 955 9091, fax:(0117) 955 6091. sales@universal-aluminium.co.uk, www.universal-aluminium.co.uk
Uponor Ltd - Hillcote Plant, PO Box 1, Berristow Lane, Blackwell, Nr Alfreton, Derbys. DE55 5JD tel:(01773) 811112, fax:(01773) 812343. enquiries.uk@uponor.com, www.uponor.co.uk
Uponor Ltd - Uponor Road, Lutterworth, Leicestershire. LE17 4DU tel:(01455) 550300, fax:(01455) 550366. enquiries.uk@uponor.com, www.uponor.co.uk
Urban Planters - PO Box 24, Skipton, North Yorkshire. BD23 9AN tel:(01274) 579331, fax:(01274) 521150. sales@urbanplanters.co.uk, www.urbanplanters.co.uk
Urbis Lighting Ltd - Telford Road, Houndmills, Basingstoke, Hants. RG21 6YW tel:(01756) 720010, fax:(0870) 056 9359. sales@urbislighting.com, www.urbislighting.com
Uretek (UK) Ltd - Mere One, Mere Grange, Elton Head Road, St Helens, Lancashire. WA9 5GG tel:(01695) 50525, fax:(01695) 555212. sales@uretek.co.uk, www.uretek.co.uk

V

V B Johnson LLP

9-11 Castleford Road, Normanton, W. Yorks. WF6 2DP tel:(01924) 897373, fax:(03332) 407363. wakefield@vbjohnson.co.uk, www.vbjohnson.co.uk

V B Johnson LLP

St John's House, 304-310 St Albans Road, Watford, Herts. WD24 6PW tel:(01923) 227236. watford@vbjohnson.co.uk, www.vbjohnson.co.uk

Vador Security Systems Ltd - 52 Lowdown, Pudsey, West Yorkshire. LS28 7AA tel:(0800) 037 5365, fax:(0870) 766 4025. info@vadorsecurity.co.uk, www.vadorsecurity.co.uk
Valmont Stainton - Teeside Industrial Estate, Thornaby, Cleveland. TS17 9LT tel:(01642) 766242, fax:(01642) 765509. stainton@valmont.com, www.stainton-metal.co.uk
Valor - Wood Lane, Erdington, Birmingham. B24 9QP tel:(0121) 373 8111, fax:(0121) 373 8181. marketing@valor.co.uk, www.valor.co.uk
Valspar Powder Coatings Ltd - 95 Aston Church Road, Birmingham. B7 5RQ tel:(0121) 322 6900, fax:(0121) 322 6901. powdercoatings@valspareurope.com, www.valspareurope.com
Valspar UK Corporation Ltd - Avenue One, Station Lane, Witeney, Oxon. OX28 4XR tel:(01993) 707400, fax:(01993) 775579. packaging@valspareurope.com, www.valspareurope.com/
Valtti Specialist Coatings Ltd - Unit 3B, South Gyle Cresent Lane, Edinburgh. EH12 9EG tel:(0131) 334 4999, fax:(0131) 334 3987. enquiries@valtti.co.uk, www.valtti.co.uk
Vantrunk Engineering Ltd - Goodard Road, Astmoor, Runcorn, Cheshire. WA7 1QF tel:(01928) 564211, fax:(01928) 580157. sales@vantrunk.co.uk, www.Vantrunk.com
Varley & Gulliver Ltd - 57-70 Alfred Street, Sparkbrook, Birmingham. B12 8JR tel:(0121) 773 2441, fax:(0121) 776 6875. sales@v-and-g.co.uk, www.v-and-g.co.uk
VBH (GB) Ltd - VBH House, Bailey Drive, Gillingham Business Park, Gillingham, Kent. ME8 0WG tel:(01634) 263 263, fax:(01634) 263 300. sales@vbhgb.com, www.vbhgb.com
Veka PLC - Farrington Road, Rossendale Road Indstrial Estate, Burnley, Lancs. BB11 5DA tel:(01282) 416611, fax:(01282) 439260. salesenquiry@veka.com, www.vekauk.com
Veka Plc - Farrington Road, Rossendale Road Industrial Estate, Burnley. BB11 5DA tel:(01282) 716611, fax:(01282) 725257. salesenquiry@veka.com, www.veka.com
VELFAC Ltd - The Old livery, Hildersham, Cambridge. CB21 6DR tel:(01223) 897100, fax:(01223) 897101. post@velfac.co.uk, www.velfac.co.uk
VELUX Co Ltd, The - Woodside Way, Glenrothes, E. Fife. KY7 4ND tel:(01592) 772211, fax:(01592) 771839. sales@velux.co.uk, www.velux.co.uk
Vent-Axia - Flemming Way, Crawley, W. Sussex. RH10 9YX tel:(01293) 526062, fax:(01293) 551188. info@vent-axia.com, www.vent-axia.com
Ventrolla Ltd - Crimple Court, Hornbeam Business Park, Harrogate, N. Yorks. HG2 8PB tel:(01423) 859323, fax:(01423) 859321. info@ventrolla.co.uk, www.ventrolla.co.uk
Ventura - Pampisford, Cambridge, Cambs. CB2 4EW tel:(01223) 837007, fax:(01223) 837215. info@ventura.uk.com, www.ventura.uk.com
Veoila Water Technologies UKElga Ltd - Windsor Court, Kingsmead Business Park, High Wycombe, Bucks. HP11 1JU tel:(01628) 897000, fax:(01628) 897001. sales.uk@veolia.com, www.elgalabwater.com
Verco Office Furniture Ltd - Chapel lane, Sands, High Wycombe, Bucks. HP12 4BG tel:(01494) 448000, fax:(01494) 464216. info@verco.co.uk, www.verco.co.uk
Vermeer United Kingdom - 45-51 Rixon Road, Wellingborough, Northants. NN8 4BA tel:(01933) 274400, fax:(01933) 274403. sales@vermeeruk.co.uk, http://vermeer-uk.co.uk/
Verna Ltd - Folds Road, Bolton, Lancs. BL1 2TX tel:(01204) 529494, fax:(01204) 521862. vcorders@vernagroup.com, www.vernagroup.com
Ves Andover Ltd - Eagle Close, Chandlers Ford Industrial Estate, Eastleigh, Hampshire. SO53 4NF tel:(08448) 15 60 60, fax:(02360) 261 204. info@ves.co.uk, www.ves.co.uk
Vescom UK Limited - Unit 3 Canada Close, Banbury, Oxfordshire. OX16 2RT tel:(01295) 273644, fax:(01295) 273646. uk@vescom.com, www.vescom.com
Vetter UK - Archway 3, Birley Fields, Greenheys Lane West, Hulme, Manchester. M15 5QJ tel:(0161) 227 6400, fax:(0161) 227 6449. enquiries@vetteruk.com, www.vetteruk.com
Vianen UK Ltd - Canary Warf, 29th Floor, 1 Canada Square, London. E14 5DY tel:(07800) 794 612, fax:(02476) 456 118. sales@vianenkvs.com, www.vianenkvs.com
Vicaima Ltd - Marlowe Avenue, Greenbridge Industrial Estate, Swindon, Wilts. SN3 3JF tel:(01793) 532333, fax:(01793) 530193. info@vicaima.com, www.vicaima.com
Vicon Industries Ltd - Brunel Way, Fareham, Hants. PO15 5TX tel:(01489) 566300, fax:(01489) 566322. sales@vicon.co.uk, www.vicon-cctv.com
Victaulic Systems - 46-48 Wibury Way, Hitchin, Herts. SG4 0UD tel:(01462) 443322, fax:(01462) 443148. info@victaulic.co.uk, www.victaulic.co.uk
Vidionics Security Systems Ltd - Systems House, Desborough Industrial Park, Desborough Park Road, High Wycombe, Bucks. HP12 3BG tel:(01494) 459606, fax:(01494) 461936. sales@vidionics.co.uk, www.vidionics.co.uk
Viking Johnson - 46-48 Wilbury Way, Hichen, Herts. SG4 0UD tel:(0121) 700 1000, fax:(0121) 700 1001. info@vikingjohnson.com, www.glynwedpipesystems-uk.com
Viking Laundry Equipment Ltd - 1 Carlisle Road, London. NW9 0HZ tel:(020) 8205 7285, fax:(020) 8200 3741. info@girbau.co.uk, www.viking1959.co.uk
Viking Security Systems Ltd - Suite C, Anchor House, School Lane, Chandlers Ford, Hampshire. SO53 4DY tel:(023) 8115 9545. www.vikingsecuritysystems.co.uk
Vink Plastics Ltd - 27 Long Wood Road, Trafford Park, Manchester. M17 1PZ tel:(01618) 737080, fax:(01618) 737079. info@vink.com, www.vinkplastics.co.uk
Viscount Catering Ltd - PO Box 16, Green Lane, Ecclesfield, Sheffield, S. Yorks. S35 9ZY tel:(0114) 257 0100, fax:(0114) 257 0251. fsuk.info@manitowoc.com, www.viscount-catering.co.uk
Visqueen Building Products - Heanor Gate, Heanor, Derbyshire. DE75 7RG tel:(0845) 302 4758, fax:(0845) 017 8663. enquiries@visqueenbuilding.co.uk, www.visqueenbuilding.co.uk
Vista Engineering Ltd - Carr Brook Works, Shallcross Mill Road, Whaley Bridge, High Peak. SK23 7JL tel:(01663) 736700, fax:(01663) 736710. sales@vistaeng.co.uk, www.vistaeng.co.uk
Vista Plan International PLC - High March, Daventry, Northants. NN11 4QE tel:(01327) 704767, fax:(01327) 300243. sales@vistaplan.com, www.vistaplan.com
Vita Liquid Polymers Ltd - Harling Road, Wythenshawe, Manchester. M22 4SZ tel:(0161) 998 3226, fax:(0161) 946 0118. info@vita-liquid.co.uk, www.vita-liquid.co.uk
Vitalighting Ltd - Unit 4, Sutherland Court, Moor Park Industrial Estate, Tolpits Lane, Watford, Herts. WD18 9NA tel:(01923) 896476, fax:(018923) 897741. sales@vitalighting.com, www.vitalighting.com
Vitra (UK) Ltd - Park 34, Collet Way, Didcot, Oxon. OX11 7WB tel:(01235) 750990, fax:(01235) 750980. info@vitra.co.uk, www.vitrauk.com
Vitra UK Ltd - 30 Clerkenwell Road, London. EC1M 5PG tel:(020) 7608 6200, fax:(020) 7608 6201. info-uk@vitra.com, www.vitra.com
Vitreous Enamel Association, The - Belfield House, Firs Lane, Appleton, Warrington. WA4 5LE tel:(0755) 55 96164. info@vea.org.uk, www.vea.org.uk
Vokes Air - Farrington Road, Farrington Road, Burnley, Lancs. BB11 5SY tel:(01282) 413131, fax:(01282) 686200. vokes@btrinc.com, www.vokesair.com
Vola - Unit 12, Ampthill Business Park, Station Road, Ampthill, Beds. MK45 2QW tel:(01525) 841155, fax:(01525) 841177. sales@vola.co.uk, www.vola.com
Volex Group PLC - Dornoch House, Kelvin Close, Birchwood, Warrington. WA3 7JX tel:(01925) 830101, fax:(01925) 830141. sales@volex.com, www.volex.com
Vortice Ltd - Beeches House, Eastern Avenue, Burton-on-Trent, Staffs. DE13 0BB tel:(01283) 492949, fax:(01283) 544121. sales@vortice.ltd.uk, www.vortice.ltd.uk
Vulcan Cladding Systems - 4 Imperial Way, Croydon. CR0 4RR tel:(020) 8681 0617, fax:(020) 8256 5977. sales@vulcansystems.co.uk, www.vulcansystems.co.uk
Vulcascot Cable Protectors Ltd - Unit 12, Norman-D-Gate, Bedford Road, Northampton. NN1 5NT tel:(0800) 0352842, fax:(01604) 632344. sales@vulcascotcableprotectors.co.uk, www.vulcascotcableprotectors.com

W

W Ward Fencing Contractors - K & M Hauliers Indust. Estate, The Aerodrome, Watnall Road, Hucknall, Notts. NG15 6EQ tel:(0115) 963 6948, fax:(0115) 963 0864. enquiries@wardfencingltd.co.uk, www.wardfencingltd.co.uk
W. Hawley & Son Ltd - Colour Works, Lichfield Road, Branston, Burton-on-Trent, Staffs. DE14 3WH tel:(01283) 714200, fax:(01283) 714201. info@concos.co.uk, www.hawley.co.uk
W.H Colt Son & Co Ltd - Unit 4 Counter Buildings, Brook Street, Woodchurch, Kent. TN26 3SP tel:(01233) 812919, fax:(01233) 740123. mail@colthouses.co.uk, www.colthouses.co.uk
W.L. West & Sons Ltd - Selham, Petworth, W. Sussex. GU28 0PJ tel:(01798) 861611, fax:(01798) 861633. sales@wlwest.co.uk, www.wlwest.co.uk
WA Skinner & Co Limited - Dorset Way, off Abbot Close, Byfleet, Surrey. KT14 7LB tel:(01932) 344228, fax:(01932) 348517
Waco UK Limnited - Catfoss Lane, Brandsburton, Nr Driffield, E. Yorks. YO25 8EJ tel:(0800) 3160888, fax:(01964) 545001. sales@waco.co.uk, www.waco.co.uk
Wade Building Services Ltd - Groveland Road, Tipton, W. Midlands. DY4 7TN tel:(0121) 520 8121, fax:(0121) 557 7061. sales@wade-bs.co.uk, www.wade-bs.co.uk
Wade Ceramics Ltd - Bessemer Drive, Stoke-on-Trent, Staffordshire. ST1 5GR tel:(01782) 577321, fax:(01782) 575195. enquiries@wade.co.uk, www.wade.co.uk
Wade International Ltd - Third Avenue, Halstead, Essex. CO9 2SX tel:(01787) 475151, fax:(01787) 475579. tech@wadedrainage.co.uk, www.wadedrainage.co.uk
Wales & Wales - The Long Barn Workshop, Muddles Green, Chiddingly, Lewes, Sussex. BN8 6HW tel:(01825) 872764, fax:(01825) 873197. info@walesandwales.com, www.walesandwales.com
Walker Timber Ltd - Carriden Sawmills, Bo'ness, W. Lothian. EH51 9SQ tel:(01506) 823331, fax:(01506) 822590. mail@walkertimber.com
Wall Tie Installers Federation - Heald House, Heald Street, Liverpool. L19 2LY tel:(0151) 494 2503. admin@wtif.org.uk, www.wtif.org.uk
Wallace & Tiernan - Chemfeed Limited, Priory Works, Five Oak Green Road, Tonbridge, Kent. TN11 0QL tel:(01732) 771777, fax:(01732) 771800. information.water@siemens.com, www.wallace-tiernan.com
Wallbank - P.O. Box 317, Wilmslow, Cheshire. SK9 5WA tel:(0161) 439 0908, fax:(0161) 439 0908. john@wallbank-lfc.co.uk, www.wallbank-lfc.co.uk
Wallcovering Manufacturers Association (now part of the British Coatings Federation) - Riverbridge House, Guildford Road, Leatherhead, Surrey. KT22 9AD tel:(01372) 365989. enquiry@bcf.co.uk, www.coatings.org.uk
Wallgate Ltd - Crow Lane, Wilton, Salisbury, Wiltshire. SP2 0HB tel:(01722) 744594, fax:(01722) 742096. sales@wallgate.com, www.wallgate.com
Wallis Conservation Limited trading as Dorothea Restorations - Unit 4, William Street, Bristol. BS2 0RG tel:(0117) 979 8397 , fax:(0117) 977 1677. john@dorothearestorations.com, www.dorothearestorations.com
Walls & Ceilings International Ltd - Tything Road, Arden Forest Industrial Estate, Alcester, Warwicks. B49 6EP tel:(01789) 763727, fax:(01789) 400312. sales@walls-and-ceilings.co.uk, www.walls-and-ceilings.co.uk
Walney UK Ltd - The Keys, Latchford Mews, Wheathampstead, Hertfordshire. AL4 8BB tel:(0870) 733 0011, fax:(0870) 733 0016. sales@walneyuk.com, www.walneyuk.com
Walraven Ltd - Thorpe Way, Banbury, Oxon. OX16 4UU tel:(01295) 75 34 00, fax:(01295) 75 34 28. sales.banbury@walraven.com, www.walraven.com/gb/en/
Wandsworth Elecrtrical Ltd - Albert Drive, Sheerwater, Woking, Surrey. GU21 5SE tel:(01483) 740740, fax:(01483) 740384. info@wandsworthgroup.com, www.wandsworth-electrical.com
Ward & Co (Letters) Ltd - Maze House, Maze Street, Barton Hill, Bristol. BS5 9TE tel:(0117) 955 5292, fax:(0117) 955 7518. web@ward-signs.co.uk, www.ward-signs.co.uk
Ward Insulated Panels Ltd - Sherburn, Malton, N. Yorks. YO17 8PQ tel:(01944) 710591, fax:(01944) 710555. wbc@wards.co.uk, www.wards.co.uk
Wardle Storeys - Grove Mill, Earby via Colne, Lancs. BB18 6UT tel:(01282) 842511, fax:(01282) 843170. sales@wardlestoreys.com, www.wardlestoreys.com
Ward's Flexible Rod Company Ltd - 22 James Carter Road, Mildenhall, Suffolk. IP28 7DE tel:(01638) 713800, fax:(01638) 716863. sales@wardsflex.co.uk, www.wardsflex.co.uk
Warmafloor (GB) Ltd - Concord House, Concord Way, Segensworth North, Fareham, Hampshire. PO15 5RL tel:(01489) 581787, fax:(01489) 57644. sales@warmafloor.co.uk, www.warmafloor.co.uk
Warner Howard Group Ltd - Warner Howard House, 2 Woodgrange Avenue, Harrow, Middx. HA3 0XD tel:(020) 8206 2900, fax:(020) 8206 1313. info@warnerhoward.co.uk, www.warnerhoward.w.uk
Watco UK Ltd - Watco House, Filmer Grove, Godalming, Surrey. GU7 3AL tel:(01483) 427373, fax:(01483) 428888. sales@watco.co.uk, www.watco.co.uk
Water Sculptures Ltd - Unit 4, White Lund, Morecambe. LA3 3PU tel:(01524) 37707. info@watersculptures.co.uk, www.watersculptures.co.uk

W (con't)

Water Technology Engineering Ltd - Unit 2, Bolton Lane, York, Yorkshire. YO41 5QX tel:(01759) 369915. info@wte-ltd.co.uk, www.wte-ltd.co.uk

Waterline Limited - Crown House, North Crawley Road, Newport Pagnell, Bucks. MK16 9TG tel:(08444) 122 524, fax:(0800) 585 531. sales@waterline.co.uk, www.waterline.co.uk

Waterstones - 203/206 Piccadilly, London. W1J 9HD tel:(0808) 118 8787. support@waterstones.com, www.waterstones.com

Watts Industries UK Ltd - Enterprise Way, Vale Business Park, Evesham, Worcs. WR11 1GA tel:(01386) 446997, fax:(01386) 41923. sales@wattsindustries.co.uk, www.wattsindustries.com

Wavin Ltd - Parsonage Way, Chippenham, Wilts. SN15 5PN tel:(01249) 766600, fax:(01249) 443286. info@wavin.co.uk, www.wavin.co.uk

Weatherley Fencing Contractors Limited - The Orchard, 135 North Cray Road, Sidcup, Kent. DA14 5HE tel:(020) 8300 6421, fax:(020) 8308 1317. weatherleyfence@btconnect.com, www.weatherleyfencing.co.uk/

Weber Building Solutions - Dickens House, Enterprise Way, Flitwick, Beds. MK45 5BY tel:(08703) 330070, fax:(01525) 718988. mail@weberbuildingsolutions.co.uk, www.weberbuildingsolutions.co.uk

Wednesbury Tube - Oxford Street, Bilston, W. Midlands. WV14 7DS tel:(01902) 491133, fax:(01902) 405838. sales@muellereurope.com, www.muellereurope.com

Weir Group - Clydesdale Bank Exchange, 20 Waterloo Street, Glasgow. G2 6DB tel:(0141) 637 7111, fax:(0141) 221 9789. pr@weir.co.uk, www.weir.co.uk

Weir Pumps Ltd - 149 Newlands Road, Cathcart, Glasgow. G44 4EX tel:(01416) 377141, fax:(01416) 377358. weirspmaberdeen@weirspm.com, www.weirclearliquid.com

Welco - Innovation Centre, 1 Devon Way, Longbridge Technology Park, Birmingham. B31 2TS tel:(0800) 954 9001, fax:(0845) 688 8900. sales@welco.co.uk, www.welco.co.uk

Wellman Robey Ltd - Newfield Road, off Dudley Road East, Oldbury, Warley, W. Midlands. B69 3ET tel:(0121) 552 3311, fax:(0121) 552 4571. sales@wellmanrobey.com, www.wellmanrobey.com

Welsh Slate Limited - Penrhyn Quarry, Bethesda, Bangor, Gwynedd. LL57 4YG tel:(01248) 600656, fax:(01248) 601171. enquiries@welshslate.com, www.welshslate.com

Wentworth Sawmills Ltd - Barrowfield Lane, Wentworth, Nr. Rotherham, S. Yorks. S62 7TP tel:(01226) 742206, fax:(01226) 742484. victoria.earnshaw@jobearnshaw.co.uk, www.wentworthsawmills.co.uk

Wernick Group Ltd - Molineux House, Russell Gardens, Wickford, Essex. SS11 8BL tel:(01268) 735544, fax:(01268) 560026. sales@wernick.co.uk, www.wernick.co.uk

Wessex Lift Co Ltd - Budds Lane, Romsey, Hants. SO51 0HA tel:(01794) 830303, fax:(01794) 512621. info@wessexlifts.co.uk, www wessexlifts.co.uk

Wessex Resins & Adhesives Ltd - Cuperham House, Cuperham Lane, Romsey, Hants. SO51 7LF tel:(01794) 521111, fax:(01794) 521271. info@wessex-resins.com, www.wessex-resins.com

Wesson Fencing - 48 High Street, Knaphill, Woking, Surrey. GU21 2PY tel:(01483) 472124, fax:(01483) 472115. wesson-fencing@hotmail.co.uk, www.wesson-fencing.co.uk

West London Security - 22-36 Paxton Place, London. SE27 9SS tel:(020) 8676 4300, fax:(020) 8676 4301. info@westlondonsecurity.com, www.westlondonsecurity.com

Westbury Filters Limited - Hall Farm Estate, Gadbrook Road, Betchworth, Surrey. RH3 7AH tel:(01306) 611611, fax:(01306) 611613. sales@westburyfiltermation.com, www.westburyfilters.co.uk

Westco Group Ltd - Penarth Road, Cardiff. CF11 8YN tel:(029) 20 376700, fax:(029) 20 383573. westco@westcodiy.co.uk, www.westcodiy.co.uk

Western Cork Ltd - Penarth Road, Cardiff. CF11 8YN tel:(029) 20 376700, fax:(029) 20 383573. customerservices@westcodiy.co.uk, www.westcofloors.co.uk

Western landscapes Ltd (trading as Wyevale) - Upper Buckover Farm, Buckover, Wotton-under-Edge, Gloucestershire. GL12 8DZ tel:(01454) 419175, fax:(01454) 412901. wyevale@wyevale-landscapes.co.uk, www.wyevale-landscapes.co.uk

Westpile Ltd - Dolphin Bridge House, Rockingham Road, Uxbridge, Middx. UB8 2UB tel:(01895) 258266, fax:(01895) 271805. estimating@westpile.co.uk, www.westpile.co.uk

WF Senate - 313-333 Rainham Road South, Dagenham, Essex. RM10 8SX tel:(0208) 984 2000, fax:(0121) 567 2638. info@rexelsenate.co.uk, www.wfsenate.co.uk

Which Building Contract - 96 Rutland Road, London. E7 8PH tel:(0844) 567 5087. malcolmb@goeshere.co.uk, www.whichbuildingcontract.co.uk

Whitecroft Lighting Limited - Burlington Street, Ashton-under-Lyne, Lancs. OL7 0AX tel:(0870) 5087087, fax:(0870) 5084210. email@whitecroftlight.com, www.whitecroftlighting.com

Whitehill Spindle Tools Ltd - 2-8 Bolton Road, Luton, Beds. LU1 3HR tel:(01582) 736881, fax:(01582) 488987. sales@whitehill-tools.com, www.whitehill-tools.com

Whitmore's Timber Co Ltd - Main Road, Claybrooke Magna, Lutterworth, Leics. LE17 5AQ tel:(01455) 209121, fax:(01455) 209041. esales@whitmores.co.uk, www.whitmores.co.uk/

Wienerberger Ltd - Wienerberger House, Brooks Drive, Cheadle Royal Business Park, Cheadle, Cheshire. SK8 3SA tel:(0161) 491 8200, fax:(0161) 491 1270. office@wienerberger.co.uk, www.wienerberger.co.uk

Willamette Europe Ltd - 10th Floor, Maitland House, Warrior Square, Southend-on-Sea, Essex. SS1 2JY tel:(01702) 619044, fax:(01702) 617162. euinfo@medite-europe.com, www.willamette-europe.com

William Hopkins & Sons Ltd - Gardine House, 147-147 Dollman Street, Nechells, Birmingham. B7 4RS tel:(0121) 333 3577, fax:(0121) 333 3480. info@william-hopkins.co.uk, www.william-hopkins.co.uk

William Wilson Ltd - SCO 14691 Registered office, Hareness Road, Altens Industrial Estate, Aberdeen. AB12 3QA tel:(01224) 877522, fax:(01224) 879650. marketing@williamwilson.co.uk, www.williamwilson.co.uk

Williamson T & R Ltd - 36 Stonebridge Gate, Ripon, N. Yorks. HG4 1TP tel:(01765) 607711, fax:(01765) 607908. enquiries@trwilliamson.co.uk, www.trwilliamson.co.uk

Wilo Samson Pumps Ltd - 2ND Avenue, Centrum 100, Burton-on-Trent, Staffs. DE14 2WJ tel:(01332) 385181, fax:(01332) 344423. sales@wilo.co.uk, www.wilo.co.uk

Wilsonart Ltd - Lambton Street Industrial Estate, Shildon, County Durham. DL4 1PX tel:+44 0 1388 774661, fax:+44 0 1388 774861. customerservices@wilsonart.co.uk, www.wilsonart.co.uk

Wincro Metal Industries Ltd - Fife Street, Wincobank, Sheffield, S. Yorks. S9 1NJ tel:(0114) 242 2171, fax:(0114) 243 4306. sales@wincro.com, www.wincro.com

Windowbuild - 55-56 Lewis Road, East Moors, Cardiff. CF24 5EB tel:(029) 20 307 200, fax:(029) 20 480 030. info@windowbuild.co.uk, www.windowbuild.co.uk

Winn & Coales (Denso) Ltd - Denso House, Chapel Road, London. SE27 0TR tel:(020) 8670 7511, fax:(020) 8761 2456. mail@denso.net, www.denso.net

Winther Browne & co Ltd - 75 Bilton Way, Enfield, London. EN3 7ER tel:(020) 8884 6000, fax:(020) 8884 6001. sales@wintherbrowne.co.uk, www.wintherbrowne.co.uk

Wirquin - Warmsworth Halt Industrial Estate, Warmsworth, Doncaster. DN4 9LS tel:(0844) 412 2029. sales@wirquin.co.uk, www.wirquin.co.uk

Witham Oil & Paint Ltd - Stanley Road, Oultoh Broad, Lowestoft, Suffolk. NR33 9ND tel:(01502) 563434, fax:(01502) 500010. enquiries@withamgroup.co.uk, www.withamoil.co.uk

Wm. Bain Fencing Ltd - Lochrin Works, 7 Limekilns Road, Blairlinn Ind. Est, Cumbernauld. G67 2RN tel:(01236) 457333, fax:(01236) 451166. sales@lochrin-bain.co.uk, www.lochrin-bain.co.uk

Wolseley UK - The Wolseley Center, Harrison Way, Spa Park, Royal Leamington Spa. CV31 3HH tel:(01923) 705 000. customerservices@wolseley.co.uk, www.wolseley.co.uk

Wood International Agency Ltd - Wood House, 16 King Edward Road, Brentwood, Essex. CM14 4HL tel:(01277) 232991, fax:(01277) 222108. woodia@msn.com, www.ourworld.compuserve.com/homepages/wood_international_agency

Wood Panels Industries Federation - Autumn Park Business Centre, Dysart Road, Grantham, Lincolnshire. NG31 7EU tel:(01476) 512381, fax:(01476) 575683. enquiries@wpif.org.uk, www.wpif.org.uk

Woodburn Engineering Ltd - Rosganna Works, Trailcock Road, Carrickfergus, Co. Antrim. BT38 7NU tel:(028) 93366404, fax:(028) 93367539. rose@woodburnengineeringltd.co.uk

Woodhouse UK Plc - Spa Park, Lemington Spa, Warwicks. CV31 3HL tel:(01926) 314313, fax:(01926) 883778. enquiries@woodhouse.co.uk, www.woodhouse.co.uk

Woodscape Ltd - 1 Sett End Road West, Shadsworth Business Park, Blackburn, Lancashire. BB1 2QJ tel:(01254) 685185, fax:(01254) 671237. sales@woodscape.co.uk, www.woodscape.co.uk

Worcester Heat Systems Ltd - Cotswold Way, Warndon, Worcester, Worcs. WR4 9SW tel:(01905) 754624, fax:(01905) 754619. sales.mailbox@uk.bosch.com, www.worcester-bosch.co.uk

Workspace Technology Ltd - Technology House, 5 Emmanuel Court, Reddicroft, Sutton Coldfield. B72 1TJ tel:(0121) 354 4894. sales@workspace-technology.com, www.workspace-technology.com

WP Metals Ltd - Westgate, Aldridge, W. Midlands. WS9 8DJ tel:(01922) 743111, fax:(01922) 743344. info@wpmetals.co.uk

Wragg Bros (Aluminium Equipment) Ltd - Robert Way, Wickford Business Park, Wickford, Essex. S11 8DQ tel:(01268) 732607, fax:(01268) 768499. wragg.bros@btclick.com, www.wraggbros.co.uk

WT Henley Ltd - Crete Hall Road, Gravesend, Kent. DA11 9DA tel:(01474) 564466, fax:(01474) 566703. sales@wt-henley.com, www.wt-henley.com

Wybone Ltd - Mason Way, Platts Common Industrial Estate, Hoyland, Nr Barnsley, S. Yorks. S74 9TF tel:(01226) 744010, fax:(01226) 350105. sales@wybone.co.uk, www.wybone.co.uk

Wyckham Blackwell Ltd - Old Station Road, Hampton in Arden, Solihull, W. Midlands. B92 0HB tel:(01675) 442233, fax:(01675) 442227. info@wyckham-blackwell.co.uk, www.wyckham-blackwell.co.uk

Wylex Ltd - Wylex Works, Wythenshaw, Manchester. M22 4RA tel:(0161) 998 5454, fax:(0161) 945 1587. wylex.sales@electrium.co.uk, www.electrium.co.uk/wylex.htm

Wymark Ltd - Runnings Road Industrial Estate, Cheltenham, Glos. G51 9NQ tel:(01242) 520966, fax:(01242) 519925. info@wymark.co.uk, www.wymark.co.uk

Wyvern Marlborough Ltd - Grove Trading Estate, Dorchester, Dorset. DT1 1SU tel:(01305) 264716, fax:(01305) 264717. mailto:info@wyvernfireplaces.com, www.wyvernfireplaces.com

X

Xella Aircrete Systems UK - PO Box 10028, Sutton Coldfield. B75 7ZF tel:(08432) 909 080, fax:(08432) 909 081. hebel-uk@xella.com, www.xella.co.uk

Xpelair Ltd - PO Box 279, Morley Way, Peterborough. PE2 9JJ tel:(01733) 456789, fax:(01733) 310606. info@redring.co.uk, www.xpelair.co.uk

Xtralis - Peoplebuilding, Ground Floor, Maylands Avenue, Hemel Hempstead, Herts. HP2 4NW tel:(01442) 242330, fax:(01442) 240327. sales@xtralis.com, http://xtralis.com

Y

Yale Security Products Ltd - Wood Street, Willenhall, W. Midlands. WV13 1LA tel:(01902) 366911, fax:(01902) 368535. info@yale.co.uk, www.yale.co.uk

York Handmade Brick Co Ltd - Winchester House, Forest Lane, Alne, York, N. Yorks. Y061 1TU tel:(01347) 838881, fax:(01347) 830065. sales@yorkhandmade.co.uk, www.yorkhandmade.co.uk

Yorkon Ltd - New Lane, Huntington, York, Yorks. YO32 9PT tel:(01904) 610990, fax:(01904) 610880. contact@yorkon.com, www.yorkon.com

Yorkshire Copper Tube - East Lancashire Road, Kirkby, Liverpool. L33 7TU tel:(0151) 545 5079, fax:(0151) 549 2139. sales@yct.com, www.yct.com

Yorkstone Products Ltd - Britannia Quarries, Morley, Leeds. LS7 0SW tel:(0113) 253 0464, fax:(0113) 252 7520. sales@woodkirkstone.co.uk, www.woodkirkstone.co.uk

Youngman Group Limited - The Causeway, Maldon, Essex. CM9 4LJ tel:(01621) 745900. youngman-sales@wernerco.com, www.youngmangroup.com

Youngs Doors Ltd - City Road Works, Norwich, Norfolk. NR1 3AN tel:(01603) 629889, fax:(01603) 764650. mail@youngs-doors.co.uk, www.youngs-doors.co.uk

Z

Zarges (UK) Ltd - 8 Holdom Avenue, Saxon Park Industrial Estate, Bletchley, Milton Keynes. MK1 1QU tel:(01908) 641118, fax:(01908) 648176. sales@zarges.co.uk, www.zargesuk.co.uk

Zehnder Ltd - B15 Armstrong Mall, Southwood Business Park, Farnborough, Hants. GU14 0NR tel:(01252) 515151, fax:(01252) 522528. sales@zehnder.co.uk, www.zehnder.co.uk

Zellweger Analytic Ltd, Sieger Division - Hatch Pond House, 4 Stinsford Road, Poole, Dorset. BH17 0RZ tel:(01202) 676161, fax:(01202) 678 011. consumer@honeywell.com, www.zelana.com

Zephyr The Visual Communicators - Midland Road, Thrapston, Northants. NN14 4LX tel:(01832) 734 484, fax:(01832) 733 064. sales@zephyr-tvc.com, www.zephyr-tvc.com

Zled (DPC) Co Ltd - Unit 9, Tonbridge Chambers, Pembury Road, Tonbridge, Kent. TN9 2HZ tel:(01732) 363 443, fax:(01732) 363 553. info@z-led.com, www.z-led.com

Zon International - PO Box 329, Edgware, Middx. HA8 6NH tel:(020) 8381 1222, fax:(020) 8381 1333. sales@zon.co.uk, www.zon.co.uk

A01   Institutions/ Associations/ Trade Organisations

Accrediting body and learned society - Chartered Institution of Building Services Engineers (CIBSE)
Aluminium trade association - Aluminium Federation Ltd
Aluminium window association - Council for Aluminium in Building
Architectural aluminium association - Council for Aluminium in Building
Asbestos training courses for management and operatives - Asbestos Removal Contractors' Association
Association - Resin Flooring Association (FeRFA)
Association and institution - Joint Council For The Building & Civil Engineering Industry (N. Ireland)
Association of concrete technologists - The Institute of Concrete Technology
Association of Concrete Industrial Flooring Contractors - Association of Concrete Industrial Flooring Contractors (ACIFC)
BCIS Dayworks Rates Online - BCIS
BCIS Online - BCIS
BCIS Rebuild Online - BCIS
BCIS Review Online - BCIS
Bookshops - Royal Incorporation of Architects in Scotland
Brick information service - Brick Development Association
Builders and Master Wrights Association, Scotland - Scottish Master Wrights and Builders Association (SMWBA)
Builders federation - Federation of Master Builders
Builders federation - National Federation of Builders
Building & Engineering Services Association - Building & Engineering Services Association
Building Cost Information Service - BCIS
Campaigns in Britain and abroad - Society for the Protection of Ancient Buildings (SPAB)
Carpet manufacturers trade association - The Carpet Foundation
Carpet Industry Trade Association - The Carpet Foundation
Carpet promotion - The Carpet Foundation
Casework to save old buildings at risk from damage or demolition - Society for the Protection of Ancient Buildings (SPAB)
Ceramic Research Association (formerly Ceram) - Lucideon
Charity for the built environment - Civic Trust
Clay pipe trade asssociation - Clay Pipe Development Association Ltd
Client services and advice - Royal Incorporation of Architects in Scotland
Concrete Society - The Institute of Concrete Technology
Conferences and events - Royal Incorporation of Architects in Scotland
Construction employers federation - Construction Employers Federation
Construction information on CD-Rom - IHS
Construction information service online - IHS
Consultancy on Timber - TRADA Technology Ltd
Consultancy services on procurement - Royal Incorporation of Architects in Scotland
Contract flooring journal - Contract Flooring Association
Copper Development Association - Copper Development Association
Cost engineers' association - Association of Cost Engineers
Damp control - The Property Care Association
Decorators trade association - Painting & Decorating Association
Design competitions administration - Royal Incorporation of Architects in Scotland
Distributor of the Arts' share of the National Lottery - Arts Council England
Draught proofing trade association - Draught Proofing Advisory Association
Dry stone walling association - Dry Stone Walling Association of Great Britain
Education and training for building professionals - Brick Development Association
Electrical association - Electrical Contractors' Association (ECA)
Estimators - John Watson & Carter
Estimators - V B Johnson LLP
Event organizing/book seller/information provider - Chartered Institution of Building Services Engineers (CIBSE)
Events including lectures and regional group visits - Society for the Protection of Ancient Buildings (SPAB)
Fencing Contractor Association - Fencing Contractors Association
Fire Safety Industry - Fire Industry Association
Flood Protection - The Property Care Association
Flood remediation - The Property Care Association
Forestry - Confor: Promoting forestry and wood

Free technical advice on the maintenance of old buildings - Society for the Protection of Ancient Buildings (SPAB)
Gallery - Royal Incorporation of Architects in Scotland
Galvanizers trade association - Galvanizers Association
Gas safety consultants - Corgi Technical Dervices
Glazing trade association - Glass and Glazing Federation
Heating and ventilating trade association - Building & Engineering Services Association
Imlementation of working rules & wage rates for the industry - Joint Industry Board for the Electrical Contracting Industry
Information & advice on craft of dry stone walling - Dry Stone Walling Association of Great Britain
Information & helpline services - Flat Roofing Alliance (FRA)
Information on Copper & Copper alloys - Copper Development Association
Information service to businesses - British Library Business Information Service
Institution of Incorporated Engineers in electronic, electrical and mechanical engineering (IIE) - The Institution of Engineering and Technology
Institutions and association - Architectural Association
Institutions and association - Asbestos Removal Contractors' Association
Institutions and association - Association of Interior Specialists (AIS)
Institutions and association - British Precast Concrete Federation
Institutions and association - British Standards Institution
Institutions and association - Catering Equipment Suppliers Association
Institutions and association - Cement Admixtures Association
Institutions and association - Craft Guild of Chefs
Institutions and association - Friends of the Earth
Institutions and association - Historic England
Institutions and association - HMSO Publications Office
Institutions and association - Institution of Electrical Engineers
Institutions and association - Institution of Mechanical Engineers
Institutions and association - Institution of Structural Engineers
Institutions and association - Mortar Industry Association
Institutions and association - Ordnance Survey
Institutions and association - Quarry Products Association
Institutions and association - Royal Institute of British Architects (RIBA)
Institutions and association - Royal Town Planning Institute
Institutions and associations - British Association for Chemical Specialities
Institutions and associations - British Association of Landscape Industries
Institutions and associations - British Refrigeration Association (BRA)
Institutions and Associations - Building Controls Industry Association (BCIA)
Institutions and associations - Civic Trust
Institutions and Associations - Cold Rolled Sections Association
Institutions and associations - European Lead Sheet Industry Association
Institutions and associations - Institution of Civil Engineers
Institutions and associations - Interpave
Institutions and associations - National Federation of Master Steeplejacks & Lightning Conductor Engineers
Institutions and associations - Precast Flooring Federation
Institutions and associations - Prestressed Concrete Association
Institutions and associations - Society for the Protection of Ancient Buildings (SPAB)
Institutions and associations - The Concrete Society
Institutions and associations - The Kitchen Bathroom Bedroom Specialists Association
Institutions/ Associations/ Trade Organisations - Design Council
Institutions/Associations - British Slate Association
Institutions/Associations - FeRFA
Institutions/Associations - Institute of Quarrying
Institutions/Associations - Leatherhead Food Research Association
Institutions/Associations - Sprayed Concrete Association
Institutions/Associations - Stone Federation Great Britain
Institutions/Associations/Trade Organisations - Be-Plas Marketing Ltd
Insulated Render & Cladding Association Ltd - Insulated Render & Cladding Association
Insulation association - National Insulation Association Ltd

Invasive weed control - The Property Care Association
Journals - Contract Journal
LIA - Lighting Industry Association, The
Lift and Escalator Industry Association - LEIA
Lighting Industry Association - Lighting Industry Association, The
Lucideon - Ceramic Research - Lucideon
Manholes - Concrete Pipeline Systems Association
National Access & Scaffolding Confederation Ltd (NASC) - National Access & Scaffolding Confederation (NASC)
National funding body for arts in England - Arts Council England
National organisation for the steel construction industry - British Constructional Steelwork Association Ltd
Nurseries - Confor: Promoting forestry and wood
Painting contractors trade association - Painting & Decorating Association
Patent glazing contractors association - Council for Aluminium in Building
Perspective artist - Green Gerald
Pipes - Concrete Pipeline Systems Association
Plant Managers' Journal - Contract Journal
Product and material information and literature - The Building Centre Group Ltd
Professional body - Chartered Institute of Building
Professional Body - Royal Institution of Chartered Surveyors (RICS)
Professional body for wastes managment - Chartered Institute of Wastes Management
Professional body registered as an educational charity - Chartered Institute of Plumbing & Heating Engineering
Professional institution - Chartered Building Company Scheme
Professional institution - Chartered Building Consultancy Scheme
Professional recognition and development for Incorporated Engineers and Engineering Technicians in electronic, electrical and mechanical engineering - The Institution of Engineering and Technology
Publications on care and repair of old buildings - Society for the Protection of Ancient Buildings (SPAB)
Publications: contract flooring association members handbook - Contract Flooring Association
Publishing house (rutland press), guides etc - Royal Incorporation of Architects in Scotland
Registration of apprentice electricians - Joint Industry Board for the Electrical Contracting Industry
Represent RIBA in Scotland - Royal Incorporation of Architects in Scotland
Scottish architectural trade association - Royal Incorporation of Architects in Scotland
Sealant Applicators Trade Organisation - Association of Sealant Applicators Ltd (ASA)
SED - The National Event For Construction - Contract Journal
Separate mills section - Society for the Protection of Ancient Buildings (SPAB)
Services to architects including PI insurance - Royal Incorporation of Architects in Scotland
Special events run by SPAB in Scotland - Society for the Protection of Ancient Buildings (SPAB)
Steel window association - Steel Window Association
Structural repair - The Property Care Association
Structural Waterproofing - The Property Care Association
TCPA Publications - Town and Country Planning Association
Technical advice information sheets - Flat Roofing Alliance (FRA)
Technical Consultancy - British Glass Manufacturers Confederation
Technical, training, health & safety representation - Flat Roofing Alliance (FRA)
Telephone enquiry point - British Library Business Information Service
Telephone help line for building professionals - Barbour Enquiry Service
Terrazzo, marble and mosaic federation - National Federation of Terrazzo, Marble & Mosaic Specialists
Testing, certification and technical approval - British Board of Agrément
Thatched roof advice, training and insurance - Thatching Advisory Services Ltd
The construction information service - NBS
The Institution of Engineering and Technology - The Institution of Engineering and Technology
The Lighting Industry Association - Lighting Industry Association, The
The welding institute - TWI
Timber preservation - The Property Care Association
Town and Country Planing Journal - Town and Country Planning Association
Trade Association - Association of Specialist Fire Protection (ASFP)
Trade Association - Bitumen Waterproofing Association

Trade association - Brick Development Association
Trade association - British Blind and Shutter Association (BBSA)
Trade Association - British Coatings Federation
Trade association - British Constructional Steelwork Association Ltd
Trade Association - British Electrotechnical and Allied Manufacturers Association
Trade Association - British Fenestration Rating Council
Trade Association - British Flue & Chimney Manufacterers Association
Trade Association - British Glass Manufacturers Confederation
Trade Association - British Plastics Federation, The
Trade Association - British Pump Manufacturers Association
Trade Association - British Security Industry Association
Trade Association - British Stainless Steel Association
Trade Association - British Structural Waterproofing Association, The
Trade Association - British Woodworking Federation
Trade Association - Builders Merchants Federation
Trade Association - Building Reserch Establishment
Trade Association - Chartered Institute of Arbitrators
Trade Association - Civil Engineering Contractors Association
Trade Association - Concrete Block Association
Trade Association - Construction Industry Research & Information Association
Trade Association - Construction Industry Training Board
Trade Association - Construction Plant-hire Association
Trade Association - Constructionline
Trade Association - Door and Hardware Federation
Trade Association - Drilling & Sawing Association, The
Trade Association - Federation of Piling Specialists
Trade Association - Federation of Plastering and Drywall Contractors
Trade Association - Garage Equipment Association Ltd
Trade Association - Guild of Architectural Ironmongers
Trade Association - Guild of Builders & Contractors, The
Trade Association - Heating Ventilating and Air Conditioning Manufacturers Association (HEVAC)
Trade Association - Institute of Ashphalt Technology, the
Trade Association - Institute of Specialist Surveyors and Engineers
Trade Association - Joint Industry Board for the Electrical Contracting Industry
Trade Association - Lift and Escalator Industry Association, The
Trade Association - National Association of Shopfitters
Trade Association - National Federation of Demolition Contractors
Trade Association - National Federation of Roofing Contractors Ltd., The
Trade Association - National House Building Council
Trade Association - National Inspection Council for Electrical Installation Contracting
Trade Association - Painting & Decorating Association
Trade Association - PRA Coatings Technology Centre
Trade association - Quarry Products Association
Trade Association - Rural Industrial Design & Building Association
Trade Association - Scottish & Northern Ireland Plumbing Employers' Federation
Trade Association - Single Ply Roofing Association
Trade Association - Solar Trade Association, The
Trade Association - The Tile Association
Trade Association - Thermal Insulation Contractors Association
Trade Association - Timber Trade Federation
Trade Association - UK Mineral Wool Association,The
Trade Association - Vitreous Enamel Association, The
Trade Association - Wall Tie Installers Federation
Trade Association - Wallcovering Manufacturers Association (now part of the British Coatings Federation)
Trade association for the ceramic tile industry - The Tile Association
Trade association for the factory produced mortar industry - Mortar Industry Association
Trade association for the lift & escalator industry - LEIA
Trade Association Service - Mastic Asphalt Council

A01 Institutions/ Associations/ Trade Organisations (con't)

Trade associations of electrical and electronics sector - Beama Ltd
Trade Body - British Adhesives and Sealants Association
Trade Organisation - Advisory, Conciliation and Arbitration Service
Trade Organisation - Association for Consultancy and Engineering (ACE)
Trade Organisation - Association of Plumbing and Heating Contractors
Trade Organisation - Bathroom Manufacturers Association, The
Trade Organisation - British Ceramic Confederation
Trade Organisation - MPA Cement
Training programs and courses for professionals, craftsmen and owners - Society for the Protection of Ancient Buildings (SPAB)
Urethane foam industry trade association - BRITISH URETHANE FOAM CONTRACTORS ASSOCIATION
Wall ties trade association - British Wall Tie Association
Water and Wastewater Association - British Water
Wood - Confor: Promoting forestry and wood
Wood Panels - Wood Panels Industries Federation

A02 Building contractors

Asbestos removal - Rhodar Limited
Association of Concrete Industrial Flooring Contractors - Association of Concrete Industrial Flooring Contractors (ACIFC)
Bespoke joinery - Pearce Construction (Barnstaple) Limited
Builders - Ogilvie Construction Ltd
Building and development - Pearce Construction (Barnstaple) Limited
Building contractor - TH Kenyon & Sons Ltd
Building contractor incorporating partnering, guaranteed max price, PFI, lease back, facilities management, site finding - Miller Construction
Building contractors - Advanced Interior Solutions Ltd
Building Contractors - Clifford Partitioning Co Ltd
Civil engineering - Freyssinet Ltd
Construction - Kier Limited
Design and build fencing contractors - Baker Fencing Ltd
Design and build surfacings contractors - Easi-Fall International Ltd
Electrical services Contractor - Kershaw Mechanical Services Ltd
Fencing contractors - Burn Fencing Limited
Housing - Kier Limited
Housing - Pearce Construction (Barnstaple) Limited
Interior contractors - Gifford Grant Ltd
Landscape contractors - Gearing F T Landscape Services Ltd
Mechcanical services Contractor - Kershaw Mechanical Services Ltd
Minor civil engineering - Chestnut Products Limited
Plumbers - Ogilvie Construction Ltd
Property development - Kier Limited
Refurbishment and Building contractors - Shaylor Group Plc
Retail, Leisure, Ditribution, manufacturing, Pharmaceutical, heathcare, education, offices and food processing sector of the industry - Simons Construction Ltd
Roofing contractor - Evans Howard Roofing Ltd
Sports field contractor - Bernhards Landscapes Ltd
Sports ground construction - AMK Fence-In Limited
Structural enhancement - Freyssinet Ltd

A10 Project particulars

Architectural and Exhibition model makers - Modelscape
Architectural presentation planning design - Comber Models
ATM enclosures - Pearce Security Systems
Business to Business Listings - Glenigan Ltd
Capital Allowance Assessors - Johnson VB LLP
Chartered quantity surveyors - John Watson & Carter
Chartered quantity surveyors - Sawyer & Fisher
Chartered quantity surveyors - V B Johnson LLP
Construction related Sector - Glenigan Ltd
Consultancy - Carter Retail Equipment Ltd
Cost Engineers - Johnson VB LLP
Design an Procurement - Robinson Lloyds Architecture
Education services design and procurement - Robinson Lloyds Architecture
Electrical services - H Pickup Mechanical & Electrical Services Ltd
Estimators - DWA Partnership
Estimators - Johnson VB LLP
General engineering - Turnbull & Scott (Engineers) Ltd
Health services design and procurement - Robinson Lloyds Architecture
Mechanical services - H Pickup Mechanical & Electrical Services Ltd
Mechanical, electrical and process services engineers - Shepherd Engineering Services Ltd
Models - Comber Models
Modular cash rooms - Pearce Security Systems
Project management - Carter Retail Equipment Ltd
Project Managers - Johnson VB LLP
Project particulars - Advanced Interior Solutions Ltd
Project particulars - British Standards Institution
Quantity Surveyors - DWA Partnership
Quantity Surveyors - Johnson VB LLP
Rising armour screens - Pearce Security Systems
Rotary transfer units - Pearce Security Systems
Sales Leads - Glenigan Ltd
Security lobbies - Pearce Security Systems
Social care services design and procurement - Robinson Lloyds Architecture
Stats & Rankings Lists - Glenigan Ltd
Structural Steelwork Design - Billington Structures Ltd
Take off, bills of quantities and estimating - EITE & Associates
Trade Association - Cast Metals Federation

A20 The Contract/ Sub-contract

Advice given to Contractors - National Federation of Builders
Expert witness - British Glass Manufacturers Confederation
Guide to sustainable procurement in the water industry - British Water

A30 Tendering/ Subletting/ Supply

Building Price Book - Laxton's Publishing Ltd

A31 Provision, content and use of documents

Books - Town and Country Planning Association
CPD seminars - DW Windsor Lighting
NBS Building - NBS
NBS Create - NBS
Online bookshop - BRE

A32 Management of the Works

Building regulations approved documents England & Wales - NBS
Management of the Works - Advanced Interior Solutions Ltd
Project management - DW Windsor Lighting
Publishers of the National Building Specification - NBS
Sub Contract Engineers - Multi Marque Production Engineering Ltd
Windows based Q.S, and civils contract administration and estimating software - CSSP (Construction Software Services Partnership)

A33 Quality standards/ control

Administration of electrotechnical certification scheme in afiliation with CSCS - Joint Industry Board for the Electrical Contracting Industry
Alarm monitoring & routine maintenance - ADT Fire and Security Plc
Analysis & Testing - British Glass Manufacturers Confederation
ATaC asbestos testing and consulting - a division of Arca - Asbestos Removal Contractors' Association
Carpet quality control - The Carpet Foundation
Civil engineering test equipment - Capco Test Equipment, A division of Castlebroom Engineering Ltd
Dilapidation Surveys - EITE & Associates
Environmental services - British Glass Manufacturers Confederation
Food safety - National Britannia Ltd
Fracture & failure - British Glass Manufacturers Confederation
Health & safety - National Britannia Ltd
Height safety training - Capital Safety Group (NE) Ltd
In-house laboratory - DW Windsor Lighting
Methodologies within the water industry - British Water
Product & Perfomance - British Glass Manufacturers Confederation
Quality standards/control - British Standards Institution
Service & Inspection Schemes for mechanical Plant - Multi Marque Production Engineering Ltd
Servicing and maintenance - Planet Communications Group Ltd
Stock testing - British Glass Manufacturers Confederation
Technical support & customer service - Hilmor
Training - National Britannia Ltd
Training Services for Gas and water and drainage installation - Uponor Ltd

A34 Security/ Safety/ Protection

Access - Magnum Ltd T/A Magnum Scaffolding
Access and Fall Arrest Systems - Safety Works & Solutions Ltd
Access control - Allegion (UK) Ltd
Access systems - Clow Group Ltd
Anti bandit & bullet resistant counters - Pearce Security Systems
Asbestos Removal - Kitsons Insulation Products Ltd
Automatic / Self-Testing Emergency Lighting - P4 Limited
Battery operated smoke & fire alarms - BRK Brands Europe Ltd
Bullet resistant gazing, windows, doors, screens - Pearce Security Systems
Cable fall arrest system - Latchways plc
Carbon monoxide - detectors - BRK Brands Europe Ltd
Cash Handling & Recycling - Gunnebo UK Limited
Door and window Locks - Securikey Ltd
Energy conservation in built environment - BRE
External Mirrors - Moravia (UK) Ltd
Fall arrest and safety - Bacon Group
Fall arrest systems - Latchways plc
Fencing Security - Darfen Durafencing
Fire research/consultancy - BRE
Fire safety - National Britannia Ltd
Fire training - Griffin and General Fire Services Ltd
Floor protection - PHS Group plc
Footwear - industrial - Briggs Industrial Footwear Ltd t/a Briggs Safety Wear
Fragile roof and rooflight covers - Latchways plc
Gaseous extinguishing - ADT Fire and Security Plc
Hardwire smoke & fire alarms - BRK Brands Europe Ltd
Industrial footwear - Briggs Industrial Footwear Ltd t/a Briggs Safety Wear
Ladder safety systems - Latchways plc
Manufacturers of Fast Tow Mixers - Multi Marque Production Engineering Ltd
Marking Tapes - Moravia (UK) Ltd
Natural gas alarms - BRK Brands Europe Ltd
Occupational health - National Britannia Ltd
Passive fire protection, Resistance fire testing, Reaction to fire testing - BM TRADA
Pedestrian barrier - Wade Building Services Ltd
Personal Protective Equipment - Briggs Industrial Footwear Ltd t/a Briggs Safety Wear
Personal Protective Equipment - Kem Edwards Ltd
Planning Supervisors - Johnson VB LLP
Portable Security - Allegion (UK) Ltd
Protection sheets - Cordek Ltd
Respiratory protection and safety helmets - MSA (Britain) Ltd
Rolled protection - Cordek Ltd
Roof safety systems - Latchways plc
Safe Deposit Lockers - Gunnebo UK Limited
Safes & Vaults - Gunnebo UK Limited
Safety access systems - Kobi Ltd
Safety Equipment - HSS Hire Service Group
Safety eyebolts - Latchways plc
Safety harnesses, lifelines etc - Capital Safety Group (NE) Ltd
Safety Nets - Richard Lees Steel Decking Ltd
Safety systems - Unistrut Limited
Safety/ security/ protection - Pilkington UK Ltd
Security / safety / protection - Initial Fire Services
Security / safety protection - Specialist Building Products
Security units - Parklines Building Ltd
Security/ Safety/ Protection - Abloy Security Ltd
Security/ Safety/ Protection - Harling Security Solutions
Security/safety - Intersolar Group Ltd
Security/safety/protection - BPT Automation Ltd
Site Safety Equipment - Sportsmark Group Ltd
Steel Hoard Site Hoarding - Darfen Durafencing
Steel security units - Ideal Building Systems Ltd
Superbolt - Furmanite International Ltd
Temporary security fences - Wade Building Services Ltd
Temporary site hoarding - Wade Building Services Ltd
Workwear - Briggs Industrial Footwear Ltd t/a Briggs Safety Wear

A36 Facilities/ Temporary works/ services

Bespoke esign services - Camlok Lifting Clamps Ltd
Builders bags - Hilton Banks Ltd
Cabins, site accommodation and event toilets - Rollalong Ltd
Demountable Flood Defense Systems - Bauer Inner City Ltd
Emergency Lighting - P4 Limited
Facilities/temporary works/services - Interface Europe Ltd
Graphic design services - Milesahead Graphic Design/Consultant
Non-mechanical access and building equipment (hire and sales) - Actavo (UK) Ltd
Palletwrap - Hilton Banks Ltd
Repairs and service of industrial doors - Avon Industrial Doors Ltd
Spill Kits - Furmanite International Ltd
Tarpaulins - Hilton Banks Ltd
Temporary fencing and site guard panels - Guardian Wire Ltd
Temporary roadways - Eve Trakway
Temporary works - Frosts Landscape Construction Ltd
Trench Cover - MCL Composites Ltd

A37 Operation/ Maintenance of the finished building

Anti-finger trap devices - Dorma Door Services Ltd
BCIS Building Running Costs Online - BCIS
Door safety products - Dorma Door Services Ltd
Engineers tools - Spear & Jackson
Hand tools - Hilmor
Lift & Escalator Maintenance / Repair / Service - Schindler Ltd
Maintenance and service - Cooper Group Ltd
Maintenance and Service of environmental systems - Kingspan Environmental Ltd
Operation/ Maintenance of the finished building - Advanced Interior Solutions Ltd
Operation/ Maintenance of the finished building - Johnson Controls Building Efficiency UK Ltd
Operational and maintenance contracts and technical advice - Doosan Power Systems Ltd
Safety, Access & Retrieval Systems - Capital Safety Group (NE) Ltd
Service - GP Burners (CIB) Ltd
Structural repairs - Proten Services Ltd

A40 Management and staff

Training - Airedale International Air Conditioning Ltd

A41 Site accommodation

Doulble Store and Mobile Tool Control - Railex Systems Ltd
Modular buildings - Wernick Group Ltd
Modular buildings hire - Waco UK Limnited
Modular buildings sale - Waco UK Limnited
Schools, Key Worker & Student Accomadation - Rollalong Ltd
Secure accommodation - Waco UK Limnited
Site accommodation - RB Farquhar Manufacturing Ltd
Site accommodation - Waco UK Limnited
Site Accomodation - Caledonian Modular
Site Accomodation - Thurston Building Systems
Site cabins - RB Farquhar Manufacturing Ltd

A42 Services and facilities

24 hour service - Servowarm
Abrasives tungsten carbide burs - PCT Group
Access Equipment hire - HSS Hire Service Group
Air quality - National Britannia Ltd
Air qulity monitoring - Zellweger Analytic Ltd, Sieger Division
Air tightness testing, Acoustic testing, Mechanical testing, Passive fire protection, Structural testing, Timber testing, Third party certification - BM TRADA
Alloy tower and aerial platform training - Turner Access Ltd
Aluminium towers, aerial platforms, scissors and booms - Turner Access Ltd
Annual Services - Servowarm
Application Tooling - Hellermann Tyton
Asbestos abatement and consultancy services - Rhodar Limited
Automative tools - Spear & Jackson
Bonded abrasives - Marcrist International Ltd
Bookshop - The Building Centre Group Ltd
Breakdown repairs - Servowarm
Bricklifter handlifting tools (bricks) - Edward Benton & Co Ltd
Buckets - Airflow (Nicoll Ventilators) Ltd
Builders tools - Spear & Jackson
Building Project Information - The Building Centre Group Ltd
Bulk handling systems - Portasilo Ltd
Cartridge tools - ITW Spit
Coal tar B.P - Legge Thompson F & Co Ltd
Conference & Seminar Facilities - The Building Centre Group Ltd
Conferences and seminars - Town and Country Planning Association
Construction equipment - Schwing Stetter (UK) Ltd
Construction research and development - BRE
Construction tools - Ingersoll-Rand European Sales Ltd
Core drills - Norton Diamond Products
Cutting machines - Norton Diamond Products
Decontamination units - Rollalong Ltd
Decorators Tools - Ciret Limited
Diamond blades - Norton Diamond Products
Disabled access door operators - Dorma Door Services Ltd
Drainage industry and plumbers tools - Monument Tools Ltd
Electric Tools - ITW Spit
Electrical and Mechanical Services Contractor - Kershaw Mechanical Services Ltd
Engineering and production tools - Spear & Jackson
Engineering Stockist - Thanet Ware Ltd
Exhibitions - building materials and products - The Building Centre Group Ltd
Exhibitions, conferences, and meeting rooms - The Building Centre Group Ltd
Global after-sales support network - FG Wilson
Guideline - The Building Centre Group Ltd
Hand & power tools - Kem Edwards Ltd
Hand and garden tools - Spear & Jackson
Hand tools - Footprint Sheffield Ltd
Hand tools, power tool accessories and abrasives - Hyde Brian Ltd
Highway maintenance contractors - Tarmac Ltd
Hot dip galvanising - Jones of Oswestry Ltd
Installation - Servowarm
Kermaster handlifting tools (Kerbs) - Edward Benton & Co Ltd
Lifting Equipment hire - HSS Hire Service Group
Local engineers (corgi regestered) - Servowarm
Lubricants - Witham Oil & Paint Ltd
Market & research schemes - The Building Centre Group Ltd
Masking tape - Ciret Limited
Metal cutting tools - Spear & Jackson
Office Equipment - Cave Tab Ltd
Open and enclosed skips, cargo containers and custom built units - Randalls Fabrications Ltd
Organisational Tools - Cave Tab Ltd
Pipe tools - Hilmor
Plant - Linatex Ltd
Power tools-pneumatic/electric/hydraulic - PCT Group
Publications - Butterworth-Heinemann
Publications and technical information on timber - TRADA Technology Ltd
Scaffold hire - HSS Hire Service Group
Services and facilities - Advanced Interior Solutions Ltd
Services and Facilities - B & D Clays & Chemicals Ltd
Services and facilities - Barlow Group
Services and facilities - Frosts Landscape Construction Ltd

A42 Services and facilities (con't)

Servocare economic insurance - Servowarm
Servocare priority insurance - Servowarm
Site services - Magnum Ltd T/A Magnum Scaffolding
Skin care products - Deb Ltd
Soft faced hammers and mallets - Thor Hammer Co Ltd
Soft soap - Liver Grease Oil & Chemical Co Ltd
Spindle tooling - Whitehill Spindle Tools Ltd
Suppliers of Bituminous Binders Road Surface Dressing Contractors - Colas Ltd
Support Services - Kier Limited
Surveying Equipment - HSS Hire Service Group
Timber merchants, Importers, Oak specialist - W.L. West & Sons Ltd
Tool & Equipment Repair - HSS Hire Service Group
Tool hire - HSS Hire Service Group
Tools - Eliza Tinsley Ltd
Tools - Footprint Sheffield Ltd
Tools - Walls & Ceilings International Ltd
Transportation of Brituminous Products, Liquid and Oils - Colas Ltd
Upgrade to existing system - Servowarm
Wall protection - Moffett Thallon & Co Ltd
Wall tie condition surveys - HCL Contracts Ltd
Wallpaper pasting machines - Ciret Limited

A43 Mechanical plant

Access platforms - Manitou (Site Lift) Ltd
Attachments/Accessories - HE Services (Plant Hire) Ltd
Backhoe Loaders, Dumpers, Excavators, and Rollers - HE Services (Plant Hire) Ltd
Boring machines, boring tools and trenchers - Vermeer United Kingdom
Builders hand tools - Benson Industries Ltd
C-hooks/coil hooks - Camlok Lifting Clamps Ltd
Compressor crankcase heaters - Flexelec(UK) Ltd
Construction equipment - Schwing Stetter (UK) Ltd
Construction equipment, compressors, rock drills and accessories, compaction and paving equipment - Ingersoll-Rand European Sales Ltd
Core drills - Harbro Supplies Ltd
Crane forks - Camlok Lifting Clamps Ltd
Creteangle bulkbag systems - Edward Benton & Co Ltd
Creteangle hopper vibratory feeder systems - Edward Benton & Co Ltd
Creteangle moulds for precast concrete products - Edward Benton & Co Ltd
Creteangle pan type mixers - Edward Benton & Co Ltd
Diesel power generators - Sandhurst Manufacturing Co Ltd
Directional drill rigs - Vermeer United Kingdom
Double vacuum impregnation plant - Protim Solignum Ltd
Drill bits - Artur Fischer (UK) Ltd
Drum handling Equipment - Camlok Lifting Clamps Ltd
Electric tools - ITW Construction Products
Excavating and earth moving plant - J C Bamford (Excavators) Ltd
Forklift truck attachments - Camlok Lifting Clamps Ltd
Gas nailers - ITW Construction Products
Gas power tools - ITW Spit
General Site Equipment - Kem Edwards Ltd
Grinding machines, surface and double ended grindes - Beacon Machine Tools Ltd
Height Safety, Controlled Access & Retrieval Systems - Capital Safety Group (NE) Ltd
Hire of Mechanical Plant - HE Services (Plant Hire) Ltd
Horizontal grinders - Vermeer United Kingdom
Horticultural and agricultural tools - Spear & Jackson
Lifting clamps and grabs - Camlok Lifting Clamps Ltd
Loading bay equipment - Amber Doors Ltd
Machine tools for the manufacture of building products - BSA Machine Tools Ltd
Machinery - The Saw Centre Group
Manual and powered hoists, winches and jacks - PCT Group
Manufacturer of Conveyors - Multi Marque Production Engineering Ltd
Manufacturers of Roller Mixer - Multi Marque Production Engineering Ltd
Material handling equipment - Harbro Supplies Ltd
Mechanical Plant - Access Industries Group Ltd
Mechanical plant - Johnson Controls Building Efficiency UK Ltd
Mechanical plant spares - O'Brian Manufacturing Ltd
Mechcanical services Contractor - Kershaw Mechanical Services Ltd
Mini excavators, dumper trucks, rollers and cement mixers - Chippindale Plant Ltd
Mobile site lighting - Sandhurst Manufacturing Co Ltd
Moles - Vermeer United Kingdom
Pan type mixing machinery - Benton Co Ltd Edward
Plant Body Shop - HE Services (Plant Hire) Ltd
Plant hire - HSS Hire Service Group
Plant hire - Signature Ltd
Plate compactors - Chippindale Plant Ltd
Pneumatic nailers - ITW Construction Products
Power actuated tools - ITW Construction Products
Power tools and accessories - Robert Bosch Ltd
Powered barrows - Bitmen Products Ltd
Rail handling Equipment - Camlok Lifting Clamps Ltd
Road Maintenance Equipment - Phoenix Engineering Co Ltd, The
Rough terrain forklifts - Manitou (Site Lift) Ltd
Sand blast equipment - Harbro Supplies Ltd
Schedule of basic plant changes - BCIS
Screwlok beam clamps - Camlok Lifting Clamps Ltd
Scrubber driers - BVC
Site dumpers, concrete and mortar mixers, sawbenches and telescopic handlers - Multi Marque Production Engineering Ltd
Snow Clearance - HE Services (Plant Hire) Ltd
Sorbents - Furmanite International Ltd
Special rigging structures - Scotia Rigging Services (Industrial Stainless Steel Rigging Specialists)
Stone processing machinery - Harbro Supplies Ltd
Telehandlers - Chippindale Plant Ltd
Telehandlers - HE Services (Plant Hire) Ltd
Telescopic loaders - Manitou (Site Lift) Ltd
Timber machining service - Brewer T. & Co Ltd
Tipping skips - Woodburn Engineering Ltd
Tools - Duffells Limited
Transport Services - HE Services (Plant Hire) Ltd
Tree equipment, woodchippers, stump cutters and tub grinders - Vermeer United Kingdom
Trenchers - Vermeer United Kingdom
Truck mounted forklift - Manitou (Site Lift) Ltd
Tube Bending Machines - Tubela Engineering Co Ltd
Vehicle turntables - British Turntable Ltd
Welding and Brazing equipment \\ consumables - Johnson Matthey PLC - Metal Joining
Welding Equipment - HSS Hire Service Group
Welding equipment - PCT Group
Wood working tools - Record Tools Ltd

A44 Temporary works

Abnormal lead support - Eve Trakway
Access platforms - Chippindale Plant Ltd
Access towers - SGB, a Brand Company
Adhesive, solid, metallic, plastic tapes - Marking Machines International
All forms of line (sports and commercial) marking machines - Marking Machines International
Aluminium towers - Youngman Group Limited
Anchor testing & installation - HCL Contracts Ltd
Bridge jacking and monitoring - Mabey Hire Ltd
Bridges - Eve Trakway
Cable grips - Albion Manufacturing
Contract scaffolding - Magnum Ltd T/A Magnum Scaffolding
Ground protection - Eve Trakway
Industrial scaffolding - Industrial Solutions - Actavo (UK) Ltd
Ladders - Titan Ladders Ltd
Ladders - Youngman Group Limited
Ladders & steps - Youngman Group Limited
Loftladders - Titan Ladders Ltd
Measuring tapes and equipment - Marking Machines International
Modular buildings - Elliott Ltd
Pest control - humane live catch and trap - Albion Manufacturing
Safety Deck Systems – Hire and Sales - Actavo (UK) Ltd
Scaffold boards - Brash John & Co Ltd
Scaffold sheeting - Icopal Limited
Scaffold towers - Titan Ladders Ltd
Scaffolding - Howard Evans Roofing and Cladding Ltd
Scaffolding - Schauman (UK) Ltd
Scaffolding - SGB, a Brand Company
Scaffolding - Wade Building Services Ltd
Scaffolding accessories - B C Barton Limited
Scaffolding restraints - HCL Contracts Ltd
Scaffolding Systems - Turner Access Ltd
Site buildings - Elliott Ltd
Staging - CPS Manufacturing Co
Stagings - Clow Group Ltd
Steps - Titan Ladders Ltd
Temporary fencing - Rom Ltd
Temporary fencing and barriers - Elliott Ltd
Temporary Road Markings - Sportsmark Group Ltd
Temporary roadways - Sommerfeld Flexboard Ltd
Temporary roadways and bridging - Mabey Hire Ltd
Temporary works - Freyssinet Ltd
Tempory protection - British Sisalkraft Ltd
Towers - Clow Group Ltd

A52 Nominated suppliers

Building supplies - McAthur Group Limited

A55 Dayworks

Building Price Book - Laxton's Publishing Ltd

A70 General specification requirements for work package

Book sales - Coutts Information Services
Book Sales - Gardners Books
Book sales - Professional Bookshops Limited
Book sales - RIBA Enterprises
Book Sales - TRADA Bookshop
Book sales - TRADA Technology Ltd
Book sales - tso shop
Book Sales - Waterstones
Book Sales - Which Building Contract
Book sellers - Construction Industry Publications
Book sellers - Dandy Booksellers
Book sellers - Hammicks Legal Bookshop
Book Sellers - RICS Book Shop
Business to Business Publications - EMAP
Online Book Suppliers - The Book Depository Ltd
Specialist Book Sellers - RICS Book Shop
Wholsale booksellers - Bertrams

B10 Prefabricated buildings/ structures

Absolut XPert & Isokern DM - block chimney systems - Schiedel Chimney Systems Ltd
Aluminium special structures - Alifabs (Woking) Ltd
Architectural fabrications - Clow Group Ltd
Barns - Purewell Timber Buildings Limited
Battery Garages - concrete - Liget Compton
Bike Shelters - Glasdon U.K. Limited
Bridges in timber or steel, boardwalks, decking, ramps, steps and bespoke structures - CTS Bridges Ltd
Builders merchant - Jewson Ltd
Builders merchant - Nicholls & Clarke Ltd
Building systems to buy or hire, self contained buildings and modular buildings - Portakabin Ltd
Buildings - Glasdon U.K. Limited
Cantilever and pigeon hole racking - Hi-Store Ltd
Car parks - Bourne Steel Ltd
Chemical stores - Parklines Building Ltd
Complete buildings - Elliott Ltd
Concrete Garages - Parklines Building Ltd
Concrete garages, sheds, workshops, garden rooms - Liget Compton
Conservatories - Alitex Ltd
Conservatories - Duraflex Limited
Conservatories - Solair Ltd
Conservatories - Spectra Glaze Services
Conservatories, garden buildings - Amdega Ltd
Corrugated structural steel plate bridges - Asset International Ltd
Covered Walkways - Fibaform Products Ltd
Cycle Shelter - The No Butts Bin Company (NBB), Trading as NBB Outdoor Shelters
Design and build of timber frame components and floor joist system - Prestoplan
Environmental trade products - Furmanite International Ltd
Equestrian buildings - Scotts of Thrapston Ltd
Export Buildings - Wernick Group Ltd
Extruded conservatories - Smart Systems Ltd
Field Shelters - Newton & Frost Fencing Ltd
Flat Pack Buildings - Wernick Group Ltd
Garages - concrete - Liget Compton
Garden buildings - Anthony de Grey Trellises + Garden Lighting
Garden buildings - Weatherley Fencing Contractors Limited
Garden rooms - concrete - Liget Compton
Gatehouses - Fibaform Products Ltd
Grp modular buildings - Dewey Waters Ltd
Hire of buildings - Wernick Group Ltd
Industrial housings - Fibaform Products Ltd
Kennels - Weatherley Fencing Contractors Limited
Log cabins - Newton & Frost Fencing Ltd
Log Cabins - Purewell Timber Buildings Limited
Log stores - Purewell Timber Buildings Limited
Magnum firechests - Schiedel Chimney Systems Ltd
Mechcanical services - Kershaw Mechanical Services Ltd
Modular and volumetric buildings - Rollalong Ltd
Modular buildings - Bell & Webster Concrete Ltd
Modular buildings - Terrapin Ltd
Modular buildings - Wernick Group Ltd
Modular portable and prefabricated buildings - Ideal Building Systems Ltd
Module based buildings - Thurston Building Systems
Outbuildings, garages, stables and accessories - Scotts of Thrapston Ltd
Outdoor Shelter - The No Butts Bin Company (NBB), Trading as NBB Outdoor Shelters
Overhead monorail systems - Arc Specialist Engineering Limited
Passenger shelters - Signature Ltd
Play Houses - Purewell Timber Buildings Limited
Play houses - Weatherley Fencing Contractors Limited
Pre engineered buildings - Terrapin Ltd
Pre engineered metal buildings - Atlas Ward Structures Ltd
Pre fabricated buildings - Brimar Plastics Ltd
Prefabricated Buildings - Elwell Buildings Ltd
Prefabricated Buildings - Fibaform Products Ltd
Prefabricated buildings - Parklines Building Ltd
Prefabricated buildings - Waco UK Limnited
Prefabricated buildings / structures - Caledonian Modular
Prefabricated buildings /structures - Rom Ltd
Prefabricated Buildings/Structures - Dartford Portable Buildings Ltd
Prefabricated Log Buildings - Norwegian Log Buildings Ltd
Prefabricated metal buildings - Elwell Buildings Ltd
Proprietary buildings - Leander Architectural
PVCu & Aluminium Conservatories - KCW Comercial Windows Ltd
Security Kiosks - Glasdon U.K. Limited
Sheds - Purewell Timber Buildings Limited
Sheds - Weatherley Fencing Contractors Limited
Sheds - concrete - Liget Compton
sheds and workshops - Newton & Frost Fencing Ltd
Sheds, Timber Buildings - Tate Fencing
Silos - CST Industries, Inc. - UK
Smartroof panels - Wyckham Blackwell Ltd
Smoking Shelter - The No Butts Bin Company (NBB), Trading as NBB Outdoor Shelters
Smoking shelters - Fibaform Products Ltd
Smoking Shelters - Glasdon U.K. Limited
Stables - Newton & Frost Fencing Ltd
Steel frame buildings - Terrapin Ltd
Steel framed buildings - Yorkon Ltd
Steel portal frame buildings - Leofric Building Systems Ltd
Summer Houses - Purewell Timber Buildings Limited
Summer houses - Weatherley Fencing Contractors Limited
Summerhouses, Clubhouses, Pavilions, Gazebos and Garden Buildings - Scotts of Thrapston Ltd
System Buildings - Wernick Group Ltd
Tailor made steel buildings - Astron Buildings Parker Hannifin PLC
Technical consultancy - BFRC Services Ltd
Timber building systems - Bullock & Driffill Ltd
Timber framed buildings - Wyckham Blackwell Ltd
Timber framed houses and structures - W.H Colt Son & Co Ltd
Timber Storage Buildings - Parklines Building Ltd
Toilets - Inter Public Urban Systems Uk Ltd
Tool Stores - Purewell Timber Buildings Limited
Units available for hire - Fibaform Products Ltd
Waiting Shelter - The No Butts Bin Company (NBB), Trading as NBB Outdoor Shelters
Weather shelters - NuLite Ltd
Workshops - Purewell Timber Buildings Limited
Workshops - concrete - Liget Compton
Workshops (Garden) - Weatherley Fencing Contractors Limited

B11 Prefabricated building units

Bathroom pods - Caledonian Modular
Boarding kennel & cattery maufacturers - Broxap Ltd
Cleanrooms - Hemsec Panel Technologies
Conservatories - BAC Ltd
Conservatories - Ultraframe PLC
Dock Leveller Pits - Ebor Concrete Ltd
Factory Prefabricated Boiler Plant Rooms - Fulton Boiler Works (Great Britain) Ltd
Glazed Extensions - Ultraframe PLC
Horizontal panel concrete buildings - Leofric Building Systems Ltd
Hotels, Timber or Steel Framed Buildings - Rollalong Ltd
Housing - Glasdon U.K. Limited
Loggia - Ultraframe PLC
Modular building units - Caledonian Modular
Modular Buildings – Building Solutions - Actavo (UK) Ltd
Orangeries - Ultraframe PLC
Pay to enter systems - Inter Public Urban Systems Uk Ltd
Pre fabricated building units - Brimar Plastics Ltd
Prefabricated building units - Ideal Building Systems Ltd
Public toilets - Inter Public Urban Systems Uk Ltd
Smoking shelters - Elwell Buildings Ltd
Stadium Terracing - Ebor Concrete Ltd
Temporary buildings - Waco UK Limnited
Vertical panel concrete buildings - Leofric Building Systems Ltd

C10 Site survey

Asbestos removal - Hawkins Insulation Ltd
Concrete repair systems - PermaRock Products Ltd
Disability access audits - Peter Cox Ltd
Site survey - Landscape Institute
Site surveys - Focal Signs Ltd
Survey anchors and markers - Earth Anchors Ltd

C11 Ground investigation

Asbestos consultancy - National Britannia Ltd
Contaminated land - National Britannia Ltd
Geotechnical consultancy and contamination assessment - Structural Soils Ltd
Ground improvement - Keller Ground Engineering
Ground investigation - May Gurney (Technical Services) Ltd - Fencing Division
Groundworks - Rom Ltd
Site investigation,soils and material testing - Structural Soils Ltd
Surveying instruments - Topcon GB Ltd

C12 Underground services survey

Cable locator - Radiodetection Ltd
Leak detection - Radiodetection Ltd
Pipe locator - Radiodetection Ltd
Underground services survey - Radiodetection Ltd

C13 Building fabric survey

Building fabric survey - Colebrand Ltd
Building Fabric Survey - Concrete Repairs Ltd
CCTV Surveys - CICO Chimney Linings Ltd
Consultancy and building survey, schedules of work - Stonewest Ltd
Glass assessment surveys - Durable Ltd
Large-scale structural testing, On-site testing and consultancy, Testing to standard or bespoke requirements, Dynamic wind loading and uplift testing, Hygrothermal performance testing - Lucideon
Toxic mould detection - Peter Cox Ltd

C14 Building services survey

Energy efficiency in lighting advisory service - Poselco Lighting
Preparation of boilers for insurance inspection - Hodgkinson Bennis Ltd
Preparation of boilers for ultrsonic testing - Hodgkinson Bennis Ltd
Structural testing and consultancy for masonry/steel/concrete/timber - Lucideon

C20 Demolition

Demolition - Parker Plant Ltd
Demolition Attachments and Accessories - HE Services (Plant Hire) Ltd
Office refurbishment - Cumberland Construction

C21 Toxic/ Hazardous material removal

Asbestos removal - Industrial Solutions - Actavo (UK) Ltd
Toxic / Hazardous material removal - Hertel Services

C30 Shoring/ Facade retention

Ground Support System - Hire and Sales - Actavo (UK) Ltd
Proping and needling - Mabey Hire Ltd
Trench strutting equipment - Mabey Hire Ltd

C40 Cleaning masonry/ concrete

Brick and stone cleaning - Gun-Point Ltd
Brick replacement - Gun-Point Ltd
Cleaning &restoration of building facades - Szerelmey Ltd
Cleaning &restoration of building stonework - Szerelmey Ltd
Cleaning masonary/concrete - TH Kenyon & Sons Ltd
Cleaning masonry/ concrete - Chichester Stoneworks Limited
Cleaning masonry/ concrete - Colebrand Ltd
Concrete cleaner - Tank Storage & Services Ltd
Concrete maintenance and repair products - HMG Paints
Concrete repair - STO Ltd
Concrete repair mortars - Hines & Sons, P E, Ltd
Designed grout anchoring system - Cintec International Ltd
Epoxy resin repairs - Gun-Point Ltd
Grouting - Gun-Point Ltd
Physical dpc insertion - Dampcoursing Limited
Pressure pointing - Gun-Point Ltd
Remedial pointing - Kingfisher Building Products Ltd
Repairing concrete/ brick/ block/ stone - Concrete Repairs Ltd
Repairs and restorations - Alpha Mosaic & Terrazzo Ltd
Repointing brickwall and stonework - Gun-Point Ltd
Resin and precision grout - Armorex Ltd
Restoration of facades in stone, brick & terracotta - Stonewest Ltd
Wall tie replacement - Timberwise (UK) Ltd

C41 Repairing/ Renovating/Conserving masonry

Chemical DPC and remedial treatments - Protim Solignum Ltd
Chemical dpc installers - Dampcoursing Limited
Chemicals - Baxenden Chemicals Ltd
Concrete repair - Mapei (UK) Ltd
Conservation works - Szerelmey Ltd
Damp-proofing - Timberwise (UK) Ltd
External refurbishment works - Szerelmey Ltd
Faience blocks - Shaws of Darwen
In-Situ Chimney & Flue Lining Systems - CICO Chimney Linings Ltd
Isokoat - Schiedel Chimney Systems Ltd
Lightweight Concrete Repair Materials - Belzona PolymericLtd
Lintel replacement - HCL Contracts Ltd
On-site sampling and structural testing - Lucideon
Polyurethane and acylic resins - Baxenden Chemicals Ltd
Remedial wall ties - Peter Cox Ltd
Repairing masonry - Flexcrete Technologies Limited
Repairing/ renovating conserving masonry - Colebrand Ltd
Repairing/ Renovating/Conserving masonry - Chichester Stoneworks Limited
Repairing/Renovating/Conserving masonry - David Ball Group Ltd
Repairing/renovating/conserving masonry - G Miccoli & Sons Limited
Repairing/Renovating/Conserving masonry - Safeguard Europe Ltd
Restoration of facades in stone, brick & terracotta - Stonewest Ltd
Stainless steel remedial wall ties - Helifix
Structural repairs - Szerelmey Ltd
Technical publications and publication brick bulletin - Brick Development Association
Terracotta blocks - Shaws of Darwen
Underground Waterproffing - Sika Ltd

C42 Repairing/ Renovating/Conserving concrete

Commercial Conncrete repair - HCL Contracts Ltd
Concrete repair - BASF plc, Construction Chemicals
Concrete repair - Sealocrete PLA Ltd
Concrete repair mortars - Fosroc Ltd
Concrete repair products - Flexcrete Technologies Limited
Concrete repair systems - Beton Construction Ltd
Concrete repairs - Freyssinet Ltd
On-site sampling and structural testing - Lucideon
Protective coatings - Fosroc Ltd
Repair Mortars, Corrosion inhibitor, Anti-carbonation/waterproof/elastomeric coatings - Ronacrete Ltd
Repairing concrete - Don Construction Products Ltd
Repairing/ renovating conserving Concrete - Colebrand Ltd
Repairing/renovating concrete - Anglo Building Products Ltd
Repairing/Renovating/Conserving Concrete - Concrete Repairs Ltd
Repairing/Renovating/Conserving concrete - David Ball Group Ltd
Repairing/Renovating/Conserving Concrete - PermaRock Products Ltd
Repairing/renovating/conserving concrete - TH Kenyon & Sons Ltd

Resinboard Concrete Repair Materials - Belzona PolymericLtd
Sprayed concrete - Freyssinet Ltd

C45 Damp proof course renewal/ insertion

Chemical dpcs to existing walls - Peter Cox Ltd
Condensation control and DPC - Kair ventilation Ltd
Damp proof course renewal/ insertion - Terminix Property Services
Damp proof course renewal/insertion - Safeguard Europe Ltd

C50 Repairing/ Renovating/Conserving metal

Architectural metalwork - renovation and restoration - Albion Manufacturing
Rail infrastructure coatings - Corroless Corrosion Control
Repairing/ renovating conserving Metal - Colebrand Ltd
Restoration work in metal - Ballantine, Bo Ness Iron Co Ltd

C51 Repairing/ Renovating/Conserving timber

Condition surveys, Historic Buildings - BM TRADA
Repairing/ renewing/ conserving timber - Terminix Property Services
Solvent based and water based preservatives - Protim Solignum Ltd
Timber Repair Product - Belzona PolymericLtd
Timber Resin Repairs - Proten Services Ltd
Timber resin repairs - Timberwise (UK) Ltd
Timber treatment and dry rot eradication - Dampcoursing Limited
Window repair - Repair care International
Wood decay repair products - Repair care International

C52 Fungus/ Beetle eradication

Dry rot and woodworm treatment - Kair ventilation Ltd
Fungus and beetle eradication - Peter Cox Ltd
Fungus/ beetle eradication - Terminix Property Services
Fungus/beetle eradication - Specialist Building Products
Timber treatment - Timberwise (UK) Ltd

C90 Alterations - spot items

Refurbishments & replicas - DW Windsor Lighting
Sash renovation - Durable Ltd

D11 Soil stabilisation

Armortec concrete erosion - Concrete Products (Lincoln) 1980 Ltd
CFA Piling - Stent Foundations Ltd
Chemicals - Baxenden Chemicals Ltd
Civil engineering systems - Sarnafil Ltd
Dynamic Compaction - Bauer Inner City Ltd
Eco Parking - Ensor Building Products Ltd
Erosion control mats - ABG Ltd
Geopolymer Injection - Uretek (UK) Ltd
Geotextiles - Don & Low Ltd
Green roof systems - ABG Ltd
Ground improvement - Keller Ground Engineering
Ground movement products - Cordek Ltd
Lime columns - Stent Foundations Ltd
Load bearing anchors - Earth Anchors Ltd
Retaining walls - ABG Ltd
Retaining walls - PHI Group Ltd
Soil stabilisation - Skanska UK Plc
Soil stabilization - Grass Concrete Ltd
Soil stabilization - Jablite Ltd
Soil stabilization - May Gurney (Technical Services) Ltd - Fencing Division
Soil stabilization - Tensar International Ltd
Soil stabilization/ ground anchors - Freyssinet Ltd
Vibro and dynamic compaction - Bauer Inner City Ltd
Vibro Concrete Columns - Bauer Inner City Ltd
Vibro Stone Columns - Bauer Inner City Ltd

D12 Site dewatering

De watering - Keller Ground Engineering
Flood Barriers - Lakeside Group
Flood Protection - Lakeside Group
Pumping systems for ground water control - Andrews Sykes Hire Ltd
Water drainage geocomposites - Geosynthetic Technology Ltd

D20 Excavating and filling

Aggregates- crushed rock, sand and gravel - Hanson UK
Excavating and Filling - HE Services (Plant Hire) Ltd
Excavating and filling - Jablite Ltd
Excavators, all sizes - Mini, Midi Large - HE Services (Plant Hire) Ltd
Land reclamation - Bernhards Landscapes Ltd

D21 Ground gas venting

Gas Barrier & Ventings Systems - Landline Ltd
Gas barrier membranes - SCP Concrete Sealing Technology Ltd
Gas drainage geocomposites - Geosynthetic Technology Ltd
Gas venting - Cordek Ltd
Gas Venting Design and Components - Landline Ltd
ground gas venting - Jablite Ltd

Ground gas venting - Keller Ground Engineering
Methane gas venting - ABG Ltd
Radon Barriers - BRC Reinforcement
Radon membranes - Visqueen Building Products

D30 Cast in place concrete piling

Bored Piling - Stent Foundations Ltd
Cased flight auger piling - Westpile Ltd
Cast in place concrete piling - Bauer Inner City Ltd
Cast in place concrete piling - Expanded Piling Ltd
Cast in place concrete piling - Freyssinet Ltd
Cast in place concrete piling - May Gurney (Technical Services) Ltd - Fencing Division
Cast in place concrete piling - Skanska UK Plc
Cast in place concrete piling - Stent Foundations Ltd
Composite Sheet Pile Wall System - Asset International Ltd
Continuous flight auger bored piling - Westpile Ltd
Ductile Piling - Bauer Inner City Ltd
Ground engineering, Piling - Keller Ground Engineering
Large diameter bored piling - Westpile Ltd
Piling - Bauer Inner City Ltd
Piling Accessories - Cordek Ltd
Piling and ground engineering - Stent Foundations Ltd
Rotary bored piling - Westpile Ltd
Tripod bored piling - Westpile Ltd

D31 Preformed concrete piling

Bored, CFA, driven precast piling - Stent Foundations Ltd
Driven precast - Westpile Ltd
Ground engineering, piling - Keller Ground Engineering
Preformed concrete piling - Expanded Piling Ltd
Preformed concrete piling - May Gurney (Technical Services) Ltd - Fencing Division
Preformed concrete piling - Skanska UK Plc
Preformed concrete piling - Stent Foundations Ltd

D32 Steel piling

Composite Sheet Pile Wall System - Asset International Ltd
Driven Tube Piling - Westpile Ltd
Ground engineering, piling - Keller Ground Engineering
Pipe piling - Asset International Ltd
Steel piling - Deepdale Engineering Co Ltd
Steel piling - Stent Foundations Ltd
Trench strutting equipment - Mabey Hire Ltd

D40 Embedded retaining walls

Diaphragm walling - Keller Ground Engineering
Diaphragm walling - Skanska UK Plc
Diaphragm Walling - Stent Foundations Ltd
Embedded retaining walls - Expanded Piling Ltd
Hydrophilk waterstop - SCP Concrete Sealing Technology Ltd
P.V.C. Waterstop - SCP Concrete Sealing Technology Ltd
Precast concrete retaining wall units safety barriers - RCC, Division of Tarmac Precast Concrete Ltd
Prestressed concrete retaining wall units - ACP Concrete Limited
Structural retaining walls and bridge abutments - Asset International Ltd

D41 Crib walls/ Gabions/ Reinforced earth

Armortec control products - Concrete Products (Lincoln) 1980 Ltd
Crib walls/ Gabions/ Reinforced earth supplies - Johnsons Wellfield Quarries Ltd
Crib walls/Gabions/Reinforced earth - RFA-Tech Ltd
Crib walls/Gabions/Reinforced earth - Skanska UK Plc
Crib walls/Gabions/Reinforced earth - Phi Group - Keller Ground Engineering
Cribwalls/Gabions/Reinforced Earth - Grass Concrete Ltd
Geotextiles - Don & Low Ltd
Ground engineering - PHI Group Ltd
Ground slabilization - PHI Group Ltd
Precast concrete products - Ruthin Precast Concrete Ltd
Reinforced earth - Tensar International Ltd
Reinforced earth retaining walls/ bridge abutments/precast arch system - Freyssinet Ltd
Retaining walls - ABG Ltd
Tee walls - Bell & Webster Concrete Ltd

E05 In situ concrete construction generally

Association of Concrete Industrial Flooring Contractors - Association of Concrete Industrial Flooring Contractors (ACIFC)
Castle brickbond SR cement - Castle Cement Ltd
Castle high alumina cement - Castle Cement Ltd
Castle hydrated lime - Castle Cement Ltd
Castle natural hydraulic lime - Castle Cement Ltd
Castle OPC cement - Castle Cement Ltd
Castle rapid hardening portland cement - Castle Cement Ltd
Castle sulfate-resisting portland cement - Castle Cement Ltd
Castle white portland cement - Castle Cement Ltd
Formwork and palcing for in situ concrete - Perform Construction (Essex) Ltd
In-situ concrete - The Concrete Society
Insulated concrete formwork system - Beco Products Ltd
Lightweight aggregate - Lytag Ltd

Packed Products - Hanson UK
Polypropylene & steel fibers for concrete & mortar - Propex Concrete Systems
Ready mixed concrete and mortar - Hanson UK
Synthetic ironoxide colouring agents - LANXESS Inorganic Pigments Group

E10 Mixing/ Casting/ Curing in situ concrete

Additives - Mapei (UK) Ltd
Admixtures - Sealocrete PLA Ltd
Aggregate suppliers - Kerridge Stone Ltd
Bagged Aggregates - Hye Oak
Cavity fill - Knauf Insulation Ltd
Cement, high alumina cement, calcium aluminate cement - Lafarge Aluminates Ltd
Colours for concrete - Procter Johnson & Co Ltd
Columns - Broadmead Cast Stone
Concrete admixtures - Beton Construction Ltd
Concrete chemicals - Grace Construction Products Ltd
Concrete plasticizers - Sika Ltd
Concrete repair materials - Maxit UK
Pier caps - Broadmead Cast Stone
Porticos - Broadmead Cast Stone
Ready Mix - London Concrete Ltd
Readymix - Tillicoulty Quarries Ltd
Slag cement - Civil & Marine Slag Cement Ltd
Structural concrete - Lytag Ltd
Underwater concrete - Armorex Ltd

E11 Sprayed in situ concrete

Concrete admixtures - Beton Construction Ltd
Institutions/Associations - Sprayed Concrete Association
Polypropylene & steel fibers for concrete & mortar - Propex Concrete Systems
Shotcrete systems - Clarke UK Ltd
Sprayed insitu concrete - Colebrand Ltd
Tunnel concrete pumps - Clarke UK Ltd

E20 Formwork for in situ concrete

Adjustible steel pipes - Wade Building Services Ltd
Concrete formwork plywood - Wood International Agency Ltd
Film faced formwork ply - Wood International Agency Ltd
Formwork - Jablite Ltd
Formwork - Perform Construction (Essex) Ltd
Formwork and Scaffolding - PERI Ltd
Formwork board - Dufaylite Developments Ltd
Formwork coatings - Hines & Sons, P E, Ltd
Formwork equipment - Mabey Hire Ltd
Formwork for in situ concrete - Rom Ltd
Formwork manufacturers - B C Barton Limited
Hy-rib permanent formwork - Expamet Building Products
Insulated concrete formwork system - Beco Products Ltd
Metal deck flooring - Caunton Engineering Ltd
Permanent Formwork - BRC Special Products - Bekaert Ltd
Permanent shuttering - Cordek Ltd
Pneumatic void former - BRC Special Products - Bekaert Ltd
Specialist formwork 3D - Cordek Ltd
Trough moulds - Cordek Ltd
Void formers & formwork - Cordek Ltd

E30 Reinforcement for in situ concrete

Contractor Detailing Services - BRC Reinforcement
Polypropylene & steel fibers for concrete & mortar - Propex Concrete Systems
Prefabrication - BRC Reinforcement
Punching Shear Reinforcement - BRC Special Products - Bekaert Ltd
Re-bar coating - Plastic Coatings Ltd
Reinforcement - Expamet Building Products
Reinforcement accessories - Rom Ltd
Reinforcement accessories and spacers - Hines & Sons, P E, Ltd
Reinforcement for in situ concrete - Perform Construction (Essex) Ltd
Reinforcement for insitu concrete - Freyssinet Ltd
Reinforcement for Insitu Concrete - Halfen Ltd
Reinforcement mesh and bar - BRC Reinforcement
Reinforcing accessories - Rainham Steel Co Ltd
Reinforcing bar and mesh - Hy-Ten Ltd
Reinforcing Bar Couplers - Ancon Building Products
Reinforcing mesh - Rainham Steel Co Ltd
Stainless steel reinforcement - Helifix
Stainless steel reinforcing rods - Helifix
Steel reinforcement, fibre, mesh and industrial wire - Rom Ltd
Steel wire fibres for concrete reinforcement - Bekaert Building Products

E31 Post tensioned reinforcement for insitu concrete

Post-tensioned floor slabs - Freyssinet Ltd
Steel wire, cables, rods and bars - MacAlloy Limited

E40 Designed joints in insitu concrete

Concrete balcony insulated connection system - BRC Special Products - Bekaert Ltd
Concrete floor jointing system - Compriband Ltd
Hydrophilic joint seals - Charcon Tunnels (Division of Tarmac plc)
Permanent joining formwork - BRC Special Products - Bekaert Ltd
Seals - Trelleborg Sealing Profiles
Structural expansion joints - Radflex Contract Services Ltd

E40 Designed joints in insitu concrete (con't)

Waterstops - Beton Construction Ltd
Waterstops - RFA-Tech Ltd

E41 Worked finishes/ Cutting into in situ concrete

Anti slip surfacing - Addagrip Terraco Ltd
Concrete airfield pavement protection - Addagrip Terraco Ltd
Floor ventilators - Airflow (Nicoll Ventilators) Ltd
Grouts - Beton Construction Ltd
Worked finishes/Cutting into in situ concrete - Safeguard Europe Ltd

E42 Accessories cast into insitu concrete

Accessories - RFA-Tech Ltd
Concrete related accessories - BRC Reinforcement
Corner protection - Fixatrad Ltd
Hydrophilic joint seals - Charcon Tunnels (Division of Tarmac plc)
Precast Accessories - BRC Special Products - Bekaert Ltd
Reinforcement couplers - Halfen Ltd
Tying wire, spacers and chemicals for concrete - Hy-Ten Ltd

E50 Precast concrete frame structures

Architectural precast cladding - Marble Mosaic Co Ltd, The
Bespoke concrete products - Bell & Webster Concrete Ltd
Bespoke concrete products - Hanson Concrete Products
Concrete Arch Bridge System - Asset International Ltd
Concrete products - Bell & Webster Concrete Ltd
Engineering castings - Ballantine, Bo Ness Iron Co Ltd
Ground Beams - Border Concrete Products Ltd
Ground Beams - Ebor Concrete Ltd
Jointing strips for precast concrete units - Winn & Coales (Denso) Ltd
Platform walls - Bell & Webster Concrete Ltd
Polypropylene & steel fibers for concrete & mortar - Propex Concrete Systems
Precast concrete fixings - Halfen Ltd
Precast concrete floor beams, staircases, sills and lintols - ACP (Concrete) Ltd
Precast concrete joists - supply & fix or supply only - C.R. Longley & Co. Ltd
Precast concrete lifting - Halfen Ltd
Precast concrete products - Ruthin Precast Concrete Ltd
Precast Concrete Stairs & Landings - Ebor Concrete Ltd
Precast concrete structural components - Hanson Concrete Products
Precast concrete tunnels,shafts, and cover slabs and bespoke products - Charcon Tunnels (Division of Tarmac plc)
Precast concrete units - Evans Concrete Products Ltd
Precast concrete walling - Rogers Concrete Ltd
Prestressed concrete retaining wall units - ACP Concrete Limited
Retaining Walls - Ebor Concrete Ltd
Retaining walls, tee walls - Bell & Webster Concrete Ltd
Stadia components - Bell & Webster Concrete Ltd
Structural frames - Bison Manufacturing Limited
Synthetic iron oxide pigments and coloured pigments - Elementis Pigments
Timberframed multi storey modular buildings - Leofric Building Systems Ltd

E60 Precast/ Composite concrete decking

Concrete flooring - Hanson Concrete Products
Concrete Floors - CEMEX
Floor beams - Stowell Concrete Ltd
Ground beams - Bell & Webster Concrete Ltd
Hollowcore Floors - Bison Manufacturing Limited
Interior flooring - Blanc de Bierges
Polypropylene, steel fibers for concrete & mortar - Propex Concrete Systems
Precast composite concrete decking - ACP (Concrete) Ltd
Precast composite concrete decking - ACP Concrete Limited
Precast concrete flooring - Marshalls Mono Ltd
Precast concrete units - Caunton Engineering Ltd
Precast concrete; stairs, terraced seating, lintols etc - Border Concrete Products Ltd
Prestressed flooring slabs - Coltman Precast Concrete Ltd
Prestressed floors - Carter Concrete Ltd
Solid Composite Floors - Bison Manufacturing Limited
Staircases, stadium terraces, columns, wall and ground beams - Coltman Precast Concrete Ltd

F10 Brick/ Block walling

Aircrete Blocks - Tarmac Building Products Ltd
Arches - Bovingdon Brickworks Ltd
Architectural facing masonry, medium and dense concrete blocks - Lignacite Ltd
Architectural masonry - Broadmead Cast Stone
Autoclaved Aerated Concrete Building Blocks - Hanson Thermalite
Berry Machine Made Bricks - Collier W H Ltd
Brick cutting specialists - M & G Brickcutters (specialist works)
Brick Slips - Hye Oak
Brick specials - Michelmersh Brick Holdings PLC
Brick trade association - Brick Development Association
Brick/ block walling - Collier W H Ltd
Brick/block walling - Broome Bros (Doncaster) Ltd
Brick/block walling - Mansfield Brick Co Ltd
Bricks - Baggeridge Brick PLC
Bricks - Carlton Main Brickworks Ltd
Bricks - Hye Oak
Bricks - Ibstock Scotish Brick
Bricks - Kingscourt Country Manor Bricks
Bricks - Wienerberger Ltd
Bricks - York Handmade Brick Co Ltd
Bricks & Blocks - Hanson Building Products
Bricks and blocks - Taperell Taylor Co
Buff Clay - Carlton Main Brickworks Ltd
Building blocks - Aggregate Industries UK Ltd
Calcium silicate facing bricks - Hye Oak
Chimney systems - Marflex Chimney Systems
Chimney systems - Poujoulat (UK) Ltd
Clay commons, facing bricks and engineering bricks - Hanson Building Products
Clay facing bricks - Ibstock Brick Ltd
Clay facing bricks - Ibstock Building Products Ltd
Commons - Northcot Brick Ltd
Concrete blocks - CEMEX
Concrete blocks - Forticrete Ltd
Concrete Blocks - Hanson Concrete Products
Concrete blocks - Sellite Blocks Ltd
Concrete blocks - Stowell Concrete Ltd
Concrete building blocks - Lignacite (North London) Ltd
Concrete Drainage System - Hepworth Building Products Ltd
Dense, lightweight, insulating solid, cellular and hollow concrete Blocks - Armstrong (Concrete Blocks), Thomas, Ltd
Engineering and facing bricks - Carlton Main Brickworks Ltd
Engineering bricks - Hye Oak
Engineering bricks - Northcot Brick Ltd
Extruded Brick Slips - Ketley Brick Co Ltd , The
Facing Brick - Carlton Main Brickworks Ltd
Facing bricks - Baggeridge Brick PLC
Facing Bricks - Hanson Building Products
Faience - Ibstock Building Products Ltd
Faience blocks - Shaws of Darwen
Fair face blocks - Thakeham Tiles Ltd
Fireplace briquettes - Michelmersh Brick Holdings PLC
Freeze/thaw testing, Compressive/shear testing, Water absorption, Dimensional tolerances - Lucideon
Glazed Bricks - Hye Oak
Glazed bricks - Smithbrook Building Products Ltd
Glazed terracotta wall facings - Smithbrook Building Products Ltd
Hand Made Bricks - Coleford Brick & Tile Co Ltd
Hand Made Bricks - Collier W H Ltd
Handmade bricks - Bovingdon Brickworks Ltd
Handmade facing bricks - Michelmersh Brick Holdings PLC
Handmade facing bricks - Northcot Brick Ltd
Insulated concrete formwork block - Beco Products Ltd
Insulation - Polypipe Building Products
Iron cement - Legge Thompson F & Co Ltd
Isokern DM block chimney systems & Isokern pumice flue linings - Schiedel Chimney Systems Ltd
Large format blocks - Tarmac Building Products Ltd
Library and matching service - Mid Essex Trading Co Ltd
Machine made bricks - Michelmersh Brick Holdings PLC
Machine made stock bricks - Bovingdon Brickworks Ltd
Reclaimed bricks - Maxit UK
reclaimed facing bricks - Northcot Brick Ltd
Red Clay - Carlton Main Brickworks Ltd
Refractory brickwork - Ancorite Surface Protection Ltd
Rustic wirecut facing bricks - Northcot Brick Ltd
Special shaped bricks - Baggeridge Brick PLC
Special Shaped Bricks - Carlton Main Brickworks Ltd
Specials - Bovingdon Brickworks Ltd
Specials - Northcot Brick Ltd
staffordshire clay engineering bricks - Ketley Brick Co Ltd , The
Staffordshire clay facing bricks - Ketley Brick Co Ltd , The
Synthetic ironoxide colouring agents - LANXESS Inorganic Pigments Group
Terracotta - Ibstock Building Products Ltd
Terracotta blocks - Shaws of Darwen
traditional & thin joint systems - Tarmac Building Products Ltd

F11 Glass block walling

Glass blocks - Shackerley (Holdings) Group Ltd incorporating Designer ceramics
Polycarbonate clip blocks - Rockwell Sheet Sales Ltd

F20 Natural stone rubble walling

Architectural masonry - Farmington Masonry LLP
Building stone - J Suttle Swanage Quarries Ltd
Customised sheet metal in all metals - Azimex Fabrications Ltd
Dressed stone - Forest of Dean Stone Firms Ltd
Drystone walling - Dunhouse Quarry Co Ltd
Heads - Bridge Quarry
Natural limestone quarried blocks - Hanson UK - Bath & Portland Stone
Natural stone - Miles-Stone, Natural Stone Merchants
Natural stone - Realstone Ltd
Natural stone rubble walling - Chichester Stoneworks Limited
Natural stone rubble walling - Stonewest Ltd
Natural stone rubble walling supplies - Johnsons Wellfield Quarries Ltd
Natural stone suppliers - A. F Jones (Stonemasons)
Natural stone suppliers - APS Masonry Ltd
Natural stone suppliers - Chichester Stoneworks Limited
Natural stone suppliers - Dunhouse Quarry Co Ltd
Natural stone suppliers - Stancliffe Stone
Natural stone suppliers - Stoneguard (London) Ltd
Paving and walling stone - The Delabole Slate Co. Ltd
Portland Stone aggregates - Hanson UK - Bath & Portland Stone
Rock face walling - Thakeham Tiles Ltd
Sandstone - Dunhouse Quarry Co Ltd
Stone suppliers - Earl's Stone Ltd
Stone suppliers - Kerridge Stone Ltd

F21 Natural stone ashlar walling/ dressings

Architectural Mouldings - Arundel Stone Ltd
Architectural Mouldings - The David Sharp Studio Ltd
Architectural terracotta and faience - Ibstock Hathernware Ltd
Ashlar Stone - Realstone Ltd
Cills - Bridge Quarry
Cleft stone and screen walling - Thakeham Tiles Ltd
Facades in stone, brick & terracotta - Stonewest Ltd
Granite and marble - design, procurement - Szerelmey Ltd
Monumental - Forest of Dean Stone Firms Ltd
Natural stone - Miles-Stone, Natural Stone Merchants
Natural stone ashlar walling/ dressings - Chichester Stoneworks Limited
Natural stone ashlar walling/dressing - G Miccoli & Sons Limited
Rapid Set Carter-Dal FTSB spot bond - Carter-Dal International
Rapid Set Façade StopBonds - 6 min - Laticrete International Inc. UK
Sandstone - Dunhouse Quarry Co Ltd
Stone suppliers - Earl's Stone Ltd
Stone suppliers - Kerridge Stone Ltd

F22 Cast stone walling/ dressings

Architectural masonry - Forticrete Ltd
Architectural precast cladding - Marble Mosaic Co Ltd, The
Architectural Stonework - Blanc de Bierges
Architectural Stonework - Cranborne Stone Ltd
Architectural stonework - Haddonstone Ltd
Artificial stone - Allen (Concrete) Ltd
Artificial stone - Procter Cast stone & Concrete Products
Artificial stone units - Northwest Pre-cast Ltd
Bespoke cast stone service - Procter Cast stone & Concrete Products
Cast stone - Border Concrete Products Ltd
Cast stone - Ibstock Building Products Ltd
Cast stone - Lignacite Ltd
Cast stone - Procter Bros Limited
Cast stone dressings - Forticrete Ltd
Cast stone walling/ dressings - Chilstone
Cast stone walling/ dressings - GB Architectural Cladding Products Ltd
Cast stone walling/ dressings - Ruthin Precast Concrete Ltd
Cast stone walling/ dressings - Stonewest Ltd
Cast stone; sills, lintols, copings, features etc - Border Concrete Products Ltd
Decorative Commercial - Aggregate Industries UK Ltd
Fencing - Procter Bros Limited
Precast concrete and cast stone products - Sangwin Concrete Products Ltd
Precast walling - Oakdale (Contracts) Ltd
Reconstructed cast stone architectural units - Broadmead Cast Stone
Reconstructed stone walling - Atlas Stone Products
Stone - Ibstock Scotish Brick
Stone walling - reconstructed - Forticrete Ltd

F30 Accessories/ Sundry items for brick/ block/ stone walling

Abrasive discs and blocks - Harbro Supplies Ltd
Accessories/ Sundry items for stone walling - Johnsons Wellfield Quarries Ltd
Accessories/sundry items for brick/block/stone - Haddonstone Ltd
Accessories/undry items for brick/ block/stone - Mansfield Brick Co Ltd
Acoustic ventilators - Ryton's Building Products Ltd
Air bricks (cast iron) - Longbottom J & J W Ltd
Aluminium copings - Alifabs (Woking) Ltd
Arch lintels - Catnic
Arches - Cavity Trays Ltd
Arches - Northcot Brick Ltd
Brick reinforcement - RFA-Tech Ltd
Brickties and windposts - Halfen Ltd
Brickwork reinforcement - BRC Special Products - Bekaert Ltd
Brickwork support - Halfen Ltd
Building products - cavity closers - Quantum Profile Systems Ltd
Cast stone, sills, lintels, copings, features etc - Border Concrete Products Ltd
Cavity and through wall ventilators - Ryton's Building Products Ltd
Cavity fixings - Artur Fischer (UK) Ltd
Cavity tanking tiles - Atlas Stone Products
Cavity trays - Cavity Trays Ltd
Cavity trays - IG Limited
Cavity Trays - SCP Concrete Sealing Technology Ltd
Cavity Trays - Zled (DPC) Co Ltd
Cavity trays, hight performance dpc - Timloc Expamet Building Products
Cavity wall tie installation - Proten Services Ltd
Chimney bird guards - Dunbrik (Yorks) Ltd
Chimney flues and linings - Rite-Vent Ltd
Chimney open fire - Dunbrik (Yorks) Ltd
Chimney pots & terminals - Dunbrik (Yorks) Ltd
Chimney service - CICO Chimney Linings Ltd
Chimney stove connection - Dunbrik (Yorks) Ltd
Chimney systems - Marflex Chimney Systems
Chimney technical services - Dunbrik (Yorks) Ltd
Chimney twin wall flue pipes - Dunbrik (Yorks) Ltd
Clay Chimney Pots - Knowles W T & Sons Ltd
Closers - Cavity Trays Ltd
Coadestone castings - Clifton Nurseries
Concrete padstones - Procter Cast stone & Concrete Products
Copings and cills - Burlington Slate Ltd
Diamond blades - Harbro Supplies Ltd
Diamond Blades - Marcrist International Ltd
Diamond core drills - Marcrist International Ltd
Enforce - composite strengthening - Weber Building Solutions
Expension joint fillers - SCP Concrete Sealing Technology Ltd
External wall insulation systems - PermaRock Products Ltd
Faience blocks - Shaws of Darwen
Fireplaces - Farmington Masonry LLP
Flue linings - Dunbrik (Yorks) Ltd
Flues and chimneys - Docherty New Vent Chimney Group
Frames, Balustrades & Columns - The Stone Catalogue Ltd
Garden Products - Aggregate Industries UK Ltd
Gas flue blocks - Dunbrik (Yorks) Ltd
Gold leaf - Harbro Supplies Ltd
Grouts - Mapei (UK) Ltd
High performance DPC and cavity trays - Visqueen Building Products
Insulated Concrete Formwork - Beco Products Ltd
Insulating cavity closers and insulated DPC - Polypipe Building Products
Insulation fixings - Artur Fischer (UK) Ltd
Internal room ventilators - Airflow (Nicoll Ventilators) Ltd
Isokern pumic flue linings - Schiedel Chimney Systems Ltd
Isolation Sleves Remedial - Redifix Ltd
Lateral restraint - HCL Contracts Ltd
Lintels - Ancon Building Products
Lintels - Cavity Trays Ltd
Lintels - CEMEX
Lintels - Stainless steel lintels - Wincro Metal Industries Ltd
Lintels and other hand made shapes - Lignacite Ltd
Manor firechests - Schiedel Chimney Systems Ltd
Masonry coatings - Jotun Henry Clark Ltd Decorative Division
Masonry Design Development - Hanson UK - Bath & Portland Stone
Masonry reinforcement - Bekaert Building Products
Masonry support systems - Ancon Building Products
Masonry support systems - Stainless steel brickwork support systems - Wincro Metal Industries Ltd
Masonry to masonry connectors - Simpson Strong-Tie®
Masonry tools - tungsten tipped chisels - Harbro Supplies Ltd
Mortars - Mapei (UK) Ltd
Movement joint - Shackerley (Holdings) Group Ltd incorporating Designer ceramics
Movement joints - Allmat (East Surrey) Ltd
Movement Joints - Illbruck Sealant Systems UK Ltd
Movement Joints - Movement Joints (UK)
Mullions - Bridge Quarry
Natural stone fixing - Halfen Ltd
Polythene damp-proof courses - Visqueen Building Products
Power tools - Harbro Supplies Ltd
Pumice stone chimney systems - Sparkes K & L
Remedial wall ties - Redifix Ltd
Repair mortars - Armorex Ltd
Retaining wall products - Forticrete Ltd
Roof ventilation, underfloor ventilation, through the wall ventilation - Timloc Expamet Building Products
Special shaped clay bricks - Ibstock Brick Ltd
Specialist waterproofing materials, admixtures & grouts - David Ball Group Ltd
Stainless steel remedial wall ties - Helifix
Steel lintels - Catnic
Steel lintels - Expamet Building Products
Steel Lintels - Harvey Steel Lintels Limited
Steel lintels - IG Limited
Steel lintels - Wade Building Services Ltd
Steel wall connectors - Allmat (East Surrey) Ltd
Straps - Expamet Building Products
Structural Frames - Beco Products Ltd
Structural Products - Aggregate Industries UK Ltd
Terracotta blocks - Shaws of Darwen
Through wall ventilators - Airflow (Nicoll Ventilators) Ltd
Timber to masonry connectors - Simpson Strong-Tie®
Twinwall flue pipe - Dunbrik (Yorks) Ltd
Ventilators - Cavity Trays Ltd
Wall connectors - Catnic
Wall insulation - Dow Building Solutions
Wall starters - Expamet Building Products

F30 Accessories/ Sundry items for brick/ block/ stone walling (con't)

Wall tie replacement - HCL Contracts Ltd
Wall ties - Catnic
Wall ties - Vista Engineering Ltd
Wall ties and restraint fixings - Ancon Building Products
Wall ties and restraint fixings - Stainless steel building components - Wincro Metal Industries Ltd
Weepholes - L. B. Plastics Ltd
Windpots Masonry support systems reinforcement for masonry - Vista Engineering Ltd

F31 Precast concrete sills/ lintels/ copings/ features

3-point bending, Rebar corrosion, Shear testing - Lucideon
Architectural Masonary - Aggregate Industries UK Ltd
Architectural Stonework - Blanc de Bierges
Balustrades - The David Sharp Studio Ltd
Bespoke - Allen (Concrete) Ltd
Cast stone, sills, lintels, copings, features etc - Border Concrete Products Ltd
Chimney systems - Marflex Chimney Systems
Cills - Allen (Concrete) Ltd
concrete placing - Perform Construction (Essex) Ltd
Coping and caps - Rogers Concrete Ltd
Copings - Allen (Concrete) Ltd
Copings Lintels - Stowell Concrete Ltd
Lintels - Allen (Concrete) Ltd
Lintels - Allmat (East Surrey) Ltd
Lintels - Anderton Concrete Products Ltd
Lintels - Procter Cast stone & Concrete Products
Plastic cavity closers - L. B. Plastics Ltd
Polypropylene, steel fibers for concrete & mortar - Propex Concrete Systems
Precast concrete - Border Concrete Products Ltd
Precast concrete and cast stone products - Sangwin Concrete Products Ltd
Precast concrete products - Allen (Concrete) Ltd
Precast concrete sills/ lintels/ copings/ features - Chilstone
Precast concrete sills/ lintels/ copings/ features - Evans Concrete Products Ltd
Precast concrete sills/lintels/copings/features - Broadmead Cast Stone
Precast concrete sills/lintels/copings/features - Haddonstone Ltd
Precast Concrete Stairs, Beams, Cills Etc - Carter Concrete Ltd
Precast deenits - Northwest Pre-cast Ltd
Precast lintels and prestressed lintels - Procter Bros Limited
Prestressed wide slab flooring - ACP (Concrete) Ltd
Prestressed wide slab flooring, precast lintels - ACP Concrete Limited
PVCu products - Howarth Windows & Doors Ltd
Reconstituted Artstone - GB Architectural Cladding Products Ltd
Retaining walls - Concrete Products (Lincoln) 1980 Ltd
Shutters and Lintels - Shellcast Security Shutters
Steel cavity fixings - Lindapter International
Structural Precast concrete - Ebor Concrete Ltd
Supergalv steel lintels - Birtley Group

G10 Structural steel framing

Aluminium and steel fabricators - Ramsay & Sons (Forfar) Ltd
Aluminium fabrications - Midland Alloy Ltd
Architectural & General Metalwork - Thanet Ware Ltd
Blasting - Bristol Metal Spraying & Protective Coatings Ltd
Cast iron metalwork - Company name: Britannia Metalwork Services Ltd
Coated steel - SSAB Swedish Steel Ltd
Cold rolled purlins and rails - Ward Insulated Panels Ltd
Cold rolled steel sections - Hadley Industries Plc
Collled rolled sections - Walls & Ceilings International Ltd
Copper wire and rods - United Wire Ltd
C-sections - Hayes Cladding Systems
Domestic and industrial tongue and grooved flooring - Egger (UK) Ltd
Eaves beams - Hayes Cladding Systems
Erection of structural steelwork - Billington Structures Ltd
Fabrication of structural steelwork - Billington Structures Ltd
Flat to pitch conversions - Alumasc Exterior Building Products Ltd
Footbridges in steel, boardwalks decking and bespoke landscape structures - CTS Bridges Ltd
Gypframe (steel framed houses) - Metsec Ltd
Heavy duty protective coatings - International Protective Coatings
Hot and cold pressings - B C Barton Limited
Industrial Coating Specialists - Bristol Metal Spraying & Protective Coatings Ltd
Laser Welded Sections - Sections & Profiles Ltd
Meshes - Cadisch MDA Ltd
Metal fabrication - OEP Furniture Group PLC
Metal Framing - British Gypsum Ltd
Metal framing - Unistrut Limited
Metsec framing - panelised light steel buildings - Metsec Ltd
Mezanine floor structures - Hadley Industries Plc
Mezzanine floor channels and CEE section floor bea - Metsec Ltd
Mezzanine Floors - Barton Storage Systems Ltd
Mezzanine Floors - Spaceway South Ltd
Open grill flooring - Norton Engineering Alloys Co Ltd
Rail fixings - Lindapter International
Roof and soffit ventilators - Ryton's Building Products Ltd
Roof Frames: Steel - NovaSpan Structures
Roof truss system - MiTek Industries Ltd
Spacejoist V webs - ITW Consutruction Products
Stainless steel fabrications - Midland Alloy Ltd
Steel - Tata Steel Europe Limited
Steel fabrication - WA Skinner & Co Limited
Steel fabrication - Woodburn Engineering Ltd
Steel fixing - Perform Construction (Essex) Ltd
Steel framed buildings complete with fixtures, fittings and cladding - Reid John & Sons (Strucsteel) Ltd
Steel framing - Hadley Industries Plc
Steel framing - Vantrunk Engineering Ltd
Steel tube stockholders, fabrications, welding and fittings - Midland Tube and Fabrications
Steel wires, cables, rods and bars - Light (Multiforms), H A
Structural frames - Bison Manufacturing Limited
Structural hollow sections - Rainham Steel Co Ltd
Structural steel - Hy-Ten Ltd
Structural steel - Woodburn Engineering Ltd
Structural steel framing - Astron Buildings Parker Hannifin PLC
Structural steel framing - Caunton Engineering Ltd
Structural steel framing - Hubbard Architectural Metalwork Ltd
Structural steel framing - The Angle Ring Company Ltd
Structural steel towers - Franklin Hodge Industries Ltd
Structural steelwork - Atlas Ward Structures Ltd
Structural steelwork - Billington Structures Ltd
Structural steelwork - Ermine Engineering Co. Ltd
Structural steelwork - South Durham Structures Limited
Structural steelwork - supply, delivery and erection - Bourne Steel Ltd
Structural steelwork and associated products - Bourne Steel Ltd
Structural steelwork, portal frames and general steelwork - Bromag Structures Ltd
Structural, mezzanine floors, conveyor support steelwork and architectural structures - Arc Specialist Engineering Limited
Support systems - Lindapter International
Universal beams & Columns - Rainham Steel Co Ltd
Wire rope assemblies and fittings - Midland Wire Cordage Co Ltd
Zed and CEE purlins - Metsec Ltd
Z-purlins - Hayes Cladding Systems

G11 Structural aluminium framing

Aluminium extrusions/ sections - Seco Aluminium Ltd
Aluminium fabrications - Midland Alloy Ltd
Curved canopies, barrel vaults and fabrications - Midland Alloy Ltd
Glasshouses - Manufacturers of Aluminium Glasshouses and Greenhouses - Hartley Botanic Ltd
Greenhouses: Large, Small, Bespoke, Made to measure, made to order - Manufacturers of Aluminium Glasshouses and Greenhouses - Hartley Botanic Ltd
Ground floor framing - Comar Architectural Aluminuim Systems
Orangeries - Manufacturers of Aluminium Glasshouses and Greenhouses - Hartley Botanic Ltd
Roof Frames: Aluminium - NovaSpan Structures
Structural Steel Framing - Blagg & Johnson Ltd

G12 Isolated structural metal members

Aluminium Architectural Products - Gooding Aluminium
Aluminium structural members - Alcoa Custom Extrudes Solutions
Architectural steelwork - Bourne Steel Ltd
Blasting - Bristol Metal Spraying & Protective Coatings Ltd
Cold roll sections/forming - Metsec Lattice Beams Ltd
Cold roll sections/forming - Metsec Ltd
Curved steel sections, tubes, roof and cambered beams - The Angle Ring Company Ltd
Industrial Coating Specialists - Bristol Metal Spraying & Protective Coatings Ltd
Isolated structural members - Hubbard Architectural Metalwork Ltd
Isolated structural members - Hy-Ten Ltd
Lattice joist and trusses - Metsec Lattice Beams Ltd
Lattice joist and trusses - Metsec Ltd
Metal post supports/shoes - Canopy Products Ltd
Metal web floor beams - Wyckham Blackwell Ltd
Steel chimneys and associated structural steelwork - Beaumont Ltd F E
Structural secondary steelwork - Arc Specialist Engineering Limited
Structural steels - Bromford Iron & Steel Co Ltd
Subcontract fabrication - Mackay Engineering
Subcontract machining - Mackay Engineering
Tension systems - Ancon Building Products
Tubes, fittings & flanges - The BSS Group
Welding and fabrication work - Griff Chains Ltd

G20 Carpentry/ Timber framing/ First fixing

Aluminuim Fascia / Soffit - Guttercrest
Bespoke joinery - Howard Bros Joinery Ltd
Carpenters Metalwork - Allmat (East Surrey) Ltd
Carpentry/ Timber framing/ First Fixing - Anderson C F & Son Ltd
Carpentry/ Timber framing/ First fixing - Polypipe Building Products
Cladding Insect Mesh - The Flyscreen Company Ltd
Composite floor joints - Bullock & Driffill Ltd
Construction & joniery softwoods - Cox Long Ltd
Constructional timber - Brewer T. & Co Ltd
Design & Supply of Laminated Timber Engineered Beams and / or Structures - Technical Timber Services Ltd
Door Kits - Premdor
Engineered Floor Systems - Scotts of Thrapston Ltd
Engineered I beams - IJK Timber Group Ltd
Engineered Timber Edge Bonders - Panel Agency Limited
Fire resisting joinery - Decorfix Ltd
Fixings to trusses - MiTek Industries Ltd
Flooring - Metsä Wood
Footbridges in timber, boardwalks, decking and bespoke landscape structures - CTS Bridges Ltd
Glue Laminated Timber - Panel Agency Limited
Glulam beams - Lilleheden Ltd
Hardwood - IJK Timber Group Ltd
Hardwood - James Latham plc
Hardwood - Metsä Wood
Hardwoods - Brewer T. & Co Ltd
High class joinery - Platonoff & Harris Plc
I Beams - Steico
Industrial pre-treatment preservatives for timber - Protim Solignum Ltd
Intumescent coatings for timber - Quelfire
Joist hangers - Expamet Building Products
Laminated beams - Bullock & Driffill Ltd
Laminated Timber Beams - Kingston Craftsmen Structural Timber Engineers Ltd
Laminated Timber Columns - Kingston Craftsmen Structural Timber Engineers Ltd
Laminated Timber Frames - Technical Timber Services Ltd
Laminated Timber Roof and Roof Frames - Technical Timber Services Ltd
Lightweight lattice joists and trusses - Metsec Lattice Beams Ltd
Load bearing and non LB site fixed panels - Metsec Ltd
Long lengths & carcassing - Brash John & Co Ltd
MDF, plywood, and engineered wood products - Willamette Europe Ltd
Nail plates - ITW Consutruction Products
Planed and sawn timber - Mason FW & Sons Ltd
Portal frames and cranked beams - Lilleheden Ltd
Re-sawn carcassing - James Donaldson Timber
Roof trusses - Armstrong (Timber), Thomas, Ltd
Roof trusses - Crendon Timber Engineering Ltd
Roof trusses - Merronbrook Ltd
Roof Trusses - Moray Timber Ltd
Roof trusses - Scotts of Thrapston Ltd
Roofing contractor - Evans Howard Roofing Ltd
Saw millers and timber merchants - Whitmore's Timber Co Ltd
Sawmilling - Moray Timber Ltd
Softwood - Metsä Wood
Softwood - Panel Agency Limited
Stainless steel timber to masonry fixings - Helifix
Stainless steel warm roof fixings - Helifix
Stress grading - Cox Long Ltd
Stress Grading - Moray Timber Ltd
Structural timber - James Donaldson Timber
Structural timber engineers - Kingston Craftsmen Structural Timber Engineers Ltd
T&G Solid Timber Decking - Technical Timber Services Ltd
Timber - Covers Timber & Builders Merchants
Timber - IJK Timber Group Ltd
Timber - Moray Timber Ltd
Timber Floor Beams - Steico
Timber frame buildings - Fleming Buildings Ltd
Timber frame design supply & fix - Cox Long Ltd
Timber frame housing - Fleming Homes Ltd
Timber framed houses - Cambridge Structures Ltd
Timber frames and roof trusses - Walker Timber Ltd
Timber framing - BM TRADA
Timber I Beams - Panel Agency Limited
Timber importers & sawmillers - Walker Timber Ltd
Timber importers, swamillers and merchants - Brewer T. & Co Ltd
Timber merchants - Derby Timber Supplies
Timber merchants - Timbmet Ltd
Timber panels - Challenge Fencing Ltd
Timber Portal Frames - Kingston Craftsmen Structural Timber Engineers Ltd
Timber preservative treatments - Kingfisher Building Products Ltd
Timber roof trusses - Dover Trussed Roof Co Ltd
Timber Roof Trusses - Kingston Craftsmen Structural Timber Engineers Ltd
Timber strength grading - Walker Timber Ltd
Timber supplies - Bullock & Driffill Ltd
Timber to masonry connectors - Simpson Strong-Tie®
Timber to timber connectors - Simpson Strong-Tie®
Timber trussed rafters - Bullock & Driffill Ltd
Treatment - Moray Timber Ltd
Trussed rafter design & manufacture - Cox Long Ltd
Trussed rafters - Crendon Timber Engineering Ltd
Trussed rafters - Wyckham Blackwell Ltd
Trusses, beams, frames, panels , timber engineering - Cowley Structural Timberwork Ltd

G30 Metal profiled sheet decking

Aluminium sheet decking - Alcoa Custom Extrudes Solutions
Cold Roll Forming - Blagg & Johnson Ltd
Composite decking - Roles Broderick Roofing Ltd
Decking - Tata Steel Europe Limited
Insulated Roof & Wall Systems - Kingspan Ltd
Metal profile sheet decking - BriggsAmasco
Profiled metal floor decking - Ward Insulated Panels Ltd
PVC decking - L. B. Plastics Ltd
Sheet metal work - B C Barton Limited
Steel Decking - Richard Lees Steel Decking Ltd
Structural Metal Decking - Omnis
Structural steelwork - supply, delivery and erection - Bourne Steel Ltd

G31 Prefabricated timber unit decking

Decking - Anthony de Grey Trellises + Garden Lighting
Sheet materials - Metsä Wood
Timber Decking - The Outdoor Deck Company
Timber frame kits - Merronbrook Ltd

G32 Edge supported/ Reinforced woodwool/rock fibre decking

Wood wool boards - Skanda Acoustics Limited

H10 Patent glazing

Aluminium and lead clothed steel patent glazing systems - Standard Patent Glazing Co Ltd
Aluminium conservatory roofs - Sapa Building Systems Ltd
aluminium/hybrid curtain walling, windows and doors - Senior Aluminium Systems plc
Anti-vandal glazing - Amari Plastics Plc
Cladding/ covering - Pilkington UK Ltd
Conservatory components - Exitex Ltd
Curtain walling - Astrofade Ltd
Curtain Walling - Duplus Architectural Systems Ltd
Curtain Walling - Freeman T R
Curtain walling - HW Architectural Ltd
Curved Glass - Novaglaze Limited
Patent glazing - Cantifix of London Limited
Patent glazing - Duplus Architectural Systems Ltd
Patent glazing - Glassolutions Saint-Gobain Ltd
Patent glazing - Ide T & W Ltd
Patent glazing - Lonsdale Metal Company
Rooflights, lantern lights and pyramid lights - Standard Patent Glazing Co Ltd

H11 Curtain walling

Aluminium curtain walling - Sapa Building Systems Ltd
Aluminium extrusions/ sections - Seco Aluminium Ltd
aluminium/hybrid curtain walling, windows and doors - Senior Aluminium Systems plc
Curtain wall systems - Comar Architectural Aluminuim Systems
Curtain walling - Barlow Group
Curtain walling - Cantifix of London Limited
Curtain Walling - CEP Claddings Ltd
Curtain walling - Colt International
Curtain walling - Crittall Steel Windows Ltd
Curtain walling - Deceuninck Ltd
Curtain walling - Duplus Architectural Systems Ltd
Curtain walling - English Architectural Glazing Ltd
Curtain Walling - Glassolutions Saint-Gobain Ltd
Curtain walling - Ide T & W Ltd
Curtain Walling - McMullen Facades
Curtain Walling - Metal Technology Ltd
Curtain walling - Panel Systems Ltd
Curtain Walling - Sapa Building Systems (monarch and glostal)
Curtain Walling - Schueco UK
Curtain walling and cladding - Schmidlin (UK) Ltd
Curtain Walling Systems - H W Architectural Ltd
Extruded curtain walling - Smart Systems Ltd
Facades - Hunter Douglas
Facades, architectural and construction solutions - Porcelanosa Group Ltd
Fire Resistant glass - SCHOTT UK Ltd
General extrusions - L. B. Plastics Ltd
Glazing and Curtain walling - Howard Evans Roofing and Cladding Ltd
Non- combustible roofing and cladding systems - Gilmour Ecometal
Patent glazing - Astrofade Ltd
Patent Glazing - Duplus Architectural Systems Ltd
Patent glazing - HW Architectural Ltd
Rainscreen - Gilmour Ecometal
Stainless Steel Cladding - Aperam Stainless Services & Solutions UK Limited
Structural wall solutions - CGL Cometec Ltd
Terracotta Cladding Systems - Telling Architectural Ltd
Terracotta Curtain Walling - Telling Architectural Ltd
Timber and Timber aluminium - Scandanavian Timber
Unitised curtain walling - English Architectural Glazing Ltd

H12 Plastic glazed vaulting/ walling

Multiwall polycarbonate - Kenilworth Daylight Centre Ltd
Opaque chemical resistant cladding - Filon Products Ltd
Plastic glazed vaulting - Duplus Architectural Systems Ltd
Plastic glazed vaulting/ walling - Astrofade Ltd
Plastic glazed vaulting/walling - Amari Plastics Plc
Polycarbonate glazing for flat and industrial roofs - Cox Building Products Ltd
Profiled plastic sheets, conservatory roofs and flat plastic sheets - Kenilworth Daylight Centre Ltd

H13 Structural glass assemblies

Architectural Glass - Creative Glass
Atria Glazing - Fenlock-Hansen Ltd
Barrelvaults, ridgelights, pyramids in glass or polycarbonate - Cox Building Products Ltd
Canopies - Glassolutions Saint-Gobain Ltd

H13 Structural glass assemblies (con't)

Glass Walling - Telling Architectural Ltd
Steel wire, cables, rods and bars - MacAlloy Limited
Structural glass assemblers - Astrofade Ltd
Structural glass assemblies - Cantifix of London Limited
Structural glass assemblies - Colt International
Structural glass assemblies - Pilkington UK Ltd
Structural Glass Assemblies - Schueco UK
Structural Glazing - Ide T & W Ltd

H14 Concrete rooflights/ pavement lights

Floor lights - Luxcrete Ltd
Glass arches - Luxcrete Ltd
Hygienic GRP Wall Cladding - Crane Composites
Pavement lights - Longbottom J & J W Ltd
Pavement lights - Luxcrete Ltd
Police cell windows and observation units - Luxcrete Ltd
Roof lights - Luxcrete Ltd
smoke outlets - Luxcrete Ltd

H20 Rigid sheet cladding

Armour sheeting - Meggitt Armour Systems
B.R.E. Structural Render Panels - Gradient Insulations (UK) Ltd
Balcony panels - Trespa UK Ltd
Building boards - Finnish Fibreboard (UK) Ltd
Cladding - Burlington Slate Ltd
Cladding - James Donaldson Timber
Cladding - wall and ceiling - Flexitallic Ltd
Cladding & Building Boards - Marley Eternit Ltd
Cladding and Fascias - Howard Evans Roofing and Cladding Ltd
Cladding panels - Blaze Signs Ltd
Composite panels - EBC UK Ltd
Composite panels - Gradient Insulations (UK) Ltd
Exterior cladding panels - Trespa UK Ltd
Exterior grade high pressure laminate - Abet Ltd
Exterior systems - Knauf UK
Glass reinforced polyester profiled and flat sheeting - Filon Products Ltd
Gypsum Rienforced Fibreboard - Panel Agency Limited
High performance building panel - Trespa UK Ltd
High performance claddings and Fire Protection Boards - Cembrit Ltd
Hygienic wall + ceiling services - Altro Ltd
Hygienic cladding - Amari Plastics Plc
Hygrothermal performance testing, Water penetration, Pull testing, Impact testing, Dynamic wind loading and uplift testing - Lucideon
Kerto LVL (laminated veneer lumber) - Metsä Wood
L.P.C.B Roof systems - Gradient Insulations (UK) Ltd
L.P.C.B Wall panels - Gradient Insulations (UK) Ltd
Laminated Boards - Steico
Metal Façade Systems - Omnis
Metal roofing and cladding - Ash & Lacy Building Products Ltd
Plastic Coated steel in flat sheet - Colorpro Systems Ltd
PVC claddings and fascias - Rockwell Sheet Sales Ltd
Rainscreen cladding - Omnis
Rigid Sheet Cladding - CEP Claddings Ltd
Rigid sheet cladding - FGF Ltd
Rigid sheet cladding - Panel Systems Ltd
Rigid sheet cladding for kitchens and walls - Respatex International Limited
Sheet materials - Mason FW & Sons Ltd
Stainless Steel cladding - Aperam Stainless Services & Solutions UK Limited
Structural Insulated Panels - EBC UK Ltd
Wall cladding - Dunhams of Norwich
Wall Cladding, roof cladding, insulated wall and roof cladding - Kingspan Ltd
Wall lining systems - Formica Ltd
Wood wool boards - Skanda Acoustics

H21 Timber weatherboarding

Cedar shakes for roofing and cladding - Brash John & Co Ltd
Cladding - James Donaldson Timber
Cladding Insect Mesh - The Flyscreen Company Ltd
Exterior systems - Knauf UK
Timber cladding - Brash John & Co Ltd
Timber cladding - LSA Projects Ltd
Timber cladding - OSMO UK
Timber weatherboarding - Brash John & Co Ltd
Timber Weatherboarding - CEP Claddings Ltd
Timber weatherboarding - Challenge Fencing Ltd
Timber weatherboarding - Vulcan Cladding Systems
Timber weatherboarding suppliers - W.L. West & Sons Ltd

H30 Fibre cement profiled sheet cladding/ covering/siding

External Wall Cladding - Caithness Flagstone Limited
External Wall Cladding - Hanson Building Products
External Wall Cladding - Paroc Panel Systems Uk Ltd
External wall cladding - Prodema Lignum
Fastenings - Cooper & Turner
Fibre cement - FGF Ltd
Fibre cement profiled sheet cladding / covering - TRC (Midlands) Ltd
Fibre cement profiled sheet cladding/ covering/siding - SLE Cladding Ltd
Fibre cement sheeting - Coverworld UK Limited
Fibre Cement Sheeting - Marley Eternit Ltd
Fundermax cladding - Telling Architectural Ltd
Profiled Sheets - Cembrit Ltd

H31 Metal profiled/ flat sheet cladding/ covering/ siding

Alternative Coatings - PVF2, Polyester - Colorpro Systems Ltd
Aluminium & steel Profiled cladding - Rigidal Systems LTD
Aluminium & steel standing seam systems - Rigidal Systems LTD
Aluminium cladding and roofing - Rigidal Industries Ltd
CA Profiles - CA Group Ltd - MR-24 Division
Cladding - Astron Buildings Parker Hannifin PLC
Cladding - Caunton Engineering Ltd
Cladding - FGF Ltd
Cladding - Tata Steel Europe Limited
Cladding and siding - Bromag Structures Ltd
Cladding Systems - Cadisch MDA Ltd
Column Cladding - Cadisch MDA Ltd
Composite aluminium and steel flat cladding systems - Booth Muirie
Composite aluminium and steel rain screens - Booth Muirie
Composite panels - Composite Panel Services Ltd
Corrugated & Cladding Sheets - Sections & Profiles Ltd
Expanded Mesh - Cadisch MDA Ltd
Facades - Hunter Douglas
Fastenings - Cooper & Turner
Galvanised Corrugated sheet - Rainham Steel Co Ltd
Industrial and commercial roofing, Cladding and Sheeting - Avonside Roofing Group T/A Letchworth Roofing
Insulated Roof & Wall Systems - Kingspan Ltd
Insulated roof panels - Composite Panel Services Ltd
Insulated wall and roof panels and cladding sheets - Ward Insulated Panels Ltd
Lightweight sheet and panel roofing systems - Britmet Tileform Limited
Metal cladding - Decra Roofing Systems
Metal cladding - LSA Projects Ltd
Metal Cladding Systems - Euroclad Ltd
Metal fabrication, ancillary cladding components - Interlink Group
Metal Façade Systems - Omnis
Metal profiled / flat sheet cladding / covering - TRC (Midlands) Ltd
Metal profiled and flat sheet cladding, covering and siding - Hayes Cladding Systems
Metal profiled/ flat sheet cladding/ covering - Hadley Industries Plc
Metal profiled/ flat sheet cladding/ covering - Reid John & Sons (Strucsteel) Ltd
Metal profiled/ flat sheet cladding/ covering/ siding - SLE Cladding Ltd
Metal profiled/ flat sheet cladding/ covering/ siding - Vulcan Cladding Systems
Metal profiled/flat sheet cladding/covering - Amari Plastics Plc
Metal Profiled/flat sheet cladding/covering - Britmet Tileform Limited
Metal roof tiles - Catnic
Metal roofing and cladding - Ash & Lacy Building Products Ltd
Metal roofing and cladding - BriggsAmasco
Metal screens and louvres - Lang+Fulton
Metal tiles - EBC UK Ltd
Perforated Sheet - Cadisch MDA Ltd
Plastic Coated Profiles Steel Roofing - Colorpro Systems Ltd
Plastic coates flashings - Colorpro Systems Ltd
Plastic Coates Profiled Steel Cladding - Colorpro Systems Ltd
Pre Crimp Meshes - Cadisch MDA Ltd
Prebonded claddding - Roles Broderick Roofing Ltd
Prefabricated bathroom structures - IPPEC Sytsems Ltd
Profiled metal sheeting - Coverworld UK Limited
Rainscreen cladding systems - Booth Muirie
Sheet Metal Work - Thanet Ware Ltd
Sheeting - Durable Contracts Ltd
Single ply flat roof system - Intergrated Polymer Systems (UK) Ltd
Standing Seam Roof Decking - Gradient Insulations (UK) Ltd
Traditional flat roofing systems - IKO PLC
Wall cladding - Drawn Metal Ltd
Waterproofing (inc. standing seam roofing) - Alumasc Exterior Building Products Ltd
Woven Wire Mesh - Cadisch MDA Ltd

H32 Plastic profiled sheet cladding/ covering/ siding

Cellular PVC fascias and soffits - Celuform Building Products
Cellular Pvc fascias, soffits, barge boards and claddings - Swish Building Products
Corrugated or flat PVC and polycarbonate clear or coloured plastic sheets - Rockwell Sheet Sales Ltd
Covers for new and existing pools of almost any shape - Grando (UK) Ltd
Decorative internal cladding system - Swish Building Products
Fascia, soffit and cladding systems - FloPlast Ltd
Fascias, Soffits & Gutters - BAC Ltd
Fastenings - Cooper & Turner
Flat roof outlets - Hunter Plastics Ltd
Glass reinforced polyesther profiled and flat sheeting - Filon Products Ltd
GRP/APC roof lights and cladding products - Hambleside Danelaw Ltd
Internal & external cladding - Amari Plastics Plc
Multiwall polycarbonate sheets - Rockwell Sheet Sales Ltd
Over-roofing - Filon Products Ltd
Plastic coated steel flashings - Azimex Fabrications Ltd
Plastic fabrication - Colman Greaves
Plastic mouldings - Invicita Plastics Ltd
Plastic profiled sheet cladding /covering - TRC (Midlands) Ltd
Plastic profiled sheet cladding and roofing - Ariel Plastics Ltd
Plastic profiled sheet cladding, facias, soffits, interior wall and ceiling panelling - Deceuninck Ltd
Plastic profiled sheet cladding/ covering/ siding - SLE Cladding Ltd
Plastic profiled sheets - Leicester Barfitting Co Ltd
PVC cladding - L. B. Plastics Ltd
PVCu sills, trims, cladding and roof line products - Duraflex Limited
PVC-V facia & Soffit - Allmat (East Surrey) Ltd
Roof sheets - profiled and structured in GRP, PVC and polycarbonate - Brett Martin Ltd
Rooflights - Hayes Cladding Systems
Soffits and fascias - Spectra Glaze Services
Translucent profiled sheeting - Coverworld UK Limited
Vinyl cladding system for fascias soffits and full elevations - Cox Building Products Ltd

H33 Bitumen & fibre profile sheet cladding/ covering

Bitumen & fibre profile sheet cladding/ covering - SLE Cladding Ltd
Bitumen + fibre profile sheet cladding - Onduline Building Products Ltd
Bitumious felt roofing systems - IKO PLC
Foundation protection - Onduline Building Products Ltd
Plastic profiled sheet cladding and roofing - Ariel Plastics Ltd

H40 Glassfibre reinforced cement panel cladding/ features

Cladding systems - BRC Special Products - Bekaert Ltd
Exterior systems - Knauf UK
Glassfibre reinforced cement panel cladding/ features - SLE Cladding Ltd
Glassfibre reinforced cement panel cladding/ features - Vetter UK
Glassfibre reinforced cement panel cladding/features - Aden Hynes Sculpture Studios
Hygrothermal performance testing, Water penetration, Pull testing, Impact testing, Dynamic wind loading and uplift testing - Lucideon
TecLite GRC based stone - Haddonstone Ltd

H41 Glassfibre reinforced plastics panel cladding/ features

Architectural Moldings GRP - Halmark Panels Ltd
Bespoke architectural mouldings and features in GRP - Design & Display Structures Ltd
Bespoke GRP Mouldings - Sarena Mfg Ltd
Columns, canopies, cornices, porticos, bay canopies - Solair GRP Architectural Products
Dormer windows - Solair GRP Architectural Products
G. R. P. Porches - Solair Ltd
Glass fibre reinforced cladding - Brimar Plastics Ltd
Glass fibre reinforced plastics cladding features - Amari Plastics Plc
Glass reinforced plastic features - Troika Contracting Ltd
Glass reinforced polyester profiled and flat sheeting - Filon Products Ltd
Glassfibre plastics reinforced cladding/features - Martello Plastics Ltd
Glassfibre reinforced plastics cladding/architectural features - Duroy Fibreglass Mouldings Ltd
Glassfibre reinforced plastics cladding/features - CEP Claddings Ltd
Glassfibre reinforced plastics panel cladding/ features - SLE Cladding Ltd
Glassfibre reinforced plastics panel cladding/ features - Vulcan Cladding Systems
Glassfibre reinforced plastics panel cladding/features - Aden Hynes Sculpture Studios
GRG internal cladding - Multitex GRP LLP
GRP - Hodkin Jones (Sheffield) Ltd
GRP (Glass Reinforced Polyester) - Gillespie (UK) Ltd
GRP canopies - Halmark Panels Ltd
GRP cladding - Adams- Hydraulics Ltd
GRP cladding and moldings - Armfibre Ltd
GRP cladding, specialist moulding, door canopies and columns - Multitex GRP LLP
GRP Columns - Production Glassfibre
GRP conservatory roofs - Halmark Panels Ltd
GRP flashings - Hambleside Danelaw Ltd
GRP housings - Precolor Sales Ltd
GRP Mouldings - Harviglass-Fibre Ltd
GRP Pilasters - Production Glassfibre
GRP porches - Halmark Panels Ltd
GRP roofing systems - Hambleside Danelaw Ltd
GRP structures & plant rooms - APA
GRP/APC roof lights and cladding products - Hambleside Danelaw Ltd
Interior and exterior laminated glassfibre components - Pyramid Plastics UK Ltd
Modular Housings - Sarena Mfg Ltd
Plastic profiled sheet cladding and roofing - Ariel Plastics Ltd
Quickfix Dormers - Multitex GRP LLP
Quickstack Chimneys - Multitex GRP LLP
RSJ Cladding Beams - Oakleaf Reproductions Ltd

H42 Precast concrete panel cladding/ features

Architectural precast cladding - Marble Mosaic Co Ltd, The
Precast & prestressed concrete panels - ACP Concrete Limited
Precast concrete and cast stone products - Sangwin Concrete Products Ltd
Precast concrete panel cladding/ features - SLE Cladding Ltd
Precast concrete panel cladding/ features - Vetter UK
Precast Concrete Panel/ Cladding - Carter Concrete Ltd
TecStone precast concrete - Haddonstone Ltd

H43 Metal Panel Cladding/ features

Façade Systems - Rheinzink UK
Metal Cladding Pannel - McMullen Facades
Metal Panel Cladding/ features - SLE Cladding Ltd
Profiled Sheet Cladding - Tata Steel Colors
Roofing - Mells Roofing Ltd
Vitreous Enamal Cladding - Trico V.E. Ltd (Signage Division)
Vitreous Enamel Panels - Trico V.E. Ltd (Signage Division)

H50 Precast concrete slab cladding/ features

Architectural precast cladding - Marble Mosaic Co Ltd, The
Brick Cladding Systems - Eurobrick Systems Ltd
Precast concrete panels - Anderton Concrete Products Ltd
Precast concrete slab cladding/ features - Evans Concrete Products Ltd
Precast concrete slab cladding/ features - Vetter UK
Precast Slab Cladding - Carter Concrete Ltd
Prestressed concrete wall panels - ACP (Concrete) Ltd
Prestressed concrete wall panels - ACP Concrete Limited
Retaining walls - Grass Concrete Ltd

H51 Natural stone slab cladding/ features

Airtec Stone - Telling Architectural Ltd
External Wall Cladding - Johnsons Wellfield Quarries Ltd
Glazed terracotta rain screen cladding - Smithbrook Building Products Ltd
Marble - The Stone Catalogue Ltd
Masonry - Bridge Quarry
Natural Limestone Masonry - Hanson UK - Bath & Portland Stone
Natural stone cladding - Kirkstone
Natural Stone Cladding - Palladio Stone
Natural Stone Slab Cladding - Carter Concrete Ltd
Natural stone slab cladding / features - Toffolo Jackson(UK) Ltd
Natural stone slab cladding features - PermaRock Products Ltd
Natural stone slab cladding/ features - Chichester Stoneworks Limited
Natural Stone Slab Cladding/ Features - Realstone Ltd
Natural stone slab cladding/ features - Stonewest Ltd
Natural stone slab cladding/ features - Vetter UK
Natural stone slab cladding/features - CEP Claddings Ltd
Natural stone slab cladding/features - G Miccoli & Sons Limited
Sandstone - Dunhouse Quarry Co Ltd
Stone faced precast cladding - Marble Mosaic Co Ltd, The
Stonework - design, procurement and installation of - Szerelmey Ltd
Yorkstone - Hard York Quarries Ltd

H52 Cast stone slab cladding/ features

Architectural precast cladding - Marble Mosaic Co Ltd, The
Architectural Precast concrete cladding - Techrete (UK) Ltd
Architectural Stonework - Cranborne Stone Ltd
Bespoke cast stone features - GB Architectural Cladding Products Ltd
Brick & stone insulated cladding systems - Eurobrick Systems Ltd
Cast stone cladding/ features - GB Architectural Cladding Products Ltd
Cast stone cladding/features - G Miccoli & Sons Limited
Cast stone slab cladding / features - Toffolo Jackson(UK) Ltd
Cast stone slab cladding /features - Haddonstone Ltd
Cast stone slab cladding/ features - Chilstone
Cast stone slab/ cladding/ features - Evans Concrete Products Ltd
Cladding - Blanc de Bierges
Mullion windows, quoins, cornices and doorways - Minsterstone Ltd

H60 Plain roof tiling

Acrylic glass tiles - Klober Ltd
Classic Handformed Tiles - Hinton Perry & Davenhill Ltd
Clay and concrete tiles - Sandtoft Roof Tiles Ltd
Clay hanging tiles - clayton - Keymer Hand Made Clay Tiles
Clay peg tiles - traditional handmade - Keymer Hand Made Clay Tiles

H60 Plain roof tiling (con't)

Clay plain tiles - traditional handmade - Keymer Hand Made Clay Tiles
Clay roof tiles - Hinton Perry & Davenhill Ltd
Clay Roof Tiles - Marley Eternit Ltd
Clay roof tiling - Imerys Roof Tiles
Clay roofing tiles - Michelmersh Brick Holdings PLC
Concrete roof tiles - Forticrete Ltd
Concrete roof tiles - Forticrete Roofing Products
Concrete Roof Tiles - Marley Eternit Ltd
Finials - Hinton Perry & Davenhill Ltd
Flat roofing - Mells Roofing Ltd
Glazed clay roof tiles - Smithbrook Building Products Ltd
Glazed roof tiles - Hye Oak
Handmade Clay, roof tiles plus under tile vent - Tudor Roof Tile Co Ltd
Integrated Photovoltaic Tiles - Imerys Roof Tiles
Ornamental Ridges - Hinton Perry & Davenhill Ltd
Ornamental Tiles and Fittings - Hinton Perry & Davenhill Ltd
Plain Clay Roof Tiles - Hinton Perry & Davenhill Ltd
Plain Roof Tiling - Britmet Tileform Limited
Plain roof tiling - Decra Roofing Systems
Plain roof tiling - PermaSeal Ltd
Plain roof tiling - SIG Roofing
Plain tiles - Monier Redland Limited
Roof Coverings - Howard Evans Roofing and Cladding Ltd
Roof space ventilating tiles - Klober Ltd
Roof Tiles - Russell Roof Tiles Ltd
Roof ventalation - Zled (DPC) Co Ltd
Roof ventilation - Hambleside Danelaw Ltd
Roofing - InstaGroup Ltd
Roofing Accessories - Monier Redland Limited
Roofing claywork - Red Bank Maufacturing Co Ltd
Roofing fittings - Monier Redland Limited
Roofing underlay felt - Newton John & Co Ltd
Technical Solutions - Monier Redland Limited
Terracotta ridges and finals - Keymer Hand Made Clay Tiles
Tileline Ventilation Products - Hinton Perry & Davenhill Ltd
Tiling - Avonside Roofing Group T/A Letchworth Roofing
Tiling - Durable Contracts Ltd
Ventilated ridge system - Klober Ltd
Ventilation system - Keymer Hand Made Clay Tiles

H61 Fibre cement slating

Fibre Cement Slates - Marley Eternit Ltd
Fibre cement slating - SIG Roofing
Polypropylene, steel fibers for concrete & mortar - Propex Concrete Systems
Roof space ventilating slates - Klober Ltd
Roof ventilation - Hambleside Danelaw Ltd
Roofing - Avonside Roofing Group T/A Letchworth Roofing
Roofing slate - Chameleon Stone Ltd
Roofing slates, ventilation products - Cembrit Ltd
Slating - Durable Contracts Ltd
Synthetic Slates and Tiles - EBC UK Ltd

H62 Natural slating

Copings and sills - Kirkstone
Flooring Slate - Chameleon Stone Ltd
Natural Roofing Slates - Ensor Building Products Ltd
Natural slate cladding - Welsh Slate Limited
Natural slate copings - Welsh Slate Limited
Natural slate counters - Welsh Slate Limited
Natural slate flooring - Welsh Slate Limited
Natural slate landscaping - Welsh Slate Limited
Natural slate roofing - Burlington Slate Ltd
Natural slate roofing - Welsh Slate Limited
Natural slate sills - Welsh Slate Limited
Natural slate walling - Welsh Slate Limited
Natural slate worktops - Welsh Slate Limited
Natural Slates - IJK Timber Group Ltd
Natural slating - SIG Roofing
Natural Stone - Bridge Quarry
Polythene under slating sheet - Visqueen Building Products
Roof Tiles & Slates - Caithness Flagstone Limited
Roof ventilation - Hambleside Danelaw Ltd
Roof ventilation/ accessories - Zled (DPC) Co Ltd
Roofing slate - Chameleon Stone Ltd
Roofing slate - The Delabole Slate Co. Ltd
Slate Repair Clips - Owens Slate Services Ltd
Slate Repair Fixings - Owen Slate Services Ltd
Slating - Avonside Roofing Group T/A Letchworth Roofing
Welsh roofing slates - Greaves Welsh Slate Co Ltd
Welsh, Spanish Brazilian and Canadian slates - Cembrit Ltd

H63 Reconstructed stone slating/ tiling

Green roofs - Alumasc Exterior Building Products Ltd
Reconstructed stone roofing tiles - Atlas Stone Products
Reconstructed stone slating/ tiling - SIG Roofing
Reconstructed stone slating/tiling - CEP Claddings Ltd
Slates - Monier Redland Limited
Slates - Sandtoft Roof Tiles Ltd
Tiles and specialist works - conglomerate marble, limestone & assimilated granite - Marble Flooring Spec. Ltd

H64 Timber shingling

Cedar shingles - Cembrit Ltd
Cedar shingles for roofing and cladding - Brash John & Co Ltd
K20 Timber Boarding etc - Challenge Fencing Ltd
Timber shingling - Brash John & Co Ltd
Timber shingling - SIG Roofing
Timber shingling suppliers - W.L. West & Sons Ltd

H65 Single lap roof tiling

Concrete Roof Tiles - Marley Eternit Ltd
Profile tiles - Monier Redland Limited
Roof Ventalation/Accessories - Zled (DPC) Co Ltd
Roofing - Avonside Roofing Group T/A Letchworth Roofing
Roofing - Mells Roofing Ltd
Roofing Systems - Sandtoft Roof Tiles Ltd
Single lap roof tiling - Sandtoft Roof Tiles Ltd
Single lap roof tiling - SIG Roofing
Single roof tiling - Decra Roofing Systems
Thatching, thatching materials and fire retardants - Thatching Advisory Services Ltd

H66 Bituminous felt shingling

Bitumen felt shingles - Cembrit Ltd
Bituminous felt shingling - PermaSeal Ltd
Bituminous felt shingling - SIG Roofing
Bituminous felt slungies - Onduline Building Products Ltd
Foundation protection - Onduline Building Products Ltd
Reinforced Bitumen Membranes - Mells Roofing Ltd
Tegola asphalt shingles - Matthew Hebden

H70 Malleable metal sheet prebonded covering/ cladding

Fully supported roofing and cladding - Roles Broderick Roofing Ltd
Malleable metal sheet prebonded covering/ cladding - Reid John & Sons (Strucsteel) Ltd
Metal cladding coatings - Jotun Henry Clark Ltd Decorative Division
Metal roofing - Omnis
Roofing - Mells Roofing Ltd

H71 Lead sheet coverings/ flashings

Aluminium flashing - George Gilmour (Metals) Ltd
Cloaks - Cavity Trays Ltd
Flashings - Cavity Trays Ltd
Hyclad matt finish - Outokumpu
Lead and sheet flashings - Royston Lead Ltd
Lead roofing, sand cast lead sheet - Anglia Lead Ltd
Lead sheet - R.M. Easdale & Co Ltd
Lead sheet and flashing - Calder Industrial Materials Ltd
Lead sheet covering and flashings - Roles Broderick Roofing Ltd
Lead Sheet Coverings/ Flashings - BLM British Lead
Lead sheet coverings/ flashings - Freeman T R
Lead sheet coverings/ flashings - Harris & Bailey Ltd
Low reflective satin finish - Outokumpu
Ternie coated lead finish - Outokumpu
White and red lead - Liver Grease Oil & Chemical Co Ltd

H72 Aluminium strip/ sheet coverings/ flashings

Aluminium covers - Franklin Hodge Industries Ltd
Aluminium & Steel Composite Roof & Wall Cladding - Rigidal Industries Ltd
Aluminium cappings & flashings - Aluminium RW Supplies Ltd
Aluminium coping - WP Metals Ltd
Aluminium coping systems - Dales Fabrications Ltd
Aluminium copings/flashings - Alcoa Custom Extrudes Solutions
Aluminium Facia and Soffit systems - Alutec
Aluminium fascias and soffits - Aluminium RW Supplies Ltd
Aluminium flashings - WP Metals Ltd
Aluminium flashings and trims - Azimex Fabrications Ltd
Aluminium glazing sheet coverings/flashings - Exitex Ltd
Aluminium roof outlets - Marley Plumbing and Drainage
Aluminium roofing and cladding sheets - Gilmour Ecometal
Aluminium sheet coverings/ flashings - Freeman T R
Aluminium sheet coverings/ flashings - HW Architectural Ltd
Aluminium sheet coverings/ flashings - Reid John & Sons (Strucsteel) Ltd
Aluminum sheet covering and flashings - Roles Broderick Roofing Ltd
Anodised Aluminium sheet - Alanod Ltd
Insulated Roof & Wall Systems - Kingspan Ltd
Metal facia systems - Alumasc Exterior Building Products Ltd
Metal roofing and cladding - Ash & Lacy Building Products Ltd
Perforated panels in aluminium - Winther Browne & co Ltd
Powder coated aluminium - Dales Fabrications Ltd

H73 Copper strip/ sheet coverings/ flashings

Copper flashings and trims - Azimex Fabrications Ltd
Copper sheet coverings and flashings - Roles Broderick Roofing Ltd
Copper sheet coverings/ flashings - Freeman T R
Copper strip - Metra Non Ferrous Metals Ltd

H74 Zinc strip/ sheet coverings/ flashings

Zinc and titanium alloy sheets and strip - Metra Non Ferrous Metals Ltd
Zinc flashings and trims - Azimex Fabrications Ltd
Zinc roofing - Rheinzink UK
Zinc sheet coverings and flashings - Roles Broderick Roofing Ltd
Zinc sheet coverings/ flashings - Freeman T R

H75 Stainless steel strip/ sheet coverings/ flashings

Insulated Roof & Wall Systems - Kingspan Ltd
Profiled Sheet Roofing - Tata Steel Colors
Stainless steel flashings and trims - Azimex Fabrications Ltd
Stainless Steel Roofing Sheet - Aperam Stainless Services & Solutions UK Limited
Stainless steel sheet and strip - Metra Non Ferrous Metals Ltd
Stainless steel sheet coverings/ flashings - Freeman T R
Stainless steel strip/ sheet coverings/ fashings - Barlow Group
Stainless steel strip/sheet coverings and flashings - Roles Broderick Roofing Ltd
Stainless steel strip/sheet coverings/flashings - Britmet Tileform Limited
Stainless steel traditional longstrip roofing and cladding material - Outokumpu

H76 Fibre bitumen thermoplastic sheet coverings/ flashings

Roofing membranes - EBC UK Ltd
Roofing Membranes - Sika Ltd

H90 Tensile fabric coverings

Architectural Fabric Structures - TA Lucas Sails
Banners - TA Lucas Sails
Breathable membranes - Fillcrete Ltd
Covers - TA Lucas Sails
Fabric Canopies - Aura
Fabric Ceilings - Architen Landrell Associates Ltd
Rigging - TA Lucas Sails
Sails - TA Lucas Sails
Tensile structure material - Mermet U.K
Tensile Structures - Architen Landrell Associates Ltd
Vapour barriers - Fillcrete Ltd

H91 Thatch roofing

Stainless steel wire - Thatching Advisory Services Ltd
Staple gun and staples for stainless steel wire - Thatching Advisory Services Ltd
Thatch Roof Tiles - Africa Roofing Uk
Thatch roofing materials - Thatching Advisory Services Ltd

H92 Rainscreen cladding

Cellular PVC cladding - Celuform Building Products
Ceramic Rainscreens - Telling Architectural Ltd
Cladding - Blanc de Bierges
Cladding - Tata Steel Europe Limited
External decorative claddings - Cembrit Ltd
Facades - Hunter Douglas
Faience blocks - Shaws of Darwen
Hygrothermal performance testing, Water penetration, Pull testing, Impact testing, Dynamic wind loading and uplift testing - Lucideon
Metal roofing and cladding - Ash & Lacy Building Products Ltd
PVC Sheeting - Power Plastics Ltd
Rainscreed Cladding stone - Johnsons Wellfield Quarries Ltd
Rainscreen - Howard Evans Roofing and Cladding Ltd
Rainscreen cladding - Baggeridge Brick PLC
Rainscreen cladding - CEP Claddings Ltd
Rainscreen cladding - English Architectural Glazing Ltd
Rainscreen cladding - FGF Ltd
Rainscreen Cladding - Marley Eternit Ltd
Rainscreen Cladding - McMullen Facades
Rainscreen cladding - Panel Systems Ltd
Rainscreen Cladding - Schueco UK
Rainscreen cladding - Vetter UK
Rainscreen cladding - Vulcan Cladding Systems
Rainscreen facade systems - CGL Cometec Ltd
Rainscreen Panels - Rheinzink UK
Screening - Power Plastics Ltd
Terracotta and faience - design, procurement and installation of - Szerelmey Ltd
Terracotta blocks - Shaws of Darwen
Terracotta Cladding - Telling Architectural Ltd

J10 Specialist waterproof rendering

Basement tanking - Safeguard Europe Ltd
Cementitious and dry line membrane tanking - Dampcoursing Limited
Cementitious waterproof systems - Kingfisher Building Products Ltd
Conductive coatings - PermaRock Products Ltd
External rendering systems - PermaRock Products Ltd
External Wall Coatings - Belzona PolymericLtd
Institutions/Associations - Sprayed Concrete Association
Polymer modified renders - Ronacrete Ltd
renders Carter-Dal Micro Coat 701 (45) colours - Carter-Dal International
Sealants for waterproofing and damp proofing - Ancorite Surface Protection Ltd
Silicone waterproofing - Bluestar Silicones
Specialist waterproof render - Flexcrete Technologies Limited
Specialist waterproof rendering - Concrete Repairs Ltd
Specialist waterproof rendering - Don Construction Products Ltd
Specialist waterproof rendering - Mapei (UK) Ltd
Specialist waterproof rendering - Peter Cox Ltd
Specialist waterproof rendering - Specialist Building Products
Water proofing systems - Beton Construction Ltd
Waterproof renders(30 colours) - Laticrete International Inc. UK
Waterproofing - BASF plc, Construction Chemicals
Waterproofing products - IKO PLC
Waterproofing Products - Sealocrete PLA Ltd
Waterproofing systems - Grace De Neef UK

J20 Mastic asphalt tanking/ damp proofing

Asphalt - Tillicoutry Quarries Ltd
Damp proof membranes - Icopal Limited
Mastic asphalt tanking / damp proof membranes - Durable Contracts Ltd
Mastic asphalt tanking/ damp proofing - North Herts Asphalte (Roofing) Ltd
Tanking - Dyke Chemicals Ltd
Tanking - Newton John & Co Ltd
Tanking - Roofing Contractors (Cambridge) Ltd
Tanking and damp proof membranes - BriggsAmasco
Waterproofing membranes - Fosroc Ltd

J21 Mastic asphalt roofing/ insulation/ finishes

Colours for Asphalt - Procter Johnson & Co Ltd
Emergency Repair Compounds - Dyke Chemicals Ltd
Green Roof systems - Axter Ltd
GRP and aluminium roof edge trims and flashings - Pitchmastic PmB Ltd
Insulation - Rockwool Ltd
Mastic asphalt mixers - Bitmen Products Ltd
Mastic asphalt roofing and car parks - BriggsAmasco
Mastic asphalt roofing/ insulation/ finishes - BFRC Services Ltd
Mastic asphalt roofing/ insulation/ finishes - Thermica Ltd
Movement Joints - Movement Joints (UK)
Porous concrete promenade tiles - Spartan Promenade Tiles Ltd
Recycled glass tiles - Spear & Jackson Interntional Ltd
Roof insulation - North Herts Asphalte (Roofing) Ltd
Roof Waterproofing Systems - Topseal Systems Ltd
Roof waterproofing systems, colour solar reflective roof finishes - Dyke Chemicals Ltd
Roofing - North Herts Asphalte (Roofing) Ltd
Roofing - Roofing Contractors (Cambridge) Ltd
Roofing contractors - Durable Contracts Ltd
Roofing insulation boards - Firestone Building Products
Tarpauling - Power Plastics Ltd
Tarpaulins - Icopal Limited
Urethane foam industry trade association - BRITISH URETHANE FOAM CONTRACTORS ASSOCIATION

J22 Proprietary roof decking with asphalt finish

Bitumen boilers - Bitmen Products Ltd
Colours for Asphalt - Procter Johnson & Co Ltd
Flat roof decking - Isocrete Floor Screeds Ltd
Pre coated chipspreaders - Bitmen Products Ltd
Proprietary roof decking with asphalt finish - Durable Contracts Ltd
Roof decking with asphalt finish - BriggsAmasco
Roof tile underlays - Icopal Limited
Roofing contractor - Evans Howard Roofing Ltd
Roofing membranes - Onduline Building Products Ltd
Torch on roofing - IKO PLC

J30 Liquid applied tanking/ damp proofing

Adhesives - Carter-Dal International
Adhesives - Laticrete International Inc. UK
Balcony waterproofing - Sika Liquid Plastics Limited
Basement tanking - Safeguard Europe Ltd
Basement water proofing - Timberwise (UK) Ltd
Basement waterproofing - Proten Services Ltd
Damp proof membrane - Ardex UK Ltd
Damp Proofing - Proten Services Ltd
Damp proofing - RFA-Tech Ltd
DPC tanking systems - IKO PLC
Flashings - Dyke Chemicals Ltd
Heldite Jointing Compound - Heldite Ltd
Liquid applied coating - Uretech TKM - Star Uretech Ltd
Liquid applied damp proof membranes - Kingfisher Building Products Ltd
Liquid applied tanking / damp proofing - Bondaglass Voss Ltd
Liquid Applied Tanking Materials - Belzona PolymericLtd
Liquid applied tanking/ damp proof membranes - Cross-Guard International Ltd
Liquid applied tanking/ damp proof membrane - Colebrand Ltd
Liquid applied tanking/ damp proof membranes - Don Construction Products Ltd

J30 Liquid applied tanking/ damp proofing (con't)

Liquid applied tanking/ damp proof membranes - Specialist Building Products
Liquid applied tanking/ damp proofing - Flexcrete Technologies Limited
Liquid applied tanking/ damp proofing - Tremco Illbruck Ltd
Liquid applied tanking/damp proof membranes - Trade Sealants Ltd
Polymer modified coatings - Ronacrete Ltd
Structural waterproofing - Alumasc Exterior Building Products Ltd
Structural waterproofing - Grace Construction Products Ltd
Tanking - RFA-Tech Ltd
Waterproof membranes - Carter-Dal International
Waterproof membranes - Laticrete International Inc. UK
Waterproofing membranes - Kenyon Group Ltd

J31 Liquid applied waterproof roof coatings

Asbsetos roof coatings - Sika Liquid Plastics Limited
Bituminous roofing - TRC (Midlands) Ltd
Cold Bonded Built-up Roofs - Sika Liquid Plastics Limited
Damp-proof membranes - Dyke Chemicals Ltd
Epoxy Resins & Adhesives - Wessex Resins & Adhesives Ltd
Fibreglass roofing - PermaSeal Ltd
Green roofs - Sika Liquid Plastics Limited
High performance roof coatings - Intergrated Polymer Systems (UK) Ltd
Inverted Roofs - Sika Liquid Plastics Limited
Liguid applied roof waterproof coatings - Flexcrete Technologies Limited
Liquid applied systems - Mells Roofing Ltd
Liquid applied waterproof coating - Brimar Plastics Ltd
Liquid applied waterproof coatings - BFRC Services Ltd
Liquid applied waterproof coatings - Brewer C & Sons Ltd
Liquid applied waterproof coatings - BriggsAmasco
Liquid applied waterproof coatings - Colebrand Ltd
Liquid applied waterproof coatings - Concrete Repairs Ltd
Liquid applied waterproof coatings - Cross-Guard International Ltd
Liquid applied waterproof coatings - Don Construction Products Ltd
Liquid applied waterproof coatings - Precolor Sales Ltd
Liquid applied waterproof coatings - Tor Coatings Ltd
liquid applied waterproof coatings - Trade Sealants Ltd
Liquid applied waterproof roof coatings - Hyflex Roofing
Liquid applied waterproof roof coatings - Sika Liquid Plastics Limited
Liquid-applied roofing and waterproofing - Granflex (Roofing) Ltd
Liuquid Applied Roof Waterproofing - Belzona PolymericLtd
Mechanically Fixed Insulation with Cold Bonded Built-up Roofs - Sika Liquid Plastics Limited
PmB Structural Waterproofing Systems - Pitchmastic PmB Ltd
Roof Renovation Paints - Kingfisher Building Products Ltd
Roof repair and protection - Sealocrete PLA Ltd
Roof repair membranes - Kingfisher Building Products Ltd
Roof repair systems - Williamson T & R Ltd
Roof Waterproofing systems - Axter Ltd
Roofing - Avonside Roofing Group T/A Letchworth Roofing
Root resistant cold liquid applied green roof coating - Sika Liquid Plastics Limited
Structural roof waterproofing - Alumasc Exterior Building Products Ltd
Waterproofing membranes - Conren Ltd
Waterproofing Membranes - Kemper System Ltd

J32 Sprayed vapour control layers

Radon and methane gas barriers - Icopal Limited
Sprayer vapour barriers - Trade Sealants Ltd
Vapour barriers and breather membranes - Icopal Limited

J33 In situ glassfibre reinforced plastics

Fibertex geotextiles - Tex Engineering Ltd
In situ glass reinforced plastic - Specialist Building Products
In situ GRP - Precolor Sales Ltd
In-situ glass reinforced plastics - Brimar Plastics Ltd
In-situ GRP - BriggsAmasco

J40 Flexible sheet tanking/ damp proofing

Basement tanking - Safeguard Europe Ltd
Basement waterproofing - Proten Services Ltd
Condensation control - Peter Cox Ltd
Damp proof courses - Cavity Trays Ltd
Damp proof membranes - Isocrete Floor Screeds Ltd
Damp proofing - Proten Services Ltd
Damp proofing - RFA-Tech Ltd
Damp-Proofing & Tanking Products - Haynes Manufacturing (UK) Ltd
Damproof membranes - Newton John & Co Ltd
Flexible sheet tanking / damp proof membranes - Durable Contracts Ltd
Flexible sheet tanking, damp proof membanes - Terminix Property Services
Flexible sheet tanking/ damp proof membranes - BFRC Services Ltd
Flexible sheet tanking/ damp proof membranes - John Davidson (Pipes) Ltd
Flexible sheet tanking/ damp proof membranes - Specialist Building Products
Flexible Sheet Tanking/Waterproofing - Peter Cox Ltd
Gas barrier membranes - Visqueen Building Products
Gas barrier membranes and pond liners - Geosynthetic Technology Ltd
Membranes - Cavity Trays Ltd
Polythene Building Films and DPC - Ensor Building Products Ltd
Polythene dampproof membranes - Visqueen Building Products
Roofing Membranes - Sarnafil Ltd
Single Ply Roofing Systems - Sarnafil Ltd
Tanking - RFA-Tech Ltd
Waterproof expansion joints - Radflex Contract Services Ltd
Waterproof membranes - Grace Construction Products Ltd
Waterproofing - Cordek Ltd
Waterproofing membranes - SCP Concrete Sealing Technology Ltd
Waterproofing sheets - ABG Ltd
Waterproofing: Breather Membranes - Icopal Limited
Waterstops - Fosroc Ltd

J41 Built up felt roof coverings

Built up felt roof covering - Brewer C & Sons Ltd
Built up felt roof coverings - Durable Contracts Ltd
Built up felt roof coverings - PermaSeal Ltd
Built up felt roof coverings - TRC (Midlands) Ltd
Built up felt roofing - BriggsAmasco
Built Up Felt Roofing - North Herts Asphalte (Roofing) Ltd
Built up felt roofing - Roofing Contractors (Cambridge) Ltd
Built-up felt roofing - Granflex (Roofing) Ltd
Copernit - Ensor Building Products Ltd
Felting and single ply - Avonside Roofing Group T/A Letchworth Roofing
Flat roofing built up felt roofing - Chesterfelt Ltd
High performance roofing - Alumasc Exterior Building Products Ltd
Membrane waterproofing - BFRC Services Ltd
Movement Joints - Movement Joints (UK)
Pitched roofing - Mells Roofing Ltd
Porous concrete promenade tiles - Spartan Promenade Tiles Ltd
Roof coverings - Howard Evans Roofing and Cladding Ltd
Roof membranes - Axter Ltd
Roofing Accessories - Ubbink (UK) Ltd
Roofing compound - Bartoline Ltd
Roofing contractor - Evans Howard Roofing Ltd
Roofing products - Cavity Trays Ltd
Roofing products - Klober Ltd
Structural Protection Membranes - Uretech TKM - Star Uretech Ltd
Vapour membranes - Klober Ltd
Weather protection sheeting and sheeting - Power Plastics Ltd

J42 Single layer polymeric roof coverings

Conservatory roof systems, vents and glazing bars - Newdawn & Sun Ltd
Grren Roof Systems - AAC Waterproofing Ltd
PVC single-ply roofing - Granflex (Roofing) Ltd
Roof Membranes - AAC Waterproofing Ltd
Roofing - Avonside Roofing Group T/A Letchworth Roofing
RubberCover - Ensor Building Products Ltd
Single layer plastic roof coverings - BFRC Services Ltd
Single layer polymeric roof coverings - Durable Contracts Ltd
Single layer polymeric roof coverings - PermaSeal Ltd
Single Layer Polymeric Sheet Roof Coverings - Safeguard Europe Ltd
Single layer PVC and CPE membrane roofing systems - RENOLIT Cramlington Limited
Single ply polymetric membranes - Roofing Contractors (Cambridge) Ltd
Single ply roofing - Firestone Building Products
Single ply roofing - ICB (International Construction Bureau) Ltd
Single Ply roofing - Mells Roofing Ltd
Single Ply Roofing - Sika Ltd
Single ply systems - BriggsAmasco
Water repellent solutions - Dyke Chemicals Ltd
Waterproofing systems - Beton Construction Ltd

J43 Proprietary roof decking with felt finish

Proprietary roof decking with felt finish - Durable Contracts Ltd
Roof decking with felt finish - BriggsAmasco
Roofing - Avonside Roofing Group T/A Letchworth Roofing
Roofing ventilators - Airflow (Nicoll Ventilators) Ltd

J44 Sheet linings for pools/ lakes/waterways

Gas barrier membranes and pond liners - Geosynthetic Technology Ltd
Geomembrane lining supply and installation - Landline Ltd
Pond / landfill lining systems - ABG Ltd
Pool liners - Firestone Building Products
Single Ply Roofing Systems - Sarnafil Ltd
Waterproofing systems - Beton Construction Ltd

K10 Plasterboard dry linings/ partitions/ceilings

Access panels ceilings - Panel & Louvre Co Ltd
Building Boards - Cembrit Ltd
Ceilings - Hunter Douglas
Ceilings - Plasterboard, Suspended - Gridpart Interiors Ltd
Clean Room Partitioning - TROAX UK Ltd
Dry Lining - Covers Timber & Builders Merchants
Dry lining systems - British Gypsum Ltd
Dry Linings - Skanda Acoustics Limited
Dry wall partitions - Optima
Drywall systems - Knauf UK
Drywall systems - Siniat Ltd
Fire resistant panels - Promat UK Ltd
Fire stopping and barriers - Firebarrier Services Ltd
Full fit out, refurbisment and partitioning service - Neslo Interiors
GRG (glass reinforced gypsum) products - design, manufacture and installation - Gillespie (UK) Ltd
Interior wall linings - Trespa UK Ltd
Linings - FGF Ltd
Mineral fibre and vermiculite board - Cryotherm Insulation Ltd
Moveable walls - Becker (SLIDING PARTITIONS) Ltd
Partitioning - Hatmet Limited
Partitioning - Welco
Partitions - Fastwall Ltd
Partitions - Glazed, Metal stud, Aluminium, Fire barriers - Gridpart Interiors Ltd
Plasterboard - British Gypsum Ltd
Plasterboard accessories - British Gypsum Ltd
Plasterboard acoustic wall products - Trim Acoustics
Plasterboard dry lining - Harris & Bailey Ltd
Plasterboard dry lining/ partitions/ ceilings - Logic Office Contracts
Plasterboards and drywall Accessories - Knauf Drywall
Self-adhesive Drywall Joint Tape (EnsorTape) - Ensor Building Products Ltd
Steel Partition - TROAX UK Ltd
Supplier of Ceilings, Partitioning, Dry lining and Interior Building products - Nevill Long Limited
Trade Association - Association of Specialist Fire Protection (ASFP)

K11 Rigid sheet flooring/ sheathing/ decking/sarking/ linings/casings

Access flooring - Nevill Long Limited
Acoustic Insulation - Hush Acoustics
Blockboard - Wood International Agency Ltd
Building boards - Finnish Fibreboard (UK) Ltd
Cladding - James Donaldson Timber
Column and beam encasement systems - Knauf Drywall
Column Cladding Systems - Jali Ltd
Cutting, drilling and machining - Boardcraft Ltd
Drylining material - Avanti Systems
Drywall systems - Knauf UK
Film Faced plywood - Wood International Agency Ltd
Fire resistant panels - Promat UK Ltd
Floating floors - Designed for Sound Ltd
Flooring accessories - Atkinson & Kirby Ltd
Flooring systems - Knauf UK
Fsc certified plywood - Wood International Agency Ltd
Glass partitions - Style Door Systems Ltd
Hardboard - Panel Agency Limited
Hardboard - Wood International Agency Ltd
Hardwood and softwood - Boardcraft Ltd
Hardwood plywood - Wood International Agency Ltd
Insulating laminates - Knauf Drywall
Insulation Boards - Skanda Acoustics Limited
Internal Walls & Partitions - Avanti Systems
Internal walls and partitions - Sunfold Systems Ltd
Laminate joinery - Decra Ltd
Laminates and MDF's - Kronospan Ltd
Lightweight composite boards and panels - QK Honeycomb Products Ltd
Linigs and castings - Cox Long Ltd
MDF sheets - Willamette Europe Ltd
Medium Density Fibrboard (MDF) - Meyer Timber Limited
Mezanine floors - Adex Interiors for industry Ltd
Moveable internal walls and partitions - Style Door Systems Ltd
Oriented Strand Board (OSB) - Meyer Timber Limited
OSB - Kronospan Ltd
OSB - Wood International Agency Ltd
Panel products - James Latham plc
Perforated MDF - Associated Perforators & Weavers Ltd
Perforated Perforations - Associated Perforators & Weavers Ltd
Perforated Products - APW Ltd
Pipe and column casings - Alumasc Interior Building Products Limited
Plain and melamine faced chipboard - Kronospan Ltd
Plain and melamine faced flooring - Kronospan Ltd
Plain, melamine faced and lacquered MDF - Kronospan Ltd
Plywood-structure + non-structure - Meyer Timber Limited
Polyisocyanurate insulation board manufacture - Celotex Ltd
rigid sheet flooring, sheathing, decking, sarking - Cox Long Ltd
Rigid sheet flooring/ sheathing/ decking/sarking/ linings/casings - Cembrit Ltd
Rigid sheet flooring/ sheathing/ decking/sarking/ linings/casings - Skanda Acoustics
Semi finished plastics and building products - Amari Plastics Plc
Sheet materials - Cox Long Ltd
Sheet Materials - Walker Timber Ltd
Softwood plywood - Wood International Agency Ltd
Sprays fire proof - Firebarrier Services Ltd
Stainless Steel Chequered plate - Aperam Stainless Services & Solutions UK Limited
Stainless steel flooring - Wincro Metal Industries Ltd
Thermal insulated panels - Panel Systems Ltd
Undercarrige Systems - Atkinson & Kirby Ltd
Underlay - Atkinson & Kirby Ltd
Veneered Boards - Meyer Timber Limited
Wet wall and showerwall panels, wet area wall lining and ceiling panels - Mermaid Panels Ltd
Wood Partical Board - Meyer Timber Limited
Woodbased panels - Egger (UK) Ltd
Woodfibre Underlays - Panel Agency Limited

K12 Under purlin/ Inside rail panel linings

Plywood, hard, fibre, chip and particle boards - Boardcraft Ltd
Under Purlin Linings - Skanda Acoustics Limited
Under purlin/Inside rail panel linings - Skanda Acoustics

K13 Rigid sheet fine linings/ panelling

Access panels walls - Panel & Louvre Co Ltd
Acoustic Baffles, Barriers - Acousticabs Industrial Noise Control Ltd
Acoustic Ceiling & wall panels - Skanda Acoustics Limited
Acoustic Ceiling Panels, Wall Panels - including perforated timber/metal - Acousticabs Industrial Noise Control Ltd
Acoustic Curtains, Attenuators, Enclosures, Panels and Louvers - Acousticabs Industrial Noise Control Ltd
Acoustic flooring and acoustic wall products - Trim Acoustics
Acoustic wall panels - Skanda Acoustics
Aquapanel Tile Backing Board - Knauf Drywall
Bespoke infil panels - Halmark Panels Ltd
Bespoke joinery - Howard Bros Joinery Ltd
Broadcast and Audio Studios - IAC Acoustics
Carved cornice, frieze and embellishments - Oakleaf Reproductions Ltd
Cladding panel - Panel Systems Ltd
Cladding panels - CEP Claddings Ltd
Composite infil pannels - VBH (GB) Ltd
Cubicle systems - Formica Ltd
Ducting systems - Decra Ltd
Facing laminates - Wilsonart Ltd
Faux book spines, mirrors and frames - Oakleaf Reproductions Ltd
Full fit out, refurbisment and partitioning service - Neslo Interiors
High pressure decorative laminates - Abet Ltd
Hygienic and railway arch lining materials - Rockwell Sheet Sales Ltd
Hygienic Wall & Ceiling Linings - Interclad (UK) Ltd
Hygienic wall linings - Filon Products Ltd
Hygienic walls and ceiling cladding systems - Advanced Hygienic Contracting Ltd
Impact resistant Acoustic Wall Panels - Knauf AMF Ceilings Ltd
Laminates - Meyer Timber Limited
Linenfold wall panelling, riven oak wall panelling - Oakleaf Reproductions Ltd
Melamine faced chipboard - Deralam Laminates Ltd
Metal + melamine laminates - Deralam Laminates Ltd
Multi-wall roofing products - Amari Plastics Plc
Panel Products - Panel Agency Limited
Panelled interiors - Hallidays UK Ltd
Partitions - Optima
Plywood, hard, fibre, chip and particle boards - Boardcraft Ltd
Real wood veneer laminates - Deralam Laminates Ltd
Rigid sheet fine linings/ panelling - Cembrit Ltd
Sheet materials - IJK Timber Group Ltd
Slat board - Smart F & G (Shopfittings)Ltd
Softwood - James Latham plc
Structural wall panels - Marshalls Mono Ltd
Timber decorative wall panels - Vicaima Ltd
Veneered boards - Atkinson & Kirby Ltd
Wall Access Panels - The Access Panel Company
Wall lining systems - Knauf Drywall
Wall Panels and Hygienic Wall Linings - Aaztec Cubicles
Wet wall and showerwall panels, wet area wall lining and ceiling panels - Mermaid Panels Ltd
Wood Based Panel Products - Meyer Timber Limited

K20 Timber board flooring/ decking/sarking/linings/casings

Acoustic Baffles, Barrierss - Acousticabs Industrial Noise Control Ltd
Acoustic Ceiling Panels, Wall Panels - including perforated timber/metal - Acousticabs Industrial Noise Control Ltd
Acoustic Curtains, Attenuators, Enclosures, Panels and Louvers - Acousticabs Industrial Noise Control Ltd
Acoustic suspended ceilings - Skanda Acoustics
Anti-slip decking - Brash John & Co Ltd
Building boards - Finnish Fibreboard (UK) Ltd
Cane woven panels - Winther Browne & co Ltd

K20 Timber board flooring/ decking/sarking/linings/casings (con't)

Decking - Anthony de Grey Trellises + Garden Lighting
Decking - James Donaldson Timber
Decorative laminated flooring - Egger (UK) Ltd
Densified wood and industrial flooring - Permali Deho Ltd
Duct walling - LSA Projects Ltd
Fire protection boards - Knauf Drywall
First and second fix timber - Brewer T. & Co Ltd
Floor ducting - Alumasc Interior Building Products Limited
Flooring Panels - Norbord
Full fit out, refurbisment and partitioning service - Neslo Interiors
Hardwood flooring - W.L. West & Sons Ltd
MDF - Metsä Wood
Mezzanine Flooring - Sperrin Metal Products Ltd
Non-slip/Anti-slip Decking in hardwood or softwood - CTS Bridges Ltd
Perimeter casing - Alumasc Interior Building Products Limited
Pipe boxing - Alumasc Interior Building Products Limited
Plywood, hard, fibre, chip and particle boards - Boardcraft Ltd
Roofing Boards - Norbord
Sanding and Sealing wood floors - Sportsmark Group Ltd
Sheet Materials - Moray Timber Ltd
Softboard - Panel Agency Limited
Timber board flooring - Atkinson & Kirby Ltd
Timber board flooring/decking - Magnet Ltd
Timber Board flooring/decking/sarking/linings/casings - Anderson C F & Son Ltd
Timber Boarding etc - Challenge Fencing Ltd
Timber Boards - Norbord
Timber cladding, mouldings and flooring - James Latham plc
Timber decking - Wyckham Blackwell Ltd
Timber Suppliers - Crowthorne Fencing

K21 Timber strip/ board fine flooring/ linings

Dance floor - Felix Design Ltd
Hardwood flooring - Atkinson & Kirby Ltd
Laminates - Egger (UK) Ltd
Movement Joints - Movement Joints (UK)
Pannel products - IJK Timber Group Ltd
Perforated panels in hardboard and MDF - Winther Browne & co Ltd
Sheet materials - Brewer T. & Co Ltd
Solid hardwood flooring - Junckers Limited
Timber decorative wall panels - Vicaima Ltd
Timber strip/board flooring/lining - Magnet Ltd
Timber strips/ board fine flooring / linings - Barlow Group
Timber Suppliers - Crowthorne Fencing

K30 Panel partitions

Access panels - manufacture and installation - Foster W H & Sons Ltd
Acoustic folding partitions - Parthos UK Ltd
Acoustic Movable Walls - Coburn Sliding Systems Ltd
Acoustic movable walls - Parthos UK Ltd
Acoustic Moveable walls/partitions - Hufcor UK Ltd
Aluminium extrusions/ sections - Seco Aluminium Ltd
Brushstrip sealing for cable management - Kleeneze Sealtech Ltd
Changing cubicles - unframed - Helmsman
Concertina Partitions - Coburn Sliding Systems Ltd
Concertina Partitions - Dividers Modernfold Ltd
Demountable partitioning - Internal Partitions Systems
Demountable partitions - HEM Interiors Group Ltd
Demountable partitions - Komfort Workspace PLC
Demountable partitions - Norwood Partition Systems Ltd
Demountable partitions and associated electrical works - Opus 4 Ltd
Drywall systems - Knauf UK
Duct panelling and frames - Amwell Systems Ltd
Duct Panels - Aaztec Cubicles
Fire rated partitions - Avanti Systems
Fire resistant panels - Promat UK Ltd
Firewalls - Composite Panel Services Ltd
Frameless glazed partitions - Avanti Systems
Full fit out, refurbisment and partitioning service - Neslo Interiors
Industrial partitioning - Barton Storage Systems Ltd
Insulated wall panels - Composite Panel Services Ltd
Interior contractors - Gifford Grant Ltd
Internal Panels - Paroc Panel Systems Uk Ltd
IPS Systems - Aaztec Cubicles
Laminate wall claddings - T B S Fabrications
Mesh partitioning - Barton Storage Systems Ltd
Mobile partitions - Dividers Modernfold Ltd
Office partitioning - Construction Group UK Ltd
Operable wall systems - Monowa Manufacturing (UK) Ltd
Panel partitions - Logic Office Contracts
Panel partitions - Optima
Partitioning - Hatmet Limited
Partitioning - New Forest Ceilings Ltd
Partitioning - Newey Ceilings
Partitioning - Optima
Partitioning and wall storage systems - Fastwall Ltd
Partitioning systems - Knauf Drywall
Partitioning systems - Nevill Long Limited
Partitions - Adex Interiors for industry Ltd
Partitions - Firco Ltd
Partitions - ITS Ceilings & Partitions
Partitions - Shaylor Group Plc
Partitions - SIG Interiors
Partitions - Glazed, Metal stud, Aluminium, Fire barriers - Gridpart Interiors Ltd
Plywood, hard, fibre, chip and particle boards - Boardcraft Ltd
Relocatable partitioning - Clestra Limited
Relocateable partitions - Fastwall Ltd
Sliding folding partition manufacturer - Building Additions Ltd
Sliding partitions - Becker (SLIDING PARTITIONS) Ltd
Solid surface products - Foster W H & Sons Ltd
Steel caging/ partitioning - Bradbury Group Ltd
Wall panels - Foster W H & Sons Ltd
Wash hand basin units - manufacture and installation - Foster W H & Sons Ltd

K32 Panel cubicles

Changing cubicles - Detlef Muller Ltd
Changing cubicles - framed - Helmsman
Cubicle systems - Panel Systems Ltd
Cubicles - Nevill Long Limited
Cubicles - Twyford Bushboard
Cubicles - manufacture and installation - Foster W H & Sons Ltd
Cubicles, Duct Panels and Vanity Units - Aaztec Cubicles
Cubicles/lockers - Trespa UK Ltd
Duct panelling and frames - Amwell Systems Ltd
Framed panel cubical partitions - HEM Interiors Group Ltd
Framed panel cubicles - T B S Fabrications
Full fit out, refurbisment and partitioning service - Neslo Interiors
Panel cubicals - Mermaid Panels Ltd
Panel cubicles - FGF Ltd
Panel Cubicles - Grant Westfield Ltd
Plywood, hard, fibre, chip and particle boards - Boardcraft Ltd
Purpose made doors, toilet cubicles and washroom systems - Moffett Thallon & Co Ltd
Steel caging/ partitioning - Bradbury Group Ltd
Toilet and shower cubicles - Decra Ltd
Toilet, Shower, Washroom and Changing cubicles - Aaztec Cubicles
Toilet/Shower/Changing Cubicles - Dunhams of Norwich
Washroom cubicle systems-Twyford Bushboard - Twyford Bushboard
WC cubicles - Detlef Muller Ltd
WC cubicles and washroom systems - Bridgman IBC Ltd

K33 Concrete/ Terrazzo partitions

Specialist contractors in mosaic and terrazzo - Alpha Mosaic & Terrazzo Ltd

K40 Demountable suspended ceilings

Acoustic ceilings - STO Ltd
Ceiling Access Panels - Howe Green Ltd
Ceiling grid systems - Hadley Industries Plc
Ceiling systems - Knauf AMF Ceilings Ltd
Ceiling systems - OWA UK Ltd
Ceiling tiles - Cep Ceilings Ltd
Ceilings - Adex Interiors for industry Ltd
Ceilings - Hunter Douglas
Ceilings - Plasterboard, Suspended - Gridpart Interiors Ltd
Ceilings (suspended) - Firco Ltd
Designer ceiling range systems - Knauf AMF Ceilings Ltd
Fire rated ceiling systems - Fire Protection Ltd
Fire resistant ceilings - Knauf AMF Ceilings Ltd
Full fit out, refurbisment and partitioning service - Neslo Interiors
Grid, Mineral, metal, wood and soft fibre suspended ceilings - Armstrong World Industries Ltd
Kitchen extraction systems - Britannia Kitchen Ventilation Ltd
Metal and 3D ceiling systems - Knauf AMF Ceilings Ltd
Metal ceiling products - OWA UK Ltd
Metal panels/tiles - Burgess Architectural Products Ltd
Mineral wool ceiling products - OWA UK Ltd
Performance ceilings - OWA UK Ltd
Plasterboard dry linings/partitions/ceilings - Skanda Acoustics
Specialist designer ceilings - OWA UK Ltd
Supply and installation of suspended ceilings - ITS Ceilings & Partitions
Suspended ceiling brackets and accessories - PRE-MET Ltd
Suspended ceiling systems - Rockfon Ltd
Suspended Ceilings - Above All
Suspended ceilings - Hatmet Limited
Suspended ceilings - HEM Interiors Group Ltd
Suspended ceilings - HT Martingale Ltd
Suspended ceilings - Internal Partitions Systems
Suspended Ceilings - Knauf UK
Suspended ceilings - Nevill Long Limited
Suspended ceilings - New Forest Ceilings Ltd
Suspended ceilings - Newey Ceilings
Suspended ceilings - Opus 4 Ltd
Suspended ceilings - Shaylor Group Plc
Suspended Ceilings - Skanda Acoustics Limited
Suspended ceilings - Walls & Ceilings International Ltd
Suspended ceilings, shaftwall Systems - Knauf Drywall

K41 Raised access floors

Access flooring - Moseley GRP Product (Division of Moseley Rubber Co Ltd)
Access Flooring Products - Metalfloor UK Ltd
Access panels floors - Panel & Louvre Co Ltd
Access Ramps - Enable Access
Adhesives for access flooring - PolyPed &: Ultra High Performance - Uretech PA1, High Performance - Uretech PA2, General Purpose - Uretech TG1, Acoustic - Uretech AA2 - Star Uretech Ltd
Cable management - Interface Europe Ltd
Flooring and access - Wincro Metal Industries Ltd
Full fit out, refurbisment and partitioning service - Neslo Interiors
Mezzanine Floors - Davicon
Panel and pedestal testing, Air plenum testing, On-site and laboratory testing - Lucideon
Plywood, hard, fibre, chip and particle boards - Boardcraft Ltd
Raised Access Flooring - SIG Interiors
Raised access flooring systems - Knauf UK
Raised access floors - Adex Interiors for industry Ltd
Raised access floors - Devar Premier Flooring
Raised Access Floors - Hatmet Limited
Raised access floors - HEM Interiors Group Ltd
Raised access floors - Interface Europe Ltd
Raised access floors - Kingspan Access Floors Ltd
Raised access floors - New Forest Ceilings Ltd
Raised access floors - Raised Floor Systems Ltd
Raised access floors - Redman Fisher Engineering
Raised Access Floors - Shaylor Group Plc
Raised access floors - Trim Acoustics
Raised access floors - Workspace Technology Ltd
Raised Flooring and accessories - Metalfloor UK Ltd
Secondary Flooring - Safety Works & Solutions Ltd
Secondary Floors - Ecotile Flooring Ltd
Secondary Floors - RDA Projects Ltd

L10 Windows/ Rooflights/ Screens/ Louvres

Acoustic Louvres - IAC Acoustics
Aluminiu clad windows - Broxwood (scotland Ltd)
Aluminium - Olsen Doors & Windows Ltd
Aluminium - Sapa Building Systems Ltd
Aluminium clad timber - Olsen Doors & Windows Ltd
Aluminium clad timber wibdows - Benlowe Group Limited
Aluminium extrusions/ sections - Seco Aluminium Ltd
Aluminium louvres - WP Metals Ltd
Aluminium windows - Andersen/ Black Millwork
Aluminium windows - Comar Architectural Aluminuim Systems
Aluminium windows - KCW Comercial Windows Ltd
Aluminium windows - Midland Alloy Ltd
Aluminium windows - Prima Systems (South East) Ltd
Aluminium windows - Sapa Building Systems Ltd
Aluminium Windows and Doors - FineLine
Aluminium.timber composite windows - Olsen Doors & Windows Ltd
aluminium/hybrid curtain walling, windows and doors - Senior Aluminium Systems plc
Anit-shatter films for bomb-blast applications - Northgate Solar Controls
Anti-Reflective Glass - SCHOTT UK Ltd
Armour plate door assemblies - Ide T & W Ltd
Atria - Lonsdale Metal Company
Automatic opening roof vents - National Domelight Company
Bespoke joinery - AS Newbould Ltd
Bespoke joinery - Howard Bros Joinery Ltd
Bespoke Timber Windows - Scotts of Thrapston Ltd
Bird screen - Timberwise (UK) Ltd
Blind Systems - Silent Gliss Ltd
Blinds - Hunter Douglas
Blinds - Luxaflex®
Blinds and Curtain Track Systems - Goelst
Brise Soleil & Security - Lang+Fulton
Brise Soleil shades - Hunter Douglas
Brush strip window seals - Kleeneze Sealtech Ltd
Building hinges - Nico Manufacturing Ltd
Capping bars - Exitex Ltd
Casement - Blair Joinery
Commercial windows - KCW Comercial Windows Ltd
Composite aluminium windows - Scandanavian Timber
Composite aluminum/wood windows & doors - VELFAC Ltd
Composite windows - Crittall Steel Windows Ltd
Conservatories - Smith & Choyce Ltd
Conservatory roofing, atria, canopies and rooflights - Ultraframe PLC
Domes and Pyramids - National Domelight Company
Door Canopies - Canopy Products Ltd
Dormer Windows/Louvres - Duroy Fibreglass Mouldings Ltd
Double glazed roof windows - Klober Ltd
Double glazed units - Ledlite Glass (Southend) Ltd
Entrance Canopies - Duroy Fibreglass Mouldings Ltd
Extruded aluminium composite windows - Smart Systems Ltd
Extruded aluminum windows - Smart Systems Ltd
Extruded shopfronts and entrance screens - Smart Systems Ltd
Extruded UPVC windows - Smart Systems Ltd
Extruded UPVC windows - Veka Plc
Filters/Screening - Associated Perforators & Weavers Ltd
Fire resistant glazing systems - Promat UK Ltd
Fly Screens - Graepel Perforators Ltd
G. R. P. Canopies - Solair Ltd
Glass & window supplies - Spectra Glaze Services
Glazed malls - Lonsdale Metal Company
Grillers/Panels - Associated Perforators & Weavers Ltd
Grilles, Registers & Diffusers - Air Diffusion Ltd
GRP Canopies - Production Glassfibre
GRP Dormers - Production Glassfibre
GRP louvers - APA
High performance windows - Sashless Window Co Ltd
Historic Rooflights - Clement Steel Windows
Industrial and Commercial Insect Screens - The Flyscreen Company Ltd
Industrial grade plastic framed roof windows for slate/tile roofs - Cox Building Products Ltd
Industrial rooflights - Kenilworth Daylight Centre Ltd
Insect screens - President Blinds Ltd
Insect Screens - The Flyscreen Company Ltd
Internal and external doors - Jeld-Wen UK Ltd
Lantern lights - Lonsdale Metal Company
Light Pipes - Sola Skylights
Louvre blind material - Mermet U.K
Louvres - Hallmark Blinds Ltd
Louvres - McKenzie-Martin Ltd
Manual control window opening systems - Clearline Architectural
Mess Screens - Cadisch MDA Ltd
Metal and glass louvers grilles, diffusers and dampers - Naco
Metal framed double glazed windows - Nova Group Ltd
Metal windows - Anglian Building Products
Metal windows - Olivand Metal Windows Ltd
Metal Windows - Thanet Ware Ltd
Metal windows - The Cotswold Casement Company
Metal windows/ rooflights/ screens/ louvers - Drawn Metal Ltd
Metal windows/ rooflights/ screens/ louvers - HW Architectural Ltd
Metal windows/ rooflights/ screens/ louvres - Reid John & Sons (Strucsteel) Ltd
North lights - Lonsdale Metal Company
Perforated Metal - Associated Perforators & Weavers Ltd
Perforated Products - Associated Perforators & Weavers Ltd
Plastic framed double glazed windows - Nova Group Ltd
Plastic profiled sheet cladding and roofing - Ariel Plastics Ltd
Plastic windows - Andersen/ Black Millwork
Plastic windows - Anglian Building Products
PVC extrusions - Nenplas
PVC roofline products - L. B. Plastics Ltd
PVC window and door systems - L. B. Plastics Ltd
PVC window profiles - Profile 22 Systems Ltd
PVC windows - Prima Systems (South East) Ltd
PVCU extrusions for window systems - Spectus Systems
PVC-U vertical sliding windows - Masterframe Windows Ltd
PVCu windows - Interframe Ltd
PVCu windows - KCW Comercial Windows Ltd
PVCu windows - Plastal (SBP Ltd)
PVC-U Windows - Magnet Ltd
PVC-U windows, doors and conservatories fabricated and installed - FineLine
Replacement window service - Spectra Glaze Services
Roller blind material - Mermet U.K
Roof lights - Lonsdale Metal Company
Roof Lights - Sola Skylights
Roof vents, ventilators - Lonsdale Metal Company
Roof windows - SIG Roofing
Roof windows - VELUX Co Ltd, The
Rooflights - Astrofade Ltd
Rooflights - Axter Ltd
Rooflights - Designer Construction Ltd
Rooflights - Duplus Architectural Systems Ltd
Rooflights - Kenilworth Daylight Centre Ltd
Rooflights - McKenzie-Martin Ltd
Rooflights - National Domelight Company
Rooflights - NuLite Ltd
Rooflights - Ubbink (UK) Ltd
Rooflights - profiled and structured in GRP, PVC and polycarbonate - Brett Martin Ltd
Rooflights (Newdome) - Novaglaze Limited
Rooflights, skylights, covered walkways, barrel vaults, deadlights, aluminium windows - Duplus Architectural Systems Ltd
Rooflights\screen\louvres - Zled (DPC) Co Ltd
Room Darkening Systems - CBS (Curtain and Blind Specialists Ltd)
Ropes and sashcords - Marlow Ropes Ltd
Sadh & case - Blair Joinery
Sarnalite rooflights - Sarnafil Ltd
Sash balances - Garador Ltd
Sash Window Renovation - Ventrolla Ltd
Sash windows - Masterframe Windows Ltd
Screening Pannels - Winther Browne & co Ltd
Screens - Enfield Speciality Doors
Screens - Komfort Workspace PLC
Screens - Project Aluminium Ltd
Screens/louvers - Allaway Acoustics Ltd
Screens/louvers - Levolux A.T. Limited
Security Grilles - HAG Shutters & Grilles Ltd
Security Grills - O'Connor Fencing Limited
Security grills - Squires Metal Fabrications Ltd
Security Solutions - Harling Security Solutions
Shades - Hunter Douglas
Shades - Luxaflex®
Shading System - CBS (Curtain and Blind Specialists Ltd)
Shaped Aluminium Windows - Midland Alloy Ltd

L10 Windows/ Rooflights/ Screens/ Louvres (con't)

Shaped Louvres - Midland Alloy Ltd
Shelters and Canopies - Elwell Buildings Ltd
Shop Fronts - Mandor Engineering Ltd
Shopfront systems - Aardee Security Shutters Ltd
Shopfronts - BF Bassett & Findley
Shopfronts - H W Architectural Ltd
Shopfronts & Entrance Screens - Metal Technology Ltd
Shopfrontsand Entrance Screens - Sapa Building Systems (monarch and glostal)
SOLA-BOOST® Solar assisted natural ventilation systems - Monodraught Ltd
Solar blinds material - Reflex-Rol (UK)
Solar control window films - The Window Film Company UK Ltd
Solar shading - Hunter Douglas
Solar shading solutions - Faber blinds UK ltd - Faber Blinds UK Ltd
Solar, safety, security, UV control, aesthetic and architectural films - Northgate Solar Controls
SOLA-VENT® Natural daylight and solar powered extract ventilation systems - Monodraught Ltd
Steal rooflights - The Rooflight company
Steel caging - Bradbury Group Ltd
Steel screens - Fitzpatrick Doors Ltd
Steel window association - Steel Window Association
Steel windows - Crittall Steel Windows Ltd
Steel windows - Monk Metal Windows Ltd
Steel windows and doors - Clement Steel Windows
Steel Windows Installation + glazing - Rea Metal Windows Ltd
Steel Windows Manufacture - Rea Metal Windows Ltd
Sun tunnels - NuLite Ltd
SUNCATCHER® Natural ventilation and daylight systems - Monodraught Ltd
SUNPIPE® Natural daylight systems - Monodraught Ltd
Suntubes - Duplus Architectural Systems Ltd
Suspended glass smoke screens - Fenlock-Hansen Ltd
Timber - Olsen Doors & Windows Ltd
Timber framed double glazed windows - Nova Group Ltd
Timber windows - Andersen/ Black Millwork
Timber windows - Anglian Building Products
Timber windows - Benlowe Group Limited
Timber windows - Broxwood (scotland Ltd)
Timber windows - Jeld-Wen UK Ltd
Timber Windows - Magnet Ltd
Timber windows - Premdor
Timber windows - Sampson Windows Ltd
Timber windows - Scandanavian Timber
Timber windows - Timber Components (UK) Ltd
Timber windows - Walker Timber Ltd
Timber windows mauufacture - Blair Joinery
Translucent rooflights - Filon Products Ltd
Unique view control films - Northgate Solar Controls
UPVC conservatories - Windowbuild
UPVC double glazed windows - Sierra Windows
UPVC window casements - Windowbuild
UPVC window systems - REHAU Ltd
Vents - Sola Skylights
VENTSAIR Louvres and penthouse louvres - Monodraught Ltd
Walkways, Porches, Entrance Canopies - Falco UK Ltd
WINDCATCHER® - Monodraught Ltd
Window Grilles, Window Screens - Bradbury Group Ltd
Window hinges - Nico Manufacturing Ltd
Window Protector - Hilton Banks Ltd
Window repair - Repair care International
Window Shutter - Continental Shutters Ltd
Window systems - Deceuninck Ltd
Window tinting - Durable Ltd
Window ventilator manufactures - Simon R W Ltd
Window/ rooflights/ screens/ Lorrles - McMullen Facades
Windows - BAC Ltd
Windows - English Architectural Glazing Ltd
Windows - Glassolutions Saint-Gobain Ltd
Windows - Moores Furniture Group Ltd
Windows - Premdor
Windows - Smith & Choyce Ltd
Windows - Solair Ltd
Windows (alum) - Duplus Architectural Systems Ltd
windows /rooflight/ screens/ louvres - Universal Components Ltd
Windows/ rooflights/ screens/ louvers - Colt International
Windows/ Rooflights/ Screens/ Louvres - Schueco UK
Windows/ Rooflights/ Screens/ Louvres - Till & Whitehead Ltd
Windows/ Rooflights/ Screens/Luvres - Harling Security Solutions
Windows/rooflights/ screens/louvers - Cantifix of London Limited
Windows/Rooflights/Screens/Louvers - Novaglaze Limited
Windows/rooflights/screens/louvres - SCS Group
Windows: Aluminium - H W Architectural Ltd
Windows: Aluminium - Metal Technology Ltd

L20 Doors/ Shutters/hatches

ABS Doors - High Impact - Regency Garage Door Services Ltd
Acoustic Doors - Amadeus
Acoustic Doors - IAC Acoustics
Acoustic doors - Jewers Doors Ltd
Acoustic Doorsets - Humphrey & Stretton PLC
Acoustic seals - Sealmaster
Aircraft hangar doors - Jewers Doors Ltd
All types of fencing & gates - WA Skinner & Co Limited
Aluminium - Project Aluminium Ltd
Aluminium Entrances and patio doors - KCW Comercial Windows Ltd
Aluminium doors - Comar Architectural Aluminuim Systems
Aluminium doors - Duplus Architectural Systems Ltd
Aluminium Doors - Mandor Engineering Ltd
Aluminium doors - Prima Systems (South East) Ltd
Aluminium doors - Sapa Building Systems (monarch and glostal)
Aluminium doors - Sapa Building Systems Ltd
aluminium/hybrid curtain walling, windows and doors - Senior Aluminium Systems plc
Apartment Entrance Doors - Vicaima Ltd
Armour doors - Meggitt Armour Systems
Automatic & manual door repair & maintenance - Dorma Door Services Ltd
Automatic door operating equipment - Kaba Garog
Automatic Doors - DORMA UK Ltd
Automatic doors - Kone
Automatic Doors - Kone PLC
Automatic Doors - TORMAX United Kingdom Ltd
Automatic doors - swing, slide and folding - Gilgen Door Systems UK Ltd
Automatic Entrance Doors - Mandor Engineering Ltd
Balcony doors - Olsen Doors & Windows Ltd
Ballistic doors - Jewers Doors Ltd
Bespoke joinery - AS Newbould Ltd
Bespoke joinery - Howard Bros Joinery Ltd
Blast doors - Jewers Doors Ltd
Brush strip door seals - Kleeneze Sealtech Ltd
Bullet Proof Doorsets - Humphrey & Stretton PLC
Bullet resistant doors and frames - Cumberland Construction
Cedarwood doors - Regency Garage Door Services Ltd
Ceiling Access Panels - The Access Panel Company
Clean room doors - Clark Door Ltd
Clossures - Simflex Grilles and Closures Ltd
Cold store doors - Ascot Doors Ltd
Collapsible Gate Shutters - Mercian Industrial Doors
Coloured GRP doors and frames - Regency Garage Door Services Ltd
Commercial and public doors - Doors & Hardware Ltd
Commercial Doors - London Shopfitters Ltd
Commercial Doors - Sunfold Systems Ltd
Commercial steel doors - Sunray Engineering Ltd
Composite doors - Halmark Panels Ltd
Composite doors - Laird Security Hardware
Composite doors - Spectra Glaze Services
Conservatories - Solair Ltd
Corrosion Resistant Sliding and Folding Door Gear - Coburn Sliding Systems Ltd
Cottage doors - AS Newbould Ltd
Curtain walling - L. B. Plastics Ltd
Custom made doors and doorsets - Bridgman IBC Ltd
Decorative doors - Foil faced, Inlay, Veneered - Vicaima Ltd
Domestic Doors - Sunfold Systems Ltd
Door - purpose made flush - Humphrey & Stretton PLC
Door & window control equipment - GEZE UK
Door damping Solutions - Coburn Sliding Systems Ltd
Door frames - Enfield Speciality Doors
Door Hardware and accessories - Strada London
Door kits - Vicaima Ltd
Door manufactures - Edmonds A & Co Ltd
Door panels - Door Panels by Design
Door panels - Halmark Panels Ltd
Door protection - Moffett Thallon & Co Ltd
Door systems - Deceuninck Ltd
Doors - Altro Ltd
Doors - BAC Ltd
Doors - Cantifix of London Limited
Doors - Crawford Hafa Ltd
Doors - Glassolutions Saint-Gobain Ltd
Doors - H E W I (UK) Ltd
Doors - IJK Timber Group Ltd
Doors - Komfort Workspace PLC
Doors - Magnet Ltd
DOORS - McMullen Facades
Doors - Metsä Wood
Doors - Project Aluminium Ltd
Doors - Scotts of Thrapston Ltd
Doors - Smith & Choyce Ltd
Doors - Walker Timber Ltd
Doors & sets - Cox Long Ltd
Doors (alum) - Duplus Architectural Systems Ltd
Doors and entrances - BF Bassett & Findley
Doors architectural - Timbmet Ltd
Doors/ Shutters/ hatches - Harling Security Solutions
Doors/ Shutters/ Hatches - Schueco UK
Doors/ Windows/ Stairs - Covers Timber & Builders Merchants
Doors/hatches - Allaway Acoustics Ltd
Doors/shutters/hatches - Hadley Industries Plc
Doors/shutters/hatches - Hart Door Systems Ltd
Doors; Commercial & Public Amenities - BIS Door Systems Ltd
Doorsets - Vicaima Ltd
Electric openers/ spare parts - Cardale Doors Ltd
Entrance doors - Andersen/ Black Millwork
Entrance doors - Olsen Doors & Windows Ltd
Entrance systems - Taskmaster Doors Ltd
External Doors - Premdor
Extruded aluminium composite doors - Smart Systems Ltd
Extruded aluminium doors - Smart Systems Ltd
Extruded UPVC doors - Smart Systems Ltd
Extruded UPVC doors - Veka Plc
Fast action doors - Neway Doors Ltd
Fibre composite residential doors - Birtley Group
Fibreglass doors - Premdor
Fire and security doors - Tyco Fire and Integrated Solutions
Fire door certification schemes, Mechanical testing - BM TRADA
Fire doors - Enfield Speciality Doors
Fire doors - Fitzpatrick Doors Ltd
Fire doors - Scotts of Thrapston Ltd
Fire Doors - Sunray Engineering Ltd
Fire doors - Vicaima Ltd
Fire doors and door sets - Shadbolt F R & Sons Ltd
Fire escape doors - Fitzpatrick Doors Ltd
Fire rated door sets - Lami Doors UK Ltd
Fire Resistant glass - SCHOTT UK Ltd
fire resistant Glazed screens and doors - Fenlock-Hansen Ltd
Fire resistant glazing systems, fire resistant panels - Promat UK Ltd
Fire resisting doors - Mann McGowan Group
Fire resisting doors and joinery - Decorfix Ltd
Fire shutters - Ascot Doors Ltd
Fire Shutters - Blount Shutters Ltd
Fire Shutters - Continental Shutters Ltd
Flush and feature doors - Vicaima Ltd
Flush doors - Enfield Speciality Doors
Folding doors - Dividers Modernfold Ltd
Folding doors - Olsen Doors & Windows Ltd
Folding doors and partitions - Parthos UK Ltd
Folding Grilles - Simflex Grilles and Closures Ltd
Food industry doors - Ascot Doors Ltd
Frames and covers - Bilco UK Ltd
G. R. P . Porches - Solair Ltd
G.R.P Doors - Regency Garage Door Services Ltd
Garage and industrial doors - Hörmann (UK) Ltd
Garage doors - Doorfit Products Ltd
Garage Doors - Rundum Meir (UK) Ltd
Garage Doors - Electric Operated - Avon Industrial Doors Ltd
Garage doors - Roller - Moffett Thallon & Co Ltd
Garage doors - Sectional - Moffett Thallon & Co Ltd
Garage doors - Up and Over - Moffett Thallon & Co Ltd
Garage Roller Shutter Doors - Rolflex Doors Ltd
Gates and barriers - automtic doors - Hart Door Systems Ltd
Glass Door Fitting - DORMA UK Ltd
Glass doors - Style Door Systems Ltd
Glazing maintenance - Spectra Glaze Services
Glazing trade association - Glass and Glazing Federation
Grilles - Bradbury Group Ltd
Grilles & Shutters - Continental Shutters Ltd
GRP door sets - Lami Doors UK Ltd
GRP Door surrounds - Production Glassfibre
GRP doors - Benlowe Group Limited
GRP hatches - Group Four Glassfibre Ltd
Hardware - Roc Secure Ltd
Hardwood doorsets to match windows - Blair Joinery
Hatches - metal - McKenzie-Martin Ltd
High integrity doors windows blast and fire protection systems - Booth Industries Ltd
High performance doors - Sashless Window Co Ltd
High speed doors - Clark Door Ltd
High speed doors - Crawford Hafa Ltd
High speed vertical folding and roll -up doors - Mandor Engineering Ltd
Hinged access panels for ceilings and walls - Profab Access Ltd
Hinged steel dorrs - Clark Door Ltd
Hinged, pivoting and sliding doors - Optima
Industrial and cold storage strip curtain systems - Flexible Reinforcements Ltd
Industrial & commercial doors, shutters, grilles and partitions - Gilgen Door Systems UK Ltd
Industrial & commercial security & fire resisting metal doors & shutters - Bolton Gate Co. Ltd
Industrial and Commercial Insect Screens - The Flyscreen Company Ltd
Industrial Doors - BIS Door Systems Ltd
Industrial Doors - Blount Shutters Ltd
Industrial Doors - HAG Shutters & Grilles Ltd
Industrial Doors - Kone
Industrial doors - Sunray Engineering Ltd
Industrial doors and loading bay equipment - Envirodoor Markus Ltd
Industrial doors, roller, folding and insulated - Avon Industrial Doors Ltd
Industrial Roller Shutter Doors - Rolflex Doors Ltd
Industrial Sliding and Folding Door Gear - Coburn Sliding Systems Ltd
Insect Screens - The Flyscreen Company Ltd
Insulated bifolding doors - Jewers Doors Ltd
insulated doors - AS Newbould Ltd
Insulated Roller Shutter Doors - Rolflex Doors Ltd
Insulated sliding doors - Jewers Doors Ltd
Insulated sliding folding doors - Jewers Doors Ltd
Internal Doors - Premdor
Laminate doors - Vicaima Ltd
Laminated Door sets - T B S Fabrications
Loft access doors - Timloc Expamet Building Products
Loft Access Hatches - Ryton's Building Products Ltd
Maintenance of industrial doors, shutters and shopfronts - Shutter Door Repair & Maintenance Ltd
Matching Side Doors - Regency Garage Door Services Ltd
Megadoors - Crawford Hafa Ltd
Metal doors - Fitzpatrick Doors Ltd
Metal doors - The Cotswold Casement Company
Metal doors, shutters - Amber Doors Ltd
Metal doors/ shutters/ hatches - HW Architectural Ltd
Metal doors/ shutters/ hatches - Reid John & Sons (Strucsteel) Ltd
Metal frames - Fitzpatrick Doors Ltd
Metal meter boxes and replacement architrave units - Ritherdon & Co Ltd
Moveable internal walls and partitions - Style Door Systems Ltd
Operable walls - Avanti Systems
Panel tracks - Goelst
Partitioning Systems - DORMA UK Ltd
Patio doors - Andersen/ Black Millwork
Patio Doors - Solair Ltd
Pine doors - Wood International Agency Ltd
Pine furniture doors - Mason FW & Sons Ltd
Plant Room Doors - Avon Industrial Doors Ltd
Purpose made flush timber doors - Youngs Doors Ltd
Purpose made panel doors - Youngs Doors Ltd
PVC and Rubber Curtains - Mercian Industrial Doors
PVC clad doors & frames - Platonoff & Harris Plc
PVC crash and strip doors - Avon Industrial Doors Ltd
PVC door profiles - Profile 22 Systems Ltd
PVC doors - Prima Systems (South East) Ltd
PVCu doors - Interframe Ltd
PVCu doors - Plastal (SBP Ltd)
PVCu Entrances and patio doors - KCW Comercial Windows Ltd
PVCU extrusions for door systems - Spectus Systems
Rapid roll doors - Ascot Doors Ltd
Remote control garage doors - Regency Garage Door Services Ltd
Repairs - Syston Rolling Shutters Ltd
Residential garage doors - Garador Ltd
Revolving doors - Gilgen Door Systems UK Ltd
Revolving doors - LMOB Electrical Contractors Ltd
Roller doors/ side-hinged doors - Cardale Doors Ltd
Roller shutter and grilles - James & Bloom Ltd
Roller Shutter Doors - BIS Door Systems Ltd
Roller Shutter Doors - Continental Shutters Ltd
Roller shutter doors, fast action doors - Priory Shutter & Door Co Ltd
Roller shutters - Ascot Doors Ltd
Roller Shutters - Blount Shutters Ltd
Roller shutters - Fitzpatrick Doors Ltd
Roller shutters - Lakeside Group
Roller shutters and industrial doors - Regency Garage Door Services Ltd
Rolling Shutters - Simflex Grilles and Closures Ltd
Rolling shutters and individual doors in steel, aluminum and timber. Service and maintenance available nationwide - Syston Rolling Shutters Ltd
Roof Access Hatches - The Access Panel Company
Ropes and sashcords - Marlow Ropes Ltd
Rubber and PVC flexible crash doors, PVC strip curtains - Neway Doors Ltd
Rubber and PVC flexible doors - Mandor Engineering Ltd
Sectional Doors - Regency Garage Door Services Ltd
Sectional overhead doors - Ascot Doors Ltd
Sectional overhead doors - Crawford Hafa Ltd
Sectional Overhead Doors - Novoferm Europe Ltd
Security door systems - Bridgman IBC Ltd
Security door systems - Vicaima Ltd
Security Doors - ASSA ABLOY Limited
Security doors - Fitzpatrick Doors Ltd
Security doors - Gilgen Door Systems UK Ltd
Security doors - Jewers Doors Ltd
Security Doorsets - Humphrey & Stretton PLC
Security Grills - O'Connor Fencing Limited
Security grills - Squires Metal Fabrications Ltd
Security metal doors and shutters - Ascot Doors Ltd
Security roller shutters - Lakeside Group
Security shutters - Aardee Security Shutters Ltd
Security Shutters - HAG Shutters & Grilles Ltd
Security shutters - Hart Door Systems Ltd
Security Solutions - Harling Security Solutions
Self Storage Doors - Rolflex Doors Ltd
Shopfronts and Entrance Screens - London Shopfitters Ltd
Shutters - Luxaflex®
Side hung and double action door sets - Lami Doors UK Ltd
Side Hung Doors - Regency Garage Door Services Ltd
Sliding and folding partitions - Building Additions Ltd
Sliding door sets - Lami Doors UK Ltd
Sliding Door Systems - Schueco UK
Sliding doors - Clark Door Ltd
Sliding Doors - Designer Construction Ltd
Sliding doors - Kaba Garog
Sliding doors - Olsen Doors & Windows Ltd
Sliding folding doors - Ascot Doors Ltd
Special timber flush doors and door sets - Bridgman IBC Ltd
Specialist steel doorsets - Accent Hansen
Specialized timber doors - AS Newbould Ltd
Stable doors - AS Newbould Ltd
Stable Doors - Scotts of Thrapston Ltd
Stainless steel doors - Fitzpatrick Doors Ltd
Stainless steel, acoustic and insulated freezer doors - Clark Door Ltd
Stainless steel, bronze and aluminium panel doors, entrance screens, shopfronts and window frames - Drawn Metal Ltd
Steel Commercial and Industrial doors - Mercian Industrial Doors
Steel Door Frames - Coburn Sliding Systems Ltd
Steel Door Installation +glazing - Rea Metal Windows Ltd

L20 Doors/ Shutters/hatches (con't)

Steel Door Manufacture - Rea Metal Windows Ltd
Steel door sets - Ascot Doors Ltd
Steel doors - Bradbury Group Ltd
Steel doors - Crittall Steel Windows Ltd
Steel doors - Gilgen Door Systems UK Ltd
Steel doors - Monk Metal Windows Ltd
Steel doors - Premdor
Steel Doors - Regency Garage Door Services Ltd
Steel doors and frames - Fitzpatrick Doors Ltd
Steel doors, Ceilings - Norwood Partition Systems Ltd
Steel doors, frames and entrance systems - Taskmaster Doors Ltd
Steel faced doors - Benlowe Group Limited
Steel Firescreen Installation + glazing - Rea Metal Windows Ltd
Steel Firescreen Manufacture - Rea Metal Windows Ltd
Steel hinged Doorsets - Accent Hansen
Steel louvred doorsets - Sunray Engineering Ltd
Steel Paneled Doors - Longden
Steel pedestrian doors - Hart Door Systems Ltd
Steel residential doors - Birtley Group
Steel Securit Doors - Steel Doors & Security Systems
Steel Security Doors - Rolflex Doors Ltd
Steel security doorsets - Sunray Engineering Ltd
Steel shutters (Mall, Shop front, Bar/Counter) - Mercian Industrial Doors
Steel window association - Steel Window Association
Steel windows and doors - Clement Steel Windows
Straight Sliding Door Gear, Folding Door Gear and Hideaway Pocket Door Kits - Coburn Sliding Systems Ltd
Synthetic rubber moulding components and ironmongery - Notcutt Ltd
Syston doors - Syston Rolling Shutters Ltd
Thermally Broken Doors - H W Architectural Ltd
Timber doors - Benlowe Group Limited
Timber doors - Broxwood (scotland Ltd)
Timber doors - JB Kind Ltd
Timber doors - Jeld-Wen UK Ltd
Timber Doors - Longden
Timber doors - Premdor
Traditional oak doors - W.L. West & Sons Ltd
Train and Platform door systems - Gilgen Door Systems UK Ltd
Transformer chamber doorsets - Sunray Engineering Ltd
Turnstiles - full and half height - Gilgen Door Systems UK Ltd
Turnstiles for access and flow control of people - Kone
Up and over doors/ sectional doors - Cardale Doors Ltd
UPVC doors - Spectra Glaze Services
UPVC french doors - Windowbuild
Vision see-through security shutters - Cooper Group Ltd
Wall access hatches - Howe Green Ltd
Wardrobe and Cupboard Sliding and Folding Door Gear - Coburn Sliding Systems Ltd
Windows - Jeld-Wen UK Ltd
X-ray Doorsets - Humphrey & Stretton PLC

L30 Stairs/ Walkways/Balustrades

Access & Escape Ladders - Baj System Design Ltd
Access Ladders - Surespan Ltd
Acrylic, glass, steel and timber stairs - Lewes Design Contracts Ltd
Aluminium Bridge parapets - Varley & Gulliver Ltd
Aluminium extrusions/ sections - Seco Aluminium Ltd
Aluminium floor outlets - Marley Plumbing and Drainage
Aluminium handrails/balustrades - Norton Engineering Alloys Co Ltd
Aluminium ladders and step ladders - Ramsay & Sons (Forfar) Ltd
Aluminium staircase fire escapes - Baj System Design Ltd
Aluminium stairways - Zarges (UK) Ltd
Architectural Glass - Creative Glass
Balconies - Squires Metal Fabrications Ltd
Balcony ladder - Baj System Design Ltd
Balconys - Woodburn Engineering Ltd
Ballstrades - Strand Hardware Ltd
Ballustrading - Challenge Fencing Ltd
Balustrade & Barrier Systems - Brass Age Ltd
Balustrade & Handrail Systems - Delta Balustrades
Balustrade and handrail coating - Plastic Coatings Ltd
Balustrades - Ermine Engineering Co. Ltd
Balustrades - Glassolutions Saint-Gobain Ltd
Balustrades - L. B. Plastics Ltd
Balustrades - Lang+Fulton
Balustrades - Squires Metal Fabrications Ltd
Balustrades - Staines Steel Gate Co
Balustrades - Steel Line Ltd
Balustrades and Handrails - Gabriel & Co Ltd
Balustrades and wallrails - H E W I (UK) Ltd
Balustraid and Handrails - Safety Works & Solutions Ltd
Bespoke joinery - AS Newbould Ltd
Cast Spiral Stairs - Cottage Craft Spirals
Curved wooden handrails - The Angle Ring Company Ltd
Customs staircases in steel, stainless steel, waxed steel, wrought iron, wood or concrete with options for glass treads/balustrades - In Steel (Blacksmiths & Fabricators) Ltd
Decking - B & M Fencing Limited
Domestic and comercial - Loft Centre Products
Drainpipe ladders - Baj System Design Ltd
Escape and electric stairway ladders, roof exit systems, spiral, spacesaver and traditional stairs - Loft Centre Products
Fabricators of open metal flooring, handrails and standards - OSF Ltd
Fibreglass ladders and step ladders - Ramsay & Sons (Forfar) Ltd
Fire escape staircases - Crescent of Cambridge Ltd
Fire escapes - In Steel (Blacksmiths & Fabricators) Ltd
Fire escapes - Squires Metal Fabrications Ltd
Fire Escapes - Staines Steel Gate Co
Fire escapes, gates, railings, balustrades, guardrails and spiral staircases - Metalcraft (Tottenham) Ltd
Flooring - Lang+Fulton
Folding escape ladders - Baj System Design Ltd
Glass Balustrade - SG System Products Ltd
GRP Handrails - Brighton (Handrails), W
GRP String course - Production Glassfibre
Guardrail - Kee Klamp Ltd
Hand rails - RB UK Ltd
Hand rails - Squires Metal Fabrications Ltd
Handrail and Balustrade Systems - Sapphire Balustrades
Handrail Covers - DC Plastic Handrails Ltd
Handrail systems in steel, stainless steel, aluminium and GRP - Lionweld Kennedy Ltd
Handrailing - Steel Line Ltd
Handrailing & Balustrading - Laidlaw Architectural Hardware Ltd
Handrails and balustrades - Drawn Metal Ltd
Handrails/balustrades - stainless - Norton Engineering Alloys Co Ltd
Helical stairs - Cambridge Structures Ltd
Helical stairs - Crescent of Cambridge Ltd
Hooped/caged ladder - Baj System Design Ltd
In floor telescopic ladder - Baj System Design Ltd
Industrial flooring, handrailing systems and stairtreads - Redman Fisher Engineering
Insulated Door panels - Ward Insulated Panels Ltd
Ladders, walkways and hand railing - Moseley GRP Product (Division of Moseley Rubber Co Ltd)
Mechanical testing - BM TRADA
Metal railings - Varley & Gulliver Ltd
Metal Staircases - Albion Manufacturing
Metal stairs - Cottage Craft Spirals
Metal stairs / walkways / balustrades - Steelway Fensecure Ltd
Metal stairs/ walkways/ balustrades - Hubbard Architectural Metalwork Ltd
Metal stairs/balustrades - NRG Fabrications
Open steel grating in steel, stainless steel and GRP - Lionweld Kennedy Ltd
Pedstrian walkways - Eve Trakway
Perforated sheets and balustrading - Associated Perforators & Weavers Ltd
Plywood, hard, fibre, chip and particle boards - Boardcraft Ltd
Pre-Cast Terrazzo Staircases - Quiligotti Terrazzo Tiles Limited
PVC handrails - Architectural Plastics (Handrail) Ltd
PVC handrails - Brighton (Handrails), W
Railing systems - Brass Age Ltd
Rails for special needs - Gabriel & Co Ltd
Roof walkway/ fall arrest system - McKenzie-Martin Ltd
Ropes and sashcords - Marlow Ropes Ltd
Spiral and helical p.c. stairs - In Steel (Blacksmiths & Fabricators) Ltd
Spiral and straight staircases - Mackay Engineering
Spiral Staircase Systems - Lewes Design Contracts Ltd
Spiral Staircases - Spiral Construction Ltd
Spiral Staircases - Titan Ladders Ltd
Spiral staircases and special precast stairs - In Steel (Blacksmiths & Fabricators) Ltd
Spiral stairs - Gabriel & Co Ltd
Spiral stairs - Pedley Furniture International Ltd
Spiral stairways - Blanc de Bierges
Stainless Steel Balustrade Infills - Navtec
Stainless steel flooring - Ancon Building Products
Stainless steel Stairs/walkways - Wincro Metal Industries Ltd
Stainless steel, aluminium, PVC covered, galvanized and brass balustrades - SG System Products Ltd
Stair edgings - Gradus Ltd
Stair/ walkways - Redman Fisher Engineering
Staircases - Company name: Britannia Metalwork Services Ltd
Staircases - E A Higginson & Co Ltd
Staircases - Metalcraft (Tottenham) Ltd
Staircases - Staines Steel Gate Co
Staircases - metal, spiral, straight - Crescent of Cambridge Ltd
Staircases, fire escapes, ladders, handrails and balustrades - Arc Specialist Engineering Limited
Stairs - Blanc de Bierges
Stairs - Kirkstone
Stairs - Magnet Ltd
Stairs - Smith & Choyce Ltd
Stairs - Squires Metal Fabrications Ltd
Stairs manufactured to order - Higginson Staircases Ltd
Stairs walkways and Balustrades - Midland Wire Cordage Co Ltd
Stairs/ walkways - Steelway Fensecure Ltd
Stairs/ walkways/ balustrades - Broxap Ltd
Stairs/ Walkways/ Balustrades - Chilstone
Stairs/ walkways/ balustrades - Glazzard (Dudley) Ltd
Stairs/ walkways/ balustrades - Quantum Profile Systems Ltd
Stairs/walkways - Ancon Building Products
Stairs/Walkways - Portia Engineering Ltd
Stairs/Walkways/ Balusters - Andy Thornton Ltd
Stairs/walkways/ balustrades - Cantifix of London Limited
Stairs/Walkways/Balustrades - Grant & Livingston Ltd
Steel bridge parapets - Varley & Gulliver Ltd
Steel fabrication railings, gates, fire escapes - SFC (Midlands) Ltd
Steel Hand Rails - The Angle Ring Company Ltd
Steel Spiral Staircases - Graepel Perforators Ltd
Steel spiral stairs - Higginson Staircases Ltd
Steel staircases - Ermine Engineering Co. Ltd
Steel Staircases - Fire Escapes and Fabrications (UK) Ltd
Steel staircases - Gabriel & Co Ltd
Steel, Stainless steel and aluminium - Speedfab Midlands
Steps - Blanc de Bierges
Straight Flights - Cottage Craft Spirals
Straight helical and spiral stairs - Lewes Design Contracts Ltd
Straight, Winder & geometric timber stairs - Higginson Staircases Ltd
Timber loft and escape ladders, roof exit systems, spiral, spacesaver and traditional stairs - Loft Centre Products
Timber Spiral Stairs - Cottage Craft Spirals
Timber spiral stairs - Higginson Staircases Ltd
Timber staircases - Benlowe Stairs
Timber staircases - Jeld-Wen UK Ltd
Timber staircases - Northern Joinery Ltd
Timber stairs - Cottage Craft Spirals
Timber stairs - Jeld-Wen UK Ltd
Timber stairs/balustrades - Timber Components (UK) Ltd
Walkways - Clow Group Ltd
Walkways - Gabriel & Co Ltd
Walkways and platforms - Woodburn Engineering Ltd
Walkways/balustrades - Levolux A.T. Limited
Wall corner & Door Protection - Gradus Ltd
Wallrails and balustrades - H E W I (UK) Ltd
Wooden ladders and step ladders - Ramsay & Sons (Forfar) Ltd
Woodturning - Spindlewood

L40 General glazing

aluminium/hybrid curtain walling, windows and doors - Senior Aluminium Systems plc
Architectural aluminium systems for glazing - Universal Components Ltd
Architectural glass - Cambridge Structures Ltd
Architectural Glass - Creative Glass
Architectural glass - Komfort Workspace PLC
Architectural glass and glazing - Avanti Systems
Atrium glazing - English Architectural Glazing Ltd
Display Enhancement Glass - SCHOTT UK Ltd
Double glazed units - Ledlite Glass (Southend) Ltd
Dry glazing systems and ventilated glazing beads etc - Exitex Ltd
Etched and sand blasted glass - Charles Lightfoot Ltd
Films for glass - Northgate Solar Controls
Films: Solar control, Tinted, Reflective, Anti-UV, Anti-Grafiti, Anti-Fog - The Window Film Company UK Ltd
Fire rated glazing - English Architectural Glazing Ltd
Fire resistant glazing - Ide T & W Ltd
Fire resistant glazing - Mann McGowan Group
Fire resistant glazing systems - Promat UK Ltd
Fire resistant safety glasses - Pyroguard UK Ltd
Ganeral glazing - Saint-Gobain Glass UK Ltd
General glazing - AGC Glass UK Ltd
General glazing - Glassolutions Saint-Gobain Ltd
General Glazing - Olsen Doors & Windows Ltd
General glazing - Pilkington UK Ltd
Glass - Fenlock-Hansen Ltd
Glass - Glassolutions Saint-Gobain Ltd
Glass - Pilkington Birmingham
Glass - The Stone Catalogue Ltd
Glass & Glazing - Clement Steel Windows
Glass melting - British Glass Manufacturers Confederation
Glass Technology Services - British Glass Manufacturers Confederation
Glazed canopies - Lonsdale Metal Company
Glaziers and glassmerchants - Ledlite Glass (Southend) Ltd
Glazing - Ide T & W Ltd
Glazing - The Abbseal Group incorporating Everseal (Thermovitrine Ltd)
Glazing & Mirrors - Roc Secure Ltd
Glazing Accessories - Reddiglaze Ltd
Glazing Accessories - Simon R W Ltd
Glazing bars - Kenilworth Daylight Centre Ltd
Glazing bars for glass sealed units - Newdawn & Sun Ltd
Glazing bars for timber roofs - Newdawn & Sun Ltd
Glazing gaskets to windows, doors and facades - Trelleborg Sealing Profiles
Glazing maintenance - Spectra Glaze Services
Glazing trade association - Glass and Glazing Federation
High performance window films - The Window Film Company UK Ltd
Insulated Fire Resistant Glass - SCHOTT UK Ltd
Laminated - Pilkington Birmingham
Laminated glass - Float Glass Industries
Laminated glass - Solaglas Laminated
Low - emissivity - Pilkington Birmingham
Manifestation graphics - Invicta Window Films Ltd
Manifestation marking of glass - Northgate Solar Controls
Opaque bullet resistant glass - Attwater & Sons Ltd
Patterned - Pilkington Birmingham
Patterned glass - Float Glass Industries
Plastic profiled sheet cladding and roofing - Ariel Plastics Ltd
Polycarbonate glazing - Cox Building Products Ltd
Privacy window films - Invicta Window Films Ltd
Putty - Adshead Ratcliffe & Co Ltd
Restoration Glass - SCHOTT UK Ltd
Roof glazing - English Architectural Glazing Ltd
Safety and security window films - Durable Ltd
Secondary glazing manufactures - Simon R W Ltd
Security/ safety window films - Invicta Window Films Ltd
Self support glazing beams - Newdawn & Sun Ltd
Selfcleaning - Pilkington Birmingham
Shopfronts - Comar Architectural Aluminuim Systems
Silvered - Pilkington Birmingham
Smart Film - Switchable between clear and frosted - Pro Display TM Limited
Smart Glass - Switchable between clear and frosted - Pro Display TM Limited
Solar control glass - Float Glass Industries
Solar control window films - Invicta Window Films Ltd
Stock glass - Float Glass Industries
Structural bolted glazing - English Architectural Glazing Ltd
Tinted - Pilkington Birmingham
Toughened - Pilkington Birmingham
Toughened glass - Float Glass Industries
Toughened, laminate, ceramic, acoustic glass and screen printing on glass - Pilkington Plyglass
Welded Wire Mesh - Multi Mesh Ltd
Window Films - 3M
Window films - Bonwyke Ltd

L41 Lead light glazing

Architectural Glass - Creative Glass
Glazing maintenance - Spectra Glaze Services
Glazing trade association - Glass and Glazing Federation
Impact resistant safety mirrors and glass - Chelsea Artisans Ltd
Leaded lights - Charles Lightfoot Ltd
Leaded lights - Ledlite Glass (Southend) Ltd
Leaded lights - The Cotswold Casement Company
Restoration/ conservation/ design and production of stained glass - Chapel Studio
Specialist glazing - Charles Lightfoot Ltd
Stained glass - Charles Lightfoot Ltd
Stained glass windows - Ide T & W Ltd

L42 Infill panels/ sheets

Architectural Glass - Creative Glass
Clear heat control films - Northgate Solar Controls
Glazing maintenance - Spectra Glaze Services
Glazing trade association - Glass and Glazing Federation
Infill panels/sheets - Panel Systems Ltd
Mirrors - Float Glass Industries
Privacy, patterned, decorative and manifestation films - Northgate Solar Controls
Screens - Glassolutions Saint-Gobain Ltd

M10 Cement: sand/ Concrete screeds/ toppings

Association of Concrete Industrial Flooring Contractors - Association of Concrete Industrial Flooring Contractors (ACIFC)
Calcium carbonate - Imerys Minerals Ltd
Cement - Hanson UK
Cement based levelling/wearing screeds" Proprietary cement-based levelling screeds - Mapei (UK) Ltd
Cement based screeds - Flowcrete Plc
Cement: sand/ concrete screeds/ toppings - Flexcrete Technologies Limited
Cement: sand/ Concrete screeds/ toppings - Vetter UK
Cements - LaFarge Cement UK
Chemical resistant screed - Addagrip Terraco Ltd
Colours for cement - Procter Johnson & Co Ltd
Concrete topping - Ardex UK Ltd
Cristobalite flour - Sibelco UK Ltd
Cristobalite sand - Sibelco UK Ltd
Floor screed - Lytag Ltd
Floor screeds and roof screeds - Isocrete Floor Screeds Ltd
Flooring - BASF plc, Construction Chemicals
Flooring - Hanson Concrete Products Ltd
Flooring surface hardeners - Armorex Ltd
Industrial flooring - Stuarts Industrial Flooring Ltd
Mezzanine floors and allied equipment - Bromag Structures Ltd
Movement Joints - Movement Joints (UK)
Permascreed - DPC Screeding Ltd
Polymer modified/ rapid drying screeds - Ronacrete Ltd
Polymer/cementitions screeds - Conren Ltd
Polypropylene & steel fibers for concrete & mortar - Propex Concrete Systems
Repair mortars - Conren Ltd
Sand cement/ Concrete/ Granolithic screeds/ flooring - Don Construction Products Ltd
Screeds and floor coatings - Addagrip Terraco Ltd
Silica Flour - Sibelco UK Ltd
Silica Sand - Sibelco UK Ltd
Specialist screeds - Carter-Dal International
Specialist screeds - Laticrete International Inc. UK
Steel framed multi storey volumetric buildings - Leofric Building Systems Ltd
Subfloor preparation products - Ardex UK Ltd
Wilsonart solid surfacing - Wilsonart Ltd

M11 Mastic asphalt flooring/ floor underlays

Colours for Asphalt - Procter Johnson & Co Ltd
Mastic asphalt flooring - North Herts Asphalte (Roofing) Ltd

M11 Mastic asphalt flooring/ floor underlays (con't)

Paving and flooring - Roofing Contractors (Cambridge) Ltd
Tiling accessories - Dural (UK) Ltd

M12 Trowelled bitumen/ resin/ rubber-latex flooring

Anti slip surfacing - Addagrip Terraco Ltd
Car Park Decking - Sika Ltd
Celicote resin coatings - Ancorite Surface Protection Ltd
Colours for Asphalt - Procter Johnson & Co Ltd
Epoxy flooring systems - Cross-Guard International Ltd
Epoxy resin coatings for floors - Watco UK Ltd
Epoxy resin repair materials - Anglo Building Products Ltd
Floor coatings - Jotun Henry Clark Ltd Decorative Division
Flooring systems - Cross-Guard International Ltd
Industrial Flooring - Sika Ltd
Resin based floor & wall finishes - Flowcrete Plc
Resin coatings - Armorex Ltd
Resin Flooring - Altro Ltd
Resinbased Floor Coatingss - Belzona PolymericLtd
Self Leveling Moread Beds Latex - Carter-Dal International
Self Leveling Moread Beds Latex - Laticrete International Inc. UK
Trowelled bitumen/ resins/ rubber latex flooring - Don Construction Products Ltd
Urethame screeds - Cross-Guard International Ltd

M13 Calcium sulphate based screeds

Architectural Mouldings - Hodkin Jones (Sheffield) Ltd
Association - Resin Flooring Association (FeRFA)
Calcium sulphate based screeds - Flowcrete Plc
Synthascreed - DPC Screeding Ltd

M20 Plastered/ Rendered/ Roughcast coatings

Building plasters - British Gypsum Ltd
Ceiling Refurbisment - Ceiling & Lighting Ltd
Cement based floor screeds, external render systems - Maxit UK
Coil mesh - Catnic
Coloured limestone, dolomite and white sand internal and external renders - Kilwaughter Chemical Co Ltd
Drywall decorating products - Siniat Ltd
Drywall finishing systems - Siniat Ltd
Drywall Joint & Flex Tapes - Siniat Ltd
Epoxy resin coatings for walls - Watco UK Ltd
High performance building boards - Knauf Drywall
Metal lath - Catnic
Mineral renders - Keim Mineral Paints Ltd
Movement Joints - Movement Joints (UK)
Plaster - Knauf Drywall
Plaster beads - Walls & Ceilings International Ltd
Plasterboard accessories - Walls & Ceilings International Ltd
Plastered / Rendered / Roughcast Coatings - Kilwaughter Chemical Co Ltd
Plastered / Rendered / Roughcast Coatings - Safeguard Europe Ltd
Plastered/ Rendered/ Roughcast coatings - Skanda Acoustics
Plastered/Rendered/Roughcast fittings - PermaRock Products Ltd
Plasterers beads - Vista Engineering Ltd
Plasterers bends - Catnic
Plasters - Kair ventilation Ltd
plastic rendering plaster and dry wall beads - Nenplas
Polymer modified renders - Ronacrete Ltd
Premix Renders - LaFarge Cement UK
Render systems - BRC Special Products - Bekaert Ltd
Renders - Weber Building Solutions
Renders and plasters - Thermica Ltd
Roughcast coatings - Andrews Coatings Ltd
Silicone Renders - Kilwaughter Chemical Co Ltd
Synthetic resin renders - STO Ltd
Textured coatings & grout - Ciret Limited

M21 Insulation with rendered finish

Expanded polystyrene insulation - Jablite Ltd
External wall insulation - STO Ltd
External wall insulation systems - Skanda Acoustics
Foamglas Cellular glass insulation for wall - Pittsburg Corning (UK) Ltd
Hydraulic lime mortars for rendering - Telling Lime Products Ltd
Insulated external render - Maxit UK
Insulated external renders/systems - Alumasc Exterior Building Products Ltd
Insulation - Rockwool Ltd
Insulation - Skanda Acoustics Limited
Insulation with rendered finish - Concrete Repairs Ltd
Insulation with rendered finish - Insulated Render & Cladding Association
Insulation with rendered finish - PermaRock Products Ltd
Insulation with rendered finish - Thermica Ltd
Render nail type M for EML lath - Reiner Fixing Devices (Ex: Hardo Fixing Devices)
Trade Association - Association of Specialist Fire Protection (ASFP)

M22 Sprayed monolithic coatings

Grout pumps - Clarke UK Ltd
Intimescent paint - Firebarrier Services Ltd
Specialist coatings - Valtti Specialist Coatings Ltd
Sprayed & trowelled on resin coatings - Ancorite Surface Protection Ltd
Sprayed monolithic coatings - Specialist Building Products
Spray-On Insulation - Aardvark Transatlantic Ltd
Trade Association - Association of Specialist Fire Protection (ASFP)

M23 Resin bound mineral coatings

Anti skid surfacing products - Clare R S & Co Ltd
Association - Resin Flooring Association (FeRFA)
Coloured sufacing products - Clare R S & Co Ltd
Decorative bound/bonded surfacing resin -TekGrip DPB &: Professional - TekGrip DSR, Flexible - TekGrip DFX, Summer - TekGrip DD2 - Star Uretech Ltd
Epoxy coatings - Cross-Guard International Ltd
Epoxy Floor Paint - Everlac (GB) Ltd
Industrial Flooring Products - Ardex UK Ltd
Manufacture of Industrial Resin Flooring in Polyurethane or Epoxy materials - Resdev Ltd
Polymer caotings - Cross-Guard International Ltd
Resin bound mineral coatings - Addagrip Terraco Ltd
Resin bound mineral coatings - Aden Hynes Sculpture Studios
Resin bound mineral coatings - Don Construction Products Ltd
Resin sealing coats - Armorex Ltd
Resins - DSM UK Ltd
Seamless resin floors - Ancorite Surface Protection Ltd
Specialist coatings - Valtti Specialist Coatings Ltd
Trade Association - Association of Specialist Fire Protection (ASFP)
Urethame coatings - Cross-Guard International Ltd

M30 Metal mesh lathing/ Anchored reinforcement for plastered coatings

Anchored reinforcement for plastered coatings - Thermica Ltd
Arch formers - Catnic
Arch formers - Expamet Building Products
Cladding systems - BRC Special Products - Bekaert Ltd
Mesh arches - Allmat (East Surrey) Ltd
Metal lathing - Expamet Building Products
Metal beads - Expamet Building Products
Metal lathing and mesh - Simpson Strong-Tie®
Metal mesh lathing - Bekaert Building Products
Metal mesh lathing/ Anchored reinforcement for plastered coatings - Skanda Acoustics
Metal wire plaster & render beads - Bekaert Building Products
PVCu plasters beads for drylining, plastering and rendering - Renderplas Ltd
Render nail for Riblath - Reiner Fixing Devices (Ex: Hardo Fixing Devices)
Resin bound mineral coatings - Specialist Building Products
Stainless Steel Reinforcement Mesh - Multi Mesh Ltd
Stainless steel surface protection - Component Developments
Welded wire mesh - Albion Manufacturing
Welded Wire Mesh - Multi Mesh Ltd

M31 Fibrous plaster

Decorative mouldings, cornices, dados, ceiling roses and coving - Copley Decor Ltd
Decorative plaster coving and ceiling roses - Hilton Banks Ltd
Fibrous plaster moldings - Troika Contracting Ltd
Glass fiber reinforced gypsum - Troika Contracting Ltd
Plaster cornices and archways - Artistic Plastercraft Ltd
Polypropylene, steel fibers for concrete & mortar - Propex Concrete Systems

M40 Stone/ Concrete/ Quarry/ Ceramic tiling/ Mosaic

Acid resistant brickwork and Vibration tile system - Ancorite Surface Protection Ltd
Air tools - Marcrist International Ltd
Anit-Fracture Membranes for Natural Stone - Carter-Dal International
Anit-Fracture Membranes for Natural Stone - Laticrete International Inc. UK
Architectural Tiles - Solus Ceramics Ltd
Association of Concrete Industrial Flooring Contractors - Association of Concrete Industrial Flooring Contractors (ACIFC)
Bespoke glazed wall tiles and terracotta floor tiles - Smithbrook Building Products Ltd
Ceramic fiber products - Combustion Linings Ltd
Ceramic floor and wall tiles - R J Stokes & Co Limited
Ceramic marble and granite tiles - Reed Harris
Ceramic tile and stone installation system - Carter-Dal International
Ceramic tile and stone installation systems - Laticrete International Inc. UK
Ceramic tiles - Bell & Co Ltd
Ceramic tiles - British Ceramic Tile Ltd
Ceramic Tiles - Porcelanosa Group Ltd
Ceramic tiles - Swedecor
Ceramic, marble and glass mosaics, swimming pool tiles and porcelain stoneware - Domus Tiles Ltd
Ceramic, marble, granite tiles - Reed Harris
Ceramic, porcelain and mosaic tiles - Arnull Bernard J & Co Ltd
Ceramics - Wade Ceramics Ltd
Ceramics and mosaics - Shackerley (Holdings) Group Ltd incorporating Designer ceramics
Cladding tiles - porcelain - Johnson Tiles
Cleaning - Natural Stone, Terrazzo, Conglomerate Marble, Marble, Granite, Limestone, Slate & Mosaic - Fieldmount Ltd
Concrete floor screeds - Armorex Ltd
Diamond Sawn -Yorkstone - Hard York Quarries Ltd
Digital transfer decoration - wall tiles - Johnson Tiles
Faience blocks - Shaws of Darwen
Floor and wall tiles - Porcelanosa Group Ltd
Floor slabs and floor tiles - The Delabole Slate Co. Ltd
Floor systems - MiTek Industries Ltd
Floor tile separators - Compriband Ltd
Floor tiles - ceramic - Johnson Tiles
Floor tiles - porcelain - Johnson Tiles
Floor tiling - Fired Earth
Flooring - Bridge Quarry
Flooring - Burlington Slate Ltd
Flooring - Farmington Masonry LLP
Flooring - Kirkstone
Flooring Marble and granite - Diespeker Marble & Terrazzo Ltd
Glass mosaic -sandstone - Reed Harris
Grouts - Fosroc Ltd
Industrial floor tiles - Swedecor
Internal Paving - Caithness Flagstone Limited
Internal Paving - Johnsons Wellfield Quarries Ltd
Internal Paving; Limwstone and Sandstone - J Suttle Swanage Quarries Ltd
Keope Tiles - Italian Porecellian - Ruabon Sales Limited
Limestone Flooring - Chameleon Stone Ltd
Limestone quarry tiles - Reed Harris
Marble Flooring - Chameleon Stone Ltd
Marble tiles - Emsworth Fireplaces Ltd
Marble, granite, limestone, precast and insitu terrazzo mosaic - Fieldmount Ltd
Marble, granite, slate and limestone - Toffolo Jackson(UK) Ltd
Marble, granite, slate, limestone & terrazzo - Diespeker Marble & Terrazzo Ltd
Mosaic tiles - Chameleon Stone Ltd
Mosaics - Porcelanosa Group Ltd
Movement Joints - Movement Joints (UK)
Natural Stone paving - Palladio Stone
Natural stone tiles - R J Stokes & Co Limited
Natural stones like marble, slate - Porcelanosa Group Ltd
New Staining Silicon Joint for String Work - Carter-Dal International
New Staining Silicon Joint for String Work - Laticrete International Inc. UK
Promenade Tiles - Slatesystem
Purbeck Limestone - Haysom (Purbeck Stone) Ltd
Quality tile adhesives - Ardex UK Ltd
Quality Tile Grout - Ardex UK Ltd
Quarry tiling - Ruabon Sales Limited
Quartz - Diespeker Marble & Terrazzo Ltd
Quartzite Flooring - Chameleon Stone Ltd
recycled glass promenade tiles - Spartan Promenade Tiles Ltd
Restoration and Renovation - Natural Stone, Terrazzo, Conglomerate Marble, Marble, Granite, Limestone, Slate & Mosaic - Fieldmount Ltd
Slate - Reed Harris
Slate Flooring - Chameleon Stone Ltd
Specialist contractors in marble and granite mosaics - Alpha Mosaic & Terrazzo Ltd
Sphinx Ttiles - Glazed Wall and Floor - Ruabon Sales Limited
Split face wall tiles - Chameleon Stone Ltd
Stone Quarries & Sandstone - Johnsons Wellfield Quarries Ltd
Stone Quarries: Sandstone - Caithness Flagstone Limited
Stone tiles - Kirkstone
Stone/ Concrete/ Quarry/ Ceramic tiling/ Mosaic - Haddonstone Ltd
Stone/ Concrete/ Quarry/ Ceramic tiling/ Mosaic - Stonewest Ltd
Stone/ Concrete/ Quarry/ Ceramic tiling/ Mosaic - Vetter UK
Stone/ concrete/ quarry/ ceramic tiling/ mosaic/ Tile installation adhesives/ grouts - Mapei (UK) Ltd
Stone/concrete/quarry/ceramic tiling/mosaic - Chilstone
Stone/concrete/quarry/ceramic tiling/mosaic - G Miccoli & Sons Limited
Stone/concrete/quarry/ceramic tiling/mosaic - Spartan Promenade Tiles Ltd
Supply and Fix - Natural Stone, Terrazzo, Conglomerate Marble, Marble, Granite, Limestone, Slate & Mosaic - Fieldmount Ltd
Surface Preparation Equipment - Marcrist International Ltd
Swimming pool tiles - Swedecor
Swimmingpool tiles - ceramic - Johnson Tiles
Terracotta - Reed Harris
Terracotta blocks - Shaws of Darwen
Terrazzo - Reed Harris
Tile cleaners - Tank Storage & Services Ltd
Tiling accessories - Dural (UK) Ltd
Tiling Products - Sealocrete PLA Ltd
Travertine Limestone Flooring - Chameleon Stone Ltd
Vessel & tank linings - Ancorite Surface Protection Ltd
Wall tile fittings - ceramic - Johnson Tiles
Wall tiles - Fired Earth
Wall tiles - ceramic - Johnson Tiles
Water jet cut murals - wall & floor tiles - Johnson Tiles
Wire Brushes - Marcrist International Ltd

M41 Terrazzo tiling/ In situ terrazzo

Association - Resin Flooring Association (FeRFA)
Bespoke Terrazzo Tiling - Quiligotti Terrazzo Tiles Limited
Cleaning - Natural Stone, Terrazzo, Conglomerate Marble, Marble, Granite, Limestone, Slate & Mosaic - Fieldmount Ltd
Customised Terrazzo Tiling - Quiligotti Terrazzo Tiles Limited
Finished Laid Terrazzo - Quiligotti Terrazzo Tiles Limited
Marble, granite, limestone, precast and insitu terrazzo mosaic - Fieldmount Ltd
Restoration and Renovation - Natural Stone, Terrazzo, Conglomerate Marble, Marble, Granite, Limestone, Slate & Mosaic - Fieldmount Ltd
Specialist contractors in terrazzo - Alpha Mosaic & Terrazzo Ltd
Stainless Grout - Carter-Dal International
Stainless Grout - Laticrete International Inc. UK
Supply and Fix - Natural Stone, Terrazzo, Conglomerate Marble, Marble, Granite, Limestone, Slate & Mosaic - Fieldmount Ltd
Terracotta floor tiles - York Handmade Brick Co Ltd
Terrazzo flooring - Diespeker Marble & Terrazzo Ltd
Terrazzo polishing and grinding - Ancorite Surface Protection Ltd
Terrazzo tile manufacturers - Toffolo Jackson(UK) Ltd
Terrazzo Tiling - Quiligotti Terrazzo Tiles Limited
Terrazzo tiling/ In situ terrazzo - Stonewest Ltd
Terrazzo tiling/ In situ terrazzo - Vetter UK
Terrazzo, marble and mosaic federation - National Federation of Terrazzo, Marble & Mosaic Specialists
Tiling accessories - Dural (UK) Ltd

M42 Wood block/ Composition block/ Parquet flooring

Ceramic Parquet - Porcelanosa Group Ltd
Commercial Wood Flooring - Tarkett Sommer Ltd
Dance floor - Felix Design Ltd
ESCO flooring pre-finished with Polyx®-Oil - OSMO UK
Flooring - Deralam Laminates Ltd
Flooring (leather) - Edelman Leather
Hardwood - Porcelanosa Group Ltd
HPL flooring - Abet Ltd
Laminated flooring - Kronospan Ltd
Parquet and laminate flooring - Western Cork Ltd
Solid & engineered parquet - Western Cork Ltd
Solid wood flooring - Fired Earth
Sports flooring systems and sprung flooring - Granwood Flooring Ltd
Sprung Floor Systems - British Harlequin Plc
Wood block/ composition block/ parquet flooring - Westco Group Ltd
Wood Flooring - Posners the floorstore
Wood flooring/ texture adhesives - Mapei (UK) Ltd

M50 Rubber/ Plastics/ Cork/ Lino/ Carpet tiling/ sheeting

Accessories - Interfloor Limited
Aluminium edge trim for vinyl flooring - Howe Green Ltd
Anti Slip Products - 3M
Anti-Fatique Matting - Kleen-Tex Industries Ltd
Anti-slip surfaces - Redman Fisher Engineering
Bonded tiles - Carpets of Worth Ltd
Carpet - Armstrong Floor Products UK Ltd
Carpet and tiles - Adam Carpets Ltd
Carpet fitting tools and accessories - Interfloor Limited
Carpet Protector - Hilton Banks Ltd
Carpet Tiles - Armstrong Floor Products UK Ltd
Carpet tiles - Brintons Ltd
Carpet Tiles - Interface Europe Ltd
Carpet tiles - Milliken Carpet
Carpet tiles - Rawson Carpets Ltd
Carpet Tiling and Sheeting - Mat.Works Flooring Solutions
Carpet, cork and vinyl tiles and laminate flooring - Western Cork Ltd
Carpet, Vinyl and Laminate flooring - Posners the floorstore
Carpets - CFS Carpets
Carpets - HEM Interiors Group Ltd
Carpets - Opus 4 Ltd
Carpets - Scotmat Flooring Services
Commercial carpet tiles, broadloom and entrance clean off zones - Checkmate Industries Ltd
Commercial Resilient Flooring - Tarkett Sommer Ltd
Commercial Textile Flooring - Tarkett Sommer Ltd
Contamination control flooring - Dycem Ltd
Contamination control mats - Dycem Ltd
Contract carpet and carpet tile manufacturer - Burmatex Ltd
Cork - Olley & Sons Ltd, C
Cork flooring - Amorim (UK) Ltd
Cork Rubber and Non slip Surfaces - Tiflex Ltd
Custom mixing - Interfloor Limited
Data installations - HEM Interiors Group Ltd
Desk chair mats - Jaymart Rubber & Plastics Ltd
Discipative matting soloutions - Kleen-Tex Industries Ltd
Duckboard matting - Jaymart Rubber & Plastics Ltd
Dust control matting soloutions - Kleen-Tex Industries Ltd
Edging - Egger (UK) Ltd
Electrical Safety Matting - Jaymart Rubber & Plastics Ltd
Electrical Safety Matting - MacLellan Rubber Ltd
Entrance barrier systems - Kleen-Tex Industries Ltd
Entrance Matting - 3M
Entrance matting - Gradus Ltd
entrance matting systems - Jaymart Rubber & Plastics Ltd

M50 Rubber/ Plastics/ Cork/ Lino/ Carpet tiling/ sheeting (con't)

Entrance matting, flooring and roof walkway matting - Plastic Extruders Ltd
Flameproof sheeting - Icopal Limited
Flexible sheets and Tiles - Forbo Flooring
Floor protection coatings - Glixtone Ltd
Floor smoothing underlay - Ball F & Co Ltd
Floor trims - Gradus Ltd
Floorcoverings - Interface Europe Ltd
Flooring - HEM Interiors Group Ltd
Flooring accessories, stairnosings and PVC accessories - Quantum Profile Systems Ltd
Flooring adhesives - Ball F & Co Ltd
Flooring coatings - Williamson T & R Ltd
Flooring products - Weber Building Solutions
Garage floor tiles - Rockwell Sheet Sales Ltd
Handrail Covers - DC Plastic Handrails Ltd
Heavy duty carpet tiles - Paragon by Heckmondwike
Heavy duty impact protection rubber sheeting - MacLellan Rubber Ltd
High pressure decorative laminates - Abet Ltd
Hygiene matting - Kleen-Tex Industries Ltd
Interlocking tiles - Jaymart Rubber & Plastics Ltd
Laminate flooring - Westco Group Ltd
Leather floor and wall tiles - Edelman Leather
Lettered & logo mats - Jaymart Rubber & Plastics Ltd
Linoleum - Armstrong Floor Products UK Ltd
Linoleum - Tarkett Sommer Ltd
Luxury Vinyl Tiles - Armstrong Floor Products UK Ltd
Natural floor covering - Fired Earth
Non-slip floor coatings - Andura Coatings Ltd
Non-slip matting soloutions - Kleen-Tex Industries Ltd
Nylon reinfoced PVC sheeting - Flexible Reinforcements Ltd
Plastic coatings - Plastic Coatings Ltd
Plastic roofing, cladding, and glazing - Amari Plastics Plc
Plastics/ cork/ lino/ carpet tilling/ sheeting - Westco Group Ltd
Polyurethane re-bond underlays - Interfloor Limited
PTFE coatings - Plastic Coatings Ltd
Publications: contract flooring association guide to contract flooring - Contract Flooring Association
PVC floorcoverings and accessories - Polyflor Ltd
PVC safety floor coverings - Polyflor Ltd
PVC sheeting - Flexible Reinforcements Ltd
Realwood HPL Decorative Laminates - Abet Ltd
Recycled flooring products - Jaymart Rubber & Plastics Ltd
Room Improvement Products - NMC (UK) Ltd
Rubber flooring - Altro Ltd
Rubber flooring systems - Jaymart Rubber & Plastics Ltd
Rubber for floors and other surfaces in a huge range of colours and textures - Dalhaus Ltd
Rubber linings - Ancorite Surface Protection Ltd
Rubber sheeting and expansion joints & bellows - MacLellan Rubber Ltd
Rubber sheeting and flooring - Linatex Ltd
Rubber stud tiles - Polyflor Ltd
Rubber/ Plastics sheeting and flooring - Renqiu Jingmei Rubber&Plastic Products Co., Ltd
Rugs - Brintons Ltd
Safety Flooring - Armstrong Floor Products UK Ltd
Safety floorings - Jaymart Rubber & Plastics Ltd
Sheeting and matting - Interfloor Limited
Slip resistant flooring - Altro Ltd
Smooth vinyl flooring - Altro Ltd
Specialised flooring materials, nosing and adhesives - Tiflex Ltd
Sponge rubber underlays (carpets/wood & laminate) - Interfloor Limited
Sports flooring - Jaymart Rubber & Plastics Ltd
Sports Surfaces - Altro Ltd
Underlay and flooring - Interfloor Limited
Vinyl - Armstrong Floor Products UK Ltd
Vinyl floorcovering sheets, tiles and accessories - Tarkett Ltd
Vinyl sheet and tile flooring - DLW Flooring
Vinyl, safety and sports flooring - Gerflor Ltd
Wall and Furniture Linoleum - Forbo Flooring
Wall furnishings - Vescom UK Limited
Wetwall - Tarkett Sommer Ltd

M51 Edge fixed carpeting

Carpet - EGE Carpets Ltd
Carpet (office/Domestic) - Evergreens Uk
Carpet for domestic and contract use - Carpets of Worth Ltd
Carpeting - Shaylor Group Plc
Carpets - CFS Carpets
Carpets - Ulster Carpet Mills Ltd
Carpets and Matting - Heckmondwicke FB
Carpets and Matting - Mat.Works Flooring Solutions
Commercial carpet tiles, broadloom and entrance clean off zones - Checkmate Industries Ltd
Contract flooring manufacturers - Rawson Carpets Ltd
Custom carpets - Ulster Carpet Mills Ltd
Domestic and contract locations - Adam Carpets Ltd
Edge fixed carpeting - Brockway Carpets Ltd
Edge fixed carpeting - Quantum Profile Systems Ltd
Edge fixed floorcoverings - Interface Europe Ltd
Entrance Matting - Heckmondwicke FB
Fibre bonded floor coverings - Rawson Carpets Ltd
Fitted Carpets - Posners the floorstore
Floorcovering accessories - Gripperrods Ltd
Heavy duty carpets - Paragon by Heckmondwike
Surface membranes and release systems for carpet - Laybond Products Ltd
Wool-rich Natural products - Ulster Carpet Mills Ltd
Woven and tufted carpet - Brintons Ltd
Woven axminster carpets - Ulster Carpet Mills Ltd
Woven wilton carpets - Ulster Carpet Mills Ltd

M52 Decorative papers/ fabrics

130cm wide fabric backed vinyls - Dixon Turner Wallcoverings
135cm wide fabric backed vinyls - Dixon Turner Wallcoverings
53cm x 10m paper backed vinyls - Dixon Turner Wallcoverings
Acrylic coatd paper - Package Products Ltd
Authentic tribal rugs - Fired Earth
Contract wall coverings - Muraspec
Contract wallcoverings - Dixon Turner Wallcoverings
Decorative finishes - Glixtone Ltd
Decorative finishes - Weber Building Solutions
Decorative papers/ fabrics - Brewer C & Sons Ltd
Decorative wallcoverings, fabrics and borders - Dixon Turner Wallcoverings
Fabric backed wall coverings - Muraspec
Fabric Linning Systems - Stretched Fabric Systems
Fabrics - Tektura Plc
Fabrics and papers - Interface Europe Ltd
Fabrics and wall papers - Today Interiors Ltd
Fabrics and wallpapers - Sanderson
Flexible sheets and Tiles - Forbo Flooring
Glass fibre - Muraspec
Hessian - Muraspec
Interior finishes - Muraspec
Metallic finishes - Muraspec
Notice board material - Forbo Flooring
Paint alternatives - Muraspec
Paint stripper, wall paper stripping machines - Ciret Limited
Paints and wallcoverings - Kalon Decorative Products
Paperbacked wall coverings - Muraspec
Primers - Tektura Plc
PVC coated fabrics - Flexible Reinforcements Ltd
Textile wallcoverings - Dixon Turner Wallcoverings
Textiles - Muraspec
Textured coatings - Glixtone Ltd
Timeless fabrics - Fired Earth
Wall coverings - Porcelanosa Group Ltd
Wallcoverings - Muraspec
Wallcoverings - Tektura Plc
Wallcoverings and adhesives - Architectural Textiles Ltd
Wholesalers of furnishing fabrics & wallpapers - CrowsonFabrics Ltd

M60 Painting/ Clear finishing

Agricultural paints - Teal & Mackrill Ltd
Anit graffiti - Kingstonian Paint Ltd
Anti - static coatings - Kenyon Group Ltd
Anti Climb, Anti Fly Poster, Anti Graffiti, paints and coatings - Kenyon Paints Limited
Anti Condensation Paints - Witham Oil & Paint Ltd
Anti condonsation - Kingstonian Paint Ltd
Anti corrosion coatings - Corroless Corrosion Control
Anti corrosion coatings - Kenyon Group Ltd
Anticlimb paint - Coo-Var Ltd
Antifouling paints - Teal & Mackrill Ltd
Antigraffiti coatings - Coo-Var Ltd
Bitumen paint - Liver Grease Oil & Chemical Co Ltd
Black varnish - Liver Grease Oil & Chemical Co Ltd
Blasting - Bristol Metal Spraying & Protective Coatings Ltd
Brushing tar, creosote, tallow and linseed oil - Legge Thompson F & Co Ltd
Cadmium pigments - Rockwood Pigments (UK) Ltd
Chrome oxide green pigments - Rockwood Pigments (UK) Ltd
Cladding paint - Kingstonian Paint Ltd
Cleaners for paving - Resiblock Ltd
Cleaning Chemicals - Tank Storage & Services Ltd
Coal tar creosote & pitch - Liver Grease Oil & Chemical Co Ltd
Coatings - Fujichem Sonneborn Ltd
Coatings - PPG Protective and Marine Coatings
Coatings for concrete floors - Everlac (GB) Ltd
Colour Schemes - Kalon Decorative Products
Complex inorganic pigments - Rockwood Pigments (UK) Ltd
Decorating - Shaylor Group Plc
Decorating materials - Brewer C & Sons Ltd
Decoration - HEM Interiors Group Ltd
Decorative ancillary products - Bartoline Ltd
Decorative coatings for timber - Protim Solignum Ltd
Decorative Finishes - ICI Woodcare
Decorative paints - Akzo Noble Coatings Ltd
Decorative paints, opaque and translucent wood stains - Jotun Henry Clark Ltd Decorative Division
Decorative Sundries - Kalon Decorative Products
Decorative surface enhancer - TekGrip DSE - Star Uretech Ltd
Decorative Walls - Altro Ltd
Elastomeric, anticarbonation coatings - Andura Coatings Ltd
Electroplating, powder coating and painting - Ritherdon & Co Ltd
Epoxy floor/wall coatings - Conren Ltd
Epoxy paints - Trade Sealants Ltd
Epoxy resin coatings - Anglo Building Products Ltd
Epoxy resins - Addagrip Terraco Ltd
Epoxy systems - Teal & Mackrill Ltd
Exterior woodcare - Blackfriar Paints Ltd
External Decorative Finishes - ICI Woodcare
Fillers - Sadolin UK Ltd
Fine and coarse textured coatings - Andura Coatings Ltd
First for finishes - Chestnut Products
Flame retardant lacquers - Bollom Fire Protection Ltd
Flame retardant paints - Bollom Fire Protection Ltd
Flame retardant varnishes - Bollom Fire Protection Ltd
Floor and line paint - Coo-Var Ltd
Floor coating systems - Corroless Corrosion Control
Floor coatings - Firwood Paints Ltd
Floor coatings, self levellers - Kenyon Group Ltd
Floor paints - Kingstonian Paint Ltd
Floor paints - Teal & Mackrill Ltd
Floor Paints - Witham Oil & Paint Ltd
Floor seals - Evans Vanodine International PLC
Fluorescent Systems - Witham Oil & Paint Ltd
Glass flake reinforced coatings - Corroless Corrosion Control
High performance coatings - Leigh's Paints
Hygiene coatings - Kenyon Group Ltd
Hygienic coatings - Corroless Corrosion Control
Hygienic coatings for walls & ceilings - Tefcote Surface Systems
Industrial Coating Specialists - Bristol Metal Spraying & Protective Coatings Ltd
Industrial paints - Firwood Paints Ltd
Industrial paits - Teal & Mackrill Ltd
Interior coatings - Jotun Henry Clark Ltd Decorative Division
Interior varnishes - Sadolin UK Ltd
Internal Special Effects - Dulux Trade
Internal/External Decorative Finishes - Dulux Trade
Intumescent coatings - Bollom Fire Protection Ltd
Lacquers and seals - Junckers Limited
Lime based paints - Telling Lime Products Ltd
Lime putty - Legge Thompson F & Co Ltd
Long life exterior wall coating - Everlac (GB) Ltd
Luminous & fluorescent paints - Coo-Var Ltd
Lump and french chalk - Legge Thompson F & Co Ltd
Maintanance Painting service - International Protective Coatings
Maintenance & Protective Coatings - ICI Woodcare
Maintenance paints - Firwood Paints Ltd
Marine & antifouling coatings - Coo-Var Ltd
Marine paints - Witham Oil & Paint Ltd
Marine paints and coatings - Teal & Mackrill Ltd
Masonry paint - STO Ltd
Metal - Bristol Metal Spraying & Protective Coatings Ltd
Mineral Paints - Keim Mineral Paints Ltd
Natural Oil Woodstain - OSMO UK
Opaque finishes - Sadolin UK Ltd
Opaque ironoxide pigments - Rockwood Pigments (UK) Ltd
Paint - Fired Earth
Paint - Sanderson
Paint and powder manufacturer - Howse Thomas Ltd
Paint brushes, rollers and decorating tools - Hamilton Acorn Ltd
Paint effects - Kingstonian Paint Ltd
Paint spraying equipment - Bambi Air Compressors Ltd.
Painting / Clear Finishing - Safeguard Europe Ltd
Painting / clear finishing - Tor Coatings Ltd
Painting /Clear Finishing - Wessex Resins & Adhesives Ltd
Painting aids - Polycell Products Ltd
Painting and decorating materials - Covers Timber & Builders Merchants
Painting/ Clear finishing - Andura Coatings Ltd
Painting/ Clear finishing - Arkema France
Painting/ Clear finishing - Bondaglass Voss Ltd
Painting/ clear finishing - Concrete Repairs Ltd
Painting/ clear finishing - Coo-Var Ltd
Painting/ Clear finishing - Don Construction Products Ltd
Painting/ clear finishing - Liquid Technology Ltd
Painting/clear finishes - TH Kenyon & Sons Ltd
Painting/Clear Finishing - International Protective Coatings
Painting/clear finishing - Tefcote Surface Systems
Paints - AkzoNobel Decorative Paints UK
Paints - Crown Paints Ltd
Paints - Emusol Products (LEICS) Ltd
Paints - HMG Paints
Paints - Manders Paint Ltd
Paints - PPG Protective and Marine Coatings
Paints - Weber Building Solutions
Paints and coatings for tennis courts and other sports surfaces - Everlac (GB) Ltd
Paints and high temperature resistant coatings - Minco Sampling Tech UK Ltd (Incorporating Fortafix)
Paints and surface coatings - Dacrylate Paints Ltd
Paints primers and specialized coatings - Coo-Var Ltd
Paints, wood finishes, solvents and chemicals - Anderson Gibb & Wilson
Partnering - Kalon Decorative Products
Pigments for the surface coatings - Rockwood Pigments (UK) Ltd
Polish and sealers - British Nova Works Ltd
Polyx®-Oil - OSMO UK
Powder and ready mixed fillers - Ciret Limited
Preparation products - Blackfriar Paints Ltd
Primers - ICI Woodcare
Protective and decorative masonry coatings - Andura Coatings Ltd
Protective coating for steelwork - International Protective Coatings
Protective coatings - Andrews Coatings Ltd
Protective Coatings - Dulux Trade
Protective maintenance coatings - Corroless Corrosion Control
Rust stabilising primers - Corroless Corrosion Control
Seals and coatings for wood floors - Everlac (GB) Ltd
Silicone waterproofing - Bluestar Silicones
Slurry Granules - LANXESS Inorganic Pigments Group
Solvent bourne coatings - Sadolin UK Ltd
Specialised paints - Teal & Mackrill Ltd
Specialist Coatings - Valtti Specialist Coatings Ltd
Specialist paint systems - Williamson T & R Ltd
Specialist Paints Varnishes and Woodcare Products - Blackfriar Paints Ltd
Specialist Primers - Witham Oil & Paint Ltd
Specialist Water Based Coattings - Witham Oil & Paint Ltd
Standard, paint quality and fair face finishes - Tarmac Building Products Ltd
Structural painting and spraying - Masonary Cleaning Services
Surface dressings - Decorating products - Sealocrete PLA Ltd
Suspended Ceiling Cleanings - Above All
Suspended Ceiling Coatings - Above All
Swimming Pool Paints - Witham Oil & Paint Ltd
Technical Specifications - Kalon Decorative Products
Tennis court spraying - Sportsmark Group Ltd
Tile transfers - Homelux Nenplas
Timber preservatives, fire retardants - decorative finishes - Arch Timber Protection
Trade Paints and coatings - Kenyon Paints Limited
Translucent finishes - Sadolin UK Ltd
Transparent ironoxide pigments - Rockwood Pigments (UK) Ltd
Twin Pack Paints - Witham Oil & Paint Ltd
Two Pack Floor Paints - Coo-Var Ltd
UV-Protection Oil - OSMO UK
Vapour corrosion inhibitors - Corroless Corrosion Control
Wall and floor coatings - Sealocrete PLA Ltd
Waterbourne coatings - Sadolin UK Ltd
Wet paint - Plastic Coatings Ltd
Wood Finishes - Dulux Trade
Wood Finishes - ICI Woodcare
Wood finishes, speciality paints, paint removers and decorating sundries - Rustins Ltd
Wood protection products - Sadolin UK Ltd
Wood Stains - Akzo Noble Coatings Ltd
Wood Wax Finish - OSMO UK
Woodfinishes - Smith & Rodger Ltd
woodworm, rot and damp treatment - Rentokil Property Care

M61 Intumescent coatings for fire protection of steelwork

Fire protection coatings for steelwork - Thermica Ltd
Fire Retart and Coatings - Bristol Metal Spraying & Protective Coatings Ltd
High temperature coatings - Kenyon Paints Limited
Hyclad matt finish - Outokumpu
Industrial Coating Specialists - Bristol Metal Spraying & Protective Coatings Ltd
Intermescant coatings for fire protection of steelwork - International Protective Coatings
Intumescant coatings - Andrews Coatings Ltd
Intumescant coatings, protecting steel from fire - Smith & Rodger Ltd
Intumescent Coatings - Brewer C & Sons Ltd
Intumescent coatings - PPG Protective and Marine Coatings
Intumescent coatings for steel - Quelfire
Nullifire intumescent range - Nullifire Ltd
Passive fire protection products - Leigh's Paints
Structural steel fire protection - Grace Construction Products Ltd
Structural steel protection - Promat UK Ltd
Trade Association - Association of Specialist Fire Protection (ASFP)

N10 General fixtures/ furnishings/ equipment

Acoustic telephone enclosure - Burgess Architectural Products Ltd
All types of window blinds - Northgate Solar Controls
Aluminium Shelving - Bedford Shelving Ltd
Aluminum shelving - RB UK Ltd
Anti fatigue mats - Dycem Ltd
Anti-Ligature Furniture - Aaztec Cubicles
Architectural antiques and Furniture - Andy Thornton Ltd
Archive research - Sanderson
Awnings - Luxaflex®
Awnings - Tidmarsh & Sons
Bamer Carpeting/Matting - PHS Group plc
Banks of post boxes - Signs Of The Times Ltd
Basket units - Smart F & G (Shopfittings)Ltd
Bathroom cabinets and tube fittings - William Hopkins & Sons Ltd
Baulstrading - Spindlewood
Bedroom & home study furniture - Ball William Ltd
Bedroom furniture - Magnet Ltd
Bedroom furniture - OEP Furniture Group PLC
Bedrooms - Moores Furniture Group Ltd
Bedspreads - Sanderson
Benches - Detlef Muller Ltd
Benching - Aaztec Cubicles
Benching - Envopak Group Ltd
Bencing - Libraco
Bespoke furniture - Hyperion Wall Furniture Ltd
Bespoke Furniture Design - Branson Leisure Ltd
Bespoke Furniture Manufacturers - Branson Leisure Ltd
Bespoke museum display cases - The Benbow Group

N10 General fixtures/ furnishings/ equipment (con't)

BioCote - Allermuir Contract Furniture Ltd
Blackout blinds - Tidmarsh & Sons
Blind and awning fabrics - Flexible Reinforcements Ltd
Blinds - Durable Ltd
Blinds - Eclipse Blind Systems Ltd
Blinds - Hallmark Blinds Ltd
Blinds - Integra Products
Blinds - Invicta Window Films Ltd
Blinds - Opus 4 Ltd
Blinds - VELUX Co Ltd, The
Blinds in sealed units - Pilkington Plyglass
Blinds, curtains, tracks - Sunrite Blinds Ltd
Block & sheet polyester, bonded products - Rawson Fillings Ltd
Boards for furniture, whiteboards, chaklboards advertising - Finnish Fibreboard (UK) Ltd
Boltless Shelving - Sperrin Metal Products Ltd
Bomb curtains and blankets - Meggitt Armour Systems
Bookcase Systems - Jali Ltd
Bracketry - Vantrunk Engineering Ltd
Broadcast and audio visual products & services - Streamtec Limited
Brush strip cable management - Kleeneze Sealtech Ltd
Brushes - Industrial Brushware Ltd
Builders merchant - First Stop Builders Merchants Ltd
Building materials and DIY products - First Stop Builders Merchants Ltd
Cabinet Hinge Manufacturers - Daro Factors Ltd
Caravan security boxes - Ritherdon & Co Ltd
Card/ magazine racks - Smart F & G (Shopfittings)Ltd
Carpet Tile - PHS Group plc
Carved stone - Caithness Flagstone Limited
Cash and security boxes - Securikey Ltd
Cast aliminium mail boxes - Signs Of The Times Ltd
Changing room furniture - Helmsman
Choir risers - Felix Design Ltd
Chrome Plated Steel Shelving - Bedford Shelving Ltd
Chrome shelving - RB UK Ltd
Clearette bins - Iles Waste Systems
Coal and log effect gas fires - BFM Europe Limited
Coir matting & logo - PHS Group plc
Conservatory Blinds - Amdega Ltd
Conservatory blinds - Tidmarsh & Sons
Contract fruniture - Taskworthy Ltd
Contract furnishers - John Pulsford Associates Ltd
Contract furniture - Metalliform Products Plc
Contract furniture - Ness Furniture Ltd
Contract office and leisure furniture - Zon International
Contract Residential Furniture - Ellis J T & Co Ltd
Contract seating - Laytrad Contracts Ltd
Contract seating & tables - Knightbridge Furniture Productions Ltd
Contract tables - Laytrad Contracts Ltd
Cord drawn tracks - Goelst
Cotton & wool felt - Rawson Fillings Ltd
Curtain fabrics - Eclipse Blind Systems Ltd
Curtain poles - Rothley Burn Ltd
Curtain poles - Silent Gliss Ltd
Curtain Tracks and Poles - Integra Products
Curtain tracks, cubicle rails and window blinds - electric & manual operation - Silent Gliss Ltd
Curtain tracks, curtains and blinds - Marlux Ltd
Curtains - Sanderson
Curtains and Blinds - Roc Secure Ltd
Custom design staging - Felix Design Ltd
Custom mouldings - Harrison Thompson & Co Ltd
Customer guidence and queue management systems - Tensator Ltd
Customised metal trims in all metals - Azimex Fabrications Ltd
Customized panel work - Ritherdon & Co Ltd
Decorative melamine faced chipboard - Egger (UK) Ltd
Designers for Working, Learning and Living Spaces - John Pulsford Associates Ltd
Desk lighting - Anglepoise Lighting Ltd
Desks and furniture in metalwork, glass, marble and granite - Howard Bros Joinery Ltd
Display Panels - Harewood Products Ltd (Adboards)
Domestic bedroom furniture - Ball William Ltd
Door Closers - DORMA UK Ltd
Drop down grab rails - Cefndy Healthcare
Electronic fly killers - P+L Systems Ltd
Entrance Matting - EMS Entrance Matting Systems Ltd
Entrance matting - Norton Engineering Alloys Co Ltd
Entrance matting - Shackerley (Holdings) Group Ltd incorporating Designer ceramics
Equestrian Arenas and Track PVC-u Fencing and Barriers - Duralock (UK) Ltd
Equipment cabinets - Arbory Group Limited
External Ashtray - The No Butts Bin Company (NBB), Trading as NBB Outdoor Shelters
External blinds - Tidmarsh & Sons
External Litter Bins - The No Butts Bin Company (NBB), Trading as NBB Outdoor Shelters
Filing and storage cabinets - Railex Systems Ltd, Elite Division
Fine surfaced furniture chip board - Egger (UK) Ltd
Finger posts - Signs Of The Times Ltd
Fire Blankets - BRK Brands Europe Ltd
Fire extinguishers - ASCO Extinguishers Co. Ltd
Fire extinguishers - BRK Brands Europe Ltd
Fire extinguishers - Fife Fire Engineers & Consultants Ltd
Fire extinguishers - Firetecnics Systems Ltd
Fire extinguishers - Griffin and General Fire Services Ltd
Fire extinguishers - Kidde Fire Protection Services Ltd
Fire extinguishers - Richard Hose Ltd
Fire fighting equipment - Eagles, William, Ltd
Fire rated mailboxes - The Safety Letterbox Company Ltd
Fire resistant fabrics - Mermet U.K
Fireplace service - CICO Chimney Linings Ltd
Fireplaces - Bridge Quarry
Fireplaces - Emsworth Fireplaces Ltd
Fireplaces - The Stone Catalogue Ltd
Fireplaces - Wyvern Marlborough Ltd
Fireplaces and fires - Bell & Co Ltd
Fireplaces, surrounds, suites and hearths - Winther Browne & co Ltd
Fitted Bedrooms - Hammonds - New Home Division
Fitted Bedrooms - Symphony Group PLC, The
Fitting & Installation - Sanderson
Fittings - Kee Klamp Ltd
Flags, flagstaffs, banners and bunting - Zephyr The Visual Communicators
Flat end grab rails - Cefndy Healthcare
Flooring solutions - Mat.Works Flooring Solutions
Fragrance dispensers - P+L Systems Ltd
Frames Historical Photographs - Francis Frith Collection
Freestanding mailboxes - The Safety Letterbox Company Ltd
Furniture - Bevan Funnell Group Limited
Furniture - Fired Earth
Furniture - Moores Furniture Group Ltd
Furniture - Roc Secure Ltd
Furniture - Sanderson
Furniture - Welco
Furniture - Bedroom, Office and School - Aaztec Cubicles
Furniture fittings - Nico Manufacturing Ltd
Furniture fittings and accessories - Hafele UK Ltd
Furniture for restuarants and catering facilities - Sandler Seating Ltd
Furniture lifting device - Interface Europe Ltd
Furniture Repairs & Restoration - Branson Leisure Ltd
Furniture, fittings and hardware - manufacturers and importers - Daro Factors Ltd
Garment rails/shoe fittings - Smart F & G (Shopfittings)Ltd
Geeral fixture/ furnishings and equipment - Hipkiss, H, & Co Ltd
General Equipment - Cottam Bros Ltd
General fixtures - Anders + Kern UK Ltd
General Fixtures and Fittings - Coverscreen UK LLP
General fixtures and furnishings - Be-Modern Ltd
General fixtures and furnishings - Elite Trade & Contract Kitchens Ltd
General fixtures/ furnishings/ equipment - AMO Blinds
General fixtures/ furnishings/ equipment - Artex Ltd
General fixtures/ furnishings/ equipment - Barlow Group
General fixtures/ furnishings/ equipment - Covers Timber & Builders Merchants
General fixtures/ furnishings/ Equipment - Laidlaw Ltd
General fixtures/ furnishings/ equipment - Libraco
General furniture - Ellis J T & Co Ltd
General purpose fittings and furnishings - Hettich UK Ltd
Glass fibre UPVC fabrics - Mermet U.K
Greenhouse accessories for Hartley Greenhouses - Hartley Botanic Ltd
GRP enclosures and kiosks - Dewey Waters Ltd
hand drawn tracks - Goelst
Handling equipment - Welco
Handmade Fireplace Briquettes - Collier W H Ltd
Hardwood worktops & bartops - W.L. West & Sons Ltd
Hinges - Cooke Brothers Ltd
Home furniture - Vitra UK Ltd
Hooks & Rails - Roc Secure Ltd
Hotel bedroom Furniture - Ellis J T & Co Ltd
Hotel bedroom furniture and contract furniture - Pedley Furniture International Ltd
Hotel furniture - Laytrad Contracts Ltd
Hotel security boxes - Ritherdon & Co Ltd
Illuminated Mirrors - Astro Lighting Ltd
Import furniture - Kush UK Ltd
Insect Screens - The Flyscreen Company Ltd
Installation Services - Moores Furniture Group Ltd
Instrument cabinet - Ruthin Precast Concrete Ltd
Interior and exterior blinds and awnings - Deans Blinds & Awnings (UK) Ltd
Interior fit outs - OEP Furniture Group PLC
Iso-design venetian blinds - Eclipse Blind Systems Ltd
Key filing, security cabinets and safes - Securikey Ltd
Kiosks - Fibaform Products Ltd
Kitchen furniture - OEP Furniture Group PLC
Leaflet holders - Smart F & G (Shopfittings)Ltd
Leather for walling/upholstry - Edelman Leather
Letter Boxes - Signs Of The Times Ltd
Lifestyle user friendly shelving kits - RB UK Ltd
Litter bins and recycling containers - Glasdon UK Limited
Lockers - Decra Ltd
Lockers - School, Sport, Office, Healthcare, MoD, Police etc - 3d Storage Systems (UK) LTD
Lockers and cubicles - Shackerley (Holdings) Group Ltd incorporating Designer ceramics
Loft hatches - Zled (DPC) Co Ltd
Made-to-measure furnishings - Sanderson
Magnetic picture hooks and coat hooks - Bunting Magnetics Europe Limited
Mailboxes - The Safety Letterbox Company Ltd
Mailing systems - The Safety Letterbox Company Ltd
Manufacturers of display & exhibition systems - Nimlok Ltd
Marble vanity tops - Emsworth Fireplaces Ltd
Material handling equipment - Moresecure Ltd
Mats and matting - PHS Group plc
Matwell frames - PHS Group plc
Mechanical extract units - Greenwood Airvac
Mirrors - Charles Lightfoot Ltd
Mirrors - Ledlite Glass (Southend) Ltd
Mobile Folding Furniture - Spaceright Europe Limited
Moulded grab rails - Cefndy Healthcare
Music Suite Equipment - Black Cat Music
Nameplates and fireplaces - The Delabole Slate Co. Ltd
Natural & polyester fillings for bedding & upholstery - Rawson Fillings Ltd
Natural ventilation products - Greenwood Airvac
Non slip bench mat - Dycem Ltd
Nostalgic gift products - Francis Frith Collection
Notice boards - Arbory Group Limited
Notice Boards - Harewood Products Ltd (Adboards)
Nylon Coated Steel Shelving - Bedford Shelving Ltd
Odour control products - P+L Systems Ltd
P.O.S support - Eclipse Blind Systems Ltd
Pallet racking, shelving systems and lockers - Sperrin Metal Products Ltd
Panel Blinds - Louver-Lite Ltd
Panel glides - Silent Gliss Ltd
Parafin pressure lamp - Tilley International Ltd
Perfect Fit Blinds - Louver-Lite Ltd
Pine beds, bunk beds and headboards - Airsprung Beds Ltd
Pine room paneling, bookcases and display cupboards - Hallidays UK Ltd
Plastic injection moulding for custom furnishing, fittings and parts - Hille Educational Products Ltd
Portable auditorium seating - Sandler Seating Ltd
Portable gas cookers - Tilley International Ltd
Press tooling - Cooke Brothers Ltd
Pressings - Cooke Brothers Ltd
Product finishing - Cooke Brothers Ltd
Public area bins, ashtrays and fire safe bins - Lesco Products Ltd
PVC strip curtains - Mandor Engineering Ltd
Radiator casings - T B S Fabrications
Ready made curtain - Eclipse Blind Systems Ltd
Reconstructed stone fireplaces - Minsterstone Ltd
Recycled Plastic Outdoor Furniture (Picknic Tables, Benches, Seating) - The No Butts Bin Company (NBB), Trading as NBB Outdoor Shelters
Refuse sack holders and fire retardant bins - Unicorn Containers Ltd
Roller blind systems - Goelst
Roller blinds - Marlux Ltd
Roller blinds - Reflex-Rol (UK)
Roller blinds - Tidmarsh & Sons
Roller/roman/venetian/vertical blinds - Silent Gliss Ltd
Roman blinds - Goelst
Room Darkening Systems - CBS (Curtain and Blind Specialists Ltd)
Ropes and twines - Kent & Co (Twines) Ltd
Rose end grab rails - Cefndy Healthcare
Safety films - The Window Film Company UK Ltd
Satallite TV - West London Security
School blinds - CBS (Curtain and Blind Specialists Ltd)
School Fit Out - West London Security
Screens & decorative panels - Jali Ltd
Sculptures - Aden Hynes Sculpture Studios
Sealer units - Pilkington Plyglass
Seating - CPS Manufacturing Co
Seating and tables for schools, waiting areas, train stations, airports, ferry terminals, hospitals and surgeries - Hille Educational Products Ltd
Seating systems - Harrison Thompson & Co Ltd
Security counters & screens - Platonoff & Harris Plc
Security mirrors - Securikey Ltd
Self retracting key reels - Securikey Ltd
Service desking - Grant Westfield Ltd
Shading System - CBS (Curtain and Blind Specialists Ltd)
Shatterproof safety mirrors - Deralam Laminates Ltd
Sheet metal enclosures - Ritherdon & Co Ltd
Shelving - Barton Storage Systems Ltd
Shelving - RB UK Ltd
Shelving - Smart F & G (Shopfittings)Ltd
Shelving systems - E-Z Rect Ltd
Shelving systems - Moresecure Ltd
Shoot bolt Locking Mechanisms - Kenrick Archibald & Sons Ltd
Showroom design - Eclipse Blind Systems Ltd
Slotted angle plastic containers and boxes - Moresecure Ltd
Slotvents - Greenwood Airvac
Soft fabrics - Eclipse Blind Systems Ltd
Solar blinds material - Reflex-Rol (UK)
Solar reflective fabrics - Mermet U.K
Solar shading - Levolux A.T. Limited
Solid surface fabrication - Grant Westfield Ltd
Special Stainless Fabrications - Pland Stainless Ltd
Sprung mattresses, divan bases, pillows and accessories - Airsprung Beds Ltd
Square tube construction systems - Moresecure Ltd
Stainless steel general fixings - Pland
Stainless steel post boxes - Signs Of The Times Ltd
Stainless steel, aluminium, PVC covered, galvanized and brass handrails - SG System Products Ltd
Steel shelving uprights and bracketry - RB UK Ltd
Steel Storage Fittings - TROAX UK Ltd
Steel tube and fittings - Rothley Burn Ltd
Storage - Rothley Burn Ltd
Storage cabinets - Richard Hose Ltd
Storage walls - Avanti Systems
Study, lounge, bedroom, fitted furniture - Hyperion Wall Furniture Ltd
Sun blinds - Tidmarsh & Sons
Sunbeds, lockers and cubicles - Dalesauna Ltd
Support rails and systems - Amilake Southern Ltd
Synthetic grass for display - Artificial Grass Ltd
Tables - Libraco
Tables & Benches - The Stone Catalogue Ltd
Themed features (any material) - Design & Display Structures Ltd
Training rooms installations - Planet Communications Group Ltd
Trolley - Unicorn Containers Ltd
Vandal resistant blinds for public buildings - CBS (Curtain and Blind Specialists Ltd)
Vending machines - Unicorn Containers Ltd
Venetian blinds - Komfort Workspace PLC
Venetian blinds - Tidmarsh & Sons
Vertical and horizontal carousels - Construction Group UK Ltd
Vertical Blind System - Goelst
Vertical blinds - Marlux Ltd
Vertical Blinds - Tidmarsh & Sons
Wall mounted mailboxes - The Safety Letterbox Company Ltd
Wall Murals - Francis Frith Collection
Wall rails - Fixatrad Ltd
Weighing scales and weigh bridges - Phoenix Scales Ltd
Window blind systems, fabrics - vertical, roller, venetian, pleated, woven wood, contract/perofrmance - Louver-Lite Ltd
Window blinds and sunscreening systems - Faber Blinds UK Ltd
Window furnishings, blinds, curtain tracks and poles - Integra Products
Wooden venetian blinds - Eclipse Blind Systems Ltd
Wooden venetian blinds, timbershades - Tidmarsh & Sons
Work benches - Fine H & Son Ltd
Work benches and cupboards - Emmerich (Berlon) Ltd
Workbenches - Barton Storage Systems Ltd
Worktops - The Delabole Slate Co. Ltd
Worktops and vanity units - Domus Tiles Ltd
Woven fabrics - Mermet U.K
Write-on, wipe-off products - Tektura Plc
Writing boards and notice boards - AudiocomPendax Ltd
Zinc Plated Steel Shelving - Bedford Shelving Ltd

N11 Domestic kitchen fittings

Appliances - Beko plc
Basin traps and wastes - Wirquin
Built -in cookers - Stoves PLC
Built-in dishwashers - Stoves PLC
Built-in ovens and hobs - Rangemaster
Built-in refrigeerators - Stoves PLC
Cookers - Stoves PLC
Cookers and electrical appliances - Electrolux Domestic Appliances
Customised work tops: stainless steel - Azimex Fabrications Ltd
Domestic kitchen appliances - De Longhi Limited
Domestic kitchen fitting - Johnson & Johnson PLC
Domestic Kitchen Fittings - Anderson C F & Son Ltd
Domestic kitchen fittings - Be-Modern Ltd
Domestic kitchen fittings - Commodore Kitchens Ltd
Domestic kitchen fittings - Covers Timber & Builders Merchants
Domestic kitchen fittings - Elite Trade & Contract Kitchens Ltd
Domestic kitchen fittings - Hipkiss, H, & Co Ltd
Domestic kitchen fittings - Jenkins Newell Dunford Ltd
Domestic kitchen fittings, Lever action - Shavrin Levatap Co Ltd
Domestic kitchen furniture - Alno (United Kingdom) Ltd
Domestic kitchen furniture - Ball William Ltd
Domestic kitchen sinks, taps and accessories - Carron Phoenix Ltd
Electrical appliances - Pifco Ltd
Fitted Kitchens - Anson Concise Ltd
Fitted kitchens - Miele Co Ltd
Free standing cookers - Stoves PLC
Granite kitchen work top - Emsworth Fireplaces Ltd
Kitchen and bathroom furniture - Dennis & Robinson Ltd
Kitchen Appliances - Waterline Limited
Kitchen fittings - Magnet Ltd
Kitchen fittings - Symphony Group PLC, The
Kitchen furniture - Ellis J T & Co Ltd
Kitchen furniture - Gower Furniture Ltd
Kitchen furniture - Waterline Limited
Kitchen sinks - Kohler UK
Kitchen units - Ball William Ltd
Kitchen worksurfaces - Twyford Bushboard
Kitchen worktops - Kronospan Ltd
Kitchens - Bell & Co Ltd
Kitchens - Moores Furniture Group Ltd
Kitchens - Smith & Choyce Ltd
Kitchens - Taskworthy Ltd
Kitchens, kitchen furniture, taps - Porcelanosa Group Ltd
Laminated + Solid Hardwood Worktops - Meyer Timber Limited
Mini Kitchens - Anson Concise Ltd
Oil stoves - Gazco Ltd
Range Cookers - Gazco Ltd
Range cookers - Rangemaster
Shelving - Bedford Shelving Ltd
Sinks - Astracast PLC

N11 Domestic kitchen fittings (con't)

Solid hardwood worktops - Junckers Limited
Solid wood + melamine worktops - Deralam Laminates Ltd
Stoves inset and free standing - Emsworth Fireplaces Ltd
Takeaway Kitchens - Jewson Ltd
Taps - kitchen - Hansgrohe
Traditional kitchen taps and Mixers - Shavrin Levatap Co Ltd
Vanitories worktops - Burlington Slate Ltd
Work surfaces - Kirkstone
Worktops - Formica Ltd
Worktops - Twyford Bushboard

N12 Catering Equipment

Automatic fill water boilers - Calomax Ltd
Bacteria Grease traps - Progressive Product Developments Ltd
Bar Tender Sinks - Pland Stainless Ltd
Bars - Smith & Choyce Ltd
Canteen tables and chairs - Q.M.P
Cast iron multifuel stoves - Valor
Catering - Welco
Catering and culinary fittings - DuPont Corian
Catering equipment - Barlow Group
Catering equipment - Electrolux Foodservice
Catering equipment - Franke Sissons Ltd
Catering equipment - Miele Co Ltd
Catering Equipment - Stamford Products Limited
Catering sinks and tables - Pland Stainless Ltd
Catering storage - Hispack Limited
Catering water boilers - Calomax Ltd
Coffee Machines - Pressure Coolers Ltd T/A Maestro Pressure Coolers
Coffee percolators - TSE Brownson
Commercial catering and bar equipment - Imperial Machine Co Ltd
Commercial catering facilities - design, supply, and installation UK wide - Lockhart Catering Equipment
Counter top water boilers - Calomax Ltd
Counters and Serveries - Viscount Catering Ltd
Deep fat fryers - Preston & Thomas Ltd
Dishwashers - Clenaware Systems Ltd
Display carts - TSE Brownson
Domestic and commercial waste disposal units - Max Appliances Ltd
Domestic kitchen foodwaste disposers - The Haigh Tweeny Co Ltd
Filtered water dispensers - PHS Group plc
Fish and chip frying ranges - Preston & Thomas Ltd
Fish and Chip Ranges & Catering equipment - Nuttall, Henry (Viscount Catering Ltd)
Food and drink vending machines - Hillday Limited
Food processing - Hemsec Panel Technologies
Freezers - Hemsec Panel Technologies
Gas stoves - Valor
Glass dryers - Clenaware Systems Ltd
Glasswashers - Clenaware Systems Ltd
Grease traps, oil and grease water separators - Progressive Product Developments Ltd
Grease treatment plants - Progressive Product Developments Ltd
Heater service counter tops - TSE Brownson
Heavy duty catering equipment - Viscount Catering Ltd
High performance kitchen furniture - Rixonway Kitchens Ltd
Insulated Urns - TSE Brownson
Kitchen equipment - Magnet Ltd
Light duty catering equipment - Viscount Catering Ltd
Medium duty cateing equipment - Viscount Catering Ltd
Milk heaters - TSE Brownson
Mini kitchens - Anson Concise Ltd
Rifrgeration - Viscount Catering Ltd
Show cases - TSE Brownson
Stainless Steel Cabinets - GEC Anderson Ltd
Stainless steel catering equipment - Moffat Ltd, E & R
Stainless steel catering equipment - Pland
Stainless Steel Shelving - GEC Anderson Ltd
Stainless steel shelving systems for kitchens and coldrooms - Bedford Shelving Ltd
Stainless Steel Sinks - GEC Anderson Ltd
Stainless Steel Worktops - GEC Anderson Ltd
Steel, Stainless steel and aluminium - Speedfab Midlands
Tray/plate dispensers - TSE Brownson
Trolleys - TSE Brownson
Undersink Chillers - Pressure Coolers Ltd T/A Maestro Pressure Coolers
Vending machines - Autobar Vending Services Limited
Vending systems - The Kenco Coffee Company
Ventilated ceiling system for commercial cooking areas - Britannia Kitchen Ventilation Ltd
Ventilation / gas interlocks - TCW Services (Control) Ltd
Wall kettles - Calomax Ltd
Waste disposal equipment, balers, shredders, drinks can crushers - Tony Team Ltd
Waste disposal units and catering cupboards - Pland Stainless Ltd
Water calcium treatment - Clenaware Systems Ltd
Water softeners - Clenaware Systems Ltd

N13 Sanitary appliances/ fittings

Accessories - Deva Tap Co Ltd
Accessories - Ideal-Standard Ltd
Acrylic baths - Vitra (UK) Ltd
Aluminium Shower outlets - Alutec
Aluminium shower outlets - Marley Plumbing and Drainage
Anti-ligature system for hospital cubicle/shower curtain tracks - President Blinds Ltd
Anti-vandal showers - Shavrin Levatap Co Ltd
Architechtural hardware - Samuel Heath & Sons plc
Architectural glass - Creative Glass
Automatic self Cleaning Toilets - Healthmatic Ltd
Bath and kitchen seals - Homelux Nenplas
Bath Screens - Lakes Bathrooms Limited
Bath screens - Majestic Shower Co Ltd
Bathing equipment for elderly and disabled - Chiltern Invadex
Bathroom accessories - H E W I (UK) Ltd
Bathroom Accessories - Marflow Eng Ltd
Bathroom accessories - Samuel Heath & Sons plc
Bathroom accessories - Triton PLC
Bathroom accessories - Vitra (UK) Ltd
Bathroom and shower accessories - Marleton Cross Ltd
Bathroom and showering equipment - Amilake Southern Ltd
Bathroom fittings - Rothley Burn Ltd
Bathroom fittings - Silverdale Bathrooms
Bathroom fittings - William Hopkins & Sons Ltd
Bathroom furniture - Ellis J T & Co Ltd
Bathroom furniture - Kohler UK
Bathroom furniture - Vitra (UK) Ltd
Bathroom Pods - Polybeam Ltd
Bathroom Pods - RB Farquhar Manufacturing Ltd
Bathroom suites - Kohler UK
Bathroom systems - Multikwik Ltd
Bathrooms - Bell & Co Ltd
Bathrooms - Fired Earth
Bathrooms - Vitra (UK) Ltd
Bathrooms, bathroom accessories and furniture, Sanitary ware, taps - Porcelanosa Group Ltd
Baths - Astracast PLC
Baths - Ideal-Standard Ltd
Baths and Bathroom suites - Silverdale Bathrooms
Baths and Shower Trays - Kaldewei
Bedroom furniture - Bell & Co Ltd
Bidets - Kohler UK
Brassware for bathrooms - Silverdale Bathrooms
Brassware, showers and accessories - Hansgrohe
Cabinet furniture - H E W I (UK) Ltd
Ceramic and acrylic sanitaryware - Armitage Shanks Ltd
Childrens Washrooms - Amwell Systems Ltd
Chrome bathroom accessories - William Hopkins & Sons Ltd
Cisterns - Multikwik Ltd
Cisterns - Wirquin
Cubicle tracks - Marlux Ltd
Curved glass bathroom bowls - Novaglaze Limited
Domestic kitchen sinks and taps - Astracast
Domestic showering products - Kohler Mira Ltd
Drinking fountains and water coolers - Pressure Coolers Ltd T/A Maestro Pressure Coolers
Duct Panelling - Amwell Systems Ltd
Duct panels - Dunhams of Norwich
Dušo sport shower column - Horne Engineering Ltd
Electric mixer and power showers, hand wash units and water heaters - Triton PLC
Electric showers - Heatrae Sadia Heating Ltd
Electric towel rails - Ecolec
Enclosures - Ideal-Standard Ltd
Ensuite shower pods - RB Farquhar Manufacturing Ltd
Fill valves - Multikwik Ltd
Filter water taps-Springflow - Astracast
Fitted Bathrooms - Symphony Group PLC, The
Fittings for disabled and aged - H E W I (UK) Ltd
Flexible plumbing fittings - Multikwik Ltd
Flush pipes - Airflow (Nicoll Ventilators) Ltd
Flushing systems - Wirquin
Flushvalves - Multikwik Ltd
Frameless shower screens - ShowerSport Ltd
Grab rails - Rothley Burn Ltd
Grabrails - William Hopkins & Sons Ltd
Group thermostatic valves - Delabie UK Ltd
Hand dryers - Warner Howard Group Ltd
Handwash units and Sanitaryware - Wallgate Ltd
Hygiene equipment - Imperial Machine Co Ltd
Hygiene equipment - P+L Systems Ltd
In Wall Frames - Wirquin
In-Build Automatic Toilets - Healthmatic Ltd
Incinerators - Combustion Linings Ltd
Intumescent mastics - Bollom Fire Protection Ltd
Jacuzzi®, Whirlpool Baths - Jacuzzi® Spa and Bath Ltd
Kick Space electric heaters - Ecolec
Kits - Ideal-Standard Ltd
Lead Sheet Coverings/ Flashings - BLM British Lead
Modern bathroom taps and mixers - Shavrin Levatap Co Ltd
Modular multi-User Toilets - Healthmatic Ltd
Nappy changing units - Magrini Ltd
Optitherm thermostatic tap - Horne Engineering Ltd
Outdoor Hot tubs - Jacuzzi® Spa and Bath Ltd
Pan connectors - Multikwik Ltd
Pillar taps - Tyco Waterworks - Samuel Booth
Plumbing fittings - Flowflex Components Ltd
Power, mixer valve and electric showers - NewTeam Ltd
Ready Plumbed Modules - Twyford Bushboard
Safety accessories for bathrooms (seats, handles, supports) - Lakes Bathrooms Limited
Sanitary appliances - ArjoHuntleigh UK
Sanitary appliances - Bathstore
Sanitary appliances & fittings - Olby H E & Co Ltd
Sanitary appliances and accessories - Secure Sanitaryware, Toilets, Basins, Showers, Vanity units and Washroom accessories - Aaztec Cubicles
Sanitary appliances and fittings - Bristan Group
Sanitary appliances and fittings - Cefndy Healthcare
Sanitary appliances and fittings - Guest (Speedfit) Ltd, John
Sanitary appliances/ fittings - Covers Timber & Builders Merchants
Sanitary appliances/ fittings - Geberit Ltd
Sanitary appliances/ fittings - Harris & Bailey Ltd
Sanitary appliances/fittings - Allgood plc
Sanitary appliances/fittings - FC Frost Ltd
Sanitary appliances/fittings - The Haigh Tweeny Co Ltd
Sanitary Appliances/Fittings - Wybone Ltd
Sanitary bins - Unicorn Containers Ltd
Sanitary disposal units, washroom and baby room products - Cannon Hygiene Ltd
Sanitary equipment and accessories - Ideal-Standard Ltd
Sanitary Fittings - Acorn Powell Ltd
Sanitary Fittings - Dart Valley Systems Ltd
Sanitary fittings - DuPont Corian
Sanitary fittings - Franke Sissons Ltd
Sanitary Fittings - Wallgate Ltd
Sanitary fittings - Wirquin
Sanitary fixings - Artur Fischer (UK) Ltd
Sanitary towel incinerator and towel rails - Consort Equipment Products Ltd
Sanitary ware and appliances - Diamond Merchants
Sanitary ware and fittings - Twyford Bathrooms
Sanitaryware, associated fittings and accessories - Shires Bathrooms
Sanitory units - Max Appliances Ltd
Sauna - Golden Coast Ltd
Saunas, steam rooms and spa baths - Dalesauna Ltd
Seats for bathrooms - Lakes Bathrooms Limited
Sharps disposal units - Inter Public Urban Systems Uk Ltd
Shower accessories - Bristan Group
Shower accessories - Delabie UK Ltd
Shower accessories - NewTeam Ltd
Shower Curtains - Roc Secure Ltd
Shower Doors - Lakes Bathrooms Limited
Shower doors - Majestic Shower Co Ltd
Shower doors and carer screens - Norton Engineering Alloys Co Ltd
Shower doors, enclosures, bath screen and shower trays - Kohler UK
Shower doors, trays bath screens and shower mixers - Matki plc
Shower enclosers - Majestic Shower Co Ltd
Shower enclosures - Contour Showers Ltd
Shower Enclosures - Lakes Bathrooms Limited
Shower enclosures and towel rails - Kermi (UK) Ltd
Shower grilles - Norton Engineering Alloys Co Ltd
Shower products - Aqualisa Products
Shower pumps - NewTeam Ltd
Shower screens - Majestic Shower Co Ltd
Shower trays, cubicles and bath screens - Coram Showers Ltd
Shower valves - Ideal-Standard Ltd
Showering systems - Kohler UK
Showers - Samuel Heath & Sons plc
Showers - Silverdale Bathrooms
Showers & Mixers - Aquacontrol
Showers and bathrooom fittings - Redring
Showers and Baths - Hansgrohe
Sinks - Franke UK Ltd
Sinks, Basins & Barhs - The Stone Catalogue Ltd
Spa baths - Astracast PLC
Specialist sanitaryware - Wallgate Ltd
Stainless Steel Basins - GEC Anderson Ltd
Stainless Steel Faucets - Vola
Stainless steel sanitary appliances and fittings - Blücher UK Ltd
Stainless Steel Sanitaryware - GEC Anderson Ltd
Stainless steel sanitaryware - Pland Stainless Ltd
Stainless steel sanitaryware and washroom equipment - Pland
Stainless Steel Sluices - GEC Anderson Ltd
Stainless Steel Urinals - GEC Anderson Ltd
Stainless Steel W.C.s - GEC Anderson Ltd
Stainless Steel Washing Troughs - GEC Anderson Ltd
Stainless steel, ceramic, synthetic sinks and shower cubicles - Rangemaster
Taps - Astracast PLC
Taps - Deva Tap Co Ltd
Taps - Franke UK Ltd
Taps - Samuel Heath & Sons plc
Taps - Shavrin Levatap Co Ltd
Taps and mixers - Ideal-Standard Ltd
Taps and valves - Opella Ltd
Taps for the disabled and elderly - Shavrin Levatap Co Ltd
Taps, mixers, and showers - Pegler Ltd
Taps, showers and thermostatic mixers - Grohe Ltd
Thermostatic mixing valves - Delabie UK Ltd
Thermostatic shower controls - Delabie UK Ltd
Thermostatic shower panels - Horne Engineering Ltd
Thermostatic shower valves - Shavrin Levatap Co Ltd
Time flow shower controls - Delabie UK Ltd
Toilet Cubicles - Amwell Systems Ltd
Toilet cubicles - Grant Westfield Ltd
Toilet Cubicles - Relcross Ltd
Toilet cubicles and washroom systems - Moffett Thallon & Co Ltd
Toilet seats - Pressalit Care
Toilet Seats - Wirquin
Toilet, shower and changing cubicles - Dunhams of Norwich
Tool hire - Jewson Ltd
Towel and soap dispensers - Kimberly-Clark Ltd
Trays - Ideal-Standard Ltd
Urinal controls - Delabie UK Ltd
Urinal filters - Cistermiser Ltd
Urinal flush controls - Cistermiser Ltd
Valves - Horne Engineering Ltd
Vandel resistant toilets - Inter Public Urban Systems Uk Ltd
Vanity units - Dunhams of Norwich
Vitreous china ceramic sanitaryware and brassware - Vitra (UK) Ltd
Vitrious Sanitryware - Relcross Ltd
W. C.'s - Ideal-Standard Ltd
Walk-in Shower Enclosures - Lakes Bathrooms Limited
Warm air hand and face dryers - Airdri Ltd
Warm air hand dryers - Wandsworth Elecrtrical Ltd
Wash basins - Ideal-Standard Ltd
Washbasins - Kohler UK
Washroom control systems - Cistermiser Ltd
Washroom Equipment - Relcross Ltd
Washroom Services - PHS Group plc
Washroom system - DuPont Corian
Washrooms - SIG Interiors
Washrooms and washroom cubicles - Amwell Systems Ltd
Wastes - Deva Tap Co Ltd
Water saving taps - Delabie UK Ltd
Water taps and mixers - Barber Wilsons & Co Ltd
Waterless Urinals - Relcross Ltd
WC flush controls - Delabie UK Ltd
WC flush valves - Cistermiser Ltd
WC suites - Kohler UK
Wetrooms - Wirquin
Whirlpool Baths - Jacuzzi® Spa and Bath Ltd
Whirlpools, steam showers - Hansgrohe

N15 Signs/ Notices

3D. Domed labels - Mockridge Labels & Nameplates Ltd
All health & safety signs - Masson Seeley & Co Ltd
Anit-vandal signage - Trico V.E. Ltd (Signage Division)
Anodised aluminium labels - Mockridge Labels & Nameplates Ltd
Architectural aluminium systems for signs - Universal Components Ltd
Architectural signage - Signs & Labels Ltd
Automatic / Self-Testing Emergency Lighting - P4 Limited
Badges - Mockridge Labels & Nameplates Ltd
Banners - GB Sign Solutions Limited
Bespoke Signage - Concept Sign & Display Ltd
Bespoke Signs - The Delabole Slate Co. Ltd
Braille signs - Signs & Labels Ltd
Canvas prints - Francis Frith Collection
Carved stone - Caithness Flagstone Limited
Cast polyester resin molded letters - Metalline Signs Ltd
Cast signs - Signs Of The Times Ltd
CCTV mandatory sinage - Hoyles Electronic Developments Ltd
Chemical Etching - Mockridge Labels & Nameplates Ltd
Coats of arms - Signs Of The Times Ltd
Commemorative plaques, coats of arms and architectural lettering in cast bronze & cast aluminum - Metalline Signs Ltd
Commercial signs - Dee-Organ Ltd
Company logos - Signs Of The Times Ltd
Contract enameling - Trico V.E. Ltd (Signage Division)
Corporate signs - Pryorsign
Custom Made signage - Stocksigns Ltd
Decorative etching - Mockridge Labels & Nameplates Ltd
Design service - Pryorsign
Digital Signage - Concept Sign & Display Ltd
Direction signs - Simplex Signs Limited
Directory systems - Simplex Signs Limited
Doors - Pearce Security Systems
Electric, illuminated and neon signs - Pearce Signs Ltd
Enameling cladding panels - Trico V.E. Ltd (Signage Division)
Engraved metals - Signs Of The Times Ltd
Engraved signs - Signs & Labels Ltd
Engraved slate - Signs Of The Times Ltd
Engravers - Abbey Nameplates Ltd
Fabricated signs - Simplex Signs Limited
Financial services signs & structures - Pryorsign
Fingerposts - GB Sign Solutions Limited
GRP structures - Pryorsign
Health and Safety Signage - Stocksigns Ltd
Heritage and conservation signs - Pryorsign
Housesigns - The Delabole Slate Co. Ltd
Illuminated signs - Signs & Labels Ltd
Instrumentation panels - Pryorsign
Interactive signs - Signature Ltd
Internally illminated signs - Signature Ltd
Large format digital printing - Signs & Labels Ltd
Large format printing - Zephyr The Visual Communicators
LED signs - Pearce Signs Ltd
Marking inks and products - Eyre & Baxter Limited
Metal lettering - Simplex Signs Limited
Nameplates - Eyre & Baxter Limited
Nameplates - Mockridge Labels & Nameplates Ltd
Nameplates & Memorial Tablets in Metal & Plastic - Abbey Nameplates Ltd
Notice board material - Forbo Flooring
Notice panels - Stamford Products Limited
Noticeboards, pinboards - Magiboards Ltd
Personnel equipment - Welco
Pinboards - Sundeala Ltd
Plaques - GB Sign Solutions Limited
Plaques - stainless steel, brass, engraved - Mockridge Labels & Nameplates Ltd
Poster displays - Fairfield Displays & Lighting Ltd
Printed banner material - Mermet U.K
Professional signs - Abacus Signs
RICS & RIBA Site Boards - Abacus Signs
Road traffic signs - Dee-Organ Ltd
Safety signs - ASCO Extinguishers Co. Ltd
Safety signs - Eyre & Baxter Limited
Safety signs - Fife Fire Engineers & Consultants Ltd
Sign manufacturers - Blaze Signs Ltd

N15 Signs/ Notices (con't)

Sign posts - Fabrikat (Nottingham) Ltd
Sign Solutions - Concept Sign & Display Ltd
Sign systems - Signs & Labels Ltd
Signage - Fairfield Displays & Lighting Ltd
Signage - Furnitubes International Ltd
Signage - Mockridge Labels & Nameplates Ltd
Signage & graphics - Komfort Workspace PLC
Signboards - Duroy Fibreglass Mouldings Ltd
Signing Systems - Concept Sign & Display Ltd
Signing Systems - Ward & Co (Letters) Ltd
Signs - Allgood plc
Signs - Dee-Organ Ltd
Signs - Doorfit Products Ltd
Signs - GB Sign Solutions Limited
Signs - Hoyles Electronic Developments Ltd
Signs - Neocrylic Signs Ltd
Signs - design - Masson Seeley & Co Ltd
Signs - manufacture - Masson Seeley & Co Ltd
Signs - site surveying - Masson Seeley & Co Ltd
Signs and notices - Arbory Group Limited
Signs and Notices - brissco signs & graphics
Signs and notices - Focal Signs Ltd
Signs and notices - Trico V.E. Ltd (Signage Division)
Signs and Plaques - Sign Industries
Signs and sign systems - Simplex Signs Limited
Signs installation - Masson Seeley & Co Ltd
Signs/ notices - Laidlaw Ltd
Signs/ Notices - Leander Architectural
Signs/ notices - Stocksigns Ltd
Signs/notices - Duroy Fibreglass Mouldings Ltd
Signs/notices - Rivermeade Signs Ltd
Stencils - Eyre & Baxter Limited
Traffic management systems - Marshalls Mono Ltd
Traffic signs - Signature Ltd
uv Stable Signage - Trico V.E. Ltd (Signage Division)
Viewpoint window display signs - Pryorsign
Vinyl signs - Invicta Window Films Ltd
Visual communication equipment - Spaceright Europe Limited
Vitreous Enamal Signs - Trico V.E. Ltd (Signage Division)
Wall plaques - Rogers Concrete Ltd
Waterproof, vandal resistant illuminated, health and safety and shop signs, nameplates and fascias labels - Pryorsign
Wayfinding schemes - Masson Seeley & Co Ltd
Whiteboards - Magiboards Ltd

N16 Bird/ Vermin control

Bait boxes - P+L Systems Ltd
Bird control - South east Bird Control
Bird Nets - P+L Systems Ltd
Bird Spikes - P+L Systems Ltd
Bird/ vermin control - Terminix Property Services
Brushstrip for pest and bird control - Kleeneze Sealtech Ltd
Cladding Insect Mesh - The Flyscreen Company Ltd
Flyscreens - P+L Systems Ltd
Industrial and Commercial Insect Screens - The Flyscreen Company Ltd
Insect Screens - The Flyscreen Company Ltd
Pest Control Services - Rentokil Specialist Hygiene

N20 Safety Equipment

Anti slip metal flooring - Graepel Perforators Ltd
Anti-slip sighting bars - Redman Fisher Engineering
Barriers - Q.M.P
Body armour - Meggitt Armour Systems
Breathing apparatus - Draeger Safety UK Ltd
Carbon monoxide detectors - Firemaster Extinguisher Ltd
Confined Space access Equipment - Didsbury Engineering Co Ltd
Confined space equipment - MSA (Britain) Ltd
Davits - Didsbury Engineering Co Ltd
Disabled toilet alarms - Hoyles Electronic Developments Ltd
Domestic smoke alarms - Firemaster Extinguisher Ltd
Emergency evacuation equipment for the mobility impaired - Evac+Chair International Ltd
Emergency Lighting Exit Signs - P4 Limited
Emergency safety showers and eye and face wash units - Pressure Coolers Ltd T/A Maestro Pressure Coolers
Environemental monitoring & measurement equipment - Casella Measurement Ltd
Evacuation Equipment - Enable Access
Explosion containment tubes - Attwater & Sons Ltd
Fall arrest equipment - Capital Safety Group (NE) Ltd
Fall arrest systems and Cable fall arrest systems - Latchways plc
Fire & Burglary Resistant Equipment - Gunnebo UK Limited
Fire blankets - Firemaster Extinguisher Ltd
Fire extinguishers - Tyco Safety Products
Fire hose and fittings - Tyco Safety Products
Fire hose reels - Kidde Fire Protection Services Ltd
Firemaster powder, water, AFFF foam, and carbon dioxide fire extinguishers - Firemaster Extinguisher Ltd
Footwear - industrial - Briggs Industrial Footwear Ltd t/a Briggs Safety Wear
Fragile roof and rooflight covers - Latchways plc
Gas detection equipment - Draeger Safety UK Ltd
Gas detection systems - Blakell Europlacer Ltd
Guards & Cages - Roc Secure Ltd
Harnesses, lanyards and fall arrest blocks - Miller by Honeywell - Honeywell Safety Products
Hoists - Didsbury Engineering Co Ltd
Horizontal Lifelines - Capital Safety Group (NE) Ltd
Industrial footwear - Briggs Industrial Footwear Ltd t/a Briggs Safety Wear
Ladder safety post - Bilco UK Ltd
Ladders and safety cages - Redman Fisher Engineering
Personal Protective Equipment - Briggs Industrial Footwear Ltd t/a Briggs Safety Wear
Personal Protective equipment - MSA (Britain) Ltd
Respiratory filter masks - Draeger Safety UK Ltd
Ropes and sashcords - Marlow Ropes Ltd
Safety & Security - HC Slingsby PLC
Safety and Security Equipment - Welco
Safety equipment - Focal Signs Ltd
Safety equipment - Hoyles Electronic Developments Ltd
Safety equipment - Interface Europe Ltd
Safety equipment - Kee Klamp Ltd
Safety equipment - Metreel Ltd
Safety equipment - MSA (Britain) Ltd
Safety equipment - Radiodetection Ltd
Safety eyebolts - Latchways plc
Safety harnesses etc - Capital Safety Group (NE) Ltd
Safety interlocking systems - Castell Safety International
Safety items - Roc Secure Ltd
Safety treads - Trimplex Safety Tread
Safety, drench, decontamination, emergency showers and eyebaths - Hughes Safety Showers Limited
Safty systems for Ladder and Roof - Latchways plc
Special purpose safety equipment - Hazard Safety Products Ltd
Sports Markings - Sportsmark Group Ltd
Steel security doorsets - Sunray Engineering Ltd
Vertical access fall safe access ladders - Miller by Honeywell - Honeywell Safety Products
Vertical Lifelines - Capital Safety Group (NE) Ltd
Water safety products - Arbory Group Limited
Welding curtains and hangers - Nederman UK
Workwear - Briggs Industrial Footwear Ltd t/a Briggs Safety Wear

N21 Storage Equipment

Adjustable Pallet Racking - Barton Storage Systems Ltd
Adjustable pallet racking - Link 51 (Storage Products)
Aluminium shelving brackets - Tebrax Ltd
Architectural products - Arbory Group Limited
Archive storage - Construction Group UK Ltd
Bespoke/Musical instrument storage - Amadeus
Cantilever Racking - Sperrin Metal Products Ltd
Chemical Storage Tanks - Titan Environmental Ltd
Coldstores - Carter Retail Equipment Ltd
Construction products - Link 51 (Storage Products)
Cupboards, lockers - Q.M.P
Desk-side storage units - Rackline Systems Storage Ltd
Document storage shelving systems - Link 51 (Storage Products)
Drum storage - Q.M.P
Filing systems - Construction Group UK Ltd
Heavy duty shelving systems - Link 51 (Storage Products)
Library shelving - Sperrin Metal Products Ltd
Lockers - Aaztec Cubicles
Lockers - Barton Storage Systems Ltd
Lockers - Detlef Muller Ltd
Lockers - Grant Westfield Ltd
Lockers - electronic lock - Helmsman
Lockers - laminate - Helmsman
Lockers - mild steel - Helmsman
Lockers - wooden - Helmsman
Lockers and cubicles - LSA Projects Ltd
Lockers and personal storage - Moresecure Ltd
Lockers, Cloakroom equipment, Cabinets - 3d Storage Systems (UK) LTD
Manufactureres of quality filing and storage systems - Railex Systems Ltd
Megyanine Floors - Hi-Store Ltd
Mobile and static shelving equipment - Railex Systems Ltd, Elite Division
Mobile shelving - Construction Group UK Ltd
Mobile shelving - Link 51 (Storage Products)
Mobile Storage Systems - Railex Systems Ltd
Moblie shelving systems - Moresecure Ltd
Pallet parking systems - Moresecure Ltd
Pallet Racking - Construction Group UK Ltd
Picking systems - Construction Group UK Ltd
Safes and Key cabinets - Duffells Limited
Shelving - Bedford Shelving Ltd
Shelving - Lesco Products Ltd
Shelving / Racking - Smart F & G (Shopfittings)Ltd
Shelving and storage racking - Construction Group UK Ltd
Small parts storage - Link 51 (Storage Products)
Small storage - Moresecure Ltd
Specialist Joinery - Taskworthy Ltd
Stainless steel cabinets IP65 - Kensal CMS
Storage - Rothley Burn Ltd
Storage & shelving equipment - Broxap Ltd
Storage cabinets - Unicorn Containers Ltd
storage equipment - Emmerich (Berlon) Ltd
Storage equipment - Envopak Group Ltd
Storage equipment - Garran Lockers Ltd
Storage equipment - Hispack Limited
Storage Equipment - Hi-Store Ltd
Storage equipment - Kee Klamp Ltd
Storage equipment - Libraco
Storage equipment - Logic Office Contracts
Storage equipment - Project Office Furniture PLC
Storage equipment - Rackline Systems Storage Ltd
Storage Equipment - SSI Schaefer Ltd
Storage equipment - Welco
Storage Equipment - Wybone Ltd
Storage equipment, Small Parts Storage - Barton Storage Systems Ltd
Storage furniture - John Pulsford Associates Ltd
Storage racks and shelving - Arc Specialist Engineering Limited
Storage systems - Komfort Workspace PLC
Storage systems - Moresecure Ltd
Task lighting - Anglepoise Lighting Ltd
Vanity units - LSA Projects Ltd
Wallstorage systems - Fastwall Ltd
Waste containment - Randalls Fabrications Ltd

N22 Office Equipment

Audiovisual Presentation Systems - AudiocomPendax Ltd
Binders - Quartet-GBC UK Ltd
Boardroom installations - Planet Communications Group Ltd
Cabinets - Railex Systems Ltd
Carousels - Railex Systems Ltd
Church seating & accessories - Knightbridge Furniture Productions Ltd
Computer document paper handling systems - Neopost Ltd
Conference and training room systems - track based - Magiboards Ltd
Conference room installations - Planet Communications Group Ltd
Desk top products - Lesco Products Ltd
Domestic and commercial office furniture - Ball William Ltd
Education and Office Storage - Stamford Products Limited
Electronic file tracking - Rackline Systems Storage Ltd
Ergonomic products - Lesco Products Ltd
Fire & Burglary Resistant Equipment - Gunnebo UK Limited
Flipcharts - Magiboards Ltd
Floor, fire vents, Smoke vents and roof hatches - Bilco UK Ltd
Folder-inserter systems - Neopost Ltd
Franking machines - Neopost Ltd
Furniture - HEM Interiors Group Ltd
Home office to the housebuilder market - Hammonds - New Home Division
Home Offices - Moores Furniture Group Ltd
Interactive whiteboards - Quartet-GBC UK Ltd
Laminators - Quartet-GBC UK Ltd
Letter openers & extractors - Neopost Ltd
Lockers, Cloakroom equipment, Cabinets - 3d Storage Systems (UK) LTD
Magnetic white boards, hooks and drawing pins - Bunting Magnetics Europe Limited
Mailing systems - Neopost Ltd
Mailroom & packaging - HC Slingsby PLC
Mailroom equipment & supplies - Neopost Ltd
Mobile Shelving - Sperrin Metal Products Ltd
Office - Welco
Office chairs - KAB Seating Ltd
Office equipment - Antocks Lairn Ltd
Office equipment - Carleton Furniture Group Ltd
Office equipment - EFG Office Furniture Ltd
Office equipment - Envopak Group Ltd
Office equipment - HC Slingsby PLC
Office equipment - Libraco
Office Equipment - SSI Schaefer Ltd
Office equipment - Vista Plan International PLC
Office filing systems - Railex Systems Ltd
Office fit out - Staverton (UK) Ltd
Office Fit Out - West London Security
Office furniture - Vitra UK Ltd
Office furniture - Fray Design Ltd
Office furniture - Girsberger London
Office furniture - Haworth UK Ltd
Office furniture - John Pulsford Associates Ltd
Office furniture - Komfort Workspace PLC
Office furniture - Laytrad Contracts Ltd
Office furniture - OEP Furniture Group PLC
Office furniture - Progress Work Place Solutions
Office furniture - Project Office Furniture PLC
Office furniture - Staverton (UK) Ltd
Office furniture and accessories - Lesco Products Ltd
Office furniture including storage cabinets and desks - Flexiform Business Furniture Ltd
Office furniture, desking, storage, boardroom and conference furniture - Hands of Wycombe
Office furniture, seating and board room tables - Steelcase Strafor Plc
Office furniture, seating and storage - Desking Systems Ltd
Office racking, storage cabinets, mobile shelving and filing systems - Rackline Systems Storage Ltd
Office seating - Evertaut Ltd
Office seating - Laytrad Contracts Ltd
Office seating and desking - Metalliform Products Plc
Office storage equipment and desking - Bisley Office Equipment
Offices - Taskworthy Ltd
Photocopiers - Oce (UK) Ltd
Pigeon Holes - Jali Ltd
Planning boards - Magiboards Ltd
Post room furniture - Lesco Products Ltd
Postroom equipment & supplies - Neopost Ltd
Printroom equipment - Envopak Group Ltd
PVC printers - Zephyr The Visual Communicators
Rail presentation systems - Quartet-GBC UK Ltd
Reception - Evertaut Ltd
Reception desks/counters - Smith & Choyce Ltd
Residential furniture - Carleton Furniture Group Ltd
Rubber stamps - Eyre & Baxter Limited
Scales, electronic postal - Neopost Ltd
Seating and executive furniture - Verco Office Furniture Ltd
Seating and tables for schools, waiting areas, train stations, airports, ferry terminals, hospitals and surgeries - Hille Educational Products Ltd
Seating equipment - Evertaut Ltd
Shelving - Bedford Shelving Ltd
Shredders - Quartet-GBC UK Ltd
Side-opening tambour cabinets - Rackline Systems Storage Ltd
Software, laserprinting management - Neopost Ltd
Software, mail & despatch management - Neopost Ltd
Staff locators - Magiboards Ltd
Storage and shelving - HC Slingsby PLC
System desking & sorage - Verco Office Furniture Ltd
Tables - offices - Verco Office Furniture Ltd
Task lights - Lesco Products Ltd
Tea stations - Anson Concise Ltd
Visual display products - Lesco Products Ltd
Work stations & work benches - Moresecure Ltd
Writing boards, Rollerboards, Notice boards, Rail systems - Spaceright Europe Limited

N23 Special purpose fixtures/ furnishings/equipment

Auditorium, theatre, multi purpose, public area and courtroom seating - Race Furniture Ltd
Automatic / Self-Testing Emergency Lighting - P4 Limited
Bar Stools - Sandler Seating Ltd
Barstool systems - Laytrad Contracts Ltd
Bespoke joinery/ counters & desks - Libraco
Binders - Quartet-GBC UK Ltd
Broadcast and audio visual products & services - Streamtec Limited
Canteen Furniture - Sandler Seating Ltd
Classroom furniture and equipment - Metalliform Products Plc
Computer furniture (education) - Klick Technology Ltd
Conference - Evertaut Ltd
Conference and training room systems - track based - Magiboards Ltd
Corporate seating & tables - Knightbridge Furniture Productions Ltd
Dance floor - Felix Design Ltd
Dining tables for schools etc - Metalliform Products Plc
Domestic car turntables - British Turntable Ltd
Education equipment - Evertaut Ltd
Educational equipment - Invicita Plastics Ltd
Educational furniture - Ellis J T & Co Ltd
Educational furniture for science and technology - S & B Ltd
Educational seating and desking - Metalliform Products Plc
Educational workshop equipment - Emmerich (Berlon) Ltd
Exhibition and display - Spaceright Europe Limited
Exhibition, Retail, Graphic and Presentation Systems - RTD Systems Ltd
Flood protection - Straight Ltd
Food technology furniture (education) - Klick Technology Ltd
Furniture design and production - Wales & Wales
Greenhouse accessories for Hartley Greenhouses - Hartley Botanic Ltd
In-Line Thermal Disinfection Unit (ILTDU) - Horne Engineering Ltd
Interactive whiteboards - Quartet-GBC UK Ltd
Laboratory tables - Metalliform Products Plc
Laminators - Quartet-GBC UK Ltd
Library furniture - Metalliform Products Plc
Nursery furniture and equipment - Metalliform Products Plc
Optitherm thermostatic tap - Horne Engineering Ltd
Padlocks and Chain - Duffells Limited
Public area seating - Evertaut Ltd
Rail presentation systems - Quartet-GBC UK Ltd
Raised slating rostra - Felix Design Ltd
Recycling bins, boxes and containers - Straight Ltd
Restaurant Furniture - Sandler Seating Ltd
Sanctuary stages - Felix Design Ltd
Screen printing on whiteboards - Magiboards Ltd
Seating - auditorium, theater, lecture theater, courtroom, conference - Race Furniture Ltd
Seating and tables for schools, waiting areas, train stations, airports, ferry terminals, hospitals and surgeries - Hille Educational Products Ltd
Shelving & Display - Libraco
Special Equipment (mixing Machines for wet-pour Rubber Crumb) - Edward Benton & Co Ltd
Special purpose equipment - Baz-Roll Products Ltd
Special Purpose Fixtures - Harling Security Solutions
Special purpose storage systems - Stamford Products Limited
Stacking chairs - Laytrad Contracts Ltd
Staging - carpet, hard deck, custom build, raised seating - Felix Design Ltd
Staging rostra - Felix Design Ltd
Stainless steel Welded Wire Products and Welded Wire Mesh - Multi Mesh Ltd
Technology furniture (education) - Klick Technology Ltd
Thermostatic mixing valves - Horne Engineering Ltd
Toys & games - Invicita Plastics Ltd
Whiteboard accessories - Magiboards Ltd

N24 Shopfitting

Ambient shelving - Carter Retail Equipment Ltd
Architectural Glass - Creative Glass
Bar railing - William Hopkins & Sons Ltd
Bespoke display cases - The Benbow Group
Bespoke shopfitting - Pearce Security Systems

N24 Shopfitting (con't)

Boards for shopfitting - Finnish Fibreboard (UK) Ltd
Cash Management systems and lobby shutters - Safetell Ltd
Checkouts - Carter Retail Equipment Ltd
Commercial shelving - Construction Group UK Ltd
Contour showcases - Stamford Products Limited
Counters and showcases - Smart F & G (Shopfittings)Ltd
Decorative tube & fittings - William Hopkins & Sons Ltd
Entrances - Comar Architectural Aluminuim Systems
Extrusions - Stamford Products Limited
Fabrics - Mermet U.K
Fast security screen counters - Safetell Ltd
Grabrails - William Hopkins & Sons Ltd
Interior Contracting - Parthos UK Ltd
Maintenance of industrial doors, shutters and shopfronts - Shutter Door Repair & Maintenance Ltd
Passive fire protection - BM TRADA
Point-of-sale, promotional and incentive products - Invicita Plastics Ltd
Premises equipment - HC Slingsby PLC
Refridgerated cabinets - Carter Retail Equipment Ltd
Retail and display fittings - DuPont Corian
Retractable glazed screen counters and secure access systems - Safetell Ltd
Rotary filing cabinets - Rackline Systems Storage Ltd
Screens - Pearce Security Systems
Shelving brackets - Tebrax Ltd
Shop fittings - Stamford Products Limited
Shopfitters - Cumberland Construction
Shopfitting - Blaze Signs Ltd
Shopfitting - Carter Retail Equipment Ltd
Shopfitting - Cil Retail Solutions Ltd
Shopfitting - Kee Klamp Ltd
Shopfitting - Smart F & G (Shopfittings)Ltd
Shopfitting - Steel Line Ltd
Shopfitting - Taskworthy Ltd
Shopfitting - The Benbow Group
Shopfitting and specialist joinery - Barlow Group
Shopfitting and specialist joinery - Barnwood Shopfitting Ltd
Shopfitting products and accessories - RB UK Ltd
Shopfronts - Comar Architectural Aluminuim Systems
Shopfronts - Project Aluminium Ltd
Shopfronts and showrooms - Glassolutions Saint-Gobain Ltd
Tube connectors - Stamford Products Limited

N25 Hospital/ Health Equipment

Anechoic Rooms - IAC Acoustics
Audiology Rooms - IAC Acoustics
Bariatric Products - Sidhil Care
Beds - Sidhil Care
Clean rooms and Containment Suites - Envair Ltd
Conduit systems - Marshall-Tufflex Ltd
Couches, Plinths & Treatment Chairs - Sidhil Care
Data management systems - Marshall-Tufflex Ltd
Disabled and elderly equipment - Nicholls & Clarke Ltd
Disposable Hospital Curtains - Marlux Ltd
Disposable hospital products - Verna Ltd
Environmental and waste equipment - Welco
Fireproff snaitary waste chutes - Inter Public Urban Systems Uk Ltd
First Aid & Treatment Room - Sidhil Care
Flipcharts, hanging rail systems and visual aids - The Visual Systems Healthcare
Floor graphics - The Visual Systems Healthcare
Grabrails - William Hopkins & Sons Ltd
Healthcare - Komfort Workspace PLC
Healthcare fittings - DuPont Corian
Hospital and Health Equipment - Antocks Lairn Ltd
Hospital and health equipment - Brandon Medical
Hospital and laboratory equipment - Pland
Hospital Equipment, Elbow control - Shavrin Levatap Co Ltd
Hospital furniture, accessories - Sidhil Care
Hospital Sinks - Pland Stainless Ltd
Hospital/ Health Equipment - Amilake Southern Ltd
Hospital/ health equipment - ArjoHuntleigh UK
Hospital/ Health Equipment - Franke Sissons Ltd
Hospital/ Health Equipment - The Kenco Coffee Company
Hospital/ Health furniture - Klick Technology Ltd
Hospital/health equipment - FC Frost Ltd
Hospital/health equipment - Miele Co Ltd
Hospital/Health Equipment - Wybone Ltd
Hospitol Health equipment - Hospital Metalcraft Ltd
Hotel & Leisure seating & tables - Knightbridge Furniture Productions Ltd
HTM71 medical cabinets - Klick Technology Ltd
Laboratory furniture - Decra Ltd
Lifting, moving and handling products for disabled - Chiltern Invadex
Living aids, commodes - Sidhil Care
Locakable and Confidential notice boards - The Visual Systems Healthcare
Lockers, Cloakroom equipment, Cabinets - 3d Storage Systems (UK) LTD
Magnetic indicators and symbols - The Visual Systems Healthcare
Mattresses - Sidhil Care
Medical and first aid furniture - Hospital Metalcraft Ltd
Medical examination lamps - Wandsworth Elecrtrical Ltd
Medical storage systems - Stamford Products Limited
Medical supplies - Kimberly-Clark Ltd
Nursing Chairs, Lifting aids - Sidhil Care
Overbed tables - Sidhil Care
Pedal bins - Unicorn Containers Ltd
Polymer radiator covers - Coverad
PVC pipe boxing - Coverad
Roller boards - The Visual Systems Healthcare
Seating and tables for schools, waiting areas, train stations, airports, ferry terminals, hospitals and surgeries - Hille Educational Products Ltd
Shower enclosures & Shower - Contour Showers Ltd
Slophers - Pland Stainless Ltd
Special needs kitchens - Anson Concise Ltd
Sprung mattresses, divan bases, pillows and accessories - Airsprung Beds Ltd
Stainless steel shelving systems for sterile areas - Bedford Shelving Ltd
Steel radiator covers (LST) - Coverad
Steel Trunking systems - Marshall-Tufflex Ltd
Support systems for elderly/disabled - Norton Engineering Alloys Co Ltd
Supportive bathroom products for disabled people - Pressalit Care
Surgeons scrub up trough - Pland Stainless Ltd
Swivel and fixed frame boards - The Visual Systems Healthcare
Trolleys - Sidhil Care
Water taps and mixers - Barber Wilsons & Co Ltd
White boards - The Visual Systems Healthcare
X-ray door sets - Lami Doors UK Ltd
X-ray Doorsets - Humphrey & Stretton PLC

N26 Gymnastic/ Sport/ Play Equipment

Activity Nets - HAGS SMP
Automatic slatted pvc swimming pool covers - Grando (UK) Ltd
Changing room equipment - Broxap Ltd
Cloakroom rails - H E W I (UK) Ltd
Dojo Matting - Fighting Films
Dojo matting - Geemat
Drama staging - Felix Design Ltd
Furniture, leisure - Allermuir Contract Furniture Ltd
Goals - Sportsmark Group Ltd
Golf tee mats & frames - Evergreens Uk
Groundmans Supplies - Sportsmark Group Ltd
GRP and plaster castings and molds - Regency Swimming Pools Ltd
Gymnastic matting - Continental Sports Ltd
Gymnastic, sport and play equipment - Carr Gymnasium Equipment Ltd
Gymnastic/sport/play equipment - Kee Klamp Ltd
Hydrotherapy pools - Aqua-Blue Ltd
Judo Matts - Fighting Films
Judo matts - Geemat
Lockers, Cloakroom equipment, Cabinets - 3d Storage Systems (UK) LTD
Maintenance service - Continental Sports Ltd
Multi-Use Games Areas and Multi-Play units - HAGS SMP
Nappy changers - LSA Projects Ltd
Nets - Sportsmark Group Ltd
Outdoor Fitness equipment - HAGS SMP
Panel pool kits - Golden Coast Ltd
Physical education equipment - Continental Sports Ltd
Playground equipment - HAGS SMP
Playground equipment - Lappset UK Ltd
Playground equipment and associated works - Kompan Ltd
Saunas, steam rooms and jacazzi spa pools - Aqua-Blue Ltd
Seating - cinema, public area - Race Furniture Ltd
Spare parts - Continental Sports Ltd
Sports Arena/Grounds PVC-u Fencing and Barriers - Duralock (UK) Ltd
Sports benches/Play equipment - Hispack Limited
Sports Covers - Power Plastics Ltd
Sports equipment - Continental Sports Ltd
Sports Equipment - Sportsmark Group Ltd
Stadia and auditorium seating - Metalliform Products Plc
Swimming Pool Designs - Aqua-Blue Ltd
Swimming pools and equipment - Golden Coast Ltd
Trampoline equipment - Continental Sports Ltd
Vanity units - Helmsman

N27 Cleaning Equipment

Brooms - Cottam Brush Ltd
Carpet cleaning machines - Nilfisk Limited
Cleaning & Waste - HC Slingsby PLC
Cleaning equipment - ArjoHuntleigh UK
Cleaning equipment - Donaldson Filtration (GB) Ltd
Cleaning equipment - Interface Europe Ltd
Cleaning equipment - Monument Tools Ltd
Cleaning maintenance equipment - Cento Engineering Company Ltd
Cleaning materials suppliers - Spontex Ltd
Cleaning Products - SC Johnson Wax
Floor polishers & burnishers - Nilfisk Limited
Hand cleaner - Ciret Limited
Industrial and commercial scrubbers, vacuum cleaners and floor maintenance equipment - Nilfisk Advance Ltd
Industrial and commercial vacuum cleaners - BVC
Maintenance, polish and cleaning chemicals - British Nova Works Ltd
Portable vacuum cleaners - Barloworld Vacuum Technology Plc
Power washers - Nilfisk Limited
Pressure washers - Industrial - Nilfisk Limited
Scrubber driers - Nilfisk Limited
Specialist brushes - Cottam Brush Ltd
Sweepers - Nilfisk Limited
Vacuum cleaners - Commercial & industrial - Nilfisk Limited
Wire Brushes - Cottam Brush Ltd

N28 Laboratory furniture/ equipment

Aluminium extrusions/ sections - Seco Aluminium Ltd
Bench top extraction - Nederman UK
Ceramic and acrylic sanitaryware - Armitage Shanks Ltd
Creteangle vibrating tables - Edward Benton & Co Ltd
Design, manufacture and installation of laboratory furniture and fume cupboards - Lab Systems Furniture Ltd
Enclosures and technical furniture - Knurr (UK) Ltd
Gas isolation and detection - TCW Services (Control) Ltd
Industrial laboratory furniture - S & B Ltd
Lab Furniture - Köttermann Ltd
Laboratory design and build project management - Fisher Scientific (UK) Ltd
Laboratory Equipment - Edward Benton & Co Ltd
Laboratory Furniture - Antocks Lairn Ltd
Laboratory furniture - Grant Westfield Ltd
Laboratory furniture - John Pulsford Associates Ltd
Laboratory Furniture - Simplylabs Limited
Laboratory furniture - Trespa UK Ltd
Laboratory furniture (education) - Klick Technology Ltd
Laboratory furniture/ equipment - Benton Co Ltd Edward
Laboratory furniture/equipment - FC Frost Ltd
Laboratory furniture/equipment - Franke Sissons Ltd
Laboratory furniture/equipment - Miele Co Ltd
Laboratory Sinks - Pland Stainless Ltd
Primary & Secondary Pharaceutical Solutions - Envair Ltd
Reprographic room equipment - Envopak Group Ltd
Shelving - Bedford Shelving Ltd

N29 Industrial Equipment

Access Equipment - Barton Storage Systems Ltd
Access equipment - HC Slingsby PLC
Aero-Engine Test Control System - IAC Acoustics
Bowl Pressings - Pland Stainless Ltd
Commercial and industrial equipment - HC Slingsby PLC
Gatehouse - Arbory Group Limited
Handling & lifting - HC Slingsby PLC
Hopper feeder systems - Benton Co Ltd Edward
Industrial batteries for standby applications - Hawker Ltd
Industrial equipment - Benton Co Ltd Edward
Industrial equipment - Edward Benton & Co Ltd
Industrial Equipment - Emmerich (Berlon) Ltd
Industrial Equipment - Harling Security Solutions
Industrial equipment - Kee Klamp Ltd
Industrial equipment - Monument Tools Ltd
Industrial equipment - bins - Unicorn Containers Ltd
Industrial instrumentation - ABB Instrumentation Ltd
Industrial materials handling equipment - B C Barton Limited
Industrial shelving - Construction Group UK Ltd
Industrial turntables - British Turntable Ltd
Industrial workbenches and storage drawer cabinets - Q.M.P
Key Blanks and Machines - Duffells Limited
Magnetic holding etc - Bunting Magnetics Europe Limited
Maintenance of industrial doors, shutters and shopfronts - Shutter Door Repair & Maintenance Ltd
Manual handling equipment - Hymo Ltd
Packaging - Welco
Paint brushes, rollers and brooms - Cottam Brush Ltd
Power plant noise control systems - IAC Acoustics
Ropes and sashcords - Marlow Ropes Ltd
Specialist brushes - Cottam Brush Ltd
Surveying instruments - Topcon GB Ltd
Trolleys and trucks - Q.M.P
Ultra violet chambers, lamps and control panels - Hanovia Ltd
Vehicle livery - Signs & Labels Ltd
Vibratory equipment and moulds - Benton Co Ltd Edward
Workshop equipment - HC Slingsby PLC
X-ray door sets - Lami Doors UK Ltd

N30 Laundry Equipment

Clothes lines - James Lever 1856 Ltd
Domestic appliances - Miele Co Ltd
Finishing equipment - Electrolux Laundry Systems
Hydro extractors - Electrolux Laundry Systems
Ironing tables - Electrolux Laundry Systems
Laundry equipment - Viking Laundry Equipment Ltd
Linen chutes - Hardall International Ltd
Roller ironers - Electrolux Laundry Systems
Tumble dryers - Electrolux Laundry Systems
Washers, dryers and ironers - Warner Howard Group Ltd
Washing machines - Electrolux Laundry Systems

N31 Entertainment/ Auditorium equipment

Acoustic Doorsets - Humphrey & Stretton PLC
Architectural dimming systems - Mode Lighting (UK) Ltd
Auditorium seating - Evertaut Ltd
AV Screens - Harkness Screens
Conference Equipment - Spaceright Europe Limited
Entertainment Equipment - Antocks Lairn Ltd
Entertainment/ Auditorium Equipment - Harkness Screens
Exhibition, Retail, Graphic and Presentation Systems - RTD Systems Ltd
Lecture theatre seating - Evertaut Ltd

P10 Sundry insulation/ proofing work/ fire stops

Acoustic design - Sound Reduction Systems Ltd
Acoustic door seals - AET.gb Ltd
Acoustic Doors - Amadeus
Acoustic flooring - Interfloor Limited
Acoustic insulation - InstaGroup Ltd
Acoustic insulation - Knauf Insulation Ltd
Acoustic insulation - The Sound Solution
Acoustic panels - AET.gb Ltd
Acoustic Treatment - Amadeus
Automatic fire barrier-curtains - Cooper Group Ltd
Automatic fire shutters - Cooper Group Ltd
Bitumen building papers, film/paper laminated papers - Package Products Ltd
Ceilings, floors and roofs - Promat UK Ltd
cold seal paper - Package Products Ltd
Expanded polystyrene insulation - Jablite Ltd
Expanded polystyrene insulation - Springvale E P S
External Wall Insulations - Skanda Acoustics Limited
Extruded polystyrene insulation - Dow Building Solutions
Fiberglass fabrics - Fothergill Engineered Fabrics Ltd
Fibreglass loft insulation - Davant Products Ltd
Fire Barriers - Gridpart Interiors Ltd
Fire collars - FloPlast Ltd
Fire doors and doorsets - Bridgman IBC Ltd
Fire insulation - B R Ainsworth (Southern)
Fire proof cladding - Firebarrier Services Ltd
Fire proof panels - Hemsec Panel Technologies
Fire proofing board - Dufaylite Developments Ltd
Fire Protection - Knauf Insulation Ltd
Fire protection casings - Cryotherm Insulation Ltd
Fire protection for ductwork, seals and joints - Promat UK Ltd
Fire protection insulation spray - Cryotherm Insulation Ltd
Fire resisting partitions - Mann McGowan Group
Fire sound and thermal soloutions - Siniat Ltd
Fire stopping systems - Cryotherm Insulation Ltd
Fire stops - FGF Ltd
Fire stops - Fire Protection Ltd
Fire stops - Quelfire
Fire stops around pipes - Walraven Ltd
Fire stops for cables pipes and ductwork - Quelfire
Fire stops for joints - Quelfire
Fire walls - Gilmour Ecometal
Fire walls and barriers - Cryotherm Insulation Ltd
Fireproofing and insulation - Thermica Ltd
Firestopping materials - PFC Corofil
Flame Retardant Boards - Meyer Timber Limited
Flat roof insulation - Isocrete Floor Screeds Ltd
Floor insulation - Springvale E P S
Foamglas Cellular glass insulation for cut to falls roof systems - Pittsburg Corning (UK) Ltd
Foamglas Cellular glass internal and external insulation for roof, wall, floor, below ground conditions and foundations - Pittsburg Corning (UK) Ltd
Full-Fire Certificate - Firebarrier Services Ltd
Hot and cold water cylinder jackets - Davant Products Ltd
Insuation panels - Cryotherm Insulation Ltd
Insulaion for Heating and Ventilation - Kingspan Industrial Ins. Ltd
Insulation - British Gypsum Ltd
Insulation - Covers Timber & Builders Merchants
Insulation - Nevill Long Limited
Insulation - Rockwool Ltd
Insulation - Steico
Insulation - Industrial Solutions - Actavo (UK) Ltd
Insulation accessories - Astron Buildings Parker Hannifin PLC
Insulation for Process industries - Kingspan Industrial Ins. Ltd
Insulation Products - Allmat (East Surrey) Ltd
Insulation, fire protection and fire stopping - Kitsons Insulation Products Ltd
Intermescant fire and smoke seals - Mann McGowan Group
Intumescent fire seals - Humphrey & Stretton PLC
Intumescent Firestop - Dow Corning Europe S.A
Nullifire fire stopping range - Nullifire Ltd
paritions and external walls - Promat UK Ltd
PIR Fire rated panels - Hemsec Panel Technologies
PIR insulated panels - Hemsec Panel Technologies
Plastic extrusions - Plasticable Ltd
Polyethylene and PVC pipe insulation - Davant Products Ltd
Polythene vapor barrier - Visqueen Building Products
Rigid insulation boards - Kingspan Insulation Ltd
Rigid thermal board - National Insulation Association Ltd
Roof insulation - Dow Building Solutions
Roof insulation - Springvale E P S
Roof Intallation/Tappered Cork - Westco Group Ltd
Seals - Mann McGowan Group
Seals and intumescents - Laidlaw Architectural Hardware Ltd
Self sealing test plugs - Walraven Ltd
Seperating Floors - Sound Service (Oxford) Ltd
Servery fire curtain-industrial/commercial/residential - Cooper Group Ltd
Sleeving and tubing - Plasticable Ltd
Solid wall insulation - National Insulation Association Ltd
Sound absorbtion panels - Henkel Ltd
Sound insulation - Sound Service (Oxford) Ltd
Sound insulation systems/containers - Mansfield Pollard & Co Ltd

P10 Sundry insulation/ proofing work/ fire stops (con't)

Sound proof Acoustic Rooms - Amadeus
Sound Proofing - InstaGroup Ltd
Sound proofing - The Sound Solution
soundproofing materials - Bicester Products Ltd
Spray Foam Insulation - Foamseal Ltd
Standard corkboard - Gradient Insulations (UK) Ltd
Sundry installation/proofing work/fire stops - Reiner Fixing Devices (Ex: Hardo Fixing Devices)
Sundry Insulation - Strand Hardware Ltd
Sundry insulation/ proofing work/ fire Stops - Polypipe Building Products
Sundry insulation/proofing work/ fire stops - Knauf Insulation Ltd
Tapered Rockwool - Gradient Insulations (UK) Ltd
Tapered, corkboard, cork/PUR composite, polystyrene and PUR - Gradient Insulations (UK) Ltd
Thatch Firewall Membrane - Thatching Advisory Services Ltd
Thermal dry lining - National Insulation Association Ltd
Thermal insulation boards & quilts - Skanda Acoustics
Trade Association - Association of Specialist Fire Protection (ASFP)
Urethane foam industry trade association - BRITISH URETHANE FOAM CONTRACTORS ASSOCIATION
Vessel,Tank and Pipe insulation - NMC (UK) Ltd
Wall insulation - Springvale E P S
Wax coated paper - Package Products Ltd

P11 Foamed/ Fibre/ Bead cavity wall insulation

Acoustic - Thermica Ltd
Bead cavity wall insulation - National Insulation Association Ltd
British wall cavity insulation - Kleeneze Sealtech Ltd
Cavity wall insulation and loft insulation - InstaGroup Ltd
Dry lining and building insulation - B R Ainsworth (Southern)
Expanded polystyrene insulation - Jablite Ltd
External wall insulation - Weber Building Solutions
Extruded polystyrene - Knauf Insulation Ltd
Fibre cavity wall insulation - Knauf Insulation Ltd
Foamed/ Fibre/ Bead cavity wall insulation - Polypipe Building Products
Insulation - FGF Ltd
Insulation Products - Allmat (East Surrey) Ltd
Mineral wool cavity wall insulation - National Insulation Association Ltd
Polyisocyanurate insulation board manufacture - Celotex Ltd
Urethane foam industry trade association - BRITISH URETHANE FOAM CONTRACTORS ASSOCIATION

P20 Unframed isolated trims/ skirtings/ sundry items

Architectural and decorative mouldings - Mason FW & Sons Ltd
Architectural trims - Rockwell Sheet Sales Ltd
Architraue - Enfield Speciality Doors
Bath and Kitchen Trims - Homelux Nenplas
Cellular PVC window trims, skirtings & architraves - celuform - Celuform Building Products
Doors and joinery - Brewer T. & Co Ltd
Finger Guards - Safety Assured Ltd
Fixings for fall arrest and safety - Bacon Group
GRP Dentil cornice - Production Glassfibre
GRP Window architrave - Production Glassfibre
Hardwood - Panel Agency Limited
Hardwood mouldings - Atkinson & Kirby Ltd
Hardwoods - James Donaldson Timber
MDF moldings - Fibercill
MDF Mouldings (Silktrim) - James Donaldson Timber
Metal Trims - Homelux Nenplas
Perimeter Casing Systems - Jali Ltd
Plastic and Foam Components - NMC (UK) Ltd
Plastic handrails - Brighton (Handrails), W
Protection rails and corner angles - Harrison Thompson & Co Ltd
PVC trims, angles & edges - Homelux Nenplas
Rope and harness - Baj System Design Ltd
Sashcords, clothes lines and building - Kent & Co (Twines) Ltd
Simulated wood mouldings, reproduction oak beams - Oakleaf Reproductions Ltd
Skirting - Enfield Speciality Doors
Softwoods: Mouldings, CLS, Clear softwoods - James Donaldson Timber
Trims/ skirtings - Jali Ltd
UPVC Plaster Beading & Trims - Homelux Nenplas
Veneered MDF Mouldings - James Donaldson Timber

P21 Door/ Window ironmongery

Anti-Ligature (or Ligature Resistant) Fittings - Roc Secure Ltd
Architectural door furniture and security hardware - Allgood plc
Architectural Hardware - Allegion (UK) Ltd
Architectural Hardware - Duffells Limited
Architectural hardware - Laidlaw Ltd
Architectural hardware including hinges, locks, levers, flush bolts and other ancilliary items including ceiling fans - James Gibbons Format Ltd
Architectural hinges, door closers and thresholds - Relcross Ltd
Architectural ironmomgery - Samuel Heath & Sons plc
Architectural ironmongerers - Moffett Thallon & Co Ltd
Architectural ironmongery - Doorfit Products Ltd
Architectural Ironmongery - Hafele UK Ltd
Architectural ironmongery, cabinet hardware and fittings - Lord Isaac Ltd
Bespoke Ironmongery Package Provision - Strada London
Black Antique Ironmongery - Kirkpatrick Ltd
Brass furniture fittings - Martin & Co Ltd
Brass hardware - Marcus M Ltd
Brass, bronze door and window hardware - Allart, Frank, & Co Ltd
Brush strip door and window seals - Kleeneze Sealtech Ltd
Brushstrip door and letterbox seals - Kleeneze Sealtech Ltd
Butt hinges - Caldwell Hardware (UK) Ltd
Catches & bolts - Caldwell Hardware (UK) Ltd
Closers and Panic Hardware - Duffells Limited
Concealed door closers - Samuel Heath & Sons plc
Cotton twine - James Lever 1856 Ltd
Cupboard fittings - Samuel Heath & Sons plc
Door & window control equipment - GEZE UK
Door & window ironmongery - Olby H E & Co Ltd
Door / window ironmongery - Brass Art Ltd
Door closers - Exidor Limited
Door Controls - Owlett - Jaton
Door entry equipment - The Safety Letterbox Company Ltd
Door furniture - ASSA Ltd
Door furniture - H E W I (UK) Ltd
Door furniture - Samuel Heath & Sons plc
Door furniture and hinges - Laidlaw Architectural Hardware Ltd
Door hardware - ASSA ABLOY Limited
Door hardware - ASSA Ltd
Door Hardware - Paddock Fabrications Ltd
Door ironmongery - Dorma Door Services Ltd
Door Locks - Bramah Security Equipment Ltd
Door Operators/Closers & Accessories - Abloy Security Ltd
Door/ Window ironmongery - Schlegel UK (2006) Ltd
Door/ Window ironmongery - Till & Whitehead Ltd
door/window iironmongery - UK Fasteners Ltd
Door/Window Ironmongery - Crowthorne Fencing
Door/Window Ironmongery - Daro Factors Ltd
Door/window ironmongery - Draught Proofing Advisory Association
Door/window ironmongery - Hart Door Systems Ltd
Door/window ironmongery - Magnet Ltd
Door/Window Ironmongery - Roto Frank Ltd
Dor plates - Fixatrad Ltd
Draught, weather, intumescent, fire and smoke seals - Sealmaster
Draught/Acoustic Seal - Ventura
Duncombe - Exidor Limited
Electric Door Operator - Relcross Ltd
Emergency exit door hardware - Exidor Limited
Emergency exit hardware - VBH (GB) Ltd
Emergency exit locks, break glass locks - Pickersgill Kaye Ltd
Espagnolette locking systems - Nico Manufacturing Ltd
Fire door motors and controls - Kaba Garog
Fire door operating equipment - Kaba Garog
Flyscreens - P+L Systems Ltd
Folding openers - Caldwell Hardware (UK) Ltd
Furniture castors and hardware - Kenrick Archibald & Sons Ltd
Galvanised water bar - Mid Essex Trading Co Ltd
Gas springs - Arvin Motion Control Ltd
Gate locks - Clarke Instruments Ltd
Glazing trade association - Glass and Glazing Federation
Gloro door furniture - Focus SB Ltd
GRP Decorative brackets - Production Glassfibre
Hardware - Reddiglaze Ltd
Hardware for windows and doors - Avocet Hardware Ltd
Hardware for windows and doors - Roto Frank Ltd
High security locks and locking systems - Abloy Security Ltd
Hinges - Cooke Brothers Ltd
Hotel locking - Allegion (UK) Ltd
Industrial locks - ASSA Ltd
Intumescent fire stopping products - Sealmaster
Ironmongery - Allegion (UK) Ltd
Ironmongery - Brewer T. & Co Ltd
Ironmongery - Caldwell Hardware (UK) Ltd
Ironmongery - Challenge Fencing Ltd
Ironmongery - Clayton Munroe
Ironmongery - Covers Timber & Builders Merchants
Ironmongery - Eliza Tinsley Ltd
Ironmongery - Grorud Industries Ltd
Ironmongery - Groupco Ltd
Ironmongery - Harris & Bailey Ltd
Ironmongery - Hipkiss, H, & Co Ltd
Ironmongery - Maco Door & Window Hardware (UK) Ltd
Ironmongery - Strand Hardware Ltd
Ironmonqery - Owlett - Jaton
Leading manufacturer for door and window furnitures - HOPPE (UK) Ltd
Locks - Aardee Security Shutters Ltd
Locks - Guardian Lock and Engineering Co Ltd
Locks - Lowe & Fletcher Ltd
Locks and Security - Duffells Limited
Locks, Door number plates - 3d Storage Systems (UK) LTD
Locksmiths - Doorfit Products Ltd
Magnetic door stops and catches - Bunting Magnetics Europe Limited
Mail boxes, door closers, door furniture, door locks and cylinders - JB Architectural Ltd
Manual and electrical systems for opening windows - Dyer Environmental Controls
Multi point locking systems - Paddock Fabrications Ltd
Multipoints and UPVC - Duffells Limited
Padlocks - Bramah Security Equipment Ltd
Padlocks - Securikey Ltd
Padlocks security products, cyclelocks and motorcycle locks - Squire Henry & Sons
Panic Exit Hardware - Abloy Security Ltd
Panic hardware - ASSA Ltd
Patio hardware - Caldwell Hardware (UK) Ltd
Pivot hinges - Caldwell Hardware (UK) Ltd
Profile cylinder master keying service - VBH (GB) Ltd
Push button and combination locks - Relcross Ltd
Restrictors - Caldwell Hardware (UK) Ltd
Rim, mortice, padlocks, cylinders and panic escape locks - Yale Security Products Ltd
Roller shutter motors and controls - Kaba Garog
Rolling Fire Shutters - Avon Industrial Doors Ltd
Roof windows and ventilators - NuLite Ltd
Ropes and sashcords - Marlow Ropes Ltd
Sah window hardware and accessories - Garador Ltd
Sash balances - Caldwell Hardware (UK) Ltd
Sash cord - James Lever 1856 Ltd
Sash Window Locks - Bramah Security Equipment Ltd
Sectional door motors and controls - Kaba Garog
Secure Letterboxes - The Safety Letterbox Company Ltd
Security Door Hardware - Exidor Limited
Security Equipment - Gunnebo UK Limited
Security Grills - O'Connor Fencing Limited
Security products - Marcus M Ltd
Security products, black ironmongery and hardware - ERA Products
Sercurity Products - Owlett - Jaton
Shelving brackets - decorative - Winther Browne & co Ltd
Shootbolt Locking Systems - Nico Manufacturing Ltd
Sliding door gear - Avon Industrial Doors Ltd
Sliding Door Gear - Lakeside Group
Smoke ventilation systems - Dyer Environmental Controls
Solenoid locks - Clarke Instruments Ltd
Specialist ironmongery - The Cotswold Casement Company
Steel Hinged Doors for Fire Exits - Avon Industrial Doors Ltd
Trickle Ventilations - Ryton's Building Products Ltd
Weatherstriping joinery seal, roof cappings, glazing bars and ridge systems - Exitex Ltd
Weatherstripping - Reddiglaze Ltd
Window and door hardware - PVCu, timber and aluminum - VBH (GB) Ltd
Window and door ironmongery - Laird Security Hardware
Window furniture - Samuel Heath & Sons plc
Window hardware - Paddock Fabrications Ltd
Window Locks - Bramah Security Equipment Ltd

P22 Sealant joints

Acoustic seals - Mann McGowan Group
Adhesives - Geocel Ltd
Bath and kitchen seals - Homelux Nenplas
Building chemicals - Geocel Ltd
Expanding foam sealant - Bondaglass Voss Ltd
Expansion joints - Fillcrete Ltd
Expansion Joints - Movement Joints (UK)
Expansion joints - Pitchmastic PmB Ltd
Fire resistant expansion joint seals - Mann McGowan Group
Joint fillers - Fosroc Ltd
Joint sealants - Fosroc Ltd
Joint sealants - SCP Concrete Sealing Technology Ltd
Jointing & tooling - The BSS Group
Jointing compounds - The BSS Group
Jointing compounds, quick repair products and solder & flux - Fernox
Jointing materials - Furmanite International Ltd
Mechanical movement joints and fire rated sealants - Compriband Ltd
Profiled sealants - Henkel Ltd
Sealant joints - Firebarrier Services Ltd
Sealants - Geocel Ltd
Sealants, solvents, wood preservatives and creosote - Bartoline Ltd
Seals and Caulking Profiles - Trelleborg Sealing Profiles
Seals for doors, windows, glazing - Schlegel UK (2006) Ltd
Silicone sealants - Bluestar Silicones
Weather draft seals - Mann McGowan Group
Weatherstrip sealing strips - Linear Ltd

P30 Trenches/ Pipeways/Pits for buried engineering services

Chambers - Tyco Water Works
Meter Boxes - Tyco Water Works
Pipeways - Polypipe Civils
Plastic Drainage System - Hepworth Building Products Ltd
Trench Cover - MCL Composites Ltd
Underground chambers - MCL Composites Ltd

P31 Holes/ Chases/ Covers/ Supports for services

Access covers - Bilco UK Ltd
Access Covers and Panels - Surespan Ltd
Access floor service outlet boxes and grommet outlets - Electrak International Ltd
Aluminium and brass recessed floor access covers - Howe Green Ltd
Builders cast iron products - Dudley Thomas Ltd
Cabinets - Glasdon U.K. Limited
Channel Supports - Kem Edwards Ltd
Electrical fixings - Walraven Ltd
Enclosures - Glasdon U.K. Limited
Gear Trays - Jerrards Plc
Inlet cabinets - Delta Fire
Low Surface temperature Covers - Zehnder Ltd
Manhole covers - MCL Composites Ltd
Mechanical Pipe Supports - Kem Edwards Ltd
Micro Trenching Compound - Uretech MTC - Star Uretech Ltd
Pipe support componants - Walraven Ltd
Pressed grating - B C Barton Limited
Radiator cabinets / covers - Winther Browne & co Ltd
Roll grooving machine - Hilmor
Stainless steel access covers - Howe Green Ltd

Q10 Kerbs/ Edgings/Channels/Paving accessories

Car Park Marking - Sportsmark Group Ltd
Channel Drainage System - Ensor Building Products Ltd
Channels - Polypipe Civils
Cills, lintols, steps - Forest of Dean Stone Firms Ltd
Combined kerb/drainage units, Grease traps, Protective skirting - ACO Technologies plc
Concrete gutter blocks - Finlock Gutters
Edging - Rogers Concrete Ltd
Edgings and copings - Oakdale (Contracts) Ltd
Floor & Line markings - Kenyon Paints Limited
Garden edgings - York Handmade Brick Co Ltd
Granite kerbs - CED Ltd
Grass concrete blocks - Concrete Products (Lincoln) 1980 Ltd
GRC surface water channels - Althon Ltd
Handcut & Masoned - Yorkstone - Hard York Quarries Ltd
Hard landscaping - Bridge Quarry
Kerbing - Rediweld Rubber & Plastics Ltd
Kerbing & Paving - Gearing F T Landscape Services Ltd
Kerbing, drainage, brickwork, road planing and white lining - AMK Fence-In Limited
Kerbs - Harris & Bailey Ltd
Kerbs / Edgings / Channels / Paving Accessories - Formpave Ltd
Kerbs and accessories - Brett Landscaping
Kerbs edgings - Stowell Concrete Ltd
Kerbs, edgings, Channels, Paving accessavial - Grass Concrete Ltd
Kerbs/ Edgings/Channels/Paving accessories - Vetter UK
Kerbs/ Edgings/Channels/Paving stone accessories - Johnsons Wellfield Quarries Ltd
Kerbs/Edgings/channels/pavind accessories - Townscape Products Ltd
Kerbs/Paving Accessories - Realstone Ltd
Landscape Suppliers - Crowthorne Fencing
Living stone paving - Concrete Products (Lincoln) 1980 Ltd
Marking - Moravia (UK) Ltd
Natural Stone Kerbs - Palladio Stone
Path edging - Atlas Stone Products
Pavers - Baggeridge Brick PLC
Paving slab supports - Compriband Ltd
Precast concrete products - Ruthin Precast Concrete Ltd
Precast concrete surface water channels - Althon Ltd
Road Blockers - Darfen Durafencing
Road Stud - Sportsmark Group Ltd
Sleeping Policemen - Sportsmark Group Ltd
Specialist Kerbs - Brett Landscaping
Stone suppliers - Earl's Stone Ltd
Stone suppliers - Kerridge Stone Ltd
Synthetic ironoxide colouring agents - LANXESS Inorganic Pigments Group
Traffic Calmings - Moravia (UK) Ltd
Traffic Regulations - Moravia (UK) Ltd
White Lining - Sportsmark Group Ltd

Q20 Granular sub-bases to roads/ pavings

Aggregates- crushed rock, sand and gravel - Hanson UK
Clay Pavers - Hanson Building Products
Crushed & graded stone - Aggregates - Hard York Quarries Ltd

Q21 Insitu concrete roads/ pavings/ bases

Insitu concrete roads/pavings/bases - Grass Concrete Ltd
Movement Joints - Movement Joints (UK)
Resin bonded surfacing, Resin bound surfacing, Repair mortars - Ronacrete Ltd

Q22 Coated macadam/ Asphalt roads/ pavings

Anti Slip Surface - Sportsmark Group Ltd
Ashphalt road reinforcement - Bekaert Building Products
Asphalt - Hanson UK
Asphalt Roads - Gearing F T Landscape Services Ltd
Asphalt, tar macadam roads/ pavings - AMK Fence-In Limited
Bitumen boilers - Bitmen Products Ltd
Bitumen macadam, hot rolled asphalt, and sub base aggregates - Breedon Aggregates
Bituminous Binders Road Surface Dressing - Colas Ltd
Cold emulsion sprayers - Bitmen Products Ltd
Colours for Asphalt - Procter Johnson & Co Ltd

Q22 Coated macadam/ Asphalt roads/ pavings (con't)

Infra ray heaters - Bitmen Products Ltd
Ironwork installation: Bedding and Locking - Uretech RR - Star Uretech Ltd
Mastic asphalt mixers - Bitmen Products Ltd
Mastic asphalt road contractors - Tarmac Ltd
Mastic Asphalt roads/ pavings - North Herts Asphalte (Roofing) Ltd
Movement Joints - Movement Joints (UK)
Paver hire - Breedon Aggregates
Permanent Pothole Repair - Uretech RR System - Star Uretech Ltd
Pre coated chipspreaders - Bitmen Products Ltd
Red macadam - Breedon Aggregates
Reflective tarffic markings - Reflecting Roadstuds Ltd
Resin bonded surfacing, Resin bound surfacing, Pot hole repair - Ronacrete Ltd
Surfacing contractors - Tarmac Ltd
Synthetic ironoxide colouring agents - LANXESS Inorganic Pigments Group

Q23 Gravel/ Hoggin/ Woodchip roads/ pavings

Aggregate & Stone - Aggregate Industries UK Ltd
Aggregates- crushed rock, sand and gravel - Hanson UK
Decorative Aggregates - Brett Landscaping
Decorative aggregates - CED Ltd
Footpath gravels & hoggin - CED Ltd
Living stone decorative aggregates - Concrete Products (Lincoln) 1980 Ltd
Loose aggregates - B & M Fencing Limited
Slate chippings - The Delabole Slate Co. Ltd

Q24 Interlocking brick/ block roads/ pavings

Block Paving - B & M Fencing Limited
Block paving - Thakeham Tiles Ltd
Block pavings - AMK Fence-In Limited
Blockpaving contractors - Gearing F T Landscape Services Ltd
Clay block paving - Oakdale (Contracts) Ltd
Clay pavers - Ibstock Brick Ltd
Clay pavers - Marshalls Mono Ltd
Concrete Block Kerbs - Brett Landscaping
Concrete block paving - CEMEX
Concrete Block Permeable Paving - Brett Landscaping
Interlocking block pavings - Formpave Ltd
Interlocking brick/ block roads/ pavings - Grass Concrete Ltd
Landscaping products - Aggregate Industries UK Ltd
Pavers - Baggeridge Brick PLC
Pavers - Ibstock Building Products Ltd
Pavers - York Handmade Brick Co Ltd
Paving - Townscape Products Ltd
Paving - block - Brett Landscaping
Paving/ hard landscaping - Blanc de Bierges
Pavings - Prismo Ltd
Pavings - York Handmade Brick Co Ltd
Q24 Concrete Block Paving - Brett Landscaping
Staffordshire Clay paving - Ketley Brick Co Ltd , The
Synthetic ironoxide colouring agents - LANXESS Inorganic Pigments Group

Q25 Slab/ Brick/ Sett/ Cobble pavings

Armorflex concrete mattresses - Concrete Products (Lincoln) 1980 Ltd
Boulders, cobbles and pebbles - CED Ltd
Clay pavers - Hanson Building Products
Clay Pavers - Ruabon Sales Limited
Cobbles - York Handmade Brick Co Ltd
Concrete Flag Paving - Brett Landscaping
Concrete slab, block paving - Marshalls Mono Ltd
Decorative precast paving - Oakdale (Contracts) Ltd
Granite paving - CED Ltd
Granite setts - CED Ltd
Grass reinforcement systems - Grass Concrete Ltd
Heavyside paving - Oakdale (Contracts) Ltd
Landscape Suppliers - Crowthorne Fencing
Landscaping - J Suttle Swanage Quarries Ltd
Limestone paving - CED Ltd
Municipal Paving - Hye Oak
Natural Stone Cobbles & Setts - Palladio Stone
Natural Stone Paving - Brett Landscaping
Natural stone paving - Marshalls Mono Ltd
Pavers - Baggeridge Brick PLC
Paving - Atlas Stone Products
Paving - Burlington Slate Ltd
Paving - Forest of Dean Stone Firms Ltd
Paving - sett - Brett Landscaping
Paving slabs - Stowell Concrete Ltd
Pavings - Prismo Ltd
Pavings - Ruthin Precast Concrete Ltd
Paviors - Kingscourt Country Manor Bricks
Porphyry setts and edging - CED Ltd
Precast concrete paving - Rogers Concrete Ltd
Promenade Tiling - North Herts Asphalte (Roofing) Ltd
Regency paving and flagstone paving - Thakeham Tiles Ltd
Roadstone - Hard York Quarries Ltd
Roofing and paving - Bridge Quarry
Slab / Brick / Sett / Cobble pavings - Formpave Ltd
Slab / brick/ sett/ cobble pavings - Harris & Bailey Ltd
Slab/ Brick/ Sett/ Cobble pavings - Chichester Stoneworks Limited
Slab/ brick/ sett/ cobble pavings - Chilstone
Slab/ Brick/ Sett/ Cobble pavings - Vetter UK
Slab/ Brick/stett/cobble pavings - Townscape Products Ltd
Slab/ Sett Stone pavings - Johnsons Wellfield Quarries Ltd
Slab/Brick/Cobble Pavings - Realstone Ltd
Slabs - B & M Fencing Limited
Staffordshire Patterned Paving - Ketley Brick Co Ltd , The
Stone suppliers - Earl's Stone Ltd
Stone suppliers - Kerridge Stone Ltd
Synthetic ironoxide colouring agents - LANXESS Inorganic Pigments Group
Yorkstone paving - CED Ltd

Q26 Special surfacings/ pavings for sport/general amenity

Anti skid surfacing products - Clare R S & Co Ltd
Anti slip metal flooring - Graepel Perforators Ltd
Anti slip surfacing - Addagrip Terraco Ltd
Antiskid surfaces - Colas Ltd
Artifical sports surfaces - En-tout-cas Tennis Courts Ltd
Artificial Grass - Evergreens Uk
Athletic tracks and synthetic grass sports pitches - Bernhards Landscapes Ltd
Bowling green equipment and channels - Sportsmark Group Ltd
Children's play areas and sports pitches - Easi-Fall International Ltd
Clay pavers - Michelmersh Brick Holdings PLC
Coloured sufacing products - Clare R S & Co Ltd
Construction, refurbishing and all types of tennis courts - En-tout-cas Tennis Courts Ltd
Decorative bound/bonded surfacing resin -TekGrip DPB &: Professional - TekGrip DSR, Flexible - TekGrip DFX, Summer - TekGrip DD2, High Friction / Anti-skid road coating - Uretech HFS - Star Uretech Ltd
Exterior Pitch Markings - Sportsmark Group Ltd
Floor & Line markings - Kenyon Paints Limited
Games area surfaces - Playtop Licensing Limited
Gauged arches - Michelmersh Brick Holdings PLC
Grass Honeycomb Rubber Matting - MacLellan Rubber Ltd
Hardwood decking - W.L. West & Sons Ltd
Indoor/outdoor wet areas - Evergreens Uk
Marking products and materials - Marking Machines International
Mulit-Use Games Areas (fencing and artificial surfaces) - En-tout-cas Tennis Courts Ltd
Outdoor matting soloutions - Kleen-Tex Industries Ltd
Permeable paving - CEMEX
Play area surfaces - Playtop Licensing Limited
Playground surfacing - HAGS SMP
Poolsurrounds - Evergreens Uk
Resin bonded surface dressing - Addagrip Terraco Ltd
Resin bonded surfacing, Resin bound surfacing - Ronacrete Ltd
Resin bound porous surfacing - Addagrip Terraco Ltd
Road Carpet - Howard Evans Roofing and Cladding Ltd
Road markings - Prismo Ltd
Rubber paving - Linatex Ltd
Rubber surfaces for play areas, paddling and swimming pools surrounds, "kick about" play areas and lockers rooms in golf clubs - Easi-Fall International Ltd
Safety surfaces - Playtop Licensing Limited
Seeding Turfing - Jack Moody Ltd
Special surfacings/pavings for span/general amenity - Grass Concrete Ltd
Specialist surfaces for raods - Conren Ltd
Specialist surfacing contractors - Tarmac Ltd
Sports court marking - Sportsmark Group Ltd
Sports ground construction - AMK Fence-In Limited
Sports Hall Markings - Sportsmark Group Ltd
Sports Surfaces - Evergreens Uk
Sports surfacing and pavings - Charles Lawrence Surfaces Ltd
Synthetic grass for residential use - Artificial Grass Ltd
Synthetic grass for sport - Artificial Grass Ltd
Tactile surfaces for visually impaired - Rediweld Rubber & Plastics Ltd

Q30 Seeding/ Turfing

External planting - Jack Moody Ltd
Grass seed and turf - Sportsmark Group Ltd
Hydraulic seedings - Hydraseeders Ltd
Landscapes - Ogilvie Construction Ltd
Seeding/ turfing - Clifton Nurseries
Seeding/ turfing - Coblands Landscapes Ltd
Seeding/ turfing - Frosts Landscape Construction Ltd
Seeding/ Turfing - Reyven (Sportsfields) Ltd
Seeding/Turfing - Gearing F T Landscape Services Ltd
Soft and hard landscaping - Western landscapes Ltd (trading as Wyevale)
Soft landscaping - AMK Fence-In Limited
Turfing - Town & Country Landscapes Ltd

Q31 External planting

External planting - Landscape Institute
External planting - Urban Planters
External planting - Wybone Ltd
General exterior works contractors - Gearing F T Landscape Services Ltd
Interiorscapes - Leaflike
Landscape contractors - Gearing F T Landscape Services Ltd
Landscape mulcher - ABG Ltd
Landscape ornaments - Haddonstone Ltd
Landscapes - Ogilvie Construction Ltd
Landscaping and grounds maintenance - Reyven (Sportsfields) Ltd
Landscaping Garden Products - Titan Environmental Ltd
Plant trellis - Lang+Fulton
Planting - Bernhards Landscapes Ltd
Planting - Clifton Nurseries
Planting - Coblands Landscapes Ltd
Planting - Frosts Landscape Construction Ltd
Soft and hard landscaping - Western landscapes Ltd (trading as Wyevale)
Soft landscaping - AMK Fence-In Limited
Tree and shrub planting - Town & Country Landscapes Ltd

Q32 Internal planting

Aftercare - Omnipex Ltd
Artificial plants and trees - Floralsilk Ltd
Bespoke design service - Omnipex Ltd
Containers - Omnipex Ltd
In house horticulture - Omnipex Ltd
Interior planting - Urban Planters
Interiorscapes - Leaflike
Interiorscapes - Omnipex Ltd
Internal planting - Frosts Landscape Construction Ltd
Internal planting - Landscape Institute
Mail order range - Leaflike
Office/ interior planting displays - Clifton Nurseries
Plastic Hardware - Titan Environmental Ltd
Professional interior landscaping - Leaflike
Rental - Omnipex Ltd
Rental & maintenance schemes - Leaflike
Replica, living & preserved displays (interior & exterior) - Leaflike
Sale - Omnipex Ltd
Thermoplastic road marking - Clare R S & Co Ltd

Q35 Landscape maintenance

Garden furniture - Clifton Nurseries
Garden timbers - Metsä Wood
Landscape design - Clifton Nurseries
Landscape featurestone - The Delabole Slate Co. Ltd
Landscape furniture - Clifton Nurseries
Landscape Maintenance - Chilstone
Landscape maintenance - Coblands Landscapes Ltd
Landscape maintenance - Frosts Landscape Construction Ltd
Landscape maintenance - Gearing F T Landscape Services Ltd
Landscape maintenance - Landscape Institute
Landscape maitenance - Jack Moody Ltd
Landscaping - Bernhards Landscapes Ltd
Light landscaping installation and maintenance - Sportsmark Group Ltd
Manufacturer of artificial trees and plant displays - The Silk Forest
Soft and hard landscaping - Western landscapes Ltd (trading as Wyevale)
Special surfaces etc - Jack Moody Ltd
Sports/Play Sands and Rootzones - Sibelco UK Ltd
Top derssings - Sibelco UK Ltd
Tree soils - Sibelco UK Ltd
Tree surrounds - Jones of Oswestry Ltd
Walling and rockery - Forest of Dean Stone Firms Ltd
Wood pole protectors - Anti-Climb Guards Ltd

Q40 Fencing

Acoustic barriers - PHI Group Ltd
Agricultural fencing - Chestnut Products Limited
Alarm fencing - Allen (Fencing) Ltd
All Types - Havering Fencing Co
All types of fencing - Dinnington Fencing Co. Limited
Architectural trellises and planters - Anthony de Grey Trellises + Garden Lighting
Argricultural & Equestrian - Metalwood Fencing (Contracts) Limited
Automated & manual gates - Metalwood Fencing (Contracts) Limited
Automated gates - Dinnington Fencing Co. Limited
Automatic Gates - Tate Fencing
Barbed wire - Anti-Climb Guards Ltd
Barbican steel fencing - Jackson H S & Son (Fencing) Ltd
BFS Pro-net Licensed - Burn Fencing Limited
Bowtop fencing - Jackson H S & Son (Fencing) Ltd
BS1722 parts 1-16 inclusive - Burn Fencing Limited
Chain - open linked for fencing or lifting - Griff Chains Ltd
Chain Link - Alltype Fencing Specialists
Chain link fencing, concrete fence posts - Procter Cast stone & Concrete Products
Chainlink - B & M Fencing Limited
Chainlink fencing - Hy-Ten Ltd
Chestnut steel palisade, chain link, railings, weld mesh, barbed and razor wire - Guardian Wire Ltd
Closeboard - B & M Fencing Limited
Closeboard - Danbury Fencing Limited
Closeboard fencing - SJ McBride Ltd T/A Warefence
Commercial fencing and gates - Tonbridge Fencing Limited
Concrete, wire fencing and gates - Challenge Fencing Ltd
Diamond trellis - Danbury Fencing Limited
Domestic - Metalwood Fencing (Contracts) Limited
Domestic fencing and gates - Tonbridge Fencing Limited
Enclosures - Fibaform Products Ltd
Environmental barrier/ acoustic fencing - Newton & Frost Fencing Ltd
Erectors for private, domestic and commercial - Danbury Fencing Limited
Euroguard mesh fencing - Jackson H S & Son (Fencing) Ltd
Fence panel manufacturers - Derby Timber Supplies
Fencepost Support System - Catnic
Fencing - Advanced Fencing Systems Ltd
Fencing - Allen (Concrete) Ltd
Fencing - Alltype Fencing Specialists
Fencing - Ambec Fencing
Fencing - AMK Fence-In Limited
Fencing - B & M Fencing Limited
Fencing - Barker & Geary Limited
Fencing - Bedford Fencing Co. Limited
Fencing - Berry Systems
Fencing - Betafence Limited
Fencing - Bowden Fencing Ltd
Fencing - Burn Fencing Limited
Fencing - Capel Fencing Contractors Ltd
Fencing - Challenge Fencing Ltd
Fencing - Chestnut Products Limited
Fencing - Claydon Architectural Metalwork Ltd
Fencing - Coates Fencing Limited
Fencing - Covers Timber & Builders Merchants
Fencing - Danbury Fencing Limited
Fencing - Dinnington Fencing Co. Limited
Fencing - Ebor Concrete Ltd
Fencing - Eve Trakway
Fencing - Fabrikat (Nottingham) Ltd
Fencing - Fencing Contractors Association
Fencing - Fernden Construction (Winchester) Ltd
Fencing - Frosts Landscape Construction Ltd
Fencing - GBG Fences Limited
Fencing - GJ Durafencing Limited
Fencing - Goodacres Fencing
Fencing - Gravesend Fencing Limited
Fencing - Hadley Industries Plc
Fencing - HAGS SMP
Fencing - Harling Security Solutions
Fencing - Havering Fencing Co
Fencing - Huntree Fencing Ltd
Fencing - J Pugh-Lewis Limited
Fencing - J. B. Corrie & Co Limited
Fencing - Jack Moody Ltd
Fencing - Kingsforth Security Fencing Limited
Fencing - Maldon Fencing Co
Fencing - May Gurney (Technical Services) Ltd - Fencing Division
Fencing - Metalwood Fencing (Contracts) Limited
Fencing - Newton & Frost Fencing Ltd
Fencing - Northern Fencing Contractors Limited
Fencing - Pearman Fencing
Fencing - Pembury Fencing Ltd
Fencing - Portia Engineering Ltd
Fencing - Pro-Fence (Midlands)
Fencing - Reyven (Sportsfields) Ltd
Fencing - Ryall & Edwards Limited
Fencing - Scott Fencing Limited
Fencing - SFC (Midlands) Ltd
Fencing - SJ McBride Ltd T/A Warefence
Fencing - Sovereign Fencing Limited
Fencing - Squires Metal Fabrications Ltd
Fencing - Termstall Limited
Fencing - Town & Country Landscapes Ltd
Fencing - W Ward Fencing Contractors
Fencing - W.L. West & Sons Ltd
Fencing - Wallbank
Fencing - Weatherley Fencing Contractors Limited
Fencing - Wentworth Sawmills Ltd
Fencing - Wesson Fencing
Fencing - all types - Anti-Climb Guards Ltd
Fencing (industrial) - Darfen Durafencing
Fencing and gates - Farefence (NW) Ltd
Fencing and gates, security fencing - Binns Fencing Ltd
Fencing and Sectional Buildings - Howe Fencing & Sectional Buildings
Fencing and security fencing - Jackson H S & Son (Fencing) Ltd
Fencing contractors - Gearing F T Landscape Services Ltd
Fencing contractors - Grange Fencing Limited
Fencing gates and environmental barriers - Baker Fencing Ltd
Fencing including steel palisade, power operated gates and metal railings - Charnwood Fencing Limited
Fencing suppliers - Crowthorne Fencing
Fencing Supply - Havering Fencing Co
Fencing, gates and barriers - Scotfen Limited
Fencing, gates, automatic gates and barriers - Tate Fencing
Fencing, gates, railings, pedestrian guard rail and crash barrier - Billericay Fencing Ltd
Fencino (metal) - Steelway Fensecure Ltd
Flow plates - Autopa Ltd
Games area fencing - En-tout-cas Tennis Courts Ltd
Gate automation - Kingsforth Security Fencing Limited
Gates - Alltype Fencing Specialists
Gates - Cannock Gates UK Ltd
Gates - Claydon Architectural Metalwork Ltd
Gates - Coates Fencing Limited
Gates - Lang+Fulton
Gates - WA Skinner & Co Limited
Gates and access control - Chestnut Products Limited
Gates, barriers, shutters, fencing and parking post - DP Sercurity
Gates, coloured concrete posts, ornamental, vertical bar and bow top railings - Procter Bros Limited
General fencing, safety and noise barrier - Chris Wheeler Construction Limited
Grating fencing - Lang+Fulton
Guard rail and barrier fencing - Valmont Stainton

Q40 Fencing (con't)

Heavy duty gates and railings - Ermine Engineering Co. Ltd
High Security - Alltype Fencing Specialists
Highway safety fencing - Chestnut Products Limited
Hinged poles - Valmont Stainton
Jakcure treated timber fencing - Jackson H S & Son (Fencing) Ltd
Local authority suppliers of PVC-u fencing - Duralock (UK) Ltd
Lochrin classic rivetless palisade - Wm. Bain Fencing Ltd
Mesh - Darfen Durafencing
NRG fabrications - NRG Fabrications
On site welding - C & W Fencing Ltd
Paladin - Alltype Fencing Specialists
Palisade - Alltype Fencing Specialists
Palisade - Procter Cast stone & Concrete Products
Palisade fencing and gates - Wm. Bain Fencing Ltd
Panels and posts - B & M Fencing Limited
Perimeter Protection - Guardall Ltd
Perimeter security - Geoquip Worldwide
Plastic battens and fencing - Rockwell Sheet Sales Ltd
Post & Rail - B & M Fencing Limited
Post and rail fencing - SJ McBride Ltd T/A Warefence
Power fencing - Allen (Fencing) Ltd
Precast concrete fencing - Anderton Concrete Products Ltd
PVC fencing - L. B. Plastics Ltd
Railings - Alltype Fencing Specialists
Railings, gates - Albion Manufacturing
Recycling green waste - Jack Moody Ltd
Rotating anti climb guards - Anti-Climb Guards Ltd
Safety fencing - Newton & Frost Fencing Ltd
Security - Metalwood Fencing (Contracts) Limited
Security and industrial fencing - Sampson & Partners Fencing
Security barriers - Bell & Webster Concrete Ltd
Security fencing - Allen (Fencing) Ltd
Security fencing - C & W Fencing Ltd
Security fencing - Chestnut Products Limited
Security fencing - Dinnington Fencing Co. Limited
Security Fencing - Sharman Fencing Ltd
Security fencing - WA Skinner & Co Limited
Security fencing and gates - SJ McBride Ltd T/A Warefence
Security fencing, gates and access systems - O'Connor Fencing Limited
Security grilles & guards - WA Skinner & Co Limited
Security palisade fencing - GJ Durafencing Limited
Security steel palisade fencing and gates - Prima Security & Fencing Products
Security, industrial fencing, railings, crash barriers, gates, palisade, posts and rails - J & M Fencing Services
Security, industrial, commercial new housing and highways - Ambec Fencing
Sentry bar steel fencing - Jackson H S & Son (Fencing) Ltd
Speed ramps - Autopa Ltd
Sports - Metalwood Fencing (Contracts) Limited
Sports pitch fencing - Chestnut Products Limited
Steel fencing and gates - C & W Fencing Ltd
Steel Gates - Simflex Grilles and Closures Ltd
Steel gates - Staines Steel Gate Co
Steel mesh system fences - SJ McBride Ltd T/A Warefence
Steel Palisade - Darfen Durafencing
Steel palisading, welded mesh, chainlink fencing - Fencelines Ltd
Supply & Fix - Havering Fencing Co
Tennis court fencing - En-tout-cas Tennis Courts Ltd
Timber fencing and gates - Tonbridge Fencing Limited
Timber garden products - OSMO UK
Timber palisade - Danbury Fencing Limited
Timber Panels - Huntree Fencing Ltd
Timber, chainlink and weldmesh fencing systems - Prima Security & Fencing Products
Trellis - Danbury Fencing Limited
Tubular construction, barriers and rials - Tubeclamps Ltd
Ultra bar (adaptable ranking fence) - Sections & Profiles Ltd
Ultra Fence (Palisade Fence) - Sections & Profiles Ltd
Vertical tube fencing - Lang+Fulton
Waney Lap - Danbury Fencing Limited
Weld mesh - Alltype Fencing Specialists
Welded Mesh - Kingsforth Security Fencing Limited
Weldmesh - Procter Cast stone & Concrete Products
Wire - Betafence Limited
Wire mesh - Betafence Limited
Wood - Alltype Fencing Specialists
Wrough iron fences - Staines Steel Gate Co

Q41 Barriers/ Guard-rails

Acoustic barriers - Chestnut Products Limited
Anti-Ram Perimeter protection - R.A.M. Perimeter Protection Ltd
Architectural railings - Chestnut Products Limited
Automated gates and barriers - Sampson & Partners Fencing
Automated traffic barriers - Dinnington Fencing Co. Limited
Automatic Barriers - Kone PLC
Balustrade - Coates Fencing Limited
Balustrade & Barrier Systems - Brass Age Ltd
Barriers - Autopa Ltd
Barriers - May Gurney (Technical Services) Ltd - Fencing Division
Barriers - Metalwood Fencing (Contracts) Limited
Barriers - Moravia (UK) Ltd
Barriers - Rediweld Rubber & Plastics Ltd
Barriers - Scotfen Limited
Barriers and guard rails - Havering Fencing Co
Barriers and guardrails - Hart Door Systems Ltd
Barriers and Guardrails - Hubbard Architectural Metalwork Ltd
Barriers and guardrails - J. B. Corrie & Co Limited
Barriers Guardrails - Steelway Fensecure Ltd
Barriers/ guardrails - Claydon Architectural Metalwork Ltd
Barriers/ Guard-rails - O'Connor Fencing Limited
Barriers/ Guards rails - Fencing Contractors Association
Barriers/Gaurdrails - Crowthorne Fencing
Barriers/guard rails - Challenge Fencing Ltd
Barriers/Guard Rails - Gearing F T Landscape Services Ltd
Barriers/guard rails - NRG Fabrications
Barriers/guardrails - Metalcraft (Tottenham) Ltd
Barriers/guard-rails - Berry Systems
Barriers/guard-rails - BPT Automation Ltd
Barriers/Guard-rails - Grant & Livingston Ltd
Barriers/Guard-rails - Pembury Fencing Ltd
Barriers/Guard-rails - Portia Engineering Ltd
Barriers/sGuard-rails - Newton & Frost Fencing Ltd
Cantilever gates - Jackson H S & Son (Fencing) Ltd
Decorative railings - Claydon Architectural Metalwork Ltd
Domestic PVC-u fencing for housing projects - Duralock (UK) Ltd
Electrically operated barriers - O'Connor Fencing Limited
Endless cords - James Lever 1856 Ltd
Fencing - Lang+Fulton
Garden twines - James Lever 1856 Ltd
Gates - May Gurney (Technical Services) Ltd - Fencing Division
Gates and barriers - Hart Door Systems Ltd
Guardrails/ Barriers - Harling Security Solutions
Handrail - Kee Klamp Ltd
Handrail Covers - DC Plastic Handrails Ltd
Hazard Warning & Protection - Moravia (UK) Ltd
Hoop Barriers - R.A.M. Perimeter Protection Ltd
Industrial and Commercial - Metalwood Fencing (Contracts) Limited
Industrial gates - Guardian Wire Ltd
Metal security fencing - Town & Country Landscapes Ltd
Ornamental Rails + Gates - Procter Cast stone & Concrete Products
Palisade - Coates Fencing Limited
Pedguard rail - Jackson H S & Son (Fencing) Ltd
Perimeter protection - APT Controls Ltd
Post & Rail - Furnitubes International Ltd
Precast Complete - RCC, Division of Tarmac Precast Concrete Ltd
Precast concrete products - Ruthin Precast Concrete Ltd
Racecourse rails and PVC-u equestrian fencing - Duralock (UK) Ltd
Railing systems - Brass Age Ltd
Railings - Coates Fencing Limited
Railings - Darfen Durafencing
Railings and gates - installers - GJ Durafencing Limited
Ropes and sashcords - Marlow Ropes Ltd
Safe,strong PVC-u fencing for Schools and Education Facilities - Duralock (UK) Ltd
Security barriers - APT Controls Ltd
Security equipment - Anti-Climb Guards Ltd
Security fencing and gates - Tonbridge Fencing Limited
Steel security palisade fencing and gates - Wm. Bain Fencing Ltd
Steel, Stainless steel and aluminium - Speedfab Midlands
Swing gates - Jackson H S & Son (Fencing) Ltd
Timber gate manufacturers - W.L. West & Sons Ltd
Traffic barriers - Kaba Garog
Traffic Barriers and Gates - Kone
Turnstiles - Kone
Turnstiles & Access Control - Clarke Instruments Ltd

Q50 Site/ Street furniture/ equipment

Access covers and frames - Jones of Oswestry Ltd
Aggregate Blocks - Hanson Building Products UK
Architectural Artwork - Broadbent
Architectural Glass - Creative Glass
Architectural trellises and planters - Anthony de Grey Trellises + Garden Lighting
Arm chairs - Gloster Furniture Ltd
Automated gates - West London Security
Automatic Bollards - APT Controls Ltd
Automatic gates and barriers - BPT Automation Ltd
Automatic Sliding and Hinged Gates - Geoquip Worldwide
Balustrade - Kee Klamp Ltd
Barrier posts - Aremco Products
Bench seating - LSA Projects Ltd
Benches - Dee-Organ Ltd
Benches - Erlau AG
Benches - Gloster Furniture Ltd
Bespoke planters - Design & Display Structures Ltd
Bi - foloding Gates - APT Controls Ltd
Bicycle Racks - Moravia (UK) Ltd
Bins - Cigarette & Litter - Glasdon U.K. Limited
Bollard shells - Signature Ltd
Bollards - Autopa Ltd
Bollards - Dee-Organ Ltd
Bollards - Erlau AG
Bollards - Furnitubes International Ltd
Bollards - Glasdon U.K. Limited
Bollards - Orchard Street Furniture
Bollards - R.A.M. Perimeter Protection Ltd
Bollards - Rediweld Rubber & Plastics Ltd
Bollards - Scotfen Limited
Bollards and airlock doors - Geoquip Worldwide
Bollards and verge marker posts - Glasdon UK Limited
Bollards, cable covers, troughs and bespoke products - Anderton Concrete Products Ltd
Bridges, boardwalks, decking and pergolas - CTS Bridges Ltd
Bus & rail shelters - Abacus Lighting Ltd
Bus and glazed shelters - Midland Alloy Ltd
Car Park Posts - Furnitubes International Ltd
Car park/Forecourt boundary PVC-u fencing - Duralock (UK) Ltd
Carved stone - Caithness Flagstone Limited
Cigarette ash bins - Swintex Ltd
Concrete bollards - Procter Cast stone & Concrete Products
Covered walkways - Q.M.P
Cycle & Motorcycle parking - Furnitubes International Ltd
Cycle Lockers - Autopa Ltd
Cycle parking - Q.M.P
Cycle parking equipment - Broxap Ltd
Cycle shelters - Dee-Organ Ltd
Cycle Shelters - Fibaform Products Ltd
Cycle Shelters - Parklines Building Ltd
Cycle stands - Dee-Organ Ltd
Cycle stands - Erlau AG
Cycle stands - HAGS SMP
Cycle stands - Rediweld Rubber & Plastics Ltd
Cycle stands and shelters - Autopa Ltd
Cycle storage systems - Elwell Buildings Ltd
Demarcation studs - brass, aluminum, stainless steel - Reflecting Roadstuds Ltd
Dining chairs - Gloster Furniture Ltd
Dog bins - Neptune Outdoor Furniture Ltd
Dog waste bins - Glasdon UK Limited
Drainage gratings - Jones of Oswestry Ltd
Electrical termination/ service pillar - Ritherdon & Co Ltd
Electrically operated gates - SJ McBride Ltd T/A Warefence
Engraving service - Gloster Furniture Ltd
Entrance Gates, Turnstiles - Gunnebo UK Limited
Exterior lighting - Abacus Lighting Ltd
Folding chairs - Gloster Furniture Ltd
Furniture - Barlow Tyrie Ltd
Garden and Entrance Gates - Danbury Fencing Limited
Garden furniture - Nova Garden Furniture Ltd
Gates - Squires Metal Fabrications Ltd
Gates timber & metal - B & M Fencing Limited
Glass fibre and aluminium flag poles - Harrision Flagpoles
Grit bins - Amberol Ltd
Grit bins and spreaders - Glasdon UK Limited
Grit/ salt bins - Balmoral Tanks
Gritbins - Titan Environmental Ltd
Ground and wall mounted banner poles - Harrision Flagpoles
Landscape components and street furniture - Woodscape Ltd
Landscape furniture, seatings, shelters, bridges, litter bins, and trim trails - Lappset UK Ltd
Landscape Suppliers - Crowthorne Fencing
Letter bins - Iles Waste Systems
Litter bins - Amberol Ltd
Litter bins - Erlau AG
Litter Bins - Furnitubes International Ltd
Litter Bins - Glasdon UK Limited
Litter bins - Swintex Ltd
Litter Bins and Ashtrays - The No Butts Bin Company (NBB), Trading as NBB Outdoor Shelters
Loungers - Erlau AG
Manufacturers - Autopa Ltd
Motorway crash barriers and pedestrian guard rails - Lionweld Kennedy Ltd
Natural Stone Fountains & Features - Palladio Stone
Occasional furniture - Gloster Furniture Ltd
Ornamental - The Stone Catalogue Ltd
Ornamental statues, pots and urns - Rogers Concrete Ltd
Outdoor litter bins - Unicorn Containers Ltd
Parking posts - Autopa Ltd
Parking posts - Rediweld Rubber & Plastics Ltd
PAS 68 crash rated products - Geoquip Worldwide
Pavior infill covers and frames - Jones of Oswestry Ltd
Pedestrian crossing equipment - Signature Ltd
Pedestrian guard rails - Fabrikat (Nottingham) Ltd
Pedestrian Turnstiles & Gates - Geoquip Worldwide
Picnic tables - HAGS SMP
Planters - Erlau AG
Play area rubber animals and spheres - Playtop Licensing Limited
Posts, Chains, Stands, Bollards - Moravia (UK) Ltd
Public seats, litter bins, benches, bollards, picnic tables and planters - Neptune Outdoor Furniture Ltd
Purpose made furniture - Orchard Street Furniture
Reconstructed stone balustrading, paving, garden ornaments and architectural dressings - Minsterstone Ltd
Recycled Plastic Outdoor Furniture (Picknic Tables, Benches, Seating) - The No Butts Bin Company (NBB), Trading as NBB Outdoor Shelters
Recycling bins - Amberol Ltd
Recycling bins - Glasdon UK Limited
Recycling bins - Iles Waste Systems
Recycling bins, boxes and containers - Straight Ltd
Recycling Bottle Banks - Titan Environmental Ltd
Recycling units & liter bins - Broxap Ltd
Restaurant chairs - Laytrad Contracts Ltd
Retractable Posts - R.A.M. Perimeter Protection Ltd
Rising Arm Barriers - Darfen Durafencing
Road Blockers - APT Controls Ltd
Road Blockers - Geoquip Worldwide
Rockery stone - The Delabole Slate Co. Ltd
Scaffolding protectors - Anti-Climb Guards Ltd
Seating - Erlau AG
Seating - Glasdon U.K. Limited
Seating and tables for schools, waiting areas, train stations, airports, ferry terminals, hospitals and surgeries - Hille Educational Products Ltd
Seating, Benches - Furnitubes International Ltd
Seats - HAGS SMP
Seats and benches - Glasdon UK Limited
Security bolders - Kerridge Stone Ltd
Security gates - Staines Steel Gate Co
Self watering baskets - Amberol Ltd
Self watering Precinct planters - decorative planters - Amberol Ltd
Self watering window boxes and barrier baskets - Amberol Ltd
Shelters - Broxap Ltd
Shelters (Smoking, Cycle, Waiting) - The No Butts Bin Company (NBB), Trading as NBB Outdoor Shelters
Shelters, buildings and housings - Glasdon UK Limited
Sign posts - Valmont Stainton
Site Equipment - Bainbridge Engineering Ltd
Site equipment - Frosts Landscape Construction Ltd
Site furniture / equipment - Glasdon UK Limited
Site furniture, car park furniture and equipment - APT Controls Ltd
Site furniture/ equipment - Broxap Ltd
Site/ Street furniture/ equipment - 3d Storage Systems (UK) LTD
Site/ street furniture/ equipment - Chilstone
Site/ street furniture/ equipment - Haddonstone Ltd
Site/ Street furniture/ equipment - Leander Architectural
Site/Street funiture/ Equipment - Berry Systems
Site/Street furniture/Equipment - Pembury Fencing Ltd
Site/Street Furniture/Equipment - Wybone Ltd
Sleeping policemen - Rediweld Rubber & Plastics Ltd
Sliding Gates - Darfen Durafencing
Smoking shelters - Autopa Ltd
Smoking shelters - Q.M.P
Speed ramp - Swintex Ltd
Speed ramps - Berry Systems
Steel Storage Buildings - Parklines Building Ltd
Steel, Stainless steel and aluminium - Speedfab Midlands
Stone features and seats - Earl's Stone Ltd
Stone street furniture - CED Ltd
Storage Systems - Toprail Systems Ltd
Street furniture - Abacus Lighting Ltd
Street furniture - Ballantine, Bo Ness Iron Co Ltd
Street furniture - Blanc de Bierges
Street furniture - Broxap Ltd
Street furniture - Claydon Architectural Metalwork Ltd
Street Furniture - CU Phosco Lighting
Street furniture - Dee-Organ Ltd
Street furniture - DW Windsor Lighting
Street furniture - Earth Anchors Ltd
Street Furniture - Ebor Concrete Ltd
Street Furniture - Falco UK Ltd
Street Furniture - Furnitubes International Ltd
Street furniture - Glasdon UK Limited
Street Furniture - Harling Security Solutions
Street furniture - Marshalls Mono Ltd
Street Furniture - Townscape Products Ltd
Street furniture - Urbis Lighting Ltd
Street furniture - Wales & Wales
Street furniture - Woodhouse UK Plc
Street Furniture Installation Services - Branson Leisure Ltd
Street Furniture Manufacturers - Branson Leisure Ltd
Street furniture, lighting, equipment installation - AMK Fence-In Limited
Street furniture, shelters and walkways - design and manufacture - Macemain + Amstad Ltd
Tables - Erlau AG
Tables - Furnitubes International Ltd
Tables - Gloster Furniture Ltd
Talking bin - Amberol Ltd
Teen shelters - HAGS SMP
Temporary flood defence systems - Bauer Inner City Ltd
Timber, steel, cast iron street and garden furniture - Orchard Street Furniture
Tower protectors - Anti-Climb Guards Ltd
Traffic calming products - Rediweld Rubber & Plastics Ltd
Traffic products - Swintex Ltd
Tree grilles and guards - Furnitubes International Ltd
Turnstyles - Darfen Durafencing
Turnstyles - Sentra Access Engineering Ltd
Vases, Urns & Pots - The Stone Catalogue Ltd
Vehicle Barriers - Geoquip Worldwide
Wase and recycling bins - Straight Ltd
Waste containers and systems - Iles Waste Systems
Water gardens and water features - Lotus Water Garden Products Ltd
Wheeled bins - Titan Environmental Ltd

R10 Rainwater pipework/ gutters

Aluminium flashings - Gilmour Ecometal
Aluminium gutter and rainwater pipe products - Marley Plumbing and Drainage

R10 Rainwater pipework/ gutters (con't)

Aluminium gutters - George Gilmour (Metals) Ltd
Aluminium gutters - Gilmour Ecometal
Aluminium pipework and gutters - Aluminium RW Supplies Ltd
Aluminium rainwater good - Alifabs (Woking) Ltd
Aluminium rainwater goods - WP Metals Ltd
Aluminium Roof Outlets - Alutec
Aluminuim Gutters, Downpipes and Hoppers - Alutec
Aluminuim rainwater goods - Guttercrest
Building Products incl window & guttering systems - Marshall-Tufflex Ltd
Building Rainwater Disposals - Geberit Ltd
Cast iron rainwater and soil goods - Longbottom J & J W Ltd
Cast iron Rainwater gutter systems to BS 460 - Saint-Gobain PAM UK
Cast iron roof outlets to BS EN 1253 - Saint-Gobain PAM UK
Cast iron soil, vent and waste and rainwater pipes to BS EN 877 - Saint-Gobain PAM UK
Channel Drainage - Wavin Ltd
Clay, Plastic and concrete drainage systems - Hepworth Building Products Ltd
Copper pipework - Servomac Limited
Copper rainwater system - Klober Ltd
Copper Rainwater Systems - Coppa Cutta Ltd
Custom fabricated in metals - Azimex Fabrications Ltd
Downpipe - Round & Square - Coppa Cutta Ltd
Fascia and soffit systems in powder coated aluminium and steel - Dales Fabrications Ltd
Galvanised steel pipework - Servomac Limited
GRP valley troughs - Timloc Expamet Building Products
Gutter heating - Findlay Irvine Ltd
Gutter leaf protection system - FloPlast Ltd
Guttering - Avonside Roofing Group T/A Letchworth Roofing
Guttering - 1/2 Round, Box & Ogee - Coppa Cutta Ltd
Insulated gutters - CGL Cometec Ltd
Pipes and fittings - Geberit Ltd
Pipes and Fittings - Interflow UK
Plastic drainage systems - Hunter Plastics Ltd
PVCu rainwater systems - Marley Plumbing and Drainage
Rain water management systems - Fullflow Group Ltd
Rainwater diverter and systems - FloPlast Ltd
Rainwater Goods - Rheinzink UK
Rainwater goods - Rockwell Sheet Sales Ltd
Rainwater goods - Roles Broderick Roofing Ltd
Rainwater outlets/ gutters - McKenzie-Martin Ltd
Rainwater pipework & gutters - Olby H E & Co Ltd
Rainwater pipework/ gutters - Harris & Bailey Ltd
Rainwater pipework/ gutters - Interlink Group
Rainwater pipework/ gutters - John Davidson (Pipes) Ltd
Rainwater recirculation systems - Kingspan Environmental
Rainwater systems - Allmat (East Surrey) Ltd
Rainwater systems - Harrison Thompson & Co Ltd
Rainwater systems in powder coated aluminium and steel - Dales Fabrications Ltd
Rainwater systems, sacias and soffits - Alumasc Exterior Building Products Ltd
Rainwater, soil and wate systems - Brett Martin Ltd
Relining of concrete gutters - Finlock Gutters
Roof Drainage Outlets - Caroflow Ltd
Roof outlets - ACO Technologies plc
Siphonic rainwater drainage systems - Dales Fabrications Ltd
Stainless steel drainage - Blücher UK Ltd
Stainless Steel Drainage systems - Aperam Stainless Services & Solutions UK Limited
Stainless steel pipework - Blucher UK
Stainless steel pipework - Servomac Limited
Steel gutters and downpipe systems - Saint-Gobain PAM UK
Surface water drainage - Hodkin Jones (Sheffield) Ltd
Syphonic drainage systems - Fullflow Group Ltd
Syphonic roof drainage - Fullflow Group Ltd
T-Pren-gutter expansion material - Matthew Hebden
Water storage containers and accessories - Straight Ltd
Zinc rainwater goods - Metra Non Ferrous Metals Ltd

R11 Foul drainage above ground

Above ground drainage - Ballantine, Bo Ness Iron Co Ltd
Above ground drainage - FloPlast Ltd
Air admittance valves - FloPlast Ltd
Cast iron above ground system to BS 416 - Saint-Gobain PAM UK
Cast iron floor and shower gullies, gratings and accessories to BS EN 1253 - Saint-Gobain PAM UK
Cast iron sanitary pipework system to BS EN 877 - Saint-Gobain PAM UK
Connectors - Wirquin
Drainage above ground - Brett Martin Ltd
Drainage systems - Alumasc Exterior Building Products Ltd
Filter beds - Milton Pipes Ltd
Foul drainage above ground - Durey Casting Ltd
Foul drainage above ground - FC Frost Ltd
Foul drainage above ground - Wavin Ltd
Gutters, drainage pipes and fittings - Longbottom J & J W Ltd
Metal pipework - Copper, Galvanised steel and Stainless steel - Servomac Limited
MUPVC waste system - Davant Products Ltd
Plastic drainage systems - Hunter Plastics Ltd
PVCu soil and waste systems - Marley Plumbing and Drainage
Rainwater pipework/gutters - Blücher UK Ltd
Retro fit above ground & built in submerged covers - Grando (UK) Ltd
Seamless cold drawn copper tube - Lawton Tube Co Ltd, The
Soil and vent terminals - Klober Ltd
Stainless steel drainage channels, gratings and gullies - Component Developments
Stainless Steel Drainage systems - Aperam Stainless Services & Solutions UK Limited
Stainless Steel Pipes - ACO Technologies plc
Stainless steel pipework - Blucher UK
Tanks (retention) - Milton Pipes Ltd
Traps and waste fittings - Opella Ltd
Waste Traps - FloPlast Ltd
Wastes - Wirquin

R12 Drainage below ground

Access covers and frames linear surface water drainage system and special fabrication - Clarksteel Ltd
Bacteria Grease traps - Progressive Product Developments Ltd
Barrier pipe for contaminated land - Wavin Ltd
Below ground drainage and maholes - Ballantine, Bo Ness Iron Co Ltd
Box culverts - Milton Pipes Ltd
Caisson shafts - Milton Pipes Ltd
Cast iron Below ground system to BS 437 - Saint-Gobain PAM UK
Cast iron, below ground system to BS EN 877 - Saint-Gobain PAM UK
Cesspools and septic tanks - Kingspan Environmental
Channel grating - Shackerley (Holdings) Group Ltd incorporating Designer ceramics
Clay Drainage - Knowles W T & Sons Ltd
Commercial Gullies, gratings, channels and traps - ACO Technologies plc
Concrete drainage systems - Hanson Concrete Products
Concrete manholes and soakaways - Milton Pipes Ltd
Cop - SUDS - British Water
Design, installation and maintenance of building services - Cofely Engineering Services Ltd
Drain and pipework cleaning - Dyno Rod PLC
Drain and pipework inspection - Dyno Rod PLC
Drain and pipework installation - Dyno Rod PLC
Drain and pipework repairs - Dyno Rod PLC
Drainage - Marshalls Mono Ltd
Drainage - Polypipe Civils
Drainage Below Ground - Aluline Ltd
Drainage below ground - CPV Ltd
Drainage below ground - FC Frost Ltd
Drainage below ground - Formpave Ltd
Drainage Below Ground - Gatic
Drainage below ground - Harris & Bailey Ltd
Drainage below ground - John Davidson (Pipes) Ltd
Drainage Below Ground - Safeguard Europe Ltd
Drainage below ground - Wavin Ltd
Drainage media - Lytag Ltd
Drainage membranes - SCP Concrete Sealing Technology Ltd
Flexible couplings - Flex-Seal Couplings Ltd
Floor Drainage Outlets - Caroflow Ltd
Floor gullies, roof outlets, access covers, grease converters and linear drainage - Wade International Ltd
Gravity drain systems - Uponor Ltd
Grease traps, oil and grease water separators - Progressive Product Developments Ltd
Grease treatment plants - Progressive Product Developments Ltd
GRP covers - Armfibre Ltd
GRP Kiosks - Adams- Hydraulics Ltd
Installation services - Uponor Ltd
Jetters cold and hot, winches, rods and accessories - Ward's Flexible Rod Company Ltd
Manhole covers and gratings - Durey Casting Ltd
Manhole covers and reservoir lid - Ruthin Precast Concrete Ltd
Monolithic road gullies - Milton Pipes Ltd
Non entry inspection chambers - Wavin Ltd
Package sewage pumping stations - KSB Ltd
Plastic drainage systems - Hunter Plastics Ltd
Pump stations - Kingspan Environmental
Pumping stations for storm water/Foul Water - Pims Pumps Ltd
PVCu underground drainage systems - Marley Plumbing and Drainage
Rainwater Harvesting System & Underground Drainage - FloPlast Ltd
Rainwater harvesting systems - Kingspan Environmental
Rectangular inspection chambers - Milton Pipes Ltd
Septic tank conversion unit - MANTAIR Ltd
Septic tanks - Balmoral Tanks
Settlement tanks and components - Armfibre Ltd
Stainless steel channels, gratings, gullies, access covers and grease separators - Blücher UK Ltd
Stainless steel manhole covers - Component Developments
Stainless steel pipework - Blucher UK
StormWater Management - Wavin Ltd
Surface Drainage - Interflow UK
Surface water drainage systems - Mea UK Ltd
Terracotta - Hepworth Building Products Ltd
Ultra rib sewer systems - Uponor Ltd
Underground drainage - Naylor Drainage Ltd
Underground drainage systems (PVC) - Brett Martin Ltd

R13 Land drainage

Cast Iron gratings and grills - Company name: Britannia Metalwork Services Ltd
Clay Drainage System - Hepworth Building Products Ltd
Inspection chambers - Hunter Plastics Ltd
Land drainage - Durey Casting Ltd
Land drainage - Flex-Seal Couplings Ltd
Land drainage - J Pugh-Lewis Limited
Land drainage - John Davidson (Pipes) Ltd
Land Drainage - Naylor Drainage Ltd
Land drainage - Polypipe Civils
Land Drainage - T-T Pumps & T-T controls
Paving Support & Drainage - Caroflow Ltd
Plumbing and Drainage - Wavin Ltd
Preformed sheets, cavity and channel drainage - ABG Ltd
PVCu quantum highway and sewer drainage - Marley Plumbing and Drainage
Rainwater harvesting - MANTAIR Ltd
Vertical Band Drains - Bauer Inner City Ltd

R14 Laboratory/ Industrial waste drainage

Bacteria Grease traps - Progressive Product Developments Ltd
Grease traps, oil and grease water separators - Progressive Product Developments Ltd
Grease treatment plants - Progressive Product Developments Ltd
Laboratory drainage systems - CPV Ltd
Laboratory Ind Waste - George Fischer Sales Ltd
Laboratory/ Industrial Waste - Naylor Drainage Ltd
Laboratory/industrial waste drainage - Satec Service Ltd
Stainless steel channels, gratings, gullies, access covers and grease separators - Blücher UK Ltd
Stainless steel drainage systems - ACO Technologies plc
Stainless Steel Drainage systems - Aperam Stainless Services & Solutions UK Limited
Stainless steel pipework - Blucher UK
Thermoplastic pipes - CPV Ltd
Underground storage tanks - Kingspan Environmental Ltd

R20 Sewage pumping

Centrifugal pumps - Calpeda Limited
Drainage and sewage package systems - Pump Technical Services Ltd
Maintenance and repair of pumping equipment - Pims Pumps Ltd
Package pump stations - Tuke & Bell Ltd
Pumps and ejectors - Tuke & Bell Ltd
Sewage pumping - Adams- Hydraulics Ltd
Sewage Pumping - Binder
Sewage Pumping - Mono Pumps Ltd
Sewage pumping - Satec Service Ltd
Sewage Pumping - T-T Pumps & T-T controls
Sewage pumps and pumping systems - New Haden Pumps Ltd
Sewage, drainage, pumps and systems - KSB Ltd

R21 Sewage treatment/ sterilisation

Cesspools and septic tanks - Kingspan Environmental
Cop - Flows & Loads - British Water
Effluent treatment - Armfibre Ltd
Fuel water seperators - Kingspan Environmental
Grease removal system - Tuke & Bell Ltd
Grease traps - Kingspan Environmental
Grit removal systems - Tuke & Bell Ltd
GRP Equipment - Adams- Hydraulics Ltd
Package sewage treatment plants - Kingspan Environmental
Package sewage treatment plants - Tuke & Bell Ltd
Packaged Sewage Treatment Plant - MANTAIR Ltd
Packaged Sewage Treatment Plant - Titan Polution Control
Pump stations - Kingspan Environmental
Rotating biological contactors - Tuke & Bell Ltd
Sample chambers - Kingspan Environmental
Separators - Kingspan Environmental
Settling tank scrapers - Tuke & Bell Ltd
Sewage purification equipment - Tuke & Bell Ltd
Sewage storage tanks - Franklin Hodge Industries Ltd
Sewage treatment - Balmoral Tanks
Sewage treatment equipment - Adams- Hydraulics Ltd
Sewage treatment systems - Kingspan Environmental Ltd
Sewage treatment/ sterilisation - Aldous & Stamp (Services) Ltd
Sewage treatment/sterilization - Satec Service Ltd
Supply of electric Sewage Treatment Plants - Water Technology Engineering Ltd
Supply of Non-electric Sewage Treatment Plants - Water Technology Engineering Ltd
Surface aerators - Tuke & Bell Ltd
Tanks (retention) - Milton Pipes Ltd
Trickling filter distributors - Tuke & Bell Ltd
Ultra violet chambers, lamps and control panels - Hanovia Ltd

R30 Centralised vacuum cleaning

Central Vacuum Systems - Barloworld Vacuum Technology Plc
Centralised vacuum cleaning - Donaldson Filtration (GB) Ltd
Centralised vacuum systems - Nilfisk Limited
Centralized vacuum cleaning systems - BVC

R31 Refuse chutes

Waste Disposal Systems - Hardall International Ltd

R32 Compactors/ Macerators

Bag compactors - Tony Team Ltd
Bin compactors - Tony Team Ltd
Compactors/ Macerators - Mono Pumps Ltd
Demountable compactors - Randalls Fabrications Ltd
Domestic lifting stations - Stuart Turner Ltd
Harpac Compactors - Hardall International Ltd
Shower Waste Pumps - Stuart Turner Ltd
Sluiceroom macerators - Verna Ltd
Static & portable compaction units - Randalls Fabrications Ltd
Waste management equipment, compactors/ macerators - Imperial Machine Co Ltd
WC Macerators - Stuart Turner Ltd

S10 Cold Water

Booster sets - Stokvis Industrial Boilers International Limited
Cold Water - British Water
Cold Water - George Fischer Sales Ltd
Cold Water - Holden Brooke Pullen
Cold water - John Davidson (Pipes) Ltd
Copper tube for water - Yorkshire Copper Tube
Design, installation and maintenance of building services - Cofely Engineering Services Ltd
Distribution Mains - Durotan Ltd
Ductwork clearing - Aquastat Ltd
External Mains - Durotan Ltd
Level gauges and transmitters - Klinger Fluid Instrumentation Ltd
Manufacturer & supplier of materials & liners for cured-in-place pipe rehabilitation - Applied Felts Ltd
MDPE pipe and fittings - FloPlast Ltd
MDPE water pipes and fittings - Brett Martin Ltd
Non portable water systems - Uponor Ltd
Packaged Cold Water Storage/Pump - Harton Heating Appliances
Pipe bending equipment - Hilmor
Pipe freezing kits - Hilmor
Piped supply systems cold water - Precolor Sales Ltd
Piping systems - George Fischer Sales Ltd
Plumbing systems - Hepworth Building Products Ltd
Polyethylene systems - Uponor Ltd
Portable water pipe fittings - Uponor Ltd
Seamless cold drawn copper tube - Lawton Tube Co Ltd, The
Tank fittings - Sarena Mfg Ltd
Threading equipment - Hilmor
Valves - Delta Fire
Valves - Sarena Mfg Ltd
Water boundary control boxes - EBCO
Water fittings and controls - Flamco UK Ltd
Water meters - SAV United Kingdom Ltd
Water meters - Switch2 Energy Solutions Ltd
Water meters and oil meters - Elster Metering Ltd
Water storage tanks - Sarena Mfg Ltd

S11 Hot Water

Boiling water products - Heatrae Sadia Heating Ltd
Cistern & Cistern - Opella Ltd
Cistern & Cistern Fittings - Opella Ltd
Commercial Water Heating - Heatrae Sadia Heating Ltd
Copper tube for water - Yorkshire Copper Tube
Design, installation and maintenance of building services - Cofely Engineering Services Ltd
Dual fired boilers - Stokvis Industrial Boilers International Limited
Heat exchangers - Stokvis Industrial Boilers International Limited
Heated towel rails - electric - Pitacs Ltd
Hot Water - Chromalox (UK) L td
Hot Water - George Fischer Sales Ltd
Hot Water - Holden Brooke Pullen
Industrial boilers - Stokvis Industrial Boilers International Limited
Pre-commisson clearing of heating & chilled water systems - Aquastat Ltd
Pre-insulated Pipework supply and installation - Durotan Ltd
Pre-insulated Pipework Supply only - Durotan Ltd
Premix boilers - Stokvis Industrial Boilers International Limited
Solar collector systems - Stokvis Industrial Boilers International Limited
Switches, valves and manifolds - Klinger Fluid Instrumentation Ltd
Temperature blending valves - Deva Tap Co Ltd
Thermal Stores and storage vessels - Lochinvar Limited
Unvented Combination Units with Unvented Hot Water Storage & Boilers - Harton Heating Appliances
Water boiling - Applied Energy Products Ltd
Water heaters - Stokvis Industrial Boilers International Limited
Water heating - Applied Energy Products Ltd

S12 Hot & cold water (self-contained specification)

Cable Duct - Hepworth Building Products Ltd
Commercial water control systems - Kohler Mira Ltd
Copper tube for water - Yorkshire Copper Tube
Copper Unvented Units - Harton Heating Appliances
Cylinder Packs With Integral Controls - Harton Heating Appliances
Direct Electric Boilers - Dimplex UK Limited
Drinking Water System - Tapworks Water Softeners
Drinking water systems - EcoWater Systems Ltd
Electric Showers - Deva Tap Co Ltd
Enamelled Steel Unvented Units - Harton Heating Appliances
Filtration equipment - EcoWater Systems Ltd
Fittings - Pegler Ltd
Flexible plastic plumbing - IPPEC Sytsems Ltd
FloFit plumbing system - FloPlast Ltd

S12 Hot & cold water (self-contained specification) (con't)

Hot & Cold Water - George Fischer Sales Ltd
Hot & cold water - Hertel Services
Hot and cold water - Guest (Speedfit) Ltd, John
Hot and cold water and central heating, polybutylene pipework - Brett Martin Ltd
Hot and cold water cylinder jackets - Davant Products Ltd
Hot and cold water systems - Equator PEX - Marley Plumbing and Drainage
Manifold Plumbing Systems - Davant Products Ltd
Mixers - Marflow Eng Ltd
Packaged boiler houses - Stokvis Industrial Boilers International Limited
Plumbing - Marflow Eng Ltd
Plumbing accessories - PTS Plumbing Trade Supplies
Plumbing systems - REHAU Ltd
Pre- insulated pipes - IPPEC Sytsems Ltd
Prefabricated Plumbing Units - Harton Heating Appliances
Presurised hot and chilled water units - Stokvis Industrial Boilers International Limited
PTFE sealant - Klinger Fluid Instrumentation Ltd
Sealed system equipment - Flamco UK Ltd
Shower Valves - Marflow Eng Ltd
Showers - Deva Tap Co Ltd
Stainless Steel Pipes - ACO Technologies plc
Swivel ferrule straps, stop taps, leadpacks and boundary boxes - Tyco Waterworks - Samuel Booth
Taps - Marflow Eng Ltd
Tube Plugs - Furmanite International Ltd
Water pumps - Stuart Turner Ltd
Water purifiers - EcoWater Systems Ltd
Water Purifiers - Tapworks Water Softeners
Water saving products - Deva Tap Co Ltd
Water softeners - EcoWater Systems Ltd
Water softeners - tapworks water softeners - Tapworks Water Softeners

S13 Pressurised water

Control valves - Auld Valves Ltd
Design, installation and maintenance of building services - Cofely Engineering Services Ltd
Expansion vessels for portable water - Flamco UK Ltd
Pressure booster pumps - Stuart Turner Ltd
Pressure boosting systems - Armstrong Integrated Ltd
Pressure sets - Stuart Turner Ltd
Pressure sprayers - Hozelock Ltd
Pressure units - Pressmain
Pressurised water - Armstrong Integrated Ltd
Pressurised water - Danfoss Ltd
Pressurised Water - Holden Brooke Pullen
Pressurized Water - British Water
Pressurized water - Calpeda Limited
Pressurized water - John Davidson (Pipes) Ltd
Pressurized water - New Haden Pumps Ltd
PVC pressure pipes - Uponor Ltd
Rainwater utilization pump systems - KSB Ltd
Shower booster pumps - Stuart Turner Ltd
Shower pumps - Deva Tap Co Ltd
Valves for high pressure pneumatic and hydraulic appliances - Hale Hamilton Valves Ltd
Water Booster - Pressmain
Water management - Verna Ltd
Water Supply Pipe - Hepworth Building Products Ltd

S14 Irrigation

Irrigation - Calpeda Limited
Irrigation - Holden Brooke Pullen
Stainless steel Drainage pipework - Blücher UK Ltd

S15 Fountains/ Water features

Architectural Glass - Creative Glass
Bespoke water features - Water Sculptures Ltd
Events water features - Water Sculptures Ltd
Fountain and Water Display Equipment - Simon Moore Water Services
Fountains - The Stone Catalogue Ltd
Fountains and Water Displays - Fountains Direct Ltd
Fountains/ Water Features - Chilstone
Fountains/ Water features - Haddonstone Ltd
Fountains/Water Features - Holden Brooke Pullen
Indoor water features - Water Sculptures Ltd
Internal Fountains - The Fountain Company
Millstone & Boulder Fountains - Fountains Direct Ltd
Sculptures in water - Water Sculptures Ltd
Stainless Steel Pipes - ACO Technologies plc
Theatrical water features - Water Sculptures Ltd
Water features - Water Sculptures Ltd
Water gardens, pumps, filters, fountains - Lotus Water Garden Products Ltd
Water storage containers and accessories - Straight Ltd

S20 Treated/ Deionised/ Distilled water

Cooling tower water treatment, cleaning and refurb - Aquastat Ltd
Filteration Equipment - Tapworks Water Softeners
Flexible tubes - United Flexibles (A Division of Senior Flex)
Limescale inhibitors - Hydropath (UK) Ltd
Treated/ Deionised/ Distilled water - Aldous & Stamp (Services) Ltd
Treated/ deionized/distilled water - Satec Service Ltd
Treated/Deionised/Distilled Water - British Water
Treated/Deionised/Distilled water - Hertel Services
Ultra violet chambers, lamps and control panels - Hanovia Ltd

Waste water treatment - The Clearwater Group
Water chlorination - Portacel
Water conditioners - Cistermiser Ltd
Water filtration and treatment - Barr & Wray Ltd
Water filtration equipment - Eden Springs (UK) Ltd
Water filtration media - Lytag Ltd
Water purification systems - Veoila Water Technologies UKElga Ltd
Water treatment - Feedwater Ltd

S21 Swimming pool water treatment

Automatic chemical control and dosing - Golden Coast Ltd
Disinfection systems - Portacel
Pool & Spa Treatment - BioLab UK
Pool covers - Amdega Ltd
Spas - Barr & Wray Ltd
Stainless Steel Pipes, Commercial Gullies, Stainless Steel hygienic channels - ACO Technologies plc
Swimming pool installations - Barr & Wray Ltd
Swimming pool kits and accessories-domestic - Fox Pool (UK) Ltd
Swimming pool water treatment - Aldous & Stamp (Services) Ltd
Swimming pool water treatment - British Water
Swimming pool water treatment - Precolor Sales Ltd
Swimming pools and accessories - Aqua-Blue Ltd
Swimming pools and equipment, spas, heat pumps, water features and chemicals - Certikin International Ltd
Swimming pools, spa baths, sauna and steam rooms - Buckingham Swimming Pools Ltd
Swimming pools, spas, saunas, steam rooms, baptisteries and mosaic murals - Regency Swimming Pools Ltd
Ultra violet chambers, lamps and control panels - Hanovia Ltd
Water features - Aqua-Blue Ltd
Water Features - Water Sculptures Ltd
Water treatment products for swimming pools & spars - BioLab UK
Watertank refurbishment - Aquastat Ltd

S30 Compressed air

Compressed air - Danfoss Ltd
Compressed air - Guest (Speedfit) Ltd, John
Compressed air - Hertel Services
Compressed air supply systems - Fluidair International Ltd
Compressors - Chippindale Plant Ltd
Hand and actuated valves - George Fischer Sales Ltd
Oil free air compressors, vacuum pumps and gas circulators - Bambi Air Compressors Ltd.
Pipework and Installation - Fluidair International Ltd
Valves and controls for presurised gas - Hale Hamilton Valves Ltd

S31 Instrument air

Valves and controls for presurised gas - Hale Hamilton Valves Ltd

S32 Natural gas

Design, installation and maintenance of building services - Cofely Engineering Services Ltd
Gas and sanitation applications - Yorkshire Copper Tube
Gas detection - Portacel
Gas Detection Equipment - TCW Services (Control) Ltd
Gas detection systems - Blakell Europlacer Ltd
Gas fires - natural - Valor
Gas isolation devices - TCW Services (Control) Ltd
Gas Isolation Equipment - TCW Services (Control) Ltd
Gas meters - Switch2 Energy Solutions Ltd
Gas pipe fittings - Uponor Ltd
Natural gas - John Davidson (Pipes) Ltd
Natural Gas - Saacke Combustion Services Ltd
Seamless cold drawn copper tube - Lawton Tube Co Ltd, The
Valves and controls for presurised gas - Hale Hamilton Valves Ltd

S33 Liquefied petroleum gas

Calor gas - Bell & Co Ltd
Cylinder & bulk LPG - Calor Gas Ltd
Gas Detection Equipment - TCW Services (Control) Ltd
Gas Isolation Equipment - TCW Services (Control) Ltd
Heating appliances - Bell & Co Ltd
Liquified petroleum gas storage - Grant & Livingston Ltd
LPG gas fires - Valor
LPG vessels - Flamco UK Ltd
Portable gas heaters - Valor
Valves and controls for presurised gas - Hale Hamilton Valves Ltd

S34 Medical/ Laboratory gas

Stainless Steel Pipes - ACO Technologies plc
Valves and controls for presurised gas - Hale Hamilton Valves Ltd

S40 Petrol/ Diesel storage/distribution

Bulk liquid/ chemical storage tanks - Balmoral Tanks
Fill Point cabinets - Metcraft Ltd
Fuel tanks - Balmoral Tanks
Insulation for Petrochemical Industries - Kingspan Industrial Ins. Ltd

Oil pollution monitoring and control systems - design, manufacture, installation and commissioning of - Aquasentry
Petrol/diesel storage - Grant & Livingston Ltd

S41 Fuel oil storage/ distribution

Adblue storage and dispensing Tanks - Kingspan Environmental Ltd
Bunded storage tanks - Cookson and Zinn (PTL) Ltd
Fuel oil storage/ distribution - John Davidson (Pipes) Ltd
Fuel oil storage/ distribution - Metcraft Ltd
Liquid storage tanks - Franklin Hodge Industries Ltd
Oil and Fuel Tanks - Kingspan Environmental Ltd
Oil meters - Switch2 Energy Solutions Ltd
Oil pollution monitoring and control systems - design, manufacture, installation and commissioning of - Aquasentry
Oil refinery - Petrochem Carless
Oil storage tanks - Davant Products Ltd
Oil tanks and storage bunkers - Titan Environmental Ltd
Refurbishment of tanks - Franklin Hodge Industries Ltd
Tank level monitoring Domestic and Commercial - Kingspan Environmental Ltd

S50 Vacuum

Vacuum - Barloworld Vacuum Technology Plc
Vacuum controls - Danfoss Ltd

S51 Steam

Steam - Golden Coast Ltd
Steam Boilerhouse Equipment - Thermal Technology Sales Ltd
Steam design packages - AEL
Steam meters - Switch2 Energy Solutions Ltd
Steam Traps - Flowserve Flow Control

S60 Fire hose reels

Fire fighting hoses, nozzles, adaptors and standpipes - Richard Hose Ltd
Fire Hose Reel - Angus Fire
Fire hose reels - Armstrong Integrated Ltd
Fire hose reels - Fire Security (Sprinkler Installations) Ltd
Fire hose reels - Initial Fire Services
Fire hose reels - Nu-Swift International Ltd
Firefighting equipment - The BSS Group
Hosereels - fixed, automatic and swinging - Tyco Safety Products
Wet risers - Tyco Safety Products

S61 Dry risers

Dry riser - Eagles, William, Ltd
Dry riser equipment - Delta Fire
Dry riser equipment - Kidde Fire Protection Services Ltd
Dry Risers - Armstrong Integrated Ltd
Dry risers - Firetecnics Systems Ltd
Dry risers - Tyco Safety Products
Flooring Installation - Tiflex Ltd
Hose reels for fire fighting - Richard Hose Ltd
Metal pipework - Copper, Galvanised steel and Stainless steel - Servomac Limited
Safety and Fire resistant flooring - Tiflex Ltd
Sprinkler Systems - Kidde Fire Protection Services Ltd

S62 Wet risers

Metal pipework - Copper, Galvanised steel and Stainless steel - Servomac Limited
Wet Risers - Armstrong Integrated Ltd
Wet risers - Holden Brooke Pullen

S63 Sprinklers

Fire sprinkler tanks - Franklin Hodge Industries Ltd
GRP sprinkler and pump suction tanks - Dewey Waters Ltd
Sprinkler systems - Firetecnics Systems Ltd
Sprinklers - Fire Security (Sprinkler Installations) Ltd
Sprinklers - Holden Brooke Pullen
Sprinklers - Initial Fire Services
Sprinklers - Richard Hose Ltd

S64 Deluge

Deluge - Fire Security (Sprinkler Installations) Ltd
Deluge - Initial Fire Services
Fire hydrant valves including dry riser equipment - Richard Hose Ltd

S65 Fire hydrants

Fire extinguishers supply and maintenance - Nu-Swift International Ltd
Fire Hydrants - Angus Fire
Fire hydrants - Bailey Birkett Sales Division

S70 Gas fire fighting

Explosion suppression systems - Kidde Fire Protection Ltd
Fire extinguishers - Britannia Fire
Fire extinguishers supply and maintenance - Nu-Swift International Ltd
Fire suppression equipment - Kidde Fire Protection Ltd
Fire suppression systems - Tyco Safety Products
Gas detection equipment - MSA (Britain) Ltd

Gas detection equipment - Zellweger Analytic Ltd, Sieger Division
Gas detection systems - Blakell Europlacer Ltd
Gas fire fighting - Initial Fire Services
Marine fire protection systems - Kidde Fire Protection Ltd
Rail fire protection systems - Kidde Fire Protection Ltd
Vehicle fire protection systems - Kidde Fire Protection Ltd

S71 Foam fire fighting

Extinguishers - Channel Safety Systems Ltd
Fire extinguisher systems - Fife Fire Engineers & Consultants Ltd
Fire extinguishers - Britannia Fire
Fire extinguishers - Richard Hose Ltd
Fire extinguishers supply and maintenance - Nu-Swift International Ltd
Foam equipment - Eagles, William, Ltd
Foam equipment and concentrate - Tyco Safety Products
Foam Fire Fighting - Angus Fire
Foam fire fighting - Fire Security (Sprinkler Installations) Ltd
Foam fire fighting - Initial Fire Services
Hose assemblies for fire fighting, industrial and drinking purposes - Richard Hose Ltd

T10 Gas/ Oil fired boilers

Biomass boilers - Broag Ltd
Boiler maintenance - H Pickup Mechanical & Electrical Services Ltd
Boiler Paint Instrumentation - SK Environmental Ltd
Boilers - Alpha Therm Ltd
Boilers - Ferroli Ltd
Boilers (commercial condensing) - AEL
Combustion equipment - Nu-Way Ltd
Condensing boilers (domestic & commercial) - Broag Ltd
Electric Boilers - Baxi Heating UK Ltd
Floor standing condensing horizontal modular boilers - Fleet H and Purewell VariHeat - Hamworthy Heating Limited
Floor standing condensing horizontal modular boilers - Fleet H and Purewell VariHeat - Hamworthy Heating Ltd
Floor standing condensing vertical modular boilers - Fleet V and Wessex ModuMax - Hamworthy Heating Limited
Floor standing condensing vertical modular boilers - Fleet V and Wessex ModuMax - Hamworthy Heating Ltd
Floor standing high efficiency steel boiler - Melbury C - Hamworthy Heating Ltd
Floor standing high efficiency steel boiler - Melbury HE - Hamworthy Heating Limited
Floor standing low temperature high efficiency steel boiler - Ensbury LT - Hamworthy Heating Limited
Floor standing low temperature high efficiency steel boiler - Ensbury LT - Hamworthy Heating Ltd
Flue and chimney systems - Flamco UK Ltd
Flue systems - Marflex Chimney Systems
Flue terminals - Tower Flue Componants Ltd
Gas and solid fuel heating appliances - Hepworth Heating Ltd
Gas boilers - AEL
Gas central heating boilers - Burco Maxol
Gas fired boilers - Fulton Boiler Works (Great Britain) Ltd
Gas Fired Hot water heaters - HRS Hevac Ltd
Gas flue systems - Ubbink (UK) Ltd
Gas, oil fired boilers - Broag Ltd
Gas, oil fired boilers - Hartley & Sugden
Gas, oil, dual fuel - GP Burners (CIB) Ltd
Gas/oil fired boilers - Docherty New Vent Chimney Group
Gas/Oil Fired Boilers - service maintenance - Hodgkinson Bennis Ltd
Gas/oil fired burners - Worcester Heat Systems Ltd
Heat transfer thermal fluid systems and heaters - Babcock Wanson Uk Ltd
Heating - Dantherm Ltd
Heating controls - Danfoss Ltd
Heating products - The BSS Group
High Efficiency Boilers - Baxi Heating UK Ltd
High Efficiency Water Heaters and Boilers - Lochinvar Limited
Instantaneous hot water systems - HRS Hevac Ltd
Metal pipework - Copper, Galvanised steel and Stainless steel - Servomac Limited
Prefabricated boilers and cylinder sets - IPPEC Sytsems Ltd
Quantec- High efficiency Boiler Range - Johnson & Starley
Quantec HR28C – Passive Flue Gas Heat Recovery - Johnson & Starley
Refractory boiler linings, bricks, castables, ceramic fibre products & incinerators - Combustion Linings Ltd
Service Terminals - Ubbink (UK) Ltd
Skid mounted Boiler Plant - Fulton Boiler Works (Great Britain) Ltd
Spare parts for burners - Nu-Way Ltd
Spares - GP Burners (CIB) Ltd
Steam and hot water boilers and pressure vessels - Wellman Robey Ltd
Steam generators & fire tube boilers - Babcock Wanson Uk Ltd
TecnoFlex flexible liner - Schiedel Chimney Systems Ltd
Thermal fluid heaters, thermal oxidizers for solids, liquids & gaseous waste - Beverley Environmental Ltd
Wall hung, condensing modular boilers - Fleet W and Stratton - Hamworthy Heating Limited

T10 Gas/ Oil fired boilers (con't)

Wall hung, condensing modular boilers - Fleet W and Stratton - Hamworthy Heating Ltd
Water boilers - TSE Brownson

T11 Coal fired boilers

Coal fired boilers - Docherty New Vent Chimney Group
Coal fired boilers repair and maintenance - Doosan Power Systems Ltd
Combustion equipment - Saacke Combustion Services Ltd
Flue and chimney systems - Poujoulat (UK) Ltd
Flue systems - Marflex Chimney Systems
Refractory boiler linings, bricks, castables, ceramic fibre products & incinerators - Combustion Linings Ltd
Solid fuel central HT boiler - Dunsley Heat Ltd
Steam and hot water boilers and pressure vessels - Wellman Robey Ltd
Stokers and associated equipment - service and maintenance - Hodgkinson Bennis Ltd
TecnoFlex flexible liner - Schiedel Chimney Systems Ltd

T12 Electrode/ Direct electric boilers

Electric boilers - Eltron Chromalox
Electric boilers - Fulton Boiler Works (Great Britain) Ltd
Electric boilers - Technical Control Systems Ltd
Electric calorifiers - Eltron Chromalox
Electrical boilers:Aztec Boilers 2-12KW - TR Engineering Ltd
Electrode Boilers - Dimplex UK Limited
Electrode/direct electric boilers - Collins Walker Ltd, Elec Steam & Hot Water Boilers
Flue systems - Marflex Chimney Systems
Hot Water - Dimplex UK Limited
Refractory boiler linings, bricks, castables, ceramic fibre products & incinerators - Combustion Linings Ltd

T13 Packaged steam generators

Bird/vermin control - Cannon Hygiene Ltd
Packaged steam generators - Collins Walker Ltd, Elec Steam & Hot Water Boilers
Packaged steam generators - Rycroft Ltd
Packaged steam turbines and generator equipment - Doosan Power Systems Ltd
Steam and hot water boilers and pressure vessels - Wellman Robey Ltd
Steam boilers - Fulton Boiler Works (Great Britain) Ltd

T14 Heat pumps

Air Source Heat Pump - Lochinvar Limited
Ground source heat pumps - Applied Energy Products Ltd
Ground Source Heat Pumps - H Pickup Mechanical & Electrical Services Ltd
Heat pumps - Holden Brooke Pullen
Heat pumps - MacMarney Refrigeration & Air Conditioning Ltd
Heat pumps - Rycroft Ltd
Heat Pumps - Stiebel Eltron Ltd
Jaga's DBE range of radiators are highly recommended by Mitsubishi and other leading heat pump manufacturers to work in conjunction with heat pumps and other renewable energy sources - Jaga Heating Products (UK) Ltd
Water chillers - Thermal Technology Sales Ltd

T15 Solar collectors

Boiler controls and accessories - Alpha Therm Ltd
Integrated Photovoltaic Tiles as replacement for clay roof tiles - Imerys Roof Tiles
Renewable Energy Solutions - Schueco UK
Solar - Stiebel Eltron Ltd
Solar colectors - Applied Energy Products Ltd
Solar collectors - Aqua-Blue Ltd
Solar collectors - Solar Trade Association, The
Solar collectors - Ubbink (UK) Ltd
Solar collectors, solar electric powered products - Intersolar Group Ltd
Solar heatin - Applied Energy Products Ltd
Solar panels - CPV Ltd
Solar panelsand cylinders - IPPEC Sytsems Ltd
Solar powered lighting - Labcraft Ltd
Solar PV - Schueco UK
Solar slats - Grando (UK) Ltd
Solar Thermal - Schueco UK
Solar thermal and photo voltaic installation - H Pickup Mechanical & Electrical Services Ltd
Solar thermal hot water - Lochinvar Limited
Solar water heating - Alpha Therm Ltd
Solar water heating systems - Atlas Solar
Trigon solar hot water systems - Hamworthy Heating Limited
Trigon solar hot water systems - Hamworthy Heating Ltd

T16 Alternative fuel boilers

Absolut XPert - Schiedel Chimney Systems Ltd
Alternative fuel boilers - Hartley & Sugden
Biomass solutions including fuel storage and automatic feed system - Hamworthy Heating Limited
Biomass, Energy from Waste - Doosan Power Systems Ltd
Boiler stoves - Gazco
Combustion burners - GP Burners (CIB) Ltd
Energy meters - Switch2 Energy Solutions Ltd
Energy saving equipment - Saacke Combustion Services Ltd
Flue and chimney systems - Poujoulat (UK) Ltd
Flue systems - Marflex Chimney Systems
ICS System Chimneys - Schiedel Chimney Systems Ltd
Multi fuel Stoves - Dunsley Heat Ltd
Multi fuel/wood burning stoves - TR Engineering Ltd
Neutralizer for multi boiler applications - Dunsley Heat Ltd
Refractory boiler linings, bricks, castables, ceramic fibre products & incinerators - Combustion Linings Ltd
Solid fuel boilers - TR Engineering Ltd
Spares - GP Burners (CIB) Ltd
Wood Pellet and Wood Chip Boilers - H Pickup Mechanical & Electrical Services Ltd

T20 Primary heat distribution

Custom made fires - Valor
Design, installation and maintenance of building services - Cofely Engineering Services Ltd
Drum heaters - Flexelec(UK) Ltd
Flue systems - Marflex Chimney Systems
Heater cables - Flexelec(UK) Ltd
Heater mats - Flexelec(UK) Ltd
Heating - Birdsall Services Ltd
Heating - PTS Plumbing Trade Supplies
Industrial and commercial heating systems - AmbiRad Ltd
Industrial heating equipment Manufacturers - Turnbull & Scott (Engineers) Ltd
Primary Heat distribution - Saacke Combustion Services Ltd
Temperature controls - Flexelec(UK) Ltd
Thermal fluid heaters - Fulton Boiler Works (Great Britain) Ltd

T30 Medium temperature hot water heating

Control equipment - The BSS Group
Decorative heating - Bristan Group
Drinking water heaters - Santon
Heating - Dunham Bush Ltd
Heating and Plumbing Design & Installation - H Pickup Mechanical & Electrical Services Ltd
Hot water steel panel radiators - Pitacs Ltd
Immersion heaters and thermostats - Heatrae Sadia Heating Ltd
LST Radiators - Myson Radiators Ltd
MTHW Heating - Rycroft Ltd
Radiators - Designer Radiators Direct
Steam and hot water boilers and pressure vessels - Wellman Robey Ltd
Thermostats - Danfoss Ltd
Valves - The BSS Group

T31 Low temperature hot water heating

Boiler and calorifier descaling - Aquastat Ltd
Boilers - HRS Hevac Ltd
Buffer and storage vessels - Flamco UK Ltd
Butterfly valves - Watts Industries UK Ltd
Cast Iron radiators - Designer Radiators Direct
Central heating - Walney UK Ltd
Central heating equipment - Hudevad Britain
Central heating programmers, water heating controllers, motorised valves, room stats and thermostats - Horstmann Controls Ltd
Cistern fed water heaters - Heatrae Sadia Heating Ltd
Compact Radiators - Barlo Radiators Limited
Cylinder Stats - Horstmann Controls Ltd
Design Radiators - Barlo Radiators Limited
Domestic and commercial central heating controls - Danfoss Ltd
Gas convector heaters - Burco Maxol
Gate valves - Watts Industries UK Ltd
Heating - Dunham Bush Ltd
Heating - PTS Plumbing Trade Supplies
Heating - Myson Radiators - Myson Radiators Ltd
Heating & Cooling Systems - Durotan Ltd
Heating and plumbing - Diamond Merchants
Heating to pool areas - Buckingham Swimming Pools Ltd
Hot water heating - Jaga Heating Products (UK) Ltd
Large unvented water heaters - Heatrae Sadia Heating Ltd
Low surface temperature radiators - Jaga Heating Products (UK) Ltd
Low Surface Tttemperature Radiators - Barlo Radiators Limited
Low temperature hot water heating - Atlas Solar
Low water content heating soloutions - Jaga Heating Products (UK) Ltd
LPHW - Turnbull & Scott (Engineers) Ltd
LTHW Heating - Rycroft Ltd
LTHW Heating - SAV United Kingdom Ltd
Manifolds - Watts Industries UK Ltd
Panel Radiators - Myson Radiators Ltd
Point of use water heaters - Heatrae Sadia Heating Ltd
Pressure reducing valves - Watts Industries UK Ltd
Radiant Ceiling Panels - Zehnder Ltd
Radiator gaurds (low surface temperature) - Alumasc Interior Building Products Limited
Radiator thermostats - Danfoss Ltd
Radiator valves - Watts Industries UK Ltd
Radiators - Designer Radiators Direct
Radiators - Zehnder Ltd
Round top Radiators - Barlo Radiators Limited
Solar heating - Solar Trade Association, The
Stainless steel designer radiators - Pitacs Ltd
Steam and hot water boilers and pressure vessels - Wellman Robey Ltd
Swing check valves - Watts Industries UK Ltd
Thermostatic mixinet valves - Watts Industries UK Ltd
Towel Radiators - Barlo Radiators Limited
Towel Rails - Designer Radiators Direct
Underfloor central heating - Polypipe Building Products Ltd
Underfloor heating - FloRad Heating and Cooling
Underfloor Heating - Wavin Ltd
Underfloor, ceiling and wall heating cooling - Warmafloor (GB) Ltd
Valves - Designer Radiators Direct
Valves - Watts Industries UK Ltd
Warm water underfloor heating - IPPEC Sytsems Ltd
Water Heating - Stiebel Eltron Ltd

T32 Low temperature hot water heating (self-contained specification)

Aluminum radiators - AEL
Cast Iron Radiators - AEL
Commissioning Valves - Marflow Eng Ltd
Corgi Appilasers Engineers - AEL
Decorative Radiators - Myson Radiators Ltd
Domestic radiators - Barlo Radiators Limited
Electric radiators - Loblite Ltd
Gas Central Heating - Servowarm
Gas fires - Robinson Willey Ltd
Heating - Beehive Coils Ltd
Heating - Morgan Hope Industries Ltd
Heating and plumbing - Diamond Merchants
Heating systems - Marflow Eng Ltd
Low surface temperature radiator gaurds - Alumasc Interior Building Products Limited
Low temperature hot water heating - Multibeton Ltd
LTHW Heating (self contained) - Rycroft Ltd
Manifolds - Marflow Eng Ltd
Panel radiators - Pitacs Ltd
Radiator valves - AEL
Radiator valves - Pegler Ltd
Radiators - K. V. Radiators
Radiators - Kermi (UK) Ltd
Small unvented water heaters - Heatrae Sadia Heating Ltd
Steel Panel Radiators - AEL
Storage water heating - Johnson & Starley
Underfloor heating - Multibeton Ltd
Water Heaters - Baxi Heating UK Ltd
Water heaters - Johnson & Starley Ltd
Water Heaters - Redring
Water Heaters - Santon
Water treatment of heating and chilled water systems - Aquastat Ltd

T33 Steam heating

Air conditioning - Dunham Bush Ltd
Clean steam boilers - Fulton Boiler Works (Great Britain) Ltd
Controls - Danfoss Ltd
Metal pipework - Copper, Galvanised steel and Stainless steel - Servomac Limited
Radiators - Designer Radiators Direct
Steam - Turnbull & Scott (Engineers) Ltd
Steam heating - Rycroft Ltd
Stream heating - Saacke Combustion Services Ltd

T40 Warm air heating

Air conditioning - Dunham Bush Ltd
Air Curtains - Robinson Willey Ltd
Air warmers - Seagoe Technologies Ltd
Electric warm air heating - Kaloric Heater Co Ltd
Fan convectors - S & P Coil Products Ltd
Finned tubing & trench heaters - Turnbull & Scott (Engineers) Ltd
Gas fired air heaters - Reznor UK Ltd
Gas-fired warm aire unit heaters - Radiant Services Ltd
Heat exchangers gasketted plate - HRS Hevac Ltd
Heating - Encon Air Systems Ltd
Heating equipment - NuAir Ltd
HPHW - Turnbull & Scott (Engineers) Ltd
Industrial & commercial, gas or oil indirect fired warm air heating systems - Elan-Dragonair
Industrial and commercial heating systems - AmbiRad Ltd
Industrial heating - Hoval Ltd
Radiant ceiling panels - S & P Coil Products Ltd
Space heating - Applied Energy Products Ltd
Thermal Fluid - Turnbull & Scott (Engineers) Ltd
Timers - Danfoss Ltd
Trench Heating - S & P Coil Products Ltd
Unit heater - Johnson & Starley Ltd
Warm air heaters - Saacke Combustion Services Ltd
Warm air heating - Consort Equipment Products Ltd
Warm air heating - Docherty New Vent Chimney Group
Warm air heating - May William (Ashton) Ltd
Warm Air heating - Nu-Way Ltd
Warm air heating - Roberts-Gordon

T41 Warm air heating (self-contained specification)

Basket fires - Valor
Controls - shearflow LTD
Convectors - shearflow LTD
Electric warm air heating - Kaloric Heater Co Ltd
fan heaters - shearflow LTD
Natural and fan assisted floor convectors - Multibeton Ltd
Oil & gas fired warm air cabinet - Radiant Services Ltd
Oil fired warm air unit heaters - Radiant Services Ltd
Unit Heaters - Turnbull & Scott (Engineers) Ltd
Warm air heating - Dimplex UK Limited
Warm Air Heating - Vortice Ltd
Warm air heating (small scale) - Consort Equipment Products Ltd
Warm air heating (small scale) - May William (Ashton) Ltd

T42 Local heating units

Anti condensation heaters - Eltron Chromalox
Central heating equipment - Hudevad Britain
Ceramic gas burners - Infraglo Limited
Convector heaters - De Longhi Limited
Custom gas burners - Infraglo Limited
Decorative gas fires & stoves - Magiglo Ltd
Electric fires - Gazco Ltd
Electric fires - Valor
Electric fuel effect fires - BFM Europe Limited
Electric radiant heater panel - Morgan Hope Industries Ltd
Electric water heating - Kaloric Heater Co Ltd
Fires and fireplaces - Gazco
Flameproof heaters, radiant panel heaters - Seagoe Technologies Ltd
Gas burners - Infraglo Limited
Gas fired air heaters - Reznor UK Ltd
Gas fired radiant heating equipment - Horizon International Ltd
Gas fired space heating - domestic/commercial - Johnson & Starley
Gas fires - Gazco Ltd
Gas fires/ Gas conductors - Robinson Willey Ltd
Gas stoves - Gazco Ltd
Gas wall heaters - Valor
Gas-fired radiant tube heater - Radiant Services Ltd
Heating products - Pressmain
Infra ray heaters - Bitmen Products Ltd
Local heating units - Consort Equipment Products Ltd
Metal fibre gas burners - Infraglo Limited
Mobile radiant gas heaters - Infraglo Limited
Portable heating - Andrews Sykes Hire Ltd
Premix gas burneres - Infraglo Limited
Products for commercial and industrial applications - Bush Nelson PLC
Radiant gas heaters - Infraglo Limited
radiant heaters - shearflow LTD
Radiant infra-red plaque gas heaters - Space - Ray UK
Radiant infra-red poultry gas brooders - Space - Ray UK
Radiant infra-red tube gas heaters - Space - Ray UK
Radiant tube and radiant heaters - Radiant Services Ltd
Radiators - Purmo-UK
Radiators (oil filled electric) - De Longhi Limited
Refractory boiler linings, bricks, castables, ceramic fibre products & incinerators - Combustion Linings Ltd
Solid Fuel Suites - Winther Browne & co Ltd
Space Heating - Stiebel Eltron Ltd
Storage and panel heaters - Seagoe Technologies Ltd
Storage gas water heaters, focal point fan convector - Burco Maxol
Stoves - Gazco
Stoves, wood burning and multifuel - BFM Europe Limited
Unit heaters - Biddle Air Systems Ltd

T50 Heat recovery

Brazed plate heat exchangers - HRS Hevac Ltd
Heat meters - Switch2 Energy Solutions Ltd
Heat pipes - S & P Coil Products Ltd
Heat recovery - Aqua-Blue Ltd
Heat recovery - Dantherm Ltd
Heat recovery - Flambeau
Heat recovery - Greenwood Airvac
Heat recovery - Kair ventilation Ltd
Heat recovery - May William (Ashton) Ltd
Heat recovery - Polypipe Ventilation
Heat recovery - Rycroft Ltd
Heat Recovery - Ubbink (UK) Ltd
Heat Recovery - Vent-Axia
Heat recovery products - Silavent Ltd
Heat recovery unit - Dunsley Heat Ltd
Heat recovery units - Thermal Technology Sales Ltd
Heat recovery ventilation units - Ves Andover Ltd
Heating and heat recovery equipment thermal wheels - AEL
Home 'n' Dry units - Johnson & Starley
Run around coils - S & P Coil Products Ltd
Steam and hot water boilers and pressure vessels - Wellman Robey Ltd

T60 Central refrigeration plant

Air conditioning - Clivet UK Ltd
Central refrigeration plant - Johnson Controls Building Efficiency UK Ltd
Central refrigeration plant - MacMarney Refrigeration & Air Conditioning Ltd
Design, installation and maintenance of building services - Cofely Engineering Services Ltd
Industrial refridgeration equipment and systems - J & E Hall Ltd
Refrigerant dryers - Fluidair International Ltd
Refrigeration - Birdsall Services Ltd
Refrigeration - Daikin Applied (UK) Ltd
Refrigeration Controllers and Data loggers - Kingspan Environmental Ltd
Refrigeration plant - Emerson Climate Technologies Retail Solutions
Refrigeration systems - Tyco Fire and Integrated Solutions

T61 Chilled water

Chilled water - J & E Hall Ltd
Chilled water - Johnson Controls Building Efficiency UK Ltd
Chilled water - Rycroft Ltd
Chilled water - TEV Limited
Data centre cooling - Eaton-Williams Group Limited
Stainless Steel Pipes - ACO Technologies plc
Underfloor cooling - FloRad Heating and Cooling

T70 Local cooling units

Absorption chillers - Birdsall Services Ltd
Close cooling units - Clivet UK Ltd
Cooling Units - Carter Environmental Engineers Ltd
Heating & Cooling Systems - Durotan Ltd
Local Cooling Units - Climate Center
multi service chilled beam - Trox (UK) Ltd
Portable coolers - London Fan Company Ltd

T71 Cold rooms

Cold rooms - Clark Door Ltd
Cold rooms - J & E Hall Ltd
Cold rooms - MacMarney Refrigeration & Air Conditioning Ltd
Cold rooms - SCHOTT UK Ltd
Cold rooms - TEV Limited
Commercial freezer environments - Hemsec Panel Technologies

T72 Ice pads

Ice pads - J & E Hall Ltd

U10 General ventilation

Access doors - Hotchkiss Air Supply
Air conditioners, dampers, air ductline - Advanced Air (UK) Ltd
Air conditioning - Birdsall Services Ltd
Air conditioning - Purified Air Ltd
Air conditioning - Trox (UK) Ltd
Air distribution equipment - Designed for Sound Ltd
Air Handing Units - Fläkt Woods Ltd
Air seperators - Watts Industries UK Ltd
Air transfer grilles - Mann McGowan Group
Airflow instrumentation - Airflow Developments Ltd
Anti-condensation for motors heater tapes - Flexelec(UK) Ltd
Design, installation and maintenance of building services - Cofely Engineering Services Ltd
Domestic and commercial extract fans - Greenwood Airvac
Domestic ventilation equipment - Silavent Ltd
Domestic ventilation systems - Airflow (Nicoll Ventilators) Ltd
Exhauster blowers - BVC
Fan convectors - Biddle Air Systems Ltd
Flanging systems - Hotchkiss Air Supply
Flatoval tube - Hotchkiss Air Supply
Flexible ducting - Hotchkiss Air Supply
Flue systems - Marflex Chimney Systems
Full freshair systems - Dantherm Ltd
General supply/ extract - Donaldson Filtration (GB) Ltd
General supply/ extract - Emsworth Fireplaces Ltd
General ventilation - Colt International
General Ventilation - Constant Air Systems Ltd
General ventilation - Dyer Environmental Controls
General ventilation - Flambeau
General ventilation - GVS Filter Technology UK
General ventilation - Intersolar Group Ltd
General ventilation - London Fan Company Ltd
General ventilation - May William (Ashton) Ltd
Industrial air filtration - GVS Filter Technology UK
Mechanical extract terminals - Klober Ltd
Modular ventilation ducting - Polypipe Ventilation
Natural ventalation Systems - Passivent Ltd
Natural Ventilation Systems - Dyer Environmental Controls
Oxygen is our energy efficient, intelligent and fully programmable heating and ventilation solution which improves Indoor Air Quality (IAQ) in classrooms and buildings - Jaga Heating Products (UK) Ltd
Roof ventilators - (ridge & slope mounted) - McKenzie-Martin Ltd
Sealed system equipment and air separation and dirt removal equipment - Flamco UK Ltd
Smoke and natural ventilation systems - SE Controls
SOLA-BOOST® - Monodraught Ltd
Spiral tube - Hotchkiss Air Supply
Splits, VRF, Packaged & Chillers - Toshiba Carrier UK
Support systems - Hotchkiss Air Supply
Trade suppliers of commercial and decorative ceiling fans - Fantasia Distribution Ltd
Underfloor ventilators - Ryton's Building Products Ltd
Ventelated Ceilings - Vianen UK Ltd
Ventialtion - Applied Energy Products Ltd
Ventilation - Encon Air Systems Ltd
Ventilation - Johnson & Starley Ltd
Ventilation - NuAir Ltd
Ventilation - Thermal Technology Sales Ltd
Ventilation - Ubbink (UK) Ltd
Ventilation - domestic, commercial and industrial - Vent-Axia
Ventilation equipment - Johnson & Starley
Ventilation Fans - Haynes Manufacturing (UK) Ltd
Ventilation of pool areas - Buckingham Swimming Pools Ltd
Ventilation products - VBH (GB) Ltd
Ventilation systems - Mansfield Pollard & Co Ltd
Ventilation systems - Polypipe Ventilation Ltd
Ventilators - Kair ventilation Ltd
VENTSAIR Louvres and penthouse louvres - Monodraught Ltd
WINDCATCHER® Natural ventilation systems - Monodraught Ltd
Window ventilator manufactures - Simon R W Ltd
Window Ventilators - Passivent Ltd

U11 Toilet ventilation

Domestic extractor fans - Polypipe Ventilation
Toilet ventilation - Flambeau
Toilet ventilation - Kair ventilation Ltd
Ventialtion - Applied Energy Products Ltd

U12 Kitchen ventilation

Design, Supply and Installation of Commercial Kitchen Ventelation - Vianen UK Ltd
Domestic extractor fans - Polypipe Ventilation
Ducting and fans - Waterline Limited
Extract fan units - Ves Andover Ltd
Extraction systems and odour control - Purified Air Ltd
Grease eliminators - Nationwide Filter Company Ltd
Grease filters - Camfil Farr Ltd
Grease filters - Westbury Filters Limited
Kitchen Canopies - Vianen UK Ltd
Kitchen extract - Fire Protection Ltd
Kitchen extraction systems - Britannia Kitchen Ventilation Ltd
Kitchen ventilation - Flambeau
Kitchen ventilation - Kair ventilation Ltd
Kitchen ventilation - London Fan Company Ltd
Kitchen ventilation - Ryton's Building Products Ltd
Kitchen Ventilation & Exhaust Systems - Interflow UK
Ventilation - domestic, commercial and industrial - Vent-Axia

U13 Car park ventilation

Air Handing Units - Fläkt Woods Ltd
Car park extract - Fire Protection Ltd
Car park Ventilation - Lang+Fulton
car park ventilation - SCS Group
Carpark ventilation - London Fan Company Ltd

U14 Smoke extract/ Smoke control

Air filtration units - PHS Group plc
Air Handing Units - Fläkt Woods Ltd
Ancillaries - Advanced Air (UK) Ltd
Automatic smoke barrier-curtains / blinds - Cooper Group Ltd
Birdguards - Brewer Metalcraft
Chimney cappers - Brewer Metalcraft
Chimney cowls - Brewer Metalcraft
Chimneys and flues - Metcraft Ltd
Commercial and domestic electrostatic air filters - Robinson Willey Ltd
Emission monitors - SK Environmental Ltd
Exhaust gas cleaning - Saacke Combustion Services Ltd
Industrial exhaust ventilation - Horizon International Ltd
Industrial fume extractors - Horizon International Ltd
Powered cowls - Brewer Metalcraft
Rain caps - Brewer Metalcraft
Smoke and natural ventilation systems - SE Controls
Smoke control systems - SCS Group
Smoke extract - Consort Equipment Products Ltd
Smoke extract - McKenzie-Martin Ltd
Smoke extract fans - London Fan Company Ltd
Smoke extract/ smoke control - Active Carbon Filters Ltd
Smoke extract/ smoke control - Colt International
Smoke extract/ smoke control - Fire Protection Ltd
Smoke extract/ smoke control - Photain Controls (part of GE security)
Smoke extract/control - Trox (UK) Ltd
Smoke management systems - Advanced Air (UK) Ltd
Spark arrestors - Brewer Metalcraft
Specials - Brewer Metalcraft
Spinning cowls - Brewer Metalcraft
Terminals - Brewer Metalcraft
Vehicle exhaust extraction - Nederman UK
Ventilation - domestic, commercial and industrial - Vent-Axia

U15 Safety cabinet/ Fume cupboard extract

Domestic and commercial flue and ventilation products - Rega Ductex Ltd
Downflow Booths - Envair Ltd
Fume cupboards - S & B Ltd
Fume Cupboards - Simplylabs Limited
Fume Extraction - Simplylabs Limited
Laminar Flow Cabinets - Envair Ltd
Pharmaceutical Isolators - Envair Ltd
Safety cabinet / fume cupboards - Envair Ltd
Safety cabinet/fume cupboard - Trox (UK) Ltd
Safety extract/ smoke control - Fire Protection Ltd

U16 Fume extract

Air pollution control equipment and technical advice - Doosan Power Systems Ltd
Boiler Flue - Alpha Therm Ltd
Dust and fume extraction - Encon Air Systems Ltd
Extractor fans - Tower Flue Componants Ltd
Flue systems - Marflex Chimney Systems
Fume extract - Donaldson Filtration (GB) Ltd
Fume extract - Emsworth Fireplaces Ltd
Fume extract - Trox (UK) Ltd
Fume extraction - London Fan Company Ltd
Fume extraction - Nederman UK
Fume extraction - PCT Group
Industrial fume extractors - Horizon International Ltd
Portable heating - Andrews Sykes Hire Ltd

U17 Anaesthetic gas extract

Anaesthetic gas extract - Active Carbon Filters Ltd

U20 Dust collection

Clean Rooms - DencoHappel UK Ltd
Dust Collection - Barloworld Vacuum Technology Plc
Dust collection - BVC
Dust collection - Camfil Farr Ltd
Dust collection - Donaldson Filtration (GB) Ltd
Dust collection - Encon Air Systems Ltd
Dust control equipment - Dantherm FiltrationLtd
Dust extraction - Nederman UK
Dust extractors - Horizon International Ltd
Industrial health & safety vacuums - Nilfisk Limited
Industrial vacuums with power tool extraction - Nilfisk Limited

U30 Low velocity air conditioning

Air conditioning - Airedale International Air Conditioning Ltd
Air conditioning - Daikin Applied (UK) Ltd
Air conditioning - Dalair Ltd
Air conditioning - Johnson & Starley Ltd
Air conditioning - Toshiba Carrier UK
Air conditioning accessories - Johnson Matthey PLC - Metal Joining
Air conditioning supply and servicing - Aircare Europe Ltd
Air conditioning systems - Mansfield Pollard & Co Ltd
Air handling units - ABB Ltd
Air handling units - Eaton-Williams Group Limited
Comfort - Airedale International Air Conditioning Ltd
Condensing units - Ves Andover Ltd
Design, installation and maintenance of building services - Cofely Engineering Services Ltd
Domestic and commercial flue and ventilation products - Rega Ductex Ltd
HVAC service and maintenance providers - Eaton-Williams Group Limited
Low velocity air conditioning - MacMarney Refrigeration & Air Conditioning Ltd
Portable heating - Andrews Sykes Hire Ltd
Spares - Airedale International Air Conditioning Ltd
Spares - Daikin Applied (UK) Ltd

U31 VAV air conditioning

Air Handing Units - Fläkt Woods Ltd
VAV air conditioning - Andrews Sykes Hire Ltd
VAV air conditioning - MacMarney Refrigeration & Air Conditioning Ltd
VAV air conditioning - May William (Ashton) Ltd
VAV Air conditioning - Trox (UK) Ltd
VAV terminal units - Advanced Air (UK) Ltd

U32 Dual-duct air conditioning

Dual-duct air conditioning - Advanced Air (UK) Ltd
Dual-duct air conditioning - MacMarney Refrigeration & Air Conditioning Ltd

U33 Multi-zone air conditioning

Multi-zone air conditioning - Andrews Sykes Hire Ltd
Multi-zone air conditioning - MacMarney Refrigeration & Air Conditioning Ltd

U41 Fan-coil air conditioning

Air conditioning - Encon Air Systems Ltd
Air Handing Units - Fläkt Woods Ltd
Dampers - Ruskin Air Management
Data centre cooling solutions - Eaton-Williams Group Limited
Fan coil air conditioning - Colt International
Fan coil air conditioning - Trox (UK) Ltd
Fan coil heater casings - Alumasc Interior Building Products Limited
Fan Coil Induction Units - Motorised Air Products Ltd
Fan coil units - Thermal Technology Sales Ltd
Fan coil valves - Watts Industries UK Ltd
Fan coils - Daikin Applied (UK) Ltd
Fan coils units - Biddle Air Systems Ltd
Fan-coil air conditioning - Advanced Air (UK) Ltd
Fan-coil air conditioning - Andrews Sykes Hire Ltd
Fan-coil air conditioning - DencoHappel UK Ltd
Fan-coil air conditioning - MacMarney Refrigeration & Air Conditioning Ltd
Fan-coil air conditioning - TEV Limited
Legionella prevention and control - Aquastat Ltd
Refrigeration equipment - Dunham Bush Ltd
Service - Airedale International Air Conditioning Ltd
Service - Daikin Applied (UK) Ltd
Split unit air conditioning - Emerson Climate Technologies Retail Solutions

U42 Terminal re-heat air conditioning

Heat exchangers gasketted plate - HRS Hevac Ltd
Terminal re-heat air conditioning - Advanced Air (UK) Ltd

U43 Terminal heat pump air conditioning

Air conditioning - Clivet UK Ltd
Chillers / Heat Pumps - DencoHappel UK Ltd
Terminal heat pump air conditioning - MacMarney Refrigeration & Air Conditioning Ltd

U50 Hybrid system air conditioning

Air heating and ventilating systems - Stokvis Industrial Boilers International Limited
COOL-PHASE® Low energy cooling and ventilation systems - Monodraught Ltd
Heat recovery - Eaton-Williams Group Limited
Hybrid system air conditioning - TEV Limited
Integrated air conditioning and HVAC systems - Tyco Fire and Integrated Solutions

U60 Air conditioning units

Air Conditioning - Climate Center
Air Conditioning - Daikin Airconditioning UK Ltd
Air Conditioning Systems - LTI Advanced Systems Technology Ltd
Air conditioning units - Andrews Sykes Hire Ltd
Air conditioning units - Colt International
Air Conditioning Units - Constant Air Systems Ltd
Air conditioning units - DencoHappel UK Ltd
Air conditioning units - Emerson Climate Technologies Retail Solutions
Air conditioning units - Encon Air Systems Ltd
Air conditioning units - Johnson Controls Building Efficiency UK Ltd
Air conditioning units - MacMarney Refrigeration & Air Conditioning Ltd
Air conditioning units - Mansfield Pollard & Co Ltd
Air conditioning units - TEV Limited
Air conditioning units - Toshiba Carrier UK
Air Cooled Condensers - Thermal Technology Sales Ltd
Air Handing Units - Fläkt Woods Ltd
Chillers - Daikin Airconditioning UK Ltd
Condensing units - Airedale International Air Conditioning Ltd
Design, installation and maintenance of building services - Cofely Engineering Services Ltd
Portable air conditioners - Thermoscreens Ltd
Refrigeration equipment - Dunham Bush Ltd
Ventialtion - Applied Energy Products Ltd

U70 Air curtains

Air curtains - Biddle Air Systems Ltd
Air curtains - DencoHappel UK Ltd
Air curtains - shearflow LTD
Air curtains - Thermoscreens Ltd
Air curtains / Convectors - S & P Coil Products Ltd
Aircurtains - AmbiRad Ltd

V10 Electricity generation plant

Automatic start generators - Cummins Power Generation Ltd
Automatic synchronising generators - Cummins Power Generation Ltd
Baseload diesel generator - Broadcrown Ltd
Central battery systems - Channel Safety Systems Ltd
Continuous and emergency standby power generator sets from 5.5kVA to 2500kVA power range - FG Wilson
General electrical products - WF Senate
General electrical products - wholesale - Jolec Electrical Supplies
Generation plant - Broadcrown Ltd
Generator sets manufactured to suit specific requirements - FG Wilson
generators - Chippindale Plant Ltd
Mobile generator sets for on-set power requirements - FG Wilson
Mobile generators - Cummins Power Generation Ltd
Peak loppingd diesel generators - Broadcrown Ltd
Power generators - Cummins Power Generation Ltd
Power plants - Cummins Power Generation Ltd
Range of generator canopies including sound attenuated - FG Wilson
Silenced generators - Cummins Power Generation Ltd
Standby diesel generators - Broadcrown Ltd
Super silenced generators - Cummins Power Generation Ltd
Transformers and packaged substations - Schneider Electric Ltd
Ventilation Products - Allmat (East Surrey) Ltd

V11 HV supply/ distribution/ public utility supply

Accessories for electrical services - Lewden Electrical Industries
Electrical distribution - Schneider Electric Ltd
Extra high voltage - Schneider Electric Ltd
HV Supply/Distribution/Public Utility Supply - Peek Traffic Ltd
Public distribution - Schneider Electric Ltd
Substation kiosks - MCL Composites Ltd

V12 LV supply/ public utility supply

Electricity meters - Switch2 Energy Solutions Ltd
Final distribution - Schneider Electric Ltd
LV Supply/Public Utility Supply - Peek Traffic Ltd
Retail Lighting - Fitzgerald Lighting Ltd

V20 LV distribution

2V Distribuition - Eaton Electric Limited
Accessories for electrical services - Lewden Electrical Industries
Cable Reel - Metreel Ltd
Circuit protection and control - Bill Switchgear
Electrical installations - HEM Interiors Group Ltd

V20 LV distribution (con't)

Electricity meter boxes - MCL Composites Ltd
Festoon Systems - Metreel Ltd
Low voltage distribution - Schneider Electric Ltd
LV Distribution - Peek Traffic Ltd
Modular desk & Screen Power, Telecom & Data Units - Electrak International Ltd
Permanent Power Distribution - 110v to 400v - Blakley Electrics Ltd
Protected sockets - Blakley Electrics Ltd
Safe Supply Units - Blakley Electrics Ltd
Small power distribution - Ackermann Ltd
Surge protection systems - Erico Europa (GB) Ltd
Temporary power Distribution - 110v to 400v - Blakley Electrics Ltd
Tranformers 100VA to 250kVA, 12v to 3300v - Blakley Electrics Ltd

V21 General lighting

Access panels to ceiling void - Saint-Gobain Ecophon Ltd
Accessories for electrical services - Lewden Electrical Industries
Acoustic suspended ceiling systems - Saint-Gobain Ecophon Ltd
Aisle and step lighting - Gradus Ltd
Architectural lighting - Optelma Lighting Ltd
Asymmetric lighting - Crescent Lighting
Ballasts: magnetic and electronic - Helvar Ltd
Bathroom Lighting - Astro Lighting Ltd
Bespoke Lighting - ALD Lighting Solutions
Cabinet and display lighting - Crescent Lighting
Cold cathode - Oldham Lighting Ltd
Commercial Lighting - Illuma Lighting
Commercial wiring systems - Metway Electrical Industries Ltd
Compact Fluorecent Downlights - ITAB Prolight UK Limited
Decorative and display lighting - Myddleton Hall Lighting, Division of Peerless Design Ltd
Design, supply and install - Oldham Lighting Ltd
Display lighting - Cooper Lighting and Security Ltd
Display Lighting - Fitzgerald Lighting Ltd
Display lighting - Sylvania Lighting International INC
Domestic lighting - Sylvania Lighting International INC
Downlights - Illuma Lighting
Electrical Accessories - Focus SB Ltd
Emergency batteries - Bernlite Ltd
Emergency modules - Bernlite Ltd
Energy efficiency in lighting - Poselco Lighting
Energy saving lighting - Anglepoise Lighting Ltd
Energy saving lighting - Sylvania Lighting International INC
Exterior lighting - DW Windsor Lighting
Exterior lighting - Fagerhult Lighting
Exterior lighting - Woodhouse UK Plc
Feature Lighting - Illuma Lighting
Fibre optic systems - ALD Lighting Solutions
Fibreoptic lighting - Crescent Lighting
Flourescent Lighting - Vitalighting Ltd
Genaral Lighting - Commercial Lighting Systems
General Lighting - Best & Lloyd Ltd
General lighting - Brandon Medical
General Lighting - Consult Lighting Ltd
General lighting - Cooper Lighting and Security Ltd
General Lighting - Designplan Lighting Ltd
General lighting - Glamox Electric (UK) Ltd
General lighting - Glamox Luxo Lighting Limited
General lighting - IBL Lighting Limited
General lighting - IKM Systems Ltd
General lighting - Jerrards Plc
General lighting - Lampways Ltd
General Lighting - Morgan Hope Industries Ltd
General lighting - Oldham Lighting Ltd
General Lighting - Peek Traffic Ltd
General lighting - Sylvania Lighting International INC
General lighting and bulbs - GE Lighting Ltd
General lighting and purpose built lighting - Quip Lighting Consultants & Suppliers Ltd
General lighting products - Whitecroft Lighting Limited
General, modular, compact Fluorescent - Lumitron Ltd
Gimbal spots - ITAB Prolight UK Limited
Green (Low enegry) Lighting - Blakley Electrics Ltd
Handlamps - Ensto Briticent Ltd
Hazardous area lighting - Glamox Electric (UK) Ltd
Hazardous area lighting - PFP Electrical Products Ltd
Healthcare lighting - Glamox Luxo Lighting Limited
High Bay Lighting - Setsquare
High frequency electronic ballasts - Bernlite Ltd
Hospital lighting - Thorlux Lighting
Ignitors - Bernlite Ltd
Industrial lighting - Cooper Lighting and Security Ltd
Industrial Lighting - Fitzgerald Lighting Ltd
Intelligent lighting - Cooper Lighting and Security Ltd
Interior and exterior lighting - Andy Thornton Ltd
Kitchen lighting - Waterline Limited
Lamps - Philips Lighting
Lamps - ceramic - BLV Licht-und Vakuumtecnik
Large lights - Signature Ltd
LED Drivers/Modules - Helvar Ltd
LED lighting - Glamox Luxo Lighting Limited
LED lighting - ITAB Prolight UK Limited
LED lighting - Labcraft Ltd
LED Lighting - NET LED Limited
LED lighting for pool installations - Golden Coast Ltd
LED luminaires - ALD Lighting Solutions
LED luminaires - Crescent Lighting
LED replacement flourescent tubes - Crescent Lighting
LED retrofit lighting - Crescent Lighting
Light bulbs, tungsten halogen and metal halide - BLV Licht-und Vakuumtecnik
Light fittings - ITAB Prolight UK Limited
Light fittings - Thorn Lighting
Light fittings - Vitalighting Ltd
Lighting - Eglo UK
Lighting - HT Martingale Ltd
Lighting - Light Engine Lighting Ltd
Lighting - Litex Design Ltd
Lighting - Optelma Lighting Ltd
Lighting - Shaylor Group Plc
Lighting control equipment - Philips Lighting Solutions
Lighting control gear inc. ballasts, chokes - Bernlite Ltd
Lighting Control Products - Eaton Electric Limited
Lighting control systems - Hamilton R & Co Ltd
Lighting controllers, sensors, relay units, interfaces, input devices and enclosures - Helvar Ltd
Lighting Controls - Fitzgerald Lighting Ltd
Lighting controls - Philips Lighting
Lighting Controls - Setsquare
Lighting equipment - Thorlux Lighting
Lighting Manufacturer and supply - Whitecroft Lighting Limited
Lighting products - Friedland Ltd
Lighting support systems - Vantrunk Engineering Ltd
low energy downlighter - Vitalighting Ltd
Low energy lighting systems - Connect Lighting Systems (UK) Ltd
Low voltage - ITAB Prolight UK Limited
low voltage downlighter - Vitalighting Ltd
Low-voltage lighting - Fairfield Displays & Lighting Ltd
Luminaires - Philips Lighting
Luminaires - office, retail, industrial, amenity, pharmaceutical, hospital etc - Whitecroft Lighting Limited
Mains voltage halogen lighting - Sylvania Lighting International INC
Marine lighting - Chalmit Lighting
Marshalling Boxes - Setsquare
Office lighting - Cooper Lighting and Security Ltd
Office lighting - Sylvania Lighting International INC
PowerX, Lighting, Timers, Chimes, Wiring Accessories, Ventilation - KingShield - Green Brook
Purpose made luminaires - ALD Lighting Solutions
Recessed Lighting - Fitzgerald Lighting Ltd
Recessed and Surface LG3 Luminaires - ITAB Prolight UK Limited
Retail and Display Lighting - Illuma Lighting
Retail Display Luminaires - ITAB Prolight UK Limited
Security PIR lighting - Timeguard Ltd
Self-learning lighting control solutions - Helvar Ltd
Site Lighting 110v Lighting - Blakley Electrics Ltd
Specialist Lighting - LPA Group PLC
Specials - ITAB Prolight UK Limited
Spotlights - Illuma Lighting
Standalone solutions - Helvar Ltd
Suspended ceiling luminaires - Burgess Architectural Products Ltd
Switches, accessories and weatherproof sockets - Loblite Ltd
Task Lighting - Glamox Luxo Lighting Limited
Track - Illuma Lighting

V22 General LV power

Dimmers, switches, sockets, on plates of brass, bronze,steel, chrome, wood polycarbonate - Hamilton R & Co Ltd
Electrical Engineering Services (Mains Distribution, Single and Three Phase Power) - H Pickup Mechanical & Electrical Services Ltd
Power - Shaylor Group Plc
Power cables - Pitacs Ltd

V30 Extra low voltage supply

Accessories for electrical services - Lewden Electrical Industries
Entry sound communoication products - Friedland Ltd
Exrta low voltage supply - Brandon Medical
General electrical products - WF Senate
General electrical products - wholesale - Jolec Electrical Supplies
LED Lighting - NET LED Limited
Low-Voltage Garden Lighting - Anthony de Grey Trellises + Garden Lighting
Miniature precision switches, push button and cord pull etc - Omeg Ltd
Transformers for low voltage lighting - Mode Lighting (UK) Ltd

V31 DC supply

Accessories for electrical services - Lewden Electrical Industries
Batteries - MJ Electronics Services (International) Ltd
Battery chargers - MJ Electronics Services (International) Ltd
DC Supplies - Peek Traffic Ltd
Direct current supply - Brandon Medical

V32 Uninterruptible power supply

Central battery systems - Emergi-Lite Safety Systems
Electrical accessories - Electrak International Ltd
Industrial batteries for standby applications - Hawker Ltd
Power electronic repair and service - MJ Electronics Services (International) Ltd
Uninterrupted power supply - Emerson Network Power Ltd
Uninterruptible power supply - ITM Communications LtdLtd
Uninterrupted Power Supply - Eaton Electric Limited
Uniterupted power supply - Channel Safety Systems Ltd
UPS services - MJ Electronics Services (International) Ltd
UPS Systems - Peek Traffic Ltd
UPS systems and inverters - MJ Electronics Services (International) Ltd

V40 Emergency lighting

Automatic / Self-Testing Emergency Lighting - P4 Limited
Bespoke light fitting manufacturers - Vitalighting Ltd
Dimming controls - Jerrards Plc
Electronic starter switches - Jerrards Plc
Emergency lighting - ADT Fire and Security Plc
Emergency lighting - ALD Lighting Solutions
Emergency lighting - Channel Safety Systems Ltd
Emergency lighting - Connect Lighting Systems (UK) Ltd
Emergency lighting - Cooper Lighting and Security Ltd
Emergency lighting - Cooper Safety
Emergency lighting - Designplan Lighting Ltd
Emergency lighting - Emergi-Lite Safety Systems
Emergency lighting - Fagerhult Lighting
Emergency Lighting - Fitzgerald Lighting Ltd
Emergency lighting - Gent Limited
Emergency lighting - Glamox Electric (UK) Ltd
Emergency Lighting - Illuma Lighting
Emergency lighting - Metway Electrical Industries Ltd
Emergency lighting - MJ Electronics Services (International) Ltd
Emergency Lighting - Morgan Hope Industries Ltd
Emergency lighting - Orbik Electrics Ltd
Emergency lighting - PFP Electrical Products Ltd
Emergency lighting - Poselco Lighting
Emergency lighting - Thorlux Lighting
Emergency lighting - Vitalighting Ltd
Emergency lighting systems - Cooper Lighting
Emergency lighting systems - Mackwell Electronics Ltd
Emergency lighting testing systems - Emergi-Lite Safety Systems
Emergency lighting, switch tripping and battery chargers - Peek Traffic Ltd
Emergency lighting/exit lights - Thomas & Betts Manufacturing Services Ltd
Emergency lights - Firetecnics Systems Ltd
Emergency Lights - Whitecroft Lighting Limited
General electrical products - WF Senate
General electrical products - wholesale - Jolec Electrical Supplies
Hazardous area lighting - Chalmit Lighting
High frequency ballast - Jerrards Plc
Industrial batteries for standby applications - Hawker Ltd
Integral emergency lighting - Lumitron Ltd
LED Emergency Lighting - NET LED Limited
Light fittings - Thorn Lighting
Lighting components - Jerrards Plc
Magnetic ballast - Jerrards Plc
Powdered way guidance - Hoyles Electronic Developments Ltd
Security lighting - ASCO Extinguishers Co. Ltd

V41 Street/ Area/ Flood lighting

Amenity & area lighting - DW Windsor Lighting
Amentity Lighting - Fitzgerald Lighting Ltd
Architectural interior lighting for commercial applications - Fagerhult Lighting
Architectural Lighting - Noral Scandinavia AB
Architectural lighting - Thorlux Lighting
Area/street lighting, floodlights - CU Phosco Lighting
Bollard baselights - Signature Ltd
Coloured metal hacide floodlamps - BLV Licht-und Vakuumtecnik
Emergency lighting - Chalmit Lighting
Exterior lighting - Vitalighting Ltd
Exterior uplighters - Lumitron Ltd
External Lighting - Guardall Ltd
Flood Lighting - Blakley Electrics Ltd
Flood Lighting - Fitzgerald Lighting Ltd
Floodlighting - Thorlux Lighting
Floodlighting poles - Valmont Stainton
Hazardous area lighting - Thorlux Lighting
High Mast Lighting - CU Phosco Lighting
LED Flood Lighting - NET LED Limited
LED street lighting - Glamox Luxo Lighting Limited
Light Towers - Genie Europe
Low-Voltage Garden Lighting - Anthony de Grey Trellises + Garden Lighting
Outdoor lighting - Holophane Europe Ltd
Outdorr Lights - Astro Lighting Ltd
Sign lights - Signature Ltd
Street lighting - DW Windsor Lighting
Street lighting - Urbis Lighting Ltd
Street lighting - Woodhouse UK Plc
Street lighting columns and accessories - Lampost Construction Co Ltd
Street lighting columns and high masts - Valmont Stainton
Street lighting columns and sign posts - Fabrikat (Nottingham) Ltd
Street/ area & flood lighting - Sylvania Lighting International INC
Street/ Area/ Flood lighting - Consult Lighting Ltd
Street/ area/ flood lighting - Cooper Lighting and Security Ltd
Street/ area/ flood lighting - Designplan Lighting Ltd
Street/ Area/ flood lighting - FG Wilson
Street/ Area/ Flood lighting - Glamox Luxo Lighting Limited
Street/Area/Flood Lighting - Commercial Lighting Systems
Street/Area/Flood Lighting - Whitecroft Lighting Limited
Tunnel and underpass lighting - Thorlux Lighting

V42 Studio/ Auditorium/ Arena lighting

Dimming Systems - Setsquare
Display Lighting - Jerrards Plc
DLCMS Lighting control system - Setsquare
Emergency Lighting - P4 Limited
Industrial lighting - Chalmit Lighting
Industrial lighting - Glamox Electric (UK) Ltd
Lamps - capsule - BLV Licht-und Vakuumtecnik
Lamps - dichroic - BLV Licht-und Vakuumtecnik
LED Tube and Panel Lighting - NET LED Limited
Lighting project design - DW Windsor Lighting
Scene Setting - Setsquare
Sports arera lighting - En-tout-cas Tennis Courts Ltd
Studio/ Auditorium/ Arena Lighting - Consult Lighting Ltd
Studio/ auditorium/ arena lighting - Sylvania Lighting International INC
Studio/Auditorium/Arena Lighting - Peek Traffic Ltd
Tennis court lighting - En-tout-cas Tennis Courts Ltd
Theatrical lighting - Multitex GRP LLP

V50 Electric underfloor/ ceiling heating

Ceiling Heating - Flexel International Ltd
Electric underfloor heating - Bush Nelson PLC
Electric underfloor/ ceiling heating - Dimplex UK Limited
Floor heating - Flexel International Ltd
Ramp heating - Findlay Irvine Ltd
Undefloor Heating - Begetube UK Ltd
Underfloor heating - H Pickup Mechanical & Electrical Services Ltd
Underfloor heating - Purmo-UK
Underfloor heating - Uponor Ltd
Underfloor heating controls - Danfoss Ltd
Underfloor heating systems - REHAU Ltd

V51 Local electric heating units

Ceiling heaters - Ecolec
Electric air duct heaters - Eltron Chromalox
Electric cartridge heaters - Eltron Chromalox
Electric convector heaters - Eltron Chromalox
Electric fan heaters - Eltron Chromalox
Electric fire manufacturer - Dimplex UK Limited
Electric Fires - Winther Browne & co Ltd
Electric heater batteries - Ves Andover Ltd
Electric heaters - Eltron Chromalox
Electric Heating - Applied Energy Products Ltd
Electric heating - Kaloric Heater Co Ltd
Electric heating - Robinson Willey Ltd
Electric panel heaters - Valor
Electric Stoves - Gazco Ltd
Electric stripring heaters - Eltron Chromalox
Electrical heating units - Consort Equipment Products Ltd
Fan heaters - Ecolec
General electrical products - WF Senate
General electrical products - wholesale - Jolec Electrical Supplies
Glass radiators - Ecolec
Heated towel rails - hot water - Pitacs Ltd
Infra red heating - Robinson Willey Ltd
Local electric heating units - Bush Nelson PLC
Local electric heating units - Dimplex UK Limited
Local heating units - Dimplex UK Limited
Local Heating Units - Morgan Hope Industries Ltd
Low energy electric panel heaters - Ecolec
Mirror radiators - Ecolec
Radiators - Designer Radiators Direct
Towel rails - Ecolec

V90 Electrical installation (self-contained spec)

Accessories for electrical services - Lewden Electrical Industries
Design of electrical services - Aercon Consultants Ltd
Domestic wiring accessories and circuit protection equipment - Contactum Ltd
Electrcity supply: switches & accessories - Hamilton R & Co Ltd
Electrical accessories - Marcus M Ltd
Electrical distribution blocks - Erico Europa (GB) Ltd
Electrical Installation - West London Security
Electrical installation equipment - Legrand Power Centre
Electrical installation products - Barton Engineering
Electrical Instillation - Barton Engineering
Electrical wiring accessories - Eaton Electric Limited
Electrical works - Shaylor Group Plc
Electrically operated tracks - Goelst
Floor laid flexible cable protectors for every application - Vulcascot Cable Protectors Ltd
General electrical products - WF Senate
General electrical products - wholesale - Jolec Electrical Supplies
Mains lighting - Cooper Lighting
Ropes and sashcords - Marlow Ropes Ltd
Specialsist electrical connectors - LPA Group PLC
Wire accessories - Hellermann Tyton
Workplace systems - Ackermann Ltd

W10 Telecommunications

Audio visual hire - Planet Communications Group Ltd
Audio visual solutions & sales - Planet Communications Group Ltd
Cables - Nexans UK Ltd
Communications - Ackermann Ltd
Communications - Ogilvie Construction Ltd
Hospital communications systems - Tunstal Healthcare (UK) Ltd
Industrial batteries for standby applications - Hawker Ltd
Telecommunications - Connaught Communications Systems Ltd
Telecommunications - ITM Communications LtdLtd
Telephone enclosures - Midland Alloy Ltd
TV/ Telephone systems for hospitals - Wandsworth Elecrtrical Ltd
Video conferencing - Planet Communications Group Ltd
Voice recording, processing and training enhancement software - Computertel Ltd

W11 Paging/ Emergency call

Care communications - ADT Fire and Security Plc
Data paging systems - Channel Safety Systems Ltd
Deaf emergency alerter - Deaf Alerter plc
Dispersed alarms - Telelarm care Ltd
Doctor call systems - Channel Safety Systems Ltd
Intercoms - PEL Services Limited
Nurse call and on-site paging systems - Blick Communication Systems Ltd
Nurse call systems - Channel Safety Systems Ltd
Paging - Wandsworth Elecrtrical Ltd
Paging/ Emergency call - C-TEC (Computionics) Ltd
Signalling and warning equipment - Castle Care-Tech
Warden call systems - Telelarm care Ltd

W12 Public address/ Conference audio facilities

Background music - PEL Services Limited
Broadcast and audio visual products & services - Streamtec Limited
Conference equipment - Anders + Kern UK Ltd
Public address - Blick Communication Systems Ltd
Public address - PEL Services Limited
Public address system - AET.gb Ltd
Sound masking systems - AET.gb Ltd

W20 Radio/ TV/ CCTV

advanced digital video recording - Xtralis
Audio visual - Anders + Kern UK Ltd
Broadcast and audio visual products & services - Streamtec Limited
C.C.T.V - Sampson & Partners Fencing
CCTV - Abel Alarm Co Ltd
CCTV - ADT Fire and Security Plc
CCTV - Allen (Fencing) Ltd
CCTV - Allgood plc
CCTV - Connaught Communications Systems Ltd
CCTV - Guardall Ltd
CCTV - PEL Services Limited
CCTV - Prima Security & Fencing Products
CCTV - Ward's Flexible Rod Company Ltd
CCTV - West London Security
CCTV and CCTV monitoring - Crimeguard Protection Ltd
CCTV systems - Interphone Limited
CCTV systems - Viking Security Systems Ltd
CCTV's - Chubb Systems Ltd
CCTV's - Initial Electronic Security Systems Ltd
Closed circuit television - Crimeguard Protection Ltd
Closed circuit television - Vicon Industries Ltd
Closed circuit television - Vidionics Security Systems Ltd
Design, manufacture and installation of audio visual systems - Spaceright Europe Limited
Design, Supply & Installation - Vidionics Security Systems Ltd
Ethernet video surveillance - Vidionics Security Systems Ltd
Home Automation & AV - West London Security
Plasma/LCD Systems - Spaceright Europe Limited
Presentation and conference equipment - Anders + Kern UK Ltd
Radio / TV / CCTV - ITM Communications LtdLtd
Radio telemetry - Eaton's Security Business
remote video transmition - Xtralis
Social care alarms - Eaton's Security Business
Transmission poles - Valmont Stainton
TV distribution - Blick Communication Systems Ltd
Video entry - Crimeguard Protection Ltd
Visual communications systems - Quartet-GBC UK Ltd

W21 Projection

Big screens (indoor/outdoor) - Pro Display TM Limited
Electronic displays - Pro Display TM Limited
Plasma monitors/TVs - Pro Display TM Limited
Projection equipment - Spaceright Europe Limited
Projection Screen Systems - Harkness Screens
Projection screens - Magiboards Ltd
Projectors and projection screens - Pro Display TM Limited
Touch screens - Pro Display TM Limited
TV feature walls - Gridpart Interiors Ltd
Video walls - Pro Display TM Limited

W22 Information/ Advertising display

Broadcast and audio visual products & services - Streamtec Limited
Electronic information display systems - Techspan Systems
Information cases - GB Sign Solutions Limited
Information kiosks - Pryorsign
Information/ advertising display - Focal Signs Ltd
Information/ advertizing display - Stocksigns Ltd
Interpretation Boards - GB Sign Solutions Limited
Light boxes - Simplex Signs Limited
Poster/ menu cases/cabinets - Simplex Signs Limited
Showcases - Fairfield Displays & Lighting Ltd
Suspend display systems - Fairfield Displays & Lighting Ltd
TV feature walls - Gridpart Interiors Ltd
Urban traffic management - Techspan Systems
Variable message signs - Techspan Systems

W23 Clocks

Clock towers - Scotts of Thrapston Ltd
Clocks - General Time Europe
Clocks - Lesco Products Ltd
Time Recording Equipment - Isgus
Timer switches for central heating and controls for central heating - Tower Flue Componants Ltd
Timers - Elkay Electrical Manufacturing Co Ltd

W30 Data transmission

Cables - Nexans UK Ltd
Car Park Guidance and Occupany - Techspan Systems
Data cabling - Shaylor Group Plc
Data transmission - C-TEC (Computionics) Ltd
Data Transmission - ITM Communications LtdLtd
Datacoms - MK (MK Electric Ltd)
Electric information systems - Ferrograph Limited
Software for the construction industry - Masterbill Micro Systems Ltd
Urban traffic management - Techspan Systems

W40 Access control

Access - Doorfit Products Ltd
Access control - Abel Alarm Co Ltd
Access control - ADT Fire and Security Plc
Access control - Allen (Fencing) Ltd
Access control - Allgood plc
Access control - BPT Automation Ltd
Access control - Chubb Systems Ltd
Access Control - Connaught Communications Systems Ltd
Access control - Crimeguard Protection Ltd
Access control - Dinnington Fencing Co. Limited
Access Control - Directional Data Systems Ltd
Access control - Dorma Door Services Ltd
Access Control - Duffells Limited
Access Control - Eaton's Security Business
Access Control - Focus SB Ltd
Access Control - Guardall Ltd
Access control - H E W I (UK) Ltd
Access control - Honeywell Control Systems Ltd
Access control - Hoyles Electronic Developments Ltd
Access control - Initial Electronic Security Systems Ltd
Access control - Initial Fire Services
Access control - ITM Communications LtdLtd
Access control - Kaba Garog
Access control - Laidlaw Ltd
Access control - PEL Services Limited
Access Control - Pike Signals Ltd
Access control - Siemens Building Technologies
Access Control - Tate Fencing
Access control - Trend Control Systems Ltd
Access control - Vidionics Security Systems Ltd
Access control - Viking Security Systems Ltd
Access control & security products - Clarke Instruments Ltd
Access control equipment - Hafele UK Ltd
Access control products - ASSA ABLOY Limited
Access control products - ASSA Ltd
Access Control Security Solutions - Laidlaw Architectural Hardware Ltd
Access control system design - Clarke Instruments Ltd
Access control systems - CEM Systems Ltd
Access control systems - Interphone Limited
Access control systems - Pac International Ltd
Access control systems - Sampson & Partners Fencing
Access control systems - Securikey Ltd
Access control systems - Time and Data Systems International Ltd (TDSI)
Access control systems - Wandsworth Elecrtrical Ltd
Access control systems - Yale Security Products Ltd
Access control systems / equipment - APT Controls Ltd
Access control systems, ID and access card production and registered key systems - Kaba Ltd
Access Control, Biometric Access Systems, Door Entry Systems - West London Security
Access Systems, exit devices, electromagnetic locks and electric strikes - Relcross Ltd
Auto sliding gates - Allen (Fencing) Ltd
Automatic gates, barriers and railings - Parsons Brothers Gates
Automatic Rising Bollards - BPT Automation Ltd
Barriers and fully automated equipment - Prima Security & Fencing Products
Bird control - Timberwise (UK) Ltd
Call systems - Wandsworth Elecrtrical Ltd
Card Access - Crimeguard Protection Ltd
Card readers - CEM Systems Ltd
Door & window control equipment - GEZE UK
Door controls and locking devices - Laidlaw Architectural Hardware Ltd
Door entry systems - Aardee Security Shutters Ltd
Electric openers/ spare parts - Cardale Doors Ltd
Electric releases - Clarke Instruments Ltd
Electrical garage door accessories - Garador Ltd
Electronic remote control window opening systems - Clearline Architectural
Gate automation - Clarke Instruments Ltd
High speed vertical folding and roll -up doors - Mandor Engineering Ltd
ID badging systems - CEM Systems Ltd
Installation service - Clarke Instruments Ltd
Intercom - Wandsworth Elecrtrical Ltd
Maintenance service - Clarke Instruments Ltd
Mechanical and electrical locks - ASSA ABLOY Limited
Mechanical and electrical locks - ASSA Ltd
Proximity access control systems - BPT Security Systems (UK) Ltd
Security access and control - Hart Door Systems Ltd
Security detection and alarm control equipment - Castle Care-Tech
Sliding Door gear - Hafele UK Ltd
Standalone Access control panels - Abloy Security Ltd
Turnstiles - Allen (Fencing) Ltd
Turnstiles - Clarke Instruments Ltd
Vehicle access control - Autopa Ltd
Video and audio door entry systems - BPT Security Systems (UK) Ltd
Video Entry systems - Tate Fencing

W41 Security detection and alarm

Access Control - Sentra Access Engineering Ltd
Access Control - Vador Security Systems Ltd
Access Control Systems - Intime Fire and Security
Anti-terrorist devices - Castle Care-Tech
Automated gates and barriers - Vador Security Systems Ltd
Burglar alarms - Chubb Systems Ltd
Burglar alarms - Initial Electronic Security Systems Ltd
CCTV Systems - Intime Fire and Security
Digital image recorders - Optex (Europe) Ltd
Disabled toilet alarm systems - Channel Safety Systems Ltd
Door security locks - Hafele UK Ltd
Electric fence system for commercial & industrial sites - Advanced Perimeter Systems Ltd
Electronic security services, Intruder Alarms - West London Security
Fence Intrusion Detection System - Advanced Perimeter Systems Ltd
Gates and barriers - high speed - Hart Door Systems Ltd
Integrated security systems - Vidionics Security Systems Ltd
Intruder alarm and CCTV - PEL Services Limited
Intruder alarms - ADT Fire and Security Plc
Intruder alarms - Crimeguard Protection Ltd
Intruder alarms - Eaton's Security Business
Intruder Alarms - Intime Fire and Security
Intruder Detection - Guardall Ltd
passive infared detectors - Xtralis
Passive infra red detectors and photoelectric beams - Optex (Europe) Ltd
PC based security control and monitoring system - Advanced Perimeter Systems Ltd
Perimeter protection - Vidionics Security Systems Ltd
Personal security system - Wandsworth Elecrtrical Ltd
Security systems - Pac International Ltd
security and surveillance systemas - Xtralis
Security detection - Friedland Ltd
Security detection - Honeywell Control Systems Ltd
Security detection and alarm - Initial Fire Services
Security detection and alarm - Johnson Controls Building Efficiency UK Ltd
Security detection and alarm - Tower Manufacturing
Security detection and alarm control equipment - Castle Care-Tech
Security detection and alarm systems - Abel Alarm Co Ltd
Security detection and alarm systems - Interphone Limited
Security detection and alarms - Geoquip Worldwide
Security detection equipment - Hoyles Electronic Developments Ltd
Security detection screening - Specialist Building Products
Security doors - Jewers Doors Ltd
Security equipment and alarms - ADT Fire and Security Plc
Security equipment and alarms - Advanced Perimeter Systems Ltd
Security fencing - C & W Fencing Ltd
Security management systems and high security locks - Relcross Ltd
Security mirrors - Smart F & G (Shopfittings)Ltd
Security products - Viking Security Systems Ltd
Security Systems - Cooper Lighting
Security, CCTV - WF Senate
Sercurity detection & alarms - Siemens Building Technologies
Sounders and strobes - KAC Alarm Company Ltd
Vibration sensors - Optex (Europe) Ltd
video intrusion detection - Xtralis
Voice alarms - PEL Services Limited
Warehouse intrusion detection system - Advanced Perimeter Systems Ltd

W50 Fire detection and alarm

Battery & Mains Heat Alarms - BRK Brands Europe Ltd
Break glass and call points - KAC Alarm Company Ltd
Emergency Lighting - P4 Limited
Fire alarm equipment - Hoyles Electronic Developments Ltd
Fire Alarm Systems - Intime Fire and Security
Fire alarms - Chubb Systems Ltd
Fire alarms - Fife Fire Engineers & Consultants Ltd
Fire alarms - Griffin and General Fire Services Ltd
Fire alarms - Initial Electronic Security Systems Ltd
Fire alarms - Viking Security Systems Ltd
Fire alarms and smoke detectors - ASCO Extinguishers Co. Ltd
Fire Alarms, Data Wiring, Call Systems - H Pickup Mechanical & Electrical Services Ltd
Fire and gas control systems - Zellweger Analytic Ltd, Sieger Division
Fire detection - Abel Alarm Co Ltd
Fire detection - ADT Fire and Security Plc
Fire detection - Honeywell Control Systems Ltd
Fire detection - Kidde Fire Protection Ltd
Fire detection - Thomas & Betts Manufacturing Services Ltd
Fire detection and alarm - Cooper Lighting and Security Ltd
Fire detection and alarm - Initial Fire Services
Fire detection and alarm - Johnson Controls Building Efficiency UK Ltd
Fire detection and alarm - PEL Services Limited
Fire detection and alarm - Photain Controls (part of GE security)
Fire detection and alarm systems - Blick Communication Systems Ltd
Fire detection and alarm systems - BRK Brands Europe Ltd
Fire detection and alarm systems - Channel Safety Systems Ltd
Fire detection and alarm systems - Firetecnics Systems Ltd
Fire detection and alarm systems - Gent Limited
Fire detection and alarm systems - Hochiki Europe (UK) Ltd
Fire detection and alarm systems - Interphone Limited
Fire detection and alarm systems - Kidde Fire Protection Services Ltd
Fire detection and alarms - C-TEC (Computionics) Ltd
Fire detection and suppression - Tyco Fire and Integrated Solutions
Fire detection equipment - Nittan (UK) Ltd
Fire detection Systems - Cooper Lighting
Fire detection systems - Cooper Safety
Fire detection systems - Emergi-Lite Safety Systems
Fire protection and alarm - Siemens Building Technologies
Fire protection products for doors and building openings - Moffett Thallon & Co Ltd
Fire systems - West London Security
Gas detection - Gas Measurement Instruments Ltd
Gates and barriers - Hart Door Systems Ltd
Lifts - SCS Group
Radio fire alarm systems - Channel Safety Systems Ltd
Smoke and heat detectors - Apollo Fire Detectors Ltd
Smoke detection, carbon monoxide and heat alarms - BRK Brands Europe Ltd
Vesda aspirating smoke detection equipment - Xtralis

W51 Earthing and bonding

Earth rods and earthing systems - Erico Europa (GB) Ltd
Earthing & Bonding - Tower Manufacturing
Earthing and bonding - Cutting R C & Co
Earthing systems - Furse
Lighting protection engineers - J W Gray Lightning Protection Ltd

W52 Lightning protection

Earthing & bonding installations - J W Gray Lightning Protection Ltd
Lightning conductor systems - Erico Europa (GB) Ltd
Lightning protection - Cutting R C & Co
Lightning protection - Furse
Lightning Protection - R. C. Cutting & Company Ltd
Lightning protection, earthing and electronic protection - Bacon Group
Transient overvoltage protection - Furse

W53 Electromagnetic screening

Patient tagging system - Wandsworth Elecrtrical Ltd

W54 Liquid detection alarm

Gas detection and alarm - Tyco Fire and Integrated Solutions
Liquid detection alarm - Specialist Building Products
Oil pollution monitoring and control systems - design, manufacture, installation and commissioning of - Aquasentry
Water leak detection equipment - Aquasentry

W60 Central control/ Building management

Access Control - Allegion (UK) Ltd
Automatic / Self-Testing Emergency Lighting - P4 Limited
Automation and control - Schneider Electric Ltd
Blind/Shade control systems - Hunter Douglas
Broadcast and audio visual products & services - Streamtec Limited
Building and energy management - ADT Fire and Security Plc
Building automation - Guthrie Douglas Ltd
Building control and management systems - Siemens Building Technologies
Building management systems - Honeywell Control Systems Ltd
Building management systems - Trend Control Systems Ltd
CCTV - Vador Security Systems Ltd
Central control software - Telelarm care Ltd
Central control systems - Honeywell Control Systems Ltd
Central control/ Building management - C-TEC (Computionics) Ltd
Central control/ Building management - Johnson Controls Building Efficiency UK Ltd
Central control/Building management - SCS Group
Community care alarm and monitering equipment - Tunstal Healthcare (UK) Ltd
Control equipment - George Fischer Sales Ltd
Control systems - Eltron Chromalox
Control systems - Technical Control Systems Ltd
Electric radiator controls - Loblite Ltd
Energy management systems - EnergyICT Ltd
Energy monitoring & targeting - Trend Control Systems Ltd
Home control products - Friedland Ltd
Indicator panels - Hoyles Electronic Developments Ltd
Industrial controls, air compressor controls and building management systems - Rapaway Energy Ltd
Instrumentation for boiler plant, incinerators and furnaces - SK Environmental Ltd
Integrated security management systems (SMS) - CEM Systems Ltd
Led's & colour mixing - Mode Lighting (UK) Ltd
Lighting control systems - Electrak International Ltd
Lighting control systems - Whitecroft Lighting Limited
Monitoring - Chubb Systems Ltd
Monitoring - Initial Electronic Security Systems Ltd
Remote management and monitoring - Guardall Ltd
Safety control systems - Tyco Fire and Integrated Solutions
Voice recording, processing and training enhancement software - Computertel Ltd
Wireless Nursecall system - Telelarm care Ltd

X10 Lifts

Access Lifts - Phoenix Lifting Systems Ltd
Access wheelchair platforms lifts - Stannah Lifts Ltd
Access, hydraulic and towable lifts - Denka International
Bathlifts - Bison Bede Ltd
Brush sealing for lifts and glass doors - Kleeneze Sealtech Ltd
Disabled access lifts - Hymo Ltd
Disabled access lifts and ramps - Chase Equipment Ltd
Dissabld Access Lifts - ThyssenKrupp Encasa, A Division of ThyssenKrupp Access Ltd
Dumb waiter - Platform Lift Company Ltd, The
Elevators - Kone
Good lifts - Stannah Lifts Ltd
Goods lifts - Hymo Ltd
Goods lifts - Platform Lift Company Ltd, The
Home lifts - Wessex Lift Co Ltd
Hydraulic lifting equipment - Hymo Ltd
Hydraulic passenger lifts - Stannah Lifts Ltd
Lift manufacturers, installers service and maintenance - Pickerings Ltd
Lifting equipment - Hymo Ltd
Lifting platforms - Wessex Lift Co Ltd
Lifts - Gartec Ltd
Lifts - Kone PLC
Lifts - Otis Ltd
Lifts - Schindler Ltd
Lifts - SSL Access
Lifts, ramps and loading bay equipment - specialist manufacturer - Chase Equipment Ltd
Lifts: Passenger, Platform, Goods and Vehicle - Kone
Mechanical Screw Lifts - Hymo Ltd
Mobile scissors lifts - Hymo Ltd
Pallet lifts - Hymo Ltd
Passenger lifts - Platform Lift Company Ltd, The
Platform lifts - Platform Lift Company Ltd, The
Platform lifts - ThyssenKrupp Encasa, A Division of ThyssenKrupp Access Ltd
Pool lifts - Platform Lift Company Ltd, The
Portable material lifts - Genie Europe
Portable personal lifts - Genie Europe
Public access wheelchair - ThyssenKrupp Encasa, A Division of ThyssenKrupp Access Ltd
Pushbuttons and controls for lifts - Dewhurst plc
Scissor lift tables - Hymo Ltd
Scissor lifts - Platform Lift Company Ltd, The
Scissors lifts - Hymo Ltd
Stair - Brooks Stairlifts Ltd
Stair lifts - Stannah Lifts Ltd
Stairlifts - Gartec Ltd
Step lifts - Wessex Lift Co Ltd
Wheelchair lifts - Stannah Lifts Ltd
Wheelchair stairlifts - Wessex Lift Co Ltd

X11 Escalators

Brush deflection for escalators - Kleeneze Sealtech Ltd
Escalators - Kone
Escalators - Kone PLC
Escalators - Otis Ltd
Escalators - Schindler Ltd
Escalators and passenger conveyers, service and maintenance - Kone Escalators Ltd

X12 Moving pavements

Moving pavements - Schindler Ltd
Moving walkways and autowalks - Kone
Passenger conveyors - Kone PLC
Travolators - Otis Ltd

X13 Powered stairlifts

Domestic stairlifts - ThyssenKrupp Encasa, A Division of ThyssenKrupp Access Ltd
Indoor, Straight and Curved Stairlifts - Brooks Stairlifts Ltd
Outdoor - Brooks Stairlifts Ltd
Perch and/or Stand - Brooks Stairlifts Ltd
Stair lifts - Bison Bede Ltd
Stairlifts - ThyssenKrupp Encasa, A Division of ThyssenKrupp Access Ltd
Stairlifts - Wessex Lift Co Ltd

X14 Fire escape chutes/ slings

Fire escape chutes/slings - NRG Fabrications
Ropes and sashcords - Marlow Ropes Ltd

X20 Hoists

Access lifting equipment - Genie Europe
Access platforms - Kobi Ltd
Confined Space access Equipment - Didsbury Engineering Co Ltd
Davits - Didsbury Engineering Co Ltd
Hoists - Didsbury Engineering Co Ltd
Hoists - Metreel Ltd
Hoists - Morris Material Handling Ltd
Hoists - Wessex Lift Co Ltd
Hoists & handling equipment - Randalls Fabrications Ltd
Hoists and trolleys - PCT Group
Hydraulic movement systems - Hydratight Sweeney Limited
Manhole cover lifter - Didsbury Engineering Co Ltd
Permanent access cradles - Kobi Ltd
Personnel and material transport hoists - SGB, a Brand Company
Ropes and sashcords - Marlow Ropes Ltd
Slings - Wessex Lift Co Ltd
Telehanders - Genie Europe
Tripods - Didsbury Engineering Co Ltd
Wheelchair Platform Lifts - ThyssenKrupp Encasa, A Division of ThyssenKrupp Access Ltd
Woven lifting slings - Albion Manufacturing

X21 Cranes

Boom lifts (Telescopic articulating) - Genie Europe
Crane support - Eve Trakway
Cranes - Metreel Ltd
Cranes - Morris Material Handling Ltd
Cranes - PCT Group
Grade 8 chain slings - PCT Group
Webbing slings - PCT Group

X22 Travelling cradles/ gantries/ladders

Access cradles - Cento Engineering Company Ltd
Access ladders - Clow Group Ltd
Cradle runways - Cento Engineering Company Ltd
Cradles - Clow Group Ltd
Dock levelers - Crawford Hafa Ltd
Dock shelters - Crawford Hafa Ltd
Façade access equipment - Cento Engineering Company Ltd
Façade hoists - Cento Engineering Company Ltd
Gantries - Clow Group Ltd
Gantry - Cento Engineering Company Ltd
Glazing trade association - Glass and Glazing Federation
Hoists and trolleys - PCT Group
Latchways - Cento Engineering Company Ltd
Steps - Clow Group Ltd
Suspended access equipment - Cento Engineering Company Ltd
Travelling Cradles/ Gantrie/ Ladders - Metreel Ltd
Travelling gantries - Kobi Ltd
Travelling ladders - Cento Engineering Company Ltd
Travelling ladders - Kobi Ltd
Trestles - Clow Group Ltd
Window cleaning cradles - Cento Engineering Company Ltd

X23 Goods distribution/ Mechanised warehousing

Conveyors - Arc Specialist Engineering Limited
Dock levelers - Chase Equipment Ltd
Dock shelters - Chase Equipment Ltd
Loading bay equipment - Crawford Hafa Ltd
Material handling equipment - Barton Storage Systems Ltd
Mechanised Warehousing - Chiorino UK Ltd
Mobile food conveyors - TSE Brownson
Mobile yard ramps - Chase Equipment Ltd
Modular loading dock - Chase Equipment Ltd
Road haulage - Eve Trakway
Scissor lifts - Genie Europe
Steel conyor plates for sugar industry - manufacturer - Wm. Bain Fencing Ltd

X30 Mechanical document conveying

Despatch equipment - Envopak Group Ltd

X31 Pneumatic document conveying

Air tube conveying systems - Barloworld Vacuum Technology Plc
Mailroom equipment - Envopak Group Ltd
Pneumatic air tube conveying systems:- cash, sample, hospital and document - Barloworld Vacuum Technology Plc
Pneumatic conveying systems - BVC
Transport systems - Hardall International Ltd

X32 Automatic document filing and retrieval

Sorting equipment - Envopak Group Ltd

Y10 Pipelines

Cost Forecasting and Reporting - Procurement - British Water
DuraFuse - Viking Johnson
Durapipe ABS - Viking Johnson
Friatherm - Viking Johnson
GRE GRP pipes, tubes and fabrications - Deva Composites Limited
Mechanical joints - Victaulic Systems
Metal pipework - Copper, Galvanised steel and Stainless steel - Servomac Limited
Philmac - Viking Johnson
Pipe clips - Unifix Ltd
Pipe couplings - Viking Johnson
Pipelines - CPV Ltd
Pipelines - Deepdale Engineering Co Ltd
Pipelines - Grant & Livingston Ltd
Pipelines - Insituform Technologies® Ltd
Pipelines - John Davidson (Pipes) Ltd
Pipelines - Polypipe Civils
Pipelines - Tata Steel Europe Limited
Pipelines - Yorkshire Copper Tube
Pipework - Woodburn Engineering Ltd
Pipework Fabrication - Turnbull & Scott (Engineers) Ltd
Plastic pipelines - IPPEC Sytsems Ltd
Plumbing accessories - PTS Plumbing Trade Supplies
Polybutylene pipe and fittings - Davant Products Ltd
Polyethylene plastic pipes - Uponor Ltd
Posiflex - Viking Johnson
Preinsulated Piping - Uponor Ltd
Pressure Pipes - Wavin Ltd
Proctect-line - Viking Johnson
Service fittings - Tyco Water Works
Spiral weld steel tube - Tex Engineering Ltd
Stainless steel fittings, flanges and fabrications - Lancashire Fittings Ltd
Stainless steel tubes and sections - Avesta Sheffield Ltd
Stainless steel valves - Lancashire Fittings Ltd
Victaulic - Viking Johnson
Wask - Viking Johnson

Y11 Pipeline ancillaries

Accessories - Tyco Water Works
Air release valves - Delta Fire
Balancing valves - Crane Fluid Systems
Balancing valves - Static, DPCV, PICV - Hattersley
Ball valves - Crane Fluid Systems
Boiler House Equipment - Flowserve Flow Control
Bursting (rupture) discs - Bailey Birkett Sales Division
Butterfly valves - Crane Fluid Systems
Check valves - Crane Fluid Systems
Check Valves - Flowserve Flow Control
Control Valves - Flowserve Flow Control
Diverters - Deva Tap Co Ltd
Fire and gas safety valves - Teddington Controls Ltd
Fitttings, valves and tanks - CPV Ltd
Flame arresters - Bailey Birkett Sales Division
Gate valves - Crane Fluid Systems
Gate valves - Delta Fire
Globe valves - Crane Fluid Systems
Hand- Hydraulic -Electric - Tube & Pipe Manuipulation Service - Tubela Engineering Co Ltd
Heating manifolds - SAV United Kingdom Ltd
Hydrant valves and pressure valves - Eagles, William, Ltd
Industrial Flow Control Equipment - Flowserve Flow Control
Inlet breeching - Delta Fire
In-Line Thermal Disinfection Unit (ILTDU) - Horne Engineering Ltd
Isolation Ball Valves - Flowserve Flow Control
Isolation valves and expansion vessels - SAV United Kingdom Ltd
Main Fittings - Tyco Water Works
Malleable iron pipe fittings - Crane Fluid Systems
Mixer valves - Bristan Group
Pipe fixings - Ellis Patents
Pipeline ancillaries - Deepdale Engineering Co Ltd
Pipeline ancillaries - Flex-Seal Couplings Ltd
Pipeline ancillaries - Isolated Systems Ltd
Pipeline ancillaries - John Davidson (Pipes) Ltd
Pipeline ancillaries - Polypipe Civils
Plug valves, control valves, gate valves, butterfly valves, check valves and consistency transmitters - Dezurik International Ltd
Plumbing accessories - PTS Plumbing Trade Supplies
Plumbing products and pipe fixings - Conex Universal Limited
Pressure reducing valves - Auld Valves Ltd
Pressure surplus valves - Auld Valves Ltd
Pushfit systems - Davant Products Ltd
Radiator valves - Crane Fluid Systems
Relief valves - Crane Fluid Systems
Safety & relief valves - Auld Valves Ltd
Sealed system equipment, air vents, pressure reducers, mixing valves, flow switches etc - Altecnic Ltd
Strainers - Crane Fluid Systems
Temperature control valves - Horne Engineering Ltd
Thermostatic and manual shower controls - Bristan Group
Valves - Pegler Ltd
Valves - Ball, Butterfly, Check, Gate, Press-Fit and Strainers - Hattersley
Water valves and plumbing equipment - EBCO
Welding and Brazing equipment \\ consumables - Johnson Matthey PLC - Metal Joining
White petroleum jelly - Legge Thompson F & Co Ltd

Y20 Pumps

AP1610 - Weir Pumps Ltd
Centrifugal pumps - Weir Pumps Ltd
Diaphragm hand pumps - Pump International Ltd
Fixing systems for building services - Erico Europa (GB) Ltd
Gear and centrifugal pumps - Albany Standard Pumps
Grout pumps - Clarke UK Ltd
GRP packaged pumping stations - Pims Pumps Ltd
H.V Pumps - Pressmain
Industrial, chemical, oil & gas, and petrochemical pumps - Weir Pumps Ltd
Mixers - Hansgrohe
Oil pollution monitoring and control systems - design, manufacture, installation and commissioning of - Aquasentry
Packaged pump sets - Armstrong Integrated Ltd
Process pumps - Clarke UK Ltd
Pump repair - Pumps & Motors UK Ltd
Pumps - Albany Engineering Co Ltd
Pumps - Andrews Sykes Hire Ltd
Pumps - Armstrong Integrated Ltd
Pumps - Barr & Wray Ltd
Pumps - Calpeda Limited
Pumps - Goodwin HJ Ltd
Pumps - Grosvenor Pumps Ltd
Pumps - Ingersoll-Rand European Sales Ltd
Pumps - JLC Pumps & Engineering Co Ltd
Pumps - Mono Pumps Ltd
Pumps - New Haden Pumps Ltd
Pumps - Polypipe Civils
Pumps - Pumps & Motors UK Ltd
Pumps - The BSS Group
Pumps - T-T Pumps & T-T controls
Pumps - Wilo Samson Pumps Ltd
Pumps and Pump stations - Kingspan Environmental Ltd
Pumps, mixers and aerators - Sulzer Pumps
Submersible electric pumps and mixers - ITT Flygt Ltd
Submersible pumps, servicing, maintenance and technical back up - Pump Technical Services Ltd
Sump drainage pumps - Stuart Turner Ltd
Trays for the elderly & special needs - Contour Showers Ltd
Tunnel concrete pumps - Clarke UK Ltd
Variable pumpdrives - Armstrong Integrated Ltd
Water leak detection equipment - Aquasentry
Water pumps - Stuart Turner Ltd

Y21 Water tanks/ cisterns

Carbon and stainless steel storage tanks - Cookson and Zinn (PTL) Ltd
Cistern floats - Airflow (Nicoll Ventilators) Ltd
Cold watertank upgrading, refurbishment & replacement Y21 - Nicholson Plastics Ltd
Custom moulded GRP products - Nicholson Plastics Ltd
Epoxy Tanks - CST Industries, Inc. - UK
Fibreglass water storage tanks - Parton Fibreglass Ltd
G R P Tanks - Pressmain
Glass-Fused-to-Steel-Tanks - CST Industries, Inc. - UK
GRP conical settlement tanks, covers and effluent treatment - Armfibre Ltd
GRP one piece and sectional water storage tanks - Dewey Waters Ltd
GRP storage cisterns - Precolor Sales Ltd
GRP storage tanks - APA
GRP tank housings - Group Four Glassfibre Ltd
GRP water storage tanks - Group Four Glassfibre Ltd
Hot water cylinders and tanks - Gledhill Water Storage Ltd
Liquid Storage Tanks - CST Industries, Inc. - UK
One piece/ semi sectional/ sectional GRP water storage tanks - AquaTech Ltd
Plastic cisterns - Dudley Thomas Ltd
Plastic water storage tanks - Titan Environmental Ltd
Rainwater harvesting - Kingspan Environmental
Refurbishment of tanks - Franklin Hodge Industries Ltd
Rust stabilisation coatings - Corroless Corrosion Control
Sectional and one piece GRP cold water tanks - Nicholson Plastics Ltd
Sectional Steel Tanks and foundations - Goodwin Tanks Ltd
Silage storage tanks - Kingspan Environmental
Silage Storage tanks - Kingspan Environmental
Sprinkler Tanks - CST Industries, Inc. - UK
Storage tanks - Franklin Hodge Industries Ltd
Storage tanks and vessels - Metcraft Ltd

Y21 Water tanks/ cisterns (con't)

Tanks - Arbory Group Limited
Tanks - Balmoral Tanks
Tanks - Braithaite Engineers Ltd
Tanks - Fibaform Products Ltd
Water storage containers and accessories - Straight Ltd
Water storage tanks - Balmoral Tanks
Water storage tanks - Brimar Plastics Ltd
Water storage tanks - CST Industries, Inc. - UK
Water tanks - Harris & Bailey Ltd
Water tanks/ cisterns - Grant & Livingston Ltd

Y22 Heat exchangers

Coil heat exchangers - S & P Coil Products Ltd
Corrugated shell and tube heat exchanger - HRS Hevac Ltd
Heat exchangers - Armstrong Integrated Ltd
Heat exchangers - Atlas Solar
Heat exchangers - Grant & Livingston Ltd
Heat exchangers - Rycroft Ltd
Heat Exchangers - SAV United Kingdom Ltd
Heat exchangers - Turnbull & Scott (Engineers) Ltd
Manufacturer of heating elements - Electric Elements Co, The
Plate heat exchangers - AEL

Y23 Storage cylinders/ Calorifiers

Carbon and stainless steel pressure vessels - Cookson and Zinn (PTL) Ltd
Cold water storage cisterns - Davant Products Ltd
Corrugated shell and tube heat exchanger - HRS Hevac Ltd
Mains pressure thermal storage - Gledhill Water Storage Ltd
Powerstock Calorifiers and Storage Tanks - Hamworthy Heating Limited
Powerstock Calorifiers and Storage Tanks - Hamworthy Heating Ltd
Storage cylinders / calorifiers - Atlas Solar
Storage cylinders/ calorifiers - Grant & Livingston Ltd
Storage cylinders/ calorifiers - Harris & Bailey Ltd
Storage cylinders/ Calorifiers - Rycroft Ltd
Vented hot water storage cylinders - Range Cylinders
Water heaters/calorifiers - Flamco UK Ltd

Y24 Trace heating

Electric space heating - Bush Nelson PLC
Electric surface heating systems manufacturer - Pentair Thermal Management
Heating jackets and panels manufacturer - Pentair Thermal Management
Resistance heating equipment manufacturer - Pentair Thermal Management
Self- regulating cut-to-length heater tapes - Flexelec(UK) Ltd
Zone parallel constant wattage heater tapes - Flexelec(UK) Ltd

Y25 Cleaning and chemical treatment

Central heating system corrosion protectors, cleaners and leak sealers - Fernox
Central heating system corrosion protectors, cleansers + leak sealers - Fernox
Chemical photo etchings - Photofabrication Limited
Chemicals - Baxenden Chemicals Ltd
Chemicals - DSM UK Ltd
Chlorination of water services - Aquastat Ltd
Cleaners and descalers - Tank Storage & Services Ltd
Cleaning & chemical treatment - Precolor Sales Ltd
Cleaning and chemical treatment - Frosts Landscape Construction Ltd
Cleaning and chemical treatment - Hertel Services
Cleaning and chemical treatments - Deb Ltd
Cleaning materials - Evans Vanodine International PLC
Construction chemicals - Artur Fischer (UK) Ltd
Detergents - Evans Vanodine International PLC
Detergents & Disinfectants - Deb Ltd
Disinfectants - Evans Vanodine International PLC
Disinfection systems - Portacel
Grit blasting, chemical cleaning, high pressure water jetting - Masonary Cleaning Services
Jointing compounds, quick repair products and solder & flux - Fernox
Leak stopper and inhibitor for central heating systems - Dunsley Heat Ltd
Limescale remedies - Fernox
Limescale remedies, jointing compounds + quick repair products - Fernox
On site services - flue cleaning, pipework and fabrications - Hodgkinson Bennis Ltd
Polishes - Evans Vanodine International PLC
Polishes and floor cleaners - Johnson Wax Professional
Solvent Degreasers - Deb Ltd
Specialist cleaning services - Rentokil Specialist Hygiene
Specialist pigments : artist colours - W. Hawley & Son Ltd
Specialist pigments : asphalt - W. Hawley & Son Ltd
Specialist pigments : concrete paving - W. Hawley & Son Ltd
Specialist pigments : concrete roofing - W. Hawley & Son Ltd
Specialist pigments : concrete walling - W. Hawley & Son Ltd
Specialist pigments : mortar - W. Hawley & Son Ltd
Specialist pigments : paint - W. Hawley & Son Ltd
Specialist pigments : plastic - W. Hawley & Son Ltd
Vehicle Cleaners - Deb Ltd
Water treatment chemicals - Tower Flue Componants Ltd
Water treatment products and services - Aldous & Stamp (Services) Ltd

Y30 Air ductlines/ Ancillaries

Air ductlines/Ancillaries - Isolated Systems Ltd
Connection and sleves for air ducts - Westbury Filters Limited
Ducting - Alumasc Interior Building Products Limited
Ducting - MacLellan Rubber Ltd
Ductwork cleaning - Nationwide Filter Company Ltd
Fire rated ductwork - Mansfield Pollard & Co Ltd
Fire rated ductwork systems - Fire Protection Ltd
Flexible ducting - Flexible Ducting Ltd
Rubber mouldings and elastomeric bearings - MacLellan Rubber Ltd
Sheet metal ductwork - Hi-Vee Ltd

Y40 Air handling units

Air conditioning - Dantherm Ltd
Air handling unit - Daikin Applied (UK) Ltd
Air Handling Units - Aqua-Blue Ltd
Air handling units - Colt International
Air handling units - Dunham Bush Ltd
Air handling units - Hi-Vee Ltd
Air handling units - May William (Ashton) Ltd
Air handling units - Polypipe Ventilation
Air handling units - Thermal Technology Sales Ltd
Air handling units - Ves Andover Ltd
Air Handling Units, with heat recovery - SAV United Kingdom Ltd
Connect grid and air handling - Saint-Gobain Ecophon Ltd
Curve and coffers - Saint-Gobain Ecophon Ltd
Hybrid system air conditioning - Eaton-Williams Group Limited

Y41 Fans

Axial Fan - Air Control Industries Ltd
Axial, centrifugal, mixed flow fans and dust extraction units - Howden Buffalo
Axial, roof extract and centrifugal fans - Matthews & Yates Ltd
Backward Curve Fan - Air Control Industries Ltd
Chimney extractor fans - Poujoulat (UK) Ltd
Crossflow fans - Consort Equipment Products Ltd
Diagonal Fans - Air Control Industries Ltd
Domestic and commercial and industrial fans - Xpelair Ltd
Domestic and industrial fans - Airflow Developments Ltd
Dust and fume extraction - Horizon International Ltd
EC Fans - Air Control Industries Ltd
Evaporators - Portacel
Fans - BOB Stevenson Ltd
Fans - Colt International
Fans - Elta Fans Ltd
Fans - Gebhardt Kiloheat
Fans - Hotchkiss Air Supply
Fans - Mawdsleys Ber Ltd
Fans - McKenzie-Martin Ltd
Fans - Polypipe Ventilation
fans - SCS Group
Fans - Vortice Ltd
Forward Curve Fan - Air Control Industries Ltd
Industrial axial fans and impellers - London Fan Company Ltd
Jet tunnel fans - Matthews & Yates Ltd
Mixed Flow Fans - Air Control Industries Ltd
PowerX, Lighting, Timers, Chimes, Wiring Accessories, Ventilation - KingShield - Green Brook
Road fans - Matthews & Yates Ltd
Road tunnel fans - Matthews & Yates Ltd
Tangential Fans - Air Control Industries Ltd
Twin fans - Ves Andover Ltd

Y42 Air filtration

10 Year guarantees - Fluidair International Ltd
Air ailtration - Bassaire Ltd
Air cleaners - Warner Howard Group Ltd
Air cleaning equipment - Horizon International Ltd
Air filter frames & housings - Auchard Development Co Ltd
Air filters - Nationwide Filter Company Ltd
Air filters - Universal Air Products Ltd
Air filters - Vokes Air
Air filters - absolute - Auchard Development Co Ltd
Air filters - panel & bags - Auchard Development Co Ltd
Air filters - synthetic - Auchard Development Co Ltd
Air filtration - Camfil Farr Ltd
Air filtration - Daikin Applied (UK) Ltd
Air filtration - Donaldson Filtration (GB) Ltd
Air filtration - GVS Filter Technology UK
Air Filtration - Vortice Ltd
Air filtration (aircleaners) - Aircare Europe Ltd
Air Purifiers - De Longhi Limited
Air purifiers - Purified Air Ltd
Air quality consultancy - Nationwide Filter Company Ltd
Air quality testing - Westbury Filters Limited
Antimicrobial filters - Nationwide Filter Company Ltd
Carbon filters - Nationwide Filter Company Ltd
Carbon filters - Westbury Filters Limited
Dust and fume extraction - Horizon International Ltd
Equipment filters - GVS Filter Technology UK
Face masks - GVS Filter Technology UK
Filter loss gauges - Westbury Filters Limited
Filtration - Fluidair International Ltd
Filtration - Trox (UK) Ltd
Fit & disposal service - Nationwide Filter Company Ltd
Health air filters - GVS Filter Technology UK
Industrial air filtration - GVS Filter Technology UK
Medical air filters - GVS Filter Technology UK
Odour control - Britannia Kitchen Ventilation Ltd
Panel filters - Westbury Filters Limited
Pleateed filters - Westbury Filters Limited
Replacement elements - Fluidair International Ltd
Roll filters - Westbury Filters Limited
Rotary screw - Fluidair International Ltd
Rotary sliding vane - Fluidair International Ltd
Slimline filters - Westbury Filters Limited
Washable filters - Westbury Filters Limited

Y43 Heating/ Cooling coils

Chillers - Airedale International Air Conditioning Ltd
Chillers - Daikin Applied (UK) Ltd
Condensers - Airedale International Air Conditioning Ltd
Fan Coils Units - Ruskin Air Management
Heat exchangers gasketted plate - HRS Hevac Ltd
Heating and cooling coils - S & P Coil Products Ltd
Heating and cooling radiant panel manufacturer - Comyn Ching & Co (Solray) Ltd
Heating/Cooling coils - Andrews Sykes Hire Ltd

Y44 Air treatment

Air cleaners - Nationwide Filter Company Ltd
Air dryers - Fluidair International Ltd
Air treatment - Consort Equipment Products Ltd
Air treatment - Danfoss Ltd
Air treatment - Hertel Services
Condensate treatment - Fluidair International Ltd
Condensation control - Proten Services Ltd
Dehumidifiers - De Longhi Limited
Dehumidifiers - S & P Coil Products Ltd
Dehumification - Dantherm Ltd
Disinfection system - Britannia Kitchen Ventilation Ltd
Dust and fume extraction - Horizon International Ltd
Humidifiers - Eaton-Williams Group Limited
Local cooling units - Eaton-Williams Group Limited
Radon control - Proten Services Ltd
Roof extract units - Ves Andover Ltd
Ultra violet chambers, lamps and control panels - Hanovia Ltd

Y45 Silencers/ Acoustic treatment

Abuse resistant panels - Salex Acoustics Limited
Accoustic flooring - Icopal Limited
Acoustic Baffles, Barriers - Acousticabs Industrial Noise Control Ltd
Acoustic Barriers - IAC Acoustics
Acoustic Ceiling Panels, Wall Panels - including perforated timber/metal - Acousticabs Industrial Noise Control Ltd
Acoustic ceilings - Knauf AMF Ceilings Ltd
Acoustic cladding and lining products - Gilmour Ecometal
Acoustic Curtains, Attenuators, Enclosures, Panels and Louvers - Acousticabs Industrial Noise Control Ltd
Acoustic doors and frames - Fitzpatrick Doors Ltd
Acoustic enclosures - Designed for Sound Ltd
Acoustic Flooring Products - Coburn Sliding Systems Ltd
Acoustic inserts - Hotchkiss Air Supply
Acoustic insulation - B R Ainsworth (Southern)
Acoustic products - Nendle Acoustic Company Ltd
Acoustic Walling Products - Coburn Sliding Systems Ltd
Attenuators/ silencers - Allaway Acoustics Ltd
Boiler Noise Silencer - Fernox
Building regs. Part E compliances - Sound Reduction Systems Ltd
Ceiling panels - Salex Acoustics Limited
Ductwork, silencers, acoustic products - Isolated Systems Ltd
Fabric wrapped panels - Salex Acoustics Limited
Impact resistant Acoustic Wall Panels - Knauf AMF Ceilings Ltd
Noise absorbing panels - Salex Acoustics Limited
Noise control systems - Acousticabs Industrial Noise Control Ltd
Noise Limiters - AET.gb Ltd
Seamless ceilings - Salex Acoustics Limited
Silencers - Hotchkiss Air Supply
Silencers - Ves Andover Ltd
Silencers - Acoustic - Sound Service (Oxford) Ltd
Silencers/ Acoustic treatment - Polypipe Ventilation
Sound masking systems - AET.gb Ltd
Studio lining - Salex Acoustics Limited
Timber veneered panels - Salex Acoustics Limited
Wall panels - Salex Acoustics Limited

Y46 Grilles/ Diffusers/Louvres

Acoustic products - Nendle Acoustic Company Ltd
Air diffusers - HT Martingale Ltd
Anodised Aluminium coils - Alanod Ltd
Design and CAD support - Saint-Gobain Ecophon Ltd
Diffusion accessories - Panel & Louvre Co Ltd
Flange Covers - Furmanite International Ltd
Grilles - Panel & Louvre Co Ltd
Grilles & diffusers - Hotchkiss Air Supply
Grilles and diffusers - Trox (UK) Ltd
Grilles, diffusers and louvres - Grille Diffuser & Louvre Co Ltd The
Grilles/ Diffusers/Louvres - Polypipe Ventilation
Grilles/Diffusers/Louvres - Isolated Systems Ltd
Grills /diffusers/ louvres - SCS Group
Integrated lighting systems - Saint-Gobain Ecophon Ltd
Louvers - McKenzie-Martin Ltd
Louvre, hit and miss ventilators - Ryton's Building Products Ltd
Security grilles - Cooper Group Ltd
Smoke ventilation louvres - Matthews & Yates Ltd
Solar shading systems - Dales Fabrications Ltd

Y50 Thermal insulation

Acoustic Insulation - Sound Reduction Systems Ltd
Acoustic Insulation - Soundsorba Ltd
Aluzink insulated jacketing - SSAB Swedish Steel Ltd
Cork/ rubber - Olley & Sons Ltd, C
Dobelshield - SSAB Swedish Steel Ltd
Draught proofing/ loft insulation - National Insulation Association Ltd
Energy services - BM TRADA
Expanded polystyrene insulation - Jablite Ltd
Floor insulation - Springvale E P S
Foamglas Cellular glass insulation for cut to falls roof systems - Pittsburg Corning (UK) Ltd
Foamglas Cellular glass internal and external insulation for roof, wall, floor, below ground conditions and foundations - Pittsburg Corning (UK) Ltd
Garden Products - NMC (UK) Ltd
Insulation - Rockwool Ltd
Insulation - Skanda Acoustics Limited
Insulation cork - Olley & Sons Ltd, C
Insulation materials - Gradient Insulations (UK) Ltd
Pre-insulated pipework - CPV Ltd
Roof insulation - Springvale E P S
Sarnatherm insulation - Sarnafil Ltd
Thermal Insulation - Aqua-Blue Ltd
Thermal insulation - B R Ainsworth (Southern)
Thermal Insulation - Draught Proofing Advisory Association
Thermal insulation - Firebarrier Services Ltd
Thermal Insulation - Hawkins Insulation Ltd
Thermal insulation - Hertel Services
Thermal Insulation - Insulated Render & Cladding Association
Thermal Insulation - Isolated Systems Ltd
Thermal insulation - Knauf Insulation Ltd
Thermal Insulation - Modern Plan Insulation Ltd
Thermal Insulation - National Insulation Association Ltd
Thermal insulation - Polypipe Ventilation
Thermal Insulation - Reiner Fixing Devices (Ex: Hardo Fixing Devices)
Thermal insulation - Thermica Ltd
Trade Association - Association of Specialist Fire Protection (ASFP)
Urethane foam industry trade association - BRITISH URETHANE FOAM CONTRACTORS ASSOCIATION
Wall insulation - Springvale E P S

Y51 Testing and commissioning of mechanical services

Acoustic testing, Air tightness testing, Secured by design, durability and weather tightness testing - BM TRADA
Chlorine residual analysers - Portacel
Equipment for measuring, testing and commissioning of mechanical services - PCE Instruments UK Ltd
Installation & Commissioning - GP Burners (CIB) Ltd
Testing and commissioning of mechanical services - Polypipe Ventilation
Testing and maintenance for fall arrest and safety - Bacon Group
Testing facilities for GRP tanks - APA

Y52 Vibration isolation mountings

Acoustic products - Nendle Acoustic Company Ltd
Anti - vibration mountings - MacLellan Rubber Ltd
Anti vibration mounting - Designed for Sound Ltd
Anti Vibration mountings and hangers - Salex Acoustics Limited
Anti vibration products - Nendle Acoustic Company Ltd
Anti-vibration mounts - Isolated Systems Ltd
Anti-vibration pads - Olley & Sons Ltd, C
Building vibration control products - Salex Acoustics Limited
Machine Mounting Anti Vibration Materials, Structural and sliding bearings and resilient - Tiflex Ltd
Structural. Bridge & Seismic Bearings - Tiflex Ltd
Vibration control equipment/containers - Mansfield Pollard & Co Ltd
Vibration isolation - Allaway Acoustics Ltd
Vibration Isolation - Sound Service (Oxford) Ltd
Vibration isolation mountings - Christie & Grey Ltd

Y53 Control components - mechanical

Boiler controls and accessories - Alpha Therm Ltd
Close Control - Airedale International Air Conditioning Ltd
Control - Airedale International Air Conditioning Ltd
Control - T-T Pumps & T-T controls
Control components - Danfoss Ltd
Control components - mechanical - Guthrie Douglas Ltd
Control equipment, timers and relays - Thomas & Betts Manufacturing Services Ltd
Control Systems - Air Control Industries Ltd
Control systems - SK Environmental Ltd
Controls - Ves Andover Ltd
Embeded control systems - MJ Electronics Services (International) Ltd
Enclosures - Knurr (UK) Ltd
Hazardous area control gear - PFP Electrical Products Ltd

Y53 Control components - mechanical (con't)

Heating controls - SAV United Kingdom Ltd
Heating programmers - Timeguard Ltd
Lighting controls - Thorlux Lighting
Moisture meters, hygrometers - Protimeter, GE Thermometrics (UK) Ltd
Pump selection software - Armstrong Integrated Ltd
Thermometers, hygrometers, hydrometers, pressure and altitude gauges - Brannan S & Sons Ltd
Thermometers, pressure gauges and associated instruments - Brannan Thermometers & Gauges
Thermostats - Teddington Controls Ltd
Valve Actuators - Flowserve Flow Control
Valve Refurbishment - Hodgkinson Bennis Ltd
Water Controls - Setsquare

Y59 Sundry common mechanical items

Air receivers - Flamco UK Ltd
Exterior surface heating - Uponor Ltd
Flexible plumbing - Uponor Ltd
Fry's solder and flux - Fernox
Gaskets - Olley & Sons Ltd, C
Gaskets and Sealing Products - MacLellan Rubber Ltd
GRP Kiosks & modular enclosures - Group Four Glassfibre Ltd
Machines for Plastic Pipe Joinings - George Fischer Sales Ltd
Manufacturerers of: Metal Cutting Circular Saws - Spear & Jackson Interntional Ltd
O Rings, bonded seals, grease guns, grease nipples and bucket pumps - Ashton Seals Ltd
Parallel slide valves - Bailey Birkett Sales Division
Pipe cutting machines - George Fischer Sales Ltd
Pressure reducing valves - Bailey Birkett Sales Division
Pressure regulating valves - Bailey Birkett Sales Division
Pulley lines - James Lever 1856 Ltd
Resharpners of Circular saw blades - Spear & Jackson Interntional Ltd
Safety relief valves - Bailey Birkett Sales Division
Seals - Atkinson & Kirby Ltd
Smoke and natural ventilation systems - SE Controls
System 21 - the latest in enrgy efficiency - Servowarm
Welding and Brazing equipment \\ consumables - Johnson Matthey PLC - Metal Joining

Y60 Conduit and Cable trunking

Adaptable boxes - Vantrunk Engineering Ltd
All insulated plastic enclosures - Ensto Briticent Ltd
Aluminium Trunking Systems - Marshall-Tufflex Ltd
Cable management - Ackermann Ltd
Cable management - Hager Ltd
Cable management - Kem Edwards Ltd
Cable management systems - MK (MK Electric Ltd)
Cable management systems - REHAU Ltd
Cable management/Termination and Fixing accessories - Norslo - Green Brook
Cable routing - Hellermann Tyton
Cable tray - Legrand Electric Ltd
Cable trunking - Legrand Electric Ltd
Conduit and Cable Trunking - Barton Engineering
Conduit and ductwork - Polypipe Civils
Electrical cable management systems - Marshall-Tufflex Ltd
Electrical Conduit & Accessories - Kem Edwards Ltd
Flexible conduit systems - Adaptaflex Ltd
Floor laid flexible cable protectors for every application - Vulcascot Cable Protectors Ltd
Flooring systems - Legrand Electric Ltd
G R P systems - Marshall-Tufflex Ltd
gator shield, flo-coat, plasticoat, Aluminiumised tube, sculptured tube, engineered tube, NHT3,5,7 ERW 1-5 - Tyco European Tubes
L.S.F. Conduit - Aercon Consultants Ltd
Ladder rack - Legrand Electric Ltd
Lighting trunking - Legrand Electric Ltd
Metal cable trunking - Legrand Power Centre
Pipe supports - Unistrut Limited
Pre wired PVC conduit - Aercon Consultants Ltd
Specialised Applications - Marshall-Tufflex Ltd
Stainless steel conduit - Kensal CMS
Stainless Steel Enclosures - Electrix International Ltd
Stainless Steel Trunking and Conduit - Electrix International Ltd
Stainless Steel Wire Basket Tray - Electrix International Ltd
Steel and stainless steel enclosures - Ensto Briticent Ltd
Steel Cable Support System - Marshall-Tufflex Ltd
Trunking - Vantrunk Engineering Ltd

Y61 HV/ LV cables and wiring

Cable glands and connectors - Elkay Electrical Manufacturing Co Ltd
Cable protection - Hellermann Tyton
Cables - Nexans UK Ltd
Cables and wiring accessories - Ellis Patents
Cabling Components - Hubbell Premise Wiring
Electrical cables - Pitacs Ltd
Electrical Insulation Blankets - MacLellan Rubber Ltd
HV and LV cables and wiring - Concordia Electric Wire & Cable Co. Ltd
HV/ LV cables & wiring - Metway Electrical Industries Ltd
HV/ LV cables and wiring - Anixter (UK) Ltd
Wiring accessories - MK (MK Electric Ltd)

Y62 Busbar trunking

Busbar - Marshall-Tufflex Ltd
Busbar power distribution systems - Electrak International Ltd
Busbar trunking - Legrand Electric Ltd
Busbar, cable management systems for floors - Legrand Power Centre
Inceiling busbar trunking - Electrak International Ltd
Stainless steel enclosures IP65 - Kensal CMS
Stainless steel trunking - Kensal CMS
Underfloor busbar trunking - Electrak International Ltd

Y63 Support components-cables

Cable ladders - Vantrunk Engineering Ltd
Cable management/Termination and Fixing accessories - Norslo - Green Brook
Cable routing - Hellermann Tyton
Cable support systems - Hadley Industries Plc
Cable supports - Unistrut Limited
Cable ties and cable clips - Unifix Ltd
Cable tray - Vantrunk Engineering Ltd
Cable tray/ladder - Unistrut Limited
Cleats and cable ties - Vantrunk Engineering Ltd
Office power, data and cable management system - Staverton (UK) Ltd
Pipe and cable suppoers - LPA Group PLC
Stainless Steel Cabinets - Electrix International Ltd
Stainless Steel Cable Tray - Electrix International Ltd
Stainless steel cable tray - Kensal CMS
Stainless Steel Supports & Fasteners - Electrix International Ltd
Support Components Cables - Ellis Patents
Support systems - Legrand Electric Ltd
Wire basket cable tray - Vantrunk Engineering Ltd

Y70 HV switchgear

HV switchgear - Hager Ltd
HV switchgear - Hubbell Premise Wiring
HV Switchgear - Peek Traffic Ltd
Switchboards - Technical Control Systems Ltd
Switchgear - Hawker Siddley Switchgear

Y71 LV switchgear and distribution boards

Circuit protection - MK (MK Electric Ltd)
Domestic switch gear - Eaton Electric Limited
Earth leakage electrical protection equipment - Paslode
Earth monitoring - Paslode
Electrical switchgear - Dorman Smith Switchgear
Enclosed isolators - Ensto Briticent Ltd
Enclosures - Elkay Electrical Manufacturing Co Ltd
LV switchgear - T-T Pumps & T-T controls
LV Switchgear & Distribution Boards - Peek Traffic Ltd
LV switchgear and distribution boards - Eaton Electric Ltd
LV switchgear and distribution boards - Hager Ltd
Mains Distribution Assemblies 100A to 3000 A, 110v to 400v - Blakley Electrics Ltd
potentiometers - Omeg Ltd
RCDs - PowerBreaker - Green Brook
Residual current circuit breakers - Paslode
Residual current monitors - Paslode
Switches - Ensto Briticent Ltd
Switchgear - Green Brook
Switchgear - Hawker Siddley Switchgear

Y72 Contactors and starters

Contactors and starters - Eaton Electric Ltd
Contracts and starters - Danfoss Ltd

Y73 Luminaires and lamps

Associated accessories - Bernlite Ltd
Cable management systems - Vantrunk Engineering Ltd
Ceiling Lights - Astro Lighting Ltd
Coloured and clear covers for fluorescent lamps and fittings - Encapsulite International Ltd
Columns and brackets - DW Windsor Lighting
Converters for cold cathode - Mode Lighting (UK) Ltd
Direct and indirect luminaires - Lumitron Ltd
Downlights - Astro Lighting Ltd
Equipment and luminaires, control gear and accessories - Poselco Lighting
Flameproof EExd fluorescents - PFP Electrical Products Ltd
Fluorescent and discharge lighting - Fitzgerald Lighting Ltd
General fluorescent lighting - Fluorel Lighting Ltd
Heritage Lighting refurbishment and re-engineering - Poselco Lighting
Incandesant light bulbs and other domestic lighting - J.F. Poynter Ltd
Increased safety fluorescents - PFP Electrical Products Ltd
Lampholders - Bernlite Ltd
Lamps and electronic control gear - Osram Ltd
LED unit conversions - Poselco Lighting
Light bulbs and fittings - British Electric Lamps Ltd
Light sensors - Timeguard Ltd
Lighting designers - Thorlux Lighting
Low voltage spots - Lumitron Ltd
Luminaire & Lamp - Andy Thornton Ltd
Luminaires - Ensto Briticent Ltd
Luminaires and lamps - Brandon Medical
Luminaires and lamps - Connect Lighting Systems (UK) Ltd
Luminaires and Lamps - Consult Lighting Ltd
Luminaires and lamps - Designplan Lighting Ltd
Luminaires and lamps - Eaton Electric Ltd
Luminaires and lamps - Focal Signs Ltd
Luminares & Lamps - Glamox Electric (UK) Ltd
OEM and bespoke LED and lighting luminaire assembly - Poselco Lighting
Project management services - Saint-Gobain Ecophon Ltd
Refurbishment and upgrading service for existing luminaires - Poselco Lighting
Spotlights - Astro Lighting Ltd
Stair components - Spindlewood
Wall Lights - Astro Lighting Ltd
Zone 1 Bulkheads - PFP Electrical Products Ltd
Zone 1 floodlights - PFP Electrical Products Ltd
Zone 1 wellglasses - PFP Electrical Products Ltd

Y74 Accessories for electrical services

Accessories for electrical services - Anixter (UK) Ltd
Accessories for electrical services - Brass Art Ltd
Accessories for electrical services - Tower Manufacturing
Dimming switches - Mode Lighting (UK) Ltd
Din-rail time switches and surge protectors - Timeguard Ltd
Domestic electrical accessories and dry cell batteries and torches - J.F. Poynter Ltd
Electrical Accessories - Legrand Electric Ltd
Electrical fittings and wire accessories - Legrand Electric Ltd
Electrical products - Panasonic UK Ltd
Electrical swithces - Wandsworth Elecrtrical Ltd
General Electrical Products - WF Senate
General electrical products - wholesale - Jolec Electrical Supplies
Invisible light switches and sockets - Forbes & Lomax Ltd
Light switches - Redditch Plastic Products
Modular wiring systems - Thorlux Lighting
Painted, nickel silver, brass and frosted acrylic stainless steel switches and sockets - Forbes & Lomax Ltd
Pipeline Gaskets & Seals WRAS Approved - MacLellan Rubber Ltd
Residual current protected skt outlets - Paslode
Secure connection systems - Hager Ltd
Switch and socket boxes - Legrand Power Centre
Switches - Teddington Controls Ltd
Time controllers, security light switches and RCD switched sockets - Timeguard Ltd
Track Support Materials - Tiflex Ltd
Ultrasonic sensors - Thomas & Betts Manufacturing Services Ltd
Wire accessories - Hellermann Tyton
Wiring accessories - Eaton Electric Ltd
Wiring accessories - Hager Ltd

Y80 Earthing and bonding components

Earth leakage electrical protection equipment - Paslode
Earth monitoring - Paslode
Earth Monitoring Equipment - Blakley Electrics Ltd

Y81 Testing & commissioning of electrical services

Appliance Testing - Hawkesworth Appliance Testing
Electrical safety testing - National Britannia Ltd
Emergency Lighting - P4 Limited
Energy management - National Britannia Ltd
Equipment for measuring, testing & commissioning of electrical services - PCE Instruments UK Ltd
Hygrometers - Protimeter, GE Thermometrics (UK) Ltd
Moisture meters - Protimeter, GE Thermometrics (UK) Ltd
Super Rod, Tools - ToolShield - Green Brook
Testing & commissioning of electrical services - Eaton Electric Ltd
Testing and maintenance of lightning protection - Bacon Group

Y82 Identification - electrical

Breezair Evaporative Coolers - Radiant Services Ltd
Cable identification - Hellermann Tyton
Identification labels - Mockridge Labels & Nameplates Ltd

Y89 Sundry common electrical items

Automatic Pool covers - Golden Coast Ltd
Clip in conduit fittings - PFP Electrical Products Ltd
Custom build control panels - Golden Coast Ltd
Electrical Distribution Equipment - WT Henley Ltd
Electronic accessories - VELUX Co Ltd, The
Filter and poulton - PFP Electrical Products Ltd
MV primary and secondary products - Schneider Electric Ltd
Stainless steel isolators - Kensal CMS
Sundry common electrical items - Metway Electrical Industries Ltd
Sundry common electrical items - Tower Manufacturing

Y90 Fixing to building fabric

Cable tie and fixing assemblies - Hellermann Tyton
Copper Tacks - Crown Nail Co Ltd
Cut Tacks - Crown Nail Co Ltd
Fixings and fastenings - Tower Manufacturing
Frame cramps, restraint straps, sliding anchor stems - Vista Engineering Ltd
Insulation fixings and stainless steel helexial nails - Reiner Fixing Devices (Ex: Hardo Fixing Devices)
Mild steel and non-ferrous nails, tacks and pins - John Reynolds & Sons (Birmingham) Ltd
Nails - Stone Fasterners Ltd
Nuts and bolts - Owlett - Jaton
Remedial fixing systems - Redifix Ltd
Rotary impact and masonry drills - Unifix Ltd
Screws, nuts and bolts - Nettlefolds Ltd
Stainless steel building components - Wincro Metal Industries Ltd
Tacks: Cut / Copper - Crown Nail Co Ltd
Through bolt fixuings - Tucker Fasteners Ltd
Wall ties - remedial - Redifix Ltd
Wood, coach, self tapping, fastbrolley screws, frame anchors, hammer and expansion plugs - Unifix Ltd

Y91 Off-site painting/ Anti-corrosion treatments

Anti-corrosion and sealing systems - Winn & Coales (Denso) Ltd
Anti-rust coating - Bondaglass Voss Ltd
Corrosion Inhibitors - Kenyon Paints Limited
Hot dip galvanizing - Arkinstall Galvanizing Ltd
Hot dip galvinizing - Joseph Ash Ltd
Inks and metal coatings - Valspar UK Corporation Ltd
Paint and powder coatings - Trimite Ltd
Potable water coatings - Corroless Corrosion Control
Powder coatings - H.B. Fuller
Primers Rust Stabilising - Kenyon Paints Limited
Printing solutions - Hellermann Tyton
Thermoplastic coating powders - Plascoat Systems Ltd

Y92 Motor drives - electric

Automatic gates - Jackson H S & Son (Fencing) Ltd
Drive Systems - Air Control Industries Ltd
Electric motor repairs - Pumps & Motors UK Ltd
Electrical motors - Eaton Electric Ltd
Motor drives (electric) - T-T Pumps & T-T controls
Motor drives-electric - Danfoss Ltd
Tube motors - Kaba Garog

Z10 Purpose made joinery

Architectural joinery manufacturers - Edmonds A & Co Ltd
Bespoke - Evertaut Ltd
Bespoke joinery - Howard Bros Joinery Ltd
Bespoke Joinery - Taskworthy Ltd
Bespoke joinery products - Alumasc Interior Building Products Limited
Bespoke laminated components - T B S Fabrications
Bespoke museum display cases - The Benbow Group
Bespoke Study, lounge, bedroom, fitted furniture - Hyperion Wall Furniture Ltd
Brass - Aalco Metals Ltd
Bronze - Aalco Metals Ltd
Copper - Aalco Metals Ltd
Decorative wood moulding and carvings, cornice, picture rail, dado rail, skirting, architrave and pediments, radiator cabinet grilles - Winther Browne & co Ltd
Door joinery manufacturer - Humphrey & Stretton PLC
Fire resisting joinery - Decorfix Ltd
Hardwood mouldings - W.L. West & Sons Ltd
Joinery - Firco Ltd
Joinery solutions - Taskworthy Ltd
Machined Hardwoods - Moray Timber Ltd
Machined softwood and machined MDF mouldings - Metsä Wood
Machined softwoods - Moray Timber Ltd
Pine room paneling, bookcases and display cupboards - Hallidays UK Ltd
Propose Made Jointery - AS Newbould Ltd
Purpose made joinery - The Benbow Group
Purpose made Joinery - Canopy Products Ltd
Purpose made joinery - Channel Woodcraft Ltd
Purpose made joinery - Decorfix Ltd
Purpose made joinery - Enfield Speciality Doors
purpose made joinery - Jali Ltd
Purpose made joinery - Jeld-Wen UK Ltd
Purpose made joinery - Smith & Choyce Ltd
Purpose made joinery - Spindlewood
Radiator cabinets - Jali Ltd
Specialist joinery - Barlow Group
Specialist joinery - Barnwood Shopfitting Ltd
Specialist joinery - Cumberland Construction
Stainless steel - Aalco Metals Ltd
Stockholders of aluminium - Aalco Metals Ltd
Timber engineering services - Wyckham Blackwell Ltd
Timber specialist - Challenge Fencing Ltd
Window & door Specials - Blair Joinery

Z11 Purpose made metalwork

Aluminium and steel fabricators - Ramsay & Sons (Forfar) Ltd
Aluminium extrusions, powder coating, anodising and fabrication - Kaye Aluminium plc
Aluminium fabrication - Panel Systems Ltd
Aluminium special structures - Alifabs (Woking) Ltd
Aluminum and steel standing seam roofing system - CA Group Ltd - MR-24 Division
Architectural metalwork - Broxap Ltd
Architectural metalwork - Andy Thornton Ltd
Architectural Metalwork - Chris Brammall
architectural metalwork - Ermine Engineering Co. Ltd
Architectural metalwork - Mackay Engineering
Architectural metalwork - Multitex GRP LLP
Architectural Metalwork - Wallis Conservation Limited trading as Dorothea Restorations

Z11 Purpose made metalwork (con't)

Architectural metalwork - Woodburn Engineering Ltd
Architectural metalworkers - Edmonds A & Co Ltd
Architectural metalworks - Glazzard (Dudley) Ltd
Bespoke metal products - Alumasc Interior Building Products Limited
Bespoke museum display cases - The Benbow Group
Brass and bronze heritage products - Brass Art Ltd
Brick bond grilles - Priory Shutter & Door Co Ltd
Cast iron fittings and finishings - Company name: Britannia Metalwork Services Ltd
Cast stanchions - Gabriel & Co Ltd
Cast steelwork - Gabriel & Co Ltd
Ceiling tiles & Systems - British Gypsum Ltd
CNC milling and turning of specialist ports in metal and plastic - Griff Chains Ltd
Copper tubing - Wednesbury Tube
Custom sheet metal - Azimex Fabrications Ltd
Decorative castings - Andy Thornton Ltd
Design, manufacture and painting/ anodising - Seco Aluminium Ltd
Engineering castings - Ballantine, Bo Ness Iron Co Ltd
Galvanising Services - Birtley Group
GRP, aluminum, mild and stainless steel - Redman Fisher Engineering
Hot dip galvanizing - Corbett & Co (Galvanising) W Ltd
Laser Cutting - Coates LSF Ltd
Lead ingots - R.M. Easdale & Co Ltd
Manufacturers of aluminum extrusion - Alcoa Custom Extrudes Solutions
Marine Craft - Thanet Ware Ltd
Mesh Screens - Cadisch MDA Ltd
Metal fabrications - B C Barton Limited
Metal fabrications - Squires Metal Fabrications Ltd
Metal framing - Unistrut Limited
Metal roofing and cladding - Ash & Lacy Building Products Ltd
Metal Spraying - Thanet Ware Ltd
Metalwork - Doorfit Products Ltd
On site welding - C & W Fencing Ltd
Perforated sheet metal (mild steel, galvanised, aluminium, stainless steel and special alloys) - Multi Mesh Ltd
Plastic coates Steel Coil - Colorpro Systems Ltd
Punch & fold machine - Coates LSF Ltd
Purpose made joinery - Barlow Group
Purpose made metal work - Metalcraft (Tottenham) Ltd
Purpose Made Metalwork - BLM British Lead
Purpose made metalwork - Canopy Products Ltd
Purpose made metalwork - Claydon Architectural Metalwork Ltd
Purpose made metalwork - Consort Equipment Products Ltd
Purpose Made Metalwork - Deepdale Engineering Co Ltd
Purpose made metalwork - Grant & Livingston Ltd
Purpose made metalwork - Hipkiss, H, & Co Ltd
Purpose made metalwork - Hubbard Architectural Metalwork Ltd
Purpose made metalwork - Leander Architectural
Purpose made metalwork - NRG Fabrications
Purpose made metalwork - The Benbow Group
Purpose made metalwork - WA Skinner & Co Limited
Purpose Made Metalworks - Blagg & Johnson Ltd
Remelt Facility - Alcoa Custom Extrudes Solutions
S.F.S. building systems (Metframe - Metsec Ltd
Seamless cold drawn copper tube - Lawton Tube Co Ltd, The
Security Grilles and Shutters - London Shopfitters Ltd
Special arichitectural metalwork - BF Bassett & Findley
Special castings in brass and bronze - Brass Art Ltd
Stainless steel and sheet metal - Kensal CMS
Stainless steel brackets - Kensal CMS
Stainless steel cleaning products - Kleen-Tex Industries Ltd
Stainless steel columns - Valmont Stainton
Stainless steel custom fabrication - Kensal CMS
Stainless steel fabrications - Component Developments
Stainless Steel Mesh - MMA Architectural Systems Ltd
Stainless Steel Rods - MMA Architectural Systems Ltd
Stainless steel wire mesh - Multi Mesh Ltd
Stainless Steel Wire products - MMA Architectural Systems Ltd
Stainless steel, angles, beams, tees, channels, flats and bar - IMS UK
Steel Cables - MMA Architectural Systems Ltd
Steel canopies - Gabriel & Co Ltd
Steel fabrication - Coates LSF Ltd
Steel fabrication, plate bending & rolling, welding, drilling, turning, milling etc - Stevenson & Cheyne (1983)
Steel lintels - Jones of Oswestry Ltd
Steel plates, angles, RSJ's, Flats - Rainham Steel Co Ltd
Steel, Stainless steel and aluminium fabrications - Speedfab Midlands
Structural fixings - Expamet Building Products
Structures - Kee Klamp Ltd
Wedge Wire - Cadisch MDA Ltd
Welding and Brazing equipment \\ consumables - Johnson Matthey PLC - Metal Joining
Welding engineers - Woodburn Engineering Ltd
Wire Products - Owlett - Jaton
Wirecloth, screens, filters, vent mesh and welded mesh - United Wire Ltd
Woven wirecloth, welded wiremesh, perforated metal, quarry screening, wire products - Potter & Soar Ltd

Z12 Preservative/ Fire retardant treatments for timber

Creosote - Legge Thompson F & Co Ltd
Passive fire protection - Industrial Solutions - Actavo (UK) Ltd
Preservation - Cox Long Ltd
Preservative / fire retardent treatment - Brewer C & Sons Ltd
Preservative and fire retardant treatments for timber - Dover Trussed Roof Co Ltd
Preservative/Fire retardant treatments for timber - BM TRADA
Preservative/fire retardant treatments for timber - Brash John & Co Ltd
Preservative/fire retardent treatments - Arch Timber Protection
Preservative/Fireretardant treatments for timber - Smith & Rodger Ltd
Stockholm tar - Legge Thompson F & Co Ltd
Timber treatment - Proten Services Ltd
Timber treatment - Walker Timber Ltd
Trade Association - Association of Specialist Fire Protection (ASFP)

Z20 Fixings/ Adhesives

Adhesive - Loctite UK Ltd
Adhesive - Tremco Illbruck Ltd
Adhesive for access flooring/Laminating/Panel: Ultra High Performance - Uretech PA1, High Performance - Uretech PA2, Acoustic - Uretech AA2, Laminating / Panel adhesive - Uretech LA5 - Star Uretech Ltd
Adhesive products - Henkel Consumer Adhesives
Adhesive tapes - Advance Tapes International Ltd
Adhesive Tapes - Hilton Banks Ltd
Adhesives - Atkinson & Kirby Ltd
Adhesives - Building Adhesives Ltd
Adhesives - Dunlop Adhesives
Adhesives - Evode Ltd
Adhesives - H.B. Fuller
Adhesives - Henkel Ltd
Adhesives - Kenyon Group Ltd
Adhesives - Kenyon Performance Adhesives
Adhesives - Mapei (UK) Ltd
Adhesives - Sealocrete PLA Ltd
Adhesives - Tarkett Ltd
Adhesives - Tektura Plc
Adhesives - manufacturer - National Starch & Chemical Ltd
Adhesives and fillers - Polycell Products Ltd
Adhesives and grouts - Shackerley (Holdings) Group Ltd incorporating Designer ceramics
Adhesives and pastes - Clam Brummer Ltd
Adhesives Design Toolkit - TWI Ltd
Adhesives for wall covering and wood - Ciret Limited
Adhesives selector - TWI Ltd
Adhesives, wood filler and glue - Humbrol
Admixtures - BASF plc, Construction Chemicals
Ahesives - Combustion Linings Ltd
Airtight adhesives and tapes - Fillcrete Ltd
Blind fastening systems - Tucker Fasteners Ltd
Blue Polypropylene Ropes - James Lever 1856 Ltd
Cable clips - Artur Fischer (UK) Ltd
Cast-in fixing channel - Halfen Ltd
Cavity fixings - Ramset Fasteners Ltd
Chemical anchors - ITW Spit
Chemical anchors - Ramset Fasteners Ltd
Clips - Airflow (Nicoll Ventilators) Ltd
Coal, barn, wood and stockholm tar - Liver Grease Oil & Chemical Co Ltd
Connectors - Simpson Strong-Tie®
Contractors tapes - Airflow (Nicoll Ventilators) Ltd
Copper Tacks - Crown Nail Co Ltd
Cut steel and leak-proof nails, tacks, pins and screws - Reynolds John & Son (Birmingham) Ltd
Cut Tacks - Crown Nail Co Ltd
Cyanoacrylate adhesives - Kenyon Group Ltd
Cyanoarylate adhesives - Kenyon Performance Adhesives
Drill bits - Ramset Fasteners Ltd
Drop in anchors - Ramset Fasteners Ltd
Elastic Bonding - Sika Ltd
Engineering and construction fasteners - Allman Fasteners Ltd
Epoxy resins - Hatfield Ltd, Roy
Expanding fre rated foam - Ramset Fasteners Ltd
Fastener systems for the roofing , cladding and construction industry - EJOT UK Limited
Fasteners - UK Fasteners Ltd
Fasteners, fixings, tools, engineering supplies - Arrow Supply Company Ltd
Fixing - Owlett - Jaton
Fixing and adhesives - BASF plc, Construction Chemicals
Fixing for masonry - Vista Engineering Ltd
Fixing systems - remedial - Redifix Ltd
Fixings - Airflow (Nicoll Ventilators) Ltd
Fixings - British Gypsum Ltd
Fixings - Broen Valves Ltd
Fixings - Canopy Products Ltd
Fixings - Hipkiss, H, & Co Ltd
Fixings - ITW Construction Products
Fixings - ITW Spit
Fixings - Power Plastics Ltd
Fixings - RFA-Tech Ltd
Fixings & adhesives - Olby H E & Co Ltd
Fixings / Adhesives - Interface Europe Ltd
Fixings and adhesives - Caswell & Co Ltd
Fixings and fastenings - Kem Edwards Ltd
Fixings/ adhesives - Bondaglass Voss Ltd
Fixings/ Adhesives - Nettlefolds Ltd
Fixings/ Adhesives - Till & Whitehead Ltd
Fixings/ Adhesives - Tower Manufacturing
Fixings/Adhesives - Crowthorne Fencing
Flat roofing fasteners - SFS Intec LTD
Floor Covering Adhesives - Ardex UK Ltd
Flooring adhesives - Ball F & Co Ltd
Flooring adhesives, smoothing compounds - Laybond Products Ltd
For Flat roofing built up felt roofing - Chesterfelt Ltd
Frame fixings - Artur Fischer (UK) Ltd
Freyssinet Asphaltic plug joints - Pitchmastic PmB Ltd
General fixings - Artur Fischer (UK) Ltd
Glue guns - Kenyon Performance Adhesives
Glues, cleaners for PVCu window construction - VBH (GB) Ltd
Heavy duty anchors - Ramset Fasteners Ltd
Heavy Duty fixings - Artur Fischer (UK) Ltd
High Temperature Adhesives - Minco Sampling Tech UK Ltd (Incorporating Fortafix)
Hot melt adhesives - Kenyon Group Ltd
Hot melt ahesives - Kenyon Performance Adhesives
Hot melt guns - Kenyon Performance Adhesives
Industrial fasterners & raw materials - Montrose Fasteners
Large perforated headed, steel and stainless steel anchorages - BigHead Bonding Fasteners Ltd
Lateral Restraint Fixings - Redifix Ltd
Light & heavyduty anchors - ITW Construction Products
Lightweight fixings - Artur Fischer (UK) Ltd
Lindapter - Owlett - Jaton
Lintel Reinforcing - Redifix Ltd
Metal roofing and cladding - Ash & Lacy Building Products Ltd
Nailing Machines and Nails - Atkinson & Kirby Ltd
Nylon devices up to 300mm - Reiner Fixing Devices (Ex: Hardo Fixing Devices)
Packaging and security fasteners - Hellermann Tyton
Polyurethane adhesives - Dow Building Solutions
Polyurethane foam sealant - Dow Building Solutions
Polyurethane foam spray systems - Dow Building Solutions
Powder acurated cartridge tools - Ramset Fasteners Ltd
Rayon Twines - James Lever 1856 Ltd
Resin Bonded Steel Female Sockets - Redifix Ltd
Resins and grouts - Helifix
Rivets and Tools - Allman Fasteners Ltd
Roof Threaded Inserts - Allman Fasteners Ltd
Screned anchors - ITW Construction Products
Screws - Owlett - Jaton
Screws and frame fixings - VBH (GB) Ltd
Sealants - Kenyon Performance Adhesives
Security Fixings - Roc Secure Ltd
Self drilling fastener - SFS Intec LTD
Self tapping fasteners - SFS Intec LTD
Shear load connectors - Ancon Building Products
Shield anchors - Ramset Fasteners Ltd
Sisal Twines - James Lever 1856 Ltd
Sleeve anchors - Ramset Fasteners Ltd
Special length studs BZP and S/Steel - Redifix Ltd
Stainless steel channel bolts & screws fixings etc - Ancon Building Products
Stainless steel channel bolts & screws fixings etc -
Stainless steel fixings/fasteners - Wincro Metal Industries Ltd
Stainless steel fasteners - SFS Intec LTD
Stainless steel fixings/fasteners - Wincro Metal Industries Ltd
Stainless steel timber to masonry fixings - Helifix
Stainless steel warm roof fixings - Helifix
Steelwork fixings - Lindapter International
Structural adhesives - Kenyon Group Ltd
Structural steel connectors - Unistrut Limited
Stud anchors - Ramset Fasteners Ltd
Styreen Free Resin - Redifix Ltd
Surface Mount Adhesives - BigHead Bonding Fasteners Ltd
Tacks: Cut / Copper - Crown Nail Co Ltd
Technical adhesive tapes - Beiersdorf (UK) Ltd
Threaded fasteners - Arnold Wragg Ltd
Tile adhesives - IKO PLC
Tiling adhesives and grout - Weber Building Solutions
Tying wire, spacers and chemicals for concrete - Hy-Ten Ltd
Valves - Cottam & Preedy Ltd
Various types of metal and plastic fasteners - Righton Ltd
Wall Paper Paste - Ciret Limited
Wallpaper Adhesives - Bartoline Ltd
Water based adhesives - Kenyon Group Ltd
Wood glue and resins - Wessex Resins & Adhesives Ltd
Woodfillers - Bondaglass Voss Ltd

Z21 Mortars

Admixtures - Building Adhesives Ltd
Admixtures - Fosroc Ltd
Building solutions and compounds - IKO PLC
Cement - Combustion Linings Ltd
Cement admixtures - IKO PLC
Colour hardeners - Hatfield Ltd, Roy
Colouring admixtures - Hatfield Ltd, Roy
Grouts and resins - Helifix
High Temperature Cements - Minco Sampling Tech UK Ltd (Incorporating Fortafix)
Industrial hardeners - Hatfield Ltd, Roy
Manufacturers of mortars & Resin systems - Ancorite Surface Protection Ltd
Mortars - BASF plc, Construction Chemicals
Mortars - Flexcrete Technologies Limited
Mortars - LaFarge Cement UK
Natural Hydraulic Lime - Telling Lime Products Ltd
Powder - LANXESS Inorganic Pigments Group
Ready mixed concrete and mortar - Hanson UK
Ready mixed fillers - Bartoline Ltd
Slag cement - Civil & Marine Slag Cement Ltd
Thin bed mortars, Repair Mortars - Ronacrete Ltd

Z22 Sealants

Block and patio sealers - Kingfisher Building Products Ltd
Butyl sealants and foam fillers - Scapa Tapes UK Ltd
Ceramic tiling tools & accessories - Building Adhesives Ltd
Chemical fixings - ITW Construction Products
Concrete products - SFS Intec LTD
Concrete Repair - Sika Ltd
Façade Sealants - Sika Ltd
Fillers - Loctite UK Ltd
Fire rated sealants - Compriband Ltd
Floor fixings - Lindapter International
Grouts - BASF plc, Construction Chemicals
Grouts - Building Adhesives Ltd
Heldite Jointing Compound - Heldite Ltd
High Temperature Sealants - Minco Sampling Tech UK Ltd (Incorporating Fortafix)
Impregnated foam sealants - Compriband Ltd
Insulating Glass Silicone Sealant - Dow Corning Europe S.A
Intumescent sealant - Mann McGowan Group
Linseed oils - Liver Grease Oil & Chemical Co Ltd
Mastic sealant - Trade Sealants Ltd
Metal roofing and cladding - Ash & Lacy Building Products Ltd
Movement joint materials - Building Adhesives Ltd
Natural stone & facade sealant - Dow Corning Europe S.A
Oakum - Liver Grease Oil & Chemical Co Ltd
Plaster Accessories - British Gypsum Ltd
Polyurathane sealers - Hatfield Ltd, Roy
Sealant - Olby H E & Co Ltd
Sealants - Adshead Ratcliffe & Co Ltd
Sealants - Amari Plastics Plc
Sealants - BASF plc, Construction Chemicals
Sealants - Bondaglass Voss Ltd
Sealants - Brewer C & Sons Ltd
Sealants - Evode Ltd
Sealants - Exitex Ltd
Sealants - Firebarrier Services Ltd
Sealants - Fosroc Ltd
Sealants - Henkel Ltd
Sealants - IKO PLC
Sealants - Loctite UK Ltd
Sealants - Mapei (UK) Ltd
Sealants - Ogilvie Construction Ltd
Sealants - Sealocrete PLA Ltd
Sealants - Smith & Rodger Ltd
Sealants - Thermica Ltd
Sealants - Till & Whitehead Ltd
Sealants - Tower Manufacturing
Sealants - Tremco Illbruck Ltd
Sealants & Adhesives - Illbruck Sealant Systems UK Ltd
Sealants & Primers - Ciret Limited
Sealants, adhesives and building chemicals - Kalon Decorative Products
Sealers - Hatfield Ltd, Roy
Sealers for paving-Resiblock Ltd - Resiblock Ltd
Sealing systems - silicones and impregnated tape - VBH (GB) Ltd
Self-smoothing compounds & surface treatments - Building Adhesives Ltd
Silicone - Unifix Ltd
Silicone products - JAMAK Fabrication Europe
Silicone sealants - Compriband Ltd
Silicone weatherseal sealant - Dow Corning Europe S.A
Structural Glazing Silicone Sealant - Dow Corning Europe S.A
Waterproof expansion joints - Radflex Contract Services Ltd
Weatherproofing silicone sealant - Dow Corning Europe S.A
Woodfillers and woodcare - Clam Brummer Ltd

Z31 Powder coatings

aluminium/hybrid curtain walling, windows and doors - Senior Aluminium Systems plc
Powder coating - Consort Equipment Products Ltd
Powder coatings - Arkema France
Powder coatings - H. B. Fuller Powder Coatings Ltd
Powder Coatings - Howse Thomas Ltd
Powder coatings - Plastic Coatings Ltd
Powder coatings - Valspar Powder Coatings Ltd
Power Coatings - Coates LSF Ltd

Z32 Liquid coatings

Anti-carbonation coatings - Glixtone Ltd
Anti-graffiti coatings - Andura Coatings Ltd
Anti-graffiti coatings - Glixtone Ltd
Anti-microbial liquid coatings - Liquid Technology Ltd
Automotive lubricants and industrial paints - Witham Oil & Paint Ltd
Chemical stains - Hatfield Ltd, Roy
Clear protective coatings - Andura Coatings Ltd
Fungicidal coatings - Glixtone Ltd
Hi-tech Paint Finishes - Bristol Metal Spraying & Protective Coatings Ltd
Liquid coatings - Andrews Coatings Ltd
Liquid coatings - Arkema France
Liquid coatings - Conren Ltd
Liquid coatings - Flexcrete Technologies Limited
Liquid coatings - Smith & Rodger Ltd
Liquid coatings - Tor Coatings Ltd
Liquid Coatings - Wessex Resins & Adhesives Ltd
Nylon coating - Plastic Coatings Ltd
Paint - Fired Earth
Polymer modified coatings - Ronacrete Ltd

Z32   Liquid coatings (con't)

Protective coating - Industrial Solutions - Actavo (UK) Ltd
Roofing - Mells Roofing Ltd

Specialist decorative protective surface paints and coatings - Glixtone Ltd
Wet Coatings - Coates LSF Ltd
Wood hardener - Bondaglass Voss Ltd

Z33   Anodising

Anodised metal - Alanod Ltd
Hot dip galvanizing - Lionweld Kennedy Ltd

PRODUCTS AND SERVICES

PRODUCTS AND SERVICES